Get Connected to ConnectED

connectED.mcgraw-hill.com

- 100% Online
- One-Stop Shop, One Personalized Password
- Easy, Intuitive Navigation
- Resources, Resources, Resources

For Students

Leave your books at school. Now you can go online and interact with your **StudentWorks™ *Plus*** digital Student Edition from any place, any time!

For Teachers

ConnectED is your one-stop online center for everything you need to teach using ***Oklahoma Geometry***, including: **TeacherWorks™ *Plus*** digital Teacher Edition, lesson planning and scheduling tools, pacing, and assessment.

For Parents

Get homework help, help your student prepare for testing, and review math topics.

Get Connected to the way students learn and teachers teach TODAY!

Glencoe McGraw-Hill

Oklahoma Geometry

Authors

Carter • Cuevas • Day • Malloy • Cummins

McGraw Hill Glencoe

About the Cover

A batter swings hoping for a home run. When a bat hits a baseball, the contact can be represented two-dimensionally by a tangent to a circle. The portion of the bat that strikes the ball and the location of the tangent on the ball determine the path of the baseball. You will learn more about circles and tangents in Chapter 10.

Photo Credits:
0 (t)Richard Hutchings, (c)Getty Images, (b)Jupiter Images/Thinkstock/Alamy; **iv–v** The McGraw-Hill Companies; **vi–vii** Kristy White/Workbook Stock/Getty Images; **ix** The Garden Picture Library/Alamy Images; **x** Alaska Stock LLC/Alamy Images; **xi** Frank Krahmer/Masterfile; **xii** Jeremy Woodhouse/Masterfile; **xiii** Alan Becker/Getty Images; **xiv** Graham Henderson; Elizabeth Whiting & Associates/CORBIS; **xv** image100/age fotostock; **xvi** Andrew Hasson/Alamy Images; **xvii** Kwame Zikomo/SuperStock; **xviii** Werner Bollmann/age fotostock; **xix** Charles O'Rear/CORBIS; **xx** Jim West/PhotoEdit; **xxi** age fotostock/SuperStock; **xxii** Hill Street Studios/Blend Images/CORBIS; **xxiv–xxv** Panoramic Images/Getty Images; **OK0** Michael Newman/PhotoEdit; **P1** The Garden Picture Library/Alamy Images

The McGraw·Hill Companies

Glencoe

Send all inquiries to:
Glencoe/McGraw-Hill
8787 Orion Place
Columbus, OH 43240-4027

ISBN: 978-0-07-892256-5
MHID: 0-07-892256-9

Printed in the United States of America.

2 3 4 5 6 7 8 9 10 DOW 17 16 15 14 13 12 11 10

Designed for Oklahoma Success

Contents in Brief

Authors

Our lead authors ensure that the Macmillan/McGraw-Hill and Glencoe/McGraw-Hill mathematics programs are truly vertically aligned by beginning with the end in mind–success in Geometry and beyond. By "backmapping" the content from the high school programs, all of our mathematics programs are well articulated in their scope and sequence.

Lead Authors

John A. Carter, Ph.D.
Assistant Principal for Teaching and Learning
Adlai E. Stevenson High School
Lincolnshire, Illinois

Areas of Expertise: Using technology and manipulatives to visualize concepts; mathematics achievement of English-language learners

Gilbert J. Cuevas, Ph.D.
Professor of Mathematics Education
Texas State University–San Marcos
San Marcos, Texas

Areas of Expertise: Applying concepts and skills in mathematically rich contexts; mathematical representations

Roger Day, Ph.D., NBCT
Mathematics Department Chairperson
Pontiac Township High School
Pontiac, Illinois

Areas of Expertise: Understanding and applying probability and statistics; mathematics teacher education

Carol Malloy, Ph.D.
Associate Professor
University of North Carolina at Chapel Hill
Chapel Hill, North Carolina

Areas of Expertise: Representations and critical thinking; student success in Algebra 1

OK Math Online Meet the authors at glencoe.com.

Program Author

Jerry Cummins
Mathematics Consultant
Former President, National Council of Supervisors of Mathematics
Western Springs, Illinois

Areas of Expertise: Graphing Technology and Mathematics

Contributing Author

Dinah Zike FOLDABLES
Educational Consultant
Dinah-Might Activities, Inc.
San Antonio, Texas

Consultants and Reviewers

These professionals were instrumental in providing valuable input and suggestions for improving the effectiveness of the mathematics instruction.

Consultants

Mathematical Content

Viken Hovsepian
Professor of Mathematics
Rio Hondo College
Whittier, California

Grant A. Fraser, Ph.D.
Professor of Mathematics
California State University, Los Angeles
Los Angeles, California

Arthur K. Wayman, Ph.D.
Professor of Mathematics Emeritus
California State University, Long Beach
Long Beach, California

Gifted and Talented

Shelbi K. Cole
Research Assistant
University of Connecticut
Storrs, Connecticut

College Readiness

Robert Lee Kimball, Jr.
Department Head, Math and Physics
Wake Technical Community College
Raleigh, North Carolina

English-Language Learners

Susana Davidenko
State University of New York
Cortland, New York

Alfredo Gómez
Mathematics/ESL teacher
George W. Fowler High School
Syracuse, New York

Graphing Calculator

Ruth M. Casey
T^3 National Instructor
Frankfort, Kentucky

Mathematical Fluency

Robert M. Capraro
Associate Professor
Texas A&M University
College Station, Texas

Pre-AP

Dixie Ross
Lead Teacher for Advanced Placement Mathematics
Pflugerville High School
Pflugerville, Texas

Reading and Writing

ReLeah Cossett Lent
Author and Educational Consultant
Morganton, Georgia

Lynn T. Havens
Director of Project CRISS
Kalispell, Montana

Windmill on prairie

Reviewers

Corey Andreasen
Mathematics Teacher
North High School
Sheboygan, Michigan

Mark B. Baetz
Mathematics Coordinating Teacher
Salem City Schools
Salem, Virginia

Kathryn Ballin
Mathematics Supervisor
Newark Public Schools
Newark, New Jersey

Kevin C. Barhorst
Mathematics Department Chair
Independence High School
Columbus, Ohio

Brenda S. Berg
Mathematics Teacher
Carbondale Community High School
Carbondale, Illinois

Sheryl Pernell Clayton
Mathematics Teacher
Hume Fogg Magnet School
Nashville, Tennessee

Bob Coleman
Mathematics Teacher
Cobb Middle School
Tallahassee, Florida

Jane E. Cotts
Mathematics Teacher
O'Fallon Township High School
O'Fallon, Illinois

Michael D. Cuddy
Mathematics Instructor
Zypherhills High School
Zypherhills, Florida

Melissa M. Dalton, NBCT
Mathematics Instructor
Rural Retreat High School
Rural Retreat, Virginia

Tina S. Dohm
Mathematics Teacher
Naperville Central High School
Naperville, Illinois

Laurie L.E. Ferrari
Teacher
L'Anse Creuse High School–North
Macomb, Michigan

Patricia R. Frazier
Mathematics Department Chair/Instructor
Celina High School
Celina, Ohio

Steve Freshour
Mathematics Teacher
Parkersburg South High School
Parkersburg, West Virginia

Shirley D. Glover
Mathematics Teacher
TC Roberson High School
Asheville, North Carolina

Caroline W. Greenough
Mathematics Teacher
Cape Fear Academy
Wilmington, North Carolina

Michelle Hanneman
Mathematics Teacher
Moore High School
Moore, Oklahoma

Susan Hack, NBCT
Mathematics Teacher
Oldham County High School
Buckner, Kentucky

Theresalynn Haynes
Mathematics Teacher
Glenbard East High School
Lombard, Illinois

Sandra Hester
Mathematics Teacher/AIG Specialist
North Henderson High School
Hendersonville, North Carolina

Jacob K. Holloway
Mathematics Teacher
Capitol Heights Junior High School
Montgomery, Alabama

Robert Hopp
Mathematics Teacher
Harrison High School
Harrison, Michigan

Eileen Howanitz
Mathematics Teacher/Department Chairperson
Valley View High School
Archbald, Pennsylvania

Charles R. Howard, NBCT
Mathematics Teacher
Tuscola High School
Waynesville, North Carolina

Sue Hvizdos
Mathematics Department Chairperson
Wheeling Park High School
Wheeling, West Virginia

Elaine Keller
Mathematics Teacher
Mathematics Curriculum Director K–12
Northwest Local Schools
Canal Fulton, Ohio

Sheila A. Kotter
Mathematics Educator
River Ridge High School
New Port Richey, Florida

Frank Lear
Mathematics Department Chair
Cleveland High School
Cleveland, Tennessee

Jennifer Lewis
Mathematics Teacher
Triad High School
Troy, Illinois

Catherine McCarthy
Mathematics Teacher
Glen Ridge High School
Glen Ridge, New Jersey

Jacqueline Palmquist
Mathematics Department Chair
Waubonsie Valley High School
Aurora, Illinois

Thom Schacher
Mathematics Teacher
Otsego High School
Otsego, Michigan

Laurie Shappee
Teacher/Mathematics Coordinator
Larson Middle School
Troy, Michigan

Jennifer J. Southers
Mathematics Teacher
Hillcrest High School
Simpsonville, South Carolina

Sue Steinbeck
Mathematics Department Chair
Parkersburg High School
Parkersburg, West Virginia

Kathleen D. Van Sise
Mathematics Teacher
Mandarin High School
Jacksonville, Florida

Karen Wiedman
Mathematics Teacher
Taylorville High School
Taylorville, Illinois

CHAPTER 0

Preparing for Geometry

Chapter 0 Support

Helping You Learn

- **Vocabulary** P8, P9, P15, P17, P18, P19
- **Exercises** P5, P7, P9, P10, P12, P14, P16, P18, P20

Math Online

- **Personal Tutor** P4, P5, P6, P7, P8, P9, P10, P11, P12, P13, P14, P15, P16, P17, P18, P19, P20
- **Self-Check Quizzes** P4, P6, P8, P10, P11, P13, P15, P17, P19
- **Extra Examples** P4, P6, P8, P10, P11, P13, P15, P17, P19
- **Homework Help** P4, P6, P8, P10, P11, P13, P15, P17, P19

Unit 1
Geometric Structure

CHAPTER 1

Tools of Geometry

OK PASS

Chapter 1 Support

Helping You Learn

- **New Vocabulary** 5, 14, 25, 36, 46, 56, 67, 68
- **Key Concepts** 5, 15, 16, 25, 26, 27, 38, 46, 47, 48, 49, 56, 58, 67, 69
- **Guided Practice** 6, 7, 14, 15, 16, 17, 22, 23, 25, 26, 27, 28, 29, 37, 38, 40, 47, 48, 49, 50, 57, 58, 59, 60, 68, 69, 70
- **Check Your Understanding** 8, 18, 30, 41, 50, 61, 70
- **Multiple Representations** 11, 20, 34, 43, 53, 73
- **H.O.T. Problems** 11, 20, 34, 43, 53, 63, 73
- **Skills Review** 12, 21, 35, 44, 54, 64, 74

Math Online

- **Get** **Animated** 2, 13, 29, 37, 39, 40, 55, 74, 75
- **Personal Tutor** 5, 6, 7, 14, 15, 16, 17, 22, 23, 25, 26, 27, 28, 29, 36, 37, 38, 40, 46, 47, 48, 49, 50, 56, 57, 58, 59, 60, 67, 68, 69, 70
- **Graphing Technology Personal Tutor** 65

Preparing for Testing

- **Extended Response** 54, 85
- **Multiple Choice** 12, 21, 35, 44, 45, 54, 59, 61, 64, 81
- **Short/Gridded Response** 12, 21, 35, 44, 64, 74, 84, 85
- **Worked-Out Examples** 59, 83

CHAPTER 2

Reasoning and Proof

Chapter 2 Support

Helping You Learn

Math Online

Preparing for Testing

CHAPTER 3

Parallel and Perpendicular Lines

Chapter 3 Support

Helping You Learn

Math Online

Preparing for Testing

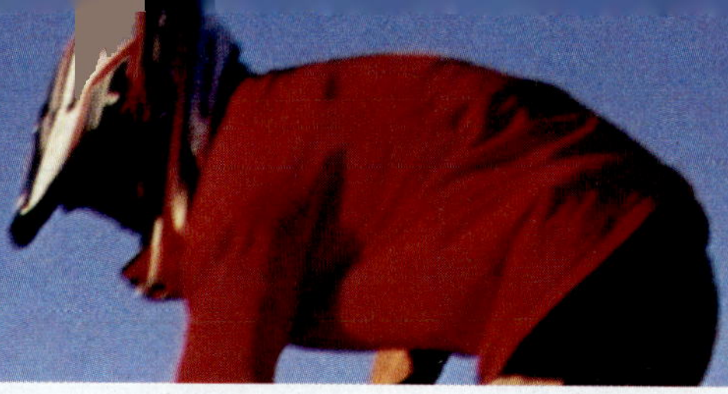

Unit 2
Congruence

CHAPTER 4

Congruent Triangles

Chapter 4 Support

Helping You Learn

- **New Vocabulary** 235, 244, 253, 262, 273, 283, 294, 301
- **Key Concepts** 235, 236, 244, 246, 247, 253, 255, 256, 262, 264, 273, 274, 276, 282, 283, 284, 294, 301
- **Guided Practice** 236, 237, 245, 247, 254, 255, 256, 262, 263, 265, 266, 274, 275, 276, 284, 285, 286, 295, 296, 301, 302, 303
- **Check Your Understanding** 238, 248, 256, 266, 276, 287, 290, 297, 304
- **Multiple Representations** 241, 251, 259, 290, 299
- **H.O.T. Problems** 241, 251, 260, 269, 279, 290, 299, 306
- **Skills Review** 242, 252, 261, 270, 280, 291, 300, 307

Math Online

- **Get**
Animated 232, 243, 265, 271, 276, 281, 301
- **Personal Tutor** 235, 236, 237, 244, 245, 247, 253, 254, 255, 256, 262, 263, 265, 266, 273, 274, 275, 276, 283, 284, 285, 286, 294, 295, 296, 301, 302, 303
- **Graphing Technology Personal Tutor** 292

Preparing for Testing

- **Extended Response** 263, 266, 270, 316
- **Multiple Choice** 242, 252, 261, 270, 272, 280, 291, 300, 307, 313, 317
- **Short/Gridded Response** 242, 252, 261, 280, 291, 300, 307, 316, 317
- **Worked-Out Examples** 263, 313

CHAPTER 5

Relationships in Triangles

OK PASS

Chapter 5 Support

Helping You Learn

Math Online

Preparing for Testing

CHAPTER 6

Quadrilaterals

Chapter 6 Support

Helping You Learn

Math Online

Preparing for Testing

Unit 3
Similarity

CHAPTER 7

Proportions and Similarity

OK PASS

Chapter 7 Support

Helping You Learn

Math Online

Preparing for Testing

Right Triangles and Trigonometry

CHAPTER 9

Transformations and Symmetry

OK PASS

Chapter 9 Support

Helping You Learn

Math Online

Preparing for Testing

Unit 4
Measurement

CHAPTER 10 Circles

Chapter 10 Support

Helping You Learn

Math Online

Preparing for Testing

CHAPTER 11

Areas of Polygons and Circles

Chapter 11 Support

Helping You Learn

Math Online

Preparing for Testing

CHAPTER 12 Extending Surface Area and Volume

CHAPTER 13

Probability and Measurement

Chapter 13 Support

Student Handbook

Contents

Welcome to *Oklahoma Geometry*

Master Oklahoma's Priority Academic Student Skills in 3 Easy Steps

1 Practice the Standards Daily

Each lesson addresses Oklahoma's Priority Academic Student Skills covered in that lesson.

Questions aligned to the standards in a format like those on the End-of-Instruction test provide you with ongoing opportunities to sharpen your test-taking skills.

Tulsa, Oklahoma

Oklahoma Reviewers

Each Oklahoma Reviewer reviewed at least two chapters of the Student Edition, giving feedback and suggestions for improving the effectiveness of the mathematics instruction.

Kelli Ball, NBCT
Mathematics Teacher
Owasso 7th Grade Center
Owasso, Oklahoma

Glenna L. Crockett
Mathematics Department Chair
Fairland High School
Fairland, Oklahoma

Michelle Hanneman
Mathematics Teacher
Moore High School
Moore, Oklahoma

2 Practice the Standards throughout the Chapter

Every chapter contains a completely worked-out test example to help you solve problems that are similar to those you might find on the test.

Every chapter contains two full pages of Test Practice with Test-Taking Tips.

3 Practice the Standards

Countdown PLUS to the Oklahoma EOI Use pages OK1–OK20 to practice questions that are similar to those found on the EOI. Lesson references are included should you need a little refresher.

Countdown PLUS to the Oklahoma EOI

End-of-Instruction Test

In the spring, you will take the End-of-Instruction test for Geometry. The EOI covers the Priority Academic Student Skills. The test has multiple-choice questions. You will apply the concepts and skills that you have learned throughout the year.

How Should I Study?

The good news is that you've been studying for the EOI all along a little bit every day. Here are some of the ways your textbook has been preparing you.

- **Every Day** Each lesson has practice questions that cover the PASS.
- **Every Week** The Mid-Chapter Quiz and Practice Test has several practice questions. Also, the following pages include a section called Countdown PLUS to the EOI. These pages give you practice questions that cover the PASS.
- **Every Chapter** Each chapter has two full pages of practice questions that are similar to those on the EOI.

How Should I Use the Countdown PLUS?

You can use these pages in the weeks before the test to determine if you are ready. Begin 20 weeks before the EOI. You should plan to complete one practice page each week to help you review for the End-of-Instruction test.

If you are struggling with any of the items, lesson references are provided so that you can go back and review from the pages in your textbook.

Multiple-Choice Items

For multiple-choice questions, you will select the correct response from four answer choices. Your teacher will provide you with an answer sheet to fill in your answer choices.

Week 20

Monday

1 Solve for x. (Lesson 1-5) (2.2.c)

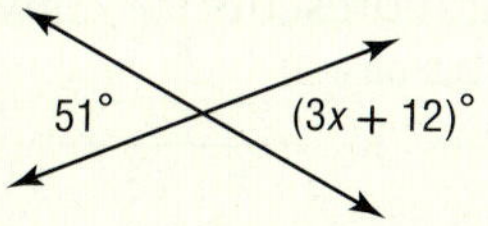

A 10
B 13
C 21
D 29

Tuesday

2 Which of the following is a convex polygon with all sides congruent and all angles congruent? (Lesson 1-6) (2.3.d)

F equiangular
G equilateral
H regular
J right

Wednesday

3 The Oklahoma state quarter was first released in 2008. It features the state bird, state flower, and 1907, the year Oklahoma became the 46th state of the union. If the radius of the coin is 12.13 millimeters, what is its circumference? (Lesson 1-6) (2.3.d)

A 38.1 mm
B 48.5 mm^2
C 76.2 mm
D 462.2 mm^2

Thursday

4 Which solid figure has two circular bases? (Lesson 1-7) (4.1.a)

F cylinder
G prism
H pyramid
J sphere

Friday

5 By drawing $\overleftrightarrow{PQ}$, what will have been constructed? (Lesson 1-5) (2.1)

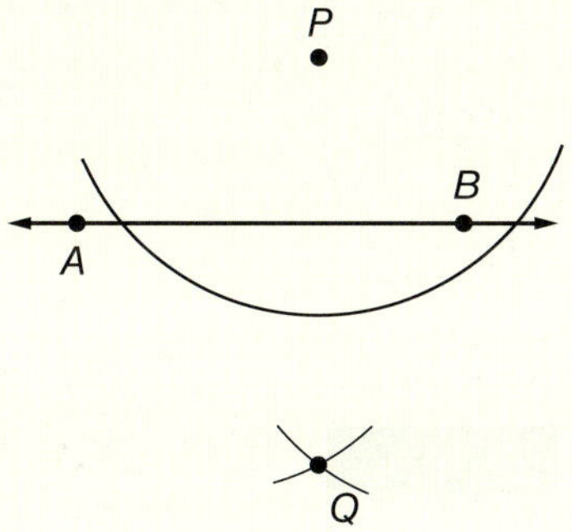

A the perpendicular bisector of $\overline{AB}$
B the angle bisector of $\angle APB$
C a line perpendicular to $\overleftrightarrow{AB}$ through P
D a segment congruent to $\overline{AB}$

Week 19

Monday

1 If a figure is a rectangle, then the figure is a parallelogram. Suppose *ABCD* is not a parallelogram. What can you conclude about the figure? (Lesson 2-3) (1.2)

A *ABCD* is not a rectangle.
B *ABCD* is a rectangle.
C *ABCD* is a parallelogram.
D *ABCD* is not a trapezoid.

Tuesday

2 Which of the following is a counterexample to the following statement? (Lesson 2-1) (1.1)

If a number is divisible by 5, then it is divisible by 10.

F 10
G 15
H 20
J 30

Wednesday

3 The conditional statement of the form if p, then q can be written as $p \to q$. Which of the following represents the converse of the statement? (Lesson 2-3) (1.2)

A $\sim p \to \sim q$
B $\sim q \to \sim p$
C $q \to p$
D $p \to \sim q$

Thursday

4 Which of the following is a valid conclusion given the figure below? (Lesson 2-8) (1.1)

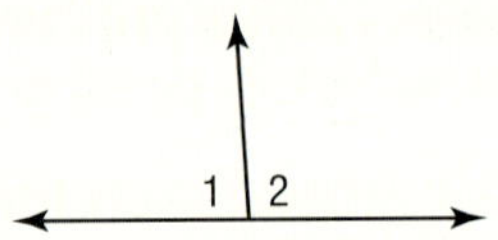

F $m\angle 1 > m\angle 2$
G $m\angle 1 < m\angle 2$
H $m\angle 1 + m\angle 2 = 180$
J $m\angle 1 + m\angle 2 = 90$

Friday

5 What is the value of a_n when $n = 6$? (Lesson 2-1) (1.1)

A 32
B 35
C 36
D 37

n	a_n
1	2
2	5
3	10
4	17
5	26

Week 18

Monday

1 Which of the following shows a construction of a perpendicular bisector of a segment? (Lesson 1-2) (2.1)

A

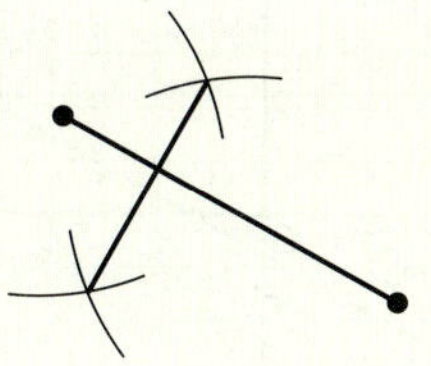

B

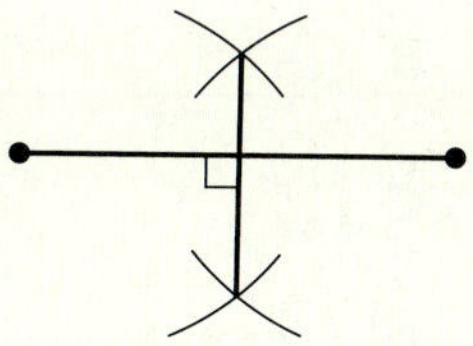

C

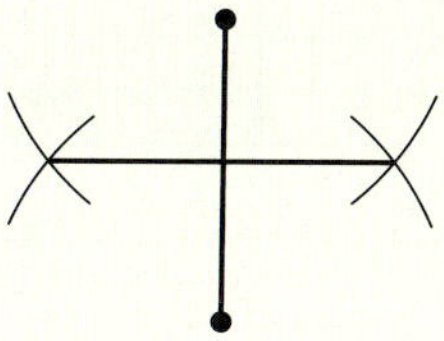

D

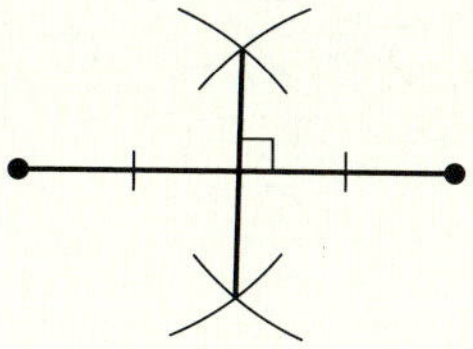

Tuesday

2 Triangle *ABC* has vertices *A*(1, 2), *B*(5, −2), and *C*(−6, −5). Classify $\triangle ABC$. (Lesson 1-6) (5.2.a)

F equilateral

G isosceles

H obtuse

J right

Wednesday

3 Solve for *x*. (Lesson 1-5) (2.2.c)

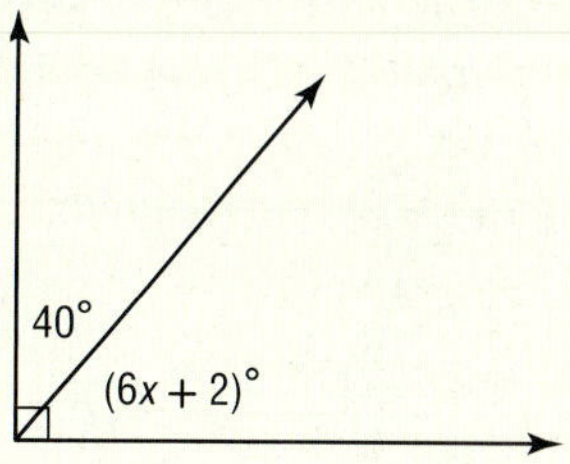

A 3

B $6\frac{1}{3}$

C 8

D $8\frac{2}{3}$

Thursday

4 What is the volume of the shoe box shown? (Lesson 1-7) (4.1.b)

F 314 in^3

G 306 in^3

H 292 in^3

J 280 in^3

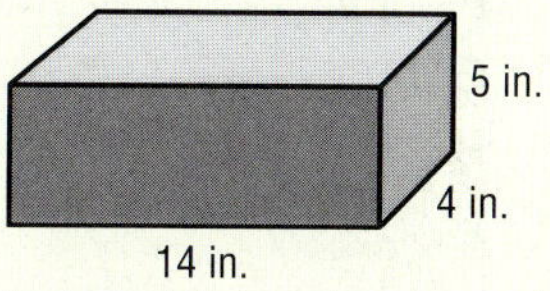

Friday

5 The Venn diagram shows the relationship between residents of Tulsa and residents of Oklahoma. Which of the following statements is true? (Lesson 2-3) (1.2)

A If you live in Oklahoma, then you live in Tulsa.

B If you do not live in Tulsa, then you do not live in Oklahoma.

C If you do not live in Oklahoma, then you do not live in Tulsa.

D If you live in Oklahoma, then you live in Oklahoma City.

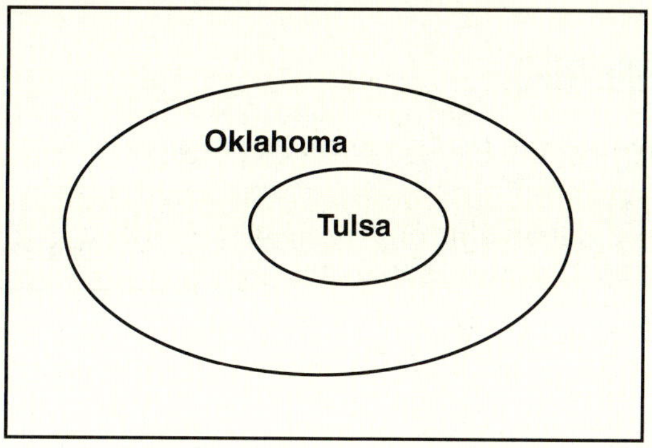

Week 17

Monday

1 Which of the following would not prove that lines a and b are parallel? (Lesson 3-5) (2.2.b)

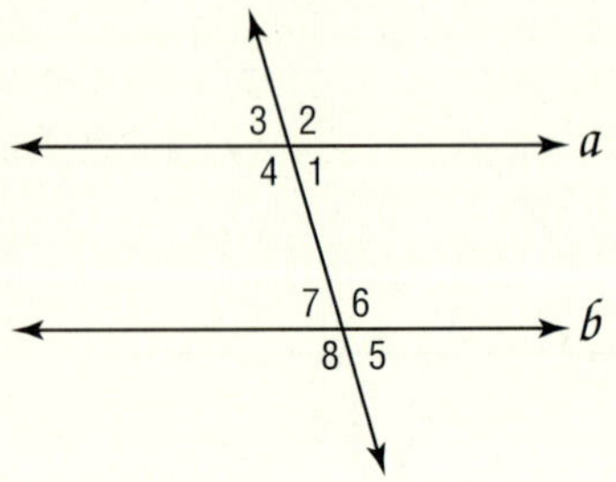

A $\angle 4 \cong \angle 6$

B $\angle 7 \cong \angle 2$

C $m\angle 3 + m\angle 6 = 180$

D $m\angle 2 + m\angle 7 = 180$

Tuesday

2 If $\ell \parallel m$, what is the value of x? (Lesson 3-5) (2.2.a)

F 15

G 16

H 18

J 19

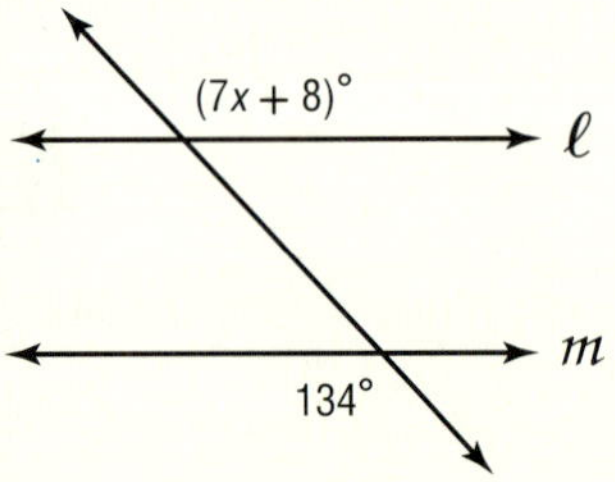

Wednesday

3 What is the slope of $\overline{CD}$? (Lesson 3-3) (5.1)

A $\frac{5}{9}$

B $\frac{3}{11}$

C $\frac{1}{6}$

D $-\frac{2}{5}$

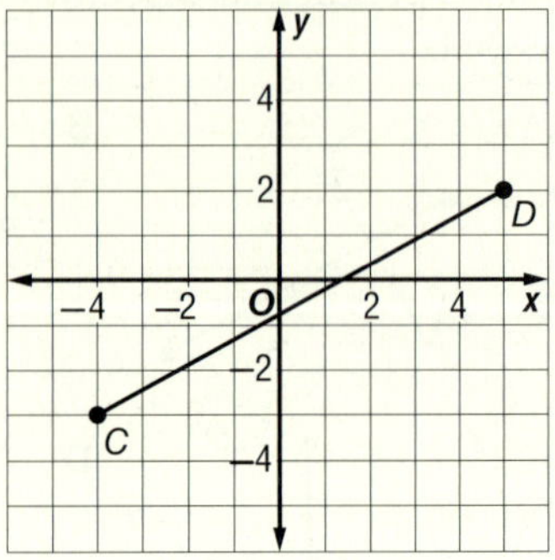

Thursday

4 Refer to the figure below. If $m\angle 5 = 40°$, what must the measure of $\angle 4$ be in order to prove that $j \parallel k$? (Lesson 3-5) (2.2.b)

F 30°

G 40°

H 50°

J 140°

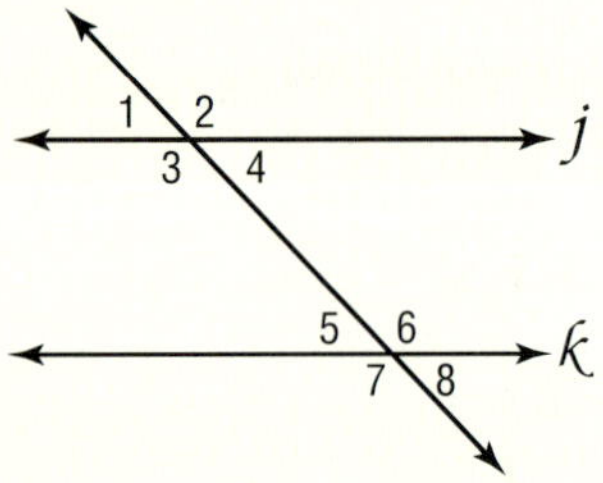

Friday

5 Solve for x in the figure below. (Lesson 3-2) (2.2.a)

A 21

B 12

C $7\frac{1}{3}$

D $\frac{1}{21}$

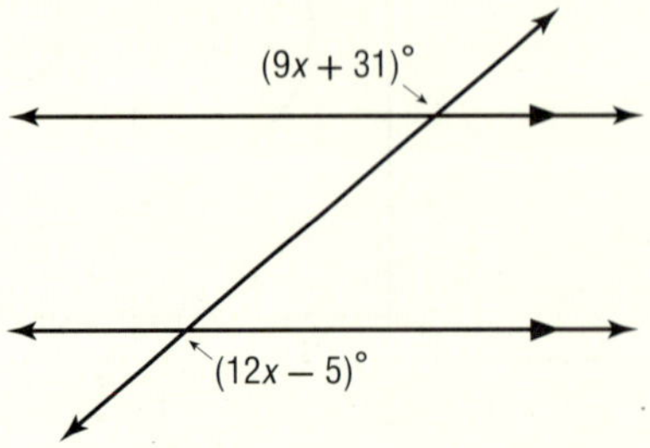

Week 16

Monday

1 What coordinates of B would form a parallelogram? (Lesson 1-6) (5.2.a)

A $B(-2, 3)$
B $B(-3, 4)$
C $B(-3, 3)$
D $B(-2, 4)$

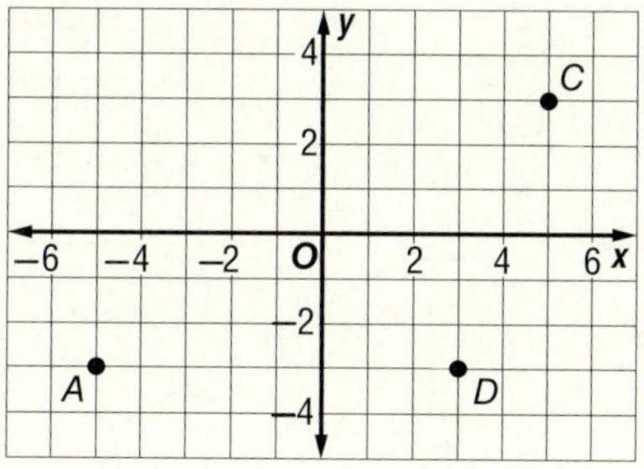

Tuesday

2 Which of the following is not a polyhedron? (Lesson 1-7) (4.1.a)

F

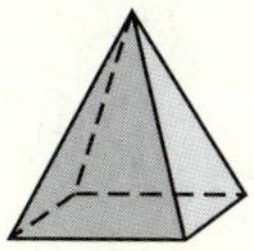

H

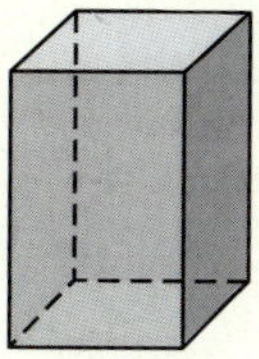

G

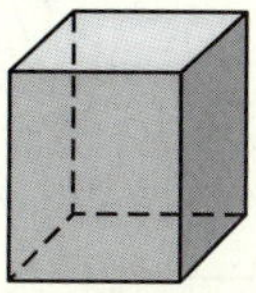

J

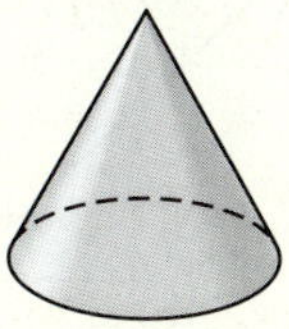

Wednesday

3 According to the Venn diagram, how many students are studying Spanish but not German? (Lesson 2-2) (1.1)

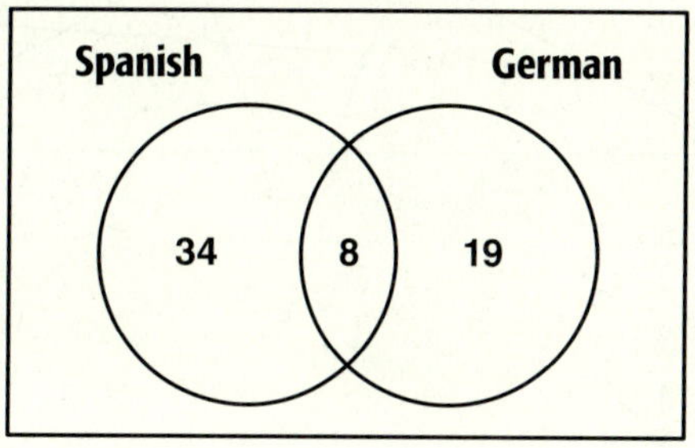

A 8
B 34
C 42
D 61

Thursday

4 What solid figure is formed by the net below? (Lesson 1-7) (4.3)

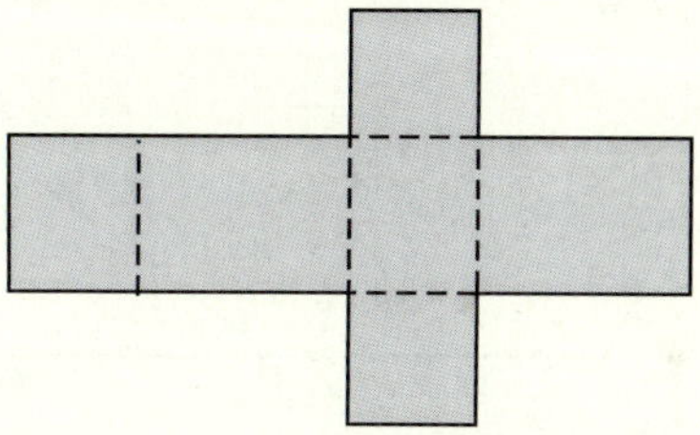

F cube
G cylinder
H prism
J pyramid

Friday

5 Roberta is using a coordinate grid to create a map of Oklahoma. She plots Oklahoma City at (3, 1) and Afton at (15, 9). If Tulsa lies at the midpoint of this segment, which coordinates represent Tulsa? (Lesson 1-3) (5.1)

A (9, 5)
B (9, 1)
C (6, 5)
D (6, 9)

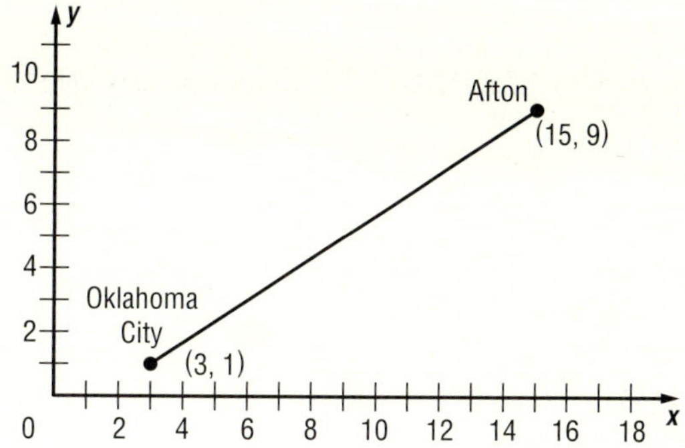

Week 15

Monday

1 If $\triangle RST \cong \triangle HKJ$, solve for x. (Lesson 4-3) (2.5.b)

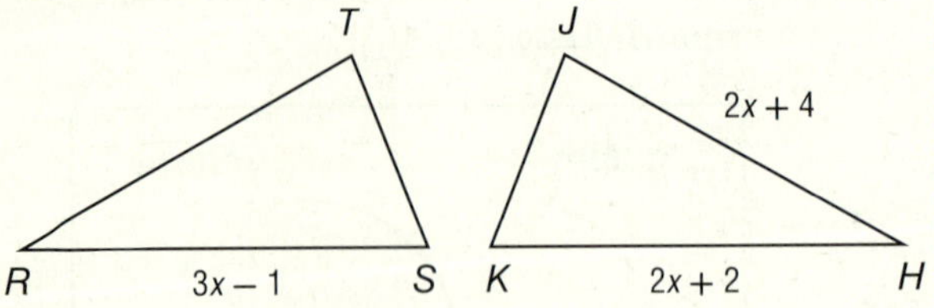

A 2
B 3
C 4
D 5

Tuesday

2 In the figure shown, $\triangle JKL \cong \triangle MNP$. What is the perimeter of $\triangle JKL$? (Lesson 4-3) (2.5.b)

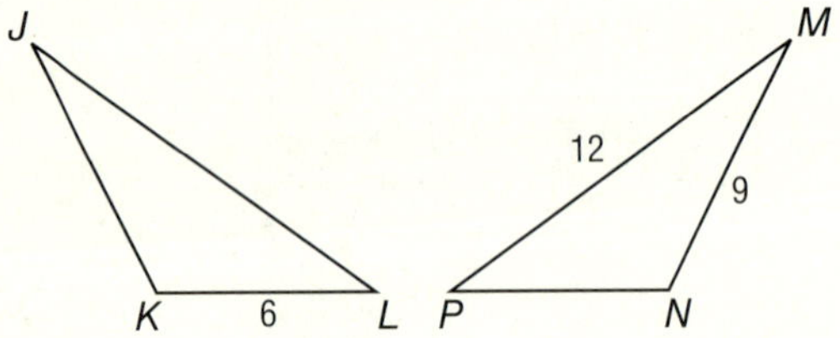

F 18
G 21
H 27
J 54

Wednesday

3 Write a congruence statement for the triangles below. (Lesson 4-5) (2.5.a)

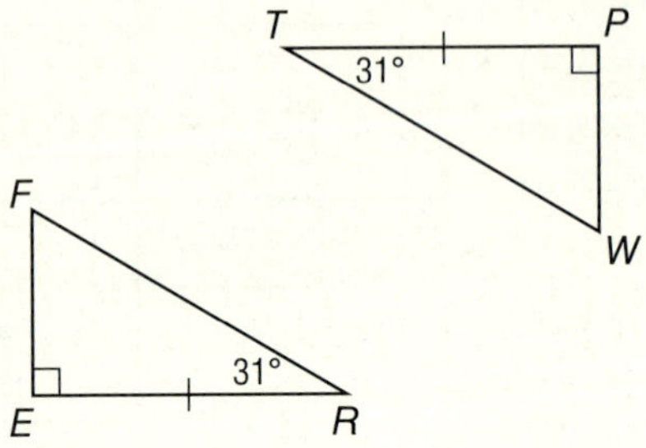

A $\triangle EFR \cong \triangle PTW$
B $\triangle EFR \cong \triangle PWT$
C $\triangle FER \cong \triangle PWT$
D $\triangle FRE \cong \triangle WPT$

Thursday

4 Given that $\triangle ABC \cong \triangle DEF$, which of the following is not true? (Lesson 4-3) (2.5.b)

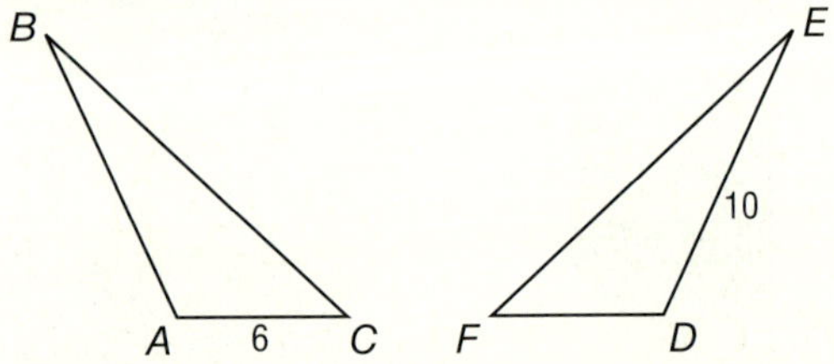

F $DF = 6$
G $m\angle B = m\angle E$
H $m\angle A = m\angle F$
J $AB = 10$

Friday

5 Which postulate could you use to prove $\triangle LMN \cong \triangle LMO$ using only the information shown? (Lesson 4-4) (2.5.a)

A ASA
B AAS
C SAS
D SSS

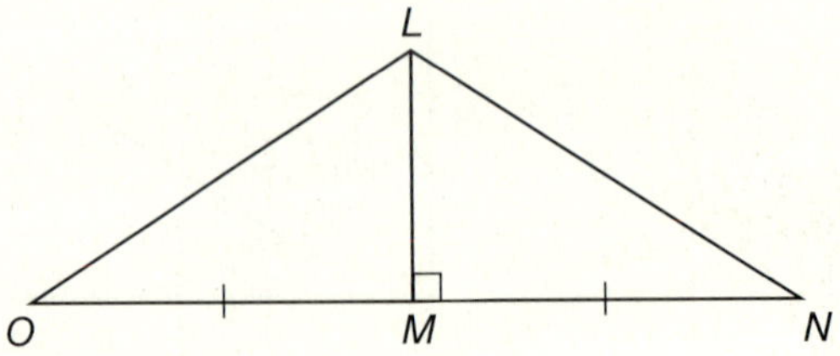

Week 14

Monday

1 What is the volume of a globe with a diameter of 40 centimeters? (Lesson 1-7) (4.1.b)

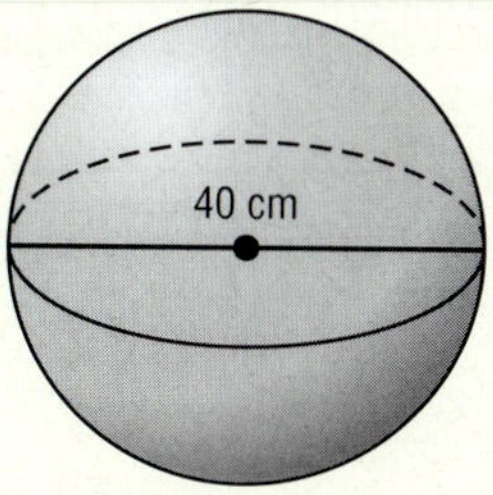

A 20,106 cm^2
B 26,357 cm^3
C 31,466 cm^2
D 33,510 cm^3

Tuesday

2 What is the distance between points $R(-4, 7)$ and $S(2, 9)$? (Lesson 3-6) (5.1)

F $\sqrt{32}$ units
G $\sqrt{40}$ units
H $\sqrt{252}$ units
J $\sqrt{260}$ units

Wednesday

3 What kind of transformation is shown on the coordinate grid? (Lesson 4-7) (5.2.b)

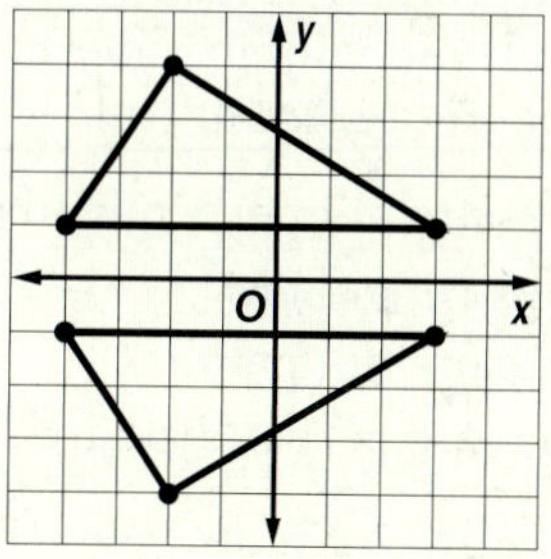

A dilation
B reflection
C rotation
D translation

Thursday

4 Solve for x. (Lesson 1-5) (2.2.c)

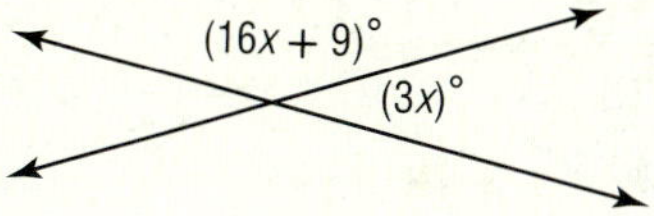

F 9
G 10
H 11
J 12

Friday

5 The area of the Oklahoma flag in Mrs. Natchez's classroom is 504 square inches. If the length of the flag is 28 inches, what is the perimeter? (Lesson 1-6) (2.3.d)

A 18 in.
B 46 in.
C 92 in.
D 100 in.

28 in.

Week 13

Monday

1 Suppose you want to prove that if two lines have slopes that are opposite reciprocals, then the lines are perpendicular. Which assumption would you begin with in order to write an indirect proof? (Lesson 5-4) (1.1)

A the lines are not perpendicular
B the lines are parallel
C the lines are not parallel
D the lines are perpendicular

Tuesday

2 Which side lengths could <u>not</u> be the sides of a triangle? (Lesson 5-5) (2.3.d)

F 8, 15, 17
G 4, 9, 13
H 15, 16, 30
J 11, 16, 21

Wednesday

3 Which of the following shows the sides of triangle *LMN* in order from shortest to longest? (Lesson 5-3) (1.1)

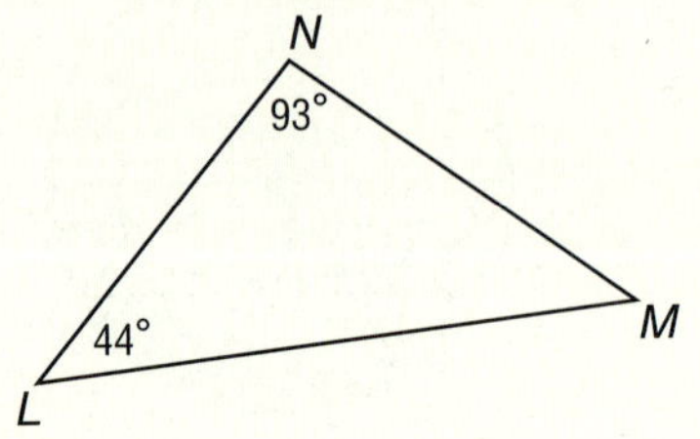

A $\overline{LM}, \overline{NM}, \overline{LN}$
B $\overline{NM}, \overline{LN}, \overline{LM}$
C $\overline{LN}, \overline{LM}, \overline{NM}$
D $\overline{LN}, \overline{NM}, \overline{LM}$

Thursday

4 What is the sum of the interior angles of this octagon? (Lesson 6-1) (2.3.a)

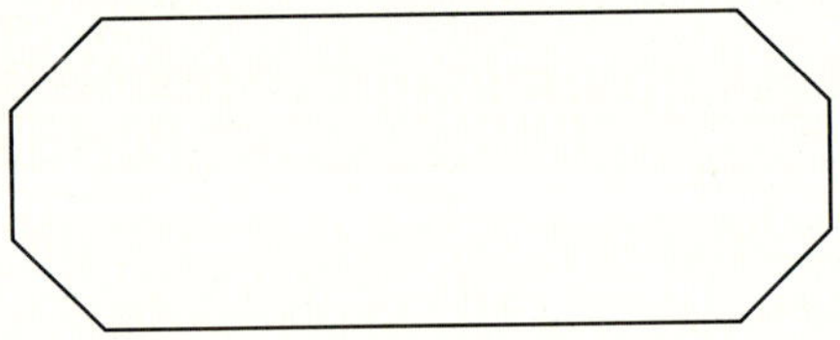

F 360°
G 720°
H 1080°
J 1440°

Friday

5 Carolyn uses a straightedge and compass to construct a median of triangle *DEF*. Which of the following is her construction? (Lesson 5-2) (2.1)

A
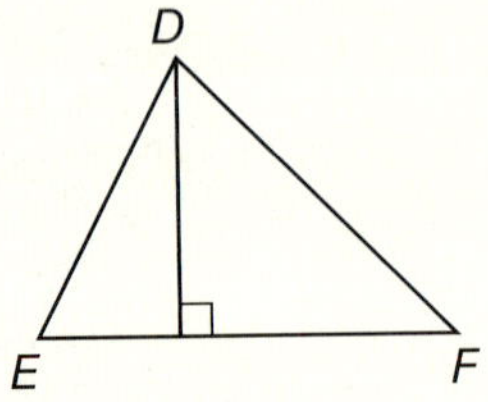

B
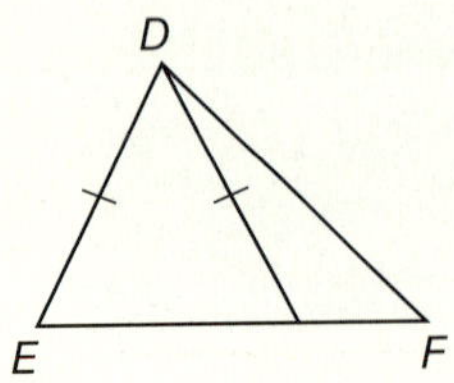

C
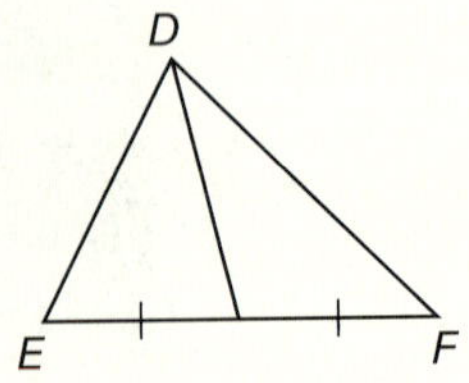

D
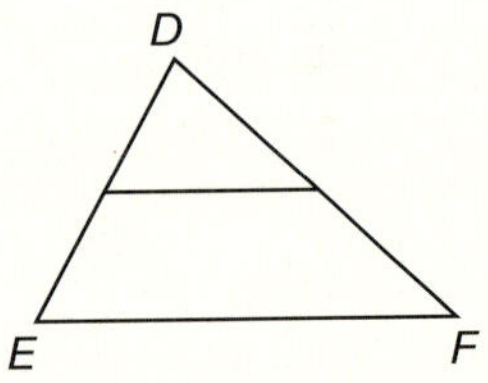

Week 12

Monday

1 Solve for x in parallelogram $RSTU$. (Lesson 6-2) (2.3.c)

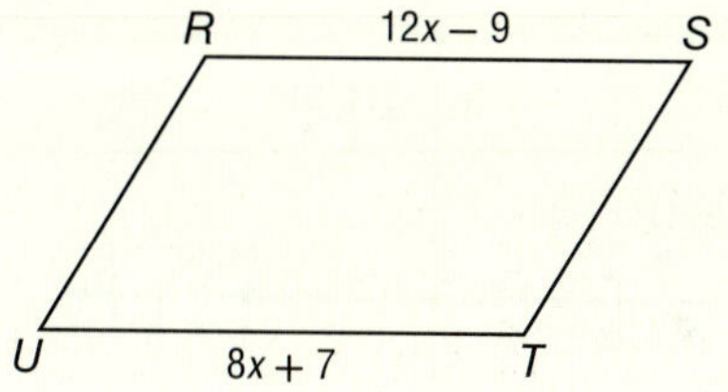

A 2
B 3
C 4
D 5

Tuesday

2 Which of the following does not describe the figure below? (Lesson 6-1) (2.3.a)

F convex
G hexagonal
H pentagonal
J regular

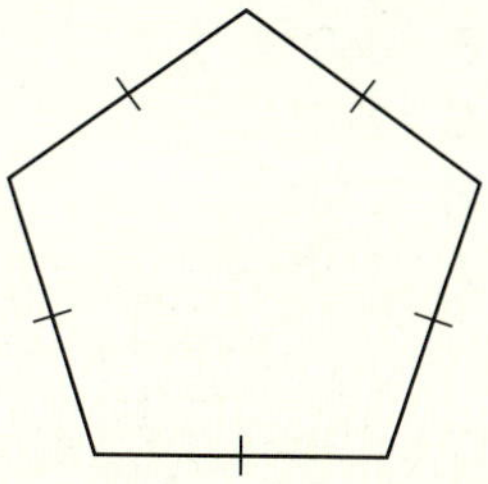

Wednesday

3 In rectangle $HIJK$, $OK = 3x + 8$ and $HJ = 8x$. Solve for x. (Lesson 6-4) (2.3.c)

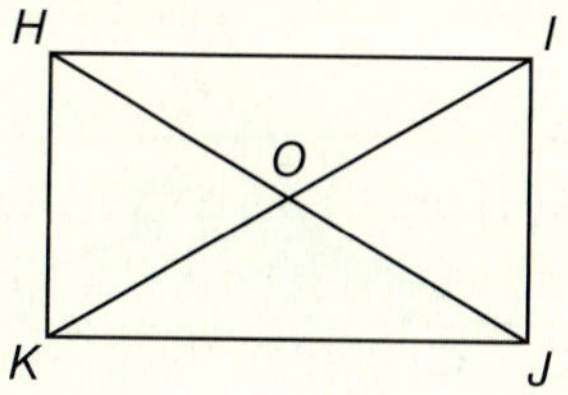

A $\frac{8}{13}$
B $1\frac{3}{5}$
C 4
D 8

Thursday

4 What is the sum of the interior angles of a hexagon? (Lesson 6-1) (2.3.b)

F 360°
G 540°
H 720°
J 1080°

Friday

5 Which of the following is not an interior angle measure of the trapezoid? (Lesson 6-1) (2.3.b)

A 60°
B 75°
C 110°
D 125°

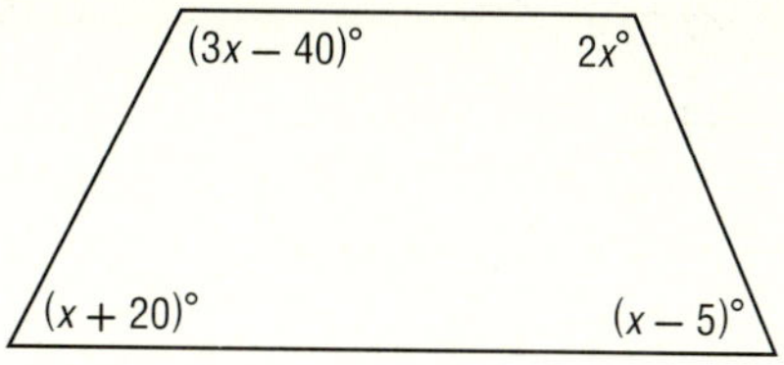

Week 11

Monday

1 What solid figure is formed by the net? (Lesson 1-7) (4.3)

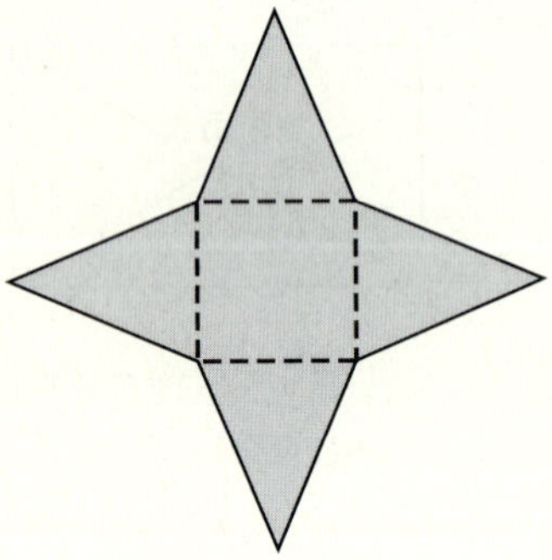

A cone
B cube
C prism
D pyramid

Tuesday

2 Which of the following is not a postulate or theorem that can be used to prove right triangle congruence? (Lesson 4-5) (2.5.a)

F Leg-Angle
G Angle-Angle
H Hypotenuse-Leg
J Leg-Leg

Wednesday

3 What type of quadrilateral is formed by the points shown? (Lesson 6-6) (2.3.c)

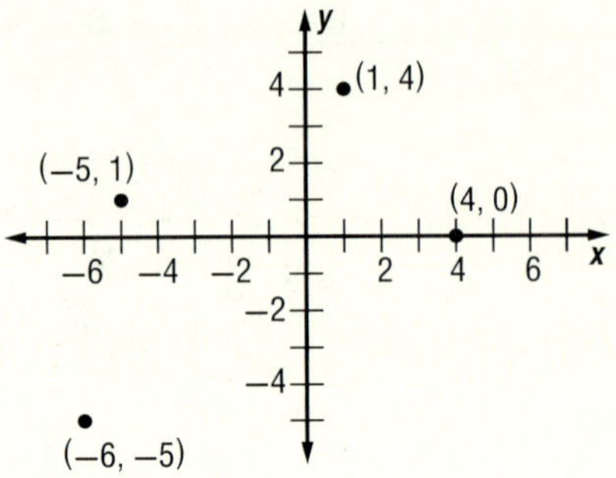

A parallelogram
B kite
C rectangle
D trapezoid

Thursday

4 What is the sum of the interior angles of the figure below? (Lesson 6-1) (2.3.b)

F 360°
G 270°
H 240°
J 180°

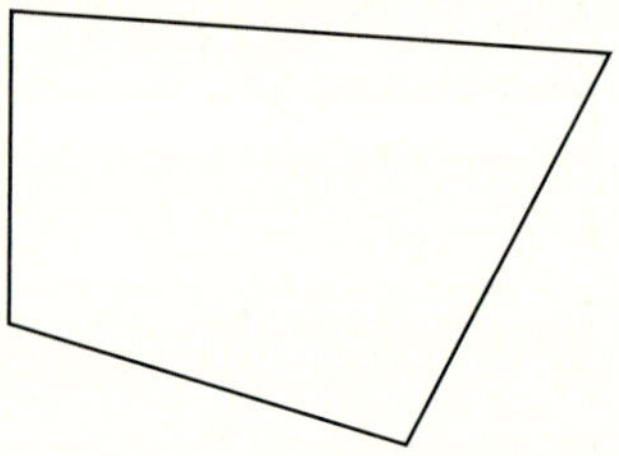

Friday

5 The diagram shows the layout of 3 highways in Oklahoma. Which conclusion can you draw about the highways? (Lesson 3-5) (2.2.b)

A I-40 is perpendicular to I-35.
B I-40 is perpendicular to the Indian Nation Turnpike.
C I-40 runs parallel to the Indian Nation Turnpike.
D I-35 runs parallel to the Indian Nation Turnpike.

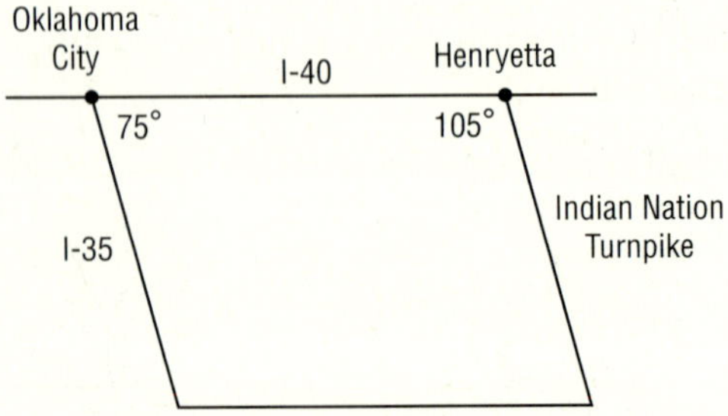

Week 10

Monday

1 Quadrilateral $QRST$ is similar to $WXYZ$ below. If $RS = 10.2$, what is the perimeter of $QRST$? (Lesson 7-2) (2.4.b)

A 45.9
B 45
C 39
D 36

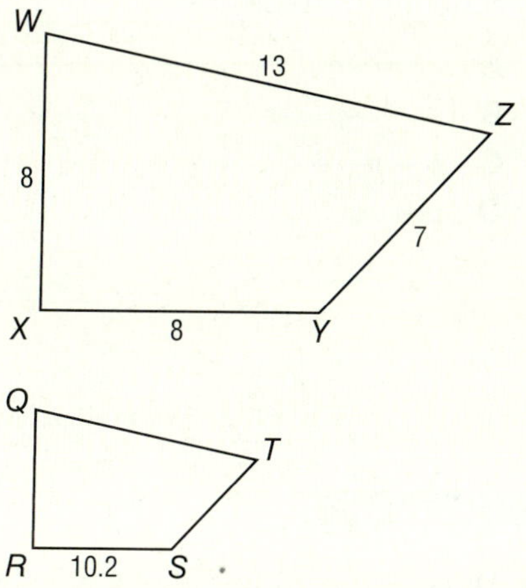

Tuesday

2 A medium pizza at an Italian restaurant has a diameter of 10 inches. A large pizza is similar to a medium with a scale factor of 1.2. How many times greater is the circumference of a large pizza than a medium pizza? (Lesson 7-6) (2.4.b)

F 0.6
G 1.2
H π
J 2π

Wednesday

3 If $\triangle NOP$ is dilated by a scale factor of 0.4, which of the following is not a side length of the resulting figure? (Lesson 7-6) (2.4.a)

A 4
B 4.8
C 5.2
D 6

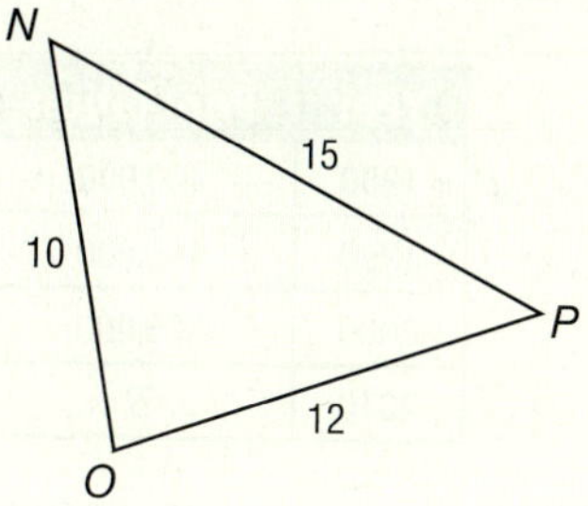

Thursday

4 If $\overline{BA} \parallel \overline{DE}$, which statement is true? (Lesson 7-4) (2.4.a)

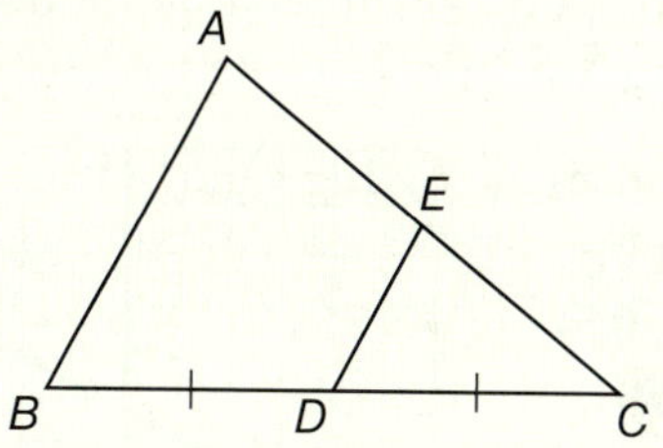

F $AE = BD$
G $DE = 0.25AB$
H $DC = EC$
J $AE = EC$

Friday

5 What is the value of x that would result in $\triangle RST \sim \triangle ABC$? (Lesson 7-3) (2.4.a)

A 3
B 3.5
C 4
D 4.5

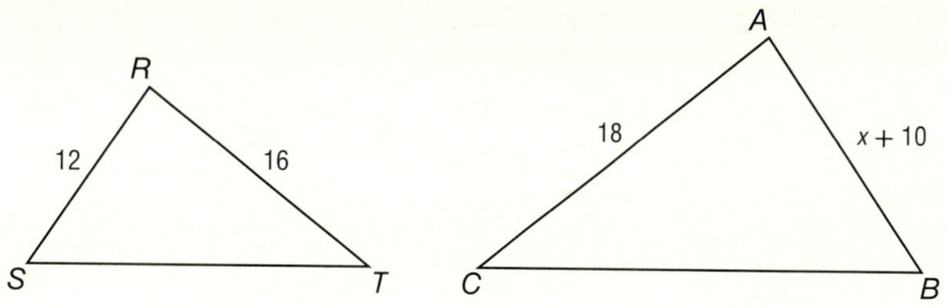

Week 9

Monday

1 The table shows the approximate population of Oklahoma City over the past several census decades. Suppose the trend continues. Predict the population in 2010. (Lesson 2-1) (1.1)

Year	Population
1980	400,000
1990	445,000
2000	506,000
2010	??

A 475,000
B 510,000
C 560,000
D 620,000

Tuesday

2 If line k is perpendicular to $\overline{MN}$ and has a y-intercept of -4, at what point do line k and $\overline{MN}$ intersect? (Lesson 3-3) (5.1)

F $(-3, -0.5)$
G $(-2, 0)$
H $(-1, 0.5)$
J $(0, 1)$

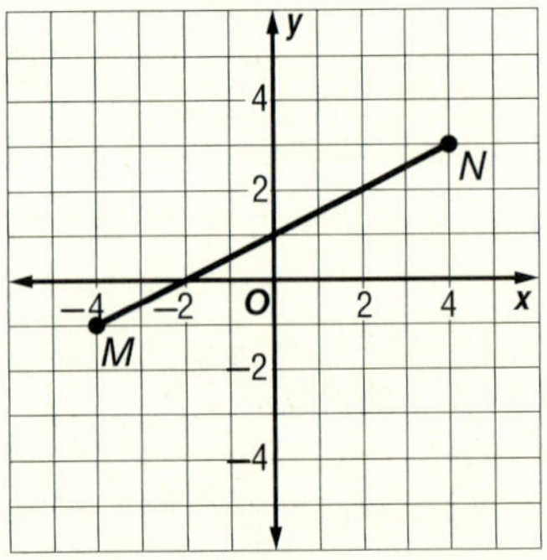

Wednesday

3 The conditional statement of the form if p, then q can be written as $p \rightarrow g$. Which of the following represents the inverse of the statement? (Lesson 2-3) (1.2)

A $\sim p \rightarrow \sim q$
B $\sim q \rightarrow \sim p$
C $q \rightarrow p$
D $p \rightarrow \sim q$

Thursday

4 What is the midpoint of $\overline{QR}$? (Lesson 1-3) (5.1)

F $(-2, -1)$
G $(-2, 1)$
H $(-1, -2)$
J $(-2, -2)$

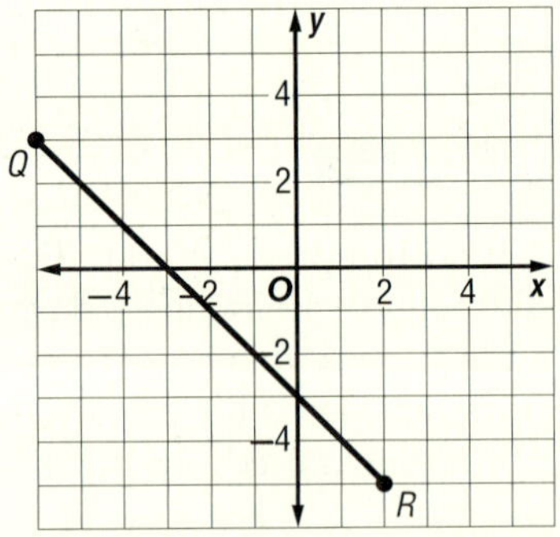

Friday

5 Which postulate or theorem can be used to prove $\triangle POM \cong \triangle NOM$? (Lesson 4-4) (2.5.a)

A AAS
B SAS
C SSS
D not possible

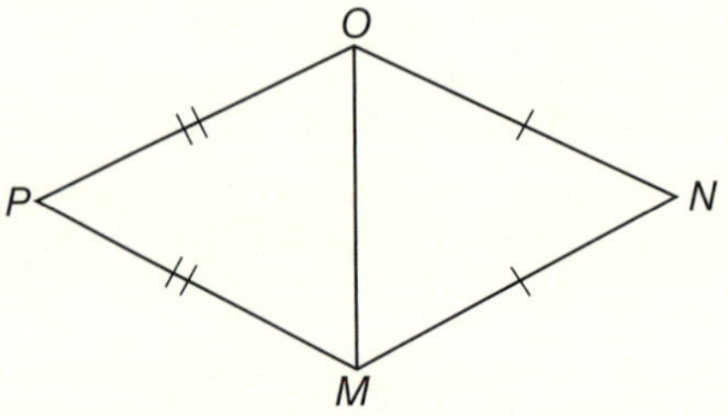

Week 8

Monday

1 Luis sets a 12-foot ladder against the side of his house. If he sets the bottom of the ladder 4 feet from the base of the wall, how high up does the ladder reach? (Lesson 8-2) (3.1)

A about 8 ft
B about 10 ft
C about 11 ft
D about 12 ft

Tuesday

2 What is the area of the triangle below? (Lesson 8-6) (3.4)

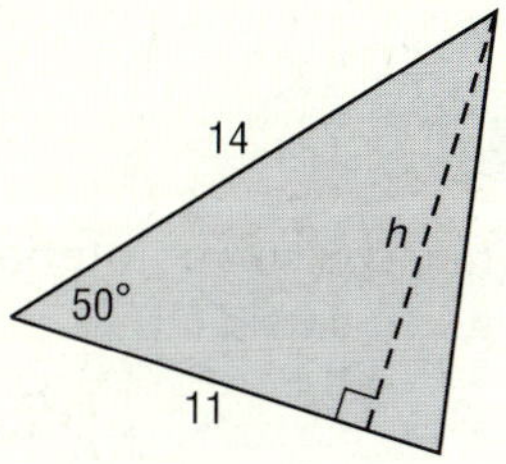

F about 51 square units
G about 54 square units
H about 59 square units
J about 62 square units

Wednesday

3 What is the value of x? (Lesson 8-3) (3.2)

A 10
B 11
C $11\sqrt{3}$
D 11R2

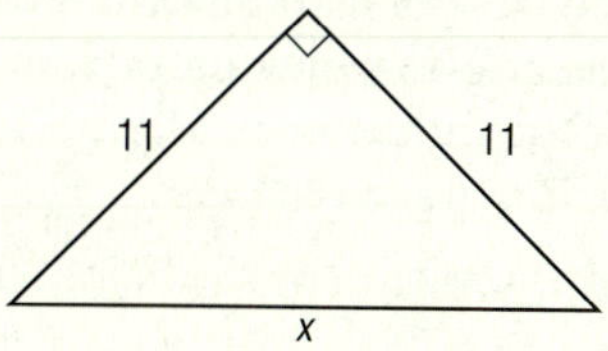

Thursday

4 Find sin A. (Lesson 8-4) (3.3)

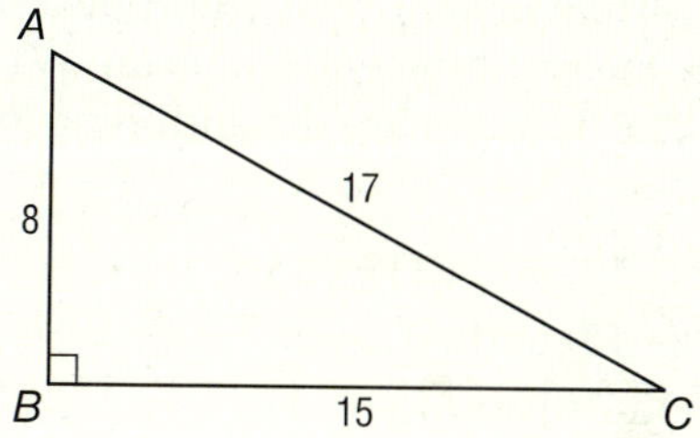

F $\frac{15}{17}$

G $\frac{8}{17}$

H $\frac{15}{8}$

J $\frac{8}{15}$

Friday

5 When a surveyor stands at a distance of 100 meters from the center of the base of the Oklahoma State Capitol building, the distance to the top of the rotunda is about 126.8 meters. What is the approximate height of the capitol building's rotunda? (Lesson 8-2) (3.1)

A 78 m
B 81 m
C 85 m
D 87 m

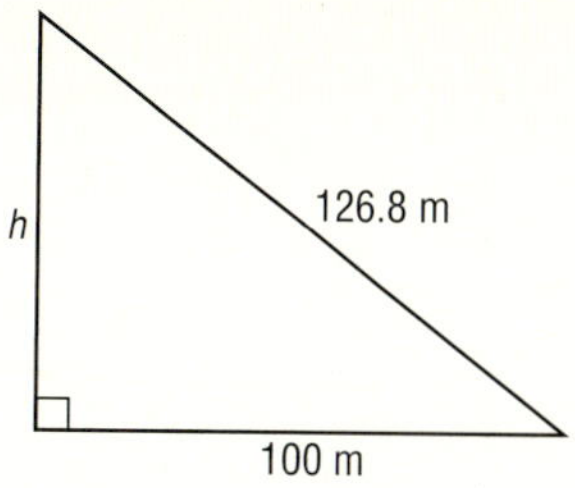

Week 7

Monday

1 A boat in a harbor is positioned 250 feet from the base of a 160-foot tall lighthouse. What is the angle of elevation from the boat to the top of the lighthouse? Round to the nearest tenth. (Lesson 8-5) (3.3)

A 29.5°
B 32.6°
C 37.1°
D 39.8°

Tuesday

2 If $\overline{WZ}$ is dilated about the origin by a scale factor of 1.25, the resulting endpoints are $W'(10, 15)$ and $Z'(25, 5)$. What are the endpoints of $\overline{WZ}$? (Lesson 7-6) (5.2.b)

F $W(6, 10)$ and $Z(18, 3)$
G $W(8, 12)$ and $Z(20, 4)$
H $W(12, 10)$ and $Z(16, 2)$
J $W(12.5, 18.75)$ and $Z(31.25, 6.25)$

Wednesday

3 What is the area of the triangle? Round to the nearest tenth. (Lesson 8-6) (3.4)

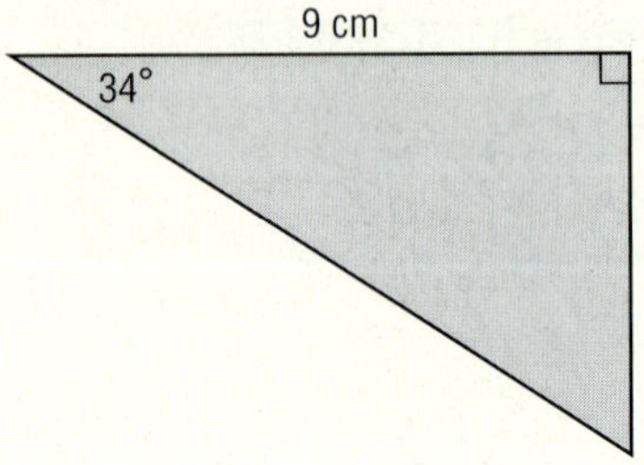

A 22.5 cm^2
B 24.9 cm^2
C 25.6 cm^2
D 27.3 cm^2

Thursday

4 What is the measure of an interior angle of a regular octagon? (Lesson 6-1) (2.3.b)

F 90°
G 120°
H 135°
J 150°

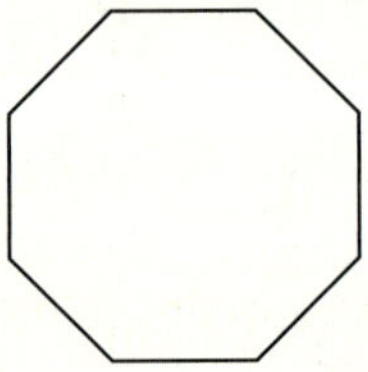

Friday

5 What is CD in the triangle shown? (Lesson 8-3) (3.2)

A 8
B $8\sqrt{3}$
C 16
D $16\sqrt{3}$

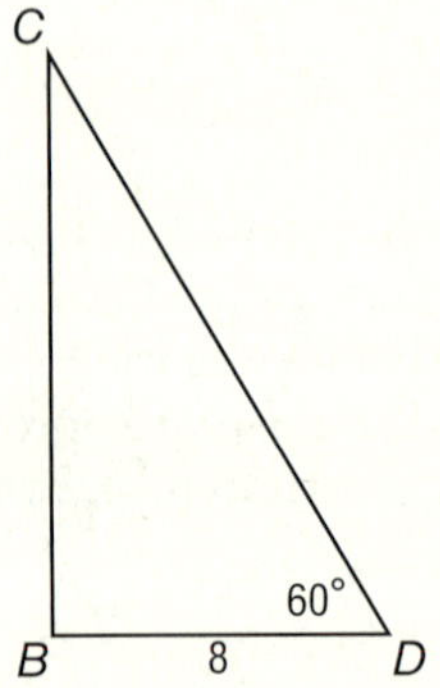

Week 6

Monday

1 How many lines of symmetry does the figure below have? (Lesson 9-5) (5.2.b)

A 0
B 1
C 2
D 4

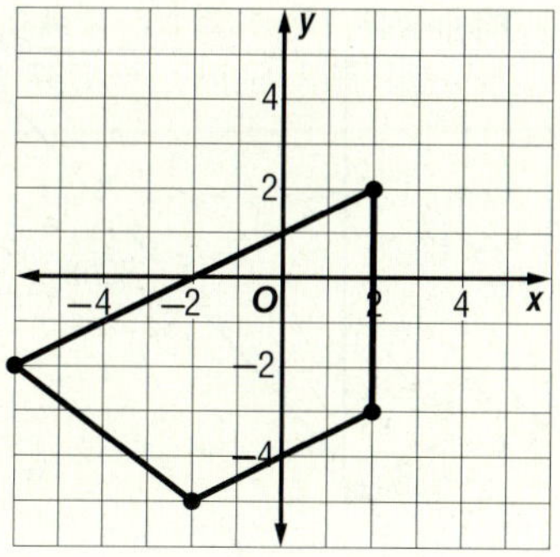

Tuesday

2 Suppose point $A(8, 5)$ is translated to create point $A'(2, 7)$. Which of the following best describes this translation? (Lesson 9-2) (5.2.b)

F 6 units right, 2 units down
G 2 units left, 6 units up
H 2 units right, 6 units down
J 6 units left, 2 units up

Wednesday

3 What is the magnitude of symmetry (or angle of rotation) of a regular pentagon? (Lesson 9-5) (5.2.b)

A 72°
B 108°
C 360°
D 540°

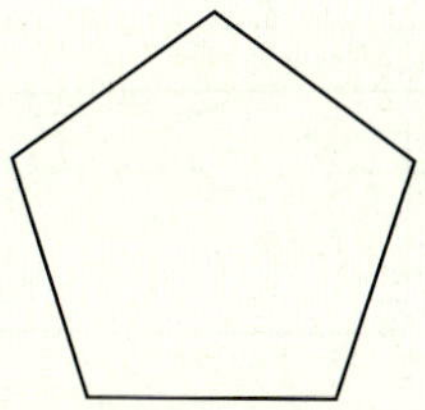

Thursday

4 For a given point (x, y), which of the following describes the image when the point is rotated 90° counterclockwise about the origin? (Lesson 9-3) (5.2.b)

F $(-x, -y)$
G $(-y, x)$
H $(y, -x)$
J (y, x)

Friday

5 Suppose Juanita constructs a triangle by reflecting triangle RST across the line $y = x$. Which of the following is <u>not</u> a vertex of the construction? (Lesson 9-1) (2.1)

A $(2, -4)$
B $(5, 2)$
C $(-1, 5)$
D $(4, -2)$

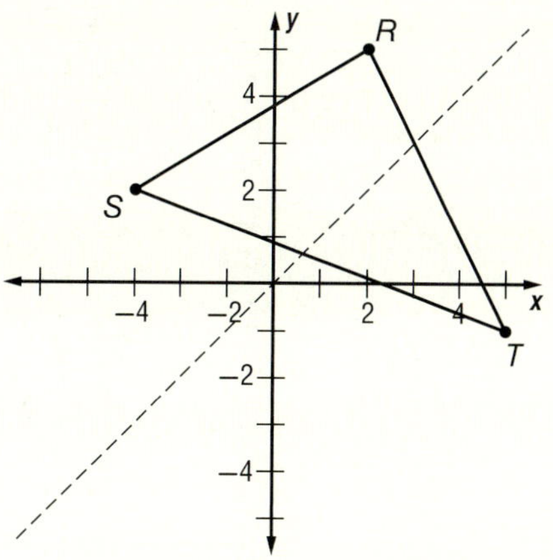

Week 5

Monday

1 What is the value of a in the right triangle below? Round to the nearest tenth if necessary. (Lesson 8-2) (3.1)

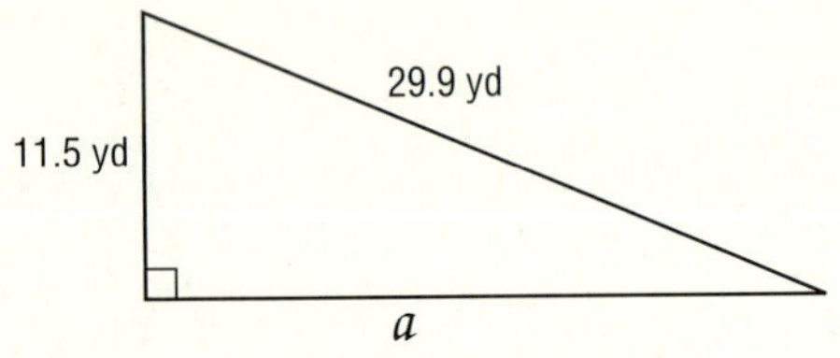

A 25.9 yd
B 27.6 yd
C 30.3 yd
D 32.0 yd

Tuesday

2 How much material is needed to cover a soccer ball with a diameter of 25 centimeters? Round to the nearest tenth. (Lesson 1-7) (4.1.b)

F 8,181.2 cm^3
G 6,518.2 cm^2
H 3,081.7 cm^2
J 1,963.5 cm^2

Wednesday

3 What is the approximate distance from Tulsa to Bartlesville along US-75? (Lesson 8-4) (3.3)

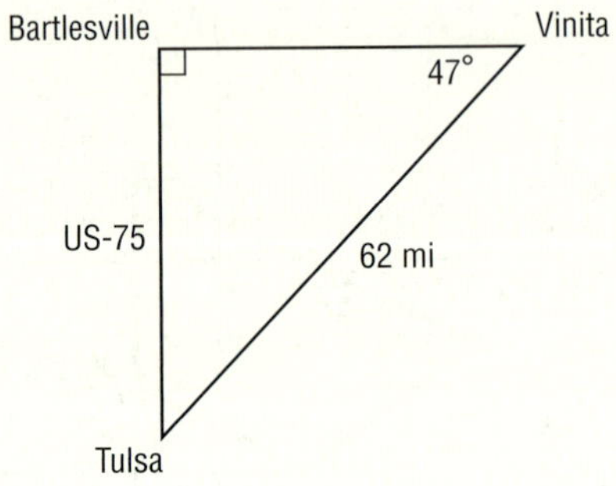

A 28 mi
B 34 mi
C 39 mi
D 45 mi

Thursday

4 Which statement is true? (Lesson 6-2) (2.3.c)

F All trapezoids are parallelograms.
G All kites are rhombi.
H All rectangles are parallelograms.
J All rectangles are squares.

Friday

5 What is the value of x given that $ABCD \sim GHJK$? (Lesson 7-2) (2.4.b)

A 4
B 6
C 7
D 9

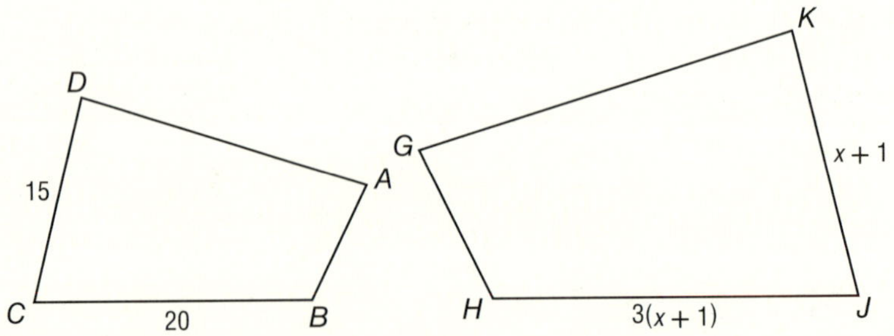

Week 4

Monday

1 Find the length of arc *MN*. Round to the nearest tenth. (Lesson 10-2) (2.6.a)

A 18.1 ft
B 21.8 ft
C 53.7 ft
D 81.3 ft

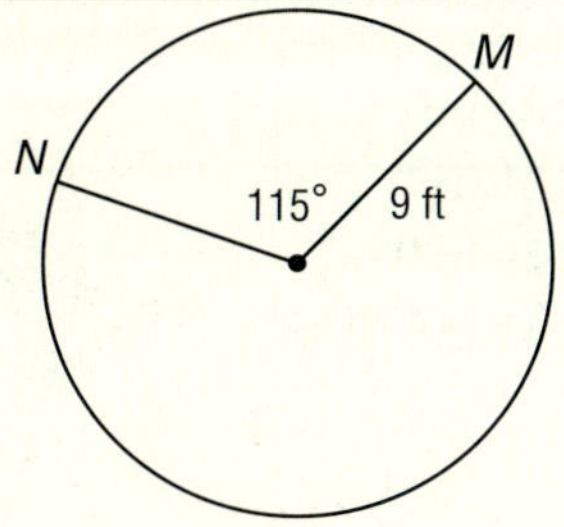

Tuesday

2 What is the measure of $\angle A$ if the measure of arc *BC* is 13° and the measure of arc *DE* is 61°? (Lesson 10-6) (2.6.b)

F 24°
G 29°
H 37°
J 48°

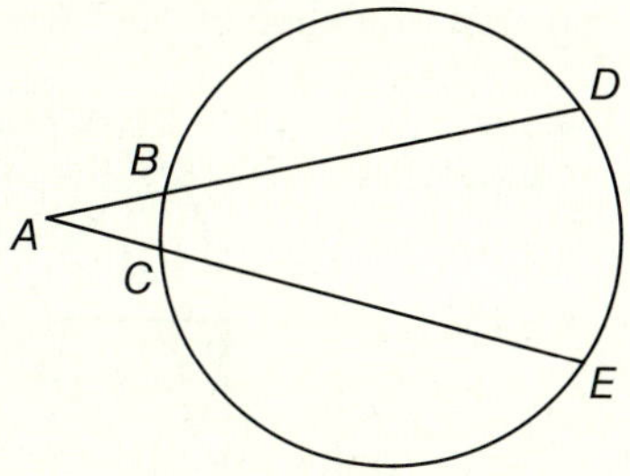

Wednesday

3 Find the value of *x*. (Lesson 10-3) (2.6.b)

A 11
B 12
C 13
D 14

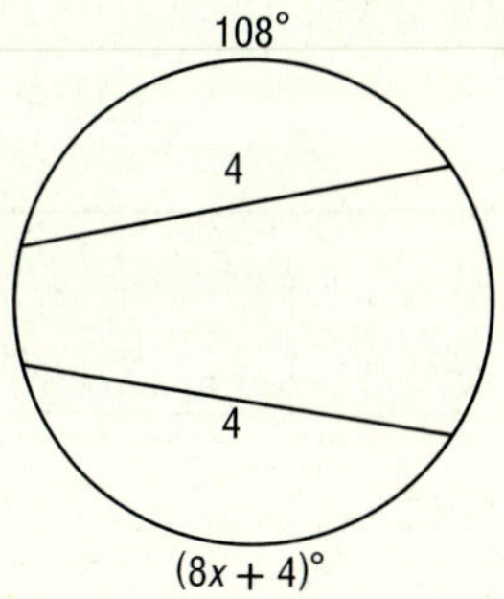

Thursday

4 Katie skates around the edge of a circular rink that is 228 feet in diameter. Her path intercepts a central angle of 105°. How far does Katie skate? Round to the nearest whole number. (Lesson 10-2) (2.6.a)

F 195 ft
G 209 ft
H 221 ft
J 236 ft

Friday

5 Tony cuts a pizza into 8 equal slices. If the diameter of the pizza is 14 inches, what is the arc length along the crust of each piece? Round to the nearest tenth. (Lesson 10-2) (2.6.a)

A 4.2 in.
B 4.7 in.
C 5.1 in.
D 5.5 in.

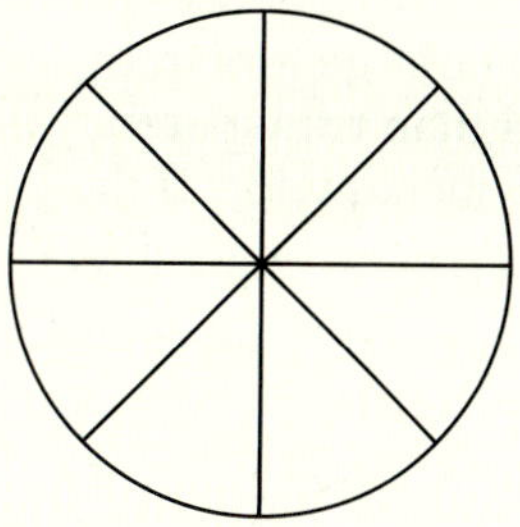

Week 3

Monday

1 What is the area of the parallelogram? (Lesson 11-1) (2.3.d)

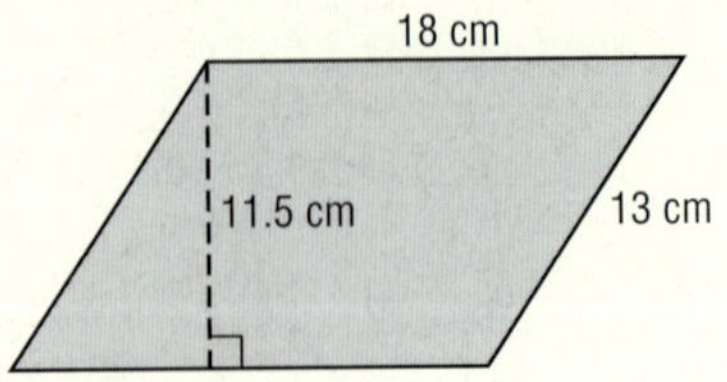

A 207 cm^2
B 185 cm^2
C 102 cm^2
D 51 cm

Tuesday

2 The area of a rhombus is 36 square feet. Which of the following could be the lengths of the diagonals of the rhombus? (Lesson 11-2) (2.3.d)

F 3 ft and 6 ft
G 4 ft and 9 ft
H 8 ft and 9 ft
J 9 ft and 16 ft

Wednesday

3 Fran built a kite with the dimensions shown. How much material did she use to make the kite? (Lesson 11-2) (2.3.d)

A 107 in^2
B 121 in^2
C 186 in^2
D 242 in^2

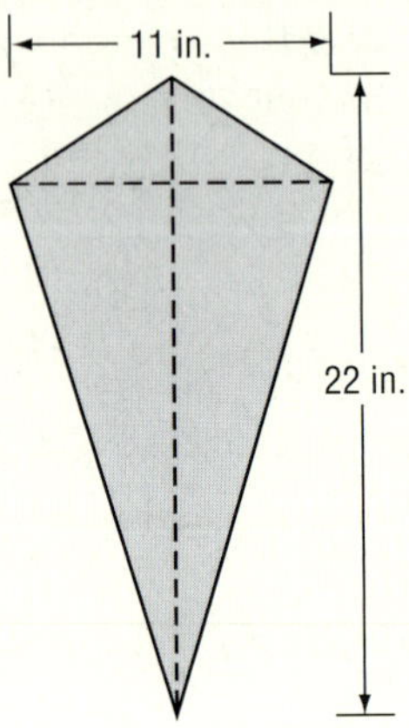

Thursday

4 Maria cuts an apple pie into 8 equal slices. What is the area of a slice of pie? (Lesson 11-3) (2.3.d)

F 4.6 in^2
G 5.8 in^2
H 6.1 in^2
J 6.3 in^2

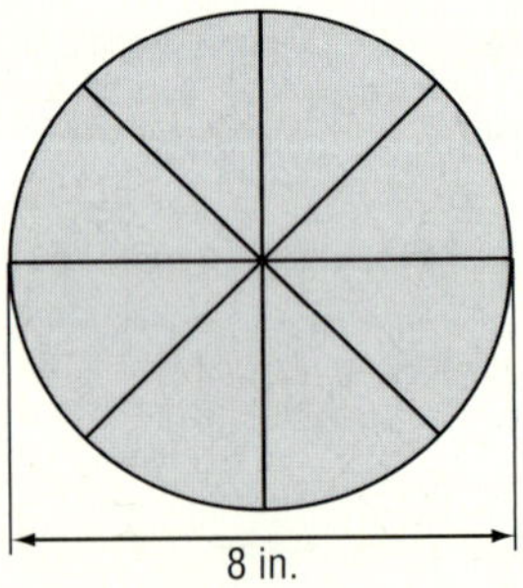

Friday

5 In her civics class, Mrs. Nichols uses an overhead projector to enlarge a map of Oklahoma. The actual map transparency is 11 inches long and 5.5 inches wide. If the area of the enlarged map is $15\frac{1}{8}$ square feet, what scale factor was applied to the length and width? (Lesson 11-5) (2.3.d)

A 5
B 5.5
C 6
D 6.25

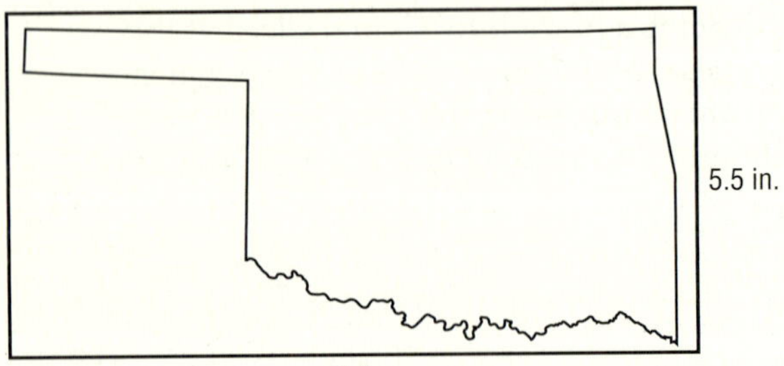

Week 2

Monday

1 Which of the following best describes a line in spherical geometry? (Lesson 12-7) (1.3)

A a diameter of the sphere
B a great circle on the sphere
C an arc of the sphere
D a radius of the sphere

Tuesday

2 If given a point in spherical geometry and a line through that point, how many lines are parallel to the line? (Lesson 12-7) (1.3)

F 0
G 1
H 2
J infinitely many

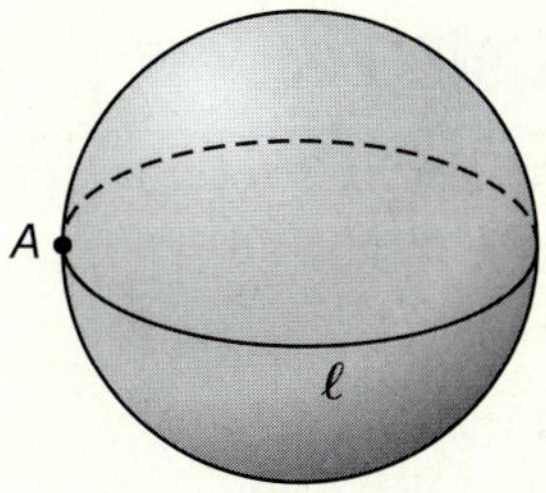

Wednesday

3 If a solid figure is scaled by a factor of 2, by what factor does the volume increase? (Lesson 12-8) (4.2.a)

A 0.5
B 2
C 4
D 8

Thursday

4 The two prisms below are congruent. What is the surface area of each prism? (Lesson 12-8) (4.2.b)

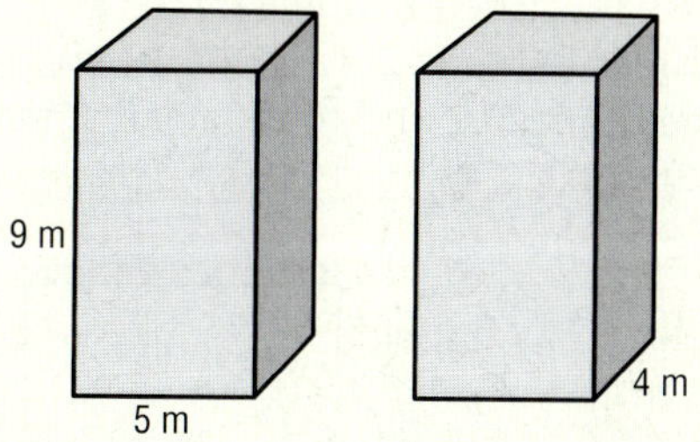

F 180 m^3
G 195 m^2
H 202 m^2
J 232 m^3

Friday

5 Which of the following is not a front, side, or top view of the block structure at the right? (Lesson 12-1) (4.3)

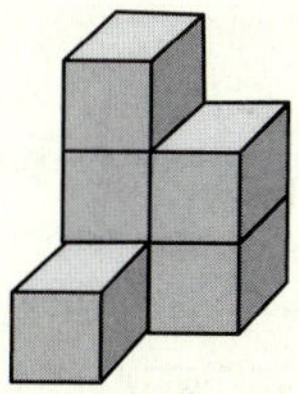

A 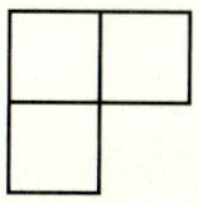

C 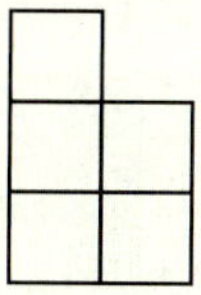

B 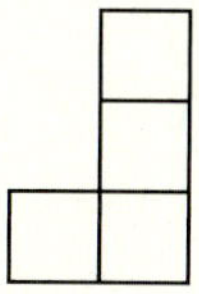

D 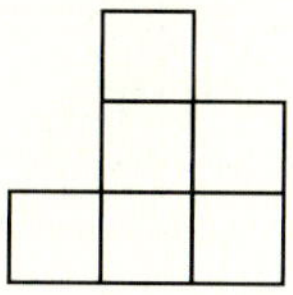

Week 1

Monday

1 The Golden Driller monument was built in Tulsa in 1953. It is the largest freestanding statue in the world. Enrique is making a model of the statue using the scale 1 inch = 12 feet. If Enrique's model is $6\frac{1}{3}$ inches tall, what is the actual height of the monument? (Lesson 12-8) (4.2.a)

A 76 ft
B 80 ft
C 88 ft
D 95 ft

Tuesday

2 A square is inscribed inside a circle with radius 12 feet. What is the area of the shaded part of the figure? Round to the nearest tenth. (Lesson 11-3) (2.3.d)

F 288 ft^2
G 194.5 ft^2
H 164.4 ft^2
J 147.3 ft^2

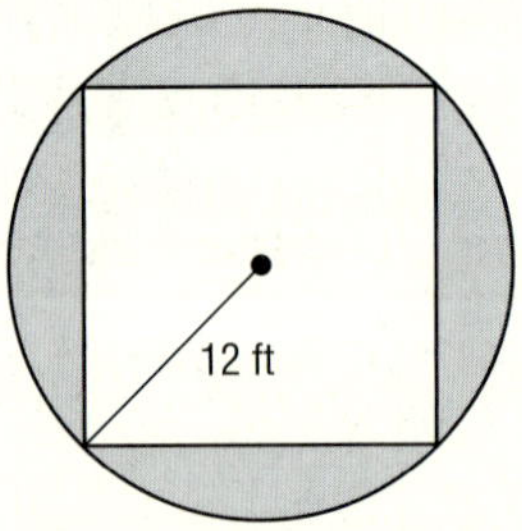

Wednesday

3 What is the area of the steel washer? Round your answer to the nearest whole number. (Lesson 11-4) (2.3.d)

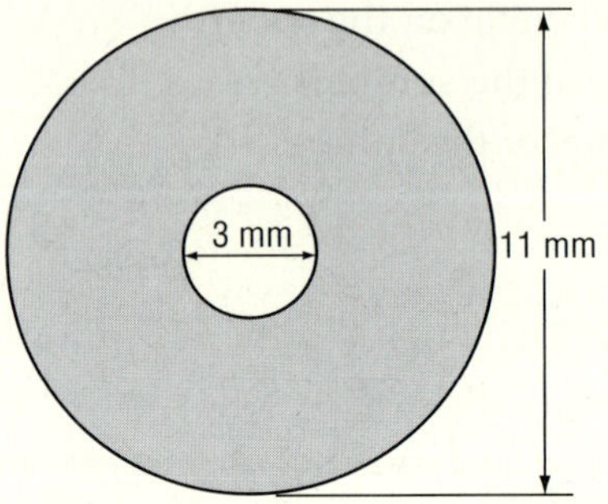

A 84 mm^2
B 88 mm^2
C 92 mm^2
D 95 mm^2

Thursday

4 What is the volume of the cone? (Lesson 12-5) (4.1.6)

F 60π cm^3
G 150π cm^3
H 225π cm^3
J 450π cm^3

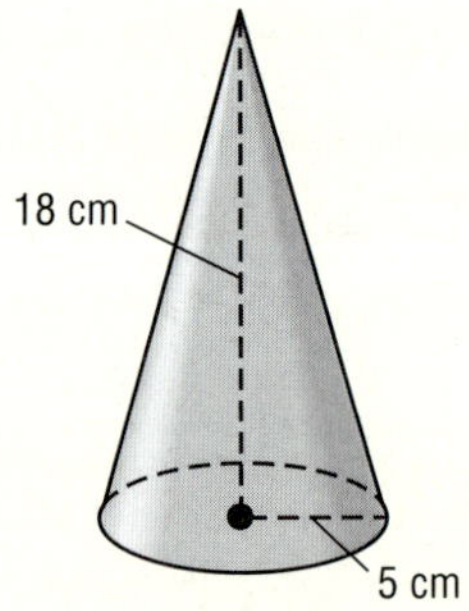

Friday

5 A farmer has two grain silos that are congruent. What is the combined volume of the silos? Round to the nearest cubic foot. (Lesson 12-8) (4.2.b)

A 42,569 ft^3
B 48,910 ft^3
C 52,182 ft^3
D 55,627 ft^3

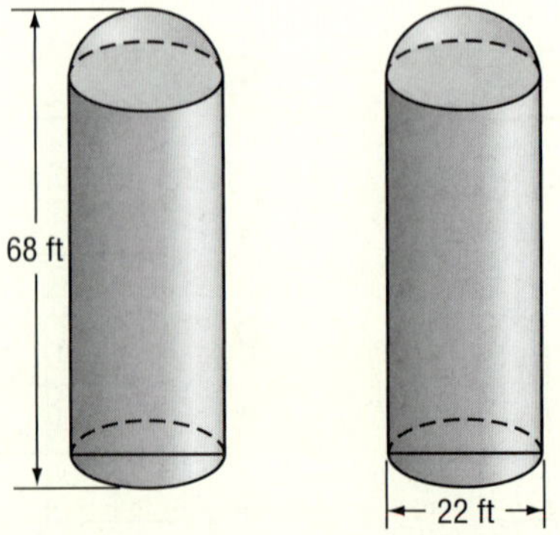

CHAPTER 0

Preparing for Geometry

Chapter 0 contains lessons on topics from previous courses. You can use this chapter in various ways.

- Begin the school year by taking the Pretest. If you need additional review, complete the lessons in this chapter. To verify that you have successfully reviewed the topics, take the Posttest.
- As you work through the text, you may find that there are topics you need to review. When this happens, complete the individual lessons that you need.
- Use this chapter for reference. When you have questions about any of these topics, flip back to this chapter to review definitions or key concepts.

Get Started on Chapter 0

You will review several concepts, skills, and vocabulary terms as you study Chapter 0. To get ready, identify important terms and organize your resources.

FOLDABLES® Study Organizer

Throughout this text, you will be invited to use Foldables to organize your notes.

Why should you use them?

- They help you organize, display, and arrange information.
- They make great study guides, specifically designed for you.
- You can use them as your math journal for recording main ideas, problem-solving strategies, examples, or questions you may have.
- They give you a chance to improve your math vocabulary.

How should you use them?

- Write general information–titles, vocabulary terms, concepts, questions, and main ideas–on the front tabs of your Foldable.
- Write specific information–ideas, your thoughts, answers to questions, steps, notes, and definitions–under the tabs.
- Use the tabs for:
  - math concepts in parts, like types of triangles,
  - steps to follow, or
  - parts of a problem, like *compare and contrast* (2 parts) or *what*, *where*, *when*, *why*, and *how* (5 parts).
- You may want to store your Foldables in a plastic zipper bag that you have three-hole punched to fit in your notebook.

When should you use them?

- Set up your Foldable as you begin a chapter, or when you start learning a new concept.
- Write in your Foldable every day.
- Use your Foldable to review for homework, quizzes, and tests.

New Vocabulary

English		Español
experiment	• p. P8 •	experimento
trial	• p. P8 •	recorrido de prueba
outcome	• p. P8 •	resultado
event	• p. P8 •	evento
probability	• p. P8 •	probabilidad
theoretical probability	• p. P9 •	probabilidad teórica
experimental probability	• p. P9 •	probababilidad experimental
ordered pair	• p. P15 •	par ordenado
x-coordinate	• p. P15 •	coordenada *x*
y-coordinate	• p. P15 •	coordenada *y*
quadrant	• p. P15 •	cuadrante
origin	• p. P15 •	origen
system of equations	• p. P17 •	sistema de ecuaciones
substitution	• p. P17 •	sustitución
elimination	• p. P18 •	eliminación
Product Property	• p. P19 •	Propriedad de Producto
Quotient Property	• p. P19 •	Propriedad de Cociente

Multilingual eGlossary glencoe.com

Math Online glencoe.com

- Study the chapter online
- Explore **Get Animated** activities
- Get extra help from your own **Personal Tutor**
- Use **Extra Examples** for additional help
- Take a **Self-Check Quiz**
- **Review Vocabulary** in fun ways

CHAPTER 0 Preparing for Geometry

State which metric unit you would probably use to measure each item.

1. length of a computer keyboard
2. mass of a large dog

Complete each sentence.

3. 4 ft = _?_ in.
4. 21 ft = _?_ yd
5. 180 g = _?_ kg
6. 3 T = _?_ lb
7. 32 g ≈ _?_ oz
8. 3 mi ≈ _?_ km
9. 35 yd ≈ _?_ m
10. 5.1 L ≈ _?_ qt

11. **TUNA** A can of tuna is 6 ounces. About how many grams is it?

12. **CRACKERS** A box of crackers is 453 grams. About how many pounds is it? Round to the nearest pound.

13. **DISTANCE** A road sign in Canada gives the distance to Toronto as 140 kilometers. What is this distance to the nearest mile?

PROBABILITY A bag contains 3 blue chips, 7 red chips, 4 yellow chips, and 5 green chips. A chip is randomly drawn from the bag. Find each probability.

14. *P*(yellow)
15. *P*(green)
16. *P*(red or blue)
17. *P*(not red)

Evaluate each expression if $r = 3$, $q = 1$, and $w = -2$.

18. $4r + q$
19. $rw - 6$
20. $\frac{r + 3q}{4r}$
21. $\frac{5w}{3r + q}$
22. $|2 - r| + 17$
23. $8 + |q - 5|$

Solve each equation.

24. $k + 3 = 14$
25. $a - 7 = 9$
26. $5c = 20$
27. $n + 2 = -11$
28. $6t - 18 = 30$
29. $4x + 7 = -1$
30. $\frac{r}{4} = -8$
31. $\frac{3}{5}b = -2$
32. $-\frac{w}{2} = -9$
33. $3y - 15 = y + 1$
34. $27 - 6d = 7 + 4d$
35. $2(m - 16) = 44$

Solve each inequality.

36. $y - 13 < 2$
37. $t + 8 \geq 19$
38. $\frac{n}{4} > -6$
39. $9a \leq 45$
40. $x + 12 > -14$
41. $-2w < 24$
42. $-\frac{n}{7} \geq 3$
43. $-\frac{b}{5} \leq -6$

Write the ordered pair for each point shown.

44. *F*
45. *H*
46. *A*
47. *D*

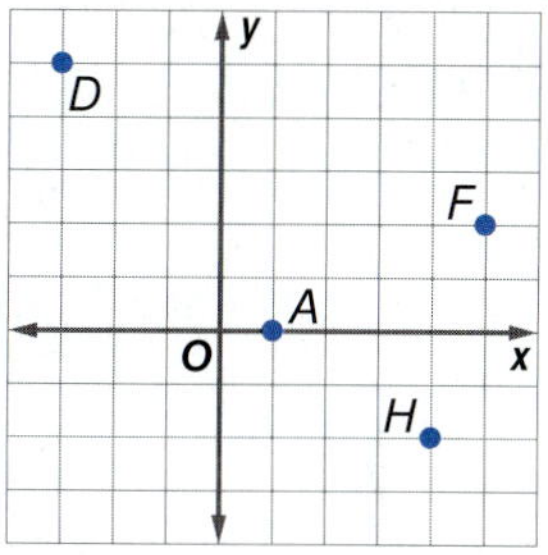

Graph and label each point on a coordinate plane.

48. $B(4, 1)$
49. $G(0, -3)$
50. $R(-2, -4)$
51. $P(-3, 3)$

52. Graph the triangle with vertices $J(1, -4)$, $K(2, 3)$, and $L(-1, 2)$.

53. Graph four points that satisfy the equation $y = 2x - 1$.

Solve each system of equations.

54. $y = 2x$
    $y = -x + 6$
55. $-3x - y = 4$
    $4x + 2y = -8$
56. $y = 2x + 1$
    $y = 3x$
57. $\frac{1}{2}x - y = -1$
    $x - 2y = 5$
58. $x + y = -6$
    $2x - y = 3$
59. $\frac{1}{3}x - 3y = -4$
    $x - 9y = -12$

Simplify.

60. $\sqrt{18}$
61. $\sqrt{\frac{25}{49}}$
62. $\sqrt{24x^2y^3}$
63. $\frac{3}{4 - \sqrt{5}}$

0-1 Changing Units of Measure Within Systems

Objective
Convert units of measure between the customary and metric systems.

Math Online
glencoe.com

EXAMPLE 1 Choose Best Unit of Measure

State which metric unit you would use to measure the length of your pen.

A pen has a small length, but not very small. The *centimeter* is the appropriate unit of measure.

Personal Tutor glencoe.com

Metric Units of Length
1 kilometer (km) = 1000 meters (m)
1 m = 100 centimeters (cm)
1 cm = 10 millimeters (mm)

Customary Units of Length
1 foot (ft) = 12 inches (in.)
1 yard (yd) = 3 ft
1 mile (mi) = 5280 ft

- To convert from larger units to smaller units, multiply.
- To convert from smaller units to larger units, divide.
- To use dimensional analysis, multiply by the ratio of the units.

EXAMPLE 2 Convert from Larger Units to Smaller Units of Length

Complete each sentence.

a. 4.2 km = __?__ m

There are 1000 meters in a kilometer.
4.2 km × 1000 = 4200 m

b. 13 yd = __?__ ft

There are 3 feet in a yard.
13 yd × 3 = 39 ft

Personal Tutor glencoe.com

EXAMPLE 3 Convert from Smaller Units to Larger Units of Length

Complete each sentence.

a. 17 mm = __?__ m

There are 100 centimeters in a meter. First change *millimeters* to *centimeters.*

17 mm = __?__ cm — **smaller unit → larger unit**
17 mm ÷ 10 = 1.7 cm — **Since 10 mm = 1 cm, divide by 10.**

Then change *centimeters* to *meters.*

1.7 cm = __?__ m — **smaller unit → larger unit**
1.7 cm ÷ 100 = 0.017 m — **Since 100 cm = 1 m, divide by 100.**

b. 6600 yd = __?__ mi

Use dimensional analysis.

$$6600\ \cancel{\text{yd}} \times \frac{3\ \cancel{\text{ft}}}{1\ \cancel{\text{yd}}} \times \frac{1\ \text{mi}}{5280\ \cancel{\text{ft}}} = 3.75\ \text{mi}$$

Personal Tutor glencoe.com

Metric Units of Capacity
1 liter (L) = 1000 milliliters (mL)

Customary Units of Capacity	
1 cup (c) = 8 fluid ounces (fl oz)	1 quart (qt) = 2 pt
1 pint (pt) = 2 c	1 gallon (gal) = 4 qt

EXAMPLE 4 Convert Units of Capacity

Complete each sentence.

a. **3.7 L = __?__ mL**

There are 1000 milliliters in a liter.

3.7 L × 1000 = 3700 mL

b. **16 qt = __?__ gal**

There are 4 quarts in a gallon.

16 qt ÷ 4 = 4 gal

c. **7 pt = __?__ fl oz**

There are 8 fluid ounces in a cup.

First change *pints* to *cups*.

7 pt = __?__ c

7 pt × 2 = 14 c

Then change *cups* to *fluid ounces*.

14 c = __?__ fl oz

14 c × 8 = 112 fl oz

d. **4 gal = __?__ pt**

There are 4 quarts in a gallon.

First change *gallons* to *quarts*.

4 gal = __?__ qt

4 gal × 4 = 16 qt

Then change *quarts* to *pints*.

16 qt = __?__ pt

16 qt × 2 = 32 pt

Personal Tutor glencoe.com

StudyTip

Dimensional Analysis You can use dimensional analysis for any conversion in this lesson.

The mass of an object is the amount of matter that it contains.

Metric Units of Mass
1 kilogram (kg) = 1000 grams (g)
1 g = 1000 milligrams (mg)

Customary Units of Weight
1 pound (lb) = 16 ounces (oz)
1 ton (T) = 2000 lb

EXAMPLE 5 Convert Units of Mass

Complete each sentence.

a. **5.47 kg = __?__ mg**

There are 1000 milligrams in a gram.

Change *kilograms* to *grams*.

5.47 kg = __?__ g

5.47 kg × 1000 = 5470 g

Then change *grams* to *milligrams*.

5470 g = __?__ mg

5470 g × 1000 = 5,470,000 mg

b. **5 T = __?__ oz**

There are 16 ounces in a pound.

Change *tons* to *pounds*.

5 T = __?__ lb

5 T × 2000 = 10,000 lb

Then change *pounds* to *ounces*.

10,000 lb = __?__ oz

10,000 lb × 16 = 160,000 oz

Personal Tutor glencoe.com

Exercises

State which metric unit you would probably use to measure each item.

1. radius of a tennis ball
2. length of a notebook
3. mass of a textbook
4. mass of a beach ball
5. liquid in a cup
6. water in a bathtub

Complete each sentence.

7. 120 in. = __?__ ft
8. 18 ft = __?__ yd
9. 10 km = __?__ m
10. 210 mm = __?__ cm
11. 180 mm = __?__ m
12. 3100 m = __?__ km
13. 90 in. = __?__ yd
14. 5280 yd = __?__ mi
15. 8 yd = __?__ ft
16. 0.62 km = __?__ m
17. 370 mL = __?__ L
18. 12 L = __?__ mL
19. 32 fl oz = __?__ c
20. 5 qt = __?__ c
21. 10 pt = __?__ qt
22. 48 c = __?__ gal
23. 4 gal = __?__ qt
24. 36 mg = __?__ g
25. 13 lb = __?__ oz
26. 130 g = __?__ kg
27. 9.05 kg = __?__ g

0-2 Changing Units of Measure Between Systems

Objective
Convert units of measure between the customary and metric systems.

Math Online
glencoe.com

The table below shows approximate equivalents between customary units of length and metric units of length.

Units of Length	
Customary → Metric	Metric → Customary
1 in. ≈ 2.5 cm	1 cm ≈ 0.4 in.
1 yd ≈ 0.9 m	1 m ≈ 1.1 yd
1 mi ≈ 1.6 km	1 km ≈ 0.6 mi

EXAMPLE 1 Convert Units of Length Between Systems

Complete each sentence.

a. 30 in. ≈ __?__ cm

There are approximately 2.5 centimeters in an inch.

30 in. × 2.5 = 75 cm

b. 5 km ≈ __?__ mi

There is approximately 0.6 mile in a kilometer.

5 km × 0.6 = 3 mi

Personal Tutor glencoe.com

EXAMPLE 2 Convert Units of Length Between Systems

Complete: 2000 yd ≈ __?__ km.

There is approximately 0.9 meter in a yard. First find the number of meters in 2000 yards.

2000 yd × 0.9 = 1800 m

Then change *meters* to *kilometers*. There are 1000 meters in a kilometer.

1800 m ÷ 1000 = 1.8 km

Personal Tutor glencoe.com

The table below shows approximate equivalents between customary units of capacity and metric units of capacity.

Units of Capacity	
Customary → Metric	Metric → Customary
1 qt ≈ 0.9 L	1 L ≈ 1.1 qt
1 pt ≈ 0.5 L	1 L ≈ 2.1 pt

EXAMPLE 3 Convert Units of Capacity Between Systems

Complete each sentence.

a. 7 qt ≈ __?__ L

There is approximately 0.9 liter in a quart.

7 qt × 0.9 = 6.3 L

b. 2 L ≈ __?__ pt

There are approximately 2.1 pints in a liter.

2 L × 2.1 = 4.2 pt

Personal Tutor glencoe.com

EXAMPLE 4 Convert Units of Capacity Between Systems

Complete: 10 L ≈ __?__ gal.

There are approximately 1.1 quarts in a liter. First find the number of quarts in 10 liters.

$10 \text{ L} \times 1.1 = 11 \text{ qt}$

Then change *quarts* to *gallons*. There are 4 quarts in a gallon.

$11 \text{ qt} \div 4 = 2.75 \text{ gal}$

You can also use dimensional analysis.

$$10 \cancel{\text{L}} \times \frac{1.1 \cancel{\text{qt}}}{1 \cancel{\text{L}}} \times \frac{1 \text{ gal}}{4 \cancel{\text{qt}}} = 2.75 \text{ gal}$$

Personal Tutor glencoe.com

StudyTip

Dimensional Analysis If the unit that you want to eliminate is in the numerator, make sure it is in the denominator of the ratio when you multiply. If it is in the denominator, make sure that it is in the numerator of the ratio.

The table below shows approximate equivalents between customary units of weight and metric units of mass.

Units of Weight/Mass	
Customary → Metric	Metric → Customary
1 oz ≈ 28.3 g	1 g ≈ 0.04 oz
1 lb ≈ 0.5 kg	1 kg ≈ 2.2 lb

EXAMPLE 5 Convert Units of Mass Between Systems

Complete each sentence.

a. 58.5 kg ≈ __?__ lb

There are approximately 2.2 pounds in a kilogram.

$58.5 \text{ kg} \times 2.2 = 128.7 \text{ lb}$

b. 14 oz ≈ __?__ g

There are approximately 28.3 grams in an ounce.

$14 \text{ oz} \times 28.3 = 396.2 \text{ g}$

Personal Tutor glencoe.com

Exercises

Complete each sentence.

1. 8 in. ≈ __?__ cm

2. 15 m ≈ __?__ yd

3. 11 qt ≈ __?__ L

4. 25 oz ≈ __?__ g

5. 10 mi ≈ __?__ km

6. 32 cm ≈ __?__ in.

7. 20 km ≈ __?__ mi

8. 9.5 L ≈ __?__ qt

9. 6 yd ≈ __?__ m

10. 4.3 kg ≈ __?__ lb

11. 10.7 L ≈ __?__ pt

12. 82.5 g ≈ __?__ oz

13. $2\frac{1}{4}$ lb ≈ __?__ kg

14. 10 ft ≈ __?__ m

15. $1\frac{1}{2}$ gal ≈ __?__ L

16. 350 g ≈ __?__ lb

17. 600 in. ≈ __?__ m

18. 2.1 km ≈ __?__ yd

19. **CEREAL** A box of cereal is 13 ounces. About how many grams is it?

20. **FLOUR** A bag of flour is 2.26 kilograms. How much does it weigh? Round to the nearest pound.

21. **SAUCE** A jar of tomato sauce is 1 pound 10 ounces. About how many grams is it?

0-3

Simple Probability

Objective
Find the probability of simple events.

New Vocabulary
experiment
trial
outcome
event
probability
theoretical probability
experimental probability

Math Online
glencoe.com

A situation involving chance such as flipping a coin or rolling a die is an **experiment**. A single performance of an experiment such as rolling a die one time is a **trial**. The result of a trial is called an **outcome**. An **event** is one or more outcomes of an experiment.

When each outcome is equally likely to happen, the **probability** of an event is the ratio of the number of favorable outcomes to the number of possible outcomes. The probability of an event is always between 0 and 1, inclusive.

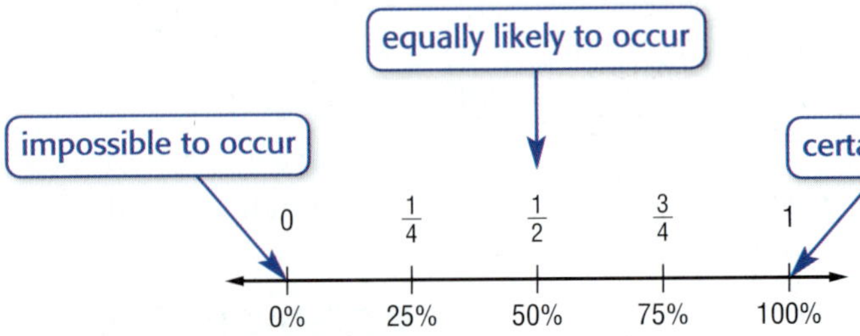

EXAMPLE 1 Find Probability

Suppose a die is rolled. What is the probability of rolling an odd number?

There are 3 odd numbers on a die: 1, 3, and 5.

There are 6 possible outcomes: 1, 2, 3, 4, 5, and 6.

$$P(\text{odd}) = \frac{\text{number of favorable outcomes}}{\text{number of possible outcomes}}$$

$$= \frac{3}{6} \text{ or } \frac{1}{2}$$

The probability of rolling an odd number is $\frac{1}{2}$ or 50%.

Personal Tutor glencoe.com

For a given experiment, the sum of the probabilities of all possible outcomes must sum to 1.

EXAMPLE 2 Find Probability

Suppose a bag contains 4 red, 3 green, 6 blue, and 2 yellow marbles. What is the probability a randomly chosen marble will not be yellow?

Since the sum of the probabilities of all of the colors must sum to 1, subtract the probability that the marble will be yellow from 1.

The probability that the marble will be yellow is $\frac{2}{15}$ because there are 2 yellow marbles and 15 total marbles.

$$P(\text{not yellow}) = 1 - P(\text{yellow})$$

$$= 1 - \frac{2}{15}$$

$$= \frac{13}{15}$$

The probability that the marble will not be yellow is $\frac{13}{15}$ or about 87%.

Personal Tutor glencoe.com

The probabilities in Examples 1 and 2 are called theoretical probabilities. The **theoretical probability** is what *should* occur. The **experimental probability** is what *actually* occurs when a probability experiment is repeated many times.

EXAMPLE 3 Find Experimental Probability

The table shows the results of an experiment in which a number cube was rolled. Find the experimental probability of rolling a 3.

Outcome	Tally	Frequency
1	𝍸𝍷	6
2	𝍷𝍷𝍷𝍷	4
3	𝍸𝍷𝍷	7
4	𝍷𝍷𝍷	3
5	𝍷𝍷𝍷𝍷	4
6	𝍷	1

$$P(3) = \frac{\text{number of times 3 occurs}}{\text{total number of outcomes}}$$

$$= \frac{7}{25}$$

The experimental probability for getting a 3 in this case is $\frac{7}{25}$ or 28%.

Personal Tutor glencoe.com

StudyTip

Experimental Probability The experimental probability of an experiment is not necessarily the same as the theoretical probability, but when an experiment is repeated many times, the experimental probability should be close to the theoretical probability.

Exercises

A die is rolled. Find the probability of each outcome.

1. P(less than 3)

2. P(even)

3. P(greater than 2)

4. P(prime)

5. P(4 or 2)

6. P(integer)

A jar contains 65 pennies, 27 nickels, 30 dimes, and 18 quarters. A coin is randomly selected from the jar. Find each probability.

7. P(penny)

8. P(quarter)

9. P(not dime)

10. P(penny or dime)

11. P(value greater than \$0.15)

12. P(not nickel)

13. P(nickel or quarter)

14. P(value less than \$0.20)

PRESENTATIONS **The students in a class are randomly drawing cards numbered 1 through 28 from a hat to determine the order in which they will give their presentations. Find each probability.**

15. P(13)

16. P(1 or 28)

17. P(less than 14)

18. P(not 1)

19. P(not 2 or 17)

20. P(greater than 16)

The table shows the results of an experiment in which three coins were tossed.

Outcome	HHH	HHT	HTH	THH	TTH	THT	HTT	TTT
Tally	𝍸	𝍸	𝍸𝍷	𝍸𝍷	𝍸𝍷𝍷	𝍸	𝍸𝍷𝍷𝍷	𝍸𝍷𝍷𝍷
Frequency	5	5	6	6	7	5	8	8

21. What is the experimental probability that all three of the coins will be heads? The theoretical probability?

22. What is the experimental probability that at least two of the coins will be heads? The theoretical probability?

23. What is the experimental probability that exactly two of the coins will be tails? The theoretical probability?

0-4 Algebraic Expressions

Objective Use the order of operations to evaluate algebraic expressions.

Math Online glencoe.com

An expression is an algebraic expression if it contains sums and/or products of variables and numbers. To evaluate an algebraic expression, replace the variable or variables with known values, and then use the order of operations.

Order of Operations	
Step 1	Evaluate expressions inside grouping symbols.
Step 2	Evaluate all powers.
Step 3	Do all multiplications and/or divisions from left to right.
Step 4	Do all additions and/or subtractions from left to right.

EXAMPLE 1 Addition/Subtraction Algebraic Expressions

Evaluate $x - 5 + y$ if $x = 15$ and $y = -7$.

$x - 5 + y = 15 - 5 + (-7)$ **Substitute.**

$= 10 + (-7)$ **Subtract.**

$= 3$ **Add.**

Personal Tutor glencoe.com

EXAMPLE 2 Multiplication/Division Algebraic Expressions

Evaluate each expression if $k = -2$, $n = -4$, and $p = 5$.

a. $\dfrac{2k + n}{p - 3}$

$\dfrac{2k + n}{p - 3} = \dfrac{2(-2) + (-4)}{5 - 3}$ **Substitute.**

$= \dfrac{-4 - 4}{5 - 3}$ **Multiply.**

$= \dfrac{-8}{2}$ or -4 **Subtract.**

b. $-3(k^2 + 2n)$

$-3(k^2 + 2n) = -3[(-2)^2 + 2(-4)]$

$= -3[4 + (-8)]$

$= -3(-4)$ or 12

Personal Tutor glencoe.com

EXAMPLE 3 Absolute Value Algebraic Expressions

Evaluate $3|a - b| + 2|c - 5|$ if $a = -2$, $b = -4$, and $c = 3$.

$3|a - b| + 2|c - 5| = 3|-2 - (-4)| + 2|3 - 5|$ **Substitute for *a*, *b*, and *c*.**

$= 3|2| + 2|-2|$ **Simplify.**

$= 3(2) + 2(2)$ **Find absolute values.**

$= 10$ **Simplify.**

Personal Tutor glencoe.com

Exercises

Evaluate each expression if $a = 2$, $b = -3$, $c = -1$, and $d = 4$.

1. $2a + c$

2. $\dfrac{bd}{2c}$

3. $\dfrac{2d - a}{b}$

4. $3d - c$

5. $\dfrac{3b}{5a + c}$

6. $5bc$

7. $2cd + 3ab$

8. $\dfrac{c - 2d}{a}$

Evaluate each expression if $x = 2$, $y = -3$, and $z = 1$.

9. $24 + |x - 4|$

10. $13 + |8 + y|$

11. $|5 - z| + 11$

12. $|2y - 15| + 7$

0-5 Linear Equations

Objective
Use algebra to solve linear equations.

Math Online
glencoe.com

If the same number is added to or subtracted from each side of an equation, the resulting equation is true.

EXAMPLE 1 Addition/Subtraction Linear Equations

Solve each equation.

a. $x - 7 = 16$

$x - 7 = 16$	Original equation
$x - 7 + 7 = 16 + 7$	Add 7 to each side.
$x = 23$	Simplify.

b. $m + 12 = -5$

$m + 12 = -5$	Original equation
$m + 12 + (-12) = -5 + (-12)$	Add -12 to each side.
$m = -17$	Simplify.

c. $k + 31 = 10$

$k + 31 = 10$	Original equation
$k + 31 - 31 = 10 - 31$	Subtract 31 from each side.
$k = -21$	Simplify.

Personal Tutor glencoe.com

If each side of an equation is multiplied or divided by the same number, the resulting equation is true.

EXAMPLE 2 Multiplication/Division Linear Equations

Solve each equation.

a. $4d = 36$

$4d = 36$	Original equation
$\frac{4d}{4} = \frac{36}{4}$	Divide each side by 4.
$x = 9$	Simplify.

b. $-\frac{t}{8} = -7$

$-\frac{t}{8} = -7$	Original equation
$-8\left(-\frac{t}{8}\right) = -8(-7)$	Multiply each side by -8.
$t = 56$	Simplify.

c. $\frac{3}{5}x = -8$

$\frac{3}{5}x = -8$	Original equation
$\frac{5}{3}\left(\frac{3}{5}\right)x = \frac{5}{3}(-8)$	Multiply each side by $\frac{5}{3}$.
$x = -\frac{40}{3}$	Simplify.

Personal Tutor glencoe.com

To solve equations with more than one operation, often called *multi-step equations*, undo operations by working backward.

EXAMPLE 3 Multi-step Linear Equations

Solve each equation.

a. $8q - 15 = 49$

$8q - 15 = 49$	Original equation
$8q = 64$	Add 15 to each side.
$q = 8$	Divide each side by 8.

b. $12y + 8 = 6y - 5$

$12y + 8 = 6y - 5$	Original equation
$12y = 6y - 13$	Subtract 8 from each side.
$6y = -13$	Subtract $6y$ from each side.
$y = -\frac{13}{6}$	Divide each side by 6.

Personal Tutor glencoe.com

Watch Out!

Order of Operations Remember that the order of operations applies when you are solving linear equations.

When solving equations that contain grouping symbols, first use the Distributive Property to remove the grouping symbols.

EXAMPLE 4 Multi-step Linear Equations

Solve $3(x - 5) = 13$.

$3(x - 5) = 13$	Original equation
$3x - 15 = 13$	Distributive Property
$3x = 28$	Add 15 to each side.
$x = \frac{28}{3}$	Divide each side by 3.

Personal Tutor glencoe.com

Exercises

Solve each equation.

1. $r + 11 = 3$
2. $n + 7 = 13$
3. $d - 7 = 8$
4. $\frac{8}{5}a = -6$
5. $-\frac{p}{12} = 6$
6. $\frac{x}{4} = 8$
7. $\frac{12}{5}f = -18$
8. $\frac{y}{7} = -11$
9. $\frac{6}{7}y = 3$
10. $c - 14 = -11$
11. $t - 14 = -29$
12. $p - 21 = 52$
13. $b + 2 = -5$
14. $q + 10 = 22$
15. $-12q = 84$
16. $5t = 30$
17. $5c - 7 = 8c - 4$
18. $2\ell + 6 = 6\ell - 10$
19. $\frac{m}{10} + 15 = 21$
20. $-\frac{m}{8} + 7 = 5$
21. $8t + 1 = 3t - 19$
22. $9n + 4 = 5n + 18$
23. $5c - 24 = -4$
24. $3n + 7 = 28$
25. $-2y + 17 = -13$
26. $-\frac{t}{13} - 2 = 3$
27. $\frac{2}{9}x - 4 = \frac{2}{3}$
28. $9 - 4g = -15$
29. $-4 - p = -2$
30. $21 - b = 11$
31. $-2(n + 7) = 15$
32. $5(m - 1) = -25$
33. $-8a - 11 = 37$
34. $\frac{7}{4}q - 2 = -5$
35. $2(5 - n) = 8$
36. $-3(d - 7) = 6$

0-6 Linear Inequalities

Objective
Use algebra to solve linear inequalities.

Math Online
glencoe.com

Statements with greater than (>), less than (<), greater than or equal to (≥), or less than or equal to (≤) are inequalities.

If any number is added or subtracted to each side of an inequality, the resulting inequality is true.

EXAMPLE 1 Addition/Subtraction Linear Inequalities

Solve each inequality.

a. $x - 17 > 12$

$x - 17 > 12$	Original inequality
$x - 17 + 17 > 12 + 17$	Add 17 to each side.
$x > 29$	Simplify.

The solution set is $\{x \mid x > 29\}$.

b. $y + 11 \leq 5$

$y + 11 \leq 5$	Original inequality
$y + 11 - 11 \leq 5 - 11$	Subtract 11 from each side.
$y \leq -6$	Simplify.

The solution set is $\{y \mid y \leq -6\}$.

Personal Tutor glencoe.com

If each side of an inequality is multiplied or divided by a positive number, the resulting inequality is true.

EXAMPLE 2 Multiplication/Division Linear Inequalities

Solve each inequality.

a. $\frac{t}{6} \geq 11$

$\frac{t}{6} \geq 11$	Original inequality
$(6)\frac{t}{6} \geq (6)11$	Multiply each side by 6.
$t \geq 66$	Simplify.

The solution set is $\{t \mid t \geq 66\}$.

b. $8p < 72$

$8p < 72$	Original inequality
$\frac{8p}{8} < \frac{72}{8}$	Divide each side by 8.
$p < 9$	Simplify.

The solution set is $\{p \mid p < 9\}$.

Personal Tutor glencoe.com

If each side of an inequality is multiplied or divided by the same negative number, the direction of the inequality symbol must be *reversed* so that the resulting inequality is true.

EXAMPLE 3 Multiplication/Division Linear Inequalities

Solve each inequality.

a. $-5c > 30$

$-5c > 30$	Original inequality
$\frac{-5c}{-5} < \frac{30}{-5}$	Divide each side by −5. Change > to <.
$c < -6$	Simplify.

The solution set is $\{c \mid c < -6\}$.

(continued on the next page)

b. $-\frac{d}{13} \leq -4$

$-\frac{d}{13} \leq -4$	Original inequality
$(-13)\left(\frac{-d}{13}\right) \geq (-13)(-4)$	Multiply each side by −13. Change ≤ to ≥.
$d \geq 52$	Simplify.

The solution set is $\{d \mid d \geq 52\}$.

Personal Tutor glencoe.com

Watch Out!

Dividing by a Negative Remember that any time you divide an inequality by a negative number you reverse the direction of the sign.

Inequalities involving more than one operation can be solved by undoing the operations in the same way you would solve an equation with more than one operation.

EXAMPLE 4 Multi-Step Linear Inequalities

Solve each inequality.

a. $-6a + 13 < -7$

$-6a + 13 < -7$	Original inequality
$-6a + 13 - 13 < -7 - 13$	Subtract 13 from each side.
$-6a < -20$	Simplify.
$\frac{-6a}{-6} > \frac{-20}{-6}$	Divide each side by −6. Change < to >.
$a > \frac{10}{3}$	Simplify.

The solution set is $\left\{a \mid a > \frac{10}{3}\right\}$.

b. $4z + 7 \geq 8z - 1$

$4z + 7 \geq 8z - 1$	Original inequality
$4z + 7 - 7 \geq 8z - 1 - 7$	Subtract 7 from each side.
$4z \geq 8z - 8$	Simplify.
$4z - 8z \geq 8z - 8 - 8z$	Subtract 8z from each side.
$-4z \geq -8$	Simplify.
$\frac{-4z}{-4} \leq \frac{-8}{-4}$	Divide each side by −4. Change ≥ to ≤.
$z \leq 2$	Simplify.

The solution set is $\{z \mid z \leq 2\}$.

Personal Tutor glencoe.com

Exercises

1. $x - 7 < 6$
2. $a + 7 \geq -5$
3. $4y < 20$
4. $-\frac{a}{8} < 5$
5. $\frac{t}{6} > -7$
6. $\frac{a}{11} \leq 8$
7. $d + 8 \leq 12$
8. $m + 14 > 10$
9. $12k \geq -36$
10. $6t - 10 \geq 4t$
11. $3z + 8 < 2$
12. $4c + 23 \leq -13$
13. $m - 21 < 8$
14. $x - 6 \geq 3$
15. $-3b \leq 48$
16. $-\frac{p}{5} \geq 14$
17. $2z - 9 < 7z + 1$
18. $-4h > 36$
19. $\frac{2}{5}b - 6 \leq -2$
20. $\frac{8}{3}t + 1 > -5$
21. $7q + 3 \geq -4q + 25$
22. $-3n - 8 > 2n + 7$
23. $-3w + 1 \leq 8$
24. $-\frac{4}{5}k - 17 > 11$

0-7 Ordered Pairs

Objective
Name and graph points in the coordinate plane.

New Vocabulary
ordered pair
x-coordinate
y-coordinate
quadrant
origin

Math Online
glencoe.com

Points in the coordinate plane are named by **ordered pairs** of the form (x, y). The first number, or **x-coordinate**, corresponds to a number on the x-axis. The second number, or **y-coordinate**, corresponds to a number on the y-axis.

EXAMPLE 1 Writing Ordered Pairs

Write the ordered pair for each point.

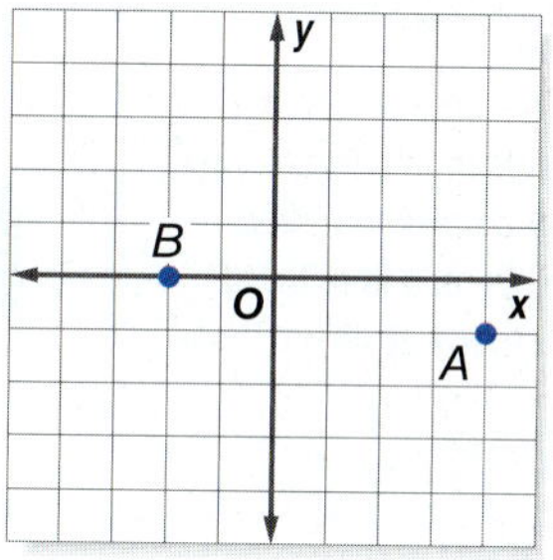

a. A

The x-coordinate is 4.

The y-coordinate is -1.

The ordered pair is $(4, -1)$.

b. B

The x-coordinate is -2.

The point lies on the x-axis, so its y-coordinate is 0.

The ordered pair is $(-2, 0)$.

Personal Tutor glencoe.com

The x-axis and y-axis separate the coordinate plane into four regions, called **quadrants**. The point at which the axes intersect is called the **origin**. The axes and points on the axes are not located in any of the quadrants.

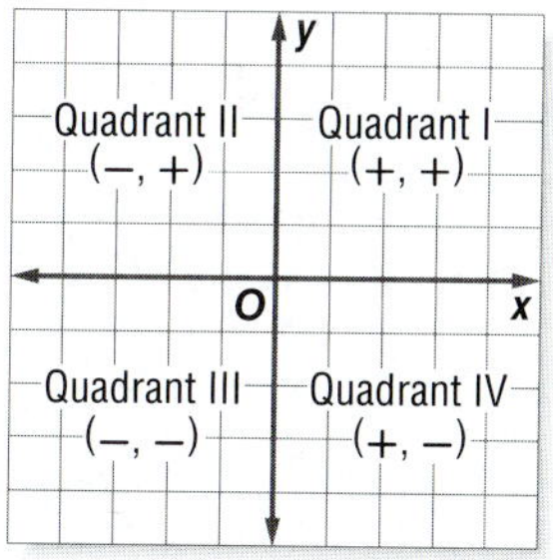

EXAMPLE 2 Graphing Ordered Pairs

Graph and label each point on a coordinate plane. Name the quadrant in which each point is located.

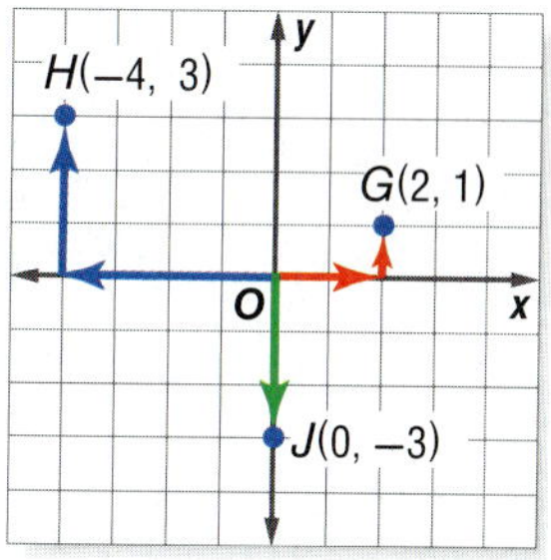

a. $G(2, 1)$

Start at the origin. Move 2 units right, since the x-coordinate is 2. Then move 1 unit up, since the y-coordinate is 1. Draw a dot, and label it G. Point $G(2, 1)$ is in Quadrant I.

b. $H(-4, 3)$

Start at the origin. Move 4 units left, since the x-coordinate is -4. Then move 3 units up, since the y-coordinate is 3. Draw a dot, and label it H. Point $H(-4, 3)$ is in Quadrant II.

c. $J(0, -3)$

Start at the origin. Since the x-coordinate is 0, the point lies on the y-axis. Move 3 units down, since the y-coordinate is -3. Draw a dot, and label it J. Because it is on one of the axes, point $J(0, -3)$ is not in any quadrant.

Personal Tutor glencoe.com

EXAMPLE 3 Graphing Multiple Ordered Pairs

Graph a polygon with vertices $A(-3, 3)$, $B(1, 3)$, $C(0, 1)$, and $D(-4, 1)$.

Graph the ordered pairs on a coordinate plane. Connect each pair of consecutive points. The polygon is a parallelogram.

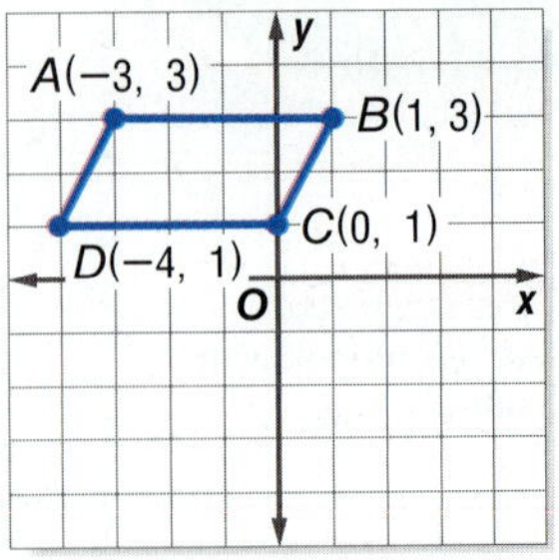

Personal Tutor glencoe.com

StudyTip

Lines There are infinitely many points on a line, so when you are asked to find points on a line, there are many answers.

EXAMPLE 4 Graphing and Solving for Ordered Pairs

Graph four points that satisfy the equation $y = 4 - x$.

Make a table.

Choose four values for x.

Evaluate each value of x for $4 - x$.

x	$4 - x$	y	(x, y)
0	4 − 0	4	(0, 4)
1	4 − 1	3	(1, 3)
2	4 − 2	2	(2, 2)
3	4 − 3	1	(3, 1)

Plot the points.

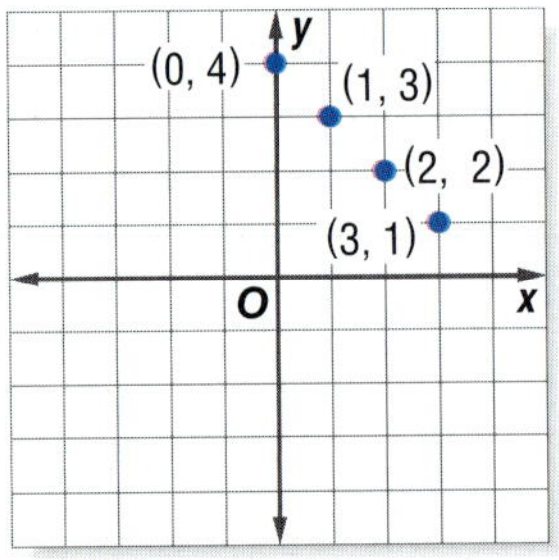

Personal Tutor glencoe.com

Exercises

Write the ordered pair for each point shown at the right.

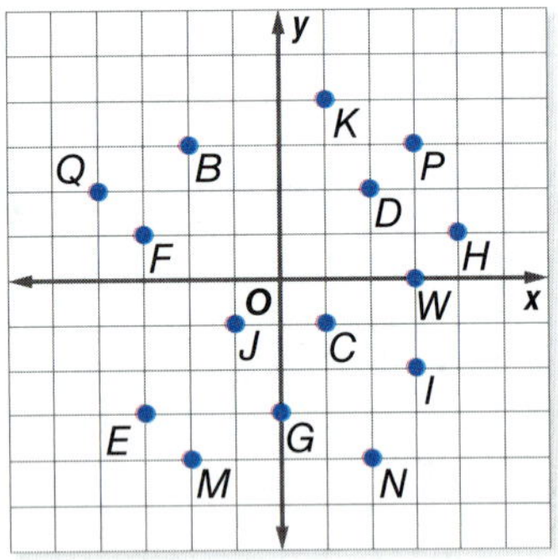

1. B **2.** C **3.** D

4. E **5.** F **6.** G

7. H **8.** I **9.** J

10. K **11.** W **12.** M

13. N **14.** P **15.** Q

Graph and label each point on a coordinate plane. Name the quadrant in which each point is located.

16. $M(-1, 3)$ **17.** $S(2, 0)$ **18.** $R(-3, -2)$ **19.** $P(1, -4)$

20. $B(5, -1)$ **21.** $D(3, 4)$ **22.** $T(2, 5)$ **23.** $L(-4, -3)$

Graph the following geometric figures.

24. a square with vertices $W(-3, 3)$, $X(-3, -1)$, $Z(1, 3)$, and $Y(1, -1)$

25. a polygon with vertices $J(4, 2)$, $K(1, -1)$, $L(-2, 2)$, and $M(1, 5)$

26. a triangle with vertices $F(2, 4)$, $G(-3, 2)$, and $H(-1, -3)$

Graph four points that satisfy each equation.

27. $y = 2x$ **28.** $y = 1 + x$ **29.** $y = 3x - 1$ **30.** $y = 2 - x$

0-8 Systems of Linear Equations

Objective
Use graphing, substitution, and elimination to solve systems of linear equations.

New Vocabulary
system of equations
substitution
elimination

Math Online
glencoe.com

Two or more equations that have common variables are called a **system of equations**. The solution of a system of equations in two variables is an ordered pair of numbers that satisfies both equations. A system of two linear equations can have zero, one, or an infinite number of solutions. There are three methods by which systems of equations can be solved: graphing, elimination, and substitution.

EXAMPLE 1 Graphing Linear Equations

Solve each system of equations by graphing. Then determine whether each system has *no* solution, *one* solution, or *infinitely many* solutions.

a. $y = -x + 3$

$y = 2x - 3$

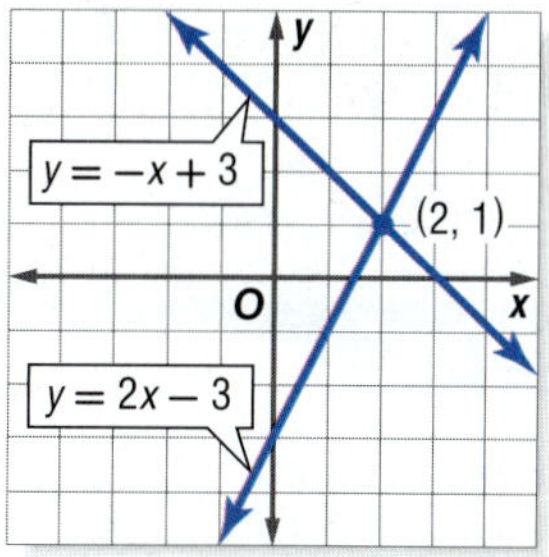

The graphs appear to intersect at (2, 1). Check this estimate by replacing x with 2 and y with 1 in each equation.

CHECK

$y = -x + 3$	$y = 2x - 3$
$1 \stackrel{?}{=} -2 + 3$	$1 \stackrel{?}{=} 2(2) - 3$
$1 = 1$ ✓	$1 = 1$ ✓

The system has one solution at (2, 1).

b. $y - 2x = 6$

$3y - 6x = 9$

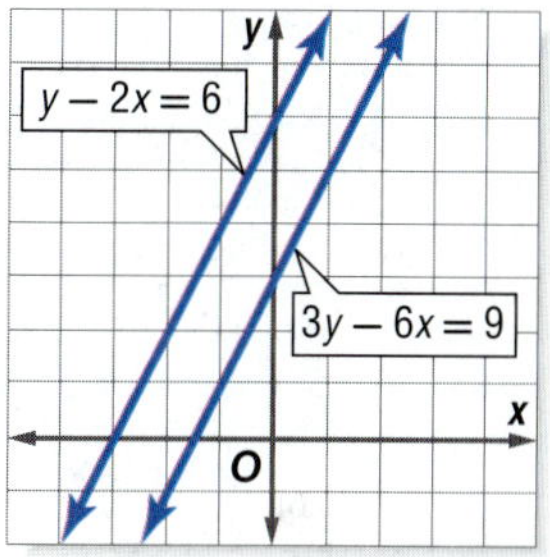

The graphs of the equations are parallel lines. Since they do not intersect, there are no solutions of this system of equations. Notice that the lines have the same slope but different y-intercepts. Equations with the same slope *and* the same y-intercepts have an infinite number of solutions.

Personal Tutor glencoe.com

It is difficult to determine the solution of a system when the two graphs intersect at noninteger values. There are algebraic methods by which an exact solution can be found. One such method is **substitution**.

EXAMPLE 2 Substitution

Use substitution to solve the system of equations.

$y = -4x$

$2y + 3x = 8$

Since $y = -4x$, substitute $-4x$ for y in the second equation.

$2y + 3x = 8$	**Second equation**
$2(-4x) + 3x = 3$	**$y = -4x$**
$-8x + 3x = 8$	**Simplify.**
$-5x = 8$	**Combine like terms.**
$\frac{-5x}{-5} = \frac{8}{-5}$	**Divide each side by −5.**
$x = -\frac{8}{5}$	**Simplify.**

Use $y = -4x$ to find the value of y.

$y = -4x$	**First equation**
$= -4\left(-\frac{8}{5}\right)$	**$x = -\frac{8}{5}$**
$= \frac{32}{5}$	**Simplify.**

The solution is $\left(-\frac{8}{5}, \frac{32}{5}\right)$.

Personal Tutor glencoe.com

Sometimes adding or subtracting two equations together will eliminate one variable. Using this step to solve a system of equations is called **elimination**.

EXAMPLE 3 Elimination

Use elimination to solve the system of equations.

$$3x + 5y = 7$$
$$4x + 2y = 0$$

Either x or y can be eliminated. In this example, we will eliminate x.

$3x + 5y = 7$ **Multiply by 4.** $12x + 20y = 28$

$4x + 2y = 0$ **Multiply by −3.** $+ (-12x) - 6y = 0$

$14y = 28$ **Add the equations.**

$\frac{14y}{14} = \frac{28}{14}$ **Divide each side by 14.**

$y = 2$ **Simplify.**

Now substitute 2 for y in either equation to find the value of x.

$4x + 2y = 0$ **Second equation**

$4x + 2(2) = 0$ **$y = 2$**

$4x + 4 = 0$ **Simplify.**

$4x + 4 - 4 = 0 - 4$ **Subtract 4 from each side.**

$4x = -4$ **Simplify.**

$\frac{4x}{4} = \frac{-4}{4}$ **Divide each side by 4.**

$x = -1$ **Simplify.**

The solution is $(-1, 2)$.

Personal Tutor glencoe.com

StudyTip

Checking Solutions You can confirm that your solutions are correct by substituting the values into both of the original equations.

Exercises

Solve by graphing.

1. $y = -x + 2$; $y = -\frac{1}{2}x + 1$
2. $y = 3x - 3$; $y = x + 1$
3. $y - 2x = 1$; $2y - 4x = 1$

Solve by substitution.

4. $-5x + 3y = 12$; $x + 2y = 8$
5. $x - 4y = 22$; $2x + 5y = -21$
6. $y + 5x = -3$; $3y - 2x = 8$

Solve by elimination.

7. $-3x + y = 7$; $3x + 2y = 2$
8. $3x + 4y = -1$; $-9x - 4y = 13$
9. $-4x + 5y = -11$; $2x + 3y = 11$

Name an appropriate method to solve each system of equations. Then solve the system.

10. $4x - y = 11$; $2x - 3y = 3$
11. $4x + 6y = 3$; $-10x - 15y = -4$
12. $3x - 2y = 6$; $5x - 5y = 5$
13. $3y + x = 3$; $-2y + 5x = 15$
14. $4x - 7y = 8$; $-2x + 5y = -1$
15. $x + 3y = 6$; $4x - 2y = -32$

0-9 Square Roots and Simplifying Radicals

Objective
Evaluate square roots and simplify radical expressions.

New Vocabulary
Product Property
Quotient Property

Math Online
glencoe.com

A radical expression is an expression that contains a square root. The expression is in simplest form when the following three conditions have been met.

- No radicands have perfect square factors other than 1.
- No radicands contain fractions.
- No radicals appear in the denominator of a fraction.

The **Product Property** states that for two numbers a and $b \geq 0$, $\sqrt{ab} = \sqrt{a} \cdot \sqrt{b}$.

EXAMPLE 1 Product Property

Simplify.

a. $\sqrt{45}$

$\sqrt{45} = \sqrt{3 \cdot 3 \cdot 5}$ Prime factorization of 45

$= \sqrt{3^2} \cdot \sqrt{5}$ Product Property of Square Roots

$= 3\sqrt{5}$ Simplify.

b. $\sqrt{6} \cdot \sqrt{15}$

$\sqrt{6} \cdot \sqrt{15} = \sqrt{6 \cdot 15}$ Product Property

$= \sqrt{3 \cdot 2 \cdot 3 \cdot 5}$ Prime factorization

$= \sqrt{3^2} \cdot \sqrt{10}$ Product Property

$= 3\sqrt{10}$ Simplify.

Personal Tutor glencoe.com

For radical expressions in which the exponent of the variable inside the radical is *even* and the resulting simplified exponent is *odd*, you must use absolute value to ensure nonnegative results.

EXAMPLE 2 Product Property

Simplify $\sqrt{20x^3y^5z^6}$.

$\sqrt{20x^3y^5z^6} = \sqrt{2^2 \cdot 5 \cdot x^3 \cdot y^5 \cdot z^6}$ Prime factorization

$= \sqrt{2^2} \cdot \sqrt{5} \cdot \sqrt{x^3} \cdot \sqrt{y^5} \cdot \sqrt{z^6}$ Product Property

$= 2 \cdot \sqrt{5} \cdot x \cdot \sqrt{x} \cdot y^2 \cdot \sqrt{y} \cdot |z^3|$ Simplify.

$= 2xy^2|z^3|\sqrt{5xy}$ Simplify.

Personal Tutor glencoe.com

The **Quotient Property** states that for any numbers a and b, where $a \geq 0$ and $b \geq 0$, $\sqrt{\frac{a}{b}} = \frac{\sqrt{a}}{\sqrt{b}}$.

EXAMPLE 3 Quotient Property

Simplify $\sqrt{\frac{25}{16}}$.

$\sqrt{\frac{25}{16}} = \frac{\sqrt{25}}{\sqrt{16}}$ Quotient Property

$= \frac{5}{4}$ Simplify.

Personal Tutor glencoe.com

Rationalizing the denominator of a radical expression is a method used to eliminate radicals from the denominator of a fraction. To rationalize the denominator, multiply the expression by a fraction equivalent to 1 such that the resulting denominator is a perfect square.

Watch Out!

Rationalizing the Denominator Don't forget to multiply both the numerator and denominator by the radical when you rationalize the denominator.

EXAMPLE 4 Rationalize the Denominator

Simplify.

a. $\frac{2}{\sqrt{3}}$

$\frac{2}{\sqrt{3}} = \frac{2}{\sqrt{3}} \cdot \frac{\sqrt{3}}{\sqrt{3}}$ Multiply by $\frac{\sqrt{3}}{\sqrt{3}}$.

$= \frac{2\sqrt{3}}{3}$ Simplify.

b. $\frac{\sqrt{13y}}{\sqrt{18}}$

$\frac{\sqrt{13y}}{\sqrt{18}} = \frac{\sqrt{13y}}{\sqrt{2 \cdot 3 \cdot 3}}$ Prime factorization

$= \frac{\sqrt{13y}}{3\sqrt{2}}$ Product Property

$= \frac{\sqrt{13y}}{3\sqrt{2}} \cdot \frac{\sqrt{2}}{\sqrt{2}}$ Multiply by $\frac{\sqrt{2}}{\sqrt{2}}$.

$= \frac{\sqrt{26y}}{6}$ Product Property

Personal Tutor glencoe.com

Sometimes, conjugates are used to simplify radical expressions. Conjugates are binomials of the form $p\sqrt{q} + r\sqrt{t}$ and $p\sqrt{q} - r\sqrt{t}$.

EXAMPLE 5 Conjugates

Simplify $\frac{3}{5 - \sqrt{2}}$.

$\frac{3}{5 - \sqrt{2}} = \frac{3}{5 - \sqrt{2}} \cdot \frac{5 + \sqrt{2}}{5 + \sqrt{2}}$ $\frac{5 + \sqrt{2}}{5 + \sqrt{2}} = 1$

$= \frac{3(5 + \sqrt{2})}{5^2 - (\sqrt{2})^2}$ $(a - b)(a + b) = a^2 - b^2$

$= \frac{15 + 3\sqrt{2}}{25 - 2}$ Multiply. $(\sqrt{2})^2 = 2$

$= \frac{15 + 3\sqrt{2}}{23}$ Simplify.

Personal Tutor glencoe.com

Exercises

Simplify.

1. $\sqrt{32}$

2. $\sqrt{75}$

3. $\sqrt{50} \cdot \sqrt{10}$

4. $\sqrt{12} \cdot \sqrt{20}$

5. $\sqrt{6} \cdot \sqrt{6}$

6. $\sqrt{16} \cdot \sqrt{25}$

7. $\sqrt{98x^3y^6}$

8. $\sqrt{56a^2b^4c^5}$

9. $\sqrt{\frac{81}{49}}$

10. $\sqrt{\frac{121}{16}}$

11. $\sqrt{\frac{63}{8}}$

12. $\sqrt{\frac{288}{147}}$

13. $\frac{\sqrt{10p^3}}{\sqrt{27}}$

14. $\frac{\sqrt{108}}{\sqrt{2q^6}}$

15. $\frac{4}{5 - 2\sqrt{3}}$

16. $\frac{7\sqrt{3}}{5 - 2\sqrt{6}}$

17. $\frac{3}{\sqrt{48}}$

18. $\frac{\sqrt{24}}{\sqrt{125}}$

19. $\frac{3\sqrt{5}}{2 - \sqrt{2}}$

20. $\frac{3}{-2 + \sqrt{13}}$

CHAPTER 0 Preparing for Geometry

State which metric unit you would probably use to measure each item.

1. mass of a book

2. length of a highway

Complete each sentence.

3. 8 in. = __?__ ft

4. 6 yd = __?__ ft

5. 24 fl oz = __?__ pt

6. 3.7 kg = __?__ lb

7. 4.2 km = __?__ m

8. 285 g = __?__ kg

9. 0.75 kg = __?__ mg

10. 1.9 L = __?__ qt

11. PROBABILITY The table shows the results of an experiment in which a number cube was rolled. Find the experimental probability of rolling a 4.

Outcome	Tally	Frequency
1	IIII	4
2	𝍸I	6
3	𝍸	5
4	III	3
5	𝍸II	7

CANDY A bag of candy contains 3 lollipops, 8 peanut butter cups, and 4 chocolate bars. A piece of candy is randomly drawn from the bag. Find each probability.

12. *P*(peanut butter cup)

13. *P*(lollipop or peanut butter cup)

14. *P*(not chocolate bar)

15. *P*(chocolate bar or lollipop)

Evaluate each expression if $x = 2$, $y = -3$, and $z = 4$.

16. $6x - z$

17. $6y + xz$

18. $3yz$

19. $\frac{6z}{xy}$

20. $\frac{y + 2x}{10z}$

21. $7 + |y - 11|$

Solve each equation.

22. $9 + s = 21$

23. $h - 8 = 12$

24. $\frac{4m}{14} = 18$

25. $\frac{2}{9}d = 10$

26. $3(20 - b) = 36$

27. $37 + w = 5w - 27$

28. $\frac{x}{6} = 7$

29. $\frac{1}{4}(n + 5) = 16$

Solve each inequality.

30. $4y - 9 > 1$

31. $-2z + 15 \geq 4$

32. $3r + 7 < r - 8$

33. $-\frac{2}{5}k - 20 \leq 10$

34. $-3(b - 4) > 33$

35. $2 - m \leq 6m - 12$

36. $8 \leq r - 14$

37. $\frac{2}{3}n < \frac{3}{9}n - 5$

Write the ordered pair for each point shown.

38. *M*

39. *N*

40. *P*

41. *Q*

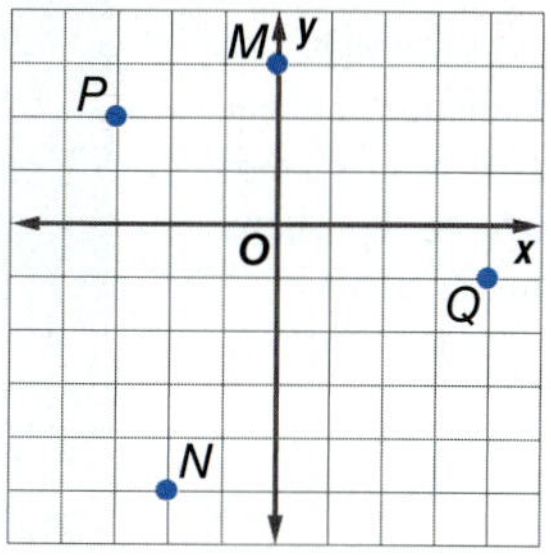

Name and label each point on a coordinate plane.

42. $A(-2, 0)$

43. $C(1, 3)$

44. $D(-4, -4)$

45. $F(3, -5)$

46. Graph the quadrilateral with vertices $R(2, 0)$, $S(4, -2)$, $T(4, 3)$, and $W(2, 5)$.

47. Graph three points that satisfy the equation $y = \frac{1}{2}x - 5$.

Solve each system of equations.

48. $2r + m = 11$
$6r - 2m = -2$

49. $2x + 4y = 6$
$7x = 4 + 3y$

50. $2c + 6d = 14$
$-\frac{7}{3} + \frac{1}{3}c = -d$

51. $5a - b = 17$
$3a + 2b = 5$

52. $6d + 3f = 12$
$2d = 8 - f$

53. $4x - 5y = 17$
$3x + 4y = 5$

Simplify.

54. $\sqrt{80}$

55. $\sqrt{\frac{128}{5}}$

56. $\sqrt{36} \cdot \sqrt{81}$

57. $\sqrt{\frac{7x^3}{3}}$

58. $\sqrt{\frac{5}{81}}$

59. $\sqrt{12x^5y^2}$

CHAPTER 1 Tools of Geometry

Then

You graphed points on the coordinate plane and evaluated mathematical expressions.

Now

In Chapter 1, you will:

- Find distances between points and midpoints of line segments.
- Identify angle relationships.
- Find perimeters, areas, surface areas, and volumes.

Why?

MAPS Geometric figures and terms can be used to represent and describe real-world situations. On a map, locations of cities can be represented by points, highways or streets by lines, and national parks by polygons that have both perimeter and area. The map itself is representative of a plane.

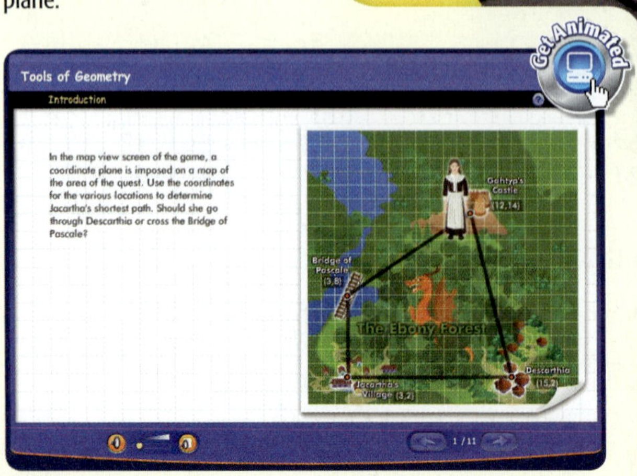

Get Ready for Chapter 1

Diagnose Readiness You have two options for checking Prerequ

Text Option

Take the Quick Check below. Refer to the Quick Review for he

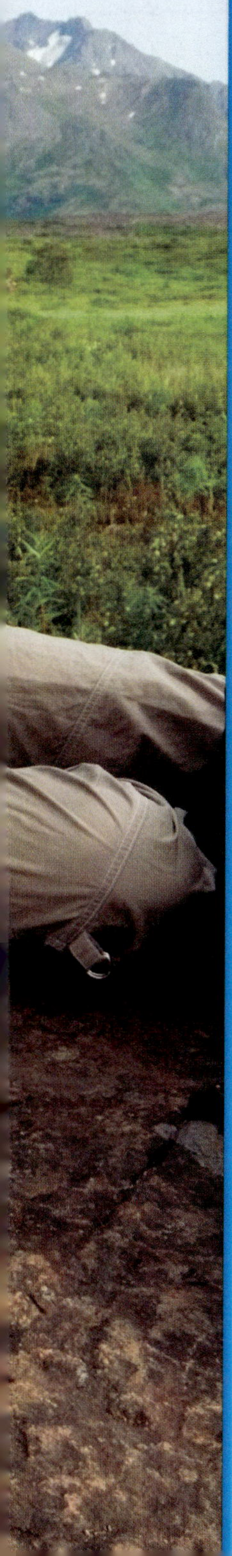

*Quick*Check

Graph and label each point in the coordinate plane. (Lesson 0-7)

1. $W(5, 2)$ **2.** $X(0, 6)$

3. $Y(-3, -1)$ **4.** $Z(4, -2)$

5. **GAMES** Carolina is using the diagram to record her chess moves. She moves her knight 2 spaces up and 1 space to the left from f3. What is the location of the knight after Carolina completes her turn?

Find each sum or difference. (Prerequisite Skill)

6. $\frac{2}{3} + \frac{5}{6}$ **7.** $2\frac{1}{18} + 4\frac{3}{4}$

8. $\frac{13}{18} - \frac{5}{9}$ **9.** $14\frac{3}{5} - 9\frac{7}{15}$

10. **FOOD** Alvin ate $\frac{1}{3}$ of a pizza for dinner and took $\frac{1}{6}$ of it for lunch the next day. How much of the pizza does he have left?

Evaluate each expression. (Lesson 0-4)

11. $(-4 - 5)^2$ **12.** $(6 - 10)^2$

13. $(8 - 5)^2 + [9 - (-3)]^2$

Solve each equation. (Lesson 0-5)

14. $6x + 5 + 2x - 11 = 90$

15. $8x - 7 = 53 - 2x$

*Quick*Review

EXAMPLE 1

Graph and label the point $Q(-3, 4)$ in the coordinate plane.

Start at the origin. Since the x-coordinate is negative, move 3 units to the left. Then move 4 units up since the y-coordinate is positive. Draw a dot and label it Q.

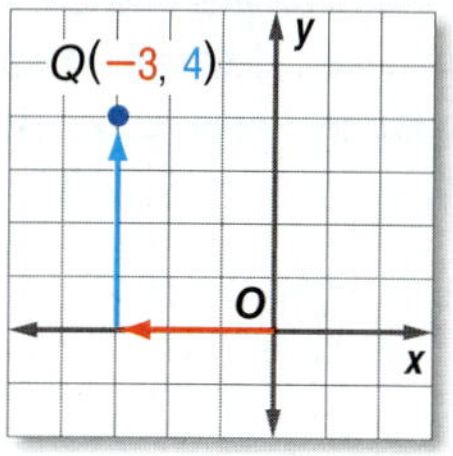

EXAMPLE 2

Find $3\frac{1}{6} + 2\frac{3}{4}$.

$3\frac{1}{6} + 2\frac{3}{4} = \frac{19}{6} + \frac{11}{4}$ **Write as improper fractions.**

$= \frac{19}{6}\left(\frac{2}{2}\right) + \frac{11}{4}\left(\frac{3}{3}\right)$ **The LCD is 12.**

$= \frac{38}{12} + \frac{33}{12}$ **Multiply.**

$= \frac{71}{12}$ or $5\frac{11}{12}$ **Simplify.**

EXAMPLE 3

Evaluate the expression $[-2 - (-7)]^2 + (1 - 8)^2$.

Follow the order of operations.

$[-2 - (-7)]^2 + (1 - 8)^2$

$= 5^2 + (-7)^2$ **Subtract.**

$= 25 + 49$ **$5^2 = 25, (-7)^2 = 49$**

$= 74$ **Add.**

Online Option

Math Online Take a self-check Chapter Readiness Quiz at glencoe.com.

t Started on Chapter 1

u will learn several new concepts, skills, and vocabulary terms as you study Chapter 1. To get ready, identify important terms and organize your resources. You may wish to refer to **Chapter 0** to review prerequisite skills.

FOLDABLES® Study Organizer

Tools of Geometry Make this Foldable to help you organize your Chapter 1 notes about points, lines, and planes; angles and angle relationships; and formulas and notes for distance, midpoint, perimeter, area, and volume. Begin with a sheet of 11″ × 17″ paper.

1. **Fold** the short sides to meet in the middle.

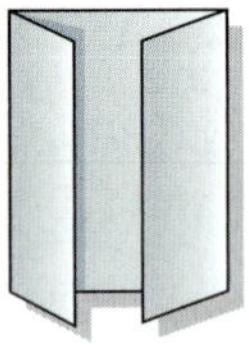

2. **Fold** the booklet in thirds lengthwise.

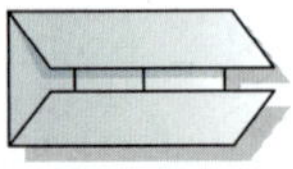

3. **Open and cut** the booklet in thirds lengthwise.

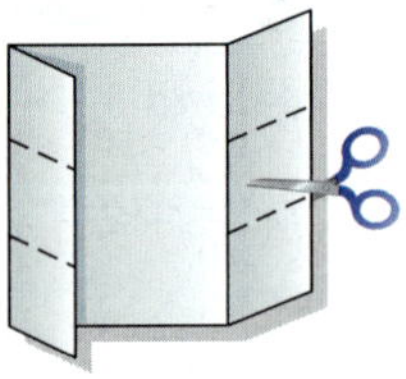

4. **Label** the tabs as shown.

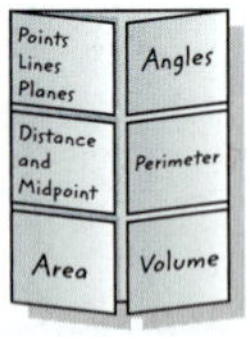

Math Online glencoe.com

- Study the chapter online
- Explore **Get Animated** activities
- Get extra help from your own **Personal Tutor**
- Use **Extra Examples** for additional help
- Take a **Self-Check Quiz**
- **Review Vocabulary** in fun ways

New Vocabulary

English		Español
collinear	• p. 5 •	colineal
coplanar	• p. 5 •	coplanar
congruent	• p. 16 •	congruente
midpoint	• p. 27 •	punto medio
segment bisector	• p. 29 •	bisectriz de segmento
angle	• p. 36 •	angulo
vertex	• p. 36 •	vertice
angle bisector	• p. 39 •	bisectriz de un angulo
perpendicular	• p. 48 •	perpendiculares
polygon	• p. 56 •	poligono
perimeter	• p. 58 •	perimetro
volume	• p. 69 •	volumen

Review Vocabulary

ordered pair • p. P15 • par ordenado a set of numbers or coordinates used to locate any point on a coordinate plane, written in the form (x, y)

origin • p. P15 • origen the point where the two axes intersect at their zero points

quadrants • p. P15 • cadrantes the four regions into which the x-axis and y-axis separate the coordinate plane

x-coordinate • p. P15 • coordenada x the first number in an ordered pair

y-coordinate • p. P15 • coordenada y the second number in an ordered pair

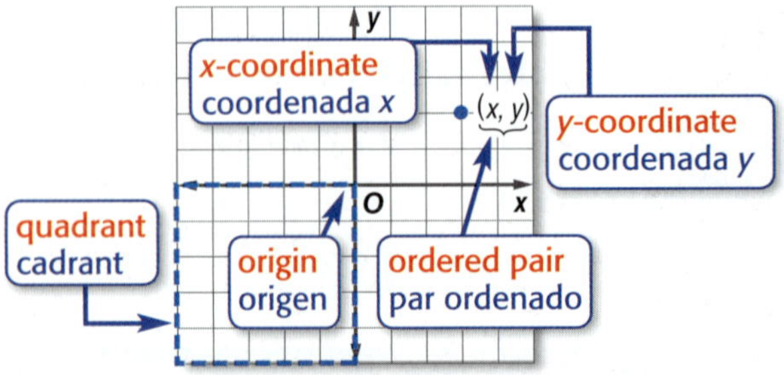

Multilingual eGlossary glencoe.com

1-1 Points, Lines, and Planes

Then

You used basic geometric concepts and properties to solve problems.

Now

- Identify and model points, lines, and planes.
- Identify intersecting lines and planes.
- Find perimeters, areas, surface areas, and volumes.

New Vocabulary

undefined term
point
line
plane
collinear
coplanar
intersection
definition
defined term
space

Math Online
glencoe.com

Why?

On a subway map, the locations of stops are represented by *points*. The route the train can take is modeled by a series of connected paths that look like *lines*. The flat surface of the map on which these points and lines lie is representative of a *plane*.

Points, Lines, and Planes Unlike the real-world objects that they model, shapes, points, lines, and planes do not have any actual size. In geometry, *point*, *line*, and *plane* are considered **undefined terms** because they are only explained using examples and descriptions.

You are already familiar with the terms point, line, and plane from algebra. You graphed on a coordinate *plane* and found ordered pairs that represented *points* on *lines*. In geometry, these terms have a similar meaning.

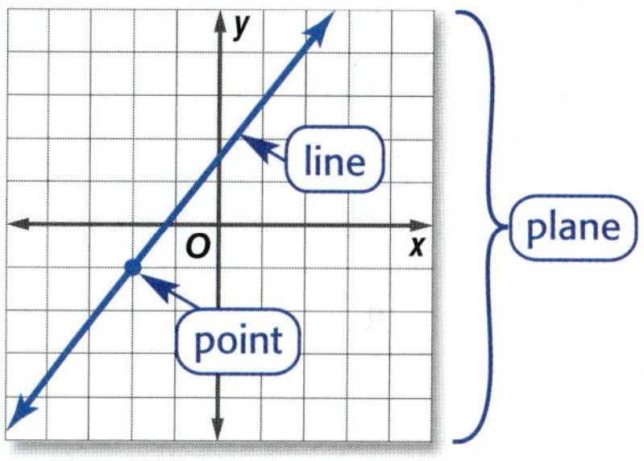

The phrase *exactly one* in a statement such as, "There is exactly one line through any two points," means that there is *one and only one.*

Key Concept — Undefined Terms

A **point** is a location. It has neither shape nor size.

Named by a capital letter

Example point A

A **line** is made up of points and has no thickness or width. There is exactly one line through any two points.

Named by the letters representing two points on the line or a lowercase script letter

Example line m, line PQ or $\overleftrightarrow{PQ}$, line QP or $\overleftrightarrow{QP}$

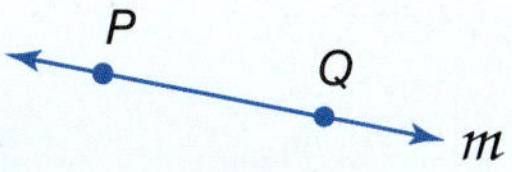

A **plane** is a flat surface made up of points that extends infinitely in all directions. There is exactly one plane through any three points not on the same line.

Named by a capital script letter or by the letters naming three points that are not all on the same line

Example plane $\mathcal{K}$, plane BCD, plane CDB, plane DCB, plane DBC, plane CBD, plane BDC

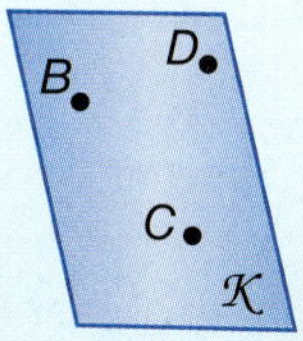

Collinear points are points that lie on the same line. *Noncollinear* points do not lie on the same line. **Coplanar** points are points that lie in the same plane. *Noncoplanar* points do not lie in the same plane.

EXAMPLE 1 Name Lines and Planes

Use the figure to name each of the following.

a. a line containing point *W*

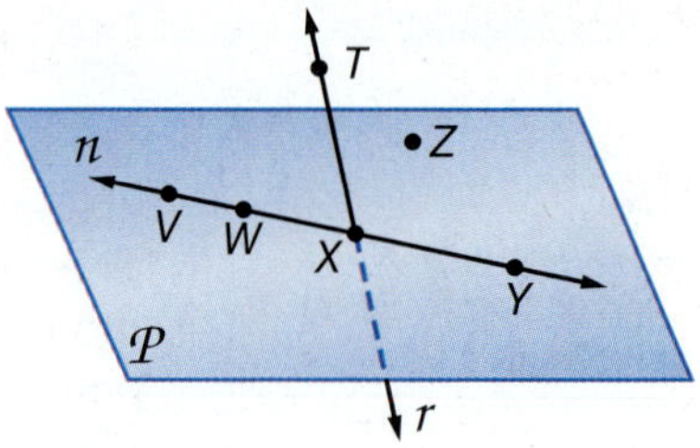

The line can be named as line n, or any two of the four points on the line can be used to name the line.

$\overleftrightarrow{VW}$ $\overleftrightarrow{WV}$ $\overleftrightarrow{VX}$ $\overleftrightarrow{XV}$ $\overleftrightarrow{VY}$ $\overleftrightarrow{YV}$

$\overleftrightarrow{WX}$ $\overleftrightarrow{XW}$ $\overleftrightarrow{WY}$ $\overleftrightarrow{YW}$ $\overleftrightarrow{XY}$ $\overleftrightarrow{YX}$

b. a plane containing point *X*

One plane that can be named is plane $\mathcal{P}$. You can also use the letters of any three *noncollinear* points to name this plane.

plane *XZY*	plane *VZW*	plane *VZX*
plane *VZY*	plane *WZX*	plane *WZY*

The letters of each of these names can be reordered to create other acceptable names for this plane. For example, *XZY* can also be written as *XYZ*, *ZXY*, *ZYX*, *YXZ*, and *YZX*. In all, there are 36 different three-letter names for this plane.

StudyTip

Additional Planes Although not drawn in Example 1b, there is another plane that contains point *X*. Since points *W*, *T*, and *X* are noncollinear, point *X* is also in plane *WTX*.

Guided Practice

1A. a plane containing points *T* and *Z* **1B.** a line containing point *T*

Personal Tutor glencoe.com

Real-World EXAMPLE 2 Model Points, Lines, and Planes

MESSAGE BOARD **Name the geometric terms modeled by the objects in the picture.**

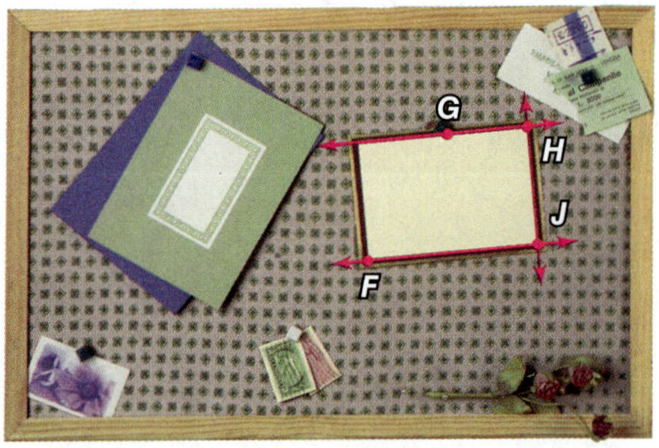

The push pin models point *G*.

The maroon border on the card models line *GH*.

The edge of the card models line *HJ*.

The card itself models plane *FGJ*.

Guided Practice

Name the geometric term modeled by each object.

2A. stripes on a sweater **2B.** the corner of a box

Personal Tutor glencoe.com

Intersections of Lines and Planes The **intersection** of two or more geometric figures is the set of points they have in common. Two lines intersect in a point. Lines can intersect planes, and planes can intersect each other.

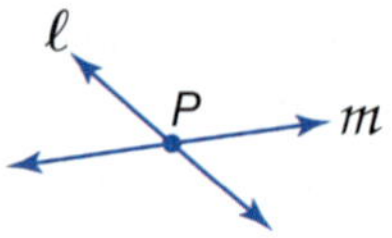

***P* represents the intersection of lines ℓ and *m*.**

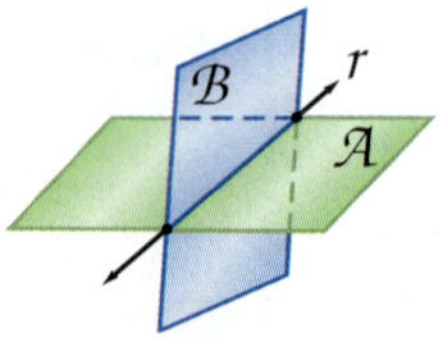

Line *r* represents the intersection of planes $\mathcal{A}$ and $\mathcal{B}$.

EXAMPLE 3 Draw Geometric Figures

Draw and label a figure for each relationship.

a. ALGEBRA Lines AB and CD intersect at E for $A(-2, 4)$, $B(0, -2)$, $C(-3, 0)$, and $D(3, 3)$ on a coordinate plane. Point F is coplanar with these points, but not collinear with $\overleftrightarrow{AB}$ or $\overleftrightarrow{CD}$.

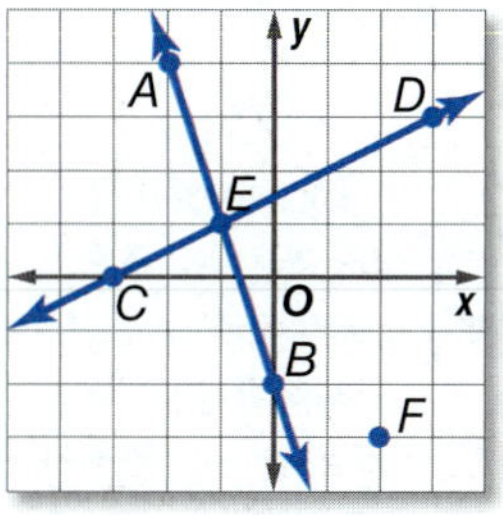

Graph each point and draw $\overleftrightarrow{AB}$ and $\overleftrightarrow{CD}$.

Label the intersection point as E.

An infinite number of points are coplanar with A, B, C, D and E but not collinear with $\overleftrightarrow{AB}$ and $\overleftrightarrow{CD}$. In the graph, one such point is $F(2, -3)$.

b. $\overleftrightarrow{QR}$ intersects plane $\mathcal{T}$ at point S.

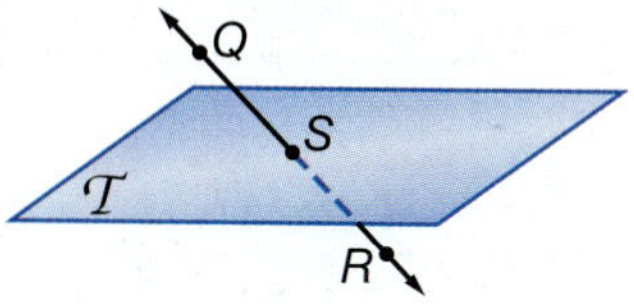

Draw a surface to represent plane $\mathcal{T}$ and label it.

Draw a dot for point S anywhere on the plane and a dot that is not on plane $\mathcal{T}$ for point Q.

Draw a line through points Q and S. Dash the line to indicate the portion hidden by the plane. Then draw another dot on the line and label it R.

Guided Practice

3A. Points $J(-4, 2)$, $K(3, 2)$, and L are collinear.

3B. Line p lies in plane $\mathcal{N}$ and contains point L.

Personal Tutor glencoe.com

> **StudyTip**
>
> **Three-Dimensional Drawings** Because it is impossible to show an entire plane in a figure, edged shapes with different shades of color are used to represent planes.

Definitions or **defined terms** are explained using undefined terms and/or other defined terms. **Space** is defined as a boundless, three-dimensional set of all points. Space can contain lines and planes.

EXAMPLE 4 Interpret Drawings

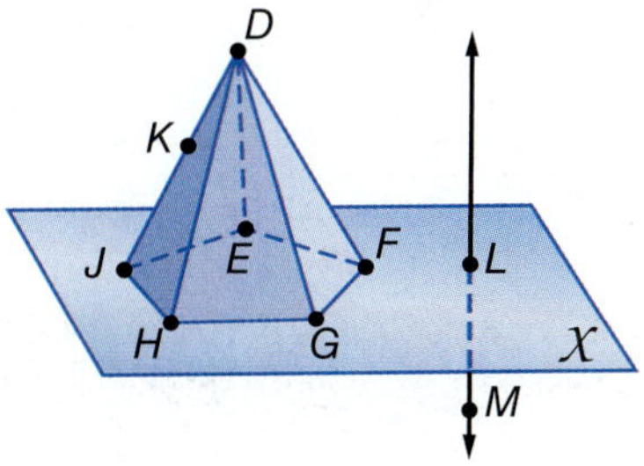

a. How many planes appear in this figure?

Six: plane $\mathcal{X}$, plane JDH, plane JDE, plane EDF, plane FDG, and plane HDG.

b. Name three points that are collinear.

Points J, K, and D are collinear.

c. Name the intersection of plane HDG with plane $\mathcal{X}$.

Plane HDG intersects plane $\mathcal{X}$ in $\overleftrightarrow{HG}$.

d. At what point do $\overleftrightarrow{LM}$ and $\overleftrightarrow{EF}$ intersect? Explain.

It does not appear that these lines intersect. $\overleftrightarrow{EF}$ lies in plane $\mathcal{X}$, but only point L of $\overleftrightarrow{LM}$ lies in $\mathcal{X}$.

Guided Practice

Explain your reasoning.

4A. Are points E, D, F, and G coplanar?

4B. At what point or in what line do planes JDH, JDE, and EDF intersect?

Personal Tutor glencoe.com

> **StudyTip**
>
> **Dimension** A point has no dimension. A line exists in one dimension. However, a circle is two-dimensional, and a pyramid is three-dimensional.

Check Your Understanding

Example 1 p. 6

Use the figure to name each of the following.

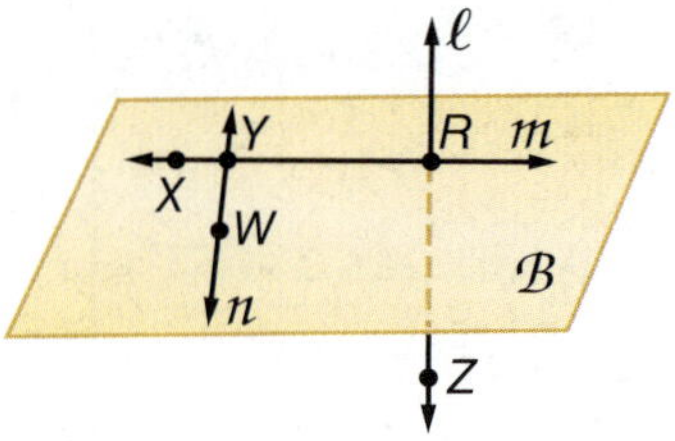

1. a line containing point X
2. a line containing point Z
3. a plane containing points W and R

Example 2 p. 6

Name the geometric term modeled by each object.

4. a beam from a laser
5. a floor

Example 3 p. 7

Draw and label a figure for each relationship.

6. A line in a coordinate plane contains $A(0, -5)$ and $B(3, 1)$ and a point C that is not collinear with $\overleftrightarrow{AB}$.
7. Plane $\mathcal{Z}$ contains lines x, y, w. Lines x and y intersect at point V and lines x and w intersect at point P.

Example 4 p. 7

Refer to the figure.

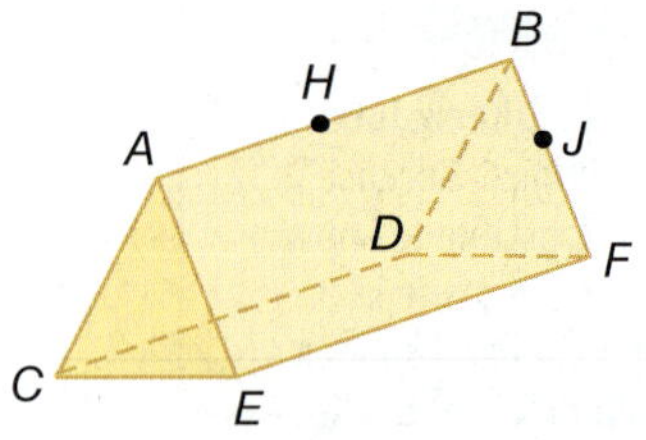

8. How many planes are shown in the figure?
9. Name three points that are collinear.
10. Are points A, H, J, and D coplanar? Explain.
11. Are points B, D, and F coplanar? Explain.
12. **ASTRONOMY** Ursa Minor, or the Little Dipper, is a constellation made up of seven stars in the northern sky including the star Polaris.
    a. What geometric figures are modeled by the stars?
    b. Are Star 1, Star 2, and Star 3 collinear on the constellation map? Explain.
    c. Are Polaris, Star 2, and Star 6 coplanar on the map?

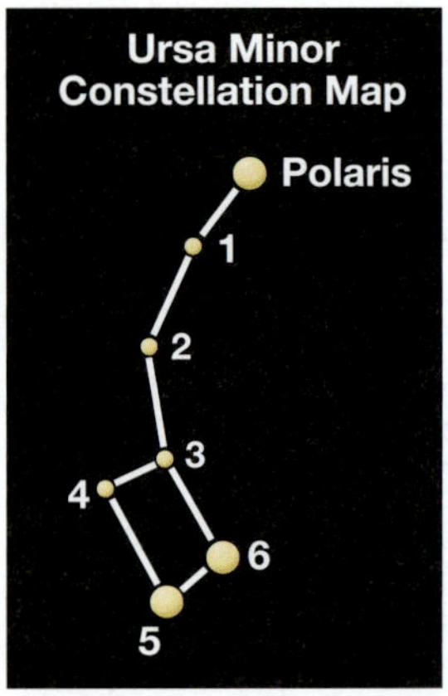

Practice and Problem Solving

= **Step-by-Step Solutions** begin on page R20.
Extra Practice begins on page 969.

Example 1 p. 6

Refer to the figure.

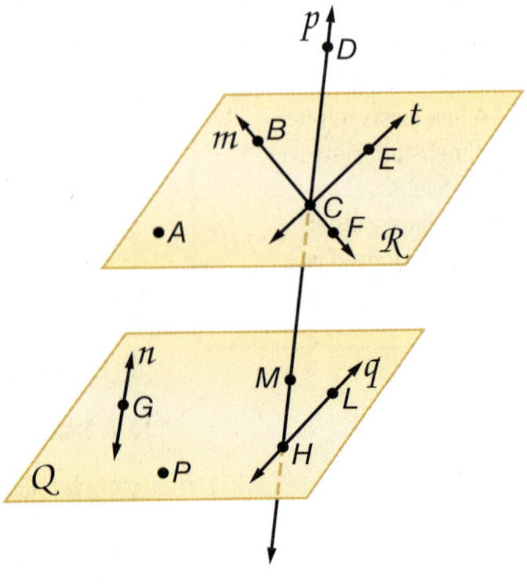

13. Name the lines that are only in plane $\mathcal{Q}$.
14. How many planes are labeled in the figure?
15. Name the plane containing the lines m and t.
16. Name the intersection of lines m and t.
17. Name a point that is not coplanar with points A, B, and C.
18. Are points F, M, G, and P coplanar? Explain.
19. Name the points not contained in a line shown.
20. What is another name for line t?
21. Does line n intersect line q? Explain.

Example 2
p. 6

Name the geometric term(s) modeled by each object.

22.
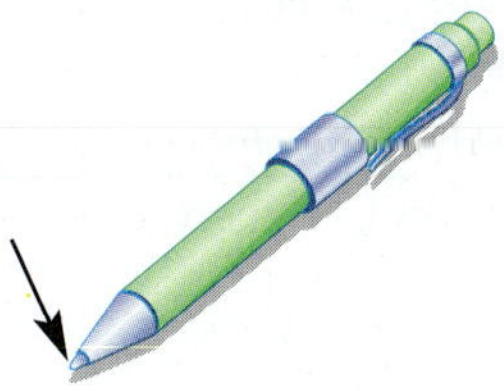

23.

24.
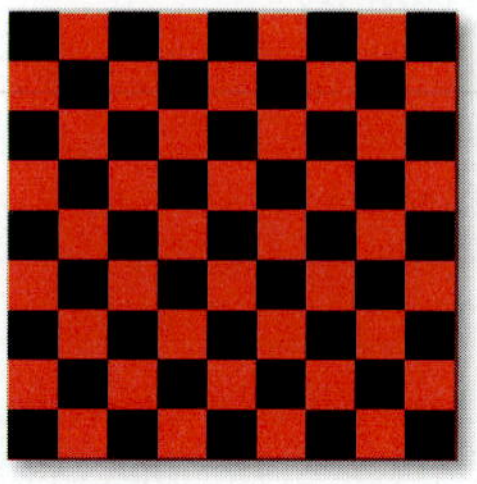

25.

26. a blanket
27. a knot in a rope
28. a telephone pole
29. the edge of a desk
30. two connected hallways
31. a partially opened folder

Example 3
p. 7

Draw and label a figure for each relationship.

32. Line m intersects plane $\mathcal{R}$ at a single point.

33. Two planes do not intersect.

34. Points X and Y lie on $\overleftrightarrow{CD}$.

35. Three lines intersect at point J but do not all lie in the same plane.

36. Points $A(2, 3)$, $B(2, -3)$, C and D are collinear, but A, B, C, D, and F are not.

37. Lines $\overleftrightarrow{LM}$ and $\overleftrightarrow{NP}$ are coplanar but do not intersect.

38. $\overleftrightarrow{FG}$ and $\overleftrightarrow{JK}$ intersect at $P(4, 3)$, where point F is at $(-2, 5)$ and point J is at $(7, 9)$.

39. Lines s and t intersect, and line v does not intersect either one.

Example 4
p. 7

PACKING When packing breakable objects such as glasses, movers frequently use boxes with inserted dividers like the one shown.

40. How many planes are modeled in the picture?

41. What parts of the box model lines?

42. What parts of the box model points?

Refer to the figure at the right.

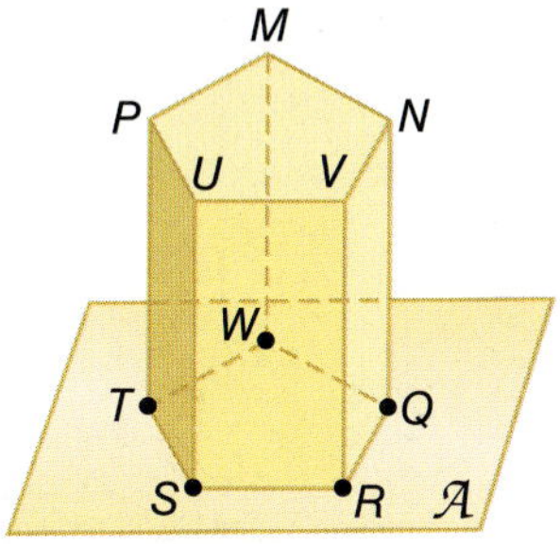

43. Name two collinear points.

44. How many planes appear in the figure?

45. Do plane $\mathcal{A}$ and plane MNP intersect? Explain.

46. In what line do planes $\mathcal{A}$ and QRV intersect?

47. Are points T, S, R, Q, and V coplanar? Explain.

48. Are points T, S, R, Q, and W coplanar? Explain.

49. FINITE PLANES A *finite plane* is a plane that has boundaries, or does not extend indefinitely. The street signs shown are finite planes.

a. If the pole models a line, name the geometric term that describes the intersection between the signs and the pole.

b. What geometric term(s) describes the intersection between the two finite planes? Explain your answer with a diagram if necessary.

Real-World Career

Drafter
Drafters use perspective to create drawings to build everything from toys to school buildings. Drafters need skills in math and computers. They get their education at trade schools, community colleges, and some 4-year colleges.

50. ONE-POINT PERSPECTIVE One-point perspective drawings use lines to convey depth. Lines representing horizontal lines in the real object can be extended to meet at a single point called the *vanishing point*. Suppose you want to draw a tiled ceiling in the room below with eight tiles across.

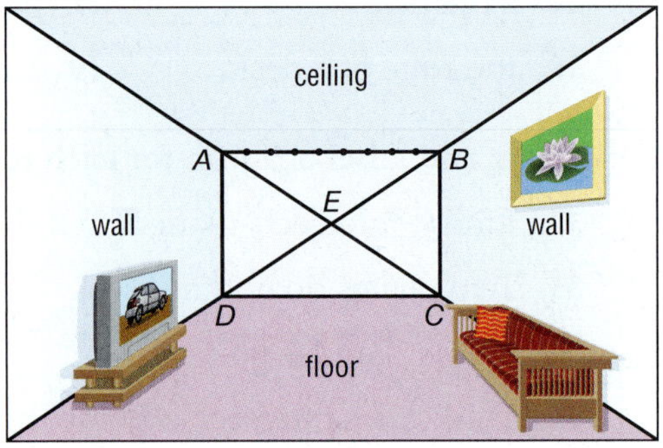

a. What point represents the vanishing point in the drawing?

b. Trace the figure. Then draw lines from the vanishing point through each of the eight points between *A* and *B*. Extend these lines to the top edge of the drawing.

c. How could you change the drawing to make the back wall of the room appear farther away?

51. TWO-POINT PERSPECTIVE Two-point perspective drawings use two vanishing points to convey depth.

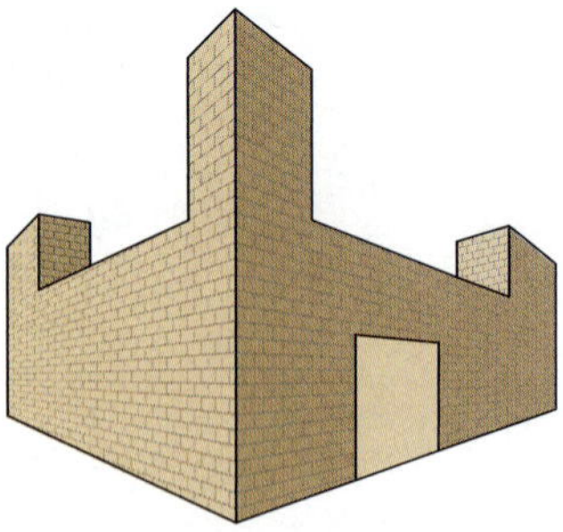

a. Trace the drawing of the castle shown. Draw five of the vertical lines used to create the drawing.

b. Draw and extend the horizontal lines to locate the vanishing points and label them.

c. What do you notice about the vertical lines as they get closer to the vanishing point?

d. Draw a two-point perspective of a home or a room in a home.

52. OPTICAL ILLUSION Name two points on the same line in the figure. How can you support your assertion?

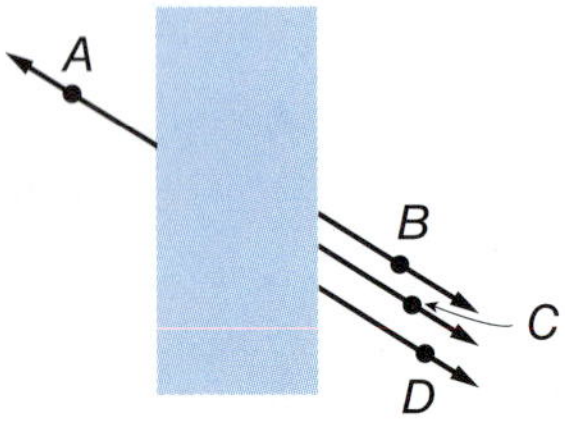

53. TRANSPORTATION When two cars enter an intersection at the same time on opposing paths, one of the cars must adjust its speed or direction to avoid a collision. Two airplanes, however, can cross paths while traveling in different directions without colliding. Explain how this is possible.

Real-World Link

Most insurance companies require teen drivers to complete a "30 and 6" driver's education course before they will offer premium discounts. Such a course consists of at least 30 hours of classroom training and 6 hours behind the wheel.

Source: Georgia Department of Driver Services

54. MULTIPLE REPRESENTATIONS Another way to describe a group of points is called a locus. A **locus** is a set of points that satisfy a particular condition. In this problem, you will explore the locus of points that satisfy an equation.

a. **TABULAR** Represent the locus of points satisfying the equation $2 + x = y$ using a table of at least five values.

b. **GRAPHICAL** Represent this same locus of points using a graph.

c. **VERBAL** Describe the geometric figure that the points suggest.

55. PROBABILITY Three of the labeled points are chosen at random.

a. What is the probability that the points chosen are collinear?

b. What is the probability that the points chosen are coplanar?

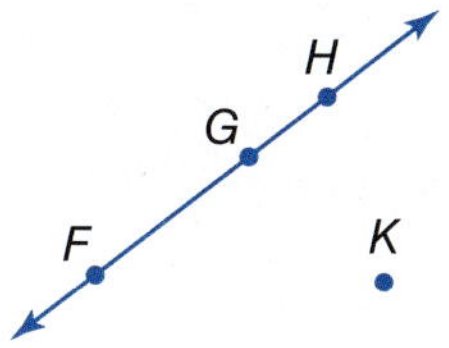

56. MULTIPLE REPRESENTATIONS In this problem, you will explore the locus of points that satisfy an inequality.

a. **TABULAR** Represent the locus of points satisfying the inequality $y < -3x - 1$ using a table of at least ten values.

b. **GRAPHICAL** Represent this same locus of points using a graph.

c. **VERBAL** Describe the geometric figure that the points suggest.

H.O.T. Problems Use Higher-Order Thinking Skills

57. OPEN ENDED Sketch three planes that intersect in a line.

58. ERROR ANALYSIS Camille and Hiroshi are trying to determine how many lines can be drawn between four points. Is either of them correct? Explain your reasoning.

Camille	Hiroshi
Since there are four points, $4 \cdot 3$ or 12 lines can be drawn between the points.	You can draw $3 \cdot 2 \cdot 1$ or 6 lines between the points.

59. CHALLENGE Points A, B, C, and D are noncollinear. What is the greatest number of planes that can be determined by points A, B, and C?

60. REASONING Is it possible for two points on the surface of a prism to be neither collinear nor coplanar? Justify your answer.

61. WRITING IN MATH Refer to Exercise 49. Give a real-life example of a finite plane. Is it possible to have a real-life object that is an infinite plane? Explain your reasoning.

Standardized Test Practice

62. Which statement about the figure below is *not* true?

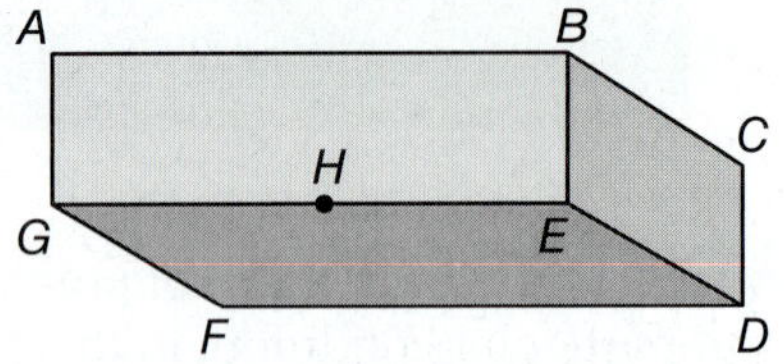

A Point H lies in planes AGE and GED.
B Planes GAB, GFD and BED intersect at point E.
C Points F, E, and B are coplanar.
D Points A, H, and D are collinear.

63. ALGEBRA What is the value of x if $3x + 2 = 8$?

F -2 **G** 0 **H** 2 **J** 6

64. GRIDDED RESPONSE An ice chest contains 3 types of drinks: 10 apple juices, 15 grape juices, and 15 bottles of water. What is the probability that a drink selected randomly from the ice chest does *not* contain fruit juice?

65. SAT/ACT A certain school's enrollment increased 6% this year over last year's enrollment. If the school now has 1378 students enrolled, how many students were enrolled last year?

A 1295 **B** 1300 **C** 1350 **D** 1460

Spiral Review

Simplify. (Lesson 0-9)

66. $\sqrt{72}$

67. $\sqrt{18} \cdot \sqrt{14}$

68. $\sqrt{44x^4y^3}$

69. $\frac{3}{\sqrt{18}}$

70. $\sqrt{\frac{28}{75}}$

71. $\frac{\sqrt{8a^6}}{\sqrt{108}}$

72. $\frac{5}{4 - \sqrt{2}}$

73. $\frac{4\sqrt{3}}{2 + \sqrt{5}}$

74. FINANCIAL LITERACY Suppose you buy 3 shirts and 2 pairs of slacks on sale at a clothing store for \$72. The next day, a friend buys 2 shirts and 4 pairs of slacks for \$96. If the shirts you each bought were all the same price and the slacks were also all the same price, then the following system of equations represents this situation.

$3x + 2y = 72$
$2x + 4y = 96$

What was the cost of each shirt and each pair of slacks? (Lesson 0-8)

Graph the following geometric figures. (Lesson 0-7)

75. a triangle with vertices $A(-5, 3)$, $B(3, -4)$, and $C(-2, -3)$

76. a rectangle with vertices $P(-2, 1)$, $Q(3, 4)$, $R(5, 1)$, and $S(0, -2)$

GROCERIES Find an approximate metric weight for each item. (Lesson 0-2)

77.

Net Wt: 15 oz

78.

Net Wt: 8.2 oz

79.

Net Wt: 2.5 lb

Skills Review

Replace each ● with >, <, or = to make a true statement.

80. $\frac{1}{4}$ in. ● $\frac{1}{2}$ in.

81. $\frac{3}{4}$ in. ● $\frac{5}{8}$ in.

82. $\frac{3}{8}$ in. ● $\frac{6}{16}$ in.

83. 18 mm ● 2 cm

84. 32 mm ● 3.2 cm

85. 0.8 m ● 8 cm

EXTEND
1-1

Geometry Lab
Describing What You See

When you are learning geometric concepts, it is critical to have accurate drawings to represent the information. It is helpful to know what words and phrases can be used to describe figures. Likewise, it is important to know how to read a geometric description and be able to draw the figure it describes.

The figures and descriptions below help you visualize and write about points, lines, and planes.

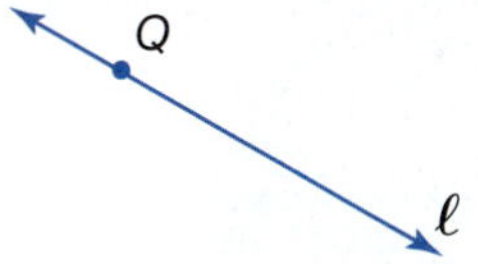

Point Q is on ℓ.
Line ℓ contains Q.
Line ℓ passes through Q.

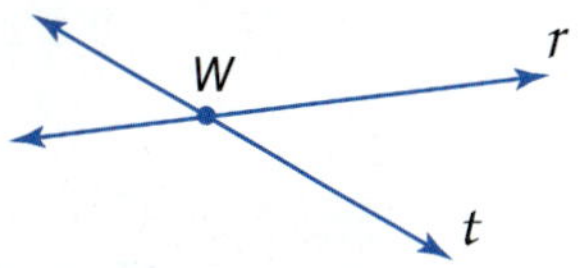

Lines r and t intersect at W.
Point W is the intersection of r and t.
Point W is on r. Point W is on t.

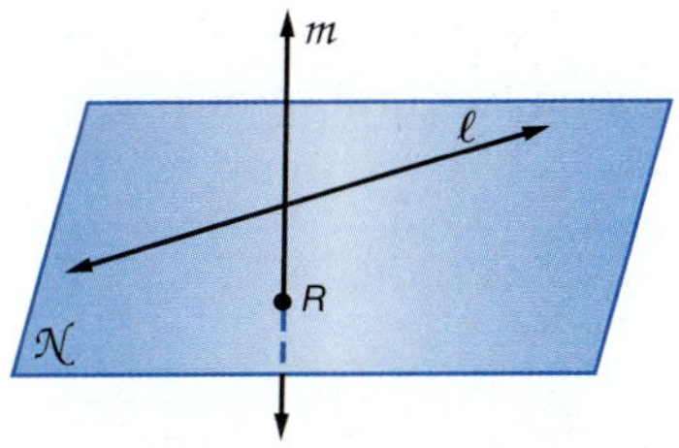

Line ℓ and point R are in $\mathcal{N}$.
Point R lies in $\mathcal{N}$.
Plane $\mathcal{N}$ contains R and ℓ.
Line m intersects $\mathcal{N}$ at R.
Point R is the intersection of m with $\mathcal{N}$.
Lines ℓ and m do not intersect.

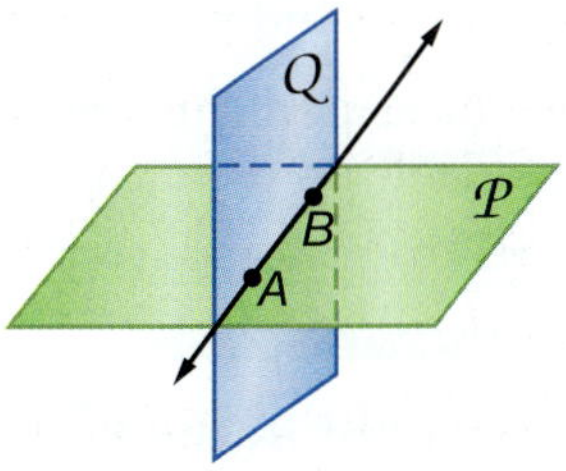

$\overleftrightarrow{AB}$ is in $\mathcal{P}$ and $\mathcal{Q}$.
Points A and B lie in both $\mathcal{P}$ and $\mathcal{Q}$.
Planes $\mathcal{P}$ and $\mathcal{Q}$ both contain $\overleftrightarrow{AB}$.
Planes $\mathcal{P}$ and $\mathcal{Q}$ intersect in $\overleftrightarrow{AB}$.
$\overleftrightarrow{AB}$ is the intersection of $\mathcal{P}$ and $\mathcal{Q}$.

Exercises

Write a description for each figure.

1.

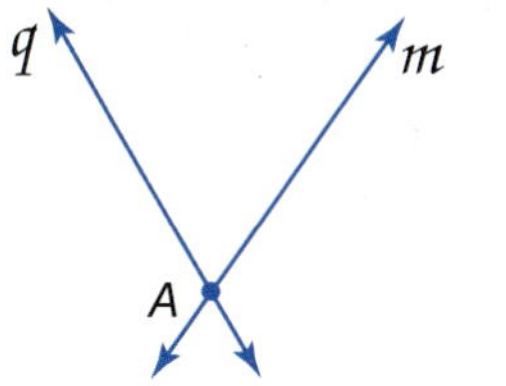

2.

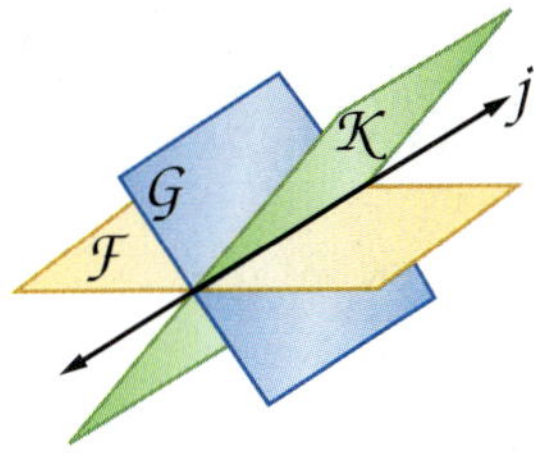

3.

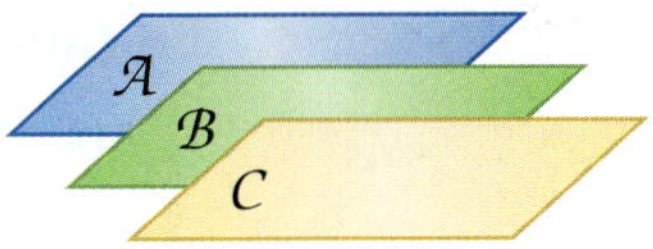

4. Draw and label a figure for the statement Planes $\mathcal{N}$ and $\mathcal{P}$ contain line a.

1-2 Linear Measure

Then

You identified and modeled points, lines, and planes. (Lesson 1-1)

Now

- Measure segments.
- Calculate with measures.

New Vocabulary

line segment
betweenness of points
between
congruent segments
construction

Math Online
glencoe.com

Why?

When the ancient Egyptians found a need for a measurement system, they used the human body as a guide. The cubit was the length of an arm from the elbow to the fingertips. Eventually the Egyptians standardized the length of a cubit, with ten *royal cubits* equivalent to one *rod*.

Measure Line Segments Unlike a line, a **line segment**, or *segment*, can be measured because it has two endpoints. A segment with endpoints A and B can be named as $\overline{AB}$ or $\overline{BA}$. The *measure* of $\overline{AB}$ is written as AB. The length or measure of a segment always includes a unit of measure, such as meter or inch.

All measurements are approximations dependent upon the smallest unit of measure available on the measuring instrument.

EXAMPLE 1 Length in Metric Units

Find the length of $\overline{AB}$ using each ruler.

a.

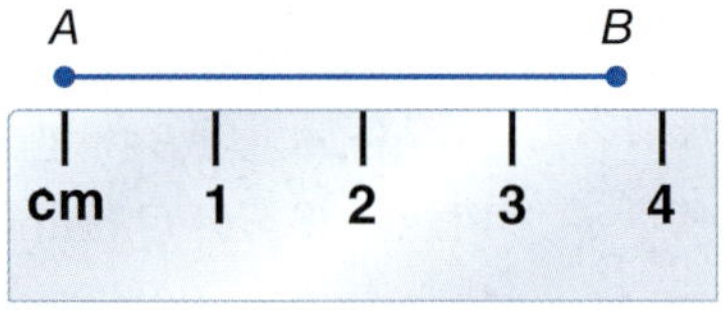

The ruler is marked in centimeters. Point B is closer to the 4-centimeter mark than to 3 centimeters.

Thus, $\overline{AB}$ is about 4 centimeters long.

b.

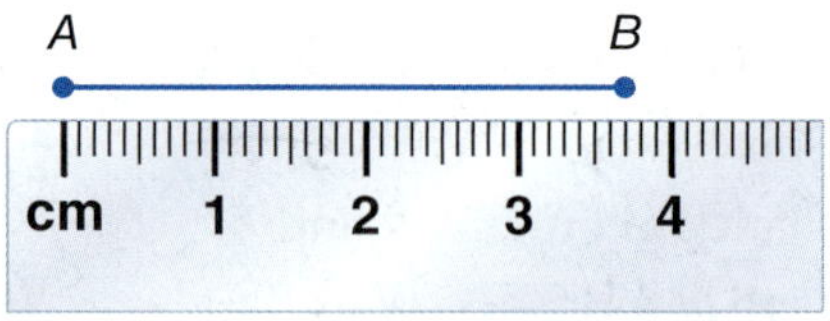

The long marks are centimeters, and the shorter marks are millimeters. There are 10 millimeters for each centimeter.

Thus, $\overline{AB}$ is about 3.7 centimeters long.

Guided Practice

1A. Measure the length of a dollar bill in centimeters.

1B. Measure the length of a pencil in millimeters.

1C. Find the length of $\overline{CD}$.

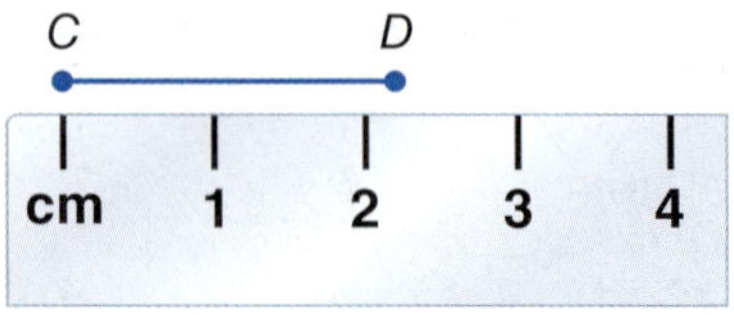

Personal Tutor glencoe.com

StudyTip

Using a Ruler The zero point on a ruler may not be clearly marked. For some rulers, zero is the left edge of the ruler. On others, it may be a fine line farther in on the scale. If it is not clear where the zero is, align one endpoint on 1 and subtract 1 from the measurement at the other endpoint.

EXAMPLE 2 Length in Standard Units

Find the length of $\overline{CD}$ using each ruler.

a.

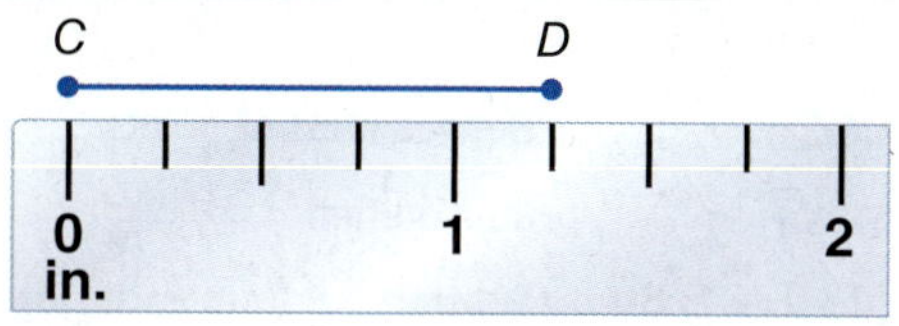

Each inch is divided into fourths.

Point D is closer to the $1\frac{1}{4}$-inch mark.

$\overline{CD}$ is about $1\frac{1}{4}$ inches long.

b.

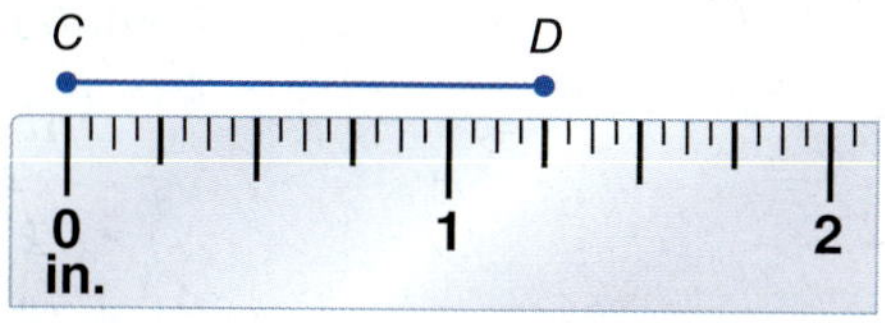

Each inch is divided into sixteenths.

Point D is closer to the $1\frac{4}{16}$-inch mark.

$\overline{CD}$ is about $1\frac{4}{16}$ or $1\frac{1}{4}$ inches long.

Guided Practice

2A. Measure the length of a dollar bill in inches.

2B. Measure the length of a pencil in inches.

Personal Tutor glencoe.com

Calculate Measures Recall that for any two real numbers a and b, there is a real number n that is *between* a and b such that $a < n < b$. This relationship also applies to points on a line and is called **betweenness of points**. In the figure, point N is between points A and B, but points R and P are not.

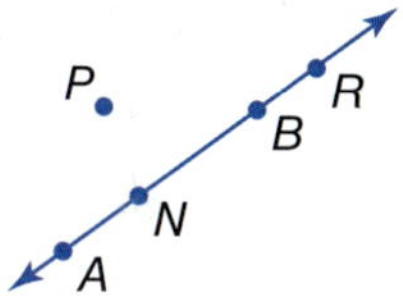

Measures are real numbers, so all arithmetic operations can be used with them. You know that the whole usually equals the sum of its parts. That is also true of line segments in geometry.

Key Concept Betweenness of Points

Words

Point M is **between** points P and Q if and only if P, Q, and M are collinear and $PM + MQ = PQ$.

Model

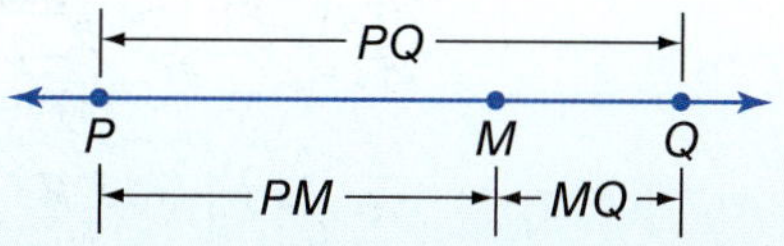

StudyTip

Comparing Measures Because measures are real numbers, you can compare them. If points X, Y, and Z are collinear in that order, then one of these statements is true: $XY = YZ$, $XY > YZ$, or $XY < YZ$.

EXAMPLE 3 Find Measurements by Adding

Find EG. Assume that the figure is not drawn to scale.

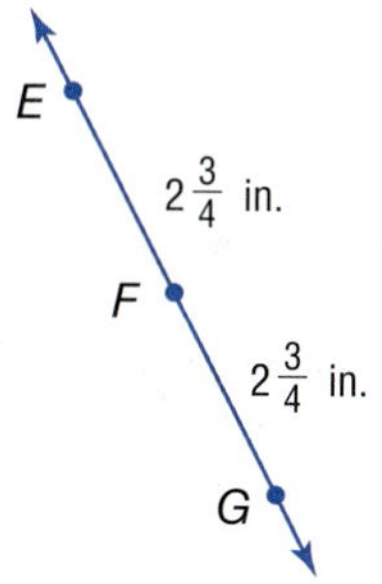

EG is the measure of $\overline{EG}$. Point F is between E and G. Find EG by adding EF and FG.

$EF + FG = EG$ **Betweenness of points**

$2\frac{3}{4} + 2\frac{3}{4} = EG$ **Substitution**

$5\frac{1}{2}$ in. $= EG$ **Add.**

Guided Practice

3. Find JL. Assume that the figure is not drawn to scale.

Personal Tutor glencoe.com

EXAMPLE 4 Find Measurements by Subtracting

Find AB. Assume that the figure is not drawn to scale.

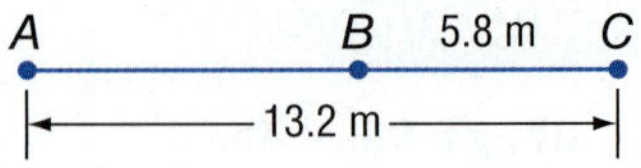

Point B is between A and C.

$AB + BC = AC$	**Betweenness of points**
$AB + 5.8 = 13.2$	**Substitution**
$AB + 5.8 - 5.8 = 13.2 - 5.8$	**Subtract 5.8 from each side.**
$AB = 7.4$ m	**Simplify.**

Guided Practice

4. Find QR. Assume that the figure is not drawn to scale.

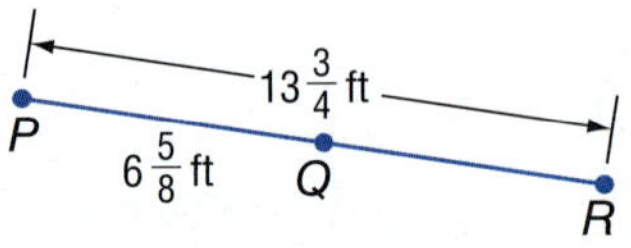

Personal Tutor glencoe.com

EXAMPLE 5 Write and Solve Equations to Find Measurements

ALGEBRA Find the value of a and XY if Y is between X and Z, $XY = 3a$, $XZ = 5a - 4$, and $YZ = 14$.

Draw a figure to represent this information.

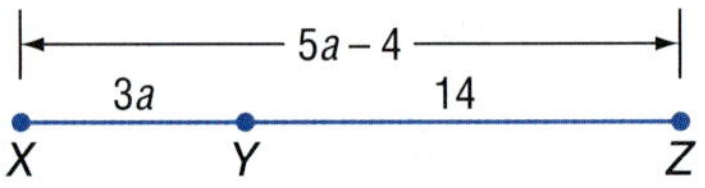

$XZ = XY + YZ$	**Betweenness of points**
$5a - 4 = 3a + 14$	**Substitution**
$5a - 4 - 3a = 3a + 14 - 3a$	**Subtract 3a from each side.**
$2a - 4 = 14$	**Simplify.**
$2a - 4 + 4 = 14 + 4$	**Add 4 to each side.**
$2a = 18$	**Simplify.**
$\frac{2a}{2} = \frac{18}{2}$	**Divide each side by 2.**
$a = 9$	**Simplify.**

Now find XY.

$XY = 3a$	**Given**
$= 3(9)$ or 27	**$a = 9$**

Guided Practice

5. Find x and BC if B is between A and C, $AC = 4x - 12$, $AB = x$, and $BC = 2x + 3$.

Personal Tutor glencoe.com

Watch Out!

Equal vs. Congruent Lengths are equal and segments are congruent. It is correct to say that $AB = CD$ and $\overline{AB} \cong \overline{CD}$. However, it is *not* correct to say that $\overline{AB} = \overline{CD}$ or that $AB \cong CD$.

Segments that have the same measure are called **congruent segments**.

Key Concept Congruent Segments

Words Congruent segments have the same measure.

Symbols $\cong$ is read *is congruent to.* Red slashes on the figure also indicate congruence.

Example $\overline{AB} \cong \overline{CD}$

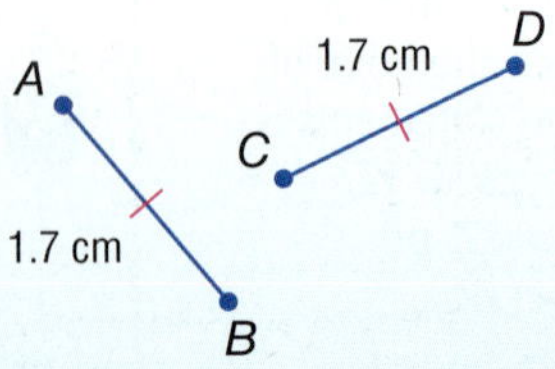

Real-World EXAMPLE 6 Congruent Segments

SKATE PARKS **In the graph, suppose a segment was drawn along the top of each bar. Which states would have segments that are congruent? Explain.**

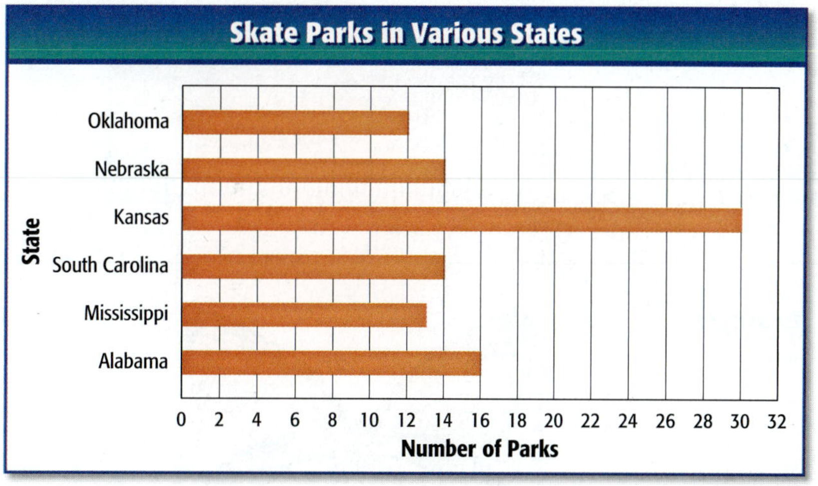

Source: SITE Design Group, Inc.

The segments on the bars for Nebraska and South Carolina would be congruent because they both represent the same number of skate parks.

Real-World Link

The first commercial skateboard was introduced in 1959. The sport has grown in popularity since then, especially after Tony Hawk won the first X-Games skateboarding competition in 1995. Now there are more than 500 skate parks in the United States.

Source: *Encyclopaedia Britannica*

Guided Practice

6A. Suppose Oklahoma added another skate park. The segment drawn along the bar representing Oklahoma would be congruent to which other segment?

6B. Name the congruent segments in the sign shown.

Personal Tutor glencoe.com

Drawings of geometric figures are created using measuring tools such as a ruler and protractor. **Constructions** are methods of creating these figures without the benefit of measuring tools. Generally, only a pencil, straightedge, and compass are used in constructions. *Sketches* are created without the use of any of these tools.

You can construct a segment that is congruent to a given segment.

Construction Copy a Segment

Step 1 Draw a segment $\overline{JK}$. Elsewhere on your paper, draw a line and a point on the line. Label the point Q.

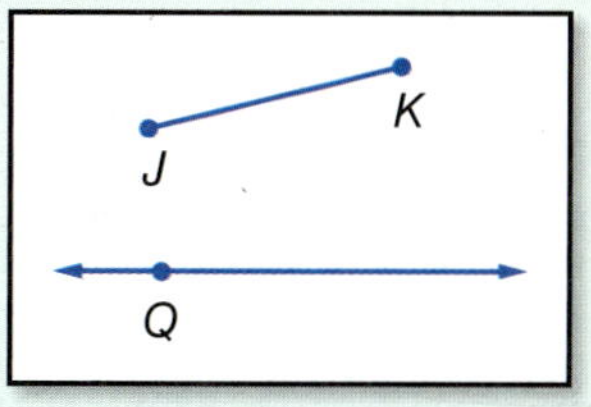

Step 2 Place the compass at point J and adjust the compass setting so that the pencil is at point K.

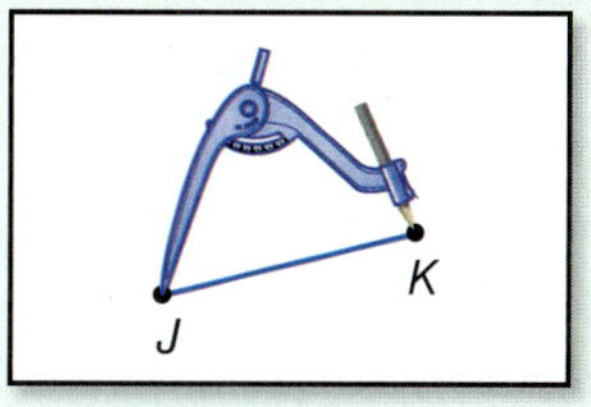

Step 3 Using that setting, place the compass point at Q and draw an arc that intersects the line. Label the point of intersection R.

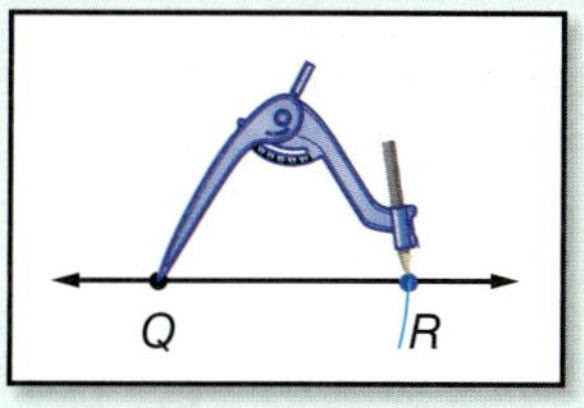

Check Your Understanding

Example 1 p. 14

Find the length of each line segment or object.

1.

2.

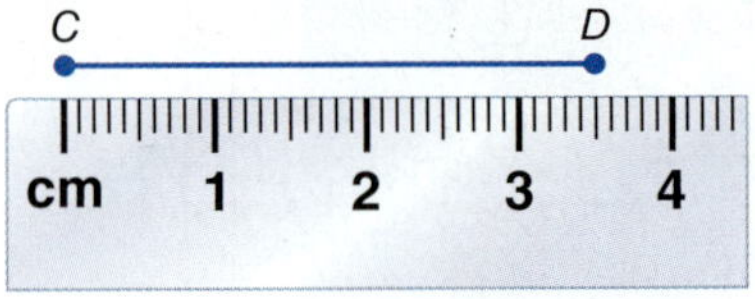

Example 2 p. 15

3.

4.

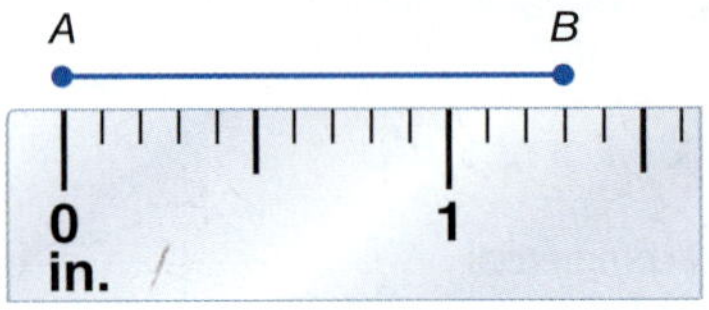

Examples 3 and 4 pp. 15–16

Find the measurement of each segment. Assume that each figure is not drawn to scale.

5. $\overline{CD}$

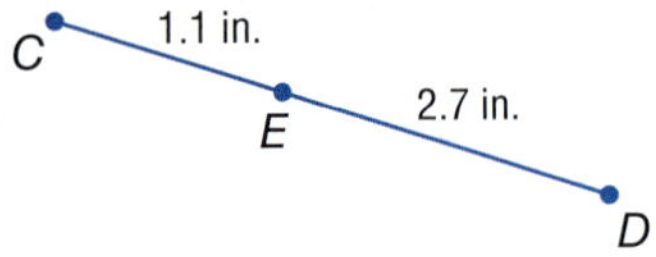

6. $\overline{RS}$

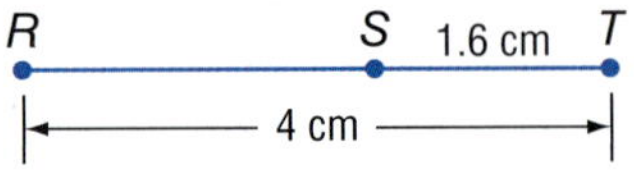

Example 5 p. 16

ALGEBRA **Find the value of x and BC if B is between C and D.**

7. $CB = 2x$, $BD = 4x$, and $BD = 12$

8. $CB = 4x - 9$, $BD = 3x + 5$, and $CD = 17$

Example 6 p. 17

9. **STATE FLAG** The Indiana State Flag was adopted in 1917. The measures of the segments between the stars and the flame are shown on the diagram in inches. List all of the congruent segments in the figure.

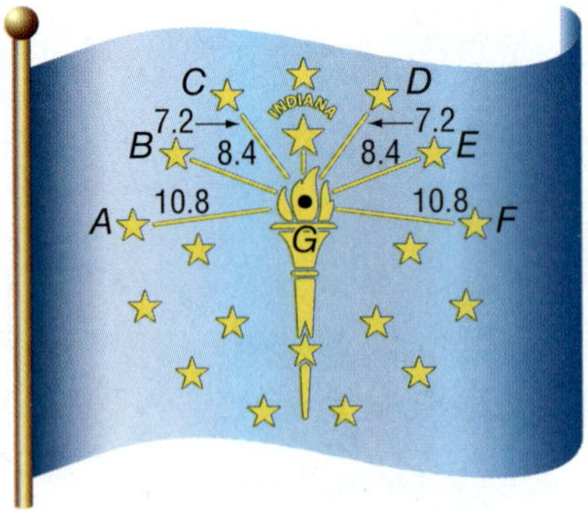

● = **Step-by-Step Solutions** begin on page R20.
Extra Practice begins on page 969.

Practice and Problem Solving

Examples 1 and 2 pp. 14–15

Find the length of each line segment.

10.

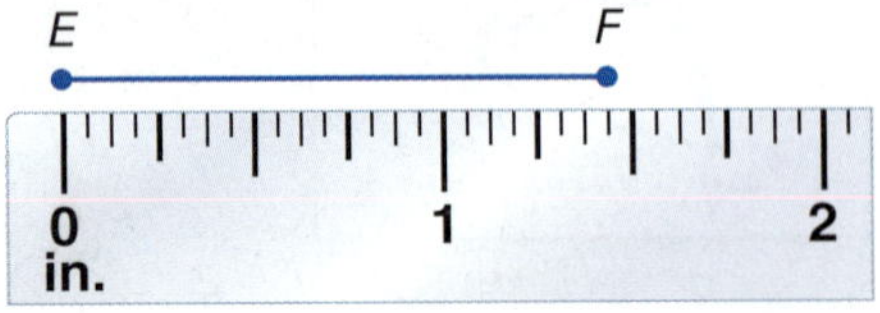

11.

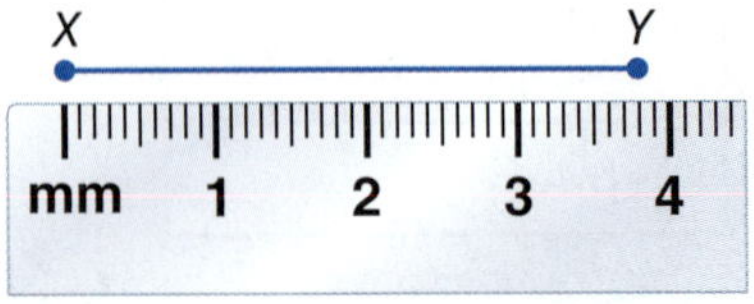

12.

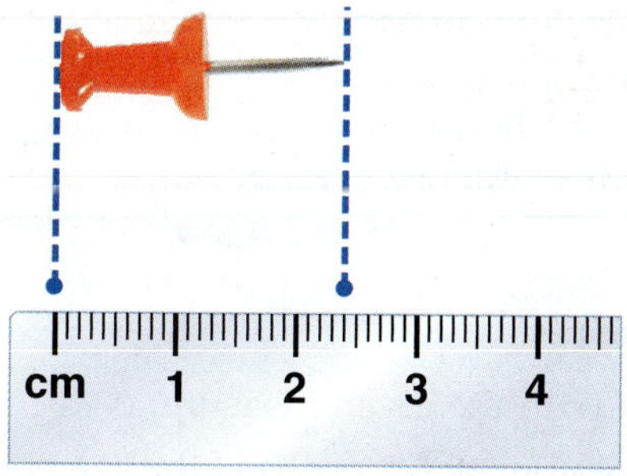

13.

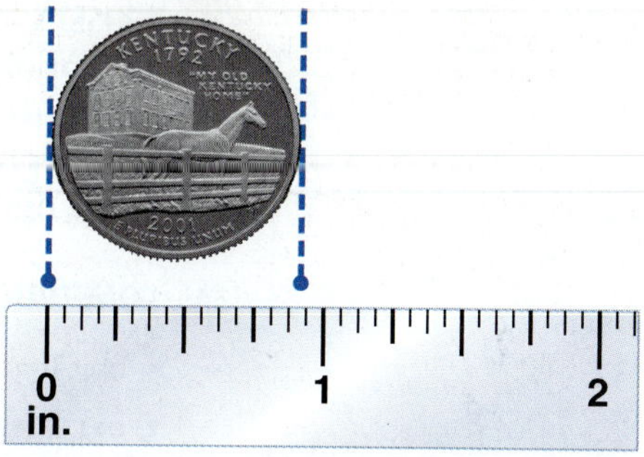

Examples 3 and 4
pp. 15–16

Find the measurement of each segment. Assume that each figure is not drawn to scale.

14. $\overline{EF}$

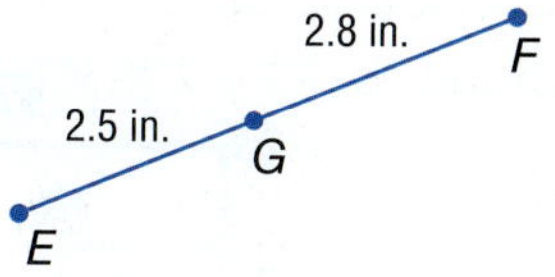

15. $\overline{JL}$

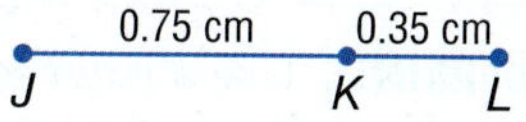

16. $\overline{PR}$

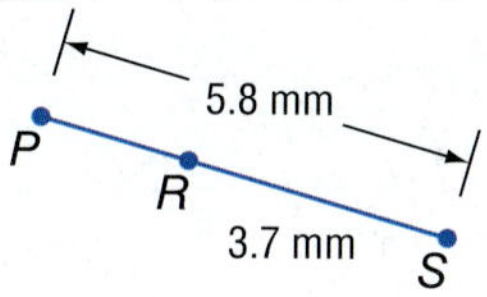

17. $\overline{SV}$

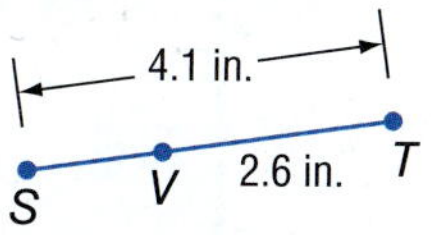

18. $\overline{WY}$

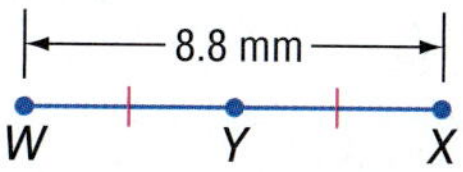

19. $\overline{FG}$

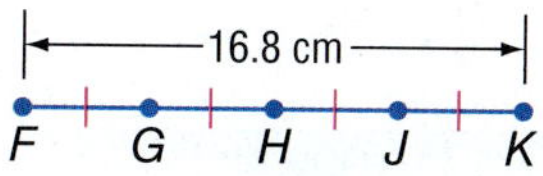

20. DONATIONS The stacked bar graph shows the number of canned food items donated by the girls and the boys in a homeroom class over the last three years. Use the concept of betweenness of points to find the number of cans donated by the boys for each year. Explain your method.

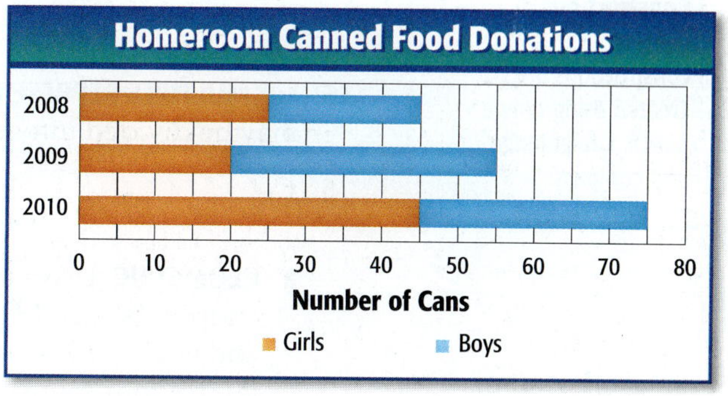

Example 5
p. 16

ALGEBRA Find the value of the variable and YZ if Y is between X and Z.

21. $XY = 11$, $YZ = 4c$, $XZ = 83$

22. $XY = 6b$, $YZ = 8b$, $XZ = 175$

23. $XY = 7a$, $YZ = 5a$, $XZ = 6a + 24$

24. $XY = 11d$, $YZ = 9d - 2$, $XZ = 5d + 28$

25. $XY = 4n + 3$, $YZ = 2n - 7$, $XZ = 22$

26. $XY = 3a - 4$, $YZ = 6a + 2$, $XZ = 5a + 22$

Example 6
p. 17

Determine whether each pair of segments is congruent.

27. $\overline{KJ}, \overline{HL}$

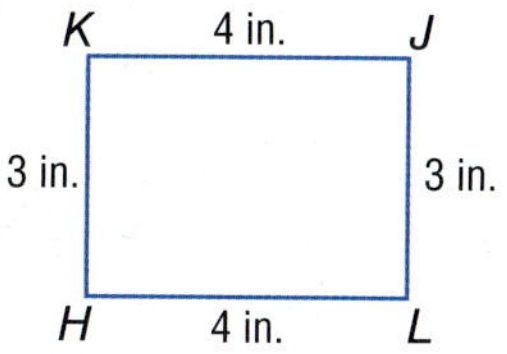

28. $\overline{AC}, \overline{BD}$

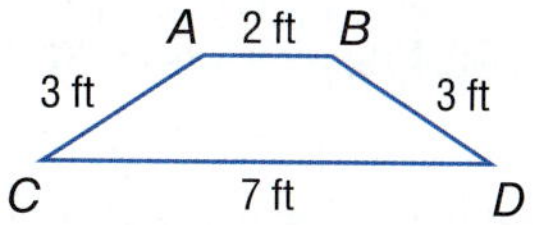

29. $\overline{EH}, \overline{FG}$

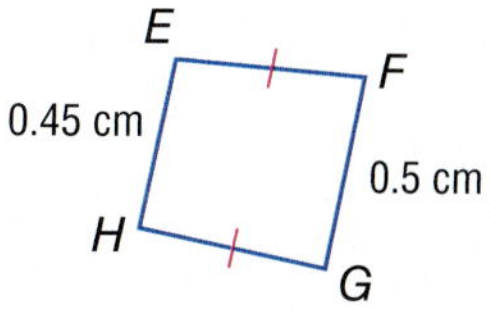

30. $\overline{VW}, \overline{UZ}$

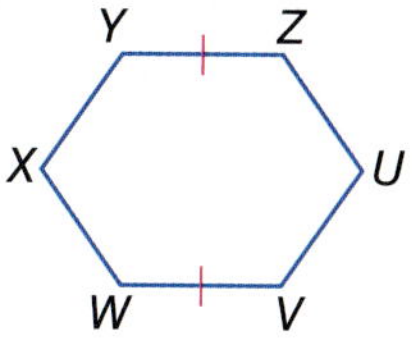

31. $\overline{MN}, \overline{RQ}$

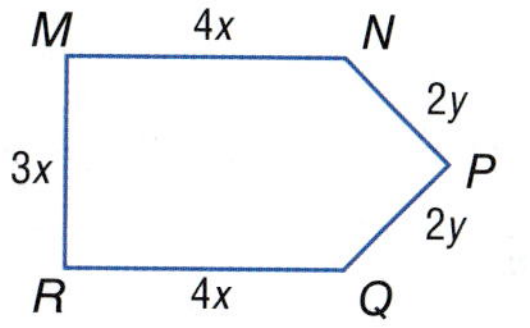

32. $\overline{SU}, \overline{VT}$

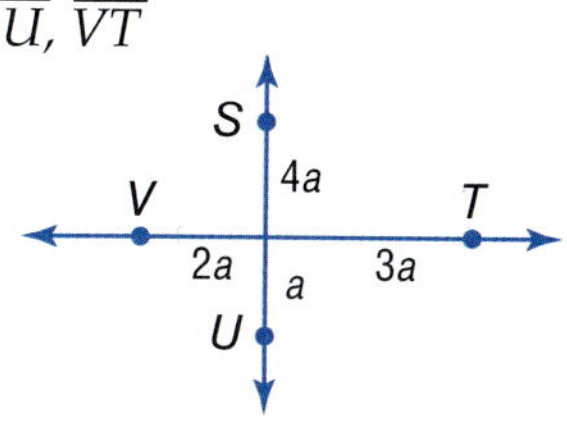

33. **TRUSSES** A truss is a structure used to support a load over a span, such as a bridge or the roof of a house. List all of the congruent segments in the figure.

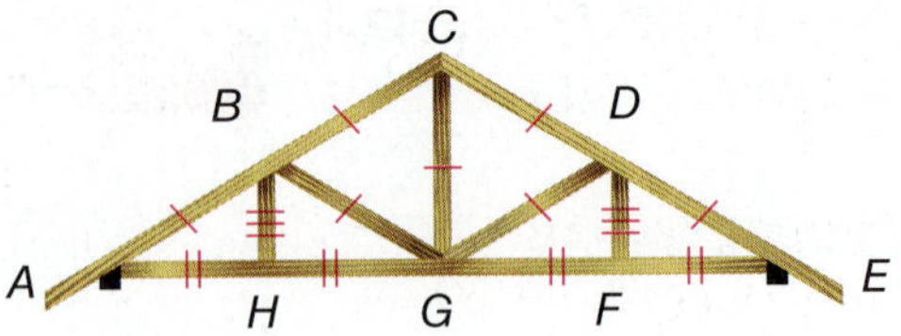

34. **CONSTRUCTION** For each expression:

- construct a segment with the given measure,
- explain the process you used to construct the segment, and
- verify that the segment you constructed has the given measure.

a. $2(XY)$ **b.** $6(WZ) - XY$

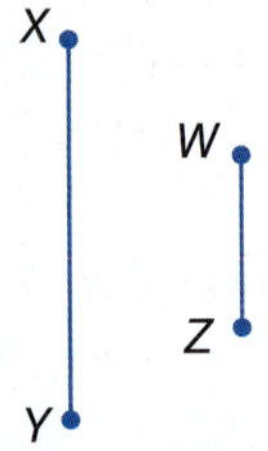

35. **BLUEPRINTS** Use a ruler to determine at least five pairs of congruent segments with labeled endpoints in the blueprint at the right.

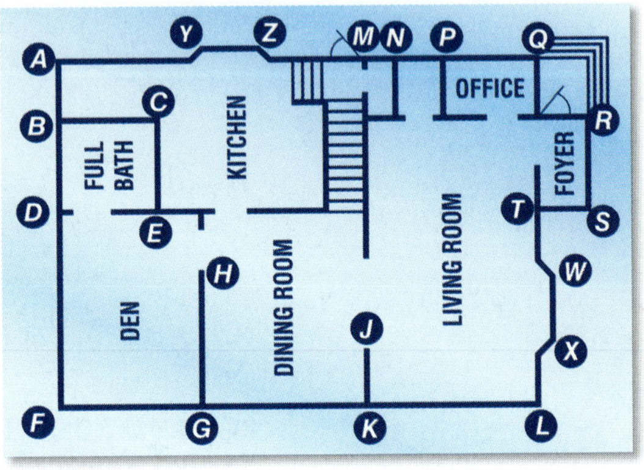

StudyTip

Constructions Remember that constructions are created using only a pencil, straightedge, and compass.

36. **MULTIPLE REPRESENTATIONS** Betweenness of points ensures that a line segment may be divided into an infinite number of line segments.

a. **GEOMETRIC** Use a ruler to draw a line segment 3 centimeters long. Label the endpoints A and D. Draw two more points along the segment and label them B and C. Draw a second line segment 6 centimeters long. Label the endpoints K and P. Add four more points along the line and label them L, M, N, and O.

b. **TABULAR** Use a ruler to measure the length of the line segment between each of the points you have drawn. Organize the lengths of the segments in $\overline{AD}$ and $\overline{KP}$ into a table. Include a column in your table to record the sum of these measures.

c. **ALGEBRAIC** Give an equation that could be used to find the lengths of $\overline{AD}$ and $\overline{KP}$. Compare the lengths determined by your equation to the actual length, 3 centimeters.

H.O.T. Problems

Use **H**igher-**O**rder **T**hinking Skills

37. **WRITING IN MATH** If point B is between points A and C, explain how you can find AC if you know AB and BC. Explain how you can find BC if you know AB and AC.

38. **OPEN ENDED** Draw a segment $\overline{AB}$ that measures between 2 and 3 inches long. Then sketch a segment $\overline{CD}$ congruent to $\overline{AB}$, draw a segment $\overline{EF}$ congruent to $\overline{AB}$, and construct a segment $\overline{GH}$ congruent to $\overline{AB}$. Compare your methods.

39. **CHALLENGE** Point K is between points J and L. If $JK = x^2 - 4x$, $KL = 3x - 2$, and $JL = 28$, write and solve an equation to find the lengths of JK and KL.

40. **REASONING** Determine whether the statement *If point M is between points C and D, then CD is greater than either CM or MD* is *sometimes*, *never*, or *always* true. Explain.

41. **WRITING IN MATH** Why is it important to have a standard of measure? Refer to page 14, and include an advantage and disadvantage to the builders of the pyramids.

Standardized Test Practice

42. SHORT RESPONSE A 36-foot-long ribbon is cut into three pieces. The first piece of ribbon is half as long as the second piece of ribbon. The third piece is 1 foot longer than twice the length of the second piece of ribbon. How long is the longest piece of ribbon?

43. In the figure, points A, B, C, D, and E are collinear. If $AE = 38$, $BD = 15$, and $\overline{BC} \cong \overline{CD} \cong \overline{DE}$, what is the length of $\overline{AD}$?

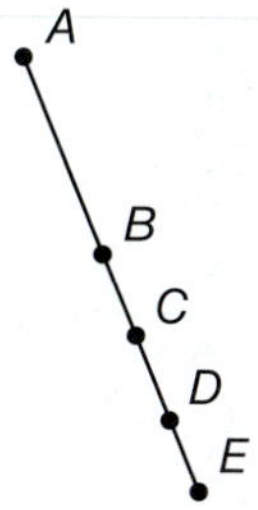

A 7.5
B 15
C 22.5
D 30.5

44. SAT/ACT If $f(x) = 7x^2 - 4x$, what is the value of $f(2)$?

F −8
G 2
H 20
J 28

45. ALGEBRA
Simplify $(3x^2 - 2)(2x + 4) - 2x^2 + 6x + 7$.

A $4x^2 + 14x - 1$
B $4x^2 - 14x + 15$
C $6x^3 + 12x^2 + 2x - 1$
D $6x^3 + 10x^2 + 2x - 1$

Spiral Review

Refer to the figure. (Lesson 1-1)

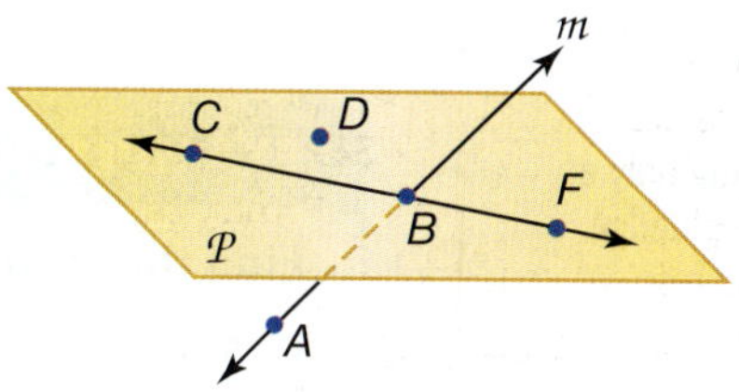

46. What are two other names for $\overleftrightarrow{AB}$?

47. Give another name for plane $\mathcal{P}$.

48. Name the intersection of plane $\mathcal{P}$ and $\overleftrightarrow{AB}$.

49. Name three collinear points.

50. Name two points that are not coplanar.

51. CLOCKS The period of a pendulum is the time required for it to make one complete swing back and forth. The formula of the period P in seconds of a pendulum is $P = 2\pi\sqrt{\frac{\ell}{32}}$, where ℓ is the length of the pendulum in feet. (Lesson 0-9)

a. What is the period of the pendulum in the clock shown to the nearest tenth of a second?

b. About how many inches long should the pendulum be in order for it to have a period of 1 second?

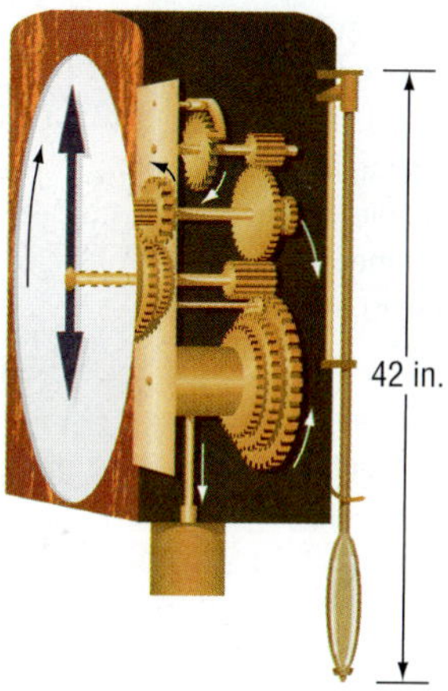

Solve each inequality. (Lesson 0-6)

52. $-14n \geq 42$

53. $p + 6 > 15$

54. $-2a - 5 < 20$

55. $5x \leq 3x - 26$

Skills Review

Evaluate each expression if $a = -7$, $b = 4$, $c = -3$, and $d = 5$. (Lessons 0-4 and 0-9)

56. $b - c$

57. $|a - d|$

58. $|d - c|$

59. $\frac{b - a}{2}$

60. $(a - c)^2$

61. $\sqrt{(a - b)^2 + (c - d)^2}$

EXTEND 1-2

Extension Lesson
Precision and Accuracy

Objectives
- Determine precision of measurements.
- Determine accuracy of measurements.

As stated in Lesson 1-2, all measurements are approximations. Two main factors are considered when determining the quality of such an approximation.

- How *precise* is the measure?
- How *accurate* is the measure?

Precision **Precision** refers to the clustering of a group of measurements. It depends only on the smallest unit of measure available on a measuring tool. Suppose you are told that a segment measures 8 centimeters. The length, to the nearest centimeter, of each segment shown below is 8 centimeters.

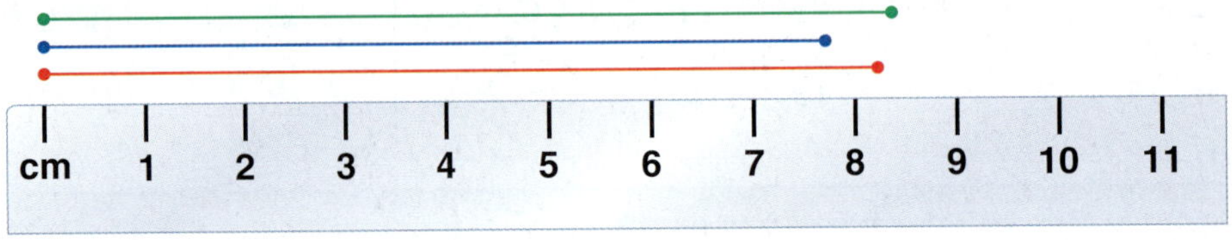

Notice that the exact length of each segment above is between 7.5 and 8.5 centimeters, or within 0.5 centimeter of 8 centimeters. The **absolute error** of a measurement is equal to one half the unit of measure. The smaller the unit of measure, the more precise the measurement.

StudyTip

Precision The absolute error of a measurement in customary units is determined before reducing the fraction. For example, if you measure the length of an object to be $1\frac{4}{16}$ inches, then the absolute error measurement is precise to within $\frac{1}{32}$ inch.

EXAMPLE 1 Find Absolute Error

Find the absolute error of each measurement. Then explain its meaning.

a. 6.4 centimeters

The measure is given to the nearest 0.1 centimeter, so the absolute error of this measurement is $\frac{1}{2}(0.1)$ or 0.05 centimeter. Therefore, the exact measurement could be between 6.35 and 6.45 centimeters. The two segments below measure 6.4 ± 0.05 centimeters.

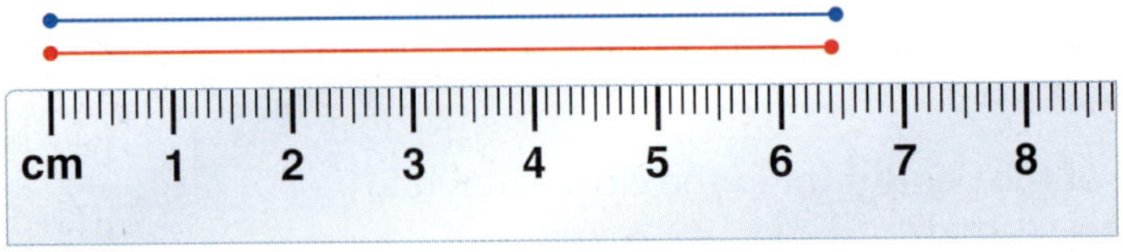

b. $2\frac{1}{4}$ inches

The measure is given to the nearest $\frac{1}{4}$ inch, so the absolute error of this measurement is $\frac{1}{2}\left(\frac{1}{4}\right)$ or $\frac{1}{8}$ inch. Therefore, the exact measurement could be between $2\frac{1}{8}$ and $2\frac{3}{8}$ inches. The two segments below measure $2\frac{1}{4} \pm \frac{1}{8}$ inches.

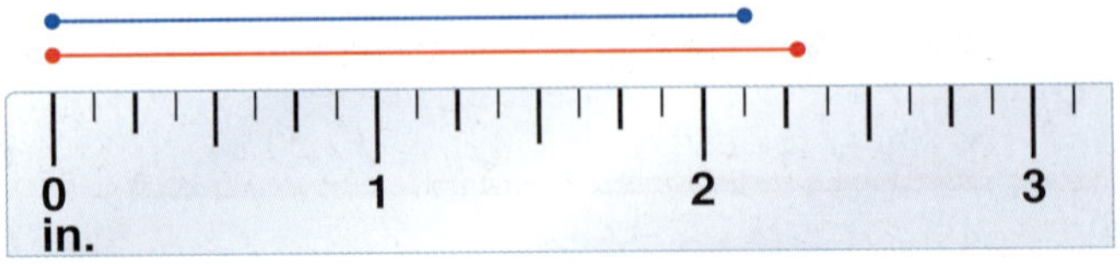

Guided Practice

1A. $1\frac{1}{2}$ inches

1B. 4 centimeters

Precision in a measurement is usually expressed by the number of **significant digits** reported. Reporting that the measure of $\overline{AB}$ is 4 centimeters is *less precise* than reporting that the measure of $\overline{AB}$ is 4.1 centimeters.

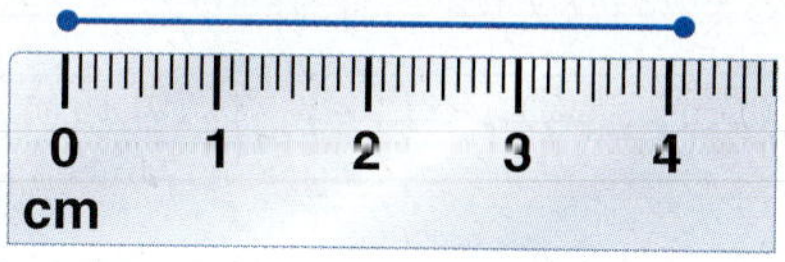

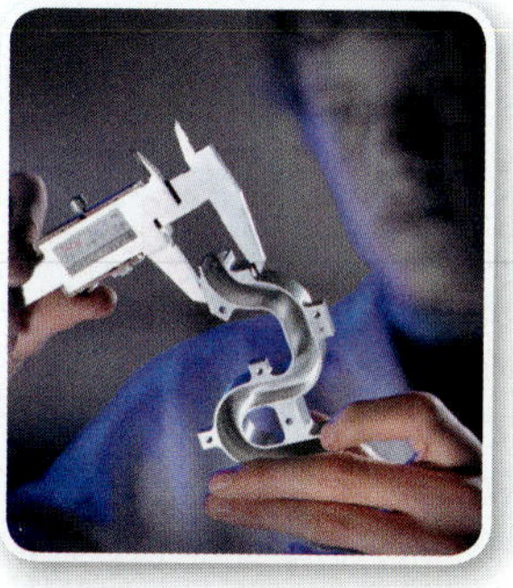

To determine whether digits are considered significant, use the following rules.

- Nonzero digits are always significant.
- In whole numbers, zeros are significant if they fall between nonzero digits.
- In decimal numbers greater than or equal to 1, every digit is significant.
- In decimal numbers less than 1, the first nonzero digit and every digit to the right are significant.

Real-World Link

Precision in measurement in the real world usually comes at a price.

- Precision in a process to 3 significant digits, commercial quality, can cost $100.
- Precision in a process to 4 significant digits, industrial quality, can cost $500.
- Precision in a process to 5 significant digits, scientific quality, can cost $2500.

Source: Southwest Texas Junior College

EXAMPLE 2 Significant Digits

Determine the number of significant digits in each measurement.

a. 430.008 meters

Since this is a decimal number greater than 1, every digit is significant. So, this measurement has 6 significant digits.

b. 0.00750 centimeter

This is a decimal number less than 1. The first nonzero digit is 7, and there are two digits to the right of 7, 5 and 0. So, this measurement has 3 significant digits.

Guided Practice

2A. 779,000 mi **2B.** 50,008 ft **2C.** 230.004500 m

Accuracy **Accuracy** refers to how close a measured value comes to the actual or desired value. The accuracy of a measurement should not be confused with its precision. Consider the target practice results shown below.

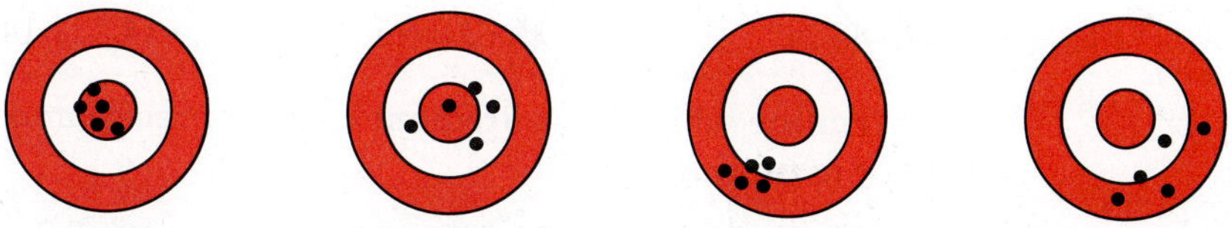

accurate and precise — accurate but not precise — precise but not accurate — not accurate and not precise

The relative error of a measure is the ratio of the absolute error to the expected measure. A measurement with a smaller relative error is said to be more accurate.

StudyTip

Accuracy The accuracy or relative error of a measurement depends on both the absolute error and the size of the object being measured.

EXAMPLE 3 Find Relative Error

MANUFACTURING **A manufacturer measures each part for a piece of equipment to be 23 centimeters in length. Find the relative error of this measurement.**

$$\text{relative error} = \frac{\text{absolute error}}{\text{expected measure}} = \frac{0.5 \text{ cm}}{23 \text{ cm}} \approx 0.022 \text{ or } 2.2\%$$

Guided Practice

Find the relative error of each measurement.

3A. 3.2 mi **3B.** 1 ft **3C.** 26 ft

Practice and Problem Solving

Find the absolute error of each measurement. Then explain its meaning.

1. 12 yd **2.** $50\frac{4}{16}$ in. **3.** 3.28 ft **4.** 2.759 cm

5. ERROR ANALYSIS In biology class, Manuel and Jocelyn measure a beetle as shown. Manuel says that the beetle measures between $1\frac{5}{8}$ and $1\frac{3}{4}$ inches. Jocelyn says that it measures between $1\frac{9}{16}$ and $1\frac{5}{8}$ inches. Is either of their statements about the beetle's measure correct? Explain your reasoning.

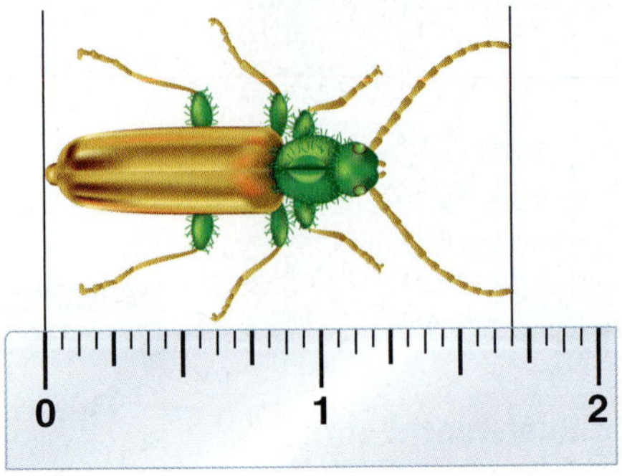

6. PYRAMIDS Refer to the information about royal cubits on page 14 and in the caption at the left. What are the greatest and least possible lengths of the sides of the Great Pyramid of Giza?

Real-World Link

Research suggests that the design dimensions of the Great Pyramid of Giza in Egypt were 440 by 440 royal cubits. The sides of the pyramid are precise within 0.05%.

Source: *Encyclopaedia Britannica*

Determine the number of significant digits in each measurement.

7. 4.05 in. **8.** 53,000 mi **9.** 0.0005 mm **10.** 750,001 ft

11. VOLUME When multiplying or dividing measures, the product or quotient should have only as many significant digits as the multiplied or divided measurement showing the least number of significant digits. To how many significant digits should the volume of the rectangle prism shown be reported? Report the volume to this number of significant digits.

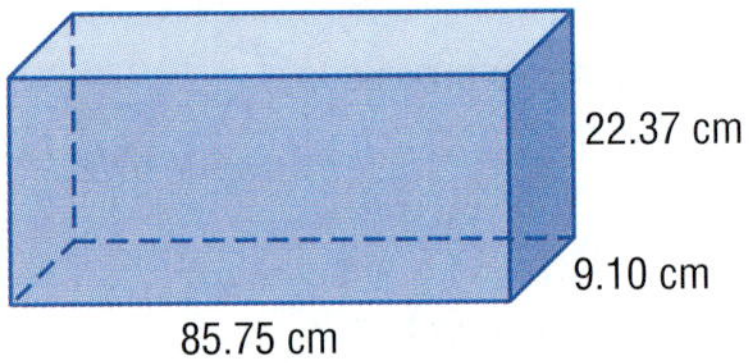

Find the relative error of each measurement.

12. 48 in. **13.** 2.0 mi **14.** 11.14 cm **15.** 0.6 m

Determine which measurement is more precise and which is more accurate. Explain your reasoning.

16. 22.4 ft; 5.82 ft **17.** 25 mi; 8 mi **18.** 9.2 cm; 42 mm **19.** $18\frac{1}{4}$ in.; 125 yd

For each situation, determine the level of accuracy needed. Explain.

20. You are estimating the height of a person. Which unit of measure should you use: 1 foot, 1 inch, or $\frac{1}{16}$ inch?

21. You are estimating the height of a mountain. Which unit of measure should you use: 1 foot, 1 inch, or $\frac{1}{16}$ inch?

22. PERIMETER The *perimeter* of a geometric figure is the sum of the lengths of its sides. Jermaine uses a ruler divided into inches and measures the sides of a rectangle to be $2\frac{1}{4}$ inches and $4\frac{3}{4}$ inches.

a. What is the least possible perimeter of the rectangle? Explain.

b. What is the greatest possible perimeter of the rectangle? Explain.

23. WRITING IN MATH Give an example of a situation in which having a measure with an absolute error of 0.5 mile would likely not cause a problem. Give an example of a situation in which this same absolute error would likely cause a problem.

1-3 Distance and Midpoints

Then
You graphed points on the coordinate plane. (Lesson 0-2)

Now
- Find the distance between two points.
- Find the midpoint of a segment.

New Vocabulary
distance
midpoint
segment bisector

Math Online
glencoe.com

Why?

The location of a city on a map is given in degrees of latitude and longitude. For short distances, the Pythagorean Theorem can be used to approximate the distance between two locations.

Distance Between Two Points The **distance** between two points is the length of the segment with those points as its endpoints. The coordinates of the points can be used to find this length. Because $\overline{PQ}$ is the same as $\overline{QP}$, the order in which you name the endpoints is not important when calculating distance.

Key Concept

Distance Formula (on Number Line)

Words The distance between two points is the absolute value of the difference between their coordinates.

Symbols If P has coordinate x_1 and Q has coordinate x_2, $PQ = |x_2 - x_1|$ or $|x_1 - x_2|$.

EXAMPLE 1 Find Distance on a Number Line

Use the number line to find BE.

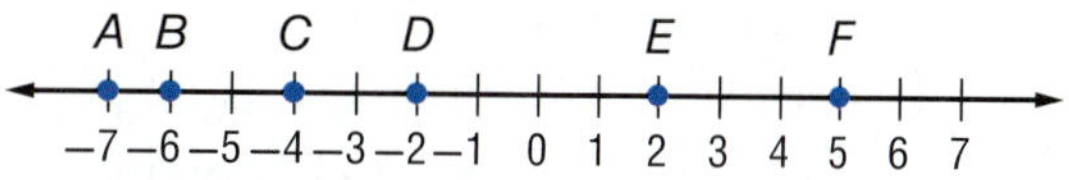

The coordinates of B and E are −6 and 2.

$BE = |x_2 - x_1|$ **Distance Formula**

$= |2 - (-6)|$ **$x_1 = -6$ and $x_2 = 2$**

$= 8$ **Simplify.**

Guided Practice

Use the number line above to find each measure.

1A. AC **1B.** CF **1C.** FB

Personal Tutor glencoe.com

StudyTip

Pythagorean Theorem Recall that the Pythagorean Theorem is often expressed as $a^2 + b^2 = c^2$, where a and b are the measures of the shorter sides (legs) of a right triangle, and c is the measure of the longest side (hypotenuse). You will prove and learn about other applications of the Pythagorean Theorem in Lesson 8-2.

To find the distance between two points A and B in the coordinate plane, you can form a right triangle with $\overline{AB}$ as its hypotenuse and point C as its vertex as shown. Then use the Pythagorean Theorem to find AB.

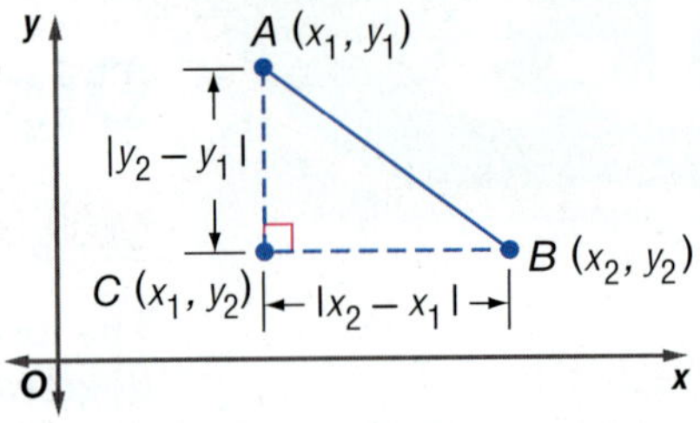

$(CB)^2 + (AC)^2 = (AB)^2$ — **Pythagorean Theorem**

$(|x_2 - x_1|)^2 + (|y_2 - y_1|)^2 = (AB)^2$ — **$CB = |x_2 - x_1|$, $AC = |y_2 - y_1|$**

$(x_2 - x_1)^2 + (y_2 - y_1)^2 = (AB)^2$ — **The square of a number is always positive.**

$\sqrt{(x_2 - x_1)^2 + (y_2 - y_1)^2} = AB$ — **Take the positive square root of each side.**

This gives us a Distance Formula for points in the coordinate plane.

Key Concept — Distance Formula (in Coordinate Plane)

If P has coordinates (x_1, y_1) and Q has coordinates (x_2, y_2), then

$$PQ = \sqrt{(x_2 - x_1)^2 + (y_2 - y_1)^2}.$$

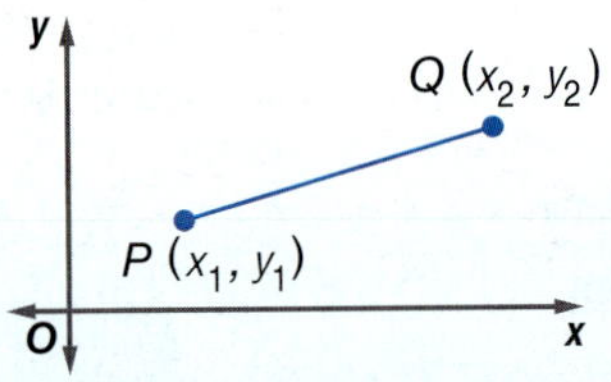

As with a coordinate on a number line, the order of the x- and y-coordinates in each set of parentheses is not important.

EXAMPLE 2 Find Distance on a Coordinate Plane

Find the distance between $C(-4, -6)$ and $D(5, -1)$.

$CD = \sqrt{(x_2 - x_1)^2 + (y_2 - y_1)^2}$ — **Distance Formula**

$= \sqrt{[5 - (-4)]^2 + [-1 - (-6)]^2}$ — **$(x_1, y_1) = (-4, -6)$ and $(x_2, y_2) = (5, -1)$**

$= \sqrt{9^2 + 5^2}$ — **Subtract.**

$= \sqrt{81 + 25}$ or $\sqrt{106}$ — **Simplify.**

The distance between C and D is $\sqrt{106}$ units. Use a calculator to find that $\sqrt{106}$ units is approximately 10.3 units.

CHECK Graph the ordered pairs and check by using the Pythagorean Theorem.

$(CD)^2 \stackrel{?}{=} (EC)^2 + (ED)^2$

$(CD)^2 \stackrel{?}{=} 5^2 + 9^2$

$(CD)^2 \stackrel{?}{=} 106$

$CD = \sqrt{106}$ ✓

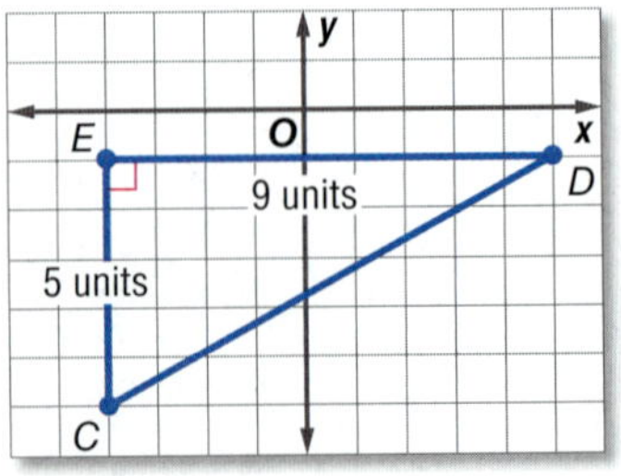

Guided Practice

Find the distance between each pair of points.

2A. $E(-5, 6)$ and $F(8, -4)$

2B. $J(4, 3)$ and $K(-3, -7)$

Personal Tutor glencoe.com

Midpoint of a Segment The **midpoint** of a segment is the point halfway between the endpoints of the segment. If X is the midpoint of $\overline{AB}$, then $AX = XB$ and $\overline{AX} \cong \overline{XB}$. You can find the midpoint of a segment on a number line by finding the *mean*, or the average, of the coordinates of its endpoints.

Key Concept — Midpoint Formula (on Number Line)

If $\overline{AB}$ has endpoints at x_1 and x_2 on a number line, then the midpoint M of $\overline{AB}$ has coordinate

$$\frac{x_1 + x_2}{2}.$$

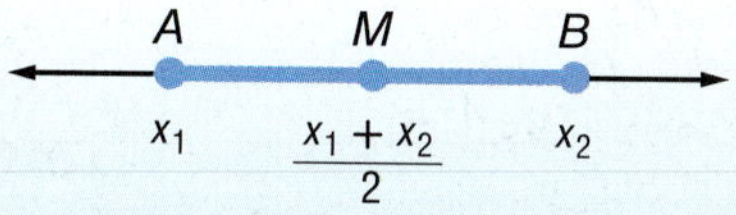

Real-World EXAMPLE 3 — Find Midpoint on a Number Line

DECORATING Jacinta hangs a picture 15 inches from the left side of a wall. How far from the edge of the wall should she mark the location for the nail the picture will hang on if the right edge is 37.5 inches from the wall's left side?

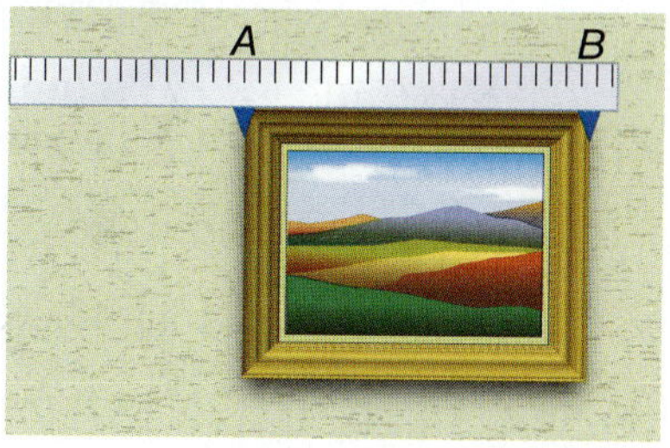

The coordinates of the endpoints of the top of the picture frame are 15 inches and 37.5 inches. Let M be the midpoint of $\overline{AB}$.

$M = \frac{x_1 + x_2}{2}$ — **Midpoint Formula**

$= \frac{15 + 37.5}{2}$ — **$x_1 = 15$, $x_2 = 37.5$**

$= \frac{52.5}{2}$ or 26.25 — **Simplify.**

The midpoint is located at 26.25 or $26\frac{1}{4}$ inches from the left edge of the wall.

StudyTip

Alternative Method In Example 3, the coordinate of the midpoint could also have been located by first finding the length of AB, which is $37.5 - 15$ or 22.5 inches. Half of this measure is the distance from one endpoint to the point midway between A and B, $\frac{22.5}{2}$ or 11.25. Add this distance to point A's distance from the left wall. So the midpoint between A and B is $15 + 11.25$ or 26.25 inches from the left wall.

Guided Practice

3. **TEMPERATURE** The temperature on a thermometer dropped from a reading of 25° to −8°. Find the midpoint of these temperatures.

Personal Tutor glencoe.com

You can find the midpoint of a segment on the coordinate plane by finding the average of the x-coordinates and of the y-coordinates of the endpoints.

Key Concept — Midpoint Formula (in Coordinate Plane)

If $\overline{PQ}$ has endpoints at $P(x_1, y_1)$ and $Q(x_2, y_2)$ in the coordinate plane, then the midpoint M of $\overline{PQ}$ has coordinates

$$M\left(\frac{x_1 + x_2}{2}, \frac{y_1 + y_2}{2}\right).$$

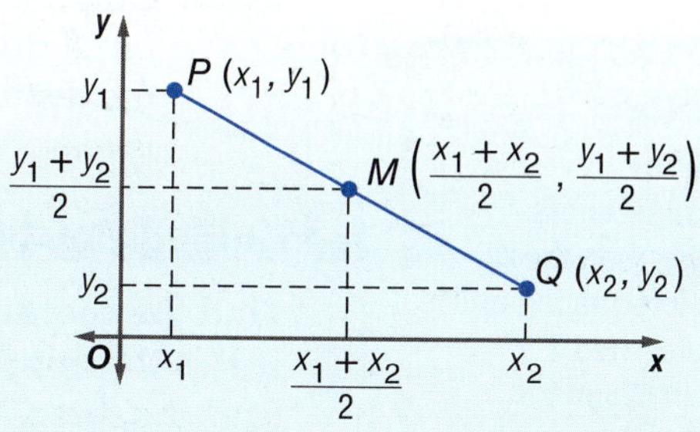

When finding the midpoint of a segment, the order of the coordinates of the endpoints is not important.

EXAMPLE 4 Find Midpoint in Coordinate Plane

Find the coordinates of M, the midpoint of $\overline{ST}$, for $S(-6, 3)$ and $T(1, 0)$.

$$M = \left(\frac{x_1 + x_2}{2}, \frac{y_1 + y_2}{2}\right) \quad \textbf{Midpoint Formula}$$

$$= \left(\frac{-6 + 1}{2}, \frac{3 + 0}{2}\right) \quad (x_1, y_1) = S(-6, 3),\ (x_2, y_2) = T(1, 0)$$

$$= \left(\frac{-5}{2}, \frac{3}{2}\right) \text{ or } M\left(-2\frac{1}{2}, 1\frac{1}{2}\right) \quad \textbf{Simplify.}$$

CHECK Graph S, T, and M. The distance from S to M does appear to be the same as the distance from M to T, so our answer is reasonable.

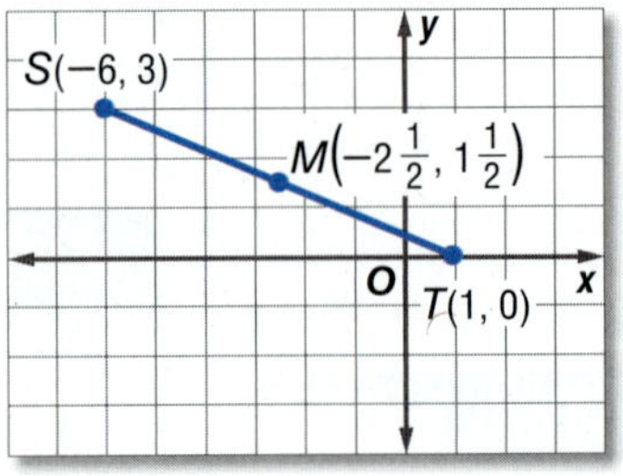

Guided Practice

Find the coordinates of the midpoint of a segment with the given coordinates.

4A. $A(5, 12)$, $B(-4, 8)$

4B. $C(-8, -2)$, $D(5, 1)$

Personal Tutor glencoe.com

You can also find the coordinates of the endpoint of a segment if you know the coordinates of its other endpoint and its midpoint.

EXAMPLE 5 Find the Coordinates of an Endpoint

Find the coordinates of J if $K(-1, 2)$ is the midpoint of $\overline{JL}$ and L has coordinates $(3, -5)$.

Step 1 Let J be (x_1, y_1) and L be (x_2, y_2) in the Midpoint Formula.

$$K\left(\frac{x_1 + 3}{2}, \frac{y_1 + (-5)}{2}\right) = K(-1, 2) \quad (x_2, y_2) = (3, -5)$$

Step 2 Write two equations to find the coordinates of J.

$\frac{x_1 + 3}{2} = -1$ **Midpoint Formula**

$x_1 + 3 = -2$ **Multiply each side by 2.**

$x_1 = -5$ **Subtract 3 from each side.**

$\frac{y_1 + (-5)}{2} = 2$ **Midpoint Formula**

$y_1 - 5 = 4$ **Multiply each side by 2.**

$y_1 = 9$ **Add 5 to each side.**

The coordinates of J are $(-5, 9)$.

CHECK Graph J, K, and L. The distance from J to K does appear to be the same as the distance from K to L, so our answer is reasonable.

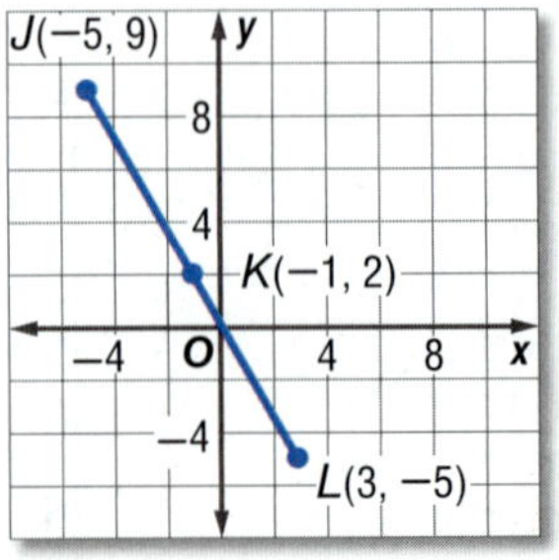

Guided Practice

Find the coordinates of the missing endpoint if P is the midpoint of $\overline{EG}$.

5A. $E(-8, 6)$, $P(-5, 10)$

5B. $P(-1, 3)$, $G(5, 6)$

Personal Tutor glencoe.com

StudyTip

Check for Reasonableness Always graph the given information and the calculated coordinates of the third point to check the reasonableness of your answer.

You can use algebra to find a missing measure or value in a figure that involves the midpoint of a segment.

EXAMPLE 6 Use Algebra to Find Measures

ALGEBRA **Find the measure of $\overline{PQ}$ if Q is the midpoint of $\overline{PR}$.**

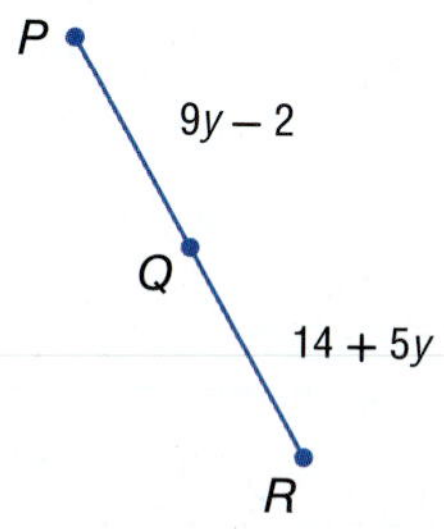

Understand You know that Q is the midpoint of $\overline{PR}$. You are asked to find the measure of $\overline{PQ}$.

Plan Because Q is the midpoint, you know that $PQ = QR$. Use this equation to find a value for y.

Solve

$PQ = QR$	Definition of midpoint
$9y - 2 = 14 + 5y$	$PQ = 9y - 2$, $QR = 14 + 5y$
$4y - 2 = 14$	Subtract $5y$ from each side.
$4y = 16$	Add 2 to each side.
$y = 4$	Divide each side by 4.

Now substitute 4 for y in the expression for PQ.

$PQ = 9y - 2$	Original measure
$= 9(4) - 2$	$y = 4$
$= 36 - 2$ or 34	Simplify.

The measure of $\overline{PQ}$ is 34.

Check Since $PQ = QR$, when the expression for QR is evaluated for 4, it should also be 34.

$QR = 14 + 5y$	Original measure
$\stackrel{?}{=} 14 + 5(4)$	$y = 4$
$= 34$ ✓	Simplify.

Guided Practice

6A. Find the measure of $\overline{YZ}$ if Y is the midpoint of $\overline{XZ}$ and $XY = 2x - 3$ and $YZ = 27 - 4x$.

6B. Find the value of x if C is the midpoint of $\overline{AB}$, $AC = 4x + 5$, and $AB = 78$.

Personal Tutor glencoe.com

StudyTip

Segment Bisectors There can be an infinite number of bisectors and each must contain the midpoint of the segment.

Any segment, line, or plane that intersects a segment at its midpoint is called a **segment bisector**. In the figure at the right, M is the midpoint of $\overline{PQ}$. Plane $\mathcal{A}$, $\overline{MJ}$, $\overleftrightarrow{KM}$, and point M are all bisectors of $\overline{PQ}$. We say that they *bisect* $\overline{PQ}$.

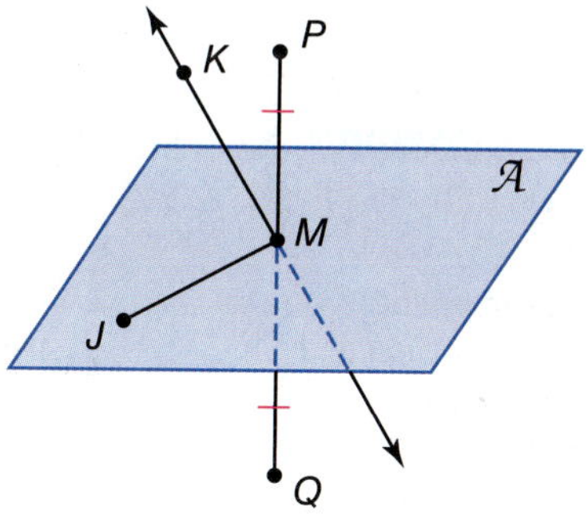

The construction on the following page shows how to construct a line that bisects a segment to find the midpoint of a given segment.

Construction — Bisect a Segment

Step 1 Draw a segment and name it $\overline{AB}$. Place the compass at point A. Adjust the compass so that its width is greater than $\frac{1}{2}AB$. Draw arcs above and below $\overline{AB}$.

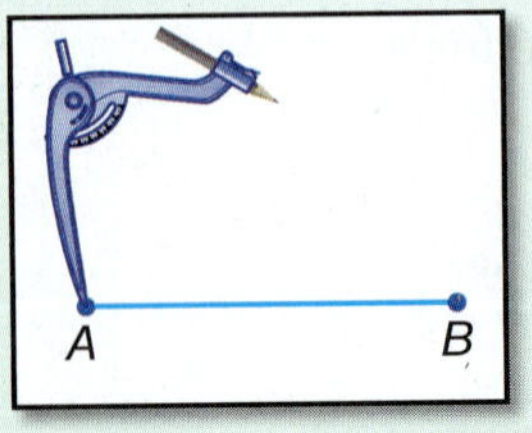

Step 2 Using the same compass setting, place the compass at point B and draw arcs above and below $\overline{AB}$ so that they intersect the two arcs previously drawn. Label the points of the intersection of the arcs as C and D.

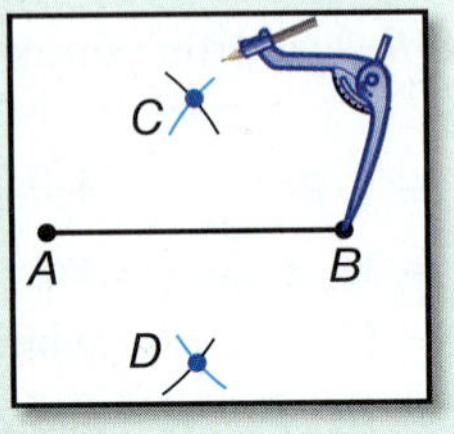

Step 3 Use a straightedge to draw $\overline{CD}$. Label the point where it intersects $\overline{AB}$ as M. Point M is the midpoint of $\overline{AB}$, and $\overline{CD}$ is a bisector of $\overline{AB}$.

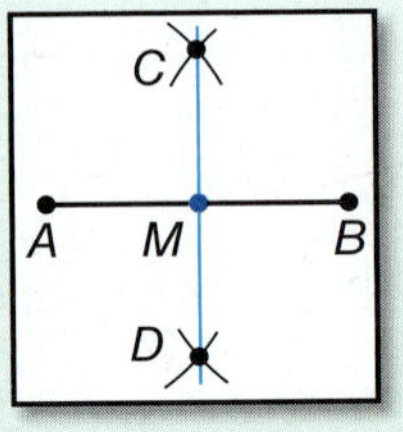

Check Your Understanding

Example 1 p. 25

Use the number line to find each measure.

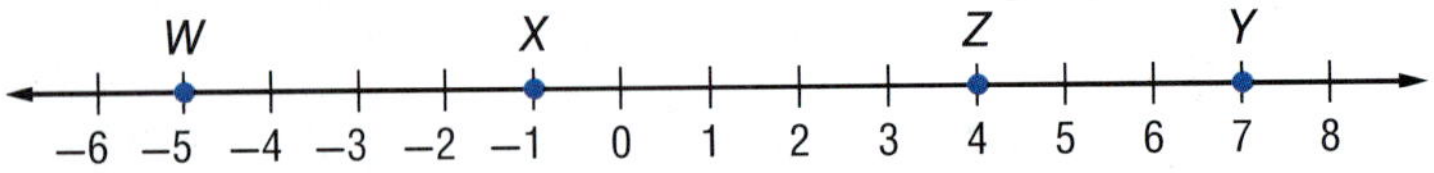

1. XY

2. WZ

Example 2 p. 26

TIME CAPSULE Graduating classes have buried time capsules on the campus of East Side High School for over twenty years. The points on the diagram show the position of three time capsules. Find the distance between each pair of time capsules.

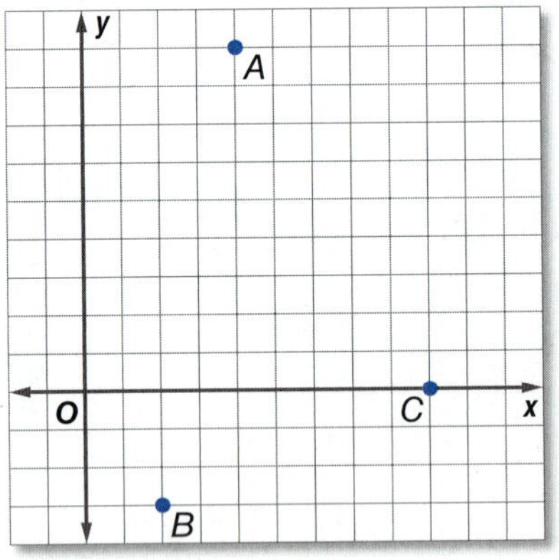

3. $A(4, 9)$, $B(2, -3)$

4. $A(4, 9)$, $C(9, 0)$

5. $B(2, -3)$, $C(9, 0)$

6. Which two time capsules are the closest to each other? Which are farthest apart?

Example 3 p. 27

Use the number line to find the coordinate of the midpoint of each segment.

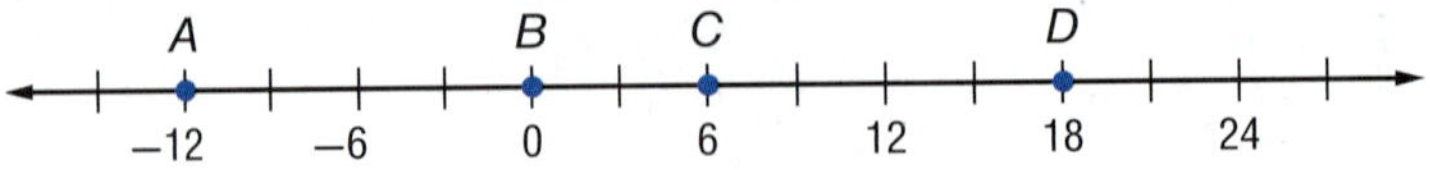

7. $\overline{AC}$

8. $\overline{BD}$

Example 4 p. 28

Find the coordinates of the midpoint of a segment with the given endpoints.

9. $J(5, -3)$, $K(3, -8)$

10. $M(7, 1)$, $N(4, -1)$

Example 5 p. 28

11 Find the coordinates of G if $F(1, 3.5)$ is the midpoint of $\overline{GJ}$ and J has coordinates $(6, -2)$.

Example 6 p. 29

12. ALGEBRA Point M is the midpoint of $\overline{CD}$. What is the value of a in the figure?

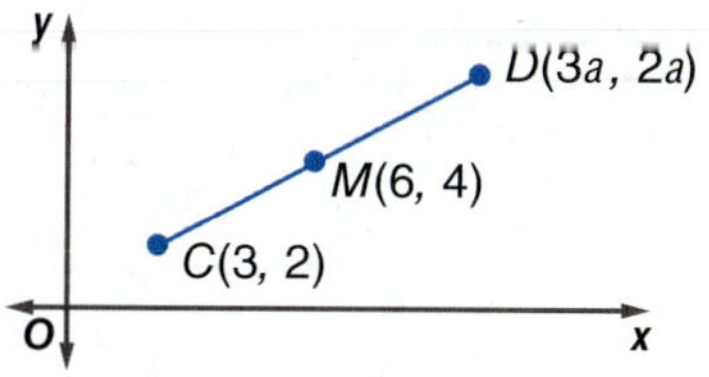

Practice and Problem Solving

● = **Step-by-Step Solutions** begin on page R20.
Extra Practice begins on page 969.

Example 1 p. 25

Use the number line to find each measure.

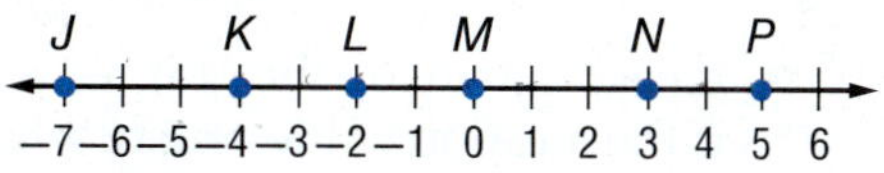

13. JL **14.** JK **15.** KP

16. NP **17.** JP **18.** LN

Example 2 p. 26

Find the distance between each pair of points.

19.

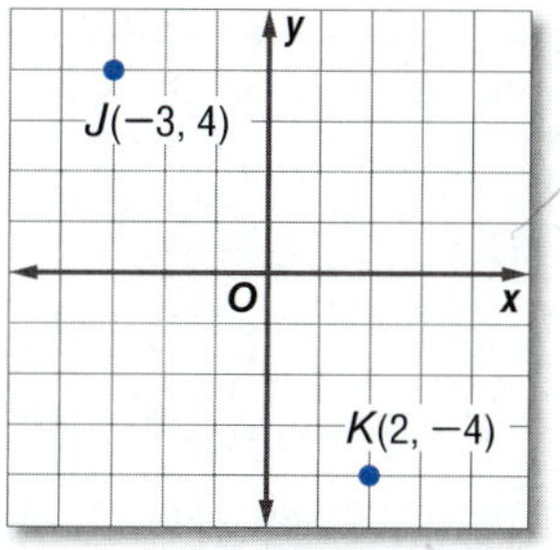

20.

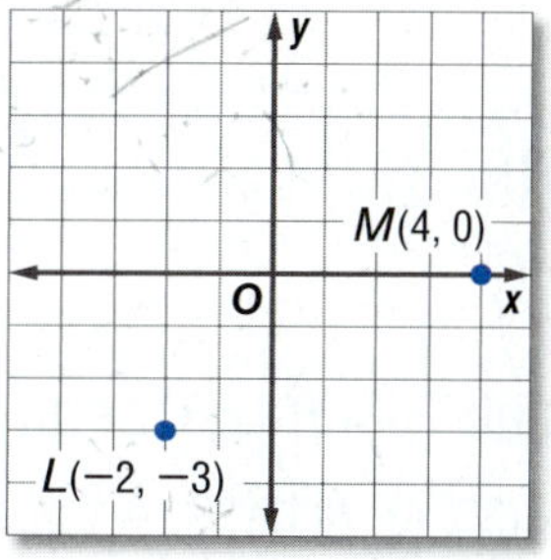

21.

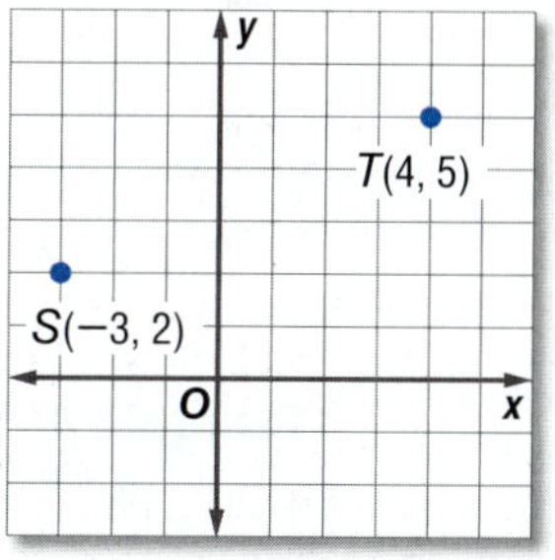

22.

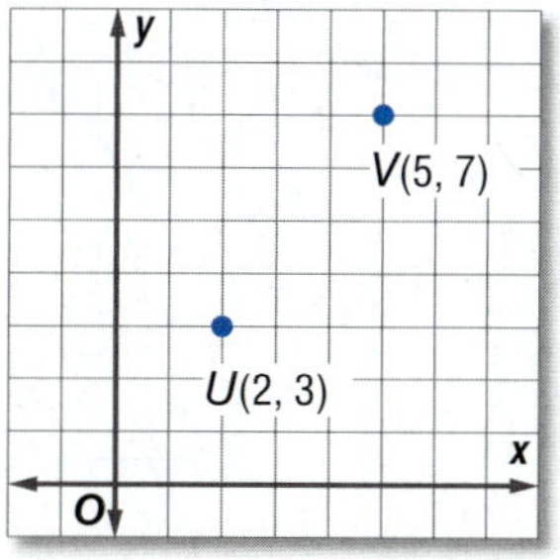

23.

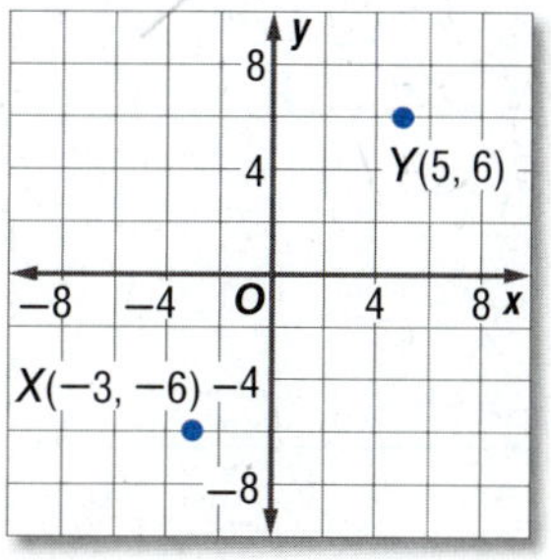

24.

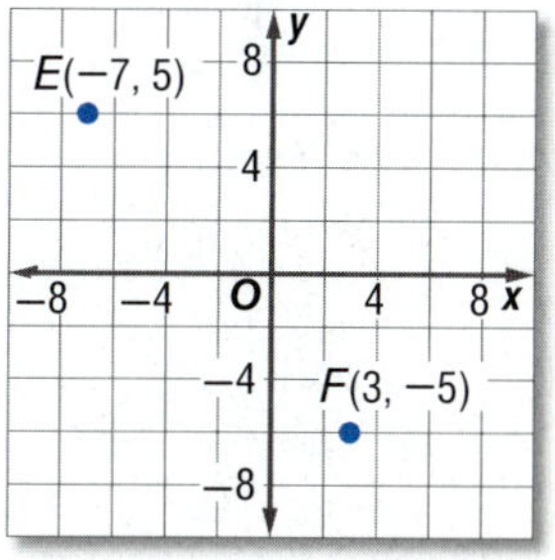

25. $X(1, 2), Y(5, 9)$ **26.** $P(3, 4), Q(7, 2)$ **27.** $M(-3, 8), N(-5, 1)$

28. $Y(-4, 9), Z(-5, 3)$ **29.** $A(2, 4), B(5, 7)$ **30.** $C(5, 1), D(3, 6)$

Real-World Link

Traveling at high altitudes (greater than 8000 feet) can cause altitude sickness because of the decreasing oxygen density in the air. It is important to let your body acclimate to new altitudes before physical activity.

Source: *Encyclopaedia Britannica*

31. HIKING Vivian is planning to hike to the top of Humphreys Peak on her family vacation. The coordinates of the peak of the mountain and of the base of the trail are shown. If the trail can be approximated by a straight line, estimate the length of the trail. (*Hint:* 1 mi = 5280 ft)

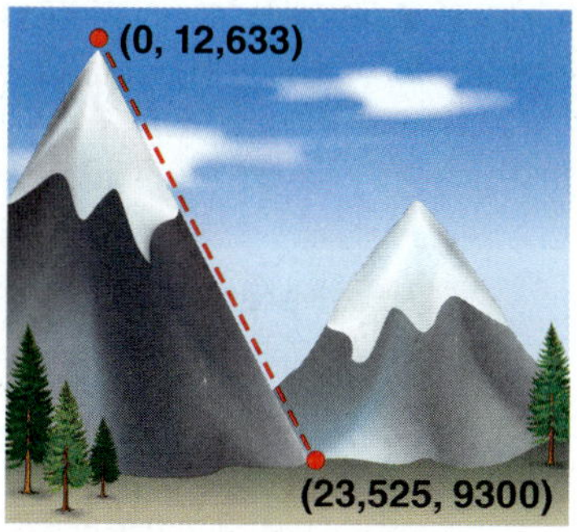

32. NEIGHBORHOODS Penny and Akiko live in the locations shown on the map below.

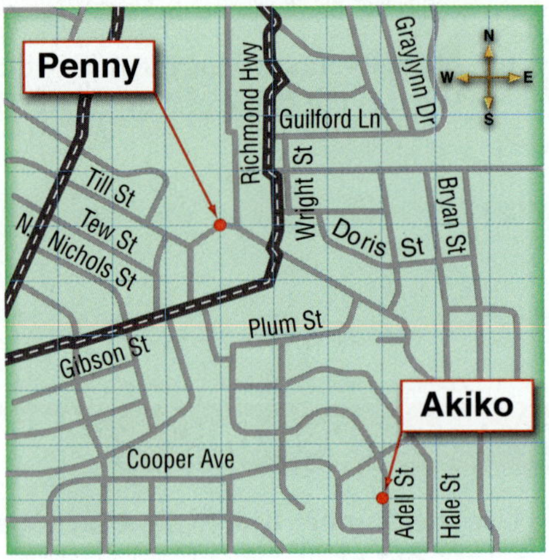

a. If each square on the grid represents one block and the bottom left corner of the grid is the location of the origin, what is the distance from Penny's house to Akiko's?

b. If Penny moves three blocks to the north and Akiko moves 5 blocks to the west, how far apart will they be?

Example 3
p. 27

Use the number line to find the coordinate of the midpoint of each segment.

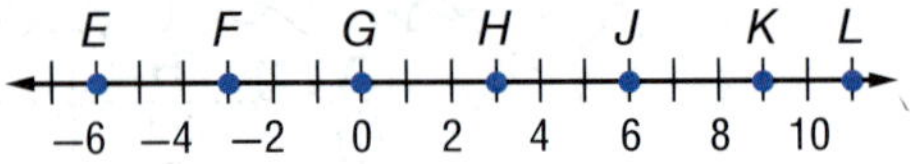

33. $\overline{HK}$ **34.** $\overline{JL}$ **35.** $\overline{EF}$

36. $\overline{FG}$ **37.** $\overline{FK}$ **38.** $\overline{EL}$

Example 4
p. 28

Find the coordinates of the midpoint of a segment with the given endpoints.

39. $C(22, 4), B(15, 7)$

40. $W(12, 2), X(7, 9)$

41. $D(-15, 4), E(2, -10)$

42. $V(-2, 5), Z(3, -17)$

43. $X(-2.4, -14), Y(-6, -6.8)$

44. $J(-11.2, -3.4), K(-5.6, -7.8)$

45.

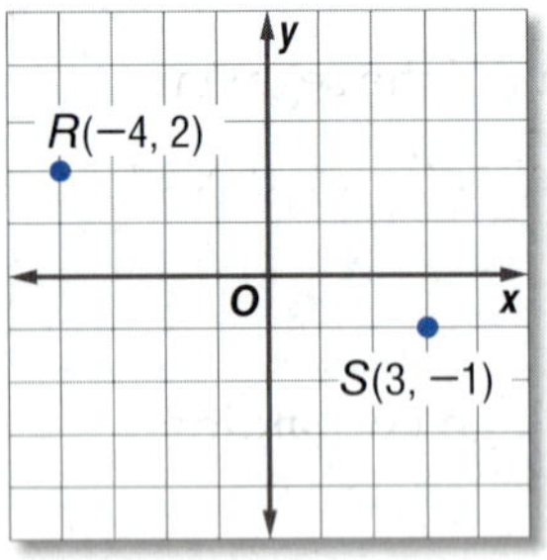

46.

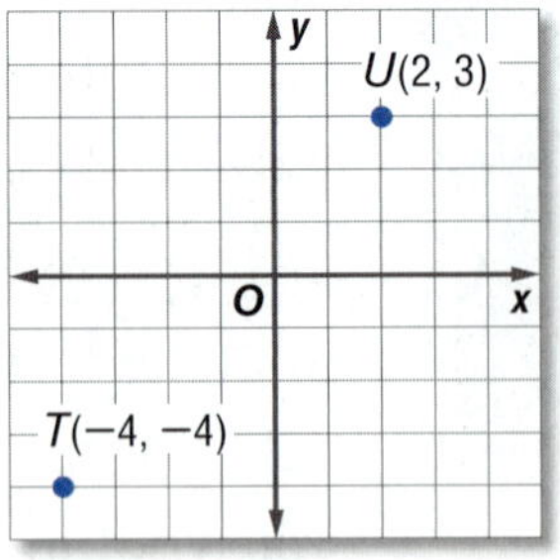

Example 5
p. 28

Find the coordinates of the missing endpoint if B is the midpoint of $\overline{AC}$.

47. $C(-5, 4), B(-2, 5)$ **48.** $A(1, 7), B(-3, 1)$ **49.** $A(-4, 2), B(6, -1)$

50. $C(-6, -2), B(-3, -5)$ **51.** $A(4, -0.25), B(-4, 6.5)$ **52.** $C\left(\frac{5}{3}, -6\right), B\left(\frac{8}{3}, 4\right)$

Example 6
p. 29

ALGEBRA **Suppose M is the midpoint of $\overline{FG}$. Use the given information to find the missing measure or value.**

53. $FM = 3x - 4, MG = 5x - 26, FG = ?$

54. $FM = 5y + 13, MG = 5 - 3y, FG = ?$

55. $MG = 7x - 15, FG = 33, x = ?$

56. $FM = 8a + 1, FG = 42, a = ?$

57 **BASKETBALL** The dimensions of a basketball court are shown below. Suppose a player throws the ball from a corner to a teammate standing at the center of the court.

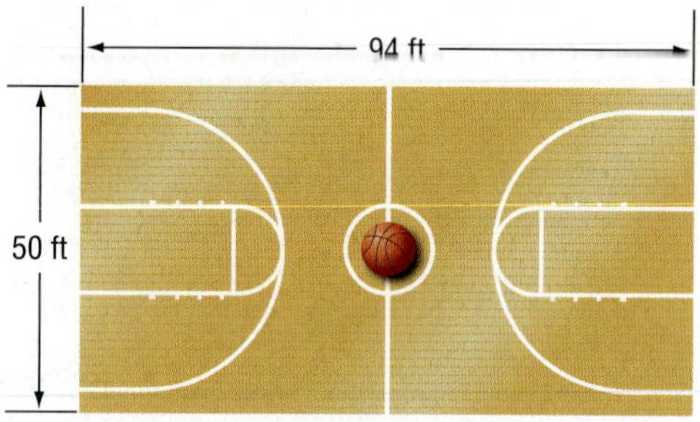

a. If center court is located at the origin, find the ordered pair that represents the location of the player in the bottom right corner.

b. Find the distance that the ball travels.

StudyTip

Spreadsheets Spreadsheets often use special commands to perform operations. For example, $\sqrt{x_2 - x_1}$ would be written as SQRT(A2–C2). To find the average of the numbers in a range of cells, use AVERAGE (range of cells). Use the symbol ^ to raise a number to a power.

SPREADSHEETS Spreadsheets can be used to perform calculations quickly. The spreadsheet below can be used to calculate the distance between two points. Values are used in formulas by using a specific cell name. The value of x_1 is used in a formula using its cell name, A2.

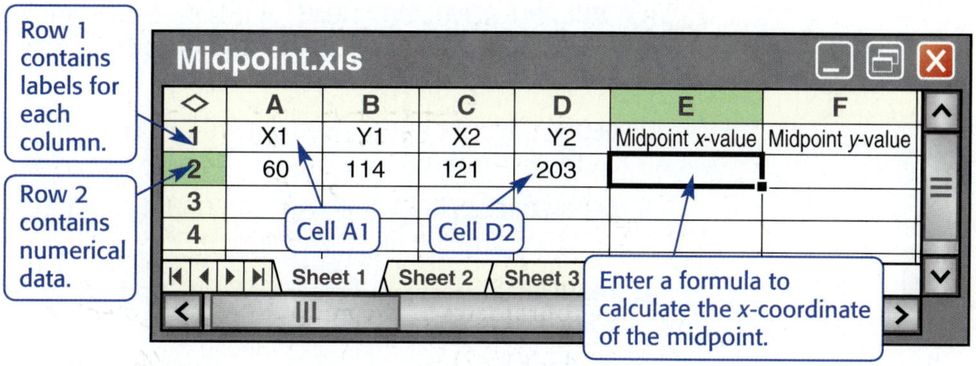

Use the information at the left to write a formula for the indicated cell that could be used to calculate the indicated value using the coordinates (x_1, y_1) and (x_2, y_2) as the endpoint of a segment.

58. E2; the x-value of the midpoint of the segment

59. F2; the y-value of the midpoint of the segment

60. G2; the length of the segment

Name the point(s) that satisfy the given condition.

61. two points on the x-axis that are 10 units from (1, 8)

62. two points on the y-axis that are 25 units from (–24, 3)

63. **COORDINATE GEOMETRY** Find the coordinates of B if B is halfway between $\overline{AC}$ and C is halfway between $\overline{AD}$.

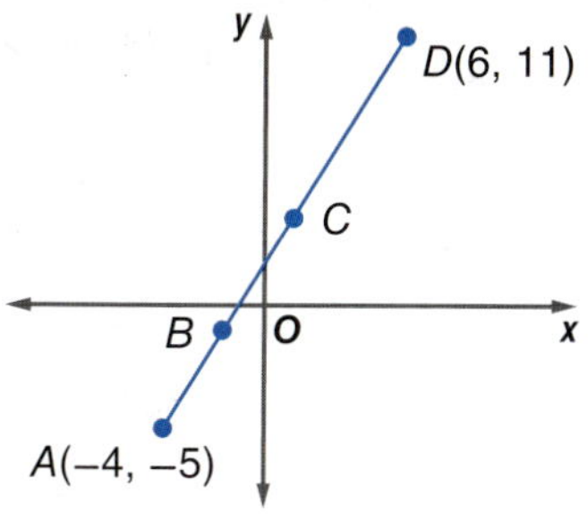

ALGEBRA **Determine the value(s) of n.**

64. $J(n, n + 2), K(3n, n - 1), JK = 5$

65. $P(3n, n - 7), Q(4n, n + 5), PQ = 13$

Real-World Link

Latitude lines are parallel to the Equator and measure distance north and south. Longitude lines measure distance east and west of the Prime Meridian running through Greenwich, England.

Source: *National Geographic*

66. GEOGRAPHY Wilmington, North Carolina, is located at (34.3°, 77.9°), which represents north latitude and west longitude. Winston-Salem is in the northern part of the state at (36.1°, 80.2°).

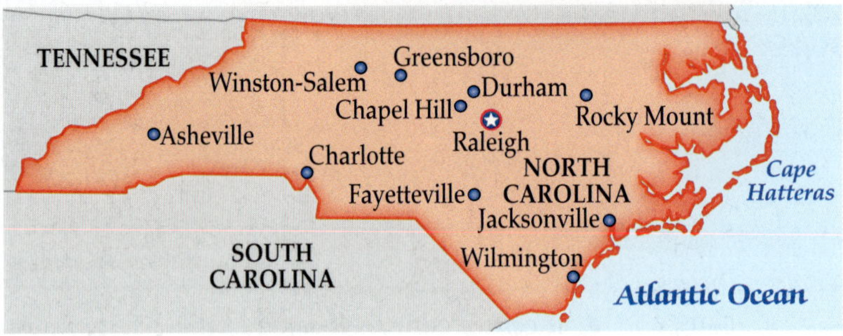

a. Find the latitude and longitude of the midpoint of the segment between Wilmington and Winston-Salem.

b. Use an atlas or the Internet to find a city near the location of the midpoint.

c. If Winston-Salem is the midpoint of the segment with one endpoint at Wilmington, find the latitude and longitude of the other endpoint.

d. Use an atlas or the Internet to find a city near the location of the other endpoint.

67. MULTIPLE REPRESENTATIONS In this problem, you will explore the relationship between a midpoint of a segment and the midpoint between the endpoint and the midpoint.

a. GEOMETRIC Use a straightedge to draw three different line segments. Label the endpoints *A* and *B*.

b. GEOMETRIC On each line segment, find the midpoint of $\overline{AB}$ and label it *C*. Then find the midpoint of $\overline{AC}$ and label it *D*.

c. TABULAR Measure and record *AB*, *AC*, and *AD* for each line segment. Organize your results into a table.

d. ALGEBRAIC If $AB = x$, write an expression for the measures *AC* and *AD*.

e. VERBAL Make a conjecture about the relationship between *AB* and each segment if you were to continue to find the midpoints of a segment and a midpoint you previously found.

H.O.T. Problems

Use Higher-Order Thinking Skills

68. WRITING IN MATH Explain how the Pythagorean Theorem and the Distance Formula are related.

69. REASONING Is the point one third of the way from (x_1, y_1) to (x_2, y_2) *sometimes*, *always*, or *never* the point $\left(\frac{x_1 + x_2}{3}, \frac{y_1 + y_2}{3}\right)$? Explain.

70. CHALLENGE Point *P* is located on the segment between point *A*(1, 4) and point *D*(7, 13). The distance from *A* to *P* is twice the distance from *P* to *D*. What are the coordinates of point *P*?

71. OPEN ENDED Draw a segment and name it $\overline{AB}$. Using only a compass and a straightedge, construct a segment $\overline{CD}$ such that $CD = 5\frac{1}{4}AB$. Explain and then justify your construction.

72. WRITING IN MATH Describe a method of finding the midpoint of a segment that has one endpoint at (0, 0). Give an example using your method, and explain why your method works.

Standardized Test Practice

73. Which of the following best describes the first step in bisecting $\overline{AB}$?

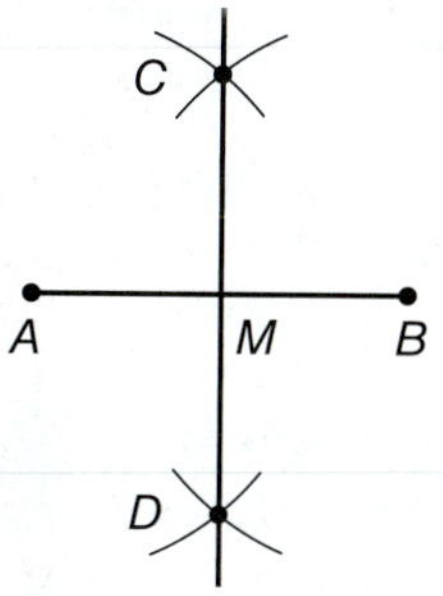

A From point A, draw equal arcs on $\overline{CD}$ using the same compass width.

B From point A, draw equal arcs above and below $\overline{AB}$ using a compass width of $\frac{1}{3}\overline{AB}$.

C From point A, draw equal arcs above and below $\overline{AB}$ using a compass width greater than $\frac{1}{2}\overline{AB}$.

D From point A, draw equal arcs above and below $\overline{AB}$ using a compass width less than $\frac{1}{2}\overline{AB}$.

74. ALGEBRA Beth paid \$74.88 for 3 pairs of jeans. All 3 pairs of jeans were the same price. How much did each pair of jeans cost?

F \$24.96

G \$37.44

H \$74.88

J \$224.64

75. SAT/ACT If $5^{2x-3} = 1$, then $x =$

A 0.4   C 1.6

B 1.5   D 2

76. GRIDDED RESPONSE One endpoint of $\overline{AB}$ has coordinates $(-3, 5)$. If the coordinates of the midpoint of $\overline{AB}$ are $(2, -6)$, what is the approximate length of $\overline{AB}$?

Spiral Review

Find the length of each object. (Lesson 1-2)

77.

78.

Draw and label a figure for each relationship. (Lesson 1-1)

79. $\overleftrightarrow{FG}$ lies in plane M and contains point H.

80. Lines r and s intersect at point W.

81. TRUCKS A sport-utility vehicle has a maximum load limit of 75 pounds for its roof. You want to place a 38-pound cargo carrier and 4 pieces of luggage on top of the roof. Write and solve an inequality to find the average allowable weight for each piece of luggage. (Lesson 0-6)

Skills Review

Solve each equation. (Lesson 0-5)

82. $8x - 15 = 5x$

83. $5y - 3 + y = 90$

84. $16a + 21 = 20a - 9$

85. $9k - 7 = 21 - 3k$

86. $11z - 13 = 3z + 17$

87. $15 + 6n = 4n + 23$

1-4 Angle Measure

Then

You measured line segments. (Lesson 1-2)

Now

- Measure and classify angles.
- Identify and use congruent angles and the bisector of an angle.

New Vocabulary

ray
opposite rays
angle
side
vertex
interior
exterior
degree
right angle
acute angle
obtuse angle
angle bisector

Math Online

glencoe.com

Why?

One of the skills Dale must learn in carpentry class is how to cut a *miter* joint. This joint is created when two boards are cut at an angle to each other. He has learned that one miscalculation in angle measure can result in mitered edges that do not fit together.

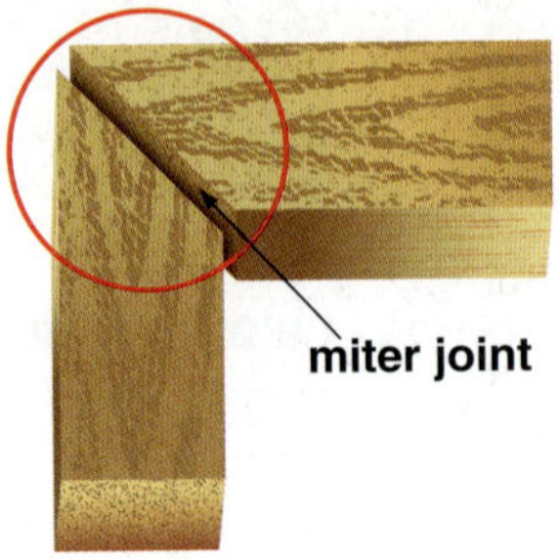

Measure and Classify Angles A **ray** is a part of a line. It has one endpoint and extends indefinitely in one direction. Rays are named by stating the endpoint first and then any other point on the ray. The ray shown cannot be named as $\overrightarrow{OM}$ because O is not the endpoint of the ray.

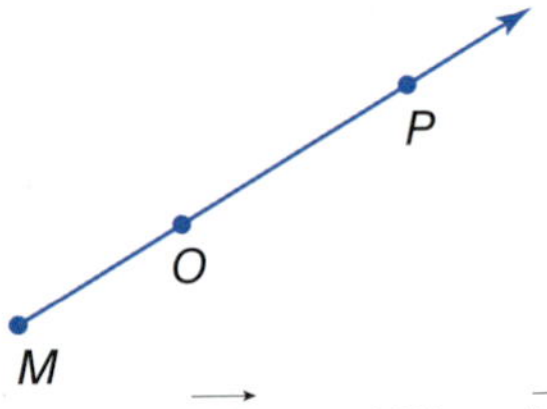

ray MP, $\overrightarrow{MP}$, ray MO, or $\overrightarrow{MO}$

If you choose a point on a line, that point determines exactly two rays called **opposite rays**. Since both rays share a common endpoint, opposite rays are collinear.

$\overrightarrow{JH}$ and $\overrightarrow{JK}$ are opposite rays.

An **angle** is formed by two *noncollinear* rays that have a common endpoint. The rays are called **sides** of the angle. The common endpoint is the **vertex**.

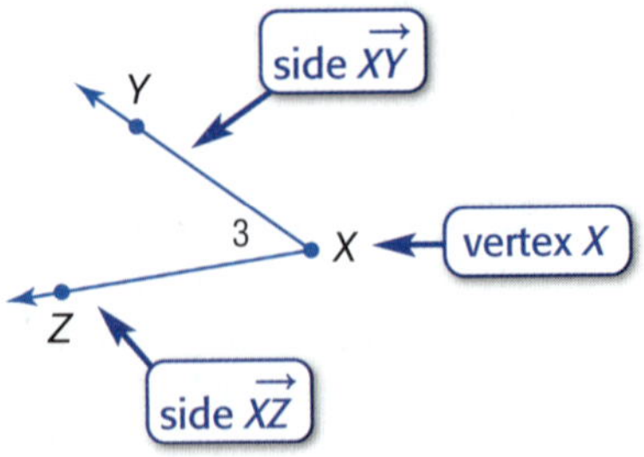

When naming angles using three letters, the vertex must be the second of the three letters. You can name an angle using a single letter only when there is exactly one angle located at that vertex. The angle shown can be named as $\angle X$, $\angle YXZ$, $\angle ZXY$, or $\angle 3$.

An angle divides a plane into three distinct parts.

- Points Q, M, and N lie on the angle.
- Points S and R lie in the **interior** of the angle.
- Points P and O lie in the **exterior** of the angle.

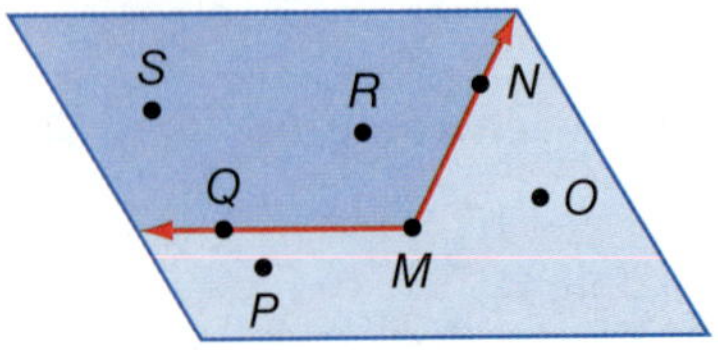

Real-World EXAMPLE 1 Angles and Their Parts

MAPS **Use the map of a high school shown.**

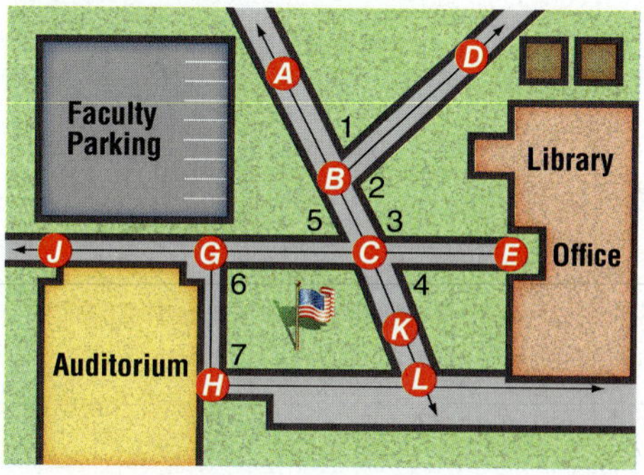

a. Name all angles that have B as a vertex.

$\angle 1$, $\angle 2$, $\angle ABD$, and $\angle DBC$

b. Name the sides of $\angle 3$.

$\overrightarrow{CA}$ and $\overrightarrow{CE}$ or $\overrightarrow{CB}$ and $\overrightarrow{CE}$

c. What is another name for $\angle GHL$?

$\angle 7$, $\angle H$, or $\angle LHG$

d. Name a point in the interior of $\angle DBK$.

Point E

> **StudyTip**
>
> **Segments as Sides** Because a ray can contain a line segment, the side of an angle can be a segment.

Guided Practice

1A. What is the vertex of $\angle 5$?

1B. Name the sides of $\angle 5$.

1C. Write another name for $\angle ECL$.

1D. Name a point in the exterior of $\angle CLH$.

Personal Tutor glencoe.com

Angles are measured in units called **degrees**. The degree results from dividing the distance around a circle into 360 parts.

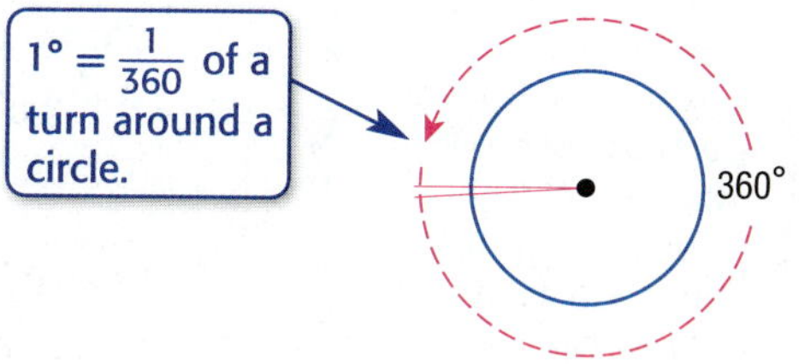

To measure an angle, you can use a *protractor*. Angle DEF below is a 50 degree (50°) angle. We say that the *degree measure* of $\angle DEF$ is 50, or $m\angle DEF = 50$.

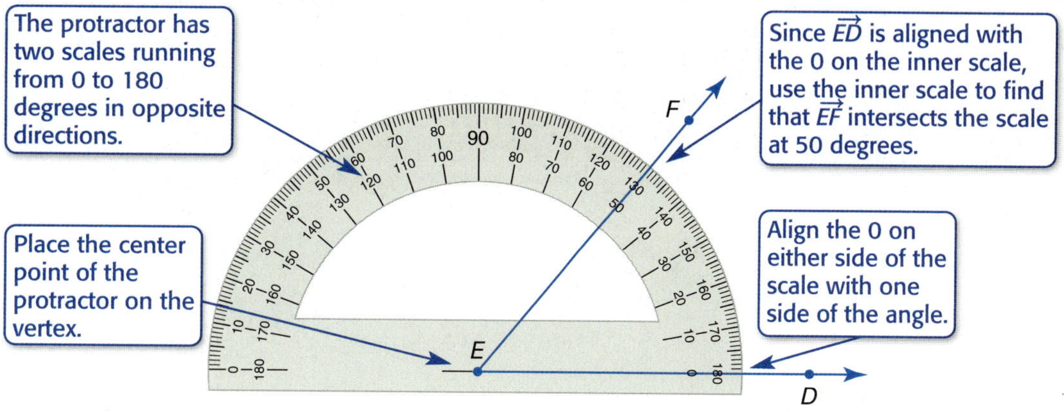

Angles can be classified by their measures as shown below.

Key Concept — Classify Angles

right angle

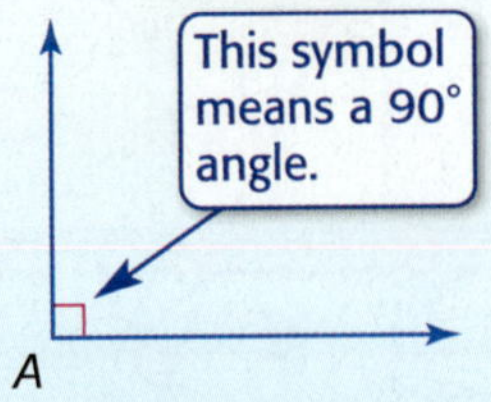

$m\angle A = 90$

acute angle

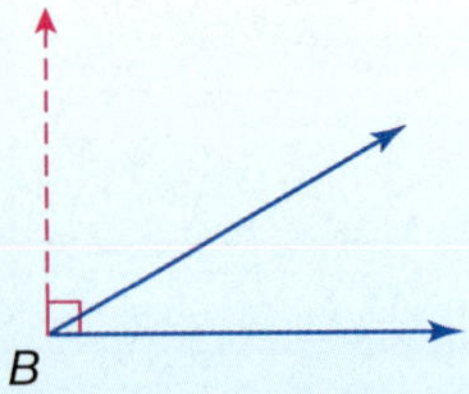

$m\angle B < 90$

obtuse angle

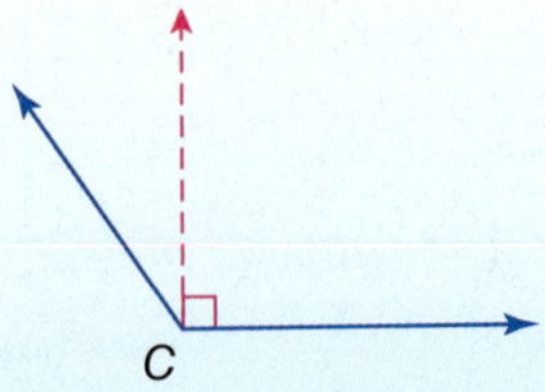

$180 > m\angle C > 90$

ReadingMath

Straight Angle Opposite rays with the same vertex form a *straight angle*. Its measure is 180. Unless otherwise specified in this book, however, the term *angle* means a nonstraight angle.

EXAMPLE 2 Measure and Classify Angles

Copy the diagram below, and extend each ray. Classify each angle as *right*, *acute*, or *obtuse*. Then use a protractor to measure the angle to the nearest degree.

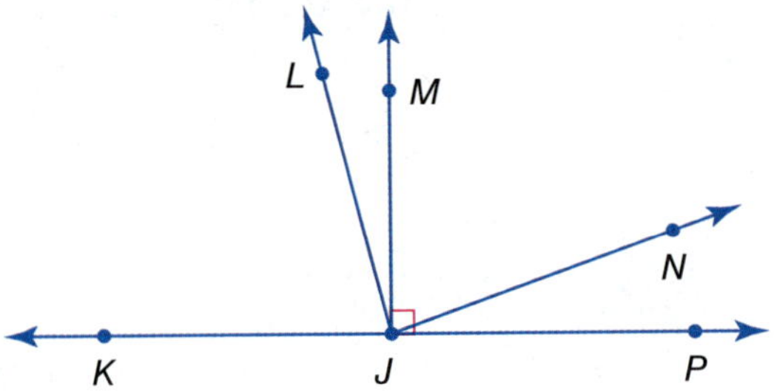

a. $\angle MJP$

$\angle MJP$ is marked as a right angle, so $m\angle MJP = 90$.

b. $\angle LJP$

Point L on angle $\angle LJP$ lies on the exterior of right angle $\angle MJP$, so $\angle LJP$ is an obtuse angle. Use a protractor to find that $m\angle LJP = 105$.

CHECK Since $105 > 90$, $\angle LJP$ is an obtuse angle. ✓

c. $\angle NJP$

Point N on angle $\angle NJP$ lies on the interior of right angle $\angle MJP$, so $\angle NJP$ is an acute angle. Use a protractor to find that $m\angle NJP = 20$.

CHECK Since $20 < 90$, $\angle NJP$ is an acute angle. ✓

Watch Out!

Classify Before Measuring Classifying an angle before measuring it can prevent you from choosing the wrong scale on your protractor. In Example 2b, you must decide whether $\angle LJP$ measures 75 or 105. Since $\angle LJP$ is an obtuse angle, you can reason that the correct measure must be 105.

Guided Practice

2A. $\angle AFB$

2B. $\angle CFA$

2C. $\angle AFD$

2D. $\angle CFD$

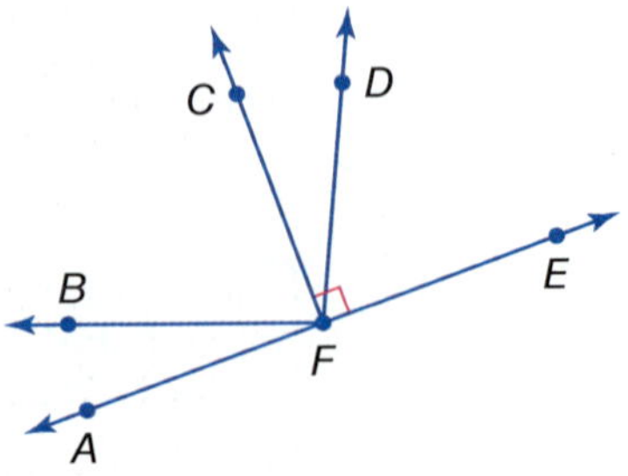

Personal Tutor glencoe.com

Congruent Angles Just as segments that have the same measure are congruent segments, angles that have the same measure are *congruent angles*.

In the figure, since $m\angle ABC = m\angle FED$, then $\angle ABC \cong \angle FED$. Matching numbers of arcs on a figure also indicate congruent angles, so $\angle CBE \cong \angle DEB$.

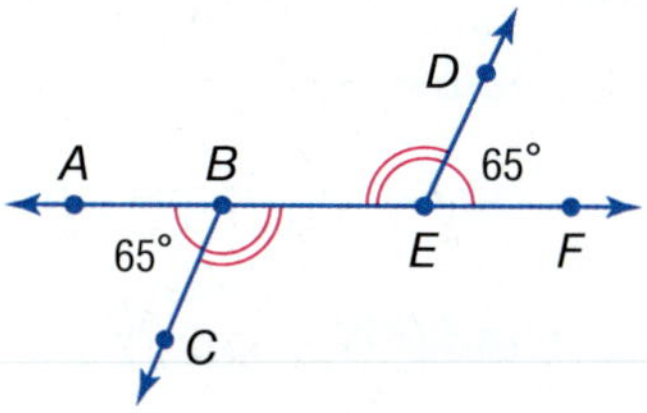

You can produce an angle congruent to a given angle using a construction.

Construction Copy an Angle

Step 1 Draw an angle like $\angle B$ on your paper. Use a straightedge to draw a ray on your paper. Label its endpoint G.

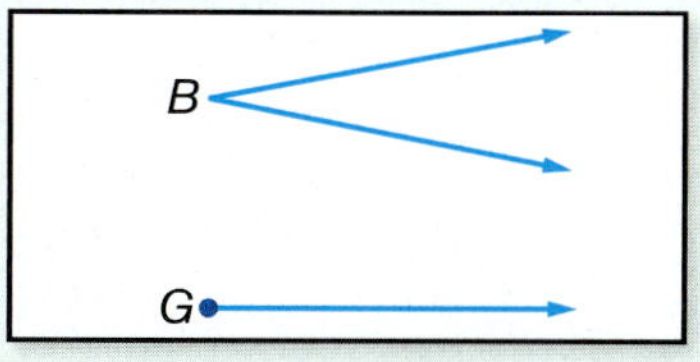

Step 2 Place the tip of the compass at point B and draw a large arc that intersects both sides of $\angle B$. Label the points of intersection A and C.

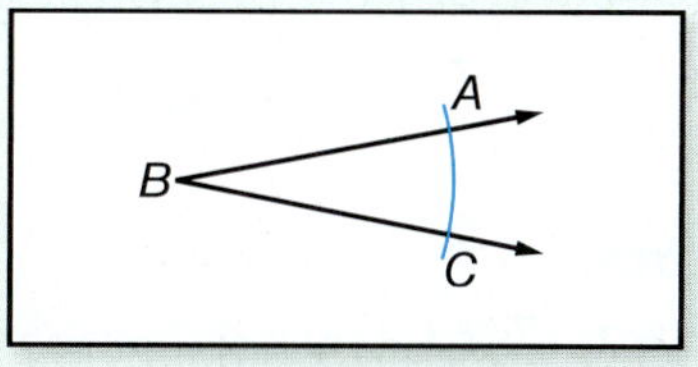

Step 3 Using the same compass setting, put the compass at point G and draw a large arc that starts above the ray and intersects the ray. Label the point of intersection H.

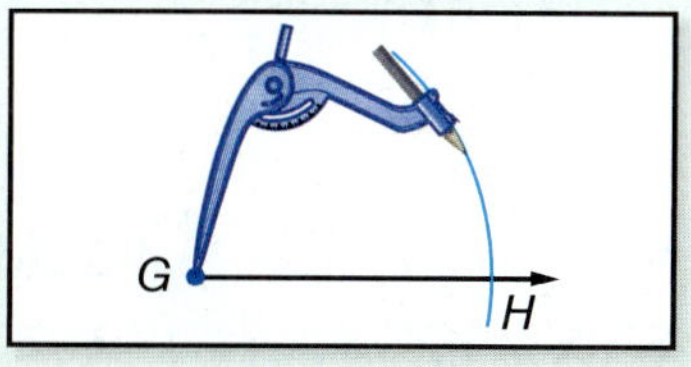

Step 4 Place the point of your compass on C and adjust so that the pencil tip is on A.

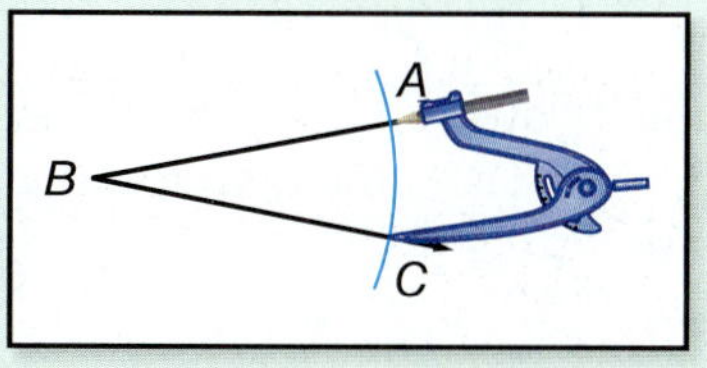

Step 5 Without changing the setting, place the compass at point H and draw an arc to intersect the larger arc you drew in Step 4. Label the point of intersection F.

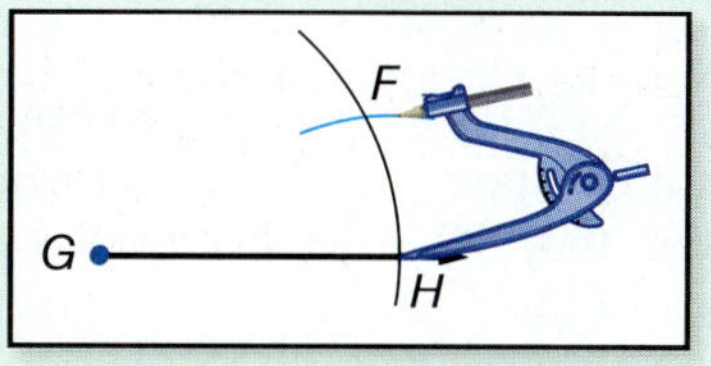

Step 6 Use a straightedge to draw $\overrightarrow{GF}$.

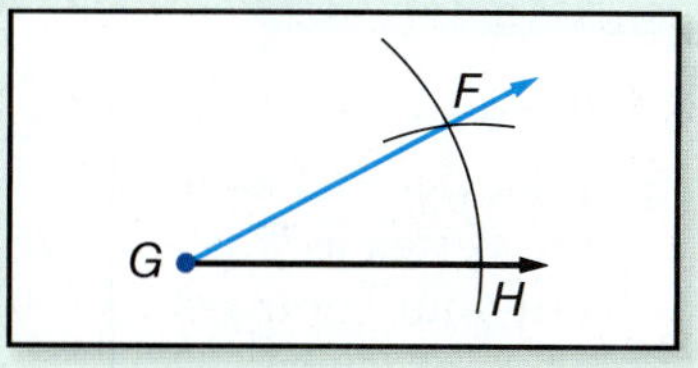

StudyTip

Segments A line segment can also bisect an angle.

A ray that divides an angle into two congruent angles is called an **angle bisector**. If $\overrightarrow{YW}$ is the angle bisector of $\angle XYZ$, then point W lies in the interior of $\angle XYZ$ and $\angle XYW \cong \angle WYZ$.

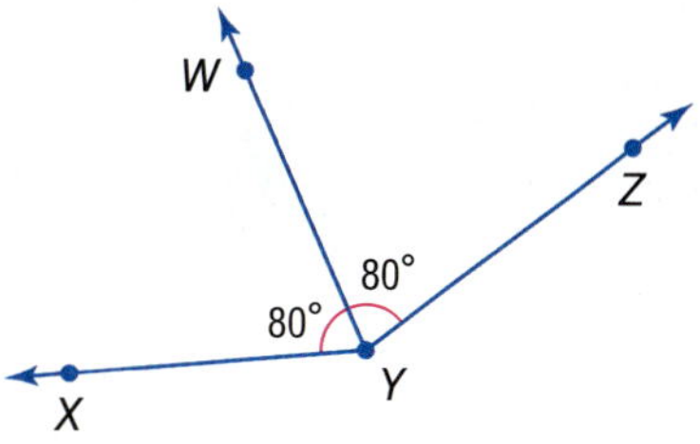

Just as with segments, when a line, segment, or ray divides an angle into smaller angles, the sum of the measures of the smaller angles equals the measure of the largest angle. So in the figure, $m\angle XYW + m\angle WYZ = m\angle XYZ$.

EXAMPLE 3 Measure and Classify Angles

ALGEBRA **In the figure, $\overrightarrow{KJ}$ and $\overrightarrow{KM}$ are opposite rays, and $\overrightarrow{KN}$ bisects $\angle JKL$. If $m\angle JKN = 8x - 13$ and $m\angle NKL = 6x + 11$, find $m\angle JKN$.**

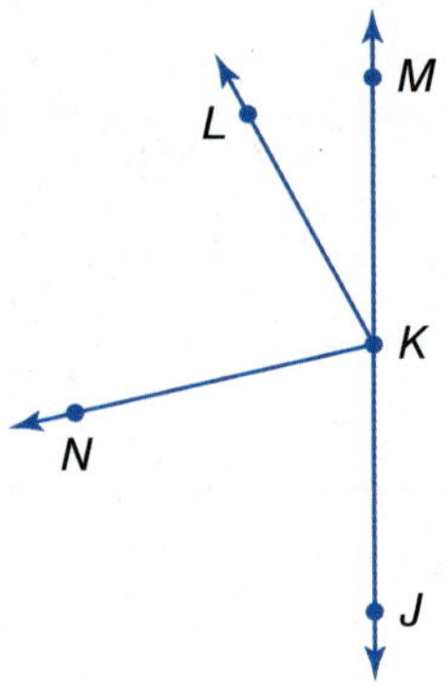

Step 1 Solve for x.

Since $\overrightarrow{KN}$ bisects $\angle JKL$, $\angle JKN \cong \angle NKL$.

$m\angle JKN = m\angle NKL$	**Definition of congruent angles**
$8x - 13 = 6x + 11$	**Substitution**
$8x = 6x + 24$	**Add 13 to each side.**
$2x = 24$	**Subtract 6x from each side.**
$x = 12$	**Divide each side by 2.**

Step 2 Use the value of x to find $m\angle JKN$.

$m\angle JKN = 8x - 13$	**Given**
$= 8(12) - 13$	**$x = 12$**
$= 96 - 13$ or 83	**Simplify.**

Guided Practice

3. Suppose $m\angle JKL = 9y + 15$ and $m\angle JKN = 5y + 2$. Find $m\angle JKL$.

Personal Tutor glencoe.com

StudyTip

Checking Solutions Check that you have computed the value of x correctly by substituting the value into the expression for $\angle JKN$. If you don't get the same measure as $\angle NKL$, you have made an error.

You can produce the angle bisector of any angle without knowing the measure of the angle.

Construction Bisect an Angle

Get Animated

Step 1 Draw an angle on your paper. Label the vertex as P. Put your compass at point P and draw a large arc that intersects both sides of $\angle P$. Label the points of intersection Q and R.

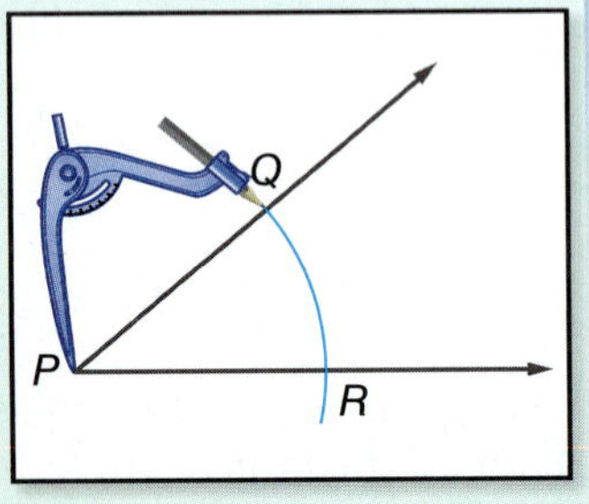

Step 2 With the compass at point Q, draw an arc in the interior of the angle.

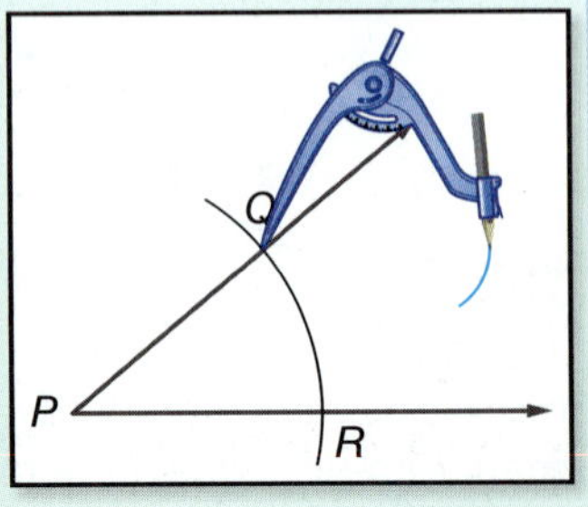

Step 3 Keeping the same compass setting, place the compass at point R and draw an arc that intersects the arc drawn in Step 2. Label the point of intersection T.

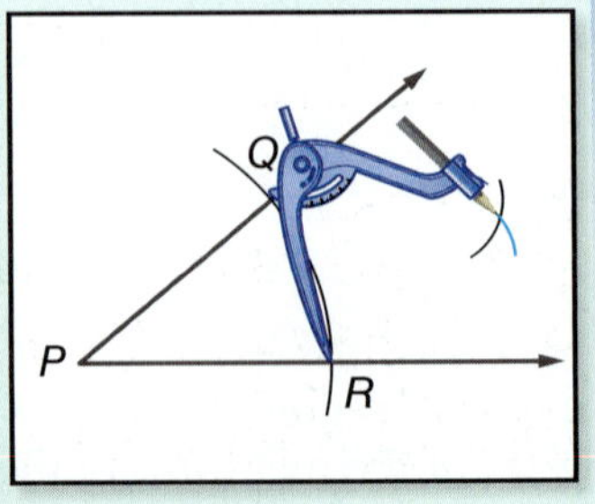

Step 4 Draw $\overrightarrow{PT}$. $\overrightarrow{PT}$ is the bisector of $\angle P$.

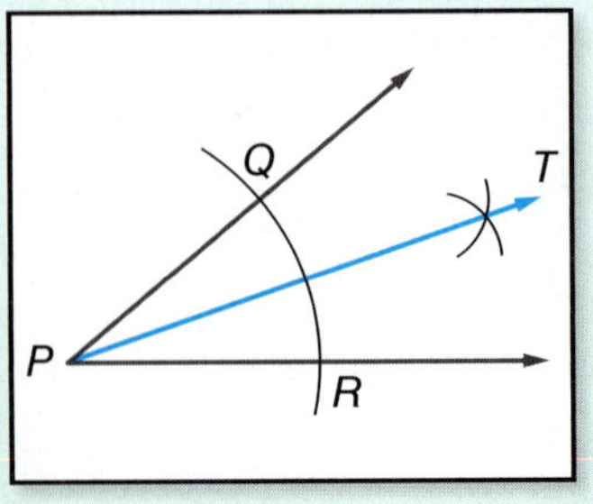

☑ Check Your Understanding

Example 1
p. 37

Use the figure at the right.

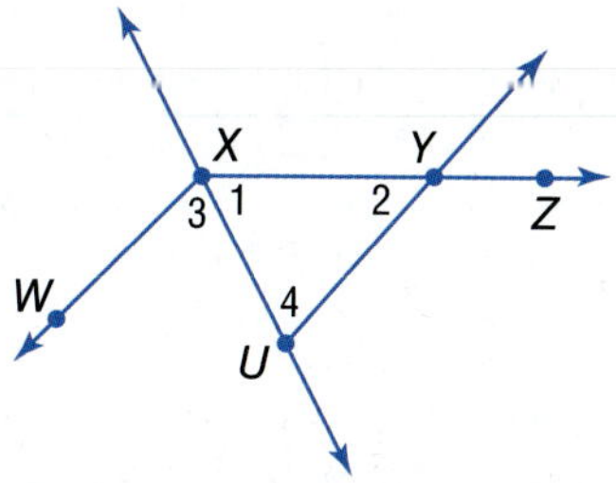

1. Name the vertex of ∠4.

2. Name the sides of ∠3.

3. What is another name for ∠2?

4. What is another name for ∠*UXY*?

Example 2
p. 38

Copy the diagram shown, and extend each ray. Classify each angle as *right*, *acute*, or *obtuse*. Then use a protractor to measure the angle to the nearest degree.

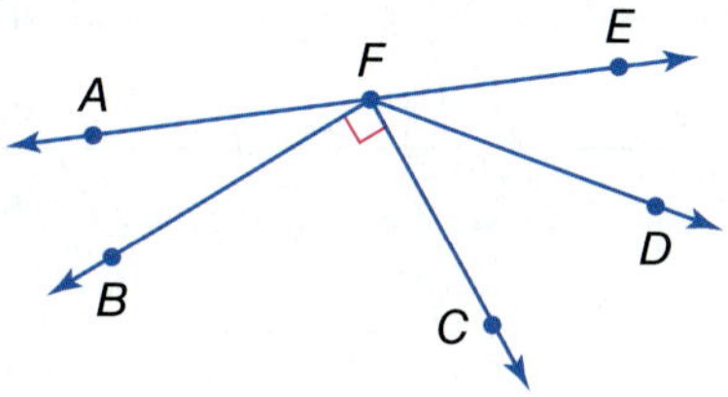

5. ∠*CFD* **6.** ∠*AFD*

7. ∠*BFC* **8.** ∠*AFB*

Example 3
p. 40

ALGEBRA **In the figure, $\overrightarrow{KJ}$ and $\overrightarrow{KL}$ are opposite rays. $\overrightarrow{KN}$ bisects ∠*LKM*.**

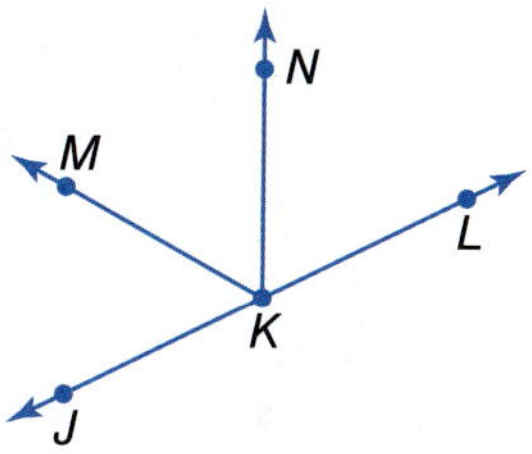

9. If $m\angle LKM = 7x - 5$ and $m\angle NKM = 3x + 9$, find $m\angle LKM$.

10. If $m\angle NKL = 7x - 9$ and $m\angle JKM = x + 3$, find $m\angle JKN$.

11. **CARPENTRY** A miter cut is used to build picture frames with corners that meet at right angles.

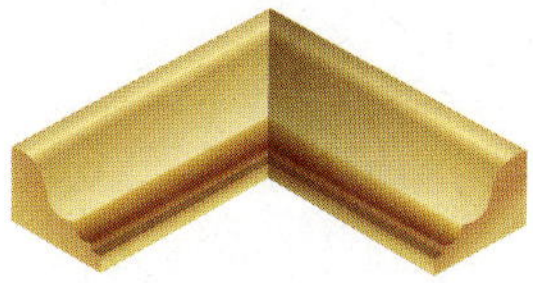

a. José miters the ends of some wood for a picture frame at congruent angles. What is the degree measure of his cut? Explain and classify the angle.

b. What does the joint represent in relation to the angle formed by the two pieces?

Practice and Problem Solving

● = **Step-by-Step Solutions** begin on page R20.
Extra Practice begins on page 969.

Example 1
p. 37

For Exercises 12–29, use the figure at the right.

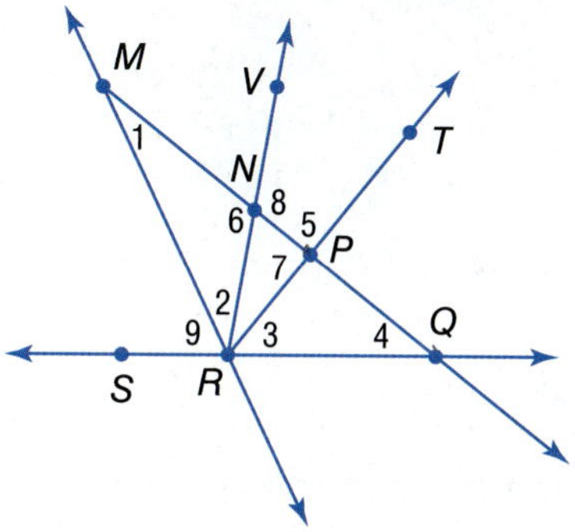

Name the vertex of each angle.

12. ∠4 **13.** ∠7 **14.** ∠2 **15.** ∠1

Name the sides of each angle.

16. ∠*TPQ* **17.** ∠*VNM* **18.** ∠6 **19.** ∠3

Write another name for each angle.

20. ∠9 **21.** ∠*QPT* **22.** ∠*MQS* **23.** ∠5

24. Name an angle with vertex *N* that appears obtuse.

25. Name an angle with vertex *Q* that appears acute.

26. Name a point in the interior of ∠*VRQ*.

27. Name a point in the exterior of ∠*MRT*.

28. Name a pair of angles that share exactly one point.

29. Name a pair of angles that share more than one point.

Example 2 p. 38

Copy the diagram shown, and extend each ray. Classify each angle as *right*, *acute*, or *obtuse*. Then use a protractor to measure the angle to the nearest degree.

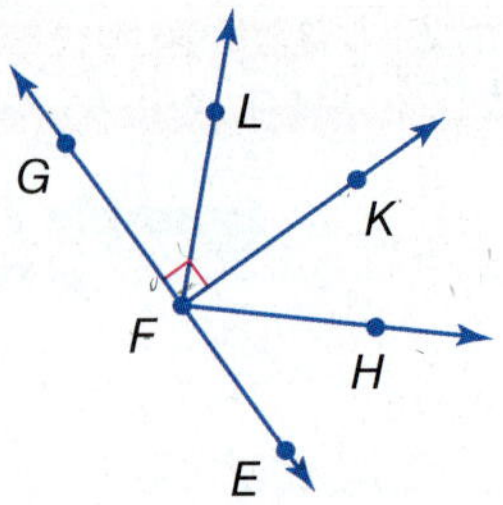

30. $\angle GFK$

31. $\angle EFK$

32. $\angle LFK$

33. $\angle EFH$

34. $\angle GFH$

35. $\angle EFL$

36. **CLOCKS** Determine at least three different times during the day when the hands on a clock form each of the following angles. Explain.

a. right angle

b. obtuse angle

c. congruent acute angles

Example 3 p. 40

ALGEBRA In the figure, $\overrightarrow{BA}$ and $\overrightarrow{BC}$ are opposite rays. $\overrightarrow{BH}$ bisects $\angle EBC$.

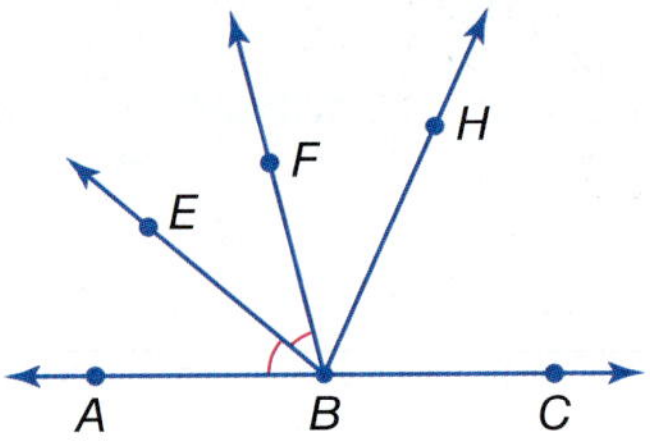

37 If $m\angle ABE = 2n + 7$ and $m\angle EBF = 4n - 13$, find $m\angle ABE$.

38. If $m\angle EBH = 6x - 20$ and $m\angle HBC = 8x - 30$, find $m\angle EBH$.

39. If $m\angle ABF = 7b - 24$ and $m\angle ABE = 2b$, find $m\angle EBF$.

40. If $m\angle EBC = 31a - 2$ and $m\angle EBH = 4a + 45$, find $m\angle HBC$.

41. If $m\angle ABF = 8s - 6$ and $m\angle ABE = 2(s + 11)$, find $m\angle EBF$.

42. If $m\angle EBC = 3r + 10$ and $m\angle ABE = 2r - 20$, find $m\angle EBF$.

43. **MAPS Estimate the measure of the angle formed by each city or location listed, the North Pole, and the Prime Meridian.**

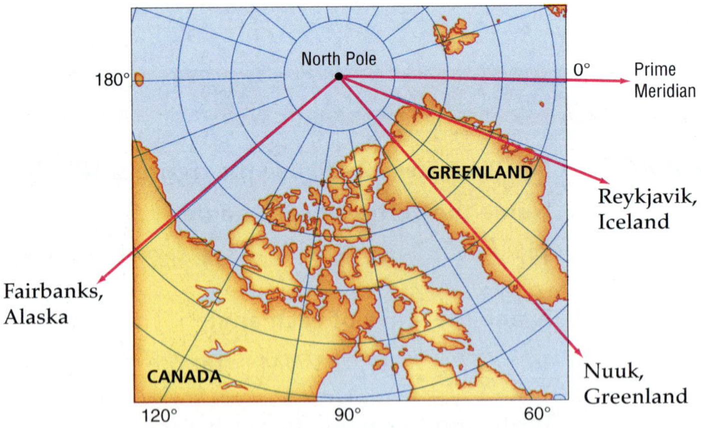

a. Nuuk, Greenland

b. Fairbanks, Alaska

c. Reykjavik, Iceland

d. Prime Meridian

Real-World Link

A compass rose is a design on a map that shows directions. In addition to the directions of north, south, east, and west, a compass rose can have as many as 32 markings.

Source: *Encyclopaedia Britannica*

44. **NAVIGATION** Use the photo and the information at the left.

a. With the center of the compass as its vertex, what is the measure of the angle between due west and due north?

b. What is the measure of the angle between due north and north-west?

c. How does the north-west ray relate to the angle in part **a**?

Plot the points in a coordinate plane and sketch $\angle XYZ$. Then classify it as *right*, *acute*, or *obtuse*.

45. $X(5, -3), Y(4, -1), Z(6, -2)$

46. $X(6, 7), Y(2, 3), Z(4, 1)$

Real-World Link

Light is also refracted when it passes through other substances, such as glass. When refracted through a prism, a white light laser beam is dispersed into the colors of the rainbow.

Source: *Encyclopaedia Britannica*

47. **PHYSICS** When you look at a pencil in water, it looks bent. This illusion is due to *refraction*, or the bending of light when it moves from one substance to the next.

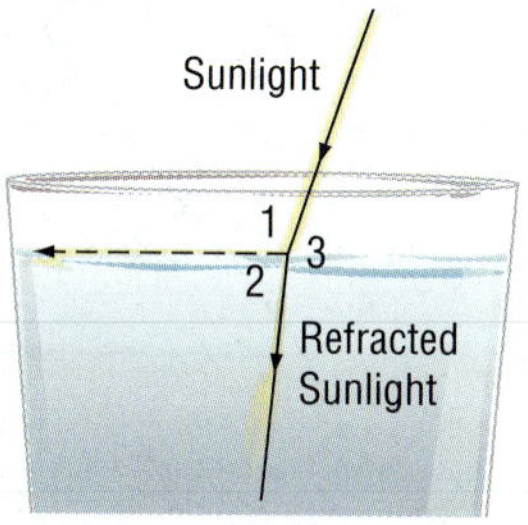

a. What is $m\angle 1$? Classify this angle as *acute*, *right*, or *obtuse*.

b. What is $m\angle 2$? Classify this angle as *acute*, *right*, or *obtuse*.

c. Without measuring, determine how many degrees the path of the light changes after it enters the water. Explain your reasoning.

48. **MULTIPLE REPRESENTATIONS** In this problem, you will explore the relationship of angles that compose opposite rays.

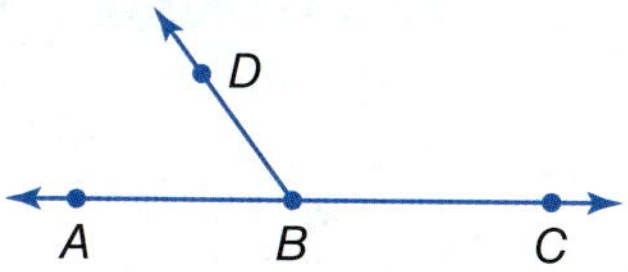

a. **GEOMETRIC** Draw four lines, each with points A, B, and C. Draw $\overrightarrow{BD}$ for each line, varying the placement of point D. Use a protractor to measure $\angle ABD$ and $\angle DBC$ for each figure.

b. **TABULAR** Organize the measures for each figure into a table. Include a column in your table to record the sum of these measures.

c. **VERBAL** Make a conjecture about the sum of the measures of the two angles. Explain your reasoning.

d. **ALGEBRAIC** If x is the measure of $\angle ABD$ and y is the measure of $\angle DBC$, write an equation that relates the two angle measures.

H.O.T. Problems

Use **H**igher-**O**rder **T**hinking Skills

49. **OPEN ENDED** Draw an obtuse angle named ABC. Measure $\angle ABC$. Construct an angle bisector $\overrightarrow{BD}$ of $\angle ABC$. Explain the steps in your construction and justify each step. Classify the two angles formed by the angle bisector.

50. **CHALLENGE** Describe how you would use a protractor to measure the angle shown.

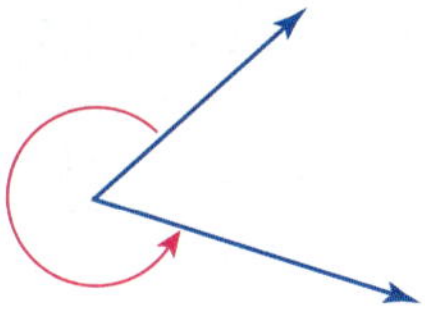

51. **REASONING** The sum of two acute angles is *sometimes*, *always*, or *never* an obtuse angle. Explain.

52. **CHALLENGE** $\overrightarrow{MP}$ bisects $\angle LMN$, $\overrightarrow{MQ}$ bisects $\angle LMP$, and $\overrightarrow{MR}$ bisects $\angle QMP$. If $m\angle RMP = 21$, find $m\angle LMN$. Explain your reasoning.

53. **WRITING IN MATH** Rashid says that he can estimate the measure of an acute angle using a piece of paper to within six degrees of accuracy. Explain how this would be possible. Then use this method to estimate the measure of the angle shown.

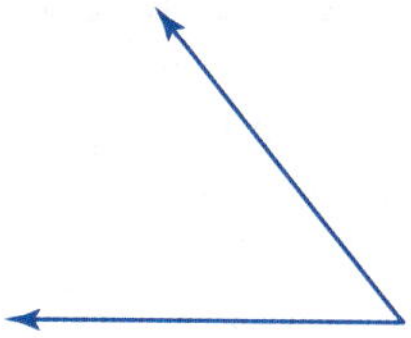

Standardized Test Practice

54. Which of the following angles measures closest to 60°?

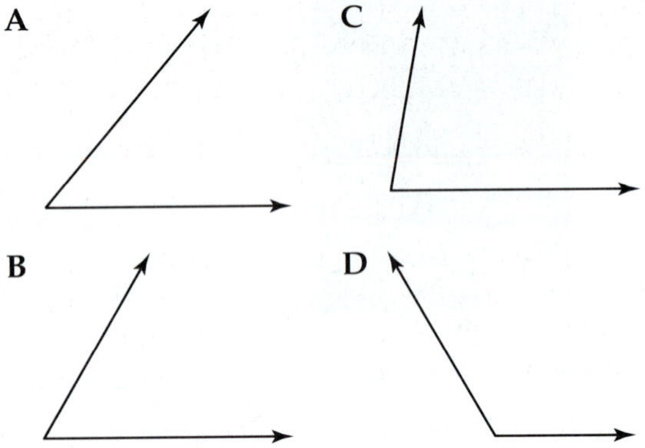

55. SHORT RESPONSE Leticia surveyed 50 English majors at a university to see if the school should play jazz music in the cafeteria during lunch. The school has 75 different majors and a total of 2000 students. Explain why the results of Leticia's survey are or are not representative of the entire student body.

56. In the figure below, if $m\angle BAC = 38$, what must be the measure of $\angle BAD$ in order for $\overrightarrow{AC}$ to be an angle bisector?

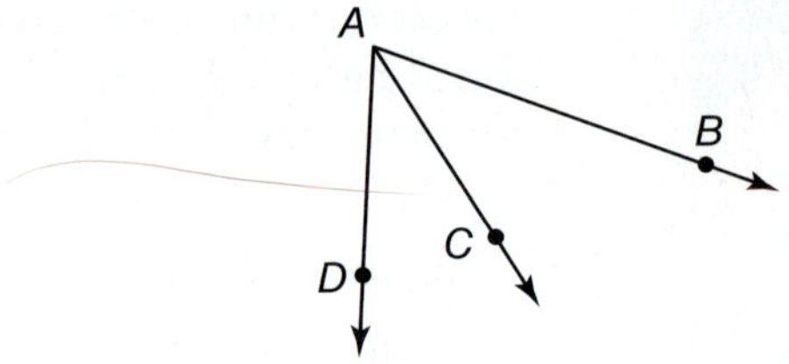

F 142
G 76
H 52
J 38

57. SAT/ACT If n is divisible by 2, 5, and 14, which of the following is also divisible by these numbers?

A $n + 7$ **C** $n + 14$
B $n + 10$ **D** $n + 70$

Spiral Review

Find the distance between each pair of points. Round to the nearest hundredth. (Lesson 1-3)

58. $A(-1, -8), B(3, 4)$ **59.** $C(0, 1), D(-2, 9)$ **60.** $E(-3, -12), F(5, 4)$

61. $G(4, -10), H(9, -25)$ **62.** $J\left(1, \frac{1}{4}\right), K\left(-3, \frac{7}{4}\right)$ **63.** $L\left(-5, \frac{8}{5}\right), M\left(5, \frac{2}{5}\right)$

Find the value of the variable and *ST* if *S* is between *R* and *T*. (Lesson 1-2)

64. $RS = 7a, ST = 12a, RT = 76$ **65.** $RS = 12, ST = 2x, RT = 34$

66. PHOTOGRAPHY Photographers often place their cameras on tripods. In the diagram, the tripod is placed on an inclined surface, and the length of each leg is adjusted so that the camera remains level with the horizon. Are the feet of the tripod coplanar? Explain your reasoning. (Lesson 1-1)

Complete each sentence. (Lesson 0-1)

67. 54 in. = __?__ ft **68.** 275 mm = __?__ m **69.** 7 gal = __?__ pt

Skills Review

Solve each equation. (Lesson 0-5)

70. $(90 - x) - x = 18$ **71.** $(5x + 3) + 7x = 180$ **72.** $(13x + 10) + 2x = 90$

73. $(180 - x) - 4x = 56$ **74.** $(4n + 17) + (n - 2) = 180$ **75.** $(8a - 23) + (9 - 2a) = 90$

CHAPTER 1

Mid-Chapter Quiz

Lessons 1-1 through 1-4

Use the figure to complete each of the following. (Lesson 1-1)

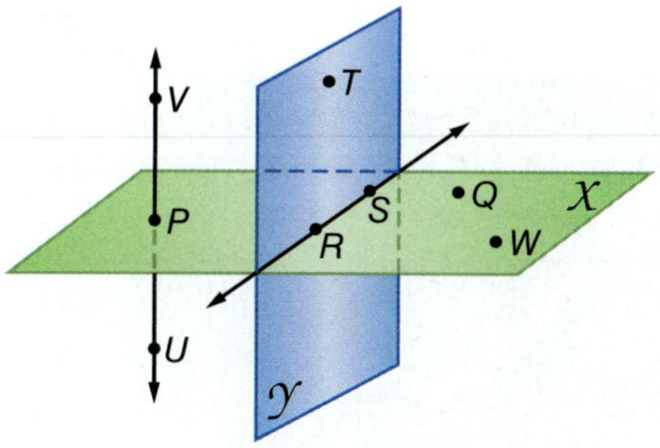

1. Name another point that is collinear with points U and V.

2. What is another name for plane $\mathcal{Y}$?

3. Name a line that is coplanar with points P, Q, and W.

Find the value of x and AC if B is between points A and C. (Lesson 1-2)

4. $AB = 12$, $BC = 8x - 2$, $AC = 10x$

5. $AB = 5x$, $BC = 9x - 2$, $AC = 11x + 7.6$

6. Find CD and the coordinate of the midpoint of $\overline{CD}$.

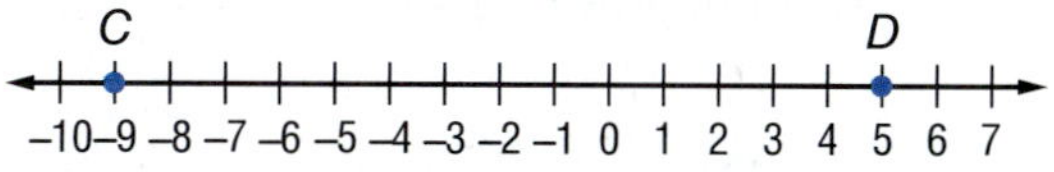

Find the coordinates of the midpoint of each segment. Then find the length of each segment. (Lesson 1-3)

7.

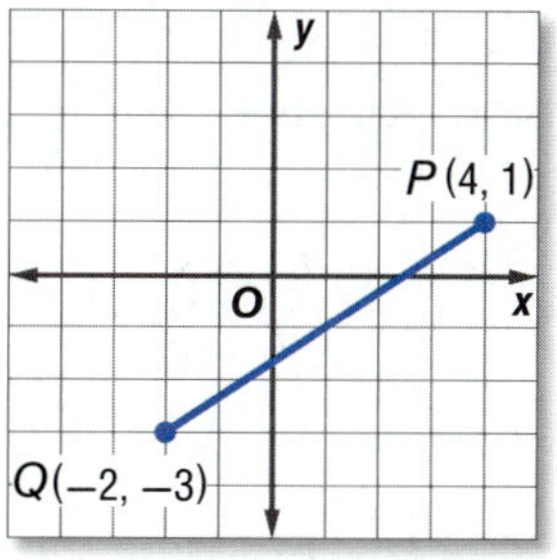

8.

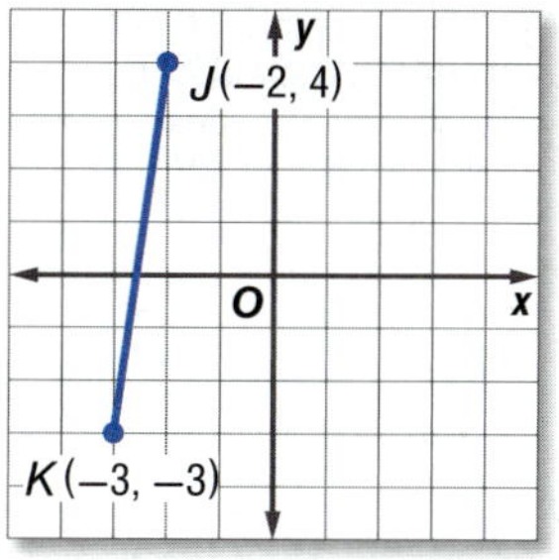

Find the coordinates of the midpoint of a segment with the given endpoints. Then find the distance between each pair of points. (Lesson 1-3)

9. $P(26, 12)$ and $Q(8, 42)$

10. $M(6, -41)$ and $N(-18, -27)$

11. **MAPS** A map of a town is drawn on a coordinate grid. The high school is found at point (3, 1) and town hall is found at (–5, 7). (Lesson 1-3)

    a. If the high school is at the midpoint between the town hall and the town library, at which ordered pair should you find the library?

    b. If one unit on the grid is equivalent to 50 meters, how far is the high school from town hall?

12. **MULTIPLE CHOICE** The vertex of $\angle ABC$ is located at the origin. Point A is located at (5, 0) and Point C is located at (0, 2). How can $\angle ABC$ be classified?

    A acute **C** right

    B obtuse **D** scalene

In the figure, $\overrightarrow{XA}$ and $\overrightarrow{XE}$ are opposite rays, and $\angle AXC$ is bisected by $\overrightarrow{XB}$. (Lesson 1-4)

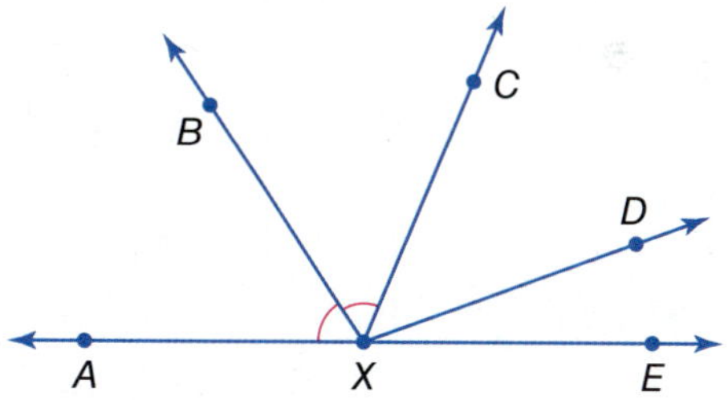

13. If $m\angle AXC = 8x - 7$ and $m\angle AXB = 3x + 10$, find $m\angle AXC$.

14. If $m\angle CXD = 4x + 6$, $m\angle DXE = 3x + 1$, and $m\angle CXE = 8x - 2$, find $m\angle DXE$.

Classify each angle as *acute*, *right*, or *obtuse*. (Lesson 1-4)

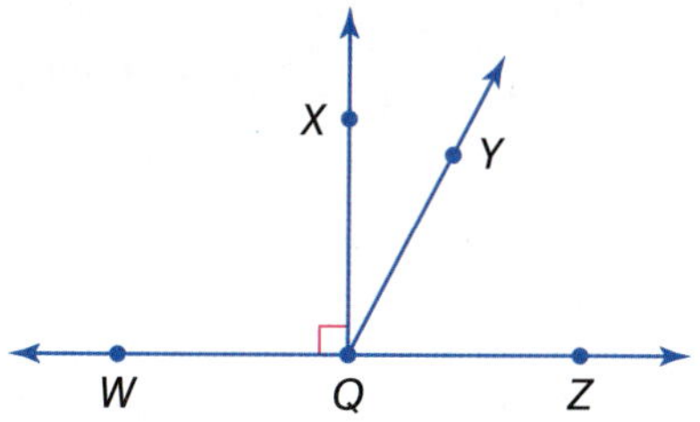

15. $\angle WQY$

16. $\angle YQZ$

1-5 Angle Relationships

Why?

Cheerleaders position their arms and legs at specific angles to create various formations when performing at games and at competitions. Certain pairs of angles have special names and share specific relationships.

Then

You measured and classified angles. (Lesson 1-4)

Now

- Identify and use special pairs of angles.
- Identify perpendicular lines.

New Vocabulary

adjacent angles
linear pair
vertical angles
complementary angles
supplementary angles
perpendicular

Math Online
glencoe.com

Pairs of Angles Some pairs of angles are special because of how they are positioned in relationship to each other. Three of these angle pairs are described below.

Key Concept Special Angle Pairs

Adjacent angles are two angles that lie in the same plane and have a common vertex and a common side, but no common interior points.

Examples ∠1 and ∠2 are adjacent angles.

Nonexamples ∠3 and ∠ABC are nonadjacent angles.

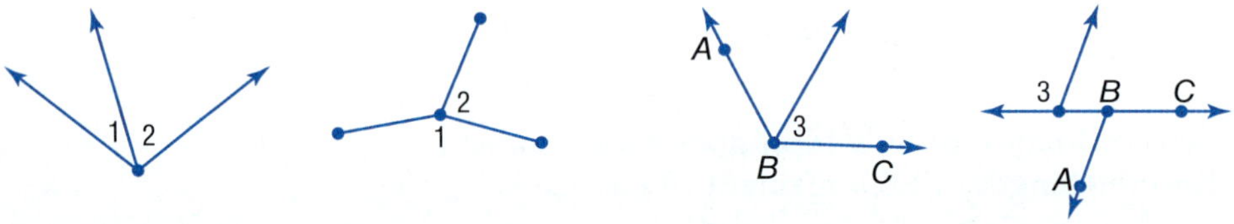

A **linear pair** is a pair of adjacent angles with noncommon sides that are opposite rays.

Example ∠1 and ∠2

Nonexample ∠ADB and ∠ADC

Vertical angles are two nonadjacent angles formed by two intersecting lines.

Examples ∠1 and ∠2; ∠3 and ∠4

Nonexample ∠AEB and ∠DEC

Real-World EXAMPLE 1 Identify Angle Pairs

CHEERLEADING **Name an angle pair that satisfies each condition.**

a. **two acute adjacent angles**

∠*HJK*, ∠*LJM*, ∠*MJN*, and ∠*NJO* are acute angles.

∠*LJM* and ∠*MJN* are acute adjacent angles, and ∠*MJN* and ∠*NJO* are acute adjacent angles.

b. **two obtuse vertical angles**

∠*HJN* and ∠*KJM* are obtuse vertical angles.

Guided Practice

1A. a linear pair

1B. two acute vertical angles

Personal Tutor glencoe.com

Some pairs of angles are special because of the relationship between their angle measures.

Key Concepts Angle Pair Relationships

Vertical angles are congruent.

Examples ∠*ABC* ≅ ∠*DBE* and ∠*ABD* ≅ ∠*CBE*

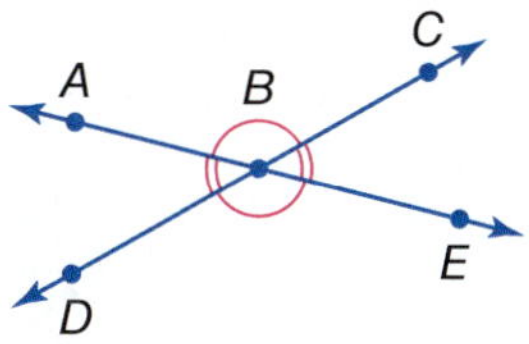

Complementary angles are two angles with measures that have a sum of 90.

Examples ∠1 and ∠2 are complementary.
∠*A* is complementary to ∠*B*.

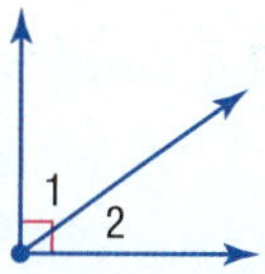

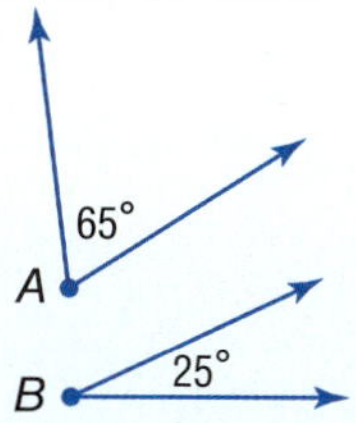

Supplementary angles are two angles with measures that have a sum of 180.

Examples ∠3 and ∠4 are supplementary.
∠*P* and ∠*Q* are supplementary.

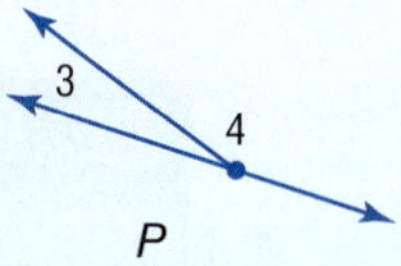

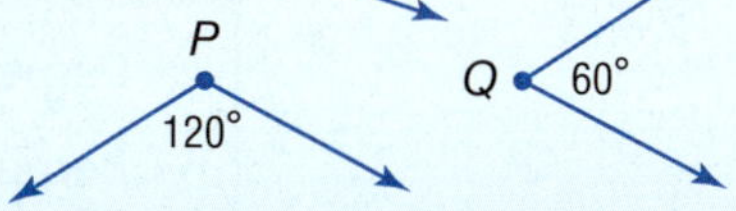

The angles in a linear pair are supplementary.

Example $m\angle 1 + m\angle 2 = 180$

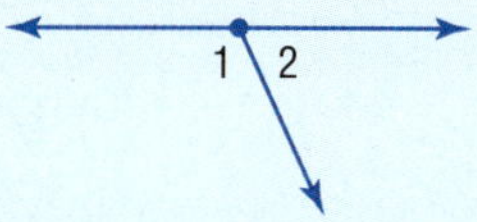

StudyTip

Linear Pair vs. Supplementary Angles While the angles in a linear pair are always supplementary, some supplementary angles do not form a linear pair.

Remember that angle measures are real numbers. So the operations for real numbers and algebra can be used with angle measures.

EXAMPLE 2 Angle Measure

ALGEBRA Find the measures of two supplementary angles if the difference in the measures of the two angles is 18.

Problem-Solving Tip

Write an Equation While you could use the guess-and-check strategy to find two measures with a sum of 180 and a difference of 18, writing an equation is a more efficient approach to this problem.

Understand The problem relates the measures of two supplementary angles. You know that the sum of the measures of supplementary angles is 180. You need to find the measure of each angle.

Plan Draw two figures to represent the angles. Let the measure of one angle be x. If $m\angle A = x$, then because $\angle A$ and $\angle B$ are supplementary, $m\angle B + x = 180$ or $m\angle B = 180 - x$.

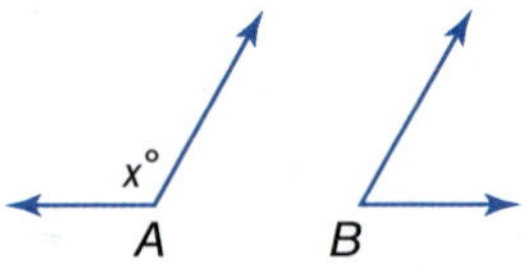

The problem states that the difference of the two angle measures is 18, or $m\angle B - m\angle A = 18$.

Solve

$m\angle B - m\angle A = 18$	Given
$(180 - x) - x = 18$	$m\angle A = x$, $m\angle B = 180 - x$
$180 - 2x = 18$	Simplify.
$-2x = -162$	Subtract 180 from each side.
$x = 81$	Divide each side by -2.

Use the value of x to find each angle measure.

$m\angle A = x$
$= 81$

$m\angle B = 180 - x$
$= 180 - 81$ or 99

Check Add the angle measures to verify that the angles are supplementary.

$m\angle A + m\angle B \stackrel{?}{=} 180$

$81 + 99 = 180$ ✓

Guided Practice

2. Find the measures of two complementary angles if the measure of the larger angle is 12 more than twice the measure of the smaller angle.

Personal Tutor glencoe.com

Perpendicular Lines Lines, segments, or rays that form right angles are **perpendicular**.

Key Concept Perpendicular Lines

- Perpendicular lines intersect to form four right angles.
- Perpendicular lines intersect to form congruent adjacent angles.
- Segments and rays can be perpendicular to lines or other line segments and rays.
- The right angle symbol in the figure indicates that the lines are perpendicular.

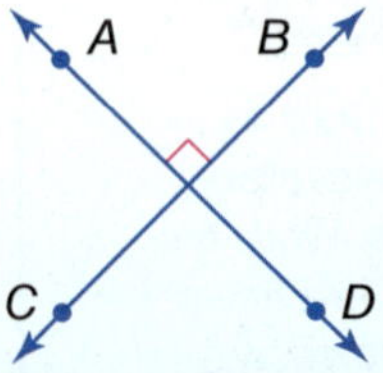

Symbol $\perp$ is read *is perpendicular to*. **Example** $\overleftrightarrow{AD} \perp \overleftrightarrow{CB}$

EXAMPLE 3 Perpendicular Lines

ALGEBRA Find x and y so that $\overleftrightarrow{PR}$ and $\overleftrightarrow{SQ}$ are perpendicular.

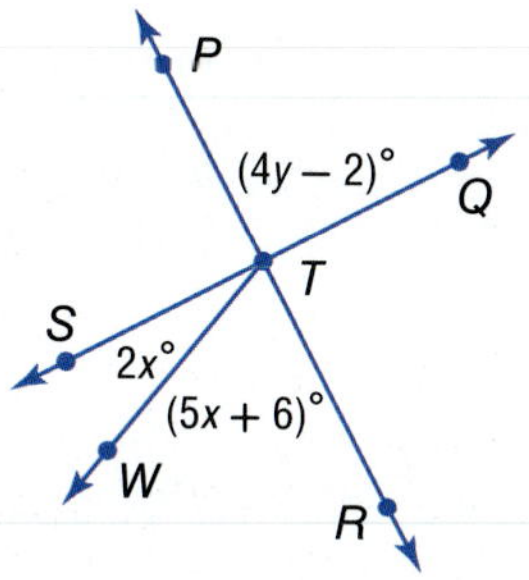

If $\overleftrightarrow{PR} \perp \overleftrightarrow{SQ}$, then $m\angle STR = 90$ and $m\angle PTQ = 90$.

To find x, use $\angle STW$ and $\angle WTR$.

$m\angle STR = m\angle STW + m\angle WTR$	**Sum of parts = whole**
$90 = 2x + (5x + 6)$	**Substitution**
$90 = 7x + 6$	**Combine like terms.**
$84 = 7x$	**Subtract 6 from each side.**
$12 = x$	**Divide each side by 7.**

To find y, use $m\angle PTQ$.

$m\angle PTQ = 4y - 2$	**Given**
$90 = 4y - 2$	**Substitution**
$92 = 4y$	**Add 2 to each side.**
$23 = y$	**Divide each side by 4.**

Guided Practice

3. Suppose $m\angle D = 3x - 12$. Find x so that $\angle D$ is a right angle.

Personal Tutor glencoe.com

StudyTip

Additional Information Additional information for a figure may be given using congruent angle markings, congruent segment markings, or right angle symbols.

In the figure at the right, it *appears* that $\overleftrightarrow{FG} \perp \overleftrightarrow{JK}$. However, you cannot assume this is true unless other information, such as $m\angle FHJ = 90$, is given.

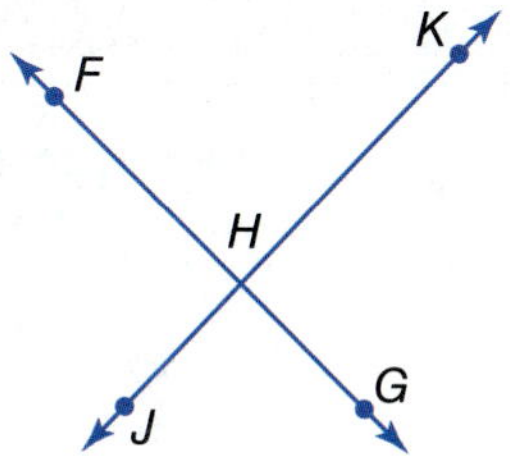

In geometry, figures are sketches used to depict a situation. They are not drawn to reflect total accuracy. There are certain relationships that you can assume to be true, but others you cannot. Study the figure and the lists below.

Key Concept Interpreting Diagrams

CAN be Assumed

All points shown are coplanar.

G, H, and J are collinear.

$\overrightarrow{HM}$, $\overrightarrow{HL}$, $\overrightarrow{HK}$, and $\overleftrightarrow{GJ}$ intersect at H.

H is between G and J.

L is in the interior of $\angle MHK$.

$\angle GHM$ and $\angle MHL$ are adjacent angles.

$\angle GHL$ and $\angle LHJ$ are a linear pair.

$\angle JHK$ and $\angle KHG$ are supplementary.

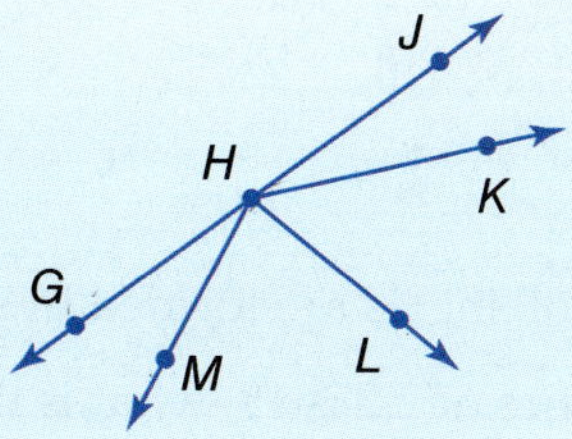

CANNOT be Assumed

Perpendicular lines: $\overrightarrow{HM} \perp \overrightarrow{HL}$

Congruent angles: $\angle JHK \cong \angle GHM$

$\angle JHK \cong \angle KHL$

$\angle KHL \cong \angle LHM$

Congruent segments: $\overline{GH} \cong \overline{HJ}$

$\overline{HJ} \cong \overline{HK}$

$\overline{HK} \cong \overline{HL}$

$\overline{HL} \cong \overline{HG}$

The list of statements that can be assumed is not a complete list. There are more special pairs of angles than those listed.

EXAMPLE 4 Interpret Figures

Determine whether each statement can be assumed from the figure. Explain.

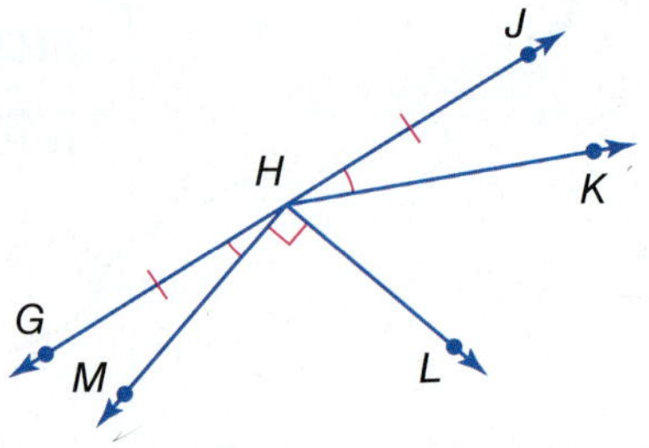

a. ∠KHJ and ∠GHM are complementary.

No; they are congruent, but we do not know anything about their exact measures.

b. ∠GHK and ∠JHK are a linear pair.

Yes; they are adjacent angles whose noncommon sides are opposite rays.

c. $\overrightarrow{HL}$ is perpendicular to $\overrightarrow{HM}$.

Yes; the right angle symbol in the figure indicates that $\overrightarrow{HL} \perp \overrightarrow{HM}$.

Guided Practice

4A. ∠GHL and ∠LHJ are supplementary.

4B. ∠GHM and ∠MHK are adjacent angles.

Personal Tutor glencoe.com

Check Your Understanding

Example 1 p. 47

Name an angle pair that satisfies each condition.

1. two acute vertical angles

2. two obtuse adjacent angles

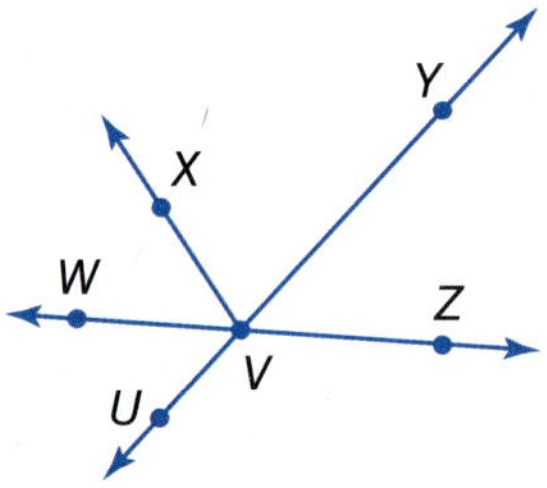

Examples 1 and 2 pp. 47–48

3. CAMERAS Cameras use lenses and light to capture images.

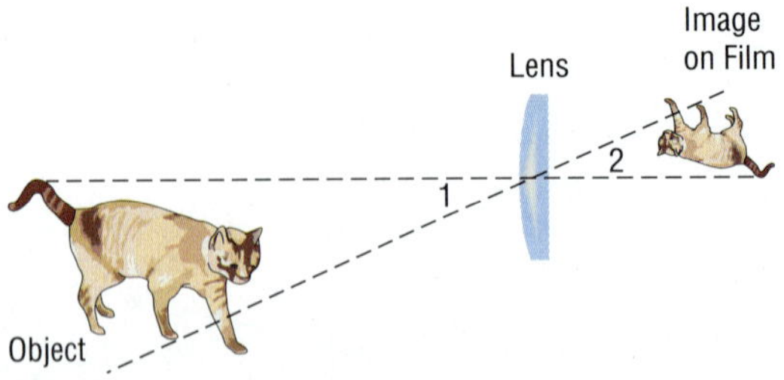

a. What type of angles are formed by the object and its image?

b. If the measure of ∠2 is 15, what is the measure of ∠1?

Examples 2 and 3 pp. 48–49

4. ALGEBRA The measures of two complementary angles are $7x + 17$ and $3x - 20$. Find the measures of the angles.

5. ALGEBRA Lines x and y intersect to form adjacent angles 2 and 3. If $m\angle 2 = 3a - 27$ and $m\angle 3 = 2b + 14$, find the values of a and b so that x is perpendicular to y.

Example 4
p. 50

Determine whether each statement can be assumed from the figure. Explain.

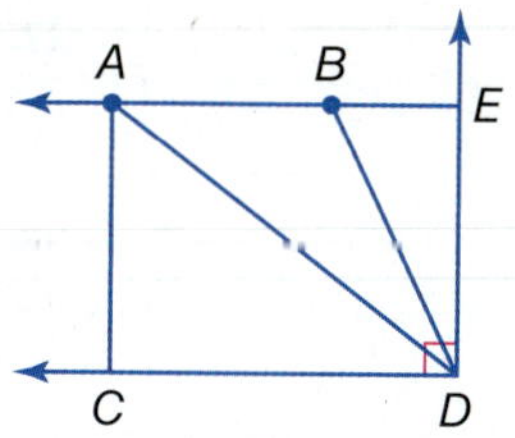

6. $\angle CAD$ and $\angle DAB$ are complementary.
7. $\angle EDB$ and $\angle BDA$ are adjacent, but they are neither complementary nor supplementary.

Practice and Problem Solving

= **Step-by-Step Solutions** begin on page R20.
Extra Practice begins on page 969.

Examples 1 and 2
pp. 47–48

Name an angle or angle pair that satisfies each condition.

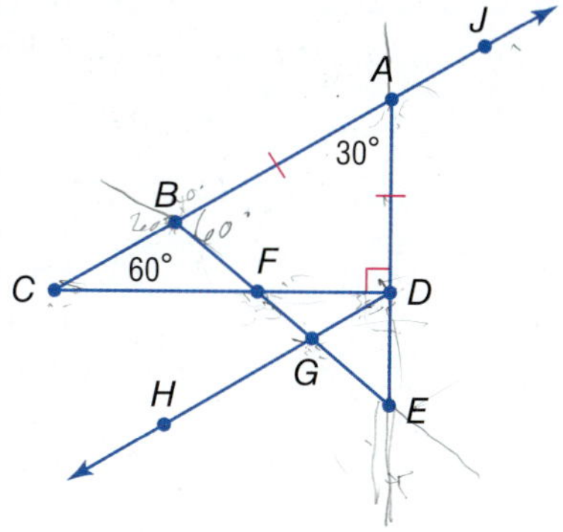

8. two adjacent angles
9. two acute vertical angles
10. two obtuse vertical angles
11. two complementary adjacent angles
12. two complementary nonadjacent angles
13. two supplementary adjacent angles
14. a linear pair whose vertex is F
15. an angle complementary to $\angle FDG$
16. an angle supplementary to $\angle CBF$
17. an angle supplementary to $\angle JAE$

18. **MAPS** You are using a compass to drive 23° east of north. Express your direction in another way using an acute angle and two of the four directions: north, south, east, and west. Explain your reasoning.

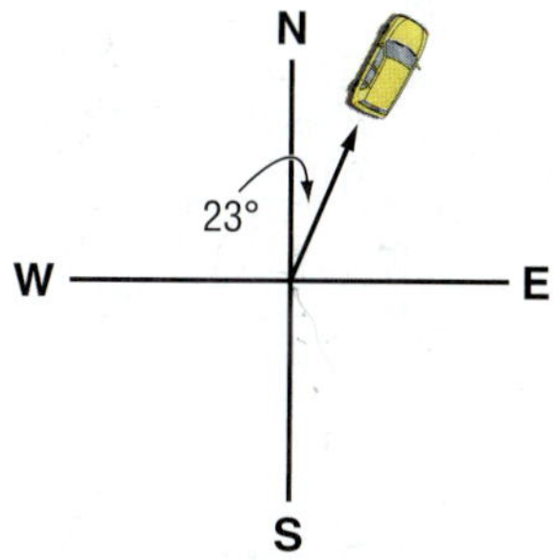

Example 2
p. 48

Find the value of each variable.

19. 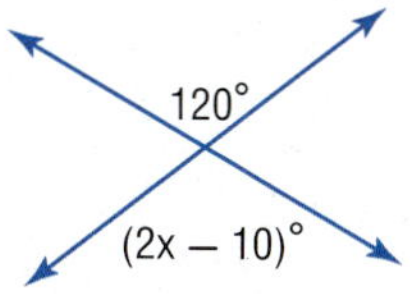

20. 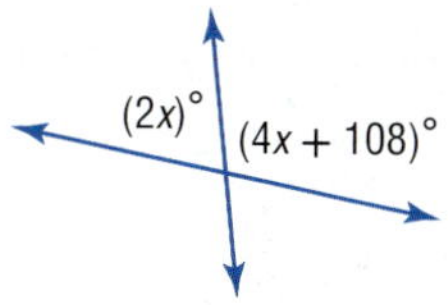

21. 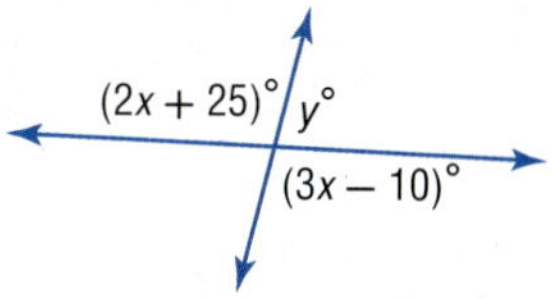

22. 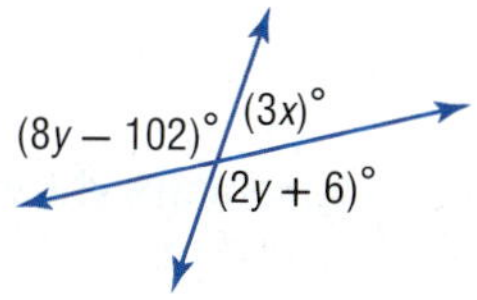

23. 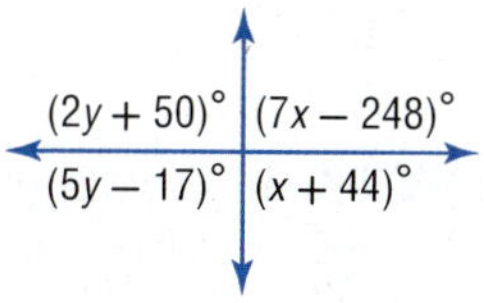

24. 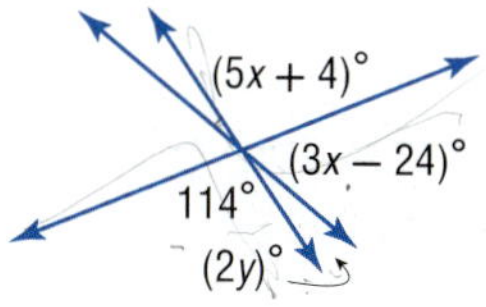

25. **ALGEBRA** $\angle E$ and $\angle F$ are supplementary. The measure of $\angle E$ is 54 more than the measure of $\angle F$. Find the measures of each angle.

26. **ALGEBRA** The measure of an angle's supplement is 76 less than the measure of the angle. Find the measure of the angle and its supplement.

27. **ALGEBRA** The measure of the supplement of an angle is 40 more than two times the measure of the complement of the angle. Find the measure of the angle.

28. **ALGEBRA** $\angle 3$ and $\angle 4$ form a linear pair. The measure of $\angle 3$ is four more than three times the measure of $\angle 4$. Find the measure of each angle.

Example 3
p. 49

ALGEBRA Use the figure at the right.

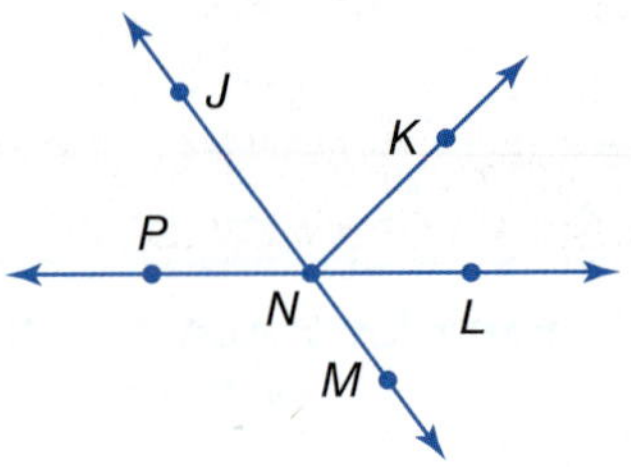

29. If $m\angle KNL = 6x - 4$ and $m\angle LNM = 4x + 24$, find the value of x so that $\angle KNM$ is a right angle.

30. If $m\angle JNP = 3x - 15$ and $m\angle JNL = 5x + 59$, find the value of x so that $\angle JNP$ and $\angle JNL$ are supplements of each other.

31. If $m\angle LNM = 8x + 12$ and $m\angle JNL = 12x - 32$, find $m\angle JNP$.

32. If $m\angle JNP = 2x + 3$, $m\angle KNL = 3x - 17$, and $m\angle KNJ = 3x + 34$, find the measure of each angle.

33. **PHYSICS** As a ray of light meets a mirror, the light is reflected. The angle at which the light strikes the mirror is the *angle of incidence*. The angle at which the light is reflected is the *angle of reflection*. The angle of incidence and the angle of reflection are congruent. In the diagram at the right, if $m\angle RMI = 106$, find the angle of reflection and $m\angle RMJ$.

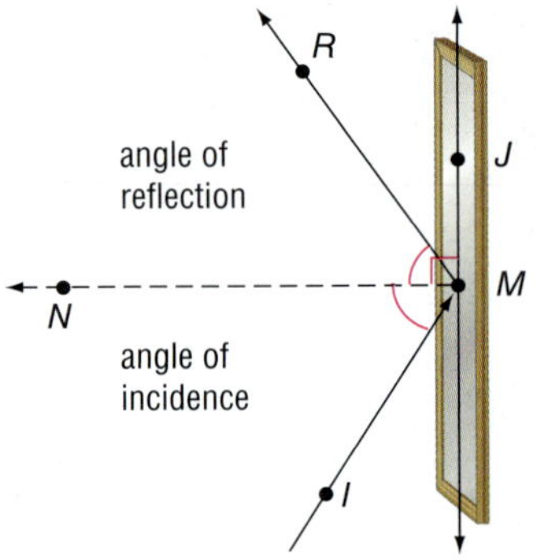

34. **ALGEBRA** Rays AB and BC are perpendicular. Point D lies in the interior of $\angle ABC$. If $m\angle ABD = 3r + 5$ and $m\angle DBC = 5r - 27$, find $m\angle ABD$ and $m\angle DBC$.

35. **ALGEBRA** $\overleftrightarrow{WX}$ and $\overleftrightarrow{YZ}$ intersect at point V. If $m\angle WVY = 4a + 58$ and $m\angle XVY = 2b - 18$, find the values of a and b so that $\overleftrightarrow{WX}$ is perpendicular to $\overleftrightarrow{YZ}$.

Example 4
p. 50

Determine whether each statement can be assumed from the figure. Explain.

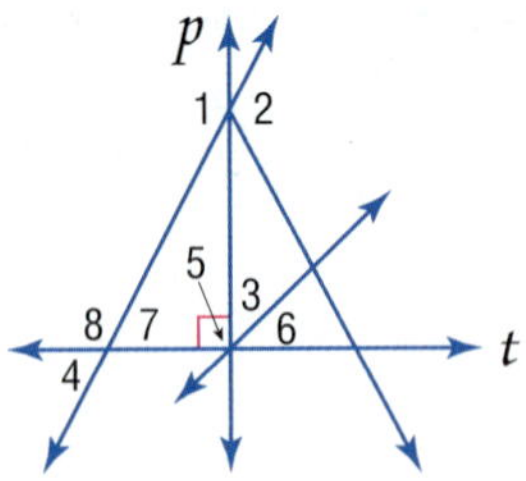

36. $\angle 4$ and $\angle 7$ are vertical angles.

37. $\angle 4$ and $\angle 8$ are supplementary.

38. $p \perp t$

39. $\angle 3 \cong \angle 6$

40. $\angle 5 \cong \angle 3 + \angle 6$

41. $\angle 5$ and $\angle 7$ form a linear pair.

42. **GARDENING** In the diagram of the pruning shears shown, $m\angle 1 = m\angle 3$. What conclusion can you reach about the relationship between $\angle 4$ and $\angle 2$? Explain.

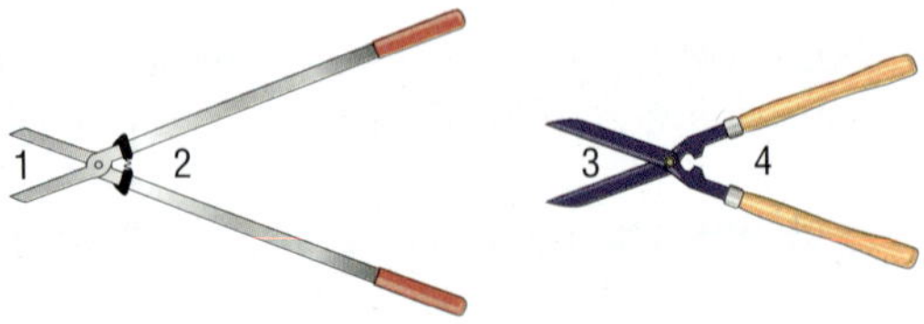

Real-World Link

The Ames-Dryden aircraft was designed to research the use of pivoting wings on aircraft to improve the efficiency of the plane. As the plane's speed increased, the wings could be pivoted up to 60° in either direction.

Source: NASA

FLIGHT Use the diagram of the Ames-Dryden aircraft shown.

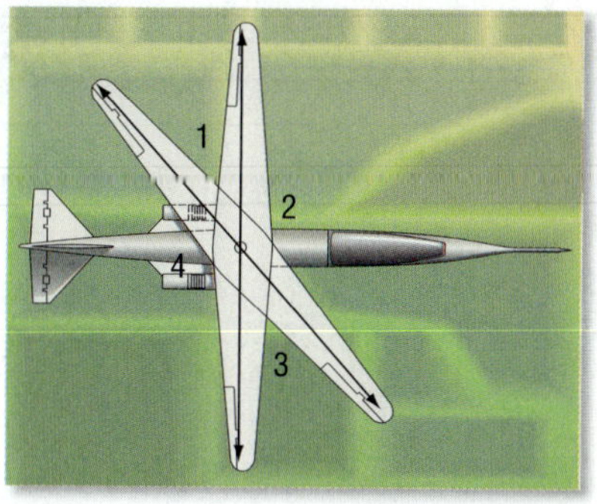

43. Identify a pair of vertical angles.

44. Identify two pairs of supplementary angles.

45. If $m\angle 1 = 110$, what is $m\angle 3$? $m\angle 4$?

46. What is the minimum possible value for $m\angle 2$? the maximum?

47. Is there a wing position in which none of the angles are obtuse? Explain.

48. **MULTIPLE REPRESENTATIONS** In this problem, you will explore the relationship between the sum of the interior angles of a triangle and the angles vertical to them.

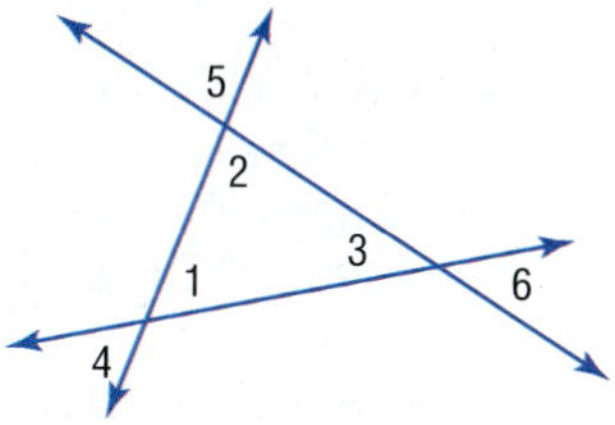

a. **GEOMETRIC** Draw three sets of three intersecting lines and label each as shown.

b. **TABULAR** For each set of lines, measure and record $m\angle 1$, $m\angle 2$, and $m\angle 3$ in a table. Record $m\angle 1 + m\angle 2 + m\angle 3$ in a separate column.

c. **VERBAL** Explain how you can find $m\angle 4$, $m\angle 5$, and $m\angle 6$ when you know $m\angle 1$, $m\angle 2$, and $m\angle 3$.

d. **ALGEBRAIC** Write an equation that relates $m\angle 1 + m\angle 2 + m\angle 3$ to $m\angle 4 + m\angle 5 + m\angle 6$. Then use substitution to write an equation that relates $m\angle 4 + m\angle 5 + m\angle 6$ to an integer.

H.O.T. Problems

Use Higher-Order Thinking Skills

49. **REASONING** Are there angles that do not have a complement? Explain.

50. **OPEN ENDED** Draw a pair of intersecting lines that forms a pair of complementary angles. Explain your reasoning.

51. **CHALLENGE** If a line, line segment, or ray is perpendicular to a plane, it is perpendicular to every line, line segment, or ray in the plane that intersects it.

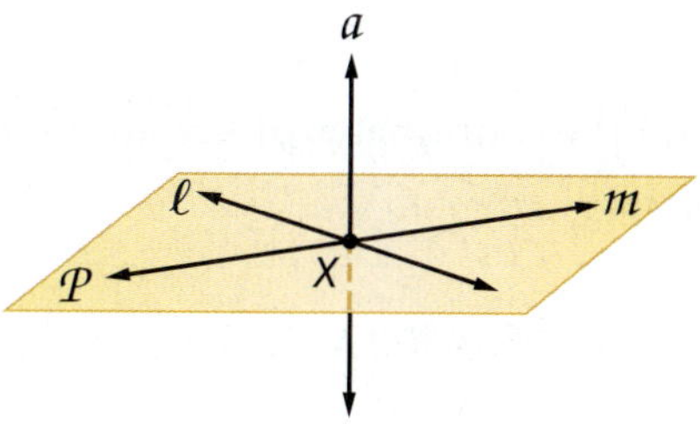

a. If a line is perpendicular to each of two intersecting lines at their point of intersection, then the line is perpendicular to the plane determined by them. If line a is perpendicular to line ℓ and line m at point X, what must also be true?

b. If a line is perpendicular to a plane, then any line perpendicular to the given line at the point of intersection with the given plane is in the given plane. If line a is perpendicular to plane $\mathcal{P}$ and line m at point X, what must also be true?

c. If a line is perpendicular to a plane, then every plane containing the line is perpendicular to the given plane. If line a is perpendicular to plane $\mathcal{P}$, what must also be true?

52. **WRITING IN MATH** Describe three different ways you can determine that an angle is a right angle.

Standardized Test Practice

53. What is $m\angle RMS$ in the figure below?

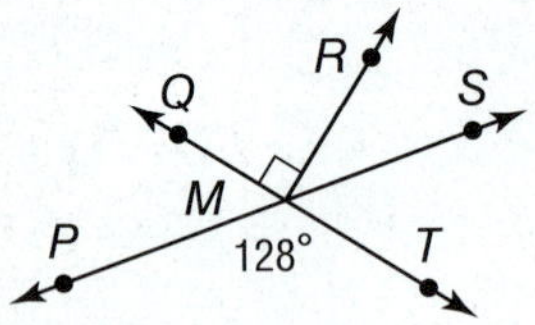

A 26
B 38
C 52
D 128

54. EXTENDED RESPONSE For a fundraiser, a theater club is making 400 cookies. They want to make twice as many chocolate chip as peanut butter cookies and three times as many peanut butter as oatmeal raisin cookies. Determine how many of each type of cookie the theater club will make. Show your work.

55. ALGEBRA Which inequality is graphed below?

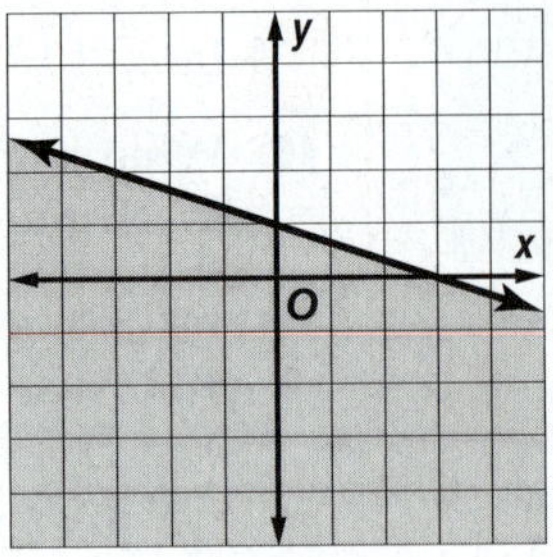

F $y > -\frac{1}{3}x + 1$ **H** $y \geq -\frac{1}{3}x + 1$

G $y < -\frac{1}{3}x + 1$ **J** $y \leq -\frac{1}{3}x + 1$

56. SAT/ACT One third of a number is three more than one fourth the same number. What is the number?

A 12 **C** 42
B 36 **D** 48

Spiral Review

Copy the diagram shown and extend each ray. Classify each angle as *right*, *acute*, or *obtuse*. Then use a protractor to measure the angle to the nearest degree. (Lesson 1-4)

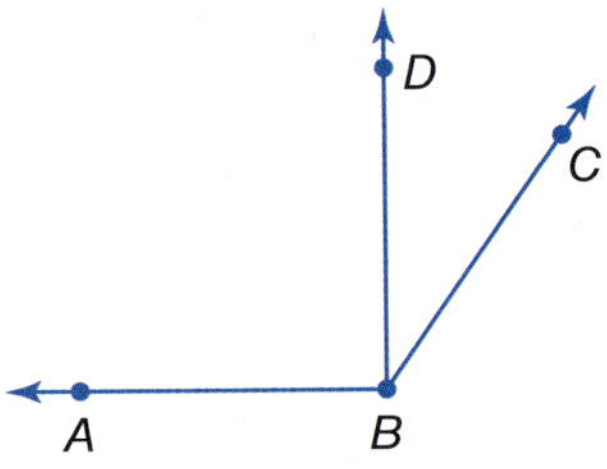

57. $\angle ABC$

58. $\angle DBC$

59. $\angle ABD$

Find the coordinates of the midpoint of a segment with the given endpoints. (Lesson 1-3)

60. $P(3, -7), Q(9, 6)$ **61.** $A(-8, -5), B(1, 7)$ **62.** $J(-7, 4), K(3, 1)$

63. SNOWBOARDING In the design on the snowboard shown, $\overline{BD}$ bisects $\overline{SN}$ at R. If $SN = 163$ centimeters, find RN. (Lesson 1-2)

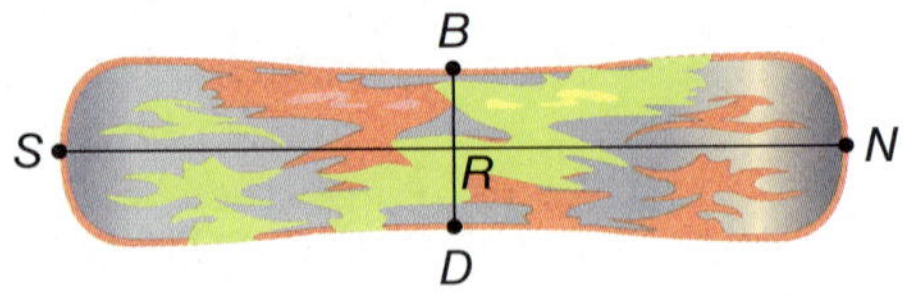

Skills Review

Name the congruent sides and angles in each figure. (Lesson 1-2)

64.

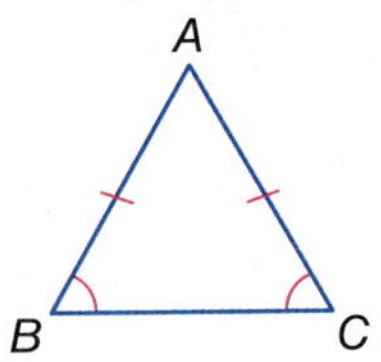

65.

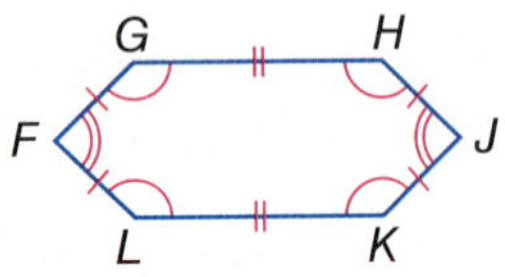

66.

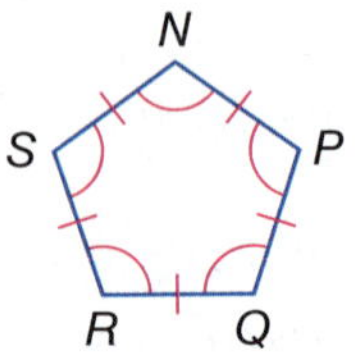

67.

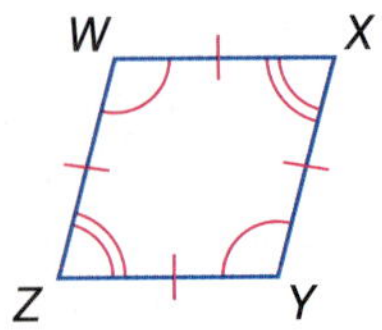

EXTEND
1-5

Geometry Lab
Constructing Perpendiculars

You can use a compass and a straightedge to construct a line perpendicular to a given line through a point on the line, or through a point *not* on the line.

ACTIVITY Construct a Perpendicular

a. Construct a line perpendicular to line ℓ and passing through point P on ℓ.

Step 1

Place the compass at P. Draw arcs to the right and left of P that intersect line ℓ using the same compass setting. Label the points of intersection A and B.

Step 2

With compass at A, draw an arc above line ℓ using a setting greater than AP. Using same compass setting, draw an arc from B that intersects previous arc. Label the intersection Q.

Step 3

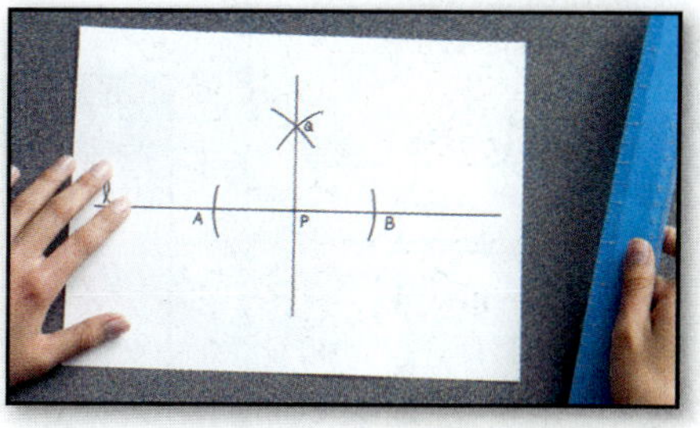

Use a straightedge to draw $\overleftrightarrow{QP}$.

b. Construct a line perpendicular to line k and passing through point P *not* on k.

Step 1

Place the compass at P. Draw an arc that intersects line k in two different places. Label the points of intersection C and D.

Step 2

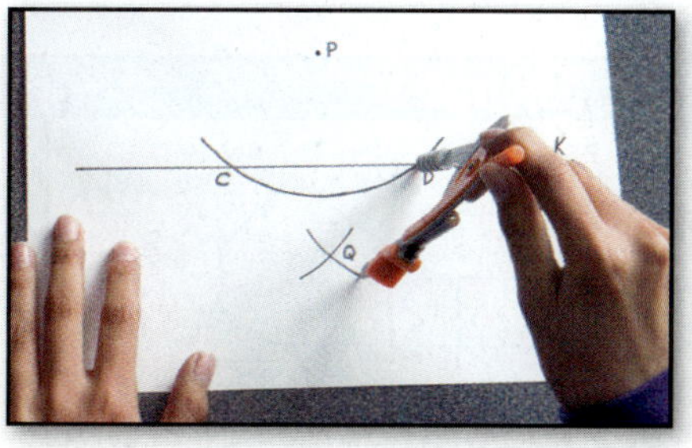

With compass at C, draw an arc below line k using a setting greater than $\frac{1}{2}CD$. Using same compass setting, draw an arc from D that intersects previous arc. Label the intersection Q.

Step 3

Use a straightedge to draw $\overleftrightarrow{PQ}$.

Model and Analyze the Results

1. Draw a line and construct a line perpendicular to it through a point on the line.
2. Draw a line and construct a line perpendicular to it through a point not on the line.
3. How is the second construction similar to the first one?

1-6 Two-Dimensional Figures

Then
You measured one-dimensional figures. (Lesson 1-2)

Now
- Identify and name polygons.
- Find perimeter, circumference, and area of two-dimensional figures.

New Vocabulary
polygon
vertex of a polygon
concave
convex
n-gon
equilateral polygon
equiangular polygon
regular polygon
perimeter
circumference
area

Math Online
glencoe.com

Why?

Mosaics are patterns or pictures created using small bits of colored glass or stone. They are usually set into a wall or floor and often make use of polygons.

Identify Polygons Most of the closed figures shown in the mosaic are polygons. The term *polygon* is derived from a Greek word meaning *many angles.*

Key Concept: Polygons

A **polygon** is a closed figure formed by a finite number of coplanar segments called *sides* such that

- the sides that have a common endpoint are noncollinear, and
- each side intersects exactly two other sides, but only at their endpoints.

The vertex of each angle is a **vertex of the polygon**. A polygon is named by the letters of its vertices, written in order of consecutive vertices.

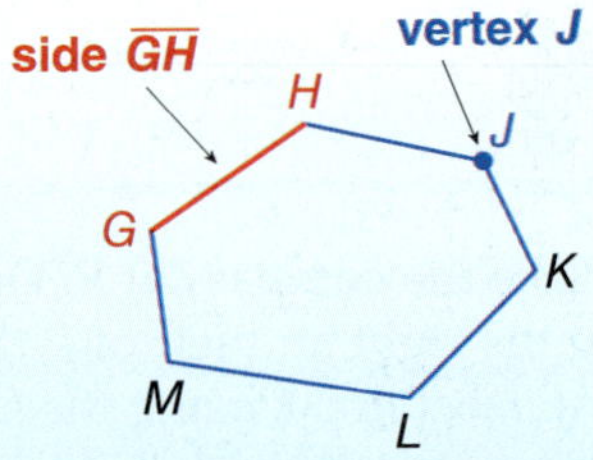

polygon *GHJKLM*

The table below shows some additional examples of polygons and some examples of figures that are not polygons.

Polygons	Not Polygons
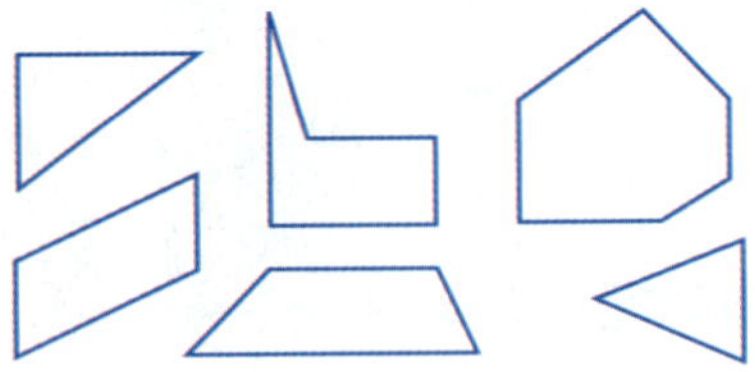	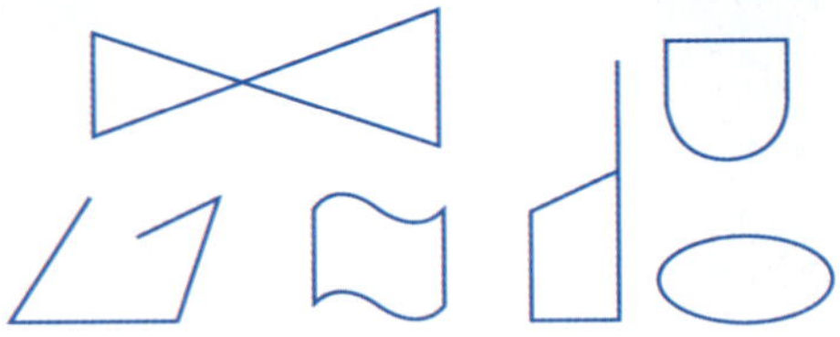

Polygons can be **concave** or **convex**. Suppose the line containing each side is drawn. If any of the lines contain any point in the interior of the polygon, then it is concave. Otherwise it is convex.

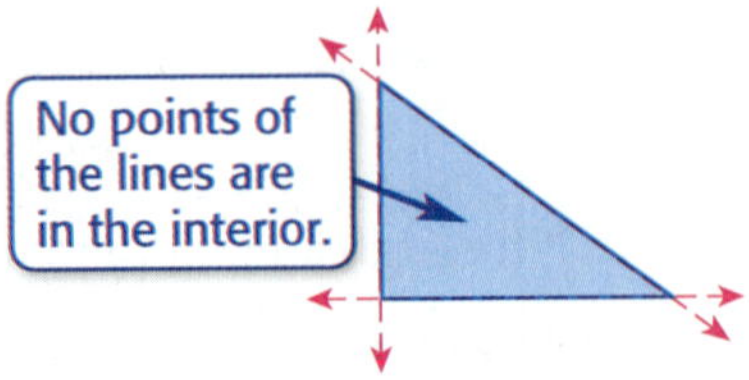

convex polygon

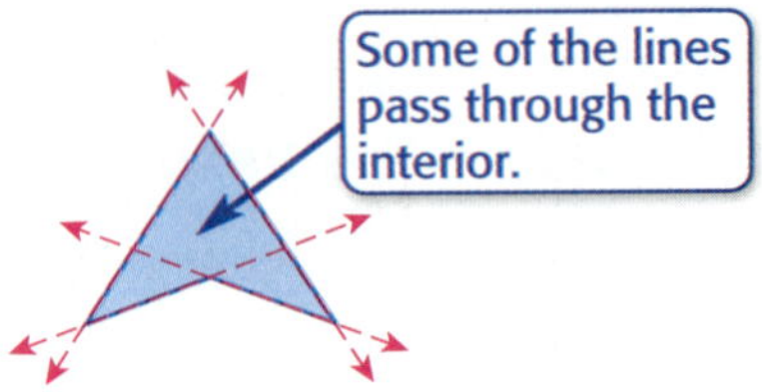

concave polygon

StudyTip

Naming Polygons The Greek prefixes used to name polygons are also used to denote number. For example a *bi*cycle has two wheels, and a *tri*pod has three legs.

In general, a polygon is classified by its number of sides. The table lists some common names for various categories of polygon. A polygon with *n* sides is an **n-gon**. For example, a polygon with 15 sides is a 15-gon.

Number of Sides	Polygon
3	triangle
4	quadrilateral
5	pentagon
6	hexagon
7	heptagon
8	octagon
9	nonagon
10	decagon
11	hendecagon
12	dodecagon
n	n-gon

An **equilateral polygon** is a polygon in which all sides are congruent. An **equiangular polygon** is a polygon in which all angles are congruent.

A convex polygon that is both equilateral and equiangular is called a **regular polygon**. An *irregular polygon* is a polygon that is *not* regular.

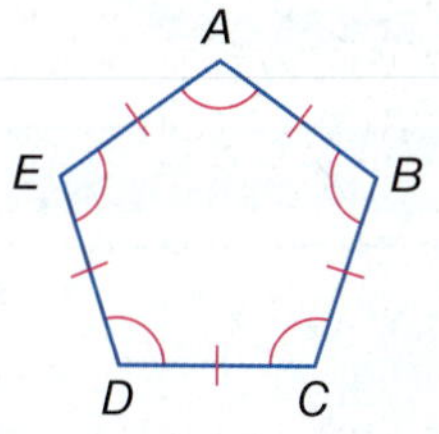

regular pentagon *ABCDE*

EXAMPLE 1 Name and Classify Polygons

Name each polygon by its number of sides. Then classify it as *convex* or *concave* and *regular* or *irregular*.

a.

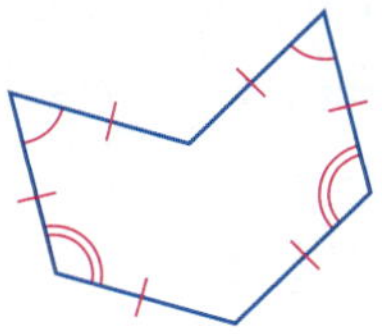

The polygon has 6 sides, so it is a hexagon.

Two of the lines containing the sides of the polygon will pass through the interior of the hexagon, so it is concave.

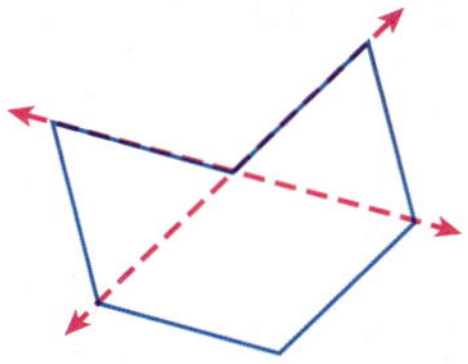

Only convex polygons can be regular, so this is an irregular hexagon.

b.

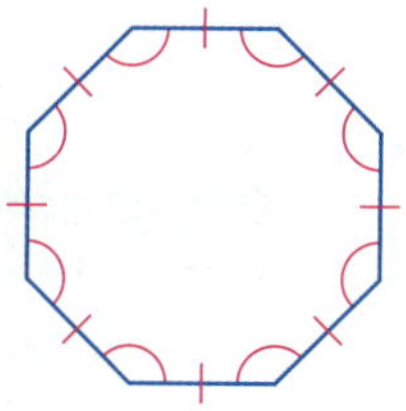

There are 8 sides, so this is an octagon.

No line containing any of the sides will pass through the interior of the octagon, so it is convex.

All of the sides are congruent, so it is equilateral. All of the angles are congruent, so it is equiangular.

Since the polygon is convex, equilateral, and equiangular, it is regular. So this is a regular octagon.

ReadingMath

Simple Closed Curves Polygons and circles are examples of *simple closed curves*. Such a curve begins and ends at the same point without crossing itself. The figures below are *not* simple closed curves.

Guided Practice

1A.

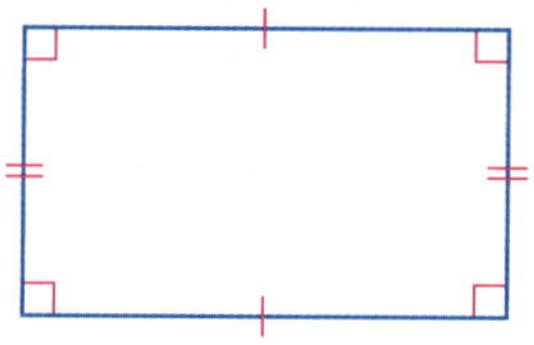

1B.

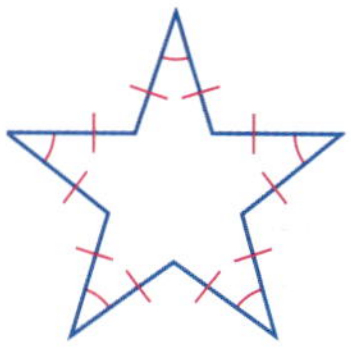

1C.

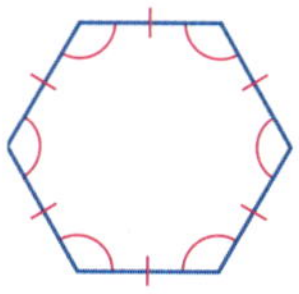

Personal Tutor glencoe.com

> **ReadingMath**
>
> **Pi** The symbol π is read *pi*. This is not a variable but an irrational number. The most accurate way to perform a calculation with π is to use a calculator. If no calculator is available, 3.14 is a good estimate for π.

Perimeter, Circumference, and Area The **perimeter** of a polygon is the sum of the lengths of the sides of the polygon. Some shapes have special formulas for perimeter, but all are derived from the basic definition of perimeter. You will derive these formulas in Chapter 11. The **circumference** of a circle is the distance around the circle.

The **area** of a figure is the number of square units needed to cover a surface. Review the formulas for the perimeter and area of three common polygons and circle given below.

Key Concepts — Perimeter, Circumference, and Area

Triangle	Square	Rectangle	Circle
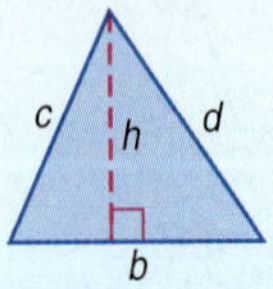	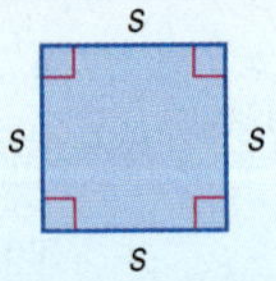	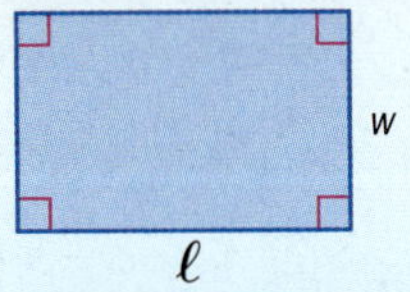	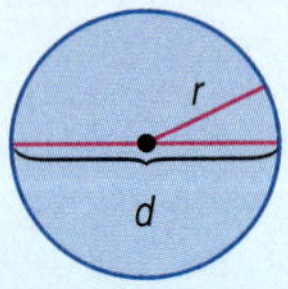
$P = b + c + d$	$P = s + s + s + s$ $= 4s$	$P = \ell + w + \ell + w$ $= 2\ell + 2w$	$C = 2\pi r$ or $C = \pi d$
$A = \frac{1}{2}bh$	$A = s^2$	$A = \ell w$	$A = \pi r^2$

P = perimeter of polygon; b = base, h = height; A = area of figure; ℓ = length, w = width; C = circumference; r = radius, d = diameter

EXAMPLE 2 Find Perimeter and Area

Find the perimeter or circumference and area of each figure.

a.

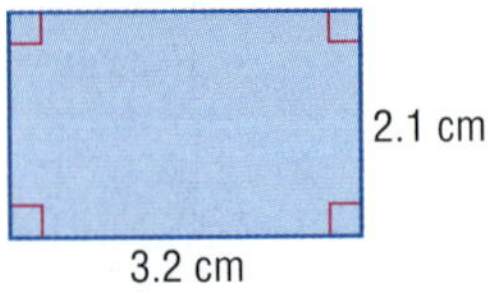

$P = 2\ell + 2w$ — **Perimeter of rectangle**

$= 2(3.2) + 2(2.1)$ — **$\ell = 3.2$, $w = 2.1$**

$= 10.6$ — **Simplify.**

The perimeter is 10.6 centimeters.

$A = \ell w$ — **Area of rectangle**

$= (3.2)(2.1)$ — **$\ell = 3.2$, $w = 2.1$**

$= 6.72$ — **Simplify.**

The area is about 6.7 square centimeters.

b.

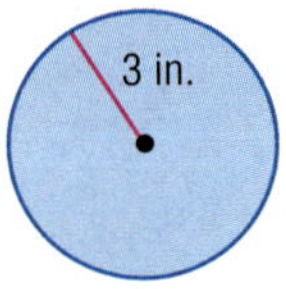

$C = 2\pi r$ — **Circumference**

$= 2\pi(3)$ — **$r = 3$**

≈ 18.85 — **Use a calculator.**

The circumference is about 18.9 inches.

$A = \pi r^2$ — **Area of circle**

$= \pi(3)^2$ — **$r = 3$**

≈ 28.3 — **Use a calculator.**

The area is about 28.3 square inches.

Guided Practice

2A.

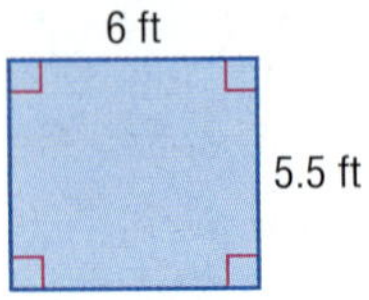

2B.

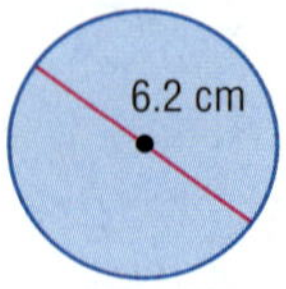

2C.

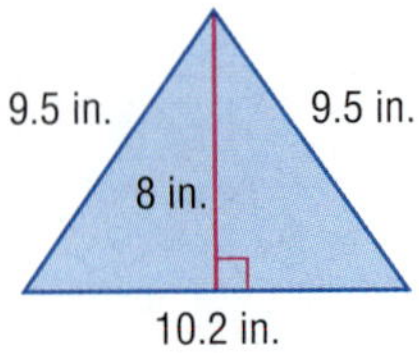

Personal Tutor glencoe.com

Test-Taking Tip

Mental Math When you are asked to compare measures for varying figures, it can be helpful to use mental math. Estimate the perimeter or area of each figure, and then check your calculations.

STANDARDIZED TEST EXAMPLE 3

Yolanda has 26 centimeters of cording to frame a photograph in her scrapbook. Which of these shapes would use *most* or all of the cording and enclose the *largest* area?

A right triangle with each leg about 7 centimeters long

B circle with a radius of about 4 centimeters

C rectangle with a length of 8 centimeters and a width of 4.5 centimeters

D square with a side length of 6 centimeters

Read the Test Item

You are asked to compare the area and perimeter of four different shapes.

Solve the Test Item

Find the perimeter and area of each shape.

Right Triangle

Use the Pythagorean Theorem to find the length of the hypotenuse.

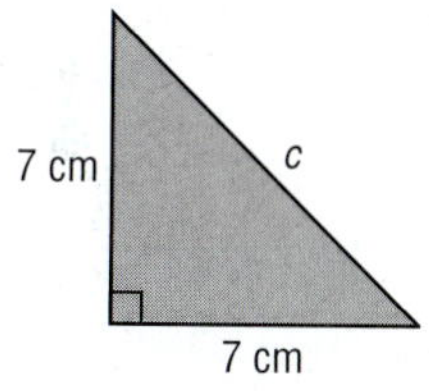

$c^2 = a^2 + b^2$ — Pythagorean Theorem

$c^2 = 7^2 + 7^2$ or 98 — $a = 7, b = 7$

$c = \sqrt{98}$ or about 9.9 — Simplify.

$P = a + b + c$ — Perimeter of a triangle

$\approx 7 + 7 + 9.9$ or about 23.9 cm — Substitution

$A = \frac{1}{2}bh$ — Area of a triangle

$= \frac{1}{2}(7)(7)$ or 24.5 cm^2 — Substitution

Circle	Rectangle	Square
$C = 2\pi r$	$P = 2\ell + 2w$	$P = 4s$
$= 2\pi(4)$	$= 2(8) + 2(4.5)$	$= 4(6)$
≈ 25.1 cm	$= 25$ cm	$= 24$ cm
$A = \pi r^2$	$A = \ell w$	$A = s^2$
$= \pi(4)^2$	$= (8)(4.5)$	$= 6^2$
≈ 50.3 cm^2	$= 36$ cm^2	$= 36$ cm^2

The shape that uses all of the cording and encloses the largest area is the circle. The answer is B.

Guided Practice

3. Dasan has 32 feet of fencing to fence in a play area for his dog. Which shape play area uses *most* or all of the fencing and encloses the *largest* area?

F circle with radius of about 5 feet

G rectangle with length 5 feet and width 10 feet

H right triangle with legs of length 10 feet

J square with side length 8 feet

Personal Tutor glencoe.com

You can use the Distance Formula to find the perimeter of a polygon graphed on a coordinate plane.

EXAMPLE 4 Perimeter and Area on the Coordinate Plane

COORDINATE GEOMETRY Find the perimeter and area of $\triangle PQR$ with vertices $P(-1, 3)$, $Q(-3, -1)$, and $R(4, -1)$.

Step 1 Find the perimeter of $\triangle PQR$.

Graph $\triangle PQR$.

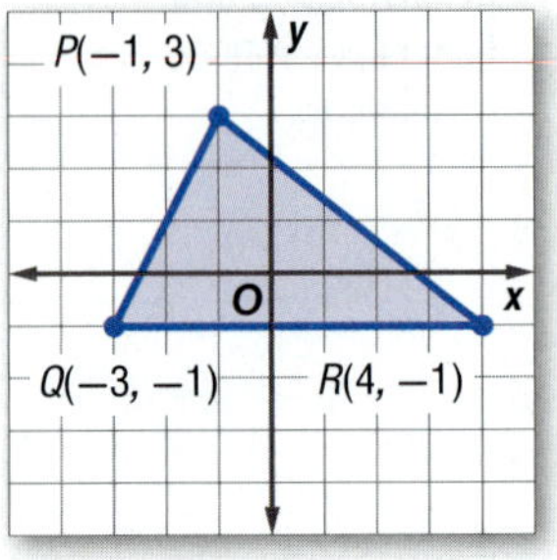

To find the perimeter of $\triangle PQR$, first find the lengths of each side. Counting the squares on the grid, we find that $QR = 7$ units. Use the Distance Formula to find the lengths of $\overline{PQ}$ and $\overline{PR}$.

$\overline{PQ}$ has endpoints at $P(-1, 3)$ and $Q(-3, -1)$.

$$PQ = \sqrt{(x_2 - x_1)^2 + (y_2 - y_1)^2}$$ **Distance Formula**

$$= \sqrt{[-1 - (-3)]^2 + [3 - (-1)]^2}$$ **Substitute.**

$$= \sqrt{2^2 + 4^2}$$ **Subtract.**

$$= \sqrt{20} \text{ or about } 4.5$$ **Simplify.**

$\overline{PR}$ has endpoints at $P(-1, 3)$ and $R(4, -1)$.

$$PR = \sqrt{(x_2 - x_1)^2 + (y_2 - y_1)^2}$$ **Distance Formula**

$$= \sqrt{(-1 - 4)^2 + [3 - (-1)]^2}$$ **Substitute.**

$$= \sqrt{(-5)^2 + 4^2}$$ **Subtract.**

$$= \sqrt{41} \text{ or about } 6.4$$ **Simplify.**

The perimeter of $\triangle PQR$ is $7 + \sqrt{20} + \sqrt{41}$ or about 17.9 units.

Step 2 Find the area of $\triangle PQR$.

To find the area of the triangle, find the lengths of the height and base. The height is the perpendicular distance from P to $\overline{QR}$. Counting squares on the graph, the height is 4 units. The length of $\overline{QR}$ is 7 units.

$$A = \frac{1}{2}bh$$ **Area of a triangle**

$$= \frac{1}{2}(7)(4) \text{ or } 14$$ **Substitute and simplify.**

The area of $\triangle PQR$ is 14 square units.

StudyTip

Linear and Square Units Remember to use linear units with perimeter and square units with area.

Guided Practice

4. Find the perimeter and area of $\triangle ABC$ with vertices $A(-1, 4)$, $B(-1, -1)$, and $C(6, -1)$.

Personal Tutor glencoe.com

Check Your Understanding

Example 1 p. 57 — **Name each polygon by its number of sides. Then classify it as *convex* or *concave* and *regular* or *irregular*.**

1.

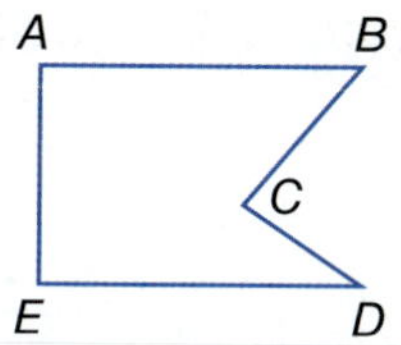

2.

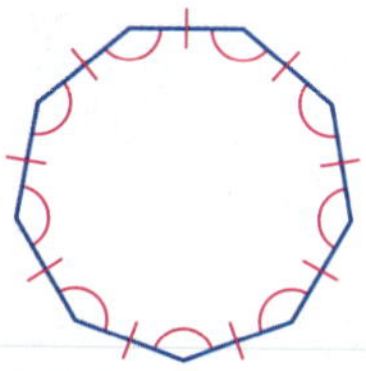

SIGNS **Identify the shape of each traffic sign and classify it as *regular* or *irregular*.**

3. stop

4. caution or warning

5. slow moving vehicle

Example 2 p. 58 — **Find the perimeter or circumference and area of each figure. Round to the nearest tenth.**

6.

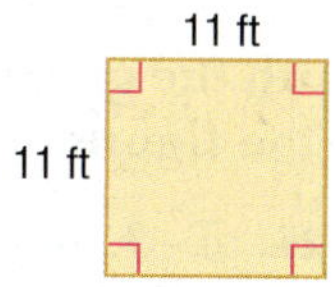

7.

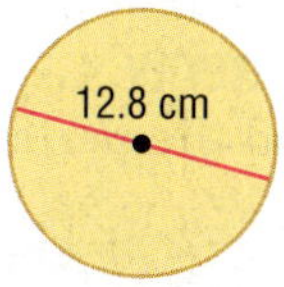

8.

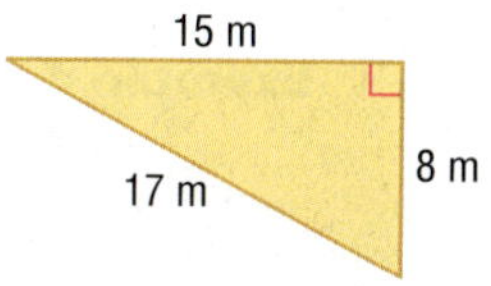

Example 3 p. 59

9. **MULTIPLE CHOICE** Vanesa is making a banner for the game. She has 20 square feet of fabric. What shape will use *most* or all of the fabric?

A a square with a side length of 4 feet

B a rectangle with a length of 4 feet and a width of 3.5 feet

C a circle with a radius of about 2.5 feet

D a right triangle with legs of about 5 feet

Example 4 p. 60

10. **COORDINATE GEOMETRY** Find the perimeter and area of $\triangle ABC$ with vertices $A(-1, 2)$, $B(3, 6)$, and $C(3, -2)$.

Practice and Problem Solving

● = **Step-by-Step Solutions** begin on page R20.
Extra Practice begins on page 969.

Example 1 p. 57 — **Name each polygon by its number of sides. Then classify it as *convex* or *concave* and *regular* or *irregular*.**

11.

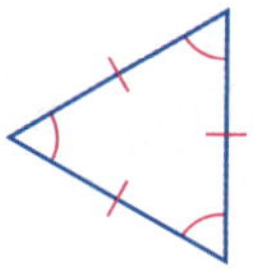

12.

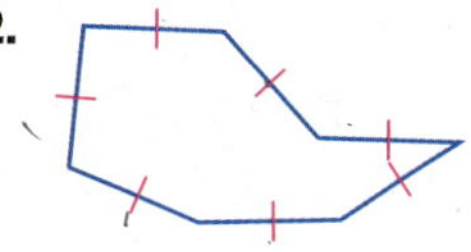

13.

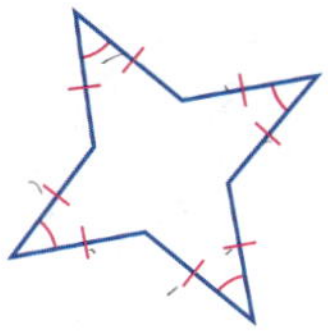

14.

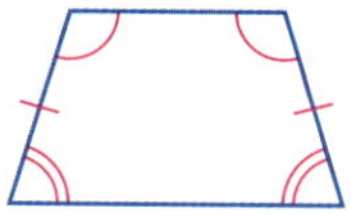

15.

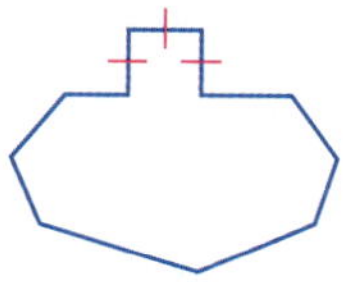

16.

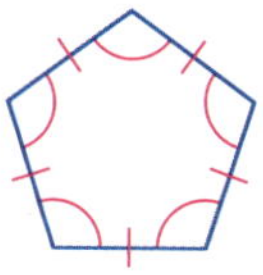

Examples 2 and 3
pp. 58–59

Find the perimeter or circumference and area of each figure. Round to the nearest tenth.

17.
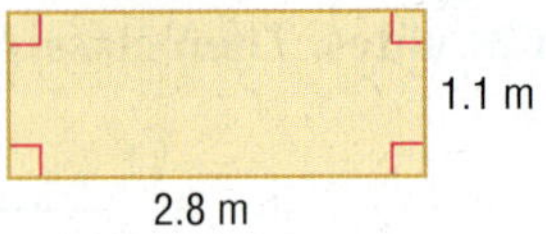

18.
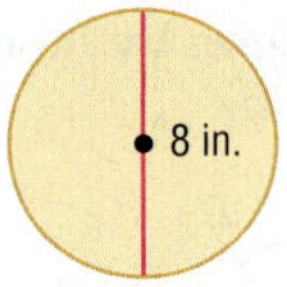

19.
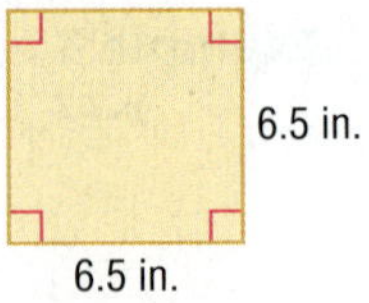

20.
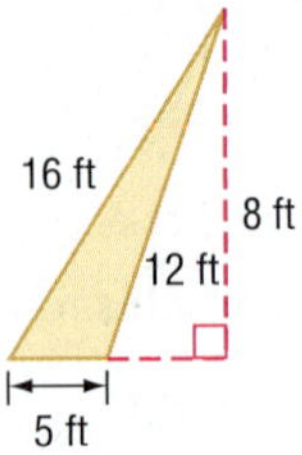

21
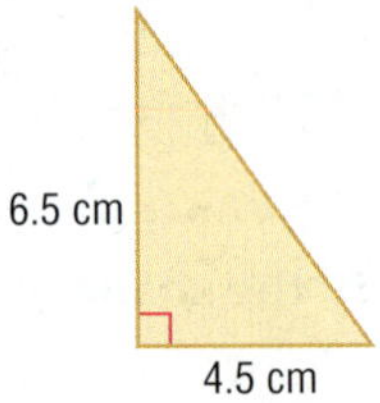

22.
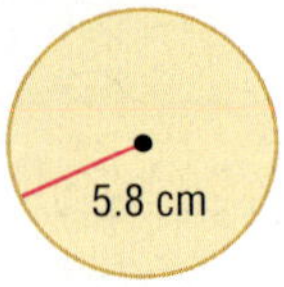

23. CRAFTS Joy has a square picture that is 4 inches on each side. The picture is framed with a length of ribbon. She wants to use the same piece of ribbon to frame a circular picture. What is the maximum radius of the circular frame?

24. LANDSCAPING Mr. Jackson has a circular garden with a diameter of 10 feet surrounded by edging. Using the same length of edging, he is going to create a square garden. What is the maximum side length of the square?

Example 4
p. 60

COORDINATE GEOMETRY Graph each figure with the given vertices and identify the figure. Then find the perimeter and area of the figure.

25. $D(-2, -2)$, $E(-2, 3)$, $F(2, -1)$

26. $J(-3, -3)$, $K(3, 2)$, $L(3, -3)$

27. $P(-1, 1)$, $Q(3, 4)$, $R(6, 0)$, $S(2, -3)$

28. $T(-2, 3)$, $U(1, 6)$, $V(5, 2)$, $W(2, -1)$

29. CHANGING DIMENSIONS Use the rectangle at the right.

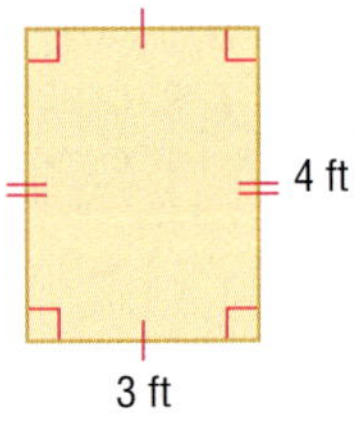

a. Find the perimeter of the rectangle.

b. Find the area of the rectangle.

c. Suppose the length and width of the rectangle are doubled. What effect would this have on the perimeter? the area? Justify your answer.

d. Suppose the length and width of the rectangle are halved. What effect does this have on the perimeter? the area? Justify your answer.

30. CHANGING DIMENSIONS Use the triangle at the right.

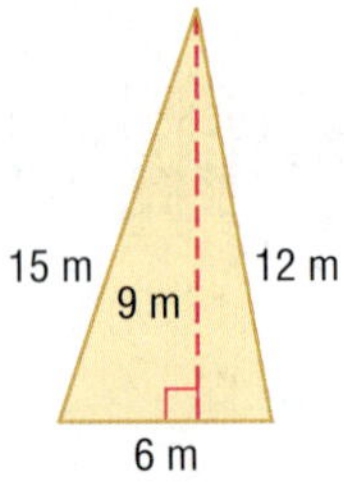

a. Find the perimeter of the triangle.

b. Find the area of the triangle.

c. Suppose the side lengths and height of the triangle were doubled. What effect would this have on the perimeter? the area? Justify your answer.

d. Suppose the side lengths and height of the triangle were divided by three. What effect would this have on the perimeter? the area? Justify your answer.

31. ALGEBRA A rectangle of area 360 square yards is 10 times as long as it is wide. Find its length and width.

32. ALGEBRA A rectangle of area 350 square feet is 14 times as wide as it is long. Find its length and width.

Real-World Link

In disc golf, a disc is thrown from a tee area to a "hole"—usually an elevated metal basket. A player must make consecutive shots from the spot where the previous throw landed in the fewest number of throws. The Professional Disc Golf Association has over 16,000 members.

Source: PDGA

33. DISC GOLF The diameter of the most popular brand of flying disc used in disc golf measures between 8 and 10 inches.

a. Find the minimum and maximum circumference of these flying discs to the nearest tenth.

b. Describe the range of possible areas for these flying discs to the nearest tenth.

ALGEBRA Find the perimeter or circumference for each figure with the given information.

34. The area of a square is 36 square units.

35. The length of a rectangle is half the width. The area is 25 square meters.

36. The area of a circle is 25π square units.

37. The area of a circle is 32π square units.

38. A rectangle's length is 3 times its width. The area is 27 square inches.

39. A rectangle's length is twice its width. The area is 48 square inches.

ALGEBRA Find the length of each side of the polygon for the given perimeter.

40. $P = 33$ in.

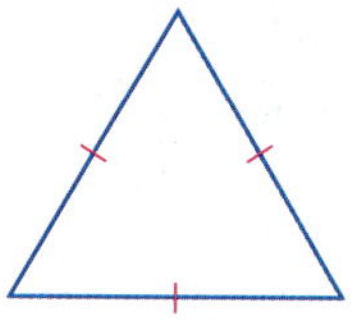

41. $P = 50$ cm

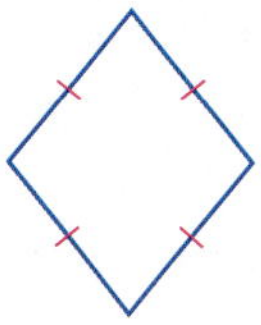

42. $P = 28$ ft

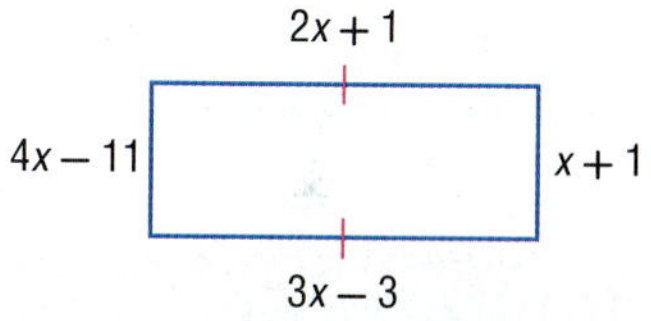

43. SWIMMING POOLS Ms. Neus has a circular swimming pool that is 20 feet in diameter. She wants to build a square fence around the pool so that the fence is a minimum of 6 feet from any edge of the pool.

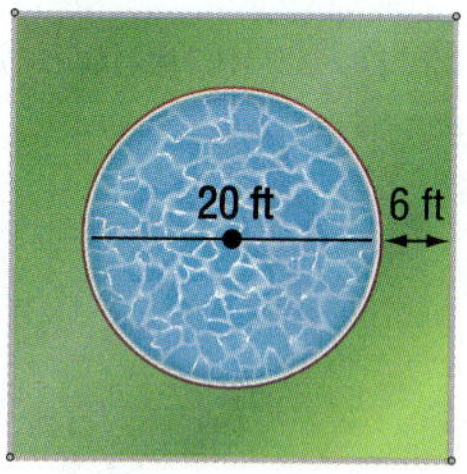

a. How much fencing will she need?

b. What percent of the area enclosed by the fence is not occupied by the pool?

H.O.T. Problems

Use **H**igher-**O**rder **T**hinking Skills

44. WHICH ONE DOESN'T BELONG? Identify the term that does not belong with the other three. Explain your reasoning.

square	circle	triangle	pentagon

45. CHALLENGE The vertices of a rectangle with side lengths of 10 and 24 units are on a circle of radius 13 units. Find the area between the figures.

46. REASONING Name a polygon that is always regular and a polygon that is sometimes regular. Explain your reasoning.

47. OPEN ENDED Draw a pentagon. Is your pentagon *convex* or *concave*? Is your pentagon *regular* or *irregular*? Justify your answers.

48. CHALLENGE A rectangular room measures 20 feet by 12.5 feet. How many 5-inch square tiles will it take to cover the floor of this room? Explain.

49. WRITING IN MATH Describe two possible ways that a polygon can be equiangular but not a regular polygon.

Standardized Test Practice

50. Find the perimeter of the figure.

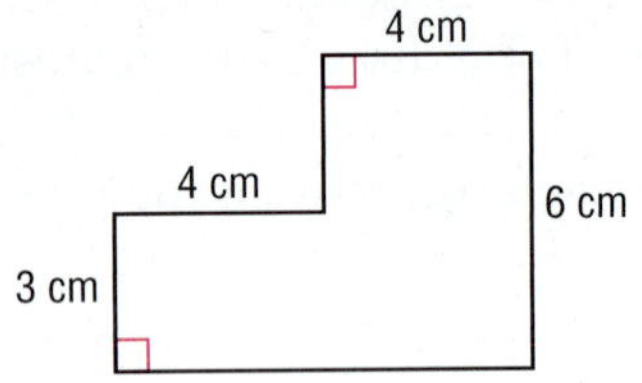

A 17 cm **C** 28 cm
B 25 cm **D** 31 cm

51. PROBABILITY In three successive rolls of a fair number cube, Matt rolls a 6. What is the probability of Matt rolling a 6 if the number cube is rolled a fourth time?

F $\frac{1}{6}$ **H** $\frac{1}{3}$
G $\frac{1}{4}$ **J** 1

52. SHORT RESPONSE Miguel is planning a party for 80 guests. According to the pattern in the table, how many gallons of ice cream should Miguel buy?

Number of Guests	Gallons of Ice Cream
8	2
16	4
24	6
32	8

53. SAT/ACT A frame 2 inches wide surrounds a painting that is 18 inches wide and 14 inches tall. What is the area of the frame?

A 84 in^2 **C** 252 in^2
B 144 in^2 **D** 396 in^2

Spiral Review

Determine whether each statement can be assumed from the figure. Explain. (Lesson 1-5)

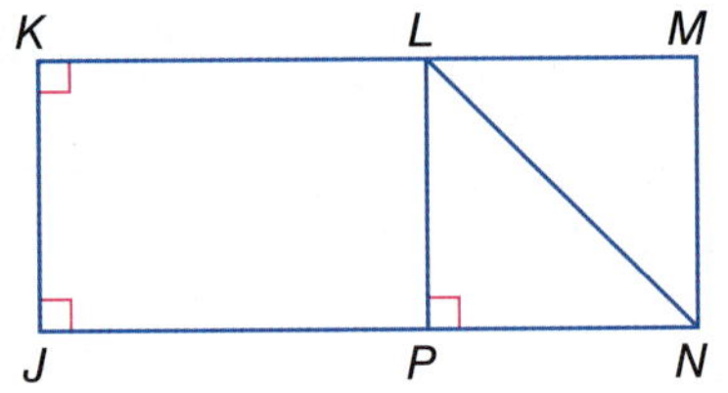

54. $\angle KJN$ is a right angle.

55. $\angle PLN \cong \angle NLM$

56. $\angle PNL$ and $\angle MNL$ are complementary.

57. $\angle KLN$ and $\angle MLN$ are supplementary.

58. TABLE TENNIS The diagram shows the angle of play for a table tennis player. If a right-handed player has a strong forehand, he should stand to the left of the center line of his opponent's angle of play. (Lesson 1-4)

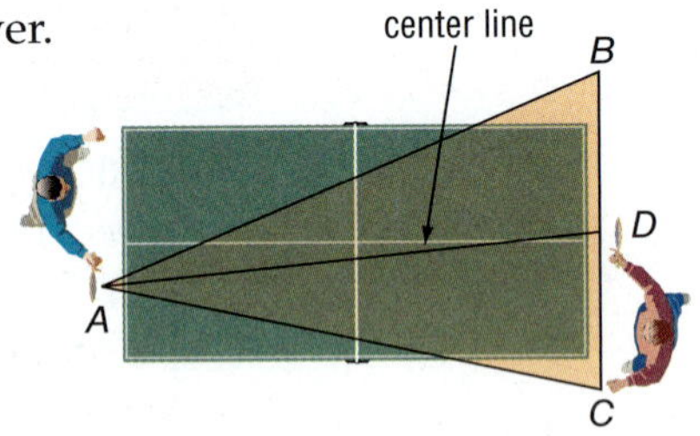

a. What geometric term describes the center line?

b. If the angle of play shown in the diagram measures 43°, what is $m\angle BAD$?

Name an appropriate method to solve each system of equations. Then solve the system. (Lesson 0-7)

59. $-5x + 2y = 13$
$2x + 3y = -9$

60. $y = -5x + 7$
$y = 3x - 17$

61. $x - 8y = 16$
$7x - 4y = -18$

Skills Review

Evaluate each expression if $P = 10$, $B = 12$, $h = 6$, $r = 3$, and $\ell = 5$. Round to the nearest tenth, if necessary. (Lesson 0-3)

62. $\frac{1}{2}P\ell + B$

63. $\frac{1}{3}Bh$

64. $\frac{1}{3}\pi r^2 h$

65. $2\pi rh + 2\pi r^2$

EXTEND 1-6

Geometry Software Lab: Two-Dimensional Figures

Math Online glencoe.com
- Other Calculator Keystrokes
- Graphing Technology Personal Tutor

You can use The Geometer's Sketchpad® to draw and investigate polygons.

ACTIVITY 1 Draw a Polygon

Draw $\triangle XYZ$.

Step 1 Select the segment tool from the toolbar, and click to set the first endpoint X of side $\overline{XY}$. Then drag the cursor, and click again to set the other endpoint Y.

Step 2 Click on point Y to set the endpoint of $\overline{YZ}$. Drag the cursor and click to set point Z.

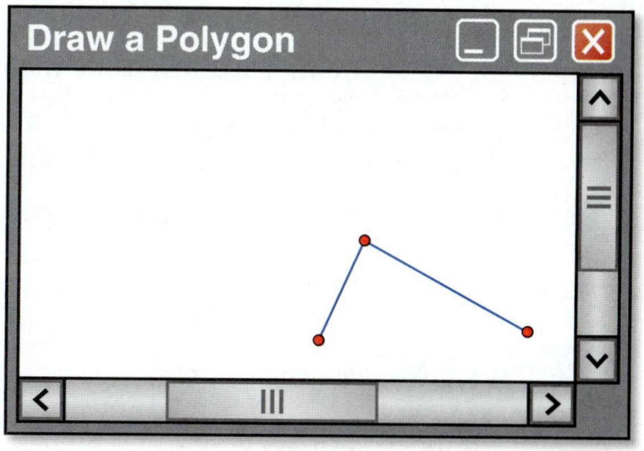

Step 3 Click on point Z to set the endpoint of $\overline{ZX}$. Then move the cursor to highlight point X. Click on X to draw $\overline{ZX}$.

Step 4 Use the pointer tool to click on points X, Y, and Z. Under the **Display** menu, select **Show Labels** to label the vertices of your triangle.

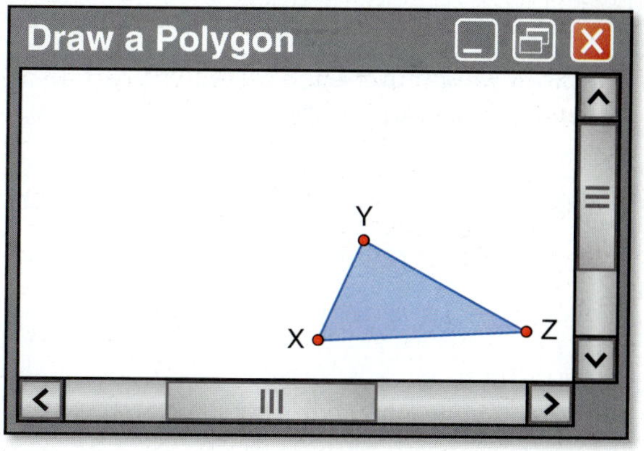

ACTIVITY 2 Measure Sides

Find XY, YZ, and ZX.

Step 1 Use the pointer tool to select $\overline{XY}$, $\overline{YZ}$, and $\overline{ZX}$.

Step 2 Select the **Length** command under the **Measure** menu to display the lengths of $\overline{XY}$, $\overline{YZ}$, and $\overline{ZX}$.

$XY = 1.79$ cm

$YZ = 3.11$ cm

$ZX = 3.48$ cm

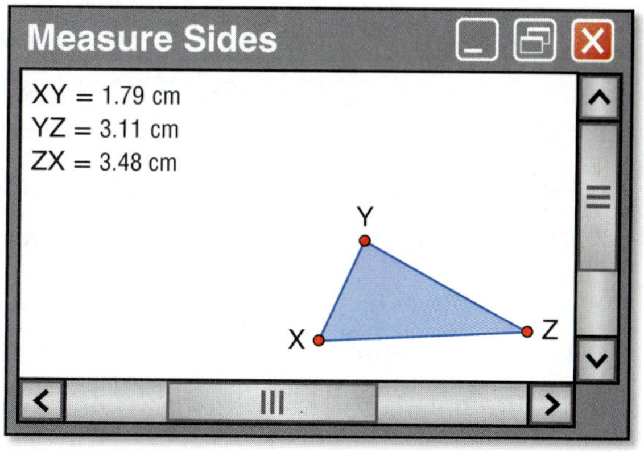

(continued on the next page)

ACTIVITY 3 Find Perimeter

Find the perimeter of $\triangle XYZ$.

Step 1 Use the pointer tool to select points X, Y, and Z.

Step 2 Under the **Construct** menu, select **Triangle Interior**. The triangle will now be shaded.

Step 3 Select the triangle interior using the pointer.

Step 4 Choose the **Perimeter** command under the **Measure** menu to find the perimeter of $\triangle XYZ$.

The perimeter of $\triangle XYZ$ is 8.38 centimeters.

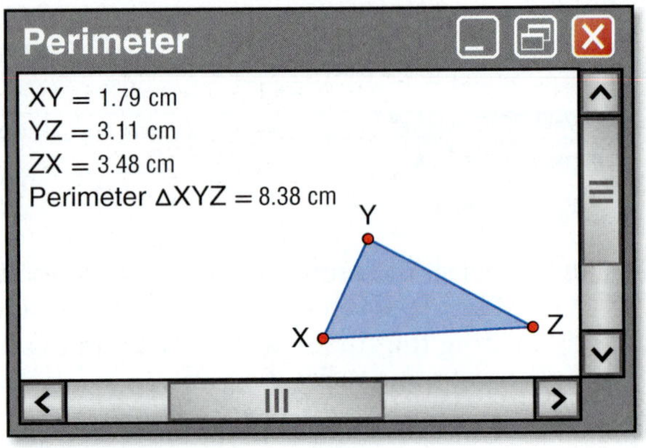

ACTIVITY 4 Measure Angles

Find $m\angle X$, $m\angle Y$, and $m\angle Z$.

Step 1 Recall that $\angle X$ can also be named $\angle YXZ$ or $\angle ZXY$. Use the pointer to select points Y, X, and Z in order.

Step 2 Select the **Angle** command from the **Measure** menu to find $m\angle X$.

Step 3 Select points X, Y, and Z. Find $m\angle Y$.

Step 4 Select points X, Z, and Y. Find $m\angle Z$.

$m\angle X = 63.16$, $m\angle Y = 86.05$, and $m\angle Z = 30.8$.

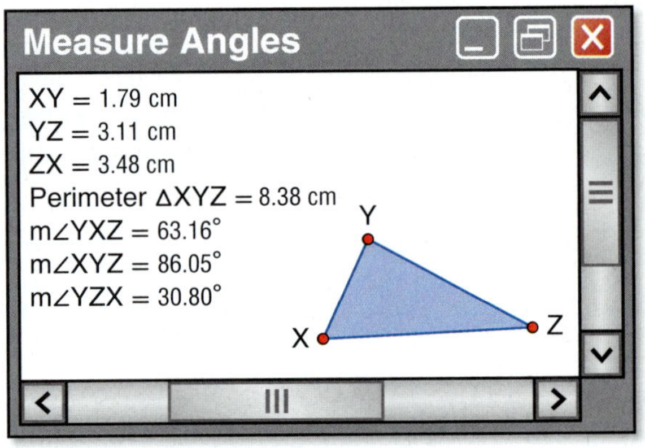

Analyze the Results

1. Add the side measures from Activity 2. How does this compare to the result in Activity 3?
2. What is the sum of the angle measures of $\triangle XYZ$?
3. Repeat the activities for each figure.
   a. irregular quadrilateral **b.** square **c.** pentagon **d.** hexagon
4. Draw another quadrilateral and find its perimeter. Then enlarge your figure using the **Dilate** command. How does changing the sides affect the perimeter?
5. Compare your results with those of your classmates.
6. Make a conjecture about the sum of the measures of the angles in any triangle.
7. What is the sum of the measures of the angles of a quadrilateral? pentagon? hexagon?
8. How are the sums of the angles of polygons related to the number of sides?
9. Test your conjecture on other polygons. Does your conjecture hold? Explain.
10. When the sides of a polygon are changed by a common factor, does the perimeter of the polygon change by the same factor as the sides? Explain.

1-7 Three-Dimensional Figures

Then

You identified and named two-dimensional figures. (Lesson 1-6)

Now

- Identify and name three-dimensional figures.
- Find surface area and volume.

New Vocabulary

polyhedron
face
edge
vertex
prism
base
pyramid
cylinder
cone
sphere
regular polyhedron
Platonic solid
surface area
volume

Math Online
glencoe.com

Why?

Architects often provide three-dimensional models of their ideas to clients. These models give their clients a better idea of what the completed structure will look like than a two-dimensional drawing. Three-dimensional figures, or *solids*, are made up of flat or curved surfaces.

Identify Three-Dimensional Figures A solid with all flat surfaces that enclose a single region of space is called a **polyhedron**. Each flat surface or **face** is a polygon. The line segments where the faces intersect are called **edges**. The point where three or more edges intersect is called a **vertex**.

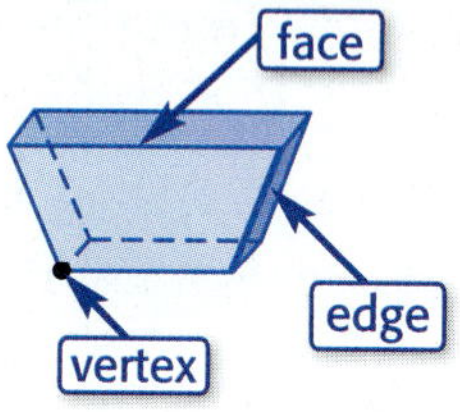

Below are examples and definitions of polyhedrons and other types of solids.

Key Concept Types of Solids

Polyhedrons

A **prism** is a polyhedron with two parallel congruent faces called **bases** connected by parallelogram faces.

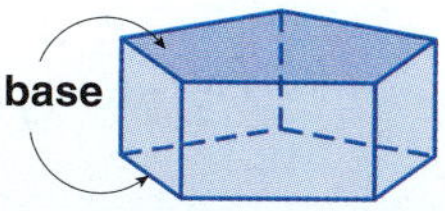

A **pyramid** is a polyhedron that has a polygonal base and three or more triangular faces that meet at a common vertex.

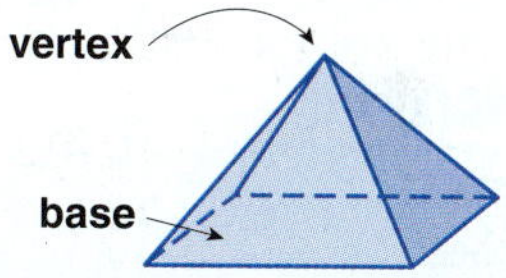

Not Polyhedrons

A **cylinder** is a solid with congruent parallel circular bases connected by a curved surface.

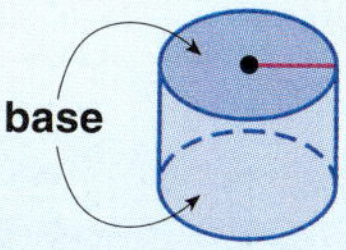

A **cone** is a solid with a circular base connected by a curved surface to a single vertex.

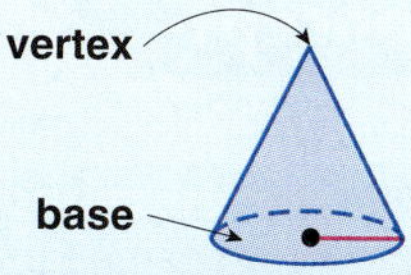

A **sphere** is a set of points in space that are the same distance from a given point. A sphere has no faces, edges, or vertices.

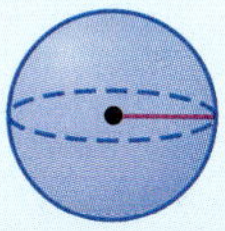

Polyhedrons or *polyhedra* are named by the shape of their bases.

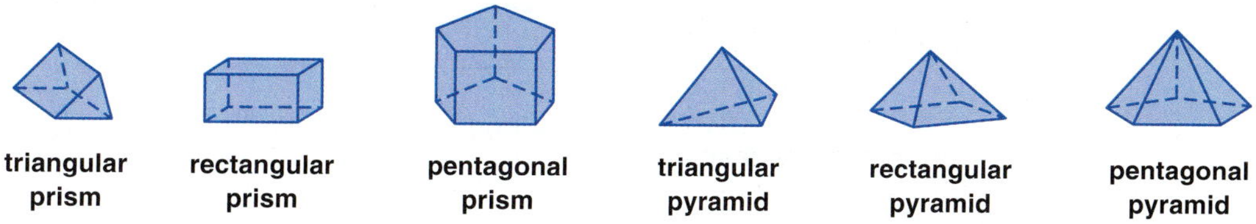

EXAMPLE 1 Identify Solids

Determine whether each solid is a polyhedron. Then identify the solid. If it is a polyhedron, name the bases, faces, edges, and vertices.

a.

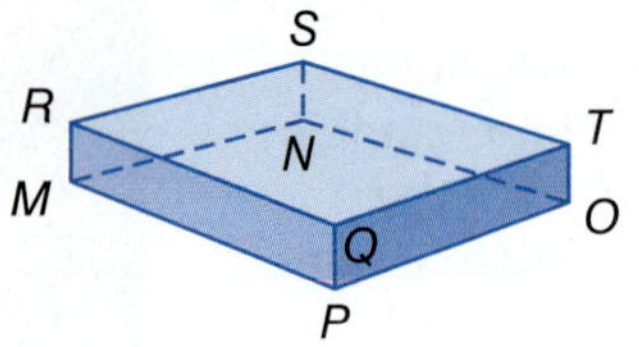

The solid is formed by polygonal faces, so it is a polyhedron. There are two parallel congruent rectangular bases, so it is a rectangular prism.

Bases: ▭*MNOP*, ▭*RSTQ*

Faces: ▭*RQPM*, ▭*RSNM*, ▭*STON*, ▭*QTOP*, ▭*RSTQ*, ▭*MNOP*

Edges: $\overline{MN}$, $\overline{NO}$, $\overline{OP}$, $\overline{PM}$, $\overline{RS}$, $\overline{ST}$, $\overline{TQ}$, $\overline{QR}$, $\overline{RM}$, $\overline{SN}$, $\overline{TO}$, $\overline{QP}$

Vertices: *M*, *N*, *O*, *P*, *Q*, *R*, *S*, *T*

b.

The solid has a curved surface, so it is not a polyhedron. It has two congruent circular bases, so it is a cylinder.

c.

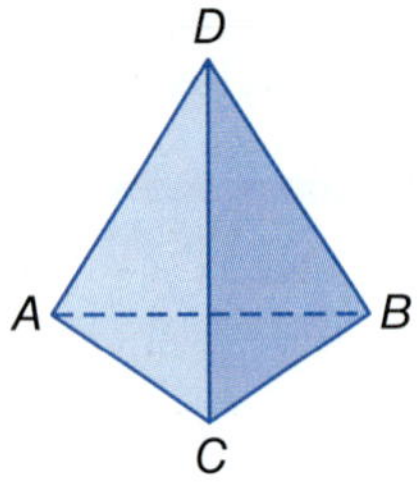

The solid is formed by polygonal faces, so it is a polyhedron. The base is a triangle, and the three faces meet in a vertex, so it is a triangular pyramid.

Bases: $\triangle ABC$

Faces: $\triangle ABC$, $\triangle ADC$, $\triangle CDB$, $\triangle BDA$

Edges: $\overline{AB}$, $\overline{BC}$, $\overline{CA}$, $\overline{DA}$, $\overline{DB}$, $\overline{DC}$

Vertices: *A*, *B*, *C*, *D*

Guided Practice

1A.

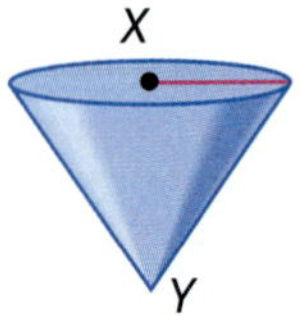

1B.

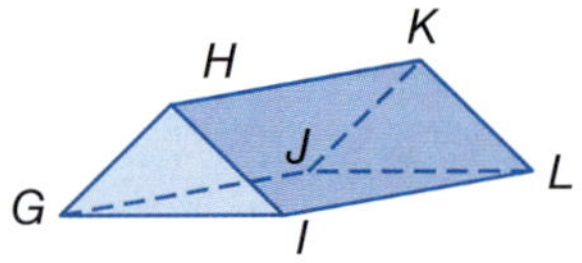

Personal Tutor glencoe.com

ReadingMath

Symbols Symbols can be used in naming the focus of polyhedra. The symbol ▭ means rectangle. The symbol △ means triangle. The symbol ⊙ means circle.

Math History Link

Plato (427–347 B.C.) Plato, a philosopher, mathematician, and scientist, lived in Athens, Greece. He is best known for founding a school known as "The Academy." In mathematics, he was concerned with the idea of proofs, and he insisted that definitions must be accurate and hypotheses must be clear.

A polyhedron is a **regular polyhedron** if all of its faces are regular congruent polygons and all of the edges are congruent. There are exactly five types of regular polyhedrons, called **Platonic Solids** because Plato used them extensively.

Key Concept Platonic Solids

Tetrahedron	Hexahedron or Cube	Octahedron	Dodecahedron	Icosahedron
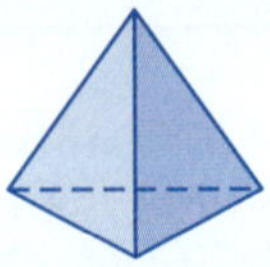	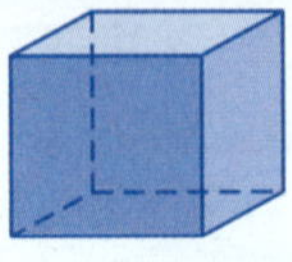	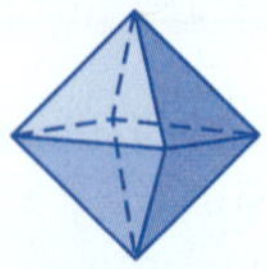	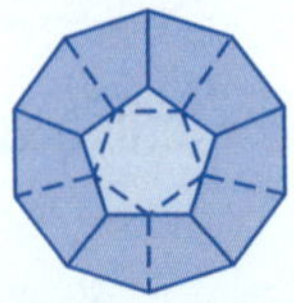	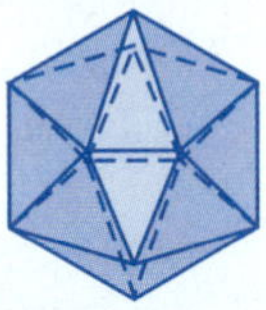
4 equilateral triangle faces	6 square faces	8 equilateral triangular faces	12 regular pentagonal faces	20 equilateral triangular faces

Surface Area and Volume

Surface area is a two-dimensional measurement of the surface of a solid figure. The surface area of a polyhedron is the sum of the areas of each face. **Volume** is the measure of the amount of space enclosed by a solid figure.

Review the formulas for the surface area and volume of five common solids given below. You will derive these formulas in Chapter 12.

StudyTip

Surface Area of a Rectangular Prism

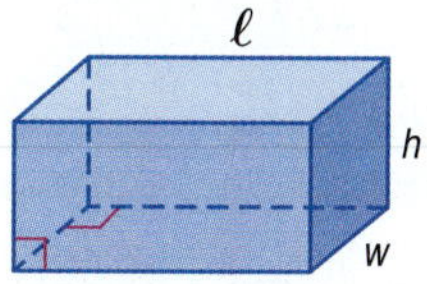

Another expression that represents the surface area of a rectangular prism is $2(\ell h + hw + \ell w)$.

Key Concept: Surface Area and Volume

Prism	Regular Pyramid	Cylinder	Cone	Sphere
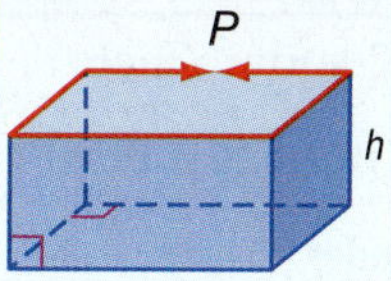	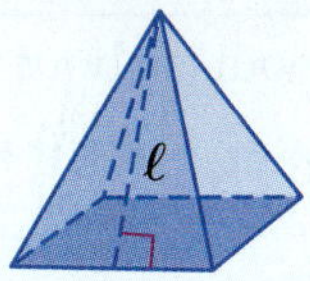	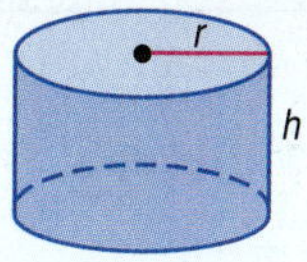	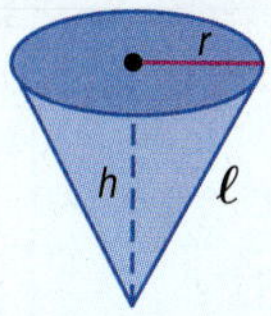	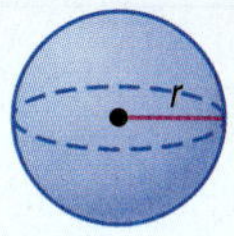
$T = Ph + 2B$	$T = \frac{1}{2}P\ell + B$	$T = 2\pi rh + 2\pi r^2$	$T = \pi r\ell + \pi r^2$	$T = 4\pi r^2$
$V = Bh$	$V = \frac{1}{3}Bh$	$V = \pi r^2 h$	$V = \frac{1}{3}\pi r^2 h$	$V = \frac{4}{3}\pi r^3$

T = total surface area; V = volume; h = height of a solid

P = perimeter of the base; B = area of base; ℓ = slant height, r = radius

EXAMPLE 2 Find Surface Area and Volume

Find the surface area and volume of the square pyramid.

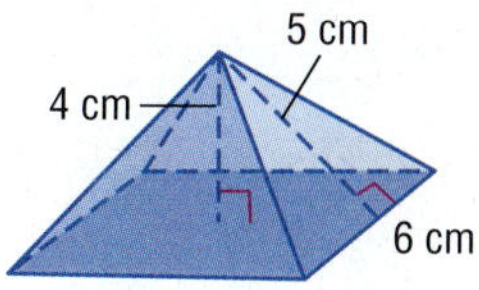

Surface Area

Since the base of the pyramid is a square, the perimeter P of the base is $4 \cdot 6$ or 24 centimeters. The area of the base B is $6 \cdot 6$ or 36 square centimeters. The slant height is 5 centimeters.

$T = \frac{1}{2}P\ell + B$ — **Surface area of pyramid**

$= \frac{1}{2}(24)(5) + 36$ or 96 — **$P = 24$ cm, $\ell = 5$ cm, $B = 36$ cm^2**

The surface area of the square pyramid is 96 square centimeters.

Volume

The height of the pyramid is 4 centimeters.

$V = \frac{1}{3}Bh$ — **Volume of pyramid**

$= \frac{1}{3}(36)(4)$ or 48 — **$B = 36$ cm^2, $h = 4$ cm**

The volume is 48 cubic centimeters.

Watch Out!

Height vs. Slant Height The *height* of a pyramid or cone is not the same as its slant height.

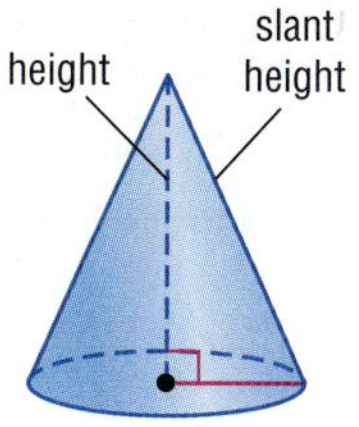

Guided Practice

Find the surface area and volume of each solid to the nearest tenth.

2A.

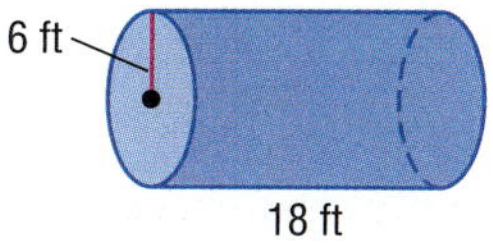

2B.

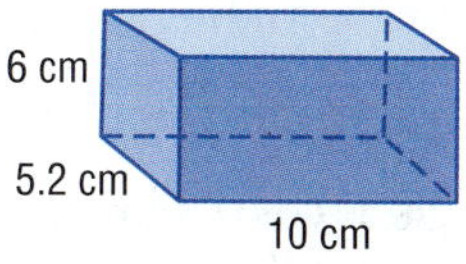

2C.

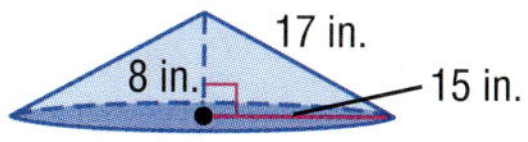

Personal Tutor glencoe.com

StudyTip

Units Be sure that you have converted all units of measure to be consistent before you begin volume or surface area calculations.

Real-World EXAMPLE 3 Surface Area and Volume

POOLS **The diameter of the pool Mr. Sato purchased is 8 feet. The height of the pool is 20 inches. Find each measure to the nearest tenth.**

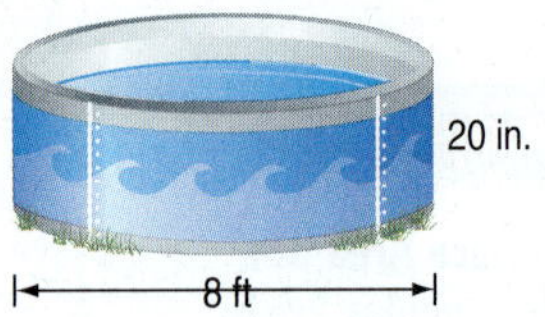

a. surface area of the pool

The pool is a cylinder.

$A = 2\pi rh + \pi r^2$	Surface area of cylinder with one base
$= 2\pi(4)\left(1\frac{2}{3}\right) + \pi(4)^2$	$r = 4$ ft, $h = 20$ in. or $1\frac{2}{3}$ ft
≈ 92.2	Use a calculator.

The surface area of the pool is about 92.2 square feet.

b. the volume of water needed to fill the pool to a depth of 16 inches

$V = \pi r^2 h$	Volume of cylinder
$= \pi(4)^2\left(1\frac{1}{3}\right)$	$r = 4$ ft, $h = 16$ in. or $1\frac{1}{3}$ ft
≈ 67.0	Use a calculator.

The volume of water needed is approximately 67.0 cubic feet.

Guided Practice

3. CRAFTS Jessica is making spherical candles using a mold that is 10 centimeters in diameter. Find each measure to the nearest tenth.

A. the volume of wax needed to fill the mold

B. the surface area of the finished candle

Personal Tutor glencoe.com

Check Your Understanding

Example 1 p. 68

Determine whether the solid is a polyhedron. Then identify the solid. If it is a polyhedron, name the bases, faces, edges, and vertices.

1.

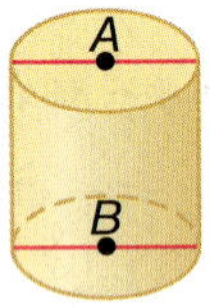

2.

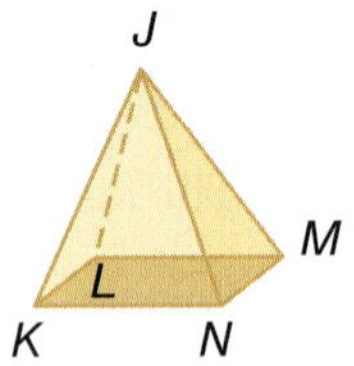

Example 2 p. 69

Find the surface area and volume of each solid to the nearest tenth.

3

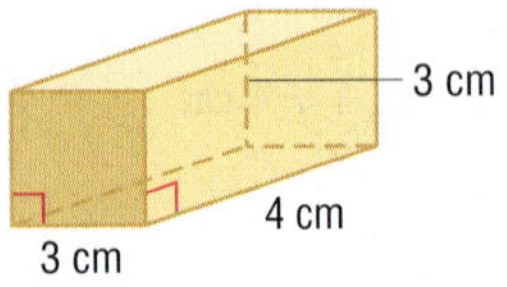

4.

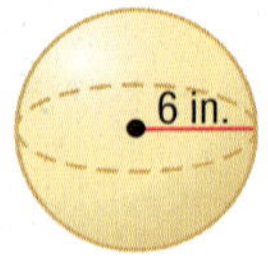

Example 3 p. 70

5. PARTY FAVORS Lawana is making cone-shaped hats 4 inches in diameter, 6.5 inches tall, with a slant height of 6.8 inches for party favors. Find each measure to the nearest tenth.

a. the volume of candy that will fill each cone

b. the area of material needed to make each hat assuming there is no overlap of material

Practice and Problem Solving

● = **Step-by-Step Solutions** begin on page R20.
Extra Practice begins on page 969.

Example 1
p. 68

Identify the solid modeled by each object. State whether the solid modeled is a polyhedron.

6. 

7. 

8. 

9. 

10. 

11. 

Determine whether the solid is a polyhedron. Then identify the solid. If it is a polyhedron, name the bases, faces, edges, and vertices.

12. 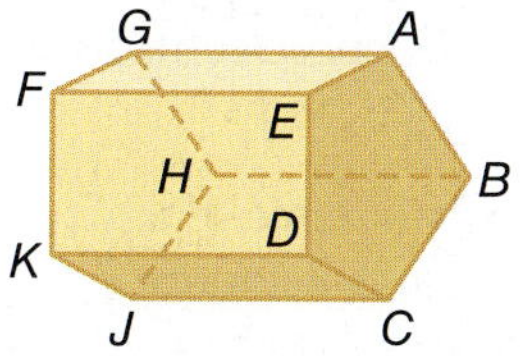

13. 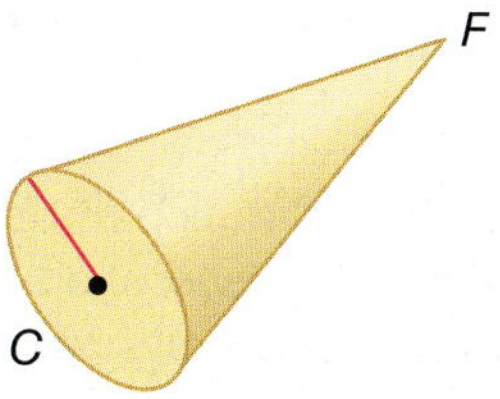

14. 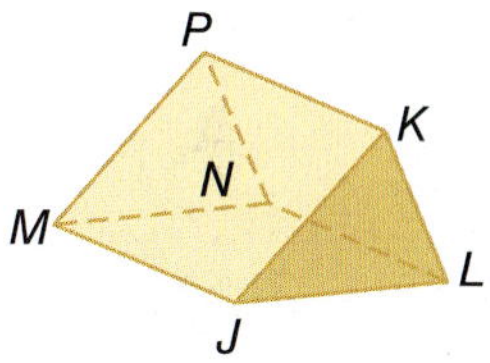

15. 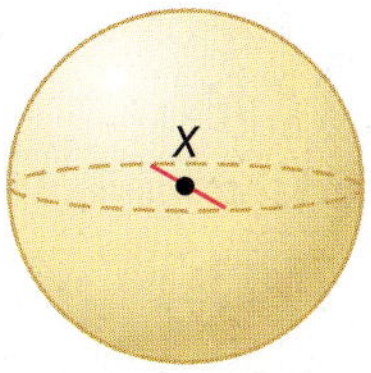

16. 

17. 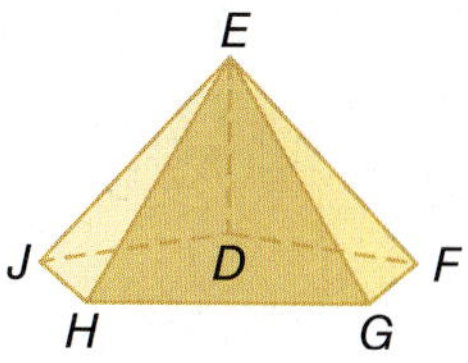

Example 2
p. 69

Find the surface area and volume of each solid to the nearest tenth.

18. 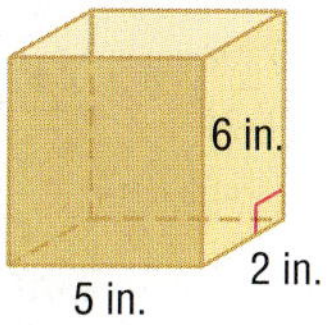

19. 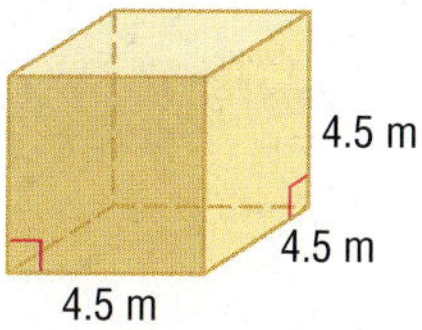

20. 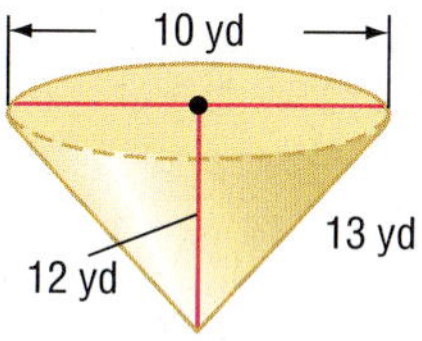

21. 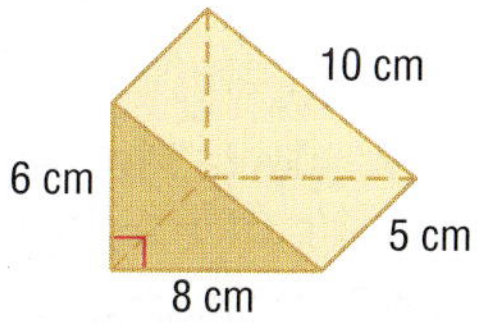

22. 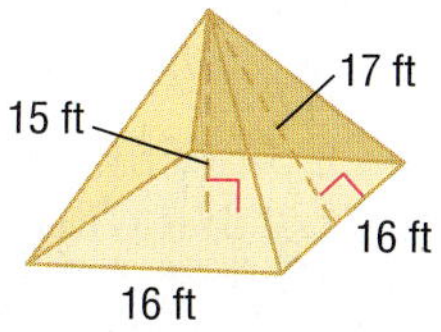

23. 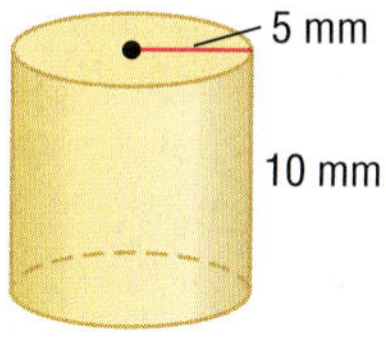

Example 3
p. 70

24. **SANDBOX** A rectangular sandbox is 3 feet by 4 feet. The depth of the box is 8 inches, but the depth of the sand is $\frac{3}{4}$ of the depth of the box. Find each measure to the nearest tenth.

a. the surface area of the sandbox assuming there is no lid

b. the volume of sand in the sandbox

Real-World Link

The Campana brothers started a design workshop in which participants create art using objects that can be inflated, such as tires or balloons.

Source: Arango Design

25 **ART** Fernando and Humberto Campana designed the Inflating Table shown at the left. The diameter of the table is $15\frac{1}{2}$ inches. Suppose the height of the cylinder is $11\frac{3}{4}$ inches. Find each measure to the nearest tenth. Assume that the sides of the table are perpendicular to the bases of the table.

a. the volume of air that will fully inflate the table

b. the surface area of the table when fully inflated

26. **FOOD** In 1999, Marks & Spencer, a British department store, created the biggest sandwich ever made. The tuna and cucumber sandwich was in the form of a triangular prism. Suppose each slice of bread was 8 inches thick. Find each measure to the nearest tenth.

a. the surface area in square feet of the sandwich when filled

b. the volume of filling in cubic feet to the nearest tenth

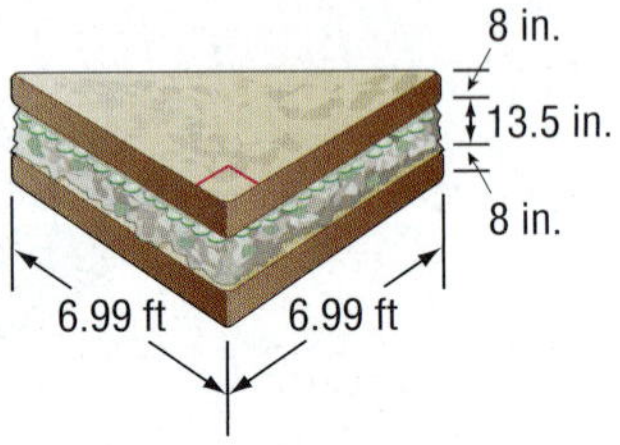

27. **ALGEBRA** The surface area of a cube is 54 square inches. Find the length of each edge.

28. **ALGEBRA** The volume of a cube is 729 cubic centimeters. Find the length of each edge.

29. **PAINTING** Tara is painting her family's fence. Each post is composed of a square prism and a square pyramid. The height of the pyramid is 4 inches. Determine the surface area and volume of each post.

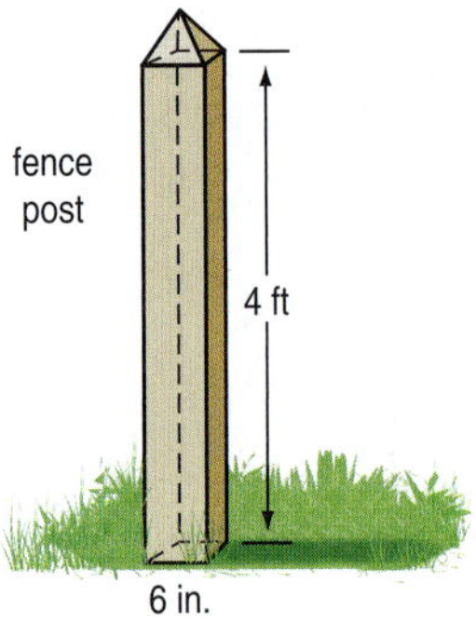

30. **COLLECT DATA** Use a ruler or tape measure and what you have learned in this lesson to find the surface area and volume of a soup can.

31. **CAKES** Cakes come in many shapes and sizes. Often they are stacked in two or more layers, like those in the diagrams shown below.

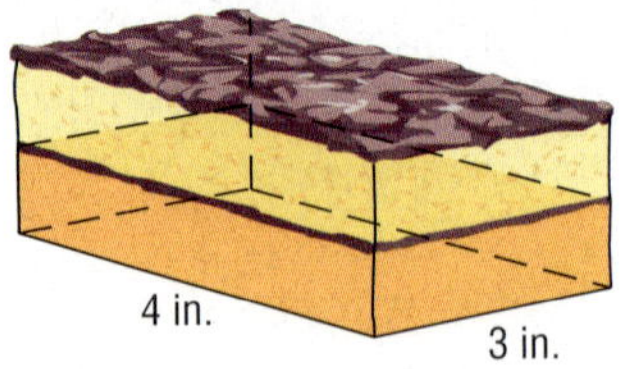

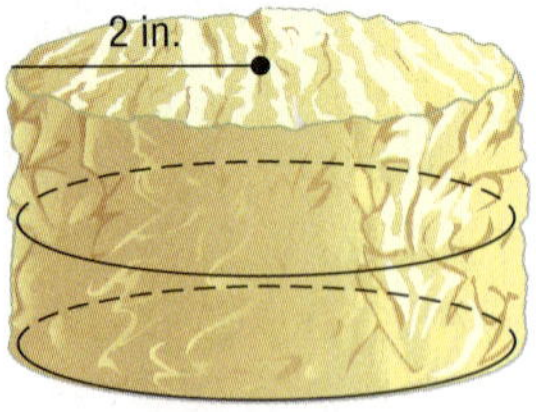

a. If each layer of the rectangular prism cake is 3 inches high, calculate the area of the cake that will be frosted assuming there is no frosting between layers.

b. Calculate the area of the cylindrical cake that will be frosted, if each layer is 4 inches in height.

c. If one can of frosting will cover 50 square inches of cake, how many cans of frosting will be needed for each cake?

d. If the height of each layer of cake is 5 inches, what does the radius of the cylindrical cake need to be, so the same amount of frosting is used for both cakes? Explain your reasoning.

32. **CHANGING UNITS** A gift box has a surface area of 6.25 square feet. What is the surface area of the box in square inches?

33 **CHANGING UNITS** A square pyramid has a volume of 4320 cubic inches. What is the volume of this pyramid in cubic feet?

ReadingMath

Euler's Formula The formula $F + V = E + 2$ is interpreted to mean the number of vertices and faces of a polyhedron together is exactly two more than the polyhedron's number of edges.

34. **EULER'S FORMULA** The number of faces F, vertices V, and edges E of a polyhedron are related by Euler's (OY luhrz) Formula: $F + V = E + 2$. Determine whether Euler's Formula is true for each of the figures in Exercises 18–23.

35. **CHANGING DIMENSIONS** A rectangular prism has a length of 12 centimeters, width of 18 centimeters, and height of 22 centimeters. Describe the effect on the volume of a rectangular prism when each dimension is doubled.

36. **MULTIPLE REPRESENTATIONS** In this problem, you will investigate how changing the length of the radius of a cone affects the cone's volume.

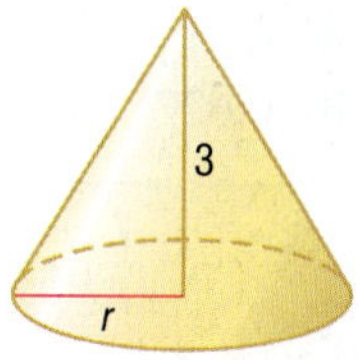

a. **TABULAR** Create a table showing the volume of a cone when doubling the radius. Use radius values between 1 and 8.

b. **GRAPHICAL** Use the values from your table to create a graph of radius versus volume.

c. **VERBAL** Make a conjecture about the effect of doubling the radius of a cone on the volume. Explain your reasoning.

d. **ALGEBRAIC** If r is the radius of a cone, write an expression showing the effect doubling the radius has on the cone's volume.

H.O.T. Problems Use Higher-Order Thinking Skills

37. **ERROR ANALYSIS** Alex and Emily are calculating the surface area of the rectangular prism shown. Is either of them correct? Explain your reasoning.

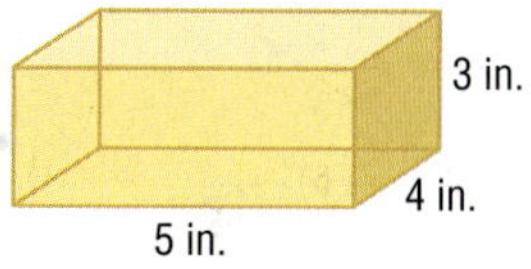

Alex	Emily
$(5 \cdot 3) \cdot 6$ faces	$2(5 \cdot 4 \cdot 3)$
$= 90\ in^2$	$= 120\ in^2$

38. **REASONING** Is a cube a regular polyhedron? Explain.

39. **CHALLENGE** Describe the solid that results if the number of sides of each base increases infinitely. The bases of each solid are regular polygons inscribed in a circle.

a. pyramid

b. prism

40. **OPEN ENDED** Draw an irregular 14-sided polyhedron in which all of the sides are congruent and all of the angles are congruent.

41. **CHALLENGE** Find the volume of a cube that has a total surface area of 54 square millimeters.

42. **WRITING IN MATH** Write a description of a prism in your school or home that another person could use to recreate the prism. Explain your reasoning for choosing each characteristic in your description.

Standardized Test Practice

43. GRIDDED RESPONSE What is the surface area of the triangular prism in square centimeters?

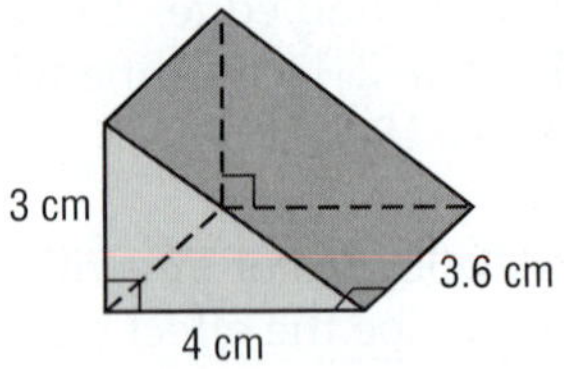

44. ALGEBRA What is the value of $(-0.8)^2 + (-0.3)^3$?

A 0.613
B 0.627
C 0.370
D 0.327

45. The length of each side of a cube is multiplied by 5. What is the change in the volume of the cube?

F The volume is 125 times the original volume.
G The volume is 25 times the original volume.
H The volume is 10 times the original volume.
J The volume is 5 times the original volume.

46. SAT/ACT What is the difference in surface area between a cube with an edge length of 7 inches and a cube with edge length of 4 inches?

A 33 in^2
B 66 in^2
C 99 in^2
D 198 in^2

Spiral Review

Name each polygon by its number of sides. Then classify it as *convex* or *concave* and *regular* or *irregular*. (Lesson 1-6)

47.

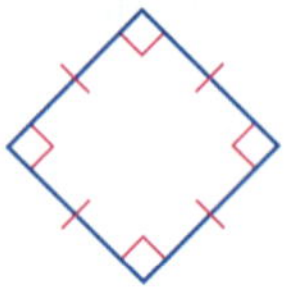

48.

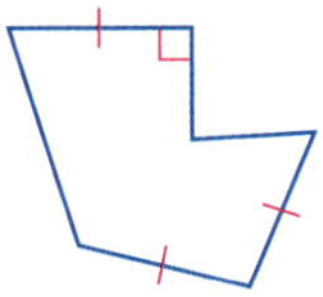

49.

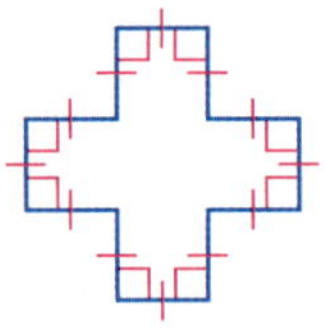

Find the value of each variable. (Lesson 1-5)

50.

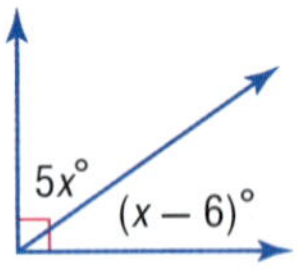

51.

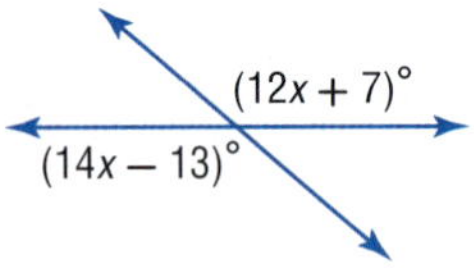

52.

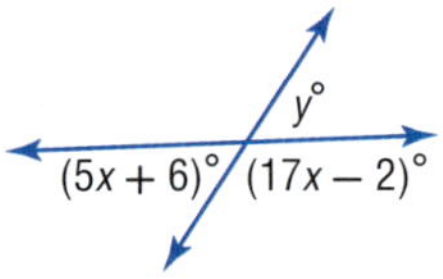

GAMES What type of geometric intersection is modeled in each photograph? (Lesson 1-1)

53.

54.

55.

Skills Review

Sketch the next two figures in each pattern.

56.

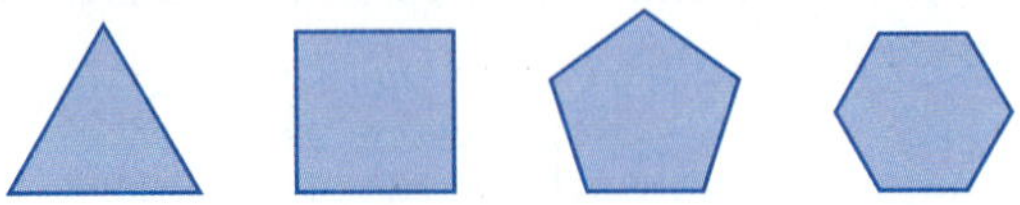

57.

58.

59.

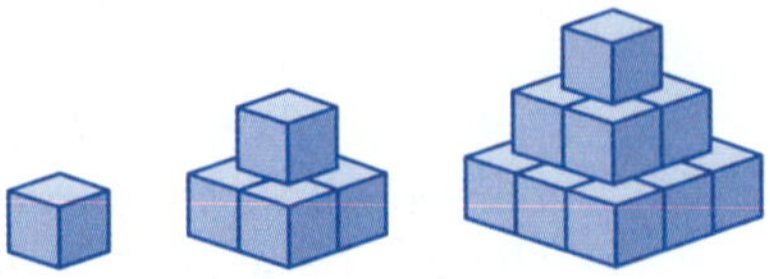

EXTEND 1-7

Geometry Lab
Orthographic Drawings and Nets

- Use orthographic views and nets to represent and construct three-dimensional figures.

If you see a three-dimensional object from only one viewpoint, you may not know its true shape. Here are four views of a square pyramid.

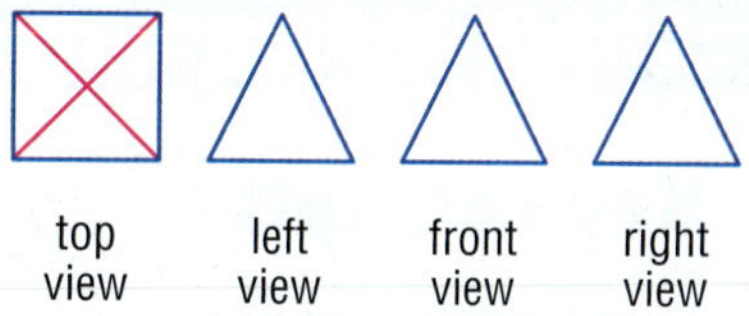

The two-dimensional views of the top, left, front, and right sides of an object are called an **orthographic drawing**.

ACTIVITY 1

Make a model of a figure given the orthographic drawing.

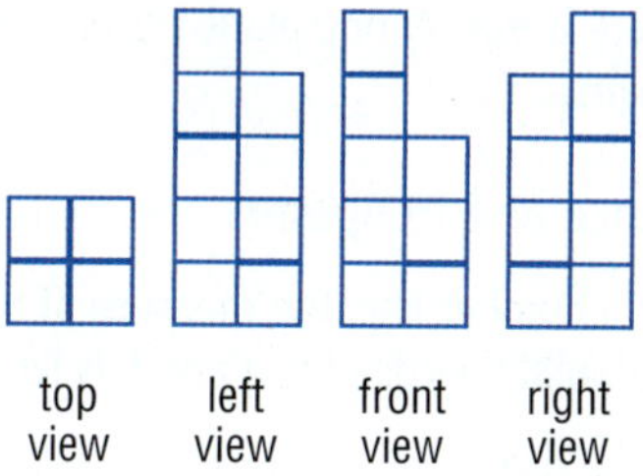

- The top view indicates two rows and two columns of different heights.
- The front view indicates that the left side is 5 blocks high and the right side is 3 blocks high. The dark segments indicate breaks in the surface.
- The right view indicates that the right front column is only one block high. The left front column is 4 blocks high. The right back column is 3 blocks high.
- Check the left side of your model. All of the blocks should be flush.

If you cut a cardboard box at the edges and lay it flat, you will have a pattern, or **net**, for the three-dimensional solid.

ACTIVITY 2

Make a model of a figure given the net.

This is the net of a triangular prism. Use a large sheet of paper, a ruler, scissors, and tape. Measure the dimensions on the paper. Cut around the edges. Fold the pattern on the solid lines and secure the edges with tape.

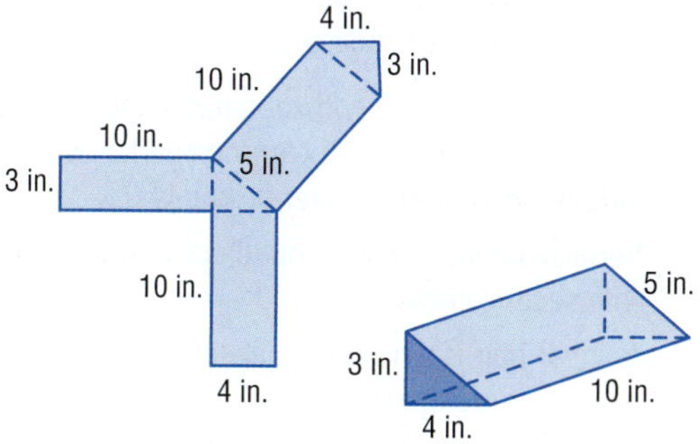

Model and Analyze

1. Make a model of a figure given the orthographic drawing. Then find the volume of the model.

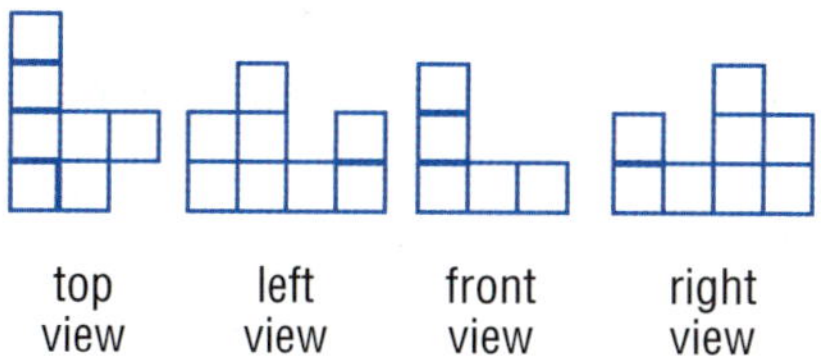

2. Make a model of a figure given the net. Then find the surface area of the model.

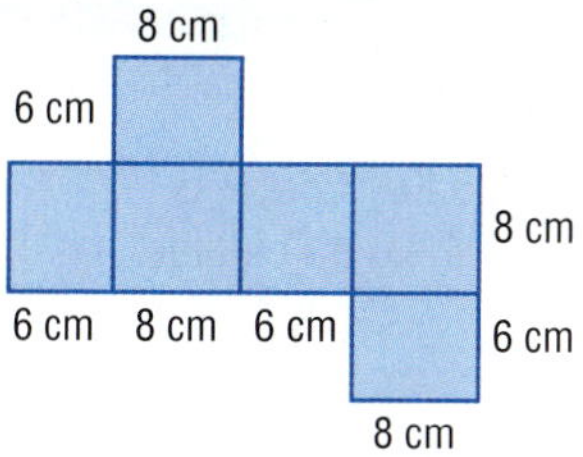

Math Online glencoe.com
- STUDY TO GO
- Vocabulary Review

Chapter Summary

Key Concepts

Points, Lines, and Planes (Lesson 1-1)

- There is exactly one line through any two points.
- There is exactly one plane through any three noncollinear points.

Distance and Midpoints (Lesson 1-3)

- On a number line, the measure of a segment with endpoint coordinates a and b is $|a - b|$.
- In the coordinate plane, the distance between two points (x_1, y_1) and (x_2, y_2) is given by $d = \sqrt{(x_2 - x_1)^2 + (y_2 - y_1)^2}$.
- On a number line, the coordinate of the midpoint of a segment with endpoints a and b is $\frac{a+b}{2}$.
- In the coordinate plane, the coordinates of the midpoint of a segment with endpoints that are (x_1, y_1) and (x_2, y_2) are $\left(\frac{x_1 + x_2}{2}, \frac{y_1 + y_2}{2}\right)$.

Angles (Lessons 1-3, 1-4, and 1-5)

- An angle is formed by two noncollinear rays that have a common endpoint, called its vertex. Angles can be classified by their measures.
- Adjacent angles are two coplanar angles that lie in the same plane and have a common vertex and a common side but no common interior points.
- Vertical angles are two nonadjacent angles formed by two intersecting lines.
- A linear pair is a pair of adjacent angles with noncommon sides that are opposite rays.
- Complementary angles are two angles with measures that have a sum of 90.
- Supplementary angles are two angles with measures that have a sum of 180.

FOLDABLES® Study Organizer

Be sure the Key Concepts are noted in your Foldable.

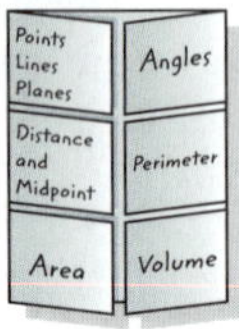

Key Vocabulary

acute angle (p. 38)
adjacent angles (p. 46)
angle (p. 36)
angle bisector (p. 39)
area (p. 58)
base (p. 67)
between (p. 15)
circumference (p. 58)
collinear (p. 5)
complementary angles (p. 47)
concave (p. 56)
cone (p. 67)
congruent (p. 16)
construction (p. 17)
convex (p. 56)
coplanar (p. 5)
cylinder (p. 67)
degree (p. 37)
distance (p. 25)
edge (p. 67)
equilateral polygon (p. 57)
equiangular polygon (p. 57)
exterior (p. 36)
face (p. 67)
interior (p. 36)
intersection (p. 6)
line (p. 5)
linear pair (p. 46)
line segment (p. 14)
midpoint (p. 27)
***n*-gon** (p. 57)
obtuse angle (p. 38)
opposite rays (p. 36)
perimeter (p. 58)
perpendicular (p. 48)
plane (p. 5)
Platonic solid (p. 68)
point (p. 5)
polygon (p. 56)
polyhedron (p. 67)
prism (p. 67)
pyramid (p. 67)
ray (p. 36)
regular polygon (p. 57)
regular polyhedron (p. 68)
right angle (p. 38)
segment bisector (p. 29)
side (p. 36)
space (p. 7)
sphere (p. 67)
supplementary angles (p. 47)
surface area (p. 69)
undefined term (p. 5)
vertex (pp. 36, 67)
vertex of a polygon (p. 56)
vertical angles (p. 46)
volume (p. 69)

Vocabulary Check

Fill in the blank in each sentence with the vocabulary term that best completes the sentence.

1. A _____ is a flat surface made up of points that extends infinitely in all directions.
2. A set of points that all lie on the same line are said to be _____.
3. If two lines intersect to form four right angles, the lines are called _____.
4. If the sum of the measures of two angles is 180, then the angles are called _____ angles.

Lesson-by-Lesson Review

1-1 Points, Lines, and Planes (pp. 5–12)

Use the figure to complete each of the following.

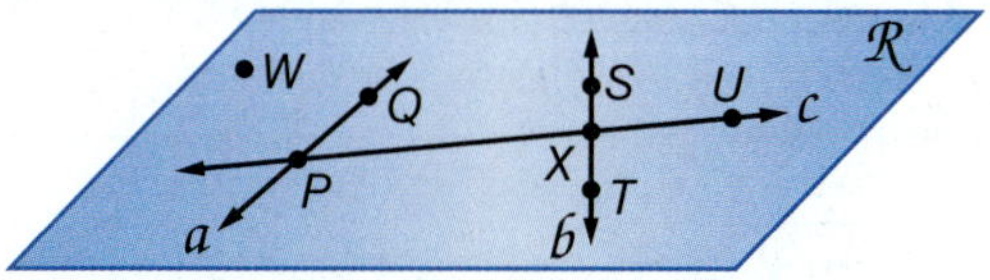

5. Name the intersection of lines *a* and *c*.
6. Give another name for line *b*.
7. Name a point that is not contained in any of the three lines *a*, *b*, or *c*.
8. Give another name for plane *WPX*.

Name the geometric term that is best modeled by each item.

9. 

10. 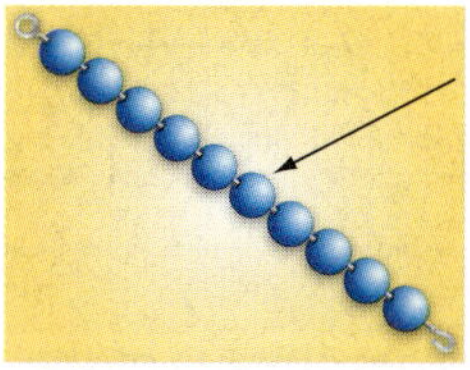

EXAMPLE 1

Draw and label a figure for the relationship below.

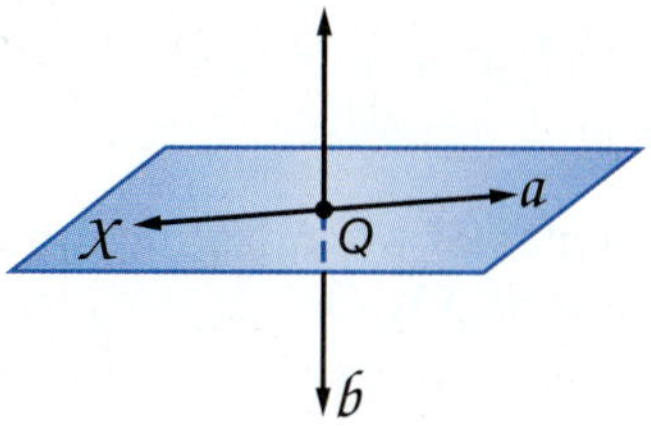

Plane $\mathcal{X}$ contains line *a*, line *b* intersects line *a* at point *Q*, but line *b* is not in plane $\mathcal{X}$.

Draw a surface to represent plane $\mathcal{X}$ and label it.

Draw a line in plane $\mathcal{X}$ and label it line *a*.

Draw a line *b* intersecting both the plane and line *a* and label the point of intersection *Q*.

1-2 Linear Measure (pp. 14–21)

Find the value of the variable and *XP*, if *X* is between *P* and *Q*.

11. $XQ = 13$, $XP = 5x - 3$, $PQ = 40$
12. $XQ = 3k$, $XP = 7k - 2$, $PQ = 6k + 16$

Determine whether each pair of segments is congruent.

13. $\overline{AB}$, $\overline{CD}$

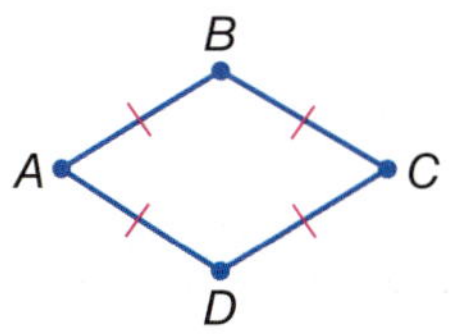

14. $\overline{XY}$, $\overline{YZ}$

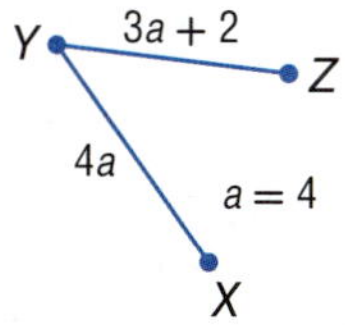

15. **DISTANCE** The distance from Salvador's job to his house is 3 times greater than the distance from his house to school. If his house is between his job and school and the distance from his job to school is 6 miles, how far is it from Salvador's house to school?

EXAMPLE 2

Use the figure to find the value of the variable and the length of $\overline{YZ}$.

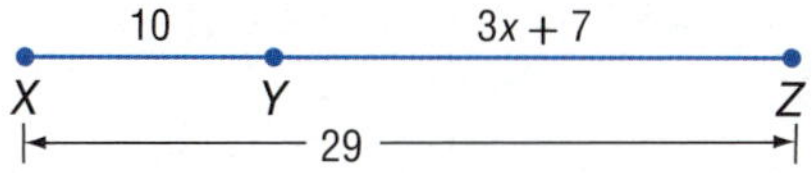

$XZ = XY + YZ$	**Betweenness of points**
$29 = 10 + 3x + 7$	**Substitution**
$29 = 3x + 17$	**Simplify.**
$12 = 3x$	**Subtract 17 from each side.**
$4 = x$	**Divide each side by 3.**
$YZ = 3x + 7$	**Given**
$= 3(4) + 7$ or 19	**Substitution**

So, $x = 4$ and $YZ = 19$.

CHAPTER 1 Study Guide and Review

1-3 Distance and Midpoints (pp. 25–35)

Find the distance between each pair of points.

16. $A(-3, 1)$, $B(7, 13)$

17. $P(2, -1)$, $Q(10, -7)$

Find the coordinates of the midpoint of a segment with the given endpoints.

18. $L(-3, 16)$, $M(17, 4)$

19. $C(32, -1)$, $D(0, -12)$

Find the coordinates of the missing endpoint if M is the midpoint of $\overline{XY}$.

20. $X(-11, -6)$, $M(15, 4)$

21. $M(-4, 8)$, $Y(19, 0)$

22. **HIKING** Carol and Marita are hiking in a state park and decide to take separate trails. The map of the park is set up on a coordinate grid. Carol's location is at the point (7, 13) and Marita is at (3, 5).

a. Find the distance between them.

b. Find the coordinates of the point midway between their locations.

EXAMPLE 3

Find the distance between $X(5, 7)$ and $Y(-7, 2)$.

Let $(x_1, y_1) = (5, 7)$ and $(x_2, y_2) = (-7, 2)$.

$$d = \sqrt{(x_2 - x_1)^2 + (y_2 - y_1)^2}$$
$$= \sqrt{(-7 - 5)^2 + (2 - 7)^2}$$
$$= \sqrt{(-12)^2 + (-5)^2}$$
$$= \sqrt{169} \text{ or } 13$$

The distance from X to Y is 13 units.

EXAMPLE 4

Find the coordinates of the midpoint between $P(-4, 13)$ and $Q(6, 5)$.

Let $(x_1, y_1) = (-4, 13)$ and $(x_2, y_2) = (6, 5)$.

$$M\left(\frac{x_1 + x_2}{2}, \frac{y_1 + y_2}{2}\right) = M\left(\frac{-4 + 6}{2}, \frac{13 + 5}{2}\right)$$
$$= M(1, 9)$$

The coordinates of the midpoint are (1, 9).

1-4 Angle Measure (pp. 36–44)

For Exercises 23–26, refer to the figure below.

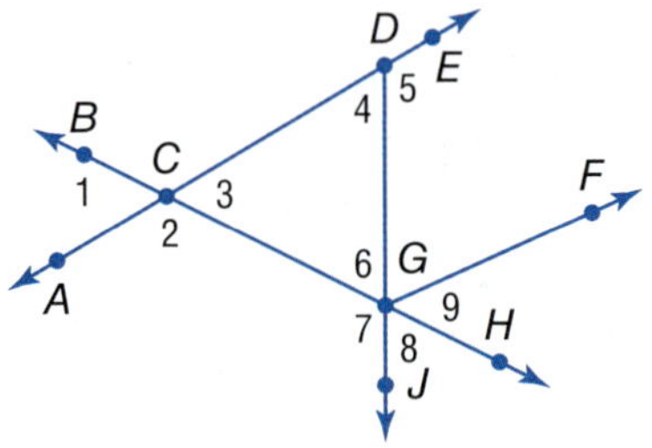

23. Name the vertex of $\angle 7$.

24. Write another name for $\angle 4$.

25. Name the sides of $\angle 2$.

26. Name a pair of opposite rays.

27. **SIGNS** A sign at West High School has the shape shown. Measure each of the angles and classify them as *right*, *acute*, or *obtuse*.

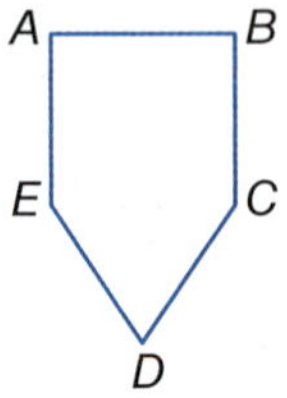

EXAMPLE 5

Refer to the figure below. Name all angles that have Q as a vertex.

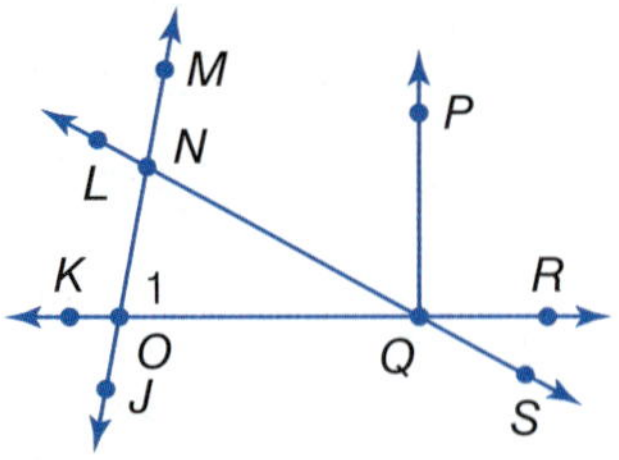

$\angle OQN$, $\angle NQP$, $\angle PQR$, $\angle RQS$, $\angle SQO$, $\angle OQP$, $\angle NQR$, $\angle PQS$, $\angle OQR$, $\angle NQS$

EXAMPLE 6

In the figure above, list all other names for $\angle 1$.

$\angle NOQ$, $\angle QON$, $\angle MOQ$, $\angle QOM$, $\angle MOR$, $\angle ROM$, $\angle NOR$, $\angle RON$

MIXED PROBLEM SOLVING
For mixed problem-solving practice, see page 999.

1-5 Angle Relationships (pp. 46–54)

For Exercises 28–30, refer to the figure below.

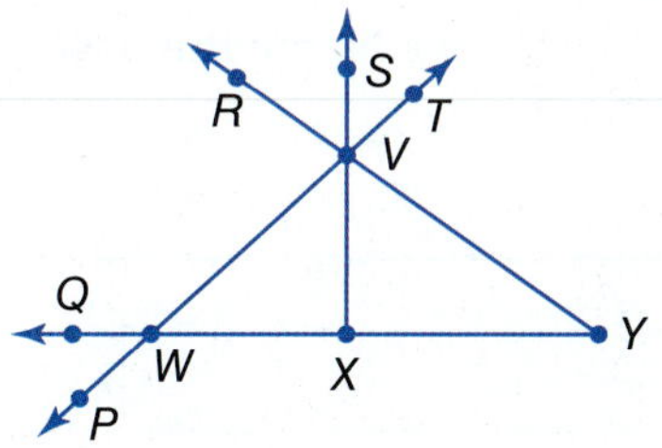

28. Name an angle supplementary to $\angle TVY$.

29. Name a pair of vertical angles with vertex W.

30. If $m\angle SXW = 5x - 16$, find the value of x so that $\overline{SX} \perp \overline{WY}$.

31. **PARKING** The parking arm shown below rests in a horizontal position and opens to a vertical position. After the arm has moved 24°, how many more degrees does it have to move so that it is vertical?

EXAMPLE 7

Name a pair of supplementary angles and a pair of complementary angles in the figure below.

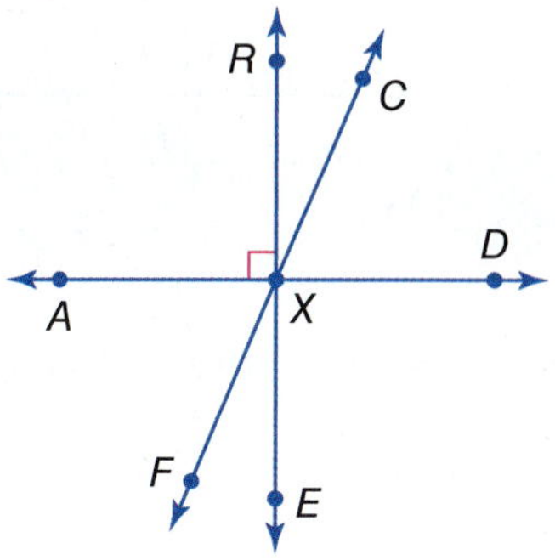

Sample answers:

Supplementary angles: $\angle RXA$ and $\angle RXD$

Complementary angles: $\angle RXC$ and $\angle CXD$

1-6 Two-Dimensional Figures (pp. 56–64)

Name each polygon by its number of sides. Then classify it as *convex* or *concave* and *regular* or *irregular*.

32.

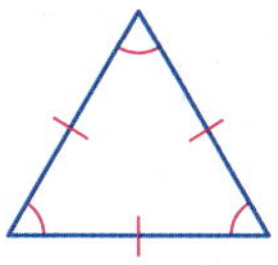

33.

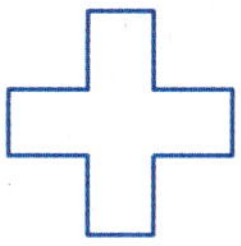

34. Find the perimeter of quadrilateral $ABCD$ with vertices $A(-3, 5)$, $B(0, 5)$, $C(2, 0)$, and $D(-5, 0)$.

35. **PARKS** Westside Park received 440 feet of chain-link fencing as a donation to build an enclosed play area for dogs. The park administrators need to decide what shape the area should have. They have three options: (1) a rectangle with length of 100 feet and width of 120 feet, (2) a square with sides of length 110 feet, or (3) a circle with radius of approximately 70 feet. Find the areas of all three enclosures and determine which would provide the largest area for the dogs.

EXAMPLE 8

Name the polygon by its number of sides. Then classify it as *convex* or *concave* and *regular* or *irregular*.

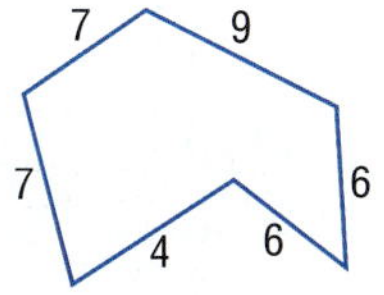

There are 6 sides, so this is a hexagon. If two of the sides are extended to make lines, they will pass through the interior of the hexagon, so it is concave. Since it is concave, it cannot be regular.

EXAMPLE 9

Find the perimeter of the polygon in the figure above.

$P = s_1 + s_2 + s_3 + s_4 + s_5 + s_6$	**Definition of perimeter**
$= 7 + 7 + 9 + 6 + 6 + 4$	**Substitution**
$= 39$	**Simplify.**

The perimeter of the polygon is 39 units.

1-7 Three-Dimensional Figures (pp. 67–74)

Identify each solid. Name the bases, faces, edges, and vertices.

36.

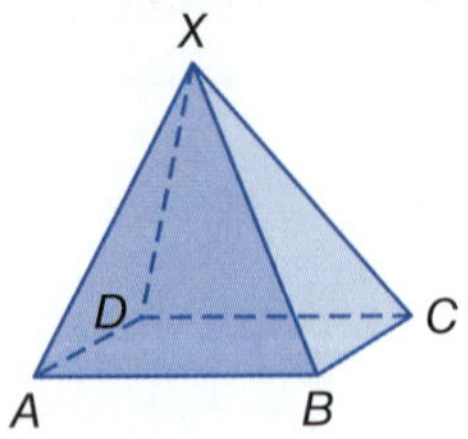

37.

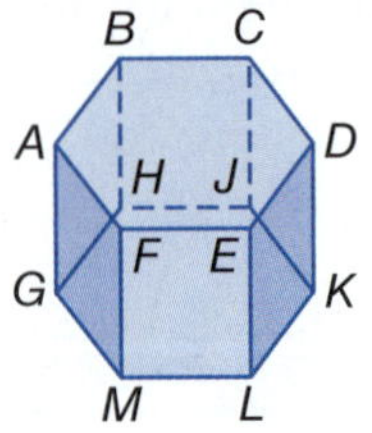

Find the surface area and volume of each solid.

38.

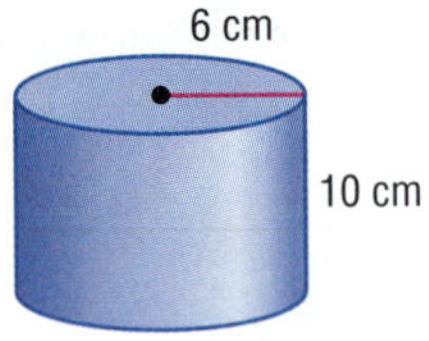

39.

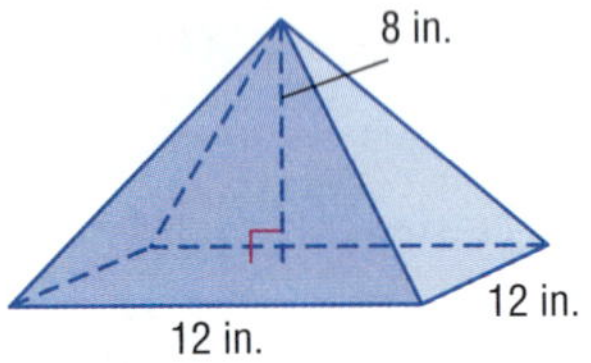

40.

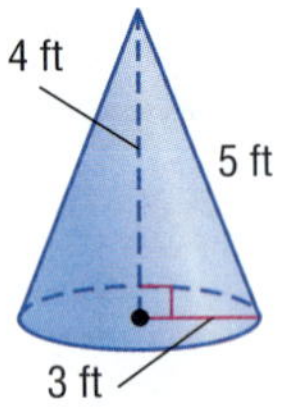

41.

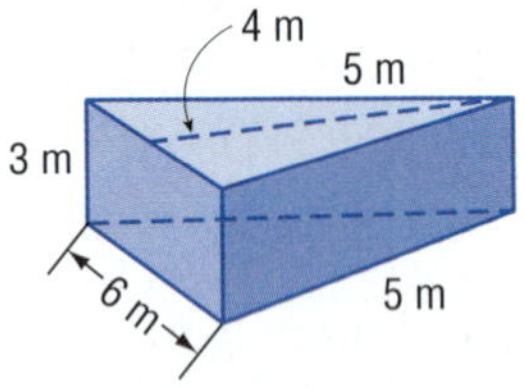

42. BUILDING Chris is building a trunk like the one shown below. His design is a square prism. What is the volume of the trunk?

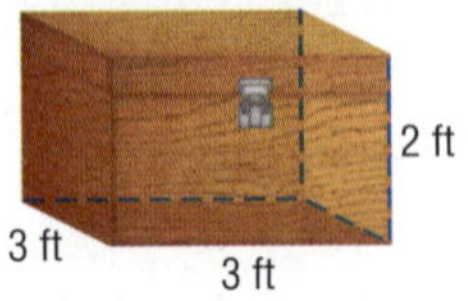

43. HOCKEY A regulation hockey puck is a cylinder made of vulcanized rubber 1 inch thick and 3 inches in diameter. Find the surface area and volume of a hockey puck.

EXAMPLE 10

Identify the solid below. Name the bases, faces, edges, and vertices.

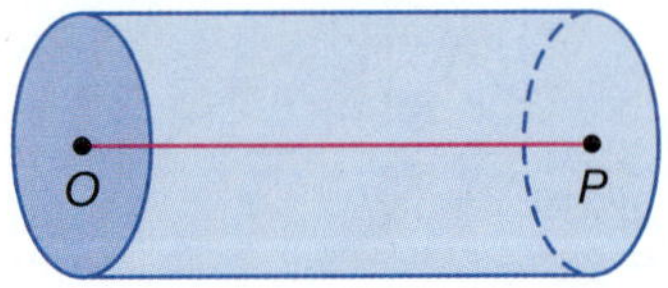

This solid has congruent circular bases in a pair of parallel planes. So, it is a cylinder.

Bases: circle O and circle P

A cylinder has no faces, edges, or vertices.

EXAMPLE 11

Find the surface area and volume of the rectangular prism below.

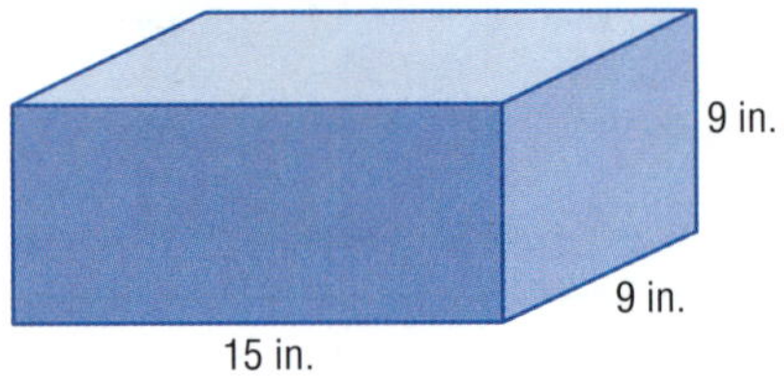

$T = Ph + 2B$ **Surface area of a prism**

$= (48)(9) + 2(135)$ **Substitution**

$= 702$ **Simplify.**

The surface area is 702 square inches.

$V = Bh$ **Volume of a prism**

$= (135)(9)$ **Substitution**

$= 1215$ **Simplify.**

The volume is 1215 cubic inches.

Practice Test

Math Online glencoe.com
Chapter Test

Use the figure to name each of the following.

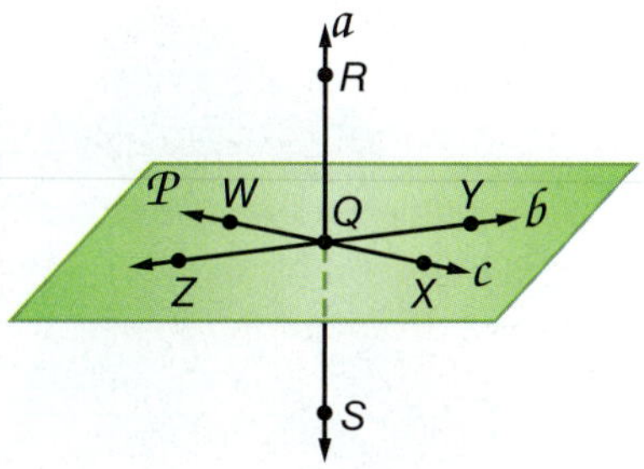

1. the line that contains points Q and Z
2. two points that are coplanar with points W, X, and Y
3. the intersection of lines a and b

Find the value of the variable if P is between J and K.

4. $JP = 2x$, $PK = 7x$, $JK = 27$
5. $JP = 3y + 1$, $PK = 12y - 4$, $JK = 75$
6. $JP = 8z - 17$, $PK = 5z + 37$, $JK = 17z - 4$

Find the coordinates of the midpoint of a segment with the given endpoints.

7. (16, 5) and (28, −13)
8. (−11, 34) and (47, 0)
9. (−4, −14) and (−22, 9)

Find the distance between each pair of points.

10. (43, −15) and (29, −3)
11. (21, 5) and (28, −1)
12. (0, −5) and (18, −10)

13. **ALGEBRA** The measure of $\angle X$ is 18 more than three times the measure of its complement. Find the measure of $\angle X$.

14. Find the value of x that will make lines a and b perpendicular in the figure below.

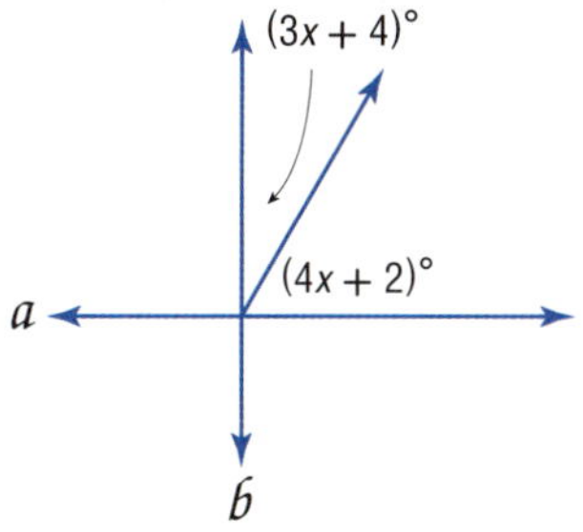

For Exercises 15–18, use the figure below.

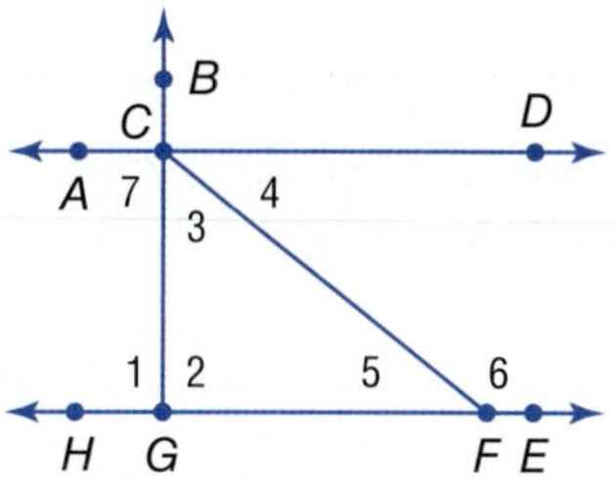

15. Name the vertex of $\angle 3$.
16. Name the sides of $\angle 1$.
17. Write another name for $\angle 6$.
18. Name a pair of angles that share exactly one point.

19. **MULTIPLE CHOICE** If $m\angle 1 = m\angle 2$, which of the following statements is true?

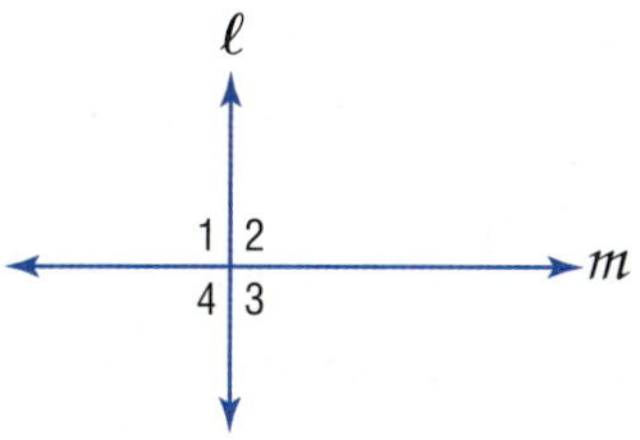

A $\angle 2 \cong \angle 4$

B $\angle 2$ is a right angle.

C $\ell \perp m$

D All of the above

Find the perimeter of each polygon.

20. triangle XYZ with vertices $X(3, 7)$, $Y(-1, -5)$, and $Z(6, -4)$
21. rectangle $PQRS$ with vertices $P(0, 0)$, $Q(0, 7)$, $R(12, 7)$, and $S(12, 0)$

22. **SAFETY** A severe weather siren in a local city can be heard within a radius of 1.3 miles. If the mayor of the city wants a new siren that will cover double the area of the old siren, what should the radius of the new siren be? Round to the nearest tenth of a mile.

Refer to the figure at the right.

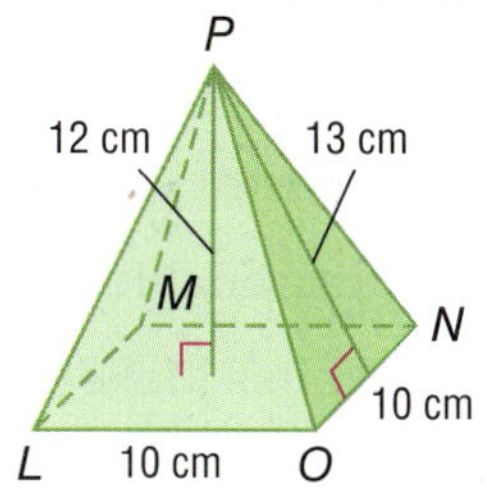

23. Name the base.
24. Find the surface area.
25. Find the volume.

Solving Math Problems

Strategies for Solving Math Problems

The first step to solving any math problem is to read the problem. When reading a math problem to get the information you need to solve, it is helpful to use special reading strategies.

Step 1

Read the problem to determine what information is given.

- **Analyze:** Determine what exactly the problem is asking you to solve.
- **Underline:** If you are able to write in your test book, underline any important information.

Step 2

Reread the problem to determine what information is needed to solve the problem.

- **Think:** How does the information fit together?
- **Key Words:** Are there any key words, variables or mathematical terms in the problem?
- **Diagrams:** Do you need to use a diagram, list or table?
- **Formulas:** Do you need a formula or an equation to solve the problem?

Step 3

Devise a plan and solve the problem. Use the information you found in Steps 1 and 2.

- **Question:** What problem are you solving?
- **Estimate:** Estimate an answer.
- **Eliminate:** Eliminate all answers that do not make sense and/or vary greatly from your estimate.

Step 4

Check your answer.

- **Reread:** Quickly reread the problem to make sure you solved the whole problem.
- **Reasonableness:** Is your answer reasonable?
- **Units:** Make sure your answer has the correct units of measurement.

EXAMPLE

Read the problem. Identify what you need to know. Then use the information in the problem to solve.

Carmen is using a coordinate grid to make a map of her backyard. She plots the swing set at point $S(2, 5)$ and the big oak tree at point $O(-3, -6)$. If each unit on the grid represents 5 feet, what is the distance between the swing set and the oak tree? Round your answer to the nearest whole foot.

A 12 ft **B** 25 ft **C** 60 ft **D** 74 ft

Determine what exactly the problem is asking you to solve. Underline any important information.

Carmen is using a coordinate grid to make a map of her backyard. She plots the swing set at point $S(2, 5)$ and the big oak tree at point $O(-3, -6)$. If each unit on the grid represents 5 feet, what is the distance between the swing set and the oak tree? Round your answer to the nearest whole foot.

The problem is asking for the distance between the swing set and the oak tree. The key word is distance, so you know you will need to use the Distance Formula.

$d = \sqrt{(x_2 - x_1)^2 + (y_2 - y_1)^2}$ **Distance Formula**

$= \sqrt{(-3 - 2)^2 + (-6 - 5)^2}$ **$(x_1, y_1) = (2, 5)$ and $(x_2, y_2) = (-3, -6)$**

$= \sqrt{(-5)^2 + (-11)^2}$ **Subtract.**

$= \sqrt{25 + 121}$ or $\sqrt{146}$ **Simplify.**

The distance between swing set and the oak tree is $\sqrt{146}$ units. Use a calculator to find that $\sqrt{146}$ units is approximately 12.08 units.

Since each unit on the grid represents 5 feet, the distance is $(12.08) \cdot (5)$ or 60.4 ft. Therefore, the correct answer is C.

Check your answer to make sure it is reasonable, and that you have used the correct units.

Exercises

Read each question. Then fill in the correct answer on the answer document provided by your teacher or on a sheet of paper.

1. A regular pentagon has a perimeter of 24 inches. What is the measure of each side?

A 3 inches **C** 4 inches

B 3.8 inches **D** 4.8 inches

2. What is the value of x in the figure at the right?

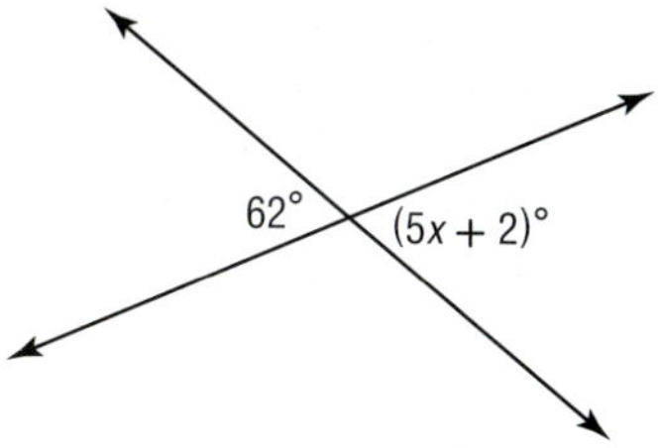

F 10

G 12

H 14

J 15

CHAPTER 1 Standardized Test Practice

Multiple Choice

Read each question. Then fill in the correct answer on the answer document provided by your teacher or on a sheet of paper.

1. If the dimensions of the prism below were doubled, by what factor would the volume of the prism increase?

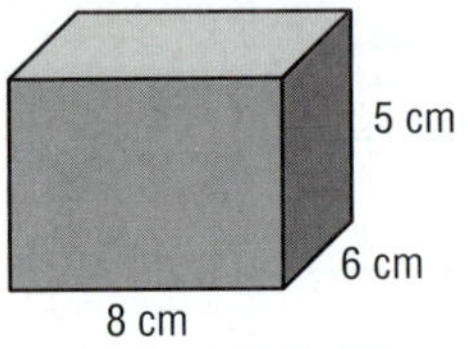

A 2 **C** 8

B 4 **D** 16

2. Find the distance between $M(-3, 1)$ and $N(2, 8)$ on a coordinate plane.

F 6.1 units

G 6.9 units

H 7.3 units

J 8.6 units

3. Which of the following terms best describes points *F*, *G*, and *H*?

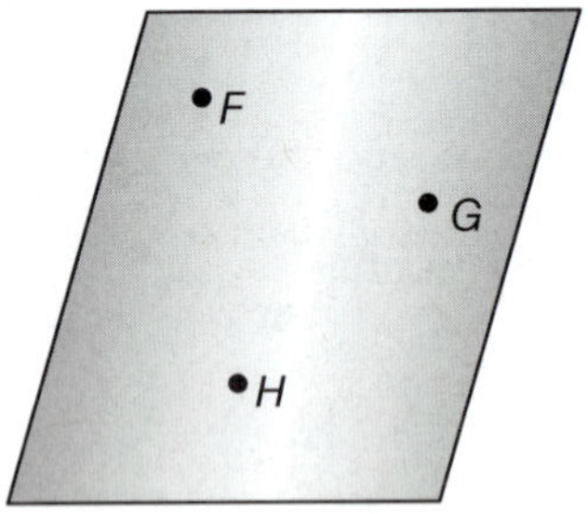

A collinear **C** coplanar

B congruent **D** skew

Test-Taking Tip

Question 3 Understanding the terms of geometry can help you solve problems. The term *congruent* refers to geometric figures, and *skew* refers to lines, therefore both answers can be eliminated.

4. What is the length of segment *BD*?

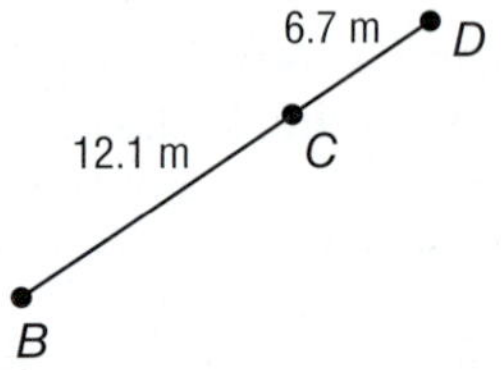

F 17.4 m **H** 18.8 m

G 18.3 m **J** 19.1 m

5. In the figure below, what is the measure of angle *CDN*?

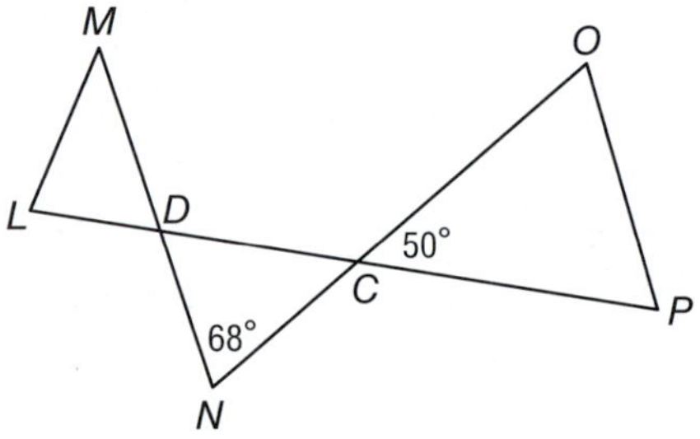

A 58° **C** 68°

B 62° **D** 70°

6. Find the perimeter of the figure below.

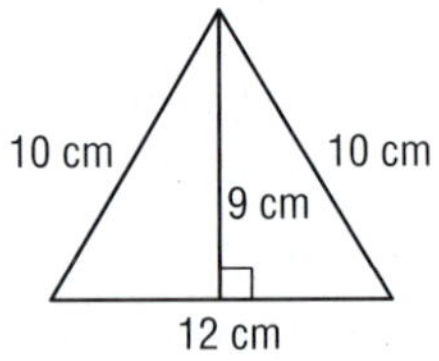

F 20 cm **H** 32 cm

G 29 cm **J** 41 cm

7. What is the relationship of ∠1 and ∠2?

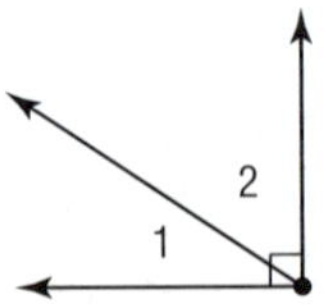

A complementary angles

B linear pair

C supplementary angles

D vertical angles

Short Response/Gridded Response

Record your answers on the answer sheet provided by your teacher or on a sheet of paper.

8. Find the distance between points R and S on the coordinate grid below. Round to the nearest tenth.

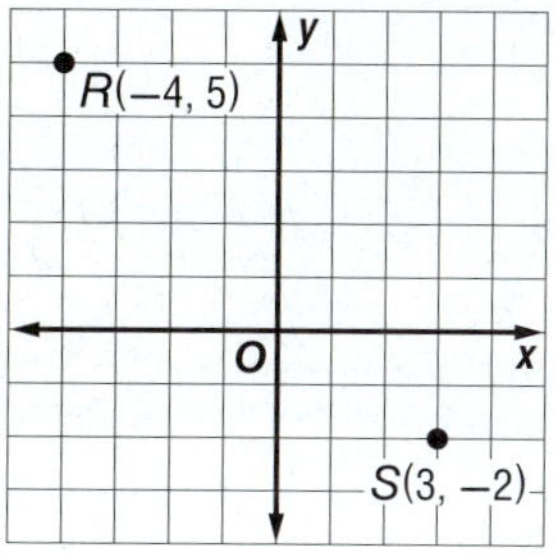

9. SHORT RESPONSE Find the value of x and AB if B is between A and C, $AB = 2x$, $AC = 6x - 5$, and $BC = 7$.

10. Suppose two lines intersect in a plane.

a. What do you know about the two pairs of vertical angles formed?

b. What do you know about the pairs of adjacent angles formed?

11. GRIDDED RESPONSE How many planes are shown in the figure below?

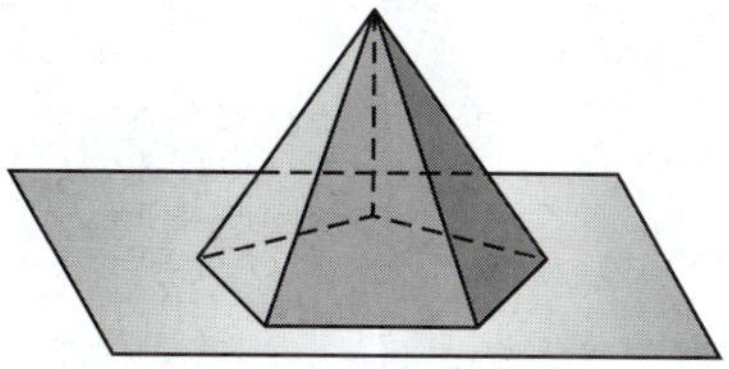

12. GRIDDED RESPONSE What is the total surface area of the cone? Round your answer to the nearest square centimeter.

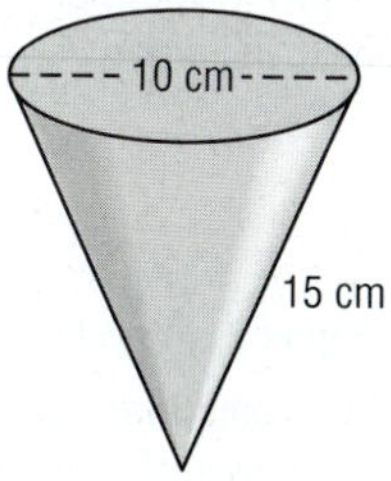

13. GRIDDED RESPONSE What is the value of x in the figure?

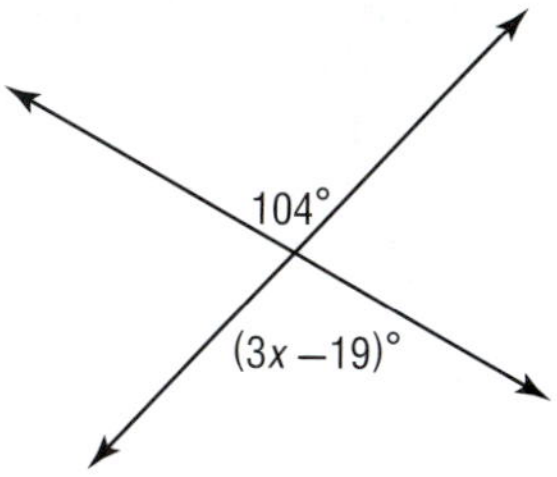

Extended Response

Record your answers on a sheet of paper. Show your work.

14. Julie's room has the dimensions shown in the figure.

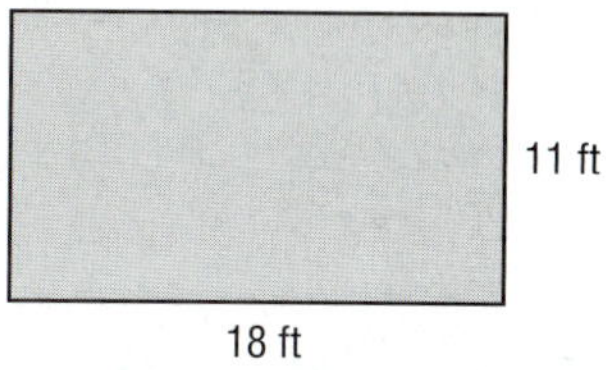

a. Find the perimeter of her room.

b. Find the area of her room.

c. If the length and width doubled, what effect would it have on the perimeter?

d. What effect would it have on the area?

Need Extra Help?														
If you missed Question...	1	2	3	4	5	6	7	8	9	10	11	12	13	14
Go to Lesson or Page...	1-7	1-3	1-1	1-2	1-5	1-6	1-5	1-3	1-2	1-5	1-1	1-7	1-5	1-6

CHAPTER 2 Reasoning and Proof

Then
In Chapter 1, you used segment and angle relationships.

Now
In Chapter 2, you will:

- Make conjectures and find counterexamples for statements.
- Use deductive reasoning to reach valid conclusions.
- Write proofs involving segment and angle theorems.

Why?
SCIENCE AND NATURE
Biologists and other scientists use inductive and deductive reasoning to make decisions and draw logical conclusions about animal populations.

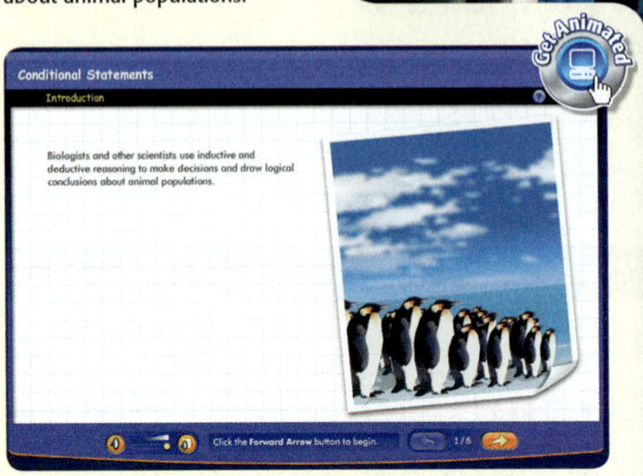

Get Ready for Chapter 2

Diagnose Readiness You have two options for checking Prerequisite Skills.

Text Option

Take the Quick Check below. Refer to the Quick Review for help.

QuickCheck

Evaluate each expression for the given value of x. (Lesson 0–4)

1. $4x + 7; x = 6$
2. $(x - 2)180; x = 8$
3. $5x^2 - 3x; x = 2$
4. $\frac{x(x-3)}{2}; x = 5$
5. $x + (x + 1) + (x + 2); x = 3$

Write each verbal expression as an algebraic expression.

6. eight less than five times a number
7. three more than the square of a number

Solve each equation. (Lesson 0–5)

8. $8x - 10 = 6x$
9. $18 + 7x = 10x + 39$
10. $3(11x - 7) = 13x + 25$
11. $3x + 8 = \frac{1}{2}x + 35$
12. $\frac{2}{3}x + 1 = 5 - 2x$
13. **CLOTHING** Nancy bought 4 shirts at the mall for \$52. Write and solve an equation to find the average cost of one shirt.

Refer to the figure in Example 3. (Lessons 1–4 and 1–5)

14. Identify a pair of vertical angles that appear to be obtuse.
15. Identify a pair of adjacent angles that appear to be complementary.
16. Identify a linear pair.
17. If $m\angle DXB = 116$ and $m\angle EXA = 3x + 2$, find x.
18. If $m\angle BXC = 90$, $m\angle CXD = 6x - 13$, and $m\angle DXE = 10x + 7$, find x.

QuickReview

EXAMPLE 1

Evaluate $x^2 - 2x + 11$ for $x = 6$.

$x^2 - 2x + 11$	Original expression
$= (6)^2 - 2(6) + 11$	Substitute 6 for x.
$= 36 - 2(6) + 11$	Evaluate the exponent.
$= 36 - 12 + 11$	Multiply.
$= 35$	Simplify.

EXAMPLE 2

Solve $36x - 14 = 16x + 58$.

$36x - 14 = 16x + 58$	Original equation
$36x - 14 - 16x = 16x + 58 - 16x$	Subtract $16x$ from each side.
$20x - 14 = 58$	Simplify.
$20x - 14 + 14 = 58 + 14$	Add 14 to each side.
$20x = 72$	Simplify.
$\frac{20x}{20} = \frac{72}{20}$	Divide each side by 20.
$x = 3.6$	Simplify.

EXAMPLE 3

If $m\angle BXA = 3x + 5$ and $m\angle DXE = 56$, find x.

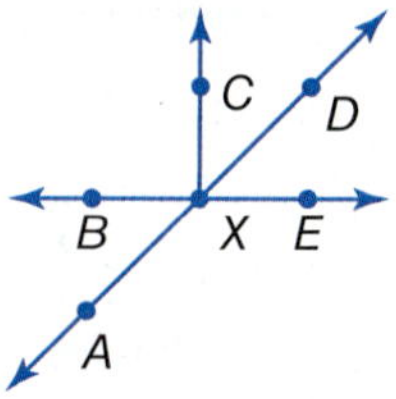

$m\angle BXA = m\angle DXE$	Vertical ∡ are ≅.
$3x + 5 = 56$	Substitution
$3x = 51$	Subtract 5 from each side.
$x = 17$	Divide each side by 3.

Online Option

Math Online Take a self-check Chapter Readiness Quiz at glencoe.com.

Get Started on Chapter 2

You will learn several new concepts, skills, and vocabulary terms as you study Chapter 2. To get ready, identify important terms and organize your resources. You may wish to refer to **Chapter 0** to review prerequisite skills.

FOLDABLES® Study Organizer

Reasoning and Proof Make this Foldable to help you organize your Chapter 2 notes about logic, reasoning, and proof. Begin with one sheet of notebook paper.

1. **Fold** lengthwise to the holes.

2. **Cut** five tabs in the top sheet.

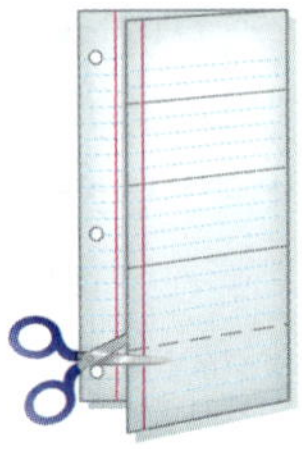

3. **Label** the tabs as shown.

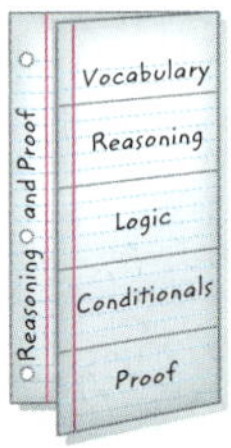

New Vocabulary

English		Español
inductive reasoning	• p. 89 •	razonamiento inductivo
conjecture	• p. 89 •	conjetura
counterexample	• p. 92 •	contraejemplo
negation	• p. 97 •	negación
if-then statement	• p. 105 •	enunciado si-entonces
hypothesis	• p. 105 •	hipótesis
conclusion	• p. 105 •	conclusión
converse	• p. 107 •	recíproco
inverse	• p. 107 •	inversa
postulate	• p. 125 •	postulado
proof	• p. 126 •	demostración
theorem	• p. 127 •	teorema

Review Vocabulary

complementary angles • p. 47 • ángulos complementarios two angles whose sum is 90 degrees

supplementary angles • p. 47 • ángulos suplementarios two angles whose sum is 180 degrees

vertical angles • p. 46 • ángulos opuestos por el vértice two nonadjacent angles formed by intersecting lines

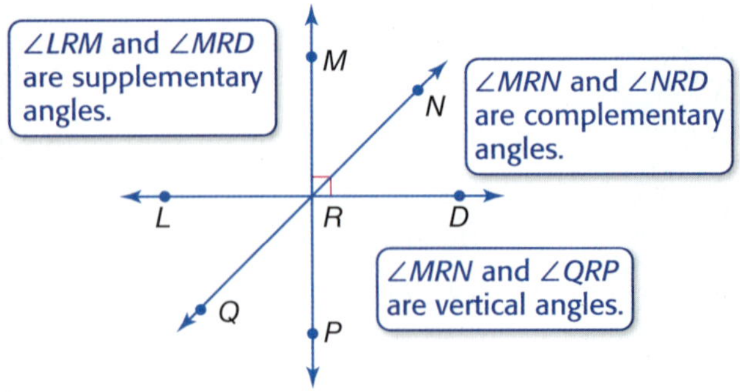

Multilingual eGlossary glencoe.com

Math Online glencoe.com

- Study the chapter online
- Explore **Get Animated** activities
- Get extra help from your own **Personal Tutor**
- Use **Extra Examples** for additional help
- Take a **Self-Check Quiz**
- **Review Vocabulary** in fun ways

2-1 Inductive Reasoning and Conjecture

Then
You used data to find patterns and make predictions.

Now
- Make conjectures based on inductive reasoning.
- Find counterexamples.

New Vocabulary
inductive reasoning
conjecture
counterexample

Math Online
glencoe.com

Why?

Market research is conducted by an analyst to answer specific questions about products. Research analysts use focus groups to collect data and to look for patterns in the success or failure of a product. The process of using patterns to analyze the effectiveness of a product involves inductive reasoning.

Make Conjectures **Inductive reasoning** is reasoning that uses a number of specific examples to arrive at a conclusion. When you assume that an observed pattern will continue, you are applying inductive reasoning. A concluding statement reached using inductive reasoning is called a **conjecture**.

EXAMPLE 1 Patterns and Conjecture

Write a conjecture that describes the pattern in each sequence. Then use your conjecture to find the next item in the sequence.

a. Movie show times: 8:30 A.M., 9:45 A.M., 11:00 A.M., 12:15 P.M., . . .

Step 1 Look for a pattern.

8:30 A.M., 9:45 A.M., 11:00 A.M., 12:15 P.M., . . .

+1 hr 15 min +1 hr 15 min +1 hr 15 min

Step 2 Make a conjecture.

The show time is 1 hour and fifteen minutes greater than the previous show time. The next show time will be 12:15 P.M. + 1:15 or 1:30 P.M.

b.

4 10 18 28 40 . . .

Step 1 4, 10, 18, 28, 40

+6 +8 +10 +12

The numbers increase by 6, 8, 10, and 12.

Step 2 The next figure will increase by 12 + 2 or 14 segments. So, the next figure will have 40 + 14 or 54 segments.

CHECK Draw the next figure to check your conjecture. ✓

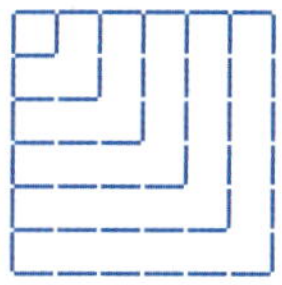

54

Review Vocabulary

sequence an arrangement of numbers or items in a particular order

Guided Practice

Write a conjecture that describes the pattern in each sequence. Then use your conjecture to find the next item in the sequence.

1A. Follow-up visits: Dec., May, Oct., Mar., . . .

1B. 10, 4, −2, −8, . . .

1C.

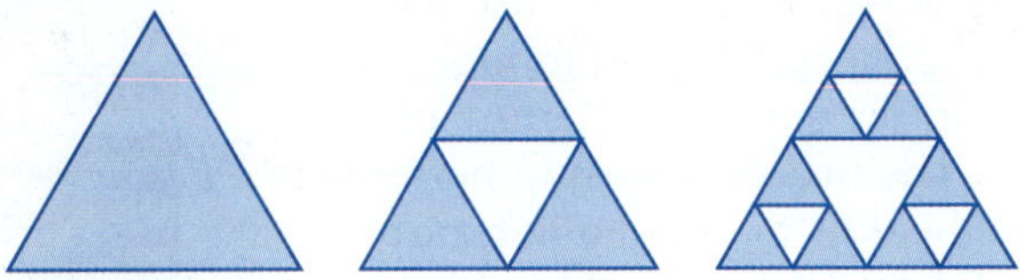

Personal Tutor glencoe.com

To make some algebraic and geometric conjectures, you will need to provide examples.

StudyTip

Supporting Examples and Proofs Examples that support a conjecture are not enough to show that a conjecture is true. To show that an algebraic or geometric conjecture is true, you must offer a logical argument called a proof. You will learn more about proofs in Lesson 2-5.

EXAMPLE 2 Algebraic and Geometric Conjectures

Make a conjecture about each value or geometric relationship. List or draw some examples that support your conjecture.

a. the sum of two odd numbers

Step 1 List examples.

$1 + 3 = 4$ $1 + 5 = 6$ $3 + 5 = 8$ $7 + 9 = 16$

Step 2 Look for a pattern.

Notice that the sums 4, 6, 8, and 16 are all even numbers.

Step 3 Make a conjecture.

The sum of two odd numbers is an even number.

b. segments joining opposite vertices of a rectangle

Step 1

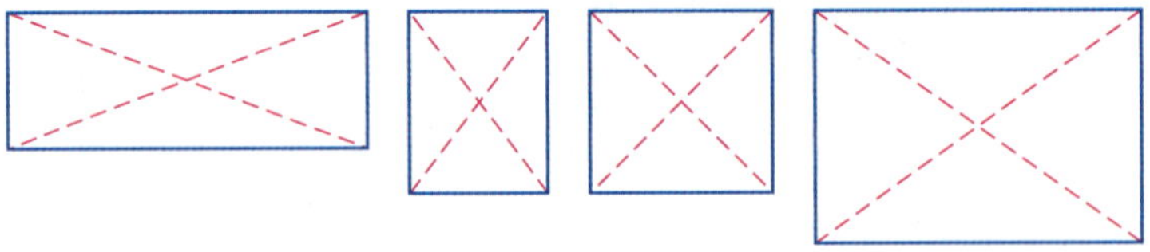

Step 2 Notice that the segments joining opposite vertices of each rectangle appear to have the same measure. Use a ruler or compass to confirm this.

Step 3 Conjecture: the segments joining opposite vertices of a rectangle are congruent.

Guided Practice

2A. the sum of two even numbers

2B. the relationship between AB and EF, if $AB = CD$ and $CD = EF$

2C. the sum of the squares of two consecutive natural numbers

Personal Tutor glencoe.com

Real-World Career

Hair Stylist Hair stylists work in salons where various services, including skin and nail treatments, may be provided in addition to hair care. About 48% of hair stylists are self-employed and own their own businesses. Hair stylists must attend cosmetology school and obtain a license.

Real-world conjectures are often made based on data gathered about a specific topic of interest.

Real-World EXAMPLE 3 Make Conjectures from Data

BUSINESS The owner of a hair salon collected data on the number of customers her salon had each Friday, Saturday, and Sunday for 6 months to decide whether she should increase the number of stylists working each weekend. The data she collected are shown below.

Number of Customers on the Weekend

Day	Month 1	Month 2	Month 3	Month 4	Month 5	Month 6
Friday	225	255	321	406	540	450
Saturday	603	658	652	712	746	832
Sunday	552	635	642	692	685	705
Total	1380	1548	1615	1810	1971	1987

a. Make a statistical graph that best displays the data.

Since you want to look for a pattern over time, use a scatter plot to display the data. Label the horizontal axis with the months and the vertical axis with the number of customers. Plot each set of data using a different color and include a legend.

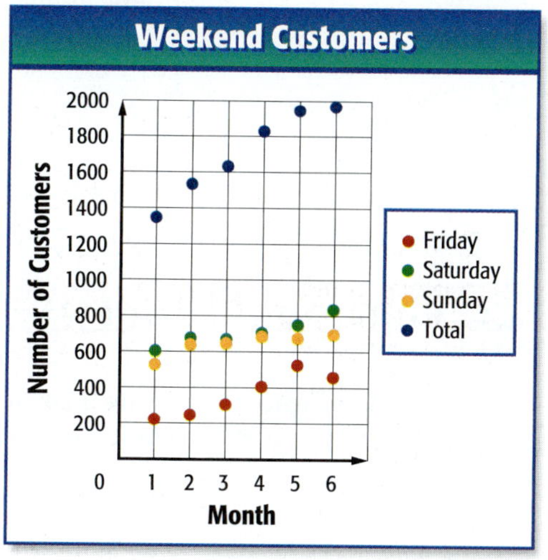

b. Make a conjecture based on the data and explain how this conjecture is supported by your graph.

Look for patterns in the data. The number of customers on each day usually increases each month, and the total number of customers increases every single month.

Survey data supports a conjecture that the amount of business on the weekends has increased, so the owner should schedule more stylists to work on those days.

Guided Practice

3. **POSTAGE** The table at the right shows the price of postage for the years 1982 through 2007.

   A. Construct a statistical graph that best displays the data.

   B. Predict the postage rate in 2013 based on the graph.

   C. Does it make sense that the pattern of the data will continue over time? If not, how will it change? Explain your reasoning.

Year	Rate (cents)
1982	20
1987	22
1992	29
1997	32
2002	37
2007	41

Personal Tutor glencoe.com

Vocabulary Link

Counterexample
Everyday Use The prefix *counter-* means *the opposite of.*
Math Use A counterexample is the opposite of an example.

Find Counterexamples To show that a conjecture is true for all cases, you must prove it. It takes only one false example, however, to show that a conjecture is not true. This false example is called a **counterexample**, and it can be a number, a drawing, or a statement.

EXAMPLE 4 Find Counterexamples

Find a counterexample to show that each conjecture is false.

a. If n is a real number, then $n^2 > n$.

When n is 1, the conjecture is false, since $1^2 \not> 1$.

b. If $JK = KL$, then K is the midpoint of $\overline{JL}$.

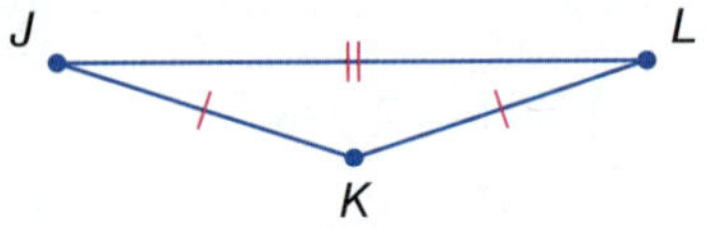

When J, K, and L are noncollinear, the conjecture is false. In the figure, $JK = KL$, but K is not the midpoint of $\overline{JL}$.

Guided Practice

4A. If n is a real number, then $-n$ is a negative.

4B. If $\angle ABC \cong \angle DBE$, then $\angle ABC$ and $\angle DBE$ are vertical angles.

Personal Tutor glencoe.com

Check Your Understanding

Example 1 p. 89

Write a conjecture that describes the pattern in each sequence. Then use your conjecture to find the next item in the sequence.

1. Costs: \$4.50, \$6.75, \$9.00 . . .

2. Appointment times: 10:15 A.M., 11:00 A.M., 11:45 A.M., . . .

3.

4.

5 3, 3, 6, 9, 15, . . .

6. 2, 6, 14, 30, 62, . . .

Example 2 p. 90

Make a conjecture about each value or geometric relationship.

7. the product of two even numbers

8. the relationship between a and b if $a + b = 0$

9. the relationship between the set of points in a plane equidistant from point A

10. the relationship between $\overline{AP}$ and $\overline{PB}$ if M is the midpoint of $\overline{AB}$ and P is the midpoint of $\overline{AM}$

Example 3 p. 91

11. **CELL PHONES** Refer to the table of the number of cell phone subscriptions in the United States by year.

    a. Make a graph that shows U.S. cell phone use from 2000 to 2005.

    b. Make a conjecture about U.S. cell phone use in 2010.

U.S. Cell Phone Subscriptions

Year	Subscriptions (Millions)
2000	109.5
2001	128.4
2002	140.8
2003	158.7
2004	182.1
2005	207.9

Source: CTIA – The Wireless Association

Example 4 p. 92

Find a counterexample to show that each conjecture is false.

12. If $\angle A$ and $\angle B$ are complementary angles, then they share a common side.

13. If a ray intersects a segment at its midpoint, then the ray is perpendicular to the segment.

Practice and Problem Solving

= **Step-by-Step Solutions** begin on page R20. **Extra Practice** begins on page 969.

Example 1 p. 89

Write a conjecture that describes the pattern in each sequence. Then use your conjecture to find the next item in the sequence.

14. 0, 2, 4, 6, 8

15. 3, 6, 9, 12, 15

16. 4, 8, 12, 16, 20

17. 2, 22, 222, 2222

18. 1, 4, 9, 16

19. $1, \frac{1}{2}, \frac{1}{4}, \frac{1}{8}$

20. Arrival times: 3:00 P.M., 12:30 P.M., 10:00 A.M., . . .

21. Percent humidity: 100%, 93%, 86%, . . .

22. Work-out days: Sunday, Tuesday, Thursday, . . .

23. Club meetings: January, March, May, . . .

24. 

25. 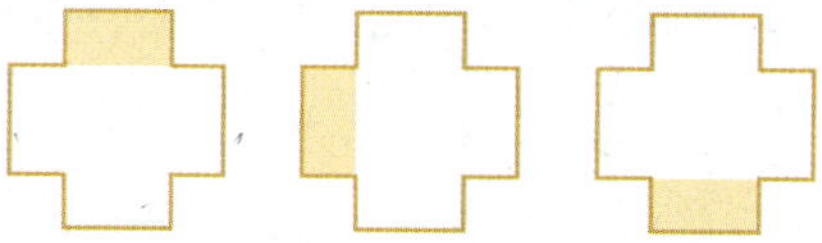

26. 

27. 

28. **FITNESS** Gabriel started training with the track team five weeks ago. During the first week, he ran 0.5 mile at each practice. The next three weeks he ran 0.75 mile, 1 mile, and 1.25 miles at each practice. If he continues this pattern, how many miles will he be running at each practice during the 7th week?

Real-World Link

Nearly two thirds of Earth is covered in water, but only a fraction of that water is drinkable. Almost 97% of Earth's water is salt water, and 2% is in glaciers and ice caps, which leaves only about 1% for human use.

Source: Earth 911

29. **CONSERVATION** When there is a shortage of water, some municipalities limit the amount of water each household is allowed to consume. Most cities that experience water restrictions are in the western and southern parts of the United States. Make a conjecture about why water restrictions occur in these areas.

30. **VOLUNTEERING** Carrie collected canned food for a homeless shelter in her area each day for one week. On day one, she collected 7 cans of food. On day two, she collected 8 cans. On day three she collected 10 cans. On day four, she collected 13 cans. If Carrie wanted to give at least 100 cans of food to the shelter and this pattern of can collecting continued, did she meet her goal?

Example 2 p. 90

Make a conjecture about each value or geometric relationship.

31. the product of two odd numbers

32. the product of two and a number, plus one

33. the relationship between a and c if $ab = bc, b \neq 0$

34. the relationship between a and b if $ab = 1$

35. the relationship between $\overline{AB}$ and the set of points equidistant from A and B

36. the relationship between the angles of a triangle with all sides congruent

37. the relationship between the areas of a square with side x and a rectangle with sides x and $2x$

38. the relationship between the volume of a prism and a pyramid with the same base

Example 3 p. 91

39. SPORTS Refer to the table of Americans over the age of 7 that played hockey.

Year	Number of Participants (millions)
2000	1.9
2002	2.1
2004	2.4
2006	2.6

a. Construct a statistical graph that best displays the data.

b. Make a conjecture based on the data and explain how this conjecture is supported by your graph.

Example 4 p. 92

Determine whether each conjecture is *true* or *false*. Give a counterexample for any false conjecture.

40. If n is a prime number, then $n + 1$ is not prime.

41. If x is an integer, then $-x$ is positive.

42. If $\angle 2$ and $\angle 3$ are supplementary angles, then $\angle 2$ and $\angle 3$ form a linear pair.

43. If you have three points A, B, and C, then A, B, C are noncollinear.

44. If in $\triangle ABC$, $(AB)^2 + (BC)^2 = (AC)^2$, then $\triangle ABC$ is a right triangle.

45. If the area of a rectangle is 20 square meters, then the length is 10 meters and the width is 2 meters.

Real-World Career

Natural Science Manager Natural science managers direct research projects and coordinate the activities of other scientists, such as biologists. Many begin their careers as scientists or mathematicians and have bachelor's degrees in the type of work they supervise.

46. NATURE Refer to the table and find a counterexample for each statement.

Fishing

State	Number of Youth Anglers	Percent of Total Anglers in State
California	1,099,000	31
Florida	543,000	15
Michigan	452,000	25
North Carolina	353,000	21.5

Source: American Sportfishing Association

a. The number of youth anglers in a state is less than one fourth of the total anglers in that state.

b. Each state listed has at least 3,000,000 total anglers.

Math History Link

Sophie Germain (1776–1831)

Sophie Germain was born in Paris, France. Like Goldbach, she studied relationships involving prime numbers, assuming a man's identity in order to pursue her passion.

47. GOLDBACH'S CONJECTURE Goldbach's conjecture states that every even number greater than 2 can be written as the sum of two primes. For example, $4 = 2 + 2$, $6 = 3 + 3$, and $8 = 3 + 5$.

a. Show that the conjecture is true for the even numbers from 10 to 20.

b. Given the conjecture *All odd numbers greater than 2 can be written as the sum of two primes,* is the conjecture *true* or *false*? Give a counterexample if the conjecture is false.

48. SEGMENTS Two collinear points form one segment, as shown for $\overline{AB}$. If a collinear point is added to $\overline{AB}$, the three collinear points form three segments.

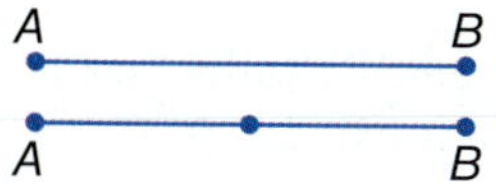

a. How many distinct segments are formed by four collinear points? by five collinear points?

b. Make a conjecture about the number of distinct segments formed by n collinear points.

c. Test your conjecture by finding the number of distinct segments formed by six points.

49. CHECKUPS Nora's doctor measures her height each year when she goes in for her annual checkup. When she was 3 years old, she was 33 inches tall. At 5, 7, and 9 years old, she was 42, 49, and 53 inches tall, respectively.

a. Create a table of Nora's height data.

b. Graph Nora's height based on her age.

c. Nora is 11 this year. Make a conjecture about her height this year.

d. Does it make sense that this pattern of data will continue over time? If not, how will it change? Explain your reasoning.

H.O.T. Problems
Use **H**igher-**O**rder **T**hinking Skills

50. CHALLENGE If you draw points on a circle and connect every pair of points, the circle is divided into regions. For example, two points form two regions and three points form four regions.

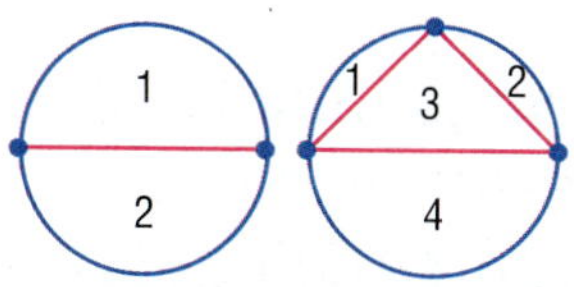

a. Make a conjecture about the relationship between the number of points on a circle and the number of regions formed in the circle.

b. Does your conjecture hold true when there are six points? Support your answer with a diagram.

51. ERROR ANALYSIS Juan and Jack are discussing prime numbers. Juan states a conjecture that all prime numbers are odd. Jack disagrees with the conjecture and states not all prime numbers are odd. Is either of them correct? Explain.

52. OPEN ENDED Write a number sequence that can be generated by two different patterns. Explain your patterns.

53. REASONING Consider the conjecture *If two points are equidistant from a third point, then the three points are collinear.* Is the conjecture *true* or *false*? If false, give a counterexample.

54. WRITING IN MATH Suppose you are conducting a survey. Choose a topic and write three questions you would include in your survey. How would you use inductive reasoning with your responses?

Standardized Test Practice

55. Look at the pattern below.

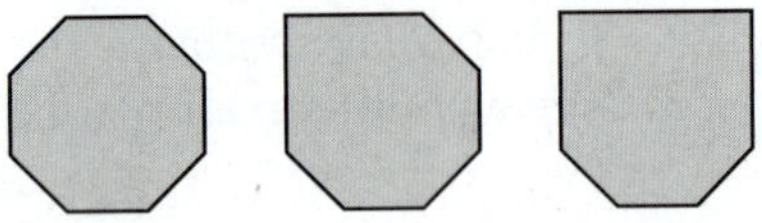

If the pattern continues, what will be the next shape?

A

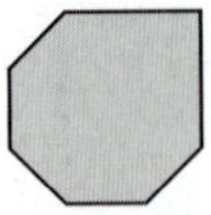

B

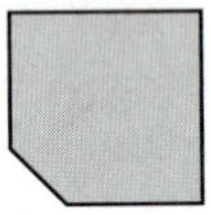

C

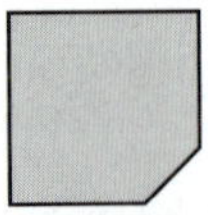

D

56. GRIDDED RESPONSE What is the value of the expression below if $a = 10$ and $b = 1$?

$$2b + ab \div (a + b)$$

57. ALGEBRA A chemistry student mixed some 30% copper sulfate with some 40% copper sulfate solution to obtain 100 mL of a 32% copper sulfate solution. How much of the 30% copper sulfate solution did the student use in the mixture?

F 90 mL
G 80 mL
H 60 mL
J 20 mL

58. SAT/ACT Which of the following is equal to $2x$?

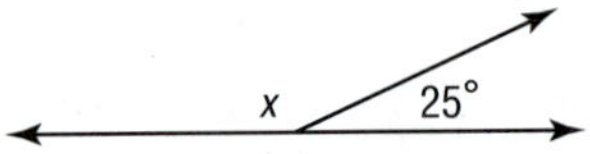

A 50°
B 155°
C 310°
D 360°

Spiral Review

Find the surface area and volume of each solid. (Lesson 1-7)

59.

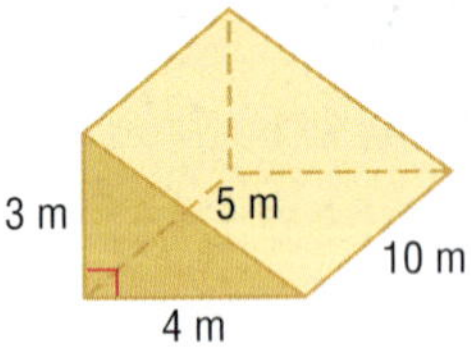

60.

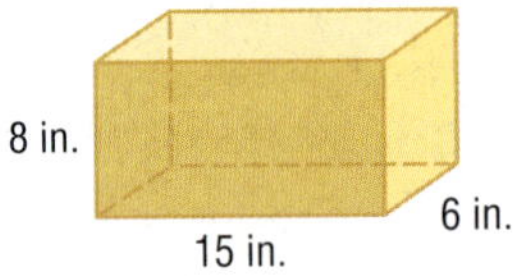

61.

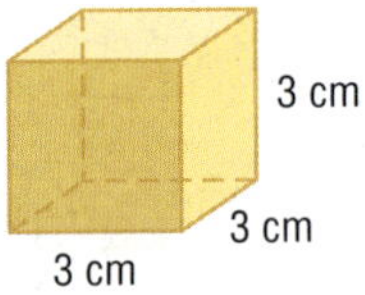

Find the perimeter of $\triangle ABC$ to the nearest hundredth, given the coordinates of its vertices. (Lesson 1-6)

62. $A(1, 6), B(1, 2), C(3, 2)$

63. $A(-3, 2), B(2, -9), C(0, -10)$

64. ALGEBRA The measures of two complementary angles are $16z - 9$ and $4z + 3$. Find the measures of the angles. (Lesson 1-5)

65. FLAGS The Wyoming state flag is shown at the right. Name the geometric term modeled by this flag: point, line, or plane. (Lesson 1-1)

66. ALGEBRA Evaluate $5|x + y| - 3|2 - z|$ if $x = 3, y = -4$, and $z = -5$. (Lesson 0-4)

Skills Review

ALGEBRA Determine which values in the replacement set make the inequality true. (Lesson 0-6)

67. $x - 3 > 12$
$\{6, 10, 14, 18\}$

68. $6 + x > 9$
$\{8, 6, 4, 2\}$

69. $2x - 4 > 10$
$\{5, 6, 7, 8\}$

2-2 Logic

Then
You found counterexamples for false conjectures. (Lesson 2-1)

Now
- Determine truth values of negations, conjunctions, and disjunctions, and represent them using Venn diagrams.
- Find counterexamples.

New Vocabulary
statement
truth value
negation
compound statement
conjunction
disjunction
truth table

Math Online
glencoe.com

Why?

Many electrical circuits operate by evaluating a series of tests that are either true or false. For example, a single light can be controlled by two different switches connected on a circuit. The positions of both switches, either up or down, determine whether the light is on or off.

Determine Truth Values A **statement** is a sentence that is either true or false. The **truth value** of a statement is either true (T) or false (F). Statements are often represented using a letter such as p or q.

p: **A rectangle is a quadrilateral.** Truth value: T

The **negation** of a statement has the opposite meaning, as well as an opposite truth value. For example, the negation of the statement above is *not p* or $\sim p$.

$\sim p$: **A rectangle is** not **a quadrilateral.** Truth value: F

Two or more statements joined by the word *and* or *or* form a **compound statement**. A compound statement using the word *and* is called a **conjunction**. A conjunction is true only when both statements that form it are true.

p: **A rectangle is a quadrilateral.** Truth value: T

q: **A rectangle is convex.** Truth value: T

p and *q*: **A rectangle is a quadrilateral,** and **a rectangle is convex.**

Since both p and q are true, the conjunction p and q, also written $p \wedge q$, is true.

EXAMPLE 1 Truth Values of Conjunctions

Use the following statements to write a compound statement for each conjunction. Then find its truth value. Explain your reasoning.

p: **The figure is a triangle.**

q: **The figure has two congruent sides.**

r: **The figure has three acute angles.**

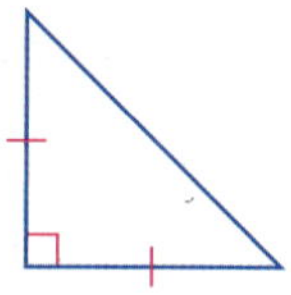

a. p **and** r

p and r: The figure is a triangle, and the figure has three acute angles.

Although p is true, r is false. So p and r is false.

b. $q \wedge \sim r$

$q \wedge \sim r$: The figure has two congruent sides, and the figure does not have three acute angles.

Both q and $\sim r$ are true, so $q \wedge \sim r$ is true.

Guided Practice

1A. $p \wedge q$ **1B.** not p and not r

Personal Tutor glencoe.com

Watch Out!

Negation Just as the opposite of an integer is not always negative, the negation of a statement is not always false. The negation of a statement has the opposite truth value of the original statement.

A compound statement that uses the word *or* is called a **disjunction**.

p: **Malik studies geometry.**

q: **Malik studies chemistry.**

p or *q*: **Malik studies geometry,** or **Malik studies chemistry.**

A disjunction is true if at least one of the statements is true. If Malik studies either geometry or chemistry or both subjects, the disjunction *p* or *q*, also written as $p \vee q$, is true. If Malik studies neither geometry nor chemistry, *p* or *q* is false.

EXAMPLE 2 Truth Values of Disjunctions

Use the following statements to write a compound statement for each disjunction. Then find its truth value. Explain your reasoning.

***p*: January is a fall month.**

***q*: January has only 30 days.**

***r*: January 1 is the first day of a new year.**

a. ***q* or *r***

q or *r*: January has only 30 days, or January 1 is the first day of a new year.

q or *r* is true because *r* is true. It does not matter that *q* is false.

b. $p \vee q$

$p \vee q$: January is a fall month, or January has only 30 days.

Since both *p* and *q* are false, $p \vee q$ is false.

c. $\sim p \vee r$

$\sim p \vee r$: January is *not* a fall month, or January 1 is the first day of a new year.

Not *p* or *r* is true, because not *p* is true and *r* is true.

Guided Practice

2A. *r* or *p* **2B.** $q \vee \sim r$ **2C.** $p \vee \sim q$

Personal Tutor glencoe.com

Concept Summary Negation, Conjunction, Disjunction

Statement	Words	Symbols
Negation	a statement that has the opposite meaning and truth value of an original statement	$\sim p$, read *not p*
Conjunction	a compound statement formed by joining two or more statements using the word *and*	$p \wedge q$, read *p and q*
Disjunction	a compound statement formed by joining two or more statements using the word *or*	$p \vee q$, read *p or q*

StudyTip

Truth Tables The truth tables for a conjunction and a disjunction are easier to recall if you remember the following.

- A conjunction is true only when both statements are true.
- A disjunction is false only when both statements are false.

A convenient method for organizing the truth values of statements is to use a **truth table**. Truth tables can be used to determine truth values of negations and compound statements.

Negation

p	$\sim p$
T	F
F	T

Conjunction

p	q	$p \wedge q$
T	T	T
T	F	F
F	T	F
F	F	F

Disjunction

p	q	$p \vee q$
T	T	T
T	F	T
F	T	T
F	F	F

You can use the truth values for negation, conjunction, and disjunction to construct truth tables for more complex compound statements.

EXAMPLE 3 Construct Truth Tables

Construct a truth table for $\sim p \vee q$.

❶ Make columns with headings that include each original statement, any negations of these statements, and the compound statement itself.

p	q	$\sim p$	$\sim p \vee q$
T	T	F	T
T	F	F	F
F	T	T	T
F	F	T	T

❷ List the possible combinations of truth values.

❸ Use the truth values of p to determine the truth values of its negation.

❹ Use the truth values for each part of the compound statement to determine the truth value of the statement.

Guided Practice

3. Construct a truth table for $\sim p \wedge \sim q$.

Personal Tutor glencoe.com

Vocabulary Link

Intersection

Everyday Use the point at which two or more objects overlap

Math Use The intersection of two sets is the set of elements that are common to both.

Venn Diagrams Conjunctions can be illustrated with Venn diagrams. Consider the conjunction given at the beginning of the lesson.

p and q: **A rectangle is a quadrilateral,** and **a rectangle is convex.**

The Venn diagram shows that a rectangle (R) is located in the *intersection* of the set of quadrilaterals and the set of convex polygons. In other words, rectangles must be in the set containing quadrilaterals *and* in the set of convex polygons.

All Polygons

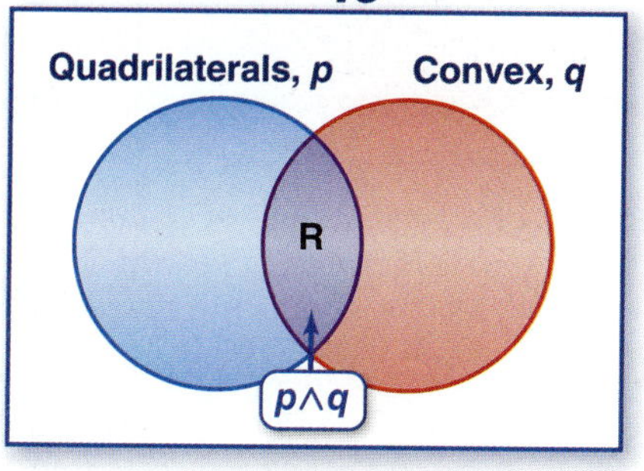

Vocabulary Link

Union
Everyday Use the joining of two or more objects
Math Use The union of two sets is the set of elements that appear in either of the sets.

A disjunction can also be illustrated with a Venn diagram. Consider the following statements.

p: A figure is a quadrilateral.
q: A figure is convex.
p or q: A figure is a quadrilateral or convex.

In the Venn diagram, the disjunction is represented by the *union* of the two sets. The union includes all polygons that are quadrilaterals, convex, *or* both.

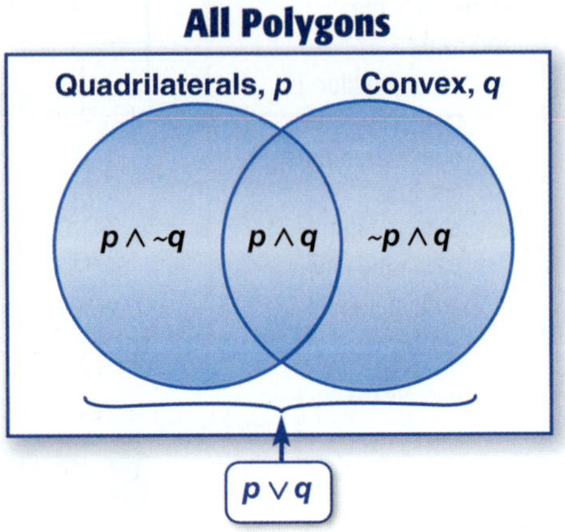

The disjunction includes these three regions:

$p \wedge {\sim}q$ quadrilaterals that are *not* convex
${\sim}p \wedge q$ convex polygons that are *not* quadrilaterals
$p \wedge q$ polygons that are both quadrilaterals and convex

Real-World EXAMPLE 4 Use Venn Diagrams

SCHEDULING The Venn diagram shows the number of people who can or cannot attend the May or the June Spanish Club meetings.

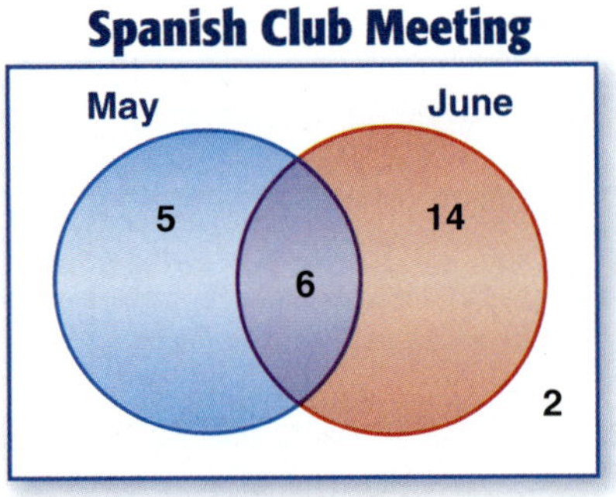

a. How many people can attend the May or the June meeting?

The people who can attend either the May meeting or the June meeting are represented by the union of the sets. There are 5 + 6 + 14 or 25 people who can attend either night.

b. How many people can attend both the May and the June meetings?

The people who can attend both the May and the June meetings are represented by the intersection of the two sets. There are 6 people who can attend both meetings.

c. Describe the meetings that the 14 people located in the nonintersecting portion of the June region can attend.

These 14 people can attend the June meeting but not the May meeting.

Guided Practice

4. PROM The Venn diagram shows the number of graduates last year who did or did not attend their junior or senior prom.

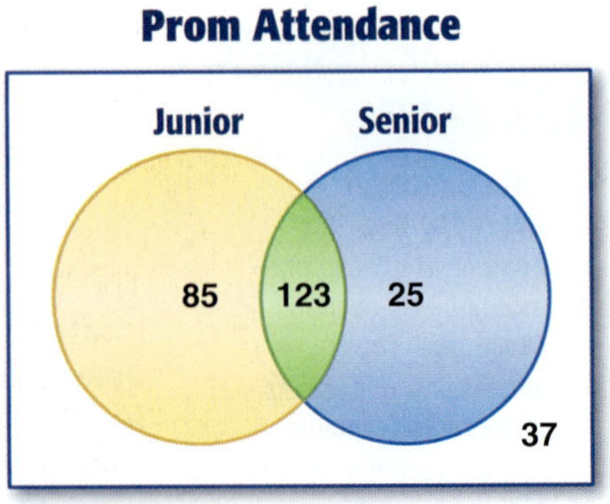

A. How many graduates attended their senior but not their junior prom?
B. How many graduates attended their junior and senior proms?
C. How many graduates did not attend either of their proms?
D. How many students graduated last year? Explain your reasoning.

Personal Tutor glencoe.com

Check Your Understanding

Examples 1 and 2 pp. 97–98

Use the following statements to write a compound statement for each conjunction or disjunction. Then find its truth value. Explain your reasoning.

p: A week has seven days.

q: There are 20 hours in a day.

r: There are 60 minutes in an hour.

1. p and r
2. $p \wedge q$
3. $q \vee r$
4. $\sim p$ or q
5. $p \vee r$
6. $\sim p \wedge \sim r$

Example 3 p. 99

7. Copy and complete the truth table at the right.

p	q	$\sim q$	$p \vee \sim q$
T	T	F	
T	F		
F	T		
F	F		

Construct a truth table for each compound statement.

8. $p \wedge q$
9. $\sim p \vee \sim q$

Example 4 p. 100

10. **CLASSES** Refer to the Venn diagram that represents the foreign language classes students selected in high school.
    a. How many students chose only Spanish?
    b. How many students chose Spanish and French?
    c. Describe the class(es) the three people in the nonintersecting portion of the French region chose.

Foreign Language Classes Selected

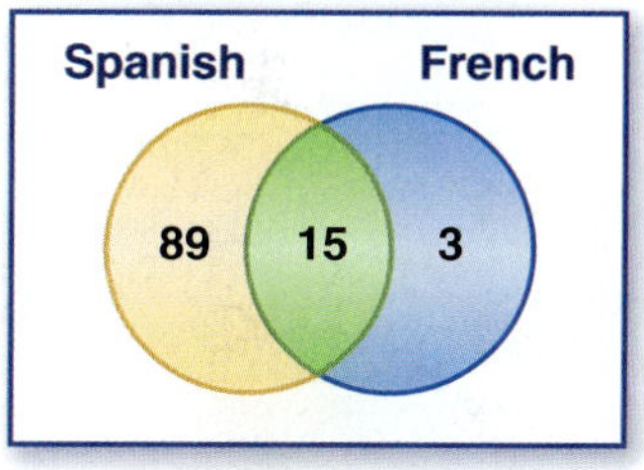

Practice and Problem Solving

● = **Step-by-Step Solutions** begin on page R20.
Extra Practice begins on page 969.

Example 1 p. 97

Use the following statements and figure to write a compound statement for each conjunction or disjunction. Then find its truth value. Explain your reasoning.

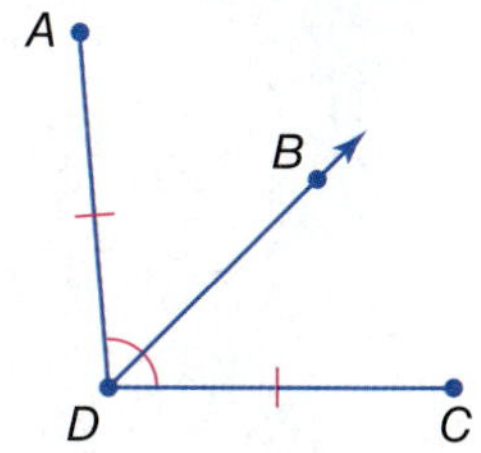

p: $\overrightarrow{DB}$ is the angle bisector of $\angle ADC$.

q: Points C, D, and B are collinear.

r: $\overline{AD} \cong \overline{DC}$

11. p and r
12. q or p
13. r or $\sim p$
14. r and q
15. $\sim p$ or $\sim r$
16. $\sim p$ and $\sim r$

Example 2 p. 98

Use the following statements to write a compound statement for each conjunction or disjunction. Then find its truth value. Explain your reasoning.

p: Springfield is the capital of Illinois.

q: Illinois borders the Atlantic Ocean.

r: Illinois shares a border with Kentucky.

s: Illinois is to the west of Missouri.

17. $p \wedge r$
18. $p \wedge q$
19. $\sim r \vee s$
20. $r \vee q$
21. $\sim p \wedge \sim r$
22. $\sim s \vee \sim p$

Example 3
p. 99

Copy and complete each truth table.

23.

p	q	$\sim p$	$\sim p \wedge q$
T		F	
T		F	
F		T	
F		T	

24.

p	q	$\sim p$	$\sim q$	$\sim p \vee \sim q$
T			F	
T			T	
F			F	
F			T	

Construct a truth table for each compound statement.

25. $p \wedge r$

26. $r \wedge q$

27. $p \vee r$

28. $q \vee r$

29. $\sim p \wedge r$

30. $\sim q \vee \sim r$

Example 4
p. 100

31. **WATER SPORTS** Refer to the Venn diagram that represents the number of students who swim and dive at a high school.

a. How many students dive?

b. How many students participate in swimming or diving or both?

c. How many students swim and dive?

Swimming and Diving

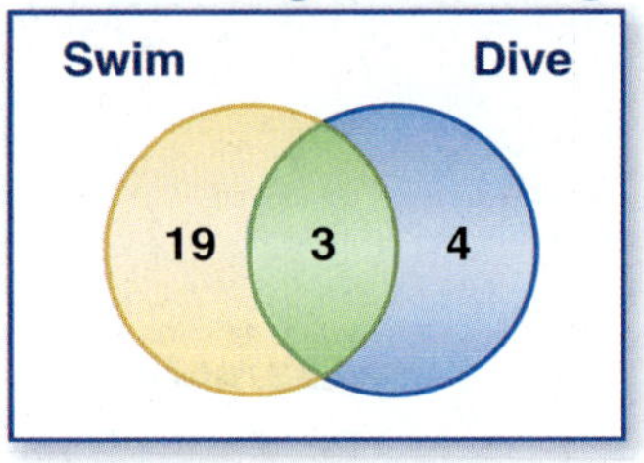

32. **ELECTRICITY** Venus has switches at the top and bottom of her stairs to control the light for the stairwell. She notices that when the upstairs switch is up and the downstairs switch is down, the light is turned on.

a. Copy and complete the truth table.

Position of Switch		Light On
Upstairs	**Downstairs**	
up		
up	down	T

b. If both the upstairs and downstairs switches are in the up position, will the light be on? Explain your reasoning.

c. If the upstairs switch is in the down position and the downstairs switch is in the up position, will the light be on?

d. In general, how should the two switches be positioned so that the light is on?

Real-World Link

About 60% of American teenagers own a cell phone and talk on them an average of an hour each day.

Source: U.S. Cellular

33. **ELECTRONICS** A group of 330 teens were surveyed about what type of electronics they used. They chose from a cell phone, an MP3 player, and a DVR. The results are shown in the Venn diagram.

a. How many teens used only an MP3 player and DVR?

b. How many said they used all three types of electronics?

c. How many said they used only a cell phone?

d. How many teens said they used only an MP3 player and a cell phone?

e. Describe the electronics that the 10 teens outside of the regions own.

Type of Electronics Used

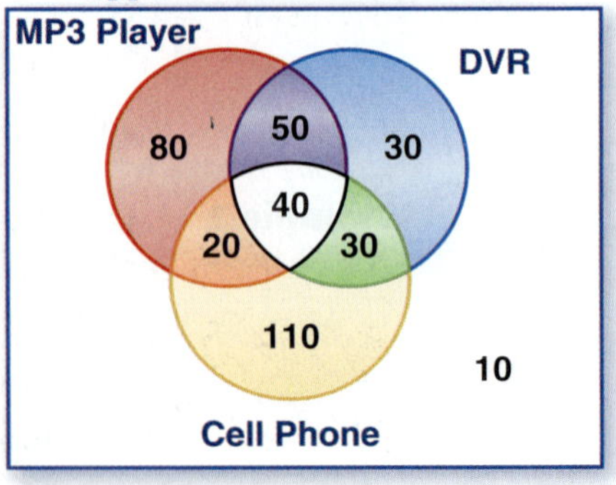

Real-World Link

In 2004, out of all the U.S. international travelers, 43% traveled to Europe. England and France were the top two international destinations visited in Europe.

Source: U.S. Dept. of Commerce

Construct a truth table for each compound statement. Determine the truth value of each compound statement if the given statements are true.

34. $p \wedge (q \wedge r)$; p, q

35 $p \wedge (\sim q \vee r)$; p, r

36. $(\sim p \vee q) \wedge r$; q, r

37. $p \wedge (\sim q \wedge \sim r)$; p, q, r

38. $\sim p \wedge (\sim q \wedge \sim r)$; p, q, r

39. $(\sim p \vee q) \vee \sim r$; p, q

40. **TOURISM** A travel agency surveyed 70 of their clients who had visited Europe about international travel. Of the 70 clients who had visited Europe, 60 had traveled to England, France, or both. Of those 60 clients, 45 had visited England, and 50 had visited France.

a. Make a Venn diagram to show the results of the survey.

b. If p represents a client who has visited England and q represents a client who has visited France, write a compound statement to represent each area of the Venn diagram. Include the compound statements on your Venn diagram.

c. What is the probability that a randomly chosen participant in the survey will have visited both England and France? Explain your reasoning.

H.O.T. Problems

Use **H**igher-**O**rder **T**hinking Skills

41. **REASONING** Irrational numbers and integers both belong to the set of real numbers (R). Based upon the Venn diagram, is it *sometimes, always,* or *never* true that integers (Z) are irrational numbers (I)? Explain your reasoning.

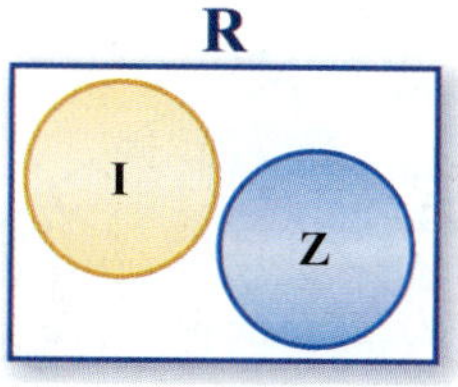

CHALLENGE To negate a statement containing the words *all* or *for every*, you can use the phrase *at least one* or *there exists*. To negate a statement containing the phase *there exists*, use the phrase *for all* or *for every*.

p: *All* polygons are convex.

$\sim p$: *At least one* polygon is *not* convex.

q: *There exists* a problem that has no solution.

$\sim q$: *For every* problem, there is a solution.

Sometimes these phrases may be implied. For example, *The square of a real number is nonnegative* implies the following conditional and its negation.

p: *For every* real number x, $x^2 \geq 0$.

$\sim p$: *There exists* a real number x such that $x^2 < 0$.

Write the negation of each statement.

42. Every student at Hammond High School has a locker.

43. All squares are rectangles.

44. There exists a real number x such that $x^2 = x$.

45. There exists a student who has at least one class in C-Wing.

46. Every real number has a real square root.

47. There exists a segment that has no midpoint.

48. **WRITING IN MATH** Describe a situation that might be depicted using the Venn diagram shown.

49. **OPEN ENDED** Write a compound statement that results in a true conjunction.

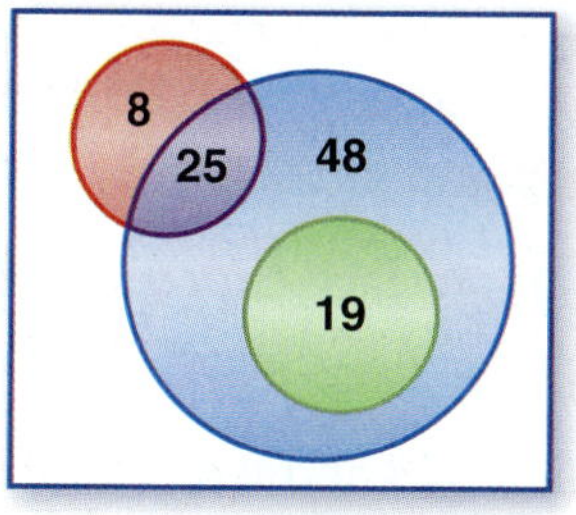

Standardized Test Practice

50. Which statement about $\triangle ABC$ has the same truth value as $AB = BC$?

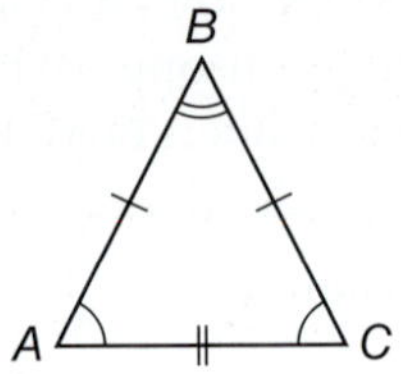

A $m\angle A = m\angle C$

B $m\angle A = m\angle B$

C $AC = BC$

D $AB = AC$

51. EXTENDED RESPONSE What is the area of the triangle shown below? Explain how you found your answer.

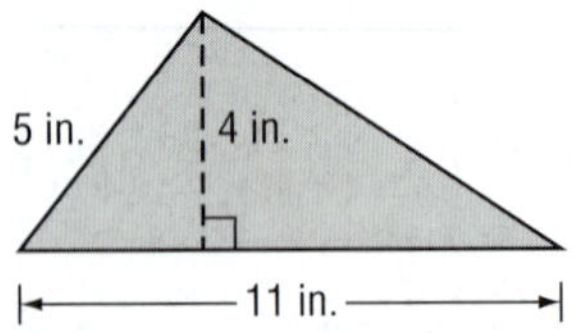

52. STATISTICS The box-and-whisker plot below represents the height of 9th graders at a certain high school. How much greater was the median height of the boys than the median height of the girls?

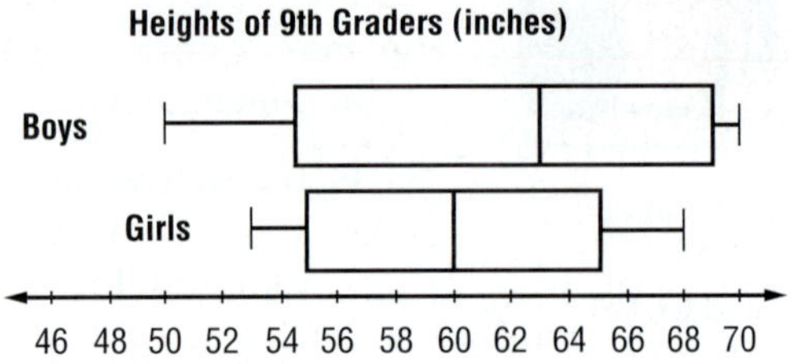

F 3 inches

G 4 inches

H 5 inches

J 6 inches

53. SAT/ACT Heather, Teresa, and Nina went shopping for new clothes. Heather spent twice as much as Teresa, and Nina spent three times what Heather spent. If they spent a total of \$300, how much did Teresa spend?

A \$33.33

B \$66.33

C \$100.00

D \$104.33

Spiral Review

54. LUNCH For the past four Tuesdays, Jason's school has served chicken sandwiches for lunch. Jason assumes that chicken sandwiches will be served for lunch on the next Tuesday. What type of reasoning did he use? Explain. (Lesson 2-1)

Identify each solid. Name the bases, faces, edges, and vertices. (Lesson 1-7)

55.

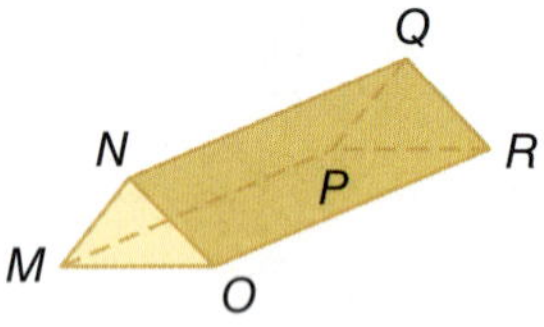

56.

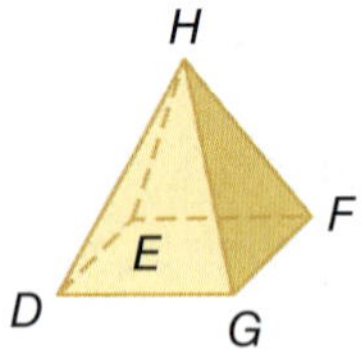

57.

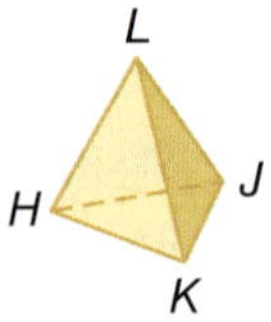

ALGEBRA Solve each equation. (Lesson 0-5)

58. $\frac{y}{2} - 7 = 5$

59. $3x + 9 = 6$

60. $4(m - 5) = 12$

61. $6(w + 7) = 0$

62. $2x - 7 = 11$

63. $\frac{y}{5} + 4 = 9$

Skills Review

ALGEBRA Evaluate each expression for the given values. (Lesson 0-4)

64. $2y + 3x$ if $y = 3$ and $x = -1$

65. $4d - c$ if $d = 4$ and $c = 2$

66. $m^2 + 7n$ if $m = 4$ and $n = -2$

67. $ab - 2a$ if $a = -2$ and $b = -3$

2-3 Conditional Statements

Then
You used logic and Venn diagrams to determine truth values of negations, conjunctions, and disjunctions. (Lesson 2-2)

Now
- Analyze statements in if-then form.
- Write the converse, inverse, and contrapositive of if-then statements.

New Vocabulary
conditional statement
if-then statement
hypothesis
conclusion
related conditionals
converse
inverse
contrapositive
logically equivalent

Math Online
glencoe.com

Why?
Call centers route calls to the appropriate departments using menus that allow callers to choose from a number of options. The recorded directions are frequently in the form of conditional statements.

If-Then Statements A **conditional statement** is a statement that can be written in *if-then form*. The direction given above is an example of a conditional statement.

If you would like to speak to a representative, **then** you will press 0 now.

Key Concept — Conditional Statement

Words	Symbols	Model
An **if-then statement** is of the form *if p, then q*.	$p \rightarrow q$ read *if p then q*, or *p implies q*	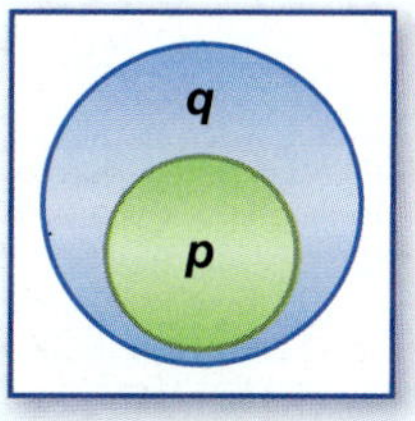 $p \rightarrow q$
The **hypothesis** of a conditional statement is the phrase immediately following the word *if*.	p	
The **conclusion** of a conditional statement is the phrase immediately following the word *then*.	q	

When a conditional statement is written as an if-then statement, you can quickly identify its hypothesis and conclusion.

EXAMPLE 1 Identify the Hypothesis and Conclusion

Identify the hypothesis and conclusion of each conditional statement.

a. If the forecast is rain, then I will take an umbrella.

Hypothesis: The forecast is rain.

Conclusion: I will take an umbrella.

b. A number is divisible by 10 if its last digit is a 0.

Hypothesis: The last digit of a number is zero.

Conclusion: The number is divisible by 10.

Guided Practice

1A. If a polygon has six sides, then it is a hexagon.

1B. Another performance will be scheduled if the first one is sold out.

Personal Tutor glencoe.com

ReadingMath

If and Then The word *if* is not part of the hypothesis. The word *then* is not part of the conclusion.

Many conditional statements are written without using the words *if* and *then*. To write these statements in if-then form, identify the hypothesis and conclusion.

Points will be deducted from any **paper turned in after Wednesday's deadline.**

Conclusion (Points will be deducted) — **Hypothesis** (paper turned in after Wednesday's deadline)

If **a paper is turned in after Wednesday's deadline**, then **points will be deducted.**

Remember, the conclusion depends upon the hypothesis.

EXAMPLE 2 Write a Conditional in If-Then Form

Identify the hypothesis and conclusion for each conditional statement. Then write the statement in if-then form.

a. A mammal is a warm-blooded animal.

Hypothesis: An animal is a mammal.

Conclusion: It is warm-blooded.

If an animal is a mammal, then it is warm-blooded.

b. A prism with bases that are regular polygons is a regular prism.

Hypothesis: A prism has bases that are regular polygons.

Conclusion: It is a regular prism.

If a prism has bases that are regular polygons, then it is a regular prism.

Guided Practice

2A. Four quarters can be exchanged for a $1 bill.

2B. The sum of the measures of two supplementary angles is 180.

Personal Tutor glencoe.com

The hypothesis and the conclusion of a conditional statement can have a truth value of true or false, as can the conditional statement itself. Consider the following conditional.

If **Tom finishes his homework**, then **he will clean his room**.

Hypothesis	Conclusion	Conditional	
Tom finishes his homework.	**Tom cleans his room.**	If **Tom finishes his homework,** then **he will clean his room.**	
T	T	T	If Tom *does* finish his homework and he *does* clean his room, then the conditional is true.
T	F	F	If Tom does *not* clean his room after he *does* finish his homework, then he has not fulfilled his promise and the conditional is false.
F	T	?	The conditional only indicates what will happen if Tom *does* finish his homework. He could clean his room or not clean his room if he does *not* finish his homework.
F	F	?	

ReadingMath

Not False If a statement is *not false*, logic dictates that it must be *true*.

When the hypothesis of a conditional is not met, the truth of a conditional cannot be determined. When the truth of a conditional statement cannot be determined, it is considered true by default.

Watch Out!

Analyzing Conditionals
When analyzing a conditional, do not try to determine whether the argument makes sense. Instead, analyze the form of the argument to determine whether the conclusion follows logically from the hypothesis.

The results from the previous page can be used to create a truth table for conditional statements.

Conditional Statements

p	q	$p \rightarrow q$
T	T	T
T	F	F
F	T	T
F	F	T

Notice that a conditional is false *only* when its hypothesis is true and its conclusion is false.

Notice too that when a hypothesis is false, the conditional will *always* be considered true, regardless of whether the conclusion is true or false.

To show that a conditional is true, you must show that for each case when the hypothesis is true, the conditional is also true. To show that a conditional is false, you need only to find one counterexample.

EXAMPLE 3 Truth Values of Conditionals

Determine the truth value of each conditional statement. If *true*, explain your reasoning. If *false*, give a counterexample.

a. If you divide an integer by another integer, the result is also an integer.

Counterexample: When you divide 1 by 2, the result is 0.5.
Since 0.5 is not an integer, the conclusion is false.
Since you can find a counterexample, the conditional statement is false.

b. If next month is August, then this month is July.

When the hypothesis is true, the conclusion is also true, since August is the month that follows July. So, the conditional statement is true.

c. If a triangle has four sides, then it is concave.

The hypothesis is false, since a triangle can never have four sides. A conditional with a false hypothesis is always true.

Guided Practice

3A. If $\angle A$ is an acute angle, then $m\angle A$ is 35. **3B.** If $\sqrt{x} = -1$, then $(-1)^2 = -1$.

Personal Tutor glencoe.com

Related Conditionals There are other statements that are based on a given conditional statement. These are known as **related conditionals**.

Key Concept Related Conditionals

Words	Symbols	Examples
A conditional statement is a statement that can be written in the form *if p, then q*.	$p \rightarrow q$	If $m\angle A$ is 35, then $\angle A$ is an acute angle.
The **converse** is formed by exchanging the hypothesis and conclusion of the conditional.	$q \rightarrow p$	If $\angle A$ is an acute angle, then $m\angle A$ is 35.
The **inverse** is formed by negating both the hypothesis and conclusion of the conditional.	$\sim p \rightarrow \sim q$	If $m\angle A$ is *not* 35, then $\angle A$ is *not* an acute angle.
The **contrapositive** is formed by negating both the hypothesis and the conclusion of the converse of the conditional.	$\sim q \rightarrow \sim p$	If $\angle A$ is *not* an acute angle, then $m\angle A$ is *not* 35.

A conditional and its contrapositive are either both true or both false. Similarly, the converse and inverse of a conditional are either both true or both false. Statements with the same truth values are said to be **logically equivalent**.

Key Concept — Logically Equivalent Statements

- A conditional and its contrapositve are logically equivalent.
- The converse and inverse of a conditional are logically equivalent.

If a conditional is true, the converse may or may not be true.

You can use logical equivalence to check the true value of statements. Notice that in Example 4, both the conditional and contrapositive are true. Also, both the converse and inverse are false.

Real-World EXAMPLE 4 — Related Conditionals

NATURE Write the converse, inverse, and contrapositive of the following true statement. Then use the information at the left to determine whether each related conditional is *true* or *false*. If a statement is false, find a counterexample.

Lions are cats that can roar.

Conditional: First, rewrite the conditional in if-then form.

If an animal is a lion, then it is a cat that can roar.

Based on the information at the left, this statement is true.

Converse: If an animal is a cat that can roar, then it is a lion.

Counterexample: A tiger is a cat that can roar, but it is not a lion.

Therefore, the converse is false.

Inverse: If an animal is not a lion, then it is not a cat that can roar.

Counterexample: A tiger is not a lion, but it is a cat that can roar.

Therefore, the inverse is false.

Contrapositive: If an animal is not a cat that can roar, then it is not a lion.

Based on the information at the left, this statement is true.

CHECK Check to see that logically equivalent statements have the same truth value.

Both the conditional and contrapositive are true. ✓

Both the converse and inverse are false. ✓

Real-World Link

Cats in the genus *Panthera* include the leopard, jaguar, lion, and tiger. These are the only cats that can roar. They cannot, however, purr.

Source: *Encyclopaedia Britannica*

Guided Practice

Write the converse, inverse, and contrapositive of each true conditional statement. Determine whether each related conditional is *true* or *false*. If a statement is false, find a counterexample.

4A. Two angles that have the same measure are congruent.

4B. A hamster is a rodent.

Personal Tutor glencoe.com

Check Your Understanding

Example 1 p. 105 — **Identify the hypothesis and conclusion of each conditional statement.**

1. If today is Friday, then tomorrow is Saturday.
2. If $2x + 5 > 7$, then $x > 1$.
3. If two angles are supplementary, then the sum of the measures of the angles is 180.
4. If two lines form right angles, then the lines are perpendicular.

Example 2 p. 106 — **Write each statement in if-then form.**

5. Sixteen-year-olds are eligible to drive.
6. Cheese contains calcium.
7. The measure of an acute angle is between 0 and 90.
8. Equilateral triangles are equiangular.
9. **WEATHER** Various kinds of precipitation form under different conditions. Write the three conditionals below in if-then form.
   a. Moisture in the air condenses and falls to form rain.
   b. Supercooled moisture in cumulonimbus clouds forms hail.
   c. When the temperature is freezing in all or most of the atmosphere, precipitation falls as snow.

Example 3 p. 107 — **Determine the truth value of each conditional statement. If *true*, explain your reasoning. If *false*, give a counterexample.**

10. If $x^2 = 16$, then $x = 4$.
11. If you live in Charlotte, then you live in North Carolina.
12. If tomorrow is Friday, then today is Thursday.
13. If an animal is spotted, then it is a Dalmatian.
14. If the measure of a right angle is 95, then bees are lizards.
15. If pigs can fly, then $2 + 5 = 7$.

Example 4 p. 108 — **Write the converse, inverse, and contrapositive of each true conditional statement. Determine whether each related conditional is *true* or *false*. If a statement is false, find a counterexample.**

16. If a number is divisible by 2, then it is divisible by 4.
17. All whole numbers are integers.

Practice and Problem Solving

● = **Step-by-Step Solutions** begin on page R20.
Extra Practice begins on page 969.

Example 1 p. 105 — **Identify the hypothesis and conclusion of each conditional statement.**

18. If two angles are adjacent, then they have a common side.
19. If you lead, then I will follow.
20. If $3x - 4 = 11$, then $x = 5$.
21. If two angles are vertical, then they are congruent.

Identify the hypothesis and conclusion of each conditional statement.

22. If the degree measure of an angle is between 90 and 180, then the angle is obtuse.

23. "If there is no struggle, there is no progress." (Frederick Douglass)

24. If a quadrilateral has four congruent sides, then it is a square.

25. If a convex polygon has five sides, then it is a pentagon.

Example 2
p. 106

Write each statement in if-then form.

26. Get a free water bottle with a one-year membership.

27. Everybody at the party received a gift.

28. The intersection of two planes is a line.

29. The area of a circle is πr^2.

30. Collinear points lie on the same line.

31. A right angle measures 90 degrees.

32. MUSIC Different instruments are emphasized in different types of music. Write each statement in if-then form.

- Jazz music often incorporates trumpet or saxophone.
- Rock music emphasizes guitar and drums.
- In hip-hop music, the bass is featured.

Real-World Link

The Andy Warhol Museum has over 4000 works of art including prints, paintings, films, photographs, and sculpture. About 500 of these pieces are on display at a time. The museum also has works by colleagues of Andy Warhol.

Source: The Andy Warhol Museum

33. ART Write the following statement in if-then form: At the Andy Warhol Museum in Pittsburgh, Pennsylvania, most of the collection is Andy Warhol's artwork.

34. SCIENCE The water on Earth is constantly changing through a process called the *water cycle*. Write the three conditionals below in if-then form.

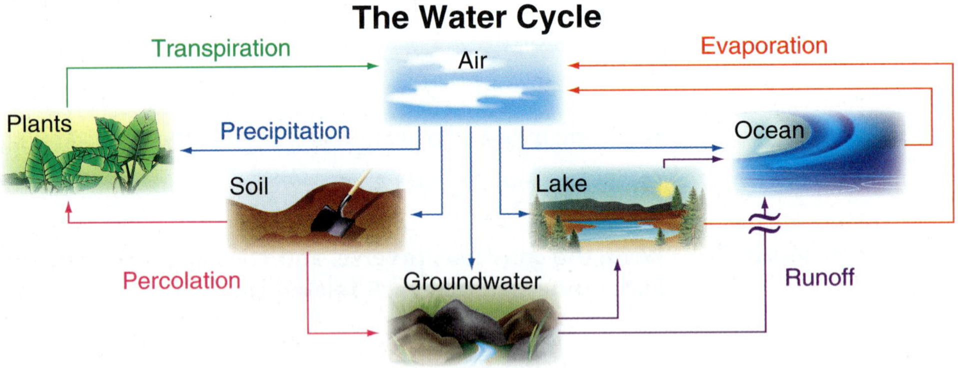

a. As runoff, water flows into bodies of water.

b. Plants return water to the air through transpiration.

c. Water bodies return water to the air through evaporation.

Example 3
p. 107

Determine the truth value of each conditional statement. If *true*, explain your reasoning. If *false*, give a counterexample.

35 If a number is odd, then it is divisible by 5.

36. If a dog is an amphibian, then the season is summer.

37. If an angle is acute, then it has a measure of 45.

38. If a polygon has six sides, then it is a regular polygon.

Determine the truth value of each conditional statement. If *true*, explain your reasoning. If *false*, give a counterexample.

39. If an angle's measure is 25, then the measure of the angle's complement is 65.

40. If North Carolina is south of Florida, then the capital of Ohio is Columbus.

41. If red paint and blue paint mixed together make white paint, then $3 - 2 = 0$.

42. If two angles are congruent, then they are vertical angles.

43. If an animal is a bird, then it is an eagle.

44. If two angles are acute, then they are supplementary.

45. If two lines intersect, then they form right angles.

46. If a banana is blue, then an apple is a vegetable.

Example 4
p. 108

Write the converse, inverse, and contrapositive of each true conditional statement. Determine whether each related conditional is *true* or *false*. If a statement is false, find a counterexample.

47. If you live in Chicago, you live in Illinois.

48. If a bird is an ostrich, then it cannot fly.

49. If two angles have the same measure, then the angles are congruent.

50. All squares are rectangles.

51. All congruent segments have the same length.

52. A right triangle has an angle measure of 90.

Real-World Link

Duikers are small antelopes that live in Africa. They can range from about one foot tall to a little over two feet tall. The duiker shown in the photo is a banded duiker.

Source: *Encyclopaedia Britannica*

NATURE Write the statement indicated, and use the information at the left to determine the truth value of each statement. If a statement is false, give a counterexample.

Animals with stripes are zebras.

53. conditional

54. converse

55 inverse

56. contrapositive

57. **SCIENCE** Chemical compounds are grouped and described by the elements that they contain. Acids contain hydrogen (H). Bases contain hydroxide (OH). Hydrocarbons contain only hydrogen (H) and carbon (C).

Compound	Example	Chemical Formula
Acid	Hydrochloric Acid	HCl
Base	Sodium Hydroxide	NaOH
Hydrocarbon	Methane	CH_4

a. Write three conditional statements in if-then form for classifying chemical compounds.

b. Write the converse of the three true conditional statements. State whether each is *true* or *false*. If a statement is false, find a counterexample.

58. **SPORTS** In football, touchdowns are worth 6 points, extra point conversions are worth 2 points, and safeties are worth 2 points.

a. Write three conditional statements in if-then form for scoring in football.

b. Write the converse of the three true conditional statements. State whether each is *true* or *false*. If a statement is false, find a counterexample.

Use the Venn diagrams below to determine the truth value of each conditional. Explain your reasoning.

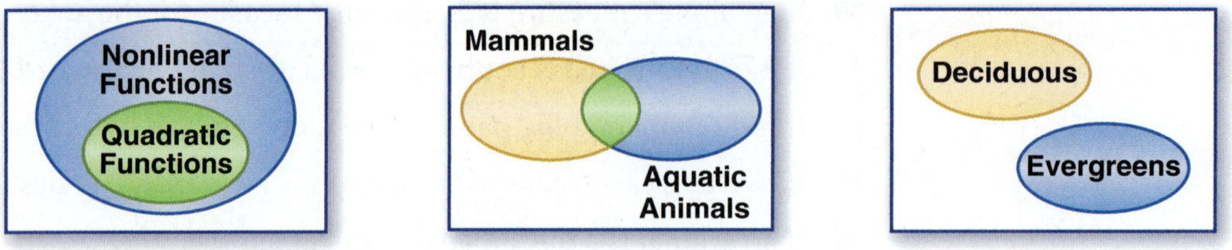

59. If a function is nonlinear, then it is quadratic.

60. If an animal is a mammal, then it cannot be aquatic.

61. If a tree is deciduous, then it is not an evergreen.

62. **MULTIPLE REPRESENTATIONS** In this problem, you will investigate a law of logic by using conditionals.

a. **LOGICAL** Write three true conditional statements, using each consecutive conclusion as the hypothesis for the next statement.

b. **GRAPHICAL** Create a Venn diagram to model your series of statements.

c. **LOGICAL** Write a conditional using the hypothesis of your first conditional and the conclusion of your third conditional. Is the conditional true if the hypothesis is true?

d. **VERBAL** Given two conditionals *If a, then b* and *If b, then c,* make a conjecture about the truth value of c when a is true. Explain your reasoning.

StudyTip

Venn Diagrams The enclosing rectangle of a Venn diagram represents the *universe*. A Venn diagram with 2 circular sets can divide this universe into at most 4 regions. One with 3 circular sets divides this universe into at most 8 regions. It can be shown that a Venn diagram with n circular sets divides this universe into at most 2^n regions.

H.O.T. Problems Use Higher-Order Thinking Skills

63. **ERROR ANALYSIS** Nicole and Kiri are evaluating the conditional *If 15 is a prime number, then 20 is divisible by 4.* Both think that the conditional is true, but their reasoning differs. Is either of them correct? Explain.

Nicole	Kiri
The conclusion is true, because 20 is divisible by 4, so the conditional is true.	*The hypothesis is false, because 15 is not a prime number, so the conditional is true.*

64. **CHALLENGE** You have learned that statements with the same truth value are logically equivalent. Use logical equivalence to create a truth table that summarizes the conditional, converse, inverse, and contrapositive for the statements p and q.

65. **REASONING** You are evaluating a conditional statement in which the hypothesis is true, but the conclusion is false. Is the inverse of the statement true or false? Explain your reasoning.

66. **OPEN ENDED** Write a conditional statement in which the converse, inverse, and contrapositive are all true. Explain your reasoning.

67. **CHALLENGE** The inverse of conditional A is given below. Write conditional A, its converse, and its contrapositive. Explain your reasoning.

If I received a detention, then I did not arrive at school on time.

68. **WRITING IN MATH** Describe the relationship between a conditional, its converse, its inverse, and its contrapositive.

Standardized Test Practice

69. *If the sum of the measures of two angles is 90, then the angles are complementary angles.*

Which of the following is the converse of the conditional above?

A If the angles are complementary angles, then the sum of the measures of two angles is 90.

B If the angles are not complementary angles, then the sum of the measures of the angles is 90.

C If the angles are complementary angles, then the sum of the measures of the angles is not 90.

D If the angles are not complementary angles, then the sum of the measures of two angles is not 90.

70. ALGEBRA What is $\frac{10a^2 - 15ab}{4a^2 - 9b^2}$ reduced to lowest terms?

F $\frac{5a}{2a - 2b}$

G $\frac{5a}{2a + 3b}$

H $\frac{a}{2a + 3b}$

J $\frac{a}{2a - 3b}$

71. SHORT RESPONSE What is the standard notation for the following expression?

$$4.62 \times 10^{-3}$$

72. SAT/ACT What is the greatest common prime factor of 18 and 33?

A 1

B 2

C 3

D 11

Spiral Review

Construct a truth table for each compound statement. (Lesson 2-2)

73. p and q

74. p or $\sim q$

75. $\sim p \wedge q$

76. $\sim p \wedge \sim q$

Make a conjecture based on the given information. Draw a figure to illustrate your conjecture. (Lesson 2-1)

77. Points H, J, and K are each located on different sides of a triangle.

78. Collinear points X, Y, and Z; Z is between X and Y.

79. $R(3, -4)$, $S(-2, -4)$, and $T(0, -4)$

80. $A(-1, -7)$, $B(4, -7)$, $C(4, -3)$, and $D(-1, -3)$

81. KITES Kite making has become an art form. The kite shown is known as a diamond kite. The measures are in inches. Name all of the congruent segments in the figure. (Lesson 1-2)

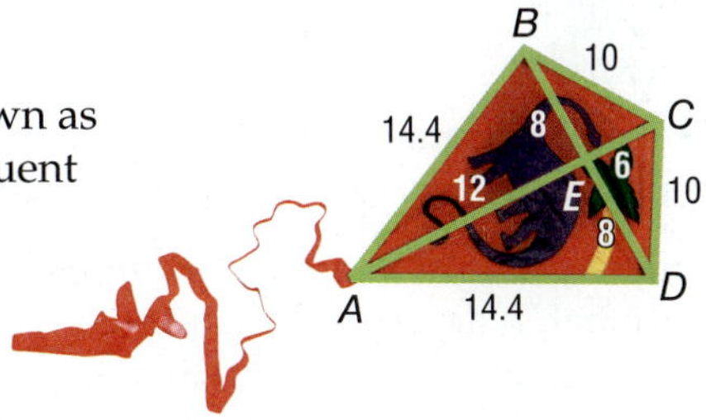

Refer to the conversion charts inside the back cover of your textbook and in Lesson 0-2. (Lesson 0-2)

82. RUNNING Ling is participating in a 5-kilometer charity run next weekend. About how many miles is the race?

83. NATURE An African elephant weighs about 9 tons. About how many kilograms is this?

84. SPORTS A football field is 100 yards long from one end zone to the other. How many feet long is a football field?

Skills Review

ALGEBRA Identify the operation used to change Equation (1) to Equation (2).

85. (1) $8(y - 11) = 32$

(2) $y - 11 = 4$

86. (1) $x + 9 = 4 - 3x$

(2) $4x + 9 = 4$

87. (1) $\frac{1}{3}m = 2$

(2) $m = 6$

EXTEND **2-3**

Geometry Lab
Biconditional Statements

Samantha is the starting pitcher for her high school softball team. If she is elected by the district coaches, she will be named to the All-Star Team. If named to the All-Star Team, she has been elected by the district coaches.

p: Samantha is elected by the district coaches
q: Samantha is named to the All-Star Team

$p \rightarrow q$: If Samantha is elected by the district coaches, then she is named to the All-Star Team.

$q \rightarrow p$: If Samantha is named to the All-Star Team, then she was elected by the district coaches.

In this case, both the conditional and its converse are true. The conjunction of the two statements is called a **biconditional**.

Key Concept — Biconditional Statement

Words A biconditional statement is the conjunction of a conditional and its converse.

Symbols $(p \rightarrow q) \wedge (q \rightarrow p) \rightarrow (p \leftrightarrow q)$, read *p if and only if q*

If and only if **can be abbreviated *iff*.**

So, the biconditional statement is as follows.

$p \leftrightarrow q$: Samantha is named to the All-Star Team if and only if she is elected by the district coaches.

EXAMPLES

Write each biconditional as a conditional and its converse. Then determine whether the biconditional is *true* or *false*. If false, give a counterexample.

a. An angle is a right angle if and only if its measure is 90.
Conditional: If an angle measures 90, then the angle is right.
Converse: If an angle is right, then the angle measures 90.
Both the conditional and the converse are true, so the biconditional is true.

b. $x > -2$ iff x is positive.
Conditional: If x is positive, then $x > -2$.
Converse: If $x > -2$, then x is positive.
Let $x = -1$. Then $-1 > -2$, but -1 is not positive. So, the biconditional is false.

Exercises

Write each biconditional as a conditional and its converse. Then determine whether the biconditional is *true* or *false*. If false, give a counterexample.

1. Two angles are complements if and only if their measures have a sum of 90.
2. There is no school if and only if it is Saturday.
3. Two lines intersect if and only if they are not horizontal.
4. $|2x| = 4$ iff $x = 2$.

2-4 Deductive Reasoning

Then

You used inductive reasoning to analyze patterns and make conjectures. (Lesson 2-1)

Now

- Use the Law of Detachment.
- Use the Law of Syllogism.

New Vocabulary

deductive reasoning
valid
Law of Detachment
Law of Syllogism

Math Online
glencoe.com

Why?

When detectives are trying to solve a case, they use techniques like fingerprinting to analyze evidence. Then they use this evidence to eliminate suspects and eventually identify the person responsible for the crime.

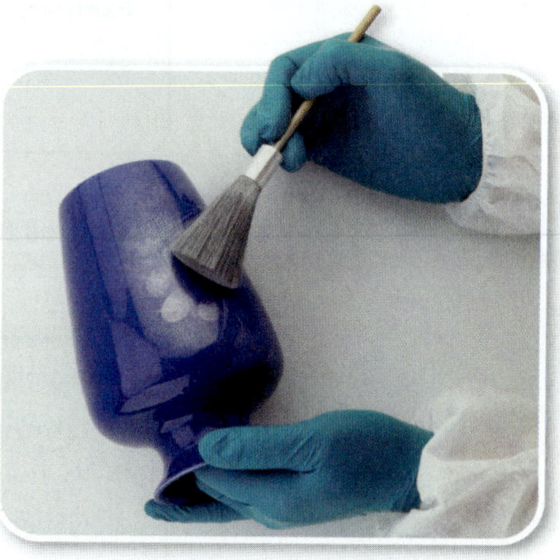

Law of Detachment The process that detectives use to identify who is most likely responsible for a crime is called deductive reasoning. Unlike inductive reasoning, which uses a pattern of examples or observations to make a conjecture, **deductive reasoning** uses facts, rules, definitions, or properties to reach logical conclusions from given statements.

Real-World EXAMPLE 1 Inductive and Deductive Reasoning

Determine whether each conclusion is based on *inductive* or *deductive* reasoning.

a. Every time Katie has worn her favorite socks to a softball game, she has gotten at least one hit. Katie is wearing her favorite socks to a game tonight, so she concludes that she will get at least one hit.

Katie is basing her conclusion on a pattern of observations, so she is using inductive reasoning.

b. If John is late making his car insurance payment, he will be assessed a late fee of $50. John's payment is late this month, so he concludes that he will be assessed a late fee of $50.

John is basing his conclusion on facts provided to him by his insurance company, so he is using deductive reasoning.

Guided Practice

1A. All of the signature items on the restaurant's menu shown are noted with a special symbol. Kevin orders a menu item that has this symbol next to it, so he concludes that the menu item that he has ordered is a signature item.

1B. None of the students who ride Raul's bus own a car. Ebony rides a bus to school, so Raul concludes that Ebony does not own a car.

Personal Tutor glencoe.com

While one counterexample is enough to disprove a conjecture reached using inductive reasoning, it is not a logically correct, or **valid**, method of proving a conjecture. To prove a conjecture requires deductive reasoning. One valid form of deductive reasoning is the **Law of Detachment**.

Key Concept — Law of Detachment

Words If $p \rightarrow q$ is a true statement and p is true, then q is true.

Example *Given:* If **a car is out of gas**, then **it will not start**.
Sarah's **car is out of gas**.

Valid Conclusion: Sarah's **car will not start**.

As long as the facts given are true, the conclusion reached using deductive reasoning will also be true.

EXAMPLE 2 Use the Law of Detachment

Determine whether each conclusion is valid based on the given information. If not, write *invalid*. Explain your reasoning.

a. Given: If two angles form a linear pair, then their noncommon sides are opposite rays.
$\angle AED$ and $\angle AEB$ form a linear pair.

Conclusion: $\overrightarrow{ED}$ and $\overrightarrow{EB}$ are opposite rays.

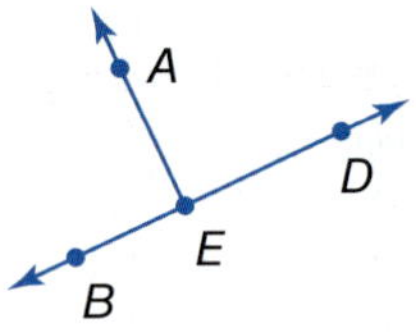

Step 1 Identify the hypothesis p and the conclusion q of the true conditional.

p: **Two angles form a linear pair.**

q: **Their noncommon sides are opposite rays.**

Step 2 Analyze the conclusion.

The given statement *$\angle AED$ and $\angle AEB$ form a linear pair* satisfies the hypothesis, so p is true. By the Law of Detachment, *$\overrightarrow{ED}$ and $\overrightarrow{EB}$ are opposite rays*, which matches q, is a true or valid conclusion.

b. Given: If Mika goes to the beach, she will wear sunscreen.
Mika is wearing sunscreen.

Conclusion: Mika is at the beach.

Step 1 p: **Mika goes to the beach.**

q: **Mika wears sunscreen.**

Step 2 The given statement *Mika is wearing sunscreen* satisfies the conclusion q of the true conditional. However, knowing that a conditional statement and its conclusion are true does not make the hypothesis true. Mika could be wearing sunscreen because she is at the pool. The conclusion is invalid.

StudyTip

Given Information From this point forward in this text, all given information can be assumed true.

Guided Practice

2A. Given: If three points are noncollinear, they determine a plane.
Points A, B, and C lie in plane G.

Conclusion: Points A, B, and C are noncollinear.

2B. Given: If a student turns in a permission slip, then the student can go on the field trip.
Felipe turned in his permission slip.

Conclusion: Felipe can go on the field trip.

Personal Tutor glencoe.com

You can also use a Venn diagram to test the validity of a conclusion.

Real-World Link

The easiest way to distinguish monkeys from other primates is to look for a tail. Most monkey species have tails, but apes do not.

Source: *Encyclopaedia Britannica*

EXAMPLE 3 Judge Conclusions Using Venn Diagrams

NATURE Determine whether each conclusion is valid based on the given information. If not, write *invalid*. Explain your reasoning using a Venn diagram.

Given: **If a primate is an ape, then it does not have a tail.**

Koko is a primate who does not have a tail.

Conclusion: **Koko is an ape.**

Understand Draw a Venn diagram. According to the conditional, an ape does not have a tail, so draw a circle for apes that does not intersect the circle for primates with tails.

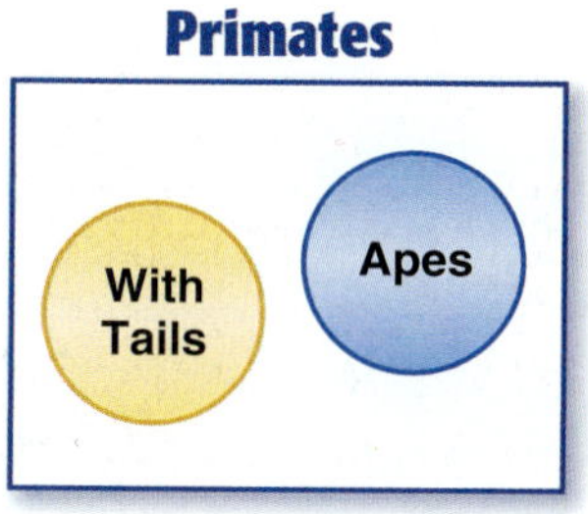

Plan Since we are only given that Koko does not have a tail, we can only conclude that Koko belongs outside the circle for primates with tails.

Solve This could put her in the area inside or outside of the Apes circle, so the conclusion is invalid.

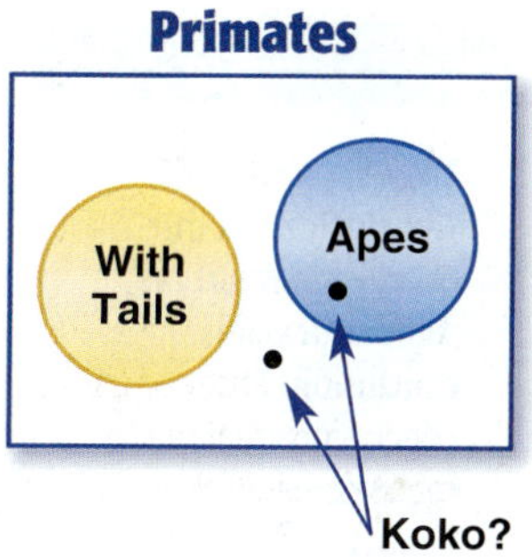

Check From the given information, we know that apes are primates that do not have tails. We also know that Koko is a primate that does not have a tail. It is possible for Koko to be a primate without a tail and still not be an ape. Therefore, the conclusion *is* invalid. ✔

Guided Practice

3. Given: If a figure is a square, then it is a polygon.
Figure A is a square.

Conclusion: Figure A is a polygon.

Personal Tutor glencoe.com

StudyTip

A logical argument such as the one shown is supported by the rules of logic. This is different from a statistical argument, which is supported by examples or data.

Law of Syllogism The **Law of Syllogism** is another valid form of deductive reasoning. This law allows you to draw conclusions from two true conditional statements when the conclusion of one statement is the hypothesis of the other.

Key Concept Law of Syllogism

Words If $p \to q$ and $q \to r$ are true statements, then $p \to r$ is a true statement.

Example *Given:* If **you get a job**, then **you will earn money**.
If **you earn money**, then **you will buy a car**.

Valid Conclusion: If **you get a job**, then **you will buy a car**.

It is important to remember that if the conclusion of the first statement is *not* the hypothesis of the second statement, no valid conclusion can be drawn.

STANDARDIZED TEST EXAMPLE 4

Determine which statement follows logically from the given statements.

(1) If you like musicals, then you enjoy theater productions.
(2) If you are an actor, then you enjoy theater productions.

A If you are an actor, then you like musicals.
B If you like musicals, then you are an actor.
C If you do not enjoy musicals, then you are not an actor.
D There is no valid conclusion.

Read the Test Item

Let p, q, and r represent the parts of the given conditional statements.

p: You like musicals.
q: You enjoy theater productions.
r: You are an actor.

Solve the Test Item

Analyze the logic of the given conditional statement using symbols.

Statement (1): $p \rightarrow q$

Statement (2): $r \rightarrow q$

Both statements are considered true. However, the Law of Syllogism does not apply since q, the conclusion of the Statement (1), is not the hypothesis of the second statement. While choices A, B, and C may be true, the logic used to draw these conclusions is not valid. Therefore, choice D is correct.

> **Test-Taking Tip**
>
> **True vs. Valid Conclusions** A true conclusion is not the same as a valid conclusion. True conclusions that are reached using invalid deductive reasoning are still invalid.

Guided Practice

4. Determine which statement follows logically from the given statements.
(1) If you do not get enough sleep, then you will be tired.
(2) If you are tired, then you will not do well on the test.

F If you are tired, then you will not get enough sleep.
G If you do not get enough sleep, then you will not do well on the test.
H If you do not do well on the test, then you did not get enough sleep.
J There is no valid conclusion.

Personal Tutor glencoe.com

EXAMPLE 5 Apply Laws of Deductive Reasoning

Draw a valid conclusion from the given statements, if possible. Then state whether your conclusion was drawn using the Law of Detachment or the Law of Syllogism. If no valid conclusion can be drawn, write *no valid conclusion* and explain your reasoning.

Given: If you are 16 years old, then you can apply for a driver's license. Nate is 16 years old.

p: You are 16 years old.
q: You can apply for a driver's license.

Since *Nate is 16 years old* satisfies the hypothesis p is true. By the Law of Detachment, a valid conclusion is *Nate can apply for a driver's license.*

Guided Practice

5. Given: The midpoint divides a segment into two congruent segments. If two segments are congruent, then their measures are equal. M is the midpoint of $\overline{AB}$.

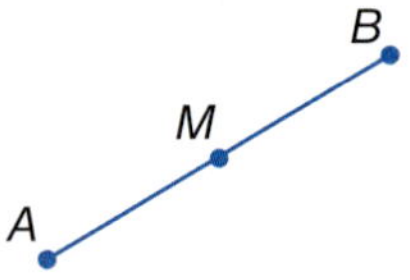

Personal Tutor glencoe.com

Check Your Understanding

Example 1 p. 115

Determine whether each conclusion is based on *inductive* or *deductive* reasoning.

1. Students at Olivia's high school must have a B average in order to participate in sports. Olivia has a B average, so she concludes that she can participate in sports at school.

2. Holly notices that every Saturday, her neighbor mows his lawn. Today is Saturday. Holly concludes her neighbor will mow his lawn.

Example 2 p. 116

Determine whether the stated conclusion is valid based on the given information. If not, write *invalid*. Explain your reasoning.

3. **Given:** If a number is divisible by 4, then the number is divisible by 2. 12 is divisible by four.

   Conclusion: 12 is divisible by 2.

4. **Given:** If Elan stays up late, he will be tired the next day. Elan is tired.

   Conclusion: Elan stayed up late.

Example 3 p. 117

Determine whether the stated conclusion is valid based on the given information. If not, write *invalid*. Explain your reasoning using a Venn diagram.

5. **Given:** If a beach is public, then it does not have a lifeguard. Bayview does not have a lifeguard.

   Conclusion: Bayview is a public beach.

6. **Given:** If students pass an entrance exam, they will be accepted into college. Latisha passed the entrance exam.

   Conclusion: Latisha will be accepted into college.

Example 4 p. 118

7. **MULTIPLE CHOICE** Determine which statement follows logically from the given statements.

   (1) If a triangle is a right triangle, then it has an angle that measures 90.

   (2) If a triangle has an angle that measures 90, then its acute angles are complementary.

   A If a triangle is not a right triangle, then it has an angle that measures 90.

   B If an angle of a triangle measures 90, then its acute angles are not complementary.

   C If a triangle is a right triangle, then its acute angles are complementary.

   D If a triangle has an angle that measures 90, then it is not a right triangle.

Example 5 p. 118

Draw a valid conclusion from the given statements, if possible. Then state whether your conclusion was drawn using the Law of Detachment or the Law of Syllogism. If no valid conclusion can be drawn, write *no valid conclusion* and explain your reasoning.

8. **Given:** If Dalila finishes her chores, she will receive her allowance.

   If Dalila receives her allowance, she will buy a CD.

9. **Given:** Vertical angles are congruent.

   $\angle 1 \cong \angle 2$

Practice and Problem Solving

● = **Step-by-Step Solutions** begin on page R20.
Extra Practice begins on page 969.

Example 1 p. 115

Determine whether each conclusion is based on *inductive* or *deductive* reasoning.

10. At Fumio's school if you are late five times, you will receive a detention. Fumio has been late to school five times; therefore he will receive a detention.

11. A dental assistant notices a patient has never been on time for an appointment. She concludes the patient will be late for her next appointment.

12. A person must have a membership to work out at a gym. Jesse is working out at a gym. Jesse has a membership to the gym.

13. If Eduardo decides to go to a concert tonight, he will miss football practice. Tonight, Eduardo went to a concert. Eduardo missed football practice.

14. Every Wednesday Lucy's mother calls. Today is Wednesday, so Lucy concludes her mother will call.

15. Whenever Juanita has attended a tutoring session she notices that her grades have improved. Juanita attends a tutoring session and she concludes her grades will improve.

Example 2 p. 116

Determine whether the stated conclusion is valid based on the given information. If not, write *invalid*. Explain your reasoning.

16. Given: Right angles are congruent. $\angle 1$ and $\angle 2$ are right angles.

Conclusion: $\angle 1 \cong \angle 2$

17. Given: If a figure is a square, it has four right angles. Figure $ABCD$ has four right angles.

Conclusion: Figure $ABCD$ is a square.

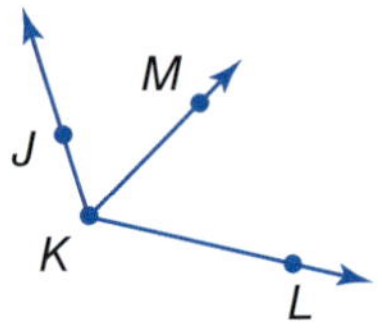

18. Given: An angle bisector divides an angle into two congruent angles. $\overrightarrow{KM}$ is an angle bisector of $\angle JKL$.

Conclusion: $\angle JKM \cong \angle MKL$

19. Given: If you leave your lights on while your car is off, your battery will die. Your battery is dead.

Conclusion: You left your lights on while the car was off.

20. Given: If Dante obtains a part-time job, he can afford a car payment. Dante can afford a car payment.

Conclusion: Dante obtained a part-time job.

21. Given: If 75% of the prom tickets are sold, the prom will be held at the country club. 75% of the prom tickets were sold.

Conclusion: The prom will be held at the country club.

22. COMPUTER GAMES One of the tasks of the Entertainment Software Rating Board, a nonprofit organization, is to assign ratings for interactive entertainment software. Determine whether the stated conclusion is valid based on the given information. If not, write *invalid*. Explain your reasoning.

Given: If a title is rated E, then it has content that may be suitable for ages 6 and older. Cesar buys a computer game that he believes is suitable for his little sister, who is 7.

Conclusion: The game Cesar purchased has a rating of E.

EVERYONE ® E CONTENT RATED BY ESRB

Real-World Link

Possible ratings assigned to computer games include:

- **EC (Early Childhood)** for ages 3 and older,
- **E (Everyone)** for ages 6 and older,
- **E10+ (Everyone 10 and older)**,
- **T (Teen)** for ages 13 and older, and
- **M (Mature)** for ages 17 and older.

Source: Entertainment Software Rating Board

Example 3
p. 117

Determine whether the stated conclusion is valid based on the given information. If not, write *invalid*. Explain your reasoning using a Venn diagram.

23. **Given:** If the temperature drops below 32°F, it may snow. The temperature did not drop below 32°F on Monday.

Conclusion: It did not snow on Monday.

24. **Given:** If a person is a Missouri resident, he or she does not live by a beach. Michelle does not live by the beach.

Conclusion: Michelle is a Missouri resident.

25. **Given:** Some nurses wear blue uniforms. Sabrina is a nurse.

Conclusion: Sabrina wears a blue uniform.

26. **Given:** All vegetarians do not eat meat. Theo is a vegetarian.

Conclusion: Theo does not eat meat.

27 **TRANSPORTATION** There are many types of vehicles and they are classified using different sets of criteria. Determine whether the stated conclusion is valid based on the given information. If not, write *invalid*. Explain your reasoning using a Venn diagram.

Given: If a vehicle is a sport-utility vehicle, then it is a four-wheel-drive car built on a truck chassis. Ms. Rodriguez has just purchased a vehicle that has four-wheel drive.

Conclusion: Ms. Rodriguez has just purchased a sport-utility vehicle.

Examples 4 and 5
p. 118

Real-World Link

Jack Nicklaus holds the record for the most Masters Tournament wins at 6, followed by Tiger Woods, who has won 4 as of 2007.

Source: The Masters

28. **GOLF** Zach Johnson won the Masters Tournament in 2007. Use the Law of Syllogism to draw a valid conclusion from each set of statements, if possible. If no valid conclusion can be drawn, write *no valid conclusion* and explain your reasoning.

(1) If Zach Johnson's score is lower than the other golfers at the end of the tournament, then he wins the tournament.

(2) If a golfer wins the Masters Tournament, then he gets a green jacket.

Use the Law of Syllogism to draw a valid conclusion from each set of statements, if possible. If no valid conclusion can be drawn, write *no valid conclusion* and explain your reasoning.

29. If you interview for a job, then you wear a suit.
If you interview for a job, then you will update your resume.

30. If Tina has a grade point average of 3.0 or greater, she will be on the honor roll.
If Tina is on the honor roll, then she will have her name in the school paper.

31. If two lines are perpendicular, then they intersect to form right angles.
Lines r and s form right angles.

32. If the measure of an angle is between 90 and 180, then it is obtuse.
If an angle is obtuse, then it is not acute.

33. If two lines in a plane are not parallel, then they intersect.
If two lines intersect, then they intersect in a point.

34. If a number ends in 0, then it is divisible by 2.
If a number ends in 4, then it is divisible by 2.

Draw a valid conclusion from the given statements, if possible. Then state whether your conclusion was drawn using the Law of Detachment or the Law of Syllogism. If no valid conclusion can be drawn, write *no valid conclusion* and explain your reasoning.

35. Given: If a figure is a square, then all the sides are congruent.
Figure $ABCD$ is a square.

36. Given: If two angles are complementary, the sum of the measures of the angles is 90.
$\angle 1$ and $\angle 2$ are complements of each other.

37. Given: Ballet dancers like classical music.
If you like classical music, then you enjoy the opera.

38. Given: If you are athletic, then you enjoy sports.
If you are competitive, then you enjoy sports.

39. Given: If a polygon is regular, then all of its sides are congruent.
All sides of polygon $WXYZ$ are congruent.

40. Given: If Bob completes a course with a grade of C, then he will not receive credit.
If Bob does not receive credit, he will have to take the course again.

Review Vocabulary

scatter plot a graph that represents two sets of data as ordered pairs in the coordinate plane used to investigate the relationship between two quantities

41. DATA ANALYSIS The table shows the number of at bats and hits for some of the members of the Florida Marlins in a recent season.

At Bats	Hits
13	6
576	195
240	79
502	139
157	36
64	11

Source: ESPN

a. Construct a scatter plot to represent the data.

b. Predict the number of hits a player with 300 at bats would get. Identify and explain your reasoning.

c. Did the player with 157 at bats or the player with 240 at bats get more hits? What type of reasoning did you use? Explain.

H.O.T. Problems
Use Higher-Order Thinking Skills

42. WRITING IN MATH Explain why the Law of Syllogism cannot be used to draw a conclusion from these conditionals.

If you wear winter gloves, then you will have warm hands.
If you do not have warm hands, then your gloves are too thin.

43. CHALLENGE Use the symbols from Lesson 2-2 for *conjunction* and *disjunction*, and the symbol for implies from Lesson 2-3 to represent the Law of Detachment and the Law of Syllogism symbolically. Let p represent the hypothesis, and let q represent the conclusion.

44. OPEN ENDED Write a pair of statements in which the Law of Syllogism can be used to reach a valid conclusion. Specify the conclusion that can be reached.

45. REASONING Students in Mr. Kendrick's class are divided into two groups for an activity. Students in group A must always tell the truth. Students in group B must always lie. Jonah and Janeka are in Mr. Kendrick's class. When asked if he and Janeka are in group A or B, Jonah says, "We are both in Group B." To which group does each student belong? Explain your reasoning.

46. WRITING IN MATH Compare and contrast the Law of Syllogism and the Transitive Property.

Standardized Test Practice

47. Determine which statement follows logically from the given statements.

If you order two burritos, then you also get nachos.
Michael ordered two burritos.

A Michael ordered one burrito.
B Michael will order two burritos.
C Michael ordered nachos.
D Michael got nachos.

48. ALGEBRA Solve for x:

$$4(x + 2) = x - 1$$

F -3
G -5
H -6
J -8

49. SHORT RESPONSE If the perimeter of the figure shown is 52 units, what is the value of x?

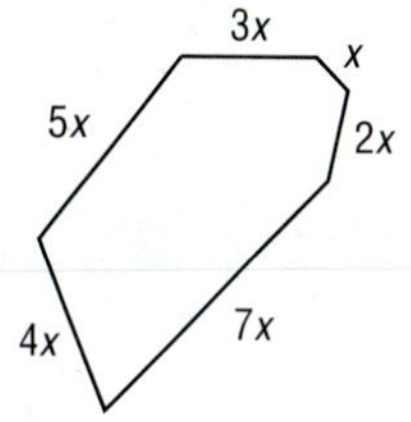

50. SAT/ACT If 30% of x is 50, then 60% of x is

A 25
B 100
C 150
D 250

Spiral Review

51. TIME All states in the United States observe daylight savings time except for Arizona and Hawaii. (Lesson 2-3)

a. Write a true conditional statement in if-then form for daylight savings time.
b. Write the converse of the true conditional statement. State whether the statement is *true* or *false*. If false, find a counterexample.

Construct a truth table for each compound statement. (Lesson 2-2)

52. a and b **53.** $\sim p$ or $\sim q$ **54.** k and $\sim m$ **55.** $\sim y$ or z

ALGEBRA Find x. (Lesson 1-5)

56.
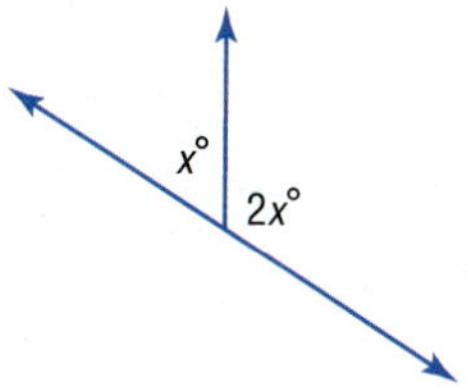

57.
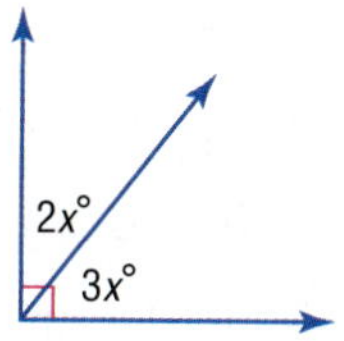

58.
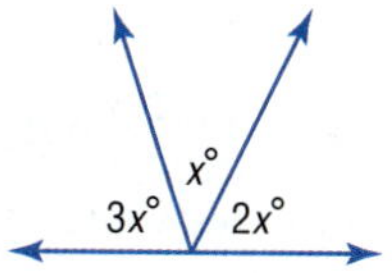

Skills Review

Determine whether each statement can be assumed from the figure. Explain. (Lesson 1-5)

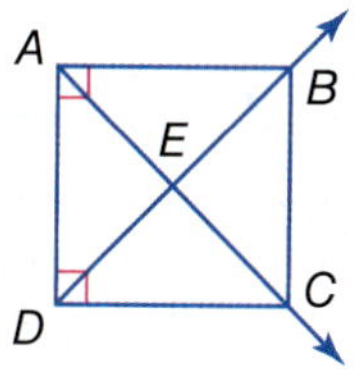

59. $\angle DAB$ is a right angle.

60. $\angle AEB \cong \angle DEC$

61. $\angle ADB$ and $\angle BDC$ are complementary.

62. $\angle DAE \cong \angle ADE$

63. $\overline{AB} \perp \overline{BC}$

64. $\angle AEB$ and $\angle BEC$ are supplementary.

EXPLORE

2-5 Geometry Lab
Necessary and Sufficient Conditions

We all know that water is a *necessary* condition for plants to survive. However, it is not a *sufficient* condition. For example, plants also need sunlight to survive.

Necessary and sufficient conditions are important in mathematics. Consider the property of having four sides. While *having four sides* is a necessary condition for something being a square, that single condition is not, by itself, a sufficient condition to guarantee that it is a square. Trapezoids are four-sided figures that are not squares.

Condition	Definition	Examples
necessary	A condition *A* is said to be *necessary* for a condition *B*, if and only if the falsity or nonexistence of *A* guarantees the falsity or nonexistence of *B*.	Having opposite sides parallel is a necessary condition for something being a square.
sufficient	A condition *A* is said to be *sufficient* for a condition *B*, if and only if the truth or existence of *A* guarantees the truth or existence of *B*.	Being a square is a sufficient condition for something being a rectangle.

Exercises

Determine whether each statement is *true* or *false*. If false, give a counterexample.

1. Being a square is a necessary condition for being a rectangle.
2. Being a rectangle is a necessary condition for being a square.
3. Being greater than 5 is a necessary condition for being less than 10.
4. Being less than 18 is a sufficient condition for being less than 25.
5. Walking on four legs is a sufficient condition for being a dog.
6. Breathing air is a necessary condition for being a human being.
7. Being an equilateral rectangle is both a necessary and sufficient condition for being a square.

Determine whether I is a *necessary* condition for II, a *sufficient* condition for II, or *both*. Explain.

8. I. Two points are given.
   II. An equation of a line can be written.
9. I. Two planes are parallel.
   II. Two planes do not intersect.
10. I. Two angles are acute.
    II. Two angles are complementary.

2-5 Postulates and Paragraph Proofs

Then

You used deductive reasoning by applying the Law of Detachment and the Law of Syllogism. (Lesson 2-4)

Now

- Identify and use basic postulates about points, lines, and planes.
- Write paragraph proofs.

New Vocabulary

postulate
axiom
proof
theorem
deductive argument
paragraph proof
informal proof

Math Online
glencoe.com

Why?

The experiment shown, of a feather and an apple falling at the same rate in a vacuum chamber, demonstrates Sir Isaac Newton's laws on gravity and inertia. These laws are accepted as fundamental truths of physics. Some laws in geometry also must be assumed or accepted as true.

Points, Lines, and Planes A **postulate** or **axiom** is a statement that is accepted as true without proof. In Chapter 1, you studied basic ideas about points, lines, and planes. These ideas can be stated as postulates.

Postulates — Points, Lines, and Planes

Words	Example	
2.1 Through any two points, there is exactly one line.	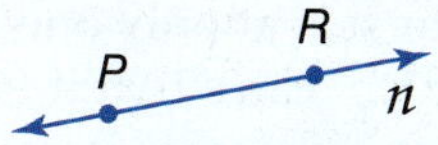	Line n is the only line through points P and R.
2.2 Through any three noncollinear points, there is exactly one plane.	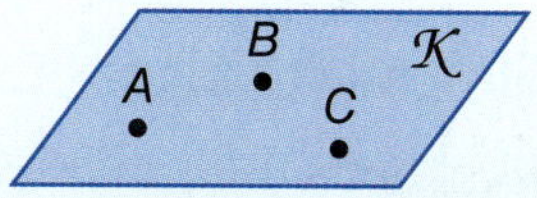	Plane $\mathcal{K}$ is the only plane through noncollinear points A, B, and C.
2.3 A line contains at least two points.	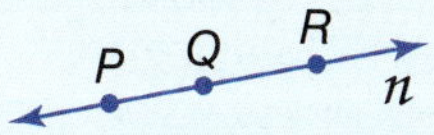	Line n contains points P, Q, and R.
2.4 A plane contains at least three noncollinear points.	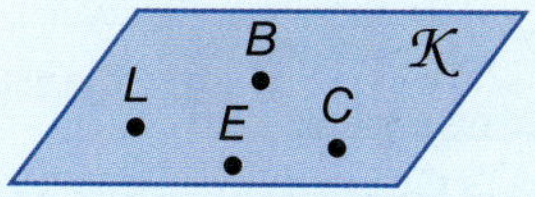	Plane $\mathcal{K}$ contains noncollinear points L, B, C, and E.
2.5 If two points lie in a plane, then the entire line containing those points lies in that plane.	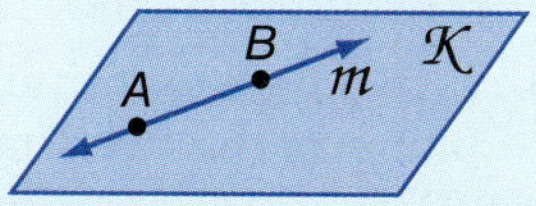	Points A and B lie in plane $\mathcal{K}$, and line m contains points A and B, so line m is in plane $\mathcal{K}$.

Other postulates relate the intersections of lines and planes.

Key Concept — Intersections of Lines and Planes

Words	Example	
2.6 If two lines intersect, then their intersection is exactly one point.	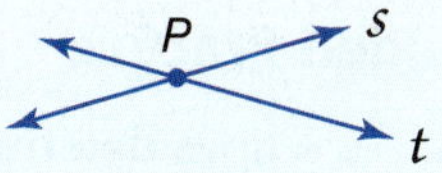	Lines s and t intersect at point P.
2.7 If two planes intersect, then their intersection is a line.	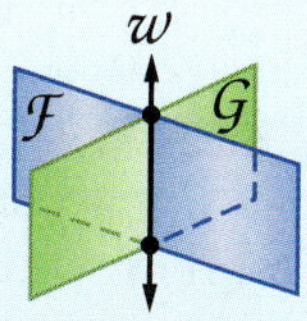	Planes $\mathcal{F}$ and $\mathcal{G}$ intersect in line w.

StudyTip

Undefined Terms Recall from Lesson 1-1 that points, lines, and planes are *undefined terms*. The postulates that you have learned in this lesson describe special relationships between them.

These additional postulates form a foundation for proofs and reasoning about points, lines, and planes.

Real-World EXAMPLE 1 Identifying Postulates

ARCHITECTURE Explain how the picture illustrates that each statement is true. Then state the postulate that can be used to show each statement is true.

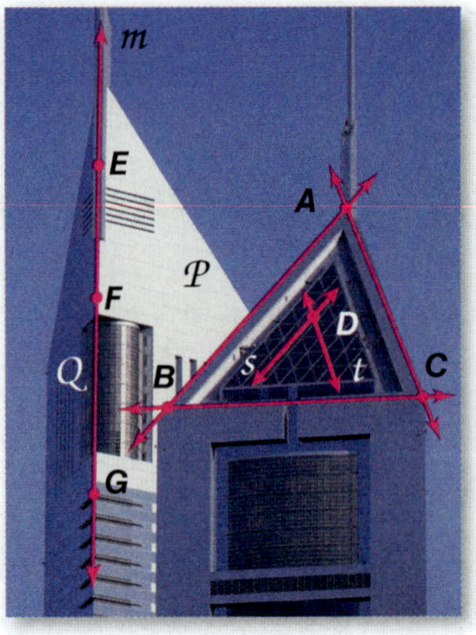

a. Line *m* contains points *F* and *G*. Point *E* can also be on line *m*.

The edge of the building is a straight line *m*. Points *E*, *F*, and *G* lie along this edge, so they lie along a line *m*. Postulate 2.3, which states that a line contains at least two points, shows that this is true.

b. Lines *s* and *t* intersect at point *D*.

The lattice on the window of the building forms intersecting lines. Lines *s* and *t* of this lattice intersect at only one location, point *D*. Postulate 2.6, which states that if two lines intersect, then their intersection is exactly one point, shows that this is true.

Guided Practice

1A. Points *A*, *B*, and *C* determine a plane. **1B.** Planes $\mathcal{P}$ and *Q* intersect in line *m*.

Personal Tutor glencoe.com

You can use postulates to explain your reasoning when analyzing statements.

EXAMPLE 2 Analyze Statements Using Postulates

Determine whether each statement is *always*, *sometimes*, or *never* true. Explain your reasoning.

a. If two coplanar lines intersect, then the point of intersection lies in the same plane as the two lines.

Always; Postulate 2.5 states that if two points lie in a plane, then the entire line containing those points lies in that plane. So, since both points lie in the plane, any point on those lines, including their point of intersection, also lies in the plane.

b. Four points are noncollinear.

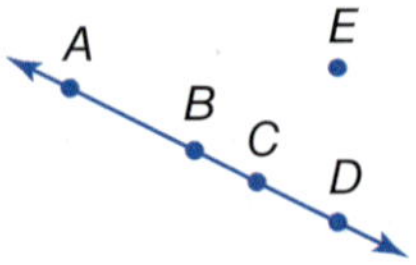

Sometimes; Postulate 2.3 states that a line contains at least two points. This means that a line can contain two *or more* points. So four points can be noncollinear, like *A*, *E*, *C*, and *D*, or collinear, like points *A*, *B*, *C*, and *D*.

Guided Practice

2A. Two lines determine a plane. **2B.** Three lines intersect in two points.

Personal Tutor glencoe.com

StudyTip

Axiomatic System An axiomatic system is a set of axioms, from which some or all axioms can be used to logically derive theorems.

Paragraph Proofs To prove a conjecture, you use deductive reasoning to move from a hypothesis to the conclusion of the conjecture you are trying to prove. This is done by writing a **proof**, which is a logical argument in which each statement you make is supported by a statement that is accepted as true.

StudyTip

Proposition A *proposition* is a statement that makes an assertion that is either false or true. In mathematics, a proposition is usually used to mean a true assertion and can be synonymous with theorem.

Once a statement or conjecture has been proven, it is called a **theorem**, and it can be used as a reason to justify statements in other proofs.

Key Concept **The Proof Process**

Step 1 List the given information and, if possible, draw a diagram to illustrate this information.

Step 2 State the theorem or conjecture to be proven.

Step 3 Create a **deductive argument** by forming a logical chain of statements linking the given to what you are trying to prove.

Step 4 Justify each statement with a reason. Reasons include definitions, algebraic properties, postulates, and theorems.

Step 5 State what it is that you have proven.

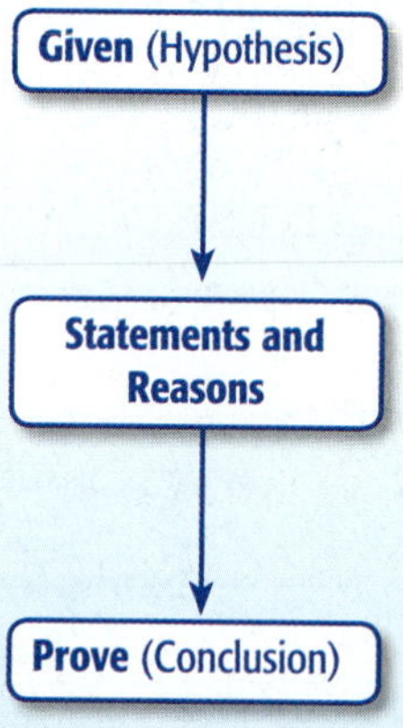

One method of proving statements and conjectures, a **paragraph proof**, involves writing a paragraph to explain why a conjecture for a given situation is true. Paragraph proofs are also called **informal proofs**, although the term *informal* is not meant to imply that this form of proof is any less valid than any other type of proof.

EXAMPLE 3 **Write a Paragraph Proof**

Given that M is the midpoint of $\overline{XY}$ write a paragraph proof to show that $\overline{XM} \cong \overline{MY}$.

Steps 1 and 2 →

Given: M is the midpoint of $\overline{XY}$.

Prove: $\overline{XM} \cong \overline{MY}$

Steps 3 and 4 →

If M is the midpoint of $\overline{XY}$, then from the definition of midpoint of a segment, we know that $XM = MY$. This means that $\overline{XM}$ and $\overline{MY}$ have the same measure. By the definition of congruence, if two segments have the same measure, then they are congruent.

Step 5 →

Thus, $\overline{XM} \cong \overline{MY}$.

Problem-SolvingTip

Work Backward One strategy for writing a proof is to *work backward*. Start with what you are trying to prove, and work backward step by step until you reach the given information.

Guided Practice

3. Given that C is between A and B and $\overline{AC} \cong \overline{CB}$, write a paragraph proof to show that C is the midpoint of $\overline{AB}$.

Personal Tutor glencoe.com

Once a conjecture has been proven true, it can be stated as a theorem and used in other proofs. The conjecture in Example 3 is known as the Midpoint Theorem.

Theorem 2.1 **Midpoint Theorem**

If M is the midpoint of $\overline{AB}$, then $\overline{AM} \cong \overline{MB}$.

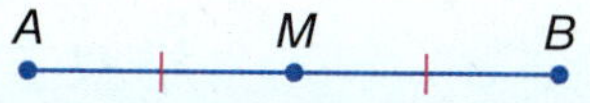

Check Your Understanding

Example 1 p. 126

Explain how the figure illustrates that each statement is true. Then state the postulate that can be used to show each statement is true.

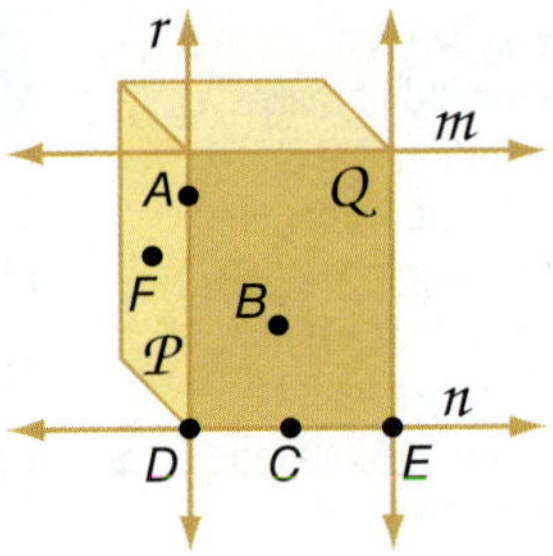

1. Planes $\mathcal{P}$ and $\mathcal{Q}$ intersect in line r.
2. Lines r and n intersect at point D.
3. Line n contains points C, D, and E.
4. Plane $\mathcal{P}$ contains the points A, F, and D.
5. Line n lies in plane $\mathcal{Q}$.
6. Line r is the only line through points A and D.

Example 2 p. 126

Determine whether each statement is *always*, *sometimes*, or *never* true. Explain your reasoning.

7. The intersection of three planes is a line.
8. Line r contains only point P.
9. Through two points, there is exactly one line.

In the figure, $\overrightarrow{AK}$ is in plane $\mathcal{P}$ and M is on $\overleftrightarrow{NE}$. State the postulate that can be used to show each statement is true.

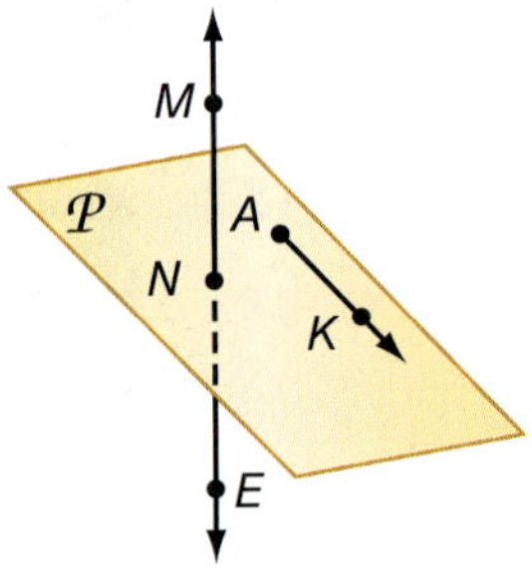

10. M, K, and N are coplanar.
11. $\overleftrightarrow{NE}$ contains points N and M.
12. N and K are collinear.
13. Points N, K, and A are coplanar.

14. **SPORTS** Each year, Jennifer's school hosts a student vs. teacher basketball tournament to raise money for charity. This year, there are eight teams participating in the tournament. During the first round, each team plays all of the other teams.
    a. How many games will be played in the first round?
    b. Draw a diagram to model the number of first round games. Which postulate can be used to justify your diagram?
    c. Find a numerical method that you could use regardless of the number of the teams in the tournament to calculate the number of games in the first round.

Example 3 p. 127

15. **PROOF** In the figure at the right, $\overline{AE} \cong \overline{DB}$ and C is the midpoint of $\overline{AE}$ and $\overline{DB}$. Write a paragraph proof to show that $AC = CB$.

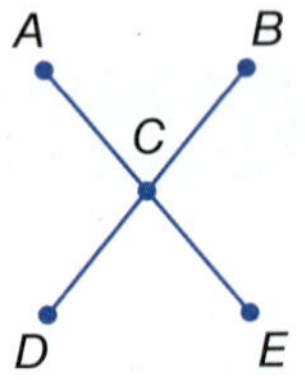

Practice and Problem Solving

● = **Step-by-Step Solutions** begin on page R20. **Extra Practice** begins on page 969.

Example 1 p. 126

CAKES **Explain how the picture illustrates that each statement is true. Then state the postulate that can be used to show each statement is true.**

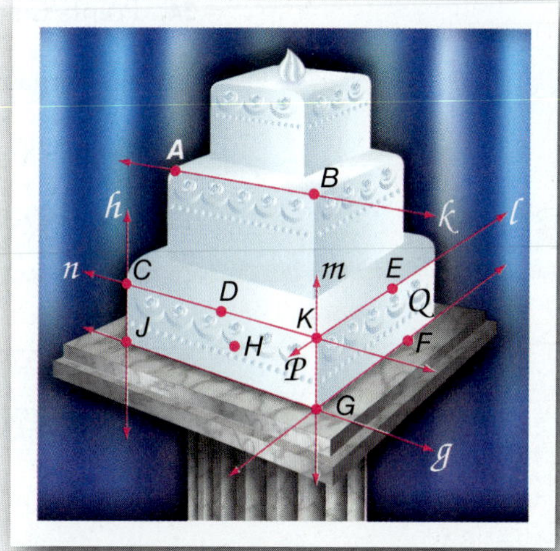

16. Lines n and ℓ intersect at point K.

17. Planes $\mathcal{P}$ and $\mathcal{Q}$ intersect in line m.

18. Points D, K, and H determine a plane.

19. Point D is also on the line n through points C and K.

20. Points D and H are collinear.

21. Points E, F, and G are coplanar.

22. $\overleftrightarrow{EF}$ lies in plane $\mathcal{Q}$.

23. Lines h and g intersect at point J.

Example 2 p. 126

Determine whether each statement is *always*, *sometimes*, or *never* true. Explain.

24. There is exactly one plane that contains noncollinear points A, B, and C.

25. There are at least three lines through points J and K.

26. If points M, N, and P lie in plane X, then they are collinear.

27. Points X and Y are in plane Z. Any point collinear with X and Y is in plane Z.

28. The intersection of two planes can be a point.

29. Points A, B, and C determine a plane.

Example 3 p. 127

30. **PROOF** Point Y is the midpoint of $\overline{XZ}$. Z is the midpoint of $\overline{YW}$. Prove that $\overline{XY} \cong \overline{ZW}$.

31. **PROOF** Point L is the midpoint of $\overline{JK}$. $\overline{JK}$ intersects $\overline{MK}$ at K. If $\overline{MK} \cong \overline{JL}$, prove that $\overline{LK} \cong \overline{MK}$.

32. **RECREATION** Last weekend, Emilio and his friends spent Saturday afternoon at the park. There were several people there with bikes and skateboards. There were a total of 11 bikes and skateboards that had a total of 36 wheels. Use a paragraph proof to show how many bikes and how many skateboards there were.

33. **DRIVING** Keisha is traveling from point A to point B. Two possible routes are shown on the map. Assume that the speed limit on Southside Boulevard is 55 miles per hour and the speed limit on I–295 is 70 miles per hour.

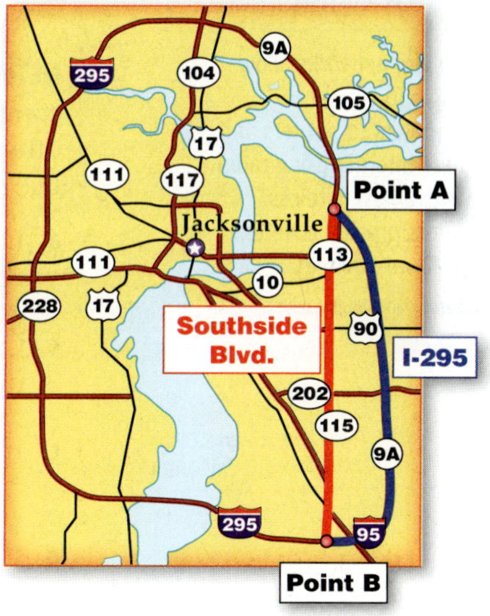

a. Which of the two routes covers the shortest distance? Explain your reasoning.

b. If the distance from point A to point B along Southside Boulevard is 10.5 miles and the distance along I-295 is 11.6 miles, which route is faster, assuming that Keisha drives the speed limit?

In the figure at the right, $\overleftrightarrow{CD}$ and $\overleftrightarrow{CE}$ lie in plane $\mathcal{P}$ and $\overleftrightarrow{DH}$ and $\overleftrightarrow{DJ}$ lie in plane Q. State the postulate that can be used to show each statement is true.

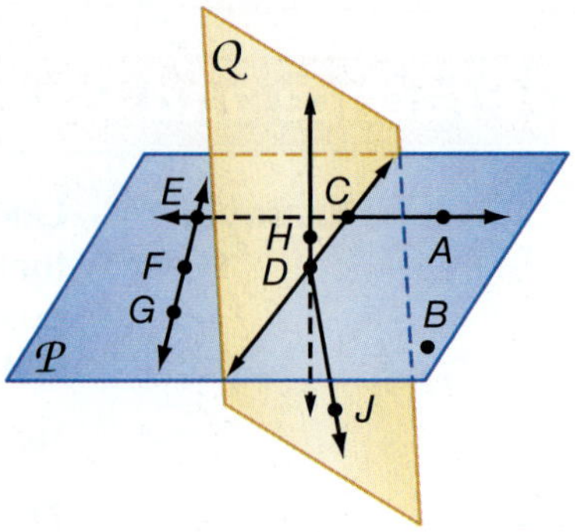

34. Points C and B are collinear.

35. $\overleftrightarrow{EG}$ contains points E, F, and G.

36. $\overleftrightarrow{DA}$ lies in plane $\mathcal{P}$.

37. Points D and F are collinear.

38. Points C, D, and B are coplanar.

39. Plane Q contains the points C, H, D, and J.

40. $\overleftrightarrow{AC}$ and $\overleftrightarrow{FG}$ intersect at point E.

41. Plane $\mathcal{P}$ and plane Q intersect at $\overleftrightarrow{CD}$.

42. **ARCHITECTURE** Roofs are designed based on the materials used to ensure that water does not leak into the buildings they cover. Some roofs are constructed from waterproof material, and others are constructed for watershed, or gravity removal of water. The pitch of a roof is the rise over the run, which is generally measured in rise per foot of run. Use the statements below to write a paragraph proof justifying the following statement: The pitch of the roof in Den's design is not steep enough.

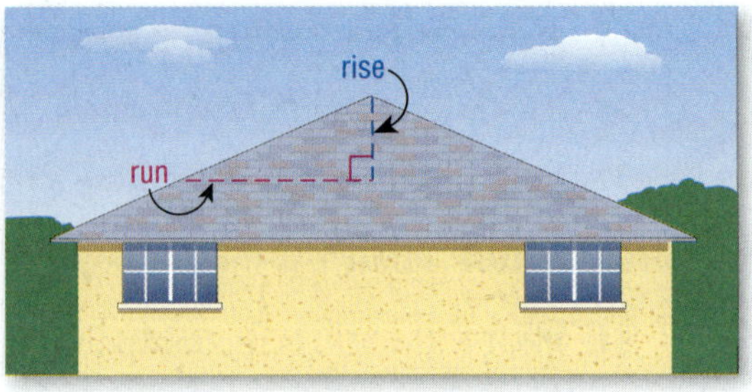

- Waterproof roofs should have a minimum slope of $\frac{1}{4}$ inch per foot.
- Watershed roofs should have a minimum slope of 4 inches per foot.
- Den is designing a house with a watershed roof.
- The pitch in Den's design is 2 inches per foot.

Real-World Link

Wireless networks allow people to connect computers to printers and to the Internet using radio waves. Most wireless routers provide coverage for about 100 feet.

Source: How Stuff Works

43. **NETWORKS** Diego is setting up a network of multiple computers so that each computer is connected to every other. The diagram at the right illustrates this network if Diego has 5 computers.

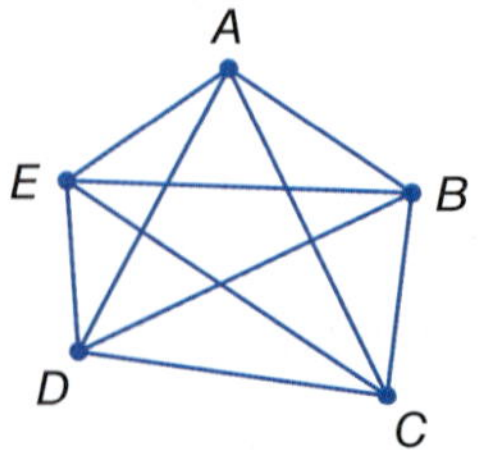

a. Draw diagrams of the networks if Diego has 2, 3, 4, or 6 computers.

b. Create a table with the number of computers and the number of connections for the diagrams you drew.

c. If there are n computers in the network, write an expression for the number of computers to which each of the computers is connected.

d. If there are n computers in the network, write an expression for the number of connections there are.

44. BUILDINGS The photo is of the rotunda in the capital building in St. Paul, Minnesota. A rotunda is a round building, usually covered by a dome. Use Postulate 1.1 to help you answer parts **a–c**.

a. If you were standing in the middle of the rotunda, which arched exit is the closest to you?

b. What information did you use to formulate your answer?

c. What term describes the shortest distance from the center of a circle to a point on the circle?

H.O.T. Problems Use Higher-Order Thinking Skills

45. ERROR ANALYSIS Omari and Lisa were working on a paragraph proof to prove that if $\overline{AB}$ is congruent to $\overline{BD}$ and A, B, and D are collinear, then B is the midpoint $\overline{AD}$. Each student started his or her proof in a different way. Is either of them correct? Explain your reasoning.

Omari	Lisa
If B is the midpoint of $\overline{AB}$, then B divides $\overline{AD}$ into two congruent segments.	$\overline{AB}$ is congruent to $\overline{BD}$ and A, B, and D are collinear.

46. OPEN ENDED Draw a figure that satisfies five of the seven postulates you have learned. Explain which postulates you chose and how your figure satisfies each postulate.

47. CHALLENGE Use the following true statement and the definitions and postulates you have learned to answer each question.

Two planes are perpendicular if and only if one plane contains a line perpendicular to the second plane.

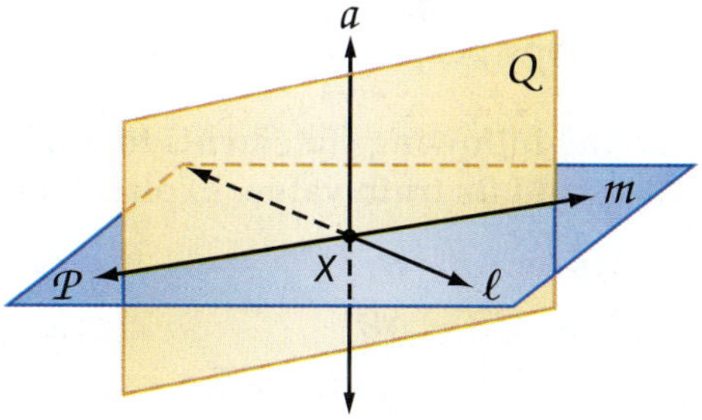

a. Through a given point, there passes one and only one plane perpendicular to a given line. If plane Q is perpendicular to line ℓ at point X and line ℓ lies in plane $\mathcal{P}$, what must also be true?

b. Through a given point, there passes one and only one line perpendicular to a given plane. If plane Q is perpendicular to plane $\mathcal{P}$ at point X and line a lies in plane Q, what must also be true?

Real-World Link

You can apply the statement in Exercise 47 when assembling furniture. A shelf is perpendicular to the back of a bookcase if one side of the shelf is perpendicular to the back of the bookcase.

REASONING Determine if each statement is *sometimes*, *always*, or *never* true. Explain your reasoning or provide a counterexample.

48. Through any three points, there is exactly one plane.

49. Three coplanar lines have two points of intersection.

50. WRITING IN MATH Explain how undefined terms, definitions, postulates, and theorems are alike and how they are different.

Standardized Test Practice

51. ALGEBRA Which is one of the solutions of the equation $3x^2 - 5x + 1 = 0$?

A $\frac{5 + \sqrt{13}}{6}$

B $\frac{-5 - \sqrt{13}}{6}$

C $\frac{5}{6} - \sqrt{13}$

D $-\frac{5}{6} + \sqrt{13}$

52. GRIDDED RESPONSE Steve has 20 marbles in a bag, all the same size and shape. There are 8 red, 2 blue, and 10 yellow marbles in the bag. He will select a marble from the bag at random. What is the probability that the marble Steve selects will be yellow?

53. Which statement *cannot* be true?

F Three noncollinear points determine a plane.

G Two lines intersect in exactly one point.

H At least two lines can contain the same two points.

J A midpoint divides a segment into two congruent segments.

54. SAT/ACT What is the greatest number of regions that can be formed if 3 distinct lines intersect a circle?

A 4

B 5

C 6

D 7

Spiral Review

Determine whether a valid conclusion can be reached from the two true statements using the Law of Detachment or the Law of Syllogism. If a valid conclusion is possible, state it and the law that is used. If a valid conclusion does not follow, write *no conclusion*. (Lesson 2-4)

55. (1) If two angles are vertical, then they do not form a linear pair.

(2) If two angles form a linear pair, then they are not congruent.

56. (1) If an angle is acute, then its measure is less than 90.

(2) $\angle EFG$ is acute.

Write each statement in if-then form. (Lesson 2-3)

57. Happy people rarely correct their faults.

58. A champion is afraid of losing.

Use the following statements to write a compound statement for each conjunction. Then find its truth value. Explain your reasoning. (Lesson 2-2)

***p*: *M* is on $\overline{AB}$.**

***q*: $AM + MB = AB$**

***r*: *M* is the midpoint of $\overline{AB}$.**

59. $p \wedge q$

60. $\sim p \vee \sim r$

61. GARDENING A landscape designer is putting black plastic edging around a rectangular flower garden that has length 5.7 meters and width 3.8 meters. The edging is sold in 5-meter lengths. Find the perimeter of the garden and determine how much edging the designer should buy. (Lesson 1-6)

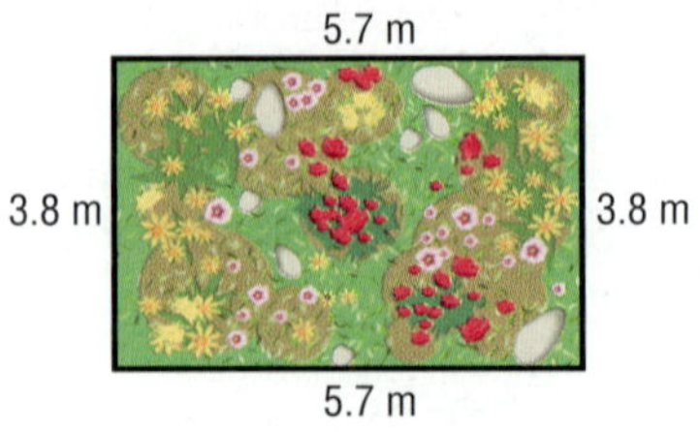

62. HEIGHT Taylor is 5 feet 8 inches tall. How many inches tall is Taylor? (Lesson 0-1)

Skills Review

ALGEBRA Solve each equation. (Lesson 0-5)

63. $4x - 3 = 19$

64. $\frac{1}{3}x + 6 = 14$

65. $5(x^2 + 2) = 30$

Mid-Chapter Quiz

Lessons 2-1 through 2-5

Write a conjecture that describes the pattern in each sequence. Then use your conjecture to find the next item in the sequence. (Lesson 2-1)

1. 5, 5, 10, 15, 25, . . .

2. 

Find a counterexample to show that each conjecture is false. (Lesson 2-1)

3. If $AB = BC$, then B is the midpoint of $\overline{AC}$.

4. If n is a real number, then $n^3 > n$.

Use the following statements to write a compound statement for each conjunction or disjunction. Then find its truth value. Explain your reasoning. (Lesson 2-2)

p: A dollar is equal to 100 cents.

q: There are 4 quarters in a dollar.

r: February is the month before January.

5. $p \wedge r$

6. p and q

7. $p \wedge \sim r$

8. Copy and complete the truth table. (Lesson 2-2)

p	q	$\sim q$	$p \vee \sim q$
T	F		
F	T		
F	F		
T	T		

Identify the hypothesis and conclusion of each conditional statement. (Lesson 2-3)

9. If a polygon has five sides, then it is a pentagon.

10. If $4x - 6 = 10$, then $x = 4$.

11. An angle with a measure less than 90 is an acute angle.

Determine the truth value of each conditional statement. If *true*, explain your reasoning. If *false*, give a counterexample. (Lesson 2-3)

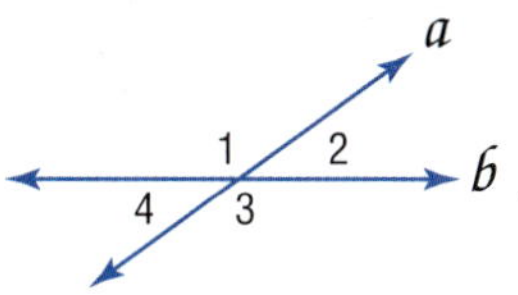

12. $\angle 1$ and $\angle 2$ are supplementary angles.

13. $\angle 1$ and $\angle 4$ are congruent angles.

Use the Venn diagrams below to determine the truth value of each conditional. Explain your reasoning. (Lesson 2-3)

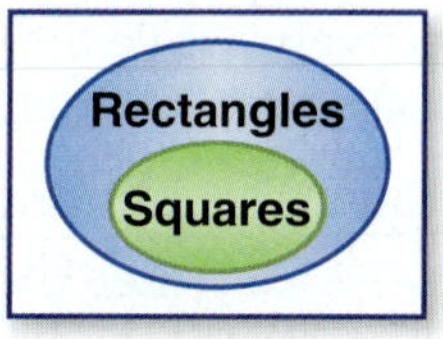

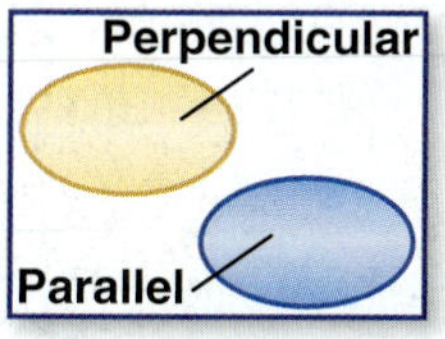

14. If a polygon is a square, then it is a rectangle.

15. If two lines are perpendicular, then they cannot be parallel.

16. **FOOTBALL** The Indianapolis Colts played the Chicago Bears in the 2007 Super Bowl. Determine whether the stated conclusion is valid based on the given information. If not, write *invalid*. Explain your reasoning. (Lesson 2-4)

Given: The Super Bowl winner has the highest score at the end of the game. The Colts had a score of 29 and the Bears had a score of 17.
Conclusion: The Colts won the Super Bowl.

17. **MULTIPLE CHOICE** Determine which statement follows logically from the given statements. (Lesson 2-4)

(1) If you are a junior in high school, then you are at least 16 years old.

(2) If you are at least 16 years old, then you are old enough to drive.

A If you are old enough to drive, then you are a junior in high school.

B If you are not old enough to drive, then you are a sophomore in high school.

C If you are a junior in high school, then you are old enough to drive.

D No valid conclusion possible.

Determine whether each statement is *always*, *sometimes*, or *never* true. Explain your reasoning. (Lesson 2-5)

18. Points J, K, L, and N are noncollinear and lie in the same plane $\mathcal{M}$.

19. There is exactly one line through points R and S.

20. Line a contains only point Q.

2-6

Algebraic Proof

Why?

Some cars have displays that show the outside air temperature in Fahrenheit or Celsius. The Fahrenheit scale sets the freezing and boiling points of water at 32° and 212°, respectively, while the Celsius scale sets them at 0° and 100°.

You can use an algebraic proof to show that if these scales are related by the formula $C = \frac{5}{9}(F - 32)$, then they are also related by the formula $F = \frac{9}{5}C + 32$.

Then

You used postulates about points, lines, and planes to write paragraph proofs. (Lesson 2-5)

Now

- Use algebra to write two-column proofs.
- Use properties of equality to write geometric proofs.

New Vocabulary

algebraic proof
two-column proof
formal proof

Math Online
glencoe.com

Algebraic Proof Algebra is a system with sets of numbers, operations, and properties that allow you to perform algebraic operations. The following table summarizes several properties of real numbers that you studied in algebra.

Key Concept — Properties of Real Numbers

The following properties are true for any real numbers a, b, and c.

Addition Property of Equality	If $a = b$, then $a + c = b + c$.
Subtraction Property of Equality	If $a = b$, then $a - c = b - c$.
Multiplication Property of Equality	If $a = b$, then $a \cdot c = b \cdot c$.
Division Property of Equality	If $a = b$ and $c \neq 0$, then, $\frac{a}{c} = \frac{b}{c}$.
Reflexive Property of Equality	$a = a$
Symmetric Property of Equality	If $a = b$, then $b = a$.
Transitive Property of Equality	If $a = b$ and $b = c$, then $a = c$.
Substitution Property of Equality	If $a = b$, then a may be replaced by b in any equation or expression.
Distributive Property	$a(b + c) = ab + ac$

An **algebraic proof** is a proof that is made up of a series of algebraic statements. The properties of equality provide justification for many statements in algebraic proofs.

EXAMPLE 1 Justify Each Step When Solving an Equation

Prove that if $-5(x + 4) = 70$, then $x = -18$. Write a justification for each step.

$-5(x + 4) = 70$	Original equation or Given
$-5 \cdot x + (-5) \cdot 4 = 70$	Distributive Property
$-5x - 20 = 70$	Substitution Property of Equality
$-5x - 20 + 20 = 70 + 20$	Addition Property of Equality
$-5x = 90$	Substitution Property of Equality
$\frac{-5x}{-5} = \frac{90}{-5}$	Division Property of Equality
$x = -18$	Substitution Property of Equality

Guided Practice

State the property that justifies each statement.

1A. If $4 + (-5) = -1$, then $x + 4 + (-5) = x - 1$.

1B. If $5 = y$, then $y = 5$.

1C. Prove that if $2x - 13 = -5$, then $x = 4$. Write a justification for each step.

Personal Tutor glencoe.com

StudyTip

Algorithms An *algorithm* is a series of steps for carrying out a procedure or solving a problem. Proofs can be considered a type of algorithm because they go step by step.

Example 1 is a proof of the conditional statement *If* $-5(x + 4) = 70$, *then* $x = -18$. Notice that the column on the left is a step-by-step process that leads to a solution. The column on the right contains the reason for each statement.

In geometry, a similar format is used to prove conjectures and theorems. A **two-column proof** or **formal proof** contains *statements* and *reasons* organized in two columns.

Real-World EXAMPLE 2 Write an Algebraic Proof

SCIENCE If the formula to convert a Fahrenheit temperature to a Celsius temperature is $C = \frac{5}{9}(F - 32)$, then the formula to convert a Celsius temperature to a Fahrenheit temperature is $F = \frac{9}{5}C + 32$. Write a two-column proof to verify this conjecture.

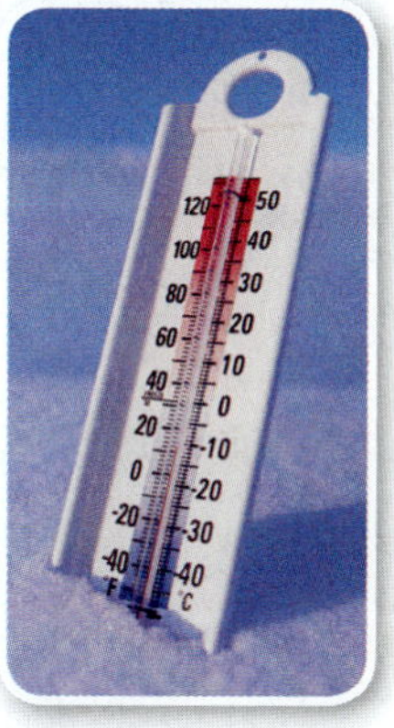

Begin by stating what is given and what you are to prove.

Given: $C = \frac{5}{9}(F - 32)$

Prove: $F = \frac{9}{5}C + 32$

Proof:

Statements	Reasons
1. $C = \frac{5}{9}(F - 32)$	**1.** Given
2. $\frac{9}{5}C = \frac{9}{5} \cdot \frac{5}{9}(F - 32)$	**2.** Multiplication Property of Equality
3. $\frac{9}{5}C = F - 32$	**3.** Substitution Property of Equality
4. $\frac{9}{5}C + 32 = F - 32 + 32$	**4.** Addition Property of Equality
5. $\frac{9}{5}C + 32 = F$	**5.** Substitution Property of Equality
6. $F = \frac{9}{5}C + 32$	**6.** Symmetric Property of Equality

StudyTip

Mental Math If your teacher permits you to do so, some steps may be eliminated by performing mental calculations. For example, steps 2 and 4 in Example 2 could be omitted. Then the reason for statement 3 would be Multiplication Property of Equality and the reason for statement 5 would be Addition Property of Equality.

Guided Practice

Write a two-column proof to verify that each conjecture is true.

2A. If $\frac{5x + 1}{2} - 8 = 0$, then $x = 3$.

2B. **PHYSICS** If the distance d moved by an object with initial velocity u and final velocity v in time t is given by $d = t \cdot \frac{u + v}{2}$, then $u = \frac{2d}{t} - v$.

Personal Tutor glencoe.com

StudyTip

Commutative and Associative Properties Throughout this text we shall assume that if a, b, and c are real numbers, then the following properties are true.

Commutative Property of Addition
$a + b = b + a$

Commutative Property of Multiplication
$a \cdot b = b \cdot a$

Associative Property of Addition
$(a + b) + c = a + (b + c)$

Associative Property of Multiplication
$(a \cdot b) \cdot c = a \cdot (b \cdot c)$

Geometric Proof Since geometry also uses variables, numbers, and operations, many of the properties of equality used in algebra are also true in geometry. For example, segment measures and angle measures are real numbers, so properties from algebra can be used to discuss their relationships as shown in the table below.

Property	Segments	Angles
Reflexive	$AB = AB$	$m\angle 1 = m\angle 1$
Symmetric	If $AB = CD$, then $CD = AB$.	If $m\angle 1 = m\angle 2$, then $m\angle 2 = m\angle 1$.
Transitive	If $AB = CD$ and $CD = EF$, then $AB = EF$.	If $m\angle 1 = m\angle 2$ and $m\angle 2 = m\angle 3$, then $m\angle 1 = m\angle 3$.

These properties can be used to write geometric proofs.

EXAMPLE 3 Write a Geometric Proof

If $\angle FGJ \cong \angle JGK$ and $\angle JGK \cong \angle KGH$, then $x = 6$. Write a two-column proof to verify this conjecture.

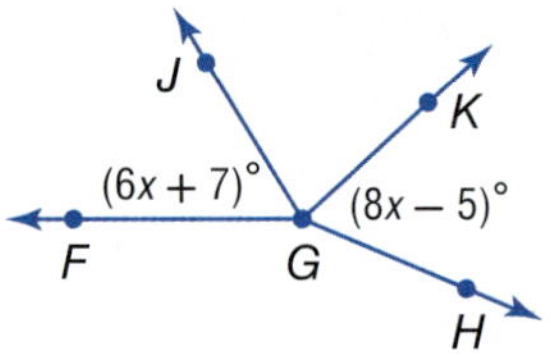

Given: $\angle FGJ \cong \angle JGK$, $\angle JGK \cong \angle KGH$, $m\angle FGJ = 6x + 7$, $m\angle KGH = 8x - 5$

Prove: $x = 6$

Proof:

Statements	Reasons
1. $\angle FGJ \cong \angle JGK$; $\angle JGK \cong \angle KGH$	**1.** Given
2. $m\angle FGJ = m\angle JGK$; $m\angle JGK = m\angle KGH$	**2.** Definition of congruent angles
3. $m\angle FGJ = m\angle KGH$	**3.** Transitive Property of Equality
4. $6x + 7 = 8x - 5$	**4.** Substitution Property of Equality
5. $6x + 7 + 5 = 8x - 5 + 5$	**5.** Addition Property of Equality
6. $6x + 12 = 8x$	**6.** Substitution Property of Equality
7. $6x + 12 - 6x = 8x - 6x$	**7.** Subtraction Property of Equality
8. $12 = 2x$	**8.** Substitution Property of Equality
9. $\frac{12}{2} = \frac{2x}{2}$	**9.** Division Property of Equality
10. $6 = x$	**10.** Substitution Property of Equality
11. $x = 6$	**11.** Symmetric Property of Equality

Guided Practice

Write a two-column proof to verify each conjecture.

3A. If $\angle A \cong \angle B$ and $m\angle A = 37$, then $m\angle B = 37$.

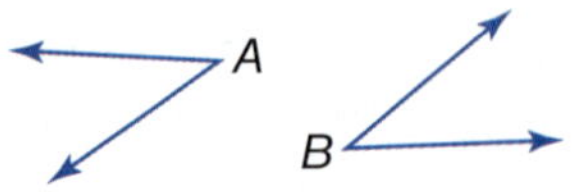

3B. If $\overline{CD} \cong \overline{EF}$, then $y = 8$.

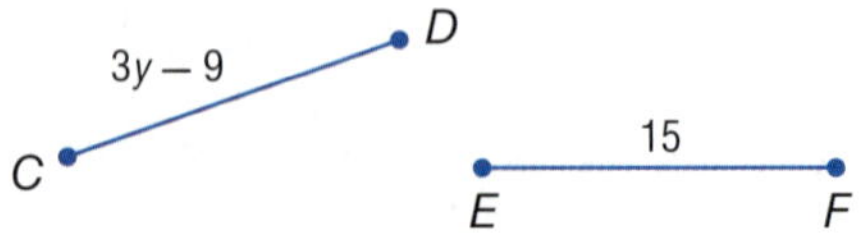

Personal Tutor glencoe.com

Check Your Understanding

Example 1
p. 134

State the property that justifies each statement.

1. If $m\angle 1 = m\angle 2$ and $m\angle 2 = m\angle 3$, then $m\angle 1 = m\angle 3$.

2. $XY = XY$

3. If $5 = x$, then $x = 5$.

4. If $2x + 5 = 11$, then $2x = 6$.

Example 2
p. 135

5. Complete the following proof.

Given: $\frac{y+2}{3} = 3$

Prove: $y = 7$

Proof:

Statements	Reasons
a. ___?___	**a.** Given
b. $3\left(\frac{y+2}{3}\right) = 3(3)$	**b.** ___?___
c. ___?___	**c.** ___?___
d. $y = 7$	**d.** Subtraction Property

Examples 2 and 3
pp. 135–136

PROOF **Write a two-column proof to verify each conjecture.**

6. If $-4(x - 3) + 5x = 24$, then $x = 12$.

7. If $\overline{AB} \cong \overline{CD}$, then $x = 7$.

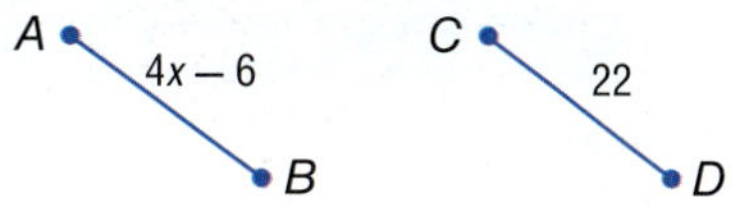

8. **FITNESS** Mai-Lin measures her heart rate whenever she exercises and tries to make sure that she is staying in her target heart rate zone. The American Heart Association suggests a target heart rate of $T = 0.75(220 - a)$, where T is a person's target heart rate and a is his or her age.

a. Prove that given a person's target heart rate, you can calculate his or her age using the formula $a = 220 - \frac{T}{0.75}$.

b. If Mai-Lin's target heart rate is 153, then how old is she? What property justifies your calculation?

Practice and Problem Solving

● = **Step-by-Step Solutions** begin on page R20.
Extra Practice begins on page 969.

Example 1
p. 134

State the property that justifies each statement.

9. If $a + 10 = 20$, then $a = 10$.

10. If $\frac{x}{3} = -15$, then $x = -45$.

11. If $4x - 5 = x + 12$, then $4x = x + 17$.

12. If $\frac{1}{5}BC = \frac{1}{5}DE$, then $BC = DE$.

State the property that justifies each statement.

13. If $5(x + 7) = -3$, then $5x + 35 = -3$.

14. If $m\angle 1 = 25$ and $m\angle 2 = 25$, then $m\angle 1 = m\angle 2$.

15. If $AB = BC$ and $BC = CD$, then $AB = CD$.

16. If $3\left(x - \frac{2}{3}\right) = 4$, then $3x - 2 = 4$.

Example 2 p. 135

Complete each proof.

17. Given: $\frac{8 - 3x}{4} = 32$

Prove: $x = -40$

Proof:

Statements	Reasons
a. $\frac{8 - 3x}{4} = 32$	**a.** Given
b. $4\left(\frac{8 - 3x}{4}\right) = 4(32)$	**b.** ___?___
c. $8 - 3x = 128$	**c.** ___?___
d. ___?___	**d.** Subtraction Property
e. $x = -40$	**e.** ___?___

Real-World Link

The Formula Society of Automotive Engineers at University of California, Berkeley, holds a competition each year for the design and construction of a race car. The cars are judged on many factors, including acceleration.

Source: Formula Society of Automotive Engineers

18. Given: $\frac{1}{5}x + 3 = 2x - 24$

Prove: $x = 15$

Proof:

Statements	Reasons
a. ___?___	**a.** Given
b. ___?___	**b.** Multiplication Property
c. $x + 15 = 10x - 120$	**c.** ___?___
d. ___?___	**d.** Subtraction Property
e. $135 = 9x$	**e.** ___?___
f. ___?___	**f.** Division Property
g. ___?___	**g.** Symmetric Property

Example 3 p. 136

PROOF Write a two-column proof to verify each conjecture.

19. If $-\frac{1}{3}n = 12$, then $n = -36$.

20. If $-3r + \frac{1}{2} = 4$, then $r = -\frac{7}{6}$.

21 **SCIENCE** Acceleration a in feet per second squared, distance traveled d in feet, velocity v in feet per second, and time t in seconds are related in the formula $d = vt + \frac{1}{2}at^2$.

a. Prove that if the values for distance, velocity, and time are known, then the acceleration of an object can be calculated using the formula $a = \frac{2d - 2vt}{t^2}$.

b. If an object travels 2850 feet in 30 seconds with an initial velocity of 50 feet per second, what is the acceleration of the object? What property justifies your calculation?

22. **SCIENCE** The Ideal Gas Law is given by the formula $PV = nRT$, where P = pressure in atmospheres, V = volume in liters, n = the amount of gas in moles, R is a constant value, and T = temperature in degrees Kelvin.

   a. Prove that if the pressure, volume, and amount of the gas are known, then the formula $T = \frac{PV}{nR}$ gives the temperature of the gas.

   b. If you have 1 mole of oxygen with a volume of 25 liters at a pressure of 1 atmosphere, what is the temperature of the gas? The value of R is 0.0821. What property justifies your calculation?

PROOF **Write a two-column proof.**

23. If $\overline{DF} \cong \overline{EG}$, then $x = 10$.

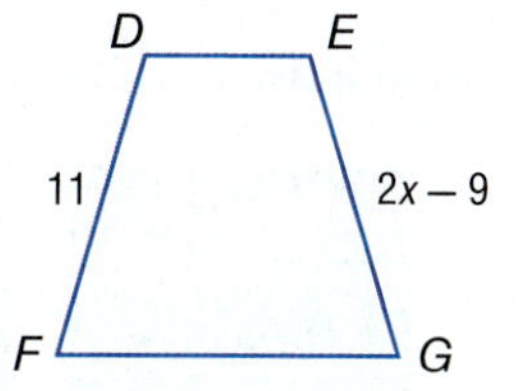

24. If $\overline{AB} \cong \overline{AC}$, then $x = 4$.

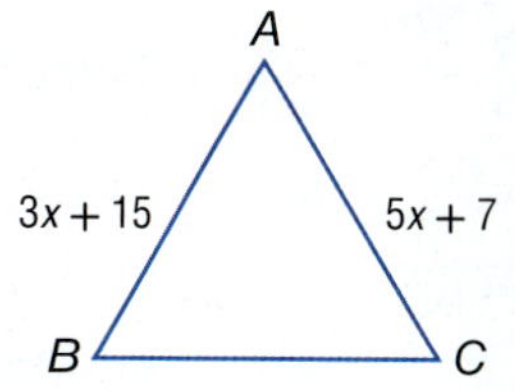

25. If $\angle Y \cong \angle Z$, then $x = 100$.

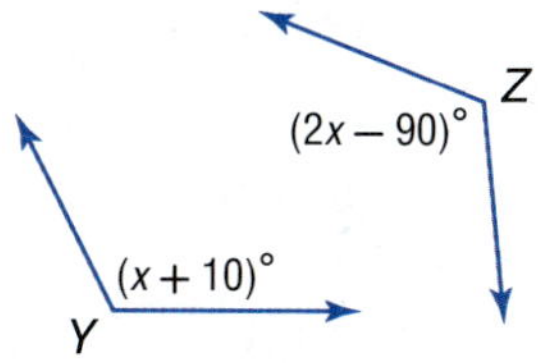

26. If $\angle MPN \cong \angle QPN$, then $x = 16$.

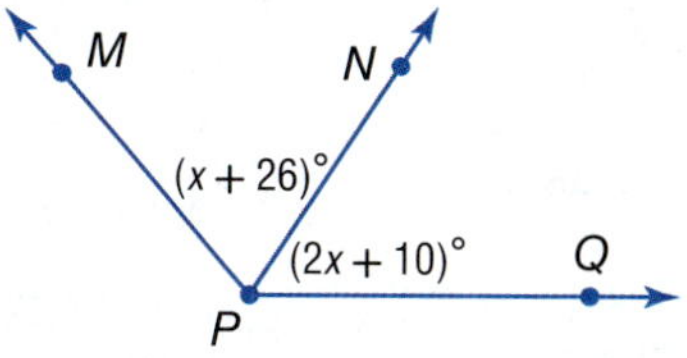

Real-World Link

Lightning results from a build-up of charge in a cloud. A strike lasts less than 1 second and can generate 100 million to 1 billion volts of electricity. Compare this to a household outlet which supplies 120 volts of electricity.

Source: NOAA

27. **ELECTRICITY** The voltage V of a circuit can be calculated using the formula $V = \frac{P}{I}$, where P is the power and I is the current of the circuit.

   a. Write a proof to show that when the power is constant, the voltage is halved when the current is doubled.

   b. Write a proof to show that when the current is constant, the voltage is doubled when the power is doubled.

28. **MULTIPLE REPRESENTATIONS** Consider a cube with a side length of s.

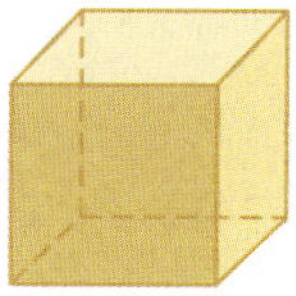

s units

   a. CONCRETE Sketch or build a model of cubes with side lengths of 2, 4, 8, and 16 units.

   b. TABULAR Find the volume of each cube. Organize your results into a table like the one shown.

Side Length (s)	Volume (V)
2	
4	
8	
16	

   c. VERBAL Use your table to make a conjecture about the change in volume when the side length of a cube is doubled. Express your conjecture in words.

   d. ANALYTICAL Write your conjecture as an algebraic equation.

   e. LOGICAL Write a proof of your conjecture. Be sure to write the *Given* and *Prove* statements at the beginning of your proof.

29. **PYTHAGOREAN THEOREM** The Pythagorean Theorem states that in a right triangle ABC, the sum of the squares of the measures of the lengths of the legs, a and b, equals the square of the measure of the hypotenuse c, or $a^2 + b^2 = c^2$. Write a two-column proof to verify that $a = \sqrt{c^2 - b^2}$. Use the Square Root Property of Equality, which states that if $a^2 = b^2$, then $a = \pm\sqrt{b^2}$.

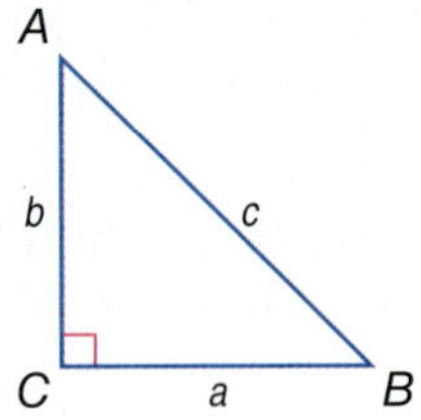

An *equivalence relation* is any relationship that satisfies the Reflexive, Symmetric, and Transitive Properties. For real numbers, equality is one type of equivalence relation. Determine whether each relation is an equivalence relation. Explain your reasoning.

30. "has the same birthday as," for the set of all human beings

31. "is taller than," for the set of all human beings

32. "is bluer than" for all the paint colors with blue in them

33. $\neq$, for the set of real numbers

34. $\geq$, for the set of real numbers

35. $\approx$, for the set of real numbers

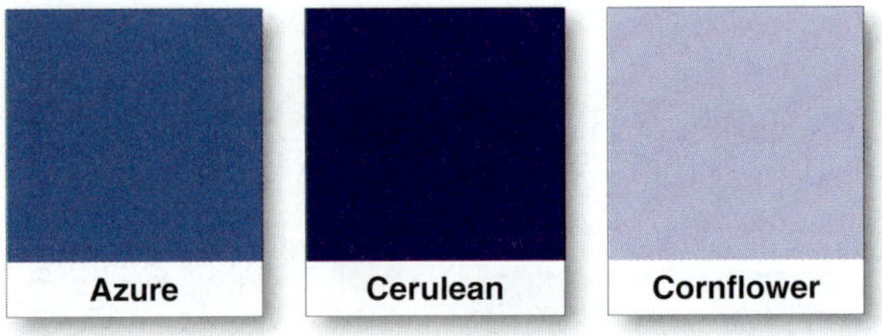

Real-World Link

In a room of 23 people, there is about a 50% chance that at least two of those people will have the same birthday. This is known as the *birthday paradox.*

H.O.T. Problems

Use **H**igher-**O**rder **T**hinking Skills

36. **OPEN ENDED** Give one real-world *example* and one real-world *non-example* of the Symmetric, Transitive, and Substitution properties.

37. **CHALLENGE** Point P is located on $\overline{AB}$. The length of $\overline{AP}$ is $2x + 3$, and the length of $\overline{PB}$ is $\frac{3x + 1}{2}$. Segment AB is 10.5 units long. Draw a diagram of this situation, and prove that point P is located two thirds of the way between point A and point B.

REASONING **Classify each statement below as *sometimes*, *always*, or *never* true. Explain your reasoning.**

38. If a and b are real numbers and $a + b = 0$, then $a = -b$.

39. If a and b are real numbers and $a^2 = b$, then $a = \sqrt{b}$.

40. **CHALLENGE** Ayana makes a conjecture that the sum of two odd integers is an even integer.

   a. List information that supports this conjecture. Then explain why the information you listed does not prove that this conjecture is true.

   b. Two odd integers can be represented by the expressions $2n - 1$ and $2m - 1$. Give information that supports this statement.

   c. If a number is even, then it is a multiple of what number? Explain in words how you could use the expressions in part **a** and your answer to part **b** to prove Ayana's conjecture.

   d. Write an algebraic proof that the sum of two odd integers is an even integer.

41. **WRITING IN MATH** Compare and contrast informal or paragraph proofs with formal or two-column proofs. Which type of proof do you find easier to write? Justify your answer.

Standardized Test Practice

42. In the diagram, $m\angle CFE = 90$ and $\angle AFB \cong \angle CFD$. Which of the following conclusions does not have to be true?

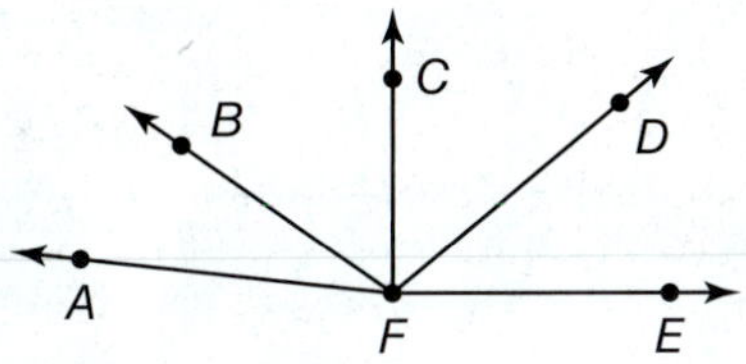

A $m\angle BFD = m\angle BFD$
B $\overline{BF}$ bisects $\angle AFD$.
C $m\angle CFD = m\angle AFB$
D $\angle CFE$ is a right angle.

43. SHORT RESPONSE Find the measure of $\angle B$ when $m\angle A = 55$ and $m\angle C = 42$.

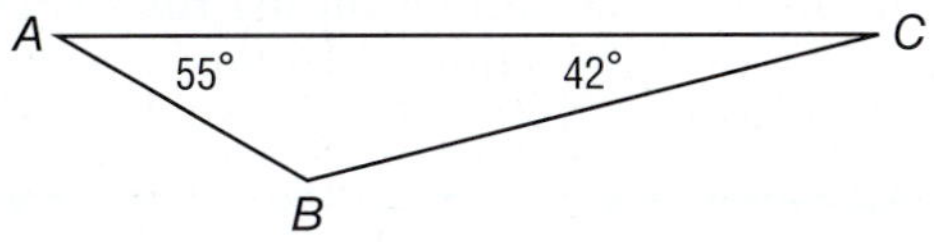

44. ALGEBRA Kendra's walk-a-thon supporters have pledged \$30 plus \$7.50 for each mile she walks. Rebecca's supporters have pledged \$45 plus \$3.75 for each mile she walks. After how many miles will Kendra and Rebecca have raised the same amount of money?

F 10
G 8
H 5
J 4

45. SAT/ACT When 17 is added to $4m$, the result is $15z$. Which of the following equations represents the statement above?

A $17 + 15z = 4m$
B $(4m)(15z) = 17$
C $4m - 15z = 17$
D $4m + 17 = 15z$

Spiral Review

Determine whether the following statements are *always*, *sometimes*, or *never* true. Explain. (Lesson 2-5)

46. Four points will lie in one plane.

47. Two obtuse angles will be supplementary.

48. Planes $\mathcal{P}$ and $\mathcal{Q}$ intersect in line m. Line m lies in both plane $\mathcal{P}$ and plane $\mathcal{Q}$.

49. ADVERTISING An ad for Speedy Delivery Service says *When it has to be there fast, it has to be Speedy.* Catalina needs to send a package fast. Does it follow that she should use Speedy? Explain. (Lesson 2-4)

Write the ordered pair for each point shown. (Lesson 0-7)

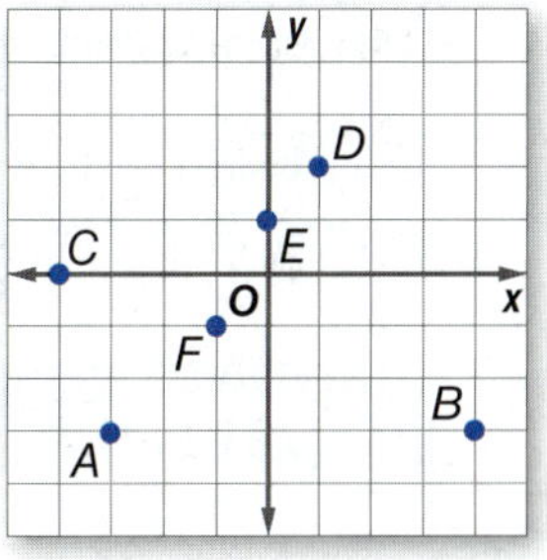

50. *A*

51. *B*

52. *C*

53. *D*

54. *E*

55. *F*

Skills Review

Find the measurement of each segment. Assume that each figure is not drawn to scale. (Lesson 1-2)

56. $\overline{ST}$

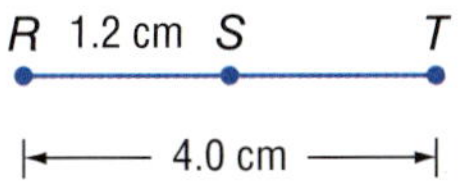

57. $\overline{WX}$

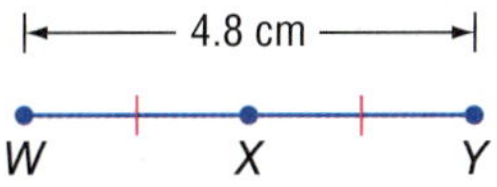

58. $\overline{BC}$

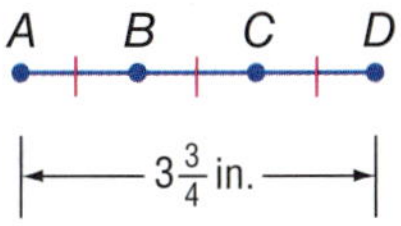

2-7 Proving Segment Relationships

Then

You wrote algebraic and two-column proofs. (Lesson 2-6)

Now

- Write proofs involving segment addition.
- Write proofs involving segment congruence.

Math Online

glencoe.com

Why?

Emma works at a fabric store after school. She measures a length of fabric by holding the straight edge of the fabric against a yardstick. To measure lengths such as 39 inches, which is longer than the yardstick, she marks a length of 36 inches. From the end of that mark, she measures an additional length of 3 inches. This ensures that the total length of fabric is 36 + 3 inches or 39 inches.

Ruler Postulate In Lesson 1-2, you measured segments with a ruler by matching the mark for zero with one endpoint and then finding the number on the ruler that corresponded to the other endpoint. This illustrates the Ruler Postulate.

Postulate 2.8

Ruler Postulate

Words The points on any line or line segment can be put into one-to-one correspondence with real numbers.

Example Given any two points A and B on a line, if A corresponds to zero, then B corresponds to a positive real number.

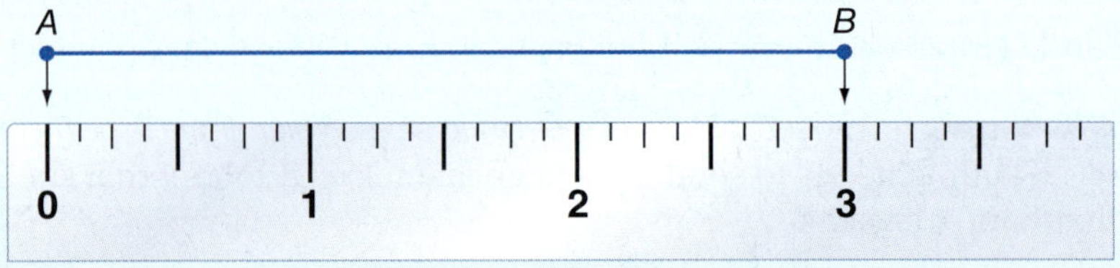

In Lesson 1-2, you also learned about what it means for a point to be *between* two other points. This relationship can be expressed as the Segment Addition Postulate.

Postulate 2.9

Segment Addition Postulate

Words If A, B, and C are collinear, then point B is between A and C if and only if $AB + BC = AC$.

Model

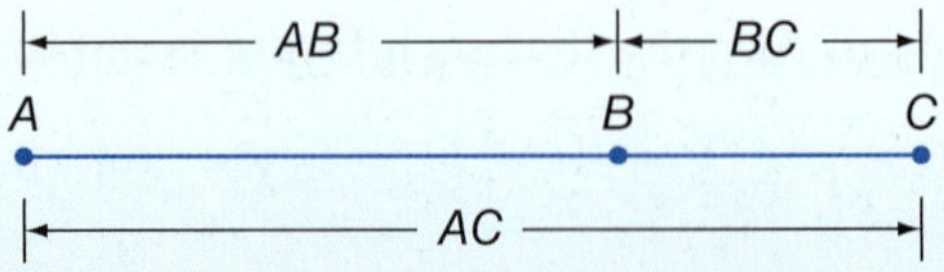

The Segment Addition Postulate is used as a justification in many geometric proofs.

EXAMPLE 1 Use the Segment Addition Postulate

Prove that if $\overline{CE} \cong \overline{FE}$ and $\overline{ED} \cong \overline{EG}$ then $\overline{CD} \cong \overline{FG}$.

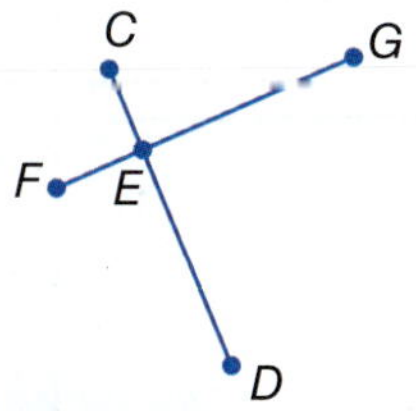

Given: $\overline{CE} \cong \overline{FE}$; $\overline{ED} \cong \overline{EG}$

Prove: $\overline{CD} \cong \overline{FG}$

Proof:

Statements	Reasons
1. $\overline{CE} \cong \overline{FE}$; $\overline{ED} \cong \overline{EG}$	**1.** Given
2. $CE = FE$; $ED = EG$	**2.** Definition of congruence
3. $CE + ED = CD$	**3.** Segment Addition Postulate
4. $FE + EG = CD$	**4.** Substitution (Steps 2 & 3)
5. $FE + EG = FG$	**5.** Segment Addition Postulate
6. $CD = FG$	**6.** Substitution (Steps 4 & 5)
7. $\overline{CD} \cong \overline{FG}$	**7.** Definition of congruence

ReadingMath

Substitution Property The Substitution Property of Equality is often just written as *Substitution*.

Guided Practice

Copy and complete the proof.

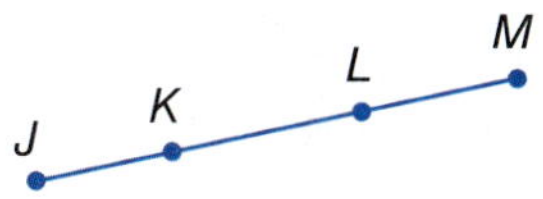

1. Given: $\overline{JL} \cong \overline{KM}$

Prove: $\overline{JK} \cong \overline{LM}$

Proof:

Statements	Reasons
a. $\overline{JL} \cong \overline{KM}$	**a.** Given
b. $JL = KM$	**b.** ___?___
c. $JK + KL =$ ___?___; $KL + LM =$ ___?___	**c.** Segment Addition Postulate
d. $JK + KL = KL + LM$	**d.** ___?___
e. $JK + KL - KL = KL + LM - KL$	**e.** Subtraction Property of Equality
f. ___?___	**f.** Substitution
g. $\overline{JK} \cong \overline{LM}$	**g.** Definition of congruence

Personal Tutor glencoe.com

Segment Congruence In Lesson 2-6, you saw that segment measures are reflexive, symmetric, and transitive. Since segments with the same measure are congruent, congruence of segments is also reflexive, symmetric, and transitive.

Vocabulary Link

Symmetric
Everyday Use balanced or proportional
Math Use If $a = b$, then $b = a$.

Theorem 2.2 Properties of Segment Congruence

Reflexive Property of Congruence	$\overline{AB} \cong \overline{AB}$
Symmetric Property of Congruence	If $\overline{AB} \cong \overline{CD}$, then $\overline{CD} \cong \overline{AB}$.
Transitive Property of Congruence	If $\overline{AB} \cong \overline{CD}$ and $\overline{CD} \cong \overline{EF}$, then $\overline{AB} \cong \overline{EF}$.

You will prove the Reflexive and Symmetric Properties in Exercises 6 and 7, respectively.

Proof **Transitive Property of Congruence**

Given: $\overline{AB} \cong \overline{CD}$; $\overline{CD} \cong \overline{EF}$

Prove: $\overline{AB} \cong \overline{EF}$

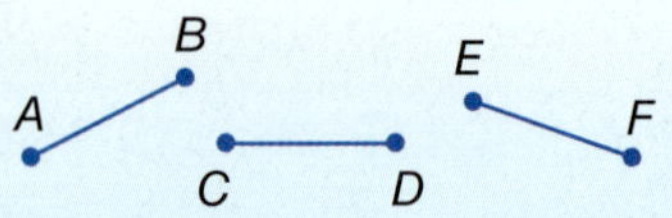

Paragraph Proof:

Since $\overline{AB} \cong \overline{CD}$ and $\overline{CD} \cong \overline{EF}$, $AB = CD$ and $CD = EF$ by the definition of congruent segments. By the Transitive Property of Equality, $AB = EF$. Thus, $\overline{AB} \cong \overline{EF}$ by the definition of congruence.

Real-World Link

According to a recent poll, 70% of teens who volunteer began doing so before age 12. Others said they would volunteer if given more opportunities to do so.

Source: Youth Service America

Real-World EXAMPLE 2 Proof Using Segment Congruence

VOLUNTEERING **The route for a charity fitness run is shown. Checkpoints *X* and *Z* are the midpoints between the starting line and Checkpoint *Y* and Checkpoint *Y* and the finish line *F*, respectively. If Checkpoint *Y* is the same distance from Checkpoints *X* and *Z*, prove that the route from Checkpoint *Z* to the finish line is congruent to the route from the starting line to Checkpoint *X*.**

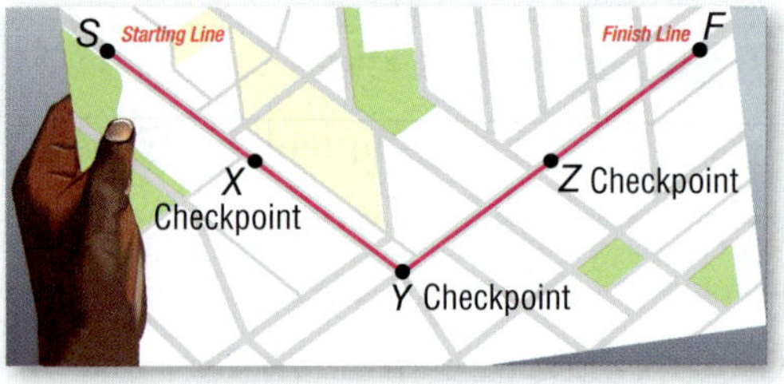

Given: X is the midpoint of $\overline{SY}$. Z is the midpoint of $\overline{YF}$. $XY = YZ$

Prove: $\overline{ZF} \cong \overline{SX}$

Two-Column Proof:

Statements	Reasons
1. X is the midpoint of $\overline{SY}$. Z is the midpoint of $\overline{YF}$. $XY = YZ$	**1.** Given
2. $\overline{SX} \cong \overline{XY}$; $\overline{YZ} \cong \overline{ZF}$	**2.** Definition of midpoint
3. $\overline{XY} \cong \overline{YZ}$	**3.** Definition of congruence
4. $\overline{SX} \cong \overline{YZ}$	**4.** Transitive Property of Congruence
5. $\overline{SX} \cong \overline{ZF}$	**5.** Transitive Property of Congruence
6. $\overline{ZF} \cong \overline{SX}$	**6.** Symmetric Property of Congruence

Guided Practice

2. **CARPENTRY** A carpenter cuts a 2″ × 4″ board to a desired length. He then uses this board as a pattern to cut a second board congruent to the first. Similarly, he uses the second board to cut a third board and the third board to cut a fourth board. Prove that the last board cut has the same measure as the first.

Personal Tutor glencoe.com

Check Your Understanding

Example 1 p. 143

1. Copy and complete the proof.

Given: $\overline{LK} \cong \overline{NM}$, $\overline{KJ} \cong \overline{MJ}$

Prove: $\overline{LJ} \cong \overline{NJ}$

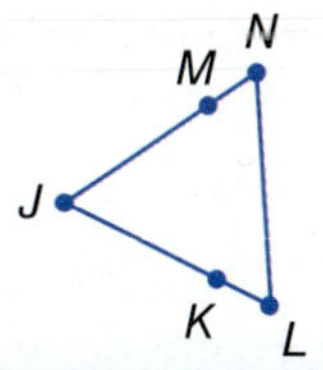

Proof:

Statements	Reasons
a. $\overline{LK} \cong \overline{NM}$, $\overline{KJ} \cong \overline{MJ}$	**a.** __?__
b. __?__	**b.** Def. of congruent segments
c. $LK + KJ = NM + MJ$	**c.** __?__
d. __?__	**d.** Segment Addition Postulate
e. $LJ = NJ$	**e.** __?__
f. $\overline{LJ} \cong \overline{NJ}$	**f.** __?__

Example 2 p. 144

2. **PROOF** Prove the following.

Given: $\overline{WX} \cong \overline{YZ}$

Prove: $\overline{WY} \cong \overline{XZ}$

3. **SCISSORS** Refer to the diagram shown. $\overline{AR}$ is congruent to $\overline{CR}$. $\overline{DR}$ is congruent to $\overline{BR}$. Prove that $AR + DR = CR + BR$.

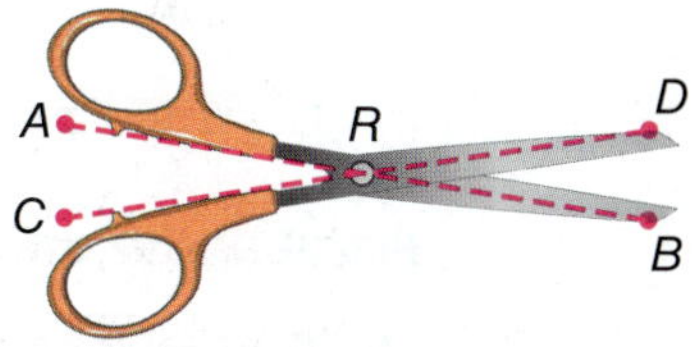

Practice and Problem Solving

= **Step-by-Step Solutions** begin on page R20.
Extra Practice begins on page 969.

Example 1 p. 143

4. Copy and complete the proof.

Given: C is the midpoint of $\overline{AE}$.
C is the midpoint of $\overline{BD}$.
$\overline{AE} \cong \overline{BD}$

Prove: $\overline{AC} \cong \overline{CD}$

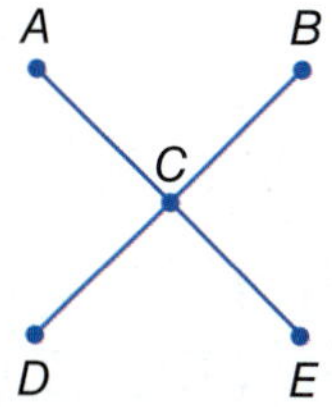

Proof:

Statements	Reasons
a. __?__	**a.** Given
b. $AC = CE$, $BC = CD$	**b.** __?__
c. $AE = BD$	**c.** __?__
d __?__	**d.** Segment Addition Postulate
e. $AC + CE = BC + CD$	**e.** __?__
f. $AC + AC = CD + CD$	**f.** __?__
g. __?__	**g.** Substitution
h. __?__	**h.** Division Property
i. $\overline{AC} \cong \overline{CD}$	**i.** __?__

Example 2
p. 144

Real-World Career

Tile Setters Tile setters install tile on floors, walls, ceilings, and countertops. They use measuring devices and levels to place the tiles and arrange them in specific patterns. Most tile setters receive on-the-job training.

Source: *Occupational Outlook Handbook*

5. **TILING** A tile setter cuts a piece of tile to a desired length. He then uses this tile as a pattern to cut a second tile congruent to the first. He uses the first two tiles to cut a third tile whose length is the sum of the measures of the first two tiles. Prove that the measure of the third tile is twice the measure of the first tile.

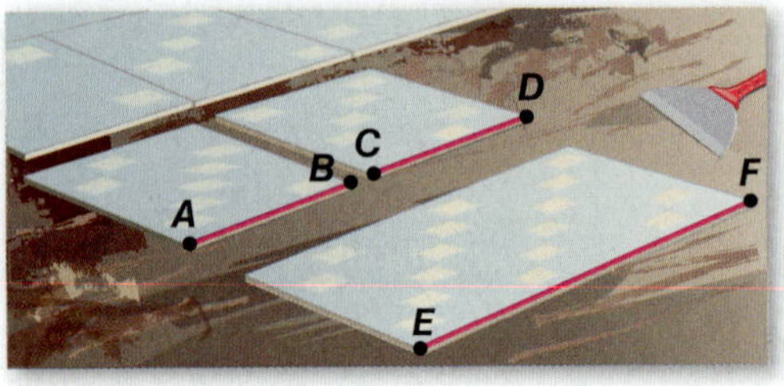

Prove each theorem.

6. Symmetric Property of Congruence (Theorem 2.2)

7 Reflexive Property of Congruence (Theorem 2.2)

8. **TRAVEL** Buffalo, Utica, Albany, and Syracuse, New York, are all connected by Interstate 90.

- Albany is 126 miles from Syracuse and 263 miles from Buffalo.
- Buffalo is 137 miles from Syracuse and 184 miles from Utica.

a. Draw a diagram to represent the locations of the cities in relation to each other and the distances between each city. Assume that Interstate 90 is straight.

b. Write a paragraph proof to support your conclusion.

PROOF **Prove the following.**

9. If $\overline{SC} \cong \overline{HR}$ and $\overline{HR} \cong \overline{AB}$, then $\overline{SC} \cong \overline{AB}$.

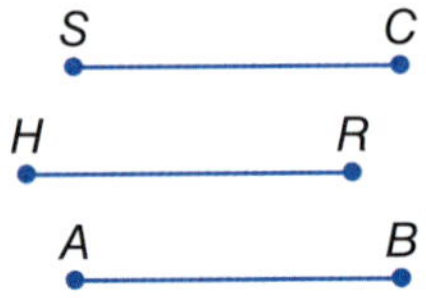

10. If $\overline{VZ} \cong \overline{VY}$ and $\overline{WY} \cong \overline{XZ}$, then $\overline{VW} \cong \overline{VX}$.

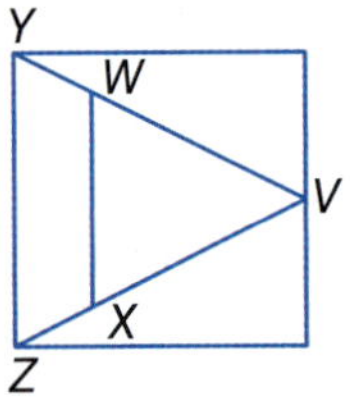

11. If E is the midpoint of $\overline{DF}$ and $\overline{CD} \cong \overline{FG}$, then $\overline{CE} \cong \overline{EG}$.

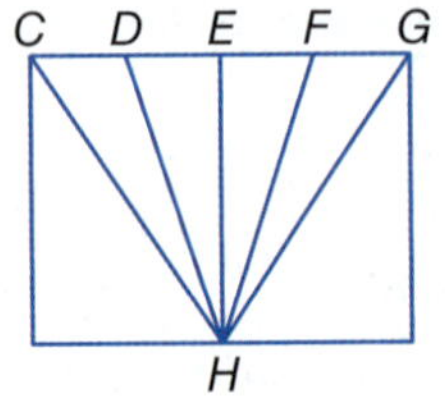

12. If B is the midpoint of $\overline{AC}$, D is the midpoint of $\overline{CE}$, and $\overline{AB} \cong \overline{DE}$, then $AE = 4AB$.

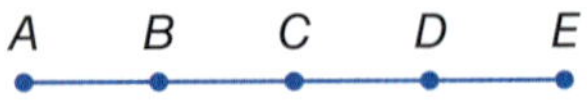

13. **OPTICAL ILLUSION** $\overline{AC} \cong \overline{GI}$, $\overline{FE} \cong \overline{LK}$, and $AC + CF + FE = GI + IL + LK$.

a. Prove that $\overline{CF} \cong \overline{IL}$.

b. Justify your proof using measurement. Explain your method.

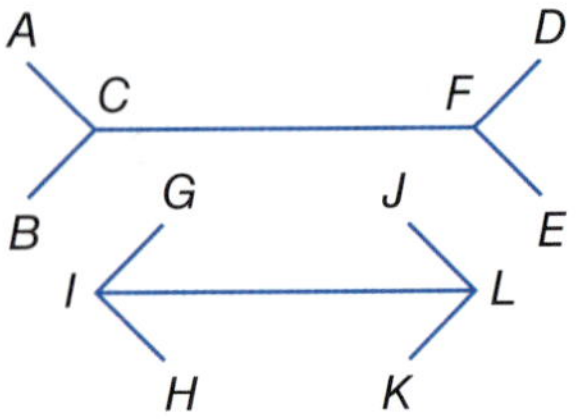

14. **CONSTRUCTION** Construct a segment that is twice as long as $\overline{PQ}$. Explain how the Segment Addition Postulate can be used to justify your construction.

Real-World Link

More than 14.3 million boys participated in school sports at 18,500 schools in the 2006–2007 school year. At 15,458 of these schools, baseball was the most popular sport.

Source: *USA TODAY*

15. **BASEBALL** Use the diagram of a baseball diamond shown.

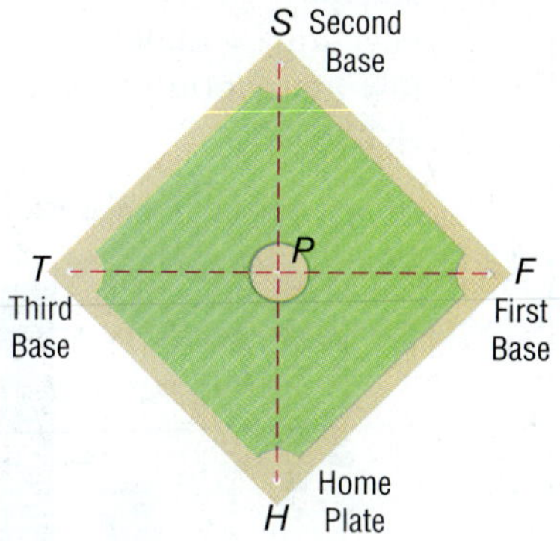

a. On a baseball field, $\overline{SH} \cong \overline{TF}$. P is the midpoint of $\overline{SH}$ and $\overline{TF}$. Using a two-column proof, prove that $\overline{SP} \cong \overline{TP}$.

b. The distance from home plate to second base is 127.3 feet. What is the distance from first base to second base?

16. **MULTIPLE REPRESENTATIONS** A is the midpoint of $\overline{PQ}$, B is the midpoint of $\overline{PA}$, and C is the midpoint of $\overline{PB}$.

a. **GEOMETRIC** Make a sketch to represent this situation.

b. **ALGEBRAIC** Make a conjecture as to the algebraic relationship between PC and PQ.

c. **GEOMETRIC** Copy segment $\overline{PQ}$ from your sketch. Then construct points B and C on $\overline{PQ}$. Explain how you can use your construction to support your conjecture.

d. **CONCRETE** Use a ruler to draw a segment congruent to $\overline{PQ}$ from your sketch and to draw points B and C on $\overline{PQ}$. Use your drawing to support your conjecture.

e. **LOGICAL** Prove your conjecture.

H.O.T. Problems

Use **H**igher-**O**rder **T**hinking Skills

17. **ERROR ANALYSIS** In the diagram, $\overline{AB} \cong \overline{CD}$ and $\overline{CD} \cong \overline{BF}$. Examine the conclusions made by Leslie and Shantice. Is either of them correct?

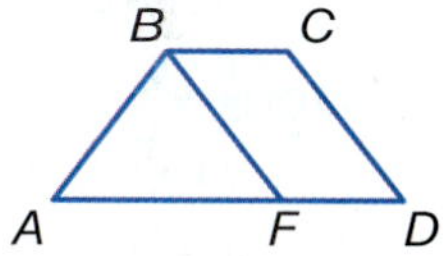

Leslie

Since $\overline{AB} \cong \overline{CD}$ and $\overline{CD} \cong \overline{BF}$, then $AB \cong AF$ by the Transitive Property of Congruence

Shantice

Since $\overline{AB} \cong \overline{CD}$ and $\overline{CD} \cong \overline{BF}$, then $\overline{AB} \cong \overline{BF}$ by the Reflexive Property of Congruence.

18. **CHALLENGE** $ABCD$ is a square. Prove that $\overline{AC} \cong \overline{BD}$.

19. **WRITING IN MATH** Does there exist an Addition Property of Congruence? Explain.

20. **REASONING** Classify the following statement as *true* or *false*. If false, provide a counterexample.

If A, B, C, D, and E are collinear with B between A and C, C between B and D, and D between C and E, and AC = BD = CE, then AB = BC = DE.

21. **OPEN ENDED** Draw a representation of the Segment Addition Postulate in which the segment is two inches long, contains four collinear points, and contains no congruent segments.

22. **WRITING IN MATH** Compare and contrast paragraph proofs and two-column proofs.

Standardized Test Practice

23. ALGEBRA The chart below shows annual recycling by material in the United States. About how many pounds of aluminum are recycled each year?

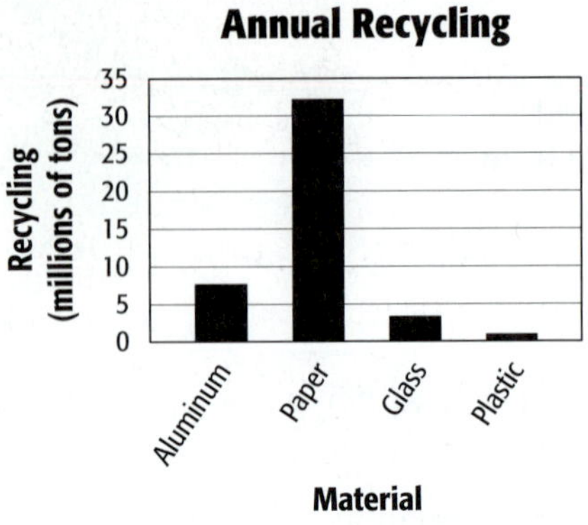

A 7.5
B 15,000
C 7,500,000
D 15,000,000,000

24. ALGEBRA Which expression is equivalent to $\frac{12x^{-4}}{4x^{-8}}$?

F $\frac{1}{3x^4}$
G $3x^4$
H $8x^2$
J $\frac{x^4}{3}$

25. SHORT RESPONSE The measures of two complementary angles are in the ratio 4 : 1. What is the measure of the smaller angle?

26. SAT/ACT Julie can word process 40 words per minute. How many minutes will it take Julie to word process 200 words?

A 0.5
B 2
C 5
D 12

Spiral Review

27. PROOF Write a two-column proof. (Lesson 2-6)
Given: $AC = DF$
$AB = DE$
Prove: $BC = EF$

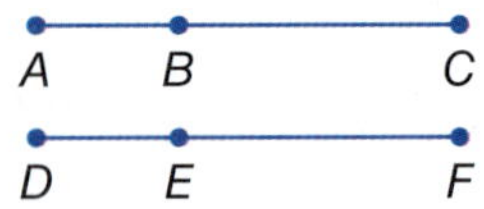

28. MODELS Brian is using six squares of cardboard to form a rectangular prism. What geometric figure do the pieces of cardboard represent, and how many lines will be formed by their intersections? (Lesson 2-5)

29. PATTERN BLOCKS Pattern blocks can be arranged to fit in a circular pattern without leaving spaces. Remember that the measurement around a full circle is 360°. Determine the degree measure of the numbered angles shown below. (Lesson 1-4)

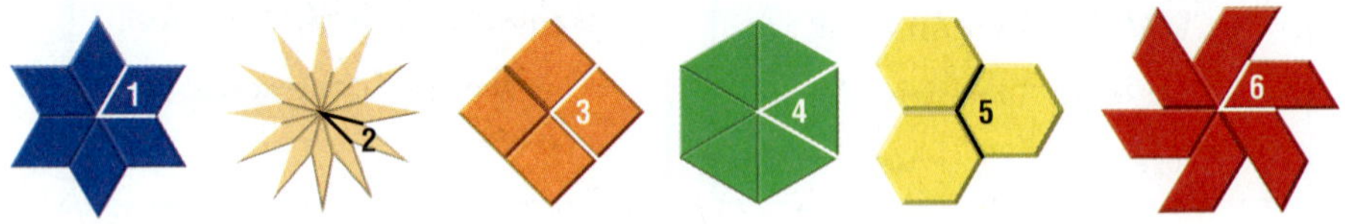

Simplify. (Lesson 0-9)

30. $\sqrt{48}$ **31.** $\sqrt{162}$ **32.** $\sqrt{25a^6b^4}$ **33.** $\sqrt{45xy^8}$

Skills Review

ALGEBRA Find x. (Lesson 1-5)

34.

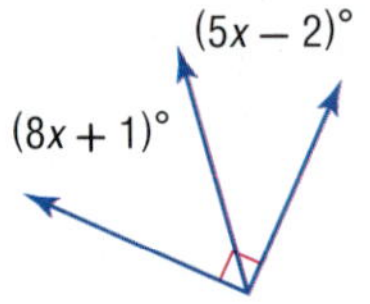

35.

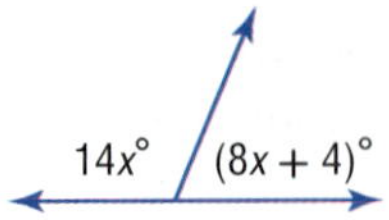

36.

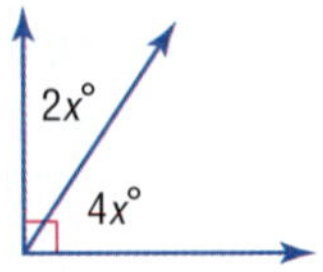

2-8 Proving Angle Relationships

Then

You identified and used special pairs of angles. (Lesson 1-5)

Now

- Write proofs involving supplementary and complementary angles.
- Write proofs involving congruent and right angles.

Math Online

glencoe.com

Why?

Jamal's school has a walkway that includes bricks with the names of graduates from each class. All of the bricks are rectangular, so when the bricks are laid, all of the angles form linear pairs.

Supplementary and Complementary Angles The Protractor Postulate illustrates the relationship between angle measures and real numbers.

Postulate 2.10 Protractor Postulate

Words Given any angle, the measure can be put into one-to-one correspondence with real numbers between 0 and 180.

Example

If $\overrightarrow{BA}$ is placed along the protractor at 0°, then the measure of $\angle ABC$ corresponds to a positive real number.

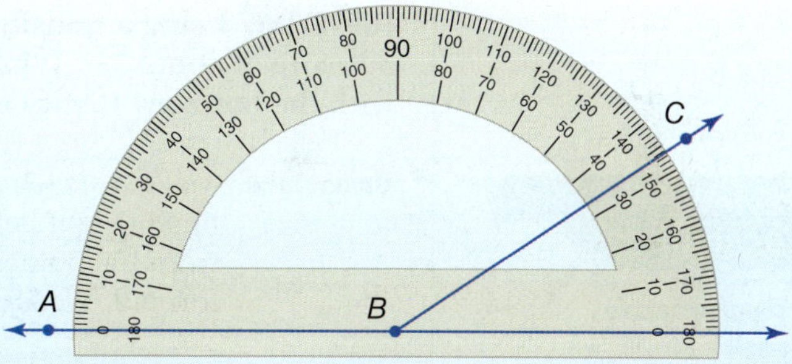

In Lesson 2-7, you learned about the Segment Addition Postulate. A similar relationship exists between the measures of angles.

Postulate 2.11 Angle Addition Postulate

D is in the interior of $\angle ABC$ if and only if $m\angle ABD + m\angle DBC = m\angle ABC$.

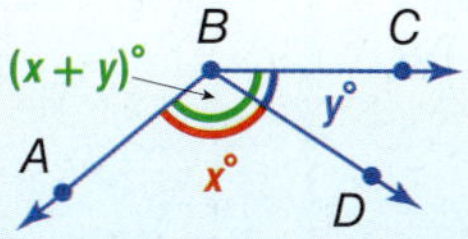

EXAMPLE 1 Use the Angle Addition Postulate

Find $m\angle 1$ if $m\angle 2 = 56$ and $m\angle JKL = 145$.

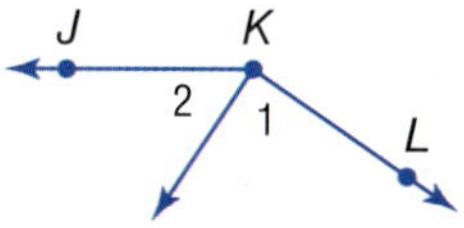

$m\angle 1 + m\angle 2 = m\angle JKL$	Angle Addition Postulate
$m\angle 1 + 56 = 145$	$m\angle 2 = 56$ $m\angle JKL = 145$
$m\angle 1 + 56 - 56 = 145 - 56$	Subtraction Property of Equality
$m\angle 1 = 89$	Substitution

Guided Practice

1. If $m\angle 1 = 23$ and $m\angle ABC = 131$, find the measure of $\angle 3$. Justify each step.

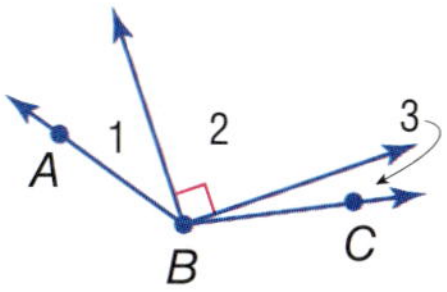

Personal Tutor glencoe.com

The Angle Addition Postulate can be used with other angle relationships to provide additional theorems relating to angles.

Theorems

2.3 Supplement Theorem If two angles form a linear pair, then they are supplementary angles.

Example $m\angle 1 + m\angle 2 = 180$

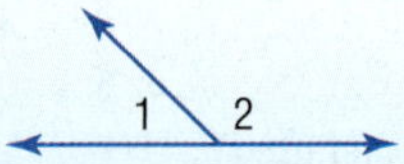

2.4 Complement Theorem If the noncommon sides of two adjacent angles form a right angle, then the angles are complementary angles.

Example $m\angle 1 + m\angle 2 = 90$

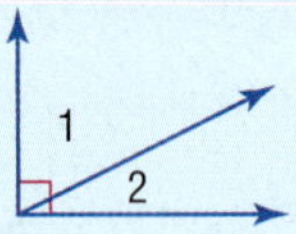

You will prove Theorems 2.3 and 2.4 in Exercises 16 and 17, respectively.

Real-World EXAMPLE 2 Use Supplement or Complement

SURVEYING Using a transit, a surveyor sights the top of a hill and records an angle measure of about 73°. What is the measure of the angle the top of the hill makes with the horizon? Justify each step.

Understand Make a sketch of the situation. The surveyor is measuring the angle of his line of sight below the vertical. Draw a vertical ray and a horizontal ray from the point where the surveyor is sighting the hill, and label the angles formed. We know that the vertical and horizontal rays form a right angle.

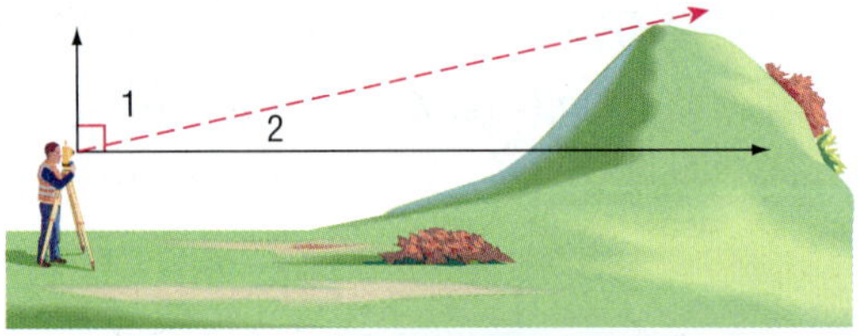

Plan Since $\angle 1$, and $\angle 2$ form a right angle, you can use the Complement Theorem.

Solve

$m\angle 1 + m\angle 2 = 90$	**Complement Theorem**
$73 + m\angle 2 = 90$	**$m\angle 1 = 73$**
$73 + m\angle 2 - 73 = 90 - 73$	**Subtraction Property of Equality**
$m\angle 2 = 17$	**Substitution**

The top of the hill makes a 17° angle with the horizon.

Check Since we know that the sum of the angles should be 90, check your math. The sum of 17 and 73 is 90. ✓

Guided Practice

2. $\angle 6$ and $\angle 7$ form linear pair. If $m\angle 6 = 3x + 32$ and $m\angle 7 = 5x + 12$, find x, $m\angle 6$, and $m\angle 7$. Justify each step.

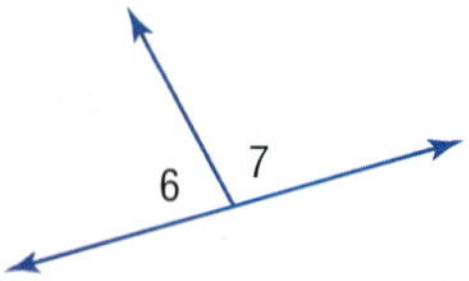

Personal Tutor glencoe.com

> **Review Vocabulary**
>
> **supplementary angles** two angles with measures that add to 180 (Lesson 1-5)
>
> **complementary angles** two angles with measures that add to 90 (Lesson 1-5)
>
> **linear pair** a pair of adjacent angles with noncommon sides that are opposite rays (Lesson 1-5)

Congruent Angles The properties of algebra that applied to the congruence of segments and the equality of their measures also hold true for the congruence of angles and the equality of their measures.

Theorem 2.5 Properties of Angle Congruence

Reflexive Property of Congruence
$\angle 1 \cong \angle 1$

Symmetric Property of Congruence
If $\angle 1 \cong \angle 2$, then $\angle 2 \cong \angle 1$.

Transitive Property of Congruence
If $\angle 1 \cong \angle 2$ and $\angle 2 \cong \angle 3$, then $\angle 1 \cong \angle 3$.

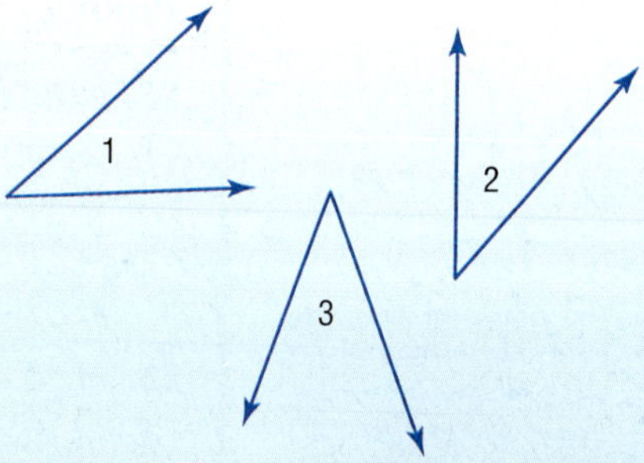

You will prove the Reflexive and Transitive Properties of Congruence in Exercises 18 and 19, respectively.

Proof Symmetric Property of Congruence

Given: $\angle A \cong \angle B$

Prove: $\angle B \cong \angle A$

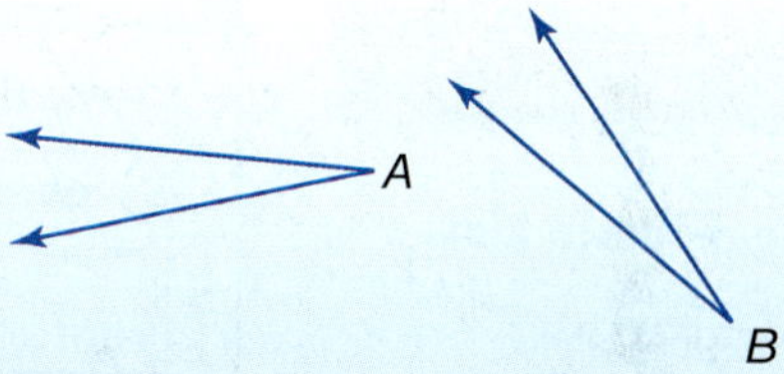

Paragraph Proof:

We are given $\angle A \cong \angle B$. By the definition of congruent angles, $m\angle A = m\angle B$. Using the Symmetric Property of Equality, $m\angle B = m\angle A$. Thus, $\angle B \cong \angle A$ by the definition of congruent angles.

Algebraic properties can be applied to prove theorems for congruence relationships involving supplementary and complementary angles.

Theorems

2.6 Congruent Supplements Theorem
Angles supplementary to the same angle or to congruent angles are congruent.

Abbreviation $\measuredangle$ suppl. to same $\angle$ or $\cong$ $\measuredangle$ are $\cong$.

Example If $m\angle 1 + m\angle 2 = 180$ and $m\angle 2 + m\angle 3 = 180$, then $\angle 1 \cong \angle 3$.

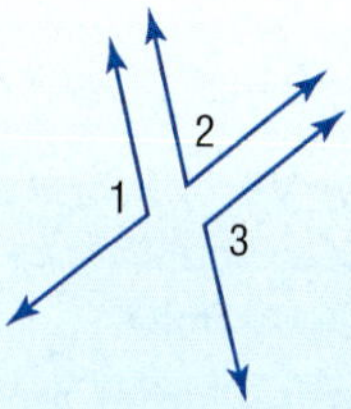

2.7 Congruent Complements Theorem
Angles complementary to the same angle or to congruent angles are congruent.

Abbreviation $\measuredangle$ compl. to same $\angle$ or $\cong$ $\measuredangle$ are $\cong$.

Example: If $m\angle 4 + m\angle 5 = 90$ and $m\angle 5 + m\angle 6 = 90$, then $\angle 4 \cong \angle 6$.

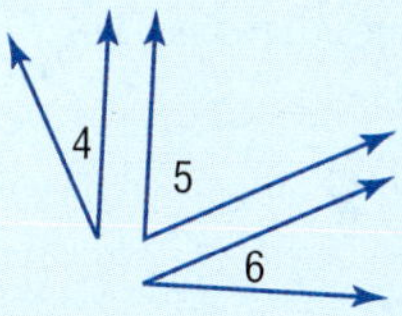

You will prove one case of Theorem 2.6 in Exercise 6.

ReadingMath

Abbreviations and Symbols The notation $\measuredangle$ means *angles*.

Proof: One Case of the Congruent Supplements Theorem

Given: ∠1 and ∠3 are supplementary.
∠2 and ∠3 are supplementary.

Prove: ∠1 ≅ ∠2

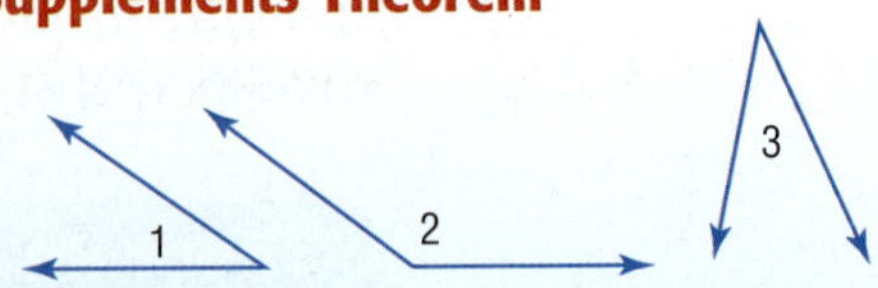

Proof:

Statements	Reasons
1. ∠1 and ∠3 are supplementary. ∠2 and ∠3 are supplementary.	**1.** Given
2. $m\angle 1 + m\angle 3 = 180$; $m\angle 2 + m\angle 3 = 180$	**2.** Definition of supplementary angles
3. $m\angle 1 + m\angle 3 = m\angle 2 + m\angle 3$	**3.** Substitution
4. $m\angle 3 = m\angle 3$	**4.** Reflexive Property
5. $m\angle 1 = m\angle 2$	**5.** Subtraction Property
6. ∠1 ≅ ∠2	**6.** Definition of congruent angles

Real-World Link

The 100-story John Hancock Building uses huge X-braces in its design. These diagonals are connected to the exterior columns, making it possible for strong wind forces to be carried from the braces to the exterior columns and back.

Source: PBS

EXAMPLE 3 Proofs Using Congruent Comp. or Suppl. Theorems

Prove that vertical angles 2 and 4 in the photo at the left are congruent.

Given: ∠2 and ∠4 are vertical angles.
Prove: ∠2 ≅ ∠4

Proof:

Statements	Reasons
1. ∠2 and ∠4 are vertical angles.	**1.** Given
2. ∠2 and ∠4 are nonadjacent angles formed by intersecting lines.	**2.** Definition of vertical angles
3. ∠2 and ∠3 from a linear pair. ∠3 and ∠4 form a linear pair.	**3.** Definition of a linear pair
4. ∠2 and ∠3 are supplementary. ∠3 and ∠4 are supplementary.	**4.** Supplement Theorem
5. ∠2 ≅ ∠4	**5.** ∡ suppl. to same ∠ or ≅ ∡ are ≅.

Guided Practice

3. In the figure, ∠*ABE* and ∠*DBC* are right angles. Prove that ∠*ABD* ≅ ∠*EBC*.

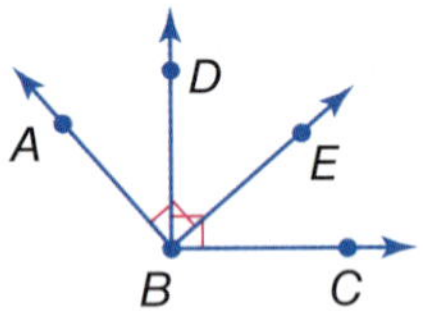

Personal Tutor glencoe.com

Review Vocabulary

Vertical Angles two nonadjacent angles formed by intersecting lines (Lesson 1-5)

Note that in Example 3, ∠1 and ∠3 are vertical angles. The conclusion in the example supports the following Vertical Angles Theorem.

Theorem 2.8 Vertical Angles Theorem

If two angles are vertical angles, then they are congruent.

Abbreviation Vert. ∡ are ≅.

Example ∠1 ≅ ∠3 and ∠2 ≅ ∠4

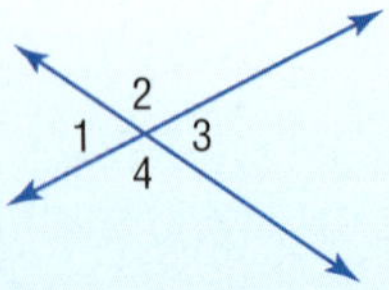

You will prove Theorem 2.8 in Exercise 28.

EXAMPLE 4 Use Vertical Angles

Prove that if $\overrightarrow{DB}$ bisects $\angle ADC$, then $\angle 2 \cong \angle 3$.

Given: $\overrightarrow{DB}$ bisects $\angle ADC$.

Prove: $\angle 2 \cong \angle 3$

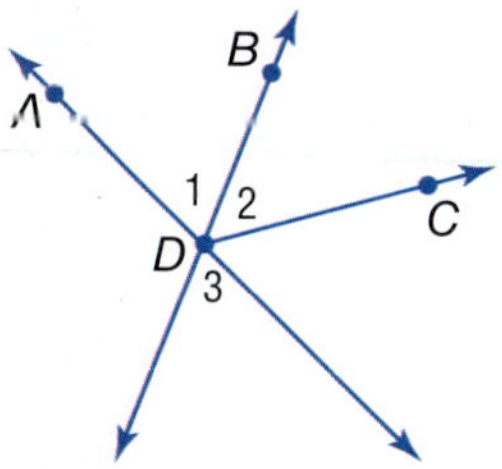

Proof:

Statements	Reasons
1. $\overrightarrow{DB}$ bisects $\angle ADC$.	**1.** Given
2. $\angle 1 \cong \angle 2$	**2.** Definition of angle bisector
3. $\angle 1$ and $\angle 3$ are vertical angles.	**3.** Definition of vertical angles
4. $\angle 3 \cong \angle 1$	**4.** Vert. ∠s are ≅.
5. $\angle 3 \cong \angle 2$	**5.** Transitive Property of Congruence
6. $\angle 2 \cong \angle 3$	**6.** Symmetric Property of Congruence

Guided Practice

4. If $\angle 3$ and $\angle 4$ are vertical angles, $m\angle 3 = 6x + 2$, and $m\angle 4 = 8x - 14$, find $m\angle 3$ and $m\angle 4$. Justify each step.

Personal Tutor glencoe.com

The theorems in this lesson can be used to prove the following right angle theorems.

Theorems Right Angle Theorems

Theorem	Example
2.9 Perpendicular lines intersect to form four right angles. **Example** If $\overleftrightarrow{AC} \perp \overleftrightarrow{DB}$, then $\angle 1$, $\angle 2$, $\angle 3$, and $\angle 4$ are rt. ∠s.	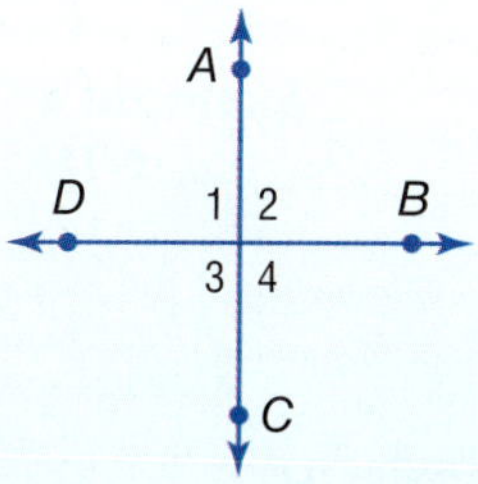
2.10 All right angles are congruent. **Example** If $\angle 1$, $\angle 2$, $\angle 3$, and $\angle 4$ are rt. ∠s, then $\angle 1 \cong \angle 2 \cong \angle 3 \cong \angle 4$.	
2.11 Perpendicular lines form congruent adjacent angles. **Example** If $\overleftrightarrow{AC} \perp \overleftrightarrow{DB}$, then $\angle 1 \cong \angle 2$, $\angle 2 \cong \angle 4$, $\angle 3 \cong \angle 4$, and $\angle 1 \cong \angle 3$.	
2.12 If two angles are congruent and supplementary, then each angle is a right angle. **Example** If $\angle 5 \cong \angle 6$ and $\angle 5$ is suppl. to $\angle 6$, then $\angle 5$ and $\angle 6$ are rt. ∠s.	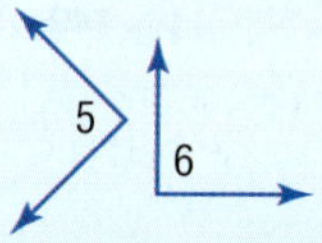
2.13 If two congruent angles form a linear pair, then they are right angles. **Example** If $\angle 7$ and $\angle 8$ form a linear pair, then $\angle 7$ and $\angle 8$ are rt. ∠s.	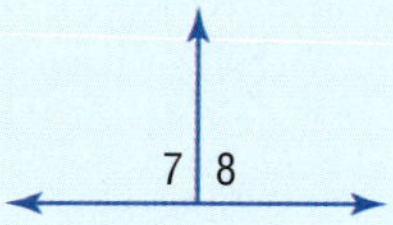

You will prove these theorems in Exercises 22–26.

ReadingMath

Perpendicular Recall from Lesson 1-5 that the symbol ⊥ means *is perpendicular to.*

Check Your Understanding

Example 1
p. 149

Find the measure of each numbered angle, and name the theorems that justify your work.

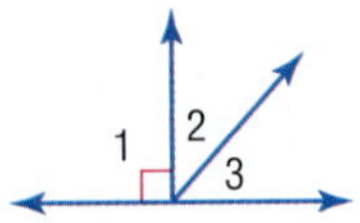

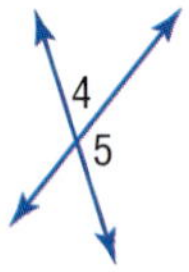

1. $m\angle 2 = 26$

2. $m\angle 2 = x, m\angle 3 = x - 16$

3. $m\angle 4 = 2x, m\angle 5 = x + 9$

4. $m\angle 4 = 3(x - 1), m\angle 5 = x + 7$

Example 2
p. 150

5. **PARKING** Refer to the diagram of the parking lot at the right. Given that $\angle 2 \cong \angle 6$, prove that $\angle 4 \cong \angle 8$.

Example 3
p. 152

6. **PROOF** **Copy and complete the proof of one case of Theorem 2.6.**

Given: $\angle 1$ and $\angle 3$ are complementary.
$\angle 2$ and $\angle 3$ are complementary.

Prove: $\angle 1 \cong \angle 2$

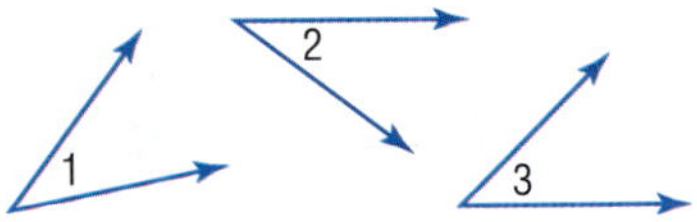

Proof:

Statements	Reasons
a. $\angle 1$ and $\angle 3$ are complementary. $\angle 2$ and $\angle 3$ are complementary.	**a.** ___?___
b. $m\angle 1 + m\angle 3 = 90$; $m\angle 2 + m\angle 3 = 90$	**b.** ___?___
c. $m\angle 1 + m\angle 3 = m\angle 2 + m\angle 3$	**c.** ___?___
d. ___?___	**d.** Reflexive Property
e. $m\angle 1 = m\angle 2$	**e.** ___?___
f. $\angle 1 \cong \angle 2$	**f.** ___?___

Example 4
p. 153

7. **PROOF** Write a two-column proof.

Given: $\angle 4 \cong \angle 7$
Prove: $\angle 5 \cong \angle 6$

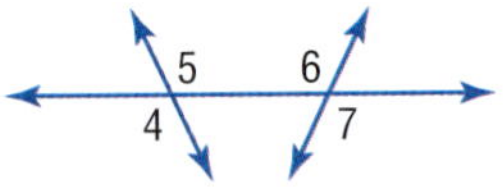

Practice and Problem Solving

● = **Step-by-Step Solutions** begin on page R20.
Extra Practice begins on page 969.

Examples 1–3
pp. 149–152

Find the measure of each numbered angle, and name the theorems used that justify your work.

8. $m\angle 5 = m\angle 6$

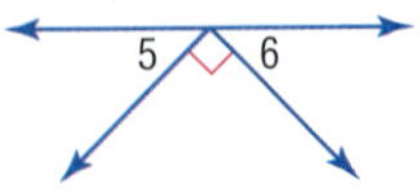

9. $\angle 2$ and $\angle 3$ are complementary. $\angle 1 \cong \angle 4$ and $m\angle 2 = 28$

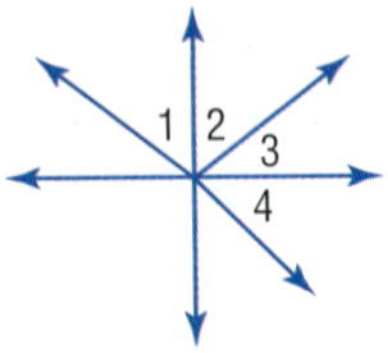

10. $\angle 2$ and $\angle 4$ and $\angle 4$ and $\angle 5$ are supplementary. $m\angle 4 = 105$

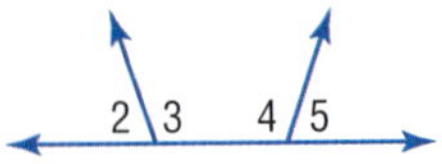

Find the measure of each numbered angle and name the theorems used that justify your work.

11. $m\angle 9 = 3x + 12$
$m\angle 10 = x - 24$

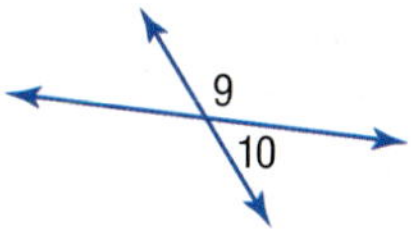

12. $m\angle 3 = 2x + 23$
$m\angle 4 = 5x - 112$

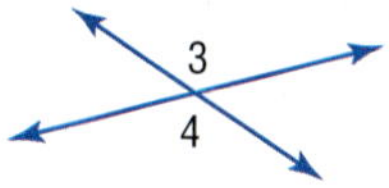

13. $m\angle 6 = 2x - 21$
$m\angle 7 = 3x - 34$

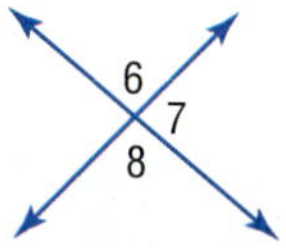

Example 4
p. 153

PROOF **Write a two-column proof.**

14. Given: $\angle ABC$ is a right angle.
Prove: $\angle ABD$ and $\angle CBD$ are complementary.

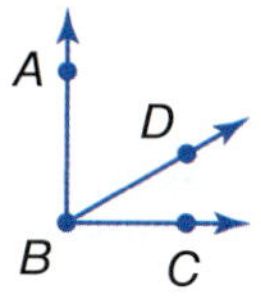

15. Given: $\angle 5 \cong \angle 6$
Prove: $\angle 4$ and $\angle 6$ are supplementary.

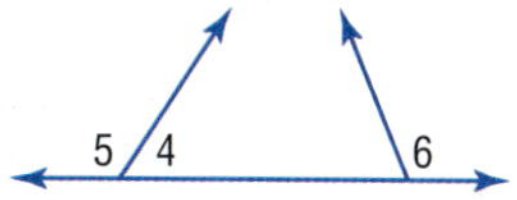

Write a proof for each theorem.

16. Supplement Theorem

17. Complement Theorem

18. Reflexive Property of Angle Congruence

19. Transitive Property of Angle Congruence

20. FLAGS Refer to the Florida state flag at the right. Prove that the sum of the four angle measures is 360.

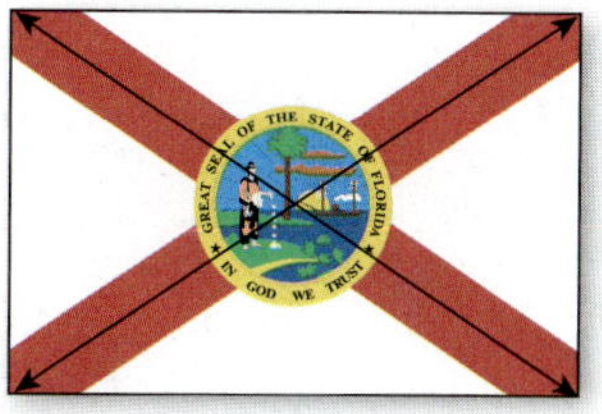

Real-World Link

A rattlesnake's fangs can reach lengths of up to six inches and fold into the snake's mouth parallel to the roof when the mouth is closed.

Source: *Encyclopaedia Britannica*

21. NATURE The diamondback rattlesnake is a pit viper with a diamond pattern on its back. An enlargement of the skin of the snake at the left is shown below. If $\angle 1 \cong \angle 4$, prove that $\angle 2 \cong \angle 3$.

PROOF **Use the figure to write a proof of each theorem.**

22. Theorem 2.9

23. Theorem 2.10

24. Theorem 2.11

25. Theorem 2.12

26. Theorem 2.13

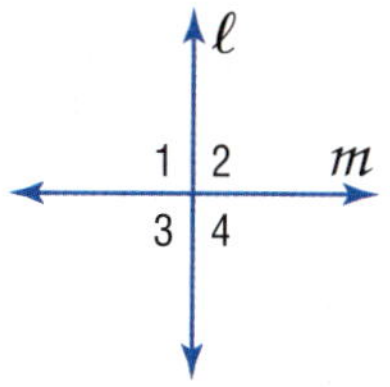

27. **MUSIC** To mark a specific tempo, the weight on the pendulum of a metronome is adjusted so that it swings at a specific rate. Suppose $\angle ABC$ in the photo is a right angle. If $m\angle 1 = 45$, write a paragraph proof to show that $\overrightarrow{BR}$ bisects $\angle ABC$.

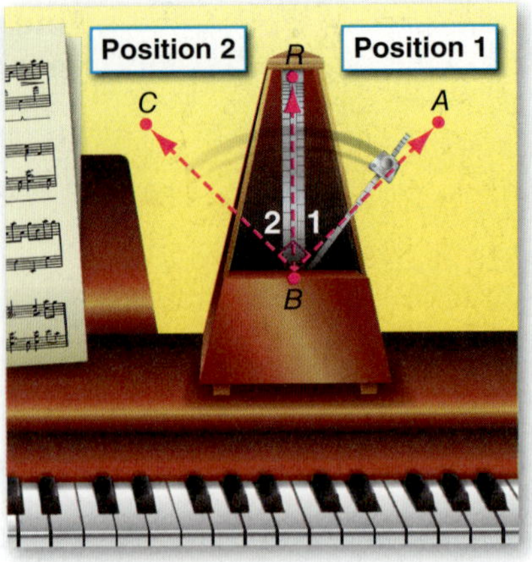

Real-World Link

The Four Corners Monument is located west of US Highway 160, 40 miles southwest of Cortez, Colorado.

Source: Utah Travel Industry

28. **PROOF** Write a proof of Theorem 2.8.

29. **GEOGRAPHY** Utah, Colorado, Arizona, and New Mexico all share a common point on their borders called Four Corners. This is the only place where four states meet in a single point. If $\angle 2$ is a right angle, prove that lines ℓ and m are perpendicular.

30. **MULTIPLE REPRESENTATIONS** In this problem, you will explore angle relationships.

a. **GEOMETRIC** Draw a right angle ABC. Place point D in the interior of this angle and draw $\overrightarrow{BD}$. Draw $\overrightarrow{KL}$ and construct $\angle JKL$ congruent to $\angle ABD$.

b. **VERBAL** Make a conjecture as to the relationship between $\angle JKL$ and $\angle DBC$.

c. **LOGICAL** Prove your conjecture.

H.O.T. Problems Use Higher-Order Thinking Skills

31. **OPEN ENDED** Draw an angle WXZ such that $m\angle WXZ = 45$. Construct $\angle YXZ$ congruent to $\angle WXZ$. Make a conjecture as to the measure of $\angle WXY$, and then prove your conjecture.

32. **WRITING IN MATH** Write the steps that you would use to complete the proof below.

Given: $\overline{BC} \cong \overline{CD}$, $AB = \frac{1}{2}BD$

Prove: $\overline{AB} \cong \overline{CD}$

33. **CHALLENGE** In this lesson, one case of the Congruent Supplements Theorem was proven. In Exercise 6, you proved the same case for the Congruent Complements Theorem. Explain why there is another case for each of these theorems. Then write a proof of this second case for each theorem.

34. **REASONING** Determine whether the following statement is *sometimes*, *always*, or *never* true. Explain your reasoning.

If one of the angles formed by two intersecting lines is acute, then the other three angles formed are also acute.

35. **WRITING IN MATH** Explain how you can use your protractor to quickly find the measure of the supplement of an angle.

Standardized Test Practice

36. GRIDDED RESPONSE What is the mode of this set of data?

4, 3, −2, 1, 4, 0, 1, 4

37. Find the measure of $\angle CFD$.

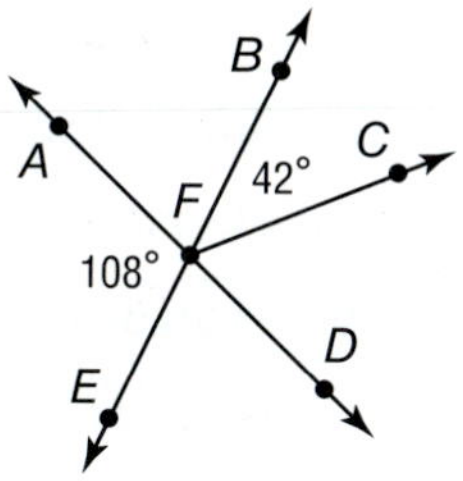

A 66° **C** 108°
B 72° **D** 138°

38. ALGEBRA Simplify.

$$4(3x - 2)(2x + 4) + 3x^2 + 5x - 6$$

F $9x^2 + 3x - 14$
G $9x^2 + 13x - 14$
H $27x^2 + 37x - 38$
J $27x^2 + 27x - 26$

39. SAT/ACT On a coordinate grid where each unit represents 1 mile, Isabel's house is located at (3, 0) and a mall is located at (0, 4). What is the distance between Isabel's house and the mall?

A 3 miles
B 5 miles
C 12 miles
D 13 miles

Spiral Review

40. MAPS On a U.S. map, there is a scale that lists kilometers on the top and miles on the bottom.

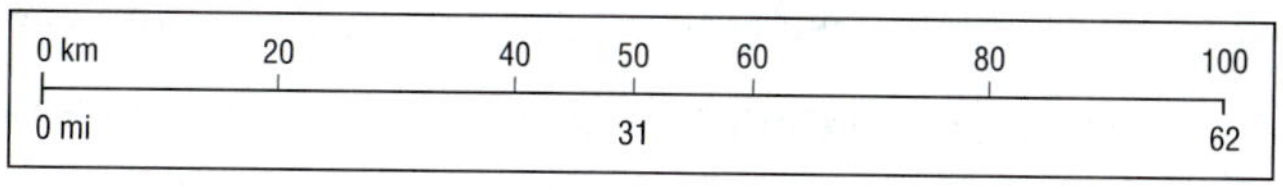

Suppose $\overline{AB}$ and $\overline{CD}$ are segments on this map. If $AB = 100$ kilometers and $CD = 62$ miles, is $\overline{AB} \cong \overline{CD}$? Explain. (Lesson 2-7)

State the property that justifies each statement. (Lesson 2-6)

41. If $y + 7 = 5$, then $y = -2$.

42. If $MN = PQ$, then $PQ = MN$.

43. If $a - b = x$ and $b = 3$, then $a - 3 = x$.

44. If $x(y + z) = 4$, then $xy + xz = 4$.

Determine the truth value of the following statement for each set of conditions.
If you have a fever, then you are sick. (Lesson 2-3)

45. You do not have a fever, and you are sick.

46. You have a fever, and you are not sick.

47. You do not have a fever, and you are not sick.

48. You have a fever, and you are sick.

Skills Review

Refer to the figure. (Lesson 1-1)

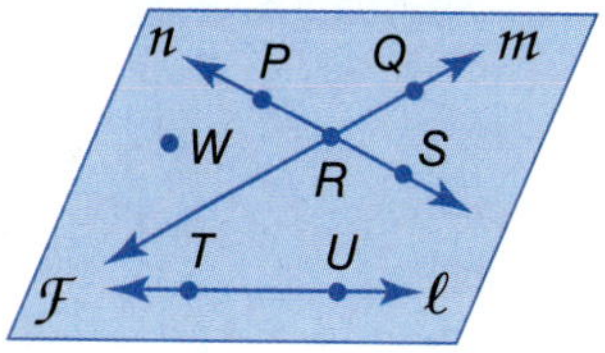

49. Name a line that contains point P.

50. Name the intersection of lines n and m.

51. Name a point not contained in lines ℓ, m, or n.

52. What is another name for line n?

53. Does line ℓ intersect line m or line n? Explain.

CHAPTER 2

Study Guide and Review

Math Online glencoe.com
- STUDY TO GO
- Vocabulary Review

Chapter Summary

Key Concepts

Inductive Reasoning and Logic (Lessons 2-1 and 2-2)

- Inductive reasoning: a conjecture is reached based on observations of a previous pattern
- Counterexample: an example that proves a conjecture is false
- Negation of statement *p: not p*
- Conjunction: a compound statement formed with the word *and*
- Disjunction: a compound statement formed with the word *or*

Conditional Statements (Lesson 2-3)

- An if-then statement is written in the form if p, then q in which p is the hypothesis and q is the conclusion.

statement	$p \rightarrow q$
converse	$q \rightarrow p$
inverse	not $p \rightarrow$ not q
contrapositive	not $q \rightarrow$ not p

Deductive Reasoning (Lesson 2-4)

- Law of Detachment: If $p \rightarrow q$ is true and p is true, then q is also true.
- Law of Syllogism: If $p \rightarrow q$ and $q \rightarrow r$ are true, then $p \rightarrow r$ is also true.

Proof (Lessons 2-5 through 2-8)

- **Step 1** List the given information and draw a diagram, if possible.
- **Step 2** State what is to be proved.
- **Step 3** Create a deductive argument.
- **Step 4** Justify each statement with a reason.
- **Step 5** State what you have proved.

FOLDABLES® Study Organizer

Be sure the Key Concepts are noted in your Foldable.

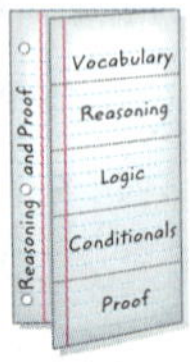

Key Vocabulary

algebraic proof (p. 134)
axiom (p. 125)
compound statement (p. 97)
conclusion (p. 105)
conditional statement (p. 105)
conjecture (p. 89)
conjunction (p. 97)
contrapositive (p. 107)
converse (p. 107)
counterexample (p. 92)
deductive argument (p. 127)
deductive reasoning (p. 115)
disjunction (p. 98)
formal proof (p. 135)
hypothesis (p. 105)
if-then statement (p. 105)
inductive reasoning (p. 89)
informal proof (p. 127)
inverse (p. 107)
logically equivalent (p. 108)
negation (p. 97)
paragraph proof (p. 127)
postulate (p. 125)
proof (p. 126)
related conditionals (p. 107)
statement (p. 97)
theorem (p. 127)
truth table (p. 99)
truth value (p. 97)
two-column proof (p. 135)

Vocabulary Check

State whether each sentence is *true* or *false*. If *false*, replace the underlined term to make a true sentence.

1. A postulate is a statement that requires proof.
2. The first part of an if-then statement is the conjecture.
3. Deductive reasoning uses the laws of mathematics to reach logical conclusions from given statements.
4. The contrapositive is formed by negating the hypothesis and conclusion of a conditional.
5. A conjunction is formed by joining two or more statements with the word *and*.
6. A theorem is a statement that is accepted as true without proof.
7. The converse is formed by exchanging the hypothesis and conclusion of a conditional.
8. To show that a conjecture is false, you would provide a disjunction.
9. The inverse of a statement p would be written in the form *not p*.
10. In a two-column proof, the properties that justify each step are called reasons.

Lesson-by-Lesson Review

2-1 Inductive Reasoning and Conjecture (pp. 89–96)

Determine whether each conjecture is *true* or *false*. If false, give a counterexample.

11. If $\angle 1$ and $\angle 2$ are supplementary angles, then $\angle 1$ and $\angle 2$ form a linear pair.

12. If $W(-3, 2)$, $X(-3, 7)$, $Y(6, 7)$, $Z(6, 2)$, then quadrilateral $WXYZ$ is a rectangle.

13. PARKS Jacinto enjoys hiking with his dog in the forest at his local park. While on vacation in Smoky Mountain National Park in Tennessee, he was disappointed that dogs were not allowed on most hiking trails. Make a conjecture about why his local park and the national park have differing rules with regard to pets.

EXAMPLE 1

Determine whether each conjecture is *true* or *false*. If false, give a counterexample.

a. **$c = d$, $d = c$ is an example of a property of real numbers.**

$c = d$, $d = c$ is an example of the Symmetric Property of real numbers, so the conjecture is true.

b. **If $AB + CD = AD$, then B and C are between A and D.**

This conjecture is false. In the figure below, $AB + CD = AD$, but B and C are not between A and D.

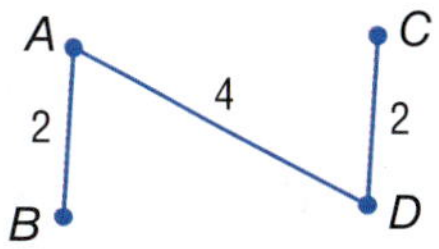

2-2 Logic (pp. 97–104)

Use the following statements to write a compound statement for each conjunction or disjunction. Then find its truth value. Explain.

p: A plane contains at least three noncollinear points.
q: A square yard is equivalent to three square feet.
r: The sum of the measures of two complementary angles is 180.

14. $\sim q \vee r$ **15.** $p \wedge \sim r$ **16.** $\sim p \vee q$

17. PETS The Venn diagram shows the results of a pet store survey to determine the pets customers owned.

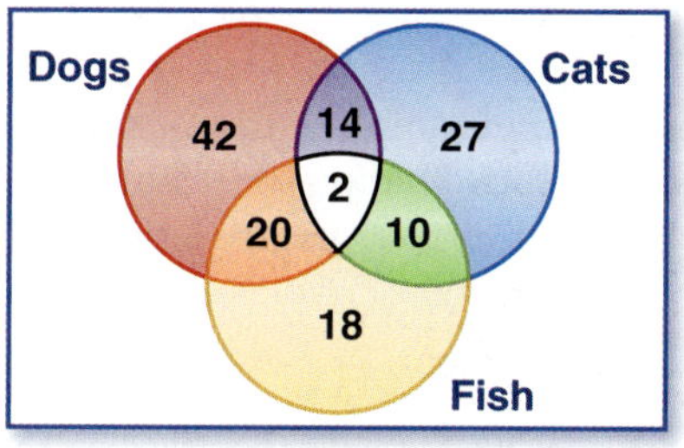

a. How many customers had only fish?

b. How many had only cats and dogs?

c. How many had dogs as well as fish?

EXAMPLE 2

Use the following statements to write a compound statement for each conjunction or disjunction. Then find its truth value. Explain.

p: x^2 is a nonnegative number.

q: Adjacent angles lie in the same plane.

r: A negative number is not a real number.

a. $\sim q \wedge r$

$\sim q \wedge r$: Adjacent angles do not lie in the same plane, and a negative number is not a real number.

Since both $\sim q$ and r are false, $\sim q \wedge r$ is false.

b. p or r

p or r: x^2 is a nonnegative number, or a negative number is not a real number.

p or r is true because p is true. It does not matter that r is false.

Study Guide and Review

2-3 Conditional Statements (pp. 105–113)

Determine the truth value of each conditional statement. If *true*, explain your reasoning. If *false*, give a counterexample.

18. If you square an integer, then the result is a positive integer.

19. If a hexagon has eight sides, then all of its angles will be obtuse.

20. Write the converse, inverse, and contrapositive of the following true conditional. Then, determine whether each related conditional is *true* or *false*. If a statement is false, find a counterexample.

    If two angles are congruent, then they have the same degree measure.

EXAMPLE 3

Write the *converse*, *inverse*, and *contrapositive* of the following true conditional.

If a figure is a square, then it is a parallelogram.

Converse: If a figure is a parallelogram, then it is a square.

Inverse: If a figure is not a square, then it is not a parallelogram.

Contrapositive: If a figure is not a parallelogram, then it is not a square.

2-4 Deductive Reasoning (pp. 115–123)

Draw a valid conclusion from the given statements, if possible. Then state whether your conclusion was drawn using the Law of Detachment or the Law of Syllogism. If no valid conclusion can be drawn, write *no valid conclusion* and explain your reasoning.

21. **Given:** If a quadrilateral has diagonals that bisect each other, then it is a parallelogram.

    The diagonals of quadrilateral *PQRS* bisect each other.

22. **Given:** If Liana struggles in science class, then she will receive tutoring.

    If Liana stays after school on Thursday, then she will receive tutoring.

23. **EARTHQUAKES** Determine whether the stated conclusion is valid based on the given information. If not, write *invalid*. Explain.

    Given: If an earthquake measures a 7.0 or higher on the Richter scale, then it is considered a major earthquake that could cause serious damage. The 1906 San Francisco earthquake measured 8.0 on the Richter scale.

    Conclusion: The 1906 San Francisco earthquake was a major earthquake that caused serious damage.

EXAMPLE 4

Use the Law of Syllogism to determine whether a valid conclusion can be reached from the following statements.

(1) If the measure of an angle is greater than 90, then it is an obtuse angle.

(2) If an angle is an obtuse angle, then it is not a right angle.

p: the measure of an angle is greater than 90

q: the angle is an obtuse angle

r: the angle is not a right angle

Statement (1): $p \rightarrow q$

Statement (2): $q \rightarrow r$

Since the given statements are true, use the Law of Syllogism to conclude that $p \rightarrow r$. That is, *If the measure of an angle is greater than 90, then it is not a right angle.*

MIXED PROBLEM SOLVING
For mixed problem-solving practice, see page 1000.

2-5 Postulates and Paragraph Proof (pp. 125–132)

Determine whether each statement is *always, sometimes,* or *never* true. Explain.

24. Two planes intersect at a point.

25. Three points are contained in more than one plane.

26. If line m lies in plane X and line m contains a point Q, then point Q lies in plane X.

27. If two angles are complementary, then they form a right angle.

28. **NETWORKING** Six people are introduced at a business convention. If each person shakes hands with each of the others, how many handshakes will be exchanged? Include a model to support your reasoning.

EXAMPLE 5

Determine whether each statement is *always, sometimes,* or *never* true. Explain.

a. If points X, Y, and Z lie in plane R, then they are not collinear.

Sometimes; the fact that X, Y, and Z are contained in plane R has no bearing on whether those points are collinear or not.

b. For any two points A and B, there is exactly one line that contains them.

Always; according to Postulate 2-1, there is exactly one line through any two points.

2-6 Algebraic Proof (pp. 134–141)

State the property that justifies each statement.

29. If $7(x - 3) = 35$, then $35 = 7(x - 3)$.

30. If $2x + 19 = 27$, then $2x = 8$.

31. $5(3x + 1) = 15x + 5$

32. $7x - 2 = 7x - 2$

33. If $12 = 2x + 8$ and $2x + 8 = 3y$, then $12 = 3y$.

34. Copy and complete the following proof.
Given: $6(x - 4) = 42$
Prove: $x = 11$

Statements	Reasons
a. $6(x - 4) = 42$	**a.** ?
b. $6x - 24 = 42$	**b.** ?
c. $6x = 66$	**c.** ?
d. $x = 11$	**d.** ?

35. Write a two-column proof to show that if $PQ = RS$, $PQ = 5x + 9$, and $RS = x - 31$, then $x = -10$.

36. **GRADES** Jerome received the same quarter grade as Paula. Paula received the same quarter grade as Heath. Which property would show that Jerome and Heath received the same grade?

EXAMPLE 6

Write a two-column proof.

Given: $\frac{5x - 3}{6} = 2x + 1$

Prove: $x = -\frac{9}{7}$

Proof:

Statements	Reasons
1. $\frac{5x - 3}{6} = 2x + 1$	**1.** Given
2. $5x - 3 = 6(2x + 1)$	**2.** Multiplication Property of Equality
3. $5x - 3 = 12x + 6$	**3.** Distributive Property of Equality
4. $-3 = 7x + 6$	**4.** Subtraction Property of Equality
5. $-9 = 7x$	**5.** Subtraction Property of Equality
6. $-\frac{9}{7} = x$	**6.** Division Property of Equality
7. $x = -\frac{9}{7}$	**7.** Symmetric Property of Equality

Study Guide and Review

2-7 Proving Segment Relationships (pp. 142–148)

Write a two-column proof.

37. **Given:** X is the midpoint of $\overline{WY}$ and $\overline{VZ}$.

Prove: $VW = ZY$

38. **Given:** $AB = DC$

Prove: $AC = DB$

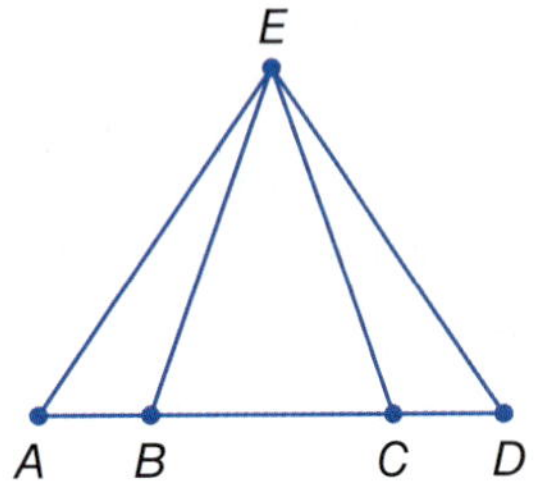

39. **GEOGRAPHY** Leandro is planning to drive from Kansas City to Minneapolis along Interstate 35. The map he is using gives the distance from Kansas City to Des Moines as 194 miles and from Des Moines to Minneapolis as 243 miles. What allows him to conclude that the distance he will be driving is 437 miles from Kansas City to Minneapolis? Assume that Interstate 35 forms a straight line.

EXAMPLE 7

Write a two-column proof.

Given: B is the midpoint of $\overline{AC}$.

C is the midpoint of $\overline{BD}$.

Prove: $\overline{AB} \cong \overline{CD}$

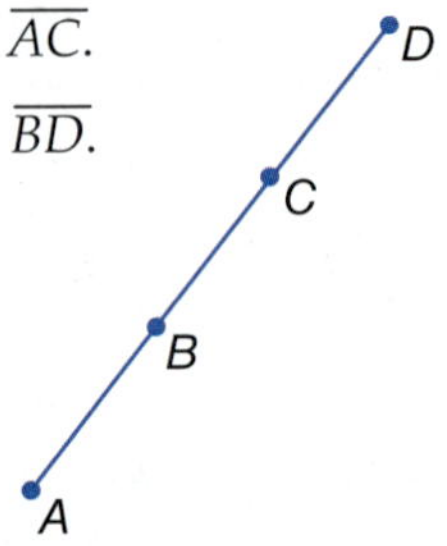

Proof:

Statements	Reasons
1. B is the midpoint of $\overline{AC}$.	**1.** Given
2. $\overline{AB} \cong \overline{BC}$	**2.** Definition of midpoint
3. C is the midpoint of $\overline{BD}$.	**3.** Given
4. $\overline{BC} \cong \overline{CD}$	**4.** Definition of midpoint
5. $\overline{AB} \cong \overline{CD}$	**5.** Transitive Property of Equality

2-8 Proving Angle Relationships (pp. 149–157)

Find the measure of each angle.

40. $\angle 5$

41. $\angle 6$

42. $\angle 7$

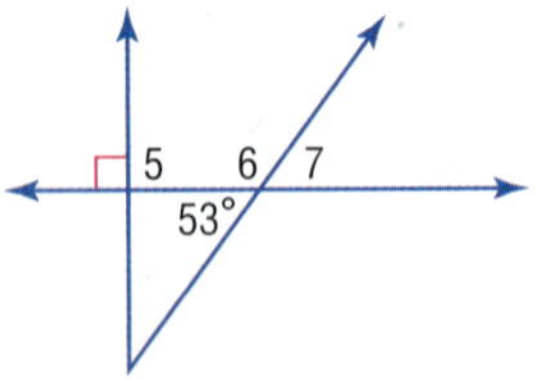

43. **PROOF** Write a two-column proof.

Given: $\angle 1 \cong \angle 4$, $\angle 2 \cong \angle 3$

Prove: $\angle AFC \cong \angle EFC$

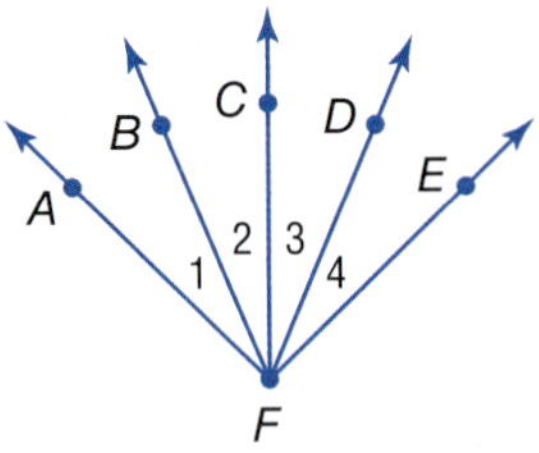

EXAMPLE 8

Find the measure of each numbered angle if $m\angle 1 = 72$ and $m\angle 3 = 26$.

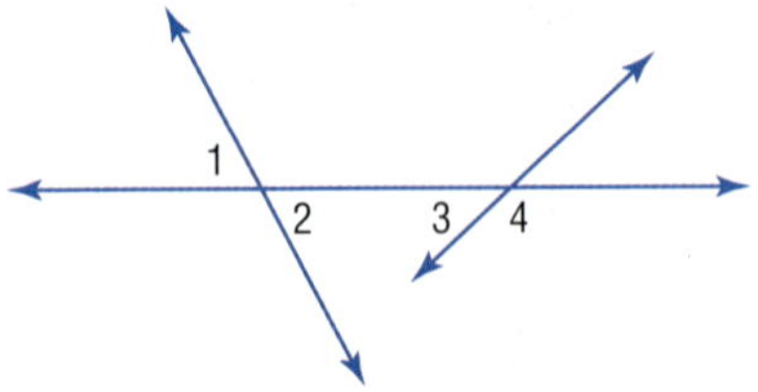

$m\angle 2 = 72$, since $\angle 1$ and $\angle 2$ are vertical angles.

$\angle 3$ and $\angle 4$ form a linear pair and must be supplementary angles.

$26 + m\angle 4 = 180$	**Definition of supplementary angles**
$m\angle 4 = 154$	**Subtract 26 from each side.**

CHAPTER 2 Practice Test

Write a conjecture that describes the pattern in each sequence. Then use your conjecture to find the next item in the sequence.

1. 15, 30, 45, 60

2.

Use the following statements to write a compound statement for each conjunction or disjunction. Then find its truth value.

p: $5 < -3$

q: **All vertical angles are congruent.**

r: **If $4x = 36$, then $x = 9$.**

3. p and q

4. $(p \vee q) \wedge r$

5. PROOF Write a paragraph proof.

Given: $\overline{JK} \cong \overline{CB}$, $\overline{KL} \cong \overline{AB}$

Prove: $\overline{JL} \cong \overline{AC}$

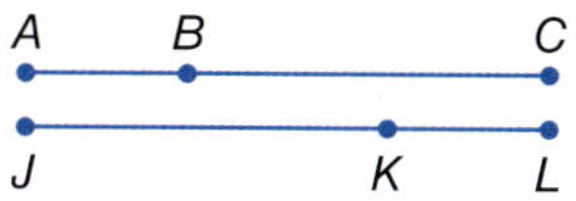

6. SPORTS Refer to the Venn diagram that represents the sports students chose to play at South High School last year.

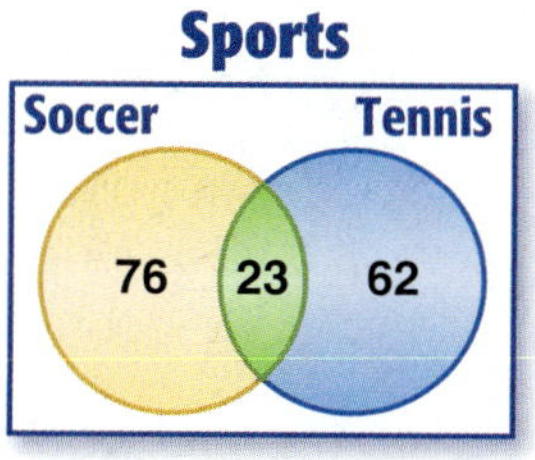

a. Describe the sports that the students in the nonintersecting portion of the tennis region chose.

b. How many students played soccer and tennis?

7. Determine whether the stated conclusion is valid based on the given information. If not, write *invalid*. Explain your reasoning.

Given: If a lawyer passes the bar exam, then he or she can practice law. Candice passed the bar exam.

Conclusion: Candice can practice law.

8. PROOF Copy and complete the following proof.

Given: $3(x - 4) = 2x + 7$

Prove: $x = 19$

Proof:

Statements	Reasons
a. $3(x - 4) = 2x + 7$	**a.** Given
b. $3x - 12 = 2x + 7$	**b.** ?
c. ?	**c.** Subtraction Property
d. $x = 19$	**d.** ?

Determine whether each statement is *always*, *sometimes*, or *never* true.

9. Two angles that are supplementary form a linear pair.

10. If B is between A and C, then $AC + AB = BC$.

11. If two lines intersect to form congruent adjacent angles, then the lines are perpendicular.

Find the measure of each numbered angle, and name the theorems that justify your work.

12. $m\angle 1 = x$,
$m\angle 2 = x - 6$

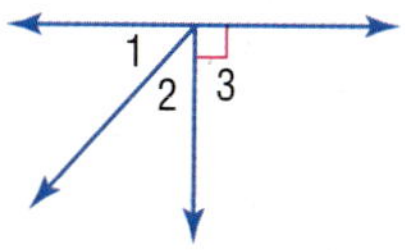

13. $m\angle 7 = 2x + 15$,
$m\angle 8 = 3x$

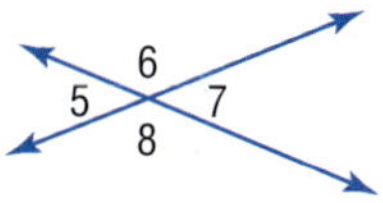

Write each statement in if-then form.

14. An acute angle measures less than 90.

15. Two perpendicular lines intersect to form right angles.

16. *If a triangle has one obtuse angle, then it is an obtuse triangle.*

Which of the following statements is the contrapositive of the conditional above?

A If a triangle is not obtuse, then it has one obtuse angle.

B If a triangle does not have one obtuse angle, then it is not an obtuse triangle.

C If a triangle is not obtuse, then it does not have one obtuse angle.

D If a triangle is obtuse, then it has one obtuse angle.

Preparing for Standardized Tests

Logical Reasoning

Solving geometry problems frequently requires the use of logical reasoning. You can use the fundamentals of logical reasoning to help you solve problems on standardized tests.

Strategies for Using Logical Reasoning

Step 1

Read the problem to determine what information you are given and what you need to find out in order to answer the question.

Step 2

Determine if you can apply one of the principles of logical reasoning to the problem.

- **Counterexample:** A counterexample contradicts a statement that is known to be true.

  Identify any answer choices that contradict the problem statement and eliminate them.

- **Postulates:** A postulate is a statement that describes a fundamental relationship in geometry.

  Determine if you can apply a postulate to draw a logical conclusion.

Step 3

If you cannot reach a conclusion using only the principles in Step 2, determine if one of the tools below would be helpful.

- **Patterns:** Look for a pattern to make a conjecture.
- **Truth Tables:** Use a truth table to organize the truth values of the statement provided in the problem.
- **Venn Diagrams:** Use a Venn Diagram to clearly represent the relationships between members of groups.
- **Proofs:** Use deductive and inductive reasoning to reach a conclusion in the form of a proof.

Step 4

If you still cannot reach a conclusion using the tools in Step 3, make a **conjecture,** or educated guess, about which answer choice is most reasonable. Then mark the problem so that you can return to it if you have extra time at the end of the exam.

EXAMPLE

Read the problem. Identify what you need to know. Then use the information in the problem to solve.

> In a school of 292 students, 94 participate in sports, 122 participate in academic clubs, and 31 participate in both. How many students at the school do not participate in sports or academic clubs?
>
> **A** 95 **C** 122
>
> **B** 107 **D** 138

Read the problem carefully. There are no clear counterexamples, and a postulate cannot be used to draw a logical conclusion. Therefore, consider the tools that you can use to organize the information.

A Venn diagram can be used to show the intersection of two sets. Make a Venn diagram with the information provided in the problem statement.

Determine how many students participate in only sports or academic clubs.

Only sports: $94 - 31 = 63$

Only academic clubs: $122 - 31 = 91$

Use the information to calculate the number of students who do not participate in either sports or academic clubs.

$292 - 63 - 91 - 31 = 107$

There are 107 students who do not participate in either sports or academic clubs. The correct answer is B.

School Participation

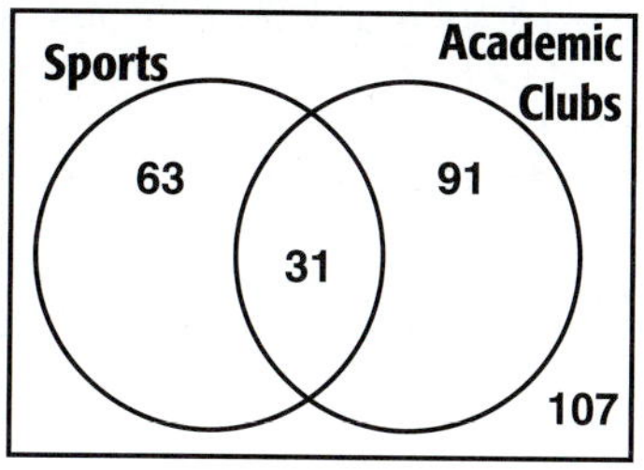

Exercises

Read each question. Then fill in the correct answer on the answer document provided by your teacher or on a sheet of paper.

1. Determine the truth of the following statement. If the statement is false, give a counterexample.

The product of two even numbers is even.

A false; $8 \times 4 = 32$

B false; $7 \times 6 = 42$

C false; $3 \times 10 = 30$

D true

2. Find the next item in the pattern.

F

H

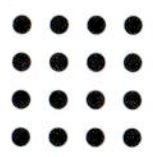

G

J

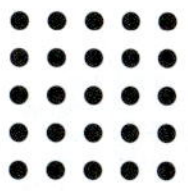

CHAPTER 2

Standardized Test Practice

Cumulative, Chapters 1 and 2

Multiple Choice

Read each question. Then fill in the correct answer on the answer document provided by your teacher or on a sheet of paper.

1. Which conjunction is true for statements p and q below?

p: There are four letters in MATH.
q: There are two vowels in MATH.

A $\sim p \wedge \sim q$

B $p \wedge q$

C $p \wedge \sim q$

D $\sim p \wedge q$

2. In the diagram below, $\angle 1 \cong \angle 3$.

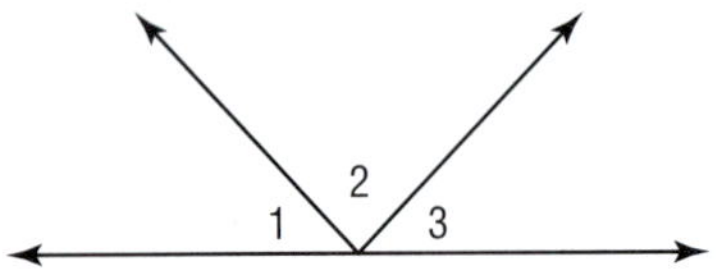

Which of the following conclusions does not have to be true?

F $m\angle 1 - m\angle 2 + m\angle 3 = 90$

G $m\angle 1 + m\angle 2 + m\angle 3 = 180$

H $m\angle 1 + m\angle 2 = m\angle 2 + m\angle 3$

J $m\angle 2 - m\angle 1 = m\angle 2 - m\angle 3$

3. *Two supplementary angles always form a linear pair.*

Which of the following best describes a *counterexample* to the assertion above?

A two acute angles

B two nonadjacent angles

C two obtuse angles

D two right angles

Test-TakingTip

Question 3 A *counterexample* is an example used to show that a given statement is not always true.

4. Determine which statement follows logically from the given statements.

If it rains today, the game will be cancelled.
Cancelled games are made up on Saturdays.

F If a game is cancelled, it was because of rain.

G If it rains today, the game will be made up on Saturday.

H Some cancelled games are not made up on Saturdays.

J If it does not rain today, the game will not be made up on Saturday.

5. In the diagram, $\overline{BD}$ intersects $\overline{AE}$ at C. Which of the following conclusions does *not* have to be true?

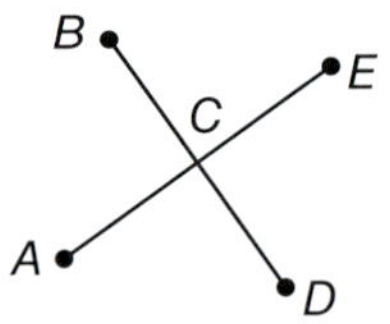

A $\angle ACB \cong \angle ECD$

B $\angle ACB$ and $\angle ACD$ form a linear pair.

C $\angle BCE$ and $\angle ACD$ are vertical angles.

D $\angle BCE$ and $\angle ECD$ are complementary angles.

6. A farmer needs to make a 1000-square-foot rectangular enclosure for her cows. She wants to save money by purchasing the least amount of fencing possible to enclose the area. What whole-number dimensions will require the least amount of fencing?

F 8 ft by 125 ft

G 10 ft by 100 ft

H 20 ft by 50 ft

J 25 ft by 40 ft

Short Response/Gridded Response

Record your answers on the answer sheet provided by your teacher or on a sheet of paper.

7. Points A, B, C, and D are collinear, with point B between points A and C and point C between points B and D. Complete the statement.

$AB +$ __?__ $= AD$

8. GRIDDED RESPONSE Suppose line m contains points D, E, and F. If $DE = 12$ millimeters, $EF = 15$ millimeters, and point D is between points E and F, what is the length of $\overline{DF}$? Express your answer in millimeters.

9. Use the proof to answer the question.

Given: $\angle A$ is the complement of $\angle B$.
$m\angle B = 46$

Prove: $m\angle A = 44$

Proof:

Statements	Reasons
1. A is the complement of $\angle B$; $m\angle B = 46$.	**1.** Given
2. $m\angle A + m\angle B = 90$	**2.** Def. of comp. angles
3. $m\angle A + 46 = 90$	**3.** Substitution Prop.
4. $m\angle A + 46 - 46 = 90 - 46$	**4.** __?__
5. $m\angle A = 44$	**5.** Substitution Prop.

What reason can be given to justify Statement 4?

10. Write the contrapositive of the statement.

If an angle measures greater than 90°, then it is obtuse.

11. GRIDDED RESPONSE Point E is the midpoint of $\overline{DF}$. If $DE = 8x - 3$ and $EF = 3x + 7$, what is x?

Extended Response

Record your answers on a sheet of paper. Show your work.

12. Consider the pattern.

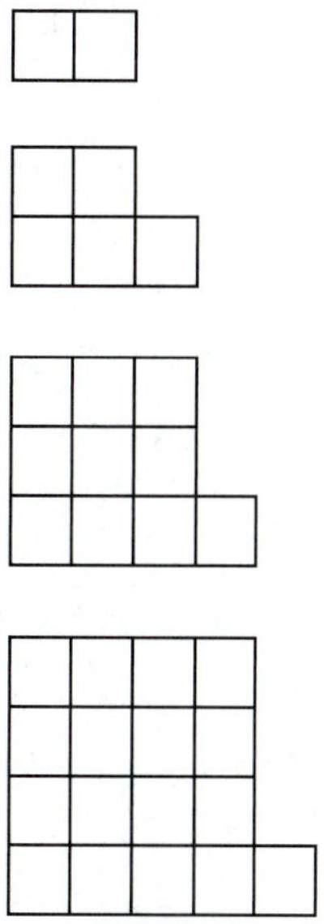

Figure 1
Figure 2
Figure 3
Figure 4

a. Make a conjecture about the number of squares in each figure.

b. Write an algebraic expression that can be used to find the number of squares in the nth figure in the pattern.

c. How many squares will be needed to make the 6th figure of the pattern?

Need Extra Help?

If you missed Question...	1	2	3	4	5	6	7	8	9	10	11	12
Go to Lesson or Page...	2-2	2-8	2-1	2-4	2-7	1-6	1-2	1-2	2-8	2-3	1-3	2-1

CHAPTER 3

Parallel and Perpendicular Lines

Then

In Chapters 1 and 2, you learned about lines and angles and writing geometric proofs.

Now

In Chapter 3, you will:

- Identify and prove angle relationships that occur with parallel lines and a transversal.
- Use slope to analyze a line and to write its equation.
- Find the distance between a point and a line and between two parallel lines.

Why?

CONSTRUCTION and ENGINEERING Architects, carpenters, and engineers, use parallel and perpendicular lines to design buildings, furniture, and machines.

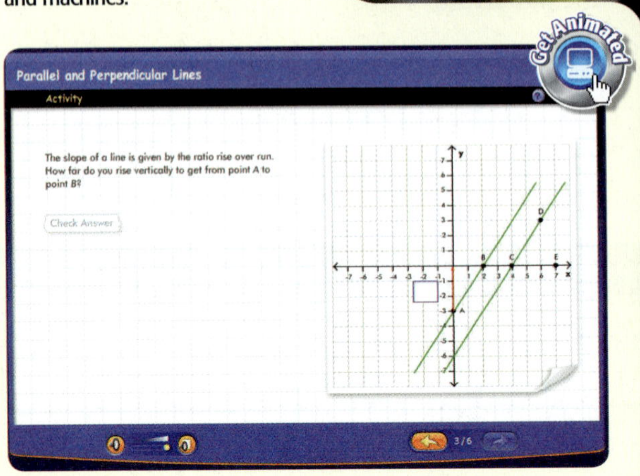

Get Ready for Chapter 3

Diagnose Readiness You have two options for checking Prerequisite Skills.

Text Option

Take the Quick Check below. Refer to the Quick Review for help.

QuickCheck

Refer to the figure to identify each of the following. (Lesson 1-1)

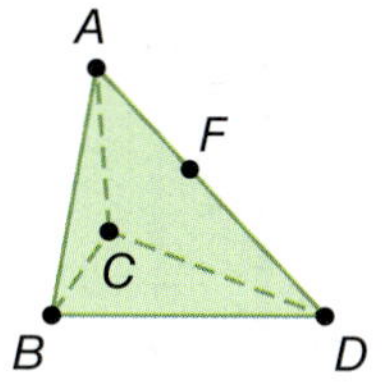

1. How many planes are shown in this figure?
2. Name three points that are collinear.
3. Are points C and D coplanar? Explain.
4. **PHOTOGRAPHY** Tina is taking a picture of her friends. If she sets a tripod level on the ground, will the bottom of each of the three legs of the tripod be coplanar? (Lesson 2-5)

Find each angle measure. (Lesson 1-5)

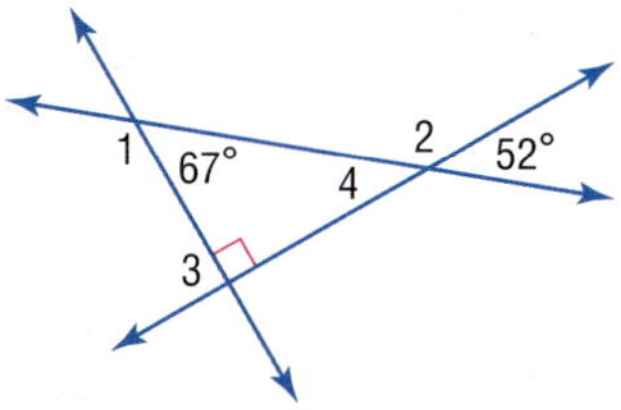

5. $\angle 1$
6. $\angle 2$
7. $\angle 3$
8. $\angle 4$

For each equation, find the value of x for the given values of a and b. (Lesson 0-4)

9. $a + 8 = -4(x - b)$, for $a = 8$ and $b = 3$
10. $b = 3x + 4a$, for $a = -9$ and $b = 12$
11. $\frac{a + 2}{b + 13} = 5x$, for $a = 18$ and $b = -1$
12. **MINIATURE GOLF** A miniature golf course offers a $1 ice cream cone with each round of golf purchased. If five friends each had a cone after golfing and spend a total of $30, how much does one round of golf cost? (Lesson 0-5)

QuickReview

EXAMPLE 1

Refer to the figure.

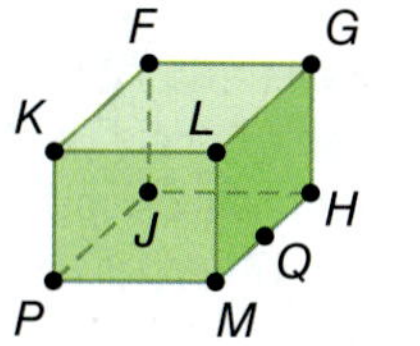

a. How many planes are shown in this figure?
Six: plane $FGLK$, plane $JHMP$, plane $FKPJ$, plane $GLMH$, plane $FGHJ$, and plane $KLMP$

b. Name three points that are collinear.
Points M, Q, and H are collinear.

c. Are points F, K, and J coplanar? Explain.
Yes. Points F, K, and J all lie in plane $FKPJ$.

EXAMPLE 2

Find $m\angle 1$.

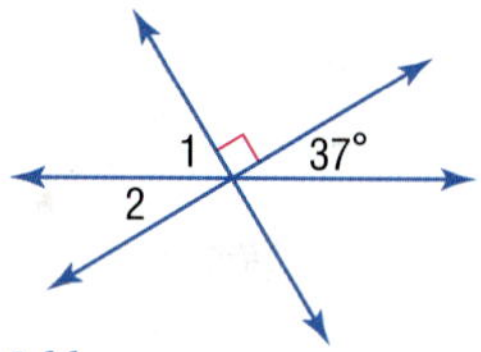

$m\angle 1 + 37 + 90 = 180$	**Add.**
$m\angle 1 = 53$	**Simplify.**

EXAMPLE 3

Find x in $a + 8 = b(x - 7)$ if $a = 12$ and $b = 10$.

$a + 8 = b(x - 7)$	**Write the equation.**
$12 + 8 = 10(x - 7)$	**$a = 12$ and $b = 10$**
$20 = 10x - 70$	**Simplify.**
$90 = 10x$	**Add.**
$x = 9$	**Divide.**

Online Option

Math Online Take a self-check Chapter Readiness Quiz at glencoe.com.

Get Started on Chapter 3

You will learn several new concepts, skills, and vocabulary terms as you study Chapter 3. To get ready, identify important terms and organize your resources. You may wish to refer to **Chapter 0** to review prerequisite skills.

FOLDABLES® Study Organizer

Parallel and Perpendicular Lines Make this Foldable to help you organize your Chapter 3 notes about relationships between lines. Begin with a sheet of 11″ × 17″ paper and six index cards.

1. **Fold** lengthwise about 3″ from the bottom.

2. **Fold** the paper in thirds.

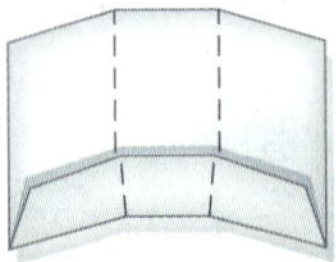

3. **Open** and staple the edges on either side to form three pockets.

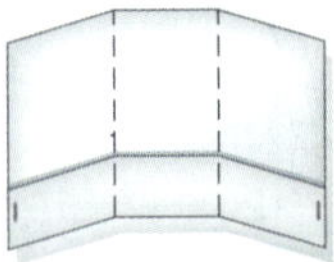

4. **Label** the pockets as shown. Place two index cards in each pocket.

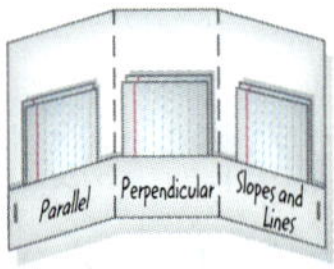

Math Online glencoe.com

- Study the chapter online
- Explore **Get Animated** activities
- Get extra help from your own **Personal Tutor**
- Use **Extra Examples** for additional help
- Take a **Self-Check Quiz**
- **Review Vocabulary** in fun ways

New Vocabulary

English		Español
parallel lines	• p. 171 •	rectas paralelas
skew lines	• p. 171 •	rectas alabeadas
parallel planes	• p. 171 •	planos paralelos
transversal	• p. 172 •	transversal
interior angles	• p. 172 •	ángulos internos
exterior angles	• p. 172 •	ángulos externos
corresponding angles	• p. 172 •	ángulos correspondientes
slope	• p. 186 •	pendiente
rate of change	• p. 187 •	tasa de cambio
slope-intercept form	• p. 196 •	forma pendiente-intersección
point-slope form	• p. 196 •	forma punto-pendiente
equidistant	• p. 216 •	equidistante

Review Vocabulary

congruent angles • p. 139 • ángulos congruentes two angles that have the same degree measure

perpendicular • p. 148 • perpendicular two lines, segments, or rays that intersect to form right angles

vertical angles • p. 46 • ángulos opusetos por el vértice two nonadjacent angles formed by intersecting lines

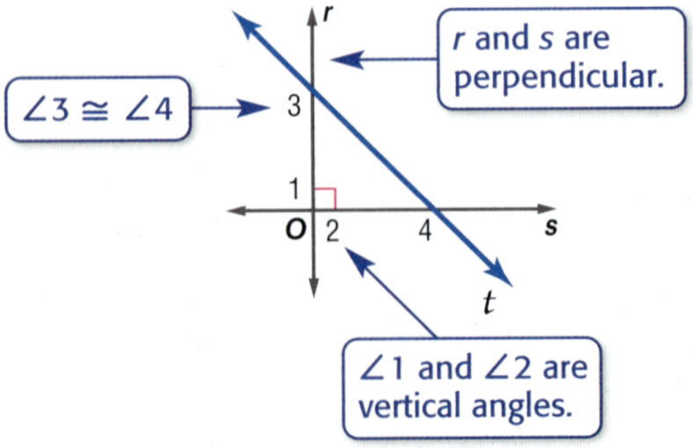

Multilingual eGlossary glencoe.com

3-1

Parallel Lines and Transversals

Then

You used angle and line segment relationships to prove theorems. (Lesson 2-8)

Now

- Identify the relationships between two lines or two planes.
- Name angle pairs formed by parallel lines and transversals.

New Vocabulary

parallel lines
skew lines
parallel planes
transversal
interior angles
exterior angles
consecutive interior angles
alternate interior angles
alternate exterior angles
corresponding angles

Math Online
glencoe.com

Why?

An *Ames room* creates the illusion that a person standing in the right corner is much larger than a person standing in the left corner.

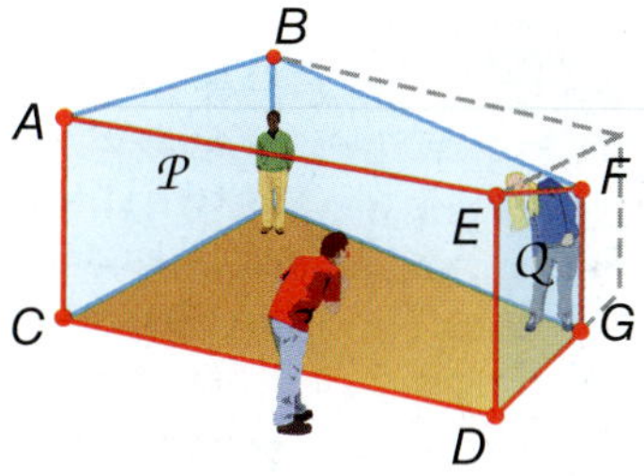

From a front viewing hole the front and back walls appear parallel, when in fact they are slanted. The ceiling and floor appear horizontal, but are actually tilted.

Relationships Between Lines and Planes The construction of the Ames room above makes use of intersecting, parallel, and skew lines, as well as intersecting and parallel planes, to create an optical illusion.

Key Concepts — Parallel and Skew

Parallel lines are coplanar lines that do not intersect.

Example $\overleftrightarrow{JK} \parallel \overleftrightarrow{LM}$

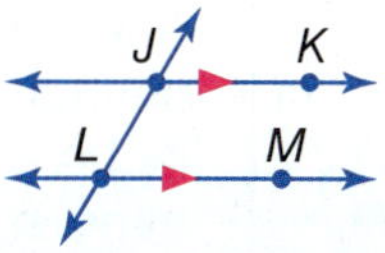

Arrows are used to indicate that lines are parallel.

Skew lines are lines that do not intersect and are not coplanar.

Example Lines ℓ and m are skew.

Parallel planes are planes that do not intersect.

Example Planes $\mathcal{A}$ and $\mathcal{B}$ are parallel.

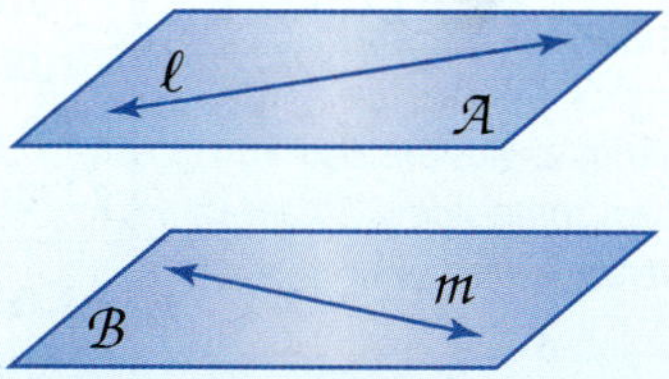

$\overleftrightarrow{JK} \parallel \overleftrightarrow{LM}$ is read as *line JK is parallel to line LM.*

If segments or rays are contained within lines that are parallel or skew, then the segments or rays are parallel or skew.

Real-World EXAMPLE 1 Identify Parallel and Skew Relationships

Identify each of the following using the wedge of cheese below.

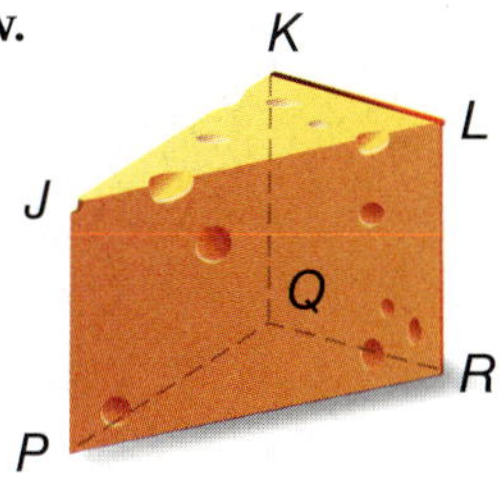

a. all segments parallel to $\overline{JP}$

$\overline{KQ}$ and $\overline{LR}$

b. a segment skew to $\overline{KL}$

$\overline{JP}$, $\overline{PQ}$, or $\overline{PR}$

c. a plane parallel to plane *PQR*

Plane *JKL* is the only plane parallel to plane *PQR*.

Watch Out!

Parallel vs. Skew In Guided Practice 1A, $\overleftrightarrow{FE}$ is *not* skew to $\overleftrightarrow{BC}$. Instead, these lines are parallel in plane *BCF*.

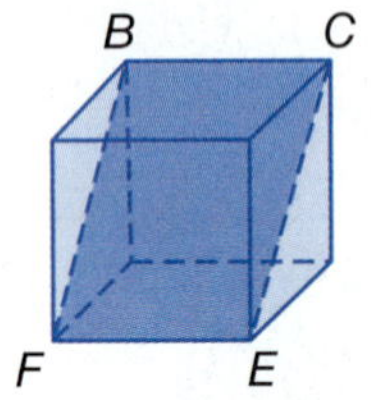

Guided Practice

Identify each of the following using the cube shown.

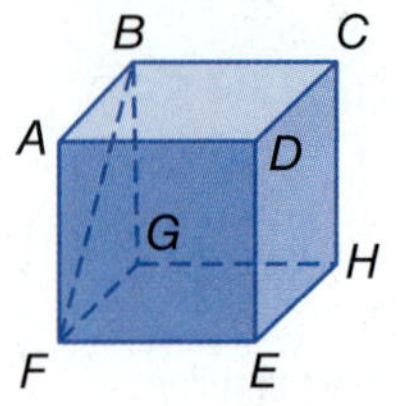

1A. all segments skew to $\overleftrightarrow{BC}$

1B. a segment parallel to $\overleftrightarrow{EH}$

1C. all planes parallel to plane *DCH*

Personal Tutor glencoe.com

Transversal Angle Pair Relationships A line that intersects two or more coplanar lines at two different points is called a **transversal**. In the diagram below, line *t* is a transversal of lines *q* and *r*. Notice that line *t* forms a total of eight angles with lines *q* and *r*. These angles, and specific pairings of these angels, are given special names.

Key Concepts **Transversal Angle Pair Relationships**

Four **interior angles** lie in the region between lines *q* and *r*.	∠3, ∠4, ∠5, ∠6
Four **exterior angles** lie in the two regions that are not between lines *q* and *r*.	∠1, ∠2, ∠7, ∠8
Consecutive interior angles are interior angles that lie on the same side of transversal *t*.	∠4 and ∠5, ∠3 and ∠6
Alternate interior angles are nonadjacent interior angles that lie on opposite sides of transversal *t*.	∠3 and ∠5, ∠4 and ∠6
Alternate exterior angles are nonadjacent exterior angles that lie on opposite sides of transversal *t*.	∠1 and ∠7, ∠2 and ∠8
Corresponding angles lie on the same side of transversal *t* and on the same side of lines *q* and *r*.	∠1 and ∠5, ∠2 and ∠6 ∠3 and ∠7, ∠4 and ∠8

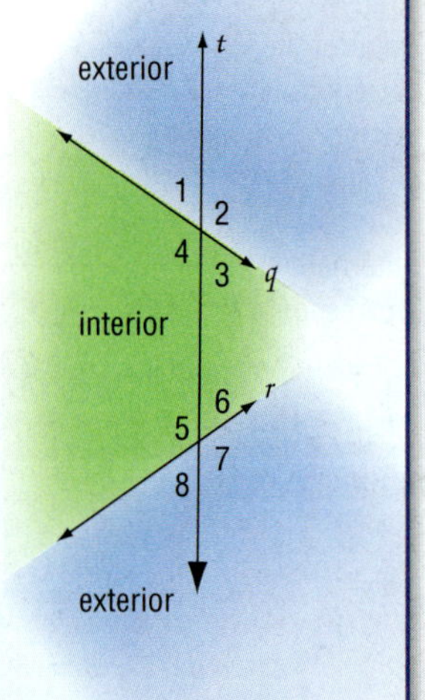

ReadingMath

Same-Side Interior Angles Consecutive interior angles are also called *same-side interior angles*.

EXAMPLE 2 Classify Angle Pair Relationships

Refer to the figure below. Classify the relationship between each pair of angles as ***alternate interior, alternate exterior, corresponding,*** **or** ***consecutive interior*** **angles.**

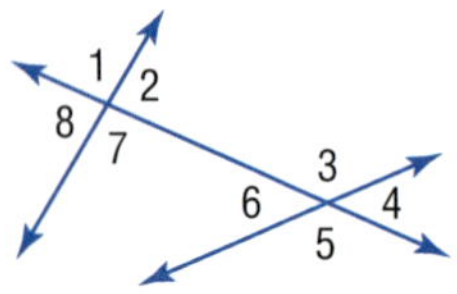

a. **∠1 and ∠5**
alternate exterior

b. **∠6 and ∠7**
consecutive interior

c. **∠2 and ∠4**
corresponding

d. **∠2 and ∠6**
alternate interior

Guided Practice

2A. ∠3 and ∠7 **2B.** ∠5 and ∠7 **2C.** ∠4 and ∠8 **2D.** ∠2 and ∠3

Personal Tutor glencoe.com

When more than one line can be considered a transversal, first identify the transversal for a given angle pair by locating the line that connects the vertices of the angles.

StudyTip

Nonexample In the figure below, line c is *not* a transversal of lines a and b, since line c intersects lines a and b in only one point.

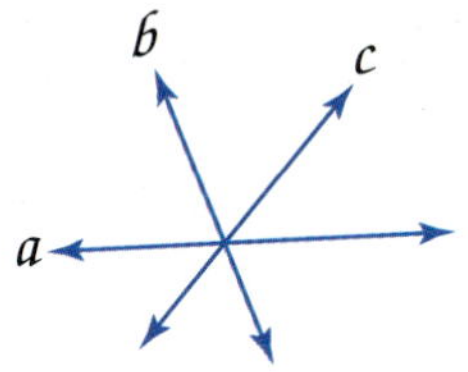

EXAMPLE 3 Identify Transversals and Classify Angle Pairs

Identify the transversal connecting each pair of angles in the photo. Then classify the relationship between each pair of angles.

a. **$\angle 1$ and $\angle 3$**

The transversal connecting $\angle 1$ and $\angle 3$ is line h. These are alternate exterior angles.

b. **$\angle 5$ and $\angle 6$**

The transversal connecting $\angle 5$ and $\angle 6$ is line k. These are consecutive interior angles.

c. **$\angle 2$ and $\angle 6$**

The transversal connecting $\angle 2$ and $\angle 6$ is line ℓ. These are corresponding angles.

Guided Practice

3A. $\angle 3$ and $\angle 5$

3B. $\angle 2$ and $\angle 8$

3C. $\angle 5$ and $\angle 7$

3D. $\angle 2$ and $\angle 9$

Personal Tutor glencoe.com

Check Your Understanding

Example 1 p. 171

Refer to the figure at the right to identify each of the following.

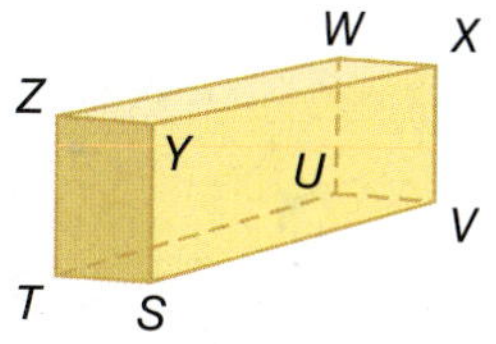

1. a plane parallel to plane ZWX
2. a segment skew to $\overline{TS}$ that contains point W
3. all segments parallel to $\overline{SV}$

4. **CONSTRUCTION** Use the diagram of the partially framed storage shed shown to identify each of the following.

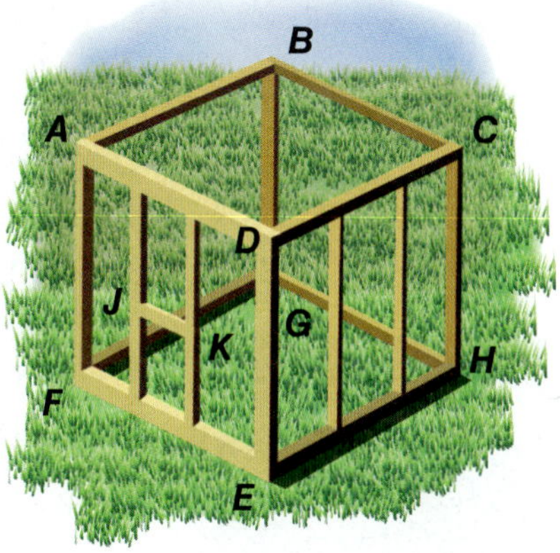

   a. Name three pairs of parallel planes.

   b. Name three segments parallel to $\overline{DE}$.

   c. Name two segments parallel to $\overline{FE}$.

   d. Name two pairs of skew segments.

Example 2 p. 172

Classify the relationship between each pair of angles as *alternate interior, alternate exterior, corresponding,* or *consecutive interior* angles.

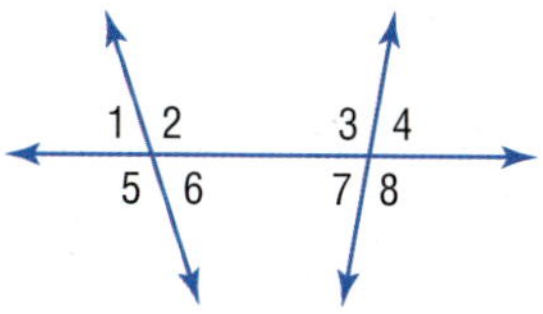

5. $\angle 1$ and $\angle 8$
6. $\angle 2$ and $\angle 4$
7. $\angle 3$ and $\angle 6$
8. $\angle 6$ and $\angle 7$

Example 3
p. 173

Identify the transversal connecting each pair of angles. Then classify the relationship between each pair of angles.

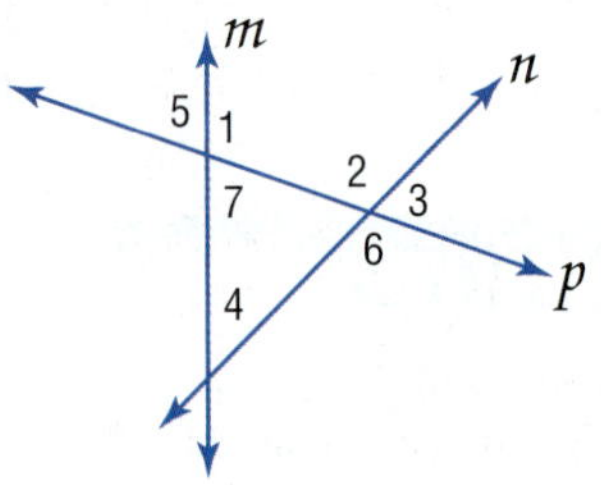

9. ∠2 and ∠4

10. ∠5 and ∠6

11. ∠4 and ∠7

12. ∠2 and ∠7

Practice and Problem Solving

● = **Step-by-Step Solutions** begin on page R20.
Extra Practice begins on page 969.

Example 1
p. 171

Refer to the figure to identify each of the following.

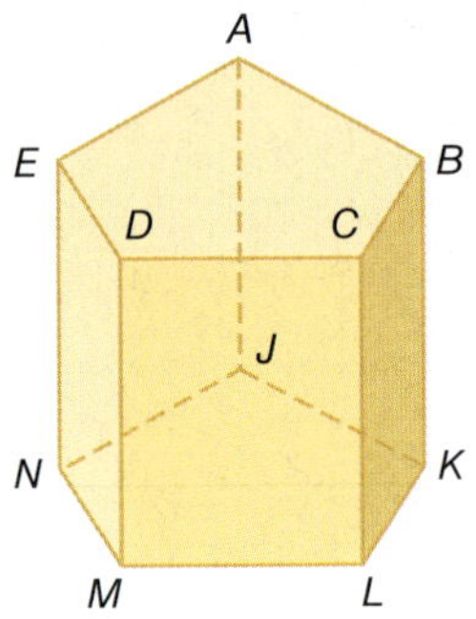

13. all segments parallel to $\overline{DM}$

14. a plane parallel to plane *ACD*

15. a segment skew to $\overline{BC}$

16. all planes intersecting plane *EDM*

17. all segments skew to $\overline{AE}$

18. a segment parallel to $\overline{EN}$

19. a segment parallel to $\overline{AB}$ through point *J*

20. a segment skew to $\overline{CL}$ through point *E*

Example 2
p. 172

Identify the transversal connecting each pair of angles. Then classify the relationship between each pair of angles as *alternate interior*, *alternate exterior*, *corresponding*, or *consecutive interior* angles.

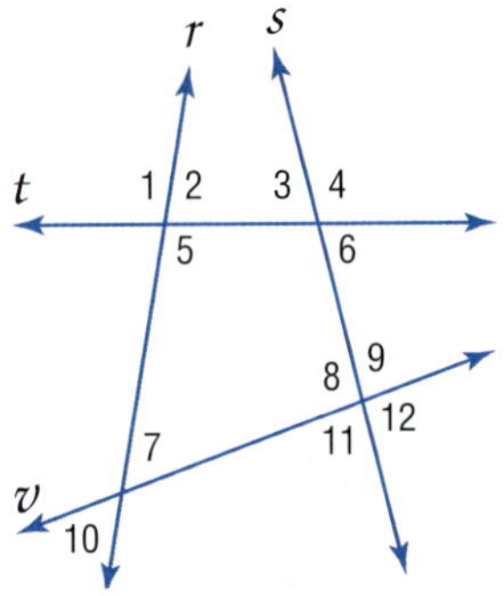

21. ∠4 and ∠9

22. ∠5 and ∠7

23. ∠3 and ∠5

24. ∠10 and ∠11

25. ∠1 and ∠6

26. ∠6 and ∠8

27. ∠2 and ∠3

28. ∠9 and ∠10

29. ∠4 and ∠11

30. ∠7 and ∠11

Example 3
p. 173

SAFETY **Identify the transversal connecting each pair of angles in the photo of a fire escape shown. Then classify the relationship between each pair of angles.**

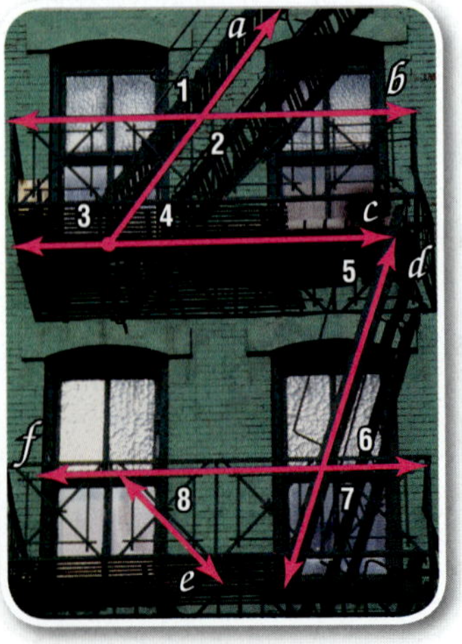

31. ∠1 and ∠2

32. ∠2 and ∠4

33. ∠4 and ∠5

34. ∠6 and ∠7

35. ∠7 and ∠8

36. ∠2 and ∠3

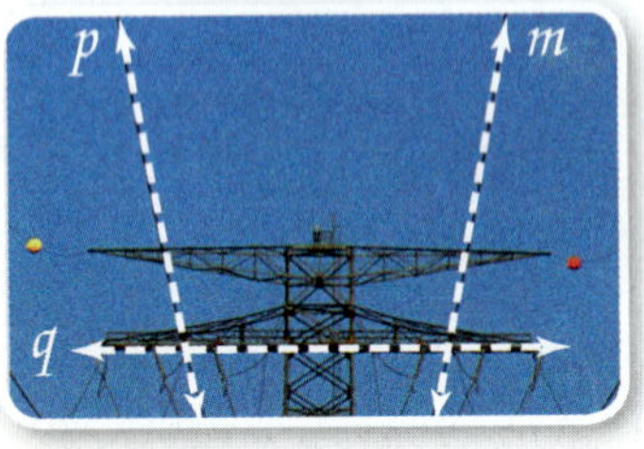

Real-World Link

To avoid electrical shorts and fires, power lines are not allowed to intersect.

37. **POWER** Use the photo and information at the left.

a. What must be the relationship between power lines *p* and *m*? Explain your reasoning.

b. What is the relationship between line *q* and lines *p* and *m*?

Real-World Link

Escalators are more efficient than elevators for moving people short distances because of their high loading capacity. Large escalators can transport 6000 passengers per hour.

Source: Otis Elevator Co.

Describe the relationship between each pair of segments as *parallel, skew,* or *intersecting.*

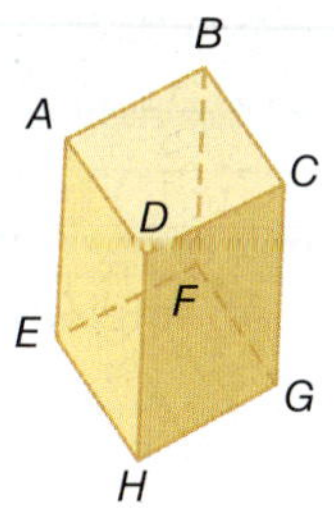

38. $\overline{FG}$ and $\overline{BC}$

39. $\overline{AB}$ and $\overline{CG}$

40. $\overline{DH}$ and $\overline{HG}$

41. $\overline{DH}$ and $\overline{BF}$

42. $\overline{EF}$ and $\overline{BC}$

43. $\overline{CD}$ and $\overline{AD}$

44. OPTICAL ILLUSION The illusion at the right is created using squares and straight lines.

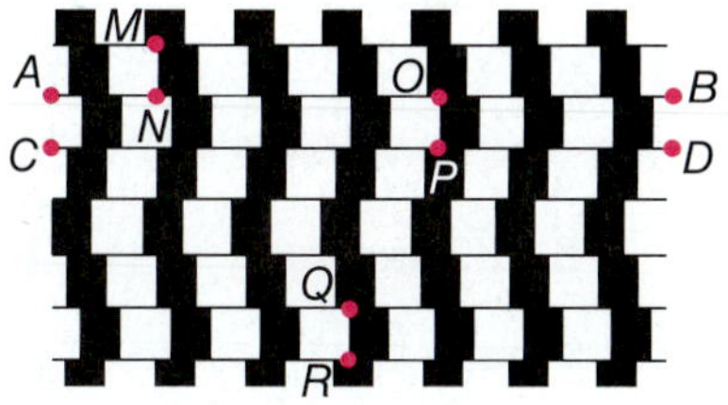

a. How are $\overline{AB}$ and $\overline{CD}$ related? Justify your reasoning.

b. How are $\overline{MN}$ and $\overline{QR}$ related? $\overline{AB}$, $\overline{CD}$, and $\overline{OP}$?

45. ESCALATORS Escalators consist of steps on a continuous loop that is driven by a motor. At the top and bottom of the platform, the steps collapse to provide a level surface for entrance and exit.

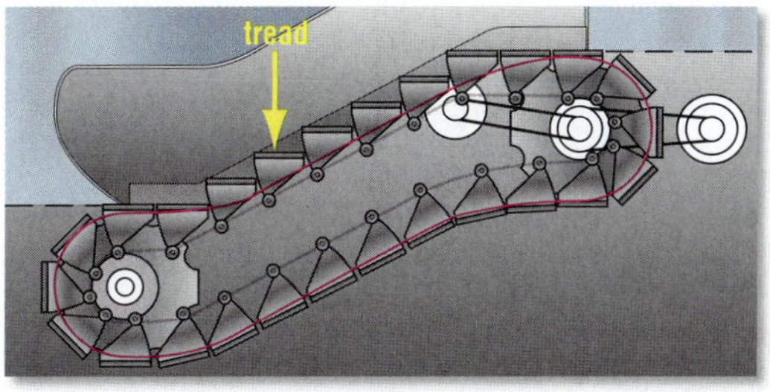

a. What is the relationship between the treads of the ascending stairs?

b. What is the relationship between the treads of the two steps at the top of the incline?

c. How do the treads of the steps on the incline of the escalator relate to the treads of the steps on the bottom of the conveyor?

H.O.T. Problems

Use Higher-Order Thinking Skills

46. OPEN ENDED Plane $\mathcal{P}$ contains lines a and b. Line c intersects plane $\mathcal{P}$ at point J. Lines a and b are parallel, lines a and c are skew, and lines b and c are not skew. Draw a figure based upon this description.

47. CHALLENGE Suppose points A, B, and C lie in plane $\mathcal{P}$, and points D, E, and F lie in plane Q. Line m contains points D and F and does not intersect plane $\mathcal{P}$. Line n contains points A and E.

a. Draw a diagram to represent the situation.

b. What is the relationship between planes $\mathcal{P}$ and Q?

c. What is the relationship between lines m and n?

REASONING Plane $\mathcal{X}$ and plane $\mathcal{Y}$ are parallel and plane $\mathcal{Z}$ intersects plane $\mathcal{X}$. Line $\overleftrightarrow{AB}$ is in plane $\mathcal{X}$, line $\overleftrightarrow{CD}$ is in plane $\mathcal{Y}$, and line $\overleftrightarrow{EF}$ is in plane $\mathcal{Z}$. Determine whether each statement is *always*, *sometimes*, or *never* true. Explain.

48. $\overleftrightarrow{AB}$ is skew to $\overleftrightarrow{CD}$.

49. $\overleftrightarrow{AB}$ intersects $\overleftrightarrow{EF}$.

50. WRITING IN MATH Explain why planes cannot be skew.

Standardized Test Practice

51. Which of the following angle pairs are alternate exterior angles?

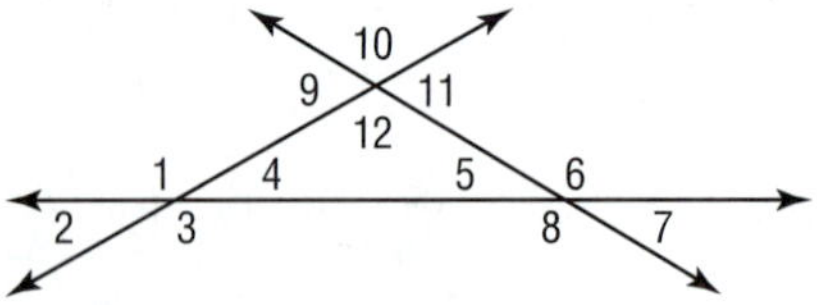

A $\angle 1$ and $\angle 5$ C $\angle 2$ and $\angle 10$
B $\angle 2$ and $\angle 6$ D $\angle 5$ and $\angle 9$

52. What is the measure of $\angle XYZ$?

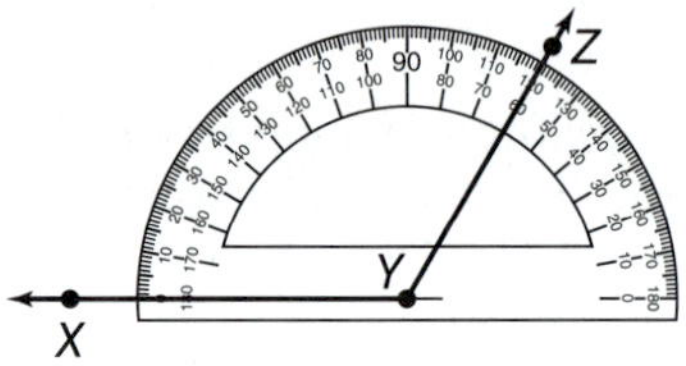

F 30° H 120°
G 60° J 150°

53. SHORT RESPONSE Name the coordinates of the points representing the x- and y-intercepts of the graph shown below.

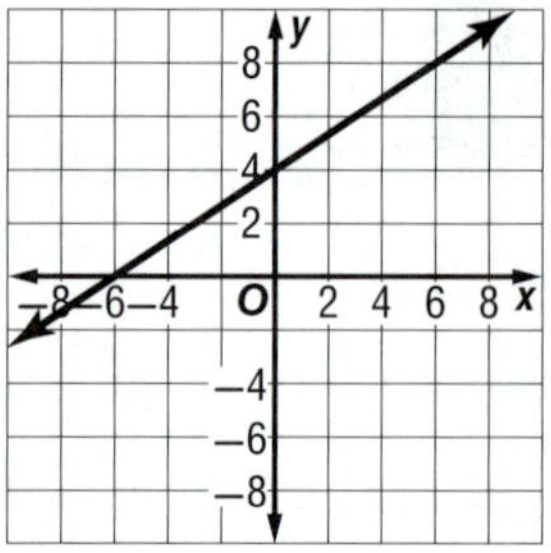

54. SAT/ACT Of the following, the one that is *not* equivalent to 485 is:

A $(3 \times 100) + (4 \times 10) + 145$
B $(3 \times 100) + (18 \times 10) + 5$
C $(4 \times 100) + (8 \times 10) + 15$
D $(4 \times 100) + (6 \times 10) + 25$

Spiral Review

Find the measure of each numbered angle. (Lesson 2-8)

55. $m\angle 9 = 2x - 4$, $m\angle 10 = 2x + 4$

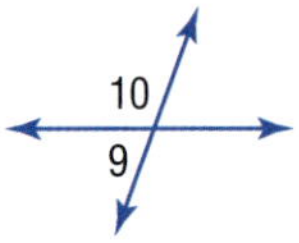

56. $m\angle 11 = 4x$, $m\angle 12 = 2x - 6$

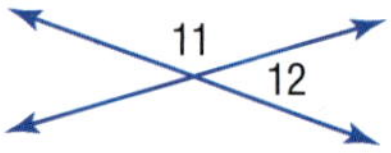

57. $m\angle 19 = 100 + 20x$, $m\angle 20 = 20x$

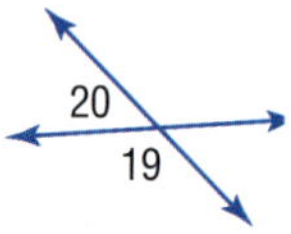

58. PROOF Prove the following. (Lesson 2-7)

Given: $\overline{WY} \cong \overline{ZX}$
A is the midpoint of $\overline{WY}$.
A is the midpoint of $\overline{ZX}$.
Prove: $\overline{WA} \cong \overline{ZA}$

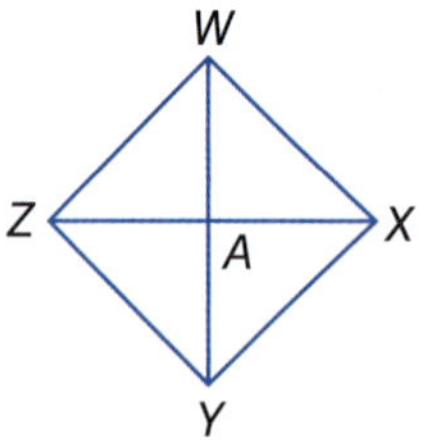

ALGEBRA Use the figure at the right. (Lesson 1-5)

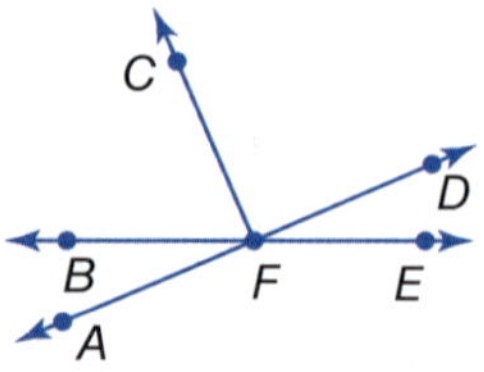

59. If $m\angle CFD = 12a + 45$, find a so that $\overrightarrow{FC} \perp \overrightarrow{FD}$.

60. If $m\angle AFB = 8x - 6$ and $m\angle BFC = 14x + 8$, find the value of x so that $\angle AFC$ is a right angle.

Skills Review

Find x. (Lesson 1-5)

61.

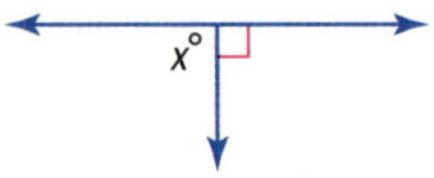

62.

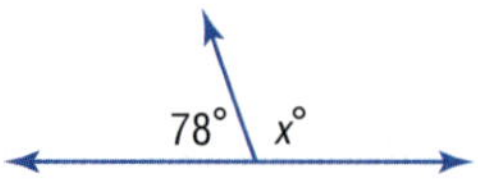

63.

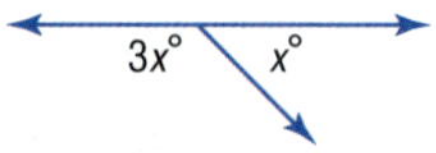

EXPLORE 3-2

Geometry Software Lab
Angles and Parallel Lines

Math Online glencoe.com
- Other Calculator Keystrokes
- Graphing Technology Personal Tutor

You can use The Geometer's Sketchpad® to explore the angles formed by two parallel lines and a transversal.

ACTIVITY Parallel Lines and a Transversal

Step 1 **Draw a line.**

Draw and label points F and G. Then use the line tool to draw $\overleftrightarrow{FG}$.

Step 2 **Draw a parallel line.**

Draw a point that is not on $\overleftrightarrow{FG}$ and label it J. Select $\overleftrightarrow{FG}$ and point J, and then choose **Parallel Line** from the **Construct** menu. Draw and label a point K on this parallel line.

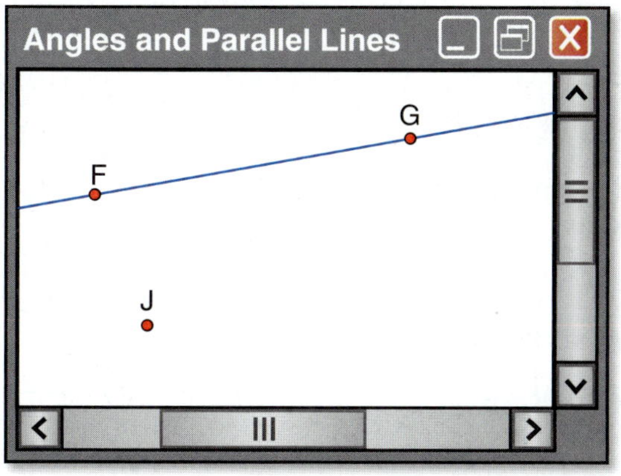

Step 3 **Draw a transversal.**

Draw and label point A on $\overleftrightarrow{FG}$ and point B on $\overleftrightarrow{JK}$. Select A and B and then choose **Line** from the **Construct** menu to draw transversal $\overleftrightarrow{AB}$. Then draw and label points C and D on $\overleftrightarrow{AB}$ as shown.

Step 4 **Measure each angle.**

Measure all eight angles formed by these lines. For example, select points F, A, then C, and choose **Angle** from the **Measure** menu to find $m\angle FAC$.

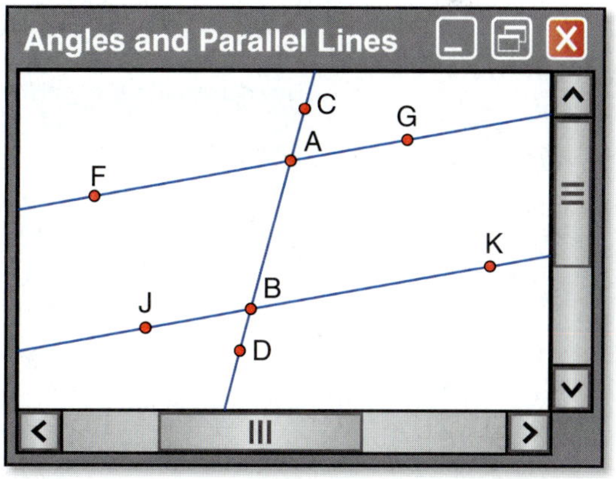

Analyze the Results

1. Record the measures from Step 4 in a table like this one. Which angles have the same measure?

Angles	∠FAC	∠CAG	∠GAB	∠FAB	∠JBA	∠ABK	∠KBD	∠JBD
1st Measure								

2. Drag point C or F to move transversal $\overleftrightarrow{AB}$ so that it intersects the two parallel lines at a different angle. Add a row **2nd Measure** to your table and record the new measures. Repeat these steps until your table has 3rd, 4th, and 5th Measure rows of data.

3. Using the angles listed in the table, identify and describe the relationship between all angle pairs that have the following special names. Then write a conjecture in if-then form about each angle pair when formed by any two parallel lines cut by a transversal.
   a. corresponding **b.** alternate interior **c.** alternate exterior **d.** consecutive interior

4. Drag point C or F so that the measure of any of the angles is 90.
   a. What do you notice about the measures of the other angles?
   b. Make a conjecture about a transversal that is perpendicular to one of two parallel lines.

3-2 Angles and Parallel Lines

Then
You named angle pairs formed by parallel lines and transversals. (Lesson 3-1)

Now
- Use theorems to determine the relationships between specific pairs of angles.
- Use algebra to find angle measurements.

Math Online
glencoe.com

Why?

Those who work in construction and maintenance often use an access scaffold. This structure provides support and access to elevated areas. The transversal t shown provides structural support to the two parallel working areas.

Parallel Lines and Angle Pairs In the photo, line t is a transversal of lines a and b, and $\angle 1$ and $\angle 2$ are corresponding angles. Since lines a and b are parallel, there is a special relationship between corresponding angle pairs.

Postulate 3.1 Corresponding Angles Postulate

If two parallel lines are cut by a transversal, then each pair of corresponding angles is congruent.

Examples $\angle 1 \cong \angle 3$, $\angle 2 \cong \angle 4$, $\angle 5 \cong \angle 7$, $\angle 6 \cong \angle 8$

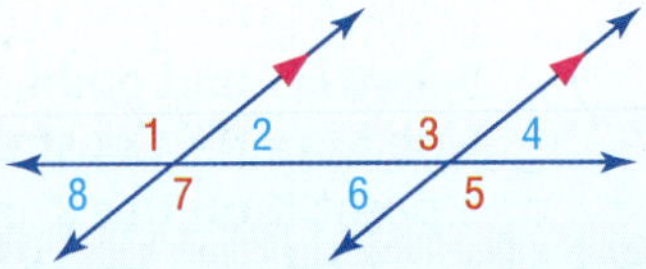

EXAMPLE 1 Use Corresponding Angles Postulate

In the figure, $m\angle 5 = 72$. Find the measure of each angle. Tell which postulates (or theorems) you used.

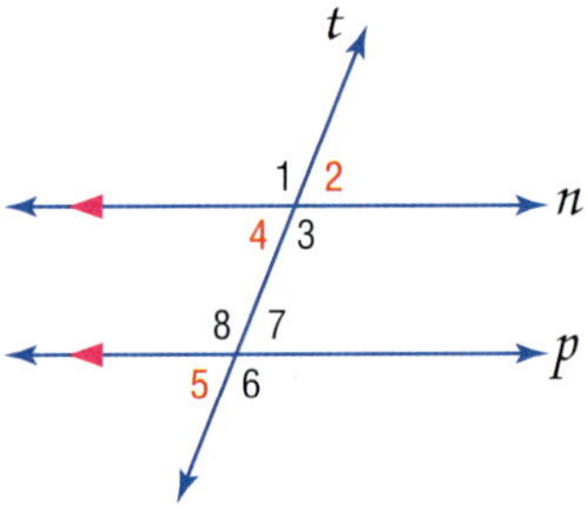

a. $\angle 4$

$\angle 4 \cong \angle 5$	Corresponding Angles Postulate
$m\angle 4 = m\angle 5$	Definition of congruent angles
$m\angle 4 = 72$	Substitution

b. $\angle 2$

$\angle 2 \cong \angle 4$	Vertical Angles Theorem
$\angle 4 \cong \angle 5$	Corresponding Angles Postulate
$\angle 2 \cong \angle 5$	Transitive Property of Congruence
$m\angle 2 = m\angle 5$	Definition of congruent angles
$m\angle 2 = 72$	Substitution

Guided Practice

In the figure, suppose that $m\angle 8 = 105$. Find the measure of each angle. Tell which postulate(s) or theorem(s) you used.

1A. $\angle 1$ **1B.** $\angle 2$ **1C.** $\angle 3$

Personal Tutor glencoe.com

In Example 1, $\angle 2$ and $\angle 5$ are congruent alternate exterior angles. This and other examples suggest the following theorems about the other angle pairs formed by two parallel lines cut by a transversal.

StudyTip

Angle Relationships These theorems generalize the relationships between specific pairs of angles. If you get confused about the relationships, you can verify them with the methods you used in Example 1, using only corresponding, vertical, and supplementary angles.

Theorems — Parallel Lines and Angle Pairs

3.1 Alternate Interior Angles Theorem If two parallel lines are cut by a transversal, then each pair of alternate interior angles is congruent.

Examples $\angle 1 \cong \angle 3$ and $\angle 2 \cong \angle 4$

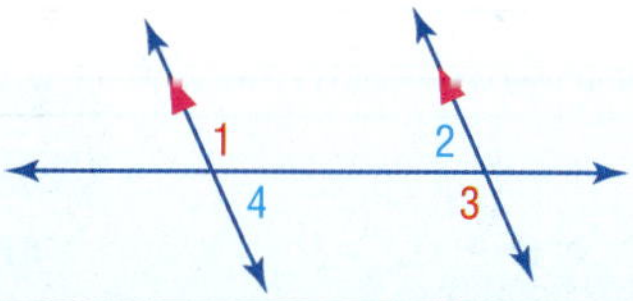

3.2 Consecutive Interior Angles Theorem If two parallel lines are cut by a transversal, then each pair of consecutive interior angles is supplementary.

Examples $\angle 1$ and $\angle 2$ are supplementary.
$\angle 3$ and $\angle 4$ are supplementary.

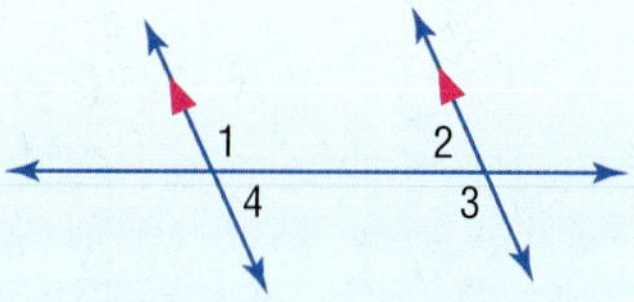

3.3 Alternate Exterior Angles Theorem If two parallel lines are cut by a transversal, then each pair of alternate exterior angles is congruent.

Examples $\angle 5 \cong \angle 7$ and $\angle 6 \cong \angle 8$

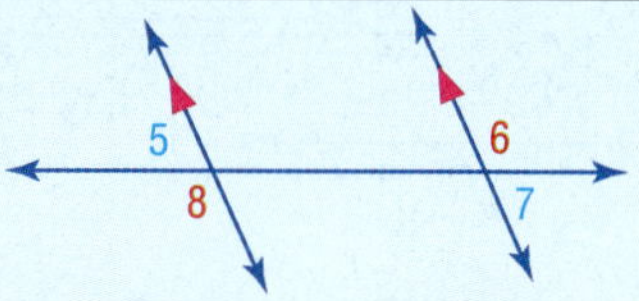

You will prove Theorems 3.2 and 3.3 in Exercises 30 and 35, respectively.

Since postulates are accepted without proof, you can use the Corresponding Angles Postulate to prove each of the theorems above.

Proof — Alternate Interior Angles Theorem

Given: $a \parallel b$
t is a transversal of a and b.

Prove: $\angle 4 \cong \angle 5$, $\angle 3 \cong \angle 6$

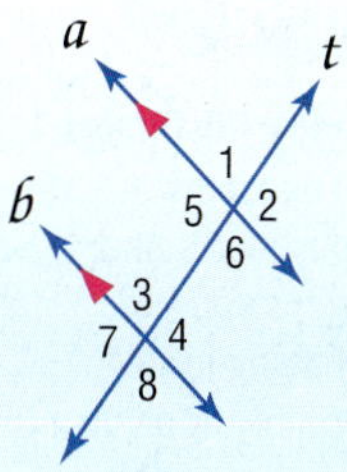

Paragraph Proof:
We are given that $a \parallel b$ with a transversal t. By the Corresponding Angles Postulate, corresponding angles are congruent. So, $\angle 2 \cong \angle 4$ and $\angle 6 \cong \angle 8$. Also, $\angle 5 \cong \angle 2$ and $\angle 8 \cong \angle 3$ because vertical angles are congruent. Therefore, $\angle 5 \cong \angle 4$ and $\angle 3 \cong \angle 6$ since congruence of angles is transitive.

Real-World Link

Some cities require that streets in newly planned subdivisions intersect at no less than a 60° angle.

Real-World EXAMPLE 2 — Use Theorems about Parallel Lines

COMMUNITY PLANNING Redding Lane and Creek Road are parallel streets that intersect Park Road along the west side of Wendell Park. If $m\angle 1 = 118$, find $m\angle 2$.

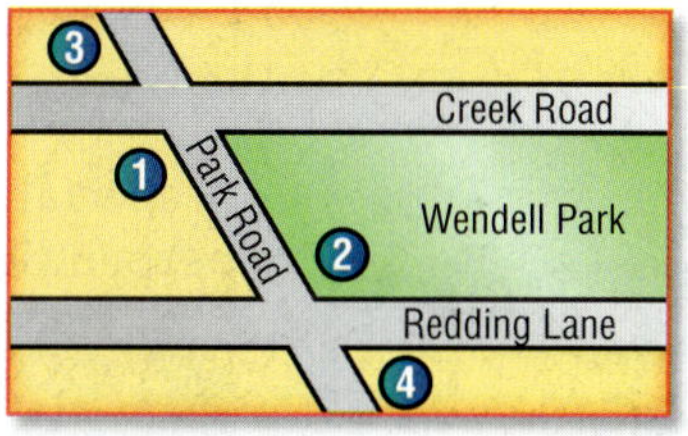

$\angle 2 \cong \angle 1$	**Alternate Interior Angles Postulate**
$m\angle 2 = m\angle 1$	**Definition of congruent angles**
$m\angle 2 = 118$	**Substitution**

Guided Practice

COMMUNITY PLANNING Refer to the diagram above to find each angle measure. Tell which postulate(s) or theorem(s) you used.

2A. If $m\angle 1 = 100$, find $m\angle 4$.

2B. If $m\angle 3 = 70$, find $m\angle 4$.

Personal Tutor glencoe.com

Algebra and Angle Measures The special relationships between the angles formed by two parallel lines and a transversal can be used to find unknown values.

EXAMPLE 3 Find Values of Variables

ALGEBRA Use the figure at the right to find the indicated variable. Explain your reasoning.

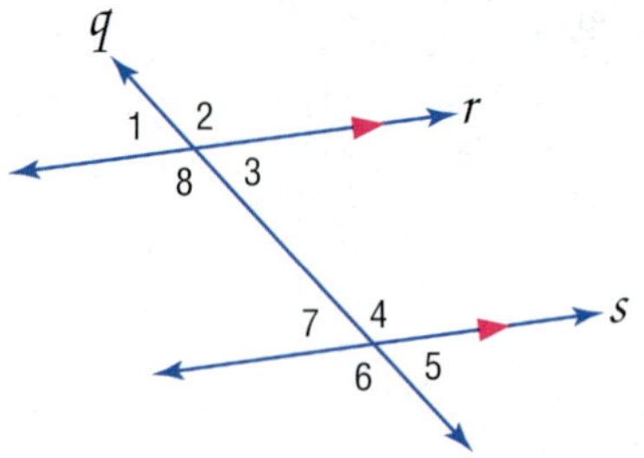

a. If $m\angle 4 = 2x - 17$ and $m\angle 1 = 85$, find x.

$\angle 3 \cong \angle 1$	**Vertical Angles Theorem**
$m\angle 3 = m\angle 1$	**Definition of congruent angles**
$m\angle 3 = 85$	**Substitution**

Since lines r and s are parallel, $\angle 4$ and $\angle 3$ are supplementary by the Consecutive Interior Angles Theorem.

$m\angle 3 + m\angle 4 = 180$	**Definition of supplementary angles**
$85 + 2x - 17 = 180$	**Substitution**
$2x + 68 = 180$	**Simplify.**
$2x = 112$	**Subtract 68 from each side.**
$x = 56$	**Divide each side by 2.**

b. Find y if $m\angle 3 = 4y + 30$ and $m\angle 7 = 7y + 6$.

$\angle 3 \cong \angle 7$	**Alternate Interior Angles Theorem**
$m\angle 3 = m\angle 7$	**Definition of congruent angles**
$4y + 30 = 7y + 6$	**Substitution**
$30 = 3y + 6$	**Subtract $4y$ from each side.**
$24 = 3y$	**Subtract 6 from each side.**
$8 = y$	**Divide each side by 3.**

Guided Practice

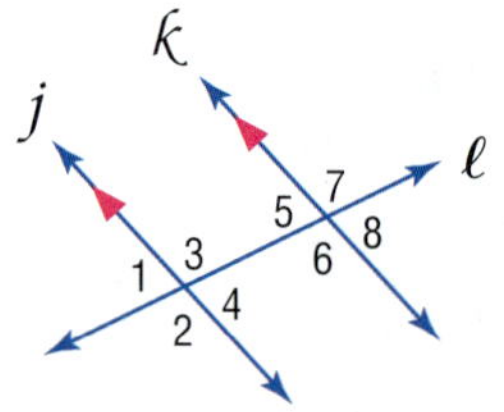

3A. If $m\angle 2 = 4x + 7$ and $m\angle 7 = 5x - 13$, find x.

3B. Find y if $m\angle 5 = 68$ and $m\angle 3 = 3y - 2$.

Personal Tutor glencoe.com

StudyTip

Applying Postulates and Theorems The postulates and theorems you will be studying in this lesson only apply to *parallel* lines cut by a transversal. You should assume that lines are parallel only if the information is given or the lines are marked with parallel arrows.

A special relationship occurs when the transversal of two parallel lines is a perpendicular line.

ReadingMath

Perpendicular Recall from Lesson 1-5 that line $b \perp$ line t is read as *Line b is perpendicular to line t.*

Theorem 3.4 Perpendicular Transversal Theorem

In a plane, if a line is perpendicular to one of two parallel lines, then it is perpendicular to the other.

Example If line $a \parallel$ line b and line $a \perp$ line t, then line $b \perp$ line t.

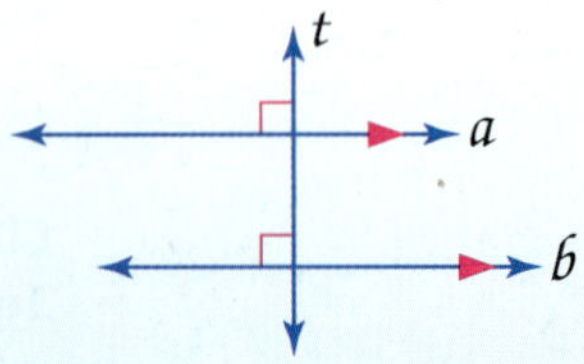

You will prove Theorem 3.4 in Exercise 37.

Check Your Understanding

Example 1 p. 178

In the figure, $m\angle 1 = 94$. Find the measure of each angle. Tell which postulate(s) or theorem(s) you used.

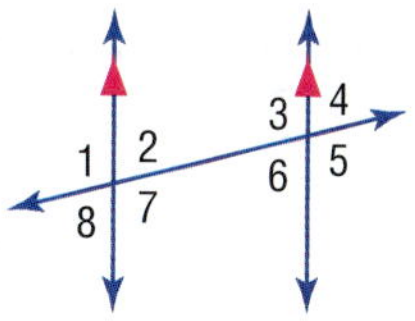

1. $\angle 3$ **2.** $\angle 5$ **3.** $\angle 4$

Example 2 p. 179

In the figure, $m\angle 4 = 101$. Find the measure of each angle. Tell which postulate(s) or theorem(s) you used.

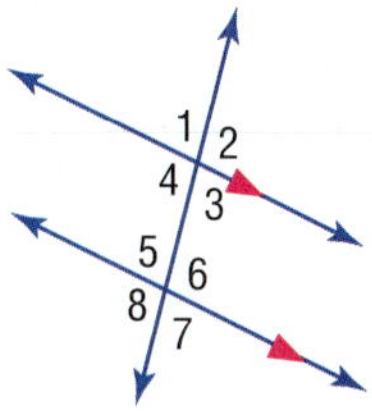

4. $\angle 6$ **5.** $\angle 7$ **6.** $\angle 5$

7. **ROADS** In the diagram, the guard rail is parallel to the surface of the roadway and the vertical supports are parallel to each other. Find the measures of angles 2, 3, and 4.

Example 3 p. 180

Find the value of the variable(s) in each figure. Explain your reasoning.

8.

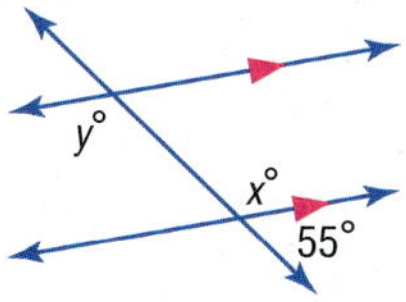

9.

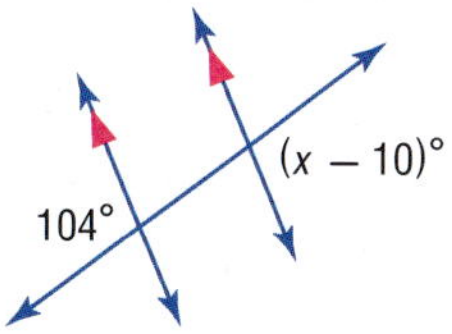

10.

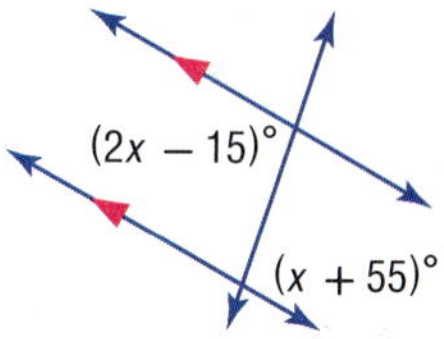

Practice and Problem Solving

● = **Step-by-Step Solutions** begin on page R20.
Extra Practice begins on page 969.

Examples 1 and 2 pp. 178–179

In the figure, $m\angle 11 = 22$ and $m\angle 14 = 18$. Find the measure of each angle. Tell which postulate(s) or theorem(s) you used.

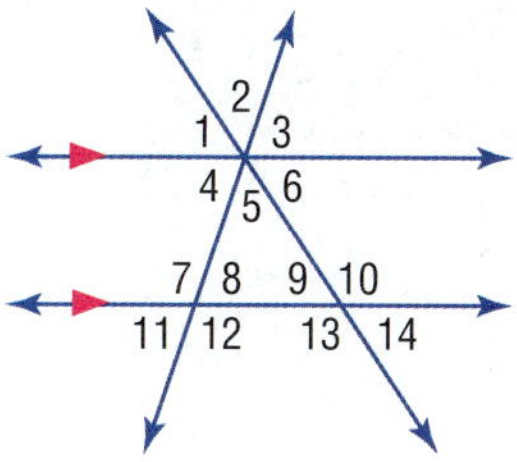

11. $\angle 4$ **12.** $\angle 3$ **13.** $\angle 12$

14. $\angle 8$ **15.** $\angle 6$ **16.** $\angle 2$

17. $\angle 10$ **18.** $\angle 5$ **19.** $\angle 1$

Example 3 p. 180

SOLAR ENERGY A solar dish collects energy by directing radiation from the sun to a receiver located at the focal point of the dish. Assume that the radiation rays are parallel. Determine the relationship between each pair of angles and explain your reasoning.

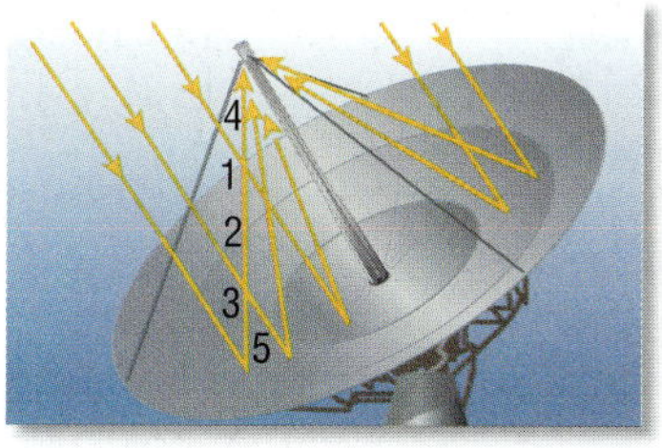

20. $\angle 1$ and $\angle 2$ **21.** $\angle 1$ and $\angle 3$ **22.** $\angle 4$ and $\angle 5$ **23.** $\angle 3$ and $\angle 4$

Find the value of the variable(s) in each figure. Explain your reasoning.

24.

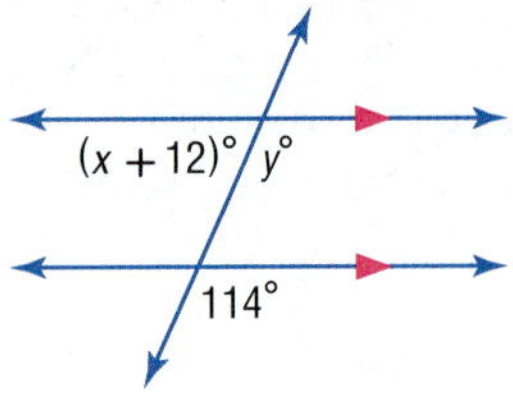

25

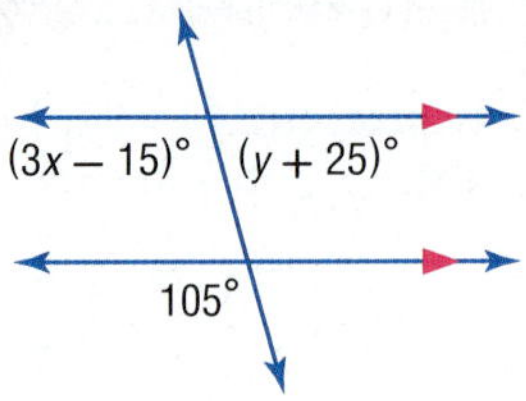

26.

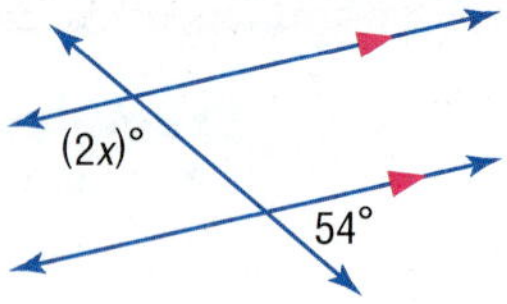

27.

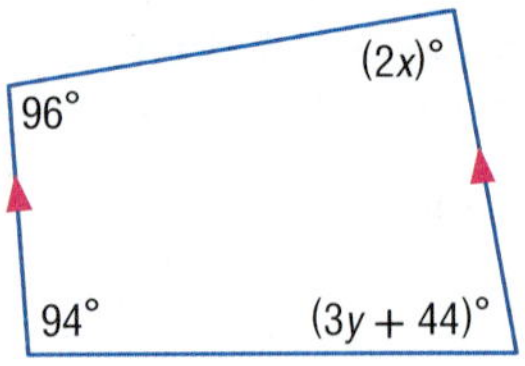

28.

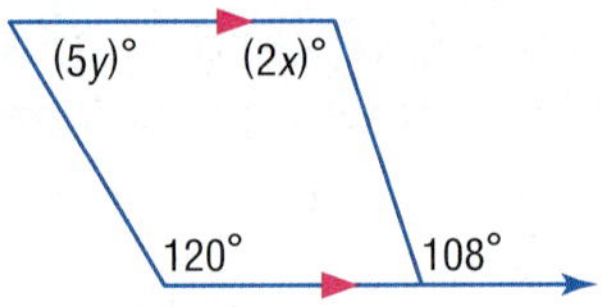

29.

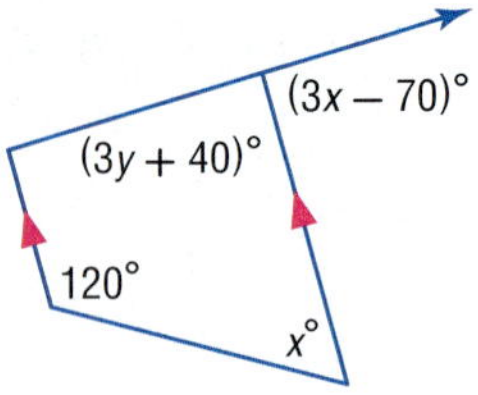

30. PROOF Copy and complete the proof of Theorem 3.2.

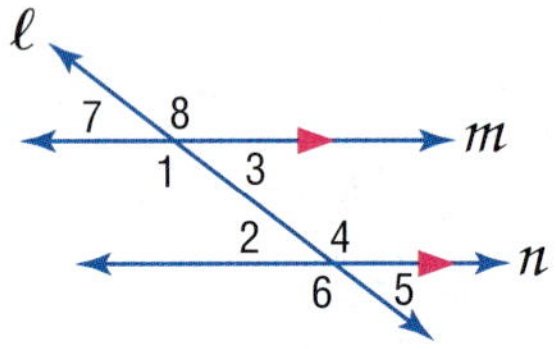

Given: $m \parallel n$; ℓ is a transversal.

Prove: $\angle 1$ and $\angle 2$ are supplementary;
$\angle 3$ and $\angle 4$ are supplementary.

Proof:

Statements	Reasons
a. ___?___	**a.** Given
b. $\angle 1$ and $\angle 3$ form a linear pair; $\angle 2$ and $\angle 4$ form a linear pair.	**b.** ___?___
c. ___?___	**c.** If two angles form a linear pair, then they are supplementary.
d. $\angle 1 \cong \angle 4$, $\angle 2 \cong \angle 3$	**d.** ___?___
e. $m\angle 1 = m\angle 4$, $m\angle 2 = m\angle 3$	**e.** Definition of Congruence
f. ___?___	**f.** ___?___

Real-World Link

The Michigan Avenue Bridge is the first bridge of its kind ever built. It is a drawbridge with a counterweight that balances Michigan Ave. through an upward swing while boat traffic passes through the Chicago River.

Source: Bridgehouse and Chicago River Museum

STORAGE When industrial shelving needs to be accessible from either side, additional support is provided on the side by transverse members. Determine the relationship between each pair of angles and explain your reasoning.

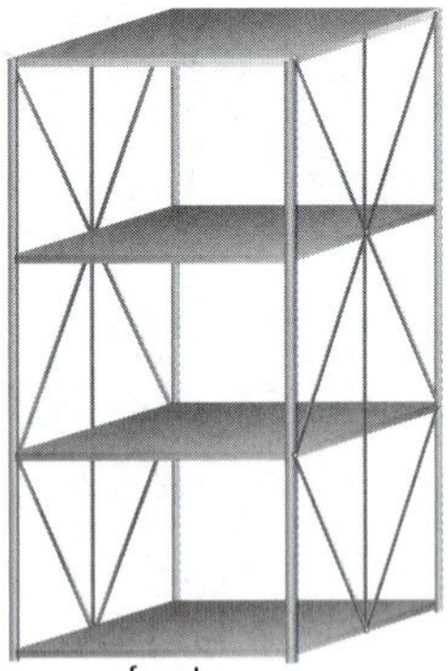

front

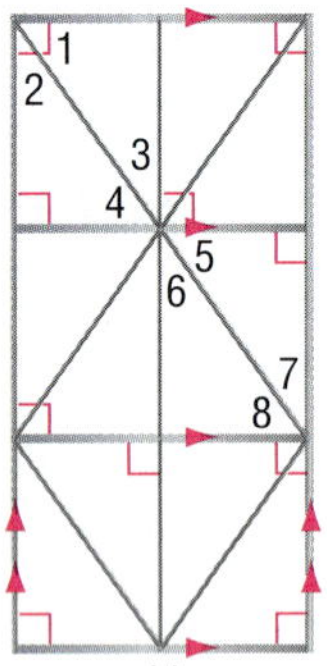

side

31. $\angle 1$ and $\angle 8$

32. $\angle 1$ and $\angle 5$

33. $\angle 3$ and $\angle 6$

34. $\angle 1$ and $\angle 2$

35. PROOF Write a two-column proof of the Alternate Exterior Angles Theorem. (Theorem 3.3)

36. BRIDGES Refer to the diagram of the double decker Michigan Avenue Bridge in Chicago, Illinois. The two levels of the bridge are parallel.

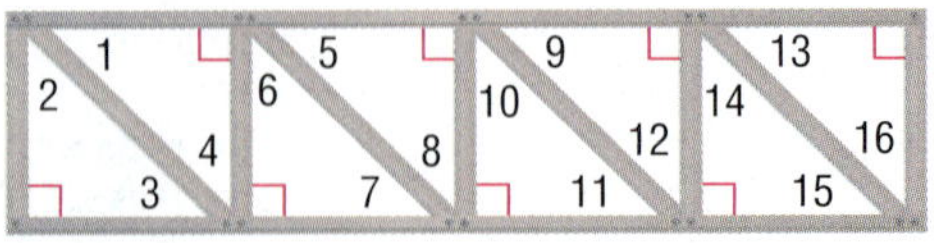

a. How are the measures of the odd-numbered angles related? Explain.

b. How are the measures of the even-numbered angles related? Explain.

c. How are any pair of angles in which one is odd and the other is even related?

d. What geometric term(s) can be used to relate the two roadways contained by the bridge?

37. **PROOF** In a plane, prove that if a line is perpendicular to one of two parallel lines, then it is perpendicular to the other. (Theorem 3.4)

Find x. (*Hint*: Draw an auxiliary line.)

38.

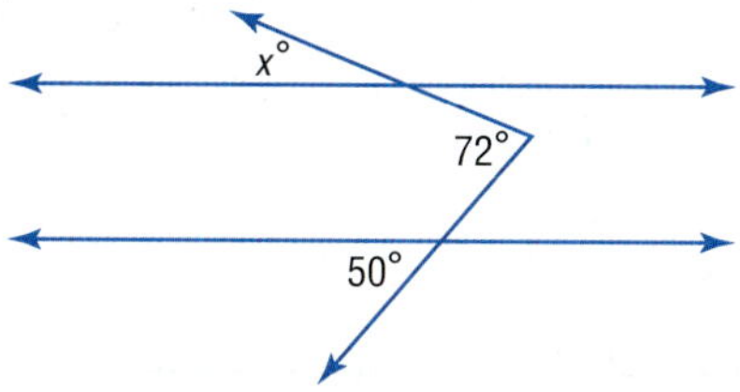

39

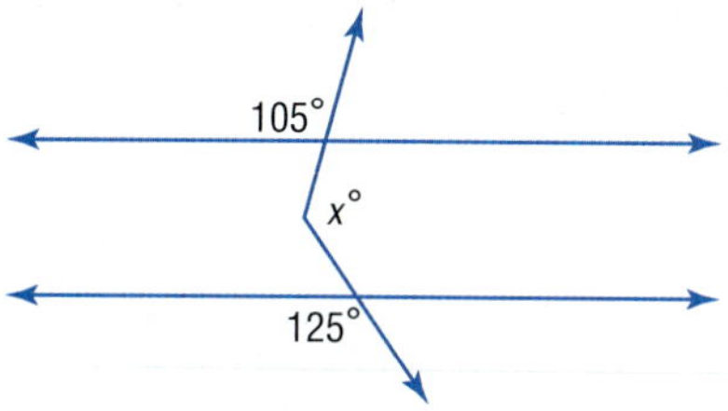

Review Vocabulary

probability the ratio of the number of favorable outcomes to the number of possible outcomes (Lesson 0-3)

40. **PROBABILITY** Suppose you were to pick any two angles in the figure below.

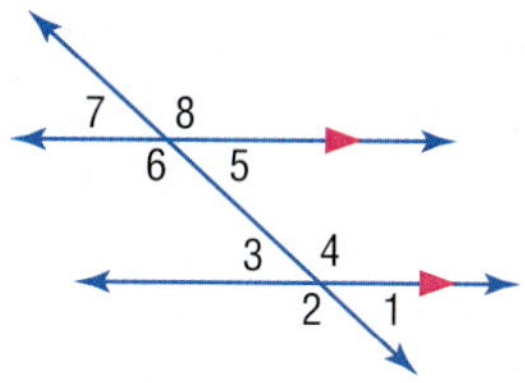

a. How many possible angle pairings are there? Explain.

b. Describe the possible relationships between the measures of the angles in each pair. Explain.

c. Describe the likelihood of randomly selecting a pair of congruent angles. Explain your reasoning.

41. **MULTIPLE REPRESENTATIONS** In this problem, you will investigate the relationship between same-side exterior angles.

a. **GEOMETRY** Draw five pairs of parallel lines, m and n, a and b, r and s, j and k, and x and y, cut by a transversal t, and measure each angle pair.

b. **TABULAR** Record your data in a table.

c. **VERBAL** Make a conjecture about the relationship between the pair of angles formed on the exterior of parallel lines and on the same side of the transversal.

d. **LOGICAL** What type of reasoning did you use to form your conjecture? Explain.

e. **PROOF** Write a proof of your conjecture.

H.O.T. Problems

Use Higher-Order Thinking Skills

42. **WRITING IN MATH** If line a is parallel to line b and $\angle 1 \cong \angle 2$, describe the relationship between lines b and c. Explain your reasoning.

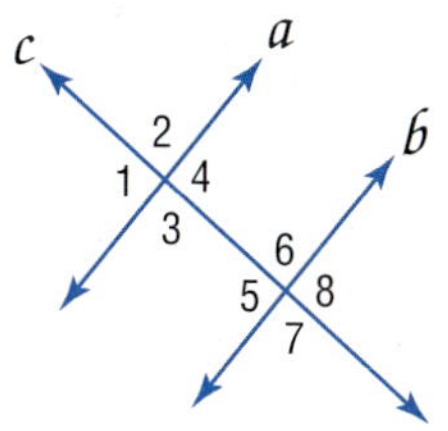

43. **WRITING IN MATH** Compare and contrast the Alternate Interior Angles Theorem and the Consecutive Interior Angles Theorem.

44. **OPEN ENDED** Draw a pair of parallel lines cut by a transversal and measure the two exterior angles on the same side of the transversal. Include the measures on your drawing. Based on the pattern you have seen for naming other pairs of angles, what do you think the name of the pair you measured would be?

45. **CHALLENGE** Find x and y.

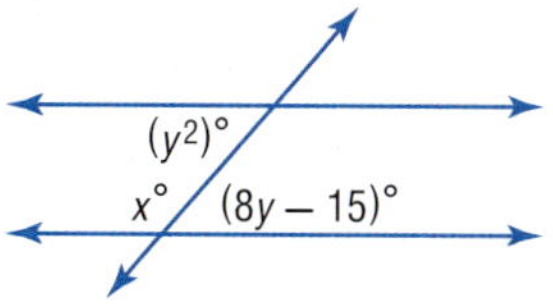

46. **REASONING** Determine the minimum number of angle measures you would have to know to find the measures of all the angles formed by two parallel lines cut by a transversal. Explain.

Standardized Test Practice

47. Suppose $\angle 4$ and $\angle 5$ form a linear pair. If $m\angle 1 = 2x$, $m\angle 2 = 3x - 20$, and $m\angle 3 = x - 4$, what is $m\angle 3$?

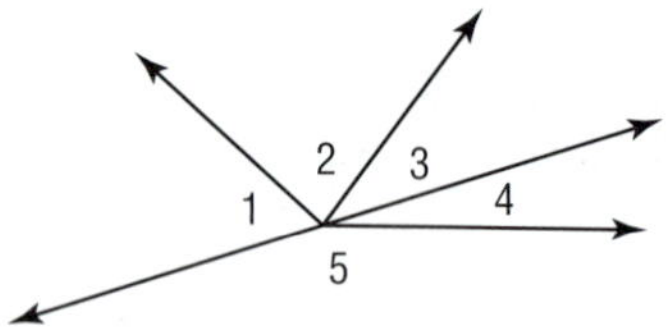

A $26°$
B $28°$
C $30°$
D $32°$

48. **SAT/ACT** A farmer raises chickens and pigs. If his animals have a total of 120 heads and a total of 300 feet, how many chickens does the farmer have?

F 60
G 70
H 80
J 90

49. **SHORT RESPONSE** If $m \parallel n$, then which of the following statements must be true?

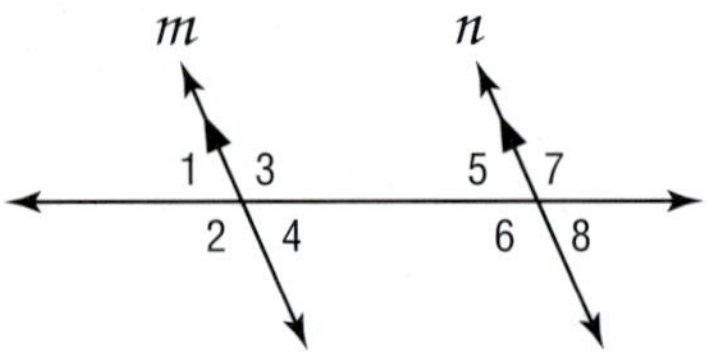

I. $\angle 3$ and $\angle 6$ are Alternate Interior Angles.
II. $\angle 4$ and $\angle 6$ are Consecutive Interior Angles.
III. $\angle 1$ and $\angle 7$ are Alternate Exterior Angles.

50. **ALGEBRA** If $-2 + x = -6$, then $-17 - x =$

A -13
B -4
C 13
D 21

Spiral Review

51. **AVIATION** Airplanes are assigned an altitude level based on the direction they are flying. If one airplane is flying northwest at 34,000 feet and another airplane is flying east at 25,000 feet, describe the type of lines formed by the paths of the airplanes. Explain your reasoning. (Lesson 3-1)

Use the given statement to find the measure of each numbered angle. (Lesson 2-8)

52. $\angle 1$ and $\angle 2$ form a linear pair and $m\angle 2 = 67$.

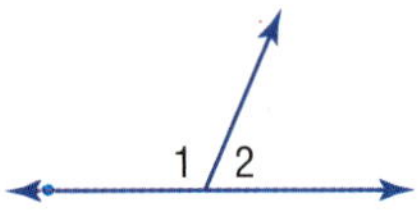

53. $\angle 6$ and $\angle 8$ are complementary; $m\angle 8 = 47$.

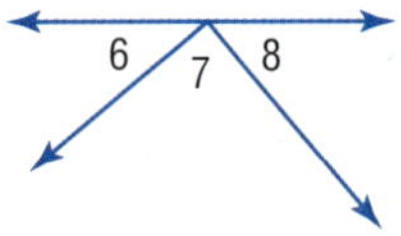

54. $m\angle 4 = 32$

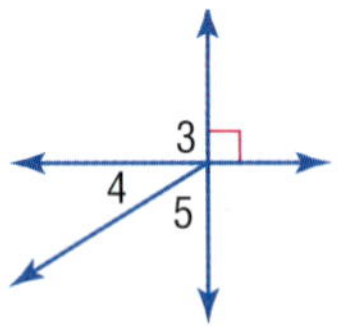

55. **TRAINS** A train company wants to provide routes to New York City, Dallas, Chicago, Los Angeles, San Francisco, and Washington, D.C. An engineer draws lines between each pair of cities on a map. No three of the cities are collinear. How many lines did the engineer draw? (Lesson 2-5)

Skills Review

Simplify each expression.

56. $\frac{6 - 5}{4 - 2}$

57. $\frac{-5 - 2}{4 - 7}$

58. $\frac{-11 - 4}{12 - (-9)}$

59. $\frac{16 - 12}{15 - 11}$

60. $\frac{10 - 22}{8 - 17}$

61. $\frac{8 - 17}{12 - (-3)}$

EXPLORE
3-3

Graphing Technology Lab
Investigating Slope

The rate of change of the steepness of a line is called the *slope*. Slope can be used to investigate the relationship between real-world quantities.

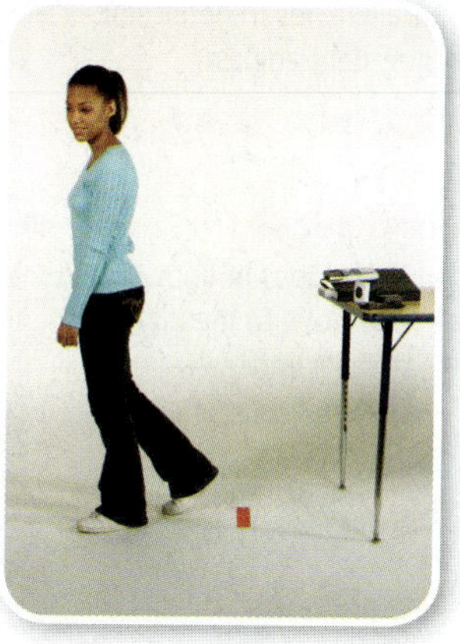

Set Up the Lab

- Connect a data collection device to a graphing calculator. Place on a desk or table so that the data collection device can read the motion of a walker.
- Mark the floor at distances of 1 meter and 6 meters from the device.

ACTIVITY

Step 1 Have one group member stand at the 1-meter mark. When another group member presses the button to begin collecting data, the walker begins to walk away from the device. Walk at a slow, steady pace.

Step 2 Stop collecting data when the walker passes the 6-meter mark. Save the data as Trial 1.

Step 3 Repeat the experiment, walking more quickly. Save the data as Trial 2.

Step 4 For Trial 3, repeat the experiment by walking toward the data collection device slowly.

Step 5 Repeat the experiment, walking quickly toward the device. Save the data as Trial 4.

Analyze the Results

1. Compare and contrast the graphs for Trials 1 and 2. How do the graphs for Trials 1 and 3 compare?
2. Use the **TRACE** feature of the calculator to find the coordinates of two points on each graph. Record the coordinates in a table like the one shown. Then use the points to find the slope of the line.

Trial	Point *A* (x_1, y_1)	Point *B* (x_2, y_2)	Slope $= \frac{y_2 - y_1}{x_2 - x_1}$
1			
2			
3			
4			

3. Compare and contrast the slopes for Trials 1 and 2. How do the slopes for Trials 1 and 2 compare to the slopes for Trials 3 and 4?
4. The slope of a line describes the rate of change of the quantities represented by the *x*- and *y*-values. What is represented by the rate of change in this experiment?
5. **MAKE A CONJECTURE** What would the graph look like if you were to collect data while the walker was standing still? Use the data collection device to test your conjecture.

3-3 Slopes of Lines

Then
You used the properties of parallel lines to determine congruent angles.
(Lesson 3-2)

Now
- Find slopes of lines.
- Use slope to identify parallel and perpendicular lines.

New Vocabulary
slope
rate of change

Math Online
glencoe.com

Why?

Ski resorts assign ratings to their ski trails according to their difficulty. A primary factor in determining this rating is a trail's steepness or *slope gradient.* A trail with a 6% or $\frac{6}{100}$ grade falls 6 feet vertically for every 100 feet traveled horizontally.

The easiest trails, labeled ●, have slopes ranging from 6% to 25%, while more difficult trails, labeled ♦ or ♦♦, have slopes of 40% or greater.

Slope of a Line The steepness or slope of a hill is described by the ratio of the hill's vertical rise to its horizontal run. In algebra, you learned that the slope of a line in the coordinate plane can be calculated using any two points on the line.

Key Concept — Slope of a Line

In a coordinate plane, the **slope** of a line is the ratio of the change along the y-axis to the change along the x-axis between any two points on the line.

The slope m of a line containing two points with coordinates (x_1, y_1) and (x_2, y_2) is given by the formula

$$m = \frac{y_2 - y_1}{x_2 - x_1}, \text{ where } x_1 \neq x_2.$$

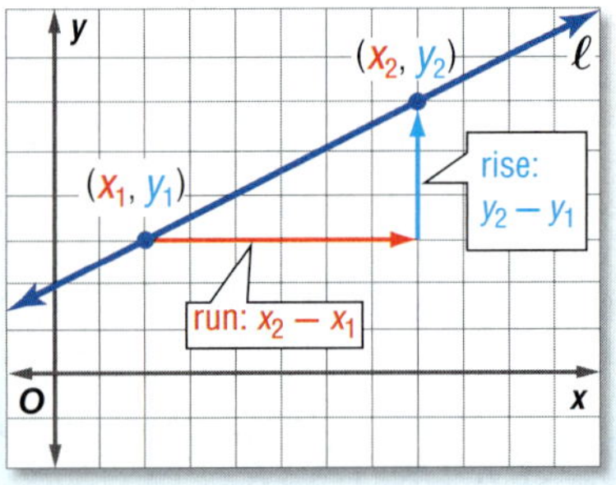

$m = \frac{\text{rise}}{\text{run}} = \frac{y_2 - y_1}{x_2 - x_1}$

EXAMPLE 1 Find the Slope of a Line

Find the slope of each line.

a.

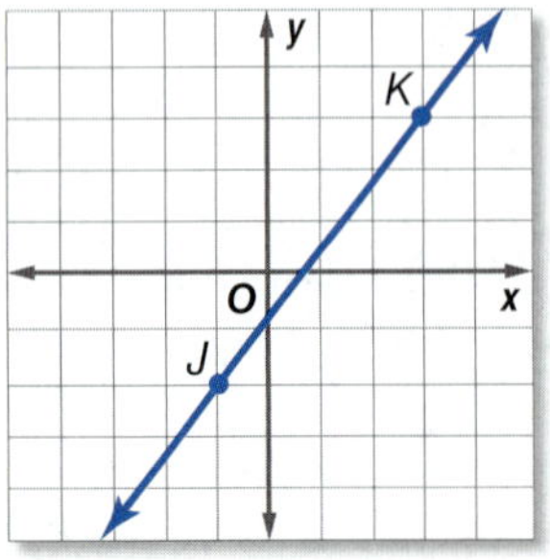

Substitute $(-1, -2)$ for (x_1, y_1) and $(3, 3)$ for (x_2, y_2).

$m = \frac{y_2 - y_1}{x_2 - x_1}$ **Slope Formula**

$= \frac{3 - (-2)}{3 - (-1)}$ **Substitution**

$= \frac{5}{4}$ **Simplify.**

b.

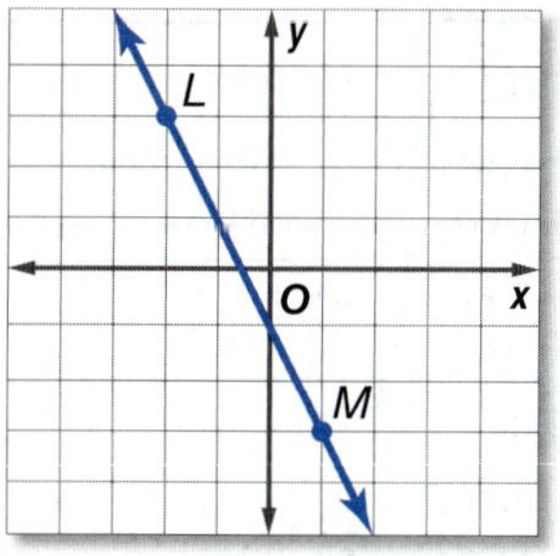

Substitute (−2, 3) for (x_1, y_1) and (1, −3) for (x_2, y_2).

$$m = \frac{y_2 - y_1}{x_2 - x_1} \quad \textbf{Slope Formula}$$

$$= \frac{-3 - 3}{1 - (-2)} \quad \textbf{Substitution}$$

$$= -2 \quad \textbf{Simplify.}$$

c.

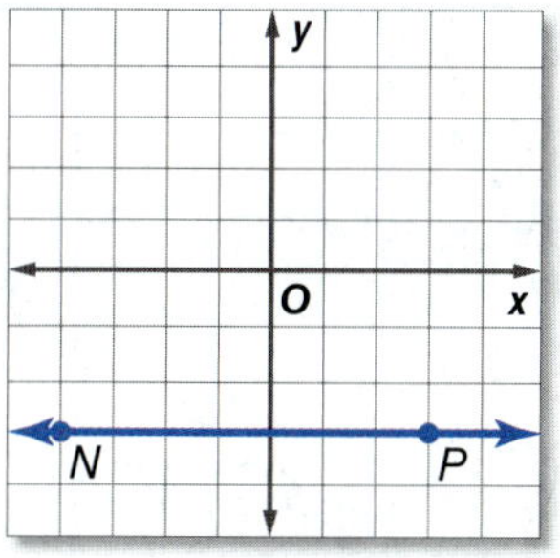

Substitute (−4, −3) for (x_1, y_1) and (3, −3) for (x_2, y_2).

$$m = \frac{y_2 - y_1}{x_2 - x_1} \quad \textbf{Slope Formula}$$

$$= \frac{-3 - (-3)}{3 - (-4)} \quad \textbf{Substitution}$$

$$= \frac{0}{7} \text{ or } 0 \quad \textbf{Simplify.}$$

d.

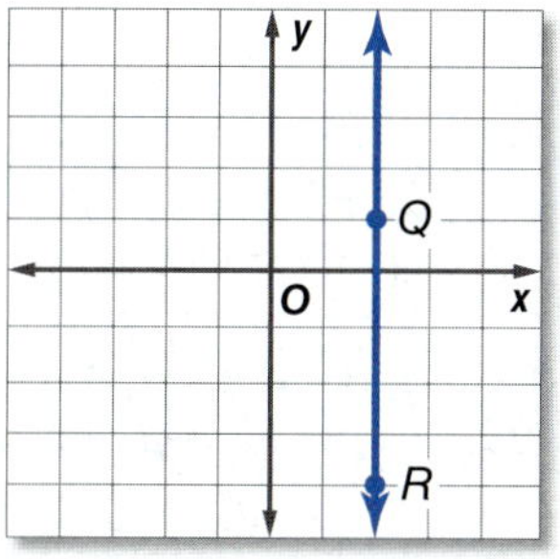

Substitute (2, 1) for (x_1, y_1) and (2, −4) for (x_2, y_2).

$$m = \frac{y_2 - y_1}{x_2 - x_1} \quad \textbf{Slope Formula}$$

$$= \frac{-4 - 1}{2 - 2} \quad \textbf{Substitution}$$

$$= \frac{-5}{0} \quad \textbf{Simplify.}$$

This slope is **undefined**.

StudyTip

Dividing by 0

The slope $\frac{-5}{0}$ is undefined because there is no number that you can multiply by 0 and get −5. Since this is true for any number, all numbers divided by 0 will have an undefined slope. All vertical lines have undefined slopes.

Guided Practice

1A. the line containing (6, −2) and (−3, −5)

1B. the line containing (8, −3) and (−6, −2)

1C. the line containing (4, 2) and (4, −3)

1D. the line containing (−3, 3) and (4, 3)

Personal Tutor glencoe.com

Example 1 illustrates the four different types of slopes.

Concept Summary — Classifying Slopes

Get Animated

Positive Slope

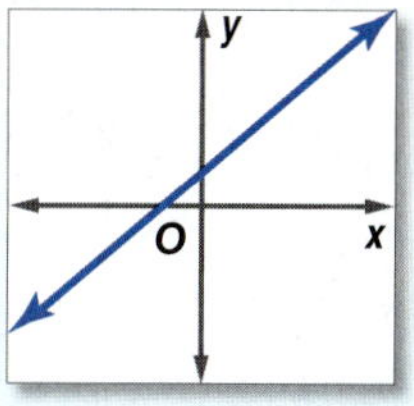

Negative Slope

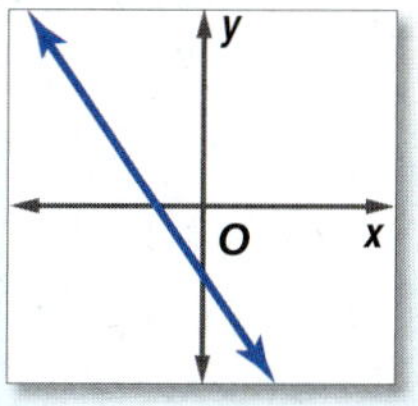

Zero Slope

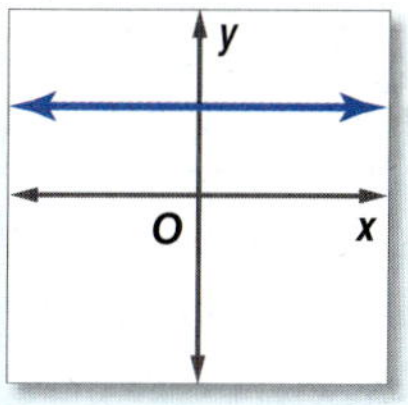

Undefined Slope

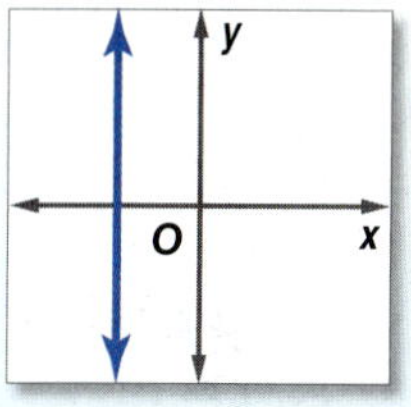

Slope can be interpreted as a **rate of change**, describing how a quantity y changes in relationship to quantity x. The slope of a line can also be used to identify the coordinates of any point on the line.

Real-World EXAMPLE 2 Use Slope as Rate of Change

TRAVEL **A pilot flies a plane from Columbus, Ohio, to Orlando, Florida. After 0.5 hour, the plane reaches its cruising altitude and is 620 miles from Orlando. Half an hour later, the plane is 450 miles from Orlando. How far was the plane from Orlando 1.25 hours after takeoff?**

Understand Use the data given to graph the line that models the distance flown y in miles as a function of time x in hours.

Assume that speed is constant. Plot the points (0.5, 620) and (1.0, 450) and draw a line through them.

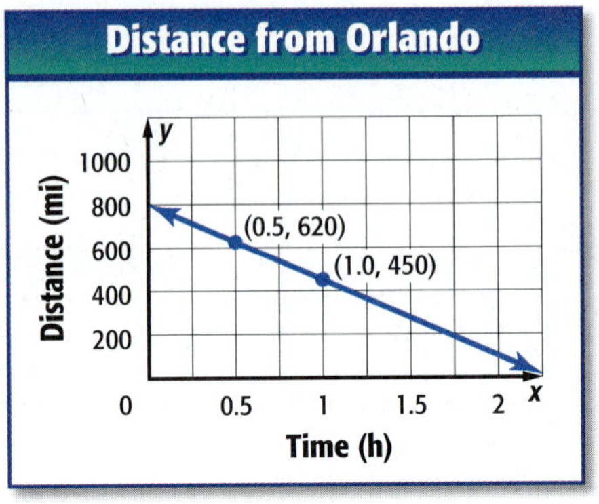

You want to find the distance from Orlando after 1.25 hours.

Plan Find the slope of the line graphed. Use this rate of change in the plane's distance from Orlando per hour to find the distance from Orlando after 1.25 hours.

Solve Use the Slope Formula to find the slope of the line.

$$m = \frac{y_2 - y_1}{x_2 - x_1} = \frac{(450 - 620) \text{ miles}}{(1.0 - 0.5) \text{ hours}} = \frac{-170 \text{ miles}}{0.5 \text{ hour}} \text{ or } -\frac{340 \text{ miles}}{1 \text{ hour}}$$

The plane traveled at an average speed of 340 miles per hour. The negative sign indicates a *decrease* in distance over time.

Use the slope of the line and one known point on the line to calculate the distance y when the time x is 1.25.

$m = \dfrac{y_2 - y_1}{x_2 - x_1}$	**Slope Formula**
$-340 = \dfrac{y_2 - 620}{1.25 - 0.5}$	**$m = -340$, $x_1 = 0.5$, $y_1 = 620$, and $x_2 = 1.25$**
$-340 = \dfrac{y_2 - 620}{0.75}$	**Simplify.**
$-255 = y_2 - 620$	**Multiply each side by 0.75.**
$365 = y_2$	**Add 620 to each side.**

Thus, the distance traveled after 1.25 hours is 365 miles.

Check From the graph we can estimate that after 1.25 hours, the distance from Orlando was a little less than 400 miles. Since 365 is close to this estimate, our answer is reasonable. ✓

Real-World Career

Flight Attendants
Flight attendants check tickets, assist passengers with boarding and carry-ons, and provide an overview of emergency equipment and procedures. A high school diploma is required, but airlines increasingly favor bi- or multi-lingual candidates with college degrees.

Guided Practice

2. **DOWNLOADS** In 2006, 500 million songs were legally downloaded from the Internet. In 2004, 200 million songs were legally downloaded.

A. Use the data given to graph the line that models the number of songs legally downloaded y as a function of time x in years.

B. Find the slope of the line and interpret its meaning.

C. If this trend continues at the same rate, how many songs will be legally downloaded in 2010?

Personal Tutor glencoe.com

Parallel and Perpendicular Lines You can use the slopes of two lines to determine whether the lines are parallel or perpendicular. Lines with the same slope are parallel.

Postulates — Parallel and Perpendicular Lines

3.2 Slopes of Parallel Lines Two nonvertical lines have the same slope if and only if they are parallel. All vertical lines are parallel.

Example Parallel lines ℓ and m have the same slope, 4.

3.3 Slopes of Perpendicular Lines Two nonvertical lines are perpendicular if and only if the product of their slopes is -1. Vertical and horizontal lines are perpendicular.

Example line $m \perp$ line p

product of slopes $= 4 \cdot -\frac{1}{4}$ or -1

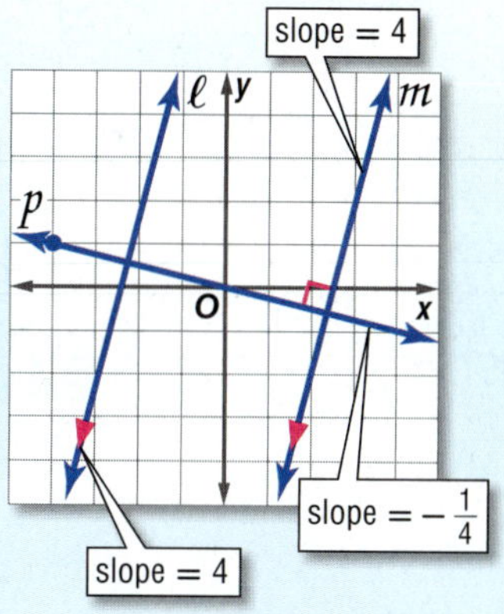

EXAMPLE 3 Determine Line Relationships

Determine whether $\overleftrightarrow{AB}$ and $\overleftrightarrow{CD}$ are *parallel, perpendicular,* or *neither* for $A(1, 1)$, $B(-1, -5)$, $C(3, 2)$, and $D(6, 1)$. Graph each line to verify your answer.

Step 1 Find the slope of each line.

slope of $\overleftrightarrow{AB} = \frac{-5-1}{-1-1} = \frac{-6}{-2}$ or 3 slope of $\overleftrightarrow{CD} = \frac{1-2}{6-3}$ or $\frac{-1}{3}$

Step 2 Determine the relationship, if any, between the lines.

The two lines do not have the same slope, so they are *not* parallel. To determine if the lines are perpendicular, find the product of their slopes.

$3\left(-\frac{1}{3}\right) = -1$ **Product of slopes for $\overleftrightarrow{AB}$ and $\overleftrightarrow{CD}$**

Since the product of their slopes is -1, $\overleftrightarrow{AB}$ is perpendicular to $\overleftrightarrow{CD}$.

CHECK When graphed, the two lines appear to intersect in right angles. ✓

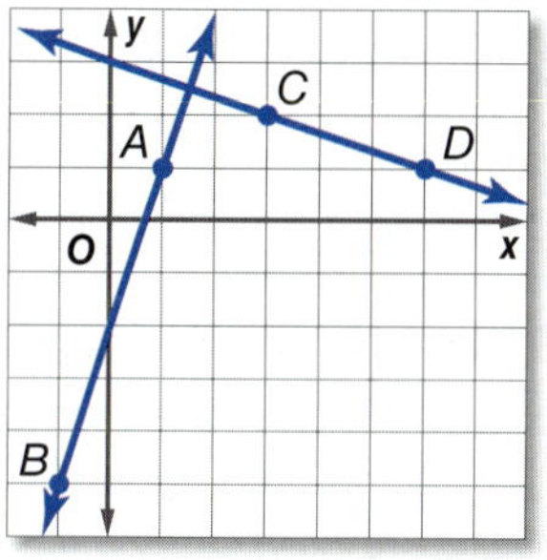

StudyTip

Slopes of Perpendiculars If a line ℓ has a slope of $\frac{a}{b}$, then the slope of a line perpendicular to line ℓ is the opposite reciprocal, $-\frac{b}{a}$, since $\frac{a}{b}\left(-\frac{b}{a}\right) = -1$.

Guided Practice

Determine whether $\overleftrightarrow{AB}$ and $\overleftrightarrow{CD}$ are *parallel, perpendicular,* or *neither.*

3A. $A(14, 13)$, $B(-11, 0)$, $C(-3, 7)$, $D(-4, -5)$

3B. $A(3, 6)$, $B(-9, 2)$, $C(5, 4)$, $D(2, 3)$

Personal Tutor glencoe.com

EXAMPLE 4 Use Slope to Graph a Line

Graph the line that contains $A(-3, 0)$ and is perpendicular to $\overleftrightarrow{CD}$ with $C(-2, -3)$ and $D(2, 0)$.

The slope of $\overleftrightarrow{CD}$ is $\frac{0-(-3)}{2-(-2)}$ or $\frac{3}{4}$.

Since $\frac{3}{4}\left(\frac{4}{-3}\right) = -1$, the slope of the line perpendicular to $\overleftrightarrow{CD}$ through A is $-\frac{4}{3}$ or $\frac{-4}{3}$.

To graph the line, start at A. Move down 4 units and then right 3 units. Label the point B and draw $\overleftrightarrow{AB}$.

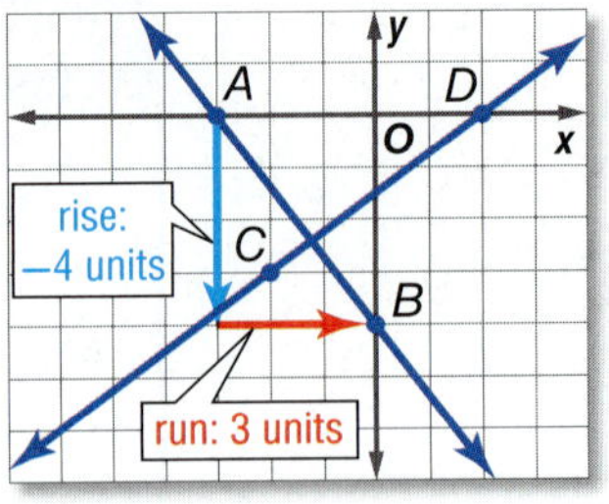

Guided Practice

4. Graph the line that contains $P(0, 1)$ and is perpendicular to $\overleftrightarrow{QR}$ with $Q(-6, -2)$ and $R(0, -6)$.

Personal Tutor glencoe.com

Check Your Understanding

Example 1 pp. 186–187

Find the slope of each line.

1.

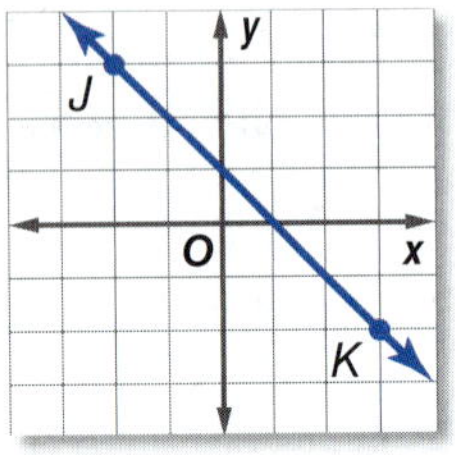

2.

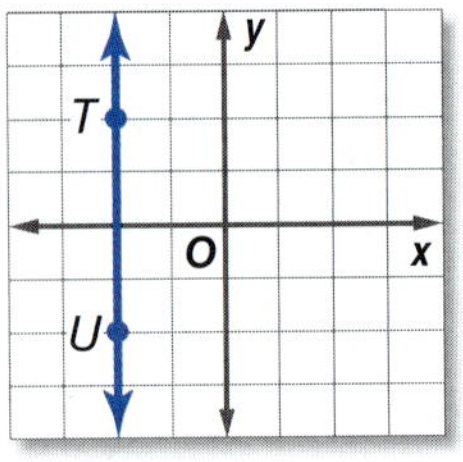

3.

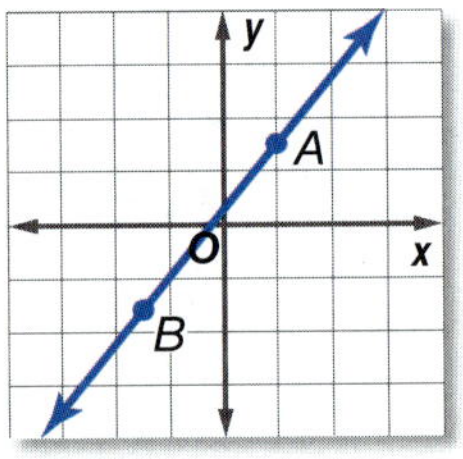

Example 2 p. 188

4. **BOTANY** Kudzu is a fast-growing vine found in the southeastern United States. An initial measurement of the length of a kudzu vine was 0.5 meter. Seven days later the plant was 4 meters long.

a. Graph the line that models the length of the plant over time.

b. What is the slope of your graph? What does it represent?

c. Assuming that the growth rate of the plant continues, how long will the plant be after 15 days?

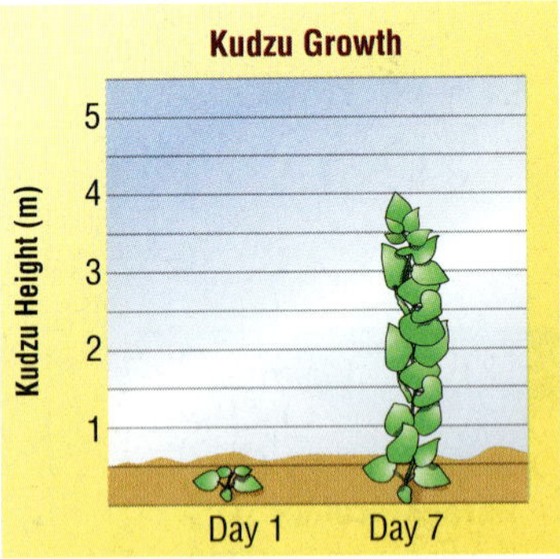

Example 3 p. 189

Determine whether $\overleftrightarrow{WX}$ and $\overleftrightarrow{YZ}$ are *parallel*, *perpendicular*, or *neither*. Graph each line to verify your answer.

5. $W(2, 4)$, $X(4, 5)$, $Y(4, 1)$, $Z(8, -7)$

6. $W(1, 3)$, $X(-2, -5)$, $Y(-6, -2)$, $Z(8, 3)$

7. $W(-7, 6)$, $X(-6, 9)$, $Y(6, 3)$, $Z(3, -6)$

8. $W(1, -3)$, $X(0, 2)$, $Y(-2, 0)$, $Z(8, 2)$

Example 4 p. 190

Graph the line that satisfies each condition.

9. passes through $A(3, -4)$, parallel to $\overleftrightarrow{BC}$ with $B(2, 4)$ and $C(5, 6)$

10. slope $= 3$, passes through $A(-1, 4)$

11. passes through $P(7, 3)$, perpendicular to $\overleftrightarrow{LM}$ with $L(-2, -3)$ and $M(-1, 5)$

Practice and Problem Solving

● = **Step-by-Step Solutions** begin on page R20.
Extra Practice begins on page 969.

Example 1
pp. 186–187

Find the slope of each line.

12.
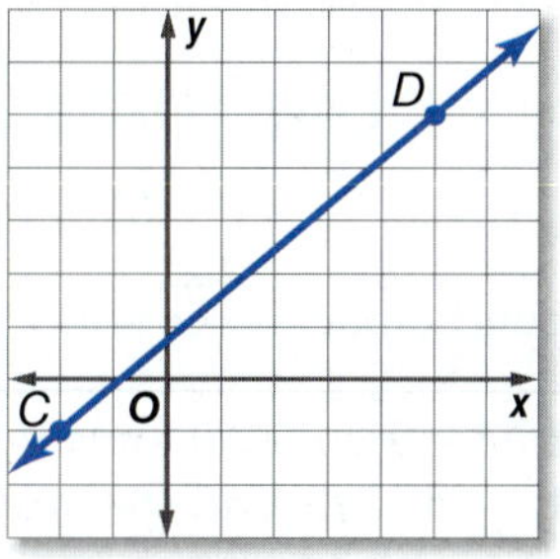

13
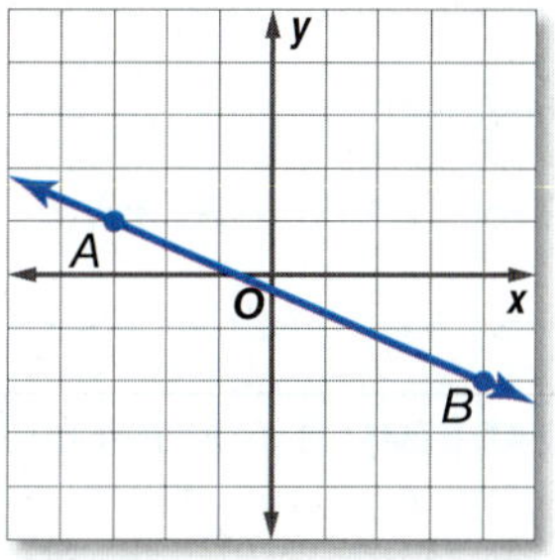

14.
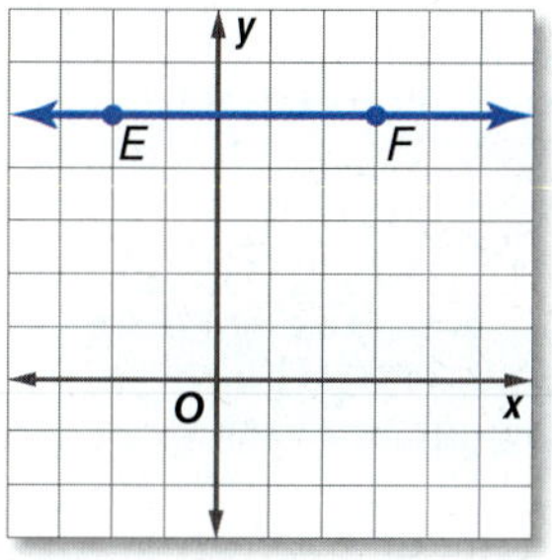

15.
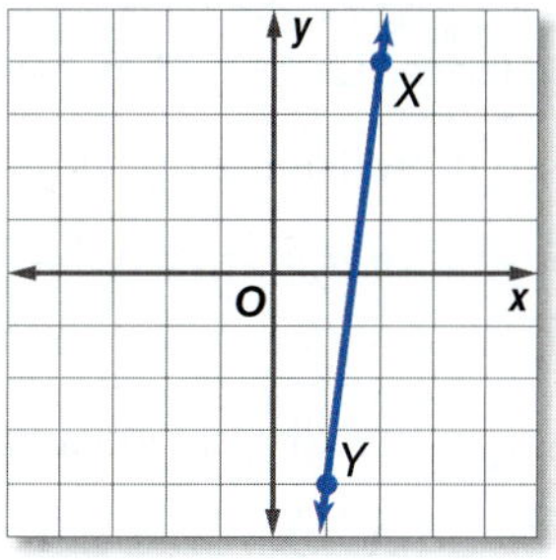

16.
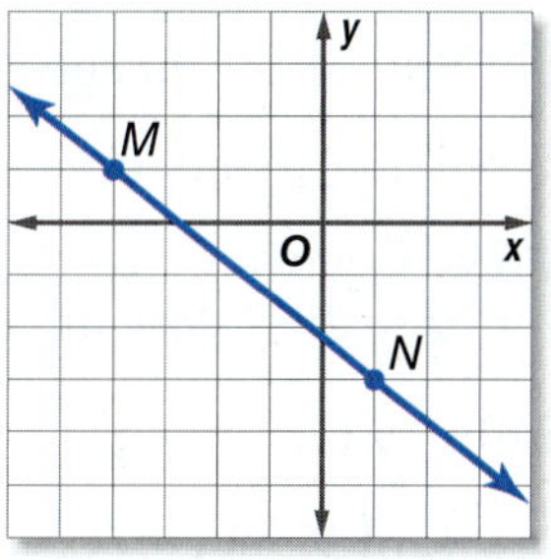

17.
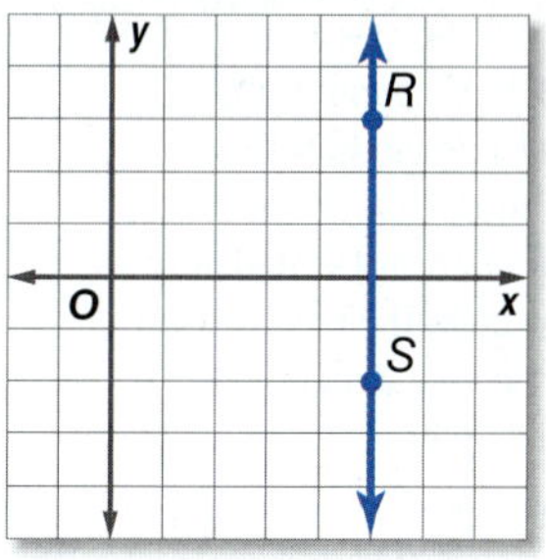

Example 2
p. 188

Determine the slope of the line that contains the given points.

18. $C(3, 1), D(-2, 1)$

19. $E(5, -1), F(2, -4)$

20. $G(-4, 3), H(-4, 7)$

21. $J(7, -3), K(-8, -3)$

22. $L(8, -3), M(-4, -12)$

23. $P(-3, -5), Q(-3, -1)$

24. $R(2, -6), S(-6, 5)$

25. $T(-6, -11), V(-12, -10)$

26. **MOUNTAIN BIKING** In 2004, 8 million Americans over the age of 7 participated in mountain biking, and in 2006, 8.5 million participated.

a. Create a graph to show the number of participants in mountain biking based on the change in participation from 2004 to 2006.

b. Based on the data, what is the growth per year of the sport?

c. If participation continues at the same rate, what will be the participation in 2013 to the nearest 10,000?

27. **FINANCIAL LITERACY** Suppose an MP3 player cost \$499 in 2003 and \$249.99 in 2007.

a. Graph a trend line to predict the price of the MP3 player for 2003 through 2007.

b. Based on the data, how much does the price drop per year?

c. If the trend continues, what will be the cost of an MP3 player in 2010?

Example 3
p. 189

Determine whether $\overleftrightarrow{AB}$ and $\overleftrightarrow{CD}$ are *parallel, perpendicular,* or *neither.* Graph each line to verify your answer.

28. $A(1, 5), B(4, 4), C(9, -10), D(-6, -5)$

29. $A(-6, -9), B(8, 19), C(0, -4), D(2, 0)$

30. $A(4, 2), B(-3, 1), C(6, 0), D(-10, 8)$

31. $A(8, -2), B(4, -1), C(3, 11), D(-2, -9)$

32. $A(8, 4), B(4, 3), C(4, -9), D(2, -1)$

33. $A(4, -2), B(-2, -8), C(4, 6), D(8, 5)$

Example 4
p. 190

Graph the line that satisfies each condition.

34. passes through $A(2, -5)$, parallel to $\overleftrightarrow{BC}$ with $B(1, 3)$ and $C(4, 5)$

35. slope $= -2$, passes through $H(-2, -4)$

36. passes through $K(3, 7)$, perpendicular to $\overleftrightarrow{LM}$ with $L(-1, -2)$ and $M(-4, 8)$

37. passes through $X(1, -4)$, parallel to $\overleftrightarrow{YZ}$ with $Y(5, 2)$ and $Z(-3, -5)$

38. slope $= \frac{2}{3}$, passes through $J(-5, 4)$

39. passes through $D(-5, -6)$, perpendicular to $\overleftrightarrow{FG}$ with $F(-2, -9)$ and $G(1, -5)$

Real-World Link

The Indianapolis Colts played in the RCA Dome for more than 20 years before Lucas Oil Stadium was built. Their first season at the new stadium was 2008-2009.

Source: Stadiums of the NFL

40. **STADIUMS** Before it was demolished, the RCA Dome was home to the Indianapolis Colts. The attendance in 2001 was 450,746, and the attendance in 2005 was 457,373.

a. What is the approximate rate of change in attendance from 2001 to 2005?

b. If this rate of change continues, predict the attendance for 2012.

c. Will the attendance continue to increase indefinitely? Explain.

d. The Colts have now built a new, larger stadium. Do you think their decision was reasonable? Why or why not?

Determine which line passing through the given points has a steeper slope.

41. Line 1: $(0, 5)$ and $(6, 1)$
Line 2: $(-4, 10)$ and $(8, -5)$

42. Line 1: $(0, -4)$ and $(2, 2)$
Line 2: $(0, -4)$ and $(4, 5)$

43 Line 1: $(-9, -4)$ and $(7, 0)$
Line 2: $(0, 1)$ and $(7, 4)$

44. Line 1: $(-6, 7)$ and $(9, -3)$
Line 2: $(-9, 9)$ and $(3, 5)$

45. **ENDANGERED SPECIES** Michigan provides habitat for two endangered species, the bald eagle and the gray wolf. The graph shows the Michigan population of each species in 1992 and 2006.

a. Which species experienced a greater rate of change in population?

b. Make a line graph showing the growth of both populations.

c. If both species continue to grow at their respective rates, what will the population of each species be in 2012?

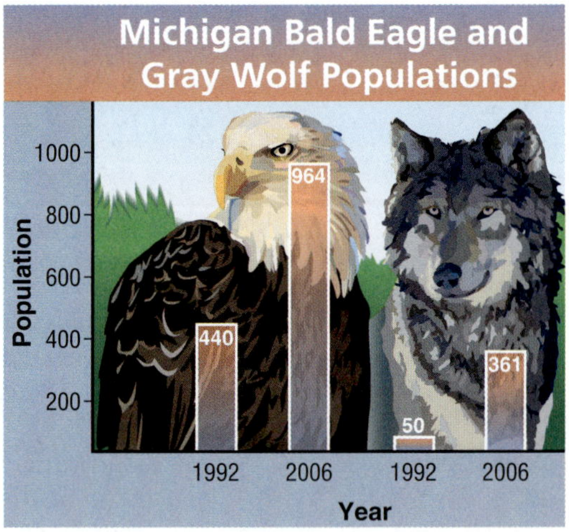

Find the value of x or y that satisfies the given conditions. Then graph the line.

46. The line containing $(4, -1)$ and $(x, -6)$ has a slope of $-\frac{5}{2}$.

47. The line containing $(-4, 9)$ and $(4, 3)$ is parallel to the line containing $(-8, 1)$ and $(4, y)$.

48. The line containing $(8, 7)$ and $(7, -6)$ is perpendicular to the line containing $(2, 4)$ and $(x, 3)$.

49. The line containing $(1, -3)$ and $(3, y)$ is parallel to the line containing $(5, -6)$ and $(9, y)$.

50. **SCHOOLS** In 2000, Jefferson High School had 1125 students. By 2006, the student body had increased to 1425 students. When Fairview High School was built in 2001, it had 1275 students. How many students did Fairview High School have in 2006 if the student body grew at the same rate as Jefferson High School?

51 **MUSIC** Maggie and Mikayla want to go to the music store near Maggie's house after school. They can walk 3.5 miles per hour and ride their bikes 10 miles per hour.

a. Create a table to show how far Maggie and Mikayla can travel walking and riding their bikes. Include distances for 0, 1, 2, 3, and 4 hours.

b. Create a graph to show how far Maggie and Mikayla can travel based on time for both walking and riding their bikes. Be sure to label the axes of your graph.

c. What does the slope represent in your graph?

d. Maggie's mom says they can only go if they can make it to the music store and back in less than two hours. If they want to spend at least 30 minutes in the music store and it is four miles away, can they make it? Should they walk or ride their bikes? Explain your reasoning.

H.O.T. Problems

Use Higher-Order Thinking Skills

52. **WRITE A QUESTION** A classmate says that all lines have positive or negative slope. Write a question that would challenge his conjecture.

53. **ERROR ANALYSIS** Terrell and Hale calculated the slope of the line passing through the points $Q(3, 5)$ and $R(-2, 2)$. Is either of them correct? Explain your reasoning.

Terrell	Hale
$m = \frac{5-2}{3-(-2)}$ $= \frac{3}{5}$	$m = \frac{5-2}{-2-3}$ $= -\frac{3}{5}$

54. **REASONING** Draw a square $ABCD$ with vertices at $A(2, -4)$ and $C(10, 4)$.

a. Find the other two vertices of the square and label them B and D.

b. Show that $\overline{AD} \parallel \overline{BC}$ and $\overline{AB} \parallel \overline{DC}$.

c. Show that the measure of each angle inside the square is equal to 90°.

Real-World Link

With a tilt of 3.97° from vertical, the Leaning Tower of Pisa is one of Italy's major tourist attractions.

Source: Reuters

55. **WRITING IN MATH** Describe the slopes of the Sears Tower and the Leaning Tower of Pisa.

Sears Tower

Leaning Tower of Pisa

56. **CHALLENGE** In this lesson you learned that $m = \frac{y_2 - y_1}{x_2 - x_1}$. Use an algebraic proof to show that the slope can also be calculated using the equation $m = \frac{y_1 - y_2}{x_1 - x_2}$.

57. **WRITING IN MATH** Find two additional points that lie along the same line as $X(3, -1)$ and $Y(-1, 7)$. Generalize a method you can use to find additional points on the line from any given point.

Standardized Test Practice

58. The graph of which equation passes through $(-3, -2)$ and is perpendicular to the graph of $y = \frac{3}{4}x + 8$?

A $y = -\frac{4}{3}x - 6$

B $y = -\frac{4}{3}x + 5$

C $y = \frac{3}{4}x + \frac{1}{4}$

D $y = -\frac{3}{4}x - 5$

59. SHORT RESPONSE A set of 25 cards is randomly placed face down on a table. 15 cards have only the letter A written on the face, and 10 cards have only the letter B. Patrick turned over 1 card. What are the odds of this card having the letter B written on its face?

60. ALGEBRA Jamie is collecting money to buy an $81 gift for her teacher. She has already contributed $24. She will collect $3 from each contributing student. From how many students must Jamie collect money?

F 3 students

G 9 students

H 12 students

J 19 students

61. SAT/ACT The area of a circle is 20π square centimeters. What is its circumference?

A $\sqrt{5}\pi$ cm

B $2\sqrt{5}\pi$ cm

C $4\sqrt{5}\pi$ cm

D 20π cm

Spiral Review

In the figure, $a \parallel b$, $c \parallel d$, and $m\angle 4 = 57$. Find the measure of each angle. (Lesson 3-2)

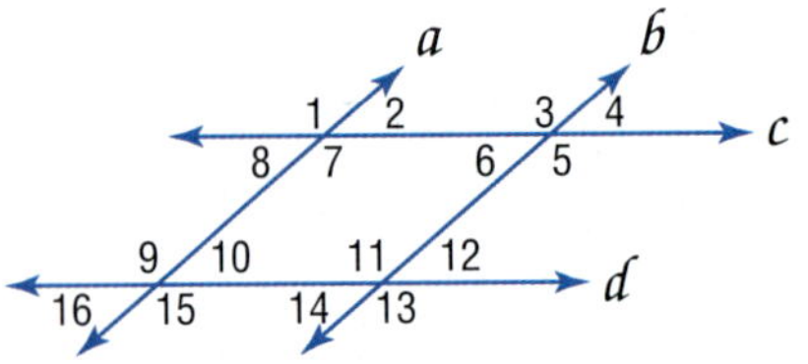

62. $\angle 5$

63. $\angle 1$

64. $\angle 8$

65. $\angle 10$

Refer to the diagram at the right. (Lesson 3-1)

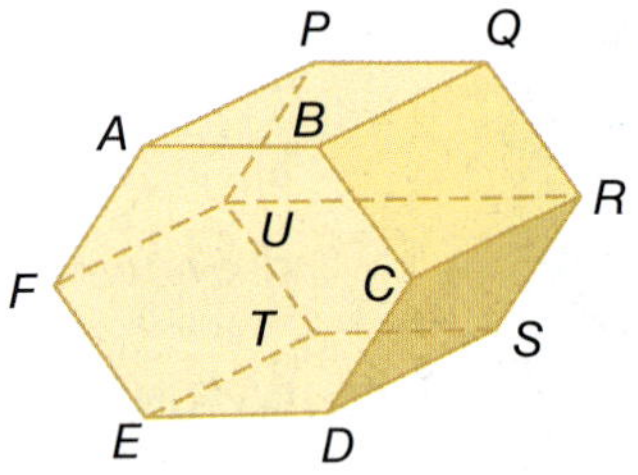

66. Name all segments parallel to $\overline{TU}$.

67. Name all planes intersecting plane *BCR*.

68. Name all segments skew to $\overline{DE}$.

Determine whether the stated conclusion is valid based on the given information. If not, write *invalid*. Explain your reasoning. (Lesson 2-4)

69. Given: $\angle B$ and $\angle C$ are vertical angles.
Conclusion: $\angle B \cong \angle C$

70. Given: $\angle W \cong \angle Y$
Conclusion: $\angle W$ and $\angle Y$ are vertical angles.

71. CONSTRUCTION There are four buildings on the Mansfield High School Campus, no three of which stand in a straight line. How many sidewalks need to be built so that each building is directly connected to every other building? (Lesson 1-1)

Skills Review

Solve for *y*.

72. $3x + y = 5$

73. $4x + 2y = 6$

74. $4y - 3x = 5$

CHAPTER 3 Mid-Chapter Quiz
Lessons 3-1 through 3-3

Identify the transversal connecting each pair of angles. Then classify the relationship between each pair of angles as *alternate interior, alternate exterior, corresponding,* or *consecutive interior* angles. (Lesson 3-1)

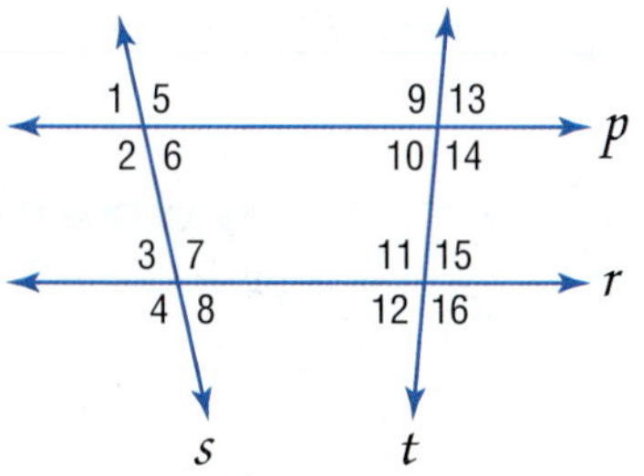

1. ∠6 and ∠3
2. ∠1 and ∠14
3. ∠10 and ∠11
4. ∠5 and ∠7

Refer to the figure to identify each of the following. (Lesson 3-1)

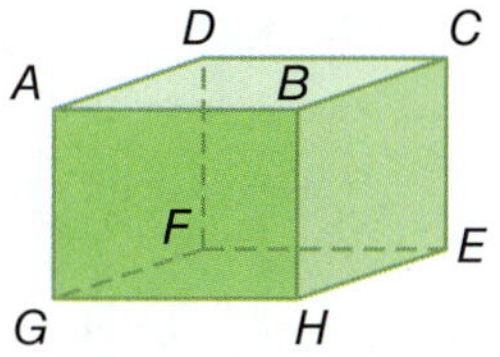

5. a plane parallel to plane *ABCD*
6. a segment skew to $\overline{GH}$ that contains point *D*
7. all segments parallel to $\overline{HE}$
8. **MULTIPLE CHOICE** Which term best describes ∠4 and ∠8? (Lesson 3-1)

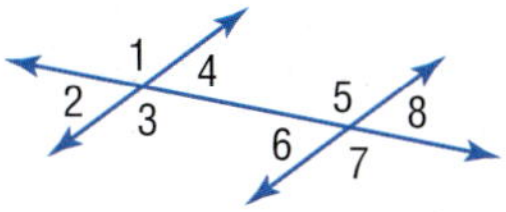

A corresponding
B alternate exterior
C alternate interior
D consecutive interior

In the figure, $m\angle 4 = 104$, $m\angle 14 = 118$. Find the measure of each angle. Tell which postulate(s) or theorem(s) you used. (Lesson 3-2)

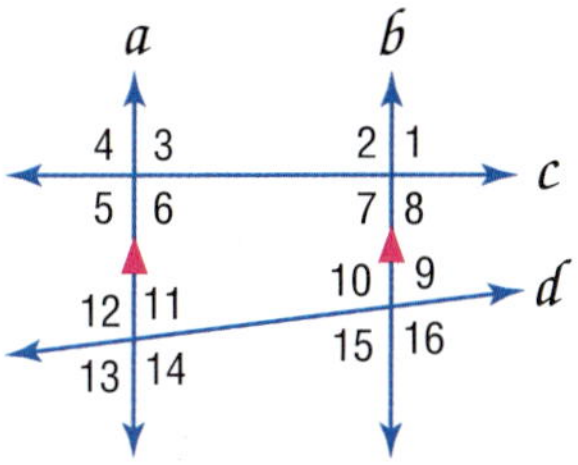

9. ∠2
10. ∠9
11. ∠10
12. ∠7
13. Find *x*. (Lesson 3-2)

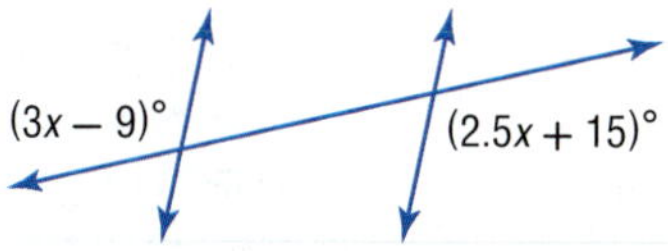

14. **MODEL TRAINS** Amy is going to set up two parallel train tracks with a third track running diagonally across the first two. To properly place a switch, she needs the angle between the diagonal and the top of the second track to be twice as large as the angle between the diagonal and bottom of the first track. What is the value of *x*? (Lesson 3-2)

Determine whether $\overleftrightarrow{AB}$ and $\overleftrightarrow{XY}$ are *parallel, perpendicular,* or *neither*. Graph each line to verify your answer. (Lesson 3-3)

15. $A(2, 0)$, $B(4, -5)$, $X(-3, 3)$, $Y(-5, 8)$
16. $A(1, 1)$, $B(6, -9)$, $X(4, -10)$, $Y(7, -4)$

Find the slope of each line. (Lesson 3-3)

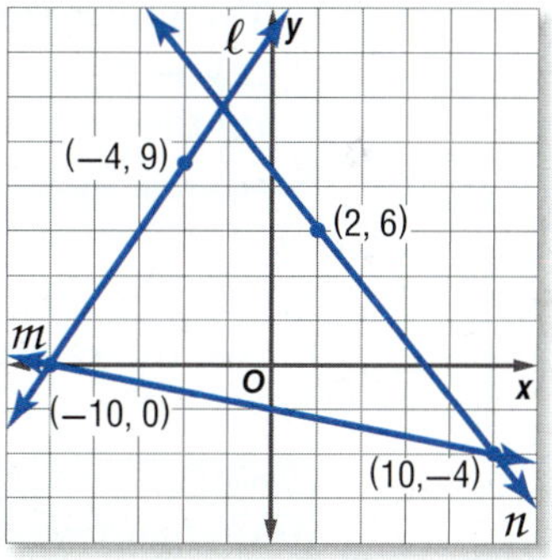

17. line ℓ
18. a line parallel to *m*
19. a line perpendicular to *n*
20. **SALES** The 2003 and 2006 sales figures for Vaughn Electronics are in the table below. (Lesson 3-3)

Year	Approximate Sales ($)
2003	240,000
2006	330,000

a. What is the rate of change in approximate sales from 2003 to 2006?

b. If this rate of change continues, predict the approximate sales for the year 2010.

3-4 Equations of Lines

Then

You found the slopes of lines. (Lesson 3-3)

Now

- Write an equation of a line given information about the graph.
- Solve problems by writing equations.

New Vocabulary

slope-intercept form
point-slope form

Math Online
glencoe.com

Why?

On an interstate near Lauren's hometown, the minimum fine for speeding ten or fewer miles per hour over the speed limit of 65 miles per hour is \$42.50. There is an additional charge of \$2 for each mile per hour over this initial ten miles per hour. The total charge, not including court costs, can be represented by the equation $C = 42.5 + 2m$.

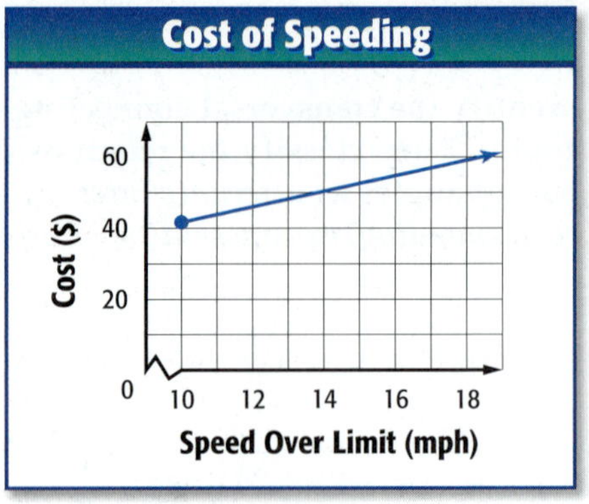

Write Equations of Lines You may remember from algebra that an equation of a nonvertical line can be written in different but equivalent forms.

Key Concept **Nonvertical Line Equations**

The **slope-intercept form** of a linear equation is $y = mx + b$, where m is the slope of the line and b is the y-intercept.

$y = mx + b$ $y = 3x + 8$ (slope; y-intercept)

The **point-slope form** of a linear equation is $y - y_1 = m(x - x_1)$, where (x_1, y_1) is any point on the line and m is the slope of the line.

$y - 5 = -2(x - 3)$ (point on line (3, 5); slope)

When given the slope and either the y-intercept or a point on a line, you can use these forms to write the equation of the line.

EXAMPLE 1 Slope and y-intercept

Write an equation in slope-intercept form of the line with slope 3 and y-intercept of -2. Then graph the line.

$y = mx + b$	Slope-intercept form
$y = 3x + (-2)$	$m = 3, b = -2$
$y = 3x - 2$	Simplify.

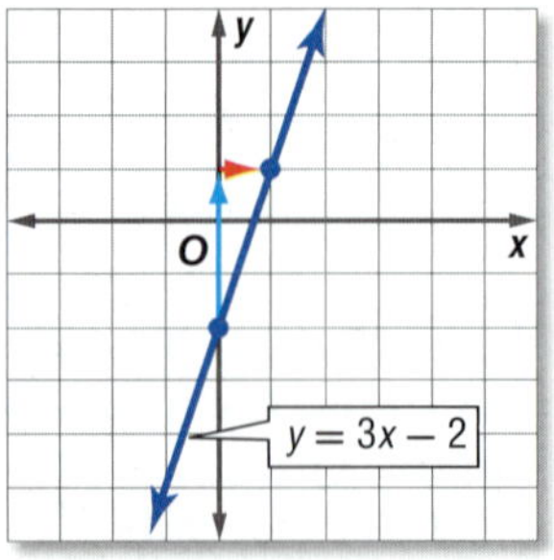

Plot a point at the y-intercept, -2. Use the slope of 3 or $\frac{3}{1}$ to find another point 3 units up and 1 unit to the right of the y-intercept. Then draw the line through these two points.

Guided Practice

1. Write an equation in slope-intercept form of the line with slope $\frac{1}{2}$ and y-intercept of 8. Then graph the line.

Personal Tutor glencoe.com

EXAMPLE 2 Slope and a Point on the Line

Write an equation in point-slope form of the line with slope $-\frac{3}{4}$ that contains (−2, 5). Then graph the line.

$y - y_1 = m(x - x_1)$ **Point-Slope form**

$y - 5 = -\frac{3}{4}[x - (-2)]$ **$m = -\frac{3}{4}$, $(x_1, y_1) = (-2, 5)$**

$y - 5 = -\frac{3}{4}(x + 2)$ **Simplify.**

Graph the given point (−2, 5). Use the slope $-\frac{3}{4}$ or $\frac{-3}{4}$ to find another point 3 units down and 4 units to the right. Then draw the line through these two points.

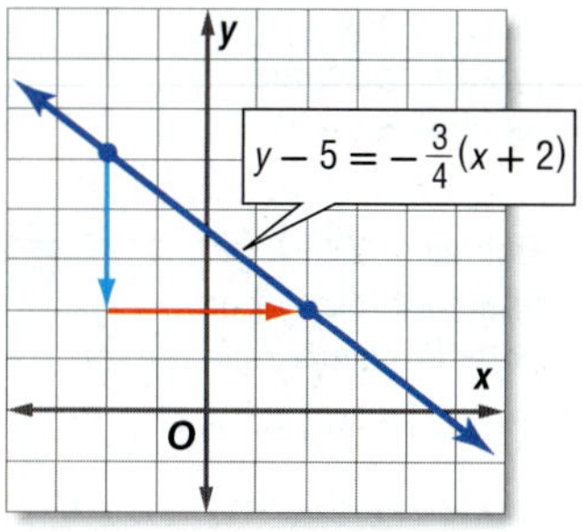

Guided Practice

2. Write an equation in point-slope form of the line with slope of 4 that contains (−3, −6). Then graph the line.

Personal Tutor glencoe.com

Watch Out!

Substituting Negative Coordinates When substituting negative coordinates, use parentheses to avoid making errors with the signs.

When the slope of a line is not given, use two points on the line to calculate the slope. Then use the point-slope or slope-intercept form to write an equation of the line.

EXAMPLE 3 Two Points

Write an equation of the line through each pair of points in slope-intercept form.

a. (0, 3) and (−2, −1)

Step 1 Find the slope of the line through the points.

$m = \frac{y_2 - y_1}{x_2 - x_1} = \frac{-1 - 3}{-2 - 0} = \frac{-4}{-2}$ or 2 **Use the Slope Formula.**

Step 2 Write an equation of the line.

$y = mx + b$ **Slope-Intercept form**

$y = 2x + 3$ **$m = 2$; (0, 3) is the y-intercept.**

b. (−7, 4) and (9, −4)

Step 1 $m = \frac{y_2 - y_1}{x_2 - x_1} = \frac{-4 - 4}{9 - (-7)} = \frac{-8}{16}$ or $-\frac{1}{2}$ **Use the Slope Formula.**

Step 2 $y - y_1 = m(x - x_1)$ **Point-Slope form**

$y - 4 = -\frac{1}{2}[x - (-7)]$ **$m = -\frac{1}{2}$, $(x_1, y_1) = (-7, 4)$**

$y - 4 = -\frac{1}{2}(x + 7)$ **Simplify.**

$y - 4 = -\frac{1}{2}x - \frac{7}{2}$ **Distribute.**

$y = -\frac{1}{2}x + \frac{1}{2}$ **Add 4 to each side: $\frac{7}{2} + 4 = -\frac{7}{2} + \frac{8}{2} = \frac{1}{2}$**

StudyTip

Alternative Method In Example 3b, you could also use the slope-intercept form and one point to find the y-intercept and write the equation.

$y = mx + b$

$4 = -\frac{1}{2}(-7) + b$

$4 = \frac{7}{2} + b$

$4 - \frac{7}{2} = b$

$b = \frac{1}{2}$

So, $y = -\frac{1}{2}x + \frac{1}{2}$.

Guided Practice

3A. (−2, 4) and (8, 10)

3B. (−1, 3) and (7, 3)

Personal Tutor glencoe.com

EXAMPLE 4 Horizontal Line

Write an equation of the line through (−2, 6) and (5, 6) in slope-intercept form.

Step 1 $m = \frac{y_2 - y_1}{x_2 - x_1} = \frac{6 - 6}{5 - (-2)} = \frac{0}{7}$ or 0 This is a horizontal line.

Step 2 $y - y_1 = m(x - x_1)$ **Point-Slope form**

$y - 6 = 0[x - (-2)]$ **$m = -\frac{1}{2}$, $(x_1, y_1) = (-2, 6)$**

$y - 6 = 0$ **Simplify.**

$y = 6$ **Add 6 to each side.**

Guided Practice

4. Write an equation of the line through (5, 0) and (0, −1) in slope-intercept form.

Personal Tutor glencoe.com

StudyTip

Writing Equations Note that the point-slope form of an equation is different for each point used. In Example 3b, another equation in point-slope form would be $y + 4 = -\frac{1}{2}(x - 9)$. However, the slope-intercept form of an equation is unique.

The equations of horizontal and vertical lines involve only one variable.

Key Concept Horizontal and Vertical Line Equations

The equation of a horizontal line is $y = b$, where b is the y-intercept of the line.

Example $y = -3$

The equation of a vertical line is $x = a$, where a is the x-intercept of the line.

Example $x = -2$

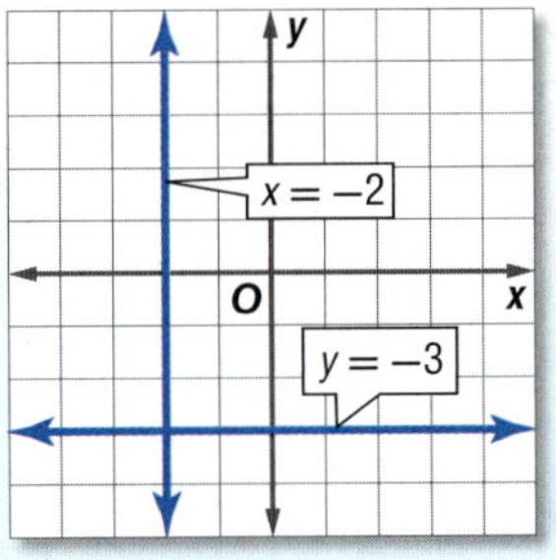

Parallel lines that are not vertical have equal slopes. Two nonvertical lines are perpendicular if the product of their slope is −1. Vertical and horizontal lines are always perpendicular to one another.

Math History Link

Gaspard Monge (1746–1818) Monge presented the point-slope form of an equation of a line in a paper published in 1784.

EXAMPLE 5 Write Parallel or Perpendicular Equations of Lines

Write an equation in slope-intercept form for a line perpendicular to the line $y = -3x + 2$ through (4, 0).

The slope of $y = -3x + 2$ is −3, so the slope of a line perpendicular to it is $\frac{1}{3}$.

$y = mx + b$ **Slope-Intercept form**

$0 = \frac{1}{3}(4) + b$ **$m = \frac{1}{3}$ and $(x, y) = (4, 0)$**

$0 = \frac{4}{3} + b$ **Simplify.**

$-\frac{4}{3} = b$ **Subtract $\frac{4}{3}$ from each side.**

So the equation is $y = \frac{1}{3}x + \left(-\frac{4}{3}\right)$ or $y = \frac{1}{3}x - 1\frac{1}{3}$.

Guided Practice

5. Write an equation in slope-intercept form for a line parallel to $y = -\frac{3}{4}x + 3$ and containing (−3, 6).

Personal Tutor glencoe.com

ReadingMath

Linear The word *linear* indicates a line. The graph of a linear equation is a line.

Problem-SolvingTip

Draw a Graph In Example 6, although the alternative plan has a lower monthly fee, the charge per text message is higher. This makes the plans more difficult to compare. A graph can often give you a better comparison of two linear situations.

Write Equations to Solve Problems

Many real-world situations can be modeled using a linear equation.

Real-World EXAMPLE 6 Write Linear Equations

FINANCIAL LITERACY Benito's current wireless phone plan, Plan X, costs $39.95 per month for unlimited calls and $0.05 per text message. He is considering switching to a Plan Y that costs $35 per month for unlimited calls plus $0.10 for each text message. Which plan offers him the better rate?

Understand Plan X costs $39.95 per month plus $0.05 per text message. Plan Y costs $35 per month plus $0.10 per text message. You want to compare the two plans to determine when the cost of one plan is less than the other.

Plan Write an equation to model the total monthly cost C of each plan for t text messages sent or received. Then graph the equations in order to compare the two plans.

Solve The rates of increase, or slopes m, in the total costs are 0.05 for Plan X and 0.10 for Plan Y. When the number of text messages is 0, the total charge is just the monthly fee. So the y-intercept b is 39.95 for Plan X and 35 for Plan Y.

Plan X		Plan Y
$C = mt + b$	Slope-intercept form	$C = mt + b$
$C = 0.05t + 39.95$	Substitute for m and b.	$C = 0.10t + 35$

Graph the two equations on the same coordinate plane.

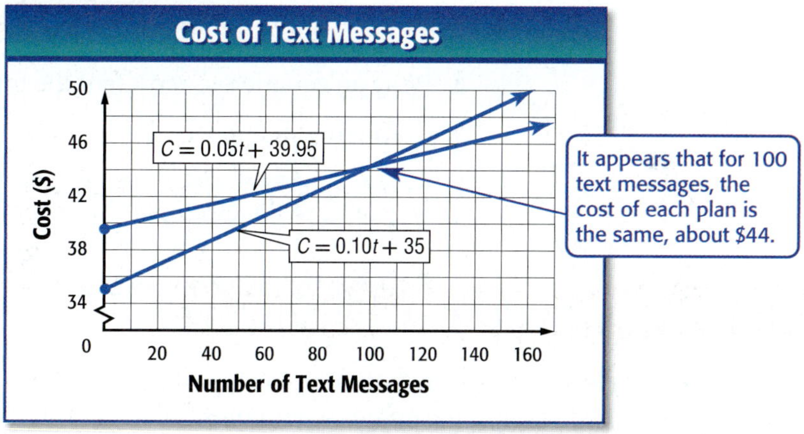

From the graph, it appears that if Benito sends or receives less than about 100 text messages, Plan Y offers the lower rate. For more than 100 messages, Plan X is lower.

Check Check your estimate. For 100 text messages, Plan X costs 0.05(100) + 39.95 or 44.95, and Plan Y costs 0.1(100) + 35 or 45. Adjusting our estimate, we find that when the number of messages is 99, both plans cost $44.90. ✓

Guided Practice

6. Suppose the rate for Plan Y was $44 a month and $0.02 per text message. Which plan offers him the better rate? Justify your answer.

Personal Tutor glencoe.com

Check Your Understanding

Example 1 p. 196

Write an equation in slope-intercept form of the line having the given slope and y-intercept. Then graph the line.

1. m: 4, y-intercept: -3 **2.** m: $\frac{1}{2}$, y-intercept: -1 **3.** m: $-\frac{2}{3}$, y-intercept: 5

Example 2 p. 197

Write an equation in point-slope form of the line having the given slope that contains the given point. Then graph the line.

4. $m = 5$, $(3, -2)$ **5.** $m = \frac{1}{4}$, $(-2, -3)$ **6.** $m = -4.25$, $(-4, 6)$

Examples 3 and 4 pp. 197–198

Write an equation of the line through each pair of points in slope-intercept form.

7. $(0, -1)$ and $(4, 4)$ **8.** $(4, 3)$ and $(1, -6)$ **9.** $(6, 5)$ and $(-1, -4)$

Example 5 p. 198

10. Write an equation in slope-intercept form for a line containing $(3, 2)$ that is perpendicular to the line with equation $y = -2x + 6$.

11. Write an equation in slope-intercept form for a line containing $(-1, 5)$ that is parallel to the line with equation $y = 4x - 5$.

Example 6 p. 199

12. **MUSIC** Kameko currently subscribes to Ace Music, an online music service, but she is considering switching to another online service, Orange Tunes. The plan for each online music service is described below.

a. Write an equation to represent the total monthly cost for each plan.

b. Graph the equations.

c. If Kameko downloads 15 songs per month, should she keep her current plan, or change to the other plan? Explain.

Practice and Problem Solving

= **Step-by-Step Solutions** begin on page R20.
Extra Practice begins on page 969.

Example 1 p. 196

Write an equation in slope-intercept form of the line having the given slope and y-intercept or given points. Then graph the line.

13. m: -5, y-intercept: -2 **14.** m: -7, b: -4 **15.** m: 9, b: 2

16. m: 12, y-intercept: $\frac{4}{5}$ **17.** m: $-\frac{3}{4}$, $(0, 4)$ **18.** m: $\frac{5}{11}$, $(0, -3)$

Example 2 p. 197

Write an equation in point-slope form of the line having the given slope that contains the given point. Then graph the line.

19 $m = 2$, $(3, 11)$ **20.** $m = 4$, $(-4, 8)$ **21.** $m = -7$, $(1, 9)$

22. $m = \frac{5}{7}$, $(-2, -5)$ **23.** $m = -\frac{4}{5}$, $(-3, -6)$ **24.** $m = -2.4$, $(14, -12)$

Examples 3 and 4 pp. 197–198

Write an equation of the line through each pair of points in slope-intercept form.

25. $(-1, -4)$ and $(3, -4)$ **26.** $(2, -1)$ and $(2, 6)$

27. $(-3, -2)$ and $(-3, 4)$ **28.** $(0, 5)$ and $(3, 3)$

29. $(-12, -6)$ and $(8, 9)$ **30.** $(2, 4)$ and $(-4, -11)$

Write an equation in slope-intercept form for each line shown or described.

31. $\overleftrightarrow{EF}$

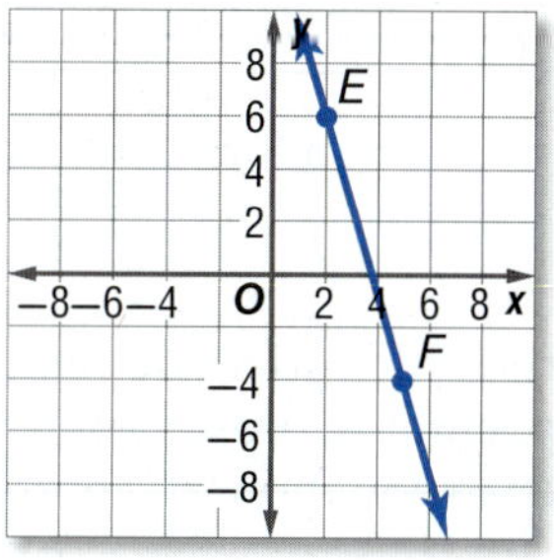

32. $\overleftrightarrow{MN}$

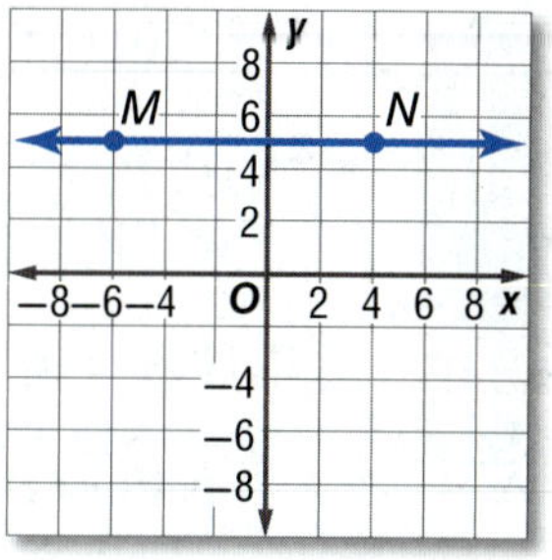

33. contains $(-1, -2)$ and $(3, 4)$

34. contains $(-4, -5)$ and $(-8, -13)$

35. x-intercept $= 3$, y-intercept $= -2$

36. x-intercept $= -\frac{1}{2}$, y-intercept $= 4$

Example 5 p. 198

Write an equation in slope-intercept form for each line described.

37. passes through $(-7, -4)$, perpendicular to $y = \frac{1}{2}x + 9$

38. passes through $(-1, -10)$, parallel to $y = 7$

39. passes through $(6, 2)$, parallel to $y = -\frac{2}{3}x + 1$

40. passes through $(-2, 2)$, perpendicular to $y = -5x - 8$

Example 6 p. 199

41. **PLANNING** Karen is planning a graduation party for the senior class. She plans to rent a meeting room at the convention center that costs \$400. There is an additional fee of \$5.50 for each person who attends the party.

a. Write an equation to represent the cost y of the party if x people attend.

b. Graph the equation.

c. There are 285 people in Karen's class. If $\frac{2}{3}$ of these people attend, how much will the party cost?

d. If the senior class has raised \$2000 for the party, how many people can attend?

Real-World Link

Whereas FM radio signals can only broadcast about 30 to 40 miles, satellite radio can broadcast its signal more than 22,000 miles.

Source: How Stuff Works

42. **SAVINGS** Victor is saving his money to buy a new satellite radio for his car. He wants to save enough money for the radio and one year of satellite radio service before he buys it. He started saving for the radio with \$50 that he got for his birthday. Since then, he has been adding \$15 every week after he cashes his paycheck.

a. Write an equation to represent Victor's savings y after x weeks.

b. Graph the equation.

c. How long will it take Victor to save \$150?

d. A satellite radio costs \$180. Satellite radio service costs \$10 per month. If Victor started saving two weeks ago, how much longer will it take him to save enough money? Explain.

Name the line(s) on the graph shown that match each description.

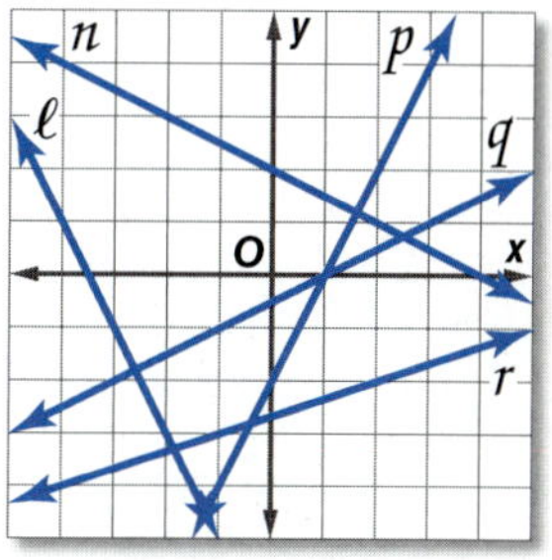

43. parallel to $y = 2x - 3$

44. perpendicular to $y = \frac{1}{2}x + 7$

45. intersecting but not perpendicular to $y = \frac{1}{2}x - 5$

Determine whether the lines are *parallel*, *perpendicular*, or *neither*.

46. $y = 2x + 4$, $y = 2x - 10$

47. $y = -\frac{1}{2}x - 12$, $y = 2x + 7$

48. $y - 4 = 3(x + 5)$, $y + 3 = -\frac{1}{3}(x + 1)$

49. $y - 3 = 6(x + 2)$, $y + 3 = -\frac{1}{3}(x - 4)$

Real-World Link

When heated to a temperature of 500°C, raw clay is transformed into a porous pottery.

Source: *The Dictionary of World Pottery and Porcelain*

50. Write an equation in slope-intercept form for a line containing (4, 2) that is parallel to the line $y - 2 = 3(x + 7)$.

51 Write an equation for a line containing (−8, 12) that is perpendicular to the line containing the points (3, 2) and (−7, 2).

52. Write an equation in slope-intercept form for a line containing (5, 3) that is parallel to the line $y + 11 = \frac{1}{2}(4x + 6)$.

53. POTTERY A local community arts center offers pottery classes. A $40 enrollment fee covers supplies and materials, including one ten-pound bag of clay. Any extra bags of clay cost $15 each. Write an equation to represent the cost of the class and x bags of clay.

54. MULTIPLE REPRESENTATIONS Ms. Smith has asked Dan to work on Friday night valet parking cars for a birthday party. Ms. Smith has given Dan two payment options: he can choose to be paid $3 per car or a $75 fee plus $1.50 per car.

a. TABULAR Create tables to show how much Dan would be paid for 20, 50, and 100 cars for both options.

b. NUMERICAL Write an equation to represent Dan's earnings for each payment option.

c. GRAPHICAL Graph the equations for both options.

d. ANALYTICAL If 35 people attend the party, which option is more profitable? If 75 people attend? Explain.

e. VERBAL Write a statement to describe which payment option is more profitable based on the number of cars that Dan parks.

f. LOGICAL If Ms. Smith sent out 50 invitations, which option should Dan choose? Explain your reasoning.

H.O.T. Problems

Use Higher-Order Thinking Skills

55. CHALLENGE Find the value of n so that the line perpendicular to the line with the equation $-2y + 4 = 6x + 8$ passes through the points at $(n, -4)$ and (2, −8).

56. REASONING Determine whether the points at (−2, 2), (2, 5), and (6, 8) are collinear. Justify your answer.

57. OPEN ENDED Write equations for two different pairs of perpendicular lines that intersect at the point at (−3, −7).

58. ERROR ANALYSIS Mark and Josefina wrote an equation of a line with slope −5 that passes through the point at (−2, 4). Is either of them correct? Explain your reasoning.

Mark

$y - 4 = -5(x - (-2))$

$y - 4 = -5(x + 2)$

$y - 4 = -5x - 10$

$y = -5x - 6$

Josefina

$y - 4 = -5(x - (-2))$

$y - 4 = -5(x + 2)$

59. WRITING IN MATH When is it easier to use the point-slope form to write an equation of a line and when is it easier to use the slope-intercept form?

Standardized Test Practice

60. Which graph best represents a line passing through the point (−2, −3)?

A
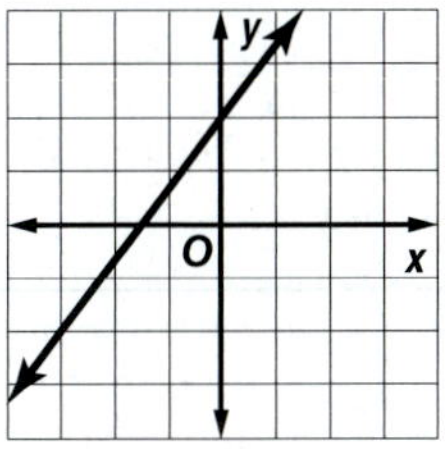

C
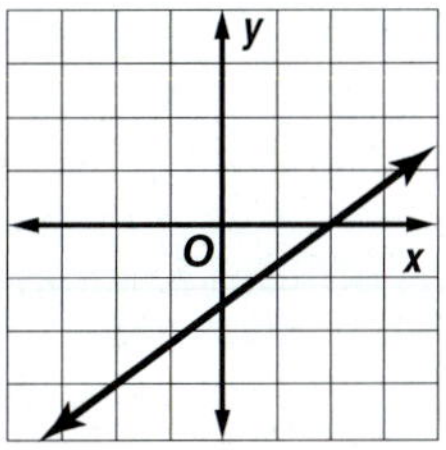

B
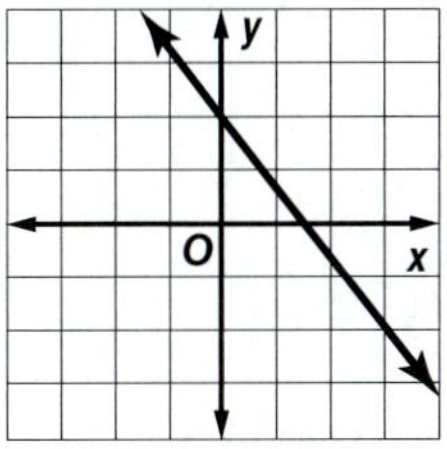

D
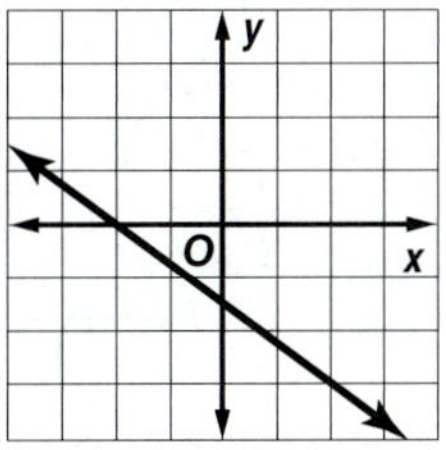

61. Which equation describes the line that passes through the point at (−2, 1) and is perpendicular to the line $y = \frac{1}{3}x + 5$?

F $y = 3x + 7$ H $y = -3x - 5$

G $y = \frac{1}{3}x + 7$ J $y = -\frac{1}{3}x - 5$

62. GRIDDED RESPONSE At Jefferson College, 80% of students have cell phones. Of the students who have cell phones, 70% have computers. What percent of the students at Jefferson College have both a cell phone and a computer?

63. SAT/ACT Which expression is equivalent to $4(x - 6) - \frac{1}{2}(x^2 + 8)$?

A $4x^2 + 4x - 28$ C $-\frac{1}{2}x^2 + 6x - 24$

B $-\frac{1}{2}x^2 + 4x - 28$ D $3x - 20$

Spiral Review

Determine the slope of the line that contains the given points. (Lesson 3-3)

64. $J(4, 3)$, $K(5, -2)$ **65.** $X(0, 2)$, $Y(-3, -4)$ **66.** $A(2, 5)$, $B(5, 1)$

Find x and y in each figure. (Lesson 3-2)

67.
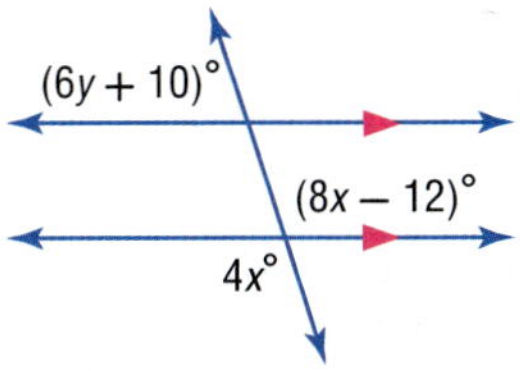

68.
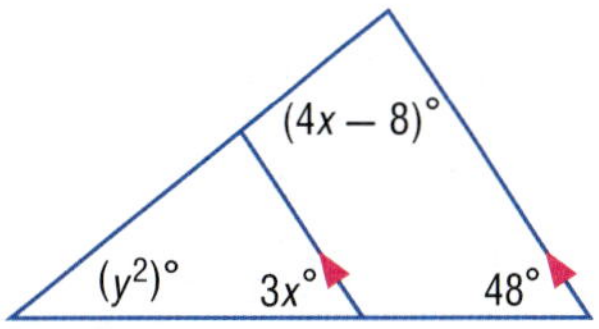

69. DRIVING Lacy's home is located at the midpoint between Newman's Gas Station and Gas-O-Rama. Newman's Gas Station is a quarter mile away from Lacy's home. How far away is Gas-O-Rama from Lacy's home? How far apart are the two gas stations? (Lesson 1-3)

Skills Review

Determine the relationship between each pair of angles. (Lesson 3-1)

70. $\angle 1$ and $\angle 12$

71. $\angle 7$ and $\angle 10$

72. $\angle 4$ and $\angle 8$

73. $\angle 2$ and $\angle 11$

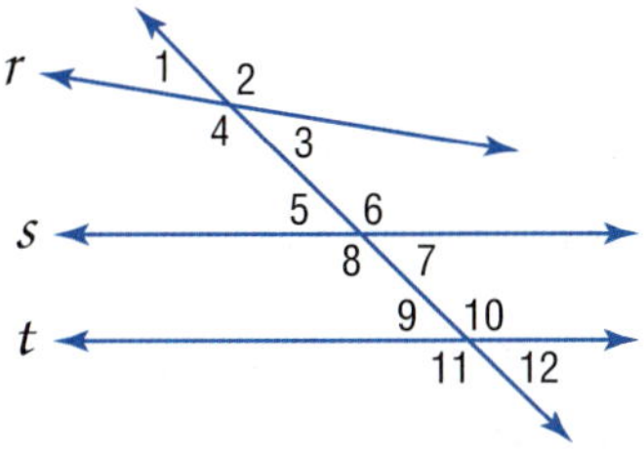

EXTEND 3-4

Geometry Lab
Equations of Perpendicular Bisectors

You can apply what you have learned about slope and equations of lines to geometric figures on a plane.

ACTIVITY

Find the equation of a line that is a perpendicular bisector of a segment AB with endpoints $A(-3, 3)$ and $B(4, 0)$.

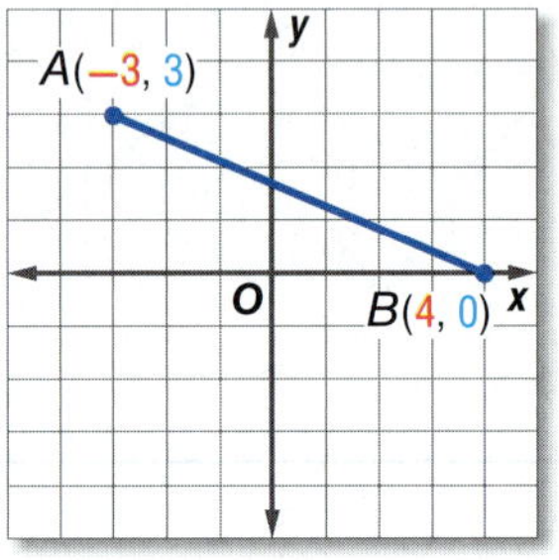

Step 1 A segment bisector contains the midpoint of the segment. Use the Midpoint Formula to find the midpoint M of $\overline{AB}$.

$$M\left(\frac{x_1 + x_2}{2}, \frac{y_1 + y_2}{2}\right) = M\left(\frac{-3 + 4}{2}, \frac{3 + 0}{2}\right)$$
$$= M\left(\frac{1}{2}, \frac{3}{2}\right)$$

Step 2 A perpendicular bisector is perpendicular to the segment through the midpoint. In order to find the slope of the bisector, first find the slope of $\overline{AB}$.

$m = \frac{y_2 - y_1}{x_2 - x_1}$ **Slope Formula**

$= \frac{0 - 3}{4 - (-3)}$ **$x_1 = -3, x_2 = 4, y_1 = 3, y_2 = 0$**

$= -\frac{3}{7}$ **Simplify.**

Step 3 Now use the point-slope form to write the equation of the line. The slope of the bisector is $\frac{7}{3}$ since $-\frac{3}{7}\left(\frac{7}{3}\right) = -1$.

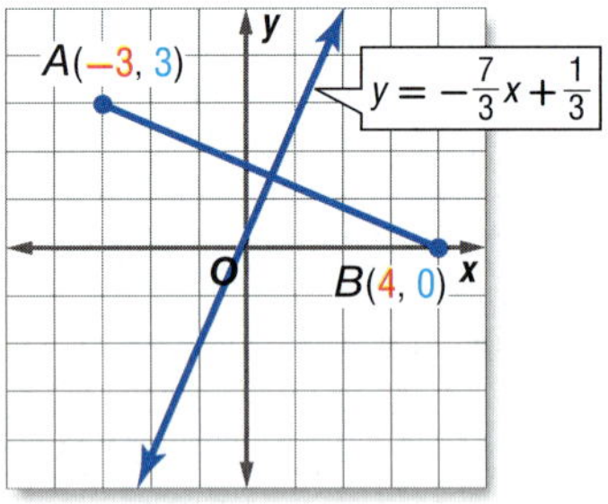

$y - y_1 = m(x - x_1)$ **Point-slope form**

$y - \frac{3}{2} = \frac{7}{3}\left(x - \frac{1}{2}\right)$ **$m = \frac{7}{3}, (x_1, y_1) = \left(\frac{1}{2}, \frac{3}{2}\right)$**

$y - \frac{3}{2} = \frac{7}{3}x - \frac{7}{6}$ **Distributive Property**

$y = \frac{7}{3}x + \frac{1}{3}$ **Add $\frac{3}{2}$ to each side.**

Exercises

Find the equation of the perpendicular bisector $\overline{PQ}$ for the given endpoints.

1. $P(5, 2), Q(7, 4)$
2. $P(-3, 9), Q(-1, 5)$
3. $P(-6, -1), Q(8, 7)$
4. $P(-2, 1), Q(0, -3)$
5. $P(0, 1.6), Q(0.5, 2.1)$
6. $P(-7, 3), Q(5, 3)$
7. Extend what you have learned to find the equations of the lines that contain the sides of $\triangle XYZ$ with vertices $X(-2, 0)$, $Y(1, 3)$, and $Z(3, -1)$.

3-5

Proving Lines Parallel

Then

You used slopes to identify parallel and perpendicular lines. (Lesson 3-3)

Now

- Recognize angle pairs that occur with parallel lines.
- Prove that two lines are parallel.

Math Online
glencoe.com

Why?

When you see a roller coaster track, the two sides of the track are always the same distance apart, even though the track curves and turns. The tracks are carefully constructed to be parallel at all points so that the car is secure on the track.

Identify Parallel Lines The two sides of the track of a roller coaster are parallel, and all of the supports along the track are also parallel. Each of the angles formed between the track and the supports are corresponding angles. We have learned that corresponding angles are congruent when lines are parallel. The converse of this relationship is also true.

Postulate 3.4

Converse of Corresponding Angles Postulate

If two lines are cut by a transversal so that corresponding angles are congruent, then the lines are parallel.

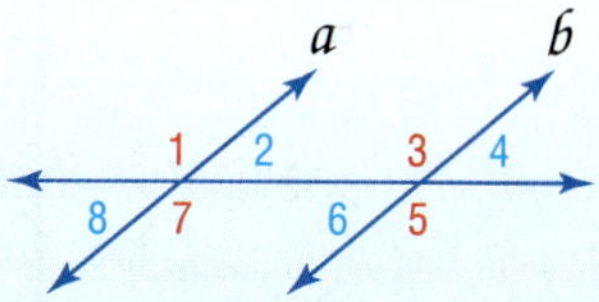

Examples If $\angle 1 \cong \angle 3$, $\angle 2 \cong \angle 4$, $\angle 5 \cong \angle 7$, $\angle 6 \cong \angle 8$, then $a \parallel b$.

The Converse of the Corresponding Angles Postulate can be used to construct parallel lines.

Construction — Parallel Line Through a Point Not on the Line

Step 1 Use a straightedge to draw $\overleftrightarrow{AB}$. Draw a point C that is not on $\overleftrightarrow{AB}$. Draw $\overleftrightarrow{CA}$.

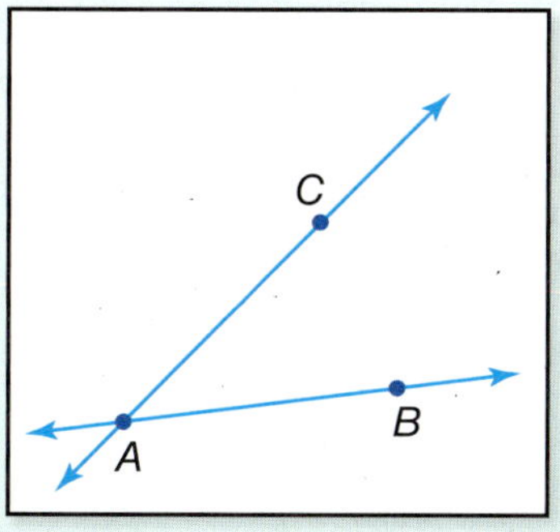

Step 2 Copy $\angle CAB$ so that C is the vertex of the new angle. Label the intersection points D and E.

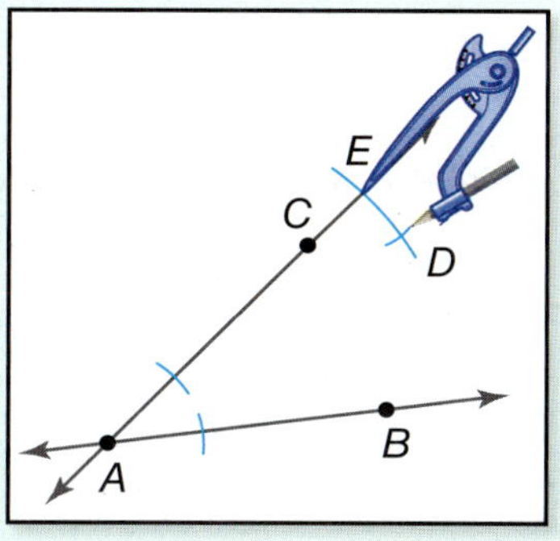

Step 3 Draw CD. Because $\angle ECD \cong \angle CAB$ by construction and they are corresponding angles, $\overleftrightarrow{AB} \parallel \overleftrightarrow{CD}$.

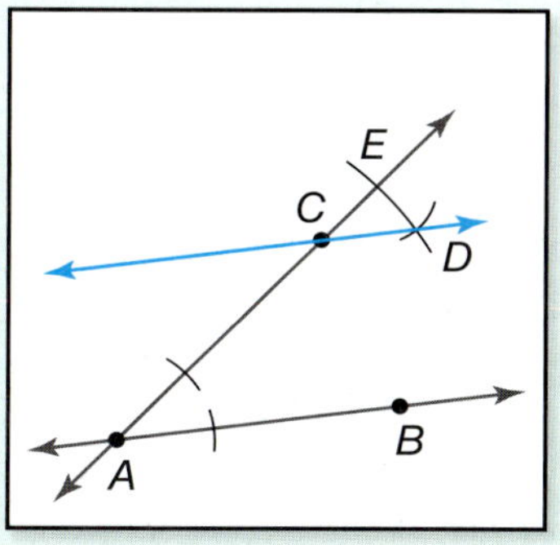

StudyTip

Euclid's Postulates The father of modern geometry, Euclid (c. 300 B.C.) realized that only a few postulates were needed to prove the theorems in his day. Postulate 3.5 is one of Euclid's five original postulates. Postulate 2.1 and Theorem 2.10 also reflect two of Euclid's postulates.

The construction establishes that there is *at least* one line through C that is parallel to $\overleftrightarrow{AB}$. The following postulate guarantees that this line is the *only* one.

Postulate 3.5 Parallel Postulate

If given a line and a point not on the line, then there exists exactly one line through the point that is parallel to the given line.

Parallel lines that are cut by a transversal create several pairs of congruent angles. These special angle pairs can also be used to prove that a pair of lines are parallel.

Theorems Proving Lines Parallel

3.5 Alternate Exterior Angles Converse
If two lines in a plane are cut by a transversal so that a pair of alternate exterior angles is congruent, then the two lines are parallel.

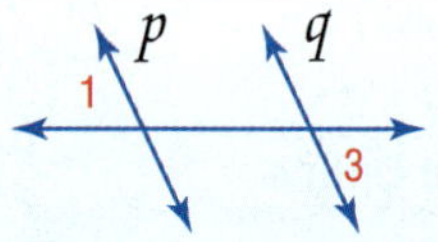

If $\angle 1 \cong \angle 3$, then $p \parallel q$.

3.6 Consecutive Interior Angles Converse
If two lines in a plane are cut by a transversal so that a pair of consecutive interior angles is supplementary, then the lines are parallel.

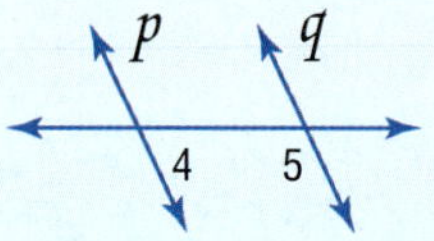

If $\angle 4 + \angle 5 = 180$, then $p \parallel q$.

3.7 Alternate Interior Angles Converse
If two lines in a plane are cut by a transversal so that a pair of alternate interior angles is congruent, then the lines are parallel.

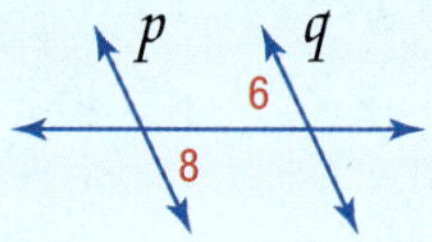

If $\angle 6 \cong \angle 8$, then $p \parallel q$.

3.8 Perpendicular Transversal Converse
In a plane, if two lines are perpendicular to the same line, then they are parallel.

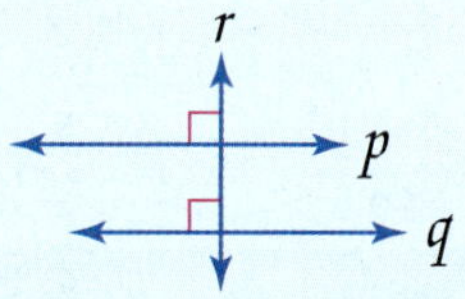

If $p \perp r$ and $q \perp r$, then $p \parallel q$.

You will prove Theorems 3.5, 3.6, 3.7, and 3.8 in Exercises 6, 23, 31, and 30, respectively.

EXAMPLE 1 Identify Parallel Lines

Given the following information, is it possible to prove that any of the lines shown are parallel? If so, state the postulate or theorem that justifies your answer.

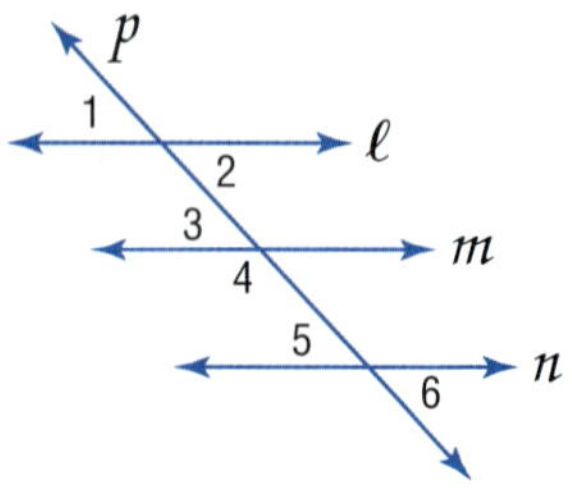

a. $\angle 1 \cong \angle 6$

$\angle 1$ and $\angle 6$ are alternate exterior angles of lines ℓ and n.

Since $\angle 1 \cong \angle 6$, $\ell \parallel n$ by the Converse of the Alternate Exterior Angles Theorem.

b. $\angle 2 \cong \angle 3$

$\angle 2$ and $\angle 3$ are alternate interior angles of lines ℓ and m.

Since $\angle 2 \cong \angle 3$, $\ell \parallel m$ by the Converse of the Alternate Interior Angles Theorem.

Guided Practice

1A. $\angle 2 \cong \angle 8$

1B. $\angle 3 \cong \angle 11$

1C. $\angle 12 \cong \angle 14$

1D. $\angle 1 \cong \angle 15$

1E. $m\angle 8 + m\angle 13 = 180$

1F. $\angle 8 \cong \angle 6$

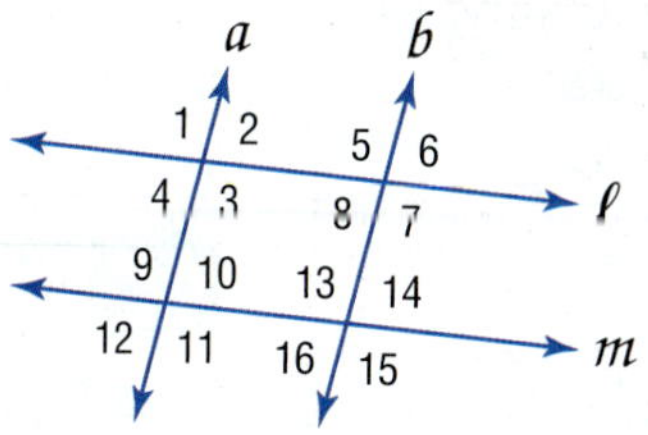

Personal Tutor glencoe.com

Angle relationships can be used to solve problems involving unknown values.

STANDARDIZED TEST EXAMPLE 2

OPEN ENDED **Find $m\angle MRQ$ so that $a \parallel b$. Show your work.**

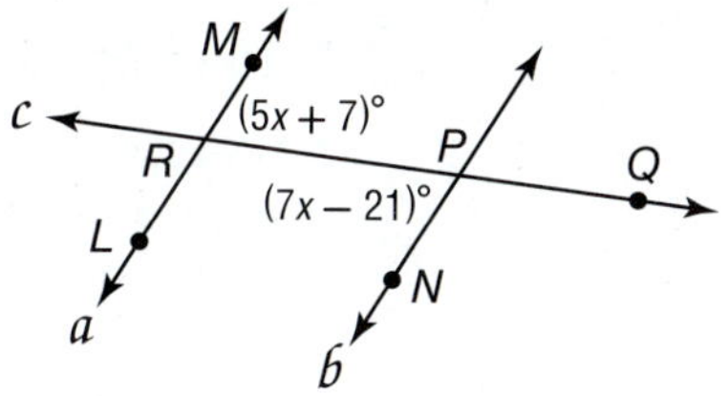

Read the Test Item

From the figure, you know that $m\angle MRQ = 5x + 7$ and $m\angle RPN = 7x - 21$. You are asked to find the measure of $\angle MRQ$.

Solve the Test Item

$\angle MRQ$ and $\angle RPN$ are alternate interior angles. For lines a and b to be parallel, alternate interior angles must be congruent, so $\angle MRQ \cong \angle RPN$. By the definition of congruence, $m\angle MRQ = m\angle RPN$. Substitute the given angle measures into this equation and solve for x.

$m\angle MRQ = m\angle RPN$	**Alternate interior angles**
$5x + 7 = 7x - 21$	**Substitution**
$7 = 2x - 21$	**Subtract 5x from each side.**
$28 = 2x$	**Add 21 to each side.**
$14 = x$	**Divide each side by 2.**

Now, use the value of x to find $\angle MRQ$.

$m\angle MRQ = 5x + 7$	**Substitution**
$= 5(14) + 7$	**$x = 14$**
$= 77$	**Simplify.**

CHECK Check your answer by using the value of x to find $m\angle RPN$.

$m\angle RP = 7x - 21$

$= 7(14) - 21$ or 77 ✓

Since $m\angle MRQ = m\angle RPN$, $\angle MRQ \cong \angle RPN$ and $a \parallel b$. ✓

Test-TakingTip

Finding What Is Asked For
Be sure to reread test questions carefully to be sure you are answering the question that was asked. In Example 2, a common error would be to stop after you have found the value of x and say that the solution of the problem is 14.

Guided Practice

2. Find y so that $e \parallel f$. Show your work.

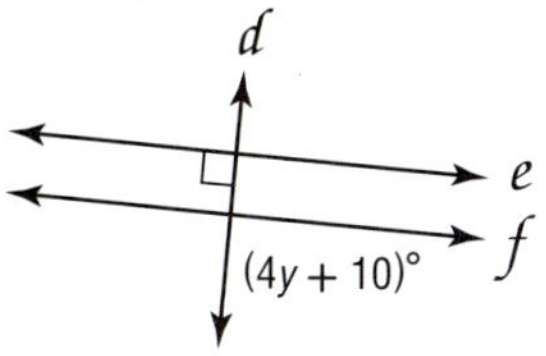

Personal Tutor glencoe.com

StudyTip

Proving Lines Parallel When two parallel lines are cut by a transversal, the angle pairs formed are either congruent or supplementary. When a pair of lines forms angles that do not meet this criterion, the lines cannot possibly be parallel.

Prove Lines Parallel The angle pair relationships formed by a transversal can be used to prove that two lines are parallel.

Real-World EXAMPLE 3 Prove Lines Parallel

HOME FURNISHINGS In the ladder shown, each rung is perpendicular to the two rails. Is it possible to prove that the two rails are parallel and that all of the rungs are parallel? If so, explain how. If not, explain why not.

Since both rails are perpendicular to each rung, the rails are parallel by the Perpendicular Transversal Converse. Since any pair of rungs is perpendicular to the rails, they are also parallel.

Guided Practice

3. **ROWING** In order to move in a straight line with maximum efficiency, rower's oars should be parallel. Refer to the photo at the right. Is it possible to prove that any of the oars are parallel? If so, explain how. If not, explain why not.

Personal Tutor glencoe.com

Check Your Understanding

Example 1 p. 206

Given the following information, determine which lines, if any, are parallel. State the postulate or theorem that justifies your answer.

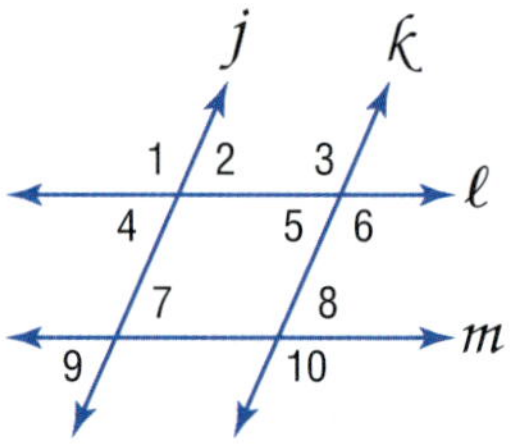

1. $\angle 1 \cong \angle 3$
2. $\angle 2 \cong \angle 5$
3. $\angle 3 \cong \angle 10$
4. $m\angle 6 + m\angle 8 = 180$

Example 2 p. 207

5. **SHORT RESPONSE** Find x so that $m \parallel n$. Show your work.

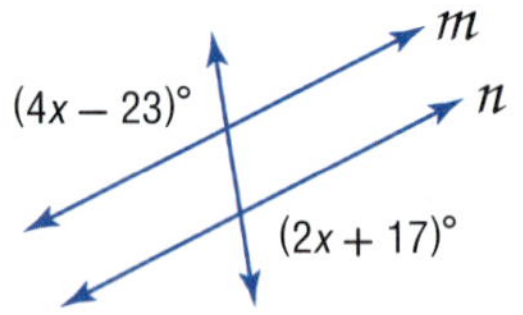

Example 3 p. 208

6. **PROOF** Copy and complete the proof of Theorem 3.5.

Given: $\angle 1 \cong \angle 2$

Prove: $\ell \parallel m$

Proof:

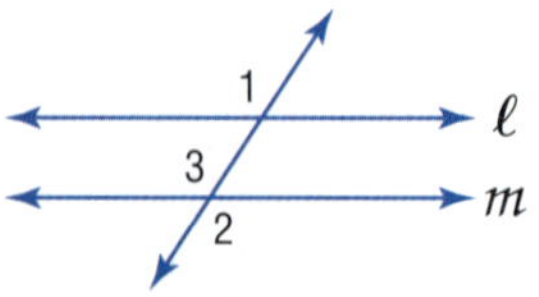

Statements	Reasons
a. $\angle 1 \cong \angle 2$	**a.** Given
b. $\angle 2 \cong \angle 3$	**b.** ___?___
c. $\angle 1 \cong \angle 3$	**c.** Transitive Property
d. ___?___	**d.** ___?___

7. **RECREATION** Is it possible to prove that the backrest and footrest of the lounging beach chair are parallel? If so, explain how. If not, explain why not.

Practice and Problem Solving

= **Step-by-Step Solutions** begin on page R20.
Extra Practice begins on page 969.

Example 1 p. 206

Given the following information, determine which lines, if any, are parallel. State the postulate or theorem that justifies your answer.

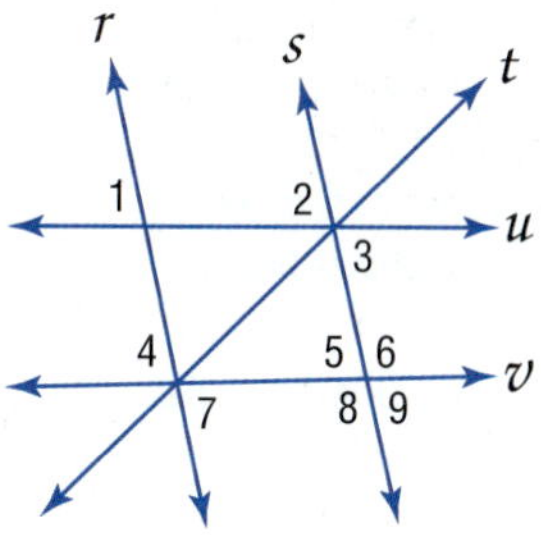

8. $\angle 1 \cong \angle 2$
9. $\angle 2 \cong \angle 9$
10. $\angle 5 \cong \angle 7$
11. $m\angle 7 + m\angle 8 = 180$
12. $m\angle 3 + m\angle 6 = 180$
13. $\angle 3 \cong \angle 5$
14. $\angle 3 \cong \angle 7$
15. $\angle 4 \cong \angle 5$

Example 2 p. 207

Find x so that $m \parallel n$. Identify the postulate or theorem you used.

16.

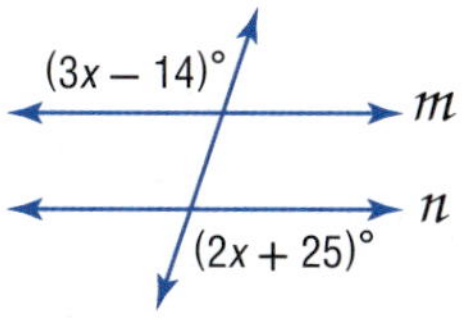

17.

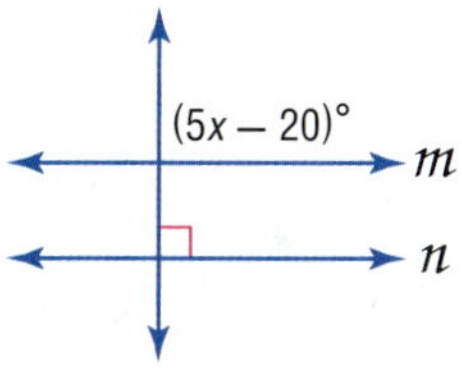

18.

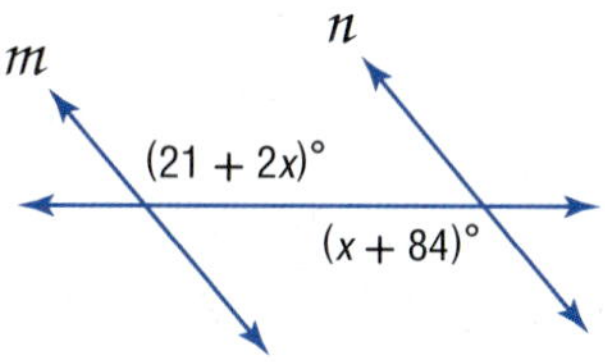

19.

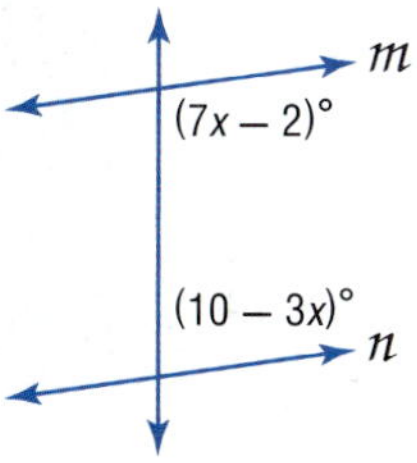

20.

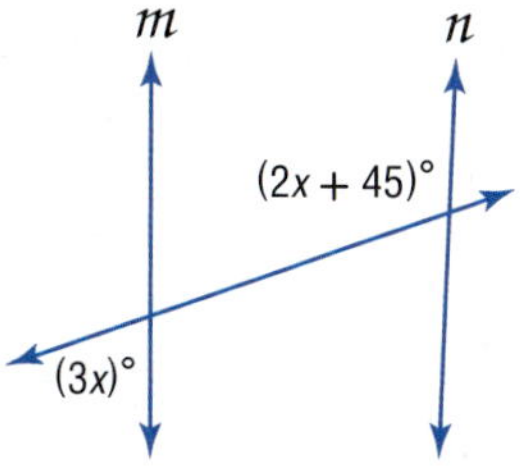

21.

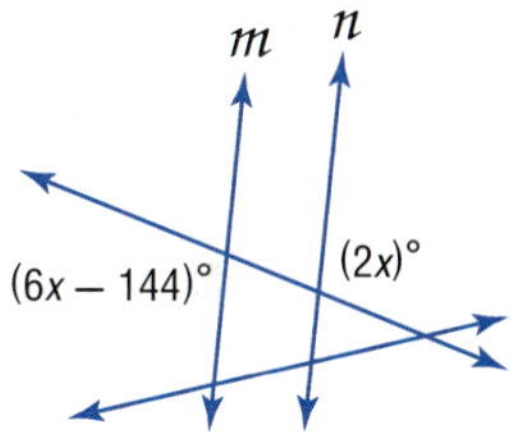

22. **FRAMING** Wooden picture frames are often constructed using a miter box or miter saw. These tools allow you to cut at an angle of a given size. If each of the four pieces of framing material is cut at a 45° angle, will the sides of the frame be parallel? Explain your reasoning.

Example 3 p. 208

23. **PROOF** Copy and complete the proof of Theorem 3.6.

Given: $\angle 1$ and $\angle 2$ are supplementary.

Prove: $\ell \parallel m$

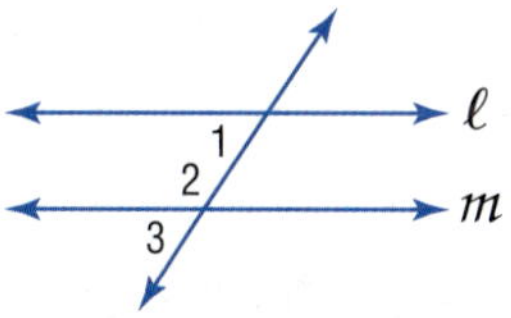

Proof:

Statements	Reasons
a. ___?___	**a.** Given
b. $\angle 2$ and $\angle 3$ form a linear pair.	**b.** ___?___
c. ___?___	**c.** ___?___
d. $\angle 1 \cong \angle 3$	**d.** ___?___
e. $\ell \parallel m$	**e.** ___?___

Real-World Link

The intensity and color of natural daylight can result in variations in the appearance of stained glass. The effect changes throughout the day.

Source: *Encyclopaedia Britannica*

24. CRAFTS Jacqui is making a stained glass piece. She cuts the top and bottom pieces at a 30° angle. If the corners are right angles, explain how Jacqui knows that each pair of opposite sides are parallel.

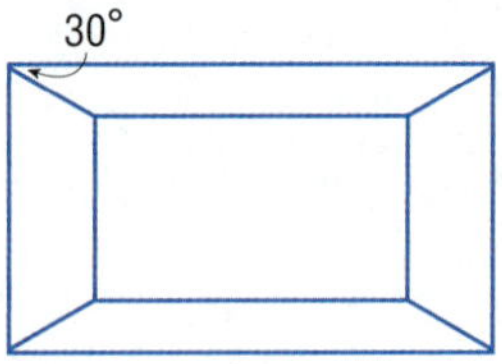

PROOF **Write a two-column proof for each of the following.**

25. **Given:** $\angle 1 \cong \angle 3$
$\overline{AC} \parallel \overline{BD}$
Prove: $\overline{AB} \parallel \overline{CD}$

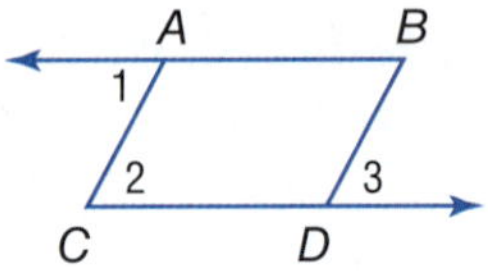

26. **Given:** $\overline{WX} \parallel \overline{YZ}$
$\angle 2 \cong \angle 3$
Prove: $\overline{WY} \parallel \overline{XZ}$

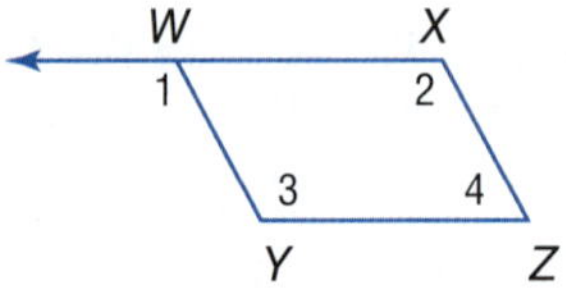

27. **Given:** $\angle ABC \cong \angle ADC$
$m\angle A + m\angle ABC = 180$
Prove: $\overline{AB} \parallel \overline{CD}$

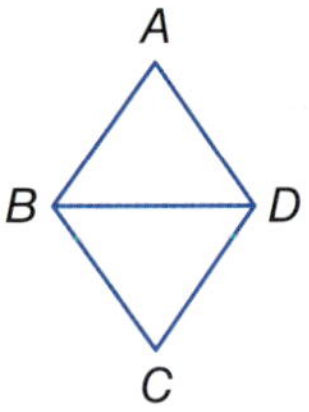

28. **Given:** $\angle 1 \cong \angle 2$
$\overline{LJ} \perp \overline{ML}$
Prove: $\overline{KM} \perp \overline{ML}$

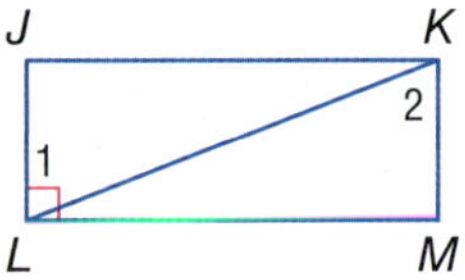

29 **MAILBOXES** Mail slots are used to make the organization and distribution of mail easier. In the mail slots shown, each slot is perpendicular to each of the sides. Explain why you can conclude that the slots are parallel.

30. PROOF Write a paragraph proof of Theorem 3.8.

31. PROOF Write a two-column proof of Theorem 3.7.

32. STAIRS Based upon the information given in the photo of the staircase at the right, what is the relationship between each step? Explain your answer.

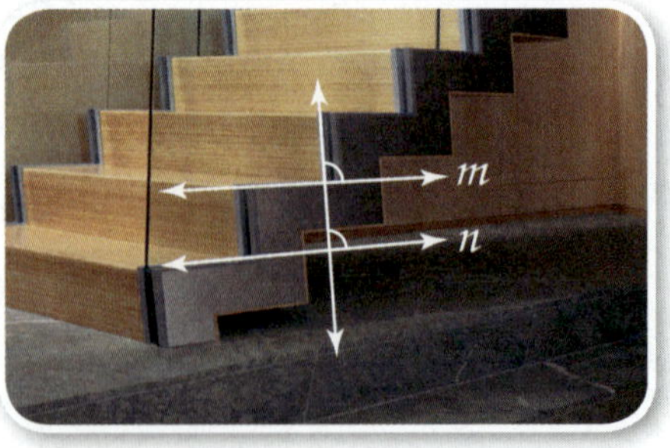

Determine whether lines r and s are parallel. Justify your answer.

33.

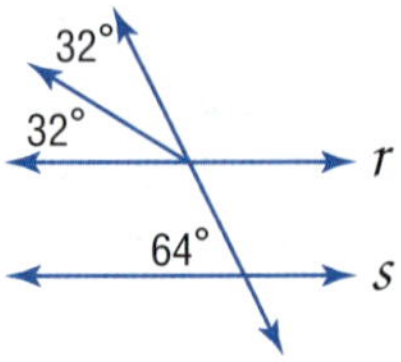

34.

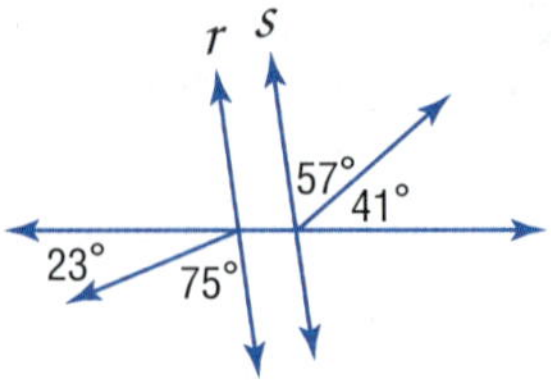

35.

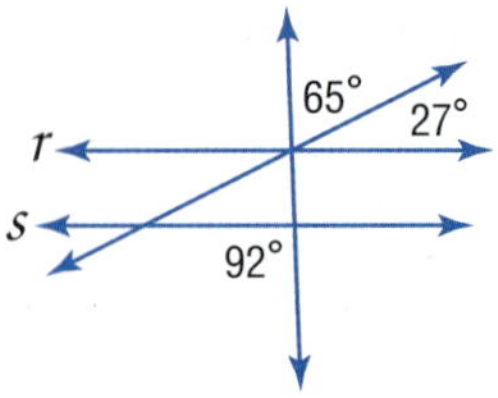

36. **MULTIPLE REPRESENTATIONS** In this problem, you will explore the shortest distance between two parallel lines.

a. **GEOMETRIC** Draw three sets of parallel lines k and ℓ, s and t, and x and y. Draw the shortest segment $\overline{BC}$ and label points A and D as shown below.

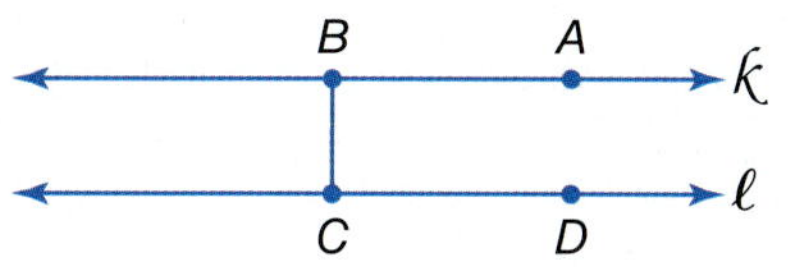

b. **TABULAR** Copy the table below, measure $\angle ABC$ and $\angle BCD$, and complete the table.

Set of Parallel Lines	$m\angle ABC$	$m\angle BCD$
k and ℓ		
s and t		
x and y		

c. **VERBAL** Make a conjecture about the angle the shortest segment forms with both parallel lines.

H.O.T. Problems Use Higher-Order Thinking Skills

37. **ERROR ANALYSIS** Sumi and Daniela are determining which lines are parallel in the figure at the right. Sumi says that since $\angle 1 \cong \angle 2$, $\overline{WY} \parallel \overline{XZ}$. Daniela disagrees and says that since $\angle 1 \cong \angle 2$, $\overline{WX} \parallel \overline{YZ}$. Is either of them correct? Explain.

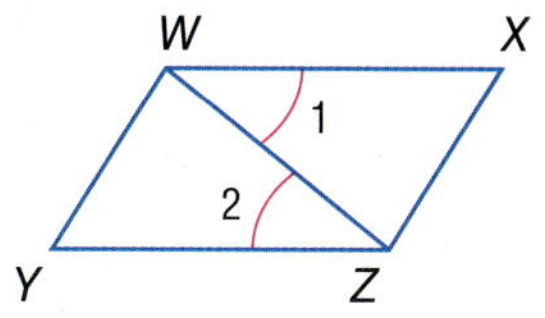

38. **REASONING** Given the statement, *In a plane, if two lines are perpendicular to the same line, then they are parallel* (Theorem 3.8), is the statement still true if the lines are not coplanar? Draw a figure to justify your answer.

39. **OPEN ENDED** Draw a triangle ABC.

a. Construct the line parallel to $\overline{BC}$ through point A.

b. Use measurement to justify that the line you constructed is parallel to $\overline{BC}$.

c. Use mathematics to justify this construction.

40. **REASONING** Determine whether parallel lines are *reflexive, symmetric,* or *transitive.* Do parallel lines form an equivalence relation? Explain.

Review Vocabulary

equivalence relation any relationship that satisfies the reflexive, symmetric, and transitive properties. (Lesson 2-6)

41. **CHALLENGE** Refer to the figure at the right.

a. If $m\angle 1 + m\angle 2 = 180$, prove that $a \parallel c$.

b. Given that $a \parallel c$, if $m\angle 1 + m\angle 3 = 180$, prove that $t \perp c$.

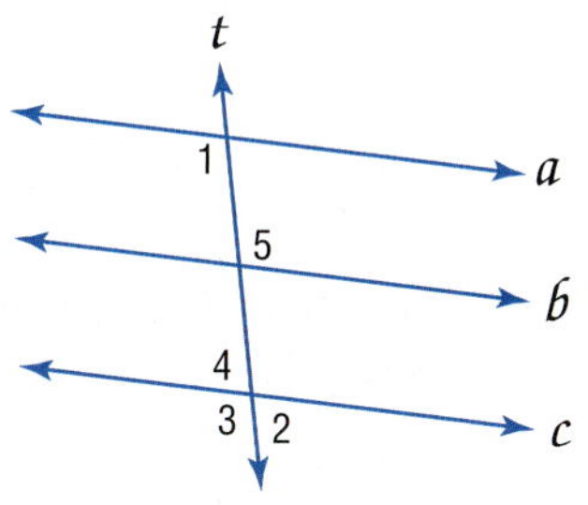

42. **WRITING IN MATH** Summarize the five methods used in this lesson to prove that two lines are parallel.

43. **WRITING IN MATH** Describe a situation in which two parallel lines are cut by a transversal and a pair of consecutive interior angles are both supplementary and congruent. Explain.

Standardized Test Practice

44. Which of the following facts would be sufficient to prove that line d is parallel to $\overline{XZ}$?

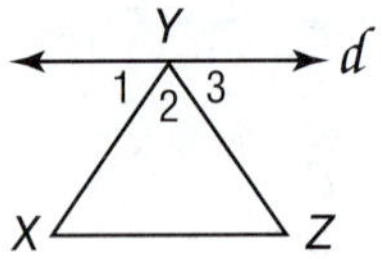

A $\angle 1 \cong \angle 3$
B $\angle 3 \cong \angle Z$
C $\angle 1 \cong \angle Z$
D $\angle 2 \cong \angle X$

45. **ALGEBRA** The expression $\sqrt{52} + \sqrt{117}$ is equivalent to

F 13
G $5\sqrt{13}$
H $6\sqrt{13}$
J $13\sqrt{13}$

46. What is the approximate surface area of the figure?

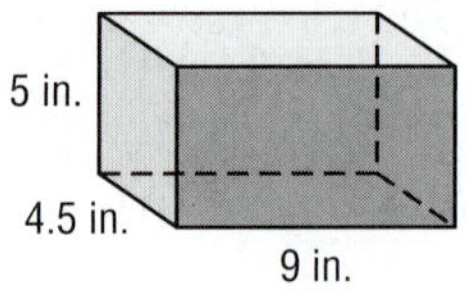

A 101.3 in^2
B 108 in^2
C 202.5 in^2
D 216 in^2

47. **SAT/ACT** If $x^2 = 25$ and $y^2 = 9$, what is the greatest possible value of $(x - y)^2$?

F 4
G 58
H 64
J 70

Spiral Review

Write an equation in slope-intercept form of the line having the given slope and y-intercept. (Lesson 3-4)

48. m: 2.5, (0, 0.5)

49. m: $\frac{4}{5}$, (0, −9)

50. m: $-\frac{7}{8}$, $\left(0, -\frac{5}{6}\right)$

51. **ROAD TRIP** Anne is driving 400 miles to visit Niagara Falls. She manages to travel the first 100 miles of her trip in two hours. If she continues at this rate, how long will it take her to drive the remaining distance? (Lesson 3-3)

Find a counterexample to show that each conjecture is false. (Lesson 2-1)

52. **Given:** $\angle 1$ and $\angle 2$ are complementary angles.
Conjecture: $\angle 1$ and $\angle 2$ form a right angle.

53. **Given:** points W, X, Y, and Z
Conjecture: W, X, Y, and Z are noncollinear.

Find the perimeter or circumference and area of each figure. Round to the nearest tenth. (Lesson 1-6)

54.

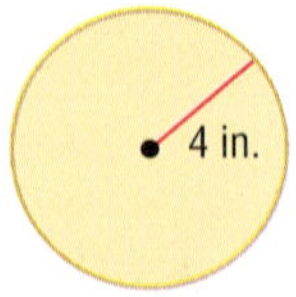

55.

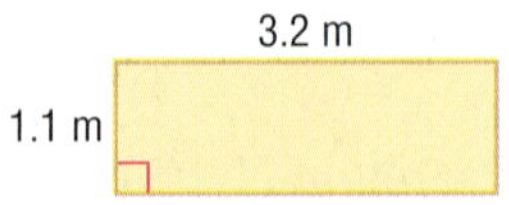

56.

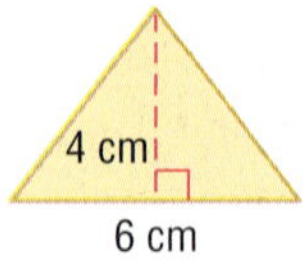

Skills Review

57. Find x and y so that $\overline{BE}$ and $\overline{AD}$ are perpendicular. (Lesson 1-5)

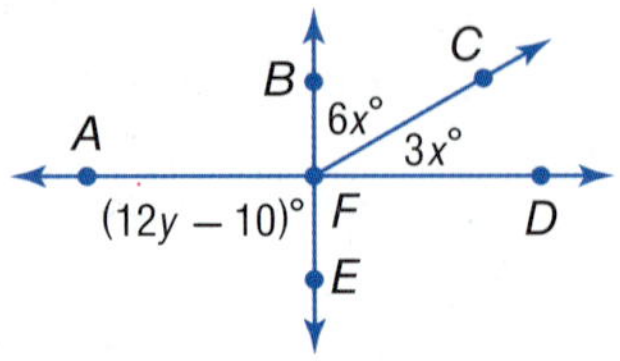

3-6 Perpendiculars and Distance

Then

You proved that two lines are parallel using angle relationships. (Lesson 3-3)

Now

- Find the distance between a point and a line.
- Find the distance between parallel lines.

New Vocabulary

equidistant

Why?

A *plumb bob* is made of string with a specially designed weight. When the weight is suspended and allowed to swing freely, the point of the bob is precisely below the point to which the string is fixed.

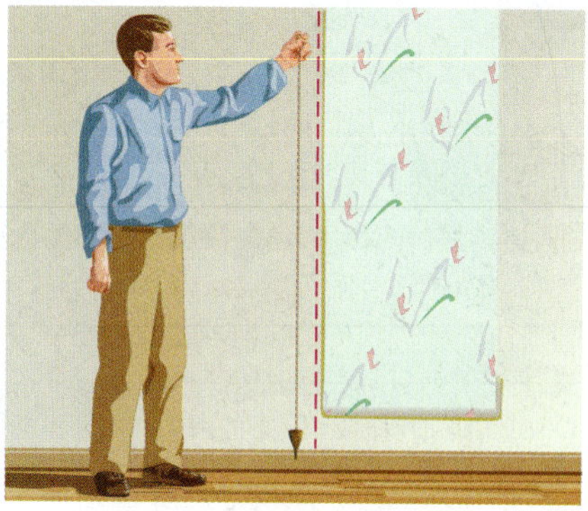

The plumb bob is useful in establishing what is the true vertical or *plumb* when constructing a wall or when hanging wallpaper.

Distance From a Point to a Line The plumb bob also indicates the shortest distance between the point at which it is attached on the ceiling and a level floor below. This perpendicular distance between a point and a line is the shortest in all cases.

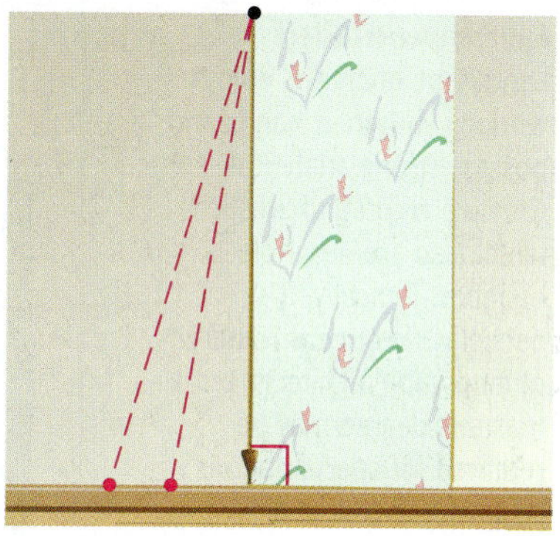

Key Concept — Distance Between a Point and a Line

Words The distance between a line and a point not on the line is the length of the segment perpendicular to the line from the point.

Model

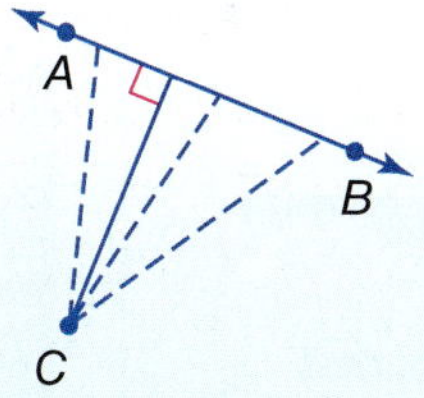

The construction of a line perpendicular to an existing line through a point not on the existing line in Extend Lesson 1-5 establishes that there is *at least one* line through a point P that is perpendicular to a line AB. The following postulate states that this line is the *only* line through P perpendicular to $\overleftrightarrow{AB}$.

Postulate 3.6 — Perpendicular Postulate

Words If given a line and a point not on the line, then there exists exactly one line through the point that is perpendicular to the given line.

Model

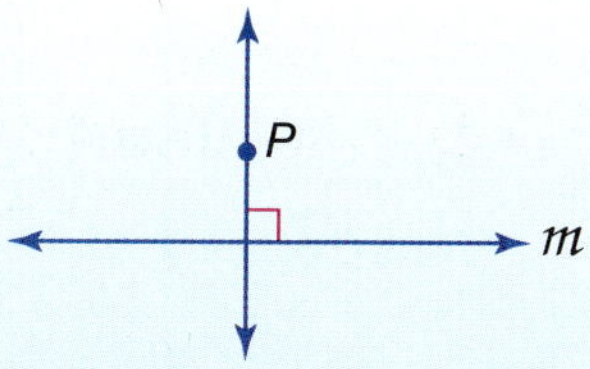

Real-World Career

Landscape Architect
Landscape architects enjoy working with their hands and possess strong analytical skills. Creative vision and artistic talent also are desirable qualities. Typically, a bachelor's degree is required of landscape architects, but a master's degree may be required for specializations such as golf course design.

Real-World EXAMPLE 1 Construct Distance From Point to a Line

LANDSCAPING A landscape architect notices that one part of a yard does not drain well. She wants to tap into an existing underground drain represented by line *m*. Construct and name the segment with the length that represents the shortest amount of pipe she will need to lay to connect this drain to point *A*.

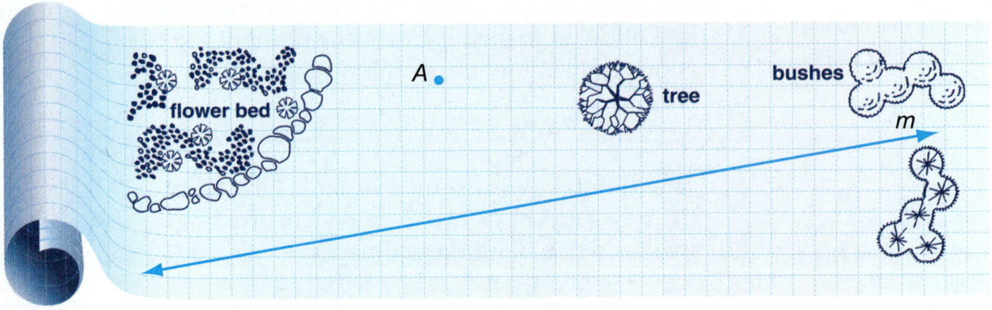

The distance from a line to a point not on the line is the length of the segment perpendicular to the line from the point. Locate points B and C on line m equidistant from point A.

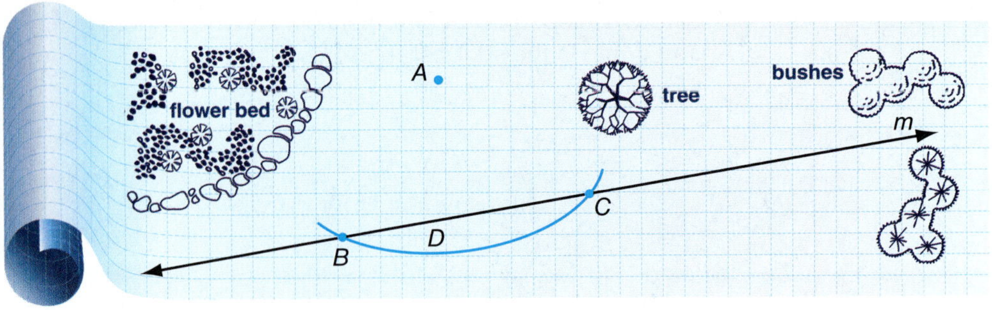

Locate a second point not on line m equidistant from A and B. Construct $\overleftrightarrow{AD}$ so that $\overleftrightarrow{AD} \perp \overleftrightarrow{BC}$.

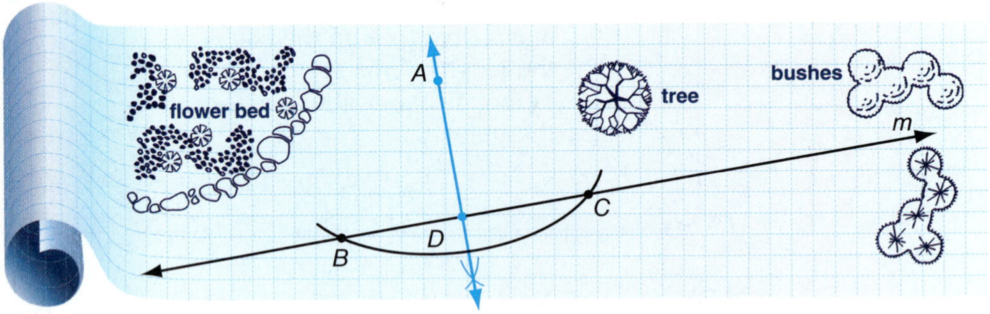

The measure of $\overline{AD}$ represents the shortest amount of pipe the architect will need to lay to connect the drain to point A.

StudyTip

Drawing the Shortest Distance You can use tools like the corner of a piece of paper to help you draw a perpendicular segment from a point to a line, but only a compass and a straightedge can be used to construct this segment.

Guided Practice

1. Copy the figure. Then construct and name the segment that represents the distance from Q to $\overleftrightarrow{PR}$.

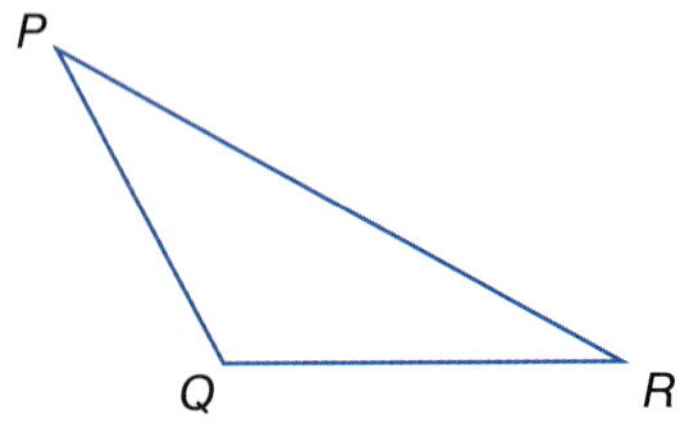

Personal Tutor glencoe.com

EXAMPLE 2 Distance from a Point to a Line on Coordinate Plane

COORDINATE GEOMETRY **Line ℓ contains points at (−5, 3) and (4, −6). Find the distance between line ℓ and point $P(2, 4)$.**

Step 1 Find the equation of the line ℓ.

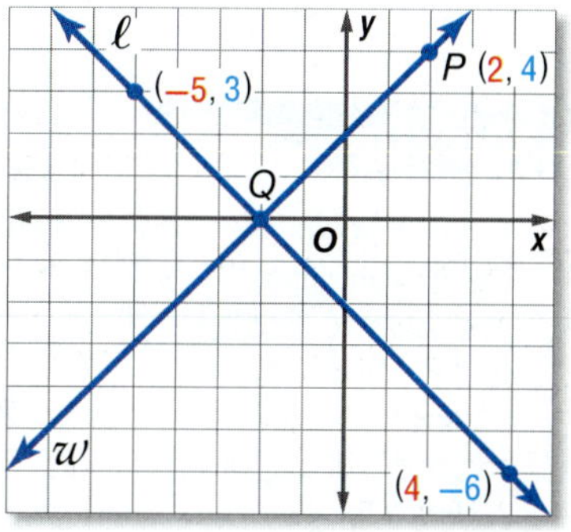

Begin by finding the slope of the line through points (−5, 3) and (4, −6).

$$m = \frac{y_2 - y_1}{x_2 - x_1} = \frac{-6 - 3}{4 - (-5)} = \frac{-9}{9} \text{ or } -1$$

Then write the equation of this line using the point (4, −6) on the line.

$y = mx + b$	**Slope-intercept form**
$-6 = -1(4) + b$	**$m = -1$, $(x, y) = (4, -6)$**
$-6 = -4 + b$	**Simplify.**
$-2 = b$	**Add 4 to each side.**

The equation of line ℓ is $y = -x + (-2)$ or $y = -x - 2$.

Step 2 Write an equation of the line w perpendicular to line ℓ through $P(2, 4)$.

Since the slope of line ℓ is −1, the slope of a line p is 1. Write the equation of line w through $P(2, 4)$ with slope 1.

$y = mx + b$	**Slope-intercept form**
$4 = 1(2) + b$	**$m = -1$, $(x, y) = (2, 4)$**
$4 = 2 + b$	**Simplify.**
$2 = b$	**Subtract 2 from each side.**

The equation of line w is $y = x + 2$.

Step 3 Solve the system of equations to determine the point of intersection.

line ℓ: $\quad y = -x - 2$

line w: $(+)\ y = x + 2$

$2y = 0$	**Add the two equations.**
$y = 0$	**Divide each side by 2.**

Solve for x.

$0 = x + 2$	**Substitute 0 for y in the second equation.**
$-2 = x$	**Subtract 2 from each side.**

The point of intersection is (−2, 0). Let this be point Q.

Step 4 Use the Distance Formula to determine the distance between $P(2, 4)$ and $Q(-2, 0)$.

$d = \sqrt{(x_2 - x_1)^2 + (y_2 - y_1)^2}$	**Distance formula**
$= \sqrt{(-2 - 2)^2 + (0 - 4)^2}$	**$x_2 = -2$, $x_1 = 2$, $y_2 = 0$, $y_1 = 4$**
$= \sqrt{32}$	**Simplify.**

The distance between the point and the line is $\sqrt{32}$ or about 5.66 units.

StudyTip

Distance to Axes Note that the distance from a point to the x-axis can be determined by looking at the y-coordinate, and the distance from a point to the y-axis can be determined by looking at the x-coordinate.

StudyTip

Elimination Method To review **solving systems of equations** using the elimination method, see p. P18.

Guided Practice

2. Line ℓ contains points at (1, 2) and (5, 4). Construct a line perpendicular to ℓ through $P(1, 7)$. Then find the distance from P to ℓ.

Personal Tutor glencoe.com

StudyTip

Equidistant You will use this concept of *equidistant* to describe special points and lines relating to the sides and angles of triangles in Lesson 5-1.

Distance Between Parallel Lines By definition, parallel lines do not intersect. An alternate definition states that two lines in a plane are parallel if they are everywhere **equidistant**. Equidistant means that the distance between two lines measured along a perpendicular line to the lines is always the same.

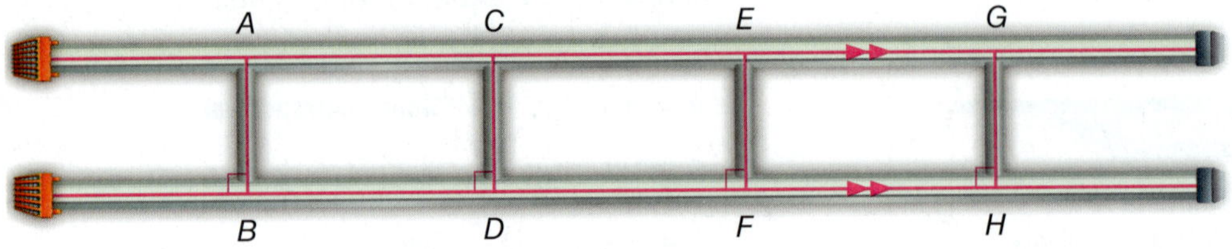

$AB = CD = EF = GH$

This leads us to the definition of the distance between two parallel lines.

Key Concept — Distance Between Parallel Lines

The distance between two parallel lines is the perpendicular distance between one of the lines and any point on the other line.

Recall from Lesson 1-1 that a *locus* is the set of all points that satisfy a given condition. Parallel lines can be described as the locus of points in a plane equidistant from a given line.

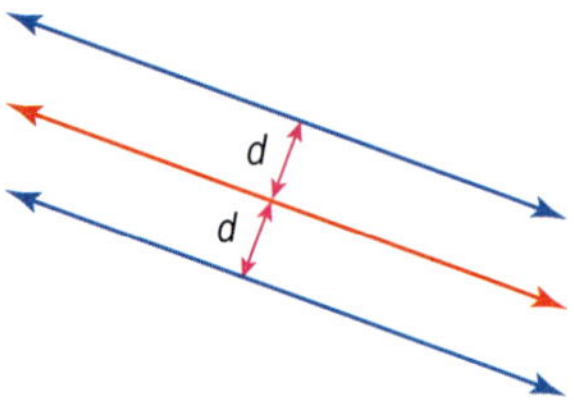

Theorem 3.9 — Two Lines Equidistant from a Third

In a plane, if two lines are each equidistant from a third line, then the two lines are parallel to each other.

You will prove Theorem 3.9 in Exercise 30.

EXAMPLE 3 Distance Between Parallel Lines

Find the distance between the parallel lines ℓ and m with equations $y = 2x + 1$ and $y = 2x - 3$, respectively.

You will need to solve a system of equations to find the endpoints of a segment that is perpendicular to both ℓ and m. From their equations, we know that the slope of line ℓ and line m is 2.

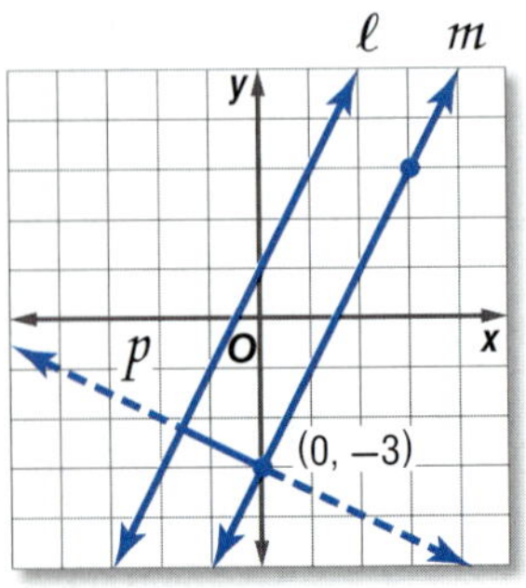

Sketch line p through the y-intercept of line m, (0, −3), perpendicular to lines m and ℓ.

Step 1 Write an equation of line p. The slope of p is the opposite reciprocal of 2, or $-\frac{1}{2}$. Use the y-intercept of line m, (0, −3), as one of the endpoints of the perpendicular segment.

$(y - y_1) = m(x - x_1)$ — **Point-slope form**

$[y - (-3)] = -\frac{1}{2}(x - 0)$ — **$x_1 = 0$, $y_1 = 3$, and $m = -\frac{1}{2}$**

$y + 3 = -\frac{1}{2}x$ — **Simplify.**

$y = -\frac{1}{2}x - 3$ — **Subtract 3 from each side.**

Step 2 Use a system of equations to determine the point of intersection of lines ℓ and p.

ℓ: $y = 2x + 1$

p: $y = -\frac{1}{2}x - 3$

$2x + 1 = -\frac{1}{2}x - 3$ — **Substitute $2x + 1$ for y in the second equation.**

$2x + \frac{1}{2}x = -3 - 1$ — **Group like terms on each side.**

$\frac{5}{2}x = -4$ — **Simplify on each side.**

$x = -\frac{8}{5}$ — **Multiply each side by $\frac{2}{5}$.**

$y = -\frac{1}{2}\left(-\frac{8}{5}\right) - 3$ — **Substitute $-\frac{8}{5}$ for x in the equation for p.**

$= -\frac{11}{5}$ — **Simplify.**

The point of intersection is $\left(-\frac{8}{5}, -\frac{11}{5}\right)$ or (−1.6, −2.2).

Step 3 Use the Distance Formula to determine the distance between (0, −3) and (−1.6, −2.2).

$d = \sqrt{(x_2 - x_1)^2 + (y_2 - y_1)^2}$ — **Distance Formula**

$= \sqrt{(-1.6 - 0)^2 + [-2.2 - (-3)]^2}$ — **$x_2 = -1.6$, $x_1 = 0$, $y_2 = -2.2$, and $y_1 = -3$**

≈ 1.8 — **Simplify using a calculator.**

The distance between the lines is about 1.8 units.

StudyTip

Substitution Method To review **solving systems of equations** using the substitution method, see p. P17.

Guided Practice

3A. Find the distance between the parallel lines r and s whose equations are $y = -3x - 5$ and $y = -3x + 6$, respectively.

3B. Find the distance between parallel lines a and b with equations $x + 3y = 6$ and $x + 3y = -14$, respectively.

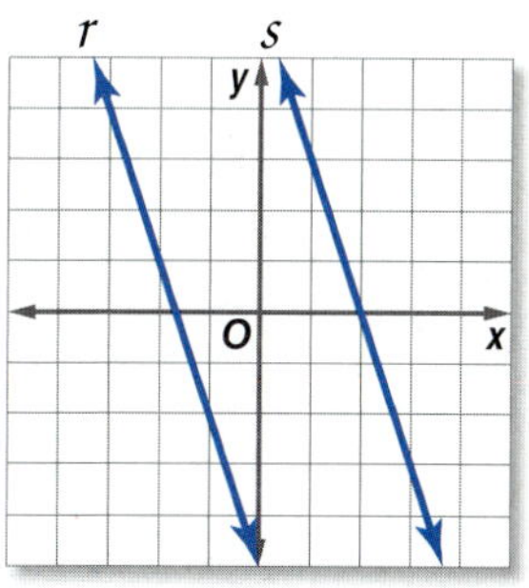

Personal Tutor glencoe.com

Check Your Understanding

Example 1 p. 214

Copy each figure. Construct the segment that represents the distance indicated.

1. Y to $\overleftrightarrow{TS}$

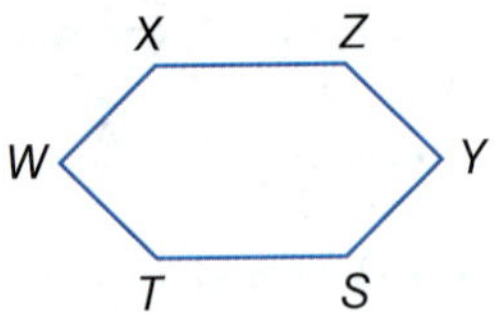

2. C to $\overleftrightarrow{AB}$

3. MARCHING BAND After forming a line, every even member of a marching band turns to face the home team's end zone and marches 5 paces straight forward. At the same time, every odd member turns in the opposite direction and marches 5 paces straight forward. Assuming that each band member covers the same distance, what formation should result? Justify your answer.

Example 2 p. 215

COORDINATE GEOMETRY Find the distance from P to ℓ.

4. Line ℓ contains points (4, 3) and (−2, 0). Point P has coordinates (3, 10).

5. Line ℓ contains points (−6, 1) and (9, −4). Point P has coordinates (4, 1).

6. Line ℓ contains points (4, 18) and (−2, 9). Point P has coordinates (−9, 5).

Example 3 pp. 216–217

Find the distance between each pair of parallel lines with the given equations.

7 $y = -2x + 4$
$y = -2x + 14$

8. $y = 7$
$y = -3$

Practice and Problem Solving

= **Step-by-Step Solutions** begin on page R20.
Extra Practice begins on page 969.

Example 1 p. 214

Copy each figure. Construct the segment that represents the distance indicated.

9. Q to $\overline{RS}$

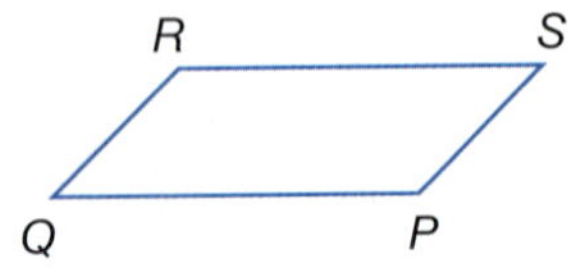

10. A to $\overline{BC}$

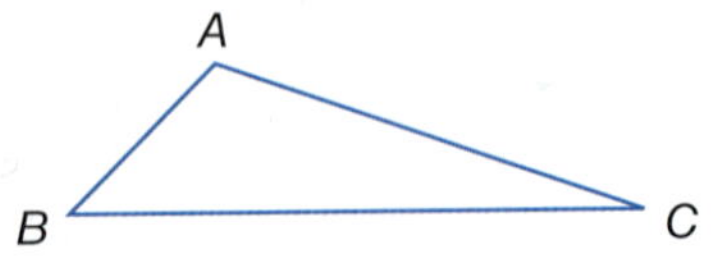

11. H to $\overline{FG}$

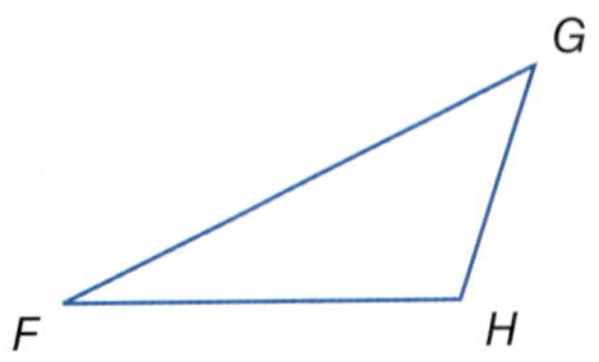

12. K to $\overline{LM}$

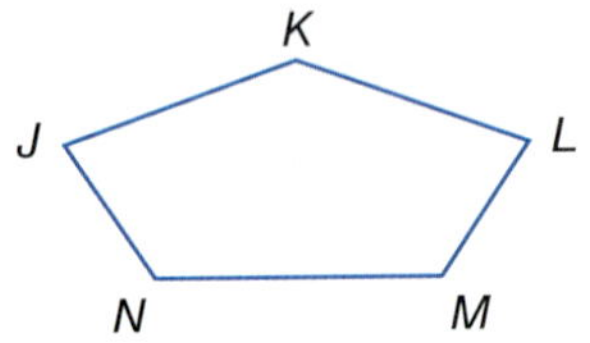

13. **DRIVEWAYS** In the diagram at the right, is the driveway shown the shortest possible one from the house to the road? Explain why or why not.

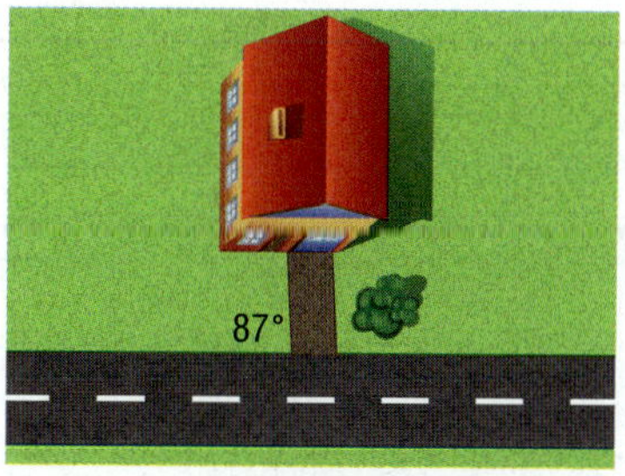

14. **SCHOOL** Rondell is crossing the courtyard in front of his school. Three possible paths are shown in the diagram at the right. Which of the three paths shown is the shortest? Explain your reasoning.

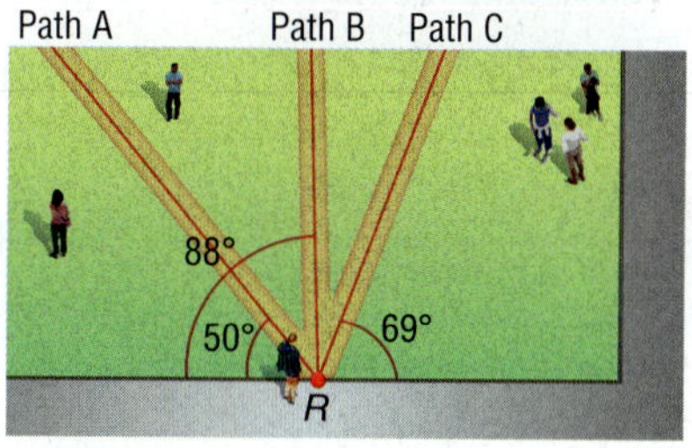

Example 2
p. 215

COORDINATE GEOMETRY **Find the distance from P to ℓ.**

15. Line ℓ contains points (0, −3) and (7, 4). Point P has coordinates (4, 3).
16. Line ℓ contains points (11, −1) and (−3, −11). Point P has coordinates (−1, 1).
17. Line ℓ contains points (−2, 1) and (4, 1). Point P has coordinates (5, 7).
18. Line ℓ contains points (4, −1) and (4, 9). Point P has coordinates (1, 6).
19. Line ℓ contains points (1, 5) and (4, −4). Point P has coordinates (−1, 1).
20. Line ℓ contains points (−8, 1) and (3, 1). Point P has coordinates (−2, 4).

Example 3
pp. 216–217

Find the distance between each pair of parallel lines with the given equations.

21. $y = -2$
    $y = 4$
22. $x = 3$
    $x = 7$
23. $y = 5x - 22$
    $y = 5x + 4$
24. $y = \frac{1}{3}x - 3$
    $y = \frac{1}{3}x + 2$
25. $x = 8.5$
    $x = -12.5$
26. $y = 15$
    $y = -4$
27. $y = \frac{1}{4}x + 2$
    $4y - x = -60$
28. $3x + y = 3$
    $y + 17 = -3x$
29. $y = -\frac{5}{4}x + 3.5$
    $4y + 10.6 = -5x$

30. **PROOF** Write a two-column proof of Theorem 3.9.

Find the distance from the line to the given point.

31. $y = -3$, (5, 2)
32. $y = \frac{1}{6}x + 6$, (−6, 5)
33. $x = 4$, (−2, 5)

34. **POSTERS** Alma is hanging two posters on the wall in her room as shown. How can Alma use perpendicular distances to confirm that the posters are parallel?

Real-World Link

To promote school spirit, schools can designate a bulletin board on campus on which to display pictures of students participating in school activities.

Source: ASB Director

35. **SCHOOL SPIRIT** Brock is decorating a hallway bulletin board to display pictures of students demonstrating school spirit. He cuts off one length of border to match the board's width, and then uses that strip as a template to cut a second strip that is exactly the same length for the bottom.

When stapling the bottom border in place, he notices that the strip he cut is about a quarter of an inch too short. Describe what he can conclude about the bulletin board. Explain your reasoning.

CONSTRUCTION **Line ℓ contains points at $(-4, 3)$ and $(2, -3)$. Point P at $(-2, 1)$ is on line ℓ. Complete the following construction.**

Step 1
Graph line ℓ and point P, and put the compass at point P. Using the same compass setting, draw arcs to the left and right of P. Label these points A and B.

Step 2
Open the compass to a setting greater than AP. Put the compass at point A and draw an arc above line ℓ.

Step 3
Using the same compass setting, put the compass at point B and draw an arc above line ℓ. Label the point of intersection Q. Then draw $\overleftrightarrow{PQ}$.

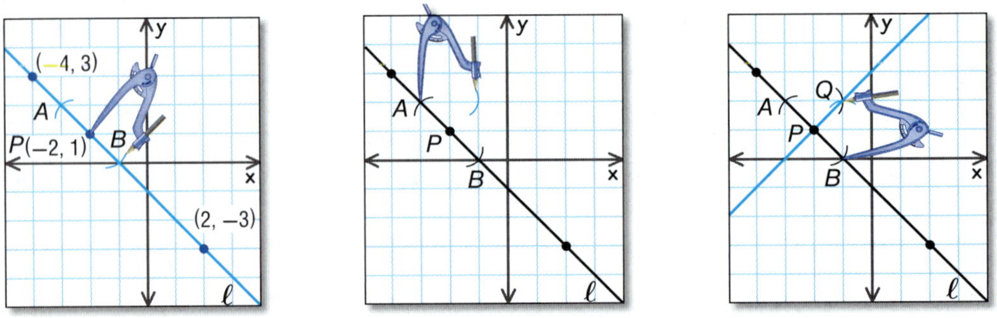

36. What is the relationship between line ℓ and $\overleftrightarrow{PQ}$? Verify your conjecture using the slopes of the two lines.

37. Repeat the construction above using a different line and point on that line.

38. **COORDINATE GEOMETRY** $\overline{AB}$ has a slope of 2 and midpoint $M(3, 2)$. A segment perpendicular to $\overline{AB}$ has midpoint $P(4, -1)$ and shares endpoint B with $\overline{AB}$.

a. Graph the segments.

b. Find the coordinates of A and B.

39. **MULTIPLE REPRESENTATIONS** In this problem, you will explore the areas of triangles formed by points on parallel lines.

a. **GEOMETRIC** Draw two parallel lines and label them as shown.

b. **VERBAL** Where would you place point C on line m to ensure that triangle ABC would have the largest area? Explain your reasoning.

c. **ANALYTICAL** If $AB = 11$ inches, what is the maximum area of $\triangle ABC$?

40. **PERPENDICULARITY AND PLANES** Make a copy of the diagram below to answer each question, marking the diagram with the given information.

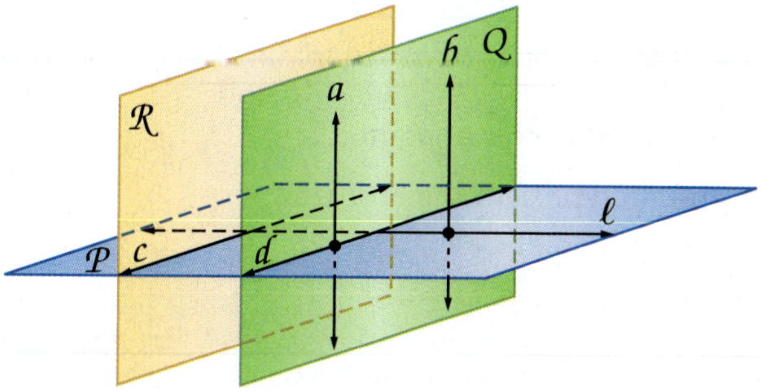

a. If two lines are perpendicular to the same plane, then they are coplanar. If both line a and line b are perpendicular to plane $\mathcal{P}$, what must also be true?

b. If a plane intersects two parallel planes, then the intersections form two parallel lines. If planes $\mathcal{R}$ and $\mathcal{Q}$ are parallel and they interesct plane $\mathcal{P}$, what must also be true?

c. If two planes are perpendicular to the same line, then they are parallel. If both plane $\mathcal{Q}$ and plane $\mathcal{R}$ are perpendicular to line ℓ, what must also be true?

H.O.T. Problems Use Higher-Order Thinking Skills

41. **ERROR ANALYSIS** Han draws the segments $\overline{AB}$ and $\overline{CD}$ shown below using a straightedge. He claims that these two lines, if extended, will never intersect. Shenequa claims that they will. Is either of them correct? Justify your answer.

42. **WRITING IN MATH** Describe a method that could be used to find the distance between a point and a plane.

43. **CHALLENGE** Suppose a line perpendicular to a pair of parallel lines intersects the lines at the points $(a, 4)$ and $(0, 6)$. If the distance between the parallel lines is $\sqrt{5}$, find the value of a and the equations of the parallel lines.

44. **REASONING** Determine whether the following statement is *sometimes*, *always*, or *never* true. Explain.

The distance between a line and a plane can be found.

45. **OPEN ENDED** Draw an irregular convex polygon using a straightedge.

a. Construct the distance between one vertex and a nonadjacent side.

b. Use measurement to justify that the angle you constructed is perpendicular to the side chosen. That is, justify that the line you constructed through the vertex is perpendicular to the side chosen.

c. Use mathematics to justify this construction.

Review Vocabulary

convex polygon a polygon for which no line contains both a side of the polygon and a point in the interior of the polygon (Lesson 1-6)

46. **CHALLENGE** Rewrite Theorem 3.9 in terms of two planes that are equidistant from a third plane. Sketch an example.

47. **WRITING IN MATH** Summarize the steps necessary to find the distance between a pair of parallel lines given the equations of the two lines.

Standardized Test Practice

48. EXTENDED RESPONSE Segment AB is perpendicular to segment CD. Segment AB and segment CD bisect each other at point X.

a. Draw a figure to represent the problem.

b. Find $\overline{BD}$ if $AB = 12$ and $CD = 16$.

c. Find $\overline{BD}$ if $AB = 24$ and $CD = 18$.

49. A city park is square and has an area of 81,000 square feet. Which of the following is the closest to the length of one side of the park?

F 100 ft　　**H** 300 ft
G 200 ft　　**J** 400 ft

50. ALGEBRA Pablo bought a sweater on sale for 25% off the original price and another 40% off the discounted price. If the sweater originally cost \$48, what was the final price of the sweater?

A \$14.40　　**C** \$31.20
B \$21.60　　**D** \$36.00

51. SAT/ACT After N cookies are divided equally among 8 children, 3 remain. How many would remain if $(N + 6)$ cookies were divided equally among the 8 children?

F 0　　**H** 2
G 1　　**J** 4

Spiral Review

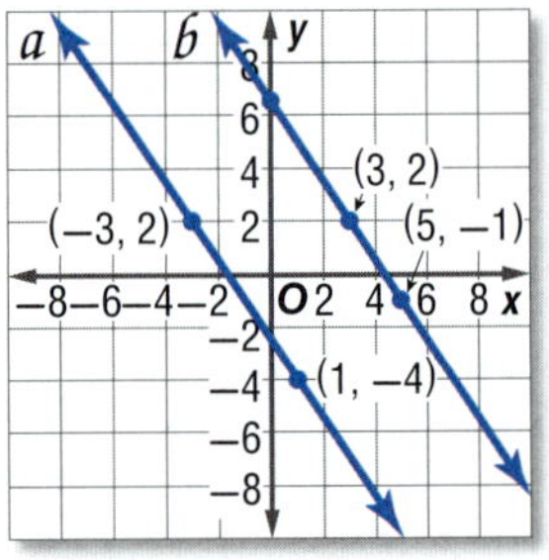

52. Refer to the figure at the right. Determine whether $a \parallel b$. Justify your answer. (Lesson 3-5)

Write an equation in point-slope form of the line having the given slope that contains the given point. (Lesson 3-4)

53. $m: \frac{1}{4}, (3, -1)$

54. $m: 0, (-2, 6)$

55. $m: -1, (-2, 3)$

56. $m: -2, (-6, -7)$

Prove the following. (Lesson 2-7)

57. If $AB = BC$, then $AC = 2BC$.

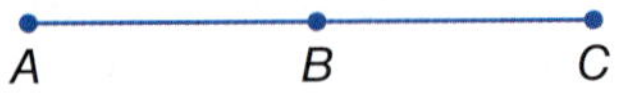

58. Given: $\overline{JK} \cong \overline{KL}$, $\overline{HJ} \cong \overline{GH}$, $\overline{KL} \cong \overline{HJ}$

Prove: $\overline{GH} \cong \overline{JK}$

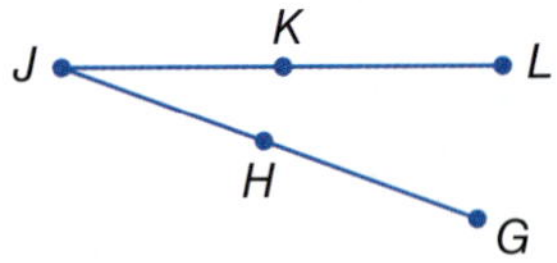

59. MAPS Darnell sketched a map for his friend of the cross streets nearest to his home. Describe two different angle relationships between the streets. (Lesson 1-5)

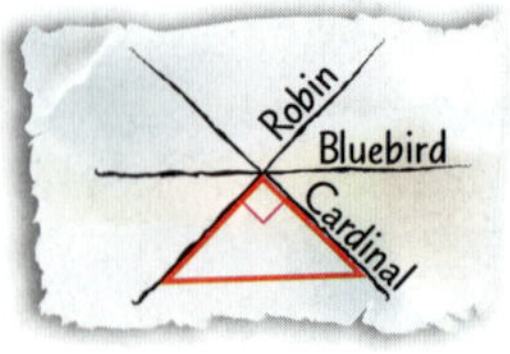

Skills Review

Use the Distance Formula to find the distance between each pair of points. (Lesson 1-3)

60. $A(0, 0), B(15, 20)$

61. $O(-12, 0), P(-8, 3)$

62. $C(11, -12), D(6, 2)$

63. $R(-2, 3), S(3, 15)$

64. $M(1, -2), N(9, 13)$

65. $Q(-12, 2), T(-9, 6)$

Math Online glencoe.com
- STUDY TO GO
- Vocabulary Review

Chapter Summary

Key Concepts

Transversals (Lessons 3-1 and 3-2)

- When a transversal intersects two lines, the following types of angles are formed: exterior, interior, consecutive interior, alternate interior, alternate exterior, and corresponding.
- If two parallel lines are cut by a transversal, then:
  - each pair of corresponding angles is congruent,
  - each pair of alternate interior angles is congruent,
  - each pair of consecutive interior angles is supplementary, and
  - each pair of alternate exterior angles is congruent.

Slope (Lessons 3-3 and 3-4)

- The slope m of a line containing two points with coordinates (x_1, y_1) and (x_2, y_2) is $m = \frac{y_2 - y_1}{x_2 - x_1}$, where $x_1 \neq x_2$.

Proving Lines Parallel (Lesson 3-5)

- If two lines in a plane are cut by a transversal so that any one of the following is true, then the two lines are parallel: a pair of corresponding angles is congruent, a pair of alternate exterior angles is congruent, a pair of alternate interior angles is congruent, or a pair of consecutive interior angles is supplementary.
- In a plane, if two lines are perpendicular to the same line, then they are parallel.

Distance (Lesson 3-6)

- The distance from a line to a point not on the line is the length of the segment perpendicular to the line from the point.
- The distance between two parallel lines is the distance between one of the lines and any point on the other line.

FOLDABLES® Study Organizer

Be sure the Key Concepts are noted in your Foldable.

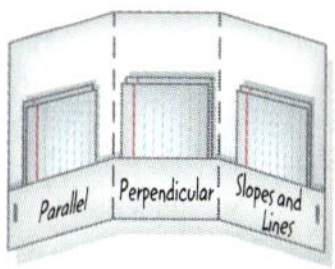

Key Vocabulary

alternate exterior angles (p. 172)
alternate interior angles (p. 172)
consecutive interior angles (p. 172)
corresponding angles (p. 172)
equidistant (p. 216)
parallel lines (p. 171)
parallel planes (p. 171)
point-slope form (p. 196)
rate of change (p. 187)
skew lines (p. 171)
slope (p. 186)
slope-intercept form (p. 196)
transversal (p. 172)

Vocabulary Check

State whether each sentence is *true* or *false*. If *false*, replace the underlined word or number to make a true sentence.

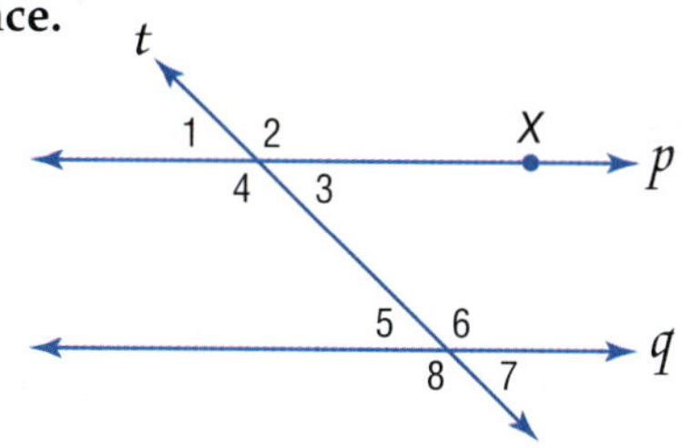

1. If $\angle 1 \cong \angle 5$, then lines p and q are <u>skew</u> lines.

2. Angles 4 and 6 are <u>alternate</u> interior angles.

3. Angles 1 and 7 are alternate <u>exterior</u> angles.

4. If lines p and q are parallel, then angles 3 and 6 are <u>congruent</u>.

5. The distance from point X to line q is the length of the segment <u>perpendicular</u> to line q from X.

6. Line t is called the <u>transversal</u> for lines p and q.

7. If $p \parallel q$, then $\angle 2$ and $\angle 8$ are <u>supplementary</u>.

8. Angles 4 and 8 are <u>corresponding</u> angles.

CHAPTER 3
Study Guide and Review

Lesson-by-Lesson Review

3-1 Parallel Lines and Transversals (pp. 171–176)

Classify the relationship between each pair of angles as *alternate interior, alternate exterior, corresponding,* or *consecutive interior* angles.

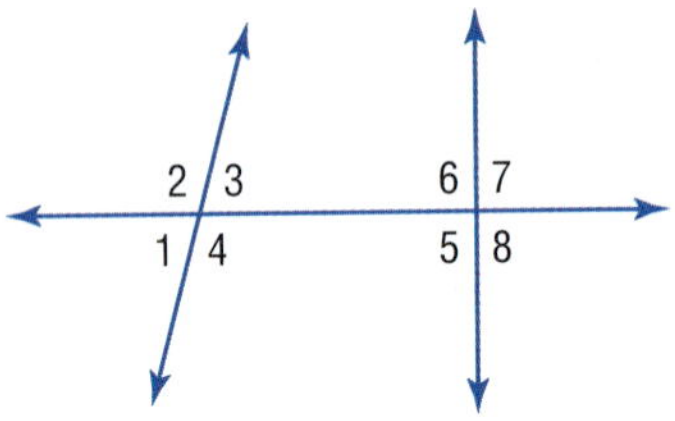

9. $\angle 1$ and $\angle 5$

10. $\angle 4$ and $\angle 6$

11. $\angle 2$ and $\angle 8$

12. $\angle 4$ and $\angle 5$

13. **BRIDGES** The Roebling Suspension Bridge extends over the Ohio River connecting Cincinnati, Ohio, to Covington, Kentucky. Describe the type of lines formed by the bridge and the river.

EXAMPLE 1

Refer to the figure below. Classify the relationship between each pair of angles as *alternate interior, alternate exterior, corresponding,* or *consecutive interior* angles.

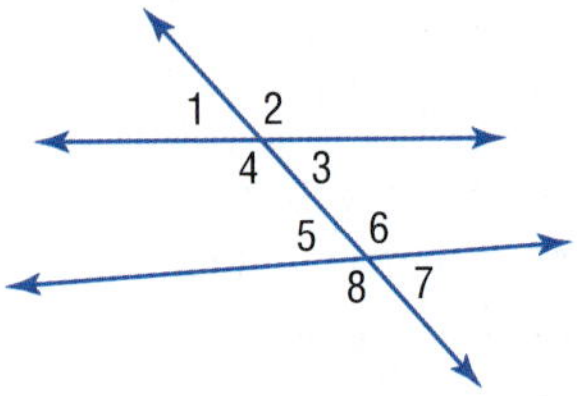

a. **$\angle 3$ and $\angle 6$**
consecutive interior

b. **$\angle 2$ and $\angle 6$**
corresponding

c. **$\angle 1$ and $\angle 7$**
alternate exterior

d. **$\angle 3$ and $\angle 5$**
alternate interior

3-2 Angles and Parallel Lines (pp. 178–184)

In the figure, $m\angle 1 = 123$. Find the measure of each angle. Tell which postulate(s) or theorem(s) you used.

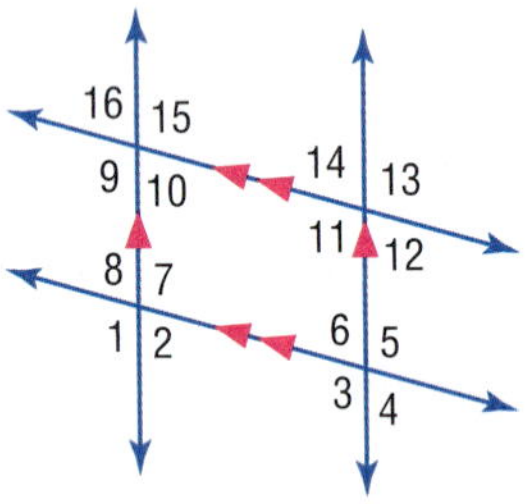

14. $\angle 5$

15. $\angle 14$

16. $\angle 16$

17. $\angle 11$

18. $\angle 4$

19. $\angle 6$

20. **MAPS** The diagram shows the layout of Elm, Plum, and Oak streets. Find the value of x.

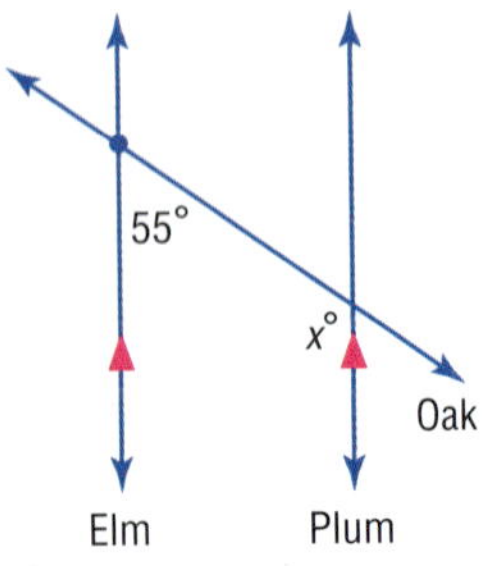

EXAMPLE 2

ALGEBRA **Use the figure below to find the indicated variable. Explain your reasoning.**

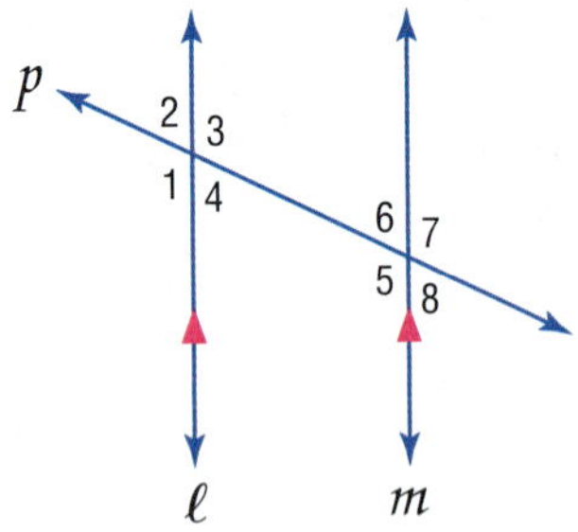

If $m\angle 5 = 7x - 5$ and $m\angle 4 = 2x + 23$, find x.

$m\angle 4 + m\angle 5 = 180$	**Def. of Supp. ∠s**
$(7x - 5) + (2x + 23) = 180$	**Substitution**
$9x + 18 = 180$	**Simplify.**
$9x = 162$	**Subtract.**
$x = 18$	**Divide.**

Since lines ℓ and m are parallel, $\angle 4$ and $\angle 5$ are supplementary by the Consecutive Interior Angles Theorem.

MIXED PROBLEM SOLVING
For mixed problem-solving practice, see page 1001.

3-3 Slopes of Lines (pp. 186–194)

Determine whether $\overleftrightarrow{AB}$ and $\overleftrightarrow{XY}$ are *parallel*, *perpendicular*, or *neither*. Graph each line to verify your answer.

21. $A(5, 3)$, $B(8, 0)$, $X(-7, 2)$, $Y(1, 10)$
22. $A(-3, 9)$, $B(0, 7)$, $X(4, 13)$, $Y(-5, 7)$
23. $A(8, 1)$, $B(-2, 7)$, $X(-6, 2)$, $Y(-1, -1)$

Graph the line that satisfies each condition.

24. contains $(-3, 4)$ and is parallel to $\overleftrightarrow{AB}$ with $A(2, 5)$ and $B(9, 2)$
25. contains $(1, 3)$ and is perpendicular to $\overleftrightarrow{PQ}$ with $P(4, -6)$ and $Q(6, -1)$
26. **AIRPLANES** Two Oceanic Airlines planes are flying at the same altitude. Using satellite imagery, each plane's position can be mapped onto a coordinate plane. Flight 815 was mapped at (23, 17) and (5, 11) while Flight 44 was mapped at (3, 15) and (9, 17). Determine whether their paths are *parallel*, *perpendicular*, or *neither*.

EXAMPLE 3

Graph the line that contains $C(0, -4)$ and is perpendicular to $\overleftrightarrow{AB}$ with $A(5, -4)$ and $B(0, -2)$.

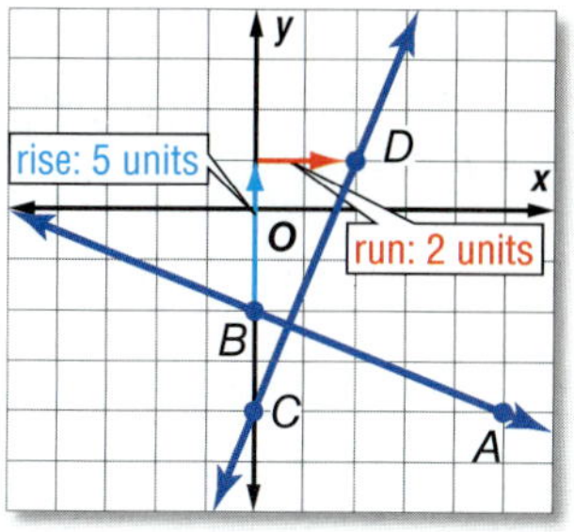

The slope of $\overleftrightarrow{AB}$ is $\frac{-2-(-4)}{0-5}$ or $-\frac{2}{5}$.

Since $-\frac{2}{5}\left(\frac{5}{2}\right) = -1$, the slope of the line perpendicular to $\overleftrightarrow{AB}$ through C is $\frac{5}{2}$.

To graph the line, start at C. Move up 5 units and then right 2 units. Label the point D and draw $\overleftrightarrow{CD}$.

3-4 Equations of Lines (pp. 196–203)

Write an equation in point-slope form of the line having the given slope that contains the given point.

27. $m = 2$, $(4, -9)$
28. $m = -\frac{3}{4}$, $(8, -1)$

Write an equation in slope-intercept form of the line having the given slope and y-intercept.

29. m: 5, y-intercept: -3
30. m: $\frac{1}{2}$, y-intercept: 4

Write an equation in slope-intercept form for each line.

31. $(-3, 12)$ and $(15, 0)$
32. $(-7, 2)$ and $(5, 8)$
33. **WINDOW CLEANING** Ace Window Cleaning Service charges \$50 for the service call and \$20 for each hour spent on the job. Write an equation in slope-intercept form that represents the total cost C in terms of the number of hours h.

EXAMPLE 4

Write an equation of the line through (2, 5) and (6, 3) in slope-intercept form.

Step 1 Find the slope of the line through the points.

$$m = \frac{y_2 - y_1}{x_2 - x_1} = \frac{3-5}{6-2} \quad \textbf{Slope Formula}$$

$$= -\frac{2}{4} \text{ or } -\frac{1}{2}$$

Step 2 Write an equation of the line.

$$y - y_1 = m(x - x_1) \quad \textbf{Point-slope form}$$

$$y - 5 = -\frac{1}{2}[x - (2)] \quad m = -\frac{1}{2}, (x_1, y_1) = (2, 5)$$

$$y - 5 = -\frac{1}{2}x + 1 \quad \textbf{Simplify.}$$

$$y = -\frac{1}{2}x + 6 \quad \textbf{Add 5 to each side.}$$

Study Guide and Review

3-5 Proving Lines Parallel (pp. 205–212)

Given the following information, determine which lines, if any, are parallel. State the postulate or theorem that justifies your answer.

34. $\angle 7 \cong \angle 10$

35. $\angle 2 \cong \angle 10$

36. $\angle 1 \cong \angle 3$

37. $\angle 3 \cong \angle 11$

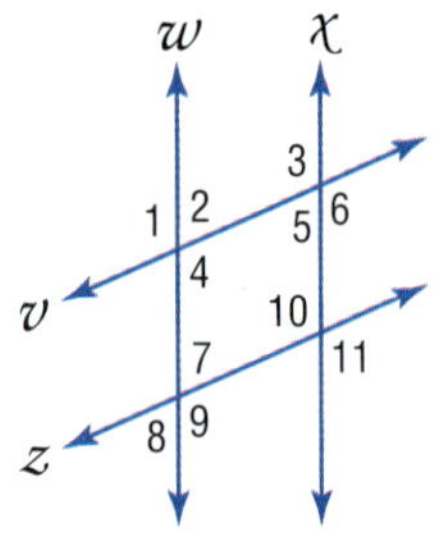

38. Find x so that $p \parallel q$. Identify the postulate or theorem you used.

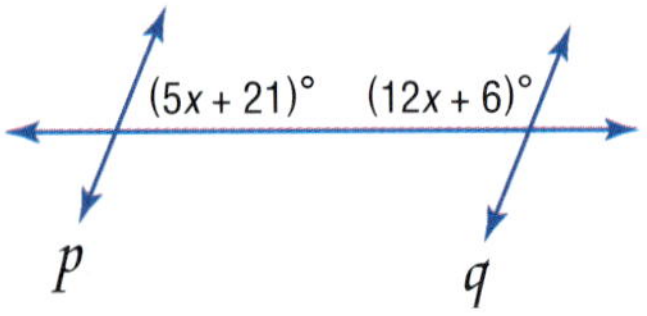

39. **LANDSCAPING** Find the measure needed for $m\angle ADC$ that will make $\overline{AB} \parallel \overline{CD}$ if $m\angle BAD = 45$.

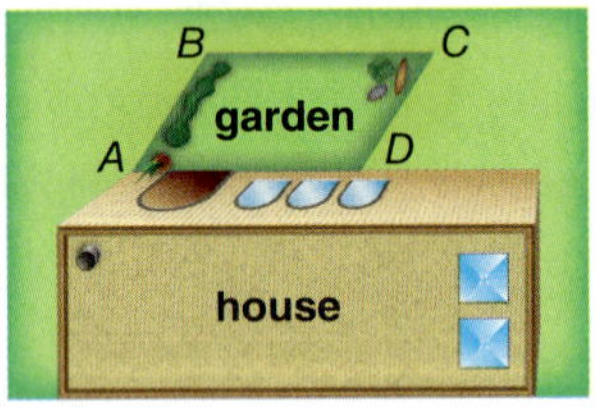

EXAMPLE 5

Given the following information, determine which lines, if any, are parallel. State the postulate or theorem that justifies your answer.

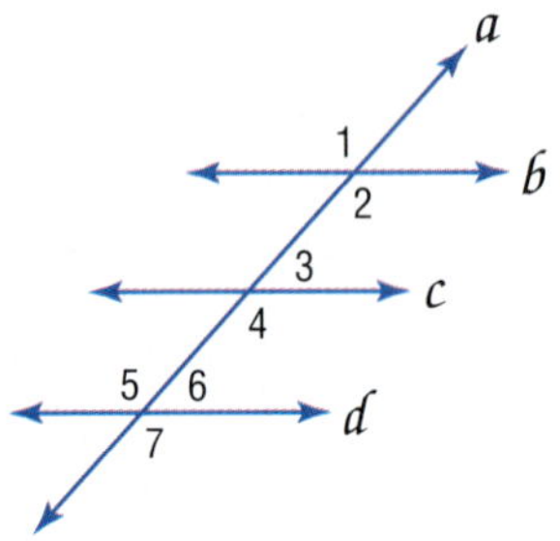

a. $\angle 1 \cong \angle 7$

$\angle 1$ and $\angle 7$ are alternate exterior angles of lines b and d.

Since $\angle 1 \cong \angle 7$, $b \parallel d$ by the Converse of the Alternate Exterior Angles Theorem.

b. $\angle 4 \cong \angle 5$

$\angle 4$ and $\angle 5$ are alternate interior angles of lines c and d.

Since $\angle 4 \cong \angle 5$, $c \parallel d$ by the Converse of the Alternate Interior Angles Theorem.

3-6 Perpendiculars and Distance (pp. 213–222)

Copy each figure. Draw the segment that represents the distance indicated.

40. X to $\overline{VW}$

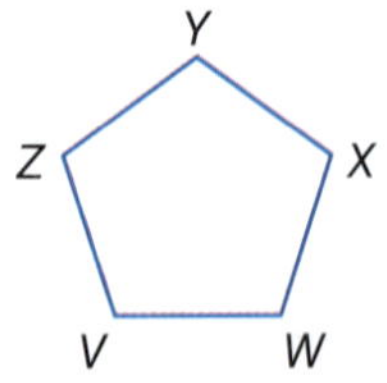

41. L to $\overline{JK}$

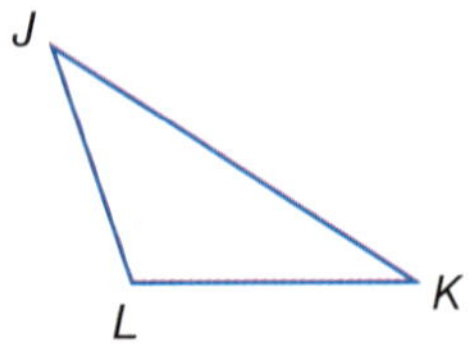

42. **HOME DÉCOR** Scott wants to hang two rows of framed pictures in parallel lines on his living room wall. He first spaces the nails on the wall in a line for the top row. Next, he hangs a weighted plumb line from each nail and measures an equal distance below each nail for the second row. Why does this ensure that the two rows of pictures will be parallel?

EXAMPLE 6

Copy the figure. Draw the segment that represents the distance from point A to $\overline{CD}$.

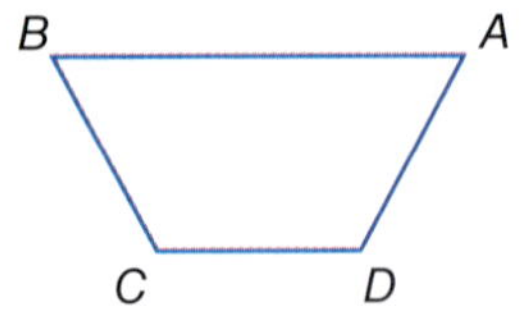

The distance from a line to a point not on the line is the length of the segment perpendicular to the line that passes through the point.

Extend $\overline{CD}$ and draw the segment perpendicular to $\overline{CD}$ from A.

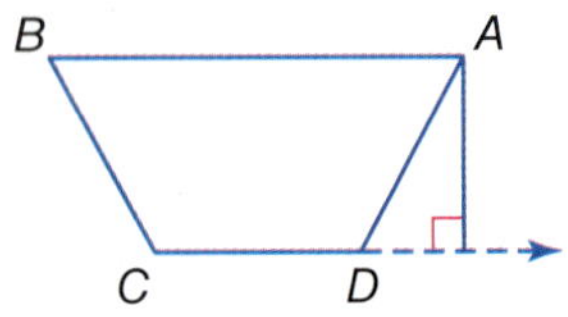

CHAPTER 3

Practice Test

Math Online glencoe.com
Chapter Test

Classify the relationship between each pair of angles as *alternate interior, alternate exterior, corresponding,* or *consecutive interior* angles.

1. $\angle 6$ and $\angle 3$
2. $\angle 4$ and $\angle 7$
3. $\angle 5$ and $\angle 4$

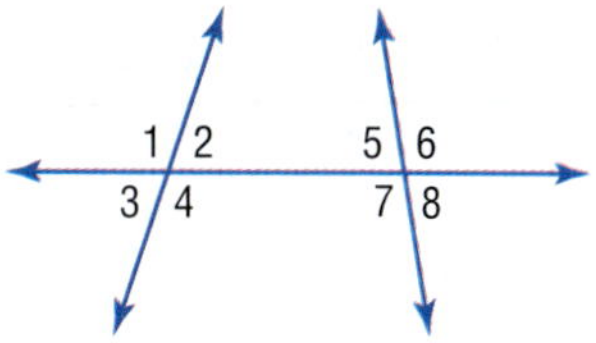

Determine the slope of the line that contains the given points.

4. $G(8, 1)$, $H(8, -6)$
5. $A(0, 6)$, $B(4, 0)$
6. $E(6, 3)$, $F(-6, 3)$
7. $E(5, 4)$, $F(8, 1)$

In the figure, $m\angle 8 = 96$ and $m\angle 12 = 42$. Find the measure of each angle. Tell which postulate(s) or theorem(s) you used.

8. $\angle 9$
9. $\angle 11$
10. $\angle 6$

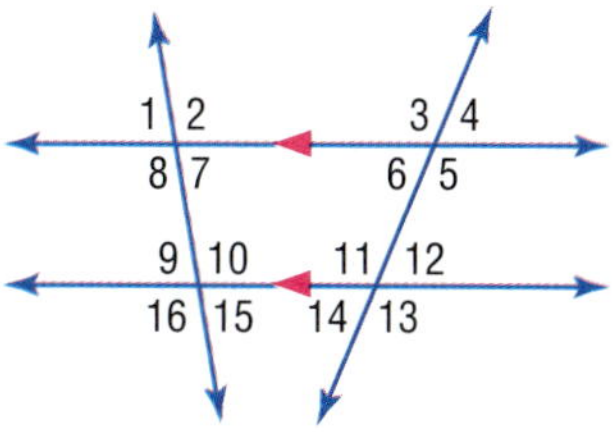

11. Find the value of the variable in the figure below.

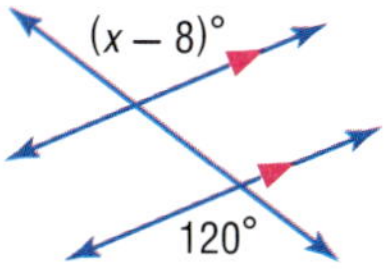

12. **FITNESS** You would like to join a fitness center. Fit-N-Trim charges \$80 per month. Fit-For-Life charges a one-time membership fee of \$75 and \$55 per month.
   a. Write and graph two equations in slope-intercept form to represent the cost y to attend each fitness center for x months.
   b. Are the lines you graphed in part **a** parallel? Explain why or why not.
   c. Which fitness center offers the better rate? Explain.

Write an equation in slope-intercept form for each line described.

13. passes through $(-8, 1)$, perpendicular to $y = 2x - 17$
14. passes through $(0, 7)$, parallel to $y = 4x - 19$
15. passes through $(-12, 3)$, perpendicular to $y = -\frac{2}{3}x - 11$

Find the distance between each pair of parallel lines with the given equations.

16. $y = x - 11$
    $y = x - 7$
17. $y = -2x + 1$
    $y = -2x + 16$

18. **MULTIPLE CHOICE** Which segment is skew to $\overline{CD}$?

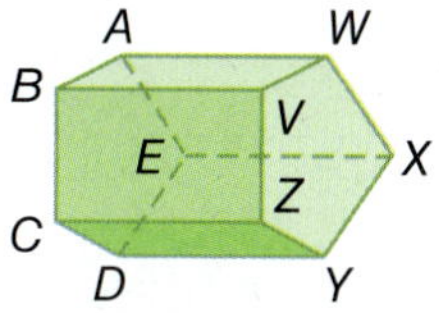

A $\overline{ZY}$ C $\overline{DE}$
B $\overline{AB}$ D $\overline{VZ}$

19. Find x so that $a \parallel b$. Identify the postulate or theorem you used.

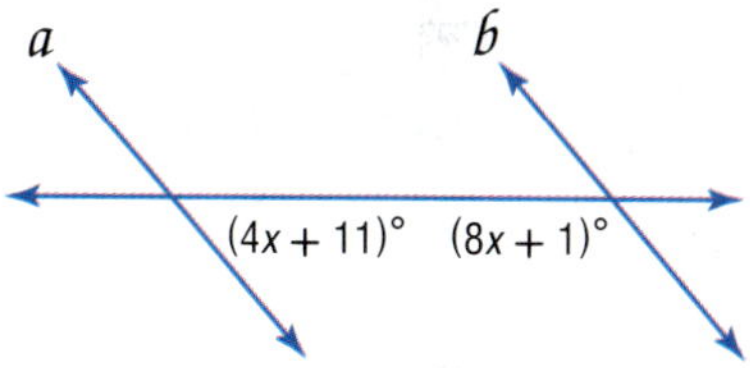

COORDINATE GEOMETRY Find the distance from P to ℓ.

20. Line ℓ contains points $(-4, 2)$ and $(3, -5)$. Point P has coordinates $(1, 2)$.
21. Line ℓ contains points $(6, 5)$ and $(2, 3)$. Point P has coordinates $(2, 6)$.

Given the following information, determine which lines, if any, are parallel. State the postulate or theorem that justifies your answer.

22. $\angle 4 \cong \angle 10$
23. $\angle 9 \cong \angle 6$
24. $\angle 7 \cong \angle 11$

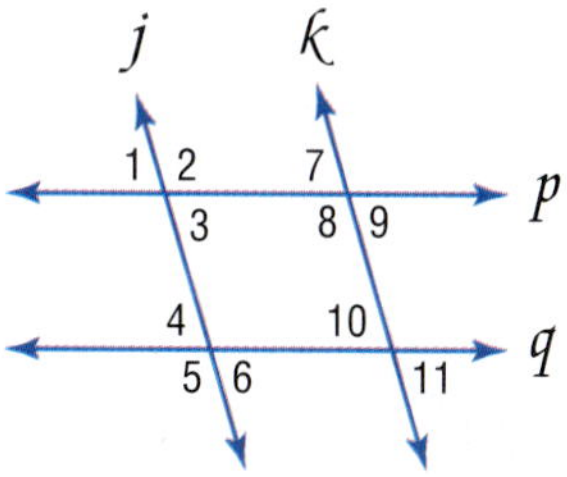

25. **JOBS** Hailey works at a gift shop after school. She is paid \$10 per hour plus a 15% commission on merchandise she sells. Write an equation that represents her earnings in a week if she sold \$550 worth of merchandise.

CHAPTER 3

Gridded Response Questions

In addition to multiple-choice, short-answer, and extended-response questions, you will likely encounter gridded-response questions on standardized tests. After solving a gridded-response question, you must print your answer on an answer sheet and mark in the correct circles on the grid to match your answer. Answers to gridded-response questions may be whole numbers, decimals, or fractions.

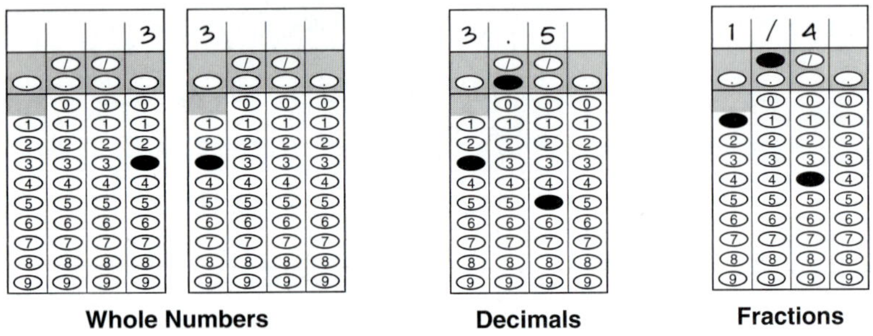

Strategies for Solving Gridded-Response Questions

Step 1

Read the problem carefully and solve.

- Be sure your answer makes sense.
- If time permits, check your answer.

Step 2

Print your answer in the answer boxes.

- Print only one digit or symbol in each answer box.
- Do not write any digits or symbols outside the answer boxes.
- Write answer as a whole number, decimal, or fraction.

Step 3

Fill in the grid.

- Fill in only one bubble for every answer box that you have written in. Be sure not to fill in a bubble under a blank answer box.
- Fill in each bubble completely and clearly.

EXAMPLE

Read the problem. Identify what you need to know. Then use the information in the problem to solve.

GRIDDED RESPONSE In the figure below, $\angle ABC$ is intersected by parallel lines ℓ and m. What is the measure of $\angle ABC$? Express your answer in degrees.

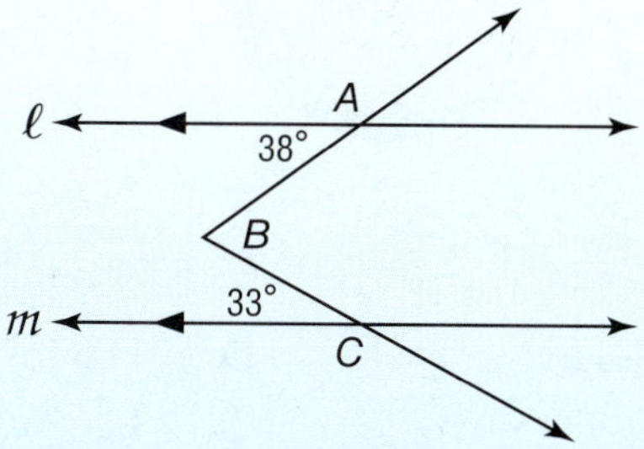

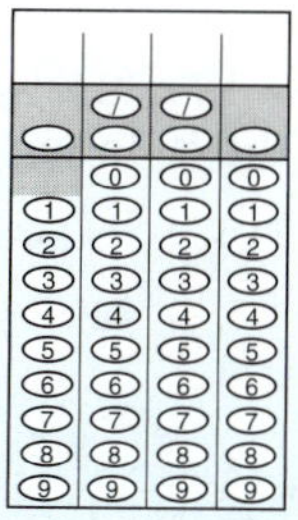

Redraw the figure and add a third line parallel to lines ℓ and m through point B. Find the angle measures using alternate interior angles.

Solve the Problem

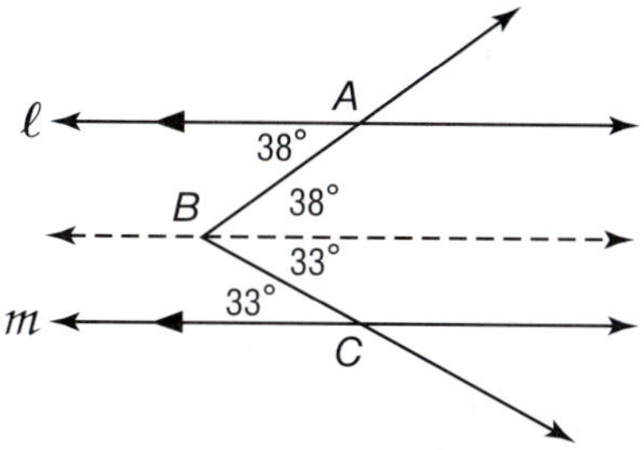

$$m\angle ABC = 38 + 33 = 71$$

Fill in the Grid

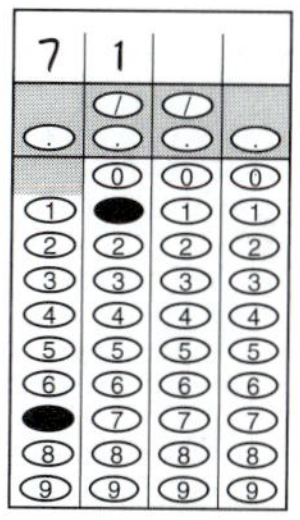

Print your answer in the answer box and fill in the grid.

Exercises

Read each question. Then fill in the correct answer on the answer document provided by your teacher or on a sheet of paper.

1. **GRIDDED RESPONSE** What is the slope of the line that contains the points $R(-2, 1)$ and $S(10, 6)$? Express your answer as a fraction.

2. **GRIDDED RESPONSE** Solve for x in the figure below.

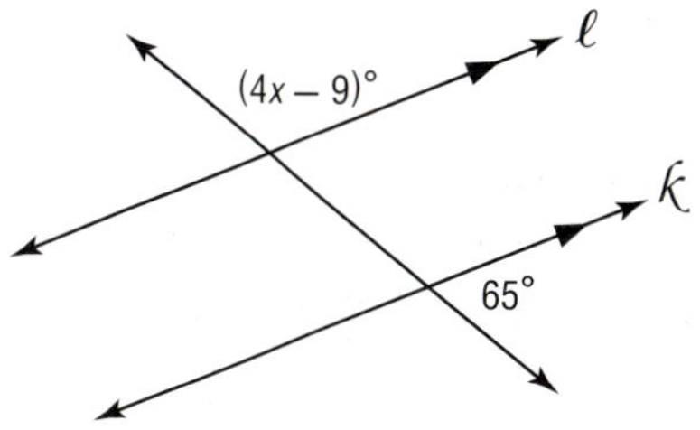

CHAPTER 3 Standardized Test Practice

Cumulative, Chapters 1 through 3

Multiple Choice

Read each question. Then fill in the correct answer on the answer document provided by your teacher or on a sheet of paper.

1. If $a \parallel b$ in the diagram below, which of the following may *not* be true?

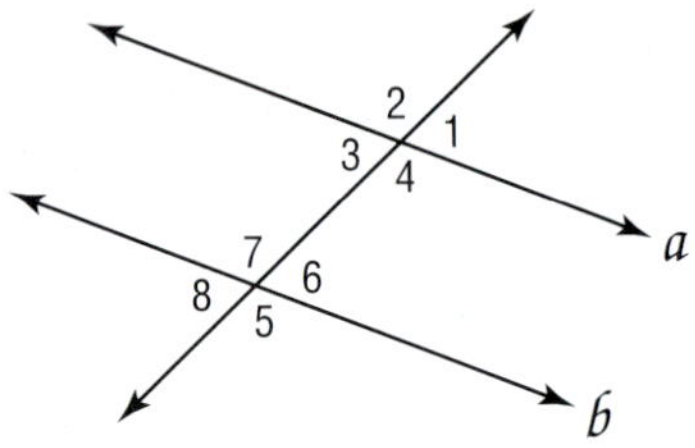

A $\angle 1 \cong \angle 3$

B $\angle 4 \cong \angle 7$

C $\angle 2 \cong \angle 5$

D $\angle 8 \cong \angle 2$

2. Which of the following best describes a *counterexample* to the assertion above?

The sum of two odd numbers is odd.

F $3 + 3 = 6$

G $5 + 4 = 9$

H $6 + 2 = 8$

J $4 + 9 = 13$

3. What is the slope of the line?

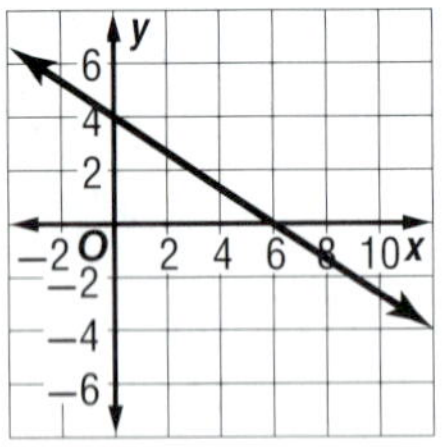

A $-\frac{2}{3}$

B $-\frac{1}{2}$

C $-\frac{2}{5}$

D $-\frac{1}{6}$

4. Line k contains points at (4, 1) and (−5, −5). Find the distance between line k and point $F(-4, 0)$.

F 3.3 units

G 3.6 units

H 4.0 units

J 4.2 units

5. The globe has a diameter of 22 inches. What is the volume of the globe?

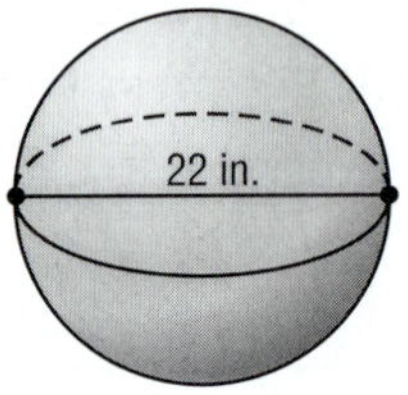

A 1520.5 in^3

B 1741.4 in^3

C 5575.3 in^3

D 6014.8 in^3

6. What is $m\angle 1$ in the figure below?

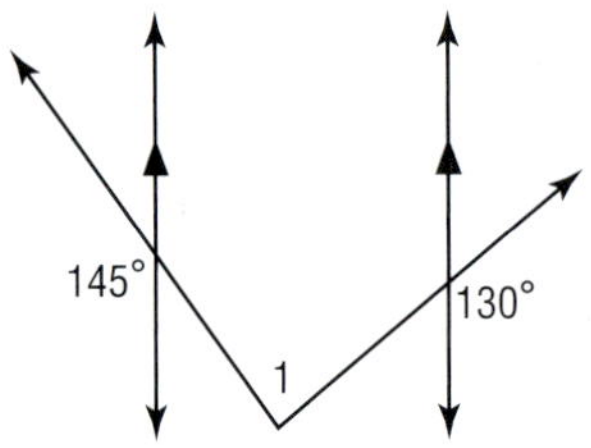

F 85

G 90

H 95

J 100

7. Jason is saving money to buy a car stereo. He has \$45 saved, and he can save \$15 per week. If the stereo that he wants is \$210, how many weeks will it take Jason to buy the stereo?

A 10

B 11

C 12

D 13

Test-TakingTip

Question 6 *Drawing a diagram* can help you solve problems. Draw a third parallel line through the vertex of angle 1. Then use the properties of parallel lines and transversals to solve the problem.

Short Response/Gridded Response

Record your answers on the answer sheet provided by your teacher or on a sheet of paper.

8. GRIDDED RESPONSE For a given line and a point not on the line, how many lines exist that pass through the point and are parallel to the given line?

9. GRIDDED RESPONSE Find the slope of the line that contains the points (4, 3) and (−2, −5).

10. Use the proof to answer the question.

Given: $\angle 1 \cong \angle 2$

Prove: $a \parallel b$

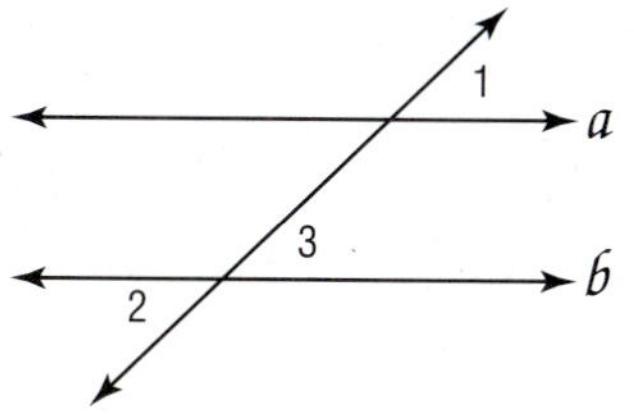

Proof:

Statements	Reasons
1. $\angle 1 \cong \angle 2$	**1.** Given
2. $\angle 2 \cong \angle 3$	**2.** ___?___
3. $\angle 1 \cong \angle 3$	**3.** Transitive Prop.
4. $a \parallel b$	**4.** If corresponding angles are congruent, then the lines are parallel.

11. Write the contrapositive of the statement.

If a figure is a square, then the figure is a parallelogram.

Extended Response

Record your answers on a sheet of paper. Show your work.

12. Refer to the figure to identify each of the following.

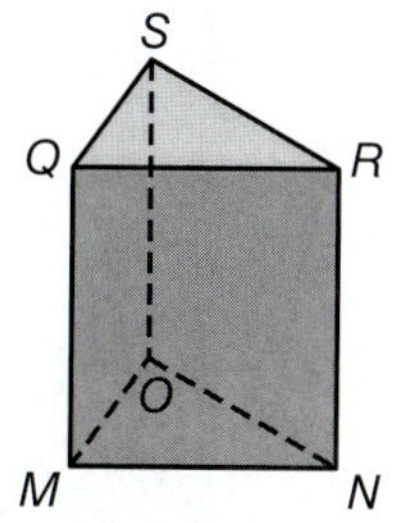

a. all segments parallel to $\overline{MQ}$

b. all planes intersecting plane SRN

c. a segment skew to $\overline{ON}$

13. Use this graph to answer each question.

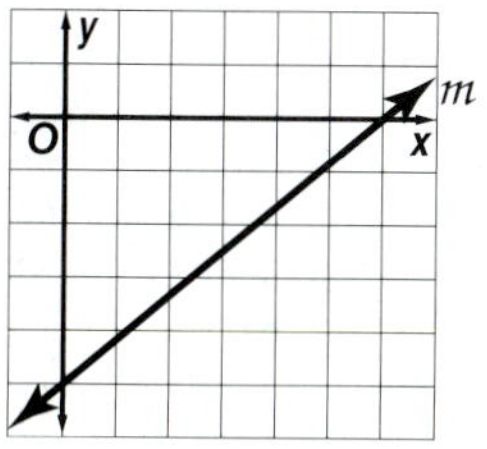

a. What is the equation of line m?

b. What is the slope of a line that is parallel to line m?

c. What is the slope of a line that is perpendicular to line m?

Need Extra Help?													
If you missed Question...	1	2	3	4	5	6	7	8	9	10	11	12	13
Go to Lesson or Page...	3-2	2-1	3-3	3-6	1-7	3-2	3-4	3-6	3-3	3-1	2-3	3-1	3-4

CHAPTER 4

Congruent Triangles

Then

In Chapters 1–3, you learned about segments and angles and discovered relationships between their measures.

Now

In Chapter 4, you will:

- Apply special relationships about the interior and exterior angles of triangles.
- Identify corresponding parts of congruent triangles and prove triangles congruent.
- Learn about the special properties of isosceles and equilateral triangles

Why?

FITNESS Triangles are used to add strength to many structures, including fitness equipment such as bike frames.

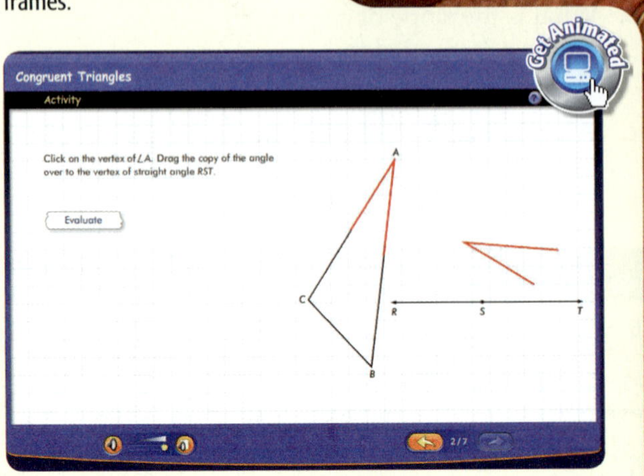

Get Ready for Chapter 4

Diagnose Readiness You have two options for checking Prerequisite Skills.

Text Option

Take the Quick Check below. Refer to the Quick Review for help.

QuickCheck

Classify each angle as *right, acute,* or *obtuse.* (Lesson 1-4)

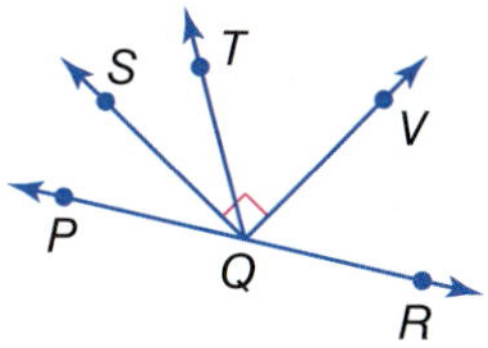

1. $m\angle VQS$
2. $m\angle TQV$
3. $m\angle PQV$
4. **ORIGAMI** The origami fold involves folding a strip of paper so that the lower edge of the strip forms a right angle with itself. Identify each angle as *right, acute,* or *obtuse.*

ALGEBRA Use the figure to find the indicated variable(s). Explain your reasoning. (Lesson 3-2)

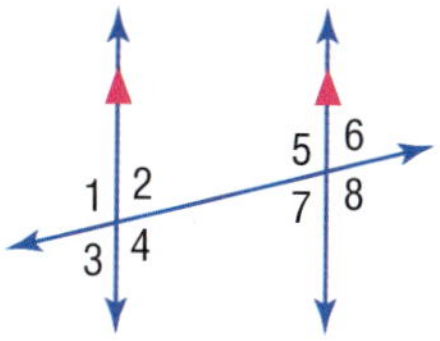

5. Find x if $m\angle 3 = x - 12$ and $m\angle 6 = 72$.
6. If $m\angle 4 = 2y + 32$ and $m\angle 5 = 3y - 3$, find y.

Find the distance between each pair of points. (Lesson 1-3)

7. $F(3, 6), G(7, -4)$
8. $X(-2, 5), Y(1, 11)$
9. $R(8, 0), S(-9, 6)$
10. $A(14, -3), B(9, -9)$
11. **MAPS** Miranda laid a coordinate grid on a map of a state where each 1 unit is equal to 10 miles. If her city is located at $(-8, -12)$ and the state capital is at $(0, 0)$, find the distance from her city to the capital to the nearest tenth of a mile.

QuickReview

EXAMPLE 1

Classify each angle as *right, acute,* or *obtuse.*

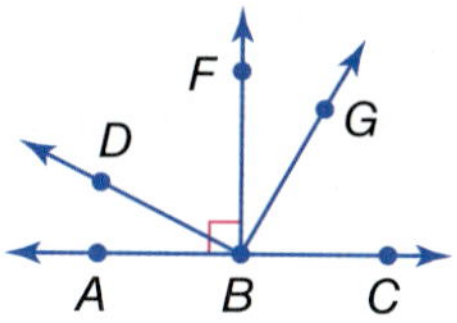

a. $m\angle ABG$
Point G on angle $\angle ABG$ lies on the exterior of right angle $\angle ABF$, so $\angle ABG$ is an obtuse angle.

b. $m\angle DBA$
Point D on angle $\angle DBA$ lies on the interior of right angle $\angle FBA$, so $\angle DBA$ is an acute angle.

EXAMPLE 2

In the figure, $m\angle 4 = 42°$. Find $m\angle 7$.

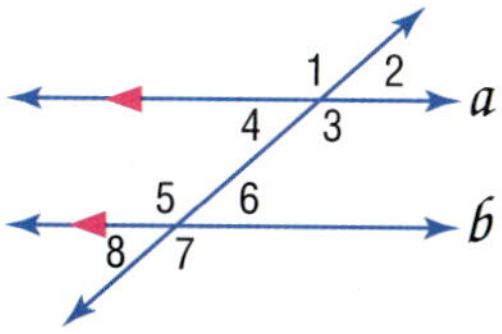

$\angle 7$ and $\angle 1$ are alternate interior angles, so they are congruent. $\angle 1$ and $\angle 4$ are a linear pair, so they are supplementary. Therefore, $\angle 7$ is supplementary to $\angle 1$. The measure of $\angle 7$ is $180 - 42$ or 138.

EXAMPLE 3

Find the distance between $J(5, 2)$ and $K(11, -7)$.

$$JK = \sqrt{(x_2 - x_1)^2 + (y_2 - y_1)^2} \quad \text{Distance Formula}$$
$$= \sqrt{(11 - 5)^2 + [(-7) - 2]^2} \quad \text{Substitute.}$$
$$= \sqrt{6^2 + (-9)^2} \quad \text{Subtract.}$$
$$= \sqrt{36 + 81} \text{ or } \sqrt{117} \quad \text{Simplify.}$$

Online Option

Math Online Take a self-check Chapter Readiness Quiz at glencoe.com.

Get Started on Chapter 4

You will learn several new concepts, skills, and vocabulary terms as you study Chapter 4. To get ready, identify important terms and organize your resources. You may wish to refer to **Chapter 0** to review prerequisite skills.

FOLDABLES® Study Organizer

Congruent Triangles Make this Foldable to help you organize your Chapter 4 notes about congruent triangles. Begin with a sheet of $8\frac{1}{2}'' \times 11''$ paper.

1. **Fold** into a taco forming a square. Cut off the excess paper strip formed by the square.

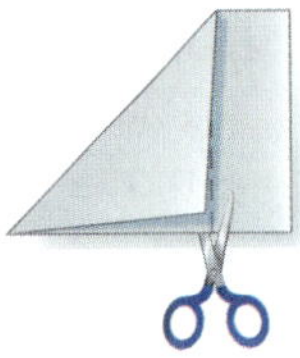

2. **Open** and fold the corners toward the center point of the X forming a small square.

3. **Trace** the square on another sheet of paper. Cut and glue it to the inside of the envelope.

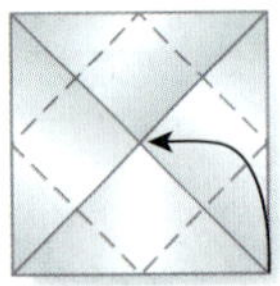

4. **Label** the flaps as shown.

Math Online glencoe.com

- Study the chapter online
- Explore **Get Animated** activities
- Get extra help from your own **Personal Tutor**
- Use **Extra Examples** for additional help
- Take a **Self-Check Quiz**
- **Review Vocabulary** in fun ways

New Vocabulary

English		Español
equiangular triangle	• p. 235 •	triángulo equiangular
equilateral triangle	• p. 236 •	triángulo equilátero
isosceles triangle	• p. 236 •	triángulo isósceles
scalene triangle	• p. 236 •	triángulo escaleno
auxiliary line	• p. 244 •	auxiliares de línea
congruent	• p. 253 •	congruente
congruent polygons	• p. 253 •	polígono congruentes
corresponding parts	• p. 253 •	partes correspondientes
included angle	• p. 264 •	ángulo incluido
included side	• p. 273 •	lado incluido
base angle	• p. 283 •	ángulo de la base
transformation	• p. 294 •	transformación
preimage	• p. 294 •	preimagen
image	• p. 294 •	imagen
reflection	• p. 294 •	reflexión
translation	• p. 294 •	traslación
rotation	• p. 294 •	rotación

Review Vocabulary

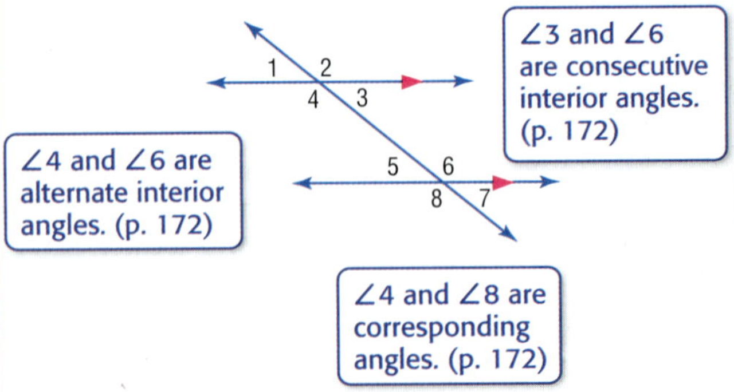

Multilingual eGlossary glencoe.com

4-1 Classifying Triangles

Then

You measured and classified angles. (Lesson 1-4)

Now

- Identify and classify triangles by angle measures.
- Identify and classify triangles by side measures.

New Vocabulary

acute triangle
equiangular triangle
obtuse triangle
right triangle
equilateral triangle
isosceles triangle
scalene triangle

Math Online
glencoe.com

Why?

Radio transmission towers are designed to support antennas for broadcasting radio or television signals. The structure of the tower shown reveals a pattern of triangular braces.

Classify Triangles by Angles Recall that a triangle is a three-sided polygon. Triangle *ABC*, written $\triangle ABC$, has parts that are named using *A*, *B*, and *C*.

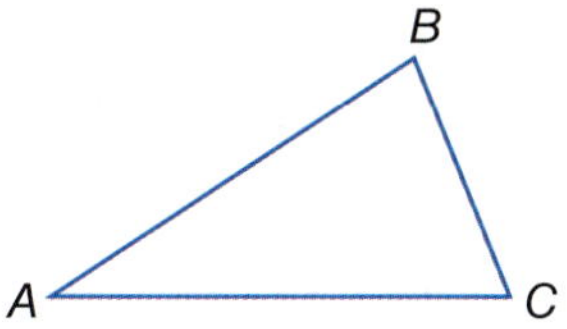

The sides of $\triangle ABC$ are $\overline{AB}$, $\overline{BC}$, and $\overline{CA}$.

The vertices are points *A*, *B*, and *C*.

The angles are $\angle BAC$ or $\angle A$, $\angle ABC$ or $\angle B$, and $\angle BCA$ or $\angle C$.

Triangles can be classified in two ways—by their angles or by their sides. All triangles have at least two acute angles, but the third angle is used to classify the triangle.

Key Concept — Classifications of Triangles by Angles

acute triangle

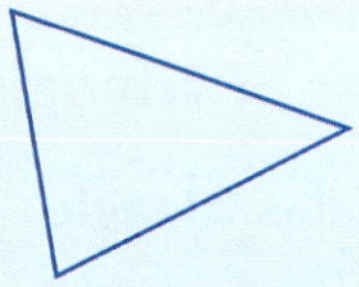

3 acute angles

equiangular triangle

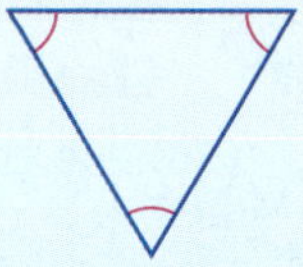

3 congruent acute angles

obtuse triangle

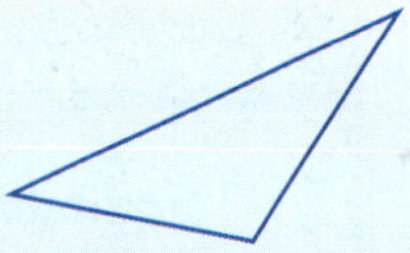

1 obtuse angle

right triangle

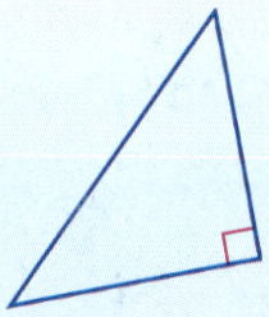

1 right angle

Each of the classifications above is a distinct group. A common mistake is to place triangles into more than one of these angle classifications. For example, a right triangle cannot be classified as an acute right triangle.

EXAMPLE 1 Classify Triangles by Angles

Classify each triangle as *acute*, *equiangular*, *obtuse*, or *right*.

a.

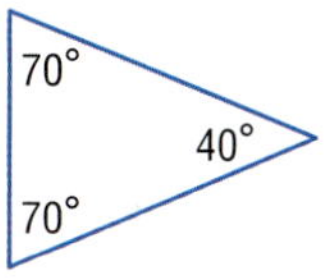

The triangle has three acute angles that are not all equal. It is an acute triangle.

b.

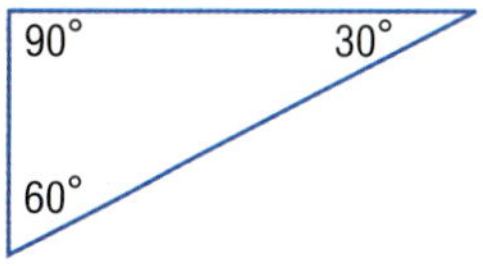

One angle of the triangle measures 90, so it is a right angle. Since the triangle has a right angle, it is a right triangle.

Review Vocabulary

acute angle an angle with a degree measure less than 90

right angle an angle with a degree measure of 90

obtuse angle an angle with a degree measure greater than 90 (Lesson 1-4)

Guided Practice

Classify each triangle as *acute, equiangular, obtuse,* or *right.*

1A.

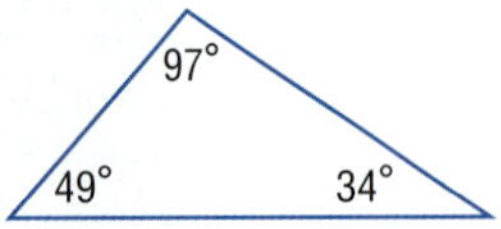

1B.

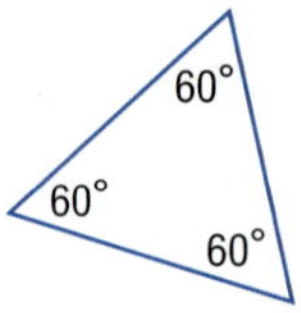

Personal Tutor glencoe.com

EXAMPLE 2 Classify Triangles by Angles Within Figures

Classify △*PQR* as *acute, equiangular, obtuse,* or *right*. Explain your reasoning.

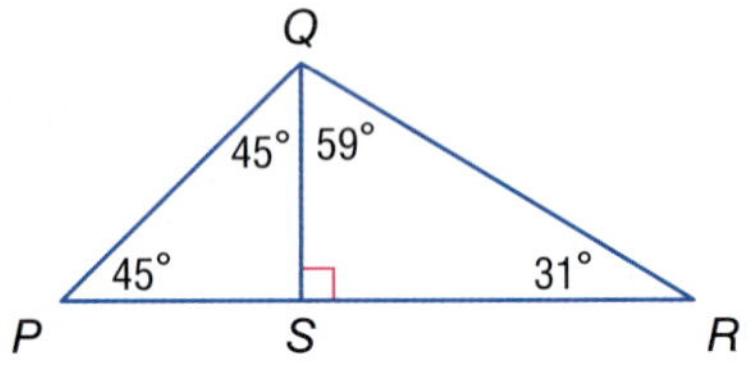

Point S is in the interior of $\angle PQR$, so by the Angle Addition Postulate, $m\angle PQR = m\angle PQS + m\angle SQR$. By substitution, $m\angle PQR = 45 + 59$ or 104.

Since $\triangle PQR$ has one obtuse angle, it is an obtuse triangle.

Guided Practice

2. Use the diagram to classify $\triangle PQS$ as *acute, equiangular, obtuse* or *right.* Explain your reasoning.

Personal Tutor glencoe.com

Classify Triangles by Sides Triangles can also be classified according to the number of congruent sides they have. To indicate that sides of a triangle are congruent, an equal number of hash marks is drawn on the corresponding sides.

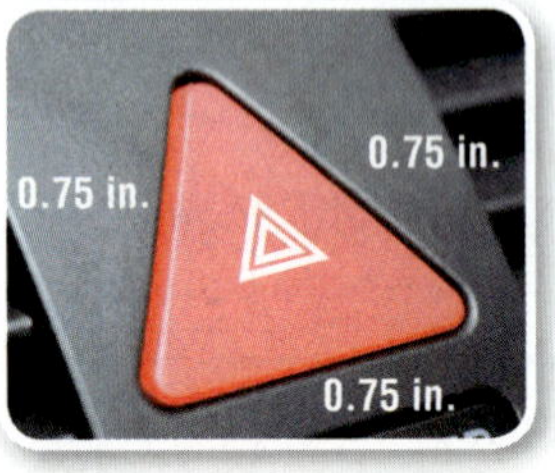

Key Concept — Classifications of Triangles by Sides

equilateral triangle

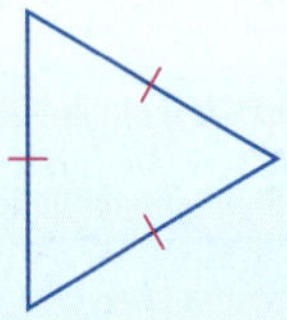

3 congruent sides

isosceles triangle

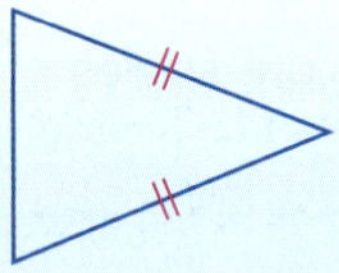

at least 2 congruent sides

scalene triangle

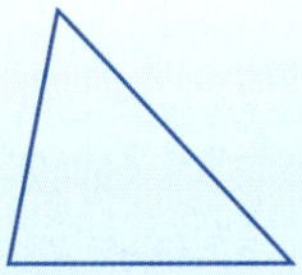

no congruent sides

An equilateral triangle is a special kind of isosceles triangle.

Real-World Link

In many cars, hazard lights are activated by pushing a small button located near the steering column. The switch is usually a small triangular red or orange icon. When the hazard lights are on, all of the turn signals on the vehicle flash in a distinct pattern to increase the vehicle's visibility.

Source: General Motors

Real-World EXAMPLE 3 Classify Triangles by Sides

MUSIC Classify the sound box of the Russian lute below as *equilateral, isosceles,* or *scalene.*

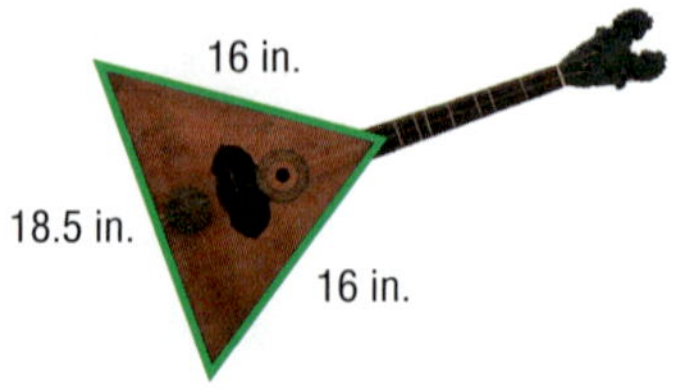

Two sides have the same measure, 16 inches, so the triangle has two congruent sides. The triangle is isosceles.

Guided Practice

3. DRIVING SAFETY Classify the button in the picture at the left by its sides.

Personal Tutor glencoe.com

EXAMPLE 4 Classify Triangles by Sides Within Figures

If point M is the midpoint of $\overline{JL}$, classify $\triangle JKM$ as *equilateral*, *isosceles*, or *scalene*. Explain your reasoning.

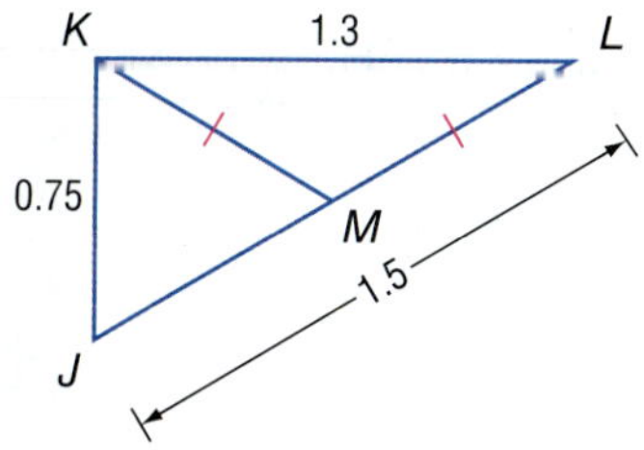

By the definition of midpoint, $JM = ML$.

$JM + ML = JL$	**Segment Addition Postulate**
$ML + ML = 1.5$	**Substitution**
$2ML = 1.5$	**Simplify.**
$ML = 0.75$	**Divide each side by 2.**

$JM = ML$ or 0.75. Since $\overline{KM} \cong \overline{ML}$, $KM = ML$ or 0.75.

Since $KJ = JM = KM = 0.75$, the triangle has three sides with the same measure. Therefore, the triangle has three congruent sides, so it is equilateral.

Guided Practice

4. Classify $\triangle KML$ as *equilateral*, *isosceles*, or *scalene*. Explain your reasoning.

Personal Tutor glencoe.com

You can also use the properties of isosceles and equilateral triangles to find missing values.

EXAMPLE 5 Finding Missing Values

ALGEBRA Find the measures of the sides of isosceles triangle ABC.

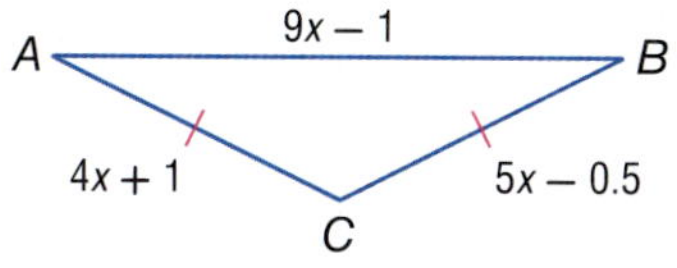

Step 1 Find x.

$AC = CB$	**Given**
$4x + 1 = 5x - 0.5$	**Substitution**
$1 = x - 0.5$	**Subtract 4x from each side.**
$1.5 = x$	**Add 0.5 to each side.**

Step 2 Substitute to find the length of each side.

$AC = 4x + 1$	**Given**
$= 4(1.5) + 1$ or 7	**$x = 1.5$**
$CB = AC$	**Given**
$= 7$	**$AC = 7$**
$AB = 9x - 1$	**Given**
$= 9(1.5) - 1$	**$x = 1.5$**
$= 12.5$	**Simplify.**

StudyTip

Check In Example 5, to check your answer, test to see if $CB = AC$ when 1.5 is substituted for x in the expression for CB, $5x - 0.5$.

$CB = 5x - 0.5$
$= 5(1.5) - 0.5$ or 7 ✓

Guided Practice

5. Find the measures of the sides of equilateral triangle FGH.

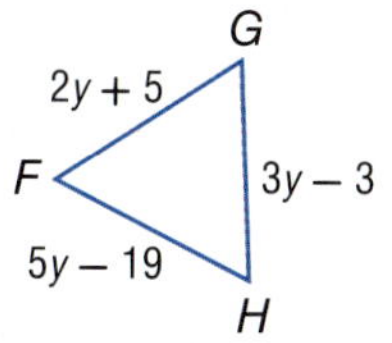

Personal Tutor glencoe.com

Check Your Understanding

Example 1
p. 235

ARCHITECTURE Classify each triangle as *acute, equiangular, obtuse,* or *right.*

1.
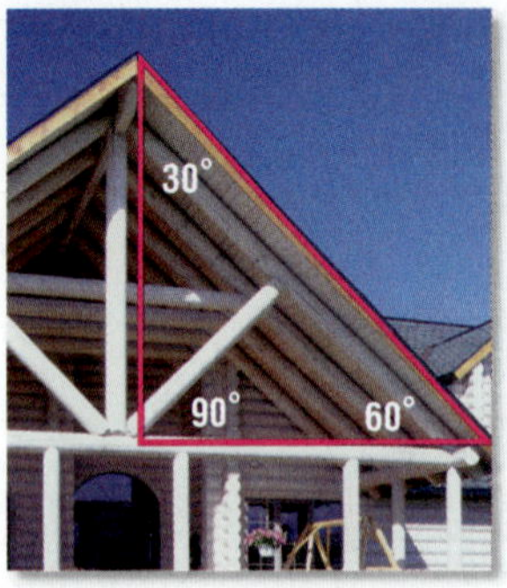

2.
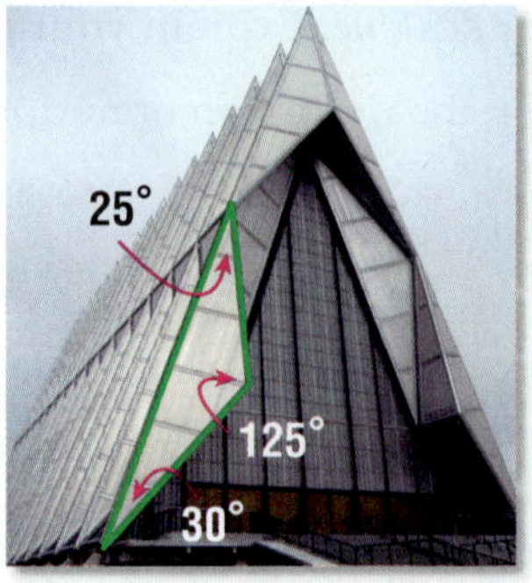

3.

Example 2
p. 236

Classify each triangle as *acute, equiangular, obtuse,* or *right.* Explain your reasoning.

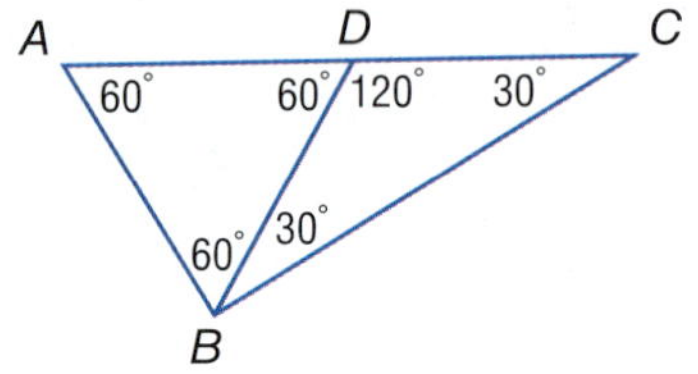

4. $\triangle ABD$

5. $\triangle BDC$

6. $\triangle ABC$

Example 3
p. 236

Classify each triangle as *equilateral, isosceles,* or *scalene.*

7.

8.
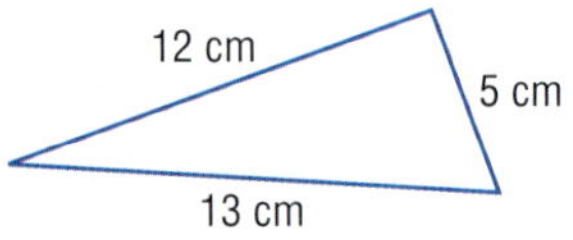

Example 4
p. 237

If point K is the midpoint of $\overline{FH}$, classify each triangle in the figure at the right as *equilateral, isosceles,* or *scalene.*

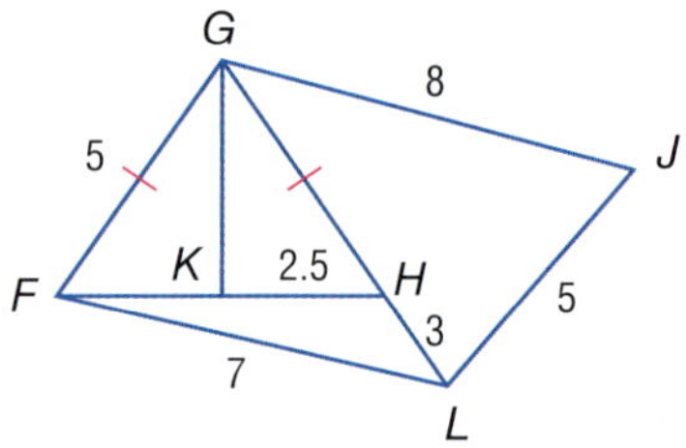

9. $\triangle FGH$

10. $\triangle GJL$

11. $\triangle FHL$

Example 5
p. 237

ALGEBRA Find x and the measures of the unknown sides of each triangle.

12.
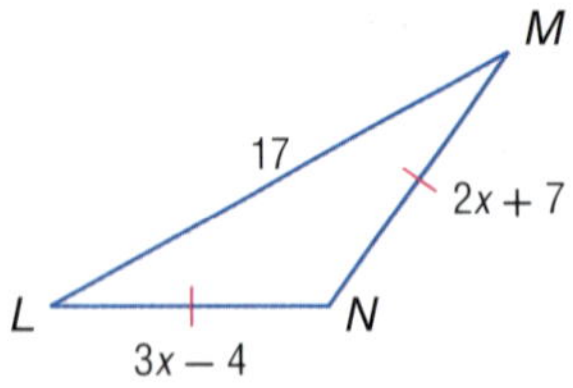

13.
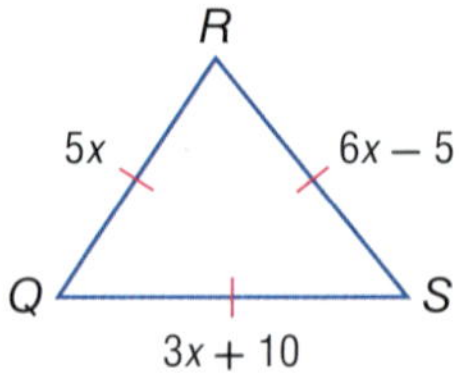

14. **JEWELRY** Suppose you are bending stainless steel wire to make the earring shown. The triangular portion of the earring is an isosceles triangle. If 1.5 centimeters are needed to make the hook portion of the earring, how many earrings can be made from 45 centimeters of wire? Explain your reasoning.

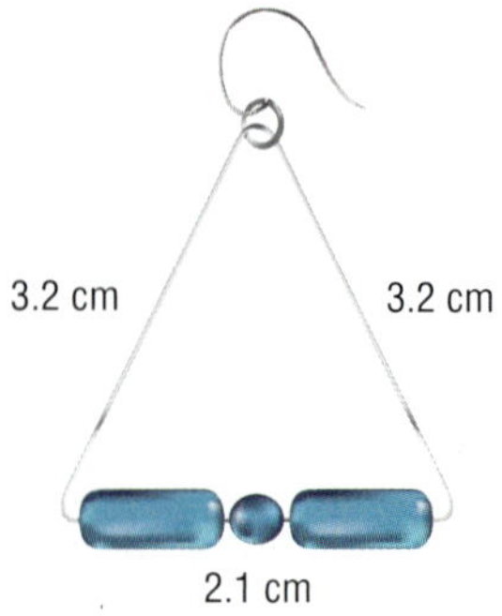

Practice and Problem Solving

= **Step-by-Step Solutions** begin on page R20.
Extra Practice begins on page 969.

Example 1
p. 235

Classify each triangle as *acute, equiangular, obtuse,* or *right*.

15.
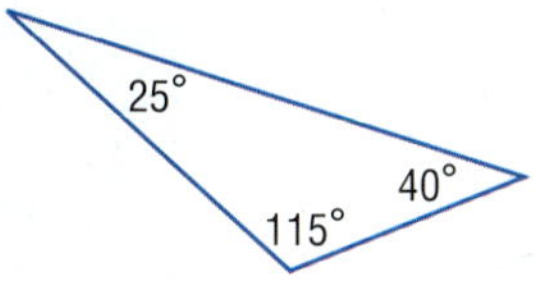

16.
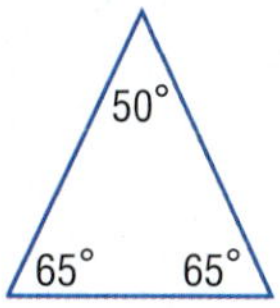

17.
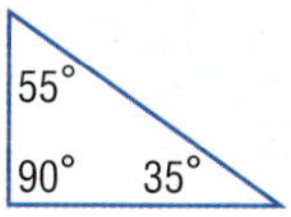

18.
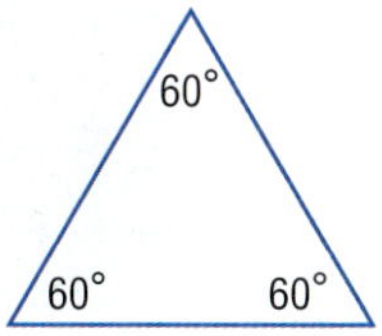

19.
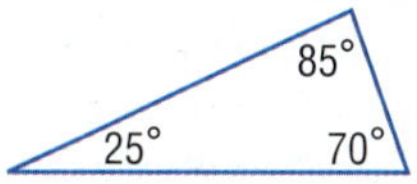

20.
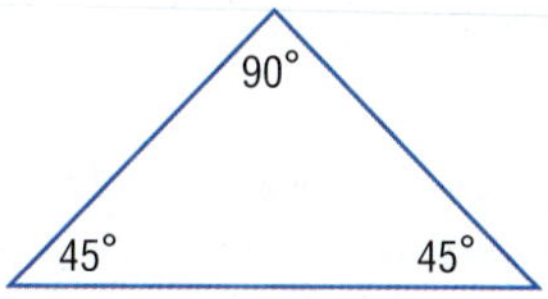

Example 2
p. 236

Classify each triangle as *acute, equiangular, obtuse,* or *right*.

21. $\triangle UYZ$

22. $\triangle BCD$

23. $\triangle ADB$

24. $\triangle UXZ$

25. $\triangle UWZ$

26. $\triangle UXY$

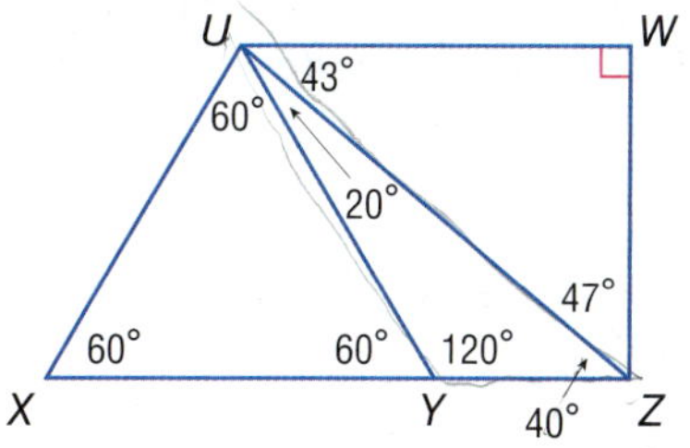

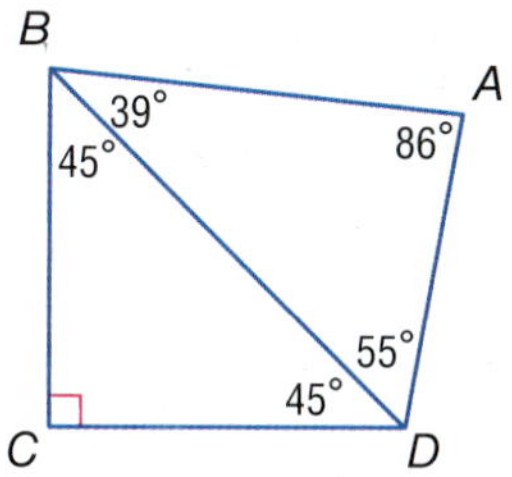

Example 3
p. 236

Classify each triangle as *equilateral, isosceles,* or *scalene*.

27.

28.

29.
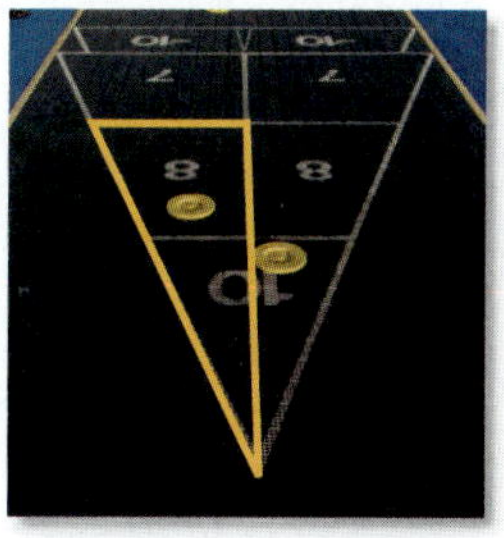

Example 4
p. 237

If point *C* is the midpoint of $\overline{BD}$ and point *E* is the midpoint of $\overline{DF}$, classify each triangle as *equilateral, isosceles,* or *scalene*.

30. $\triangle ABC$

31. $\triangle AEF$

32. $\triangle ADF$

33. $\triangle ACD$

34. $\triangle AED$

35. $\triangle ABD$

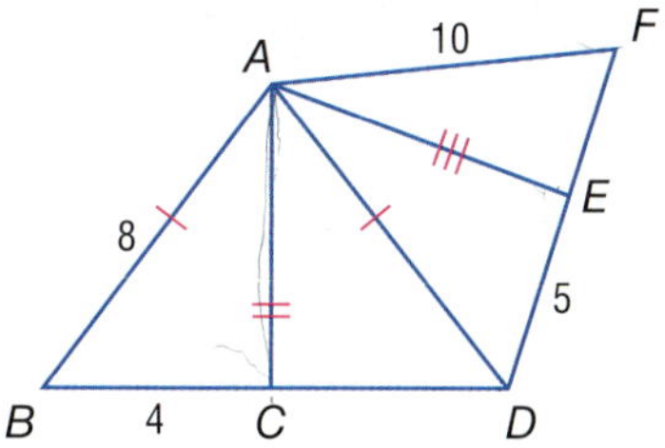

Example 5
p. 237

36. ALGEBRA Find x and the length of each side if $\triangle ABC$ is an isosceles triangle with $\overline{AB} \cong \overline{BC}$.

37. ALGEBRA Find x and the length of each side if $\triangle FGH$ is an equilateral triangle.

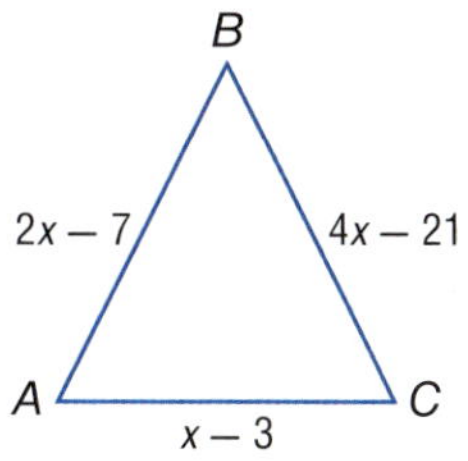

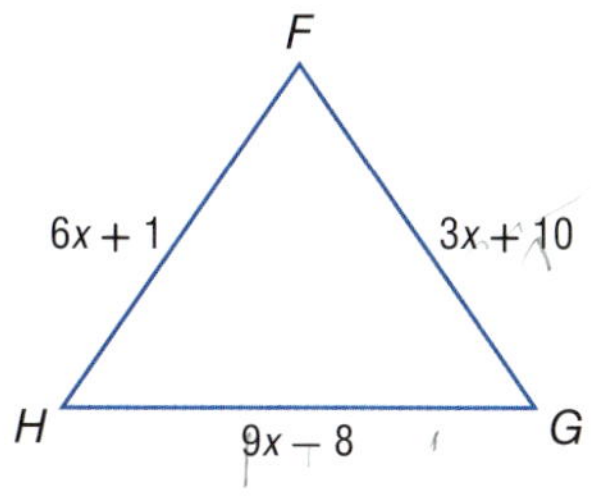

Real-World Link

A kaleidoscope consists of little pieces of colored glass and mirrors that reflect geometric images through a viewer, forming the appearance of four symmetrically placed objects. The pattern changes as the kaleidoscope is rotated.

Source: *Encyclopaedia Britannica*

38. GRAPHIC ART Classify each numbered triangle in *Kat* by its angles and by its sides. Use the corner of a sheet of notebook paper to classify angle measures and a ruler to measure sides.

39. KALEIDOSCOPE Josh is building a kaleidoscope using PVC pipe, cardboard, bits of colored paper, and a 12-inch square mirror tile. The mirror tile is to be cut into strips and arranged to form an open prism with a base like that of an equilateral triangle. Make a sketch of the prism, giving its dimensions. Explain your reasoning.

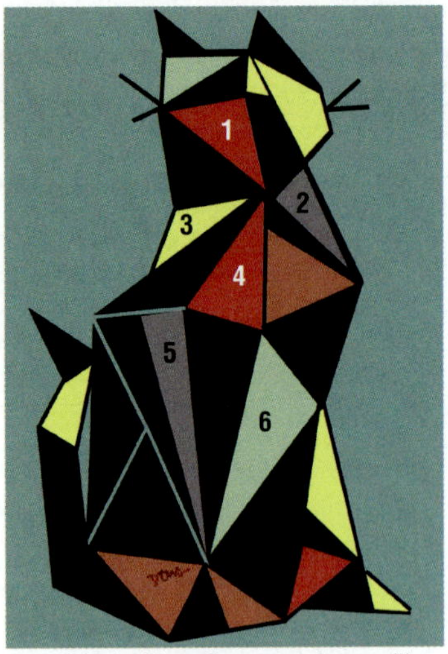

Kat, 2002, by Diana Ong, computer graphic

Classify each triangle in the figure by its angles and sides.

40. $\triangle ABE$

41. $\triangle EBC$

42. $\triangle BDC$

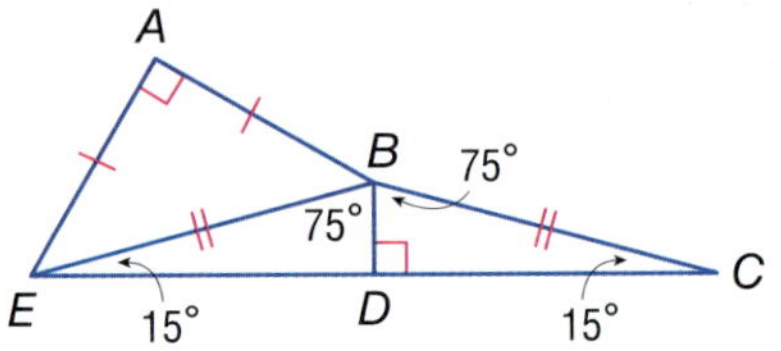

COORDINATE GEOMETRY Find the measures of the sides of $\triangle XYZ$ and classify each triangle by its sides.

43. $X(-5, 9), Y(2, 1), Z(-8, 3)$

44. $X(7, 6), Y(5, 1), Z(9, 1)$

45. $X(3, -2), Y(1, -4), Z(3, -4)$

46. $X(-4, -2), Y(-3, 7), Z(4, -2)$

47. PROOF Write a paragraph proof to prove that $\triangle DBC$ is an acute triangle if $m\angle ADC = 120$ and $\triangle ABC$ is acute.

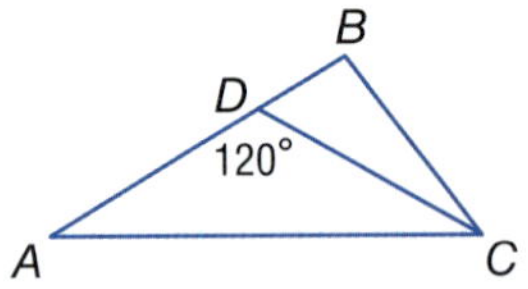

48. PROOF Write a two-column proof to prove that $\triangle BCD$ is equiangular if $\triangle ACE$ is equiangular and $\overline{BD} \parallel \overline{AE}$.

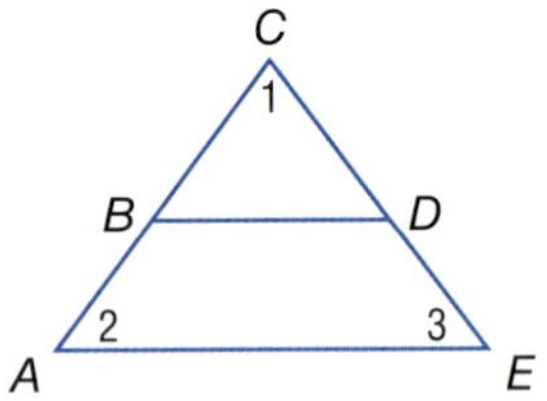

ALGEBRA For each triangle, find x and the measure of each side.

49. $\triangle FGH$ is an equilateral triangle with $FG = 3x - 10$, $GH = 2x + 5$, and $HF = x + 20$.

50. $\triangle JKL$ is isosceles with $\overline{JK} \cong \overline{KL}$, $JK = 4x - 1$, $KL = 2x + 5$, and $LJ = 2x - 1$.

51. $\triangle MNP$ is isosceles with $\overline{MN} \cong \overline{NP}$. MN is two less than five times x, NP is seven more than two times x, and PM is two more than three times x.

52. $\triangle RST$ is equilateral. RS is three more than four times x, ST is seven more than two times x, and TR is one more than five times x.

53. CONSTRUCTION Construct an equilateral triangle. Verify your construction using measurement and justify it using mathematics. (*Hint:* Use the construction for copying a segment.)

Real-World Link

The New York Stock Exchange has the capacity to process about 10 billion shares a day.

Source: NYSE

54. STOCKS Technical analysts use charts to identify patterns that can suggest future activity in stock prices. Triangle charts are most useful when the fluctuation in the price of a stock is decreasing over time.

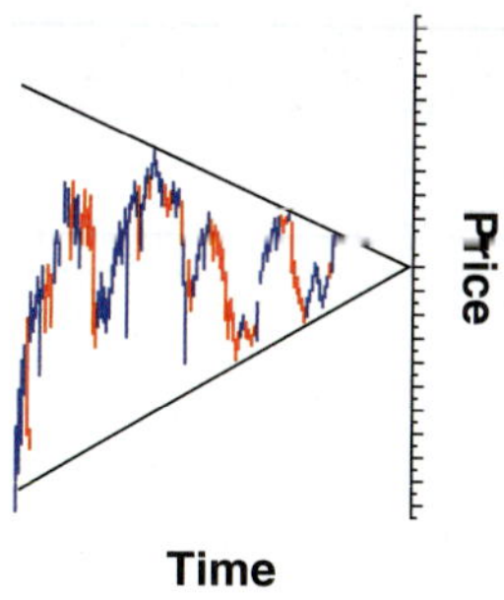

a. Classify by its sides and angles the triangle formed if a vertical line is drawn at any point on the graph.

b. How would the price have to fluctuate in order for the data to form an obtuse triangle? Draw an example to support your reasoning.

55 **MULTIPLE REPRESENTATIONS** In the diagram, the vertex *opposite* side $\overline{BC}$ is $\angle A$.

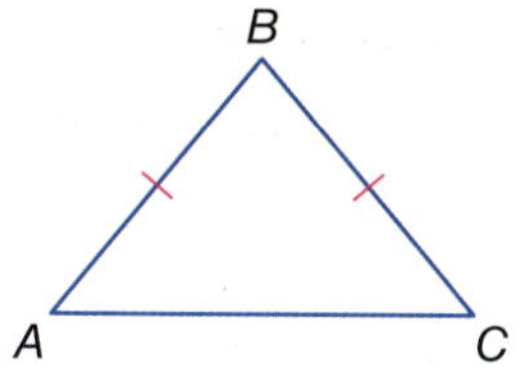

a. GEOMETRIC Draw four isosceles triangles, including one acute, one right, and one obtuse isosceles triangle. Label the vertices opposite the congruent sides as *A* and *C*. Label the remaining vertex *B*. Then measure the angles of each triangle and label each angle with its measure.

b. TABULAR Measure all the angles of each triangle. Organize the measures for each triangle into a table. Include a column in your table to record the sum of these measures.

c. VERBAL Make a conjecture about the measures of the angles that are opposite the congruent sides of an isosceles triangle. Then make a conjecture about the sum of the measures of the angles of an isosceles triangle.

d. ALGEBRAIC If x is the measure of one of the angles opposite one of the congruent sides in an isosceles triangle, write expressions for the measures of each of the other two angles in the triangle. Explain.

H.O.T. Problems

Use **H**igher-**O**rder **T**hinking Skills

56. ERROR ANALYSIS Elaina says that $\triangle DFG$ is obtuse. Ines disagrees, explaining that the triangle has more acute angles than obtuse angles so it must be acute. Is either of them correct? Explain your reasoning.

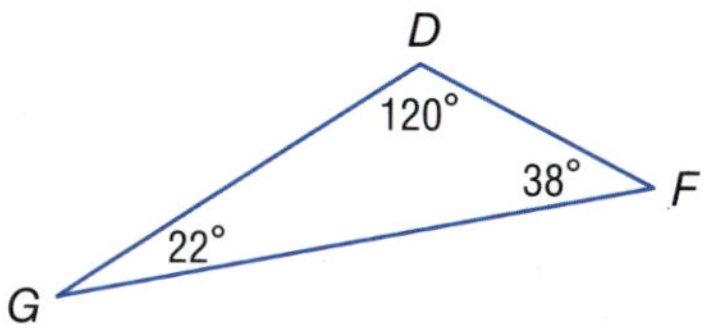

REASONING Determine whether the statements below are *sometimes*, *always*, or *never* true. Explain your reasoning.

57. Equiangular triangles are also right triangles.

58. Equilateral triangles are isosceles.

59. Right triangles are equilateral.

60. CHALLENGE An equilateral triangle has sides that measure $5x + 3$ units and $7x - 5$ units. What is the perimeter of the triangle? Explain.

OPEN ENDED Draw an example of each type of triangle below using a protractor and a ruler. Label the sides and angles of each triangle with their measures. If not possible, explain why not.

61. scalene right

62. isosceles obtuse

63. equilateral obtuse

64. WRITING IN MATH Explain why classifying an equiangular triangle as an *acute* equiangular triangle is unnecessary.

Standardized Test Practice

65. Which type of triangle can serve as a counterexample to the conjecture below?

If two angles of a triangle are acute, then the measure of the third angle must be greater than or equal to 90.

A equilateral
B obtuse
C right
D scalene

66. ALGEBRA A baseball glove originally cost $84.50. Kenji bought it at 40% off. How much was deducted from the original price?

F $50.70
G $44.50
H $33.80
J $32.62

67. GRIDDED RESPONSE Jorge is training for a 20-mile race. Jorge runs 7 miles on Monday, Tuesday, and Friday, and 12 miles on Wednesday and Saturday. After 6 weeks of training, Jorge will have run the equivalent of how many races?

68. SAT/ACT What is the slope of the line determined by the equation $2x + y = 5$?

A 2
B $\frac{5}{2}$
C -1
D -2

Spiral Review

Find the distance between each pair of parallel lines with the given equations. (Lesson 3-6)

69. $x = -2$
$x = 5$

70. $y = -6$
$y = 1$

71. $y = 2x + 3$
$y = 2x - 7$

72. $y = x + 2$
$y = x - 4$

73. FOOTBALL When striping the practice football field, Mr. Hawkins first painted the sidelines. Next he marked off 10-yard increments on one sideline. He then constructed lines perpendicular to the sidelines at each 10-yard mark. Why does this guarantee that the 10-yard lines will be parallel? (Lesson 3-5)

Identify the hypothesis and conclusion of each conditional statement. (Lesson 2-3)

74. If three points lie on a line, then they are collinear.

75. If you are a teenager, then you are at least 13 years old.

76. If $2x + 6 = 10$, then $x = 2$.

77. If you have a driver's license, then you are at least 16 years old.

Refer to the figure at the right. (Lesson 1-1)

78. How many planes appear in this figure?

79. Name the intersection of plane AEB with plane $\mathcal{N}$.

80. Name three points that are collinear.

81. Are points D, E, C, and B coplanar?

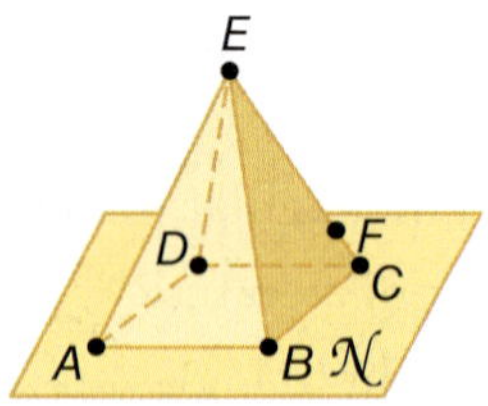

Skills Review

Identify each pair of angles as ***alternate interior, alternate exterior, corresponding,*** **or** ***consecutive interior angles.*** (Lesson 3-1)

82. $\angle 5$ and $\angle 3$

83. $\angle 9$ and $\angle 4$

84. $\angle 11$ and $\angle 13$

85. $\angle 1$ and $\angle 11$

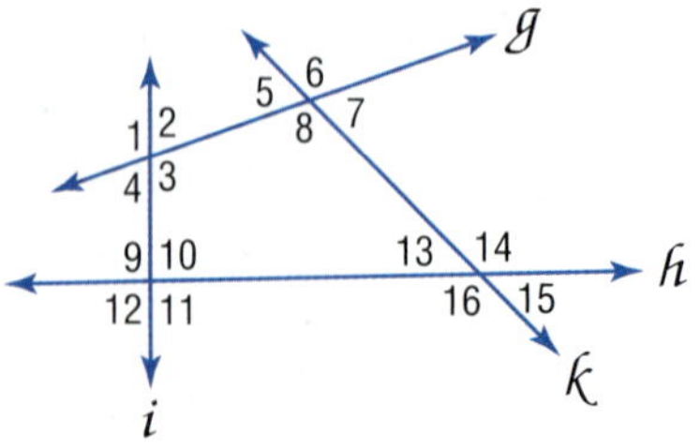

EXPLORE
4-2
Geometry Lab
Angles of Triangles

In this lab, you will find special relationships among the angles of a triangle.

ACTIVITY 1 Interior Angles of a Triangle

Step 1

Draw and cut out several different triangles. Label the vertices *A*, *B*, and *C*.

Step 2

For each triangle, fold vertex *B* down so that the fold line is parallel to $\overline{AC}$. Relabel as vertex *B*.

Step 3

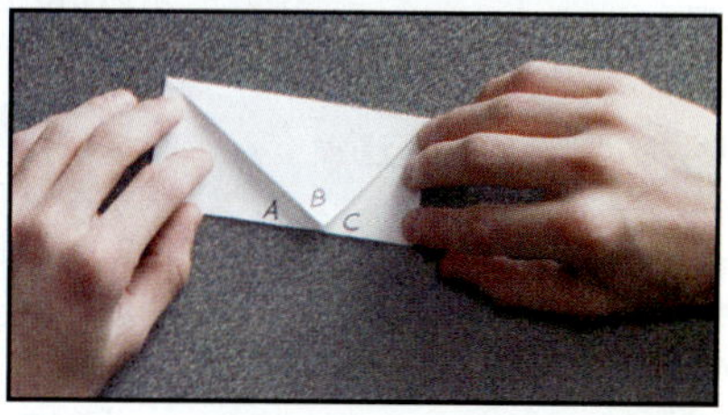

Then fold vertices *A* and *C* so that they meet vertex *B*. Relabel as vertices *A* and *C*.

Analyze the Results

1. Angles *A*, *B*, and *C* are called *interior angles* of triangle *ABC*. What type of figure do these three angles form when joined together in Step 3?
2. **Make a conjecture** about the sum of the measures of the interior angles of a triangle.

Get Animated

ACTIVITY 2 Exterior Angles of a Triangle

Step 1

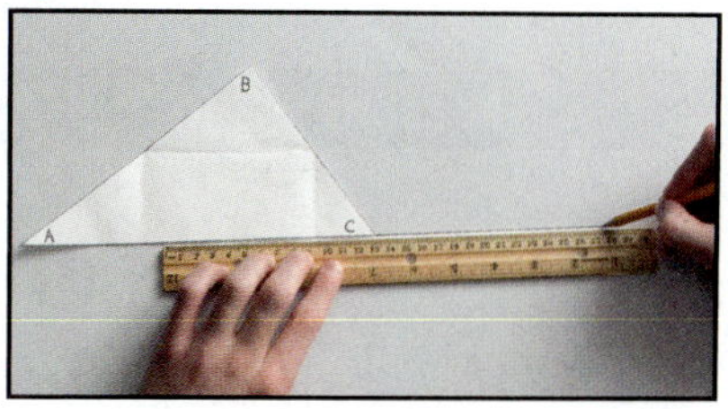

Unfold each triangle from Activity 1 and place each on a separate piece of paper. Extend $\overline{AC}$ as shown.

Step 2

For each triangle, tear off $\angle A$ and $\angle B$.

Step 3

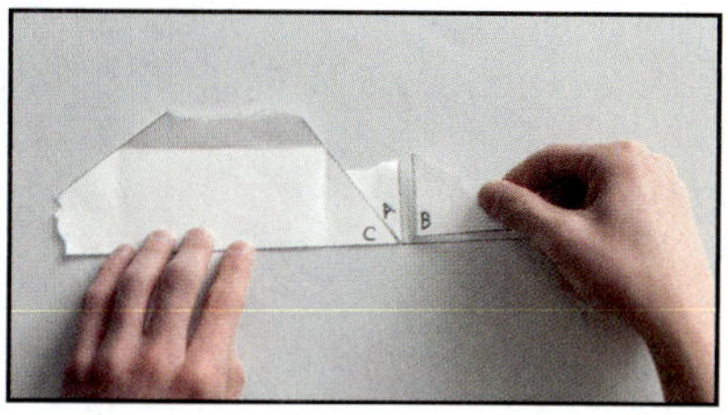

Arrange $\angle A$ and $\angle B$ so that they fill the angle adjacent to $\angle C$ as shown.

Model and Analyze the Results

3. The angle adjacent to $\angle C$ is called an *exterior angle* of triangle *ABC*. **Make a conjecture** about the relationship among $\angle A$, $\angle B$, and the exterior angle at *C*.
4. Repeat the steps in Activity 2 for the exterior angles of $\angle A$ and $\angle B$ in each triangle.
5. **Make a conjecture** about the measure of an exterior angle and the sum of the measures of its nonadjacent interior angles.

4-2 Angles of Triangles

Then

You classified triangles by their side or angle measures. (Lesson 4-1)

Now

- Apply the Triangle Angle-Sum Theorem.
- Apply Exterior Angle Theorem.

New Vocabulary

auxiliary line
exterior angle
remote interior angles
flow proof
corollary

Math Online
glencoe.com

Why?

Massachusetts Institute of Technology (MIT) sponsors the annual *Design 2.007* contest in which students design and build a robot.

One test of a robot's movements is to program it to move in a triangular path. The sum of the measures of the pivot angles through which the robot must turn will always be the same.

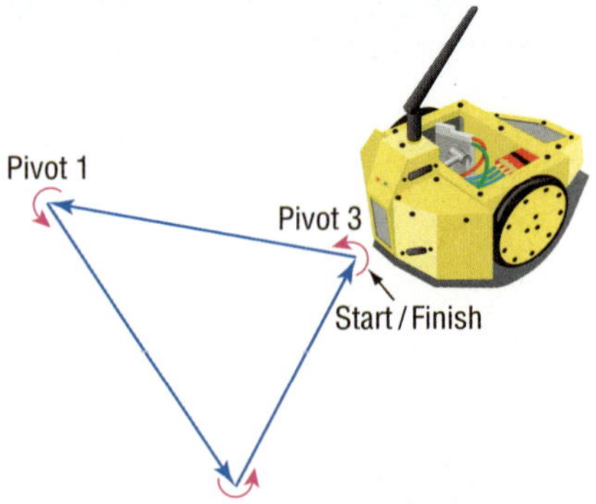

Triangle Angle-Sum Theorem The Triangle Angle-Sum Theorem gives the relationship among the interior angle measures of any triangle.

Theorem 4.1 Triangle Angle-Sum Theorem

Words The sum of the measures of the angles of a triangle is 180.

Example $m\angle A + m\angle B + m\angle C = 180$

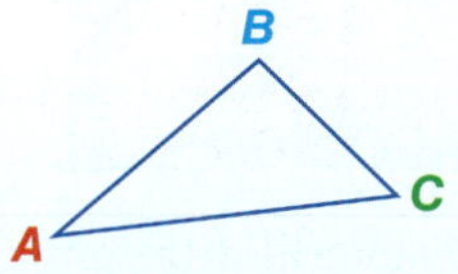

The proof of the Triangle Angle-Sum Theorem requires the use of an auxiliary line. An **auxiliary line** is an extra line or segment drawn in a figure to help analyze geometric relationships. As with any statement in a proof, you must justify any properties of an auxiliary line that you have drawn.

Proof Triangle Angle-Sum Theorem

Given: $\triangle ABC$

Prove: $m\angle 1 + m\angle 2 + m\angle 3 = 180$

Proof:

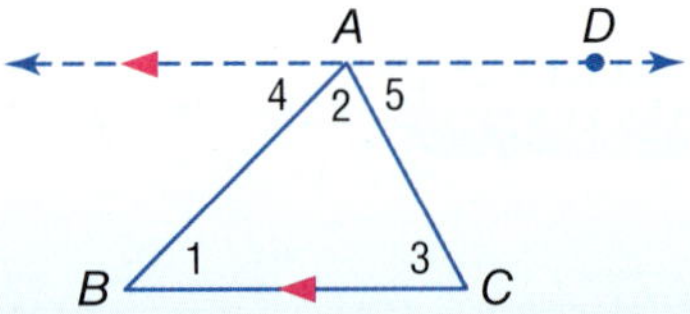

Statements	Reasons
1. $\triangle ABC$	**1.** Given
2. Draw $\overleftrightarrow{AD}$ through A parallel to $\overline{BC}$.	**2.** Parallel Postulate
3. $\angle 4$ and $\angle BAD$ form a linear pair.	**3.** Def. of a linear pair
4. $\angle 4$ and $\angle BAD$ are supplementary.	**4.** If 2 ∠s form a linear pair, they are supplementary.
5. $m\angle 4 + m\angle BAD = 180$	**5.** Def. of suppl. ∠s
6. $m\angle BAD = m\angle 2 + m\angle 5$	**6.** Angle Addition Postulate
7. $m\angle 4 + m\angle 2 + m\angle 5 = 180$	**7.** Substitution
8. $\angle 4 \cong \angle 1$, $\angle 5 \cong \angle 3$	**8.** Alt. Int. ∠s Theorem
9. $m\angle 4 = m\angle 1$, $m\angle 5 = m\angle 3$	**9.** Def. of ≅ ∠s
10. $m\angle 1 + m\angle 2 + m\angle 3 = 180$	**10.** Substitution

The Triangle Angle-Sum Theorem can be used to determine the measure of the third angle of a triangle when the other two angle measures are known.

Real-World Link

The pass-and-move soccer drill incorporates several fundamental aspects of passing. All passes in this drill are made in a triangle, which is the basis of all ball movement. Additionally, the players are forced to move immediately after passing the ball.

Real-World EXAMPLE 1 Use the Triangle Angle-Sum Theorem

SOCCER The diagram shows the path of the ball in a passing drill created by four friends. Find the measure of each numbered angle.

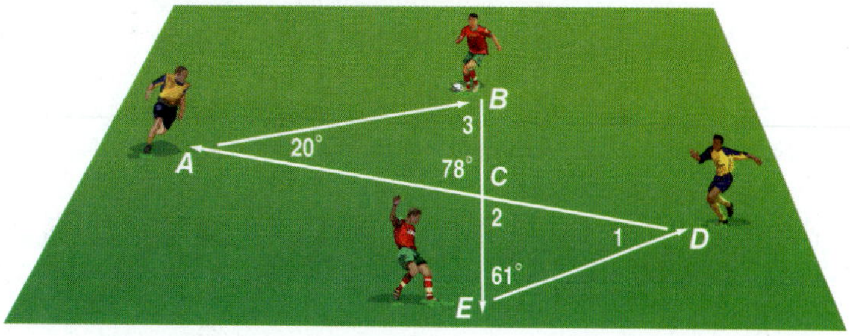

Understand Examine the information given in the diagram. You know the measures of two angles of one triangle and only one measure of another. You also know that $\angle ACB$ and $\angle 2$ are vertical angles.

Plan Find $m\angle 3$ using the Triangle Angle-Sum Theorem, because the measures of two angles of $\angle ABC$ are known. Use the Vertical Angles Theorem to find $m\angle 2$. Then you will have enough information to find the measure of $\angle 1$ in $\triangle CDE$.

Solve

$m\angle 3 + m\angle BAC + m\angle ACB = 180$ **Triangle Angle-Sum Theorem**

$m\angle 3 + 20 + 78 = 180$ **Substitution**

$m\angle 3 + 98 = 180$ **Simplify.**

$m\angle 3 = 82$ **Subtract 98 from each side.**

$\angle ACB$ and $\angle 2$ are congruent vertical angles. So, $m\angle 2 = 78$.

Use $m\angle 2$ and $\angle CED$ of $\triangle CDE$ to find $m\angle 1$.

$m\angle 1 + m\angle 2 + m\angle CED = 180$ **Triangle Angle-Sum Theorem**

$m\angle 1 + 78 + 61 = 180$ **Substitution**

$m\angle 1 + 139 = 180$ **Simplify.**

$m\angle 1 = 41$ **Subtract 139 from each side.**

Check The sums of the measures of the angles of $\triangle ABC$ and $\triangle CDE$ should be 180.

$\triangle ABC$: $m\angle 3 + m\angle BAC + m\angle ACB = 82 + 20 + 78$ or 180 ✓

$\triangle CDE$: $m\angle 1 + m\angle 2 + m\angle CED = 41 + 78 + 61$ or 180 ✓

Problem-Solving Tip

Break It Up Often a complex problem can be more easily solved if you first break it into more manageable parts. In Example 1, before you can find $m\angle 1$, you must first find $m\angle 2$.

Guided Practice

Find the measures of each numbered angle.

1A.

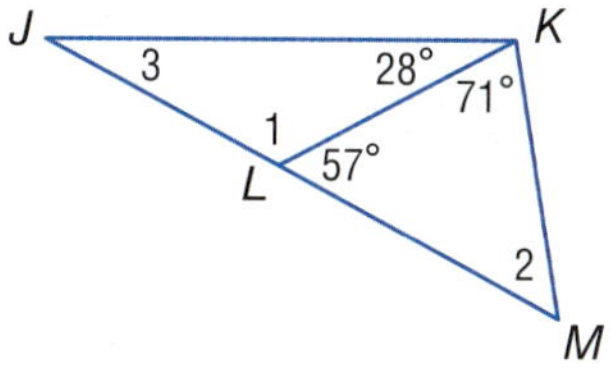

1B.

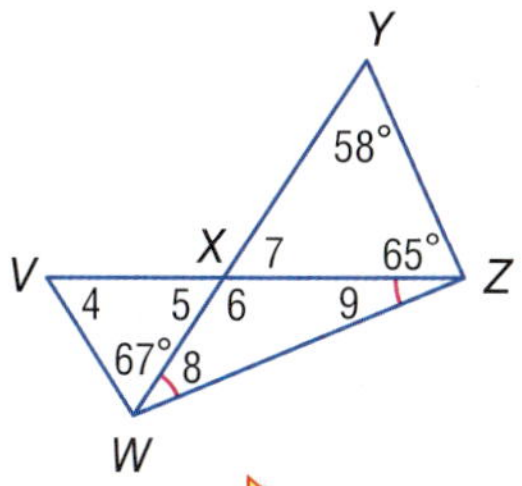

Personal Tutor glencoe.com

Exterior Angle Theorem In addition to its three interior angles, a triangle can have **exterior angles** formed by one side of the triangle and the extension of an adjacent side. Each exterior angle of a triangle has two **remote interior angles** that are not adjacent to the exterior angle.

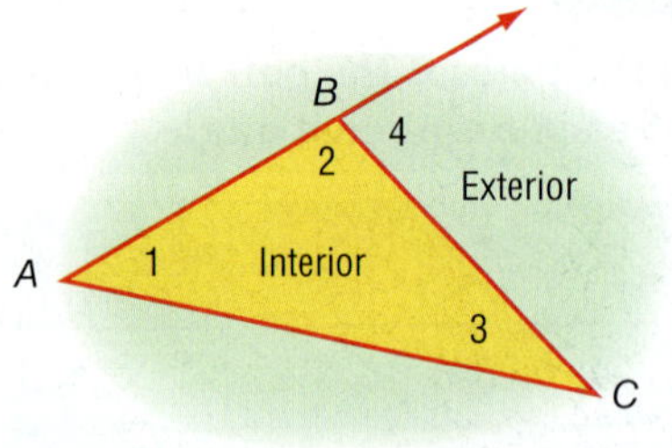

$\angle 4$ is an exterior angle of $\triangle ABC$. Its two remote interior angles are $\angle 1$ and $\angle 3$.

Theorem 4.2 Exterior Angle Theorem

The measure of an exterior angle of a triangle is equal to the sum of the measures of the two remote interior angles.

Example $m\angle A + m\angle B = m\angle 1$

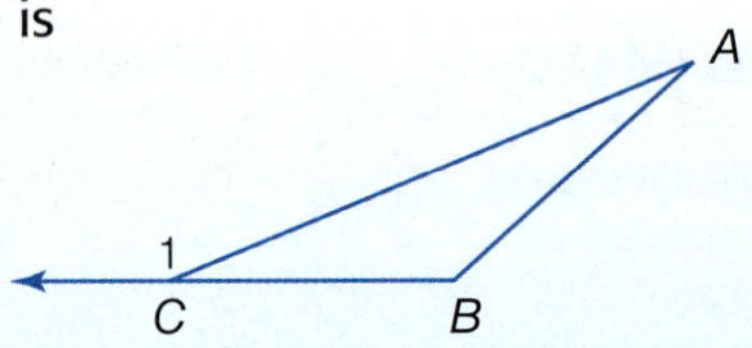

ReadingMath

Flowchart Proof A flow proof is sometimes called a *flowchart* proof.

A **flow proof** uses statements written in boxes and arrows to show the logical progression of an argument. The reason justifying each statement is written below the box. You can use a flow proof to prove the Exterior Angle Theorem.

Proof Exterior Angle Theorem

Given: $\triangle ABC$

Prove: $m\angle A + m\angle B + = m\angle 1$

Flow Proof:

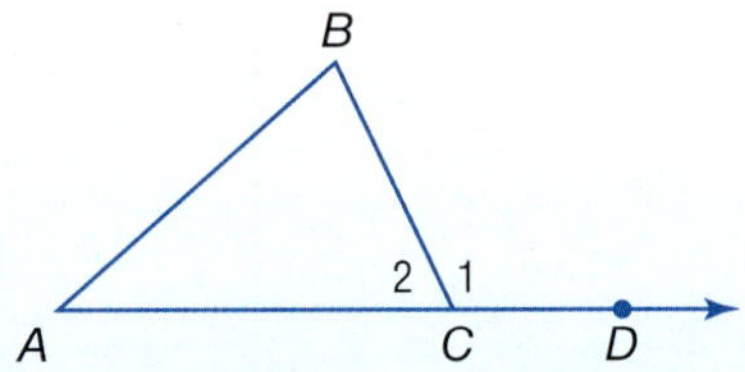

- $\triangle ABC$ — Given
- $\angle 2$ and $\angle 1$ form a linear pair. — Definition of a linear pair
- $\angle 2$ and $\angle 1$ are supplementary. — If 2 ∠s form a linear pair, they are supplementary.
- $m\angle A + m\angle B + m\angle 2 = 180$ — Triangle Angle-Sum Theorem
- $m\angle 2 + m\angle 1 = 180$ — Definition of supplementary
- $m\angle A + m\angle B + m\angle 2 = m\angle 2 + m\angle 1$ — Substitution
- $m\angle A + m\angle B = m\angle 1$ — Subtraction Property of Equality

StudyTip

Flow Proofs Flow proofs can be written vertically or horizontally.

The Exterior Angle Theorem can also be used to find missing measures.

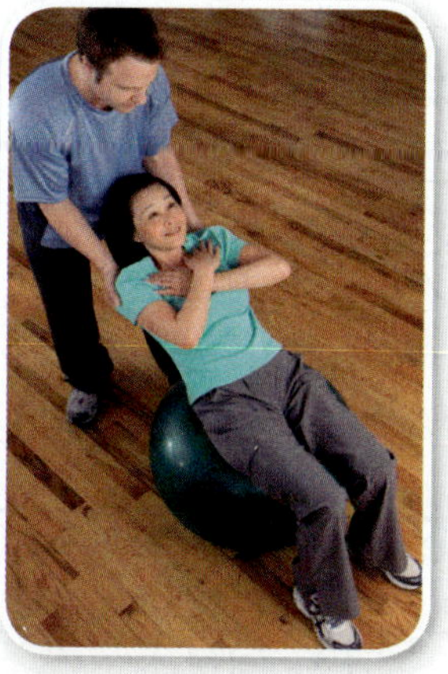

Real-World Career

Personal Trainer
Personal trainers instruct and motivate individuals in exercise activities. They demonstrate various exercises and help clients improve their exercise techniques. Personal trainers must obtain certification in the fitness field.

Real-World EXAMPLE 2 Use the Exterior Angle Theorem

FITNESS **Find the measure of $\angle JKL$ in the Triangle Pose shown.**

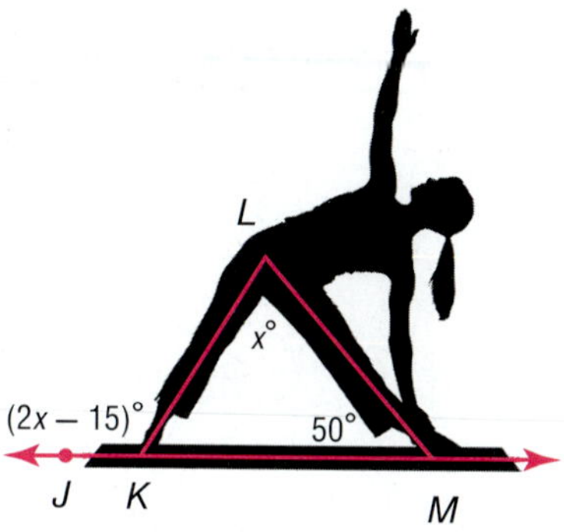

$m\angle KLM + m\angle LMK = m\angle JKL$	**Exterior Angle Theorem**
$x + 50 = 2x - 15$	**Substitution**
$50 = x - 15$	**Subtract *x* from each side.**
$65 = x$	**Add 15 to each side.**

So, $m\angle JKL = 2(65) - 15$ or 115.

Guided Practice

2. CLOSET ORGANIZING Tanya mounts the shelving bracket shown to the wall of her closet. What is the measure of $\angle 1$, the angle that the bracket makes with the wall?

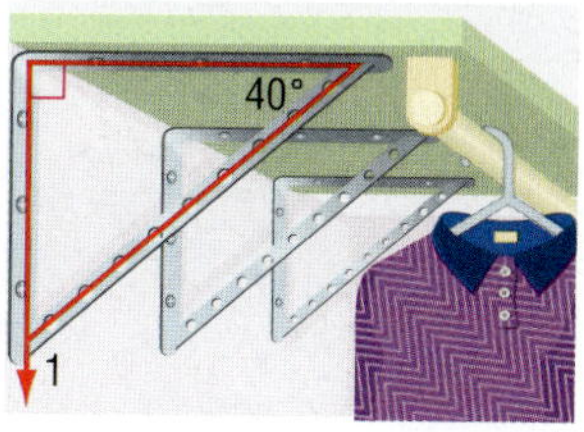

Personal Tutor glencoe.com

A **corollary** is a theorem with a proof that follows as a direct result of another theorem. As with a theorem, a corollary can be used as a reason in a proof. The corollaries below follow directly from the Triangle Angle-Sum Theorem.

Corollaries Triangle Angle-Sum Corollaries

4.1 The acute angles of a right triangle are complementary.

Abbreviation *Acute ∡ of a rt. △ are comp.*

Example If $\angle C$ is a right angle, then $\angle A$ and $\angle B$ are complementary.

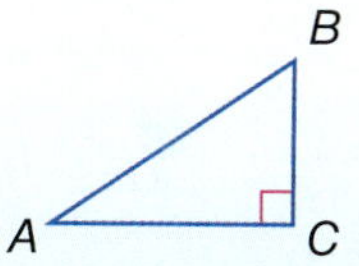

4.2 There can be at most one right or obtuse angle in a triangle.

Example If $\angle L$ is a right or an obtuse angle, then $\angle J$ and $\angle K$ must be acute angles.

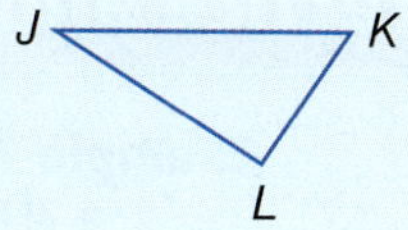

You will prove Corollaries 4.1 and 4.2 in Exercises 34 and 35.

StudyTip

Check for Reasonableness
When you are solving for the measure of one or more angles of a triangle, always check to make sure that the sum of the angle measures is 180.

EXAMPLE 3 Find Angle Measures in Right Triangles

Find the measures of each numbered angle.

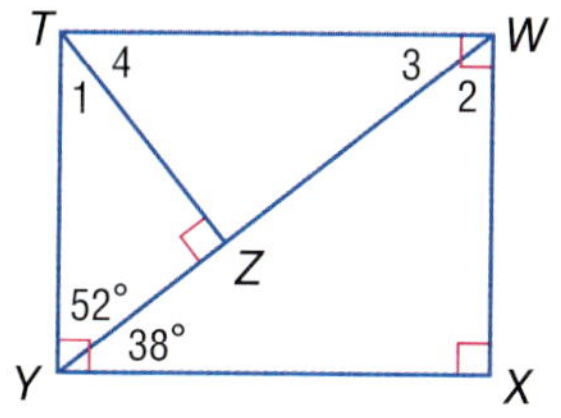

$m\angle 1 + m\angle TYZ = 90$	**Acute ∡ of a rt. △ are comp.**
$m\angle 1 + 52 = 90$	**Substitution**
$m\angle 1 = 38$	**Subtract 52 from each side.**

Guided Practice

3A. $\angle 2$ **3B.** $\angle 3$ **3C.** $\angle 4$

Personal Tutor glencoe.com

Check Your Understanding

Example 1 p. 245 — Find the measures of each numbered angle.

1. 

2. 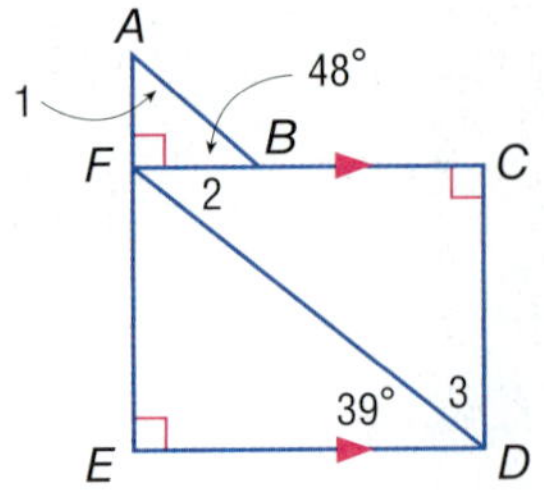

Example 2 p. 247 — Find each measure.

3. $m\angle 2$

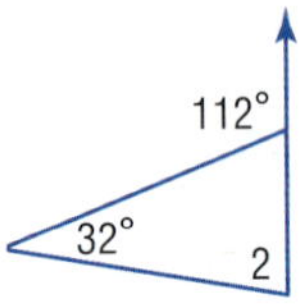

4. $m\angle MPQ$

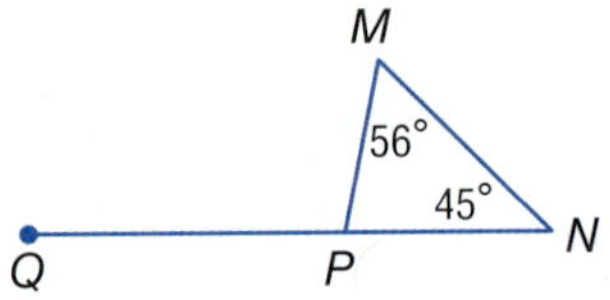

DECK CHAIRS The brace of this deck chair forms a triangle with the rest of the chair's frame as shown. If $m\angle 1 = 102$ and $m\angle 3 = 53$, find each measure.

5. $m\angle 4$

6. $m\angle 6$

7. $m\angle 2$

8. $m\angle 5$

Example 3 p. 247 — Find each measure.

9. $m\angle 1$

10. $m\angle 3$

11. $m\angle 2$

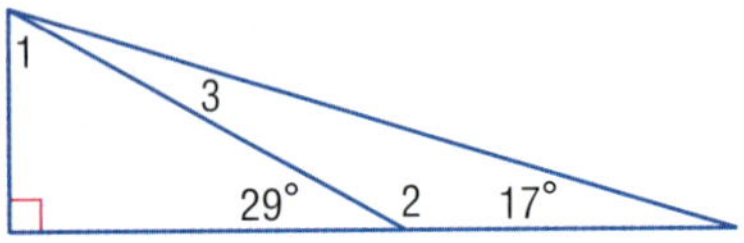

Practice and Problem Solving

● = **Step-by-Step Solutions** begin on page R20.
Extra Practice begins on page 969.

Example 1 p. 245 — Find the measure of each numbered angle.

12. 

13. 

14. 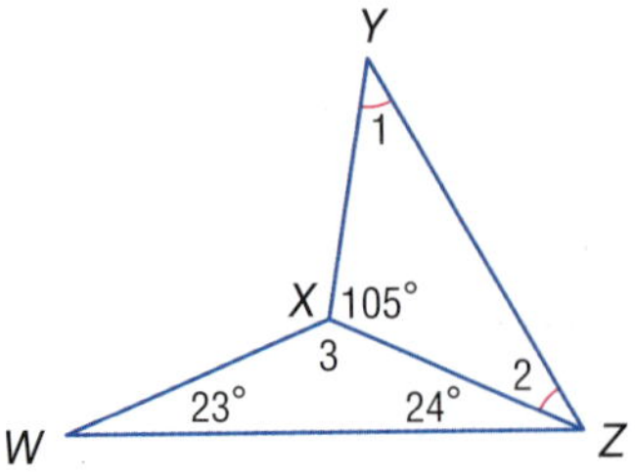

15 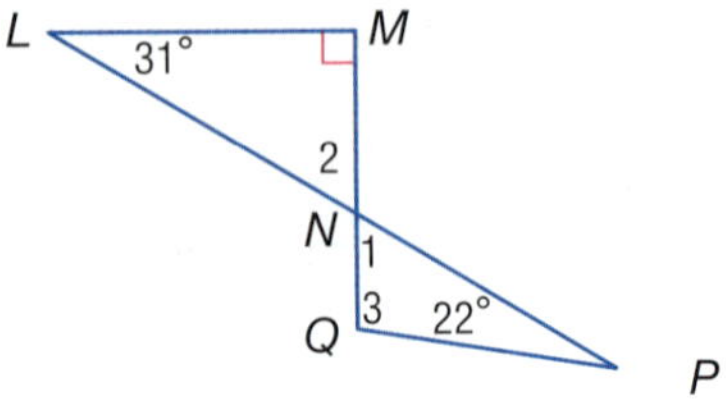

16. **AIRPLANES** The path of an airplane can be modeled using two sides of a triangle as shown. The distance covered during the plane's ascent is equal to the distance covered during its descent.

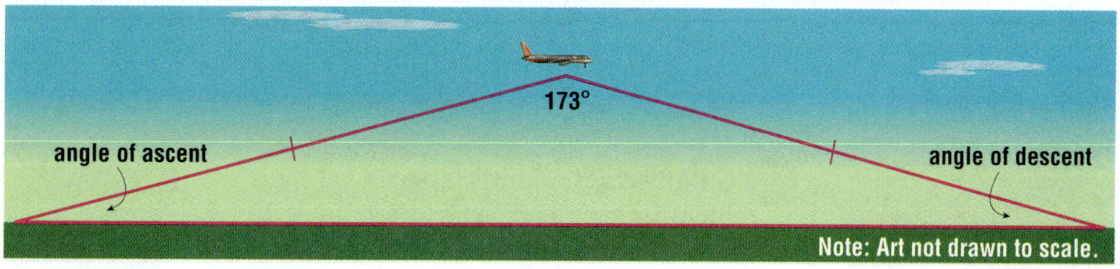

a. Classify the model using its sides and angles.

b. The angles of ascent and descent are congruent. Find their measures.

Example 2 p. 247

Find each measure.

17. $m\angle 1$

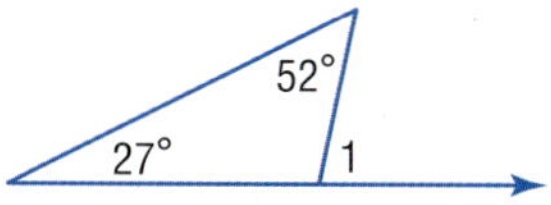

18. $m\angle 3$

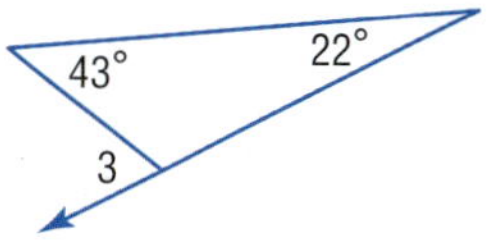

19. $m\angle 2$

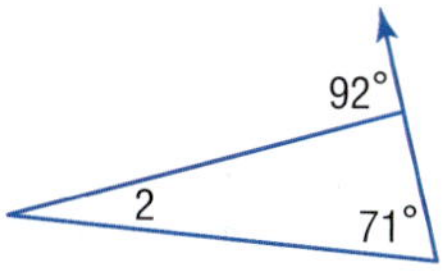

20. $m\angle 4$

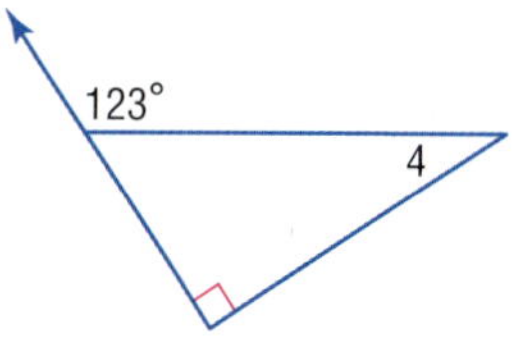

21. $m\angle ABC$

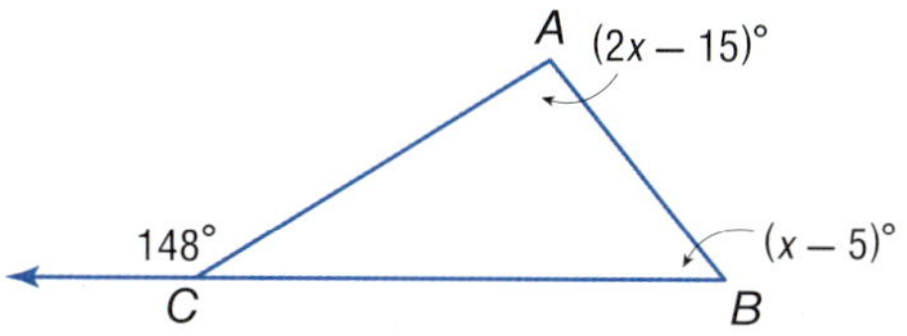

22. $m\angle JKL$

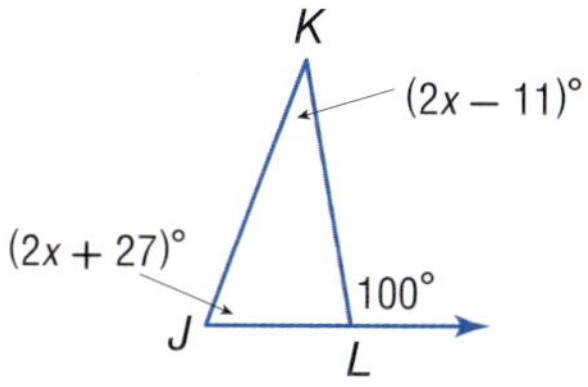

Example 3 p. 247

23. **WHEELCHAIR RAMP** Suppose the wheelchair ramp shown makes a 12° angle with the ground. What is the measure of the angle the ramp makes with the van door?

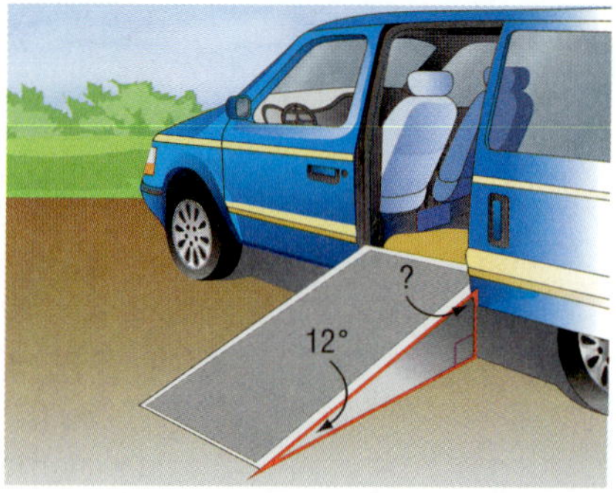

Find each measure.

24. $m\angle 1$

25. $m\angle 2$

26. $m\angle 3$

27. $m\angle 4$

28. $m\angle 5$

29. $m\angle 6$

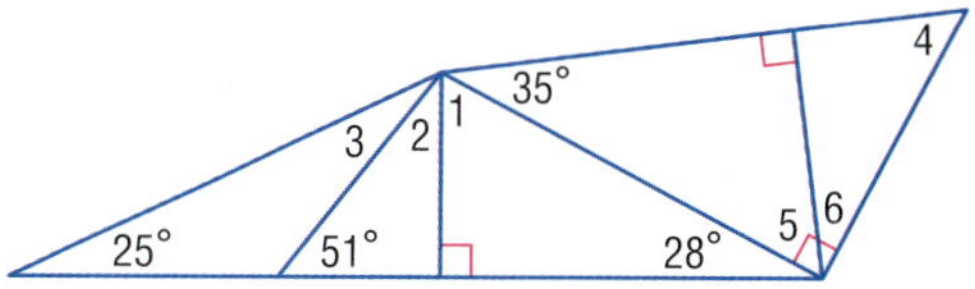

ALGEBRA **Find the value of x. Then find the measure of each angle.**

30.

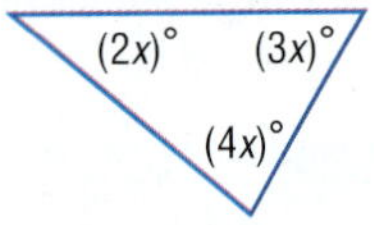

31.

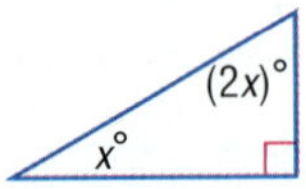

32.

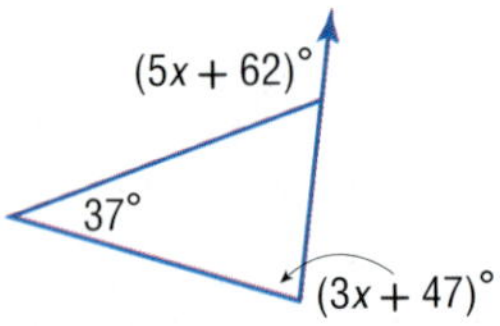

Real-World Link

Standard mums may be grown to produce a stem of 30 inches and are divided into 13 classes by their bloom forms.

Source: National Chrysanthemum Society

33. **GARDENING** A landscaper is forming an isosceles triangle in a flowerbed using chrysanthemums. She wants $m\angle A$ to be three times the measure of $\angle B$ and $\angle C$. What should the measure of each angle be?

PROOF **Write the specified type of proof.**

34. flow proof of Corollary 4.1

35. paragraph proof of Corollary 4.2

Find the measure of each numbered angle.

36.

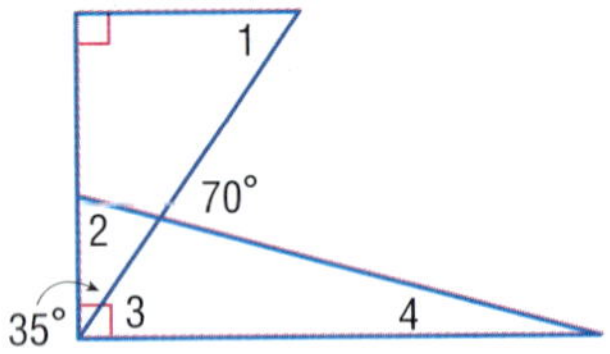

37.

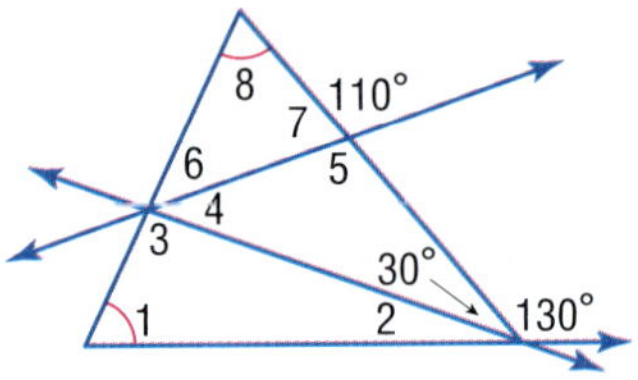

38. **ALGEBRA** Classify the triangle shown by its angles. Explain your reasoning.

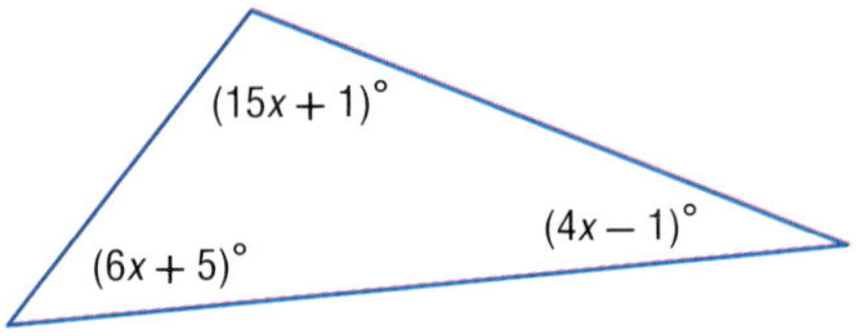

39. **ALGEBRA** The measure of the larger acute angle in a right triangle is two degrees less than three times the measure of the smaller acute angle. Find the measure of each angle.

40. Determine whether the following statement is *true* or *false*. If false, give a counterexample. If true, give an argument to support your conclusion.

If the sum of two acute angles of a triangle is greater than 90, then the triangle is acute.

41. **ALGEBRA** In $\triangle XYZ$, $m\angle X = 157$, $m\angle Y = y$, and $m\angle Z = z$. Write an inequality to describe the possible measures of $\angle Z$. Explain your reasoning.

42. **CARS** Refer to the photo at the right.

a. Find $m\angle 1$ and $m\angle 2$.

b. If the support for the hood were shorter than the one shown, how would $m\angle 1$ change? Explain.

c. If the support for the hood were shorter than the one shown, how would $m\angle 2$ change? Explain.

PROOF **Write the specified type of proof.**

43. two-column proof

Given: *RSTUV* is a pentagon.
Prove: $m\angle S + m\angle STU + m\angle TUV + m\angle V + m\angle VRS = 540$

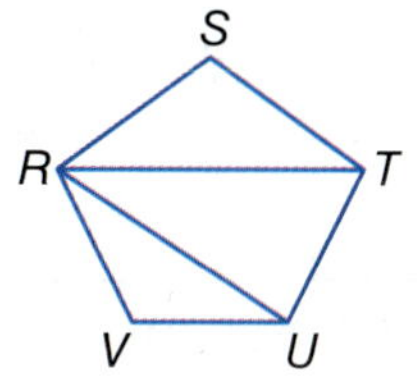

44. flow proof

Given: $\angle 3 \cong \angle 5$
Prove: $m\angle 1 + m\angle 2 = m\angle 6 + m\angle 7$

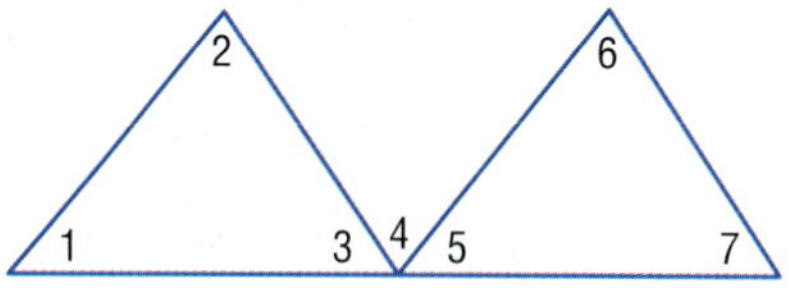

Watch Out!

Measuring Angles When measuring angles, remember to first align the 0 on either side of the protractor with one side of the angle. If the 0 you chose is on the outside scale, be sure to read the number on the outside scale where the other side of the angle intersects the protractor.

45. **MULTIPLE REPRESENTATIONS** In this problem, you will explore the sum of the measures of the exterior angles of a triangle.

a. **GEOMETRIC** Draw five different triangles, extending the sides and labeling the angles as shown. Be sure to include at least one obtuse, one right, and one acute triangle.

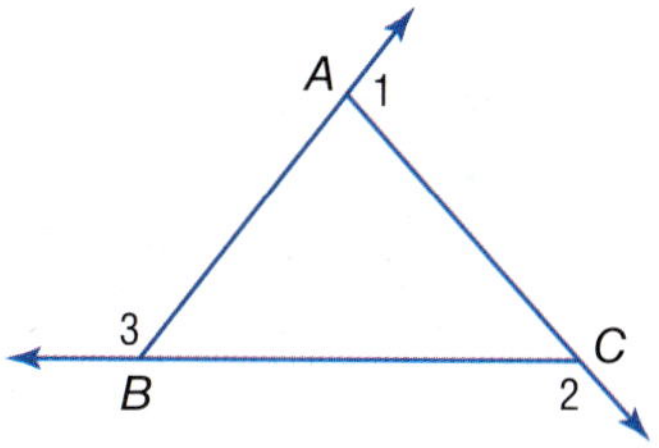

b. **TABULAR** Measure the exterior angles of each triangle. Record the measures for each triangle and the sum of these measures in a table.

c. **VERBAL** Make a conjecture about the sum of the exterior angles of a triangle. State your conjecture using words.

d. **ALGEBRAIC** State the conjecture you wrote in part **c** algebraically.

e. **ANALYTICAL** Write a paragraph proof of your conjecture.

H.O.T. Problems Use Higher-Order Thinking Skills

46. **ERROR ANALYSIS** Curtis measured and labeled the angles of the triangle as shown. Arnoldo says that at least one of his measures is incorrect. Explain in at least two different ways how Arnoldo knows that this is true.

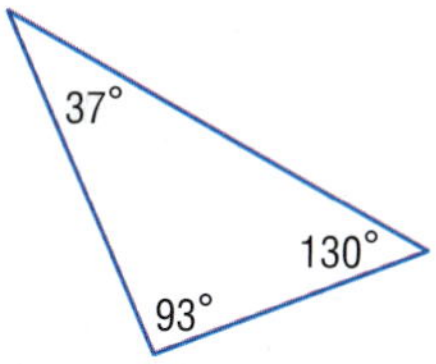

47. **WRITING IN MATH** Explain how you would find the missing measures in the figure shown.

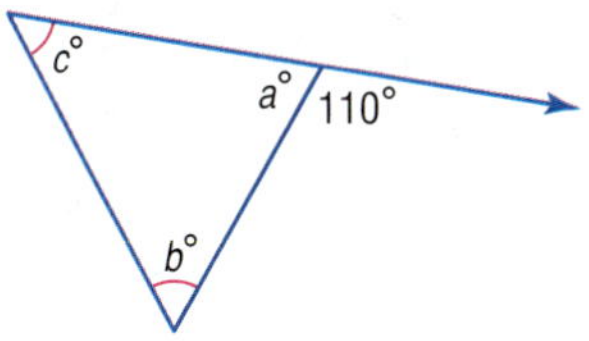

48. **OPEN ENDED** Construct a right triangle and measure one of the acute angles. Find the measure of the second acute angle using calculation and explain your method. Confirm your result using a protractor.

49. **CHALLENGE** Find the values of y and z in the figure at the right.

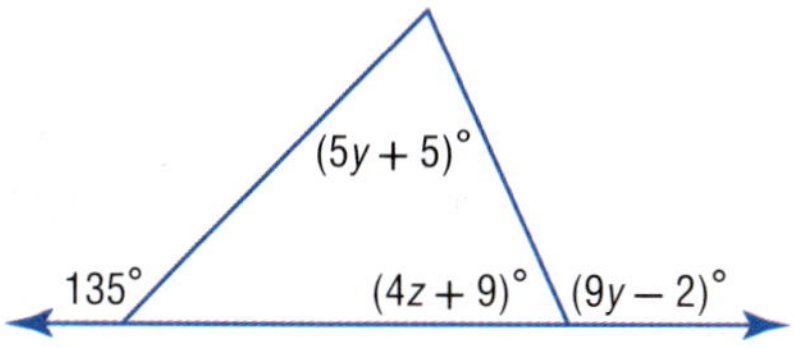

50. **REASONING** If an exterior angle adjacent to $\angle A$ is acute, is $\triangle ABC$ acute, right, obtuse, or can its classification not be determined? Explain your reasoning.

51. **WRITING IN MATH** Explain why a triangle cannot have an obtuse, acute, and a right exterior angle.

Standardized Test Practice

52. PROBABILITY Mr. Glover owns a video store and wants to survey his customers to find what type of movies he should buy. Which of the following options would be the best way for Mr. Glover to get accurate survey results?

A surveying customers who come in from 9 P.M. until 10 P.M.

B surveying customers who come in on the weekend

C surveying the male customers

D surveying at different times of the week and day

53. SHORT RESPONSE Two angles of a triangle have measures of 35° and 80°. Find the values of the exterior angle measures of the triangle.

54. ALGEBRA Which equation is equivalent to $7x - 3(2 - 5x) = 8x$?

F $2x - 6 = 8$

G $22x - 6 = 8x$

H $-8x - 6 = 8x$

J $22x + 6 = 8x$

55. SAT/ACT Joey has 4 more video games than Solana and half as many as Melissa. If together they have 24 video games, how many does Melissa have?

A 7

B 9

C 12

D 14

Spiral Review

Classify each triangle as *acute, equiangular, obtuse,* or *right.* (Lesson 4-1)

56.

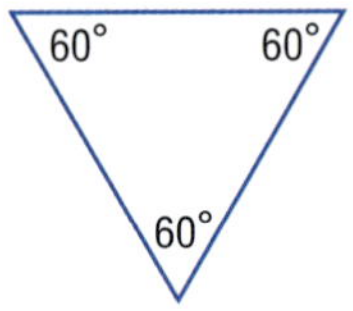

57.

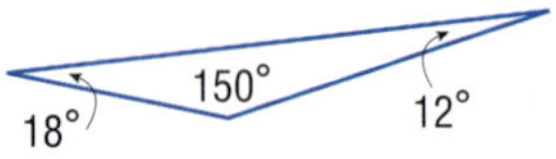

58.

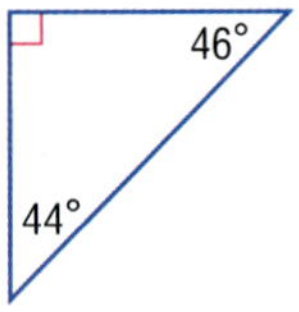

COORDINATE GEOMETRY **Find the distance from P to ℓ.** (Lesson 3-6)

59. Line ℓ contains points $(0, -2)$ and $(1, 3)$. Point P has coordinates $(-4, 4)$.

60. Line ℓ contains points $(-3, 0)$ and $(3, 0)$. Point P has coordinates $(4, 3)$.

Write a conjecture that describes the pattern in each sequence. Then use your conjecture to find the next item in the sequence. (Lesson 2-1)

61.

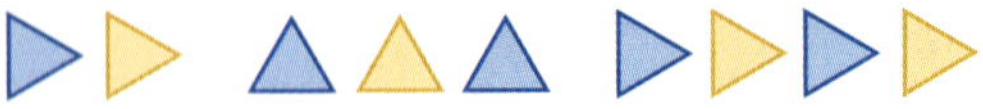

62.

Skills Review

State the property that justifies each statement. (Lesson 2-6)

63. If $\frac{x}{2} = 7$, then $x = 14$.

64. If $x = 5$ and $b = 5$, then $x = b$.

65. If $XY - AB = WZ - AB$, then $XY = WZ$.

66. If $m\angle A = m\angle B$ and $m\angle B = m\angle C$, $m\angle A = m\angle C$.

67. If $m\angle 1 + m\angle 2 = 90$ and $m\angle 2 = m\angle 3$, then $m\angle 1 + m\angle 3 = 90$.

4-3 Congruent Triangles

Then

You identified and used congruent angles. (Lesson 1-4)

Now

- Name and use corresponding parts of congruent polygons.
- Prove triangles congruent using the definition of congruence.

New Vocabulary

congruent
congruent polygons
corresponding parts

Why?

As an antitheft device, many manufacturers make car stereos with removable faceplates. The shape and size of the faceplate and of the space where it fits must be exactly the same for the faceplate to properly attach to the car's dashboard.

Congruence and Corresponding Parts

If two geometric figures have exactly the same shape and size, they are **congruent**.

Congruent	Not Congruent
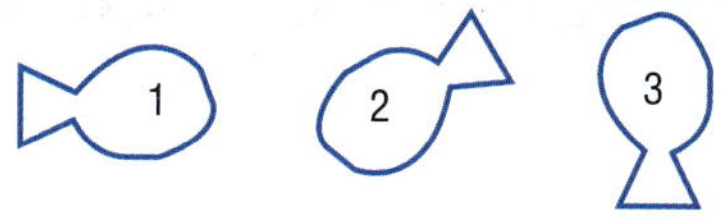	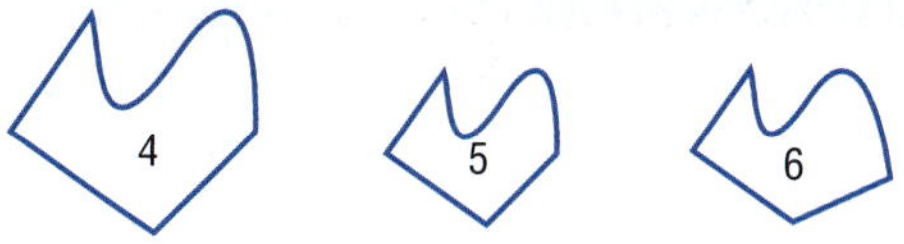
While positioned differently, Figures 1, 2, and 3 are exactly the same shape and size.	Figures 4 and 5 are exactly the same shape but not the same size. Figures 5 and 6 are the same size but not exactly the same shape.

In two **congruent polygons**, all of the parts of one polygon are congruent to the **corresponding parts** or matching parts of the other polygon. These corresponding parts include *corresponding angles* and *corresponding sides*.

Key Concept — Definition of Congruent Polygons

Words Two polygons are congruent if and only if their corresponding parts are congruent.

Example Corresponding Angles

$\angle A \cong \angle H$ $\angle B \cong \angle J$ $\angle C \cong \angle K$

Corresponding Sides

$\overline{AB} \cong \overline{HJ}$ $\overline{BC} \cong \overline{JK}$ $\overline{AC} \cong \overline{HK}$

Congruence Statement

$\triangle ABC \cong \triangle HJK$

Model

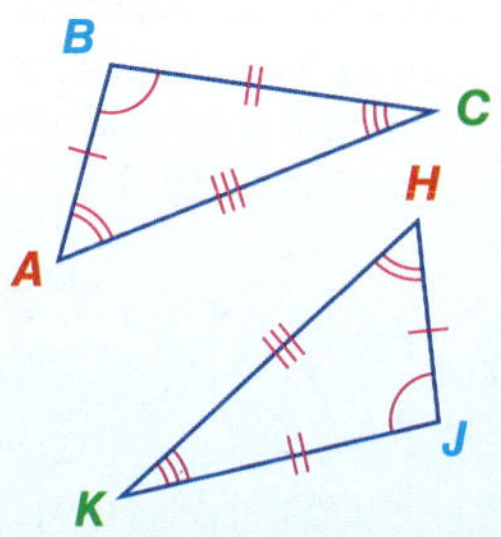

Other congruence statements for the triangles above exist. Valid congruence statements for congruent polygons list corresponding vertices in the same order.

Valid Statement	Not a Valid Statement
$\triangle BCA \cong \triangle JKH$	$\triangle ABC \cong \triangle HKJ$

Math History Link

Johann Carl Friedrich Gauss (1777–1855) Gauss developed the congruence symbol to show that two sides of an equation were the same even if they weren't equal. He made many advances in math and physics, including a proof of the fundamental theorem of algebra.

EXAMPLE 1 Identify Corresponding Congruent Parts

Show that the polygons are congruent by identifying all the congruent corresponding parts. Then write a congruence statement.

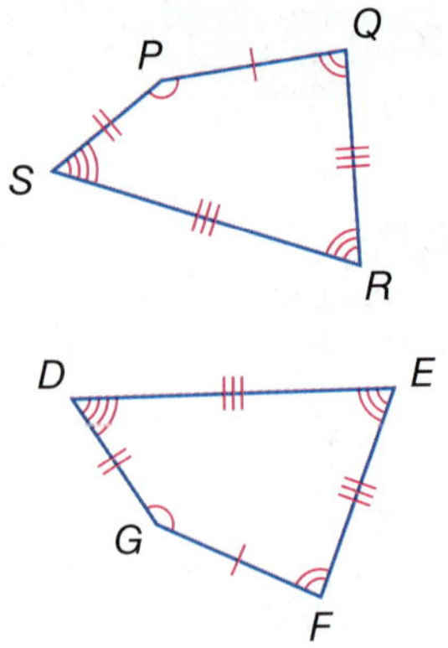

Angles: $\angle P \cong \angle G$, $\angle Q \cong \angle F$,
$\angle R \cong \angle E$, $\angle S \cong \angle D$

Sides: $\overline{PQ} \cong \overline{GF}$, $\overline{QR} \cong \overline{FE}$,
$\overline{RS} \cong \overline{ED}$, $\overline{SP} \cong \overline{DG}$

All corresponding parts of the two polygons are congruent. Therefore, polygon $PQRS \cong$ polygon $GFED$.

Guided Practice

1A.

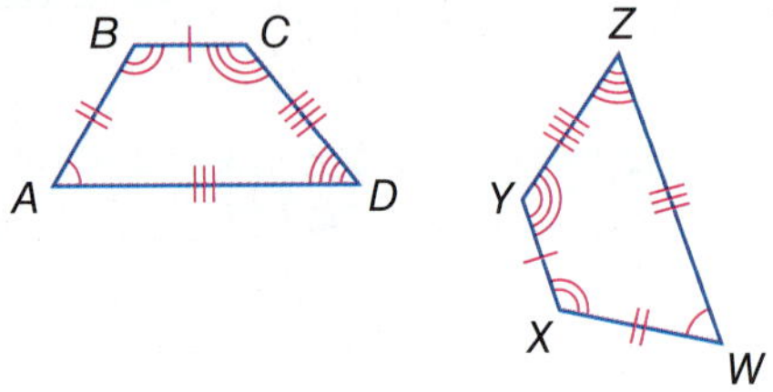

1B.

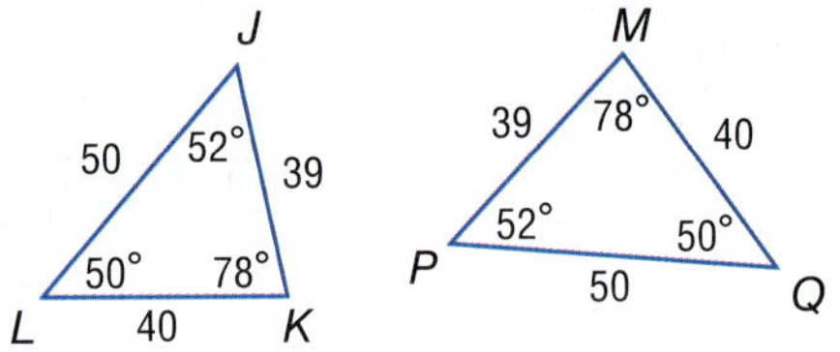

Personal Tutor glencoe.com

The phrase "if and only if" in the congruent polygon definition means that both the conditional and its converse are true. So, if two polygons are congruent, then their corresponding parts are congruent. For triangles, we say *Corresponding parts of congruent triangles are congruent*, or CPCTC.

EXAMPLE 2 Use Corresponding Parts of Congruent Triangles

In the diagram, $\triangle ABC \cong \triangle DFE$. Find the values of x and y.

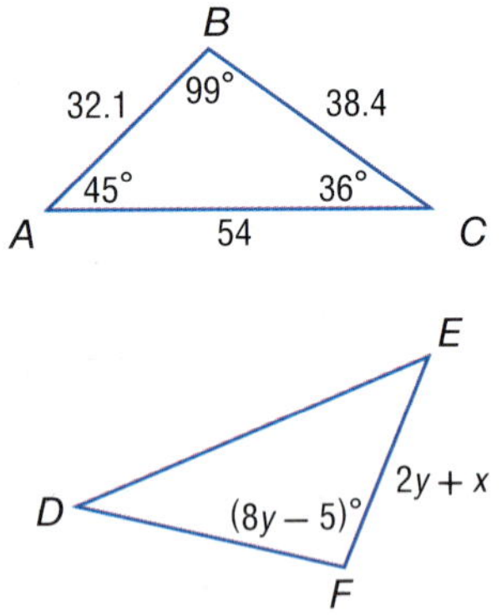

$\angle F \cong \angle B$	**CPCTC**
$m\angle F = m\angle B$	**Definition of congruence**
$8y - 5 = 99$	**Substitution**
$8y = 104$	**Add 5 to each side.**
$y = 13$	**Divide each side by 8.**
$\overline{FE} \cong \overline{BC}$	**CPCTC**
$FE = BC$	**Definition of congruence**
$2y + x = 38.4$	**Substitution**
$2(13) + x = 38.4$	**Substitution**
$26 + x = 38.4$	**Simplify.**
$x = 12.4$	**Subtract 26 from each side.**

StudyTip

Using a Congruence Statement You can use a congruence statement to help you correctly identify corresponding sides.
$\triangle ABC \cong \triangle DFE$
$\overline{BC} \cong \overline{FE}$

Guided Practice

2. In the diagram, $\triangle RSV \cong \triangle TVS$. Find the values of x and y.

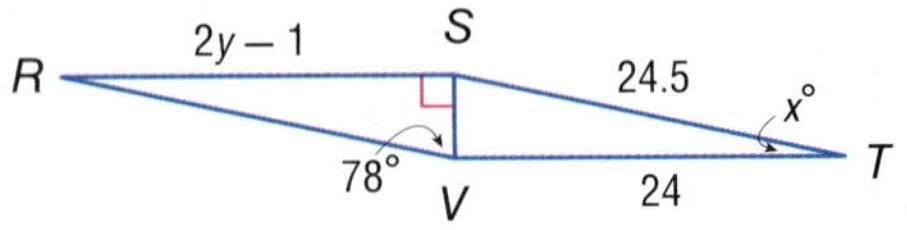

Personal Tutor glencoe.com

Prove Triangles Congruent The Triangle Angle-Sum Theorem you learned in Lesson 4-2 leads to another theorem about the angles in two triangles.

Theorem 4.3 — Third Angles Theorem

Words If two angles of one triangle are congruent to two angles of a second triangle, then the third angles of the triangles are congruent.

Example If $\angle C \cong \angle K$ and $\angle B \cong \angle J$, then $\angle A \cong \angle L$.

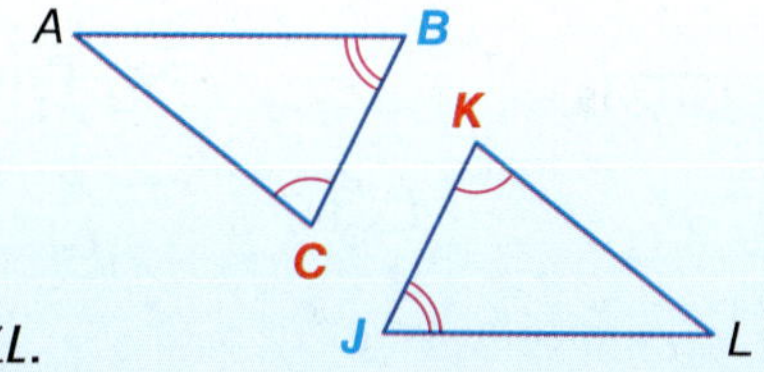

You will prove this theorem in Exercise 21.

Real-World Link

Using some basic skills with napkin folding can add an elegant touch to any party. Many of the folds use triangles.

Real-World EXAMPLE 3 Use the Third Angles Theorem

PARTY PLANNING The planners of the Senior Banquet decide to fold the dinner napkins using the Triangle Pocket Fold so that they can place a small gift in the pocket. If $\angle NPQ \cong \angle RST$, and $m\angle NPQ = 40$, find $m\angle SRT$.

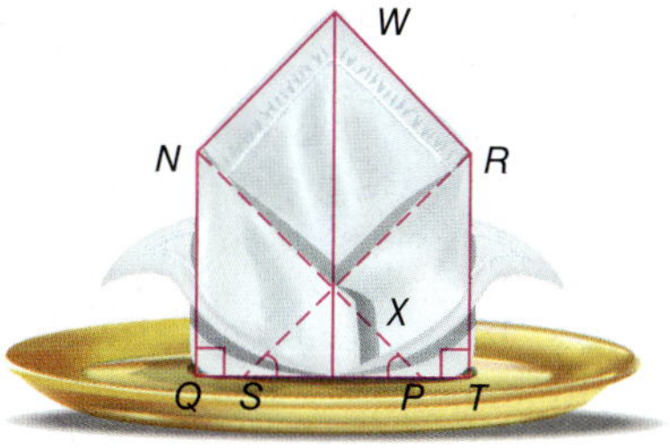

$\angle NPQ \cong \angle RST$, and since all right angles are congruent, $\angle NQP \cong \angle RTS$. So by the Third Angles Theorem, $\angle QNP \cong \angle SRT$. By the definition of congruence, $m\angle QNP = m\angle TRS$.

$m\angle QNP + m\angle NPQ = 90$ — The acute angles of a right triangle are complementary.

$m\angle QNP + 40 = 90$ — Substitution

$m\angle QNP = 50$ — Subtract 40 from each side.

By substitution, $m\angle SRT = m\angle QNP$ or 50.

Guided Practice

3. In the diagram above, if $\angle WNX \cong \angle WRX$, $\overline{WX}$ bisects $\angle NXR$, $m\angle WNX = 88$, and $m\angle NXW = 49$, find $m\angle NWR$. Explain your reasoning.

Personal Tutor glencoe.com

EXAMPLE 4 Prove That Two Triangles are Congruent

Write a two-column proof.

Given: $\overline{DE} \cong \overline{GE}$, $\overline{DF} \cong \overline{GF}$, $\angle D \cong \angle G$, $\angle DFE \cong \angle GFE$

Prove: $\triangle DEF \cong \triangle GEF$

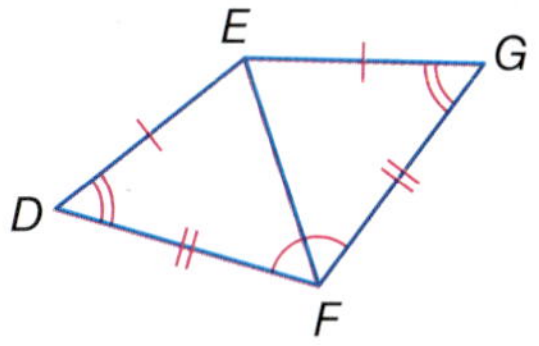

Proof:

Statements	Reasons
1. $\overline{DE} \cong \overline{GE}$, $\overline{DF} \cong \overline{GF}$	**1.** Given
2. $\overline{EF} \cong \overline{EF}$	**2.** Reflexive Property of Congruence
3. $\angle D \cong \angle G$, $\angle DFE \cong \angle GFE$	**3.** Given
4. $\angle DEF \cong \angle GEF$	**4.** Third Angles Theorem
5. $\triangle DEF \cong \triangle GEF$	**5.** Definition of Congruent Polygons

StudyTip

Reflexive Property When two triangles share a common side, use the Reflexive Property of Congruence to establish that the common side is congruent to itself.

Guided Practice

4. Write a two column proof.

Given: $\angle J \cong \angle P$, $\overline{JK} \cong \overline{PM}$, $\overline{JL} \cong \overline{PL}$, and L bisects $\overline{KM}$.

Prove: $\triangle JLK \cong \triangle PLM$

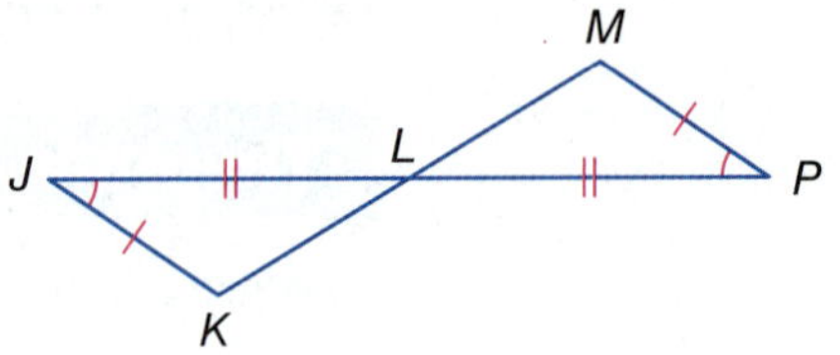

Personal Tutor glencoe.com

Like congruence of segments and angles, congruence of triangles is reflexive, symmetric, and transitive.

Theorem 4.4 Properties of Triangle Congruence

Reflexive Property of Triangle Congruence

$\triangle ABC \cong \triangle ABC$

Symmetric Property of Triangle Congruence

If $\triangle ABC \cong \triangle EFG$, then $\triangle EFG \cong \triangle ABC$.

Transitive Property of Triangle Congruence

If $\triangle ABC \cong \triangle EFG$ and $\triangle EFG \cong \triangle JKL$, then $\triangle ABC \cong \triangle JKL$.

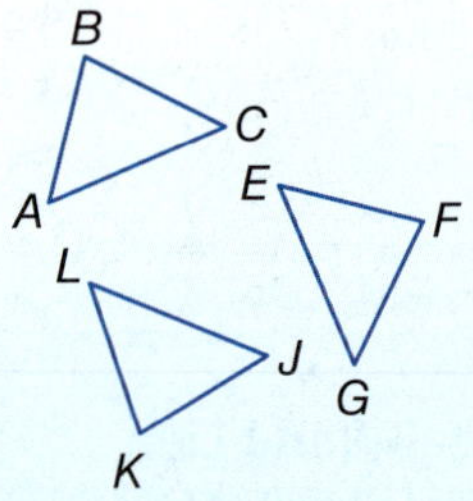

You will prove the reflexive, symmetric, and transitive parts of Theorem 4.4 in Exercises 22, 26, and 27, respectively.

Check Your Understanding

Example 1 p. 254

Show that polygons are congruent by identifying all congruent corresponding parts. Then write a congruence statement.

1.

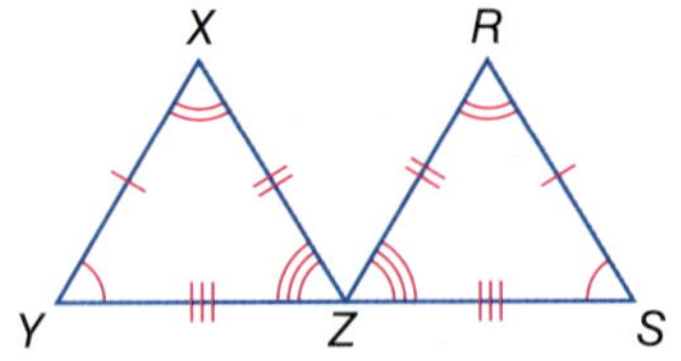

2.

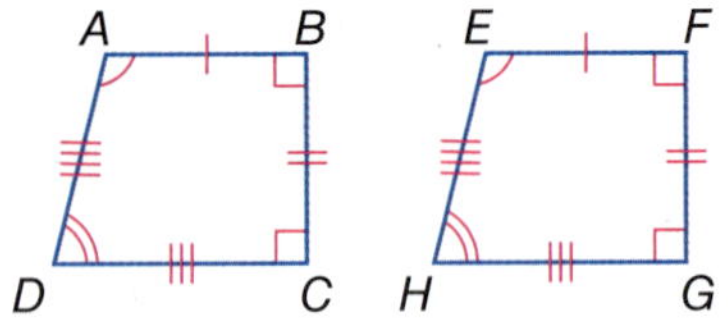

3. TOOLS Sareeta is changing the tire on her bike and the nut securing the tire looks like the one shown. Which of the sockets below should she use with her wrench to remove the tire? Explain your reasoning.

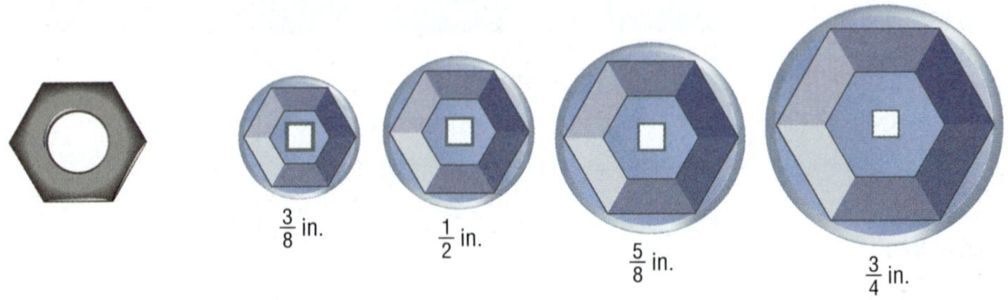

Example 2
p. 254

In the figure, $\triangle LMN \cong \triangle QRS$.

4. Find x.

5 Find y.

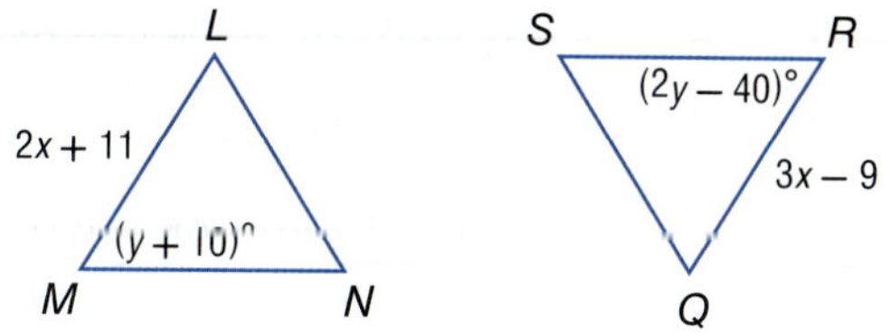

Example 3
p. 255

Find x. Explain your reasoning.

6.

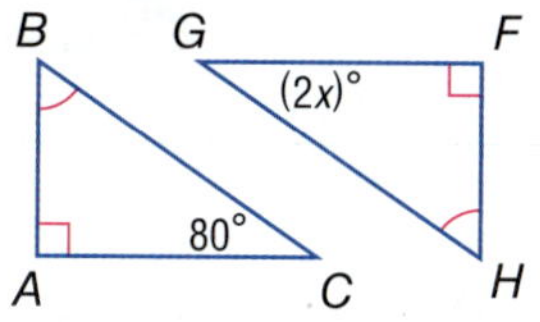

7.

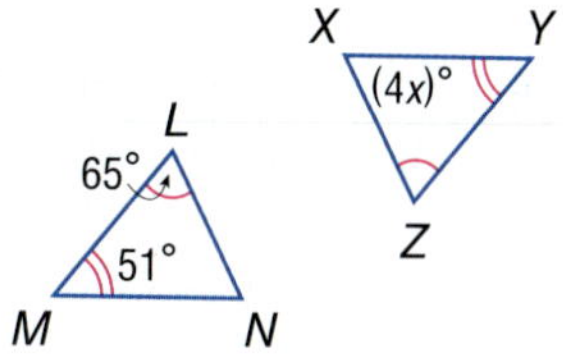

Example 4
p. 255

8. PROOF Write a paragraph proof.

Given: $\angle WXZ \cong \angle YXZ$, $\angle XZW \cong \angle XZY$, $\overline{WX} \cong \overline{YX}$, $\overline{WZ} \cong \overline{YZ}$

Prove: $\triangle WXZ \cong \triangle YXZ$

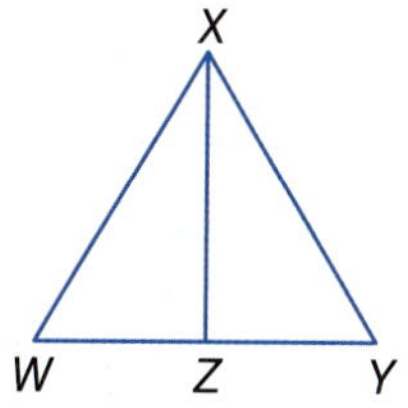

Practice and Problem Solving

● = **Step-by-Step Solutions** begin on page R20.
Extra Practice begins on page 969.

Example 1
p. 254

Show that polygons are congruent by identifying all congruent corresponding parts. Then write a congruence statement.

9.

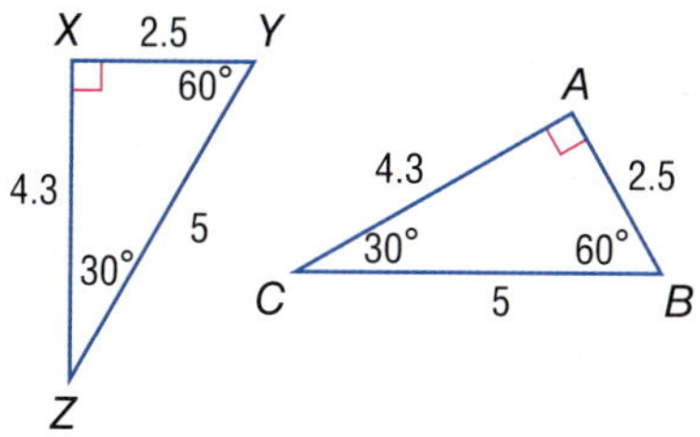

10.

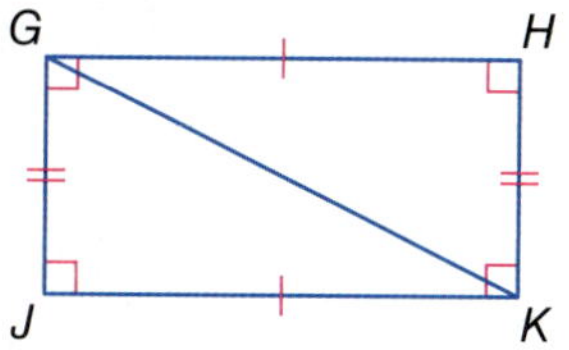

11.

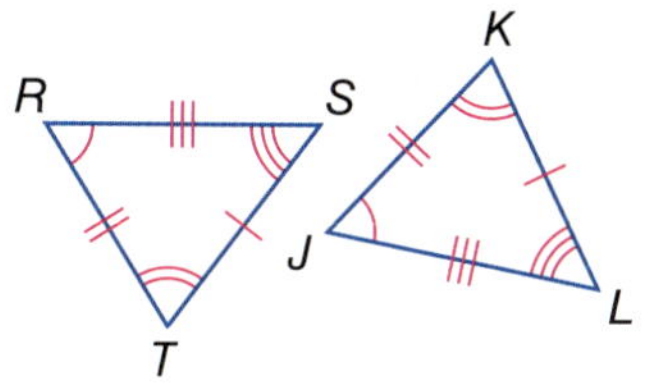

12.

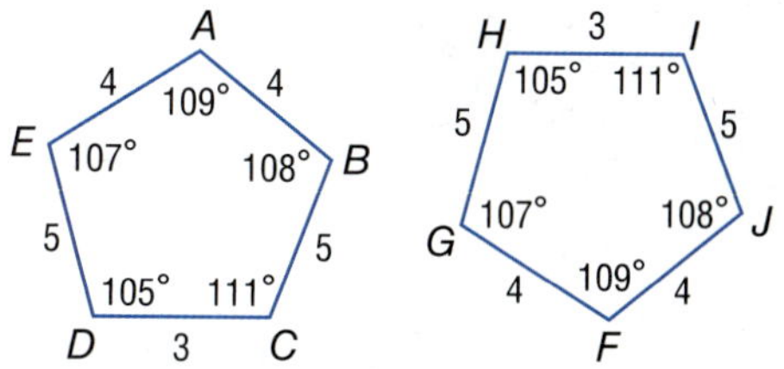

Example 2
p. 254

Polygon $BCDE \cong$ polygon $RSTU$. Find each value.

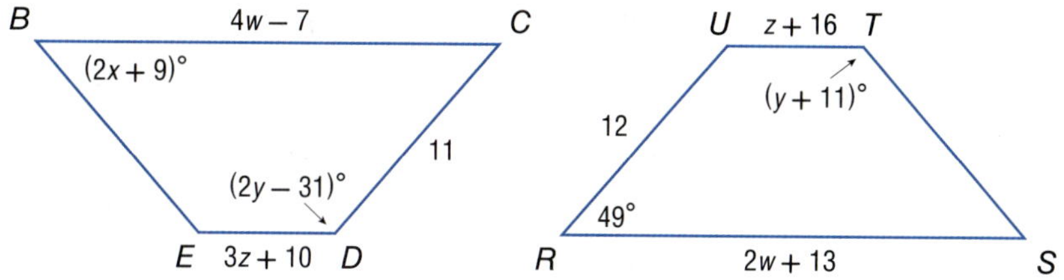

13. x

14. y

15 z

16. w

17. **SAILING** To ensure that sailboat races are fair, the boats and their sails are required to be the same size and shape.

a. Write a congruence statement relating the triangles in the photo.

b. Name six pairs of congruent segments.

c. Name six pairs of congruent angles.

Example 3
p. 255

Find x and y.

18.

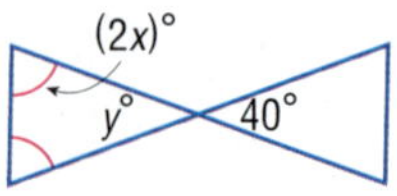

19.

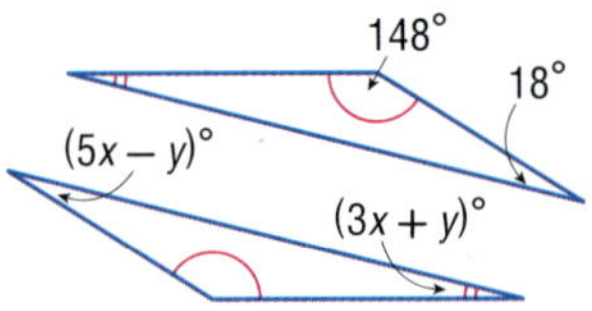

20.

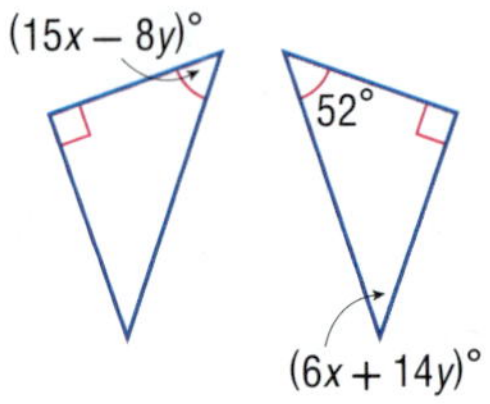

Example 4
p. 255

21. **PROOF** Write a two-column proof of Theorem 4.3.

22. **PROOF** Put the statements used to prove the statement below in the correct order. Provide the reasons for each statement.

Congruence of triangles is symmetric. (Theorem 4.4)

Given: $\triangle RST \cong \triangle XYZ$

Prove: $\triangle XYZ \cong \triangle RST$

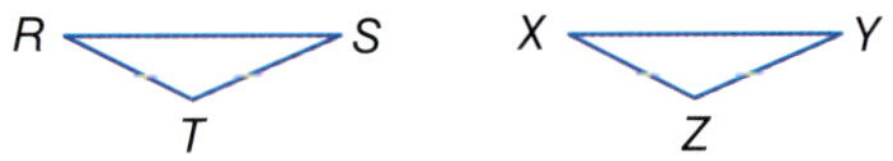

Proof:

$\angle X \cong \angle R$, $\angle Y \cong \angle S$, $\angle Z \cong \angle T$, $\overline{XY} \cong \overline{RS}$, $\overline{YZ} \cong \overline{ST}$, $\overline{XZ} \cong \overline{RT}$	$\angle R \cong \angle X$, $\angle S \cong \angle Y$, $\angle T \cong \angle Z$, $\overline{RS} \cong \overline{XY}$, $\overline{ST} \cong \overline{YZ}$, $\overline{RT} \cong \overline{XZ}$	$\triangle RST \cong \triangle XYZ$	$\triangle XYZ \cong \triangle RST$
?	?	?	?

PROOF **Write a two-column proof.**

23. **Given:** $\overline{BD}$ bisects $\angle B$.
$\overline{BD} \perp \overline{AC}$

Prove: $\angle A \cong \angle C$

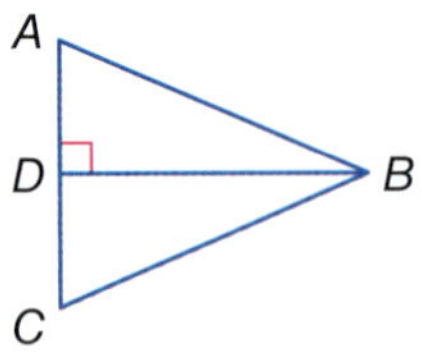

24. **Given:** $\angle P \cong \angle T$, $\angle S \cong \angle Q$
$\overline{TR} \cong \overline{PR}$, $\overline{RP} \cong \overline{RQ}$,
$\overline{RT} \cong \overline{RS}$
$\overline{PQ} \cong \overline{TS}$

Prove: $\triangle PRQ \cong \triangle TRS$

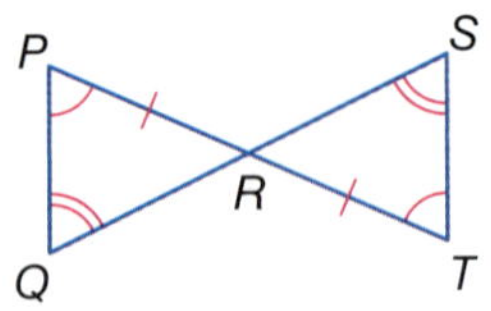

Real-World Link

Scrapbooking and other memory crafts were the most popular selling crafts in 2006, accounting for about $2.6 billion in sales.

Source: Craft & Hobby Association

25. **SCRAPBOOKING** Lanie is using a flower-shaped corner decoration punch for a scrapbook she is working on. If she punches the corners of two pages as shown, what property guarantees that the punched designs are congruent? Explain.

PROOF **Write the specified type of proof of the indicated part of Theorem 4.4.**

26. Congruence of triangles is transitive. (paragraph proof)

27. Congruence of triangles is reflexive. (flow proof)

ALGEBRA **Draw and label a figure to represent the congruent triangles. Then find x and y.**

28. $\triangle ABC \cong \triangle DEF$, $AB = 7$, $BC = 9$, $AC = 11 + x$, $DF = 3x - 13$, and $DE = 2y - 5$

29. $\triangle LMN \cong \triangle RST$, $m\angle L = 49$, $m\angle M = 10y$, $m\angle S = 70$, and $m\angle T = 4x + 9$

30. $\triangle JKL \cong \triangle MNP$, $JK = 12$, $LJ = 5$, $PM = 2x - 3$, $m\angle L = 67$, $m\angle K = y + 4$ and $m\angle N = 2y - 15$

31. **PENNANTS** Scott is in charge of roping off an area of 100 square feet for the band to use during a pep rally. He is using a string of pennants that are congruent isosceles triangles.

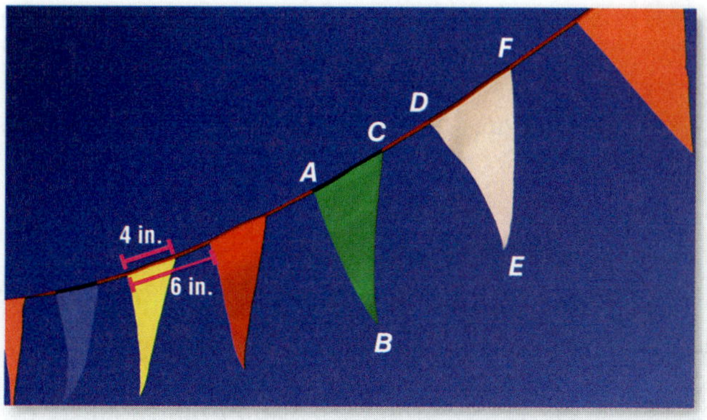

a. List seven pairs of congruent segments in the photo.

b. If the area he ropes off for the band is a square, how long will the pennant string need to be?

c. How many pennants will be on the string?

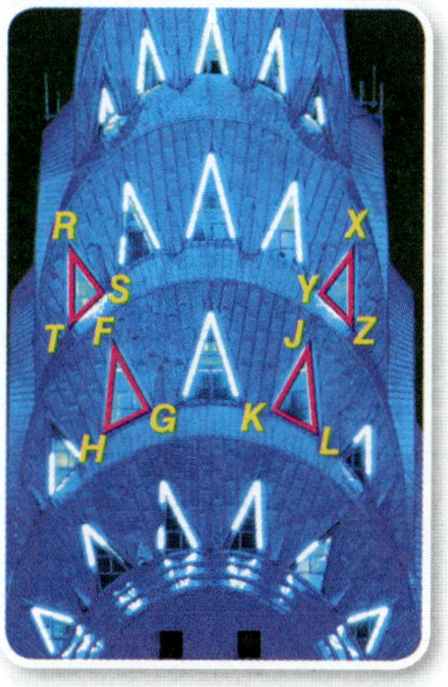

Real-World Link

The art deco design of the Chrysler Building in New York City makes extensive use of geometric figures.

32. **ARCHITECTURE** In the photo of the Chrysler Building at the left, $\overline{TS} \cong \overline{ZY}$, $\overline{XY} \cong \overline{RS}$, $\overline{TR} \cong \overline{ZX}$, $\angle X \cong \angle R$, $\angle T \cong \angle Z$, $\angle Y \cong \angle S$, and $\triangle HGF \cong \triangle LKJ$.

a. Which triangle, if any, is congruent to $\triangle YXZ$? Explain your reasoning.

b. Which side(s) are congruent to $\overline{JL}$? Explain your reasoning.

c. Which angle(s) are congruent to $\angle G$? Explain your reasoning.

33. **MULTIPLE REPRESENTATIONS** In this problem, you will explore the following statement.

The areas of congruent triangles are equal.

a. **VERBAL** Write a conditional statement to represent the relationship between the areas of a pair of congruent triangles.

b. **VERBAL** Write the converse of your conditional statement. Is the converse *true* or *false*? Explain your reasoning.

c. **GEOMETRIC** If possible, draw two equilateral triangles that have the same area but are not congruent. If not possible, explain why not.

d. **GEOMETRIC** If possible, draw two rectangles that have the same area but are not congruent. If not possible, explain why not.

e. **GEOMETRIC** If possible, draw two squares that have the same area but are not congruent. If not possible, explain why not.

f. **VERBAL** For which polygons will the following conditional and its converse both be true? Explain your reasoning.

If a pair of __________ are congruent, then they have the same area.

34. PATTERNS The pattern shown is created using regular polygons.

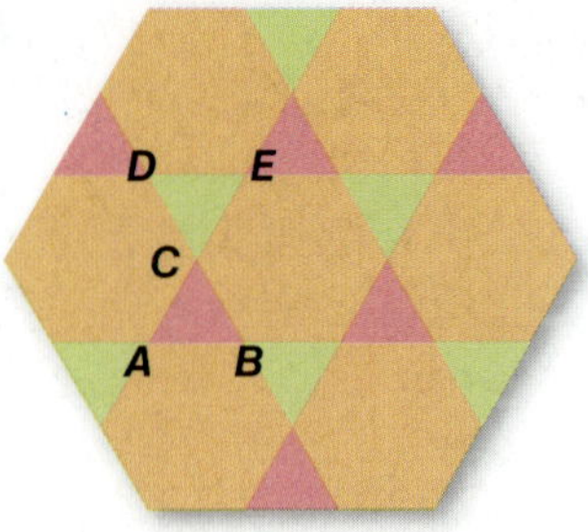

a. What two polygons are used to create the pattern?

b. Name a pair of congruent triangles.

c. Name a pair of corresponding angles.

d. If $CB = 2$ inches, what is AE? Explain.

e. What is the measure of $\angle D$? Explain.

Real-World Link

The longest verified time of continuous fitness hoop twirling is 72 hours set in October of 1984.

Source: *Guinness Book of World Records*

35. FITNESS A fitness instructor is starting a new aerobics class using fitness hoops. She wants to confirm that all of the hoops are the same size. What measure(s) can she use to prove that all of the hoops are congruent? Explain your reasoning.

H.O.T. Problems Use Higher-Order Thinking Skills

36. WRITING IN MATH Explain why the order of the vertices is important when naming congruent triangles. Give an example to support your answer.

37. ERROR ANALYSIS Jasmine and Will are evaluating the congruent figures below. Jasmine says that $\triangle CAB \cong \triangle ZYX$ and Will says that $\triangle ABC \cong \triangle YXZ$. Is either of them correct? Explain.

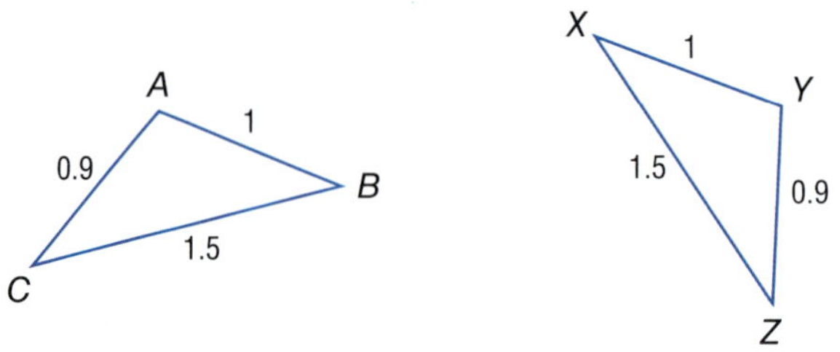

38. WRITE A QUESTION A classmate is using the Third Angles Theorem to show that if 2 corresponding pairs of the angles of two triangles are congruent, then the third pair is also congruent. Write a question to help him decide if he can use the same strategy for quadrilaterals.

39. CHALLENGE Find x and y if $\triangle PQS \cong \triangle RQS$.

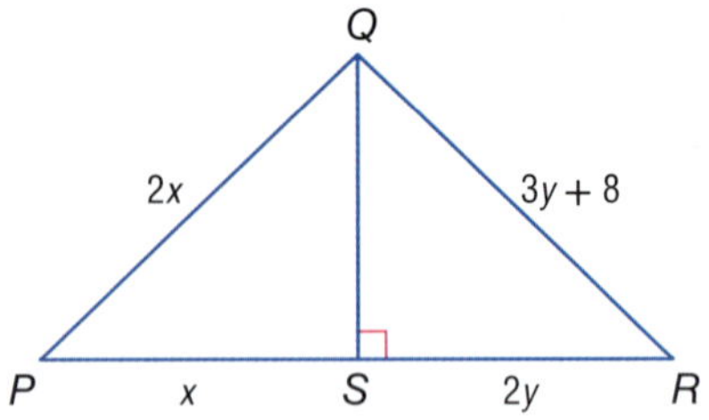

REASONING **Determine whether each statement is *true* or *false*. If false, give a counterexample. If true, explain your reasoning.**

40. Two triangles with two pairs of congruent corresponding angles and three pairs of congruent corresponding sides are congruent.

41. Two triangles with three pairs of corresponding congruent angles are congruent.

42. CHALLENGE Write a paragraph proof to prove polygon $ABED \cong$ polygon $FEBC$.

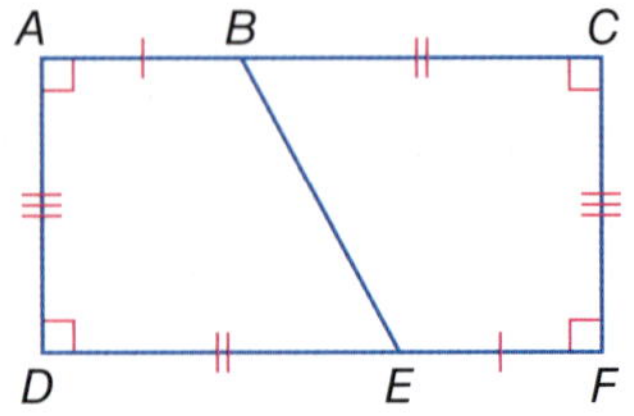

43. WRITING IN MATH Determine whether the following statement is *always*, *sometimes*, or *never* true. Explain your reasoning.

Equilateral triangles are congruent.

Standardized Test Practice

44. Barrington cut four congruent triangles off the corners of a rectangle to make an octagon as shown below. What is the area of the octagon?

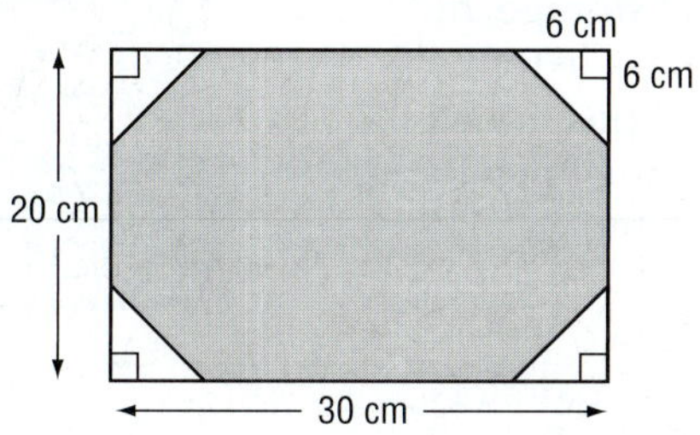

A 456 cm^2
B 528 cm^2
C 552 cm^2
D 564 cm^2

45. GRIDDED RESPONSE Triangle ABC is congruent to $\triangle HIJ$. The vertices of $\triangle ABC$ are $A(-1, 2)$, $B(0, 3)$ and $C(2, -2)$. What is the measure of side $\overline{HJ}$?

46. ALGEBRA Which is a factor of $x^2 + 19x - 42$?

F $x + 14$
G $x + 2$
H $x - 2$
J $x - 14$

47. SAT/ACT Mitsu travels a certain distance at 30 miles per hour and returns the same route at 65 miles per hour. What is his average speed in miles per hour for the round trip?

A 35.0
B 41.0
C 47.5
D 55.3

Spiral Review

Find each measure. (Lesson 4-2)

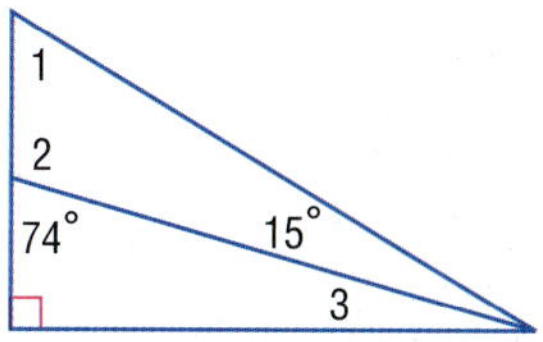

48. $m\angle 2$

49. $m\angle 1$

50. $m\angle 3$

COORDINATE GEOMETRY Find the measures of the sides of $\triangle JKL$ and classify each triangle by the measures of its sides. (Lesson 4-1)

51. $J(-7, 10)$, $K(15, 0)$, $L(-2, -1)$

52. $J(9, 9)$, $K(12, 14)$, $L(14, 6)$

53. $J(4, 6)$, $K(4, 11)$, $L(9, 6)$

54. $J(16, 14)$, $K(7, 6)$, $L(-5, -14)$

Determine whether each statement is *always, sometimes,* or *never* true. (Lesson 1-5)

55. Two angles that form a linear pair are supplementary.

56. If two angles are supplementary, then one of the angles is obtuse.

57. CARPENTRY A carpenter must cut two pieces of wood at angles so that they fit together to form the corner of a picture frame. What type of angles must he use to make sure that a 90° corner results? (Lesson 1-5)

Skills Review

58. Copy and complete the proof. (Lesson 2-7)

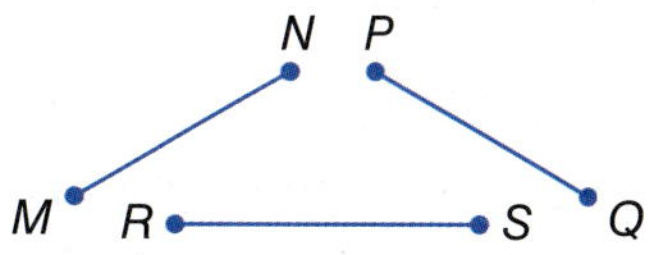

Given: $\overline{MN} \cong \overline{PQ}$, $\overline{PQ} \cong \overline{RS}$
Prove: $\overline{MN} \cong \overline{RS}$
Proof:

Statements	Reasons
a. ___?___	**a.** Given
b. $MN = PQ$, $PQ = RS$	**b.** ___?___
c. ___?___	**c.** ___?___
d. $\overline{MN} \cong \overline{RS}$	**d.** Definition of congruent segments

4-4 Proving Triangles Congruent—SSS, SAS

Then
You proved triangles congruent using the definition of congruence. (Lesson 4-3)

Now
- Use the SSS Postulate to test for triangle congruence.
- Use the SAS Postulate to test for triangle congruence.

New Vocabulary
included angle

Math Online
glencoe.com

Why?

An A-frame sandwich board is a convenient way to display information. Not only does it fold flat for easy storage, but with each sidearm locked into place, the frame is extremely sturdy. With the sidearms the same length and positioned the same distance from the top on either side, the open frame forms two congruent triangles—$\triangle ABC$ and $\triangle XYZ$.

SSS Postulate In Lesson 4-3, you proved that two triangles were congruent by showing that all six pairs of corresponding parts were congruent. It is possible to prove two triangles congruent using fewer pairs.

The sandwich board demonstrates that if two triangles have the same three side lengths, then they are congruent. This is expressed in the postulate below.

Postulate 4.1 Side-Side-Side (SSS) Congruence

If three sides of one triangle are congruent to three sides of a second triangle, then the triangles are congruent.

Example If **S**ide $\overline{AB} \cong \overline{DE}$,
Side $\overline{BC} \cong \overline{EF}$, and
Side $\overline{AC} \cong \overline{DF}$,
then $\triangle ABC \cong \triangle DEF$.

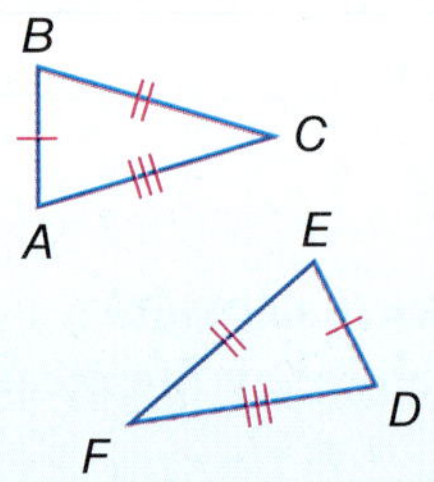

EXAMPLE 1 Use SSS to Prove Triangles Congruent

Write a flow proof.

Given: $\overline{GH} \cong \overline{KJ}$, $\overline{HL} \cong \overline{JL}$, and L is the midpoint of $\overline{GK}$.

Prove: $\triangle GHL \cong \triangle KJL$

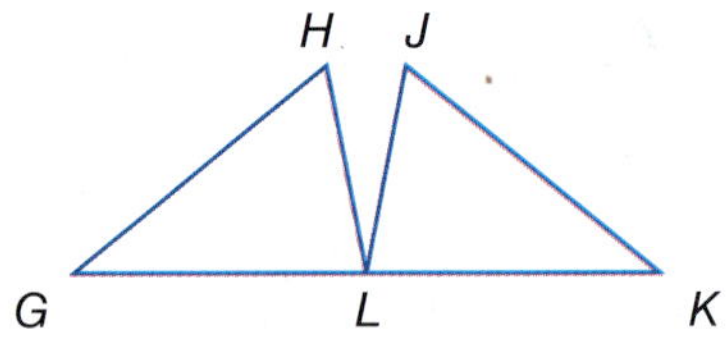

Flow Proof:

$\overline{GH} \cong \overline{KJ}$ (Given) → $\triangle GHL \cong \triangle KHL$ (SSS)

$\overline{HL} \cong \overline{JL}$ (Given) → $\triangle GHL \cong \triangle KHL$ (SSS)

L is the midpoint of $\overline{GK}$. (Given) → $\overline{GL} \cong \overline{KL}$ (Midpoint Theorem) → $\triangle GHL \cong \triangle KHL$ (SSS)

Guided Practice

1. **Write a flow proof.**

Given: $\triangle QRS$ is isosceles with $\overline{QR} \cong \overline{SR}$. $\overline{RT}$ bisects $\overline{QS}$ at point T.

Prove: $\triangle QRT \cong \triangle SRT$

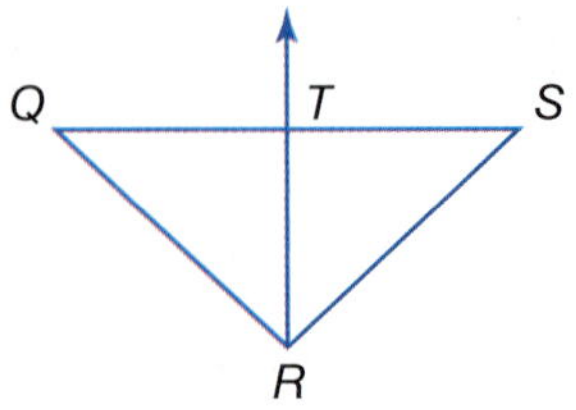

Personal Tutor glencoe.com

STANDARDIZED TEST EXAMPLE 2

EXTENDED RESPONSE **Triangle *ABC* has vertices *A*(1, 1), *B*(0, 3), and *C*(2, 5). Triangle *EFG* has vertices *E*(1, −1), *F*(2, −5), and *G*(4, −4).**

a. Graph both triangles on the same coordinate plane.

b. Use your graph to make a conjecture as to whether the triangles are congruent. Explain your reasoning.

c. Write a logical argument using coordinate geometry to support the conjecture you made in part b.

Test-Taking Tip

Coordinate Plane When you are solving problems using the coordinate plane, remember to use tools like the Distance, Midpoint, and Slope Formulas to solve problems and to check your solutions.

Read the Test Item

You are asked to do three things in this problem. In part **a**, you are to graph $\triangle ABC$ and $\triangle EFG$ on the same coordinate plane. In part **b**, you should make a conjecture that $\triangle ABC \cong \triangle EFG$ or $\triangle ABC \ncong \triangle EFG$ based on your graph. Finally, in part **c**, you are asked to prove your conjecture.

Solve the Test Item

a.

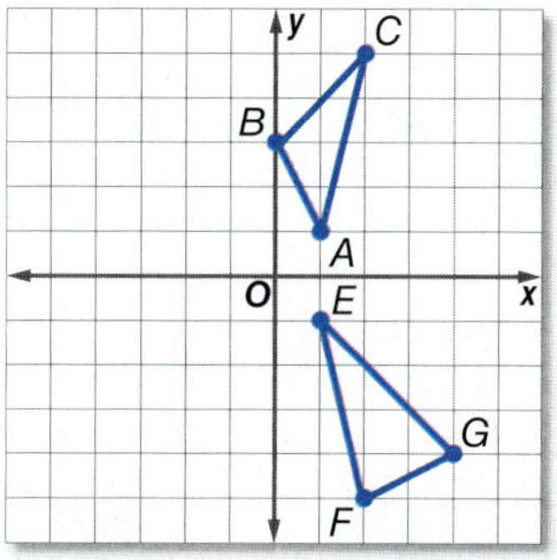

b. From the graph, it appears that the triangles do not have the same shape, so we can conjecture that they are not congruent.

c. Use the Distance Formula to show that not all corresponding sides have the same measure.

$$AB = \sqrt{(0-1)^2 + (3-1)^2} = \sqrt{1+4} \text{ or } \sqrt{5}$$

$$EF = \sqrt{(2-1)^2 + [-5-(-1)]^2} = \sqrt{1+16} \text{ or } \sqrt{17}$$

$$BC = \sqrt{(2-0)^2 + (5-3)^2} = \sqrt{4+4} \text{ or } \sqrt{8}$$

$$FG = \sqrt{(4-2)^2 + [-4-(-5)]^2} = \sqrt{4+1} \text{ or } \sqrt{5}$$

$$AC = \sqrt{(2-1)^2 + (5-1)^2} = \sqrt{1+16} \text{ or } \sqrt{17}$$

$$EG = \sqrt{(4-1)^2 + [-4-(-1)]^2} = \sqrt{9+9} \text{ or } \sqrt{18}$$

While $AB = FG$ and $AC = EF$, $BC \neq EG$. Since SSS congruence is not met, $\triangle ABC \ncong \triangle EFG$.

Reading Math

Symbols $\triangle ABC \ncong \triangle EFG$ is read as *triangle ABC is not congruent to triangle EFG.*

Guided Practice

2. Triangle *JKL* has vertices *J*(2, 5), *K*(1, 1), and *L*(5, 2). Triangle *NPQ* has vertices *N*(−3, 0), *P*(−7, 1), and *Q*(−4, 4).

A. Graph both triangles on the same coordinate plane.

B. Use your graph to make a conjecture as to whether the triangles are congruent. Explain your reasoning.

C. Write a logical argument using coordinate geometry to support the conjecture you made in part **b**.

Personal Tutor glencoe.com

Construction — Congruent Triangles Using Sides

Draw a triangle and label it $\triangle ABC$. Then use the SSS Postulate to construct $\triangle XYZ \cong \triangle ABC$.

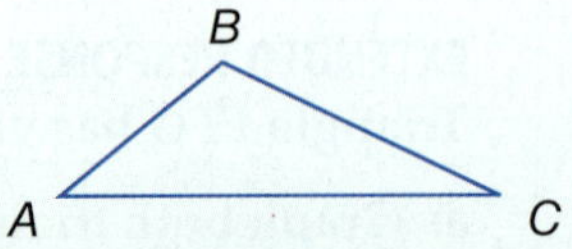

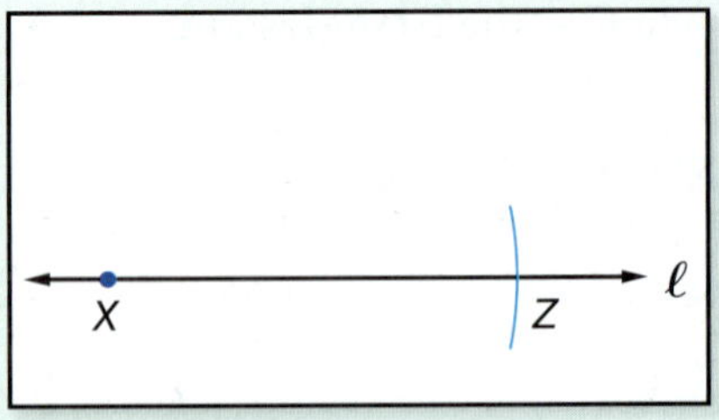

Step 1 Draw point X on a line ℓ. Then construct $\overline{XZ} \cong \overline{AC}$ on line ℓ

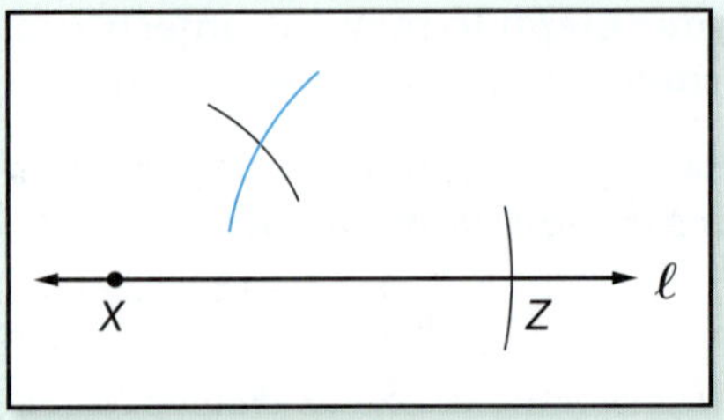

Step 2 Construct one arc with radius AB centered at point X and another arc with radius BC centered at point Z.

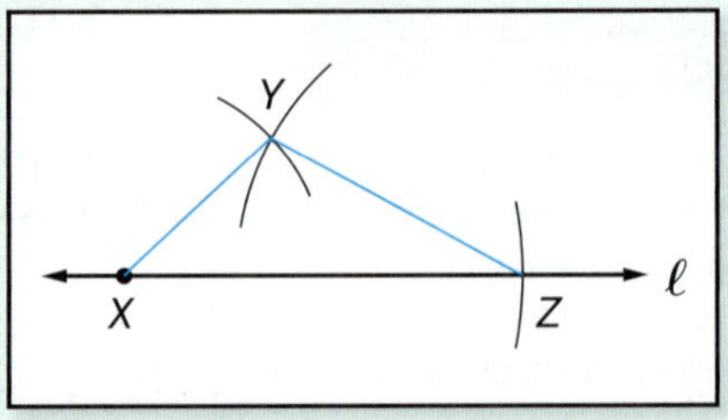

Step 3 Label the point of intersection of the two arcs Y. Draw $\overline{XY}$ and $\overline{ZY}$ to form $\triangle XYZ$.

SAS Postulate The angle formed by two adjacent sides of a polygon is called an **included angle**. Consider included angle JKL formed by the hands on the first clock shown below. Any time the hands form an angle with the same measure, the distance between the ends of the hands $\overline{JL}$ and $\overline{PR}$ will be the same.

$\triangle PKR \cong \triangle JKL$

Any two triangles formed using the same side lengths and included angle measure will be congruent. This illustrates the following postulate.

Postulate 4.2 — Side-Angle-Side (SAS) Congruence

Words If two sides and the included angle of one triangle are congruent to two sides and the included angle of a second triangle, then the triangles are congruent.

Example If **S**ide $\overline{AB} \cong \overline{DE}$,
Angle $\angle B \cong \angle E$, and
Side $\overline{BC} \cong \overline{EF}$,
then $\triangle ABC \cong \triangle DEF$.

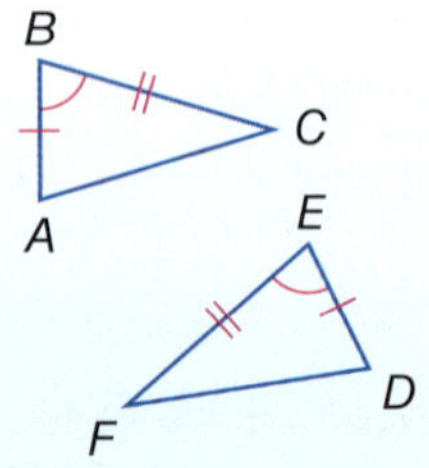

Real-World Career

Lighting Technicians
In the motion picture industry, gaffers, or lighting technicians, place the lighting required for a film. Gaffers make sure the angles the lights form are in the correct positions. They may have college or technical school degrees, or they may have completed a formal training program.

Real-World EXAMPLE 3 Use SAS to Prove Triangles are Congruent

LIGHTING **The scaffolding for stage lighting shown appears to be made up of congruent triangles. If $\overline{WZ} \cong \overline{YX}$ and $\overline{WX} \parallel \overline{ZY}$, write a two-column proof to prove that $\triangle WXZ \cong \triangle YZX$.**

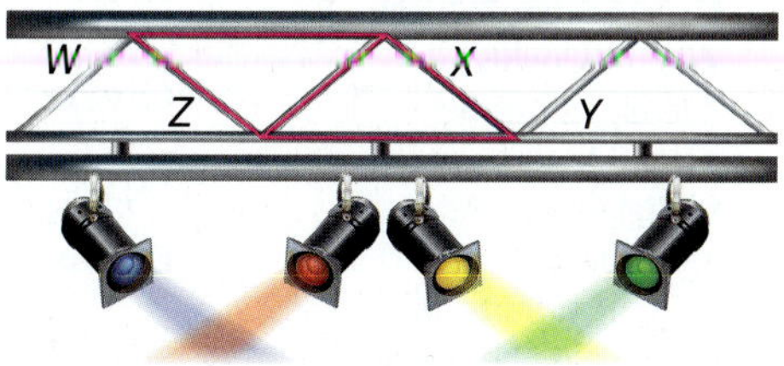

Proof:

Statements	Reasons
1. $\overline{WZ} \cong \overline{YX}$	**1.** Given
2. $\overline{WX} \parallel \overline{ZY}$	**2.** Given
3. $\angle WXZ \cong \angle XZY$	**3.** Alternate Interior Angle Theorem
4. $\overline{XZ} \cong \overline{ZX}$	**4.** Reflexive Property of Congruence
5. $\triangle WXZ \cong \triangle YZX$	**5.** SAS

Guided Practice

3. **EXTREME SPORTS** The wings of the hang glider shown appear to be congruent triangles. If $\overline{FG} \cong \overline{GH}$ and $\overline{JG}$ bisects $\angle FGH$, prove that $\triangle FGJ \cong \triangle HGJ$.

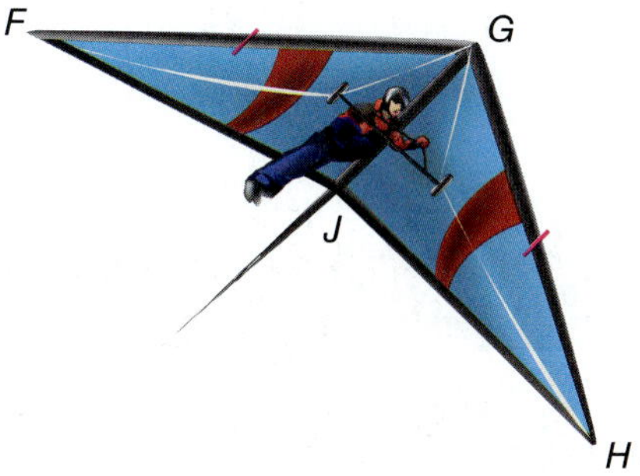

Personal Tutor glencoe.com

You can also construct congruent triangles given two sides and the included angle.

Construction Congruent Triangles Using Two Sides and the Included Angle

Draw a triangle and label it $\triangle ABC$. Then use the SAS Postulate to construct $\triangle RST \cong \triangle ABC$.

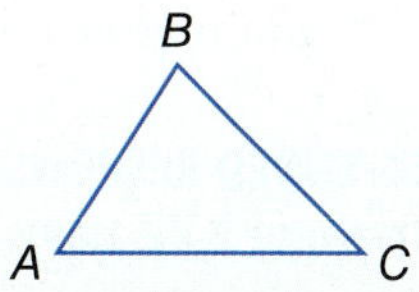

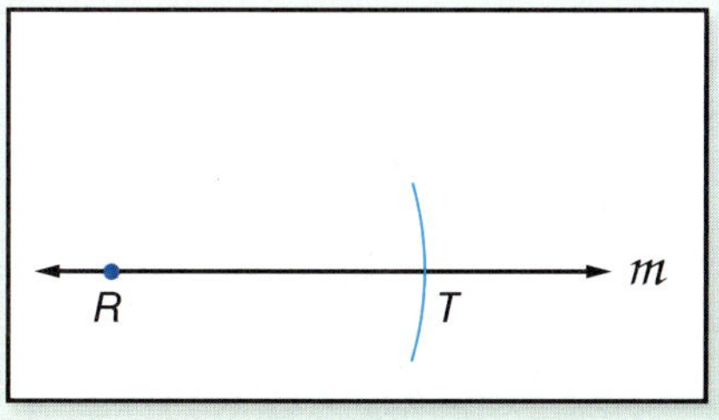

Step 1 Draw point R on a line m. Then construct $\overline{RT} \cong \overline{AC}$ on line m.

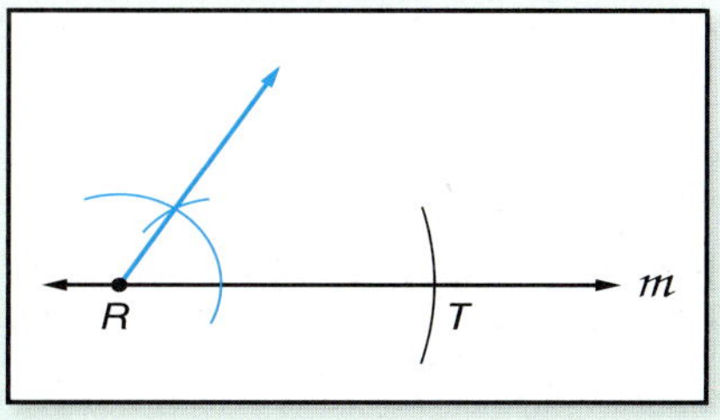

Step 2 Construct $\angle R \cong \angle A$ using $\overline{RT}$ as a side of the angle and point R.

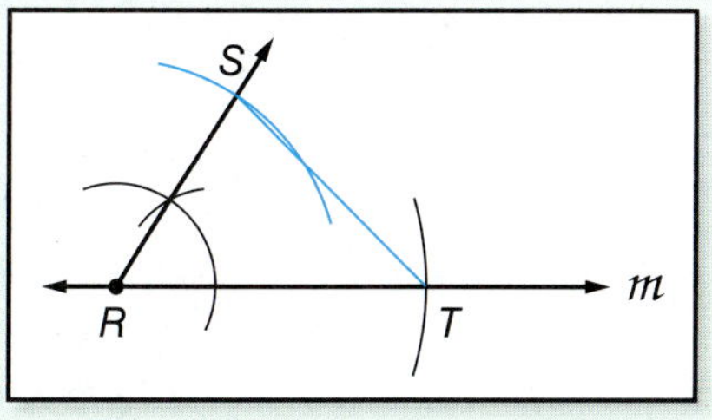

Step 3 Construct $\overline{RS} \cong \overline{AB}$. Then draw $\overline{ST}$ to form $\triangle RST$.

StudyTip

Overlapping Figures When triangles overlap, it can be helpful to draw each triangle separately and label the congruent parts. In Example 4, the figure could have been separated as shown.

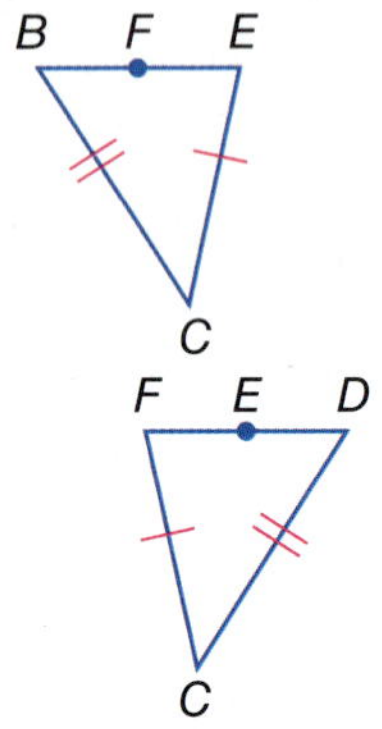

EXAMPLE 4 Use SAS or SSS in Proofs

Write a paragraph proof.

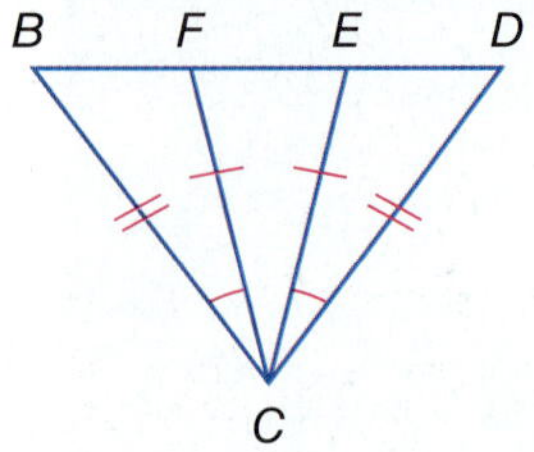

Given: $\overline{BC} \cong \overline{DC}$, $\angle BCF \cong \angle DCE$, $\overline{FC} \cong \overline{EC}$

Prove: $\angle CFD \cong \angle CEB$

Proof:

Since $\overline{BC} \cong \overline{DC}$, $\angle BCF \cong \angle DCE$, and $\overline{FC} \cong \overline{EC}$, then $\triangle BCF \cong \triangle DCE$ by SAS. By CPCTC, $\angle CFB \cong \angle CED$. $\angle CFD$ forms a linear pair with $\angle CFB$, and $\angle CEB$ forms a linear pair with $\angle CED$. By the Congruent Supplements Theorem, $\angle CFD$ is supplementary to $\angle CFB$ and $\angle CEB$ is supplementary to $\angle CED$. Since angles supplementary to the same angle or congruent angles are congruent, $\angle CFD \cong \angle CEB$.

Guided Practice

4. Write a two-column proof.

Given: $\overline{MN} \cong \overline{PN}$, $\overline{LM} \cong \overline{LP}$

Prove: $\angle LNM \cong \angle LNP$

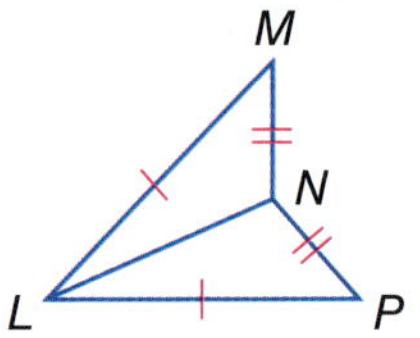

Personal Tutor glencoe.com

Check Your Understanding

Example 1 p. 262

1. **OPTICAL ILLUSION** $ABCD$ is a square with $AB = CD$ and $DA = BC$. $ABCD$ is congruent to the three other squares that form the pattern.

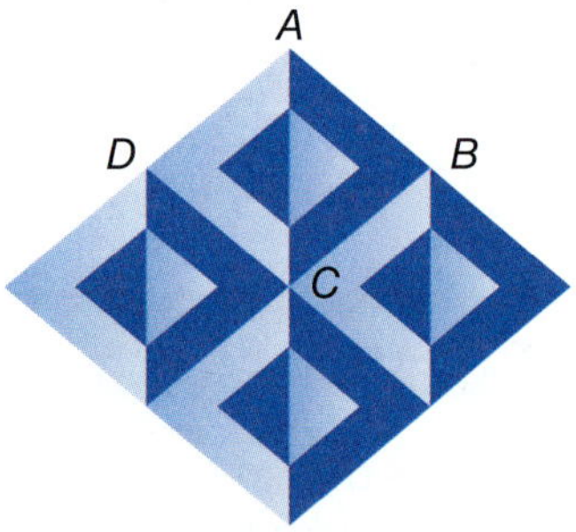

a. How many different-sized triangles are used to create the illusion?

b. Use the Side-Side-Side Congruence Postulate to prove that $\triangle ABC \cong \triangle CDA$.

c. What is the relationship between the lines formed by the bases of the triangles? Explain your reasoning.

Example 2 p. 263

2. **EXTENDED RESPONSE** Triangle ABC has vertices $A(-3, -5)$, $B(-1, -1)$, and $C(-1, -5)$. Triangle XYZ has vertices $X(5, -5)$, $Y(3, -1)$, and $Z(3, -5)$.

a. Graph both triangles on the same coordinate plane.

b. Use your graph to make a conjecture as to whether the triangles are congruent. Explain your reasoning.

c. Write a logical argument using coordinate geometry to support your conjecture.

Example 3 p. 265

3. **EXERCISE** In the exercise diagram, if $\overline{LP} \cong \overline{NO}$, $\angle LPM \cong \angle NOM$, and $\triangle MOP$ is equilateral, write a paragraph proof to show that $\triangle LMP \cong \triangle NMO$.

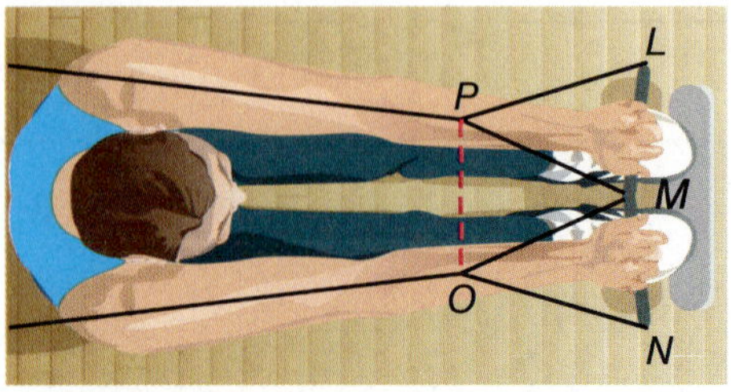

Example 4 p. 266

4. **Write a two-column proof.**

Given: $\overline{BA} \cong \overline{DC}$, $\angle BAC \cong \angle DCA$
Prove: $\overline{BC} \cong \overline{DA}$

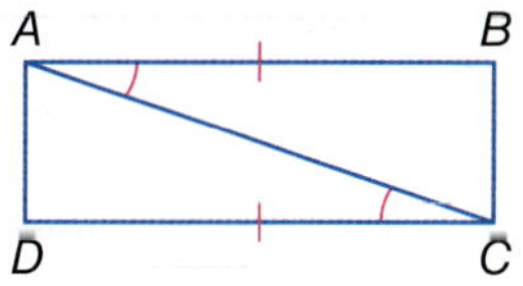

Practice and Problem Solving

= **Step-by-Step Solutions** begin on page R20.
Extra Practice begins on page 969.

Example 1 p. 262

PROOF **Write the specified type of proof.**

5. paragraph proof

Given: $\overline{QR} \cong \overline{SR}$, $\overline{ST} \cong \overline{QT}$
Prove: $\triangle QRT \cong \triangle SRT$

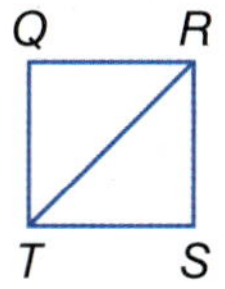

6. two-column proof

Given: $\overline{AB} \cong \overline{ED}$, $\overline{CA} \cong \overline{CE}$; $\overline{AC}$ bisects $\overline{BD}$.
Prove: $\triangle ABC \cong \triangle EDC$

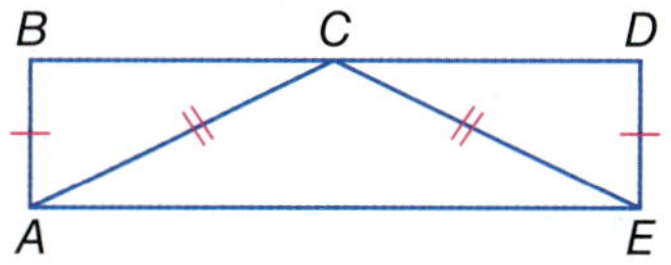

7. **BRIDGES** The Sunshine Skyway Bridge in Florida is the world's longest cable-stayed bridge, spanning 4.1 miles of Tampa Bay. It is supported using steel cables suspended from two concrete supports. If the supports are the same height above the roadway and perpendicular to the roadway, and the topmost cables meet at a point midway between the supports, prove that the two triangles shown in the photo are congruent.

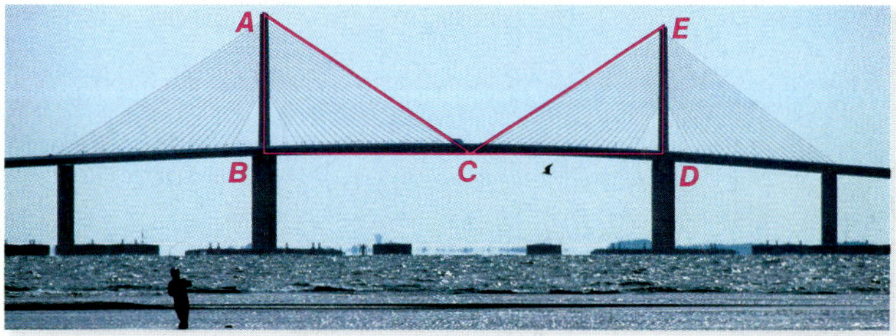

Example 2 p. 263

Determine whether $\triangle MNO \cong \triangle QRS$. Explain.

8. $M(2, 5), N(5, 2), O(1, 1), Q(-4, 4), R(-7, 1), S(-3, 0)$
9. $M(0, -1), N(-1, -4), O(-4, -3), Q(3, -3), R(4, -4), S(3, 3)$
10. $M(0, -3), N(1, 4), O(3, 1), Q(4, -1), R(6, 1), S(9, -1)$
11. $M(4, 7), N(5, 4), O(2, 3), Q(2, 5), R(3, 2), S(0, 1)$

Example 3 p. 265

PROOF **Write the specified type of proof.**

12. two-column proof

Given: $\overline{BD} \perp \overline{AC}$, $\overline{BD}$ bisects $\overline{AC}$.
Prove: $\triangle ABD \cong \triangle CBD$

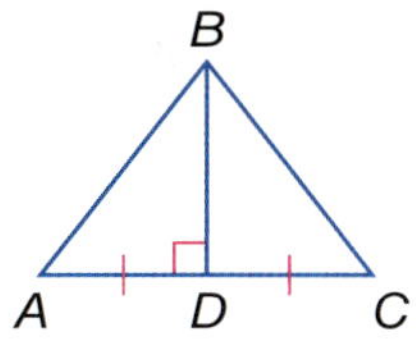

13. paragraph proof

Given: R is the midpoint of $\overline{QS}$ and $\overline{PT}$.
Prove: $\triangle PRQ \cong \triangle TRS$

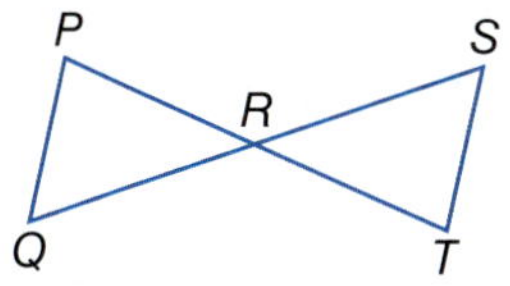

Example 4
p. 266

PROOF **Write the specified type of proof.**

14. flow proof

Given: $\overline{JM} \cong \overline{NK}$; L is the midpoint of $\overline{JN}$ and $\overline{KM}$.

Prove: $\angle MJL \cong \angle KNL$

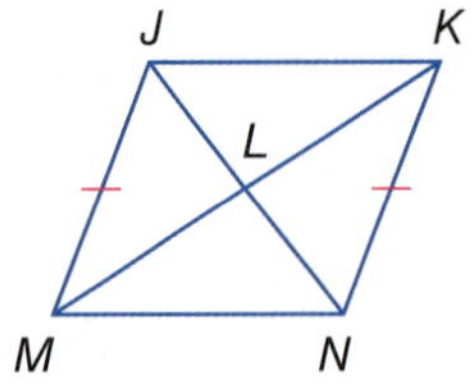

15. paragraph proof

Given: $\triangle XYZ$ is equilateral. $\overline{WY}$ bisects $\angle Y$.

Prove: $\overline{XW} \cong \overline{ZW}$

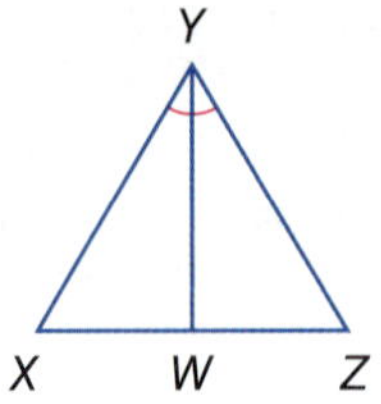

Determine which postulate can be used to prove that the triangles are congruent. If it is not possible to prove congruence, write *not possible*.

16.

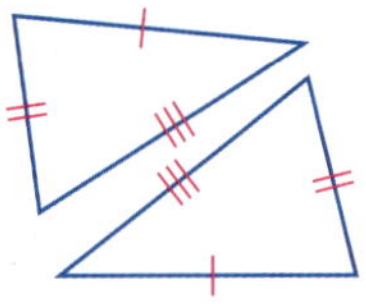

17

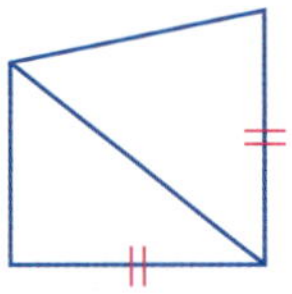

18.

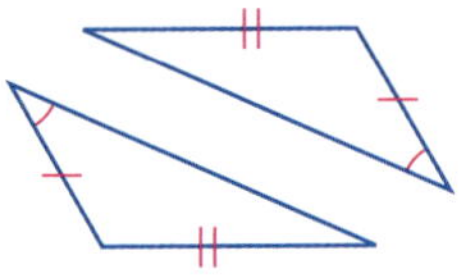

19.

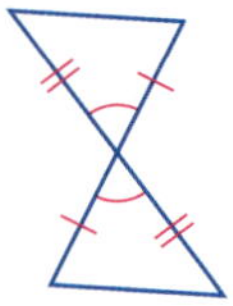

20. **SIGNS** Refer to the diagram at the right.

a. Identify the three-dimensional figure represented by the wet floor sign.

b. If $\overline{AB} \cong \overline{AD}$ and $\overline{CB} \cong \overline{DC}$, prove that $\triangle ACB \cong \triangle ACD$.

c. Why do the triangles not look congruent in the diagram?

PROOF **Write a flow proof.**

21. **Given:** $\overline{MJ} \cong \overline{ML}$; K is the midpoint of $\overline{JL}$.

Prove: $\triangle MJK \cong \triangle MLK$

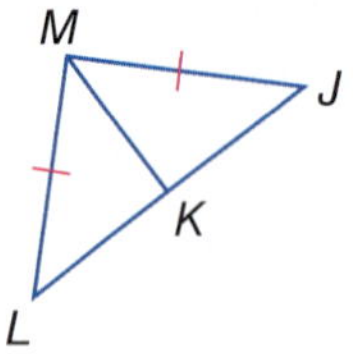

22. **Given:** $\triangle TPQ \cong \triangle SPR$
$\angle TQR \cong \angle SRQ$

Prove: $\triangle TQR \cong \triangle SRQ$

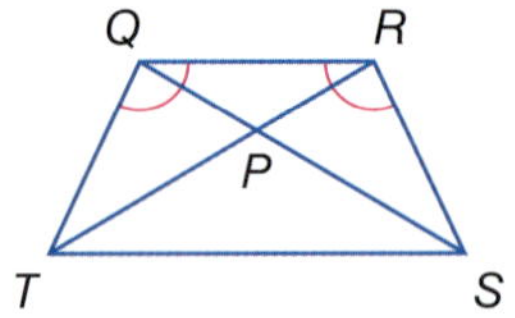

Real-World Link

The three major types of softball are fast-pitch, slow-pitch, and modified-pitch. The bases in fast-pitch and modified-pitch softball are 60 feet (18.3 m) apart. In slow-pitch, they are 65 feet (19.8 m) apart, compared to 90 feet apart in regulation baseball.

Source: MSN Encarta

23. **SOFTBALL** Use the diagram of a fast-pitch softball diamond shown.

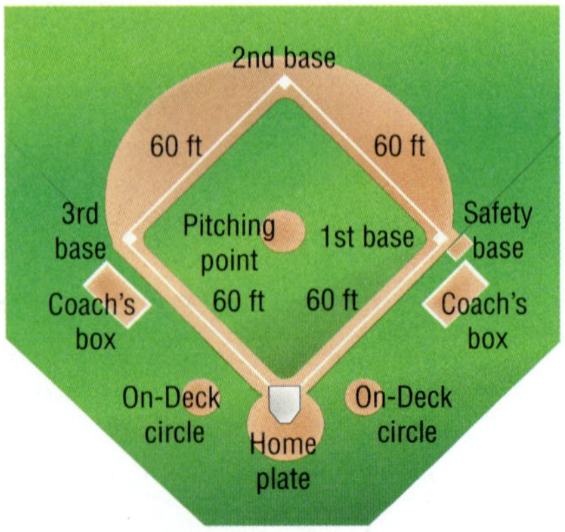

a. Write a two-column proof to prove that the distance from first base to third base is the same as the distance from home plate to second base.

b. Write a two-column proof to prove that the angle formed between second base, home plate, and third base is the same as the angle formed between second base, home plate, and first base.

Study Tip

Diagrams When writing proofs or solving problems involving congruent triangles, it is helpful to draw your own diagram and mark the congruent sides and angles you find.

PROOF **Write a two-column proof.**

24. **Given:** $\overline{YX} \cong \overline{WZ}$, $\overline{YX} \parallel \overline{ZW}$
Prove: $\triangle YXZ \cong \triangle WZX$

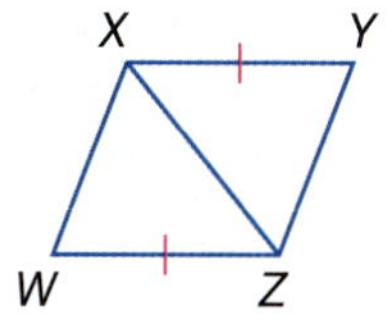

25. **Given:** $\triangle EAB \cong \triangle DCB$
Prove: $\triangle EAD \cong \triangle DCE$

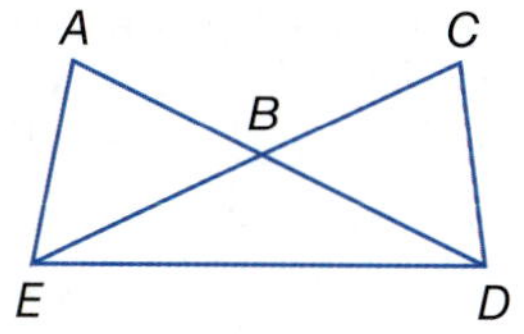

26. Write a paragraph proof.
Given: $\overline{HL} \cong \overline{HM}$, $\overline{PM} \cong \overline{KL}$, $\overline{PG} \cong \overline{KJ}$, $\overline{GH} \cong \overline{JH}$
Prove: $\angle G \cong \angle J$

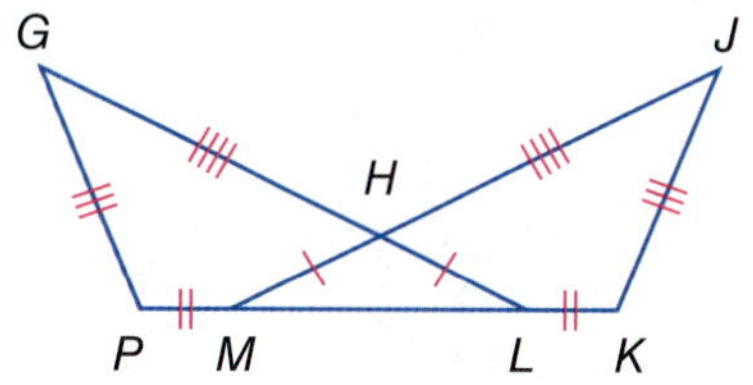

ALGEBRA **Find the value of the variable that yields congruent triangles. Explain.**

27 $\triangle WXY \cong \triangle WXZ$

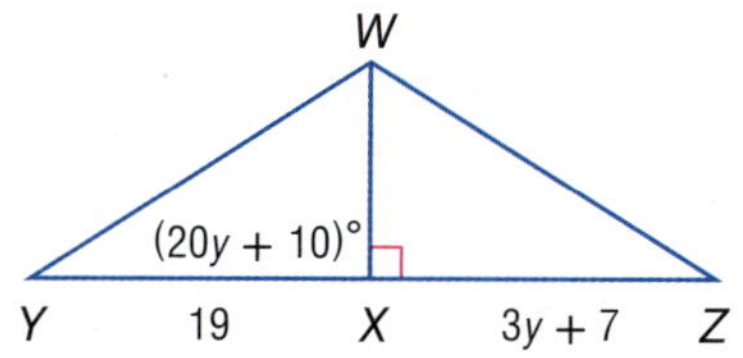

28. $\triangle ABC \cong \triangle FGH$

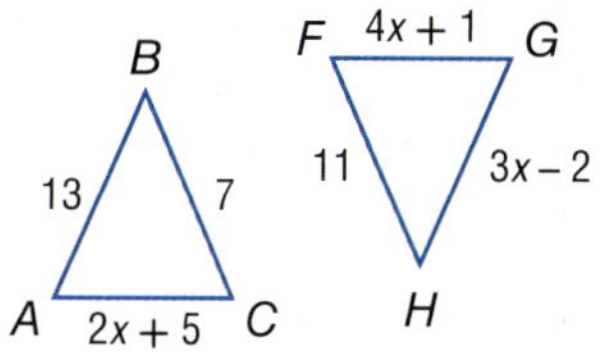

H.O.T. Problems Use Higher-Order Thinking Skills

29. **CHALLENGE** **Refer to the graph shown.**

a. Describe two methods you could use to prove that $\triangle WYZ$ is congruent to $\triangle WYX$. You may not use a ruler or a protractor. Which method do you think is more efficient? Explain.

b. Are $\triangle WYZ$ and $\triangle WYX$ congruent? Explain your reasoning.

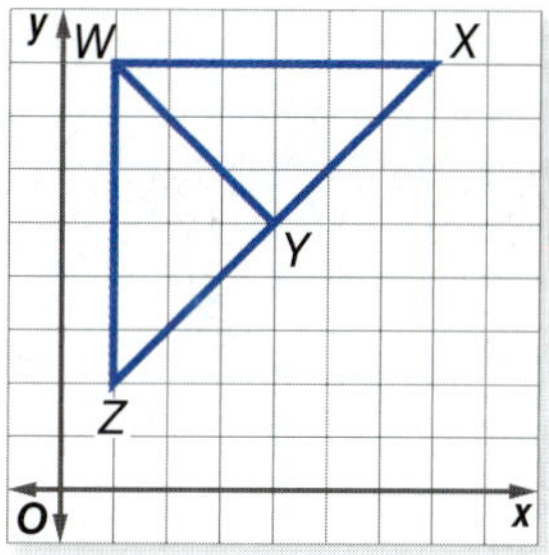

30. **REASONING** Determine whether the following statement is *true* or *false*. If true, explain your reasoning. If *false*, provide a counterexample.

If the congruent sides in one isosceles triangle have the same measure as the congruent sides in another isosceles triangle, then the triangles are congruent.

31. **ERROR ANALYSIS** Bonnie says that $\triangle PRQ \cong \triangle XYZ$ by SAS. Shada disagrees. She says that there is not enough information to prove that the two triangles are congruent. Is either of them correct? Explain.

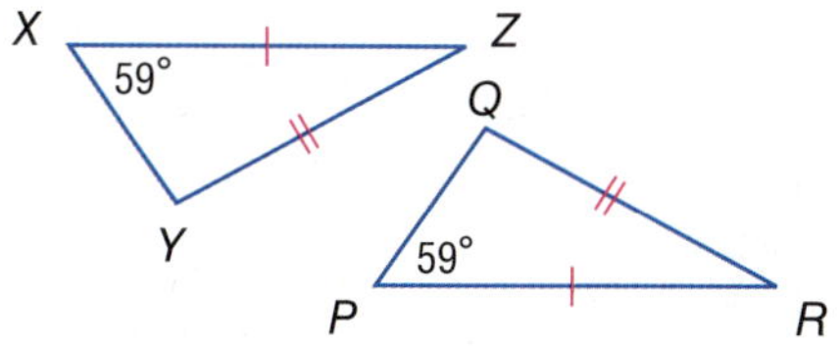

32. **OPEN ENDED** Use a straightedge to draw obtuse triangle ABC. Then construct $\triangle XYZ$ so that it is congruent to $\triangle ABC$ using either SSS or SAS. Justify your construction mathematically and verify it using measurement.

33. **WRITING IN MATH** Two pairs of corresponding sides of two right triangles are congruent. Are the triangles congruent? Explain your reasoning.

Standardized Test Practice

34. ALGEBRA The Ross Family drove 300 miles to visit their grandparents. Mrs. Ross drove 70 miles per hour for 65% of the trip and 35 miles per hour or less for 20% of the trip that was left. Assuming that Mrs. Ross never went over 70 miles per hour, how many miles did she travel at a speed between 35 and 70 miles per hour?

A 195 **C** 21
B 84 **D** 18

35. In the figure, $\angle C \cong \angle Z$ and $\overline{AC} \cong \overline{XZ}$.

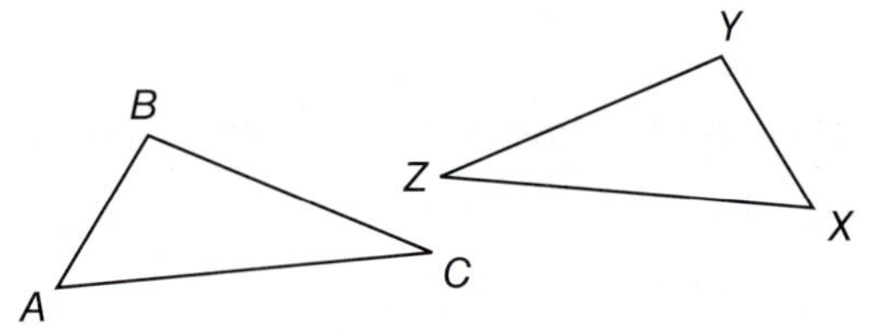

What additional information could be used to prove that $\triangle ABC \cong \triangle XYZ$?

F $\overline{BC} \cong \overline{YZ}$
G $\overline{AB} \cong \overline{XY}$
H $\overline{BC} \cong \overline{XZ}$
J $\overline{XZ} \cong \overline{XY}$

36. EXTENDED RESPONSE The graph below shows the eye colors of all of the students in a class. What is the probability that a student chosen at random from this class will have blue eyes? Explain your reasoning.

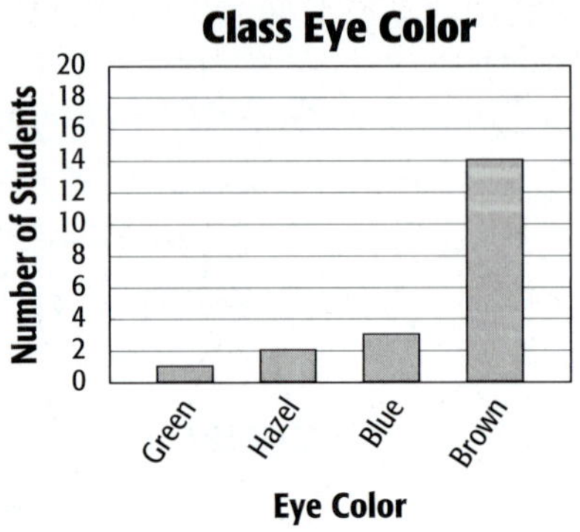

37. SAT/ACT If $4a + 6b = 6$ and $-2a + b = -7$, what is the value of a?

A -1
B 2
C 3
D 4

Spiral Review

In the diagram, $\square LMNP \cong \square QRST$. (Lesson 4-3)

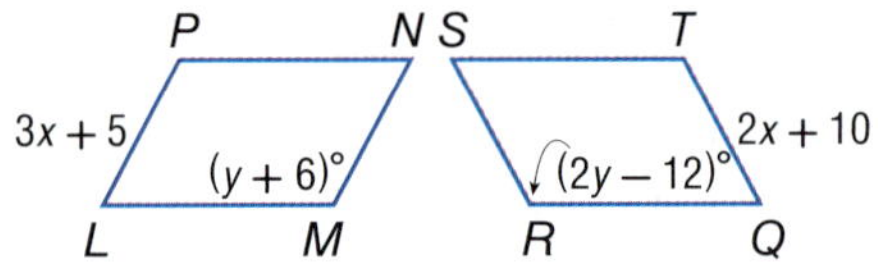

38. Find x.

39. Find y.

40. ASTRONOMY The Big Dipper is a part of the larger constellation Ursa Major. Three of the brighter stars in the constellation form $\triangle RSA$. If $m\angle R = 41$ and $m\angle S = 109$, find $m\angle A$. (Lesson 4-2)

Write an equation in slope-intercept form for each line. (Lesson 3-4)

41. $(-5, -3)$ and $(10, -6)$

42. $(4, -1)$ and $(-2, -1)$

43. $(-4, -1)$ and $(-8, -5)$

Determine the truth value of each conditional statement. If *true*, explain your reasoning. If *false*, give a counterexample. (Lesson 2-3)

44. If $x^2 = 25$, then $x = 5$.

45. If you are 16, you are a junior in high school.

Skills Review

State the property that justifies each statement. (Lesson 2-6)

46. $AB = AB$

47. If $EF = GH$ and $GH = JK$, then $EF = JK$.

48. If $a^2 = b^2 - c^2$, then $b^2 - c^2 = a^2$.

49. If $XY + 20 = YW$ and $XY + 20 = DT$, then $YW = DT$.

EXTEND 4-4

Geometry Lab
Proving Constructions

When you perform a construction using a straightedge and compass, you assume that segments constructed using the same compass setting are congruent. You can use this information, along with definitions, postulates, and theorems to prove constructions.

ACTIVITY **Follow the steps below to bisect an angle. Then prove the construction.**

Step 1

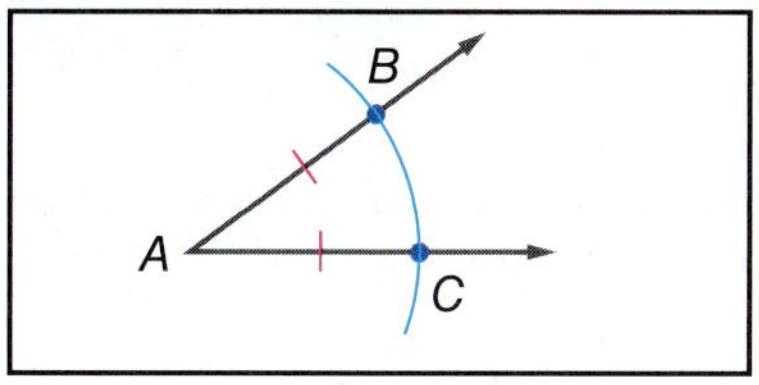

Draw any angle with vertex A. Place the compass point at A and draw an arc that intersects both sides of $\angle A$. Label the points B and C. Mark the congruent segments.

Step 2

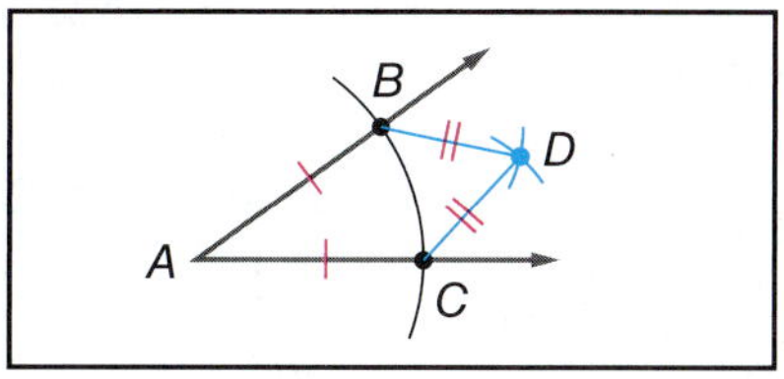

With the compass point at B, draw an arc in the interior of $\angle A$. With the same radius, draw an arc from C intersecting the first arc at D. Draw the segments $\overline{BD}$ and $\overline{CD}$. Mark the congruent segments.

Step 3

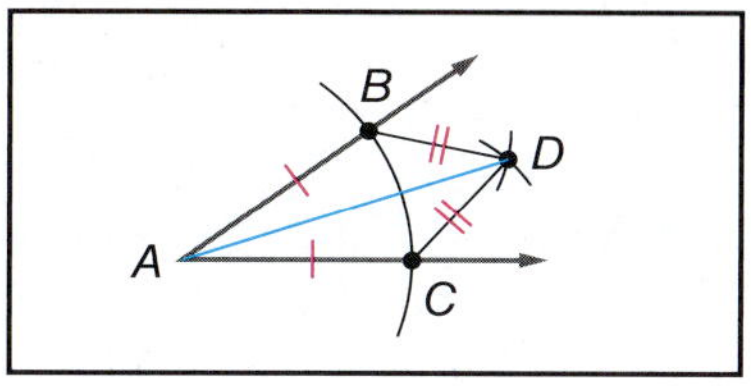

Draw $\overline{AD}$.

Given: Description of steps and diagram of construction

Prove: $\overline{AD}$ bisects $\angle BAC$.

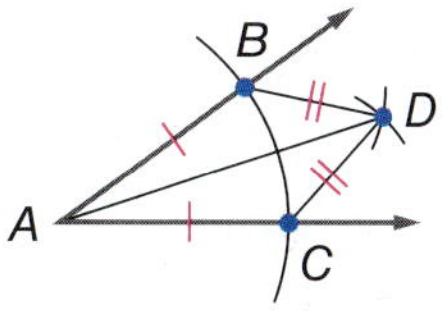

Proof:

Statements	Reasons
1. $\overline{AB} \cong \overline{AC}$	**1.** The same compass setting was used from point A to construct points B and C.
2. $\overline{BD} \cong \overline{CD}$	**2.** The same compass setting was used from points B and C to construct point D.
3. $\overline{AD} \cong \overline{AD}$	**3.** Reflexive Property
4. $\triangle ABD \cong \triangle ACD$	**4.** SSS Postulate
5. $\angle BAD \cong \angle CAD$	**5.** CPCTC
6. $\overline{AD}$ bisects $\angle BAC$.	**6.** Definition of angle bisector

Exercises

1. Construct a line parallel to a given line through a given point. Write a two-column proof of your construction.
2. Construct an equilateral triangle. Write a paragraph proof of your construction.
3. **CHALLENGE** Construct the bisector of a segment that is also perpendicular to the segment and write a two-column proof of your construction. (*Hint:* You will need to use more than one pair of congruent triangles.).

CHAPTER 4 Mid-Chapter Quiz

Lessons 4-1 through 4-4

1. **COORDINATE GEOMETRY** Classify $\triangle ABC$ with vertices $A(-2, -1)$, $B(-1, 3)$, and $C(2, 0)$ as *scalene*, *equilateral*, or *isosceles*. (Lesson 4-1)

2. **MULTIPLE CHOICE** Which of the following are the measures of the sides of isosceles triangle QRS? (Lesson 4-1)

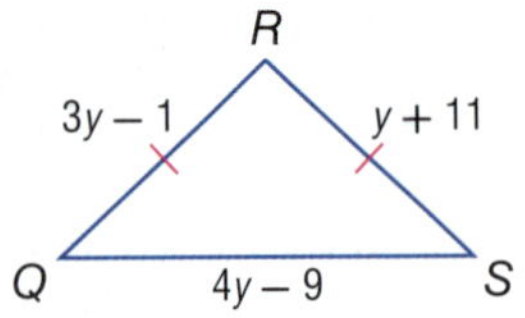

A 17, 17, 15

B 15, 15, 16

C 14, 15, 14

D 14, 14, 16

3. **ALGEBRA** Find x and the length of each side if $\triangle WXY$ is an equilateral triangle with sides $\overline{WX} = 6x - 12$, $\overline{XY} = 2x + 10$, and $\overline{WY} = 4x - 1$. (Lesson 4-1)

Find the measure of each angle indicated. (Lesson 4-2)

4. $m\angle 1$

5. $m\angle 2$

6. $m\angle 3$

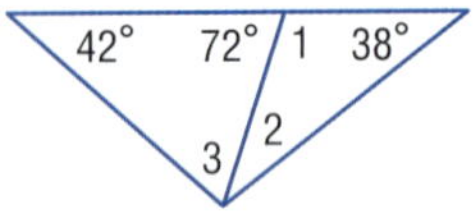

7. **ASTRONOMY** Leo is a constellation that represents a lion. Three of the brighter stars in the constellation form $\triangle LEO$. If the angles have measures as shown in the figure, find $m\angle OLE$. (Lesson 4-2)

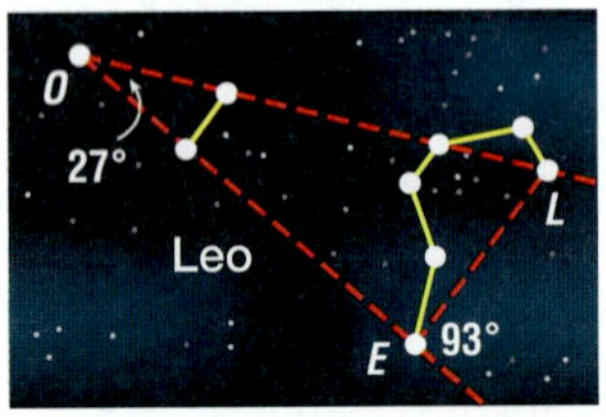

Find the measure of each numbered angle. (Lesson 4-2)

8. $m\angle 4$

9. $m\angle 5$

10. $m\angle 6$

11. $m\angle 7$

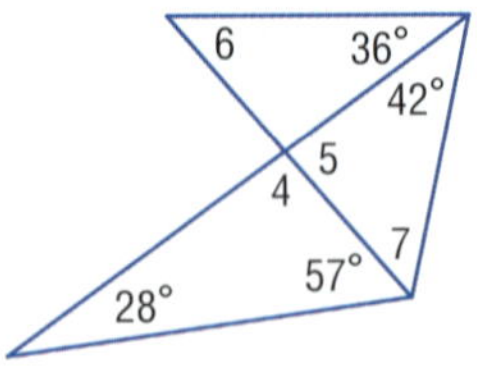

In the diagram, $\triangle RST \cong \triangle ABC$. (Lesson 4-3)

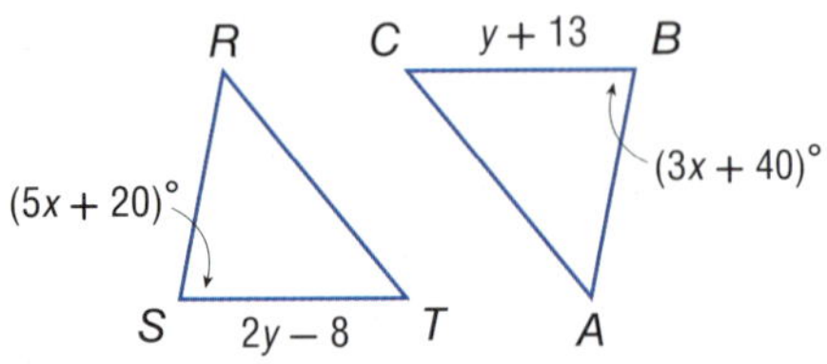

12. Find x.

13. Find y.

14. **ARCHITECTURE** The diagram shows an A-frame house with various points labeled. Assume that segments and angles that appear to be congruent in the diagram are congruent. Indicate which triangles are congruent. (Lesson 4-3)

15. **MULTIPLE CHOICE** Determine which statement is true given that $\triangle CBX \cong \triangle SML$. (Lesson 4-3)

F $\overline{MO} \cong \overline{SL}$

G $\overline{XC} \cong \overline{ML}$

H $\angle X \cong \angle S$

J $\angle XCB \cong \angle LSM$

16. **BRIDGES** A bridge truss is shown in the diagram below, where $\overline{AC} \perp \overline{BD}$ and B is the midpoint of $\overline{AC}$. What method can be used to prove that $\triangle ABD \cong \triangle CBD$? (Lesson 4-4)

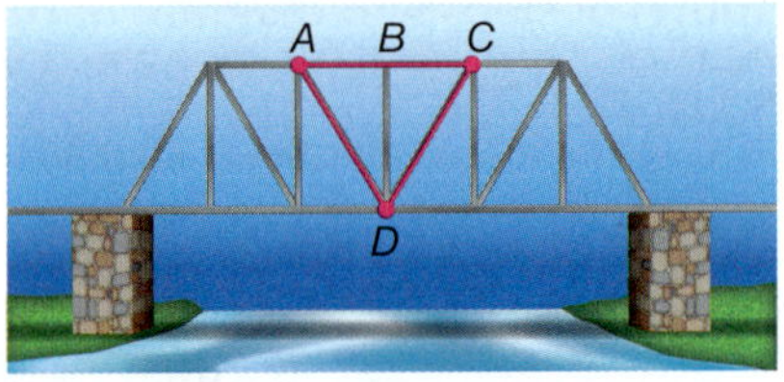

Determine whether $\triangle PQR \cong \triangle XYZ$. (Lesson 4-4)

17. $P(3, -5)$, $Q(11, 0)$, $R(1, 6)$, $X(5, 1)$, $Y(13, 6)$, $Z(3, 12)$

18. $P(-3, -3)$, $Q(-5, 1)$, $R(-2, 6)$, $X(2, -6)$, $Y(3, 3)$, $Z(5, -1)$

19. $P(8, 1)$, $Q(-7, -15)$, $R(9, -6)$, $X(5, 11)$, $Y(-10, -5)$, $Z(6, 4)$

20. **Write a two-column proof.** (Lesson 4-4)

Given: $\triangle LMN$ is isos. with $\overline{LM} \cong \overline{NM}$, and $\overline{MO}$ bisects $\angle LMN$.

Prove: $\triangle MLO \cong \triangle MNO$

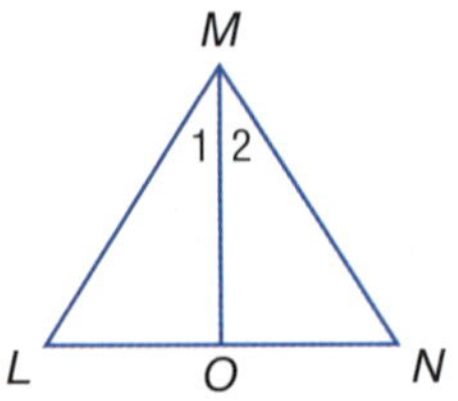

4-5 Proving Triangles Congruent—ASA, AAS

Then

You proved triangles congruent using SSS and SAS. (Lesson 4-4)

Now

- Use the ASA Postulate to test for congruence.
- Use the AAS Theorem to test for congruence.

New Vocabulary

included side

Math Online

glencoe.com

Why?

Competitive sweep rowing, also called *crew*, involves two or more people who sit facing the stern of the boat, with each rower pulling one oar. In high school competitions, a race, called a *regatta*, usually requires a body of water that is more than 1500 meters long. Congruent triangles can be used to measure distances that are not easily measured directly, like the length of a regatta course.

ASA Postulate An **included side** is the side located between two consecutive angles of a polygon. In $\triangle ABC$ at the right, $\overline{AC}$ is the included side between $\angle A$ and $\angle C$.

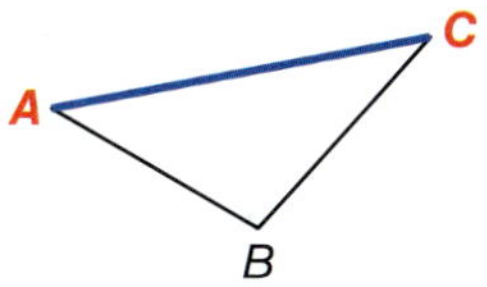

Postulate 4.3 **Angle-Side-Angle (ASA) Congruence**

If two angles and the included side of one triangle are congruent to two angles and the included side of another triangle, then the triangles are congruent.

Example If **A**ngle $\angle A \cong \angle D$,
Side $\overline{AB} \cong \overline{DE}$, and
Angle $\angle B \cong \angle E$,
then $\triangle ABC \cong \triangle DEF$.

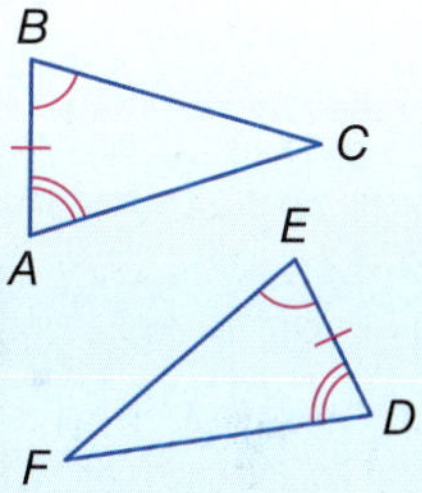

Construction **Congruent Triangles Using Two Angles and Included Side**

Draw a triangle and label it $\triangle ABC$. Then use the ASA Postulate to construct $\triangle XYZ \cong \triangle ABC$.

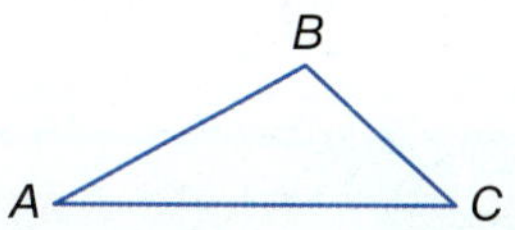

Step 1

Draw a line ℓ and select a point X. Construct $\overline{XZ}$ such that $\overline{XZ} \cong \overline{AC}$.

Step 2

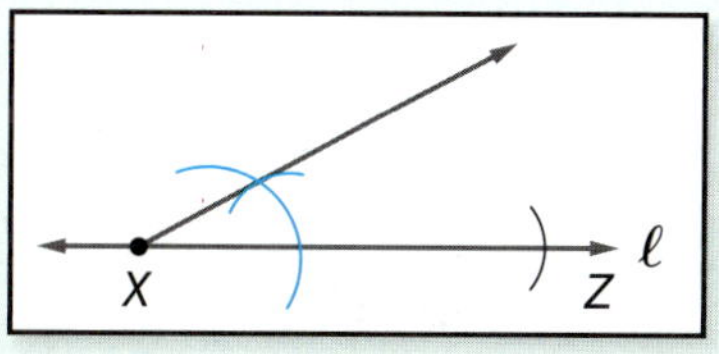

Construct an angle congruent to $\angle A$ at X using $\overleftrightarrow{XZ}$ as a side of the angle.

Step 3

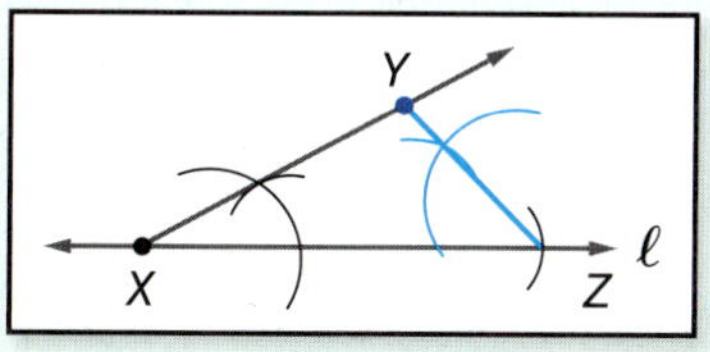

Construct an angle congruent to $\angle C$ at Z using $\overleftrightarrow{XZ}$ as a side of the angle. Label the point where the new sides of the angles meet as Y.

EXAMPLE 1 Use ASA to Prove Triangles Congruent

Write a two-column proof.

Given: $\overline{QS}$ bisects $\angle PQR$; $\angle PSQ \cong \angle RSQ$.

Prove: $\triangle PQS \cong \triangle RQS$

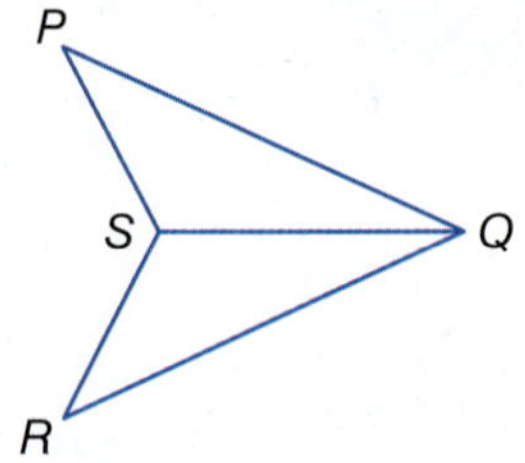

Proof:

Statements	Reasons
1. $\overline{QS}$ bisects $\angle PQR$; $\angle PSQ \cong \angle RSQ$.	**1.** Given
2. $\angle PQS \cong \angle RQS$	**2.** Definition of Angle Bisector
3. $\overline{QS} \cong \overline{QS}$	**3.** Reflexive Property of Congruence
4. $\triangle PQS \cong \triangle RQS$	**4.** ASA

Guided Practice

1. Write a flow proof.

Given: $\overline{ZX}$ bisects $\angle WZY$; $\overline{XZ}$ bisects $\angle YXW$.

Prove: $\triangle WXZ \cong \triangle XZY$

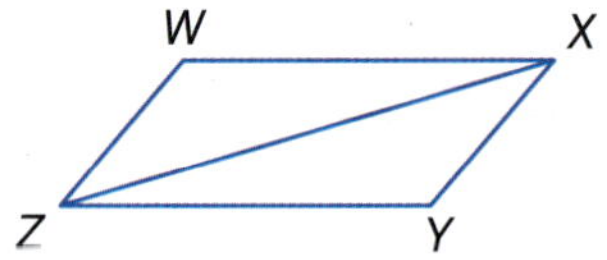

Personal Tutor glencoe.com

AAS Theorem The congruence of two angles and a nonincluded side are also sufficient to prove two triangles congruent. This congruence relationship is a theorem because it can be proved using the Third Angles Theorem.

Theorem 4.5 Angle-Angle-Side (AAS) Congruence

If two angles and the nonincluded side of one triangle are congruent to the corresponding two angles and side of a second triangle, then the two triangles are congruent.

Example If **A**ngle $\angle A \cong \angle D$,
Angle $\angle B \cong \angle E$, and
Side $\overline{BC} \cong \overline{EF}$,
then $\triangle ABC \cong \triangle DEF$.

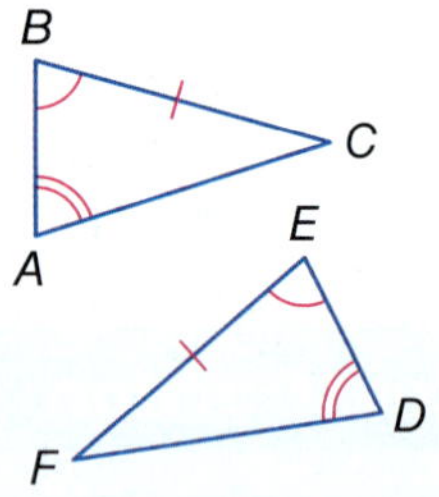

StudyTip

Side-Side-Angle The measures of two sides and a nonincluded angle are not sufficient to prove two triangles congruent.

Proof Angle-Angle-Side Theorem

Given: $\angle L \cong \angle Q$, $\angle M \cong \angle R$, $\overline{MN} \cong \overline{RS}$

Prove: $\triangle LMN \cong \triangle QRS$

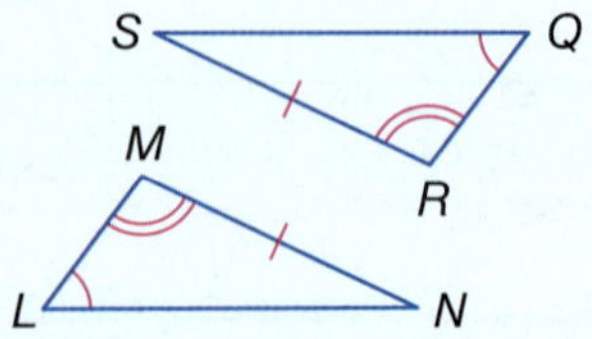

Proof:

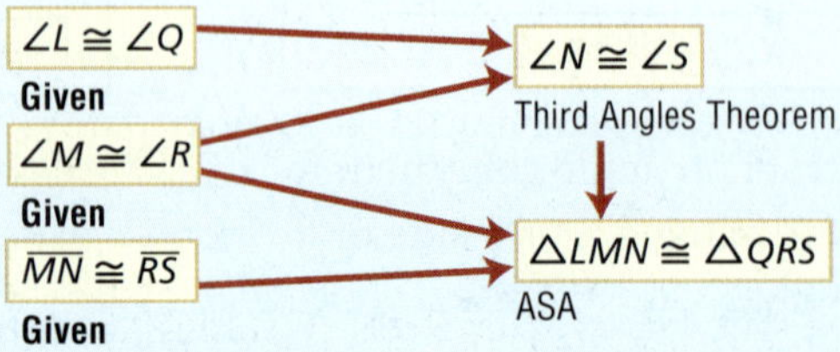

EXAMPLE 2 Use AAS to Prove Triangles Congruent

Write a paragraph proof.

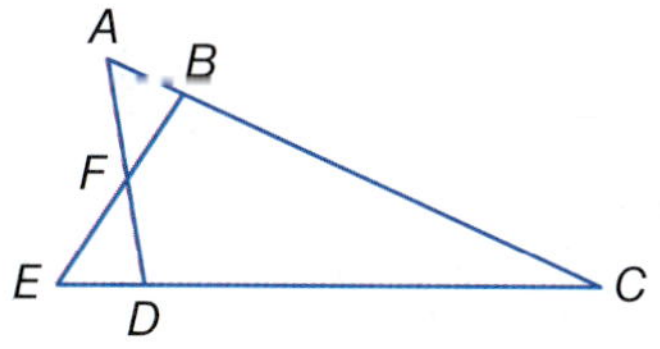

Given: $\angle DAC \cong \angle BEC$
$\overline{DC} \cong \overline{BC}$

Prove: $\triangle ACD \cong \triangle ECB$

Proof: We are given that $\angle DAC \cong \angle BEC$ and $\overline{DC} \cong \overline{BC}$. $\angle C \cong \angle C$ by the Reflexive Property. By AAS, $\triangle ACD \cong \triangle ECB$.

Guided Practice

2. Write a flow proof.

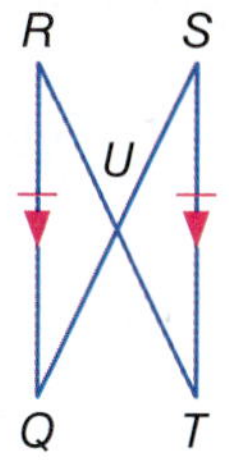

Given: $\overline{RQ} \cong \overline{ST}$ and $\overline{RQ} \parallel \overline{ST}$

Prove: $\triangle RUQ \cong \triangle TUS$

Personal Tutor glencoe.com

You can use congruent triangles to measure distances that are difficult to measure directly.

Real-World EXAMPLE 3 Apply Triangle Congruence

COMMUNITY SERVICE **Jeremias is working with a community service group to build a bridge across a creek at a local park. The bridge will span the creek between points *C* and *B*. Jeremias located a fixed point *D* to use as a reference point so that the segments have the relationships shown. *DE* is 15 feet. How long does the bridge need to be?**

In order to determine the length of $\overline{CB}$, we must first prove that the two triangles Jeremias has created are congruent.

- Since $\overline{CD}$ is perpendicular to both $\overline{CB}$ and $\overline{DE}$, the segments form right angles as shown on the diagram.
- All right angles are congruent, so $\angle BCA \cong \angle EDA$.
- Point A is the midpoint of $\overline{CD}$, so $\overline{CA} \cong \overline{AD}$.
- $\angle BAC$ and $\angle EAD$ are vertical angles, so they are congruent.

Therefore, by ASA, $\triangle BAC \cong \triangle EAD$.

Since $\triangle BAC \cong \triangle EAD$, $\overline{DE} \cong \overline{CB}$ by CPCTC. Since the measure of $\overline{DE}$ is 15 feet, the measure of $\overline{CB}$ is also 15 feet. Therefore, the bridge needs to be 15 feet long.

StudyTip

Angle-Angle-Angle In Example 3, $\angle B$ and $\angle E$ are congruent by the Third Angles Theorem. Congruence of all three corresponding angles is not sufficient, however, to prove two triangles congruent.

Guided Practice

3. In the sign scaffold shown at the right, $\overline{BC} \perp \overline{AC}$ and $\overline{DE} \perp \overline{CE}$. $\angle BAC \cong \angle DCE$, and $\overline{AB} \cong \overline{CD}$. Write a paragraph proof to show that $\overline{BC} \cong \overline{DE}$.

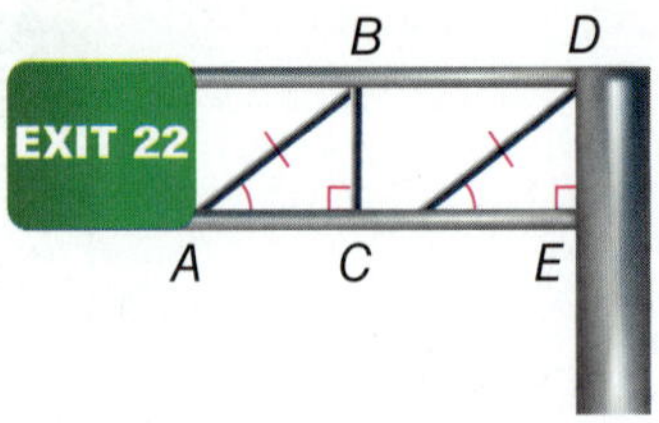

Personal Tutor glencoe.com

You have learned several methods for proving triangle congruence.

Concept Summary — Proving Triangles Congruent

SSS

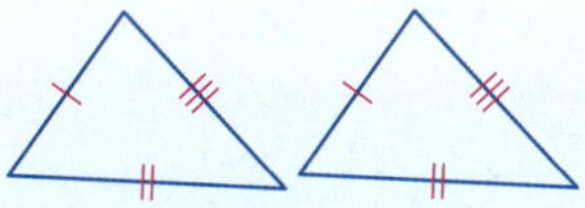

Three pairs of corresponding sides are congruent.

SAS

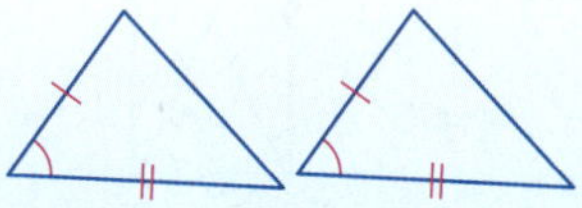

Two pairs of corresponding sides and their included angles are congruent.

ASA

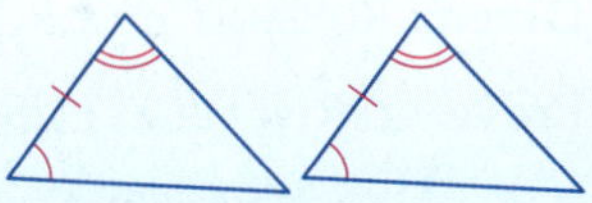

Two pairs of corresponding angles and their included sides are congruent.

AAS

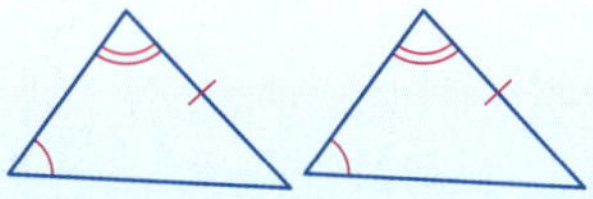

Two pairs of corresponding angles and the corresponding nonincluded sides are congruent.

Check Your Understanding

Example 1 p. 274

PROOF **Write the specified type of proof.**

1. two-column proof
Given: $\overline{CB}$ bisects $\angle ABD$ and $\angle ACD$.
Prove: $\triangle ABC \cong \triangle DBC$

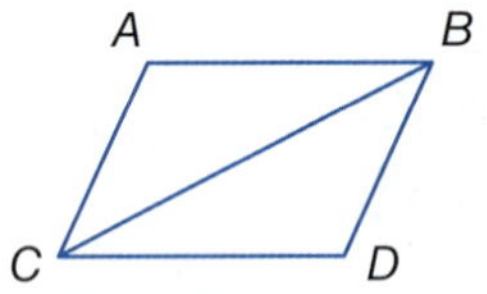

2. flow proof
Given: $\overline{JK} \parallel \overline{LM}$, $\overline{JL} \parallel \overline{KM}$
Prove: $\triangle JML \cong \triangle MJK$

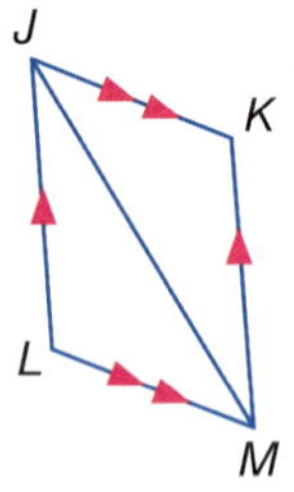

Example 2 p. 275

3. paragraph proof
Given: $\angle K \cong \angle M$, $\overline{JK} \cong \overline{JM}$, $\overline{JL}$ bisects $\angle KLM$.
Prove: $\triangle JKL \cong \triangle JML$

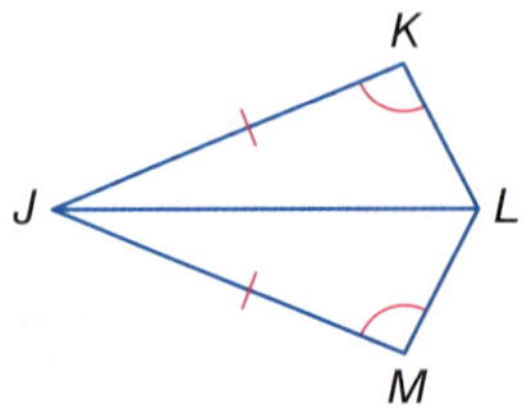

4. two-column proof
Given: $\overline{GH} \parallel \overline{FJ}$
$m\angle G = m\angle J = 90$
Prove: $\triangle HJF \cong \triangle FGH$

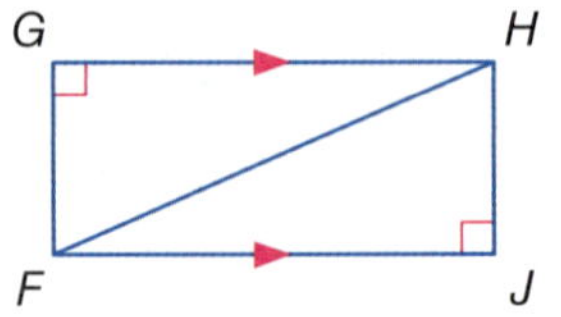

Example 3
p. 275

5 **BRIDGE BUILDING** A surveyor needs to find the distance from point A to point B across a canyon. She places a stake at A, and a coworker places a stake at B on the other side of the canyon. The surveyor then locates C on the same side of the canyon as A such that $\overline{CA} \perp \overline{AB}$. A fourth stake is placed at E, the midpoint of $\overline{CA}$. Finally, a stake is placed at D such that $\overline{CD} \perp \overline{CA}$ and D, E, and B are sited as lying along the same line.

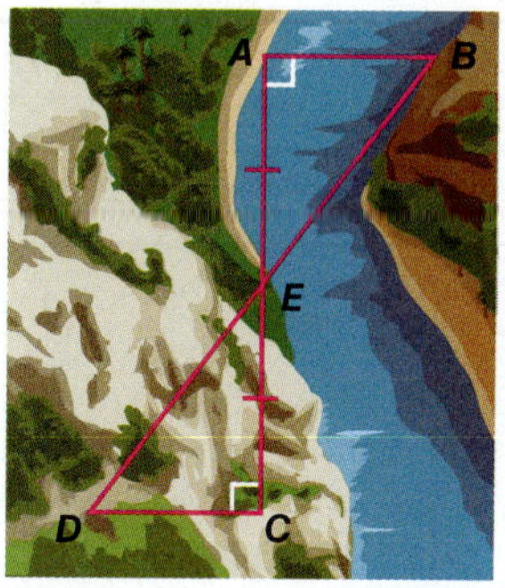

a. Explain how the surveyor can use the triangles formed to find AB.

b. If $AC = 1300$ meters, $DC = 550$ meters, and $DE = 851.5$ meters, what is AB? Explain your reasoning.

Practice and Problem Solving

= **Step-by-Step Solutions** begin on page R20.
Extra Practice begins on page 969.

Example 1
p. 274

PROOF **Write a paragraph proof.**

6. **Given:** $\overline{CE}$ bisects $\angle BED$; $\angle BCE$ and $\angle ECD$ are right angles.
Prove: $\triangle ECB \cong \triangle ECD$

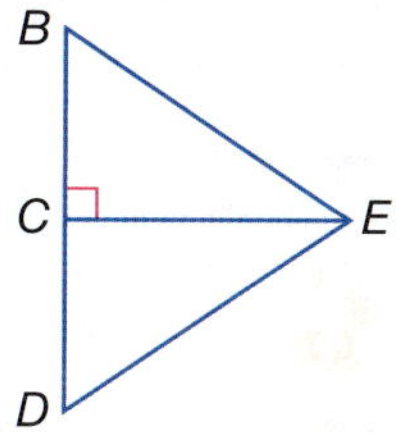

7. **Given:** $\angle W \cong \angle Y$, $\overline{WZ} \cong \overline{YZ}$, $\overline{XZ}$ bisects $\angle WZY$.
Prove: $\triangle XWZ \cong \triangle XYZ$

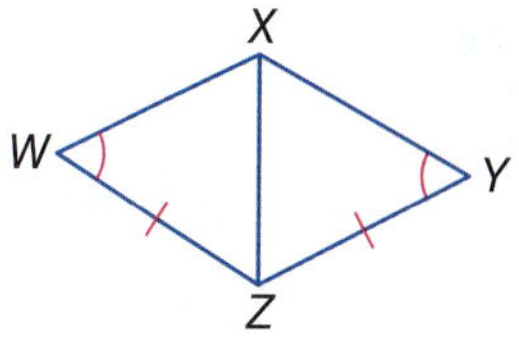

8. **TOYS** The object of the toy shown is to make the two spheres meet and strike each other repeatedly on one side of the wand and then again on the other side. If $\angle JKL \cong \angle MLK$ and $\angle JLK \cong \angle MKL$, prove that $\overline{JK} \cong \overline{ML}$.

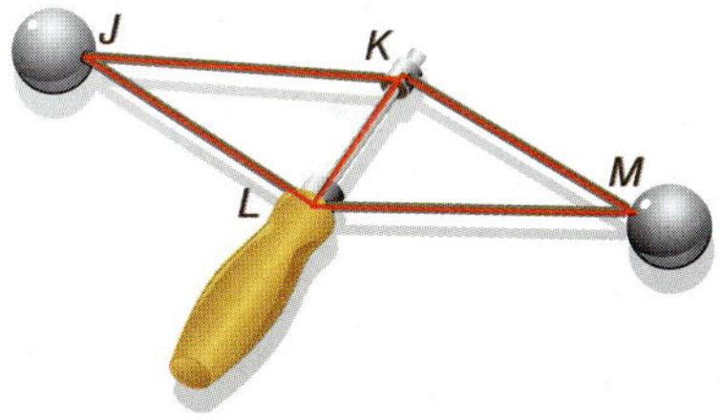

Example 2
p. 275

PROOF **Write a two-column proof.**

9 **Given:** V is the midpoint of $\overline{YW}$; $\overline{UY} \parallel \overline{XW}$.
Prove: $\triangle UVY \cong \triangle XVW$

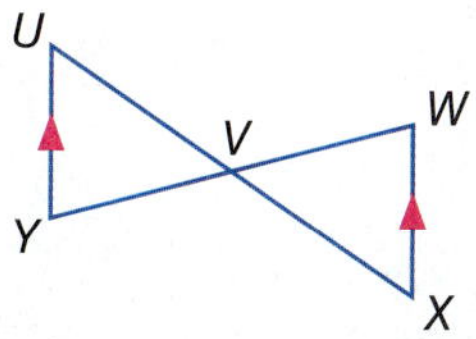

10. **Given:** $\overline{MS} \cong \overline{RQ}$, $\overline{MS} \parallel \overline{RQ}$
Prove: $\triangle MSP \cong \triangle RQP$

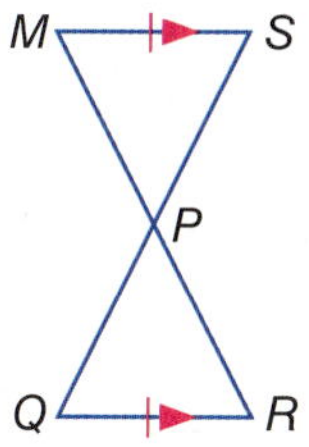

11. **PROOF** Write a flow proof.

Given: $\angle A$ and $\angle C$ are right angles. $\angle ABE \cong \angle CBD$, $\overline{AE} \cong \overline{CD}$
Prove: $\overline{BE} \cong \overline{BD}$

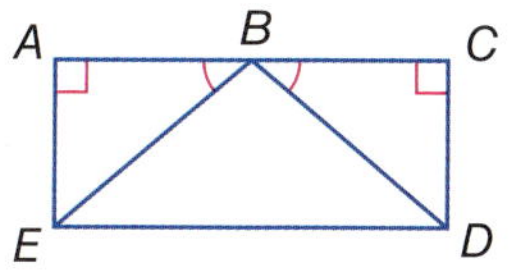

12. **PROOF** Write a flow proof.

Given: $\overline{KM}$ bisects $\angle JML$; $\angle J \cong \angle L$.
Prove: $\overline{JM} \cong \overline{LM}$

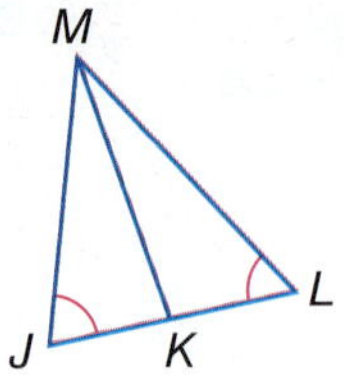

Example 3 p. 275

Real-World Link

In 2006, more than 3 million girls participated in high school sports. More girls participated in competitive rowing than boys.

Source: National Federation of State High School Associations

13. **FITNESS** A high school wants to hold a 1500-meter regatta on Lake Powell but is unsure if the lake is long enough. To measure the distance across the lake, the crew members locate the vertices of the triangles below and find the measures of the lengths of $\triangle HJK$ as shown below.

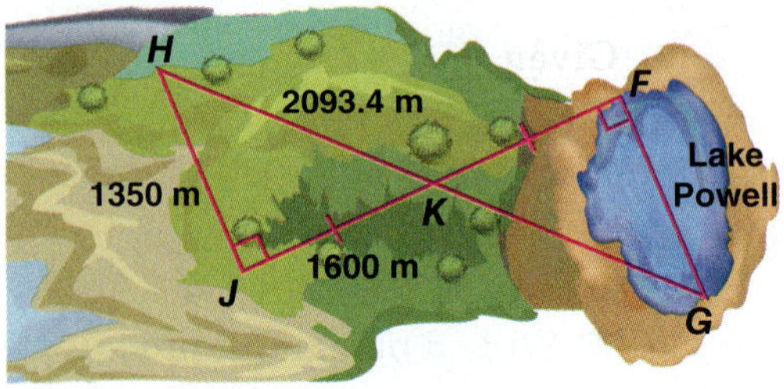

a. Explain how the crew team can use the triangles formed to estimate the distance FG across the lake.

b. Using the measures given, is the lake long enough for the team to use as the location for their regatta? Explain your reasoning.

ALGEBRA **Find the value of the variable that yields congruent triangles.**

14. $\triangle BCD \cong \triangle WXY$

15. $\triangle MHJ \cong \triangle PQJ$

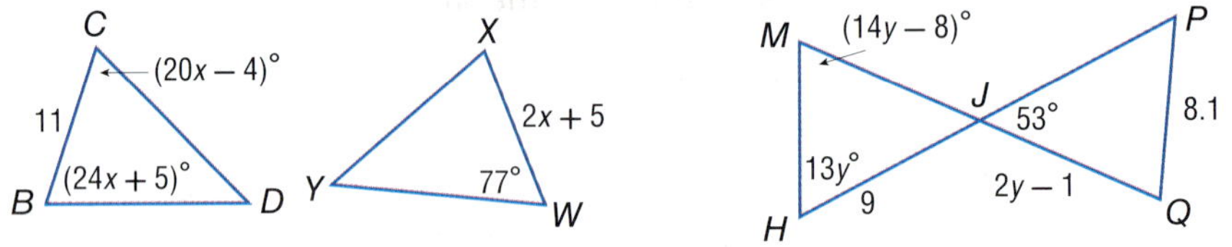

16. **THEATER DESIGN** The trusses of the roof of the outdoor theater shown below appear to be several different pairs of congruent triangles. Assume that trusses that appear to lie on the same line actually lie on the same line.

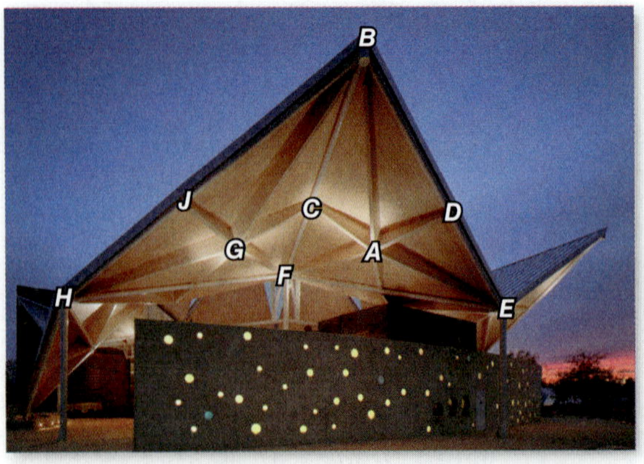

a. If $\overline{AB}$ bisects $\angle CBD$ and $\angle CAD$, prove that $\triangle ABC \cong \triangle ABD$.

b. If $\triangle ABC \cong \triangle ABD$ and $\angle FCA \cong \angle EDA$, prove that $\triangle CAF \cong \triangle DAE$.

c. If $\overline{HB} \cong \overline{EB}$, $\angle BHG \cong \angle BEA$, $\angle HGJ \cong \angle EAD$, and $\angle JGB \cong \angle DAB$, prove that $\triangle BHG \cong \triangle BEA$.

PROOF Write a paragraph proof.

17. **Given:** $\overline{AE} \perp \overline{DE}$, $\overline{EA} \perp \overline{AB}$, C is the midpoint of $\overline{AE}$.
Prove: $\overline{CD} \cong \overline{CB}$

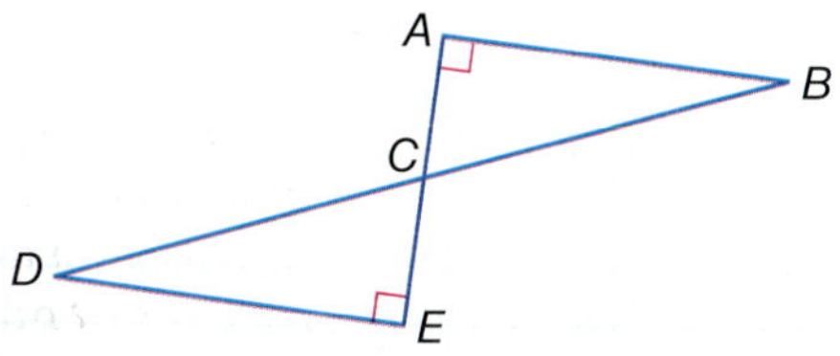

18. **Given:** $\angle F \cong \angle J$, $\overline{FH} \parallel \overline{GJ}$
Prove: $\overline{FH} \cong \overline{JG}$

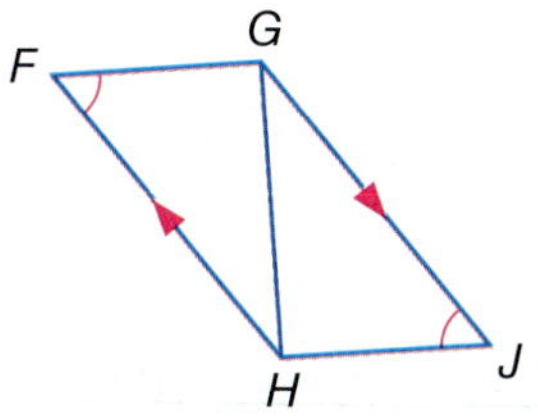

PROOF Write a two-column proof.

19. **Given:** $\angle K \cong \angle M$, $\overline{KP} \perp \overline{PR}$, $\overline{MR} \perp \overline{PR}$
Prove: $\angle KPL \cong \angle MRL$

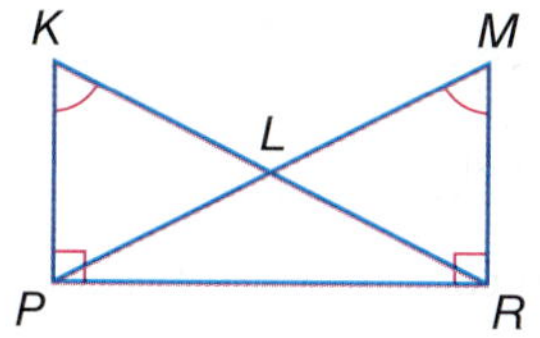

20. **Given:** $\overline{QR} \cong \overline{SR} \cong \overline{WR} \cong \overline{VR}$
Prove: $\overline{QT} \cong \overline{WU}$

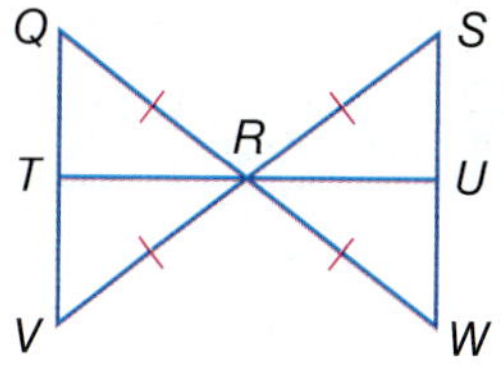

Real-World Link

Bicycles are sized based on the length of the seat tube. Standard bicycles for adults normally range from 12 to 26 inches and are considered to fit a rider when, with both feet placed flat on the ground, he or she can straddle the top tube with about 1 inch of clearance.

Source: Encarta

21. **FITNESS** The seat tube of a bicycle forms a triangle with each seat and chain stay as shown. If each seat stay makes a 44° angle with its corresponding chain stay and each chain stay makes a 68° angle with the seat tube, show that the two seat stays are the same length.

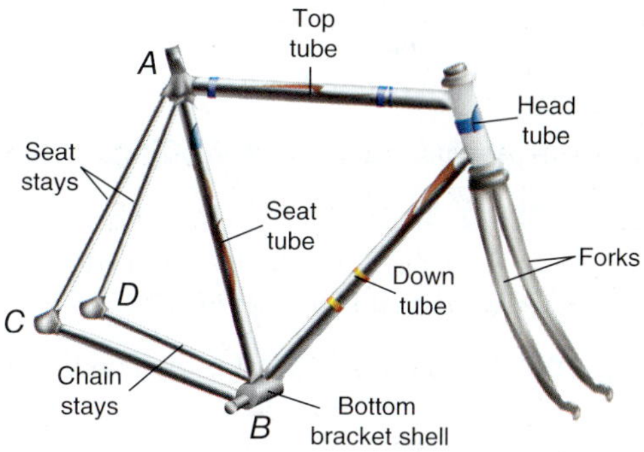

H.O.T. Problems

Use **H**igher-**O**rder **T**hinking Skills

22. **OPEN ENDED** Draw and label two triangles that could be proved congruent by ASA.

23. **ERROR ANALYSIS** Tyrone says it is not possible to show that $\triangle ADE \cong \triangle ACB$. Lorenzo disagrees, explaining that since $\angle ADE \cong \angle ACB$, and $\angle A \cong \angle A$ by the Reflexive Property, $\triangle ADE \cong \triangle ACB$. Is either of them correct? Explain.

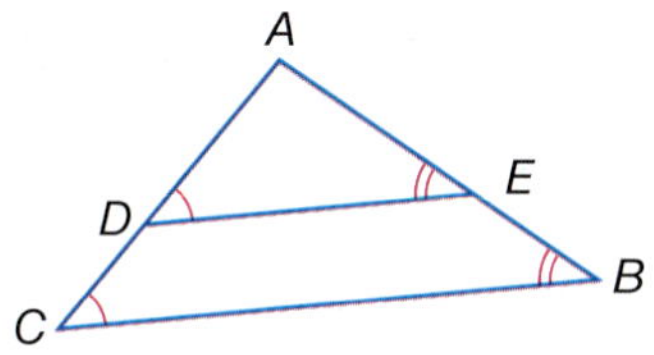

24. **REASONING** Find a counterexample to show why SSA (Side-Side-Angle) cannot be used to prove the congruence of two triangles.

25. **CHALLENGE** Using the information given in the diagram, write a flow proof to show that $\triangle PVQ \cong \triangle SVT$.

26. **WRITING IN MATH** Summarize the methods described in Lessons 4-3, 4-4, and 4-5 for proving triangle congruence into a chart that explains when to use each method.

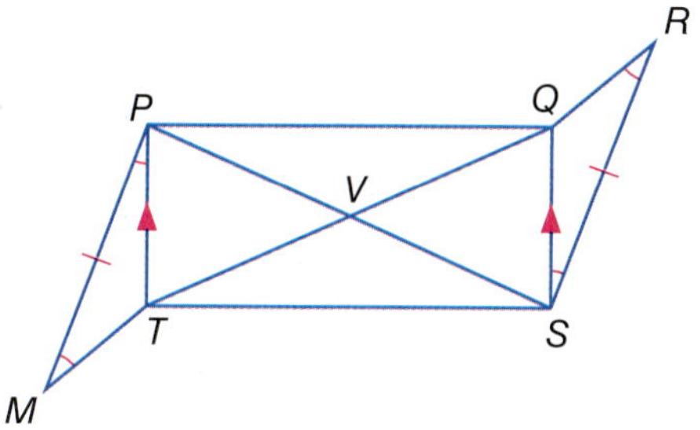

Standardized Test Practice

27. Given: $\overline{BC}$ is perpendicular to $\overline{AD}$; $\angle 1 \cong \angle 2$.

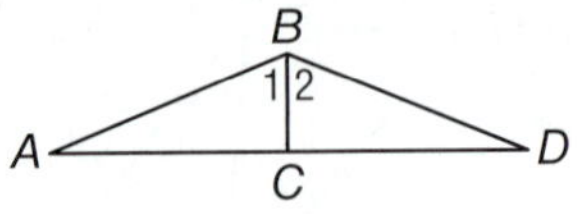

Which theorem or postulate could be used to prove $\triangle ABC \cong \triangle DBC$?

A AAS
B ASA
C SAS
D SSS

28. SHORT RESPONSE Write an expression that can be used to find the values of $s(n)$ in the table.

n	−8	−4	−1	0	1
$s(n)$	1.00	2.00	2.75	3.00	3.25

29. ALGEBRA If −7 is multiplied by a number greater than 1, which of the following describes the result?

F a number greater than 7
G a number between −7 and 7
H a number greater than −7
J a number less than −7

30. SAT/ACT $\sqrt{121 + 104} = ?$

A 15
B 21
C 125
D 225

Spiral Review

Determine whether $\triangle ABC \cong \triangle XYZ$. Explain. (Lesson 4-4)

31. $A(6, 4), B(1, -6), C(-9, 5), X(0, 7), Y(5, -3), Z(15, 8)$

32. $A(0, 5), B(0, 0), C(-2, 0), X(4, 8), Y(4, 3), Z(6, 3)$

33. ALGEBRA If $\triangle RST \cong \triangle JKL$, $RS = 7$, $ST = 5$, $RT = 9 + x$, $JL = 2x - 10$, and $JK = 4y - 5$, draw and label a figure to represent the congruent triangles. Then find x and y. (Lesson 4-3)

34. FINANCIAL LITERACY Maxine charges \$5 to paint a mailbox and \$4 per hour to mow a lawn. Write an equation to represent the amount of money Maxine can earn from a homeowner who has his or her mailbox painted and lawn mowed. (Lesson 3-4)

Copy and complete each truth table. (Lesson 2-2)

35.

p	q	$\sim p$	$\sim p \vee q$
F	T		
T	T		
F	F		
T	F		

36.

p	q	$\sim q$	$\sim q \wedge p$
F		F	
T		T	
T		F	
F		T	

Skills Review

PROOF Write a two-column proof for each of the following. (Lesson 3-5)

37. Given: $\angle 2 \cong \angle 1$
$\angle 1 \cong \angle 3$
Prove: $\overline{AB} \parallel \overline{DE}$

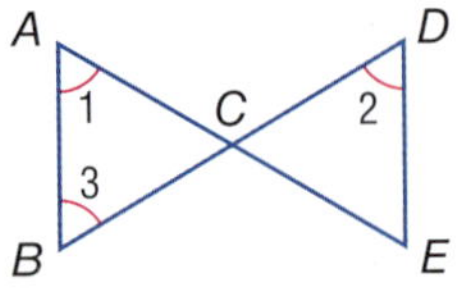

38. Given: $\angle MJK \cong \angle KLM$
$\angle LMJ$ and $\angle KLM$ are supplementary.
Prove: $\overline{KJ} \parallel \overline{LM}$

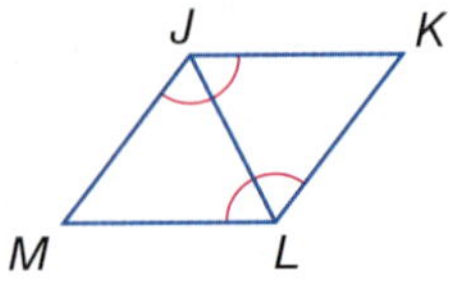

EXTEND 4-5

Geometry Lab
Congruence in Right Triangles

In Lessons 4-4 and 4-5, you learned theorems and postulates to prove triangles congruent. How do these theorems and postulates apply to right triangles?

Study each pair of right triangles.

a.

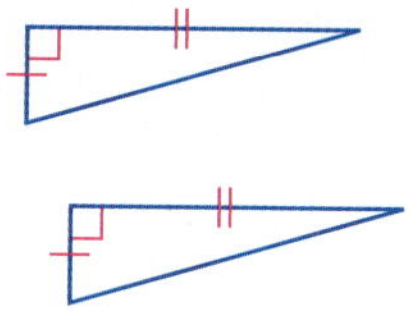

b.

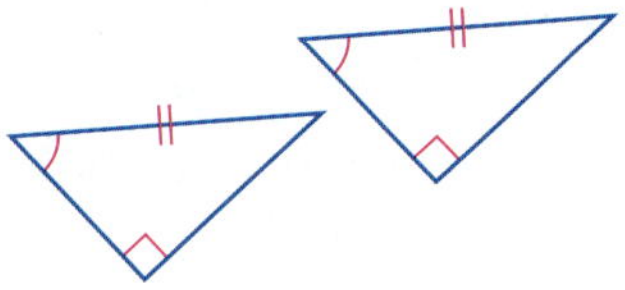

c.

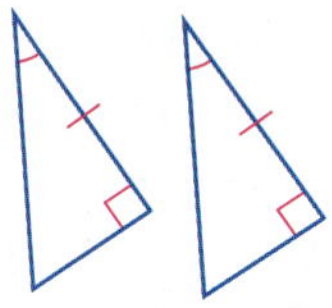

Analyze

1. Is each pair of triangles congruent? If so, which congruence theorem or postulate applies?

2. Rewrite the congruence rules from Exercise 1 using *leg*, (L), or *hypotenuse*, (H), to replace *side*. Omit the *A* for any right angle since we know that all right triangles contain a right angle and all right angles are congruent.

3. **MAKE A CONJECTURE** If you know that the corresponding legs of two right triangles are congruent, what other information do you need to declare the triangles congruent? Explain.

In Lesson 4-5, you learned that SSA is not a valid test for determining triangle congruence. Can SSA be used to prove right triangles congruent?

Get Animated ACTIVITY SSA and Right Triangles

Step 1

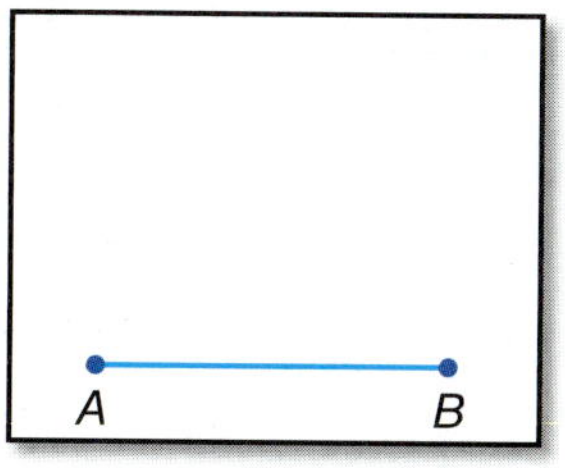

Draw $\overline{AB}$ so that $AB = 6$ centimeters.

Step 2

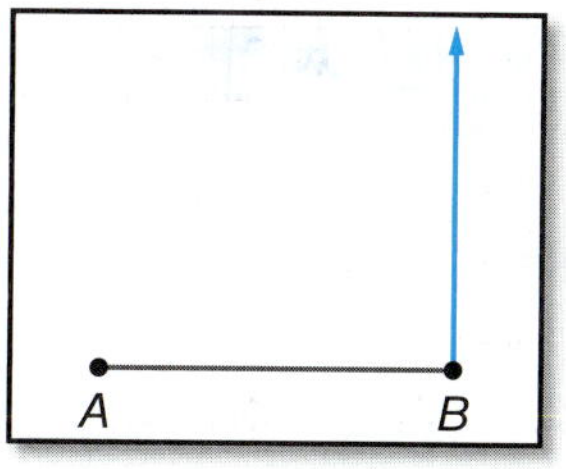

Use a protractor to draw a ray from *B* that is perpendicular to $\overline{AB}$.

Step 3

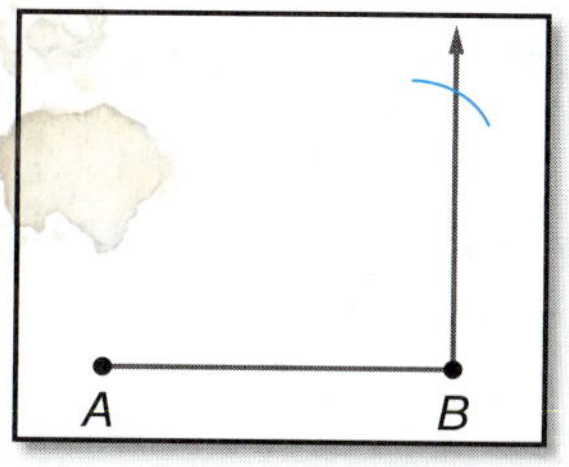

Open your compass to a width of 8 centimeters. Place the point at *A* and draw an arc to intersect the ray.

Step 4

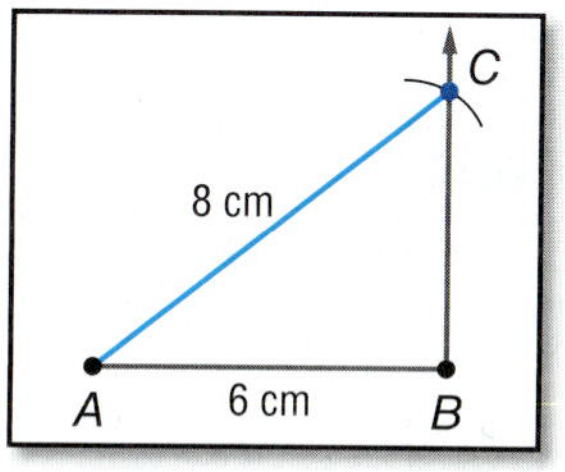

Label the intersection *C* and draw $\overline{AC}$ to complete $\triangle ABC$.

Analyze

4. Does the model yield a unique triangle?

5. Can you use the lengths of the hypotenuse and a leg to show right triangles are congruent?

6. **Make a conjecture** about the case of SSA that exists for right triangles.

(continued on the next page)

Activities 1 and 2 provide evidence for four ways to prove right triangles congruent.

Theorems — Right Triangle Congruence

Theorem 4.6 Leg-Leg Congruence
If the legs of one right triangle are congruent to the corresponding legs of another right triangle, then the triangles are congruent.
Abbreviation *LL*

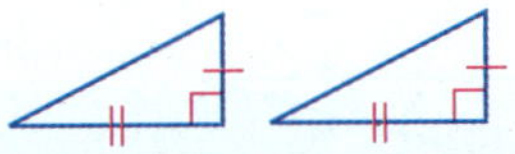

Theorem 4.7 Hypotenuse-Angle Congruence
If the hypotenuse and acute angle of one right triangle are congruent to the hypotenuse and corresponding acute angle of another right triangle, then the two triangles are congruent.
Abbreviation *HA*

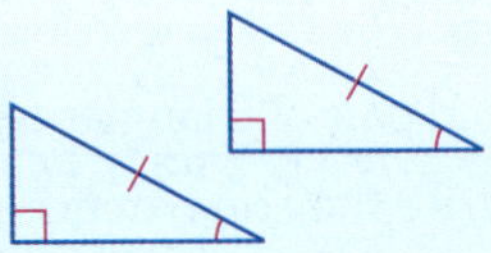

Theorem 4.8 Leg-Angle Congruence
If one leg and an acute angle of one right triangle are congruent to the corresponding leg and acute angle of another right triangle, then the triangles are congruent.
Abbreviation *LA*

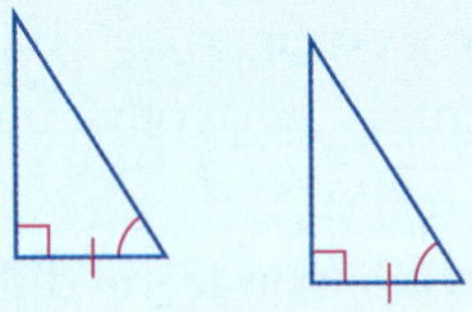

Theorem 4.9 Hypotenuse-Leg Congruence
If the hypotenuse and a leg of one right triangle are congruent to the hypotenuse and corresponding leg of another right triangle, then the triangles are congruent.
Abbreviation *HL*

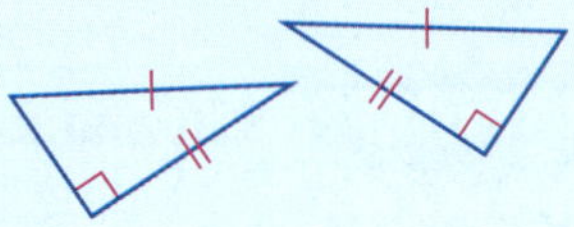

Exercises

Determine whether each pair of triangles is congruent. If yes, tell which postulate or theorem applies.

7.

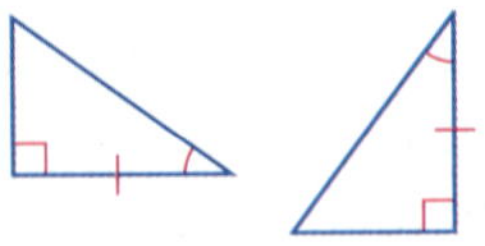

8.

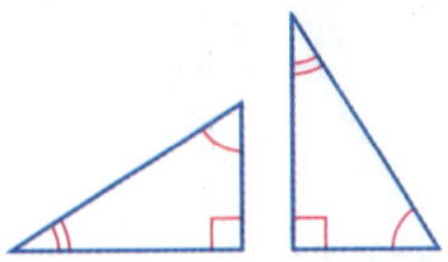

9.

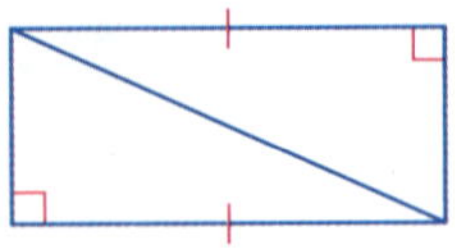

PROOF **Write a proof for each of the following.**

10. Theorem 4.6

11. Theorem 4.7

12. Theorem 4.8 (*Hint*: There are two possible cases.)

13. Theorem 4.9 (*Hint*: Use the Pythagorean Theorem.)

Use the figure at the right.

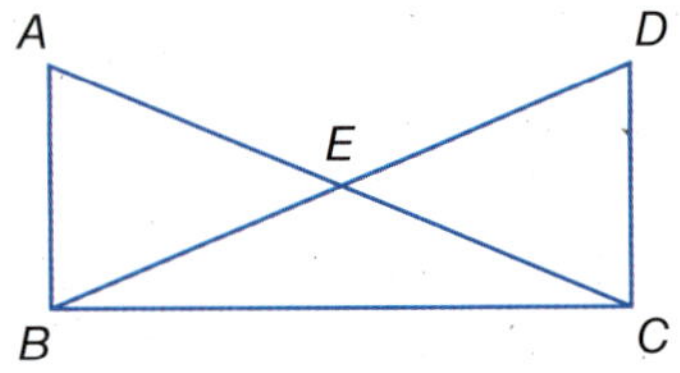

14. Given: $\overline{AB} \perp \overline{BC}$, $\overline{DC} \perp \overline{BC}$
$\overline{AC} \cong \overline{BD}$
Prove: $\overline{AB} \cong \overline{DC}$

15. Given: $\overline{AB} \parallel \overline{DC}$, $\overline{AB} \perp \overline{BC}$
E is the midpoint of $\overline{AC}$ and $\overline{BD}$.
Prove: $\overline{AC} \cong \overline{DB}$

4-6 Isosceles and Equilateral Triangles

Then
You identified isosceles and equilateral triangles. (Lesson 4-1)

Now
- Use properties of isosceles triangles.
- Use properties of equilateral triangles.

New Vocabulary
legs of an isosceles triangle
vertex angle
base angles

Math Online
glencoe.com

Why?

The tracks on the roller coaster have triangular reinforcements between the tracks for support and stability. The triangle supports in the photo are isosceles triangles.

Properties of Isosceles Triangles Recall from Lesson 4-1 that isosceles triangles have at least two congruent sides. The parts of an isosceles triangle have special names.

The two congruent sides are called the **legs**, and the angle with sides that are the legs is called the **vertex angle**. The side of the triangle opposite the vertex angle is called the *base*. The two angles formed by the base and the congruent sides are called the **base angles**.

$\angle 1$ is the vertex angle.
$\angle 2$ and $\angle 3$ are the base angles.

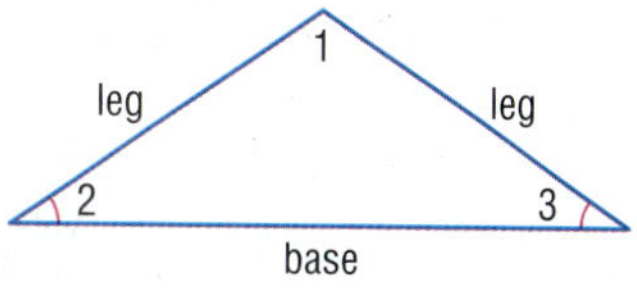

Theorems Isosceles Triangle

4.10 Isosceles Triangle Theorem
If two sides of a triangle are congruent, then the angles opposite those sides are congruent.

Example If $\overline{AC} \cong \overline{BC}$, then $\angle 2 \cong \angle 1$.

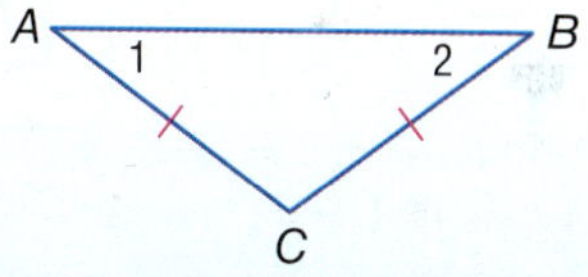

4.11 Converse of Isosceles Triangle Theorem
If two angles of a triangle are congruent, then the sides opposite those angles are congruent.

Example If $\angle 1 \cong \angle 2$, then $\overline{FE} \cong \overline{DE}$.

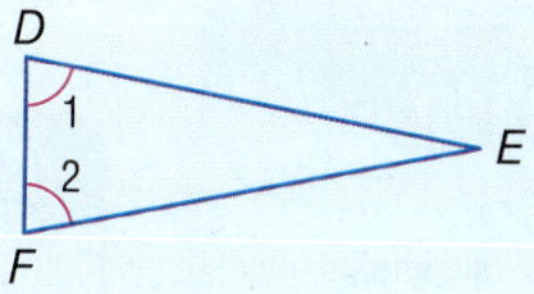

You will prove Theorem 4.11 in Exercise 37.

EXAMPLE 1 Congruent Segments and Angles

a. Name two unmarked congruent angles.

$\angle ACB$ is opposite $\overline{AB}$ and $\angle B$ is opposite $\overline{AC}$, so $\angle ACB \cong \angle B$.

b. Name two unmarked congruent segments.

$\overline{AD}$ is opposite $\angle ACD$ and $\overline{AC}$ is opposite $\angle D$, so $\overline{AD} \cong \overline{AC}$.

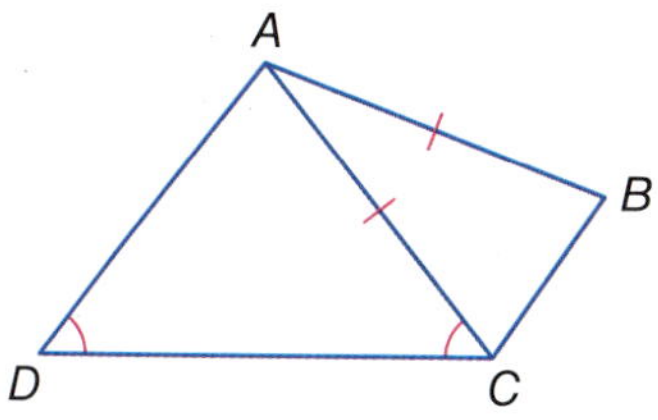

Guided Practice

1A. Name two unmarked congruent angles.

1B. Name two unmarked congruent segments.

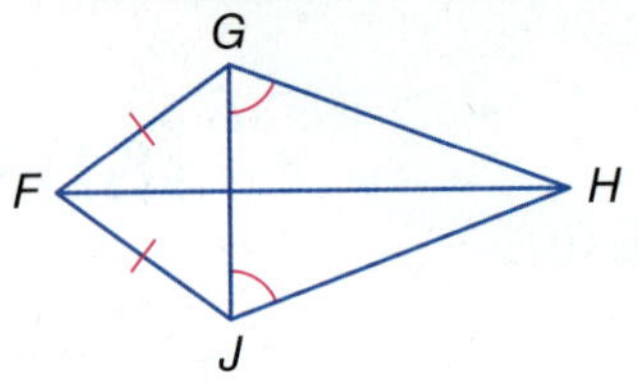

Personal Tutor glencoe.com

To prove the Isosceles Triangle Theorem, draw an auxiliary line and use the two triangles formed.

PROOF Isosceles Triangle Theorem

Given: $\triangle LMP$; $\overline{LM} \cong \overline{LP}$

Prove: $\angle M \cong \angle P$

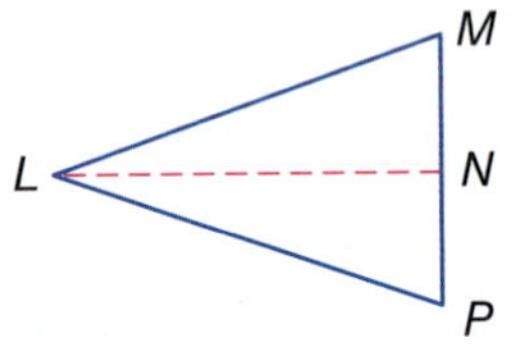

Proof:

Statements	Reasons
1. Let N be the midpoint of $\overline{MP}$.	**1.** Every segment has exactly one midpoint.
2. Draw an auxiliary segment $\overline{LN}$.	**2.** Two points determine a line.
3. $\overline{MN} \cong \overline{PN}$	**3.** Midpoint Theorem
4. $\overline{LN} \cong \overline{LN}$	**4.** Reflexive Property of Congruence
5. $\overline{LM} \cong \overline{LP}$	**5.** Given
6. $\triangle LMN \cong \triangle LPN$	**6.** SSS
7. $\angle M \cong \angle P$	**7.** CPCTC

Properties of Equilateral Triangles The Isosceles Triangle Theorem leads to two corollaries about the angles of an equilateral triangle.

Review Vocabulary

Equilateral Triangle a triangle with three congruent sides (Lesson 4-1)

Corollaries Equilateral Triangle

4.3 A triangle is equilateral if and only if it is equiangular.

Example If $\angle A \cong \angle B \cong \angle C$, then $\overline{AB} \cong \overline{BC} \cong \overline{CA}$.

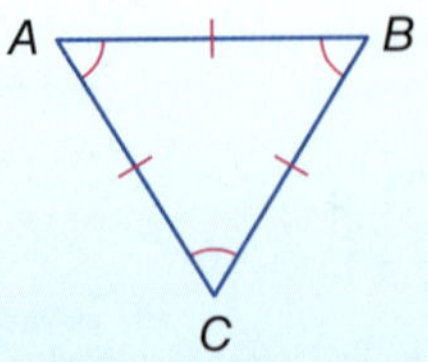

4.4 Each angle of an equilateral triangle measures 60.

Example If $\overline{DE} \cong \overline{EF} \cong \overline{FE}$, then $m\angle A = m\angle B = m\angle C = 60$.

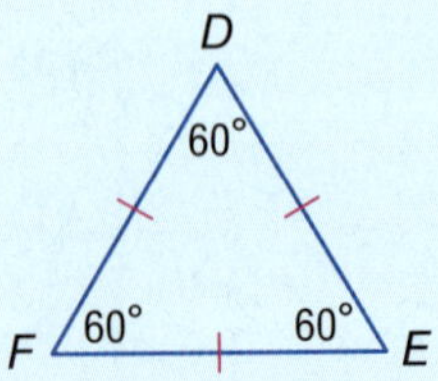

You will prove Corollaries 4.3 and 4.4 in Exercises 35 and 36.

EXAMPLE 2 Find Missing Measures

Find each measure.

a. $m\angle Y$

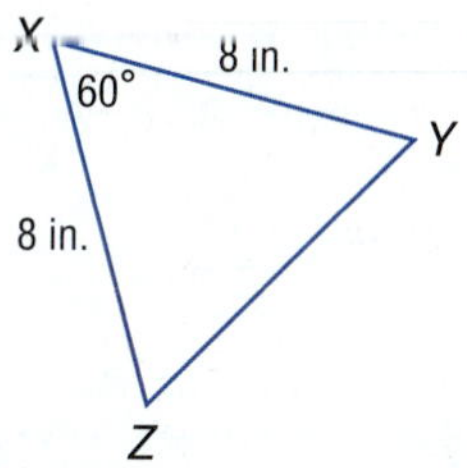

Since $XY = XZ$, $\overline{XY} \cong \overline{XZ}$. By the Isosceles Triangle Theorem, base angles Z and Y are congruent, so $m\angle Z = m\angle Y$. Use the Triangle Sum Theorem to write and solve an equation to find $m\angle Y$.

$m\angle X + m\angle Y + m\angle Z = 180$	**Triangle Sum Theorem**
$60 + m\angle Y + m\angle Y = 180$	**$m\angle X = 60$, $m\angle Z = m\angle Y$**
$60 + 2(m\angle Y) = 180$	**Simplify.**
$2(m\angle Y) = 120$	**Subtract 60 from each side.**
$m\angle Y = 60$	**Divide each side by 2.**

b. YZ

$m\angle Z = m\angle Y$, so $m\angle Z = 60$ by substitution. Since $m\angle X = 60$, all three angles measure 60, so the triangle is equiangular. Because an equiangular triangle is also equilateral, $XY = XZ = ZY$. Since $XY = 8$ inches, $YZ = 8$ inches by substitution.

Guided Practice

2A. $m\angle M$ **2B.** PN

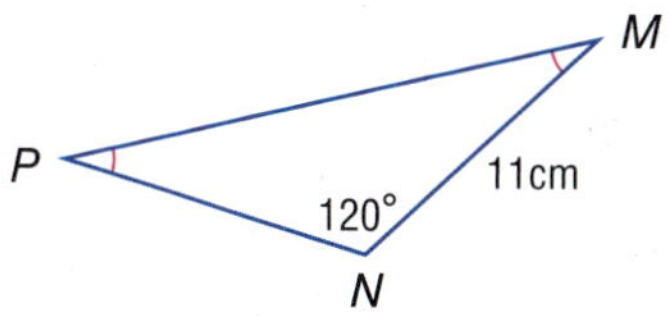

Personal Tutor glencoe.com

> **StudyTip**
>
> **Isosceles Triangles** As you discovered in Example 2, any isosceles triangle that has one 60° angle must be an equilateral triangle.

You can use the properties of equilateral triangles and algebra to find missing values.

EXAMPLE 3 Find Missing Values

ALGEBRA Find the value of each variable.

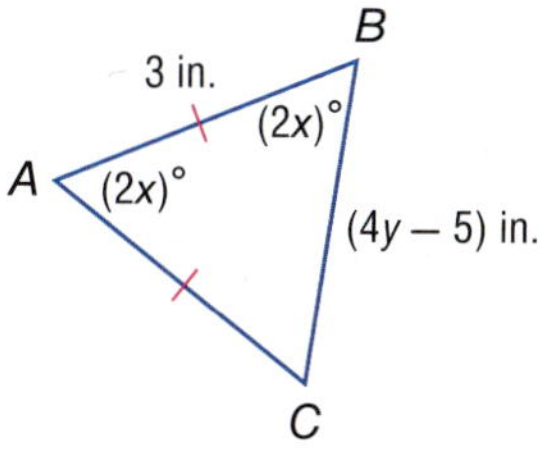

Since $\angle B = \angle A$, $\overline{AC} \cong \overline{BC}$ by the Converse of the Isosceles Triangle Theorem. All of the sides of the triangle are congruent, so the triangle is equilateral. Each angle of an equilateral triangle measures 60°, so $2x = 60$ and $x = 30$.

The triangle is equilateral, so all of the sides are congruent, and the lengths of all of the sides are equal.

$AB = BC$	**Definition of equilateral triangle**
$3 = 4y - 5$	**Substitution**
$8 = 4y$	**Add 5 to each side.**
$2 = y$	**Divide each side by 4.**

Guided Practice

3. Find the value of each variable.

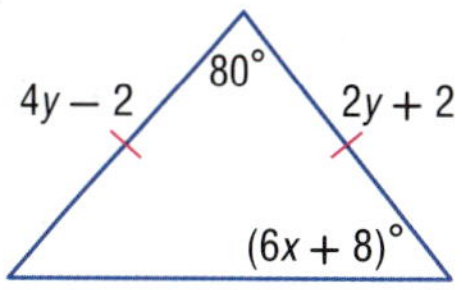

Personal Tutor glencoe.com

Real-World Link

Biosphere II is the largest totally enclosed ecosystem ever built, covering 3.14 acres in Oracle, Arizona. The controlled-environment facility is 91 feet at its highest point, and it has 6500 windows that enclose a volume of 7.2 million cubic feet.

Source: University of Arizona

Real-World EXAMPLE 4 Apply Triangle Congruence

ENVIRONMENT **Refer to the photo of Biosphere II at the right. $\triangle ACE$ is an equilateral triangle. F is the midpoint of $\overline{AE}$, D is the midpoint of $\overline{EC}$, and B is the midpoint of $\overline{CA}$. Prove that $\triangle FBD$ is also equilateral.**

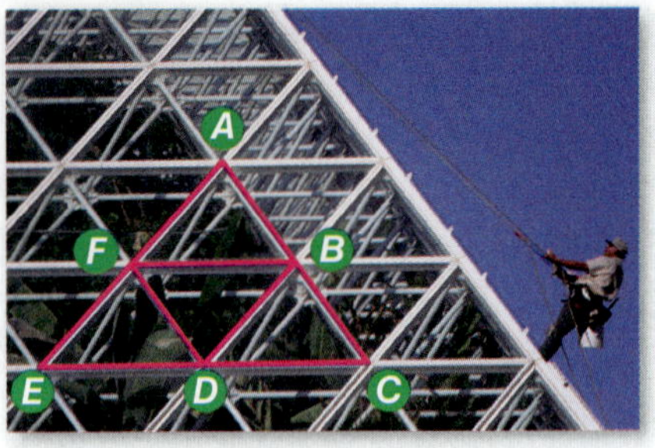

Given: $\triangle ACE$ is equilateral. F is the midpoint of $\overline{AE}$, D is the midpoint of $\overline{EC}$, and B is the midpoint of $\overline{CA}$.

Prove: $\triangle FBD$ is equilateral.

Proof:

Statements	Reasons
1. $\triangle ACE$ is equilateral.	**1.** Given
2. F is the midpoint of $\overline{AE}$, D is the midpoint of $\overline{EC}$, and B is the midpoint of $\overline{CA}$.	**2.** Given
3. $m\angle A = 60$, $m\angle C = 60$, $m\angle E = 60$	**3.** Each angle of an equilateral triangle measures 60.
4. $\angle A \cong \angle C \cong \angle E$	**4.** Definition of congruence and substitution
5. $\overline{AE} \cong \overline{EC} \cong \overline{CA}$	**5.** Definition of equilateral triangle
6. $AE = EC = CA$	**6.** Definition of congruence
7. $\overline{AF} \cong \overline{FE}$, $\overline{ED} \cong \overline{DC}$, $\overline{CB} \cong \overline{BA}$	**7.** Midpoint Theorem
8. $AF = FE$, $ED = DC$, $CB = BA$	**8.** Definition of congruence
9. $AF + FE = AE$, $ED + DC = EC$, $CB + BA = CA$	**9.** Segment Addition Postulate
10. $AF + AF = AE$, $FE + FE = AE$, $ED + ED = EC$, $DC + DC = EC$, $CB + CB = CA$, $BA + BA = CA$	**10.** Substitution
11. $2AF = AE$, $2FE = AE$, $2ED = EC$, $2DC = EC$, $2CB = CA$, $2BA = CA$	**11.** Addition Property
12. $2AF = AE$, $2FE = AE$, $2ED = AE$, $2DC = AE$, $2CB = AE$, $2BA = AE$	**12.** Substitution Property
13. $2AF = 2ED = 2CB$, $2FE = 2DC = 2BA$	**13.** Transitive Property
14. $AF = ED = CB$, $FE = DC = BA$	**14.** Division Property
15. $\overline{AF} \cong \overline{ED} \cong \overline{CB}$, $\overline{FE} \cong \overline{DC} \cong \overline{BA}$	**15.** Definition of congruence
16. $\triangle AFB \cong \triangle EDF \cong \triangle CBD$	**16.** SAS
17. $\overline{DF} \cong \overline{FB} \cong \overline{BD}$	**17.** CPCTC
18. $\triangle FBD$ is equilateral.	**18.** Definition of equilateral triangle

Guided Practice

4. Given that $\triangle ACE$ is equilateral, $\overline{FB} \parallel \overline{EC}$, $\overline{FD} \parallel \overline{BC}$, $\overline{BD} \parallel \overline{EF}$, and D is the midpoint of $\overline{EC}$, prove that $\triangle FED \cong \triangle BDC$.

Personal Tutor glencoe.com

Check Your Understanding

Example 1 p. 283

Refer to the figure at the right.

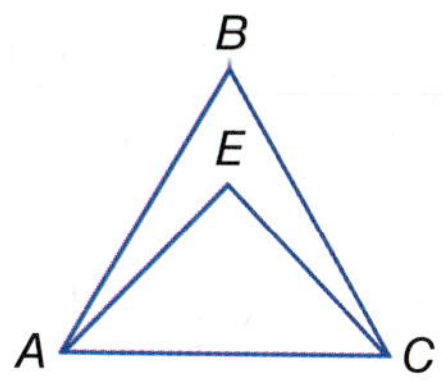

1. If $\overline{AB} \cong \overline{CB}$, name two congruent angles.

2. If $\angle EAC \cong \angle ECA$, name two congruent segments.

Example 2 p. 285

Find each measure.

3. *FH*

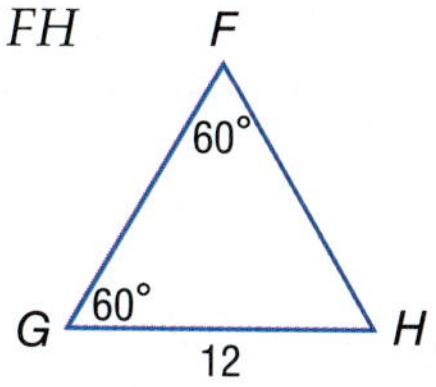

4. $m\angle MRP$

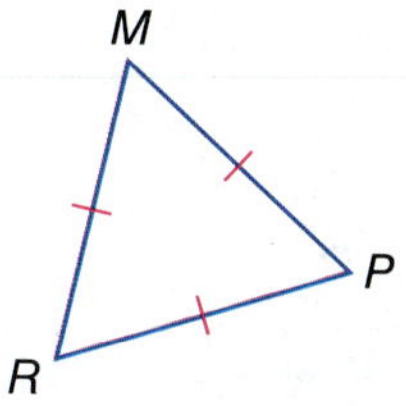

Example 3 p. 285

ALGEBRA **Find the value of each variable.**

5.

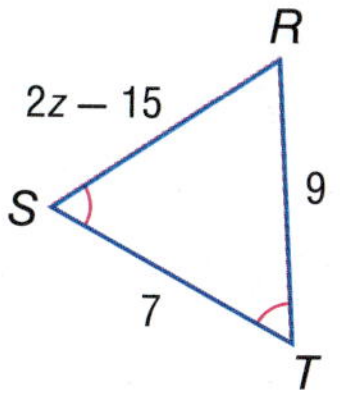

6.

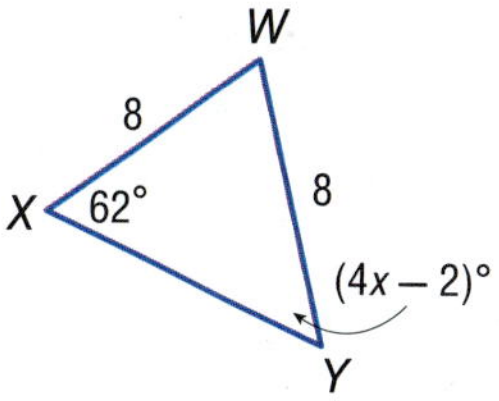

Example 4 p. 286

7. PROOF **Write a two-column proof.**

Given: $\triangle ABC$ is isosceles; $\overline{EB}$ bisects $\angle ABC$.

Prove: $\triangle ABE \cong \triangle CBE$

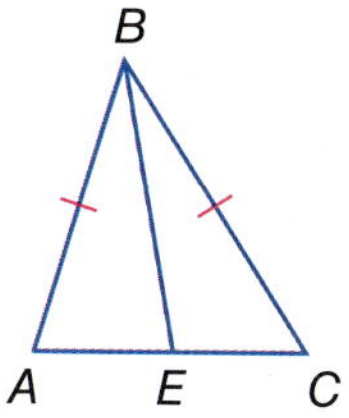

8. ROLLER COASTERS The roller coaster track shown in the photo on page 283 appears to be composed of congruent triangles. A portion of the track is shown.

a. If $\overline{QR}$ and $\overline{ST}$ are perpendicular to $\overline{QT}$, $\triangle VSR$ is isosceles with base $\overline{SR}$, and $\overline{QT} \parallel \overline{SR}$, prove that $\triangle RQV \cong \triangle STV$.

b. If $VR = 2.5$ meters and $QR = 2$ meters, find the distance between $\overline{QR}$ and $\overline{ST}$. Explain your reasoning.

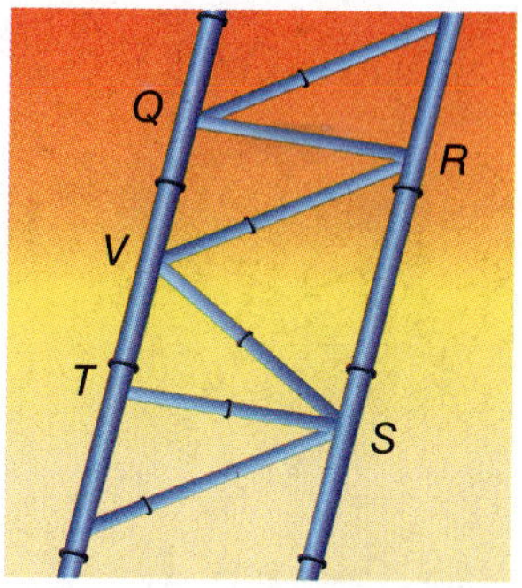

Practice and Problem Solving

● = **Step-by-Step Solutions** begin on page R20.
Extra Practice begins on page 969.

Example 1 p. 283

Refer to the figure at the right.

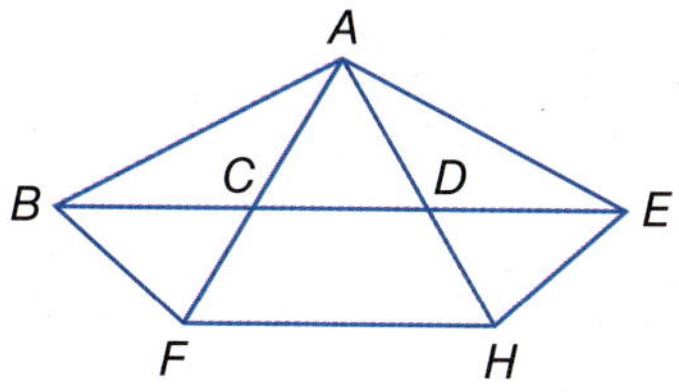

9. If $\overline{AB} \cong \overline{AE}$, name two congruent angles.

10. If $\angle ABF \cong \angle AFB$, name two congruent segments.

11. If $\overline{CA} \cong \overline{DA}$, name two congruent angles.

12. If $\angle DAE \cong \angle DEA$, name two congruent segments.

13. If $\angle BCF \cong \angle BFC$, name two congruent segments.

14. If $\overline{FA} \cong \overline{AH}$, name two congruent angles.

Example 2 p. 285

Find each measure.

15. $m\angle BAC$

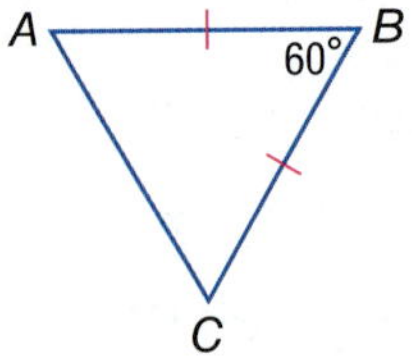

16. $m\angle SRT$

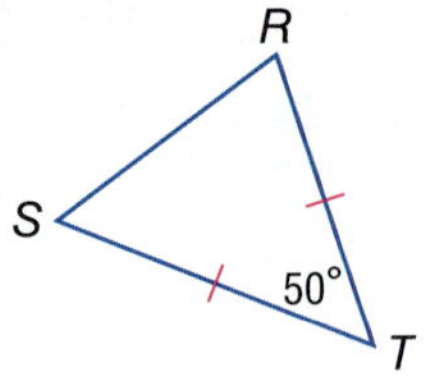

17. TR

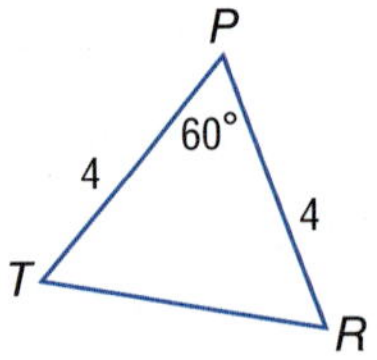

18. CB

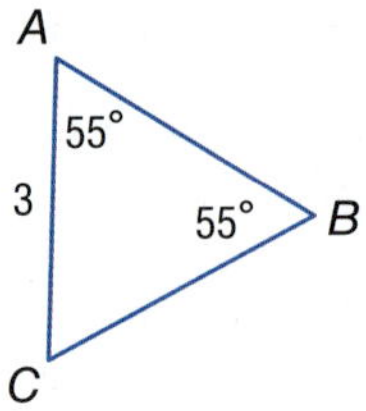

Example 3 p. 285

ALGEBRA **Find the value of each variable.**

19.

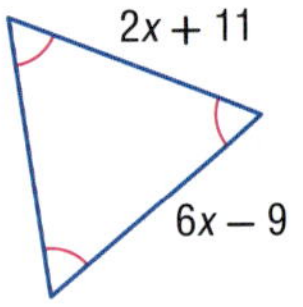

20.

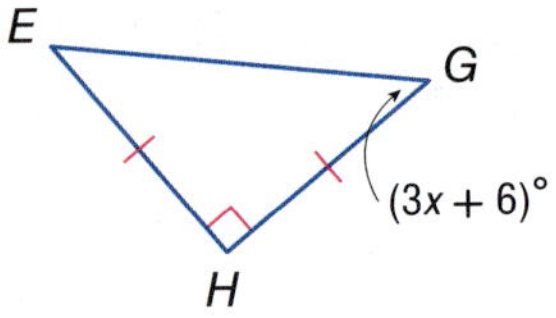

21.

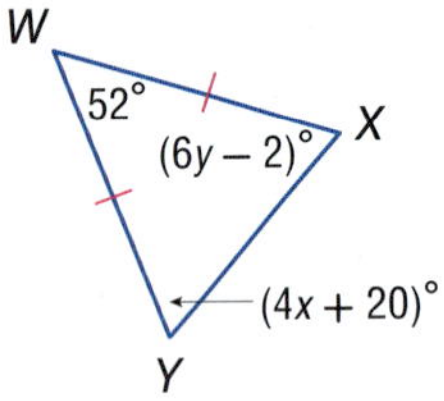

22.

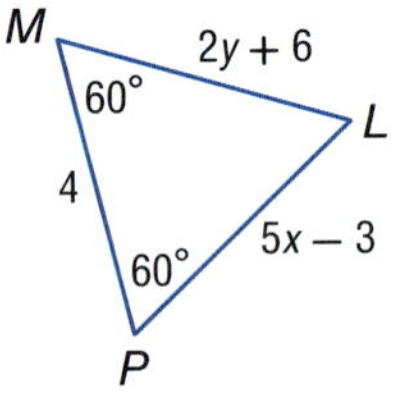

Example 4 p. 286

PROOF **Write a paragraph proof.**

23. Given: $\triangle HJM$ is an isosceles triangle, and $\triangle HKL$ is an equilateral triangle. $\angle JKH$ and $\angle HKL$ and $\angle HLK$ and $\angle MLH$ are supplementary.

Prove: $\angle JHK \cong \angle MHL$

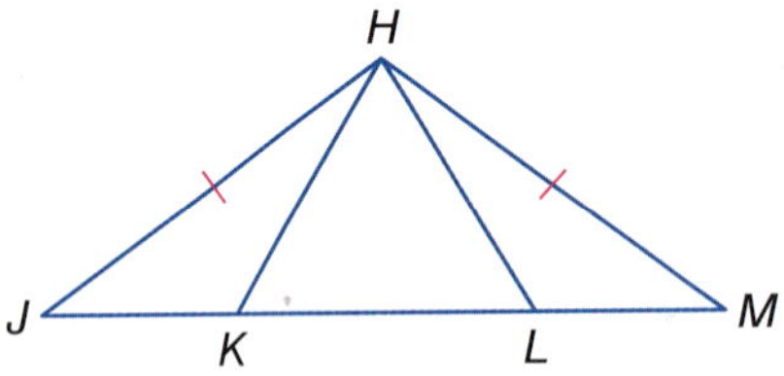

24. Given: $\overline{XY} \cong \overline{XZ}$
W is the midpoint of $\overline{XY}$.
Q is the midpoint of $\overline{XZ}$.

Prove: $\overline{WZ} \cong \overline{QY}$

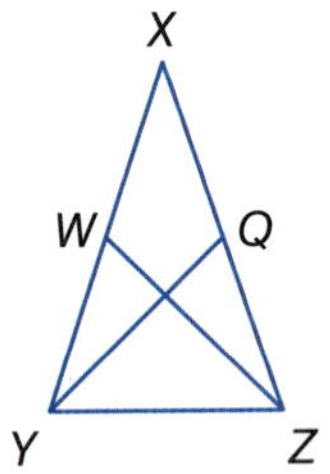

Real-World Link

Babysitter training courses are available that include safety and first aid, choosing age-appropriate toys and games, and diapering and feeding techniques.

Source: American Red Cross

25. BABYSITTING While babysitting her neighbor's children, Elisa observes that the supports on either side of a park swing set form two sets of triangles. Using a jump rope to measure, Elisa is able to determine that $\overline{AB} \cong \overline{AC}$, but $\overline{BC} \not\cong \overline{AB}$.

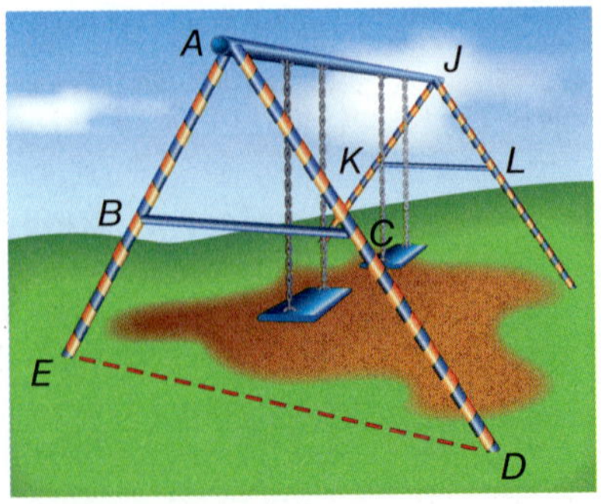

a. Elisa estimates $m\angle BAC$ to be 50. Based on this estimate, what is $m\angle ABC$? Explain.

b. If $\overline{BE} \cong \overline{CD}$, show that $\triangle AED$ is isosceles.

c. If $\overline{BC} \parallel \overline{ED}$ and $\overline{ED} \cong \overline{AD}$, show that $\triangle AED$ is equilateral.

d. If $\triangle JKL$ is isosceles, what is the minimum information needed to prove that $\triangle ABC \cong \triangle JLK$? Explain your reasoning.

26. CHIMNEYS In the picture, $\overline{BD} \perp \overline{AC}$ and $\triangle ABC$ is an isosceles triangle with base $\overline{AC}$. Show that the chimney of the house, represented by $\overline{BD}$, bisects the angle formed by the sloped sides of the roof, $\angle ABC$.

27. CONSTRUCTION Construct three different isosceles right triangles. Explain your method. Then verify your constructions using measurement and mathematics.

28. PROOF Based on your construction in Exercise 27, make and prove a conjecture about the relationship between the base angles of an isosceles right triangle.

Find each measure.

29 $m\angle CAD$

30. $m\angle ACD$

31. $m\angle ACB$

32. $m\angle ABC$

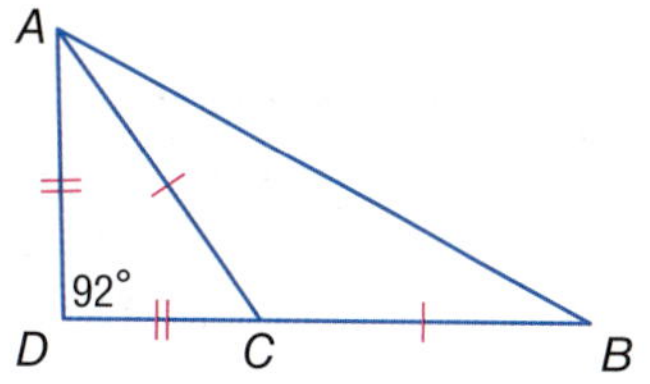

Real-World Link

In bike trials, a rider uses a special bicycle to maneuver and balance to complete a specially designed course composed of 3–10 artificial or natural *sections*. A *dab* or penalty point is charged against a rider if he or she puts a foot down during a run through a section.

Source: *North American Bike Trials Regulations*

33. FITNESS In the diagram, the rider will use his bike to hop across the tops of each of the concrete solids shown. If each triangle is isosceles with vertex angles G, H, and J, and $\overline{BG} \cong \overline{HC}$, $\overline{HD} \cong \overline{JF}$, $\angle G \cong \angle H$, and $\angle H \cong \angle J$, show that the distance from B to F is three times the distance from D to F.

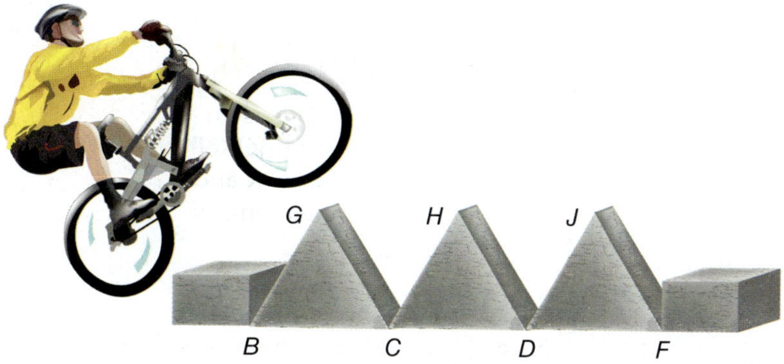

34. Given: $\triangle XWV$ is isosceles; $\overline{ZY} \perp \overline{YV}$.
Prove: $\angle X$ and $\angle YZV$ are complementary.

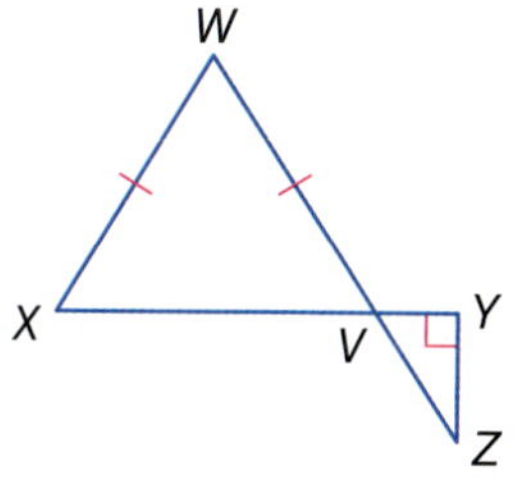

PROOF **Write a two-column proof of each corollary or theorem.**

35. Corollary 4.3

36. Corollary 4.4

37. Theorem 4.11

Find the value of each variable.

38.

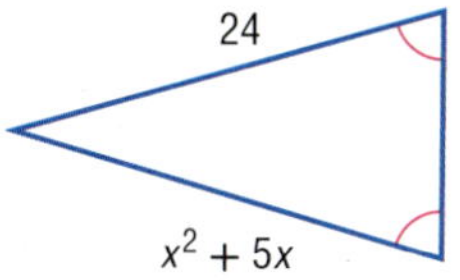

39.

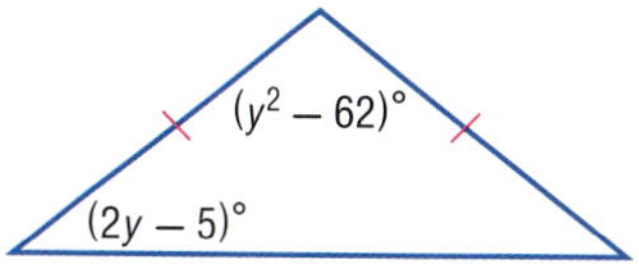

GAMES **Use the diagram of a game timer shown to find each measure.**

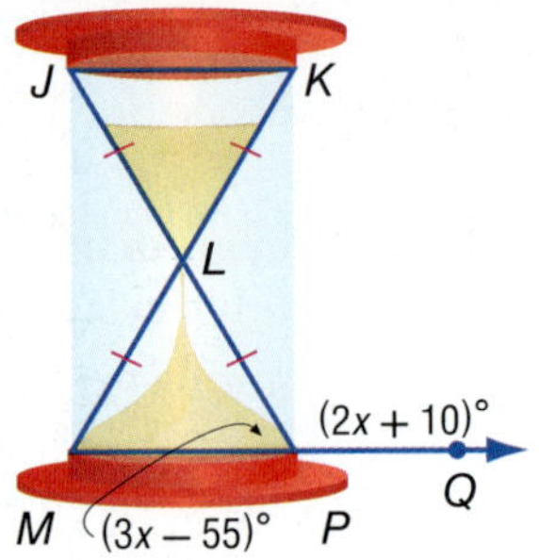

40. $m\angle LPM$

41. $m\angle LMP$

42. $m\angle JLK$

43. $m\angle JKL$

Real-World Link

In order to be accurate, an hourglass game timer must have a flow rate that does not fluctuate. This is dependent on the ratio of hole or tube width to the diameter of the sand particles used.

Source: Library of Congress

44. **MULTIPLE REPRESENTATIONS** In this problem, you will explore possible measures of the interior angles of an isosceles triangle given the measure of one exterior angle.

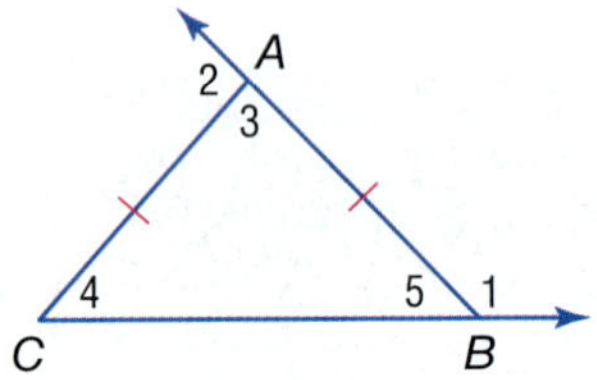

a. **GEOMETRIC** Use a ruler and a protractor to draw three different isosceles triangles, extending one of the sides adjacent to the vertex angle and to one of the base angles, and labeling as shown.

b. **TABULAR** Use a protractor to measure and record $m\angle 1$ for each triangle. Use $m\angle 1$ to calculate the measures of $\angle 3$, $\angle 4$, and $\angle 5$. Then find and record $m\angle 2$ and use it to calculate these same measures. Organize your results in two tables.

c. **VERBAL** Explain how you used $m\angle 1$ to find the measures of $\angle 3$, $\angle 4$, and $\angle 5$. Then explain how you used $m\angle 2$ to find these same measures.

d. **ALGEBRAIC** If $m\angle 1 = x$, write an expression for the measures of $\angle 3$, $\angle 4$, and $\angle 5$. Likewise, if $m\angle 2 = x$, write an expression for these same angle measures.

H.O.T. Problems Use Higher-Order Thinking Skills

45. **CHALLENGE** In the figure at the right, if $\triangle WJZ$ is equilateral and $\angle ZWP \cong \angle WJM \cong \angle JZL$, prove that $\overline{WP} \cong \overline{ZL} \cong \overline{JM}$.

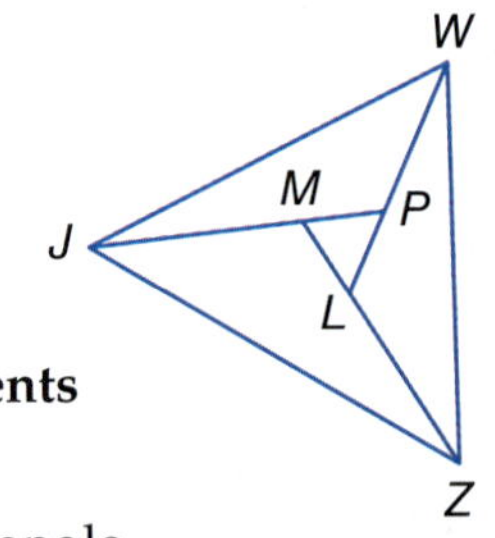

REASONING **Determine whether the following statements are *sometimes*, *always*, or *never* true. Explain.**

46. If the measure of the vertex angle of an isosceles triangle is an integer, then the measure of each base angle is an integer.

47. If the measures of the base angles of an isosceles triangle are integers, then the measure of its vertex angle is odd.

48. **ERROR ANALYSIS** Alexis and Miguela are finding $m\angle G$ in the figure shown. Alexis says that $m\angle G = 35$, while Miguela says that $m\angle G = 60$. Is either of them correct? Explain your reasoning.

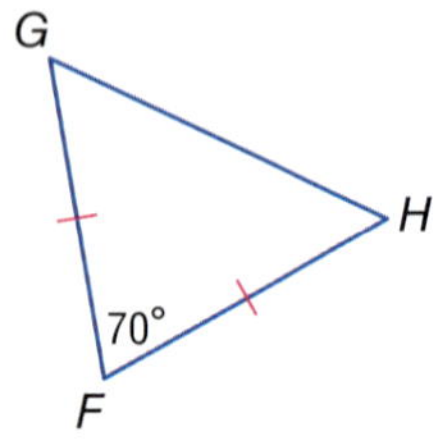

49. **OPEN ENDED** If possible, draw an isosceles triangle with base angles that are obtuse. If it is not possible, explain why not.

50. **REASONING** In isosceles $\triangle ABC$, $m\angle B = 90$. Draw the triangle. Indicate the congruent sides and label each angle with its measure.

51. **WRITING IN MATH** Explain how you can use the measure of a base angle of an isosceles triangle to find the measure of the vertex angle.

Standardized Test Practice

52. ALGEBRA What quantity should be added to both sides of this equation to complete the square?

$$x^2 - 10x = 3$$

A -25
B -5
C 5
D 25

53. SHORT RESPONSE In a school of 375 students, 150 students play sports and 70 students are involved in the community service club. 30 students play sports and are involved in the community service club. How many students are *not* involved in either sports or the community service club?

54. In the figure below, AE and BD bisect each other at point C.

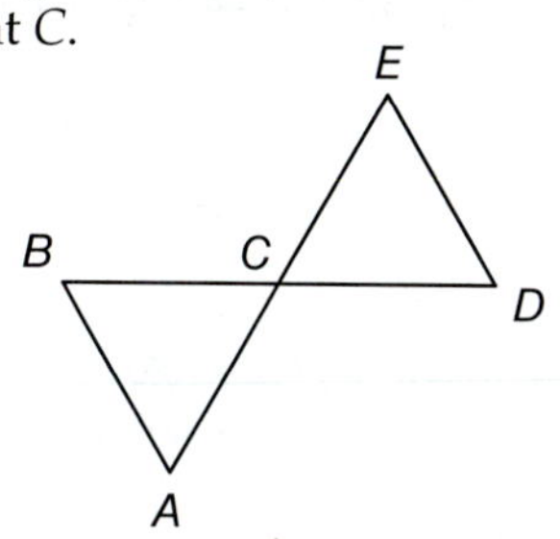

Which additional piece of information would be enough to prove that $\overline{DE} \cong \overline{DC}$?

F $\angle A \cong \angle BCA$
G $\angle B \cong \angle D$
H $\angle ACB \cong \angle EDC$
J $\angle A \cong \angle B$

55. ACT/SAT If $x = -3$, then $4x^2 - 7x + 5 =$

A 2
B 20
C 42
D 62

Spiral Review

56. If $m\angle ADC = 35$, $m\angle ABC = 35$, $m\angle DAC = 26$, and $m\angle BAC = 26$, determine whether $\triangle ADC \cong \triangle ABC$. (Lesson 4-5)

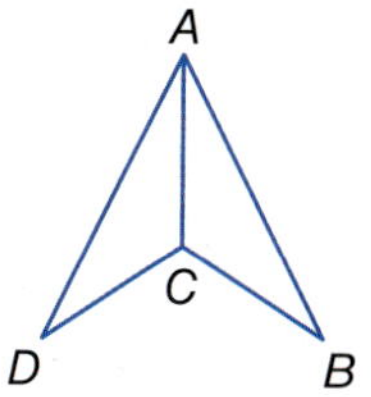

Determine whether $\triangle STU \cong \triangle XYZ$. Explain. (Lesson 4-4)

57. $S(0, 5)$, $T(0, 0)$, $U(1, 1)$, $X(4, 8)$, $Y(4, 3)$, $Z(6, 3)$

58. $S(2, 2)$, $T(4, 6)$, $U(3, 1)$, $X(-2, -2)$, $Y(-4, 6)$, $Z(-3, 1)$

59. PHOTOGRAPHY Film is fed through a traditional camera by gears that catch the perforation in the film. The distance from A to C is the same as the distance from B to D. Show that the two perforated strips are the same width. (Lesson 2-7)

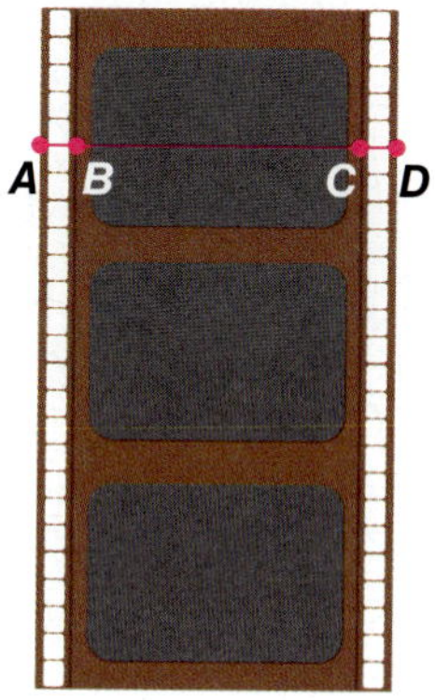

State the property that justifies each statement. (Lesson 2-6)

60. If $x(y + z) = a$, then $xy + xz = a$.

61. If $n - 17 = 39$, then $n = 56$.

62. If $m\angle P + m\angle Q = 110$ and $m\angle R = 110$, then $m\angle P + m\angle Q = m\angle R$.

63. If $cv = md$ and $md = 15$, then $cv = 15$.

Refer to the figure at the right. (Lesson 1-1)

64. How many planes appear in this figure?

65. Name three points that are collinear.

66. Are points A, C, D, and J coplanar?

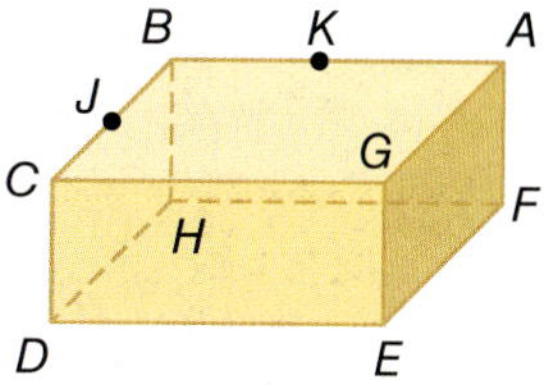

Skills Review

67. PROOF If $\angle ACB \cong \angle ABC$, then $\angle XCA \cong \angle YBA$. (Lesson 2-8)

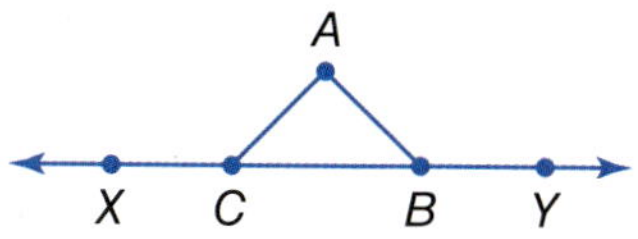

Graphing Technology Lab
Congruence Transformations

Math Online glencoe.com
- Other Calculator Keystrokes
- Graphing Technology Personal Tutor

You can use TI-Nspire™ or TI-Nspire™ CAS technology to perform *transformations* on triangles in the coordinate plane and test for congruence.

ACTIVITY 1 Translate a Triangle and Test for Congruence

Step 1 Open a new **Graphs and Geometry** page. Select **Show Grid** from the **View** menu.

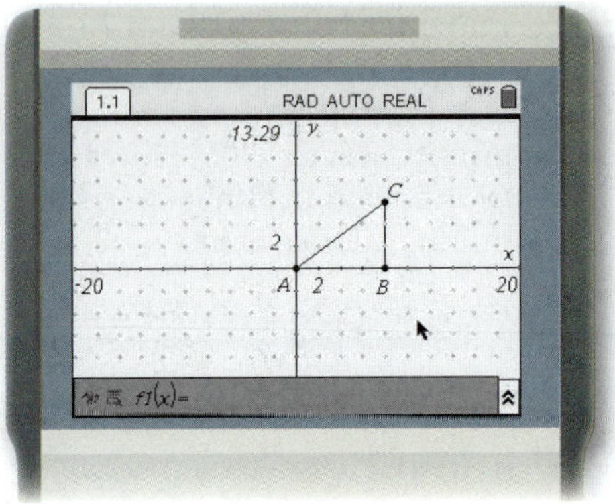

Step 2 Select **Triangle** from the **Shapes** menu and draw a right triangle with legs measuring 6 units and 8 units as shown by placing the first point at (8, 6), the second point at (8, 0), and the third point at (0, 0). Use the Text tool to label the vertices of the triangle as *A*, *B*, and *C*.

Step 3 Select **Translation** from the **Transformation** menu. Then select $\triangle ABC$ and translate or *slide* it 8 units down and 14 units left. Label the corresponding vertices of the image as A', B', and C'.

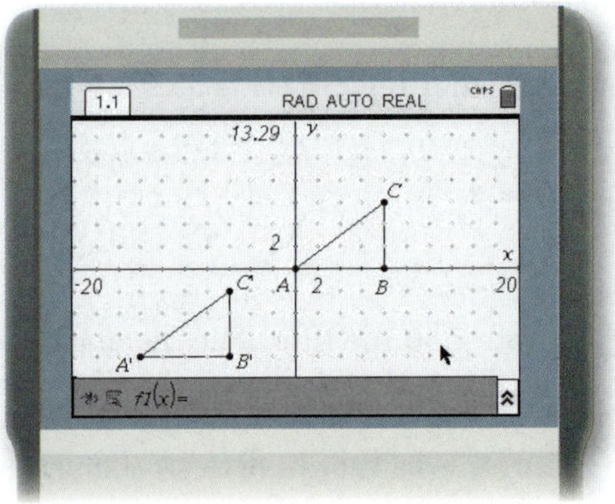

Step 4 To verify that $\triangle A'B'C'$ is congruent to $\triangle ABC$, select **Length** from the **Measurement** menu. Then select the endpoints of each side of each triangle to determine their measures.

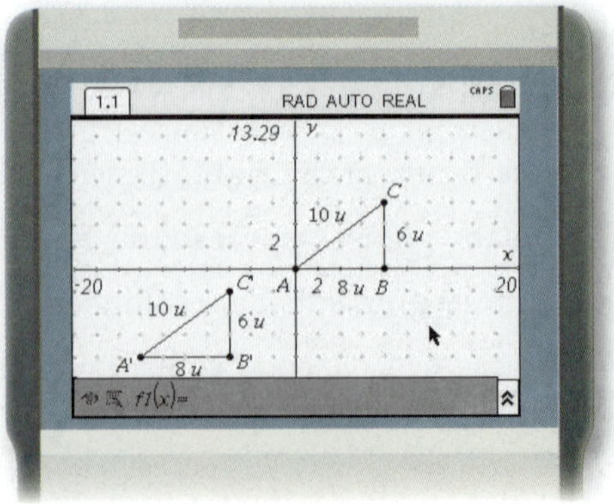

In addition to measuring lengths, the TI-Nspire™ can also be used to measure angles. This will allow you to use other tests for triangle congruence that involve angle measure.

ACTIVITY 2 Reflect a Triangle and Test for Congruence

Step 1 Open a new **Graphs and Geometry** page, and redraw $\triangle ABC$ from Activity 1.

Step 2 Select **Reflection** from the **Transformation** menu. Then select $\triangle ABC$ and the y-axis to reflect or *flip* $\triangle ABC$ in the y-axis. Label the corresponding vertices of the image as A', B', and C'.

Step 3 Use the **Angle** measurement tool to find $m\angle A$ and $m\angle A'$. Use the **Length** measurement tool to find AB, $A'B'$, AC, and $A'C'$.

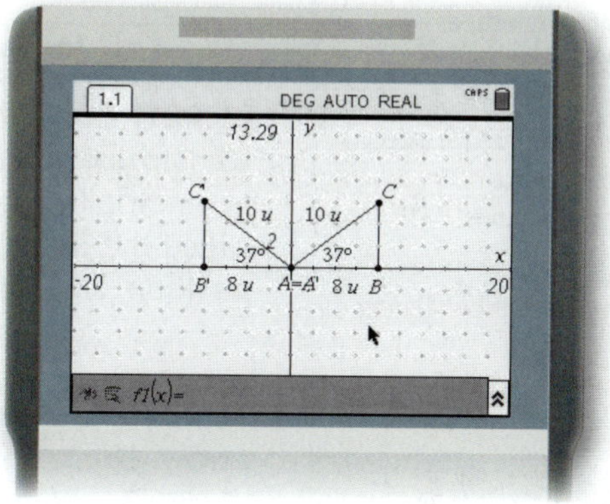

To rotate a figure about the origin using the TI-Nspire, use the Rotation tool to select the figure, then the point (0, 0), then draw an angle of rotation.

ACTIVITY 3 Rotate a Triangle and Test for Congruence

Step 1 Open a new **Graphs and Geometry** page and redraw $\triangle ABC$ from Activity 1.

Step 2 Select **Rotation** from the **Transformation** menu. Then select $\triangle ABC$ and rotate or *turn* it counterclockwise about the point (0, 0) through any acute angle.

Step 3 Use the **Angle** measurement tool to find $m\angle A$, $m\angle A'$, $m\angle C$, and $m\angle C'$. Use the **Length** measurement tool to find AC and $A'C'$.

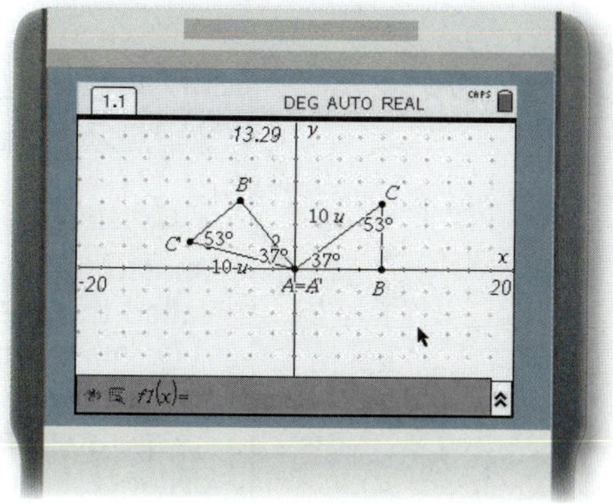

Analyze the Results

Determine whether $\triangle ABC$ and $\triangle A'B'C'$ are congruent. Explain your reasoning.

1. Activity 1

2. Activity 2

3. Activity 3

4. Explain why $\triangle A'B'C'$ in Activity 3 does not appear to be congruent to $\triangle ABC$.

5. **MAKE A CONJECTURE** Repeat Activities 1–3 using a different triangle XYZ. Analyze your results and compare them to those found in Exercises 1–3. Make a conjecture as to the relationship between a triangle and its transformed image under a translation, reflection, or a rotation.

4-7 Congruence Transformations

Why?

The fashion industry often uses prints that display patterns. Many of these patterns are created by taking one figure and sliding it to create another figure in a different location, flipping the figure to create a mirror image of the original, or turning the original figure to create a new one.

Then

You proved whether two triangles were congruent. (Lessons 4-3, 4-4, and 4-5)

Now

- Identify reflections, translations, and rotations.
- Verify congruence after a congruence transformation.

New Vocabulary

transformation
congruence transformation
preimage
image
isometry
reflection
translation
rotation

Math Online
glencoe.com

Identify Congruence Transformations A **transformation** is an operation that maps an original geometric figure, the **preimage**, onto a new figure called the **image**. A transformation can change the position, size, or shape of a figure.

A transformation can be noted using an arrow. The transformation statement $\triangle ABC \rightarrow \triangle XYZ$ tells you that A is mapped to X, B is mapped to Y, and C is mapped to Z.

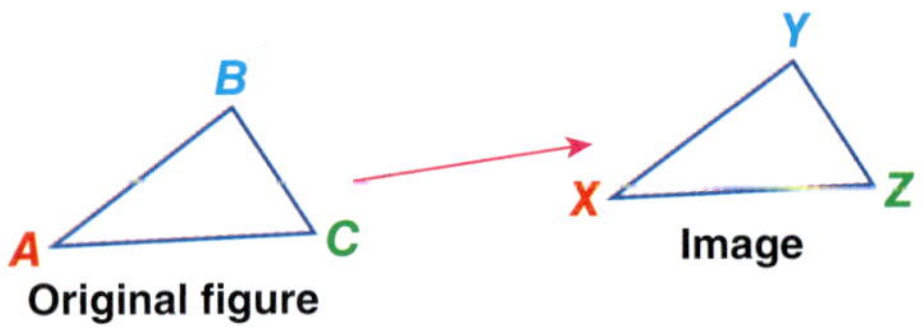

A **congruence transformation**, also called a *rigid transformation* or an **isometry**, is one in which the position of the image may differ from that of the preimage, but the two figures remain congruent. The three main types of congruence transformations are shown below.

Key Concept Reflections, Translations, and Rotations

A **reflection** or *flip* is a transformation over a line called the *line of reflection*. Each point of the preimage and its image are the same distance from the line of reflection.

Example

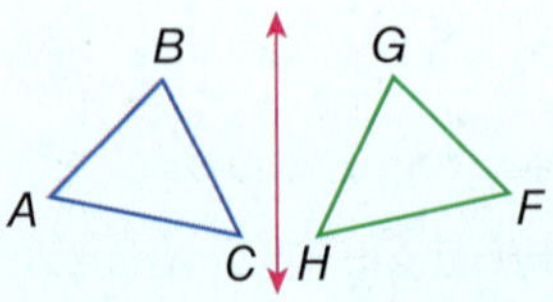

$\triangle ABC \longrightarrow \triangle FGH$

A **translation** or *slide* is a transformation that moves all points of the original figure the same distance in the same direction.

Example

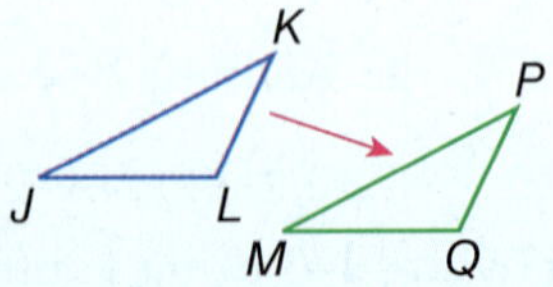

$\triangle JKL \longrightarrow \triangle MPQ$

A **rotation** or *turn* is a transformation around a fixed point called the *center of rotation*, through a specific angle, and in a specific direction. Each point of the original figure and its image are the same distance from the center.

Example

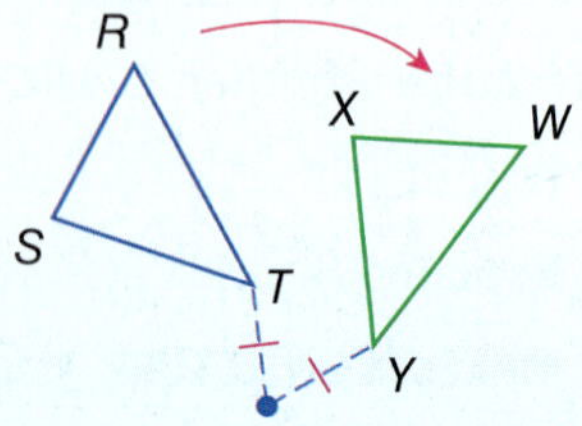

$\triangle RST \longrightarrow \triangle WXY$

StudyTip

Transformations Not all transformations preserve congruence. Only transformations that do not change the size or shape of the figure are congruence transformations. You will learn more about transformations in Chapter 9.

EXAMPLE 1 Identify Congruence Transformations

Identify the type of congruence transformation shown as a *reflection, translation,* or *rotation.*

a.

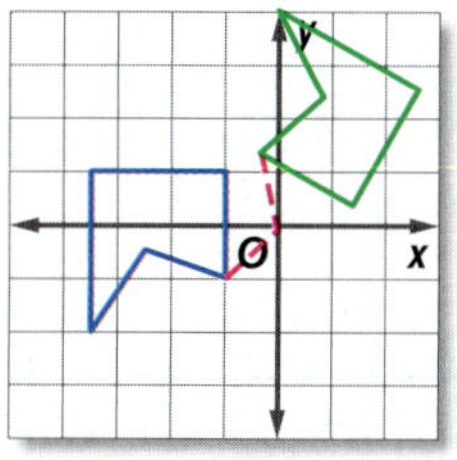

Each vertex and its image are the same distance from the origin. The angles formed by each pair of corresponding points and the origin are congruent. This is a rotation.

b.

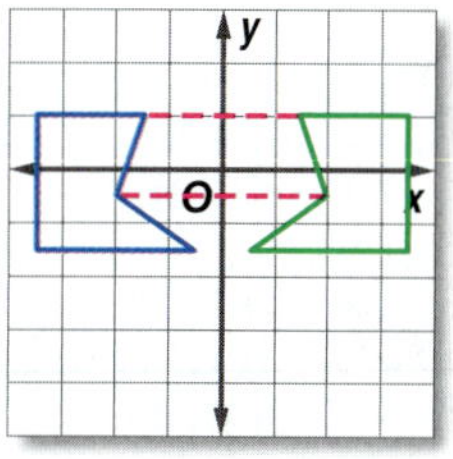

Each vertex and its image are the same distance from the *y*-axis. This is a reflection.

c.

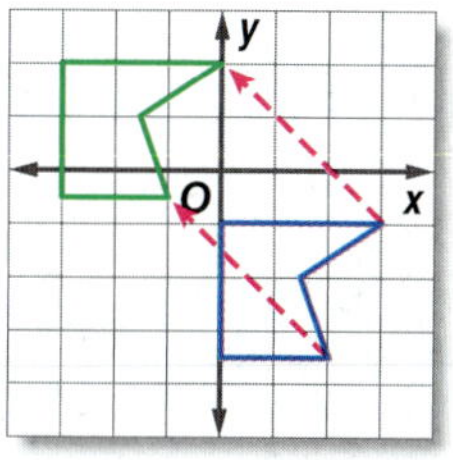

Each vertex and its image are in the same position, just 3 units left and 3 units up. This is a translation.

Guided Practice

1A.

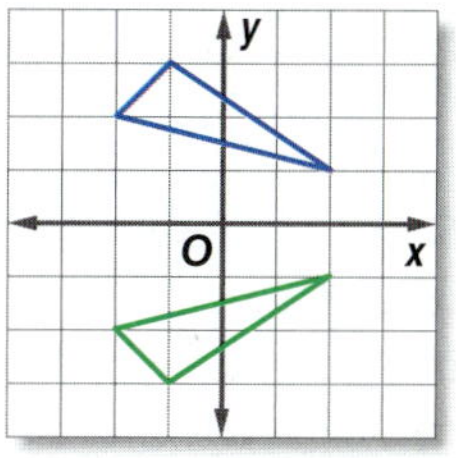

1B.

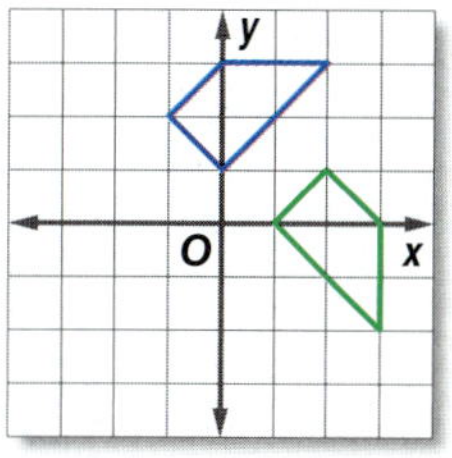

1C.

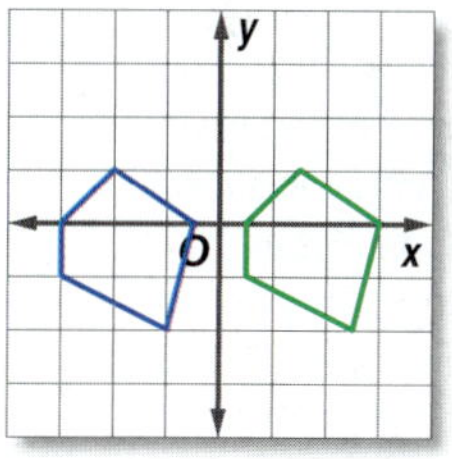

Personal Tutor glencoe.com

Some real-world motions or objects can be represented by transformations.

Real-World Link

The game shown above involves a weight attached by a bar to a ring that you can place around your ankle. As the bar passes in front of your other foot, you skip over it.

Real-World EXAMPLE 2 Identify a Real-World Transformation

GAMES **Refer to the information at the left. Identify the type of congruence transformation shown in the diagram as a *reflection, translation,* or *rotation.***

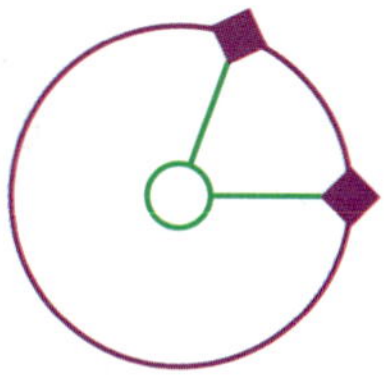

The position of the weight at different times is an example of a rotation. The center of rotation is the person's ankle.

Guided Practice

Identify the type of congruence transformation shown as a *reflection, translation,* or *rotation.*

2A.

2B.

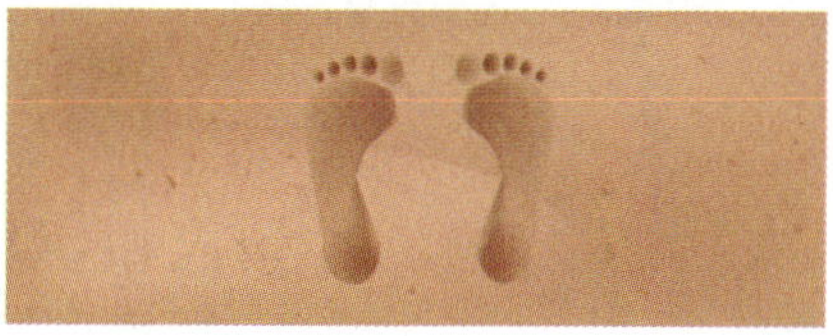

Personal Tutor glencoe.com

Verify Congruence You can verify that reflections, translations, and rotations of triangles produce congruent triangles using SSS.

EXAMPLE 3 Verify Congruence after a Transformation

Triangle XZY with vertices $X(2, -8)$, $Z(6, -7)$, and $Y(4, -2)$ is a transformation of $\triangle ABC$ with vertices $A(2, 8)$, $B(6, 7)$, and $C(4, 2)$. Graph the original figure and its image. Identify the transformation and verify that it is a congruence transformation.

Understand You are asked to identify the type of transformation—reflection, translation, or rotation. Then, you need to show that the two figures are congruent.

Plan Use the Distance Formula to find the measure of each side. Then show that the two triangles are congruent by SSS.

Solve Graph each figure. The transformation appears to be a reflection over the *x*-axis. Find the measures of the sides of each triangle.

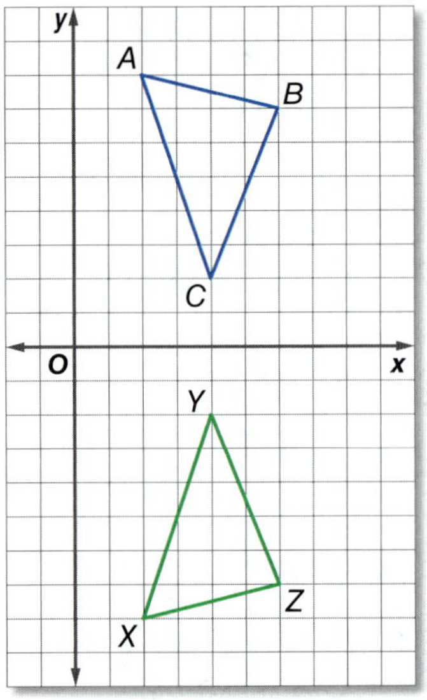

$AB = \sqrt{(6-2)^2 + (7-8)^2}$ or $\sqrt{17}$

$BC = \sqrt{(6-4)^2 + (7-2)^2}$ or $\sqrt{29}$

$AC = \sqrt{(4-2)^2 + (2-8)^2}$ or $\sqrt{40}$

$XZ = \sqrt{(6-2)^2 + [-7-(-8)]^2}$ or $\sqrt{17}$

$ZY = \sqrt{(6-4)^2 + [-7-(-2)]^2}$ or $\sqrt{29}$

$XY = \sqrt{(2-4)^2 + [-8-(-2)]^2}$ or $\sqrt{40}$

Since $AB = XZ$, $BC = ZY$, and $AC = XY$, $\overline{AB} \cong \overline{XZ}$, $\overline{BC} \cong \overline{ZY}$, and $\overline{AC} \cong \overline{XY}$. By SSS, $\triangle ABC \cong \triangle XZY$.

Check Use the definition of a reflection. Use a ruler to measure and compare the segments connecting each vertex and its image to the line of symmetry. These segments are congruent, so the triangles are congruent. ✓

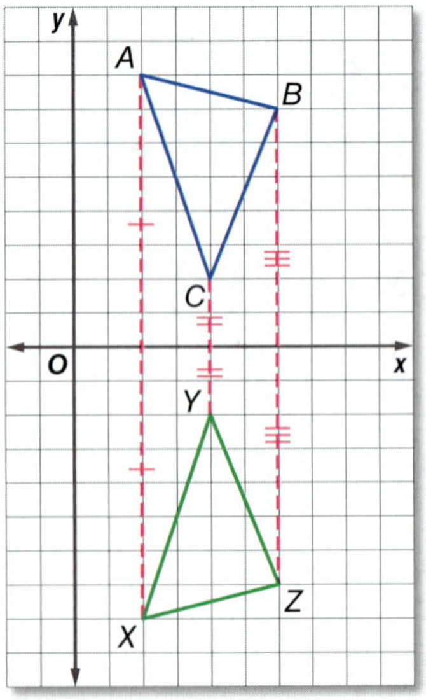

StudyTip

Isometry While an isometry preserves congruence, a *direct isometry* also preserves orientation or order of lettering. An *indirect* or *opposite isometry* changes this order, such as from clockwise to counterclockwise. The reflection shown in Example 3 is an example of an indirect isometry.

Guided Practice

3. Triangle JKL with vertices $J(-2, 2)$, $K(-8, 5)$, and $L(-4, 6)$ is a transformation of $\triangle PQR$ with vertices $P(2, -2)$, $Q(8, -5)$, and $R(4, -6)$. Graph the original figure and its image. Identify the transformation and verify that it is a congruence transformation.

Personal Tutor glencoe.com

Check Your Understanding

Example 1 p. 295

Identify the type of congruence transformation shown as a *reflection, translation,* or *rotation.*

1.

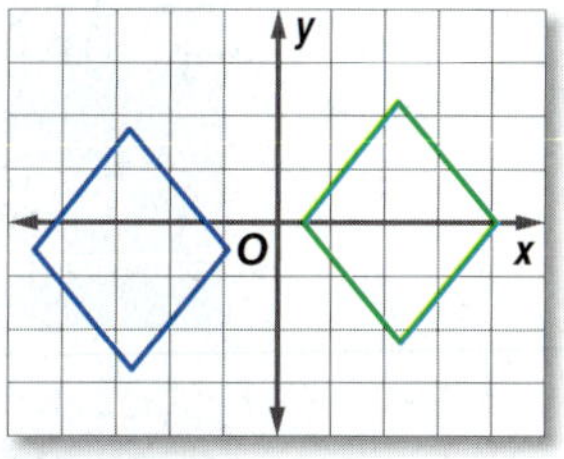

2.

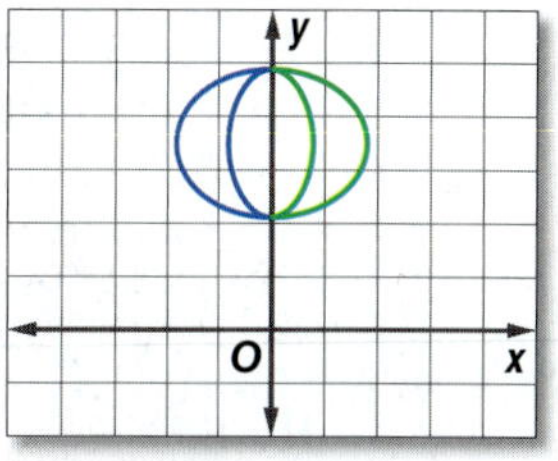

Example 2 p. 295

3.

4.

Example 3 p. 296

COORDINATE GEOMETRY Identify each transformation and verify that it is a congruence transformation.

5.

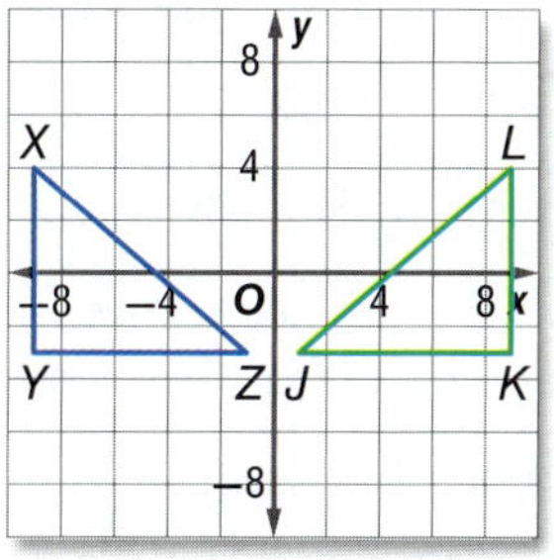

6.

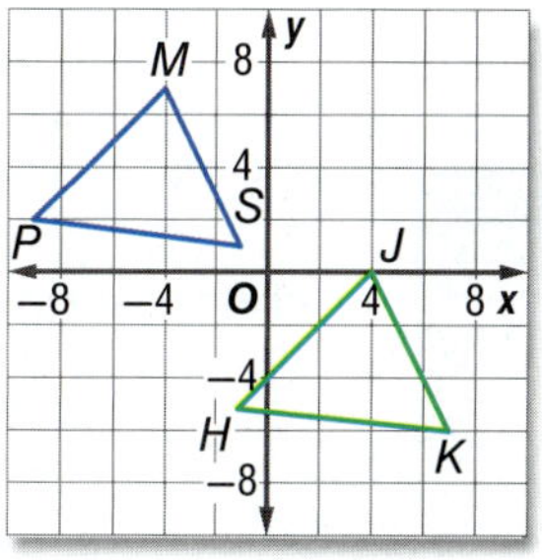

Practice and Problem Solving

= **Step-by-Step Solutions** begin on page R20.
Extra Practice begins on page 969.

Example 1 p. 295

Identify the type of congruence transformation shown as a *reflection, translation,* or *rotation.*

7.

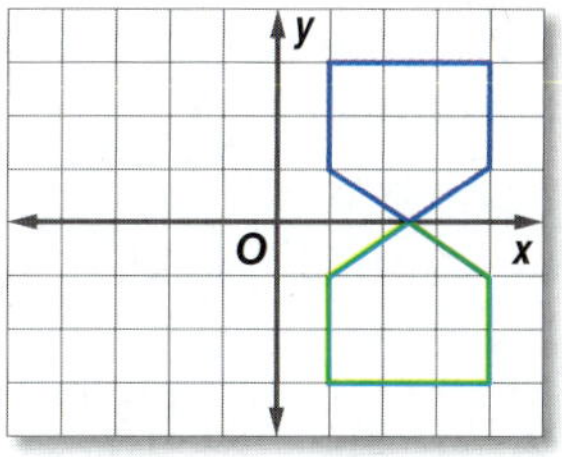

8.

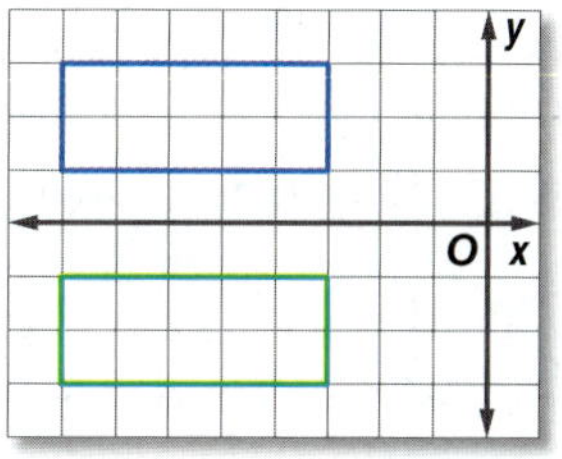

9.

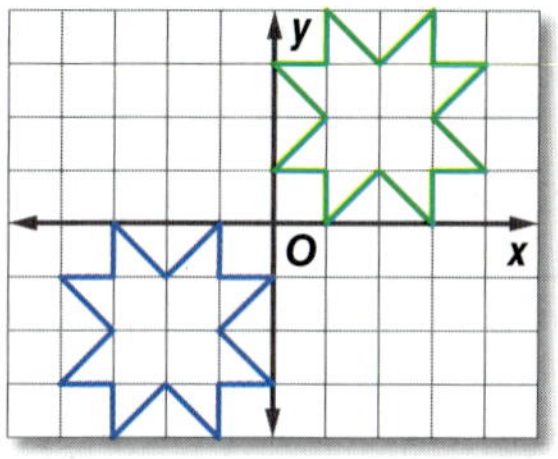

10.

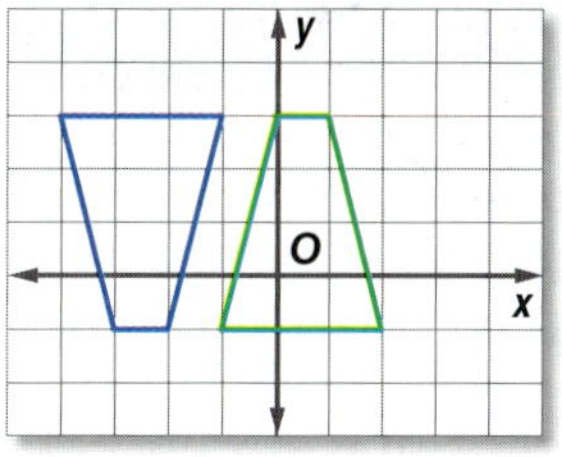

11.

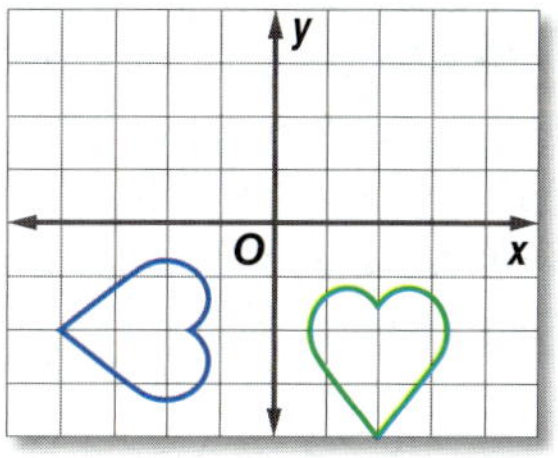

12.

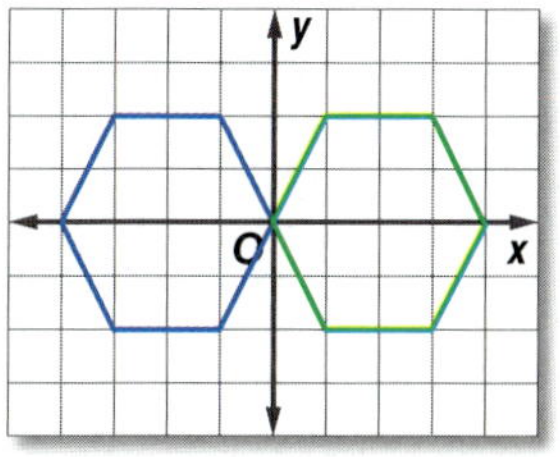

Example 2
p. 295

Identify the type of congruence transformation shown in each picture as a *reflection*, *translation*, or *rotation*.

13. 

14. 

15. 

16. 

Example 3
p. 296

COORDINATE GEOMETRY **Graph each pair of triangles with the given vertices. Then, identify the transformation, and verify that it is a congruence transformation.**

17. $M(-7, -1), P(-7, -7), R(-1, -4)$; $T(7, -1), V(7, -7), S(1, -4)$

18. $A(3, 9), B(3, 7), C(7, 7)$; $S(3, 5), T(3, 3), R(7, 3)$

19. $A(-4, 5), B(0, 2), C(-4, 2)$; $X(-5, -4), Y(-2, 0), Z(-2, -4)$

20. $A(2, 2), B(4, 7), C(6, 2)$; $D(2, -2), F(4, -7), G(6, -2)$

Real-World Link

A truss is a structure that uses triangles to give shape and strength to roofs. Standard roof trusses have a triangular shape, but they can have many different configurations.

CONSTRUCTION **Identify the type of congruence transformation performed on each given triangle to generate the other triangle in the truss shown below.**

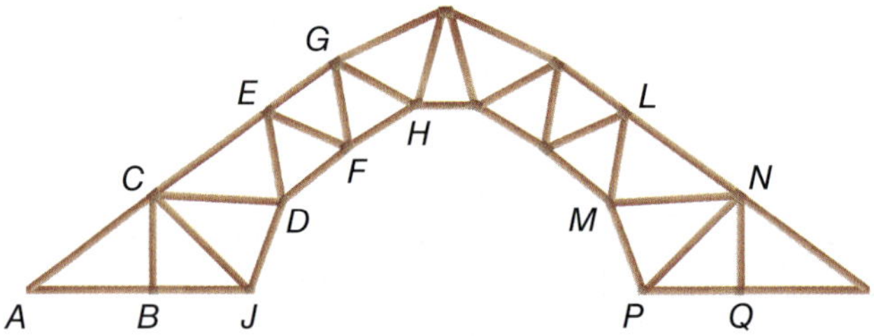

21. $\triangle NML$ to $\triangle CDE$

22. $\triangle EFD$ to $\triangle GHF$

23. $\triangle CBJ$ to $\triangle NQP$

AMUSEMENT RIDES **Identify the type of congruence transformation shown in each picture as a *reflection*, *translation*, or *rotation*.**

24. 

25. 

26. 

27. **SCHOOL** Identify the transformations that are used to open a combination lock on a locker. If appropriate, identify the line of symmetry or center of rotation.

28. **ALPHABET** Determine which letters of the alphabet have vertical and/or horizontal lines of reflection.

29 **DECORATING** Tionne is redecorating her bedroom. She can use stencils or a stamp to create the design shown.

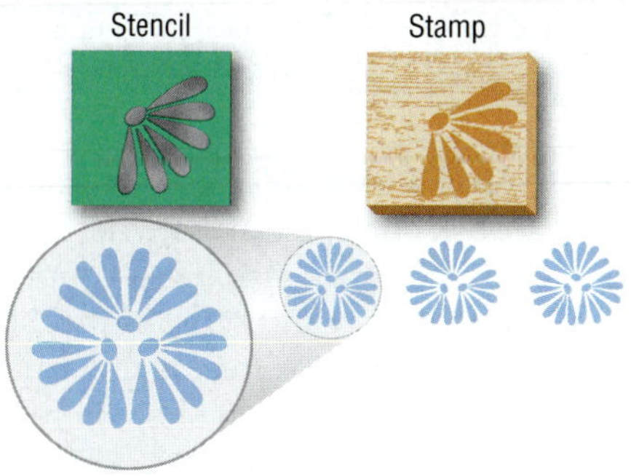

a. If Tionne used the stencil, what type of transformation was used to produce each flower in the design?

b. What type of transformation was used if she used the stamp to produce each flower in the design?

30. **MULTIPLE REPRESENTATIONS** In this problem, you will investigate the relationship between the ordered pairs of a figure and its translated image.

a. **GEOMETRIC** Draw congruent rectangles *ABCD* and *WXYZ* on a coordinate plane.

b. **VERBAL** How do you get from a vertex on *ABCD* to the corresponding vertex on *WXYZ* using only horizontal and vertical movement?

c. **TABULAR** Copy the table shown. Use your rectangles to fill in the *x*-coordinates, the *y*-coordinates, and the unknown value in the transformation column.

Rectangle *ABCD*	Transformation	Rectangle *WXYZ*
A(?, ?)	$(x_1 + ?, y_1 + ?)$	*W*(?, ?)
B(?, ?)	$(x_1 + ?, y_1 + ?)$	*X*(?, ?)
C(?, ?)	$(x_1 + ?, y_1 + ?)$	*Y*(?, ?)
D(?, ?)	$(x_1 + ?, y_1 + ?)$	*Z*(?, ?)

d. **ALGEBRAIC** Complete the following notation that represents the rule for the translation $ABCD \rightarrow WXYZ$: $(x, y) \rightarrow (x + a, y + b)$.

StudyTip

Function Notation The notation $(x, y) \rightarrow (x + a, y + b)$, where *a* and *b* are real numbers, represents a mapping from one set of coordinates onto another. This notation is also referred to as function notation.

H.O.T. Problems Use Higher-Order Thinking Skills

31. **CHALLENGE** Use the diagram at the right.

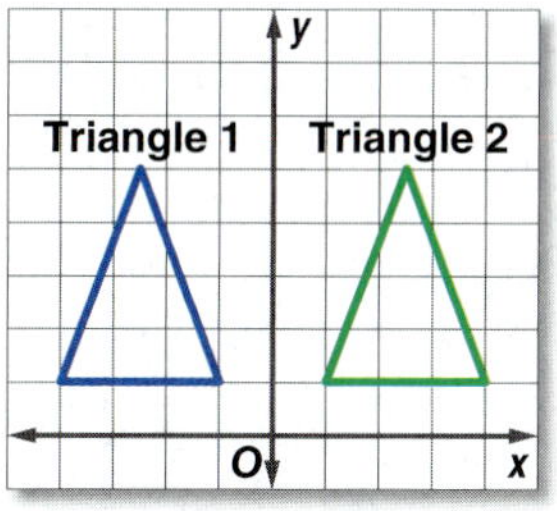

a. Identify two transformations of Triangle 1 that can result in Triangle 2.

b. What must be true of the triangles in order for more than one transformation on a preimage to result in the same image? Explain your reasoning.

32. **REASONING** A *dilation* is another type of transformation. In the diagram, a small paper clip has been dilated to produce a larger paper clip. Explain why dilations are not a congruence transformation.

OPEN ENDED **Describe a real-world example of each of the following transformations, other than those given in this lesson.**

33. reflection **34.** translation **35.** rotation

36. **WRITING IN MATH** In the diagram at the right $\triangle DEF$ is called a *glide reflection* of $\triangle ABC$. Based on the diagram, define a glide reflection. Is a glide reflection a congruence transformation? Explain your reasoning.

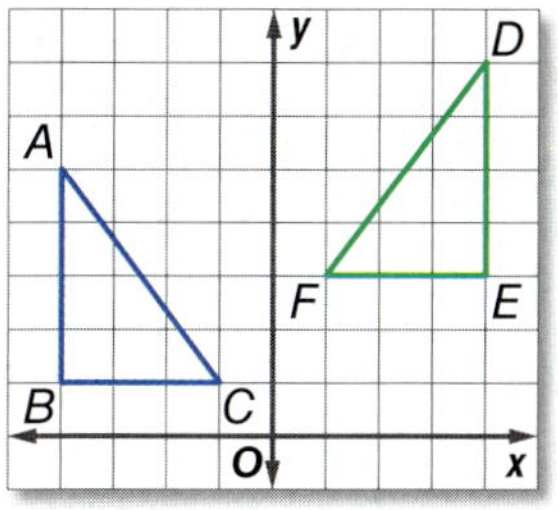

Standardized Test Practice

37. SHORT RESPONSE Cindy is shopping for a new desk chair at a store where the desk chairs are 50% off. She also has a coupon for 50% off any one item. Cindy thinks that she can now get the desk chair for free. Is this true? If not, what will be the percent off she will receive with both the sale and the coupon?

38. Identify the congruence transformation shown.

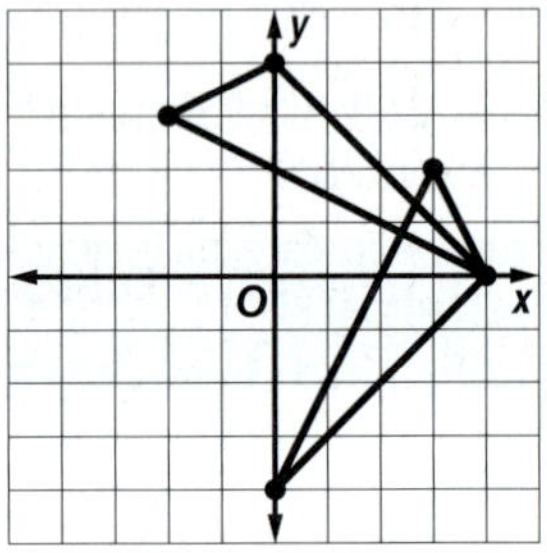

A dilation
B reflection
C rotation
D translation

39. Look at the graph below. What is the slope of the line shown?

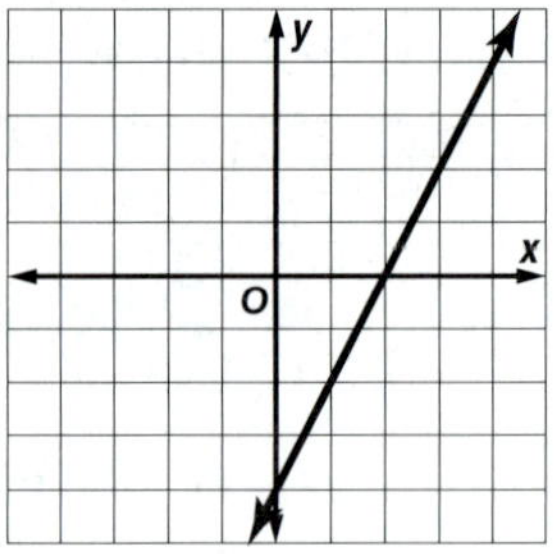

F −2
G −1
H 1
J 2

40. ACT/SAT What is the y-intercept of the line determined by the equation $3x - 4 = 12y - 3$?

A −12
B $-\frac{1}{12}$
C $\frac{1}{12}$
D 12

Spiral Review

Find each measure. (Lesson 4-6)

41. YZ

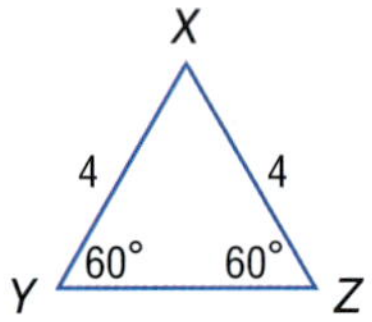

42. $m\angle JLK$

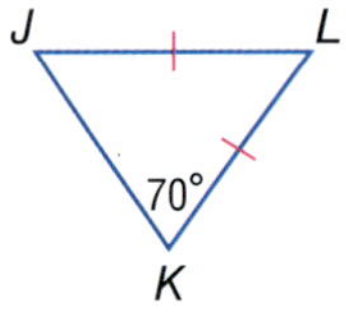

43. AB

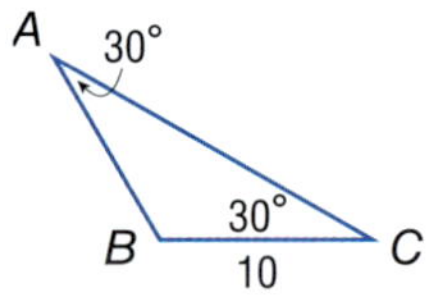

44. PROOF Write a paragraph proof. (Lesson 4-5)

Given: $\angle YWZ \cong \angle XZW$ and $\angle YZW \cong \angle XWZ$

Prove: $\triangle WXZ \cong \triangle ZYW$

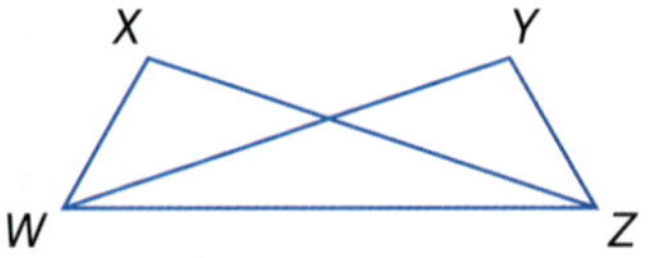

45. ROLLER COASTERS The sign in front of the Electric Storm roller coaster states that all riders must be at least 54 inches tall to ride. If Andy is 5 feet 8 inches tall, can he ride the Electric Storm? Which law of logic leads you to this conclusion? (Lesson 2-4)

Skills Review

Find the coordinates of the midpoint of a segment with the given endpoints. (Lesson 1-3)

46. $A(10, -12)$, $C(5, -6)$

47. $A(13, 14)$, $C(3, 5)$

48. $A(-28, 8)$, $C(-10, 2)$

49. $A(-12, 2)$, $C(-3, 5)$

50. $A(0, 0)$, $C(3, -4)$

51. $A(2, 14)$, $C(0, 5)$

4-8 Triangles and Coordinate Proof

Why?

A global positioning system (GPS) receives transmissions from satellites that allow the exact location of a car to be determined. The information can be used with navigation software to provide driving directions.

Then
You used coordinate geometry to prove triangle congruence. (Lesson 4-4)

Now
- Position and label triangles for use in coordinate proofs.
- Write coordinate proofs.

New Vocabulary
coordinate proof

Math Online
glencoe.com

Position and Label Triangles As with global positioning systems, knowing the coordinates of a figure in a coordinate plane allows you to explore its properties and draw conclusions about it. **Coordinate proofs** use figures in the coordinate plane and algebra to prove geometric concepts. The first step in a coordinate proof is placing the figure on the coordinate plane.

EXAMPLE 1 Position and Label a Triangle

Position and label right triangle MNP on the coordinate plane so that leg $\overline{MN}$ is a units long and leg $\overline{NP}$ is b units long.

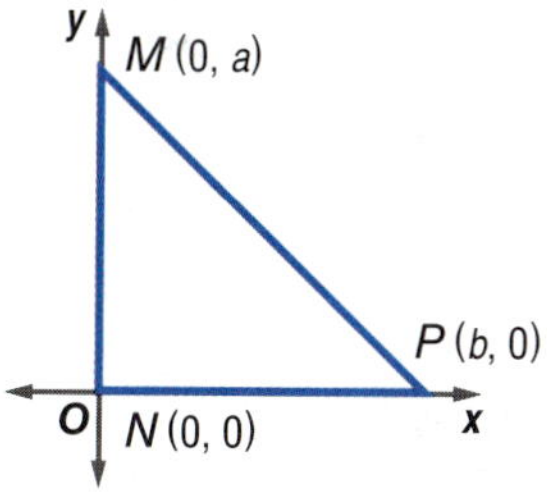

- The length(s) of the side(s) that are along the axes will be easier to determine than the length(s) of side(s) that are not along an axis. Since this is a right triangle, two sides can be located on an axis.
- Placing the right angle of the triangle, $\angle N$, at the origin will allow the two legs to be along the x- and y-axes.
- Position the triangle in the first quadrant.
- Since M is on the y-axis, its x-coordinate is 0. Its y-coordinate is a because the leg is a units long.
- Since P is on the x-axis, its y-coordinate is 0. Its x-coordinate is b because the leg is b units long.

Guided Practice

1. Position and label isosceles triangle JKL on the coordinate plane so that its base $\overline{JL}$ is a units long, vertex K is on the y-axis, and the height of the triangle is b units.

Personal Tutor glencoe.com

Key Concept Placing Triangles on Coordinate Plane

Get Animated

Step 1 Use the origin as a vertex or center of the triangle.
Step 2 Place at least one side of a triangle on an axis.
Step 3 Keep the triangle within the first quadrant if possible.
Step 4 Use coordinates that make computations as simple as possible.

EXAMPLE 2 Identify Missing Coordinates

Name the missing coordinates of isosceles triangle *XYZ*.

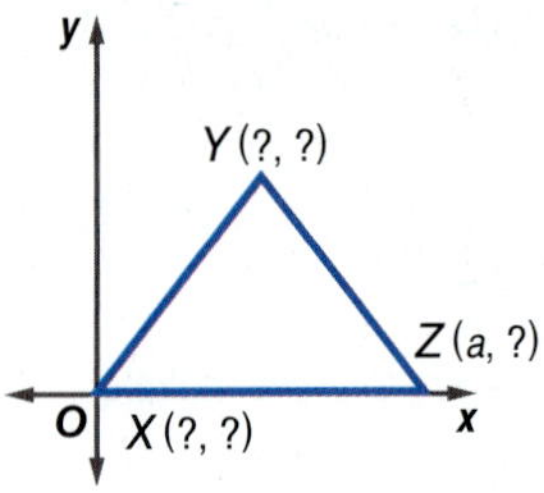

Vertex X is positioned at the origin; its coordinates are (0, 0).

Vertex Z is on the x-axis, so its y-coordinate is 0. The coordinates of vertex Z are $(a, 0)$.

$\triangle XYZ$ is isosceles, so the x-coordinate of Y is located halfway between 0 and a or $\frac{a}{2}$. We cannot write the y-coordinate in terms of a, so call it b. The coordinates of point Y are $\left(\frac{a}{2}, b\right)$.

StudyTip

Right Angle The intersection of the x- and y-axis forms a right angle, so it is a convenient place to locate the right angle of a figure such as a right triangle.

Guided Practice

2. Name the missing coordinates of isosceles right triangle ABC.

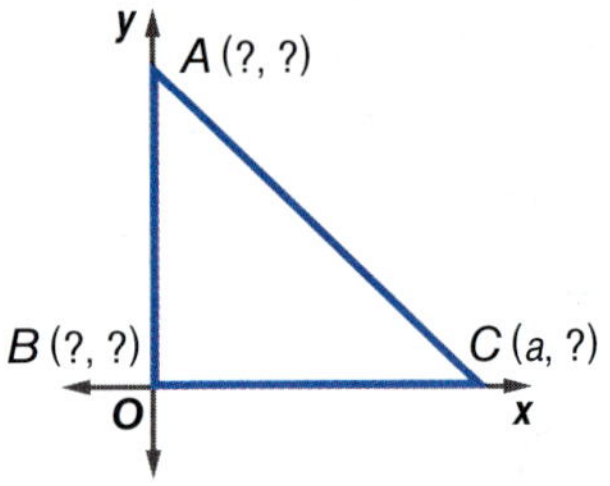

Personal Tutor glencoe.com

Write Coordinate Proofs After a triangle is placed on the coordinate plane and labeled, we can use coordinate proofs to verify properties and to prove theorems.

EXAMPLE 3 Write a Coordinate Proof

Write a coordinate proof to show that a line segment joining the midpoints of two sides of a triangle is parallel to the third side.

StudyTip

Coordinate Proof The guidelines and methods used in this lesson apply to all polygons, not just triangles.

Place a vertex at the origin and label it A. Use coordinates that are multiples of 2 because the Midpoint Formula involves dividing the sum of the coordinates by 2.

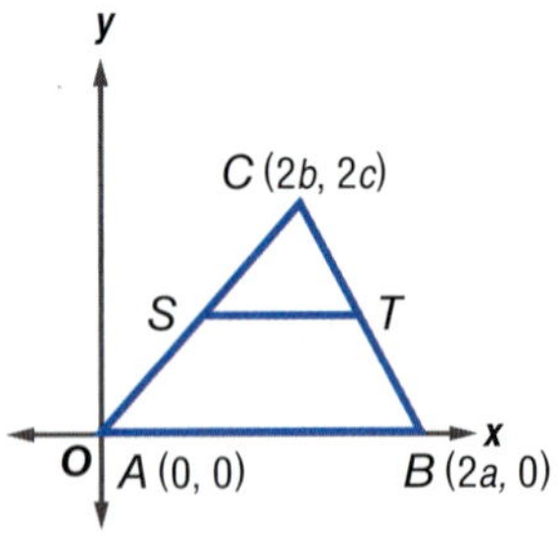

Given: $\triangle ABC$
S is the midpoint of $\overline{AC}$.
T is the midpoint of $\overline{BC}$.

Prove: $\overline{ST} \parallel \overline{AB}$

Proof:

By the Midpoint Formula, the coordinates of S are $\left(\frac{2b+0}{2}, \frac{2c+0}{2}\right)$ or (b, c) and the coordinates of T are $\left(\frac{2a+2b}{2}, \frac{0+2c}{2}\right)$ or $(a + b, c)$.

By the Slope Formula, the slope of $\overline{ST}$ is $\frac{c-c}{a+b-b}$ or 0 and the slope of $\overline{AB}$ is $\frac{0-0}{2a-0}$ or 0.

Since $\overline{ST}$ and $\overline{AB}$ have the same slope, $\overline{ST} \parallel \overline{AB}$.

Guided Practice

3. Write a coordinate proof to show that $\triangle ABX \cong \triangle CDX$.

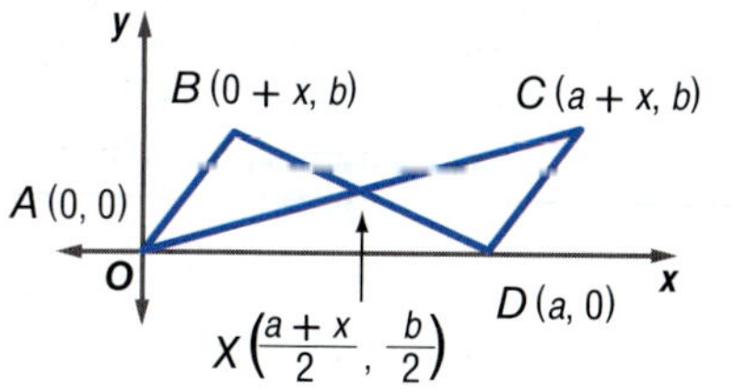

Personal Tutor glencoe.com

The techniques used for coordinate proofs can be used to solve real-world problems.

Real-World Link

More than 50 ships and 20 airplanes have mysteriously disappeared from a section of the North Atlantic Ocean off of North America commonly referred to as the Bermuda Triangle.

Source: *Encyclopaedia Britannica*

Real-World EXAMPLE 4 Classify Triangles

GEOGRAPHY The Bermuda Triangle is a region formed by Miami, Florida, San Jose, Puerto Rico, and Bermuda. The approximate coordinates of each location, respectively, are 25.8°N 80.27°W, 18.48°N 66.12°W, and 33.37°N 64.68°W. Write a coordinate proof to prove that the Bermuda Triangle is scalene.

The first step is to label the coordinates of each location. Let M represent Miami, B represent Bermuda, and P represent Puerto Rico.

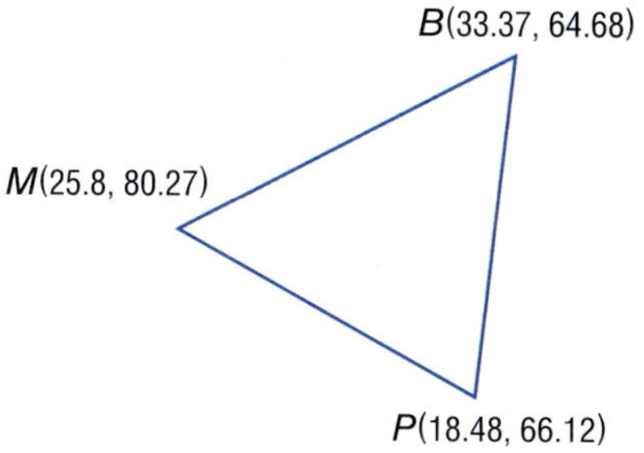

If no two sides of $\triangle MPB$ are congruent, then the Bermuda Triangle is scalene. Use the Distance Formula and a calculator to find the distance between each location.

$$MB = \sqrt{(33.37 - 25.8)^2 + (64.68 - 80.27)^2}$$
$$\approx 17.33$$

$$MP = \sqrt{(25.8 - 18.48)^2 + (80.27 - 66.12)^2}$$
$$\approx 15.93$$

$$PB = \sqrt{(33.37 - 18.48)^2 + (64.68 - 66.12)^2}$$
$$\approx 14.96$$

Since each side is a different length, $\triangle MPB$ is scalene. Therefore, the Bermuda Triangle is scalene.

Guided Practice

4. **GEOGRAPHY** In 2006, a group of art museums collaborated to form the West Texas Triangle to promote their collections. This region is formed by the cities of Odessa, Albany, and San Angelo. The approximate coordinates of each location, respectively, are 31.9°N 102.3°W, 32.7°N 99.3°W, and 31.4°N 100.5°W. Write a coordinate proof to prove that the West Texas Triangle is approximately isosceles.

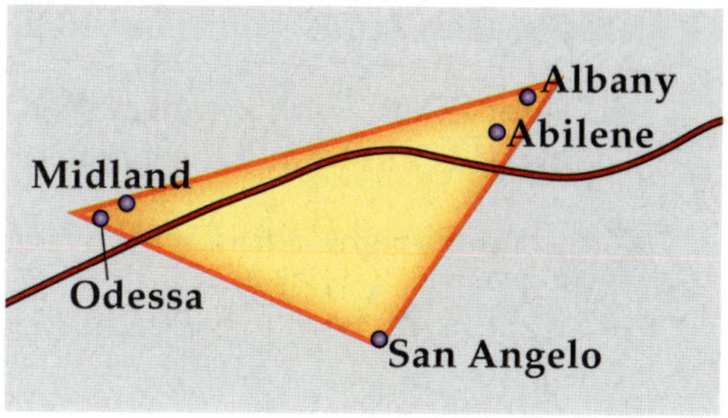

Personal Tutor glencoe.com

Check Your Understanding

Example 1
p. 301

Position and label each triangle on the coordinate plane.

1. right $\triangle ABC$ with legs $\overline{AC}$ and $\overline{AB}$ so that $\overline{AC}$ is $2a$ units long and leg $\overline{AB}$ is $2b$ units long

2. isosceles $\triangle FGH$ with base $\overline{FG}$ that is $2a$ units long

Example 2
p. 302

Name the missing coordinate(s) of each triangle.

3.

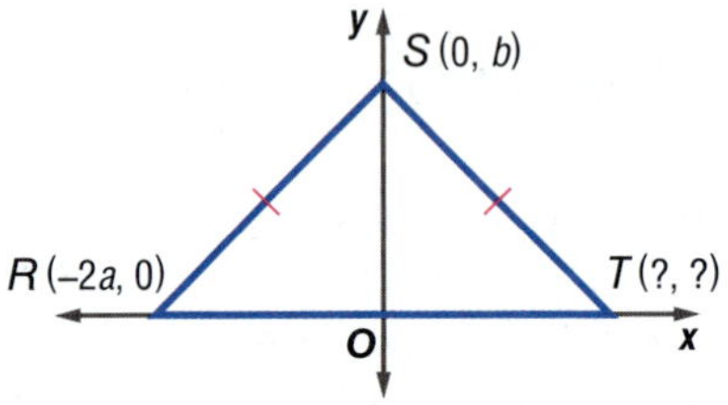

4.

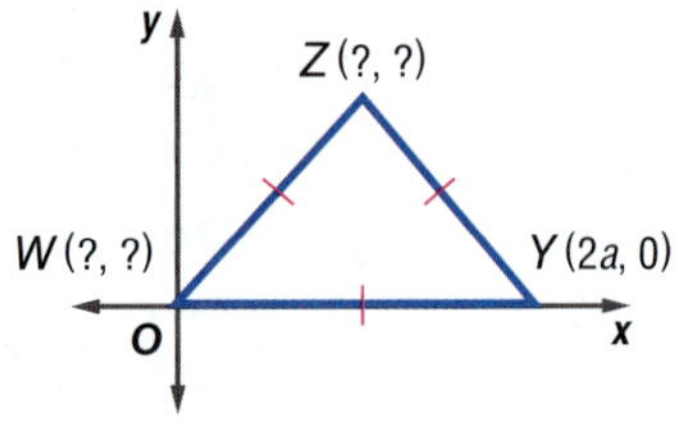

Example 3
p. 302

5. Write a coordinate proof to show that $\triangle FGH \cong \triangle FDC$.

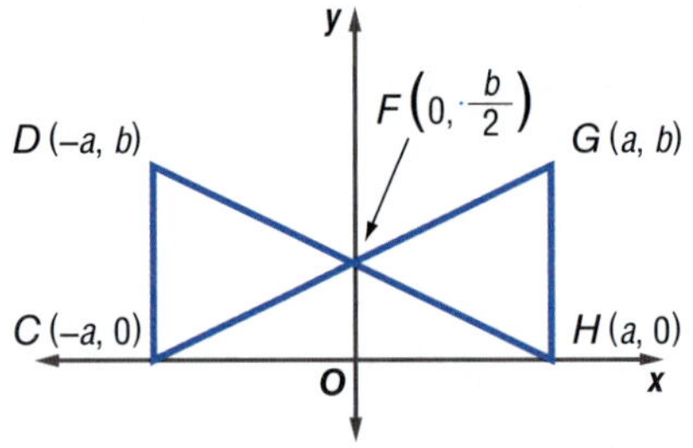

Example 4
p. 303

6. **FLAGS** Write a coordinate proof to prove that the large triangle in the center of the flag is isosceles. The dimensions of the flag are 4 feet by 6 feet and point B of the triangle bisects the bottom of the flag.

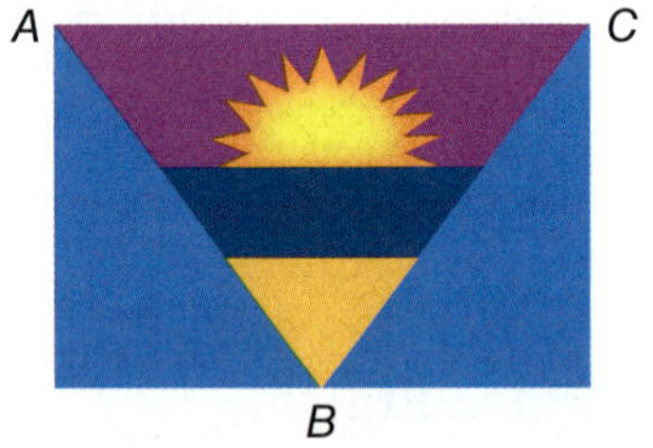

Practice and Problem Solving

● = **Step-by-Step Solutions** begin on page R20.
Extra Practice begins on page 969.

Example 1
p. 301

Position and label each triangle on the coordinate plane.

7. isosceles $\triangle ABC$ with base $\overline{AB}$ that is a units long

8. right $\triangle XYZ$ with hypotenuse $\overline{YZ}$, the length of $\overline{XY}$ is b units long, and the length of $\overline{XZ}$ is three times the length of $\overline{XY}$

9. isosceles right $\triangle RST$ with hypotenuse $\overline{RS}$ and legs $3a$ units long

10. right $\triangle JKL$ with legs $\overline{JK}$ and $\overline{KL}$ so that $\overline{JK}$ is a units long and leg $\overline{KL}$ is $4b$ units long

11. equilateral $\triangle GHJ$ with sides $\frac{1}{2}a$ units long

12. equilateral $\triangle DEF$ with sides $4b$ units long

Example 2
p. 302

Name the missing coordinate(s) of each triangle.

13.

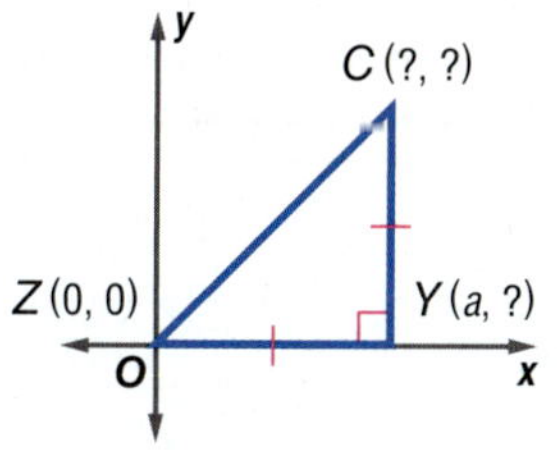

14.

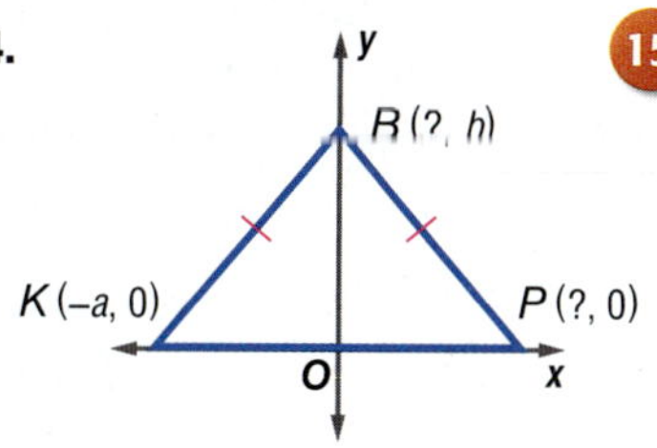

15.

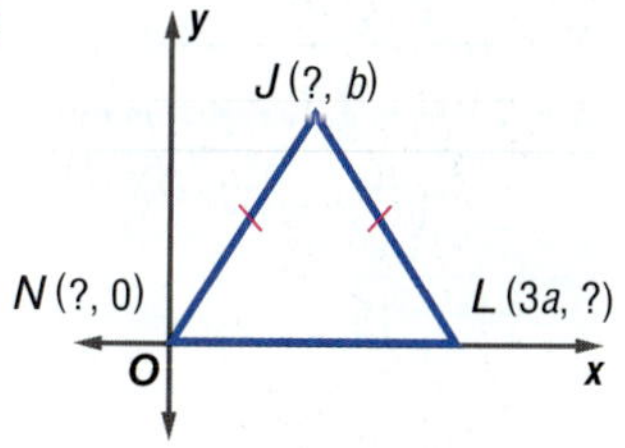

16.

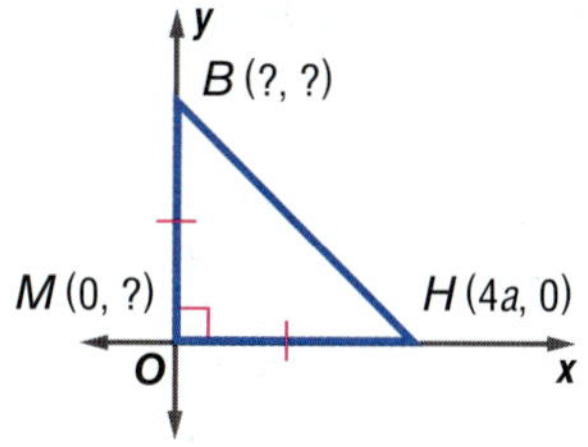

17.

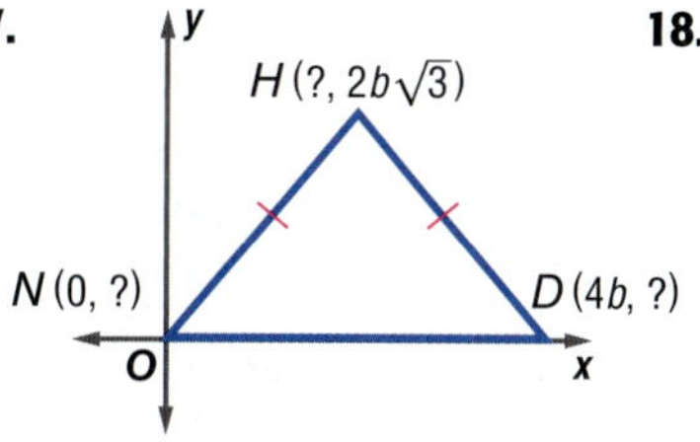

18.

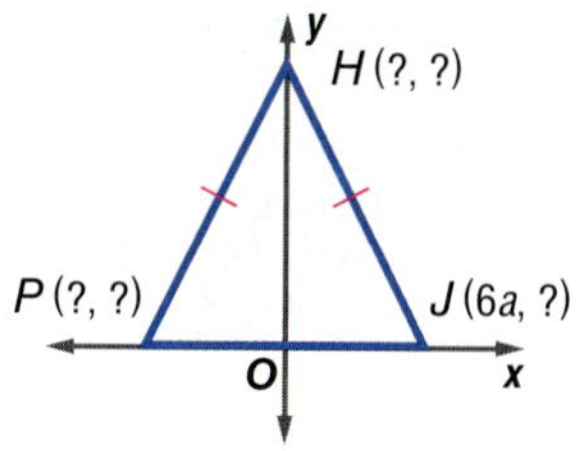

Example 3
p. 302

PROOF **Write a coordinate proof for each statement.**

19. The segments joining the base vertices to the midpoints of the legs of an isosceles triangle are congruent.

20. The three segments joining the midpoints of the sides of an isosceles triangle form another isosceles triangle.

Example 4
p. 303

PROOF **Write a coordinate proof for each statement.**

21. The measure of the segment that joins the vertex of the right angle in a right triangle to the midpoint of the hypotenuse is one-half the measure of the hypotenuse.

22. If a line segment joins the midpoints of two sides of a triangle, then its length is equal to one half the length of the third side.

Real-World Link

Duke University, the University of North Carolina at Chapel Hill, and North Carolina State University are all located in the Research Triangle.

23. RESEARCH TRIANGLE The cities of Raleigh, Durham, and Chapel Hill, North Carolina, form what is known as the Research Triangle. The approximate latitude and longitude of Raleigh are 35.82°N 29.9°W, of Durham are 35.97°N 78.9°W, and of Chapel Hill are 35.92°N 79.03°W. Show that the triangle formed by these three cities is scalene.

24. PARTY PLANNING Three friends live in houses with backyards adjacent to a neighborhood bike path. They decide to have a round-robin party using their three homes, inviting their friends to start at one house and then move to each of the other two. If one friend's house is centered at the origin, then the location of the other homes are (5, 12) and (13, 0). Write a coordinate proof to prove that the triangle formed by these three homes is isosceles.

Draw $\triangle XYZ$ and find the slope of each side of the triangle. Determine whether the triangle is a right triangle. Explain.

25. $X(0, 0)$, $Y(2h, 2h)$, $Z(4h, 0)$

26. $X(0, 0)$, $Y(1, h)$, $Z(2h, 0)$

27. CAMPING Two families set up tents at a state park. If the ranger's station is located at (0, 0), and the locations of the tents are (0, 25) and (12, 9), write a coordinate proof to prove that the figure formed by the locations of the ranger's station and the two tents is a right triangle.

28. PROOF Write a coordinate proof to prove that $\triangle ABC$ is an isosceles triangle if the vertices are $A(0, 0)$, $B(a, b)$, and $C(2a, 0)$.

Real-World Link

Personal watercraft are less than 13 feet in length. In 2005, there were 1,550,000 personal watercraft in use.

Source: U.S. Department of Transportation

29 **WATER SPORTS** Three personal watercraft vehicles launch from the same dock. The first vehicle leaves the dock traveling due northeast, while the second vehicle travels due northwest. Meanwhile, the third vehicle leaves the dock traveling due north.

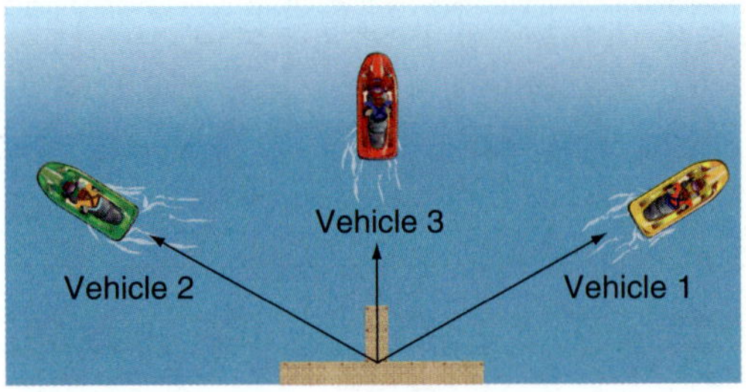

The first and second vehicles stop about 300 yards from the dock, while the third stops about 212 yards from the dock.

a. If the dock is located at (0, 0), sketch a graph to represent this situation. What is the equation of the line along which the first vehicle lies? What is the equation of the line along which the second vehicle lies? Explain your reasoning.

b. Write a coordinate proof to prove that the dock, the first vehicle, and the second vehicle form an isosceles right triangle.

c. Find the coordinates of the locations of all three watercrafts. Explain your reasoning.

d. Write a coordinate proof to prove that the positions of all three watercrafts are approximately collinear and that the third watercraft is at the midpoint between the other two.

H.O.T. Problems Use Higher-Order Thinking Skills

30. **REASONING** The midpoints of the sides of a triangle are located at $(a, 0)$, $(2a, b)$ and (a, b). If one vertex is located at the origin, what are the coordinates of the other vertices? Explain your reasoning.

CHALLENGE **Find the coordinates of point L so $\triangle JKL$ is the indicated type of triangle. Point J has coordinates (0, 0) and point K has coordinates $(2a, 2b)$.**

31. scalene triangle **32.** right triangle **33.** isosceles triangle

34. **OPEN ENDED** Draw an isosceles right triangle on the coordinate plane so that the midpoint of its hypotenuse is the origin. Label the coordinates of each vertex.

35. **CHALLENGE** Use a coordinate proof to show that if you add n units to each x-coordinate of the vertices of a triangle and m to each y-coordinate, the resulting figure is congruent to the original triangle.

36. **REASONING** A triangle has vertex coordinates (0, 0) and $(a, 0)$. If the coordinates of the third vertex are in terms of a, and the triangle is isosceles, identify the coordinates and position the triangle on the coordinate plane.

37. **WRITING IN MATH** Explain why following each guideline below for placing a triangle on the coordinate plane is helpful in proving coordinate proofs.

a. Use the origin as a vertex of the triangle.

b. Place at least one side of the triangle on the x- or y-axis.

c. Keep the triangle within the first quadrant if possible.

Standardized Test Practice

38. GRIDDED RESPONSE In the figure below, $m\angle B = 76$. The measure of $\angle A$ is half the measure of $\angle B$. What is $m\angle C$?

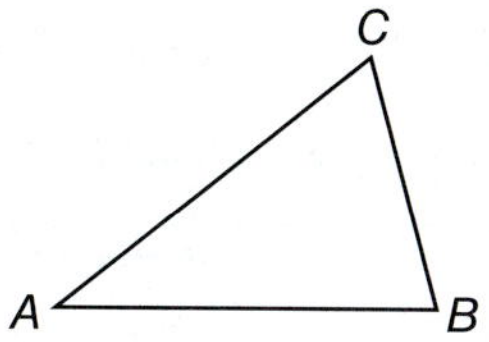

39. ALGEBRA What is the x-coordinate of the solution to the system of equations shown below?

$$\begin{cases} 2x - 3y = 3 \\ -4x + 2y = -18 \end{cases}$$

A -6 **C** 3

B -3 **D** 6

40. What are the coordinates of point R in the triangle?

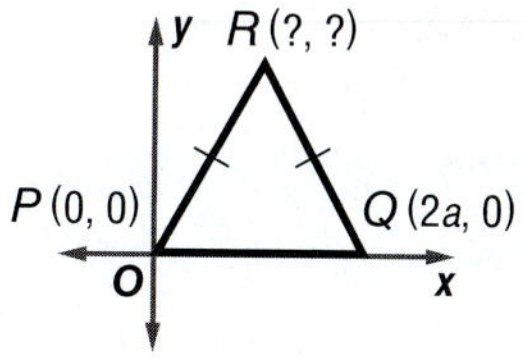

F $\left(\frac{a}{2}, a\right)$ **H** $\left(\frac{b}{2}, a\right)$

G (a, b) **J** $\left(\frac{b}{2}, \frac{a}{2}\right)$

41. SAT/ACT For all x,
$17x^5 + 3x^2 + 2 - (-4x^5 + 3x^3 - 2) =$

A $13x^5 + 3x^3 + 3x^2$

B $21x^5 - 3x^3 + 3x^2 + 4$

C $13x^5 + 6x^2 + 4$

D $21x^5 + 3x^2 + 3x^3$

Spiral Review

Identify the type of congruence transformation shown as a *reflection, translation,* or *rotation.* (Lesson 4-7)

42.

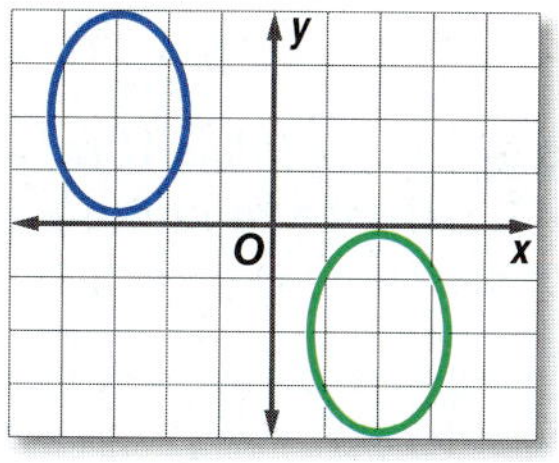

43.

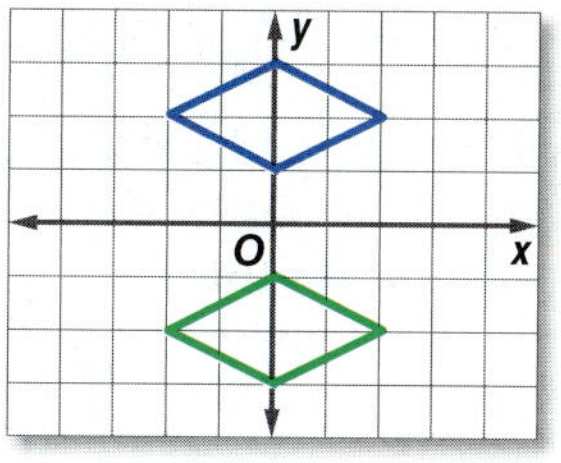

44.

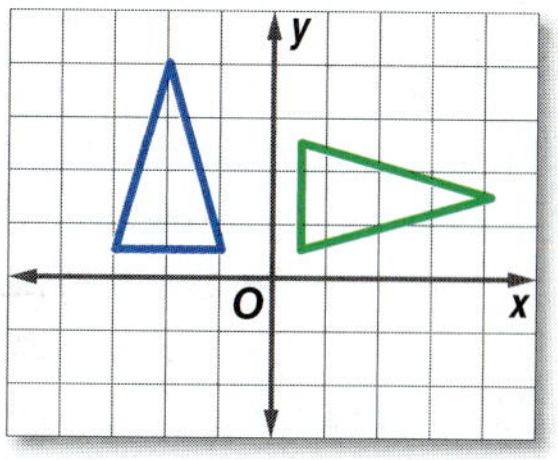

Refer to the figure at the right. (Lesson 4-6)

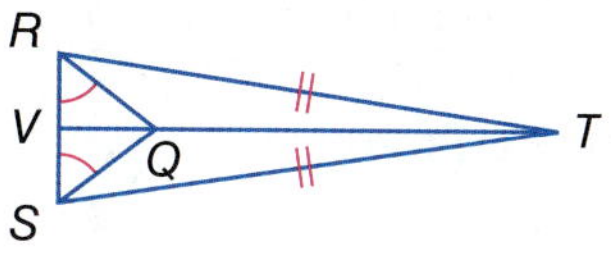

45. Name two congruent angles.

46. Name two congruent segments.

47. Name a pair of congruent triangles.

48. RAMPS The Americans with Disabilities Act requires that wheelchair ramps have at least a 12-inch run for each rise of 1 inch. (Lesson 3-3)

a. Determine the slope represented by this requirement.

b. The maximum length that the law allows for a ramp is 30 feet. How many inches tall is the highest point of this ramp?

Skills Review

Find the distance between each pair of points. Round to the nearest tenth. (Lesson 1-3)

49. $X(5, 4)$ and $Y(2, 1)$ **50.** $A(1, 5)$ and $B(-2, -3)$ **51.** $J(-2, 6)$ and $K(1, 4)$

CHAPTER 4

Study Guide and Review

Math Online glencoe.com
- STUDY TO GO
- Vocabulary Review

Chapter Summary

Key Concepts

Classifying Triangles (Lesson 4-1)

- Triangles can be classified by their angles as acute, obtuse, or right, and by their sides as scalene, isosceles, or equilateral.

Angles of Triangles (Lesson 4-2)

- The measure of an exterior angle is equal to the sum of its two remote interior angles.

Congruent Triangles (Lesson 4-3 through 4-5)

- SSS: If all of the corresponding sides of two triangles are congruent, then the triangles are congruent.
- SAS: If two pairs of corresponding sides of two triangles and the included angles are congruent, then the triangles are congruent.
- ASA: If two pairs of corresponding angles of two triangles and the included sides are congruent, then the triangles are congruent.
- AAS: If two pairs of corresponding angles of two triangles are congruent, and a corresponding pair of nonincluded sides is congruent, then the triangles are congruent.

Isosceles Triangles (Lesson 4-6)

- The base angles of an isosceles triangle are congruent and a triangle is equilateral if it is equiangular.

Transformations and Coordinate Proofs (Lessons 4-7 and 4-8)

- In a congruence transformation, the position of the image may differ from the preimage, but the two figures remain congruent.
- Coordinate proofs use algebra to prove geometric concepts.

FOLDABLES® Study Organizer

Be sure the Key Concepts are noted in your Foldable.

Key Vocabulary

acute triangle (p. 235)
auxiliary line (p. 244)
base angles (p. 283)
congruence transformation (p. 294)
congruent polygons (p. 253)
coordinate proof (p. 301)
corollary (p. 247)
corresponding parts (p. 253)
equiangular triangle (p. 235)
equilateral triangle (p. 236)
exterior angle (p. 246)
flow proof (p. 246)
included angle (p. 264)
included side (p. 273)
isosceles triangle (p. 236)
obtuse triangle (p. 235)
reflection (p. 294)
remote interior angles (p. 246)
right triangle (p. 235)
rotation (p. 294)
scalene triangle (p. 236)
translation (p. 294)
vertex angle (p. 283)

Vocabulary Check

State whether each sentence is *true* or *false*. If *false*, replace the underlined word or phrase to make a true sentence.

1. An equiangular triangle is also an example of an acute triangle.
2. A triangle with an angle that measures greater than 90° is a right triangle.
3. An equilateral triangle is always equiangular.
4. A scalene triangle has at least two congruent sides.
5. The vertex angles of an isosceles triangle are congruent.
6. An included side is the side located between two consecutive angles of a polygon.
7. The three types of congruence transformations are rotation, reflection, and translation.
8. A rotation moves all points of a figure the same distance and in the same direction.
9. A flow proof uses figures in the coordinate plane and algebra to prove geometric concepts.
10. The measure of an exterior angle of a triangle is equal to the sum of the measures of its two remote interior angles.

Lesson-by-Lesson Review

4-1 Classifying Triangles (pp. 235–242)

Classify each triangle as *acute*, *equiangular*, *obtuse*, or *right*.

11. $\triangle ADB$

12. $\triangle BCD$

13. $\triangle ABC$

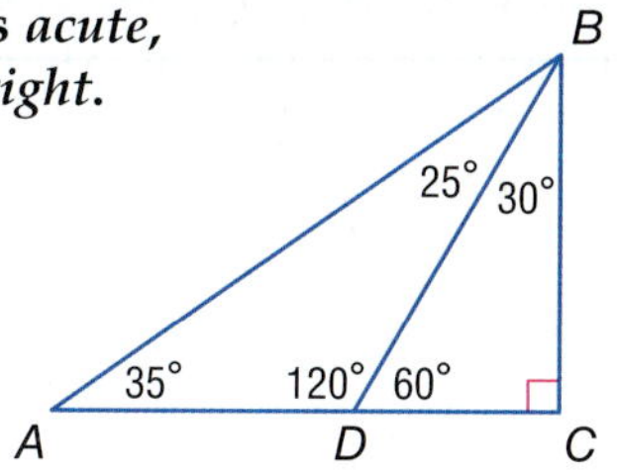

ALGEBRA **Find x and the measures of the unknown sides of each triangle.**

14.

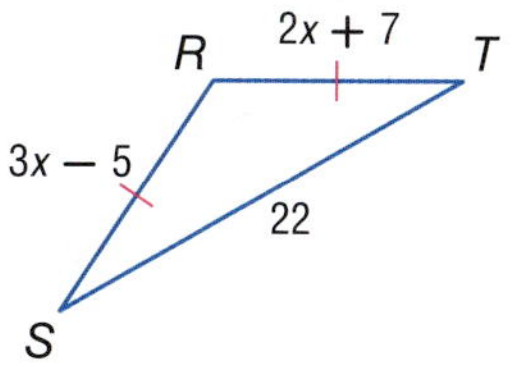

15.

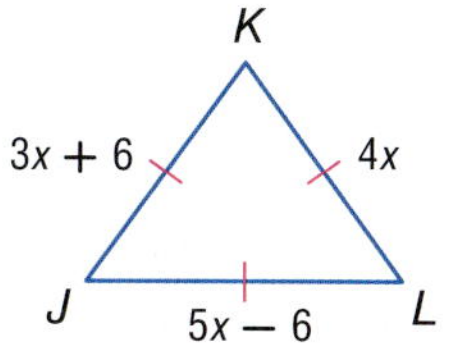

16. MAPS The distance from Chicago to Cleveland to Cincinnati and back to Chicago is 900 miles. The distance from Chicago to Cleveland is 50 miles more than the distance from Cincinnati to Chicago, and the distance from Cleveland to Cincinnati is 50 miles less than the distance from Cincinnati to Chicago. Find each distance and classify the triangle formed by the three cities.

EXAMPLE 1

Classify each triangle as *acute*, *equiangular*, *obtuse*, or *right*.

a.

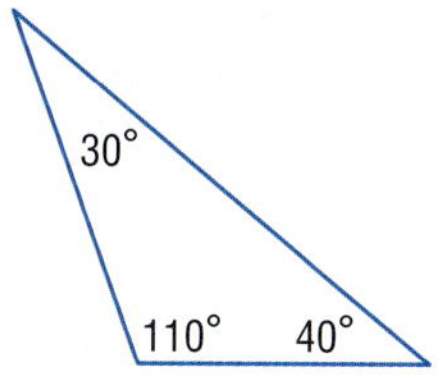

Since the triangle has one obtuse angle, it is an obtuse triangle.

b.

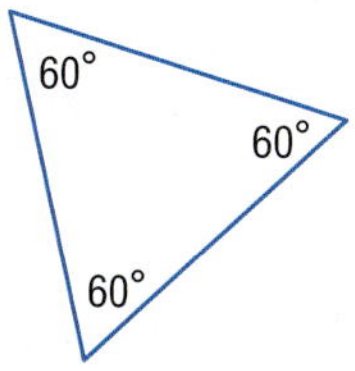

The triangle has three acute angles that are all equal. It is an equiangular triangle.

4-2 Angles of Triangles (pp. 244–252)

Find the measure of each numbered angle.

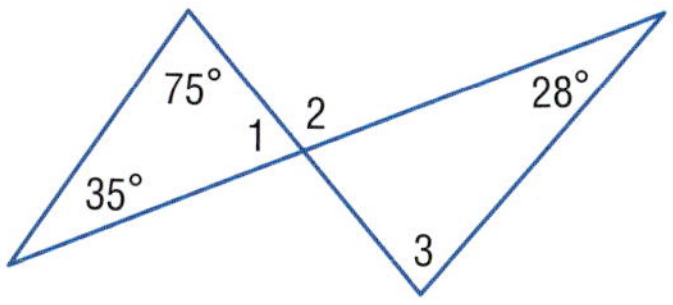

17. $\angle 1$

18. $\angle 2$

19. $\angle 3$

20. HOUSES The roof support on Lamar's house is in the shape of an isosceles triangle with base angles of 38°. Find x.

EXAMPLE 2

Find the measure of each numbered angle.

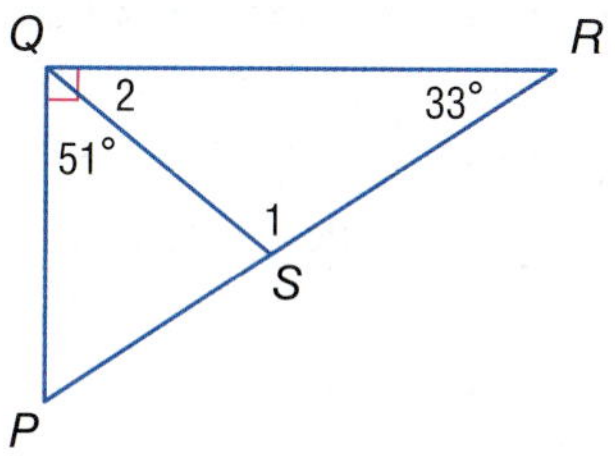

$m\angle 2 + m\angle PQS = 90$	
$m\angle 2 + 51 = 90$	**Substitution**
$m\angle 2 = 39$	**Subtract 51 from each side.**
$m\angle 1 + m\angle 2 + 33 = 180$	**Triangle Sum Theorem**
$m\angle 1 + 39 + 33 = 180$	**Substitution**
$m\angle 1 + 72 = 180$	**Simplify.**
$m\angle 1 = 108$	**Subtract.**

Study Guide and Review

4-3 Congruent Triangles (pp. 253–261)

Show that the polygons are congruent by identifying all congruent corresponding parts. Then write a congruence statement.

21.

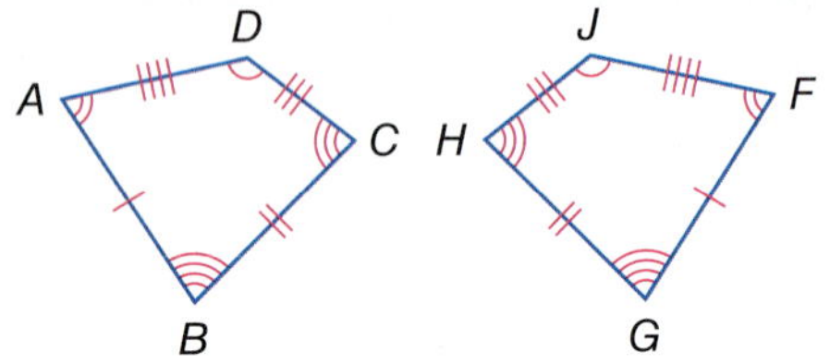

22.

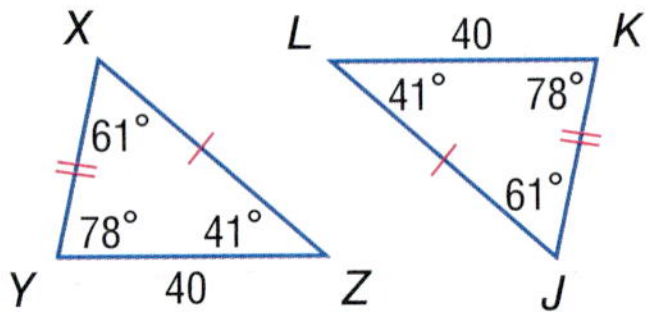

23. **MOSAIC TILING** A section of a mosaic tiling is shown. Name the triangles that appear to be congruent.

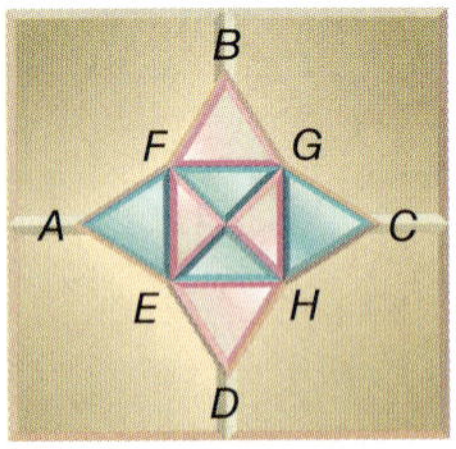

EXAMPLE 3

Show that the polygons are congruent by identifying all the congruent corresponding parts. Then write a congruence statement.

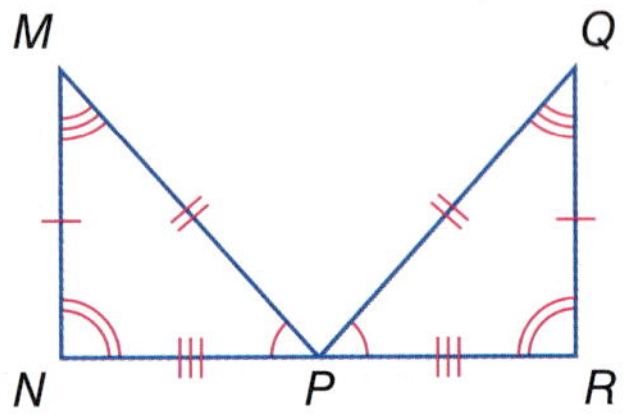

Angles: $\angle N \cong \angle R$, $\angle M \cong \angle Q$, $\angle MPN \cong \angle QPR$

Sides: $\overline{MN} \cong \overline{QR}$, $\overline{MP} \cong \overline{QP}$, $\overline{NP} \cong \overline{RP}$

All corresponding parts of the two triangles are congruent. Therefore, $\triangle MNP \cong \triangle QRP$.

4-4 Proving Triangles Congruent–SSS, SAS (pp. 262–270)

Determine whether $\triangle ABC \cong \triangle XYZ$. Explain.

24. $A(5, 2)$, $B(1, 5)$, $C(0, 0)$, $X(-3, 3)$, $Y(-7, 6)$, $Z(-8, 1)$

25. $A(3, -1)$, $B(3, 7)$, $C(7, 7)$, $X(-7, 0)$, $Y(-7, 4)$, $Z(1, 4)$

Determine which postulate can be used to prove that the triangles are congruent. If it is not possible to prove that they are congruent, write *not possible*.

26.

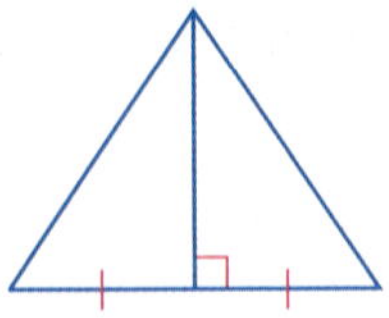

27.

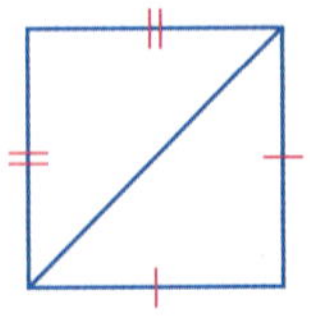

28. **PARKS** The diagram shows a park in the shape of a pentagon with five sidewalks of equal length leading to a central point. If all the angles at the central point have the same measure, how could you prove that $\triangle ABX \cong \triangle DCX$?

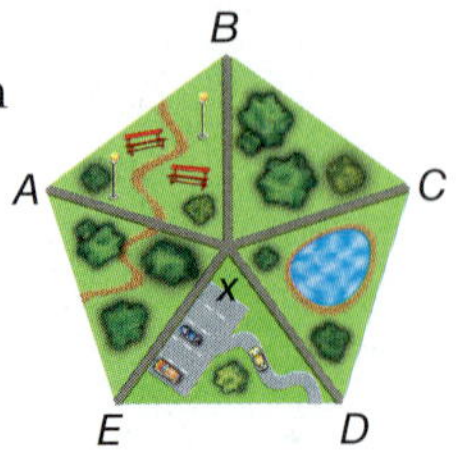

EXAMPLE 4

Write a two-column proof.

Given: $\triangle KPL$ is equilateral.
$\overline{JP} \cong \overline{MP}$,
$\angle JPK \cong \angle MPL$

Prove: $\triangle JPK \cong \triangle MPL$

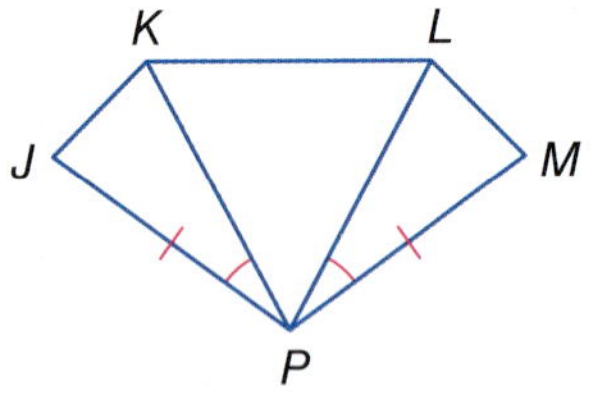

Statements	Reasons
1. $\triangle KPL$ is equilateral.	**1.** Given
2. $\overline{PK} \cong \overline{PL}$	**2.** Def. of Equilateral $\triangle$
3. $\overline{JP} \cong \overline{MP}$	**3.** Given
4. $\angle JPK \cong \angle MPL$	**4.** Given
5. $\triangle JPK \cong \triangle MPL$	**5.** SAS

MIXED PROBLEM SOLVING
For mixed problem-solving practice, see page 1002.

4-5 Proving Triangles Congruent–ASA, AAS (pp. 273–280)

Write a two-column proof.

29. **Given:** $\overline{AB} \parallel \overline{DC}$, $\overline{AB} \cong \overline{DC}$
Prove: $\triangle ABE \cong \triangle CDE$

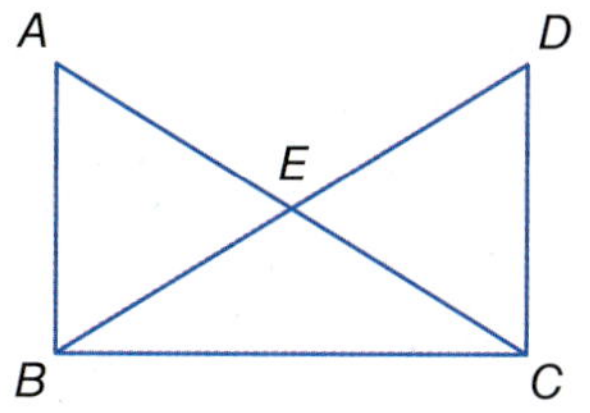

30. KITES Denise's kite is shown in the figure at the right. Given that $\overline{WY}$ bisects both $\angle XWZ$ and $\angle XYZ$, prove that $\triangle WXY \cong \triangle WZY$.

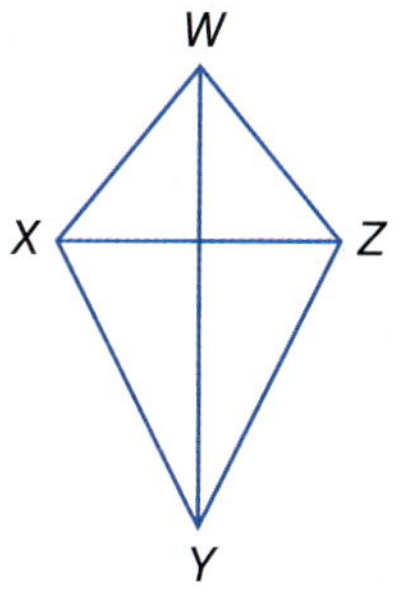

EXAMPLE 5

Write a flow proof.

Given: $\overline{PQ}$ bisects $\angle RPS$.
$\angle R \cong \angle S$

Prove: $\triangle RPQ \cong \triangle SPQ$

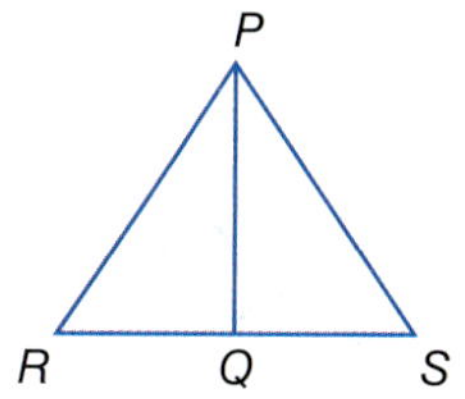

Flow Proof:

$\overline{PQ}$ bisects $\angle RPS$. — **Given** → $\angle RPQ \cong \angle SPQ$ — Def. of ∠ Bisector

$\angle R \cong \angle S$ — **Given**

$\overline{PQ} \cong \overline{PQ}$ — **Reflexive**

→ $\triangle RPQ \cong \triangle SPQ$ — AAS

4-6 Isosceles and Equilateral Triangles (pp. 283–291)

Find the value of each variable.

31.

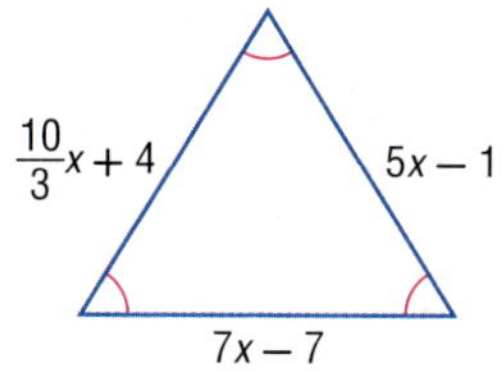

32.

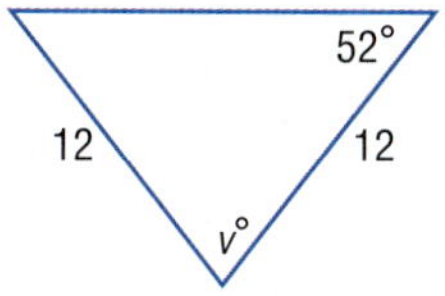

33. PAINTING Pam is painting using a wooden easel. The support bar on the easel forms an isosceles triangle with the two front supports. According to the figure below, what are the measures of the base angles of the triangle?

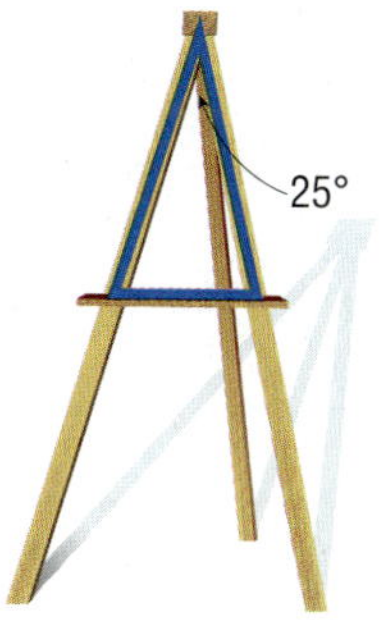

EXAMPLE 6

Find each measure.

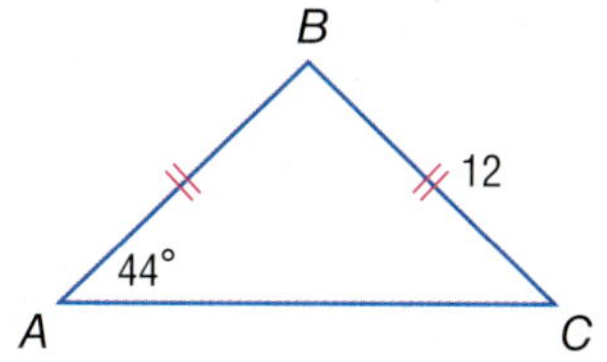

a. $m\angle B$

Since $AB = BC$, $\overline{AB} \cong \overline{BC}$. By the Isosceles Triangle Theorem, base angles A and C are congruent, so $m\angle A = m\angle C$. Use the Triangle Sum Theorem to write and solve an equation to find $m\angle B$.

$m\angle A + m\angle B + m\angle C = 180$	**△ Sum Theorem**
$44 + m\angle B + 44 = 180$	$m\angle A = m\angle C = 44$
$88 + m\angle B = 180$	**Simplify.**
$m\angle B = 92$	**Subtract.**

b. AB

$AB = BC$, so $\triangle ABC$ is isosceles. Since $BC = 12$, $AB = 12$ by substitution.

CHAPTER 4

Study Guide and Review

4-7 Congruence Transformations (pp. 294–300)

Identify the type of congruence transformation shown as a *reflection*, *translation*, or *rotation*.

34.

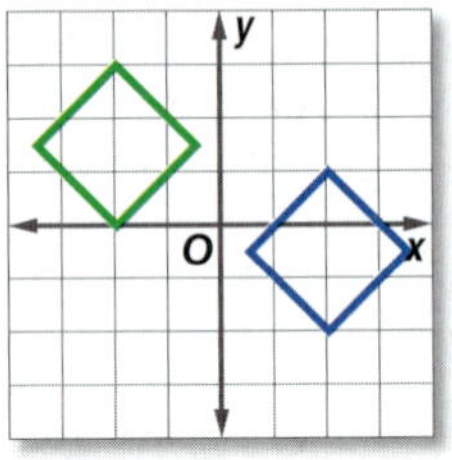

35.

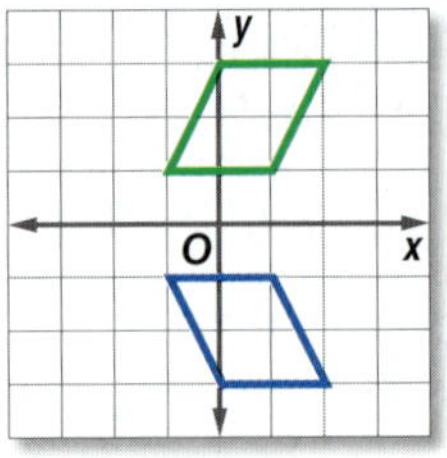

36.

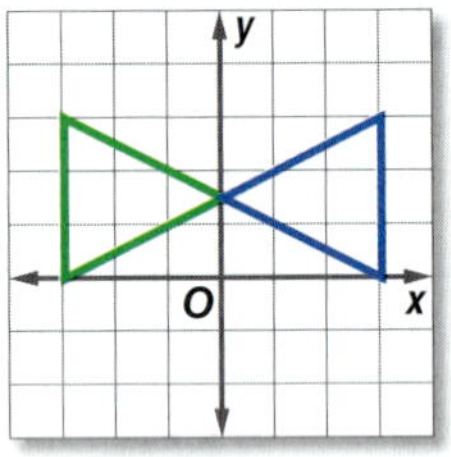

37.

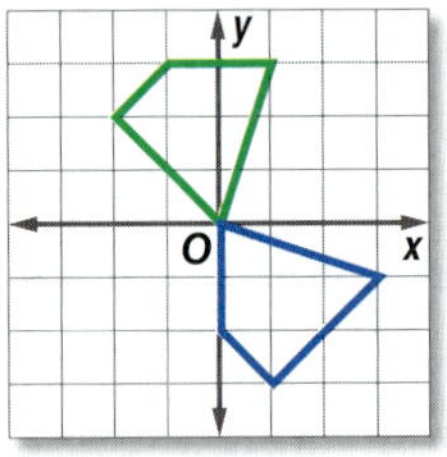

38. Triangle ABC with vertices $A(1, 1)$, $B(2, 3)$, and $C(3, -1)$ is a transformation of $\triangle MNO$ with vertices $M(-1, 1)$, $N(-2, 3)$, and $O(-3, -1)$. Graph the original figure and its image. Identify the transformation and verify that it is a congruence transformation.

EXAMPLE 7

Triangle RST with vertices $R(4, 1)$, $S(2, 5)$, and $T(-1, 0)$ is a transformation of $\triangle CDF$ with vertices $C(1, -3)$, $D(-1, 1)$, and $F(-4, -4)$. Identify the transformation and verify that it is a congruence transformation.

Graph each figure. The transformation appears to be a translation. Find the lengths of the sides of each triangle.

$$RS = \sqrt{(4-2)^2 + (1-5)^2} \text{ or } \sqrt{20}$$
$$TS = \sqrt{(-1-2)^2 + (0-5)^2} \text{ or } \sqrt{34}$$
$$RT = \sqrt{(-1-4)^2 + (0-1)^2} \text{ or } \sqrt{26}$$
$$CD = \sqrt{(-1-1)^2 + [1-(-3)]^2} \text{ or } \sqrt{20}$$
$$DF = \sqrt{[-4-(-1)]^2 + (-4-1)^2} \text{ or } \sqrt{34}$$
$$CF = \sqrt{(-4-1)^2 + [-4-(-3)]^2} \text{ or } \sqrt{26}$$

Since each vertex of $\triangle CDF$ has undergone a transformation 3 units to the right and 4 units up, this is a translation.

Since $RS = CD$, $TS = DF$, and $RT = CF$, $\overline{RS} \cong \overline{CD}$, $\overline{TS} \cong \overline{DF}$, and $\overline{RT} \cong \overline{CF}$. By SSS, $\triangle RST \cong \triangle CDF$.

4-8 Triangles and Coordinate Proof (pp. 301–307)

Position and label each triangle on the coordinate plane.

39. right $\triangle MNO$ with right angle at point M and legs of lengths a and $2a$.

40. isosceles $\triangle WXY$ with height h and base $\overline{WY}$ with length $2a$.

41. GEOGRAPHY Jorge plotted the cities of Dallas, San Antonio, and Houston as shown. Write a coordinate proof to show that the triangle formed by these cities is scalene.

EXAMPLE 8

Position and label an equilateral triangle $\triangle XYZ$ with side lengths of $2a$.

- Use the origin for one of the three vertices of the triangle.
- Place one side of the triangle along the positive side of the x-axis.
- The third point should be located above the midpoint of the base of the triangle.

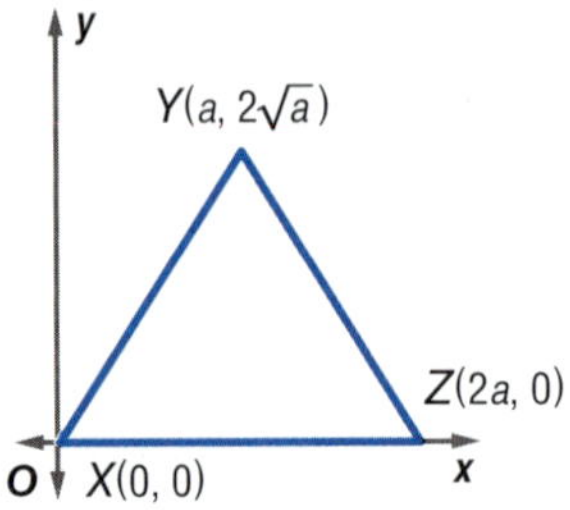

Practice Test

Math Online glencoe.com
Chapter Test

Classify each triangle as *acute, equilateral, obtuse,* or *right.*

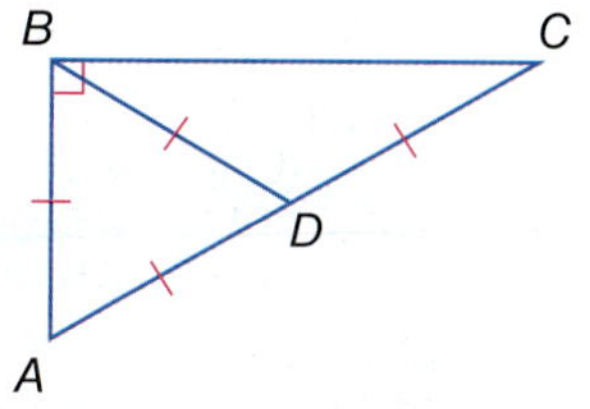

1. $\triangle ABD$
2. $\triangle ABC$
3. $\triangle BDC$

Find the measure of each numbered angle.

4. $\angle 1$
5. $\angle 2$
6. $\angle 3$
7. $\angle 4$

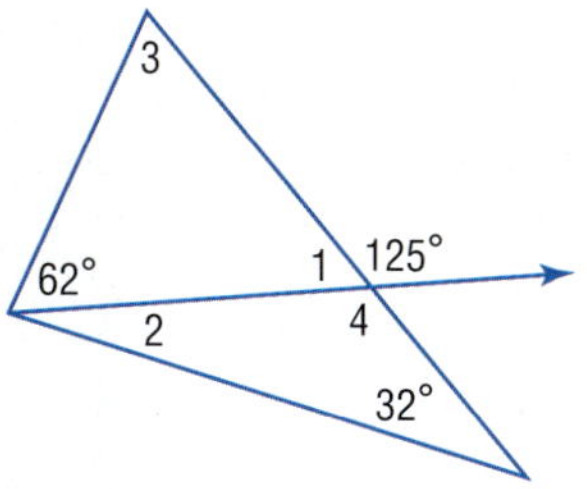

In the diagram, $\triangle RST \cong \triangle XYZ$.

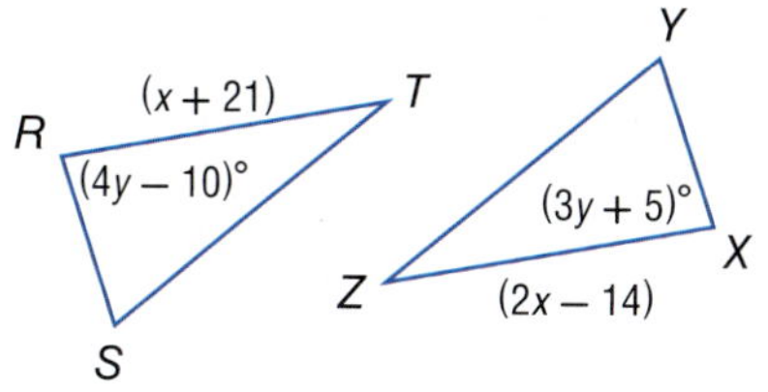

8. Find x.
9. Find y.
10. **PROOF** Write a flow proof.
    Given: $\overline{XY} \parallel \overline{WZ}$ and $\overline{XW} \parallel \overline{YZ}$
    Prove: $\triangle XWZ \cong \triangle ZYX$

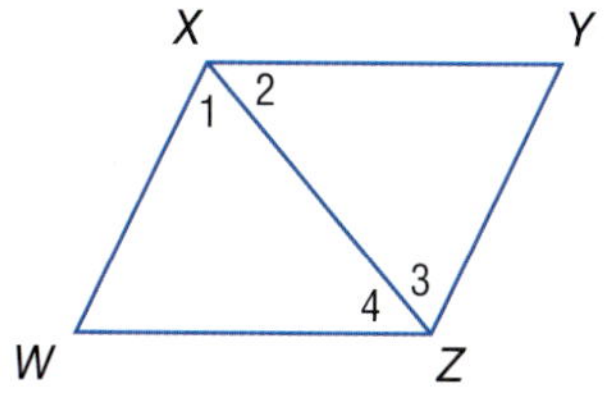

11. **MULTIPLE CHOICE** Find x.

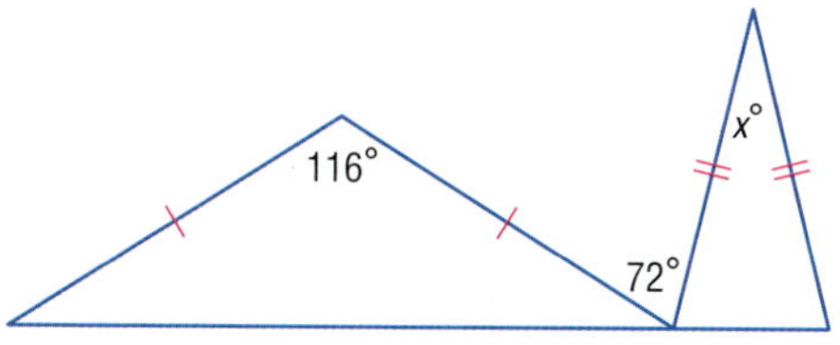

A 36
B 32
C 28
D 22

12. Determine whether $\triangle TJD \cong \triangle SEK$ given $T(-4, -2)$, $J(0, 5)$, $D(1, -1)$, $S(-1, 3)$, $E(3, 10)$, and $K(4, 4)$. Explain.

Determine which postulate or theorem can be used to prove each pair of triangles congruent. If it is not possible to prove them congruent, write *not possible*.

13. 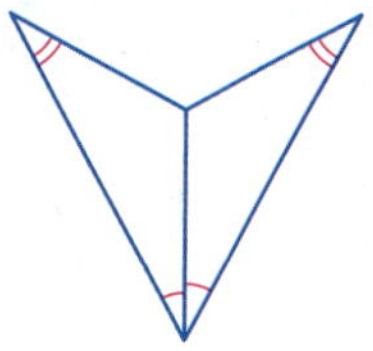

14. 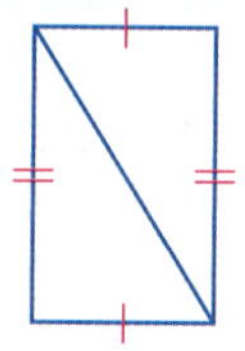

15. 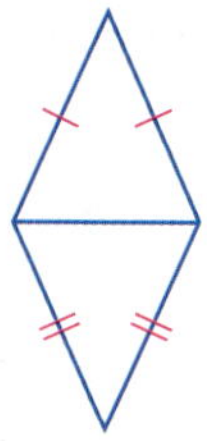

16. 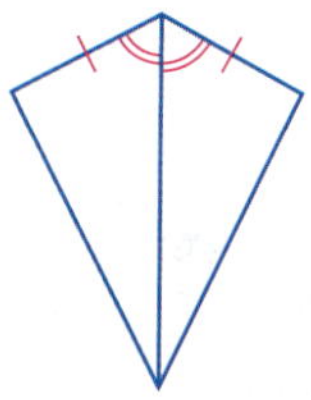

17. **LANDSCAPING** Angie has laid out a design for a garden consisting of two triangular areas as shown below. The points are $A(0, 0)$, $B(0, 5)$, $C(3, 5)$, $D(6, 5)$, and $E(6, 0)$. Name the type of congruence transformation for the preimage $\triangle ABC$ to $\triangle EDC$.

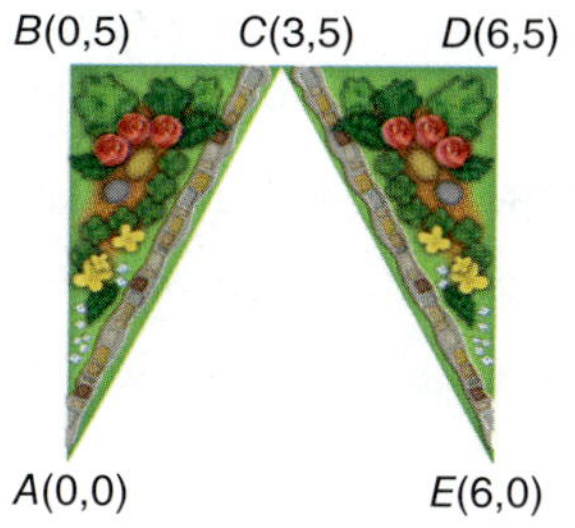

Find the measure of each numbered angle.

18. $\angle 1$
19. $\angle 2$

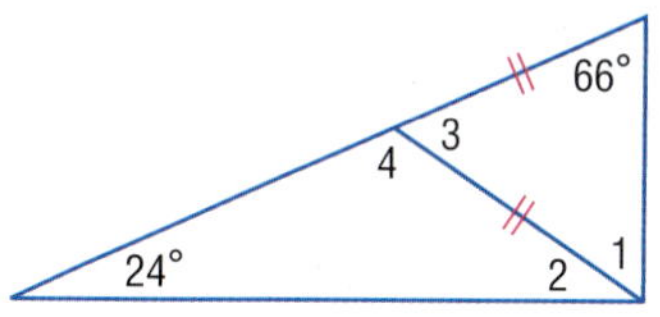

20. **PROOF** $\triangle ABC$ is a right isosceles triangle with hypotenuse $\overline{AB}$. M is the midpoint of $\overline{AB}$. Write a coordinate proof to show that $\overline{CM}$ is perpendicular to $\overline{AB}$.

Short-Answer Questions

Short-answer questions require you to provide a solution to the problem, along with a method, explanation, and/or justification used to arrive at the solution.

Short-answer questions are typically graded using a **rubric**, or a scoring guide.

The following is an example of a short-answer question scoring rubric.

Scoring Rubric		
Criteria		Score
Full Credit	The answer is correct and a full explanation is provided that shows each step.	2
Partial Credit	• The answer is correct, but the explanation is incomplete. • The answer is incorrect, but the explanation is correct.	1 1
No Credit	Either an answer is not provided or the answer does not make sense.	0

Strategies for Solving Short-Answer Questions

Step 1

Read the problem to gain an understanding of what you are trying to solve.

- Identify relevant facts.
- Look for key words and mathematical terms.

Step 2

Make a plan and solve the problem.

- Explain your reasoning or state your approach to solving the problem.
- Show all of your work or steps.
- Check your answer if time permits.

EXAMPLE

Read the problem. Identify what you need to know. Then use the information in the problem to solve. Show your work.

Triangle *ABC* is an isosceles triangle with base $\overline{BC}$. What is the perimeter of the triangle?

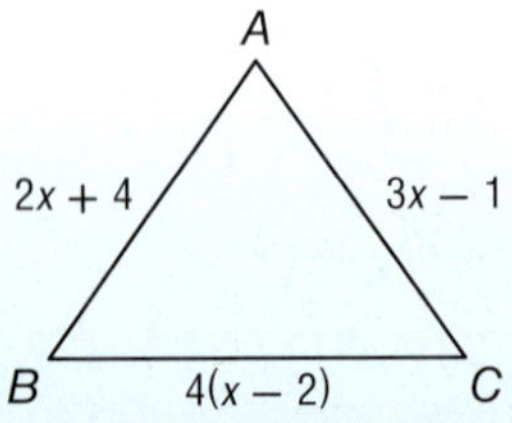

Read the problem carefully. You are told that $\triangle ABC$ is isosceles with base $\overline{BC}$. You are asked to find the perimeter of the triangle.

Make a plan and solve the problem.

The legs of an isosceles triangle are congruent.
So, $\overline{AB} \cong \overline{AC}$ or $AB = AC$. Solve for x.

$$AB = AC$$
$$2x + 4 = 3x - 1$$
$$2x - 3x = -1 - 4$$
$$-x = -5$$
$$x = 5$$

Next, find the length of each side.

$\overline{AB}$: $2(5) + 4 = 14$ units
$\overline{AC}$: $3(5) - 1 = 14$ units
$\overline{BC}$: $4(5 - 2) = 12$ units

The perimeter of $\triangle ABC$ is $14 + 14 + 12 = 40$ units.

The steps, calculations, and reasoning are clearly stated. The student also arrives at the correct answer. So, this response is worth the full 2 points.

Exercises

Read each problem. Identify what you need to know. Then use the information in the problem to solve. Show your work.

1. Classify $\triangle DEF$ according to its angle measures.

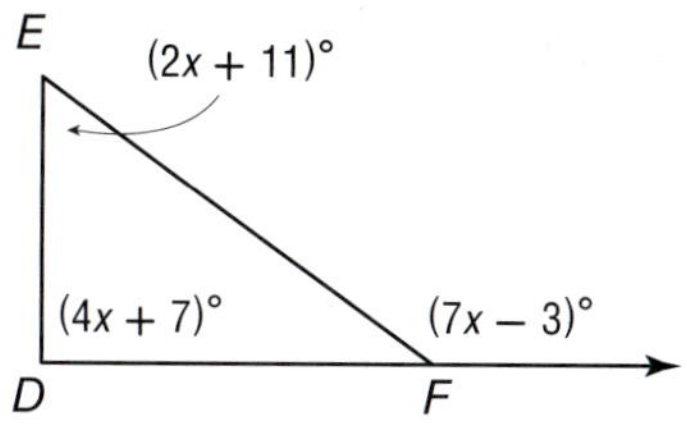

2. In the figure below, $\triangle RST \cong \triangle VUT$. What is the area of $\triangle RST$?

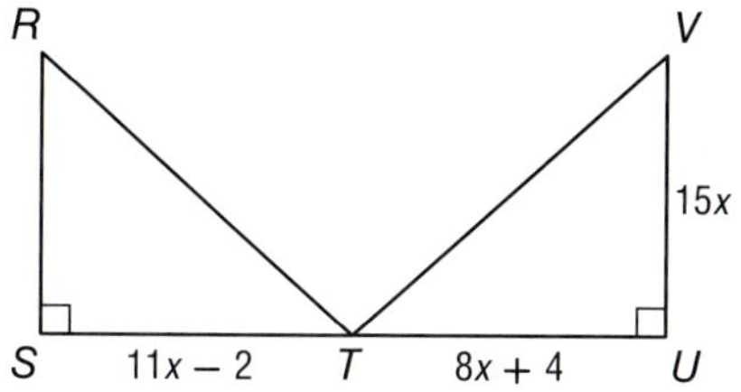

3. A farmer needs to make a 1000-square-foot rectangular enclosure for her cattle. She wants to save money by purchasing the least amount of fencing possible to enclose the area. What whole-number dimensions will require the least amount of fencing?

4. What is the $m\angle 1$ in degrees?

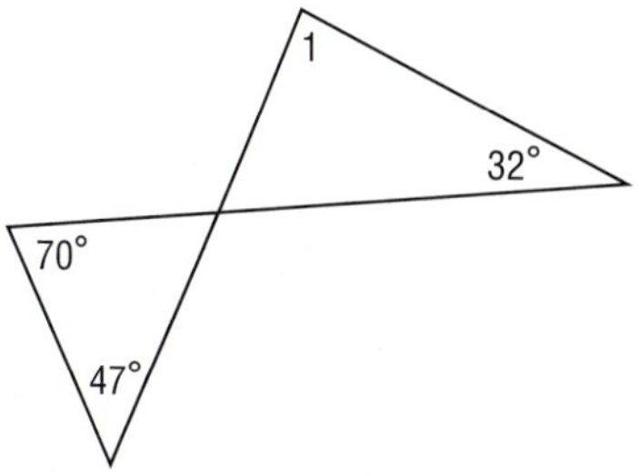

5. Write an equation of the line containing the points (2, 4) and (0, –2)

CHAPTER 4

Standardized Test Practice

Cumulative, Chapters 1 through 4

Multiple Choice

Read each question. Then fill in the correct answer on the answer document provided by your teacher or on a sheet of paper.

1. If $m\angle 1 = 110°$, what must $m\angle 2$ equal for lines x and z to be parallel?

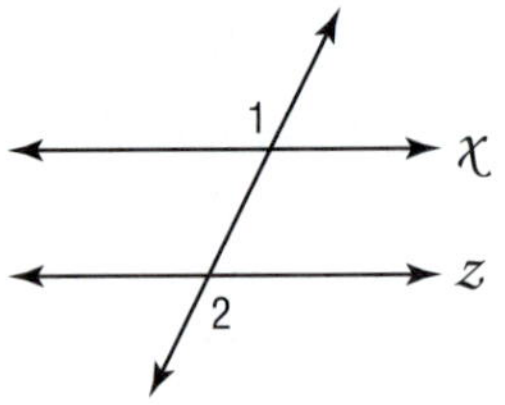

A 30° **B** 60° **C** 70° **D** 110°

2. Which of the following terms best describes the transformation below?

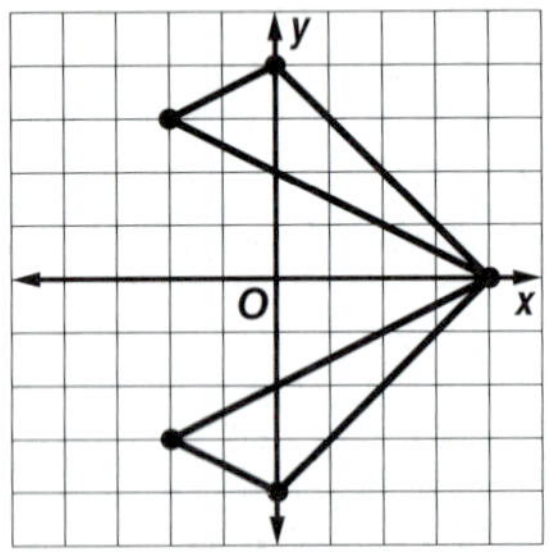

F dilation

G reflection

H rotation

J translation

3. Classify the triangle below according to its side lengths.

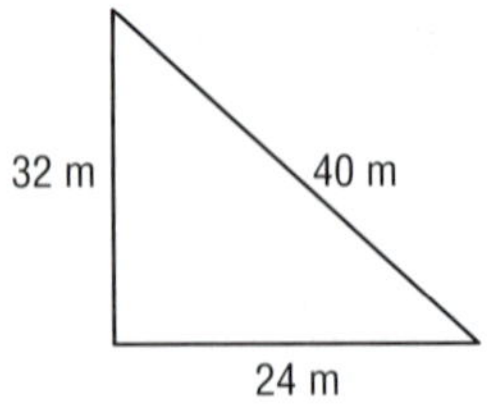

A equilateral

B isosceles

C right

D scalene

Test-Taking Tip

Question 3 Read the problem statement carefully to make sure you select the correct answer.

4. Given: $\overline{WX} \cong \overline{JK}$, $\overline{YX} \cong \overline{IK}$, $\angle X \cong \angle K$

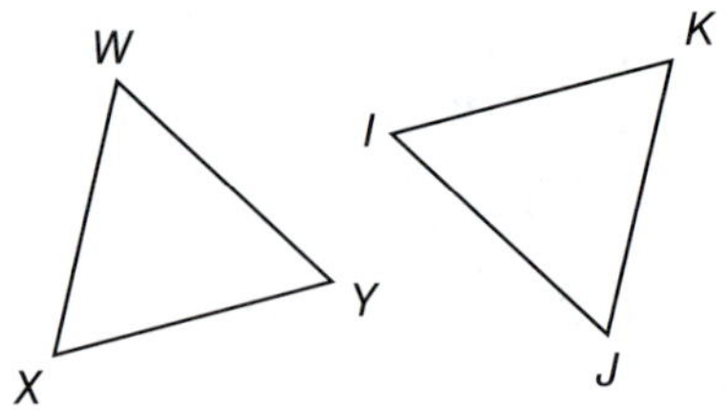

Which of the following lists the correct triangle congruence?

F $\triangle WXY \cong \triangle KIJ$

G $\triangle WXY \cong \triangle IKJ$

H $\triangle WXY \cong \triangle JKI$

J $\triangle WXY \cong \triangle IJK$

5. Suppose the dimensions of the prism below are tripled. By what factor will the volume of the prism increase?

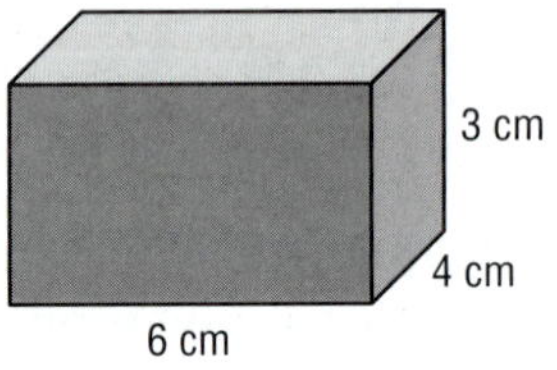

A 3 **B** 9 **C** 27 **D** 81

6. What is the measure of angle R below?

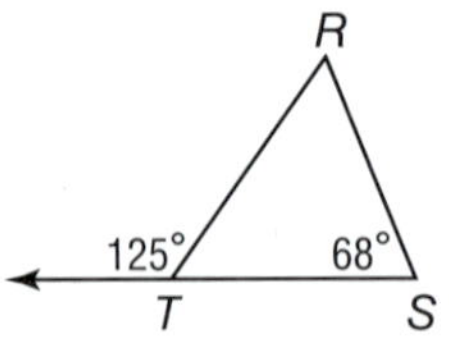

F 57° **G** 59° **H** 65° **J** 68°

7. Suppose one base angle of an isosceles triangle has a measure of 44°. What is the measure of the vertex angle?

A 108°

B 92°

C 56°

D 44°

Short Response/Gridded Response

Record your answers on the answer sheet provided by your teacher or on a sheet of paper.

8. **GRIDDED RESPONSE** In the figure below, $\triangle NDG \cong \triangle LGD$. What is the value of x?

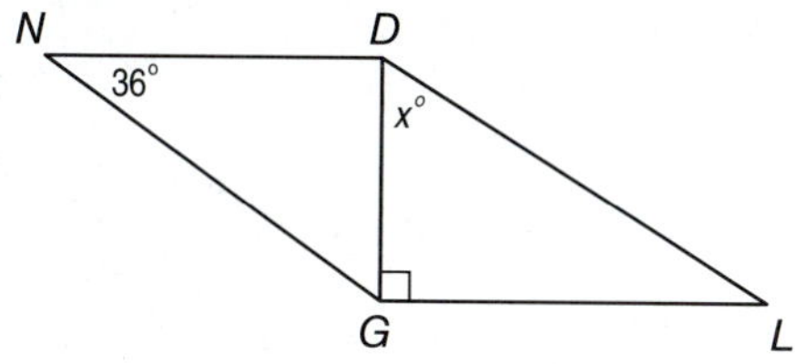

9. **GRIDDED RESPONSE** Suppose line ℓ contains points A, B, and C. If $AB = 7$ inches, $AC = 32$ inches, and point B is between points A and C, what is the length of $\overline{BC}$? Express your answer in inches.

10. Write the converse of the statement.

    If you are the winner, then I am the loser.

11. Use the figure and the given information below.

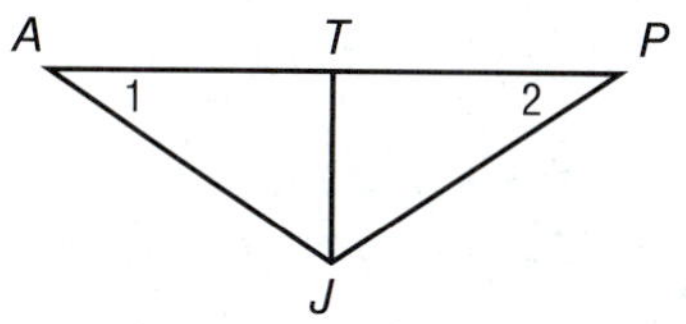

Given: $\overline{JT} \perp \overline{AP}$
$\angle 1 \cong \angle 2$

Which congruence theorem could you use to prove $\triangle PTJ \cong \triangle ATJ$ with only the information given? Explain.

12. Write an equation in slope intercept form for the line which goes through the points (0, 3) and (4, −5).

13. **GRIDDED RESPONSE** Find $m\angle TUV$ in the figure.

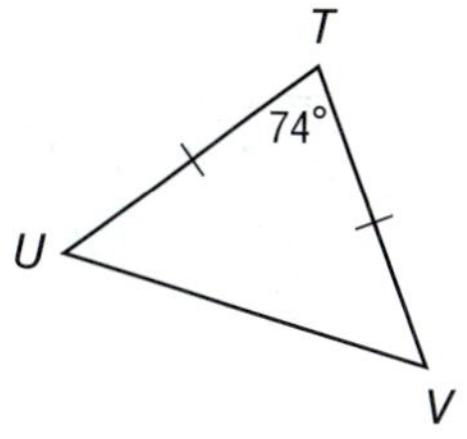

14. Suppose two sides of triangle ABC are congruent to two sides of triangle MNO. Also, suppose one of the nonincluded angles of $\triangle ABC$ is congruent to one of the nonincluded angles of $\triangle MNO$. Are the triangles congruent? If so, write a paragraph proof showing the congruence. If not, sketch a counterexample.

Extended Response

Record your answers on a sheet of paper. Show your work.

15. Use a coordinate grid to write a coordinate proof of the following statement.

    If the vertices of a triangle are A(0, 0), B(2a, b), and C(4a, 0), then the triangle is isosceles.

    a. Plot the vertices on a coordinate grid to model the problem.

    b. Use the Distance Formula to write an expression for AB.

    c. Use the Distance Formula to write an expression for BC.

    d. Use your results from parts **b** and **c** to draw a conclusion about $\triangle ABC$.

Need Extra Help?

If you missed Question...	1	2	3	4	5	6	7	8	9	10	11	12	13	14	15
Go to Lesson or Page...	3-2	4-7	4-1	4-3	1-7	4-2	4-6	4-3	1-2	2-3	4-5	3-4	4-6	4-4	4-8

CHAPTER 5 Relationships in Triangles

Then

In Chapter 4, you learned how to classify triangles.

Now

In Chapter 5, you will:

- Learn about special segments and points related to triangles.
- Learn about relationships between the sides and angles of triangles.
- Learn to write indirect proofs.

Why?

INTERIOR DESIGN Triangle relationships are used to find and compare angle measures and distances. Interior designers use the relationships in triangles to maximize efficiency and create balance in their designs.

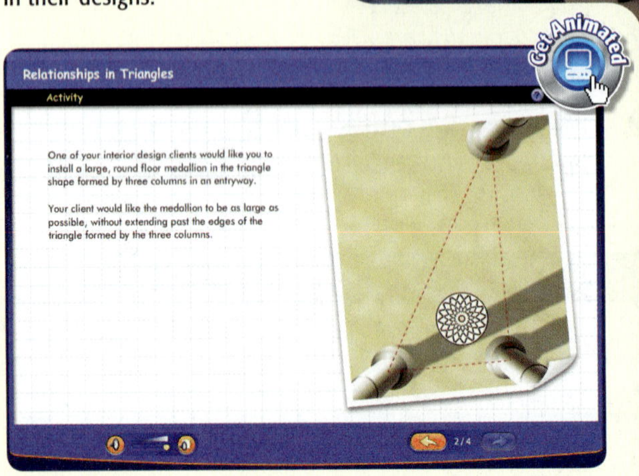

Get Ready for Chapter 5

Diagnose Readiness You have two options for checking Prerequisite Skills.

Text Option

Take the Quick Check below. Refer to the Quick Review for help.

QuickCheck

Find each measure. (Lesson 4-6)

1. BC

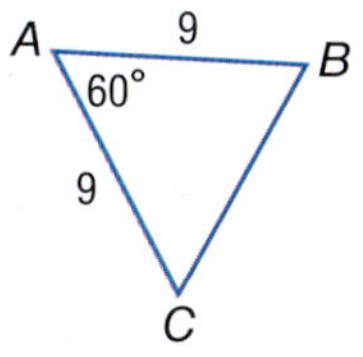

2. $m\angle RST$

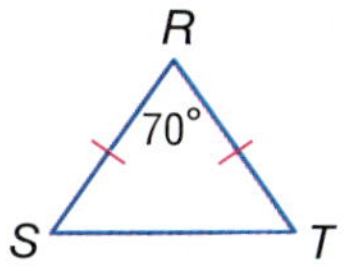

3. GARDENS Bronson is creating a right triangular flower bed. If two of the sides of the flower bed are 7 feet long each, what is the length of the third side to the nearest foot? (Prerequisite Skill)

Make a conjecture based on the given information. (Lesson 2-1)

4. $\angle 3$ and $\angle 4$ are a linear pair.

5. $JKLM$ is a square.

6. $\overrightarrow{BD}$ is an angle bisector of $\angle ABC$.

7. REASONING Determine whether the following conjecture is *always*, *sometimes*, or *never* true based on the given information. Justify your reasoning.

Given: collinear points D, E, and F

Conjecture: $DE + EF = DF$

Solve each inequality. (Lesson 0-6)

8. $x + 13 < 41$

9. $x - 6 > 2x$

10. $6x + 9 < 7x$

11. $8x + 15 > 9x - 26$

12. MUSIC Nina added 15 more songs to her MP3 player, resulting in a total of more than 120 songs. How many songs were originally on the player?

QuickReview

EXAMPLE 1

Find each measure.

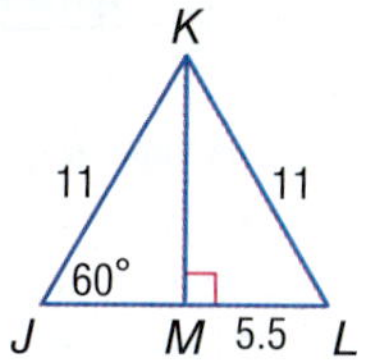

a. JM

$m\angle J = m\angle L$, so $m\angle L = 60$ and $\overline{JM} \cong \overline{LM}$ by the Converse of the Isosceles Triangle Theorem. Since $LM = 5.5$, $JM = 5.5$ by substitution.

b. $m\angle JKL$

$m\angle J + m\angle JKL + m\angle L = 180$ — **△ Sum Theorem**

$60 + m\angle JKL + 60 = 180$ — $m\angle J = m\angle L = 60$

$120 + m\angle JKL = 180$ — **Simplify.**

$m\angle JKL = 60$ — **Subtract.**

EXAMPLE 2

K is the midpoint of $\overline{JL}$. Make a conjecture based on the given information and draw a figure to illustrate your conjecture.

Given: K is the midpoint of $\overline{JL}$
J, K, and L are collinear points, and K lies an equal distance between J and L.

Conjecture: $\overline{JK} \cong \overline{KL}$

Check: Draw $\overline{JL}$. This illustrates the conjecture.

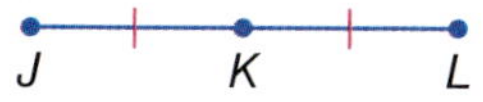

EXAMPLE 3

Solve $3x + 5 > 2x$.

$3x + 5 > 2x$ — **Given**

$3x - 3x + 5 > 2x - 3x$ — **Subtract.**

$5 > -x$ — **Simplify.**

$-5 < x$ — **Divide.**

Online Option

Math Online Take a self-check Chapter Readiness Quiz at glencoe.com.

Get Started on Chapter 5

You will learn several new concepts, skills, and vocabulary terms as you study Chapter 5. To get ready, identify important terms and organize your resources. You may wish to refer to **Chapter 0** to review prerequisite skills.

FOLDABLES® Study Organizer

Relationships in Triangles Make this Foldable to help you organize your Chapter 5 notes about relationships in triangles. Begin with seven sheets of grid paper.

1. **Stack** the sheets. Fold the top right corner to the bottom edge to form an isosceles right triangle.

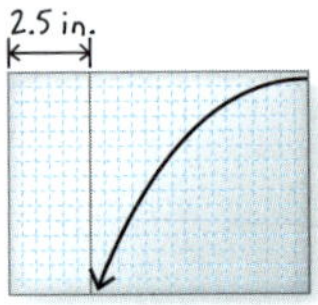

2. **Fold** the rectangular part in half.

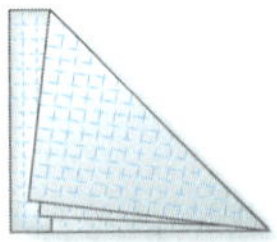

3. **Staple** the sheets along the rectangular fold in four places.

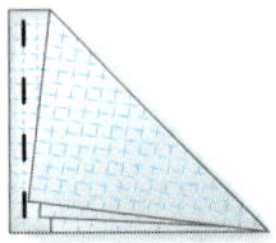

4. **Label** each sheet with a lesson number and the rectangular tab with the chapter title.

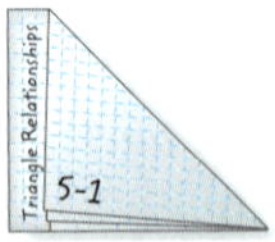

Math Online glencoe.com

- Study the chapter online
- Explore **Get Animated** activities
- Get extra help from your own **Personal Tutor**
- Use **Extra Examples** for additional help
- Take a **Self-Check Quiz**
- **Review Vocabulary** in fun ways

New Vocabulary

English		Español
perpendicular bisector	• p. 322 •	mediatriz
circumcenter	• p. 323 •	circuncentro
concurrent lines	• p. 323 •	rectas concurrentes
point of concurrency	• p. 323 •	punto de concurrencia
incenter	• p. 326 •	incentro
centroid	• p. 333 •	centriode
median	• p. 333 •	mediana
altitude	• p. 335 •	altura
orthocenter	• p. 335 •	ortocentro
indirect proof	• p. 351 •	demostración indirecta
indirect reasoning	• p. 351 •	razonamiento indirecto
proof by contradiction	• p. 351 •	demostración por contradicción

Review Vocabulary

angle bisector • p. 39 • bisectriz de un ángulo a ray that divides an angle into two congruent angles (Lesson 1-4)

midpoint • p. 27 • punto medio the point on a segment exactly halfway between the endpoints of the segment (Lesson 1-3)

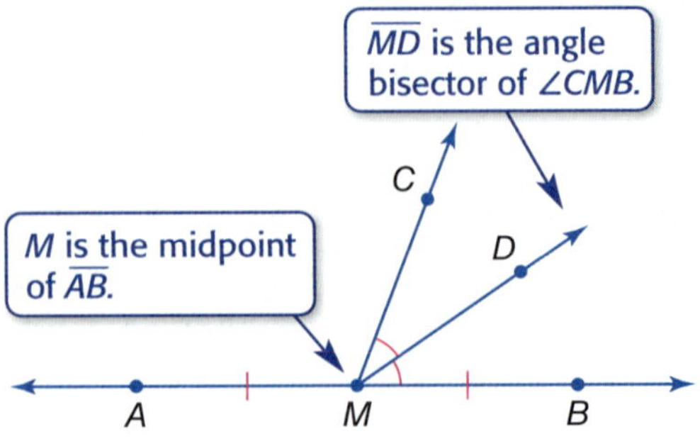

Multilingual eGlossary glencoe.com

EXPLORE

5-1 Geometry Lab
Constructing Bisectors

You can use the constructions for perpendiculars and angle bisectors to construct special segments in triangles.

Construction: Perpendicular Bisector

Construct a perpendicular bisector of the side of a triangle.

Step 1

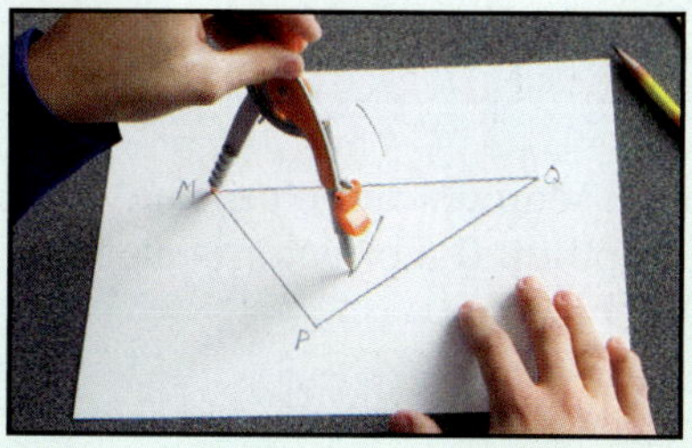

Draw $\triangle MPQ$. Adjust the compass to an opening greater than $\frac{1}{2}MQ$. From vertex M, draw an arc above and below $\overline{MQ}$.

Step 2

Use the same compass settings. From Q, draw an arc above and below $\overline{MQ}$. Label the points of intersection of the arcs A and B.

Step 3

Use a straightedge to draw $\overleftrightarrow{AB}$. Label the point where $\overleftrightarrow{AB}$ intersects $\overline{MQ}$ as C.

An angle bisector in a triangle is a line containing a vertex of the triangle and bisecting that angle.

Construction: Angle Bisector

Construct an angle bisector of a triangle.

Step 1

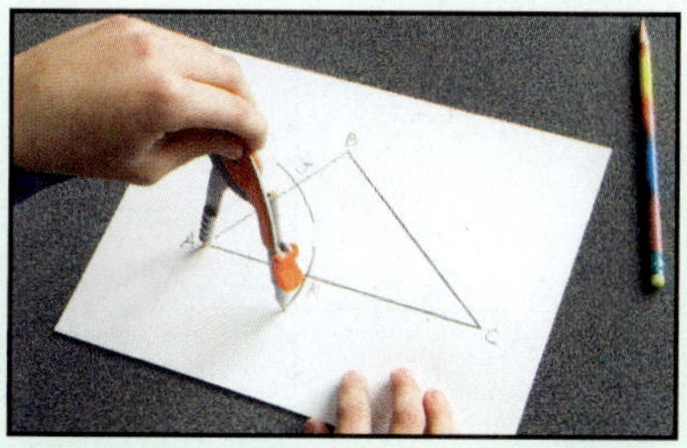

Place the compass on vertex A and draw an arc through $\overline{AB}$ and an arc through $\overline{AC}$. Label the intersections J and K.

Step 2

Place the compass on J and draw an arc. From K, draw an arc intersecting the first arc. Label the intersection L.

Step 3

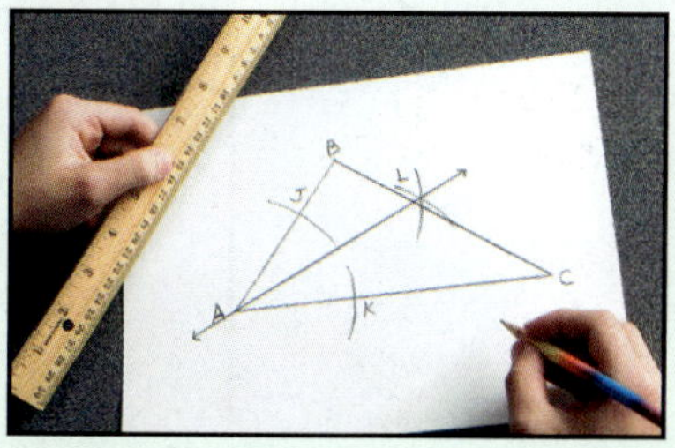

Use a straightedge to draw $\overleftrightarrow{AL}$. $\overline{AL}$ is an angle bisector of $\triangle ABC$.

Model and Analyze

1. Construct the perpendicular bisectors and angle bisectors of the other two sides of $\triangle MPQ$. What do you notice about their intersections?

Repeat the two constructions for each type of triangle.

2. acute
3. obtuse
4. right

5-1 Bisectors of Triangles

Then

You used segment and angle bisectors. (Lessons 1-3 and 1-4)

Now

- Identify and use perpendicular bisectors in triangles.
- Identify and use angle bisectors in triangles.

New Vocabulary

perpendicular bisector
concurrent lines
point of concurrency
circumcenter
incenter

Math Online
glencoe.com

Why?

Creating a work triangle in a kitchen can make food preparation more efficient by cutting down on the number of steps you have to take. If the triangle is too large, an island work station can be added in the middle. To locate the point that is equidistant from the sink, stove, and refrigerator, you can use the perpendicular bisectors of the triangle.

Perpendicular Bisectors In Lesson 1-3, you learned that a segment bisector is any segment, line, or plane that intersects a segment at its midpoint. If a bisector is also perpendicular to the segment, it is called a **perpendicular bisector**.

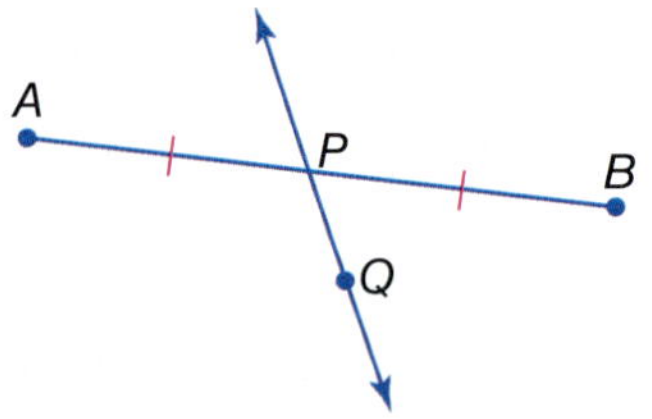

$\overleftrightarrow{PQ}$ is a bisector of $\overline{AB}$.

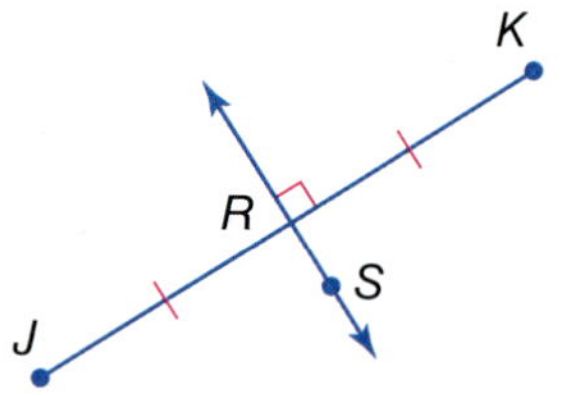

$\overleftrightarrow{RS}$ is a perpendicular bisector of $\overline{JK}$.

Recall that a *locus* is a set of points that satisfies a particular condition. The perpendicular bisector of a segment is the locus of points in a plane equidistant from the endpoints of the segment. This leads to the following theorems.

Theorems — Perpendicular Bisectors

5.1 Perpendicular Bisector Theorem

If a point is on the perpendicular bisector of a segment, then it is equidistant from the endpoints of the segment.

Example If $\overline{CD}$ is a $\perp$ bisector of $\overline{AB}$, then $AC = BC$.

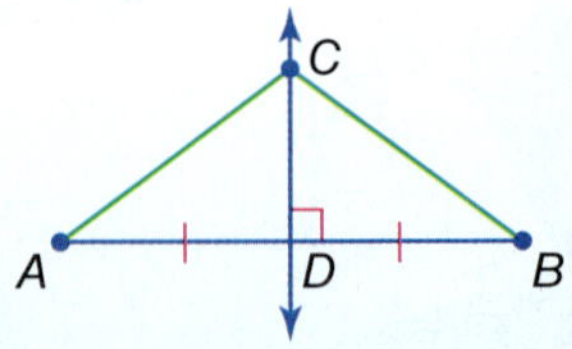

5.2 Converse of the Perpendicular Bisector Theorem

If a point is equidistant from the endpoints of a segment, then it is on the perpendicular bisector of the segment.

Example If $AE = BE$, then E lies on $\overline{CD}$, the $\perp$ bisector of $\overline{AB}$.

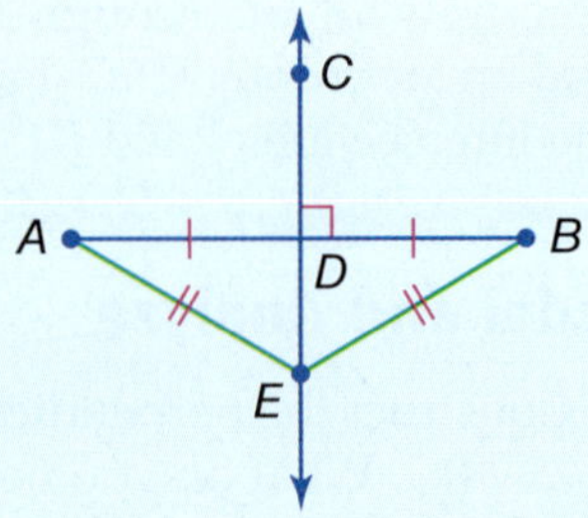

You will prove Theorems 5.1 and 5.2 in Exercises 39 and 37.

EXAMPLE 1 Use the Perpendicular Bisector Theorems

Find each measure.

a. AB

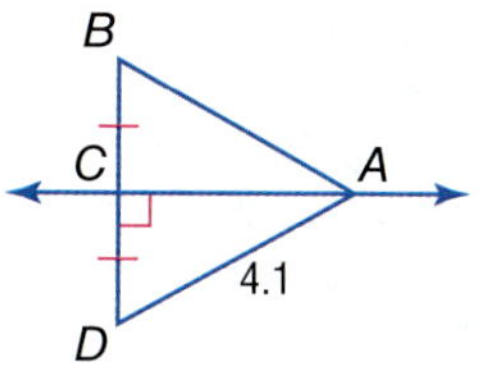

From the information in the diagram, we know that $\overleftrightarrow{CA}$ is the perpendicular bisector of $\overline{BD}$.

$AB = AD$	**Perpendicular Bisector Theorem**
$AB = 4.1$	**Substitution**

b. WY

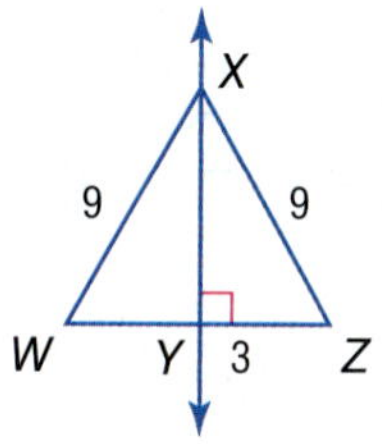

Since $WX = ZX$ and $\overleftrightarrow{XY} \perp \overline{WZ}$, $\overleftrightarrow{XY}$ is the perpendicular bisector of $\overline{WZ}$ by the Converse of the Perpendicular Bisector Theorem. By the definition of segment bisector, $WY = YZ$. Since $YZ = 3$, $WY = 3$.

c. RT

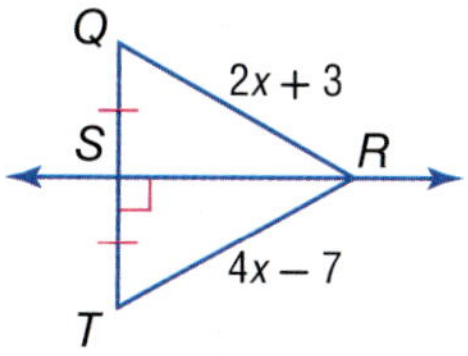

$\overleftrightarrow{SR}$ is the perpendicular bisector of $\overline{QT}$.

$RT = RQ$	**Perpendicular Bisector Theorem**
$4x - 7 = 2x + 3$	**Substitution**
$2x - 7 = 3$	**Subtract 2*x* from each side.**
$2x = 10$	**Add 7 to each side.**
$x = 5$	**Divide each side by 2.**

So $RT = 4(5) - 7$ or 13.

Guided Practice

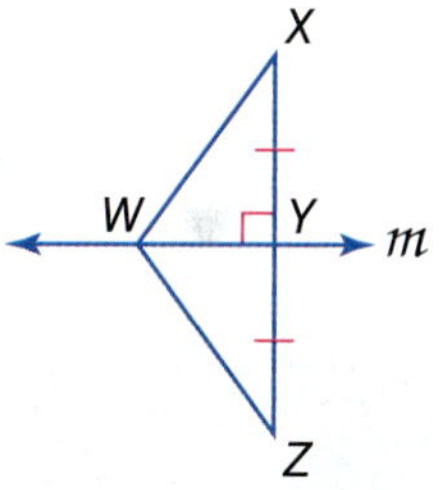

1A. If $WX = 25.3$, $YZ = 22.4$, and $WZ = 25.3$, find XY.

1B. If m is the perpendicular bisector of XZ and $WZ = 14.9$, find WX.

1C. If m is the perpendicular bisector of XZ, $WX = 4a - 15$, and $WZ = a + 12$, find WX.

Personal Tutor glencoe.com

StudyTip

Perpendicular Bisectors The perpendicular bisector of a side of a triangle does not necessarily pass through a vertex of the triangle. For example, in $\triangle XYZ$ below, the perpendicular bisector of $\overline{XY}$ does not pass through point Z.

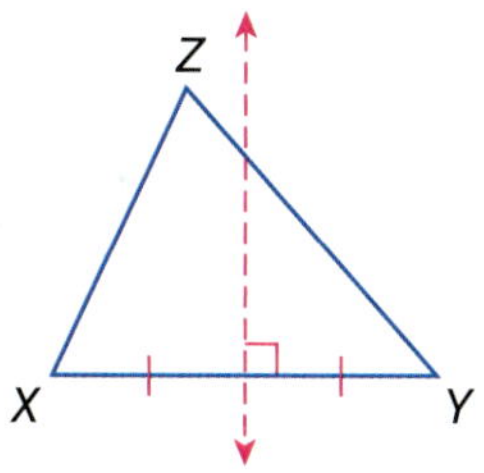

When three or more lines intersect at a common point, the lines are called **concurrent lines**. The point where concurrent lines intersect is called the **point of concurrency**.

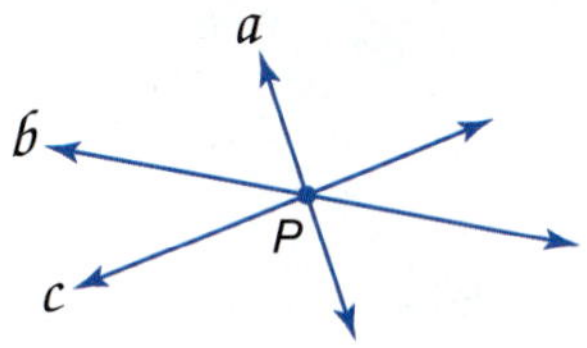

Lines a, b, and c are concurrent at P.

A triangle has three sides, so it also has three perpendicular bisectors. These bisectors are concurrent lines. The point of concurrency of the perpendicular bisectors is called the **circumcenter** of the triangle.

Theorem 5.3 Circumcenter Theorem

Words The perpendicular bisectors of a triangle intersect at a point called the *circumcenter* that is equidistant from the vertices of the triangle.

Example If P is the circumcenter of $\triangle ABC$, then $PB = PA = PC$.

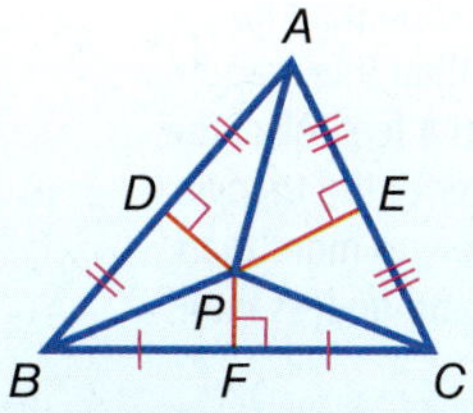

ReadingMath

Circum–
The prefix *circum–* means *about* or *around.* The circumcenter is the center of a circle around a triangle that contains the vertices of the triangle.

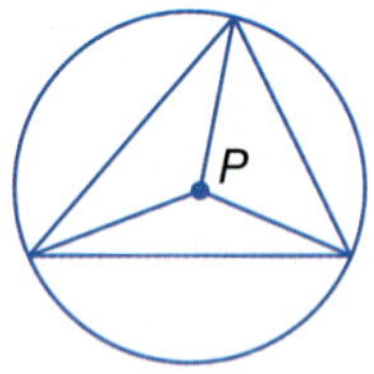

The circumcenter can be on the interior, exterior, or side of a triangle.

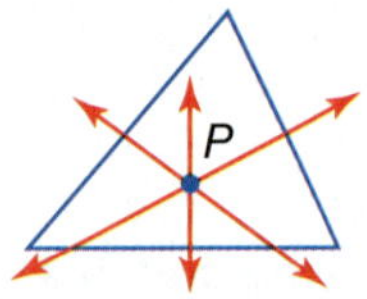

acute triangle

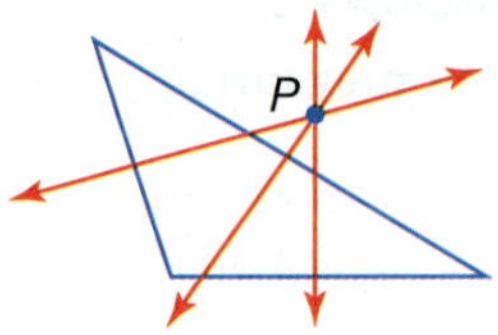

obtuse triangle

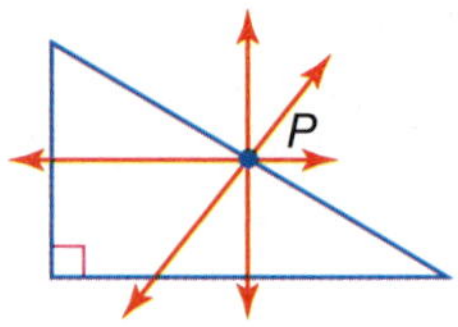

right triangle

PROOF Circumcenter Theorem

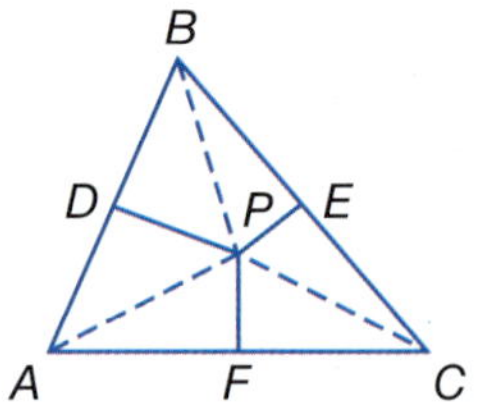

Given: $\overline{PD}$, $\overline{PF}$, and $\overline{PE}$ are perpendicular bisectors of $\overline{AB}$, $\overline{AC}$, and $\overline{BC}$, respectively.

Prove: $AP = CP = BP$

Paragraph Proof:

Since P lies on the perpendicular bisector of $\overline{AC}$, it is equidistant from A and C. By the definition of equidistant, $AP = CP$. The perpendicular bisector of $\overline{BC}$ also contains P. Thus, $CP = BP$. By the Transitive Property of Equality, $AP = BP$. Thus, $AP = CP = BP$.

Real-World EXAMPLE 2 Use the Circumcenter Theorem

INTERIOR DESIGN A stove *S*, sink *K*, and refrigerator *R* are positioned in a kitchen as shown. Find the location for the center of an island work station so that it is the same distance from these three points.

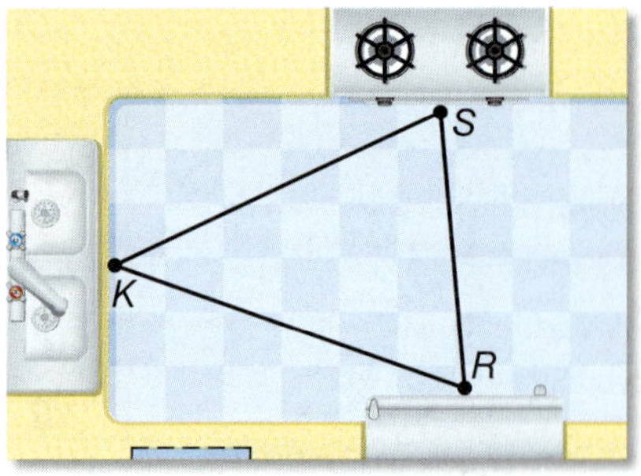

By the Circumcenter Theorem, a point equidistant from three points is found by using the perpendicular bisectors of the triangle formed by those points.

Copy $\triangle SKR$, and use a ruler and protractor to draw the perpendicular bisectors. The location for the center of the island is C, the circumcenter of $\triangle SKR$.

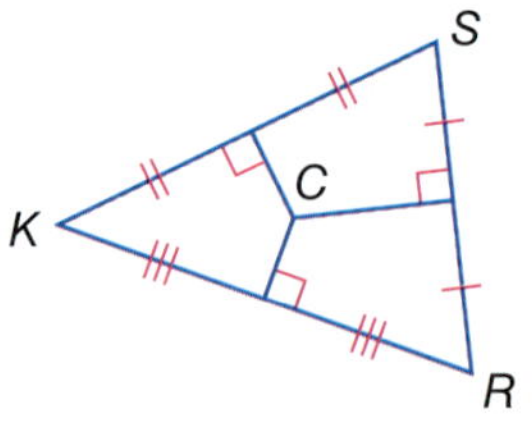

Guided Practice

2. To water his triangular garden, Alex needs to place a sprinkler equidistant from each vertex. Where should Alex place the sprinkler?

Personal Tutor glencoe.com

Real-World Link

Some basic rules of thumb for a kitchen work triangle are that the sides of the triangle should be no greater than 9 feet and no less than 4 feet. Also, the perimeter of the triangle should be no more than 26 feet and no less than 12 feet.

Source: Merillat

Angle Bisectors Recall from Lesson 1-4 that an angle bisector divides an angle into two congruent angles. The angle bisector can be a line, segment, or ray.

The bisector of an angle can be described as the locus of points in the interior of the angle equidistant from the sides of the angle. This description leads to the following theorems.

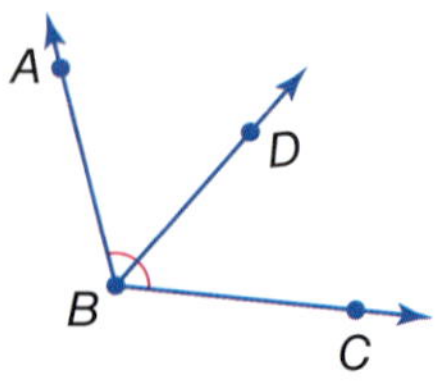

$\overrightarrow{BD}$ is the angle bisector of $\angle ABC$.

Theorems — Angle Bisectors

5.4 Angle Bisector Theorem

If a point is on the bisector of an angle, then it is equidistant from the sides of the angle.

Example If $\overrightarrow{BF}$ bisects $\angle DBE$, $\overline{FD} \perp \overrightarrow{BD}$, and $\overline{FE} \perp \overrightarrow{BE}$, then $DF = FE$.

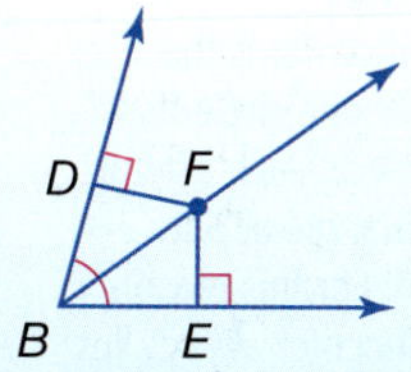

5.5 Converse of the Angle Bisector Theorem

If a point in the interior of an angle is equidistant from the sides of the angle, then it is on the bisector of the angle.

Example If $\overline{FD} \perp \overrightarrow{BD}$, $\overline{FE} \perp \overrightarrow{BE}$, and $DF = FE$, then $\overrightarrow{BF}$ bisects $\angle DBE$.

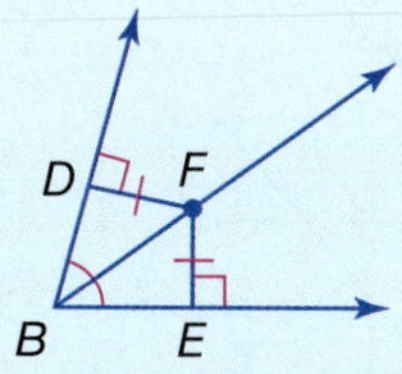

You will prove Theorems 5.4 and 5.5 in Exercises 43 and 40.

EXAMPLE 3 Use the Angle Bisector Theorems

Find each measure.

a. XY

$XY = XW$	**Angle Bisector Theorem**
$XY = 7$	**Substitution**

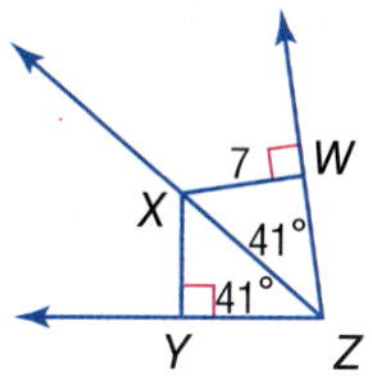

b. $m\angle JKL$

Since $\overline{LJ} \perp \overrightarrow{KJ}$, $\overline{LM} \perp \overrightarrow{KM}$, $\overline{LJ} \cong \overline{LM}$, L is equidistant from the sides of $\angle JKM$. By the Converse of the Angle Bisector Theorem, $\overrightarrow{KL}$ bisects $\angle JKM$.

$\angle JKL \cong \angle LKM$	**Definition of angle bisector**
$m\angle JKL = m\angle LKM$	**Definition of congruent angles**
$m\angle JKL = 37$	**Substitution**

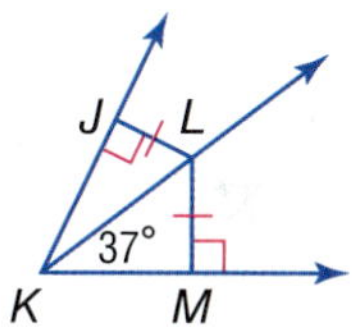

> **StudyTip**
>
> **Angle Bisector** For part b, only having the information that $JL = LM$ would not be enough information to conclude that $\overrightarrow{KL}$ bisects $\angle JKM$.

c. SP

$SP = SM$	**Angle Bisector Theorem**
$6x - 7 = 3x + 5$	**Substitution**
$3x - 7 = 5$	**Subtract 3x from each side.**
$3x = 12$	**Add 7 to each side.**
$x = 4$	**Divide each side by 3.**

So, $SP = 6(4) - 7$ or 17.

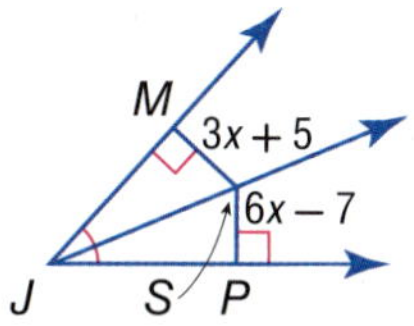

Guided Practice

3A. If $m\angle BAC = 38$, $BC = 5$, and $DC = 5$, find $m\angle DAC$.

3B. If $m\angle BAC = 40$, $m\angle DAC = 40$, and $DC = 10$, find BC.

3C. If $\overrightarrow{AC}$ bisects $\angle DAB$, $BC = 4x + 8$, and $DC = 9x - 7$, find BC.

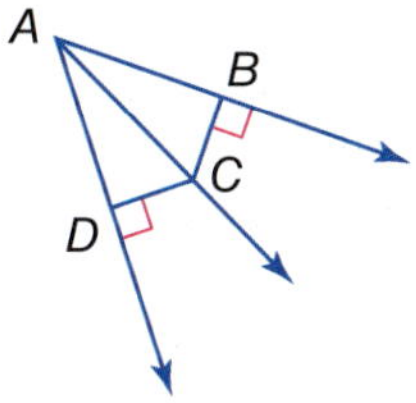

Personal Tutor glencoe.com

ReadingMath

Incenter
The incenter is the center of a circle that intersects each side of the triangle at one point. For this reason, the incenter always lies in the interior of a triangle.

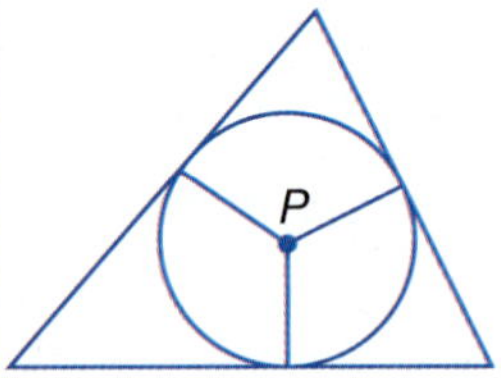

Similar to perpendicular bisectors, since a triangle has three angles, it also has three angle bisectors. The angle bisectors of a triangle are concurrent, and their point of concurrency is called the **incenter** of a triangle.

Theorem 5.6 Incenter Theorem

Words The angle bisectors of a triangle intersect at a point called the *incenter* that is equidistant from each side of the triangle.

Example If P is the incenter of $\triangle ABC$, then $PD = PE = PF$.

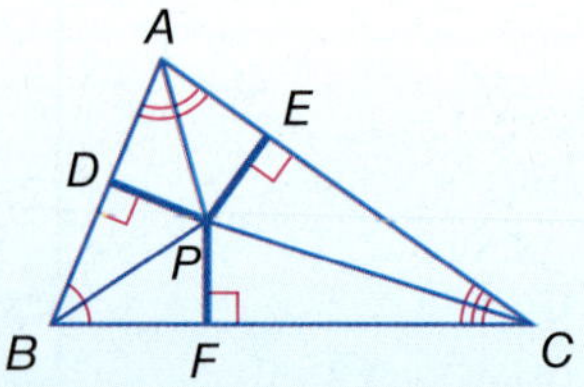

You will prove Theorem 5.6 in Exercise 38.

EXAMPLE 4 Use the Incenter Theorem

Find each measure if J is the incenter of $\triangle ABC$.

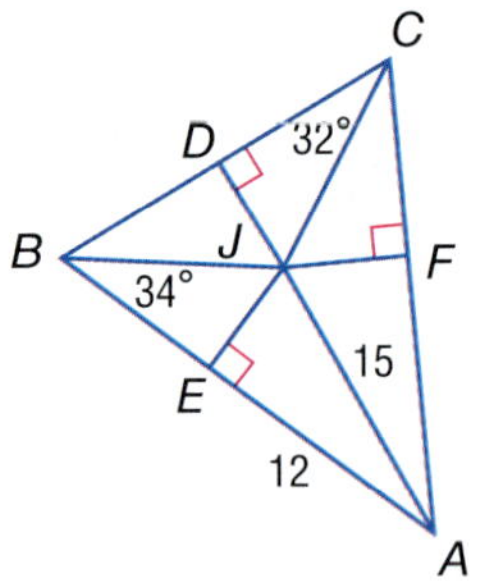

a. JF

By the Incenter Theorem, since J is equidistant from the sides of $\triangle ABC$, $JF = JE$. Find JF by using the Pythagorean Theorem.

$a^2 + b^2 = c^2$	**Pythagorean Theorem**
$JE^2 + 12^2 = 15^2$	**Substitution**
$JE^2 + 144 = 225$	**$12^2 = 144$ and $15^2 = 225$.**
$JE^2 = 81$	**Subtract 144 from each side.**
$JE = \pm 9$	**Take the square root of each side.**

Since length cannot be negative, use only the positive square root, 9. Since $JE = JF$, $JF = 9$.

b. $m\angle JAC$

Since $\overrightarrow{BJ}$ bisects $\angle CBE$, $m\angle CBE = 2m\angle JBE$. So $m\angle CBE = 2(34)$ or 68. Likewise, $m\angle DCF = 2m\angle DCJ$, so $m\angle DCF = 2(32)$ or 64.

$m\angle CBE + m\angle DCF + m\angle FAE = 180$	**Triangle Angle Sum Theorem**
$68 + 64 + m\angle FAE = 180$	**$m\angle CBE = 68$, $m\angle DCF = 64$**
$132 + m\angle FAE = 180$	**Simplify.**
$m\angle FAE = 48$	**Subtract 132 from each side.**

Since $\overrightarrow{AJ}$ bisects $\angle FAE$, $2m\angle JAC = m\angle FAE$. This means that $m\angle JAC = \frac{1}{2}m\angle FAE$, so $m\angle JAC = \frac{1}{2}(48)$ or 24.

Guided Practice

If P is the incenter of $\triangle XYZ$, find each measure.

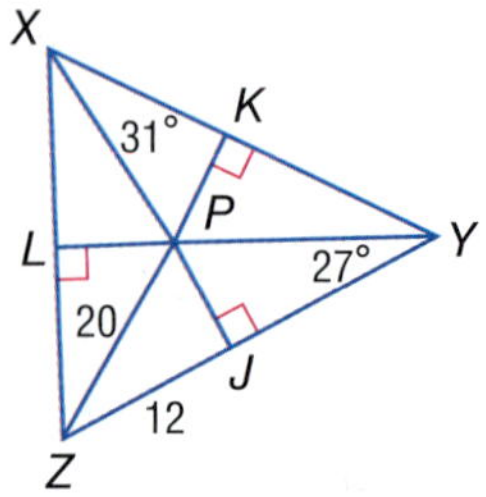

4A. PK

4B. $m\angle LZP$

Personal Tutor glencoe.com

Check Your Understanding

Example 1 p. 323

Find each measure.

1. XW

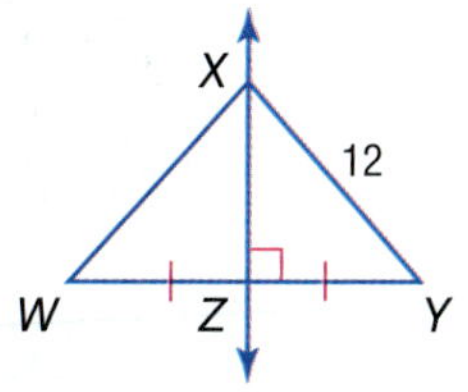

2. AC

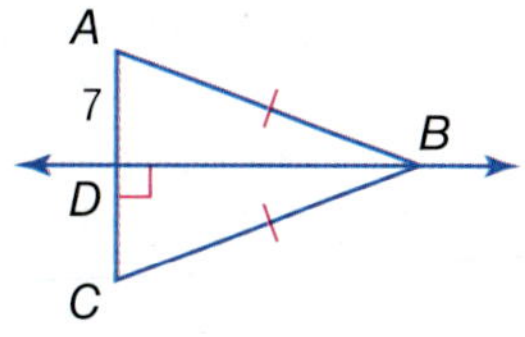

3. LP

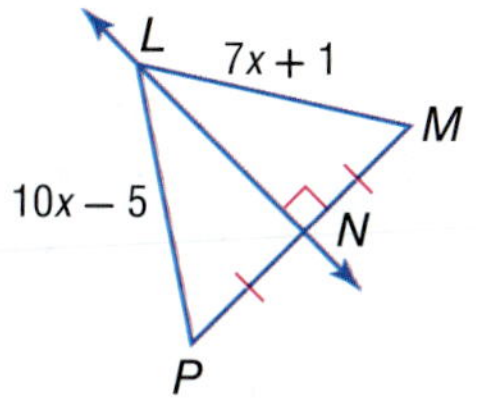

Example 2 p. 324

4. ADVERTISING Four friends are passing out flyers at a mall food court. Three of them take as many flyers as they can and position themselves as shown. The fourth one keeps the supply of additional flyers. Copy the positions of points A, B, and C. Then position the fourth friend at D so that she is the same distance from each of the other three friends.

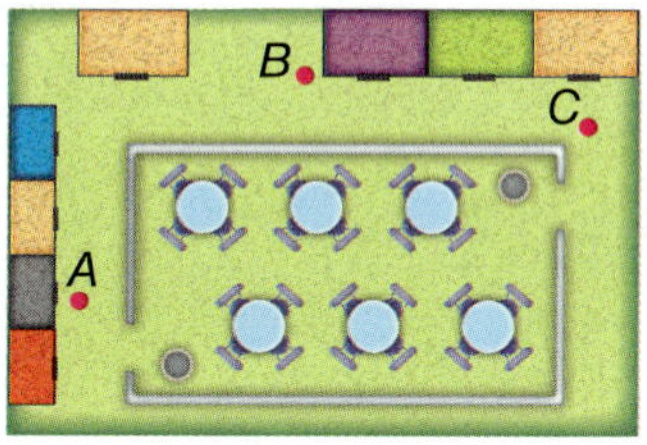

Example 3 p. 325

Find each measure.

5. CP

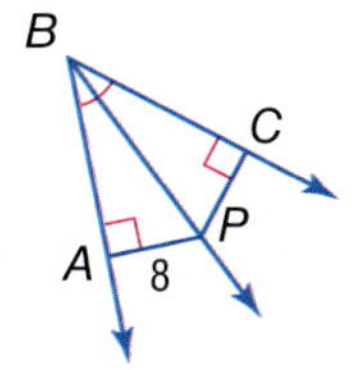

6. $m\angle WYZ$

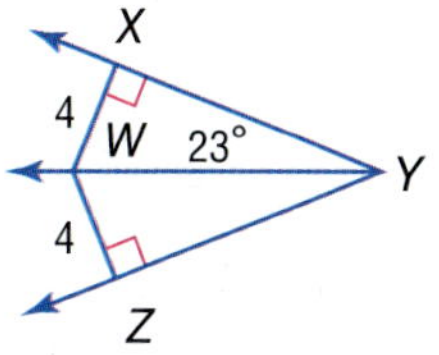

7. QM

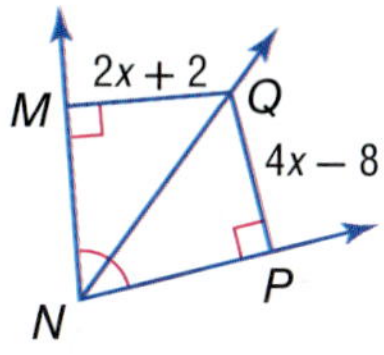

Example 4 p. 326

8. Find JQ if Q is the incenter of $\triangle JLN$.

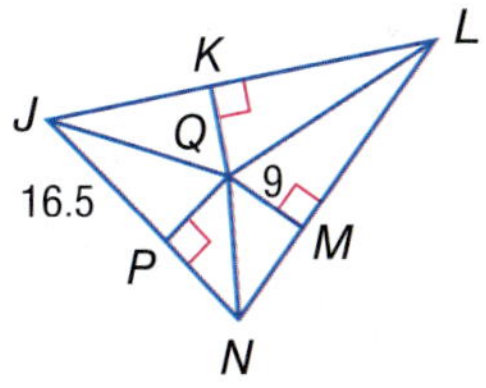

Practice and Problem Solving

● = **Step-by-Step Solutions** begin on page R20.
Extra Practice begins on page 969.

Example 1 p. 323

Find each measure.

9. NP

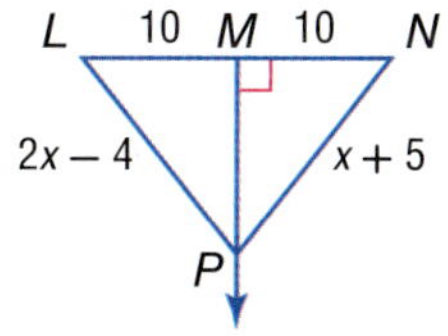

10. PS

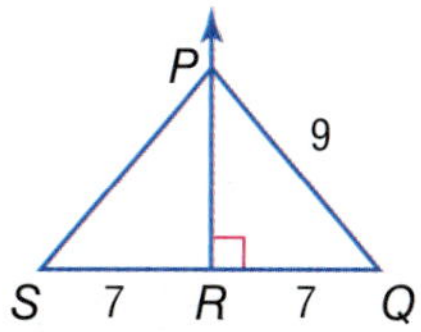

11. KL

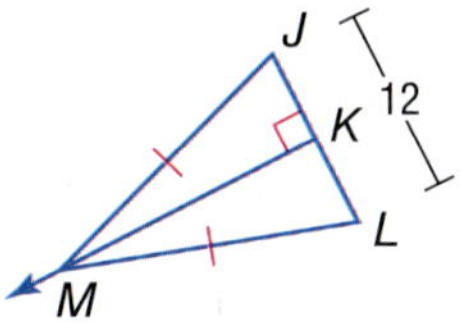

12. EG

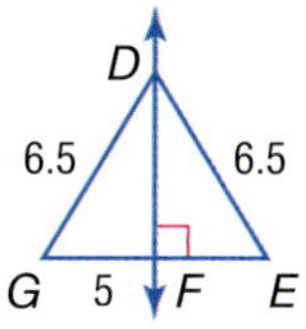

13. CD

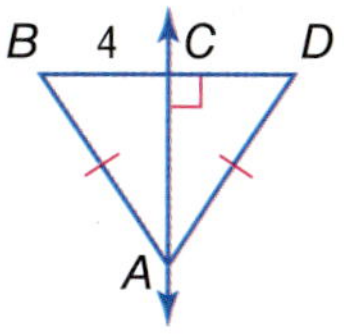

14. SW

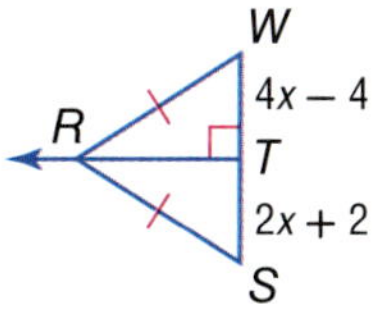

Example 2
p. 324

15. STATE FAIR The state fair has set up the location of the midway, livestock competition, and food vendors. The fair planners decide that they want to locate the portable restrooms the same distance from each location. Copy the positions of points M, L, and F. Then find the location for the restrooms and label it R.

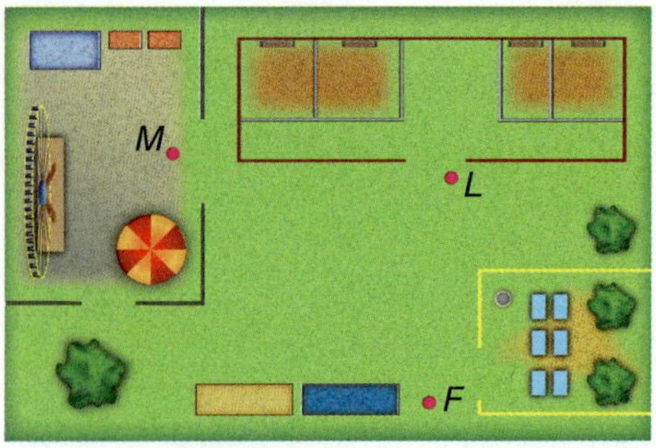

16. SCHOOL A school system has built an elementary, middle, and high school at the locations shown in the diagram. Copy the positions of points E, M, and H. Then find the location for the bus yard B that will service these schools so that it is the same distance from each school.

Point D is the circumcenter of $\triangle ABC$. List any segment(s) congruent to each segment.

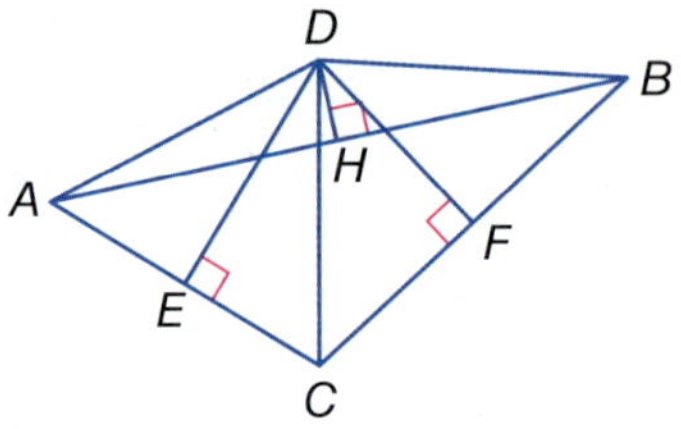

17. $\overline{AD}$

18. $\overline{BF}$

19. $\overline{AH}$

20. $\overline{DC}$

Example 3
p. 325

Find each measure.

21. AF

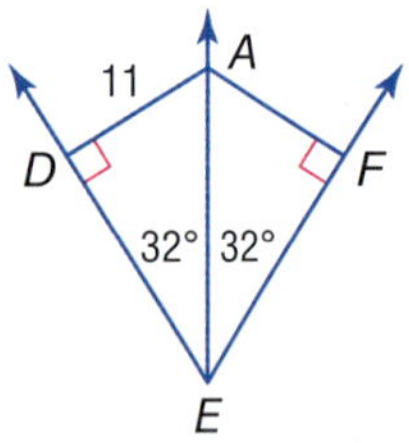

22. $m\angle DBA$

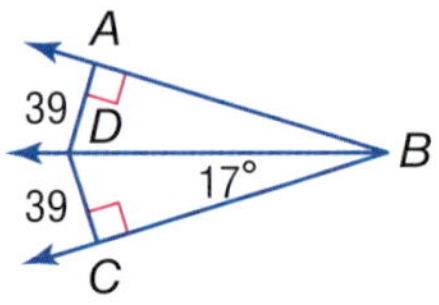

23 $m\angle PNM$

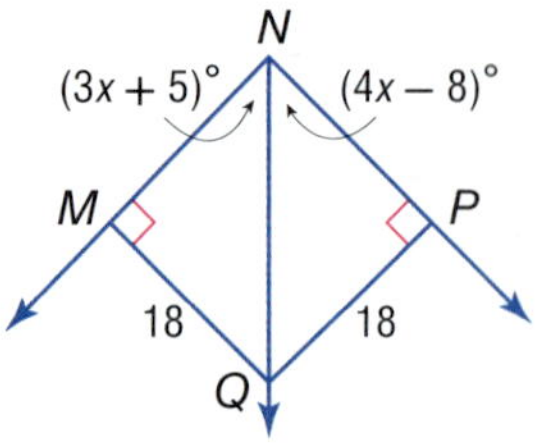

24. XA

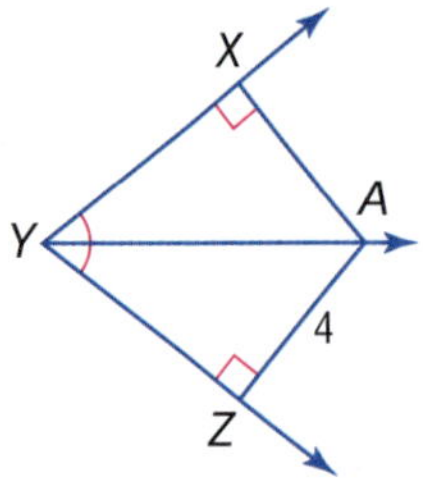

25. $m\angle PQS$

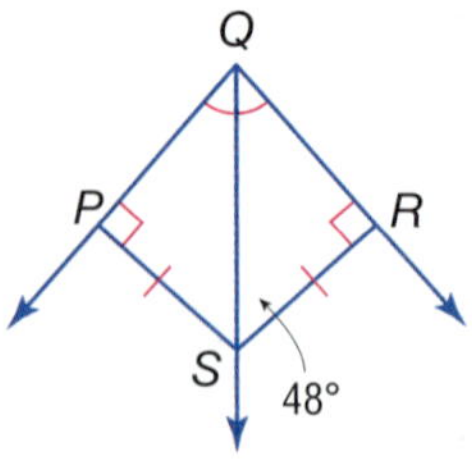

26. PN

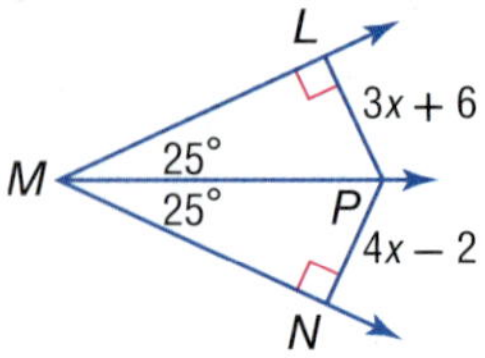

Example 4
p. 326

Point P is the incenter of $\triangle AEC$. Find each measure below.

27. PB

28. DE

29. $m\angle DAC$

30. $m\angle DEP$

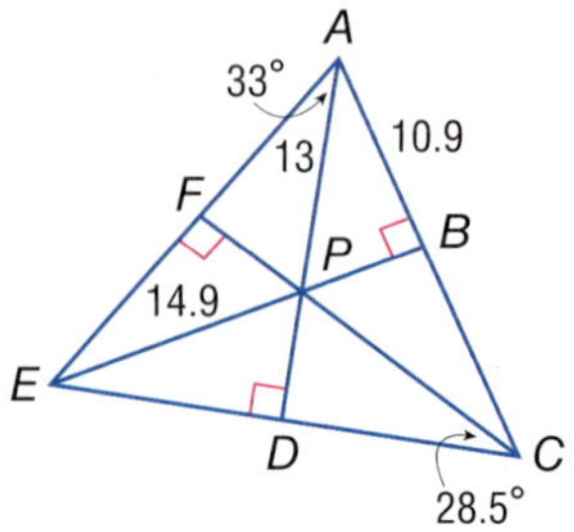

Real-World Career

Interior Designer
An interior designer decorates a space so that it is visually pleasing and comfortable for people to live or work in. Designers must know color and paint theory, lighting design, and space planning. A bachelor's degree is recommended for entry-level positions. Graduates usually enter a 1- to 3-year apprenticeship before taking a licensing exam.

31. INTERIOR DESIGN You want to place a centerpiece on a corner table so that it is located the same distance from each edge of the table. Make a sketch to show where you should place the centerpiece. Explain your reasoning.

Determine whether there is enough information given in each diagram to find the value of x. Explain your reasoning.

32.

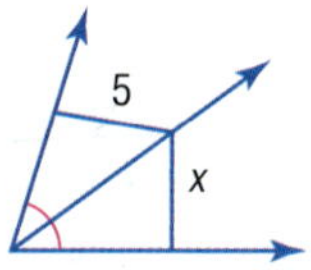

33.

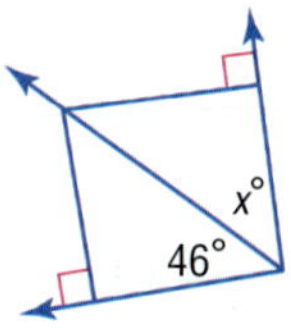

34.

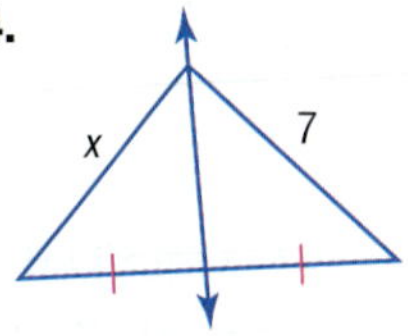

35.

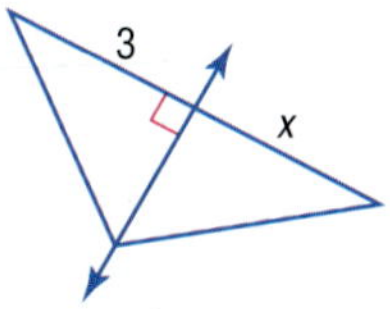

36. SOCCER A soccer player P is approaching the opposing team's goal as shown in the diagram. To make the goal, the player must kick the ball between the goal posts at L and R. The goalkeeper faces the kicker. He then tries to stand so that if he needs to dive to stop a shot, he is as far from the left-hand side of the shot angle as the right-hand side.

a. Describe where the goalkeeper should stand. Explain your reasoning.

b. Copy $\triangle PRL$. Use a compass and a straightedge to locate point G, the desired place for the goalkeeper to stand.

c. If the ball is kicked so it follows the path from P to R, construct the shortest path the goalkeeper should take to block the shot. Explain your reasoning.

PROOF Write a two-column proof.

37. Theorem 5.2

Given: $\overline{CA} \cong \overline{CB}$, $\overline{AD} \cong \overline{BD}$

Prove: C and D are on the perpendicular bisector of $\overline{AB}$.

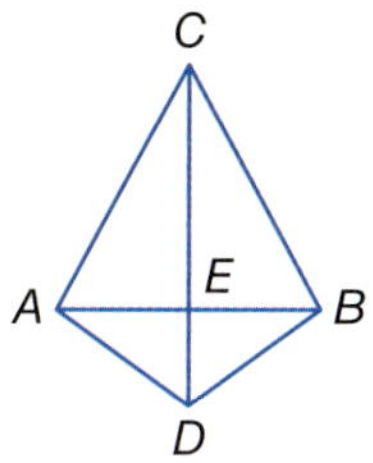

38. Theorem 5.6

Given: $\triangle ABC$, angle bisectors $\overline{AD}$, $\overline{BE}$, and $\overline{CF}$; $\overline{KP} \perp \overline{AB}$, $\overline{KQ} \perp \overline{BC}$, $\overline{KR} \perp \overline{AC}$

Prove: $KP = KQ = KR$

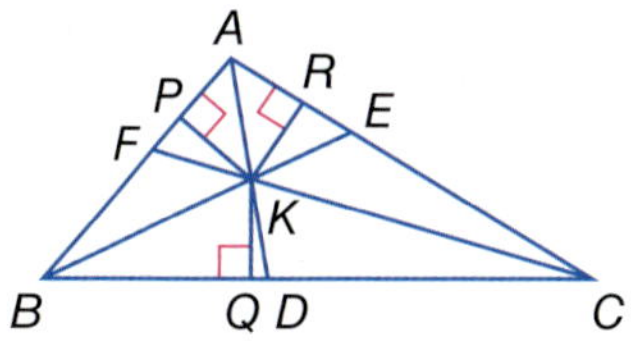

PROOF Write a paragraph proof of each theorem.

39. Theorem 5.1

40. Theorem 5.5

COORDINATE GEOMETRY Write an equation in slope-intercept form for the perpendicular bisector of the segment with the given endpoints. Justify your answer.

41. $A(-3, 1)$ and $B(4, 3)$

42. $C(-4, 5)$ and $D(2, -2)$

43. **PROOF** Write a two-column proof of Theorem 5.4.

44. **GRAPHIC DESIGN** Mykia is designing a pennant for her school. She wants to put a picture of the school mascot inside a circle on the pennant. Copy the outline of the pennant and locate the point where the center of the circle should be to create the largest circle possible. Justify your drawing.

COORDINATE GEOMETRY **Find the coordinates of the circumcenter of the triangle with the given vertices. Explain.**

Review Vocabulary

locus a set of points that satisfy a particular condition

45. $A(0, 0)$, $B(0, 6)$, $C(10, 0)$

46. $J(5, 0)$, $K(5, -8)$, $L(0, 0)$

47. **LOCI** Consider $\overline{CD}$. Describe the set of all points in space that are equidistant from C and D.

H.O.T. Problems

Use Higher-Order Thinking Skills

48. **ERROR ANALYSIS** Claudio says that from the information supplied in the diagram, he can conclude that K is on the perpendicular bisector of $\overline{LM}$. Caitlyn disagrees. Is either of them correct? Explain your reasoning.

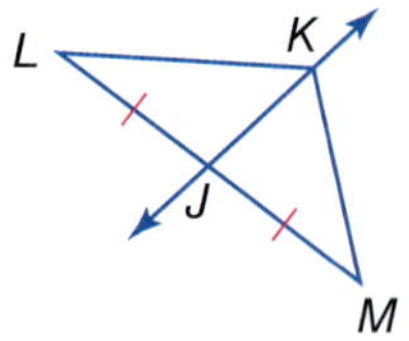

49. **OPEN ENDED** Draw a triangle with an incenter located inside the triangle but a circumcenter located outside. Justify your drawing by using a straightedge and a compass to find both points of concurrency.

REASONING **Determine whether each statement is *sometimes, always,* or *never* true. Justify your reasoning using a counterexample or proof.**

50. The angle bisectors of a triangle intersect at a point that is equidistant from the vertices of the triangle.

51. In an isosceles triangle, the perpendicular bisector of the base is also the angle bisector of the opposite vertex.

CHALLENGE **Write a two-column proof for each of the following.**

52. **Given:** Plane $\mathcal{Y}$ is a perpendicular bisector of $\overline{DC}$.
Prove: $\angle ADB \cong \angle ACB$

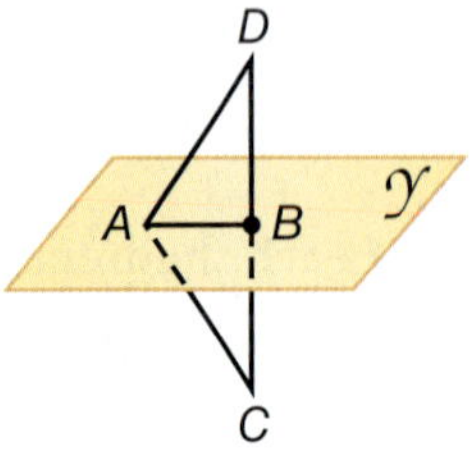

53. **Given:** Plane $\mathcal{Z}$ is an angle bisector of $\angle KJH$.
$\overline{KJ} \cong \overline{HJ}$
Prove: $\overline{MH} \cong \overline{MK}$

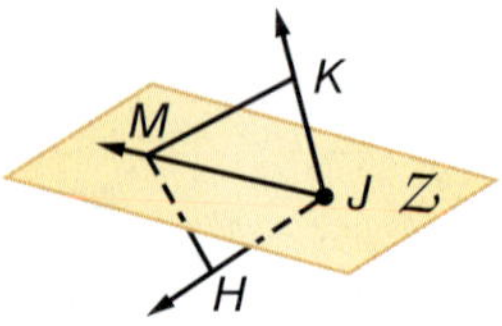

54. **WRITING IN MATH** Compare and contrast the perpendicular bisectors and angle bisectors of a triangle. How are they alike? How are they different? Be sure to compare their points of concurrency.

Standardized Test Practice

55. ALGEBRA An object is projected straight upward with initial velocity v meters per second from an initial height of s meters. The height h in meters of the object after t seconds is given by $h = -10t^2 + vt + s$. Sherise is standing at the edge of a balcony 54 meters above the ground and throws a ball straight up with an initial velocity of 12 meters per second. After how many seconds will it hit the ground?

A 3 seconds
B 4 seconds
C 6 seconds
D 9 seconds

56. SHORT RESPONSE Write an equation in slope-intercept form that describes the line containing the points (−1, 0) and (2, 4).

57. A line drawn through which of the following points would be a perpendicular bisector of $\triangle JKL$?

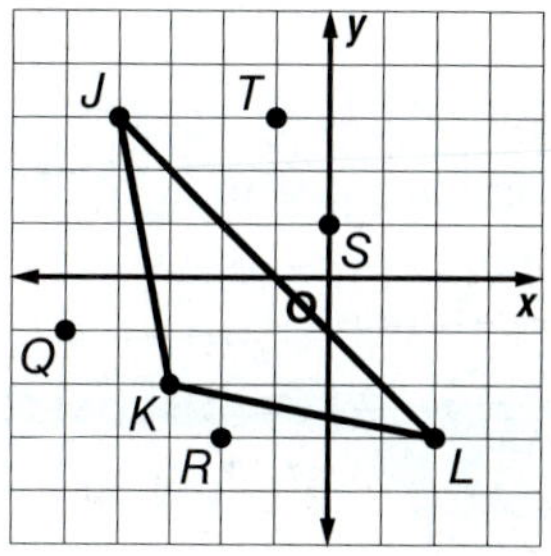

F T and K
G L and Q
H J and R
J S and K

58. SAT/ACT For $x \neq -3$, $\frac{3x+9}{x+3} =$

A $x + 9$
B $x + 3$
C x
D 3

Spiral Review

Name the missing coordinate(s) of each triangle. (Lesson 4-8)

59.

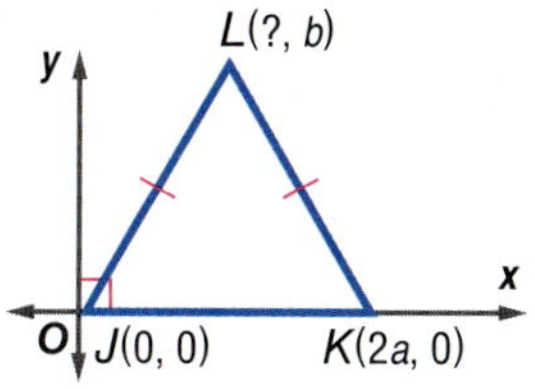

60.

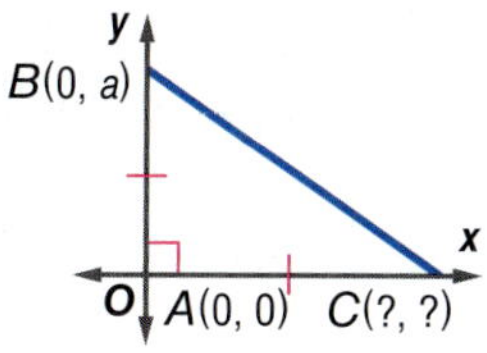

61.

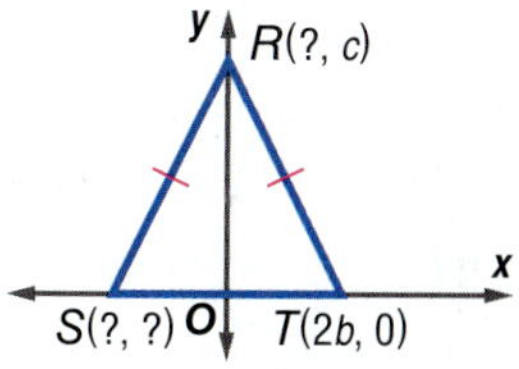

COORDINATE GEOMETRY **Graph each pair of triangles with the given vertices. Then identify the transformation and verify that it is a congruence transformation.** (Lesson 4-7)

62. $A(-2, 4)$, $B(-2, -2)$, $C(4, 1)$; $R(12, 4)$, $S(12, -2)$, $T(6, 1)$

63. $J(-3, 3)$, $K(-3, 1)$, $L(1, 1)$; $X(-3, -1)$, $Y(-3, -3)$, $Z(1, -3)$

Find the distance from the line to the given point. (Lesson 3-6)

64. $y = 5$, $(-2, 4)$

65. $y = 2x + 2$, $(-1, -5)$

66. $2x - 3y = -9$, $(2, 0)$

67. AUDIO ENGINEERING A studio engineer charges a flat fee of \$450 for equipment rental and \$42 an hour for recording and mixing time. Write the equation that shows the cost to hire the studio engineer as a function of time. How much would it cost to hire the studio engineer for 17 hours? (Lesson 3-4)

Skills Review

PROOF **Write a two-column proof for each of the following.** (Lesson 4-6)

68. Given: $\triangle XKF$ is equilateral.
$\overline{XJ}$ bisects $\angle X$.
Prove: J is the midpoint of $\overline{KF}$.

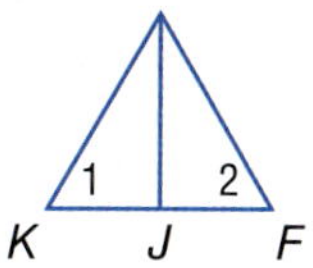

69. Given: $\triangle MLP$ is isosceles.
N is the midpoint of $\overline{MP}$.
Prove: $\overline{LN} \perp \overline{MP}$

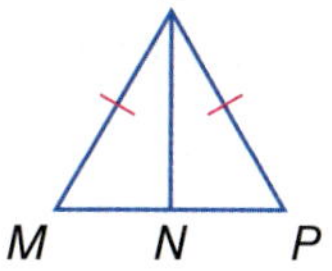

EXPLORE 5-2

Geometry Lab: Constructing Medians and Altitudes

A *median* of a triangle is a segment with endpoints that are a vertex and the midpoint of the side opposite that vertex. You can use the construction for the midpoint of a segment to construct a median.

Construction 1 Median of a Triangle

Step 1

Draw intersecting arcs above and below $\overline{DE}$. Label the points of intersection R and S.

Step 2

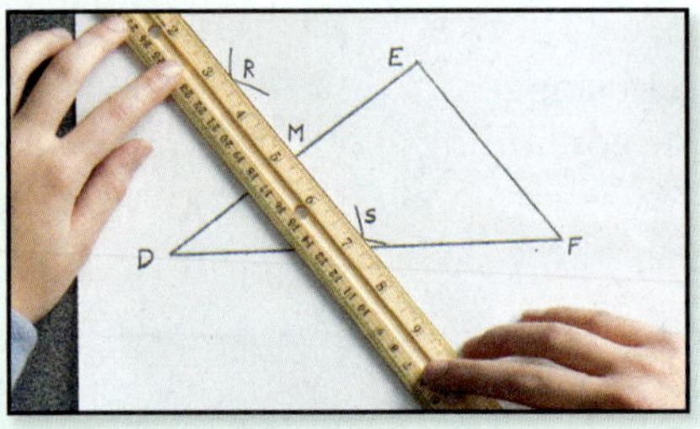

Use a straightedge to find the point where $\overline{RS}$ intersects $\overline{DE}$. Label the midpoint M.

Step 3

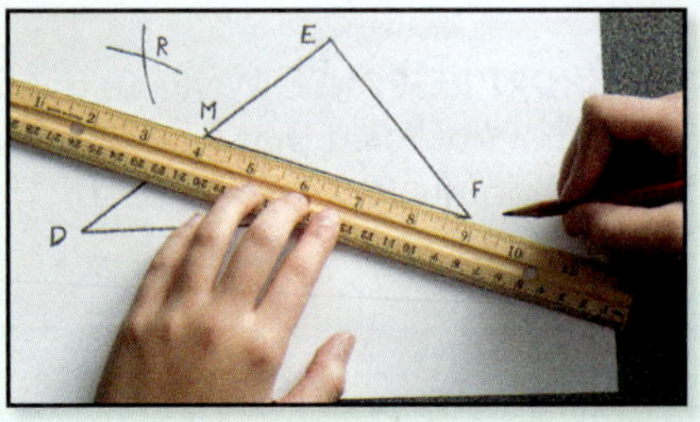

Draw a line through F and M. $\overline{FM}$ is a median of $\triangle DEF$.

An *altitude* of a triangle is a segment from a vertex of the triangle to the line containing the opposite side and is perpendicular to the line containing that side.

Construction 2 Altitude of a Triangle

Step 1

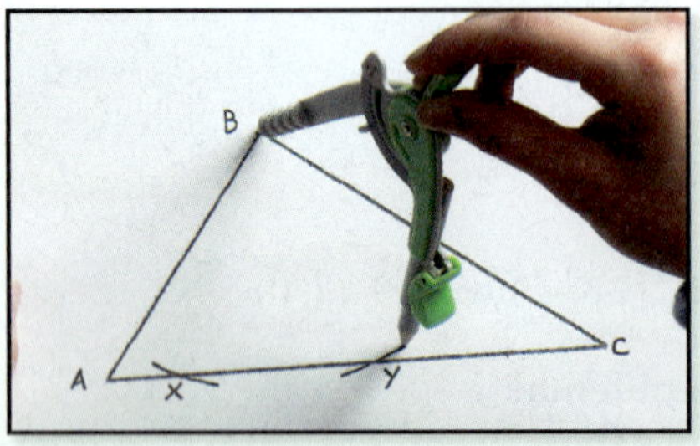

Place the compass at vertex B and draw two arcs intersecting $\overline{AC}$. Label the points where the arcs intersect the side X and Y.

Step 2

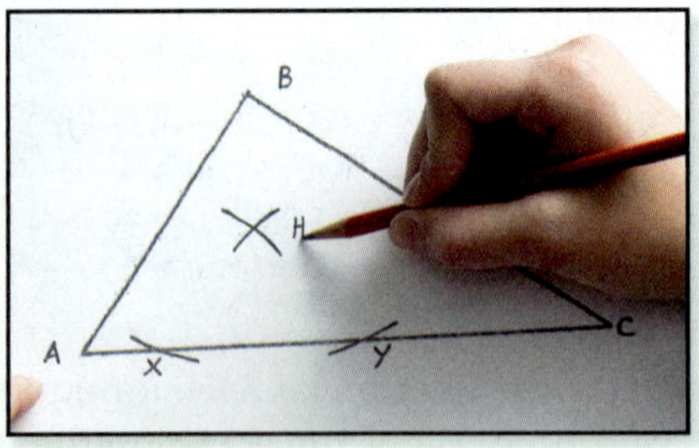

Adjust the compass to an opening greater than $\frac{1}{2}XY$. Place the compass on X and draw an arc above $\overline{AC}$. Use the same setting to draw an arc from Y. Label the point of intersection of the arcs H.

Step 3

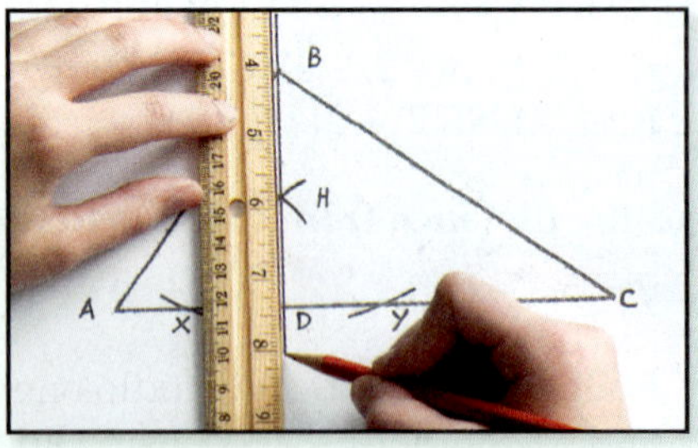

Use a straightedge to draw $\overleftrightarrow{BH}$. Label the point where $\overleftrightarrow{BH}$ intersects $\overline{AC}$ as D. $\overline{BD}$ is an altitude of $\triangle ABC$ and is perpendicular to $\overline{AC}$.

Model and Analyze

1. Construct the medians of the other two sides of $\triangle DEF$. What do you notice about the medians of a triangle?
2. Construct the altitudes to the other two sides of $\triangle ABC$. What do you observe?

5-2 Medians and Altitudes of Triangles

Then
You identified and used perpendicular and angle bisectors in triangles. (Lesson 5-1)

Now
- Identify and use medians in triangles.
- Identify and use altitudes in triangles.

New Vocabulary
median
centroid
altitude
orthocenter

Math Online
glencoe.com

Why?

A mobile is a *kinetic* or moving sculpture that uses the principles of balance and equilibrium. Simple mobiles consist of several rods attached by strings from which objects of varying weights hang. The hanging objects balance each other and can rotate freely. To ensure that a triangle in a mobile hangs parallel to the ground, artists have to find the triangle's balancing point.

Medians A **median** of a triangle is a segment with endpoints being a vertex of a triangle and the midpoint of the opposite side.

Every triangle has three medians that are concurrent. The point of concurrency of the medians of a triangle is called the **centroid** and is always inside the triangle.

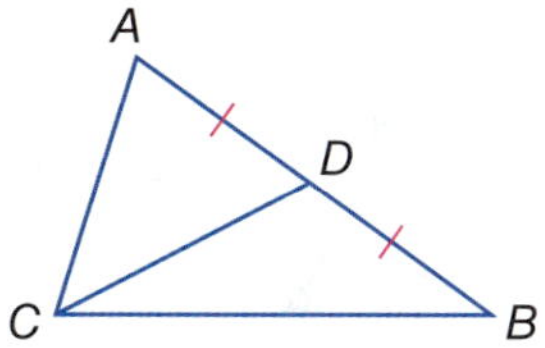

$\overline{CD}$ is a median of $\triangle ABC$.

Theorem 5.7 Centroid Theorem

The medians of a triangle intersect at a point called the centroid that is two thirds of the distance from each vertex to the midpoint of the opposite side.

Example If P is the centroid of $\triangle ABC$, then

$AP = \frac{2}{3}AK$, $BP = \frac{2}{3}BL$, and $CP = \frac{2}{3}CJ$.

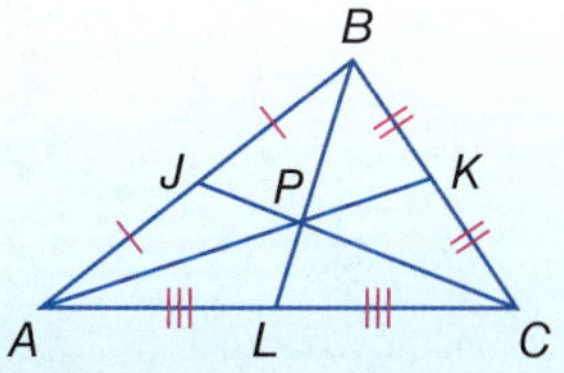

You will prove Theorem 5.7 in Exercise 36.

EXAMPLE 1 Use the Centroid Theorem

In $\triangle ABC$, Q is the centroid and $BE = 9$. Find BQ and QE.

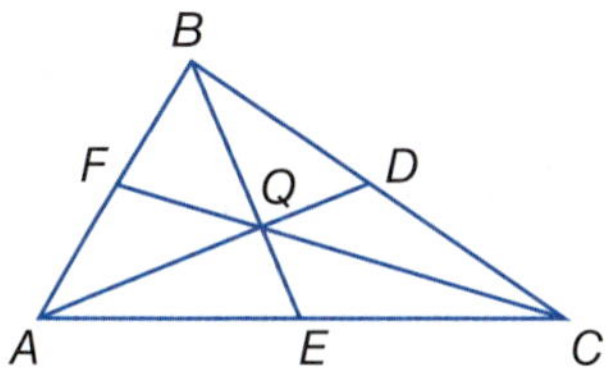

$BQ = \frac{2}{3}BE$	Centroid Theorem
$= \frac{2}{3}(9)$ or 6	$BE = 9$
$BQ + QE = 9$	Segment Addition
$6 + QE = 9$	$BQ = 6$
$QE = 3$	Subtract 6 from each side.

Guided Practice

In $\triangle ABC$ above, $FC = 15$. Find each length.

1A. FQ **1B.** QC

Personal Tutor glencoe.com

EXAMPLE 2 Use the Centroid Theorem

In $\triangle JKL$, $PT = 2$. Find KP.

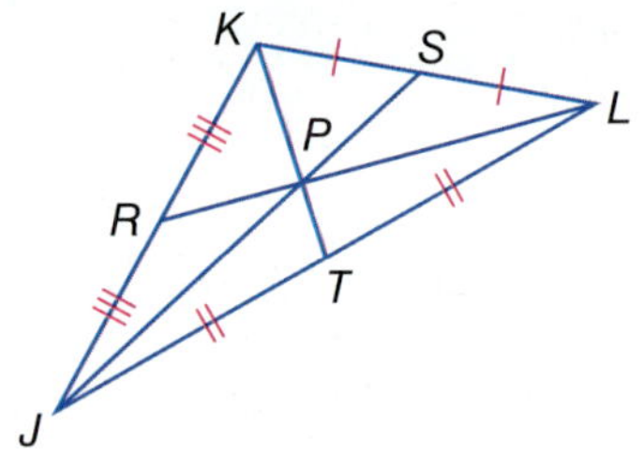

Since $\overline{JR} \cong \overline{RK}$, R is the midpoint of $\overline{JK}$ and $\overline{LR}$ is a median of $\triangle JKL$. Likewise, S and T are the midpoints of $\overline{KL}$ and $\overline{LJ}$ respectively, so $\overline{JS}$ and $\overline{KT}$ are also medians of $\triangle JKL$. Therefore, point P is the centroid of $\triangle JKL$.

$KP = \frac{2}{3}KT$	**Centroid Theorem**
$KP = \frac{2}{3}(KP + PT)$	**Segment Addition and Substitution**
$KP = \frac{2}{3}(KP + 2)$	**$PT = 2$**
$KP = \frac{2}{3}KP + \frac{4}{3}$	**Distributive Property**
$\frac{1}{3}KP = \frac{4}{3}$	**Subtract $\frac{2}{3}KP$ from each side.**
$KP = 4$	**Multiply each side by 3.**

StudyTip

Use Number Sense In Example 2, you can also use number sense to find KP. Since $KP = \frac{2}{3}KT$, $PT = \frac{1}{3}KT$, and $KP = 2PT$. Therefore, if $PT = 2$, then $KP = 2(2)$ or 4.

Guided Practice

In $\triangle JKL$ above, $RP = 3.5$ and $JP = 9$. Find each measure.

2A. PL

2B. PS

Personal Tutor glencoe.com

All polygons have a balance point or centroid. The centroid is also the balancing point or *center of gravity* for a triangular region. The center of gravity is the point at which the region is stable under the influence of gravity.

Real-World EXAMPLE 3 Find the Centroid on Coordinate Plane

PERFORMANCE ART **A performance artist plans to balance triangular pieces of metal during her next act. When one such triangle is placed on the coordinate plane, its vertices are located at (1, 10), (5, 0), and (9, 5). What are the coordinates of the point where the artist should support the triangle so that it will balance?**

Understand You need to find the centroid of the triangle with the given coordinates. This is the point at which the triangle will balance.

Plan Graph and label the triangle with vertices $A(1, 10)$, $B(5, 0)$, and $C(9, 5)$. Since the centroid is the point of concurrency of the medians of a triangle, use the Midpoint Theorem to find the midpoint of one of the sides of the triangle. The centroid is two-thirds the distance from the opposite vertex to that midpoint.

Solve Graph $\triangle ABC$.

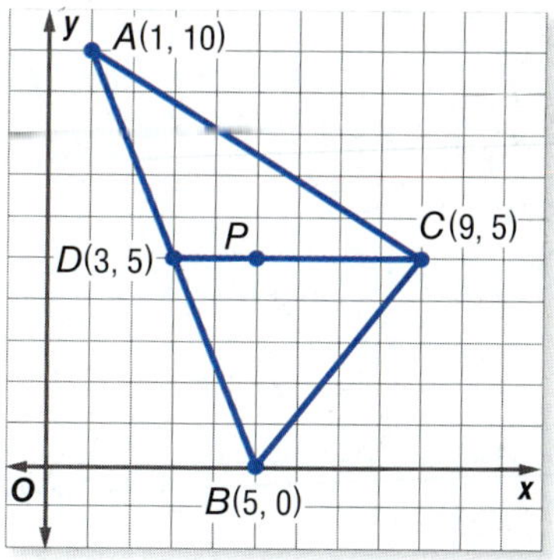

Find the midpoint D of side $\overline{AB}$ with endpoints $A(1, 10)$ and $B(5, 0)$.

$$D\left(\frac{1+5}{2}, \frac{10+0}{2}\right) = D(3, 5)$$

Graph point D. Notice that $\overline{DC}$ is a horizontal line. The distance from $D(3, 5)$ to $C(9, 5)$ is $9 - 3$ or 6 units.

If P is the centroid of $\triangle ABC$, then $PC = \frac{2}{3}DC$. So the centroid is $\frac{2}{3}(6)$ or 4 units to the left of C. The coordinates of P are $(9 - 4, 5)$ or $(5, 5)$.

The performer should balance the triangle at the point (5, 5).

Check Use a different median to check your answer. The midpoint F of side $\overline{AC}$ is $F\left(\frac{1+9}{2}, \frac{10+5}{2}\right)$ or $F(5, 7.5)$. $\overline{BF}$ is a vertical line, so the distance from B to F is $7.5 - 0$ or 7.5. $PB = \frac{2}{3}(7.5)$ or 5, so P is 5 units up from B. The coordinates of P are $(5, 0 + 5)$ or $(5, 5)$. ✓

> **ReadingMath**
>
> **Height of a Triangle** The length of an altitude is known as the *height* of the triangle. The height of a triangle is used to calculate the triangle's area.

Guided Practice

3. A second triangle has vertices at (0, 4), (6, 11.5), and (12, 1). What are the coordinates of the point where the artist should support the triangle so that it will balance? Explain your reasoning.

Personal Tutor glencoe.com

Altitudes An **altitude** of a triangle is a segment from a vertex to the line containing the opposite side and perpendicular to the line containing that side. An altitude can lie in the interior, exterior, or on the side of a triangle.

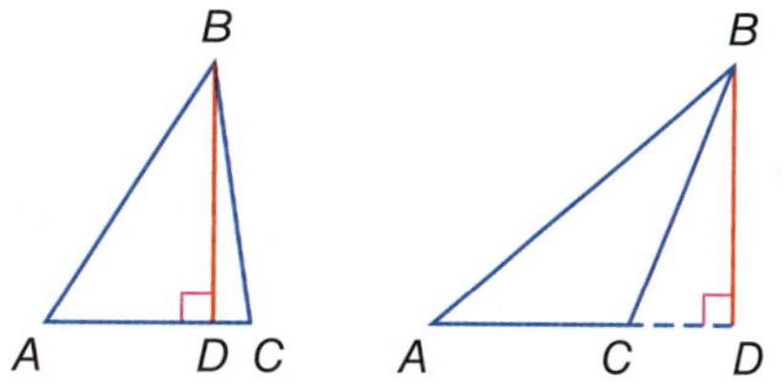

$\overline{BD}$ is an altitude from B to $\overline{AC}$.

Every triangle has three altitudes. If extended, the altitudes of a triangle intersect in a common point.

Key Concept Orthocenter

The lines containing the altitudes of a triangle are concurrent, intersecting at a point called the **orthocenter**.

Example The lines containing altitudes $\overline{AF}$, $\overline{CD}$, and $\overline{BG}$ intersect at P, the orthocenter of $\triangle ABC$.

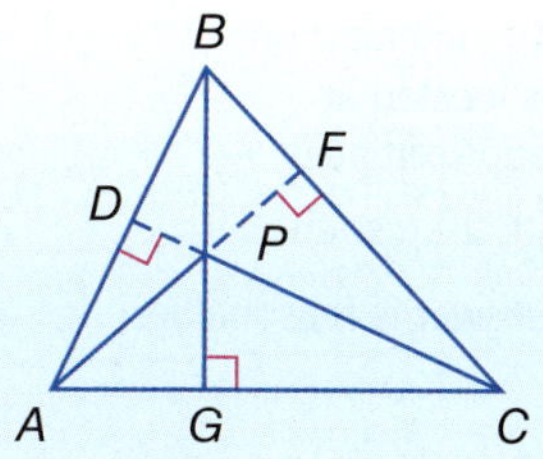

EXAMPLE 4 Find the Orthocenter on a Coordinate Plane

COORDINATE GEOMETRY **The vertices of $\triangle FGH$ are $F(-2, 4)$, $G(4, 4)$, and $H(1, -2)$. Find the coordinates of the orthocenter of $\triangle FGH$.**

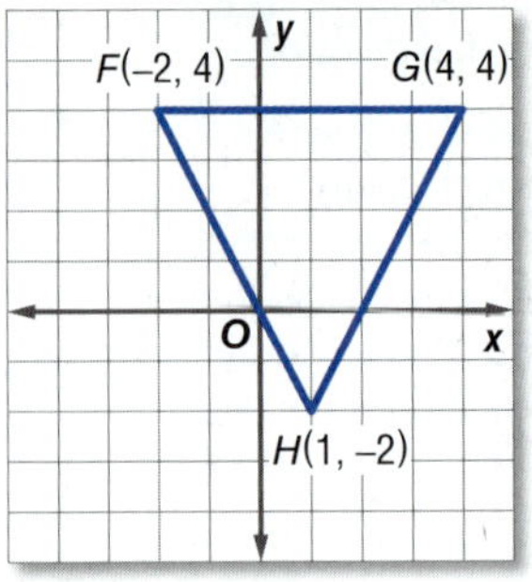

Step 1 Graph $\triangle FGH$. To find the orthocenter, find the point where two of the three altitudes intersect.

Step 2 Find an equation of the altitude from F to $\overline{GH}$. The slope of $\overline{GH}$ is $\frac{4-(-2)}{4-1}$ or 2, so the slope of the altitude, which is perpendicular to $\overline{GH}$, is $-\frac{1}{2}$.

$y - y_1 = m(x - x_1)$ **Point-slope form**

$y - 4 = -\frac{1}{2}[x - (-2)]$ **$m = -\frac{1}{2}$ and $(x_1, y_1) = F(-2, 4)$.**

$y - 4 = -\frac{1}{2}(x + 2)$ **Simplify.**

$y - 4 = -\frac{1}{2}x - 1$ **Distributive Property**

$y = -\frac{1}{2}x + 3$ **Add 4 to each side.**

Find an equation of the altitude from G to $\overline{FH}$. The slope of $\overline{FH}$ is $\frac{-2-4}{1-(-2)}$ or -2, so the slope of the altitude is $\frac{1}{2}$.

$y - y_1 = m(x - x_1)$ **Point-slope form**

$y - 4 = \frac{1}{2}(x - 4)$ **$m = \frac{1}{2}$ and $(x_1, y_1) = G(4, 4)$**

$y - 4 = \frac{1}{2}x - 2$ **Distributive Property**

$y = \frac{1}{2}x + 2$ **Add 4 to each side.**

Step 3 Solve the resulting system of equations $\begin{cases} y = -\frac{1}{2}x + 3 \\ y = \frac{1}{2}x + 2 \end{cases}$ to find the point of intersection of the altitudes.

Adding the two equations to eliminate x results in $2y = 5$ or $y = \frac{5}{2}$.

$y = \frac{1}{2}x + 2$ **Equation of altitude from G**

$\frac{5}{2} = \frac{1}{2}x + 2$ **$y = \frac{5}{2}$**

$\frac{1}{2} = \frac{1}{2}x$ **Subtract $\frac{4}{2}$ or 2 from each side.**

$1 = x$ **Multiply each side by 2.**

The coordinates of the orthocenter of $\triangle JKL$ are $\left(1, \frac{5}{2}\right)$ or $\left(1, 2\frac{1}{2}\right)$.

StudyTip

Check for Reasonableness Use the corner of a sheet of paper to draw the altitudes of each side of the triangle.

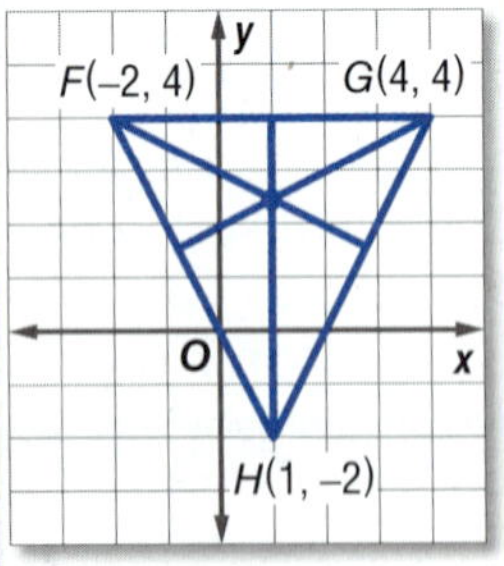

The intersection is located at approximately $\left(1, 2\frac{1}{2}\right)$, so the answer is reasonable.

Guided Practice

4. Find the coordinates of the orthocenter of $\triangle ABC$ graphed at the right.

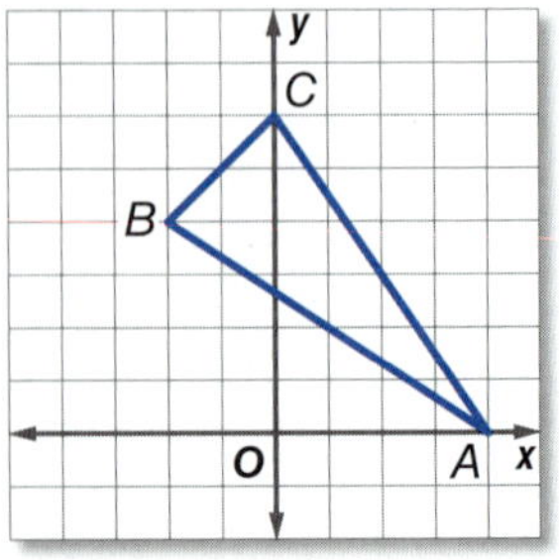

Personal Tutor glencoe.com

Concept Summary — Special Segments and Points in Triangles

Name	Example	Point of Concurrency	Special Property	Example
perpendicular bisector	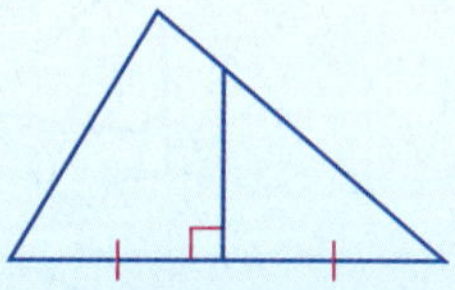	circumcenter	The circumcenter P of $\triangle ABC$ is equidistant from each vertex.	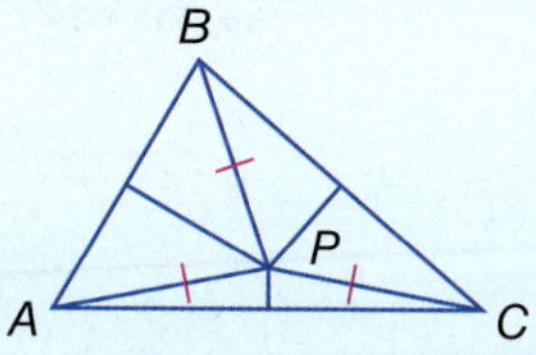
angle bisector	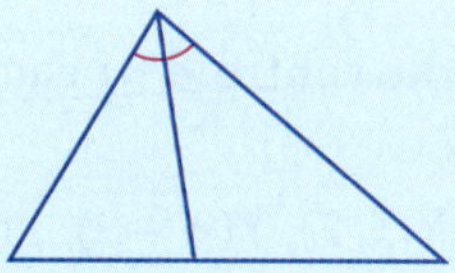	incenter	The incenter Q of $\triangle ABC$ is equidistant from each side of the triangle.	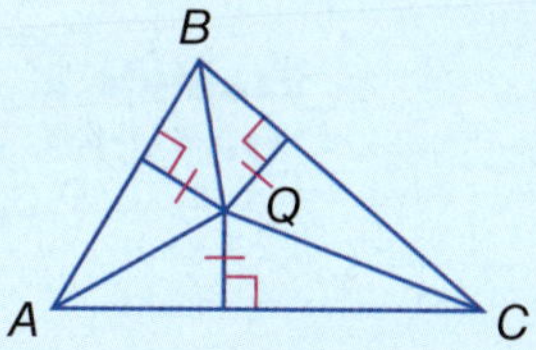
median	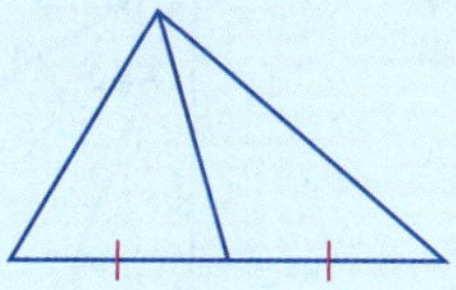	centroid	The centroid R of $\triangle ABC$ is two thirds of the distance from each vertex to the midpoint of the opposite side.	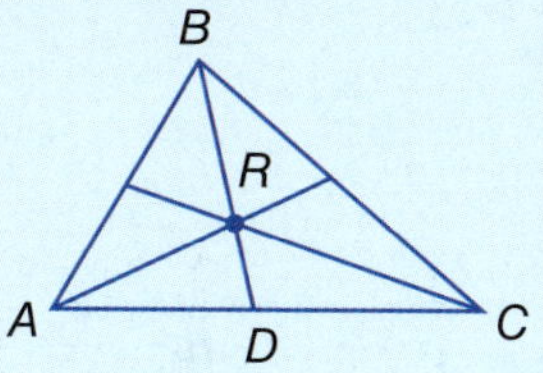
altitude	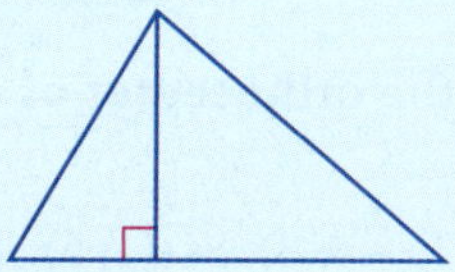	orthocenter	The lines containing the altitudes of $\triangle ABC$ are concurrent at the orthocenter S.	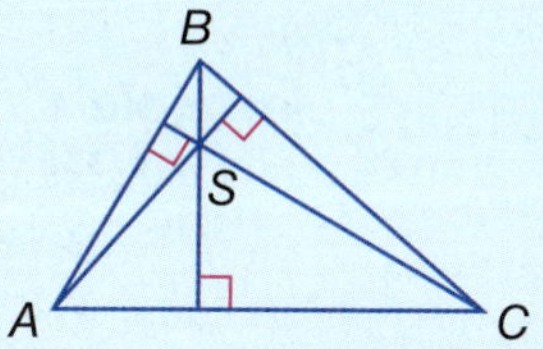

Check Your Understanding

Examples 1 and 2
pp. 333–334

In $\triangle ACE$, P is the centroid, $PF = 6$, and $AD = 15$. Find each measure.

1. PC
2. AP

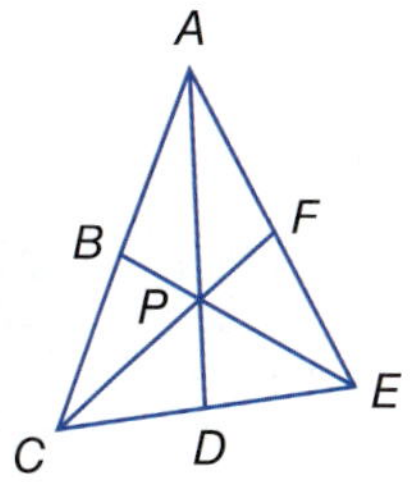

Example 3
pp. 334–335

3. **INTERIOR DESIGN** An interior designer is creating a custom coffee table for a client. The top of the table is a glass triangle that needs to balance on a single support. If the coordinates of the vertices of the triangle are at (3, 6), (5, 2), and (7, 10), at what point should the support be placed?

Example 4
p. 336

4. **COORDINATE GEOMETRY** Find the coordinates of the orthocenter of $\triangle ABC$ with vertices $A(-3, 3)$, $B(-1, 7)$, and $C(3, 3)$.

Practice and Problem Solving

● = **Step-by-Step Solutions** begin on page R20.
Extra Practice begins on page 969.

Examples 1 and 2 pp. 333–334

In $\triangle SZU$, $UJ = 9$, $VJ = 3$, and $ZT = 18$. Find each length.

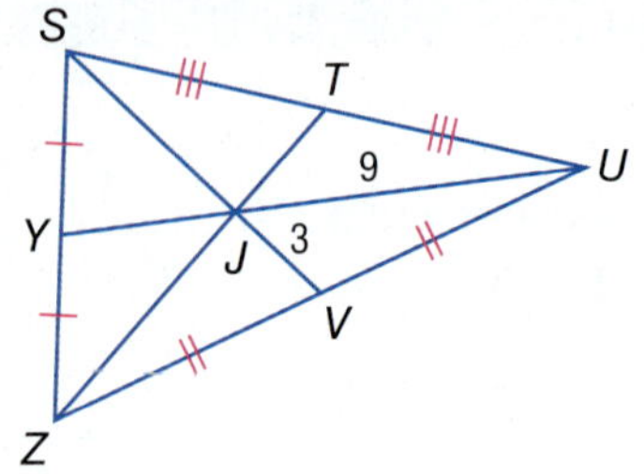

5. YJ
6. SJ
7. YU
8. SV
9. JT
10. ZJ

Example 3 pp. 334–335

COORDINATE GEOMETRY **Find the coordinates of the centroid of each triangle with the given vertices.**

11. $A(-1, 11)$, $B(3, 1)$, $C(7, 6)$
12. $X(5, 7)$, $Y(9, -3)$, $Z(13, 2)$

13. **INTERIOR DESIGN** Emilia made a collage with pictures of her friends. She wants to hang the collage from the ceiling in her room so that it is parallel to the ceiling. A diagram of the collage is shown in the graph at the right. At what point should she place the string?

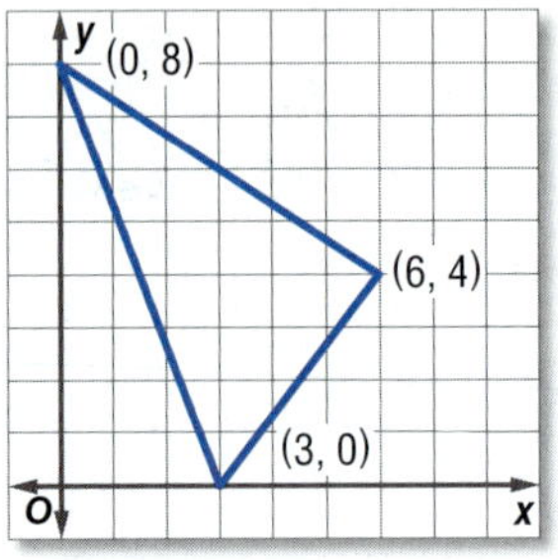

Example 4 p. 336

COORDINATE GEOMETRY **Find the coordinates of the orthocenter of each triangle with the given vertices.**

14. $J(3, -2)$, $K(5, 6)$, $L(9, -2)$
15. $R(-4, 8)$, $S(-1, 5)$, $T(5, 5)$

Identify each segment $\overline{BD}$ as a(n) altitude, median, or perpendicular bisector.

16.
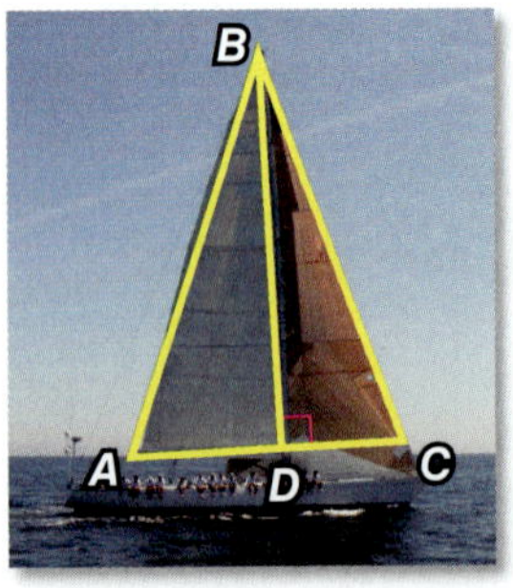

17.
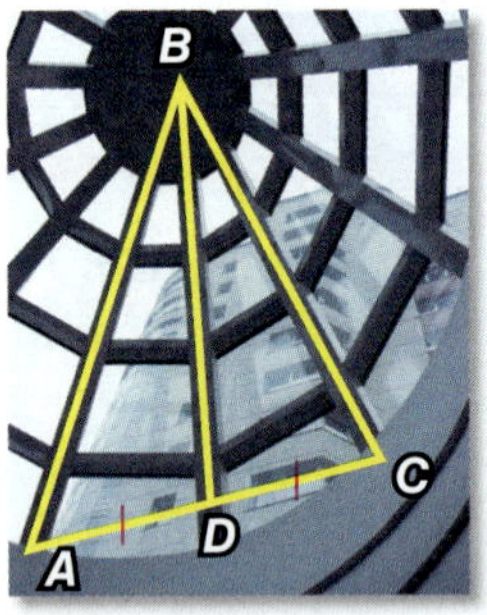

18.

19.
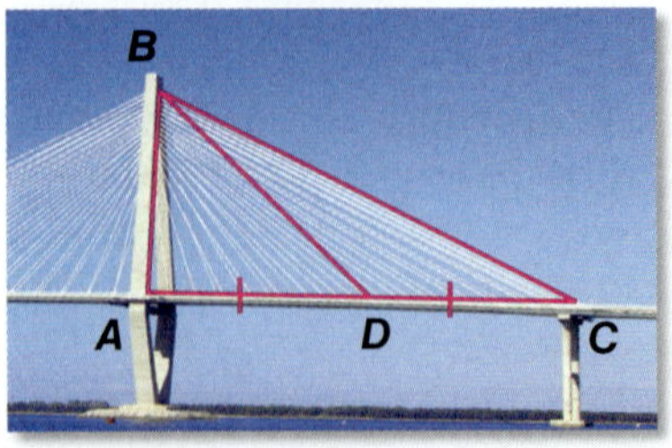

20. **ALGEBRA** In the figure at the right, if J, P, and L are the midpoints of $\overline{KH}$, $\overline{HM}$, and $\overline{MK}$, respectively, find x, y, and z.

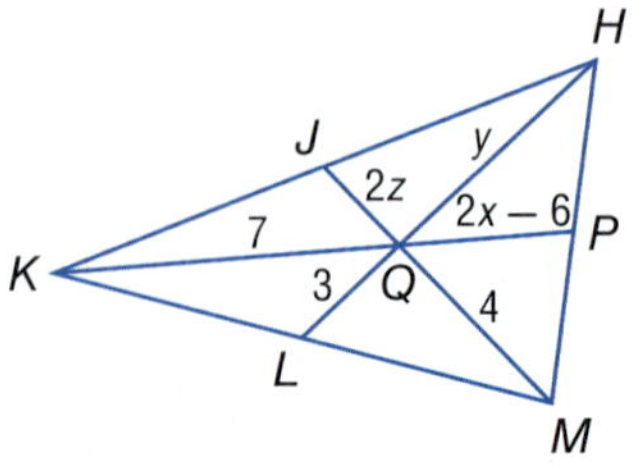

Copy and complete each statement for $\triangle RST$ for medians $\overline{RM}$, $\overline{SL}$ and $\overline{TK}$, and centroid J.

21. $SL = x(JL)$

22. $JT = x(TK)$

23. $JM = x(RJ)$

ALGEBRA **Use the figure at the right.**

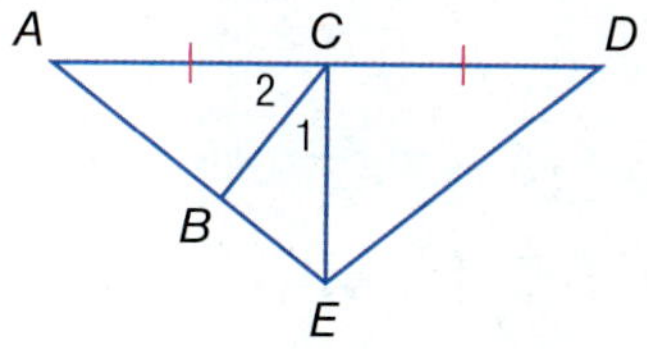

24. If $\overline{EC}$ is an altitude of $\triangle AED$, $m\angle 1 = 2x + 7$, and $m\angle 2 = 3x + 13$, find $m\angle 1$ and $m\angle 2$.

25 Find the value of x if $\overline{AC} = 4x - 3$, $\overline{DC} = 2x + 9$, $m\angle ECA = 15x + 2$, and $\overline{EC}$ is a median of $\triangle AED$. Is $\overline{EC}$ also an altitude of $\triangle AED$? Explain.

Real-World Link

The 5×5 triangular peg game shown above has 15 holes and is played with 14 pegs. The object is to eliminate all but one peg. The game has 179,124 possible solutions for leaving just 1 peg in the hole that was initially empty.

Source: Ventura Manufacturing

26. **GAMES** The game board shown is shaped like an equilateral triangle and has indentations for game pieces. The game's objective is to remove pegs by jumping over them until there is only one peg left. Copy the game board's outline and determine which of the points of concurrency the blue peg represents: *circumcenter, incenter, centroid,* or *orthocenter*. Explain your reasoning.

Use the given information to determine whether $\overline{LM}$ is a *perpendicular bisector, median,* and/or an *altitude* of $\triangle JKL$.

27. $\overline{LM} \perp \overline{JK}$

28. $\triangle JLM \cong \triangle KLM$

29. $\overline{JM} \cong \overline{KM}$

30. $\overline{LM} \perp \overline{JK}$ and $\overline{JL} \cong \overline{KL}$

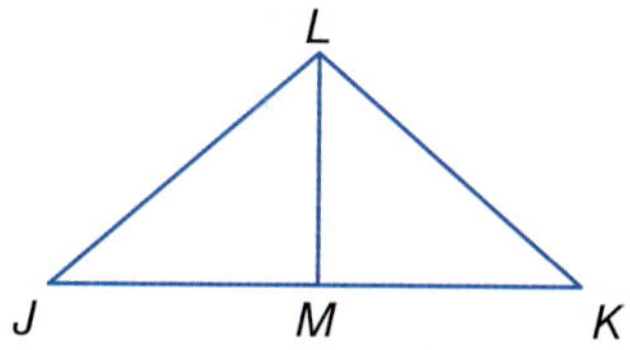

31. **PROOF** Write a paragraph proof.

Given: $\triangle XYZ$ is isosceles.
$\overline{WY}$ bisects $\angle Y$.

Prove: $\overline{WY}$ is a median.

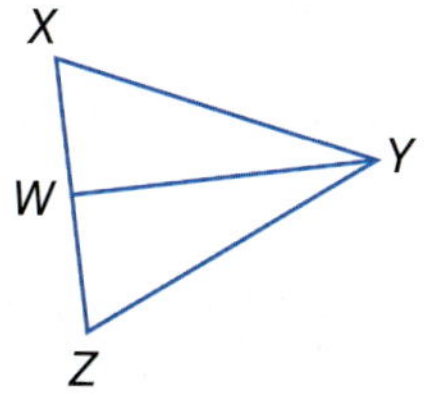

32. **PROOF** Write an algebraic proof.

Given: $\triangle XYZ$ with medians $\overline{XR}$, $\overline{YS}$, $\overline{ZQ}$

Prove: $\frac{XP}{PR} = 2$

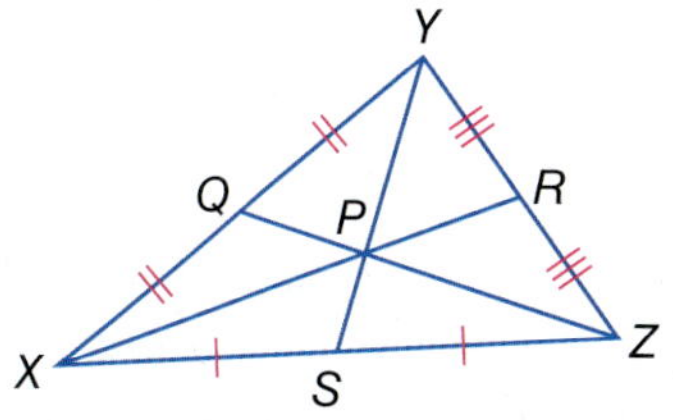

33. **MULTIPLE REPRESENTATIONS** In this problem, you will investigate the location of the points of concurrency for any equilateral triangle.

a. **CONCRETE** Construct three different equilateral triangles on patty paper and cut them out. Fold each triangle to locate the circumcenter, incenter, centroid, and orthocenter.

b. **VERBAL** Make a conjecture about the relationships among the four points of concurrency of any equilateral triangle.

c. **GRAPHICAL** Position an equilateral triangle and its circumcenter, incenter, centroid, and orthocenter on the coordinate plane using variable coordinates. Determine the coordinates of each point of concurrency.

ALGEBRA **In $\triangle JLP$, $m\angle JMP = 3x - 6$, $JK = 3y - 2$, and $LK = 5y - 8$.**

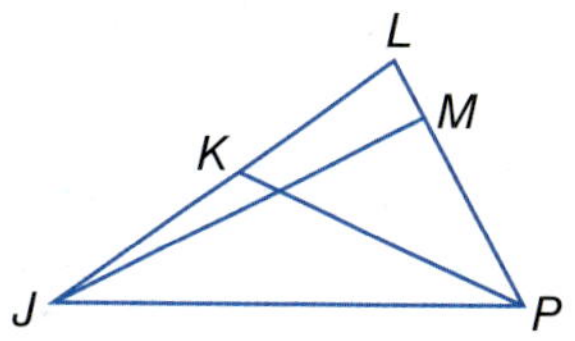

34. If $\overline{JM}$ is an altitude of $\triangle JLP$, find x.

35. Find LK if $\overline{PK}$ is a median.

The Granger Collection, New York

Math History Link

Pierre de Fermat
(1601–1665)

Another triangle center is the Fermat point, which minimizes the sum of the distances from the three vertices. Fermat is one of the best-known mathematicians for writing proofs.

36. PROOF Write a coordinate proof to prove the Centroid Theorem.

Given: $\triangle ABC$, medians $\overline{AR}$, $\overline{BS}$, and $\overline{CQ}$

Prove: The medians intersect at point P and P is two thirds of the distance from each vertex to the midpoint of the opposite side.

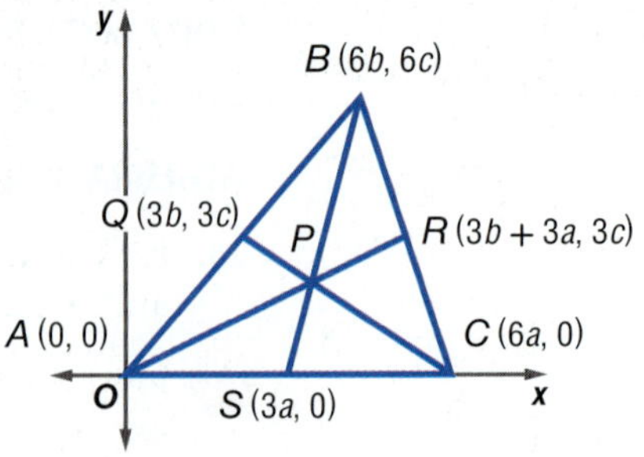

(*Hint*: First, find the equations of the lines containing the medians. Then find the coordinates of point P and show that all three medians intersect at point P. Next, use the Distance Formula and multiplication to show $AP = \frac{2}{3}AR$, $BP = \frac{2}{3}BS$, and $CP = \frac{2}{3}CQ$.)

H.O.T. Problems

Use **H**igher-**O**rder **T**hinking Skills

37. ERROR ANALYSIS Based on the figure at the right, Luke says that $\frac{2}{3}AP = AD$. Kareem disagrees. Is either of them correct? Explain your reasoning.

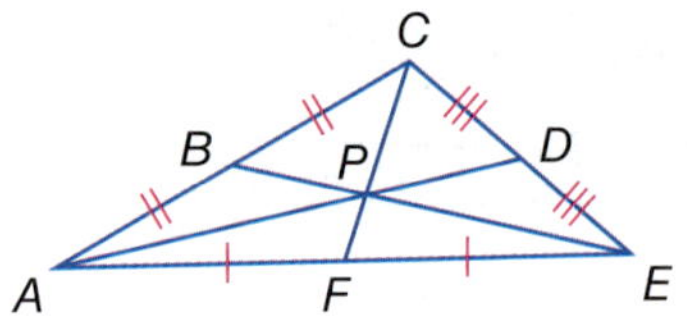

38. REASONING Determine whether the following statement is *true* or *false*. If true, explain your reasoning. If false, provide a counterexample.

The orthocenter of a right triangle is always located at the vertex of the right angle.

39. CHALLENGE $\triangle ABC$ has vertices $A(-3, 3)$, $B(2, 5)$, and $C(4, -3)$. What are the coordinates of the centroid of $\triangle ABC$? Explain the process you used to reach your conclusion.

40. WRITING IN MATH Compare and contrast the perpendicular bisectors, medians, and altitudes of a triangle.

41. CHALLENGE In the figure at the right, segments $\overline{AD}$ and $\overline{CE}$ are medians of $\triangle ACB$, $\overline{AD} \perp \overline{CE}$, $AB = 10$, and $CE = 9$. Find CA.

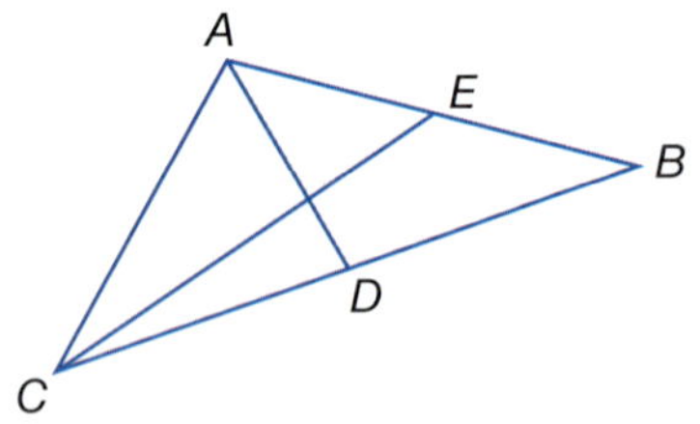

42. OPEN ENDED In this problem, you will investigate the relationships among three points of concurrency in a triangle.

a. Draw an acute triangle and find the circumcenter, centroid, and orthocenter.

b. Draw an obtuse triangle and find the circumcenter, centroid, and orthocenter.

c. Draw a right triangle and find the circumcenter, centroid, and orthocenter.

d. Make a conjecture about the relationships among the circumcenter, centroid, and orthocenter.

43. WRITING IN MATH Use area to explain why the centroid of a triangle is its center of gravity. Then use this explanation to describe the location for the balancing point for a rectangle.

Standardized Test Practice

44. In the figure below, $\overline{GJ} \cong \overline{HJ}$. Which must be true?

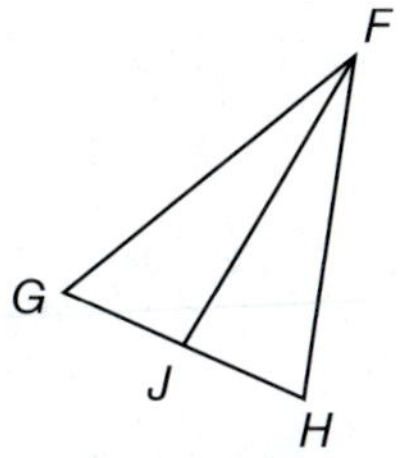

A $\overline{FJ}$ is an altitude of $\triangle FGH$.

B $\overline{FJ}$ is an angle bisector of $\triangle FGH$.

C $\overline{FJ}$ is a median of $\triangle FGH$.

D $\overline{FJ}$ is a perpendicular bisector of $\triangle FGH$.

45. GRIDDED RESPONSE What is the x-intercept of the graph of $4x - 6y = 12$?

46. ALGEBRA Four students have volunteered to fold pamphlets for a local community action group. Which student is the fastest?

Student	Folding Speed
Neiva	1 page every 3 seconds
Sarah	2 pages every 10 seconds
Quinn	30 pages per minute
Deron	45 pages in 2 minutes

F Deron H Quinn

G Neiva J Sarah

47. SAT/ACT 80 percent of 42 is what percent of 16?

A 240 C 150

B 210 D 50

Spiral Review

Find each measure. (Lesson 5-1)

48. LM

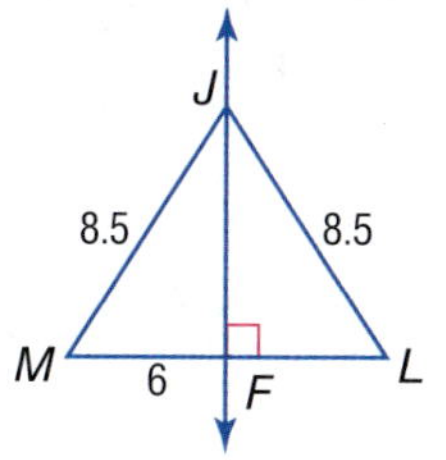

49. DF

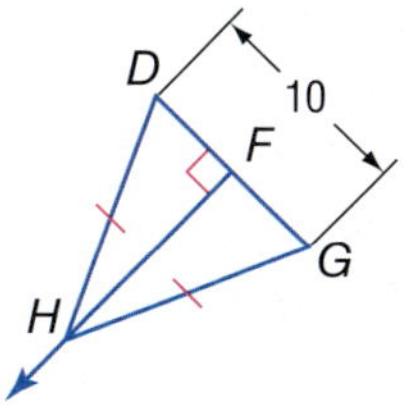

50. TQ

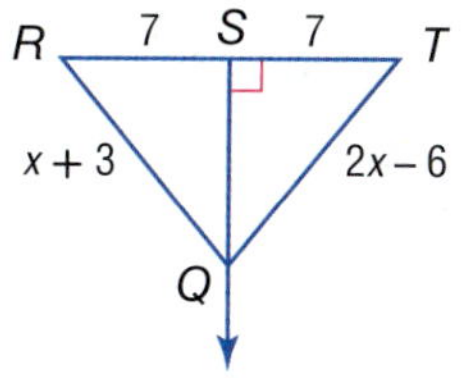

Position and label each triangle on the coordinate plane. (Lesson 4-8)

51. right $\triangle XYZ$ with hypotenuse $\overline{XZ}$, ZY is twice XY, and $\overline{XY}$ is b units long

52. isosceles $\triangle QRT$ with base $\overline{QR}$ that is b units long

Determine whether $\overleftrightarrow{RS}$ and $\overleftrightarrow{JK}$ are *parallel*, *perpendicular*, or *neither*. Graph each line to verify your answer. (Lesson 3-3)

53. $R(5, -4)$, $S(10, 0)$, $J(9, -8)$, $K(5, -13)$

54. $R(1, 1)$, $S(9, 8)$, $J(-6, 1)$, $K(2, 8)$

55. HIGHWAYS Near the city of Hopewell, Virginia, Route 10 runs perpendicular to Interstate 95 and Interstate 295. Show that the angles at the intersections of Route 10 with Interstate 95 and Interstate 295 are congruent. (Lesson 2-8)

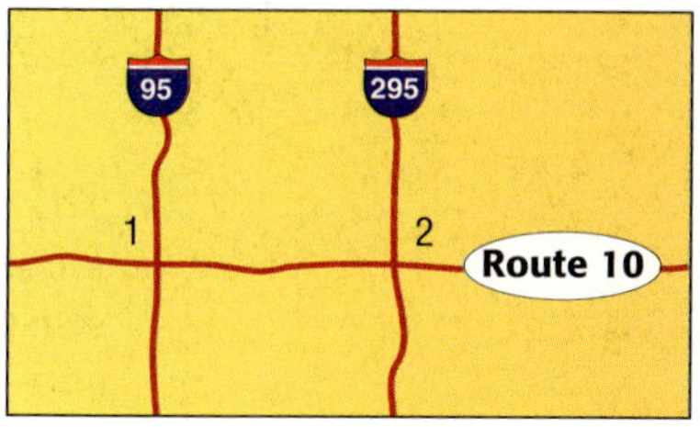

Skills Review

PROOF Write a flow proof of the Exterior Angle Theorem. (Lesson 4-2)

56. Given: $\triangle XYZ$

Prove: $m\angle X + m\angle Z = m\angle 1$

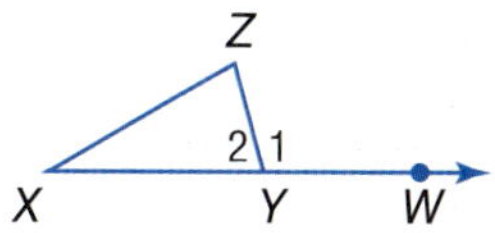

5-3 Inequalities in One Triangle

Then

You found the relationship between the angle measures of a triangle. (Lesson 4-2)

Now

- Recognize and apply properties of inequalities to the measures of the angles of a triangle.
- Recognize and apply properties of inequalities to the relationships between the angles and sides of a triangle.

Math Online
glencoe.com

Why?

To create the appearance of depth in a room, interior designers use a technique called *triangulation*. A basic example of this technique is the placement of an end table on each side of a sofa with a painting over the sofa.

The measures of the base angles of the triangle should be less than the measure of the other angle.

Angle Inequalities In algebra, you learned about the inequality relationship between two real numbers. This relationship is often used in proofs.

Key Concept — Definition of Inequality

Words For any real numbers a and b, $a > b$ if and only if there is a positive number c such that $a = b + c$.

Example If $5 = 2 + 3$, then $5 > 2$ and $5 > 3$.

The table below lists some of the properties of inequalities you studied in algebra.

Key Concept — Properties of Inequality for Real Numbers

The following properties are true for any real numbers a, b, and c.

Comparison Property of Inequality	$a < b$, $a = b$, or $a > b$
Transitive Property of Inequality	**1.** If $a < b$ and $b < c$, then $a < c$. **2.** If $a > b$ and $b > c$, then $a > c$.
Addition Property of Inequality	**1.** If $a > b$, then $a + c > b + c$. **2.** If $a < b$, then $a + c < b + c$.
Subtraction Property of Inequality	**1.** If $a > b$, then $a - c > b - c$. **2.** If $a < b$, then $a - c < b - c$.

The definition of inequality and the properties of inequalities can be applied to the measures of angles and segments, since these are real numbers. Consider $\angle 1$, $\angle 2$, and $\angle 3$ in the figure shown.

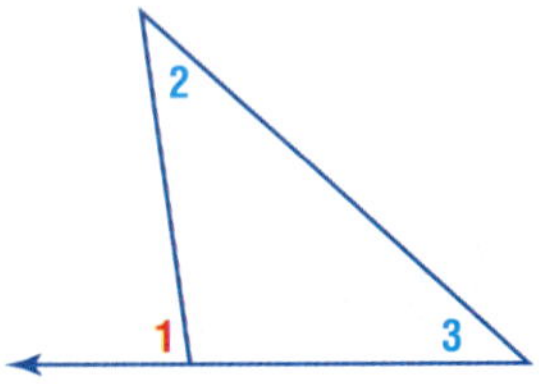

By the Exterior Angle Theorem, you know that $m\angle 1 = m\angle 2 + m\angle 3$.

Since the angle measures are positive numbers, we can also say that

$$m\angle 1 > m\angle 2 \quad \text{and} \quad m\angle 1 > m\angle 3$$

by the definition of inequality. This result suggests the following theorem.

Theorem 5.8 Exterior Angle Inequality

The measure of an exterior angle of a triangle is greater than the measure of either of its corresponding remote interior angles.

Example $m\angle 1 > m\angle A$
$m\angle 1 > m\angle B$

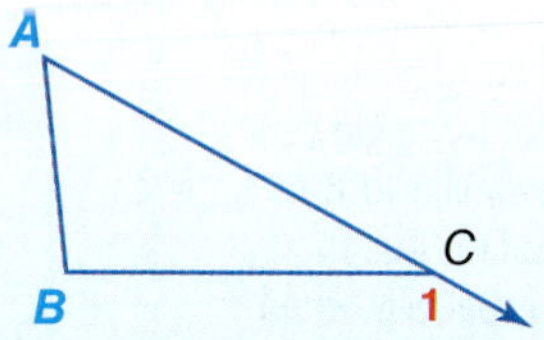

The proof of Theorem 5.8 is in Lesson 5-4.

Review Vocabulary

Remote Interior Angle Each exterior angle of a triangle has two *remote interior angles* that are not adjacent to the exterior angle. (Lesson 4-2)

EXAMPLE 1 Use the Exterior Angle Inequality Theorem

Use the Exterior Angle Inequality Theorem to list all of the angles that satisfy the stated condition.

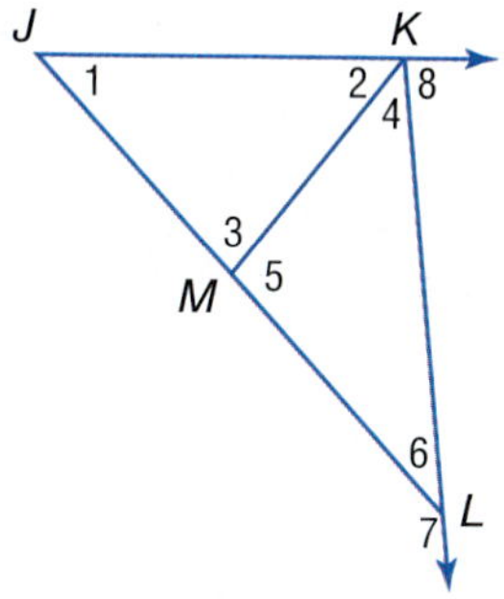

a. measures less than $m\angle 7$

$\angle 7$ is an exterior angle to $\triangle KML$, with $\angle 4$ and $\angle 5$ as corresponding remote interior angles. By the Exterior Angle Inequality Theorem, $m\angle 7 > m\angle 4$ and $m\angle 7 > m\angle 5$.

$\angle 7$ is also an exterior angle to $\triangle JKL$, with $\angle 1$ and $\angle JKL$ as corresponding remote interior angles. So, $m\angle 7 > m\angle 1$ and $m\angle 7 > m\angle JKL$. Since $m\angle JKL = m\angle 2 + m\angle 4$, by substitution $m\angle 7 > m\angle 2 + m\angle 4$. Therefore, $m\angle 7 > m\angle 2$.

So, the angles with measures less than $m\angle 7$ are $\angle 1$, $\angle 2$, $\angle 4$, $\angle 5$.

b. measures greater than $m\angle 6$

$\angle 3$ is an exterior angle to $\triangle KLM$. So by the Exterior Angle Inequality Theorem, $m\angle 3 > m\angle 6$. Because $\angle 8$ is an exterior angle to $\triangle JKL$, $m\angle 8 > m\angle 6$. Thus, the measures of $\angle 3$ and $\angle 8$ are greater than $m\angle 6$.

Guided Practice

1A. measures less than $m\angle 1$

1B. measures greater than $m\angle 8$

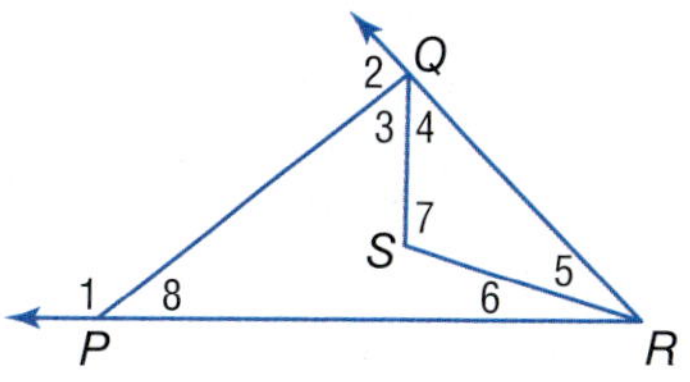

Personal Tutor glencoe.com

Watch Out!

Identifying Side Opposite Be careful to correctly identify the side opposite an angle. The sides that form the angle cannot be the sides opposite the angle.

Angle-Side Inequalities In Lesson 4-6, you learned that if two sides of a triangle are congruent, or the triangle is isosceles, then the angles opposite those sides are congruent. What relationship exists if the sides are not congruent? Examine the longest and shortest sides and smallest and largest angles of a scalene obtuse triangle.

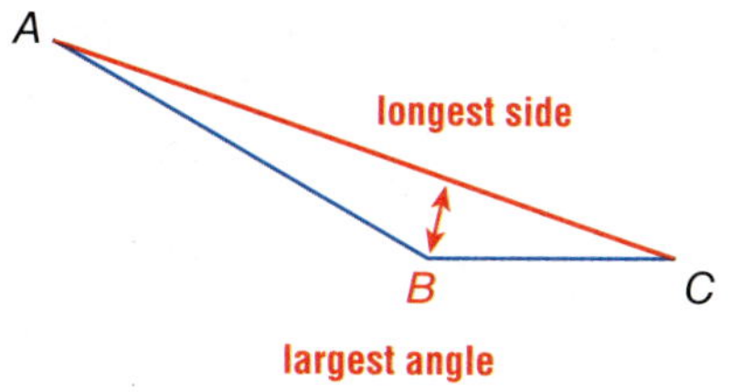

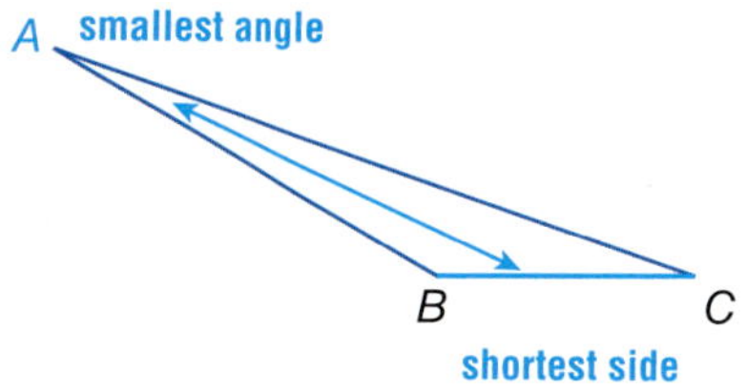

Notice that the longest side and largest angle of $\triangle ABC$ are opposite each other. Likewise, the shortest side and smallest angle are opposite each other.

Watch Out!

Symbols for Angles and Inequalities The symbol for angle ($\angle$) looks similar to the symbol for less than ($<$), especially when handwritten. Be careful to write the symbols correctly in situations where both are used.

The side-angle relationships in an obtuse scalene triangle are true for all triangles, and are stated using inequalities in the theorems below.

Theorems — Angle-Side Relationships in Triangles

5.9 If one side of a triangle is longer than another side, then the angle opposite the longer side has a greater measure than the angle opposite the shorter side.

Example $XY > YZ$, so $m\angle Z > m\angle X$.

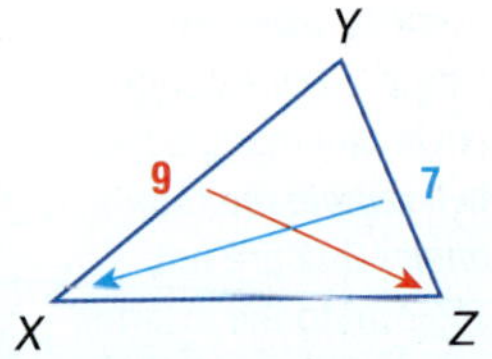

5.10 If one angle of a triangle has a greater measure than another angle, then the side opposite the greater angle is longer than the side opposite the lesser angle.

Example $m\angle J > m\angle K$, so $KL > JL$.

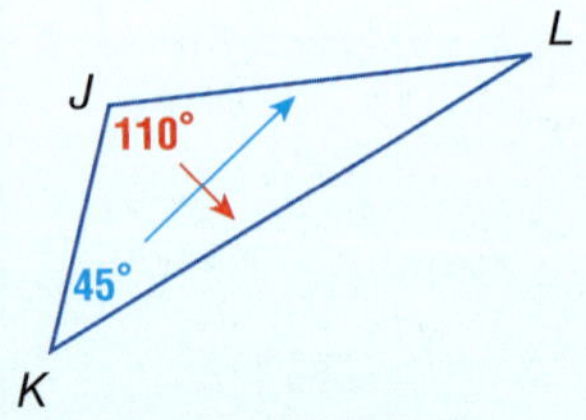

Proof — Theorem 5.9

Given: $\triangle ABC$, $AB > BC$

Prove: $m\angle BCA > m\angle A$

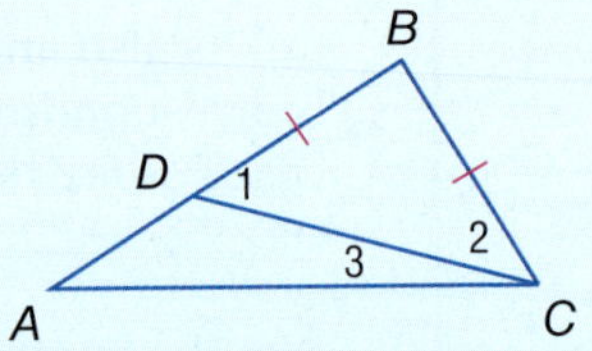

Proof:

Since $AB > BC$ in the given $\triangle ABC$, there exists a point D on $\overline{AB}$ such that $BD = BC$. Draw $\overline{CD}$ to form isosceles $\triangle BCD$. By the Isosceles Triangle Theorem, $\angle 1 \cong \angle 2$, so $m\angle 1 = m\angle 2$ by the definition of congruent angles.

By the Angle Addition Postulate, $m\angle BCA = m\angle 2 + m\angle 3$, so $m\angle BCA > m\angle 2$ by the definition of inequality. By substitution, $m\angle BCA > m\angle 1$.

By the Exterior Angle Inequality Theorem, $m\angle 1 > m\angle A$. Therefore, because $m\angle BCA > m\angle 1$ and $m\angle 1 > m\angle A$, by the Transitive Property of Inequality, $m\angle BCA > m\angle A$.

You will prove Theorem 5.10 in Lesson 5-4, Exercise 31.

EXAMPLE 2 Order Triangle Angle Measures

List the angles of $\triangle PQR$ in order from smallest to largest.

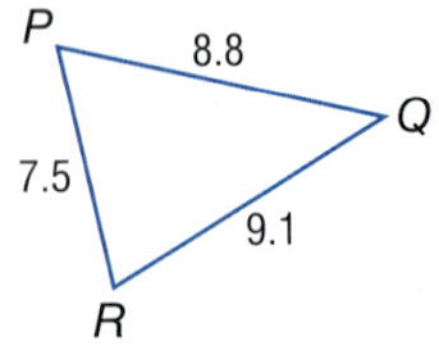

The sides from shortest to longest are $\overline{PR}$, $\overline{PQ}$, $\overline{QR}$. The angles opposite these sides are $\angle Q$, $\angle R$, and $\angle P$, respectively. So the angles from smallest to largest are $\angle Q$, $\angle R$, and $\angle P$.

Guided Practice

2. List the angles and sides of $\triangle ABC$ in order from smallest to largest.

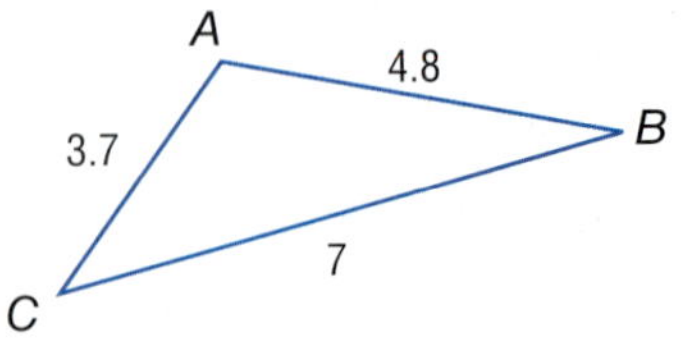

Personal Tutor glencoe.com

EXAMPLE 3 Order Triangle Side Lengths

List the sides of $\triangle FGH$ in order from shortest to longest.

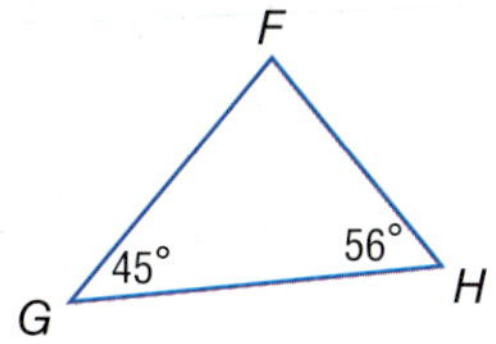

First find the missing angle measure using the Triangle Angle Sum Theorem.

$m\angle F = 180 - (45 + 56)$ or 79

So, the angles from smallest to largest are $\angle G$, $\angle H$, and $\angle F$. The sides opposite these angles are $\overline{FH}$, $\overline{FG}$, and $\overline{GH}$, respectively. So, the sides from shortest to longest are $\overline{FH}$, $\overline{FG}$, $\overline{GH}$.

Guided Practice

3. List the angles and sides of $\triangle WXY$ in order from smallest to largest.

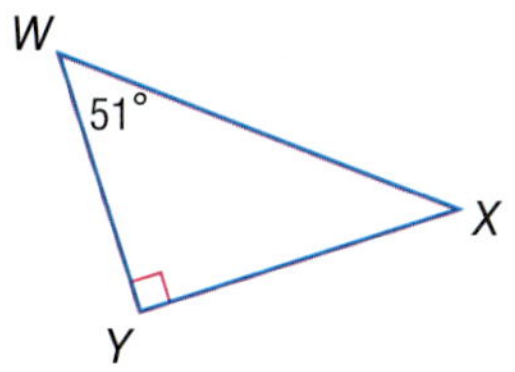

Personal Tutor glencoe.com

You can use angle-side relationships in triangles to solve real-world problems.

Real-World EXAMPLE 4 Angle-Side Relationships

INTERIOR DESIGN An interior designer uses triangulation to create depth in a client's living room. If $m\angle B$ is to be less than $m\angle A$, which distance should be longer—the distance between the two lamps or the distance from the lamp at B to the midpoint of the top of the artwork? Explain.

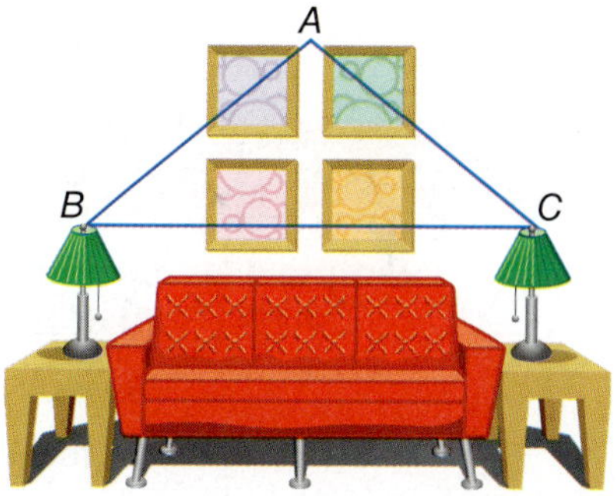

According to Theorem 5.10, in order for $m\angle B < m\angle A$, the length of the side opposite $\angle B$ must be less than the length of the side opposite $\angle A$. Since $\overline{AC}$ is opposite $\angle B$, and $\overline{BC}$ is opposite $\angle A$, then $AC < BC$ and $BC > AC$. So BC, the distance between the lamps, must be greater than the distance from the lamp at B to the midpoint of the top of the artwork.

Real-World Link

Lifeguard preparation courses include training in surveillance, rescue, and first aid. Course lengths vary from 30 to 37 hours based on the type of water environment such as pools, waterfront, or waterpark.

Source: American Red Cross

Guided Practice

4. LIFEGUARDING During lifeguard training, an instructor simulates a person in distress so that trainees can practice their rescue skills. If the instructor, Trainee 1, and Trainee 2 are located in the positions shown on the diagram, which of the two trainees is closest to the instructor?

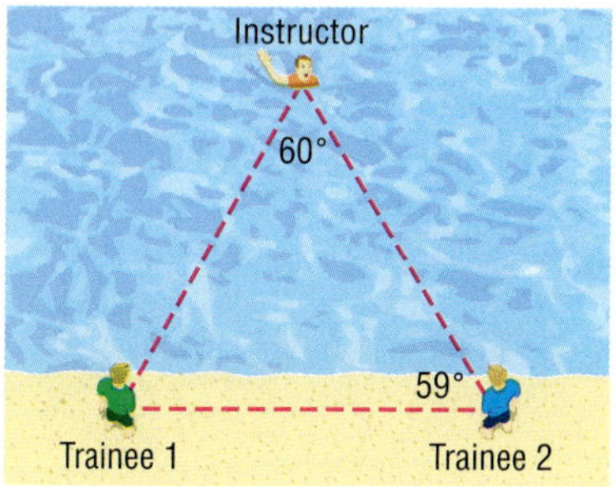

Personal Tutor glencoe.com

Check Your Understanding

Example 1
p. 343

Use the Exterior Angle Inequality Theorem to list all of the angles that satisfy the stated condition.

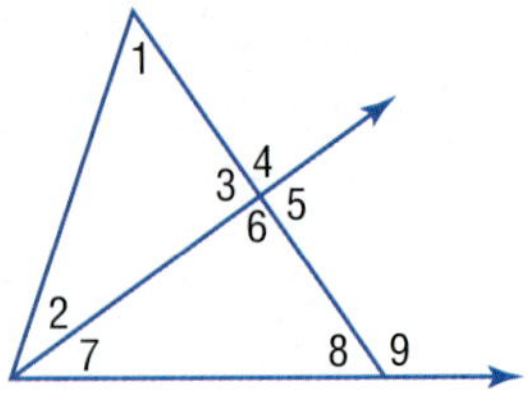

1. measures less than $m\angle 4$
2. measures greater than $m\angle 7$
3. measures greater than $m\angle 2$
4. measures less than $m\angle 9$

Examples 2 and 3
pp. 344–345

List the angles and sides of each triangle in order from smallest to largest.

5. 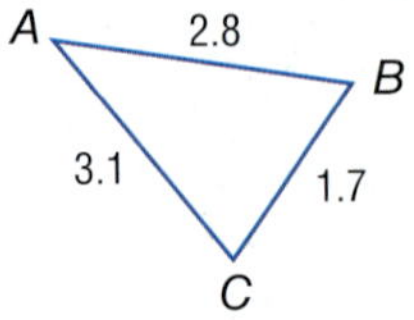

6. 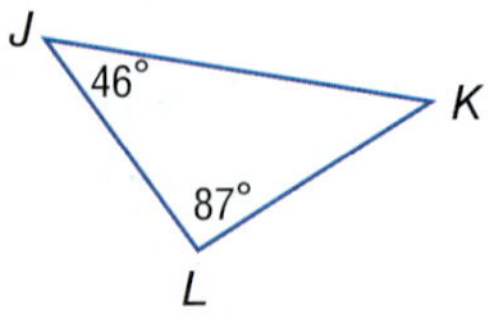

Example 4
p. 345

7. **HANG GLIDING** The supports on a hang glider form triangles like the one shown. Which is longer—the support represented by $\overline{AC}$ or the support represented by $\overline{BC}$? Explain your reasoning.

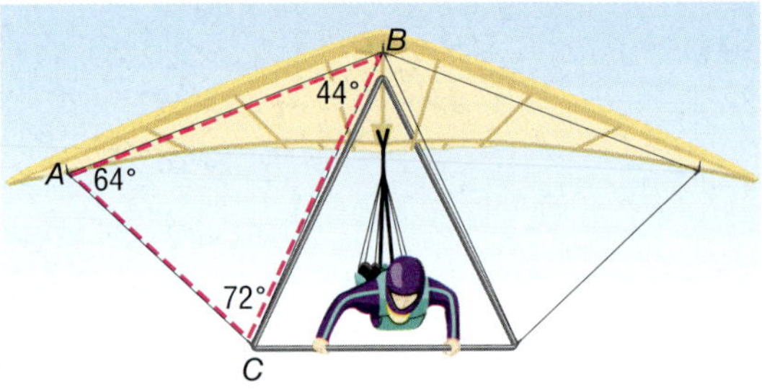

Practice and Problem Solving

● = **Step-by-Step Solutions** begin on page R20.
Extra Practice begins on page 969.

Example 1
p. 343

Use the Exterior Angle Inequality Theorem to list all of the angles that satisfy the stated condition.

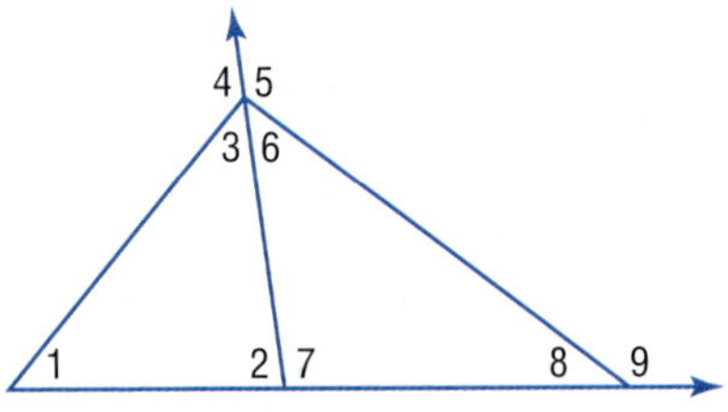

8. measures greater than $m\angle 2$
9. measures less than $m\angle 4$
10. measures less than $m\angle 5$
11. measures less than $m\angle 9$
12. measures greater than $m\angle 8$
13. measures greater than $m\angle 7$

Examples 2 and 3
pp. 344–345

List the angles and sides of each triangle in order from smallest to largest.

14. 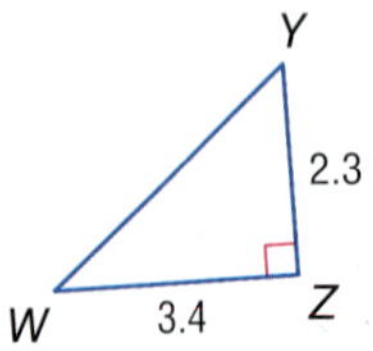

15. 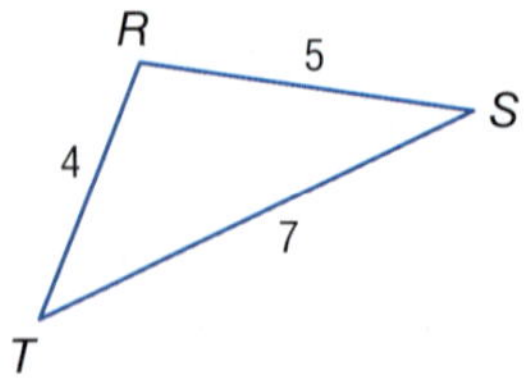

16. 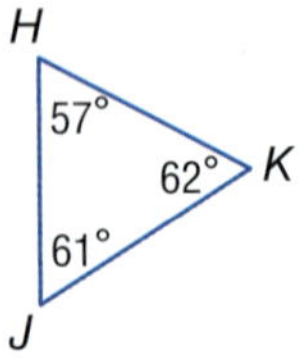

17. 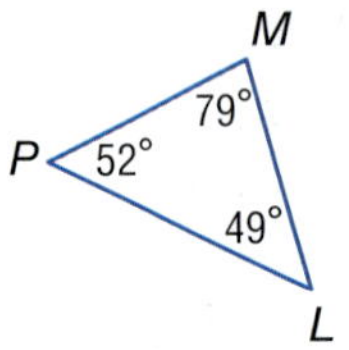

18. 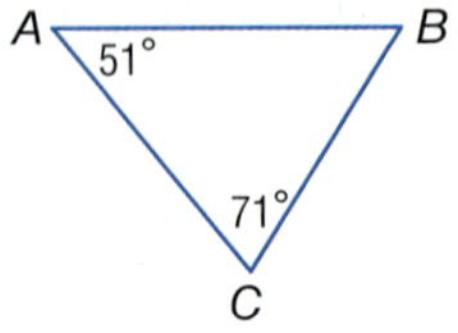

19. 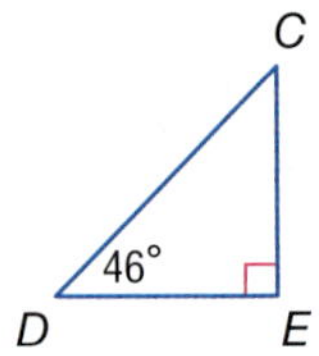

Example 4
p. 345

Real-World Link

Ultimate shares characteristics with soccer, basketball, and football. Seven-player teams score points by throwing a disc to a team member in the opponents' end zone. When holding the disc, players can only pivot in place until they pass the disc. Games are usually played to 15 points and last about 2 hours.

Source: Ultimate Players Association

20. SPORTS Ben, Gilberto, and Hannah are playing Ultimate. Hannah is trying to decide if she should pass to Ben or Gilberto. Which player should she choose in order to have the shorter passing distance? Explain your reasoning.

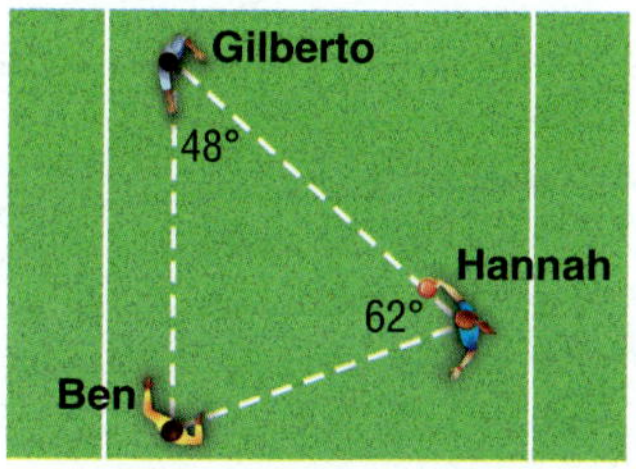

21. RAMPS The wedge below represents a bike ramp. Which is longer, the length of the ramp $\overline{XZ}$ or the length of the top surface of the ramp $\overline{YZ}$? Explain your reasoning using Theorem 5.9.

List the angles and sides of each triangle in order from smallest to largest.

22.

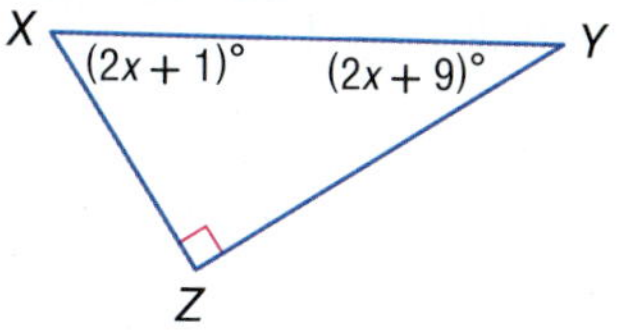

23.

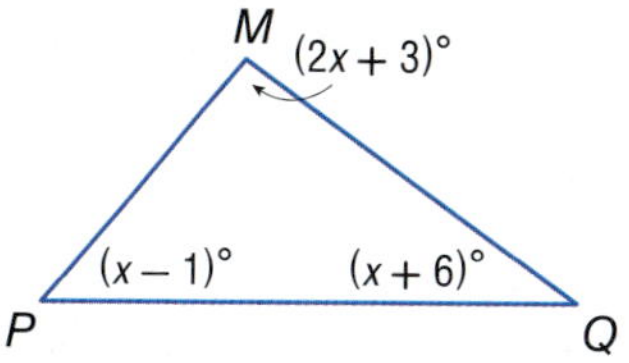

Use the figure at the right to determine which angle has the greatest measure.

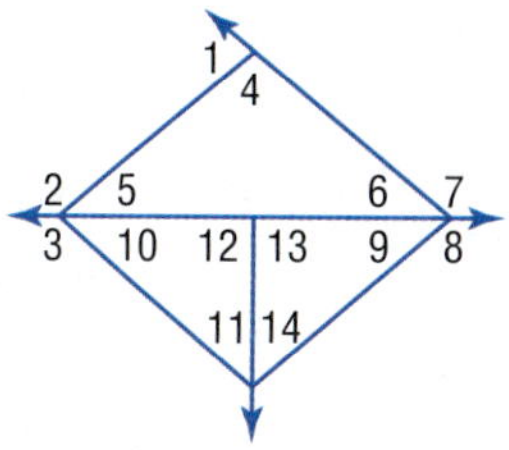

24. $\angle 1, \angle 5, \angle 6$

25. $\angle 2, \angle 4, \angle 6$

26. $\angle 7, \angle 4, \angle 5$

27. $\angle 3, \angle 11, \angle 12$

28. $\angle 3, \angle 9, \angle 14$

29. $\angle 8, \angle 10, \angle 11$

Use the figure at the right to determine the relationship between the measures of the given angles.

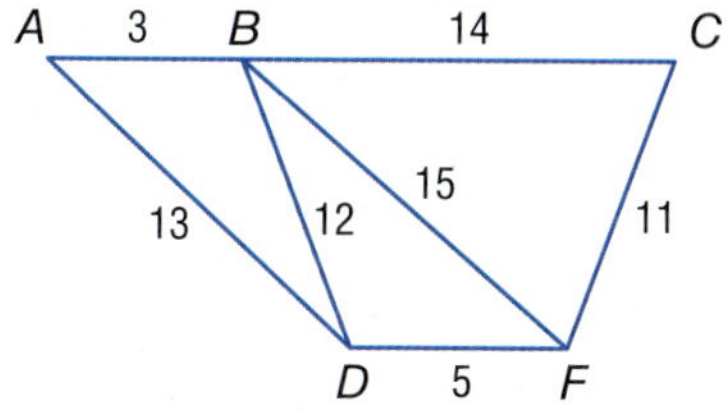

30. $\angle ABD, \angle BDA$

31. $\angle BCF, \angle CFB$

32. $\angle BFD, \angle BDF$

33. $\angle DBF, \angle BFD$

Use the figure at the right to determine the relationship between the lengths of the given sides.

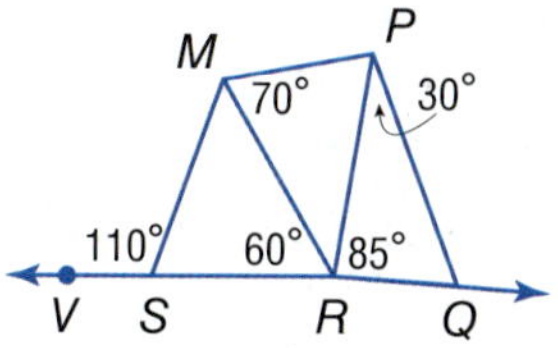

34. $\overline{SM}, \overline{MR}$

35. $\overline{RP}, \overline{MP}$

36. $\overline{RQ}, \overline{PQ}$

37. $\overline{SM}, \overline{RQ}$

38. HIKING Justin and his family are hiking around a lake as shown in the diagram at the right. Order the angles of the triangle formed by their path from largest to smallest.

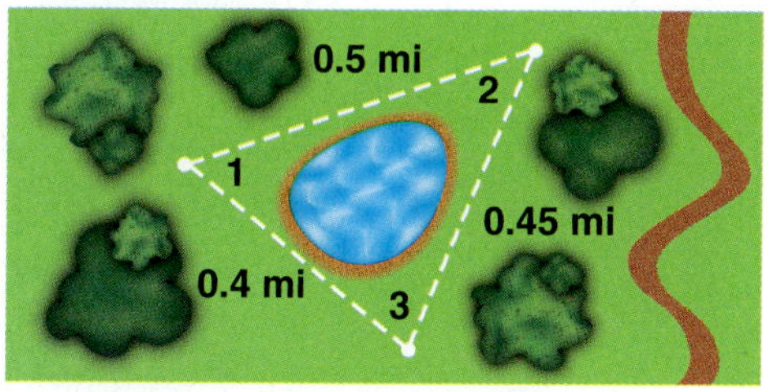

COORDINATE GEOMETRY **List the angles of each triangle with the given vertices in order from smallest to largest. Justify your answer.**

39 $A(-4, 6), B(-2, 1), C(5, 6)$

40. $X(-3, -2), Y(3, 2), Z(-3, -6)$

41. List the side lengths of the triangles in the figure from shortest to longest. Explain your reasoning.

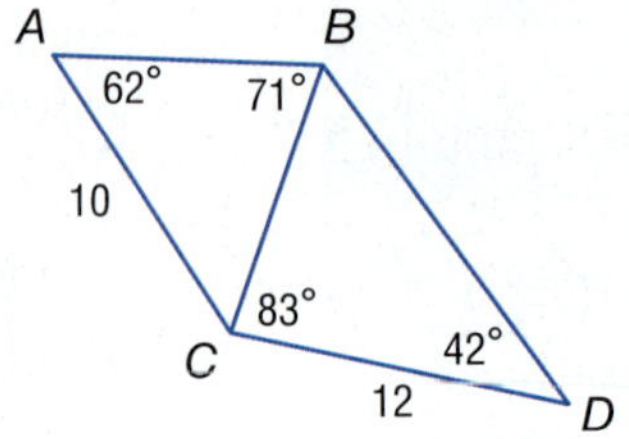

42. **MULTIPLE REPRESENTATIONS** In this problem, you will explore the relationship between the sides of a triangle.

a. **GEOMETRIC** Draw three triangles, including one acute, one obtuse, and one right angle. Label the vertices of each triangle A, B, and C.

b. **TABULAR** Measure the length of each side of the three triangles. Then copy and complete the table.

Triangle	AB	BC	$AB + BC$	CA
Acute				
Obtuse				
Right				

c. **TABULAR** Create two additional tables like the one above, finding the sum of BC and CA in one table and the sum of AB and CA in the other.

d. **ALGEBRAIC** Write an inequality for each of the tables you created relating the measure of the sum of two of the sides to the measure of the third side of a triangle.

e. **VERBAL** Make a conjecture about the relationship between the measure of the sum of two sides of a triangle and the measure of the third side.

H.O.T. Problems

Use Higher-Order Thinking Skills

43. **WRITING IN MATH** Analyze the information given in the diagram and explain why the markings must be incorrect.

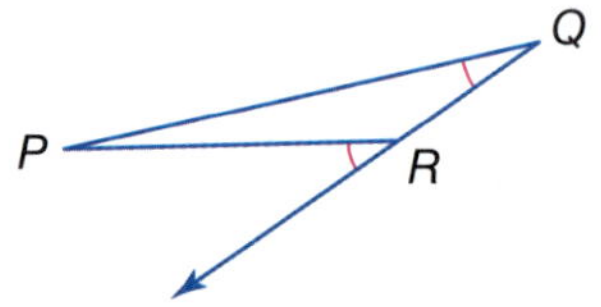

44. **CHALLENGE** Using only a ruler, draw $\triangle ABC$ such that $m\angle A > m\angle B > m\angle C$. Justify your drawing.

StudyTip

Justify Drawings To justify a drawing, use the measurements taken from your drawing to construct a logical argument.

45. **OPEN ENDED** Give a possible measure for $\overline{AB}$ in $\triangle ABC$ shown. Explain your reasoning.

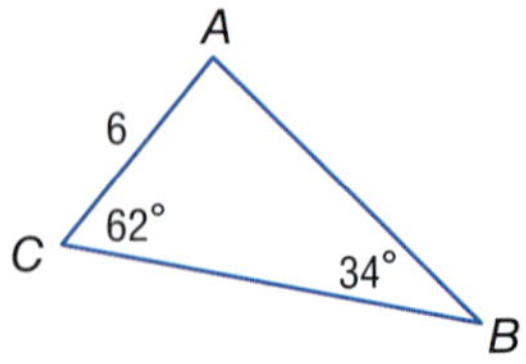

46. **REASONING** Is the base of an isosceles triangle *always*, *sometimes*, or *never* the longest side of the triangle? Explain.

47. **CHALLENGE** Use the side lengths in the figure to list the numbered angles in order from smallest to largest given that $m\angle 2 = m\angle 5$. Explain your reasoning.

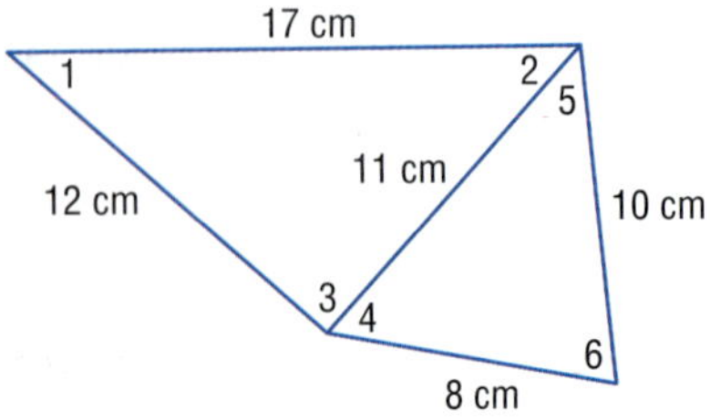

48. **WRITING IN MATH** Explain why the hypotenuse of a right triangle is always the longest side of the triangle.

Standardized Test Practice

49. STATISTICS The chart shows the number and types of DVDs sold at three stores.

DVD Type	Store 1	Store 2	Store 3
Comedy	75	80	92
Action	54	37	65
Horror	30	48	62
Science Fiction	21	81	36
Total	180	246	255

According to the information in the chart, which of these statements is true?

A The mean number of DVDs sold per store was 56.
B Store 1 sold twice as many action and horror films as store 3 sold of science fiction.
C Store 2 sold fewer comedy and science fiction than store 3 sold.
D The mean number of science fiction DVDs sold per store was 46.

50. Two angles of a triangle have measures 45° and 92°. What type of triangle is it?

F obtuse scalene
G obtuse isosceles
H acute scalene
J acute isosceles

51. EXTENDED RESPONSE At a five-star restaurant, a waiter earns a total of t dollars for working h hours in which he receives \$198 in tips and makes \$2.50 per hour.

a. Write an equation to represent the total amount of money the waiter earns.
b. If the waiter earned a total of \$213, how many hours did he work?
c. If the waiter earned \$150 in tips and worked for 12 hours, what is the total amount of money he earned?

52. SAT/ACT Which expression has the *least* value?

A $|45|$
B $|15|$
C $|-28|$
D $|-39|$

Spiral Review

In $\triangle XYZ$, P is the centroid, $KP = 3$, and $XJ = 8$. Find each length. (Lesson 5-2)

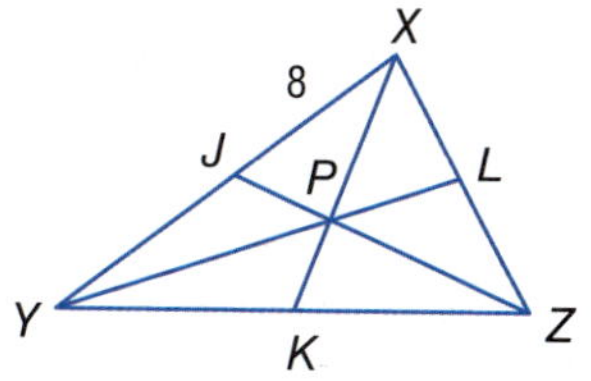

53. XK

54. YJ

COORDINATE GEOMETRY Write an equation in slope-intercept form for the perpendicular bisector of the segment with the given endpoints. Justify your answer. (Lesson 5-1)

55. $D(-2, 4)$ and $E(3, 5)$

56. $D(-2, -4)$ and $E(2, 1)$

57. JETS The United States Navy Flight Demonstration Squadron, the Blue Angels, flies in a formation that can be viewed as two triangles with a common side. Write a two-column proof to prove that $\triangle SRT \cong \triangle QRT$ if T is the midpoint of $\overline{SQ}$ and $\overline{SR} \cong \overline{QR}$. (Lesson 4-4)

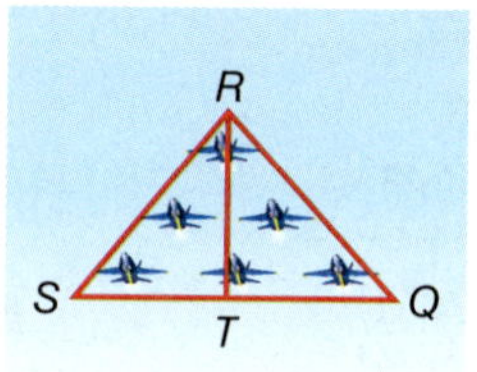

58. POOLS A rectangular pool is 20 feet by 30 feet. The depth of the pool is 60 inches, but the depth of the water is $\frac{3}{4}$ of the depth of the pool. Find each measure to the nearest tenth. (Lesson 1-7)

a. the surface area of the pool

b. the volume of water in the pool

Skills Review

Determine whether each statement is true or false if $x = 8$, $y = 2$, and $z = 3$.

59. $z(x - y) = 13$

60. $2x = 3yz$

61. $x + y > z + y$

CHAPTER 5 Mid-Chapter Quiz

Lessons 5-1 through 5-3

Find each measure. (Lesson 5-1)

1. *AB*

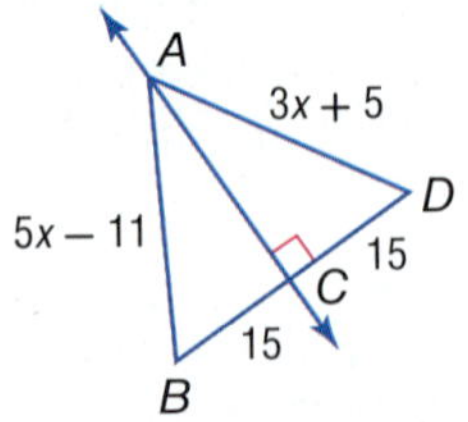

2. *JL*

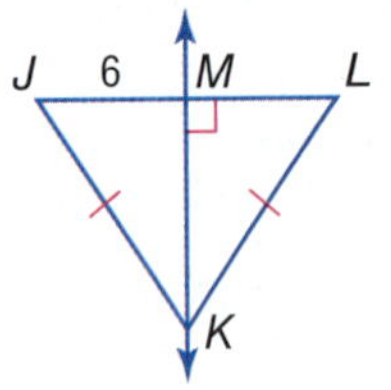

3. CAMP Camp Onawatchi ends with a game of capture the flag. If the starting locations of three teams are shown in the diagram below, with the flag at a point equidistant from each team's base, how far from each base is the flag in feet? (Lesson 5-1)

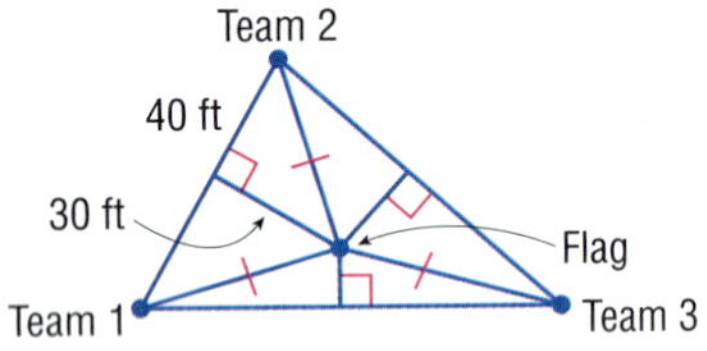

Find each measure. (Lesson 5-1)

4. $\angle MNP$

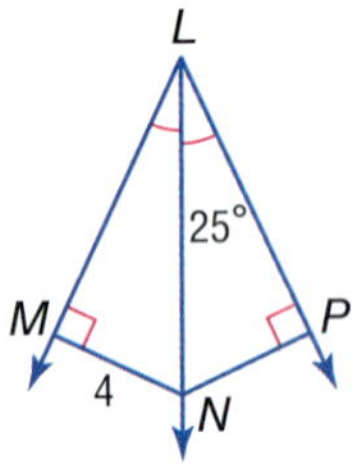

5. *XY*

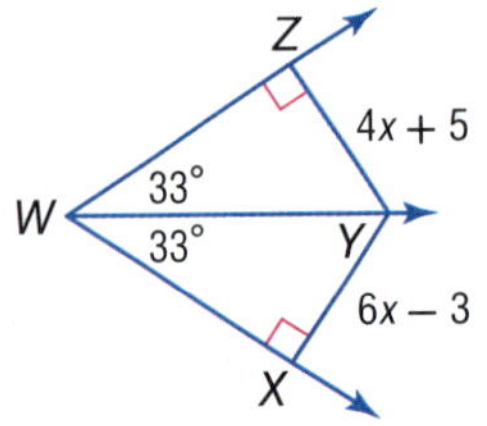

In $\triangle RST$, Z is the centroid and $RZ = 18$. Find each length. (Lesson 5-2)

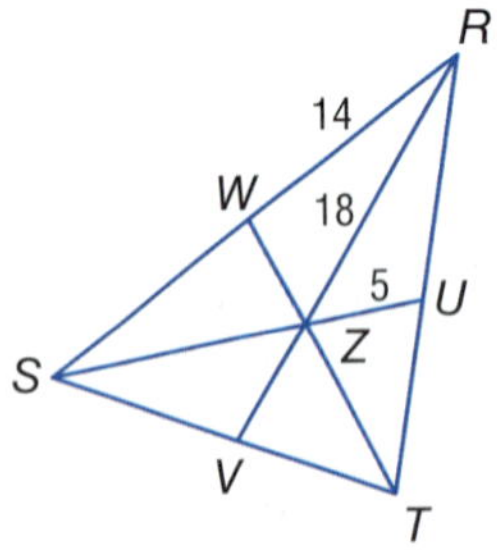

6. *ZV* **7.** *SZ* **8.** *SR*

COORDINATE GEOMETRY Find the coordinates of the centroid of each triangle with the given vertices. (Lesson 5-2)

9. $A(1, 7), B(4, 2), C(7, 7)$

10. $X(-11, 0), Y(-11, -8), Z(-1, -4)$

11. $R(-6, 4), S(-2, -2), T(2, 4)$

12. $J(-5, 5), K(-5, -1), L(1, 2)$

13. ARCHITECTURE An architect is designing a high school building. Describe how to position the central office so that it is at the intersection of each hallway connected to the three entrances to the school. (Lesson 5-2)

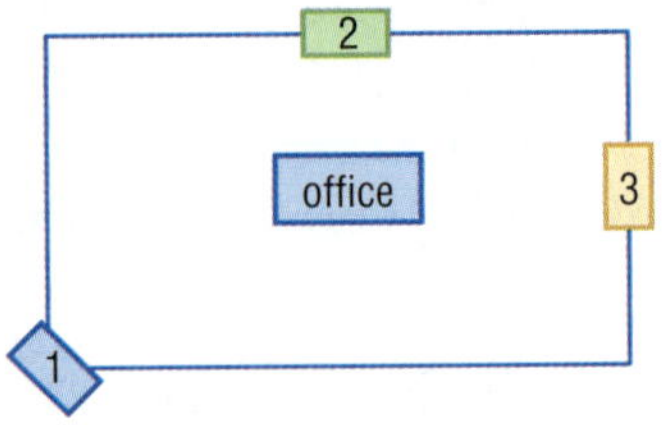

List the angles and sides of each triangle in order from smallest to largest. (Lesson 5-3)

14.

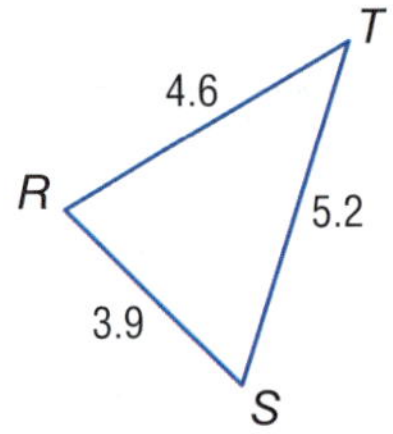

15.

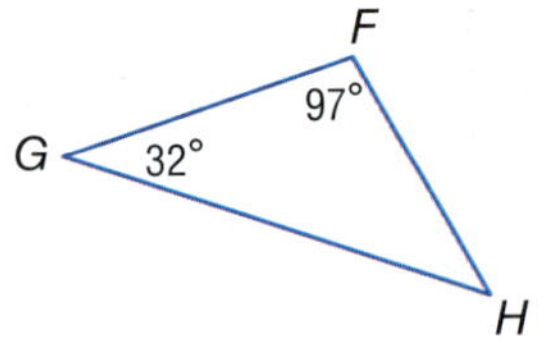

16. VACATION Kailey plans to fly over the route marked on the map of Hawaii below. (Lesson 5-3)

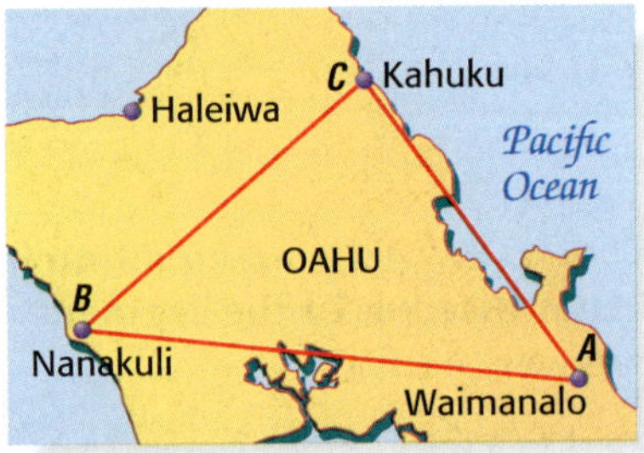

a. If $m\angle A = 2 + m\angle B$ and $m\angle C = 2(m\angle B) - 14$, what are the measures of the three angles?

b. What are the lengths of Kailey's trip in order of least to greatest?

c. The length of the entire trip is about 68 miles. The middle leg is 11 miles greater than one-half the length of the shortest leg. The longest leg is 12 miles greater than three-fourths of the shortest leg. What are the lengths of the legs of the trip?

Use the Exterior Angle Inequality Theorem to list all of the angles that satisfy the stated condition. (Lesson 5-3)

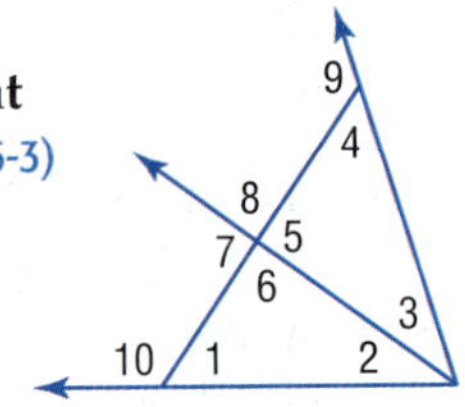

17. measures less than $m\angle 8$

18. measures greater than $m\angle 3$

19. measures less than $m\angle 10$

5-4 Indirect Proof

Then
You wrote paragraph, two-column, and flow proofs. (Lessons 2-5, 2-6, and 4-2)

Now
- Write indirect algebraic proofs.
- Write indirect geometric proofs.

New Vocabulary
indirect reasoning
indirect proof
proof by contradiction

Math Online
glencoe.com

Why?

Matthew: "I'm almost positive Friday is not a teacher work day, but I can't prove it."

Kim: "Let's assume that Friday *is* a teacher work day. What day is our next Geometry test?"

Mi-Ling: "Hmmm . . . according to the syllabus, it's this Friday. But we don't have tests on teacher work days—we're not in school."

Domingo: "Exactly—so that proves it! This Friday can't be a teacher work day."

Indirect Algebraic Proof The proofs you have written have been *direct proofs*—you started with a true hypothesis and proved that the conclusion was true. In the example above, the students used **indirect reasoning**, by assuming that a conclusion was false and then showing that this assumption led to a contradiction.

In an **indirect proof** or **proof by contradiction**, you temporarily assume that what you are trying to prove is false. By showing this assumption to be logically impossible, you prove your assumption false and the original conclusion true.

Key Concept — How to Write an Indirect Proof

Step 1 Identify the conclusion you are asked to prove. Make the assumption that this conclusion is false by assuming that the opposite is true.

Step 2 Use logical reasoning to show that this assumption leads to a contradiction of the hypothesis, or some other fact, such as a definition, postulate, theorem, or corollary.

Step 3 Point out that since the assumption leads to a contradiction, the original conclusion, what you were asked to prove, must be true.

EXAMPLE 1 State the Assumption for Starting an Indirect Proof

State the assumption necessary to start an indirect proof of each statement.

a. $\angle ABC \ncong \angle XYZ$

$\angle ABC \cong \angle XYZ$

b. If 6 is a factor of *n*, then 2 is a factor of *n*.

The conclusion of the conditional statement is *2 is a factor of n*. The negation of the conclusion is *2 is not a factor of n*.

c. ∠3 is an obtuse angle.

If *∠3 is an obtuse angle* is false, then *∠3 is not an obtuse angle* must be true. This means that *∠3 is an acute or right angle* must be true.

Guided Practice

1A. $x > 5$

1B. Points *J*, *K*, and *L* are collinear.

1C. $\triangle XYZ$ is an equilateral triangle.

Personal Tutor glencoe.com

ReadingMath

Contradiction
A contradiction is a principle of logic stating that an assumption cannot be both *A* and the opposite of *A* at the same time.

Indirect proofs can be used to prove algebraic concepts.

EXAMPLE 2 Write an Indirect Algebraic Proof

Write an indirect proof to show that if $-3x + 4 > 16$, then $x < -4$.

Given: $-3x + 4 > 16$

Prove: $x < -4$

Step 1 **Indirect Proof:**
The negation of $x < -4$ is $x \geq -4$. So, assume that $x > -4$ or $x = -4$ is true.

Step 2 Make a table with several possibilities for x assuming $x > -4$ or $x = -4$.

x	-4	-3	-2	-1	0
$-3x + 4$	16	13	10	7	4

When $x > -4$, $-3x + 4 < 16$ and when $x = -4$, $-3x + 4 = 16$.

Step 3 In both cases, the assumption leads to the contradiction of the given information that $-3x + 4 > 16$. Therefore, the assumption that $x \geq -4$ must be false, so the original conclusion that $x < -4$ must be true.

Guided Practice

Write an indirect proof of each statement.

2A. If $7x > 56$, then $x > 8$.

2B. If $-c$ is positive, then c is negative.

Personal Tutor glencoe.com

Indirect reasoning and proof can be used in everyday situations.

Real-World Link

\$100–\$300
the range in price of a girl's prom dress

\$75–\$125
the range in cost for a tuxedo rental

around \$150
the cost of a fancy dinner for two

\$100–\$200
the range in cost of prom tickets per couple

Source: PromSpot

Real-World EXAMPLE 3 Indirect Algebraic Proof

PROM COSTS Javier asked his friend Christopher the cost of his meal and his date's meal when he went to dinner for prom. Christopher could not remember the individual costs, but he did remember that the total bill, not including tip, was over \$60. Use indirect reasoning to show that at least one of the meals cost more than \$30.

Let the cost of one meal be x and the cost of the other meal be y.

Step 1 **Given:** $x + y > 60$

Prove: $x > 30$ or $y > 30$

Indirect Proof:
Assume that $x \leq 30$ and $y \leq 30$.

Step 2 If $x \leq 30$ and $y \leq 30$, then $x + y \leq 30 + 30$ or $x + y \leq 60$. This is a contradiction because we know that $x + y > 60$.

Step 3 Since the assumption that $x \leq 30$ and $y \leq 30$ leads to a contradiction of a known fact, the assumption must be false. Therefore, the conclusion that $x > 30$ or $y > 30$ must be true. Thus, at least one of the meals had to cost more than \$30.

Guided Practice

3. **TRAVEL** Cleavon traveled over 360 miles on his trip, making just two stops. Use indirect reasoning to prove that he traveled more than 120 miles on one leg of his trip.

Personal Tutor glencoe.com

Indirect proofs are often used to prove concepts in number theory. In such proofs, it is helpful to remember that you can represent an even number with the expression $2k$ and an odd number with the expression $2k + 1$ for any integer k.

EXAMPLE 4 Indirect Proofs in Number Theory

Write an indirect proof to show that if $x + 2$ is an even integer, then x is an even integer.

Step 1 **Given:** $x + 2$ is an even integer.

Prove: x is an even integer.

Indirect Proof:
Assume that x is an odd integer. This means that $x = 2k + 1$ for some integer k.

Step 2 $x + 2 = (2k + 1) + 2$ Substitution of assumption

$= (2k + 2) + 1$ Commutative Property

$= 2(k + 1) + 1$ Distributive Property

Now determine whether $2(k + 1) + 1$ is an even or odd integer. Since k is an integer, $k + 1$ is also an integer. Let m represent the integer $k + 1$.

$2(k + 1) + 1 = 2m + 1$ Substitution

So, $x + 2$ can be represented by $2m + 1$, where m is an integer. But this representation means that $x + 2$ is an odd integer, which contradicts the given statement that $x + 2$ is an even integer.

Step 3 Since the assumption that x is an odd integer leads to a contradiction of the given statement, the original conclusion that x is an even integer must be true.

Guided Practice

4. Write an indirect proof to show that if the square of an integer is odd, then the integer is odd.

Personal Tutor glencoe.com

Indirect Proof with Geometry Indirect reasoning can be used to prove statements in geometry, such as the Exterior Angle Inequality Theorem.

EXAMPLE 5 Geometry Proof

If an angle is an exterior angle of a triangle, prove that its measure is greater than the measure of either of its corresponding remote interior angles.

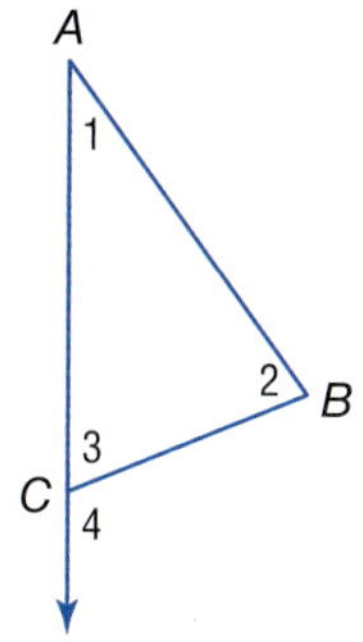

Step 1 Draw a diagram of this situation. Then identify what you are given and what you are asked to prove.

Given: $\angle 4$ is an exterior angle of $\triangle ABC$.

Prove: $m\angle 4 > m\angle 1$ and $m\angle 4 > m\angle 2$.

Indirect Proof:
Assume that $m\angle 4 \not> m\angle 1$ or $m\angle 4 \not> m\angle 2$.
In other words, $m\angle 4 \leq m\angle 1$ or $m\angle 4 \leq m\angle 2$.

(continued on the next page)

Watch Out!

Proof by Contradiction vs. Counterexample
Proof by contradiction and using a counterexample are not the same. A counterexample helps you disprove a conjecture. It cannot be used to prove a conjecture.

Step 2 You need only show that the assumption $m\angle 4 \leq m\angle 1$ leads to a contradiction. The argument for $m\angle 4 \leq m\angle 2$ follows the same reasoning.

$m\angle 4 \leq m\angle 1$ means that either $m\angle 4 = m\angle 1$ or $m\angle 4 < m\angle 1$.

Case 1: $m\angle 4 = m\angle 1$

$m\angle 4 = m\angle 1 + m\angle 2$ — **Exterior Angle Theorem**

$m\angle 4 = m\angle 4 + m\angle 2$ — **Substitution**

$0 = m\angle 2$ — **Subtract $m\angle 4$ from each side.**

This contradicts the fact that the measure of an angle is greater than 0, so $m\angle 4 \neq m\angle 1$.

Case 2: $m\angle 4 < m\angle 1$

By the Exterior Angle Theorem, $m\angle 4 = m\angle 1 + m\angle 2$. Since angle measures are positive, the definition of inequality implies that $m\angle 4 > m\angle 1$. This contradicts the assumption that $m\angle 4 < m\angle 1$.

Step 3 In both cases, the assumption leads to the contradiction of a theorem or definition. Therefore, the original conclusion that $m\angle 4 > m\angle 1$ and $m\angle 4 > m\angle 2$ must be true.

StudyTip

Recognizing Contradictions Remember that the contradiction in an indirect proof is not always of the given information or the assumption. It can be of a known fact or definition, such as in Case 1 of Example 5—the measure of an angle must be greater than 0.

Guided Practice

5. **Write an indirect proof.**

Given: $\overline{MO} \cong \overline{ON}$, $\overline{MP} \not\cong \overline{NP}$

Prove: $\angle MOP \not\cong \angle NOP$

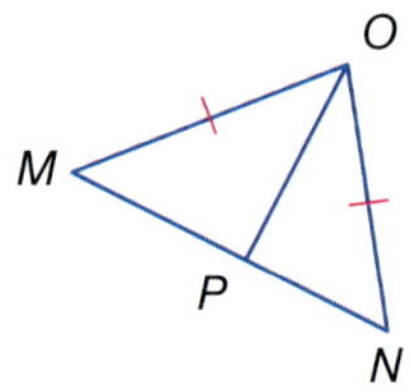

Personal Tutor glencoe.com

Check Your Understanding

Example 1 p. 351

State the assumption you would make to start an indirect proof of each statement.

1. $\overline{AB} \cong \overline{CD}$

2. $\triangle XYZ$ is a scalene triangle.

3. If $4x < 24$, then $x < 6$.

4. $\angle A$ is not a right angle.

Example 2 p. 352

Write an indirect proof of each statement.

5. If $2x + 3 < 7$, then $x < 2$.

6. If $3x - 4 > 8$, then $x > 4$.

Example 3 p. 352

7. **LACROSSE** Christina scored 13 points for her high school lacrosse team during the last six games. Prove that her average points per game was less than 3.

Example 4 p. 353

8. Write an indirect proof to show that if $5x - 2$ is an odd integer, then x is an odd integer.

Example 5 pp. 353–354

Write an indirect proof of each statement.

9. The hypotenuse of a right triangle is the longest side.

10. If two angles are supplementary, then they both cannot be obtuse angles.

Practice and Problem Solving

● = **Step-by-Step Solutions** begin on page R20.
Extra Practice begins on page 969.

Example 1 p. 351

State the assumption you would make to start an indirect proof of each statement.

11. If $2x > 16$, then $x > 8$.

12. $\angle 1$ and $\angle 2$ are not supplementary angles.

13. If two lines have the same slope, the lines are parallel.

14. If the consecutive interior angles formed by two lines and a transversal are supplementary, the lines are parallel.

15. If a triangle is not equilateral, the triangle is not equiangular.

16. An odd number is not divisible by 2.

Example 2 p. 352

Write an indirect proof of each statement.

17. If $2x - 7 > -11$, then $x > -2$.

18. If $5x + 12 < -33$, then $x < -9$.

19. If $-3x + 4 < 7$, then $x > -1$.

20. If $-2x - 6 > 12$, then $x < -9$.

Example 3 p. 352

21. **COMPUTER GAMES** Kwan-Yong bought two computer games for just over \$80 before tax. A few weeks later, his friend asked how much each game cost. Kwan-Yong could not remember the individual prices. Use indirect reasoning to show that at least one of the games cost more than \$40.

22. **FUNDRAISING** Jamila's school is having a Fall Carnival to raise money for a local charity. The cost of an adult ticket to the carnival is \$6 and the cost of a child's ticket is \$2.50. If 375 total tickets were sold and the profit was more than \$1460, prove that at least 150 adult tickets were sold.

Examples 4 and 5 pp. 353–354

Write an indirect proof of each statement.

23. **Given:** xy is an odd integer.
Prove: x and y are both odd integers.

24. **Given:** n^2 is even.
Prove: n^2 is divisible by 4.

25. **Given:** x is an odd number.
Prove: x is not divisible by 4.

26. **Given:** xy is an even integer.
Prove: x or y is an even integer.

27. **Given:** $XZ > YZ$
Prove: $\angle X \not\cong \angle Y$

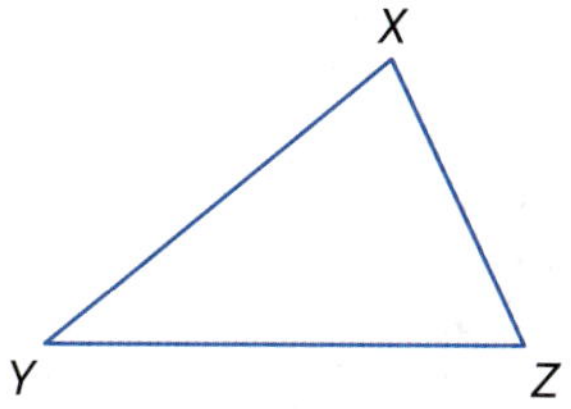

28. **Given:** $\triangle ABC$ is equilateral.
Prove: $\triangle ABC$ is equiangular.

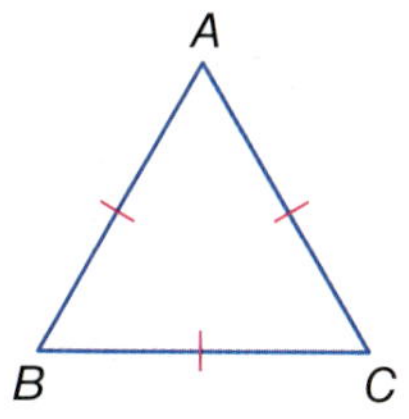

29. In an isosceles triangle neither of the base angles can be a right angle.

30. A triangle can have only one right angle.

31. Write an indirect proof for Theorem 5.10.

32. Write an indirect proof to show that if $\frac{1}{b} < 0$, then b is negative.

Real-World Link

In basketball, there are three possible ways to score three points in a single possession. A player can make a basket from behind the three-point line, a player may be fouled while scoring a two-point shot and be allowed to shoot one free throw, or a player may be fouled behind the three-point line and be allowed to shoot three free throws.

33. BASKETBALL When Katsu left to get in the concession line before the end of the first half of a basketball game, her school's team was winning 28 to 26. When she returned during halftime, the opposing team's score was 29 and Katsu's team's score was 28. Katsu concluded that a player on the visiting team had made a three-point basket. Prove or disprove her assumption using an indirect proof and the information at the left.

34. GAMES A computer game involves a knight on a quest for treasure. At the end of the journey, the knight approaches the two doors shown below.

A servant tells the knight that one of the signs is true and the other is false. Use indirect reasoning to determine which door the knight should choose. Explain your reasoning.

35. SURVEYS Luisa's local library conducted an online poll of teens to find out what activities teens participate in to preserve the environment. The results of the poll are shown in the graph below.

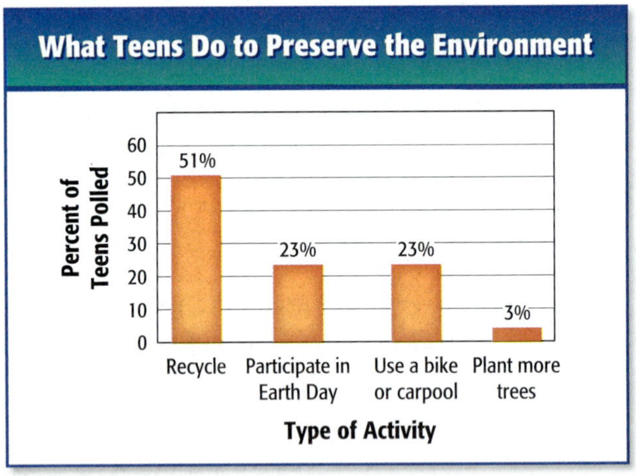

a. Prove the following statement. *More than half of teens polled said that they recycle to preserve the environment.*

b. If 400 teenagers were polled, verify that 92 of them said that they participate in Earth Day.

36. CARS James, Hector, and Mandy all have different color cars. Only one of the statements below is true. Use indirect reasoning to determine which statement is true. Explain.

(1) James has a red car.

(2) Hector does not have a red car.

(3) Mandy does not have a blue car.

Determine whether each statement about the shortest distance between a point and a line or plane can be proved using a direct or indirect proof. Then write a proof of each statement.

37. **Given:** $\overline{AB} \perp$ line p

Prove: $\overline{AB}$ is the shortest segment from A to line p.

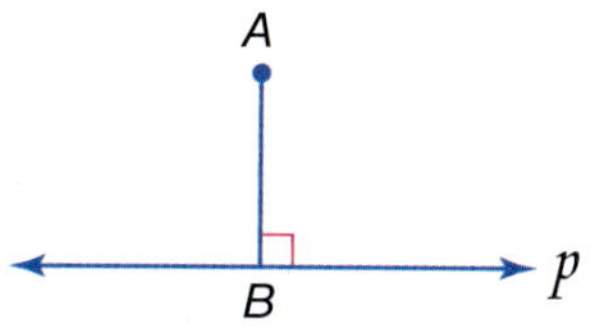

38. **Given:** $\overline{PQ} \perp$ plane $\mathcal{M}$

Prove: $\overline{PQ}$ is the shortest segment from P to plane $\mathcal{M}$.

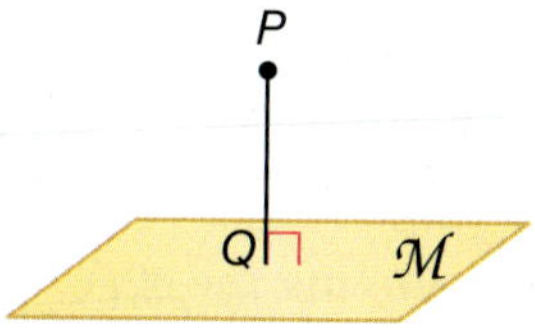

39. **NUMBER THEORY** In this problem, you will make and prove a conjecture about a number theory relationship.

a. Write an expression for *the sum of the cube of a number and three.*

b. Create a table that includes the value of the expression for 10 different values of n. Include both odd and even values of n.

c. Write a conjecture about n when the value of the expression is even.

d. Write an indirect proof of your conjecture.

H.O.T. Problems Use Higher-Order Thinking Skills

40. **WRITING IN MATH** Explain the procedure for writing an indirect proof.

41. **OPEN ENDED** Write a statement that can be proven using indirect proof. Include the indirect proof of your statement.

42. **CHALLENGE** If x is a rational number, then it can be represented by the quotient $\frac{a}{b}$ for some integers a and b, if $b \neq 0$. An irrational number cannot be represented by the quotient of two integers. Write an indirect proof to show that the product of a nonzero rational number and an irrational number is an irrational number.

Review Vocabulary

integer the set of numbers $\{\ldots, -2, -1, 0, 1, 2, \ldots\}$

43. **ERROR ANALYSIS** Amber and Raquel are trying to verify the following statement using indirect proof. Is either of them correct? Explain your reasoning.

If the sum of two numbers is even, then the numbers are even.

Amber

The statement is true. If one of the numbers is even and the other number is zero, then the sum is even. Since the hypothesis is true even when the conclusion is false, the statement is true.

Raquel

The statement is true. If the two numbers are odd, then the sum is even. Since the hypothesis is true when the conclusion is false, the statement is true.

44. **WRITING IN MATH** Refer to Exercise 8. Write the contrapositive of the statement and write a direct proof of the contrapositive. How are the direct proof of the contrapositive of the statement and the indirect proof of the statement related?

Standardized Test Practice

45. SHORT RESPONSE Write an equation in slope-intercept form to describe the line that passes through the point (5, 3) and is parallel to the line represented by the equation $-2x + y = -4$.

46. Statement: If $\angle A \cong \angle B$ and $\angle A$ is supplementary to $\angle C$, then $\angle B$ is supplementary to $\angle C$.

Dia is proving the statement above by contradiction. She began by assuming that $\angle B$ is not supplementary to $\angle C$. Which of the following definitions will Dia use to reach a contradiction?

A definition of congruence
B definition of a linear pair
C definition of a right angle
D definition of supplementary angles

47. List the angles of $\triangle MNO$ in order from smallest to largest if $MN = 9$, $NO = 7.5$, and $OM = 12$.

F $\angle N, \angle O, \angle M$
G $\angle O, \angle M, \angle N$
H $\angle O, \angle N, \angle M$
J $\angle M, \angle O, \angle N$

48. SAT/ACT If $b > a$, which of the following must be true?

A $-a > -b$
B $3a > b$
C $a^2 < b^2$
D $a^2 < ab$

Spiral Review

49. PROOF Write a two-column proof. (Lesson 5-3)

Given: $\overline{RQ}$ bisects $\angle SRT$.

Prove: $m\angle SQR > m\angle SRQ$

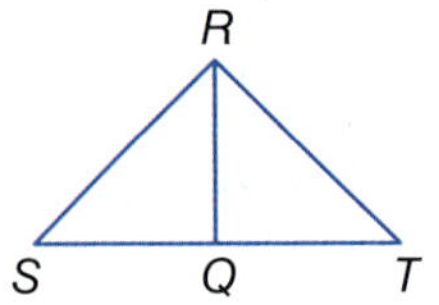

COORDINATE GEOMETRY **Find the coordinates of the circumcenter of each triangle with the given vertices.** (Lesson 5-1)

50. $D(-3, 3)$, $E(3, 2)$, $F(1, -4)$

51. $A(4, 0)$, $B(-2, 4)$, $C(0, 6)$

Find each measure. (Lesson 4-2)

52. $m\angle 1$

53. $m\angle 4$

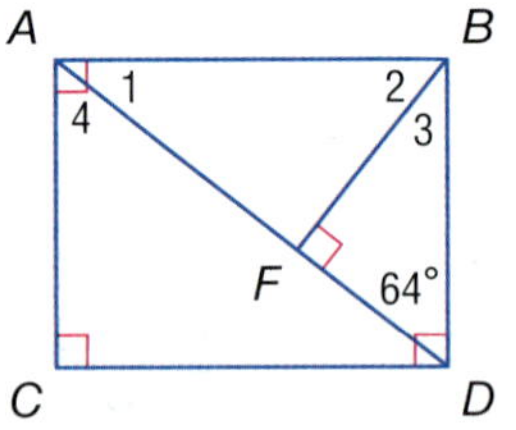

COORDINATE GEOMETRY **Find the distance between each pair of parallel lines with the given equations.** (Lesson 3-6)

54. $x + 3y = 6$
$x + 3y = -14$

55. $y = 2x + 2$
$y = 2x - 3$

56. RECYCLING Refer to the Venn diagram that represents the number of neighborhoods in a city with a curbside recycling program for paper or aluminum. (Lesson 2-2)

a. How many neighborhoods recycle aluminum?

b. How many neighborhoods recycle paper or aluminum or both?

c. How many neighborhoods recycle paper and aluminum?

Curbside Recycling

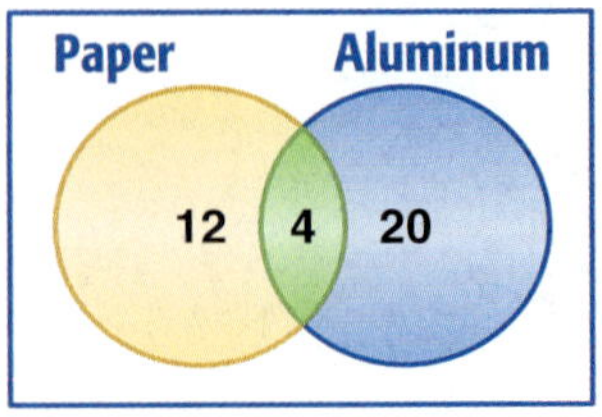

Skills Review

Determine whether each inequality is *true* or *false*.

57. $23 - 11 > 9$

58. $41 - 19 < 21$

59. $57 + 68 < 115$

EXPLORE 5-5

Graphing Technology Lab

The Triangle Inequality

Math Online glencoe.com
- Other Calculator Keystrokes
- Graphing Technology Personal Tutor

Objective
Use technology to investigate triangle inequalities.

You can use the Cabri™ Jr. application on a TI-83/84 Plus graphing calculator to discover properties of triangles.

ACTIVITY 1

Construct a triangle. Observe the relationship between the sum of the lengths of two sides and the length of the other side.

Step 1 Construct a triangle using the triangle tool on the F2 menu. Then use the Alph-Num tool on the F5 menu to label the vertices as *A*, *B*, and *C*.

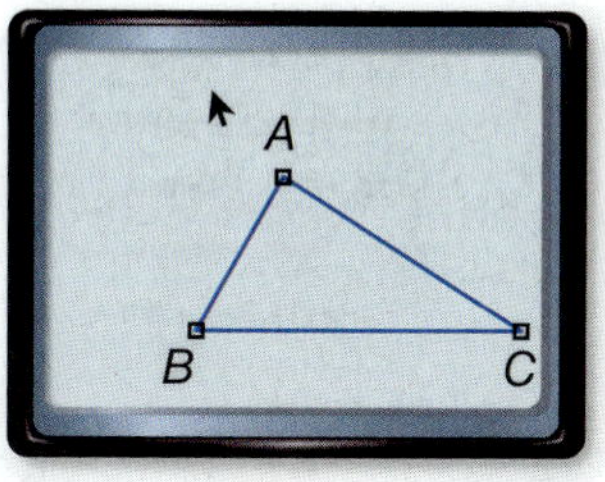

Step 1

Step 2 Access the distance & length tool, shown as D. & Length, under Measure on the F5 menu. Use the tool to measure each side of the triangle.

Step 3 Display $AB + BC$, $AB + CA$, and $BC + CA$ by using the Calculate tool on the F5 menu. Label the measures.

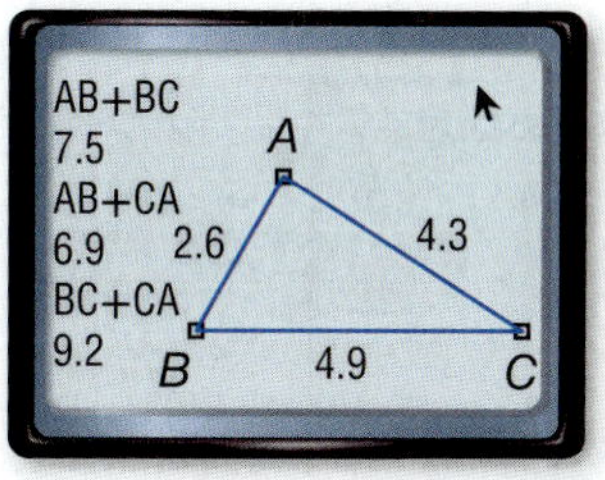

Steps 2 and 3

Step 4 Click and drag the vertices to change the shape of the triangle.

Analyze the Results

1. Replace each ● with <, >, or = to make a true statement.
   $AB + BC$ ● CA $AB + CA$ ● BC $BC + CA$ ● AB
2. Click and drag the vertices to change the shape of the triangle. Then review your answers to Exercise 1. What do you observe?
3. Click on point *A* and drag it to lie on line *BC*. What do you observe about *AB*, *BC*, and *CA*? Are *A*, *B*, and *C* the vertices of a triangle? Explain.
4. **Make a conjecture** about the sum of the lengths of two sides of a triangle and the length of the third side.
5. Replace each ● with <, >, or = to make a true statement.
   $|AB - BC|$ ● CA $|AB - CA|$ ● BC $|BC - CA|$ ● AB

   Then click and drag the vertices to change the shape of the triangle and review your answers. What do you observe?
6. How could you use your observations to determine the possible lengths of the third side of a triangle if you are given the lengths of the other two sides?

5-5 The Triangle Inequality

Why?

On a home improvement show, a designer wants to use scrap pieces of cording from another sewing project to decorate the triangular throw pillows that she and the homeowner have made. To minimize waste, she wants to use the scraps without cutting them. She selects three scraps at random and tries to form a triangle. Two such attempts are shown below.

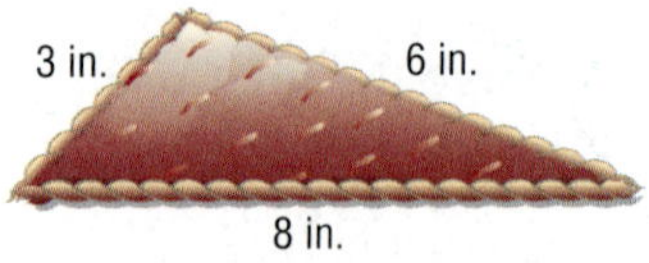

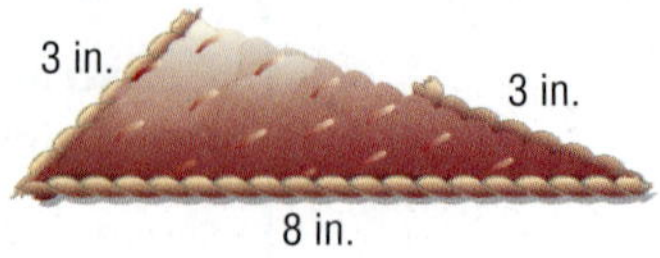

Then
You recognized and applied properties of inequalities to the relationships between the angles and sides of a triangle. (Lesson 5-3)

Now
- Use the Triangle Inequality Theorem to identify possible triangles.
- Prove triangle relationships using the Triangle Inequality Theorem.

Math Online
glencoe.com

The Triangle Inequality While a triangle is formed by three segments, a special relationship must exist among the lengths of the segments in order for them to form a triangle.

Theorem 5.11 Triangle Inequality Theorem

The sum of the lengths of any two sides of a triangle must be greater than the length of the third side.

Examples $PQ + QR > PR$
$QR + PR > PQ$
$PR + PQ > QR$

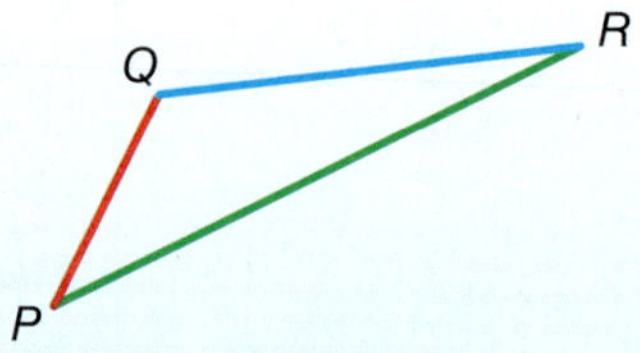

You will prove Theorem 5.11 in Exercise 23.

To show that it is not possible to form a triangle with three side lengths, you need only show that one of the three triangle inequalities is not true.

EXAMPLE 1 Identify Possible Triangles Given Side Lengths

Is it possible to form a triangle with the given side lengths? If not, explain why not.

a. **8 in., 15 in., 17 in.**

Check each inequality.

$8 + 15 \overset{?}{>} 17$
$23 > 17$ ✓

$8 + 17 \overset{?}{>} 15$
$25 > 15$ ✓

$15 + 17 \overset{?}{>} 8$
$32 > 8$ ✓

Since the sum of each pair of side lengths is greater than the third side length, sides with lengths 8, 15, and 17 inches will form a triangle.

b. **6 m, 8 m, 14 m**

$6 + 8 \overset{?}{>} 14$
$14 \not> 14$ ✗

Since the sum of one pair of side lengths is not greater than the third side length, sides with lengths 6, 8, and 14 meters will not form a triangle.

Guided Practice

1A. 15 yd, 16 yd, 30 yd

1B. 2 ft, 8 ft, 11 ft

Personal Tutor glencoe.com

When the lengths of two sides of a triangle are known, the third side can be any length in a range of values. You can use the Triangle Inequality Theorem to determine the range of possible lengths for the third side.

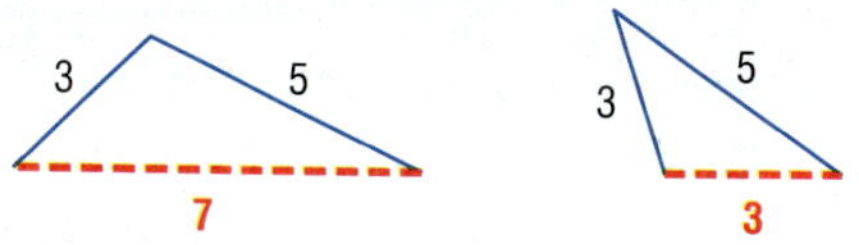

STANDARDIZED TEST EXAMPLE 2

If the measures of two sides of a triangle are 3 feet and 7 feet, which is the *least* possible whole number measure for the third side?

A 3 ft

B 4 ft

C 5 ft

D 10 ft

Test-Taking Tip

Testing Choices If you are short on time, you can test each choice to find the correct answer and eliminate any remaining choices.

Read the Test Item

You need to determine which value is the least possible measure for the third side of a triangle with sides that measure 3 feet and 7 feet.

Solve the Test Item

To determine the least possible measure from the choices given, first determine the range of possible measures for the third side.

Draw a diagram and let x represent the length of the third side.

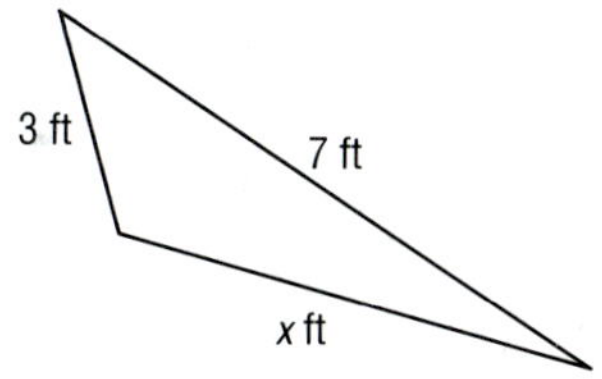

Next, set up and solve each of the three triangle inequalities.

$3 + 7 > x$	$3 + x > 7$	$x + 7 > 3$
$10 > x$ or $x < 10$	$x > 4$	$x > -4$

Reading Math

Multiple Inequality Symbols The compound inequality $4 < x < 10$ is read *x is between 4 and 10.*

Notice that $x > -4$ is always true for any whole number measure for x. Combining the two remaining inequalities, the range of values that fit both inequalities is $x > 4$ and $x < 10$, which can be written as $4 < x < 10$.

The least whole number value between 4 and 10 is 5. So the correct answer is choice C.

Guided Practice

2. Which of the following could *not* be the value of n?

F 7 **H** 10

G 13 **J** 22

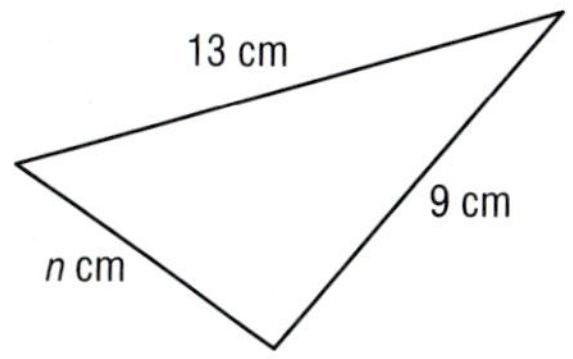

Personal Tutor glencoe.com

Proofs Using the Triangle Inequality Theorem You can use the Triangle Inequality Theorem as a reason in proofs.

Real-World EXAMPLE 3 Proof Using Triangle Inequality Theorem

Real-World Link

A direct flight is not the same as a nonstop flight. For a direct flight, passengers do not change planes, but the plane may make one or more stops before continuing to its final destination.

TRAVEL The distance from Colorado Springs, Colorado, to Abilene, Texas, is the same as the distance from Colorado Springs to Tulsa, Oklahoma. Prove that a direct flight from Colorado Springs to Tulsa is a greater distance than a nonstop flight from Colorado Springs to Abilene.

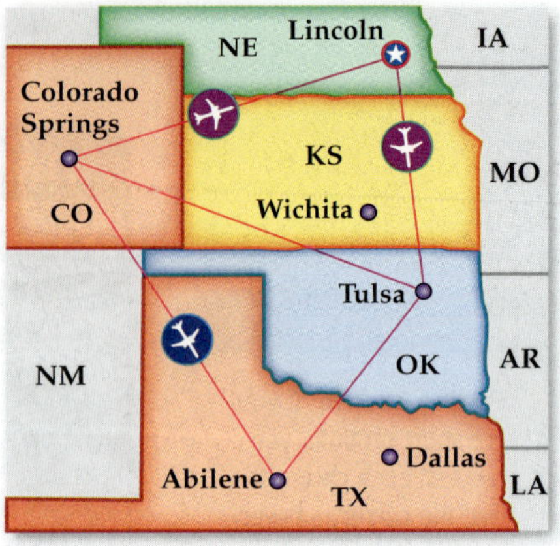

Draw a simpler diagram of the situation and label the diagram. Draw in side $\overline{LT}$ to form $\triangle CTL$.

Given: $CA = CT$

Prove: $CL + LT > CA$

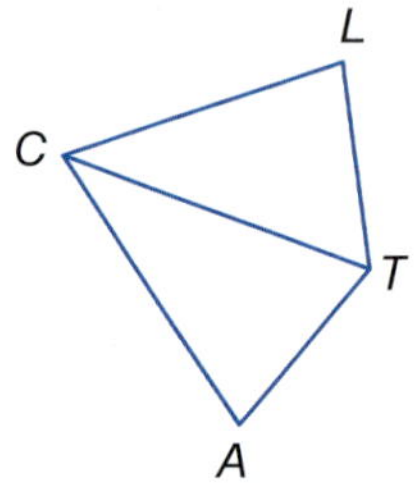

Proof:

Statements	Reasons
1. $CA = CT$	**1.** Given
2. $CL + LT > CT$	**2.** Triangle Inequality Theorem
3. $CL + LT > CA$	**3.** Substitution

Guided Practice

3. Write a two-column proof.

Given: $GL = LK$

Prove: $JH + GH > JK$

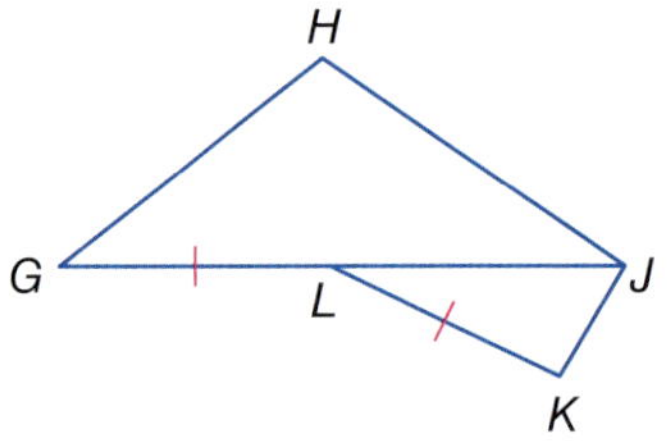

Personal Tutor glencoe.com

Check Your Understanding

Example 1 p. 360

Is it possible to form a triangle with the given side lengths? If not, explain why not.

1. 5 cm, 7 cm, 10 cm **2.** 3 in., 4 in., 8 in. **3.** 6 m, 14 m, 10 m

Example 2 p. 361

4. MULTIPLE CHOICE If the measures of two sides of a triangle are 5 yards and 9 yards, what is the least possible measure of the third side if the measure is an integer?

A 5 yd **B** 4 yd **C** 14 yd **D** 6 yd

Example 3 p. 362

5. PROOF Write a two-column proof.

Given: $\overline{XW} \cong \overline{YW}$

Prove: $YZ + ZW > XW$

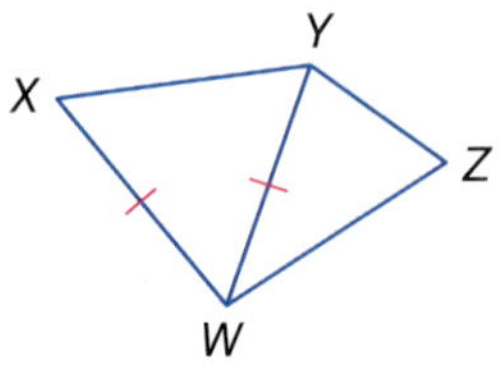

Practice and Problem Solving

= **Step-by-Step Solutions** begin on page R20.
Extra Practice begins on page 969.

Example 1 p. 360

Is it possible to form a triangle with the given side lengths? If not, explain why not.

6. 4 ft, 9 ft, 15 ft

7. 11 mm, 21 mm, 16 mm

8. 9.9 cm, 1.1 cm, 8.2 cm

9. 2.1 in., 4.2 in., 7.9 in.

10. $2\frac{1}{2}$ m, $1\frac{3}{4}$ m, $5\frac{1}{8}$ m

11. $1\frac{1}{5}$ km, $4\frac{1}{2}$ km, $3\frac{3}{4}$ km

Example 2 p. 361

Find the range for the measure of the third side of a triangle given the measures of two sides.

12. 4 ft, 8 ft

13. 5 m, 11 m

14. 2.7 cm, 4.2 cm

15. 3.8 in., 9.2 in.

16. $\frac{1}{2}$ km, $3\frac{1}{4}$ km

17. $2\frac{1}{3}$ yd, $7\frac{2}{3}$ yd

Example 3 p. 362

PROOF **Write a two-column proof.**

18. Given: $\angle BCD \cong \angle CDB$
Prove: $AB + AD > BC$

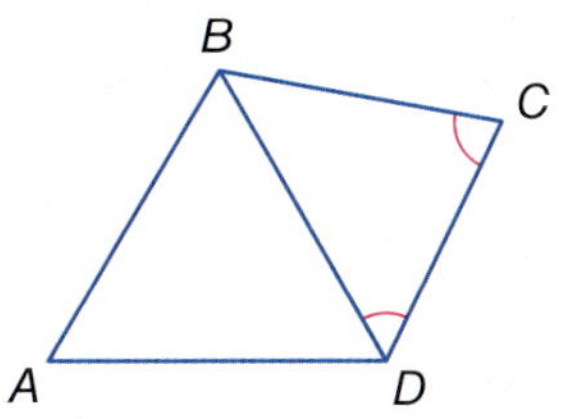

19. Given: $\overline{JL} \cong \overline{LM}$
Prove: $KJ + KL > LM$

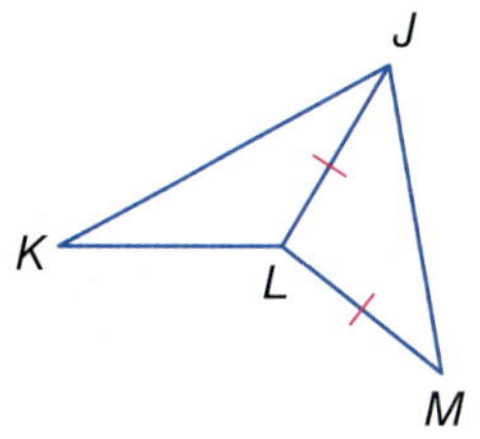

ALGEBRA **Determine the possible values of x.**

20.

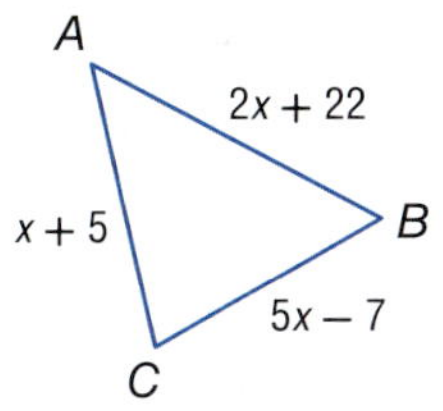

21.

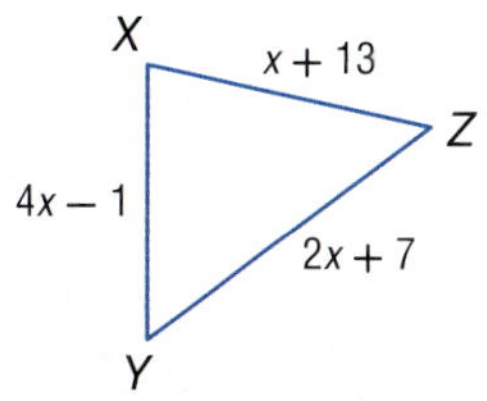

22. DRIVING Takoda wants to take the most efficient route from his house to a soccer tournament at The Sportsplex. He can take County Line Road or he can take Highway 4 and then Route 6 to the get to The Sportsplex.

a. Which of the two possible routes is the shortest? Explain your reasoning.

b. Suppose Takoda always drives below the speed limit. If the speed limit on County Line Road is 30 miles per hour and on both Highway 4 and Route 6 it is 55 miles per hour, which route will be faster? Explain.

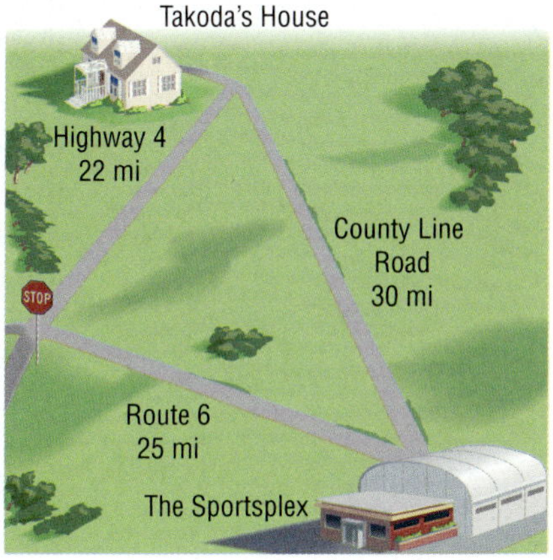

23. PROOF **Write a two-column proof.**
Given: $\triangle ABC$
Prove: $AC + BC > AB$ (Triangle Inequality Theorem)

(*Hint:* Draw auxiliary segment $\overline{CD}$, so that C is between B and D and $\overline{CD} \cong \overline{AC}$.)

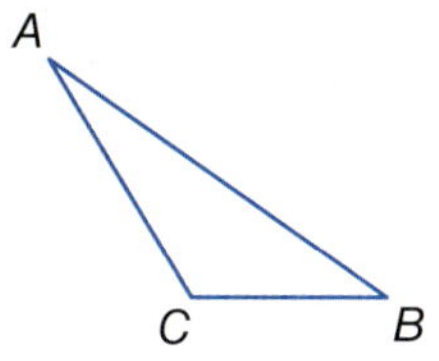

24. **SCHOOL** When Toya goes from science class to math class, she usually stops at her locker. The distance from her science classroom to her locker is 90 feet, and the distance from her locker to her math classroom is 110 feet. What are the possible distances from science class to math class if she takes the hallway that goes directly between the two classrooms?

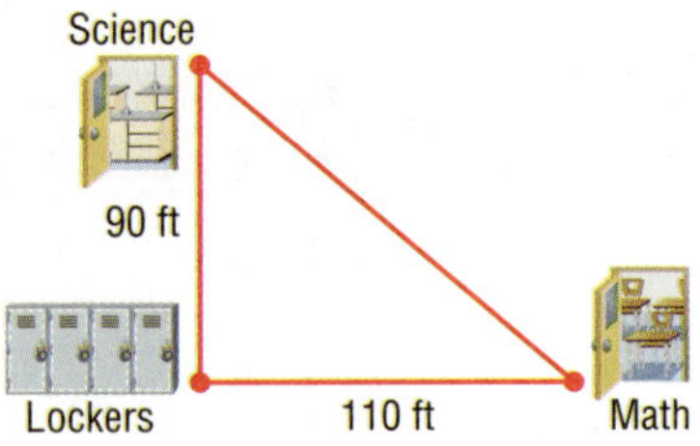

Find the range of possible measures of x if each set of expressions represents measures of the sides of a triangle.

25. $x, 4, 6$

26. $8, x, 12$

27. $x + 1, 5, 7$

28. $x - 2, 10, 12$

29. $x + 2, x + 4, x + 6$

30. $x, 2x + 1, x + 4$

Real-World Link

In addition to simple ramps, modern theaters use lifts to move actors and scenery above or below the stage floor.

Source: *Encyclopaedia Britannica*

31. **DRAMA CLUB** Anthony and Catherine are working on a ramp up to the stage for the drama club's next production. Anthony's sketch of the ramp is shown below. Catherine is concerned about the measurements and thinks they should recheck the measures before they start cutting the wood. Is Catherine's concern valid? Explain your reasoning.

32. **BIKING** Aisha is riding her bike to the park and can take one of two routes. The most direct route from her house is to take Main Street, but it is safer to take Route 3 and then turn right on Clay Road as shown. What is the maximum distance she will travel out of her way if she takes Route 3 to Clay Road?

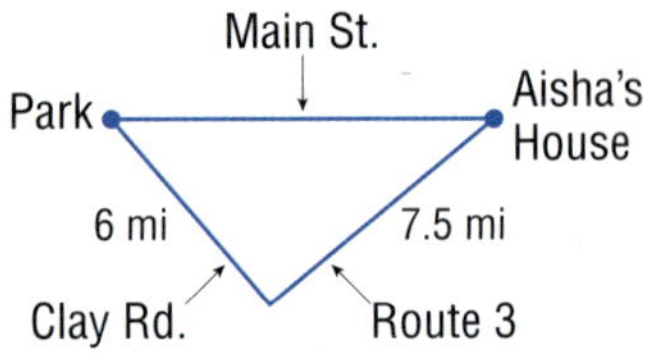

33. **DESIGN** Carlota designed an awning that she and her friends could take to the beach. Carlota decides to cover the top of the awning with material that will drape 6 inches over the front. What is the minimum length of material she should buy to use with her design? Assume that the width of the material is sufficient to cover the awning.

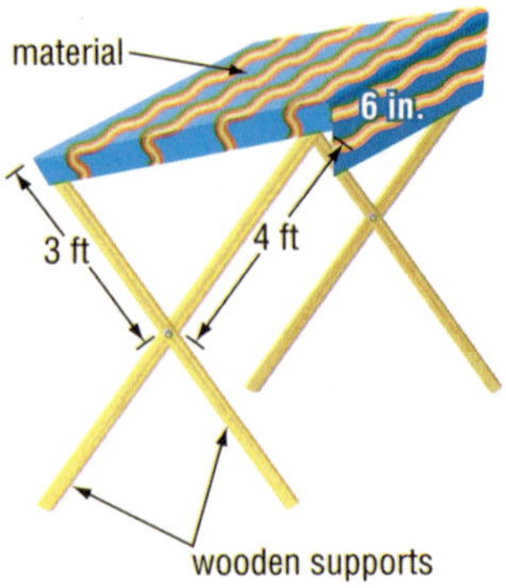

ESTIMATION **Without using a calculator, determine if it is possible to form a triangle with the given side lengths. Explain.**

34. $\sqrt{8}$ ft, $\sqrt{2}$ ft, $\sqrt{35}$ ft

35. $\sqrt{99}$ yd, $\sqrt{48}$ yd, $\sqrt{65}$ yd

36. $\sqrt{3}$ m, $\sqrt{15}$ m, $\sqrt{24}$ m

37. $\sqrt{122}$ in., $\sqrt{5}$ in., $\sqrt{26}$ in.

ALGEBRA **Determine whether the given coordinates are the vertices of a triangle. Explain.**

38. $X(1, -3), Y(6, 1), Z(2, 2)$

39 $F(-4, 3), G(3, -3), H(4, 6)$

40. $J(-7, -1), K(9, -5), L(21, -8)$

41. $Q(2, 6), R(6, 5), S(2, 6)$

42. **MULTIPLE REPRESENTATIONS** In this problem, you will use inequalities to make comparisons between the sides and angles of two triangles.

a. GEOMETRIC Draw three pairs of triangles that have two pairs of congruent sides and one pair of sides that is not congruent. Mark each pair of congruent sides. Label each triangle pair ABC and DEF, where $\overline{AB} \cong \overline{DE}$ and $\overline{AC} \cong \overline{DF}$.

b. TABULAR Copy the table below. Measure and record the values of BC, $m\angle A$, EF, and $m\angle D$ for each triangle pair.

Triangle Pair	BC	$m\angle A$	EF	$m\angle D$
1				
2				
3				

c. VERBAL Make a conjecture about the relationship between the angles opposite the noncongruent sides of a pair of triangles that have two pairs of congruent legs.

H.O.T. Problems Use Higher-Order Thinking Skills

43. **CHALLENGE** What is the range of possible perimeters for figure $ABCDE$ if $AC = 7$ and $DC = 9$? Explain your reasoning.

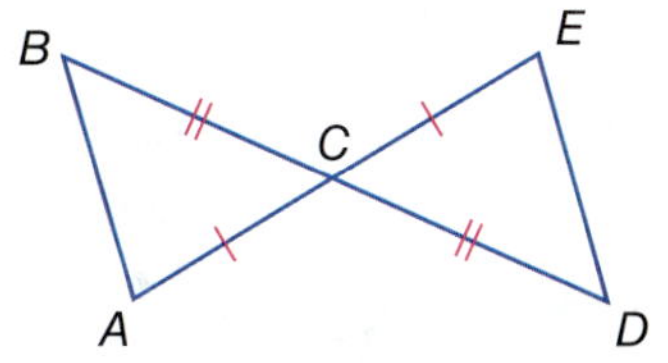

44. **REASONING** What is the range of lengths of each leg of an isosceles triangle if the measure of the base is 6 inches? Explain.

45. **WRITING IN MATH** Explain the process you use to find the minimum and maximum values of a side of a triangle if you are given the measures of the other two sides.

StudyTip

Probability Recall from Lesson 0-3 that the probability of an event is the ratio of the number of favorable outcomes to the number of possible outcomes.

46. **CHALLENGE** The sides of an isosceles triangle are whole numbers, and its perimeter is 30 units. What is the probability that the triangle is equilateral?

47. **OPEN ENDED** The length of one side of a triangle is 2 inches. Draw a triangle in which the 2-inch side is the shortest side and one in which the 2-inch side is the longest side. Include side and angle measures on your drawing.

48. **WRITING IN MATH** Suppose your house is $\frac{3}{4}$ mile from a park and the park is 1.5 miles from a shopping center.

a. If your house, the park, and the shopping center are noncollinear, what do you know about the distance from your house to the shopping center? Explain your reasoning.

b. If the three locations are collinear, what do you know about the distance from your house to the shopping center? Explain your reasoning.

Standardized Test Practice

49. If $\overline{DC}$ is a median of $\triangle ABC$ and $m\angle 1 > m\angle 2$, which of the following statements is *not* true?

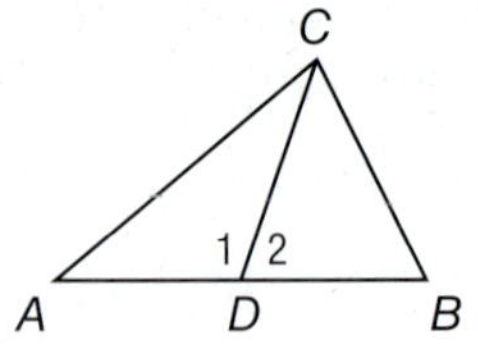

A $AD = BD$

B $m\angle ADC = m\angle BDC$

C $AC > BC$

D $m\angle 1 > m\angle B$

50. SHORT RESPONSE A high school soccer team has a goal of winning at least 75% of their 15 games this season. In the first three weeks, the team has won 5 games. How many more games must the team win to meet their goal?

51. Which of the following is a logical conclusion based on the statement and its converse below?

Statement: If a polygon is a rectangle, then it has four sides.

Converse: If a polygon has four sides, then it is a rectangle.

F The statement and its converse are both true.

G The statement and its converse are both false.

H The statement is true; the converse is false.

J The statement is false; the converse is true.

52. SAT/ACT When 7 is subtracted from $14w$, the result is z. Which of the following equations represents this statement?

A $7 - 14w = z$

B $z = 14w + 7$

C $7 - z = 14w$

D $z = 14w - 7$

Spiral Review

State the assumption you would make to start an indirect proof of each statement. (Lesson 5-4)

53. If $4y + 17 = 41$, then $y = 6$.

54. If two lines are cut by a transversal and a pair of alternate interior angles are congruent, then the two lines are parallel.

55. GEOGRAPHY The distance between San Jose, California, and Las Vegas, Nevada, is about 375 miles. The distance from Las Vegas to Carlsbad, California, is about 243 miles. Use the Triangle Inequality Theorem to find the possible distance between San Jose and Carlsbad. (Lesson 5-3)

Find x so that $m \parallel n$. Identify the postulate or theorem you used. (Lesson 3-5)

56.

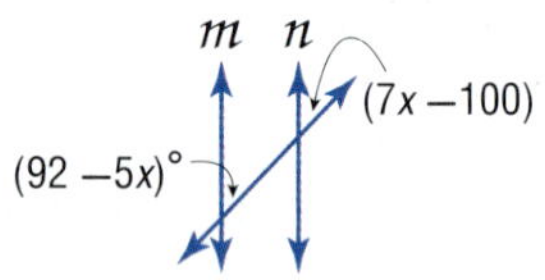

57.

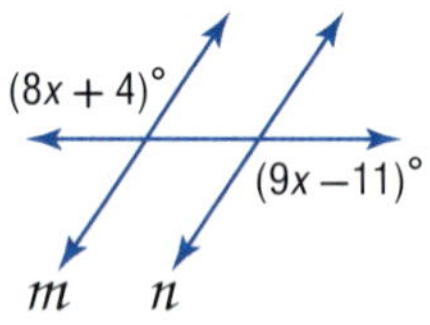

58.

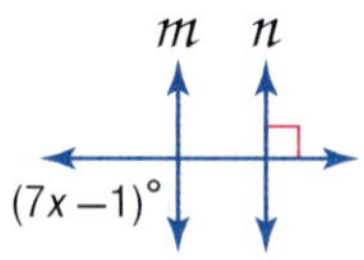

ALGEBRA **Find x and JK if J is between K and L.** (Lesson 1-2)

59. $KJ = 3x$, $JL = 6x$, and $KL = 12$

60. $KJ = 3x - 6$, $JL = x + 6$, and $KL = 24$

Skills Review

Find x and the measures of the unknown sides of each triangle. (Lesson 4-6)

61.

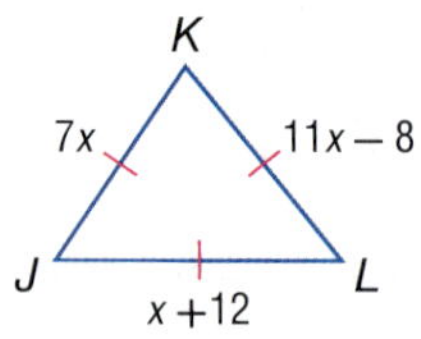

62.

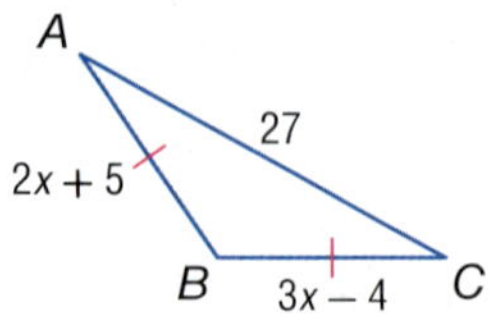

63.

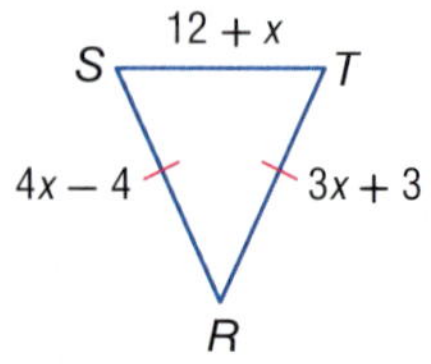

5-6 Inequalities in Two Triangles

Why?

A car jack is used to lift a car. The jack shown below is one of the simplest still in use today. Notice that as the jack is lowered, the legs of isosceles $\triangle ABC$ remain congruent, but the included angle A widens and $\overline{BC}$, the side opposite $\angle A$, lengthens.

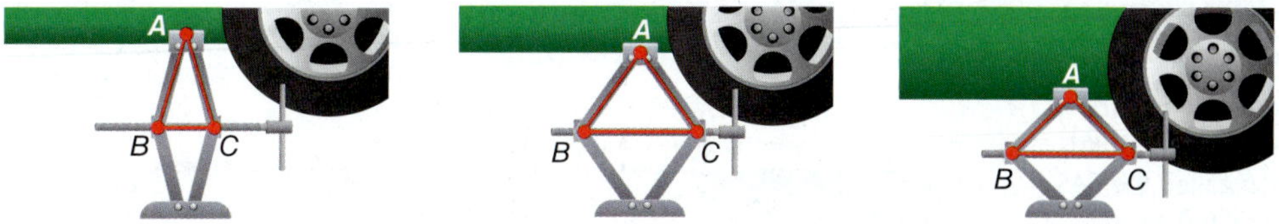

Hinge Theorem The observation in the example above is true of any type of triangle and illustrates the following theorems.

Theorems — Inequalities in Two Triangles

5.13 Hinge Theorem
If two sides of a triangle are congruent to two sides of another triangle, and the included angle of the first is larger than the included angle of the second triangle, then the third side of the first triangle is longer than the third side of the second triangle.

Example If $\overline{AB} \cong \overline{FG}$, $\overline{AC} \cong \overline{FH}$, and $m\angle A > m\angle F$, then $BC > GH$.

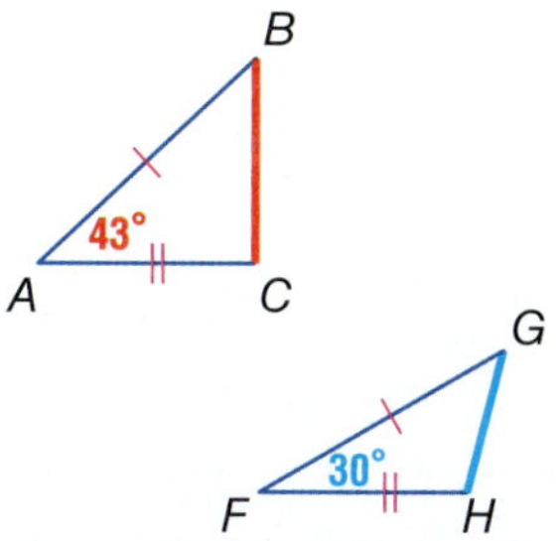

5.14 Converse of the Hinge Theorem
If two sides of a triangle are congruent to two sides of another triangle, and the third side in the first is longer than the third side in the second triangle, then the included angle measure of the first triangle is greater than the included angle measure in the second triangle.

Example If $\overline{JL} \cong \overline{PR}$, $\overline{KL} \cong \overline{QR}$, and $PQ > JK$, then $m\angle R > m\angle L$.

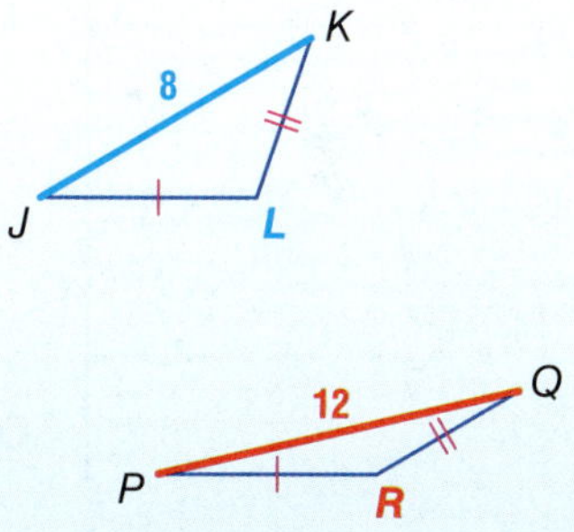

The proof of Theorem 5.13 is on p. 368. You will prove Theorem 5.14 in Exercise 28.

EXAMPLE 1 Use the Hinge Theorem and its Converse

Compare the given measures.

a. WX and XY

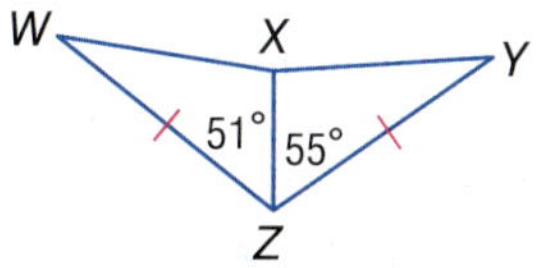

In $\triangle WXZ$ and $\triangle YXZ$, $\overline{WZ} \cong \overline{YZ}$, $\overline{XZ} \cong \overline{XZ}$, and $\angle YZX > \angle WZX$. By the Hinge Theorem, $m\angle WZX < m\angle YZX$, so $WX < XY$.

b. $m\angle FCD$ and $m\angle BFC$

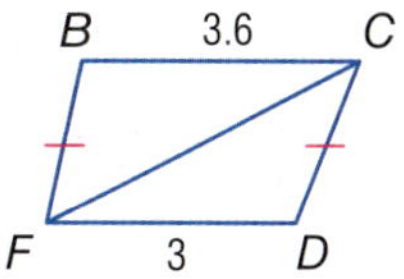

In $\triangle BCF$ and $\triangle DFC$, $\overline{BF} \cong \overline{DC}$, $\overline{FC} \cong \overline{CF}$, and $BC > FD$. By the Converse of the Hinge Theorem, $\angle BFC > \angle DCF$.

Then
You used inequalities to make comparisons in one triangle. (Lesson 5-3)

Now
- Apply the Hinge Theorem or its converse to make comparisons in two triangles.
- Prove triangle relationships using the Hinge Theorem or its converse.

Guided Practice

Compare the given measures.

1A. JK and MQ

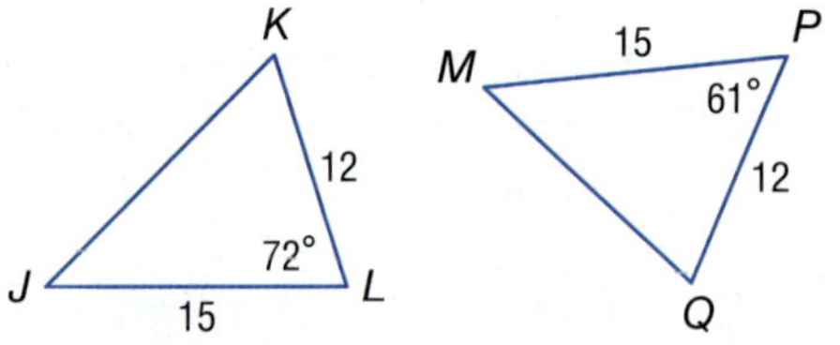

1B. $m\angle SRT$ and $m\angle VRT$

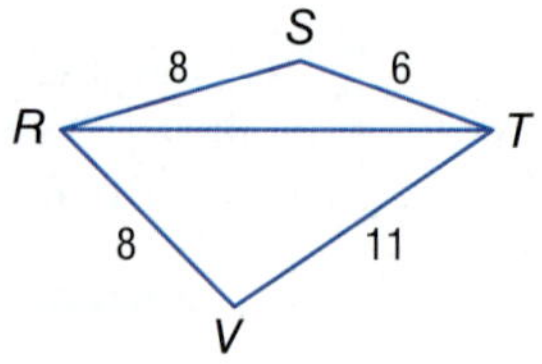

Personal Tutor glencoe.com

StudyTip

SAS and SSS Inequality Theorem The Hinge Theorem is also called the SAS Inequality Theorem. The Converse of the Hinge Theorem is also called the SSS Inequality Theorem.

PROOF Hinge Theorem

Given: $\triangle ABC$ and $\triangle DEF$
$\overline{AC} \cong \overline{DF}$, $\overline{BC} \cong \overline{EF}$
$m\angle F > m\angle C$

Prove: $DE > AB$

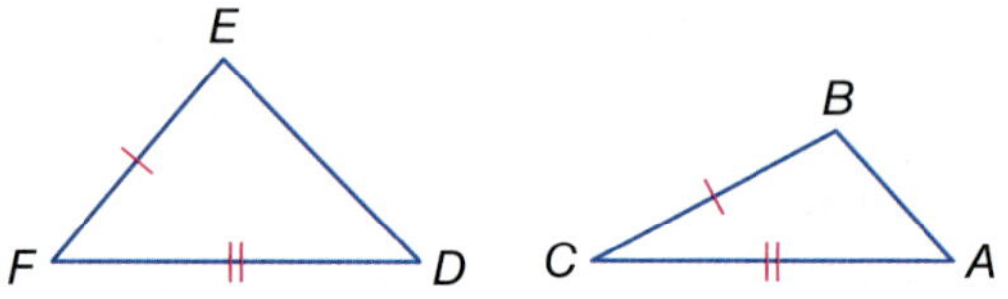

Proof:

We are given that $\overline{AC} \cong \overline{DF}$ and $\overline{BC} \cong \overline{EF}$. We also know that $m\angle F > m\angle C$.

Draw auxiliary ray FP such that $m\angle DFP = m\angle C$ and that $\overline{PF} \cong \overline{BC}$. This leads to two cases.

Case 1 P lies on $\overline{DE}$.

Then $\triangle FPD \cong \triangle CBA$ by SAS. Thus, $PD = BA$ by CPCTC and the definition of congruent segments.

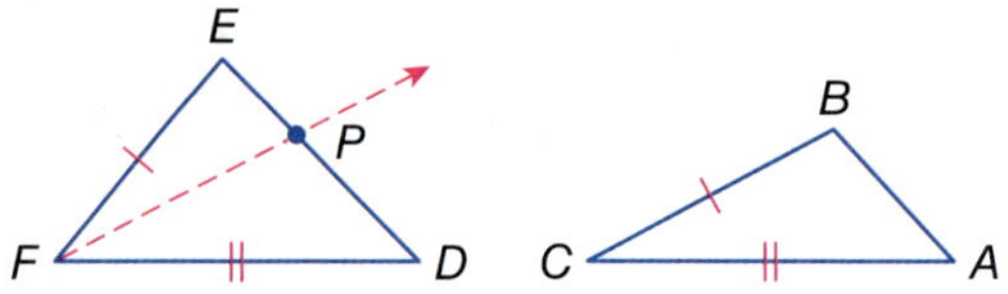

By the Segment Addition Postulate, $DE = EP + PD$. Also, $DE > PD$ by the definition of inequality. Therefore, $DE > AB$ by substitution.

Case 2 P does not lie on $\overline{DE}$.

Then let the intersection of $\overline{FP}$ and $\overline{ED}$ be point T, and draw another auxiliary segment $\overline{FQ}$ such that Q is on $\overline{DE}$ and $\angle EFQ \cong \angle QFP$. Then draw auxiliary segments $\overline{PD}$ and $\overline{PQ}$.

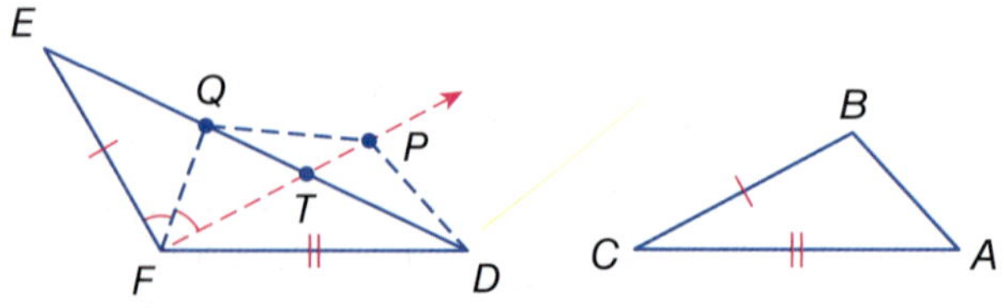

Since $\overline{FP} \cong \overline{BC}$ and $\overline{BC} \cong \overline{EF}$, we have $\overline{FP} \cong \overline{EF}$ by the Transitive Property. Also $\overline{QF}$ is congruent to itself by the Reflexive Property. Thus, $\triangle EFQ \cong \triangle PFQ$ by SAS. By CPCTC, $\overline{EQ} \cong \overline{PQ}$ or $EQ = PQ$. Also, $\triangle FPD \cong \triangle CBA$ by SAS. So, $\overline{PD} \cong \overline{BA}$ by CPCTC and $PD = BA$.

In $\triangle QPD$, $QD + PQ > PD$ by the Triangle Inequality Theorem. By substitution, $QD + EQ > PD$. Since $ED = QD + EQ$ by the Segment Addition Postulate, $ED > PD$. Using substitution, $ED > BA$ or $DE > AB$.

Real-World Link

There are over 225,000 miles of groomed and marked snowmobile trails in North America.

Source: International Snowmobile Manufacturers Association

Problem-SolvingTip

Draw a Diagram Draw a diagram to help you see and correctly interpret a problem that has been described in words.

You can use the Hinge Theorem to solve real-world problems.

Real-World EXAMPLE 2 Use the Hinge Theorem

SNOWMOBILING Two groups of snowmobilers leave from the same base camp. Group A goes 7.5 miles due west and then turns 35° north of west and goes 5 miles. Group B goes 7.5 miles due east and then turns 40° north of east and goes 5 miles. At this point, which group is farther from the base camp? Explain your reasoning.

Understand Using the sets of directions given in the problem, you need to determine which snowmobile group is farther from the base camp. A turn of 35° north of west is correctly interpreted as shown.

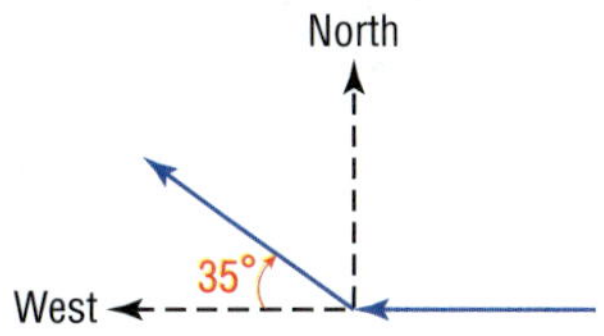

Plan Draw a diagram of the situation.

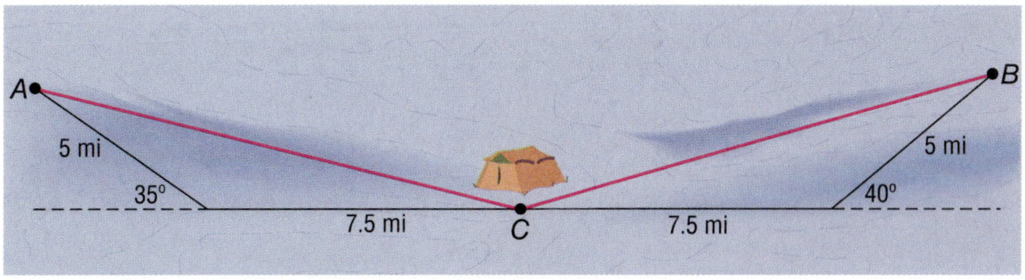

The paths taken by each group and the straight-line distance back to the camp form two triangles. Each group goes 7.5 miles and then turns and goes 5 miles.

Use linear pairs to find the measures of the included angles. Then apply the Hinge Theorem to compare the distance each group is from base camp.

Solve The included angle for the path made by Group A measures 180 − 35 or 145. The included angle for the path made by Group B is 180 − 40 or 140.

Since 145 > 140, $AC > BC$ by the Hinge Theorem. So Group A is farther from the base camp.

Check Group B turned 5° more than Group A did back toward base camp, so they should be closer to base camp than Group A. Thus, Group A should be farther from the base camp. ✓

Guided Practice

2A. SKIING Two groups of skiers leave from the same lodge. Group A goes 4 miles due east and then turns 70° north of east and goes 3 miles. Group B goes 4 miles due west and then turns 75° north of west and goes 3 miles. At this point, which group is *farther* from the lodge? Explain your reasoning.

2B. SKIING In problem 2A, suppose Group A instead went 4 miles west and then turned 45° north of west and traveled 3 miles. Which group would be *closer* to the lodge? Explain your reasoning.

Personal Tutor glencoe.com

When the included angle of one triangle is greater than the included angle in a second triangle, the Converse of the Hinge Theorem is used.

EXAMPLE 3 Apply Algebra to the Relationships in Triangles

ALGEBRA **Find the range of possible values for x.**

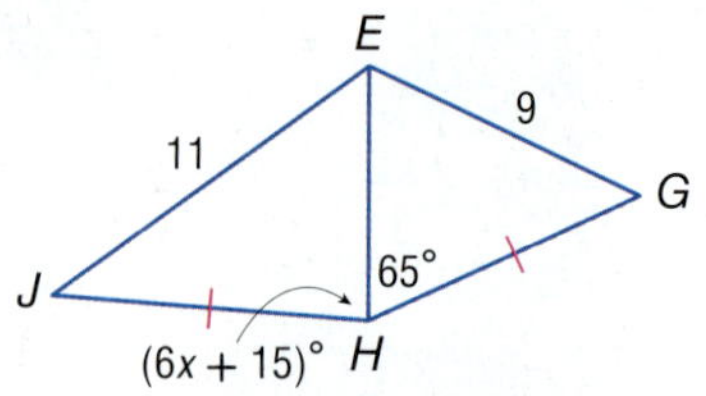

Step 1 From the diagram, we know that $\overline{JH} \cong \overline{GH}$, $\overline{EH} \cong \overline{EH}$, and $JE > EG$.

$m\angle JHE > m\angle EHG$ — **Converse of the Hinge Theorem**

$6x + 15 > 65$ — **Substitution**

$x > 8\frac{1}{3}$ — **Solve for x.**

Step 2 Use the fact that the measure of any angle in a triangle is less than 180 to write a second inequality.

$m\angle JHE < 180$

$6x + 15 < 180$ — **Substitution**

$x < 27.5$ — **Solve for x.**

Step 3 Write $x > 8\frac{1}{3}$ and $x < 27.5$ as the compound inequality $8\frac{1}{3} < x < 27.5$.

StudyTip

Using Additional Facts When finding a range for the possible values for x, you may need to use one of the following facts.

- The measure of any angle is always greater than 0 and less than 180.
- The measure of any segment is always greater than 0.

Guided Practice

3. Find the range of possible values for x.

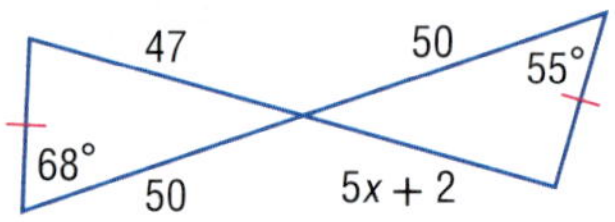

Personal Tutor glencoe.com

Prove Relationships In Two Triangles You can use the Hinge Theorem and its converse to prove relationships in two triangles.

EXAMPLE 4 Prove Triangle Relationships Using Hinge Theorem

Write a two-column proof.

Given: $\overline{AB} \cong \overline{AD}$

Prove: $EB > ED$

Proof:

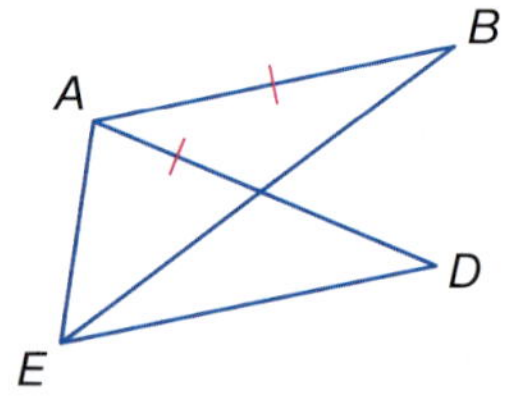

Statements	Reasons
1. $\overline{AB} \cong \overline{AD}$	**1.** Given
2. $\overline{AE} \cong \overline{AE}$	**2.** Reflexive Property
3. $m\angle EAB = m\angle EAD + m\angle DAB$	**3.** Angle Addition Postulate
4. $m\angle EAB > m\angle EAD$	**4.** Definition of Inequality
5. $EB > ED$	**5.** Hinge Theorem

Guided Practice

4. Write a two-column proof.

Given: $\overline{RQ} \cong \overline{ST}$

Prove: $RS > TQ$

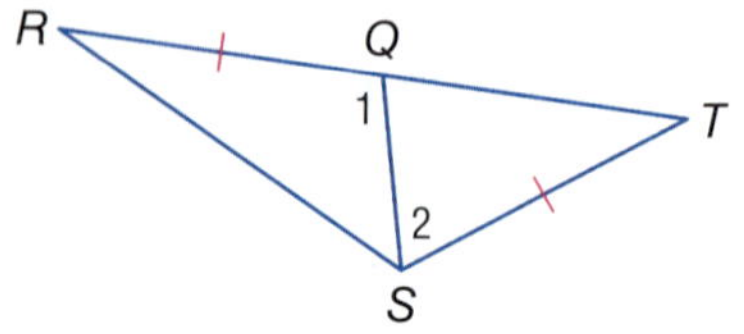

Personal Tutor glencoe.com

EXAMPLE 5 Prove Relationships Using Converse of Hinge Theorem

Write a flow proof.

Given: T is the midpoint of $\overline{ZX}$.
$\overline{ST} \cong \overline{WT}$
$SZ > WX$

Prove: $m\angle XTR > m\angle ZTY$

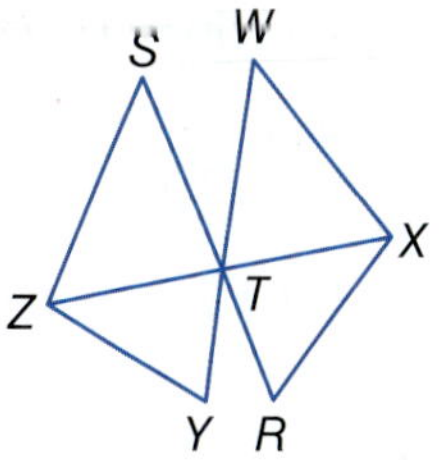

Flow Proof:

T is the midpoint of $\overline{ZX}$. — Given → $\overline{ZT} \cong \overline{TX}$ — Def. of Midpt.

$\overline{ST} \cong \overline{WT}$ — Given

$\overline{SZ} > \overline{WX}$ — Given

$\overline{ZT} \cong \overline{TX}$, $\overline{ST} \cong \overline{WT}$, $\overline{SZ} > \overline{WX}$ → $m\angle STZ > m\angle WTX$ — Converse of Hinge Thm.

$\angle STZ \cong \angle XTR$, $\angle WTX \cong \angle ZTY$ — Vert. ∠s are ≅. → $m\angle STZ = m\angle XTR$, $m\angle WTX = m\angle ZTY$ — Def. of ≅ ∠s

$m\angle STZ > m\angle WTX$ and $m\angle STZ = m\angle XTR$, $m\angle WTX = m\angle ZTY$ → $m\angle XTR > m\angle ZTY$ — Substitution

Guided Practice

5. Write a two-column proof.

Given: $\overline{NK}$ is a median of $\triangle JMN$.
$JN > NM$

Prove: $m\angle 1 > m\angle 2$

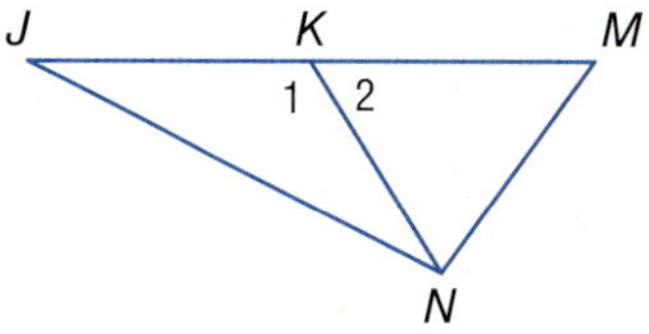

Personal Tutor glencoe.com

Check Your Understanding

Example 1 p. 367

Compare the given measures.

1. $m\angle ACB$ and $m\angle GDE$

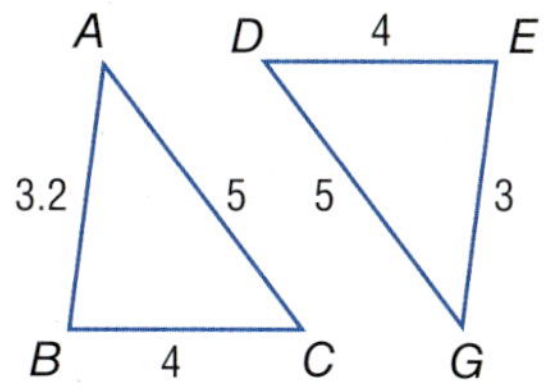

2. JL and KM

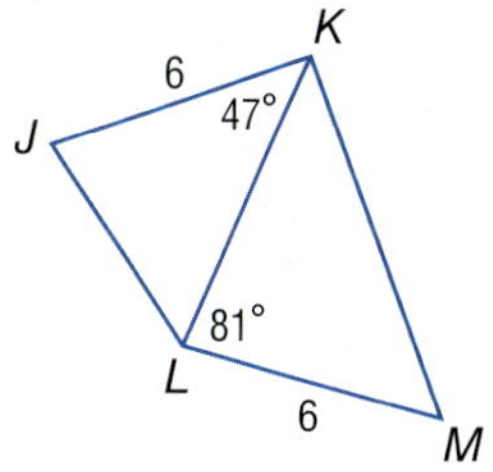

3 QT and ST

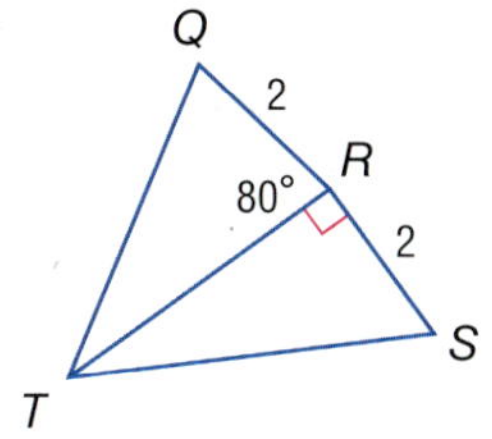

4. $m\angle XWZ$ and $m\angle YZW$

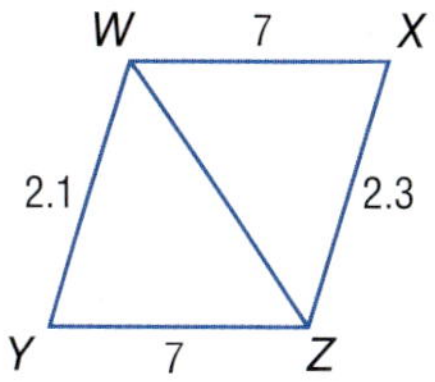

Example 2
p. 369

5. **SWINGS** The position of the swing changes based on how hard the swing is pushed.

a. Which pairs of segments are congruent?

b. Is the measure of $\angle A$ or the measure of $\angle D$ greater? Explain.

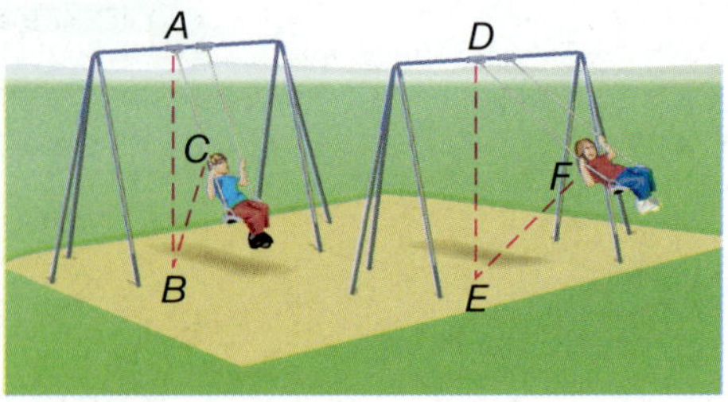

Example 3
p. 370

Find the range of possible values for x.

6.

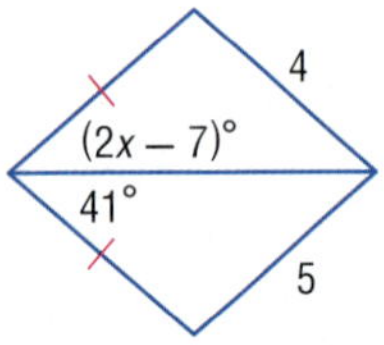

7.

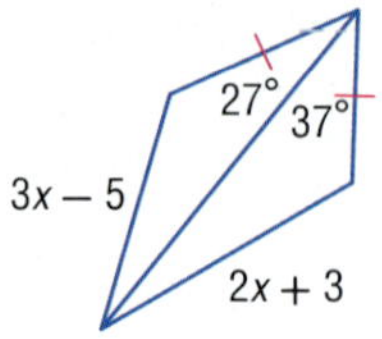

Examples 4 and 5
pp. 370–371

PROOF **Write a two-column proof.**

8. **Given:** $\triangle YZX$
$\overline{YZ} \cong \overline{XW}$

Prove: $ZX > YW$

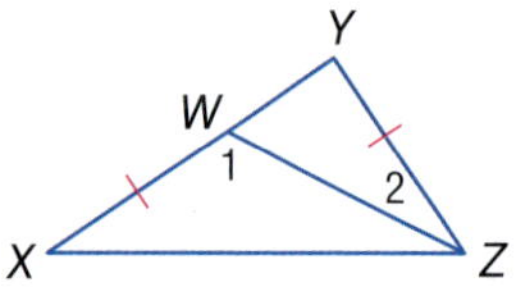

9. **Given:** $\overline{AD} \cong \overline{CB}$
$DC < AB$

Prove: $m\angle CBD < m\angle ADB$

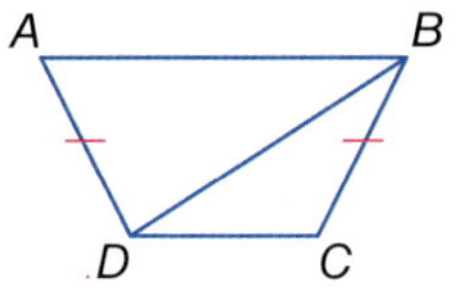

Practice and Problem Solving

● = **Step-by-Step Solutions** begin on page R20.
Extra Practice begins on page 969.

Example 1
p. 367

Compare the given measures.

10. $m\angle BAC$ and $m\angle DGE$

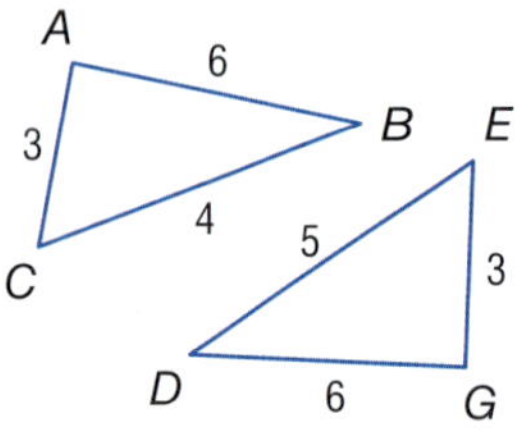

11. $m\angle MLP$ and $m\angle TSR$

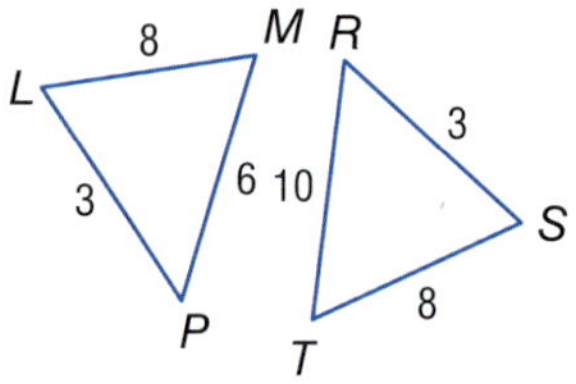

12. SR and XY

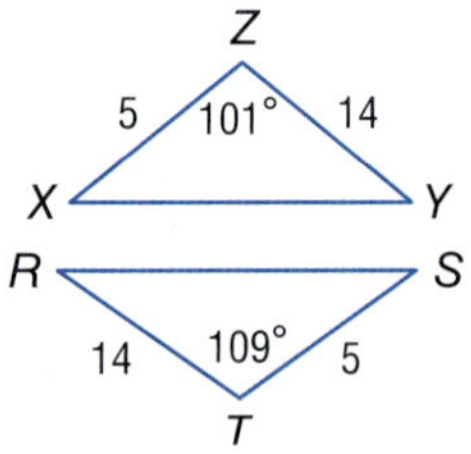

13. $m\angle TUW$ and $m\angle VUW$

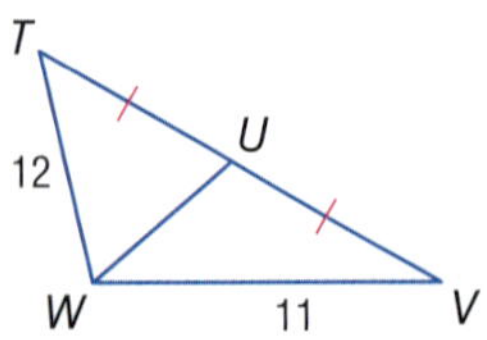

14. PS and SR

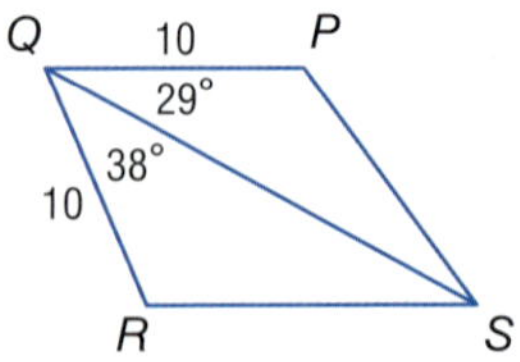

15. JK and HJ

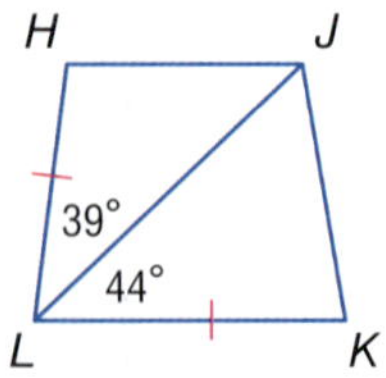

Example 2
p. 369

16. **CAMPING** Pedro and Joel are camping in a national park. One morning, Pedro decides to hike to the waterfall. He leaves camp and goes 5 miles east then turns 15° south of east and goes 2 more miles. Joel leaves the camp and travels 5 miles west, then turns 35° north of west and goes 2 miles to the lake for a swim.

a. When they reach their destinations, who is closer to the camp? Explain your reasoning. Include a diagram.

b. Suppose instead of turning 35° north of west, Joel turned 10° south of west. Who would then be farther from the camp? Explain your reasoning. Include a diagram.

Example 3
p. 370

Find the range of possible values for x.

17.

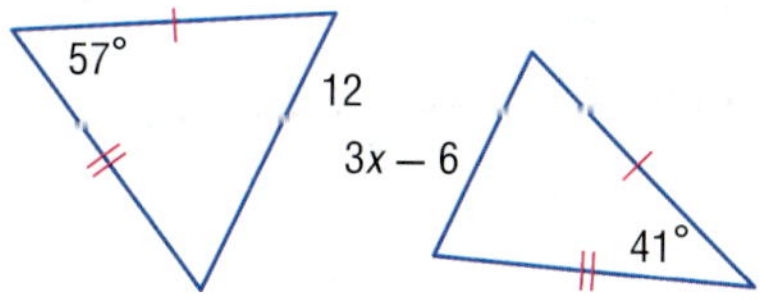

18.

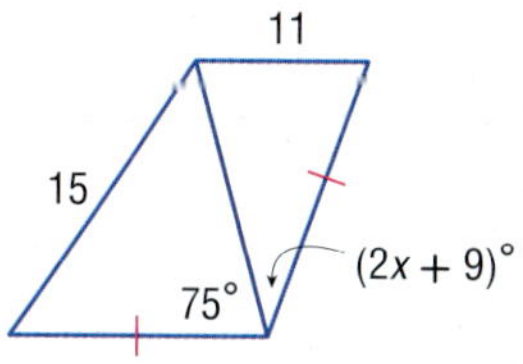

19

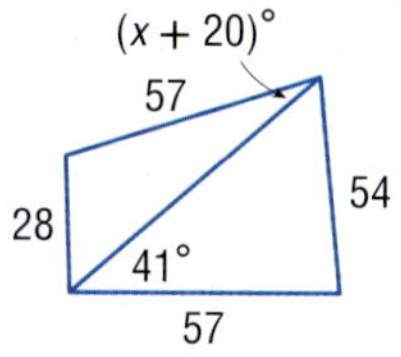

20.

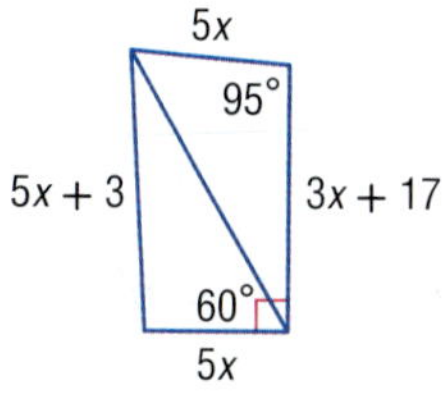

21. CRANES In the diagram, a crane is shown lifting an object to two different heights. The length of the red arm of the crane is fixed. Is $\overline{MN}$ or $\overline{RS}$ shorter? Explain your reasoning.

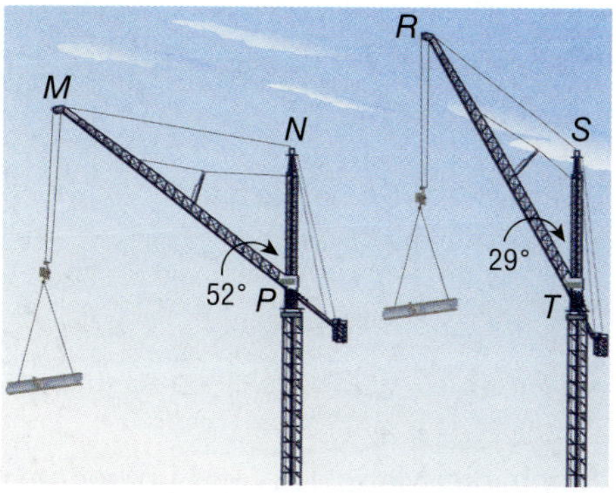

22. LOCKERS Neva and Shawn both have their lockers open as shown in the diagram. Whose locker forms a larger angle? Explain your reasoning.

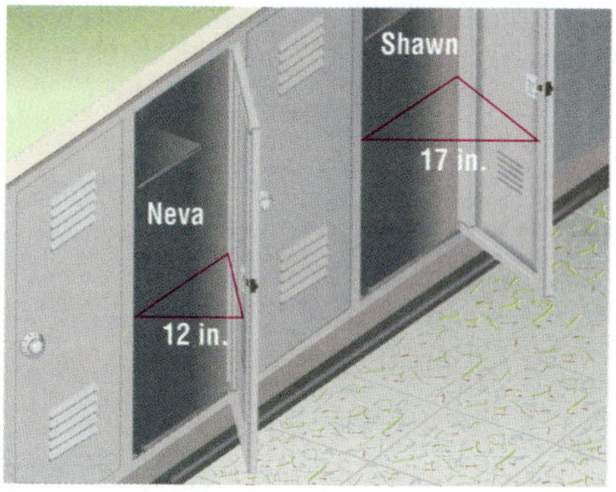

Examples 4 and 5
pp. 370–371

PROOF **Write a two-column proof.**

23. Given: $\overline{LK} \cong \overline{JK}$, $\overline{RL} \cong \overline{RJ}$
K is the midpoint of $\overline{QS}$.
$m\angle SKL > m\angle QKJ$
Prove: $RS > QR$

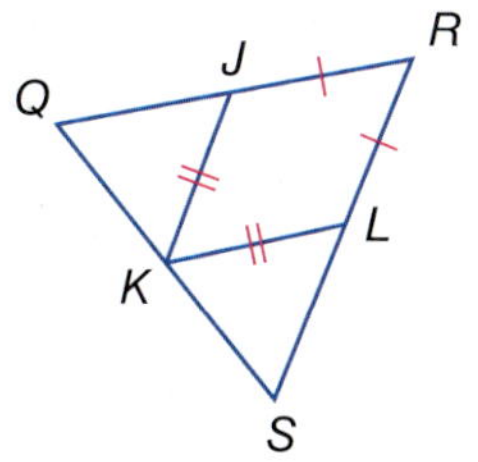

24. Given: $\overline{VR} \cong \overline{RT}$, $\overline{WV} \cong \overline{WT}$
$m\angle SRV > m\angle QRT$
R is the midpoint of $\overline{SQ}$.
Prove: $WS > WQ$

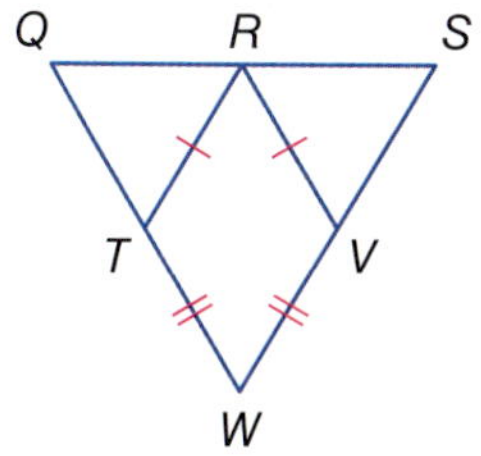

25. Given: $\overline{XU} \cong \overline{VW}$, $VW > XW$
$\overline{XU} \parallel \overline{VW}$
Prove: $m\angle XZU > m\angle UZV$

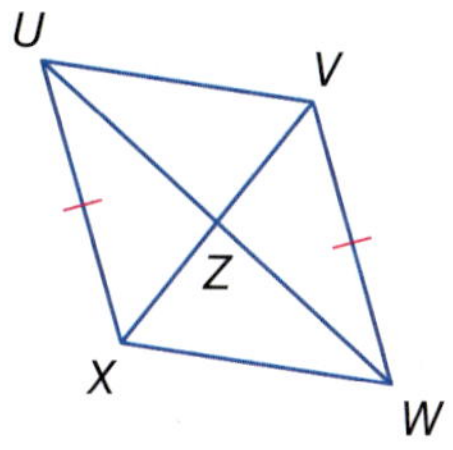

26. Given: $\overline{AF} \cong \overline{DJ}$, $\overline{FC} \cong \overline{JB}$
$AB > DC$
Prove: $m\angle AFC > m\angle DJB$

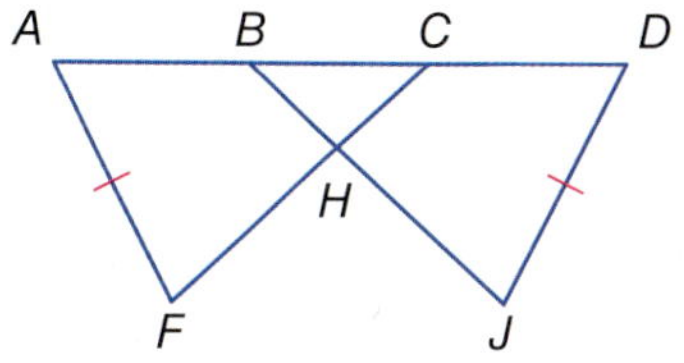

Real-World Link

Strength training increases endurance and strength for sports and fitness activities. If you are just starting out, most fitness experts recommend three strength training sessions a week, ranging from 20 minutes to 1 hour (including warm-up and cool-down), allowing at least a day off between work-outs.

Source: KidsHealth

27. EXERCISE Anica is doing knee-supported bicep curls as part of her strength training.

a. Is the distance from Anica's fist to her shoulder greater in Position 1 or Position 2? Justify your answer using measurement.

b. Is the measure of the angle formed by Anica's elbow greater in Position 1 or Position 2? Explain your reasoning.

28. PROOF Use an indirect proof to prove the SSS Inequality Theorem (Theorem 5.14).

Given: $\overline{RS} \cong \overline{UW}$
$\overline{ST} \cong \overline{WV}$
$RT > UV$

Prove: $m\angle S > m\angle W$

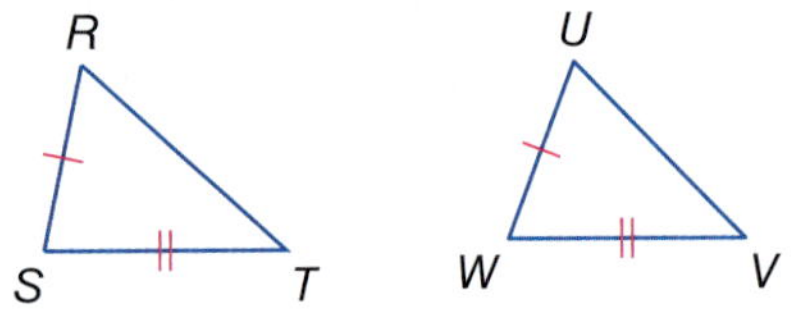

29. PROOF If $\overline{PR} \cong \overline{PQ}$ and $SQ > SR$, write a two-column proof to prove $m\angle 1 < m\angle 2$.

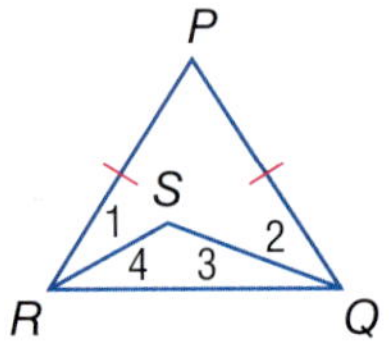

30. SCAVENGER HUNT Stephanie, Mario, Lee, and Luther are participating in a scavenger hunt as part of a geography lesson. Their map shows that the next clue is 50 feet due east and then 75 feet 35° east of north starting from the fountain in the school courtyard. When they get ready to turn and go 75 feet 35° east of north, they disagree about which way to go, so they split up and take the paths shown in the diagram below.

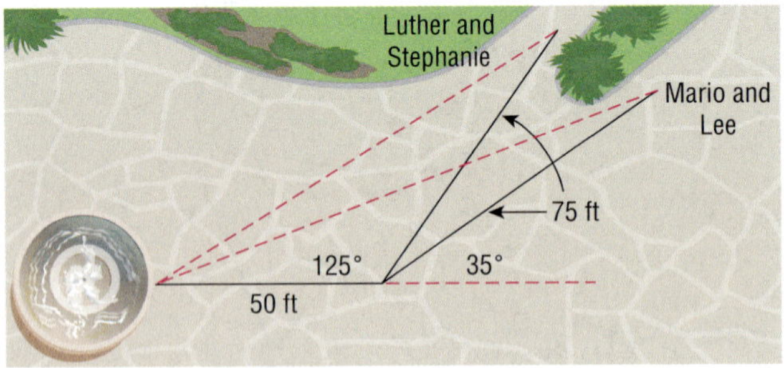

a. Which pair chose the correct path? Explain your reasoning.

b. Which pair is closest to the fountain when they stop? Explain your reasoning.

Use the figure at the right to write an inequality relating the given pair of angles or segment measures.

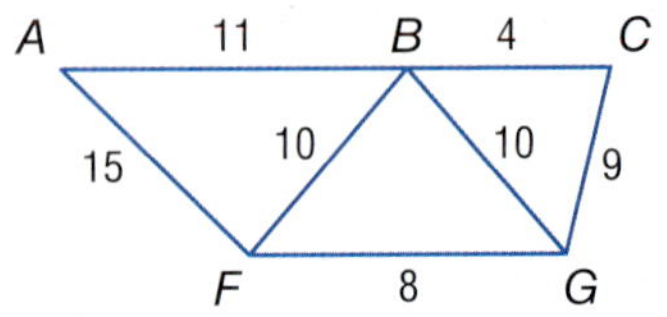

31. CB and AB

32. $\angle FBG$ and $\angle CGB$

33. $\angle BGC$ and $\angle BAF$

Use the figure at the right to write an inequality relating the given pair of angles or segment measures.

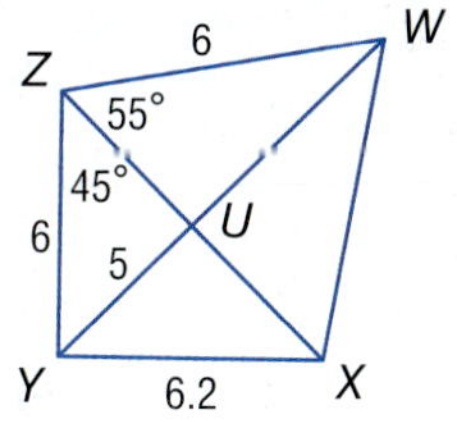

34. $m\angle ZUY$ and $m\angle YUX$

35. WU and YU

36. $m\angle WUX$ and $m\angle ZUW$

37. **MULTIPLE REPRESENTATIONS** In this problem, you will investigate properties of polygons.

a. GEOMETRIC Draw a three-sided, a four-sided, and a five-sided polygon. Label the 3-sided polygon ABC, the four-sided polygon $FGHJ$, and the five-sided polygon $PQRST$. Use a protractor to measure and label each angle.

b. TABULAR Copy and complete the table below.

Number of Sides	Angle Measures				Sum of Angles
3	$m\angle A$		$m\angle C$		
	$m\angle B$				
4	$m\angle F$		$m\angle H$		
	$m\angle G$		$m\angle J$		
5	$m\angle P$		$m\angle S$		
	$m\angle Q$		$m\angle T$		
	$m\angle R$				

c. VERBAL Make a conjecture about the relationship between the number of sides of a polygon and the sum of the measures of the angles of the polygon.

d. LOGICAL What type of reasoning did you use in part **c**? Explain.

e. ALGEBRAIC Write an algebraic expression for the sum of the measures of the angles for a polygon with n sides.

H.O.T. Problems

Use **H**igher-**O**rder **T**hinking Skills

38. **CHALLENGE** If $m\angle LJN > m\angle KJL$, $KJ \cong JN$, and $JN \perp NL$, which is greater, $m\angle LKN$ or $m\angle LNK$? Explain your reasoning.

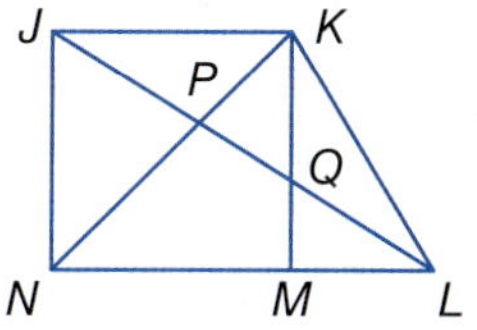

39. **OPEN ENDED** Give a real-world example of an object that uses a hinge. Draw two sketches in which the hinge on your object is adjusted to two different positions. Use your sketches to explain why Theorem 5.13 is called the Hinge Theorem.

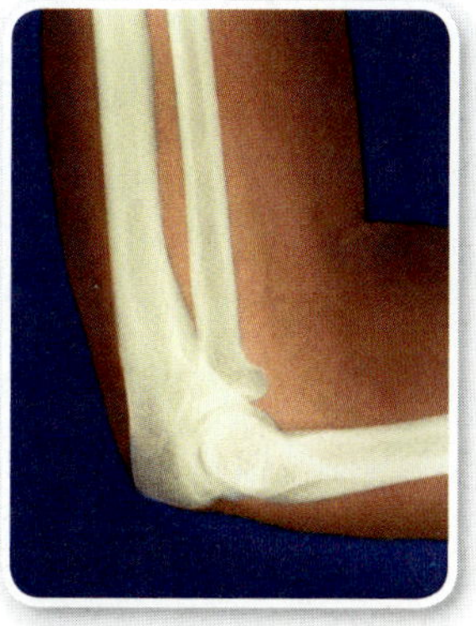

Real-World Link

Some of your joints use a hinge for the extension or retraction of an appendage such as an arm or leg.

40. **CHALLENGE** Given $\triangle RST$ with median $\overline{RQ}$, if RT is greater than or equal to RS, what are the possible classifications of $\triangle RQT$? Explain your reasoning.

41. **REASONING** If $\overline{BD}$ is a median and $AB < BC$, then $\angle BDC$ is *always, sometimes,* or *never* an acute angle. Explain.

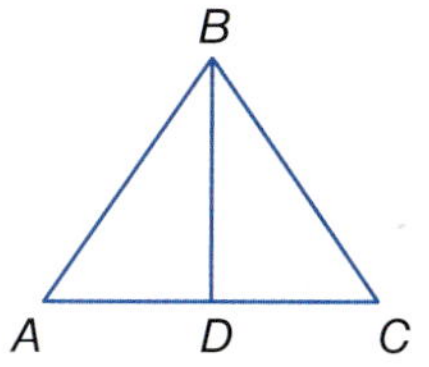

42. **WRITING IN MATH** Compare and contrast the Hinge Theorem to the SAS Postulate for triangle congruence.

Standardized Test Practice

43. SHORT RESPONSE Write an inequality to describe the possible range of values for x.

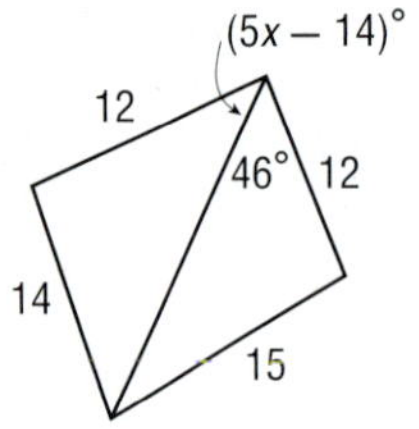

44. Which of the following is the inverse of the statement *If it is snowing, then Steve wears his snow boots?*

A If Steve wears his snow boots, then it is snowing.

B If it is not snowing, then Steve does not wear his snow boots.

C If it is not snowing, then Steve wears his snow boots.

D If it never snows, then Steve does not own snow boots.

45. ALGEBRA Which linear function best describes the graph shown?

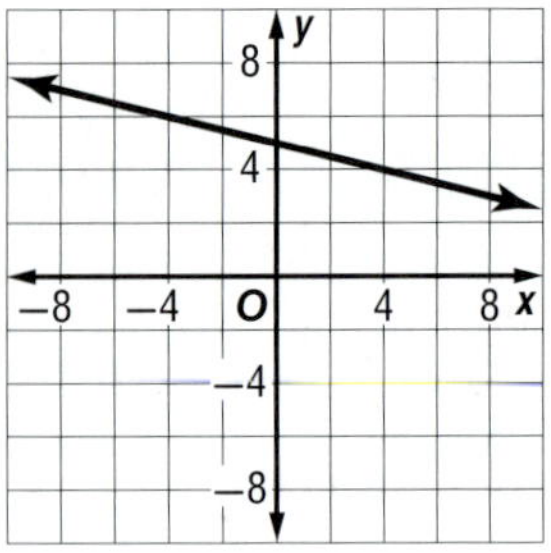

F $y = -\frac{1}{4}x + 5$

G $y = -\frac{1}{4}x - 5$

H $y = \frac{1}{4}x + 5$

J $y = \frac{1}{4}x - 5$

46. SAT/ACT If the side of a square is $x + 3$, then the diagonal of the square is

A $x^2 + 1$

B $x\sqrt{2} + 3\sqrt{2}$

C $2x + 6$

D $x^2\sqrt{2} + 6$

Spiral Review

Find the range for the measure of the third side of a triangle given the measures of two sides. (Lesson 5-5)

47. 3.2 cm, 4.4 cm

48. 5 ft, 10 ft

49. 3 m, 9 m

50. CRUISES Ally asked Tavia the cost of a cruise she and her best friend went on after graduation. Tavia could not remember how much it cost per person, but she did remember that the total cost was over \$500. Use indirect reasoning to show that the cost for one person was more than \$250. (Lesson 5-4)

Draw and label a figure to represent the congruent triangles. Then find x. (Lesson 4-3)

51. $\triangle QRS \cong \triangle GHJ$, $RS = 12$, $QR = 10$, $QS = 6$, and $HJ = 2x - 4$.

52. $\triangle ABC \cong \triangle XYZ$, $AB = 13$, $AC = 19$, $BC = 21$, and $XY = 3x + 7$.

Use the figure at the right. (Lesson 1-4)

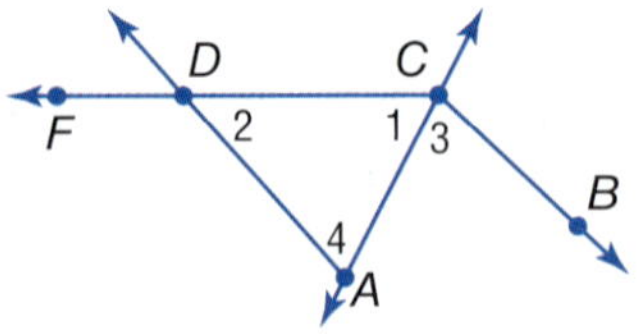

53. Name the vertex of $\angle 4$.

54. What is another name for $\angle 2$?

55. What is another name for $\angle BCA$?

Skills Review

Find the value of the variable(s) in each figure. Explain your reasoning. (Lesson 3-2)

56.

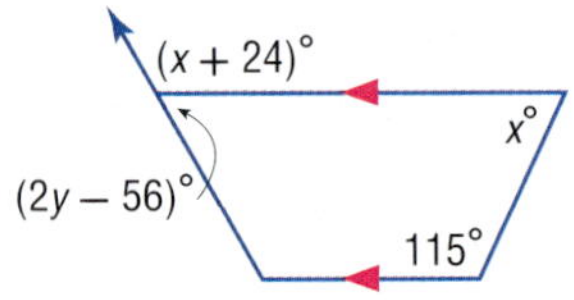

57.

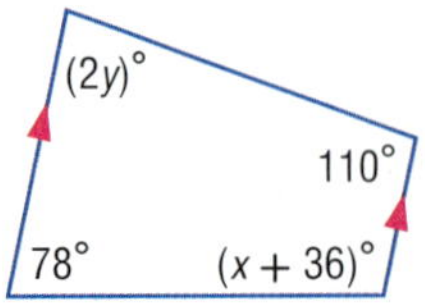

58.

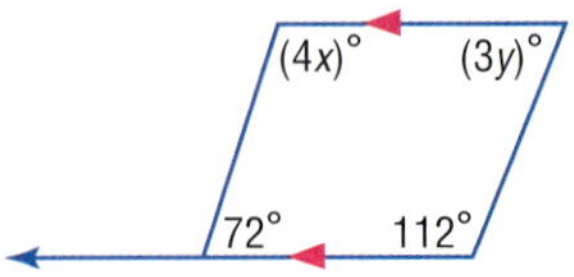

CHAPTER 5

Study Guide and Review

Math Online glencoe.com
- STUDY TO GO
- Vocabulary Review

Chapter Summary

Key Concepts

Special Segments in Triangles (Lessons 5-1 and 5-2)

- The special segments of triangles are perpendicular bisectors, angle bisectors, medians, and altitudes.
- The intersection points of each of the special segments of a triangle are called the points of concurrency.
- The points of concurrency for a triangle are the circumcenter, incenter, centroid, and orthocenter.

Indirect Proof (Lesson 5-4)

- Writing an Indirect Proof:
  1. Assume that the conclusion is false.
  2. Show that this assumption leads to a contradiction.
  3. Since the false conclusion leads to an incorrect statement, the original conclusion must be true.

Triangle Inequalities (Lessons 5-3, 5-5, and 5-6)

- The largest angle in a triangle is opposite the longest side, and the smallest angle is opposite the shortest side.
- The sum of the lengths of any two sides of a triangle is greater than the length of the third side.
- **SAS Inequality** (Hinge Theorem): In two triangles, if two sides are congruent, then the measure of the included angle determines which triangle has the longer third side.
- **SSS Inequality**: In two triangles, if two corresponding sides of each triangle are congruent, then the length of the third side determines which triangle has the included angle with the greater measure.

FOLDABLES® Study Organizer

Be sure the Key Concepts are noted in your Foldable.

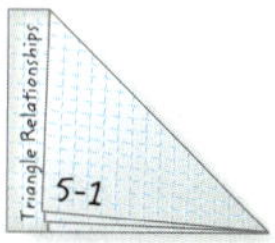

Key Vocabulary

altitude (p. 335)
centroid (p. 333)
circumcenter (p. 323)
concurrent lines (p. 323)
incenter (p. 326)
indirect proof (p. 351)
indirect reasoning (p. 351)
median (p. 333)
orthocenter (p. 335)
perpendicular bisector (p. 322)
point of concurrency (p. 323)
proof by contradiction (p. 351)

Vocabulary Check

State whether each sentence is *true* or *false*. If *false*, replace the underlined term to make a true sentence.

1. The centroid is the point at which the altitudes of a triangle intersect.
2. The point of concurrency of the medians of a triangle is called the incenter.
3. The point of concurrency is the point at which three or more lines intersect.
4. The circumcenter of a triangle is equidistant from the vertices of the triangle.
5. To find the centroid of a triangle, first construct the angle bisectors.
6. The perpendicular bisectors of a triangle are concurrent lines.
7. To start a proof by contradiction, first assume that what you are trying to prove is true.
8. A proof by contradiction uses indirect reasoning.
9. A median of a triangle connects the midpoint of one side of the triangle to the midpoint of another side of the triangle.
10. The incenter is the point at which the angle bisectors of a triangle intersect.

Study Guide and Review

Lesson-by-Lesson Review

5-1 Bisectors of Triangles (pp. 322–331)

11. Find EG if G is the incenter of $\triangle ABC$.

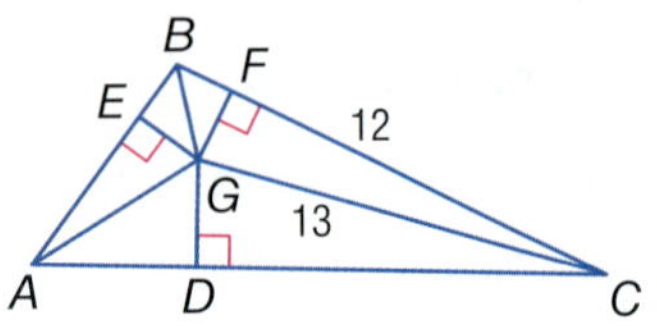

Find each measure.

12. RS

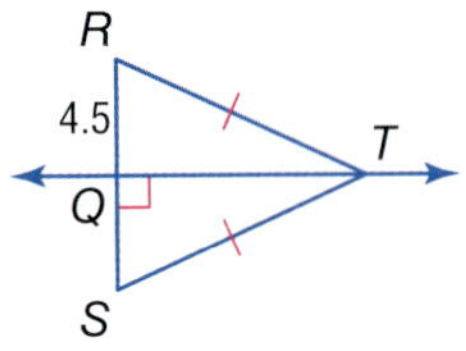

13. XZ

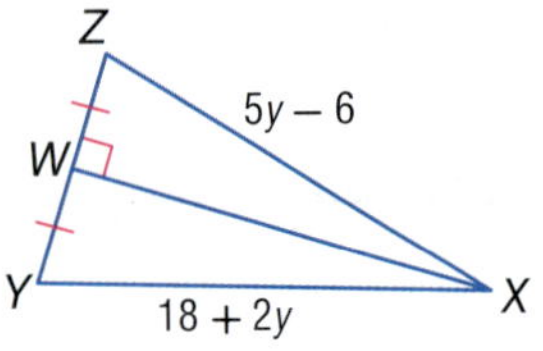

14. **BASEBALL** Jackson, Trevor, and Scott are warming up before a baseball game. One of their warm-up drills requires three players to form a triangle, with one player in the middle. Where should the fourth player stand so that he is the same distance from the other three players?

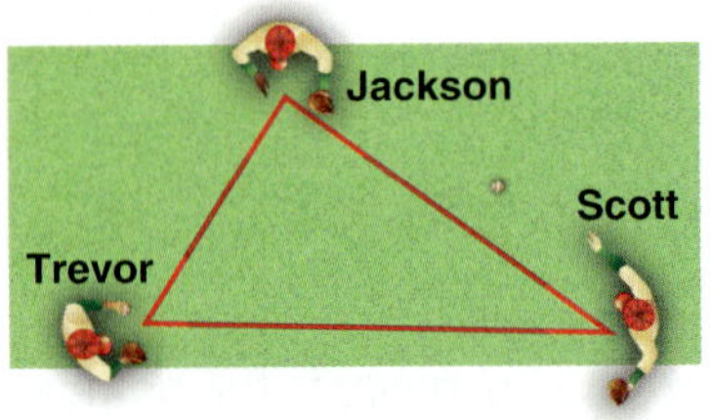

EXAMPLE 1

Find each measure if Q is the incenter of $\triangle JKL$.

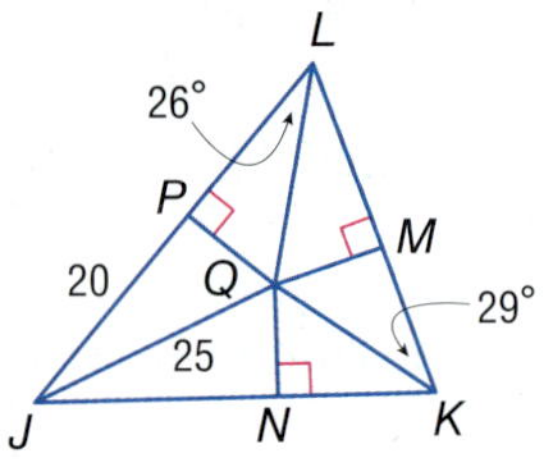

a. $\angle QJK$

$m\angle KLP + m\angle MKN + m\angle NJP = 180$	**△ Sum Theorem**
$2(26) + 2(29) + m\angle NJP = 180$	**Substitution**
$110 + m\angle NJP = 180$	**Simplify.**
$m\angle NJP = 70$	**Subtract.**

Since $\overrightarrow{JQ}$ bisects $\angle NJP$, $2m\angle QJK = m\angle NJP$. So, $m\angle QJK = \frac{1}{2}m\angle NJP$, so $m\angle QJK = \frac{1}{2}70$ or 35.

b. QP

$a^2 + b^2 = c^2$	**Pythagorean Theorem**
$(QP)^2 + 20^2 = 25^2$	**Substitution**
$(QP)^2 + 400 = 625$	**$20^2 = 400$ and $25^2 = 625$**
$(QP)^2 = 225$	**Subtract.**
$QP = 15$	**Simplify.**

5-2 Medians and Altitudes of Triangles (pp. 333–341)

15. The vertices of $\triangle DEF$ are $D(0, 0)$, $E(0, 7)$, and $F(6, 3)$. Find the coordinates of the orthocenter of $\triangle DEF$.

16. **PROM** Georgia is on the prom committee. She wants to hang a dozen congruent triangles from the ceiling so that they are parallel to the floor. She sketched out one triangle on a coordinate plane with coordinates (0, 4), (3, 8), and (6, 0). If each triangle is to be hung by one chain, what are the coordinates of the point where the chain should attach to the triangle?

EXAMPLE 2

In $\triangle EDF$, T is the centroid and $FT = 12$. Find TQ.

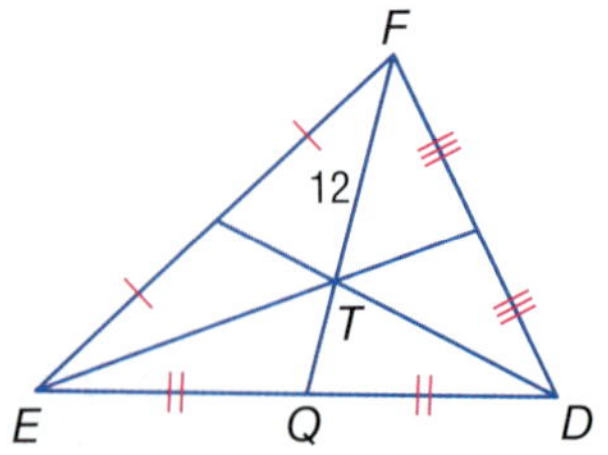

$FT = \frac{2}{3}FQ$

$FT = \frac{2}{3}(FT + TQ)$

$12 = \frac{2}{3}(12 + TQ)$	**$FT = 12$**
$12 = 8 + \frac{2}{3}TQ$	**Distributive Property**
$4 = \frac{2}{3}TQ$	**Subtract.**
$6 = TQ$	**Multiply.**

MIXED PROBLEM SOLVING
For mixed problem-solving practice, see page 1003.

5-3 Inequalities in One Triangle (pp. 342–349)

List the angles and sides of each triangle in order from smallest to largest.

17.

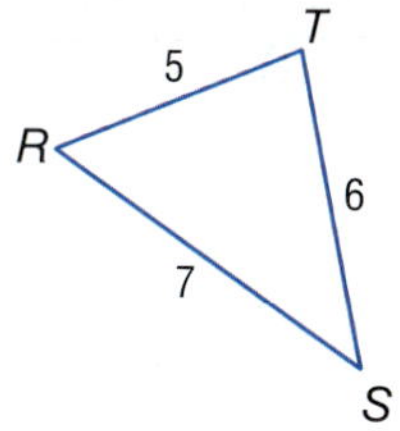

18.

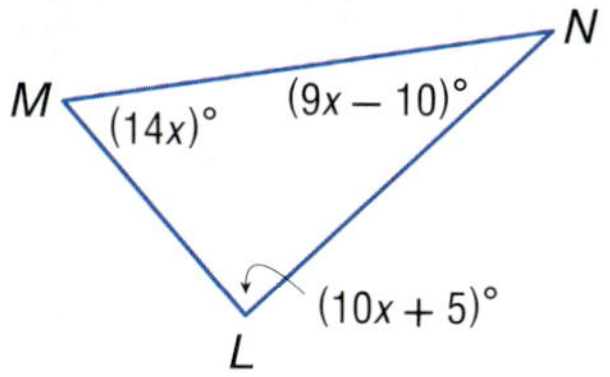

19. **NEIGHBORHOODS** Anna, Sarah, and Irene live at the intersections of the three roads that make the triangle shown. If the girls want to spend the afternoon together, is it a shorter path for Anna to stop and get Sarah and go onto Irene's house, or for Sarah to stop and get Irene and then go on to Anna's house?

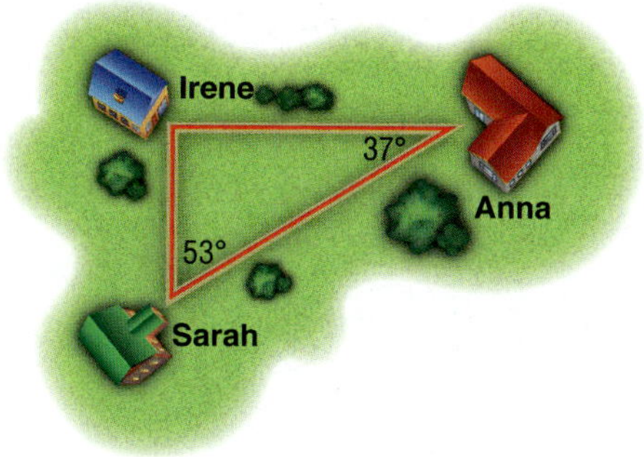

EXAMPLE 3

List the angles and sides of $\triangle ABC$ in order from smallest to largest.

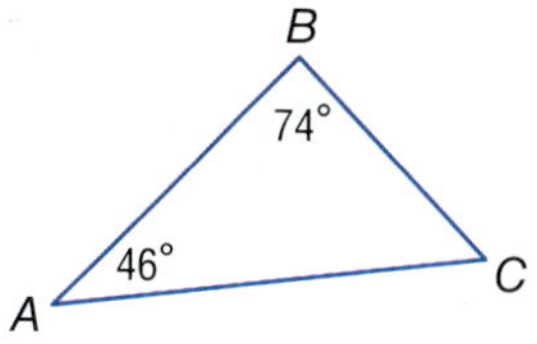

a. First, find the missing angle measure using the Triangle Sum Theorem.

$m\angle C = 180 - (46 + 74)$ or 60

So, the angles from smallest to largest are $\angle A$, $\angle C$, and $\angle B$.

b. The sides from shortest to longest are $\overline{BC}$, $\overline{AB}$, and $\overline{AC}$.

5-4 Indirect Proof (pp. 351–358)

State the assumption you would make to start an indirect proof of each statement.

20. $m\angle A \geq m\angle B$

21. $\triangle FGH \cong \triangle MNO$

22. $\triangle KLM$ is a right triangle.

23. If $3y < 12$, then $y < 4$.

24. Write an indirect proof to show that if two angles are complementary, neither angle is a right angle.

25. **MOVIES** Isaac bought two DVD's and spent over \$50. Use indirect reasoning to show that at least one of the DVD's he purchased was over \$25.

EXAMPLE 4

State the assumption necessary to start an indirect proof of each statement.

a. $\overline{XY} \ncong \overline{JK}$

$\overline{XY} \cong \overline{JK}$

b. **If $3x < 18$, then $x < 6$**

The conclusion of the conditional statement is $x < 6$. The negation of the conclusion is $x > 6$.

c. **$\angle 2$ is an acute angle.**

If *$\angle 2$ is an acute angle* is false, then *$\angle 2$ is not an acute angle* must be true. This means that *$\angle 2$ is an obtuse or right angle* must be true.

CHAPTER 5 Study Guide and Review

5-5 The Triangle Inequality (pp. 360–366)

Is it possible to form a triangle with the given lengths? If not, explain why not.

26. 5, 6, 9

27. 3, 4, 8

Find the range for the measure of the third side of a triangle given the measure of two sides.

28. 5 ft, 7 ft

29. 10.5 cm, 4 cm

30. BIKES Leonard rides his bike to visit Josh. Since High Street is closed, he has to travel 2 miles down Main Street and turn to travel 3 miles farther on 5th Street. If the three streets form a triangle with Leonard and Josh's house as two of the vertices, find the range of the possible distance between Leonard and Josh's houses when traveling straight down High Street.

EXAMPLE 5

Is it possible to form a triangle with the lengths 7, 10, and 9 feet? If not, explain why not.

Check each inequality.

$7 + 10 > 9$	$7 + 9 > 10$	$10 + 9 > 7$
$17 > 9$ ✓	$16 > 10$ ✓	$19 > 7$ ✓

Since the sum of each pair of side lengths is greater than the third side length, sides with lengths 7, 10, and 9 feet will form a triangle.

5-6 Inequalities in Two Triangles (pp. 367–376)

Compare the given measures.

31. $m\angle ABC$, $m\angle DEF$

32. QT and RS

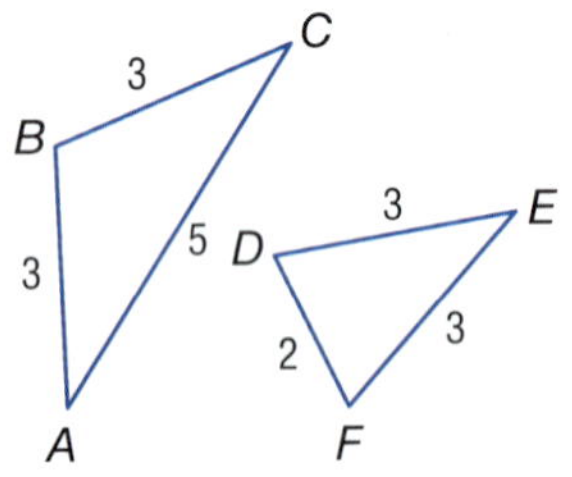

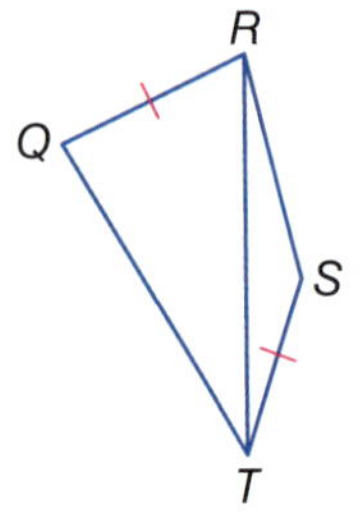

33. BOATING Rose and Connor each row across a pond heading to the same point. Neither of them has rowed a boat before, so they both go off course as shown in the diagram. After two minutes, they have each traveled 50 yards. Who is closer to their destination?

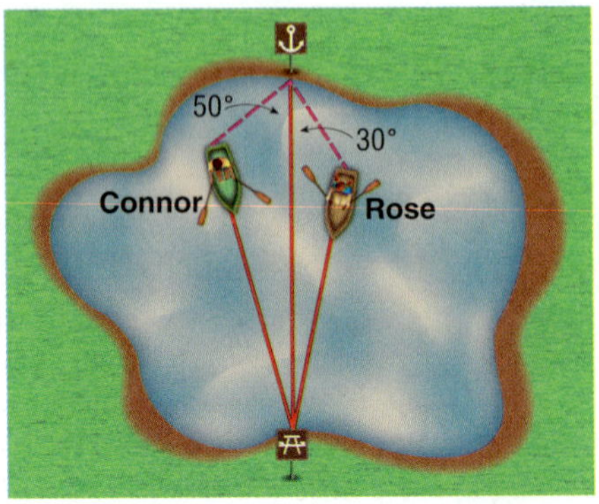

EXAMPLE 6

Compare the given measures.

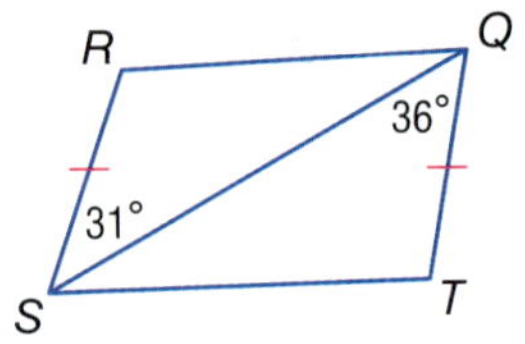

a. ***RQ* and *ST***

In $\triangle QRS$ and $\triangle STQ$, $\overline{RS} \cong \overline{TQ}$, $\overline{QS} \cong \overline{QS}$, and $\angle SQT > \angle RSQ$. By the Hinge Theorem, $m\angle SQT < m\angle RSQ$, so $RQ < ST$.

b. $m\angle JKM$ **and** $m\angle LKM$

In $\triangle JKM$ and $\triangle LKM$, $\overline{JM} \cong \overline{LM}$, $\overline{KM} \cong \overline{KM}$, and $LK > JK$. By the Converse of the Hinge Theorem, $\angle LKM > \angle JKM$.

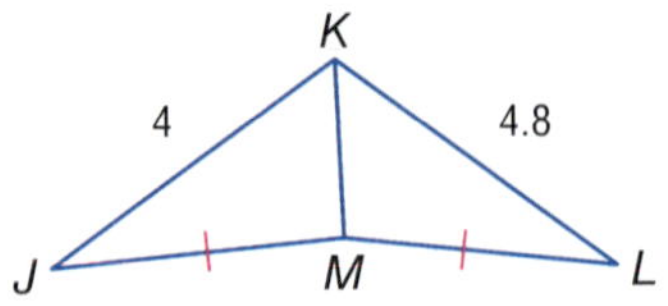

Practice Test

Math Online glencoe.com
Chapter Test

1. **GARDENS** Maggie wants to plant a circular flower bed within a triangular area set off by three pathways. Which point of concurrency related to triangles would she use for the center of the largest circle that would fit inside the triangle?

In $\triangle CDF$, K is the centroid and $DK = 16$. Find each length.

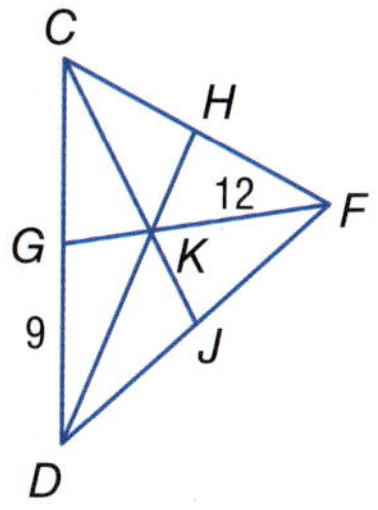

2. KH
3. CD
4. FG

5. **PROOF** Write an indirect proof.

Given: $5x + 7 \geq 52$

Prove: $x \geq 9$

Find each measure.

6. $m\angle TQR$

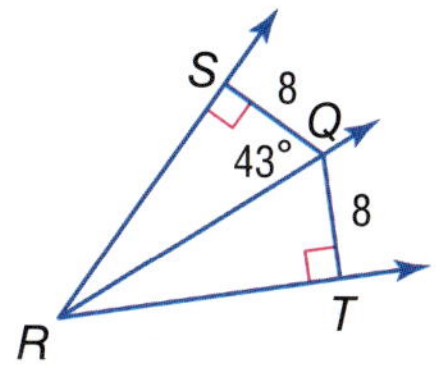

7. XZ

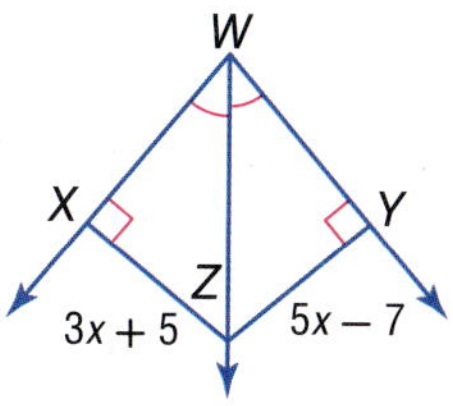

8. **GEOGRAPHY** The distance from Tonopah to Round Mountain is equal to the distance from Tonopah to Warm Springs. The distance from Tonopah to Hawthorne is the same as the distance from Tonopah to Beatty. Determine which distance is greater, Round Mountain to Hawthorne or Warm Springs to Beatty.

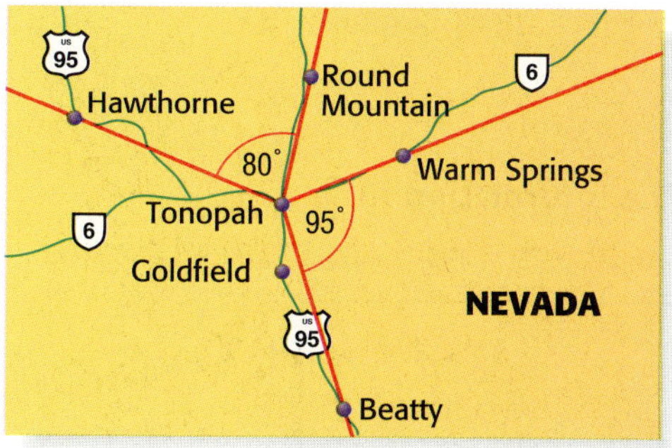

9. **MULTIPLE CHOICE** If the measures of two sides of a triangle are 3.1 feet and 4.6 feet, which is the *least* possible whole number measure for the third side?

   A 1.6 feet **C** 7.5 feet

   B 2 feet **D** 8 feet

Point H is the incenter of $\triangle ABC$. Find each measure.

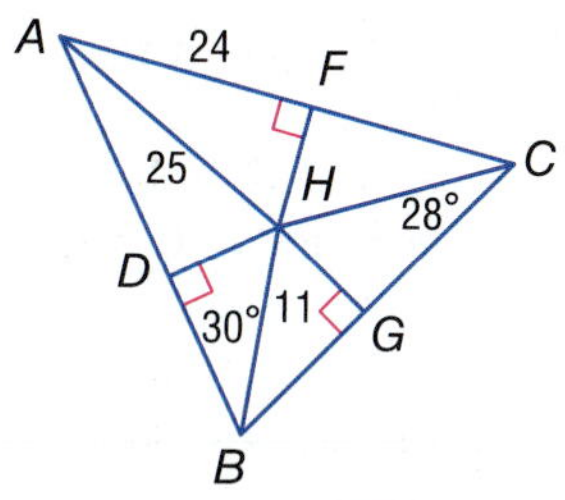

10. DH
11. BD
12. $m\angle HAC$
13. $m\angle DHG$

14. **MULTIPLE CHOICE** If the lengths of two sides of a triangle are 5 and 11, what is the range of possible lengths for the third side?

    F $6 < x < 10$ **H** $6 < x < 16$

    G $5 < x < 11$ **J** $x < 5$ or $x > 11$

Compare the given measures.

15. AB and BC

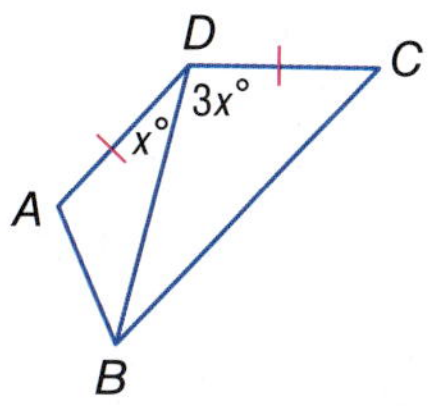

16. $\angle RST$ and $\angle JKL$

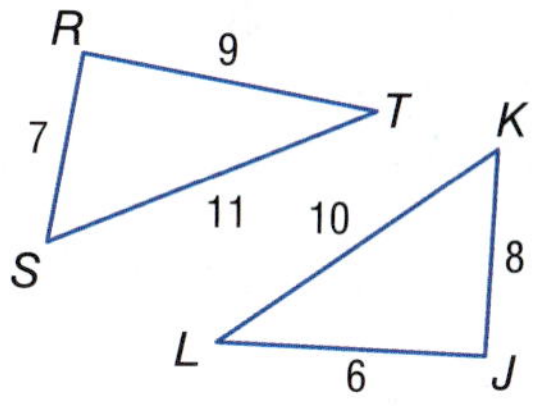

State the assumption necessary to start an indirect proof of each statement.

17. If 8 is a factor of n, then 4 is a factor of n.
18. $m\angle M > m\angle N$
19. If $3a + 7 \leq 28$, then $a \leq 7$.

Use the figure to determine which angle has the greatest measure.

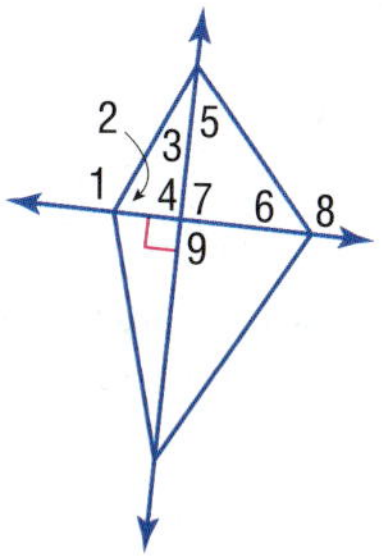

20. $\angle 1, \angle 5, \angle 6$
21. $\angle 9, \angle 8, \angle 3$
22. $\angle 4, \angle 3, \angle 2$

23. **PROOF** Write a two-column proof.

    Given: $\overline{RQ}$ bisects $\angle SRT$.

    Prove: $m\angle SQR > m\angle SRQ$

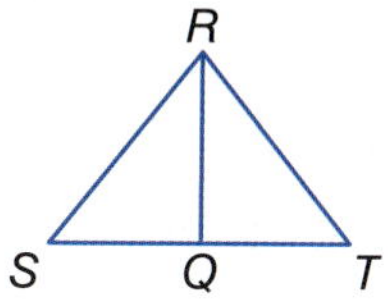

Find the range for the measure of the third side of a triangle given the measures of the two sides.

24. 10 ft, 16 ft
25. 23 m, 39 m

CHAPTER 5

Preparing for Standardized Tests

Eliminate Unreasonable Answers

You can eliminate unreasonable answers to determine the correct answer when solving multiple choice test items.

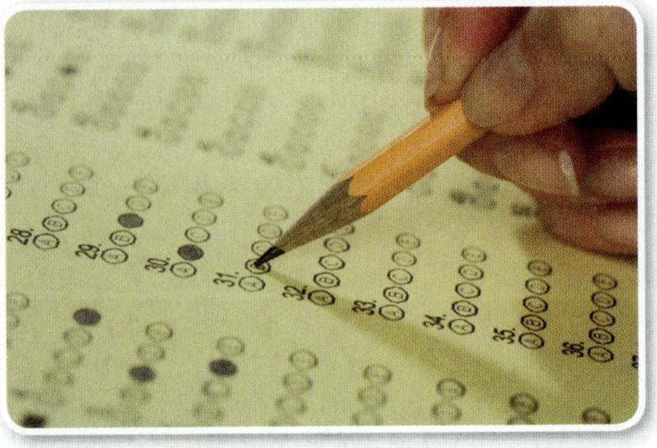

Strategies for Eliminating Unreasonable Answers

Step 1

Read the problem statement carefully to determine exactly what you are being asked to find.

- What am I being asked to solve?
- Is the correct answer a whole number, fraction, or decimal?
- Do I need to use a graph or table?
- What units (if any) will the correct answer have?

Step 2

Carefully look over each possible answer choice and evaluate for reasonableness. Do not write any digits or symbols outside the answer boxes.

- Identify any answer choices that are clearly incorrect and eliminate them.
- Eliminate any answer choices that are not in the proper format.
- Eliminate any answer choices that do not have the correct units.

Step 3

Solve the problem and choose the correct answer from those remaining. Check your answer.

EXAMPLE

Read the problem. Identify what you need to know. Then use the information in the problem to solve.

What is the measure of $\angle KLM$?

A 32

B 44

C 78

D 94

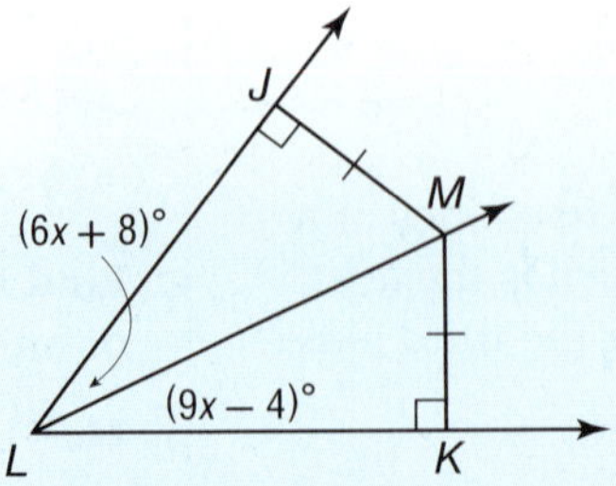

Read the problem and study the figure carefully. Triangle *KLM* is a right triangle. Since the sum of the interior angles of a triangle is 180°, $m\angle KLM + m\angle LMK$ must be equal to 90°. Otherwise, the sum would exceed 180°. Since answer choice D is an obtuse angle, it can be eliminated as unreasonable. The correct answer must be A, B, or C.

Solve the problem. According to the converse of the Angle Bisector Theorem, if a point in the interior of an angle is equidistant from the sides of the angle, then it is on the bisector of the angle. Point *M* is equidistant from rays *LJ* and *LK*, so it lies on the angle bisector of $\angle JLK$. Therefore, $\angle JLM$ must be congruent to $\angle KLM$. Set up and solve an equation for *x*.

$$6x + 8 = 9x - 4$$
$$-3x = -12$$
$$x = 4$$

So, the measure of $\angle KLM$ is $[9(4) - 4]°$, or 32°. The correct answer is A.

Exercises

Read each question. Then fill in the correct answer on the answer document provided by your teacher or on a sheet of paper.

1. Point *P* is the centroid of triangle *QUS*. If $QP =$ 14 centimeters, what is the length of $\overline{QT}$?

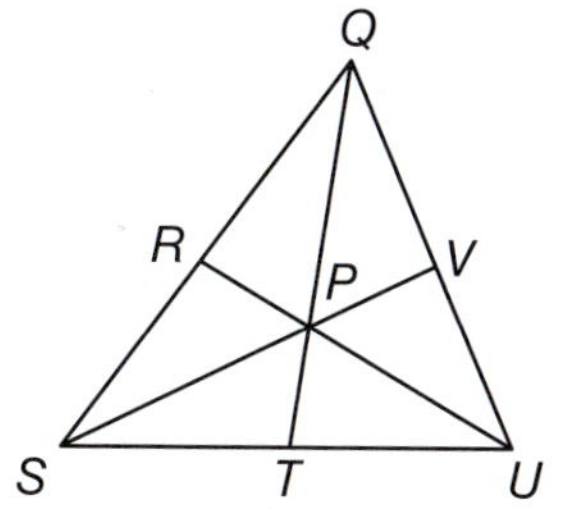

A 7 cm

B 12 cm

C 18 cm

D 21 cm

2. What is the area, in square units, of the triangle shown below?

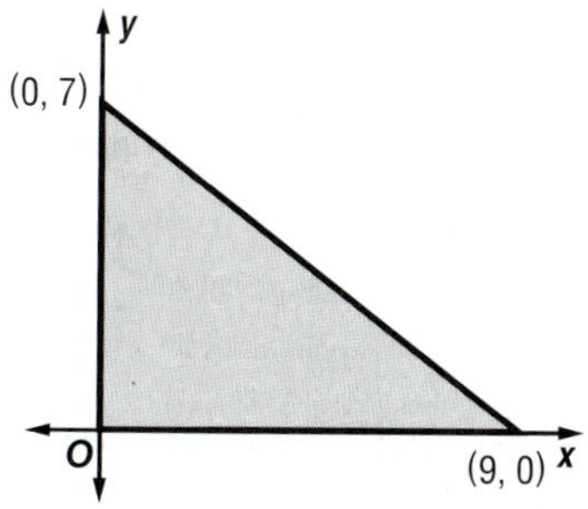

F 8

G 27.4

H 31.5

J 63

3. What are the coordinates of the orthocenter of the triangle below?

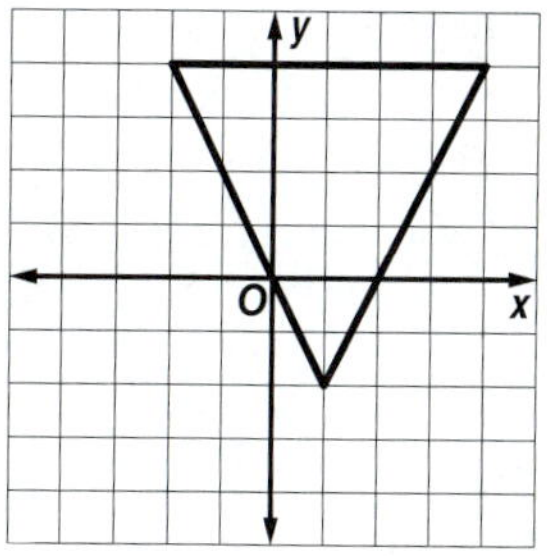

A $\left(-\frac{3}{4}, -1\right)$

B $\left(-\frac{4}{3}, 1\right)$

C $\left(1, \frac{5}{2}\right)$

D $\left(1, \frac{9}{4}\right)$

4. If $\triangle ABC$ is isosceles and $m\angle A = 94$, which of the following *must* be true?

F $m\angle B = 94$

G $m\angle B = 47$

H $AB = BC$

J $AB = AC$

5. Which of the following could *not* be the dimensions of a triangle?

A 1.9, 3.2, 4

B 1.6, 3, 4.6

C 3, 7.2, 7.5

D 2.6, 4.5, 6

Standardized Test Practice

Cumulative, Chapters 1 through 5

Multiple Choice

Read each question. Then fill in the correct answer on the answer document provided by your teacher or on a sheet of paper.

1. Solve for x.

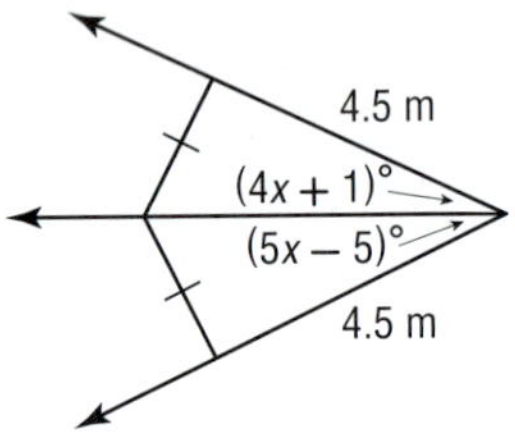

A 3
B 4
C 5
D 6

2. Which of the following could not be the value of x?

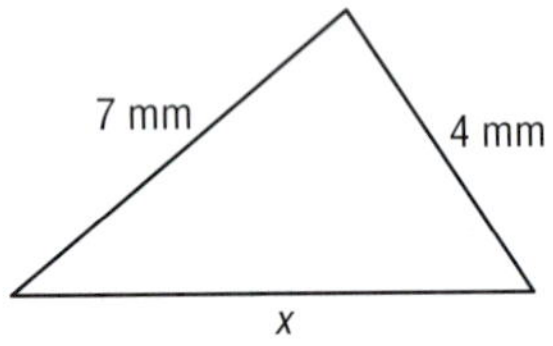

F 8 mm
G 9 mm
H 10 mm
J 11 mm

3. Jesse claims that if you live in Lexington, then you live in Kentucky. Which assumption would you need to make to form an indirect proof of this claim?

A Suppose someone lives in Kentucky, but not in Lexington.

B Suppose someone lives in Kentucky and in Lexington.

C Suppose someone lives in Lexington and in Kentucky.

D Suppose someone lives in Lexington, but not in Kentucky.

4. Which of the following best describes the shortest distance from a vertex of a triangle to the opposite side?

F altitude
G diameter
H median
J segment

5. Lin started mowing lawns. Let x represent the number of weeks after he began mowing lawns, and y represent the number of customers. Use the points (3, 4) and (9, 6) to find the equation of a line that can be used to predict how many customers Lin has by the end of a certain week.

A $y = \frac{1}{3}x$
B $y = \frac{1}{3}x + 3$
C $y = \frac{2}{3}x + 2$
D $y = \frac{2}{3}x$

6. What is the correct relationship between the angle measures of $\triangle PQR$?

F $m\angle R < m\angle Q < m\angle P$
G $m\angle R < m\angle P < m\angle Q$
H $m\angle Q < m\angle P < m\angle R$
J $m\angle P < m\angle Q < m\angle R$

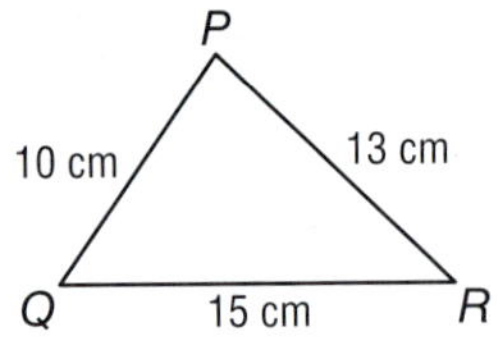

7. Which assumption would you need to make in order to start an indirect proof of the statement?

Angle S is not an obtuse angle.

A $\angle S$ is a right angle.
B $\angle S$ is an obtuse angle.
C $\angle S$ is an acute angle.
D $\angle S$ is not an acute angle.

8. Classify the triangle below according to its angle measures.

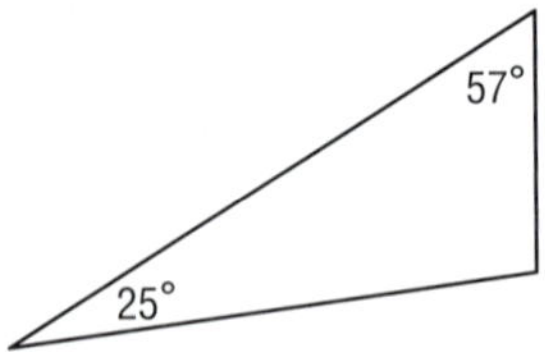

F acute
G equiangular
H obtuse
J right

Test-TakingTip

Question 2 The sum of any two sides of a triangle must be greater than the third side.

Short Response/Gridded Response

Record your answers on the answer sheet provided by your teacher or on a sheet of paper.

9. **GRIDDED RESPONSE** If the measures of two sides of a triangle are 9 centimeters and 15 centimeters, what is the least possible measure of the third side in centimeters if the measure is an integer?

10. What are the coordinates of the orthocenter of the triangle below?

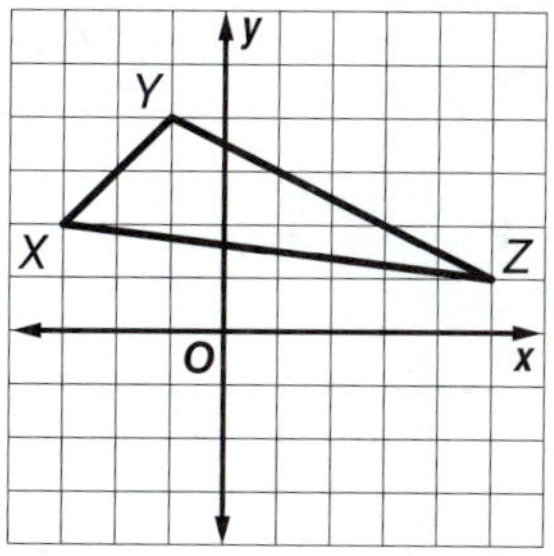

11. List the sides of the triangle below in order from shortest to longest.

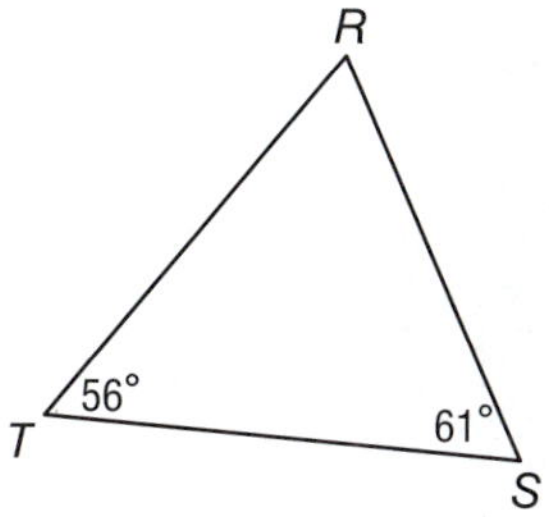

12. Suppose two lines intersect in a plane to form four angles.

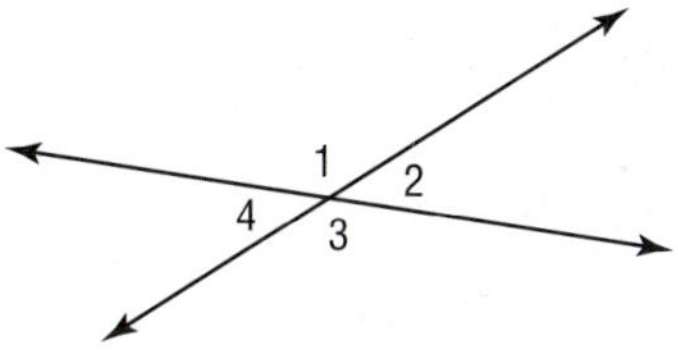

What do you know about the pairs of adjacent angles formed? Explain.

13. Eric and Heather are each taking a group of campers hiking in the woods. Eric's group leaves camp and goes 2 miles east, then turns 20° south of east and goes 4 more miles. Heather's group leaves camp and travels 2 miles west, then turns 30° north of west and goes 4 more miles. How many degrees south of east would Eric have needed to turn in order for his group and Heather's group to be the same distance from camp after the two legs of the hike?

14. **GRIDDED RESPONSE** Solve for x in the triangle below.

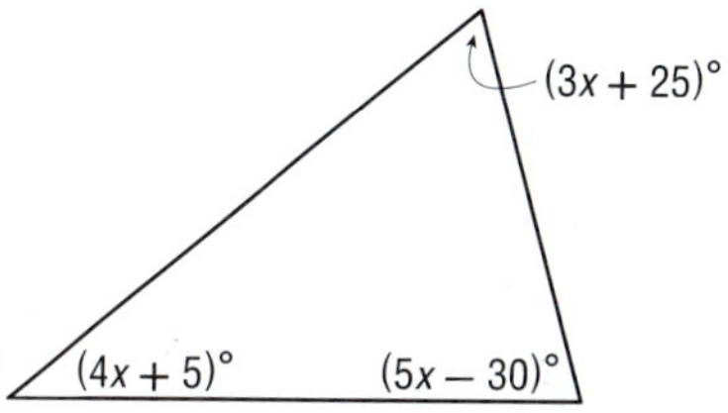

Extended Response

Record your answers on a sheet of paper. Show your work.

15. Refer to the figure to answer each question.

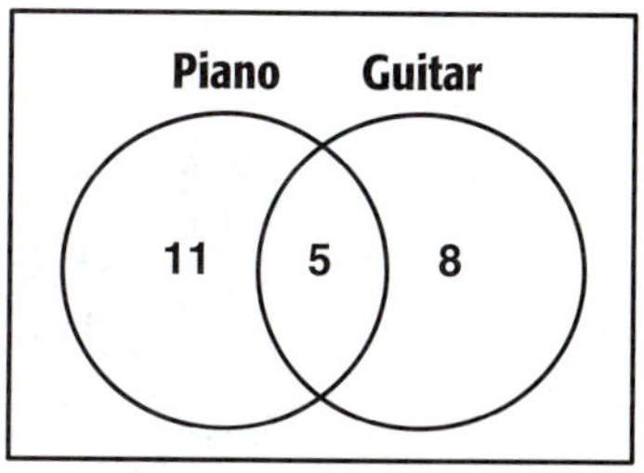

   a. How many students play the guitar?
   b. How many students play the piano?
   c. How many students play both piano and guitar?

Need Extra Help?

If you missed Question...	1	2	3	4	5	6	7	8	9	10	11	12	13	14	15
Go to Lesson or Page...	5-1	5-5	5-4	5-2	3-4	5-3	5-4	4-1	5-5	5-2	5-3	1-5	5-6	4-2	2-2

CHAPTER 6 Quadrilaterals

Then

In Chapter 1, you classified polygons. You recognized and applied properties of polygons.

Now

In Chapter 6, you will:

- Find and use the sum of the measures of the interior and exterior angles of a polygon.
- Recognize and apply properties of quadrilaterals.
- Compare quadrilaterals.

Why?

FUN AND GAMES The properties of quadrilaterals can be used to find various angle measures and side lengths such as the measures of angles in game equipment, playing fields, and game boards.

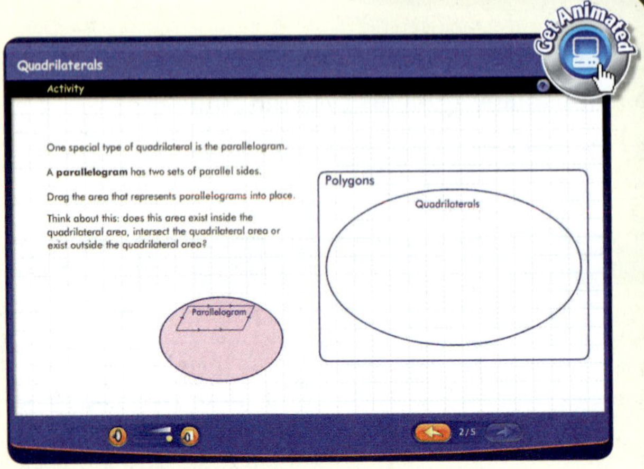

Get Ready for Chapter 6

Diagnose Readiness You have two options for checking Prerequisite Skills.

Text Option

Take the Quick Check below. Refer to the Quick Review for help.

QuickCheck

Find x to the nearest tenth. (Lesson 4-2)

1.

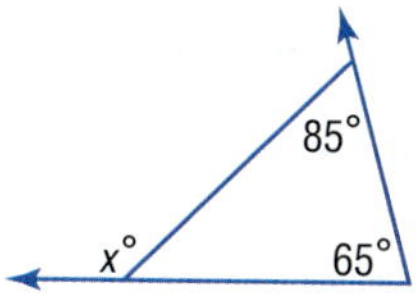

2.

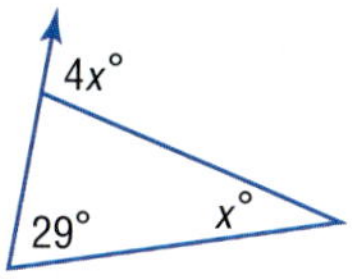

SPEED SKATING A speed skater forms at least two sets of triangles and exterior angles as she skates. Find each measure.

3. $m\angle 1$

4. $m\angle 2$

5. $m\angle 3$

6. $m\angle 4$

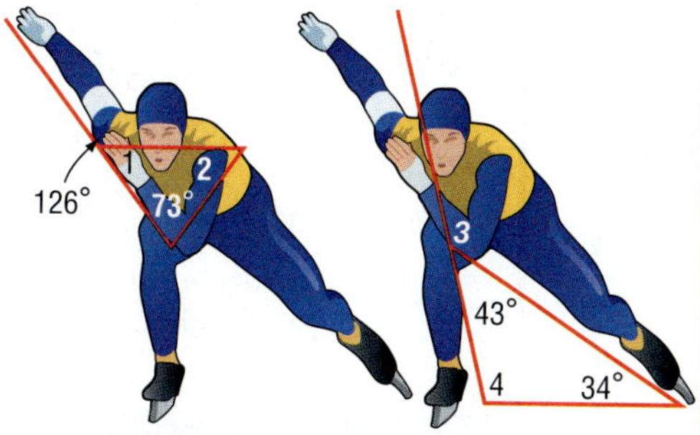

ALGEBRA **Find x and the measures of the unknown sides of each triangle.** (Lesson 4-1)

7.

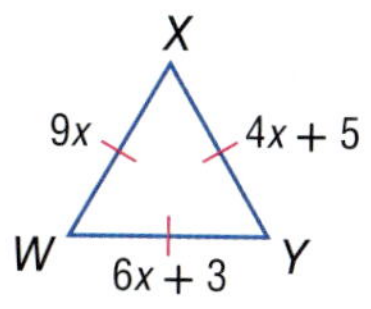

8.

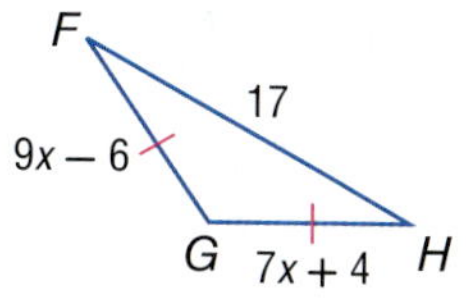

9. **TRAVEL** A plane travels from Des Moines to Phoenix, on to Atlanta, and back to Des Moines, as shown below. Find the distance in miles from Des Moines to Phoenix if the total trip was 3482 miles.

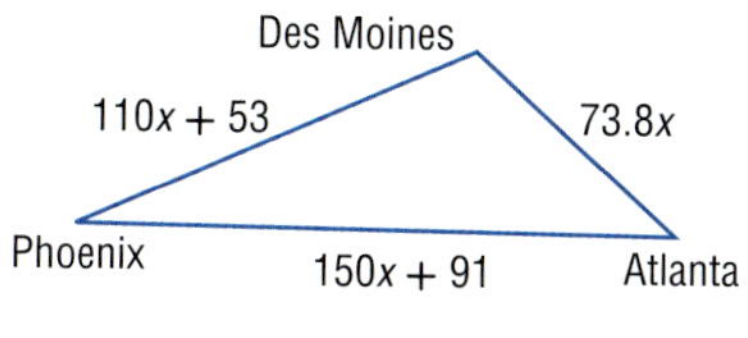

QuickReview

EXAMPLE 1

Find the measure of each numbered angle.

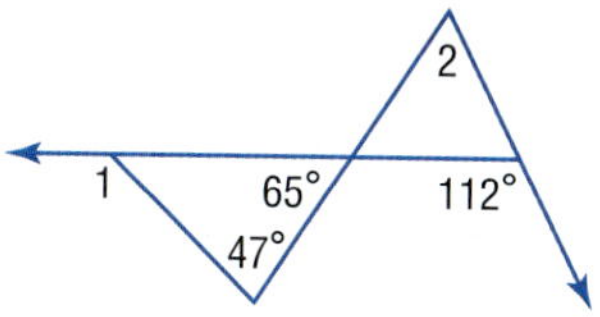

a. $m\angle 1$

$m\angle 1 = 65 + 47$	Exterior ∠ Theorem
$m\angle 1 = 112$	Add.

b. $m\angle 2$

$180 = m\angle 2 + 68 + 65$	Triangle Sum Theorem
$180 = m\angle 2 + 133$	Simplify.
$m\angle 2 = 47$	Subtract.

EXAMPLE 2

ALGEBRA **Find the measures of the sides of isosceles $\triangle XYZ$.**

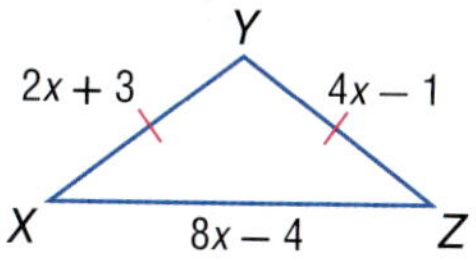

$XY = YZ$	Given
$2x + 3 = 4x - 1$	Substitution
$-2x = -4$	Subtract.
$x = 2$	Simplify.
$XY = 2x + 3$	Given
$= 2(2) + 3$ or 7	$x = 2$
$YZ = XY$	Given
$= 7$	$XY = 7$
$XZ = 8x - 4$	Given
$= 8(2) - 4$ or 12	$x = 2$

Online Option

Math Online Take a self-check Chapter Readiness Quiz at glencoe.com.

Get Started on Chapter 6

You will learn several new concepts, skills, and vocabulary terms as you study Chapter 6. To get ready, identify important terms and organize your resources. You may wish to refer to **Chapter 0** to review prerequisite skills.

FOLDABLES® Study Organizer

Quadrilaterals Make this Foldable to help you organize your Chapter 6 notes about quadrilaterals. Begin with one sheet of notebook paper.

1. **Fold** lengthwise to the holes.

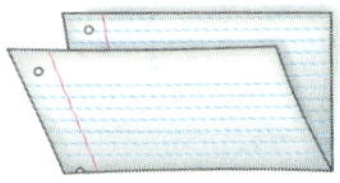

2. **Fold** along the width of the paper twice and unfold the paper.

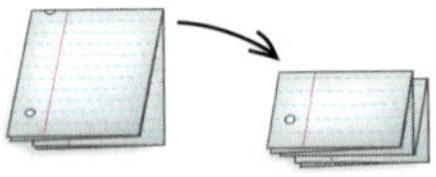

3. **Cut** along the fold marks on the left side of the paper.

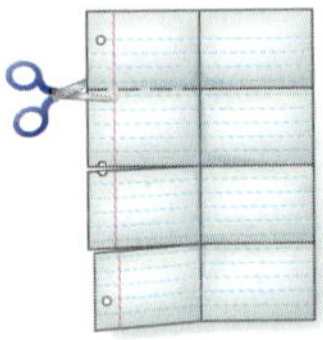

4. **Label** as shown.

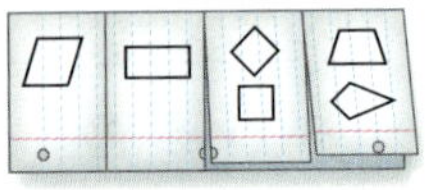

New Vocabulary

English		Español
diagonal	• p. 389 •	diagonal
parallelogram	• p. 399 •	paralelogramo
rectangle	• p. 419 •	rectángulo
rhombus	• p. 426 •	rombo
square	• p. 427 •	cuadrado
trapezoid	• p. 435 •	trapecio
base	• p. 435 •	base
legs	• p. 435 •	catetos
isosceles trapezoid	• p. 435 •	trapecio isósceles
midsegment of a trapezoid	• p. 437 •	paralela media de un trapecio

Review Vocabulary

exterior angle • p. 246 • ángulo externo an angle formed by one side of a triangle and the extension of another side

remote interior angle • p. 246 • ángulos internos no adyacentes the angles of a triangle that are not adjacent to a given exterior angle

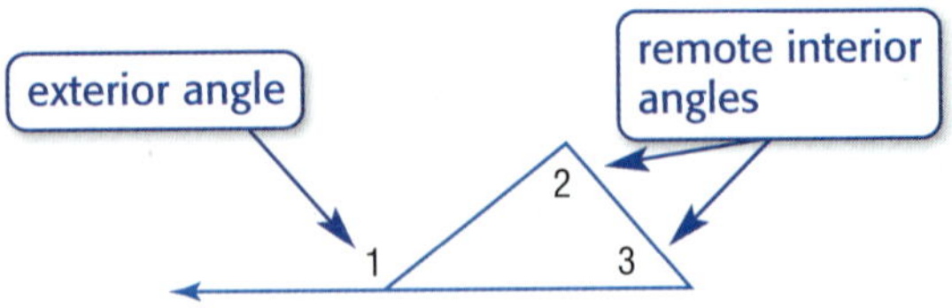

slope • p. 186 • pendiente for a (nonvertical) line containing two points (x_1, y_1) and (x_2, y_2), the number m given by the formula $m = \frac{(y_2 - y_1)}{(x_2 - x_1)}$ where $x_2 \neq x_1$

Multilingual eGlossary glencoe.com

Math Online glencoe.com

- Study the chapter online
- Explore **Get Animated** activities
- Get extra help from your own **Personal Tutor**
- Use **Extra Examples** for additional help
- Take a **Self-Check Quiz**
- **Review Vocabulary** in fun ways

6-1 Angles of Polygons

Then
You named and classified polygons. (Lesson 1-6)

Now
- Find and use the sum of the measures of the interior angles of a polygon.
- Find and use the sum of the measures of the exterior angles of a polygon.

New Vocabulary
diagonal

Math Online
glencoe.com

Why?

To create their honeycombs, young worker honeybees excrete flecks of wax that are carefully molded by other bees to form hexagonal cells. The cells are less than 0.1 millimeter thick, but they support almost 25 times their own weight. The cell walls all stand at exactly the same angle to one another. This angle is the measure of the interior angle of a regular hexagon.

Polygon Interior Angles Sum A **diagonal** of a polygon is a segment that connects any two nonconsecutive vertices.

The vertices of polygon *PQRST* that are not consecutive with vertex *P* are vertices *R* and *S*. Therefore, polygon *PQRST* has two diagonals from vertex *P*, $\overline{PR}$ and $\overline{PS}$. Notice that the diagonals from vertex *P* separate the hexagon into three triangles.

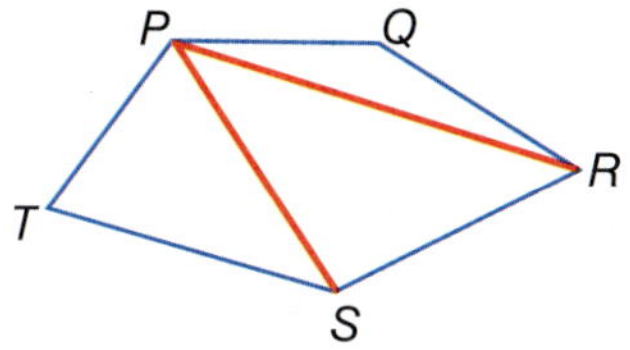

The sum of the angle measures of a polygon is the sum of the angle measures of the triangles formed by drawing all the possible diagonals from one vertex.

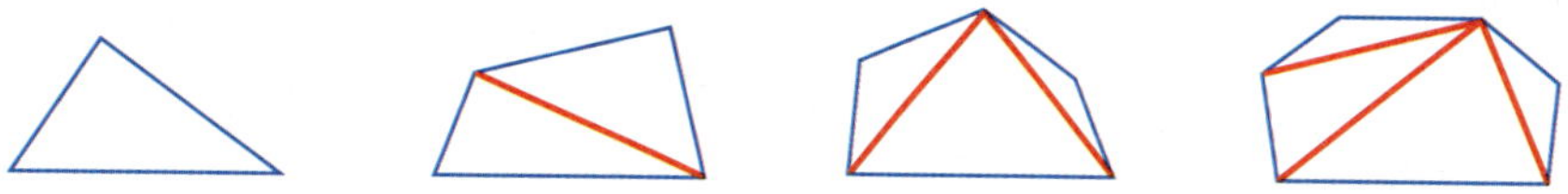

Triangle | Quadrilateral | Pentagon | Hexagon

Since the sum of the angle measures of a triangle is 180, we can make a table and look for a pattern to find the sum of the angle measures for any convex polygon.

Polygon	Number of Sides	Number of Triangles	Sum of Interior Angle Measures
Triangle	3	1	(1)180 or 180
Quadrilateral	4	2	(2)180 or 360
Pentagon	5	3	(3)180 or 540
Hexagon	6	4	(4)180 or 720
n-gon	*n*	$n - 2$	$(n - 2)180$

This leads to the following theorem.

Theorem 6.1 Polygon Interior Angles Sum

The sum of the interior angle measures of an *n*-sided convex polygon is $(n - 2) \cdot 180$.

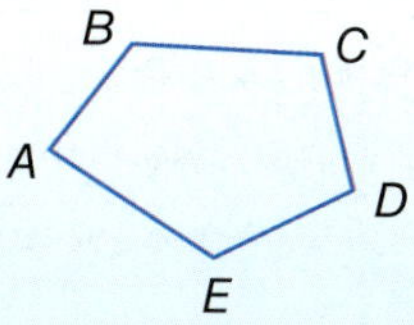

Example

$$m\angle A + m\angle B + m\angle C + m\angle D + m\angle E = (5 - 2) \cdot 180$$
$$= 540$$

You will prove Theorem 6.1 for octagons in Exercise 42.

You can use the Polygon Interior Angles Sum Theorem to find the sum of the interior angles of a polygon and to find missing measures in polygons.

StudyTip

Naming Polygons Remember, a polygon with n-sides is an n-*gon*, but several polygons have special names.

Number of Sides	Polygon
3	triangle
4	quadrilateral
5	pentagon
6	hexagon
7	heptagon
8	octagon
9	nonagon
10	decagon
11	hendecagon
12	dodecagon
n	n-gon

EXAMPLE 1 Find the Interior Angles Sum of a Polygon

a. Find the sum of the measures of the interior angles of a convex heptagon.

A heptagon has seven sides. Use the Polygon Interior Angles Sum Theorem to find the sum of its interior angle measures.

$(n - 2) \cdot 180 = (7 - 2) \cdot 180$ **$n = 7$**

$= 5 \cdot 180$ or 900 **Simplify.**

The sum of the measures is 900.

CHECK Draw a convex polygon with seven sides. Use a protractor to measure each angle to the nearest degree. Then find the sum of these measures.

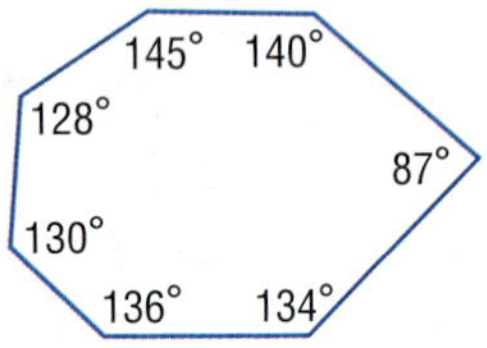

$128 + 145 + 140 + 87 + 134 + 136 + 130 = 900$ ✓

b. ALGEBRA Find the measure of each interior angle of quadrilateral *ABCD*.

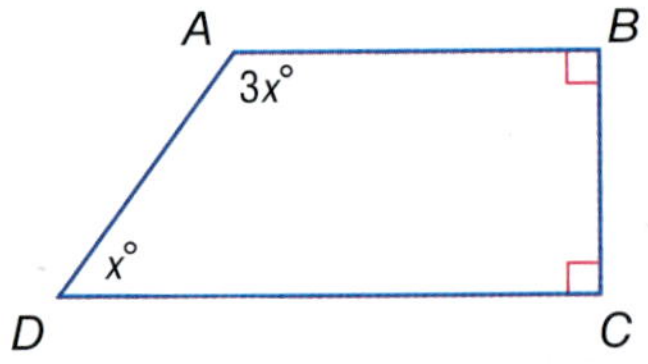

Step 1 Find x.

Since there are 4 angles, the sum of the interior angle measures is $(4 - 2) \cdot 180$ or 360.

$360 = m\angle A + m\angle B + m\angle C + m\angle D$ **Sum of interior angle measures**

$360 = 3x + 90 + 90 + x$ **Substitution**

$360 = 4x + 180$ **Combine like terms.**

$180 = 4x$ **Subtract 180 from each side.**

$45 = x$ **Divide each side by 4.**

Step 2 Use the value of x to find the measure of each angle.

$m\angle A = 3x$ $= 3(45)$ or 135

$m\angle B = 90$

$m\angle C = 90$

$m\angle D = x$ $= 45$

Guided Practice

1A. Find the sum of the measures of the interior angles of a convex octagon.

1B. Find the measure of each interior angle of pentagon *HJKLM* shown.

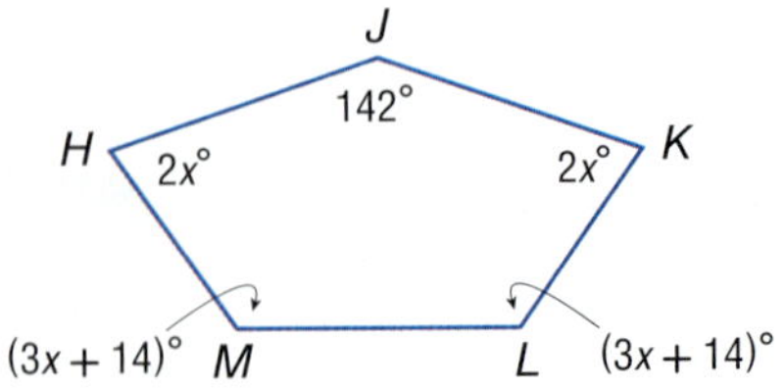

Personal Tutor glencoe.com

Recall from Lesson 1-6 that in a regular polygon, all of the interior angles are congruent. You can use this fact and the Polygon Interior Angle Sum Theorem to find the interior angle measure of any regular polygon.

Review Vocabulary

regular polygon a convex polygon in which all of the sides are congruent and all of the angles are congruent (Lesson 1-6)

Real-World Link

Susan B. Anthony was a leader of the women's suffrage movement in the late 1800s, which eventually led to the Nineteenth Amendment giving women the right to vote. In 1979, the Susan B. Anthony one-dollar coin was first minted, making her the first woman to be depicted on U.S. currency.

Source: *Encyclopaedia Britannica*

Real-World EXAMPLE 2 Interior Angle Measure of Regular Polygon

TENTS The poles for a tent form the vertices of a regular hexagon. When the poles are properly positioned, what is the measure of the angle formed at a corner of the tent?

Understand Draw a diagram of the situation.

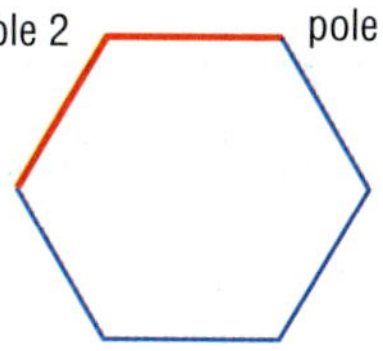

The measure of the angle formed at a corner of the tent is an interior angle of a regular hexagon.

Plan Use the Polygon Interior Angles Sum Theorem to find the sum of the measures of the angles. Since the angles of a regular polygon are congruent, divide this sum by the number of angles to find the measure of each interior angle.

Solve **Step 1** Find the sum of the interior angle measures.

$$(n - 2) \cdot 180 = (6 - 2) \cdot 180 \qquad n = 6$$
$$= 4 \cdot 180 \text{ or } 720 \qquad \text{Simplify.}$$

Step 2 Find the measure of one interior angle.

$$\frac{\text{sum of interior angle measures}}{\text{number of congruent angles}} = \frac{720}{6} \qquad \text{Substitution}$$
$$= 120 \qquad \text{Divide.}$$

The angle at a corner of the tent measures 120.

Check To verify that this measure is correct, use a ruler and a protractor to draw a regular hexagon using 120 as the measure of each interior angle. The last side drawn should connect with the beginning point of the first segment drawn. ✓

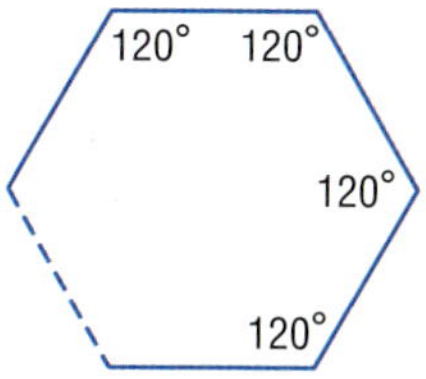

Guided Practice

2A. COINS Find the measure of each interior angle of the regular 11-gon that appears on the face of a Susan B. Anthony one-dollar coin.

2B. HOT TUBS A certain company makes hot tubs in a variety of different shapes. Find the measure of each interior angle of the nonagon model.

Personal Tutor glencoe.com

Given the interior angle measure of a regular polygon, you can also use the Polygon Interior Angles Sum Theorem to find a polygon's number of sides.

EXAMPLE 3 Find Number of Sides Given Interior Angle Measure

The measure of an interior angle of a regular polygon is 135. Find the number of sides in the polygon.

Let n = the number of sides in the polygon. Since all angles of a regular polygon are congruent, the sum of the interior angle measures is $135n$. By the Polygon Interior Angles Sum Theorem, the sum of the interior angle measures can also be expressed as $(n-2) \cdot 180$.

$135n = (n-2) \cdot 180$	Write an equation.
$135n = 180n - 360$	Distributive Property
$-45n = -360$	Subtract $180n$ from each side.
$n = 8$	Divide each side by -45.

The polygon has 8 sides.

Guided Practice

3. The measure of an interior angle of a regular polygon is 144. Find the number of sides in the polygon.

Personal Tutor glencoe.com

Review Vocabulary

exterior angle an angle formed by one side of a polygon and the extension of another side (Lesson 4-2)

Polygon Exterior Angles Sum Does a relationship exist between the number of sides of a convex polygon and the sum of its exterior angle measures? Examine the polygons below in which an exterior angle has been measured at each vertex.

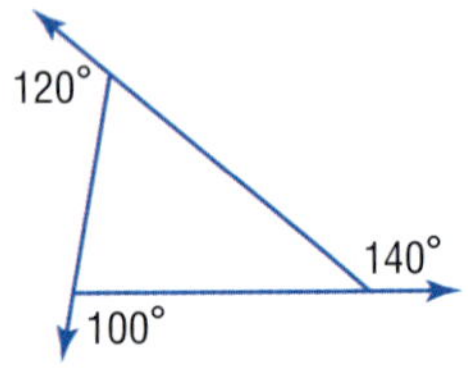

120 + 100 + 140 = 360

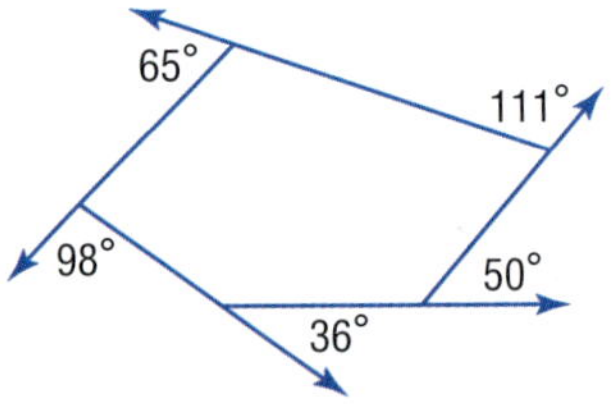

65 + 98 + 36 + 50 + 111 = 360

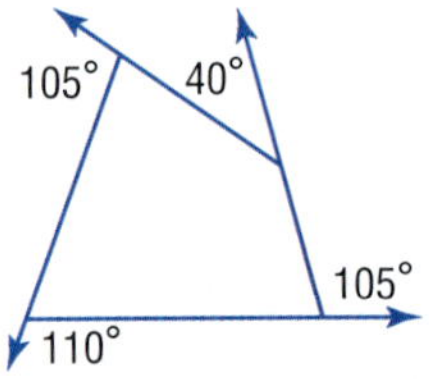

105 + 110 + 105 + 40 = 360

Notice that the sum of the exterior angle measures in each case is 360. This suggests the following theorem.

Theorem 6.2 Polygon Exterior Angles Sum

The sum of the exterior angle measures of a convex polygon, one angle at each vertex, is 360.

Example

$m\angle 1 + m\angle 2 + m\angle 3 + m\angle 4 + m\angle 5 + m\angle 6 = 360$

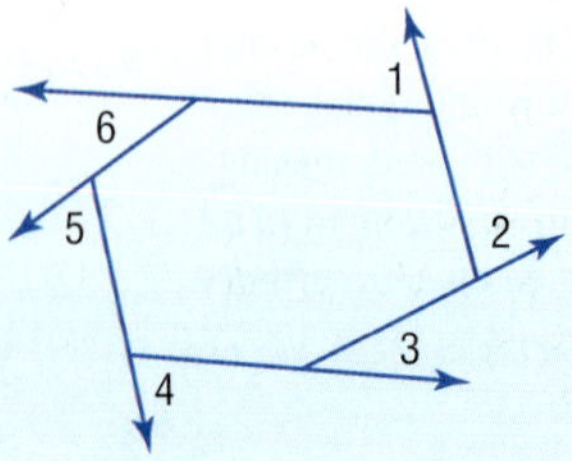

You will prove Theorem 6.2 in Exercise 43.

EXAMPLE 4 Find Exterior Angle Measures of a Polygon

a. ALGEBRA Find the value of x in the diagram.

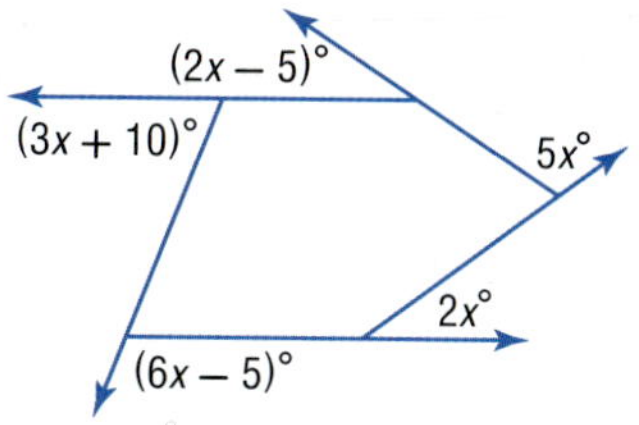

Use the Polygon Exterior Angles Sum Theorem to write an equation. Then solve for x.

$$(2x - 5) + 5x + 2x + (6x - 5) + (3x + 10) = 360$$
$$(2x + 5x + 2x + 6x + 3x) + [-5 + (-5) + 10] = 360$$
$$18x = 360$$
$$x = \frac{360}{18} \text{ or } 20$$

b. Find the measure of each exterior angle of a regular nonagon.

A regular nonagon has 9 congruent sides and 9 congruent interior angles. The exterior angles are also congruent, since angles supplementary to congruent angles are congruent. Let n = the measure of each exterior angle and write and solve an equation.

$9n = 360$ **Polygon Exterior Angles Sum Theorem**

$n = 40$ **Divide each side by 9.**

The measure of each exterior angle of a regular nonagon is 40.

StudyTip

Alternative Method To find the measure of each exterior angle of a regular polygon, you can find the measure of each interior angle and subtract this measure from 180, since an exterior angle and its corresponding interior angle are supplementary.

Guided Practice

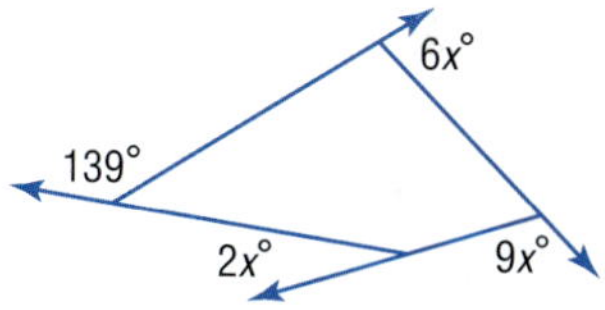

4A. Find the value of x in the diagram.

4B. Find the measure of each exterior angle of a regular dodecagon.

Personal Tutor glencoe.com

Check Your Understanding

Example 1 p. 390

Find the sum of the measures of the interior angles of each convex polygon.

1. decagon

2. pentagon

Find the measure of each interior angle.

3.

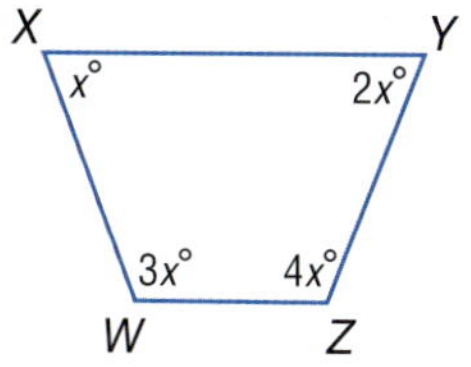

4.

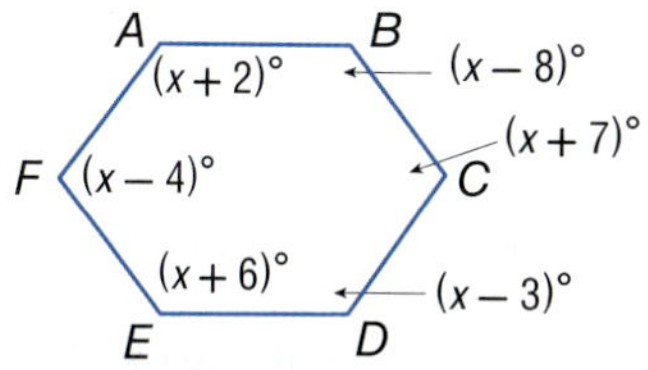

Example 2 p. 391

5. **AMUSEMENT** The Wonder Wheel at Coney Island in Brooklyn, New York, is a regular polygon with 16 sides. What is the measure of each interior angle of the polygon?

Example 3 p. 392

The measure of an interior angle of a regular polygon is given. Find the number of sides in the polygon.

6. 150

7. 170

Example 4
p. 393

Find the value of x in each diagram.

8.

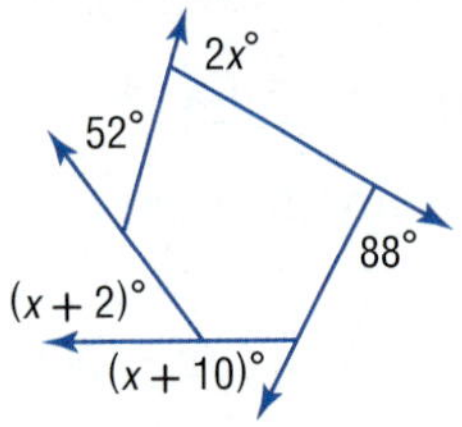

9.

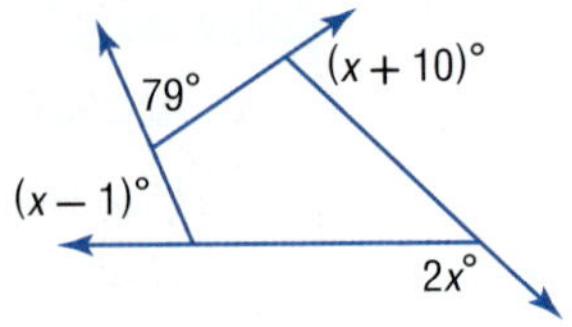

Find the measure of each exterior angle of each regular polygon.

10. quadrilateral

11. octagon

Practice and Problem Solving

● = **Step-by-Step Solutions** begin on page R20.
Extra Practice begins on page 969.

Example 1
p. 390

Find the sum of the measures of the interior angles of each convex polygon.

12. dodecagon

13. 20-gon

14. 29-gon

15. 32-gon

Find the measure of each interior angle.

16.

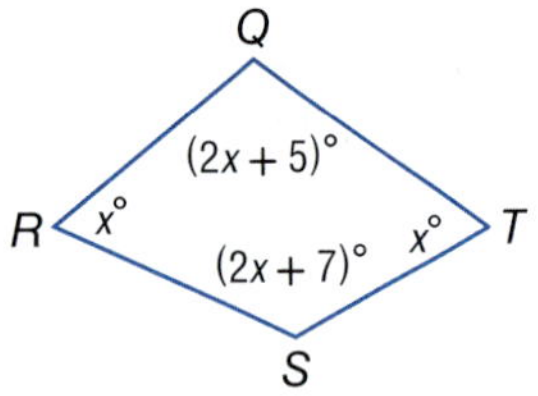

17

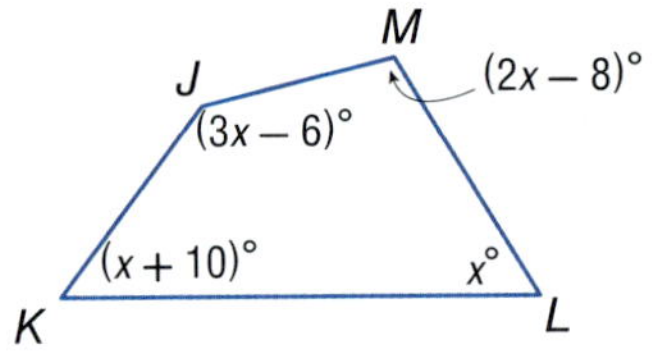

18.

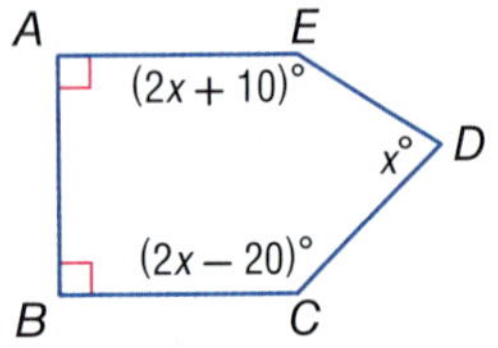

19.

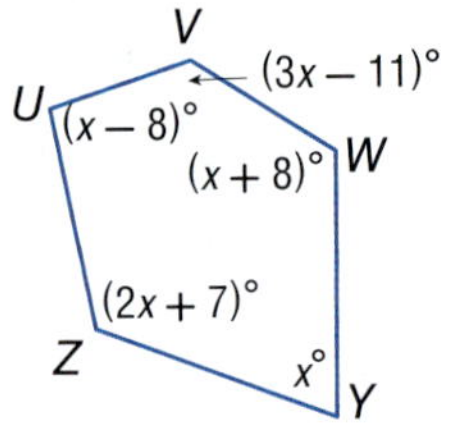

20. BASEBALL In baseball, home plate is a pentagon. The dimensions of home plate are shown. What is the sum of the measures of the interior angles of home plate?

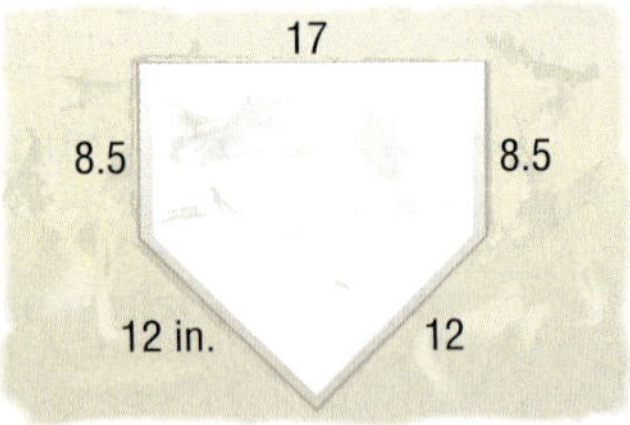

Example 2
p. 391

Find the measure of each interior angle of each regular polygon.

21. dodecagon

22. pentagon

23. decagon

24. nonagon

25. GAMES Hexagonal chess is played on a regular hexagonal board comprised of 92 small hexagons in three colors. The chess pieces are arranged so that a player can move any piece at the start of a game.

a. What is the sum of the measures of the interior angles of the chess board?

b. Does each interior angle have the same measure? If so, give the measure. Explain your reasoning.

Example 3
p. 392

The measure of an interior angle of a regular polygon is given. Find the number of sides in the polygon.

26. 60

27. 90

28. 120

29. 156

Example 4
p. 393

Find the value of x in each diagram.

30.

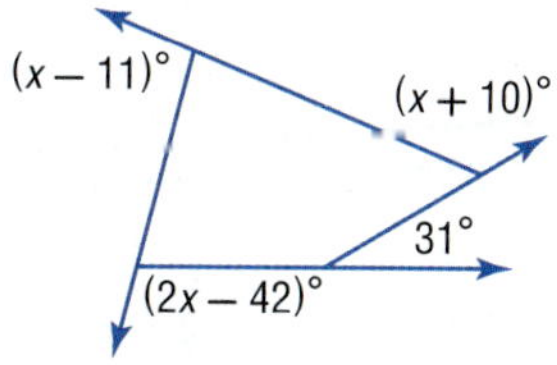

31.

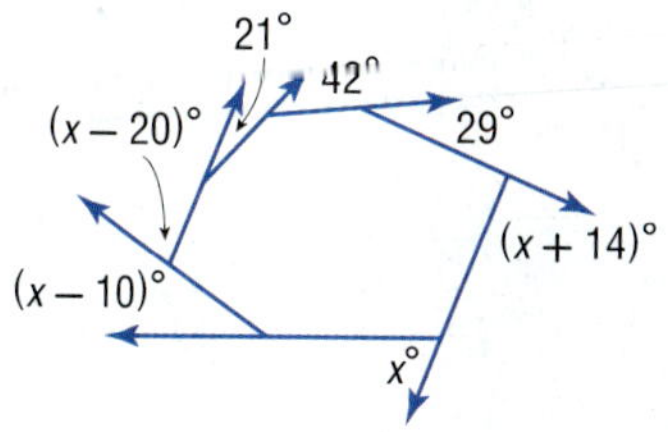

32.

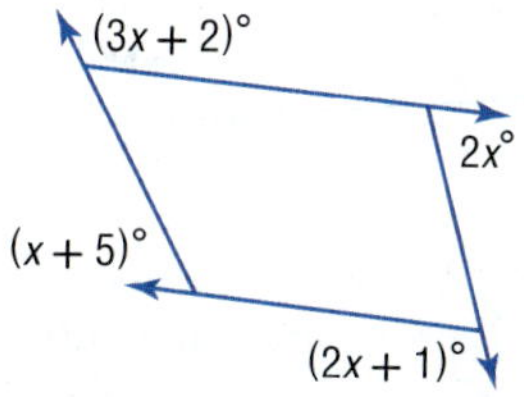

33.

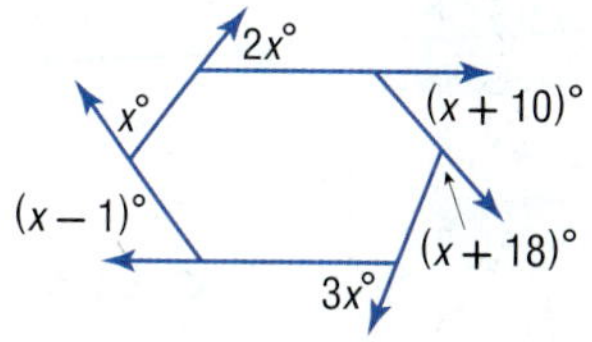

Real-World Link

A marching band's color guard provides visual interest using flag formations. Rhythm, timing, and angles are all critical to performance.

Source: Colorguard Corner

Find the measure of each exterior angle of each regular polygon.

34. decagon **35.** pentagon **36.** hexagon **37.** 15-gon

38. **COLOR GUARD** During the halftime performance for a football game, the color guard is planning a new formation in which seven members stand around a central point and stretch their flag to the person immediately to their left as shown.

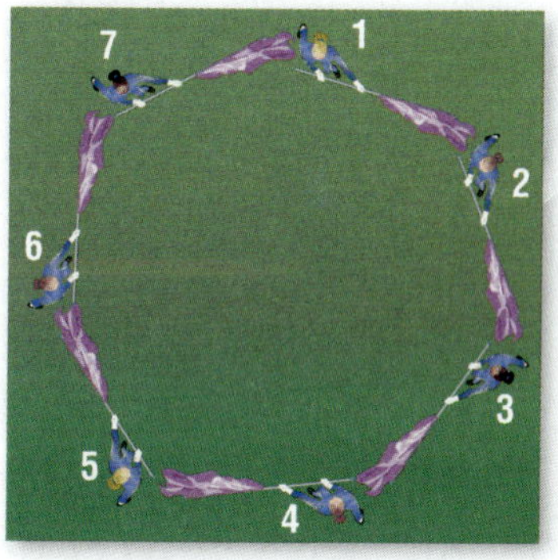

a. What is the measure of each exterior angle of the formation?

b. If the perimeter of the formation is 38.5 feet, how long is each flag?

Find the measures of an exterior angle and an interior angle given the number of sides of each regular polygon. Round to the nearest tenth, if necessary.

39. 7 **40.** 13 **41.** 14

42. **PROOF** Write a paragraph proof to prove the Polygon Interior Angles Sum Theorem for octagons.

43. **PROOF** Use algebra to prove the Polygon Exterior Angle Sum Theorem.

44. **PHOTOGRAPHY** The aperture on the camera lens shown is a regular 14-sided polygon.

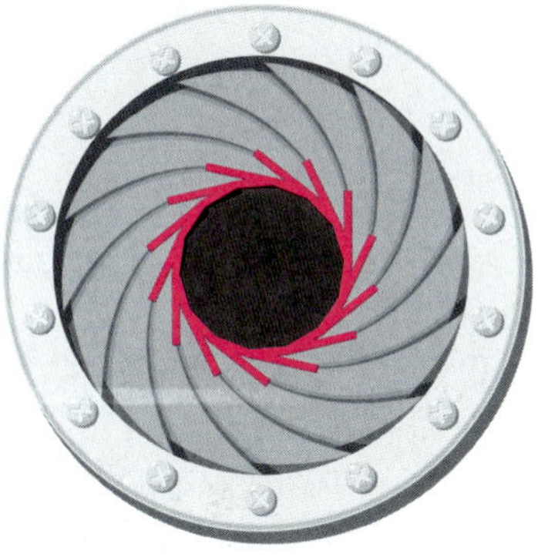

a. What is the measure of each interior angle of the polygon?

b. What is the measure of each exterior angle of the polygon?

ALGEBRA **Find the measure of each interior angle.**

45. decagon, in which the measures of the interior angles are $x + 5$, $x + 10$, $x + 20$, $x + 30$, $x + 35$, $x + 40$, $x + 60$, $x + 70$, $x + 80$, and $x + 90$

46. polygon *ABCDE,* in which the measures of the interior angles are $6x$, $4x + 13$, $x + 9$, $2x - 8$, $4x - 1$

Real-World Link

Theater in the round is staged so that the acting area is completely surrounded by the audience. The concept originated in ancient Greek theater.

Source: *Encyclopaedia Britannica*

47. **THEATER** The drama club would like to build a theater in the round for its next production.

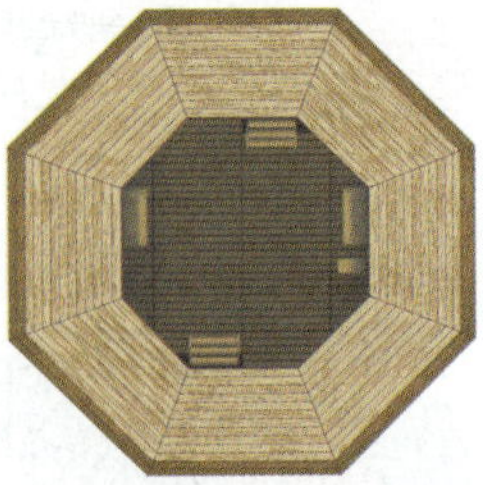

a. The stage is to be a regular octagon with a total perimeter of 60 feet. To what length should each board be cut to form the sides of the stage?

b. At what angle should each board be cut so that they will fit together as shown? Explain your reasoning.

48. **MULTIPLE REPRESENTATIONS** In this problem, you will explore angle and side relationships in special quadrilaterals.

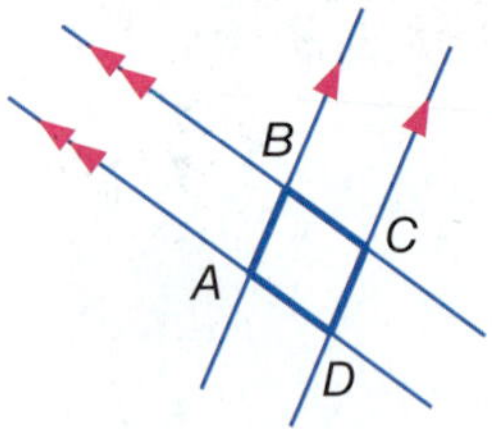

a. **GEOMETRIC** Draw two pairs of parallel lines that intersect like the ones shown. Label the quadrilateral formed by *ABCD*. Repeat these steps to form two additional quadrilaterals, *FGHJ* and *QRST*.

b. **TABULAR** Copy and complete the table below.

Quadrilateral	Lengths and Measures							
ABCD	$m\angle A$		$m\angle B$		$m\angle C$		$m\angle D$	
	AB		*BC*		*CD*		*DA*	
FGHJ	$m\angle F$		$m\angle G$		$m\angle H$		$m\angle J$	
	FG		*GH*		*HJ*		*JF*	
QRST	$m\angle Q$		$m\angle R$		$m\angle S$		$m\angle T$	
	QR		*RS*		*ST*		*TQ*	

c. **VERBAL** Make a conjecture about the relationship between the angles opposite each other in a quadrilateral formed by two pairs of parallel lines.

d. **VERBAL** Make a conjecture about the relationship between two consecutive angles in a quadrilateral formed by two pairs of parallel lines.

e. **VERBAL** Make a conjecture about the relationship between the sides opposite each other in a quadrilateral formed by two pairs of parallel lines.

H.O.T. Problems

Use Higher-Order Thinking Skills

49. **ERROR ANALYSIS** Marcus says that the sum of the exterior angles of a decagon is greater than that of a heptagon because a decagon has more sides. Liam says that the sum of the exterior angles for both polygons is the same. Is either of them correct? Explain your reasoning.

50. **CHALLENGE** Find the values of *a*, *b*, and *c* if *QRSTVX* is a regular hexagon. Justify your answer.

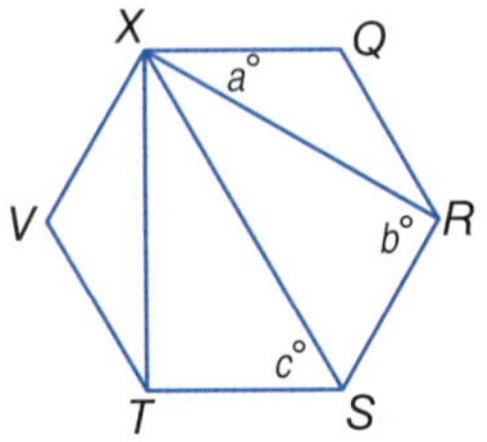

51. **REASONING** If two sides of a regular hexagon are extended to meet at a point in the exterior of the polygon, will the triangle formed *always*, *sometimes*, or *never* be equilateral? Justify your answer.

52. **OPEN ENDED** Sketch a polygon and find the sum of its interior angles. How many sides does a polygon with twice this interior angles sum have? Justify your answer.

53. **WRITING IN MATH** Explain how triangles are related to the Interior Angles Sum Theorem.

Standardized Test Practice

54. If the polygon shown is regular, what is $m\angle ABC$?

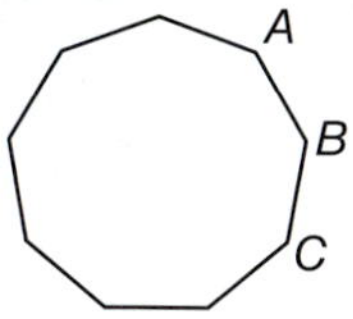

A 140
B 144
C 162
D 180

55. SHORT RESPONSE Figure *ABCDE* is a regular pentagon with line ℓ passing through side *AE*. What is $m\angle y$?

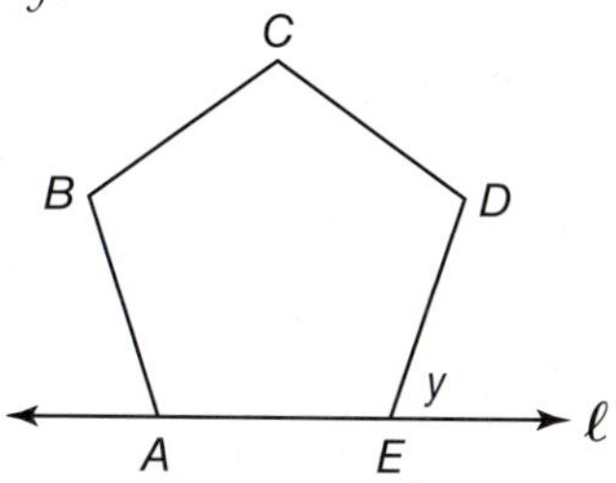

56. ALGEBRA $\frac{3^2 \cdot 4^5 \cdot 5^3}{5^3 \cdot 3^3 \cdot 4^6} =$

F $\frac{1}{60}$
G $\frac{1}{12}$
H $\frac{3}{4}$
J 12

57. SAT/ACT The sum of the measures of the interior angles of a polygon is twice the sum of the measures of its exterior angles. What type of polygon is it?

A square
B pentagon
C hexagon
D octagon

Spiral Review

Compare the given measures. (Lesson 5-6)

58. $m\angle CDE$ and $m\angle RST$

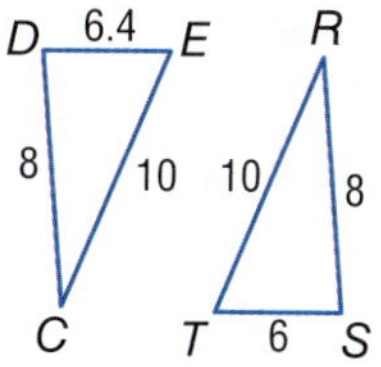

59. *JM* and *ML*

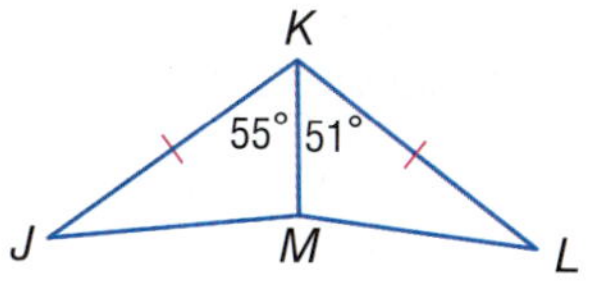

60. *WX* and *ZY*

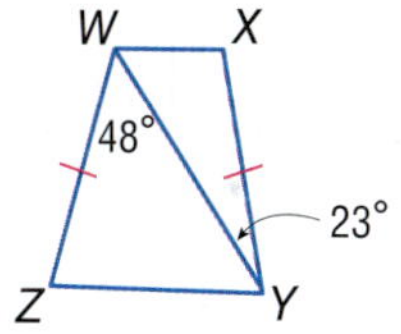

61. HISTORY The early Egyptians used to make triangles by using a rope with knots tied at equal intervals. Each vertex of the triangle had to occur at a knot. How many different triangles can be formed using the rope below? (Lesson 5-5)

Show that the triangles are congruent by identifying all congruent corresponding parts. Then write a congruence statement. (Lesson 4-3)

62.

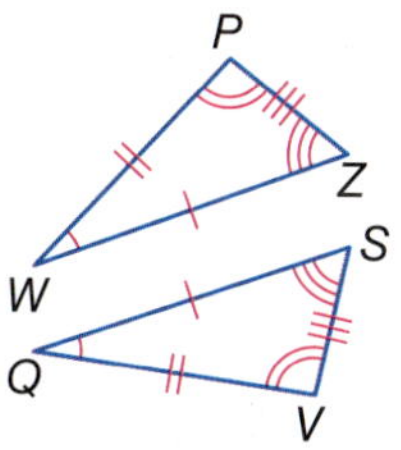

63.

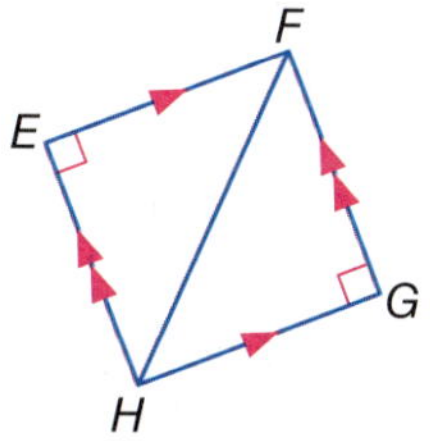

64.

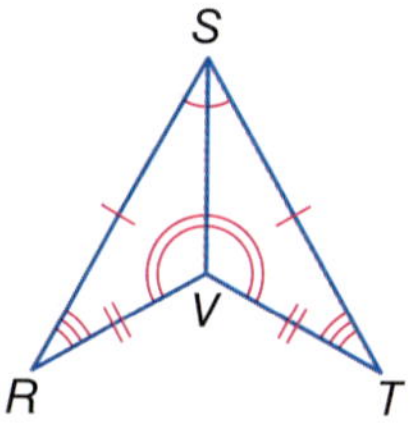

Skills Review

In the figure, $\ell \parallel m$ and $\overline{AC} \parallel \overline{BD}$. Name all pairs of angles for each type indicated. (Lesson 3-1)

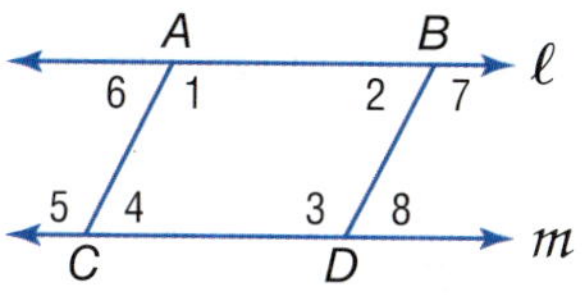

65. alternate interior angles

66. consecutive interior angles

EXTEND
6-1 Spreadsheet Lab
Angle of Polygons

Math Online glencoe.com
Graphing Technology Personal Tutor

Objective
You can use graphing technology to find the measures of the interior angles of regular polygons.

It is possible to find the interior and exterior measurements along with the sum of the interior angles of any regular polygon with n number of sides by using a spreadsheet.

ACTIVITY

Design a spreadsheet using the following steps.

- Label the columns as shown in the spreadsheet below.
- Enter the digits 3–10 in the first column.
- The number of triangles in a polygon is 2 fewer than the number of sides. Write a formula for Cell B1 to subtract 2 from each number in Cell A1.
- Enter a formula for Cell C1 so the spreadsheet will calculate the sum of the measures of the interior angles. Remember that the formula is $S = (n - 2)180$.
- Continue to enter formulas so that the indicated computation is performed. Then, copy each formula through Row 9. The final spreadsheet will appear as below.

Polygons and Angles

	A	B	C	D	E	F
1	Number of Sides	Number of Triangles	Sum of Measures of Interior Angles	Measure of Each Interior Angle	Measure of Each Exterior Angle	Measures of Exterior Angles
2	3	1	180	60	120	360
3	4	2	360	90	90	360
4	5	3	540	108	72	360
5	6	4	720	120	60	360
6	7	5	900	128.57	51.43	360
7	8	6	1080	135	45	360
8	9	7	1260	140	40	360
9	10	8	1440	144	36	360

Sheet 1 Sheet 2 Sheet 3

Exercises

1. Write the formula to find the measure of each interior angle in the polygon.
2. Write the formula to find the sum of the measures of the exterior angles.
3. What is the measure of each interior angle if the number of sides is 1? 2?
4. Is it possible to have values of 1 and 2 for the number of sides? Explain.

For Exercises 5–8, use the spreadsheet.

5. How many triangles are in a polygon with 17 sides?
6. Find the measure of an exterior angle of a regular polygon with 16 sides.
7. Find the measure of an interior angle of a regular polygon with 115 sides.
8. If the measure of the exterior angles is 0, find the measure of the interior angles. Is this possible? Explain.

6-2 Parallelograms

Then

You classified polygons with four sides as quadrilaterals. (Lesson 1-6)

Now

- Recognize and apply properties of the sides and angles of parallelograms.
- Recognize and apply properties of the diagonals of parallelograms.

New Vocabulary

parallelogram

Math Online

glencoe.com

Why?

The arm of the basketball goal shown can be adjusted to a height of 10 feet or 5 feet. Notice that as the height is adjusted, each pair of opposite sides of the quadrilateral formed by the arms remains parallel.

Sides and Angles of Parallelograms A **parallelogram** is a quadrilateral with both pairs of opposite sides parallel. To name a parallelogram, use the symbol ▱. In ▱*ABCD*, $\overline{BC} \parallel \overline{AD}$ and $\overline{AB} \parallel \overline{DC}$ by definition.

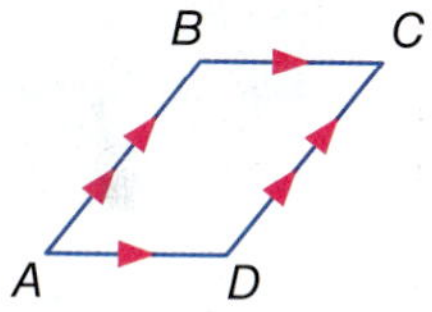

▱*ABCD*

Other properties of parallelograms are given in the theorems below.

Theorems — Properties of Parallelograms

6.3 If a quadrilateral is a parallelogram, then its opposite sides are congruent.

Abbreviation *Opp. sides of a ▱ are ≅.*

Example If *JKLM* is a parallelogram, then $\overline{JK} \cong \overline{ML}$ and $\overline{JM} \cong \overline{KL}$.

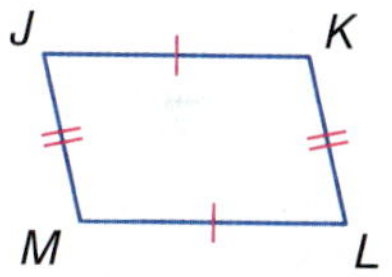

6.4 If a quadrilateral is a parallelogram, then its opposite angles are congruent.

Abbreviation *Opp. ∠s of a ▱ are ≅.*

Example If *JKLM* is a parallelogram, then $\angle J \cong \angle L$ and $\angle K \cong \angle M$.

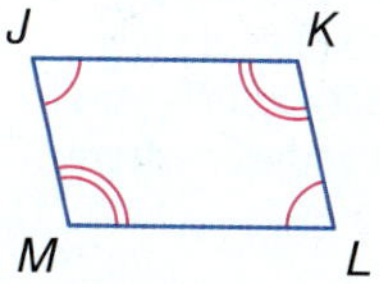

6.5 If a quadrilateral is a parallelogram, then its consecutive angles are supplementary.

Abbreviation *Cons. ∠s in a ▱ are supplementary.*

Example If *JKLM* is a parallelogram, then $x + y = 180$.

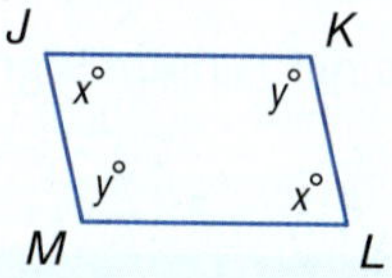

6.6 If a parallelogram has one right angle, then it has four right angles.

Abbreviation *If a ▱ has 1 rt. ∠, it has 4 rt. ∠s.*

Example In ▱*JKLM*, if $\angle J$ is a right angle, then $\angle K$, $\angle L$, and $\angle M$ are also right angles.

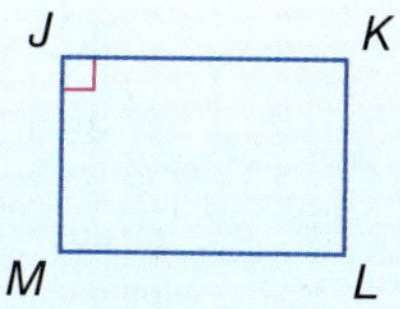

You will prove Theorems 6.3, 6.5, and 6.6 in Exercises 28, 26, and 7, respectively.

StudyTip

Including a Figure Theorems are presented in general terms. In a proof, you must include a drawing so that you can refer to segments and angles specifically.

Proof — Theorem 6.4

Write a two-column proof of Theorem 6.4.

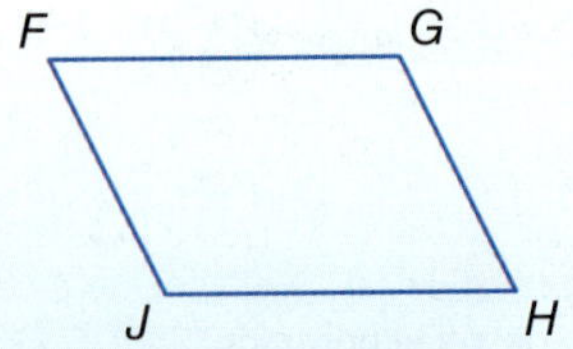

Given: $\square FGHJ$

Prove: $\angle F \cong \angle H$, $\angle J \cong \angle G$

Proof:

Statements	Reasons
1. $\square FGHJ$	**1.** Given
2. $\overline{FG} \parallel \overline{JH}$; $\overline{FJ} \parallel \overline{GH}$	**2.** Definition of parallelogram
3. $\angle F$ and $\angle J$ are supplementary. $\angle J$ and $\angle H$ are supplementary. $\angle H$ and $\angle G$ are supplementary.	**3.** If parallel lines are cut by a transversal, consecutive interior angles are supplementary.
4. $\angle F \cong \angle H$, $\angle J \cong \angle G$	**4.** Supplements of the same angles are congruent.

Real-World Career

Coach
Coaches organize amateur and professional atheletes, teaching them the fundamentals of a sport. They manage teams during both practice sessions and competitions. Additional tasks may include selecting and issuing sports equiment, materials, and supplies. Head coaches at public secondary schools usually have a bachelor's degree.

Real-World EXAMPLE 1 Use Properties of Parallelograms

BASKETBALL In $\square ABCD$, suppose $m\angle A = 55$, $AB = 2.5$ feet, and $BC = 1$ foot. Find each measure.

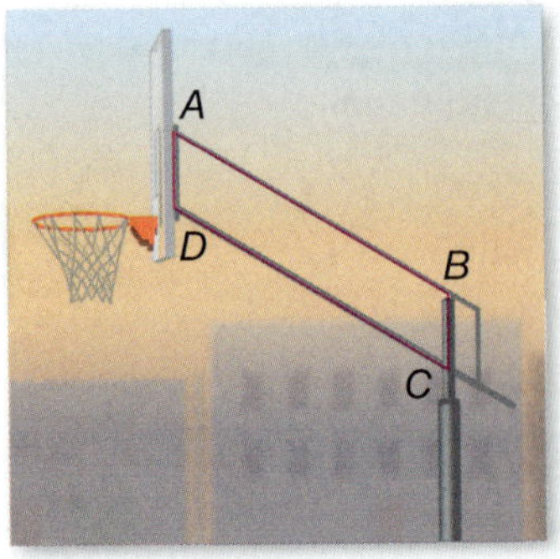

a. DC

$DC = AB$ — Opp. sides of a $\square$ are $\cong$.

$= 2.5$ ft — Substitution

b. $m\angle B$

$m\angle B + m\angle A = 180$ — Cons. ∡ in a $\square$ are supplementary.

$m\angle B + 55 = 180$ — Substitution

$m\angle B = 125$ — Subtract 55 from each side.

c. $m\angle C$

$m\angle C = m\angle A$ — Opp. ∡ of a $\square$ are $\cong$.

$= 55$ — Substitution

Guided Practice

1. MIRRORS The wall-mounted mirror shown uses parallelograms that change shape as the arm is extended. In $\square JKLM$, suppose $m\angle J = 47$. Find each measure.

A. $m\angle L$

B. $m\angle M$

C. Suppose the arm was extended further so that $m\angle J = 90$. What would be the measure of each of the other angles? Justify your answer.

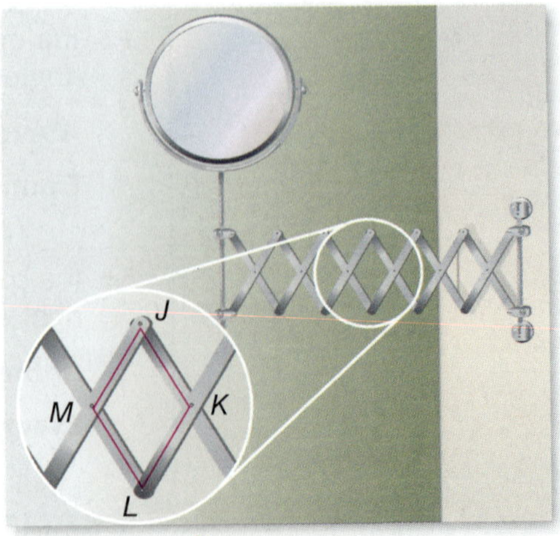

Personal Tutor glencoe.com

Diagonals of Parallelograms The diagonals of a parallelogram have special properties as well.

Theorems — Diagonals of Parallelograms

6.7 If a quadrilateral is a parallelogram, then its diagonals bisect each other.

Abbreviation *Diag. of a ▱ bisect each other.*

Example If $ABCD$ is a parallelogram, then $\overline{AP} \cong \overline{PC}$ and $\overline{DP} \cong \overline{PB}$.

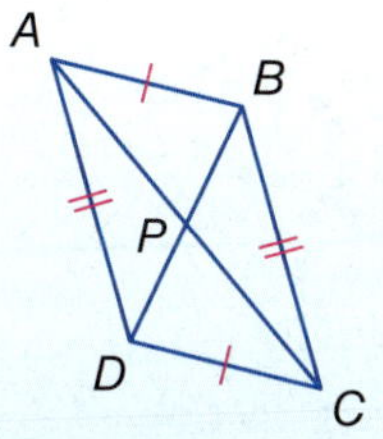

6.8 If a quadrilateral is a parallelogram, then each diagonal separates the parallelogram into two congruent triangles.

Abbreviation *Diag. separates a ▱ into 2 ≅ △s.*

Example If $ABCD$ is a parallelogram, then $\triangle ABD \cong \triangle CDB$.

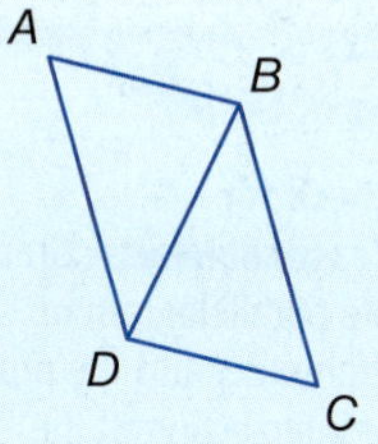

You will prove Theorems 6.7 and 6.8 in Exercises 29 and 27, respectively.

EXAMPLE 2 Use Properties of Parallelograms and Algebra

ALGEBRA If $QRST$ is a parallelogram, find the value of the indicated variable.

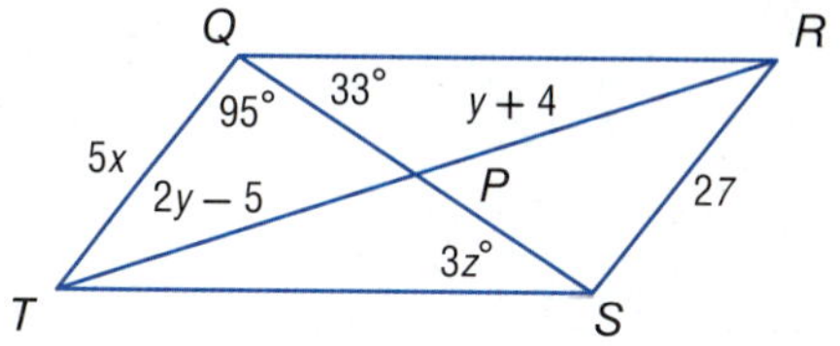

a. x

$\overline{QT} \cong \overline{RS}$	Opp. sides of a ▱ are ≅.
$QT = RS$	Definition of congruence
$5x = 27$	Substitution
$x = 5.4$	Divide each side by 5.

b. y

$\overline{TP} \cong \overline{PR}$	Diag. of a ▱ bisect each other.
$TP = PR$	Definition of congruence
$2y - 5 = y + 4$	Substitution
$y = 9$	Subtract y and add 5 to each side.

c. z

$\triangle TQS \cong \triangle RSQ$	Diag. separates a ▱ into 2 ≅ △s.
$\angle QST \cong \angle SQR$	CPCTC
$m\angle QST = m\angle SQR$	Definition of congruence
$3z = 33$	Substitution
$z = 11$	Divide each side by 3.

> **StudyTip**
> **Congruent Triangles** A parallelogram with two diagonals divides the figure into two pairs of congruent triangles.

Guided Practice

Find the value of each variable in the given parallelogram.

2A.

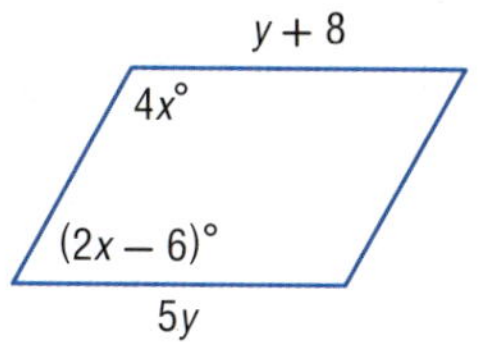

2B.

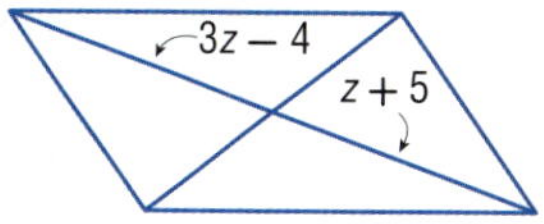

Personal Tutor glencoe.com

You can use Theorem 6.7 to determine the coordinates of the intersection of the diagonals of a parallelogram on a coordinate plane given the coordinates of the vertices.

EXAMPLE 3 Parallelograms and Coordinate Geometry

COORDINATE GEOMETRY Determine the coordinates of the intersection of the diagonals of ▱*FGHJ* with vertices *F*(−2, 4), *G*(3, 5), *H*(2, −3), and *J*(−3, −4).

Since the diagonals of a parallelogram bisect each other, their intersection point is the midpoint of $\overline{FH}$ and $\overline{GJ}$. Find the midpoint of $\overline{FH}$ with endpoints (−2, 4) and (2, −3).

$$\left(\frac{x_1 + x_2}{2}, \frac{y_1 + y_2}{2}\right) = \left(\frac{-2 + 2}{2}, \frac{4 + (-3)}{2}\right) \quad \text{Midpoint Formula}$$

$$= (0, 0.5) \quad \text{Simplify.}$$

The coordinates of the intersection of the diagonals of ▱*FGHJ* are (0, 0.5).

CHECK Find the midpoint of $\overline{GJ}$ with endpoints (3, 5) and (−3, −4).

$$\left(\frac{3 + (-3)}{2}, \frac{5 + (-4)}{2}\right) = (0, 0.5) \checkmark$$

StudyTip

Check for Reasonableness Graph the parallelogram in Example 3 and the point of intersection of the diagonals you found. Draw the diagonals. The point of intersection appears to be correct.

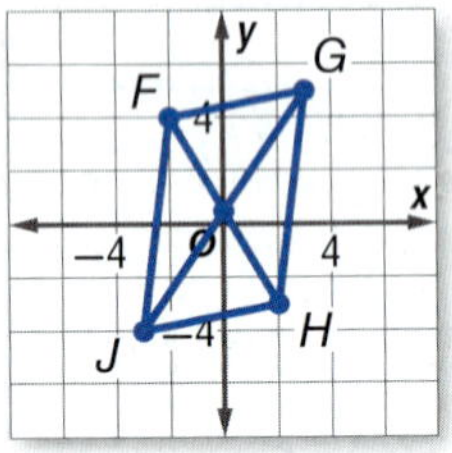

Guided Practice

3. **COORDINATE GEOMETRY** Determine the coordinates of the intersection of the diagonals of *RSTU* with vertices *R*(−8, −2), *S*(−6, 7), *T*(6, 7), and *U*(4, −2).

Personal Tutor glencoe.com

You can use the properties of parallelograms and their diagonals to write proofs.

EXAMPLE 4 Proofs Using the Properties of Parallelograms

Write a paragraph proof.

Given: ▱*ABDG*, $AF \cong CF$

Prove: $\angle BDG \cong \angle C$

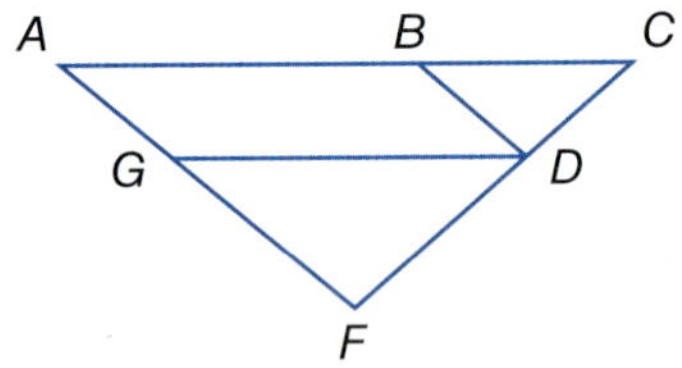

Proof:

We are given *ABDG* is a parallelogram. Since opposite angles in a parallelogram are congruent, $\angle BDG \cong \angle A$. We are also given that $\overline{AF} \cong \overline{CF}$. By the Isosceles Triangle Theorem, $\angle A \cong \angle C$. So, by the Transitive Property of Congruence, $\angle BDG \cong \angle C$.

Guided Practice

4. Write a two-column proof.

Given: ▱*HJKP* and ▱*PKLM*

Prove: $\overline{HJ} \cong \overline{ML}$

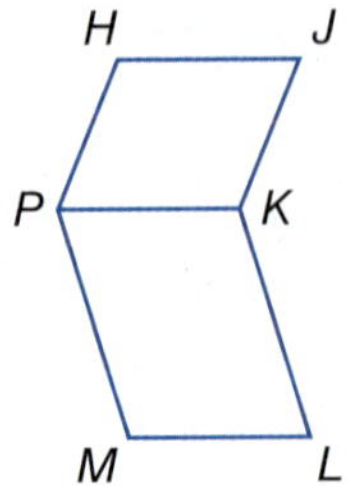

Personal Tutor glencoe.com

Check Your Understanding

Example 1
p. 400

1. **NAVIGATION** To chart a course, sailors use a *parallel ruler.* One edge of the ruler is placed along the line representing the direction of the course to be taken. Then the other ruler is moved until its edge reaches the compass rose printed on the chart. Reading the compass determines which direction to travel. The rulers and the crossbars form of the tool $\square MNPQ$.

   a. If $m\angle NMQ = 32$, find $m\angle MNP$.

   b. If $m\angle MQP = 125$, find $m\angle MNP$.

   c. If $MQ = 4$, what is NP?

Example 2
p. 401

ALGEBRA **Find the value of each variable in each parallelogram.**

2.
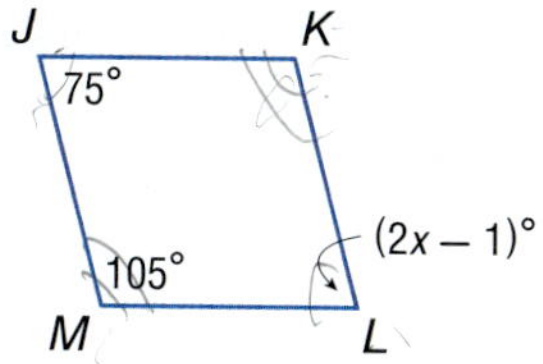

3.
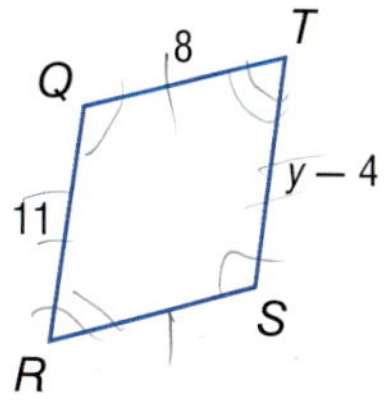

4.
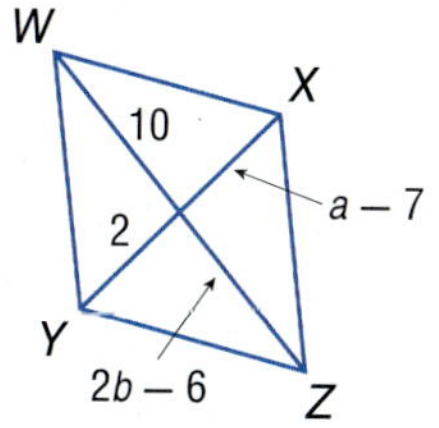

5.
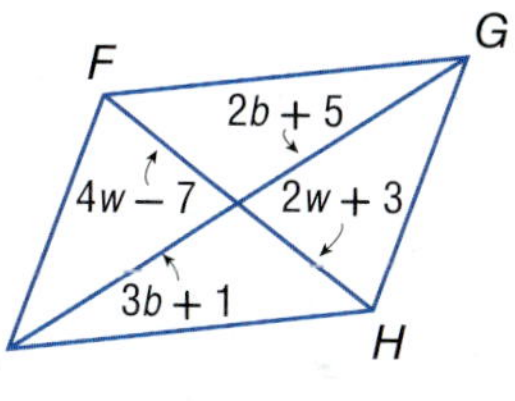

Example 3
p. 402

6. **COORDINATE GEOMETRY** Determine the coordinates of the intersection of the diagonals of $\square ABCD$ with vertices $A(-4, 6)$, $B(5, 6)$, $C(4, -2)$, and $D(-5, -2)$.

Example 4
p. 402

PROOF **Write the indicated type of proof.**

7. paragraph

   Given: $\square ABCD$, $\angle A$ is a right angle.
   Prove: $\angle B$, $\angle C$, and $\angle D$ are right angles. (Theorem 6.6)

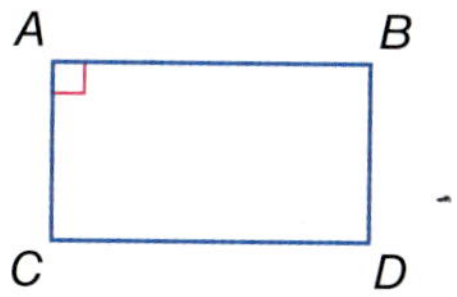

8. two-column

   Given: $ABCH$ and $DCGF$ are parallelograms.
   Prove: $\angle A \cong \angle F$

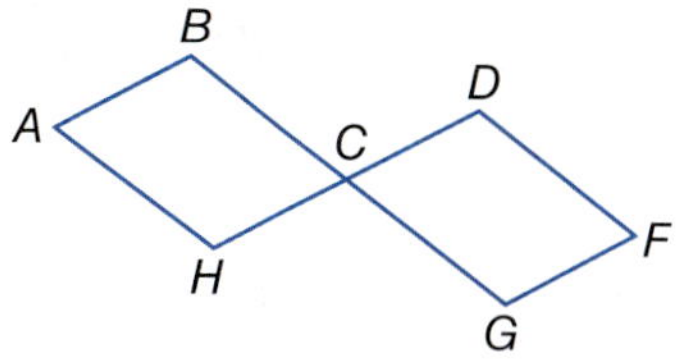

Practice and Problem Solving

● = **Step-by-Step Solutions** begin on page R20.
Extra Practice begins on page 969.

Example 1
p. 400

Use $\square PQRS$ to find each measure.

9. $m\angle R$

10. QR

11. QP

12. $m\angle S$

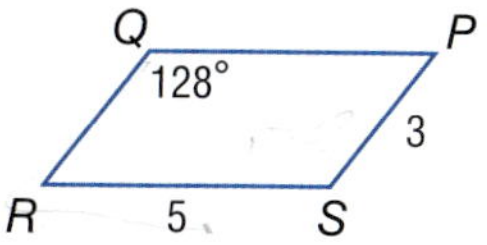

Real-World Link

Only one dog, a Smooth Fox Terrier named Champion Warren Remedy, has won Best in Show at the Westminster Kennel Club Dog Show three times.

Source: Westminster Kennel Club

13 **HOME DECOR** The slats on Venetian blinds are designed to remain parallel in order to direct the path of light coming in a widow. In ▱$FGHJ$, $FJ = \frac{3}{4}$ inch, $FG = 1$ inch, and $\angle JHG = 62$.

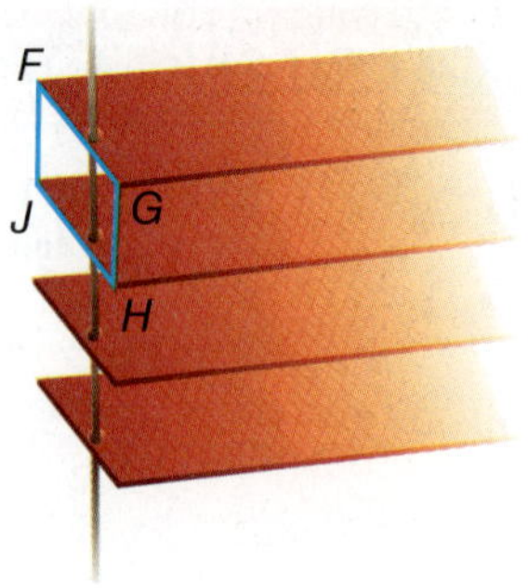

Find each measure.

a. JH

b. GH

c. $m\angle JFG$

d. $m\angle FJH$

14. **DOG SHOWS** Wesley is a member of the kennel club in his area. His club uses accordion fencing like the section shown at the right to block out areas at dog shows.

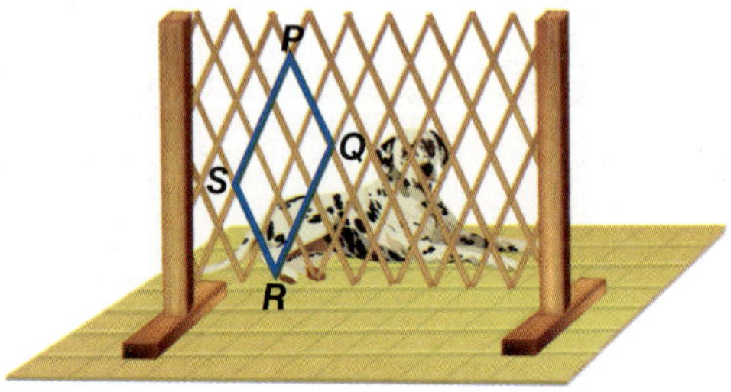

a. Identify two pairs of congruent segments.

b. Identify two pairs of supplementary angles.

Example 2
p. 401

ALGEBRA **Find the value of each variable in each parallelogram.**

15.

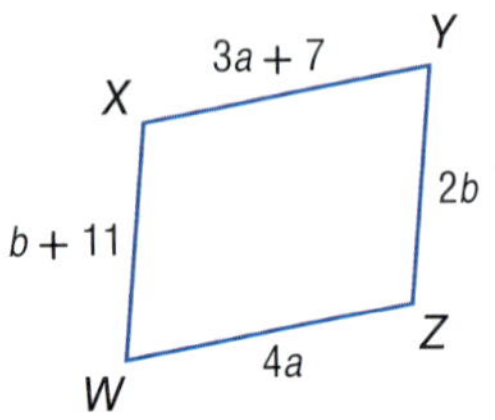

16.

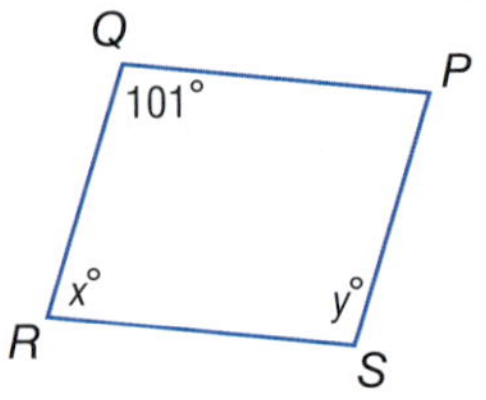

17.

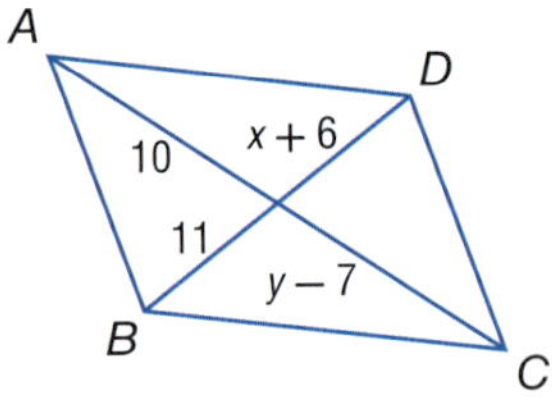

18.

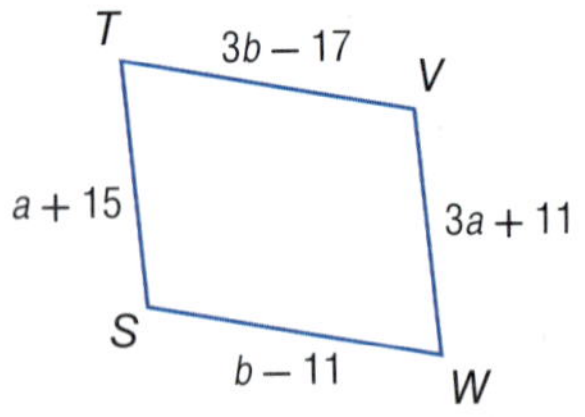

19.

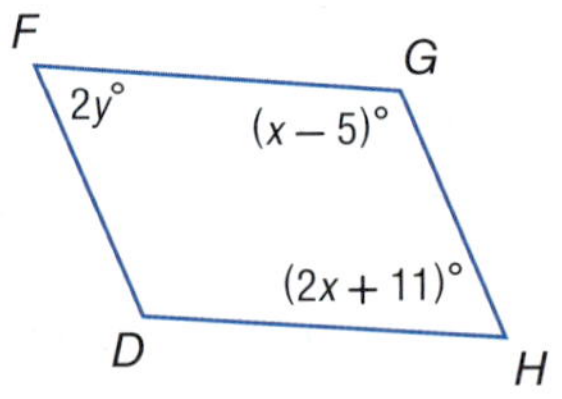

20.

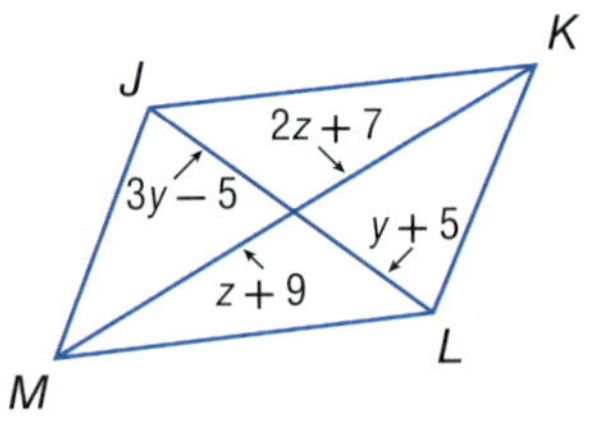

Example 3
p. 402

COORDINATE GEOMETRY **Find the coordinates of the intersection of the diagonals of ▱$WXYZ$ with the given vertices.**

21. $W(-1, 7)$, $X(8, 7)$, $Y(6, -2)$, $Z(-3, -2)$

22. $W(-4, 5)$, $X(5, 7)$, $Y(4, -2)$, $Z(-5, -4)$

Example 4
p. 402

PROOF **Write a two-column proof.**

23. **Given:** $WXTV$ and $ZYVT$ are parallelograms.
Prove: $\overline{WX} \cong \overline{ZY}$

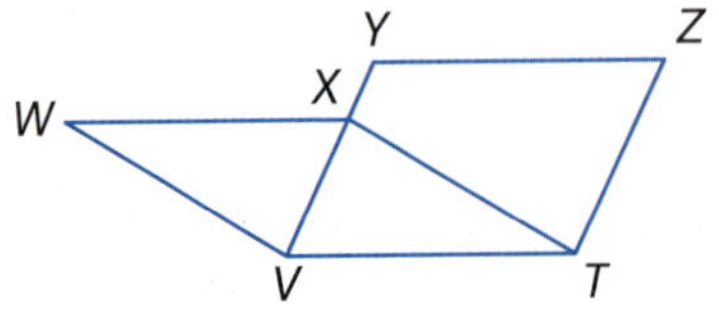

24. **Given:** ▱$BDHA$, $\overline{CA} \cong \overline{CG}$
Prove: $\angle BDH \cong \angle G$

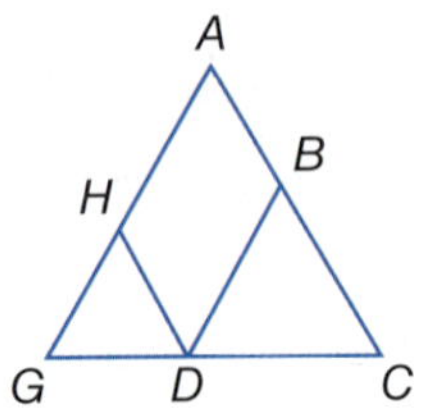

Real-World Link

Seven different national flags have flown over the state of Alabama, including the flag of the Kingdom of Spain and the flag of the British Empire.

Source: Alabama Department of Archives and History

25. FLAGS Refer to the Alabama state flag at the right.

Given: $\triangle ACD \cong \triangle CAB$

Prove: $\overline{DP} \cong \overline{PB}$

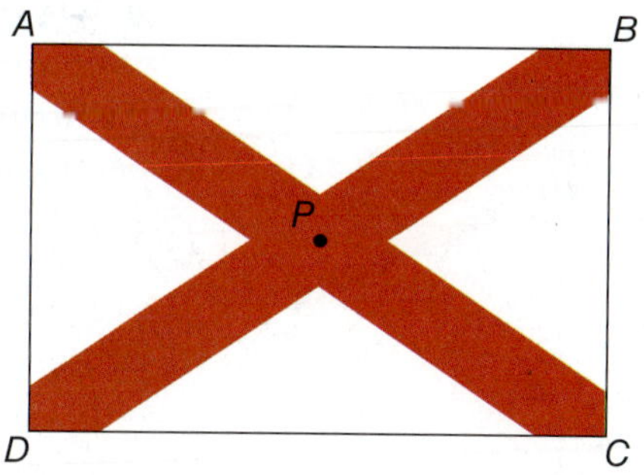

PROOF **Write the indicated type of proof.**

26. two-column
Given: $\square GKLM$
Prove: $\angle G$ and $\angle K$, $\angle K$ and $\angle L$, $\angle L$ and $\angle M$, and $\angle M$ and $\angle G$ are supplementary. (Theorem 6.5)

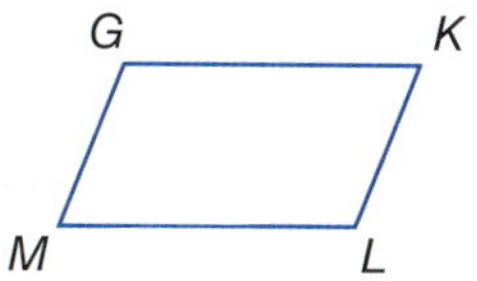

27. two-column
Given: $\square WXYZ$
Prove: $\triangle WXZ \cong \triangle YZX$ (Theorem 6.8)

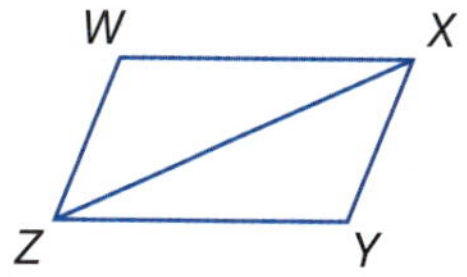

28. two-column
Given: $\square PQRS$
Prove: $\overline{PQ} \cong \overline{RS}$, $\overline{QR} \cong \overline{SP}$ (Theorem 6.3)

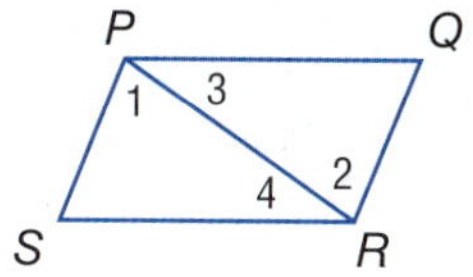

29. paragraph
Given: $\square ACDE$ is a parallelogram.
Prove: $\overline{EC}$ bisects $\overline{AD}$. (Theorem 6.7)

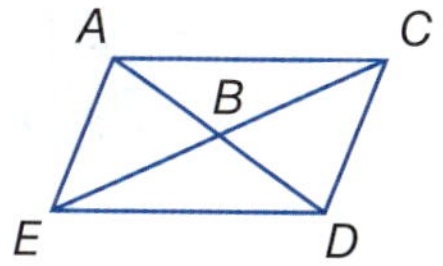

30. COORDINATE GEOMETRY Use the graph shown.

a. Use the Distance Formula to determine if the diagonals of *JKLM* bisect each other. Explain.

b. Determine whether the diagonals are congruent. Explain.

c. Use slopes to determine if the consecutive sides are perpendicular. Explain.

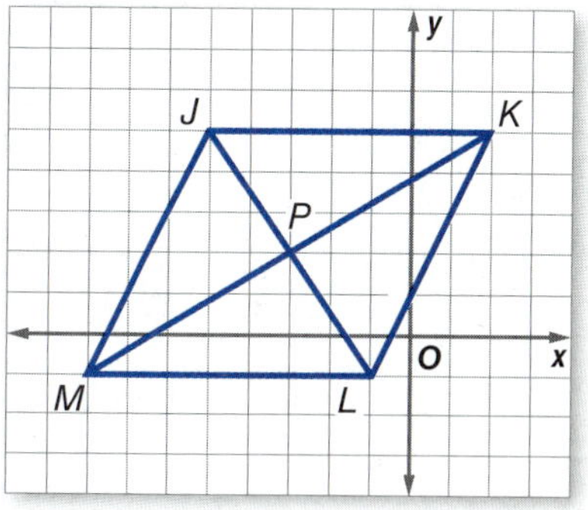

ALGEBRA **Use $\square ABCD$ to find each measure or value.**

31. x

32. y

33 $m\angle AFB$

34. $m\angle DAC$

35. $m\angle ACD$

36. $m\angle DAB$

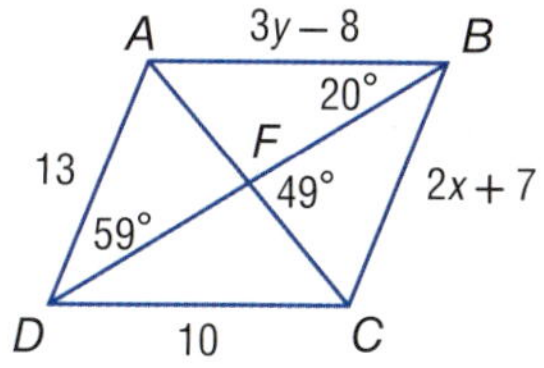

37. COORDINATE GEOMETRY $\square ABCD$ has vertices $A(-2, 5)$, $B(2, 2)$, and $C(4, -4)$. Determine the coordinates of vertex D. Explain your reasoning.

Real-World Link

Some scissor lifts can reach heights of over 100 meters. In the picture shown above, a scissor lift is used to assemble dinosaur bones.

Source: Bronto Skylift

38. **MECHANICS** Scissor lifts are variable elevation work platforms. One is shown at the right. In the diagram, *ABCD* and *DEFG* are congruent parallelograms.

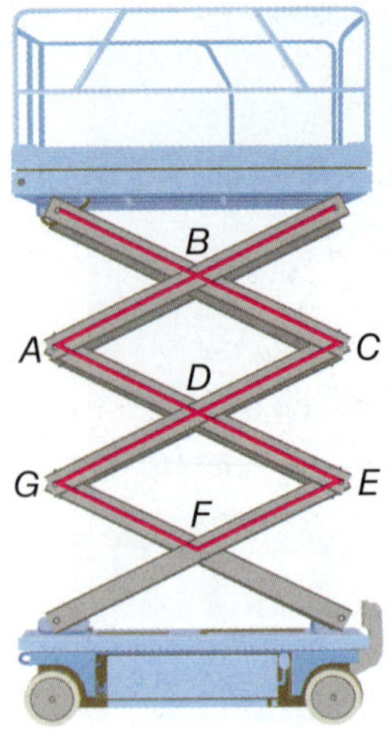

a. List the angle(s) congruent to $\angle A$. Explain your reasoning.

b. List the segment(s) congruent to $\overline{BC}$. Explain your reasoning.

c. List the angle(s) supplementary to $\angle C$. Explain your reasoning.

PROOF **Write a two-column proof.**

39. **Given:** $\square YWVZ$, $\overline{VX} \perp \overline{WY}$, $\overline{YU} \perp \overline{VZ}$
Prove: $\triangle YUZ \cong \triangle VXW$

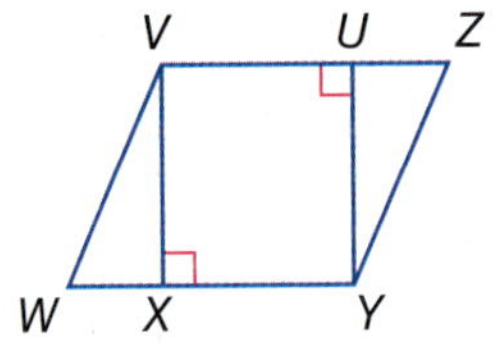

40. **MULTIPLE REPRESENTATIONS** In this problem, you will explore tests for parallelograms.

a. **GEOMETRIC** Draw three pairs of segments that are both congruent and parallel and connect the endpoints to form quadrilaterals. Label one quadrilateral *ABCD*, one *MNOP*, and one *WXYZ*. Measure and label the sides and angles of the quadrilaterals.

b. **TABULAR** Complete the table below for each quadrilateral.

Quadrilateral	Opposite Sides Congruent?	Opposite Angles Congruent?	Parallelogram
ABCD			
MNOP			
WXYZ			

c. **VERBAL** Make a conjecture about quadrilaterals with one pair of segments that are both congruent and parallel.

H.O.T. Problems

Use **H**igher-**O**rder **T**hinking Skills

41. **CHALLENGE** *ABCD* is a parallelogram with side lengths as indicated in the figure at the right. The perimeter of *ABCD* is 22. Find *AB*.

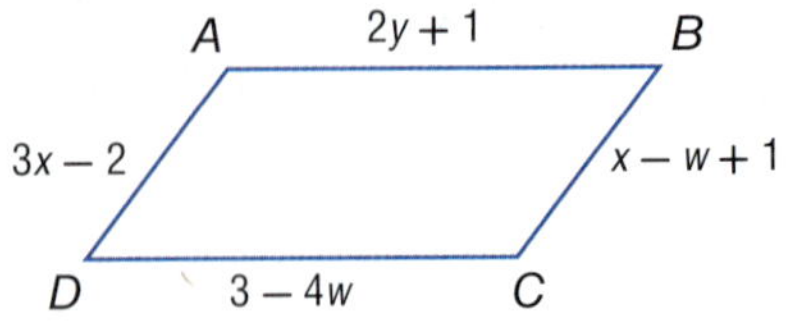

42. **WRITING IN MATH** Explain why parallelograms are *always* quadrilaterals, but quadrilaterals are *sometimes* parallelograms.

43. **OPEN ENDED** Provide a counterexample to show that parallelograms are not always congruent if their corresponding sides are congruent.

44. **REASONING** Find $m\angle 1$ and $m\angle 10$ in the figure at the right. Explain.

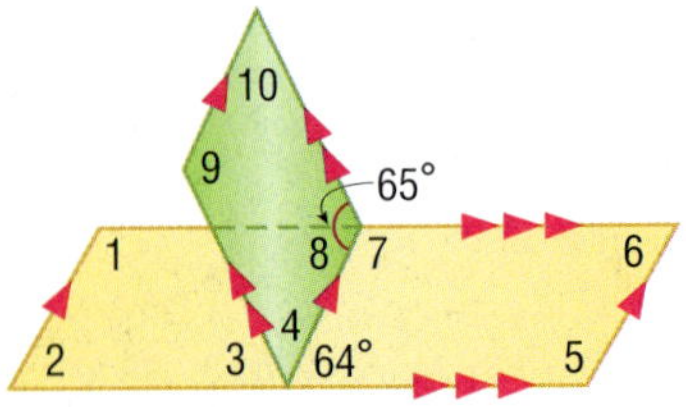

45. **WRITING IN MATH** Summarize the properties of the sides, angles, and diagonals of a parallelogram.

Standardized Test Practice

46. Two consecutive angles of a parallelogram measure $3x + 42$ and $9x - 18$. What are the measures of the angles?

A 13, 167

B 58.5, 31.5

C 39, 141

D 81, 99

47. GRIDDED RESPONSE Parallelogram *MNPQ* is shown. What is the value of *x*?

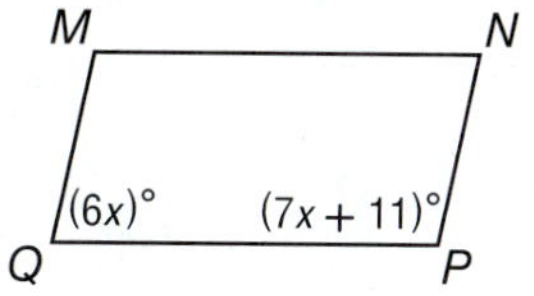

48. ALGEBRA In a history class with 32 students, the ratio of girls to boys is 5 to 3. How many more girls are there than boys?

F 2

G 8

H 12

J 15

49. SAT/ACT The table shows the heights of the tallest buildings in Kansas City, Missouri. To the nearest tenth, what is the positive difference between the median and the mean of the data?

Name	Height (m)
One Kansas City Place	193
Town Pavillion	180
Hyatt Regency	154
Power and Light Building	147
City Hall	135
1201 Walnut	130

A 5

B 6

C 7

D 8

Spiral Review

The measure of an interior angle of a regular polygon is given. Find the number of sides in the polygon. (Lesson 6-1)

50. 108 **51.** 140 **52.** 147.3 **53.** 160 **54.** 135 **55.** 176.4

56. LANDSCAPING When landscapers plant new trees, they usually brace the tree using a stake tied to the trunk of the tree. Use the SAS or SSS Inequality to explain why this is an effective method for keeping a newly planted tree perpendicular to the ground. (Lesson 5-6)

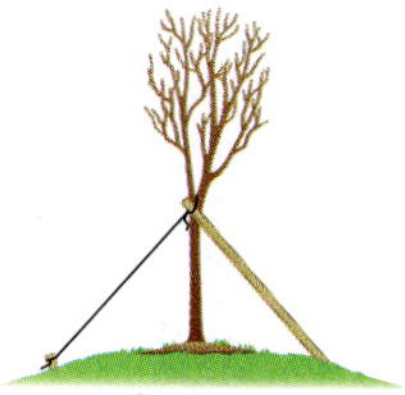

Determine whether the solid is a polyhedron. Then identify the solid. If it is a polyhedron, name the bases, faces, edges, and vertices. (Lesson 1-7)

57.

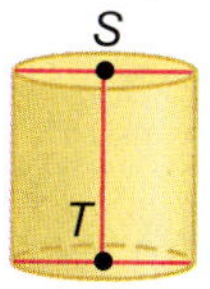

58.

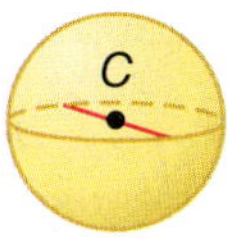

59.

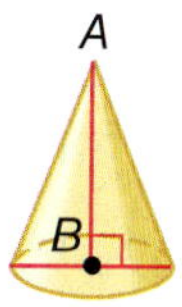

Skills Review

The vertices of a quadrilateral are $W(3, -1)$, $X(4, 2)$, $Y(-2, 3)$ and $Z(-3, 0)$. Determine whether each segment is a side or diagonal of the quadrilateral, and find the slope of each segment. (Lesson 3-3)

60. $\overline{YZ}$ **61.** $\overline{YW}$ **62.** $\overline{ZW}$

EXPLORE 6-3

Graphing Technology Lab
Parallelograms

Math Online glencoe.com
- Other Calculator Keystrokes
- Graphing Technology Personal Tutor

Objective
You can use technology to discover the properties of parallelograms.

You can use the Cabri™ Jr. application on a TI-83/84 Plus graphing calculator to discover properties of parallelograms.

ACTIVITY

Construct a quadrilateral with one pair of sides that are both parallel and congruent.

Step 1 Construct a segment using the **Segment** tool on the **F2** menu. Label the segment $\overline{AB}$. This is one side of the quadrilateral.

Step 2 Use the **Parallel** tool on the **F3** menu to construct a line parallel to the segment. Pressing ENTER will draw the line and a point on the line. Label the point C.

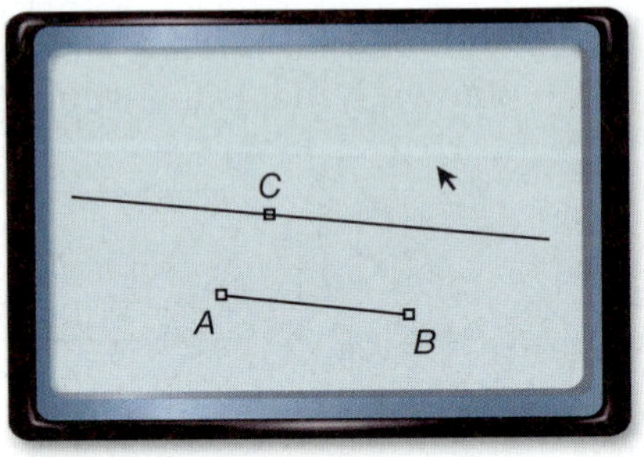

Steps 1 and 2

Step 3 Access the **Compass** tool on the **F3** menu. Set the compass to the length of $\overline{AB}$ by selecting one endpoint of the segment and then the other. Construct a circle centered at C.

Step 4 Use the **Point Intersection** tool on the **F2** menu to draw a point at the intersection of the line and the circle. Label the point D. Then use the **Segment** tool on the **F2** menu to draw $\overline{AC}$ and $\overline{BD}$.

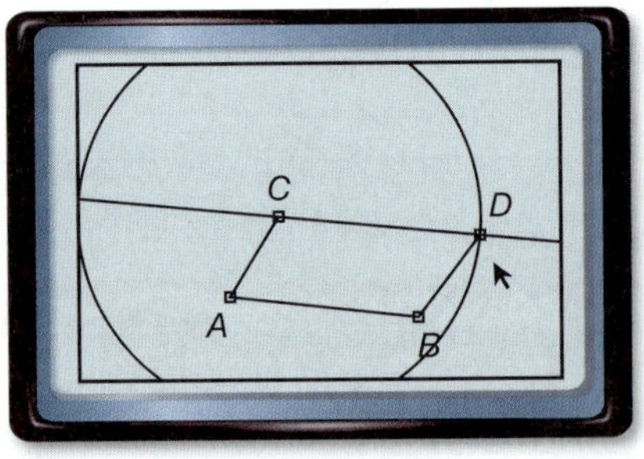

Steps 3 and 4

Step 5 Use the **Hide/Show** tool on the **F5** menu to hide the circle. Then access **Slope** tool under **Measure** on the **F5** menu. Display the slopes of $\overline{AB}$, $\overline{BD}$, $\overline{CD}$, and $\overline{AC}$.

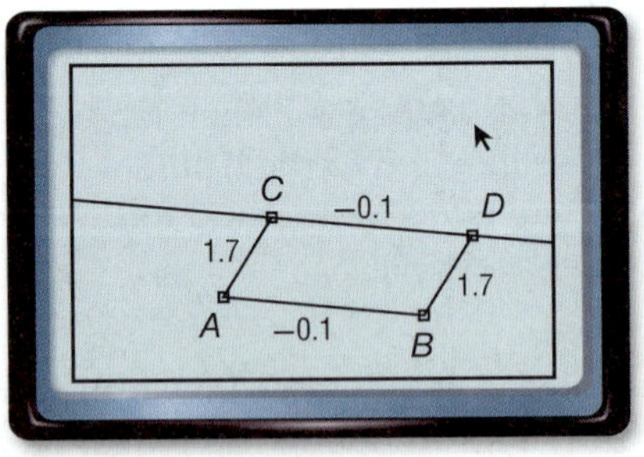

Step 5

Analyze the Results

1. What is the relationship between sides $\overline{AB}$ and $\overline{CD}$? Explain how you know.
2. What do you observe about the slopes of opposite sides of the quadrilateral? What type of quadrilateral is $ABDC$? Explain.
3. Click on point A and drag it to change the shape of $ABDC$. What do you observe?
4. Make a conjecture about a quadrilateral with a pair of opposite sides that are both congruent and parallel.
5. Use a graphing calculator to construct a quadrilateral with both pairs of opposite sides congruent. Then analyze the slopes of the sides of the quadrilateral. Make a conjecture based on your observations.

6-3 Tests for Parallelograms

Then

You recognized and applied properties of parallelograms. (Lesson 6-2)

Now

- Recognize the conditions that ensure a quadrilateral is a parallelogram.
- Prove that a set of points forms a parallelogram in the coordinate plane.

Why?

Lexi and Rosalinda cut strips of bulletin board paper at an angle to form the hallway display shown. Their friends asked them how they cut the strips so that their sides were parallel without using a protractor.

Rosalinda explained that since the left and right sides of the paper were parallel, she only needed to make sure that the sides were cut to the same length to guarantee that a strip would form a parallelogram.

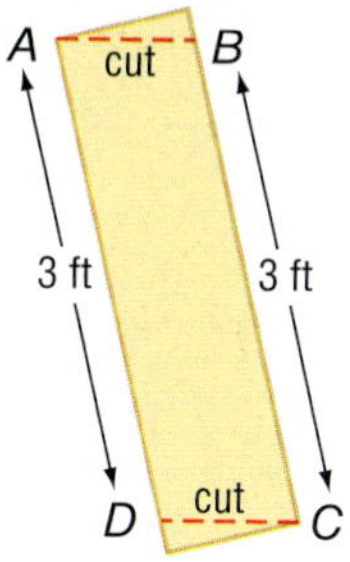

Conditions for Parallelograms If a quadrilateral has each pair of opposite sides parallel, it is a parallelogram by definition. This is not the only test, however, that can be used to determine if a quadrilateral is a parallelogram.

Theorems — Conditions for Parallelograms

6.9 If both pairs of opposite sides of a quadrilateral are congruent, then the quadrilateral is a parallelogram.

Abbreviation *If both pairs of opp. sides are ≅, then quad. is a ▱.*

Example If $\overline{AB} \cong \overline{DC}$ and $\overline{AD} \cong \overline{BC}$, then $ABCD$ is a parallelogram.

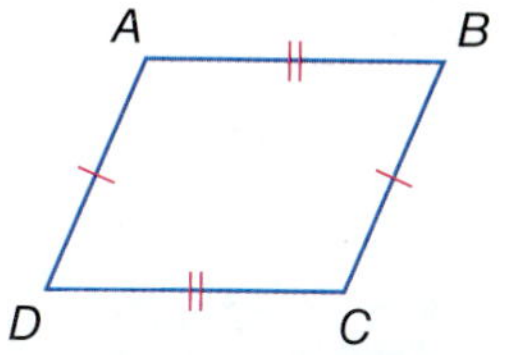

6.10 If both pairs of opposite angles of a quadrilateral are congruent, then the quadrilateral is a parallelogram.

Abbreviation *If both pairs of opp. ∠s are ≅, then quad. is a ▱.*

Example If $\angle A \cong \angle C$ and $\angle B \cong \angle D$, then $ABCD$ is a parallelogram.

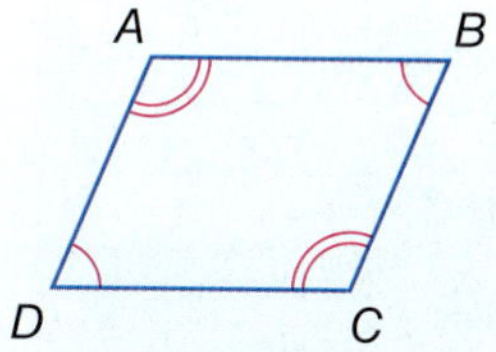

6.11 If the diagonals of a quadrilateral bisect each other, then the quadrilateral is a parallelogram.

Abbreviation *If diag. bisect each other, then quad. is a ▱.*

Example If $\overline{AC}$ and $\overline{DB}$ bisect each other, then $ABCD$ is a parallelogram.

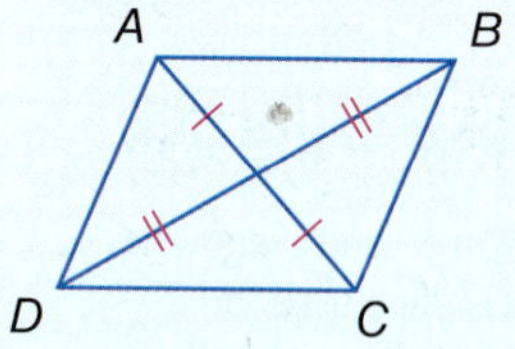

6.12 If one pair of opposite sides of a quadrilateral is both parallel and congruent, then the quadrilateral is a parallelogram.

Abbreviation *If one pair of opp. sides is ≅ and ||, then the quad. is a ▱.*

Example If $\overline{AB} \parallel \overline{DC}$ and $\overline{AB} \cong \overline{DC}$, then $ABCD$ is a parallelogram.

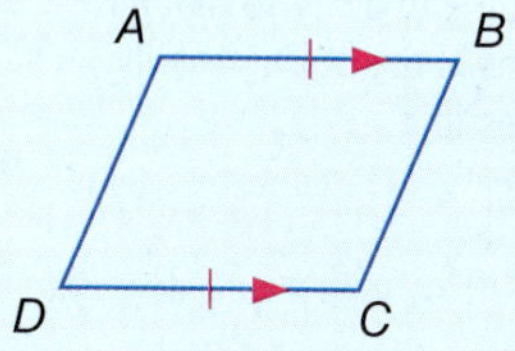

You will prove Theorems 6.10, 6.11, and 6.12 in Exercises 30, 32, and 33, respectively.

Proof **Theorem 6.9**

Write a paragraph proof of Theorem 6.9.

Given: $\overline{WX} \cong \overline{ZY}$, $\overline{WZ} \cong \overline{XY}$

Prove: $WXYZ$ is a parallelogram.

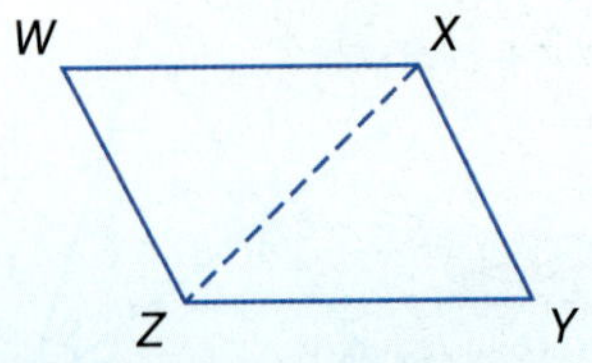

Paragraph Proof:

Two points determine a line, so we can draw auxiliary line $\overline{ZX}$ to form $\triangle ZWX$ and $\triangle XYZ$. We are given that $\overline{WX} \cong \overline{ZY}$ and $\overline{WZ} \cong \overline{XY}$. Also, $\overline{ZX} \cong \overline{XZ}$ by the Reflexive Property of Congruence. So $\triangle ZWX \cong \triangle XYZ$ by SSS. By CPCTC, $\angle WXZ \cong \angle YZX$ and $\angle WZX \cong \angle YXZ$. This means that $\overline{WX} \parallel \overline{ZY}$ and $\overline{WZ} \parallel \overline{XY}$ by the Alternate Interior Angles Converse. Opposite sides of $WXYZ$ are parallel, so by definition $WXYZ$ is a parallelogram.

EXAMPLE 1 Identify Parallelograms

Determine whether the quadrilateral is a parallelogram. Justify your answer.

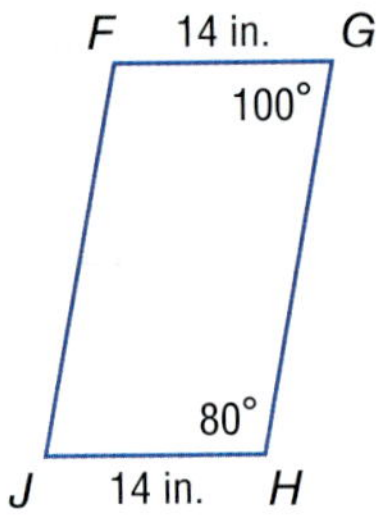

Opposite sides $\overline{FG}$ and $\overline{JH}$ are congruent because they have the same measure. Also, since $\angle FGH$ and $\angle GHJ$ are supplementary consecutive interior angles, $\overline{FG} \parallel \overline{JH}$. Therefore, by Theorem 6.12, $FGHJ$ is a parallelogram.

Guided Practice

1A.

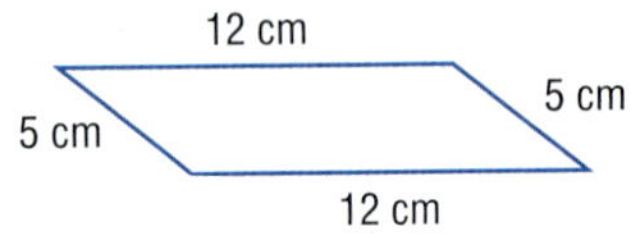

1B.

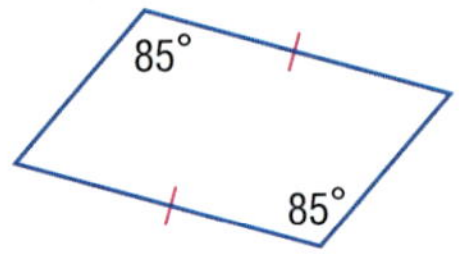

Personal Tutor glencoe.com

You can use the conditions of parallelograms to prove relationships in real-world situations.

Real-World Link

A 2- or 3-cantilever tackle box is often used to organize lures and other fishing supplies. The trays lift up and away so that all items in the box are easily accessible.

Real-World EXAMPLE 2 Use Parallelograms to Prove Relationships

FISHING The diagram shows a side view of the tackle box at the left. In the diagram, $PQ = RS$ and $PR = QS$. Explain why the upper and middle trays remain parallel no matter to what height the trays are raised or lowered.

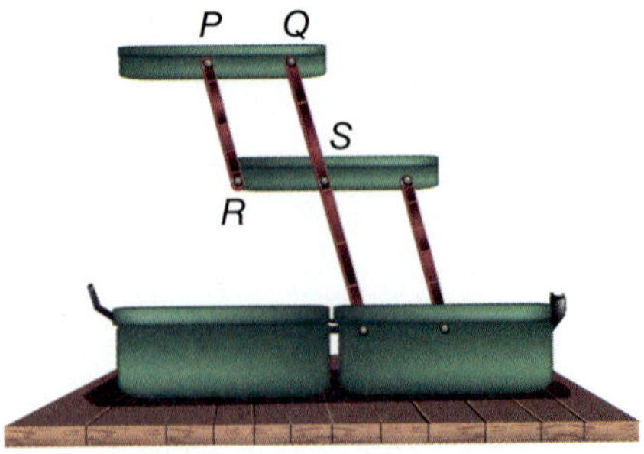

Since both pairs of opposite sides of quadrilateral $PQSR$ are congruent, $PQRS$ is a parallelogram by Theorem 6.9. By the definition of a parallelogram, opposite sides are parallel, so $\overline{PQ} \parallel \overline{RS}$. Therefore, no matter the vertical position of the trays, they will always remain parallel.

Guided Practice

2. BANNERS In the example at the beginning of the lesson, explain why the cuts made by Lexi and Rosalinda are parallel.

Personal Tutor glencoe.com

Watch Out!

Parallelograms In Example 3, if x is 4, then y must be 2.5 in order for $FGHJ$ to be a parallelogram. In other words, if x is 4 and y is 1, then $FGHJ$ is not a parallelogram.

You can also use the conditions of parallelograms along with algebra to find missing values that make a quadrilateral a parallelogram.

EXAMPLE 3 Use Parallelograms and Algebra to Find Values

If $FK = 3x - 1$, $KG = 4y + 3$, $JK = 6y - 2$, and $KH = 2x + 3$, find x and y so that the quadrilateral is a parallelogram.

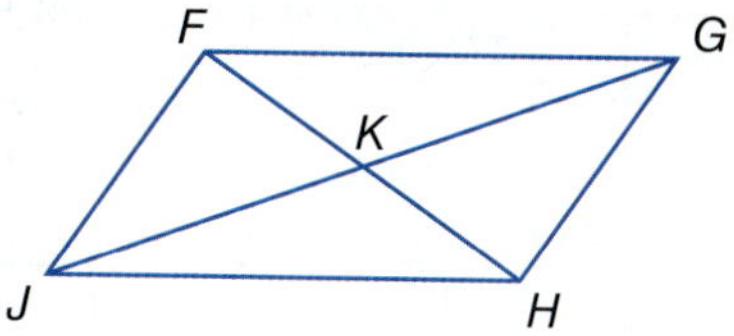

By Theorem 6.11, if the diagonals of a quadrilateral bisect each other, then it is a parallelogram. So find x such that $\overline{FK} \cong \overline{KH}$ and y such that $\overline{JK} \cong \overline{KG}$.

$FK = KH$	Definition of ≅
$3x - 1 = 2x + 3$	Substitution
$x - 1 = 3$	Subtract $2x$ from each side.
$x = 4$	Add 1 to each side.
$JK = KG$	Definition of ≅
$6y - 2 = 4y + 3$	Substitution
$2y - 2 = 3$	Subtract $4y$ from each side.
$2y = 5$	Add 2 to each side.
$y = 2.5$	Divide each side by 2.

So, when x is 4 and y is 2.5, quadrilateral $FGHJ$ is a parallelogram.

Guided Practice

Find x and y so that each quadrilateral is a parallelogram.

3A.

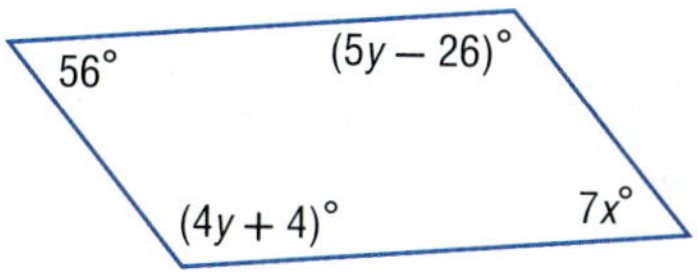

3B.

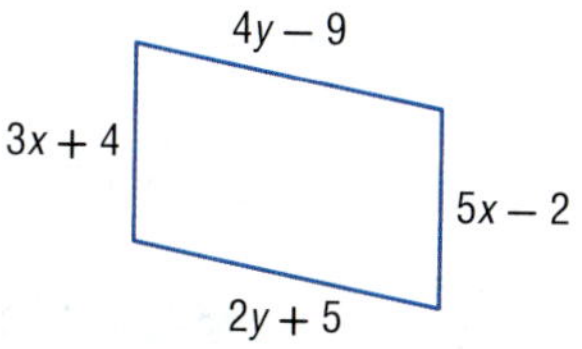

Personal Tutor glencoe.com

You have learned the conditions of parallelograms. The following list summarizes how to use the conditions to prove a quadrilateral is a parallelogram.

Concept Summary

Prove that a Quadrilateral Is a Parallelogram

1. Show that both pairs of opposite sides are parallel. (Definition)
2. Show that both pairs of opposite sides are congruent. (Theorem 6.9)
3. Show that both pairs of opposite angles are congruent. (Theorem 6.10)
4. Show that the diagonals bisect each other. (Theorem 6.11)
5. Show that a pair of opposite sides is both parallel and congruent. (Theorem 6.12)

Parallelograms on the Coordinate Plane We can use the Distance, Slope, and Midpoint Formulas to determine whether a quadrilateral in the coordinate plane is a parallelogram.

StudyTip

Midpoint Formula To show that a quadrilateral is a parallelogram, you can also use the Midpoint Formula. If the midpoint of each diagonal is the same point, then the diagonals bisect each other.

EXAMPLE 4 Parallelograms and Coordinate Geometry

COORDINATE GEOMETRY Graph quadrilateral *KLMN* with vertices *K*(2, 3), *L*(8, 4), *M*(7, −2), and *N*(1, −3). Determine whether the quadrilateral is a parallelogram. Justify your answer using the Slope Formula.

If the opposite sides of a quadrilateral are parallel, then it is a parallelogram.

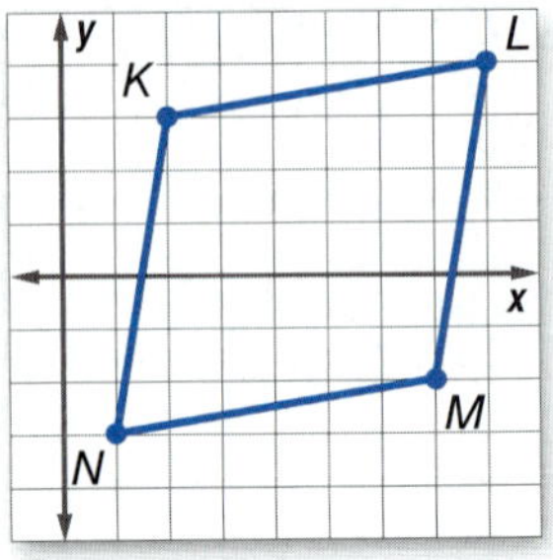

slope of $\overline{KL} = \frac{4-3}{8-2}$ or $\frac{1}{6}$

slope of $\overline{NM} = \frac{-2-(-3)}{7-1}$ or $\frac{1}{6}$

slope of $\overline{KN} = \frac{-3-3}{1-2} = \frac{-6}{-1}$ or 6

slope of $\overline{LM} = \frac{-2-4}{7-8} = \frac{-6}{-1}$ or 6

Since opposite sides have the same slope, $\overline{KL} \parallel \overline{NM}$ and $\overline{KN} \parallel \overline{LM}$. Therefore, *KLMN* is a parallelogram by definition.

Guided Practice

Determine whether the quadrilateral is a parallelogram. Justify your answer using the given formula.

4A. $A(3, 3)$, $B(8, 2)$, $C(6, -1)$, $D(1, 0)$; Distance Formula

4B. $F(-2, 4)$, $G(4, 2)$, $H(4, -2)$, $J(-2, -1)$; Midpoint Formula

Personal Tutor glencoe.com

In Chapter 4, you learned that variable coordinates can be assigned to the vertices of triangles. Then the Distance, Slope, and Midpoint Formulas were used to write coordinate proofs of theorems. The same can be done with quadrilaterals.

EXAMPLE 5 Parallelograms and Coordinate Proofs

Write a coordinate proof for the following statement.

If one pair of opposite sides of a quadrilateral is both parallel and congruent, then the quadrilateral is a parallelogram.

Review Vocabulary

coordinate proof a proof that uses figures in the coordinate plane and algebra to prove geometric concepts (Lesson 4-8)

Step 1 Position quadrilateral *ABCD* on the coordinate plane such that $\overline{AB} \parallel \overline{DC}$ and $\overline{AB} \cong \overline{DC}$.

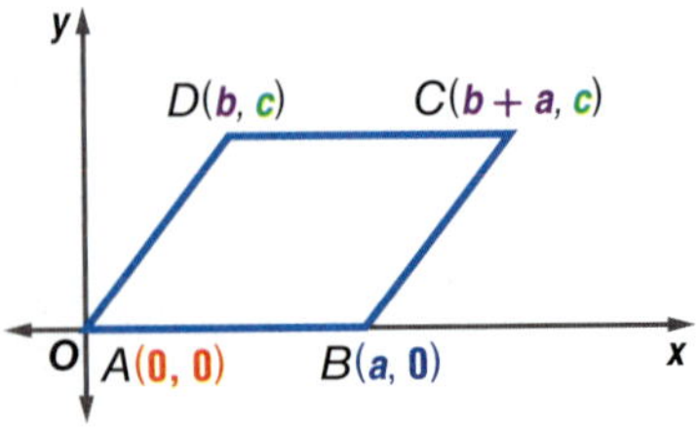

- Begin by placing the vertex *A* at the **origin**.
- Let $\overline{AB}$ have a length of a units. Then *B* has coordinates **(*a*, 0)**.
- Since horizontal segments are parallel, position the endpoints of $\overline{DC}$ so that they have the same *y*-coordinate, ***c***.
- So that the distance from *D* to *C* is also *a* units, let the *x*-coordinate of *D* be ***b*** and of *C* be ***b* + *a***.

Math History Link

René Descartes
(1596–1650)
René Descartes was a French mathematician who was the first to use a coordinate grid. It has been said that he first thought of locating a point on a plane with a pair of numbers when he was watching a fly on the ceiling, but this is a myth.

Step 2 Use your figure to write a proof.

Given: quadrilateral $ABCD$, $\overline{AB} \parallel \overline{DC}$, $\overline{AB} \cong \overline{DC}$

Prove: $ABCD$ is a parallelogram.

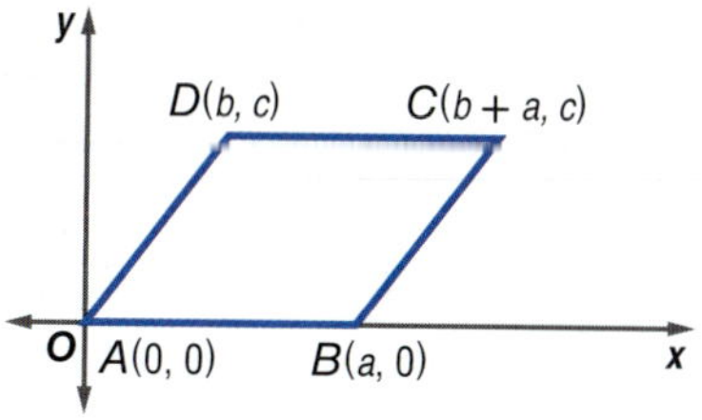

Coordinate Proof:

By definition, a quadrilateral is a parallelogram if opposite sides are parallel. We are given that $\overline{AB} \parallel \overline{DC}$, so we need only show that $\overline{AD} \parallel \overline{BC}$.

Use the Slope Formula.

slope of $\overline{AD} = \frac{c-0}{b-0} = \frac{c}{b}$ slope of $\overline{BC} = \frac{c-0}{b+a-a} = \frac{c}{b}$

Since $\overline{AD}$ and $\overline{BC}$ have the same slope, $\overline{AD} \parallel \overline{BC}$. So quadrilateral $ABCD$ is a parallelogram because opposite sides are parallel.

Guided Practice

5. Write a coordinate proof of this statement: *If a quadrilateral is a parallelogram, then opposite sides are congruent.*

Personal Tutor glencoe.com

Check Your Understanding

Example 1 p. 410

Determine whether each quadrilateral is a parallelogram. Justify your answer.

1.

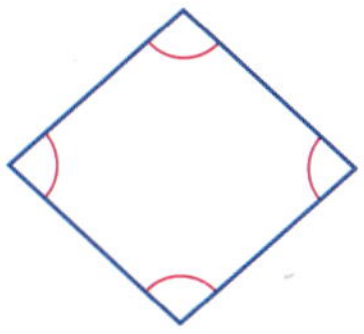

2.

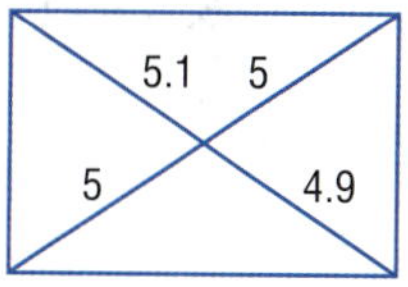

Example 2 p. 410

3. KITES Charmaine is building the kite shown below. She wants to be sure that the string around her frame forms a parallelogram before she secures the material to it. How can she use the measures of the wooden portion of the frame to prove that the string forms a parallelogram? Explain your reasoning.

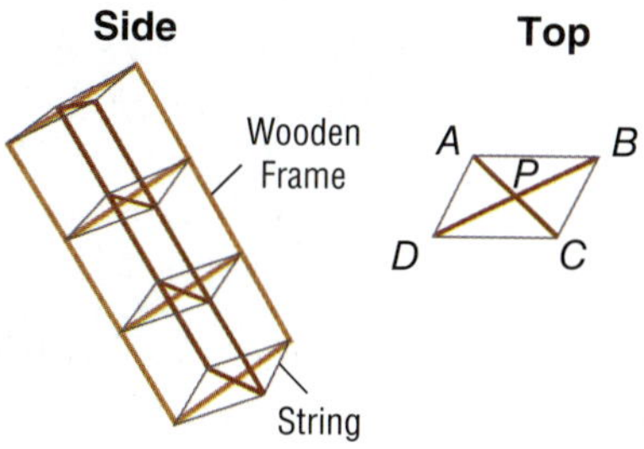

Example 3 p. 411

ALGEBRA Find x and y so that the quadrilateral is a parallelogram.

4.

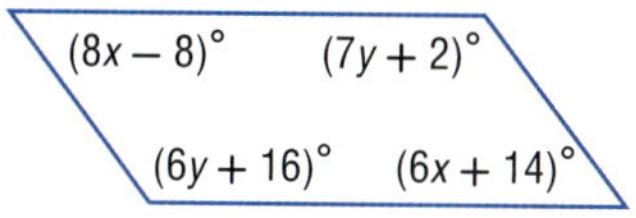

5

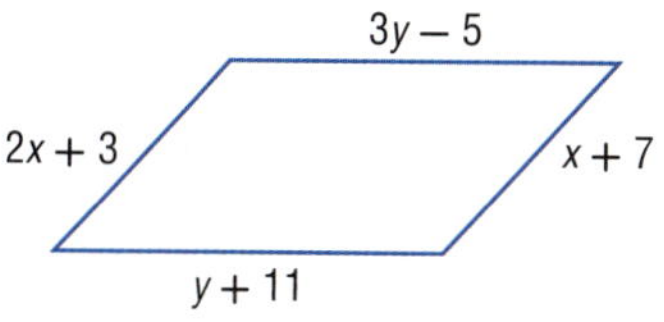

Example 4 p. 412

COORDINATE GEOMETRY **Graph each quadrilateral with the given vertices. Determine whether the figure is a parallelogram. Justify your answer with the method indicated.**

6. $A(-2, 4)$, $B(5, 4)$, $C(8, -1)$, $D(-1, -1)$; Slope Formula

7. $W(-5, 4)$, $X(3, 4)$, $Y(1, -3)$, $Z(-7, -3)$; Midpoint Formula

Example 5 pp. 412–413

8. Write a coordinate proof for the statement: *If a quadrilateral is a parallelogram, then its diagonals bisect each other.*

Practice and Problem Solving

● = **Step-by-Step Solutions** begin on page R20. **Extra Practice** begins on page 969.

Example 1 p. 410

Determine whether each quadrilateral is a parallelogram. Justify your answer.

9.
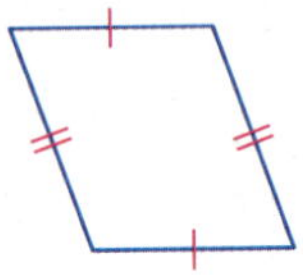

10.
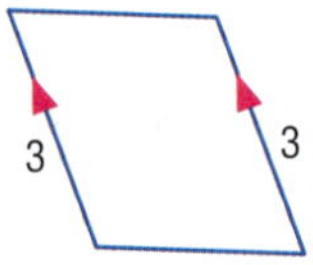

11.
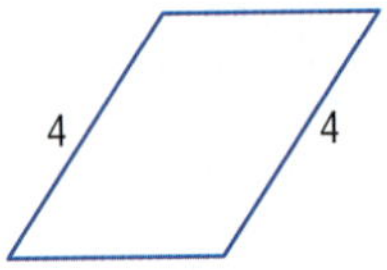

12.
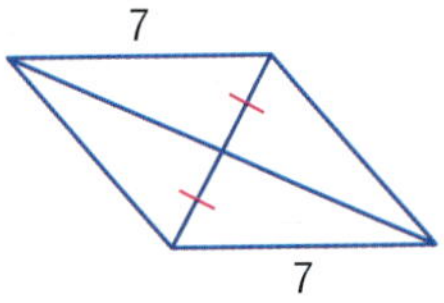

13.
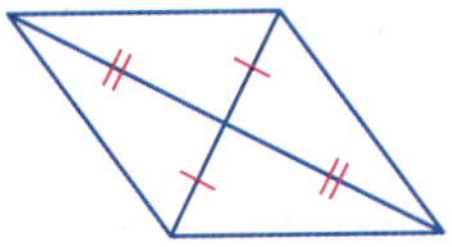

14.
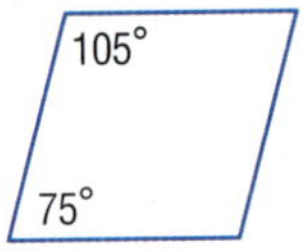

Example 2 p. 410

15. **PROOF** If *ACDH* is a parallelogram, *B* is the midpoint of $\overline{AC}$, and *F* is the midpoint of $\overline{HD}$, write a flow proof to prove that *ABFH* is a parallelogram.

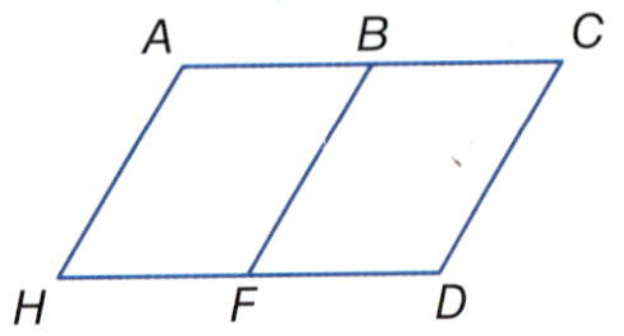

16. **PROOF** If *WXYZ* is a parallelogram, $\angle W \cong \angle X$, and *M* is the midpoint of $\overline{WX}$, write a paragraph proof to prove that *ZMY* is an isosceles triangle.

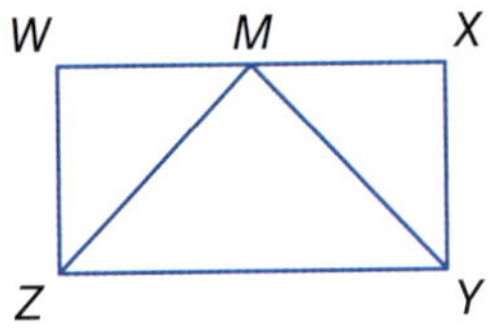

17. **REPAIR** Parallelogram lifts are used to elevate large vehicles for maintenance. In the diagram, *ABEF* and *BCDE* are parallelograms. Write a two-column proof to show that *ACDF* is also a parallelogram.

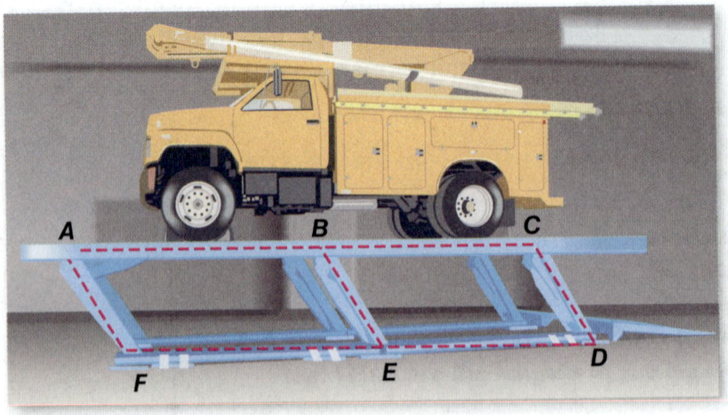

Example 3 p. 411

ALGEBRA **Find *x* and *y* so that the quadrilateral is a parallelogram.**

18.
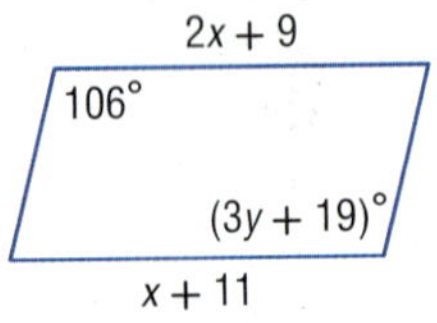

19.
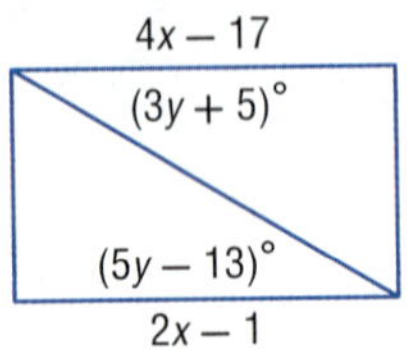

20.
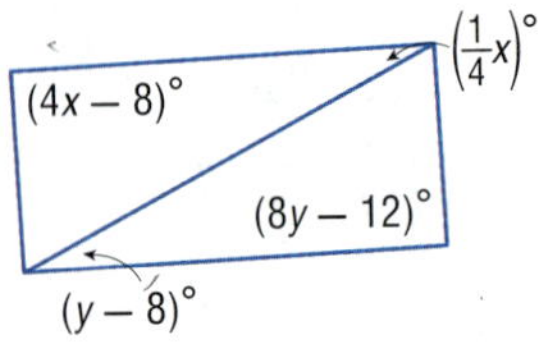

ALGEBRA **Find x and y so that the quadrilateral is a parallelogram.**

21.

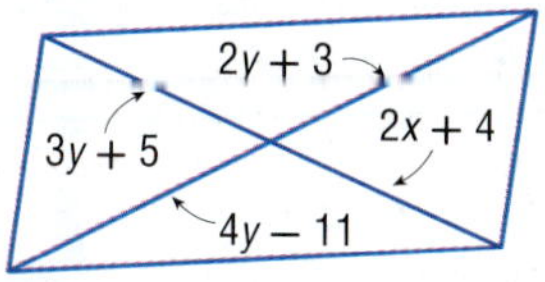

22.

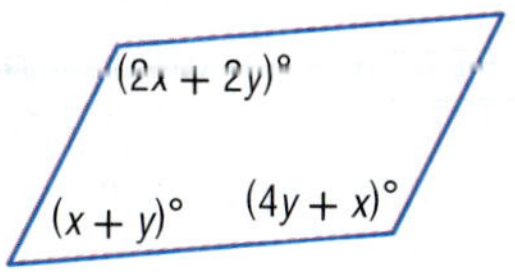

23.

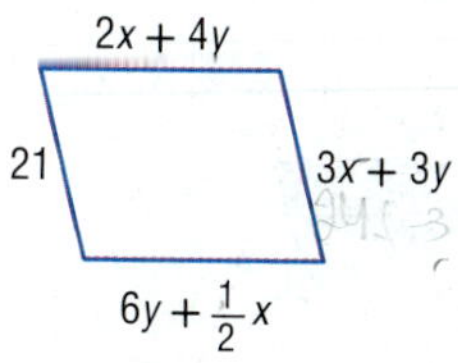

Example 4
p. 412

COORDINATE GEOMETRY **Graph each quadrilateral with the given vertices. Determine whether the figure is a parallelogram. Justify your answer with the method indicated.**

24. $A(-3, 4)$, $B(4, 5)$, $C(5, -1)$, $D(-2, -2)$; Slope Formula

25. $J(-4, -4)$, $K(-3, 1)$, $L(4, 3)$, $M(3, -3)$; Distance Formula

26. $V(3, 5)$, $W(1, -2)$, $X(-6, 2)$, $Y(-4, 7)$; Distance and Slope Formulas

27. $Q(2, -4)$, $R(4, 3)$, $S(-3, 6)$, $T(-5, -1)$; Distance and Slope Formulas

Example 5
pp. 412–413

28. Write a coordinate proof for the statement: *If both pairs of opposite sides of a quadrilateral are congruent, then the quadrilateral is a parallelogram.*

29. Write a coordinate proof for the statement: *If a parallelogram has one right angle, it has four right angles.*

30. **PROOF** Write a paragraph proof of Theorem 6.10.

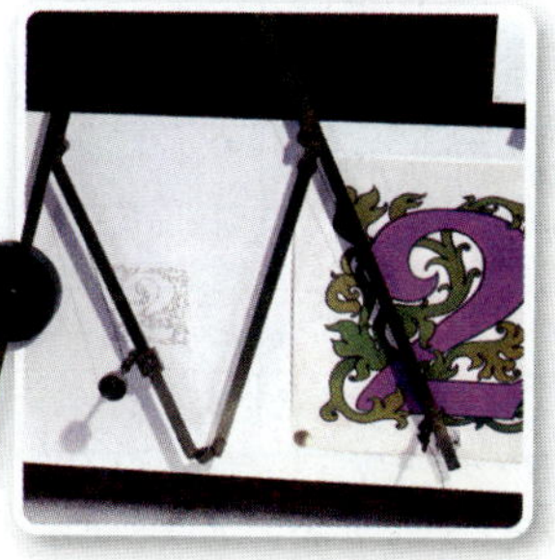

Real-World Link

A pantograph is a device that can be used to copy an object and either enlarge or reduce it based on the dimensions of the pantograph.

Source: Wolfram Math World

31. **PANTOGRAPH** Refer to the information at the left and the diagram below.

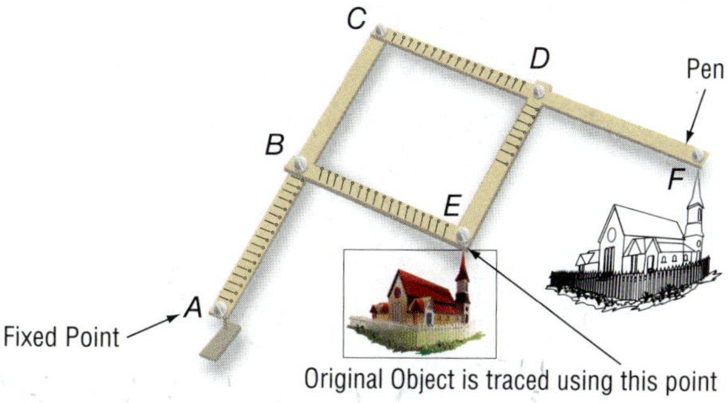

a. If $\overline{AC} \cong \overline{CF}$, $\overline{AB} \cong \overline{CD} \cong \overline{BE}$, and $\overline{DF} \cong \overline{DE}$, write a paragraph proof to show that $\overline{BE} \parallel \overline{CD}$.

b. The scale of the copied object is the ratio of CF to BE. If AB is 12 inches, DF is 8 inches, and the width of the original object is 5.5 inches, what is the width of the copy?

PROOF **Write a two-column proof.**

32. Theorem 6.11

33. Theorem 6.12

34. **CONSTRUCTION** Explain how you can use Theorem 6.11 to construct a parallelogram. Then construct a parallelogram using your method.

Name the missing coordinates for each parallelogram.

35.

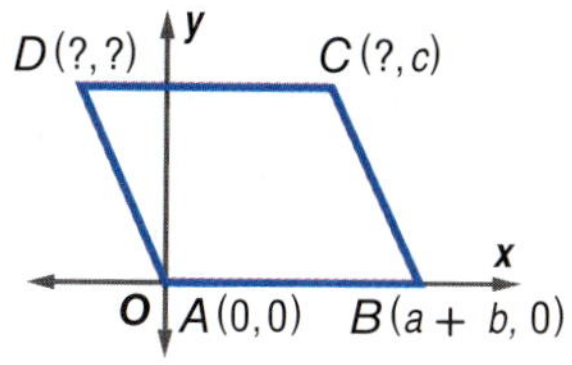

36.

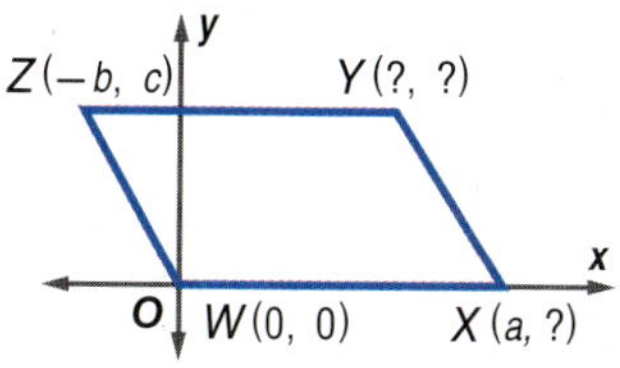

Real-World Link

Carpenter's squares can be used for many purposes, including marking patterns in framing, roofing, and stairway work and as a straightedge for determining whether or not a surface is flat.

Source: Bob Vila

37 **SERVICE** While replacing a hand rail, a contractor uses a carpenter's square to confirm that the vertical supports are perpendicular to the top step and the ground, respectively. How can the contractor prove that the two hand rails are parallel using the fewest measurements? Assume that the top step and the ground are both level.

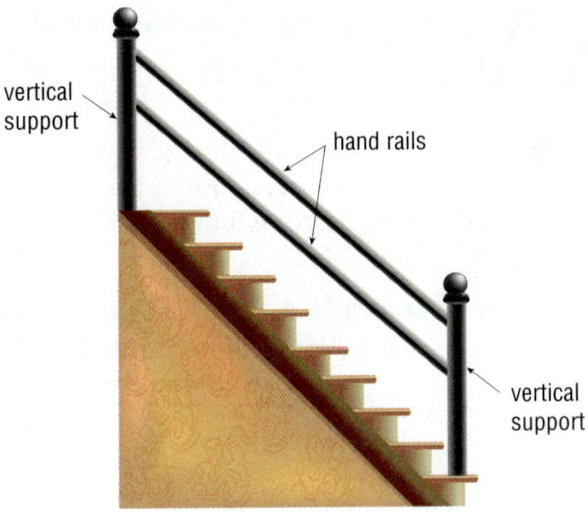

38. **PROOF** Write a coordinate proof to prove that the segments joining the midpoints of the sides of any quadrilateral form a parallelogram.

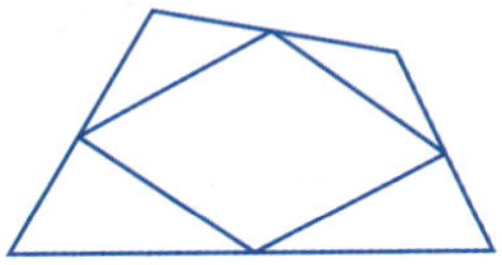

39. **MULTIPLE REPRESENTATIONS** In this problem, you will explore the properties of rectangles. A rectangle is a quadrilateral with four right angles.

a. **GEOMETRIC** Draw three rectangles with varying lengths and widths. Label one rectangle *ABCD*, one *MNOP*, and one *WXYZ*. Draw the two diagonals for each rectangle.

b. **TABULAR** Measure the diagonals of each rectangle and complete the table at the right.

c. **VERBAL** Write a conjecture about the diagonals of a rectangle.

Rectangle	Side	Length
ABCD	$\overline{AC}$	
	$\overline{BD}$	
MNOP	$\overline{MO}$	
	$\overline{NP}$	
WXYZ	$\overline{WY}$	
	$\overline{XZ}$	

H.O.T. Problems

Use **H**igher-**O**rder **T**hinking Skills

40. **CHALLENGE** The diagonals of a parallelogram meet at the point (0, 1). One vertex of the parallelogram is located at (2, 4), and a second vertex is located at (3, 1). Find the locations of the remaining vertices.

41. **WRITING IN MATH** Compare and contrast Theorem 6.9 and Theorem 6.3.

42. **REASONING** If two parallelograms have four congruent corresponding angles, are the parallelograms *sometimes*, *always*, or *never* congruent?

43. **OPEN ENDED** Position and label a parallelogram on the coordinate plane differently than shown in either Example 5, Exercise 35, or Exercise 36.

44. **CHALLENGE** If *ABCD* is a parallelogram and $\overline{AJ} \cong \overline{KC}$, show that quadrilateral *JBKD* is a parallelogram.

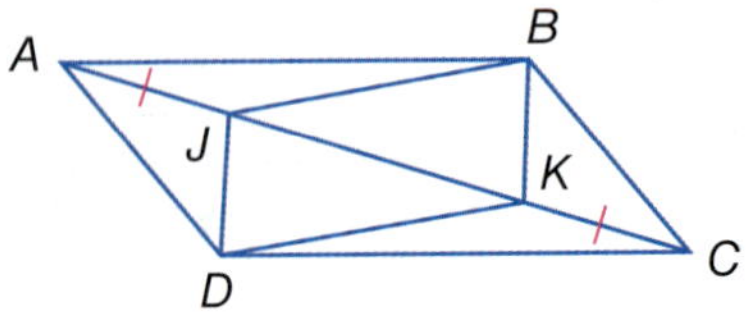

45. **WRITING IN MATH** Describe the information needed to prove that a quadrilateral is a parallelogram.

Standardized Test Practice

46. If sides AB and DC of quadrilateral $ABCD$ are parallel, which additional information would be sufficient to prove that quadrilateral $ABCD$ is a parallelogram?

A $\overline{AB} \cong \overline{AC}$ **C** $\overline{AC} \cong \overline{BD}$

B $\overline{AB} \cong \overline{DC}$ **D** $\overline{AD} \cong \overline{BC}$

47. SHORT RESPONSE Quadrilateral $ABCD$ is shown. AC is 40 and BD is $\frac{3}{5}AC$. $\overline{BD}$ bisects $\overline{AC}$. For what value of x is $ABCD$ a parallelogram?

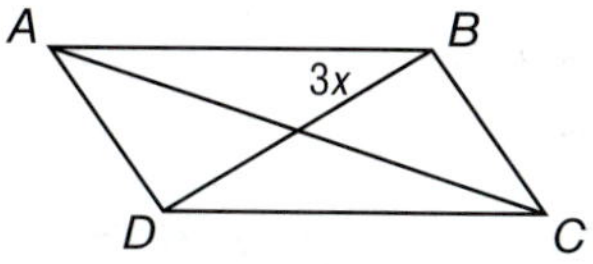

48. ALGEBRA Jarod's average driving speed for a 5-hour trip was 58 miles per hour. During the first 3 hours, he drove 50 miles per hour. What was his average speed in miles per hour for the last 2 hours of his trip?

F 70 **H** 60

G 66 **J** 54

49. SAT/ACT A parallelogram has vertices at (0, 0), (3, 5), and (0, 5). What are the coordinates of the fourth vertex?

A (0, 3) **C** (5, 0)

B (5, 3) **D** (3, 0)

Spiral Review

COORDINATE GEOMETRY **Find the coordinates of the intersection of the diagonals of ▱*ABCD* with the given vertices.** (Lesson 6-2)

50. $A(-3, 5)$, $B(6, 5)$, $C(5, -4)$, $D(-4, -4)$

51. $A(2, 5)$, $B(10, 7)$, $C(7, -2)$, $D(-1, -4)$

Find the value of x. (Lesson 6-1)

52.

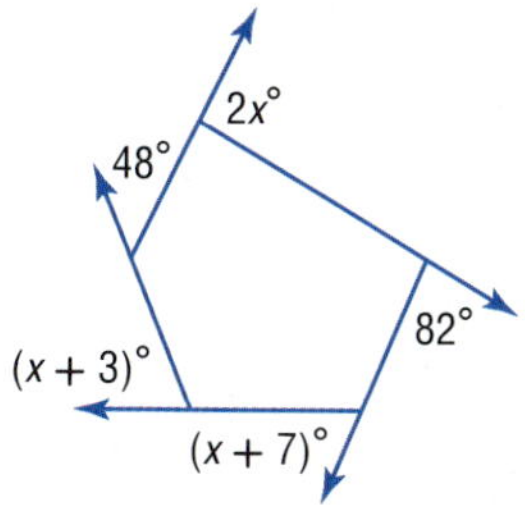

53.

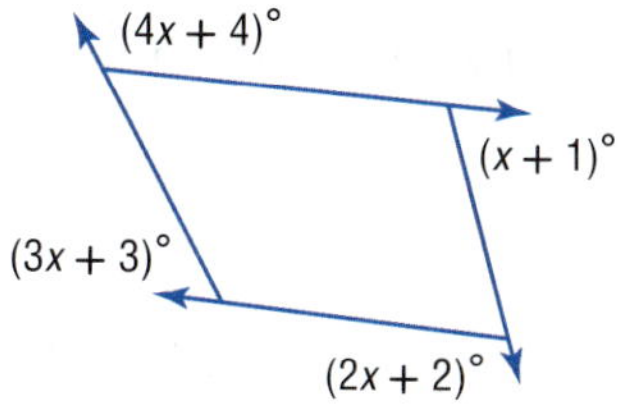

54.

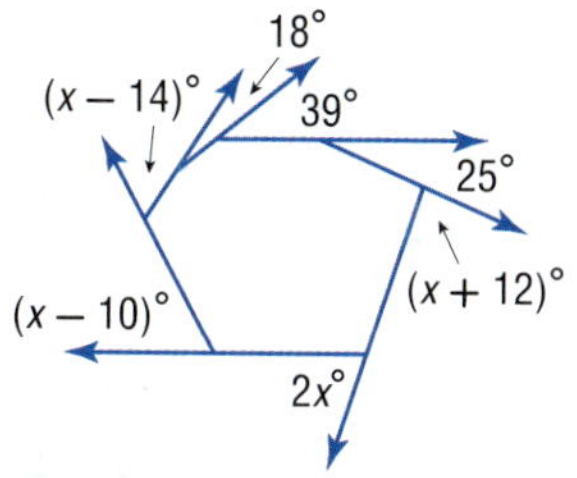

55. FITNESS Toshiro was at the gym for just over two hours. He swam laps in the pool and lifted weights. Prove that he did one of these activities for more than an hour. (Lesson 5-4)

PROOF **Write a flow proof.** (Lesson 4-5)

56. Given: $\overline{EJ} \parallel \overline{FK}$, $\overline{JG} \parallel \overline{KH}$, $\overline{EF} \cong \overline{GH}$
Prove: $\triangle EJG \cong \triangle FKH$

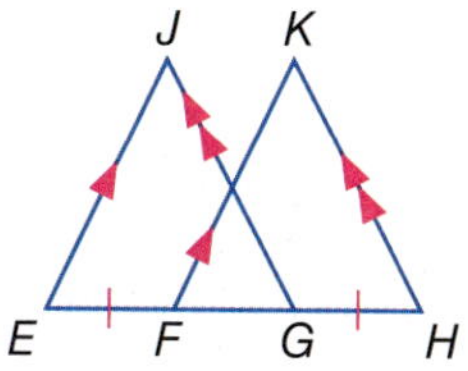

57. Given: $\overline{MN} \cong \overline{PQ}$, $\angle M \cong \angle Q$, $\angle 2 \cong \angle 3$
Prove: $\triangle MLP \cong \triangle QLN$

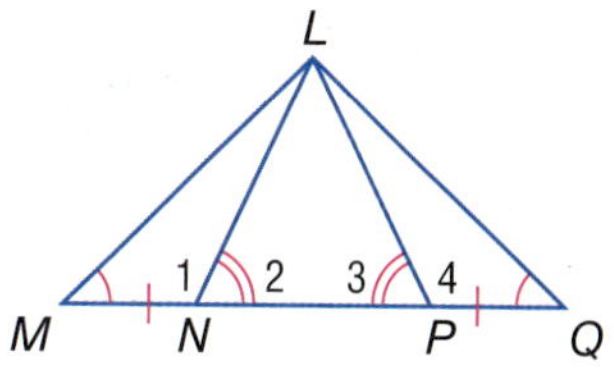

Skills Review

Use slope to determine whether XY and YZ are *perpendicular* or *not perpendicular*. (Lesson 3-3)

58. $X(-2, 2)$, $Y(0, 1)$, $Z(4, 1)$

59. $X(4, 1)$, $Y(5, 3)$, $Z(6, 2)$

CHAPTER 6 Mid-Chapter Quiz

Lessons 6-1 through 6-3

Find the sum of the measures of the interior angles of each convex polygon. (Lesson 6-1)

1. pentagon

2. heptagon

3. 18-gon

4. 23-gon

Find the measure of each interior angle. (Lesson 6-1)

5.

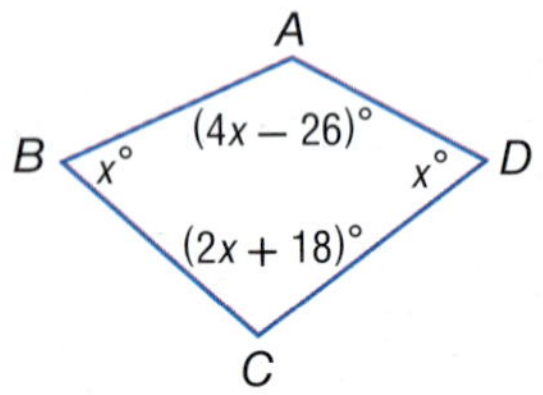

6.

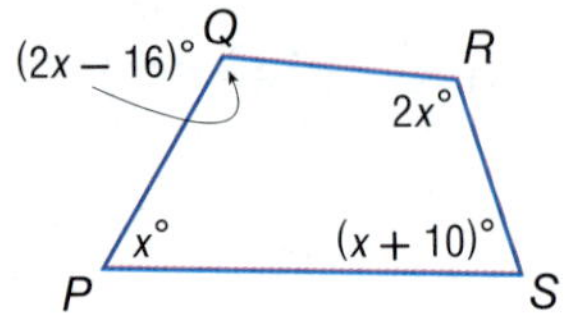

The sum of the measures of the interior angles of a regular polygon is given. Find the number of sides in the polygon. (Lesson 6-1)

7. 720

8. 1260

9. 1800

10. 4500

Find the value of x in each diagram. (Lesson 6-1)

11.

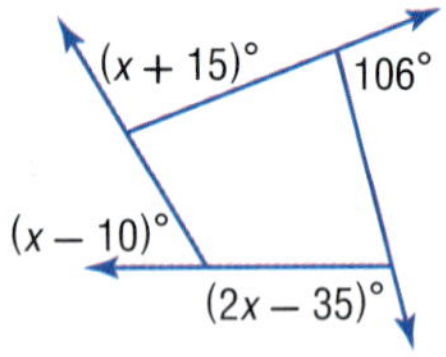

12.

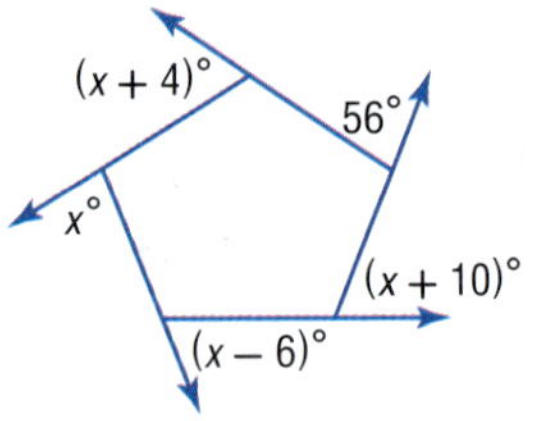

Use ▱WXYZ to find each measure. (Lesson 6-2)

13. $m\angle WZY$

14. WZ

15. $m\angle XYZ$

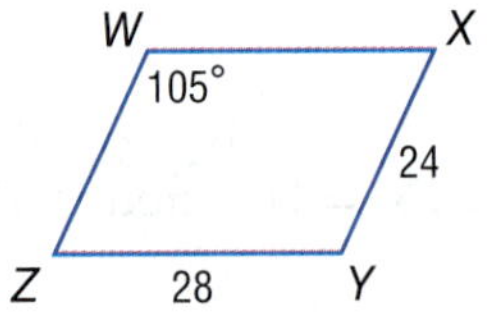

16. DESIGN Describe two ways to ensure that the pieces of the design at the right would fit properly together. (Lesson 6-2)

ALGEBRA Find the value of each variable in each parallelogram. (Lesson 6-2)

17.

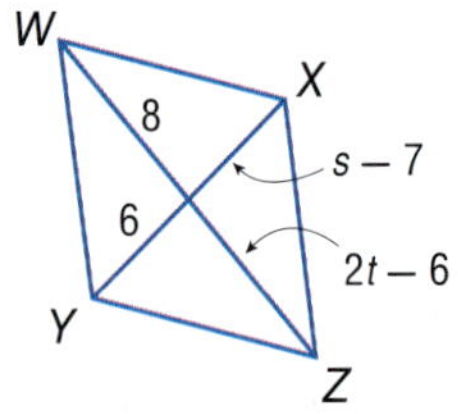

18.

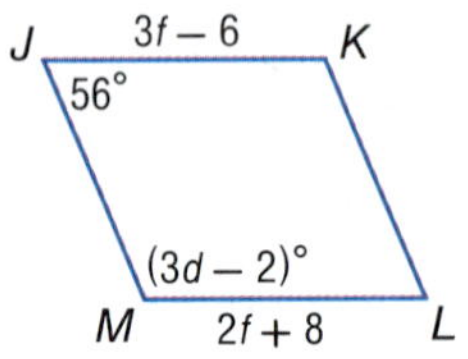

19. PROOF Write a two-column proof. (Lesson 6-2)

Given: ▱$GFBA$ and ▱$HACD$

Prove: $\angle F \cong \angle D$

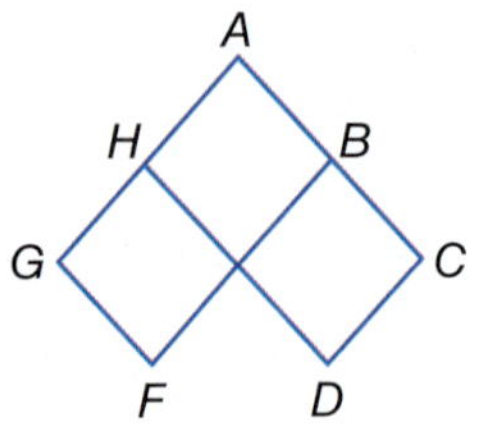

Find x and y so that each quadrilateral is a parallelogram. (Lesson 6-3)

20.

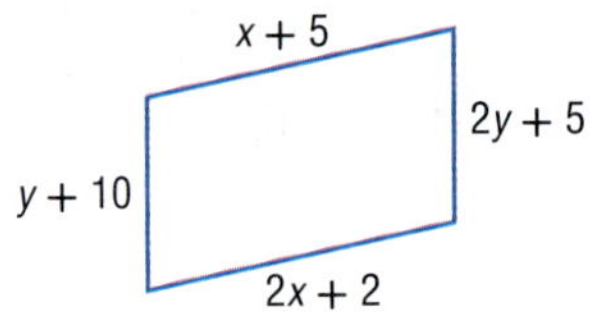

21.

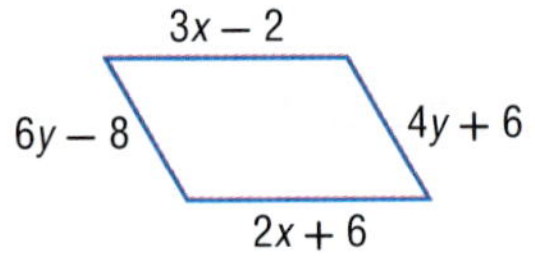

22. MUSIC Why will the keyboard stand shown below always remain parallel to the floor? (Lesson 6-3)

23. MULTIPLE CHOICE Which of the following quadrilaterals is not a parallelogram? (Lesson 6-3)

A

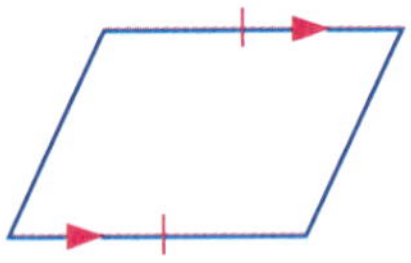

C

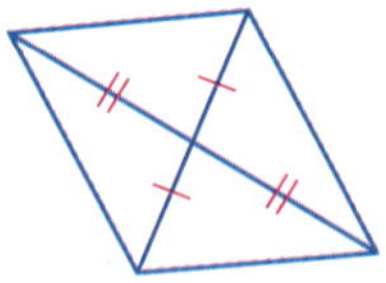

B

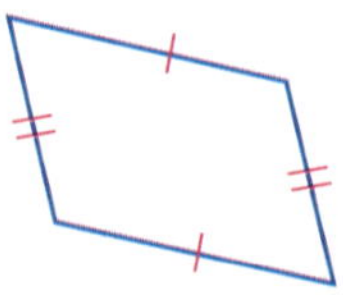

D

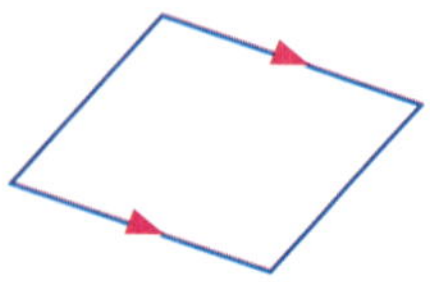

COORDINATE GEOMETRY Determine whether the figure is a parallelogram. Justify your answer with the method indicated. (Lesson 6-3)

24. $A(-6, -5)$, $B(-1, -4)$, $C(0, -1)$, $D(-5, -2)$; Distance Formula

25. $Q(-5, 2)$, $R(-3, -6)$, $S(2, 2)$, $T(-1, 6)$; Slope Formula

6-4 Rectangles

Why?

Leonardo is in charge of set design for a school play. He needs to use paint to create the appearance of a doorway on a lightweight solid wall. The doorway is to be a rectangle 36 inches wide and 80 inches tall. How can Leonardo be sure that he paints a rectangle?

Then
You used properties of parallelograms and determined whether quadrilaterals were parallelograms. (Lesson 6-2)

Now
- Recognize and apply properties of rectangles.
- Determine whether parallelograms are rectangles.

New Vocabulary
rectangle

Math Online
glencoe.com

Properties of Rectangles A **rectangle** is a parallelogram with four right angles. By definition, a rectangle has the following properties.

- All four angles are right angles.
- Opposite sides are parallel and congruent.
- Opposite angles are congruent.
- Consecutive angles are supplementary.
- Diagonals bisect each other.

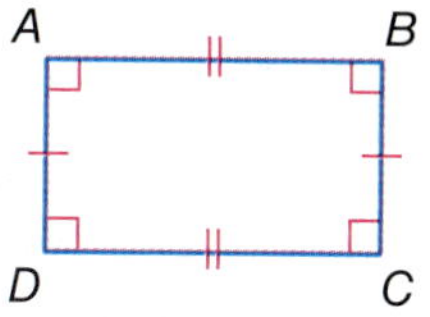

Rectangle *ABCD*

In addition, the diagonals of a rectangle are congruent.

Theorem 6.13 Diagonals of a Rectangle

If a parallelogram is a rectangle, then its diagonals are congruent.

Abbreviation *If a ▱ is a rectangle, diag. are ≅.*

Example If ▱*JKLM* is a rectangle, then $\overline{JL} \cong \overline{MK}$.

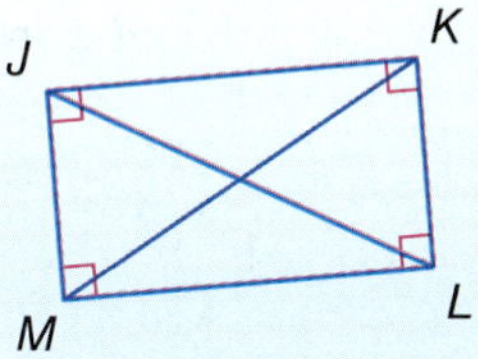

You will prove Theorem 6.13 in Exercise 33.

Real-World EXAMPLE 1 Use Properties of Rectangles

EXERCISE **A rectangular park has two walking paths as shown. If $PS = 180$ meters and $PR = 200$ meters, find QT.**

$\overline{QS} \cong \overline{PR}$	If a ▱ is a rectangle, diag. are ≅.
$QS = PR$	Definition of congruence
$QS = 200$	Substitution

Since *PQRS* is a rectangle, it is a parallelogram. The diagonals of a parallelogram bisect each other, so $QT = ST$.

$QT + ST = QS$	Segment Addition
$QT + QT = QS$	Substitution
$2QT = QS$	Simplify.
$QT = \frac{1}{2}QS$	Divide each side by 2.
$QT = \frac{1}{2}(200)$ or 100	Substitution

Guided Practice **Refer to the figure in Example 1.**

1A. If $TS = 120$ meters, find PR.

1B. If $m\angle PRS = 64$, find $m\angle SQR$.

Personal Tutor glencoe.com

StudyTip

Right Angles Recall from Theorem 6-6 that if a parallelogram has one right angle, then it has four right angles. (Lesson 6-2)

You can use the properties of rectangles along with algebra to find missing values.

EXAMPLE 2 Use Properties of Rectangles and Algebra

ALGEBRA Quadrilateral $JKLM$ is a rectangle. If $m\angle KJL = 2x + 4$ and $m\angle JLK = 7x + 5$, find x.

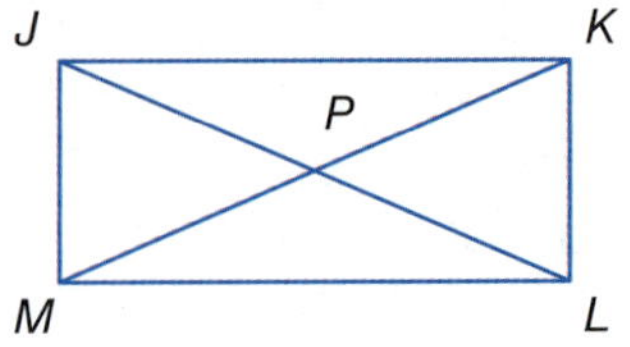

Since $JKLM$ is a rectangle, it has four right angles. So, $m\angle MLK = 90$. Since a rectangle is a parallelogram, opposite sides are parallel. Alternate interior angles of parallel lines are congruent, so $\angle JLM \cong \angle KJL$ and $m\angle JLM = m\angle KJL$.

$m\angle JLM + m\angle JLK = 90$	**Angle Addition**
$m\angle KJL + m\angle JLK = 90$	**Substitution**
$2x + 4 + 7x + 5 = 90$	**Substitution**
$9x + 9 = 90$	**Add like terms.**
$9x = 81$	**Subtract 9 from each side.**
$x = 9$	**Divide each side by 9.**

Guided Practice

2. Refer to the figure in Example 2. If $JP = 3y - 5$ and $MK = 5y + 1$, find y.

Personal Tutor glencoe.com

Prove that Parallelograms are Rectangles

The converse of Theorem 6.13 is also true.

Theorem 6.14

If the diagonals of a parallelogram are congruent, then the parallelogram is a rectangle.

Abbreviation *If diag. of a ▱ are ≅, then ▱ is a rectangle.*

Example If $\overline{WY} \cong \overline{XZ}$ in ▱$WXYZ$, then ▱$WXYZ$ is a rectangle.

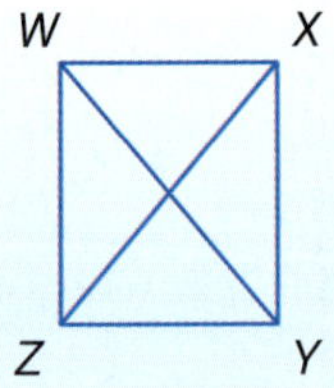

You will prove Theorem 6.14 in Exercise 34.

Real-World Link

The game of dodgeball is played on a rectangular playing field ideally 60 feet long and 30 feet wide. The field is divided into two equal sections by a centerline and attack-lines that are 3 meters (9.8 feet) from, and parallel to, the centerline.

Source: National Amateur Dodgeball Assoc.

Real-World EXAMPLE 3 Proving Rectangle Relationships

DODGEBALL A community recreation center has created an outdoor dodgeball playing field. To be sure that it meets the ideal playing field requirements, they measure the sides of the field and its diagonals. If $AB = 60$ feet, $BC = 30$ feet, $CD = 60$ feet, $AD = 30$ feet, $AC = 67$ feet, and $BD = 67$ feet, explain how the recreation center can be sure that the playing field is rectangular.

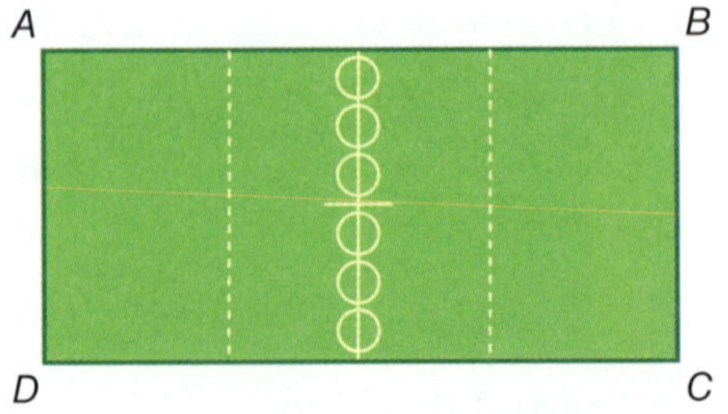

Since $AB = CD$, $BC = AD$, and $AC = BD$, $\overline{AB} \cong \overline{CD}$, $\overline{BC} \cong \overline{AD}$, and $\overline{AC} \cong \overline{BD}$. Because $\overline{AB} \cong \overline{CD}$ and $\overline{BC} \cong \overline{AD}$, $ABCD$ is a parallelogram. Since $\overline{AC}$ and $\overline{BD}$ are congruent diagonals in ▱$ABCD$, ▱$ABCD$ is a rectangle.

Real-World Link

The Mosaic Youth Theater in Detroit, Michigan, is a professional performing arts training program for young people ages 12 to 18. Students are involved in all aspects of performances, including set and lighting design, set construction, stage management, sound, and costumes.

Guided Practice

3. SET DESIGN Refer to the beginning of the lesson. Leonardo measures the sides of his figure and confirms that they have the desired measures as shown. Using a carpenter's square, he also confirms that the measure of the bottom left corner of the figure is a right angle. Can he conclude that the figure is a rectangle? Explain.

Personal Tutor glencoe.com

You can also use the properties of rectangles to prove that a quadrilateral positioned on a coordinate plane is a rectangle given the coordinates of the vertices.

EXAMPLE 4 Rectangles and Coordinate Geometry

COORDINATE GEOMETRY Quadrilateral *PQRS* has vertices *P*(−5, 3), *Q*(1, −1), *R*(−1, −4), and *S*(−7, 0). Determine whether *PQRS* is a rectangle by using the Distance Formula.

Step 1 Use the Distance Formula to determine whether *PQRS* is a parallelogram by determining if opposite sides are congruent.

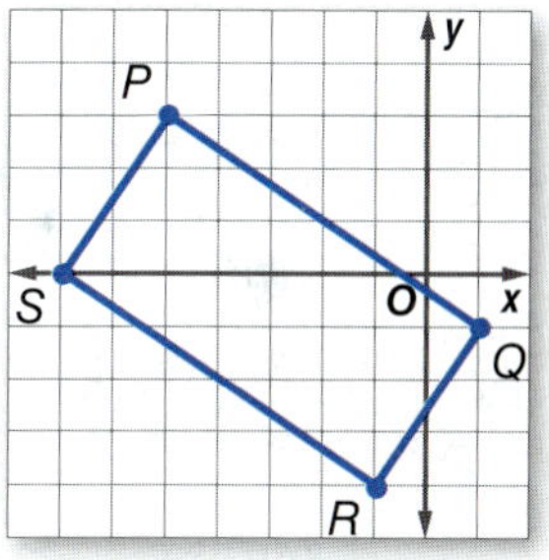

$$PQ = \sqrt{(-5-1)^2 + [3-(-1)]^2} \text{ or } \sqrt{52}$$

$$RS = \sqrt{[-1-(-7)]^2 + (-4-0)^2} \text{ or } \sqrt{52}$$

$$PS = \sqrt{[-5-(-7)]^2 + (3-0)^2} \text{ or } \sqrt{13}$$

$$QR = \sqrt{[1-(-1)^2 + [-1-(-4)]^2} \text{ or } \sqrt{13}$$

Since opposite sides of the quadrilateral have the same measure, they are congruent. So, quadrilateral *PQRS* is a parallelogram.

Step 2 Determine whether the diagonals of ▱*PQRS* are congruent.

$$PR = \sqrt{[-5-(-1)]^2 + [3-(-4)]^2} \text{ or } \sqrt{65}$$

$$QS = \sqrt{[1-(-7)]^2 + (-1-0)^2} \text{ or } \sqrt{65}$$

Since the diagonals have the same measure, they are congruent. So, ▱*PQRS* is a rectangle.

StudyTip

Rectangles and Parallelograms A rectangle is a parallelogram, but a parallelogram is not necessarily a rectangle.

Guided Practice

4. Quadrilateral *JKLM* has vertices *J*(−10, 2), *K*(−8, −6), *L*(5, −3), and *M*(2, 5). Determine whether *JKLM* is a rectangle using the Slope Formula.

Personal Tutor glencoe.com

Check Your Understanding

Example 1 p. 419

FARMING An X-brace on a barn door is both decorative and functional. It helps to prevent the door from warping over time. If $ST = 3\frac{13}{16}$ feet, $PS = 7$ feet, and $m\angle PTQ = 67$, find each measure.

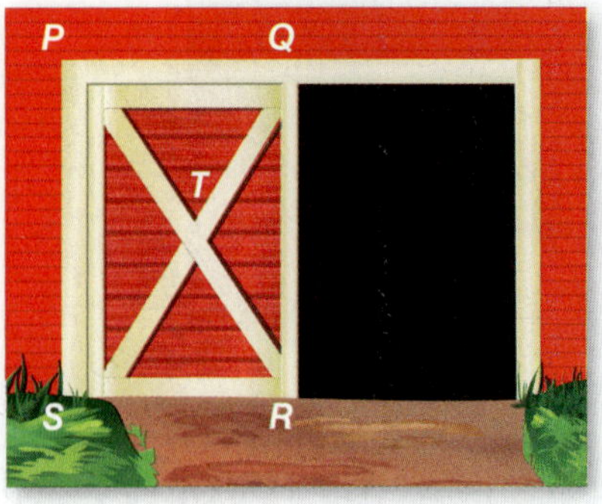

1. QR
2. SQ
3. $m\angle TQR$
4. $m\angle TSR$

Example 2 p. 420

ALGEBRA Quadrilateral $DEFG$ is a rectangle.

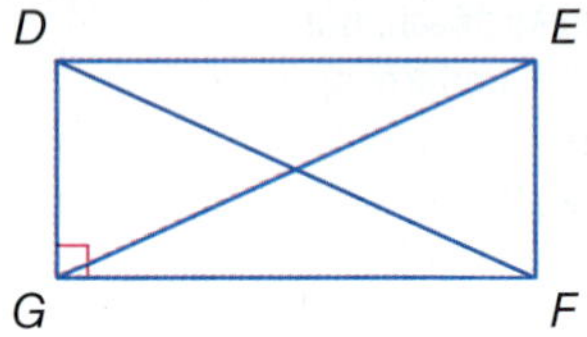

5. If $FD = 3x - 7$ and $EG = x + 5$, find EG.
6. If $m\angle EFD = 2x - 3$ and $m\angle DFG = x + 12$, find $m\angle EFD$.

Example 3 p. 420

7. **PROOF** If $ABDE$ is a rectangle and $\overline{BC} \cong \overline{DC}$, prove that $\overline{AC} \cong \overline{EC}$.

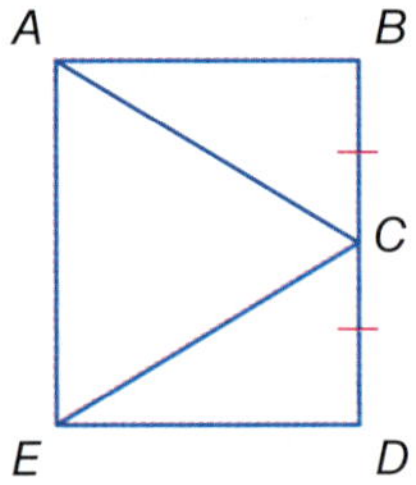

Example 4 p. 421

COORDINATE GEOMETRY Graph each quadrilateral with the given vertices. Determine whether the figure is a rectangle. Justify your answer using the indicated formula.

8. $W(-4, 3)$, $X(1, 5)$, $Y(3, 1)$, $Z(-2, -2)$; Slope Formula
9. $A(4, 3)$, $B(4, -2)$, $C(-4, -2)$, $D(-4, 3)$; Distance Formula

Practice and Problem Solving

● = **Step-by-Step Solutions** begin on page R20.
Extra Practice begins on page 969.

Example 1 p. 419

FENCING X-braces are also used to provide support in fencing. If $AB = 6$ feet, $AC = 2$ feet, and $m\angle CAE = 65$, find each measure.

10. BD
11. CB
12. $m\angle DEB$
13. $m\angle ECD$

Example 2 p. 420

ALGEBRA Quadrilateral $WXYZ$ is a rectangle.

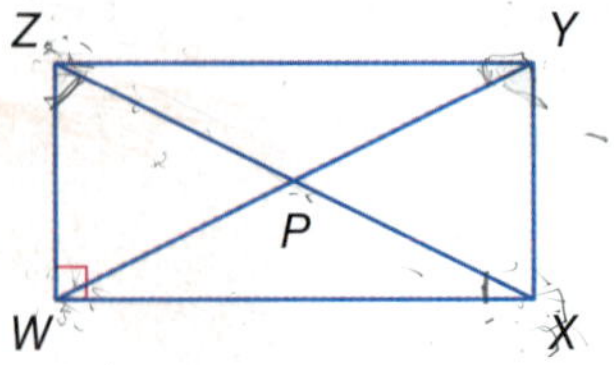

14. If $ZY = 2x + 3$ and $WX = x + 4$, find WX.
15. If $PY = 3x - 5$ and $WP = 2x + 11$, find ZP.
16. If $m\angle ZYW = 2x - 7$ and $m\angle WYX = 2x + 5$, find $m\angle ZYW$.
17. If $ZP = 4x - 9$ and $PY = 2x + 5$, find ZX.
18. If $m\angle XZY = 3x + 6$ and $m\angle XZW = 5x - 12$, find $m\angle YXZ$.
19. If $m\angle ZXW = x - 11$ and $m\angle WZX = x - 9$, find $m\angle ZXY$.

Example 3 p. 420

PROOF **Write a two-column proof.**

20. **Given:** *ABCD* is a rectangle.
Prove: $\triangle ADC \cong \triangle BCD$

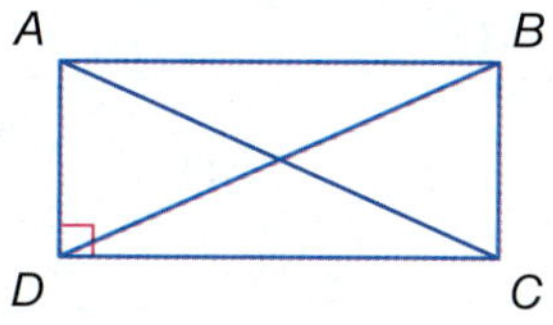

21. **Given:** *QTVW* is a rectangle.
$\overline{QR} \cong \overline{ST}$
Prove: $\triangle SWQ \cong \triangle RVT$

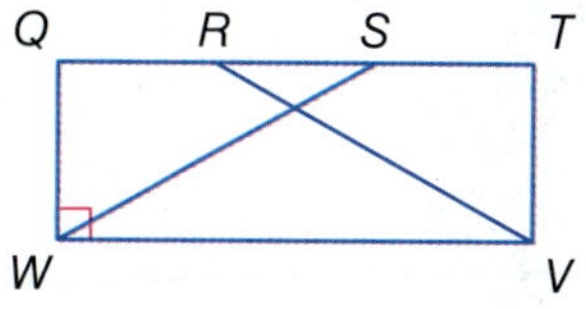

Example 4 p. 421

COORDINATE GEOMETRY **Graph each quadrilateral with the given vertices. Determine whether the figure is a rectangle. Justify your answer using the indicated formula.**

22. $W(-2, 4)$, $X(5, 5)$, $Y(6, -2)$, $Z(-1, -3)$; Slope Formula

23. $J(3, 3)$, $K(-5, 2)$, $L(-4, -4)$, $M(4, -3)$; Distance Formula

24. $Q(-2, 2)$, $R(0, -2)$, $S(6, 1)$, $T(4, 5)$; Distance Formula

25. $G(1, 8)$, $H(-7, 7)$, $J(-6, 1)$, $K(2, 2)$; Slope Formula

Quadrilateral *ABCD* is a rectangle. Find each measure if $m\angle 2 = 40$.

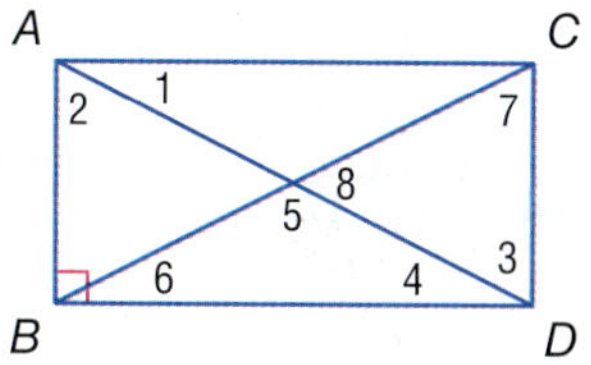

26. $m\angle 1$

27. $m\angle 7$

28. $m\angle 3$

29. $m\angle 5$

30. $m\angle 6$

31. $m\angle 8$

32. **CONSTRUCTION** Jody is building a new bookshelf using wood and metal supports like the one shown. To what length should she cut the metal supports in order for the bookshelf to be *square*, which means that the angles formed by the shelves and the vertical supports are all right angles? Explain your reasoning.

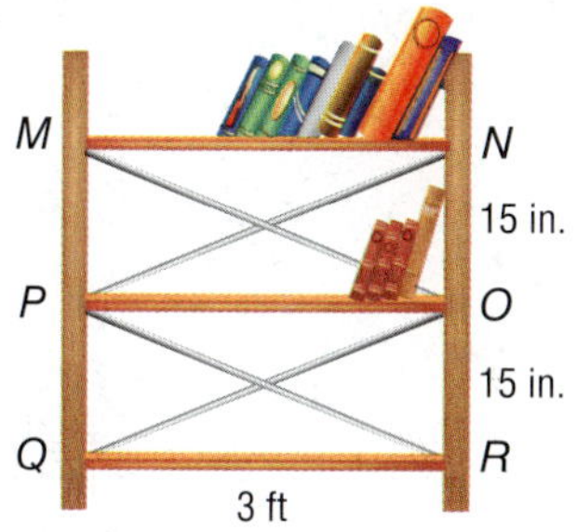

PROOF **Write a two-column proof.**

33. Theorem 6.13

34. Theorem 6.14

PROOF **Write a paragraph proof of each statement.**

35. If a parallelogram has one right angle, then it is a rectangle.

36. If a quadrilateral has four right angles, then it is a rectangle.

37. **CONSTRUCTION** Construct a rectangle using the construction for congruent segments and the construction for a line perpendicular to another line through a point on the line. Justify each step of the construction.

Real-World Link

The end zone on a football field is 160 feet wide and 30 feet long.

Source: Sports Know How

38. **SPORTS** Kyle is responsible for painting the football practice field. He has finished the end zone. Explain how Kyle can confirm that the end zone is the regulation size and be sure that it is also a rectangle using only a tape measure.

ALGEBRA **Quadrilateral *WXYZ* is a rectangle.**

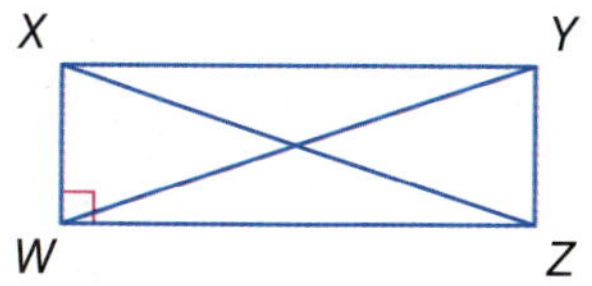

39. If $XW = 3$, $WZ = 4$, and $XZ = b$, find YW.

40. If $XZ = 2c$ and $ZY = 6$, and $XY = 8$, find WY.

Real-World Link

Many cities regulate the surface area of political signs. For example, Ft. Myers, Florida, requires that the surface area of political signs be less than 4 square feet.

Source: City of Ft. Myers, Florida

41. SIGNS The sign below is in the foyer of Nyoko's school. Based on the dimensions given, can Nyoko be sure that the sign is a rectangle? Explain your reasoning.

PROOF **Write a coordinate proof of each statement.**

42. The diagonals of a rectangle are congruent.

43 If the diagonals of a parallelogram are congruent, then it is a rectangle.

44. **MULTIPLE REPRESENTATIONS** In the problem, you will explore properties of other special parallelograms.

a. GEOMETRIC Draw three parallelograms, each with all four sides congruent. Label one parallelogram *ABCD*, one *MNOP*, and one *WXYZ*. Draw the two diagonals of each parallelogram and label the intersections *R*.

b. TABULAR Use a protractor to measure the appropriate angles and complete the table below.

Parallelogram	*ABCD*		*MNOP*		*WXYZ*	
Angle	$\angle ARB$	$\angle BRC$	$\angle MRN$	$\angle NRO$	$\angle WRX$	$\angle XRY$
Angle Measure						

c. VERBAL Make a conjecture about the diagonals of a parallelogram with four congruent sides.

H.O.T. Problems Use Higher-Order Thinking Skills

45. CHALLENGE In rectangle *ABCD*, $m\angle EAB = 4x + 6$, $m\angle DEC = 10 - 11y$, and $m\angle EBC = 60$. Find the values of x and y.

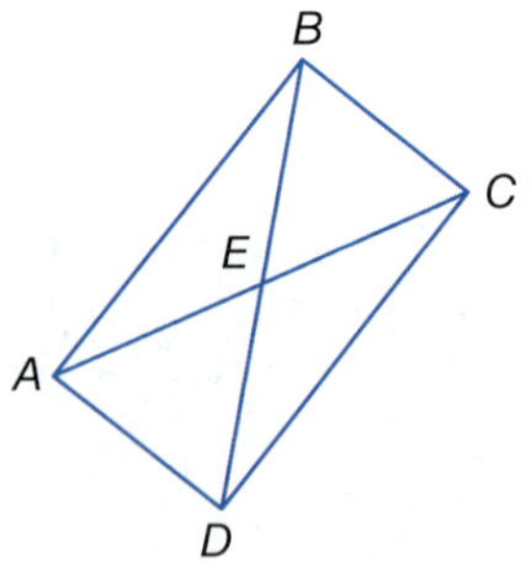

46. ERROR ANALYSIS Parker says that any two congruent acute triangles can be arranged to make a rectangle. Tamika says that only two congruent right triangles can be arranged to make a rectangle. Is either of them correct? Explain your reasoning.

47. REASONING In the diagram at the right, lines n, p, q, and r are parallel and lines ℓ and m are parallel. How many rectangles are formed by the intersecting lines?

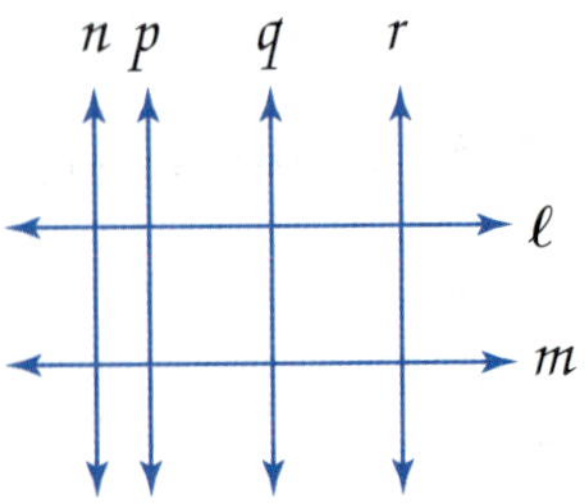

48. OPEN ENDED Write the equations of four lines having intersections that form the vertices of a rectangle. Verify your answer using coordinate geometry.

49. WRITING IN MATH Explain why all rectangles are parallelograms, but all parallelograms are not rectangles.

Standardized Test Practice

50. If $FJ = -3x + 5y$, $FM = 3x + y$, $GH = 11$, and $GM = 13$, what values of x and y make parallelogram $FGHJ$ a rectangle?

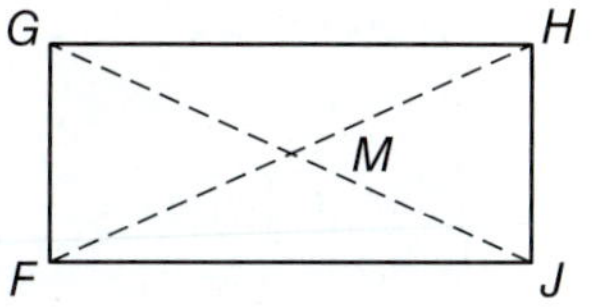

A $x = 3, y = 4$ **C** $x = 7, y = 8$

B $x = 4, y = 3$ **D** $x = 8, y = 7$

51. ALGEBRA A rectangular playground is surrounded by an 80-foot fence. One side of the playground is 10 feet longer than the other. Which of the following equations could be used to find r, the shorter side of the playground?

F $10r + r = 80$ **H** $r(r + 10) = 80$

G $4r + 10 = 80$ **J** $2(r + 10) + 2r = 80$

52. SHORT RESPONSE What is the measure of $\angle APB$?

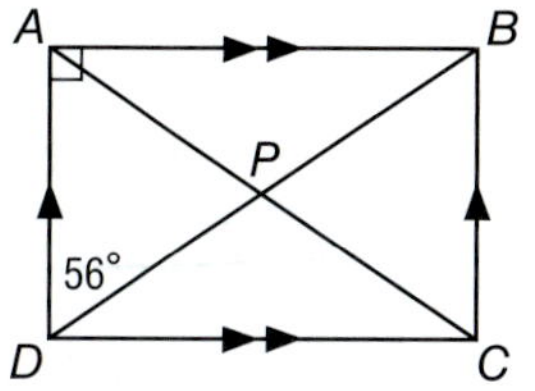

53. SAT/ACT If p is odd, which of the following must also be odd?

A $2p$

B $p + 2$

C $\frac{p}{2}$

D $2p - 2$

Spiral Review

ALGEBRA **Find x and y so that the quadrilateral is a parallelogram.** (Lesson 6-3)

54.

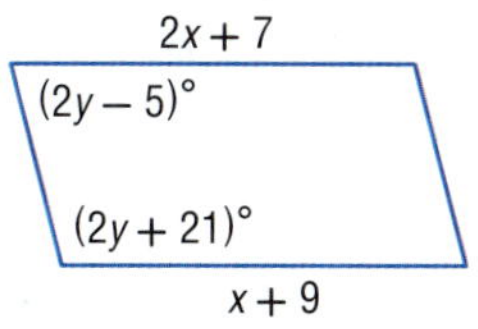

55.

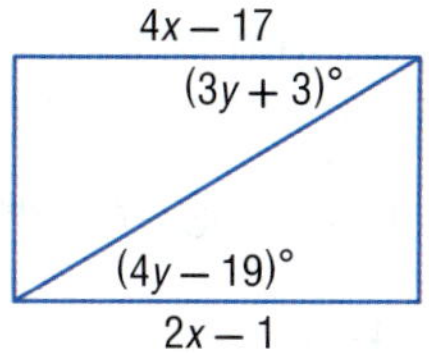

56.

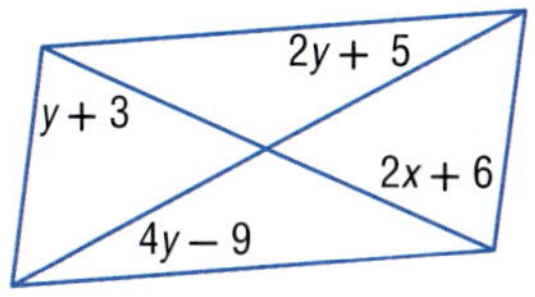

57. COORDINATE GEOMETRY Find the coordinates of the intersection of the diagonals of ▱$ABCD$ with vertices $A(1, 3)$, $B(6, 2)$, $C(4, -2)$, and $D(-1, -1)$. (Lesson 6-2)

Refer to the figure at the right. (Lesson 4-6)

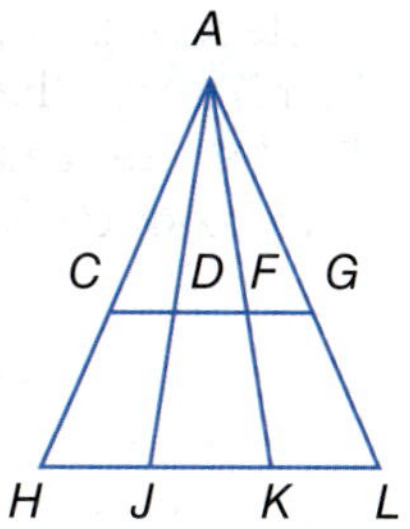

58. If $\overline{AC} \cong \overline{AF}$, name two congruent angles.

59. If $\angle AHJ \cong \angle AJH$, name two congruent segments.

60. If $\angle AJL \cong \angle ALJ$, name two congruent segments.

61. If $\overline{JA} \cong \overline{KA}$, name two congruent angles.

Skills Review

Find the distance between each pair of points. (Lesson 1-3)

62. $(4, 2), (2, -5)$

63. $(0, 6), (-1, -4)$

64. $(-4, 3), (3, -4)$

6-5 Rhombi and Squares

Then

You determined whether quadrilaterals were parallelograms and/or rectangles. (Lesson 6-4)

Now

- Recognize and apply the properties of rhombi and squares.
- Determine whether quadrilaterals are rectangles, rhombi, or squares.

New Vocabulary

rhombus
square

Math Online
glencoe.com

Why?

Some fruits, nuts, and vegetables are packaged using bags made out of rhombus-shaped tubular netting. Similar shaped nylon netting is used for goals in such sports as soccer, hockey, and football. A rhombus and a square are both types of equilateral parallelograms.

Properties of Rhombi and Squares A **rhombus** is a parallelogram with all four sides congruent. A rhombus has all the properties of a parallelogram and the two additional characteristics described in the theorems below.

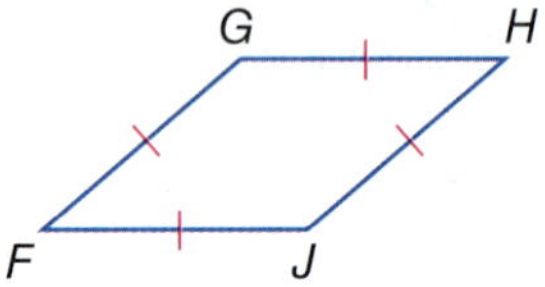

Theorems — Diagonals of a Rhombus

6.15 If a parallelogram is a rhombus, then its diagonals are perpendicular.

Example If $\square ABCD$ is a rhombus, then $\overline{AC} \perp \overline{BD}$.

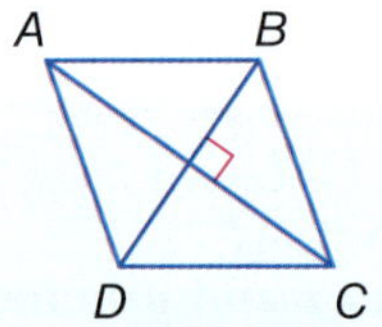

6.16 If a parallelogram is a rhombus, then each diagonal bisects a pair of opposite angles.

Example If $\square NPQR$ is a rhombus, then $\angle 1 \cong \angle 2$, $\angle 3 \cong \angle 4$, $\angle 5 \cong \angle 6$, and $\angle 7 \cong \angle 8$.

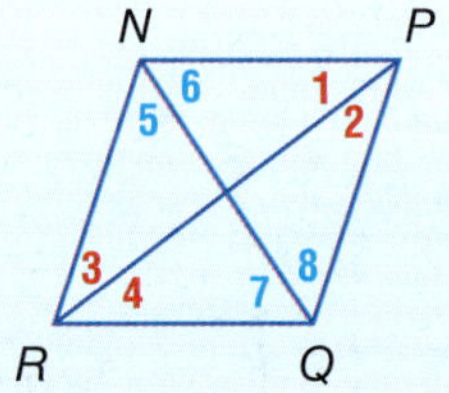

You will prove Theorem 6.16 in Exercise 34.

Proof — Theorem 6.15

Given: *ABCD* is a rhombus.

Prove: $\overline{AC} \perp \overline{BD}$

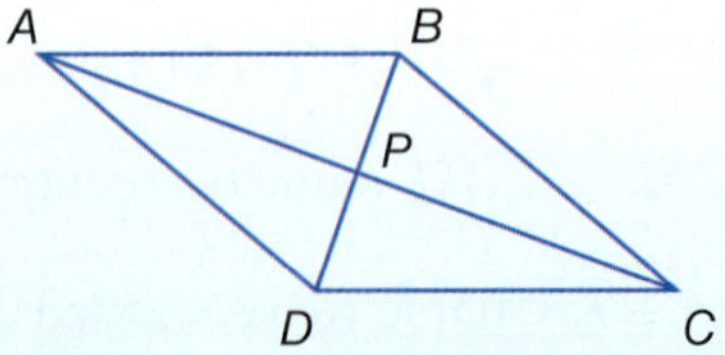

Paragraph Proof:

Since *ABCD* is a rhombus, by definition $\overline{AB} \cong \overline{BC}$. A rhombus is a parallelogram and the diagonals of a parallelogram bisect each other, so $\overline{BD}$ bisects $\overline{AC}$ at *P*. Thus, $\overline{AP} \cong \overline{PC}$. $\overline{BP} \cong \overline{BP}$ by the Reflexive Property. So, $\triangle APB \cong \triangle CPB$ by SSS. $\angle APB \cong \angle CPB$ by CPCTC. $\angle APB$ and $\angle CPB$ also form a linear pair. Two congruent angles that form a linear pair are right angles. $\angle APB$ is a right angle, so $\overline{AC} \perp \overline{BD}$ by the definition of perpendicular lines.

ReadingMath

Rhombi The plural form of rhombus is *rhombi,* pronounced ROM-bye.

EXAMPLE 1 Use Properties of a Rhombus

The diagonals of rhombus $FGHJ$ intersect at K. Use the given information to find each measure or value.

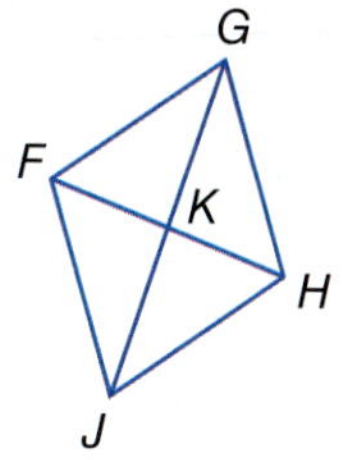

a. If $m\angle FJH = 82$, find $m\angle KHJ$.

Since $FGHJ$ is a rhombus, diagonal $\overline{JG}$ bisects $\angle FJH$. Therefore, $m\angle KJH = \frac{1}{2}m\angle FJH$. So $m\angle KJH = \frac{1}{2}(82)$ or 41. Since the diagonals of a rhombus are perpendicular, $m\angle JKH = 90$ by the definition of perpendicular lines.

$m\angle KJH + m\angle JKH + m\angle KHJ = 180$	Triangle Sum Theorem
$41 + 90 + m\angle KHJ = 180$	Substitution
$131 + m\angle KHJ = 180$	Simplify.
$m\angle KHJ = 49$	Subtract 131 from each side.

b. ALGEBRA If $GH = x + 9$ and $JH = 5x - 2$, find x.

$\overline{GH} \cong \overline{JH}$	By definition, all sides of a rhombus are congruent.
$GH = JH$	Definition of congruence
$x + 9 = 5x - 2$	Substitution
$9 = 4x - 2$	Subtract x from each side.
$11 = 4x$	Add 2 to each side.
$2.75 = x$	Divide each side by 4.

Guided Practice

Refer to rhombus $FGHJ$ above.

1A. If $FK = 5$ and $FG = 13$, find KJ.

1B. ALGEBRA If $m\angle JFK = 6y + 7$ and $m\angle KFG = 9y - 5$, find y.

Personal Tutor glencoe.com

A **square** is a parallelogram with four congruent sides and four right angles. Recall that a parallelogram with four right angles is a rectangle, and a parallelogram with four congruent sides is a rhombus. Therefore, a parallelogram that is both a rectangle and a rhombus is also a square.

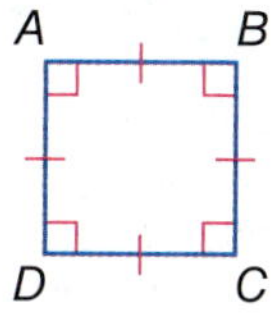

Square *ABCD*

The Venn diagram summarizes the relationships among parallelograms, rhombi, rectangles, and squares.

Concept Summary Parallelograms

Parallelograms (Opp. sides are ∥.)

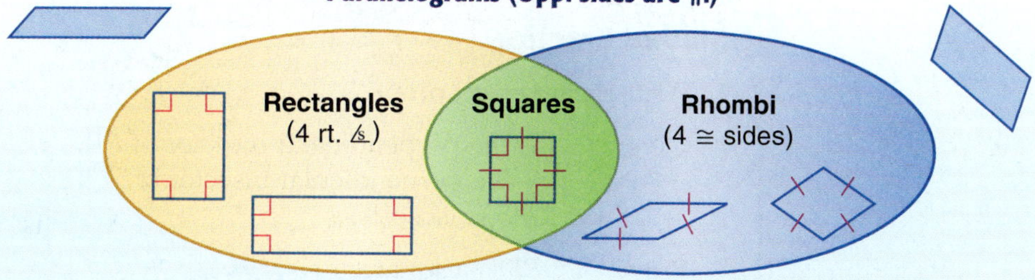

StudyTip

Common Misconception Theorems 6.17, 6.18, and 6.19 apply only if you already know that a quadrilateral is a parallelogram.

All of the properties of parallelograms, rectangles, and rhombi apply to squares. For example, the diagonals of a square bisect each other (parallelogram), are congruent (rectangle), and are perpendicular (rhombus).

Prove that Quadrilaterals are Rhombi or Squares The theorems below provide conditions for rhombi and squares.

Theorems — Conditions for Rhombi and Squares

6.17 If the diagonals of a parallelogram are perpendicular, then the parallelogram is a rhombus. (Converse of Theorem. 6.15)

Example If $\overline{JL} \perp \overline{KM}$, then ▱*JKLM* is a rhombus.

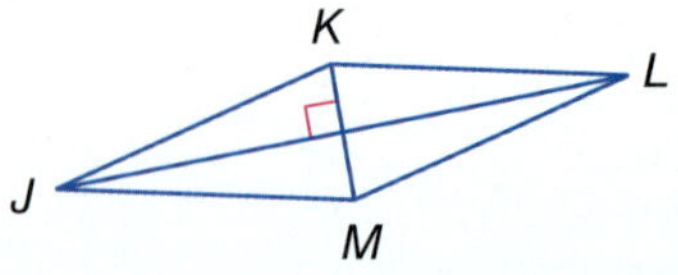

6.18 If one diagonal of a parallelogram bisects a pair of opposite angles, then the parallelogram is a rhombus. (Converse of Theorem. 6.16)

Example If $\angle 1 \cong \angle 2$ and $\angle 3 \cong \angle 4$, or $\angle 5 \cong \angle 6$ and $\angle 7 \cong \angle 8$, then ▱*WXYZ* is a rhombus.

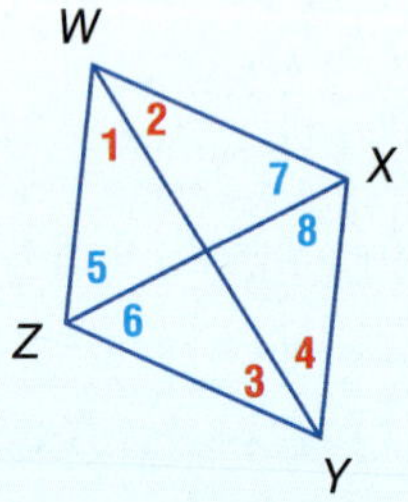

6.19 If one pair of consecutive sides of a parallelogram are congruent, the parallelogram is a rhombus.

Example If $\overline{AB} \cong \overline{BC}$, then ▱*ABCD* is a rhombus.

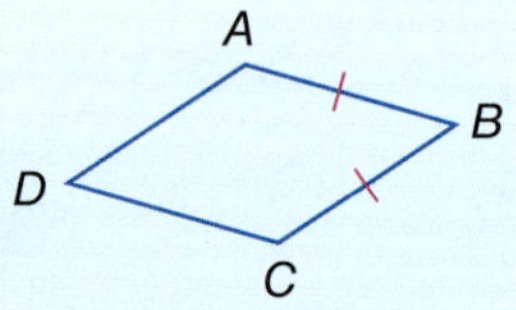

6.20 If a quadrilateral is both a rectangle and a rhombus, then it is a square.

You will prove Theorems 6.17–6.20 in Exercises 35–38, respectively.

StudyTip

Congruent Triangles Since a rhombus has four congruent sides, one diagonal separates the rhombus into two congruent isosceles triangles. Drawing two diagonals separates the rhombus into four congruent right triangles.

You can use the properties of rhombi and squares to write proofs.

EXAMPLE 2 Proofs Using Properties of Rhombi and Squares

Write a paragraph proof.

Given: *JKLM* is a parallelogram.
$\triangle JKL$ is isosceles.

Prove: *JKLM* is a rhombus.

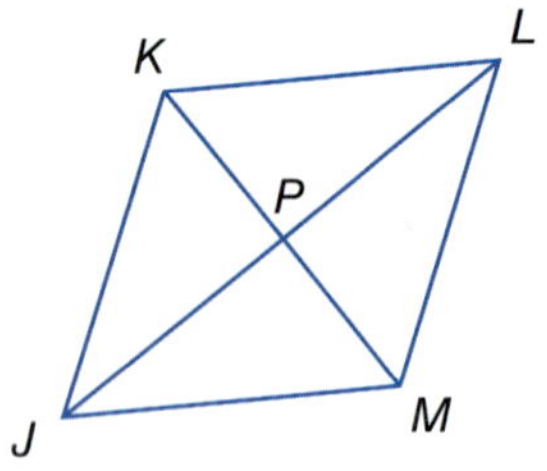

Paragraph Proof:

Since it is given that $\triangle JKL$ is isosceles, $\overline{KL} \cong \overline{JK}$ by definition. These are consecutive sides of the given parallelogram *JKLM*. So, by Theorem 6.19, *JKLM* is a rhombus.

Guided Practice

2. Write a paragraph proof.

Given: $\overline{SQ}$ is the perpendicular bisector of $\overline{PR}$.
$\overline{PR}$ is the perpendicular bisector of $\overline{SQ}$.
$\triangle RMS$ is isosceles.

Prove: *PQRS* is a square.

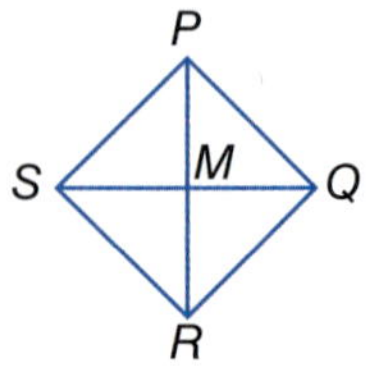

Personal Tutor glencoe.com

Real-World Link

Archaeology is the study of artifacts that provide information about human life and activities in the past. Since humans only began writing about 5000 years ago, information from periods before that time must be gathered from the objects that archeologists locate.

Source: *Encyclopaedia Britannica*

Real-World EXAMPLE 3 Use Conditions for Rhombi and Squares

ARCHAEOLOGY The key to the successful excavation of an archaeological site is accurate mapping. How can archaeologists be sure that the region they have marked off is a 1-meter by 1-meter square?

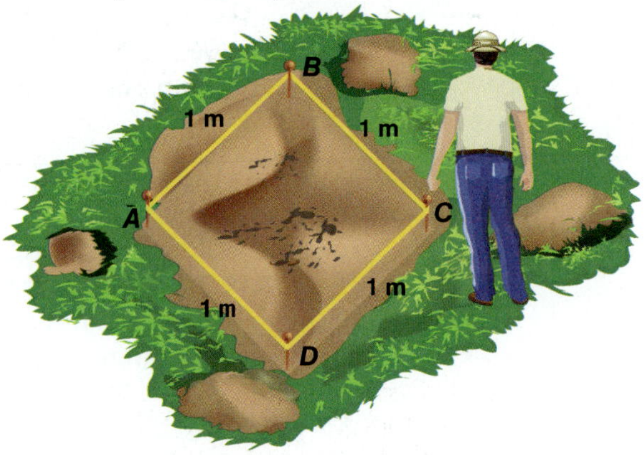

Each side of quadrilateral $ABCD$ measures 1 meter. Since opposite sides are congruent, $ABCD$ is a parallelogram. Since consecutive sides of $\square ABCD$ are congruent, it is a rhombus. If the archaeologists can show that $\square ABCD$ is also a rectangle, then by Theorem 6.20, $\square ABCD$ is a square.

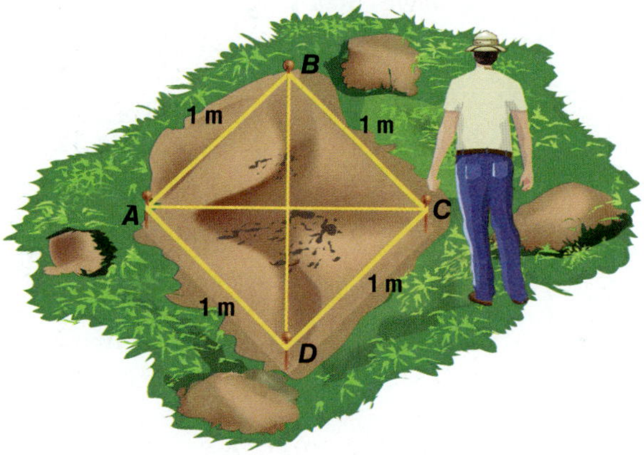

If the diagonals of a parallelogram are congruent, then the parallelogram is a rectangle. So if the archeologists measure the length of string needed to form each diagonal and find that these lengths are equal, then $ABCD$ is a square.

Guided Practice

3. QUILTING Kathy is designing a quilt with blocks like the one shown.

A. If she marks the diagonals of each yellow piece and determines that each pair of diagonals is perpendicular, can she conclude that each yellow piece is a rhombus? Explain.

B. If all four angles of the green piece have the same measure and the bottom and left sides have the same measure, can she conclude that the green piece is a square? Explain.

Personal Tutor glencoe.com

In Chapter 4, you used coordinate geometry to classify triangles. Coordinate geometry can also be used to classify quadrilaterals.

EXAMPLE 4 Classify Quadrilaterals Using Coordinate Geometry

COORDINATE GEOMETRY Determine whether ▱*JKLM* with vertices *J*(−7, −2), *K*(0, 4), *L*(9, 2), and *M*(2, −4) is a *rhombus*, a *rectangle*, or a *square*. List all that apply. Explain.

Understand Plot and connect the vertices on a coordinate plane.

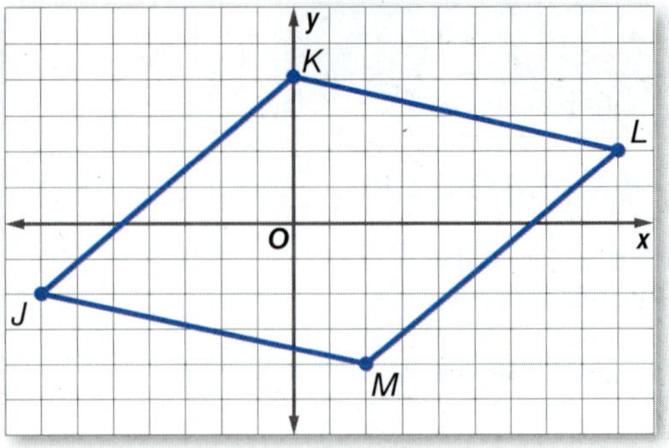

It appears from the graph that the parallelogram has four congruent sides, but no right angles. So, it appears that the figure is a rhombus, but not a square or a rectangle.

Plan If the diagonals of the parallelogram are congruent, then it is a rectangle. If they are perpendicular, then it is a rhombus. If they are both congruent and perpendicular, the parallelogram is a rectangle, a rhombus, and a square.

Solve **Step 1** Use the Distance Formula to compare the diagonal lengths.

$KM = \sqrt{(2-0)^2 + (-4-4)^2} = \sqrt{68}$ or $2\sqrt{17}$

$JL = \sqrt{[9-(-7)]^2 + [2-(-2)]^2} = \sqrt{272}$ or $4\sqrt{17}$

Since $2\sqrt{17} \neq 4\sqrt{17}$, the diagonals are not congruent. So, ▱*JKLM* is *not* a rectangle. Since the figure is not a rectangle, it also *cannot* be a square.

Step 2 Use the Slope Formula to determine whether the diagonals are perpendicular.

slope of $\overline{KM} = \frac{-4-4}{2-0} = \frac{-8}{2}$ or -4

slope of $\overline{JL} = \frac{2-(-2)}{9-(-7)} = \frac{4}{16}$ or $\frac{1}{4}$

Since the product of the slopes of the diagonals is −1, the diagonals are perpendicular, so ▱*JKLM* is a rhombus.

Check $JK = \sqrt{[4-(-2)]^2 + [0-(-7)]^2}$ or $\sqrt{85}$

$KL = \sqrt{(9-0)^2 + (2-4)^2}$ or $\sqrt{85}$

So, ▱*JKLM* is a rhombus by Theorem 6.20.

Since the slope of $\overline{JK} = \frac{4-(-2)}{0-(-7)}$ or $\frac{6}{7}$, the slope of $\overline{KL} = \frac{2-4}{9-0}$ or $-\frac{2}{9}$, and the product of these slopes is not −1, consecutive sides $\overline{JK}$ and $\overline{KL}$ are not perpendicular. Therefore, ∠*JKL* is not a right angle. So ▱*JKLM* is not a rectangle or a square. ✔

Guided Practice

4. Given *J*(5, 0), *K*(8, −11), *L*(−3, −14), *M*(−6, −3), determine whether parallelogram *JKLM* is a *rhombus*, a *rectangle*, or a *square*. List all that apply. Explain.

Personal Tutor glencoe.com

Problem-Solving Tip

Make a Graph When analyzing a figure using coordinate geometry, graph the figure to help formulate a conjecture and also to help check the reasonableness of the answer you obtain algebraically.

Study Tip

Square and Rhombus A square is a rhombus, but a rhombus is not necessarily a square.

Check Your Understanding

Example 1
p. 427

ALGEBRA **Quadrilateral *ABCD* is a rhombus. Find each value or measure.**

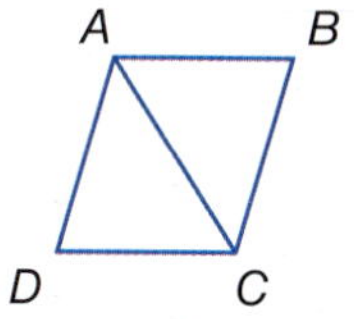

1. If $m\angle BCD = 64$, find $m\angle BAC$.

2. If $AB = 2x + 3$ and $BC = x + 7$, find CD.

Examples 2 and 3
pp. 428–429

3. **PROOF** Write a two-column proof to prove that if $ABCD$ is a rhombus with diagonal $\overline{DB}$, then $\overline{AP} \cong \overline{CP}$.

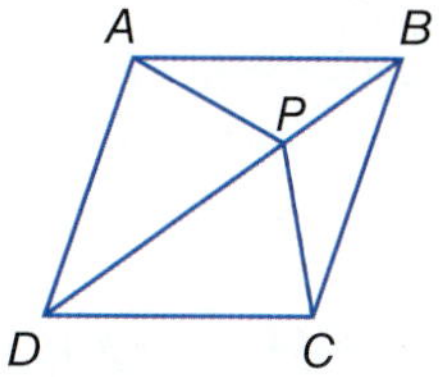

4. **GAMES** The checkerboard below is made up of 64 congruent black and red squares. Use this information to prove that the board itself is a square.

Example 4
p. 430

COORDINATE GEOMETRY **Given each set of vertices, determine whether $\square QRST$ is a *rhombus*, a *rectangle*, or a *square*. List all that apply. Explain.**

5. $Q(1, 2)$, $R(-2, -1)$, $S(1, -4)$, $T(4, -1)$

6. $Q(-2, -1)$, $R(-1, 2)$, $S(4, 1)$, $T(3, -2)$

Practice and Problem Solving

● = **Step-by-Step Solutions** begin on page R20.
Extra Practice begins on page 969.

Example 1
p. 427

ALGEBRA **Quadrilateral *ABCD* is a rhombus. Find each value or measure.**

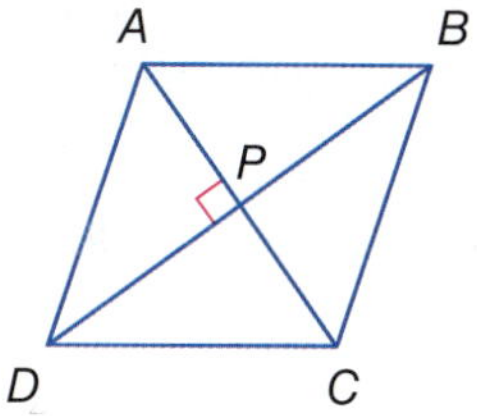

7. If $AB = 14$, find BC.

8. If $m\angle BCD = 54$, find $m\angle BAC$.

9. If $AP = 3x - 1$ and $PC = x + 9$, find AC.

10. If $DB = 2x - 4$ and $PB = 2x - 9$, find PD.

11. If $m\angle ABC = 2x - 7$ and $m\angle BCD = 2x + 3$, find $m\angle DAB$.

12. If $m\angle DPC = 3x - 15$, find x.

Example 2
p. 428

PROOF **Write a two-column proof.**

13. **Given:** $\overline{WZ} \parallel \overline{XY}$, $\overline{WX} \parallel \overline{ZY}$
$\overline{WZ} \cong \overline{ZY}$
Prove: $WXYZ$ is a rhombus.

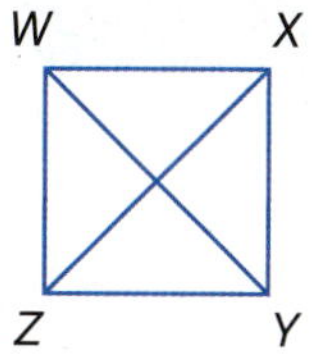

14. **Given:** $QRST$ is a parallelogram.
$\overline{TR} \cong \overline{QS}$, $m\angle QPR = 90$
Prove: $QRST$ is a square.

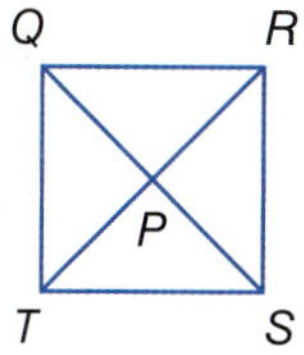

15. **Given:** $JKQP$ is a square.
$\overline{ML}$ bisects $\overline{JP}$ and $\overline{KQ}$.
Prove: $JKLM$ is a parallelogram.

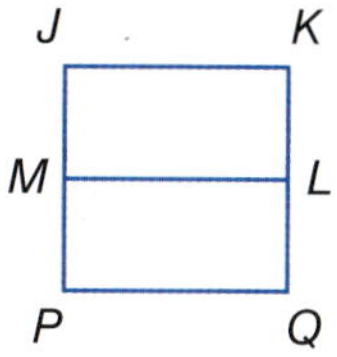

16. **Given:** $ACDH$ and $BCDF$ are parallelograms; $\overline{BF} \cong \overline{AB}$.
Prove: $ABHF$ is a rhombus.

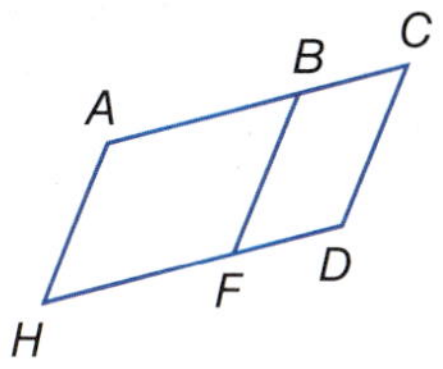

Example 3 p. 429

17. **ROADWAYS** Main Street and High Street intersect as shown in the diagram. Each of the crosswalks is the same length. Classify the quadrilateral formed by the crosswalks. Explain your reasoning.

18. **CONSTRUCTION** A landscaper has staked out the area for a square garden as shown. She has confirmed that each side of the quadrilateral formed by the stakes is congruent and that the diagonals are perpendicular. Is this information enough for the landscaper to be sure that the garden is a square? Explain your reasoning.

Example 4 p. 430

COORDINATE GEOMETRY Given each set of vertices, determine whether ▱*JKLM* is a *rhombus*, a *rectangle*, or a *square*. List all that apply. Explain.

19. $J(-4, -1), K(1, -1), L(4, 3), M(-1, 3)$
20. $J(-3, -2), K(2, -2), L(5, 2), M(0, 2)$
21. $J(-2, -1), K(-4, 3), L(1, 5), M(3, 1)$
22. $J(-1, 1), K(4, 1), L(4, 6), M(-1, 6)$

$ABCD$ is a rhombus. If $PB = 12$, $AB = 15$, and $m\angle ABD = 24$, find each measure.

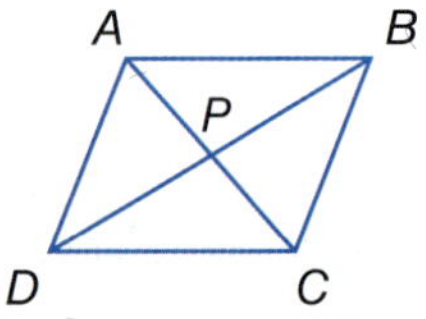

23. AP
24. CP
25. $m\angle BDA$
26. $m\angle ACB$

$WXYZ$ is a square. If $WT = 3$, find each measure.

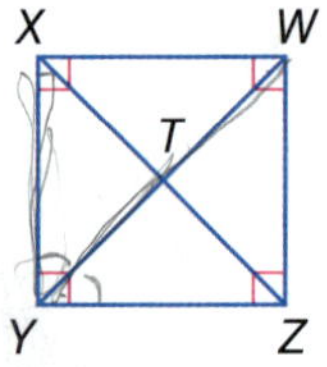

27. ZX
28. XY
29. $m\angle WTZ$
30. $m\angle WYX$

Classify each quadrilateral.

31. 

32. 

33. 

PROOF **Write a paragraph proof.**

34. Theorem 6.16
35. Theorem 6.17
36. Theorem 6.18
37. Theorem 6.19
38. Theorem 6.20

CONSTRUCTION **Use diagonals to construct each figure. Justify each construction.**

39. rhombus
40. square

PROOF **Write a coordinate proof of each statement.**

41. The diagonals of a square are perpendicular.
42. The segments joining the midpoints of the sides of a rectangle form a rhombus.

Real-World Link

Mosaics are images formed using patterns of closely set stones, glass, tile, or other material. The mosaic shown above is an early Greek pebble mosaic. By 200 B.C., the Greeks used tesserae, or regularly shaped material, instead of pebbles in their mosaics.

Source: *Encyclopaedia Britannica*

43. **DESIGN** The tile pattern below consists of regular octagons and quadrilaterals. Classify the quadrilaterals in the pattern and explain your reasoning.

44. **REPAIR** The window pane shown needs to be replaced. What are the dimensions of the replacement pane?

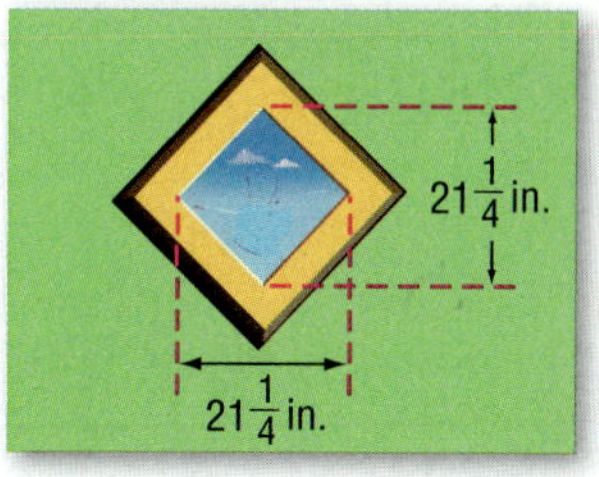

45. **MULTIPLE REPRESENTATIONS** In this problem, you will explore the properties of kites, which are quadrilaterals with exactly two distinct pairs of adjacent congruent sides.

a. **GEOMETRIC** Draw three kites with varying side lengths. Label one kite *ABCD*, one *PQRS*, and one *WXYZ*. Then draw the diagonals of each kite, labeling the point of intersection *N* for each kite.

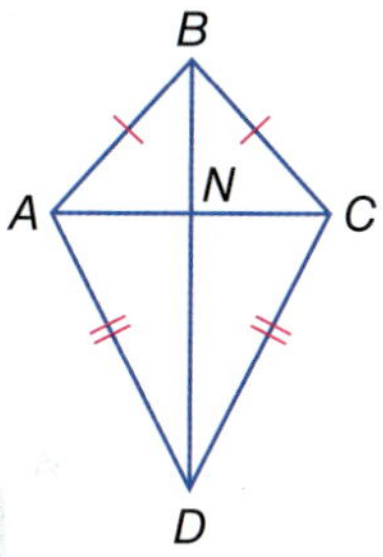

Kite *ABCD*

b. **TABULAR** Measure the distance from *N* to each vertex. Record your results in a table like the one shown.

Figure	Distance from *N* to Each Vertex Along Shorter Diagonal		Distance from *N* to Each Vertex Along Longer Diagonal	
ABCD				
PQRS				
WXYZ				

c. **VERBAL** Make a conjecture about the diagonals of a kite.

H.O.T. Problems

Use **H**igher-**O**rder **T**hinking Skills

46. **ERROR ANALYSIS** In quadrilateral *PQRS*, $\overline{PR} \cong \overline{QS}$. Lola thinks that the quadrilateral is a square, and Xavier thinks that it is a rhombus. Is either of them correct? Explain your reasoning.

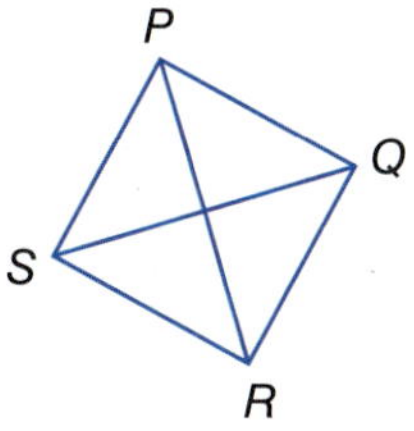

47. **REASONING** Determine whether the statement is *true* or *false*. Then write the converse, inverse, and contrapositive of the statement and determine the truth value of each. Explain your reasoning.

If a quadrilateral is a square, then it is a rectangle.

48. **CHALLENGE** The area of square *ABCD* is 36 square units and the area of $\triangle EBF$ is 20 square units. If $\overline{EB} \perp \overline{BF}$ and $\overline{AE} = 2$, find the length of $\overline{CF}$.

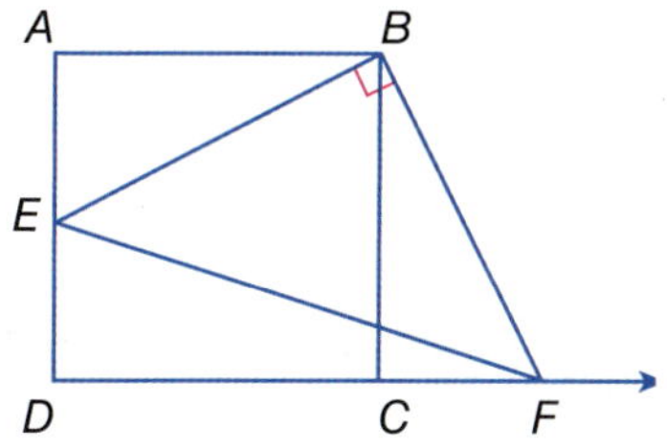

49. **OPEN ENDED** Find the vertices of a square with diagonals that are contained in the lines $y = x$ and $y = -x + 6$. Justify your reasoning.

50. **WRITING IN MATH** Compare all of the properties of the following quadrilaterals: parallelograms, rectangles, rhombi, and squares.

Standardized Test Practice

51. *JKLM* is a rhombus. If $CK = 8$ and $JK = 10$, find *JC*.

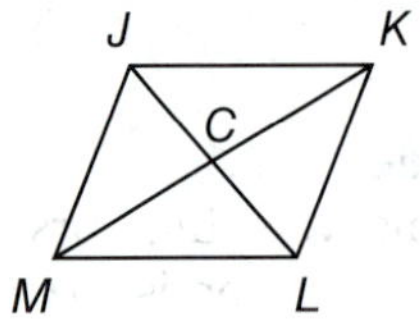

A 4
B 6
C 8
D 10

52. EXTENDED RESPONSE The sides of square *ABCD* are extended by sides of equal length to form square *WXYZ*.

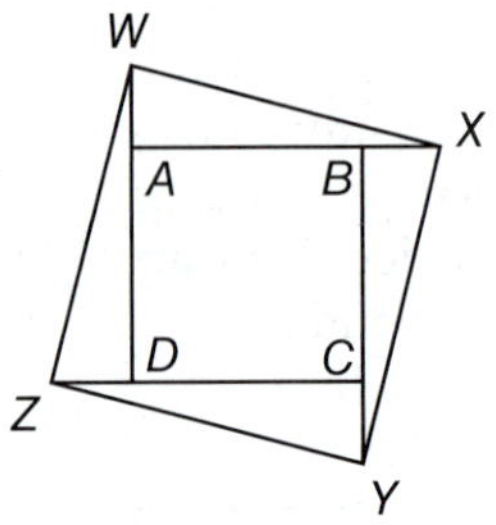

a. If $CY = 3$ cm and the area of *ABCD* is 81 cm^2, find the area of *WXYZ*.

b. If the areas of *ABCD* and *WXYZ* are 49 cm^2 and 169 cm^2 respectively, find *DZ*.

c. If $AB = 2CY$ and the area of $ABCD = m$ square meters, find the area of *WXYZ* in square meters.

53. ALGEBRA What values of x and y make quadrilateral *ABCD* a parallelogram?

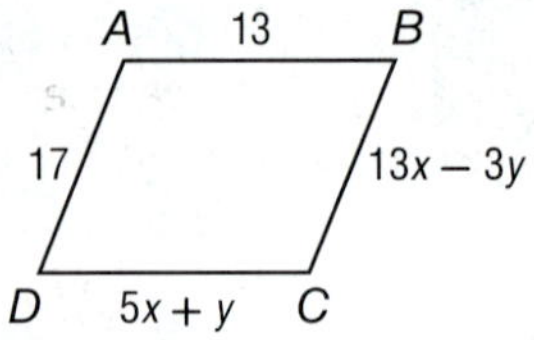

F $x = 3, y = 2$
G $x = \frac{3}{2}, y = -1$
H $x = 2, y = 3$
J $x = 3, y = -1$

54. SAT/ACT What is 6 more than the product of -3 and a certain number x?

A $-3x - 6$
B $-3x$
C $-x$
D $-3x + 6$

Spiral Review

Quadrilateral *ABDC* is a rectangle. Find each measure if $m\angle 1 = 38$. (Lesson 6-4)

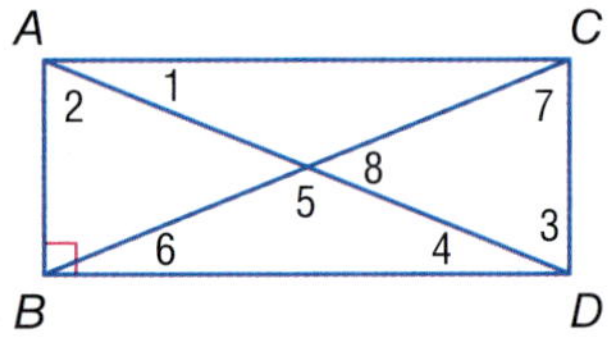

55. $m\angle 2$ **56.** $m\angle 5$ **57.** $m\angle 6$

Determine whether each quadrilateral is a parallelogram. Justify your answer. (Lesson 6-3)

58.

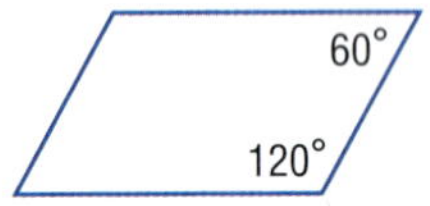

59.

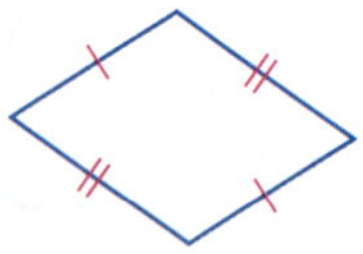

60.

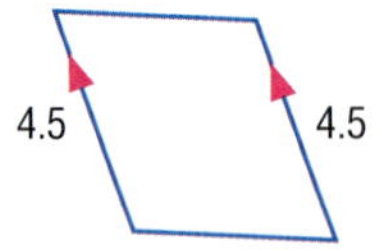

61. MEASUREMENT Monifa says that her backyard is shaped like a triangle and that the lengths of its sides are 22 feet, 23 feet, and 45 feet. Do you think these measurements are correct? Explain your reasoning. (Lesson 5-5)

62. COORDINATE GEOMETRY Identify the transformation and verify that it is a congruence transformation. (Lesson 4-7)

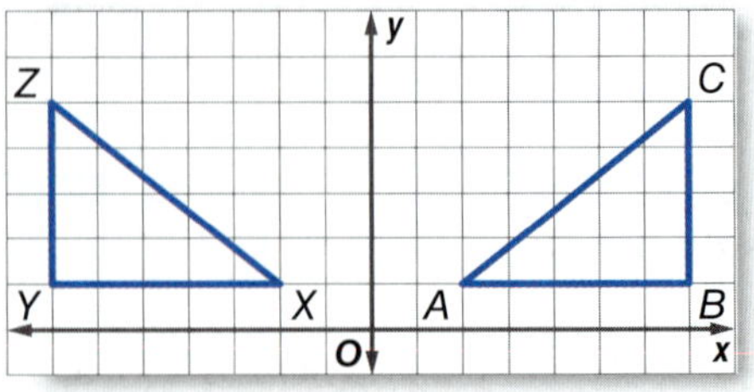

Skills Review

Solve each equation. (Lesson 0-5)

63. $\frac{1}{2}(5x + 7x - 1) = 11.5$ **64.** $\frac{1}{2}(10x + 6x + 2) = 7$ **65.** $\frac{1}{2}(12x + 6 - 8x + 7) = 9$

6-6 Trapezoids and Kites

Then
You used properties of special parallelograms. (Lesson 6-5)

Now
- Apply properties of trapezoids.
- Apply properties of kites.

New Vocabulary
trapezoid
bases
legs of a trapezoid
base angles
isosceles trapezoid
midsegment of a trapezoid
kite

Math Online
glencoe.com

Why?

In gymnastics, vaulting boxes made out of high compression foam are used as spotting platforms, vaulting horses, and steps. The left and right side of each section is a *trapezoid*.

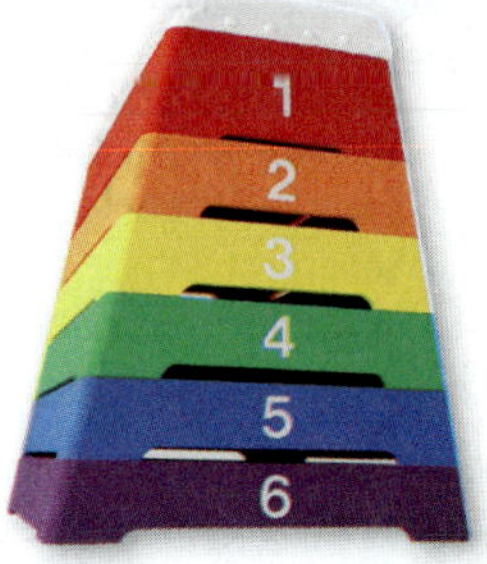

Properties of Trapezoids A **trapezoid** is a quadrilateral with exactly one pair of parallel sides. The parallel sides are called **bases**. The nonparallel sides are called **legs**. The **base angles** are formed by the base and one of the legs. In trapezoid *ABCD*, ∠*A* and ∠*B* are one pair of base angles and ∠*C* and ∠*D* are the other pair.

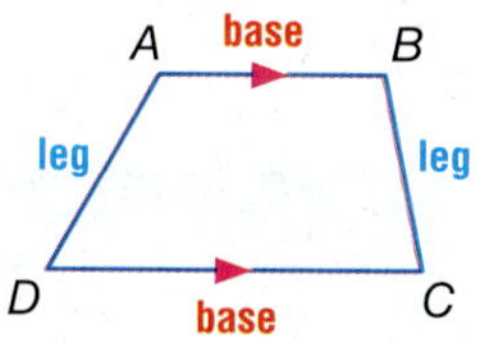

If the legs of a trapezoid are congruent, then it is an **isosceles trapezoid**.

Theorems — Isosceles Trapezoids

6.21 If a trapezoid is isosceles, then each pair of base angles is congruent.

Example If trapezoid *FGHJ* is isosceles, then ∠*G* ≅ ∠*H* and ∠*F* ≅ ∠*J*.

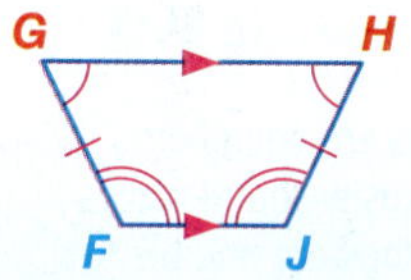

6.22 If a trapezoid has one pair of congruent base angles, then it is an isosceles trapezoid.

Example If ∠*L* ≅ ∠*M*, then trapezoid *KLMP* is isosceles.

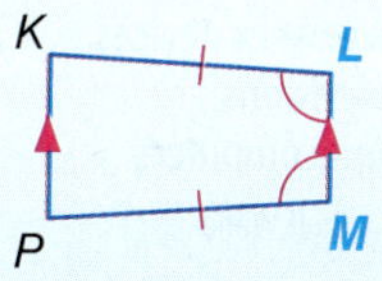

6.23 A trapezoid is isosceles if and only if its diagonals are congruent.

Example If trapezoid *QRST* is isosceles, then $\overline{QS} \cong \overline{RT}$. Likewise, if $\overline{QS} \cong \overline{RT}$, then trapezoid *QRST* is isosceles.

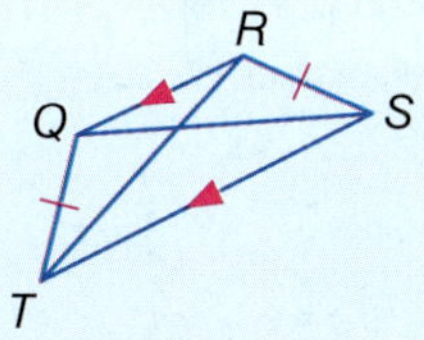

You will prove Theorem 6.21, Theorem 6.22, and the other part of Theorem 6.23 in Exercises 28, 29, and 30, respectively.

Proof — Part of Theorem 6.23

Given: *ABCD* is an isosceles trapezoid.

Prove: $\overline{AC} \cong \overline{BD}$

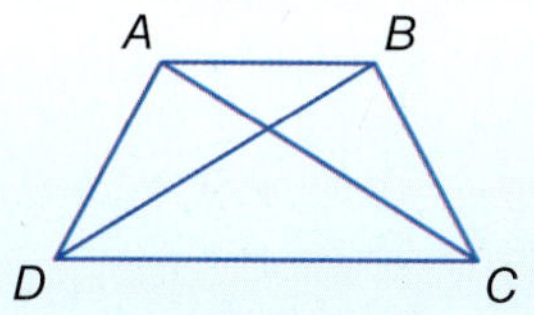

ABCD is an isosceles trapezoid. — Given

$\overline{DC} \cong \overline{CD}$ — Reflexive Property

$\overline{AD} \cong \overline{BC}$ — Def. Isos. Trapezoid

∠*ADC* ≅ ∠*BCD* — Base ∠s of trapezoid are ≅.

△*ADC* ≅ △*BCD* — SAS

$\overline{AC} \cong \overline{BD}$ — CPCTC

StudyTip

Isosceles Trapezoids The base angles of a trapezoid are only congruent if the trapezoid is isosceles.

Real-World Link

Speakers are amplifiers that intensify sound waves so that they are audible to the unaided ear. Amplifiers exist in devices such as televisions, stereos, and computers.

Source: How Stuff Works

Real-World EXAMPLE 1 Use Properties of Isosceles Trapezoids

MUSIC The speaker shown is an isosceles trapezoid. If $m\angle FJH = 85$, $FK = 8$ inches, and $JG = 19$ inches, find each measure.

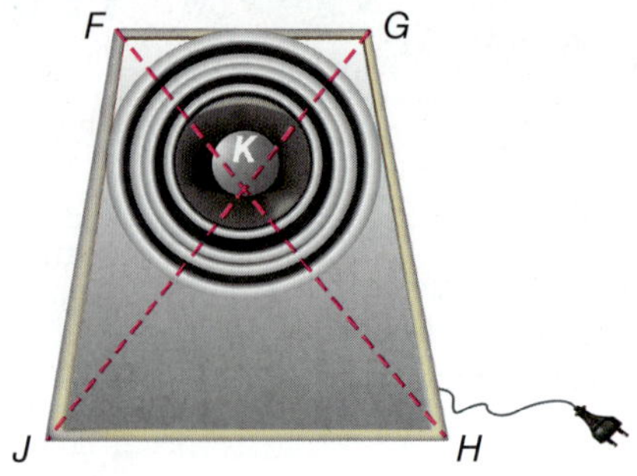

a. $m\angle FGH$

Since $FGHJ$ is an isosceles trapezoid, $\angle FJH$ and $\angle GHJ$ are congruent base angles. So, $m\angle GHJ = m\angle FJH = 85$.

Since $FGHJ$ is a trapezoid, $\overline{FG} \parallel \overline{JH}$.

$m\angle FGH + m\angle GHJ = 180$	Consecutive Interior Angles Theorem
$m\angle FGH + 85 = 180$	Substitution
$m\angle FGH = 95$	Subtract 85 from each side.

b. KH

Since $FGHJ$ is an isosceles trapezoid, diagonals $\overline{FH}$ and $\overline{JG}$ are congruent.

$FH = JG$	Definition of congruent
$FK + KH = JG$	Segment Addition
$8 + KH = 19$	Substitution
$KH = 11$ cm	Subtract 8 from each side.

Guided Practice

1. CAFETERIA TRAYS To save space at a square table, cafeteria trays often incorporate trapezoids into their design. If $WXYZ$ is an isosceles trapezoid and $m\angle YZW = 45$, $WV = 15$ centimeters, and $VY = 10$ centimeters, find each measure.

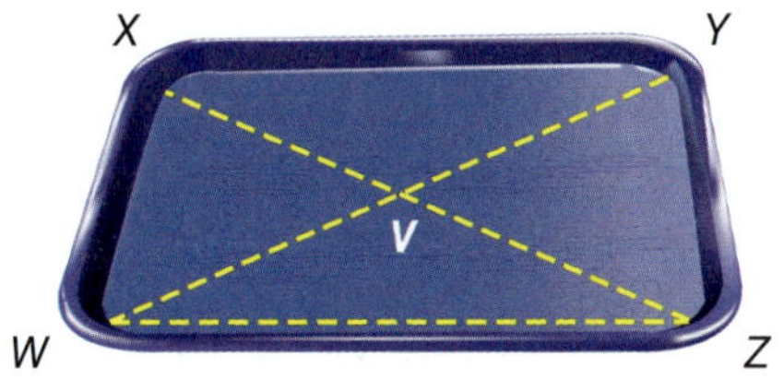

A. $m\angle XWZ$ **B.** $m\angle WXY$

C. XZ **D.** XV

Personal Tutor glencoe.com

You can use coordinate geometry to determine whether a trapezoid is an isosceles trapezoid.

EXAMPLE 2 Isosceles Trapezoids and Coordinate Geometry

COORDINATE GEOMETRY Quadrilateral $ABCD$ has vertices $A(-3, 4)$, $B(2, 5)$, $C(3, 3)$, and $D(-1, 0)$. Show that $ABCD$ is a trapezoid and determine whether it is an isosceles trapezoid.

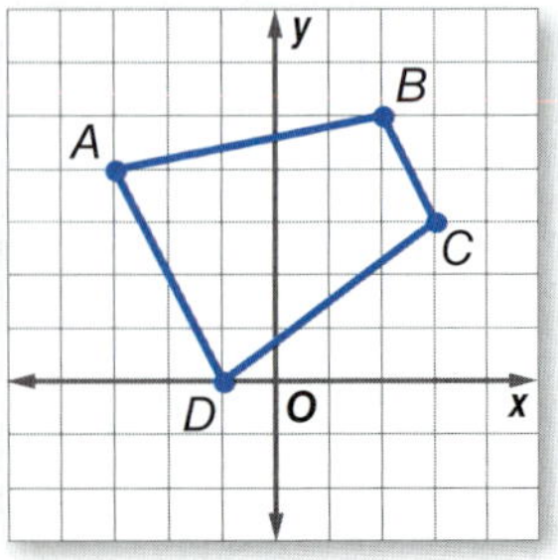

Graph and connect the vertices of $ABCD$.

Step 1 Use the Slope Formula to compare the slopes of opposite sides $\overline{BC}$ and $\overline{AD}$ and of opposite sides $\overline{AB}$ and $\overline{DC}$. A quadrilateral is a trapezoid if exactly one pair of opposite sides are parallel.

Opposite sides $\overline{BC}$ and $\overline{AD}$:

slope of $\overline{BC} = 3 - \frac{5}{3} - 2 = -\frac{2}{1}$ or -2

slope of $\overline{AD} = \frac{0 - 4}{-1 - (-3)} = \frac{-4}{2}$ or -2

Since the slopes of $\overline{BC}$ and $\overline{AD}$ are equal, $\overline{BC} \parallel \overline{AD}$.

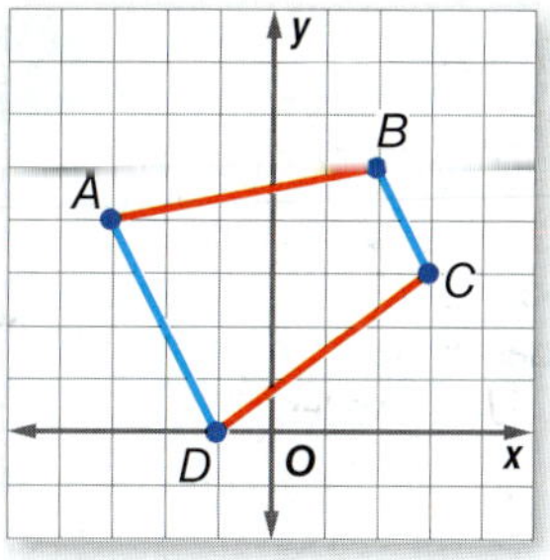

Opposite sides $\overline{AB}$ and $\overline{DC}$:

slope of $\overline{AB} = \frac{5 - 4}{2 - (-3)} = \frac{1}{5}$ slope of $\overline{DC} = \frac{0 - 3}{-1 - 3} = \frac{-3}{-4}$ or $\frac{3}{4}$

Since the slope of $\overline{AB}$ and $\overline{DC}$ are *not* equal, $\overline{BC} \nparallel \overline{AD}$. Since $ABCD$ has only one pair of parallel sides, $ABCD$ is a trapezoid.

Step 2 Use the Distance Formula to compare the lengths of legs $\overline{AB}$ and $\overline{DC}$ to determine if trapezoid $ABCD$ is isosceles.

$AB = \sqrt{(-3 - 2)^2 + (4 - 5)^2}$ or $\sqrt{26}$

$DC = \sqrt{(-1 - 3)^2 + (0 - 3)^2} = \sqrt{25}$ or 5

Since $AB \neq DC$, trapezoid $ABCD$ is not isosceles.

Guided Practice

2. Quadrilateral $QRST$ has vertices $Q(-8, -4)$, $R(0, 8)$, $S(6, 8)$, and $T(-6, -10)$. Show that $QRST$ is a trapezoid and determine whether $QRST$ is an isosceles trapezoid.

Personal Tutor glencoe.com

ReadingMath

Symbols Recall that the symbol $\nparallel$ means *is not parallel to*.

The **midsegment of a trapezoid** is the segment that connects the midpoints of the legs of the trapezoid. The theorem below relates the midsegment and the bases of a trapezoid.

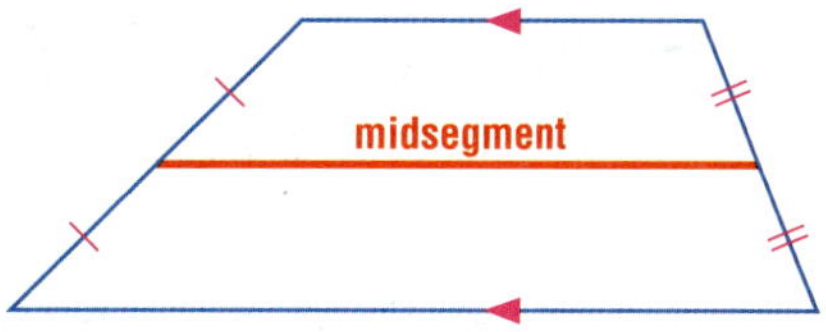

ReadingMath

Midsegment A midsegment of a trapezoid can also be called a *median*.

Theorem 6.24 Trapezoid Midsegment Theorem

The midsegment of a trapezoid is parallel to each base and its measure is one half the sum of the lengths of the bases.

Example If $\overline{BE}$ is the midsegment of trapezoid $ACDF$, then

$\overline{AF} \parallel \overline{BE}$, $\overline{CD} \parallel \overline{BE}$, and

$BE = \frac{1}{2}(AF + CD)$.

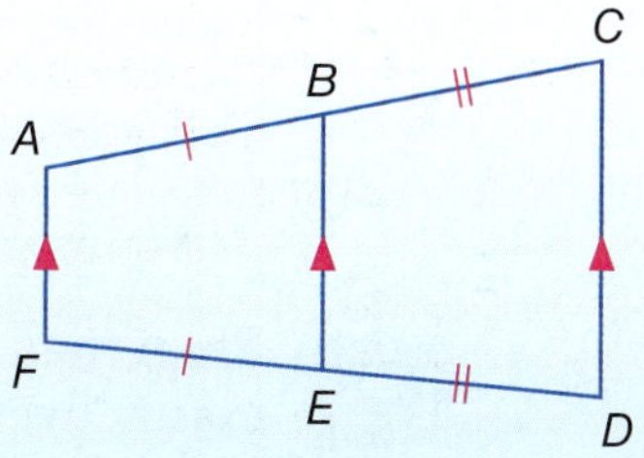

You will prove Theorem 6.24 in Exercise 33.

STANDARDIZED TEST EXAMPLE 3

GRIDDED RESPONSE In the figure, $\overline{LH}$ is the midsegment of trapezoid *FGJK*. What is the value of *x*?

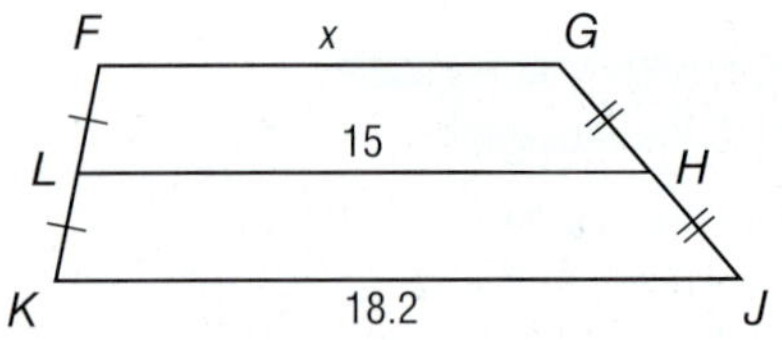

Note: The figure is not drawn to scale.

Read the Test Item

You are given the measure of the midsegment of a trapezoid and the measure of one of its bases. You are asked to find the measure of the other base.

Solve the Test Item

$LH = \frac{1}{2}(FG + KJ)$	**Trapezoid Midsegment Theorem**
$15 = \frac{1}{2}(x + 18.2)$	**Substitution**
$30 = x + 18.2$	**Multiply each side by 2.**
$11.8 = x$	**Subtract 18.2 from each side.**

Grid In Your Answer

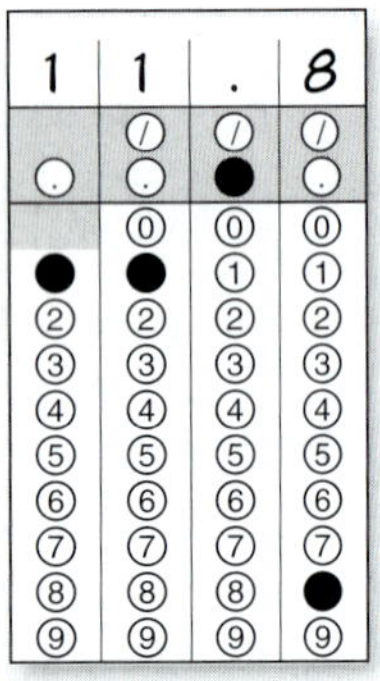

- You can align the numerical answer by placing the first digit in the left answer box or by putting the last digit in the right answer box.
- Do not leave blank boxes in the middle of an answer.
- Fill in **one** bubble for each filled answer box. Do not fill more than one bubble for an answer box. Do not fill in a bubble for blank answer boxes.

> **Test-Taking Tip**
>
> **Gridded Responses** Rational answers can often be gridded in more than one way. An answer such as $\frac{8}{5}$ could be gridded as 8/5 or 1.6, but not as 1 3/5.

Guided Practice

3. GRIDDED RESPONSE Trapezoid *ABCD* is shown below. If $\overline{FG}$ is parallel to $\overline{AD}$, what is the *x*-coordinate of point *G*?

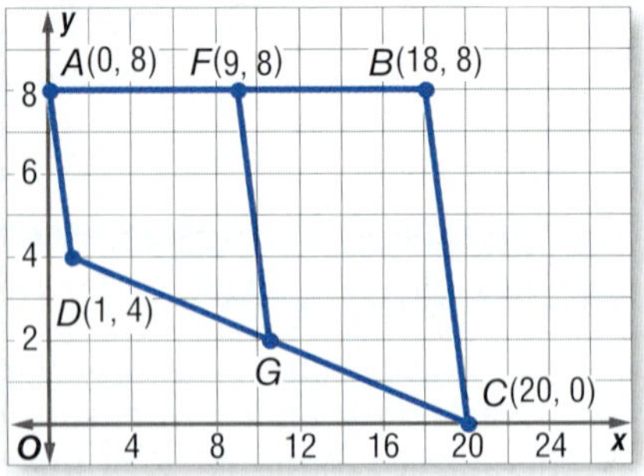

Personal Tutor glencoe.com

Properties of Kites A **kite** is a quadrilateral with exactly two pairs of consecutive congruent sides. Unlike a parallelogram, the opposite sides of a kite are not congruent or parallel.

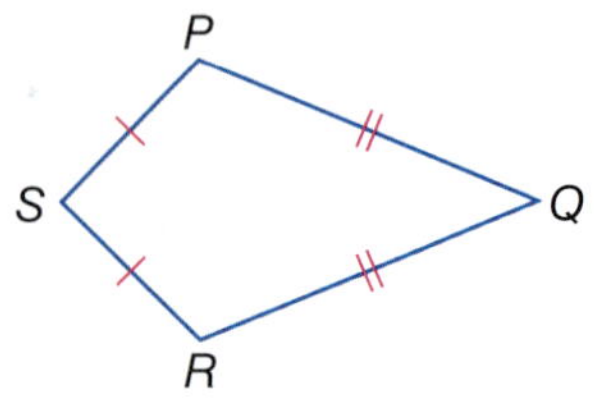

StudyTip

Kites The congruent angles of a kite are included by the non-congruent adjacent sides.

Theorems — Kites

6.25 If a quadrilateral is a kite, then its diagonals are perpendicular.

Example If quadrilateral $ABCD$ is a kite, then $\overline{AC} \perp \overline{BD}$.

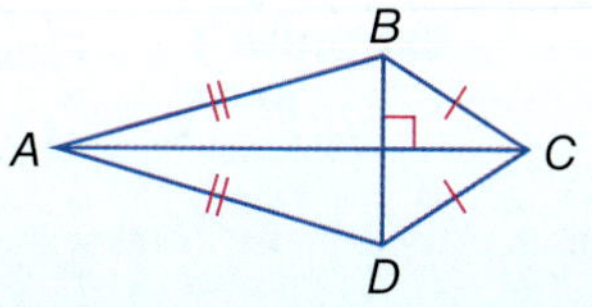

6.26 If a quadrilateral is a kite, then exactly one pair of opposite angles is congruent.

Example If quadrilateral $JKLM$ is a kite and $JK \cong KL$, then $\angle J \cong \angle L$ and $\angle K \ncong \angle M$.

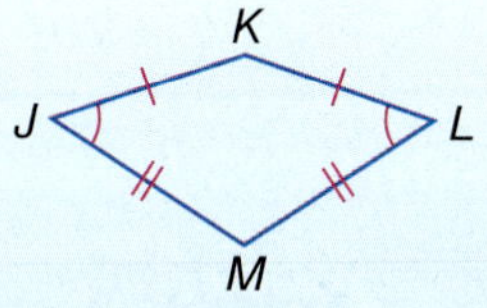

You will prove Theorems 6.25 and 6.26 in Exercises 31 and 32, respectively.

You can use the theorems above, the Pythagorean Theorem, and the Polygon Interior Angles Sum Theorem to find missing measures in kites.

Real-World Link

The fastest recorded speed of a kite is over 120 miles per hour. The record for the highest single kite flown is 12,471 feet.

Source: Borealis Kites

EXAMPLE 4 Use Properties of Kites

a. If $FGHJ$ is a kite, find $m\angle GFJ$.

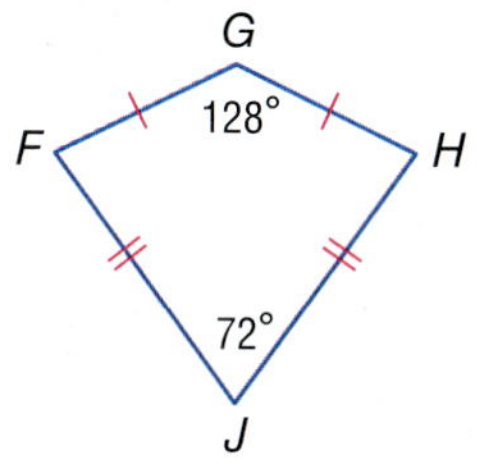

Since a kite can only have one pair of opposite congruent angles and $\angle G \ncong \angle J$, then $\angle F \cong \angle H$. So, $m\angle F = m\angle H$. Write and solve an equation to find $m\angle F$.

$m\angle F + m\angle G + m\angle H + m\angle J = 360$	**Polygon Interior Angles Sum Theorem**
$m\angle F + 128 + m\angle F + 72 = 360$	**Substitution**
$2m\angle F + 200 = 360$	**Simplify.**
$2m\angle F = 160$	**Subtract 200 from each side.**
$m\angle F = 80$	**Divide each side by 2.**

b. If $WXYZ$ is a kite, find ZY.

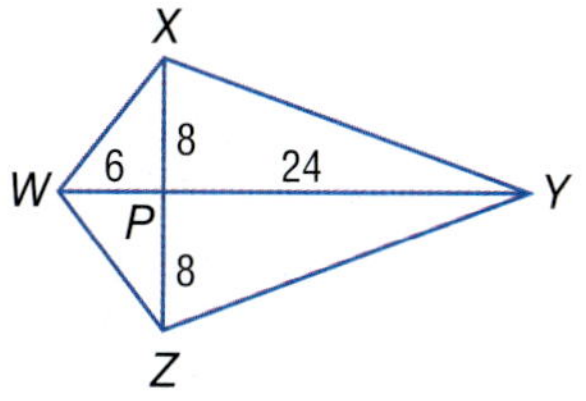

Since the diagonals of a kite are perpendicular, they divide $WXYZ$ into four right triangles. Use the Pythagorean Theorem to find ZY, the length of the hypotenuse of right $\triangle YPZ$.

$PZ^2 + PY^2 = ZY^2$	**Pythagorean Theorem**
$8^2 + 24^2 = ZY^2$	**Substitution**
$640 = ZY^2$	**Simplify.**
$\sqrt{640} = ZY$	**Take the square root of each side.**
$8\sqrt{10} = ZY$	**Simplify.**

Guided Practice

4A. If $m\angle BAD = 38$ and $m\angle BCD = 50$, find $m\angle ADC$.

4B. If $BT = 5$ and $TC = 8$, find CD.

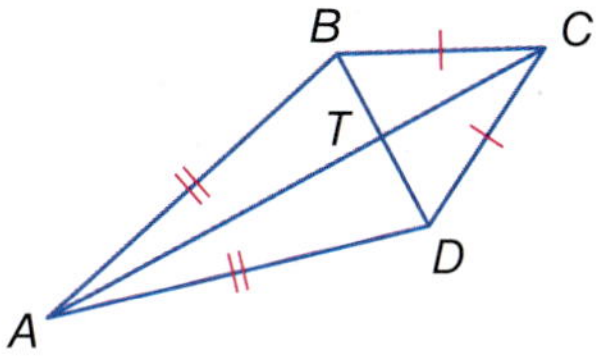

Personal Tutor glencoe.com

Check Your Understanding

Example 1 p. 436

Find each measure.

1. $m\angle D$

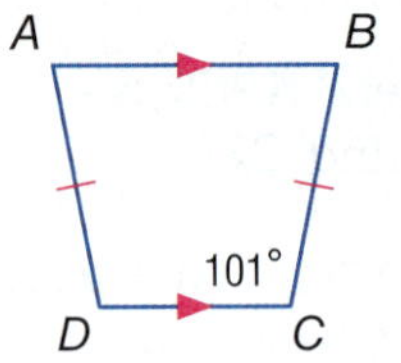

2. WT, if $ZX = 20$ and $TY = 15$

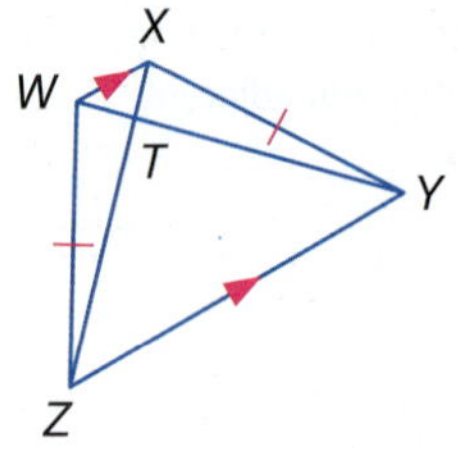

Example 2 pp. 436–437

COORDINATE GEOMETRY **Quadrilateral $ABCD$ has vertices $A(-4, -1)$, $B(-2, 3)$, $C(3, 3)$, and $D(5, -1)$.**

3. Verify that $ABCD$ is a trapezoid.

4. Determine whether $ABCD$ is an isosceles trapezoid. Explain.

Example 3 p. 438

5. **SHORT REPSONSE** In the figure at the right, $\overline{YZ}$ is the midsegment of trapezoid $TWRV$. Determine the value of x.

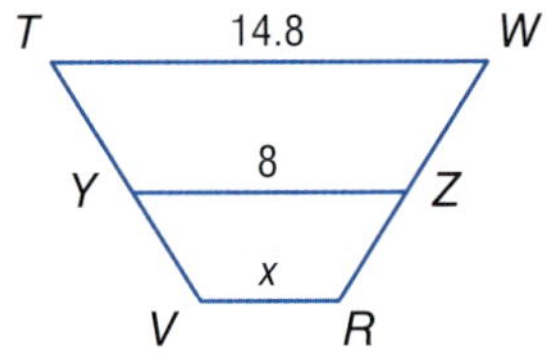

Example 4 p. 439

If $ABCD$ is a kite, find each measure.

6. AB

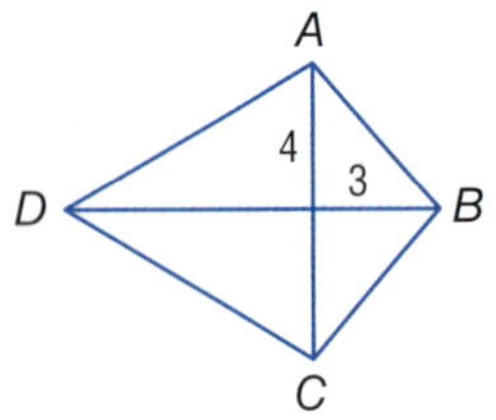

7. $m\angle C$

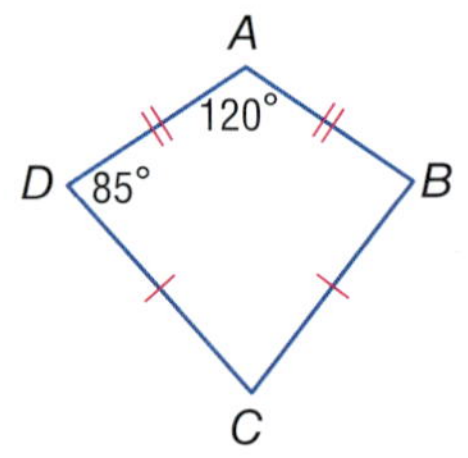

Practice and Problem Solving

● = **Step-by-Step Solutions** begin on page R20.
Extra Practice begins on page 969.

Example 1 p. 436

Find each measure.

8. $m\angle K$

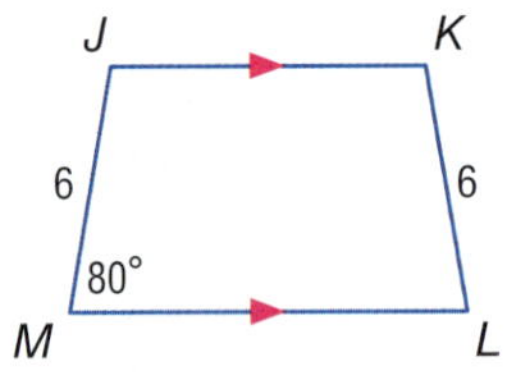

9. $m\angle Q$

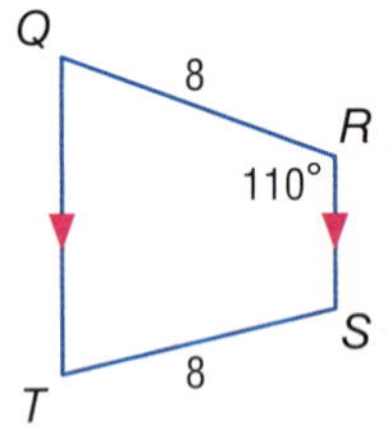

10. JL, if $KP = 4$ and $PM = 7$

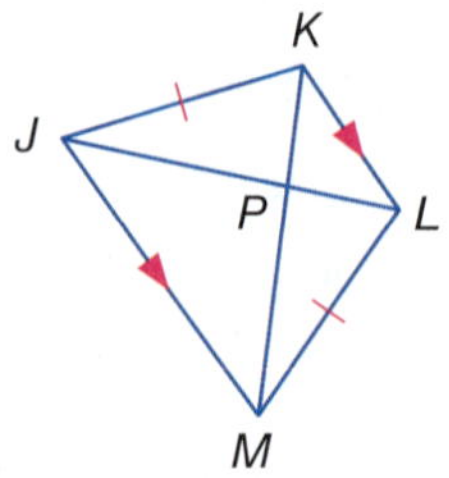

11. PW, if $XZ = 18$ and $PY = 3$

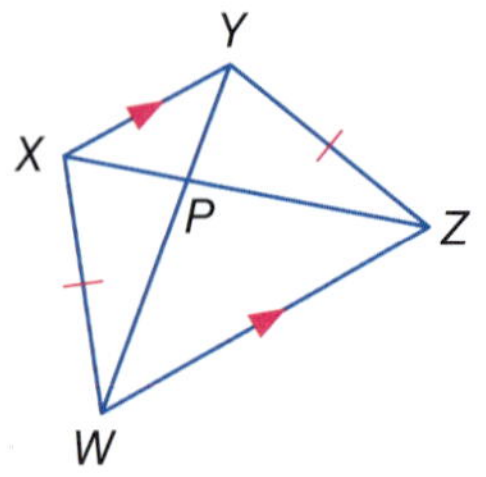

Example 2 pp. 436–437

COORDINATE GEOMETRY **For each quadrilateral with the given vertices, verify that the quadrilateral is a trapezoid and determine whether the figure is an isosceles trapezoid.**

12. $A(-2, 5)$, $B(-3, 1)$, $C(6, 1)$, $D(3, 5)$

13. $J(-4, -6)$, $K(6, 2)$, $L(1, 3)$, $M(-4, -1)$

14. $Q(2, 5)$, $R(-2, 1)$, $S(-1, -6)$, $T(9, 4)$

15. $W(-5, -1)$, $X(-2, 2)$, $Y(3, 1)$, $Z(5, -3)$

Example 3
p. 438

For trapezoid $QRTU$, V and S are midpoints of the legs.

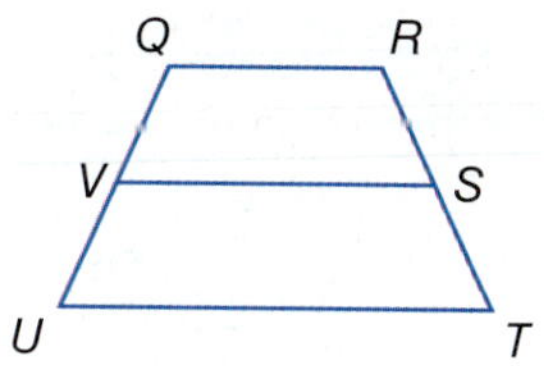

16. If $QR = 12$ and $UT = 22$, find VS.

17. If $QR = 4$ and $UT = 16$, find VS.

18. If $VS = 9$ and $UT = 12$, find QR.

19. If $TU = 26$ and $SV = 17$, find QR.

20. If $QR = 2$ and $VS = 7$, find UT.

21. If $RQ = 5$ and $VS = 11$, find UT.

Real-World Link

A window-mounted planter or window box can be any four-sided structure as long as it has two side surfaces, front and back surfaces, and one bottom surface.

Source: U.S. Patent and Trademark Office

22. **DESIGN** Juana is designing a window box. She wants the end of the box to be a trapezoid with the dimensions shown. If she wants to put a shelf in the middle for the plants to rest on, how wide should she make the shelf?

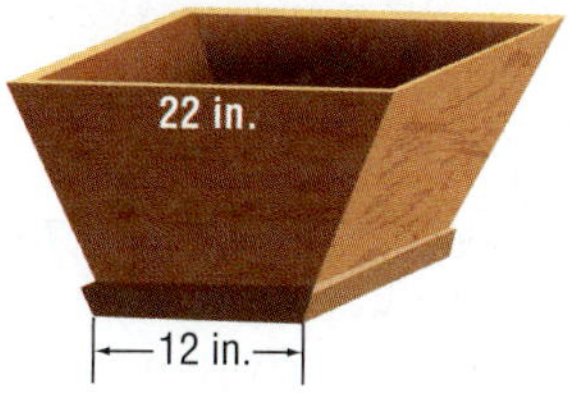

23 **MUSIC** The keys of the xylophone shown form a trapezoid. If the length of the lower pitched C is 6 inches long, and the higher pitched D is 1.8 inches long, how long is the G key?

Example 4
p. 439

If $WXYZ$ is a kite, find each measure.

24. YZ

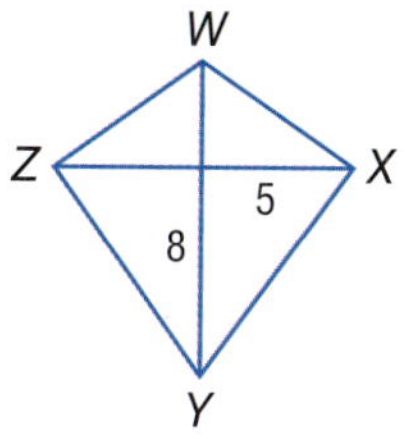

25. WP

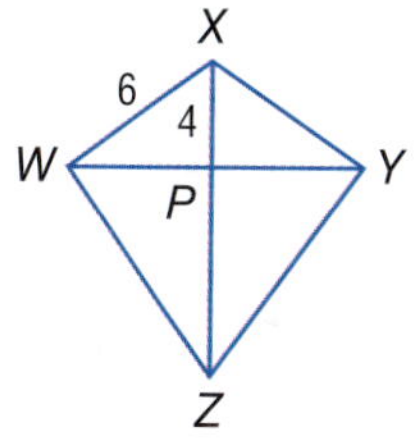

26. $m\angle X$

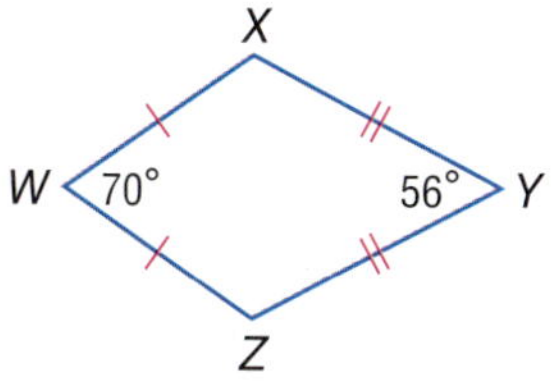

27. $m\angle Z$

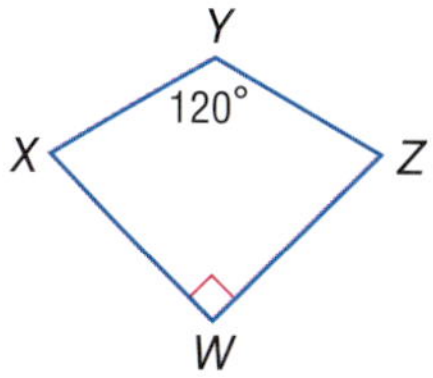

PROOF **Write a paragraph proof for each theorem.**

28. Theorem 6.21

29. Theorem 6.22

30. Theorem 6.23

31. Theorem 6.25

32. Theorem 6.26

33. **PROOF** Write a coordinate proof for Theorem 6.24.

34. **COORDINATE GEOMETRY** Refer to quadrilateral $ABCD$.

a. Determine whether the figure is a trapezoid. If so, is it isosceles? Explain.

b. Is the midsegment contained in the line with equation $y = -x + 1$? Justify your answer.

c. Find the length of the midsegment.

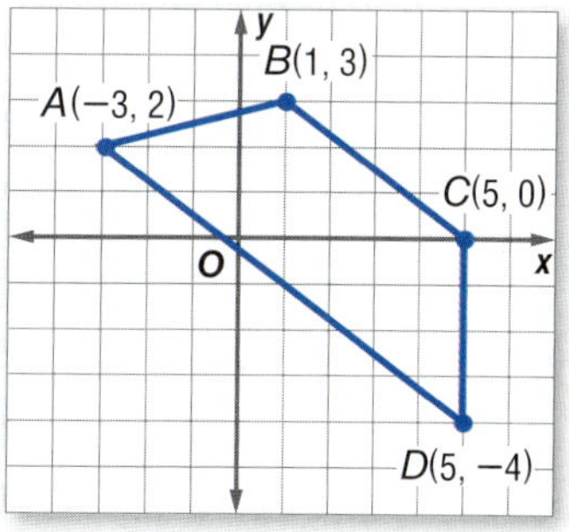

Real-World Link

Amateur and professional batters use batting cages to practice their hitting and improve their speed and accuracy. A typical batting cage is about 10 feet high, 12 to 14 feet wide, and 30 to 74 feet long.

Source: PickyGuide

ALGEBRA ***ABCD* is a trapezoid.**

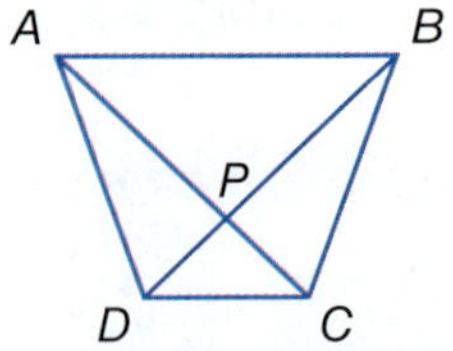

35. If $AC = 3x - 7$ and $BD = 2x + 8$, find the value of x so that $ABCD$ is isosceles.

36. If $m\angle ABC = 4x + 11$ and $m\angle DAB = 2x + 33$, find the value of x so that $ABCD$ is isosceles.

SPORTS **The end of the batting cage shown is an isosceles trapezoid. If $PT = 12$ feet, $ST = 28$ feet, and $m\angle PQR = 110$, find each measure.**

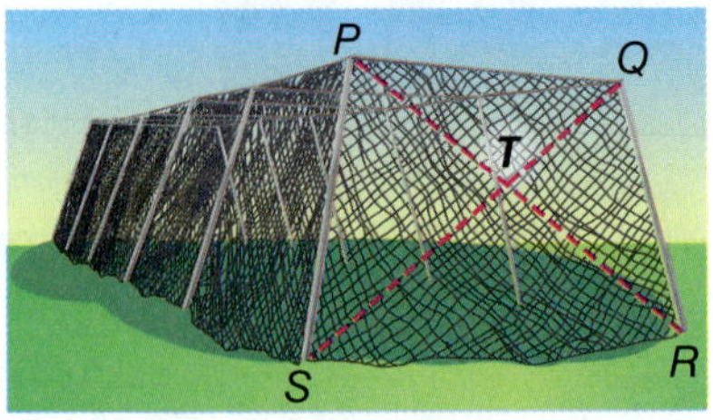

37. TR

38. SQ

39. $m\angle QRS$

40. $m\angle QPS$

ALGEBRA **For trapezoid *QRST*, *M* and *P* are midpoints of the legs.**

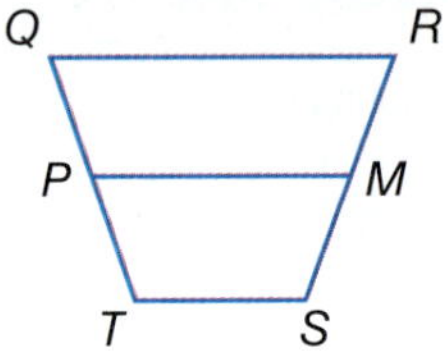

41. If $QR = 16$, $PM = 12$, and $TS = 4x$, find x.

42. If $TS = 2x$, $PM = 20$, and $QR = 6x$, find x.

43. If $PM = 2x$, $QR = 3x$, and $TS = 10$, find PM.

44. If $TS = 2x + 2$, $QR = 5x + 3$, and $PM = 13$, find TS.

SHOPPING **The side of the shopping bag shown is an isosceles trapezoid. If $EC = 9$ inches, $DB = 19$ inches, $m\angle ABE = 40$, and $m\angle EBC = 35$, find each measure.**

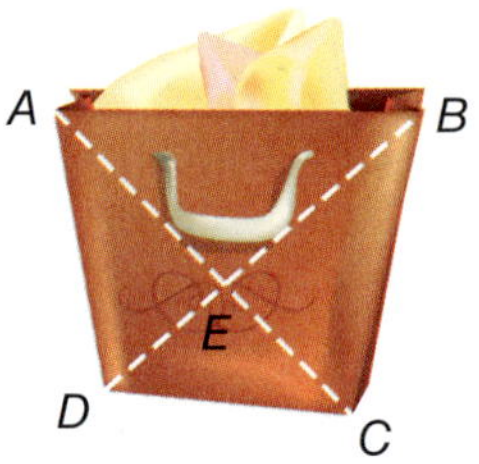

45. AE

46. AC

47. $m\angle BCD$

48. $m\angle EDC$

ALGEBRA ***WXYZ* is a kite.**

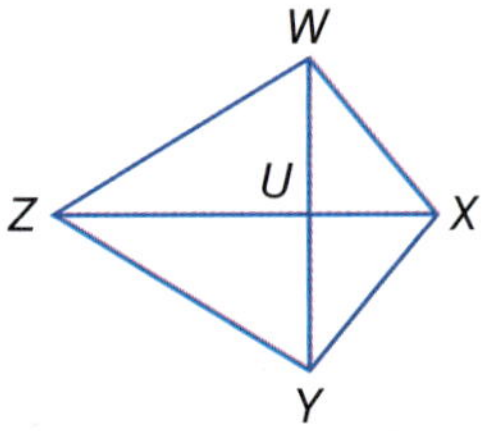

49 If $m\angle WXY = 120$, $m\angle WZY = 4x$, and $m\angle ZWX = 10x$, find $m\angle ZYX$.

50. If $m\angle WXY = 13x + 24$, $m\angle WZY = 35$, and $m\angle ZWX = 13x + 14$, find $m\angle ZYX$.

PROOF **Write a two-column proof.**

51. Given: $ABCD$ is an isosceles trapezoid.

Prove: $\angle DAC \cong \angle CBD$

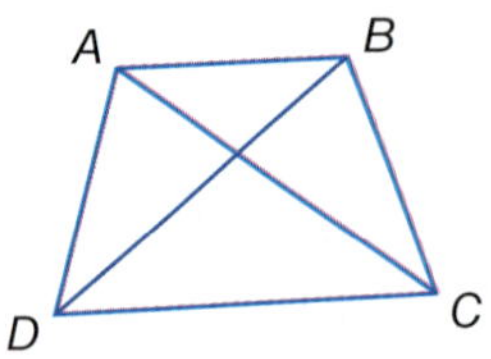

52. Given: $\overline{WZ} \cong \overline{ZV}$, $\overline{XY}$ bisects $\overline{WZ}$ and $\overline{ZV}$, and $\angle W \cong \angle ZXY$.

Prove: $WXYV$ is an isosceles trapezoid.

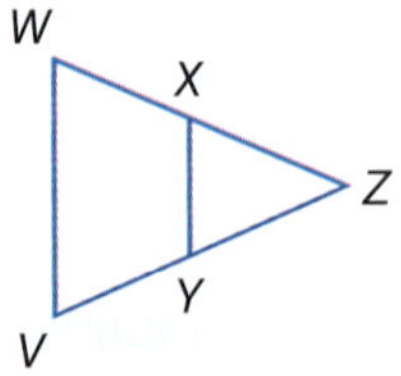

Determine whether each statement is *always*, *sometimes*, or *never* true. Explain.

53. The opposite angles of a trapezoid are supplementary.

54. One pair of opposite sides are parallel in a kite.

55. A square is a rhombus.

56. A rectangle is a square.

57. A parallelogram is a rectangle.

StudyTip

Properties of Quadrilaterals Refer to the Concept Summary box on page 427 to review the relationships among rectangles, squares, rhombi, and parallelograms.

58. KITES Refer to the kite at the right. Using the properties of kites, write a two-column proof to show that $\triangle MNR$ is congruent to $\triangle PNR$.

59. VENN DIAGRAM Create a Venn diagram that incorporates all quadrilaterals, including trapezoids, isosceles trapezoids, kites, and quadrilaterals that cannot be classified as anything other than quadrilaterals.

COORDINATE GEOMETRY **Determine whether each figure is a *trapezoid*, a *parallelogram*, a *square*, a *rhombus*, or a *quadrilateral* given the coordinates of the vertices. Choose the most specific term. Explain.**

60. $A(-1, 4), B(2, 6), C(3, 3), D(0, 1)$

61. $W(-3, 4), X(3, 4), Y(5, 3), Z(-5, 1)$

62. MULTIPLE REPRESENTATIONS In this problem, you will explore proportions in kites.

a. GEOMETRIC Draw a segment. Construct a noncongruent segment that perpendicularly bisects the first segment. Connect the endpoints of the segments to form a quadrilateral *ABCD*. Repeat the process two times. Name the additional quadrilaterals *PQRS* and *WXYZ*.

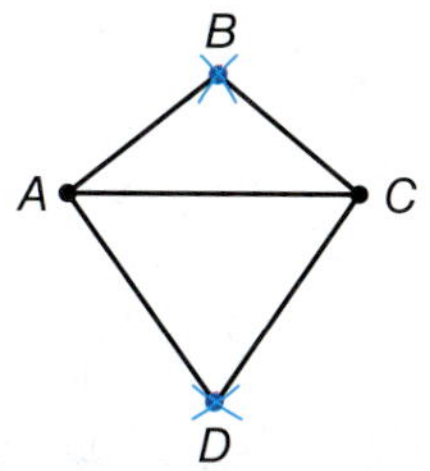

b. TABULAR Copy and complete the table below.

Figure	Side	Length	Side	Length	Side	Length	Side	Length
ABCD	*AB*		*BC*		*CD*		*DA*	
PQRS	*PQ*		*QR*		*RS*		*SP*	
WXYZ	*WX*		*XY*		*YZ*		*ZW*	

c. VERBAL Make a conjecture about a quadrilateral in which the diagonals are perpendicular, exactly one diagonal is bisected, and the diagonals are not congruent.

PROOF **Write a coordinate proof of each statement.**

63. The diagonals of an isosceles trapezoid are congruent.

64. The median of an isosceles trapezoid is parallel to the bases.

H.O.T. Problems
Use Higher-Order Thinking Skills

65. ERROR ANALYSIS Bedagi and Belinda are trying to determine $m\angle A$ in kite *ABCD* shown. Is either of them correct? Explain.

Bedagi	Belinda
$m\angle A = 45$	$m\angle A = 115$

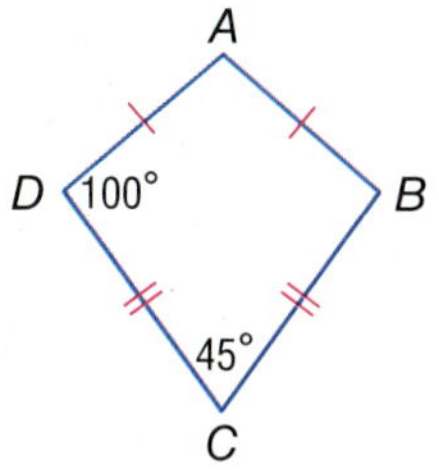

66. CHALLENGE If the parallel sides of a trapezoid are contained by the lines $y = x + 4$ and $y = x - 8$, what equation represents the line contained by the midsegment?

67. REASONING Is it *sometimes*, *always*, or *never* true that a square is also a kite? Explain.

68. OPEN ENDED Sketch two noncongruent trapezoids *ABCD* and *FGHJ* in which $\overline{AC} \cong \overline{FH}$ and $\overline{BD} \cong \overline{GJ}$.

69. WRITING IN MATH Describe the properties a quadrilateral must possess in order for the quadrilateral to be classified as a trapezoid, an isosceles trapezoid, or a kite. Compare the properties of all three quadrilaterals.

Standardized Test Practice

70. ALGEBRA All of the items on a breakfast menu cost the same whether ordered with something else or alone. Two pancakes and one order of bacon costs \$4.92. If two orders of bacon cost \$3.96, what does one pancake cost?

A \$0.96 **C** \$1.98

B \$1.47 **D** \$2.94

71. GRIDDED RESPONSE If quadrilateral *ABCD* is a kite, what is $m\angle C$?

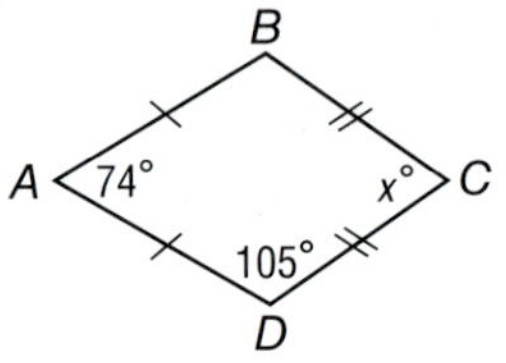

72. Which figure can serve as a counterexample to the conjecture below?

If the diagonals of a quadrilateral are congruent, then the quadrilateral is a rectangle.

F square

G rhombus

H parallelogram

J isosceles trapezoid

73. SAT/ACT In the figure below, what is the value of *x*?

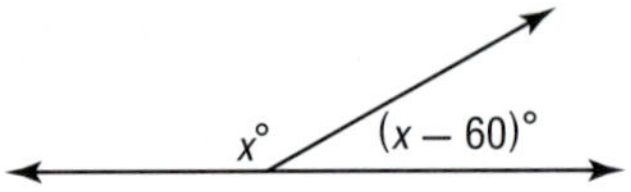

A 60 **C** 180

B 120 **D** 240

Spiral Review

ALGEBRA Quadrilateral *DFGH* is a rhombus. Find each value or measure. (Lesson 6-5)

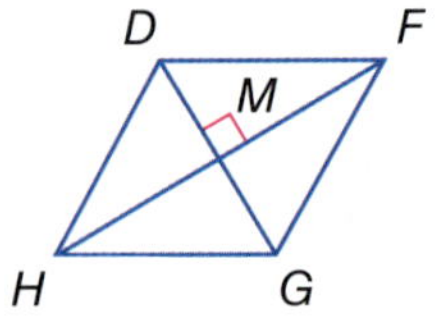

74. If $m\angle FGH = 118$, find $m\angle MHG$.

75. If $DM = 4x - 3$ and $MG = x + 6$, find DG.

76. If $DF = 10$, find FG.

77. If $HM = 12$ and $HD = 15$, find MG.

COORDINATE GEOMETRY Graph each quadrilateral with the given vertices. Determine whether the figure is a rectangle. Justify your answer using the indicated formula. (Lesson 6-4)

78. $A(4, 2)$, $B(-4, 1)$, $C(-3, -5)$, $D(5, -4)$; Distance Formula

79. $J(0, 7)$, $K(-8, 6)$, $L(-7, 0)$, $M(1, 1)$; Slope Formula

80. BASEBALL A batter hits the ball to the third baseman and begins to run toward first base. At the same time, the runner on first base runs toward second base. If the third baseman wants to throw the ball to the nearest base, to which base should he throw? Explain. (Lesson 5-3)

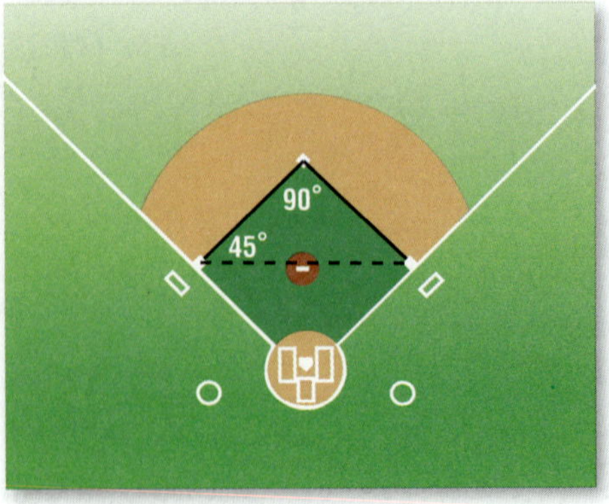

81. PROOF Write a two-column proof. (Lesson 4-5)

Given: $\angle CMF \cong \angle EMF$,
$\angle CFM \cong \angle EFM$

Prove: $\triangle DMC \cong \triangle DME$

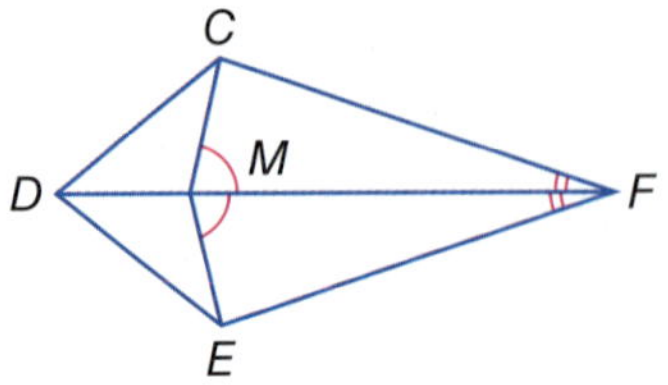

Skills Review

Write an expression for the slope of each segment given the coordinates and endpoints. (Lesson 3-3)

82. $(x, 4y)$, $(-x, 4y)$

83. $(-x, 5x)$, $(0, 6x)$

84. (y, x), (y, y)

Study Guide and Review

Math Online glencoe.com
- STUDY TO GO
- Vocabulary Review

Chapter Summary

Key Concepts

Angles of Polygons (Lesson 6-1)

- The sum of the measures of the interior angles of a polygon is given by the formula $S = 180(n - 2)$.
- The sum of the measures of the exterior angles of a convex polygon is 360.

Properties of Parallelograms (Lessons 6-2 and 6-3)

- Opposite sides are congruent and parallel.
- Opposite angles are congruent.
- Consecutive angles are supplementary.
- If a parallelogram has one right angle, it has four right angles.
- Diagonals bisect each other.

Properties of Rectangles, Rhombi, Squares, and Trapezoids (Lesson 6-4 through 6-6)

- A rectangle has all the properties of a parallelogram. Diagonals are congruent and bisect each other. All four angles are right angles.
- A rhombus has all the properties of a parallelogram. All sides are congruent. Diagonals are perpendicular. Each diagonal bisects a pair of opposite angles.
- A square has all the properties of a parallelogram, a rectangle, and a rhombus.
- In an isosceles trapezoid, both pairs of base angles are congruent and the diagonals are congruent.

FOLDABLES® Study Organizer

Be sure the Key Concepts are noted in your Foldable.

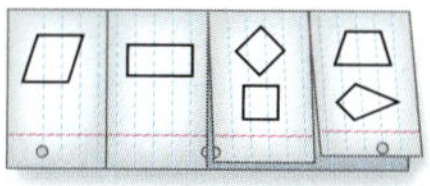

Key Vocabulary

base (p. 435)
base angle (p. 435)
diagonal (p. 389)
isosceles trapezoid (p. 435)
kite (p. 438)
legs (p. 435)
midsegment of a trapezoid (p. 437)
parallelogram (p. 399)
rectangle (p. 419)
rhombus (p. 426)
square (p. 427)
trapezoid (p. 435)

Vocabulary Check

State whether each sentence is *true* or *false*. If *false*, replace the underlined word or phrase to make a true sentence.

1. No angles in an isosceles trapezoid are congruent.
2. If a parallelogram is a rectangle, then the diagonals are congruent.
3. A midsegment of a trapezoid is a segment that connects any two nonconsecutive vertices.
4. The base of a trapezoid is one of the parallel sides.
5. The diagonals of a rhombus are perpendicular.
6. The diagonal of a trapezoid is the segment that connects the midpoints of the legs.
7. A rectangle is not always a parallelogram.
8. A quadrilateral with only one set of parallel sides is a parallelogram.
9. A rectangle that is also a rhombus is a square.
10. The leg of a trapezoid is one of the parallel sides.

Lesson-by-Lesson Review

6-1 Angles of Polygons (pp. 389–397)

Find the sum of the measures of the interior angles of each convex polygon.

11. decagon

12. 15-gon

13. **SNOWFLAKES** The snowflake decoration at the right is a regular hexagon. Find the sum of the measures of the interior angles of the hexagon.

The measure of an interior angle of a regular polygon is given. Find the number of sides in the polygon.

14. 135

15. 166.15

EXAMPLE 1

Find the sum of the measures of the interior angles of a convex 22-gon.

$m = (n - 2)180$	**Write an equation.**
$= (22 - 2)180$	**Substitution**
$= 20 \cdot 180$	**Subtract.**
$= 3600$	**Multiply.**

EXAMPLE 2

The measure of an interior angle of a regular polygon is 157.5. Find the number of sides in the polygon.

$157.5n = (n - 2)180$	**Write an equation.**
$157.5n = 180n - 360$	**Distributive Property**
$-22.5n = -360$	**Subtract.**
$n = 16$	**Divide.**

The polygon has 16 sides.

6-2 Parallelograms (pp. 399–407)

Use ▱$ABCD$ to find each measure.

16. $m\angle ADC$

17. AD

18. AB

19. $m\angle BCD$

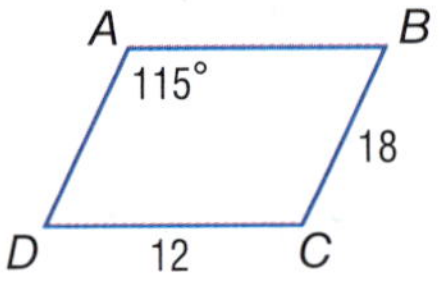

ALGEBRA Find the value of each variable in each parallelogram.

20.

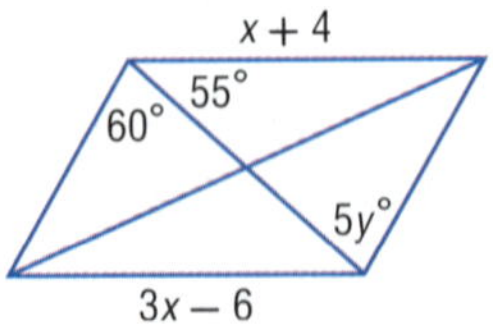

21.

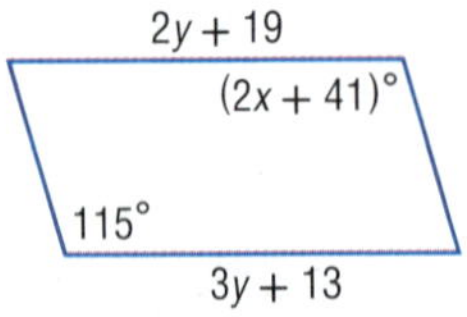

22. **DESIGN** What type of information is needed to determine whether the shapes that make up the stained glass window below are parallelograms?

EXAMPLE 3

ALGEBRA If $KLMN$ is a parallelogram, find the value of the indicated variable.

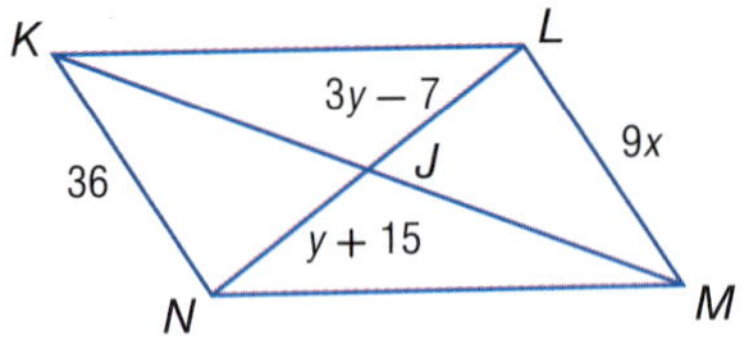

a. x

$\overline{KN} \cong \overline{LM}$	**Opp. sides of a ▱ are ≅.**
$KN = LM$	**Definition of congruence**
$36 = 9x$	**Substitution**
$4 = x$	**Divide.**

b. y

$\overline{NJ} \cong \overline{JL}$	**Diag. of a ▱ bisect each other.**
$NJ = JL$	**Definition of congruence**
$y + 15 = 3y - 7$	**Substitution**
$-2y = -22$	**Subtract.**
$y = 11$	**Divide.**

MIXED PROBLEM SOLVING
For mixed problem-solving practice, see page 1004.

6-3 Tests for Parallelograms (pp. 409–417)

Determine whether each quadrilateral is a parallelogram. Justify your answer.

23.

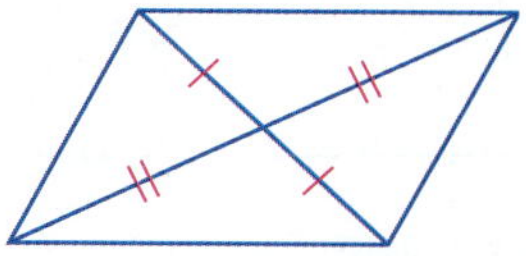

24.

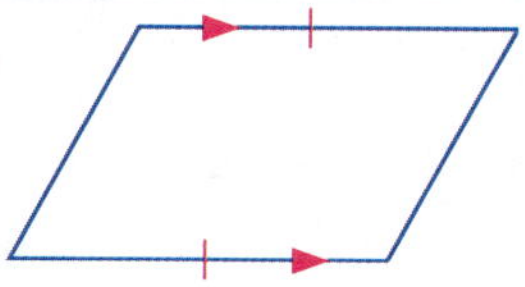

25. PROOF Write a two-column proof.

Given: $\square ABCD$, $\overline{AE} \cong \overline{CF}$

Prove: Quadrilateral *EBFD* is a parallelogram.

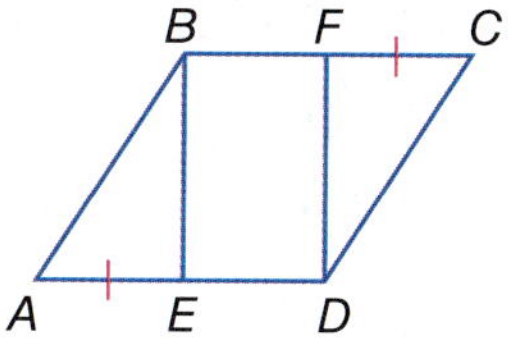

ALGEBRA Find *x* and *y* so that the quadrilateral is a parallelogram.

26.

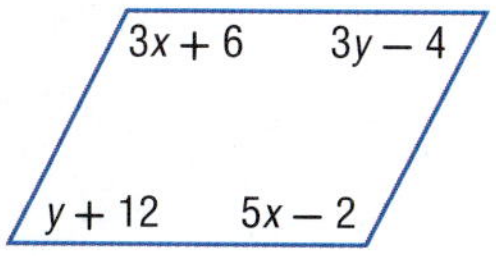

27.

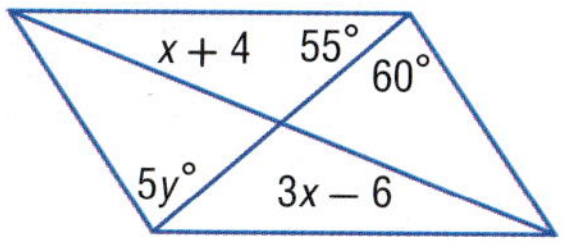

EXAMPLE 4

If $TP = 4x + 2$, $QP = 2y - 6$, $PS = 5y - 12$, and $PR = 6x - 4$, find *x* and *y* so that the quadrilateral is a parallelogram.

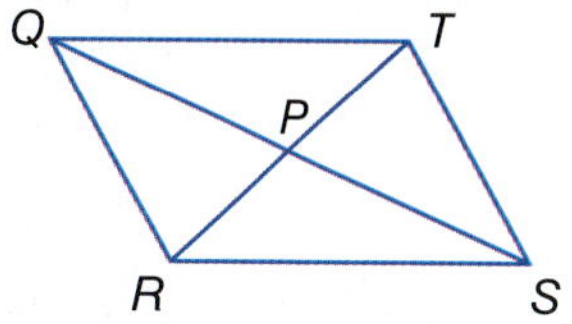

Find *x* such that $\overline{TP} \cong \overline{PR}$ and *y* such that $\overline{QP} \cong \overline{PS}$.

$TP = PR$	**Definition of ≅**
$4x + 2 = 6x - 4$	**Substitution**
$-2x = -6$	**Subtract.**
$x = 3$	**Divide.**

$QP = PS$	**Definition of ≅**
$2y - 6 = 5y - 12$	**Substitution**
$-3y = -6$	**Subtract.**
$y = 2$	**Divide.**

6-4 Rectangles (pp. 419–425)

28. PARKING The lines of the parking space shown below are parallel. How wide is the space (in inches)?

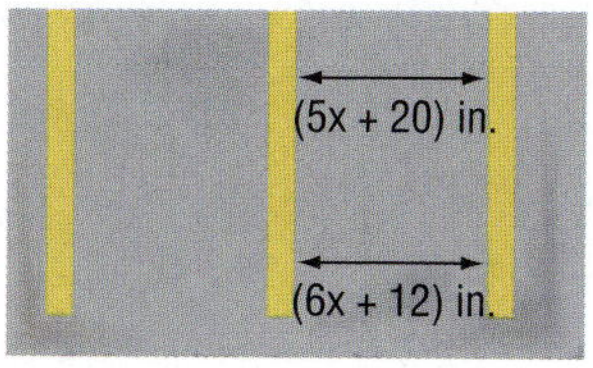

ALGEBRA Quadrilateral *EFGH* is a rectangle.

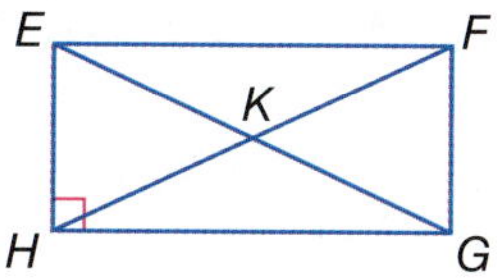

29. If $m\angle FEG = 57$, find $m\angle GEH$.

30. If $m\angle HGE = 13$, find $m\angle FGE$.

31. If $FK = 32$ feet, find EG.

32. Find $m\angle HEF + m\angle EFG$.

33. If $EF = 4x - 6$ and $HG = x + 3$, find EF.

EXAMPLE 5

ALGEBRA Quadrilateral *ABCD* is a rectangle. If $m\angle ADB = 4x + 8$ and $m\angle DBA = 6x + 12$, find *x*.

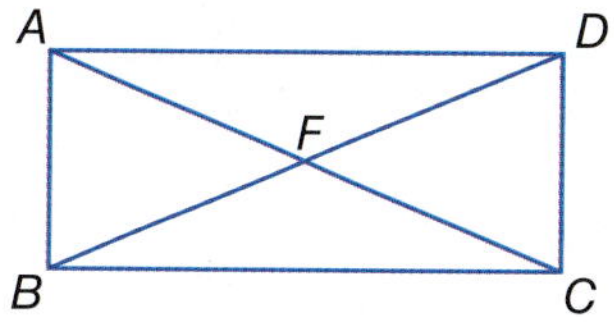

ABCD is a rectangle, so $m\angle ABC = 90$. Since the opposite sides of a rectangle are parallel, and the alternate interior angles of parallel lines are congruent, $\angle DBC \cong \angle ADB$ and $m\angle DBC = m\angle ADB$.

$m\angle DBC + m\angle DBA = 90$	**Angle Addition**
$m\angle ADB + m\angle DBA = 90$	**Substitution**
$4x + 8 + 6x + 12 = 90$	**Substitution**
$10x + 20 = 90$	**Add.**
$10x = 70$	**Subtract.**
$x = 7$	**Divide.**

CHAPTER 6 Study Guide and Review

6-5 Rhombi and Squares (pp. 426–434)

ALGEBRA *ABCD* **is a rhombus. If** $EB = 9$, $AB = 12$ **and** $m\angle ABD = 55$, **find each measure.**

34. AE

35. $m\angle BDA$

36. CE

37. $m\angle ACB$

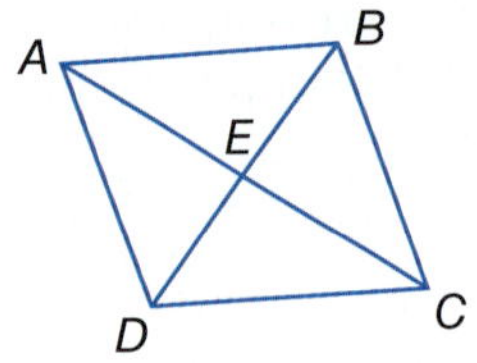

38. LOGOS A car company uses the symbol shown at the right for their logo. If the inside space of the logo is a rhombus, what is the length of FJ?

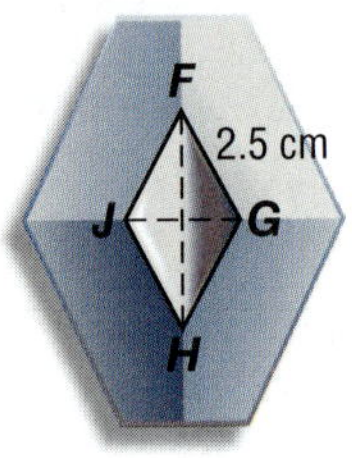

COORDINATE GEOMETRY Given each set of vertices, determine whether ▱*QRST* is a *rhombus*, a *rectangle*, or a *square*. List all that apply. Explain.

39. $Q(12, 0)$, $R(6, -6)$, $S(0, 0)$, $T(6, 6)$

40. $Q(-2, 4)$, $R(5, 6)$, $S(12, 4)$, $T(5, 2)$

EXAMPLE 6

The diagonals of rhombus *QRST* intersect at *P*. Use the information to find each measure or value.

a. ALGEBRA If $QT = x + 7$ **and** $TS = 2x - 9$, **find** x.

$\overline{QT} \cong \overline{TS}$	**Def. of rhombus**
$QT = TS$	**Def. of congruence**
$x + 7 = 2x - 9$	**Substitution**
$-x = -16$	**Subtract.**
$x = 16$	**Divide.**

b. If $m\angle QTS = 76$, **find** $m\angle TSP$.

$\overline{TR}$ bisects $\angle QTS$. Therefore, $m\angle PTS = \frac{1}{2}m\angle QTS$. So $m\angle PTS = \frac{1}{2}(76)$ or 38. Since the diagonals of a rhombus are perpendicular, $m\angle TPS = 90$.

$m\angle PTS + m\angle TPS + m\angle TSP = 180$	**△ Sum Thm.**
$38 + 90 + m\angle TSP = 180$	**Substitution**
$128 + m\angle TSP = 180$	**Add.**
$m\angle TSP = 52$	**Subtract.**

6-6 Trapezoids and Kites (pp. 435–444)

Find each measure.

41. GH

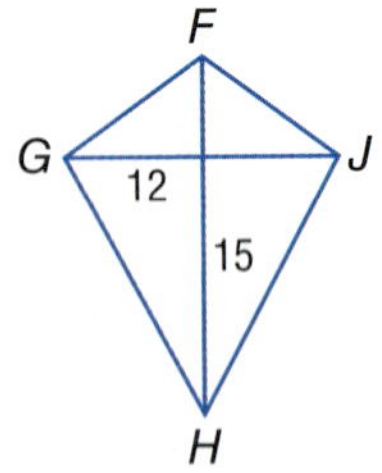

42. $m\angle Z$

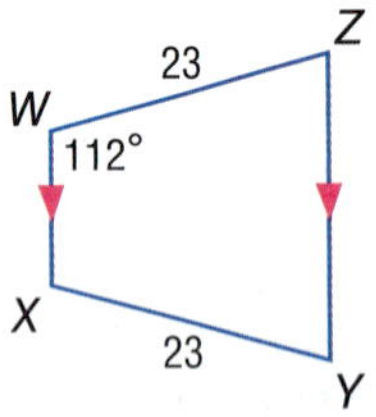

43. DESIGN Renee designed the square tile as an art project.

a. Describe a way to determine if the trapezoids in the design are isosceles.

b. If the perimeter of the tile is 48 inches and the perimeter of the red square is 16 inches, what is the perimeter of one of the trapezoids?

EXAMPLE 7

If *QRST* is a kite, find $m\angle RST$.

Since $\angle Q \cong \angle S$, $m\angle Q = m\angle S$. Write and solve an equation to find $m\angle S$.

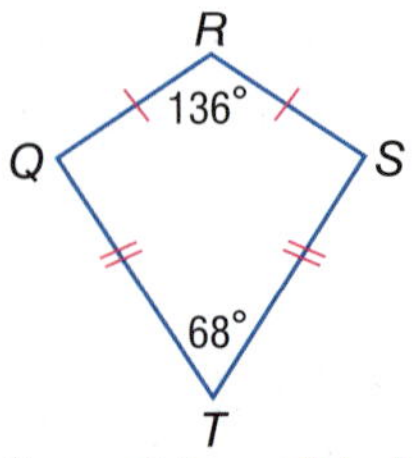

$m\angle Q + m\angle R + m\angle S + m\angle T = 360$	**Polygon Int. △ Sum Thm**
$m\angle Q + 136 + m\angle S + 68 = 360$	**Substitution**
$2m\angle S + 204 = 360$	**Simplify.**
$2m\angle S = 156$	**Subtract.**
$m\angle S = 78$	**Divide.**

CHAPTER 6 Practice Test

Find the sum of the measures of the interior angles of each convex polygon.

1. hexagon

2. 16-gon

3. ART Jen is making a frame to stretch a canvas over for a painting. She nailed four pieces of wood together at what she believes will be the four vertices of a square.

a. How can she be sure that the canvas will be a square?

b. If the canvas has the dimensions shown below, what are the missing measures?

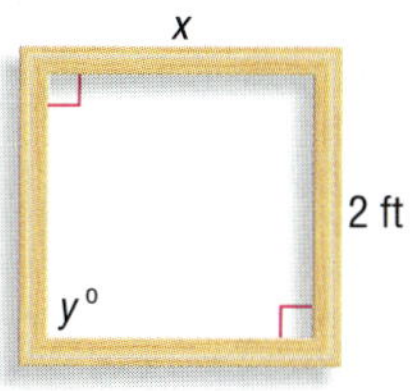

Quadrilateral *ABCD* is an isosceles trapezoid.

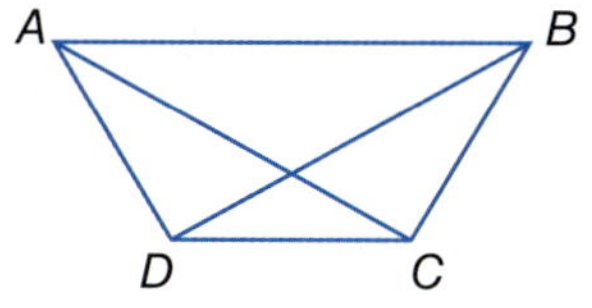

4. Which angle is congruent to $\angle C$?

5. Which side is parallel to $\overline{AB}$?

6. Which segment is congruent to $\overline{AC}$?

The measure of the interior angles of a regular polygon is given. Find the number of sides in the polygon.

7. 900

8. 1980

9. 2880

10. 5400

11. MULTIPLE CHOICE If *QRST* is a parallelogram, what is the value of x?

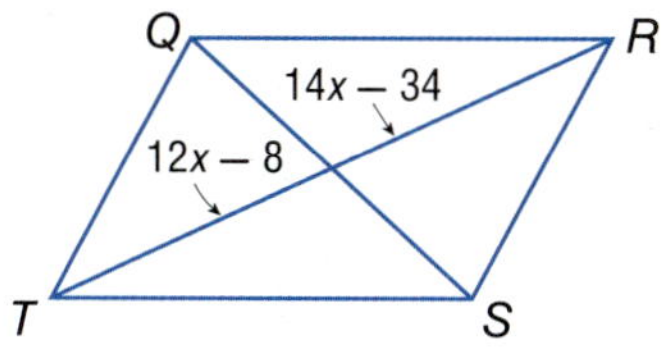

A 11

B 12

C 13

D 14

If *CDFG* is a kite, find each measure.

12. *GF*

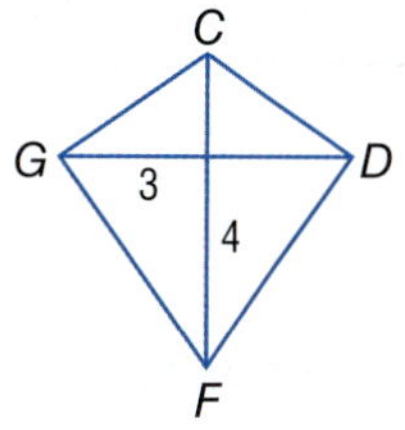

13. $m\angle D$

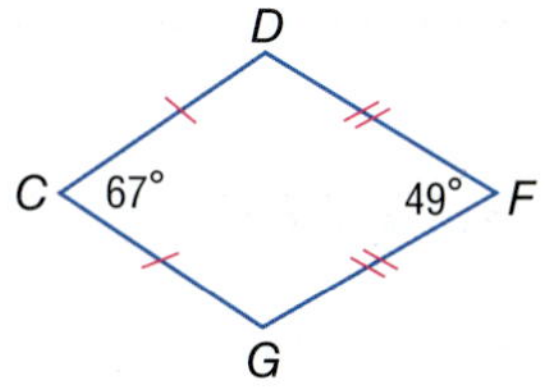

ALGEBRA Quadrilateral *MNOP* is a rhombus. Find each value or measure.

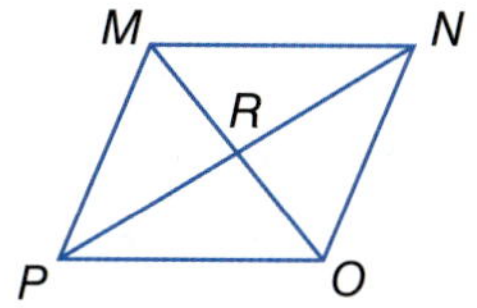

14. $m\angle MRN$

15. If $PR = 12$, find *RN*.

16. If $m\angle PON = 124$, find $m\angle POM$.

17. CONSTRUCTION The Smiths are building an addition to their house. Mrs. Smith is cutting an opening for a new window. If she measures to see that the opposite sides are congruent and that the diagonal measures are congruent, can Mrs. Smith be sure that the window opening is rectangular? Explain.

Use ▱*JKLM* to find each measure.

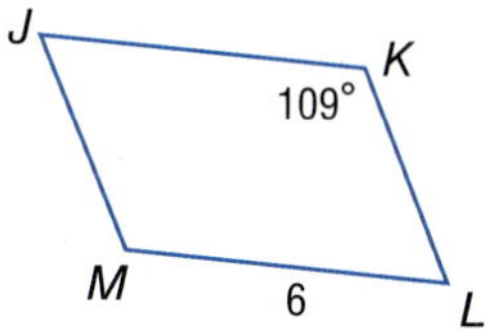

18. $m\angle JML$

19. *JK*

20. $m\angle KLM$

ALGEBRA Quadrilateral *DEFG* is a rectangle.

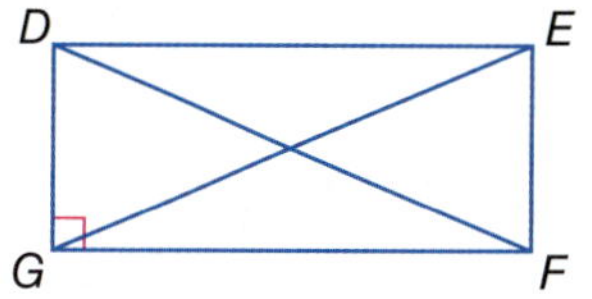

21. If $DF = 2(x + 5) - 7$ and $EG = 3(x - 2)$, find *EG*.

22. If $m\angle EDF = 5x - 3$ and $m\angle DFG = 3x + 7$, find $m\angle EDF$.

23. If $DE = 14 + 2x$ and $GF = 4(x - 3) + 6$, find *GF*.

Determine whether each quadrilateral is a parallelogram. Justify your answer.

24.

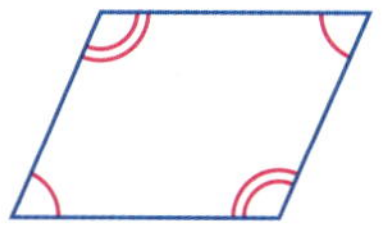

25.

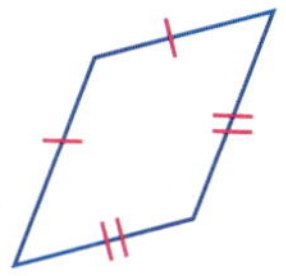

Apply Definitions and Properties

Many geometry problems on standardized tests require the application of definitions and properties in order to solve them. Use this section to practice applying definitions to help you solve extended-response test items.

Strategies for Applying Definitions and Properties

Step 1

Read the problem statement carefully.

- Determine what you are being asked to solve.
- Study any figures given in the problem.
- **Ask yourself:** What principles or properties of this figure can I apply to solve the problem?

Step 2

Solve the problem.

- Identify any definitions or geometric concepts you can use to help you find the unknowns in the problem.
- Use definitions and properties of figures to set up and solve an equation.

Step 3

- Check your answer.

EXAMPLE

Read the problem. Identify what you need to know. Then use the information in the problem to solve. Show your work.

A performing arts group is building a theater in the round for upcoming productions. The stage will be a regular octagon with a perimeter of 76 feet.

a. What length should each board be to form the sides of the stage?

b. What angle should the end of each board be cut so that they will fit together properly to form the stage? Explain.

Read the problem carefully. You are told that the boards form a regular octagon with a perimeter of 76 feet. You need to find the length of each board and the angle that they should be cut to fit together properly.

To find the length of each board, divide the perimeter by the number of boards.

$76 \div 8 = 9.5$

So, each board should be 9.5 feet, or 9 feet 6 inches, long.

Use the property of the interior angle sum of convex polygons to find the measure of an interior angle of a regular octagon. First find the sum S of the interior angles.

$$\begin{aligned} S &= (n - 2) \cdot 180 \\ &= (8 - 2) \cdot 180 \\ &= 1080 \end{aligned}$$

So, the measure of an interior angle of a regular octagon is $1080 \div 8$, or 135°. Since two boards are used to form each vertex of the stage, the end of each board should be cut at an angle of $135 \div 2$, or 67.5°.

Exercises

Read each problem. Identify what you need to know. Then use the information in the problem to solve. Show your work.

1. $\overline{RS}$ is the midsegment of trapezoid *MNOP*. What is the length of $\overline{RS}$?

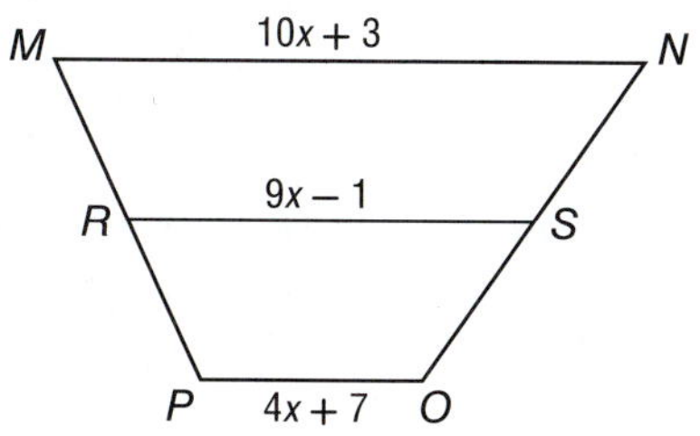

A 14 units
B 19 units
C 23 units
D 26 units

2. If $\overline{AB} \parallel \overline{DC}$, find x.

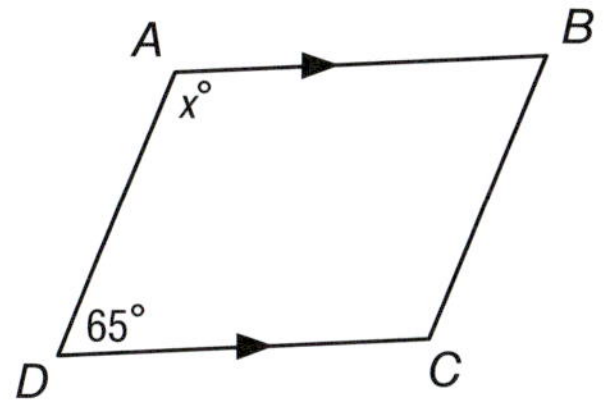

F 32.5
G 65
H 105
J 115

3. Use the graph shown below to answer each question.

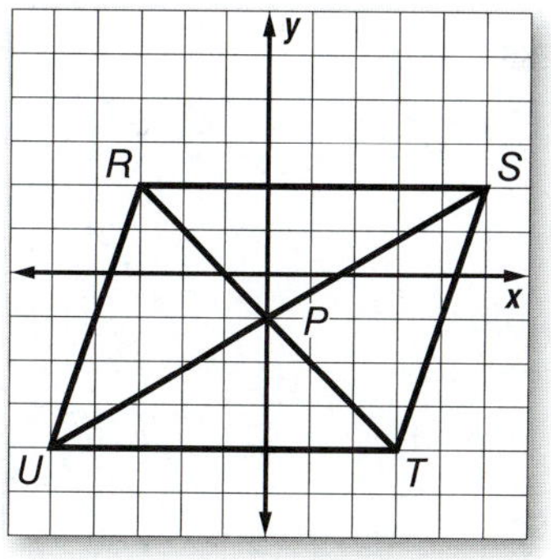

a. Do the diagonals of quadrilateral *RSTU* bisect each other? Use the Distance Formula to verify your answer.

b. What type of quadrilateral is *RSTU*? Explain using the properties and/or definitions of this type of quadrilateral.

4. What is the sum of the measures of the exterior angles of a regular octagon?

A 45
B 135
C 360
D 1080

CHAPTER 6 Standardized Test Practice

Cumulative, Chapters 1 through 6

Multiple Choice

Read each question. Then fill in the correct answer on the answer document provided by your teacher or on a sheet of paper.

1. If $a \parallel b$, which of the following is *not* true?

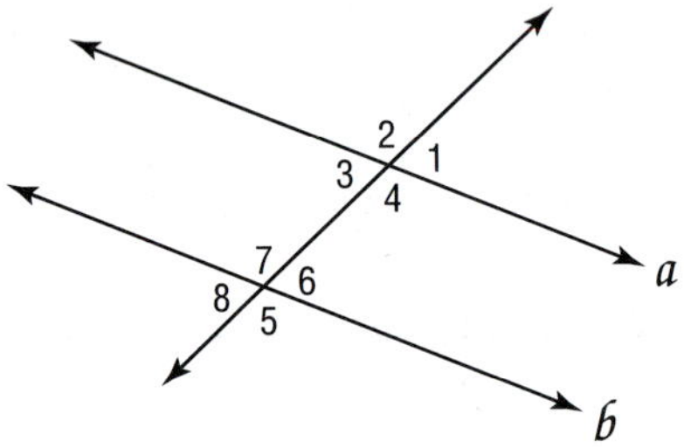

A $\angle 1 \cong \angle 3$

B $\angle 4 \cong \angle 7$

C $\angle 2 \cong \angle 5$

D $\angle 8 \cong \angle 2$

2. Classify the triangle below according to its angle measures. Choose the most appropriate term.

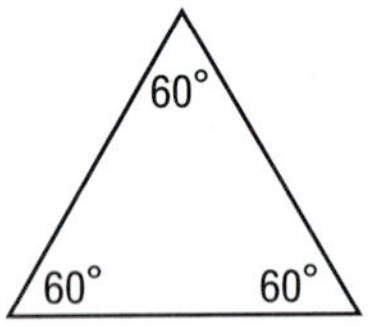

F acute

G equiangular

H obtuse

J right

3. Solve for x in parallelogram $RSTU$.

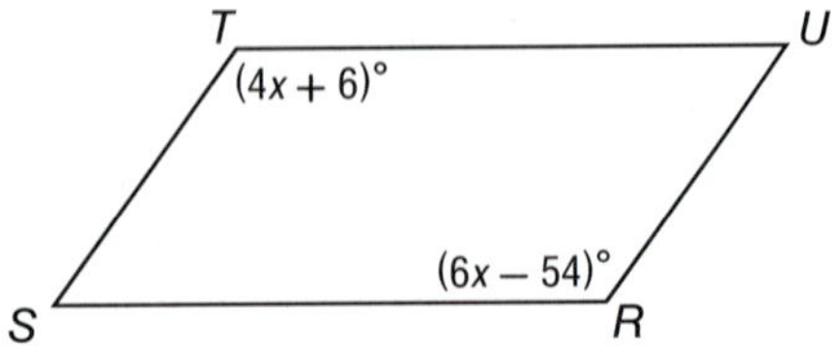

A 12

B 18

C 25

D 30

Test-TakingTip

Question 3 Use the properties of parallelograms to solve the problem. Opposite angles are congruent.

4. What is the measure of an interior angle of a regular pentagon?

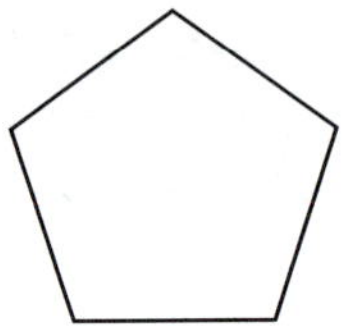

F 96

G 108

H 120

J 135

5. Quadrilateral $ABCD$ is a rhombus. If $m\angle BCD = 120$, find $m\angle DAC$.

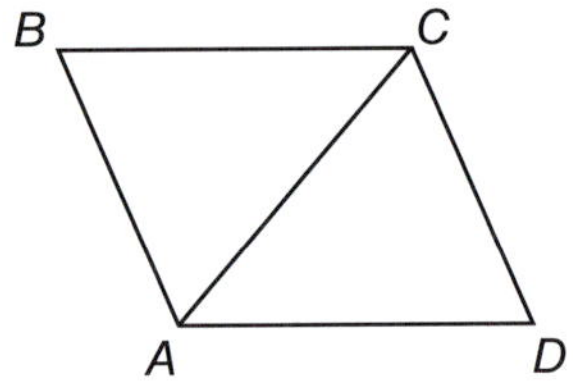

A 30

B 60

C 90

D 120

6. What is the value of x in the figure below?

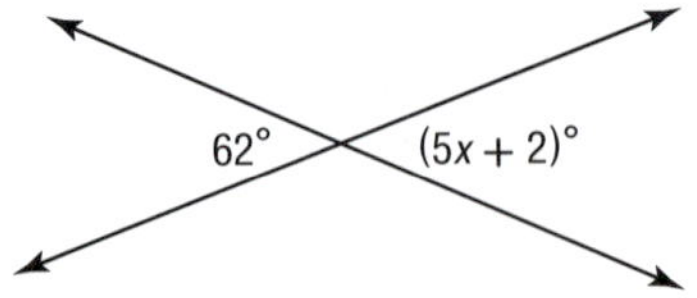

F 10

G 12

H 14

J 15

7. Which of the following statements is true?

A All rectangles are squares.

B All rhombi are squares.

C All rectangles are parallelograms.

D All parallelograms are rectangles.

Short Response/Gridded Response

Record your answers on the answer sheet provided by your teacher or a sheet of paper.

8. **GRIDDED RESPONSE** The posts for Nancy's gazebo form a regular hexagon. What is the measure of the angle formed at each corner of the gazebo?

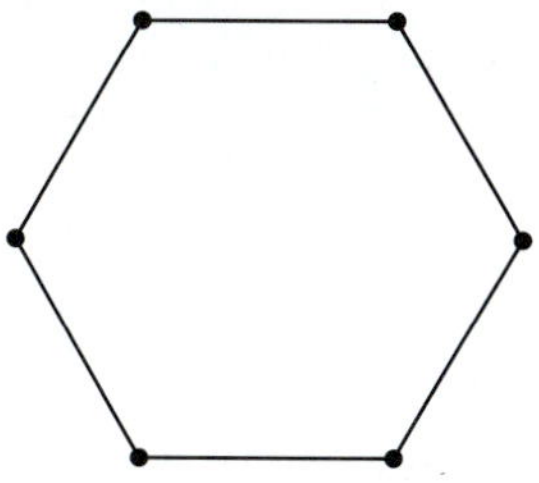

9. What are the coordinates of point O, the fourth vertex of an isosceles trapezoid? Show your work.

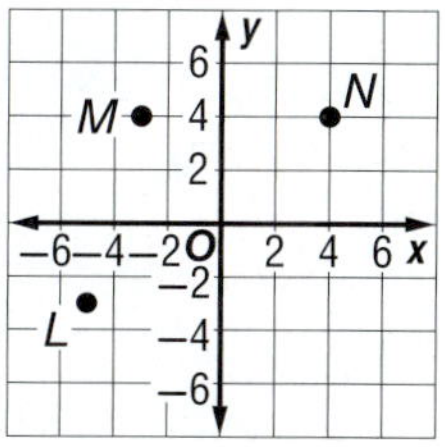

10. What do you know about a parallelogram if its diagonals are perpendicular? Explain.

11. Determine whether the stated conclusion is valid based on the given information below. If not, write *invalid*. Explain your reasoning.

Given: If a number is divisible by 9, then the number is divisible by 3. The number 144 is divisible by 9.

Conclusion: The number 144 is divisible by 3.

12. **GRIDDED RESPONSE** Solve for x in the figure below. Round to the nearest tenth if necessary.

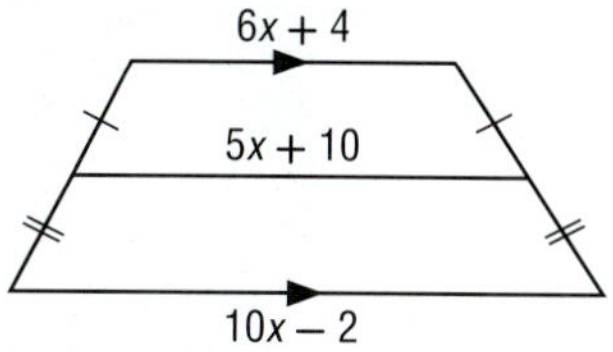

13. What are the coordinates of the circumcenter of the triangle below?

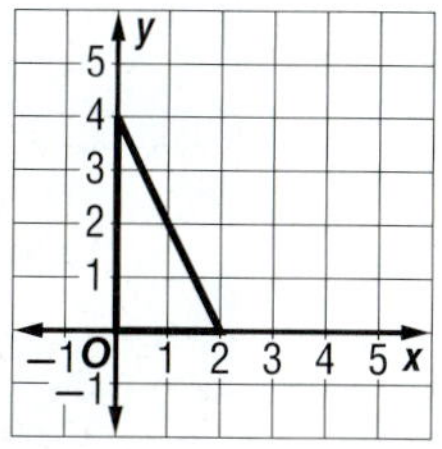

Extended Response

Record your answers on a sheet of paper. Show your work.

14. Determine whether you can prove each figure is a parallelogram. If not, tell what additional information would be needed to prove that it is a parallelogram. Explain your reasoning.

a.

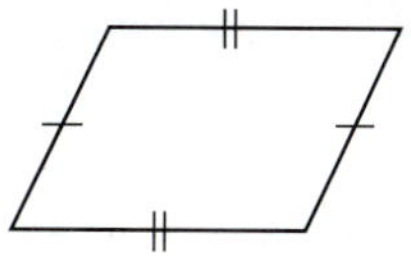

b.

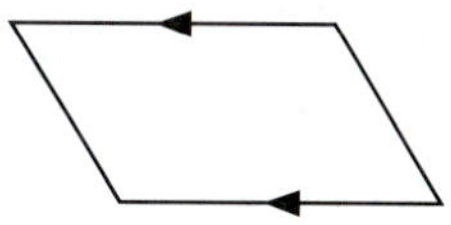

c.

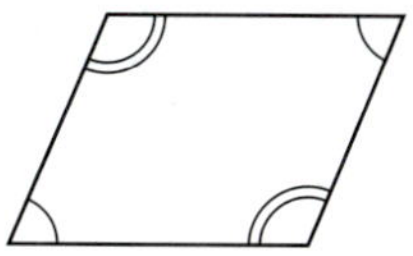

Need Extra Help?

If you missed Question...	1	2	3	4	5	6	7	8	9	10	11	12	13	14
Go to Lesson or Page...	3-2	4-1	6-2	6-1	6-5	1-5	6-5	6-1	6-6	6-5	2-4	6-6	5-1	6-3

CHAPTER 7 Proportions and Similarity

Then

In Algebra, you learned about ratios and proportions and applied them to real-world applications.

Now

In Chapter 7, you will:

- Identify similar polygons and use ratios and proportions to solve problems.
- Identify and apply similarity transformations.
- Use scale models and drawings to solve problems.

Why?

SPORTS Similar triangles can be used in sports to describe the path of a ball, such as a bounce from one person to another.

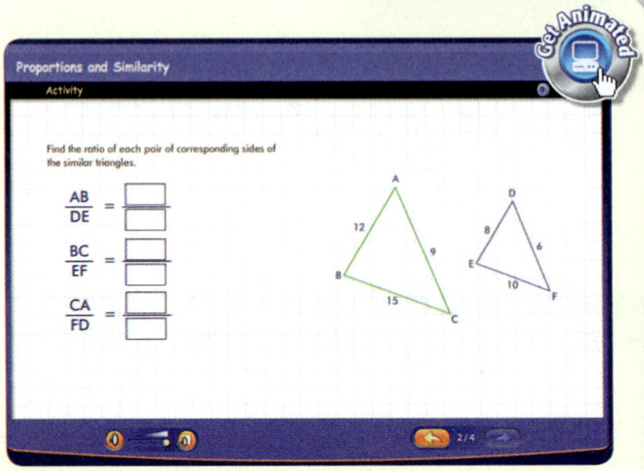

Get Ready for Chapter 7

Diagnose Readiness You have two options for checking Prerequisite Skills.

Text Option

Take the Quick Check below. Refer to the Quick Review for help.

QuickCheck

Solve each equation. (Prerequisite Skill)

1. $\frac{3x}{8} = \frac{6}{x}$

2. $\frac{7}{3} = \frac{x-4}{6}$

3. $\frac{x+9}{2} = \frac{3x-1}{8}$

4. $\frac{3}{2x} = \frac{3x}{8}$

5. **EDUCATION** The student to teacher ratio at Elder High School is 17 to 1. If there are 1088 students in the school, how many teachers are there?

ALGEBRA **In the figure, $\overrightarrow{AB}$ and $\overrightarrow{BC}$ are opposite rays and $\overrightarrow{BD}$ bisects $\angle ABF$.** (Lesson 1-5)

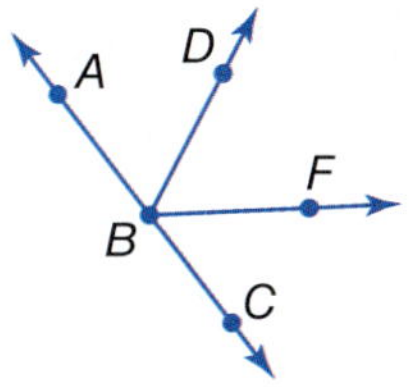

6. If $m\angle ABF = 3x - 8$ and $m\angle ABD = x + 14$, find $m\angle ABD$.

7. If $m\angle FBC = 2x + 25$ and $m\angle ABF = 10x - 1$, find $m\angle DBF$.

8. **LANDSCAPING** A landscape architect is planning to add sidewalks around a fountain as shown below. If $\overrightarrow{AB}$ and $\overrightarrow{BC}$ are opposite rays and $\overrightarrow{BD}$ bisects $\angle ABF$, find $m\angle FBC$.

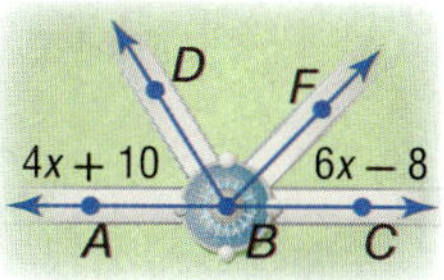

QuickReview

EXAMPLE 1

Solve $\frac{4x-3}{5} = \frac{2x+11}{3}$.

$\frac{4x-3}{5} = \frac{2x+11}{3}$	Original equation
$3(4x-3) = 5(2x+11)$	Cross multiplication
$12x - 9 = 10x + 55$	Distributive Property
$2x = 64$	Add.
$x = 32$	Simplify.

EXAMPLE 2

In the figure, $\overrightarrow{PQ}$ and $\overrightarrow{QR}$ are opposite rays, and $\overrightarrow{TQ}$ bisects $\angle SQR$. If $m\angle SQR = 6x + 8$ and $m\angle TQR = 4x - 14$, find $m\angle SQT$.

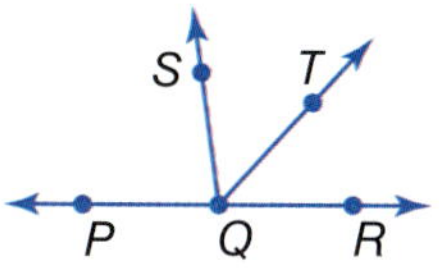

Since $\overrightarrow{TQ}$ bisects $\angle SQR$, $m\angle SQR = 2(m\angle TQR)$.

$m\angle SQR = 2(m\angle TQR)$	Def. of $\angle$ bisector
$6x + 8 = 2(4x - 14)$	Substitution
$6x + 8 = 8x - 28$	Distributive Property
$-2x = -36$	Subtract.
$x = 18$	Simplify.

Since $\overrightarrow{TQ}$ bisects $\angle SQR$, $m\angle SQT = m\angle TQR$.

$m\angle SQT = m\angle TQR$	Def. of $\angle$ bisector
$m\angle SQT = 4x - 14$	Substitution
$m\angle SQT = 58$	$x = 18$

Online Option

Math Online Take a self-check Chapter Readiness Quiz at glencoe.com.

Get Started on Chapter 7

You will learn several new concepts, skills, and vocabulary terms as you study Chapter 7. To get ready, identify important terms and organize your resources. You may wish to refer to **Chapter 0** to review prerequisite skills.

FOLDABLES® Study Organizer

Proportions and Similarity Make this Foldable to help you organize your Chapter 7 notes about proportions, similar polygons, and similarity transformations. Begin with four sheets of notebook paper.

1. **Fold** the four sheets of paper in half.

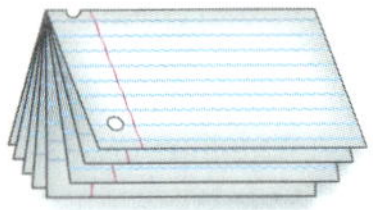

2. **Cut** along the top fold of the papers.

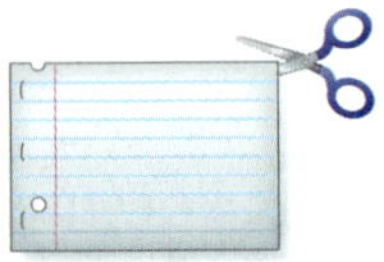

3. **Cut** the right sides of each paper to create a tab for each lesson.

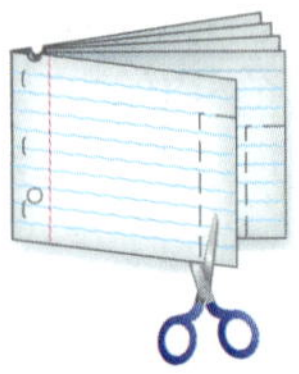

4. **Label** each tab with a lesson number, as shown.

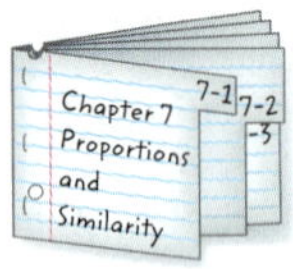

New Vocabulary

English		Español
ratio	• p. 457 •	razón
proportion	• p. 458 •	proporción
extremes	• p. 458 •	extremos
means	• p. 458 •	medios
cross products	• p. 458 •	productos cruzados
similar polygons	• p. 465 •	polígonos semejantes
scale factor	• p. 466 •	factor de escala
similarity transformation	• p. 505 •	transformación de semejanza
dilation	• p. 505 •	dilatación
enlargement	• p. 505 •	ampliación
reduction	• p. 505 •	reducción
scale model	• p. 512 •	modelo a escala
scale drawing	• p. 512 •	dibujo a escala

Review Vocabulary

altitude • p. 335 • altura a segment drawn from a vertex of a triangle perpendicular to the line containing the other side

angle bisector • p. 39 • bisectriz de un ángulo a ray that divides an angle into two congruent angles

median • p. 333 • mediana a segment drawn from a vertex of a triangle to the midpoint of the opposite side

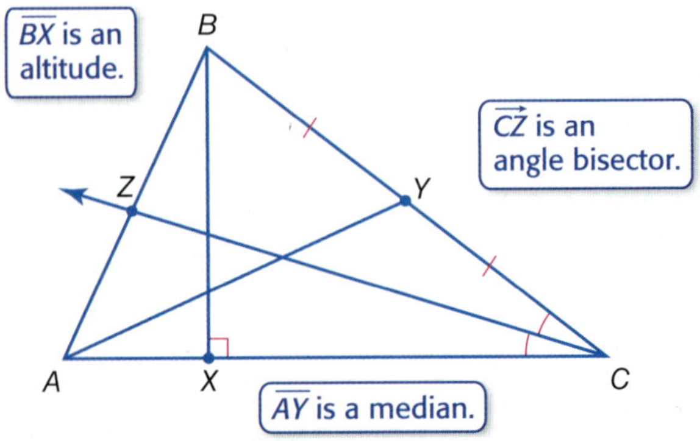

Multilingual eGlossary glencoe.com

Math Online glencoe.com

- Study the chapter online
- Explore **Get Animated** activities
- Get extra help from your own **Personal Tutor**
- Use **Extra Examples** for additional help
- Take a **Self-Check Quiz**
- **Review Vocabulary** in fun ways

7-1 Ratios and Proportions

Then

You solved problems by writing and solving equations. (Lesson 3-4)

Now

- Write ratios.
- Write and solve proportions.

New Vocabulary

ratio
extended ratios
proportion
extremes
means
cross products

Math Online
glencoe.com

Why?

The *aspect ratio* of a television or computer screen is the screen's width divided by its height. A standard television screen has an aspect ratio of $\frac{4}{3}$ or 4:3, while a high definition television screen (HDTV) has an aspect ratio of 16:9.

Write and Use Ratios A **ratio** is a comparison of two quantities using division. The ratio of quantities a and b can be expressed as a to b, $a:b$, or $\frac{a}{b}$, where $b \neq 0$. Ratios are usually expressed in simplest form.

The aspect ratios 32:18 and 16:9 are equivalent.

$$\frac{\text{width of screen}}{\text{height of screen}} = \frac{32 \text{ in.}}{18 \text{ in.}} \quad \textbf{Divide out units.}$$

$$= \frac{32 \div 2}{18 \div 2} \text{ or } \frac{16}{9} \quad \textbf{Divide out common factors.}$$

Real-World EXAMPLE 1 Write and Simplify Ratios

SPORTS A baseball player's batting average is the ratio of the number of base hits to the number of at-bats, not including walks. Minnesota Twins' Joe Mauer had the highest batting average in Major League Baseball in 2006. If he had 521 official at-bats and 181 hits, find his batting average.

Divide the number of hits by the number of at-bats.

$$\frac{\text{number of hits}}{\text{number of at-bats}} = \frac{181}{521}$$

$$\approx \frac{0.347}{1} \quad \textbf{A ratio in which the denominator is 1 is called a \textit{unit ratio}.}$$

Joe Mauer's batting average was 0.347.

Guided Practice

1. **SCHOOL** In Logan's high school, there are 190 teachers and 2650 students. What is the approximate student-teacher ratio at his school?

Personal Tutor glencoe.com

Extended ratios can be used to compare three or more quantities. The expression $a:b:c$ means that the ratio of the first two quantities is $a:b$, the ratio of the last two quantities is $b:c$, and the ratio of the first and last quantities is $a:c$.

EXAMPLE 2 Use Extended Ratios

The ratio of the measures of the angles in $\triangle ABC$ is 3:4:5. Find the measures of the angles.

Just as the ratio $\frac{3}{4}$ or 3:4 is equivalent to $\frac{3x}{4x}$ or $3x:4x$, the extended ratio 3:4:5 can be written as $3x:4x:5x$.

Sketch and label the angle measures of the triangle. Then write and solve an equation to find the value of x.

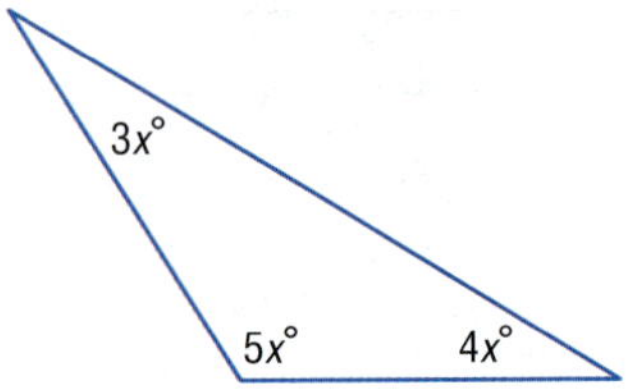

$3x + 4x + 5x = 180$	**Triangle Sum Theorem**
$12x = 180$	**Combine like terms.**
$x = 15$	**Divide each side by 12.**

So the measures of the angles are 3(15) or 45, 4(15) or 60, and 5(15) or 75.

CHECK The sum of the angle measures should be 180.

$45 + 60 + 75 = 180$ ✓

Guided Practice

2. In a triangle, the ratio of the measures of the sides is 3:3:8 and the perimeter is 392 inches. Find the length of the longest side of the triangle.

Personal Tutor glencoe.com

ReadingMath

Proportion When a proportion is written using colons, it is read using the word *to* for the colon. For example, 2:3 is read *2 to 3*. The means are the inside numbers, and the extremes are the outside numbers.

extremes

2:3 = 6:9

means

Use Properties of Proportions An equation stating that two ratios are equal is called a **proportion**. In the proportion $\frac{a}{b} = \frac{c}{d}$, the numbers a and d are called the **extremes** of the proportion, while the numbers b and c are called the **means** of the proportion.

extreme → a, mean → b in $\frac{a}{b} = \frac{c}{d}$; c ← mean, d ← extreme

The product of the extremes ad and the product of the means bc are called **cross products**.

Key Concept Cross Products Property

Words In a proportion, the product of the extremes equals the product of the means.

Symbols If $\frac{a}{b} = \frac{c}{d}$ when $b \neq 0$ and $d \neq 0$, then $ad = bc$.

Example If $\frac{4}{10} = \frac{6}{15}$, then $4 \cdot 15 = 10 \cdot 6$.

You will prove the Cross Products Property in Exercise 41.

The converse of the Cross Products Property is also true. If $ad = bc$ and $b \neq 0$ and $d \neq 0$, then $\frac{a}{b} = \frac{c}{d}$. That is, $\frac{a}{b}$ and $\frac{c}{d}$ form a proportion.

You can use the Cross Products Property to solve a proportion.

EXAMPLE 3 Use Cross Products to Solve Proportions

Solve each proportion.

a. $\frac{6}{x} = \frac{21}{31.5}$

$\frac{6}{x} = \frac{21}{31.5}$	Original proportion
$6(31.5) = x(21)$	Cross Products Property
$189 = 21x$	Simplify.
$9 = x$	Solve for x.

b. $\frac{x+3}{2} = \frac{4x}{5}$

$$\frac{x+3}{2} = \frac{4x}{5}$$
$$(x + 3)5 = 2(4x)$$
$$5x + 15 = 8x$$
$$15 = 3x$$
$$5 = x$$

StudyTip

Alternate Method Example 3b could also be solved by multiplying each side of the equation by 10, the least common denominator.

$$10\left(\frac{x+3}{2}\right) = \frac{4x}{5}(10)$$
$$5(x + 3) = 2(4x)$$
$$5x + 15 = 8x$$
$$15 = 3x$$
$$5 = x$$

Guided Practice

3A. $\frac{x}{4} = \frac{11}{-6}$ **3B.** $\frac{-4}{7} = \frac{6}{2y+5}$ **3C.** $\frac{7}{z-1} = \frac{9}{z+4}$

Personal Tutor glencoe.com

Proportions can be used to make predictions.

Real-World EXAMPLE 4 Use Proportions to Make Predictions

CAR OWNERSHIP Fernando conducted a survey of 50 students driving to school and found that 28 owned their own cars. If 755 students drive to his school, predict the total number of students with their own cars.

Write and solve a proportion that compares the number of students who own their cars to the number who drive to school.

$\frac{28}{50} = \frac{x}{755}$	← students owning their cars ← students driving to school
$28 \cdot 755 = 50 \cdot x$	Cross Products Property
$21{,}140 = 50x$	Simplify.
$422.8 = x$	Divide each side by 50.

Based on Fernando's survey, about 423 students at his school own their cars.

Real-World Link

The percent of driving-age teens (ages 15 to 20) with their own vehicles nearly doubled nationwide from 22 percent in 1985 to 42 percent in 2003.

Source: CNW Marketing Research

Guided Practice

4. BIOLOGY In an experiment, students netted butterflies, recorded the number with tags on their wings, and then released them. The students netted 48 butterflies and 3 of those had tagged wings. Predict the number of butterflies that would have tagged wings out of 100 netted.

Personal Tutor glencoe.com

The proportion shown in Example 4 is not the only correct proportion for that situation. Equivalent forms of a proportion all have identical cross products.

Key Concept Equivalent Proportions

Symbols The following proportions are equivalent.

$$\frac{a}{b} = \frac{c}{d}, \quad \frac{b}{a} = \frac{d}{c}, \quad \frac{a}{c} = \frac{b}{d}, \quad \frac{c}{a} = \frac{d}{b}$$

Examples $\frac{28}{50} = \frac{x}{755}, \frac{50}{28} = \frac{755}{x}, \frac{28}{x} = \frac{50}{755}, \frac{x}{28} = \frac{755}{50}$

Check Your Understanding

Example 1 p. 457

1. **PETS** Out of a survey of 1000 households, 460 had at least one dog or cat as a pet. What is the ratio of pet owners to households?

2. **SPORTS** Thirty girls tried out for 15 spots on the basketball team. What is the ratio of open spots to the number of girls competing?

Example 2 p. 458

3. The ratio of the measures of three sides of a triangle is 2:5:4, and its perimeter is 165 units. Find the measure of each side of the triangle.

4. The ratios of the measures of three angles of a triangle are 4:6:8. Find the measure of each angle of the triangle.

Example 3 p. 459

Solve each proportion.

5. $\frac{2}{3} = \frac{x}{24}$

6. $\frac{x}{5} = \frac{28}{100}$

7. $\frac{2.2}{x} = \frac{26.4}{96}$

8. $\frac{x-3}{3} = \frac{5}{8}$

Example 4 p. 459

9. **BAKING** Ella is baking apple muffins for the Student Council bake sale. The recipe that she is using calls for 2 eggs per dozen muffins, and she needs to make 108 muffins. How many eggs will she need?

Practice and Problem Solving

● = **Step-by-Step Solutions** begin on page R20.
Extra Practice begins on page 969.

Example 1 p. 457

MOVIES For Exercises 10 and 11, refer to the graphic below.

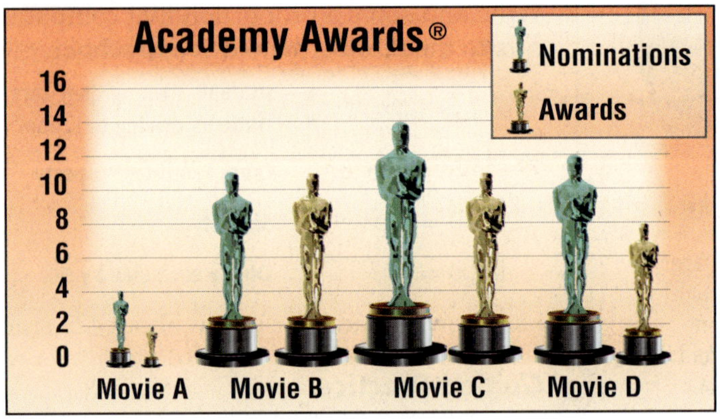

10. Of the films listed, which had the greatest ratio of Academy Awards to number of nominations?

11. Which film listed had the lowest ratio of awards to nominations?

Example 2 p. 458

12. **GAMES** A video game store has 60 games to choose from, including 40 sports games. What is the ratio of sports games to video games?

13. The ratio of the measures of the three sides of a triangle is 9:7:5. Its perimeter is 191.1 inches. Find the measure of each side

14. The ratio of the measures of the three sides of a triangle is 3:7:5, and its perimeter is 156.8 meters. Find the measure of each side.

15. The ratio of the measures of the three sides of a triangle is $\frac{1}{4}:\frac{1}{8}:\frac{1}{6}$. Its perimeter is 4.75 feet. Find the length of the longest side.

16. The ratio of the measures of the three sides of a triangle is $\frac{1}{4}:\frac{1}{3}:\frac{1}{6}$, and its perimeter is 31.5 centimeters. Find the length of the shortest side.

Find the measures of the angles of each triangle.

17. The ratio of the measures of the three angles is 3:6:1.

18. The ratio of the measures of the three angles is 7:5:8.

19. The ratio of the measures of the three angles is 10:8:6.

20. The ratio of the measures of the three angles is 5:4:7.

Example 3 p. 459

Solve each proportion.

21. $\frac{5}{8} = \frac{y}{3}$ **22.** $\frac{w}{6.4} = \frac{1}{2}$ **23.** $\frac{4x}{24} = \frac{56}{112}$ **24.** $\frac{11}{20} = \frac{55}{20x}$

25. $\frac{2x+5}{10} = \frac{42}{20}$ **26.** $\frac{a+2}{a-2} = \frac{3}{2}$ **27.** $\frac{3x-1}{4} = \frac{2x+4}{5}$ **28.** $\frac{3x-6}{2} = \frac{4x-2}{4}$

Example 4 p. 459

29. **NUTRITION** According to a recent study, 7 out of every 500 Americans aged 13 to 17 years are vegetarian. In a group of 350 13- to 17-year-olds, about how many would you expect to be vegetarian?

30. **CURRENCY** Your family is traveling to Mexico on vacation. You have saved $500 to use for spending money. If 269 Mexican pesos is equivalent to 25 United States dollars, how much money will you get when you exchange your $500 for pesos?

Real-World Link

In June of 2000, 60.2% of American teens 16 to 19 years old had summer jobs. By June of 2006, 51.6% of teens in that age group were a part of the summer work force.

Source: U.S. Department of Labor

ALGEBRA **Solve each proportion. Round to the nearest tenth.**

31. $\frac{2x+3}{3} = \frac{6}{x-1}$ **32.** $\frac{x^2+4x+4}{40} = \frac{x+2}{10}$ **33.** $\frac{9x+6}{18} = \frac{20x+4}{3x}$

34. The perimeter of a rectangle is 98 feet. The ratio of its length to its width is 5:2. Find the area of the rectangle.

35. The perimeter of a rectangle is 220 inches. The ratio of its length to its width is 7:3. Find the area of the rectangle.

36. The ratio of the measures of the side lengths of a quadrilateral is 2:3:5:4. Its perimeter is 154 feet. Find the length of the shortest side.

37. The ratio of the measures of the angles of a quadrilateral is 2:4:6:3. Find the measures of the angles of the quadrilateral.

38. **SUMMER JOBS** Refer to the information at the left.

a. Has the number of 16- to 19-year-olds with summer jobs increased or decreased since 2000? Explain your reasoning.

b. In June 2006, how many 16- to 19-year-olds would you expect to have jobs out of 700 in that age group? Explain your reasoning.

39. **GOLDEN RECTANGLES** Many artists have used golden rectangles in their work. In a golden rectangle, the ratio of the length to the width is about 1.618. This is known as the *golden ratio*.

a. A rectangle has dimensions of 19.42 feet and 12.01 feet. Determine if the rectangle is a golden rectangle. Then find the length of the diagonal.

b. Recall from page 457 that a standard television screen has an aspect ratio of 4:3, while a high definition television screen has an aspect ratio of 16:9. Is either type of screen a golden rectangle? Explain.

40. **SCHOOL ACTIVITIES** A survey of club involvement showed that, of the 36 students surveyed, the ratio of French Club members to Spanish Club members to Drama Club members was 2:3:7. How many of those surveyed participate in Spanish Club? Assume that each student is active in only one club.

41. **PROOF** Write an algebraic proof of the Cross Products Property.

Real-World Link

According to a study by the American Public Health Association, aerobic exercise such as walking or jogging done 30 minutes a day, a few days a week, can increase bone density.

Source: American Public Health Association

42. SPORTS Jane jogs the same path every day in the winter to stay in shape for track season. She runs at a constant rate, and she spends a total of 39 minutes jogging. If the ratio of the times of the four legs of the jog is 3:5:1:4, how long does the second leg of the jog take her?

43. MULTIPLE REPRESENTATIONS In this problem, you will explore proportional relationships in triangles.

a. GEOMETRIC Draw an isosceles triangle *ABC*. Measure and label the legs and the vertex angle. Draw a second triangle *MNO* with a congruent vertex angle and legs twice as long as *ABC*. Draw a third triangle *PQR* with a congruent vertex angle and legs half as long as *ABC*.

b. TABULAR Complete the table below using the appropriate measures.

Triangle	*ABC*	*MNO*	*PQR*
Leg length			
Perimeter			

c. VERBAL Make a conjecture about the change in the perimeter of an isosceles triangle if the vertex angle is held constant and the leg length is increased or decreased by a factor.

H.O.T. Problems

Use Higher-Order Thinking Skills

44. ERROR ANALYSIS Mollie and Eva have solved the proportion $\frac{x-3}{4} = \frac{1}{2}$. Is either of them correct? Explain your reasoning.

Mollie

$(x - 3)1 = 4(2)$

$x - 3 = 8$

$x = 11$

Eva

$x - 3(2) = 4(1)$

$x - 3 = 4$

$x = 7$

45. CHALLENGE The dimensions of a rectangle are y and $y^2 + 1$ and the perimeter of the rectangle is 14 units. Find the ratio of the longer side of the rectangle to the shorter side of the rectangle.

46. REASONING The ratio of the lengths of the diagonals of a quadrilateral is 1:1. The ratio of the lengths of the consecutive sides of the quadrilateral is 3:4:3:5. Classify the quadrilateral. Explain.

47. WHICH ONE DOESN'T BELONG? Identify the proportion that does not belong with the other three. Explain your reasoning.

$\frac{3}{8} = \frac{8.4}{22.4}$ $\qquad$ $\frac{2}{3} = \frac{5}{7.5}$ $\qquad$ $\frac{5}{6} = \frac{14}{16.8}$ $\qquad$ $\frac{7}{9} = \frac{19.6}{25.2}$

48. OPEN ENDED Write four ratios that are equivalent to the ratio 2:5. Explain why all of the ratios are equivalent.

49. WRITING IN MATH Compare and contrast a ratio and a proportion. Explain how you use both to solve a problem.

Standardized Test Practice

50. Solve the following proportion.

$$\frac{x}{-8} = \frac{12}{6}$$

A -12 **C** -16

B -14 **D** -18

51. What is the area of rectangle $WXYZ$?

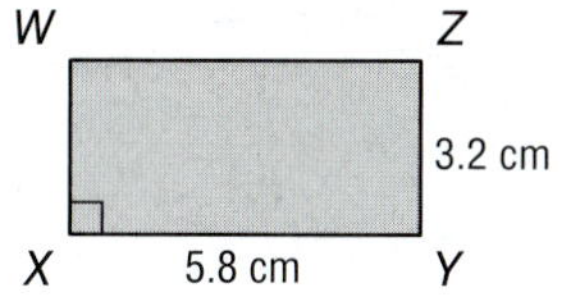

F 18.6 cm^2 **H** 21.2 cm^2

G 20.4 cm^2 **J** 22.8 cm^2

52. GRIDDED RESPONSE Mrs. Sullivan's rectangular bedroom measures 12 feet by 10 feet. She wants to purchase carpet for the bedroom that costs $2.56 per square foot, including tax. How much will it cost in dollars to carpet her bedroom?

53. SAT/ACT Kamilah has 5 more than 4 times the number of DVDs that Mercedes has. If Mercedes has x DVDs, then in terms of x, how many DVDs does Kamilah have?

A $4(x + 5)$ **C** $9x$

B $4(x + 3)$ **D** $4x + 5$

Spiral Review

For trapezoid $ABCD$, S and T are midpoints of the legs. (Lesson 6-6)

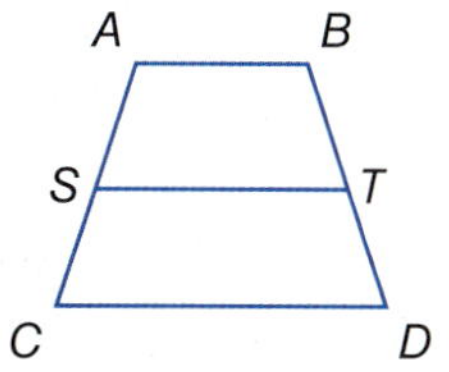

54. If $CD = 14$, $ST = 10$, and $AB = 2x$, find x.

55. If $AB = 3x$, $ST = 15$, and $CD = 9x$, find x.

56. If $AB = x + 4$, $CD = 3x + 2$, and $ST = 9$, find AB.

57. SPORTS The infield of a baseball diamond is a square, as shown at the right. Is the pitcher's mound located in the center of the infield? Explain. (Lesson 6-5)

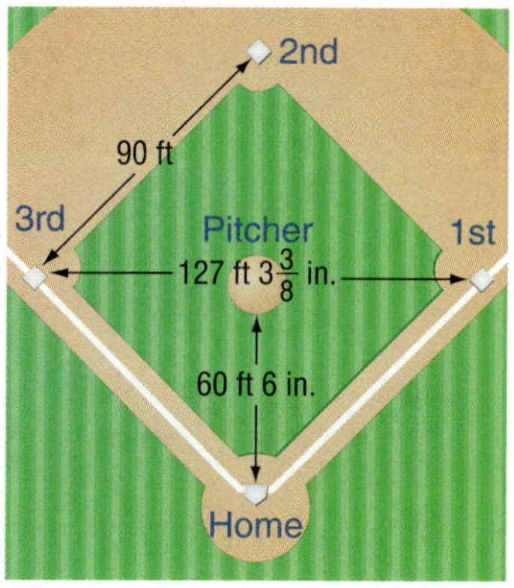

Write an inequality for the range of values for x. (Lesson 5-6)

58.

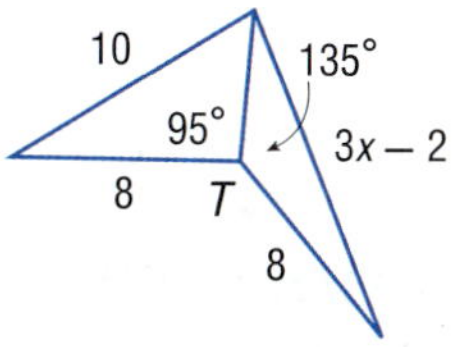

59.

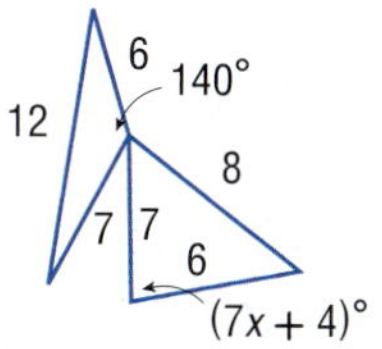

Use the Exterior Angle Inequality Theorem to list all of the angles that satisfy the stated condition. (Lesson 5-3)

60. measures less than $m\angle 5$

61. measures greater than $m\angle 6$

62. measures greater than $m\angle 10$

63. measures less than $m\angle 11$

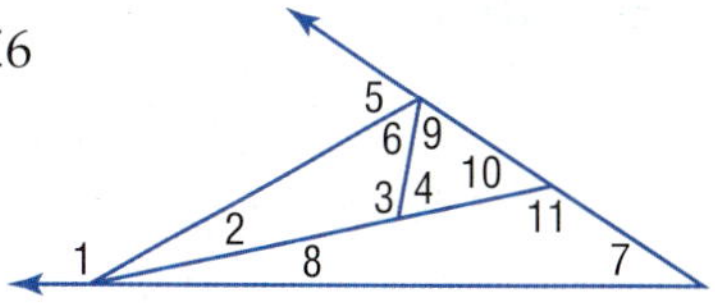

64. REASONING Find a counterexample for the following statement. (Lesson 3-5)
If lines p and m are cut by transversal t so that consecutive interior angles are congruent, then lines p and m are parallel and t is perpendicular to both lines.

Skills Review

Write a paragraph proof. (Lesson 4-3)

65. Given: $\triangle ABC \cong \triangle DEF$; $\triangle DEF \cong \triangle GHI$
Prove: $\triangle ABC \cong \triangle GHI$

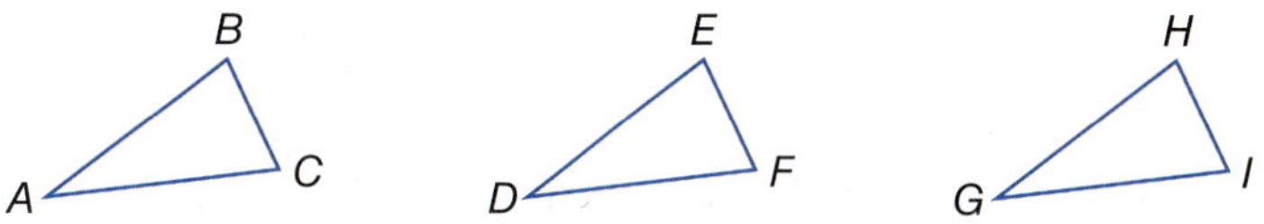

EXTEND 7-1

Graphing Technology Lab: Fibonacci Sequence and Ratios

Math Online glencoe.com
- Other Calculator Keystrokes
- Graphing Technology Personal Tutor

Leonardo Pisano (c. 1170–c. 1250), or Fibonacci, was born in Italy but educated in North Africa. As a result, his work is similar to that of other North African authors of that time. His book *Liber abaci*, published in 1202, introduced what is now called the Fibonacci sequence, in which each term after the first two terms is the sum of the two numbers before it.

Term	1	2	3	4	5	6	7
Fibonacci Number	1	1	2	3	5	8	13
			↑ 1 + 1	↑ 1 + 2	↑ 2 + 3	↑ 3 + 5	↑ 5 + 8

ACTIVITY

You can use CellSheet on a TI-83/84 Plus graphing calculator to create terms of the Fibonacci sequence. Then compare each term with its preceding term.

Step 1 Access the **CellSheet** application by pressing the APPS key. Choose the number for **CellSheet** and press ENTER.

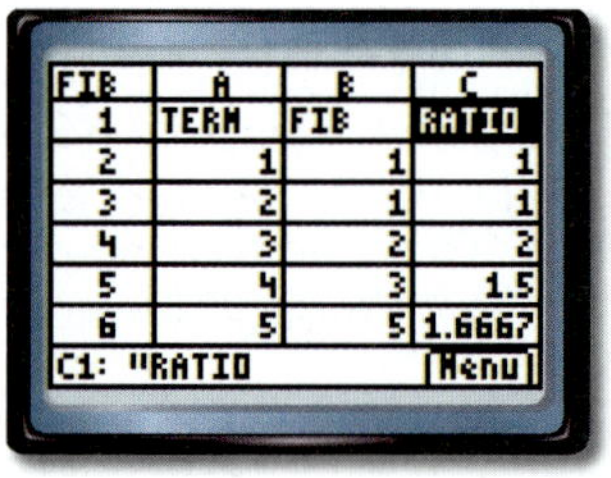

Step 2 Enter the column headings in row 1. Use the **ALPHA** key to enter letters and press ["] at the beginning of each label.

Step 3 Enter 1 into cell **A2**. Then insert the formula =A2+1 in cell **A3**. Press STO to insert the = in the formula. Then use F3 to copy this formula and use F4 to paste it in each cell in the column. This will automatically calculate the number of the term.

Step 4 In column B, we will record the Fibonacci numbers. Enter 1 in cells **B2** and **B3** since you do not have two previous terms to add. Then insert the formula =B2+B3 in cell **B4**. Copy this formula down the column.

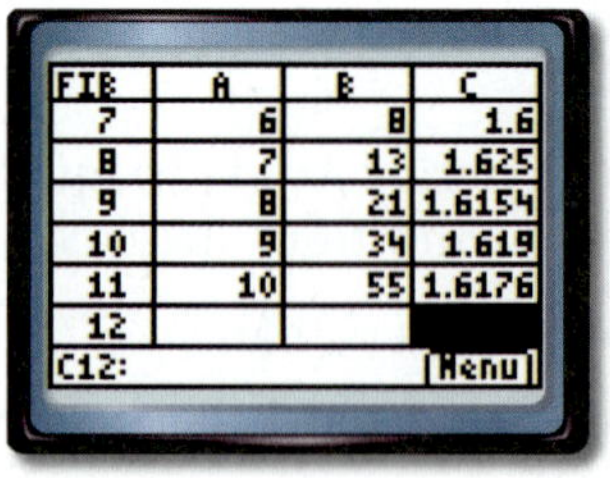

Step 5 In column C, we will find the ratio of each term to its preceding term. Enter 1 in cell **C2** since there is no preceding term. Then enter B3/B2 in cell **C3**. Copy this formula down the column. *The screens show the results for terms 1 through 10.*

Analyze the Results

1. What happens to the Fibonacci number as the number of the term increases?
2. What pattern of odd and even numbers do you notice in the Fibonacci sequence?
3. As the number of terms gets greater, what pattern do you notice in the ratio column?
4. Extend the spreadsheet to calculate fifty terms of the Fibonacci sequence. Describe any differences in the patterns you described in Exercises 1–3.
5. **MAKE A CONJECTURE** How might the Fibonacci sequence relate to the golden ratio?

7-2 Similar Polygons

Why?

People often customize their computer desktops using photos, centering the images at their original size or stretching them to fit the screen. This second method distorts the image, because the original and new images are not geometrically similar.

Then
You used proportions to solve problems. (Lesson 7-1)

Now
- Use proportions to identify similar polygons.
- Solve problems using the properties of similar polygons.

New Vocabulary
similar polygons
similarity ratio
scale factor

Math Online
glencoe.com

Identify Similar Polygons **Similar polygons** have the same shape but not necessarily the same size.

Key Concept — Similar Polygons

Two polygons are similar if and only if their corresponding angles are congruent and corresponding side lengths are proportional.

Example In the diagram below, *ABCD* is similar to *WXYZ*.

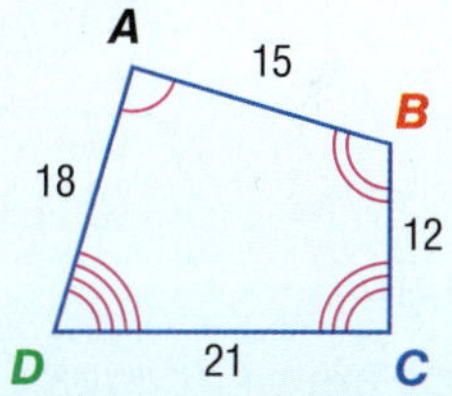

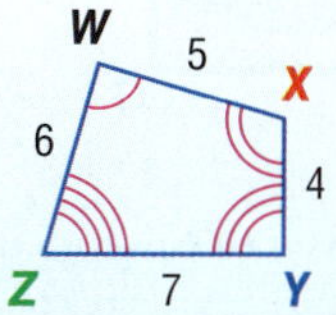

Corresponding angles

$\angle A \cong \angle W$, $\angle B \cong \angle X$, $\angle C \cong \angle Y$, and $\angle D \cong \angle Z$

Corresponding sides

$$\frac{AB}{WX} = \frac{BC}{XY} = \frac{CD}{YZ} = \frac{DA}{ZW} = \frac{3}{1}$$

Symbols $ABCD \sim WXYZ$

As with congruence statements, the order of vertices in a similarity statement like $ABCD \sim WXYZ$ is important. It identifies the corresponding angles and sides.

EXAMPLE 1 Use a Similarity Statement

If $\triangle FGH \sim \triangle JKL$, list all pairs of congruent angles and write a proportion that relates the corresponding sides.

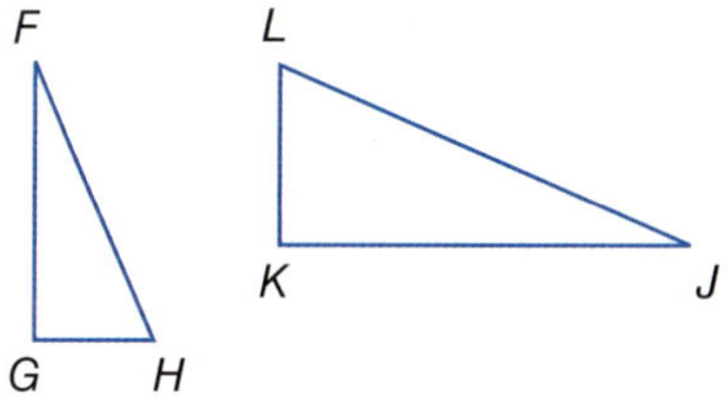

Use the similarity statement.

$\triangle FGH \sim \triangle JKL$

Congruent angles: $\angle F \cong \angle J$, $\angle G \cong \angle K$, $\angle H \cong \angle L$

Proportion: $\frac{FG}{JK} = \frac{GH}{KL} = \frac{HF}{LJ}$

Guided Practice

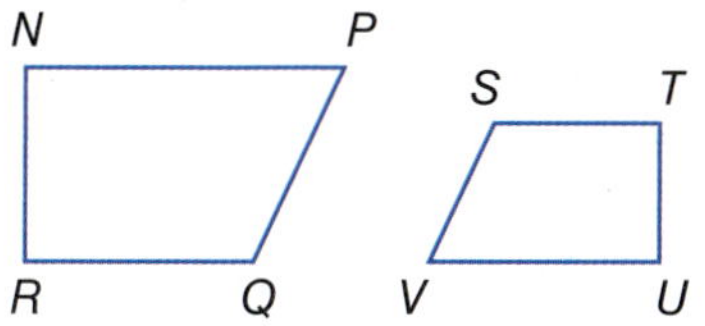

1. In the diagram, $NPQR \sim UVST$. List all pairs of congruent angles, and write a proportion that relates the corresponding sides.

Personal Tutor glencoe.com

StudyTip

Similarity Ratio The scale factor between two similar polygons is sometimes called the *similarity ratio*.

The ratio of the lengths of the corresponding sides of two similar polygons is called the **scale factor**. The scale factor depends on the order of comparison.

In the diagram, $\triangle ABC \sim \triangle XYZ$.

The scale factor of $\triangle ABC$ to $\triangle XYZ$ is $\frac{6}{3}$ or 2.

The scale factor of $\triangle XYZ$ to $\triangle ABC$ is $\frac{3}{6}$ or $\frac{1}{2}$.

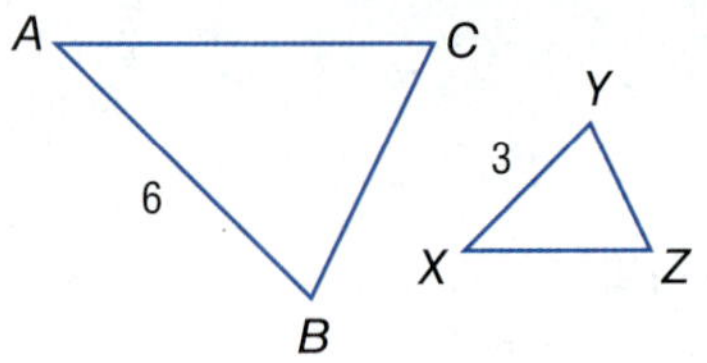

Real-World EXAMPLE 2 Identify Similar Polygons

PHOTO EDITING **Kuma wants to use the rectangular photo shown as the background for her computer's desktop, but she needs to resize it. Determine whether the following rectangular images are similar. If so, write the similarity statement and scale factor. Explain your reasoning.**

a.

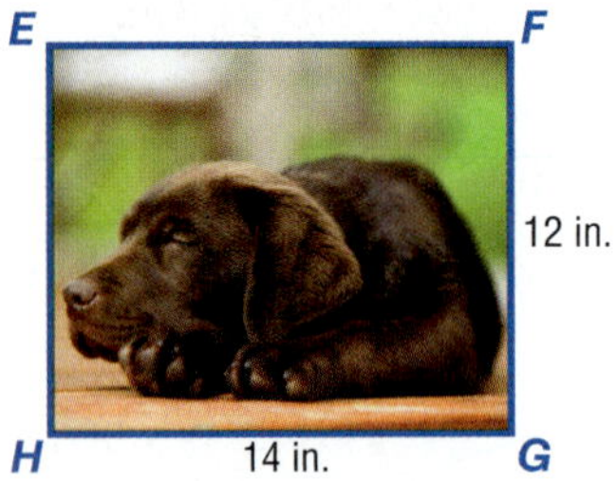

b.

a. Step 1 Compare corresponding angles.

Since all angles of a rectangle are right angles and right angles are congruent, corresponding angles are congruent.

Step 2 Compare corresponding sides.

$\frac{DC}{HG} = \frac{10}{14}$ or $\frac{5}{7}$ $\qquad \frac{BC}{FG} = \frac{8}{12}$ or $\frac{2}{3}$ $\qquad \frac{5}{7} \neq \frac{2}{3}$

Since corresponding sides are not proportional, $ABCD \nsim EFGH$. So the photos are not similar.

b. Step 1 Since *ABCD* and *JKLM* are both rectangles, corresponding angles are congruent.

Step 2 Compare corresponding sides.

$\frac{DC}{ML} = \frac{10}{15}$ or $\frac{2}{3}$ $\qquad \frac{BC}{KL} = \frac{8}{12}$ or $\frac{2}{3}$ $\qquad \frac{2}{3} = \frac{2}{3}$

Since corresponding sides are proportional, $ABCD \sim JKLM$. So the rectangles are similar with a scale factor of $\frac{2}{3}$.

ReadingMath

Similarity Symbol The symbol $\nsim$ is read as *is not similar to*.

Guided Practice

2. Determine whether the triangles shown are similar. If so, write the similarity statement and scale factor. Explain your reasoning.

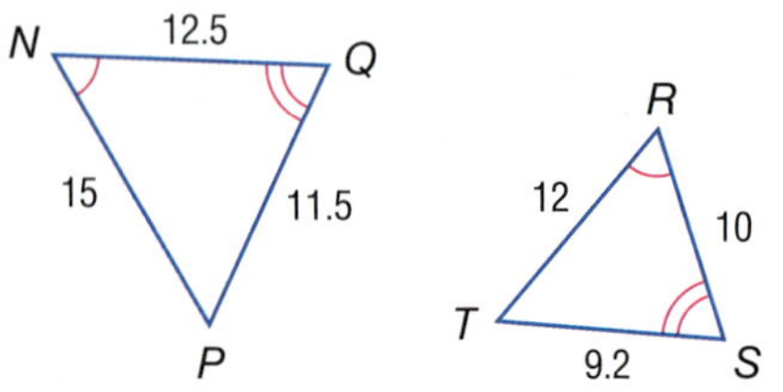

Personal Tutor glencoe.com

> **StudyTip**
>
> **Similarity and Congruence** If two polygons are congruent, they are also similar. All of the corresponding angles are congruent, and the lengths of the corresponding sides have a ratio of 1 : 1.

Use Similar Figures

You can use scale factors and proportions to solve problems involving similar figures.

EXAMPLE 3 Use Similar Figures to Find Missing Measures

In the diagram, $ACDF \sim VWYZ$.

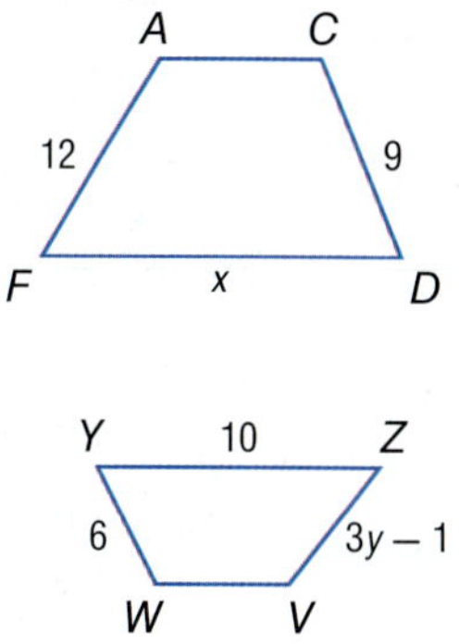

a. Find x.

Use the corresponding side lengths to write a proportion.

$\frac{CD}{WY} = \frac{DF}{YZ}$	**Similarity proportion**
$\frac{9}{6} = \frac{x}{10}$	**$CD = 9$, $WY = 6$, $DF = x$, $YZ = 10$**
$9(10) = 6(x)$	**Cross Products Property**
$90 = 6x$	**Multiply.**
$15 = x$	**Divide each side by 6.**

b. Find y.

$\frac{CD}{WY} = \frac{FA}{ZV}$	**Similarity proportion**
$\frac{9}{6} = \frac{12}{3y - 1}$	**$CD = 9$, $WY = 6$, $FA = 12$, $ZV = 3y - 1$**
$9(3y - 1) = 6(12)$	**Cross Products Property**
$27y - 9 = 72$	**Multiply.**
$27y = 81$	**Add 9 to each side.**
$y = 3$	**Divide each side by 27.**

Guided Practice

Find the value of each variable if $\triangle JLM \sim \triangle QST$.

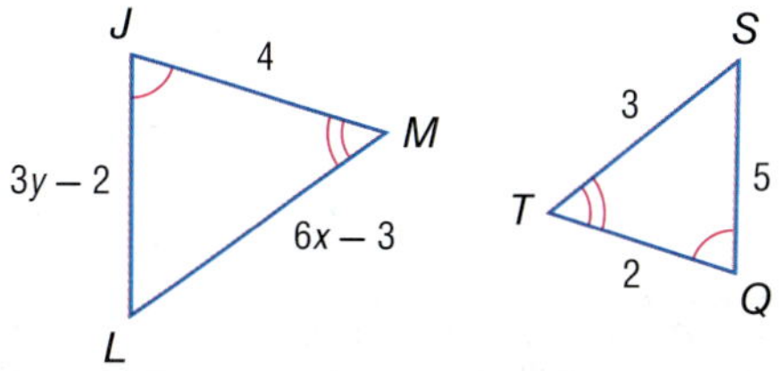

3A. x

3B. y

Personal Tutor glencoe.com

> **StudyTip**
>
> **Identifying Similar Triangles** When only two congruent angles of a triangle are given, remember that you can use the Third Angles Theorem to establish that the remaining corresponding angles are also congruent.

In similar polygons, the ratio of any two corresponding lengths is proportional to the scale factor between them. This leads to the following theorem about the perimeters of two similar polygons.

Theorem 7.1 Perimeters of Similar Polygons

If two polygons are similar, then their perimeters are proportional to the scale factor between them.

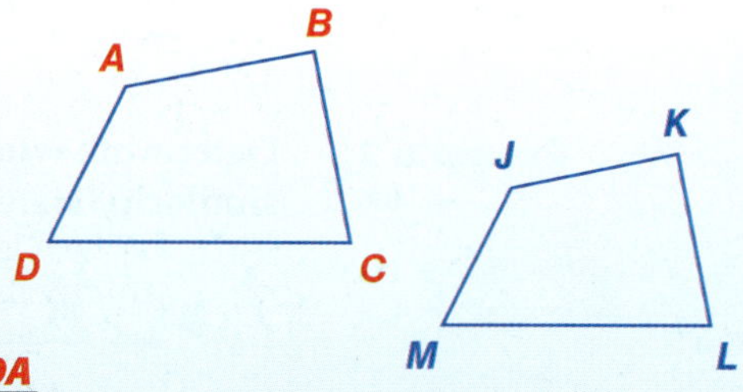

Example If $ABCD \sim JKLM$, then

$$\frac{AB + BC + CD + DA}{JK + KL + LM + MJ} = \frac{AB}{JK} = \frac{BC}{KL} = \frac{CD}{LM} = \frac{DA}{MJ}.$$

You will prove Theorem 7.1 for triangles in Exercise 46.

EXAMPLE 4 Use a Scale Factor to Find Perimeter

If $ABCDE \sim PQRST$, find the scale factor of $ABCDE$ to $PQRST$ and the perimeter of each polygon.

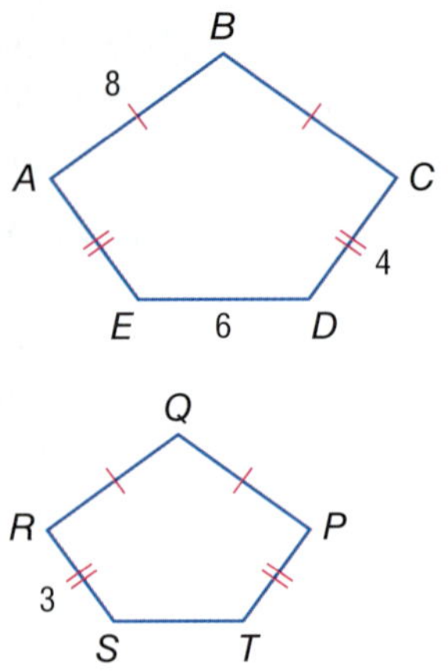

The scale factor of $ABCDE$ to $PQRST$ is $\frac{CD}{RS}$ or $\frac{4}{3}$.

Since $\overline{BC} \cong \overline{AB}$ and $\overline{AE} \cong \overline{CD}$, the perimeter of $ABCD$ is $8 + 8 + 4 + 6 + 4$ or 30.

Use the perimeter of $ABCDE$ and the scale factor to write a proportion. Let x represent the perimeter of $PQRST$.

$\frac{4}{3} = \frac{\text{perimeter of } ABCDE}{\text{perimeter of } PQRST}$ **Theorem 7.1**

$\frac{4}{3} = \frac{30}{x}$ **Substitution**

$(3)(30) = 4x$ **Cross Products Property**

$22.5 = x$ **Solve.**

So, the perimeter of $PQRST$ is 22.5.

Guided Practice

4. If $MNPQ \sim XYZW$, find the scale factor of $MNPQ$ to $XYZW$ and the perimeter of each polygon.

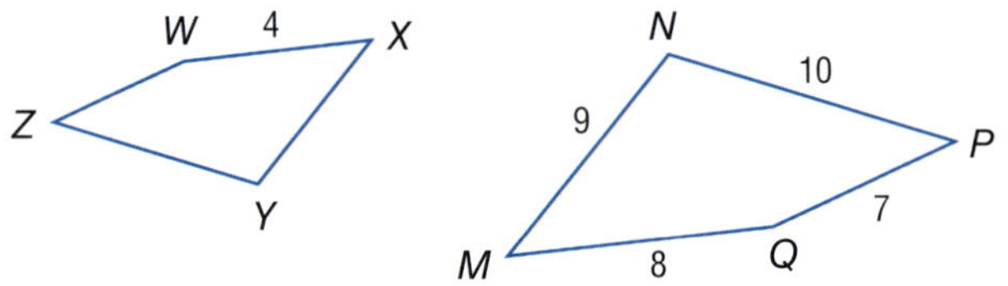

Personal Tutor glencoe.com

Check Your Understanding

Example 1 p. 465 **List all pairs of congruent angles, and write a proportion that relates the corresponding sides for each pair of similar polygons.**

1. $\triangle ABC \sim \triangle ZYX$

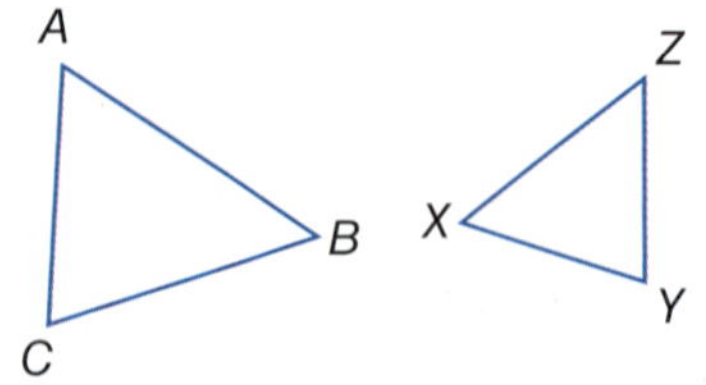

2. $JKLM \sim TSRQ$

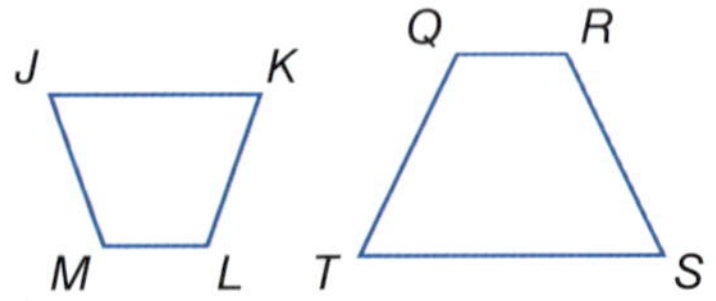

Example 2 p. 466 **Determine whether each pair of figures is similar. If so, write the similarity statement and scale factor. If not, explain your reasoning.**

3.

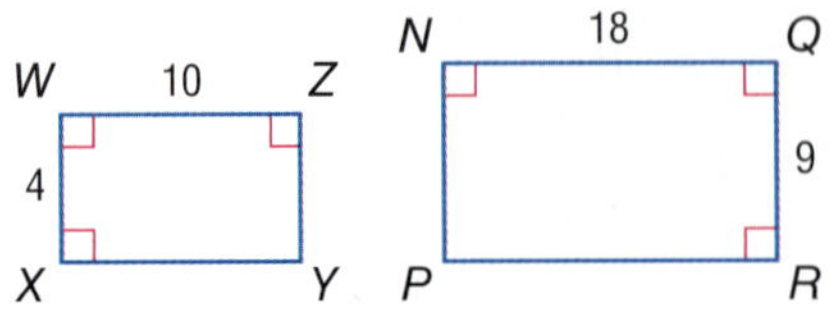

4.

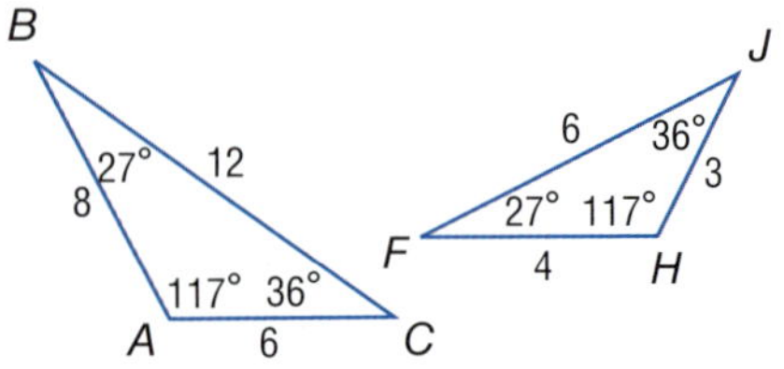

Example 3
p. 467

Each pair of polygons is similar. Find the value of x.

5.

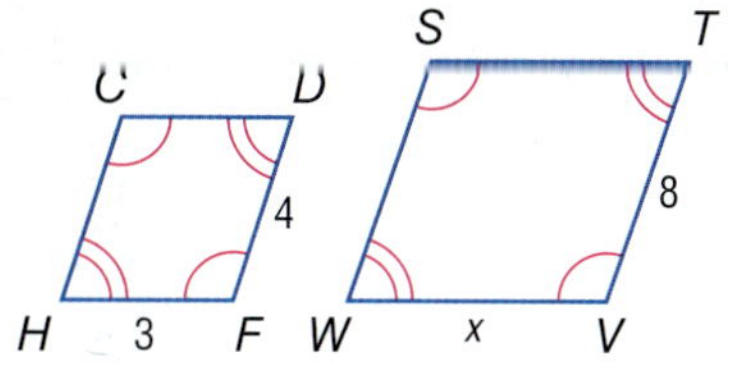

6.

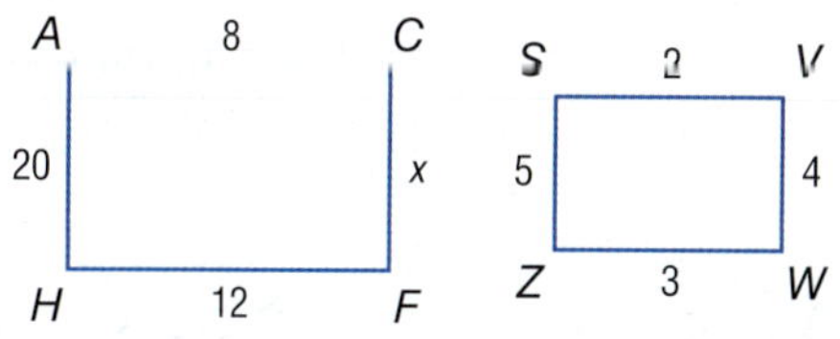

Example 4
p. 468

7. **DESIGN** On the blueprint of the apartment shown, the balcony measures 1 inch wide by 1.75 inches long. If the actual length of the balcony is 7 feet, what is the perimeter of the balcony?

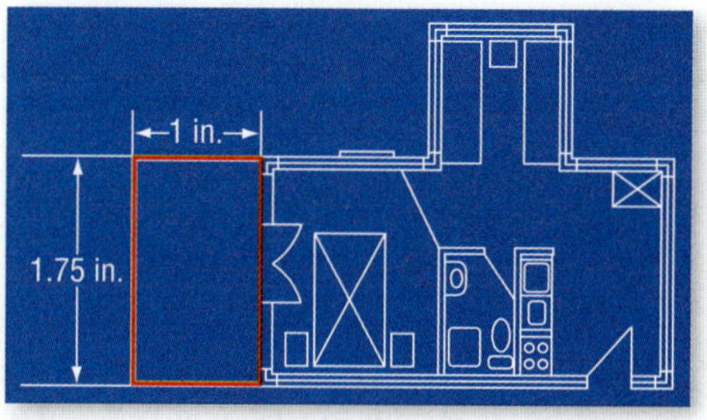

Practice and Problem Solving

= **Step-by-Step Solutions** begin on page R20.
Extra Practice begins on page 969.

Example 1
p. 465

List all pairs of congruent angles, and write a proportion that relates the corresponding sides for each pair of similar polygons.

8. $\triangle CHF \sim \triangle YWS$

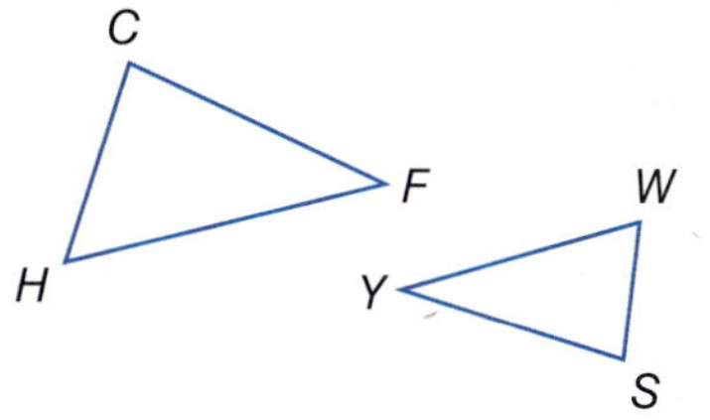

9. $JHFM \sim PQST$

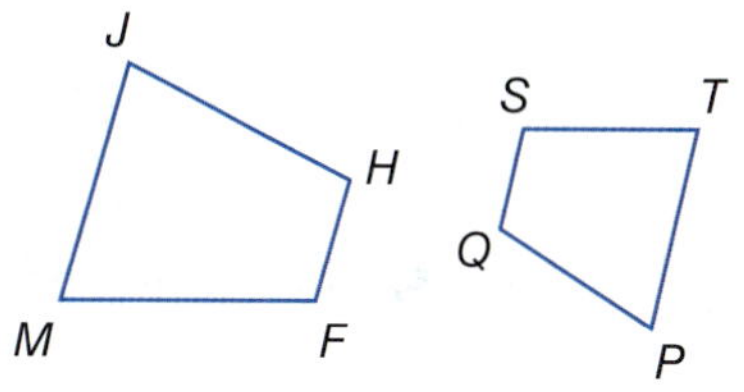

10. $ABDF \sim VXZT$

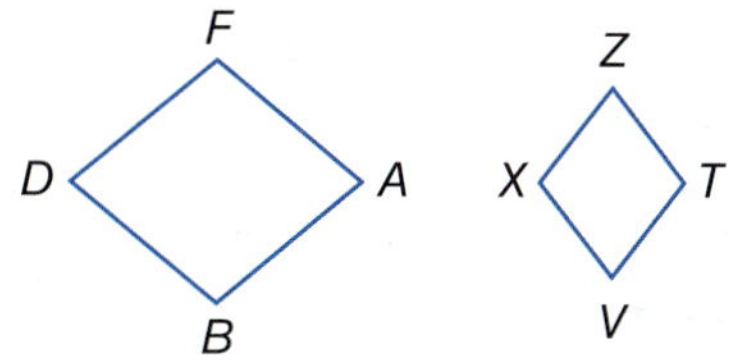

11. $\triangle DFG \sim \triangle KMJ$

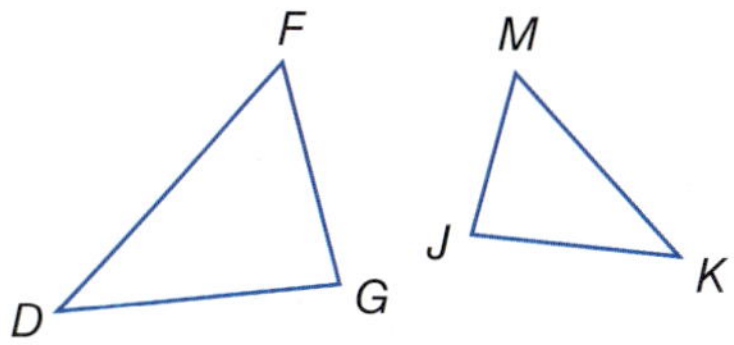

Example 2
p. 466

Determine whether each pair of figures is similar. If so, write the similarity statement and scale factor. If not, explain your reasoning.

12.

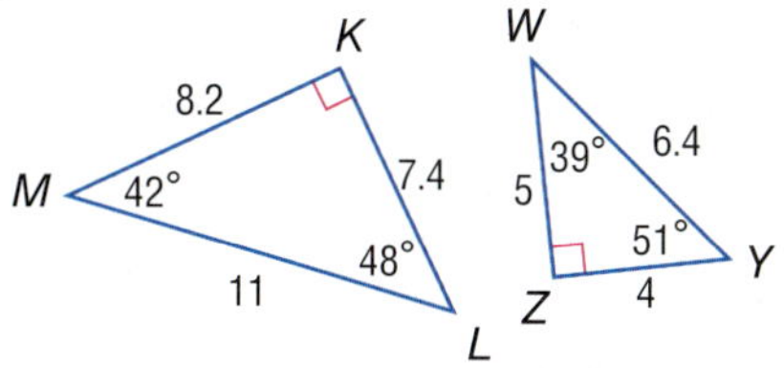

13.

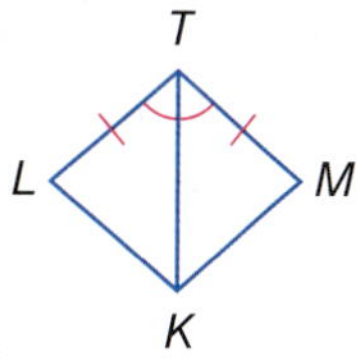

14.

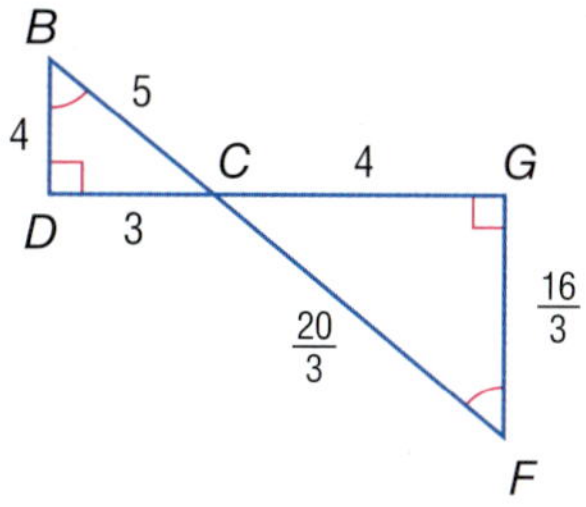

15.

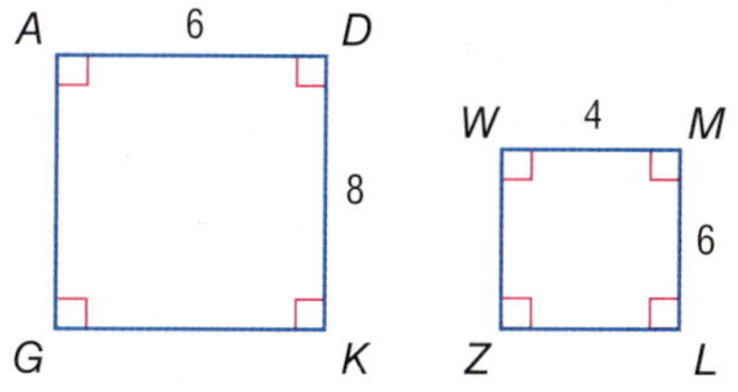

16. **GAMES** The dimensions of a hockey rink are 200 feet by 85 feet. Are the hockey rink and the air hockey table shown similar? Explain your reasoning.

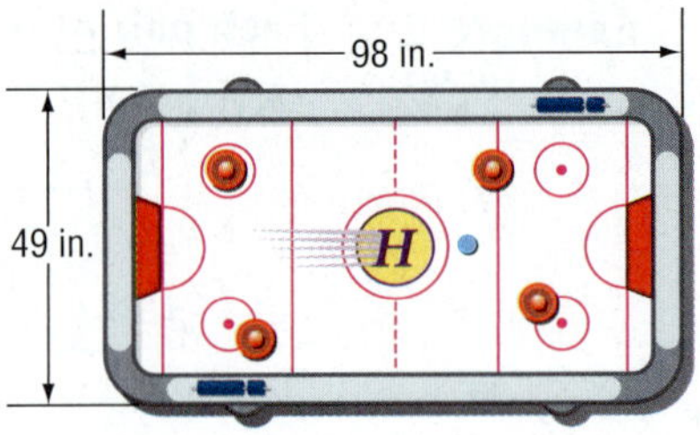

17. **COMPUTERS** The dimensions of a 17-inch flat panel computer screen are approximately $13\frac{1}{4}$ by $10\frac{3}{4}$ inches. The dimensions of a 19-inch flat panel computer screen are approximately $14\frac{1}{2}$ by 12 inches. Are the computer screens similar? Explain your reasoning.

Example 3
p. 467

Each pair of polygons is similar. Find the value of x.

18.

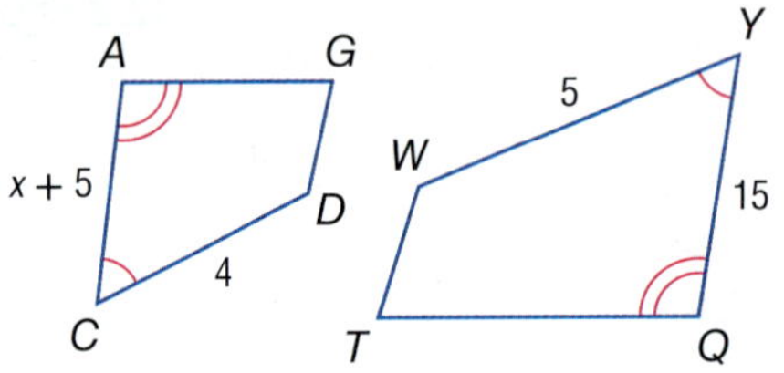

19.

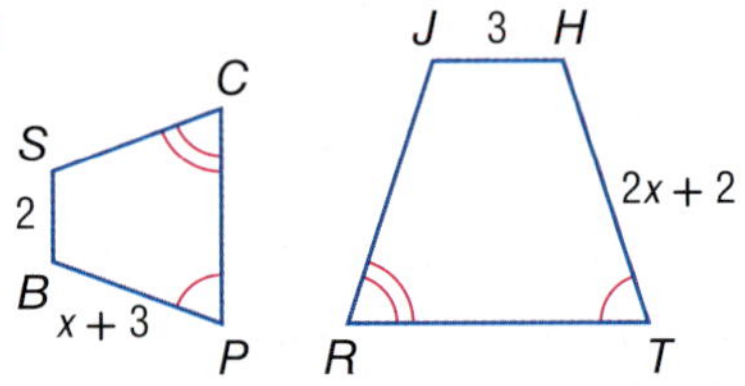

20.

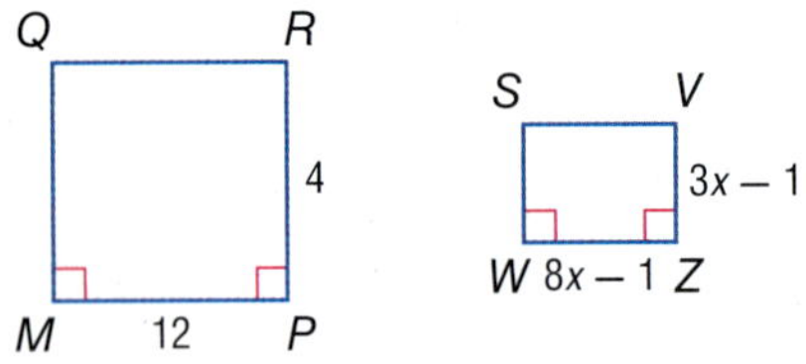

21.

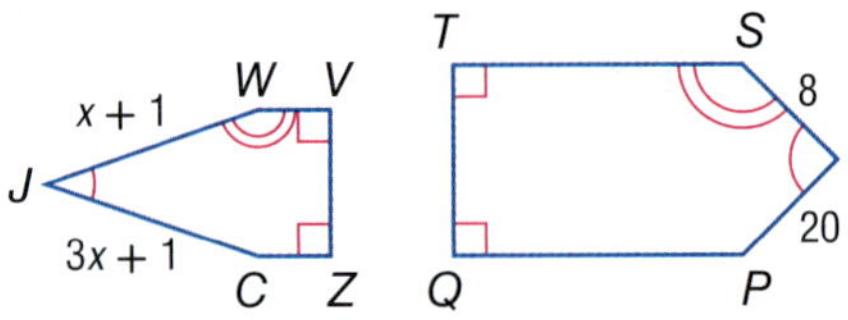

Example 4
p. 468

22. Rectangle $ABCD$ has a width of 8 yards and a length of 20 yards. Rectangle $QRST$, which is similar to rectangle $ABCD$, has a length of 40 yards. Find the scale factor of rectangle $ABCD$ to rectangle $QRST$ and the perimeter of each rectangle.

Find the perimeter of the given triangle.

23. $\triangle DEF$, if $\triangle ABC \sim \triangle DEF$, $AB = 5$, $BC = 6$, $AC = 7$, and and $DE = 3$

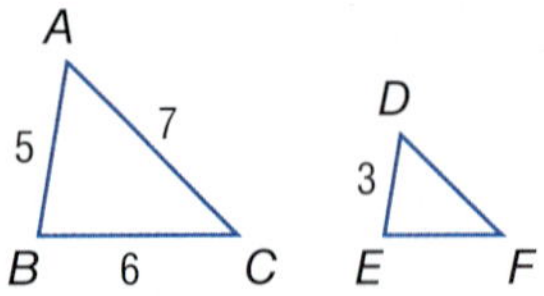

24. $\triangle WZX$, if $\triangle WZX \sim \triangle SRT$, $ST = 6$, $WX = 5$, and the perimeter of $\triangle SRT = 15$

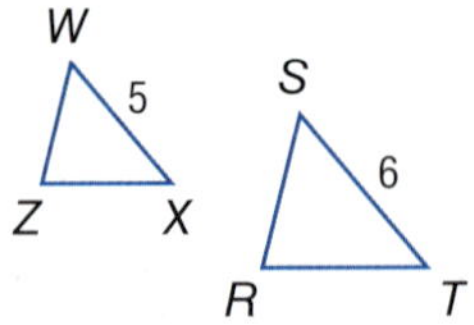

25. $\triangle CBH$, if $\triangle CBH \sim \triangle FEH$, $ADEG$ is a parallelogram, $CH = 7$, $FH = 10$, $FE = 11$, and $EH = 6$

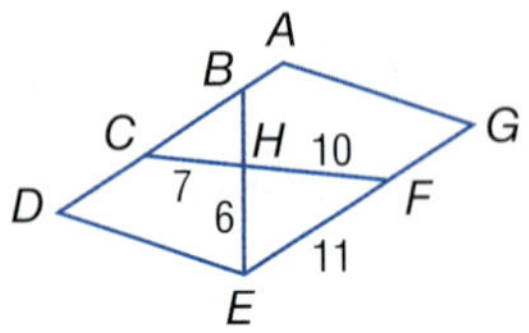

26. $\triangle DEF$, if $\triangle DEF \sim \triangle CBF$, perimeter of $\triangle CBF = 27$, $DF = 6$, $FC = 8$

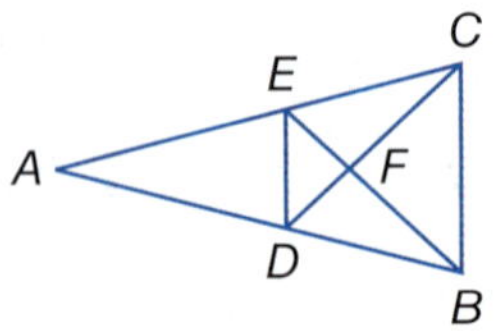

27. Two similar rectangles have a scale factor of 2:4. The perimeter of the large rectangle is 80 meters. Find the perimeter of the small rectangle.

28. Two similar squares have a scale factor of 3:2. The perimeter of the small rectangle is 50 feet. Find the perimeter of the large rectangle.

List all pairs of congruent angles, and write a proportion that relates the corresponding sides.

29.

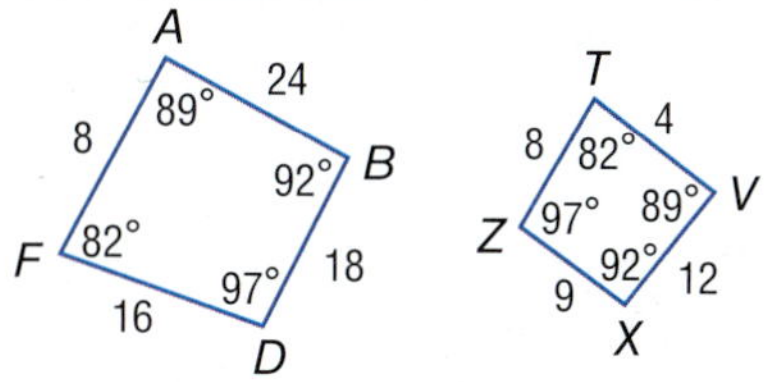

30.

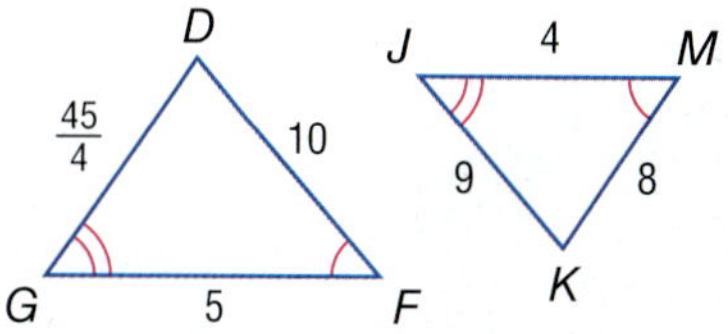

Real-World Link

In shuffleboard, players use sticks to push disks onto a numbered scoring region. The court is 52 feet long and 6 feet wide.

Source: *Columbia Encyclopedia*

SHUFFLEBOARD A shuffleboard court forms three similar triangles in which $\angle AHB \cong \angle AGC \cong \angle AFD$. Find the side(s) that correspond to the given side or angles that are congruent to the given angle.

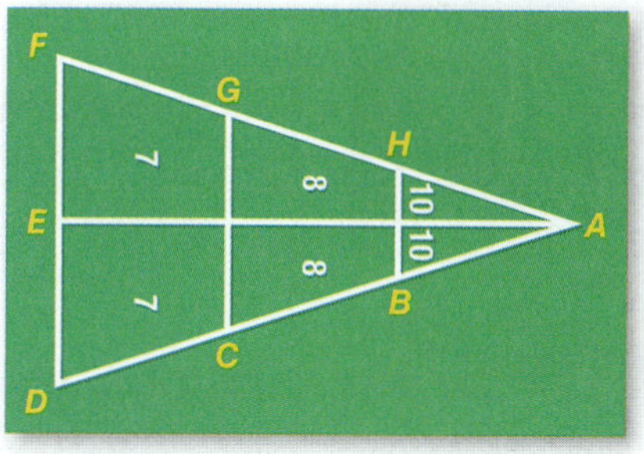

31. $\overline{AB}$

32. $\overline{FD}$

33. $\angle ACG$

34. $\angle A$

Find the value of each variable.

35 $ABCD \sim QSRP$

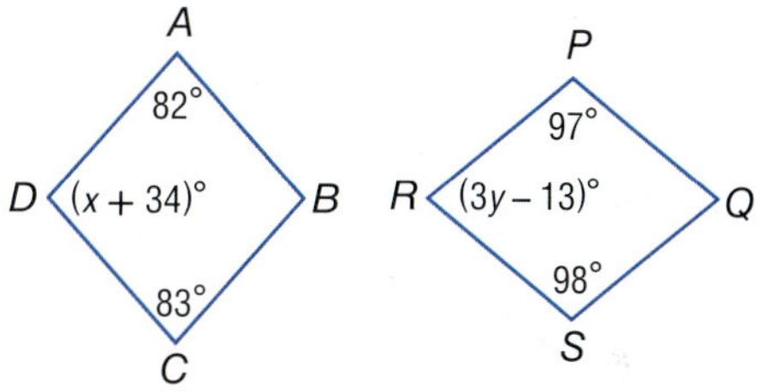

36. $\triangle JKL \sim \triangle WYZ$

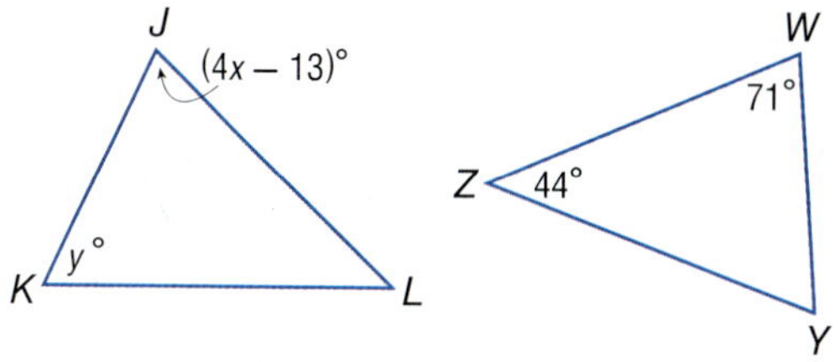

37. SLIDE SHOW You are using a digital projector for a slide show. The photos are 13 inches by $9\frac{1}{4}$ inches on the computer screen, and the scale factor of the computer image to the projected image is 1:4. What are the dimensions of the projected image?

COORDINATE GEOMETRY For the given vertices, determine whether rectangle *ABCD* is similar to rectangle *WXYZ*. Justify your answer.

38. $A(-1, 5)$, $B(7, 5)$, $C(7, -1)$, $D(-1, -1)$;
$W(-2, 10)$, $X(14, 10)$, $Y(14, -2)$, $Z(-2, -2)$

39. $A(5, 5)$, $B(0, 0)$, $C(5, -5)$, $D(10, 0)$;
$W(1, 6)$, $X(-3, 2)$, $Y(2, -3)$, $Z(6, 1)$

Determine whether the polygons are *always*, *sometimes*, or *never* similar. Explain your reasoning.

40. two obtuse triangles

41. a trapezoid and a parallelogram

42. two right triangles

43. two isosceles triangles

44. a scalene triangle and an isosceles triangle

45. two equilateral triangles

46. PROOF Write a paragraph proof of Theorem 7.1.

Given: $\triangle ABC \sim \triangle DEF$ and $\frac{AB}{DE} = \frac{m}{n}$

Prove: $\frac{\text{perimeter of } \triangle ABC}{\text{perimeter of } \triangle DEF} = \frac{m}{n}$

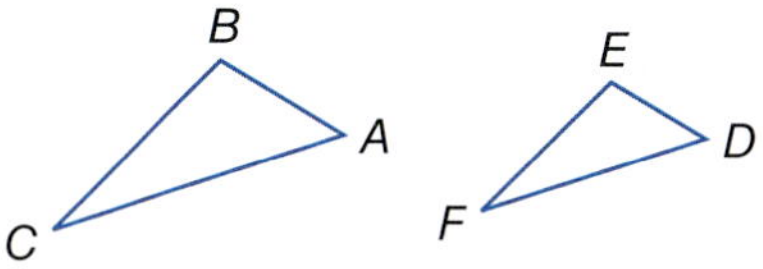

47 **PHOTOS** You are enlarging the photo shown at the right for your school yearbook. If the dimensions of the original photo are $2\frac{1}{3}$ inches by $1\frac{2}{3}$ inches and the scale factor of the old photo to the new photo is 2:3, what are the dimensions of the new photo?

Real-World Link

Most American middle schools, high schools, and colleges have traditional paper yearbooks. However, yearbooks can also exist in digital form on CD-ROM or DVD.

Source: Columbia Scholastic Press Association

48. **CHANGING DIMENSIONS** Rectangle *QRST* is similar to rectangle *JKLM* with sides in a ratio of 4:1.

a. What is the ratio of the areas of the two rectangles?

b. Suppose the dimension of each rectangle is tripled. What is the new ratio of the sides of the rectangles?

c. What is the ratio of the areas of these larger rectangles?

49. **CHANGING DIMENSIONS** In the figure shown, $\triangle FGH \sim \triangle XYZ$.

a. Show that the perimeters of $\triangle FGH$ and $\triangle XYZ$ have the same ratio as their corresponding sides.

b. If 6 units are added to the lengths of each side, are the new triangles similar? Explain.

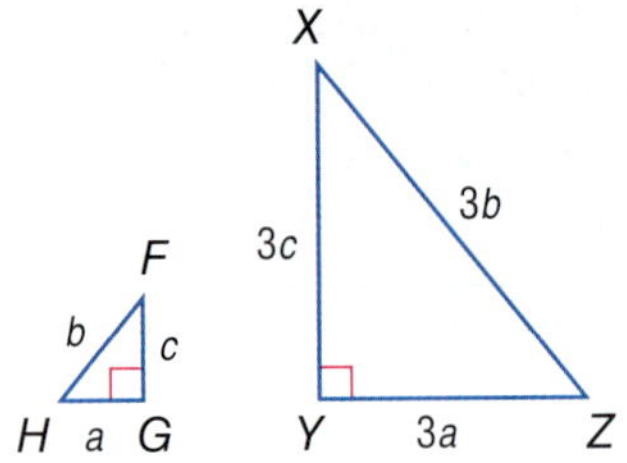

50. **MULTIPLE REPRESENTATIONS** In this problem, you will investigate similarity in squares.

a. **GEOMETRIC** Draw three different-sized squares. Label them *ABCD*, *PQRS*, and *WXYZ*. Measure and label each square with its side length.

b. **TABULAR** Calculate and record in a table the ratios of corresponding sides for each pair of squares: *ABCD* and *PQRS*, *PQRS* and *WXYZ*, and *WXYZ* and *ABCD*. Is each pair of squares similar?

c. **VERBAL** Make a conjecture about the similarity of all squares.

H.O.T. Problems

Use **H**igher-**O**rder **T**hinking Skills

51. **CHALLENGE** For what value(s) of x is $BEFA \sim EDCB$?

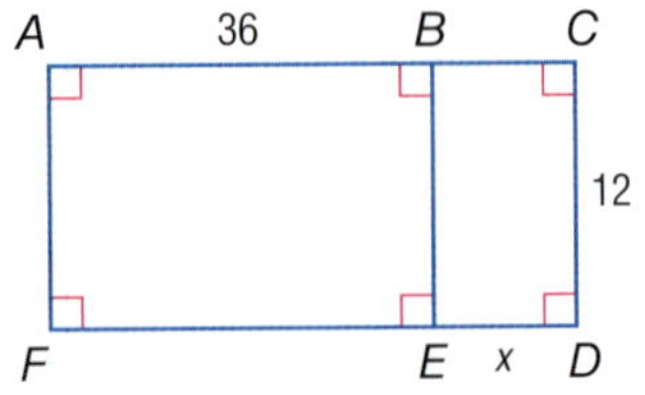

52. **REASONING** Recall that an *equivalence relation* is any relationship that satisfies the Reflexive, Symmetric, and Transitive Properties. Is similarity an equivalence relation? Explain.

53. **OPEN ENDED** Find a counterexample for the following statement.

All rectangles are similar.

54. **REASONING** Draw two regular pentagons of different sizes. Are the pentagons similar? Will any two regular polygons with the same number of sides be similar? Explain.

55. **WRITING IN MATH** Compare and contrast congruent, similar, and equal figures.

Standardized Test Practice

56. ALGEBRA If the arithmetic mean of $4x$, $3x$, and 12 is 18, then what is the value of x?

A 6
B 5
C 4
D 3

57. Two similar rectangles have a scale factor of 3:5. The perimeter of the large rectangle is 65 meters. What is the perimeter of the small rectangle?

F 29 m
G 39 m
H 49 m
J 59 m

58. SHORT RESPONSE If a jar contains 25 dimes and 7 quarters, what is the probability that a coin selected from the jar at random will be a dime?

59. SAT/ACT If the side of a square is $x + 3$, then what is the diagonal of the square?

A $x^2 + 3$
B $3x + 3$
C $x\sqrt{2} + 3\sqrt{2}$
D $x\sqrt{3} + 3\sqrt{3}$

Spiral Review

60. COMPUTERS In a survey of 5000 households, 4200 had at least one computer. What is the ratio of computers to households? (Lesson 7-1)

61. PROOF **Write a flow proof.** (Lesson 6-6)

Given: E and C are midpoints of $\overline{AD}$ and $\overline{DB}$, $\overline{AD} \cong \overline{DB}$, $\angle A \cong \angle 1$.

Prove: $ABCE$ is an isosceles trapezoid.

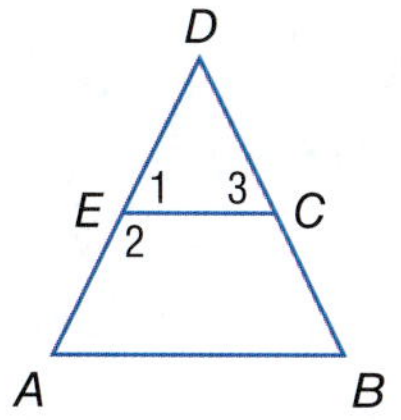

62. COORDINATE GEOMETRY Determine the coordinates of the intersection of the diagonals of $\square JKLM$ with vertices $J(2, 5)$, $K(6, 6)$, $L(4, 0)$, and $M(0, -1)$. (Lesson 6-2)

State the assumption you would make to start an indirect proof of each statement. (Lesson 5-4)

63. If $3x > 12$, then $x > 4$.

64. $\overline{PQ} \cong \overline{ST}$

65. The angle bisector of the vertex angle of an isosceles triangle is also an altitude of the triangle.

66. If a rational number is any number that can be expressed as $\frac{a}{b}$, where a and b are integers and $b \neq 0$, then 6 is a rational number.

Find the measures of each numbered angle. (Lesson 4-2)

67. $m\angle 1$

68. $m\angle 2$

69. $m\angle 3$

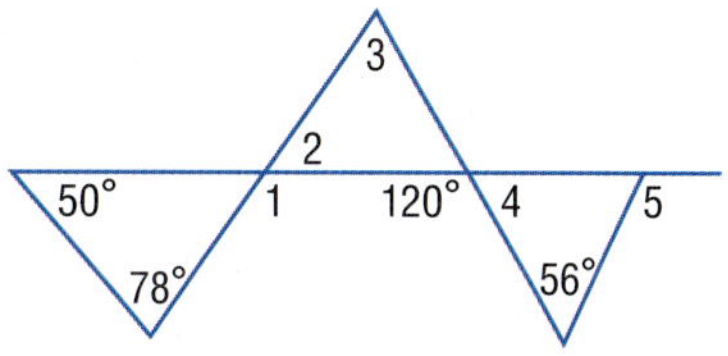

Skills Review

ALGEBRA **Find x and the unknown side measures of each triangle.** (Lesson 4-1)

70.

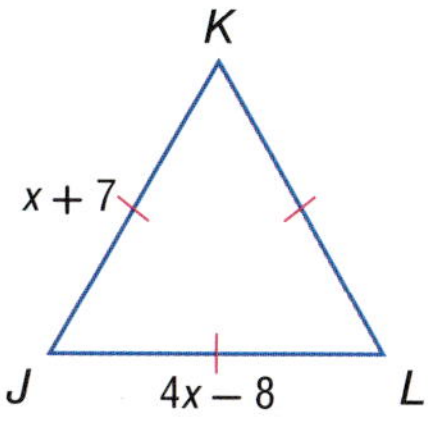

71.

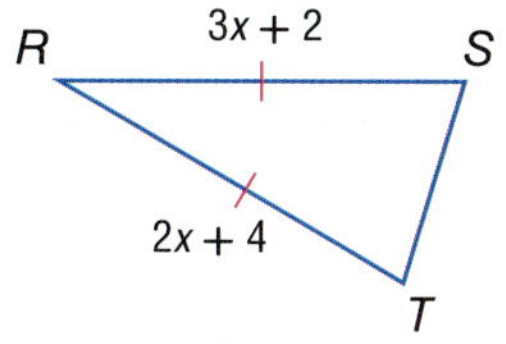

72.

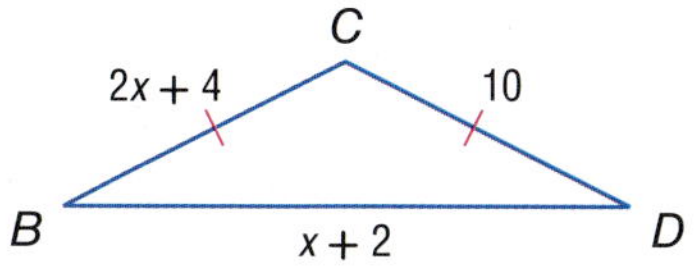

7-3 Similar Triangles

Then

You used the AAS, SSS, and SAS Congruence Theorems to prove triangles congruent. (Lesson 4-4)

Now

- Identify similar triangles using the AA Similarity Postulate and the SSS and SAS Similarity Theorems.
- Use similar triangles to solve problems.

Why?

Julian wants to draw a similar version of his skate club's logo on a poster. He first draws a line at the bottom of the poster. Next, he uses a cutout of the original triangle to copy the two bottom angles. Finally, he extends the noncommon sides of the two angles.

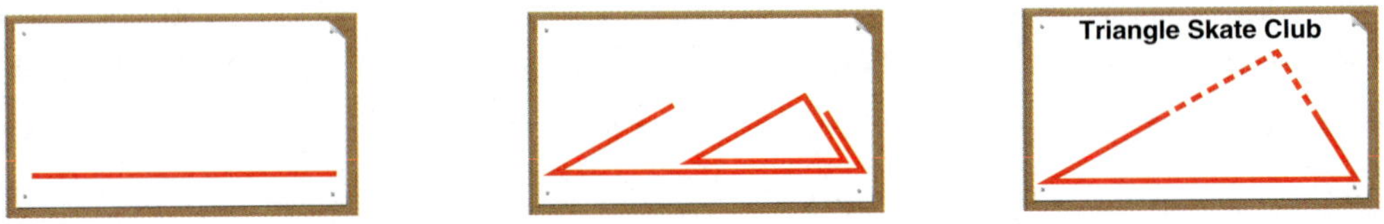

Identify Similar Triangles In Chapter 4, you explored tests for congruent triangles. There are also tests for similar triangles. The example suggests that two triangles are similar if two pairs of corresponding angles are congruent.

Postulate 7.1 Angle-Angle (AA) Similarity

If two angles of one triangle are congruent to two angles of another triangle, then the triangles are similar.

Example If $\angle A \cong \angle F$ and $\angle B \cong \angle G$, then $\triangle ABC \sim \triangle FGH$.

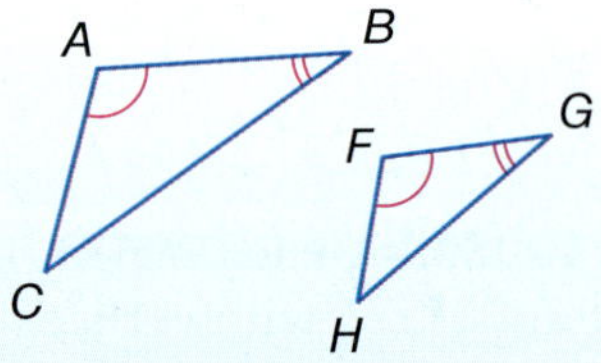

EXAMPLE 1 Use the AA Similarity Postulate

Determine whether the triangles are similar. If so, write a similarity statement. Explain your reasoning.

a.

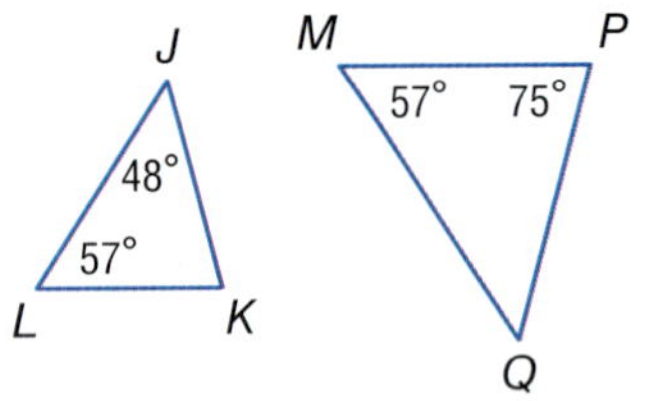

b.

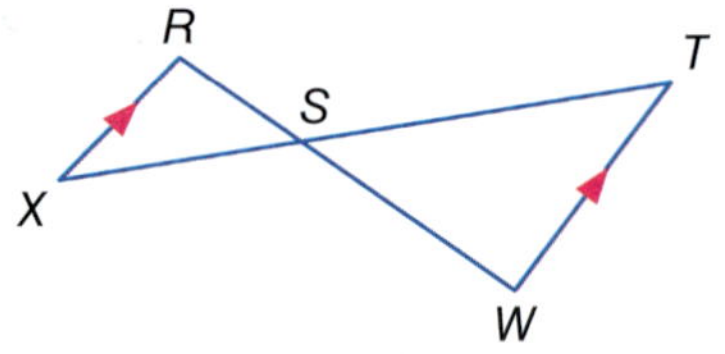

a. Since $m\angle L = m\angle M$, $\angle L \cong \angle M$. By the Triangle Sum Theorem, $57 + 48 + m\angle K = 180$, so $m\angle K = 75$. Since $m\angle P = 75$, $\angle K \cong \angle P$. So, $\triangle LJK \sim \triangle MQP$ by AA Similarity.

b. $\angle RSX \cong \angle WST$ by the Vertical Angles Theorem. Since $\overline{RX} \parallel \overline{TW}$, $\angle R \cong \angle W$. So, $\triangle RSX \sim \triangle WST$ by AA Similarity.

Guided Practice

1A.

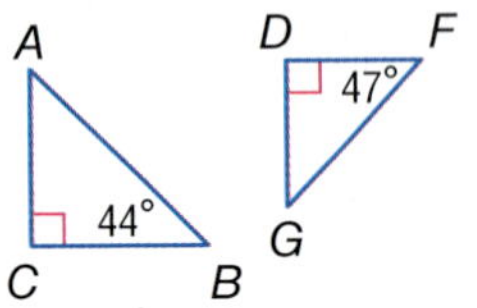

1B.

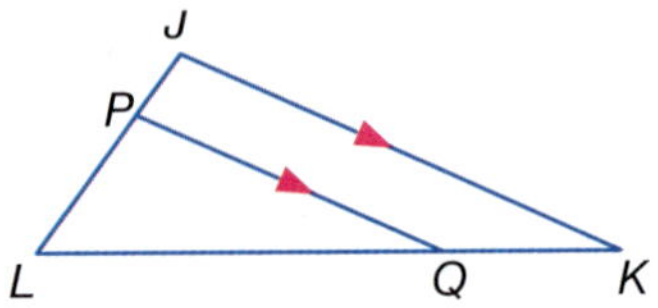

StudyTip

Draw Diagrams It is helpful to redraw similar triangles so that the corresponding side lengths have the same orientation.

You can use the AA Similarity Postulate to prove the following two theorems.

Theorems

7.2 Side-Side-Side (SSS) Similarity

If the corresponding side lengths of two triangles are proportional, then the triangles are similar.

Example If $\frac{JK}{MP} = \frac{KL}{PQ} = \frac{LJ}{QM}$, then $\triangle JKL \sim \triangle MPQ$.

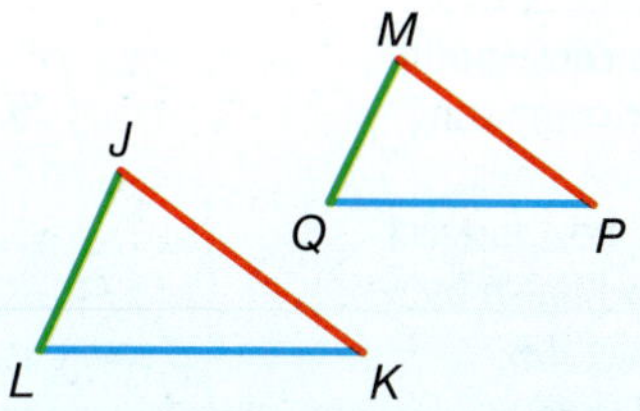

7.3 Side-Angle-Side (SAS) Similarity

If the lengths of two sides of one triangle are proportional to the lengths of two corresponding sides of another triangle and the included angles are congruent, then the triangles are similar.

Example If $\frac{RS}{XY} = \frac{ST}{YZ}$ and $\angle S \cong \angle Y$, then $\triangle RST \sim \triangle XYZ$.

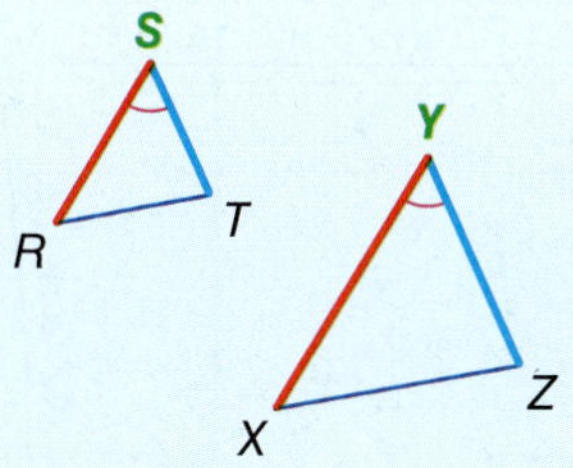

You will prove Theorem 7.3 in Exercise 25.

PROOF Theorem 7.2

Given: $\frac{AB}{FG} = \frac{BC}{GH} = \frac{AC}{FH}$

Prove: $\triangle ABC \sim \triangle FGH$

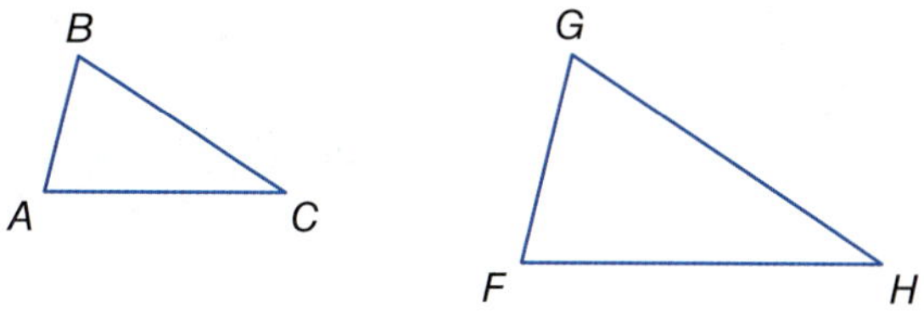

Paragraph Proof:

Locate J on $\overline{FG}$ so that $JG = AB$. Draw $\overline{JK}$ so that $\overline{JK} \parallel \overline{FH}$. Label $\angle GJK$ as $\angle 1$.

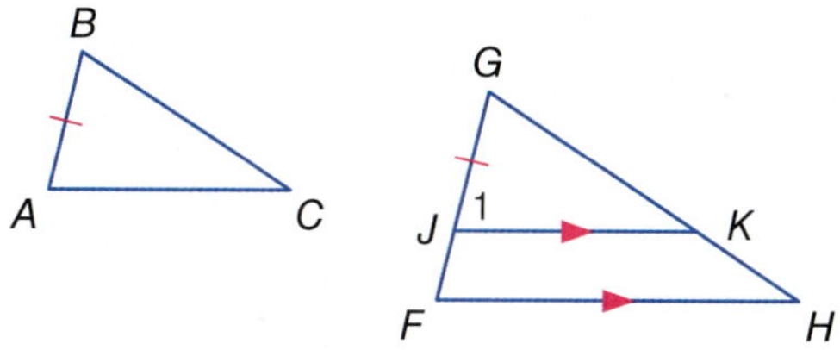

Since $\angle G \cong \angle G$ by the Reflexive Property and $\angle 1 \cong \angle F$ by the Corresponding Angles Postulate, $\triangle GJK \sim \triangle GFH$ by the AA Similarity Postulate.

By the definition of similar polygons, $\frac{JG}{FG} = \frac{GK}{GH} = \frac{JK}{FH}$. By substitution, $\frac{AB}{FG} = \frac{GK}{GH} = \frac{JK}{FH}$.

Since we are also given that $\frac{AB}{FG} = \frac{BC}{GH} = \frac{AC}{FH}$, we can say that $\frac{GK}{GH} = \frac{BC}{GH}$ and $\frac{JK}{FH} = \frac{AC}{FH}$. This means that $GK = BC$ and $JK = AC$, so $\overline{GK} \cong \overline{BC}$ and $\overline{JK} \cong \overline{AC}$. By SSS, $\triangle ABC \cong \triangle JGK$.

By CPCTC, $\angle B \cong \angle G$ and $\angle A \cong \angle 1$. Since $\angle 1 \cong \angle F$, $\angle A \cong \angle F$ by the Transitive Property. By AA Similarity, $\triangle ABC \sim \triangle FGH$.

Corresponding Sides To determine which sides of two triangles correspond, begin by comparing the longest sides, then the next longest sides, and finish by comparing the shortest sides.

EXAMPLE 2 Use the SSS and SAS Similarity Theorems

Determine whether the triangles are similar. If so, write a similarity statement. Explain your reasoning.

a.

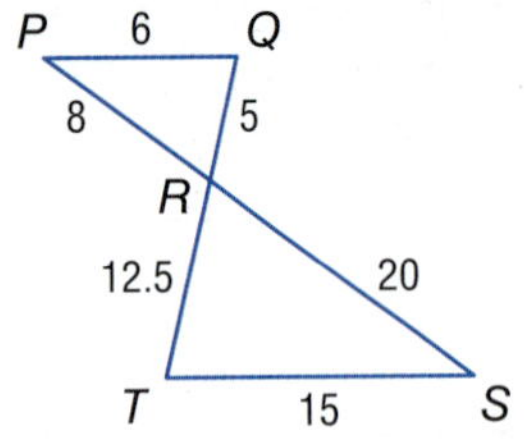

$\frac{PR}{SR} = \frac{8}{20}$ or $\frac{2}{5}$, $\frac{PQ}{ST} = \frac{6}{15}$ or $\frac{2}{5}$, and $\frac{QR}{TR} = \frac{5}{12.5} = \frac{50}{125}$ or $\frac{2}{5}$. So, $\triangle PQR \sim \triangle STR$ by the SSS Similarity Theorem.

b.

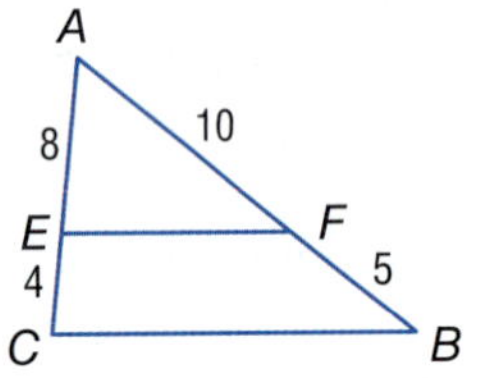

By the Reflexive Property, $\angle A \cong \angle A$.

$\frac{AF}{AB} = \frac{10}{10+5} = \frac{10}{15}$ or $\frac{2}{3}$ and $\frac{AE}{AC} = \frac{8}{8+4} = \frac{8}{12}$ or $\frac{2}{3}$.

Since the lengths of the sides that include $\angle A$ are proportional, $\triangle AEF \sim \triangle ACB$ by the SAS Similarity Theorem.

Guided Practice

2A.

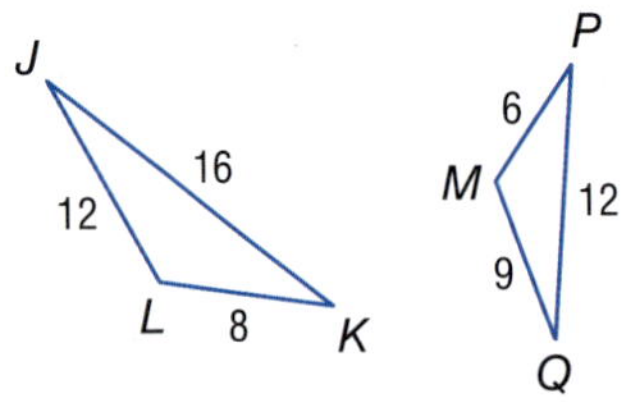

2B.

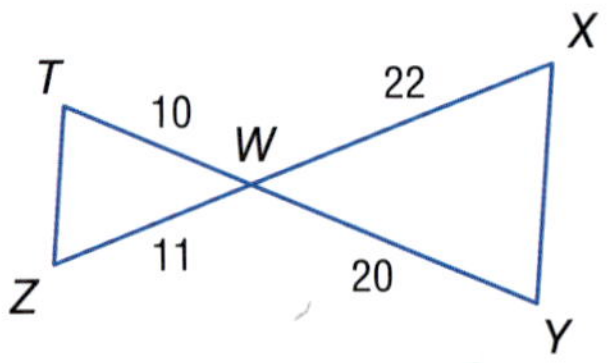

Personal Tutor glencoe.com

You can decide what is sufficient to prove that two triangles are similar.

STANDARDIZED TEST EXAMPLE 3

In the figure, $\angle ADB$ is a right angle. Which of the following would *not* be sufficient to prove that $\triangle ADB \sim \triangle CDB$?

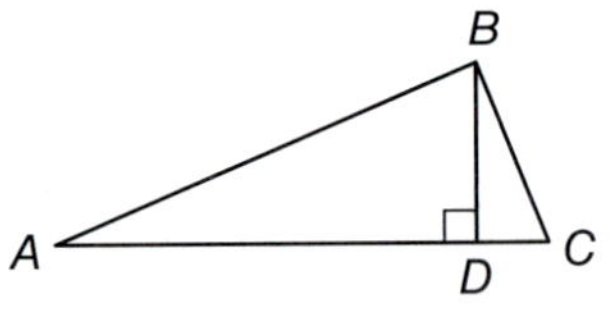

A $\frac{AD}{BD} = \frac{BD}{CD}$

B $\frac{AB}{BC} = \frac{BD}{CD}$

C $\angle ABD \cong \angle C$

D $\frac{AD}{BD} = \frac{BD}{CD} = \frac{AB}{BC}$

Read the Test Item

You are given that $\angle ADB$ is a right angle and asked to identify which additional information would not be enough to prove that $\triangle ADB \sim \triangle CDB$.

Solve the Test Item

Since $\angle ADB$ is a right angle, $\angle CDB$ is also a right angle. Since all right angles are congruent, $\angle ADB \cong \angle CDB$. Check each answer choice until you find one that does not supply a sufficient additional condition to prove that $\triangle ADB \sim \triangle CDB$.

Choice A If $\frac{AD}{BD} = \frac{BD}{CD}$ and $\angle ADB \cong \angle CDB$, then $\triangle ADB \sim \triangle CDB$ by SAS Similarity.

Choice B If $\frac{AB}{BC} = \frac{BD}{CD}$ and $\angle ADB \cong \angle CDB$, then we cannot conclude that $\triangle ADB \sim \triangle CDB$ because the included angle of side $\overline{AB}$ and $\overline{BD}$ is not $\angle ADB$. So the answer is B.

Test-TakingTip

Identifying Nonexamples Sometimes test questions require you to find a nonexample, as in this case. You must check each option until you find a valid nonexample. If you would like to check your answer, confirm that each additional option is correct.

Guided Practice

3. If $\triangle JKL$ and $\triangle FGH$ are two triangles such that $\angle J \cong \angle F$, which of the following would be sufficient to prove that the triangles are similar?

F $\frac{KL}{GH} = \frac{JL}{FH}$ **G** $\frac{JL}{JK} = \frac{FH}{FG}$ **H** $\frac{JK}{FG} = \frac{KL}{GH}$ **J** $\frac{JL}{JK} = \frac{GH}{FG}$

Personal Tutor glencoe.com

Use Similar Triangles Like the congruence of triangles, similarity of triangles is reflexive, symmetric, and transitive.

Theorem 7.4 Properties of Similarity

Reflexive Property of Similarity	$\triangle ABC \sim \triangle ABC$
Symmetric Property of Similarity	If $\triangle ABC \sim \triangle DEF$, then $\triangle DEF \sim \triangle ABC$.
Transitive Property of Similarity	If $\triangle ABC \sim \triangle DEF$, and $\triangle DEF \sim \triangle XYZ$, then $\triangle ABC \sim \triangle XYZ$.

You will prove Theorem 7.4 in Exercise 26.

EXAMPLE 4 Parts of Similar Triangles

Find *BE* and *AD*.

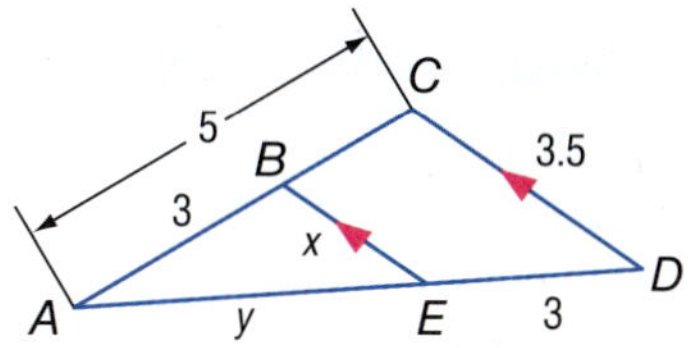

Since $\overline{BE} \parallel \overline{CD}$, $\angle ABE \cong \angle BCD$ and $\angle AEB \cong \angle EDC$ because they are corresponding angles. By AA Similarity, $\triangle ABE \sim \triangle ACD$.

$\frac{AB}{AC} = \frac{BE}{CD}$	Definition of Similar Polygons
$\frac{3}{5} = \frac{x}{3.5}$	$AC = 5$, $CD = 3.5$, $AB = 3$, $BE = x$
$3.5 \cdot 3 = 5 \cdot x$	Cross Products Property
$2.1 = x$	*BE* is 2.1.
$\frac{AC}{AB} = \frac{AD}{AE}$	Definition of Similar Polygons
$\frac{5}{3} = \frac{y+3}{y}$	$AC = 5$, $AB = 3$, $AD = y + 3$, $AE = y$
$5 \cdot y = 3(y + 3)$	Cross Products Property
$5y = 3y + 9$	Distributive Property
$2y = 9$	Subtract $3y$ from each side.
$y = 4.5$	*AD* is $y + 3$ or 7.5.

> **StudyTip**
>
> **Proportions** An additional proportion that is true for Example 4 is $\frac{AC}{CD} = \frac{AB}{BE}$.

Guided Practice

Find each measure.

4A. *QP* and *MP*

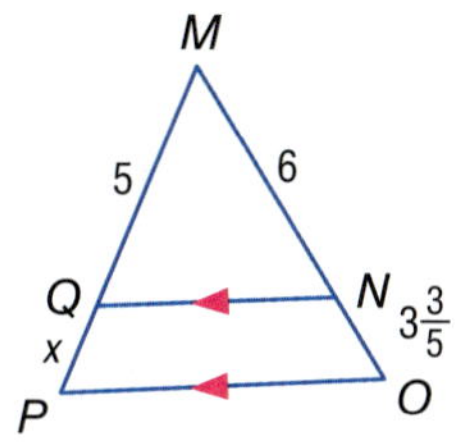

4B. *WR* and *RT*

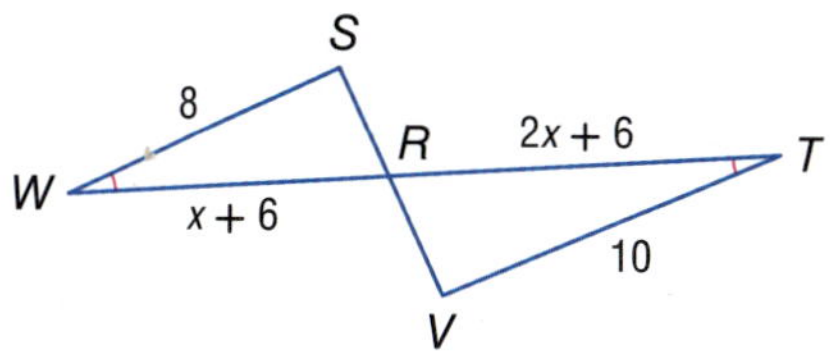

Personal Tutor glencoe.com

Real-World EXAMPLE 5 Indirect Measurement

ROLLER COASTERS Hallie is estimating the height of the Superman roller coaster in Mitchellville, Maryland. She is 5 feet 3 inches tall and her shadow is 3 feet long. If the length of the shadow of the roller coaster is 40 feet, how tall is the roller coaster?

Understand Make a sketch of the situation. 5 feet 3 inches is equivalent to 5.25 feet.

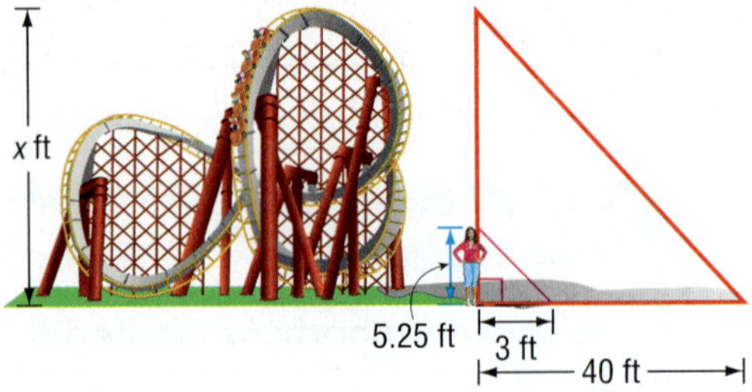

Plan In shadow problems, you can assume that the angles formed by the Sun's rays with any two objects are congruent and that the two objects form the sides of two right triangles.

Since two pairs of angles are congruent, the right triangles are similar by the AA Similarity Postulate. So, the following proportion can be written.

$$\frac{\text{Hallie's height}}{\text{coaster's height}} = \frac{\text{Hallie's shadow length}}{\text{coaster's shadow length}}$$

Solve Substitute the known values and let x = roller coaster's height.

$\frac{5.25}{x} = \frac{3}{40}$	**Substitution**
$3 \cdot x = 40(5.25)$	**Cross Products Property**
$3x = 210$	**Simplify.**
$x = 70$	**Divide each side by 3.**

The roller coaster is 70 feet tall.

Check The roller coaster's shadow length is $\frac{40 \text{ ft}}{3 \text{ ft}}$ or about 13.3 times Hallie's shadow length. Check to see that the roller coaster's height is about 13.3 times Hallie's height. $\frac{70 \text{ ft}}{5.25 \text{ ft}} \approx 13.3$ ✓

Problem-SolvingTip

Determine Reasonable Answers When you have solved a problem, check your answer for reasonableness. In this example, Hallie's shadow is a little more than half her height. The coaster's shadow is also a little more than half of the height you calculated. Therefore, the answer is reasonable.

Guided Practice

5. **BUILDINGS** Adam is standing next to the Palmetto Building in Columbia, South Carolina. He is 6 feet tall and the length of his shadow is 9 feet. If the length of the shadow of the building is 322.5 feet, how tall is the building?

Personal Tutor glencoe.com

Concept Summary Triangle Similarity

AA Similarity Postulate

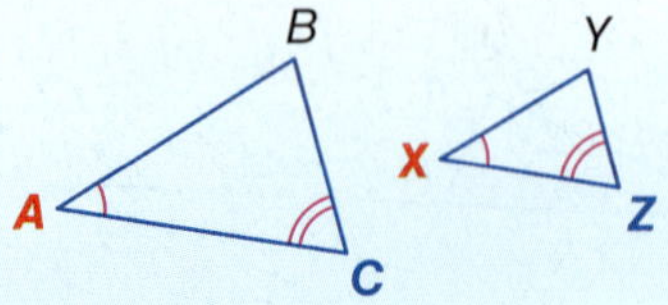

If $\angle A \cong \angle X$ and $\angle C \cong \angle Z$, then $\triangle ABC \sim \triangle XYZ$.

SSS Similarity Theorem

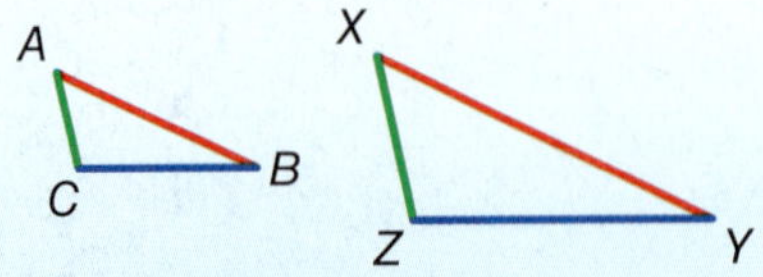

If $\frac{AB}{XY} = \frac{BC}{YZ} = \frac{CA}{ZX}$, then $\triangle ABC \sim \triangle XYZ$.

SAS Similarity Theorem

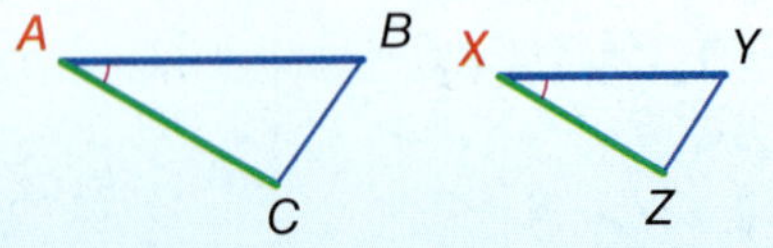

If $\angle A \cong \angle X$ and $\frac{AB}{XY} = \frac{CA}{ZX}$, then $\triangle ABC \sim \triangle XYZ$.

Check Your Understanding

Examples 1 and 2 pp. 474–476

Determine whether the triangles are similar. If so, write a similarity statement. Explain your reasoning.

1.

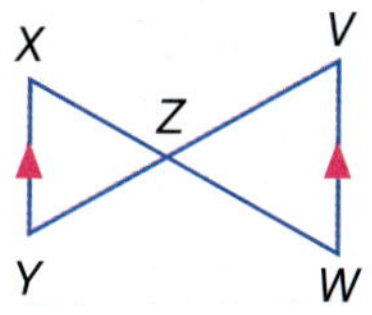

2.

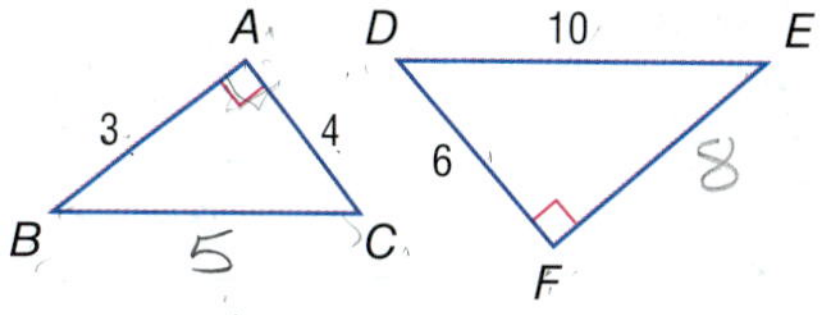

3.

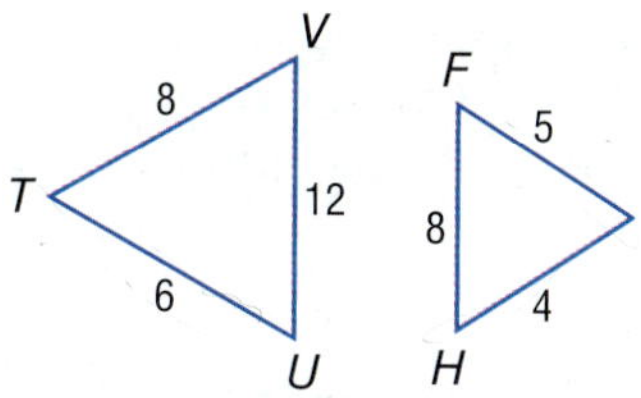

4.

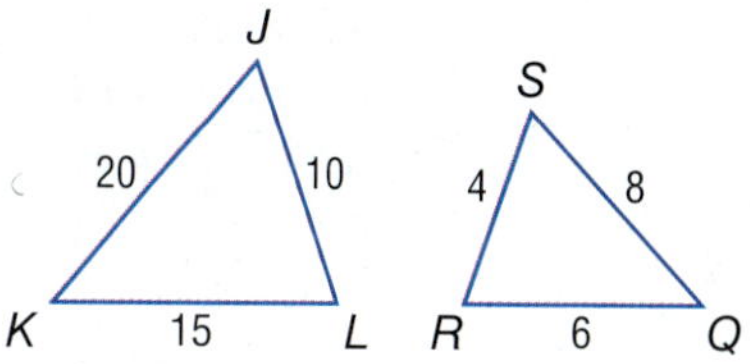

Example 3 p. 476

5. MULTIPLE CHOICE In the figure, $\overline{AB}$ intersects $\overline{DE}$ at point C. Which additional information would be enough to prove that $\triangle ADC \sim \triangle BEC$?

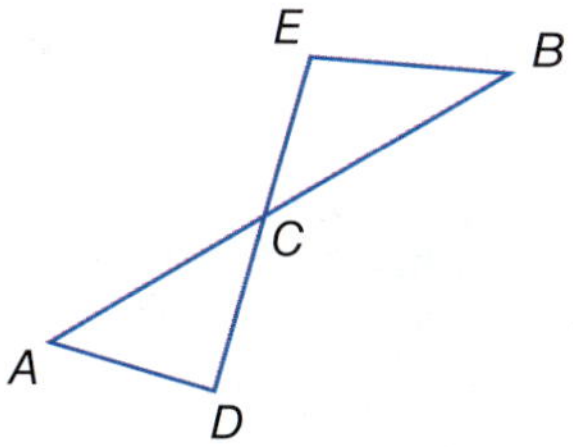

A $\angle DAC$ and $\angle ECB$ are congruent.

B $\overline{AC}$ and $\overline{BC}$ are congruent.

C $\overline{AD}$ and $\overline{EB}$ are parallel.

D $\angle CBE$ is a right angle.

Example 4 p. 477

ALGEBRA Identify the similar triangles. Find each measure.

6. KL

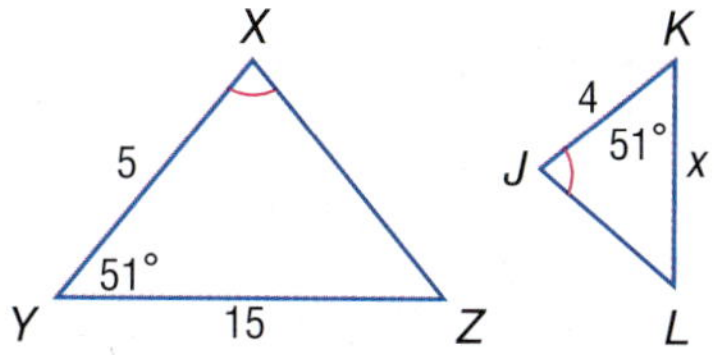

7. VS

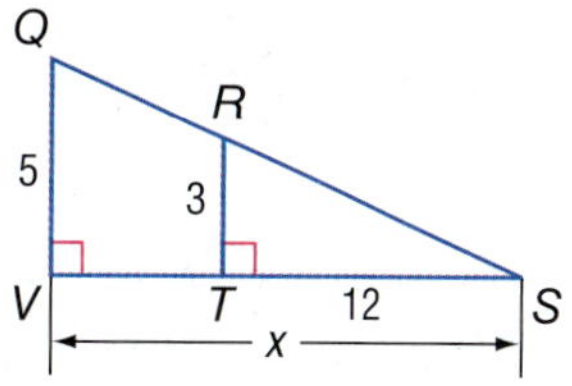

Example 5 p. 478

8. COMMUNICATION A cell phone tower casts a 100-foot shadow. At the same time, a 4-foot 6-inch post near the tower casts a shadow of 3 feet 4 inches. Find the height of the tower.

Practice and Problem Solving

= **Step-by-Step Solutions** begin on page R20.
Extra Practice begins on page 969.

Examples 1–3 pp. 474–476

Determine whether the triangles are similar. If so, write a similarity statement. If not, what would be sufficient to prove the triangles similar? Explain your reasoning.

9.

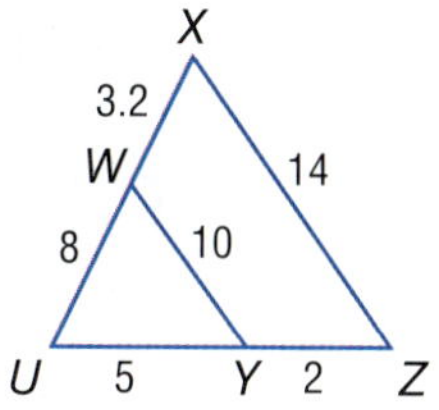

10.

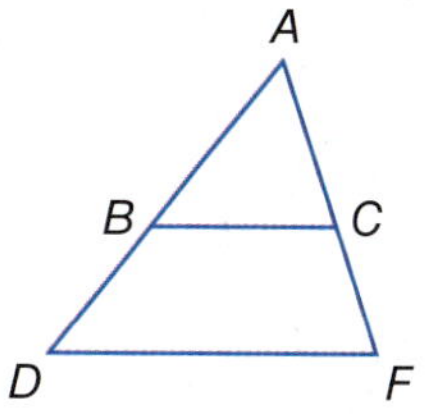

11

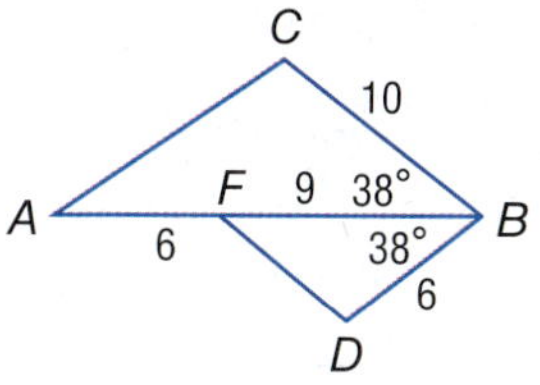

Examples 1–3
pp. 474 and 476

Determine whether the triangles are similar. If so, write a similarity statement. If not, what would be sufficient to prove the triangles similar? Explain your reasoning.

12.

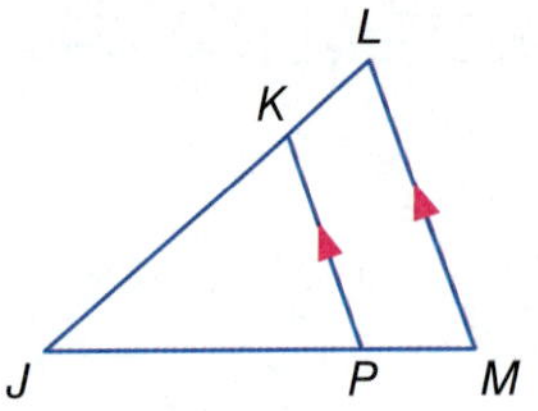

13.

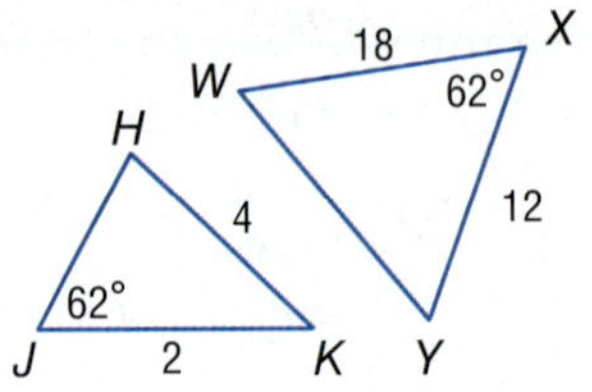

14.

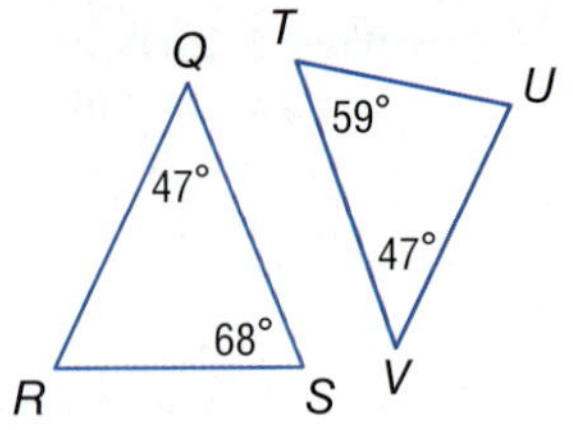

15. **VISION** When we look at an object, it is projected on the retina through the pupil. The distances from the pupil to the top and bottom of the object are congruent and the distances from the pupil to the top and bottom of the image on the retina are congruent. Are the triangles formed between the object and the pupil and the object and the image similar? Explain your reasoning.

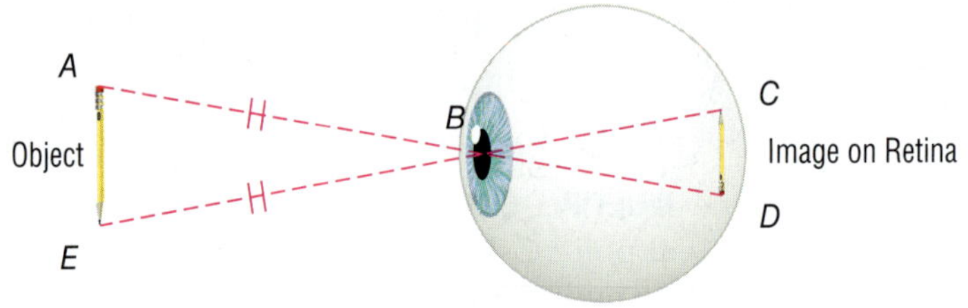

Example 4
p. 477

ALGEBRA **Identify the similar triangles. Then find each measure.**

16. *JK*

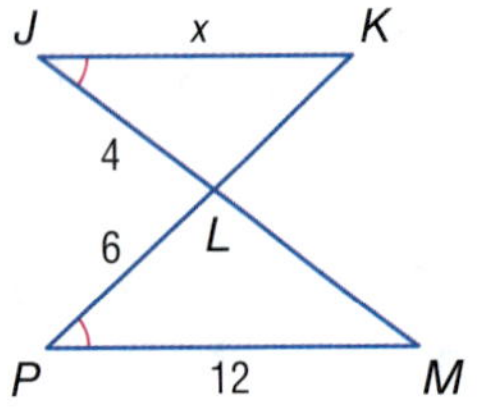

17 *ST*

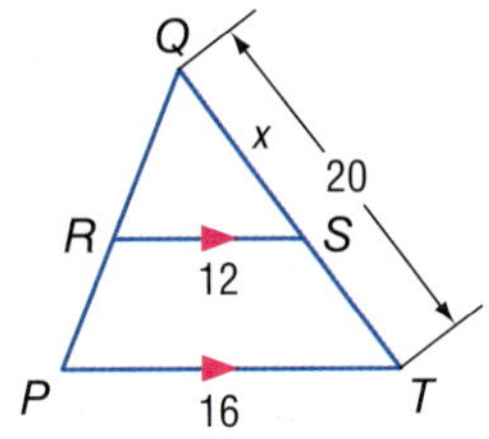

18. *WZ*, *UZ*

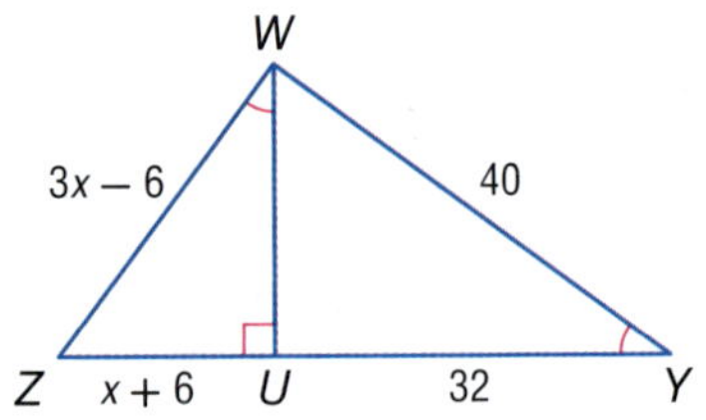

19. *HJ*, *HK*

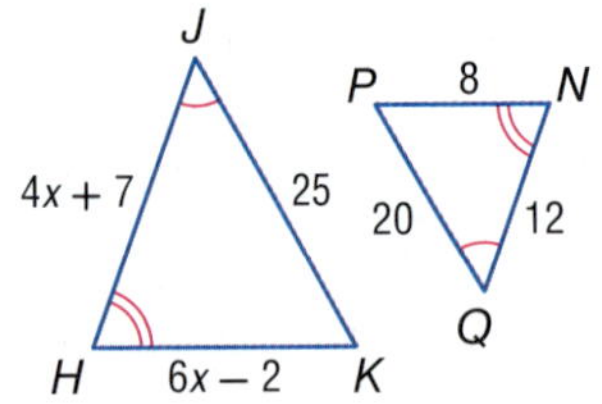

20. *DB*, *CB*

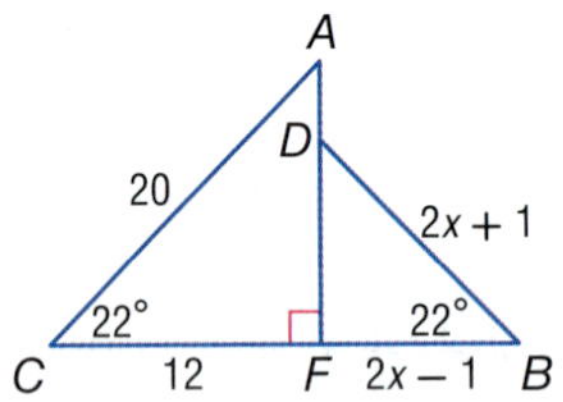

21. *GD*, *DH*

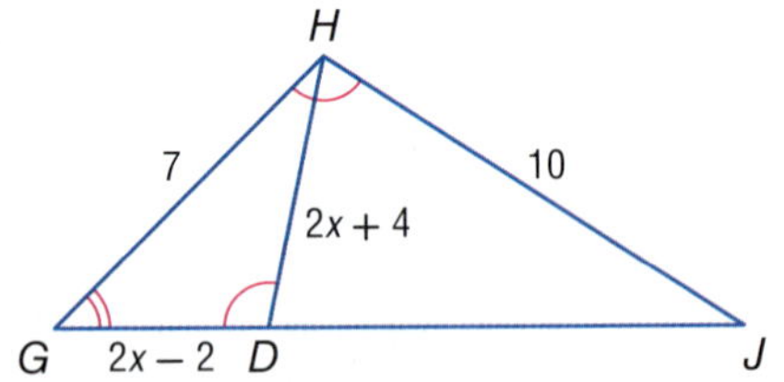

Example 5
p. 478

22. **STATUES** Mei is standing next to a statue in the park. If Mei is 5 feet tall, her shadow is 3 feet long, and the statue's shadow is $10\frac{1}{2}$ feet long, how tall is the statue?

23. **SPORTS** When Alonzo, who is 5′11″ tall, stands next to a basketball goal, his shadow is 2′ long, and the basketball goal's shadow is 4′4″ long. About how tall is the basketball goal?

24. **FORESTRY** A hypsometer, as shown, can be used to estimate the height of a tree. Bartolo looks through the straw to the top of the tree and obtains the readings given. Find the height of the tree.

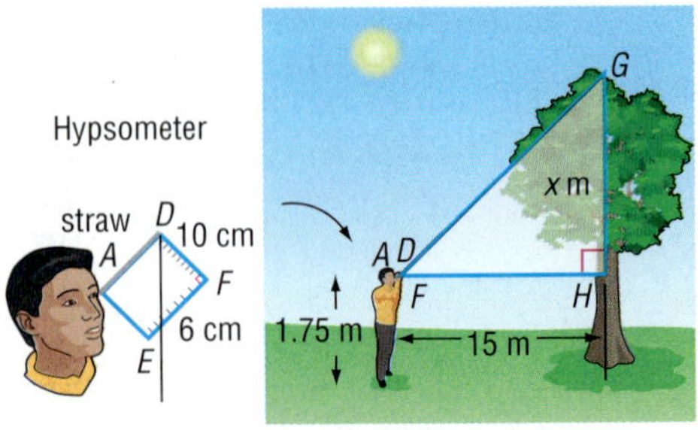

PROOF **Write a two-column proof.**

25. Theorem 7.3

26. Theorem 7.4

Real-World Link

At only 5′7″, Spud Webb is the shortest player ever to enter the NBA Slam Dunk Contest. He won the contest in 1986.

Source: National Basketball Association

PROOF **Write a two-column proof.**

27. **Given:** $\triangle XYZ$ and $\triangle ABC$ are right triangles; $\frac{XY}{AB} = \frac{YZ}{BC}$.

Prove: $\triangle YXZ \sim \triangle BAC$

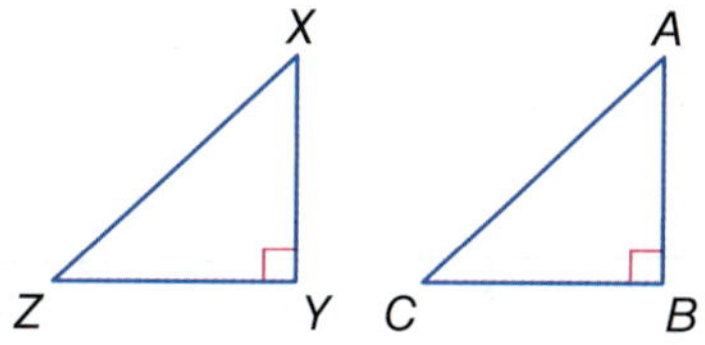

28. **Given:** $ABCD$ is a trapezoid.

Prove: $\frac{DP}{PB} = \frac{CP}{PA}$

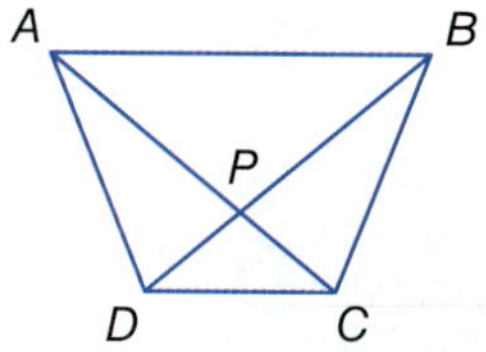

29. **SPORTS** When Luis's dad threw a bounce pass to him, the angles formed by the basketball's path were congruent. The ball landed $\frac{2}{3}$ of the way between them before it bounced back up. If Luis's dad released the ball 40 inches above the floor, at what height did Luis catch the ball?

COORDINATE GEOMETRY **$\triangle XYZ$ and $\triangle WYV$ have vertices $X(-1, -9)$, $Y(5, 3)$, $Z(-1, 6)$, $W(1, -5)$, and $V(1, 5)$.**

30. Graph the triangles, and prove that $\triangle XYZ \sim \triangle WYV$.

31 Find the ratio of the perimeters of the two triangles.

32. **BILLIARDS** When a ball is deflected off a smooth surface, the angles formed by the path are congruent. Booker hit the orange ball and it followed the path from A to B to C as shown below. What was the total distance traveled by the ball from the time Booker hit it until it came to rest at the end of the table?

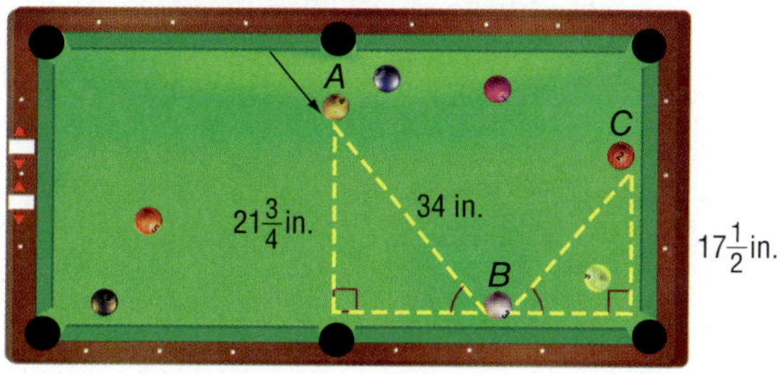

33. **PROOF** Use similar triangles to show that the slope of the line through any two points on that line is constant. That is, if points A, B, A' and B' are on line ℓ, use similar triangles to show that the slope of the line from A to B is equal to the slope of the line from A' to B'.

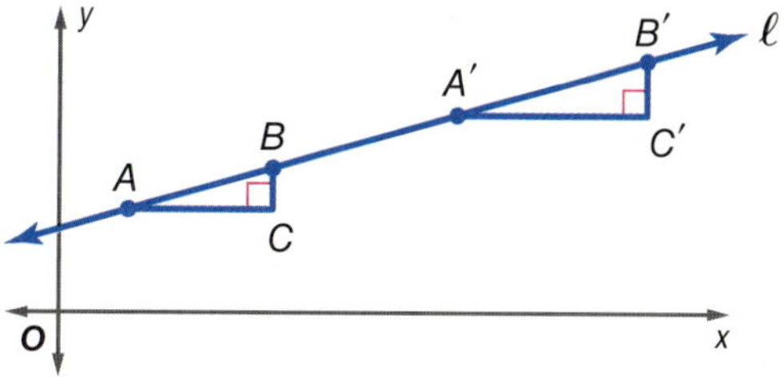

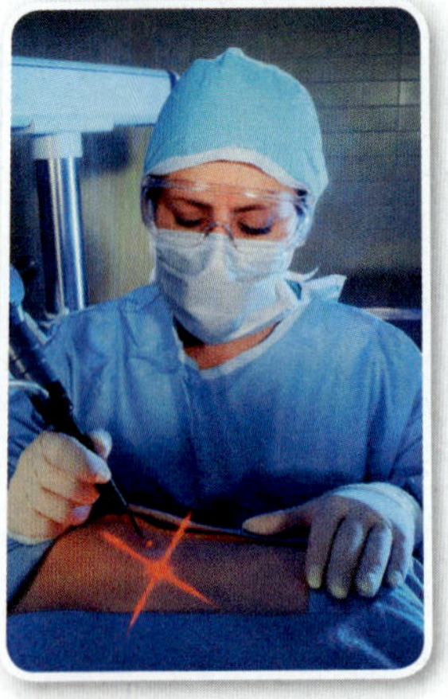

Real-World Link

Certain medical treatments involve laser beams that contact and penetrate the skin, forming similar triangles.

Source: Annenberg Media

34. CHANGING DIMENSIONS Assume that $\triangle ABC \sim \triangle JKL$.

a. If the lengths of the sides of $\triangle JKL$ are half the length of the sides of $\triangle ABC$, and the area of $\triangle ABC$ is 40 square inches, what is the area of $\triangle JKL$? How is the area related to the scale factor of $\triangle ABC$ to $\triangle JKL$?

b. If the lengths of the sides of $\triangle ABC$ are three times the length of the sides of $\triangle JKL$, and the area of $\triangle ABC$ is 63 square inches, what is the area of $\triangle JKL$? How is the area related to the scale factor of $\triangle ABC$ to $\triangle JKL$?

35. MEDICINE Refer to the information at the left and the diagram at the right. How far apart should the laser sources be placed to ensure that the areas treated by each source do not overlap?

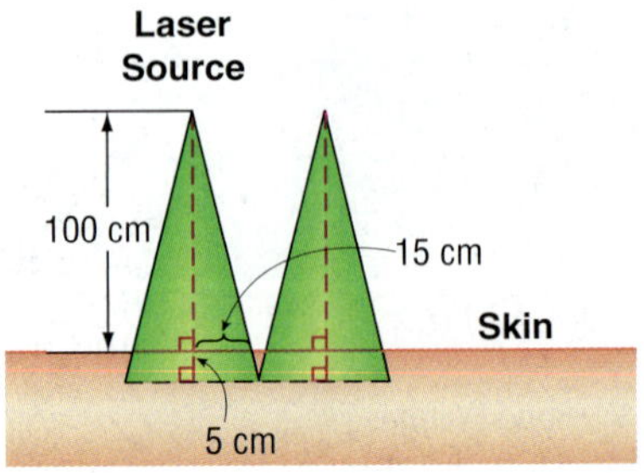

36. MULTIPLE REPRESENTATIONS In this problem, you will explore proportional parts of triangles.

a. GEOMETRIC Draw a $\triangle ABC$ with $\overline{DE}$ parallel to $\overline{AC}$ as shown at the right.

b. TABULAR Measure and record the lengths AD, DB, CD, and EB and the ratios $\frac{AD}{DB}$ and $\frac{CE}{EB}$ in a table.

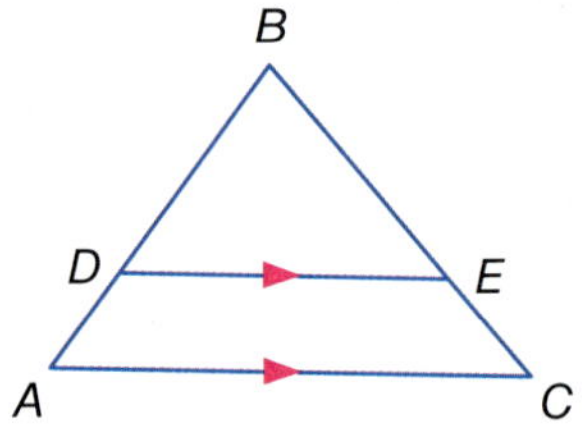

c. VERBAL Make a conjecture about the segments created by a line parallel to one side of a triangle and intersecting the other two sides.

H.O.T. Problems

Use Higher-Order Thinking Skills

37. WRITING IN MATH Compare and contrast the AA Similarity Postulate, the SSS Similarity Theorem, and the SAS similarity theorem.

38. CHALLENGE $\overline{YW}$ is an altitude of $\triangle XYZ$. Find YW.

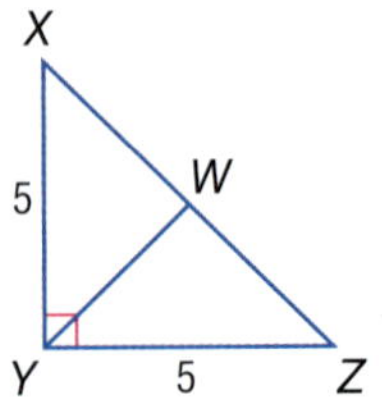

39. REASONING A pair of similar triangles has angle measures of 50°, 85°, and 45°. The sides of one triangle measure 3, 4, and 5.2 units, and the sides of the second triangle measure x, $x - 1.5$, and $x + 1.8$ units. Find the value of x.

40. OPEN ENDED Draw a triangle that is similar to $\triangle ABC$ shown. Explain how you know that it is similar.

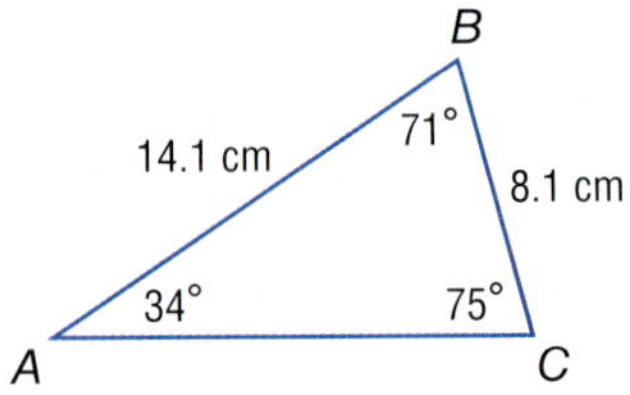

41. WRITING IN MATH Given a triangle, explain a process you can use to draw a similar triangle that is twice as large.

Standardized Test Practice

42. PROBABILITY $\frac{x!}{(x-3)!} =$

A 3.0
B 0.33
C $x^2 - 3x + 2$
D $x^3 - 3x^2 + 2x$

43. EXTENDED RESPONSE In the figure below, $\overline{EB} \parallel \overline{DC}$.

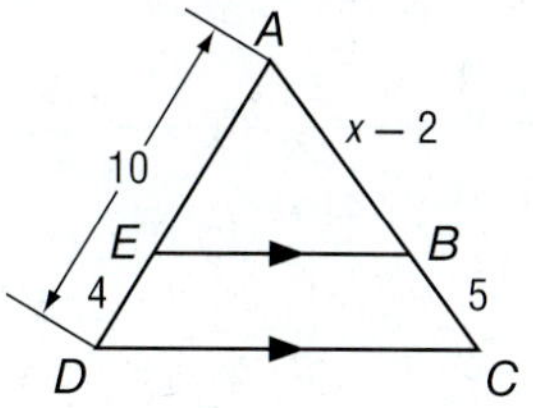

a. Write a proportion that could be used to find x.
b. Find the value of x and the measure of $\overline{AB}$.

44. ALGEBRA Which polynomial represents the area of the shaded region?

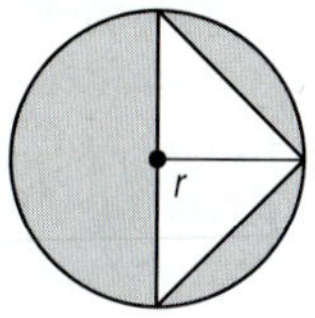

F πr^2
G $\pi r^2 + r^2$
H $\pi r^2 + r$
J $\pi r^2 - r^2$

45. SAT/ACT The volume of a certain rectangular solid is $16x$ cubic units. If the dimensions of the solid are integers x, y, and z units, what is the greatest possible value of z?

A 32
B 16
C 8
D 4

Spiral Review

List all pairs of congruent angles, and write a proportion that relates the corresponding sides for each pair of similar polygons. (Lesson 7-2)

46. $\triangle JKL \sim \triangle CDE$

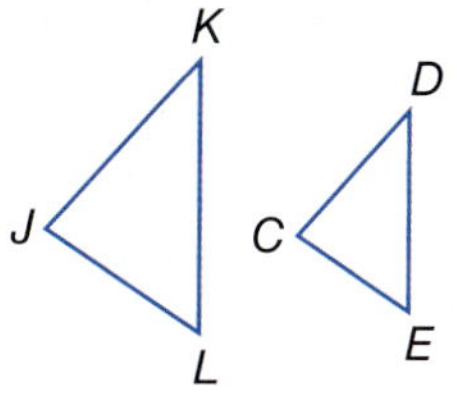

47. $WXYZ \sim QRST$

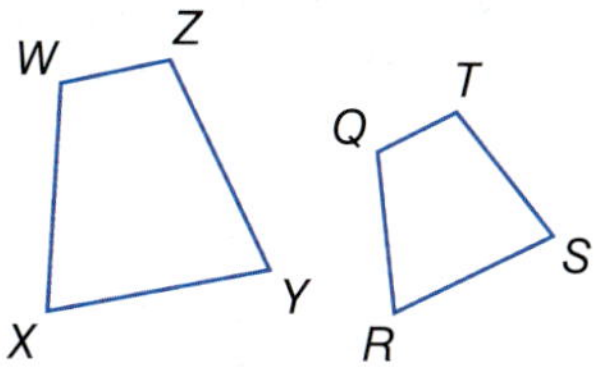

48. $FGHJ \sim MPQS$

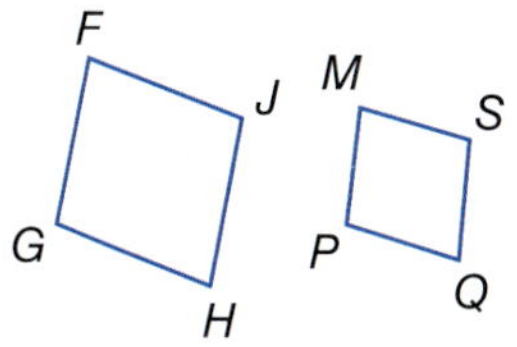

Solve each proportion. (Lesson 7-1)

49. $\frac{3}{4} = \frac{x}{16}$

50. $\frac{x}{10} = \frac{22}{50}$

51. $\frac{20.2}{88} = \frac{12}{x}$

52. $\frac{x-2}{2} = \frac{3}{8}$

53. TANGRAMS A tangram set consists of seven pieces: a small square, two small congruent right triangles, two large congruent right triangles, a medium-sized right triangle, and a quadrilateral. How can you determine the shape of the quadrilateral? Explain. (Lesson 6-3)

Determine which postulate can be used to prove that the triangles are congruent. If it is not possible to prove congruence, write *not possible*. (Lesson 4-4)

54.

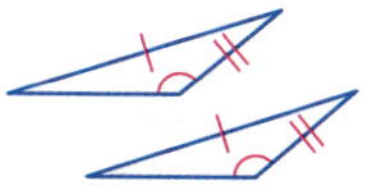

55.

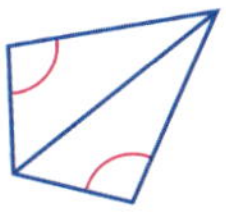

56.

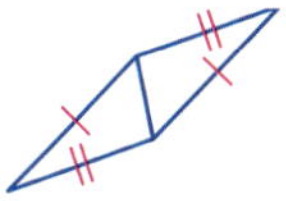

Skills Review

Write a two-column proof. (Lesson 3-5)

57. Given: $r \parallel t$; $\angle 5 \cong \angle 6$

Prove: $\ell \parallel m$

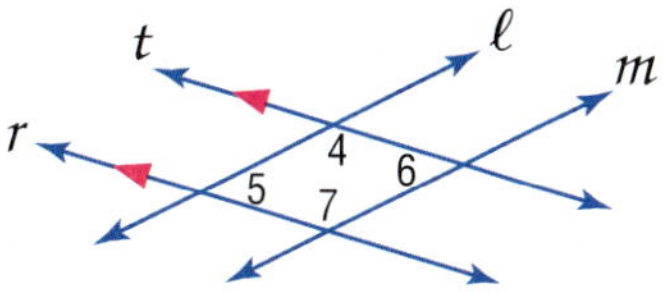

7-4 Parallel Lines and Proportional Parts

Then

You used proportions to solve problems between similar triangles. (Lesson 7-3)

Now

- Use proportional parts within triangles.
- Use proportional parts with parallel lines.

New Vocabulary

midsegment of a triangle

Math Online

glencoe.com

Why?

Cartoonists have many different techniques at their disposal to add the illusion of depth to their work. Perspective drawings make use of the fact that objects that are far away look proportionally smaller than objects that are closer to the observer. To accomplish this illusion, cartoonists make use of the Triangle Proportionality Theorem.

Proportional Parts Within Triangles When a triangle contains a line that is parallel to one of its sides, the two triangles formed can be proved similar using the Angle-Angle Similarity Postulate. Since the triangles are similar, their sides are proportional.

Theorem 7.5 Triangle Proportionality Theorem

If a line is parallel to one side of a triangle and intersects the other two sides, then it divides the sides into segments of proportional lengths.

Example If $\overline{BE} \parallel \overline{CD}$, then $\frac{AB}{BC} = \frac{AE}{ED}$.

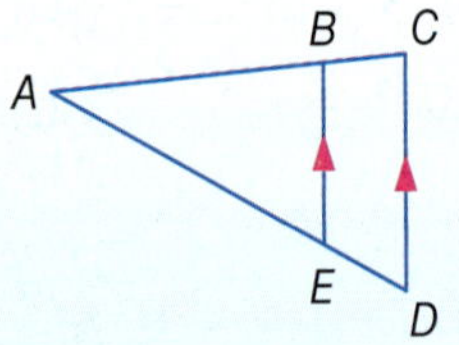

You will prove Theorem 7.5 in Exercise 30.

EXAMPLE 1 Find the Length of a Side

In $\triangle PQR$, $\overline{ST} \parallel \overline{RQ}$. If $PT = 7.5$, $TQ = 3$, and $SR = 2.5$, find PS.

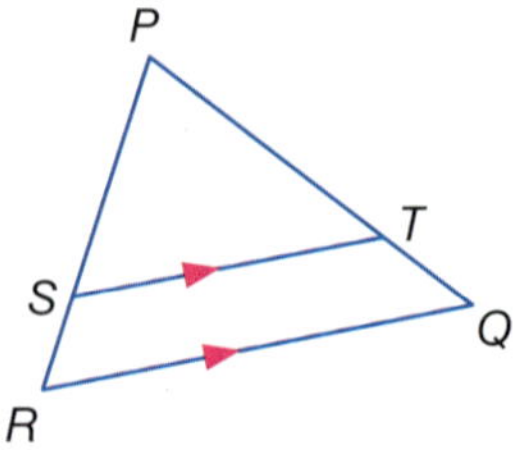

Use the Triangle Proportionality Theorem.

$\frac{PS}{SR} = \frac{PT}{TQ}$	Triangle Proportionality Theorem
$\frac{PS}{2.5} = \frac{7.5}{3}$	Substitute.
$PS \cdot 3 = (2.5)(7.5)$	Cross Products Property
$3PS = 18.75$	Multiply.
$PS = 6.25$	Divide by 3.

Guided Practice

1. If $PS = 12.5$, $SR = 5$, and $PT = 15$, find TQ.

Personal Tutor glencoe.com

The converse of Theorem 7.5 is also true and can be proved using the proportional parts of a triangle.

Theorem 7.6

Converse of Triangle Proportionality Theorem

If a line intersects two sides of a triangle and separates the sides into proportional corresponding segments, then the line is parallel to the third side of the triangle.

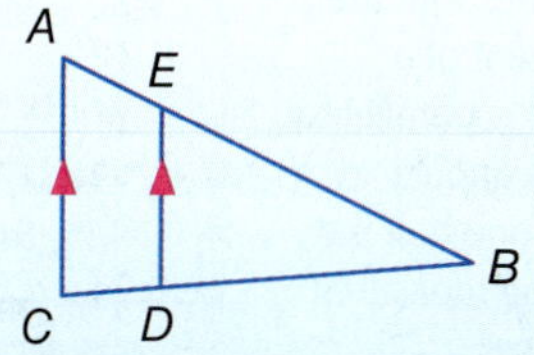

Example If $\frac{AE}{EB} = \frac{CD}{DB}$, then $\overline{AC} \parallel \overline{ED}$.

You will prove Theorem 7.6 in Exercise 31.

EXAMPLE 2 Determine if Lines are Parallel

In $\triangle DEF$, $EH = 3$, $HF = 9$, and DG is one-third the length of $\overline{GF}$. Is $\overline{DE} \parallel \overline{GH}$?

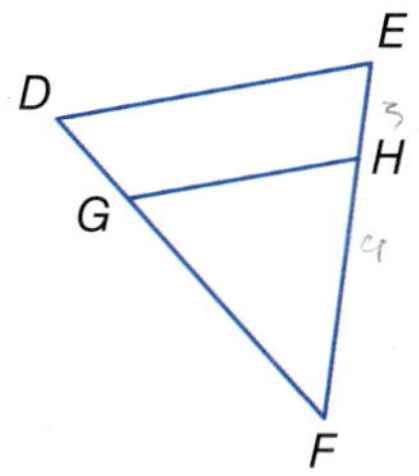

Using the converse of the Triangle Proportionality Theorem, in order to show that $\overline{DE} \parallel \overline{GH}$, we must show that $\frac{DG}{GF} = \frac{EH}{HF}$.

Find and simplify each ratio. Let $DG = x$. Since DG is one-third of GF, $GF = 3x$.

$\frac{DG}{GF} = \frac{x}{3x}$ or $\frac{1}{3}$ $\qquad$ $\frac{EH}{HF} = \frac{3}{9}$ or $\frac{1}{3}$

Since $\frac{1}{3} = \frac{1}{3}$, the sides are proportional, so $\overline{DE} \parallel \overline{GH}$.

Guided Practice

2. DG is half the length of $\overline{GF}$, $EH = 6$, and $HF = 10$. Is $\overline{DE} \parallel \overline{GH}$?

Personal Tutor glencoe.com

StudyTip

Midsegment Triangle The three midsegments of a triangle form the *midsegment triangle.*

A **midsegment of a triangle** is a segment with endpoints that are the midpoints of two sides of the triangle. Every triangle has three midsegments. The midsegments of $\triangle ABC$ are $\overline{RP}$, $\overline{PQ}$, $\overline{RQ}$.

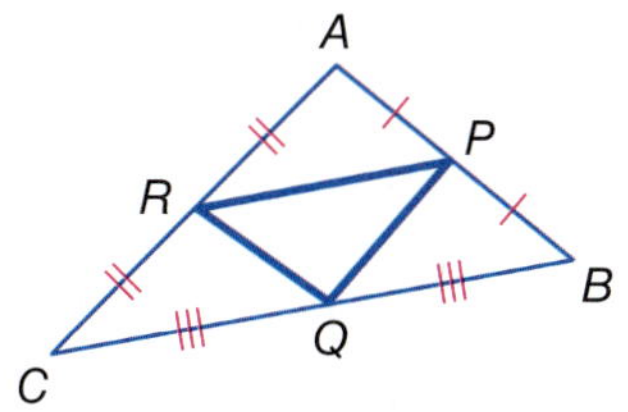

A special case of the Triangle Proportionality Theorem is the Triangle Midsegment Theorem.

Theorem 7.7 Triangle Midsegment Theorem

A midsegment of a triangle is parallel to one side of the triangle, and its length is one half the length of that side.

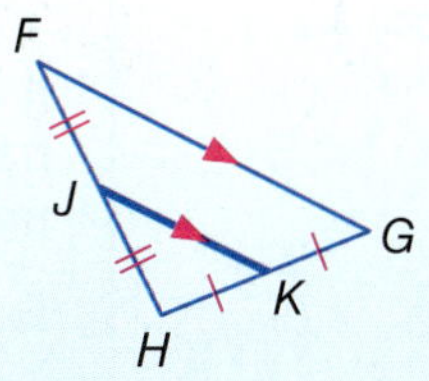

Example If J and K are midpoints of $\overline{FH}$ and $\overline{HG}$, respectively, then $\overline{JK} \parallel \overline{FG}$ and $JK = \frac{1}{2}FG$.

You will prove Theorem 7.7 in Exercise 32.

StudyTip

Midsegment The Triangle Midsegment Theorem is similar to the Trapezoid Midsegment Theorem, which states that the midsegment of a trapezoid is parallel to the bases and its length is one half the sum of the measures of the bases. (Lesson 6-6)

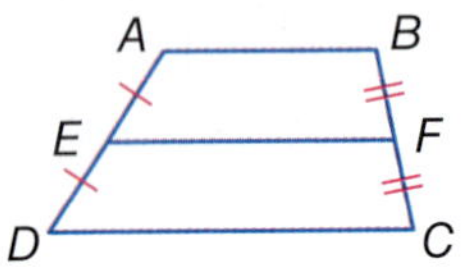

$\overline{EF} \parallel \overline{AB} \parallel \overline{DC}$

$EF = \frac{1}{2}(AB + DC)$

EXAMPLE 3 Use the Triangle Midsegment Theorem

In the figure, $\overline{XY}$ and $\overline{XZ}$ are midsegments of $\triangle RST$. Find each measure.

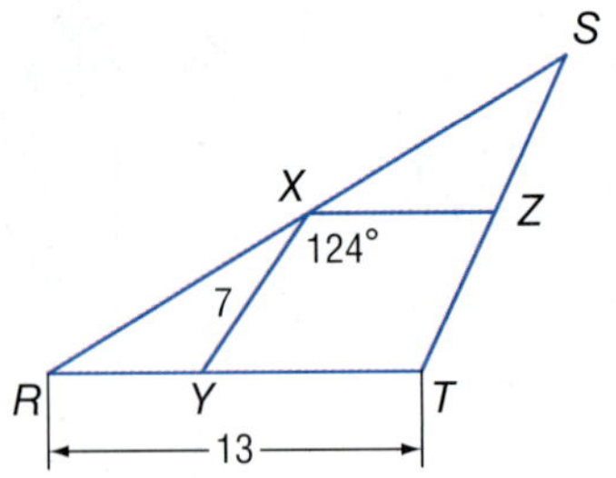

a. XZ

$XZ = \frac{1}{2}RT$ — Triangle Midsegment Theorem

$XZ = \frac{1}{2}(13)$ — Substitution

$XZ = 6.5$ — Simplify.

b. ST

$XY = \frac{1}{2}ST$ — Triangle Midsegment Theorem

$7 = \frac{1}{2}ST$ — Substitution

$14 = ST$ — Multiply each side by 2.

c. $m\angle RYX$

By the Triangle Midsegment Theorem, $\overline{XZ} \parallel \overline{RT}$.

$\angle RYX \cong \angle YXZ$ — Alternate Interior Angles Theorem

$m\angle RYX = m\angle YXZ$ — Definition of congruence

$m\angle RYX = 124$ — Substitution

Guided Practice

Find each measure.

3A. DE

3B. DB

3C. $m\angle FED$

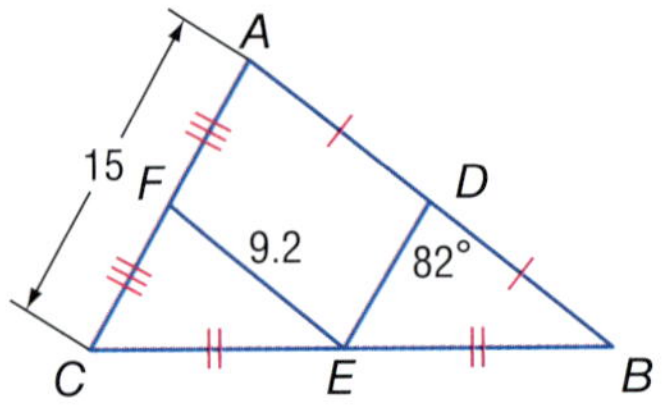

Personal Tutor glencoe.com

Proportional Parts with Parallel Lines

Another special case of the Triangle Proportionality Theorem involves three or more parallel lines cut by two transversals. Notice that if transversals a and b are extended, they form triangles with the parallel lines.

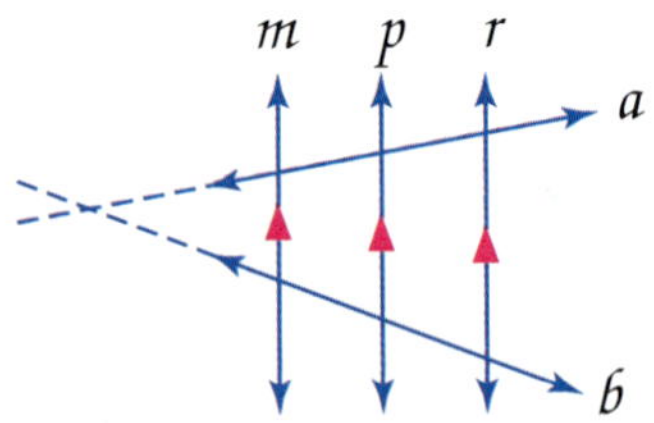

StudyTip

Other Proportions Two other proportions can be written for the example in Corollary 7.1.

$\frac{AB}{EF} = \frac{BC}{FG} = \frac{AC}{BC} = \frac{EG}{FG}$

Corollary 7.1

Proportional Parts of Parallel Lines

If three or more parallel lines intersect two transversals, then they cut off the transversals proportionally.

Example If $\overline{AE} \parallel \overline{BF} \parallel \overline{CG}$, then $\frac{AB}{BC} = \frac{EF}{FG}$.

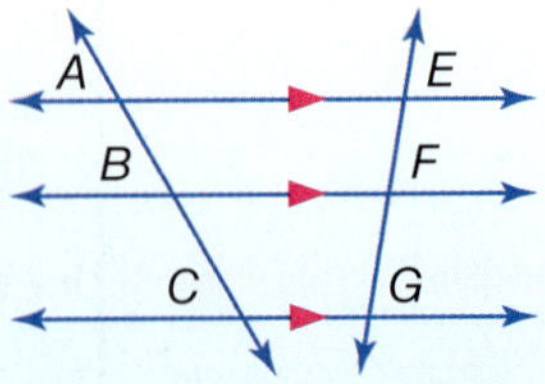

You will prove Corollary 7.1 in Exercise 28.

Real-World Link

To make a two-dimensional drawing appear three-dimensional, an artist provides several perceptual cues.

- *size* - faraway items look smaller
- *clarity* - closer objects appear more in focus
- *detail* - nearby objects have texture, while distant ones are roughly outlined

Source: Center for Media Literacy

Real-World EXAMPLE 4 Use Proportional Segments of Transversals

ART **Megan is drawing a hallway in one-point perspective. She uses the guidelines shown to draw two windows on the left wall. If segments $\overline{AD}$, $\overline{BC}$, $\overline{WZ}$, and $\overline{XY}$ are all parallel, $AB = 8$ centimeters, $DC = 9$ centimeters, and $ZY = 5$ centimeters, find WX.**

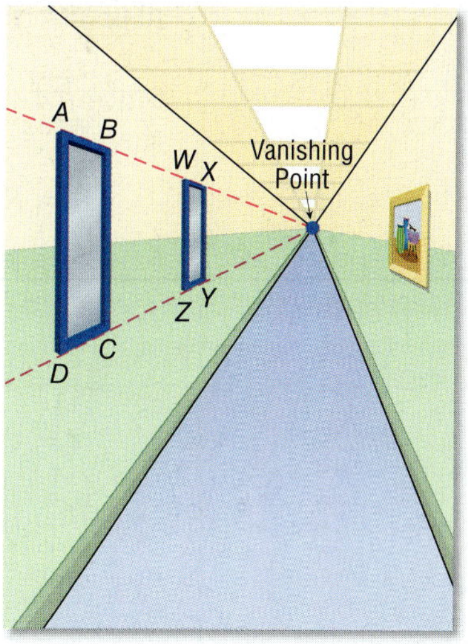

By Corollary 7.1, if $\overline{AD} \parallel \overline{BC} \parallel \overline{WZ} \parallel \overline{XY}$, then $\frac{AB}{WX} = \frac{DC}{ZY}$.

$\frac{AB}{WX} = \frac{DC}{ZY}$	**Corollary 7.1**
$\frac{8}{WX} = \frac{9}{5}$	**Substitute.**
$WX \cdot 9 = 8 \cdot 5$	**Cross Products Property**
$9WX = 40$	**Simplify.**
$WX = \frac{40}{9}$	**Divide each side by 4.**

The distance between W and X should be $\frac{40}{9}$ or about 4.4 centimeters.

CHECK The ratio of DC to ZY is 9 to 5, which is about 10 to 5 or 2 to 1. The ratio of AB to WX is 8 to 4.4 or about 8 to 4 or 2 to 1 as well, so the answer is reasonable. ✓

Guided Practice

4. **REAL ESTATE** *Frontage* is the measurement of a property's boundary that runs along the side of a particular feature such as a street, lake, ocean, or river. Find the ocean frontage for Lot A to the nearest tenth of a yard.

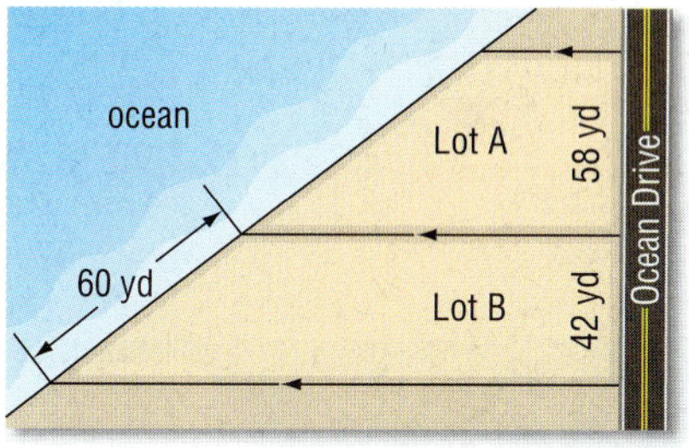

Personal Tutor glencoe.com

If the scale factor of the proportional segments is 1, they separate the transversals into congruent parts.

Corollary 7.2

Congruent Parts of Parallel Lines

If three or more parallel lines cut off congruent segments on one transversal, then they cut off congruent segments on every transversal.

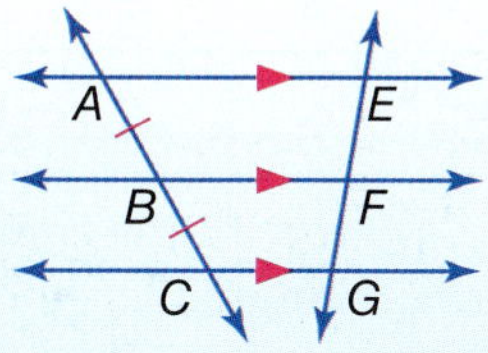

Example If $\overline{AE} \parallel \overline{BF} \parallel \overline{CG}$, and $\overline{AB} \cong \overline{BC}$, then $\overline{EF} \cong \overline{FG}$.

You will prove Corollary 7.2 in Exercise 29.

Real-World EXAMPLE 5 Use Congruent Segments of Transversals

ALGEBRA Find x and y.

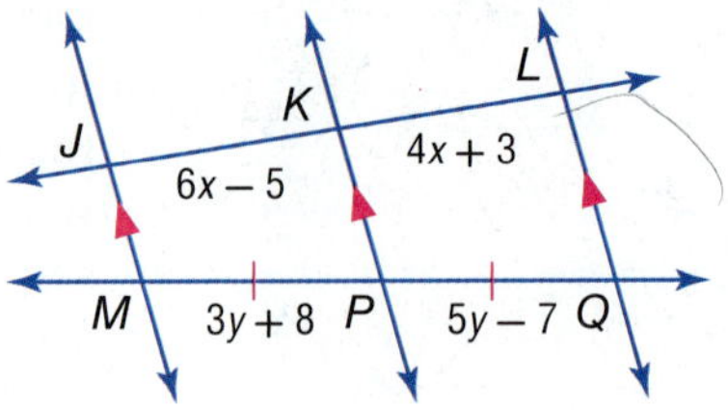

Since $\overleftrightarrow{JM} \parallel \overleftrightarrow{KP} \parallel \overleftrightarrow{LQ}$ and $\overline{MP} \cong \overline{PQ}$, then $\overline{JK} \cong \overline{KL}$ by Corollary 7.2.

$JK = KL$	**Definition of congruence**
$6x - 5 = 4x + 3$	**Substitution**
$2x - 5 = 3$	**Subtract 4x from each side.**
$2x = 8$	**Add 5 to each side.**
$x = 4$	**Divide each side by 2.**
$MP = PQ$	**Definition of congruence**
$3y + 8 = 5y - 7$	**Substitution**
$8 = 2y - 7$	**Subtract 3y from each side.**
$15 = 2y$	**Add 7 to each side.**
$7.5 = y$	**Divide each side by 2.**

Guided Practice

5A.

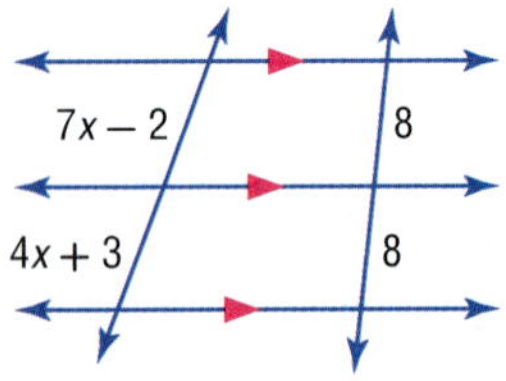

5B.

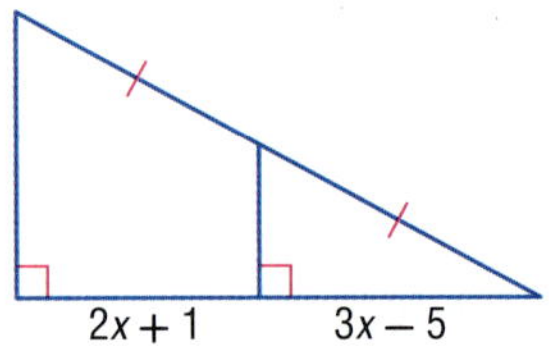

Personal Tutor glencoe.com

It is possible to separate a segment into two congruent parts by constructing the perpendicular bisector of a segment. However, a segment cannot be separated into three congruent parts by constructing perpendicular bisectors. To do this, you must use parallel lines and Corollary 7.2.

Construction Trisect a Segment

Draw a segment $\overline{AB}$. Then use Corollary 7.2 to trisect $\overline{AB}$.

Step 1

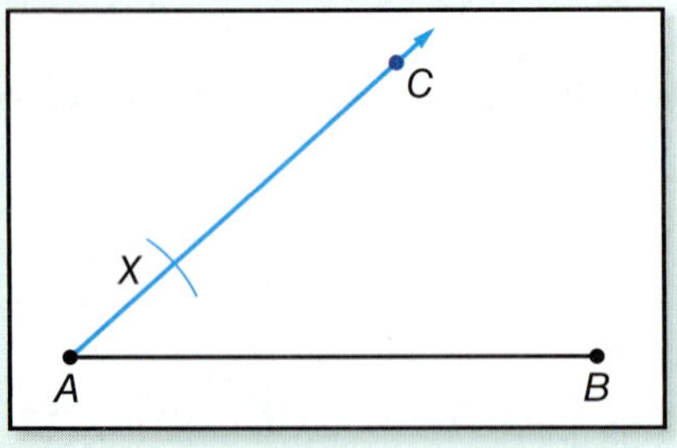

Draw $\overline{AC}$. Then with the compass at A, mark off an arc that intersects $\overline{AC}$ at X.

Step 2

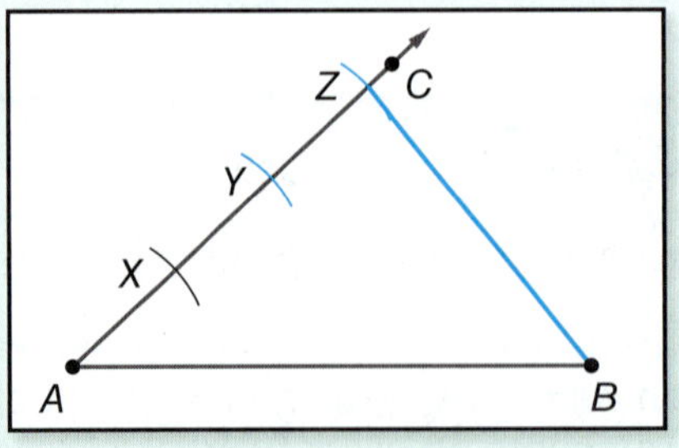

Use the same compass setting to mark off Y and Z such that $\overline{AX} \cong \overline{XY} \cong \overline{YZ}$. Then draw $\overline{ZB}$.

Step 3

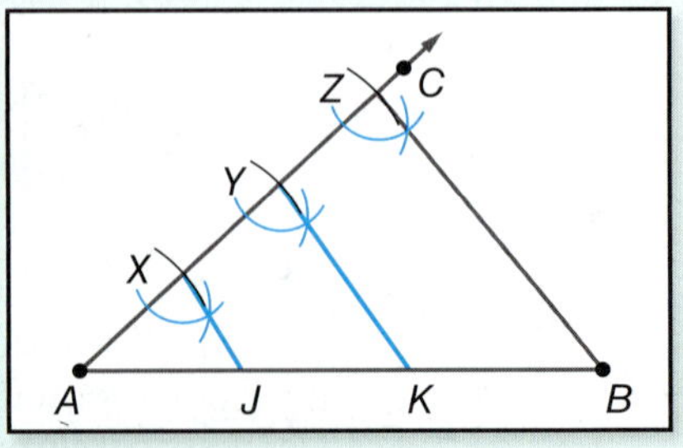

Construct lines through Y and X that are parallel to $\overline{ZB}$. Label the intersection points on $\overline{AB}$ as J and K.

Conclusion: Since parallel lines cut off congruent segments on transversals, $\overline{AJ} \cong \overline{JK} \cong \overline{KB}$.

Check Your Understanding

Example 1 p. 484

1. If $XM = 4$, $XN = 6$, and $NZ = 9$, find XY.
2. If $XN = 6$, $XM = 2$, and $XY = 10$, find NZ.

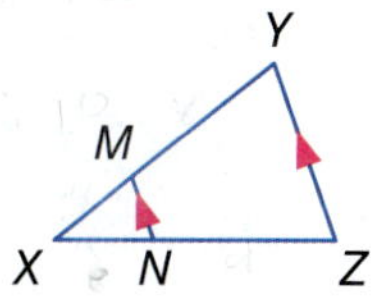

Example 2 p. 485

3. In $\triangle ABC$, $BC = 15$, $BE = 6$, $DC = 12$, and $AD = 8$. Determine whether $\overline{DE} \parallel \overline{AB}$. Justify your answer.

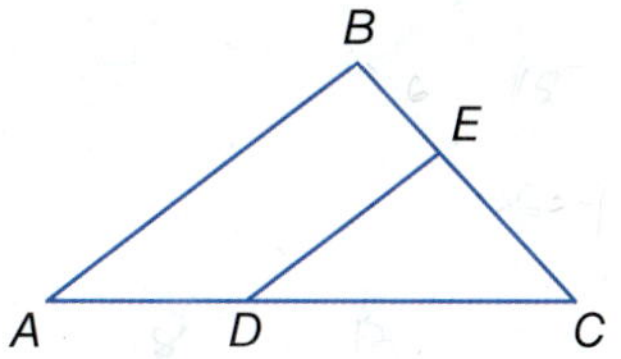

4. In $\triangle JKL$, $JK = 15$, $JM = 5$, $LK = 13$, and $PK = 9$. Determine whether $\overline{JL} \parallel \overline{MP}$. Justify your answer.

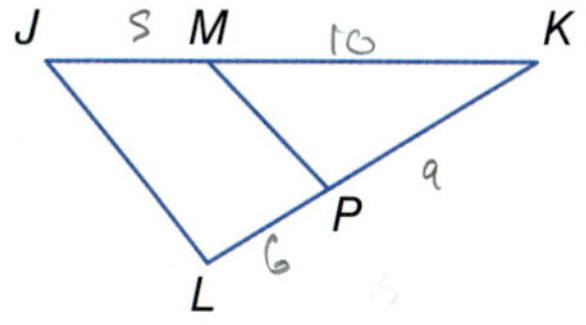

Example 3 p. 486

$\overline{JH}$ is a midsegment of $\triangle KLM$. Find the value of x.

5.

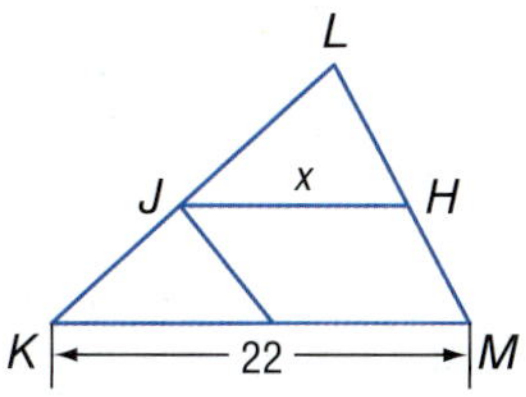

6.

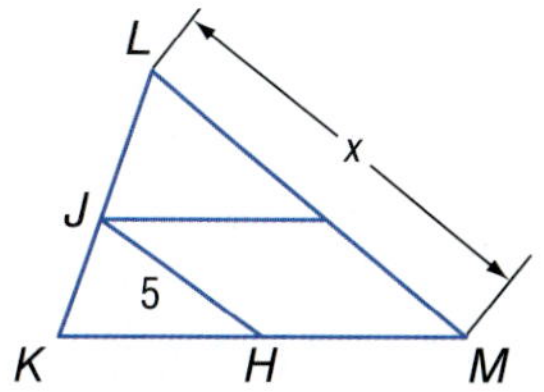

Example 4 p. 487

7. **MAPS** Refer to the map at the right. 3rd Avenue and 5th Avenue are parallel. If the distance from 3rd Avenue to City Mall along State Street is 3201 feet, find the distance between 5th Avenue and City Mall along Union Street. Round to the nearest tenth.

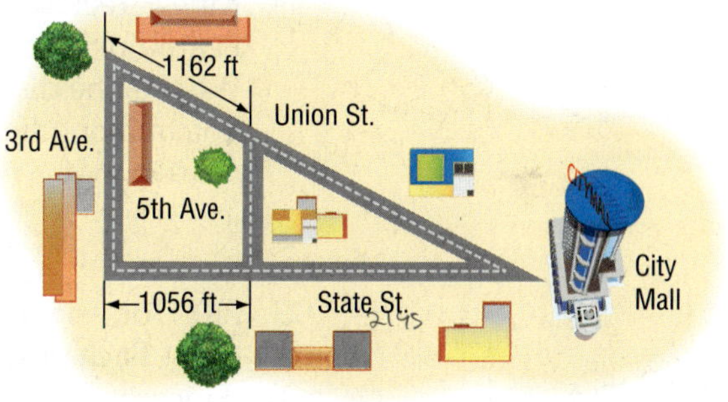

Example 5 p. 488

ALGEBRA Find x and y.

8.

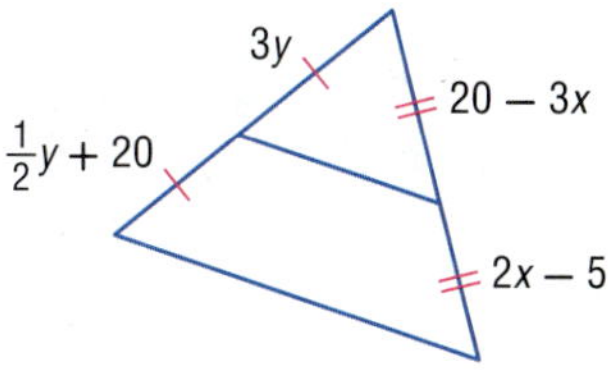

9.

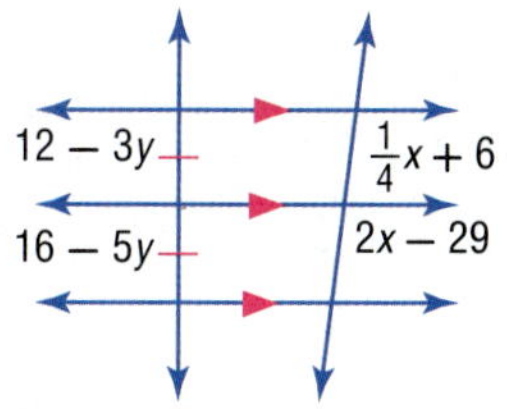

Practice and Problem Solving

● = **Step-by-Step Solutions** begin on page R20.
Extra Practice begins on page 969.

Example 1 p. 484

10. If $AB = 6$, $BC = 4$, and $AE = 9$, find ED.
11. If $AB = 12$, $AC = 16$, and $ED = 5$, find AE.
12. If $AC = 14$, $BC = 8$, and $AD = 21$, find ED.
13. If $AD = 27$, $AB = 8$, and $AE = 12$, find BC.

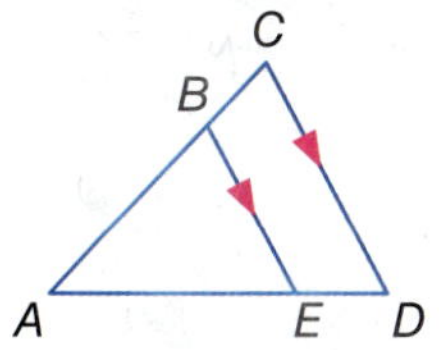

Example 2
p. 485

Determine whether $\overline{VY} \parallel \overline{ZW}$. Justify your answer.

14. $ZX = 18$, $ZV = 6$, $WX = 24$, and $YX = 16$

15. $VX = 7.5$, $ZX = 24$, $WY = 27.5$, and $WX = 40$

16. $ZV = 8$, $VX = 2$, and $YX = \frac{1}{2}WY$

17. $WX = 31$, $YX = 21$, and $ZX = 4ZV$

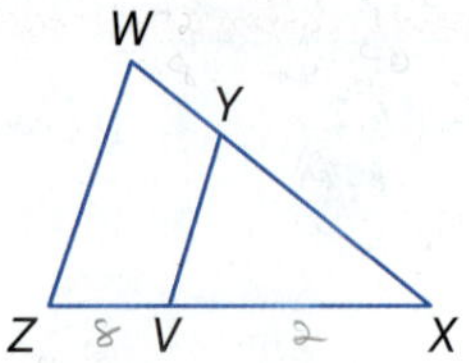

Example 3
p. 486

$\overline{JH}$, $\overline{JP}$, and $\overline{PH}$ are midsegments of $\triangle KLM$. Find the value of x.

18.

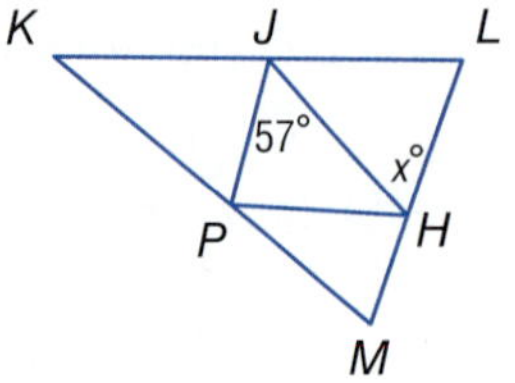

19

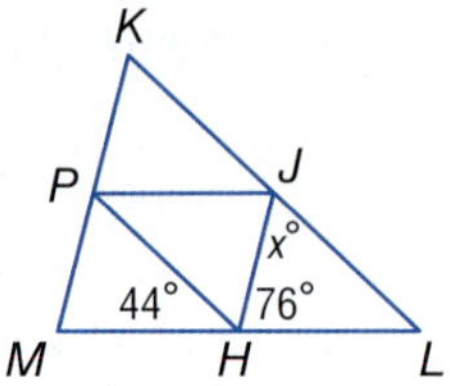

20.

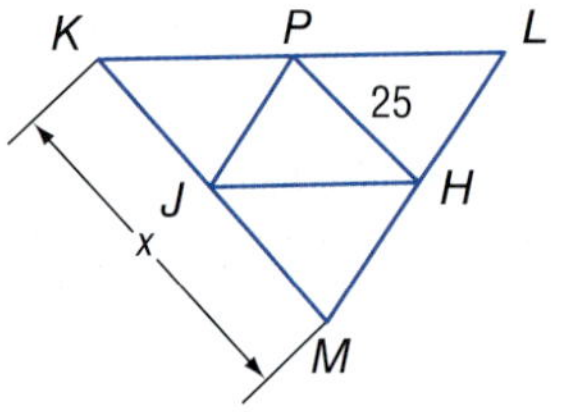

21.

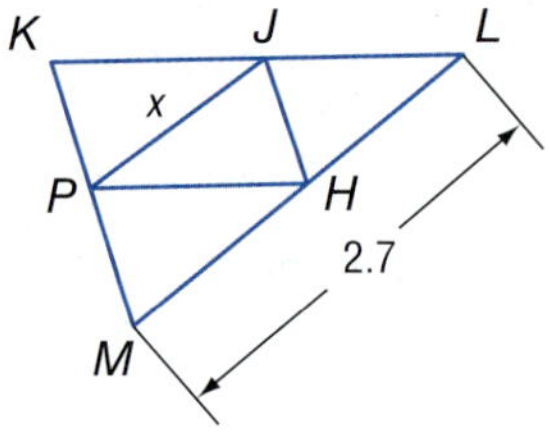

Example 4
p. 487

22. MAPS In Charleston, South Carolina, Logan Street is parallel to both King Street and Smith Street between Beaufain Street and Queen Street. What is the distance from Smith to Logan along Beaufain? Round to the nearest foot.

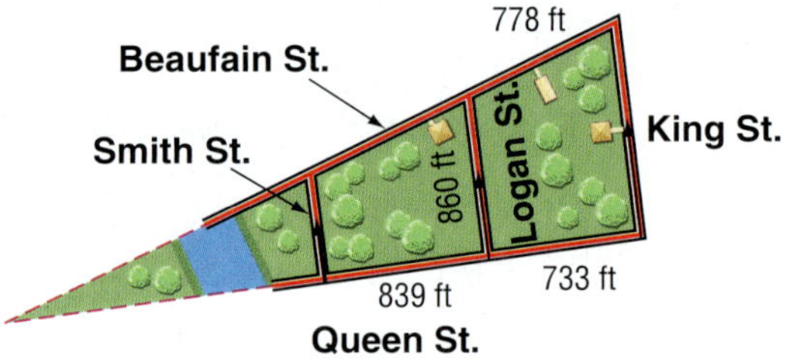

23. ART Tonisha drew the line of dancers shown below for her perspective project in art class. Each of the dancers is parallel. Find the lower distance between the first two dancers.

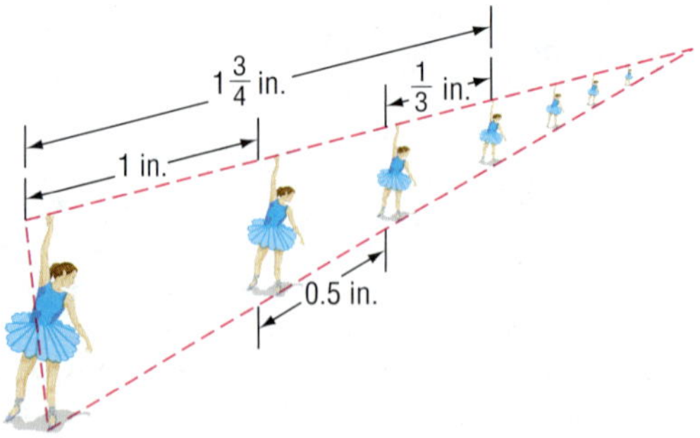

Example 5
p. 488

ALGEBRA Find x and y.

24.

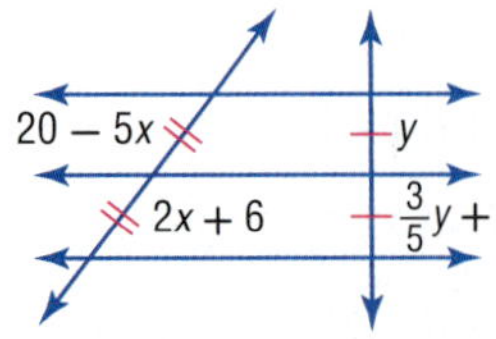

25.

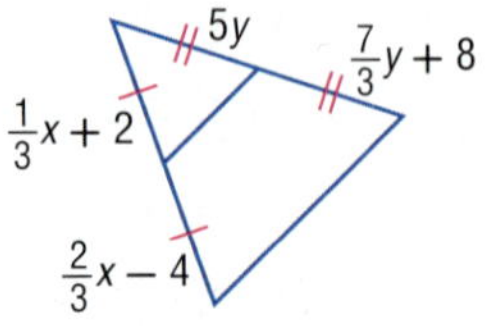

ALGEBRA **Find x and y.**

26.

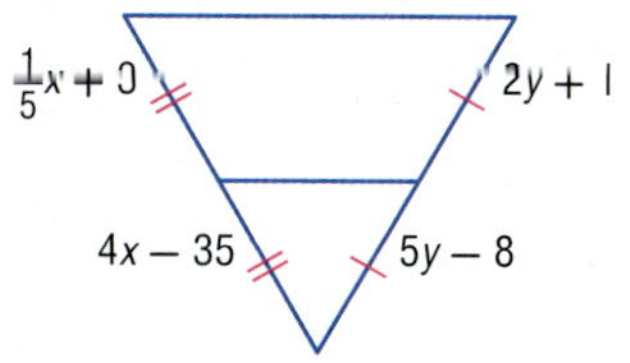

27.

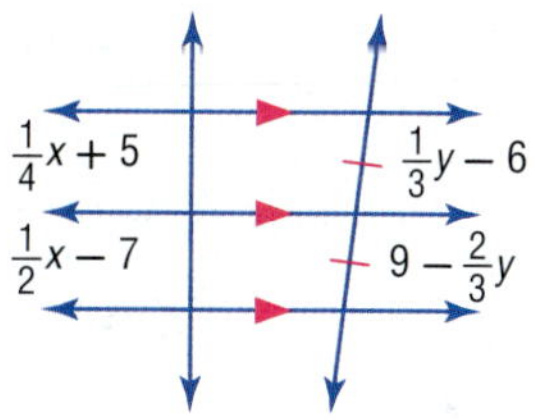

PROOF **Write a paragraph proof.**

28. Corollary 7.1

29. Corollary 7.2

30. Theorem 7.5

PROOF **Write a two-column proof.**

31. Theorem 7.6

32. Theorem 7.7

Refer to $\triangle QRS$.

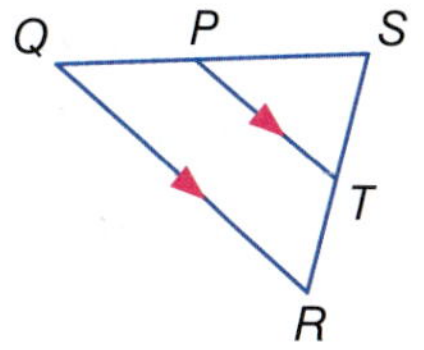

33. If $ST = 8$, $TR = 4$, and $PT = 6$, find QR.

34. If $SP = 4$, $PT = 6$, and $QR = 12$, find SQ.

35. If $CE = t - 2$, $EB = t + 1$, $CD = 2$, and $CA = 10$, find t and CE.

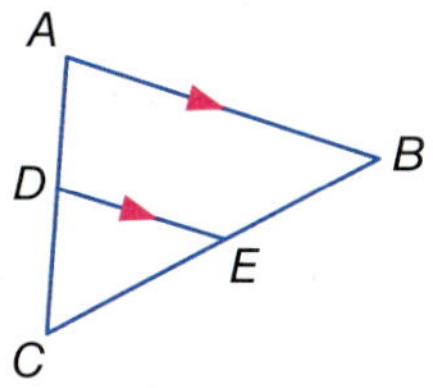

36. If $WX = 7$, $WY = a$, $WV = 6$, and $VZ = a - 9$, find WY.

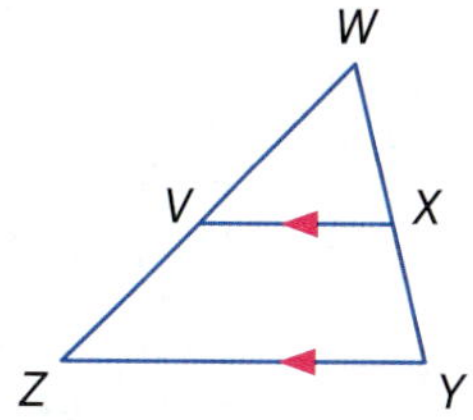

37. If $QR = 2$, $XW = 12$, $QW = 15$, and $ST = 5$, find RS and WV.

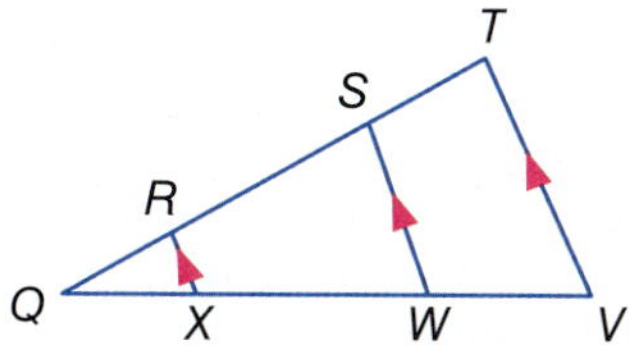

38. If $LK = 4$, $MP = 3$, $PQ = 6$, $KJ = 2$, $RS = 6$, and $LP = 2$, find ML, QR, QK, and JH.

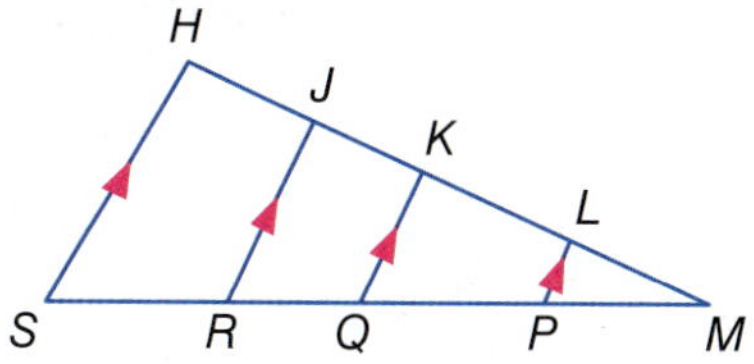

Math History Link

Galileo Galilei (1564–1642) Galileo was born in Pisa, Italy. He studied philosophy, astronomy, and mathematics. Galileo made essential contributions to all three disciplines.

Source: *Encyclopaedia Britannica*

39. **MATH HISTORY** The sector compass was a tool perfected by Galileo in the sixteenth century for measurement. To draw a segment two-fifths the length of a given segment, align the ends of the arms with the given segment. Then draw a segment at the 40 mark. Write a justification that explains why the sector compass works for proportional measurement.

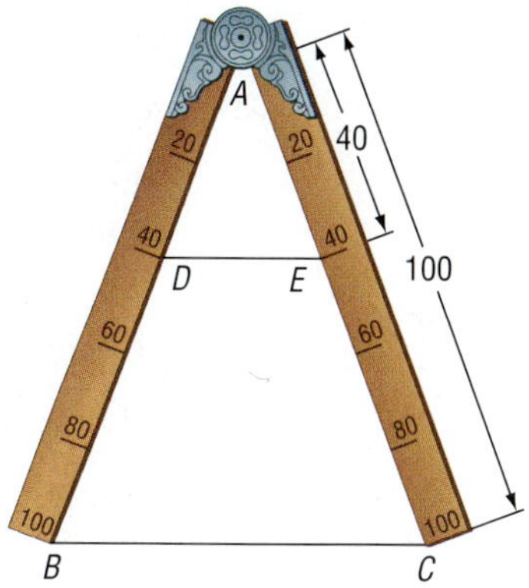

Determine the value of x so that $\overline{BC} \parallel \overline{DF}$.

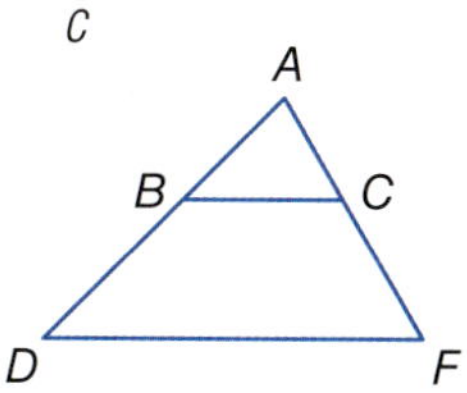

40. $AB = x + 5$, $BD = 12$, $AC = 3x + 1$, and $CF = 15$

41. $AC = 15$, $BD = 3x - 2$, $CF = 3x + 2$, and $AB = 12$

42. **COORDINATE GEOMETRY** $\triangle ABC$ has vertices $A(-8, 7)$, $B(0, 1)$, and $C(7, 5)$. Draw $\triangle ABC$. Determine the coordinates of the midsegment of $\triangle ABC$ that is parallel to $\overline{BC}$. Justify your answer.

43 **HOUSES** Refer to the diagram of the gable at the right. Each piece of siding is a uniform width. Find the lengths of $\overline{FG}$, $\overline{EH}$, and $\overline{DJ}$.

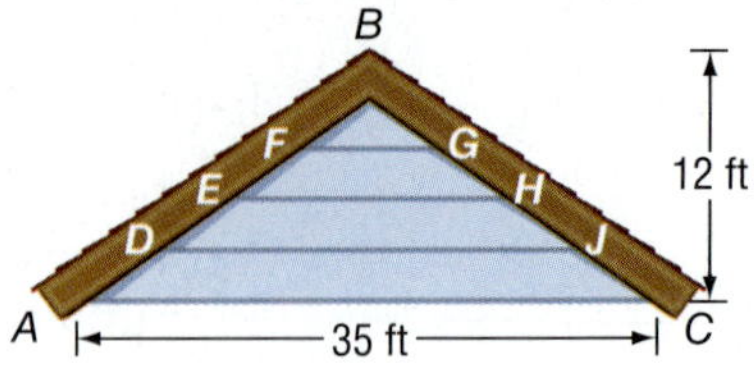

StudyTip

Constructions Remember that a compass and straightedge are the only tools used in constructions.

CONSTRUCTIONS **Construct each segment as directed.**

44. a segment separated into five congruent segments

45. a segment separated into two segments in which their lengths have a ratio of 1 to 3

46. a segment 3 inches long, separated into four congruent segments

47. **MULTIPLE REPRESENTATIONS** In this problem, you will explore angle bisectors and proportions.

a. **GEOMETRIC** Draw three triangles, one acute, one right, and one obtuse. Label one triangle ABC and draw angle bisector $\overrightarrow{BD}$. Label the second MNP with angle bisector $\overrightarrow{NQ}$ and the third WXY with angle bisector $\overrightarrow{XZ}$.

b. **TABULAR** Complete the table at the right with the appropriate values.

c. **VERBAL** Make a conjecture about the segments of a triangle created by an angle bisector.

Triangle	Length		Ratio	
ABC	AD		$\frac{AD}{CD}$	
	CD			
	AB		$\frac{AB}{CB}$	
	CB			
MNP	MQ		$\frac{MQ}{PQ}$	
	PQ			
	MN		$\frac{MN}{PN}$	
	PN			
WXY	WZ		$\frac{WZ}{YZ}$	
	YZ			
	WX		$\frac{WX}{YX}$	
	YX			

H.O.T. Problems Use Higher-Order Thinking Skills

48. **ERROR ANALYSIS** Jacob and Sebastian are finding the value of x in $\triangle JHL$. Jacob says that MP is one half of JL, so x is 4.5. Sebastian says that JL is one half of MP, so x is 18. Is either of them correct? Explain.

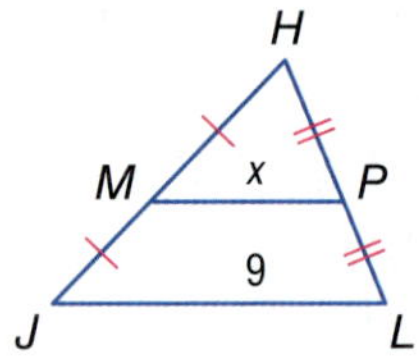

49. **REASONING** In $\triangle ABC$, $AF = FB$ and $AH = HC$. If D is $\frac{3}{4}$ of the way from A to B and E is $\frac{3}{4}$ of the way from A to C, is DE *always*, *sometimes*, or *never* $\frac{3}{4}$ of BC? Explain.

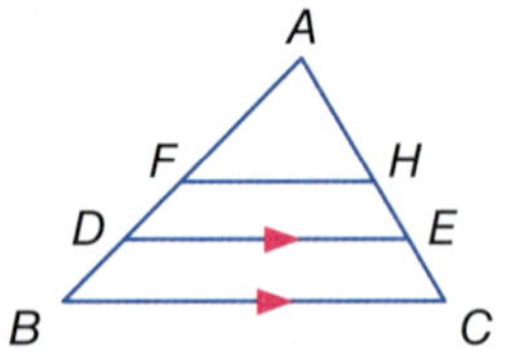

50. **CHALLENGE** Write a two-column proof.

Given: $AB = 4$ and $BC = 4$, $CD = DE$

Prove: $\overline{BD} \parallel \overline{AE}$

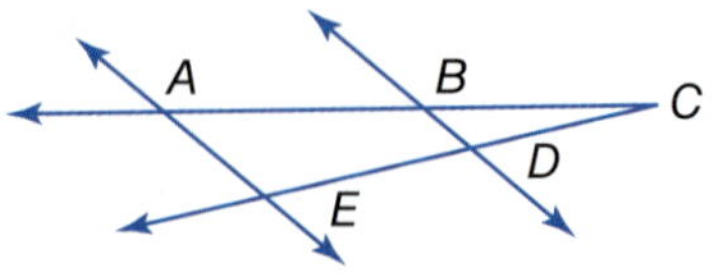

51. **OPEN ENDED** Draw three segments, a, b, and c, of all different lengths. Draw a fourth segment, d, such that $\frac{a}{b} = \frac{c}{d}$.

52. **WRITING IN MATH** Compare the Triangle Proportionality Theorem and the Triangle Midsegment Theorem.

Standardized Test Practice

53. SHORT RESPONSE What is the value of x?

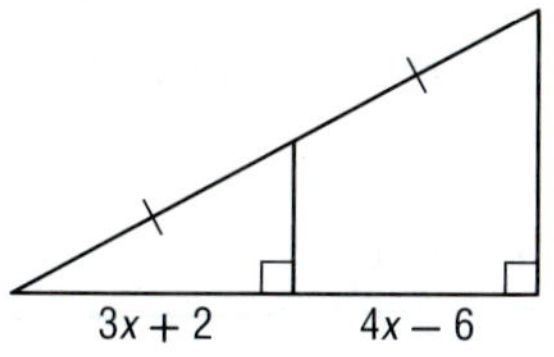

54. If the vertices of triangle JKL are (0, 0), (0, 10) and (10, 10) then the area of triangle JKL is

A 20 units^2 **C** 40 units^2
B 30 units^2 **D** 50 units^2

55. ALGEBRA A breakfast cereal contains wheat, rice, and oats in the ratio 2:4:1. If the manufacturer makes a mixture using 110 pounds of wheat, how many pounds of rice will be used?

F 120 lb **H** 240 lb
G 220 lb **J** 440 lb

56. SAT/ACT If the area of a circle is 16 square meters, what is its radius in meters?

A $\frac{8}{\pi}$ **C** 12π
B $\frac{4\sqrt{\pi}}{\pi}$ **D** $\frac{16}{\pi}$

Spiral Review

ALGEBRA Identify the similar triangles. Then find the measure(s) of the indicated segment(s). (Lesson 7-3)

57. $\overline{AB}$

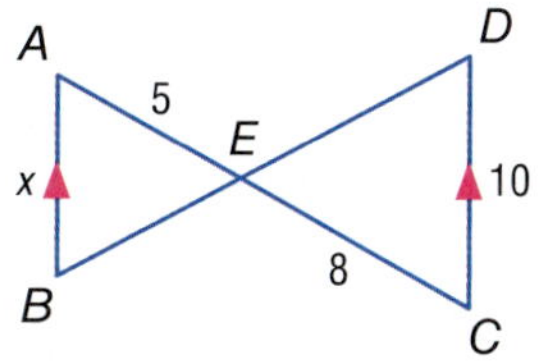

58. $\overline{RT}$, $\overline{RS}$

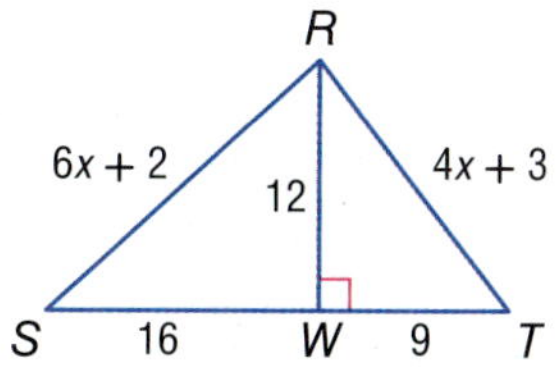

59. $\overline{TY}$

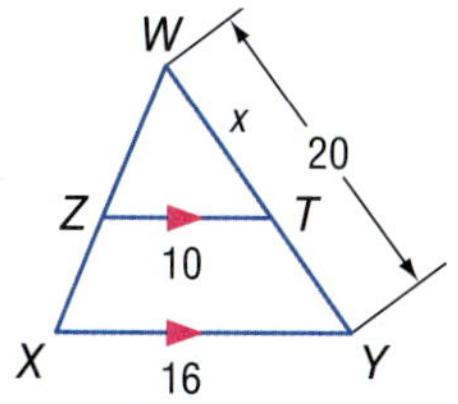

60. SURVEYING Mr. Turner uses a carpenter's square to find the distance across a stream. The carpenter's square models right angle NOL. He puts the square on top of a pole that is high enough to sight along $\overline{OL}$ to point P across the river. Then he sights along $\overline{ON}$ to point M. If MK is 1.5 feet and OK is 4.5 feet, find the distance KP across the stream. (Lesson 7-2)

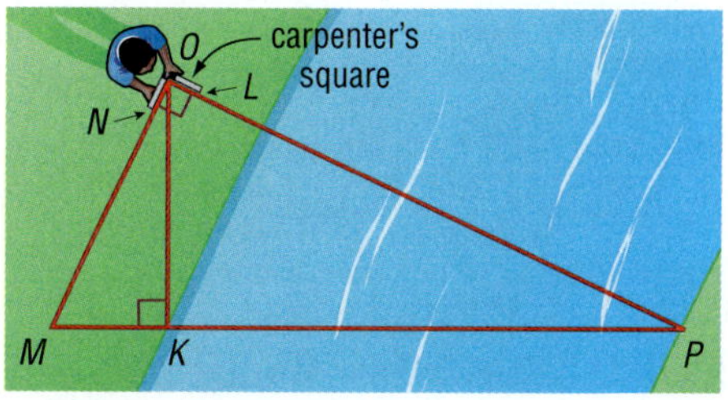

COORDINATE GEOMETRY For each quadrilateral with the given vertices, verify that the quadrilateral is a trapezoid and determine whether the figure is an isosceles trapezoid. (Lesson 6-6)

61. $Q(-12, 1)$, $R(-9, 4)$, $S(-4, 3)$, $T(-11, -4)$

62. $A(-3, 3)$, $B(-4, -1)$, $C(5, -1)$, $D(2, 3)$

Point S is the incenter of $\triangle JPL$. Find each measure. (Lesson 5-1)

63. SQ

64. QJ

65. $m\angle MPQ$

66. $m\angle SJP$

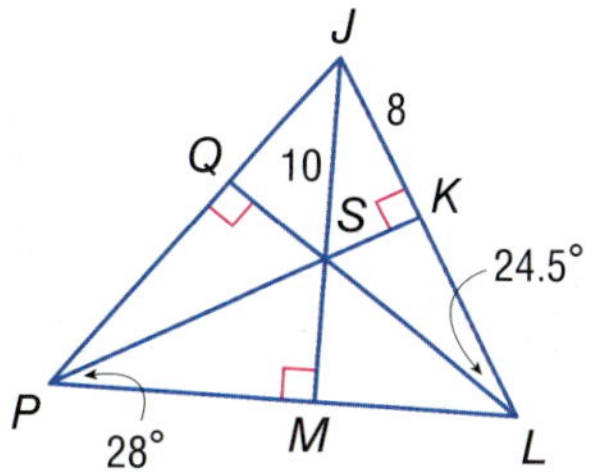

Skills Review

Solve each proportion. (Lesson 7-1)

67. $\frac{1}{3} = \frac{x}{2}$

68. $\frac{3}{4} = \frac{5}{x}$

69. $\frac{2.3}{4} = \frac{x}{3.7}$

70. $\frac{x-2}{2} = \frac{4}{5}$

71. $\frac{x}{12-x} = \frac{8}{3}$

CHAPTER 7

Mid-Chapter Quiz

Lessons 7-1 through 7-4

Solve each proportion. (Lesson 7-1)

1. $\frac{2}{5} = \frac{x}{25}$

2. $\frac{10}{3} = \frac{7}{x}$

3. $\frac{y+4}{11} = \frac{y-2}{9}$

4. $\frac{z-1}{3} = \frac{8}{z+1}$

5. BASEBALL A pitcher's earned run average or ERA is the product of 9 and the ratio of earned runs the pitcher has allowed to the number of innings pitched. During the 2007 season, Johan Santana of the Minnesota Twins allowed 81 earned runs in 219 innings pitched. Find his ERA to the nearest hundredth. (Lesson 7-1)

Each pair of polygons is similar. Find the value of x. (Lesson 7-2)

6.

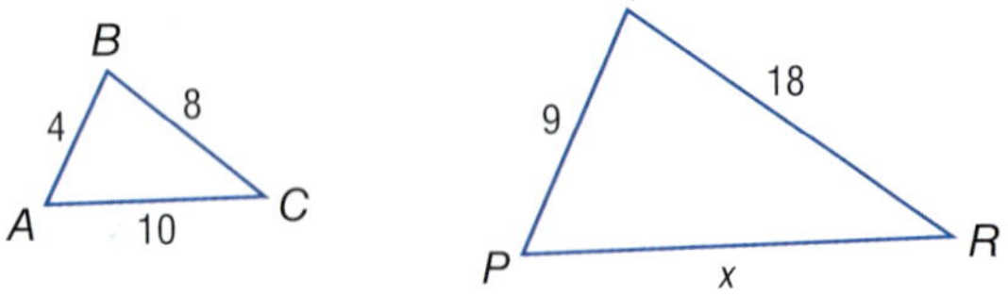

7.

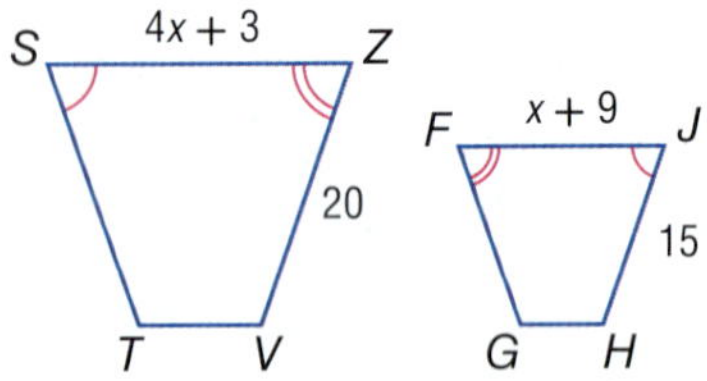

8. MULTIPLE CHOICE Two similar polygons have a scale factor of 3:5. The perimeter of the large polygon is 120 feet. Find the perimeter of the small polygon. (Lesson 7-2)

A 68 ft

B 72 ft

C 192 ft

D 200 ft

Determine whether the triangles are similar. If so, write a similarity statement. If not, what would be sufficient to prove the triangles similar? Explain your reasoning. (Lesson 7-3)

9.

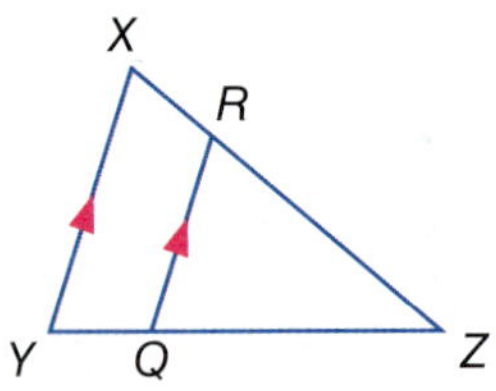

10.

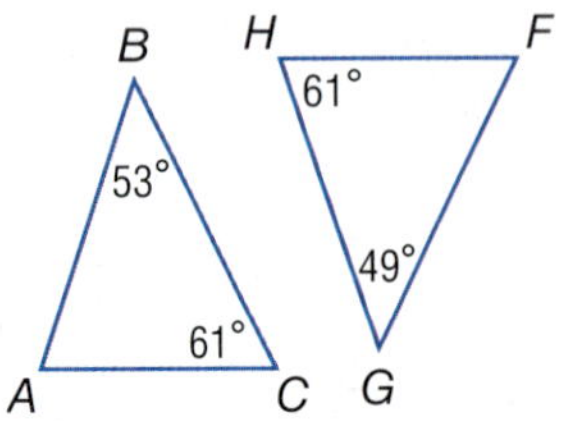

ALGEBRA Identify the similar triangles. Find each measure. (Lesson 7-3)

11. SR

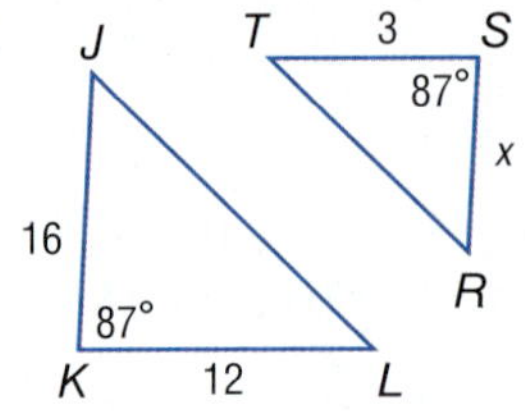

12. AF

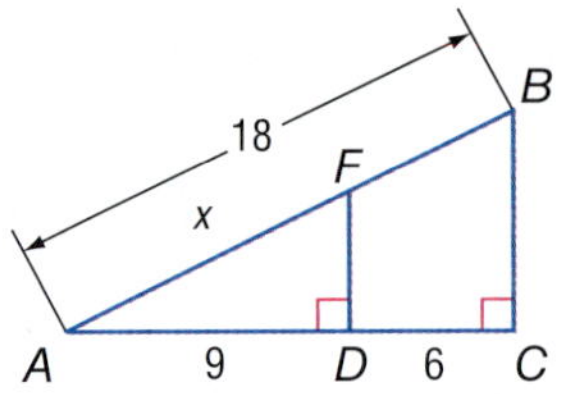

13. HISTORY In the fifteenth century, mathematicians and artists tried to construct the perfect letter. A square was used as a frame to design the letter "A," as shown below. The thickness of the major stroke of the letter was $\frac{1}{12}$ the height of the letter. (Lesson 7-4)

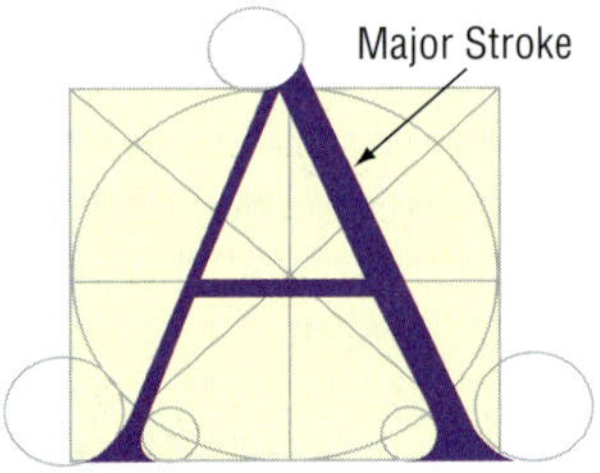

a. Explain why the bar through the middle of the *A* is half the length of the space between the outside bottom corners of the sides of the letter.

b. If the letter were 3 centimeters tall, how wide would the major stroke be?

ALGEBRA Find x and y. (Lesson 7-4)

14.

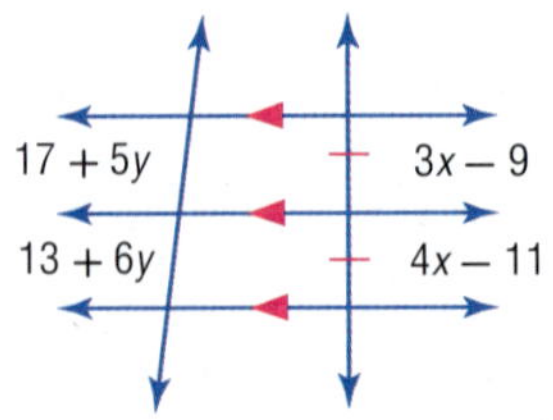

15.

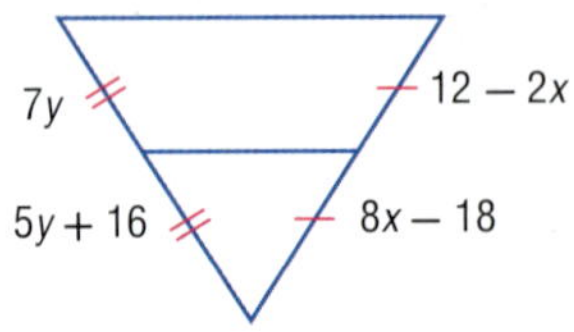

7-5 Parts of Similar Triangles

Then

You learned that corresponding sides of similar polygons are proportional. (Lesson 7-2)

Now

- Recognize and use proportional relationships of corresponding angle bisectors, altitudes, and medians of similar triangles.
- Use the Triangle Bisector Theorem.

Why?

The "Rule of Thumb" uses the average ratio of a person's arm length to the distance between his or her eyes and the altitudes of similar triangles to estimate the distance between a person and an object of approximately known width.

Special Segments of Similar Triangles You learned in Lesson 7-2 that the corresponding side lengths of similar polygons, such as triangles, are proportional. This concept can be extended to other segments in triangles.

Theorems

Special Segments of Similar Triangles

7.8 If two triangles are similar, the lengths of corresponding altitudes are proportional to the lengths of corresponding sides.

Abbreviation ~△s *have corr. altitudes proportional to corr. sides.*

Example If $\triangle ABC \sim \triangle FGH$, then $\frac{AD}{FJ} = \frac{AB}{FG}$.

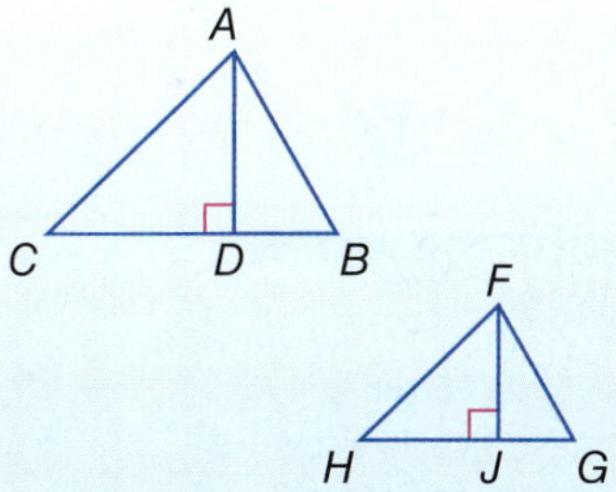

7.9 If two triangles are similar, the lengths of corresponding angle bisectors are proportional to the lengths of corresponding sides.

Abbreviation ~△s *have corr. ∠ bisectors proportional to corr. sides.*

Example If $\triangle KLM \sim \triangle QRS$, then $\frac{LP}{RT} = \frac{LM}{RS}$.

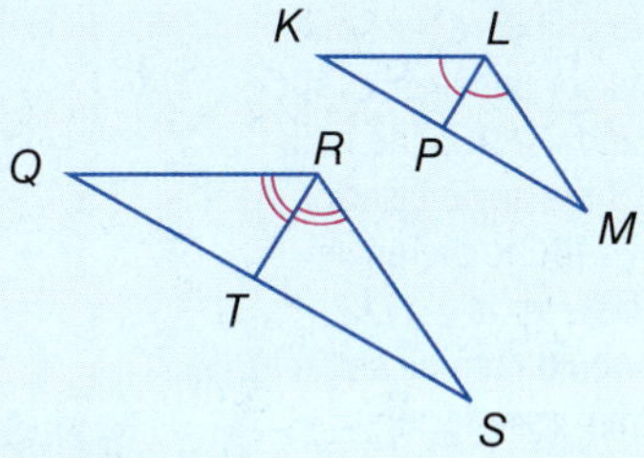

7.10 If two triangles are similar, the lengths of corresponding medians are proportional to the lengths of corresponding sides.

Abbreviation ~△s *have corr. medians proportional to corr. sides.*

Example If $\triangle ABC \sim \triangle WXY$, then $\frac{CD}{YZ} = \frac{AB}{WX}$.

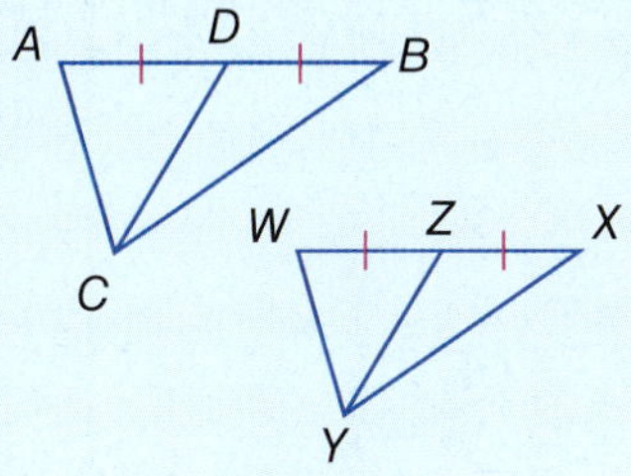

You will prove Theorems 7.9 and 7.10 in Exercises 18 and 19, respectively.

PROOF Theorem 7.8

Given: $\triangle FGH \sim \triangle KLM$
$\overline{FJ}$ and $\overline{KP}$ are altitudes.

Prove: $\frac{FJ}{KP} = \frac{HF}{MK}$

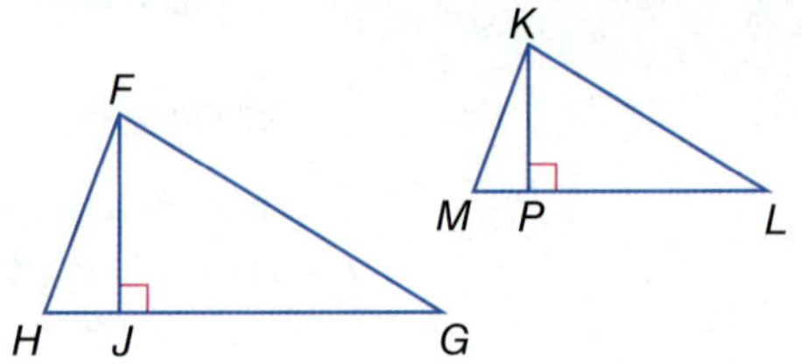

Paragraph Proof:

Since $\triangle FGH \sim \triangle KLM$, $\angle H \cong \angle M$. $\angle FJH \cong \angle KPM$ because they are both right angles created by the altitudes drawn to the opposite side and all right angles are congruent. Thus $\triangle HFJ \sim \triangle MKP$ by AA Similarity. So $\frac{FJ}{KP} = \frac{HF}{MK}$ by the definition of similar polygons.

Since the corresponding altitudes are chosen at random, we need not prove Theorem 7.8 for every pair of altitudes.

You can use special segments in similar triangles to find missing measures.

EXAMPLE 1 Use Special Segments in Similar Triangles

In the figure, $\triangle ABC \sim \triangle FDG$. Find the value of x.

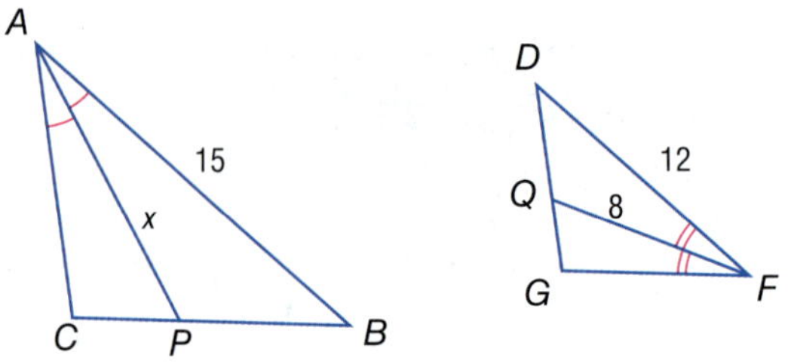

$\overline{AP}$ and $\overline{FQ}$ are corresponding angle bisectors and $\overline{AB}$ and $\overline{FD}$ are corresponding sides of similar triangles ABC and FDG.

$\frac{AP}{FQ} = \frac{AB}{FD}$	**~△s have corr. ∠ bisectors proportional to the corr. sides.**
$\frac{x}{8} = \frac{15}{12}$	**Substitution**
$8 \cdot 15 = x \cdot 12$	**Cross Products Property**
$120 = 12x$	**Simplify.**
$10 = x$	**Divide each side by 12.**

StudyTip

Use Scale Factor Example 1 could also have been solved by first finding the scale factor between $\triangle ABC$ and $\triangle FDG$. The ratio of the angle bisector in $\triangle ABC$ to the angle bisector in $\triangle FDG$ would then be equal to this scale factor.

Guided Practice

Find the value of x.

1A.

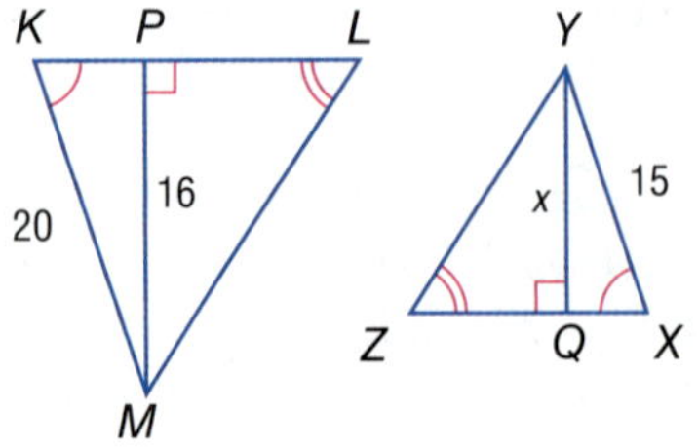

1B.

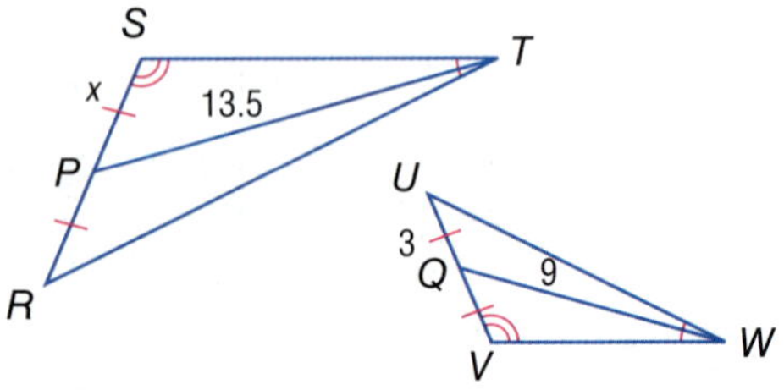

Personal Tutor glencoe.com

You can use special segments in similar triangles to solve real-world problems.

Real-World Link

Hold your outstretched hand horizontal at arm's length with your palm facing you; for each hand width the sun is above the horizon, there is one remaining hour of sunlight.

Source: Sail Island Channels

Real-World EXAMPLE 2 Use Similar Triangles to Solve Problems

ESTIMATING DISTANCES Liliana holds her arm straight out in front of her with her elbow straight and her thumb pointing up. Closing one eye, she aligns one edge of her thumb with a car she is sighting. Next she switches eyes without moving her head or her arm. The car appears to jump 4 car widths. If Liliana's arm is about 10 times longer than the distance between her eyes, and the car is about 5.5 feet wide, estimate the distance from Liliana's thumb to the car.

Understand Make a diagram of the situation labeling the given distances and the distance you need to find as x. Also, label the vertices of the triangles formed.

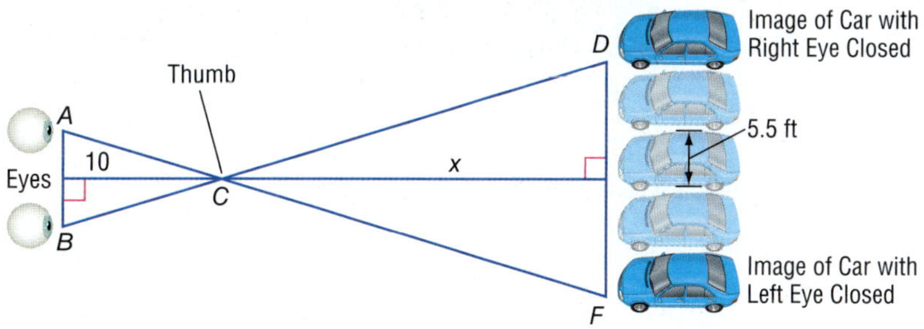

Note: Not drawn to scale.

We assume that if Liliana's thumb is straight out in front of her, then $\overline{PC}$ is an altitude of $\triangle ABC$. Likewise, $\overline{QC}$ is the corresponding altitude. We assume that $\overline{AB} \parallel \overline{DF}$.

Plan Since $\overline{AB} \parallel \overline{DF}$, $\angle BAC \cong \angle DFC$ and $\angle CBA \cong \angle CDF$ by the Alternate Interior Angles Theorem. Therefore $\triangle ABC \sim \triangle FDC$ by AA Similarity. Write a proportion and solve for x.

Solve

$\frac{PC}{QC} = \frac{AB}{DF}$ **Theorem 7.8**

$\frac{10}{x} = \frac{1}{5.5 \cdot 4}$ **Substitution**

$\frac{10}{x} = \frac{1}{22}$ **Simplify.**

$10 \cdot 22 = x \cdot 1$ **Cross Products Property**

$220 = x$ **Simplify.**

So the estimated distance to the car is 200 feet.

Check The ratio of Liliana's arm length to the width between her eyes is 10 to 1. The ratio of the distance to the car to the distance the image of the car jumped is 22 to 220 or 10 to 1. ✓

Guided Practice

2. Suppose Liliana stands at the back of her classroom and sights a clock on the wall at the front of the room. If the clock is 30 centimeters wide and appears to move 3 clock widths when she switches eyes, estimate the distance from Liliana's thumb to the clock.

Personal Tutor glencoe.com

Triangle Angle Bisector Theorem An angle bisector of a triangle also divides the side opposite the angle proportionally.

> **StudyTip**
>
> **Proportions** Another proportion that could be written using the Triangle Angle Bisector Theorem is $\frac{KM}{KJ} = \frac{LM}{LJ}$.

Theorem 7.11 Triangle Angle Bisector

An angle bisector in a triangle separates the opposite side into two segments that are proportional to the lengths of the other two sides.

Example If $\overline{JM}$ is an angle bisector of $\triangle JKL$,

then $\frac{KM}{LM} = \frac{KJ}{LJ}$. ← segments with vertex *K* / ← segments with vertex *L*

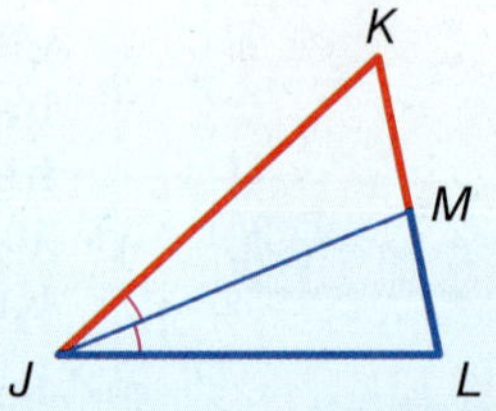

You will prove Theorem 7.11 in Exercise 25.

EXAMPLE 3 Use the Triangle Angle Bisector Theorem

Find x.

Since $\overline{RT}$ is an angle bisector of $\triangle QRS$, you can use the Triangle Angle Bisector Theorem to write a proportion.

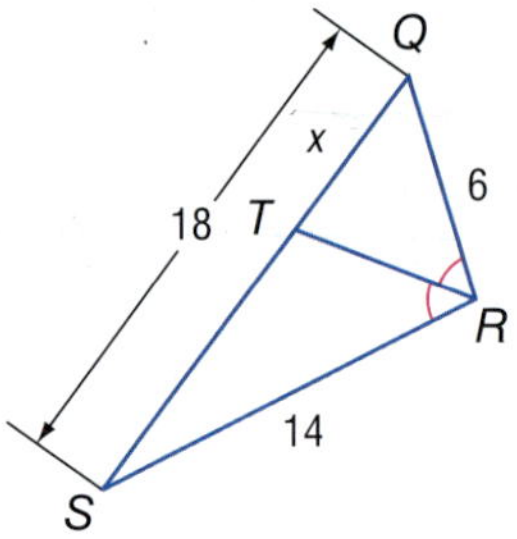

$\frac{QT}{ST} = \frac{QR}{SR}$	Triangle Angle Bisector Theorem
$\frac{x}{18 - x} = \frac{6}{14}$	Substitution
$(18 - x)(6) = x \cdot 14$	Cross Products Property
$108 - 6x = 14x$	Simplify.
$108 = 20x$	Add $6x$ to each side.
$5.4 = x$	Divide each side by 20.

Guided Practice

3A.

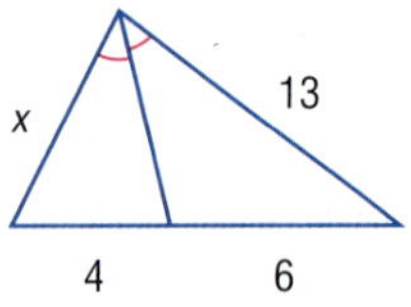

3B.

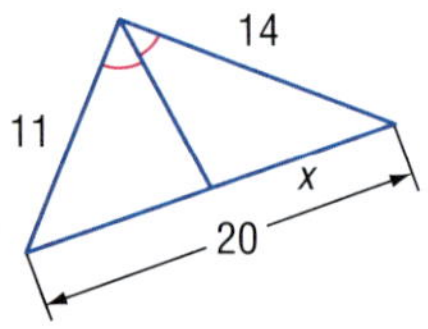

Personal Tutor glencoe.com

Check Your Understanding

Example 1 p. 496

Find x.

1

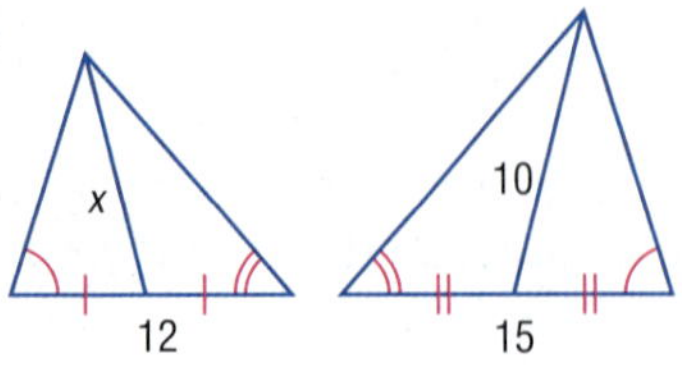

2.

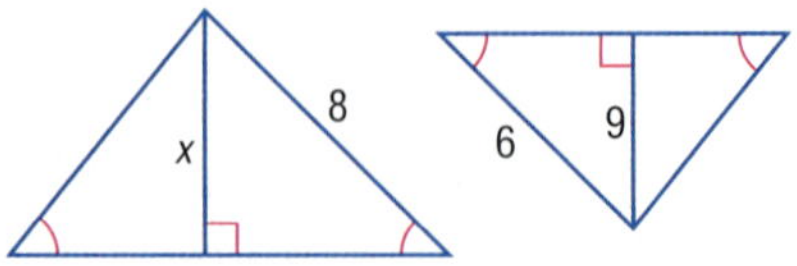

Example 2 p. 497

3. VISION A cat that is 10 inches tall forms a retinal image that is 7 millimeters tall. If $\triangle ABE \sim \triangle DBC$ and the distance from the pupil to the retina is 25 millimeters, how far away from your pupil is the cat?

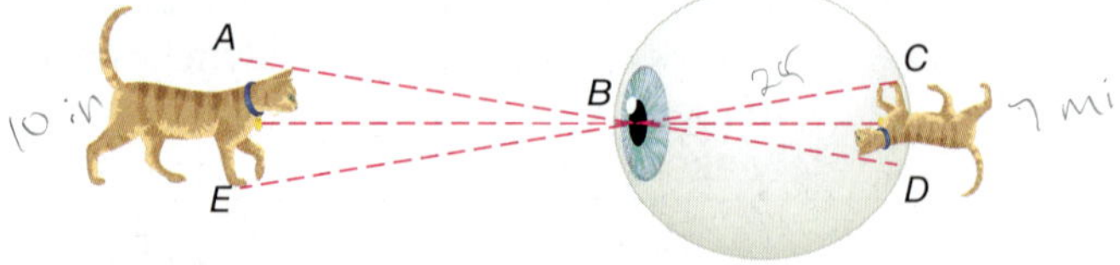

Example 3
p. 498

Find the value of each variable.Find the value of each variable.

4.

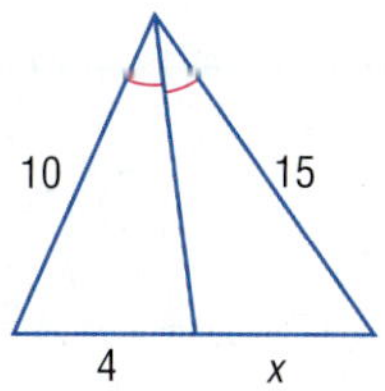

5.

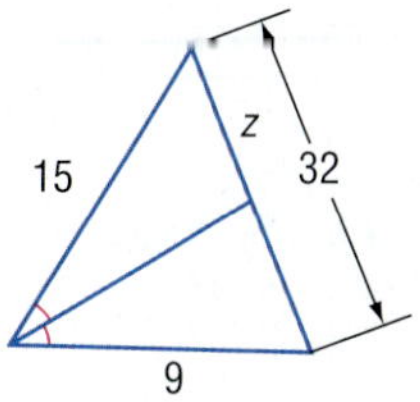

Practice and Problem Solving

● = **Step-by-Step Solutions** begin on page R20.
Extra Practice begins on page 969.

Example 1
p. 496

Find x.

6.

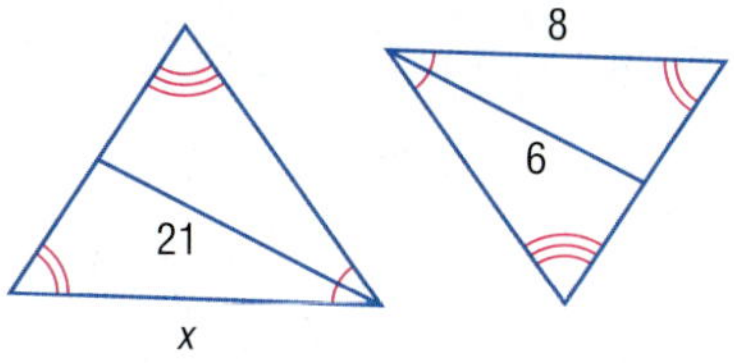

7.

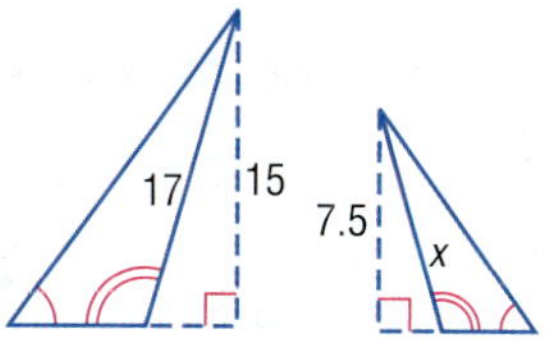

8.

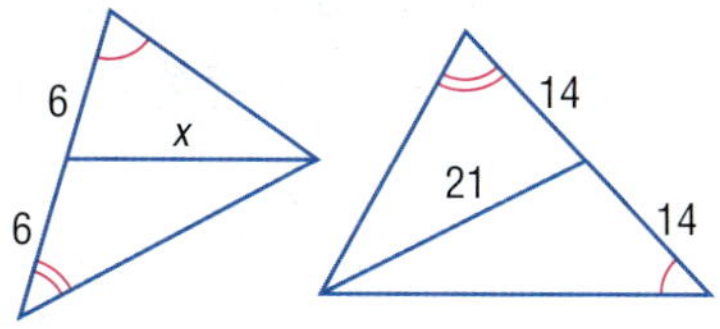

9.

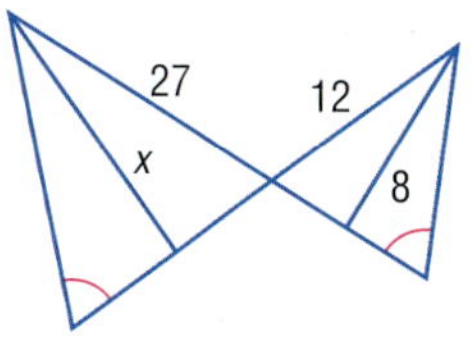

Example 2
p. 497

10. **ROADWAYS** The intersection of the two roads shown forms two similar triangles. If AC is 382 feet, MP is 248 feet, and the gas station is 50 feet from the intersection, how far from the intersection is the bank?

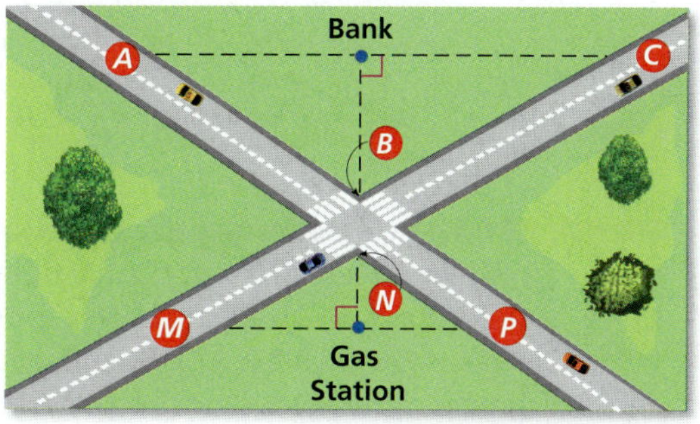

Example 3
p. 498

Find the value of each variable.

11

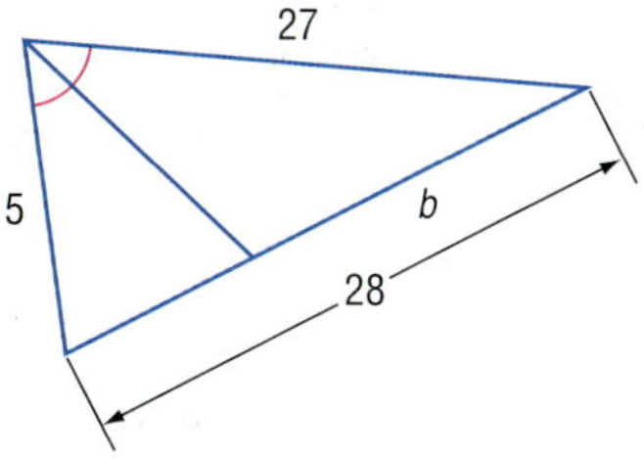

12.

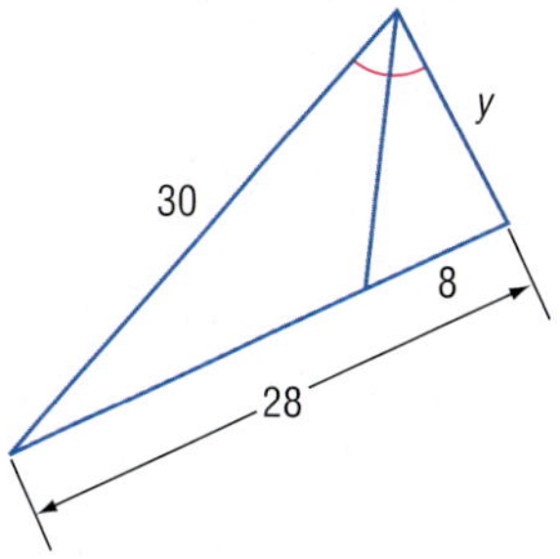

13.

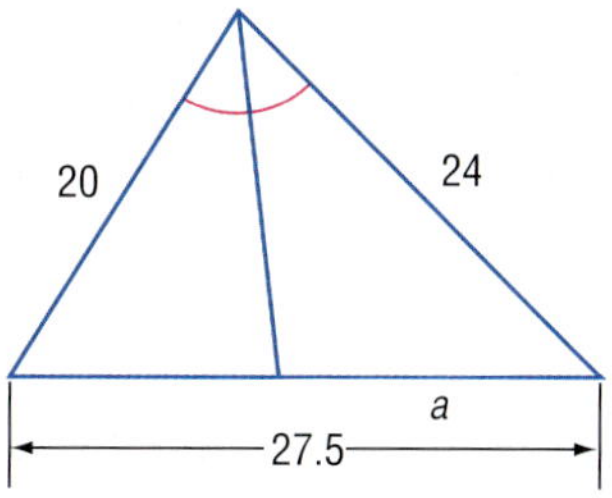

14.

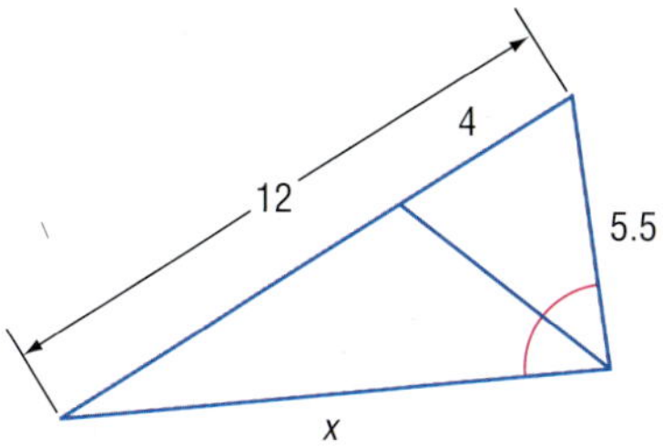

15. **ALGEBRA** If $\overline{AB}$ and $\overline{JK}$ are altitudes, $\triangle DAC \sim \triangle MJL$, $AB = 9$, $AD = 4x - 8$, $JK = 21$, $JM = 5x + 3$, find x.

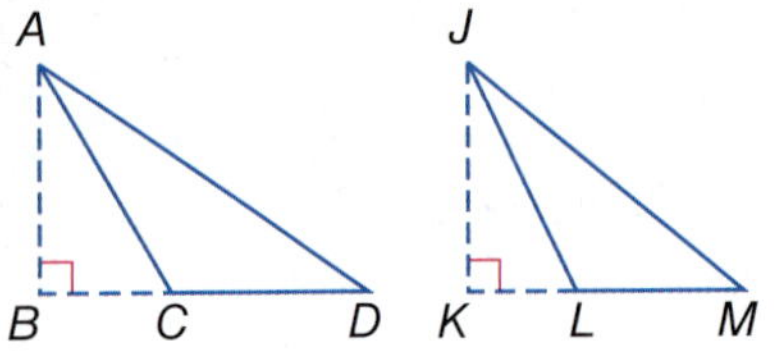

16. **ALGEBRA** If $\overline{NQ}$ and $\overline{VX}$ are medians, $\triangle PNR \sim \triangle WVY$, $NQ = 8$, $PR = 12$, $WY = 7x - 1$, and $VX = 4x + 2$, find x.

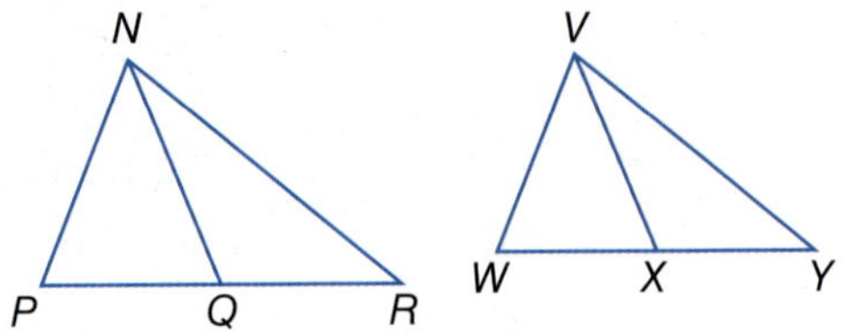

17. If $\triangle SRY \sim \triangle WXQ$, $\overline{RT}$ is an altitude of $\triangle SRY$, $\overline{XV}$ is an altitude of $\triangle WXQ$, $RT = 5$, $RQ = 4$, $QY = 6$, and $YX = 2$, find XV.

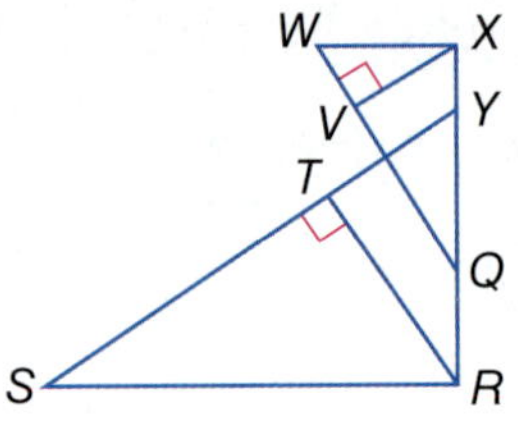

18. **PROOF** Write a paragraph proof of Theorem 7.9.

19. **PROOF** Write a two-column proof of Theorem 7.10.

ALGEBRA **Find** x.

20.

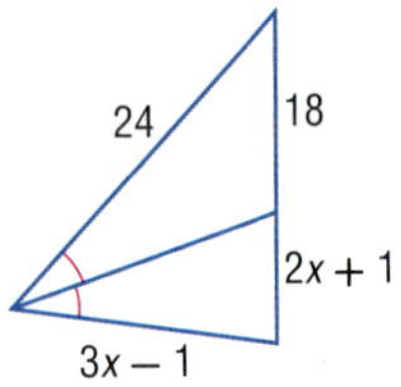

21.

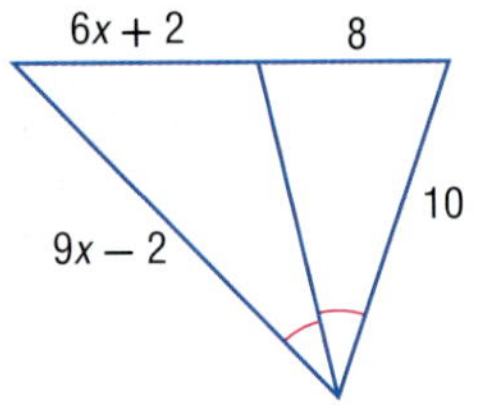

22.

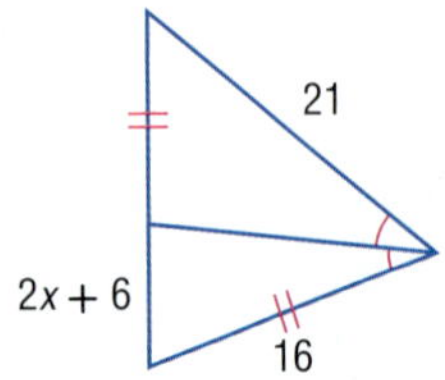

23.

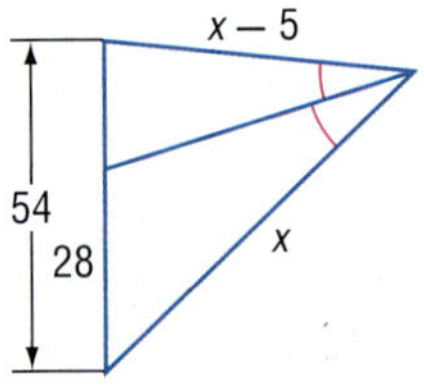

StudyTip

Proportions Given $\frac{a}{b} = \frac{c}{d}$, if $a > c$, then $b > d$. The converse is also true. If $b > d$, then $a > c$.

24. **SPORTS** Consider the triangle formed by the path between a batter, center fielder, and right fielder as shown. If the batter gets a hit that bisects the triangle at $\angle B$, is the center fielder or the right fielder closer to the ball? Explain your reasoning.

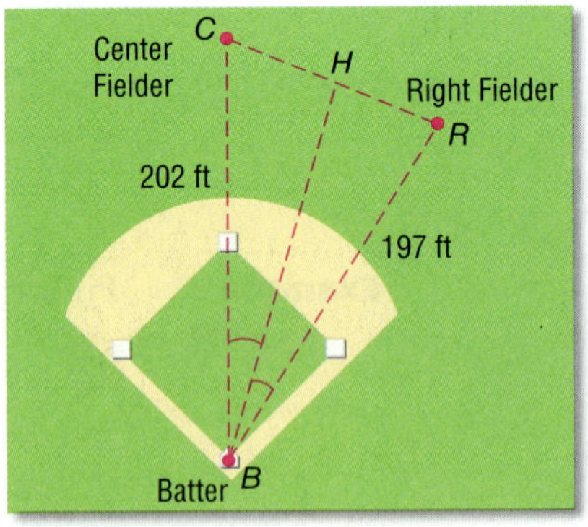

PROOF **Write a two-column proof.**

25. Theorem 7.11
Given: $\overline{CD}$ bisects $\angle ACB$.
By construction, $\overline{AE} \parallel \overline{CD}$.
Prove: $\frac{AD}{DB} = \frac{AC}{BC}$

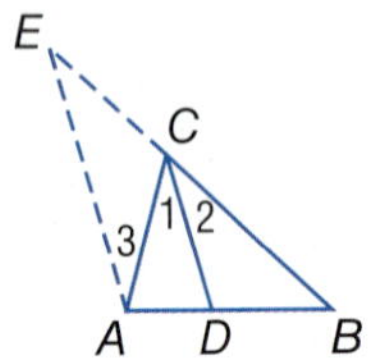

26. **Given:** $\angle H$ is a right angle.
L, K, and M are midpoints.
Prove: $\angle LKM$ is a right angle.

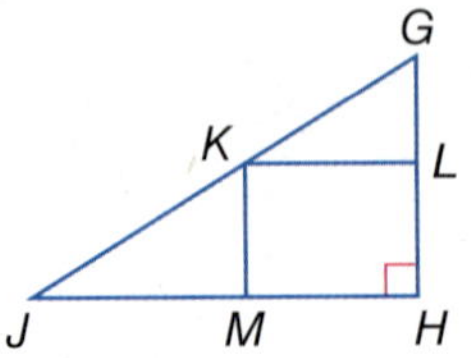

Real-World Career

Athletic Trainer
Athletic trainers help prevent and treat sports injuries. They ensure that protective equipment is used properly and that people understand safe practices that prevent injury. An athletic trainer must have a bachelor's degree to be certified. Most also have master's degrees.

PROOF Write a two-column proof.

27. Given: $\triangle QTS \sim \triangle XWZ$, $\overline{TR}$ and $\overline{WY}$ are angle bisectors.

Prove: $\frac{TR}{WY} = \frac{QT}{XW}$

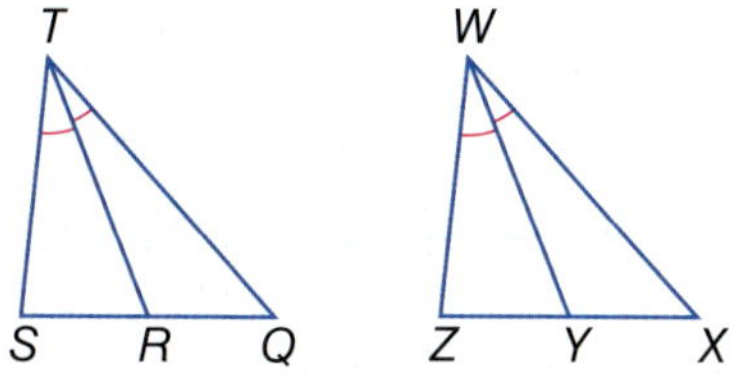

28. Given: $\overline{FD} \parallel \overline{BC}$, $\overline{BF} \parallel \overline{CD}$, $\overline{AC}$ bisects $\angle C$.

Prove: $\frac{DE}{EC} = \frac{BA}{AC}$

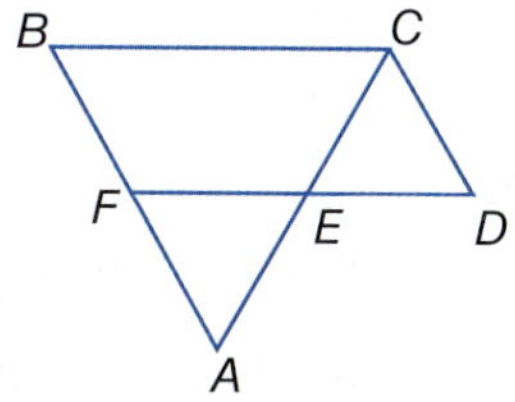

29. SPORTS During football practice, Trevor threw a pass to Ricardo as shown below. If Eli is farther from Trevor when he completes the pass to Ricardo and Craig and Eli move at the same speed, who will reach Ricardo to tackle him first?

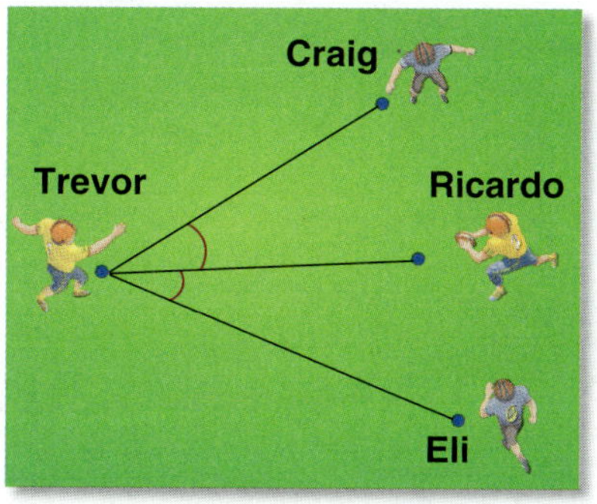

30. SHELVING In the triangular bookshelf shown, the distance between each of the shelves is 13 inches and $\overline{AK}$ is a median of $\triangle ABC$. If EF is $3\frac{1}{3}$ inches, what is BK?

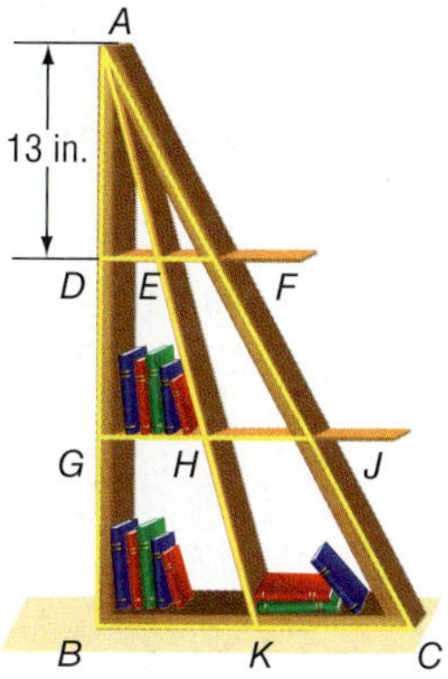

H.O.T. Problems

Use **H**igher-**O**rder **T**hinking Skills

31. ERROR ANALYSIS Chun and Traci are determining the value of x in the figure. Chun says to find x, solve the proportion $\frac{5}{8} = \frac{15}{x}$, but Traci says to find x, the proportion $\frac{5}{x} = \frac{8}{15}$ should be solved. Is either of them correct? Explain.

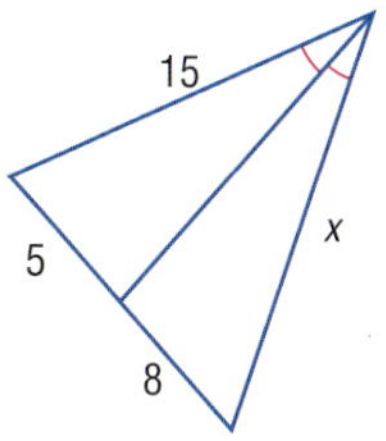

32. REASONING Find a counterexample to the following statement. Explain.

If the measure of an altitude and side of a triangle are proportional to the corresponding altitude and corresponding side of another triangle, then the triangles are similar.

33. CHALLENGE The perimeter of $\triangle PQR$ is 94 units. $\overline{QS}$ bisects $\angle PQR$. Find PS and RS.

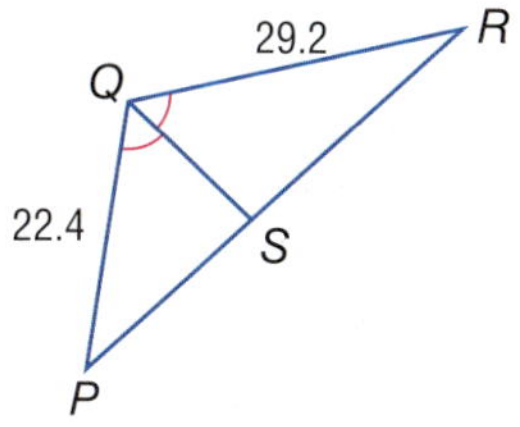

34. OPEN ENDED Draw two triangles so that the measures of corresponding medians and a corresponding side are proportional, but the triangles are not similar.

35. WRITING IN MATH Compare and contrast Theorem 7.9 and the Triangle Angle Bisector Theorem.

Standardized Test Practice

36. ALGEBRA Which shows 0.00234 written in scientific notation?

A 2.34×10^5
B 2.34×10^3
C 2.34×10^{-2}
D 2.34×10^{-3}

37. SHORT RESPONSE In the figures below, $\overline{DB} \cong \overline{BC}$ and $\overline{FH} \cong \overline{HE}$.

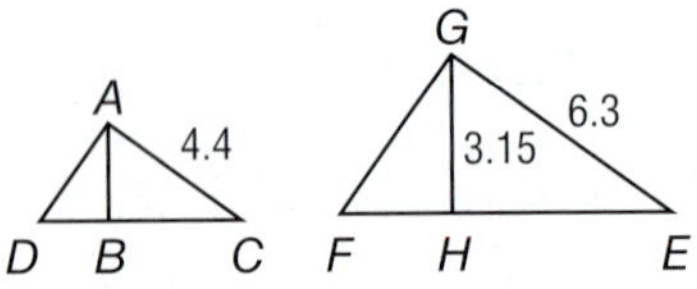

If $\triangle ACD \sim \triangle GEF$, find AB.

38. Quadrilateral $HJKL$ is a parallelogram. If the diagonals are perpendicular, which statement must be true?

F Quadrilateral $HJKL$ is a square.
G Quadrilateral $HJKL$ is a rectangle.
H Quadrilateral $HJKL$ is a rhombus.
J Quadrilateral $HJKL$ is an isosceles trapezoid.

39. SAT/ACT The sum of three numbers is 180. Two of the numbers are the same, and each of them is one third of the greatest number. What is the least number?

A 30
B 36
C 45
D 60

Spiral Review

ALGEBRA **Find x and y.** (Lesson 7-4)

40.

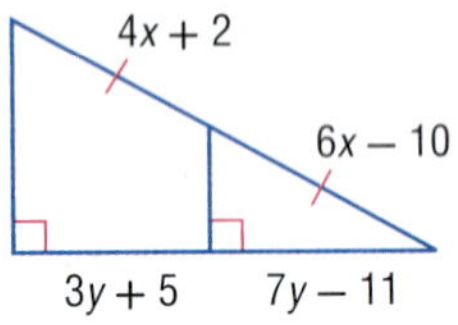

41.

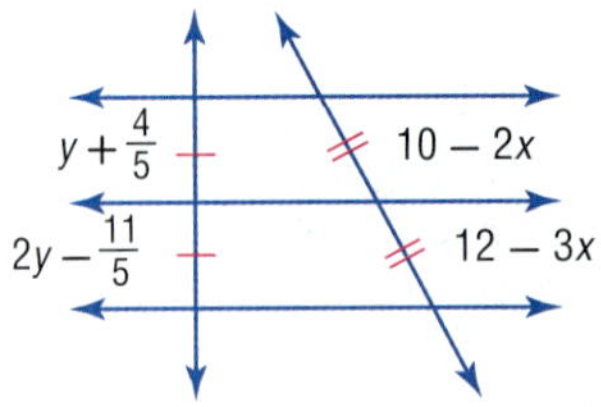

42.

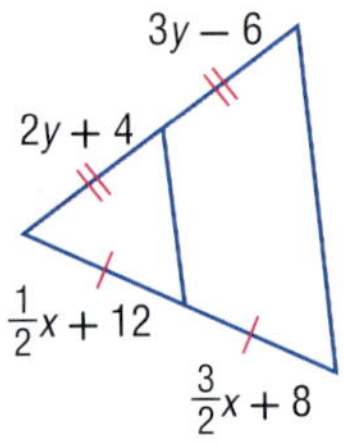

Find the indicated measure(s). (Lesson 7-3)

43. If $\overline{PR} \parallel \overline{KL}$, $KN = 9$, $LN = 16$, $PM = 2(KP)$, find KP, KM, MR, ML, MN, and PR.

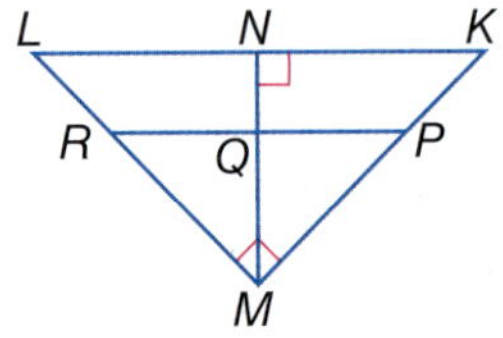

44. If $\overline{PR} \parallel \overline{WX}$, $WX = 10$, $XY = 6$, $WY = 8$, $RY = 5$, and $PS = 3$, find PY, SY, and PQ.

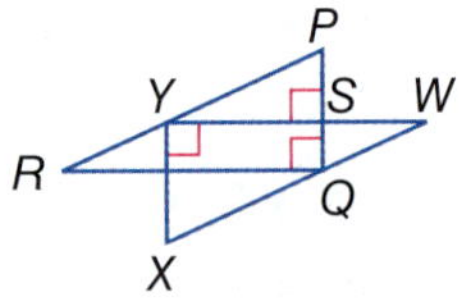

45. GEESE A flock of geese flies in formation. Prove that $\triangle EFG \cong \triangle HFG$ if $\overline{EF} \cong \overline{HF}$ and that G is the midpoint of $\overline{EH}$. (Lesson 4-4)

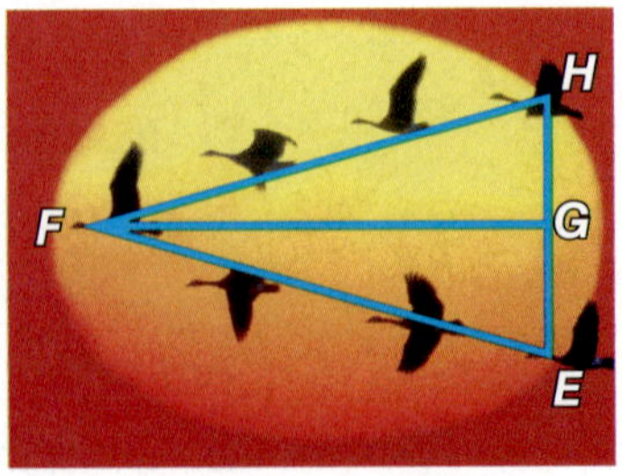

Skills Review

Find the distance between each pair of points. (Lesson 1-3)

46. $E(-3, -2)$, $F(5, 8)$

47. $A(2, 3)$, $B(5, 7)$

48. $C(-2, 0)$, $D(6, 4)$

49. $W(7, 3)$, $Z(-4, -1)$

50. $J(-4, -5)$, $K(2, 9)$

51. $R(-6, 10)$, $S(8, -2)$

EXTEND 7-5

Geometry Lab Fractals

A **fractal** is a geometric figure that is created using iteration. **Iteration** is a process of repeating the same operation over and over again. Fractals are **self-similar**, which means that the smaller details of the shape have the same geometric characteristics as the original form.

ACTIVITY 1

Stage 0 Draw an equilateral triangle on isometric dot paper in which each side is 8 units long.

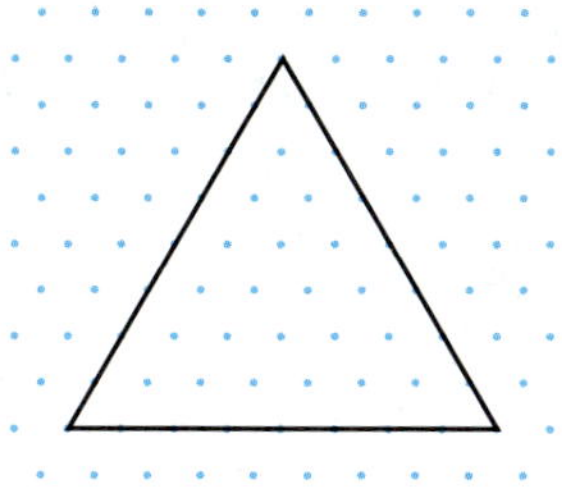

Stage 1 Connect the midpoints of the sides to form another triangle. Shade the center triangle.

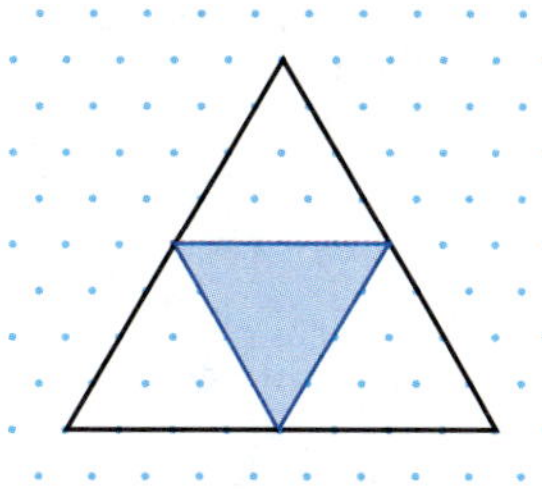

Stage 2 Repeat the process using the three unshaded triangles. Connect the midpoints of the sides to form three other triangles.

If you repeat this process indefinitely, the figure that results is called the Sierpinski Triangle.

Analyze the Results

1. If you continue the process, how many unshaded triangles will you have at Stage 3?

2. What is the perimeter of an unshaded triangle in Stage 4?

3. If you continue the process indefinitely, what will happen to the perimeters of the unshaded triangles?

4. **CHALLENGE** Complete the proof below.

Given: $\triangle KAP$ is equilateral. $D, F, M, B, C,$ and E are midpoints of $\overline{KA}, \overline{AP}, \overline{PK}, \overline{DA}, \overline{AF},$ and $\overline{FD}$, respectively.

Prove: $\triangle BAC \sim \triangle KAP$

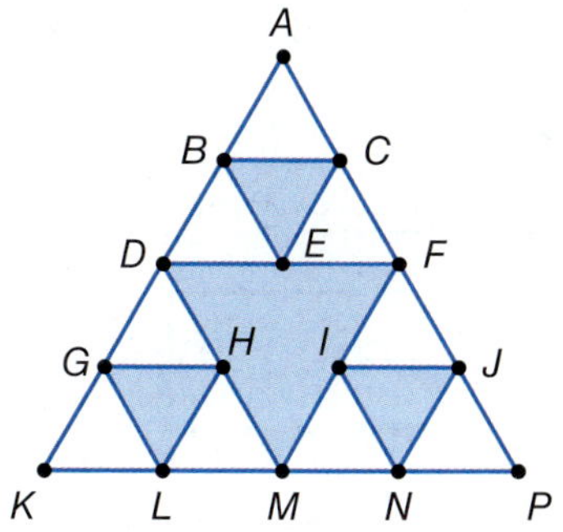

5. A *fractal tree* can be drawn by making two new branches from the endpoint of each original branch, each one-third as long as the previous branch.

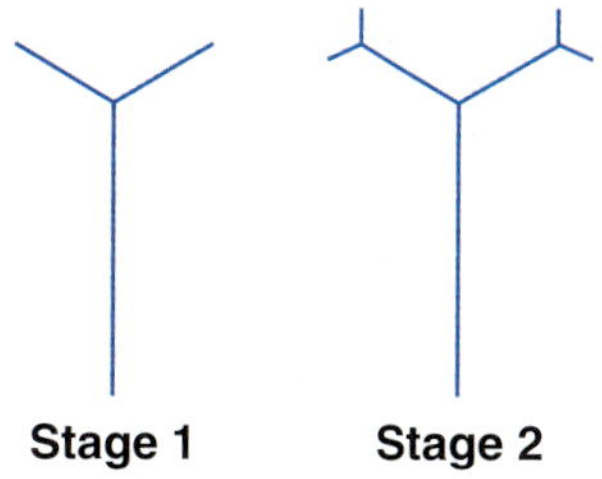

a. Draw Stages 3 and 4 of a fractal tree. How many total branches do you have in Stages 1 through 4? (Do not count the stems.)

b. Write an expression to predict the number of branches at each stage.

(continued on the next page)

Not all iterative processes involve manipulation of geometric shapes. Some iterative processes can be translated into formulas or algebraic equations, similar to the expression you wrote in Exercise 5 on the previous page. These are called **recursive formulas**.

ACTIVITY 2

Pascal's Triangle is a numerical pattern in which each row begins and ends with 1 and all other terms in the row are the sum of the two numbers above it. Find a formula in terms of the row number for any row in Pascal's Triangle.

Step 1 Draw rows 1 through 5 in Pascal's Triangle.

Step 2 Find the sum of values in each row.

Step 3 Find a pattern using the **row number** that can be used to determine the sum of any row.

Row	Pascal's Triangle	Sum	Pattern
1	1	1	$2^0 = 2^{1-1}$
2	1 1	2	$2^1 = 2^{2-1}$
3	1 2 1	4	$2^2 = 2^{3-1}$
4	1 3 3 1	8	$2^3 = 2^{4-1}$
5	1 4 6 4 1	16	$2^4 = 2^{5-1}$

Analyze the Results

6. Write a formula for the sum S of any row n in the Pascal Triangle.

7. What is the sum of the values in the eighth row of Pascal's Triangle?

Exercises

Write a recursive formula for $F(x)$.

8.

x	2	4	6	8	10
$F(x)$	3	7	11	15	19

9.

x	0	5	10	15	20
$F(x)$	0	20	90	210	380

10.

x	1	2	4	8	10
$F(x)$	1	0.5	0.25	0.125	0.1

11.

x	4	9	16	25	36
$F(x)$	5	6	7	8	9

12. CHALLENGE The pattern below represents a sequence of *triangular numbers*. How many dots will be in the 8th term in the sequence? Is it possible to write a recursive formula that can be used to determine the number of dots in the nth triangular number in the series? If so, write the formula. If not, explain why not.

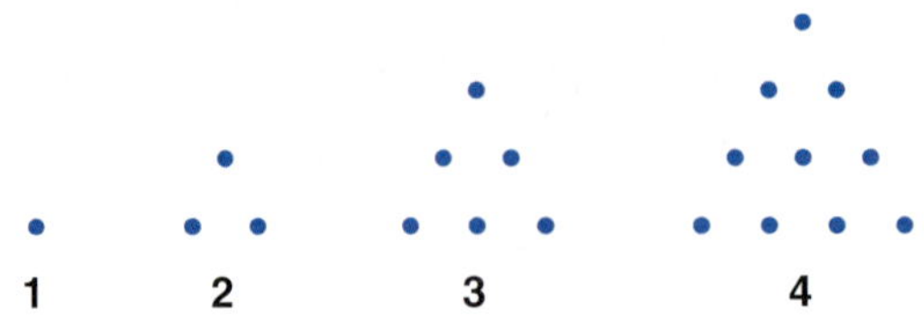

7-6 Similarity Transformations

Then

You identified congruence transformations.
(Lesson 4-7)

Now

- Identify similarity transformations.
- Verify similarity after a similarity transformation.

New Vocabulary

dilation
similarity transformation
center of dilation
scale factor of a dilation
enlargement
reduction

Math Online
glencoe.com

Why?

Adriana uses a copier to enlarge a movie ticket to use as the background for a page in her movie ticket scrapbook. She places the ticket on the glass of the copier. Then she must decide what percentage to input in order to create an image that is three times as big as her original ticket.

Identify Similarity Transformations Recall from Lesson 4-7 that a *transformation* is an operation that maps an original figure, the *preimage,* onto a new figure called the *image*.

A **dilation** is a transformation that enlarges or reduces the original figure proportionally. Since a dilation produces a similar figure, a dilation is a type of **similarity transformation**.

Dilations are performed with respect to a fixed point called the **center of dilation**.

The **scale factor of a dilation** describes the extent of the dilation. The scale factor is the ratio of a length on the image to a corresponding length on the preimage.

The letter k usually represents the scale factor of a dilation. The value of k determines whether the dilation is an enlargement or a reduction.

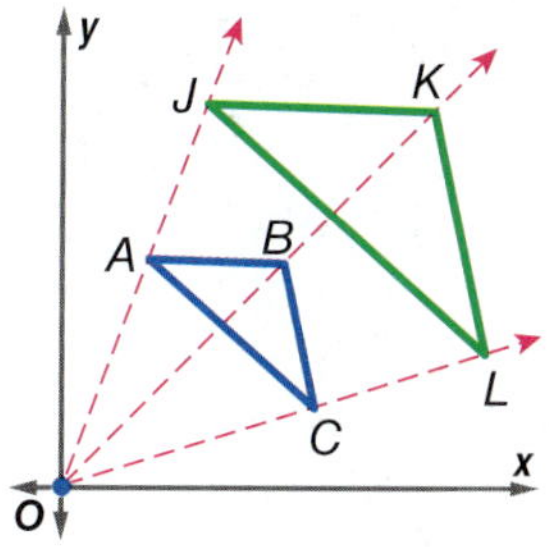

$\triangle JKL$ is a dilation of $\triangle ABC$.
Center of dilation: (0, 0)
Scale factor: $\frac{JK}{AB}$

Key Concept

Types of Dilations

A dilation with a scale factor greater than 1 produces an **enlargement**, or an image that is larger than the original figure.

Symbols If $k > 1$, the dilation is an enlargement.

Example $\triangle FGH$ is dilated by a scale factor of 3 to produce $\triangle RST$. Since $3 > 1$, $\triangle RST$ is an enlargement of $\triangle FGH$.

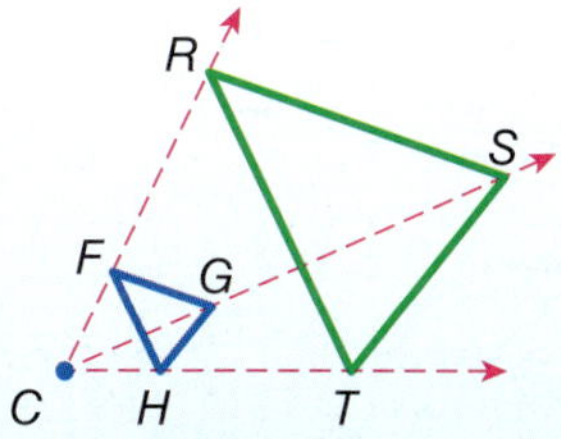

$k > 1$

A dilation with a scale factor between 0 and 1 produces a **reduction**, an image that is smaller than the original figure.

Symbols If $0 < k < 1$, the dilation is a reduction.

Example $ABCD$ is dilated by a scale factor of $\frac{1}{4}$ to produce $WXYZ$. Since $0 < \frac{1}{4} < 1$, $WXYZ$ is a reduction of $ABCD$.

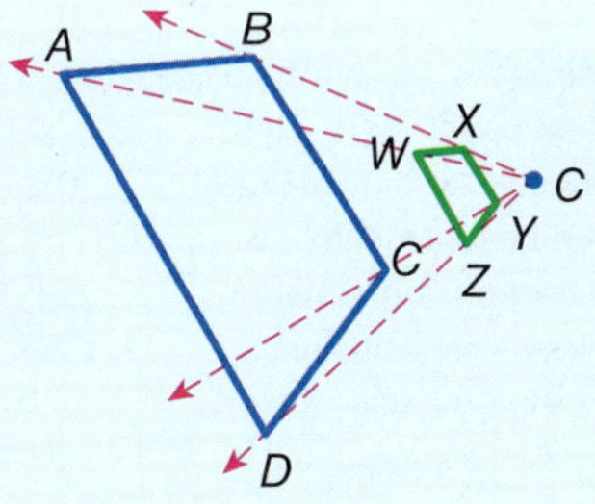

$0 < k < 1$

EXAMPLE 1 Identify a Dilation and Find Its Scale Factor

Determine whether the dilation from *A* to *B* is an *enlargement* or a *reduction*. Then find the scale factor of the dilation.

a.

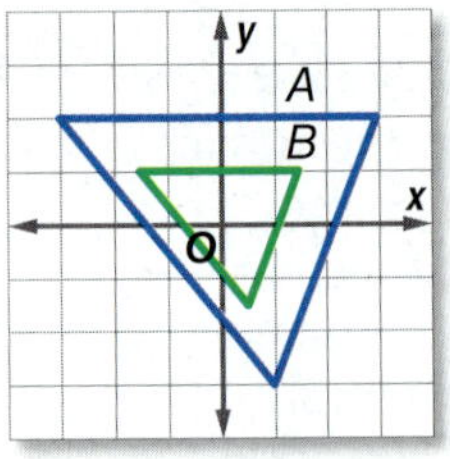

B is smaller than *A*, so the dilation is a reduction.

The distance between the vertices at (−3, 2) and (3, 2) for *A* is 6 and from the vertices at (−1.5, 1) and (1.5, 1) for *B* is 3. So the scale factor is $\frac{3}{6}$ or $\frac{1}{2}$.

b.

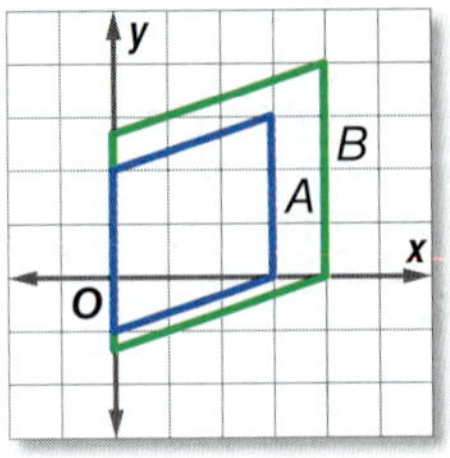

B is larger than *A*, so the dilation is an enlargement.

The distance between the vertices at (3, 3) and (3, 0) for *A* is 3 and between the vertices at (4, 4) and (4, 0) for *B* is 4. So the scale factor is $\frac{4}{3}$.

Guided Practice

1A.

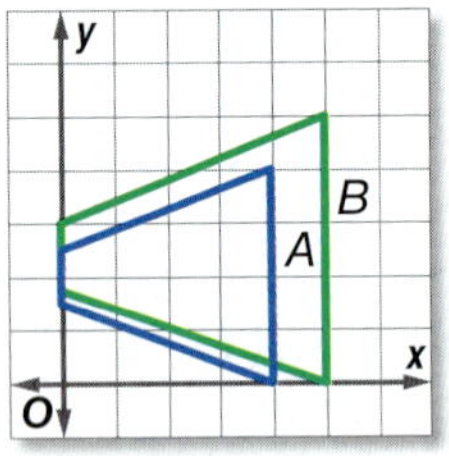

1B.

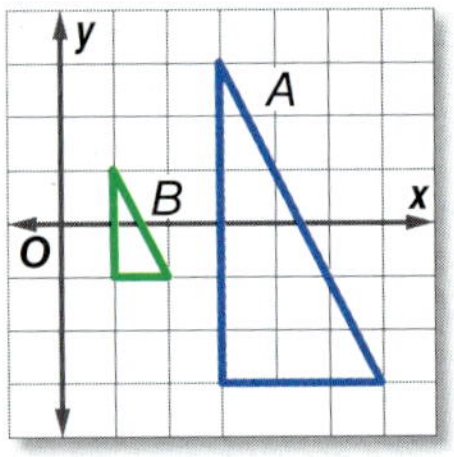

Personal Tutor glencoe.com

Real-World Link

Hew Weng Fatt accepted a contest challenge to collect the most movie stubs from a certain popular fantasy movie. He collected 6561 movie stubs in 38 days!

Source: *Youth2, Star Publications*

Dilations and their scale factors are used in many real-world situations.

Real-World EXAMPLE 2 Find and Use a Scale Factor

COLLECTING Refer to the beginning of the lesson. By what percent should Adriana enlarge the ticket stub so that the dimensions of its image are 3 times that of her original? What will be the dimensions of the enlarged image?

Adriana wants to create a dilated image of her ticket stub using the copier. The scale factor of her enlargement is 3. Written as a percent, the scale factor is (3 • 100)% or 300%. Now find the dimension of the enlarged image using the scale factor.

width: 5 cm • 300% = 15 cm   length: 6.4 cm • 300% = 19.2 cm

The enlarged ticket stub image will be 15 centimeters by 19.2 centimeters.

StudyTip

Multiple Representations The scale factor of a dilation can be represented as a fraction, a decimal, or as a percent. For example, a scale factor of $\frac{2}{5}$ can also be written as 0.4 or as 40%.

Guided Practice

2. If the resulting ticket stub image was 1.5 centimeters wide by about 1.9 centimeters long instead, what percent did Adriana mistakenly use to dilate the original image? Explain your reasoning.

Personal Tutor glencoe.com

Verify Similarity You can verify that a dilation produces a similar figure by comparing corresponding sides and angles. For triangles, you can also use SAS Similarity.

StudyTip

Center of Dilation Unless otherwise stated, all dilations on the coordinate plane use the origin as their center of dilation.

EXAMPLE 3 Verify Similarity after a Dilation

Graph the original figure and its dilated image. Then verify that the dilation is a similarity transformation.

a. original: $A(-6, -3)$, $B(3, 3)$, $C(3, -3)$; **image:** $X(-4, -2)$, $Y(2, 2)$, $Z(2, -2)$

Graph each figure. Since $\angle C$ and $\angle Z$ are both right angles, $\angle C \cong \angle Z$. Show that the lengths of the sides that include $\angle C$ and $\angle Z$ are proportional.

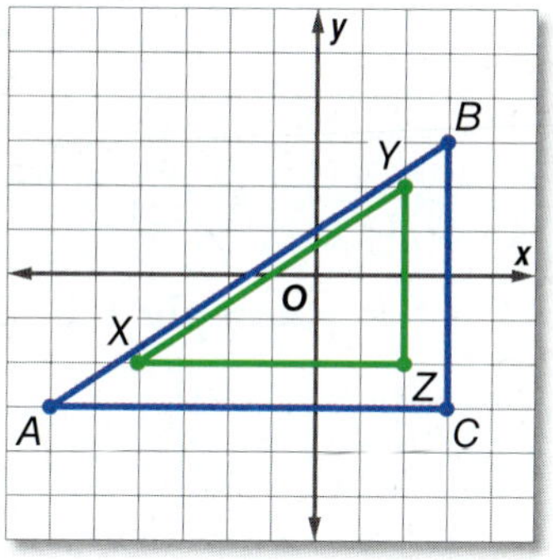

Use the coordinate grid to find the side lengths.

$\frac{XZ}{AC} = \frac{6}{9}$ or $\frac{2}{3}$, and $\frac{YZ}{BC} = \frac{4}{6}$ or $\frac{2}{3}$, so $\frac{XZ}{AC} = \frac{YZ}{BC}$.

Since the lengths of the sides that include $\angle C$ and $\angle Z$ are proportional, $\triangle XYZ \sim \triangle ABC$ by SAS Similarity.

b. original: $J(-6, 4)$, $K(6, 8)$, $L(8, 2)$, $M(-4, -2)$;

image: $P(-3, 2)$, $Q(3, 4)$, $R(4, 1)$, $S(-2, -1)$

Use the Distance Formula to find the length of each side.

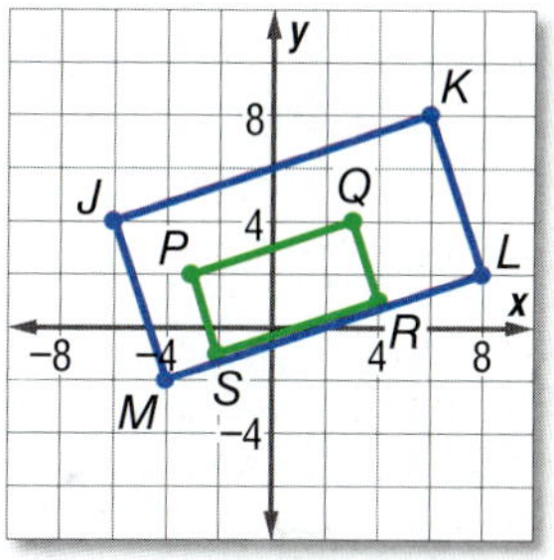

$JK = \sqrt{[6 - (-6)]^2 + (8 - 4)^2} = \sqrt{160}$ or $4\sqrt{10}$

$PQ = \sqrt{[3 - (-3)]^2 + (4 - 2)^2} = \sqrt{40}$ or $2\sqrt{10}$

$KL = \sqrt{(8 - 6)^2 + (2 - 8)^2} = \sqrt{40}$ or $2\sqrt{10}$

$QR = \sqrt{(4 - 3)^2 + (1 - 4)^2} = \sqrt{10}$

$LM = \sqrt{(-4 - 8)^2 + (-2 - 2)^2} = \sqrt{160}$ or $4\sqrt{10}$

$RS = \sqrt{(-2 - 4)^2 + (-1 - 1)^2} = \sqrt{40}$ or $2\sqrt{10}$

$MJ = \sqrt{[-6 - (-4)]^2 + [4 - (-2)]^2} = \sqrt{40}$ or $2\sqrt{10}$

$SP = \sqrt{[-3 - (-2)]^2 + [2 - (-1)]^2} = \sqrt{10}$

Find and compare the ratios of corresponding sides.

$\frac{PQ}{JK} = \frac{2\sqrt{10}}{4\sqrt{10}}$ or $\frac{1}{2}$ $\quad \frac{QR}{KL} = \frac{\sqrt{10}}{2\sqrt{10}}$ or $\frac{1}{2}$ $\quad \frac{RS}{LM} = \frac{2\sqrt{10}}{4\sqrt{10}}$ or $\frac{1}{2}$ $\quad \frac{SP}{MJ} = \frac{\sqrt{10}}{2\sqrt{10}}$ or $\frac{1}{2}$

PQRS and *JKLM* are both rectangles. This can be proved by showing that diagonals $\overline{PR} \cong \overline{SQ}$ and $\overline{JL} \cong \overline{KM}$ are congruent using the Distance Formula. Since they are both rectangles, their corresponding angles are congruent.

Since $\frac{PQ}{JK} = \frac{QR}{KL} = \frac{RS}{LM} = \frac{SP}{MJ}$ and corresponding angles are congruent, $PQRS \sim JKLM$.

Guided Practice

3A. original: $A(2, 3)$, $B(0, 1)$, $C(3, 0)$
image: $D(4, 6)$, $F(0, 2)$, $G(6, 0)$

3B. original: $H(0, 0)$, $J(6, 0)$, $K(6, 4)$, $L(0, 4)$
image: $W(0, 0)$, $X(3, 0)$, $Y(3, 2)$, $Z(0, 2)$

Personal Tutor glencoe.com

Check Your Understanding

Example 1 p. 506

Determine whether the dilation from *A* to *B* is an *enlargement* or a *reduction*. Then find the scale factor of the dilation.

1.

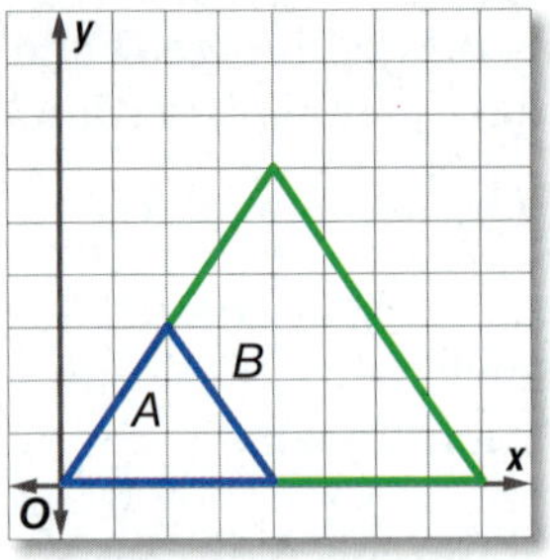

2.

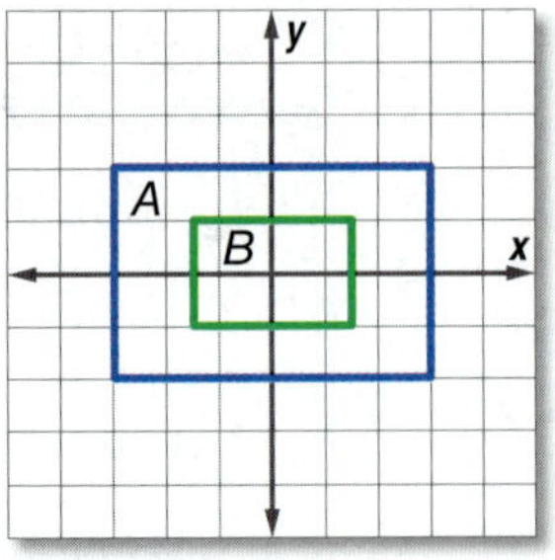

Example 2 p. 506

3 **GAMES** The dimensions of a regulation tennis court are 27 feet by 78 feet. The dimensions of a table tennis table are 152.5 centimeters by 274 centimeters. Is a table tennis table a dilation of a tennis court? If so, what is the scale factor? Explain.

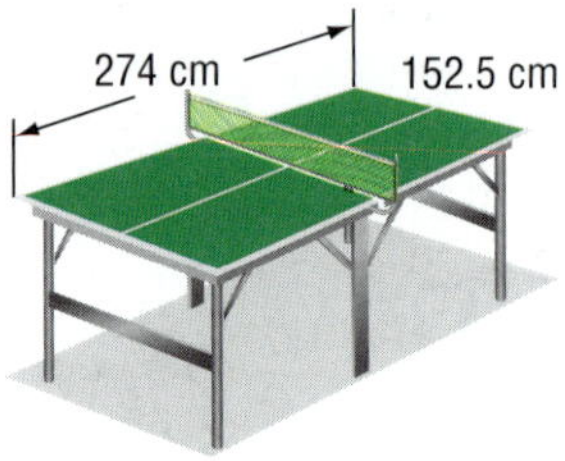

Example 3 p. 507

Verify that the dilation is a similarity transformation.

4.

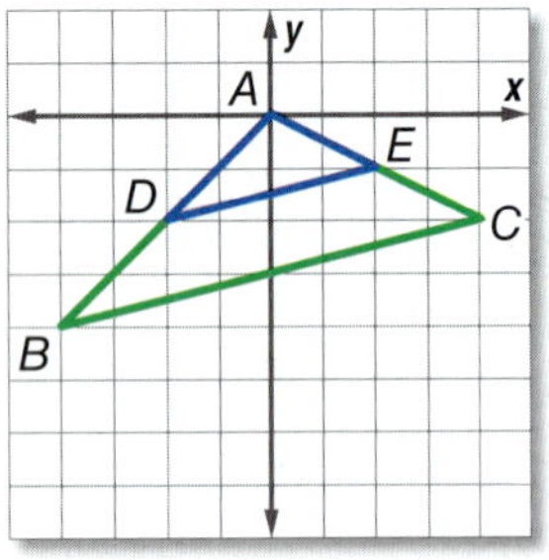

5.

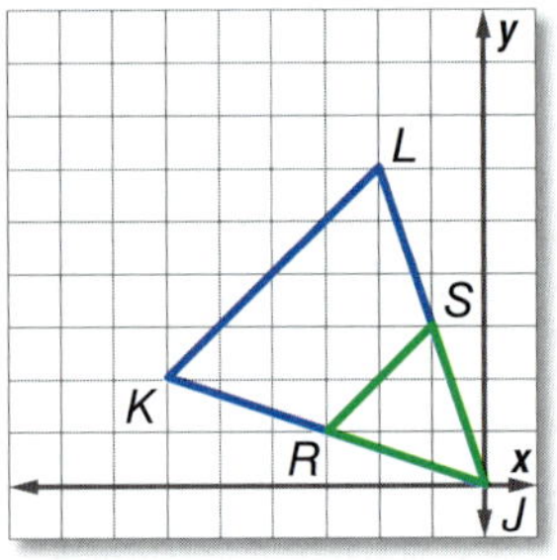

Practice and Problem Solving

● = **Step-by-Step Solutions** begin on page R20.
Extra Practice begins on page 969.

Example 1 p. 506

Determine whether the dilation from *A* to *B* is an *enlargement* or a *reduction*. Then find the scale factor of the dilation.

6.

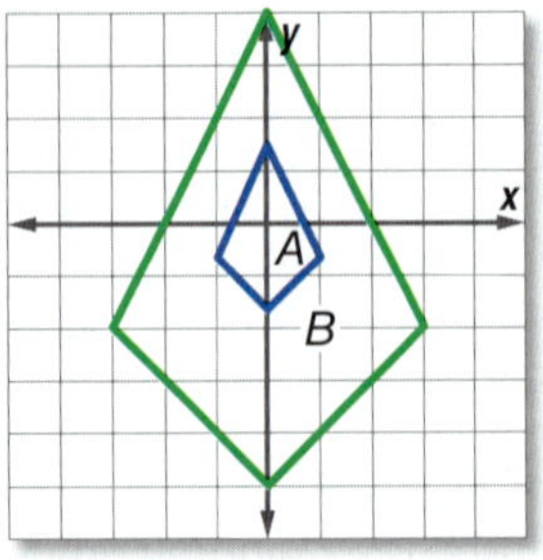

7.

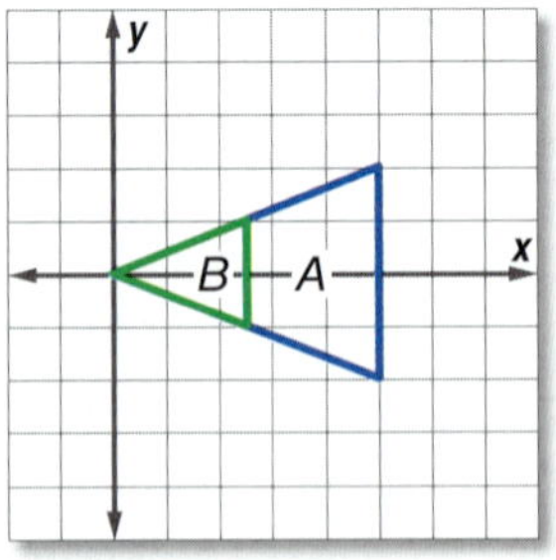

8.

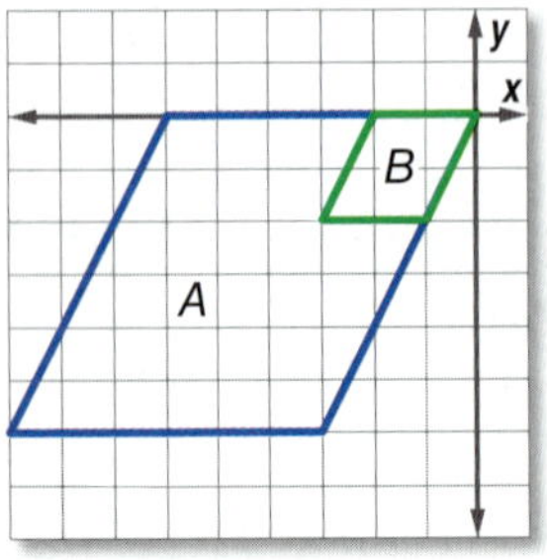

9.

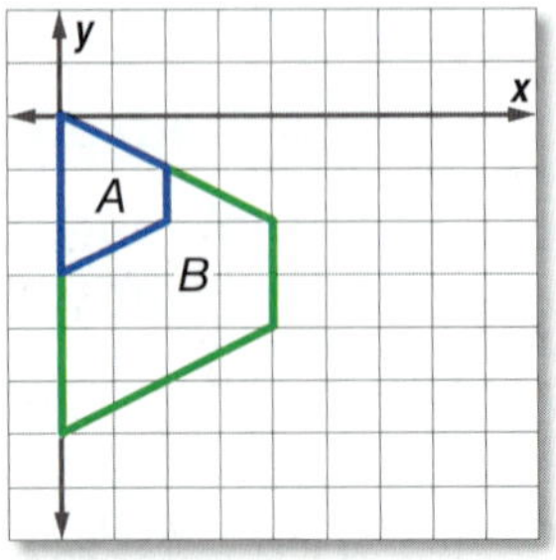

Determine whether each dilation is an *enlargement* or *reduction*.

10. Before / After

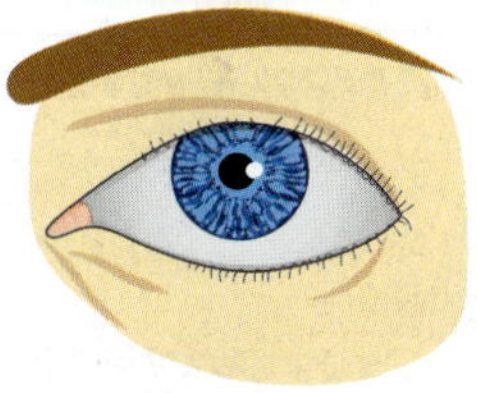

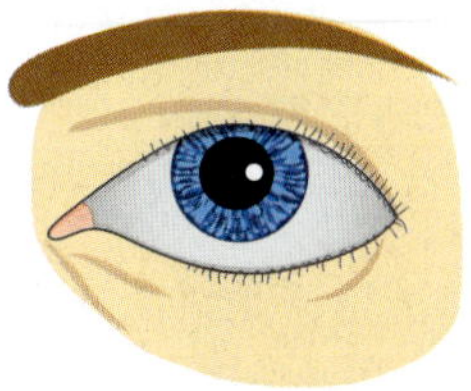

11. Painting / Postcard

Example 2
p. 506

12. **YEARBOOK** Jordan is putting a photo of the lacrosse team in a full-page layout in the yearbook. The original photo is 4 inches by 6 inches. If the photo in the yearbook is $6\frac{2}{3}$ inches by 10 inches, is the yearbook photo a dilation of the original photo? If so, what is the scale factor? Explain.

13. **SCHOOL SPIRIT** Candace created a design to be made into temporary tattoos for a homecoming game as shown. Is the temporary tattoo a dilation of the original design? If so, what is the scale factor? Explain.

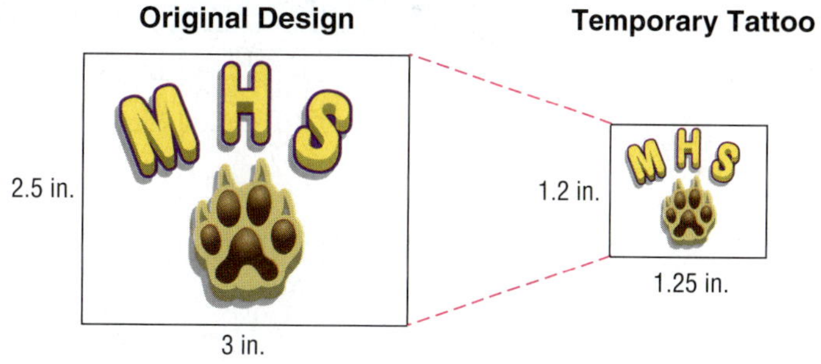

Example 3
p. 507

Graph the original figure and its dilated image. Then verify that the dilation is a similarity transformation.

14. $M(1, 4), P(2, 2), Q(5, 5); S(-3, 6), T(0, 0), U(9, 9)$

15. $A(1, 3), B(-1, 2), C(1, 1); D(-7, -1), E(1, -5)$

16. $V(-3, 4), W(-5, 0), X(1, 2); Y(-6, -2), Z(3, 1)$

17. $J(-6, 8), K(6, 6), L(-2, 4); D(-12, 16), G(12, 12), H(-4, 8)$

If $\triangle ABC \sim \triangle AYZ$, find the missing coordinate.

18.

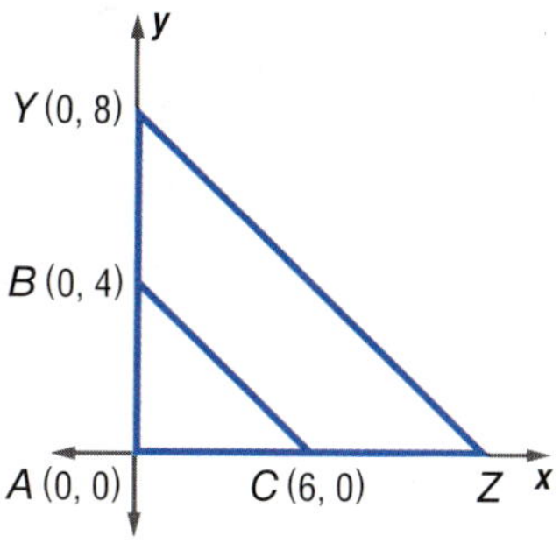

19

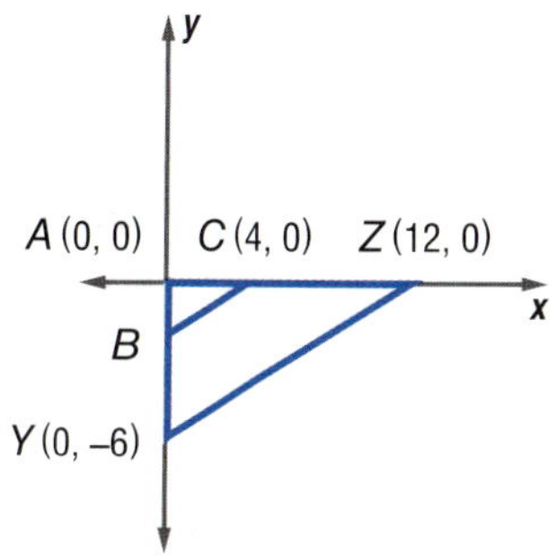

20. **GRAPHIC ART** Aimee painted the sample sign shown using $\frac{1}{2}$ bottle of glass paint. The actual sign she will paint in a shop window is to be 3 feet by $7\frac{1}{2}$ feet.

a. Explain why the actual sign is a dilation of her sample.

b. How many bottles of paint will Aimee need to complete the actual sign?

Real-World Link

In 2006 there were around 261,000 graphic designing jobs in the U.S. Many graphic designing jobs involve creative advertising for businesses, like the logo for a seafood restaurant shown above.

Source: U.S. Department of Labor

21. **MULTIPLE REPRESENTATIONS** In this problem, you will investigate similarity of triangles on the coordinate plane.

a. GEOMETRIC Draw a triangle with vertex A at the origin. Make sure that the two additional vertices B and C have whole-number coordinates. Draw a similar triangle that is twice as large as $\triangle ABC$ with its vertex also located at the origin. Label the triangle ADE.

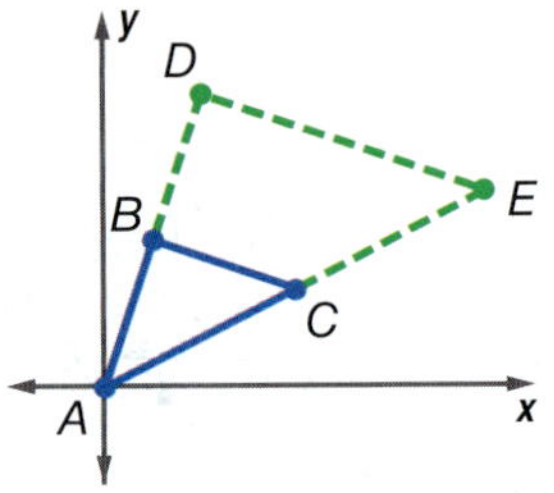

b. GEOMETRIC Repeat the process in part a two times. Label the second pair of triangles MNP and MQR and the third pair TWX and TYZ. Use different scale factors than part **a**.

c. TABULAR Complete the table below with the appropriate values.

Coordinates											
$\triangle ABC$		$\triangle ADE$		$\triangle MNP$		$\triangle MQR$		$\triangle TWX$		$\triangle TYZ$	
A		A		M		M		T		T	
B		D		N		Q		W		Y	
C		E		P		R		X		Z	

d. VERBAL Make a conjecture about how you could predict the coordinates of a dilated triangle with a scale factor of n if the two similar triangles share a corresponding vertex at the origin.

H.O.T. Problems

Use Higher-Order Thinking Skills

22. **CHALLENGE** $MNOP$ is a dilation of $ABCD$. How is the scale factor of the dilation related to the similarity ratio of $ABCD$ to $MNOP$? Explain your reasoning.

23. **REASONING** The coordinates of two triangles are provided in the table at the right. Is $\triangle XYZ$ a dilation of $\triangle PQR$? Explain.

$\triangle PQR$		$\triangle XYZ$	
P	(a, b)	X	(3a, 2b)
Q	(c, d)	Y	(3c, 2d)
R	(e, f)	Z	(3e, 2f)

OPEN ENDED Describe a real-world example of each transformation other than those given in this lesson.

24. enlargement

25. reduction

26. congruence transformation

27. **WRITING IN MATH** Explain how you can use scale factor to determine whether a transformation is an enlargement, a reduction, or a congruence transformation.

Standardized Test Practice

28. ALGEBRA Which equation describes the line that passes through (−3, 4) and is perpendicular to $3x - y = 6$?

A $y = -\frac{1}{3}x + 4$ **C** $y = 3x + 4$

B $y = -\frac{1}{3}x + 3$ **D** $y = 3x + 3$

29. SHORT RESPONSE What is the scale factor of the dilation shown below?

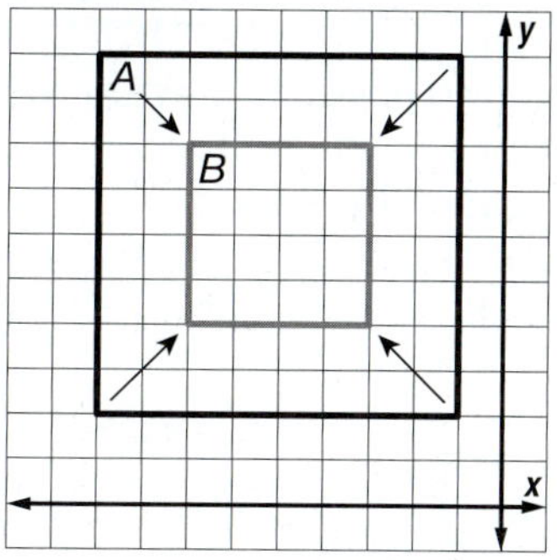

30. In the figure below, $\angle A \cong \angle C$.

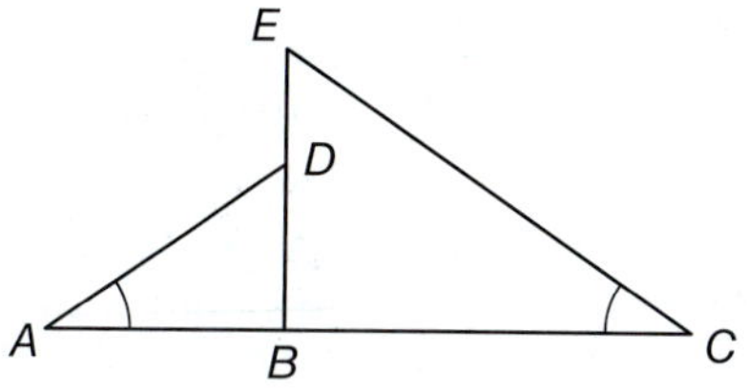

Which additional information would *not* be enough to prove that $\triangle ADB \sim \triangle CEB$?

F $\frac{AB}{DB} = \frac{CB}{EB}$

G $\angle ADB \cong \angle CEB$

H $\overline{ED} \cong \overline{DB}$

J $\overline{EB} \perp \overline{AC}$

31. SAT/ACT If $x = \frac{6}{4p + 3}$ and $xy = \frac{3}{4p + 3}$, then $y =$

A 4 **B** 2 **C** $\frac{1}{2}$ **D** $\frac{1}{4}$

Spiral Review

32. LANDSCAPING Shea is designing two gardens shaped like similar triangles. One garden has a perimeter of 53.5 feet, and the longest side is 25 feet. She wants the second garden to have a perimeter of 32.1 feet. Find the length of the longest side of this garden. (Lesson 7-5)

Determine whether $\overline{AB} \parallel \overline{CD}$. Justify your answer. (Lesson 7-4)

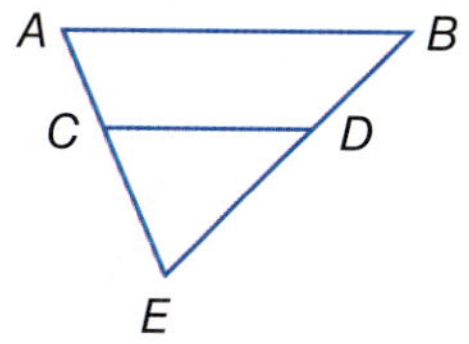

33. $AC = 8.4$, $BD = 6.3$, $DE = 4.5$, and $CE = 6$

34. $AC = 7$, $BD = 10.5$, $BE = 22.5$, and $AE = 15$

35. $AB = 8$, $AE = 9$, $CD = 4$, and $CE = 4$

If each figure is a kite, find each measure. (Lesson 6-6)

36. QR

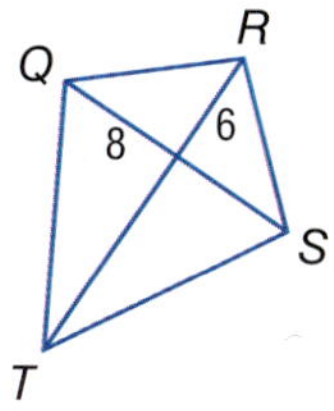

37. $m\angle K$

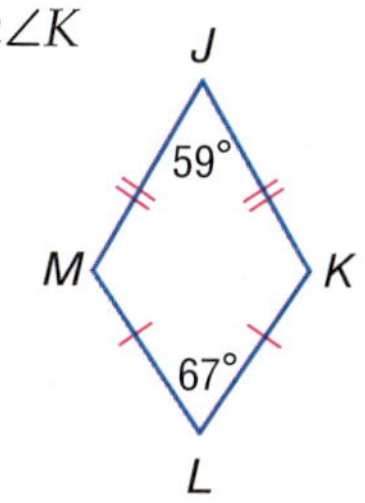

38. BC

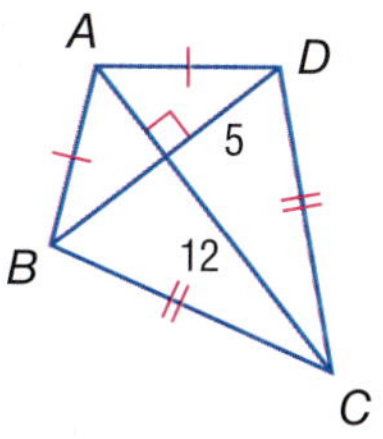

39. PROOF Write a coordinate proof for the following statement. (Lesson 4-8)
If a line segment joins the midpoints of two sides of a triangle, then it is parallel to the third side.

Skills Review

Solve each equation. (Lesson 0-5)

40. $145 = 29 \cdot t$

41. $216 = d \cdot 27$

42. $2r = 67 \cdot 5$

43. $100t = \frac{70}{240}$

44. $\frac{80}{4} = 14d$

45. $\frac{2t + 15}{t} = 92$

7-7 Scale Drawings and Models

Then

You used scale factor to solve problems with similar polygons. (Lesson 7-2)

Now

- Interpret scale models.
- Use scale factor to solve problems.

New Vocabulary

scale model
scale drawing
scale

Math Online
glencoe.com

Why?

The Big Bugs Exhibit is a collection of eight different anatomically correct insect sculptures created by artist David Rogers. The proportions of the bug sculptures are the same as the much smaller insects that inspired them. The praying mantis at the right is 18 feet long and 20 feet wide.

Scale Models A **scale model** or a **scale drawing** is an object or drawing with lengths proportional to the object it represents. The **scale** of a model or drawing is the ratio of a length on the model or drawing to the actual length of the object being modeled or drawn.

EXAMPLE 1 Use a Scale Drawing

MAPS The scale on the map shown is 0.4 inch : 40 miles. Find the actual distance from Nashville to Memphis.

Use a ruler. The distance between Nashville and Memphis is about 1.5 inches.

Method 1 Write and solve a proportion.

Let x represent the distance between Nashville and Memphis.

Scale **Nashville to Memphis**

map → / actual → $\frac{0.4 \text{ in.}}{40 \text{ mi}} = \frac{1.5 \text{ in.}}{x \text{ mi}}$ ← map / ← actual

$0.4 \cdot x = 40 \cdot 1.5$ **Cross Products Property**

$x = 150$ **Simplify.**

Method 2 Write and solve an equation.

Let a = actual distance in miles between Nashville and Memphis and m = map distance in inches. Write the scale as $\frac{40 \text{ mi}}{0.4 \text{ in.}}$, which is $40 \div 0.4$ or 100 miles per inch. So for every inch on the map, the actual distance is 100 miles.

$a = 100 \cdot m$ **Write an equation.**

$= 100 \cdot 1.5$ **$m = 1.5$ in.**

$= 150$ **Solve.**

CHECK Use dimensional analysis.

$\text{mi} = \frac{\text{mi}}{\cancel{\text{in.}}} \cdot \cancel{\text{in.}} \Rightarrow \text{mi} = \text{mi}$ ✓

The distance between Nashville and Memphis is 150 miles.

Guided Practice

1. **MAPS** Find the actual distance between Nashville and Chattanooga.

Personal Tutor glencoe.com

Use Scale Factors The scale factor of a drawing or scale model is written as a unitless ratio in simplest form. Scale factors are always written so that the model length in the ratio comes first.

EXAMPLE 2 Find the Scale

SCALE MODEL This is a miniature replica of a 1923 Checker Cab. The length of the model is 6.5 inches. The actual length of the car was 13 feet.

a. What is the scale of the model?

To find the scale, write the ratio of a model length to an actual length.

$$\frac{\text{model length}}{\text{actual length}} = \frac{6.5 \text{ in.}}{13 \text{ ft}} \text{ or } \frac{1 \text{ in.}}{2 \text{ft}}$$

The scale of the model is 1 in.:2 ft.

b. How many times as long as the actual car is the model?

To answer this question, find the scale factor of the model. Multiply by a conversion factor that relates inches to feet to obtain a unitless ratio.

$$\frac{1 \text{ in.}}{2 \text{ ft}} = \frac{1 \text{ in.}}{2 \text{ ft}} \cdot \frac{1 \text{ ft}}{12 \text{ in.}} = \frac{1}{24}$$

The scale factor is 1:24. That is, the model is $\frac{1}{24}$ as long as the actual car.

Guided Practice

2. SCALE MODEL Mrs. Alejandro's history class made a scale model of the Alamo that is 3 feet tall. The actual height of the building is 33 feet 6 inches.

A. What is the scale of the model?

B. How many times as tall as the actual building is the model? How many times as tall as the model is the actual building?

Personal Tutor glencoe.com

Real-World EXAMPLE 3 Construct a Scale Model

SCALE MODEL Suppose you want to build a model of the St. Louis Gateway Arch that is no more than 11 inches tall. Choose an appropriate scale and use it to determine the height of the model. Use the information at the left.

The actual monument is 630 feet tall. Since 630 feet ÷ 11 inches = 57.3 feet per inch, a scale of 1 inch = 60 feet is an appropriate. So for every inch on the model m, let the actual measure a be 60 feet. Write this as an equation.

$a = 60 \cdot m$ **Write an equation.**

$630 = 60 \cdot m$ **$a = 630$**

$10.5 = m$ So the height of the model would be 10.5 inches.

Guided Practice

3. SCALE DRAWING Sonya is making a scale drawing of her room on an 8.5-by-11-inch sheet of paper. If her room is 14 feet by 12 feet, find an appropriate scale for the drawing and determine the dimensions of the drawing.

Personal Tutor glencoe.com

StudyTip

Checking Results The scale factor of a model that is smaller than the original object is between 0 and 1 and the scale factor for a model that is larger than the original object is greater than 1.

Real-World Link

The St. Louis Gateway Arch is the tallest national monument in the United States at 630 feet. The span of the base is also 630 feet. The arch weighs 17,246 tons and can sway a maximum of 9 inches in each direction during high winds.

Source: Gateway Arch Facts

Check Your Understanding

Example 1 p. 512

MAPS Use the map of Maine shown and a customary ruler to find the actual distance between each pair of cities. Measure to the nearest sixteenth of an inch.

1. Bangor and Portland
2. Augusta and Houlton

Example 2 p. 513

3. **SCALE MODELS** Carlos made a scale model of a local bridge. The model spans 6 inches; the actual bridge spans 50 feet.
   a. What is the scale of the model?
   b. What scale factor did Carlos use to build his model?

Example 3 p. 513

4. **SPORTS** A volleyball court is 9 meters wide and 18 meters long. Choose an appropriate scale and construct a scale drawing of the court to fit on a 3-inch by 5-inch index card.

Practice and Problem Solving

= **Step-by-Step Solutions** begin on page R20.
Extra Practice begins on page 969.

Example 1 p. 512

MAPS Use the map of Oklahoma shown and a metric ruler to find the actual distance between each pair of cities. Measure to the nearest centimeter.

5. Guymon and Oklahoma City
6. Lawton and Tulsa
7. Enid and Tulsa
8. Ponca City and Shawnee

Example 2 p. 513

9. **SCULPTURE** A replica of *The Thinker* is 10 inches tall. A statue of *The Thinker* at the University of Louisville is 10 feet tall.
   a. What is the scale of the replica?
   b. How many times as tall as the actual sculpture is the replica?

The Thinker

10. **MAPS** The map below shows a portion of Frankfort, Kentucky.

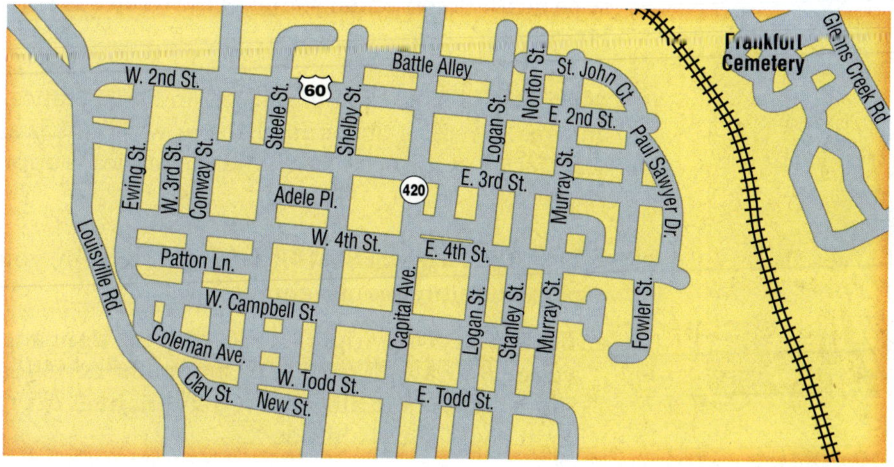

a. If the actual distance from the intersection of Conway Street and 4th Street to the intersection of Murray Street and 4th Street is 0.47 mile, use a customary ruler to estimate the scale of the map.

b. What is the approximate scale factor of the map? Interpret its meaning.

Example 3
p. 513

SPORTS **Choose an appropriate scale and construct a scale drawing of each playing area so that it would fit on an 8.5-by-11-inch sheet of paper.**

11. A baseball diamond is a square 90 feet on each side with about a 128-foot diagonal.

12. A high school basketball court is a rectangle with length 84 feet and width 50 feet.

MAPS **Use the map shown and an inch ruler to answer each question. Measure to the nearest sixteenth of an inch and assume that you can travel along any straight line.**

13. About how long would it take to drive from Valdosta, Georgia, to Daytona Beach, Florida, traveling at 65 miles per hour?

14. How long would it take to drive from Gainesville to Miami, Florida, traveling at 70 miles per hour?

Real-World Link

At forty miles from Pluto to the Sun, the Maine Solar System Model on U.S. Route 1 is the largest complete three-dimensional scale model of the Solar System in the world. Its scale factor is 1:93,000,000.

Source: Northern Maine Museum of Science

15. **SCALE MODELS** If the distance between Earth and the Sun is actually 150,000,000 kilometers, how far apart are Earth and the Sun in the model?

16. **LITERATURE** In the book, *Alice's Adventures in Wonderland*, Alice's size changes from her normal height of about 50 inches. Suppose Alice came across a door about 15 inches high and her height changed to 10 inches.

a. Find the ratio of the height of the door to Alice's height in Wonderland.

b. How tall would the door have been in Alice's normal world?

17 **ROCKETS** Peter bought a $\frac{1 \text{ in.}}{12 \text{ ft}}$ scale model of the Mercury-Redstone rocket.

a. If the height of the model is 7 inches, what is the approximate height of the rocket?

b. If the diameter of the rocket is 70 inches, what is the diameter of the model? Round to the nearest half inch.

Real-World Link

Scale replicas of the Statue of Liberty can also be found in Alabama, Colorado, and Georgia.

18. **ARCHITECTURE** A replica of the Statue of Liberty in Austin, Texas, is $16\frac{3}{4}$ feet tall. If the scale factor of the actual statue to the replica is 9:1, how tall is the statue in New York Harbor?

19. **AMUSEMENT PARK** The Eiffel Tower in Paris, France, is 986 feet tall, not including its antenna. A replica of the Eiffel Tower was built as a ride in an amusement park. If the scale factor of the actual tower to the replica is approximately 3:1, how tall is the ride?

20. **MULTIPLE REPRESENTATIONS** In this problem, you will explore the altitudes of right triangles.

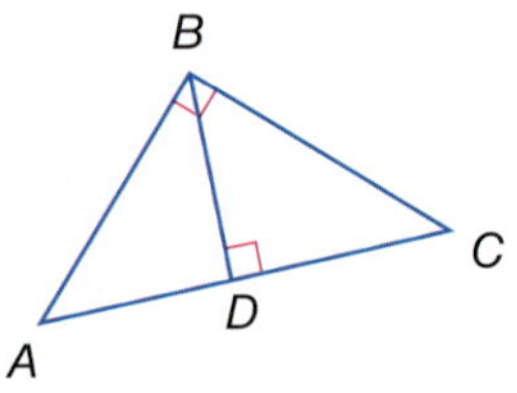

a. **GEOMETRIC** Draw right $\triangle ABC$ with the right angle at vertex B. Draw altitude $\overline{BD}$. Draw right $\triangle MNP$, with right angle N and altitude $\overline{NQ}$, and right $\triangle WXY$, with right angle X and altitude $\overline{XZ}$.

b. **TABULAR** Measure and record indicated angles in the table below.

	Angle Measure					
	$\triangle ABC$		$\triangle BDC$		$\triangle ADB$	
$\triangle ABC$	ABC		BDC		ADB	
	A		CBD		BAD	
	C		DCB		DBA	
	$\triangle MNP$		$\triangle NQP$		$\triangle MQN$	
$\triangle MNP$	MNP		NQP		MQN	
	M		PNQ		NMQ	
	P		QPN		QNM	
	$\triangle WXY$		$\triangle WZX$		$\triangle XZY$	
$\triangle WXY$	WXY		WZX		XZY	
	W		XWZ		YXZ	
	Y		ZXW		ZYX	

c. **VERBAL** Make a conjecture about the altitude of a right triangle originating at the right angle of the triangle.

H.O.T. Problems

Use Higher-Order Thinking Skills

21. **ERROR ANALYSIS** Felix and Tamara are building a replica of their high school. The high school is 75 feet tall and the replica is 1.5 feet tall. Felix says the scale factor of the actual high school to the replica is 50:1, while Tamara says the scale factor is 1:50. Is either of them correct? Explain your reasoning.

22. **CHALLENGE** You can produce a scale model of a certain object by extending each dimension by a constant. What must be true of the shape of the object? Explain your reasoning.

23. **REASONING** Sofia is making two scale drawings of the lunchroom. In the first drawing, Sofia used a scale of 1 inch = 1 foot, and in the second drawing she used a scale of 1 inch = 6 feet. Which scale will produce a larger drawing? What is the scale factor of the first drawing to the second drawing? Explain.

24. **OPEN ENDED** Draw a scale model of your classroom using any scale.

25. **WRITING IN MATH** Compare and contrast scale and scale factor.

Standardized Test Practice

26. SHORT RESPONSE If $3^x = 27^{(x-4)}$, then what is the value of x?

27. In $\triangle ABC$, $\overline{BD}$ is a median. If $AD = 3x + 5$ and $CD = 5x - 1$, find AC.

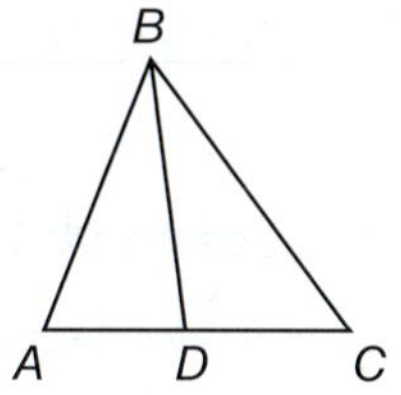

A 6 **C** 14
B 12 **D** 28

28. In a triangle, the ratio of the measures of the sides is 4 : 7 : 10, and its longest side is 40 centimeters. Find the perimeter of the triangle in centimeters.

F 37 cm **H** 84 cm
G 43 cm **J** 168 cm

29. SAT/ACT If Lydia can type 80 words in two minutes, how long will it take Lydia to type 600 words?

A 30 min **C** 10 min
B 15 min **D** 5 min

Spiral Review

30. PAINTING Aaron is painting a portrait of a friend for an art class. Since his friend doesn't have time to model, he uses a photo that is 6 inches by 8 inches. If the canvas is 24 inches by 32 inches, is the painting a dilation of the original photo? If so, what is the scale factor? Explain. (Lesson 7-6)

Find x. (Lesson 7-5)

31.

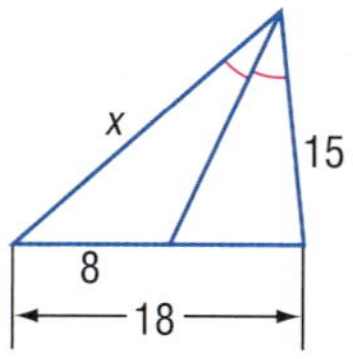

32.

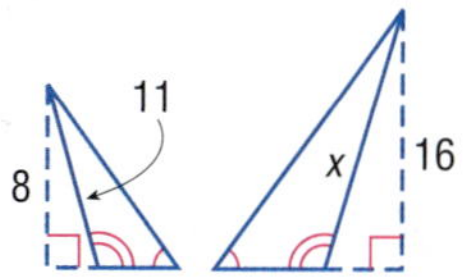

33.

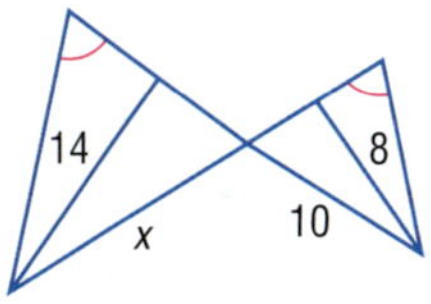

ALGEBRA Quadrilateral $JKMN$ is a rectangle. (Lesson 6-4)

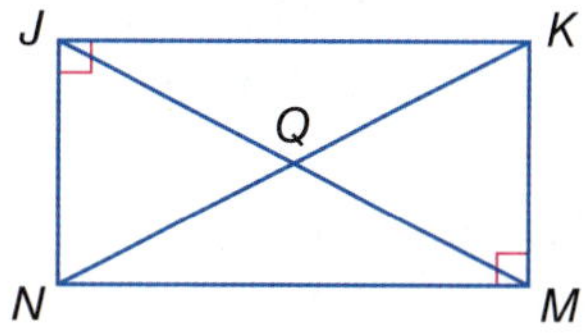

34. If $NQ = 2x + 3$ and $QK = 5x - 9$, find JQ.

35. If $m\angle NJM = 2x - 3$ and $m\angle KJM = x + 5$, find x.

36. If $NM = 8x - 14$ and $JK = x^2 + 1$, find JK.

In $\triangle ABC$, $MC = 7$, $RM = 4$, and $AT = 16$. Find each measure. (Lesson 5-2)

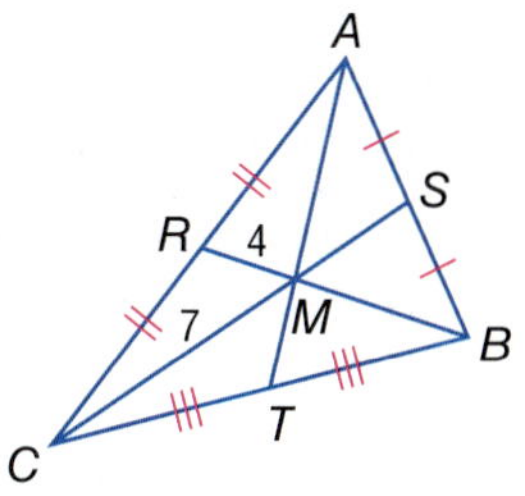

37. MS **38.** AM **39.** SC

40. RB **41.** MB **42.** TM

Determine whether $\triangle JKL \cong \triangle XYZ$. Explain. (Lesson 4-4)

43. $J(3, 9)$, $K(4, 6)$, $L(1, 5)$, $X(1, 7)$, $Y(2, 4)$, $Z(-1, 3)$

44. $J(-1, -1)$, $K(0, 6)$, $L(2, 3)$, $X(3, 1)$, $Y(5, 3)$, $Z(8, 1)$

Skills Review

Simplify each expression. (Lesson 0-9)

45. $\sqrt{4 \cdot 16}$ **46.** $\sqrt{3 \cdot 27}$ **47.** $\sqrt{32 \cdot 72}$ **48.** $\sqrt{15 \cdot 16}$ **49.** $\sqrt{33 \cdot 21}$

CHAPTER 7

Study Guide and Review

Math Online glencoe.com
- STUDY TO GO
- Vocabulary Review

Chapter Summary

Key Concepts

Proportions (Lesson 7-1)

- For any numbers a and c and any nonzero numbers b and d, $\frac{a}{b} = \frac{c}{d}$ if and only if $ad = bc$.

Similar Polygons and Triangles (Lessons 7-2 and 7-3)

- Two polygons are similar if and only if their corresponding angles are congruent and the measures of their corresponding sides are proportional.
- Two triangles are similar if:

  AA: Two angles of one triangle are congruent to two angles of the other triangle.

  SSS: The measures of the corresponding sides of the two triangles are proportional.

  SAS: The measures of two sides of one triangle are proportional to the measures of two corresponding sides of another triangle and their included angles are congruent.

Proportional Parts (Lessons 7-4 and 7-5)

- If a line is parallel to one side of a triangle and intersects the other two sides in two distinct points, then it separates these sides into segments of proportional length.
- A midsegment of a triangle is parallel to one side of the triangle and its length is one-half the length of that side.
- Two triangles are similar when each of the following are proportional in measure: their perimeters, their corresponding altitudes, their corresponding angle bisectors, and their corresponding medians.

Similarity Transformations and Scale Drawings and Models (Lessons 7-6 and 7-7)

- A **scale model** or **scale drawing** has lengths that are proportional to the corresponding lengths in the object it represents.

FOLDABLES® Study Organizer

Be sure the Key Concepts are noted in your Foldable.

Key Vocabulary

cross products (p. 458)
dilation (p. 505)
enlargement (p. 505)
extremes (p. 458)
means (p. 458)
midsegment of a triangle (p. 485)
proportion (p. 458)
ratio (p. 457)
reduction (p. 505)
scale (p. 512)
scale drawing (p. 512)
scale factor (p. 466)
scale model (p. 512)
similar polygons (p. 465)
similarity transformation (p. 505)

Vocabulary Check

Choose the letter of the word or phrase that best completes each statement.

a. ratio
b. proportion
c. means
d. extremes
e. similar
f. scale factor
g. AA Similarity Post.
h. SSS Similarity Theorem
i. SAS Similarity Theorem
j. midsegment
k. dilation
l. enlargement
m. reduction

1. A(n) ___?___ of a triangle has endpoints that are the midpoints of two sides of the triangle.
2. A(n) ___?___ is a comparison of two quantities using division.
3. If $\angle A \cong \angle X$ and $\angle C \cong \angle Z$, then $\triangle ABC \sim \triangle XYZ$ by the ___?___.
4. A(n) ___?___ is an example of a similarity transformation.
5. If $\frac{a}{b} = \frac{c}{d}$, then a and d are the ___?___.
6. The ratio of the lengths of two corresponding sides of two similar polygons is the ___?___.
7. A(n) ___?___ is an equation stating that two ratios are equivalent.
8. A dilation with a scale factor of $\frac{2}{5}$ will result in a(n) ___?___.

Lesson-by-Lesson Review

7-1 Ratios and Proportions (pp. 457–463)

Solve each proportion.

9. $\frac{x+8}{6} = \frac{2x-3}{10}$

10. $\frac{3x+9}{x} = \frac{12}{5}$

11. $\frac{x}{12} = \frac{50}{6x}$

12. $\frac{7}{x} = \frac{14}{9}$

13. The ratio of the lengths of the three sides of a triangle is 5:8:10. If its perimeter is 276 inches, find the length of the longest side of the triangle.

14. **CARPENTRY** A board that is 12 feet long must be cut into two pieces that have lengths in a ratio of 3 to 2. Find the lengths of the two pieces.

EXAMPLE 1

Solve $\frac{2x-3}{4} = \frac{x+9}{3}$.

$\frac{2x-3}{4} = \frac{x+9}{3}$	**Original proportion**
$3(2x-3) = 4(x+9)$	**Cross Products Property**
$6x - 9 = 4x + 36$	**Simplify.**
$2x - 9 = 36$	**Subtract.**
$2x = 45$	**Add.**
$x = 22.5$	**Divide.**

7-2 Similar Polygons (pp. 465–473)

Determine whether each pair of figures is similar. If so, write the similarity statement and scale factor. If not, explain your reasoning.

15.

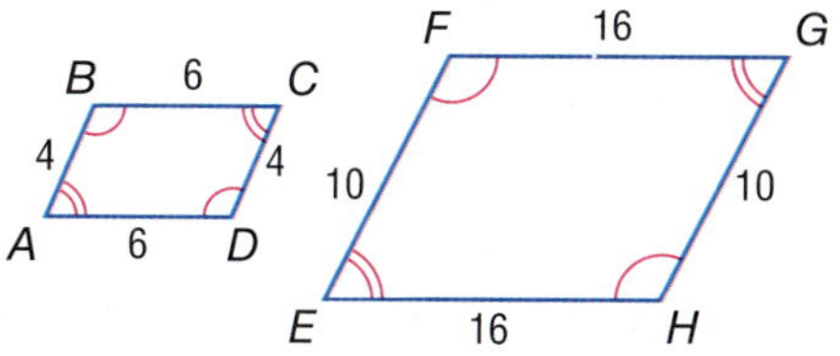

16.

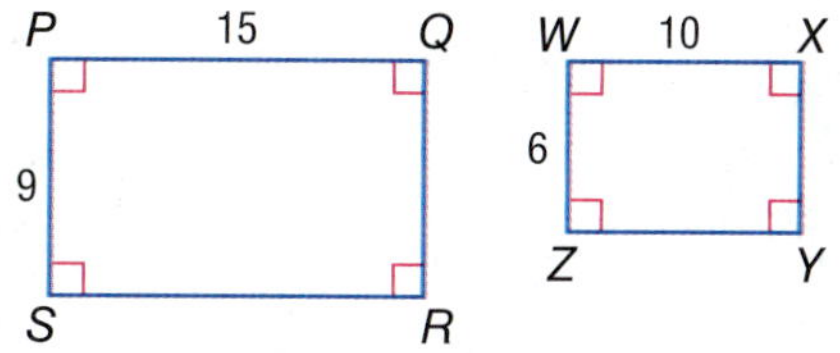

17. The two triangles in the figure below are similar. Find the value of x.

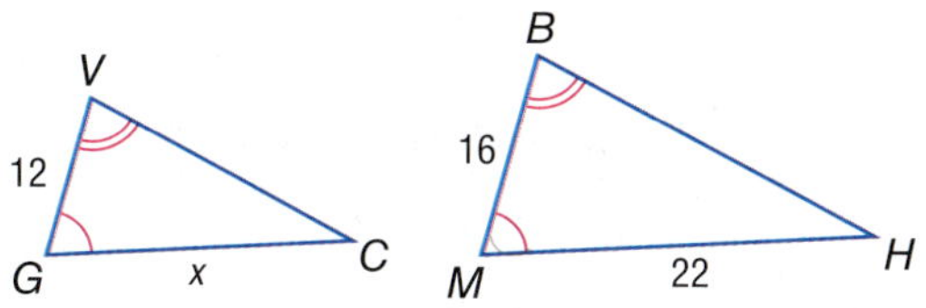

18. **MAPS** On a map of Colorado, the cities of Denver and Colorado Springs are 10.5 inches apart. If the scale of the map shows that 1.5 inches represents 10 miles, find the actual distance from Denver to Colorado Springs.

EXAMPLE 2

Determine whether the pair of triangles is similar. Justify your answer. If so, write the similarity statement and scale factor. If not, explain your reasoning.

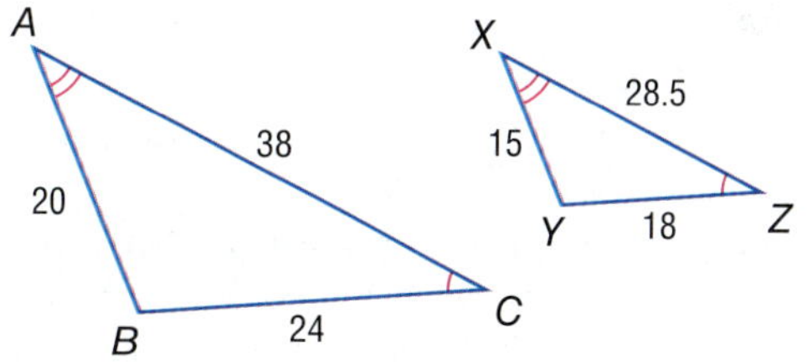

$\angle A \cong \angle X$ and $\angle C \cong \angle Z$, so by the Third Angle Theorem, $\angle B \cong \angle Y$. All of the corresponding angles are therefore congruent.

Similar polygons must also have proportional side lengths. Check the ratios of corresponding side lengths.

$\frac{AB}{XY} = \frac{20}{15}$ or $\frac{4}{3}$ $\quad \frac{BC}{YZ} = \frac{24}{18}$ or $\frac{4}{3}$ $\quad \frac{AC}{XZ} = \frac{38}{28.5}$ or $\frac{4}{3}$

Since corresponding sides are proportional, $\triangle ABC \sim \triangle XYZ$. So, the triangles are similar with a scale factor of $\frac{4}{3}$.

7-3 Similar Triangles (pp. 474–483)

Determine whether the triangles are similar. If so, write a similarity statement. Explain your reasoning.

19.

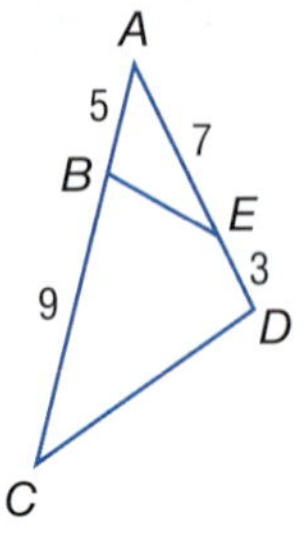

20.

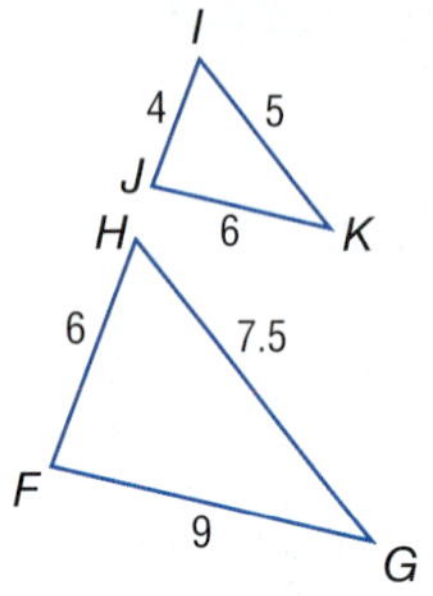

21.

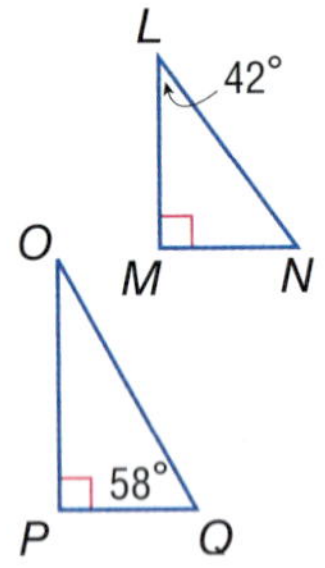

22.

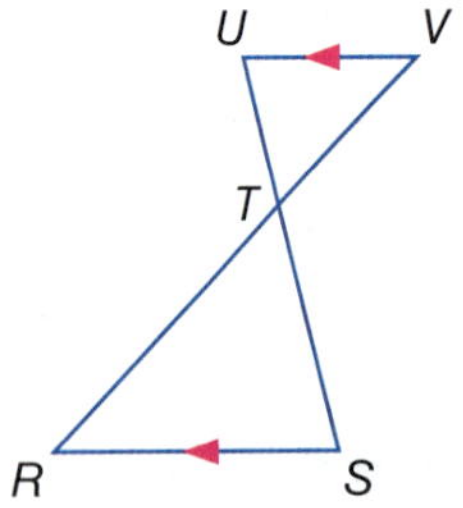

23. TREES To estimate the height of a tree, Dave stands in the shadow of the tree so that his shadow and the tree's shadow end at the same point. Dave is 6 feet 4 inches tall and his shadow is 15 feet long. If he is standing 66 feet away from the tree, what is the height of the tree?

EXAMPLE 3

Determine whether the triangles are similar. If so, write a similarity statement. Explain your reasoning.

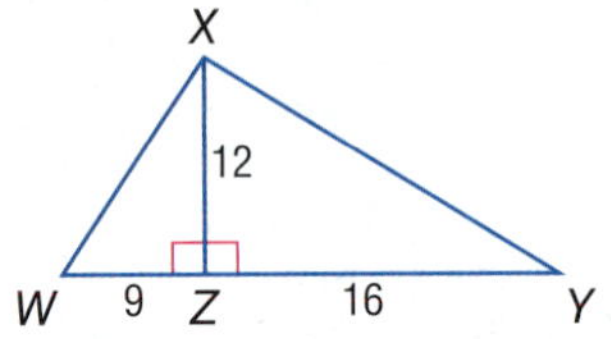

$\angle WZX \cong \angle XZY$ because they are both right angles. Now compare the ratios of the legs of the right triangles.

$$\frac{WZ}{XZ} = \frac{9}{12} = \frac{3}{4} \qquad \frac{XZ}{YZ} = \frac{12}{16} = \frac{3}{4}$$

Since two pairs of sides are proportional with the included angles congruent, $\triangle WZX \sim \triangle XZY$ by SAS Similarity.

7-4 Parallel Lines and Proportional Parts (pp. 484–493)

Find x.

24.

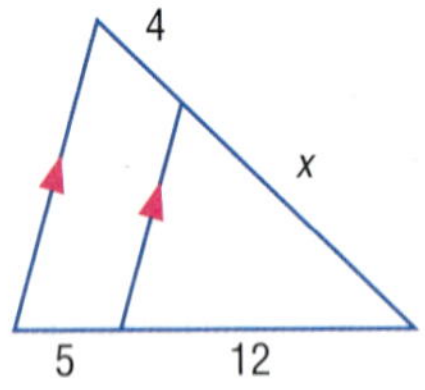

25.

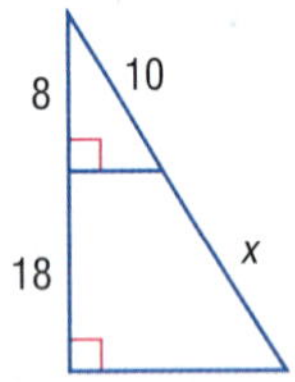

26. STREETS Find the distance along Broadway between 37th St. and 36th St.

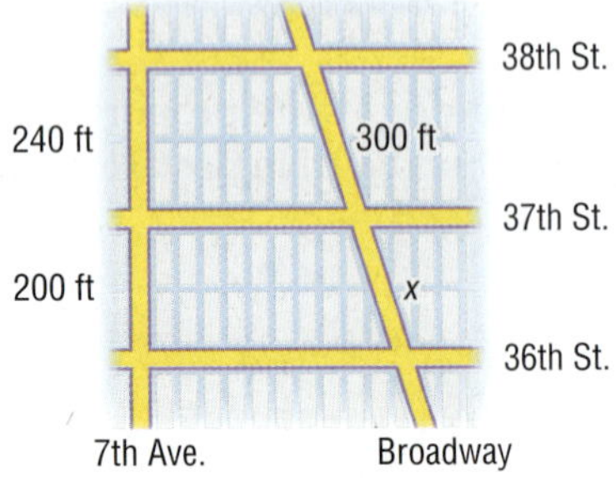

EXAMPLE 4

ALGEBRA Find x and y.

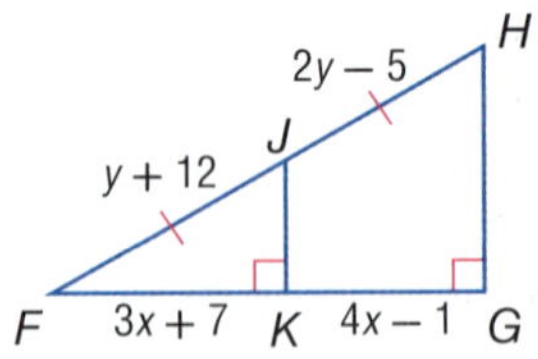

$FK = KG$

$3x + 7 = 4x - 1$

$-x = -8$

$x = 8$

$FJ = JH$	Definition of congruence
$y + 12 = 2y - 5$	Substitution
$-y = -17$	Subtract.
$y = 17$	Simplify.

MIXED PROBLEM SOLVING
For mixed problem-solving practice, see page 1005.

7-5 Parts of Similar Triangles (pp. 495–502)

Find the value of each variable.

27.

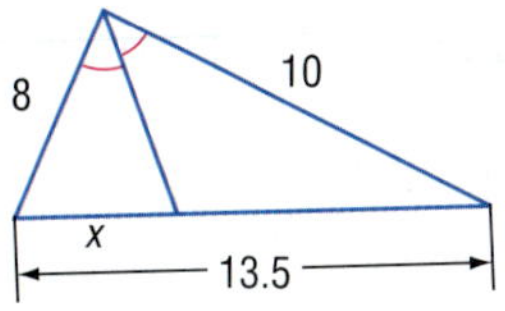

28.

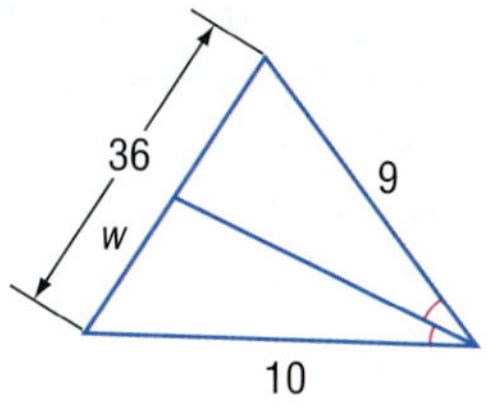

29. MAPS The scale given on a map of the state of Missouri indicates that 3 inches represents 50 miles. The cities of St. Louis, Springfield, and Kansas City form a triangle. If the measurements of the lengths of the sides of this triangle on the map are 15 inches, 10 inches, and 13 inches, find the perimeter of the actual triangle formed by these cities to the nearest mile.

EXAMPLE 5

Find x.

Use the Triangle Angle Bisector Theorem to write a proportion.

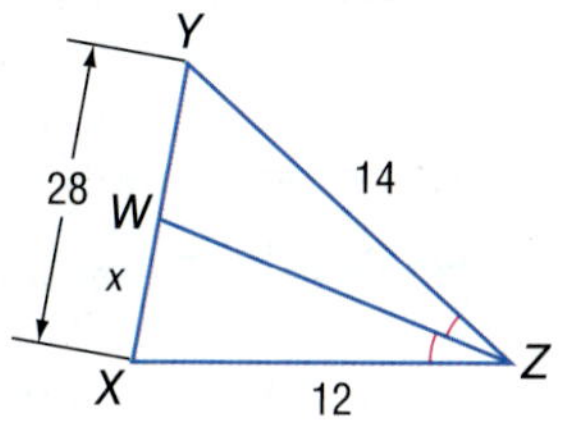

$\frac{WX}{YW} = \frac{XZ}{YZ}$	**Triangle Angle Bisector Thm.**
$\frac{x}{28 - x} = \frac{12}{14}$	**Substitution**
$(28 - x)(12) = x \cdot 14$	**Cross Products Property**
$336 - 12x = 14x$	**Simplify.**
$336 = 26x$	**Add.**
$12.9 = x$	**Simplify.**

7-6 Similarity Transformations (pp. 505–511)

Determine whether the dilation from A to B is an *enlargement* or a *reduction*. Then find the scale factor of the dilation.

30.

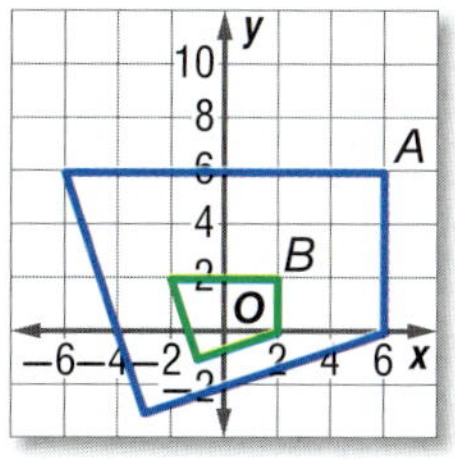

31.

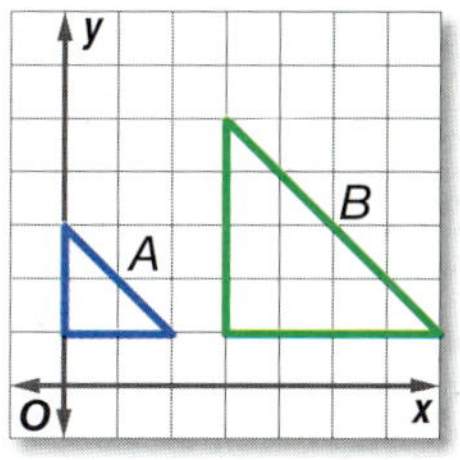

32. GRAPHIC DESIGN Jamie wants to use a photocopier to enlarge her design for the Honors Program at her school. She sets the copier to 250%. If the original drawing was 6 inches by 9 inches, find the dimensions of the enlargement.

EXAMPLE 6

Determine whether the dilation from A to B is an *enlargement* or a *reduction*. Then find the scale factor of the dilation.

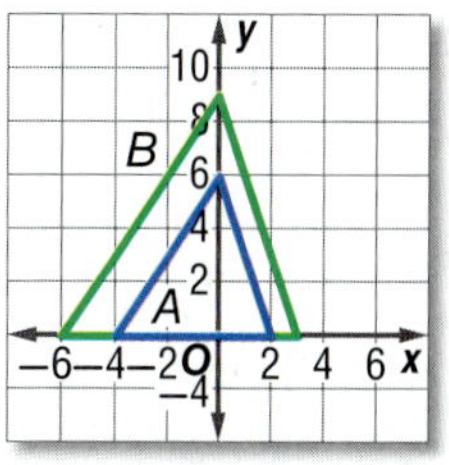

B is larger than A, so the dilation is an enlargement. The distance between the vertices at $(-4, 0)$ and $(2, 0)$ for A is 6 and the distance between the vertices at $(-6, 0)$ and $(3, 0)$ for B is 9. So the scale factor is $\frac{9}{6}$ or $\frac{3}{2}$.

Study Guide and Review

7-7 Scale Drawings and Models (pp. 512–517)

33. BUILDING PLANS In a scale drawing of a school's floor plan, 6 inches represents 100 feet. If the distance from one end of the main hallway to the other is 175 feet, find the corresponding length in the scale drawing.

34. MODEL TRAINS A popular scale for model trains is the 1:48 scale. If the actual train car had a length of 72 feet, find the corresponding length of the model in inches.

35. MAPS A map of the eastern United States has a scale where 3 inches = 25 miles. If the distance on the map between Columbia, South Carolina, and Charlotte, North Carolina, is 11.5 inches what is the actual distance between the cities?

EXAMPLE 7

In the scale of a map of the Pacific Northwest 1 inch = 20 miles. The distance on the map between Portland, Oregon, and Seattle, Washington, is 8.75 inches. Find the distance between the two cities.

$\frac{1}{20} = \frac{8.75}{x}$ **Write a proportion.**

$x = 20(8.75)$ **Cross Products Property**

$x = 175$ **Simplify.**

The distance between the two cities is 175 miles.

Practice Test

Math Online glencoe.com
Chapter Test

Solve each proportion.

1. $\frac{3}{7} = \frac{12}{x}$
2. $\frac{2x}{5} = \frac{x+3}{3}$
3. $\frac{4x}{15} = \frac{60}{x}$
4. $\frac{5x-4}{4x+7} = \frac{13}{11}$

Determine whether each pair of figures is similar. If so, write the similarity statement and scale factor. If not, explain your reasoning.

5.

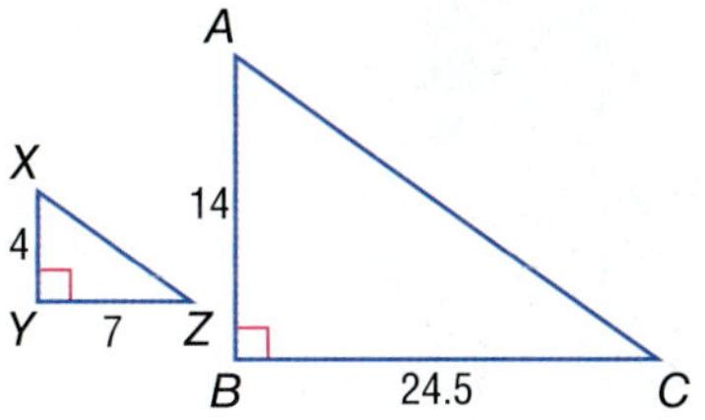

6.

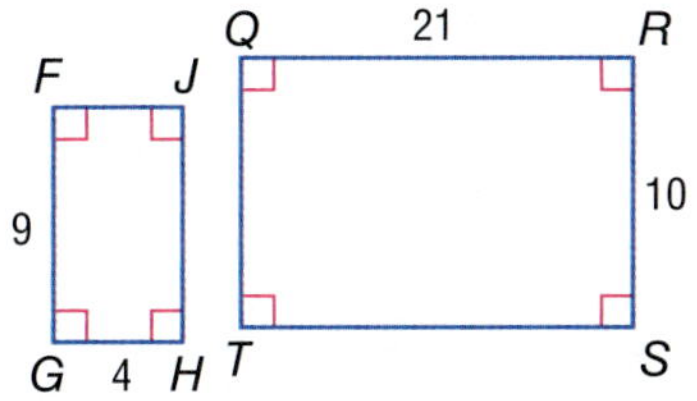

7. **CURRENCY** Jane is traveling to Europe this summer with the French Club. She plans to bring $300 to spend while she is there. If $90 in U.S. currency is equivalent to 63 euros, how many euros will she receive when she exchanges her money?

ALGEBRA Find x and y. Round to the nearest tenth if necessary.

8.

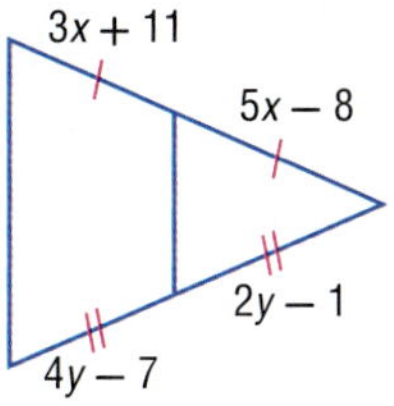

9.

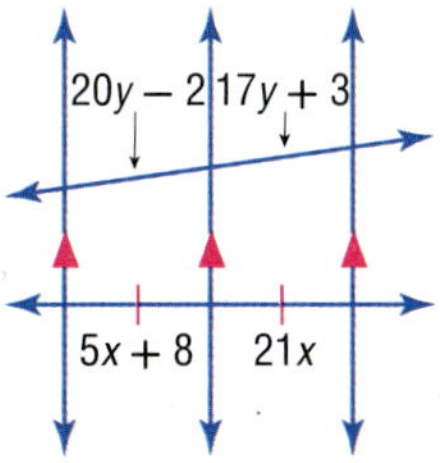

10. **ALGEBRA** Equilateral $\triangle MNP$ has perimeter $12a + 18b$. $\overline{QR}$ is a midsegment. What is QR?

11. **ALGEBRA** Right isosceles $\triangle ABC$ has hypotenuse length h. $\overline{DE}$ is a midsegment with length $4x$. What is the perimeter of $\triangle ABC$?

12. **MODELS** Jimmy has a diecast metal car that is a scale model of an actual race car. If the actual length of the car is 10 feet and 6 inches and the model has a length of 7 inches, what is the scale factor of model to actual car?

Find x.

13.

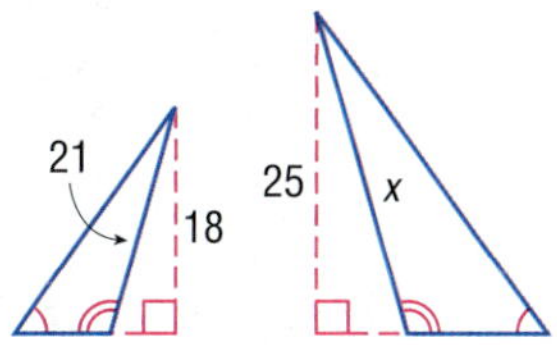

14.

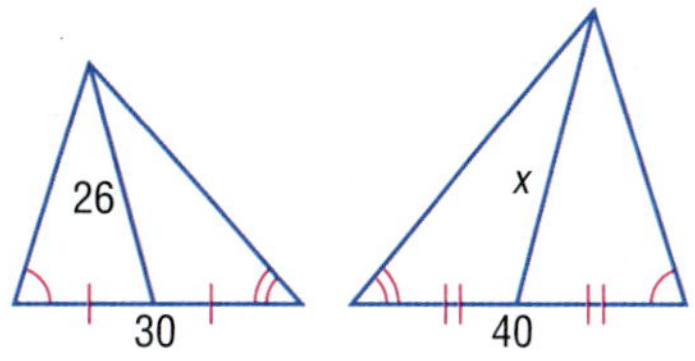

Determine whether the dilation from A to B is an *enlargement* or a *reduction*. Then find the scale factor of the dilation.

15.

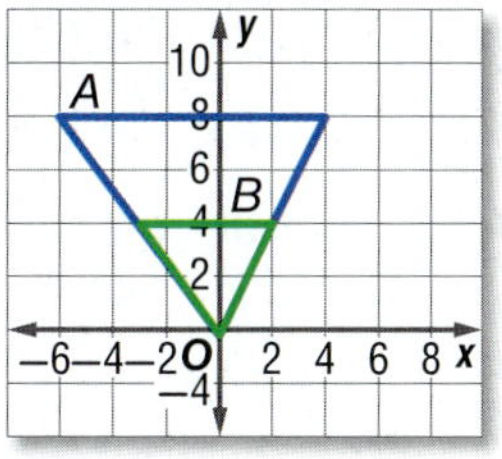

16.

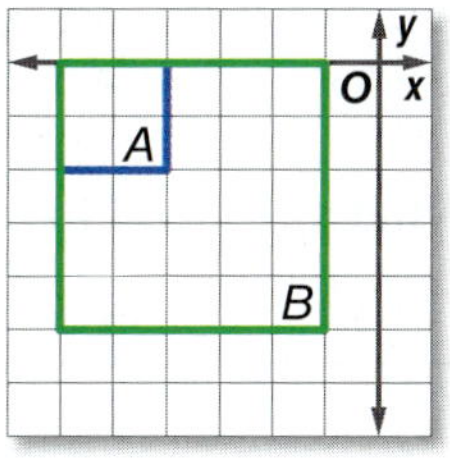

ALGEBRA Identify the similar triangles. Find each measure.

17. WZ, UZ

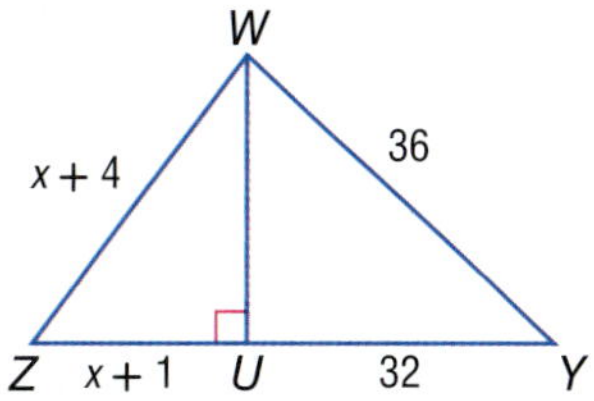

CHAPTER 7

Preparing for Standardized Tests

Identifying Non-Examples

Multiple choice items sometimes ask you to determine which of the given answer choices is a non-example. These types of problems require a different approach when solving them.

Strategies for Identifying Non-Examples

Step 1

Read and understand the problem statement.

- **Non-Example:** A non-example is an answer choice that does not satisfy the conditions of the problem statement.
- **Keywords:** Look for the word *not* (usually bold, all capital letters, or italicized) to indicate that you need to find a non-example.

Step 2

Follow the concepts and steps below to help you identify non-examples. Identify any answer choices that are clearly incorrect and eliminate them.

- Eliminate any answer choices that are not in the proper format.
- Eliminate any answer choices that do not have the correct units.

EXAMPLE

Read the problem. Identify what you need to know. Then use the information in the problem to solve.

In the adjacent triangle, you know that $\angle MQN \cong \angle RQS$. Which of the following would *not* be sufficient to prove that $\triangle QMN \sim \triangle QRS$?

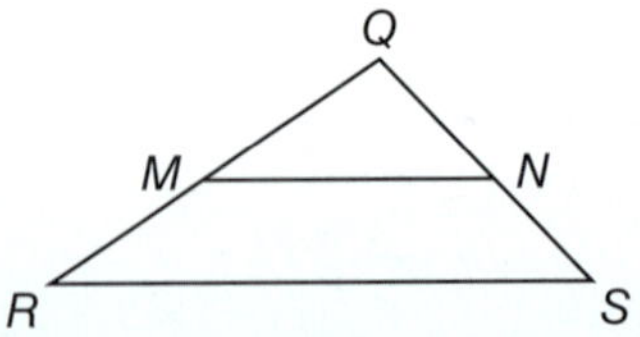

A $\angle QMN \cong \angle QRS$

B $\overline{MN} \parallel \overline{RS}$

C $\overline{QN} \cong \overline{NS}$

D $\frac{QM}{QR} = \frac{QN}{QS}$

The italicized *not* indicates that you need to find a non-example. Test each answer choice using the principles of triangle similarity to see which one would not prove $\triangle QMN \cong \triangle QRS$.

Choice A: $\angle QMN \cong \angle QRS$

If $\angle QMN \cong \angle QRS$, then $\triangle QMN \sim \triangle QRS$ by AA Similarity.

Choice B: $\overline{MN} \parallel \overline{RS}$

If $\overline{MN} \parallel \overline{RS}$, then $\angle QMN \cong \angle QRS$, because they are corresponding angles of two parallel lines cut by transversal $\overline{QR}$. Therefore, $\triangle QMN \sim \triangle QRS$ by AA Similarity.

Choice C: $\overline{QN} \cong \overline{NS}$

If $\overline{QN} \cong \overline{NS}$, we cannot conclude that $\triangle QMN \sim \triangle QRS$ because we do not know anything about $\overline{QM}$ and $\overline{MR}$. So, answer choice C is a non-example.

The correct answer is C. You should also check answer choice D to make sure it is a valid example if you have time.

Exercises

Read each problem. Identify what you need to know. Then use the information in the problem to solve.

1. The ratio of the measures of the angles of the quadrilateral below is 6:5:4:3. Which of the following is *not* an angle measure of the figure?

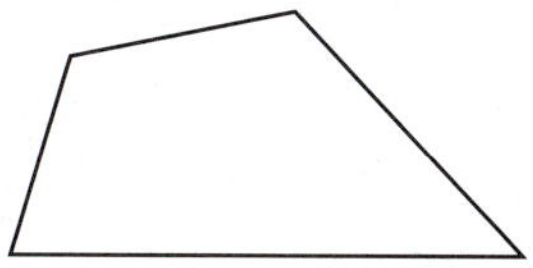

A 60° **C** 120°

B 80° **D** 140°

2. Which figure can serve as a counterexample to the conjecture below?

If all angles of a quadrilateral are right angles, then the quadrilateral is a square.

F parallelogram

G rectangle

H rhombus

J trapezoid

3. Consider the figure below. Which of the following is *not* sufficient to prove that $\triangle GIK \sim \triangle HIG$?

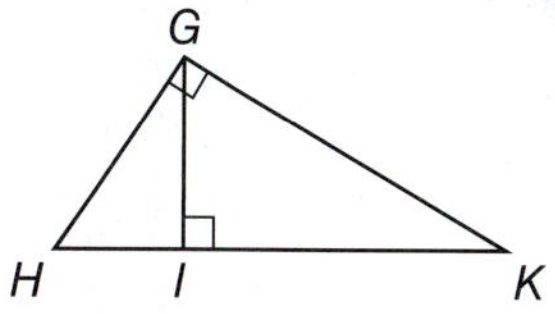

A $\angle GKI \cong \angle HGI$

B $\frac{HI}{GI} = \frac{GI}{IK}$

C $\frac{GH}{GI} = \frac{GK}{IK}$

D $\angle IGK \cong \angle IHG$

4. Which triangles are *not* necessarily similar?

F two right triangles with one angle measuring 30°

G two right triangles with one angle measuring 45°

H two isosceles triangles

J two equilateral triangles

Standardized Test Practice

Cumulative, Chapters 1 through 7

Multiple Choice

Read each question. Then fill in the correct answer on the answer document provided by your teacher or on a sheet of paper.

1. Adrian wants to measure the width of a ravine. He marks distances as shown in the diagram.

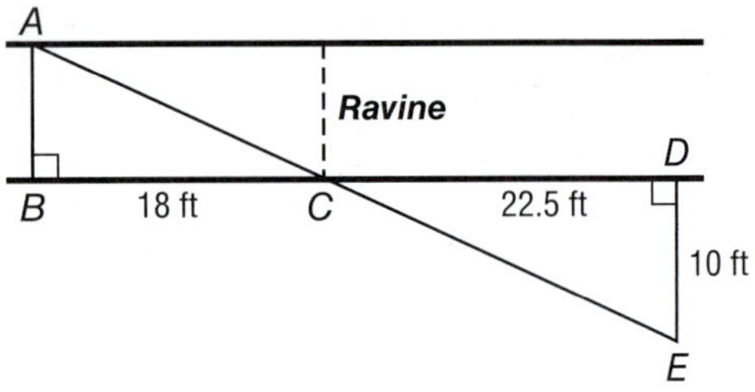

Using this information, what is the *approximate* width of the ravine?

A 5 ft

B 6 ft

C 7 ft

D 8 ft

2. Kyle and his family are planning a vacation in Cancun, Mexico. Kyle wants to convert 200 US dollars to Mexican pesos for spending money. If 278 Mexican pesos are equivalent to $25, how many pesos will Kyle get for $200?

F 2178

G 2224

H 2396

J 2504

3. Which of the following terms *best* describes the transformation below?

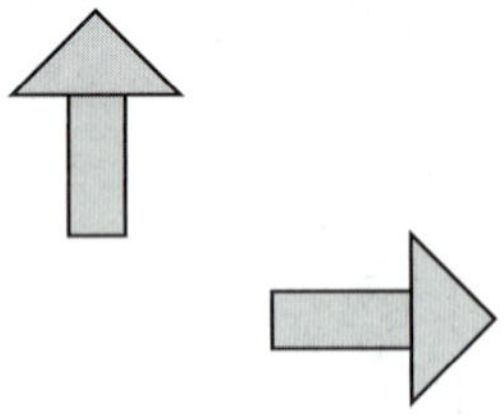

A dilation

B reflection

C rotation

D translation

Test-Taking Tip

Question 2 Set up and solve the proportion for the number of pesos. Use the ratio pesos : dollars.

4. Refer to the figures below. Which of the following terms *best* describes the transformation?

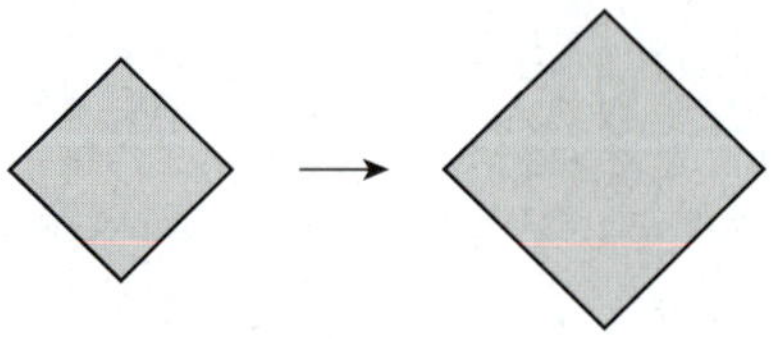

F congruent

G enlargement

H reduction

J scale

5. The ratio of North Carolina residents to Americans is about 295 to 10,000. If there are approximately 300,000,000 Americans, how many of them are North Carolina residents?

A 7,950,000

B 8,400,000

C 8,850,000

D 9,125,000

6. Solve for x.

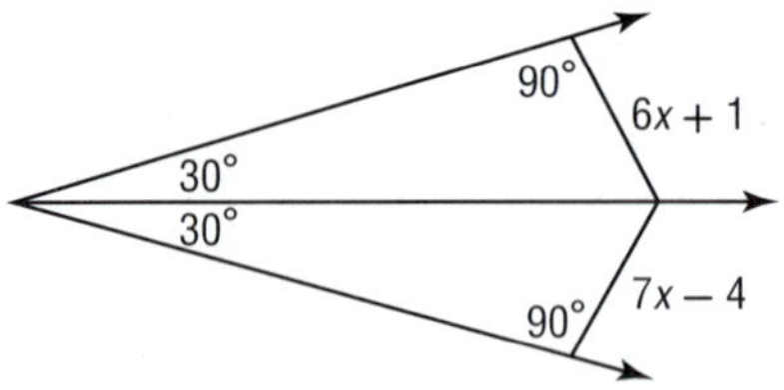

F 3

G 4

H 5

J 6

7. Two similar trapezoids have a scale factor of 3:2. The perimeter of the larger trapezoid is 21 yards. What is the perimeter of the smaller trapezoid?

A 14 yd

B 17.5 yd

C 28 yd

D 31.5 yd

Short Response/Gridded Response

Record your answers on the answer sheet provided by your teacher or on a sheet of paper.

8. **GRIDDED RESPONSE** Colleen surveyed 50 students in her school and found that 35 of them have homework at least four nights a week. If there are 290 students in the school altogether, how many of them would you expect to have homework at least four nights a week?

9. **GRIDDED RESPONSE** In the triangle below, $\overline{MN} \parallel \overline{BC}$. Solve for x.

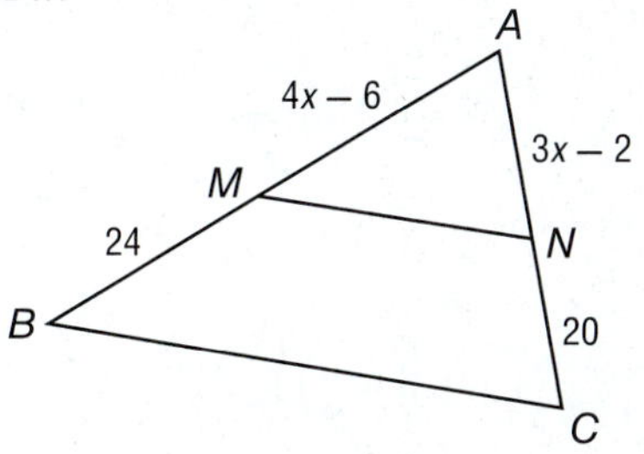

10. Quadrilateral $WXYZ$ is a rhombus. If $m\angle XYZ = 110°$, find $m\angle ZWY$.

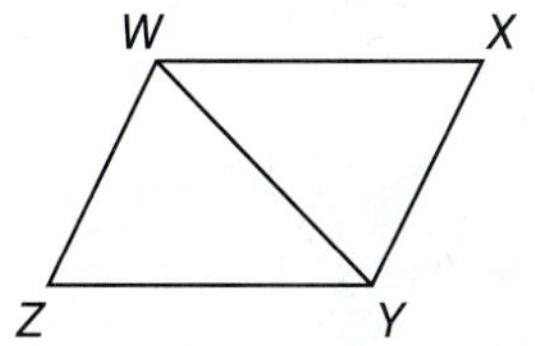

11. What is the contrapositive of the statement below?

If Tom was born in Louisville, then he was born in Kentucky.

12. **GRIDDED RESPONSE** In the triangle below, $\overline{RS}$ bisects $\angle VRU$. Solve for x.

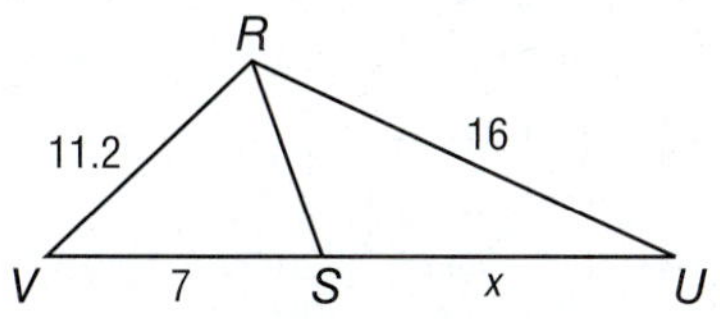

13. **GRIDDED RESPONSE** The scale of a map is 1 inch = 2.5 miles. What is the distance between two cities that are 3.3 inches apart on the map? Round to the nearest tenth, if necessary.

14. What is the value of x in the figure?

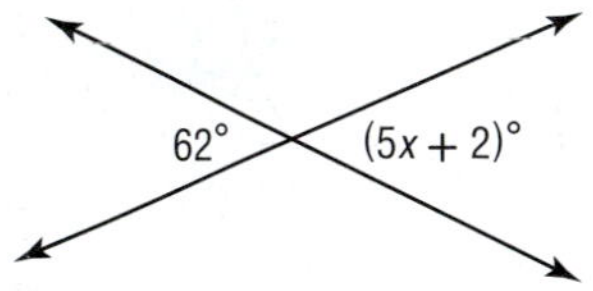

Extended Response

Record your answers on a sheet of paper. Show your work.

15. Refer to triangle XYZ to answer each question.

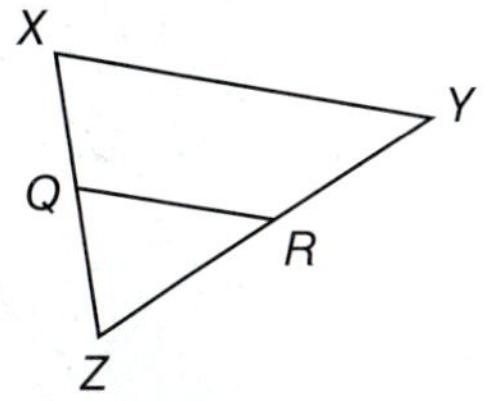

a. Suppose $\overline{QR} \parallel \overline{XY}$. What do you know about the relationship between segments XQ, QZ, YR, and RZ?

b. If $\overline{QR} \parallel \overline{XY}$, $XQ = 15$, $QZ = 12$, and $YR = 20$, what is the length of $\overline{RZ}$?

c. Suppose $\overline{QR} \parallel \overline{XY}$, $\overline{XQ} \cong \overline{QZ}$, and $QR = 9.5$ units. What is the length of $\overline{XY}$?

Need Extra Help?															
If you missed Question...	1	2	3	4	5	6	7	8	9	10	11	12	13	14	15
Go to Lesson or Page...	7-3	7-1	4-7	7-6	7-1	5-1	7-2	7-1	7-6	6-5	2-3	7-5	7-7	1-5	7-4

CHAPTER 8 Right Triangles and Trigonometry

Then

In Chapter 7, you learned to solve proportions.

Now

In Chapter 8, you will:

- Use the Pythagorean Theorem.
- Use properties of special right triangles.
- Use trigonometry to find missing measures of triangles.

Why?

Properties of triangles can be used in planning and preparation for special events including the height of decorations.

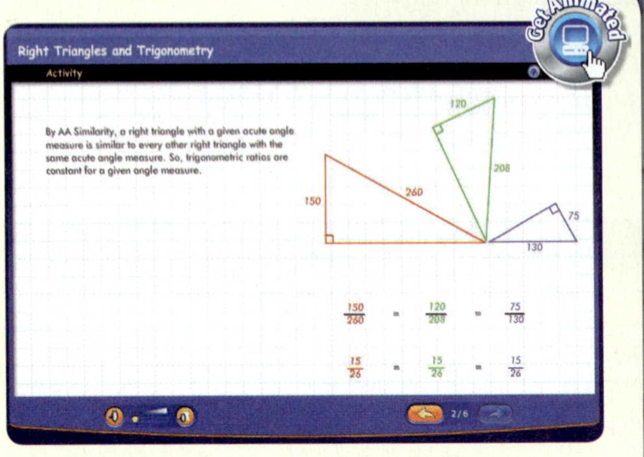

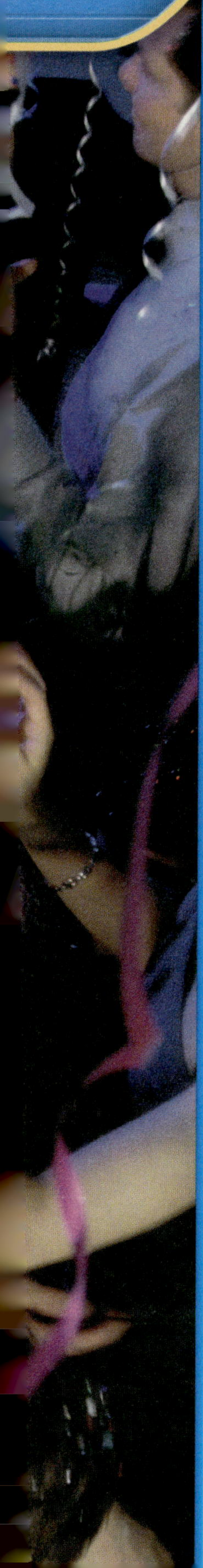

Get Ready for Chapter 8

Diagnose Readiness You have two options for checking Prerequisite Skills.

Text Option

Take the Quick Check below. Refer to the Quick Review for help.

QuickCheck

Simplify. (Lesson 0-9)

1. $\sqrt{112}$

2. $\frac{\sqrt{24}}{2\sqrt{3}}$

3. $\sqrt{15 \cdot 20}$

4. $\frac{\sqrt{6}}{\sqrt{3}} \cdot \frac{\sqrt{18}}{\sqrt{3}}$

5. $\sqrt{\frac{45}{80}}$

6. $\frac{8\sqrt{2}}{6 - 3\sqrt{8}}$

Find x. (Lesson 1-3)

7.

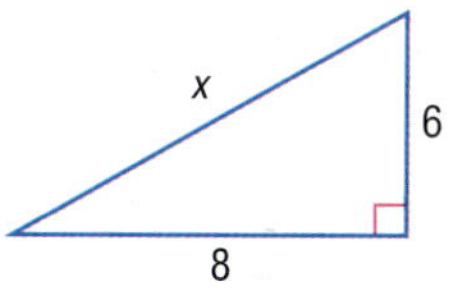

8.

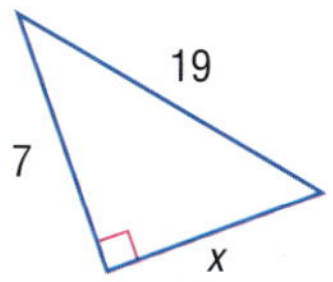

9. BANNERS Anna is making a banner out of 4 congruent triangles as shown below. How much blue trim will she need for each side?

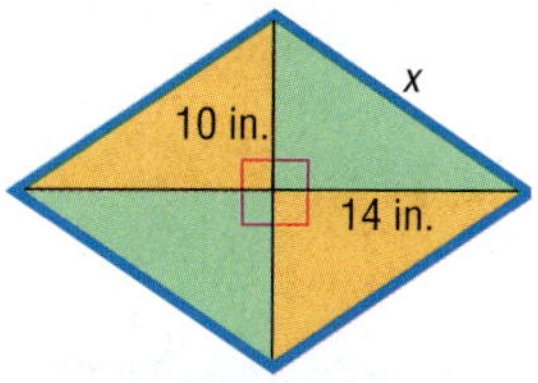

Graph the line segment with the given endpoints. (Lesson 3-4)

10. $G(3, -4)$ and $H(3, 4)$

11. $E(-3, 5)$ and $F(4, -3)$

12. COLLEGES Quinn is visiting a college campus. He notices from his map that several important buildings are located around a grassy area the students call the Quad. If the library is represented on the map by $L(6, 8)$ and the cafeteria is represented by $C(0, 0)$, graph the line segment that represents the shortest path between the two buildings.

QuickReview

EXAMPLE 1

Simplify $\frac{6}{\sqrt{3}}$.

$\frac{6}{\sqrt{3}} = \frac{6}{\sqrt{3}} \cdot \frac{\sqrt{3}}{\sqrt{3}}$ — Multiply by $\frac{\sqrt{3}}{\sqrt{3}}$.

$= \frac{6\sqrt{3}}{3}$ or $2\sqrt{3}$ — Simplify.

EXAMPLE 2

Find x.

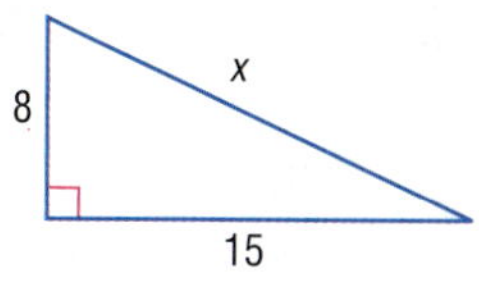

$a^2 + b^2 = c^2$	Pythagorean Theorem
$8^2 + 15^2 = x^2$	$a = 8$ and $b = 15$
$289 = x^2$	Simplify.
$\sqrt{289} = \sqrt{x^2}$	Take the positive square root of each side.
$17 = x$	Simplify.

EXAMPLE 3

Graph the line segment with endpoints $A(-4, 2)$ and $B(3, -2)$.

Plot points A and B.

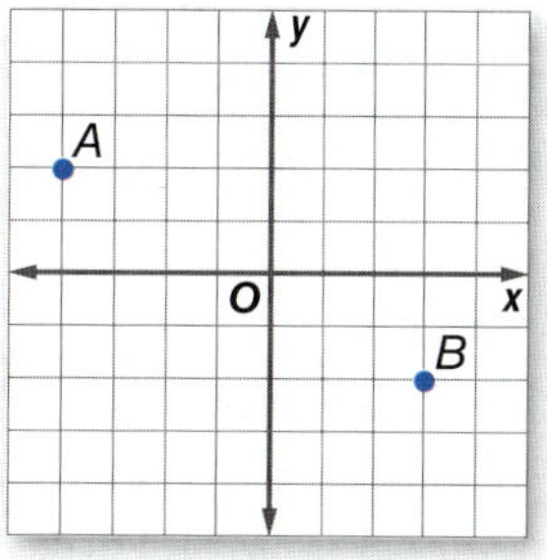

Connect the points.

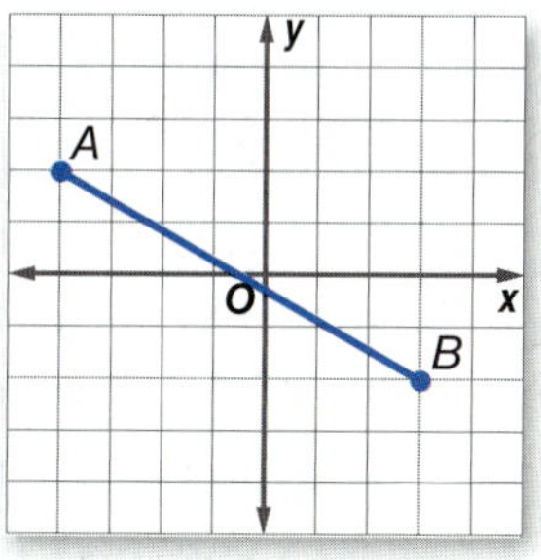

Online Option

Math Online Take a self-check Chapter Readiness Quiz at glencoe.com.

Get Started on Chapter 8

You will learn several new concepts, skills, and vocabulary terms as you study Chapter 8. To get ready, identify important terms and organize your resources. You may wish to refer to **Chapter 0** to review prerequisite skills.

FOLDABLES® Study Organizer

Right Angles and Trigonometry Make this Foldable to help you organize your Chapter 8 notes about right angles and trigonometry. Begin with three sheets of notebook paper and one sheet of construction paper.

1. **Stack** the notebook paper on the construction paper.

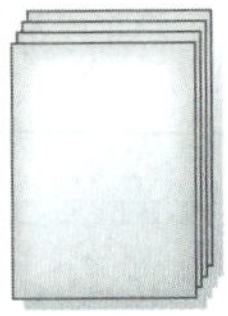

2. **Fold** the paper diagonally to form a triangle and cut off the excess.

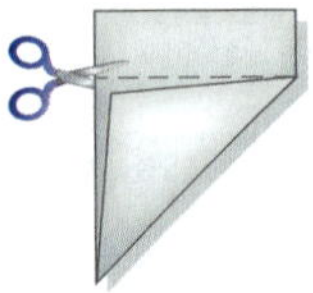

3. **Open** the paper and staple the inside fold to form a booklet.

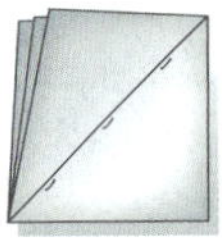

4. **Label** each page with a lesson number and title.

Math Online glencoe.com

- Study the chapter online
- Explore **Get Animated** activities
- Get extra help from your own **Personal Tutor**
- Use **Extra Examples** for additional help
- Take a **Self-Check Quiz**
- **Review Vocabulary** in fun ways

New Vocabulary

English		Español
geometric mean	p. 531	media geométrica
Pythagorean triple	p. 542	triplete de Pitágoras
cosine	p. 562	coseno
sine	p. 562	seno
tangent	p. 562	tangente
trigonometric ratio	p. 562	razón trigonométrica
trigonometry	p. 562	trigonométria
angle of depression	p. 574	ángulo de depresión
angle of elevation	p. 574	ángulo de elevación
Law of Sines	p. 582	Ley de los senos
Law of Cosines	p. 583	Ley do los cosenos
component form	p. 593	componente
magnitude	p. 593	magnitud
vector	p. 593	vector
resultant	p. 595	resultante

Review Vocabulary

altitude • p. 335 • altura a segment drawn from a vertex of a triangle perpendicular to the line containing the other side

Pythagorean Theorem • p. 26 • Teorema de Pitágoras If a and b are the measures of the legs of a right triangle and c is the measure of the hypotenuse, then $a^2 + b^2 = c^2$.

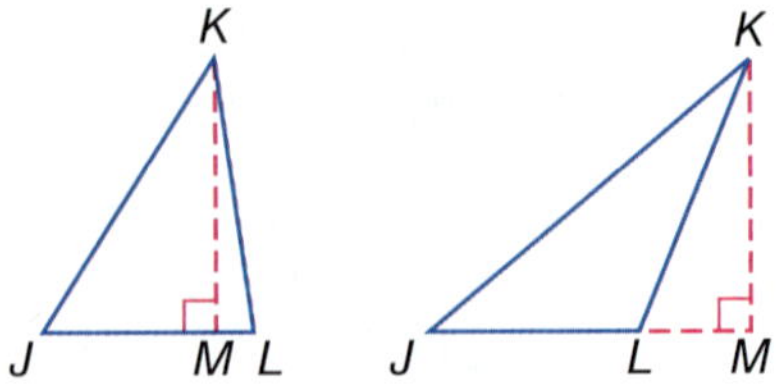

$\overline{KM}$ is an altitude from K to $\overleftrightarrow{JL}$.

Multilingual eGlossary glencoe.com

8-1 Geometric Mean

Then

You used proportional relationships of corresponding angle bisectors, altitudes, and medians of similar triangles. (Lesson 7-5)

Now

- Find the geometric mean between two numbers.
- Solve problems involving relationships between parts of a right triangle and the altitude to its hypotenuse.

New Vocabulary

geometric mean

Math Online
glencoe.com

Why?

Photographing very tall or very wide objects can be challenging. It can be difficult to include the entire object in your shot without distorting the image. If your camera is set for a vertical viewing angle of 90° and you know the height of the object you wish to photograph, you can use the geometric mean of the distance from the top of the object to your camera level and the distance from the bottom of the object to camera level.

Geometric Mean When the means of a proportion are the same number, that number is called the geometric mean of the extremes. The **geometric mean** between two numbers is the positive square root of their product.

$$\text{extreme} \rightarrow \frac{a}{x} = \frac{x}{b} \leftarrow \text{mean}$$
$$\text{mean} \rightarrow \qquad \qquad \leftarrow \text{extreme}$$

Key Concept — Geometric Mean

Words The geometric mean of two positive numbers a and b is the number x such that $\frac{a}{x} = \frac{x}{b}$. So, $x^2 = ab$ and $x = \sqrt{ab}$.

Example The geometric mean of $a = 9$ and $b = 4$ is 6, because $6 = \sqrt{9 \cdot 4}$.

EXAMPLE 1 Geometric Mean

Find the geometric mean between 8 and 10.

$x = \sqrt{ab}$	Definition of geometric mean
$= \sqrt{8 \cdot 10}$	$a = 8$ and $b = 10$
$= \sqrt{(4 \cdot 2) \cdot (2 \cdot 5)}$	Factor.
$= \sqrt{16 \cdot 5}$	Associative Property
$= 4\sqrt{5}$	Simplify.

The geometric mean between 8 and 10 is $4\sqrt{5}$ or about 8.9.

Guided Practice

Find the geometric mean between each pair of numbers.

1A. 5 and 45 **1B.** 12 and 15

Personal Tutor glencoe.com

Geometric Means in Right Triangles In a right triangle, an altitude drawn from the vertex of the right angle to the hypotenuse forms two additional right triangles. These three right triangles share a special relationship.

Theorem 8.1

If the altitude is drawn to the hypotenuse of a right triangle, then the two triangles formed are similar to the original triangle and to each other.

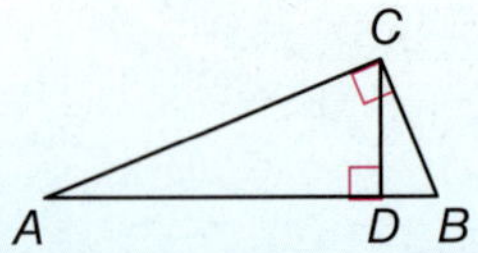

Example If $\overline{CD}$ is the altitude to hypotenuse $\overline{AB}$ of right $\triangle ABC$, then $\triangle ACD \sim \triangle ABC$, $\triangle CBD \sim \triangle ABC$, and $\triangle ACD \sim \triangle CBD$.

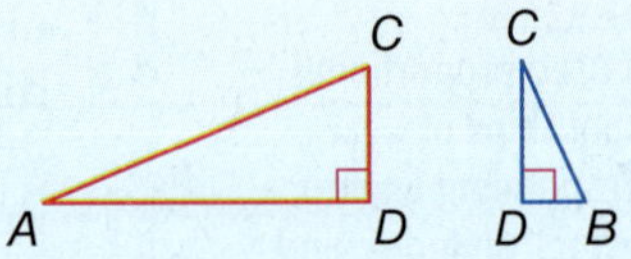

You will prove Theorem 8.1 in Exercise 39.

Review Vocabulary

altitude (of a triangle) a segment from a vertex to the line containing the opposite side and perpendicular to the line containing that side (Lesson 5-2)

EXAMPLE 2 Identify Similar Right Triangles

Write a similarity statement identifying the three similar right triangles in the figure.

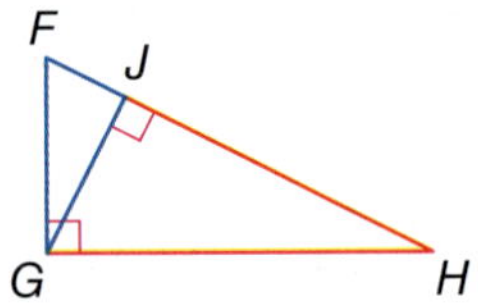

Separate the triangle into two triangles along the altitude. Then sketch the three triangles, reorienting the smaller ones so that their corresponding angles and sides are in the same positions as the original triangle.

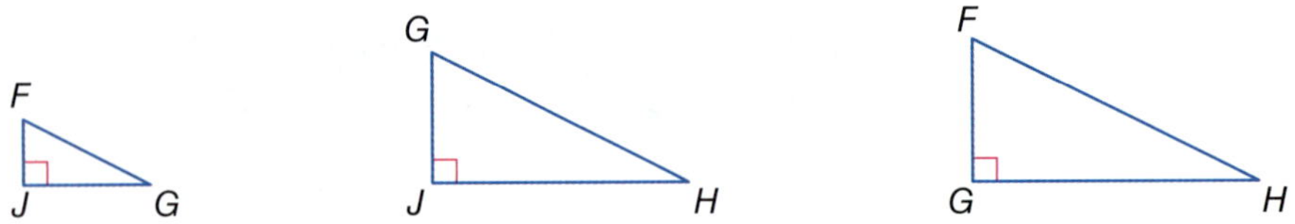

So by Theorem 8.1, $\triangle FJG \sim \triangle GJH \sim \triangle FGH$.

StudyTip

Reorienting Triangles To reorient the right triangles in Example 2, first match up the right angles. Then match up the shorter sides.

Guided Practice

2A.

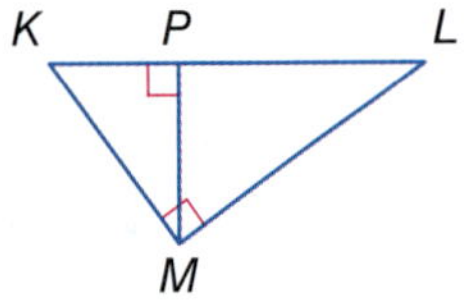

2B.

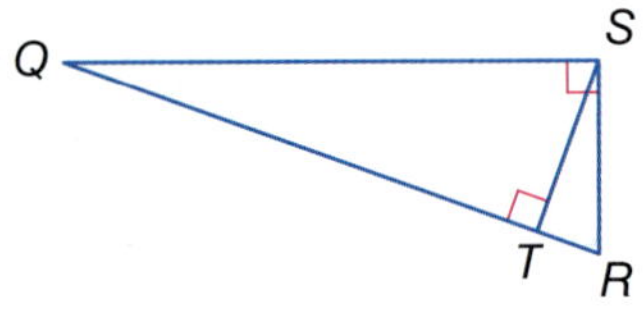

Personal Tutor glencoe.com

From Theorem 8.1, you know that altitude $\overline{CD}$ drawn to the hypotenuse of right triangle ABC forms three similar triangles: $\triangle ACB \sim \triangle ADC \sim \triangle CDB$. By the definition of similar polygons, you can write the following proportions comparing the side lengths of these triangles.

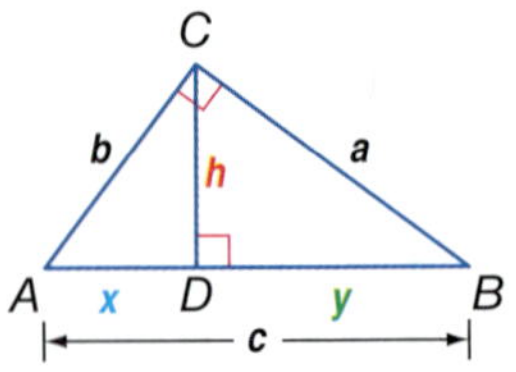

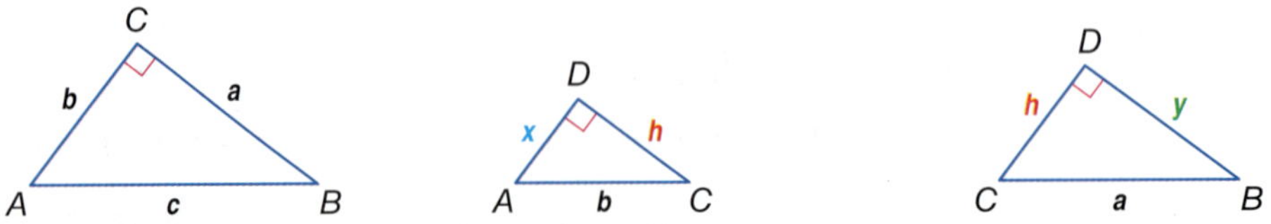

$$\frac{\text{shorter leg}}{\text{longer leg}} = \frac{b}{a} = \frac{x}{h} = \frac{h}{y} \qquad \frac{\text{hypotenuse}}{\text{shorter leg}} = \frac{c}{b} = \frac{b}{x} = \frac{a}{h} \qquad \frac{\text{hypotenuse}}{\text{longer leg}} = \frac{c}{a} = \frac{b}{h} = \frac{a}{y}$$

Notice that the circled relationships involve geometric means. This leads to the theorems at the top of the next page.

Theorems

Right Triangle Geometric Mean Theorems

8.2 Geometric Mean (Altitude) Theorem The altitude drawn to the hypotenuse of a right triangle separates the hypotenuse into two segments. The length of this altitude is the geometric mean between the lengths of these two segments.

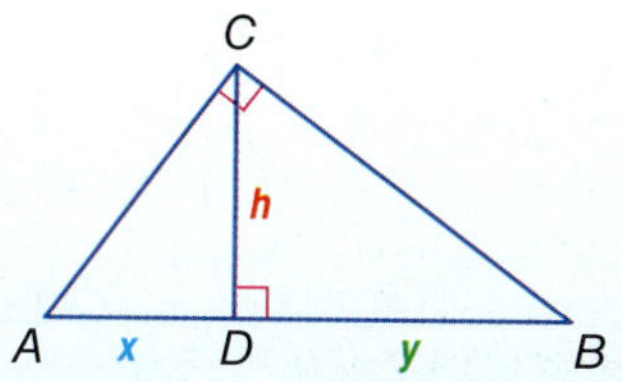

Example If $\overline{CD}$ is the altitude to hypotenuse $\overline{AB}$ of right $\triangle ABC$, then $\frac{x}{h} = \frac{h}{y}$ or $h = \sqrt{xy}$.

8.3 Geometric Mean (Leg) Theorem The altitude drawn to the hypotenuse of a right triangle separates the hypotenuse into two segments. The length of a leg of this triangle is the geometric mean between the length of the hypotenuse and the segment of the hypotenuse adjacent to that leg.

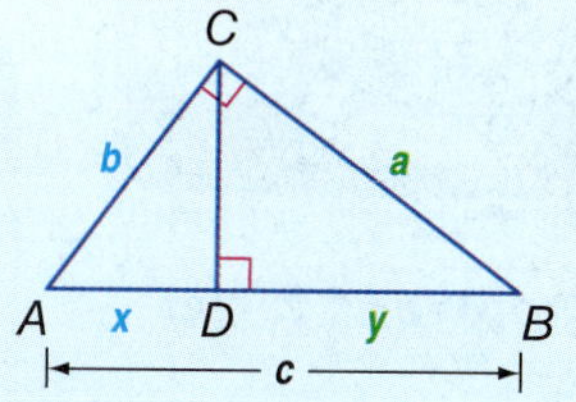

Example If $\overline{CD}$ is the altitude to hypotenuse $\overline{AB}$ of right $\triangle ABC$, then $\frac{c}{b} = \frac{b}{x}$ or $b = \sqrt{xc}$ and $\frac{c}{a} = \frac{a}{y}$ or $a = \sqrt{yc}$.

You will prove Theorems 8.2 and 8.3 in Exercises 40 and 41, respectively.

EXAMPLE 3 Use Geometric Mean with Right Triangles

Find x, y, and z.

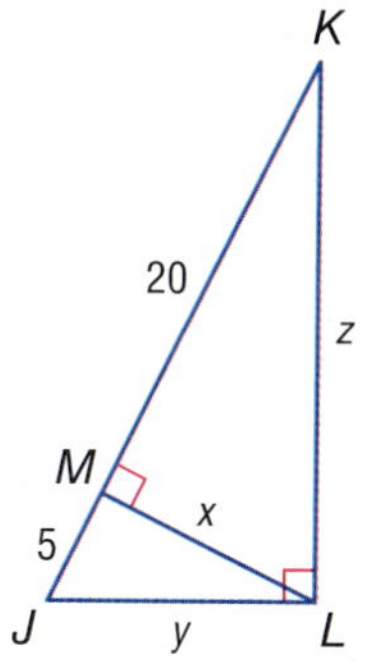

Since x is the measure of the altitude drawn to the hypotenuse of right $\triangle JKL$, x is the geometric mean of the lengths of the two segments that make up the hypotenuse, JM and MK.

$x = \sqrt{JM \cdot MK}$ **Geometric Mean (Altitude) Theorem**

$= \sqrt{5 \cdot 20}$ **Substitution**

$= \sqrt{100}$ or 10 **Simplify.**

StudyTip

Use a Proportion In Example 3, the value of x could also be found by solving the proportion $\frac{5}{x} = \frac{x}{20}$.

Since y is the measure of leg $\overline{JL}$, y is the geometric mean of JM, the measure of the segment adjacent to this leg, and the measure of the hypotenuse JK.

$y = \sqrt{JM \cdot JK}$ **Geometric Mean (Leg) Theorem**

$= \sqrt{5 \cdot (20 + 5)}$ **Substitution**

$= \sqrt{125}$ or about 11.2 **Use a calculator to simplify.**

Since z is the measure of leg $\overline{KL}$, z is the geometric mean of MK, the measure of the segment adjacent to $\overline{KL}$, and the measure of the hypotenuse JK.

$z = \sqrt{MK \cdot JK}$ **Geometric Mean (Leg) Theorem**

$= \sqrt{20 \cdot (20 + 5)}$ **Substitution**

$= \sqrt{500}$ or about 22.4 **Use a calculator to simplify.**

Guided Practice

Find x, y, and z.

3A. 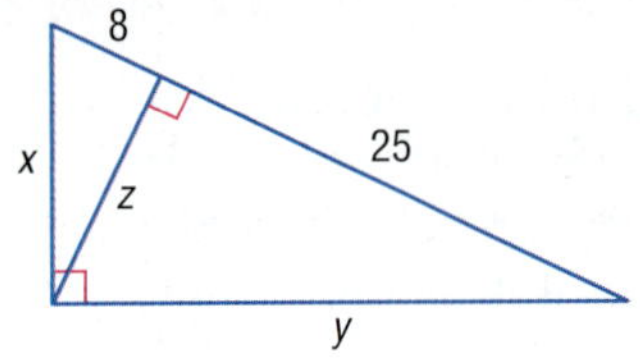

3B. 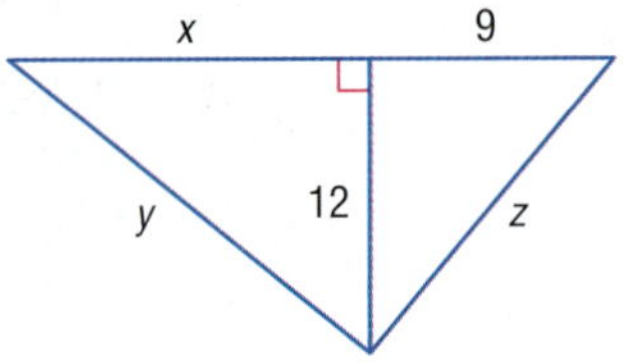

Personal Tutor glencoe.com

Real-World Career

Event Planner
Event planners organize events including choosing a location, arranging for food, and scheduling entertainment. They also coordinate services like transportation and photography.

Most of the skills required for event planning are acquired through on-the-job experience.

You can use geometric mean to measure height indirectly.

Real-World EXAMPLE 4 Indirect Measurement

ADVERTISING **Zach wants to order a banner that will hang over the side of his high school baseball stadium grandstand and reach the ground.**

To find this height, he uses a cardboard square to line up the top and bottom of the grandstand. He measures his distance from the grandstand and from the ground to his eye level. Find the height of the grandstand to the nearest foot.

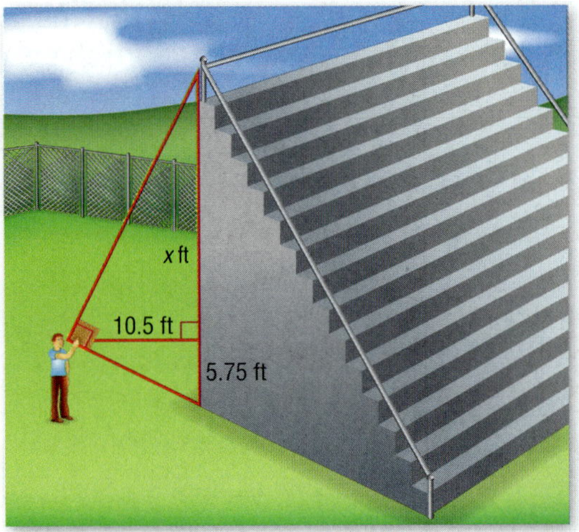

Note: Not drawn to scale.

The distance from Zach to the grandstand is the altitude to the hypotenuse of a right triangle. The length of this altitude is the geometric mean of the two segments that make up the hypotenuse. The shorter segment has the measure of 5.75 feet. Let the unknown measure be x feet.

$10.5 = \sqrt{5.75 \cdot x}$ **Geometric Mean (Altitude) Theorem**

$110.25 = 5.75x$ **Square each side.**

$19.17 \approx x$ **Divide each side by 5.75.**

The height of the grandstand is the total length of the hypotenuse, $5.75 + 19.17$, or about 25 feet.

Guided Practice

4. **SPORTS** A community center needs to estimate the cost of installing a rock climbing wall by estimating the height of the wall. Sue holds a book up to her eyes so that the top and bottom of the wall are in line with the bottom edge and binding of the cover. If her eye level is 5 feet above the ground and she stands 11 feet from the wall, how high is the wall? Draw a diagram and explain your reasoning.

Personal Tutor glencoe.com

Check Your Understanding

Example 1 p. 531

Find the geometric mean between each pair of numbers.

1. 5 and 20 **2.** 36 and 4 **3.** 40 and 15

Example 2 p. 532

4. Write a similarity statement identifying the three similar triangles in the figure.

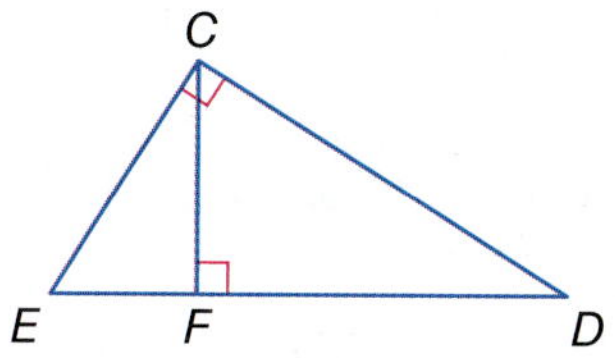

Example 3 p. 533

Find x, y, and z.

5.

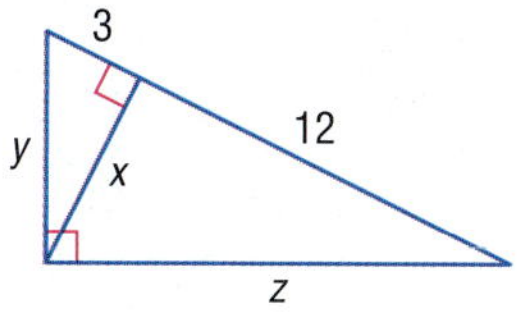

6.

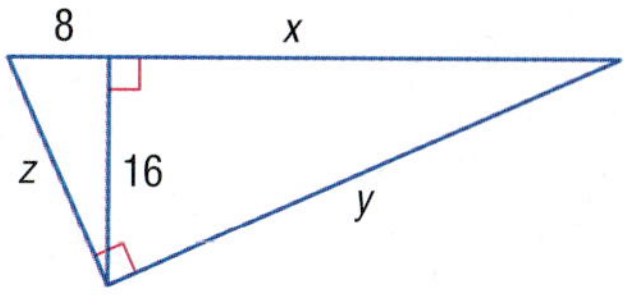

Example 4 p. 534

7. MONUMENTS Corey is visiting the Jefferson Memorial with his family. He wants to estimate the height of the statue of Thomas Jefferson. Corey stands so that his line of vision to the top and base of the statue form a right angle as shown in the diagram. About how tall is the statue?

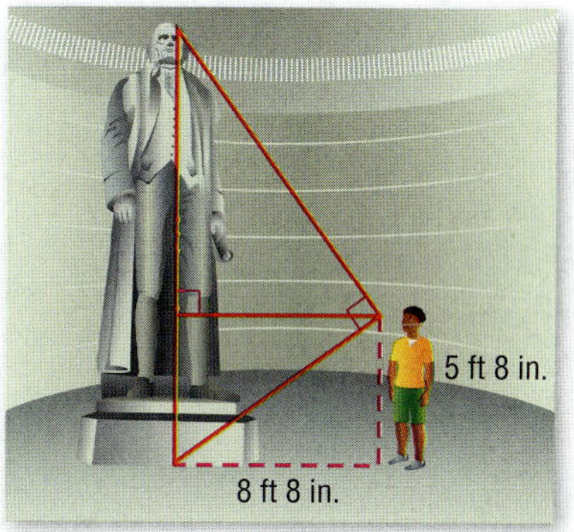

Note: Not drawn to scale.

Practice and Problem Solving

● = **Step-by-Step Solutions** begin on page R20.
Extra Practice begins on page 969.

Example 1 p. 531

Find the geometric mean between each pair of numbers.

8. 81 and 4 **9.** 25 and 16 **10.** 20 and 25

11. 36 and 24 **12.** 12 and 2.4 **13.** 18 and 1.5

Example 2 p. 532

Write a similarity statement identifying the three similar triangles in the figure.

14.

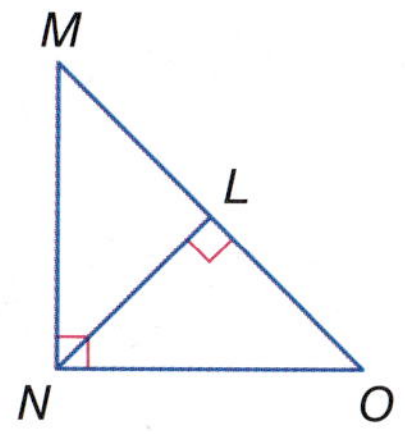

15.

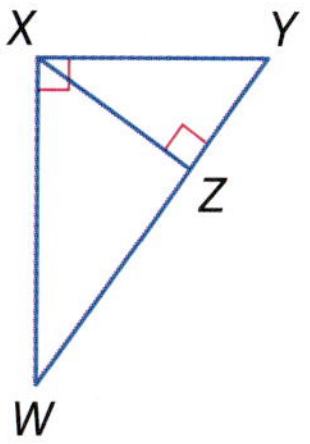

16.

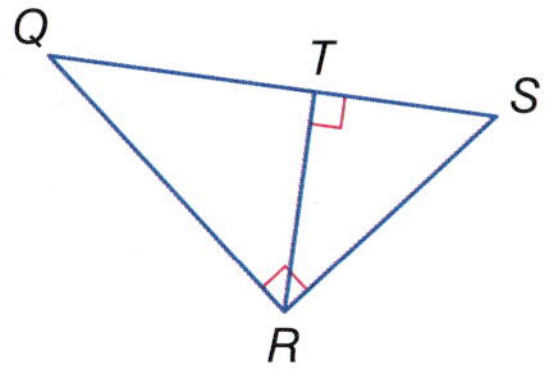

17.

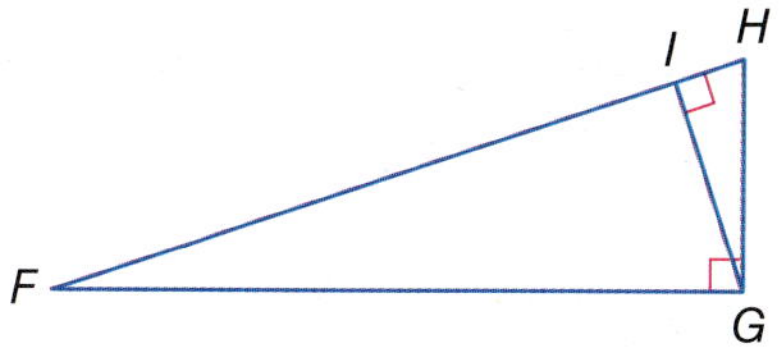

Example 3
p. 533

Find x, y, and z.

18.

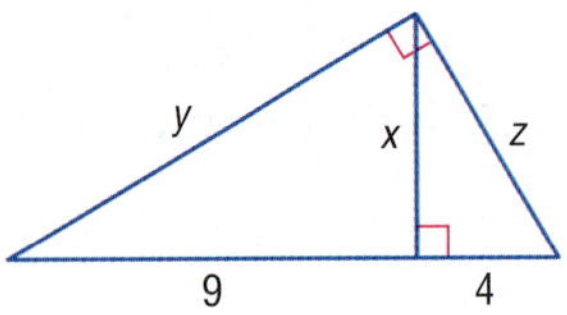

19

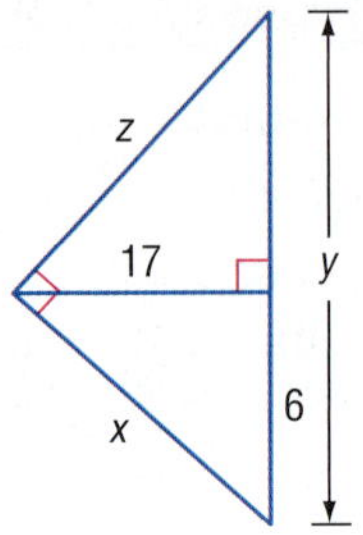

20.

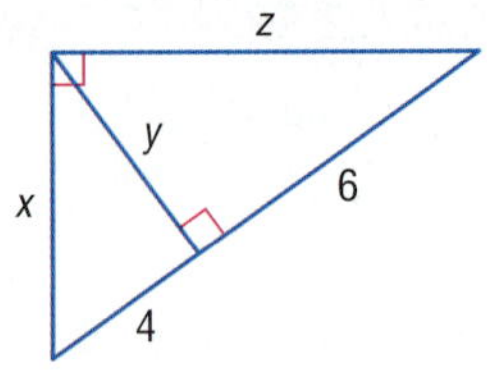

21.

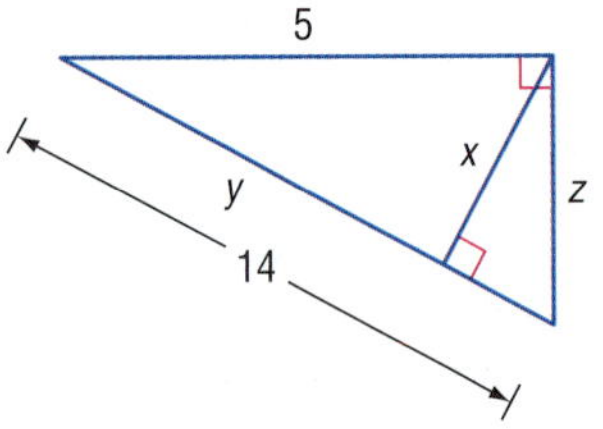

22.

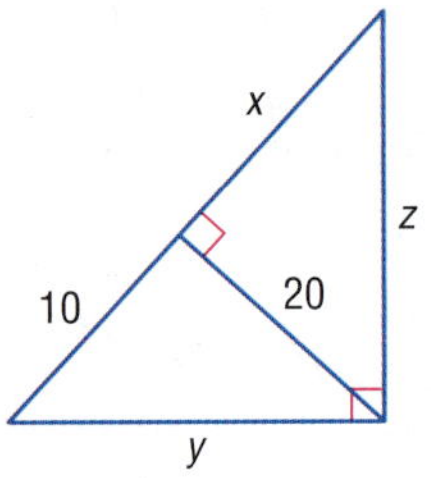

23.

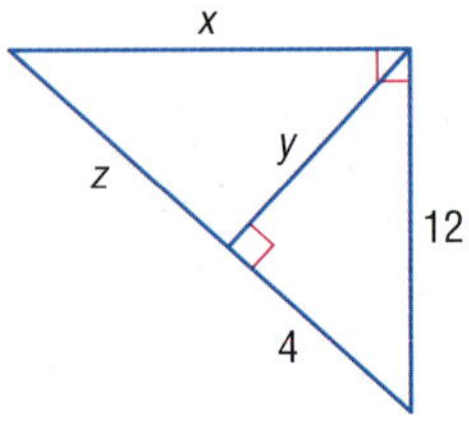

Example 4
p. 534

24. **EVENT PLANNING** Evelina is hanging silver stars from the gym ceiling using string for the homecoming dance. She wants the ends of the strings where the stars will be attached to be 7 feet from the floor. Use the diagram to determine how long she should make the strings.

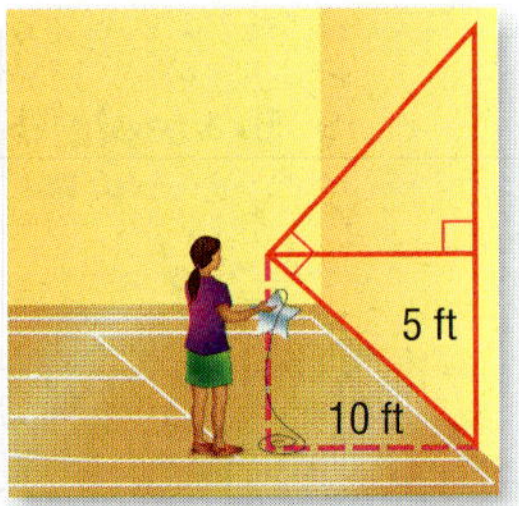

Note: Not drawn to scale.

StudyTip

Similar Triangles If you are uncertain about applying the geometric mean theorems, you can write a similarity statement and a proportion of corresponding sides to find missing lengths.

25. **WATERFALLS** Makayla is using a book to sight the top of a waterfall. Her eye level is 5 feet from the ground and she is a horizontal distance of 28 feet from the waterfall. Find the height of the waterfall to the nearest tenth of a foot.

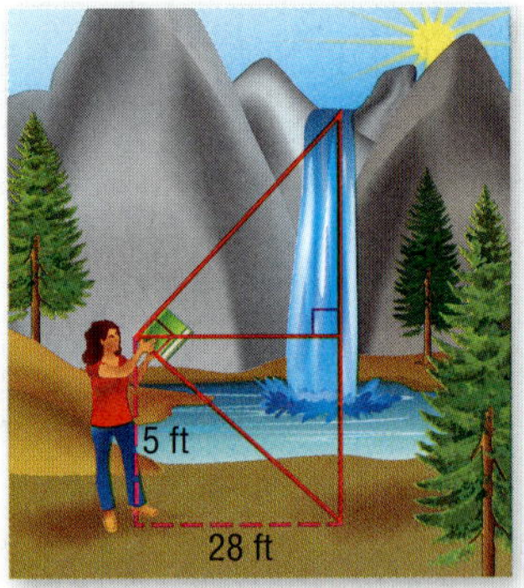

Note: Not drawn to scale.

Find the geometric mean between each pair of numbers.

26. $\frac{1}{5}$ and 60

27. $\frac{3\sqrt{2}}{7}$ and $\frac{5\sqrt{2}}{7}$

28. $\frac{3\sqrt{5}}{4}$ and $\frac{5\sqrt{5}}{4}$

Find x, y, and z.

29.

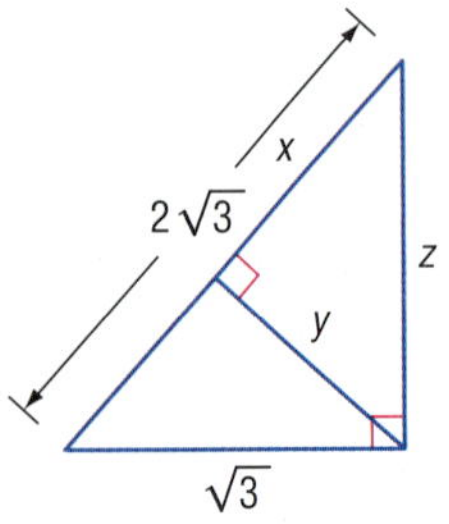

30.

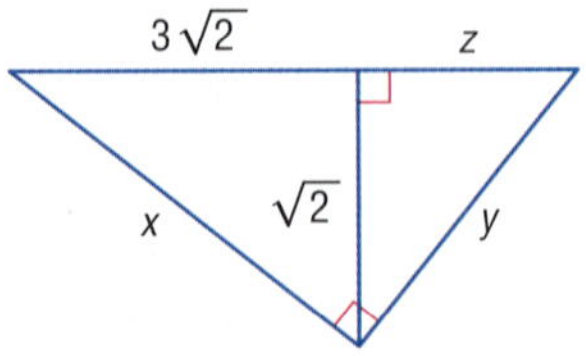

31. **ALGEBRA** The geometric mean of a number and four times the number is 22. What is the number?

Use similar triangles to find the value of x.

32.

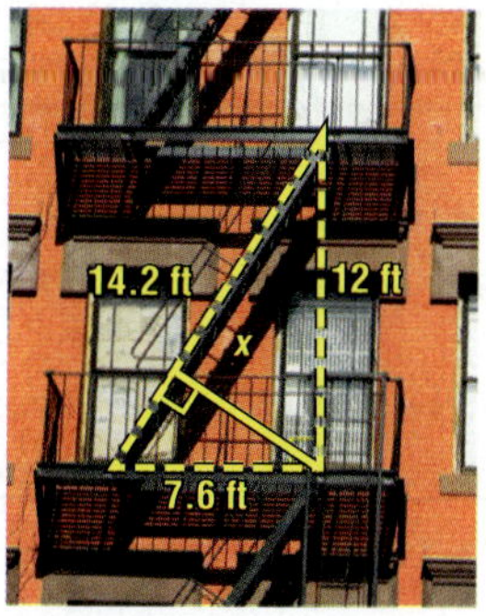

33.

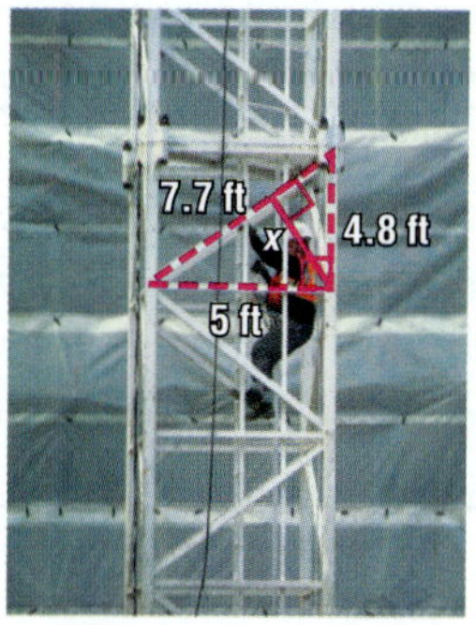

34.

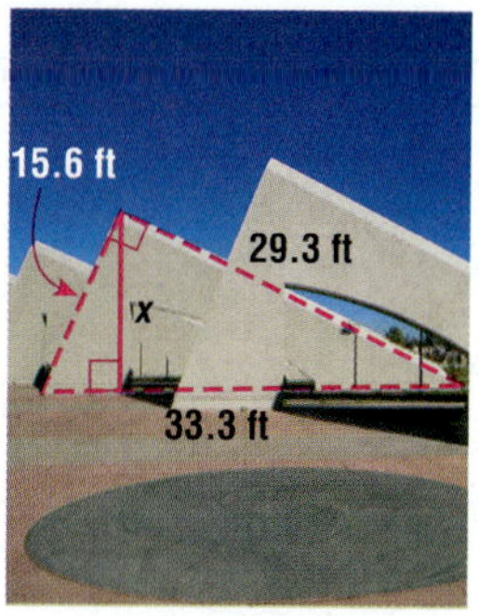

ALGEBRA **Find the value(s) of the variable.**

35.

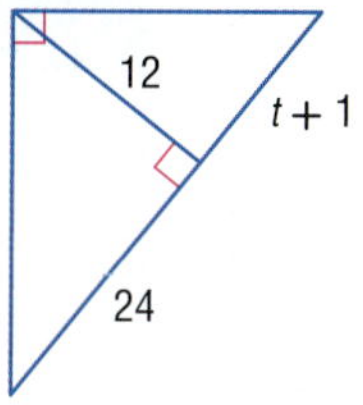

36.

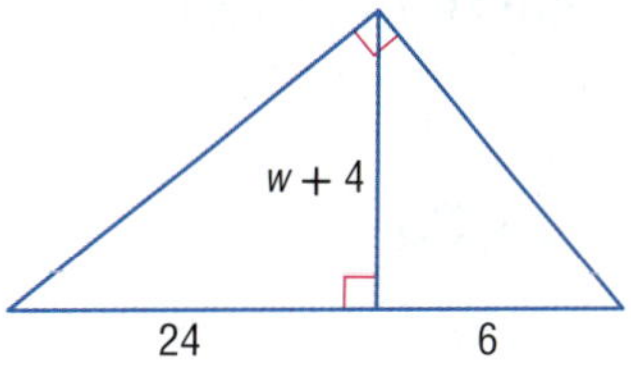

37.

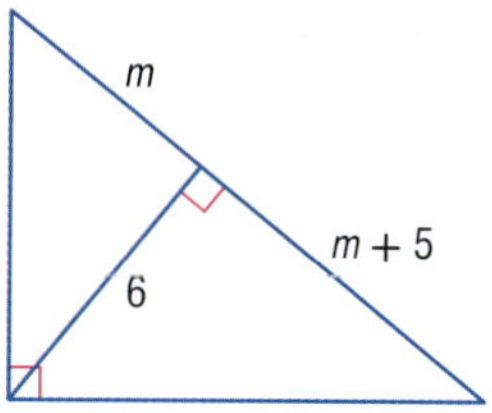

38. **CONSTRUCTION** A room-in-attic truss is a truss design that provides support while leaving area that can be enclosed as living space. In the diagram, $\angle BCA$ and $\angle EGB$ are right angles, $\triangle BEF$ is isosceles, $\overline{CD}$ is an altitude of $\triangle ABC$, and $\overline{EG}$ is an altitude of $\triangle BEF$. If $DB = 5$ feet, $CD = 6$ feet 4 inches, $BF = 10$ feet 10 inches, and $EG = 4$ feet 6 inches, what is AE?

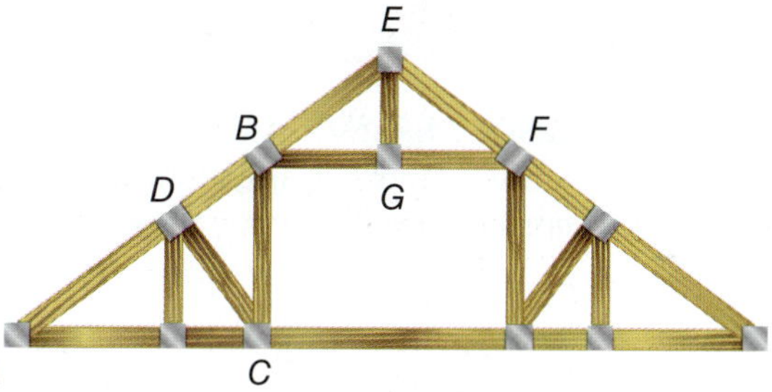

PROOF **Write a proof for each theorem.**

39. Theorem 8.1

40. Theorem 8.2

41. Theorem 8.3

Real-World Link

Standing 15 feet 6 inches tall and weighing 28,000 pounds, Bigfoot #5 is the tallest and heaviest pickup truck in the world.

Source: Bigfoot 4 × 4

42. **TRUCKS** In photography, the angle formed by the top of the subject, the camera, and the bottom of the subject is called the viewing angle, as shown at the right. Natalie is taking a picture of Bigfoot #5. She sets her camera on a tripod that is 5 feet above ground level. The vertical viewing angle of her camera is set for 90°.

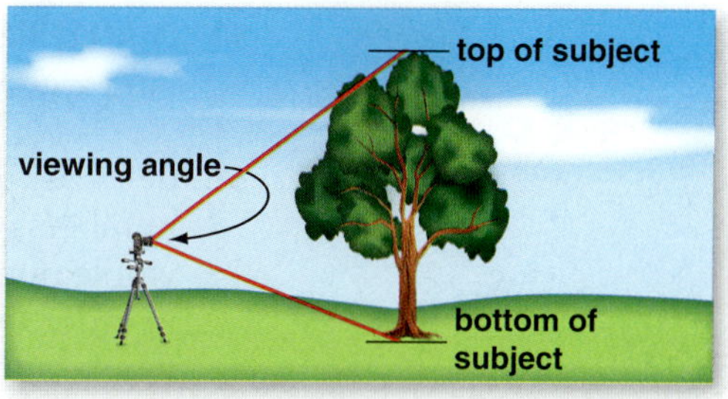

a. Sketch a diagram of this situation.

b. How far away from the truck should Natalie stand so that she perfectly frames the entire height of the truck in her shot?

43 **FINANCE** The average rate of return on an investment over two years is the geometric mean of the two annual returns. If an investment returns 12% one year and 7% the next year, what is the average rate of return on this investment over the two-year period?

44. PROOF Derive the Pythagorean Theorem using the figure at the right and the Geometric Mean (Leg) Theorem.

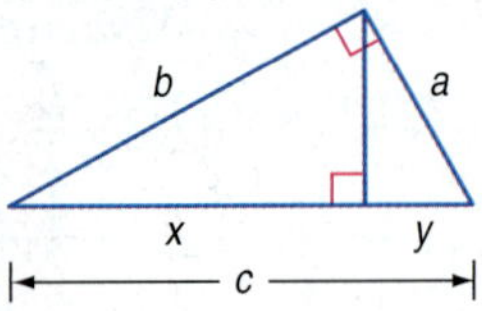

Determine whether each statement is *always*, *sometimes*, or *never* true. Explain your reasoning.

45. The geometric mean for consecutive positive integers is the mean of the two numbers.

46. The geometric mean for two perfect squares is a positive integer.

47. The geometric mean for two positive integers is another integer.

48. MULTIPLE REPRESENTATIONS In this problem, you will investigate geometric mean.

a. TABULAR Copy and complete the table of five ordered pairs (x, y) such that $\sqrt{xy} = 8$.

x	y	$\sqrt{xy}$
		8
		8
		8
		8
		8

b. GRAPHICAL Graph the ordered pairs from your table in a scatter plot.

c. VERBAL Make a conjecture as to the type of graph that would be formed if you connected the points from your scatter plot. Do you think the graph of any set of ordered pairs that results in the same geometric mean would have the same general shape? Explain your reasoning.

H.O.T. Problems

Use Higher-Order Thinking Skills

49. ERROR ANALYSIS Aiden and Tia are finding the value x in the triangle shown. Is either of them correct? Explain your reasoning.

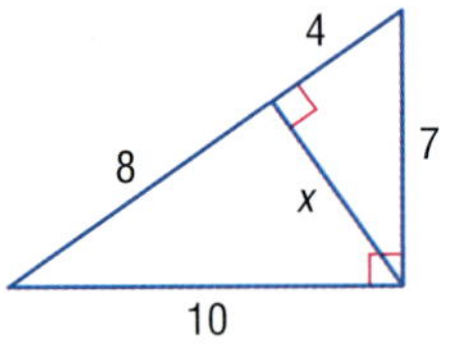

Aiden

$\frac{4}{x} = \frac{x}{7}$

$x \approx 5.3$

Tia

$\frac{4}{x} = \frac{x}{10}$

$x \approx 6.3$

50. CHALLENGE Refer to the figure at the right. Find x, y, and z.

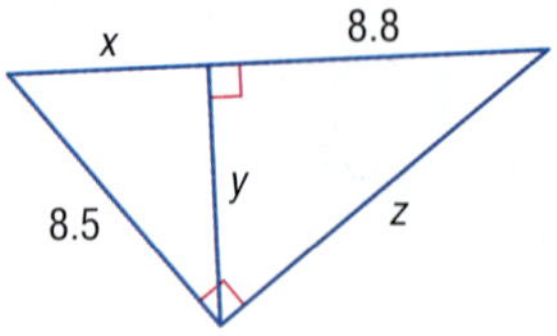

51. OPEN ENDED Find two pairs of whole numbers with a geometric mean that is also a whole number. What condition must be met in order for a pair of numbers to produce a whole-number geometric mean?

52. REASONING Refer to the figure at the right. The orthocenter of $\triangle ABC$ is located 6.4 units from point D. Find BC.

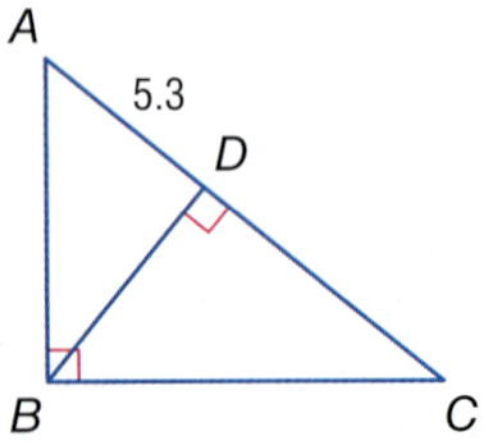

Review Vocabulary

arithmetic mean the terms between any two nonconsecutive terms of an arithmetic sequence

53. WRITING IN MATH Compare and contrast the arithmetic and geometric means of two numbers. When will the two means be equal? Justify your reasoning.

Standardized Test Practice

54. What is the geometric mean of 8 and 22 in simplest form?

A $4\sqrt{11}$ **C** $16\sqrt{11}$
B 15 **D** 176

55. SHORT RESPONSE If $\overline{MN} \parallel \overline{PQ}$, use a proportion to find the value of x. Show your work.

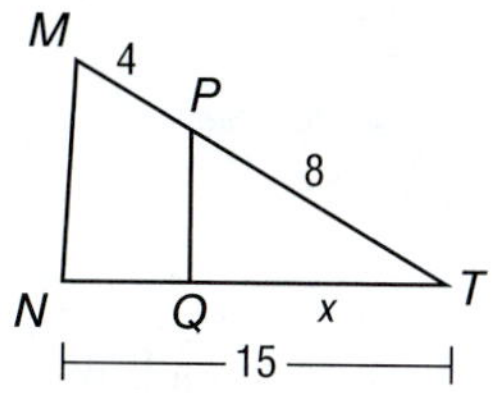

56. ALGEBRA What are the solutions of the quadratic equation $x^2 - 20 = 8x$?

F 2, 10 **H** −1, 20
G 20, 1 **J** −2, 10

57. SAT/ACT In the figure, $\overline{AD}$ is perpendicular to $\overline{BC}$, and $\overline{AB}$ is perpendicular to $\overline{AC}$. What is BC?

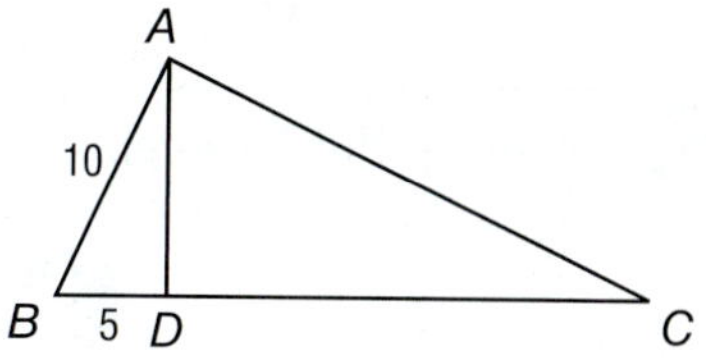

A $5\sqrt{2}$ **C** 20
B $5\sqrt{3}$ **D** 25

Spiral Review

58. MAPS Use the map to estimate how long it would take to drive from Chicago to Springfield if you averaged 65 miles per hour. (Lesson 7-7)

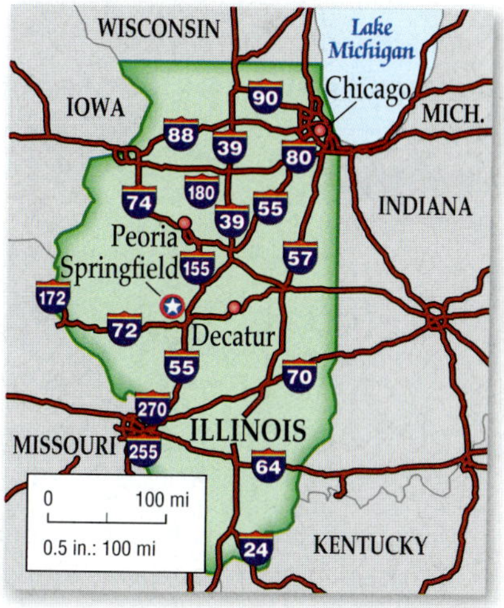

Graph the original figure and its dilated image. Then verify that the dilation is a similarity transformation. (Lesson 7-6)

59. $A(-3, 1)$, $B(9, 7)$, $C(3, -2)$; $D(-1, 1)$, $E(3, 3)$, $F(1, 0)$

60. $G(-4, -4)$, $H(-1, 2)$, $J(2, -1)$; $K(-3, -2)$, $L(1, 0)$

61. $M(7, -4)$, $N(5, -4)$, $P(7, -1)$; $Q(2, -8)$, $R(6, -8)$, $S(2, -2)$

The interior angle measure of a regular polygon is given. Identify the polygon. (Lesson 6-1)

62. 108 **63.** 135

Find x and y in each figure. (Lesson 3-2)

64.

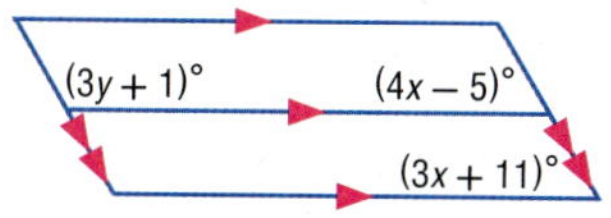

65.

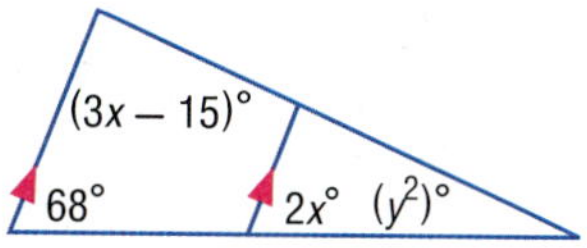

66.

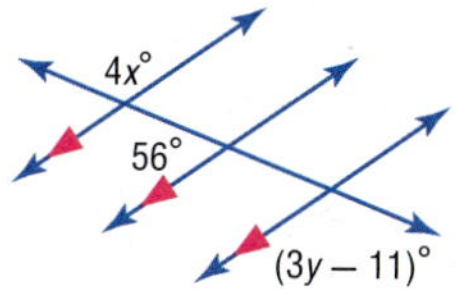

Identify each solid. Name the bases, faces, edges, and vertices. (Lesson 1-7)

67.

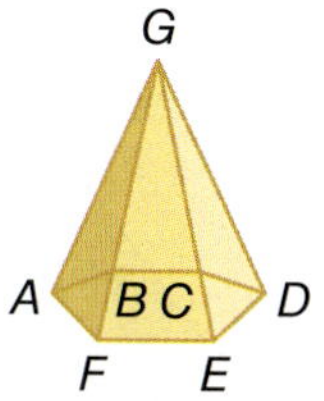

68.

69.

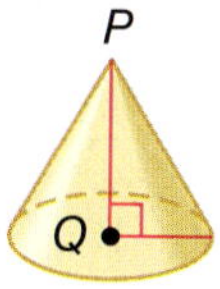

Skills Review

Simplify each expression by rationalizing the denominator. (Lesson 0-9)

70. $\frac{2}{\sqrt{2}}$ **71.** $\frac{16}{\sqrt{3}}$ **72.** $\frac{\sqrt{6}}{\sqrt{4}}$ **73.** $\frac{3\sqrt{5}}{\sqrt{11}}$ **74.** $\frac{21}{\sqrt{3}}$

EXPLORE

8-2

Geometry Lab
The Pythagorean Theorem

In Chapter 1, you learned that the Pythagorean Theorem relates the measures of the legs and the hypotenuse of a right triangle. You can use square pieces of patty paper and algebra to discover this relationship.

ACTIVITY **Develop the Pythagorean Theorem by using paper and algebra.**

Step 1

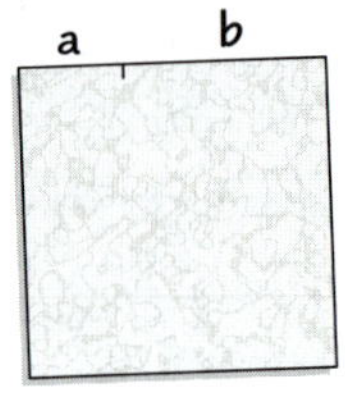

On a piece of patty paper, mark one side a and b as shown above.

Step 2

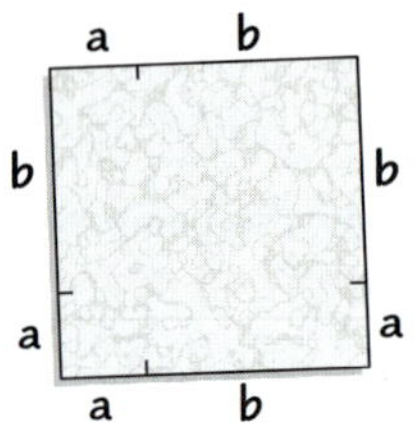

Copy these measures on each of the other sides.

Step 3

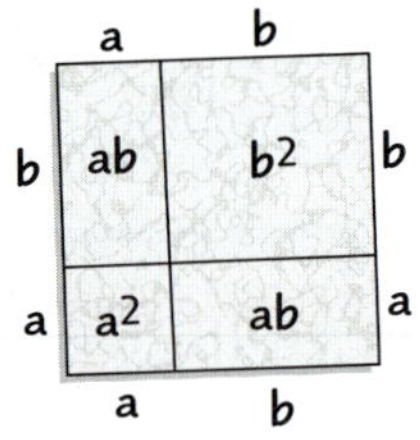

Fold the paper into four sections and label the area of each section.

Step 4

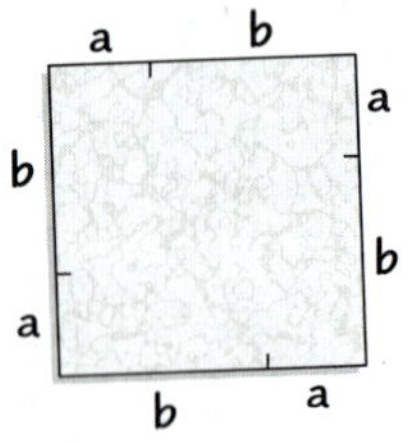

On another piece of patty paper, mark each side a and b as shown above.

Step 5

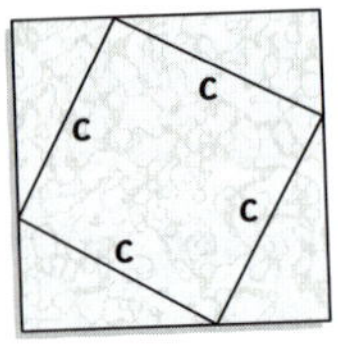

Connect the marks using a straightedge. Let c represent the length of each hypotenuse.

Step 6

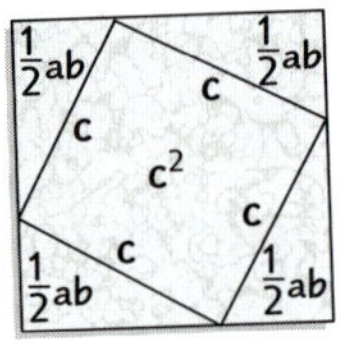

Label the area of each triangle $\frac{1}{2}ab$ and the area of each square c^2.

Step 7

Place the squares side by side and color the corresponding regions that have the same area. For example, $ab = \frac{1}{2}ab + \frac{1}{2}ab$.

The parts that are not shaded tell us that $a^2 + b^2 = c^2$.

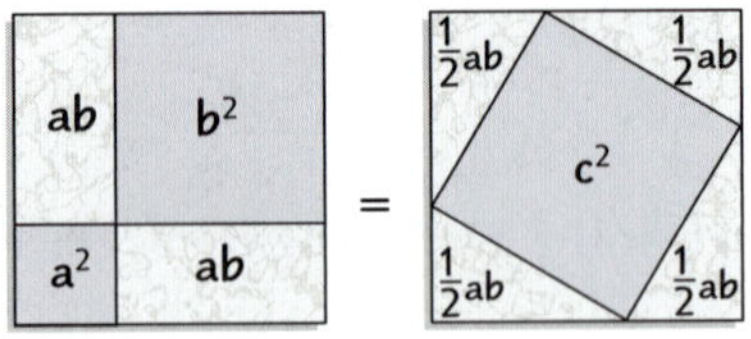

Analyze the Results

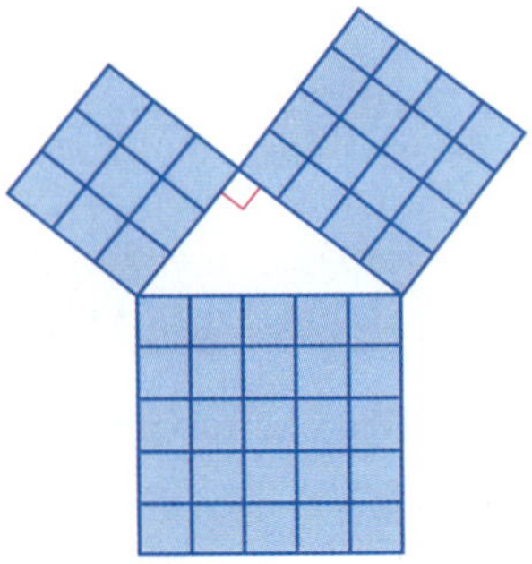

1. Use a ruler to measure a, b, and c. Do these measures confirm that $a^2 + b^2 = c^2$?
2. Repeat the activity with different a and b values. What do you notice?
3. **WRITING IN MATH** Explain why the drawing at the right is an illustration of the Pythagorean Theorem.
4. **CHALLENGE** Draw a geometric diagram to show that for any positive numbers a and b, $a + b > \sqrt{a^2 + b^2}$. Explain.

8-2 The Pythagorean Theorem and Its Converse

Then

You used the Pythagorean Theorem to develop the Distance Formula. (Lesson 1-3)

Now

- Use the Pythagorean Theorem.
- Use the Converse of the Pythagorean Theorem.

New Vocabulary

Pythagorean triple

Math Online

glencoe.com

Why?

Tether lines are used to steady an inflatable snowman. Suppose you know the height at which the tether lines are attached to the snowman and how far away you want to anchor the tether in the ground. You can use the converse of the Pythagorean Theorem to adjust the lengths of the tethers to keep the snowman perpendicular to the ground.

The Pythagorean Theorem The Pythagorean Theorem is perhaps one of the most famous theorems in mathematics. It relates the lengths of the hypotenuse (side opposite the right angle) and legs (sides adjacent to the right angle) of a right triangle.

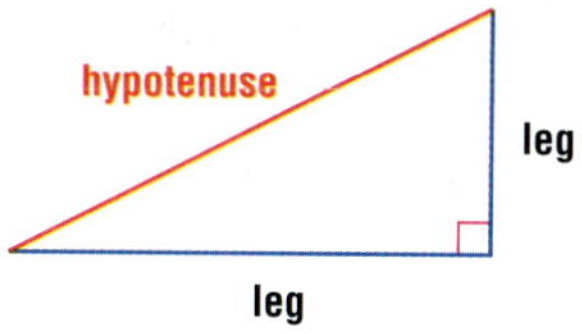

Theorem 8.4

Pythagorean Theorem

Words In a right triangle, the sum of the squares of the lengths of the legs is equal to the square of the length of the hypotenuse.

Symbols If $\triangle ABC$ is a right triangle with right angle C, then $a^2 + b^2 = c^2$.

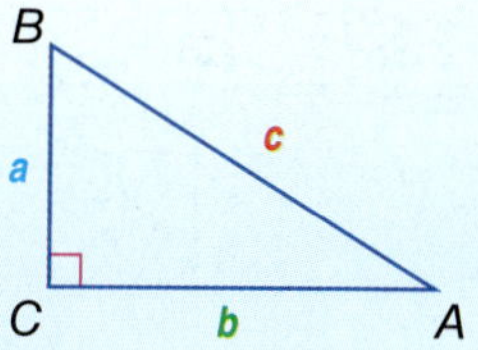

The geometric mean can be used to prove the Pythagorean Theorem.

Proof Pythagorean Theorem

Given: $\triangle ABC$ with right angle at C

Prove: $a^2 + b^2 = c^2$

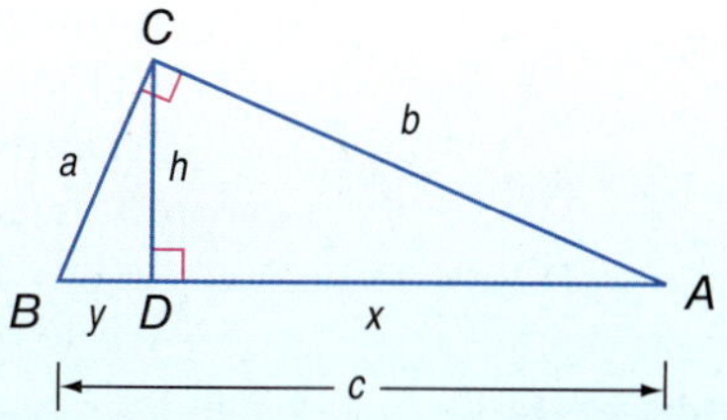

Proof:

Draw right triangle ABC so C is the right angle. Then draw the altitude from C to $\overline{AB}$. Let $AB = c$, $AC = b$, $BC = a$, $AD = x$, $DB = y$, and $CD = h$.

Two geometric means now exist.

$\frac{c}{a} = \frac{a}{y}$ and	$\frac{c}{b} = \frac{b}{x}$	Geometric Mean (Leg) Theorem
$a^2 = cy$	$b^2 = cx$	Cross products
$a^2 + b^2 = cy + cx$		Add the equations.
$a^2 + b^2 = c(y + x)$		Factor.
$a^2 + b^2 = c \cdot c$		Since $c = y + x$, substitute c for $(y + x)$.
$a^2 + b^2 = c^2$		Simplify.

You can use the Pythagorean Theorem to find the measure of any side of a right triangle given the lengths of the other two sides.

EXAMPLE 1 Find Missing Measures Using the Pythagorean Theorem

Find x.

a.

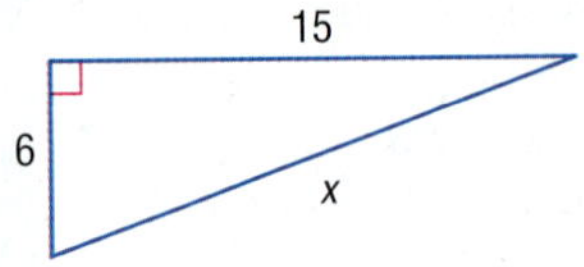

The side opposite the right angle is the hypotenuse, so $c = x$.

$a^2 + b^2 = c^2$	Pythagorean Theorem
$6^2 + 15^2 = x^2$	$a = 6$ and $b = 15$
$261 = x^2$	Simplify.
$\sqrt{261} = x$	Take the positive square root of each side.
$3\sqrt{29} = x$	Simplify.

b.

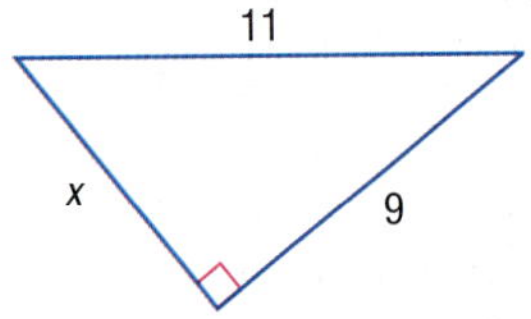

The hypotenuse is 11, so $c = 11$.

$a^2 + b^2 = c^2$	Pythagorean Theorem
$x^2 + 9^2 = 11^2$	$a = x$ and $b = 9$
$x^2 + 81 = 121$	Simplify.
$x^2 = 40$	Subtract 81 from each side.
$x = \sqrt{40}$ or $2\sqrt{10}$	Take the positive square root of each side and simplify.

Guided Practice

1A.

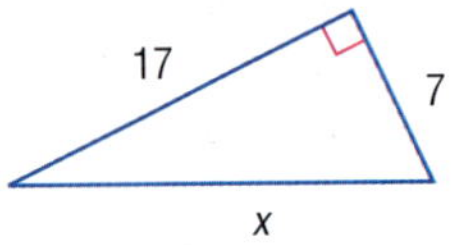

1B.

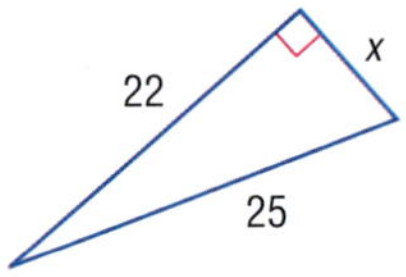

Personal Tutor glencoe.com

> **StudyTip**
>
> **Positive Square Root** When finding the length of a side using the Pythagorean Theorem, use only the positive and not the negative square root, since length cannot be negative.

A **Pythagorean triple** is a set of three nonzero whole numbers a, b, and c, such that $a^2 + b^2 = c^2$. One common Pythagorean triple is 3, 4, 5; that is, the sides of a right triangle are in the ratio 3:4:5. The most common Pythagorean triples are shown below in blue. The triples below these are found by multiplying each number in the triple by the same factor.

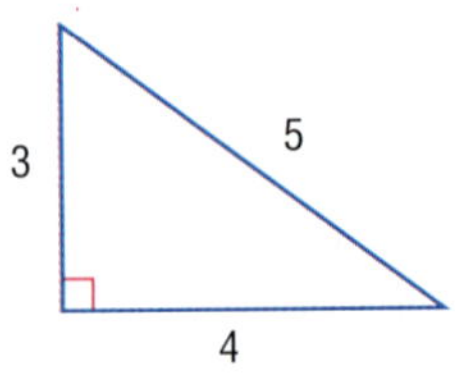

Key Concept Common Pythagorean Triples

3, 4, 5	5, 12, 13	8, 15, 17	7, 24, 25
6, 8, 10	10, 24, 26	16, 30, 34	14, 48, 50
9, 12, 15	15, 36, 39	24, 45, 51	21, 72, 75
3x, 4x, 5x	5x, 12x, 13x	8x, 15x, 17x	7x, 24x, 25x

The largest number in each triple is the length of the hypotenuse.

> **StudyTip**
>
> **Pythagorean Triples** If the measures of the sides of any right triangle are *not* whole numbers, the measures do not form a Pythagorean triple.

EXAMPLE 2 Use a Pythagorean Triple

Use a Pythagorean triple to find x. Explain your reasoning.

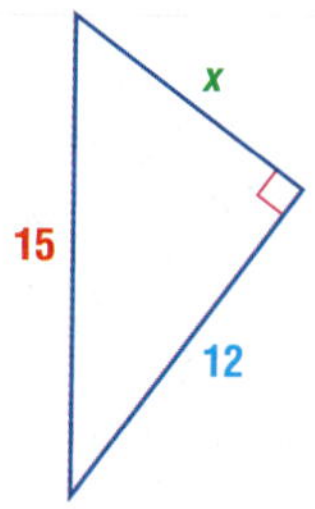

Notice that 15 and 12 are both multiples of 3: $15 = 3 \cdot 5$ and $12 = 3 \cdot 4$. Since 3, 4, 5 is a Pythagorean triple, the missing leg length x is $3 \cdot 3$ or 9.

CHECK $12^2 + 9^2 \stackrel{?}{=} 15^2$ **Pythagorean Theorem**

$225 = 225$ ✓ **Simplify.**

Guided Practice

2A.

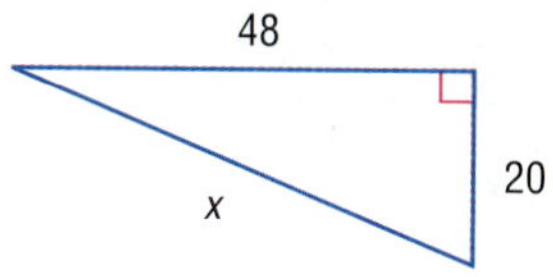

2B.

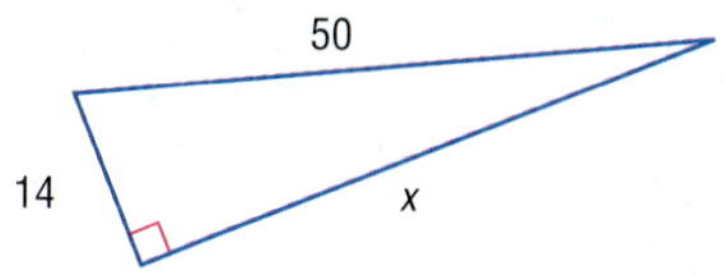

Personal Tutor glencoe.com

The Pythagorean Theorem can be used to solve many real-world problems.

STANDARDIZED TEST EXAMPLE 3

Damon is locked out of his house, with the only open window on the second floor, 12 feet above the ground. He needs to borrow a ladder from his neighbor. If he must place the ladder 5 feet from the house to avoid some bushes, what length of ladder does Damon need?

Note: Not drawn to scale.

A 7 feet

B 11 feet

C 13 feet

D 17 feet

Read the Test Item

The distance the ladder is from the house, the height the ladder reaches, and the length of the ladder itself make up the lengths of the sides of a right triangle. You need to find the length of the ladder, which is the hypotenuse.

Solve the Test Item

Method 1 Use a Pythagorean triple.

The lengths of the legs are 5 and 12. 5, 12, 13 is a Pythagorean triple, so the length of the ladder is 13 feet.

Method 2 Use the Pythagorean Theorem.

Let the length of the ladder be x.

$5^2 + 12^2 = x^2$ **Pythagorean Theorem**

$169 = x^2$ **Simplify.**

$\sqrt{169} = x$ **Take the positive square root of each side.**

$13 = x$ **Simplify.**

So the answer is choice C.

ReadingMath

3-4-5 A right triangle with side lengths 3, 4, and 5 is called a *3-4-5 right triangle*.

Test-TakingTip

Eliminate Unreasonable Answers Since the hypotenuse of a right triangle is always the longest side, the length of the ladder in Example 3 must be greater than 5 or 12 feet. Since 7 and 11 feet are both less than 12 feet, choices A and B can be eliminated.

Guided Practice

3. According to your company's safety regulations, the distance from the base of a ladder to a wall that it leans against should be at least one fourth of the ladder's total length. You are given a 20-foot ladder to place against a wall at a job site. If you follow the company's safety regulations, what is the maximum distance x up the wall the ladder will reach, to the nearest tenth?

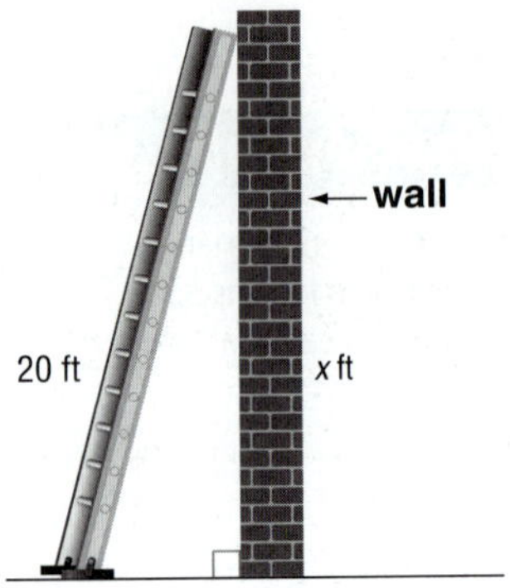

Note: Not drawn to scale.

F 12 feet **H** 20.6 feet

G 19.4 feet **J** 30.6 feet

Personal Tutor glencoe.com

Converse of the Pythagorean Theorem The converse of the Pythagorean Theorem also holds. You can use this theorem to help you determine whether a triangle is a right triangle given the measures of all three sides.

Theorem 8.5

Converse of the Pythagorean Theorem

Words If the sum of the squares of the lengths of the shortest sides of a triangle is equal to the square of the length of the longest side, then the triangle is a right triangle.

Symbols If $a^2 + b^2 = c^2$, then $\triangle ABC$ is a right triangle.

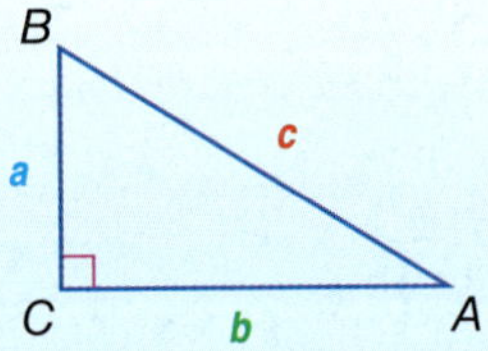

You will prove Theorem 8.5 in Exercise 35.

You can also use side lengths to classify a triangle as acute or obtuse.

Theorems

Pythagorean Inequality Theorems

8.6 If the square of the length of the longest side of a triangle is less than the sum of the squares of the lengths of the other two sides, then the triangle is an acute triangle.

Symbols If $c^2 < a^2 + b^2$, then $\triangle ABC$ is acute.

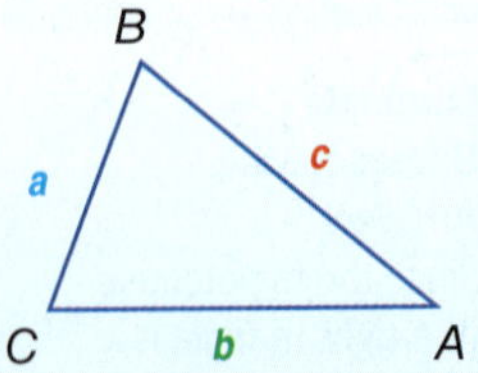

8.7 If the square of the length of the longest side of a triangle is greater than the sum of the squares of the lengths of the other two sides, then the triangle is an obtuse triangle.

Symbols If $c^2 > a^2 + b^2$, then $\triangle ABC$ is obtuse.

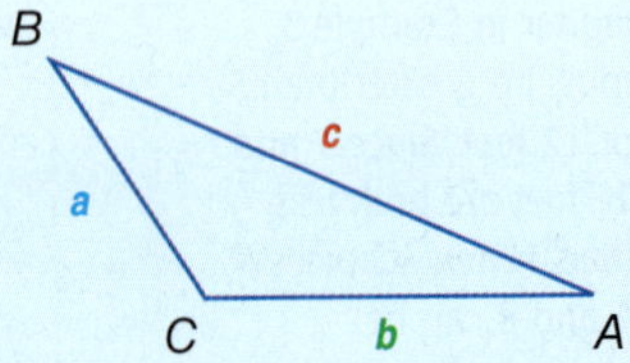

You will prove Theorems 8.6 and 8.7 in Exercises 36 and 37, respectively.

StudyTip

Determining the Longest Side If the measures of any of the sides of a triangle are expressed as radicals, you may wish to use a calculator to determine which length is the longest.

EXAMPLE 4 Classify Triangles

Determine whether each set of numbers can be the measures of the sides of a triangle. If so, classify the triangle as *acute, right,* or *obtuse.* Justify your answer.

Review Vocabulary

Triangle Inequality Theorem The sum of the lengths of any two sides of a triangle must be greater than the length of the third side. (Lesson 5-5)

a. 7, 14, 16

Step 1 Determine whether the measures can form a triangle using the Triangle Inequality Theorem.

$7 + 14 > 16$ ✓ $\quad 14 + 16 > 7$ ✓ $\quad 7 + 16 > 14$ ✓

The side lengths 7, 14, and 16 can form a triangle.

Step 2 Classify the triangle by comparing the square of the longest side to the sum of the squares of the other two sides.

$c^2 \stackrel{?}{=} a^2 + b^2$ — **Compare c^2 and $a^2 + b^2$.**

$16^2 \stackrel{?}{=} 7^2 + 14^2$ — **Substitution**

$256 > 245$ — **Simplify and compare.**

Since $c^2 > a^2 + b^2$, the triangle is obtuse.

b. 9, 40, 41

Step 1 Determine whether the measures can form a triangle.

$9 + 40 > 41$ ✓ $\quad 40 + 41 > 9$ ✓ $\quad 9 + 41 > 40$ ✓

The side lengths 9, 40, and 41 can form a triangle.

Step 2 Classify the triangle.

$c^2 \stackrel{?}{=} a^2 + b^2$ — **Compare c^2 and $a^2 + b^2$.**

$41^2 \stackrel{?}{=} 9^2 + 40^2$ — **Substitution**

$1681 = 1681$ — **Simplify and compare.**

Since $c^2 = a^2 + b^2$, the triangle is a right triangle.

Guided Practice

4A. 11, 60, 61 **4B.** $2\sqrt{3}, 4\sqrt{2}, 3\sqrt{5}$ **4C.** 6.2, 13.8, 20

Personal Tutor glencoe.com

Check Your Understanding

Example 1 p. 542 — Find x.

1.

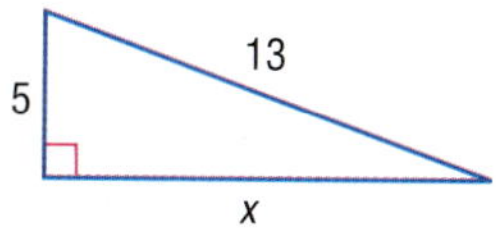

2.

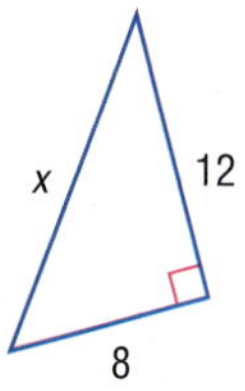

3.

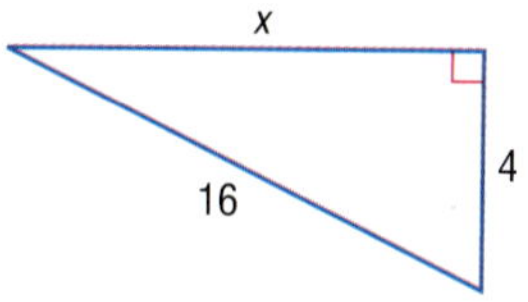

Example 2 p. 543

4. Use a Pythagorean triple to find x. Explain your reasoning.

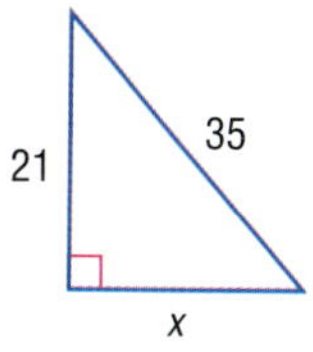

Example 3
p. 543

5. MULTIPLE CHOICE The mainsail of a boat is shown. What is the length, in feet, of $\overline{LN}$?

A 52.5 **C** 72.5

B 65 **D** 75

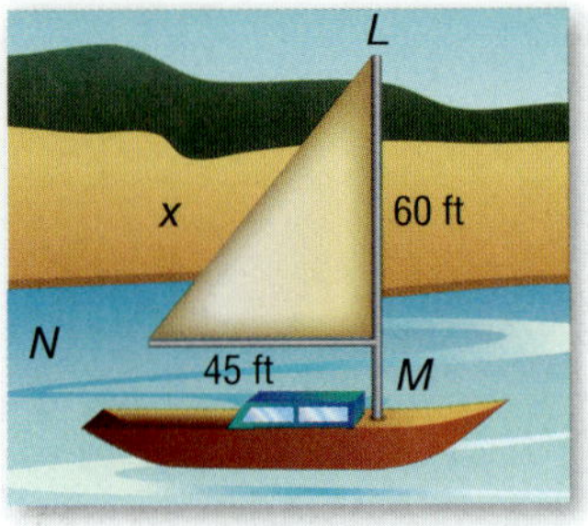

Example 4
p. 545

Determine whether each set of numbers can be the measures of the sides of a triangle. If so, classify the triangle as *acute*, *obtuse*, or *right*. Justify your answer.

6. 15, 36, 39 **7.** 16, 18, 26 **8.** 15, 20, 24

Practice and Problem Solving

= **Step-by-Step Solutions** begin on page R20. **Extra Practice** begins on page 969.

Example 1
p. 542

Find x.

9.

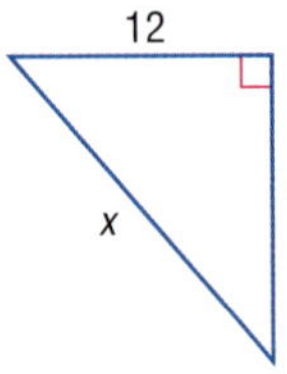

10.

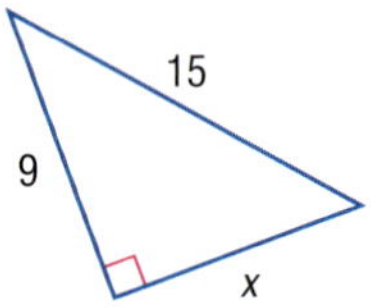

11.

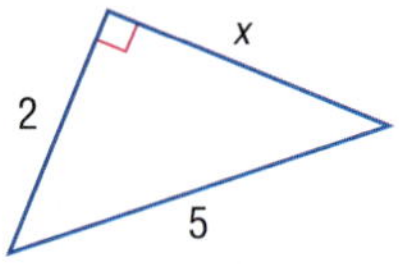

12.

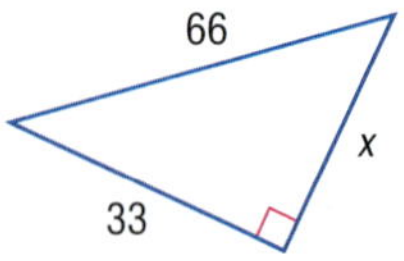

13.

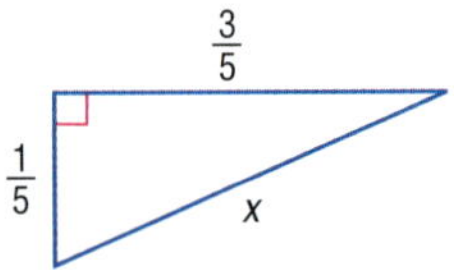

14.

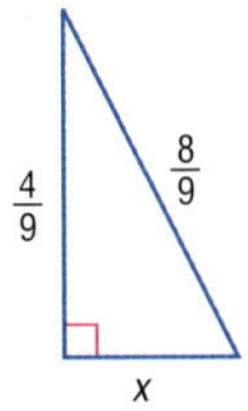

Example 2
p. 543

Use a Pythagorean Triple to find x.

15

16.

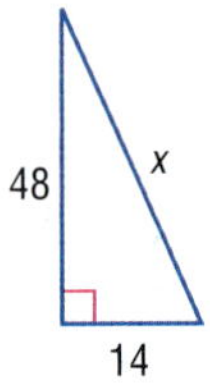

17.

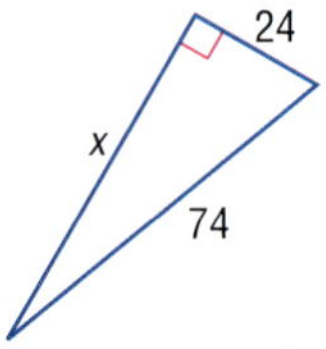

18.

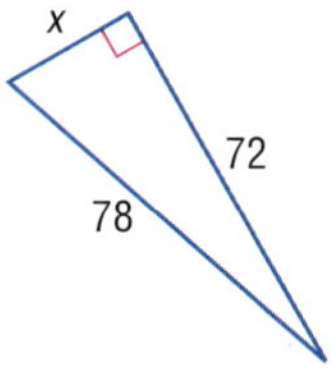

Example 3
p. 543

19. BASKETBALL The support for a basketball goal forms a right triangle as shown. What is the length x of the horizontal portion of the support?

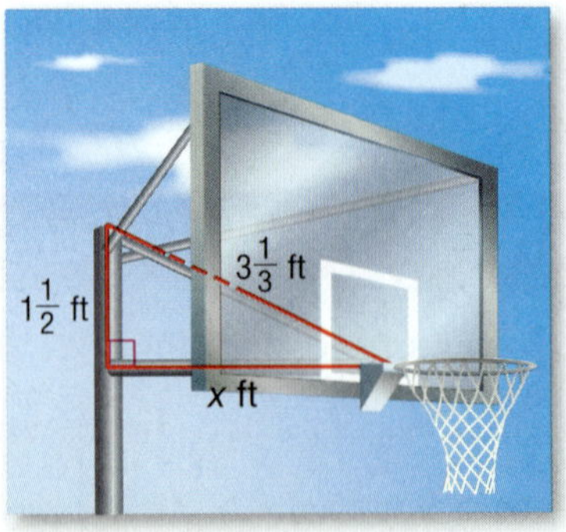

20. DRIVING The street that Khaliah usually uses to get to school is under construction. She has been taking the detour shown. If the construction starts at the point where Khaliah leaves her normal route and ends at the point where she re-enters her normal route, about how long is the stretch of road under construction?

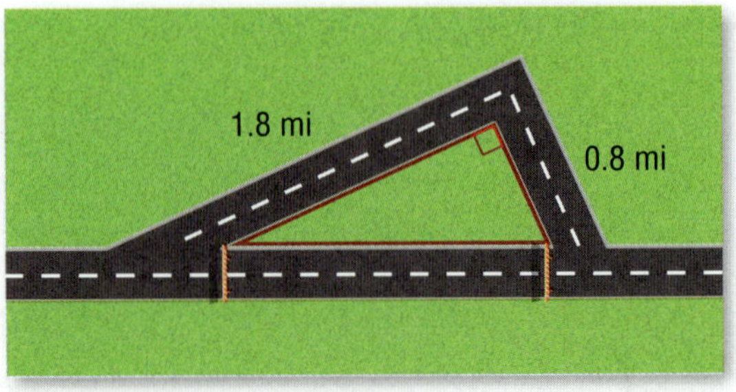

Example 4
p. 545

Determine whether each set of numbers can be the measures of the sides of a triangle. If so, classify the triangle as *acute*, *obtuse*, or *right*. Justify your answer.

21. 7, 15, 21 **22.** 10, 12, 23 **23.** 4.5, 20, 20.5

24. 44, 46, 91 **25.** 4.2, 6.4, 7.6 **26.** 4, 12, 14

Find x.

27.

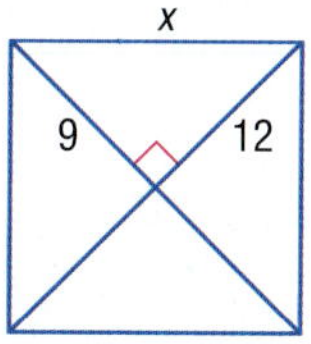

28.

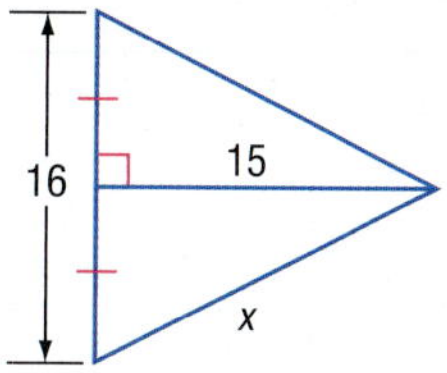

29.

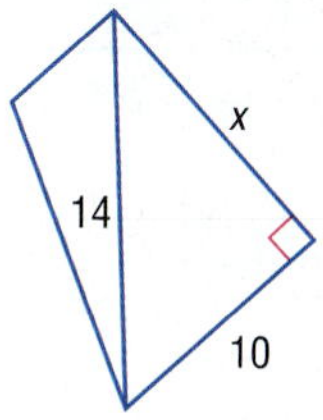

COORDINATE GEOMETRY **Determine whether $\triangle XYZ$ is an *acute*, *right*, or *obtuse* triangle for the given vertices. Explain.**

30. $X(-3, -2)$, $Y(-1, 0)$, $Z(0, -1)$

31. $X(-7, -3)$, $Y(-2, -5)$, $Z(-4, -1)$

32. $X(1, 2)$, $Y(4, 6)$, $Z(6, 6)$

33. $X(3, 1)$, $Y(3, 7)$, $Z(11, 1)$

Real-World Link

The resolution of a television is measured using pixels. An HDTV has about twice as many pixels of resolution as a standard television, so the picture is twice as clear.

Source: HDTV Info Port

34. **JOGGING** Brett jogs in the park three times a week. Usually, he takes a $\frac{3}{4}$-mile path that cuts through the park. Today, the path is closed, so he is taking the orange route shown. How much farther will he jog on his alternate route than he would have if he had followed his normal path?

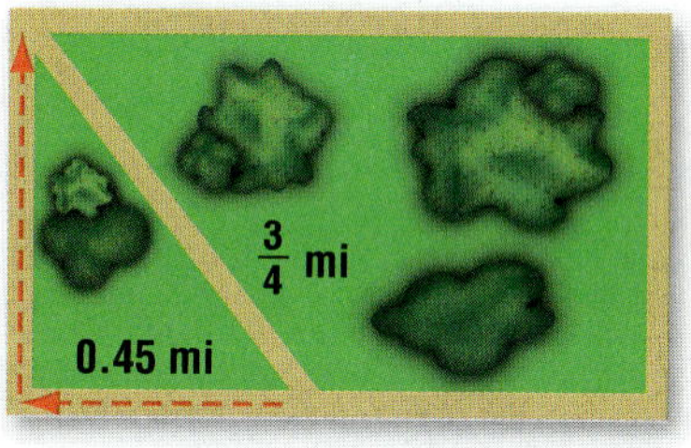

35. **PROOF** Write a paragraph proof of Theorem 8.5.

PROOF **Write a two-column proof for each theorem.**

36. Theorem 8.6 **37.** Theorem 8.7

Find the perimeter and area of each figure.

38.

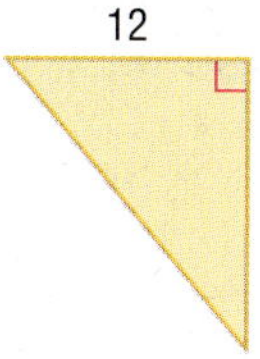

39.

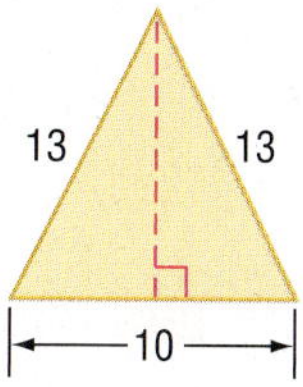

40.

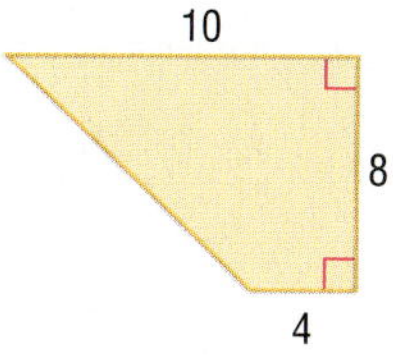

41. **ALGEBRA** The sides of a triangle have lengths x, $x + 5$, and 25. If the length of the longest side is 25, what value of x makes the triangle a right triangle?

42. **ALGEBRA** The sides of a triangle have lengths $2x$, 8, and 12. If the length of the longest side is $2x$, what values of x make the triangle acute?

43. **TELEVISION** The screen aspect ratio, or the ratio of the width to the length, of a high-definition television is 16:9. The size of a television is given by the diagonal distance across the screen. If an HDTV is 41 inches wide, what is its screen size?

Real-World Link

The Handbook for Public Playground Safety was first issued in 1981 as a guideline to help schools, day care centers, and corporations build safe playgrounds.

Source: *The Handbook for Public Playground Safety*

44. **PLAYGROUND** According to the *Handbook for Public Playground Safety*, the ratio of the vertical distance to the horizontal distance covered by a slide should not be more than about 4 to 7. If the horizontal distance allotted in a slide design is 14 feet, approximately how long should the slide be?

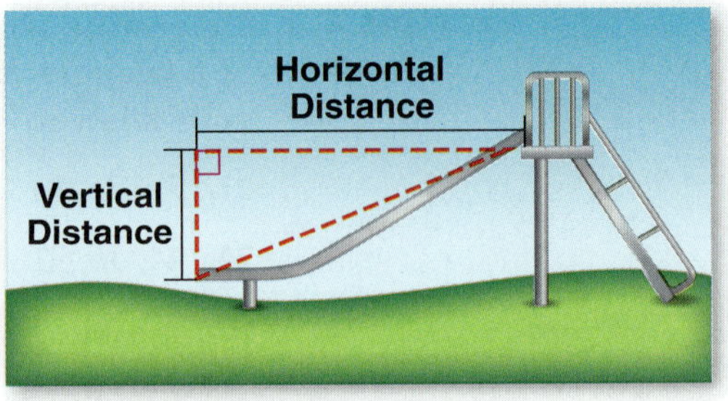

Find x.

45.

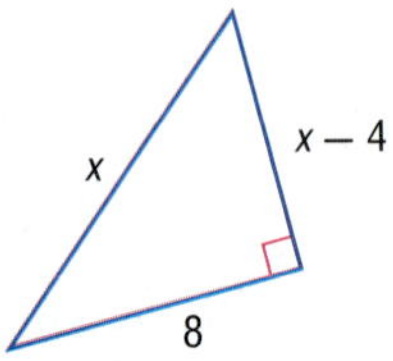

46.

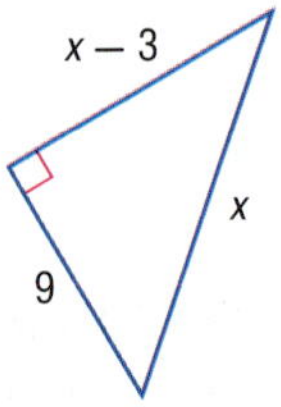

47.

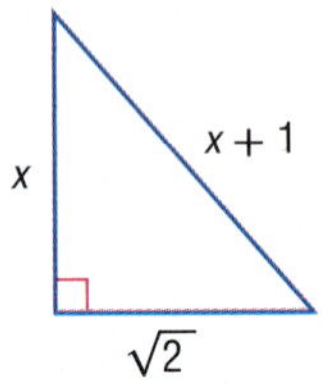

48. **MULTIPLE REPRESENTATIONS** In this problem, you will investigate special right triangles.

a. **GEOMETRIC** Draw three different isosceles right triangles that have whole-number side lengths. Label the triangles *ABC*, *MNP*, and *XYZ* with the right angle located at vertex *A*, *M*, and *X*, respectively. Label the leg lengths of each side, and find the length of the hypotenuse in simplest radical form.

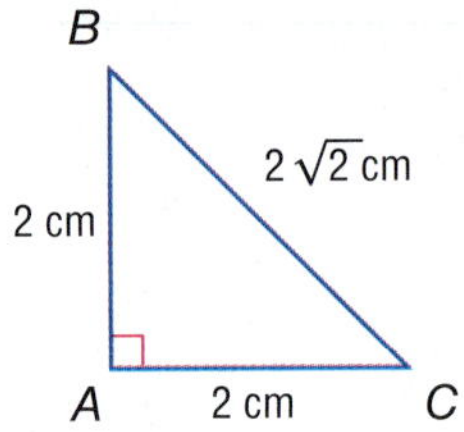

b. **TABULAR** Copy and complete the table below.

Triangle	Length				Ratio	
ABC	*BC*		*AB*		$\frac{BC}{AB}$	
MNP	*NP*		*MN*		$\frac{NP}{MN}$	
XYZ	*YZ*		*XY*		$\frac{YZ}{XY}$	

c. **VERBAL** Make a conjecture about the ratio of the hypotenuse to a leg of an isosceles right triangle.

H.O.T. Problems

Use **H**igher-**O**rder **T**hinking Skills

49. **CHALLENGE** Find the value of x in the figure at the right.

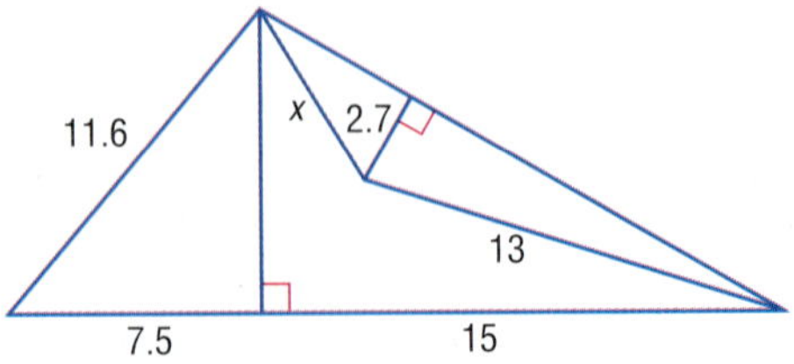

50. **REASONING** *True* or *false*? Any two right triangles with the same hypotenuse have the same area. Explain your reasoning.

51. **OPEN ENDED** Draw a right triangle with side lengths that form a Pythagorean triple. If you double the length of each side, is the resulting triangle *acute*, *right*, or *obtuse*? if you halve the length of each side? Explain.

52. **WRITING IN MATH** Explain how you can use the Pythagorean Theorem to find the length of the diagonal of a rectangle.

Standardized Test Practice

53. Which set of numbers cannot be the measures of the sides of a triangle?

A 10, 11, 20 **C** 35, 45, 75
B 14, 16, 28 **D** 41, 55, 98

54. A square park has a diagonal walkway from one corner to another. If the walkway is 120 meters long, what is the approximate length of each side of the park?

F 60 m **H** 170 m
G 85 m **J** 240 m

55. SHORT RESPONSE If the perimeter of square 2 is 200 units and the perimeter of square 1 is 150 units, what is the perimeter of square 3?

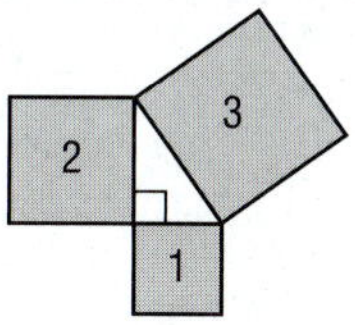

56. SAT/ACT In $\triangle ABC$, $\angle B$ is a right angle and $\angle A$ is 20° greater than $\angle C$. What is the measure of $\angle C$?

A 30 **C** 40
B 35 **D** 45

Spiral Review

Find the geometric mean between each pair of numbers. (Lesson 8-1)

57. 9 and 4 **58.** 45 and 5 **59.** 12 and 15 **60.** 36 and 48

61. SCALE DRAWING Teodoro is creating a scale model of a skateboarding ramp on a 10-by-8-inch sheet of graph paper. If the real ramp is going to be 12 feet by 8 feet, find an appropriate scale for the drawing and determine the ramp's dimensions. (Lesson 7-7)

Determine whether the triangles are similar. If so, write a similarity statement. If not, what would be sufficient to prove the triangles similar? Explain your reasoning. (Lesson 7-3)

62.

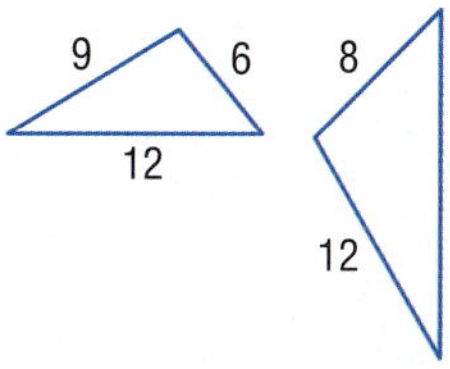

63.

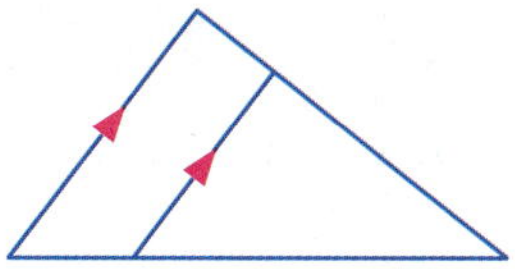

64.

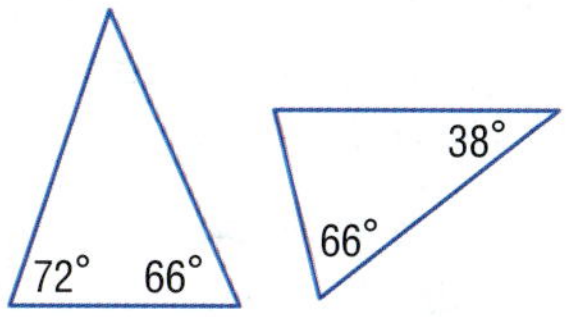

65. Write a two-column proof. (Lesson 5-3)

Given: $\overline{FG} \perp \ell$
$\overline{FH}$ is any nonperpendicular segment from F to ℓ.

Prove: $FH > FG$

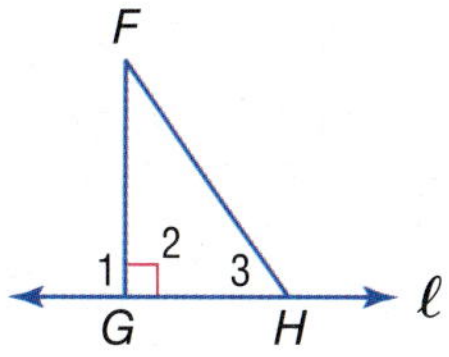

Find each measure if $m\angle DGF = 53$ and $m\angle AGC = 40$. (Lesson 4-2)

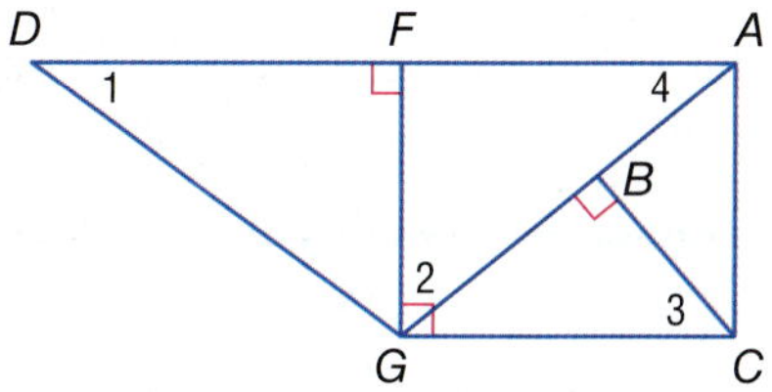

66. $m\angle 1$
67. $m\angle 2$
68. $m\angle 3$
69. $m\angle 4$

Find the distance between each pair of parallel lines with the given equations. (Lesson 3-6)

70. $y = 4x$
$y = 4x - 17$

71. $y = 2x - 3$
$2x - y = -4$

72. $y = -0.75x - 1$
$3x + 4y = 20$

Skills Review

Find the value of x. (Lesson 0-9)

73. $18 = 3x\sqrt{3}$ **74.** $24 = 2x\sqrt{2}$ **75.** $9\sqrt{2} \cdot x = 18\sqrt{2}$ **76.** $2 = x \cdot \frac{4}{\sqrt{3}}$

EXTEND 8-2

Geometry Lab: Coordinates in Space

You have used ordered pairs of two coordinates to describe the location of a point on the coordinate plane. Because space has three dimensions, a point requires three numbers, or coordinates, to describe its location in space.

A point in space is represented by an **ordered triple** of real numbers (x, y, z). In the figure at the right, the ordered triple (2, 3, 6) locates point P. Notice that a rectangular prism is used to show perspective.

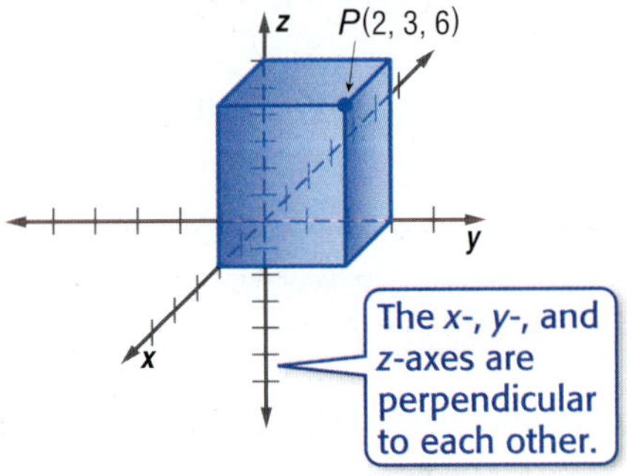

ACTIVITY 1 Graph a Rectangular Solid

Graph a rectangular solid that has two vertices, $L(4, -5, 2)$ and the origin. Label the coordinates of each vertex.

Step 1 Plot the x-coordinate first. Draw a segment from the origin 4 units in the positive direction.

Step 2 To plot the y-coordinate, draw a segment five units in the negative direction.

Step 3 Next, to plot the z-coordinate, draw a segment two units long in the positive direction.

Step 4 Label the coordinate L.

Step 5 Draw the rectangular prism and label each vertex: $L(4, -5, 2)$, $K(0, -5, 2)$, $J(0, 0, 2)$, $M(4, 0, 2)$ $Q(4, -5, 0)$, $P(0, -5, 0)$, $N(0, 0, 0)$, and $R(4, 0, 0)$.

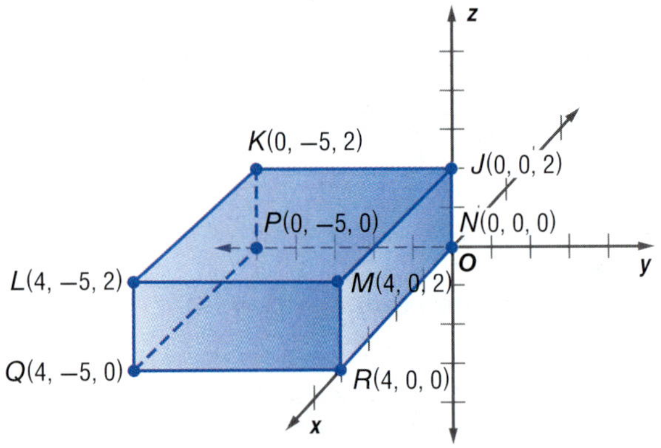

Finding the distance between points and the midpoint of a segment in space is similar to finding distance and a midpoint in the coordinate plane.

Key Concept

Distance and Midpoint Formulas in Space

If A has coordinates $A(x_1, y_1, z_1)$ and B has coordinates $B(x_2, y_2, z_2)$, then

$$AB = \sqrt{(x_2 - x_1)^2 + (y_2 - y_1)^2 + (z_2 - z_1)^2}.$$

The midpoint M of $\overline{AB}$ has coordinates

$$M\left(\frac{x_1 + x_2}{2}, \frac{y_1 + y_2}{2}, \frac{z_1 + z_2}{2}\right).$$

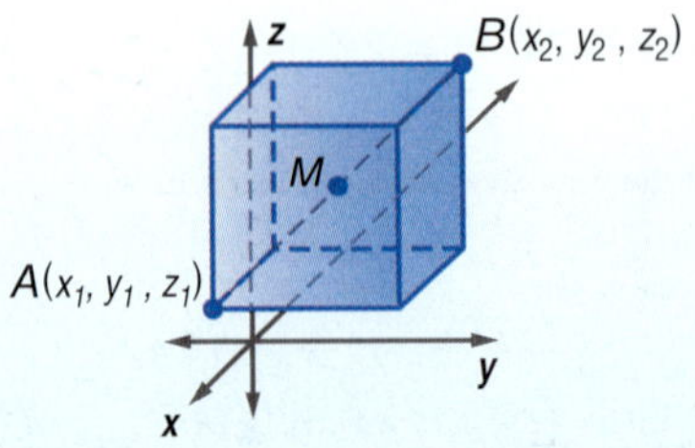

ACTIVITY 2 Distance and Midpoint Formulas in Space

Consider $J(2, 4, 9)$ and $K(-4, -5, 11)$.

a. **Find JK.**

$JK = \sqrt{(x_2 - x_1)^2 + (y_2 - y_1)^2 + (z_2 - z_1)^2}$ **Distance Formula in Space**

$= \sqrt{(-4 - 2)^2 + (-5 - 4)^2 + (11 - 9)^2}$ **Substitution**

$= \sqrt{121}$ **Simplify.**

$= 11$ **Use a calculator.**

b. **Determine the coordinates of the midpoint M of $\overline{JK}$.**

$M = \left(\frac{x_1 + x_2}{2}, \frac{y_1 + y_2}{2}, \frac{z_1 + z_2}{2}\right)$ **Midpoint Formula in Space**

$= \left(\frac{2 + (-4)}{2}, \frac{4 + (-5)}{2}, \frac{9 + 11}{2}\right)$ **Substitution**

$= \left(-1, -\frac{1}{2}, 10\right)$ **Simplify.**

Exercises

Graph a rectangular solid that contains the given point and the origin as vertices. Label the coordinates of each vertex.

1. $A(2, 1, 5)$
2. $P(-1, 4, 2)$
3. $C(-2, 2, 2)$
4. $R(3, -4, 1)$
5. $P(4, 6, -3)$
6. $G(4, 1, -3)$
7. $K(-2, -4, -4)$
8. $W(-1, -3, -6)$
9. $W(3, 3, 4)$

Determine the distance between each pair of points. Then determine the coordinates of the midpoint M of the segment joining the pair of points.

10. $D(0, 0, 0)$ and $E(1, 5, 7)$
11. $G(-3, -4, 6)$ and $H(5, -3, -5)$
12. $K(2, 2, 0)$ and $L(-2, -2, 0)$
13. $P(-2, -5, 8)$ and $Q(3, -2, -1)$
14. $A(4, 7, 9)$ and $B(-3, 8, -8)$
15. $W(-12, 8, 10)$ and $Z(-4, 1, -2)$
16. $F\left(\frac{3}{5}, 0, \frac{4}{5}\right)$ and $G(0, 3, 0)$
17. $G(1, -1, 6)$ and $H\left(\frac{1}{5}, -\frac{2}{5}, 2\right)$
18. $B(\sqrt{3}, 2, 2\sqrt{2})$ and $C(-2\sqrt{3}, 4, 4\sqrt{2})$
19. $S(6\sqrt{3}, 4, 4\sqrt{2})$ and $T(4\sqrt{3}, 5, \sqrt{2})$
20. **PROOF** Write a coordinate proof of the Distance Formula in Space.

    Given: A has coordinates $A(x_1, y_1, z_1)$, and B has coordinates $B(x_2, y_2, z_2)$.

    Prove: $AB = \sqrt{(x_2 - x_1)^2 + (y_2 - y_1)^2 + (z_2 - z_1)^2}$

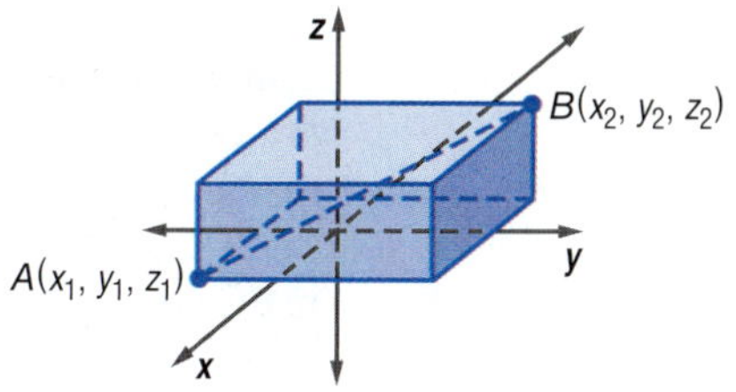

21. **WRITING IN MATH** Compare and contrast the Distance and Midpoint Formulas on the coordinate plane and in three-dimensional coordinate space.

8-3 Special Right Triangles

Why?

As part of a packet for students attending a regional student council meeting, Lyndsay orders triangular highlighters. She wants to buy rectangular boxes for the highlighters and other items, but she is concerned that the highlighters will not fit in the box she has chosen. If she knows the length of a side of the highlighter, Lyndsay can use the properties of special right triangles to determine if it will fit in the box.

Then

You used properties of isosceles and equilateral triangles. (Lesson 4-6)

Now

- Use the properties of 45°-45°-90° triangles.
- Use the properties of 30°-60°-90° triangles.

Math Online

glencoe.com

Properties of 45°-45°-90° Triangles The diagonal of a square forms two congruent isosceles right triangles. Since the base angles of an isosceles triangle are congruent, the measure of each acute angle is $90 \div 2$ or 45. Such a triangle is also known as a 45°-45°-90° triangle.

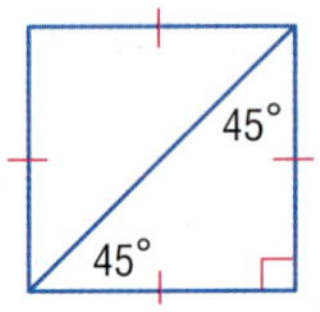

You can use the Pythagorean Theorem to find a relationship among the side lengths of a 45°-45°-90° right triangle.

$\ell^2 + \ell^2 = h^2$	**Pythagorean Theorem**
$2\ell^2 = h^2$	**Simplify.**
$\sqrt{2\ell^2} = \sqrt{h^2}$	**Take the positive square root of each side.**
$\ell\sqrt{2} = h$	**Simplify.**

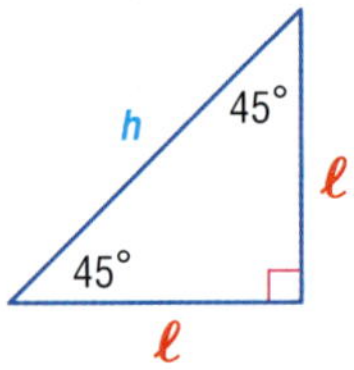

This algebraic proof verifies the following theorem.

Theorem 8.8 **45°-45°-90° Triangle Theorem**

In a 45°-45°-90° triangle, the legs ℓ are congruent and the length of the hypotenuse h is $\sqrt{2}$ times the length of a leg.

Symbols In a 45°-45°-90° triangle, $\ell = \ell$ and $h = \ell\sqrt{2}$.

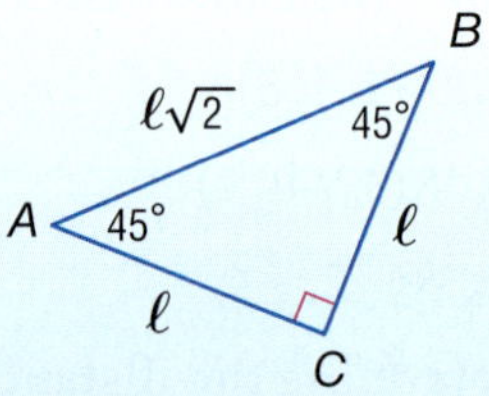

EXAMPLE 1 **Find the Hypotenuse Length in a 45°-45°-90° Triangle**

Find x.

a.

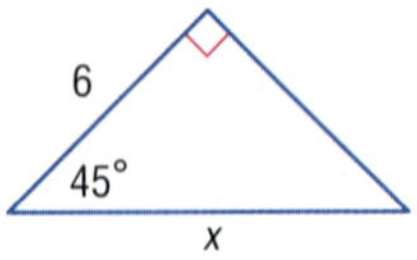

The acute angles of a right triangle are complementary, so the measure of the third angle is 90 –45 or 45. Since this is a 45°-45°-90° triangle, use Theorem 8.8.

$h = \ell\sqrt{2}$	**Theorem 8.8**
$x = 6\sqrt{2}$	**Substitution**

b.

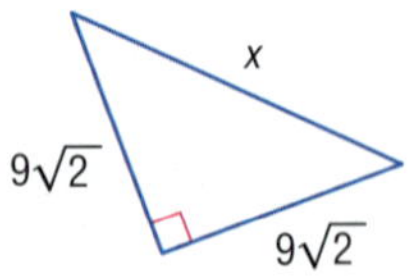

The legs of this right triangle have the same measure, so it is isosceles. Since this is a 45°-45°-90° triangle, use Theorem 8.8.

$h = \ell\sqrt{2}$	**Theorem 8.8**
$x = 9\sqrt{2} \cdot \sqrt{2}$	**Substitution**
$x = 9 \cdot 2$ or 18	$\sqrt{2} \cdot \sqrt{2} = 2$

Guided Practice

Find x.

1A.

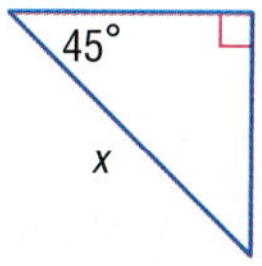

1B.

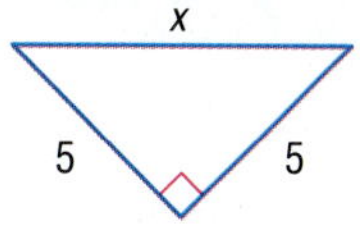

1C.

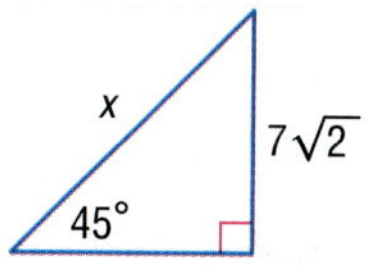

Personal Tutor glencoe.com

You can also work backward using Theorem 8.8 to find the lengths of the legs of a 45°-45°-90° triangle given the length of its hypotenuse.

EXAMPLE 2 Find the Leg Lengths in a 45°-45°-90° Triangle

Find x.

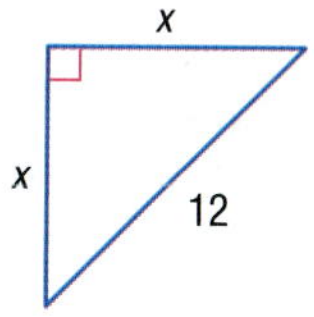

The legs of this right triangle have the same measure, x, so it is a 45°-45°-90° triangle. Use Theorem 8.8 to find x.

$h = \ell\sqrt{2}$	**45°-45°-90° Triangle Theorem**
$12 = x\sqrt{2}$	**Substitution**
$\frac{12}{\sqrt{2}} = x$	**Divide each side by $\sqrt{2}$.**
$\frac{12}{\sqrt{2}} \cdot \frac{\sqrt{2}}{\sqrt{2}} = x$	**Rationalize the denominator.**
$\frac{12\sqrt{2}}{2} = x$	**Multiply.**
$6\sqrt{2} = x$	**Simplify.**

Review Vocabulary

rationalizing the denominator a method used to eliminate radicals from the denominator of a fraction (Lesson 0-9)

Guided Practice

2A.

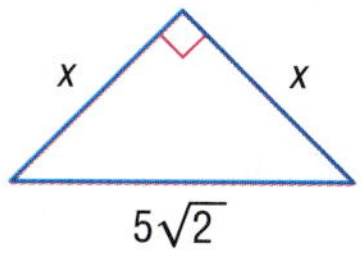

2B.

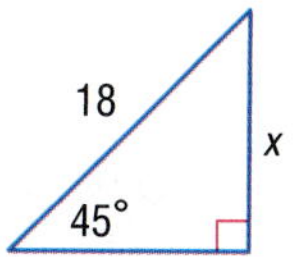

Personal Tutor glencoe.com

Properties of 30°-60°-90° Triangles A 30°-60°-90° triangle is another *special* right triangle or right triangle with side lengths that share a special relationship. You can use an equilateral triangle to find this relationship.

StudyTip

Altitudes of Isosceles Triangles Notice that an altitude of an isosceles triangle is also a median of the triangle. In the figure at the right, $\overline{BD}$ bisects $\overline{AC}$.

When an altitude is drawn from any vertex of an equilateral triangle, two congruent 30°-60°-90° triangles are formed. In the figure shown, $\triangle ABD \cong \triangle CBD$, so $\overline{AD} \cong \overline{CD}$. If $AD = x$, then $CD = x$ and $AC = 2x$. Since $\triangle ABC$ is equilateral, $AB = 2x$ and $BC = 2x$.

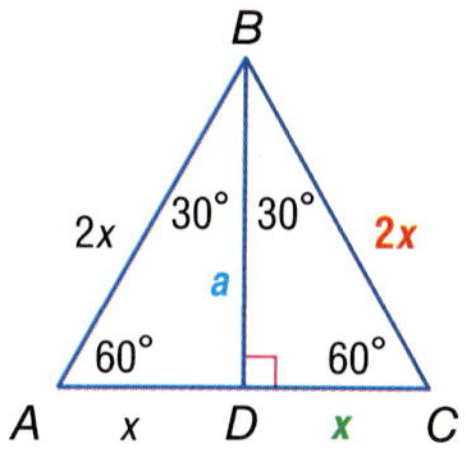

Use the Pythagorean Theorem to find a, the length of the altitude $\overline{BD}$, which is also the longer leg of $\triangle BDC$.

$a^2 + x^2 = (2x)^2$	**Pythagorean Theorem**
$a^2 + x^2 = 4x^2$	**Simplify.**
$a^2 = 3x^2$	**Subtract x^2 from each side.**
$a = \sqrt{3x^2}$	**Take the positive square root of each side.**
$a = x\sqrt{3}$	**Simplify.**

This algebraic proof verifies the following theorem.

Theorem 8.9

30°-60°-90° Triangle Theorem

In a 30°-60°-90° triangle, the length of the hypotenuse h is 2 times the length of the shorter leg s, and the length of the longer leg ℓ is $\sqrt{3}$ times the length of the shorter leg.

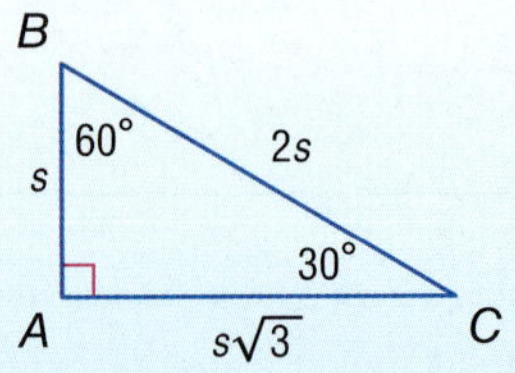

Symbols In a 30°-60°-90° triangle, $h = 2s$ and $\ell = s\sqrt{3}$.

StudyTip

Use Ratios The lengths of the sides of a 30°-60°-90° triangle are in a ratio of 1 to $\sqrt{3}$ to 2 or $1 : \sqrt{3} : 2$.

Remember, the shortest side of a triangle is opposite the smallest angle. So the shorter leg in a 30°-60°-90° triangle is opposite the 30° angle, and the longer leg is opposite the 60° angle.

EXAMPLE 3 Find Lengths in a 30°-60°-90° Triangle

Find x and y.

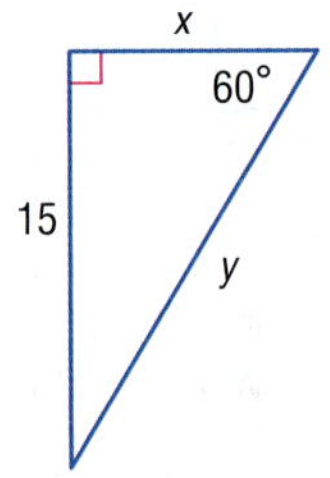

The acute angles of a right triangle are complementary, so the measure of the third angle in this triangle is 90 − 60 or 30. This is a 30°-60°-90° triangle.

Use Theorem 8.9 to find x, the length of the shorter side.

$\ell = s\sqrt{3}$ **Theorem 8.9**

$15 = x\sqrt{3}$ **Substitution**

$\frac{15}{\sqrt{3}} = x$ **Divide each side by $\sqrt{3}$.**

$\frac{15}{\sqrt{3}} \cdot \frac{\sqrt{3}}{\sqrt{3}} = x$ **Rationalize the denominator.**

$\frac{15\sqrt{3}}{\sqrt{3} \cdot \sqrt{3}} = x$ **Multiply.**

$\frac{15\sqrt{3}}{3} = x$ **$\sqrt{3} \cdot \sqrt{3} = 3$**

$5\sqrt{3} = x$ **Simplify.**

Now use Theorem 8.9 to find y, the length of the hypotenuse.

$h = 2s$ **Theorem 8.9**

$y = 2(5\sqrt{3})$ or $10\sqrt{3}$ **Substitution**

Guided Practice

Find x and y.

3A.

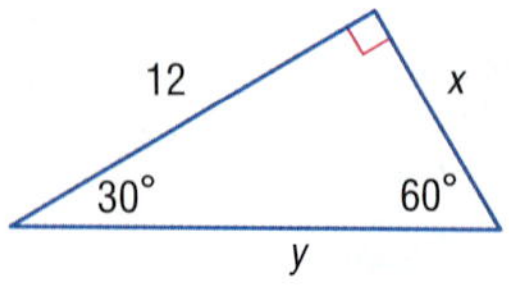

3B.

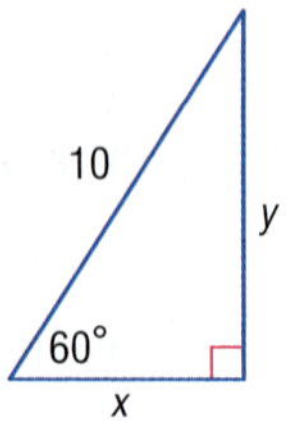

3C.

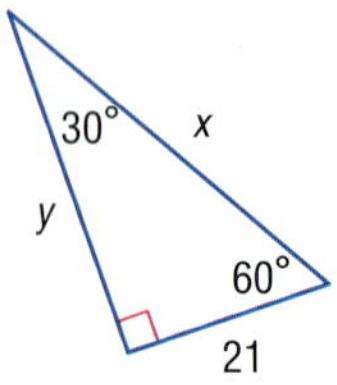

Personal Tutor glencoe.com

You can use the properties of 30°-60°-90° and 45°-45°-90° triangles to solve real-world problems.

Real-World EXAMPLE 4 Use Properties of Special Right Triangles

INVENTIONS A company makes crayons that "do not roll off tables" by shaping them as triangular prisms with equilateral bases. Sixteen of these crayons fit into a box shaped like a triangular prism that is $1\frac{1}{2}$ inches wide. The crayons stand on end in the box and the base of the box is equilateral. What are the dimensions of each crayon?

Understand You know that 16 crayons with equilateral triangular bases fit into a prism. You need to find the base length and height of each crayon.

Plan Guess and check to determine the arrangement of 16 crayons that would stack to fill the box. Find the width of one crayon and use the 30°-60°-90° Triangle Theorem to find its altitude.

Solve Make a guess that 4 equilateral crayons will fit across the base of the box. A sketch shows that the total number of crayons it takes to fill the box using 4 crayons across the base is 16. ✓

The width of the box is $1\frac{1}{2}$ inches, so the width of one crayon is $1\frac{1}{2} \div 4$ or $\frac{3}{8}$ inch.

Draw an equilateral triangle representing one crayon. Its altitude forms the longer leg of two 30°-60°-90° triangles. Use Theorem 8.9 to find the approximate length of the altitude a.

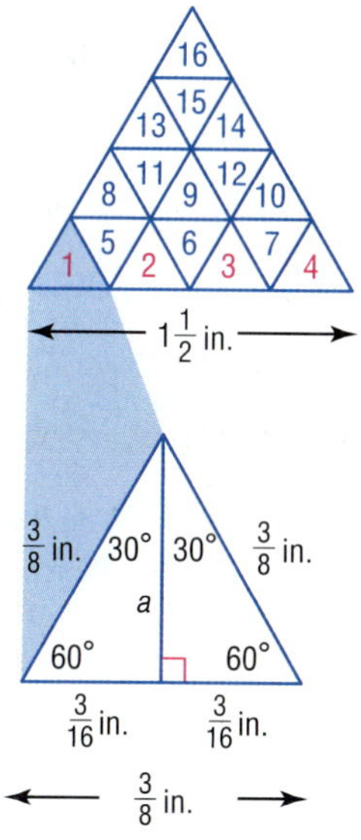

longer leg length = shorter leg length $\cdot \sqrt{3}$

$$a = \frac{3}{16} \cdot \sqrt{3} \text{ or about } 0.3$$

Each crayon is $\frac{3}{8}$ or about 0.4 inch by about 0.3 inch.

Check Find the height of the box using the 30°-60°-90° Triangle Theorem. Then divide by four, since the box is four crayons high. The result is a crayon height of about 0.3 inch. ✓

Problem-Solving Tip

Guess and Check When using the guess and check strategy, it can be helpful to keep a list of those guesses that you have already tried and know do not work.

In Example 4, suppose your first guess had been that the box was 5 crayons wide.

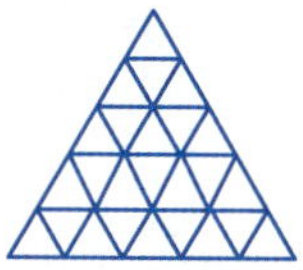

The sketch of this possibility reveals that this leads to a stack of 25, not 16 crayons.

Guided Practice

4. FURNITURE The top of the aquarium coffee table shown is an isosceles right triangle. The table's longest side, $\overline{AC}$, measures 107 centimeters. What is the distance from vertex B to side $\overline{AC}$? What are the lengths of the other two sides?

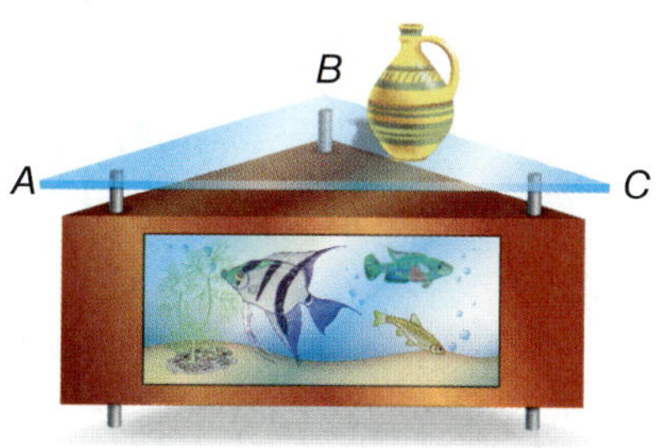

Personal Tutor glencoe.com

Check Your Understanding

Examples 1 and 2
pp. 552–553

Find x.

1.

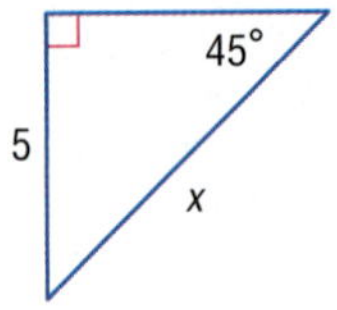

2.

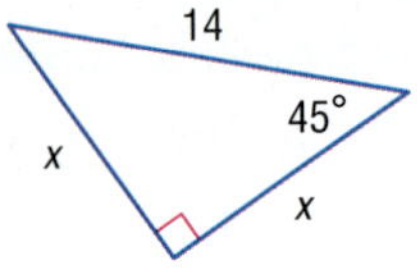

3.

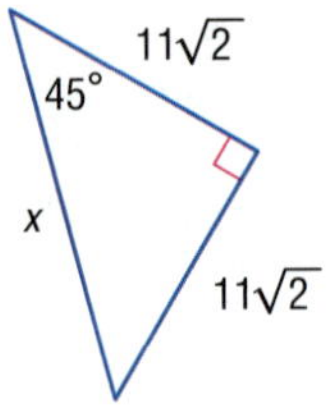

Example 3
p. 554

Find x and y.

4.

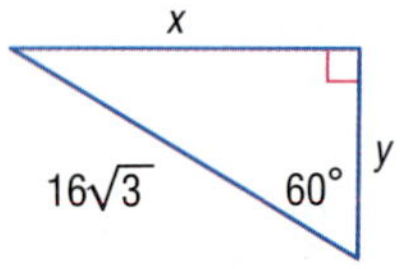

5.

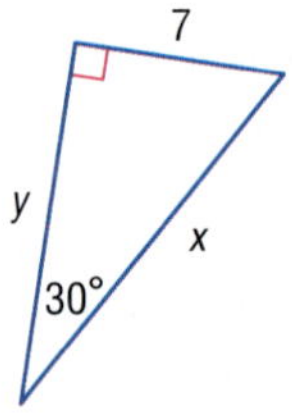

6.

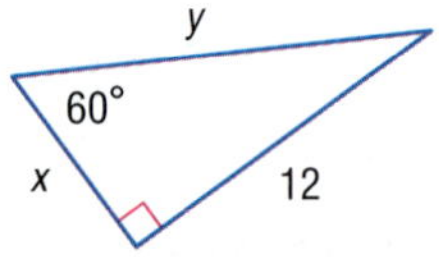

Example 4
p. 555

7. **ART** Paulo is mailing a piece of art that he has rolled into a tube. The diameter of the roll is $3\frac{1}{4}$ inches. He has a mailer that is a triangular prism with 4-inch equilateral triangle bases as shown in the diagram. Will the rolled art fit through the opening of the mailer? Explain.

Practice and Problem Solving

● = **Step-by-Step Solutions** begins on page R20.
Extra Practice begins on page 969.

Examples 1 and 2
pp. 552–553

Find x.

8.

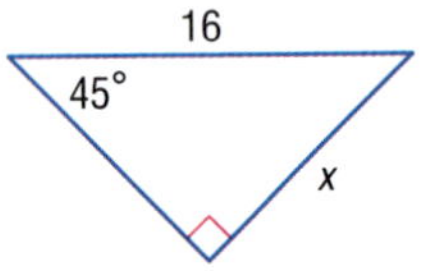

9.

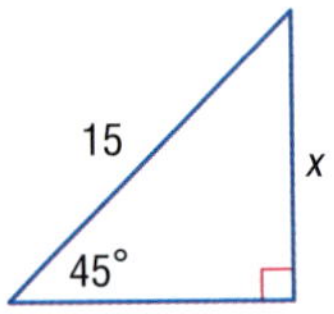

10.

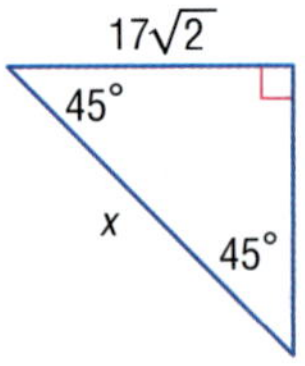

11.

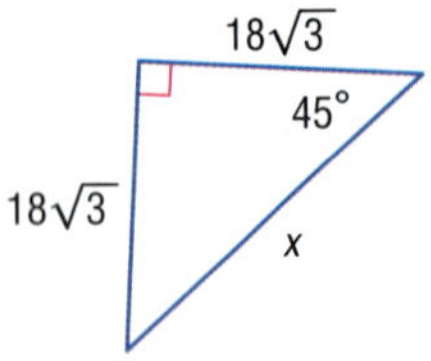

12.

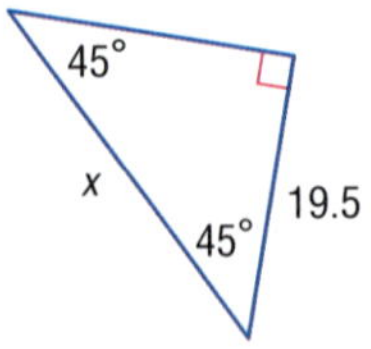

13.

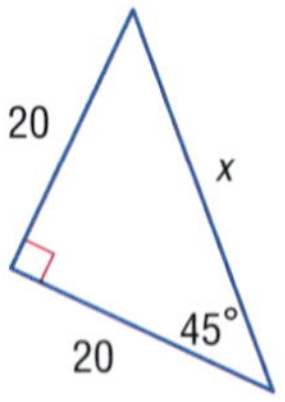

14. If a 45°-45°-90° triangle has a hypotenuse length of 9, find the leg length.

15. Determine the length of the leg of a 45°-45°-90° triangle with a hypotenuse length of 11.

16. What is the length of the hypotenuse of a 45°-45°-90° triangle if the leg length is 6 centimeters?

17. Find the length of the hypotenuse of a 45°-45°-90° triangle with a leg length of 8 centimeters.

Example 3
p. 554

Find x and y.

18.
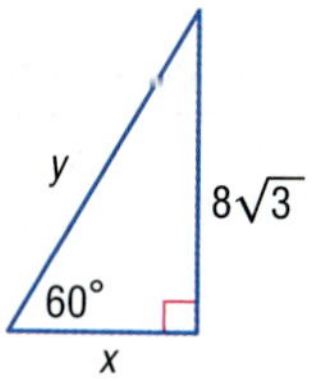

19.
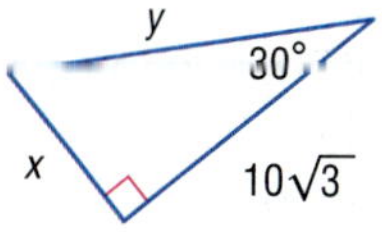

20.
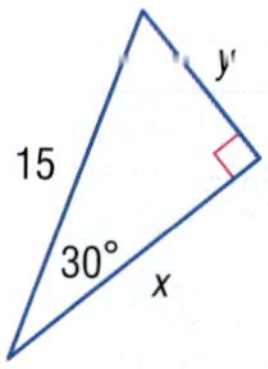

21.
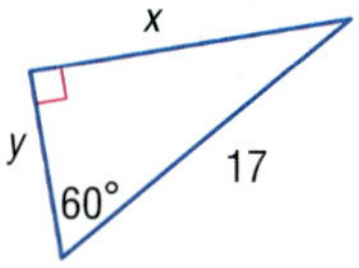

22.
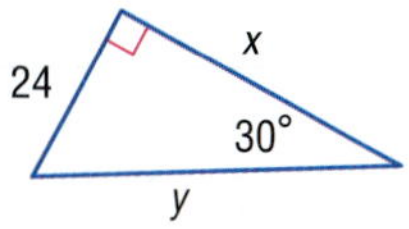

23.
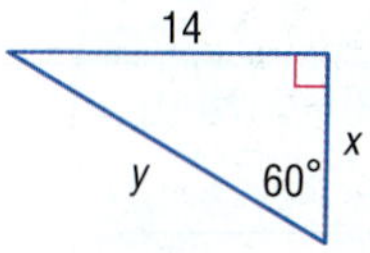

24. An equilateral triangle has an altitude length of 18 feet. Determine the length of a side of the triangle.

25. Find the length of the side of an equilateral triangle that has an altitude length of 24 feet.

Example 4
p. 555

26. PACKAGING Refer to the beginning of the lesson. Each highlighter is an equilateral triangle with 9-centimeter sides. Will the highlighter fit in a 10-inch by 7-inch rectangular box? Explain.

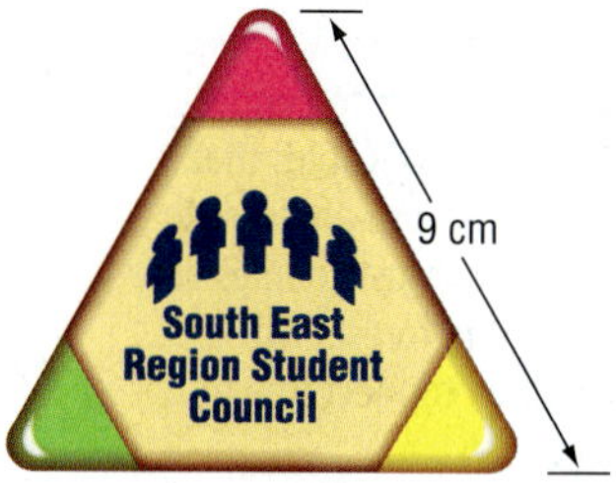

27. EVENT PLANNING Grace is having a party, and she wants to decorate the gable of the house as shown. The gable is an isosceles right triangle and she knows that the height of the gable is 8 feet. What length of lights will she need to cover the gable below the roof line?

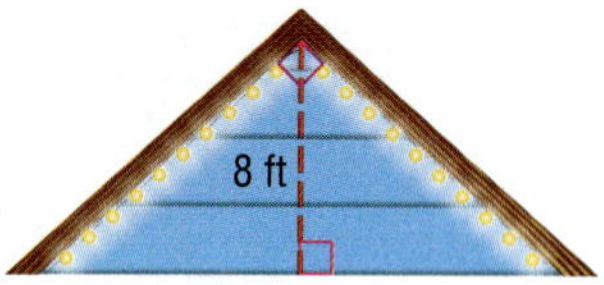

Find x and y.

StudyTip

Use Ratios The lengths of the sides of a 45°-45°-90° triangle are in a ratio of 1 to 1 to $\sqrt{2}$ or $1:1:\sqrt{2}$.

28.
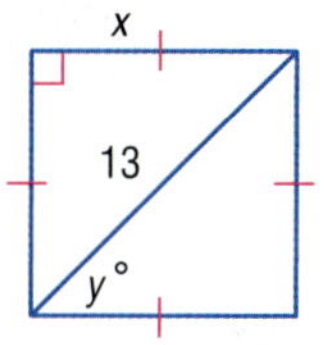

29
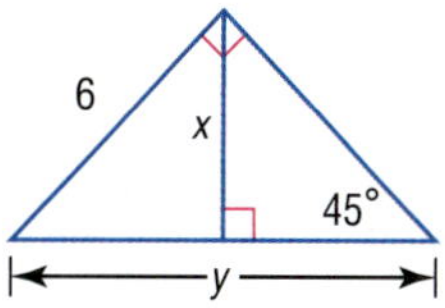

30.
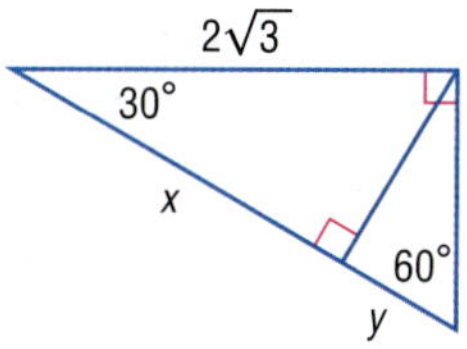

31.
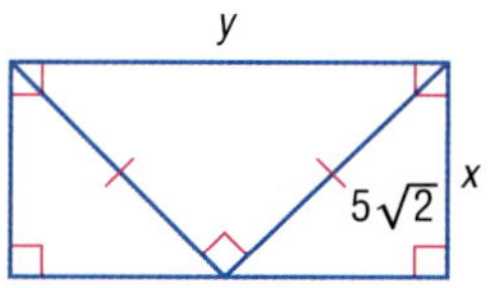

32.
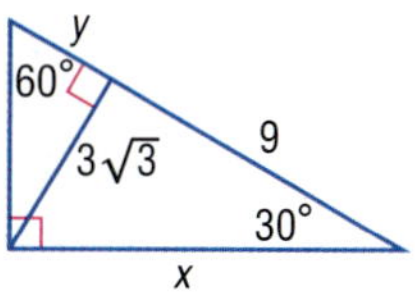

33.
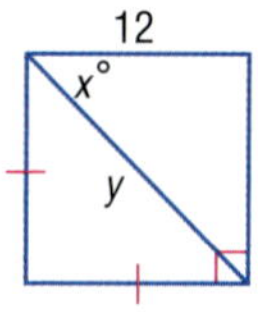

34. QUILTS The quilt block shown is made up of a square and four isosceles right triangles. What is the value of x? What is the side length of the entire quilt block?

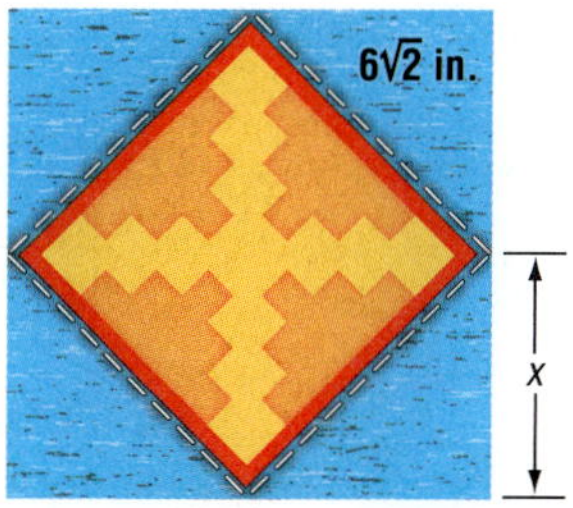

Real-World Link

Zip line tours are outdoor adventure tours in which participants use a harness and carabiner to connect to a series of cables that zip them from platform to platform. Most courses are located in forests where zip lines connect tree to tree.

Source: Adventure Travel

35. **ZIP LINE** Suppose a zip line is anchored in one corner of a course shaped like a rectangular prism. The other end is anchored in the opposite corner as shown. If the zip line makes a 60° angle with post $\overline{AF}$, find the zip line's length, AD.

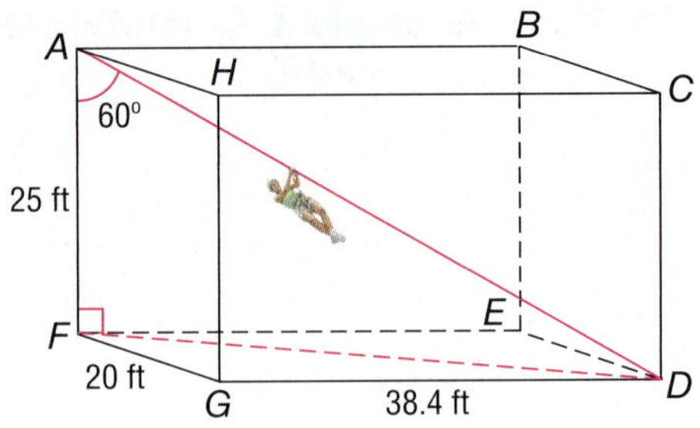

36. **GAMES** Kei is building a bean bag toss for the school carnival. He is using a 2-foot back support that is perpendicular to the ground 2 feet from the front of the board. He also wants to use a support that is perpendicular to the board as shown in the diagram. How long should he make the support?

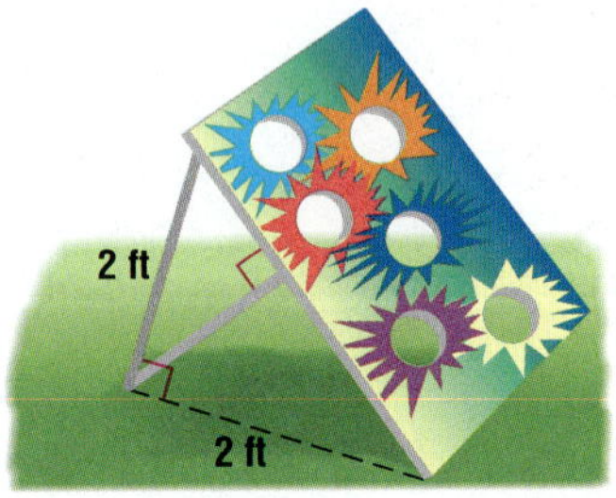

37. Find x, y, and z.

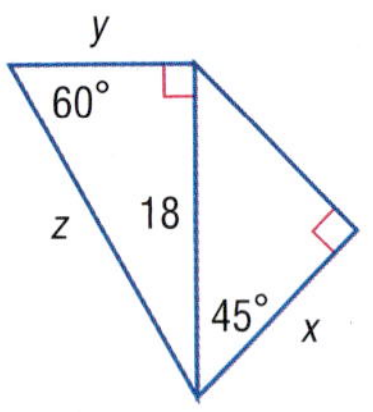

38. Each triangle in the figure is a 45°-45°-90° triangle. Find x.

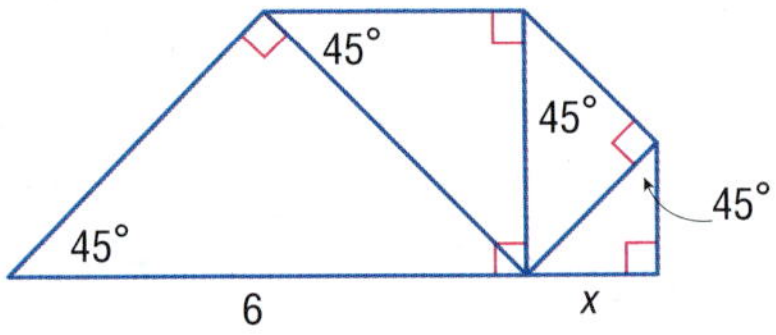

39. **MACHINERY** The dump truck shown has a 15-foot bed length. What is the height of the bed h when angle x is 30°? 45°? 60°?

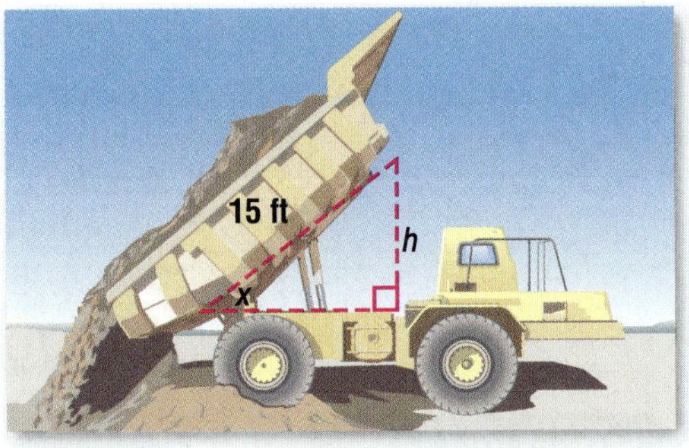

40. Find x, y, and z, and the perimeter of trapezoid $PQRS$.

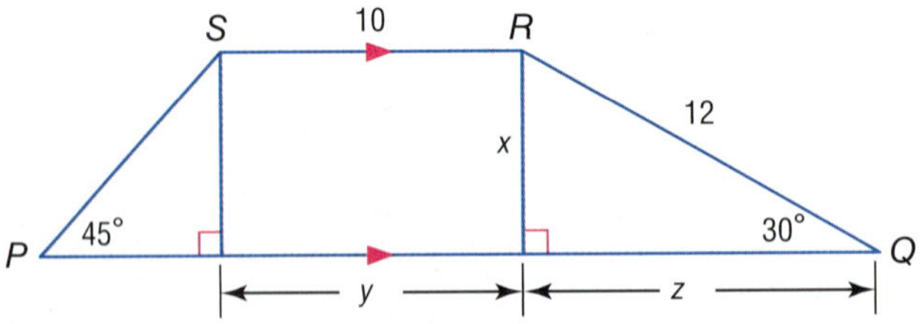

41. **COORDINATE GEOMETRY** $\triangle XYZ$ is a 45°-45°-90° triangle with right angle Z. Find the coordinates of X in Quadrant I for $Y(-1, 2)$ and $Z(6, 2)$.

42. **COORDINATE GEOMETRY** $\triangle EFG$ is a 30°-60°-90° triangle with $m\angle F = 90°$. Find the coordinates of E in Quadrant III for $F(-3, -4)$ and $G(-3, 2)$. $\overline{FG}$ is the longer leg.

43. **COORDINATE GEOMETRY** $\triangle JKL$ is a 45°-45°-90° triangle with right angle K. Find the coordinates of L in Quadrant IV for $J(-3, 5)$ and $K(-3, -2)$.

Real-World Link

When planning a party with a stand-up buffet, a host should allow 8 square feet of area for each guest.

Source: *San Antonio News*

44. EVENT PLANNING Eva has reserved a gazebo at a local park for a party. She wants to be sure that there will be enough space for her 12 guests to be in the gazebo at the same time. If the floor of the gazebo is a regular hexagon and each side is 7 feet, will there be enough room for Eva and her friends? Explain. (*Hint:* Use the Polygon Interior Angle Sum Theorem and the properties of special right triangles.)

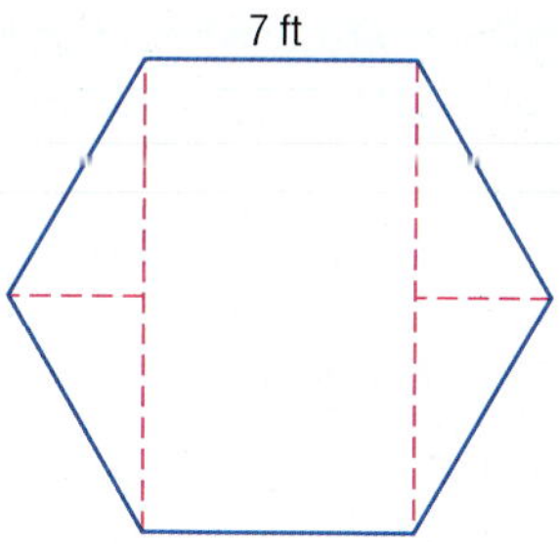

45. MULTIPLE REPRESENTATIONS In this problem, you will investigate ratios in right triangles.

a. GEOMETRIC Draw three similar right triangles with a 50° angle. Label one triangle *ABC* where angle *A* is the right angle and *B* is the 50° angle. Label a second triangle *MNP* where *M* is the right angle and *N* is the 50° angle. Label the third triangle *XYZ* where *X* is the right angle and *Y* is the 50° angle.

b. TABULAR Copy and complete the table below.

Triangle	Length				Ratio	
ABC	*AC*		*BC*		$\frac{AC}{BC}$	
MNP	*MP*		*NP*		$\frac{MP}{NP}$	
XYZ	*XZ*		*YZ*		$\frac{XZ}{YZ}$	

c. VERBAL Make a conjecture about the ratio of the leg opposite the 50° angle to the hypotenuse in any right triangle with an angle measuring 50°.

H.O.T. Problems

Use **H**igher-**O**rder **T**hinking Skills

46. ERROR ANALYSIS Carmen and Audrey want to find *x* in the triangle shown. Is either of them correct? Explain.

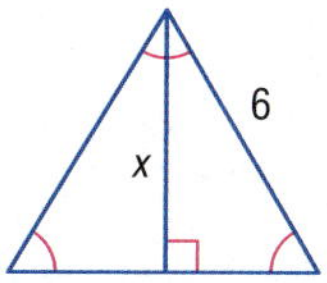

Carmen

$x = \frac{6\sqrt{3}}{2}$

$x = 3\sqrt{3}$

Audrey

$x = \frac{6\sqrt{2}}{2}$

$x = 3\sqrt{2}$

47. OPEN ENDED Draw a rectangle that has a diagonal twice as long as its width. Then write an equation to find the length of the rectangle.

48. CHALLENGE Find the perimeter of quadrilateral *ABCD*.

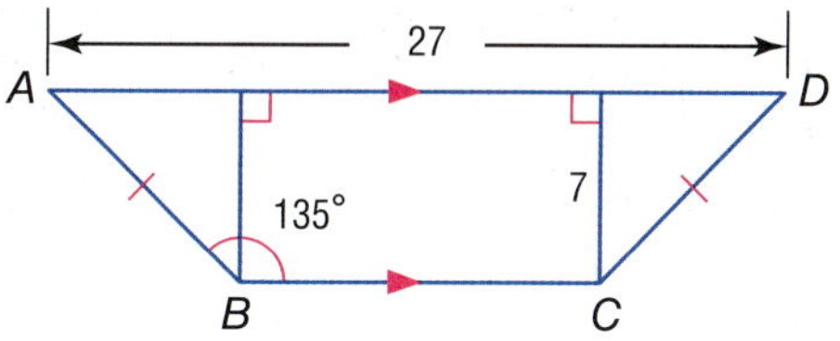

49. REASONING The ratio of the measure of the angles of a triangle is 1:2:3. The length of the shortest side is 8. What is the perimeter of the triangle?

50. WRITING IN MATH Explain how you can find the lengths of the two legs of a 30°-60°-90° triangle in radical form if you are given the length of the hypotenuse.

Standardized Test Practice

51. If the length of the longer leg in a 30°-60°-90° triangle is $5\sqrt{3}$, what is the length of the shorter leg?

A 3
B $5\sqrt{2}$
C 5
D 10

52. ALGEBRA Solve $\sqrt{5 - 4x} - 6 = 7$.

F −44
G −41
H 41
J 44

53. SHORT RESPONSE $\triangle XYZ$ is a 45°-45°-90° triangle with right angle Y. Find the coordinates of X in Quadrant III for $Y(-3, -3)$ and $Z(-3, 7)$.

54. SAT/ACT In the figure, below, square $ABCD$ is attached to $\triangle ADE$ as shown. If $m\angle EAD$ is 30° and AE is equal to $4\sqrt{3}$, then what is the area of square $ABCD$?

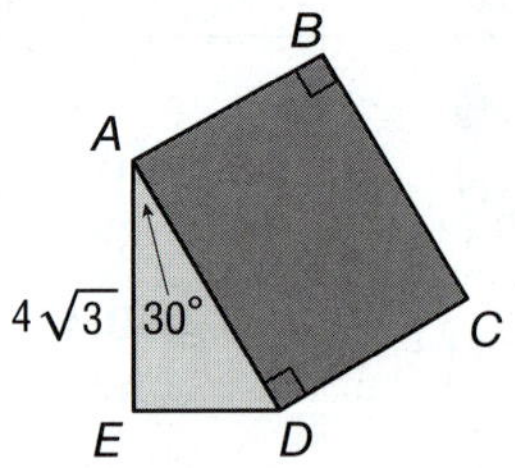

A $8\sqrt{3}$
B 64
C 72
D $64\sqrt{2}$

Spiral Review

55. SPORTS Dylan is making a ramp for bike jumps. The ramp support forms a right angle. The base is 12 feet long, and the height is 9 feet. What length of plywood does Dylan need for the ramp? (Lesson 8-2)

Find x, y, and z. (Lesson 8-1)

56.

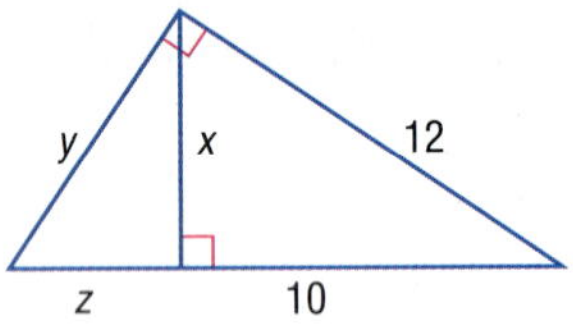

57.

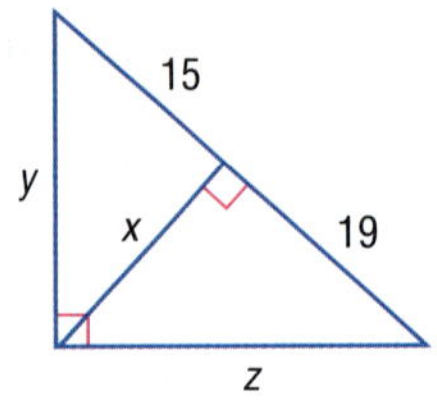

58.

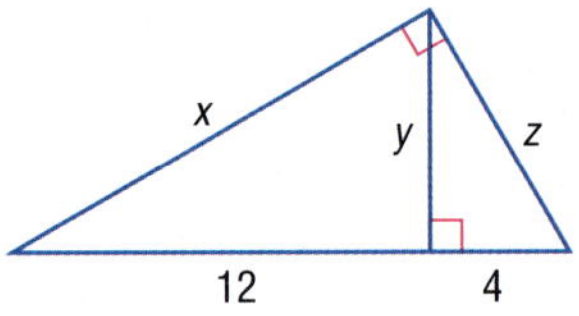

Find the measures of the angles of each triangle. (Lesson 7-1)

59. The ratio of the measures of the three angles is 2:5:3.

60. The ratio of the measures of the three angles is 6:9:10.

61. The ratio of the measures of the three angles is 5:7:8.

Use the Exterior Angle Inequality Theorem to list all of the angles that satisfy the stated condition. (Lesson 5-3)

62. measures less than $m\angle 5$

63. measures greater than $m\angle 6$

64. measures greater than $m\angle 10$

65. measures less than $m\angle 11$

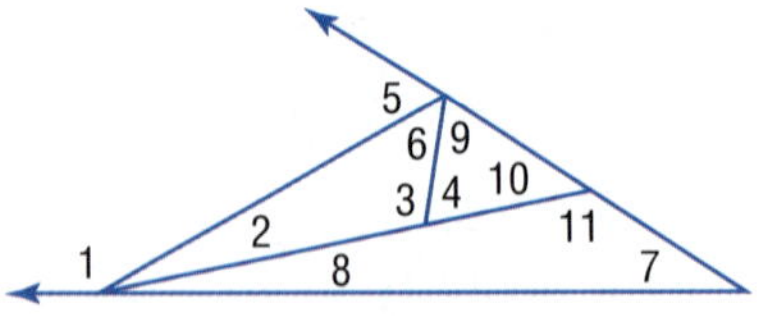

Skills Review

Find x. (Lesson 8-2)

66.

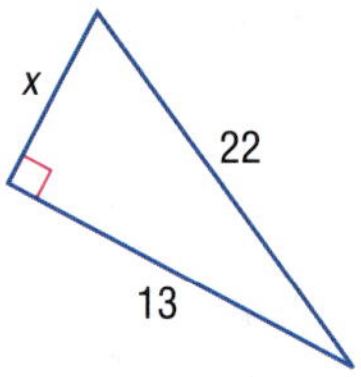

67.

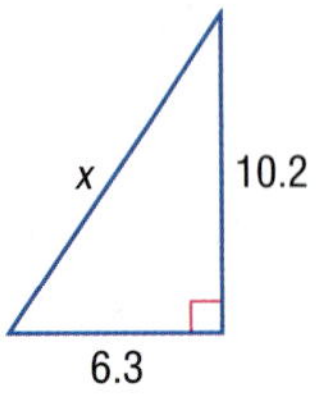

68.

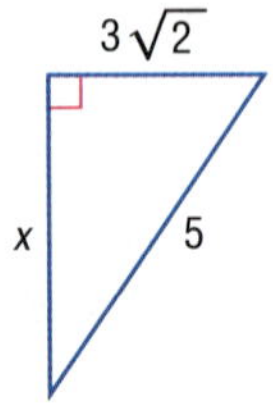

EXPLORE

8-4

Graphing Technology Lab
Trigonometry

You have investigated patterns in the measures of special right triangles. *Trigonometry* is the study of the patterns in all right triangles. You can use the Cabri™ Jr. application on a graphing calculator to investigate these patterns.

ACTIVITY Investigate Trigonometric Ratios

Step 1 Use the line tool on the **F2** menu to draw a horizontal line. Label the points on the line A and B.

Step 2 Press **F3** and choose the **Perpendicular** tool to create a perpendicular line through point B. Draw and label a point C on the perpendicular line.

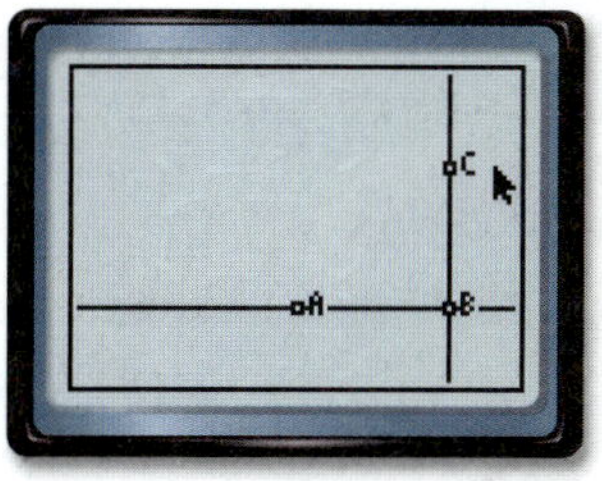

Steps 1 and 2

Step 3 Use the segment tool on the **F2** menu to draw $\overline{AC}$.

Step 4 Find and label the measures of $\overline{BC}$ and $\overline{AC}$ using the **Distance** and **Length** tool under **Measure** on the **F5** menu. Use the **Angle** tool to find the measure of $\angle A$.

Step 5 Calculate and display the ratio $\frac{BC}{AC}$ using the **Calculate** tool on the **F5** menu. Label the ratio as A/B.

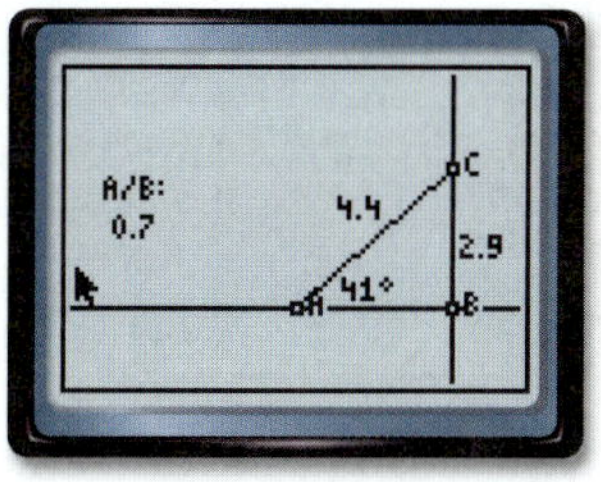

Steps 3 through 5

Step 6 Press CLEAR. Then use the arrow keys to move the cursor close to point B. When the arrow is clear, press and hold the ALPHA key. Drag B and observe the ratio.

Analyze the Results

1. Discuss the effect on $\frac{BC}{AC}$ by dragging point B on $\overline{BC}$, $\overline{AC}$, and $\angle A$.

2. Use the calculate tool to find the ratios $\frac{AB}{AC}$ and $\frac{BC}{AB}$. Then drag B and observe the ratios.

3. **MAKE A CONJECTURE** The *sine*, *cosine*, and *tangent* functions are trigonometric functions based on angle measures. Make a note of $m\angle A$. Exit Cabri Jr. and use SIN, COS, and TAN on the calculator to find *sine*, *cosine* and *tangent* for $m\angle A$. Compare the results to the ratios you found in the activity. Make a conjecture about the definitions of sine, cosine, and tangent.

8-4 Trigonometry

Why?

The steepness of a hiking trail is often expressed as a *percent of grade*. The steepest part of Bright Angel Trail in the Grand Canyon National Park has about a 15.7% grade. This means that the trail rises or falls 15.7 feet over a horizontal distance of 100 feet. You can use trigonometric ratios to determine that this steepness is equivalent to an angle of about 9°.

Then

You used the Pythagorean Theorem to find missing lengths in right triangles. (Lesson 8-2)

Now

- Find trigonometric ratios using right triangles.
- Use trigonometric ratios to find angle measures in right triangles.

New Vocabulary

trigonometry
trigonometric ratio
sine
cosine
tangent
inverse sine
inverse cosine
inverse tangent

Math Online
glencoe.com

Trigonometric Ratios The word **trigonometry** comes from two Greek terms, *trigon*, meaning triangle, and *metron*, meaning measure. The study of trigonometry involves triangle measurement. A **trigonometric ratio** is a ratio of the lengths of two sides of a right triangle. One trigonometric ratio of $\triangle ABC$ is $\frac{AC}{AB}$.

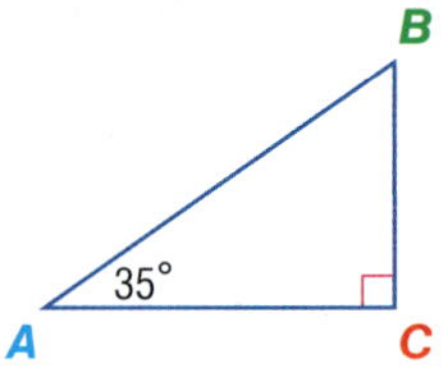

By AA Similarity, a right triangle with a given acute angle measure is similar to every other right triangle with the same acute angle measure. So, trigonometric ratios are constant for a given angle measure.

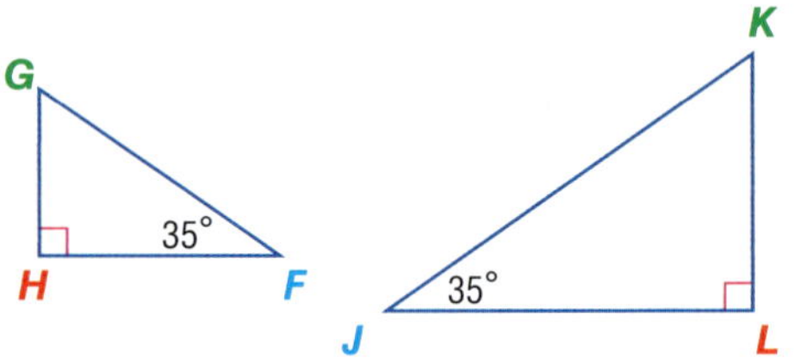

$\triangle ABC \sim \triangle FGH \sim \triangle JKL$, so $\frac{AC}{AB} = \frac{FH}{FG} = \frac{JL}{JK}$

The names of the three most common trigonometric ratios are given below.

Key Concept **Trigonometric Ratios**

Words	Symbols
If $\triangle ABC$ is a right triangle with acute $\angle A$, then the **sine** of $\angle A$ (written sin A) is the ratio of the length of the leg opposite $\angle A$ (opp) to the length of the hypotenuse (hyp).	$\sin A = \frac{\text{opp}}{\text{hyp}}$ or $\frac{a}{c}$ $\sin B = \frac{\text{opp}}{\text{hyp}}$ or $\frac{b}{c}$
If $\triangle ABC$ is a right triangle with acute $\angle A$, then the **cosine** of $\angle A$ (written cos A) is the ratio of the length of the leg adjacent $\angle A$ (adj) to the length of the hypotenuse (hyp).	$\cos A = \frac{\text{adj}}{\text{hyp}}$ or $\frac{b}{c}$ $\cos B = \frac{\text{adj}}{\text{hyp}}$ or $\frac{a}{c}$
If $\triangle ABC$ is a right triangle with acute $\angle A$, then the **tangent** of $\angle A$ (written tan A) is the ratio of the length of the leg opposite $\angle A$ (opp) to the length of the leg adjacent $\angle A$ (adj).	$\tan A = \frac{\text{opp}}{\text{adj}}$ or $\frac{a}{b}$ $\tan B = \frac{\text{opp}}{\text{adj}}$ or $\frac{b}{a}$

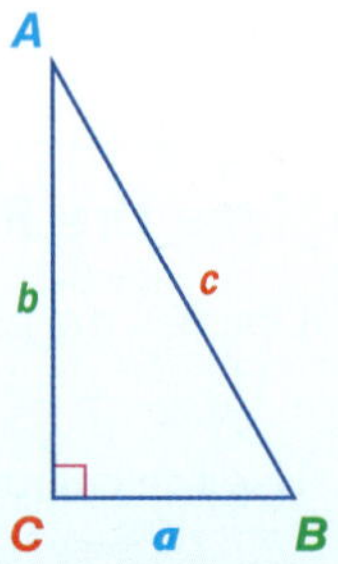

EXAMPLE 1 Find Sine, Cosine, and Tangent Ratios

Express each ratio as a fraction and as a decimal to the nearest hundredth.

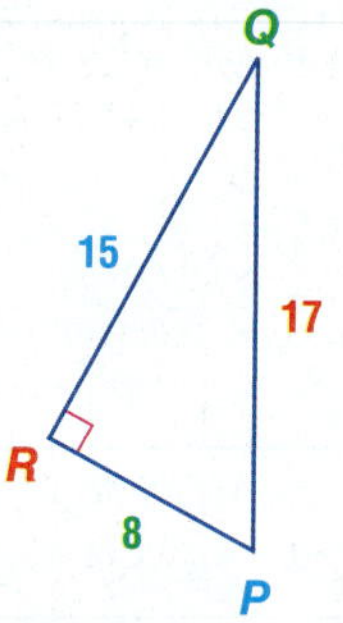

a. $\sin P$

$\sin P = \frac{\text{opp}}{\text{hyp}}$

$= \frac{15}{17}$ or about 0.88

b. $\cos P$

$\cos P = \frac{\text{adj}}{\text{hyp}}$

$= \frac{8}{17}$ or about 0.47

c. $\tan P$

$\tan P = \frac{\text{opp}}{\text{adj}}$

$= \frac{15}{8}$ or about 1.88

d. $\sin Q$

$\sin Q = \frac{\text{opp}}{\text{hyp}}$

$= \frac{8}{17}$ or about 0.47

e. $\cos Q$

$\cos Q = \frac{\text{adj}}{\text{hyp}}$

$= \frac{15}{17}$ or about 0.88

f. $\tan Q$

$\tan Q = \frac{\text{opp}}{\text{adj}}$

$= \frac{8}{15}$ or about 0.53

StudyTip

Memorizing Trigonometric Ratios SOH-CAH-TOA is a mnemonic device for learning the ratios for sine, cosine, and tangent using the first letter of each word in the ratios.

$\sin A = \frac{\text{opp}}{\text{hyp}}$

$\cos A = \frac{\text{adj}}{\text{hyp}}$

$\tan A = \frac{\text{opp}}{\text{adj}}$

Guided Practice

1. Find $\sin J$, $\cos J$, $\tan J$, $\sin K$, $\cos K$, and $\tan K$. Express each ratio as a fraction and as a decimal to the nearest hundredth.

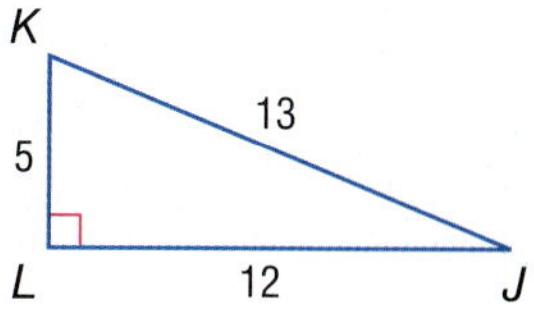

Personal Tutor glencoe.com

Special right triangles can be used to find the sine, cosine, and tangent of 30°, 60°, and 45° angles.

EXAMPLE 2 Use Special Right Triangles to Find Trigonometric Ratios

Use a special right triangle to express the tangent of 30° as a fraction and as a decimal to the nearest hundredth.

Draw and label the side lengths of a 30°-60°-90° right triangle, with x as the length of the shorter leg.

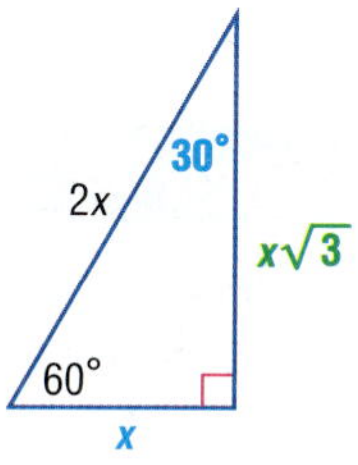

The side opposite the 30° angle has a measure of x.

The side adjacent to the 30° angle has a measure of $x\sqrt{3}$.

$\tan 30° = \frac{\text{opp}}{\text{adj}}$ **Definition of tangent ratio**

$= \frac{x}{x\sqrt{3}}$ **Substitution**

$= \frac{1}{\sqrt{3}} \cdot \frac{\sqrt{3}}{\sqrt{3}}$ **Simplify and rationalize the denominator.**

$= \frac{\sqrt{3}}{3}$ or about 0.58 **Simplify and use a calculator.**

Guided Practice

2. Use a special right triangle to express the cosine of 45° as a fraction and as a decimal to the nearest hundredth.

Personal Tutor glencoe.com

Real-World Link

The grade of a trail often changes many times. Average grade is the average of several consecutive running grades of a trail. Maximum grade is the smaller section of a trail that exceeds the trail's typical running grade. Trails often have maximum grades that are much steeper than a trail's average running grade.

Source: Federal Highway Administration

Real-World EXAMPLE 3 Estimate Measures Using Trigonometry

HIKING A certain part of a hiking trail slopes upward at about a 5° angle. After traveling a horizontal distance of 100 feet along this part of the trail, what would be the change in a hiker's vertical position? What distance has the hiker traveled along the path?

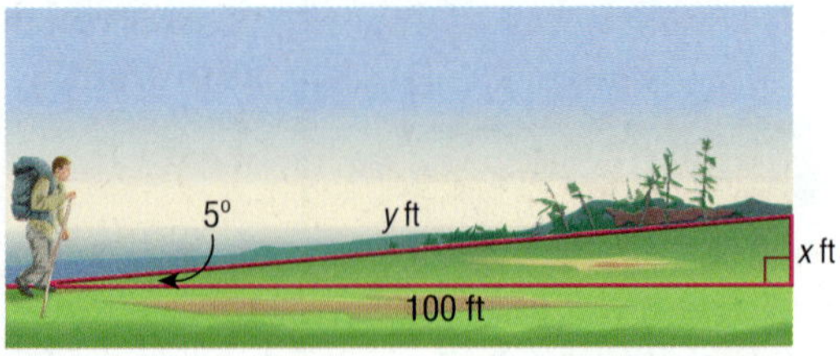

Let $m\angle A = 5$. The vertical change in the hiker's position is x, the measure of the leg opposite $\angle A$. The horizontal distance traveled is 100 feet, the measure of the leg adjacent to $\angle A$. Since the length of the leg opposite and the leg adjacent to a given angle are involved, write an equation using a tangent ratio.

$$\tan A = \frac{\text{opp}}{\text{adj}}$$ **Definition of tangent ratio**

$$\tan 5° = \frac{x}{100}$$ **Substitution**

$$100 \cdot \tan 5° = x$$ **Multiply each side by 100.**

Use a calculator to find x.

100 [TAN] 5 [ENTER] 8.748866353

The hiker is about 8.75 feet higher than when he started walking.

The distance y traveled along the path is the length of the hypotenuse, so you can use a cosine ratio to find this distance.

$$\cos A = \frac{\text{adj}}{\text{hyp}}$$ **Definition of cosine ratio**

$$\cos 5° = \frac{100}{y}$$ **Substitution**

$$y \cdot \cos 5° = 100$$ **Multiply each side by y.**

$$y = \frac{100}{\cos 5°}$$ **Divide each side by cos 5°.**

Use a calculator to find y.

100 [÷] [COS] 5 [ENTER] 100.3819838

The hiker has traveled a distance of about 100.38 feet along the path.

StudyTip

Graphing Calculator Be sure your graphing calculator is in degree mode rather than radian mode.

Guided Practice

Find x to the nearest hundredth.

3A.

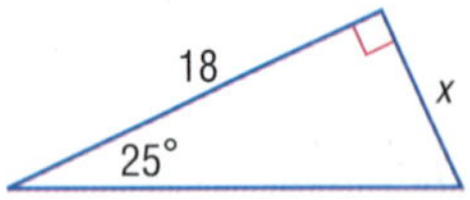

3B.

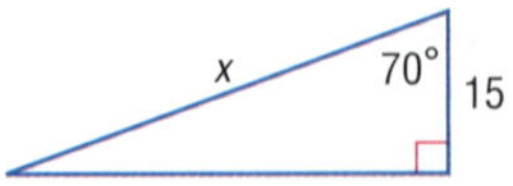

3C. ARCHITECTURE The front of the vacation cottage shown is an isosceles triangle. What is the height x of the cottage above its foundation? What is the length y of the roof? Explain your reasoning.

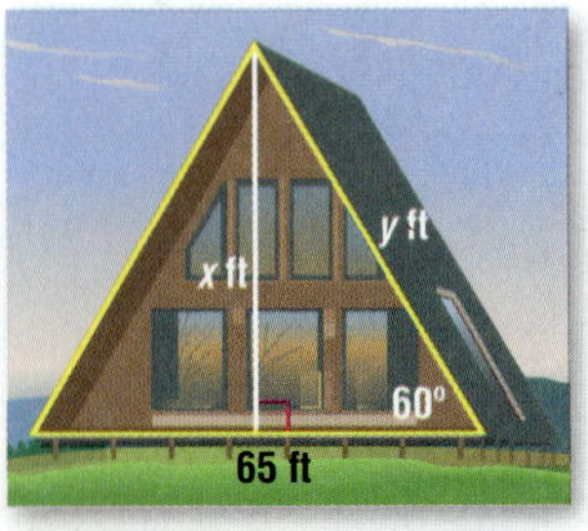

Personal Tutor glencoe.com

Use Inverse Trigonometric Ratios

In Example 2, you found that tan 30° ≈ 0.58. It follows that if the tangent of an acute angle is 0.58, then the angle measures approximately 30.

If you know the sine, cosine, or tangent of an acute angle, you can use a calculator to find the measure of the angle, which is the inverse of the trigonometric ratio.

> **ReadingMath**
>
> **Inverse Trigonometric Ratios** The expression $\sin^{-1} x$ is read *the inverse sine of x* and is interpreted as the angle with sine x. Be careful not to confuse this notation with the notation for negative exponents– $\sin^{-1} x \neq \frac{1}{\sin x}$. Instead, this notation is similar to the notation for an inverse function, $f^{-1}(x)$.

Key Concept

Inverse Trigonometric Ratios

Words If $\angle A$ is an acute angle and the sine of A is x, then the **inverse sine** of x is the measure of $\angle A$.

Symbols If $\sin A = x$, then $\sin^{-1} x = m\angle A$.

Words If $\angle A$ is an acute angle and the cosine of A is x, then the **inverse cosine** of x is the measure of $\angle A$.

Symbols If $\cos A = x$, then $\cos^{-1} x = m\angle A$.

Words If $\angle A$ is an acute angle and the tangent of A is x, then the **inverse tangent** of x is the measure of $\angle A$.

Symbols If $\tan A = x$, then $\tan^{-1} x = m\angle A$.

So if $\tan 30° \approx 0.58$, then $\tan^{-1} 0.58 \approx 30°$.

EXAMPLE 4 Find Angle Measures Using Inverse Trigonometric Ratios

Use a calculator to find the measure of $\angle A$ to the nearest tenth.

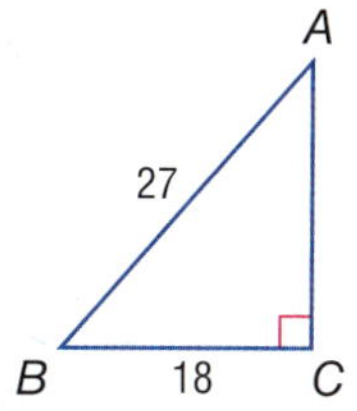

The measures given are those of the leg opposite $\angle A$ and the hypotenuse, so write an equation using the sine ratio.

$\sin A = \frac{18}{27}$ or $\frac{2}{3}$ $\quad$ **$\sin A = \frac{\text{opp}}{\text{hyp}}$**

If $\sin A = \frac{2}{3}$, then $\sin^{-1} \frac{2}{3} = m\angle A$. Use a calculator.

KEYSTROKES: [2nd] [SIN] [(] 2 [÷] 3 [)] [ENTER] 41.8103149

So, the measure of $\angle A$ is approximately 41.8°.

> **StudyTip**
>
> **Graphing Calculator** The second functions of the [SIN], [COS], and [TAN] keys are usually the inverses.

Guided Practice

Use a calculator to find the measure of $\angle A$ to the nearest tenth.

4A.

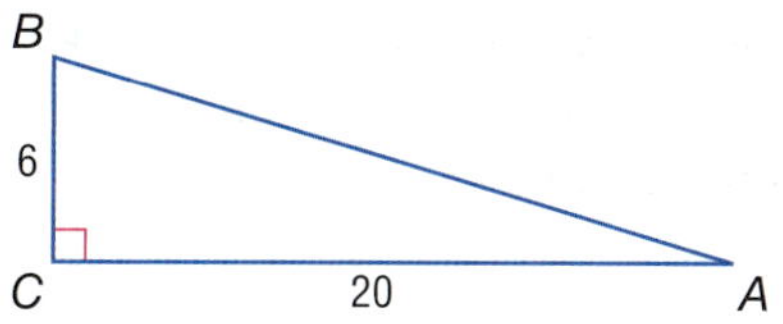

4B.

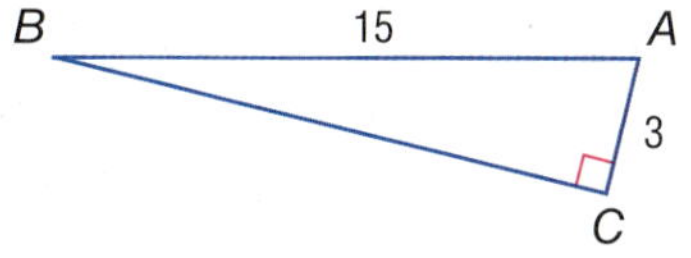

Personal Tutor glencoe.com

When you use given measures to find the unknown angle and side measures of a right triangle, this is known as *solving a right triangle*. To solve a right triangle, you need to know

- two side lengths or
- one side length and the measure of one acute angle.

EXAMPLE 5 Solve a Right Triangle

Solve the right triangle. Round side measures to the nearest tenth and angle measures to the nearest degree.

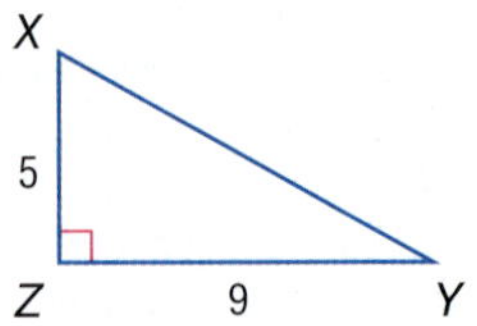

Step 1 Find $m\angle X$ by using a tangent ratio.

$$\tan X = \frac{9}{5} \qquad \textbf{tan } X = \frac{\textbf{opp}}{\textbf{adj}}$$

$$\tan^{-1}\frac{9}{5} = m\angle X \qquad \textbf{Definition of inverse tangent}$$

$$60.9453959 \approx m\angle X \qquad \textbf{Use a calculator.}$$

So the measure of $\angle X$ is about 61°.

Step 2 Find $m\angle Y$ using Corollary 4.1, which states that the acute angles of a right triangle are complementary.

$$m\angle X + m\angle Y = 90 \qquad \textbf{Corollary 4.1}$$

$$61 + m\angle Y \approx 90 \qquad m\angle X \approx 61$$

$$m\angle Y \approx 29 \qquad \textbf{Subtract 61 from each side.}$$

So the measure of $\angle Y$ is about 29°.

Step 3 Find XY by using the Pythagorean Theorem.

$$(XZ)^2 + (ZY)^2 = (XY)^2 \qquad \textbf{Pythagorean Theorem}$$

$$5^2 + 9^2 = (XY)^2 \qquad \textbf{Substitution}$$

$$106 = (XY)^2 \qquad \textbf{Simplify.}$$

$$\sqrt{106} = XY \qquad \textbf{Take the positive square root of each side.}$$

$$10.3 \approx XY \qquad \textbf{Use a calculator.}$$

So the measure of $\overline{XY}$ is about 10.3.

StudyTip

Alternative Methods Right triangles can often be solved using different methods. In Example 5, $m\angle Y$ could have been found using a tangent ratio, and $m\angle X$ and a sine ratio could have been used to find XY.

Watch Out!

Approximation If using calculated measures to find other measures in a right triangle, be careful not to round values until the last step. So in the following equation, use $\tan^{-1}\frac{9}{5}$ instead of its approximate value, 61°.

$$XY = \frac{9}{\sin X}$$

$$= \frac{9}{\sin\left(\tan^{-1}\frac{9}{5}\right)}$$

$$\approx 10.3$$

Guided Practice

Solve each right triangle. Round side measures to the nearest tenth and angle measures to the nearest degree.

5A.

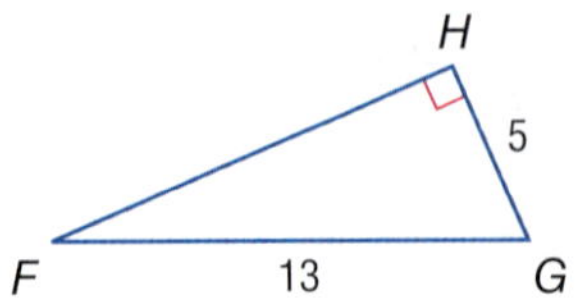

5B.

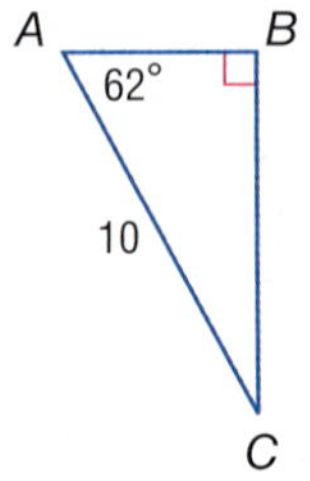

5C.

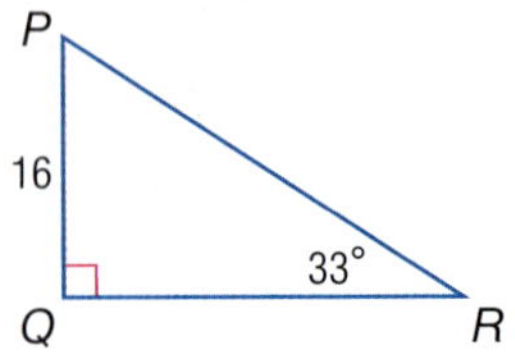

Personal Tutor glencoe.com

Check Your Understanding

Example 1 p. 563

Express each ratio as a fraction and as a decimal to the nearest hundredth.

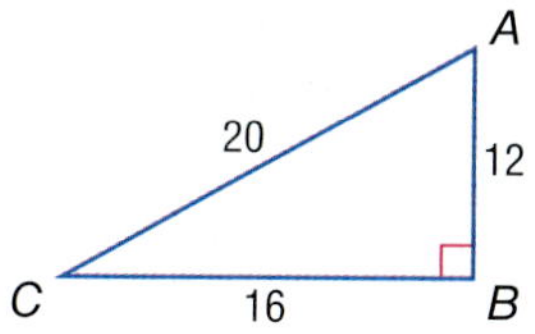

1. $\sin A$ **2.** $\tan C$ **3.** $\cos A$

4. $\tan A$ **5.** $\cos C$ **6.** $\sin C$

Example 2 p. 563

7. Use a special right triangle to express $\sin 60°$ as a fraction and as a decimal to the nearest hundredth.

Example 3 p. 564

Find x. Round to the nearest hundredth.

8.

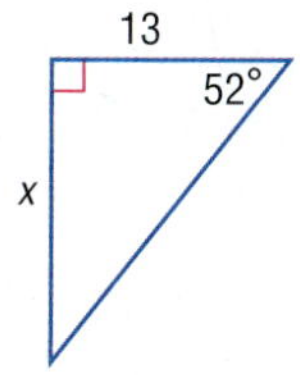

9.

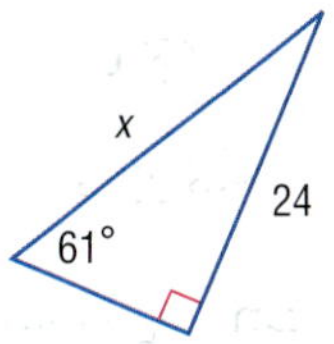

10.

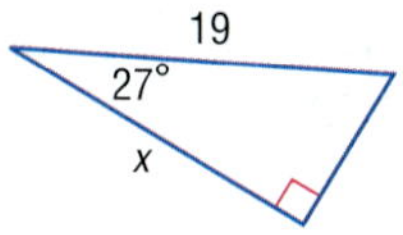

11. SPORTS David is building a bike ramp. He wants the angle that the ramp makes with the ground to be 20°. If the board he wants to use for his ramp is $3\frac{1}{2}$ feet long, about how tall will the ramp need to be at the highest point?

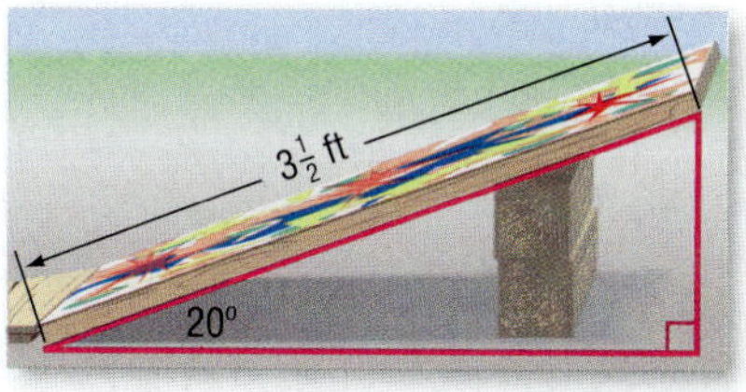

Example 4 p. 565

Use a calculator to find the measure of ∠Z to the nearest tenth.

12.

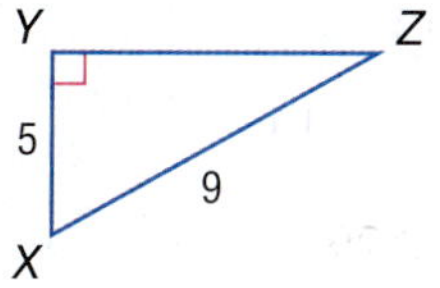

13.

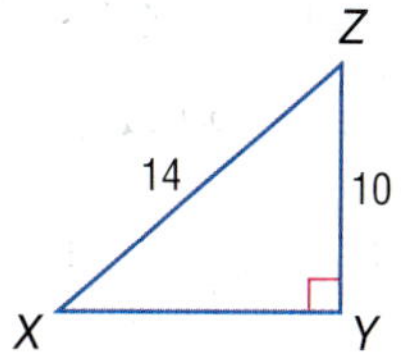

14.

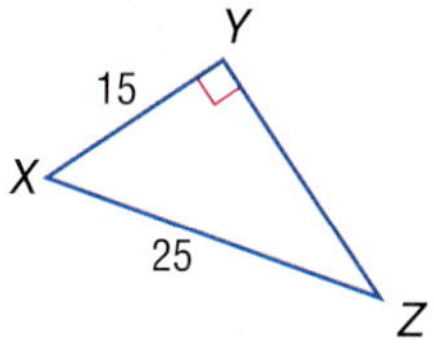

Example 5 p. 566

15. Solve the right triangle. Round side measures to the nearest tenth and angle measures to the nearest degree.

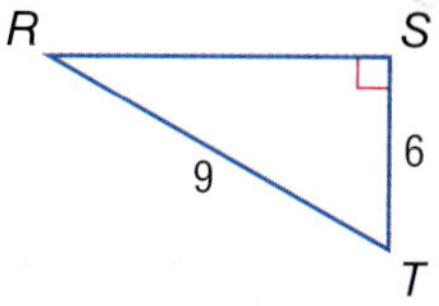

Practice and Problem Solving

● = **Step-by-Step Solutions** begin on page R20.
Extra Practice begins on page 969.

Example 1 p. 563

Find sin J, cos J, tan J, sin L, cos L, and tan L. Express each ratio as a fraction and as a decimal to the nearest hundredth.

16.

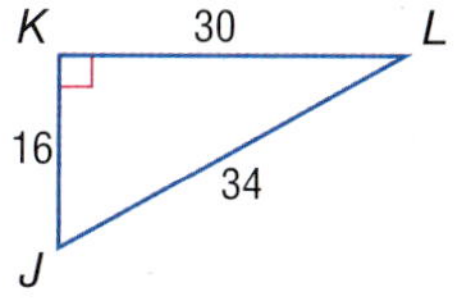

17

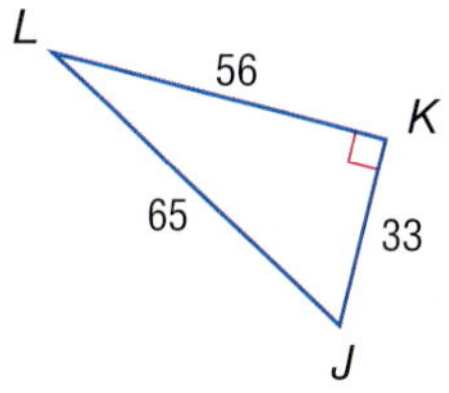

18.

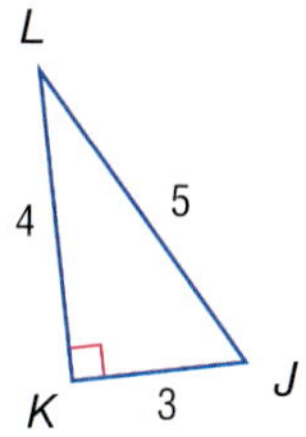

19.

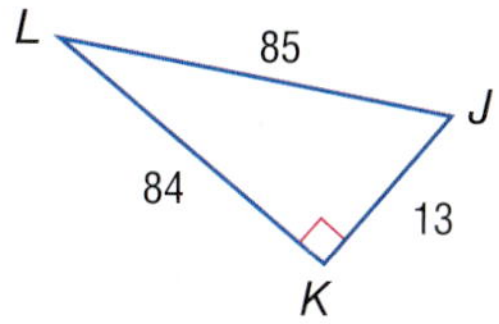

20.

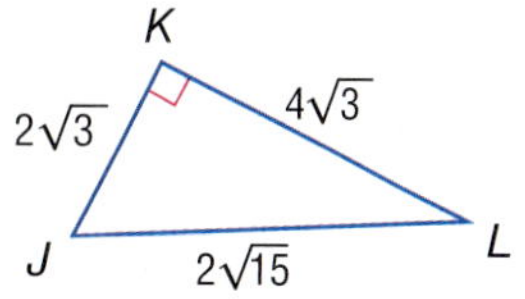

21.

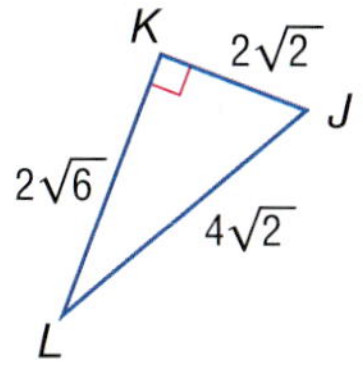

Example 2
p. 563

Use a special right triangle to express each trigonometric ratio as a fraction and as a decimal to the nearest hundredth.

22. $\tan 60°$ **23.** $\cos 30°$ **24.** $\sin 45°$

25. $\sin 30°$ **26.** $\tan 45°$ **27.** $\cos 60°$

Example 3
p. 564

Find x. Round to the nearest tenth.

28.

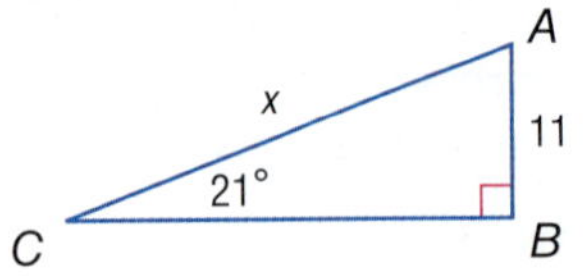

29.

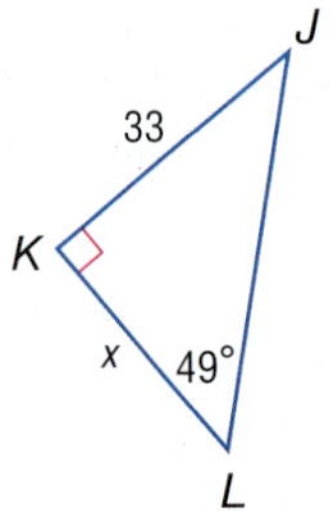

30.

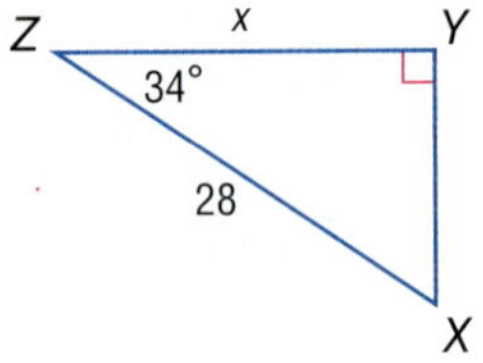

31.

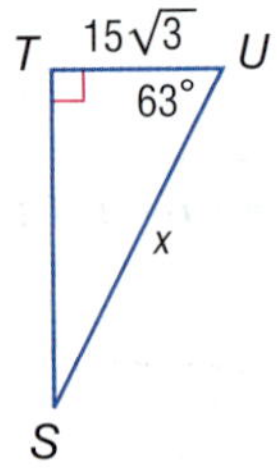

32.

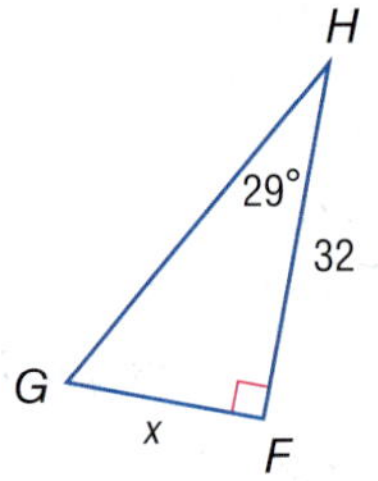

33.

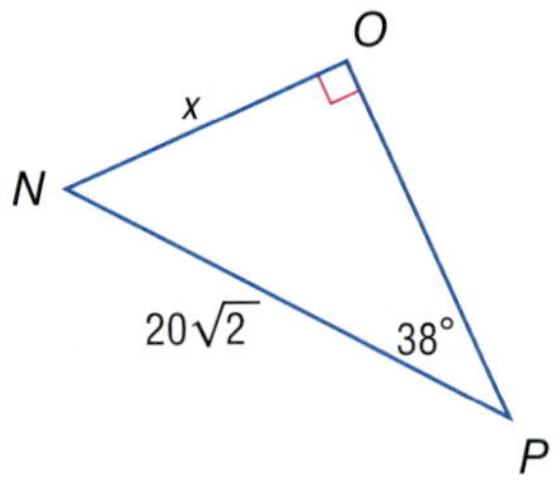

34. **GYMNASTICS** The springboard that Eric uses in his gymnastics class has 6-inch coils and forms an angle of 14.5° with the base. About how long is the springboard?

35 **ROLLER COASTERS** The angle of ascent of the first hill of a roller coaster is 55°. If the length of the track from the beginning of the ascent to the highest point is 98 feet, what is the height of the roller coaster when it reaches the top of the first hill?

Example 4
p. 565

Use a calculator to find the measure of $\angle T$ to the nearest tenth.

36.

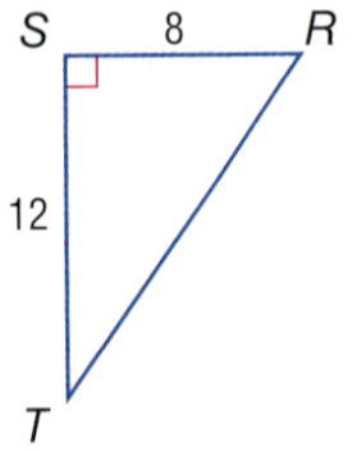

37.

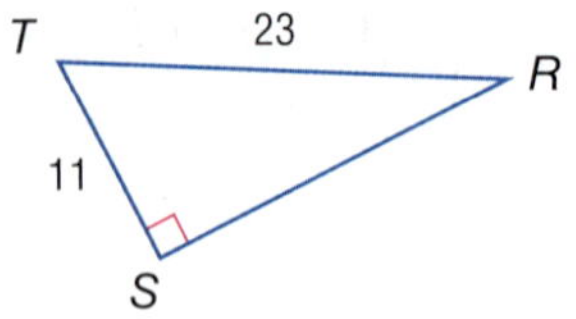

38.

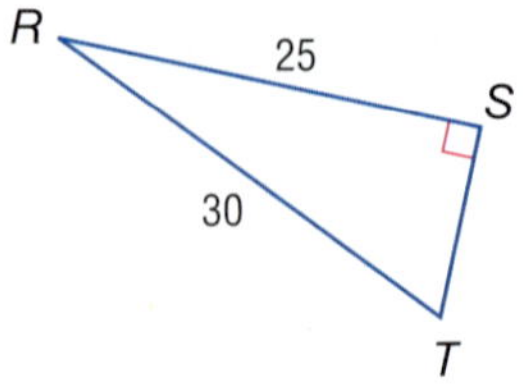

39.

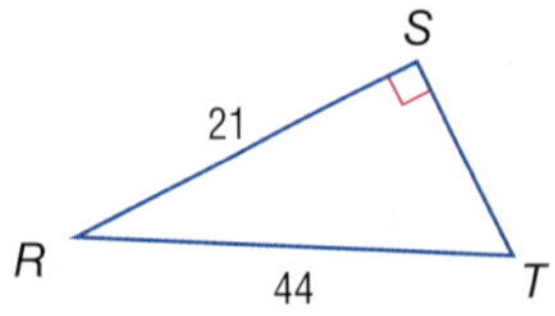

40.

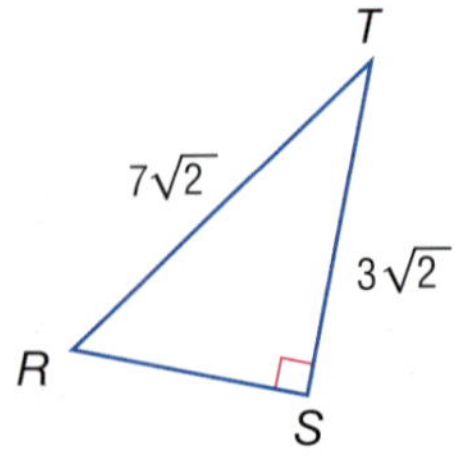

41.

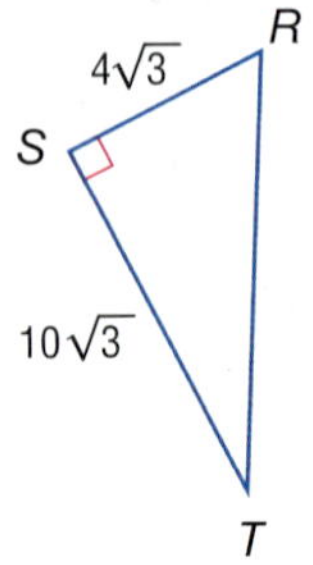

Example 5
p. 566

Real-World Link

Many orthopaedic surgeons suggest that students limit backpack weight to 20% of their body weight to prevent injuries. Newer wheeled backpacks may further prevent injuries.

Source: International Chiropractic Pediatric Association

Solve each right triangle. Round side measures to the nearest tenth and angle measures to the nearest degree.

42. 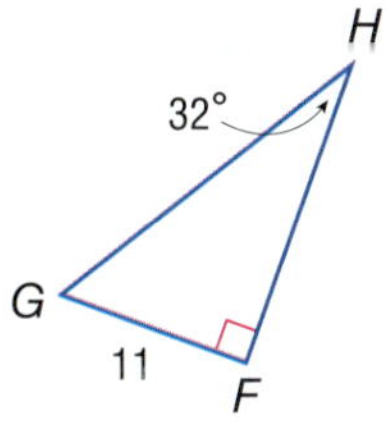

43. 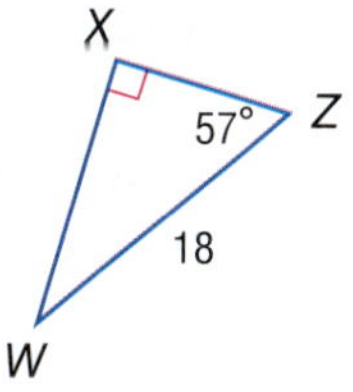

44. 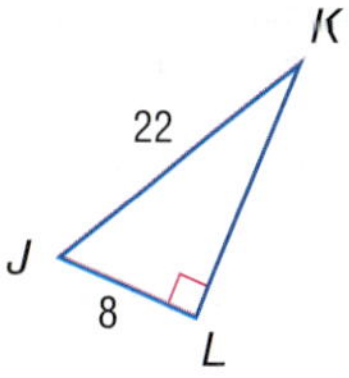

45. 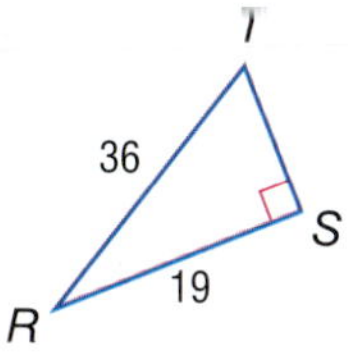

46. BACKPACKS Ramón has a rolling backpack that is $3\frac{3}{4}$ feet tall when the handle is extended. When he is pulling the backpack, Ramon's hand is 3 feet from the ground. What angle does his backpack make with the floor? Round to the nearest degree.

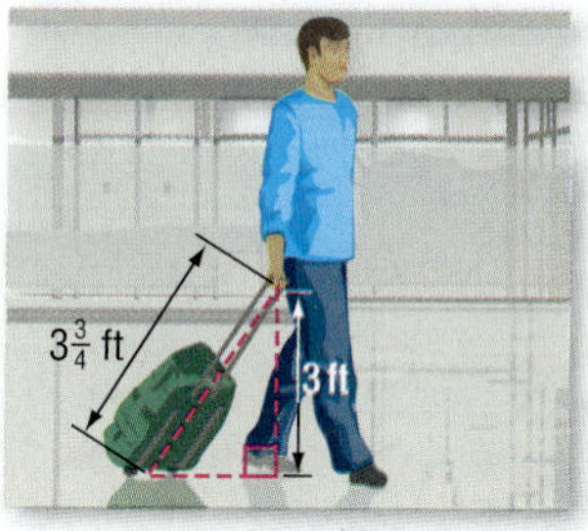

COORDINATE GEOMETRY **Find the measure of each angle to the nearest tenth of a degree using the Distance Formula and an inverse trigonometric ratio.**

47 $\angle K$ in right triangle JKL with vertices $J(-2, -3)$, $K(-7, -3)$, and $L(-2, 4)$

48. $\angle Y$ in right triangle XYZ with vertices $X(4, 1)$, $Y(-6, 3)$, and $Z(-2, 7)$

49. $\angle A$ in in right triangle ABC with vertices $A(3, 1)$, $B(3, -3)$, and $C(8, -3)$

50. SCHOOL SPIRIT Hana is making a pennant for each of the 18 girls on her basketball team. What is the total length of orange felt that Hana needs to buy in order to have enough for the borders of all of the pennants?

Find the perimeter and area of each triangle. Round to the nearest hundredth.

51. 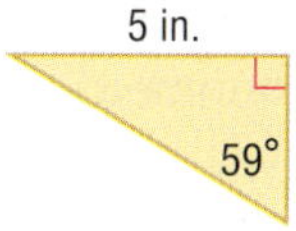

52. 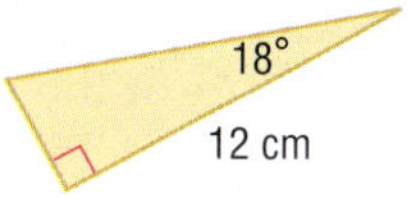

53. 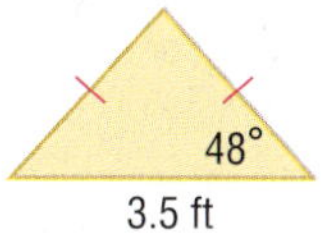

54. Find the tangent of the greater acute angle in a triangle with side lengths of 3, 4, and 5 centimeters.

55. Find the cosine of the smaller acute angle in a triangle with side lengths of 10, 24, and 26 inches.

56. ESTIMATION Ethan and Tariq want to estimate the area of the field that their team will use for soccer practice. They know that the field is rectangular, and they have paced off the width of the field as shown. They used the fence posts at the corners of the field to estimate that the angle between the length of the field and the diagonal is about 40°. If they assume that each of their steps is about 18 inches, what is the area of the practice field in square feet? Round to the nearest square foot.

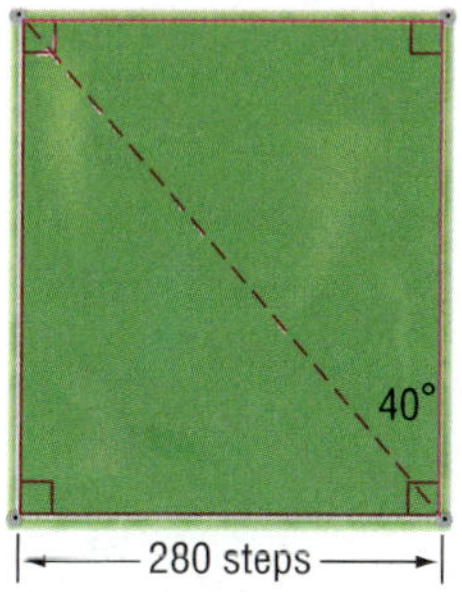

Find x and y. Round to the nearest tenth.

57.

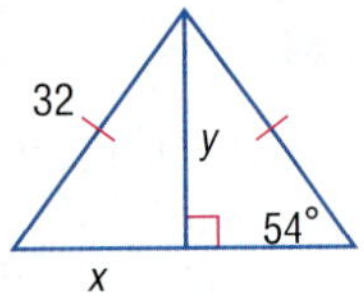

58.

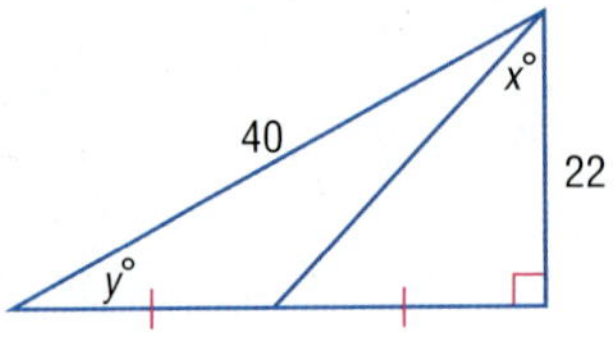

59.

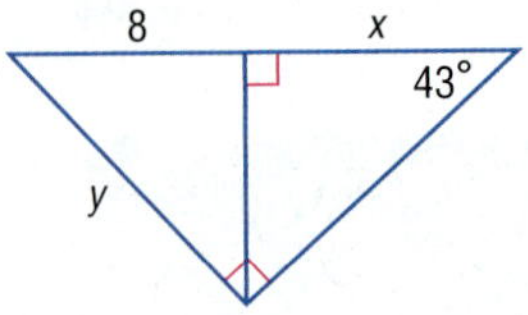

60. **COORDINATE GEOMETRY** Show that the slope of a line at 225° from the x-axis is equal to the tangent of 225°.

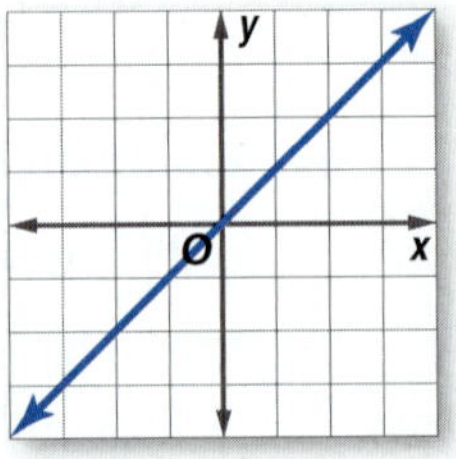

61. **MULTIPLE REPRESENTATIONS** In this problem, you will investigate an algebraic relationship between the sine and cosine ratios.

a. GEOMETRIC Draw three right triangles that are not similar to each other. Label the triangles ABC, MNP, and XYZ, with the right angles located at vertices B, N, and Y, respectively. Measure and label each side of the three triangles.

b. TABULAR Copy and complete the table below.

Triangle	Trigonometric Ratios				Sum of Ratios Squared	
ABC	$\cos A$		$\sin A$		$(\cos A)^2 + (\sin A)^2 =$	
	$\cos C$		$\sin C$		$(\cos C)^2 + (\sin C)^2 =$	
MNP	$\cos M$		$\sin M$		$(\cos M)^2 + (\sin M)^2 =$	
	$\cos P$		$\sin P$		$(\cos P)^2 + (\sin P)^2 =$	
XYZ	$\cos X$		$\sin X$		$(\cos X)^2 + (\sin X)^2 =$	
	$\cos Z$		$\sin Z$		$(\cos Z)^2 + (\sin Z)^2 =$	

c. VERBAL Make a conjecture about the sum of the squares of the cosine and sine of an acute angle of a right triangle.

d. ALGEBRAIC Express your conjecture algebraically for an angle X.

e. ANALYTICAL Show that your conjecture is valid for angle A in the figure at the right using the trigonometric functions and the Pythagorean Theorem.

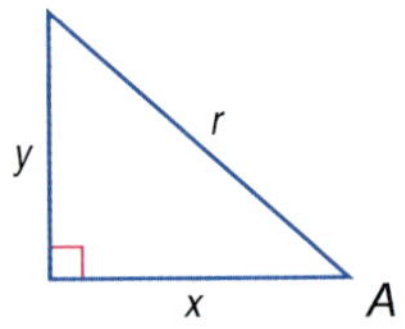

StudyTip

Trigonometric Ratios The tangent ratio $\tan A = \frac{\text{opp}}{\text{adj}}$ is equivalent to $\tan A = \frac{\text{opp}}{\text{hyp}} \cdot \frac{\text{hyp}}{\text{adj}}$ or $\tan A = \frac{\sin A}{\cos A}$.

H.O.T. Problems Use Higher-Order Thinking Skills

62. **CHALLENGE** Solve $\triangle ABC$. Round to the nearest whole number.

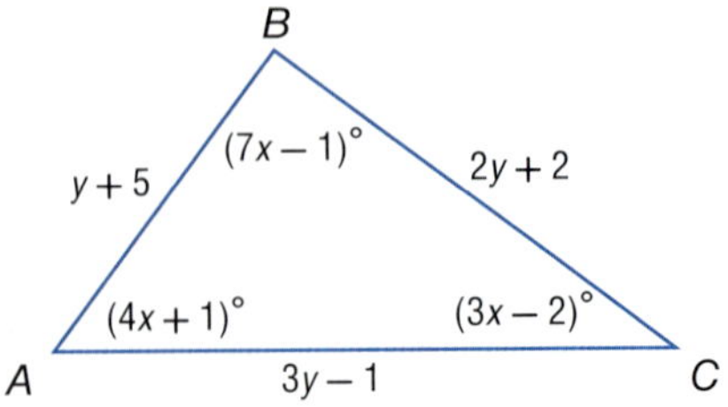

63. **REASONING** After solving several right triangle trigonometry problems, Tristan noticed that the values of sine and cosine for the acute angles of a right triangle were less than 1. Will the values of sine and cosine for an acute angle always be less than 1? Explain.

64. **OPEN ENDED** Draw right $\triangle ABC$ with the right angle at vertex B. Use a ruler to measure each side, and write the ratios for $\sin A$ and $\cos C$. How are the two ratios related?

65. **WRITING IN MATH** Explain how you can use ratios of the side lengths to find the angle measures of the acute angles in a right triangle.

Standardized Test Practice

66. What is the value of tan x?

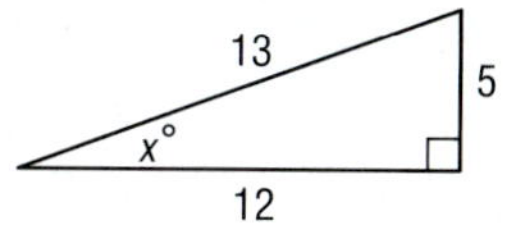

A $\tan x = \frac{13}{5}$

B $\tan x = \frac{12}{5}$

C $\tan x = \frac{5}{13}$

D $\tan x = \frac{5}{12}$

67. ALGEBRA Which of the following has the same value as $2^{-12} \times 2^{3}$?

F 2^{-36}

G 4^{-9}

H 2^{-9}

J 2^{-4}

68. GRIDDED RESPONSE If $AC = 12$ and $AB = 25$, what is the measure of $\angle B$ to the nearest tenth?

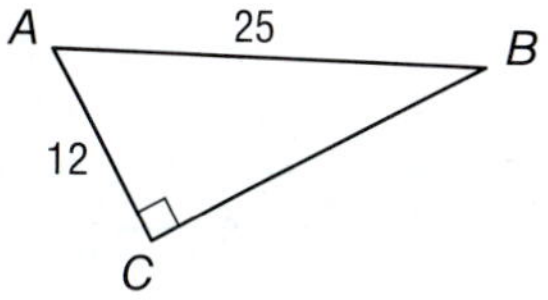

69. SAT/ACT The area of a right triangle is 240 square inches. If the base is 30 inches long, how many inches long is the hypotenuse?

A 8

B 16

C $2\sqrt{241}$

D 34

Spiral Review

Find x and y. (Lesson 8-3)

70.

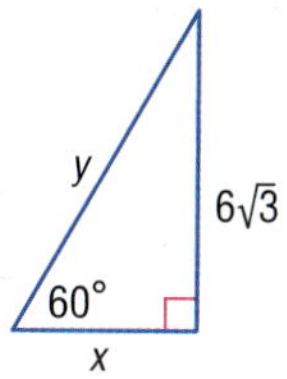

71.

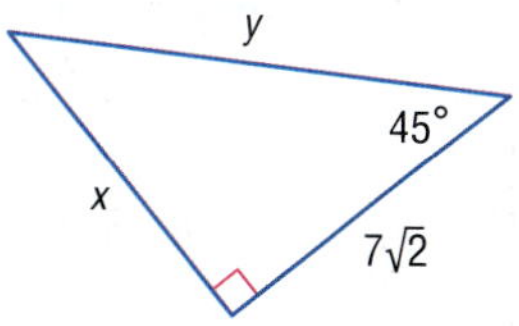

72.

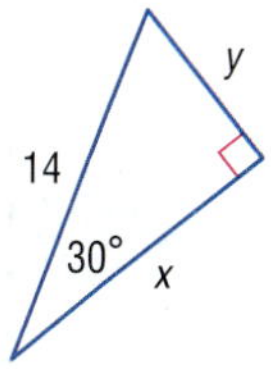

Determine whether each set of numbers can be the measures of the sides of a triangle. If so, classify the triangle as *acute*, *obtuse*, or *right*. Justify your answer. (Lesson 8-2)

73. 8, 15, 17

74. 11, 12, 24

75. 13, 30, 35

76. 18, 24, 30

77. 3.2, 5.3, 8.6

78. $6\sqrt{3}$, 14, 17

79. MAPS The scale on the map of New Mexico is 2 centimeters = 160 miles. The width of New Mexico through Albuquerque on the map is 4.1 centimeters. How long would it take to drive across New Mexico if you drove at an average of 60 miles per hour? (Lesson 7-7)

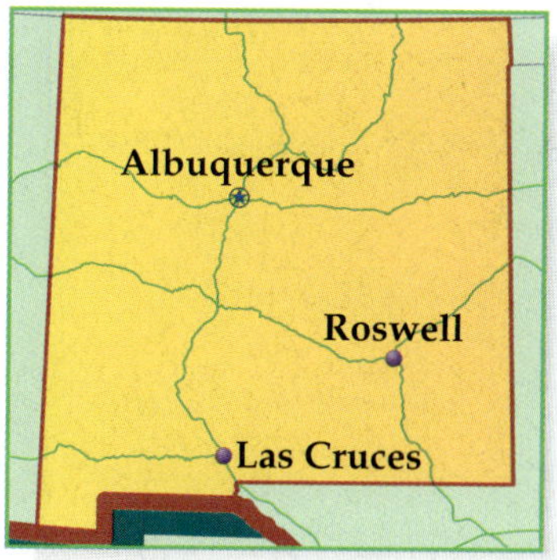

ALGEBRA Find x and y. (Lesson 7-4)

80.

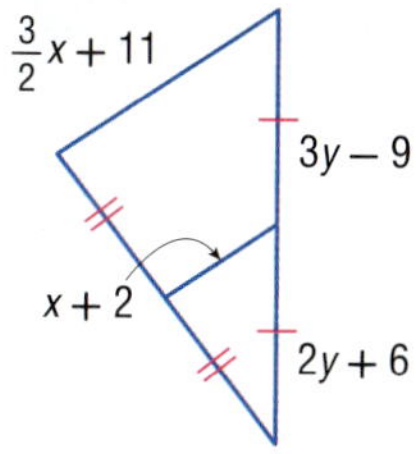

81.

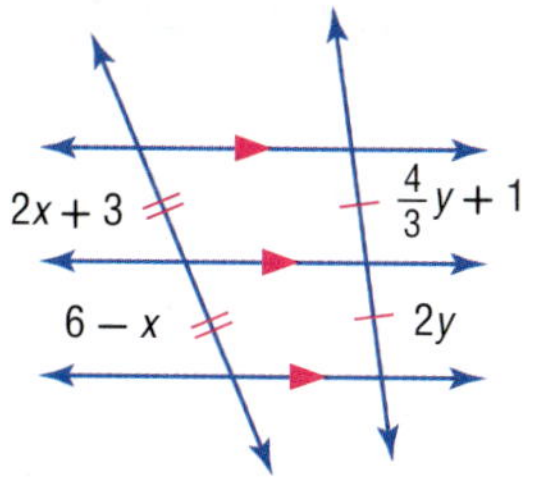

Skills Review

Solve each proportion. Round to the nearest tenth if necessary. (Lesson 0-5)

82. $2.14 = \frac{x}{12}$

83. $0.05x = 13$

84. $0.37 = \frac{32}{x}$

85. $0.74 = \frac{14}{x}$

86. $1.66 = \frac{x}{23}$

87. $0.21 = \frac{33}{x}$

EXTEND

8-4

Graphing Technology Lab
Secant, Cosecant, and Cotangent

Objective
Investigate the other trigonometric ratios using technology.

In the previous lesson, you used the trigonometric functions sine, cosine, and tangent to find angle relationships in right angles. In this activity, you will use the reciprocals of those functions, cosecant, secant, and cotangent, to explore angle and side relationships in right triangles.

Key Concept — Reciprocal Trigonometric Ratios

Words	Symbols
The **cosecant** of $\angle A$ (written csc A) is the reciprocal of sin A.	$\csc A = \frac{1}{\sin A}$ or $\frac{c}{a}$
The **secant** of $\angle A$ (written sec A) is the reciprocal of cos A.	$\sec A = \frac{1}{\cos A}$ or $\frac{c}{b}$
The **cotangent** of $\angle A$ (written cot A) is the reciprocal of tan A.	$\cot A = \frac{1}{\tan A}$ or $\frac{b}{a}$

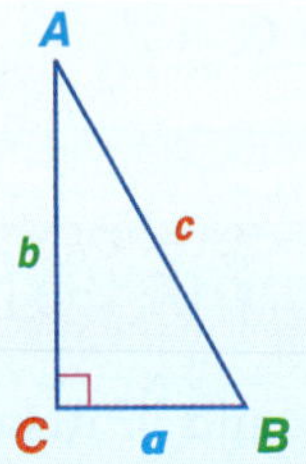

ACTIVITY Find Trigonometric Values

Step 1 Draw and label a right triangle with the dimensions shown at the right.

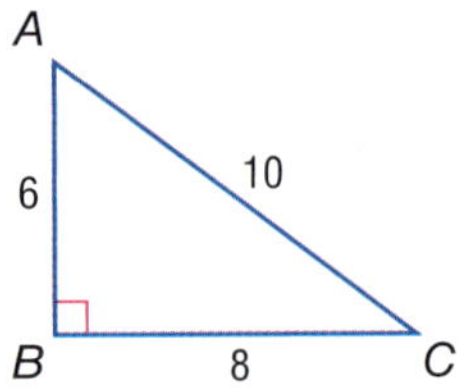

Step 2 Use your graphing calculator to find the values for sin A, cos A, and tan A.

Step 3 Next, find the value for csc A by dividing 1 by [SIN] A. Repeat step 3 to find sec A and cot A.

Step 4 Copy the table below and record your results. Next, find the value of each trigonometric function for angle C.

Angle	sin	cos	tan	csc	sec	cot
A						
C						

Exercises

1. Find the values of the six trigonometric functions for a 45° angle in a 45°-45°-90° triangle with legs that are 4 cm.

2. In $\triangle FGH$, $\tan F = \frac{5}{12}$. Find cot F and sin F if $\angle G$ is a right angle.

3. Find the values of the six trigonometric functions for angle T in $\triangle RST$ if $m\angle R = 36$ and $r = 12$. Round to the nearest hundredth.

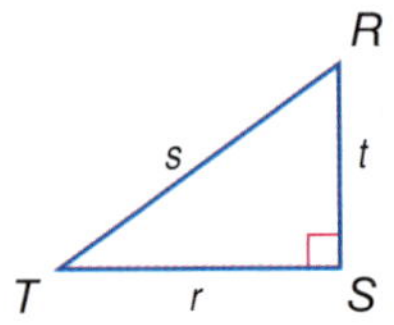

CHAPTER 8 Mid-Chapter Quiz

Lessons 8-1 through 8-4

Find the geometric mean between each pair of numbers. (Lesson 8-1)

1. 12 and 3
2. 63 and 7
3. 45 and 20
4. 50 and 10

Write a similarity statement identifying the three similar triangles in each figure. (Lesson 8-1)

5.

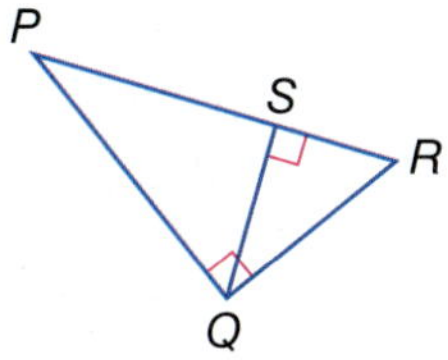

6.

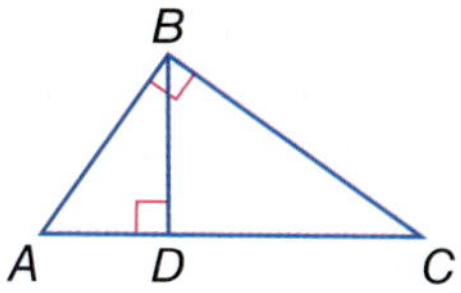

7. Find x, y, and z. (Lesson 8-1)

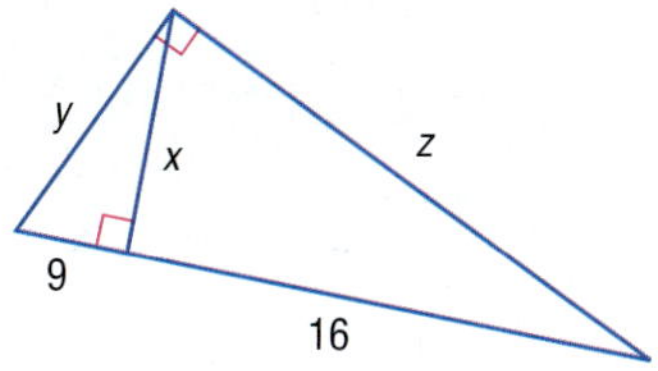

8. **PARKS** There is a small park in a corner made by two perpendicular streets. The park is 100 ft by 150 ft, with a diagonal path, as shown below. What is the length of path $\overline{AC}$? (Lesson 8-2)

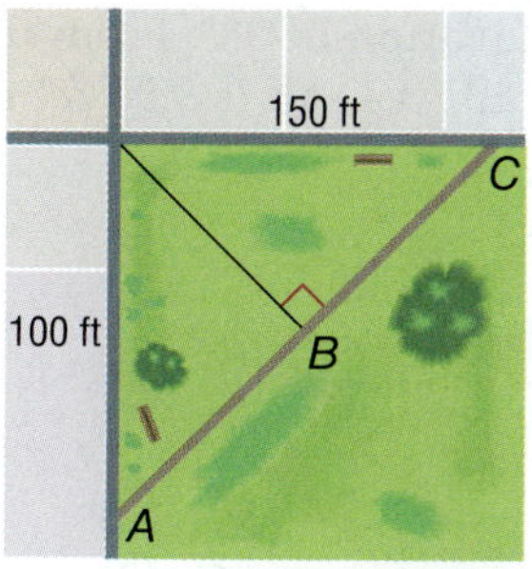

Find x. Round to the nearest hundredth. (Lesson 8-2)

9.

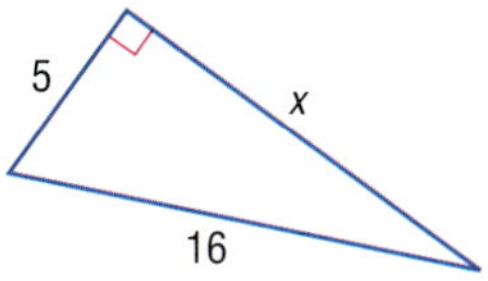

10.

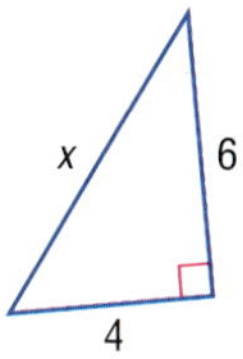

11. **MULTIPLE CHOICE** Which of the following sets of numbers is not a Pythagorean triple? (Lesson 8-2)

A 9, 12, 15
B 21, 72, 75
C 15, 36, 39
D 8, 13, 15

Find x. (Lesson 8-3)

12.

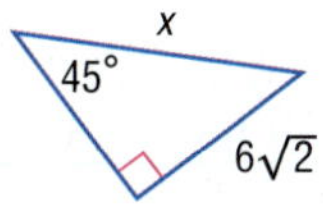

13.

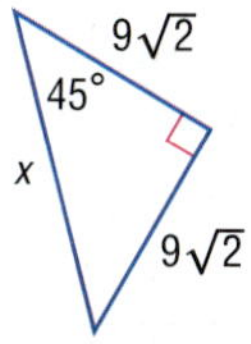

14. **DESIGN** Jamie designed a pinwheel to put in her garden. In the pinwheel, the blue triangles are congruent equilateral triangles, each with an altitude of 4 inches. The red triangles are congruent isosceles right triangles. The hypotenuse of a red triangle is congruent to a side of the blue triangle. (Lesson 8-3)

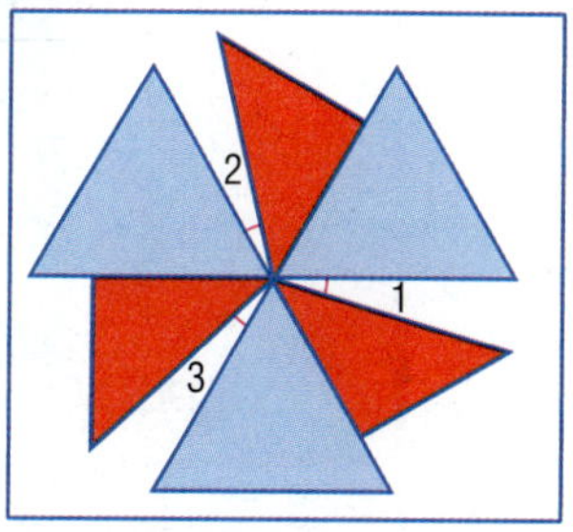

a. If angles 1, 2, and 3 are congruent, find the measure of each angle.

b. Find the perimeter of the pinwheel.

Find x and y. (Lesson 8-3)

15.

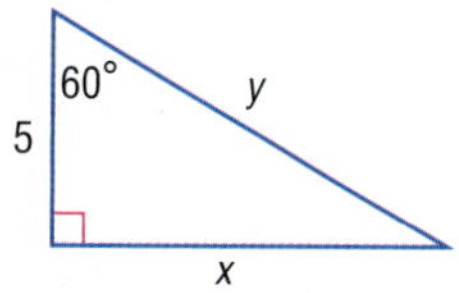

16.

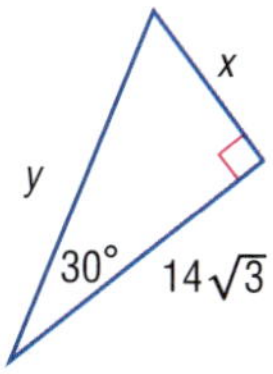

Express each ratio as a fraction and as a decimal to the nearest hundredth. (Lesson 8-4)

17. tan M
18. cos M
19. cos N
20. sin N

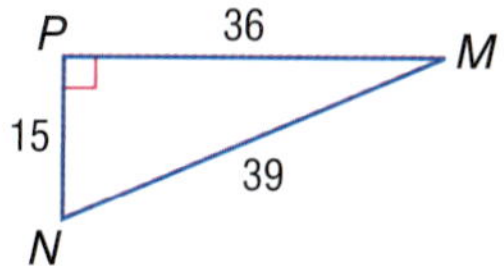

21. Solve the right triangle. Round angle measures to the nearest degree and side measures to the nearest tenth. (Lesson 8-4)

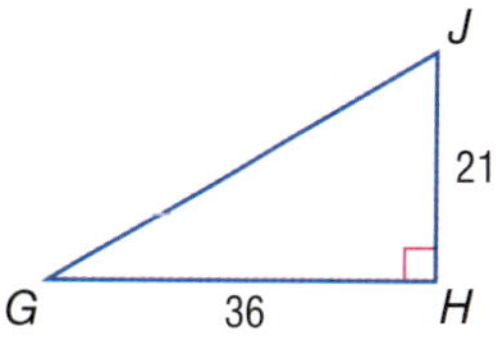

8-5 Angles of Elevation and Depression

Then

You used similar triangles to measure distances indirectly. (Lesson 7-3)

Now

- Solve problems involving angles of elevation and depression.
- Use angles of elevation and depression to find the distance between two objects.

New Vocabulary

angle of elevation
angle of depression

Math Online

glencoe.com

Why?

To make a field goal, a kicker must kick the ball with enough force and at an appropriate angle of elevation to ensure that the ball will reach the goal post at a level high enough to make it over the horizontal bar. This angle must change depending on the initial placement of the ball away from the base of the goalpost.

Angles of Elevation and Depression An **angle of elevation** is the angle formed by a horizontal line and an observer's line of sight to an object above the horizontal line. An **angle of depression** is the angle formed by a horizontal line and an observer's line of sight to an object below the horizontal line.

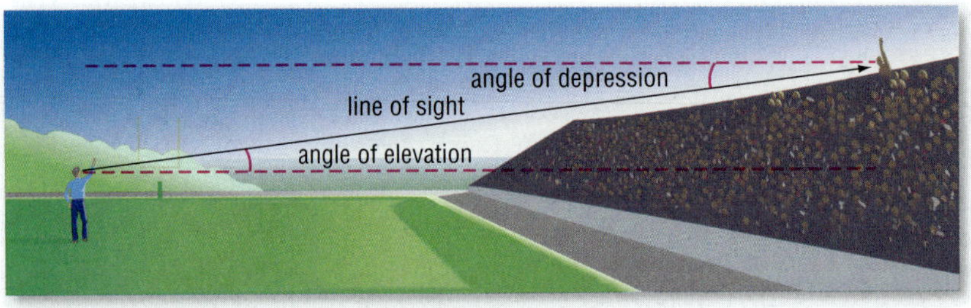

Horizontal lines are parallel, so the angle of elevation and the angle of depression in the diagram are congruent by the Alternate Interior Angles Theorem.

EXAMPLE 1 Angle of Elevation

VACATION Leah is meeting friends at the castle in the center of an amusement park. She sights the top of the castle at an angle of elevation of 38°. From the park's brochure, she knows that the castle is 190 feet tall. If Leah is 5.5 feet tall, about how far is she from the castle to the nearest foot?

Make a sketch.

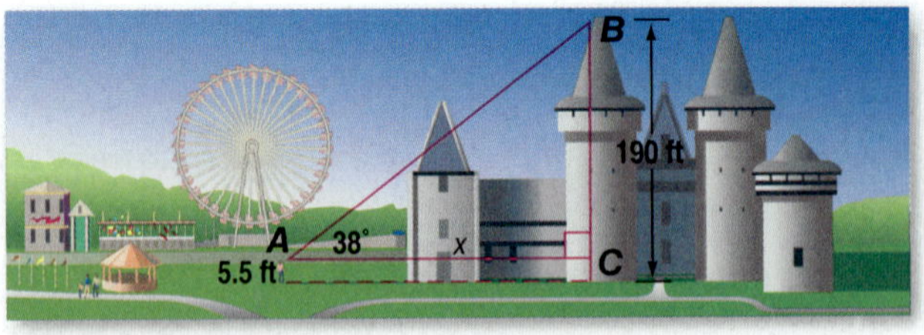

Since Leah is 5.5 feet tall, $BC = 190 - 5.5$ or 184.5 feet. Let x represent the distance from Leah to the castle, AC.

$\tan A = \frac{BC}{AC}$ — **$\tan = \frac{\text{opposite}}{\text{adjacent}}$**

$\tan 38° = \frac{184.5}{x}$ — **$m\angle A = 38$, $BC = 184.5$, $AC = x$**

$x = \frac{184.5}{\tan 38°}$ — **Solve for x.**

$x \approx 236.1$ — **Use a calculator.**

Leah is about 236 feet from the castle.

Guided Practice

1. **FOOTBALL** The cross bar of a goalpost is 10 feet high. If a field goal attempt is made 25 yards from the base of the goalpost that clears the goal by 1 foot, what is the smallest angle of elevation at which the ball could have been kicked to the nearest degree?

Personal Tutor glencoe.com

EXAMPLE 2 Angle of Depression

EMERGENCY A search and rescue team is airlifting people from the scene of a boating accident when they observe another person in need of help. If the angle of depression to this other person is 42° and the helicopter is 18 feet above the water, what is the horizontal distance from the rescuers to this person to the nearest foot?

Make a sketch of the situation.

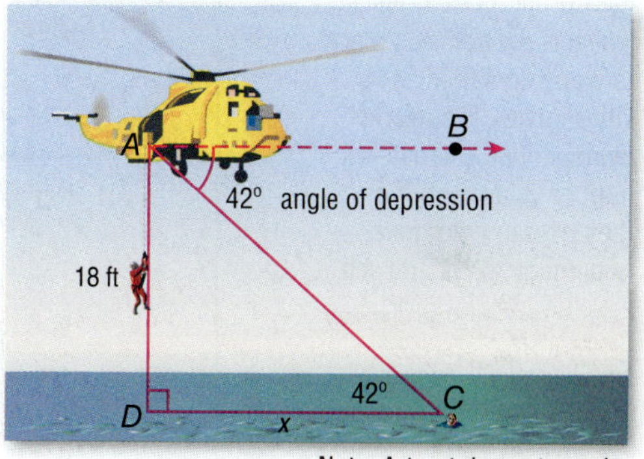

Note: Art not drawn to scale.

Since $\overrightarrow{AB}$ and $\overline{DC}$ are parallel, $m\angle BAC = m\angle ACD$ by the Alternate Interior Angles Theorem.

Let x represent the horizontal distance from the rescuers to the person DC.

$\tan C = \frac{AD}{DC}$	$\tan = \frac{\text{opposite}}{\text{adjacent}}$
$\tan 42° = \frac{18}{x}$	$C = 42$, $AD = 18$, and $DC = x$
$x \tan 42° = 18$	Multiply each side by x.
$x = \frac{18}{\tan 42°}$	Divide each side by tan 42°.
$x \approx 20.0$	Use a calculator.

The horizontal distance from the rescuers to the person is 20.0 feet.

Watch Out!

Angles of Elevation and Depression To avoid mislabeling, remember that angles of elevation and depression are always formed with a horizontal line and never with a vertical line.

Guided Practice

2. **LIFEGUARDING** A lifeguard is watching a beach from a line of sight 6 feet above the ground. She sees a swimmer at an angle of depression of 8°. How far away from the tower is the swimmer?

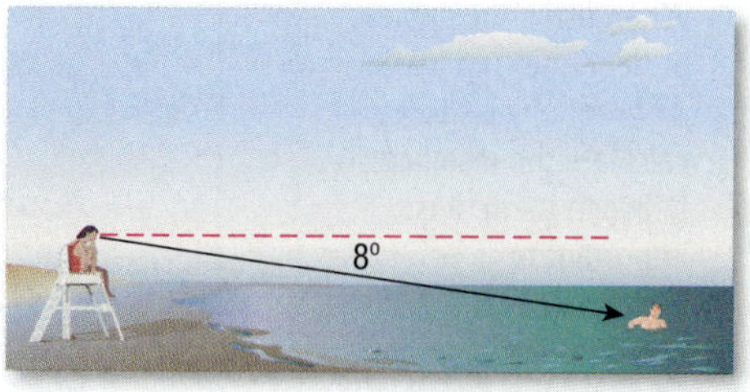

Personal Tutor glencoe.com

Two Angles of Elevation or Depression Angles of elevation or depression to two different objects can be used to estimate the distance between those objects. Similarly, the angles from two different positions of observation to the same object can be used to estimate the object's height.

Real-World Link

In the United States, lumber volume is measured in board-feet, which is defined as a piece of wood containing 144 cubic inches. Woodland owners often estimate the lumber volume of trees they own to determine how many to cut and sell.

Source: The Ohio State University School of Natural Resources

EXAMPLE 3 Use Two Angles of Elevation or Depression

TREE REMOVAL **To estimate the height of a tree she wants removed, Mrs. Long sights the tree's top at a 70° angle of elevation. She then steps back 10 meters and sights the top at a 26° angle. If Mrs. Long's line of sight is 1.7 meters above the ground, how tall is the tree to the nearest meter?**

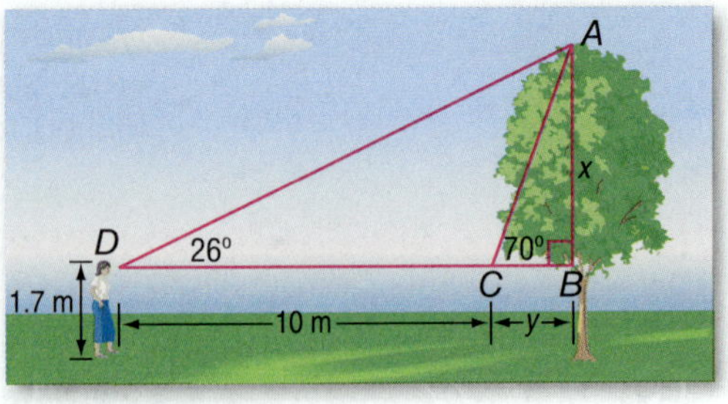

Understand $\triangle ABC$ and $\triangle ABD$ are right triangles. The height of the tree is the sum of Mrs. Long's height and AB.

Plan Since her initial distance from the tree is not given, write and solve a system of equations using both triangles. Let $AB = x$ and $CB = y$. So $DB = y + 10$ and the height of the tree is $x + 1.7$.

Solve Use $\triangle ABC$.

$\tan 70° = \frac{x}{y}$ — $\tan = \frac{\text{opposite}}{\text{adjacent}}$; $m\angle ACB = 70$

$y \tan 70° = x$ — Multiply each side by y.

Use $\triangle ABD$.

$\tan 26° = \frac{x}{y + 10}$ — $\tan = \frac{\text{opposite}}{\text{adjacent}}$; $m\angle D = 26$

$(y + 10) \tan 26° = x$ — Multiply each side by $y + 10$.

Substitute the value for x from $\triangle ABD$ in the equation for $\triangle ABC$ and solve for y.

$$y \tan 70° = x$$
$$y \tan 70° = (y + 10) \tan 26°$$
$$y \tan 70° = y \tan 26° + 10 \tan 26°$$
$$y \tan 70° - y \tan 26° = 10 \tan 26°$$
$$y(\tan 70° - \tan 26°) = 10 \tan 26°$$
$$y = \frac{10 \tan 26°}{\tan 70° - \tan 26°}$$

Use a calculator to find that $y \approx 2.16$. Using the equation from $\triangle ABC$, $x = 2.16 \tan 70°$ or about 5.9.

The height of the tree is 5.9 + 1.7 or 7.6, which is about 8 meters.

Check Substitute the value for y in the equation from $\triangle ABD$.

$x = (2.16 + 10) \tan 26°$ or about 5.9. This is the same value found using the equation from $\triangle ABC$. ✓

Guided Practice

3. **SKYSCRAPERS** Two buildings are sited from atop a 200-meter skyscraper. Building A is sited at a 35° angle of depression, while Building B is sighted at a 36° angle of depression. How far apart are the two buildings to the nearest meter?

Personal Tutor glencoe.com

StudyTip

Indirect Measurement When using the angles of depression to two different objects to calculate the distance between them, it is important to remember that the two objects must lie in the same horizontal plane. In other words, one object cannot be higher or lower than the other.

Check Your Understanding

Example 1 p. 574

1. **BIKING** Lenora wants to build the bike ramp shown. Find the length of the base of the ramp.

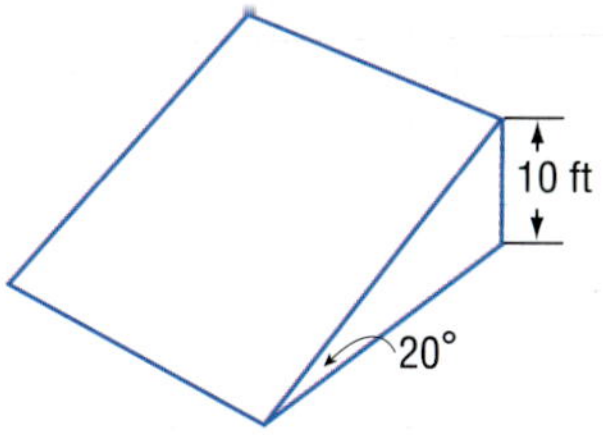

Example 2 p. 575

2. **BASEBALL** A fan is seated in the upper deck of a stadium 200 feet away from home plate. If the angle of depression to the field is 62°, at what height is the fan sitting?

Example 3 p. 576

3. **DANCES** Annabelle and Rich are setting up decorations for their school dance. Rich is standing 5 feet directly in front of Annabelle under a disco ball. If the angle of elevation from Annabelle to the ball is 40° and Rich to the ball is 50°, how high is the disco ball?

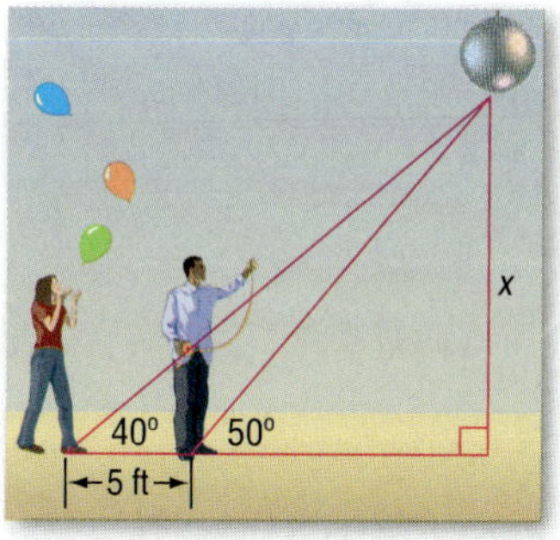

Practice and Problem Solving

= **Step-by-Step Solutions** begin on page R20.
Extra Practice begins on page 969.

Example 1 p. 574

4. **HOCKEY** A hockey player takes a shot at a distance of 20 feet away from a 5-foot goal. If the puck travels at a 15° angle of elevation toward the center of the goal, will the player score?

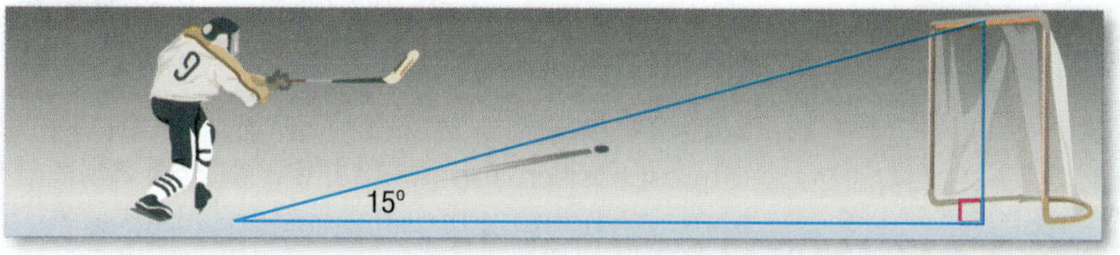

5. **MOUNTAINS** Find the angle of elevation to the peak of a mountain for an observer who is 155 meters from the mountain if the observer's eye is 1.5 meters above the ground and the mountain is 350 meters tall.

Example 2 p. 575

6. **PLANTS** Alicia hangs a plant in her bedroom window. If the sunlight comes in with a 46° angle of depression, and the length of the basket chain is 9 inches, how far is the plant from the window?

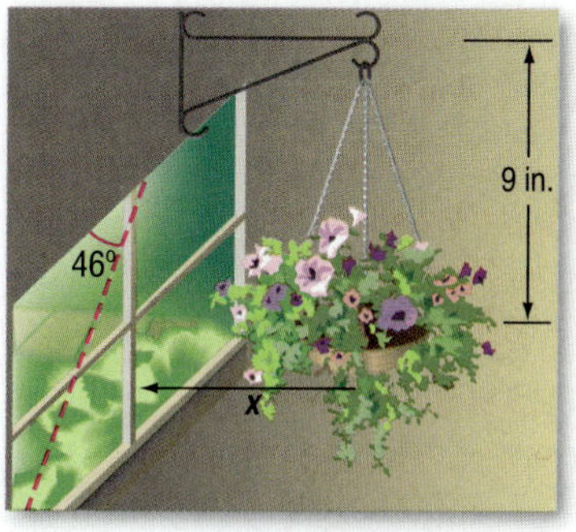

7. **AVIATION** Due to a storm, a pilot flying at an altitude of 528 feet has to land early. If he has a horizontal distance of 2000 feet to land, at what angle of depression should he land?

Example 3 p. 576

8. **PYRAMIDS** Miko and Tyler are visiting the Great Pyramid in Egypt. From where Miko is standing, the angle of elevation to the top of the pyramid is 48.6°. From Tyler's position, the angle of elevation is 50°. If they are standing 20 feet apart, how tall is the pyramid?

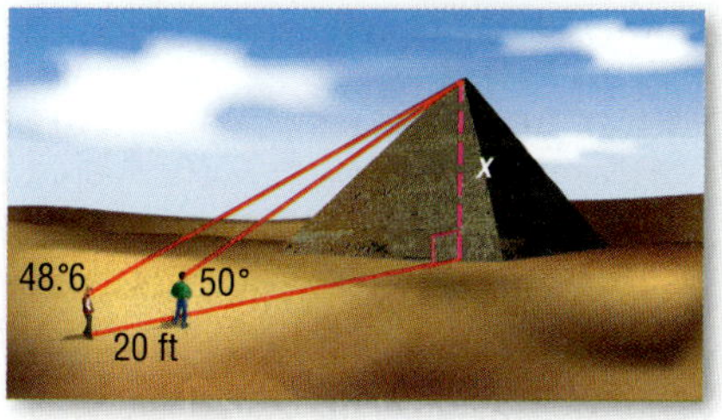

9. **DIVING** Austin is standing on the high dive at the local pool. Two of his friends are in the water on the opposite side of the pool. If the angle of depression to one of his friends is 40°, and 30° to his other friend who is 5 feet beyond the first, how tall is the platform?

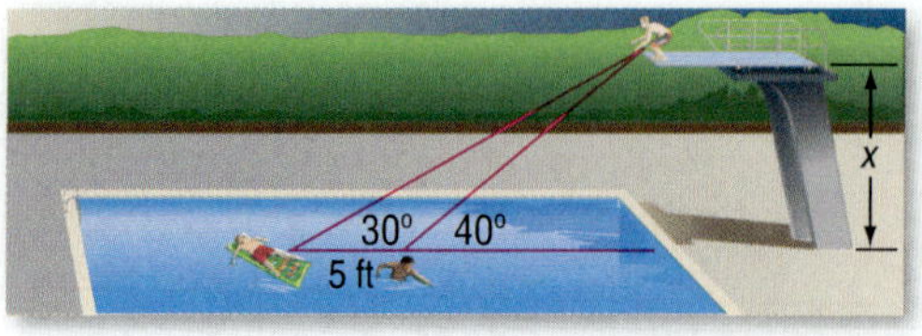

10. **BASKETBALL** Claire and Marisa are both waiting to get a rebound during a basketball game. If the height of the basketball hoop is 10 feet, the angle of elevation between Claire and the goal is 35°, and the angle of elevation between Marisa and the goal is 25°, how far apart are they standing?

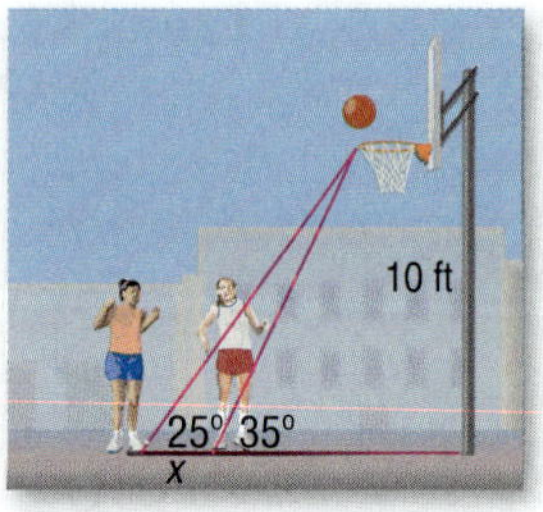

11. **RIVERS** Hugo is standing in the top of St. Louis' Gateway Arch, looking down on the Mississippi River. The angle of depression to the closer bank is 45° and the angle of depression to the farther bank is 18°. The arch is 630 feet tall. Estimate the width of the river at that point.

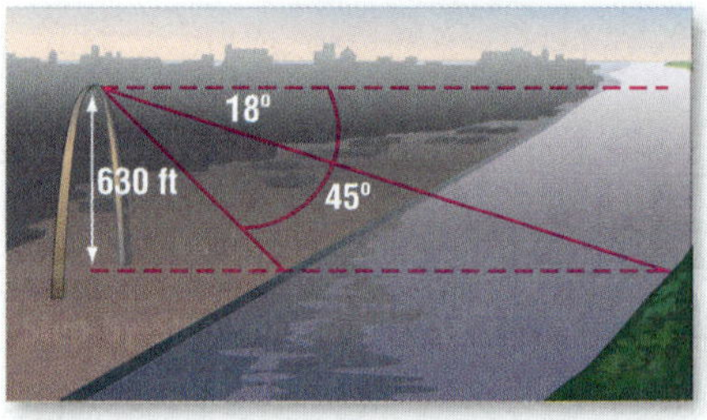

12. **VOLCANOES** The Unzen Volcano in Japan has a magma reservoir that is located 15 kilometers beneath the Chijiwa Bay. The magma rises eastward at an angle of elevation of 40°. What length of magma is below sea level?

13. **BRIDGES** Suppose you are standing in the middle of the platform of the world's longest suspension bridge, the Akashi Kaikyo Bridge. If the height from the top of the platform holding the suspension cables is 297 meters, and the length from the platform to the center of the bridge is 995 meters, what is the angle of depression from the platform to the center of the bridge?

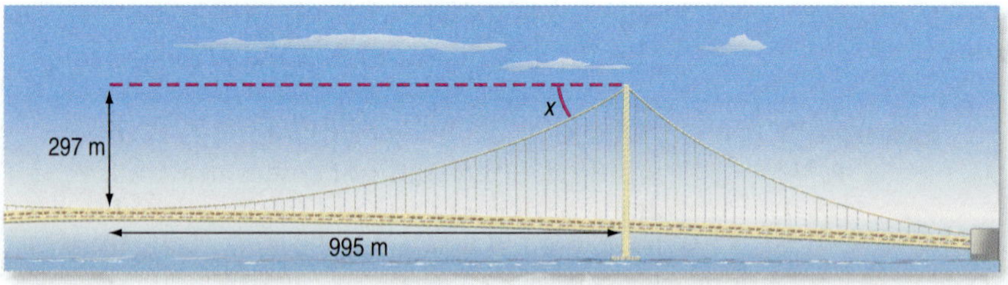

Real-World Link

The Akashi Kaikyo Bridge in Japan has the longest span of any suspension bridge in the world. The main span of the bridge is 1991 meters, 366 meters longer than the old record holder, the StoreBaelt Bridge in Denmark.

Source: U.S. Department of Transportation

14. **LIGHTHOUSES** To aid in navigation, Little Gull Island Lighthouse shines a light from a height of 91 feet with a 6° angle of depression. Plum Island Lighthouse, 1800 feet away, shines a light from a height of 34 feet with a 2° angle of depression. Which light will reach a boat that sits exactly between Little Gull Island Lighthouse and Plum Island Lighthouse?

15. **TOURISM** Hakeem is on a tour bus in Paris with his family. He sees L'arc de Triomphe at an angle of 34°. If the arc is 162 feet tall, how far away is the bus? Round to the nearest tenth.

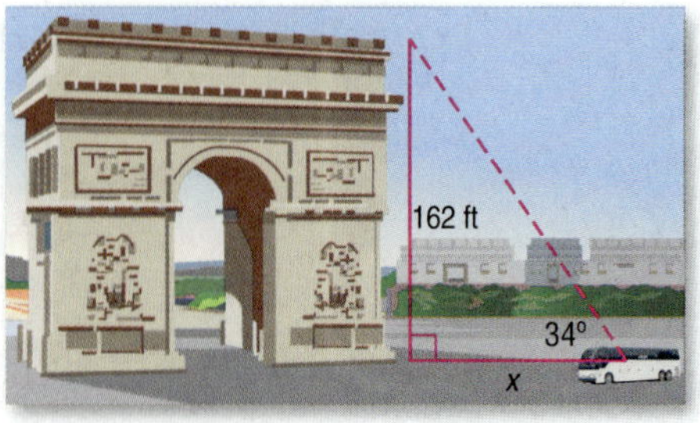

16. **MAINTENANCE** Two telephone repair workers arrive at a location to restore electricity after a power outage. One of the workers climbs up the telephone pole while the other worker stands 10 feet to left of the pole. If the terminal box is located 30 feet above ground on the pole and the angle of elevation from the truck to the repair worker is 27°, how far is the worker on the ground standing from the truck?

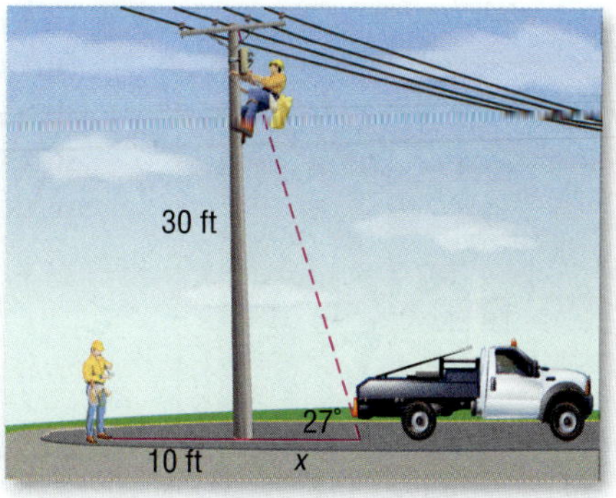

17. **PHOTOGRAPHY** A digital camera with a panoramic lens is described as having a view with an angle of elevation of 38°. If the camera is on a 3-foot tripod aimed directly at a 124-foot-tall monument, how far from the monument should you place the tripod to see the entire monument in your photograph?

Math History Link

Eratosthenes (276–194 B.C.)

Eratosthenes was a mathematician and astronomer who was born in Cyrene, which is now Libya. He used the angle of elevation of the Sun at noon in the cities of Alexandria and Syene (now Egypt) to measure the circumference of Earth.

Source: *Encyclopaedia Britannica*

18. **BALLOONING** As a part of their weather unit, Anoki's science class took a hot air balloon ride. As they passed over a fenced field, the angle of depression of the closer side of the fence was 32°, and the angle of depression of the farther side of the fence was 27°. If the height of the balloon was 800 feet, estimate the width of the field.

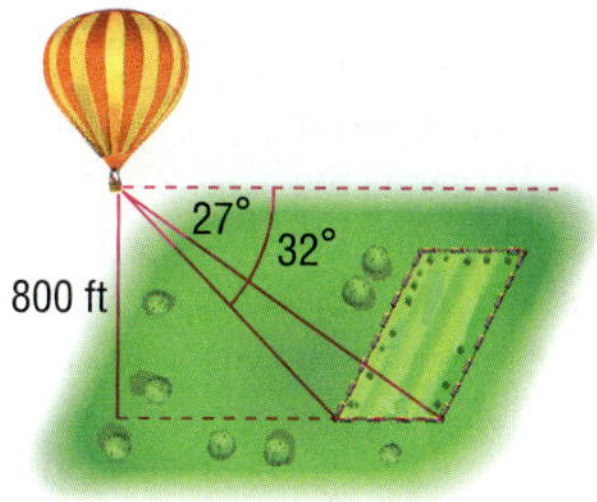

19. **MARATHONS** The Badwater Ultramarathon is a race that begins at the lowest point in North America, Death Valley, and ends at the highest point on the continent, Mount Whitney. The race starts at a depth of 86 meters from sea level and ends 4421 meters above sea level.

a. Determine the angle of elevation to Mount Whitney if the length from the base to the peak is 1200 meters.

b. If the angle of depression to Death Valley is 38°, what is the horizontal distance from sea level?

20. **AMUSEMENT PARKS** India, Enrique, and Trina went to an amusement park while visiting Japan. They went on a Ferris wheel that was 100 meters in diameter and on an 80-meter cliff-dropping slide.

a. When Enrique and Tina are at the topmost point on the Ferris wheel shown below, how far are they from India?

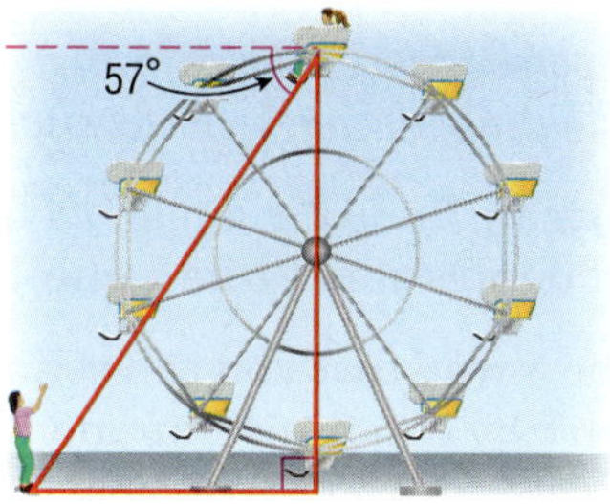

b. If the cliff-dropping ride has an angle of depression of 46°, how long is the slide?

Real-World Link

A regulation dart board is 18 inches in diameter and is divided into 20 sections. In a standard game, the board is hung so that the bull's-eye is 5 feet 8 inches from the floor.

Source: World Darts Federation

21 **DARTS** Kelsey and José are playing a very competitive game of darts. The center of the bull's-eye on the dartboard is 5.7 feet from the floor, and they throw from a distance of 8.5 feet. José throws from a height of 6 feet from the floor, and Kelsey throws 5 feet from the floor. What are the angles of elevation or depression from which each must throw to get a bull's-eye?

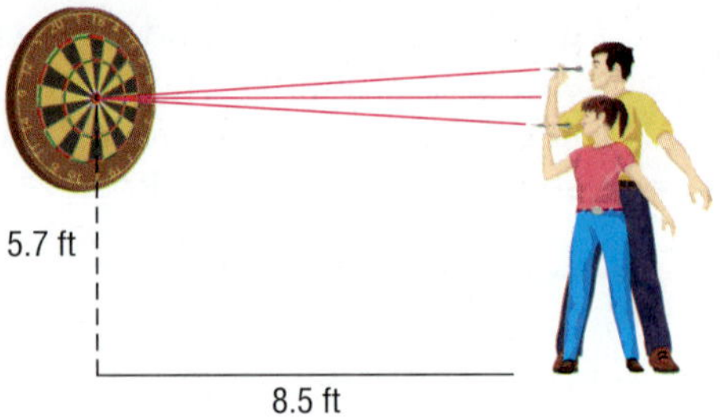

22. **MULTIPLE REPRESENTATIONS** In this problem, you will investigate relationships between the sides and angles of triangles.

a. **GEOMETRIC** Draw three triangles. Make one acute, one obtuse, and one right. Label one triangle *ABC*, a second *MNP*, and the third *XYZ*. Label the side lengths and angle measures of each triangle.

b. **TABULAR** Copy and complete the table below.

Triangle	Ratios		
ABC	$\frac{\sin A}{BC} =$	$\frac{\sin B}{CA} =$	$\frac{\sin C}{AB} =$
MNP	$\frac{\sin M}{NP} =$	$\frac{\sin N}{PM} =$	$\frac{\sin P}{MN} =$
XYZ	$\frac{\sin X}{YZ} =$	$\frac{\sin Y}{ZX} =$	$\frac{\sin Z}{XY} =$

c. **VERBAL** Make a conjecture about the ratio of the sine of an angle to the length of the leg opposite that angle for a given triangle.

H.O.T. Problems Use Higher-Order Thinking Skills

23. **ERROR ANALYSIS** Terrence and Rodrigo are trying to determine the relationship between angles of elevation and depression. Terrence says that if you are looking up at someone with an angle of elevation of 35°, then they are looking down at you with an angle of depression of 55°, which is the complement of 35°. Rodrigo disagrees and says that the other person would be looking down at you with an angle of depression equal to your angle of elevation, or 35°. Is either of them correct? Explain.

24. **CHALLENGE** Find the value of *x*. Round to the nearest tenth.

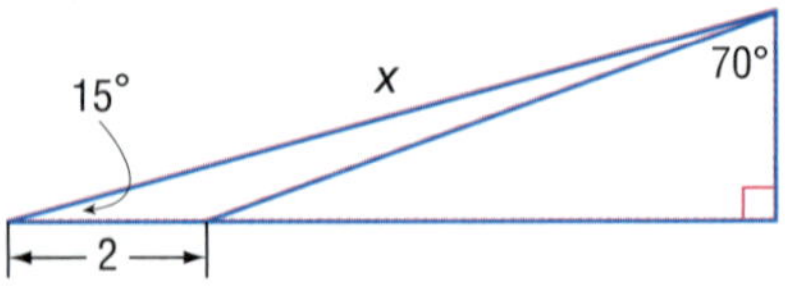

25. **REASONING** Classify the statement below as *true* or *false*. Explain.

As a person moves closer to an object he or she is sighting, the angle of elevation increases.

26. **WRITE A QUESTION** A classmate finds the angle of elevation of an object, but she is trying to find the angle of depression. Write a question to help her solve the problem.

27. **WRITING IN MATH** Describe a way that you can estimate the height of an object without using trigonometry by choosing your angle of elevation. Explain why your method will work.

Standardized Test Practice

28. Ryan wanted to know the height of a cell-phone tower neighboring his property. He walked 80 feet from the base of the tower and measured the angle of elevation to the top of the tower at 54°. If Ryan is 5 feet tall, what is the height of the cell-phone tower?

A 52 **C** 110

B 63 **D** 115

29. SHORT RESPONSE A searchlight is 6500 feet from a weather station. If the angle of elevation to the spot of light on the clouds above the station is 45°, how high is the cloud ceiling?

30. ALGEBRA What is the solution of this system of equations?

$$2x - 4y = -12$$
$$-x + 4y = 8$$

F (4, 4) **H** (−4, −4)

G (−4, 1) **J** (1, −4)

31. SAT/ACT A triangle has sides in the ratio of 5 : 12 : 13. What is the measure of the triangle's smallest angle in degrees?

A 13.34 **C** 34.14

B 22.62 **D** 42.71

Spiral Review

Express each ratio as a fraction and as a decimal to the nearest hundredth. (Lesson 8-4)

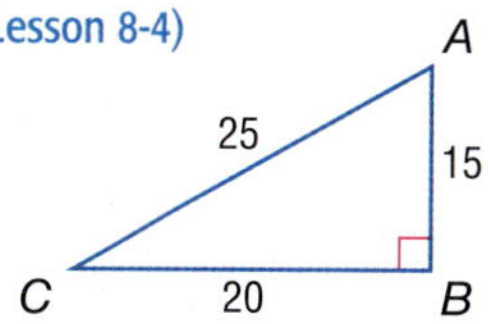

32. sin C **33.** tan A **34.** cos C

35. tan C **36.** cos A **37.** sin A

38. LANDSCAPING Imani needs to determine the height of a tree. Holding a drafter's 45° triangle so that one leg is horizontal, she sights the top of the tree along the hypotenuse, as shown at the right. If she is 6 yards from the tree and her eyes are 5 feet from the ground, find the height of the tree. (Lesson 8-3)

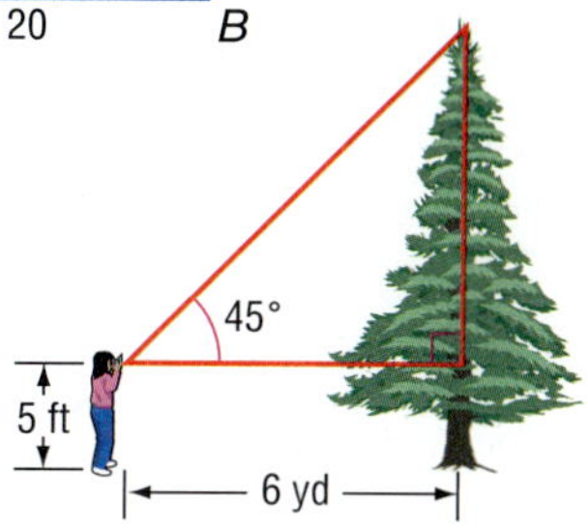

PROOF **Write a two-column proof.** (Lesson 7-5)

39. Given: $\overline{CD}$ bisects $\angle ACB$.

By construction, $\overline{AE} \parallel \overline{CD}$.

Prove: $\frac{AD}{DB} = \frac{AC}{BC}$

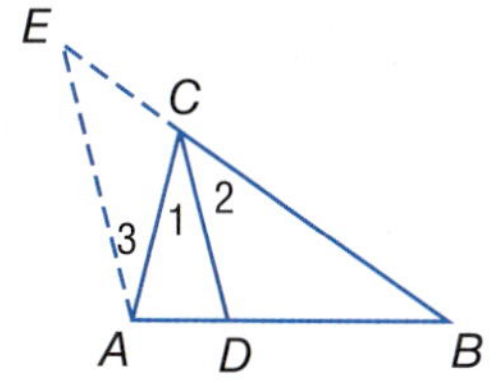

40. Given: $\overline{JF}$ bisects $\angle EFG$.

$\overline{EH} \parallel \overline{FG}$, $\overline{EF} \parallel \overline{HG}$

Prove: $\frac{EK}{KF} = \frac{GJ}{JF}$

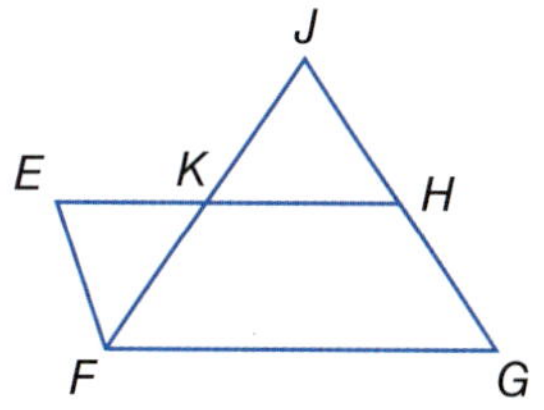

COORDINATE GEOMETRY **Find the coordinates of the centroid of each triangle.** (Lesson 5-2)

41. $A(2, 2)$, $B(7, 8)$, $C(12, 2)$

42. $X(-3, -2)$, $Y(1, -12)$, $Z(-7, -7)$

43. $A(-1, 11)$, $B(-5, 1)$, $C(-9, 6)$

44. $X(4, 0)$, $Y(-2, 4)$, $Z(0, 6)$

Skills Review

Solve each proportion. (Lesson 7-1)

45. $\frac{1}{5} = \frac{x}{10}$ **46.** $\frac{2x}{11} = \frac{3}{8}$ **47.** $\frac{4x}{16} = \frac{62}{118}$ **48.** $\frac{12}{21} = \frac{45}{10x}$

8-6 The Law of Sines and Law of Cosines

Then

You used trigonometric ratios to solve right triangles. (Lesson 8-4)

Now

- Use the Law of Sines to solve triangles.
- Use the Law of Cosines to solve triangles.

New Vocabulary

Law of Sines
Law of Cosines

Math Online glencoe.com

Why?

You have learned that the height or length of a tree can be calculated using *right triangle trigonometry* if you know the angle of elevation to the top of the tree and your distance from the tree. Some trees, however, grow at an angle or lean due to weather damage. To calculate the length of such trees, you must use other forms of trigonometry.

Law of Sines In Lesson 8-4, you used trigonometric ratios to find angle and side measures in *right* triangles. The **Law of Sines** can be used to find missing measures in nonright triangles.

Theorem 8.10 Law of Sines

If $\triangle ABC$ has lengths a, b, and c, representing the lengths of the sides opposite the angles with measures A, B, and C, then

$$\frac{\sin A}{a} = \frac{\sin B}{b} = \frac{\sin C}{c}.$$

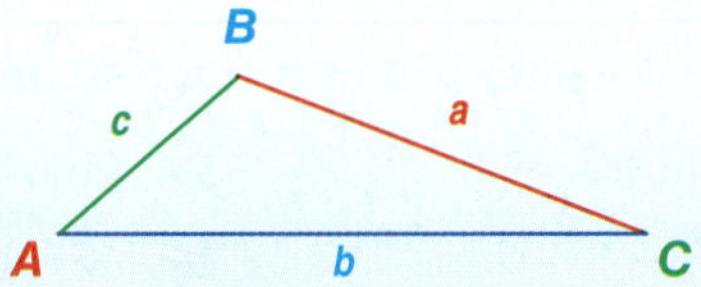

You will prove one of the proportions for Theorem 8.10 in Exercise 45.

You can use the Law of Sines to solve a triangle if you know the measures of two angles and any side (AAS or ASA). If given ASA, use the Triangle Angle Sum Theorem to first find the measure of the third angle.

EXAMPLE 1 Law of Sines (AAS or ASA)

Find x. Round to the nearest tenth.

We are given the measures of two angles and a nonincluded side, so use the Law of Sines to write a proportion.

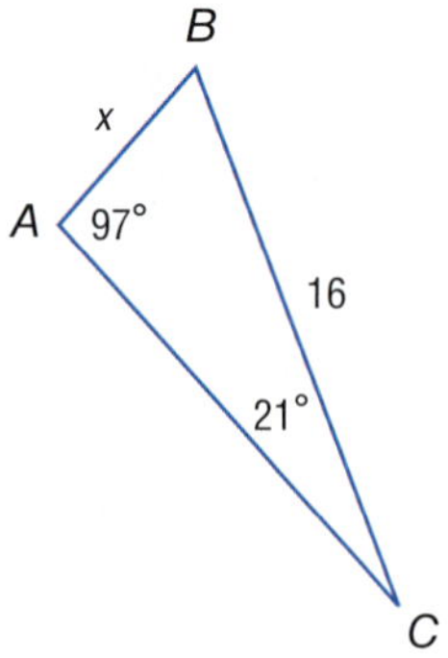

$\frac{\sin A}{a} = \frac{\sin C}{c}$	Law of Sines
$\frac{\sin 97°}{16} = \frac{\sin 21°}{x}$	$m\angle A = 97$, $a = 16$, $m\angle C = 21$, $c = x$
$x \sin 97° = 16 \sin 21°$	Cross Products Property
$x = \frac{16 \sin 21°}{\sin 97°}$	Divide each side by sin 97°.
$x \approx 5.8$	Use a calculator.

Guided Practice

1A.

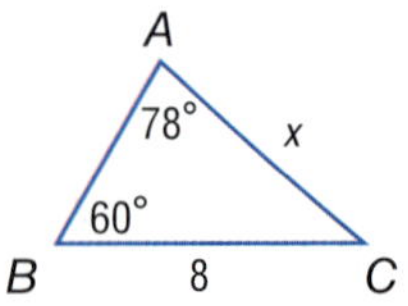

1B.

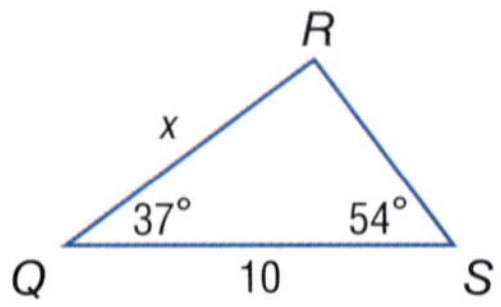

Personal Tutor glencoe.com

You can also use the Law of Sines to solve a triangle if you know the measures of two sides and a *nonincluded* angle (SSA).

EXAMPLE 2 Law of Sines (SSA)

Find x. Round to the nearest degree.

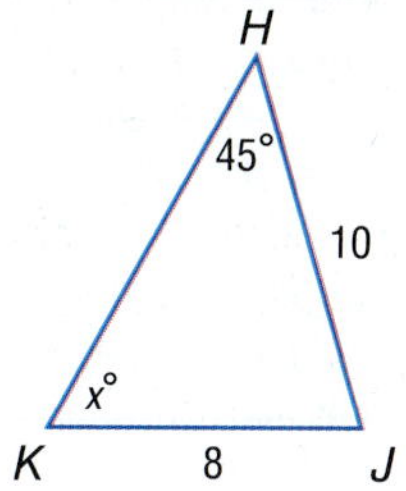

$$\frac{\sin H}{h} = \frac{\sin K}{k}$$ Law of Sines

$$\frac{\sin 45°}{8} = \frac{\sin x°}{10}$$ $m\angle H = 45$, $h = 8$, $m\angle K = x$, $k = 10$

$$10 \sin 45° = 8 \sin x°$$ Cross Products Property

$$\frac{10 \sin 45°}{8} = \sin x°$$ Divide each side by 8.

$$x = \sin^{-1} \frac{10 \sin 45°}{8}$$ Use the inverse sine ratio.

$$x \approx 62°$$ Use a calculator.

Guided Practice

2A.

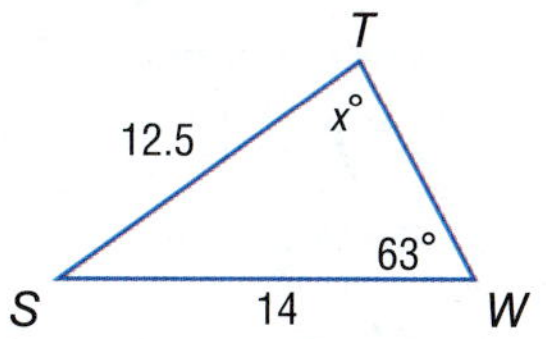

2B.

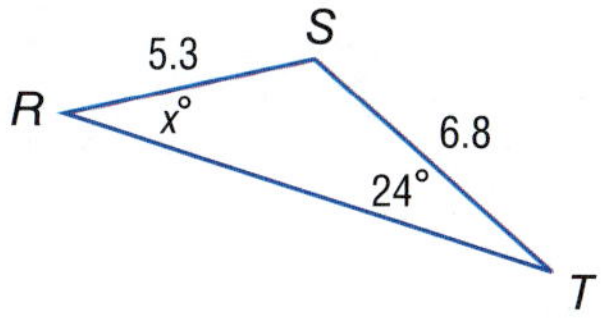

Personal Tutor glencoe.com

StudyTip

Law of Cosines Notice that the angle referenced in each equation of the Law of Cosines corresponds to the side opposite that angle on the other side of the equation.

$a^2 = b^2 + c^2 - 2bc \cos A$

$b^2 = a^2 + c^2 - 2ac \cos B$

$c^2 = a^2 + b^2 - 2ab \cos C$

Law of Cosines When the Law of Sines cannot be used to solve a triangle, the Law of Cosines may apply.

Theorem 8.11 Law of Cosines

If $\triangle ABC$ has lengths a, b, and c, representing the lengths of the sides opposite the angles with measures A, B, and C, then

$a^2 = b^2 + c^2 - 2bc \cos A$,

$b^2 = a^2 + c^2 - 2ac \cos B$, and

$c^2 = a^2 + b^2 - 2ab \cos C$.

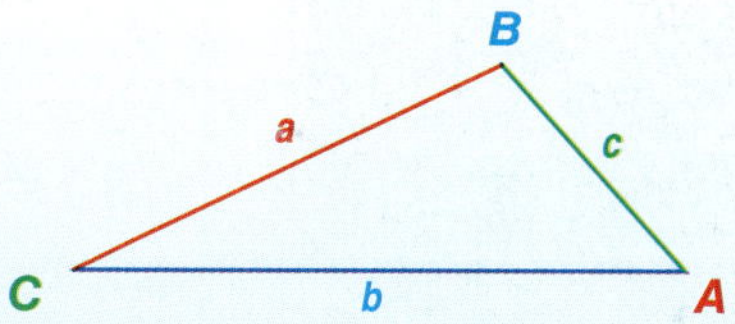

You will prove one of the equations for Theorem 8.11 in Exercise 46.

You can use the **Law of Cosines** to solve a triangle if you know the measures of two sides and the *included* angle (SAS).

Watch Out!

Order of operations Remember to follow the order of operations when simplifying expressions. Multiplications or divisions must be performed before additions or subtractions. So, $202 - 198 \cos 28°$ *cannot* be simplified to $4 \cos 28°$.

EXAMPLE 3 Law of Cosines (SAS)

Find x. Round to the nearest tenth.

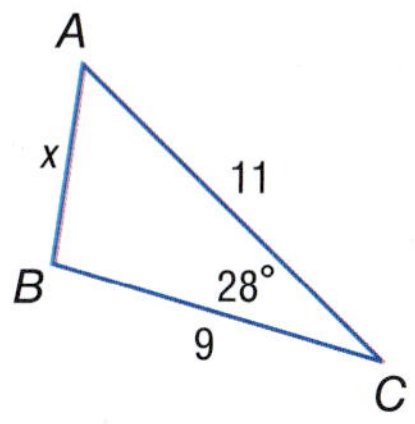

We are given the measures of two sides and their included angle, so use the Law of Cosines.

$$c^2 = a^2 + b^2 - 2ab \cos C$$ Law of Cosines

$$x^2 = 9^2 + 11^2 - 2(9)(11) \cos 28°$$ Substitution

$$x^2 = 202 - 198 \cos 28°$$ Simplify.

$$x = \sqrt{202 - 198 \cos 28°}$$ Take the square root of each side.

$$x \approx 5.2°$$ Use a calculator.

StudyTip

Obtuse Angles There are also values for sin *A*, cos *A*, and tan *A* when $A \geq 90°$. Values of the ratios for these angles will be found using the trigonometric functions on your calculator

Guided Practice

Find x. Round to the nearest tenth.

3A.

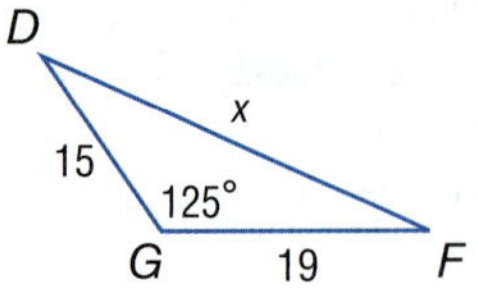

3B.

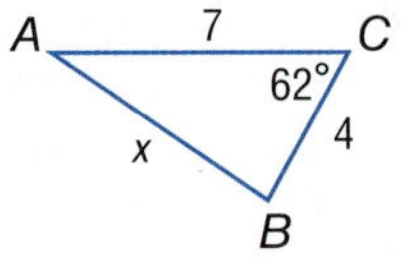

Personal Tutor glencoe.com

You can also use the Law of Cosines if you know three side measures (SSS).

EXAMPLE 4 Law of Cosines (SSS)

Find x. Round to the nearest degree.

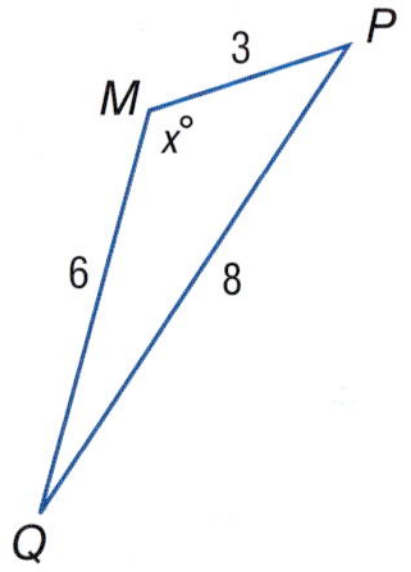

$m^2 = p^2 + q^2 - 2pq \cos M$	**Law of Cosines**
$8^2 = 6^2 + 3^2 - 2(6)(3) \cos x°$	**Substitution**
$64 = 45 - 36 \cos x°$	**Simplify.**
$19 = -36 \cos x°$	**Subtract 45 from each side.**
$\frac{19}{-36} = \cos x°$	**Divide each side by −36.**
$x = \cos^{-1} -\frac{19}{36}$	**Use the inverse cosine ratio.**
$x \approx 122$	**Use a calculator.**

Guided Practice

4A.

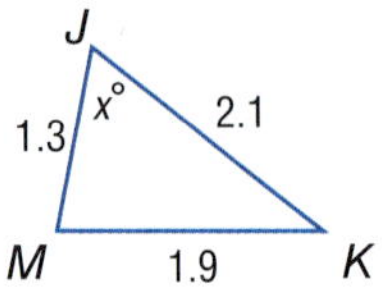

4B.

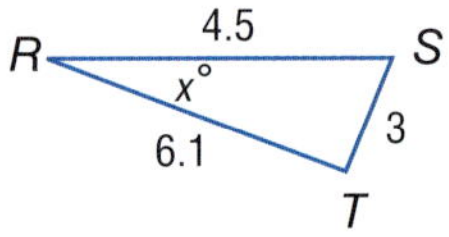

Personal Tutor glencoe.com

Real-World Link

The first game of basketball was played at a YMCA in Springfield, Massachusetts, on December 1, 1891. James Naismith, a physical education instructor, invented the sport using a soccer ball and two half-bushel peach baskets, which is how the name *basketball* came about.

Source: *Encyclopaedia Britannica*

You can use the Law of Sines and Law of Cosines to solve direct and indirect measurement problems.

Real-World EXAMPLE 5 Indirect Measurement

BASKETBALL **Drew and Hunter are playing basketball. Drew passes the ball to Hunter when he is 26 feet from the goal and 24 feet from Hunter. How far is Hunter from the goal if the angle from the goal to Drew and then to Hunter is 34°?**

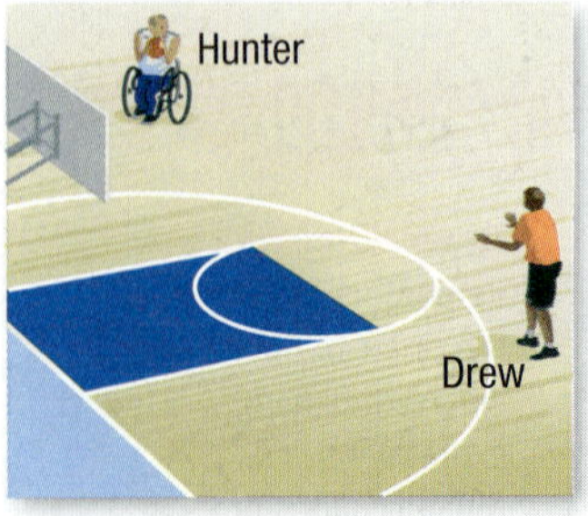

Draw a diagram. Since we know two sides of a triangle and the included angle, use the Law of Cosines.

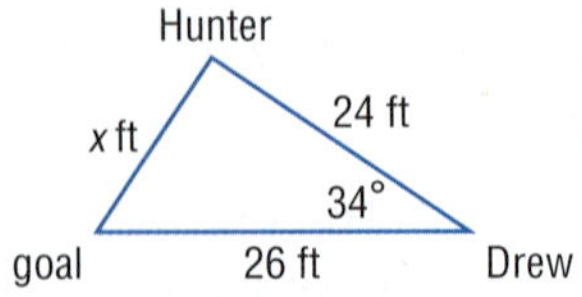

$x^2 = 24^2 + 26^2 - 2(24)(26) \cos 34°$	**Law of Cosines**
$x = \sqrt{1252 - 1248 \cos 34°}$	**Simplify and take the positive square root of each side.**
$x \approx 15$	**Use a calculator.**

Hunter is about 15 feet from the goal when he takes his shot.

Guided Practice

5. LANDSCAPING At 10 feet away from the base of a tree, the angle the top of a tree makes with the ground is 61°. If the tree grows at an angle of 78° with respect to the ground, how tall is the tree to the nearest foot?

Personal Tutor glencoe.com

When solving right triangles, you can use sine, cosine, or tangent. When solving other triangles, you can use the Law of Sines or the Law of Cosines, depending on what information is given.

EXAMPLE 6 Solve a Triangle

Solve triangle *ABC*. Round to the nearest degree.

Since $13^2 + 12^2 \neq 15^2$, this is not a right triangle. Since the measures of all three sides are given (SSS), begin by using the Law of Cosines to find $m\angle A$.

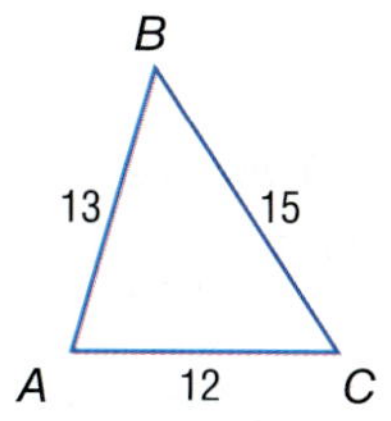

$a^2 = b^2 + c^2 - 2bc \cos A$	**Law of Cosines**
$15^2 = 12^2 + 13^2 - 2(12)(13) \cos A$	**$a = 15$, $b = 12$, and $c = 13$**
$225 = 313 - 312 \cos A$	**Simplify.**
$-88 = -312 \cos A$	**Subtract 313 from each side.**
$\frac{-88}{-312} = \cos A$	**Divide each side by −312.**
$m\angle A = \cos^{-1} \frac{88}{312}$	**Use the inverse cosine ratio.**
$m\angle A \approx 74$	**Use a calculator.**

> **ReadingMath**
>
> **Solve a Triangle** Remember that to *solve* a triangle means to find all of the missing side measures and/or angle measures.

Use the Law of Sines to find $m\angle B$.

$\frac{\sin A}{a} = \frac{\sin B}{b}$	**Law of Sines**
$\frac{\sin 74°}{15} \approx \frac{\sin B}{12}$	**$m\angle A \approx 74$, $a = 15$, and $b = 12$**
$12 \sin 74° = 15 \sin B$	**Cross Products Property**
$\frac{12 \sin 74°}{15} = \sin B$	**Divide each side by 15.**
$m\angle B = \sin^{-1} \frac{12 \sin 74°}{15}$	**Use the inverse sine ratio.**
$m\angle B \approx 50$	**Use a calculator.**

By the Triangle Angle Sum Theorem, $m\angle C \approx 180 - (74 + 50)$ or 56.

Therefore $m\angle A \approx 74$, $m\angle B \approx 50$, and $m\angle C \approx 56$.

Guided Practice

Solve triangle *ABC* using the given information. Round angle measures to the nearest degree and side measures to the nearest tenth.

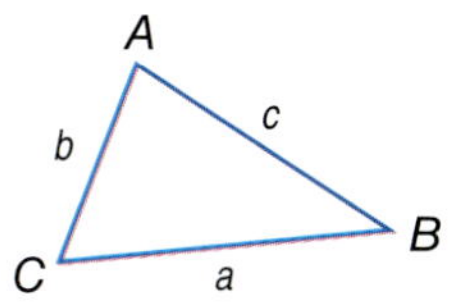

6A. $b = 10.2$, $c = 9.3$, $m\angle A = 26$

6B. $a = 6.4$, $c = 5.8$, $m\angle C = 46$

Personal Tutor glencoe.com

Watch Out!

Rounding When you round a numerical solution and then use it in later calculations, your answers may be inaccurate. Wait until after you have completed all of your calculations to round.

Concept Summary — **Solving a Triangle**

To solve . . .	Given	Begin by using . . .
a right triangle:	leg-leg (LL)	tangent ratio
	hypotenuse-leg (HL)	sine or cosine ratio
	acute angle-hypotenuse (AH)	sine or cosine ratio
	acute angle-leg (AL)	sine, cosine, or tangent ratios
any triangle:	angle-angle-side (AAS or ASA)	Law of Sines
	side-side-angle (SSA)	Law of Sines
	angle-side-angle (SAS)	Law of Cosines
	side-side-side (SSS)	Law of Cosines

Check Your Understanding

Examples 1–4
pp. 582–584

Find x. Round angle measures to the nearest degree and side measures to the nearest tenth.

1.

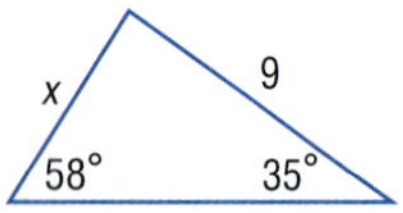

2.

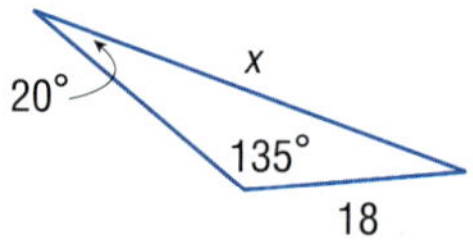

3.

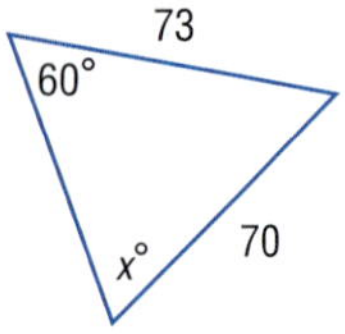

4.

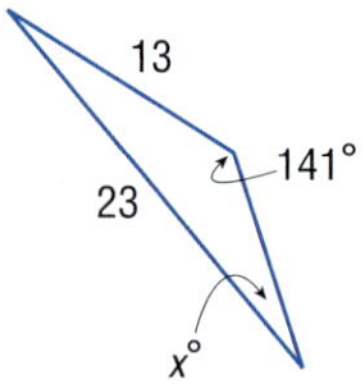

Examples 3 and 4
pp. 583–584

5.

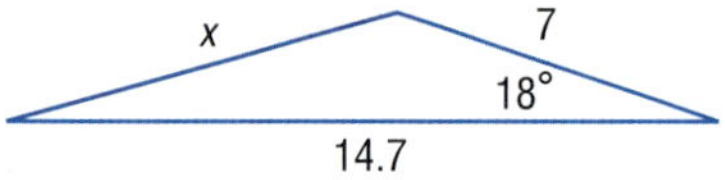

6.

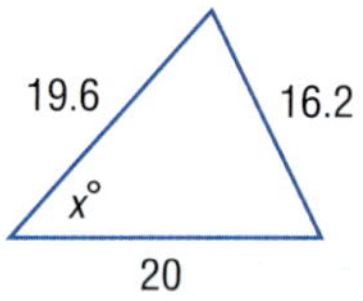

Example 5
p. 584

7. SAILING Determine the length of the bottom edge, or foot, of the sail.

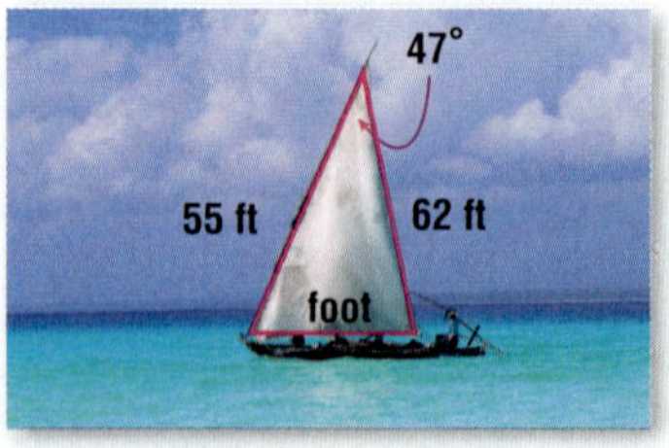

Example 6
p. 585

Solve each triangle. Round angle measures to the nearest degree and side measures to the nearest tenth.

8.

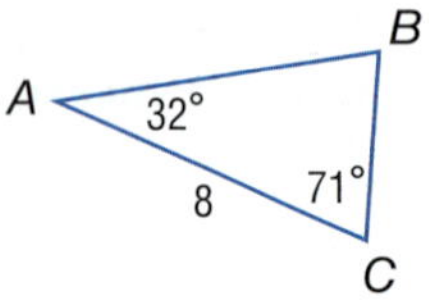

9.

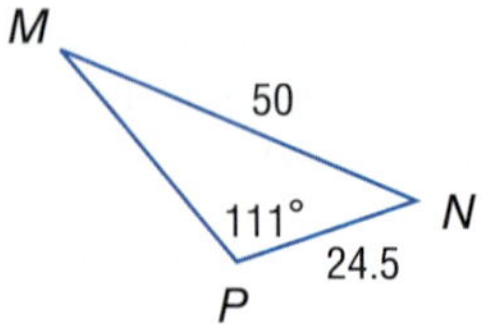

10.

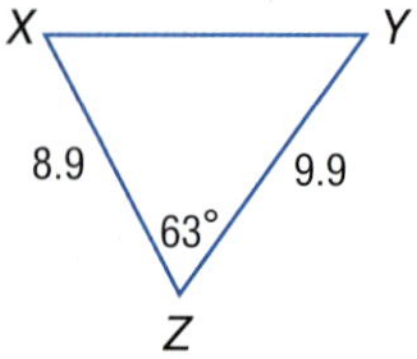

11. Solve $\triangle DEF$ if $DE = 16$, $EF = 21.6$, $FD = 20$.

Practice and Problem Solving

● = **Step-by-Step Solutions** begin on page R20.
Extra Practice begins on page 969.

Examples 1 and 2
pp. 582–583

Find x. Round angle measures to the nearest degree and side measures to the nearest tenth.

12.
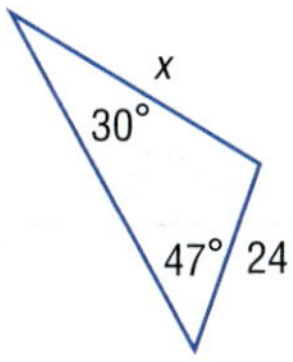

13.
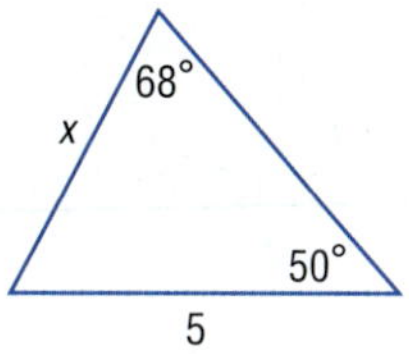

14.
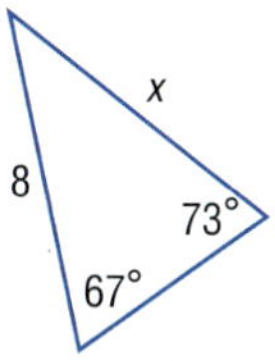

15.
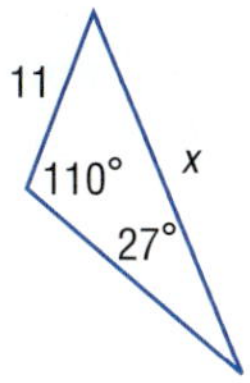

16.
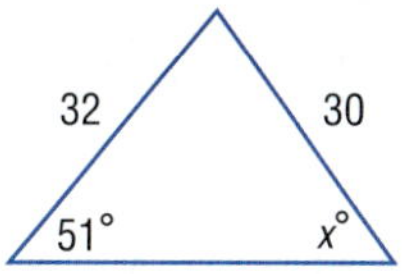

17.
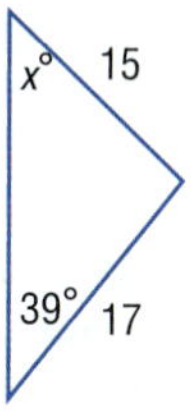

18.
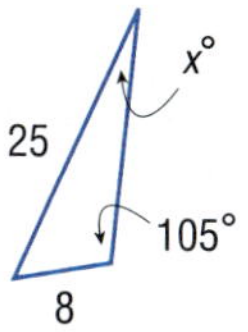

19.
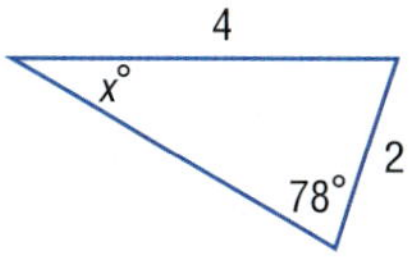

20.
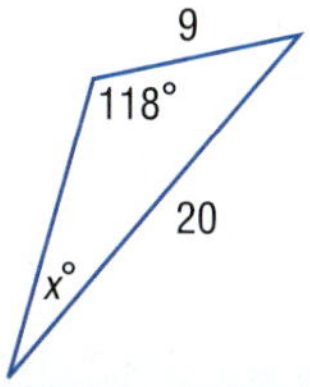

Real-World Link

The Big Dipper is part of the constellation Ursa Major, also called Great Bear. Over the next 100,000 years, stellar motions will cause a change in the configuration of the Big Dipper.

Source: NASA

21. ASTRONOMY Angelina is looking at the Big Dipper through a telescope. From her view, the cup of the constellation forms a triangle that has measurements shown on the diagram at the right. Use the Law of Sines to determine distance between A and C.

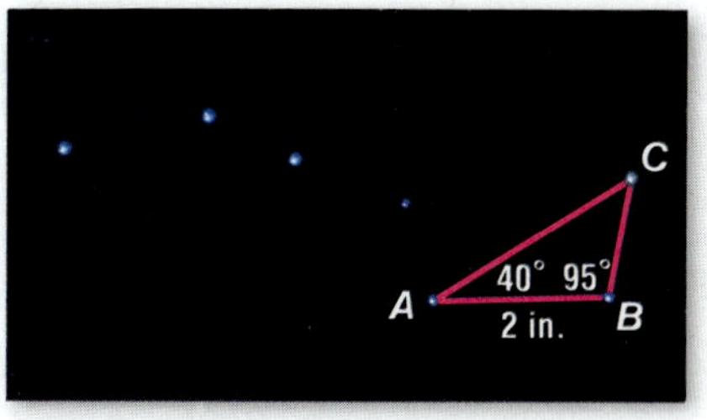

Examples 3 and 4
pp. 583–584

Find x. Round angle measures to the nearest degree and side measures to the nearest tenth.

22.
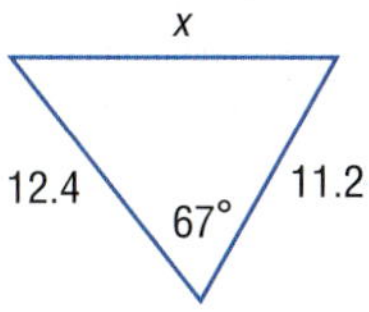

23.
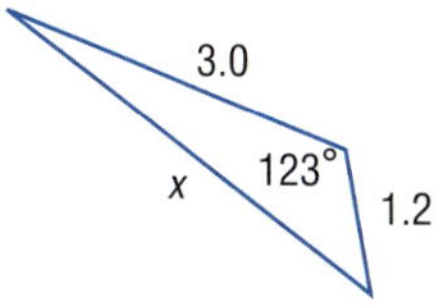

24.
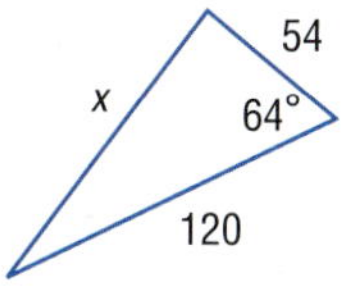

25.
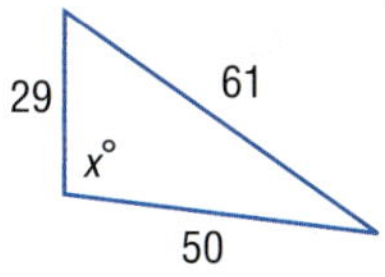

26.
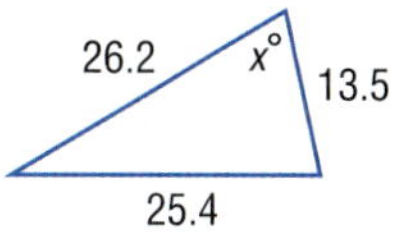

27.
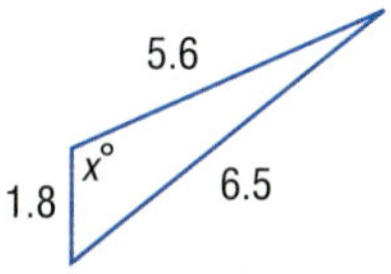

28. HIKING A group of friends who are camping decide to go on a hike. According to the map shown at the right, what is the angle between Trail 1 and Trail 2?

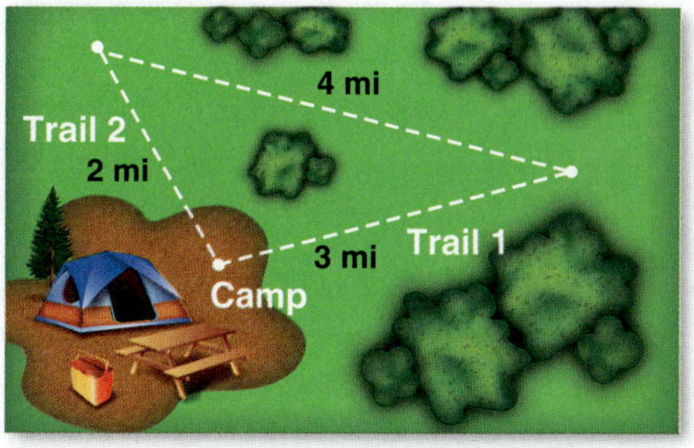

Example 5 p. 584

29. **TORNADOES** Find the width of the mouth of the tornado shown below.

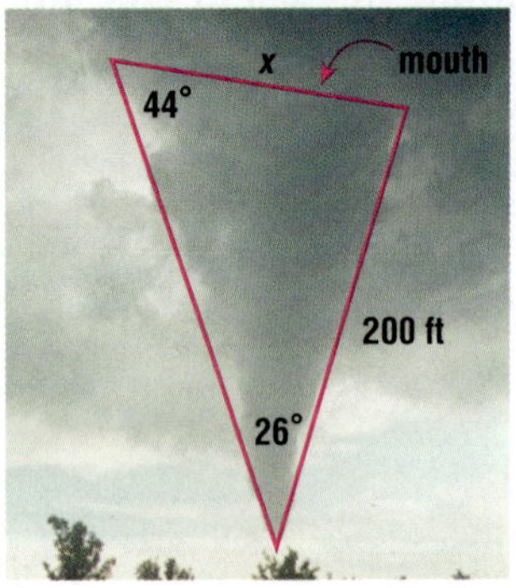

30. **TRAVEL** A pilot flies 90 miles from Memphis, Tennessee, to Tupelo, Mississippi, to Huntsville, Alabama, and finally back to Memphis. How far is Memphis from Huntsville?

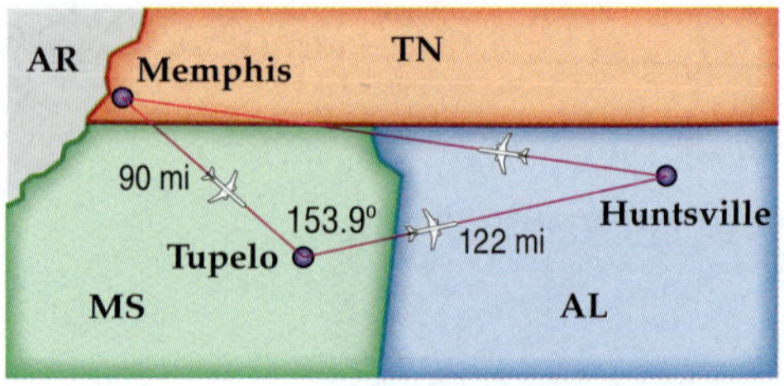

Example 6 p. 585

Solve each triangle. Round angle measures to the nearest degree and side measures to the nearest tenth.

31.

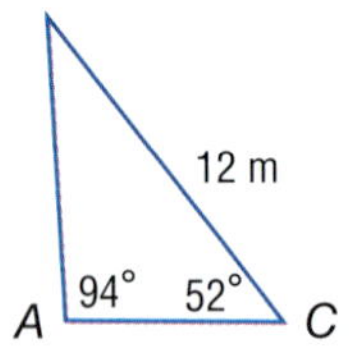

32.

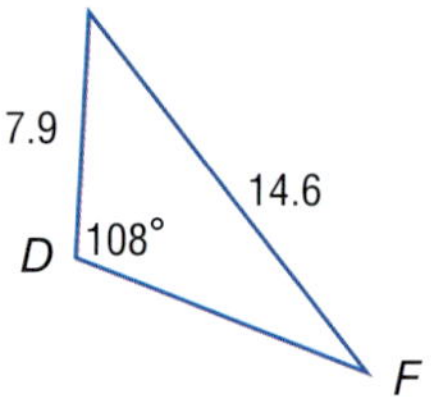

33.

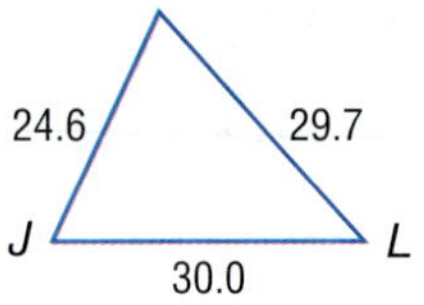

34.

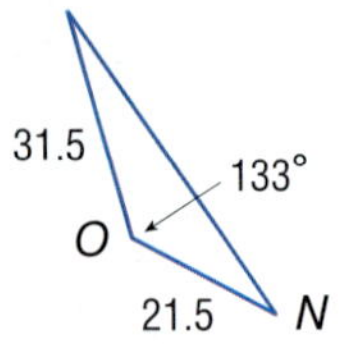

35.

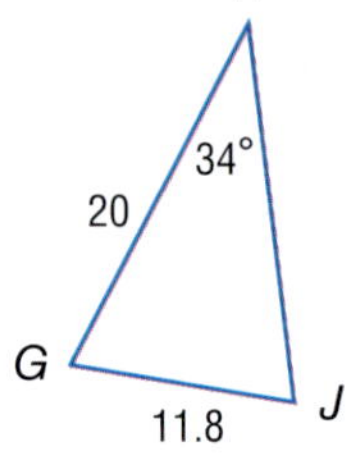

36.

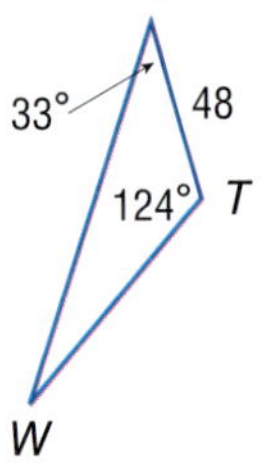

37.

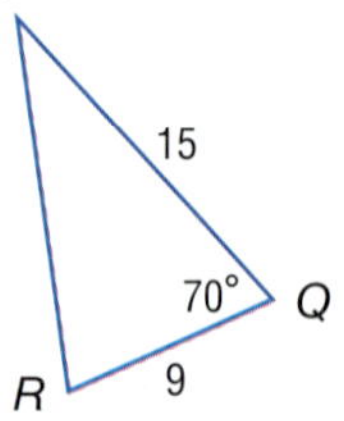

38.

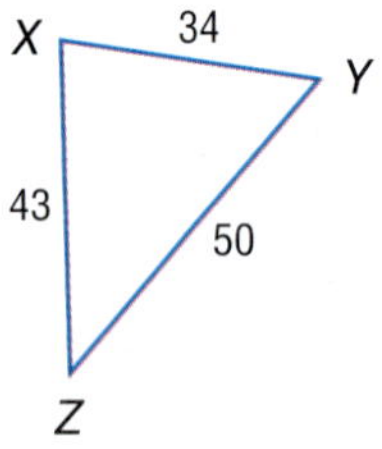

39.

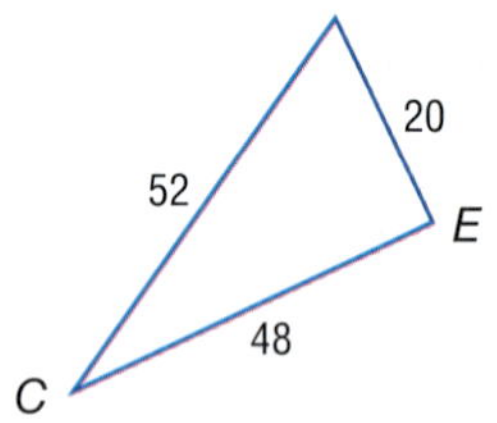

40. Solve $\triangle JKL$ if $JK = 33$, $KL = 56$, $LJ = 65$.

41. Solve $\triangle ABC$ if $m\angle B = 119$, $m\angle C = 26$, $CA = 15$.

42. Solve $\triangle XYZ$ if $XY = 190$, $YZ = 184$, $ZX = 75$.

43. **GARDENING** Crystal has an organic vegetable garden. She wants to add another triangular section so that she can start growing tomatoes. If the garden and neighboring space have the dimensions shown, find the perimeter of the new garden.

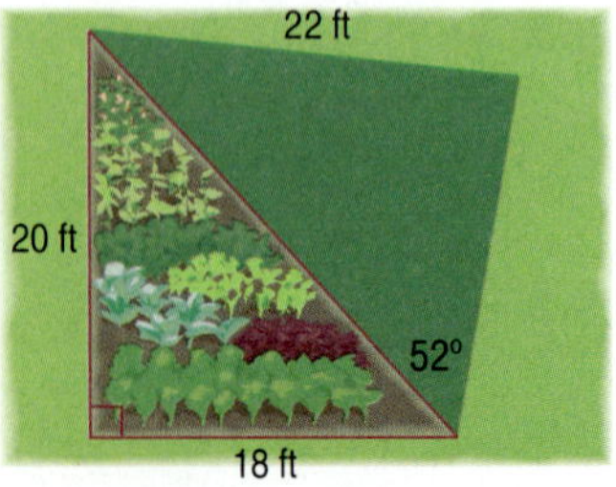

Real-World Link

The Pan American Games are held every four years during the summer before the Olympic Games. They serve as a showcase of international talent, and for the sport of field hockey, and can act as a qualifying event for the next year's Olympics.

Source: USA Field Hockey

44. FIELD HOCKEY Alyssa and Nari are playing field hockey. Alyssa is standing 20 feet from one post of the goal and 25 feet from the opposite post. Nari is standing 45 feet from one post of the goal and 38 feet from the other post. If the goal is 12 feet wide, which player has a greater chance to make a shot? What is the measure of the player's angle?

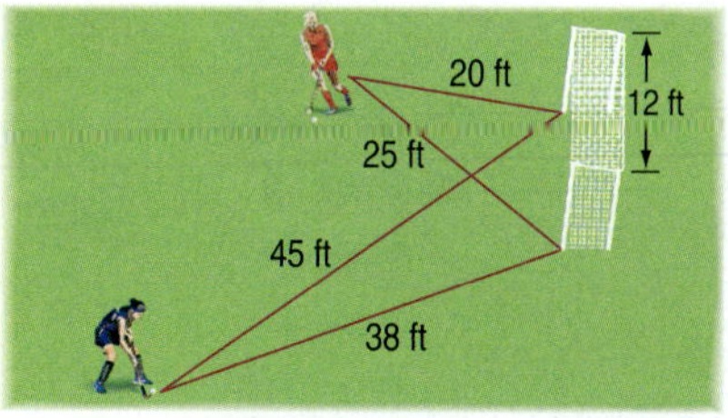

45. PROOF Justify each statement for the derivation of the Law of Sines.

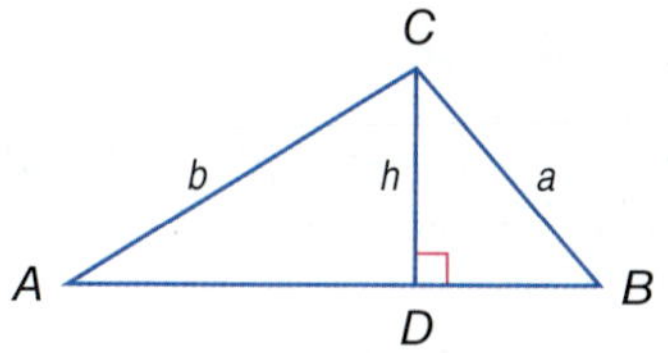

Given: $\overline{CD}$ is an altitude of $\triangle ABC$.

Prove: $\frac{\sin A}{a} = \frac{\sin B}{b}$

Proof:

Statements	Reasons
a. $\sin A = \frac{h}{b}, \sin B = \frac{h}{a}$	**a.** ___?___
b. $b \sin A = h, a \sin B = h$	**b.** ___?___
c. $b \sin A = a \sin B$	**c.** ___?___
d. $\frac{\sin A}{a} = \frac{\sin B}{b}$	**d.** ___?___

46. PROOF Justify each statement for the derivation of the Law of Cosines.

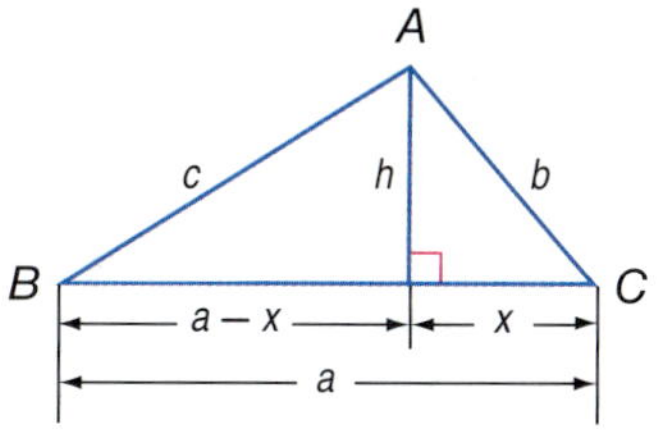

Given: h is an altitude of $\triangle ABC$.

Prove: $c^2 = a^2 + b^2 - 2ab \cos C$

Proof:

Statements	Reasons
a. $c^2 = (a - x)^2 + h^2$	**a.** ___?___
b. $c^2 = a^2 - 2ax + x^2 + h^2$	**b.** ___?___
c. $x^2 + h^2 = b^2$	**c.** ___?___
d. $c^2 = a^2 - 2ax + b^2$	**d.** ___?___
e. $\cos C = \frac{x}{b}$	**e.** ___?___
f. $b \cos C = x$	**f.** ___?___
g. $c^2 = a^2 - 2a(b \cos C) + b^2$	**g.** ___?___
h. $c^2 = a^2 + b^2 - 2ab \cos C$	**h.** ___?___

Find the perimeter of each figure. Round to the nearest tenth.

47.

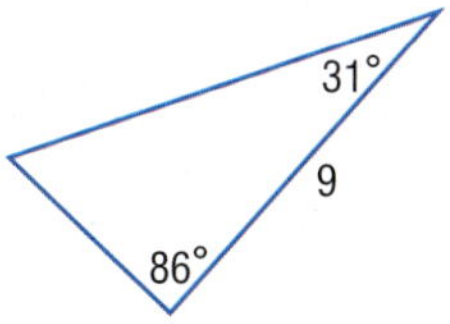

48.

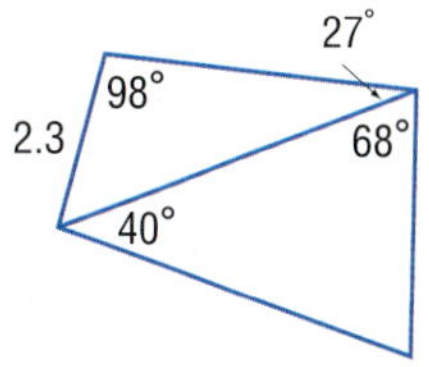

49.

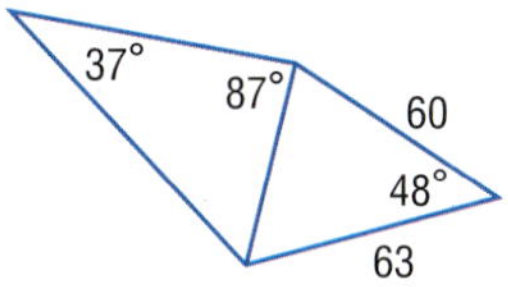

50.

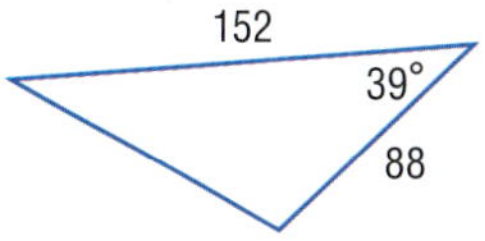

Real-World Link

Professional scale model makers often create high quality models for displays, corporations, and private collectors. Commissions can exceed over $10,000.

Source: Twenty-First Century Models

51. **MODELS** Vito is working on a model castle. Find the length of the missing side (in inches) using the diagram at the right.

52. **COORDINATE GEOMETRY** Find the measure of the largest angle in $\triangle ABC$ with coordinates $A(-3, 6)$, $B(4, 2)$, and $C(-5, 1)$. Explain your reasoning.

53. **MULTIPLE REPRESENTATIONS** In this problem, you will use trigonometry to find the area of a triangle.

a. **GEOMETRIC** Draw an acute, scalene $\triangle ABC$ including an altitude of length h originating at vertex A.

b. **ALGEBRAIC** Use trigonometry to represent h in terms of $m\angle B$.

c. **ALGEBRAIC** Write an equation to find the area of $\triangle ABC$ using trigonometry.

d. **NUMERICAL** If $m\angle B$ is 47, $AB = 11.1$, $BC = 14.1$, and $CA = 10.4$, find the area of $\triangle ABC$. Round to the nearest tenth.

e. **ANALYTICAL** Write an equation to find the area of $\triangle ABC$ using trigonometry in terms of a different angle.

H.O.T. Problems

Use **H**igher-**O**rder **T**hinking Skills

54. **ERROR ANALYSIS** Colleen and Mike are planning a party. Colleen wants to sew triangular decorations and needs to know the perimeter of one of the triangles to buy enough trim. The triangles are isosceles with angle measurements of 64° at the base and side lengths of 5 inches. Colleen thinks the perimeter is 15.7 inches and Mike thinks it is 15 inches. Is either of them correct?

55. **CHALLENGE** Find the value of x in the figure at the right.

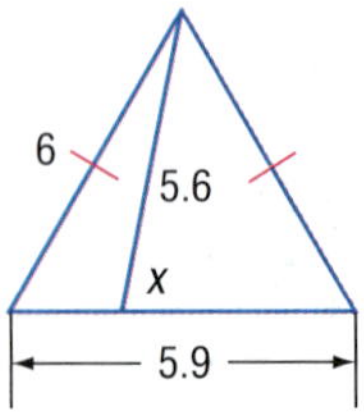

56. **REASONING** Are there any sets of three measures that cannot be used to solve a triangle? Explain.

57. **OPEN ENDED** Draw and label a triangle that can be solved:

a. using only the Law of Sines.

b. using only the Law of Cosines.

58. **WRITING IN MATH** Explain why the Law of Sines cannot be used to solve a triangle for which we know three side lengths. Include an example to illustrate your explanation.

Standardized Test Practice

59. For $\triangle ABC$, $m\angle A = 42$, $m\angle B = 74$, and $a = 3$, what is the value of b?

A 4.3
B 3.8
C 2.1
D 1.5

60. ALGEBRA Which inequality *best* describes the graph below?

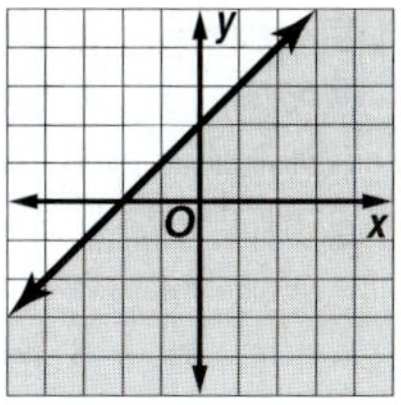

F $y \geq -x + 2$
G $y \leq x + 2$
H $y \geq -3x + 2$
J $y \leq 3x + 2$

61. SHORT RESPONSE What is the perimeter of the triangle shown below? Round to the nearest tenth.

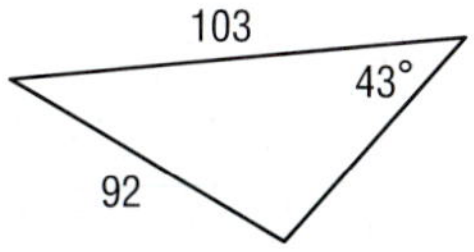

62. SAT/ACT If $\sin x = 0.6$ and $AB = 12$, what is the area of $\triangle ABC$?

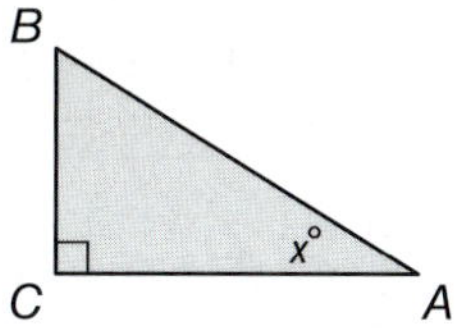

A 28.8 units^2
B 31.2 units^2
C 34.6 units^2
D 42.3 units^2

Spiral Review

63. HIKING A hiker is on top of a mountain 250 feet above sea level with a 68° angle of depression. She can see her camp from where she is standing. How far is her camp from the top of the mountain? (Lesson 8-5)

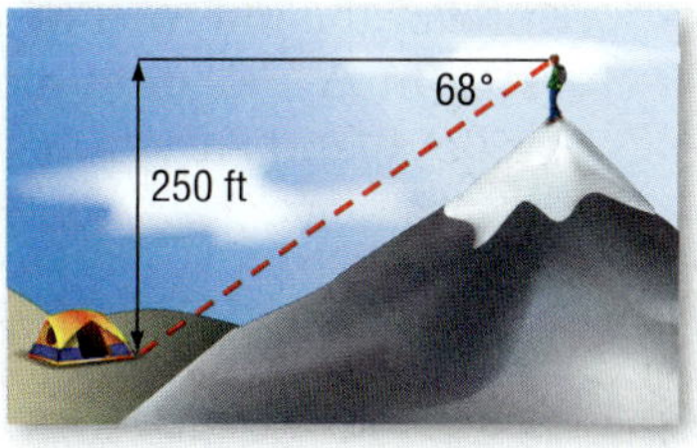

Use a calculator to find the measure of $\angle J$ to the nearest degree. (Lesson 8-4)

64.

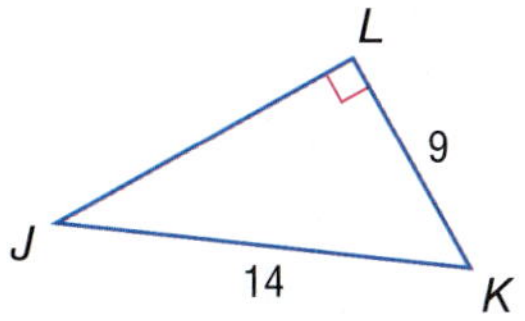

65.

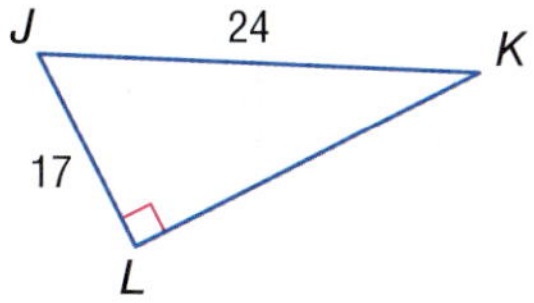

Determine whether the polygons are *always*, *sometimes*, or *never* similar. Explain your reasoning. (Lesson 7-2)

66. a right triangle and an isosceles triangle

67. an equilateral triangle and a scalene triangle

Name the missing coordinates of each triangle. (Lesson 4-8)

68.

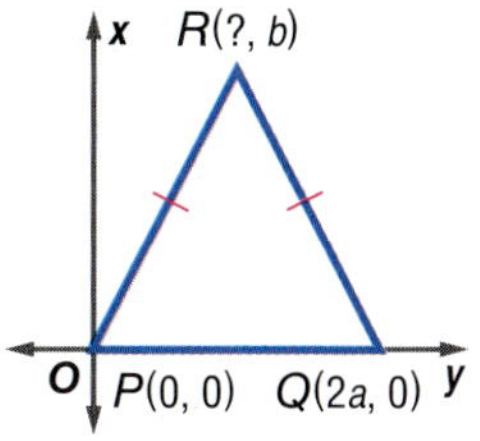

69.

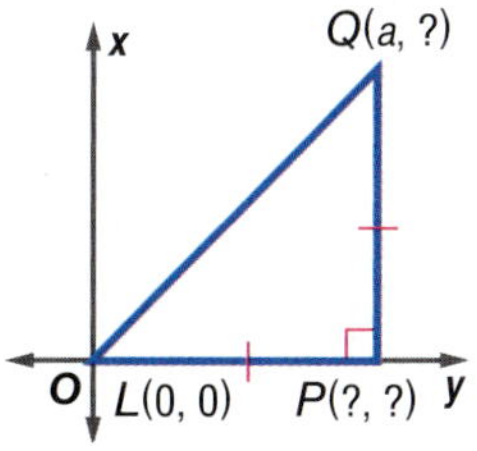

70.

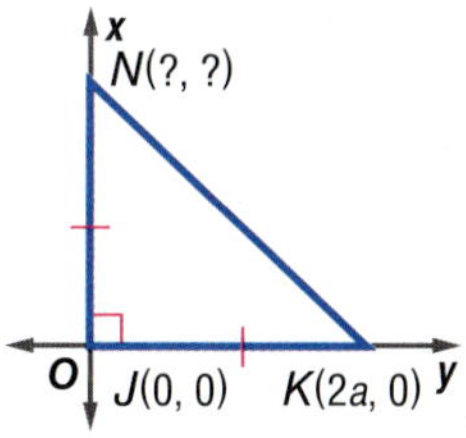

Skills Review

Find the distance between each pair of points. Round to the nearest tenth. (Lesson 1-3)

71. $A(5, 1)$ and $C(-3, -3)$

72. $J(7, 11)$ and $K(-1, 5)$

73. $W(2, 0)$ and $X(8, 6)$

EXTEND 8-6

Geometry Software Lab
The Ambiguous Case of the Laws of Sines

Math Online glencoe.com
- Other Calculator Keystrokes
- Graphing Technology Personal Tutor

In Lesson 8-6, you learned that you could solve a triangle using the Law of Sines if you know the measures of two angles and any side of the triangle (AAS or ASA). You can also solve a triangle using the Law of Sines if you know the measures of two sides and an angle opposite one of the sides (SSA). When you use SSA to solve a triangle, and the given angle is acute, sometimes it is possible to find two different triangles. You can use the Geometer's Sketchpad to explore this case, called **the ambiguous case**, of the Law of Sines.

ACTIVITY

Step 1 Construct $\overrightarrow{AB}$ and $\overrightarrow{AC}$. Construct a circle with center at B so that it intersects $\overrightarrow{AC}$ at two points. Then, construct any radius $\overline{BD}$.

Step 2 Find the measures of $\overline{BD}$, $\overline{AB}$, and $\angle A$.

Step 3 Use the rotate tool to move D so that it lies on one of the intersection points of circle B and $\overrightarrow{AC}$. In $\triangle ABD$, find the measures of $\angle ABD$, $\angle BDA$, and AD.

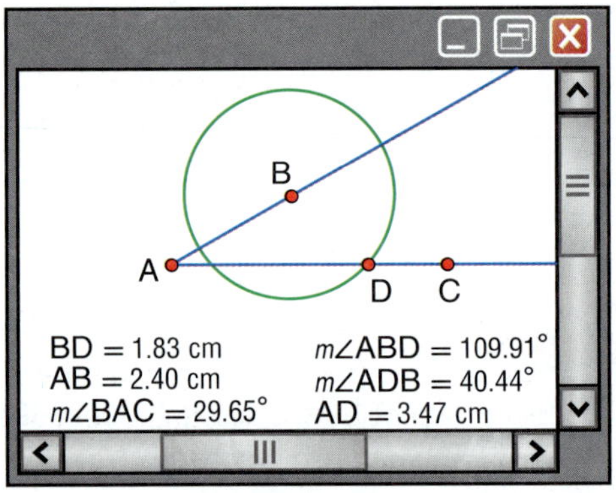

Steps 1–3

Step 4 Using the rotate tool, move D to the other intersection point of circle B and $\overrightarrow{AC}$.

Step 5 Note the measures of $\angle ABD$, $\angle BDA$, and $\overline{AD}$ in $\triangle ABD$.

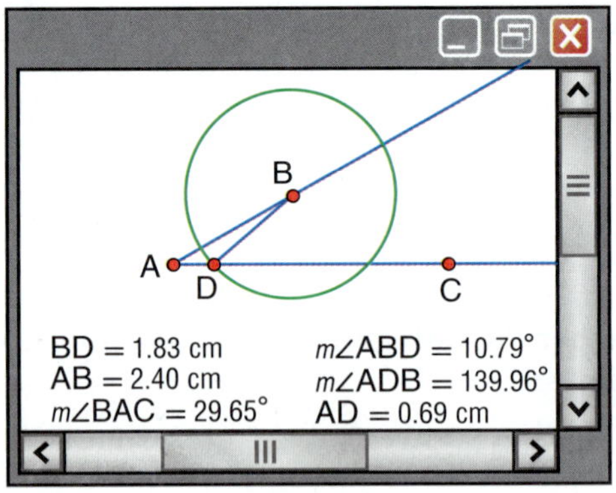

Steps 4–5

Analyze the Results

1. Which measures are the same in both triangles?
2. Repeat the activity using different measures for $\angle A$, $\overline{BD}$, and $\overline{AB}$. How do the results compare to the earlier results?

Make a Conjecture

3. Compare your results with those of your classmates. How do the results compare?
4. What would have to be true about circle B in order for there to be one unique solution? Test your conjecture by repeating the activity.
5. **WRITING IN MATH** Is it possible, given the measures of $\overline{BD}$, $\overline{AB}$, and $\angle A$, to have no solution? Test your conjecture and explain.

8-7 Vectors

Why?

A pilot must correct for environmental factors such as wind to keep a plane on course. They can use vectors to take head or tail winds into account.

Then

You used trigonometry to find side lengths and angle measures of right triangles. (Lesson 8-4)

Now

- Find magnitudes and directions of vectors.
- Add and subtract vectors.

New Vocabulary

vector
magnitude
direction
standard position
component form
resultant

Math Online
glencoe.com

Describe Vectors A **vector** is a quantity that has both *magnitude* and *direction.*

The **magnitude** of a vector is the length of the vector from its initial point to its terminal point. The **direction** of a vector is the angle that is formed with the positive x-axis or any other horizontal line.

magnitude of $\overrightarrow{AB} = |\overrightarrow{AB}|$ direction of $\overrightarrow{AB} = x°$

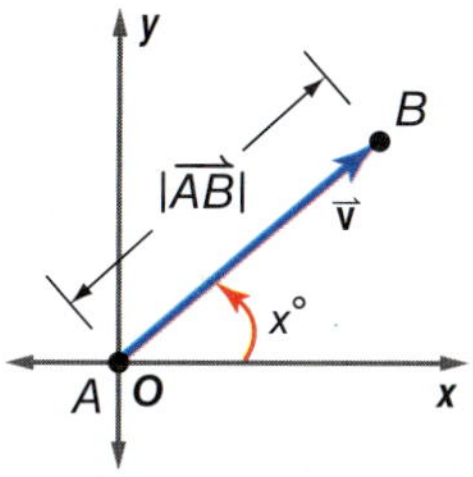

Vectors can be represented as directed line segments. The vector shown, which can be called $\overrightarrow{AB}$ or $\vec{\mathbf{v}}$, has an *initial point A* and *terminal point B.*

When placed on the coordinate plane, a vector is in **standard position** when it has its initial point at the origin. Vector w is in standard position.

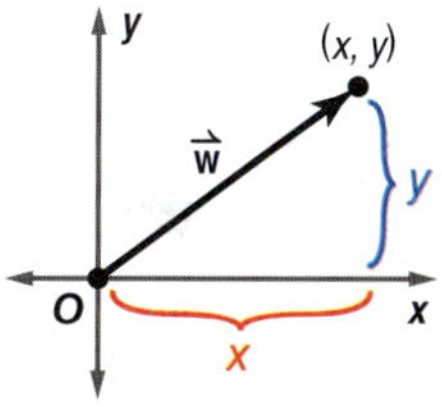

The **component form** of a vector is described in terms of ts horizontal change x and vertical change y from its initial point to its terminal point. The component form of $\vec{\mathbf{w}}$ is $\langle x, y \rangle$.

EXAMPLE 1 Write Vectors in Component Form

Write the component form of $\overrightarrow{CD}$.

Use component notation and find the change in x-values and the change in y-values.

$\overrightarrow{CD} = \langle x_2 - x_1, y_2 - y_1 \rangle$	Component form of vector
$= \langle 4 - 0, 1 - (-1) \rangle$	$x_1 = 0, x_2 = 4, y_1 = -1, y_2 = 1$
$= \langle 4, 2 \rangle$	Simplify.

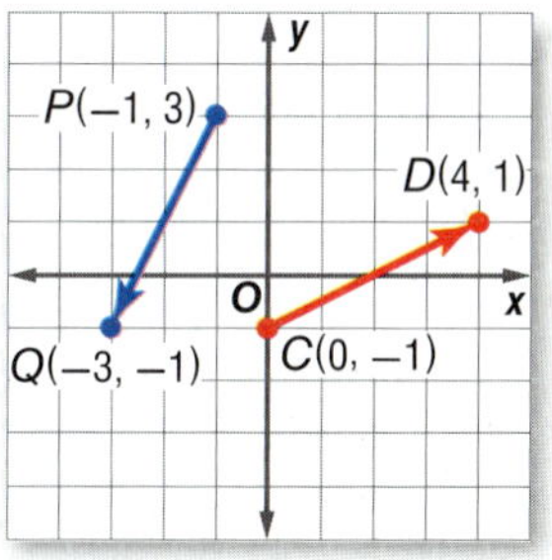

Guided Practice

1. Write the component form of $\overrightarrow{PQ}$.

Personal Tutor glencoe.com

The Distance Formula can be used to find the magnitude of a vector. You can use trigonometric ratios to find the direction of a vector.

EXAMPLE 2 Find the Magnitude and Direction of a Vector

Find the magnitude and direction of $\overrightarrow{RT}$ for $R(2, 4)$ and $T(-3, -2)$.

> **StudyTip**
>
> **Alternate Method** You can also use the Pythagorean Theorem to find the magnitude of the vector.

Step 1 Use the Distance Formula to find the vector's magnitude.

$$\overrightarrow{RT} = \sqrt{(x_2 - x_1)^2 + (y_2 - y_1)^2} \quad \textbf{Distance Formula}$$

$$= \sqrt{(-3-2)^2 + (-2-4)^2} \quad x_1 = 2, x_2 = -3, y_1 = 4, \text{ and } y_2 = -2$$

$$= \sqrt{61} \text{ or about } 7.8 \quad \textbf{Simplify.}$$

Step 2 Use trigonometry to find the vector's direction.

Graph $\overrightarrow{RT}$. Draw a right triangle with $\overrightarrow{RT}$ as its hypotenuse.

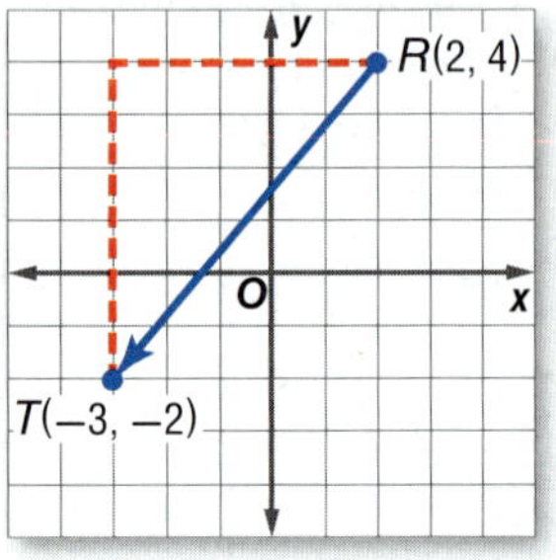

$$\tan R = \left|\frac{y_2 - y_1}{x_2 - x_1}\right| \quad \tan = \frac{\text{opp}}{\text{adj}}; \text{ length cannot be negative.}$$

$$= \left|\frac{-2-4}{-3-2}\right| \text{ or } \frac{6}{5} \quad x_1 = 2, x_2 = -3, y_1 = 4, y_2 = -2$$

$$m\angle R = \tan^{-1}\frac{6}{5} \quad \textbf{Definition of inverse tangent}$$

$$\approx 50.2 \quad \textbf{Use a calculator.}$$

> **StudyTip**
>
> **Direction Angles** Vectors that lie in the third or fourth quadrants when placed in standard position will have direction angles greater than 180°.

When repositioned so that its initial point is at the origin, $\overrightarrow{RT}$ lies in the third quadrant and forms an angle with the *negative* x-axis equal to $m\angle R$.

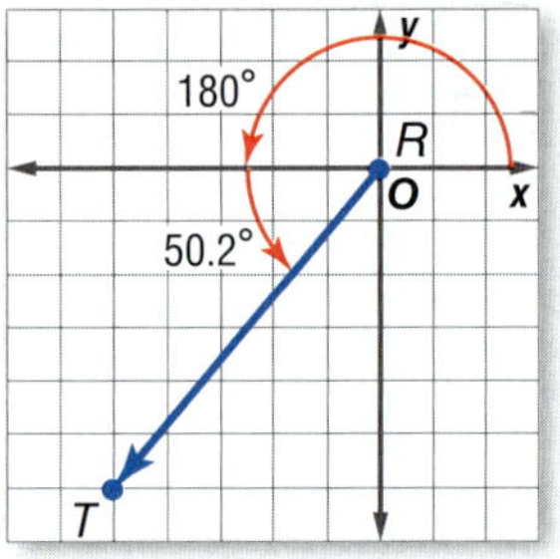

The direction of $\overrightarrow{RT}$ is the angle it makes with the *positive* x-axis, which is $m\angle R + 180$ or about 230.2.

Therefore, $\overrightarrow{RT}$ has a magnitude of about 7.8 units and a direction of about 230.2°.

Guided Practice

2. Find the magnitude and direction for $\overrightarrow{FG}$ for $F(-1, 5)$ and $G(3, -2)$.

Personal Tutor glencoe.com

Vector Addition Three types of vectors are described below.

Key Concept Equal, Parallel, and Opposite Vectors

Equal Vectors Two vectors are equal if and only if they have the same magnitude and direction.

Example $\vec{b} = \vec{c}$ **Nonexample** $\vec{b} \neq \vec{e}$

Parallel Vectors Two vectors are parallel if and only if they have the same or opposite direction.

Example $\vec{d} \parallel \vec{a}$ **Nonexample** $\vec{d} \nparallel \vec{f}$

Opposite Vectors Two vectors are opposites if they have the same magnitude and opposite directions.

Example $\vec{b}$ and $\vec{e}$ **Nonexample** $\vec{d}$ and $\vec{f}$

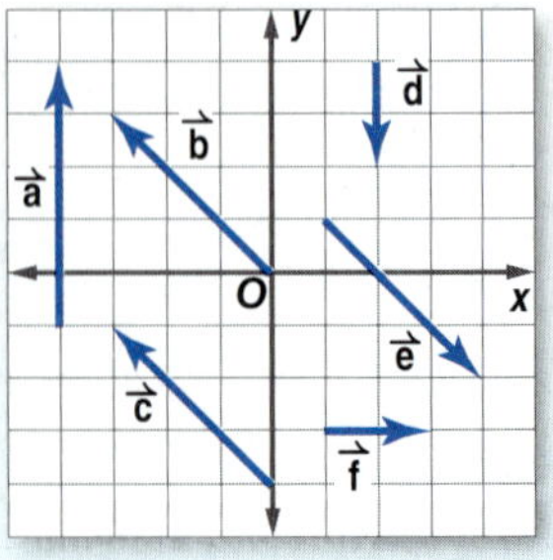

The sum of two vectors is called the resultant. The **resultant** of two vectors can be found geometrically by using two different methods.

> **StudyTip**
>
> **Triangle Method** The triangle method may also be called the *tip-to-tail method.*

Key Concept Vector Addition

Parallelogram Method

Step 1 Place both vectors at the same initial point.

Step 2 Complete the parallelogram. The resultant is the diagonal of the parallelogram.

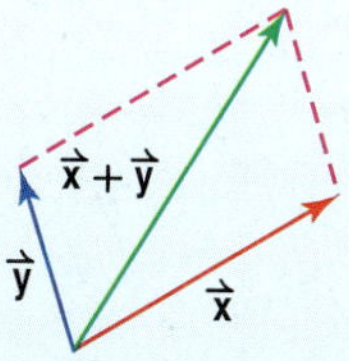

Triangle Method

Step 1 Place the initial point of the second vector at the terminal point of the first.

Step 2 The resultant connects the initial point of the first vector and the terminal point of the second.

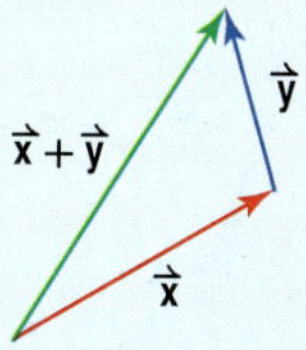

EXAMPLE 3 Vector Addition and Subtraction

Copy the vectors to find $\vec{a}-\vec{b}$.

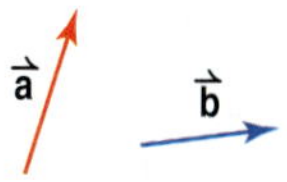

Subtracting a vector is equivalent to adding its opposite.

Method 1 Use the parallelogram method.

Step 1

Copy $\vec{a}$ and $\vec{b}$ with the same initial point.

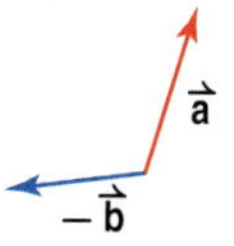

Step 2

Complete the parallelogram.

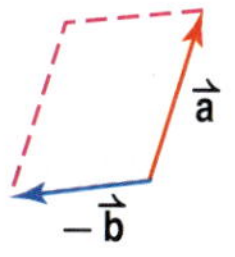

Step 3

Draw the diagonal of the parallelogram from the initial point.

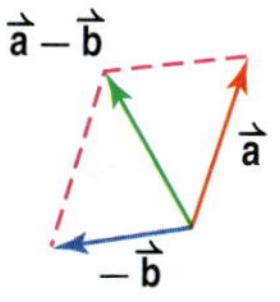

> **StudyTip**
>
> **Opposite Vectors** A vector has the same magnitude and a direction that is 180° more than its opposite vector.

Method 2 Use the triangle method.

Step 1

Copy $\vec{a}$.

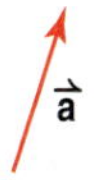

Step 2

Place the initial point of $-\vec{b}$ at the terminal point of $\vec{a}$.

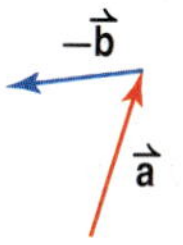

Step 3

Draw the vector from the initial point of $\vec{a}$ to the terminal point of $\vec{b}$.

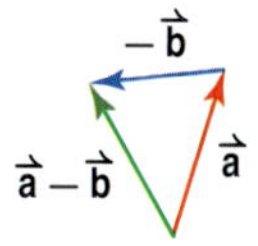

Compare the resultants from the parallelogram method and the triangle method. They are equal.

Guided Practice

3A. Find $\vec{a}+\vec{b}$.

3B. Find $\vec{b}-\vec{a}$.

Personal Tutor glencoe.com

Vectors can also be added or subtracted algebraically by adding or subtracting their horizontal and vertical components.

$$\underset{\langle x_1, y_1\rangle}{\vec{\mathbf{v}}} + \underset{\langle x_2, y_2\rangle}{\vec{\mathbf{w}}} = \underset{\langle x_1 + x_2, y_1 + y_2\rangle}{\vec{\mathbf{v}} + \vec{\mathbf{w}}}$$

Real-World EXAMPLE 4 Algebraic Vectors

Real-World Link

Launchers are used at sporting events to project T-shirts and other soft prizes into the stands. The launchers make it possible for the prizes to travel farther than if they were hand-thrown by increasing their speed and changing their angles of elevation.

EVENT PLANNING Cheerleaders are using an air launcher to project T-shirts into the stands. The wind is blowing in the same direction that shirts are launched. If a shirt leaves the launcher at 40 feet per second at an angle of 45° and the wind is blowing 7 feet per second parallel to the horizontal, what is the resultant velocity and direction of the T-shirt?

Step 1 Draw a diagram to represent each direction vector.

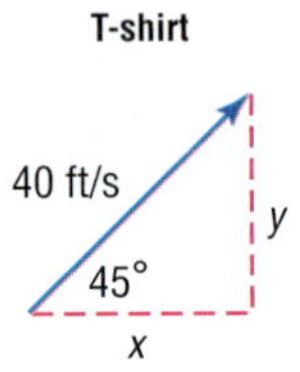

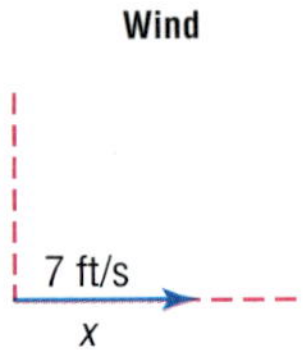

Step 2 Write the direction vector for the T-shirt in component form.

$\cos 45° = \frac{x}{40}$	**$\cos = \frac{\text{adj}}{\text{hyp}}$ and $\sin = \frac{\text{opp}}{\text{hyp}}$**	$\sin 45° = \frac{y}{40}$
$40 \cos 45° = x$	**Multiply each side by 40.**	$40 \sin 45° = y$
$28.3 \approx x$	**Use a calculator.**	$28.3 \approx y$

The component form for the T-shirt's vector is $\langle 28.3, 28.3\rangle$.

Since the wind is blowing parallel to the horizontal, the vector will only have a horizontal component. The vector describing the wind is $\langle 7, 0\rangle$.

Use the algebraic method to add the vectors for the shirt and the wind.
$\langle 28.3, 28.3\rangle + \langle 7, 0\rangle = \langle 28.3 + 7, 28.3 + 0\rangle$ or $\langle 35.3, 28.3\rangle$

Draw a diagram. Let v represent the velocity of the shirt, and let a represent the angle of direction.

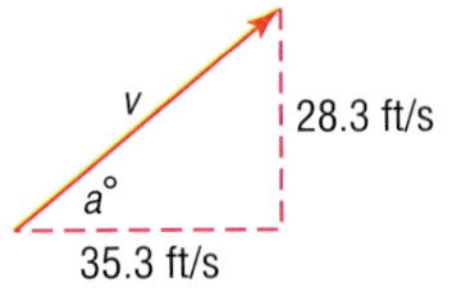

Determine the vector's magnitude.

$v^2 = x^2 + y^2$	**Pythagorean Theorem**
$v^2 \approx (35.3)^2 + (28.3)^2$	**Substitute.**
$v \approx 45.2$	**Use a calculator.**

Use trigonometry to find the vector's direction.

$\tan a = \frac{y}{x}$	**$\tan = \frac{\text{opp}}{\text{adj}}$**
$\tan a \approx \frac{28.3}{35.3}$	**Substitute.**
$a \approx \tan^{-1} \frac{28.3}{35.3}$ or about 38.7	**Definition of inverse tangent**

The resultant velocity of the T-shirt is 45.2 feet per second at an angle of 38.7° with the horizontal.

Guided Practice

4. What would be the resultant magnitude and direction of the shirt if the wind is blowing in the opposite direction?

Personal Tutor glencoe.com

Check Your Understanding

Example 1 p. 593

Write the component form of each vector.

1.

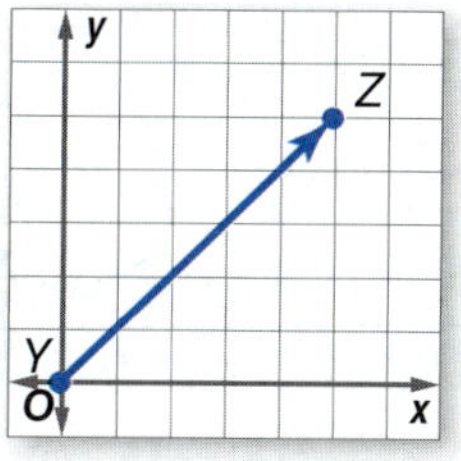

2.

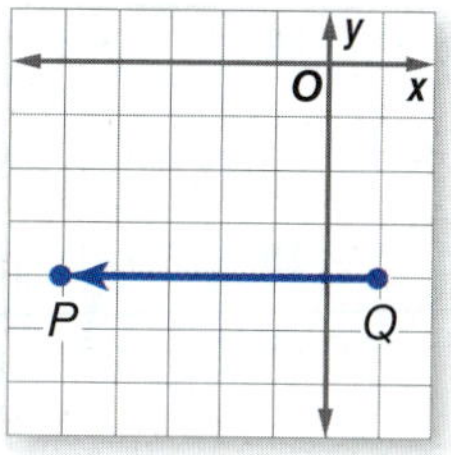

Example 2 p. 594

Find the magnitude and direction of each vector.

3. $\overrightarrow{PQ}$: $P(2, 4)$ and $Q(8, 4)$

4. $\overrightarrow{WX}$: $W(-9, -5)$ and $X(-1, -7)$

Example 3 p. 595

Copy the vectors to find each sum or difference.

5. $\vec{c} + \vec{d}$

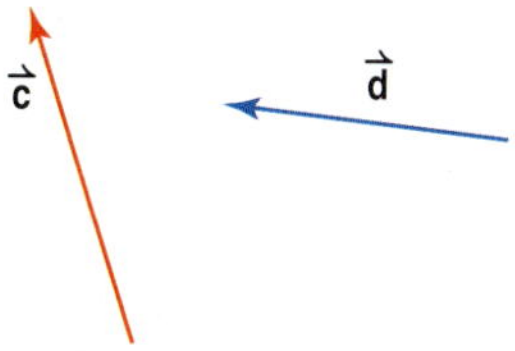

6. $\vec{y} - \vec{z}$

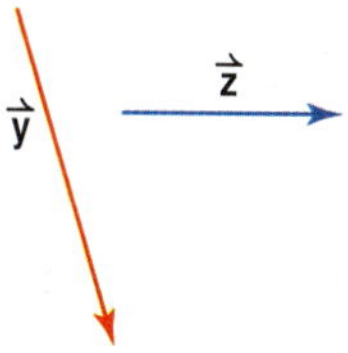

Example 4 p. 596

7. **TRAVEL** A plane takes off going east at 100 miles per hour and 8° above ground. If the wind is blowing 10 miles per hour parallel to the ground and in the same direction as the plane, what is the plane's resultant velocity and direction?

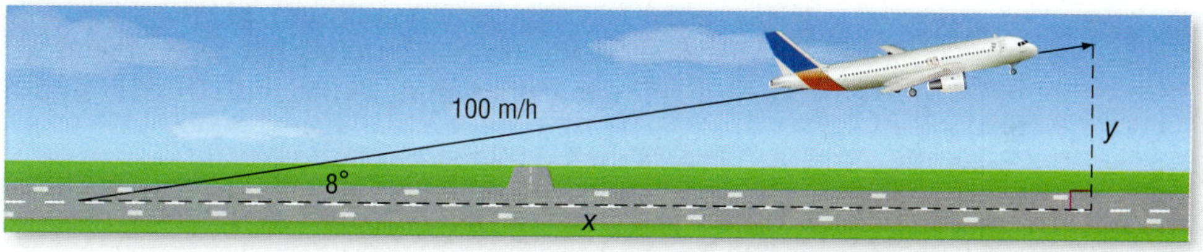

Practice and Problem Solving

● = **Step-by-Step Solutions** begin on page R20.
Extra Practice begins on page 969.

Example 1 p. 593

Write the component form of each vector.

8.

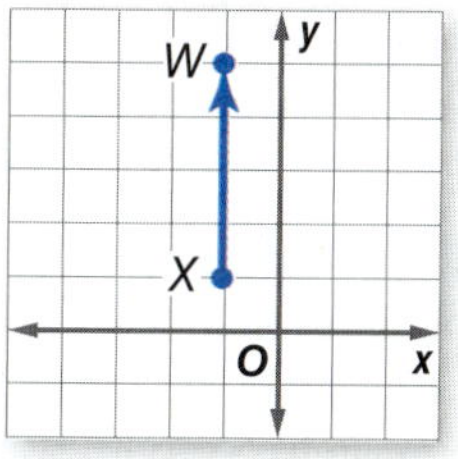

9.

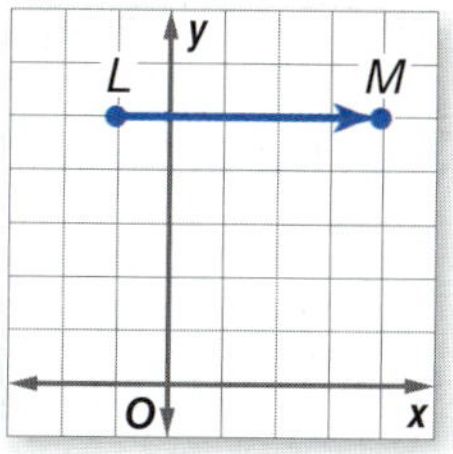

10.

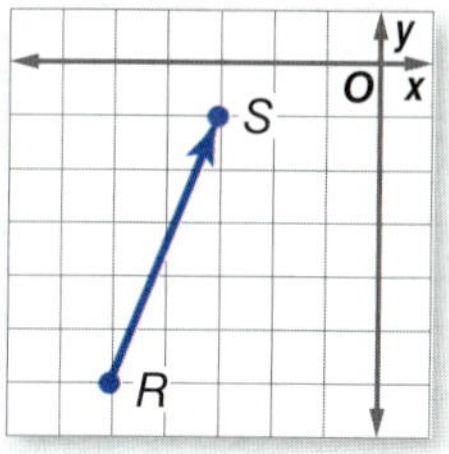

11.

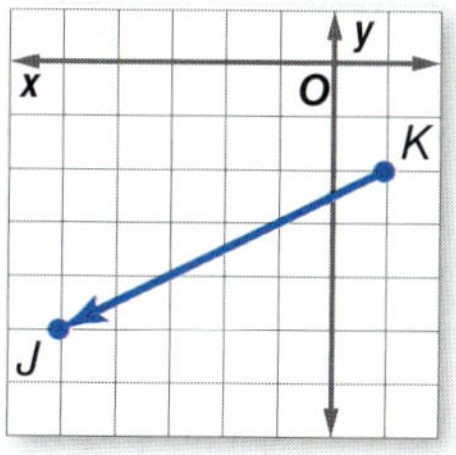

12.

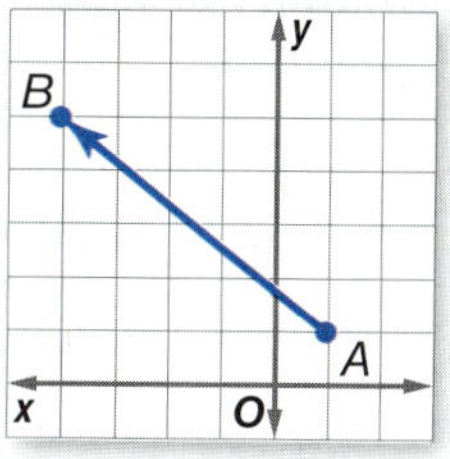

13.

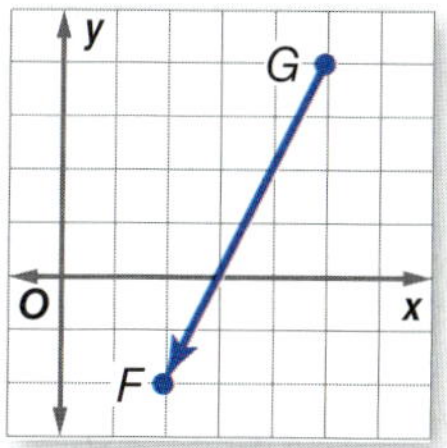

Example 2
p. 594

Find the magnitude and direction of each vector.

14. $\overrightarrow{JK}$: $J(-4, -6)$ and $K(-7, -6)$

15. $\overrightarrow{AB}$: $A(9, -2)$ and $B(9, -8)$

16. $\overrightarrow{RS}$: $R(-1, 1)$ and $S(-7, 7)$

17. $\overrightarrow{FG}$: $F(-8, -8)$ and $G(-3, -3)$

18. $\overrightarrow{CD}$: $C(10, 5)$ and $D(4, 2)$

19. $\overrightarrow{XY}$: $X(6, 7)$ and $Y(9, 11)$

Example 3
p. 595

Copy the vectors to find each sum or difference.

20. $\vec{t} - \vec{m}$

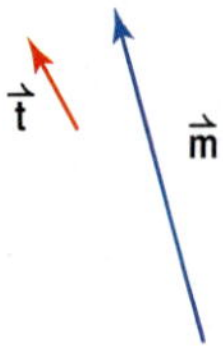

21. $\vec{j} - \vec{k}$

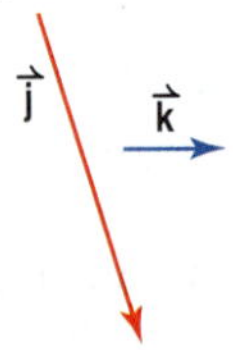

22. $\vec{w} + \vec{z}$

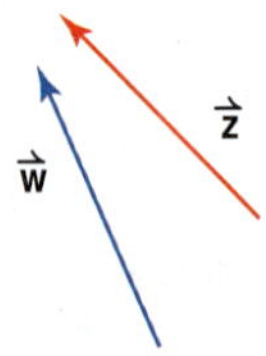

23. $\vec{c} + \vec{a}$

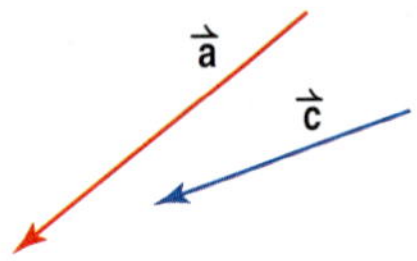

24. $\vec{d} - \vec{f}$

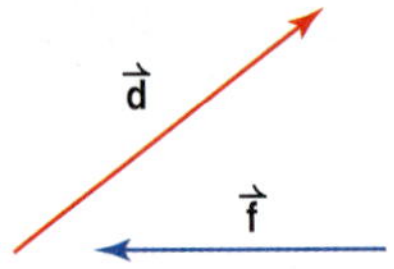

25. $\vec{t} - \vec{m}$

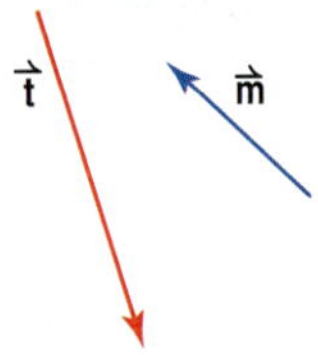

Example 4
p. 596

26. **SPORTS** Trey rents a personal watercraft and heads north at 15 meters per second. The river is moving with a velocity of 2 meters per second to the west. What is the resultant velocity and direction of the watercraft according to an observer on the shore?

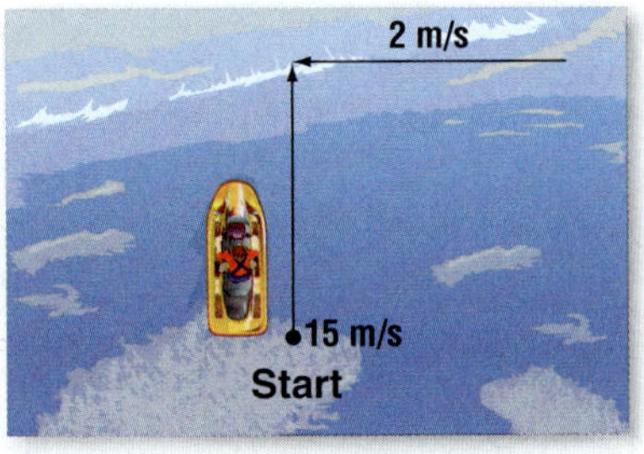

27. **TRACK** Camila is running the 100-meter dash at a speed of 6.7 meters per second parallel with the horizontal. The wind is 1.2 meters per second from behind, blowing at an angle of 30° with Camila's path.

a. What is the Camila's resultant velocity?

b. How long will it take her to complete the dash?

Find the sum of each pair of vectors.

28. $\langle 3, 5\rangle + \langle 17, 1\rangle$

29. $\langle 4, -7\rangle + \langle -6, -3\rangle$

30. $\langle -1, 12\rangle + \langle 14, 2\rangle$

31. $\langle a, b\rangle + \langle c, d\rangle$

32. $\langle x, 2y\rangle + \langle -x, 3y\rangle$

33. $\langle -3p, 3r\rangle + \langle 2p, -r\rangle$

34. **HOMECOMING** Nikki is on a committee to help plan her school's homecoming parade. The parade starts at the high school and continues as shown in the figure at the right.

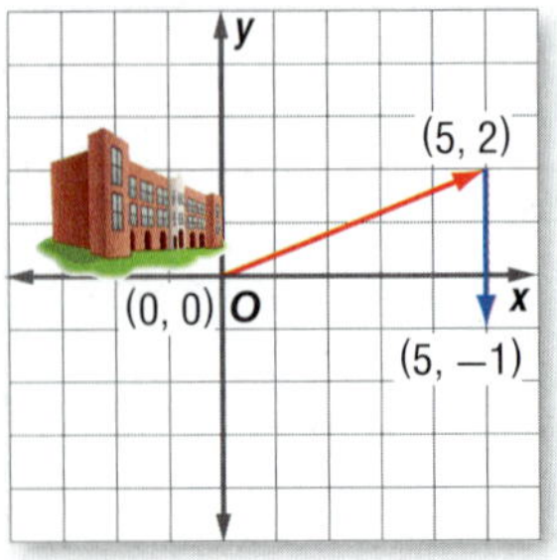

a. What is the magnitude and direction of the vector formed from the ending point of the parade to the school?

b. Find the sum of the two vectors to determine the length of the parade if 1 unit $= \frac{1}{4}$ mile.

Write each vector in component form. Round to the nearest tenth.

35 magnitude = 5, direction = 37°

36. magnitude = 12, direction = 195°

37. magnitude = 7, direction = 104°

38. magnitude = 20, direction = 308°

Real-World Link

Thunder Over Louisville is the Kentucky Derby Festival's opening ceremony. It includes a fireworks show that is the largest annual pyrotechnic display in the United States. About 60 tons of fireworks shells, 250 tons of launching tubes, and 700 miles of wire cable are used to assemble the show.

Source: Thunder Over Louisville

39. **FIREWORKS** The ascent of a firework can be modeled using a vector. Write a vector in component form to describe a firework at the point when it has traveled 70 feet horizontally and 400 feet vertically.

40. **SWIMMING** Jonas is swimming from the east bank to the west bank of a stream at a speed of 3.3 feet per second. The stream is 80 feet wide and flows south. If Jonas crosses the stream in 20 seconds, what is the speed of the current?

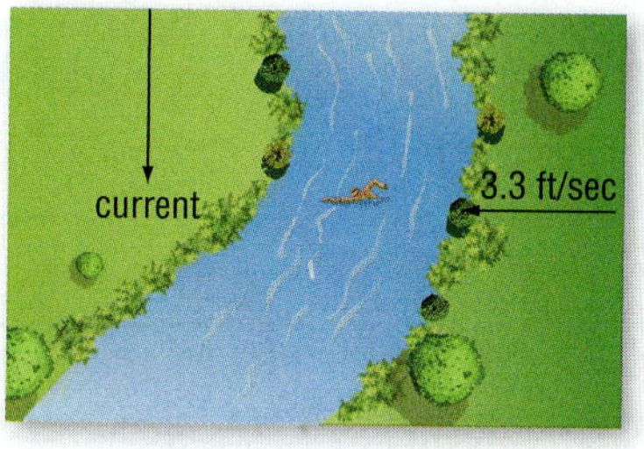

41. **MULTIPLE REPRESENTATIONS** In this problem, you will investigate multiplication of vectors by a scalar quantity.

   a. **GEOMETRIC** Draw a vector on the coordinate plane with the initial point at the origin. Label the initial point *A* and the terminal point *B*. Then draw $\overrightarrow{AC}$ twice as long as $\overrightarrow{AB}$ with the same direction.

   b. **GEOMETRIC** Draw $\overrightarrow{DE}$ with point *D* located at the origin and $\overrightarrow{DF}$ three times as long as $\overrightarrow{DE}$ in the same direction. Draw $\overrightarrow{GH}$ with point *G* located at the origin and $\overrightarrow{GJ}$ four times as long as $\overrightarrow{GH}$ in the same direction.

   c. **TABULAR** Copy and complete the table below.

Coordinates				Scalar Quantity
B		*C*		
E		*F*		
H		*J*		

   d. **VERBAL** Make a conjecture about the relationship between coordinates of the terminal point of the initial vector and the coordinates of the terminal point of the vector that has been increased by a scalar quantity.

   e. **ALGEBRAIC** If a vector has a terminal point (x, y) and it is multiplied by a scalar quantity k, write an algebraic expression for the terminal point of the new vector.

H.O.T. Problems

Use Higher-Order Thinking Skills

42. **WRITING IN MATH** Explain why it is necessary to use both magnitude and direction to describe a vector.

43. **CHALLENGE** What are the resultant velocity and direction of $\vec{x}$?

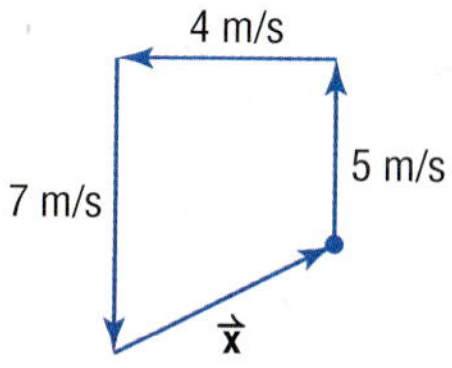

44. **REASONING** Are parallel vectors *sometimes*, *always*, or *never* opposite vectors? Explain.

45. **OPEN ENDED** Draw a pair of parallel vectors.

   a. Draw the sum of the two vectors. What is true of the direction of the vector representing the sum?

   b. Draw the difference of the two vectors. What is true of the vector representing the difference?

46. **WRITING IN MATH** Compare and contrast the parallelogram and triangle methods of adding vectors.

Standardized Test Practice

47. EXTENDED RESPONSE Sydney parked her car and hiked along two paths described by the vectors $\langle 2, 3\rangle$ and $\langle 5, -1\rangle$.

a. What vector represents her hike along both paths?

b. When she got to the end of the second path, how far is she from her car if the numbers represent miles?

48. In right triangle ABC shown below, what is the measure of $\angle A$ to the nearest tenth of a degree?

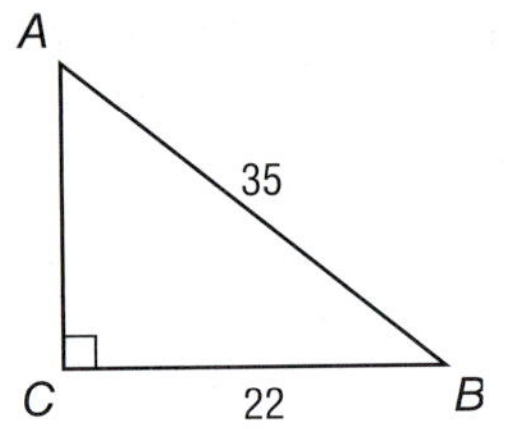

A 32.2 **C** 51.1
B 38.9 **D** 57.8

49. PROBABILITY A die is rolled. Find the probability of rolling a number greater than 4 to the nearest hundredth.

F 0.17 **H** 0.50
G 0.33 **J** 0.67

50. SAT/ACT Caleb followed the two paths shown below to get to his house C from a store S. What is the total distance of the two paths, in meters, from C to S?

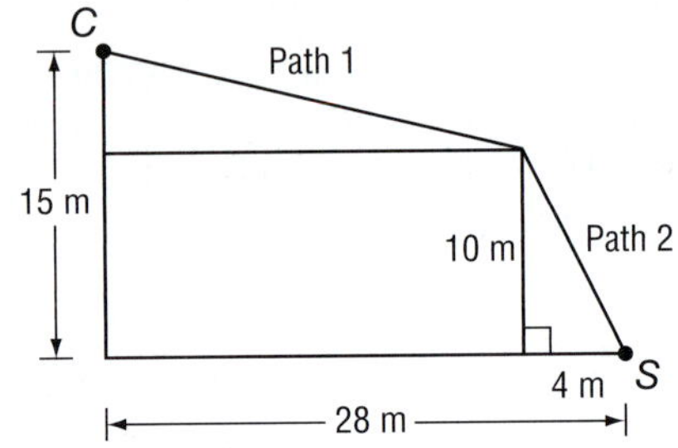

A 10.8 m **C** 31.8 m
B 24.5 m **D** 35.3 m

Spiral Review

Find x. Round angle measures to the nearest degree and side measures to the nearest tenth. (Lesson 8-6)

51.

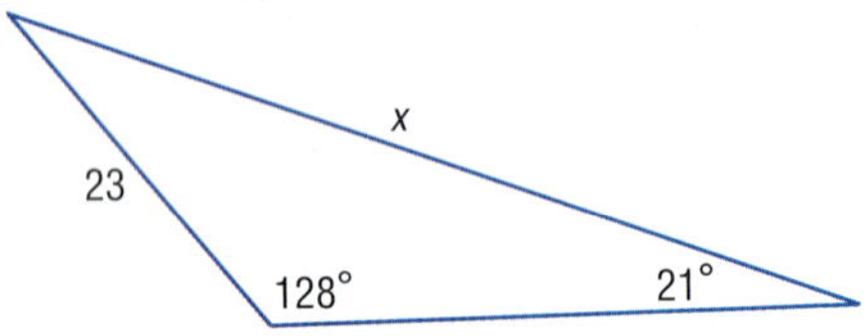

52.

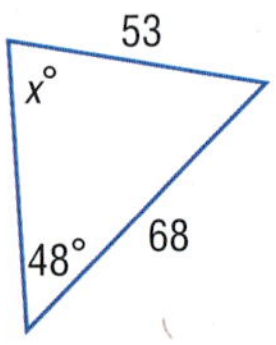

53.

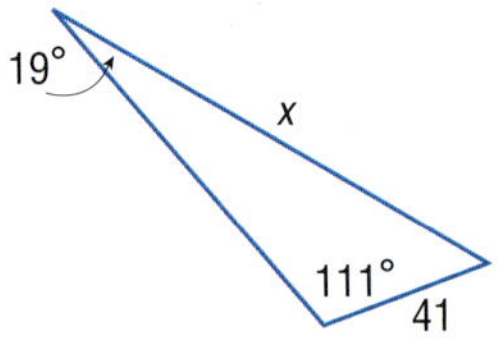

54. SOCCER Adelina is in a soccer stadium 80 feet above the field. The angle of depression to the field is 12°. What is the horizontal distance between Adelina and the soccer field? (Lesson 8-5)

Quadrilateral $WXYZ$ is a rectangle. Find each measure if $m\angle 1 = 30$. (Lesson 6-4)

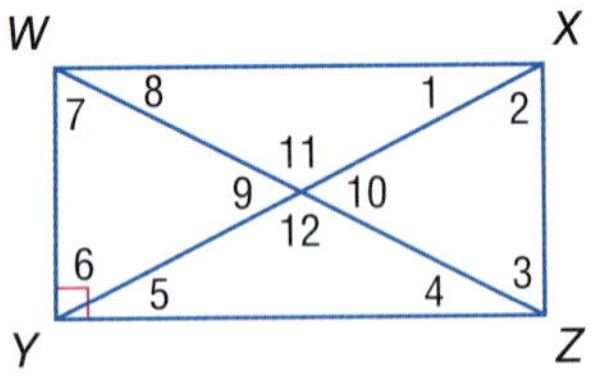

55. $m\angle 2$ **56.** $m\angle 8$ **57.** $m\angle 12$

58. $m\angle 5$ **59.** $m\angle 6$ **60.** $m\angle 3$

Skills Review

Assume that segments and angles that appear to be congruent in each figure are congruent. Indicate which triangles are congruent. (Lesson 4-3)

61.

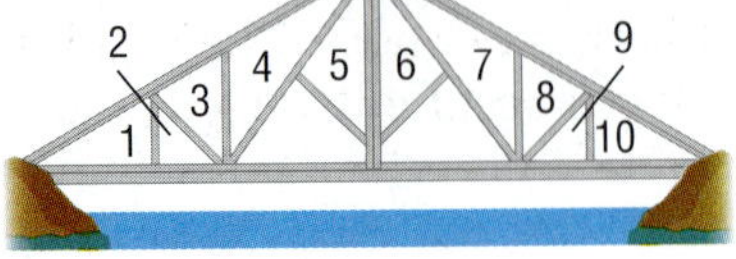

62.

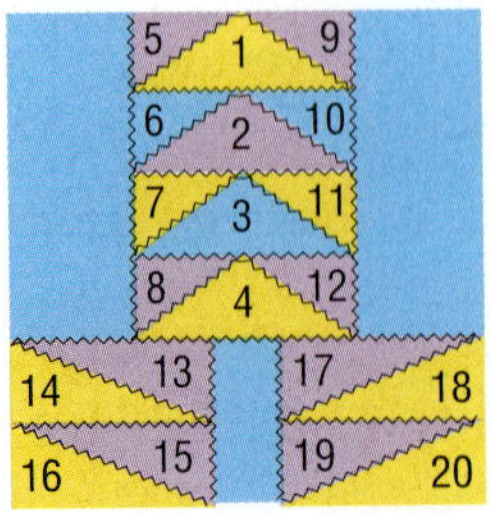

63.

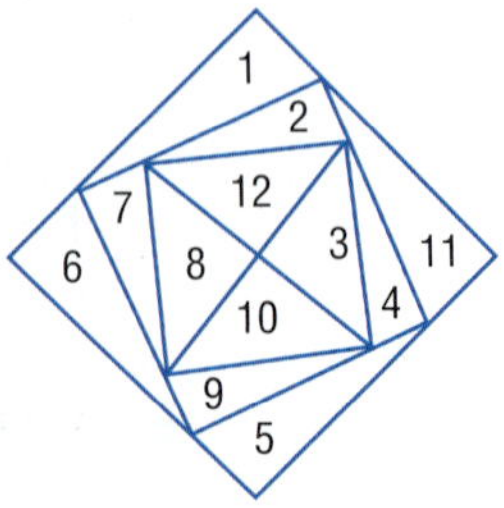

EXTEND 8-7

Geometry Lab
Adding Vectors

Objective
Use scale drawings and direct measurement to find the resultant of two vectors.

You can use scale drawings to represent vectors and solve problems.

ACTIVITY

A small aircraft flies due south at an average speed of 175 miles per hour. The wind is blowing 30° south of west at 25 miles per hour. What is the resultant velocity and direction of the plane?

Step 1 Choose a scale.

Since it is not reasonable to represent the vectors using their actual sizes, you can use a scale drawing. For this activity, let 2 inches represent 100 miles.

Step 2 Make a scale drawing.

Use a ruler and protractor to make a scale drawing of the two vectors.

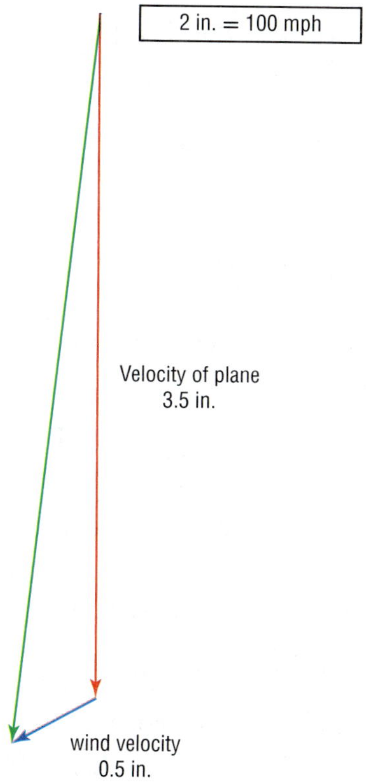

Step 3 Find the resultant.

Find the resultant of the two vectors by using the triangle method or the parallelogram method.

Step 4 Measure the resultant.

Measure the length and angle of the resultant.

The resultant length is $3\frac{3}{4}$ inches, and it makes a 7° angle with the vector representing the velocity of the plane.

Step 5 Find the magnitude and direction of the resultant.

Use the scale with the length that you measured in Step 4 to calculate the magnitude of the plane's resultant velocity.

$$3\frac{3}{4}\text{ in.} \times \frac{100\text{ mph}}{2\text{ in.}} = 187.5\text{ mph}$$

The resultant velocity of the plane is 187.5 miles per hour 7° west of south.

Exercises

Make a scale drawing to solve each problem.

1. **BIKING** Lance is riding his bike west at a velocity of 10 miles per hour. The wind is blowing 5 miles per hour 20° north of east. What is Lance's resultant velocity and direction?

2. **CANOEING** Bianca is trying to canoe directly across a river with a current of 3 miles per hour due west. If Bianca can canoe at a rate of 7 miles per hour, what is her resultant velocity and direction?

CHAPTER 8
Study Guide and Review

Math Online glencoe.com
- STUDY TO GO
- Vocabulary Review

Chapter Summary

Key Concepts

Geometric Mean (Lesson 8-1)

- For two positive numbers a and b, the geometric mean is the positive number x where $a : x = x : b$ is true.

Pythagorean Theorem (Lesson 8-2)

- Let $\triangle ABC$ be a right triangle with right angle C. Then $a^2 + b^2 = c^2$.

Special Right Triangles (Lesson 8-3)

- The measures of the sides of a 45°-45°-90° triangle are x, x, and $x\sqrt{2}$.
- The measures of the sides of a 30°-60°-90° triangle are x, $2x$, and $x\sqrt{3}$.

Trigonometry (Lesson 8-4)

- $\sin A = \dfrac{\text{opposite leg}}{\text{hypotenuse}}$
- $\cos A = \dfrac{\text{adjacent leg}}{\text{hypotenuse}}$
- $\tan A = \dfrac{\text{opposite leg}}{\text{adjacent leg}}$

Angles of Elevation and Depression (Lesson 8-5)

- An angle of elevation is the angle formed by a horizontal line and the line of sight to an object above.
- An angle of depression is the angle formed by a horizontal line and the line of sight to an object below.

Laws of Sines and Cosines (Lesson 8-6)

Let $\triangle ABC$ be any triangle.

- Law of Sines: $\dfrac{\sin A}{a} = \dfrac{\sin B}{b} = \dfrac{\sin C}{c}$
- Law of Cosines: $a^2 = b^2 + c^2 - 2bc \cos A$
  $b^2 = a^2 + c^2 - 2ac \cos B$
  $c^2 = a^2 + b^2 - 2ab \cos C$

Vectors (Lesson 8-7)

- A vector is a quantity with both magnitude and direction.

FOLDABLES® Study Organizer

Be sure the Key Concepts are noted in your Foldable.

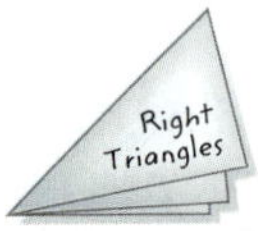

Key Vocabulary

angle of depression (p. 574)
angle of elevation (p. 574)
component form (p. 593)
cosine (p. 562)
direction (p. 593)
geometric mean (p. 531)
inverse cosine (p. 565)
inverse sine (p. 565)
inverse tangent (p. 565)
Law of Sines (p. 582)
Law of Cosines (p. 583)
magnitude (p. 593)
Pythagorean triple (p. 542)
resultant (p. 595)
sine (p. 562)
standard position (p. 593)
tangent (p. 562)
trigonometric ratio (p. 562)
trigonometry (p. 562)
vector (p. 593)

Vocabulary Check

State whether each sentence is *true* or *false*. If *false*, replace the underlined word or phrase to make a true sentence.

1. The arithmetic mean of two numbers is the positive square root of the product of the numbers.
2. Extended ratios can be used to compare three or more quantities.
3. To find the length of the hypotenuse of a right triangle, take the square root of the difference of the squares of the legs.
4. An angle of elevation is the angle formed by a horizontal line and an observer's line of sight to an object below the horizon.
5. The sum of two vectors is the resultant.
6. Magnitude is the angle a vector makes with the x-axis.
7. A vector is in standard position when the initial point is at the origin.
8. The component form of a vector describes the vector in terms of change in x and change in y.
9. The Law of Sines can be used to find an angle measure when given three side lengths.
10. A trigonometric ratio is a ratio of the lengths of two sides of a right triangle.

Lesson-by-Lesson Review

8-1 Geometric Mean (pp. 531–539)

Find the geometric mean between each pair of numbers.

11. 9 and 4 **12.** $\sqrt{20}$ and $\sqrt{80}$ **13.** $\frac{8\sqrt{2}}{3}$ and $\frac{4\sqrt{2}}{3}$

14. Find x, y, and z.

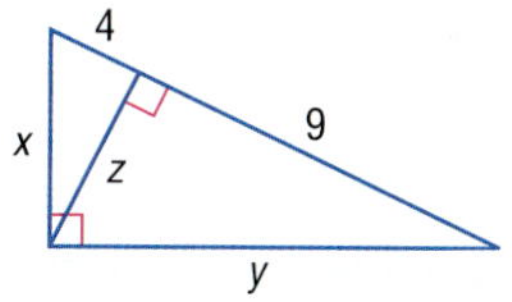

15. DANCES Mike is hanging a string of lights on his barn for a square dance. Using a book to sight the top and bottom of the barn, he can see he is 15 feet from the barn. If his eye level is 5 feet from the ground, how tall is the barn?

EXAMPLE 1

Find the geometric mean between 10 and 15.

$x = \sqrt{ab}$	**Definition of geometric mean**
$= \sqrt{10 \cdot 15}$	**$a = 10$ and $b = 15$**
$= \sqrt{(5 \cdot 2) \cdot (3 \cdot 5)}$	**Factor.**
$= \sqrt{25 \cdot 6}$	**Associative Property**
$= 5\sqrt{6}$	**Simplify.**

8-2 The Pythagorean Theorem and Its Converse (pp. 541–549)

Find x.

16.

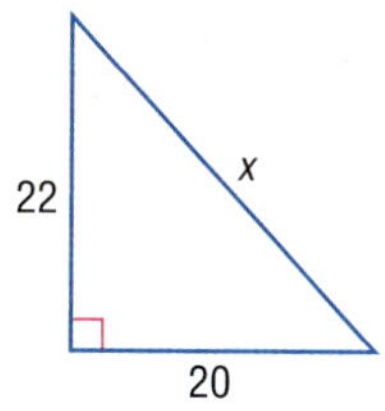

17.

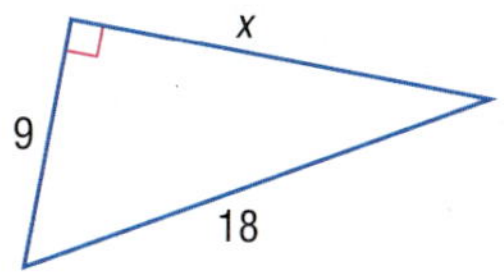

Determine whether each set of numbers can be the measures of the sides of a triangle. If so, classify the triangle as *acute*, *obtuse*, or *right*. Justify your answer.

18. 7, 24, 25 **19.** 13, 15, 16 **20.** 65, 72, 88

21. SWIMMING Alexi walks 27 meters south and 38 meters east to get around a lake. Her sister swims directly across the lake. How many meters to the nearest tenth did Alexi's sister save by swimming?

EXAMPLE 2

Find x.

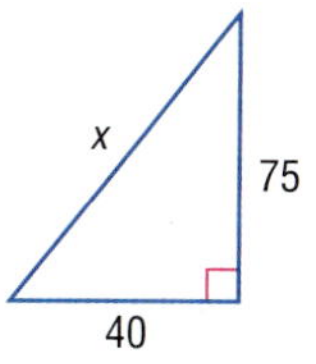

The side opposite the right angle is the hypotenuse, so $c = x$.

$a^2 + b^2 = c^2$	**Pythagorean Theorem**
$40^2 + 75^2 = x^2$	**$a = 40$ and $b = 75$**
$7225 = x^2$	**Simplify.**
$\sqrt{7225} = x$	**Take the positive square root of each side.**
$85 = x$	**Simplify.**

CHAPTER 8 Study Guide and Review

8-3 Special Right Triangles (pp. 552–560)

Find x and y.

22.

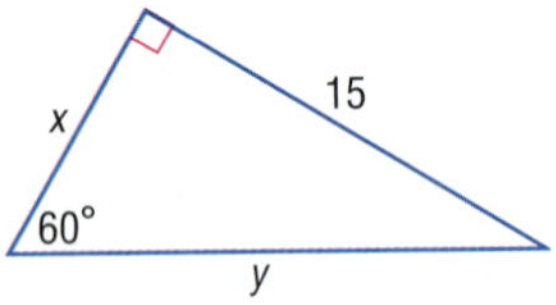

23.

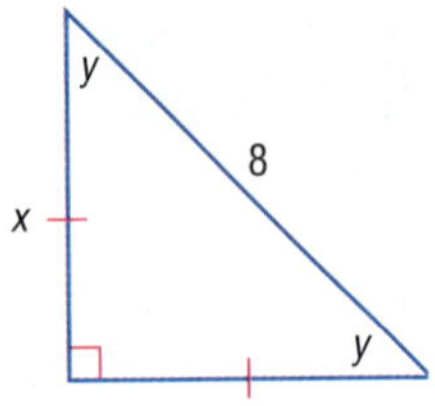

24. CLIMBING Jason is adding a climbing wall to his little brother's swing-set. If he starts building 5 feet out from the existing structure, and wants it to have a 60° angle, how long should the wall be?

EXAMPLE 3

Find x and y.

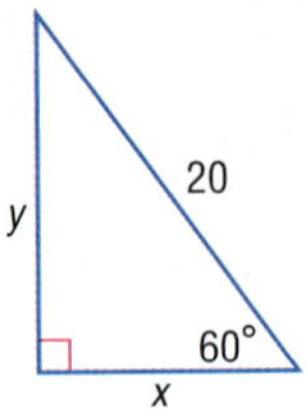

The measure of the third angle in this triangle is 90 – 60 or 30. This is a 30°-60°-90° triangle.

$h = 2s$ **30°-60°-90° Triangle Theorem**

$20 = 2x$ **Substitute.**

$10 = x$ **Divide.**

Now find y, the length of the longer leg.

$\ell = s\sqrt{3}$ **30°-60°-90° Triangle Theorem**

$y = 10\sqrt{3}$ **Substitute.**

8-4 Trigonometry (pp. 562–571)

Express each ratio as a fraction and as a decimal to the nearest hundredth.

25. $\sin A$ **26.** $\tan B$

27. $\sin B$ **28.** $\cos A$

29. $\tan A$ **30.** $\cos B$

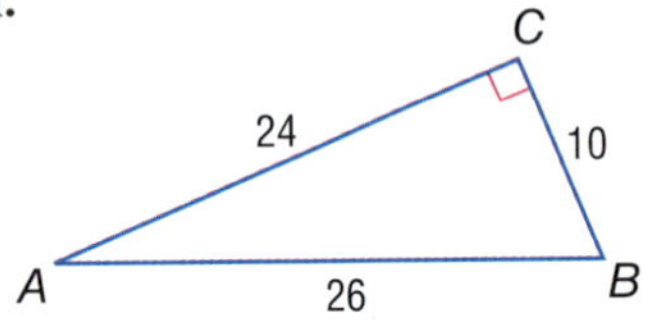

Find x.

31.

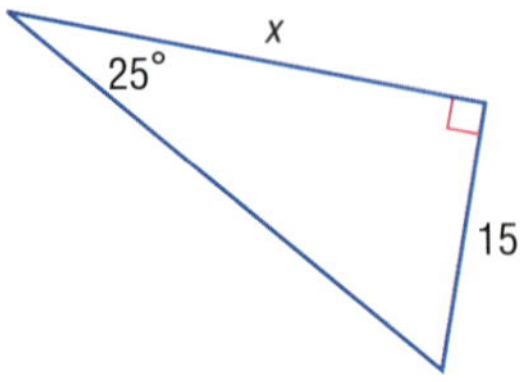

32.

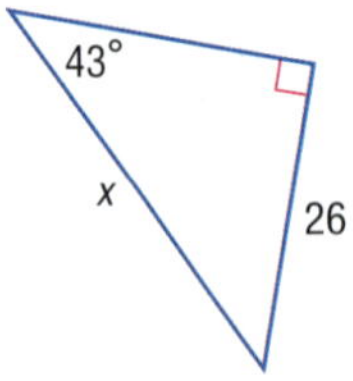

33. GARDENING Sofia wants to put a flower bed in the corner of her yard by laying a stone border that starts 3 feet from the corner of one fence and ends 6 feet from the corner of the other fence. Find the angles, x and y, the fence make with the border.

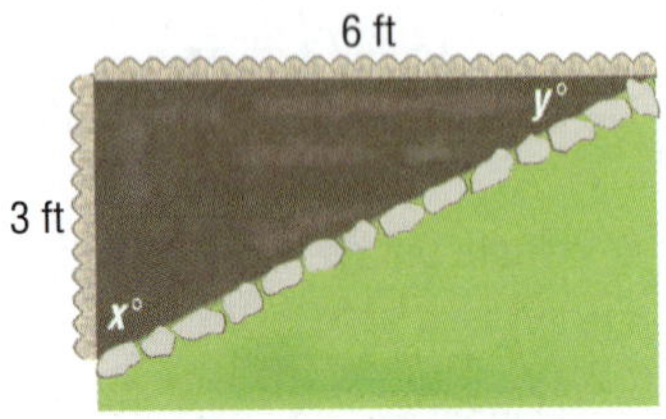

EXAMPLE 4

Express each ratio as a fraction and as a decimal to the nearest hundredth.

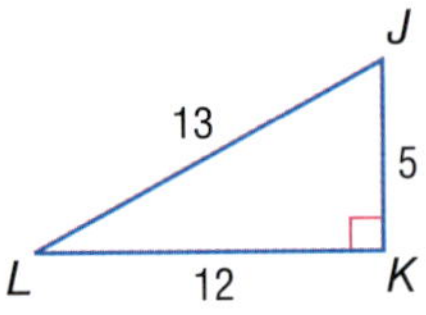

a. $\sin L$

$\sin L = \frac{5}{13}$ or about 0.38 $\quad$ **$\sin L = \frac{\text{opp}}{\text{hyp}}$**

b. $\cos L$

$\cos L = \frac{12}{13}$ or about 0.92 $\quad$ **$\cos L = \frac{\text{adj}}{\text{hyp}}$**

c. $\tan L$

$\tan L = \frac{5}{12}$ or 0.42 $\quad$ **$\tan L = \frac{\text{opp}}{\text{adj}}$**

MIXED PROBLEM SOLVING
For mixed problem-solving practice, see page 1006.

8-5 Angles of Elevation and Depression (pp. 574–581)

34. JOBS Tom delivers papers on a rural route from his car. If he throws a paper from a height of 4 feet, and it lands 15 feet from the car, at what angle of depression did he throw the paper to the nearest degree?

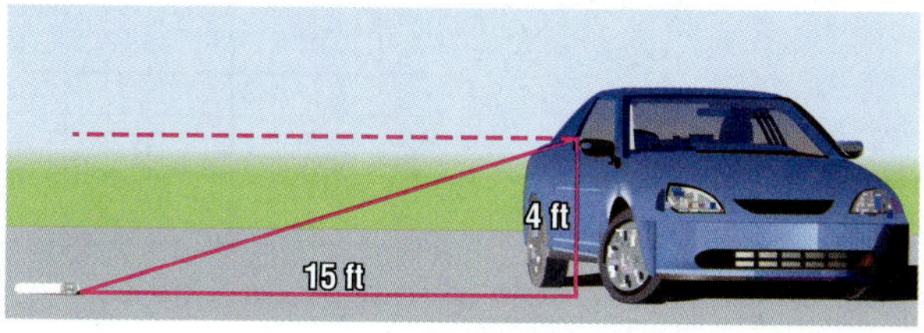

35. TOPIC There is a cell phone tower in the field across from Jen's house. If Jen walks 50 feet from the tower, and finds the angle of elevation from her position to the top of the tower to be 60°, how tall is the tower?

EXAMPLE 5

Sarah's cat climbed up a tree. If she sights her cat at an angle of elevation of 40°, and her eyes are 5 feet off the ground, how high up from the ground is her cat?

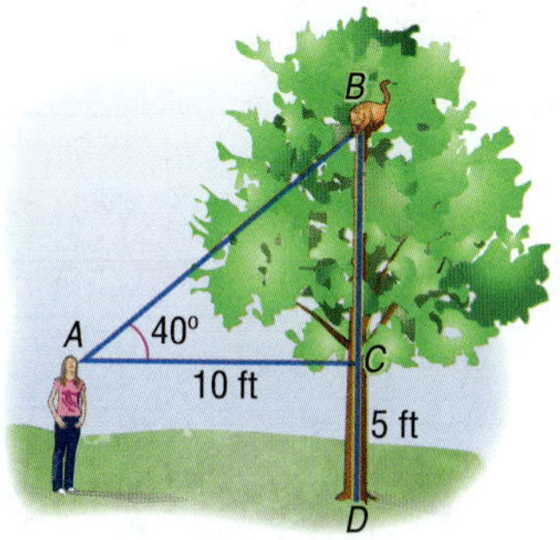

To find the how high the cat is up the tree, find CB.

$\tan 40 = \frac{CB}{10}$	**$\tan = \frac{\text{opposite}}{\text{adjacent}}$**
$10(\tan 40) = CB$	**Multiply.**
$8.4 = CB$	**Simplify.**

Since Sarah's eyes are 5 feet from the ground, add 5 to 8.4. Sarah's cat is 13.4 feet up.

8-6 The Law of Sines and Law of Cosines (pp. 582–591)

Find x. Round angle measures to the nearest degree and side measures to the nearest tenth.

36.

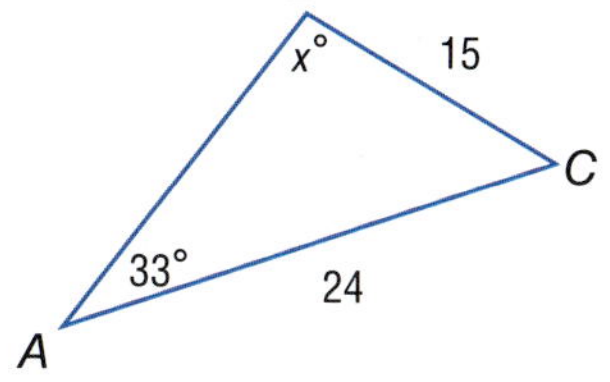

37.

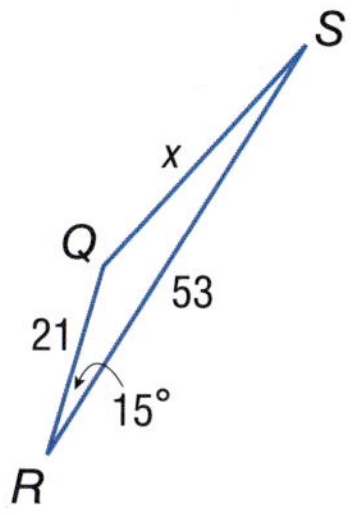

38. SKIING At Crazy Ed's Ski resort, Ed wants to put in another ski lift for the skiers to ride from the base to the summit of the mountain. The run over which the ski lift will go is represented by the figure below. The length of the lift is represented by SB. If Ed needs twice as much cable as the length of $\overline{SB}$, how much cable does he need?

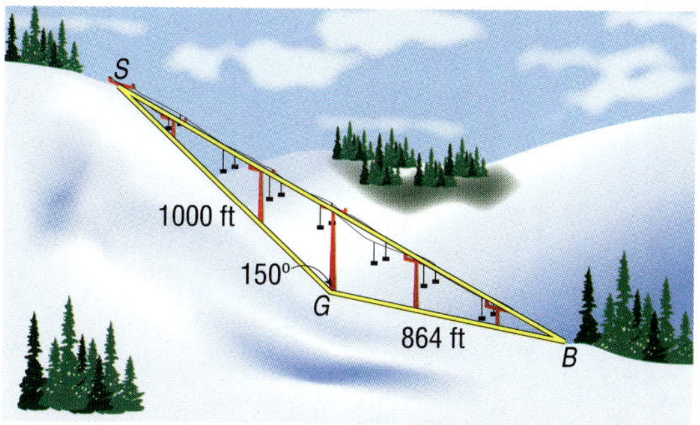

EXAMPLE 6

Find x. Round to the nearest tenth.

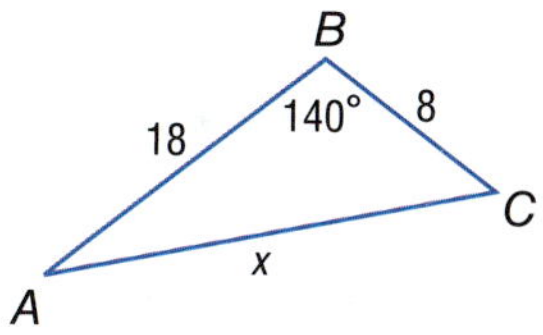

We are given the measures of two sides and their included angle, so use the Law of Cosines.

$b^2 = a^2 + c^2 - 2ac \cos B$	**Law of Cosines**
$x^2 = 8^2 + 18^2 - 2(8)(18) \cos 140°$	**Substitution**
$x^2 = 388 - 288 \cos 140°$	**Simplify.**
$x = \sqrt{388 - 288 \cos 140°} \approx 24.7$	**Take the square root of each side.**

EXAMPLE 7

Find x. Round to the nearest tenth.

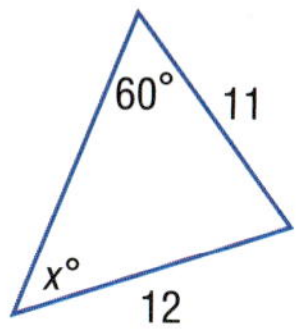

$\frac{\sin A}{a} = \frac{\sin C}{c}$	**Law of Sines**
$\frac{\sin 60}{12} = \frac{\sin x}{11}$	**Substitution**
$11 \sin 60° = 12 \sin x$	**Cross Products Property**
$\frac{11 \sin 60}{12} = \sin x$	**Divide.**

$$x = \sin^{-1} \frac{11 \sin 60}{12} \text{ or about } 52.5°$$

Study Guide and Review

8-7 Vectors (pp. 593–600)

39. Write the component form of the vector shown.

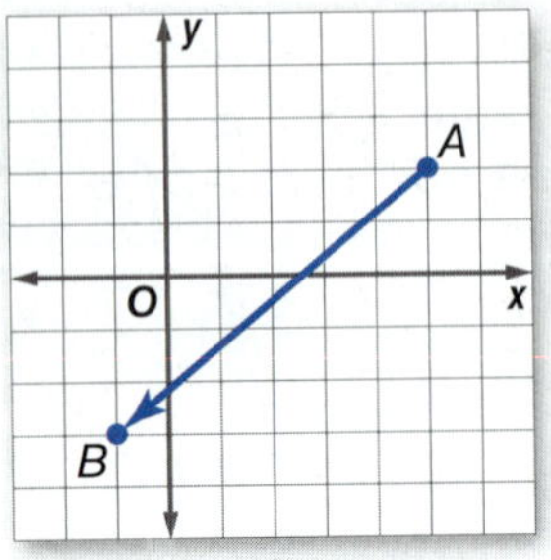

40. Copy the vectors to find $\vec{a} + \vec{b}$.

41. Given that $\vec{s}$ is $\langle 2, -6\rangle$ and $\vec{t}$ is $\langle -10, 7\rangle$, find the component form of $\vec{s} + \vec{t}$.

EXAMPLE 8

Find the magnitude and direction of $\overrightarrow{AB}$ for $A(1, 2)$ and $B(-1, 5)$.

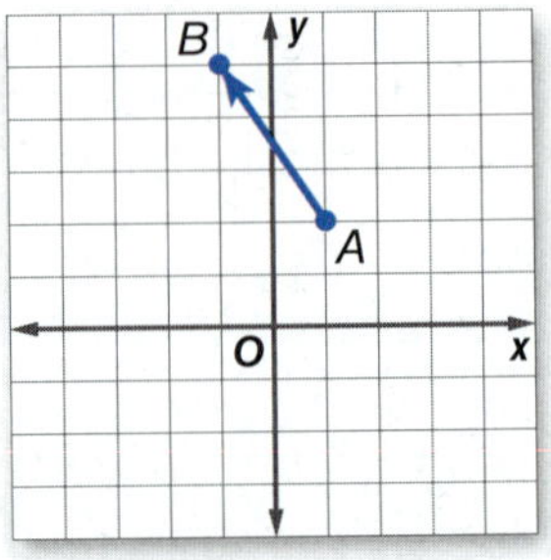

Use the Distance Formula to find the magnitude.

$|\overrightarrow{AB}| = \sqrt{(x_2 - x_1)^2 + (y_2 - y_1)^2}$ **Distance Formula**

$= \sqrt{(-1 - 1)^2 + (5 - 2)^2}$ **Substitute.**

$= \sqrt{13}$ or about 3.6 **Simplify.**

Draw a right triangle with hypotenuse $\overrightarrow{AB}$ and acute angle A.

$\tan A = \left|\frac{5-2}{-1-1}\right|$ or $\frac{3}{2}$ **$\tan = \frac{\text{opp}}{\text{adj}}$; length cannot be negative.**

$m\angle A = \tan^{-1}\left(-\frac{3}{2}\right)$ **Def. of inverse tangent**

≈ -56.3 **Use a calculator.**

The direction of $\overrightarrow{AB}$ is 180 – 56.3 or 123.7°.

Practice Test

Math Online glencoe.com
Chapter Test

Find the geometric mean between each pair of numbers.

1. 7 and 11

2. 12 and 9

3. 14 and 21

4. $4\sqrt{3}$ and $10\sqrt{3}$

5. Find x, y, and z.

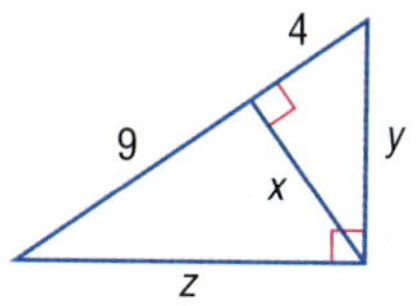

6. FAIRS Blake is setting up his tent at a renaissance fair. If the tent is 8 feet tall, and the tether can be staked no more than two feet from the tent, how long should the tether be?

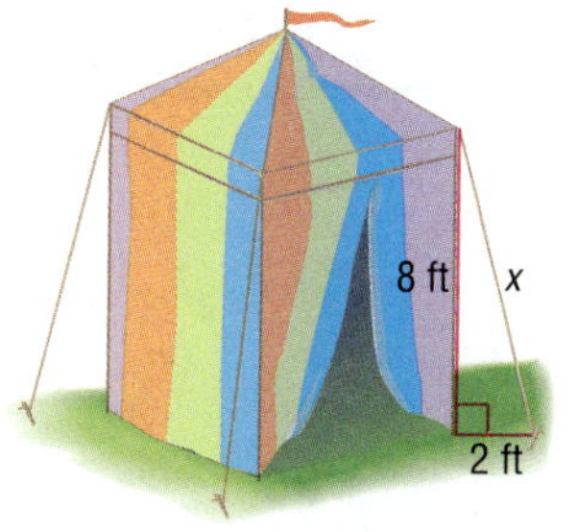

Use a calculator to find the measure of $\angle R$ to the nearest tenth.

7.

8.

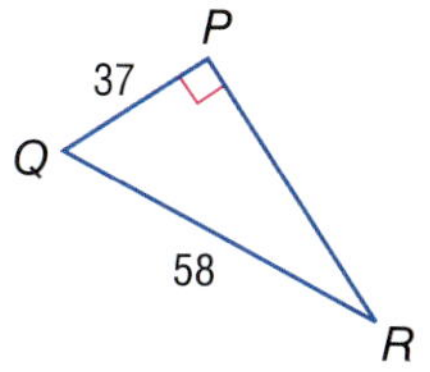

9. Find x and y.

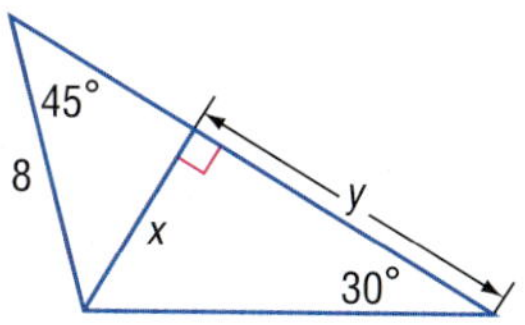

Express each ratio as a fraction and as a decimal to the nearest hundredth.

10. $\cos X$

11. $\tan X$

12. $\tan V$

13. $\sin V$

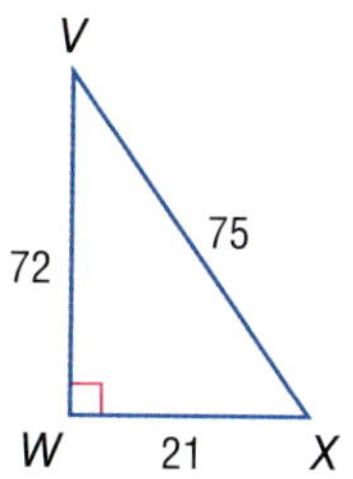

Find the magnitude and direction of each vector.

14. $\overrightarrow{JK}$: $J(-6, -4)$ and $K(-10, -4)$

15. $\overrightarrow{RS}$: $R(1, 0)$ and $S(-2, 3)$

16. SPACE Anna is watching a space shuttle launch 6 miles from Cape Canaveral in Florida. When the angle of elevation from her viewing point to the shuttle is 80°, how high is the shuttle, if it is going straight up?

Find x. Round angle measures to the nearest degree and side measures to the nearest tenth.

17.

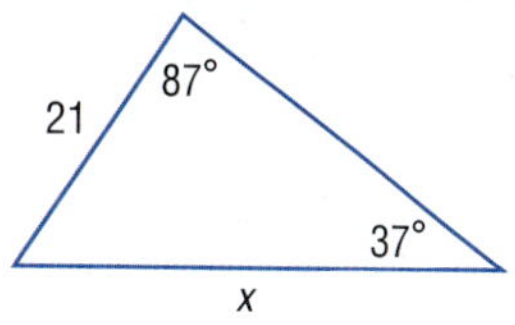

18.

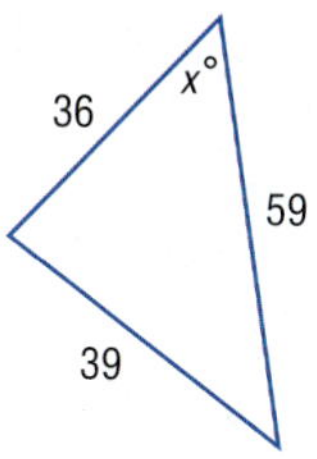

19. MULTIPLE CHOICE Which of the following is the length of the leg of a 45°-45°-90° triangle with a hypotenuse of 20?

A 10

B $10\sqrt{2}$

C 20

D $20\sqrt{2}$

Find x.

20.

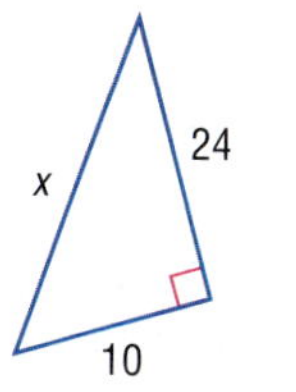

21.

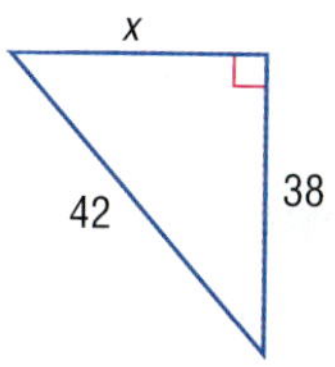

22. WHALE WATCHING Isaac is looking through binoculars on a whale watching trip when he notices a sea otter in the distance. If he is 20 feet above sea level in the boat, and the angle of depression is 30°, how far away from the boat is the otter to the nearest foot?

Write the component form of each vector.

23.

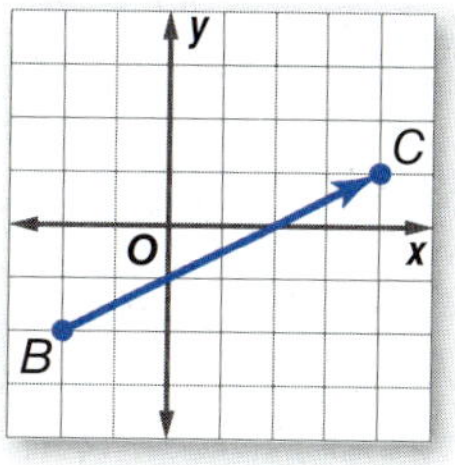

24.

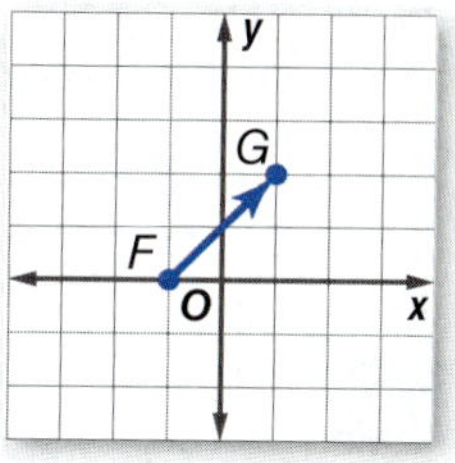

25. Solve $\triangle FGH$. Round to the nearest degree.

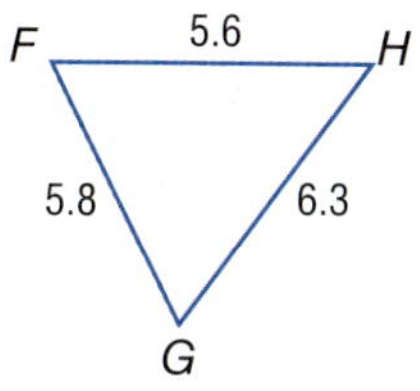

CHAPTER 8

Preparing for Standardized Tests

Use a Formula

Sometimes it is necessary to use a formula to solve problems on standardized tests. In some cases you may even be given a sheet of formulas that you are permitted to reference while taking the test.

Formulas

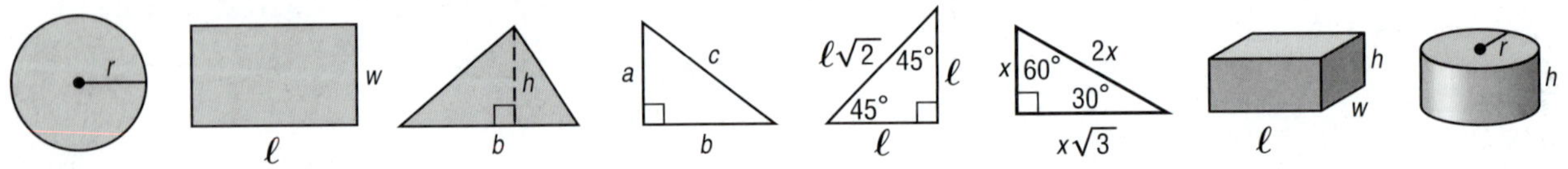

$C = 2\pi r$ $A = \pi r^2$ | $A = \ell w$ | $A = \frac{1}{2}bh$ | $a^2 + b^2 = c^2$ | Special Right Triangles | $V = \ell wh$ | $V = \pi r^2 h$

The are 360 degrees in a circle.

The sum of the measures of the angles of a triangle is 180.

Strategies for Using a Formula

Step 1

Read the problem statement carefully.

Ask yourself:

- What am I being asked to solve?
- What information is given in the problem?
- Are there any formulas that I can use to help me solve the problem?

Step 2

Solve the problem.

- Substitute the known quantities that are given in the problem statement into the formula.
- Simplify to solve for the unknown values in the formula.

Step 3

Check your solution.

- Determine a reasonable range of values for the answer.
- Check to make sure that your answer makes sense.
- If time permits, check your answer.

EXAMPLE

Read the problem. Identify what you need to know. Then use the information in the problem to solve.

The ratio of the width to the height of a high-definition television is 16 : 9. This is also called the *aspect ratio* of the television. The size of a television is given in terms of the diagonal distance across the screen. If an HD television is 25.5 inches tall, what is its screen size?

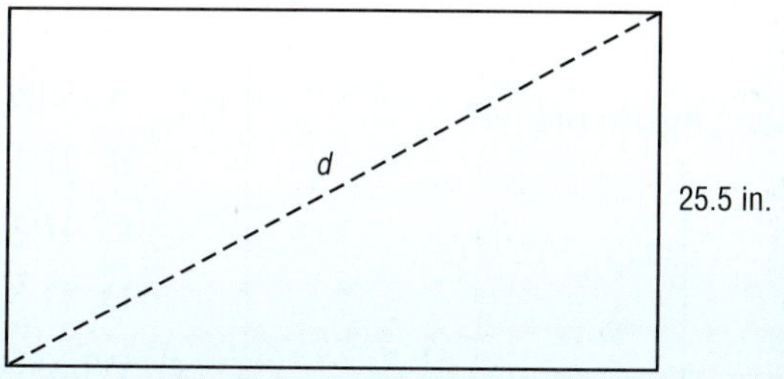

A 48 inches

B 50 inches

C 51 inches

D 52 inches

Read the problem statement carefully. You are given the height of the screen and the ratio of the width to the height. You are asked to find the diagonal distance of the screen. You can use the **Pythagorean Theorem** to solve the problem.

Find the width of the screen. Set up and solve a proportion using the aspect ratio 16 : 9.

$\frac{16}{9} = \frac{w}{25.5}$ ← **width of the screen** / ← **height of the screen**

$9w = 408$ **Cross Products Property**

$w = 45\frac{1}{3}$ **Divide each side by 9.**

So, the width of the screen is $45\frac{1}{3}$ inches. Use the Pythagorean Theorem to solve for the diagonal distance.

$c^2 = a^2 + b^2$ **Pythagorean Theorem**

$c^2 = (25.5)^2 + \left(45\frac{1}{3}\right)^2$ **Substitute for *a* and *b*.**

$c \approx 52.01$ **Simplify. Take the square root of both sides to solve for *c*.**

The diagonal distance of the screen is about 52 inches. So, the answer is D.

Exercises

Read each problem. Identify what you need to know. Then use the information in the problem to solve.

1. Christine is flying a kite on the end of a taut string. The kite is 175 feet above the ground and is a horizontal distance of 130 feet from where Christine is standing. How much kite string has Christine let out? Round to the nearest foot.

A 204 ft

B 218 ft

C 225 ft

D 236 ft

2. What is the value of x below to the nearest tenth?

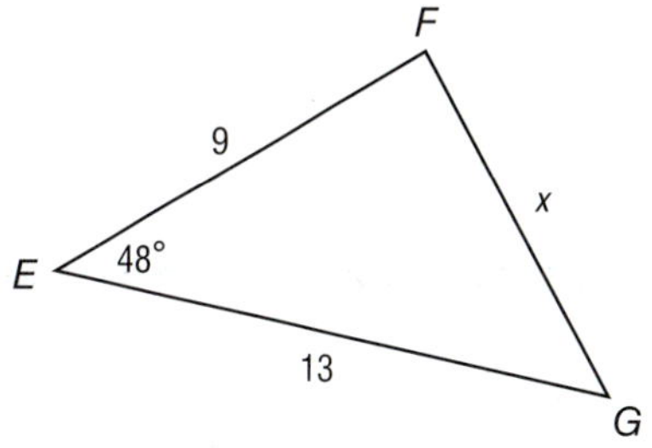

F 9.7 **G** 10.2 **H** 10.5 **J** 11.1

CHAPTER 8

Standardized Test Practice

Cumulative, Chapters 1 through 8

Multiple Choice

Read each question. Then fill in the correct answer on the answer document provided by your teacher or on a sheet of paper.

1. What is the value of x in the figure below?

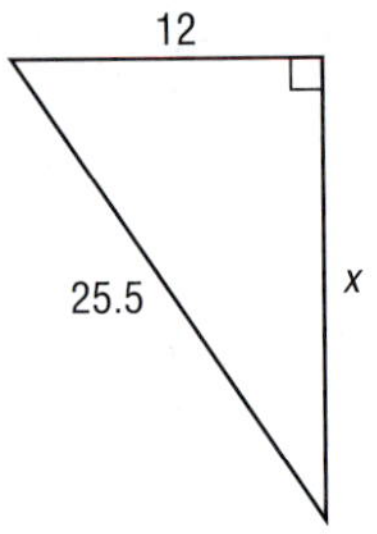

A 22.5

B 23

C 23.5

D 24

2. A baseball diamond is a square with 90-ft sides. What is the length from 3rd base to 1st base? Round to the nearest tenth.

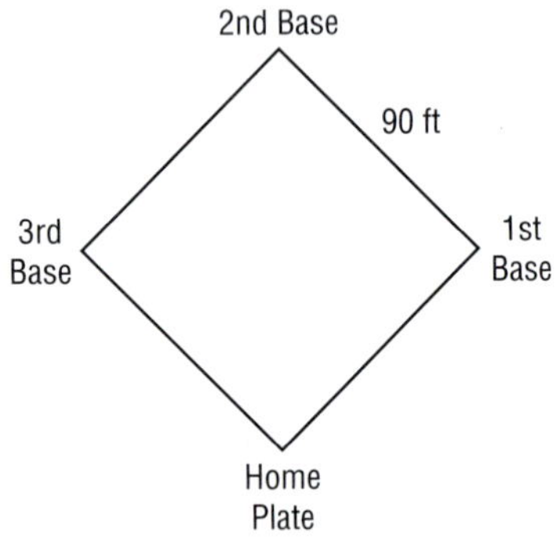

F 155.9 ft

G 141.6 ft

H 127.3 ft

J 118.2 ft

Test-Taking Tip

Question 1 Some test items require the use of a formula to solve them. Use the Pythagorean Theorem to find x.

3. The scale of a map is 1 inch = 4.5 kilometers. What is the distance between two cities that are 2.4 inches apart on the map?

A 10.8 kilometers

B 11.1 kilometers

C 11.4 kilometers

D 11.5 kilometers

4. What is the value of x in the figure below? Round to the nearest tenth.

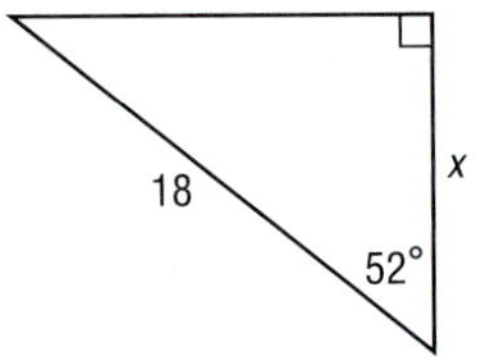

F 10.5

G 11.1

H 13.6

J 14.2

5. What type of triangle is formed by the locations of Lexington, Somerset, and Bowling Green?

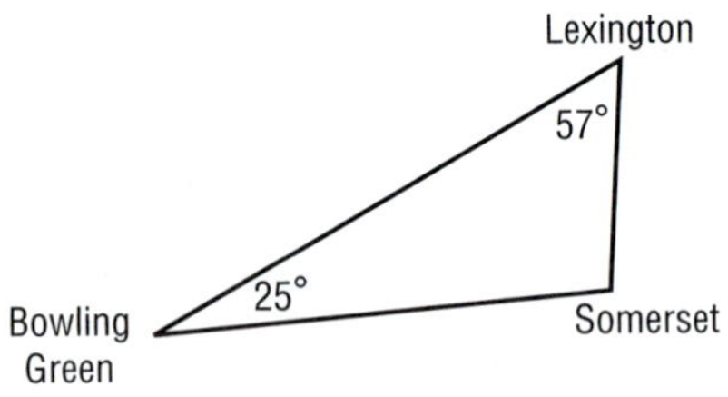

A acute

B equiangular

C obtuse

D right

6. Grant is flying a kite on the end of a string that is 350 feet long. The angle elevation from Grant to the kite is 74°. How high above the ground is the kite? Round your answer to the nearest tenth if necessary.

F 336.4ft

G 295.6 ft

H 141.2 ft

J 96.5 ft

Short Response/Gridded Response

Record your answers on the answer sheet provided by your teacher or on a sheet of paper.

7. **GRIDDED RESPONSE** Find x in the figure below. Round your answer to the nearest tenth if necessary.

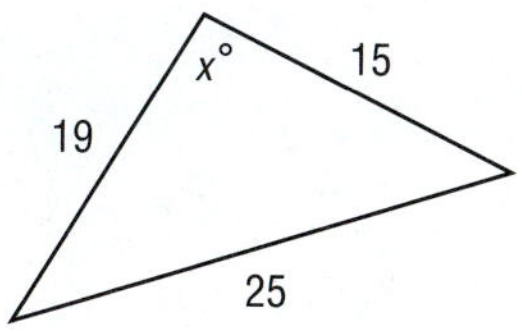

8. Amy is paddling her canoe across a lake at a speed of 10 feet per second headed due north. The wind is blowing 40° east of north with a velocity of 2.8 feet. What is Amy's resultant velocity? Express your answer as a vector. Show your work.

9. Janice used a 16-inch dowel and a 21-inch dowel to build a kite as shown below. What is the perimeter of her kite?

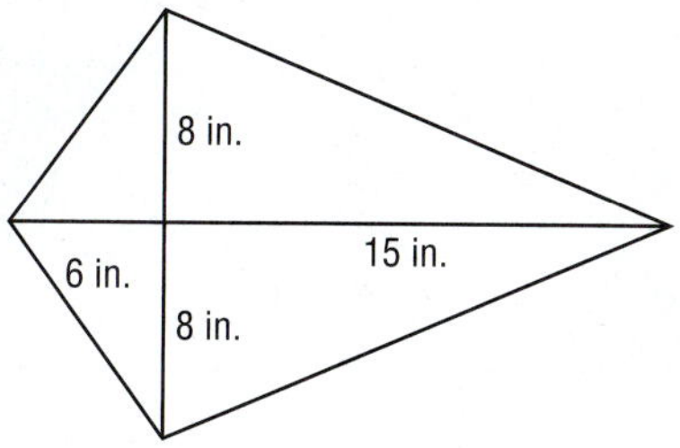

10. **GRIDDED RESPONSE** A model airplane takes off at an angle of elevation of 30°. How high will the plane be after traveling 100 feet horizontally? Round to the nearest tenth. Show your work.

11. According to the Perpendicular Bisector Theorem, what is the length of segment AB below?

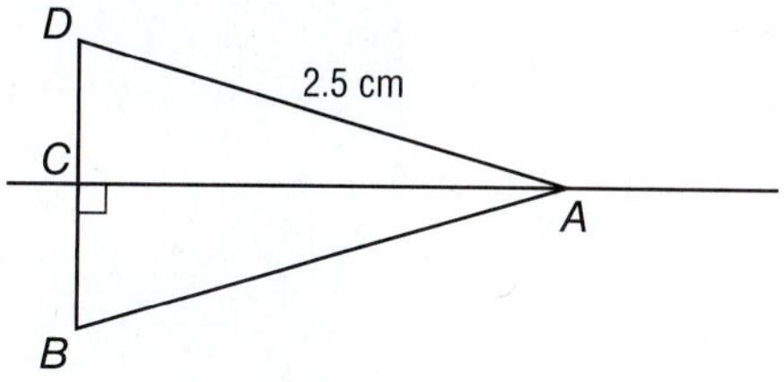

12. Find the slope of the line that contains the points (7, 2) and (3, 4).

13. If $EG = 15$ meters, what is the length of segment FG?

14. What is the contrapositive of the statement below?

If a quadrilateral is a rectangle, then it is a parallelogram.

Extended Response

Record your answers on a sheet of paper. Show your work.

15. Refer to the triangle shown below.

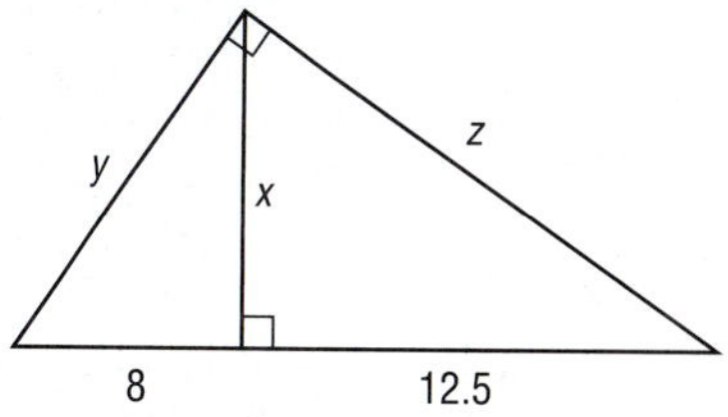

a. Find x to the nearest tenth.
b. Find y to the nearest tenth.
c. Find z to the nearest tenth.

Need Extra Help?

If you missed Question...	1	2	3	4	5	6	7	8	9	10	11	12	13	14	15
Go to Lesson or Page...	8-2	8-3	7-7	8-4	4-1	8-5	8-6	8-7	6-6	8-5	5-1	3-3	1-2	2-3	8-1

CHAPTER 9 Transformations and Symmetry

Then

In Chapter 4, you learned to identify reflections, translations, and rotations.

Now

In Chapter 9, you will:

- Name and draw figures that have been reflected, translated, rotated, or dilated.
- Recognize and draw compositions of translations.
- Identify symmetry in two- and three-dimensional figures.

Why?

PHOTOGRAPHY
Photographers use reflections, rotations, and symmetry to make photographs interesting and visually appealing.

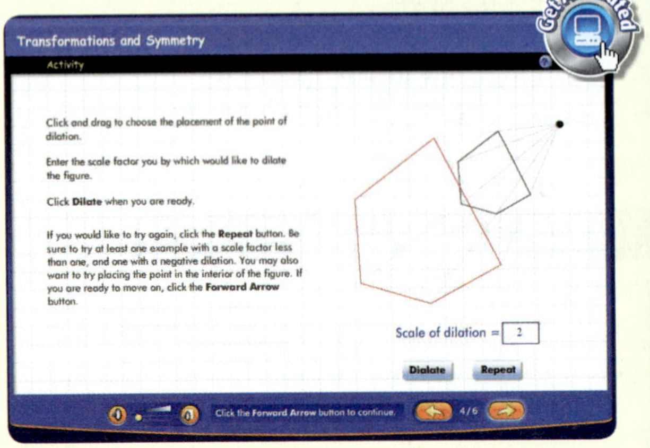

Get Ready for Chapter 9

Diagnose Readiness You have two options for checking Prerequisite Skills.

Text Option

Take the Quick Check below. Refer to the Quick Review for help.

QuickCheck

Identify the type of congruence transformation shown as a *reflection, translation,* or *rotation.* (Lesson 4-7)

1. A to B
2. D to A
3. A to C

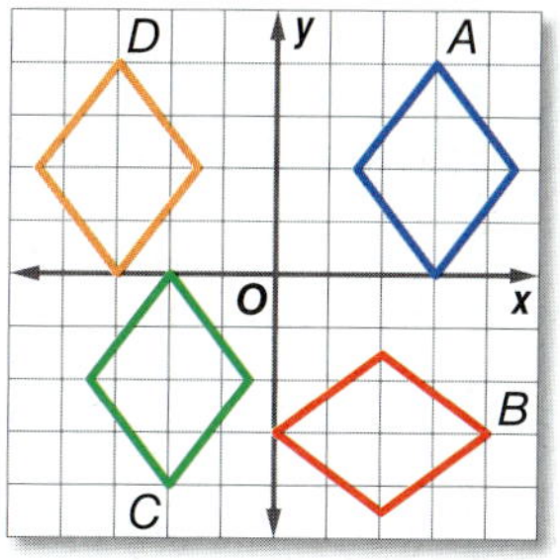

Find the sum of each pair of vectors. (Lesson 8-7)

4. $\langle 13, -4\rangle + \langle -11, 9\rangle$
5. $\langle 6, -31\rangle + \langle -22, 3\rangle$
6. **BAND** During part of a song, the drummer in a marching band moves from (1, 4) to (5, 1). Write the component form of the vector that describes his movement.

Determine whether the dilation from A to B is an *enlargement* or a *reduction.* Then find the scale factor of the dilation. (Lesson 7-6)

7.

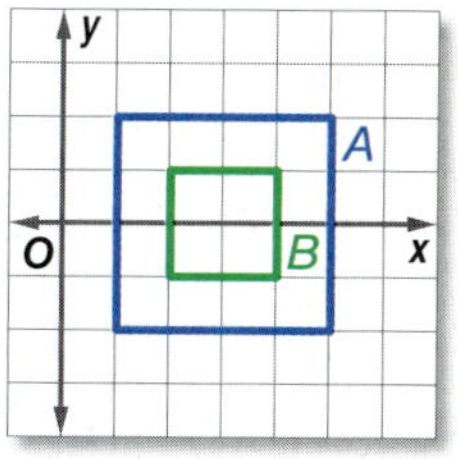

8. **PLAYS** Bob is making a model of an ant for a play. Find the scale factor of the model if the ant is $\frac{1}{2}$ inch long and the model is 1 foot long.

QuickReview

EXAMPLE 1

Identify the type of congruence transformation shown as a *reflection, translation,* or *rotation.*

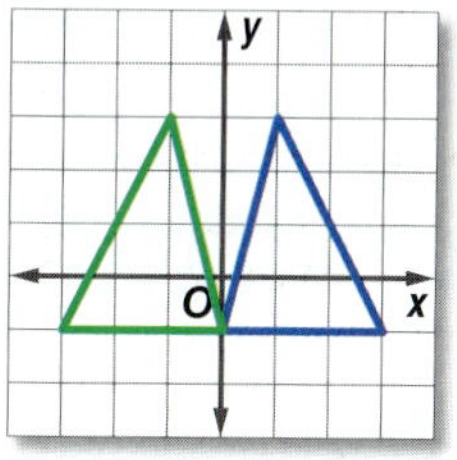

Each vertex and its image are the same distance from the y-axis. This is a reflection.

EXAMPLE 2

Write the component form of $\overrightarrow{AB}$ for $A(-1, 1)$ and $B(4, -3)$.

$\overrightarrow{AB} = \langle x_2 - x_1, y_2 - y_1\rangle$ **Component form of vector**

$= \langle 4 - (-1), -3 - 1\rangle$ **Substitute**

$= \langle 5, -4\rangle$ **Simplify.**

EXAMPLE 3

Determine whether the dilation from A to B is an *enlargement* or a *reduction.* Then find the scale factor of the dilation.

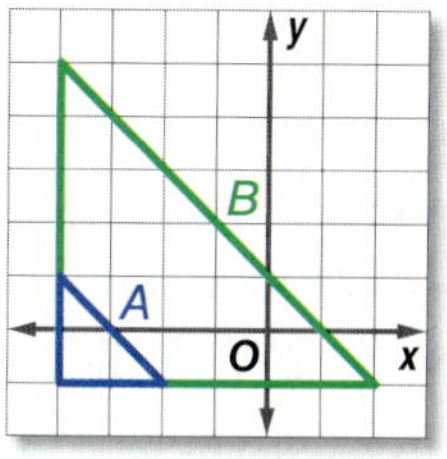

B is larger than A, so it is an enlargement.

The distance between the vertices of A is 2 and the corresponding distance for B is 6.

The scale factor is $\frac{6}{2}$ or 3.

Online Option

Math Online Take a self-check Chapter Readiness Quiz at glencoe.com.

Get Started on Chapter 9

You will learn several new concepts, skills, and vocabulary terms as you study Chapter 9. To get ready, identify important terms and organize your resources. You may wish to refer to **Chapter 0** to review prerequisite skills.

FOLDABLES® Study Organizer

Transformations and Symmetry Make this Foldable to help you organize your Chapter 9 notes about transformations and symmetry. Begin with three sheets of notebook paper.

1. **Fold** each sheet of paper in half.

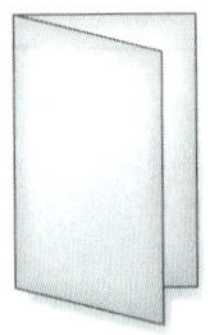

2. **Open** the folded papers and fold each paper lengthwise two inches, to form a pocket.

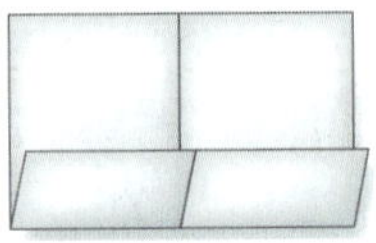

3. **Glue** the sheets side-by-side to create a booklet.

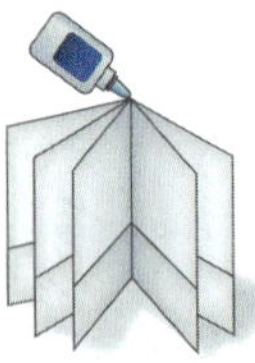

4. **Label** each of the pockets as shown.

Math Online glencoe.com

- Study the chapter online
- Explore **Get Animated** activities
- Get extra help from your own **Personal Tutor**
- Use **Extra Examples** for additional help
- Take a **Self-Check Quiz**
- **Review Vocabulary** in fun ways

New Vocabulary

English		Español
line of reflection	• p. 615 •	línea de reflexión
center of rotation	• p. 632 •	centro de rotación
angle of rotation	• p. 632 •	ángulo de rotación
composition of transformations	• p. 641 •	compasición de transformaciones
symmetry	• p. 653 •	símetria
line symmetry	• p. 653 •	símetria lineal
line of symmetry	• p. 653 •	eje de símetria

Review Vocabulary

reflection • p. 294 • reflexión a transformation representing a flip of the figure over a point, line or plane

rotation • p. 294 • rotación a transformation that turns every point of a preimage through a specified angle and direction about a fixed point

translation • p. 294 • translación a transformation that moves all points of a figure the same distance in the same direction

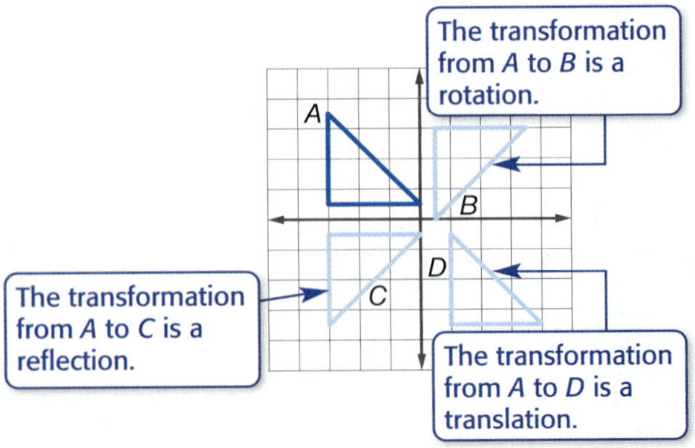

Multilingual eGlossary glencoe.com

9-1 Reflections

Then

You identified reflections and verified them as congruence transformations. (Lesson 4-7)

Now

- Draw reflections.
- Draw reflections in the coordinate plane.

New Vocabulary

line of reflection

Math Online
glencoe.com

Why?

Photographers often use reflections to add tension and more interesting composition to their photographs. Notice in this water reflection that the distance a point lies above the water line appears the same as the distance its image lies below the water.

Draw Reflections In Lesson 4-7, you learned that a reflection or *flip* is a transformation in a line called the **line of reflection**. Each point of the preimage and its corresponding point on the image are the same distance from this line.

Key Concept Reflection in a Line

A reflection in a line maps a point to its image such that

- if the point is on the line, then the image and preimage are the same point, or
- if the point does not lie on the line, the line is the perpendicular bisector of the segment joining the two points.

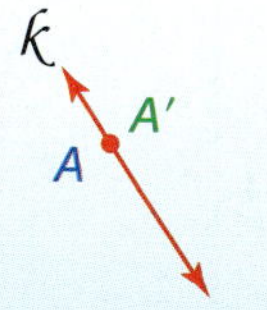

A is on line k.

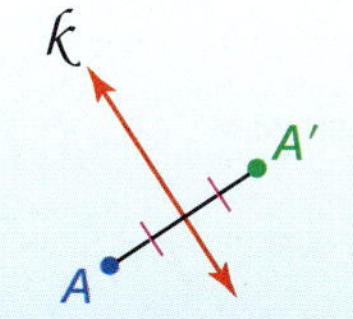

A is not on line k.

A′, A″, A‴, and so on, name corresponding points for one or more transformations.

To reflect a polygon in a line, reflect each of the polygon's vertices. Then connect these vertices to form the reflected image.

EXAMPLE 1 Reflect a Figure in a Line

Copy the figure and the given line of reflection. Then draw the reflected image in this line using a ruler.

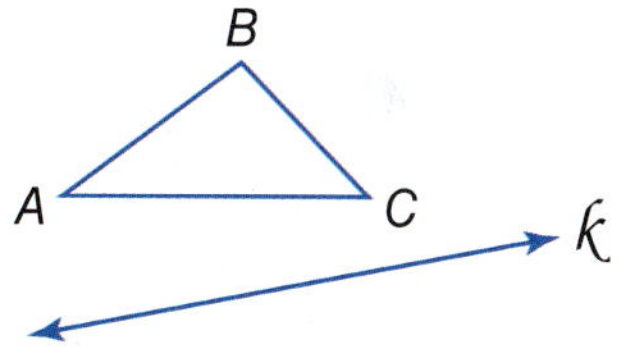

Step 1 Draw a line through each vertex that is perpendicular to line *k*.

Step 2 Measure the distance from point *A* to line *k*. Then locate *A′* the same distance from line *k* on the opposite side.

Step 3 Repeat Step 2 to locate points *B′* and *C′*. Then connect vertices *A′*, *B′*, and *C′* to form the reflected image.

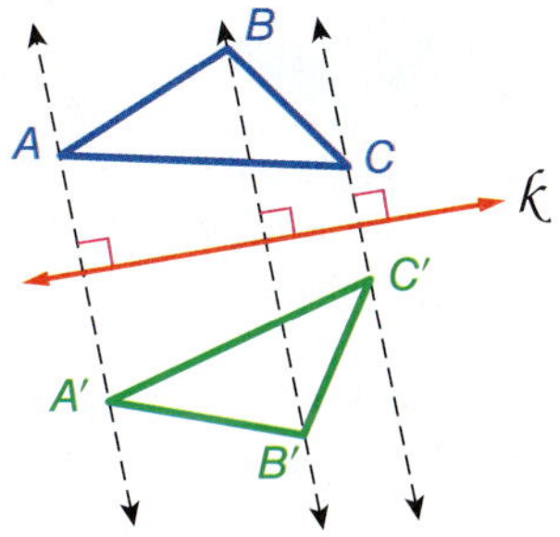

Guided Practice

1A.

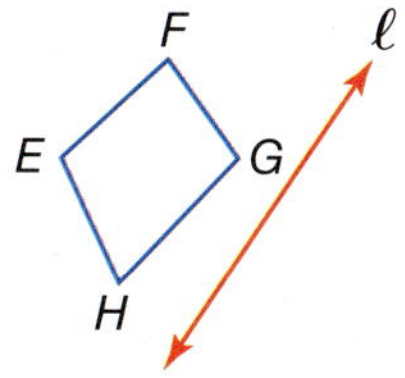

1B.

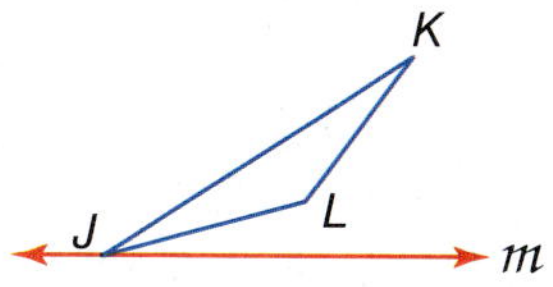

1C.

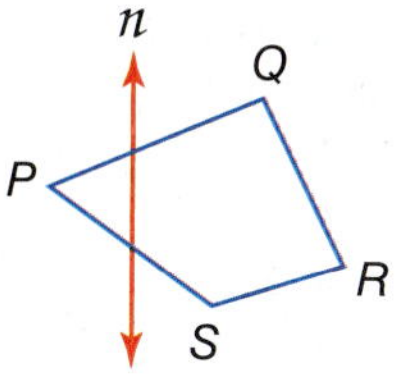

Personal Tutor glencoe.com

Recall that a reflection is a *congruence transformation* or *isometry*. In the figure in Example 1, $\triangle ABC \cong \triangle A'B'C'$.

Real-World EXAMPLE 2 Minimize Distance by Using a Reflection

SHOPPING Suppose you are going to buy clothes in Store B, return to your car, and then buy shoes at Store G. Where along line s of parking spaces should you park to minimize the distance you will walk?

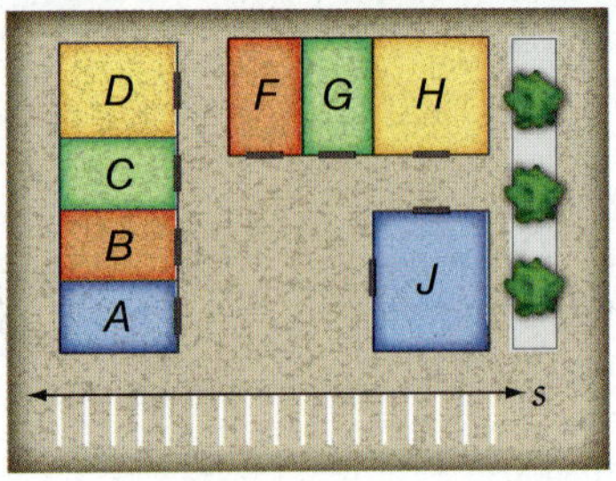

Understand You are asked to locate a point P on line s such that $BP + PG$ has the least possible value.

Plan The total distance from B to P and then from P to G is least when these three points are collinear. Use the reflection of point B in line s to find the location for point P.

Solve Draw $\overline{B'G}$. Locate P at the intersection of line s and $\overline{B'G}$.

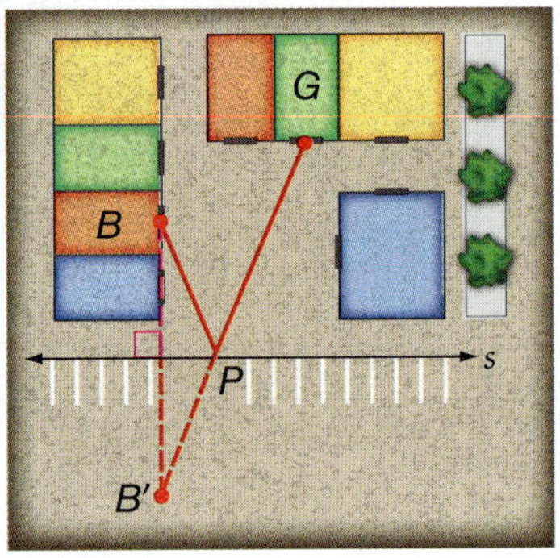

Check Guess and check other locations on line s for point P. Compare the sum $BP + PG$ for each case to verify that the location found for P minimizes this sum.

Guided Practice

2. **TICKET SALES** Joy wants to select a good location to sell tickets for a dance. Locate point P such that the distance someone would have to walk from Hallway A, to point P on the wall, and then to their next class in Hallway B is minimized.

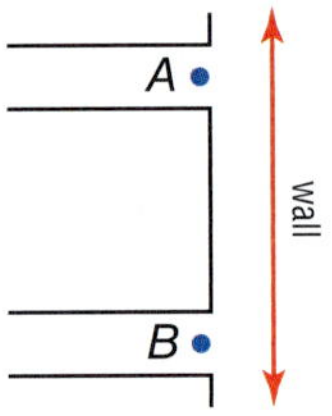

Personal Tutor glencoe.com

Draw Reflections in the Coordinate Plane Reflections can also be performed in the coordinate plane by using the techniques presented in Example 3.

StudyTip

Characteristics of a Reflection Reflections, like all isometries, preserve distance, angle measure, betweenness of points, and collinearity. The orientation of a preimage and its image, however, are reversed.

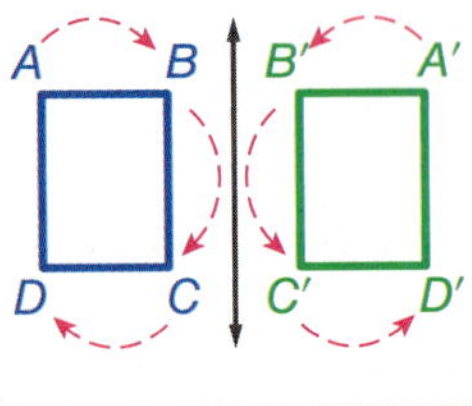

EXAMPLE 3 Reflect a Figure in a Horizontal or Vertical Line

Triangle JKL has vertices $J(0, 3)$, $K(-2, -1)$, and $L(-6, 1)$. Graph $\triangle JKL$ and its image in the given line.

a. $x = -4$

Use the horizontal grid lines to find a corresponding point for each vertex so that each vertex and its image are equidistant from the line $x = -4$.

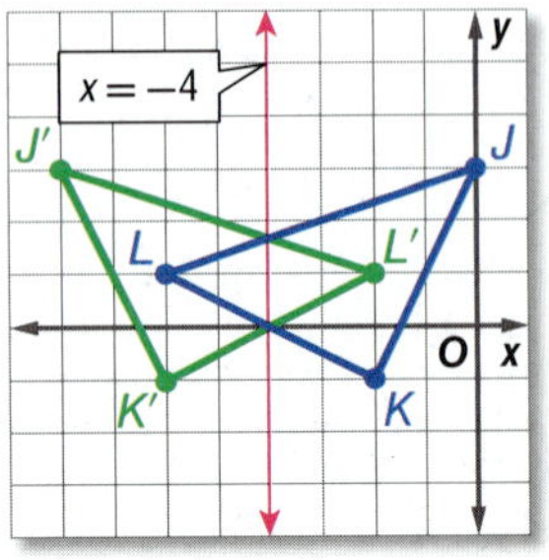

b. $y = 2$

Use the vertical grid lines to find a corresponding point for each vertex so that each vertex and its image are equidistant from the line $y = 2$.

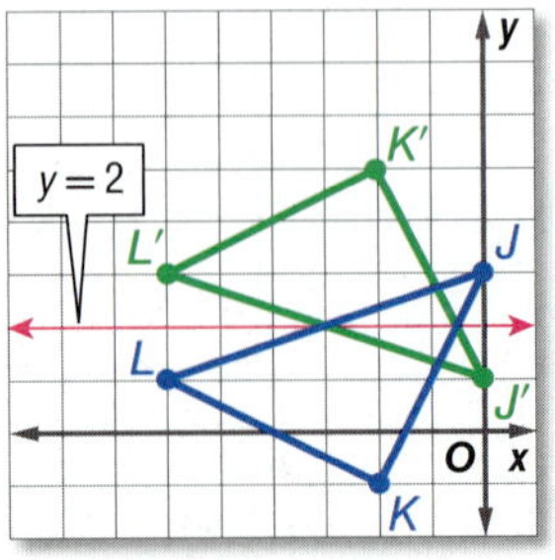

Guided Practice

Trapezoid *RSTV* has vertices $R(-1, 1)$, $S(4, 1)$, $T(4, -1)$, and $V(-1, -3)$. Graph trapezoid *RSTV* and its image in the given line.

3A. $y = -3$

3B. $x = 2$

Personal Tutor glencoe.com

When the line of reflection is the x- or y-axis, you can use the following rule.

Key Concept — Reflection in the x- or y-axis

Reflection in the x-axis

Words To reflect a point in the x-axis, multiply its y-coordinate by -1.

Symbols $(x, y) \rightarrow (x, -y)$

Example

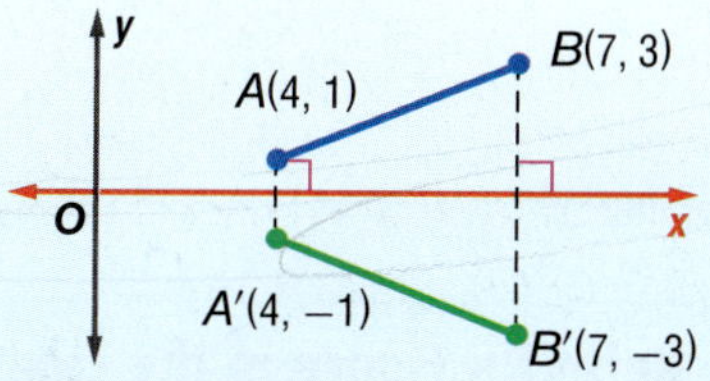

Reflection in the y-axis

Words To reflect a point in the y-axis, multiply its x-coordinate by -1.

Symbols $(x, y) \rightarrow (-x, y)$

Example

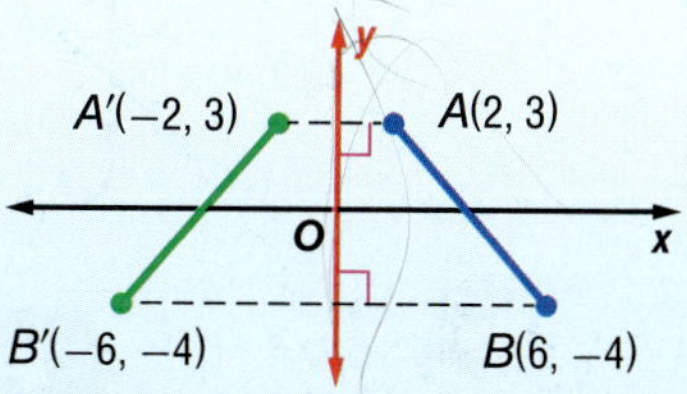

ReadingMath

Coordinate Function Notation The expression $P(a, b) \rightarrow P'(a, -b)$ can be read as *point P with coordinates a and b is mapped to new location P prime with coordinates a and negative b.*

EXAMPLE 4 Reflect a Figure in the x- or y-axis

Graph each figure and its image under the given reflection.

a. **$\triangle ABC$ with vertices $A(-5, 3)$, $B(2, 0)$, and $C(1, 2)$ in the x-axis**

Multiply the y-coordinate of each vertex by -1.

$(x, y) \rightarrow (x, -y)$

$A(-5, 3) \rightarrow A'(-5, -3)$

$B(2, 0) \rightarrow B'(2, 0)$

$C(1, 2) \rightarrow C'(1, -2)$

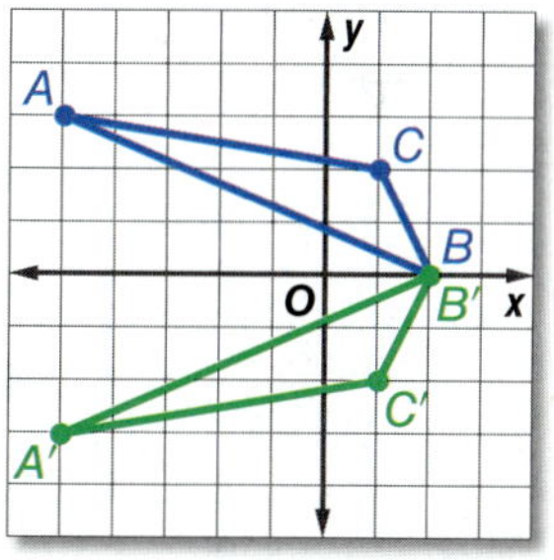

b. **parallelogram *PQRS* with vertices $P(-4, 1)$, $Q(2, 3)$, $R(2, -1)$, and $S(-4, -3)$ in the y-axis**

Multiply the x-coordinate of each vertex by -1.

$(x, y) \rightarrow (-x, y)$

$P(-4, 1) \rightarrow P'(4, 1)$

$Q(2, 3) \rightarrow Q'(-2, 3)$

$R(2, -1) \rightarrow R'(-2, -1)$

$S(-4, -3) \rightarrow S'(4, -3)$

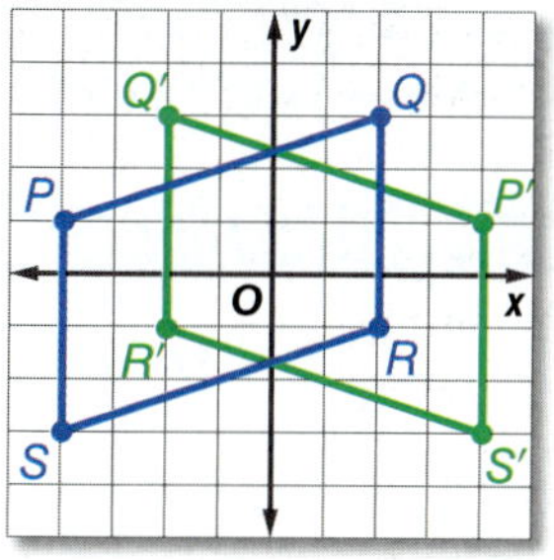

StudyTip

Invariant Points In Example 4a, point *B* is called an *invariant point* because it maps onto itself. Only points that lie on the line of reflection are invariant under a reflection.

Guided Practice

4A. rectangle with vertices $E(-4, -1)$, $F(2, 2)$, $G(3, 0)$, $H(-3, -3)$ in the x-axis

4B. $\triangle JKL$ with vertices $J(3, 2)$, $K(2, -2)$, $L(4, -5)$ in the y-axis

Personal Tutor glencoe.com

Review Vocabulary

perpendicular lines Two nonvertical lines are perpendicular if and only if the product of their slopes is −1. (Lesson 3-3)

You can also reflect an image in the line $y = x$.

The slope of $y = x$ is 1. In the graph shown, $\overline{CC'}$ is perpendicular to $y = x$, so its slope is −1. From $C(-3, 2)$, move right 2.5 units and down 2.5 units to reach $y = x$. From this point on $y = x$, move right 2.5 units and down 2.5 units to locate $C'(2, -3)$. Using a similar method, the image of $D(-3, -1)$ is found to be $D'(-1, -3)$.

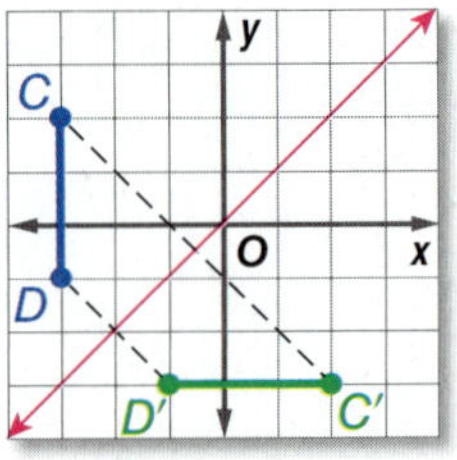

Comparing the coordinates of these and other examples leads to the following rule for reflections in the line $y = x$.

Key Concept — Reflection in Line $y = x$

Words To reflect a point in the line $y = x$, interchange the x- and y-coordinates.

Symbols $(x, y) \rightarrow (y, x)$

Example

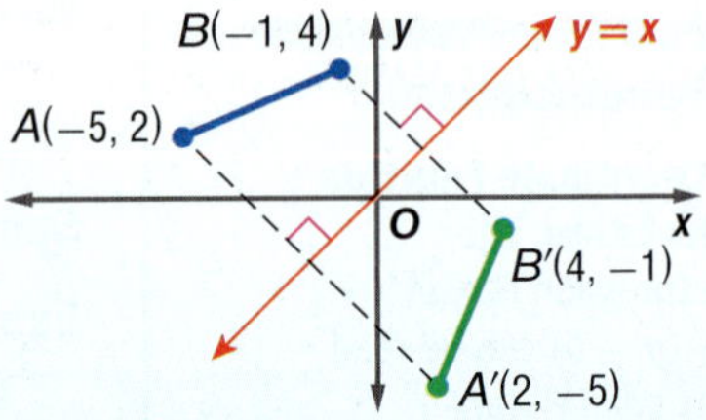

EXAMPLE 5 Reflect a Figure in the Line $y = x$

Quadrilateral $JKLM$ has vertices $J(2, 2)$, $K(4, 1)$, $L(3, -3)$, and $M(0, -4)$. Graph $JKLM$ and its image $J'K'L'M'$ in the line $y = x$.

Interchange the x- and y-coordinates of each vertex.

(x, y)	→	(y, x)
$J(2, 2)$	→	$J'(2, 2)$
$K(4, 1)$	→	$K'(1, 4)$
$L(3, -3)$	→	$L'(-3, 3)$
$M(0, -4)$	→	$M'(-4, 0)$

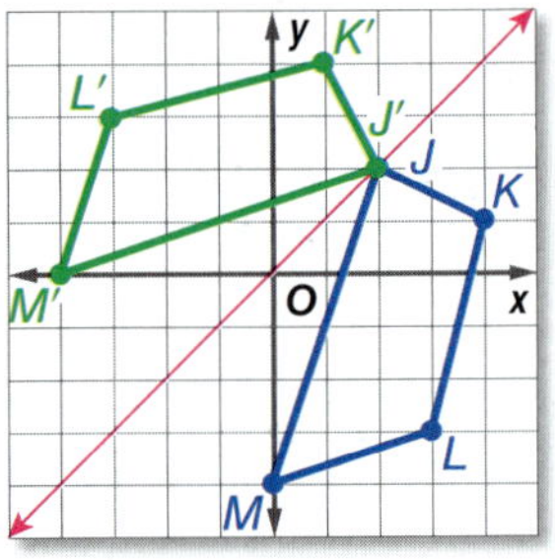

StudyTip

Preimage and Image In this book, the preimage will always be blue and the image will always be green.

Guided Practice

5. $\triangle BCD$ has vertices $B(-3, 3)$, $C(1, 4)$, and $D(-2, -4)$. Graph $\triangle BCD$ and its image in the line $y = x$.

Personal Tutor glencoe.com

Concept Summary — Reflection in the Coordinate Plane

Reflection in the x-axis

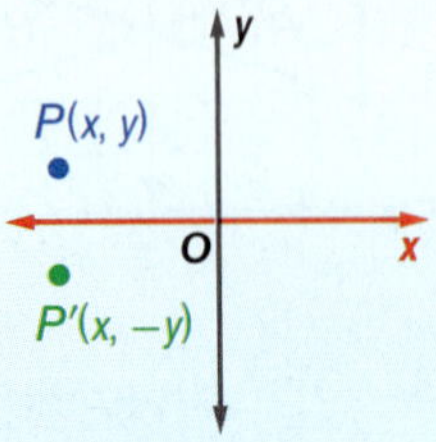

$(x, y) \rightarrow (x, -y)$

Reflection in the y-axis

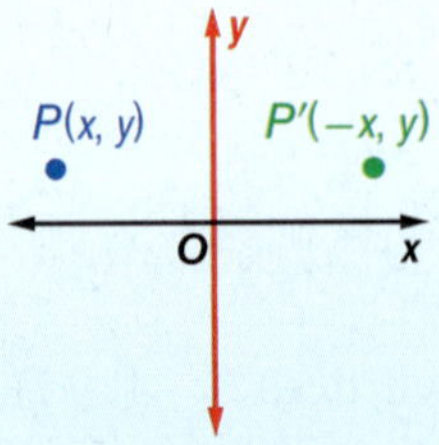

$(x, y) \rightarrow (-x, y)$

Reflection in the line $y = x$

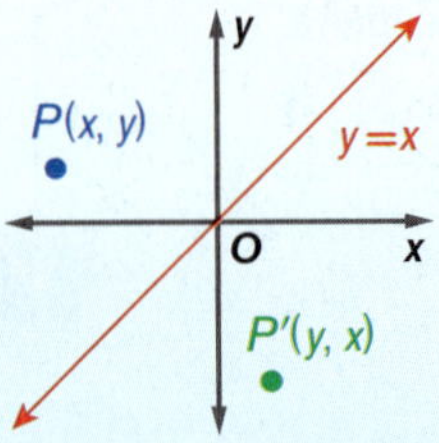

$(x, y) \rightarrow (y, x)$

Check Your Understanding

Example 1 p. 615

Copy the figure and the given line of reflection. Then draw the reflected image in this line using a ruler.

1.

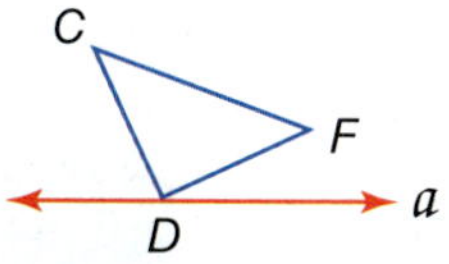

2.

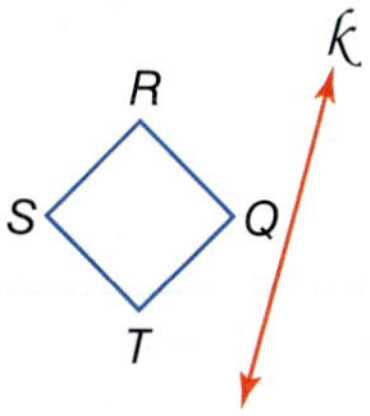

3.

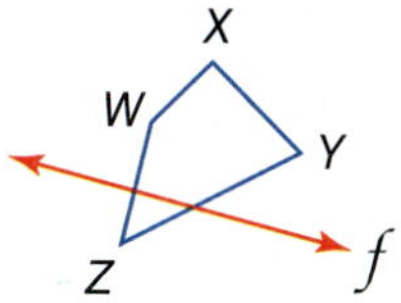

Example 2 p. 616

4. SPORTING EVENTS Toru is waiting at a café for a friend to bring him a ticket to a sold-out sporting event. At what point P along the street should the friend try to stop his car to minimize the distance Toru will have to walk from the café, to the car, and then to the arena entrance? Draw a diagram.

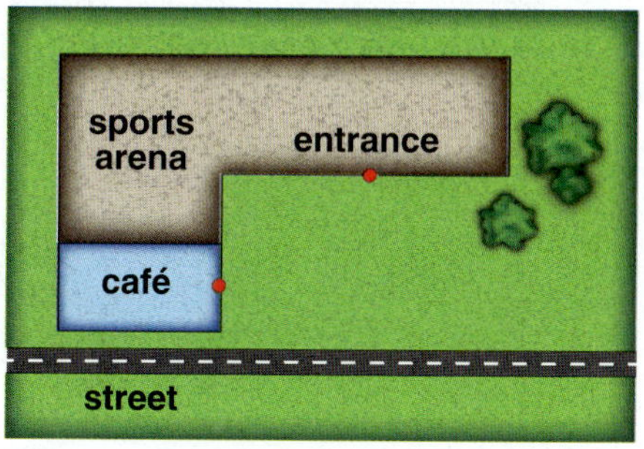

Example 3 p. 616

Graph $\triangle ABC$ and its image in the given line.

5. $y = -2$

6. $x = 3$

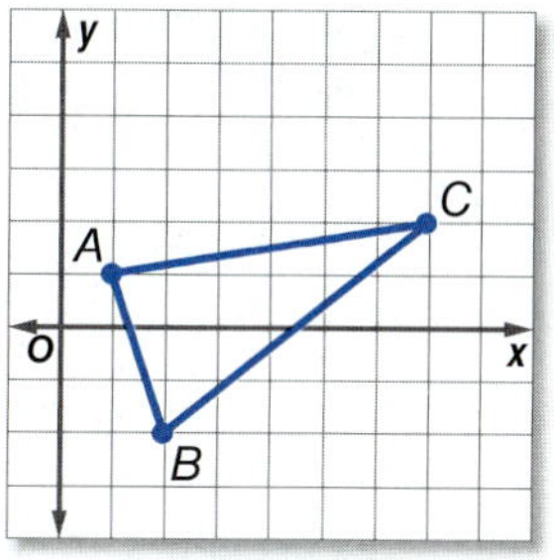

Examples 4 and 5 pp. 617–618

Graph each figure and its image under the given reflection.

7. $\triangle XYZ$ with vertices $X(0, 4)$, $Y(-3, 4)$, and $Z(-4, -1)$ in the y-axis

8. $\square QRST$ with vertices $Q(-1, 4)$, $R(4, 4)$, $S(3, 1)$, and $T(-2, 1)$ in the x-axis

9. quadrilateral $JKLM$ with vertices $J(-3, 1)$, $K(-1, 3)$, $L(1, 3)$, and $M(-3, -1)$ in the line $y = x$

Practice and Problem Solving

● = **Step-by-Step Solutions** begin on page R20.
Extra Practice begins on page 969.

Example 1 p. 615

Copy the figure and the given line of reflection. Then draw the reflected image in this line using a ruler.

10.

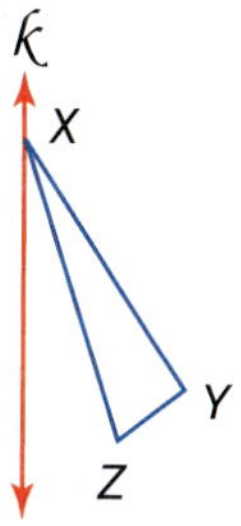

11.

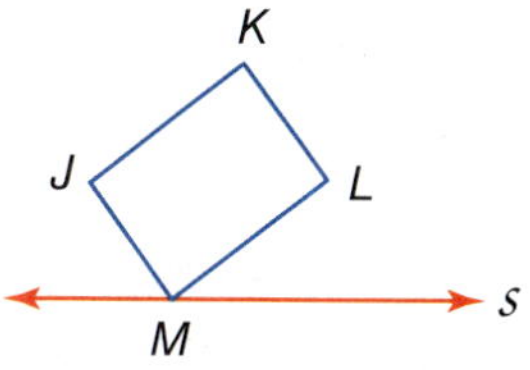

12.

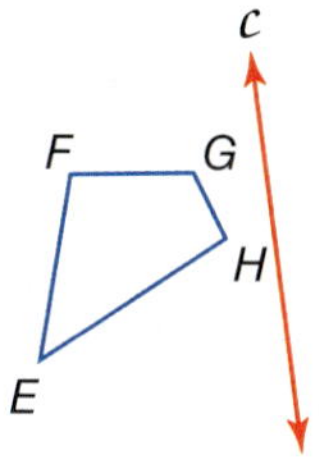

13.

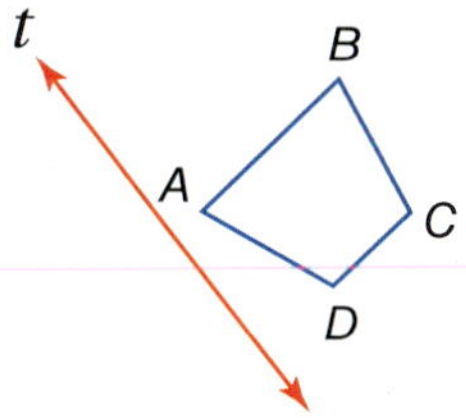

14.

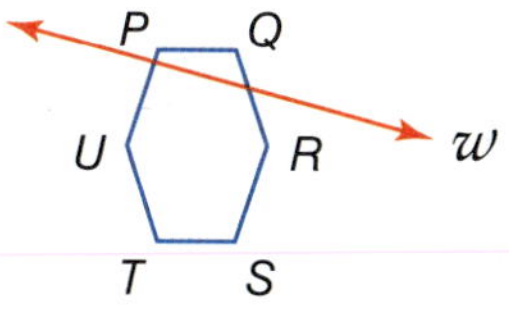

15.

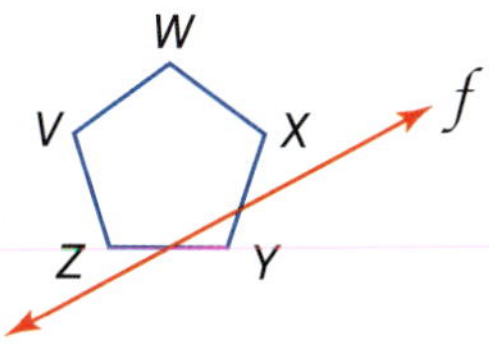

Example 2
p. 616

SPORTS When a ball is rolled or struck without spin against a wall, it bounces off the wall and travels in a ray that is the reflected image of the path of the ball if it had gone straight through the wall. Use this information in Exercises 16 and 17.

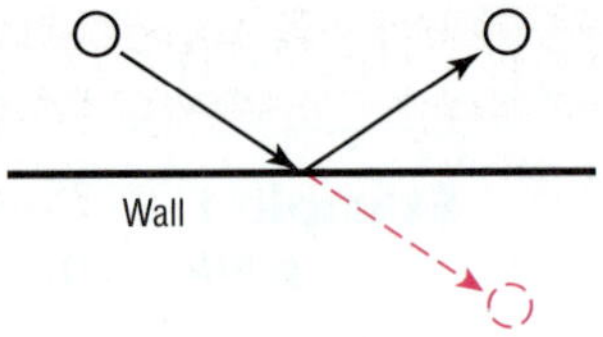

16. **BILLIARDS** Tadeo is playing billiards. He wants to strike the eight ball with the cue ball so that the eight ball bounces off the rail and rolls into the indicated pocket. If the eight ball moves with no spin, draw a diagram showing the exact point P along the right rail where the eight ball should hit after being struck by the cue ball.

17. **INDOOR SOCCER** Abby is playing indoor soccer, and she wants to hit the ball to point C, but must avoid an opposing player at point B. She decides to hit the ball at point A so that it bounces off the side wall. Draw a diagram that shows the exact point along the top wall for which Abby should aim.

Example 3
p. 616

Graph each figure and its image in the given line.

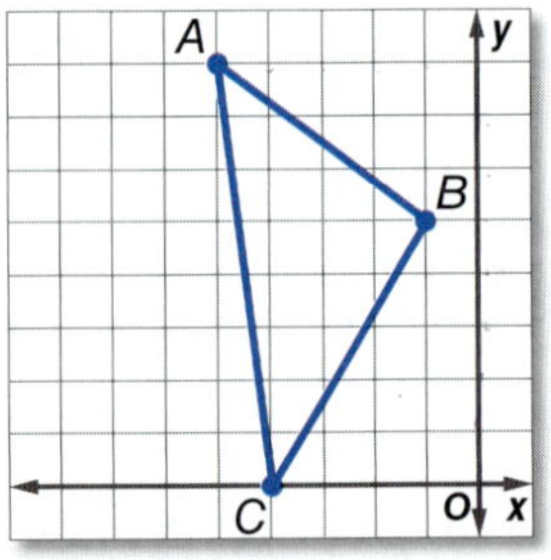

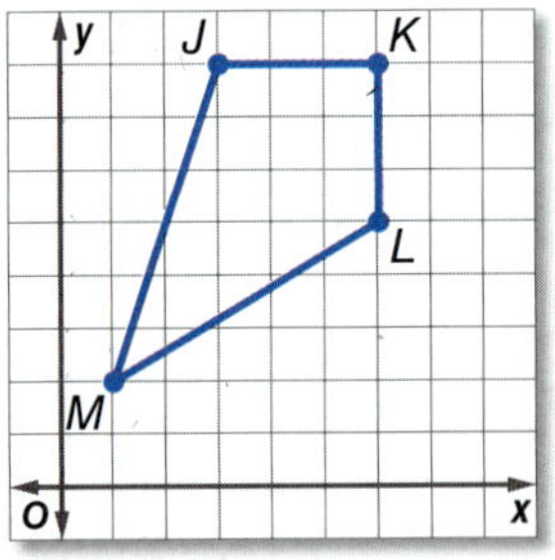

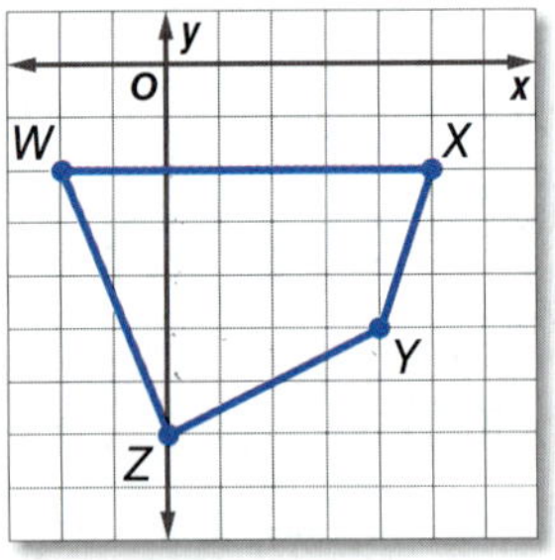

18. $\triangle ABC$; $y = 3$

19. $\triangle ABC$; $x = -1$

20. $JKLM$; $x = 1$

21. $JKLM$; $y = 4$

22. $WXYZ$; $y = -4$

23. $WXYZ$; $x = -2$

Examples 4 and 5
pp. 617–618

Graph each figure and its image under the given reflection.

24. rectangle $ABCD$ with vertices $A(-5, 2)$, $B(1, 2)$, $C(1, -1)$, and $D(-5, -1)$ in the line $y = -2$

25. square $JKLM$ with vertices $J(-4, 6)$, $K(0, 6)$, $L(0, 2)$, and $M(-4, 2)$ in the y-axis

26. $\triangle FGH$ with vertices $F(-3, 2)$, $G(-4, -1)$, and $H(-6, -1)$ in the line $y = x$

27. ▱$WXYZ$ with vertices $W(2, 3)$, $X(7, 3)$, $Y(6, -1)$, and $Z(1, -1)$ in the x-axis

28. trapezoid $PQRS$ with vertices $P(-1, 4)$, $Q(2, 4)$, $R(1, -1)$, and $S(-1, -1)$ in the y-axis

29. $\triangle STU$ with vertices $S(-3, -2)$, $T(-2, 3)$, and $U(2, 2)$ in the line $y = x$

Each figure shows a preimage and its reflected image in some line. Copy each figure and draw the line of reflection.

30.

31.

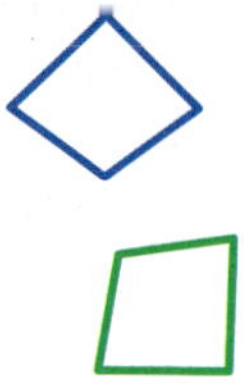

32.

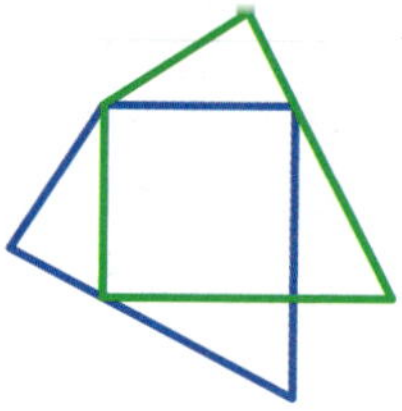

CONSTRUCTION To construct the reflection of a figure in a line using only a compass and a straightedge, you can use:

- the construction of a line perpendicular to a given line through a point not on the line (p. 55), and
- the construction of a segment congruent to a given segment (p. 17).

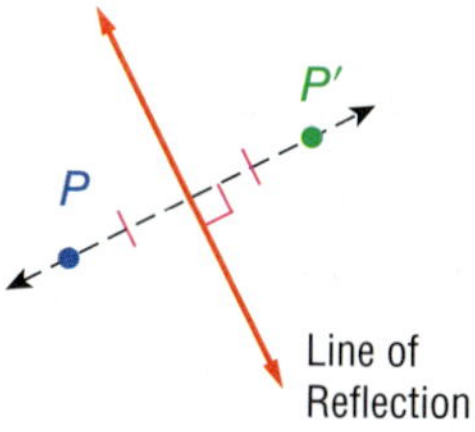

Real-World Career

Photographer
Photographers take photos for a variety of reasons such as journalism, art, to record an event, or for scientific purposes. In some photography fields such as photojournalism and scientific photography, a bachelor's degree is required. For others, such as portrait photography, technical proficiency is the only requirement.

Copy each figure and the given line of reflection. Then construct the reflected image.

33.

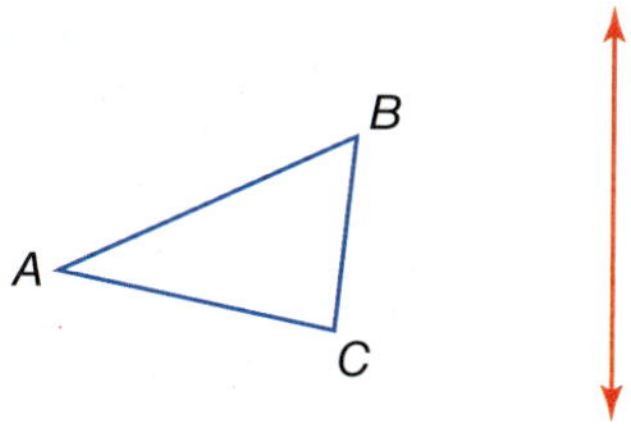

34.

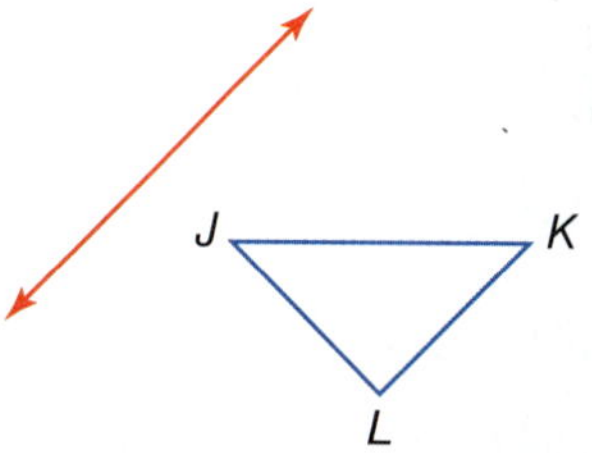

35. **PHOTOGRAPHY** Refer to the photo at the right.

a. What object separates the zebras and their reflections?

b. What geometric term can be used to describe this object?

ALGEBRA **Graph the line $y = 2x - 3$ and its reflected image in the given line. What is the equation of the reflected image?**

36. x-axis

37. y-axis

38. $y = x$

39. Reflect $\triangle CDE$ shown below in the line $y = 3x$.

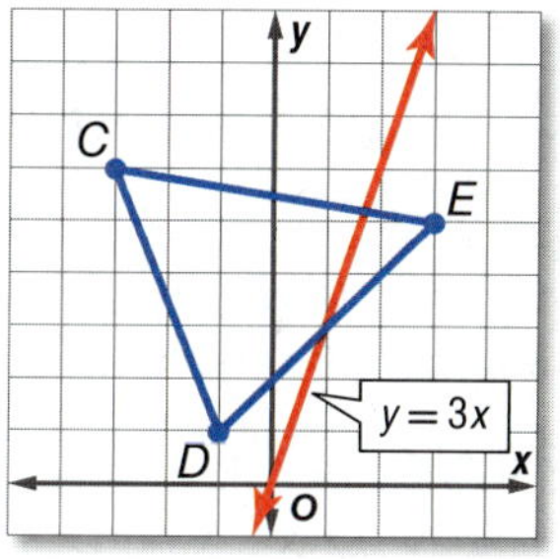

40. Relocate vertex C so that $ABCDE$ is convex, and all sides remain the same length.

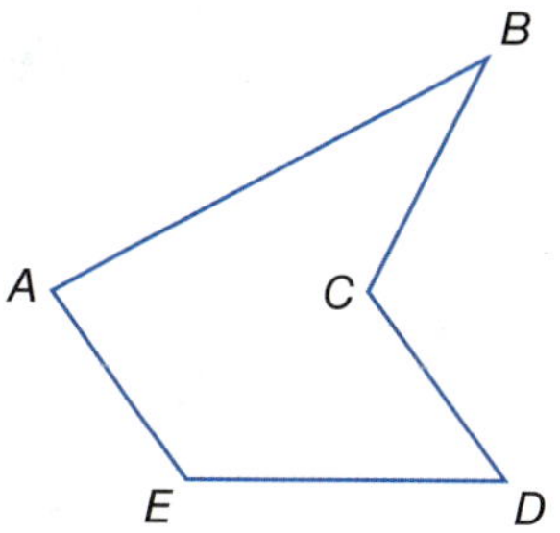

ALGEBRA **Graph the reflection of each function in the given line. Then write the equation of the reflected image.**

41 x-axis

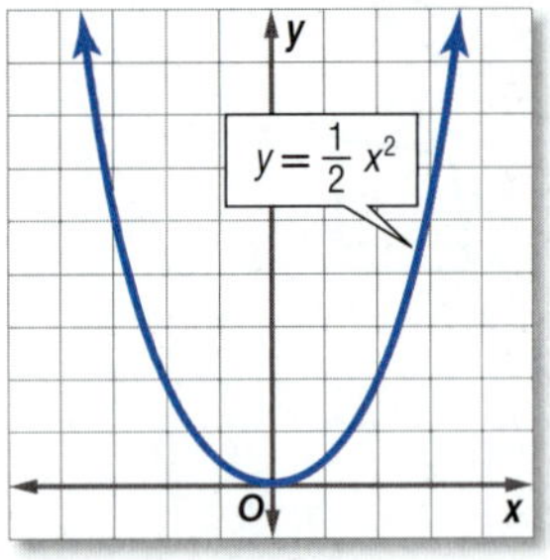

42. y-axis

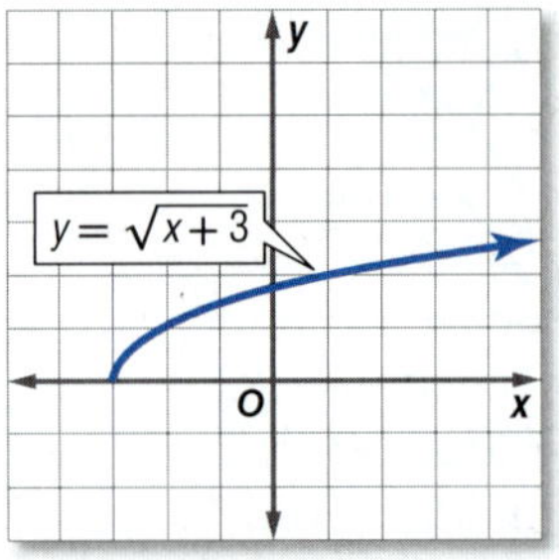

43. x-axis

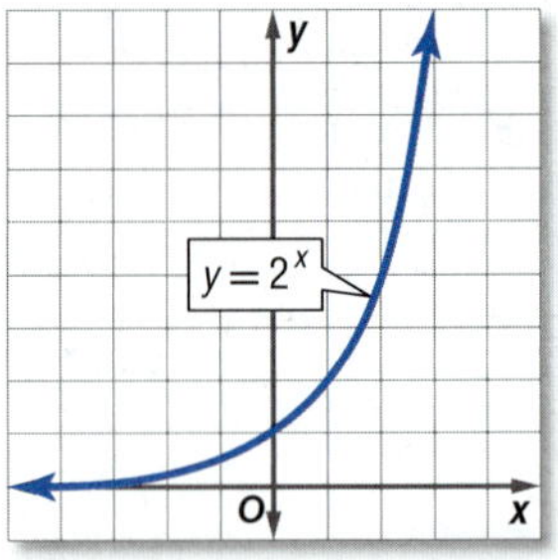

44. **MULTIPLE REPRESENTATIONS** In this problem, you will investigate a reflection in the origin.

a. **GEOMETRIC** Draw $\triangle ABC$ in the coordinate plane so that each vertex is a whole-number ordered pair.

b. **GRAPHICAL** Locate each reflected point A', B', and C' so that the reflected point, the original point, and the origin are collinear, and both the original point and the reflected point are equidistant from the origin.

c. **TABULAR** Copy and complete the table below.

	$\triangle ABC$		$\triangle A'B'C'$	
Coordinates	A		A'	
	B		B'	
	C		C'	

d. **VERBAL** Make a conjecture about the relationship between corresponding vertices of a figure reflected in the origin.

Math History Link

Felix Klein (1849–1925) Klein was a German mathematician. His definition of geometry as the study of the properties of a space that remain invariant under a group of transformations allowed for the inclusion of both Euclidean and non-Euclidean geometry.

H.O.T. Problems

Use **H**igher-**O**rder **T**hinking Skills

45. **ERROR ANALYSIS** Jamil and Ashley are finding the coordinates of the image of (2, 3) after a reflection in the x-axis. Is either of them correct? Explain.

Jamil	Ashley
$C'(2, -3)$	$C'(-2, 3)$

46. **WRITING IN MATH** Describe how to reflect a figure not on the coordinate plane across a line.

47. **CHALLENGE** A point in the second quadrant with coordinates $(-a, b)$ is reflected in the x-axis. If the reflected point is then reflected in the line $y = -x$, what are the final coordinates of the image?

48. **OPEN ENDED** Draw a polygon on the coordinate plane that when reflected in the x-axis looks exactly like the original figure.

49. **CHALLENGE** When $A(4, 3)$ is reflected in a line, its image is $A'(-1, 0)$. Find the equation of the line of reflection. Explain your reasoning.

50. **REASONING** The image of a point reflected in a line is *always*, *sometimes*, or *never* located on the other side of the line of reflection.

51. **WRITING IN MATH** Suppose points P, Q, and R are collinear, with point Q between points P and R. Describe a plan for a proof that the reflection of points P, Q, and R in a line preserves collinearity and betweenness of points.

Standardized Test Practice

52. SHORT RESPONSE If quadrilateral $WXYZ$ is reflected across the y-axis to become quadrilateral $W'X'Y'Z'$, what are the coordinates of X'?

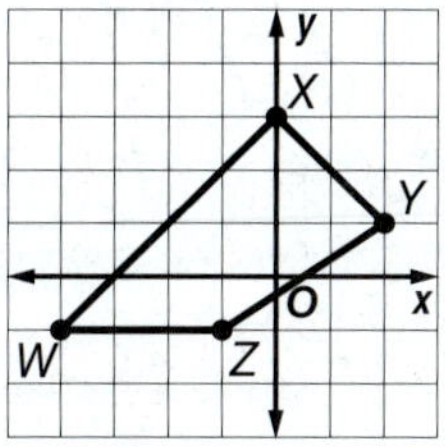

53. ALGEBRA If the arithmetic mean of $6x$, $3x$, and 27 is 18, then what is the value of x?

A 2 **C** 5
B 3 **D** 6

54. In $\triangle DEF$, $m\angle E = 108$, $m\angle F = 26$, and $f = 20$. Find d to the nearest whole number.

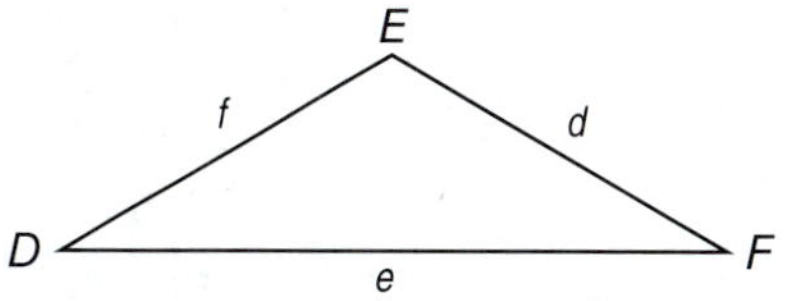

F 26 **H** 60
G 33 **J** 65

55. SAT/ACT In a coordinate plane, points A and B have coordinates $(-2, 4)$ and $(3, 3)$, respectively. What is the value of AB?

A $(1, 7)$ **C** $(5, -1)$
B $\sqrt{26}$ **D** $\sqrt{50}$

Spiral Review

Write the component form of each vector. (Lesson 8-7)

56.

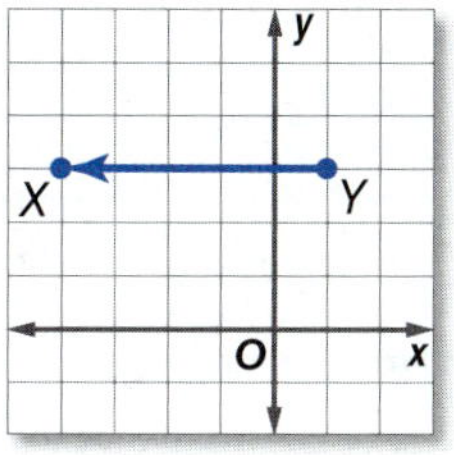

57.

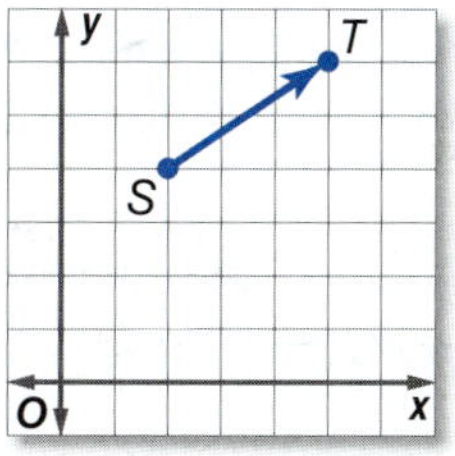

58.

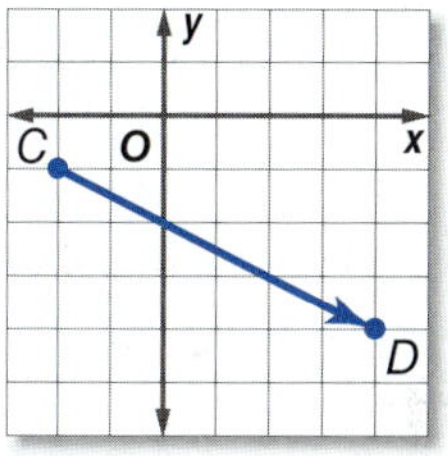

59. REAL ESTATE A house is built on a triangular plot of land. Two sides of the plot are 160 feet long, and they meet at an angle of 85°. If a fence is to be placed along the perimeter of the property, how much fencing material is needed? (Lesson 8-6)

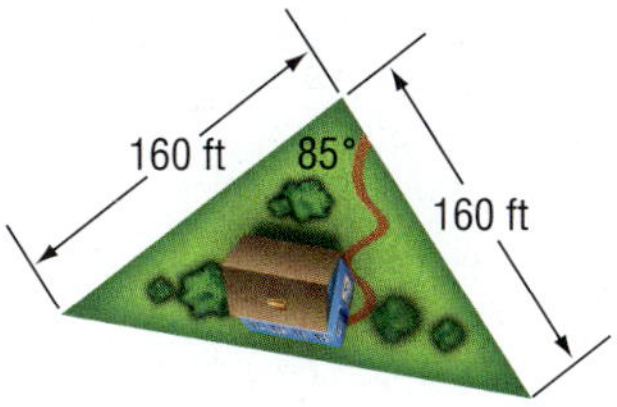

60. COORDINATE GEOMETRY In $\triangle LMN$, $\overline{PR}$ divides $\overline{NL}$ and $\overline{MN}$ proportionally. If the vertices are $N(8, 20)$, $P(11, 16)$, and $R(3, 8)$ and $\frac{LP}{PN} = \frac{2}{1}$, find the coordinates of L and M. (Lesson 7-4)

Use the figure at the right to write an inequality relating the given pair of angle or segment measures. (Lesson 5-6)

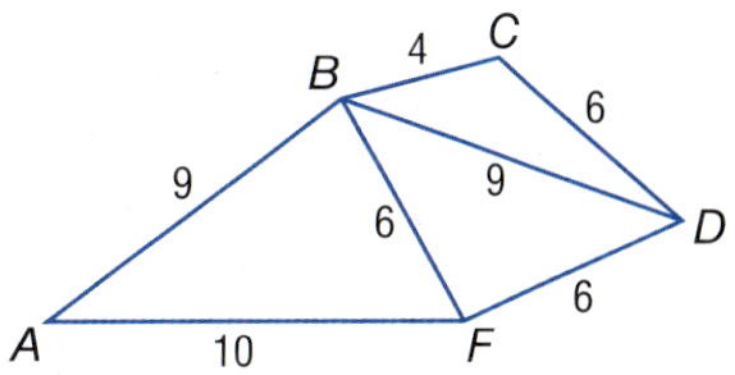

61. AB, FD

62. $m\angle BDC$, $m\angle FDB$

63. $m\angle FBA$, $m\angle DBF$

Skills Review

Find the magnitude and direction of each vector. (Lesson 8-7)

64. $\overrightarrow{RS}$: $R(-3, 3)$ and $S(-9, 9)$

65. $\overrightarrow{JK}$: $J(8, 1)$ and $K(2, 5)$

66. $\overrightarrow{FG}$: $F(-4, 0)$ and $G(-6, -4)$

67. $\overrightarrow{AB}$: $A(-1, 10)$ and $B(1, -12)$

9-2 Translations

Why?

Stop-motion animation is a technique in which an object is moved by very small amounts between individually photographed frames. When the series of frames is played as a continuous sequence, the result is the illusion of movement.

Then

You found the magnitude and direction of vectors. (Lesson 8-7)

Now

- Draw translations.
- Draw translations in the coordinate plane.

New Vocabulary

translation vector

Math Online
glencoe.com

Draw Translations In Lesson 4-7, you learned that a translation or *slide* is a transformation that moves all points of a figure the same distance in the same direction. Since vectors can be used to describe both distance and direction, vectors can be used to define translations.

Key Concept — Translation

A translation maps each point to its image along a vector, called the **translation vector**, such that

- each segment joining a point and its image has the same length as the vector, and
- this segment is also parallel to the vector.

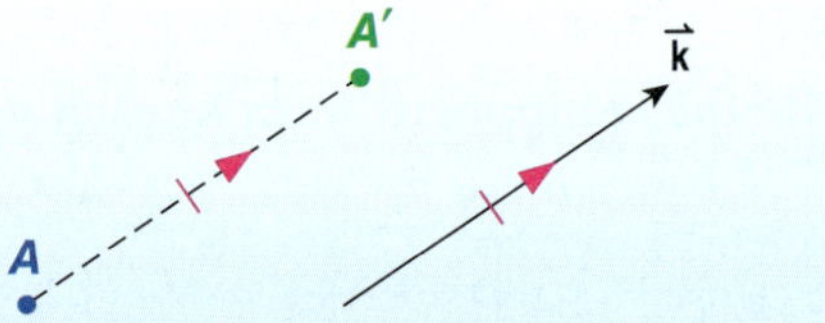

Point ***A′*** is a translation of point ***A*** along translation vector $\vec{\mathbf{k}}$.

EXAMPLE 1 Draw a Translation

Copy the figure and the given translation vector. Then draw the translation of the figure along the translation vector.

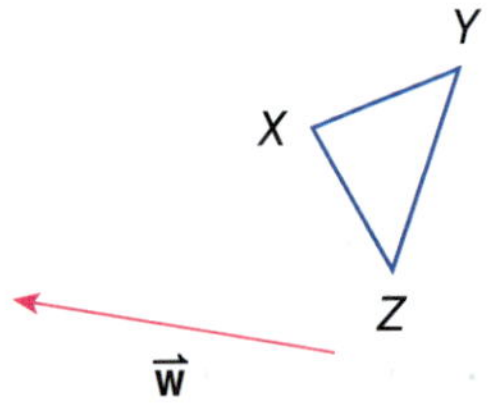

Step 1 Draw a line through each vertex parallel to vector $\vec{w}$.

Step 2 Measure the length of vector $\vec{w}$. Locate point X' by marking off this distance along the line through vertex X, starting at X and in the same direction as the vector.

Step 3 Repeat Step 2 to locate points Y' and Z'. Then connect vertices X', Y', and Z' to form the translated image.

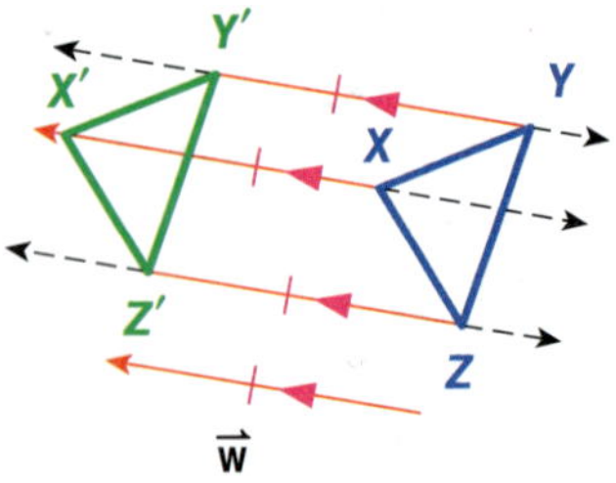

Guided Practice

1A.

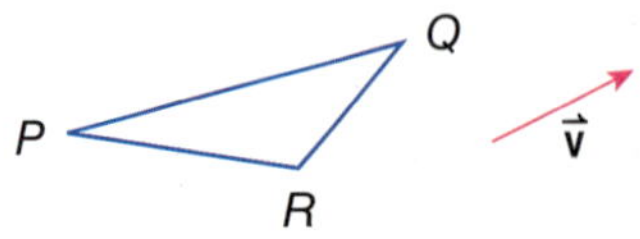

1B.

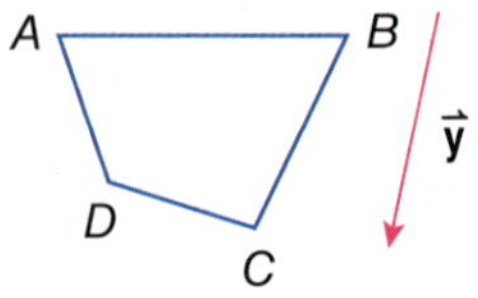

Personal Tutor glencoe.com

Draw Translations in the Coordinate Plane Recall that a vector in the coordinate plane can be written as $\langle a, b \rangle$, where a represents the horizontal change and b is the vertical change from the vector's tip to its tail. $\overline{CD}$ is represented by the ordered pair $\langle 2, -4 \rangle$.

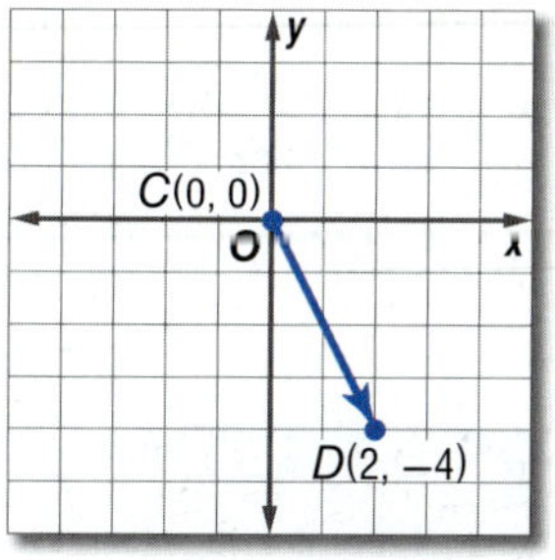

Written in this form, called the component form, a vector can be used to translate a figure in the coordinate plane.

> **ReadingMath**
>
> **Horizontal and Vertical Translations** When the translation vector is of the form $\langle a, 0 \rangle$, the translation is horizontal only. When the translation vector is of the form $\langle 0, b \rangle$, the translation is vertical only.

Key Concept — Translation in the Coordinate Plane

Words To translate a point along vector $\langle a, b \rangle$, add a to the x-coordinate and b to the y-coordinate.

Symbols $(x, y) \rightarrow (x + a, y + b)$

Example The image of $P(-2, 3)$ translated along vector $\langle 7, 4 \rangle$ is $P'(5, 7)$.

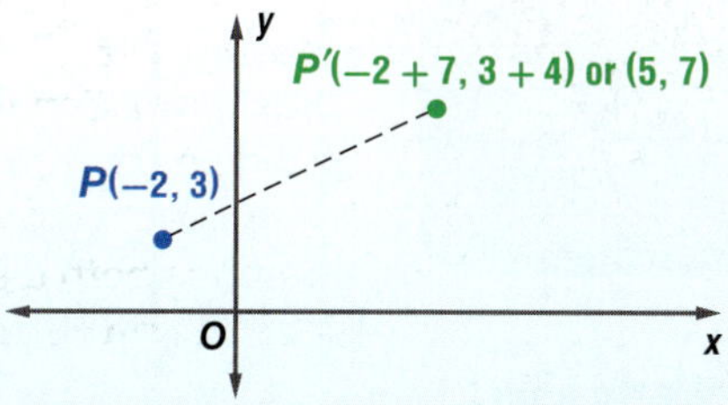

EXAMPLE 2 Translations in the Coordinate Plane

Graph each figure and its image along the given vector.

a. **$\triangle EFG$ with vertices $E(-7, -1)$, $F(-4, -4)$, and $G(-3, -1)$; $\langle 2, 5 \rangle$**

The vector indicates a translation 2 units right and 5 units up.

$(x, y) \rightarrow (x + 2, y + 5)$

$E(-7, -1) \rightarrow E'(-5, 4)$

$F(-4, -4) \rightarrow F'(-2, 1)$

$G(-3, -1) \rightarrow G'(-1, 4)$

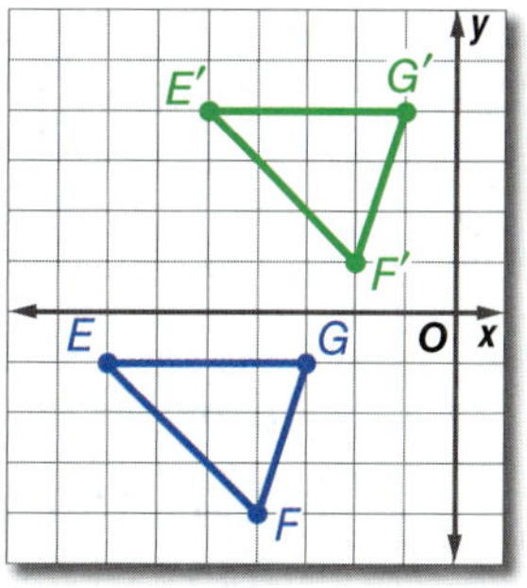

> **StudyTip**
>
> **Congruence Transformations** Recall that a translation is another type of congruence transformation or isometry. In Example 2a, $\triangle EFG \cong \triangle E'F'G'$.

b. **square $JKLM$ with vertices $J(3, 4)$, $K(5, 2)$, $L(7, 4)$, and $M(5, 6)$; $\langle -3, -4 \rangle$**

The vector indicates a translation 3 units left and 4 units down.

$(x, y) \rightarrow (x + (-3), y + (-4))$

$J(3, 4) \rightarrow J'(0, 0)$

$K(5, 2) \rightarrow K'(2, -2)$

$L(7, 4) \rightarrow L'(4, 0)$

$M(5, 6) \rightarrow M'(2, 2)$

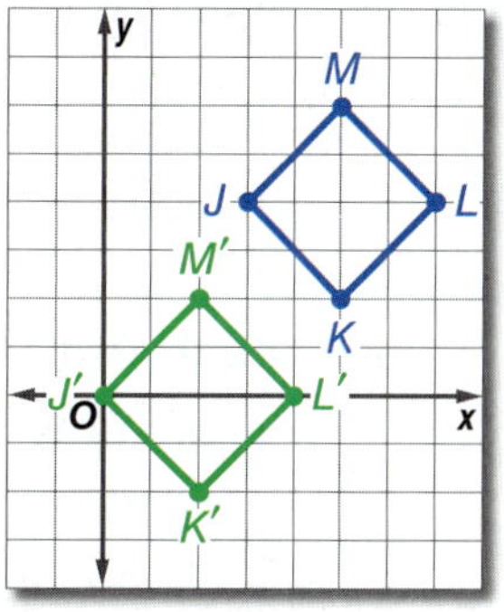

Guided Practice

2A. $\triangle ABC$ with vertices $A(2, 6)$, $B(1, 1)$, $C(7, 5)$; $\langle -4, -1 \rangle$

2B. quadrilateral $QRST$ with vertices $Q(-8, -2)$, $R(-9, -5)$, $S(-4, -7)$, $T(-4, -2)$; $\langle 7, 1 \rangle$

Personal Tutor glencoe.com

Real-World EXAMPLE 3 Describing Translations

MARCHING BAND **In one part of a marching band's performance, a line of trumpet players starts at position 1, marches to position 2, and then to position 3. Each unit on the graph represents one step.**

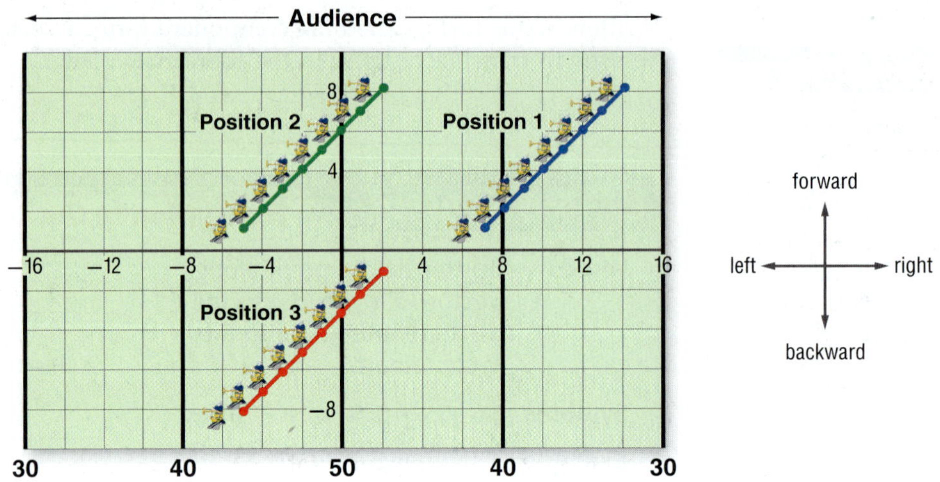

Real-World Link

Marching bands often make use of a series of formations that can include geometric shapes. Usually, each band member has an assigned position in each formation. *Floating* is the movement of a group of members together without changing the shape or size of their formation.

a. Describe the translation of the trumpet line from position 1 to position 2 in function notation and in words.

One point on the line in position 1 is (14, 8). In position 2, this point moves to (2, 8). Use the translations function $(x, y) \rightarrow (x + a, y + b)$ to write and solve equations to find a and b.

$(14 + a, 8 + b)$ or $(2, 8)$

$14 + a = 2$ $8 + b = 8$

$a = -12$ $b = 0$

function notation: $(x, y) \rightarrow (x + (-12), y + 0)$

So, the trumpet line is translated 12 steps *left* but no steps forward or backward from position 1 to position 2.

b. Describe the translation of the line from position 1 to position 3 using a translation vector.

$(14 + a, 8 + b)$ or $(2, -1)$

$14 + a = 2$ $8 + b = -1$

$a = -12$ $b = -9$

translation vector: $\langle -12, -9 \rangle$

Guided Practice

3. **ANIMATION** A coin is filmed using stop-motion animation so that it appears to move.

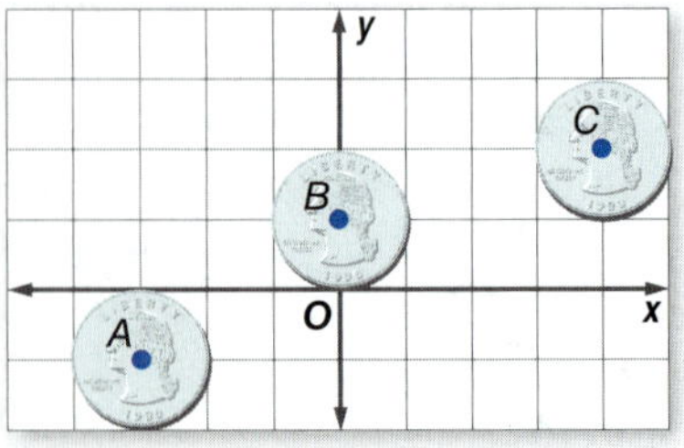

A. Describe the translation from A to B in function notation and in words.

B. Describe the translation from A to C using a translation vector.

Personal Tutor glencoe.com

Check Your Understanding

Example 1
p. 624

Copy the figure and the given translation vector. Then draw the translation of the figure along the translation vector.

1.

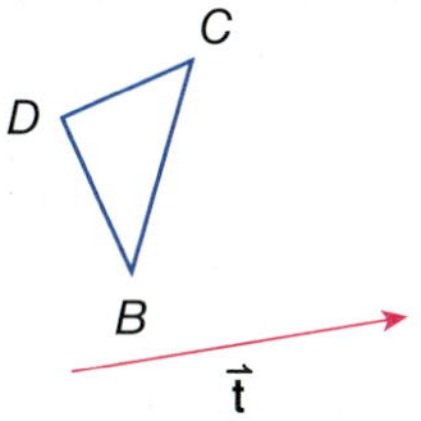

2.

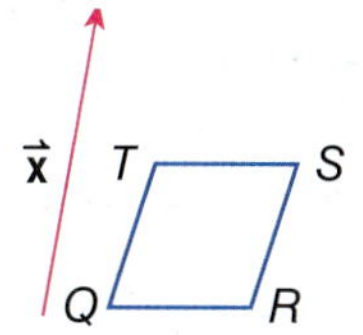

3.

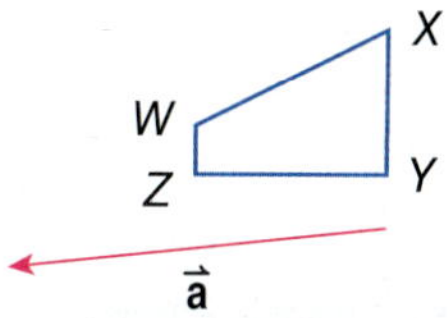

Example 2
p. 625

Graph each figure and its image along the given vector.

4. trapezoid *JKLM* with vertices $J(2, 4)$, $K(1, 1)$, $L(5, 1)$ and $M(4, 4)$; $\langle 7, 1 \rangle$

5. $\triangle DFG$ with vertices $D(-8, 8)$, $F(-10, 4)$, and $G(-7, 6)$; $\langle 5, -2 \rangle$

6. parallelogram *WXYZ* with vertices $W(-6, -5)$, $X(-2, -5)$, $Y(-1, -8)$, and $Z(-5, -8)$; $\langle -1, 4 \rangle$

Example 3
p. 626

7. **VIDEO GAMES** The object of the video game shown is to manipulate the colored tiles left or right as they fall from the top of the screen to completely fill each row without leaving empty spaces. If the starting position of the tile piece at the top of the screen is (x, y), use function notation to describe the translation that will fill the indicated row.

Practice and Problem Solving

= **Step-by-Step Solutions** begin on page R20.
Extra Practice begins on page 969.

Example 1
p. 624

Copy the figure and the given translation vector. Then draw the translation of the figure along the translation vector.

8.

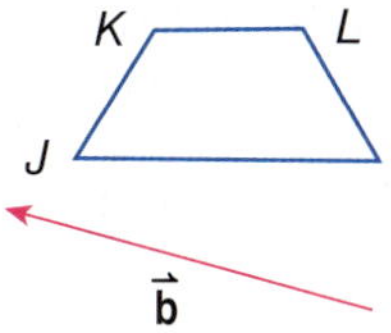

9.

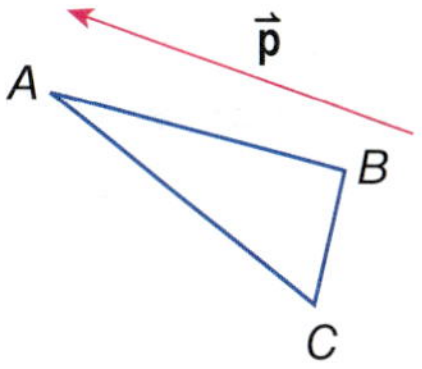

10.

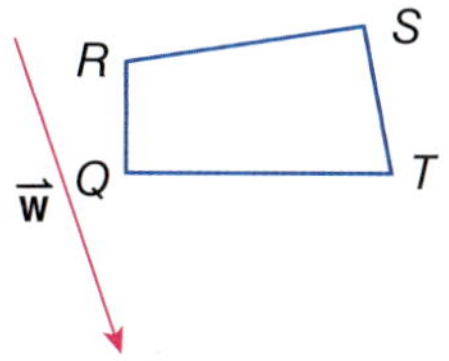

11.

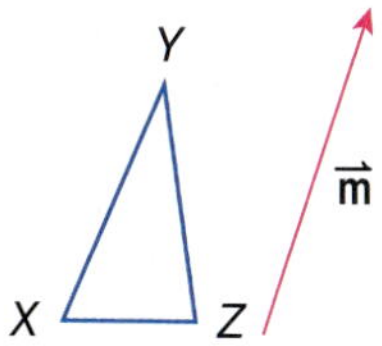

12.

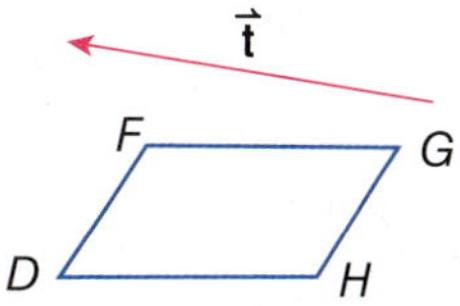

13.

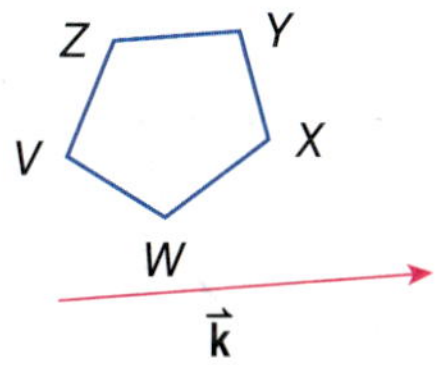

Example 2
p. 625

Graph each figure and its image along the given vector.

14. $\triangle ABC$ with vertices $A(1, 6)$, $B(3, 2)$, and $C(4, 7)$; $\langle 4, -1 \rangle$

15. $\triangle MNP$ with vertices $M(4, -5)$, $N(5, -8)$, and $P(8, -6)$; $\langle -2, 5 \rangle$

16. rectangle *QRST* with vertices $Q(-8, 4)$, $R(-8, 2)$, $S(-3, 2)$, and $T(-3, 4)$; $\langle 2, 3 \rangle$

17. quadrilateral *FGHJ* with vertices $F(-4, -2)$, $G(-1, -1)$, $H(0, -4)$, and $J(-3, -6)$; $\langle -5, -2 \rangle$

18. ▱*WXYZ* with vertices $W(-3, -1)$, $X(1, -1)$, $Y(2, -4)$, and $Z(-2, -4)$; $\langle -3, 4 \rangle$

19. trapezoid *JKLM* with vertices $J(-4, -2)$, $K(-1, -2)$, $L(0, -5)$, and $M(-5, -5)$; $\langle 6, 5 \rangle$

Example 3
p. 626

20. LOCATION Brittany's neighborhood is shown on the grid at the right.

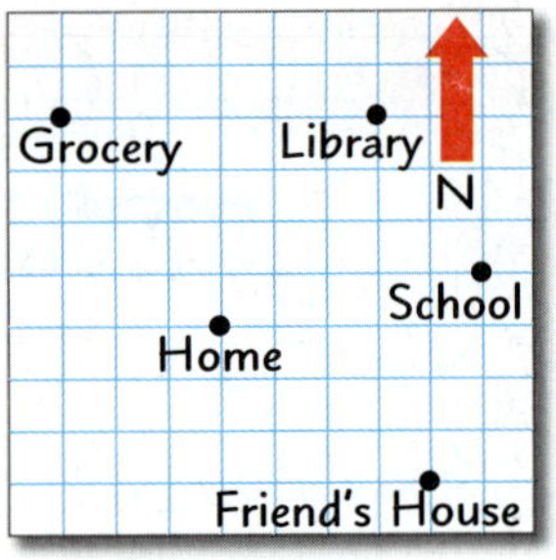

a. If she leaves home and travels 4 blocks north and 3 blocks east, what is her new location?

b. Use words to describe two possible translations that will take Brittany home from school.

Real-World Link

As of 2007, Jerry Rice has the record for the most receiving yards in a career at 22,895.

Source: National Football League

21. FOOTBALL A wide receiver starts from the 15-yard line on the right hash mark and runs a route that takes him 12 yards to the left and down field for a gain of 17 yards. Write a translation vector to describe the receiver's route.

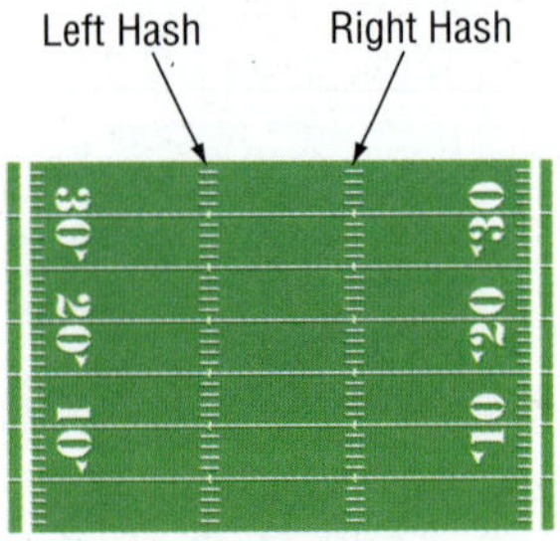

22. CHESS Each chess piece has a path that it can follow to move. The rook, which begins in square a8, can only move vertically or horizontally. The knight, which begins in square b8, can move two squares horizontally and then one square vertically, or two squares vertically and one square horizontally. The bishop, which begins in square f8, can only move diagonally.

a. The knight moves 2 squares vertically and 1 square horizontally on its first move, then two squares horizontally and 1 square vertically on its second move. What are the possible locations for the knight after two moves?

b. After two moves, the rook is in square d3. Describe a possible translation to describe the two moves.

c. Describe a translation that can take the bishop to square a1. What is the minimum number of moves that can be used to accomplish this translation?

Write each translation vector.

23.

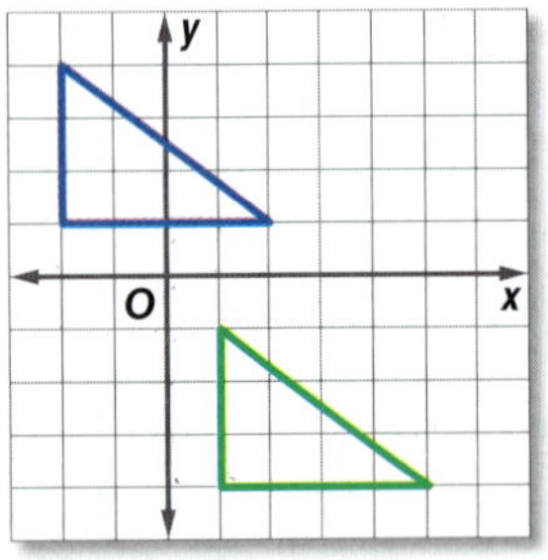

24.

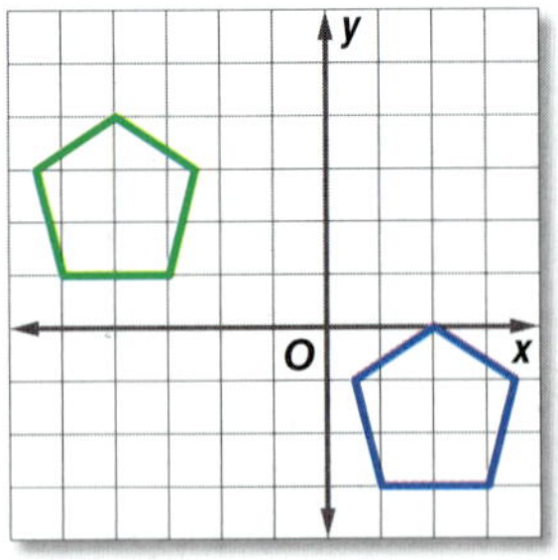

25. CONCERTS Dexter's family buys tickets every year for a concert. Last year they were in seats C3, C4, C5, and C6. This year, they will be in seats D16, D17, D18, and D19. Write a translation in words and using vector notation that can be used to describe the change in their seating.

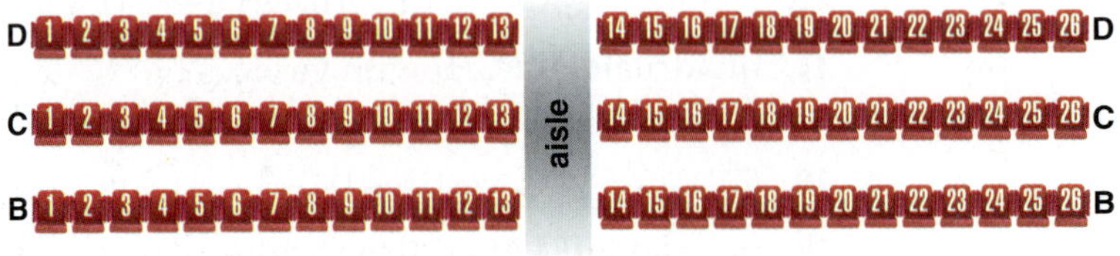

ALGEBRA **Graph the translation of each function along the given vector. Then write the equation of the translated image.**

> **StudyTip**
>
> **Translations of Continuous Functions** When a continuous function is transformed as in Exercises 26 and 27, shape is preserved as it is in congruence transformations.

26. $\langle 4, 1 \rangle$

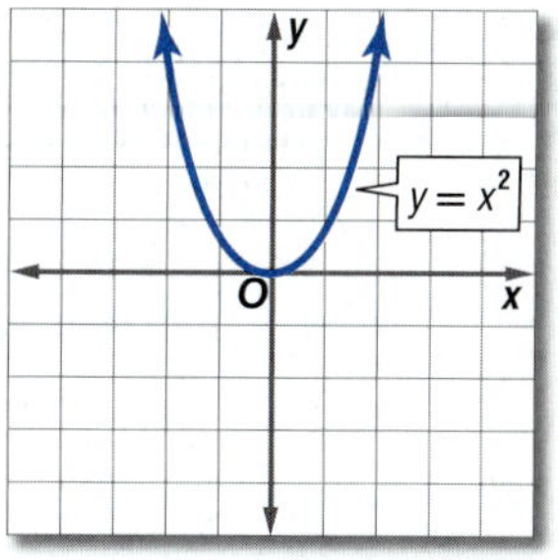

27. $\langle -2, 0 \rangle$

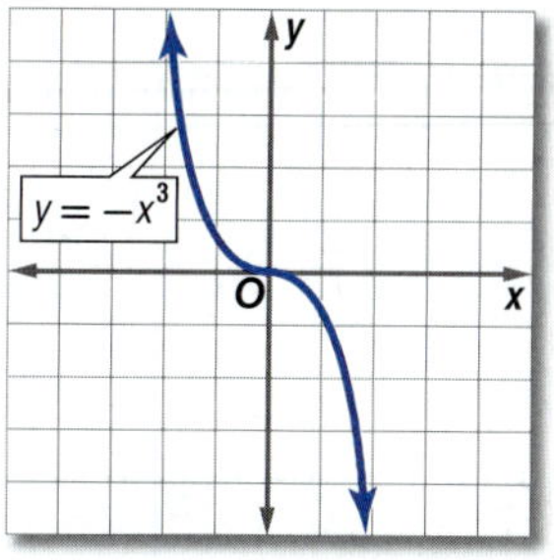

28. **ROLLER COASTERS** The length of the roller coaster track from the top of a hill to the bottom of the hill is 125 feet at a 53° angle with the vertical. If the position at the top of the hill is (x, y), use function notation to describe the translation to the bottom of the hill. Round to the nearest foot.

29. **MULTIPLE REPRESENTATIONS** In this problem, you will investigate reflections over a pair of parallel lines.

a. **GEOMETRIC** On patty paper, draw $\triangle ABC$ and a pair of vertical lines ℓ and m. Reflect $\triangle ABC$ in line ℓ by folding the patty paper. Then reflect $\triangle A'B'C'$, in line m. Label the final image $\triangle A''B''C''$.

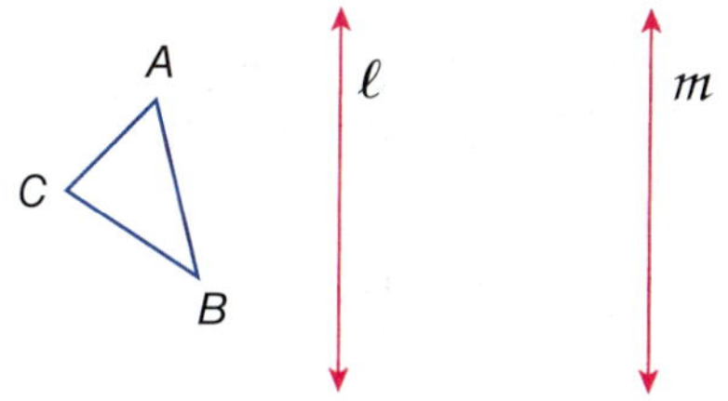

b. **GEOMETRIC** Repeat the process in part **a** for $\triangle DEF$ reflected in vertical lines n and p and $\triangle MNP$ reflected in vertical lines q and r.

c. **TABULAR** Copy and complete the table below.

Distance Between Corresponding Points (cm)		Distance Between Vertical Lines (cm)	
A and A'', B and B'', C and C''		ℓ and m	
D and D'', E and E'', F and F''		n and p	
M and M'', N and N'', P and P''		q and r	

d. **VERBAL** Describe the result of two reflections in two vertical lines using one transformation.

H.O.T. Problems Use Higher-Order Thinking Skills

30. **REASONING** Determine a rule to find the final image of a point that is translated along $\langle x + a, y + b \rangle$ and then $\langle x + c, y + d \rangle$.

31. **CHALLENGE** A line $y = mx + b$ is translated using the vector $\langle a, b \rangle$. Write the equation of the translated line. What is the value of the y-intercept?

32. **OPEN ENDED** Draw a figure on the coordinate plane so that the figure has the same orientation after it is reflected in the line $y = 1$. Explain what must be true in order for this to occur.

33. **WRITING IN MATH** Compare and contrast function notation and vector notation for translations.

34. **WRITING IN MATH** Recall from Lesson 9-1 that an invariant point maps onto itself. Can invariant points occur with translations? Explain why or why not.

Standardized Test Practice

35. Identify the location of point P under translation $(x + 3, y + 1)$.

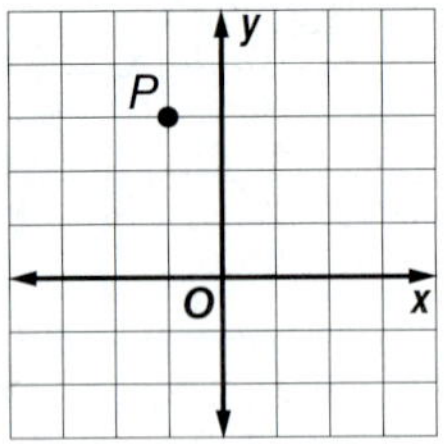

A (0, 6) **C** (2, −4)
B (0, 3) **D** (2, 4)

36. SHORT RESPONSE Which vector best describes the translation of $A(3, -5)$ to $A'(-2, -8)$?

37. ALGEBRA Over the next four days, Amanda plans to drive 160 miles, 235 miles, 185 miles, and 220 miles. If her car gets an average of 32 miles per gallon of gas, how many gallons of gas should she expect to use in all?

F 25 **G** 30 **H** 35 **J** 40

38. SAT/ACT A bag contains 5 red marbles, 2 blue marbles, 4 white marbles, and 1 yellow marble. If two marbles are chosen in a row, without replacement, what is the probability of getting 2 white marbles?

A $\frac{1}{66}$ **B** $\frac{1}{11}$ **C** $\frac{1}{9}$ **D** $\frac{5}{33}$

Spiral Review

Graph each figure and its image under the given reflection. (Lesson 9-1)

39. $\overline{DJ}$ with endpoints $D(4, 4)$, $J(-3, 2)$ in the y-axis

40. $\triangle XYZ$ with vertices $X(0, 0)$, $Y(3, 0)$, and $Z(0, 3)$ in the x-axis

41. $\triangle ABC$ with vertices $A(-3, -1)$, $B(0, 2)$, and $C(3, -2)$, in the line $y = x$

42. quadrilateral $JKLM$ with vertices $J(-2, 2)$, $K(3, 1)$, $L(4, -1)$, and $M(-2, -2)$ in the origin

Copy the vectors to find each sum or difference. (Lesson 8-7)

43. $\vec{c} - \vec{d}$

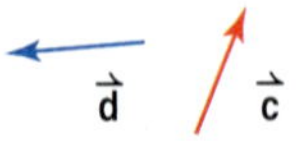

44. $\vec{w} + \vec{x}$

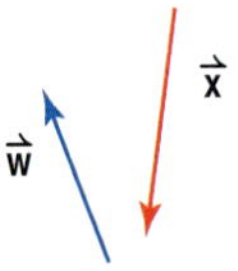

45. $\vec{n} - \vec{p}$

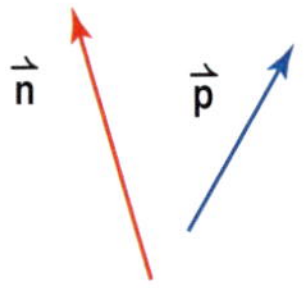

46. NAVIGATION An airplane is three miles above sea level when it begins to climb at a 3.5° angle. If this angle is constant, how far above sea level is the airplane after flying 50 miles? (Lesson 8-4)

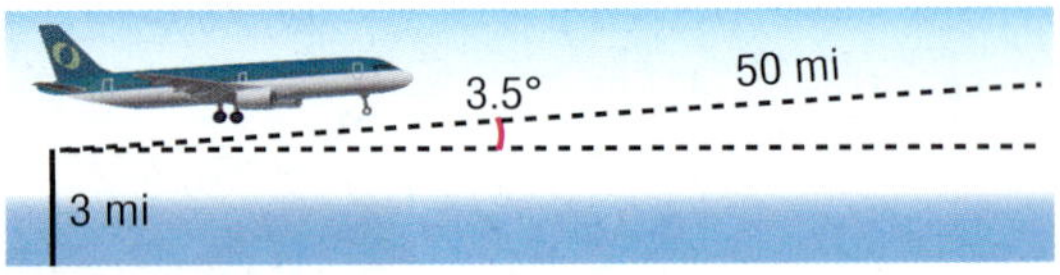

Use ▱$JKLM$ to find each measure. (Lesson 6-2)

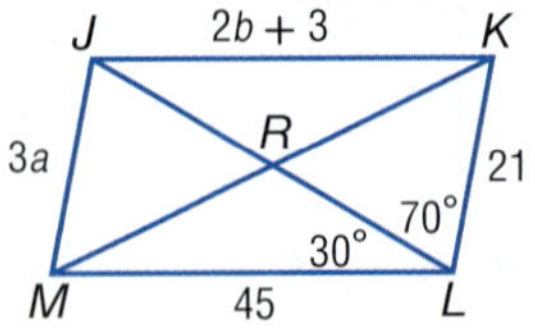

47. $m\angle MJK$ **48.** $m\angle JML$

49. $m\angle JKL$ **50.** $m\angle KJL$

Skills Review

Copy the diagram shown, and extend each ray. Classify each angle as *right*, *acute*, or *obtuse*. Then use a protractor to measure the angle to the nearest degree. (Lesson 1-4)

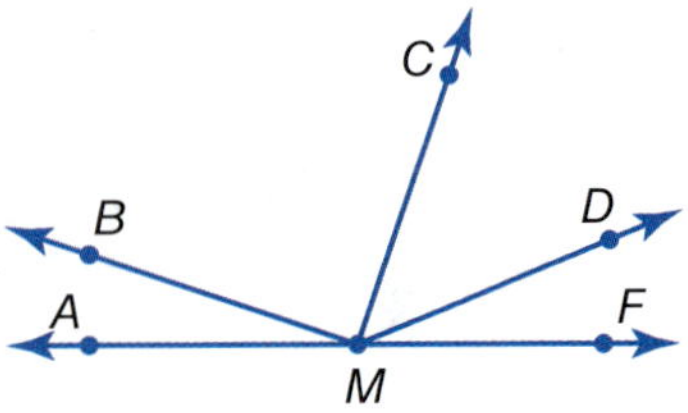

51. $\angle AMC$ **52.** $\angle FMD$

53. $\angle BMD$ **54.** $\angle CMB$

EXPLORE 9-3

Geometry Lab
Rotations

In Chapter 4, you learned that a rotation is a type of transformation that moves a figure about a fixed point, or center of rotation, through a specific angle and in a specific direction. In this activity you will use patty paper to explore the properties of rotations.

ACTIVITY Explore Rotations by Using Patty Paper

Step 1 On a piece of patty paper, draw quadrilateral *ABCD* and a point *P*.

Step 2 On another piece of patty paper, trace quadrilateral *ABCD* and point *P*. Label the new quadrilateral *A′B′C′D′* and the new point *P*.

Step 3 Position the patty paper so that both points *P* coincide. Rotate the paper so that *ABCD* and *A′B′C′D′* do not overlap. Tape the two pieces of patty paper together.

Step 4 Measure the distance between *A*, *B*, *C*, and *D* to point *P*. Repeat for quadrilateral *A′B′C′D′*. Then copy and complete the table below.

Quadrilateral	Length			
ABCD	*AP*	*BP*	*CP*	*DP*
A′B′C′D′	*A′P*	*B′P*	*C′P*	*D′P*

Step 1

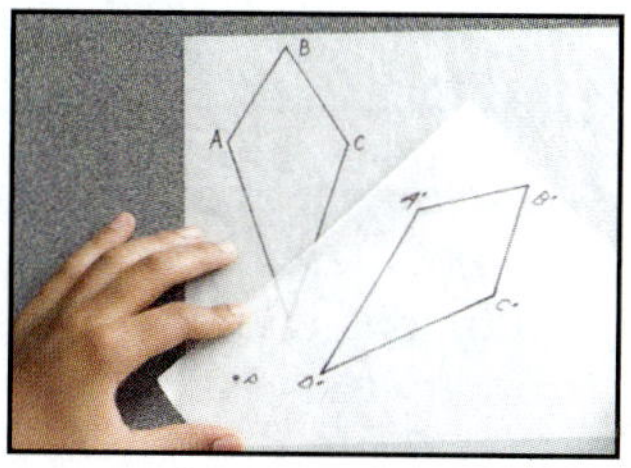

Steps 2 and 3

Exercises

1. Graph $\triangle JKL$ with vertices $J(1, 3)$, $K(2, 1)$, and $L(3, 4)$ on a coordinate plane, and then trace on patty paper.

   a. Use a protractor to rotate each vertex 90° clockwise about the origin as shown in the figure at the right. What are the vertices of the rotated image?

   b. Rotate $\triangle JKL$ 180° about the origin. What are the vertices of the rotated image?

   c. Use the Distance Formula to find the distance from points *J*, *K*, and *L* to the origin. Repeat for $J'K'L'$ and $J''K''L''$.

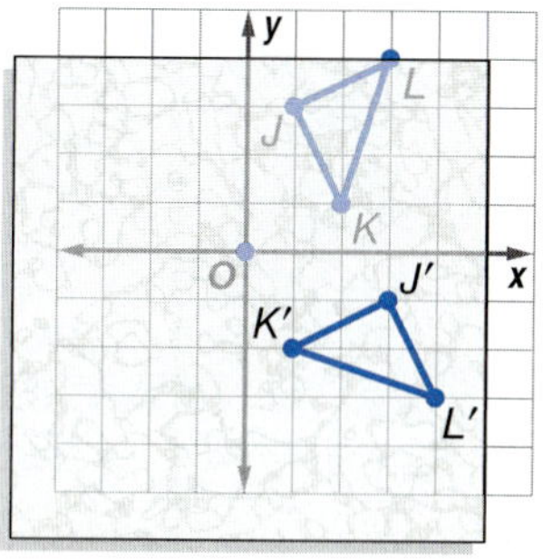

2. **WRITING IN MATH** If you rotate point (4, 2) 90° and 180° about the origin, how do the *x*- and *y*-coordinates change?

3. **MAKE A PREDICTION** What are the new coordinates of a point (x, y) that is rotated 270°?

4. **MAKE A CONJECTURE** Make a conjecture about the distances from the center of rotation *P* to each corresponding vertex of *ABCD* and *A′B′C′D′*.

9-3 Rotations

Why?

Traditionally, the energy generated by a windmill was used to pump water or grind grain into flour. Modern windmill technology may be an important alternative to fossil fuels. Windmills convert the wind's energy into electricity through the rotation of turbine blades.

Then
You identified rotations and verified them as congruence transformations. (Lesson 4-7)

Now
- Draw rotations.
- Draw rotations in the coordinate plane.

New Vocabulary
center of rotation
angle of rotation

Math Online
glencoe.com

Draw Rotations In Lesson 4-7, you learned that a rotation or *turn* moves every point of a preimage through a specified angle and direction about a fixed point.

Key Concept — Rotation

A rotation about a fixed point, called the **center of rotation**, through an angle of $x°$ maps a point to its image such that

- if the point is the center of rotation, then the image and preimage are the same point, or
- if the point is not the center of rotation, then the image and preimage are the same distance from the center of rotation and the measure of the **angle of rotation** formed by the preimage, center of rotation, and image points is x.

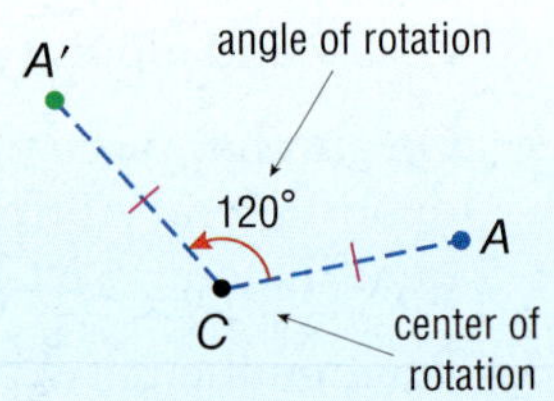

A' is the image of A after a 120° rotation about point C.

The direction of a rotation can be either clockwise or counterclockwise. From this point forward, all rotations will be counterclockwise, unless stated otherwise.

EXAMPLE 1 Draw a Rotation

Copy $\triangle ABC$ and point K. Then use a protractor and ruler to draw a 140° rotation of $\triangle ABC$ about point K.

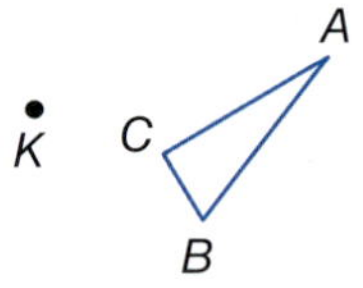

Step 1 Draw a segment from vertex A to point K.

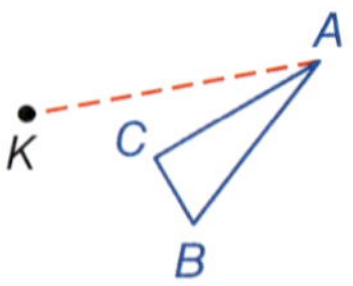

Step 2 Draw a 140° angle using $\overline{KA}$ as one side.

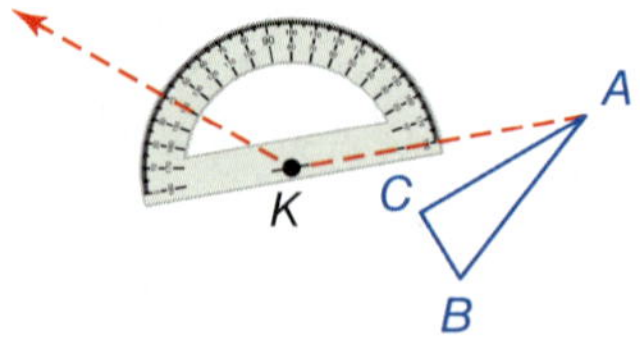

Step 3 Use a ruler to draw A' such that $KA' = KA$.

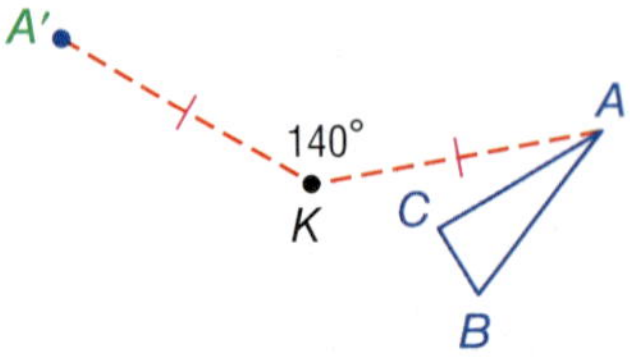

Step 4 Repeat Steps 1–3 for vertices B and C and draw $\triangle A'B'C'$.

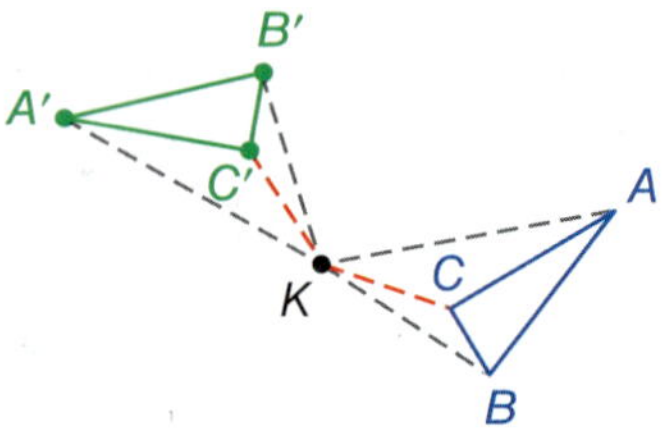

Guided Practice

Copy each figure and point K. Then use a protractor and ruler to draw a rotation of the figure the given number of degrees about K.

1A. 65°

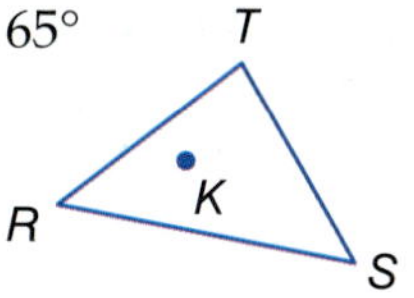

1B. 170°

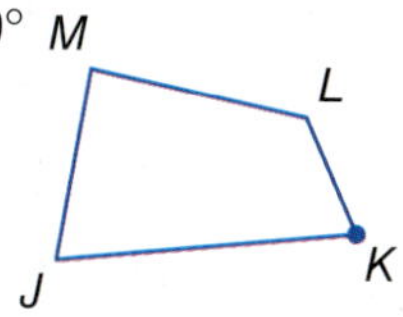

Personal Tutor glencoe.com

StudyTip

Clockwise Rotation Clockwise rotation can be designated by a negative angle measure. For example a rotation of −90° about the origin is a rotation 90° clockwise about the origin.

Draw Rotations in the Coordinate Plane When a point is rotated 90°, 180°, or 270° counterclockwise about the origin, you can use the following rules.

Key Concept — Rotations in the Coordinate Plane

90° Rotation
To rotate a point 90° counterclockwise about the origin, multiply the y-coordinate by −1 and then interchange the x- and y-coordinates.

Symbols $(x, y) \rightarrow (-y, x)$

Example

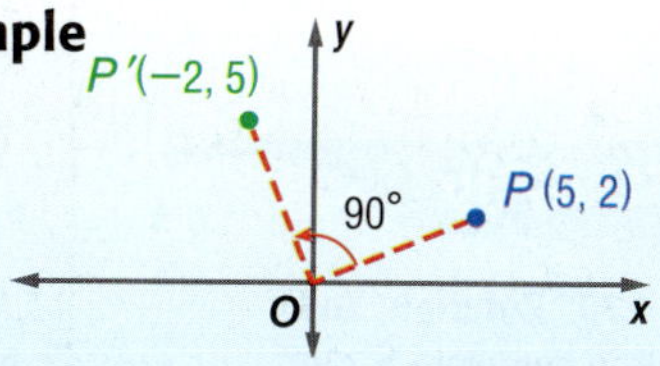

180° Rotation
To rotate a point 180° counterclockwise about the origin, multiply the x- and y-coordinates by −1.

Symbols $(x, y) \rightarrow (-x, -y)$

Example

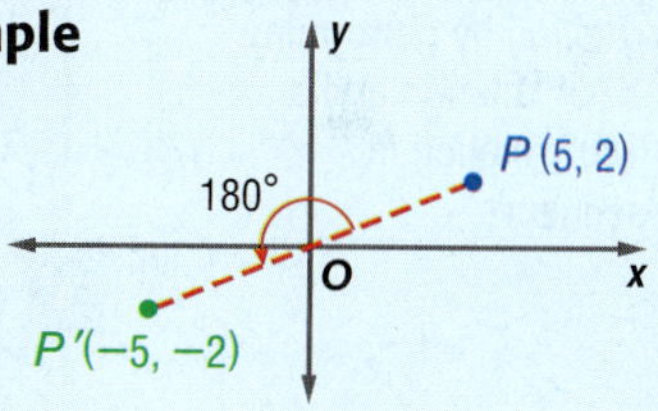

270° Rotation
To rotate a point 270° counterclockwise about the origin, multiply the x-coordinate by −1 and then interchange the x- and y-coordinates.

Symbols $(x, y) \rightarrow (y, -x)$

Example

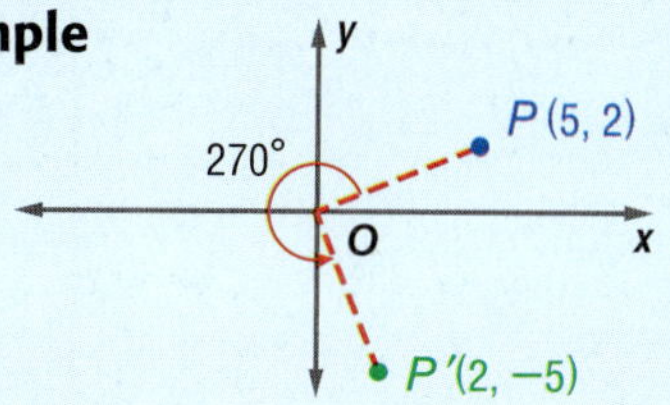

StudyTip

360° Rotation A rotation of 360° about a point returns a figure to its original position. That is, the image under a 360° rotation is equal to the preimage.

EXAMPLE 2 Rotations in the Coordinate Plane

Triangle PQR has vertices $P(1, 1)$, $Q(4, 5)$, and $R(5, 1)$. Graph $\triangle PQR$ and its image after a rotation 90° about the origin.

Multiply the y-coordinate of each vertex by −1 and interchange.

$(x, y) \rightarrow (-y, x)$

$P(1, 1) \rightarrow P'(-1, 1)$

$Q(4, 5) \rightarrow Q'(-5, 4)$

$R(5, 1) \rightarrow R'(-1, 5)$

Graph $\triangle PQR$ and its image $\triangle P'Q'R'$.

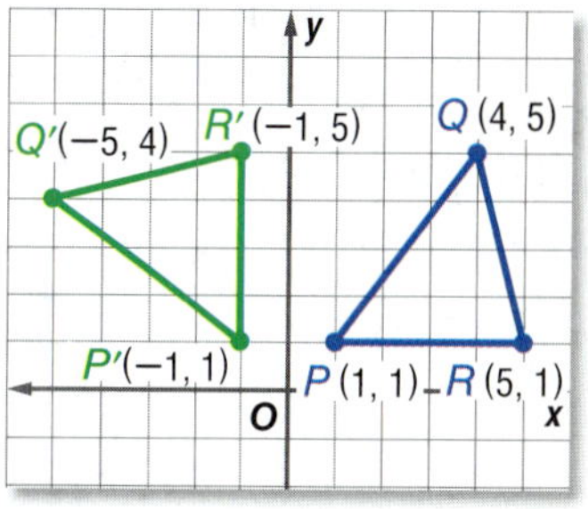

Guided Practice

2. Parallelogram $FGHJ$ has vertices $F(2, 1)$, $G(7, 1)$, $H(6, -3)$, and $J(1, -3)$. Graph $FGHJ$ and its image after a rotation 180° about the origin.

Personal Tutor glencoe.com

STANDARDIZED TEST EXAMPLE 3

Triangle *JKL* is shown at the right. What is the image of point *J* after a rotation 270° counterclockwise about the origin?

A (−3, −7)

B (−7, 3)

C (−7, −3)

D (7, −3)

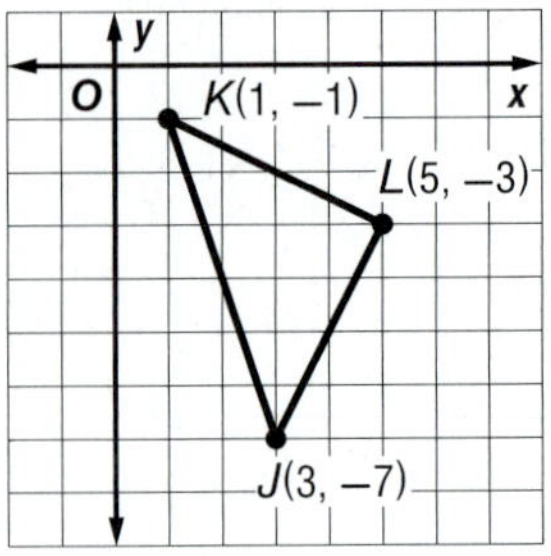

Read the Test Item

You are given that $\triangle JKL$ has coordinates $J(3, -7)$, $K(1, -1)$, and $L(5, -3)$ and are then asked to identify the coordinates of the image of point *J* after a 270° counterclockwise rotation about the origin.

Solve the Test Item

To find the coordinates of point *J* after a 270° counterclockwise rotation about the origin, multiply the *x*-coordinate by −1 and then interchange the *x*- and *y*-coordinates.

$(x, y) \rightarrow (y, -x)$ $\qquad$ $(3, -7) \rightarrow (-7, -3)$

The answer is choice C.

StudyTip

270° Rotation You can complete a 270° rotation by performing a 90° rotation and a 180° rotation in sequence.

Guided Practice

3. Parallelogram *WXYZ* is rotated 180° counterclockwise about the origin. Which of these graphs represents the resulting image?

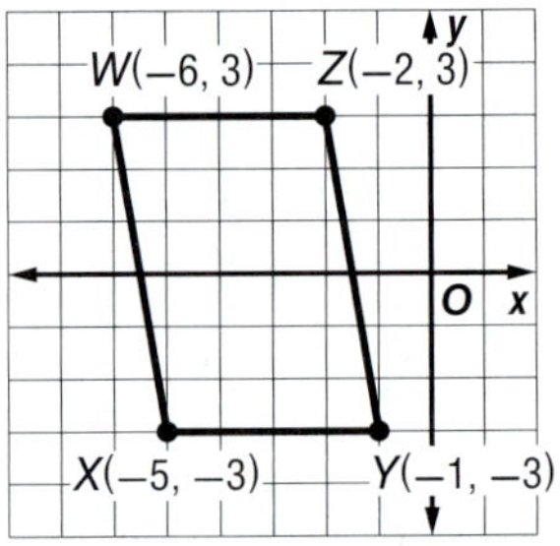

F

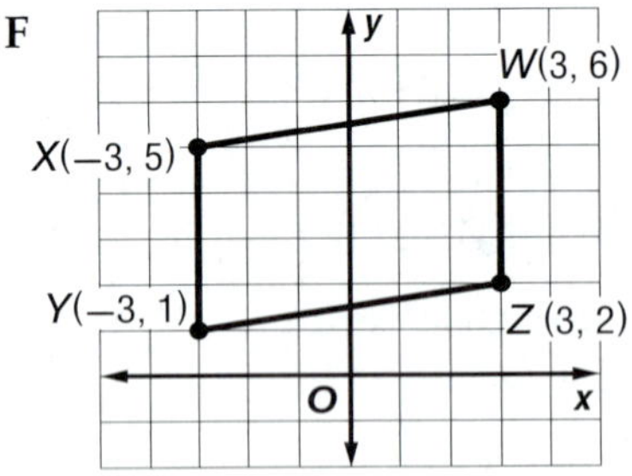

H

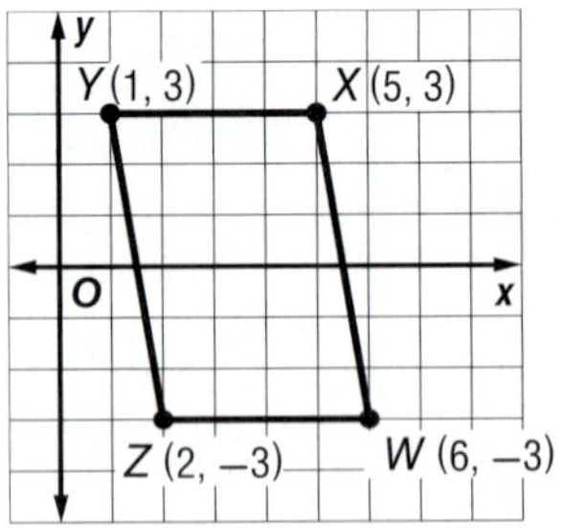

G

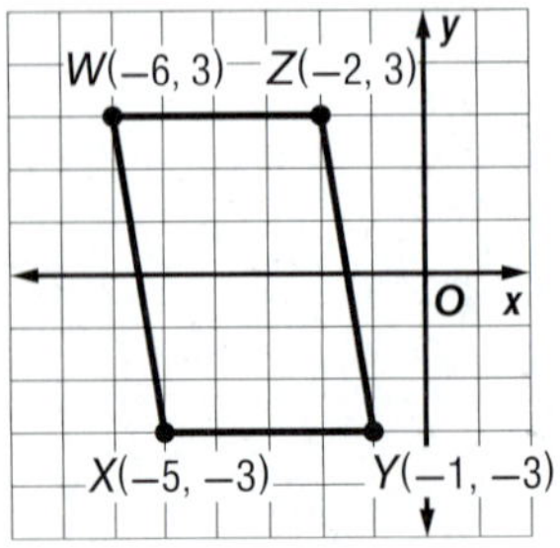

J

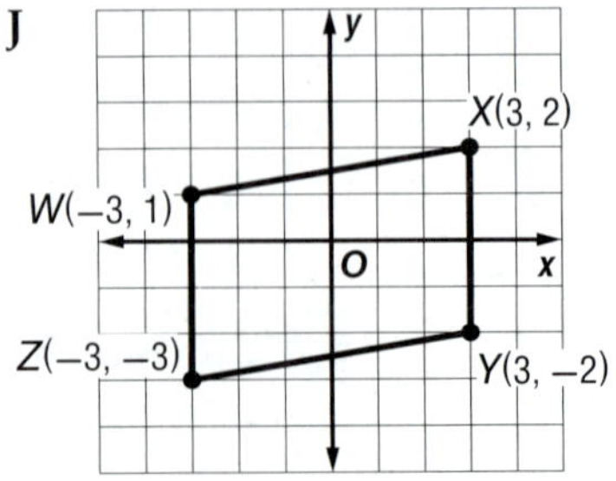

Test-TakingTip

Solve a Simpler Problem Instead of checking all four vertices of parallelogram *WXYZ* in each graph, check just one vertex, such as *X*.

Personal Tutor glencoe.com

Check Your Understanding

Example 1 p. 632

Copy each polygon and point K. Then use a protractor and ruler to draw the specified rotation of each figure about point K.

1. 45°

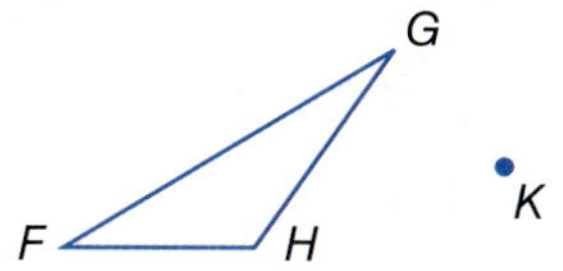

2. 120°

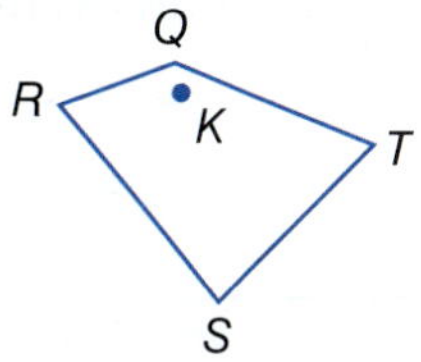

Example 2 p. 633

3 Triangle DFG has vertices $D(-2, 6)$, $F(2, 8)$, and $G(2, 3)$. Graph $\triangle DFG$ and its image after a rotation 180° about the origin.

Example 3 p. 634

4. **MULTIPLE CHOICE** For the transformation shown, what is the measure of the angle of rotation of $ABCD$ about the origin?

A 90°

B 180°

C 270°

D 360°

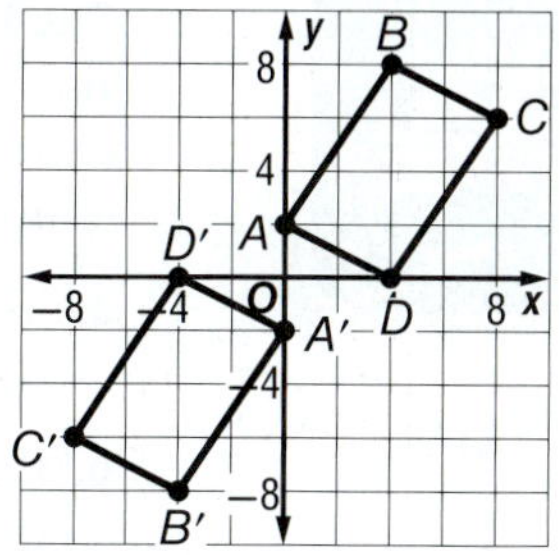

Practice and Problem Solving

= **Step-by-Step Solutions** begin on page R20.
Extra Practice begins on page 969.

Example 1 p. 632

Copy each polygon and point K. Then use a protractor and ruler to draw the specified rotation of each figure about point K.

5. 90°

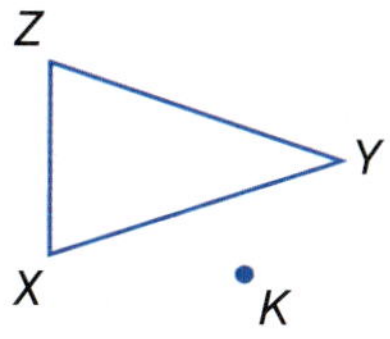

6. 15°

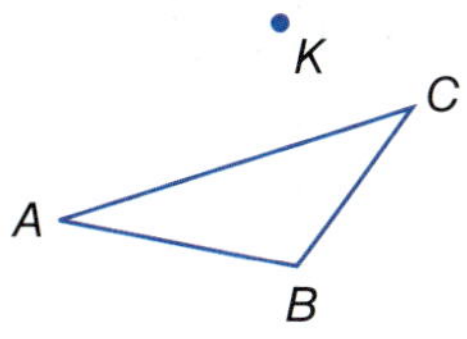

7. 145°

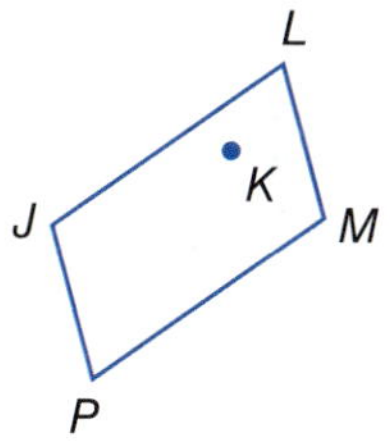

8. 30°

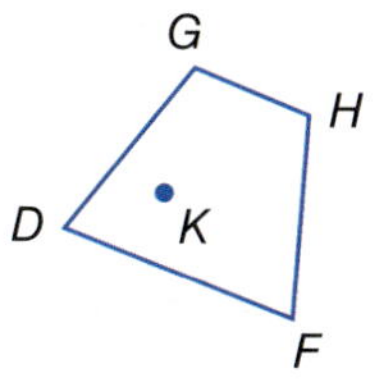

9. 260°

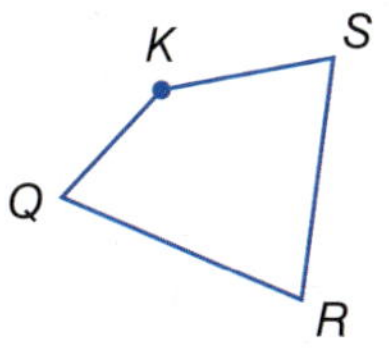

10. 50°

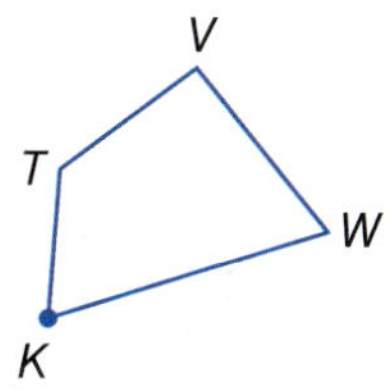

PINWHEELS **Find the angle of rotation to the nearest tenth of a degree that maps P onto P'. Explain your reasoning.**

11.

12.

13.

Examples 2 and 3
pp. 633–634

Graph each figure and its image after the specified rotation about the origin.

14. $\triangle JKL$ has vertices $J(2, 6)$, $K(5, 2)$, and $L(7, 5)$; 90°

15. rhombus $WXYZ$ has vertices $W(-3, 4)$, $X(0, 7)$, $Y(3, 4)$, and $Z(0, 1)$; 90°

16. $\triangle FGH$ has vertices $F(2, 4)$, $G(5, 6)$, and $H(7, 2)$; 180°

17. trapezoid $ABCD$ has vertices $A(-7, -2)$, $B(-6, -6)$, $C(-1, -1)$, and $D(-5, 0)$; 180°

18. $\triangle RST$ has vertices $R(-6, -1)$, $S(-4, -5)$, and $T(-2, -1)$; 270°

19. parallelogram $MPQV$ has vertices $M(-6, 3)$, $P(-2, 3)$, $Q(-3, -2)$, and $V(-7, -2)$; 270°

Real-World Link

The Grande Roue was built in Paris in 1999 for the millennium celebrations. The diameter of the wheel is 60 meters.

Source: *Speeple News*

20. WEATHER A weathervane is used to indicate the direction of the wind. If the vane is pointing northeast and rotates 270°, what is the new wind direction?

21. PHOTOGRAPHY The photograph of the Grande Roue, or Big Wheel, at the left appears blurred because of the camera's shutter speed—the length of time the camera's shutter was open.

a. Estimate the angle of rotation in the photo. (*Hint:* Use points A and A'.)

b. If the Ferris wheel makes one revolution per minute, use your estimate from part **a** to estimate the camera's shutter speed.

Each figure shows a preimage and its image after a rotation about point P. Copy each figure, locate point P, and find the angle of rotation.

22.

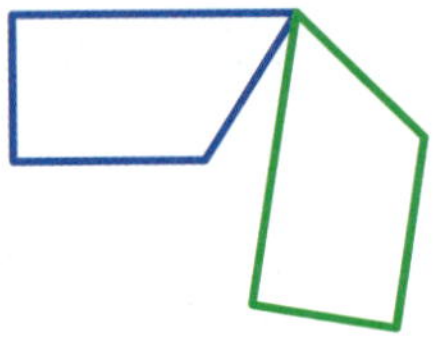

23.

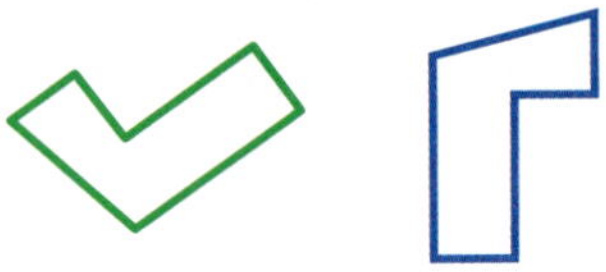

ALGEBRA Give the equation of the line $y = -x - 2$ after a rotation about the origin through the given angle. Then describe the relationship between the equations of the image and preimage.

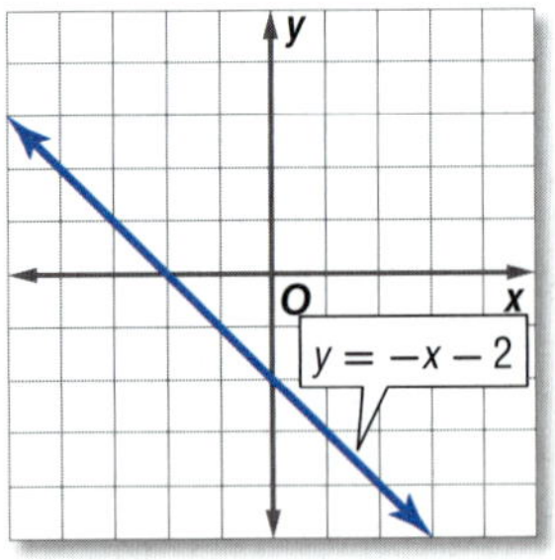

24. 90°

25. 180°

26. 270°

27. 360°

ALGEBRA Rotate the line the specified number of degrees about the x- and y-intercepts and find the equation of the resulting image.

28. $y = x - 5$; 90°

29. $y = 2x + 4$; 180°

30. $y = 3x - 2$; 270°

31. RIDES An amusement park ride consists of four circular cars. The ride rotates at a rate of 0.25 revolution per second. In addition, each car rotates 0.5 revolution per second. If Jane is positioned at point P when the ride begins, what coordinates describe her position after 31 seconds?

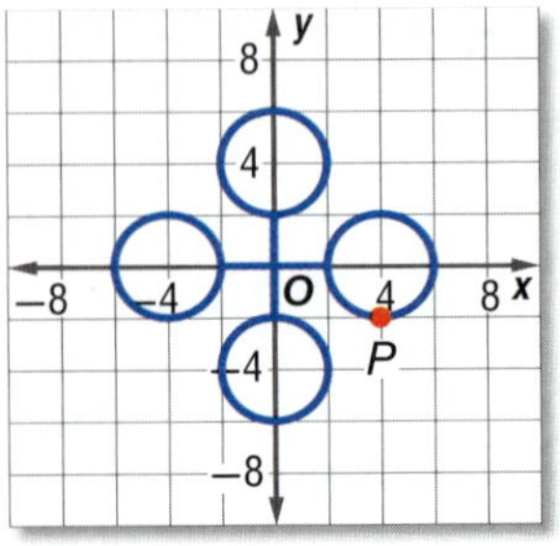

32. **BICYCLE RACING** Brandon and Nestor are participating in a bicycle race on a circular track with a radius of 200 feet.

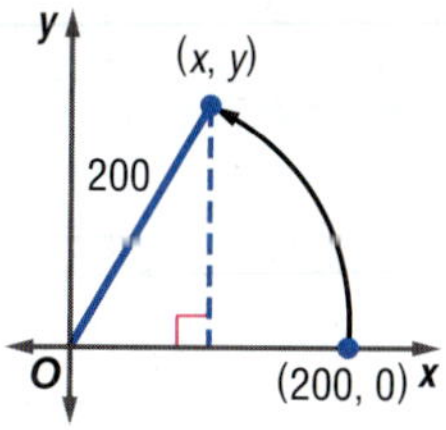

a. If the race starts at (200, 0) and both complete one rotation in 30 seconds, what are their coordinates after 5 seconds?

b. Suppose the length of race is 50 laps and Brandon continues the race at the same rate. If Nestor finishes in 26.2 minutes, who is the winner?

Real-World Link

Bicycle tires can support about 400 times their own weight and only collapse under about 700 times their own weight.

Source: Exploratorium

33. **MULTIPLE REPRESENTATIONS** In this problem, you will investigate reflections over a pair of intersecting lines.

a. **GEOMETRIC** On a coordinate plane, draw a triangle and a pair of intersecting lines. Label the triangle ABC and the lines ℓ and m. Reflect $\triangle ABC$ in the line ℓ. Then reflect $\triangle A'B'C'$ in the line m. Label the final image $A''B''C''$.

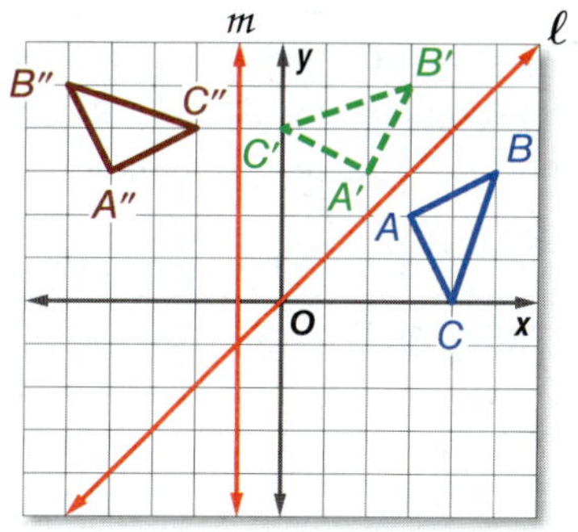

b. **GEOMETRIC** Repeat the process in part a two more times in two different quadrants. Label the second triangle DEF and reflect it in intersecting lines n and p. Label the third triangle MNP and reflect it in intersecting lines q and r.

c. **TABULAR** Measure the angle of rotation of each triangle about the point of intersection of the two lines. Copy and complete the table below.

Angle of Rotation Between Figures		Angle Between Intersecting Lines	
$\triangle ABC$ and $\triangle A''B''C''$		ℓ and m	
$\triangle DEF$ and $\triangle D''E''F$		n and p	
$\triangle MNP$ and $\triangle M''N''P$		q and r	

d. **VERBAL** Make a conjecture about the angle of rotation of a figure about the intersection of two lines after the figure is reflected in both lines.

H.O.T. Problems

Use **H**igher-**O**rder **T**hinking Skills

34. **WRITING IN MATH** Are collinearity and betweenness of points maintained under rotation? Explain.

35. **CHALLENGE** Point C has coordinates $C(5, 5)$. The image of this point after a rotation of 100° about a certain point is $C'(-5, 7.5)$. Find the coordinates of the center of this rotation. Explain.

36. **OPEN ENDED** Draw a figure on the coordinate plane. Describe a nonzero rotation that maps the image onto the preimage with no change in orientation.

37. **REASONING** Is the reflection of a figure in the x-axis equivalent to the rotation of that same figure 180° about the origin? Explain.

38. **WRITING IN MATH** Do invariant points *sometimes*, *always*, or *never* occur in a rotation? Explain your reasoning.

Standardized Test Practice

39. What rotation of trapezoid $QRST$ creates an image with point R' at (4, 3)?

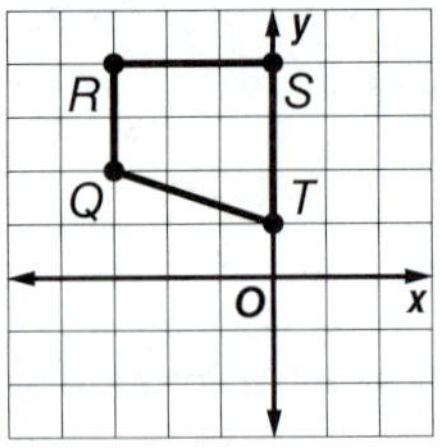

- **A** 270° counterclockwise about point T
- **B** 185° counterclockwise about point T
- **C** 180° clockwise about the origin
- **D** 90° clockwise about the origin

40. SHORT RESPONSE $\triangle XYZ$ has vertices $X(1, 7)$, $Y(0, 2)$, and $Z(-5, -2)$. What are the coordinates of X' after a rotation 270° counterclockwise about the origin?

41. ALGEBRA The population of the United States in July of 2007 was estimated to have surpassed 301,000,000. At the same time the world population was estimated to be over 6,602,000,000. What percent of the world population, to the nearest tenth, lived in the United States at this time?

F 3.1%	**H** 4.2%
G 3.5%	**J** 4.6%

42. SAT/ACT An 18-foot ladder is placed against the side of a house. The base of the ladder is positioned 8 feet from the house. How high up on the side of the house, to the nearest tenth of a foot, does the ladder reach?

A 10.0 ft	**C** 19.7 ft
B 16.1 ft	**D** 26.0 ft

Spiral Review

43. VOLCANOES A cloud of dense gas and dust from a volcano blows 40 miles west and then 30 miles north. Make a sketch to show the translation of the dust particles. Then find the distance of the shortest path that would take the particles to the same position. (Lesson 9-2)

Copy the figure and the given line of reflection. Then draw the reflected image in this line using a ruler. (Lesson 9-1)

44.

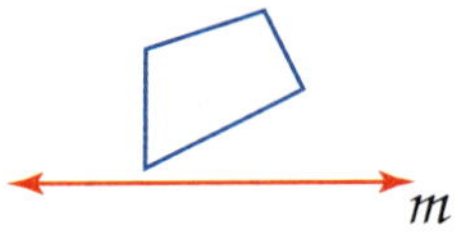

45.

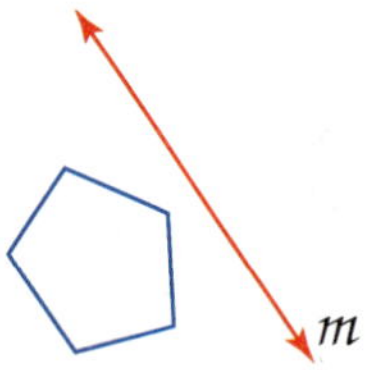

46.

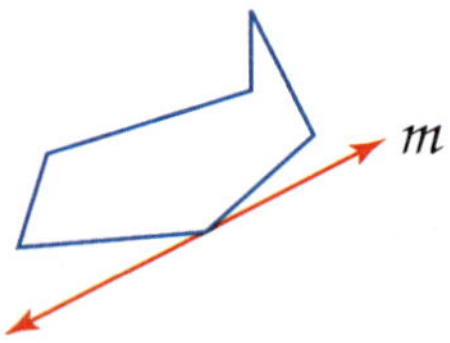

Skills Review

Identify the type of congruence transformation shown as a *reflection*, *translation*, or *rotation*. (Lesson 4-7)

47.

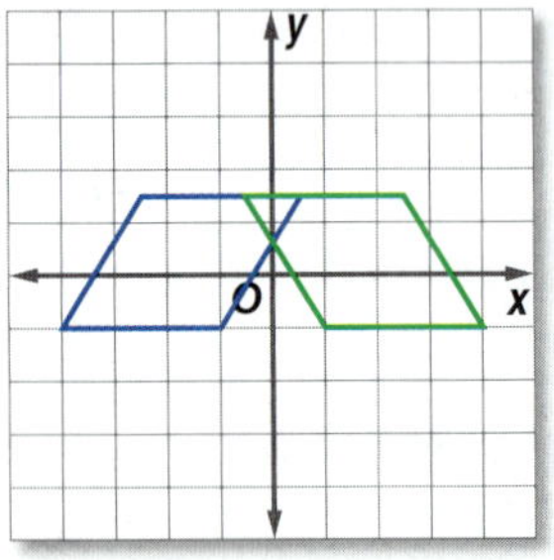

48.

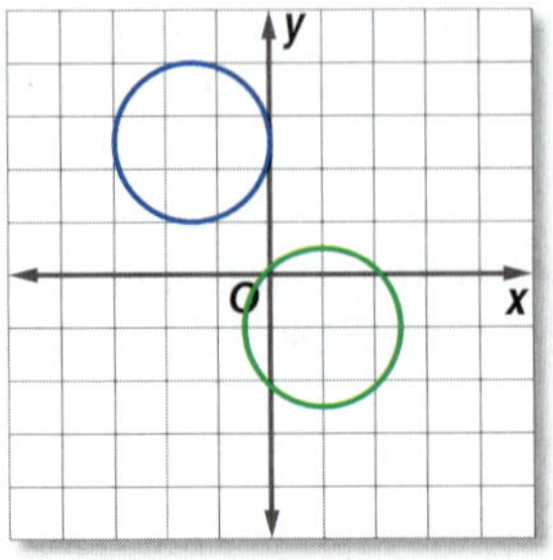

49.

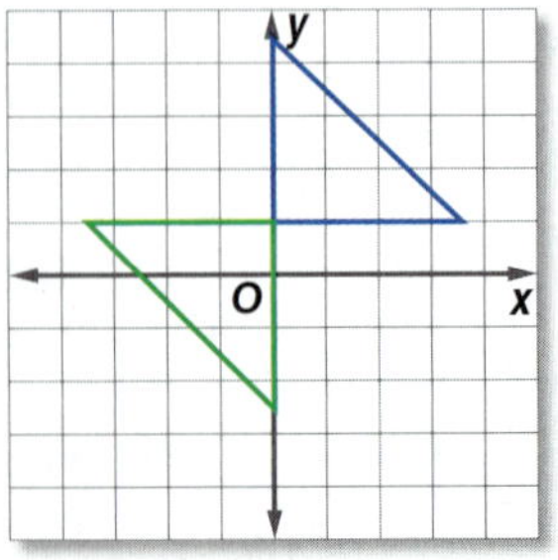

CHAPTER 9 Mid-Chapter Quiz

Lessons 9-1 through 9-3

Copy the figure and the given line of reflection. Then draw the reflected image in this line using a ruler. (Lesson 9-1)

1.

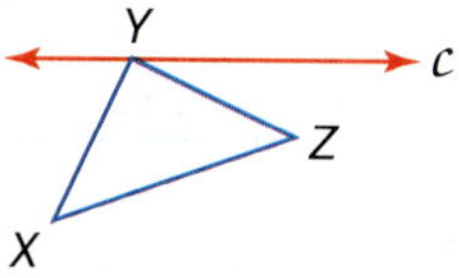

2.

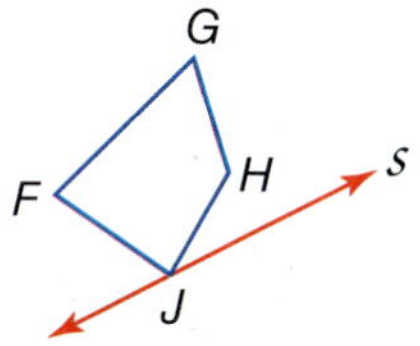

Graph each figure and its image after the specified reflection. (Lesson 9-1)

3. $\triangle FGH$ has vertices $F(-4, 3)$, $G(-2, 0)$, and $H(-1, 4)$; in the y-axis

4. rhombus $QRST$ has vertices $Q(2, 1)$, $R(4, 3)$, $S(6, 1)$, and $T(4, -1)$; in the x-axis

5. **CLUBS** The drama club is selling candy during the intermission of a school play. Locate point P along the wall to represent the candy table so that people coming from either door A or door B would walk the same distance to the table. (Lesson 9-1)

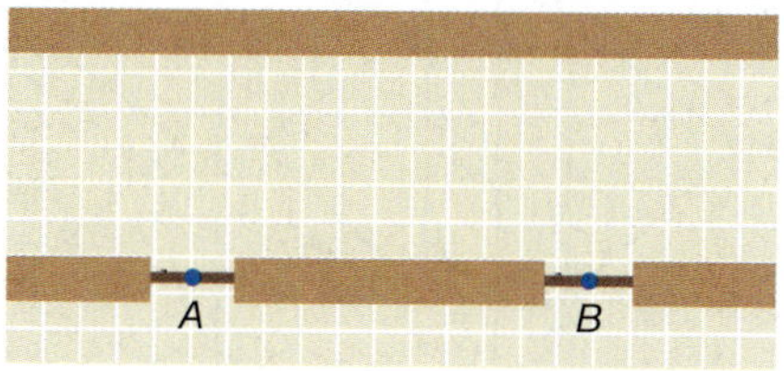

Graph each figure and its image after the specified translation. (Lesson 9-2)

6. $\triangle ABC$ with vertices $A(0, 0)$, $B(2, 1)$, $C(1, -3)$; $\langle 3, -1\rangle$

7. rectangle $JKLM$ has vertices $J(-4, 2)$, $K(-4, -2)$, $L(-1, -2)$, and $M(-1, 2)$; $\langle 5, -3\rangle$

Copy the figure and the given translation vector. Then draw the translation of the figure along the translation vector. (Lesson 9-2)

8.

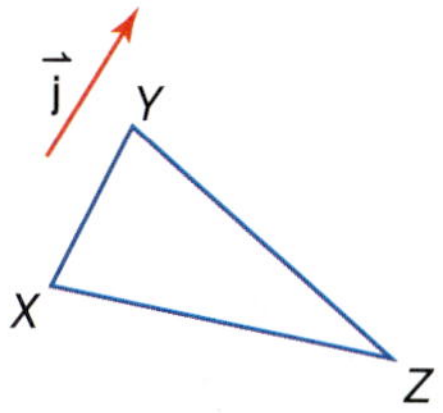

9.

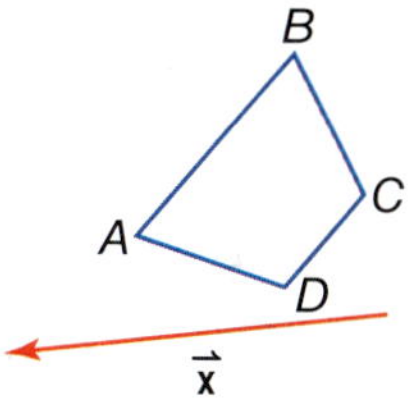

10. **COMICS** Alex is making a comic. He uses graph paper to make sure the dimensions of his drawings are accurate. If he draws a coordinate plane with two flies as shown below, what vector represents the movement from fly 1 to fly 2? (Lesson 9-2)

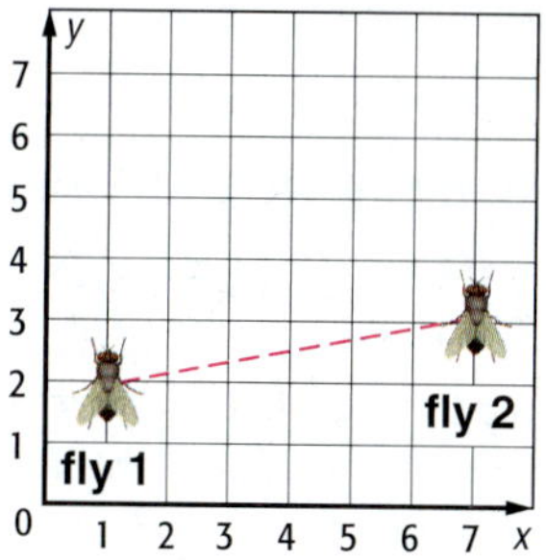

Copy each polygon and point R. Then use a protractor and ruler to draw the specified rotation of each figure about point R. (Lesson 9-3)

11. 45°

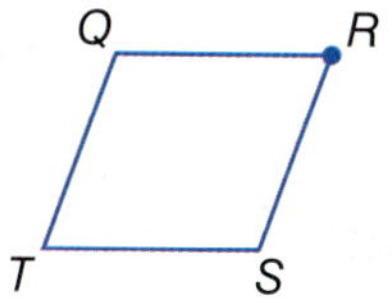

12. 60°

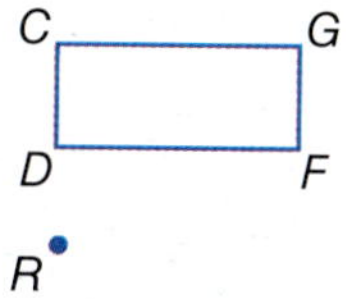

13. **MULTIPLE CHOICE** What is the image of point M after a rotation of 90° about the origin? (Lesson 9-3)

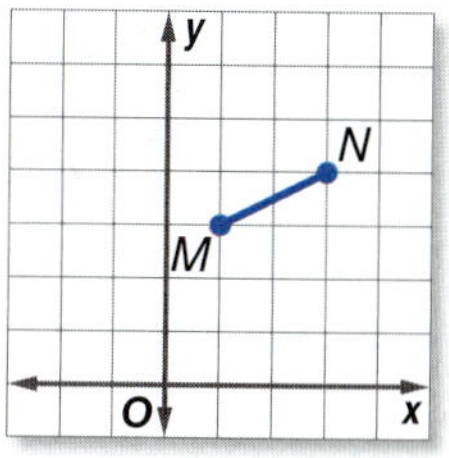

A $(-3, 1)$

B $(-3, -1)$

C $(-1, -3)$

D $(3, 1)$

Graph each figure and its image after the specified rotation. (Lesson 9-3)

14. $\triangle RST$ has vertices $R(-3, 0)$, $S(-1, -4)$, and $T(0, -1)$; 90°

15. square $JKLM$ has vertices $J(-1, 2)$, $K(-1, -2)$, $L(3, -2)$, and $M(3, 2)$; 180°

EXPLORE 9-4

Geometry Software Lab

Compositions of Transformations

Math Online glencoe.com

- Other Calculator Keystrokes
- Graphing Technology Personal Tutor

In this lab, you will use Geometer's Sketchpad to explore the effects of performing multiple transformations on a figure.

ACTIVITY

Reflect a figure in two vertical lines.

Step 1 Use the line segment tool to construct a triangle with one vertex pointing to the left so that you can easily see changes as you perform transformations. Label the triangle *ABC*.

Step 2 Insert and label a line *m* to the right of $\triangle ABC$. Insert a point so that the distance from the point to line *m* is greater than the width of $\triangle ABC$. Draw the line parallel to line *m* through the point and label the new line *r*.

Step 3 Select line *m* and choose **Mark Mirror** from the **Transform** menu. Select all sides and vertices of $\triangle ABC$ and choose **Reflect** from the **Transform** menu.

Step 4 Repeat the process you used in Step 3 to reflect the new image in line *r*.

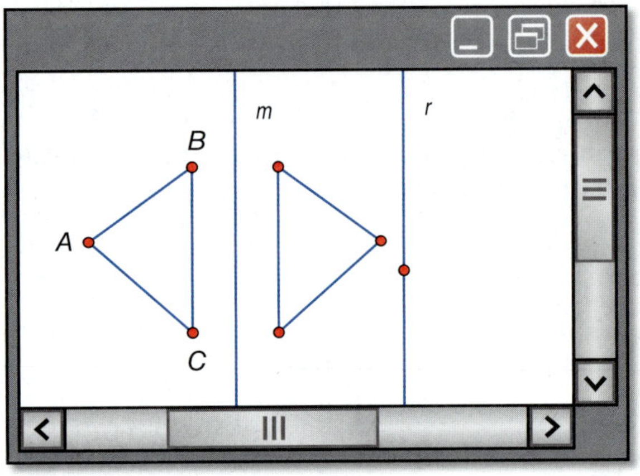

Steps 1–3

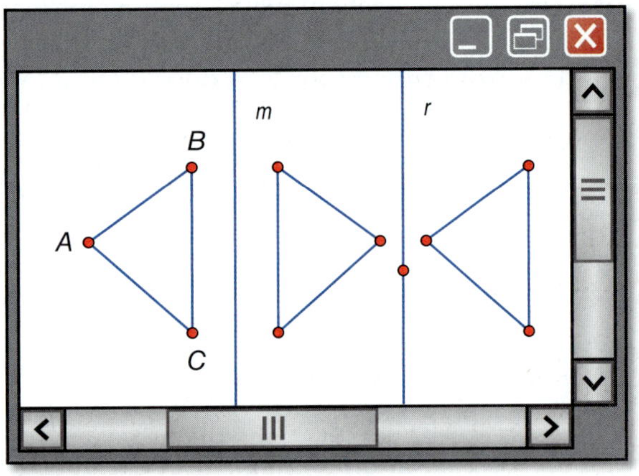

Step 4

Analyze the Results

1. How are the original figure and the final figure related?
2. What single transformation could be used to produce the final figure?
3. If you move line *m*, what happens? if you move line *r*?
4. **MAKE A CONJECTURE** If you reflected the figure in a third line, what single transformation do you think could be used to produce the final figure? Explain your reasoning.
5. Repeat the activity for a pair of perpendicular lines. What single transformation could be used to produce the same final figure?
6. **MAKE A CONJECTURE** If you reflected the figure from Exercise 5 in a third line perpendicular to the second line, what single transformation do you think could be used to produce the final figure? Explain your reasoning.

9-4 Compositions of Transformations

Then

You drew reflections, translations, and rotations. (Lessons 9-1, 9-2, and 9-3)

Now

- Draw glide reflections and other compositions of isometries in the coordinate plane.
- Draw compositions of reflections in parallel and intersecting lines.

New Vocabulary

composition of transformations
glide reflection

Math Online
glencoe.com

Why?

The pattern of footprints left in the sand after a person walks along the edge of a beach illustrates the composition of two different transformations—translations and reflections.

Glide Reflections When a transformation is applied to a figure and then another transformation is applied to its image, the result is called a **composition of transformations**. A glide reflection is one type of composition of transformations.

Key Concept: Glide Reflection

A **glide reflection** is the composition of a translation followed by a reflection in a line parallel to the translation vector.

Example
The glide reflection shown is the composition of a translation along $\vec{w}$ followed by a reflection in line ℓ.

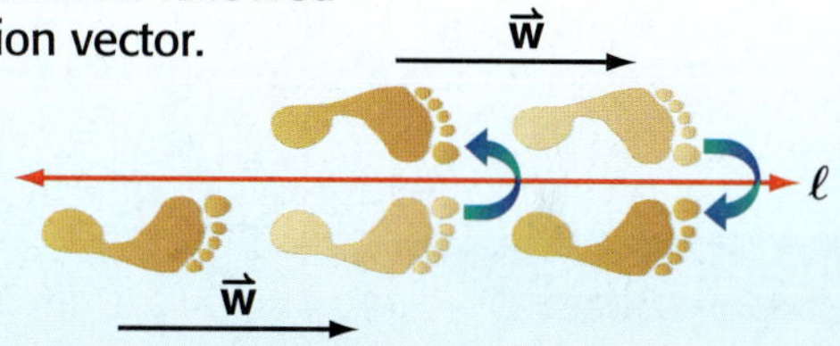

EXAMPLE 1 Graph a Glide Reflection

Triangle *JKL* has vertices *J*(6, −1), *K*(10, −2), and *L*(5, −3). Graph △*JKL* and its image after a translation along ⟨0, 4⟩ and a reflection in the *y*-axis.

Step 1 translation along ⟨0, 4⟩

(x, y)	→	$(x, y + 4)$
$J(6, -1)$	→	$J'(6, 3)$
$K(10, -2)$	→	$K'(10, 2)$
$L(5, -3)$	→	$L'(5, 1)$

Step 2 reflection in the *y*-axis

(x, y)	→	$(-x, y)$
$J'(6, 3)$	→	$J''(-6, 3)$
$K'(10, 2)$	→	$K''(-10, 2)$
$L'(5, 1)$	→	$L''(-5, 1)$

Step 3 Graph △*JKL* and its image △*J″K″L″*.

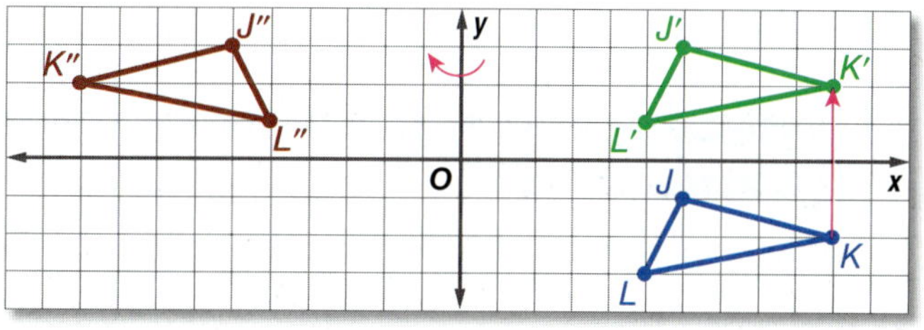

Guided Practice

Triangle *PQR* has vertices *P*(1, 1), *Q*(2, 5), and *R*(4, 2). Graph △*PQR* and its image after the indicated glide reflection.

1A. Translation: along ⟨−2, 0⟩
Reflection: in *x*-axis

1B. Translation: along ⟨−3, −3⟩
Reflection: in $y = x$

Personal Tutor glencoe.com

In Example 1, $\triangle JKL \cong \triangle J'K'L'$ and $\triangle J'K'L' \cong \triangle J''K''L''$. By the Transitive Property of Congruence, $\triangle JKL \cong \triangle J''K''L''$. This suggests the following theorem.

Theorem 9.1 **Composition of Isometries**

The composition of two (or more) isometries is an isometry.

You will prove one case of Theorem 9.1 in Exercise 30.

StudyTip

Rigid Motions Glide reflections, reflections, translations, and rotations are the only four *rigid motions* or isometries in a plane.

So, the composition of two or more isometries—reflections, translations, or rotations—results in an image that is congruent to its preimage.

EXAMPLE 2 Graph Other Compositions of Isometries

The endpoints of $\overline{CD}$ are $C(-7, 1)$ and $D(-3, 2)$. Graph $\overline{CD}$ and its image after a reflection in the x-axis and a rotation 90° about the origin.

Step 1 reflection in the x-axis

$(x, y) \rightarrow (x, -y)$

$C(-7, 1) \rightarrow C'(-7, -1)$

$D(-3, 2) \rightarrow D'(-3, -2)$

Step 2 rotation 90° about origin

$(x, y) \rightarrow (-y, x)$

$C'(-7, -1) \rightarrow C''(1, -7)$

$D'(-3, -2) \rightarrow D''(2, -3)$

Step 3 Graph $\overline{CD}$ and its image $\overline{C''D''}$.

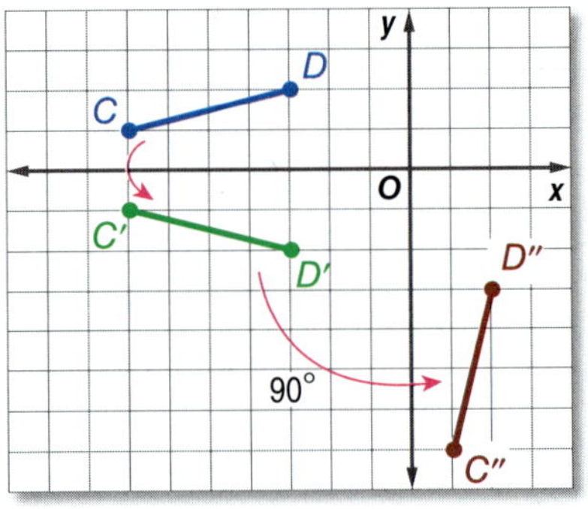

ReadingMath

Double Primes Double primes are used to indicate that a vertex is the image of a second transformation.

Guided Practice

Triangle ABC has vertices $A(-6, -2)$, $B(-5, -5)$, and $C(-2, -1)$. Graph $\triangle ABC$ and its image after the composition of transformations in the order listed.

2A. Translation: along $\langle 3, -1 \rangle$
Reflection: in y-axis

2B. Rotation: 180° about origin
Translation: along $\langle -2, 4 \rangle$

Personal Tutor glencoe.com

Compositions of Two Reflections The composition of two reflections in parallel lines is the same as a translation.

Theorem 9.2 **Reflections in Parallel Lines**

The composition of two reflections in parallel lines can be described by a translation vector that is

- perpendicular to the two lines, and
- twice the distance between the two lines.

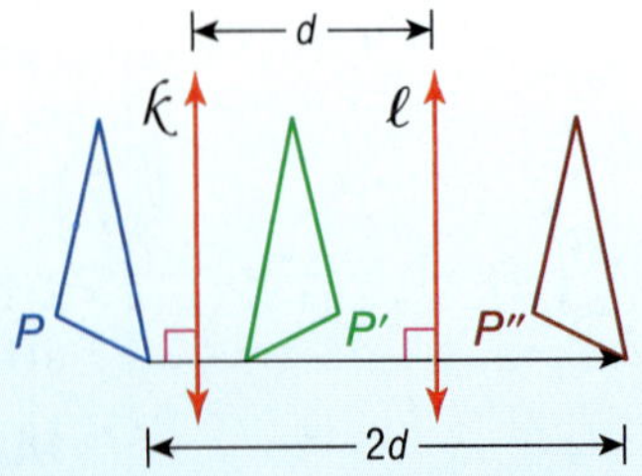

You will prove Theorem 9.2 in Exercise 36.

The composition of two reflections in intersecting lines is the same as a rotation.

Theorem 9.3 Reflections in Intersecting Lines

The composition of two reflections in intersecting lines can be described by a rotation

- about the point where the lines intersect and
- through an angle that is twice the measure of the acute or right angle formed by the lines.

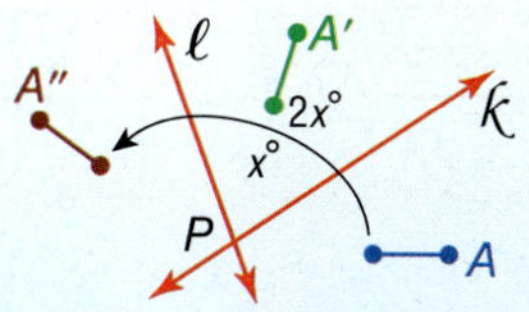

You will prove Theorem 9.3 in Exercise 37.

> **Watch Out!**
>
> **Order of Composition** Be sure to compose two transformations according to the order in which they are given.

EXAMPLE 3 Reflect a Figure in Two Lines

Copy and reflect figure *A* in line *m* and then line *p*. Then describe a single transformation that maps *A* onto *A*″.

a.

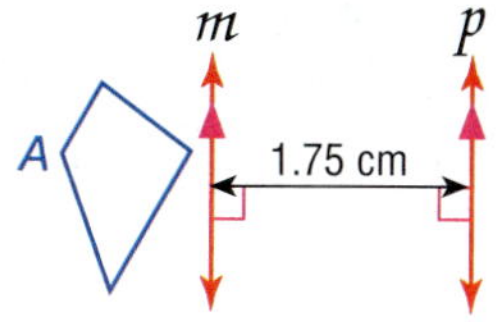

Step 1 Reflect *A* in line *m*.

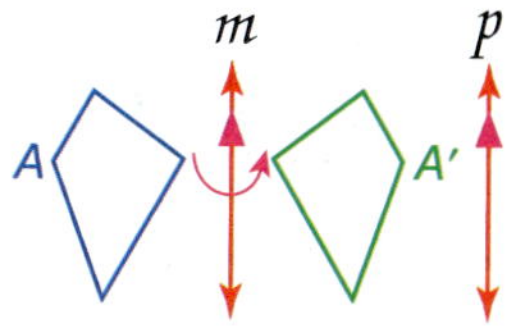

Step 2 Reflect *A* in line *p*.

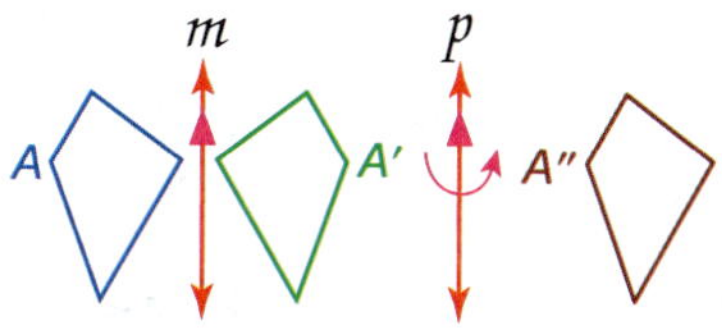

By Theorem 9.2, the composition of two reflections in parallel vertical lines *m* and *p* is equivalent to a horizontal translation right 2 • 1.75 or 3.5 centimeters.

b.

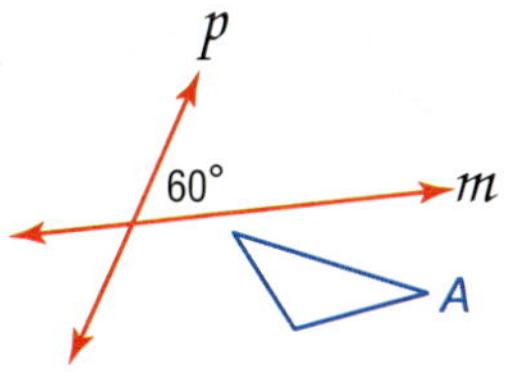

Step 1

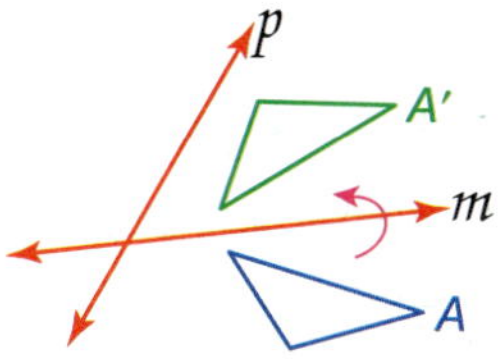

Step 2

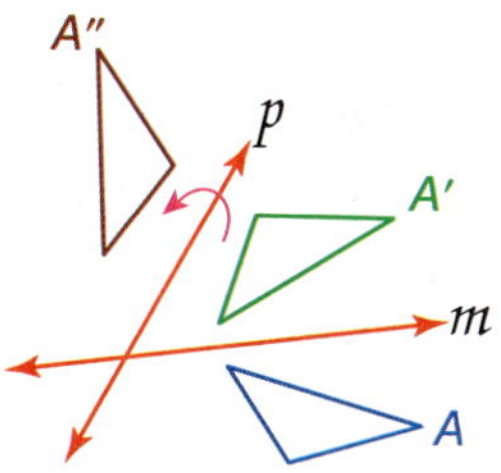

By Theorem 9.3, the composition of two reflections in intersecting lines *m* and *p* is equivalent to a 2 • 60° or 120° counterclockwise rotation about the point where lines *m* and *p* intersect.

Guided Practice

Copy and reflect figure *B* in line *n* and then line *q*. Then describe a single transformation that maps *B* onto *B*″.

3A.

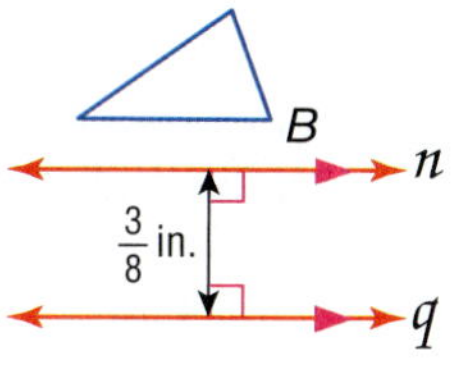

3B.

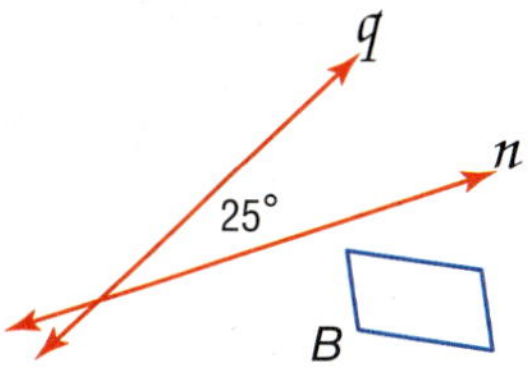

Personal Tutor glencoe.com

Many patterns in the real world are created using compositions of transformations.

Real-World Link

In carpets, border patterns result when any of several basic transformations are repeated in one direction. There are seven possible combinations: translations, horizontal reflections, vertical reflections, vertical followed by horizontal reflections, glide reflections, rotations, and reflections followed by glide reflections.

Source: The Textile Museum

Real-World EXAMPLE 4 Describe Transformations

STATIONERY BORDERS Describe the transformations that are combined to create each stationery border shown.

a.

The pattern is created by successive translations of the first four potted plants. So this pattern can be created by combining two reflections in lines m and p as shown. Notice that line m goes through the center of the preimage.

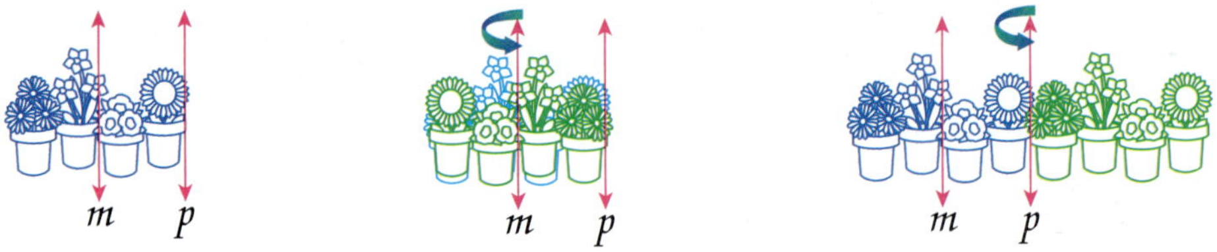

b.

The pattern is created by glide reflection. So this pattern can be created by combining a translation along translation vector $\vec{v}$ followed by a reflection over horizontal line n as shown.

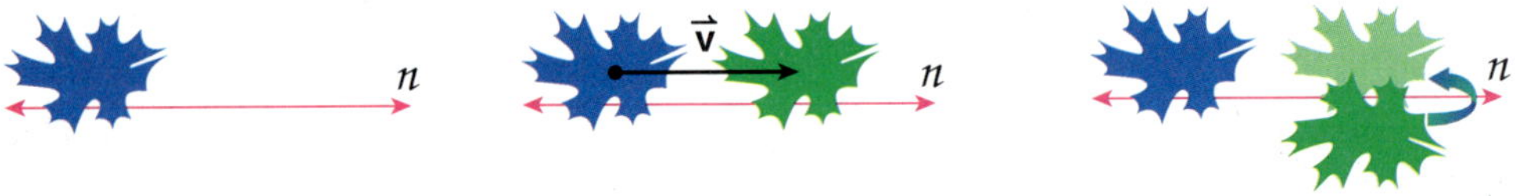

Guided Practice

4. CARPET PATTERNS Describe the transformations that are combined to create each carpet pattern shown.

A.

B.

Personal Tutor glencoe.com

Concept Summary: Compositions of Translations

Glide Reflection	Translation	Rotation
the composition of a reflection and a translation	the composition of two reflections in parallel lines	the composition of two reflections in intersecting lines

Check Your Understanding

Example 1 p. 641

Triangle CDE has vertices $C(-5, -1)$, $D(-2, -5)$, and $E(-1, -1)$. Graph $\triangle CDE$ and its image after the indicated glide reflection.

1. Translation: along $\langle 4, 0 \rangle$
Reflection: in x-axis

2. Translation: along $\langle 0, 6 \rangle$
Reflection: in y-axis

Example 2 p. 642

3. The endpoints of $\overline{JK}$ are $J(2, 5)$ and $K(6, 5)$. Graph $\overline{JK}$ and its image after a reflection in the x-axis and a rotation 90° about the origin.

Example 3 p. 643

Copy and reflect figure S in line m and then line p. Then describe a single transformation that maps S onto S''.

4.

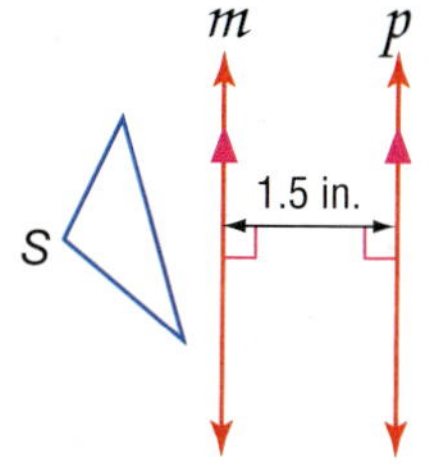

5.

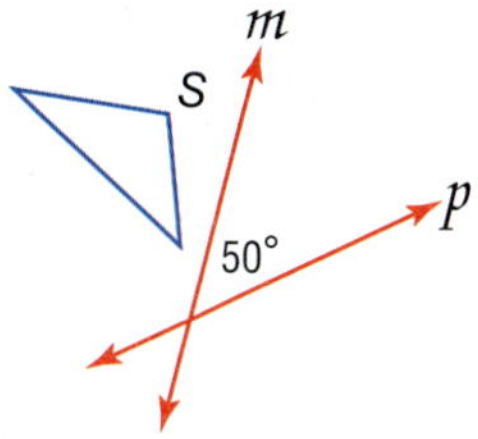

Example 4 p. 644

6. TILE PATTERNS Viviana is creating a pattern for the top of a table with tiles in the shape of isosceles triangles. Describe the transformation combination that was used to create the pattern.

Practice and Problem Solving

● = **Step-by-Step Solutions** begin on page R20.
Extra Practice begins on page 969.

Example 1 p. 641

Graph each figure with the given vertices and its image after the indicated glide reflection.

7. $\triangle RST$: $R(1, -4)$, $S(6, -4)$, $T(5, -1)$
Translation: along $\langle 2, 0 \rangle$
Reflection: in x-axis

8. $\triangle JKL$: $J(1, 3)$, $K(5, 0)$, $L(7, 4)$
Translation: along $\langle -3, 0 \rangle$
Reflection: in x-axis

9. $\triangle XYZ$: $X(-7, 2)$, $Y(-5, 6)$, $Z(-2, 4)$
Translation: along $\langle 0, -1 \rangle$
Reflection: in y-axis

10. $\triangle ABC$: $A(2, 3)$, $B(4, 7)$, $C(7, 2)$
Translation: along $\langle 0, 4 \rangle$
Reflection: in y-axis

11. $\triangle DFG$: $D(2, 8)$, $F(1, 2)$, $G(4, 6)$
Translation: along $\langle 3, 3 \rangle$
Reflection: in $y = x$

12. $\triangle MPQ$: $M(-4, 3)$, $P(-5, 8)$, $Q(-1, 6)$
Translation: along $\langle -4, -4 \rangle$
Reflection: in $y = x$

Example 2 p. 642

Graph each figure with the given vertices and its image after the indicated composition of transformations.

13. $\overline{WX}$: $W(-4, 6)$ and $X(-4, 1)$
Reflection: in x-axis
Rotation: 90° about origin

14. $\overline{AB}$: $A(-3, 2)$ and $B(3, 8)$
Rotation: 90° about origin
Translation: along $\langle 4, 4 \rangle$

15. $\overline{FG}$: $F(1, 1)$ and $G(6, 7)$
Reflection: in x-axis
Rotation: 180° about origin

16. $\overline{RS}$: $R(2, -1)$ and $S(6, -5)$
Translation: along $\langle -2, -2 \rangle$
Reflection: in y-axis

Example 3 p. 643

Copy and reflect figure D in line m and then line p. Then describe a single transformation that maps D onto D''.

17.

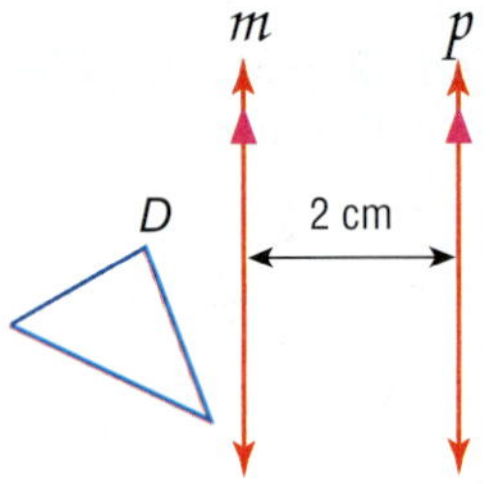

18.

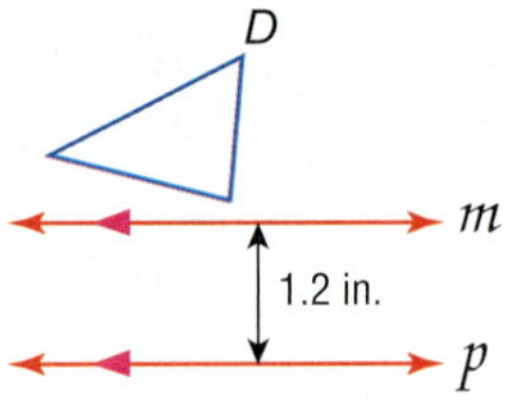

19.

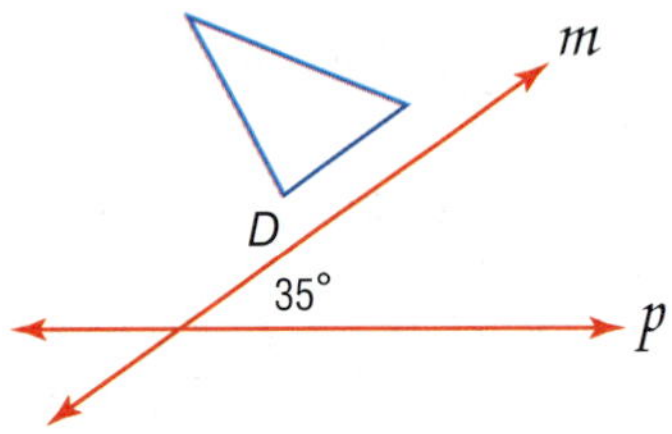

20.

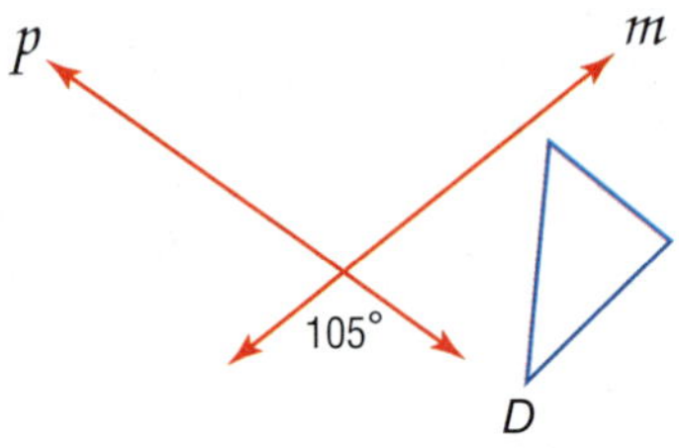

Example 4 p. 644

Describe the transformations combined to create the outlined kimono fabric pattern.

21.

22.

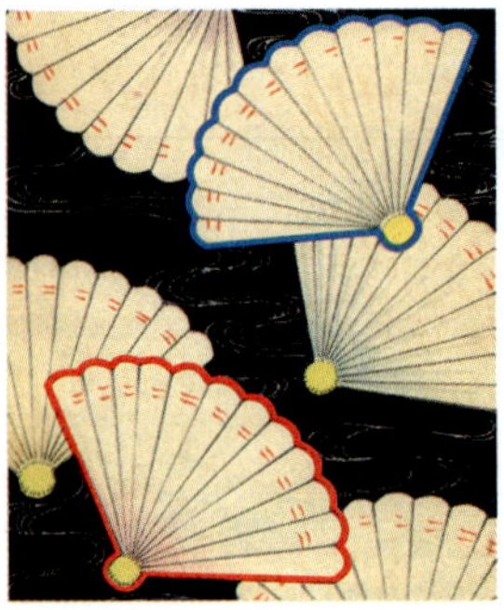

23.

24. SKATEBOARDS Elizabeth has airbrushed the pattern shown onto her skateboard. What combination of transformations did she use to create the pattern?

ALGEBRA Graph each figure and its image after the indicated transformations.

25. Rotation: 90° about the origin
Reflection: in x-axis

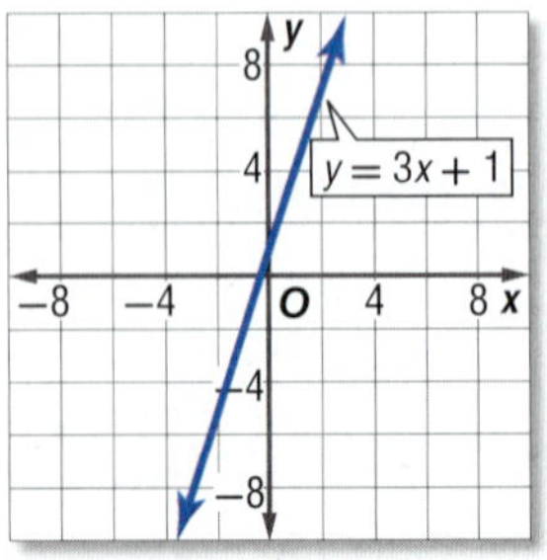

26. Reflection: in x-axis
Reflection: in y-axis

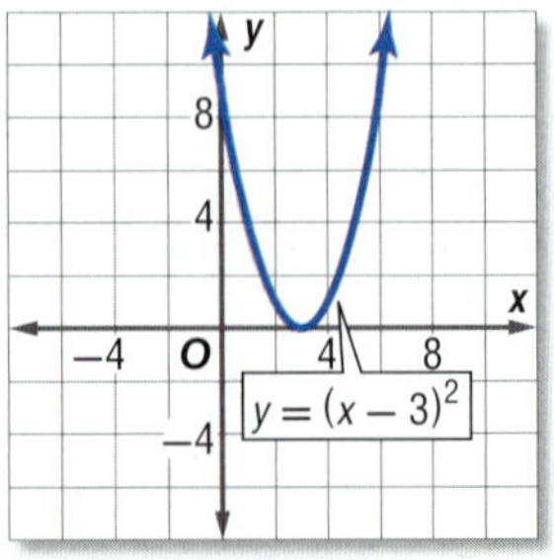

27. Find the coordinates of $\triangle A''B''C''$ after a reflection in the x-axis and a rotation of 180° about the origin if $\triangle ABC$ has vertices $A(-3, 1)$, $B(-2, 3)$, and $C(-1, 0)$.

28. FIGURE SKATING Kayla is practicing her figure skating routine. What combination of transformations is needed for Kayla to start at A, skate to A', and end up at A''?

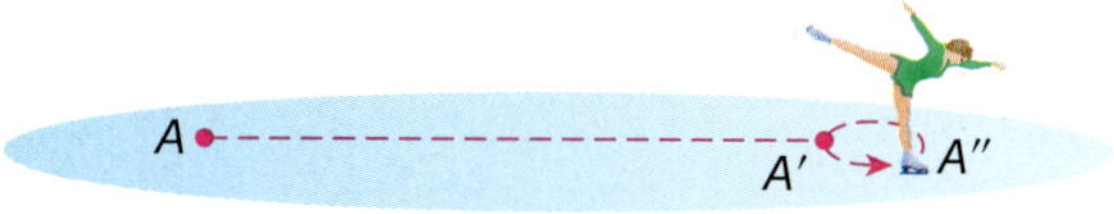

29. DANCING Describe the transformations combined to go from Step 1 to Step 3.

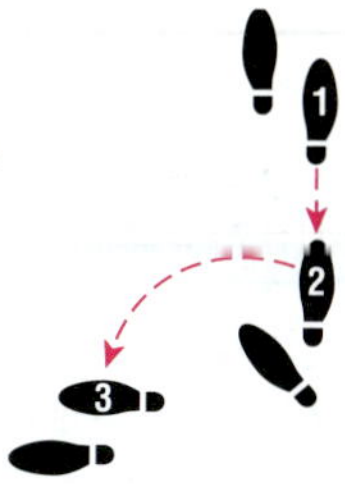

30. PROOF Write a paragraph proof for one case of the Composition of Isometries Theorem.

Given: A translation along $\langle a, b\rangle$ maps X to X' and Y to Y'. A reflection in z maps X' to X'' and Y' to Y''.

Prove: $\overline{XY} \cong \overline{X''Y''}$

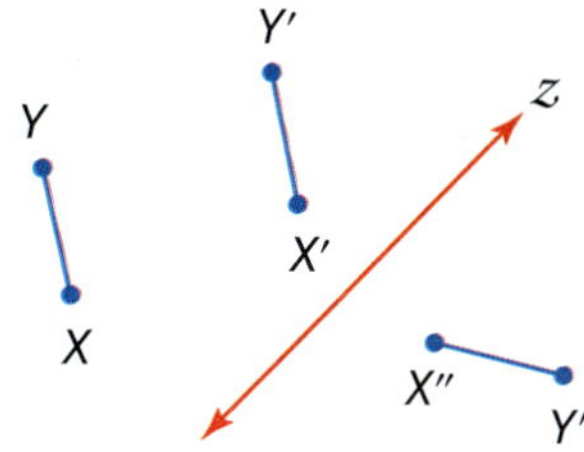

Real-World Link

The length of an animal's stride is the distance between two consecutive tracks. The average stride length of a turkey is about 11 inches, and the average stride length of a duck is about 5 inches.

Source: University of Nebraska State Museum

ANIMAL TRACKS Refer to the information at the left. Write a glide reflection that can be used to predict the location of the next track for each set of animal tracks.

31. turkey

32. duck

33. KNITTING Tonisha is knitting a scarf using the tumbling blocks pattern shown at the right. Describe the transformations combined to create the pattern.

Describe the transformations that combined to map each figure.

34.

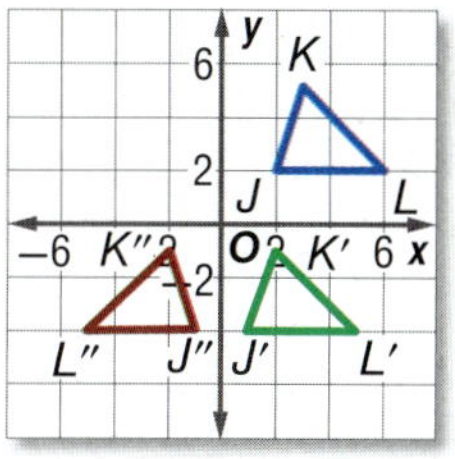

35.

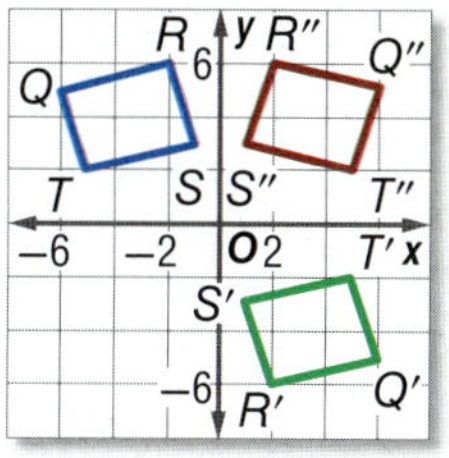

36. PROOF Write a two-column proof of Theorem 9.2.

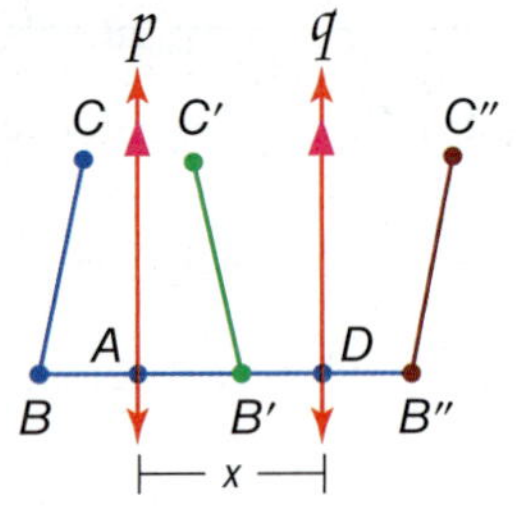

Given: A reflection in line p maps $\overline{BC}$ to $\overline{B'C'}$.
A reflection in line q maps $\overline{B'C'}$ to $\overline{B''C''}$.
$p \parallel q$, $AD = x$

Prove: **a.** $\overline{BB''} \perp p$, $\overline{BB''} \perp q$

b. $BB'' = 2x$

StudyTip

Look Back See Lesson 2-7 to review segment congruence properties.

37. PROOF Write a paragraph proof of Theorem 9.3.

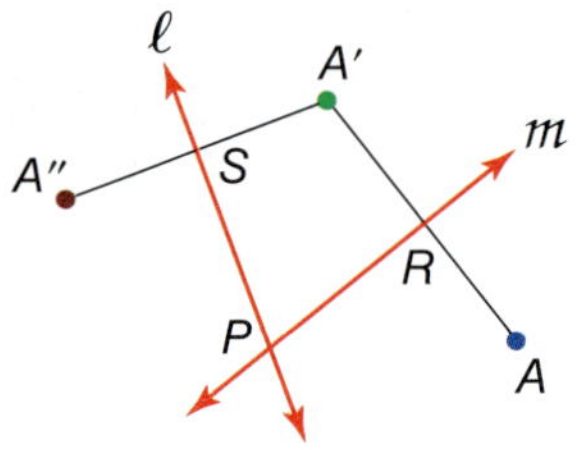

Given: Lines ℓ and m intersect at point P.
A is any point not on ℓ or m.

Prove: **a.** If you reflect point A in m, and then reflect its image A' in ℓ, A'' is the image of A after a rotation about point P.

b. $m\angle APA'' = 2(m\angle SPR)$

H.O.T. Problems Use Higher-Order Thinking Skills

38. ERROR ANALYSIS Daniel and Lolita are translating $\triangle XYZ$ along $\langle 2, 2\rangle$ and reflecting it in the line $y = 2$. Daniel says that the transformation is a glide reflection. Lolita disagrees and says that the transformation is a composition of transformations. Is either of them correct? Explain your reasoning.

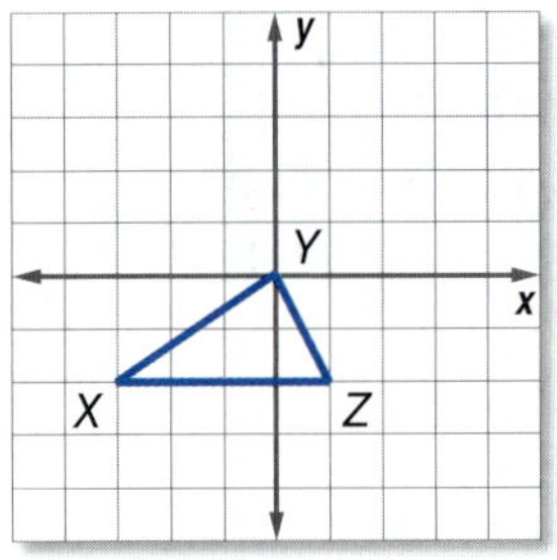

39. WRITING IN MATH Do any points remain invariant under glide reflections? under compositions of transformations? Explain.

40. CHALLENGE If $PQRS$ is translated along $\langle 3, -2\rangle$, reflected in $y = -1$, and rotated 90° about the origin, what are the coordinates of $P'''Q'''R'''S'''$?

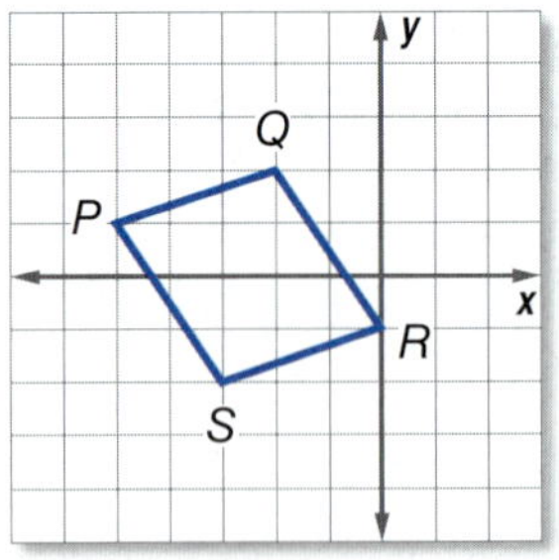

41. REASONING If an image is to be reflected in the line $y = x$ and the x-axis, does the order of the reflections affect the final image? Explain.

42. OPEN ENDED Write a glide reflection or composition of transformations that can be used to transform $\triangle ABC$ to $\triangle DEF$.

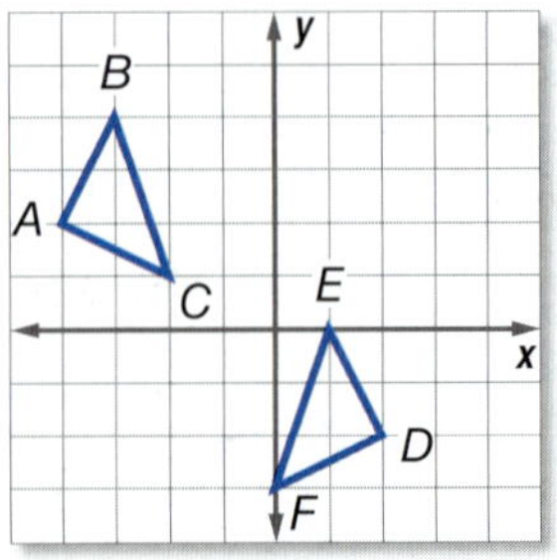

43. REASONING When two rotations are performed on a single image, does the order of the rotations *sometimes*, *always*, or *never* affect the location of the final image? Explain.

44. WRITING IN MATH Compare and contrast glide reflections and compositions of transformations.

Standardized Test Practice

45. $\triangle ABC$ is translated along the vector $\langle -2, 3\rangle$ and then reflected in the x-axis. What are the coordinates of A' after the transformation?

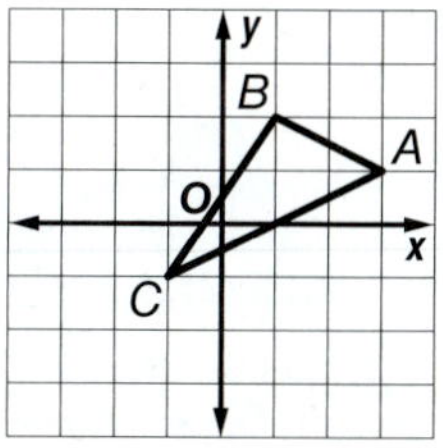

A $(1, -4)$

B $(1, 4)$

C $(-1, 4)$

D $(-1, -4)$

46. SHORT RESPONSE What are the coordinates of D'' if $\overline{CD}$ with vertices $C(2, 4)$ and $D(8, 7)$ is translated along $\langle -6, 2\rangle$ and then reflected over the y-axis?

47. ALGEBRA Write $\frac{18x^2 - 2}{3x^2 - 5x - 2}$ in simplest terms.

F $\frac{18}{3x + 1}$

G $\frac{2(3x + 1)}{x - 2}$

H $\frac{2(3x - 1)}{x - 2}$

J $2(3x - 1)$

48. SAT/ACT If $f(x) = x^3 - x^2 - x$, what is the value of $f(-3)$?

A -39

B -33

C -21

D -15

Spiral Review

Copy each polygon and point X. Then use a protractor and ruler to draw the specified rotation of each figure about point X. (Lesson 9-3)

49. $60°$

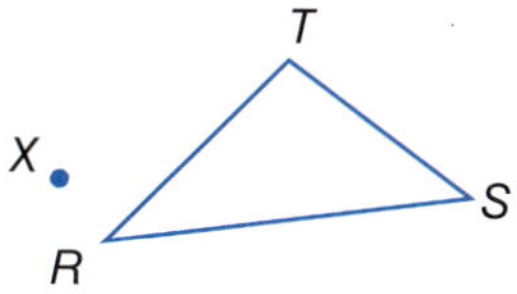

50. $120°$

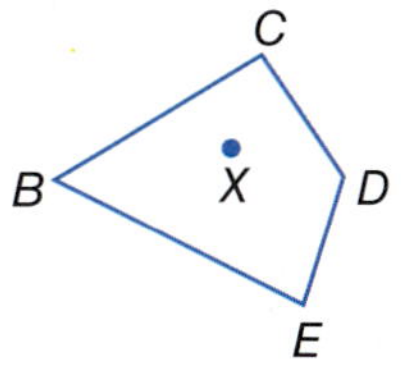

51. $180°$

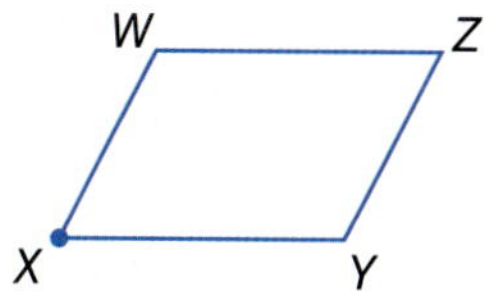

Graph each figure and its image along the given vector. (Lesson 9-2)

52. $\triangle FGH$ with vertices $F(1, -4)$, $G(3, -1)$, and $H(7, -1)$; $\langle 2, 6\rangle$

53. quadrilateral $ABCD$ with vertices $A(-2, 7)$, $B(-1, 4)$, $C(2, 3)$, and $D(2, 7)$; $\langle -3, -5\rangle$

54. AVIATION A jet is flying northwest, and its velocity is represented by $\langle -450, 450\rangle$ miles per hour. The wind is from the west, and its velocity is represented by $\langle 100, 0\rangle$ miles per hour. (Lesson 8-7)

a. Find the resultant vector for the jet in component form.

b. Find the magnitude of the resultant.

c. Find the direction of the resultant.

Skills Review

Each figure shows a preimage and its reflected image in some line. Copy each figure and draw the line of reflection. (Lesson 9-1)

55.

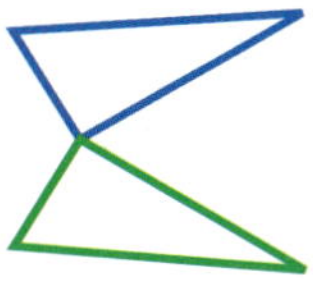

56.

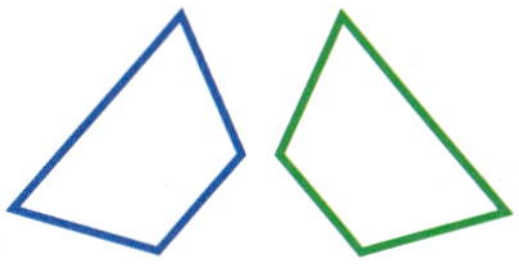

57.

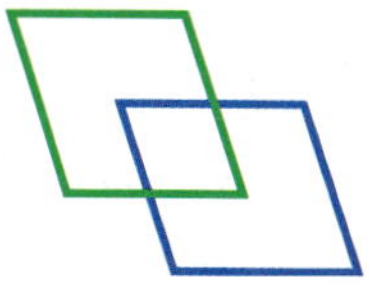

Geometry Lab
Tessellations

A **tessellation** is a pattern of one or more figures that covers a plane so that there are no overlapping or empty spaces. The sum of the angles around the vertex of a tessellation is 360°.

A **regular tessellation** is formed by only one type of regular polygon. A regular polygon will tessellate if it has an interior angle measure that is a factor of 360. A **semi-regular tessellation** is formed by two or more regular polygons.

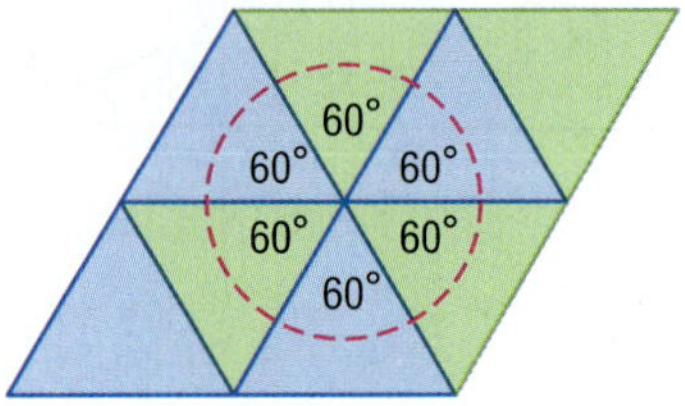

ACTIVITY 1 Regular Tessellation

Determine whether each regular polygon will tessellate in the plane. Explain.

a. hexagon

Let x represent the measure of an interior angle of a regular hexagon.

$x = \frac{180(n-2)}{n}$ **Interior Angle Formula**

$= \frac{180(6-2)}{6}$ **$n = 6$**

$= 120$ **Simplify.**

Since 120 is a factor of 360, a regular hexagon will tessellate in the plane.

b. decagon

Let x represent the measure of an interior angle of a regular decagon.

$x = \frac{180(n-2)}{n}$ **Interior Angle Formula**

$= \frac{180(10-2)}{10}$ **$n = 10$**

$= 144$ **Simplify.**

Since 144 is not a factor of 360, a regular decagon will not tessellate in the plane.

A tessellation is **uniform** if it contains the same arrangement of shapes and angles at each vertex.

Uniform

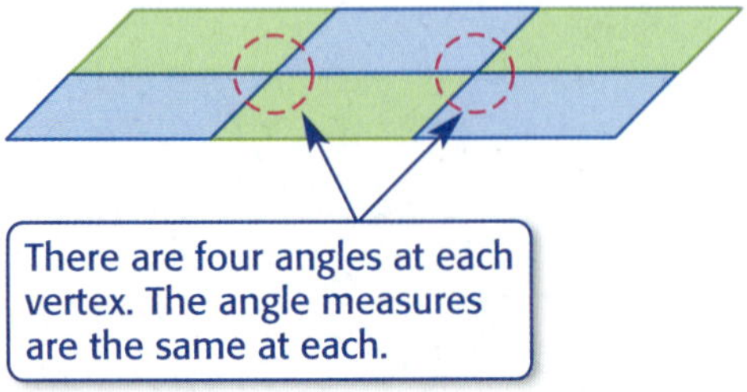

Not Uniform

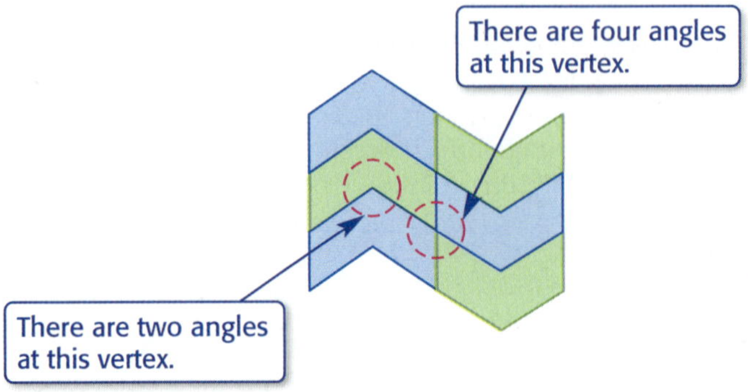

ACTIVITY 2 Classify Tessellations

Determine whether each pattern is a tessellation. If so, describe it as *regular*, *semi-regular*, or *neither* and *uniform* or *not uniform*.

a.

There is no unfilled space, and none of the figures overlap, so the pattern is a **tessellation**.

The tessellation consists of regular hexagons, squares and equilateral triangles, so it is **semi-regular**.

There are four angles around some of the vertices and five around others, so it is **not uniform**.

b.

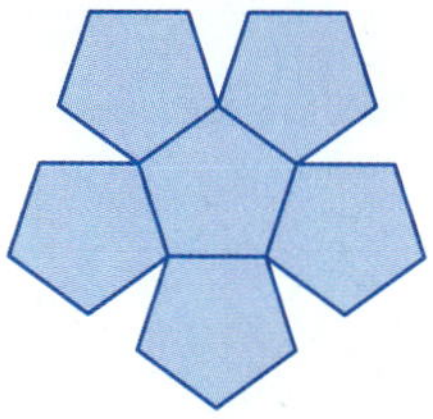

There is unfilled space, so the pattern is a **not a tessellation**.

c.

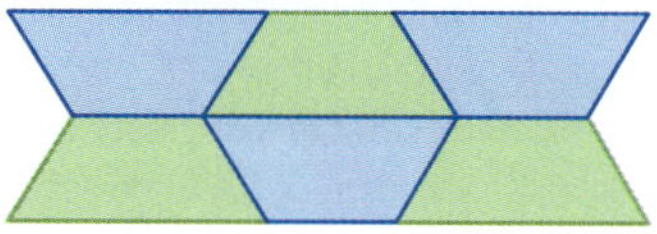

There is no unfilled space, and none of the figures overlap, so the pattern is a **tessellation**.

The tessellation consists of trapezoids, which are not regular polygons, so it is **neither** regular nor semi-regular.

There are four angles around each of the vertices and the angle measures are the same at each vertex, so it is **uniform**.

You can use the properties of tessellations to design and create tessellations.

ACTIVITY 3 Draw a Tessellation

Draw a triangle and use it to create a tessellation.

Step 1 Draw a triangle and find the midpoint of one side.

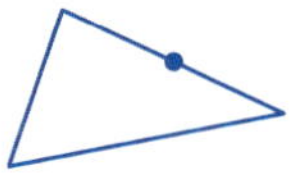

Step 2 Rotate the triangle 180° about the point.

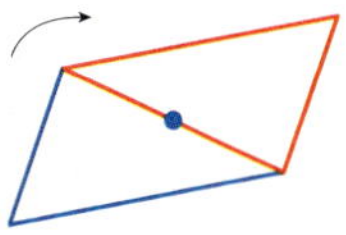

Step 3 Translate the pair of triangles to make a row.

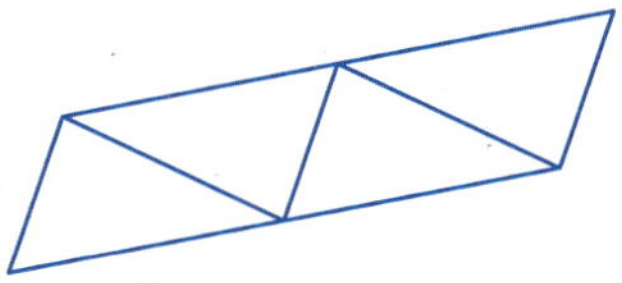

Step 4 Translate the row to make a tessellation.

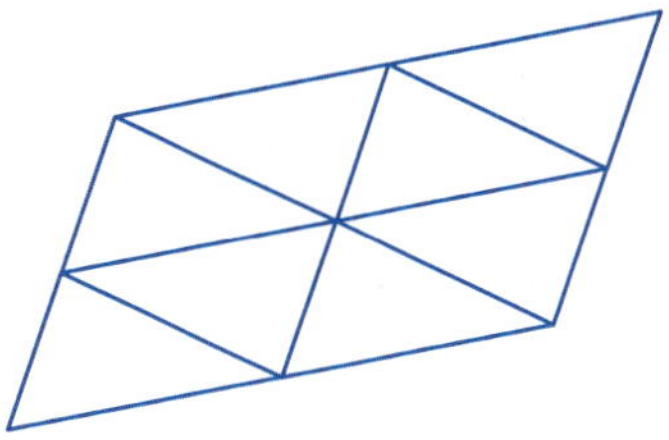

ACTIVITY 4 Tessellations using Technology

Use Geometer's Sketchpad to create a tessellation.

Step 1 Insert three points and construct a line through two of the points. Then construct the line parallel to the first line through the third point using the **Parallel Line** option from the **Construct** menu. Complete the parallelogram and label the points *A*, *B*, *C*, and *D*. Hide the lines.

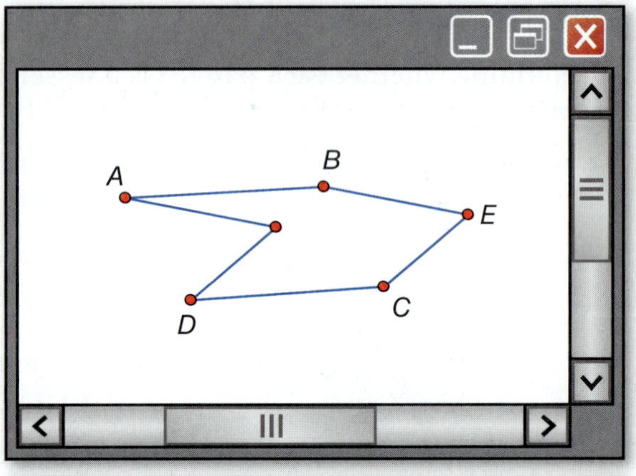

Step 2 Insert another point *E* on the exterior of the parallelogram. Draw the segments between *A* and *B*, *B* and *E*, *E* and *C*, and *C* and *D*.

Step 3 Highlight *B* and then *A*. From the **Transform** menu, choose **Mark Vector**. Select the $\overline{BE}$, $\overline{EC}$, and point *E*. From the **Transform** menu, choose **Translate**.

Step 4 Starting with *A*, select all of the vertices around the perimeter of the polygon. Choose **Hexagon Interior** from the **Construct** menu.

Step 5 Choose point *A* and then point *B* and mark the vector as you did in Step 3. Select the interior of the polygon and choose **Translate** from the **Transform** menu. Continue the tessellation by marking vectors and translating the polygon. You can choose **Color** from the **Display** menu to create a color pattern.

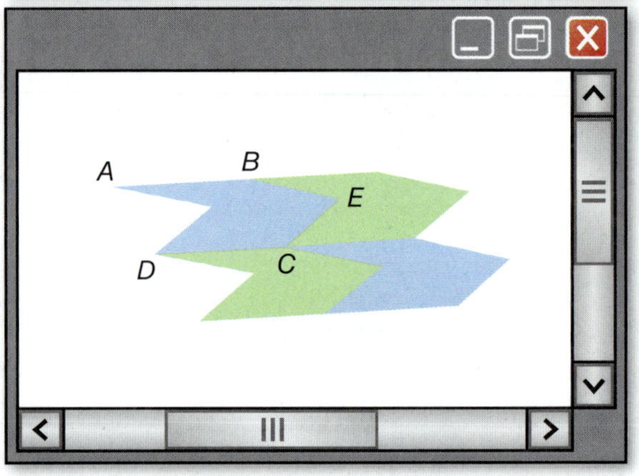

Exercises

Determine whether each regular polygon will tessellate in the plane. Write *yes* or *no*. Explain.

1. triangle

2. pentagon

3. 16-gon

Determine whether each pattern is a tessellation. Write *yes* or *no*. If so, describe it as *regular*, *semi-regular*, or *neither* and *uniform* or *not uniform*.

4.

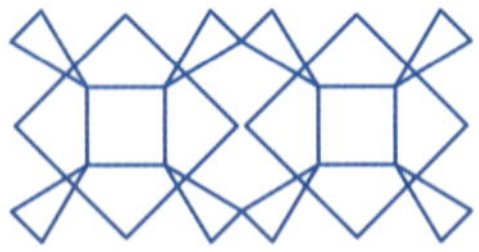

5.

6.

Draw a tessellation using the following shape(s).

7. octagon and square

8. hexagon and triangle

9. right triangle

10. trapezoid and a parallelogram

11. **WRITING IN MATH** There are only three possible regular tessellations. List the three polygons used to create these tessellations, and explain why they are the only ones possible.

12. **MAKE A CONJECTURE** Describe a figure that you think will tessellate in three-dimensional space. Explain your reasoning.

Symmetry

Then
You drew reflections and rotations of figures. (Lessons 9-1 and 9-3)

Now
- Identify line and rotational symmetries in two-dimensional figures.
- Identify line and rotational symmetries in three-dimensional figures.

New Vocabulary
symmetry
line symmetry
line of symmetry
rotational symmetry
center of symmetry
order of symmetry
magnitude of symmetry
plane symmetry
axis symmetry

Math Online
glencoe.com

Why?

In the animal kingdom, the symmetry of an animal's body is often an indication of the animal's complexity. Animals displaying line symmetry, such as insects, are usually more complex life forms than those displaying rotational symmetry, like a jellyfish.

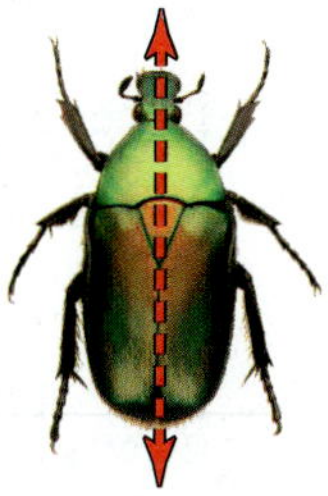

Symmetry in Two-Dimensional Figures A figure has **symmetry** if there exists a rigid motion—reflection, translation, rotation, or glide reflection—that maps the figure onto itself. One type of symmetry is line symmetry.

Key Concept — Line Symmetry

A figure in the plane has **line symmetry** (or *reflection symmetry*) if the figure can be mapped onto itself by a reflection in a line, called a **line of symmetry** (or *axis of symmetry*).

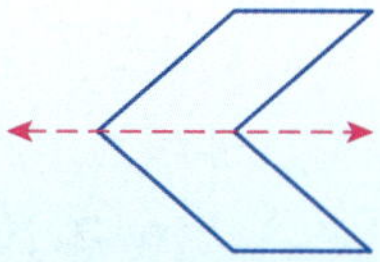

Real-World EXAMPLE 1 — Identify Line Symmetry

BEACHES **State whether the object appears to have line symmetry. Write *yes* or *no*. If so, copy the figure, draw all lines of symmetry, and state their number.**

a.

Yes; the crab has one line of symmetry.

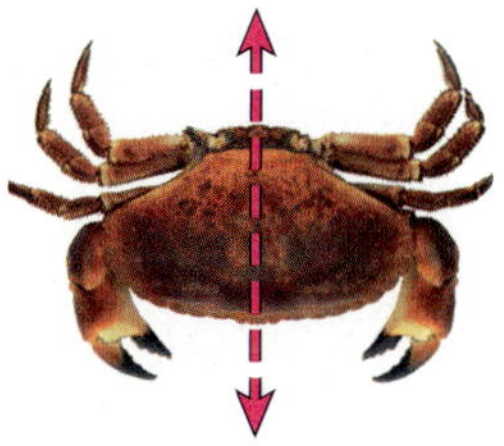

b.

Yes; the starfish has five lines of symmetry.

c.

No; there is no line in which the oyster shell can be reflected so that it maps onto itself.

Guided Practice

State whether the figure has line symmetry. Write *yes* or *no*. If so, copy the figure, draw all lines of symmetry, and state their number.

1A.

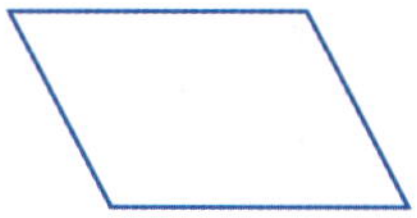

1B.

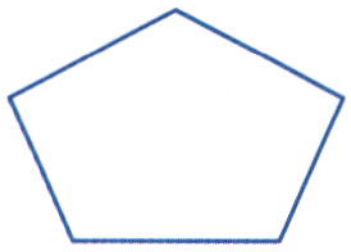

1C.

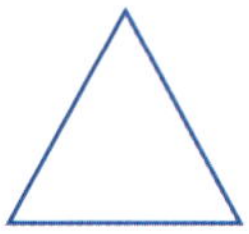

Personal Tutor glencoe.com

Another type of symmetry is rotational symmetry.

Key Concept — Rotational Symmetry

A figure in the plane has **rotational symmetry** (or *radial symmetry*) if the figure can be mapped onto itself by a rotation between 0° and 360° about the center of the figure, called the **center of symmetry** (or *point of symmetry*).

Examples The figure below has rotational symmetry because a rotation of 90°, 180°, or 270° maps the figure onto itself.

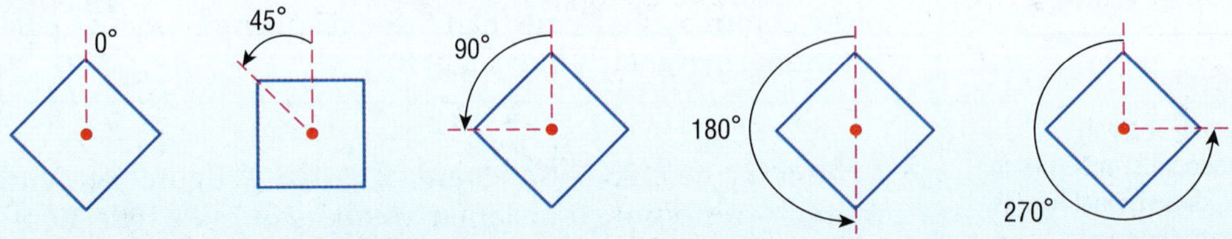

The number of times a figure maps onto itself as it rotates from 0° to 360° is called the **order of symmetry**. The **magnitude of symmetry** (or angle of rotation) is the smallest angle through which a figure can be rotated so that it maps onto itself. The order and magnitude of a rotation are related by the following equation.

$$\text{magnitude} = 360° \div \text{order}$$

The figure above has rotational symmetry of order 4 and magnitude 90°.

EXAMPLE 2 Identify Rotational Symmetry

State whether the figure has rotational symmetry. Write *yes* or *no*. If so, copy the figure, locate the center of symmetry, and state the order and magnitude of symmetry.

a.

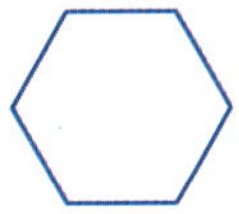

Yes; the regular hexagon has order 6 rotational symmetry and magnitude 360° ÷ 6 or 60°. The center is the intersection of the diagonals.

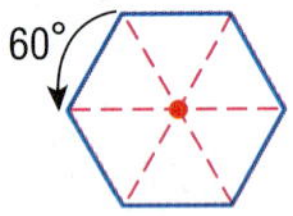

b.

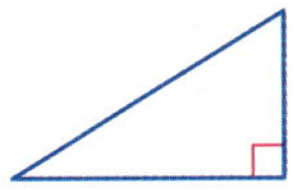

No; no rotation between 0° and 360° maps the right triangle onto itself.

c.

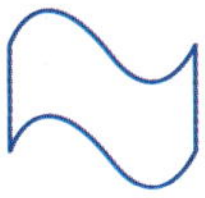

Yes; the figure has order 2 rotational symmetry and magnitude 360° ÷ 2 or 180°. The center is the intersection of the diagonals.

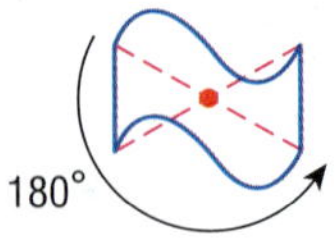

StudyTip

Point Symmetry A figure has *point symmetry* if the figure can be mapped onto itself by a rotation of 180°. A playing card exhibits point symmetry. It looks the same right-side up as upside down.

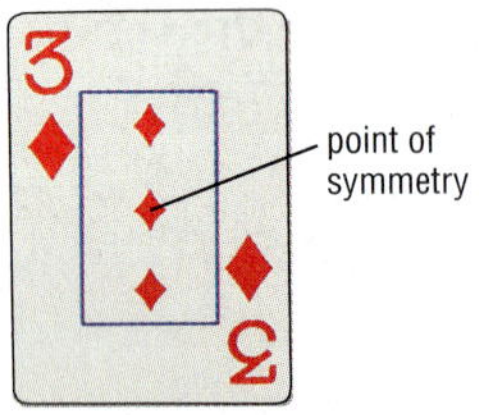

Guided Practice

FLOWERS **State whether the flower appears to have rotational symmetry. Write *yes* or *no*. If so, copy the flower, locate the center of symmetry, and state the order and magnitude of symmetry.**

2A.

2B.

2C.

Personal Tutor glencoe.com

Symmetry in Three-Dimensional Figures Three-dimensional figures can also have symmetry.

Key Concept Three-Dimensional Symmetries

Plane Symmetry

A three-dimensional figure has **plane symmetry** if the figure can be mapped onto itself by a reflection in a plane.

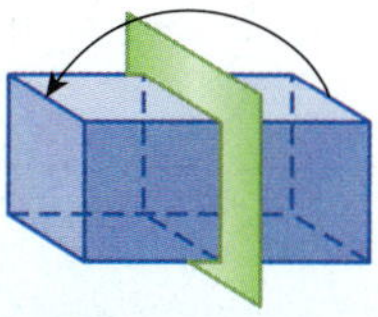

Axis Symmetry

A three-dimensional figure has **axis symmetry** if the figure can be mapped onto itself by a rotation between 0° and 360° in a line.

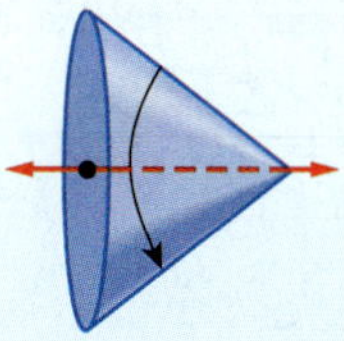

Review Vocabulary

prism a polyhedron with two parallel congruent bases connected by parallelogram faces (Lesson 1-7)

EXAMPLE 3 Three-Dimensional Symmetry

State whether the figure has *plane* symmetry, *axis* symmetry, *both*, or *neither*.

a. L-shaped prism

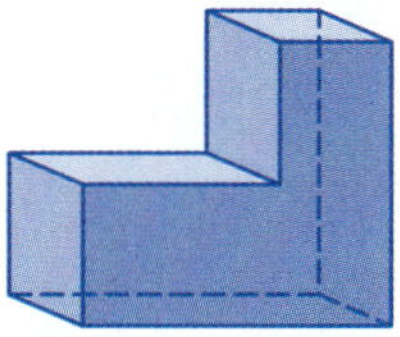

plane symmetry

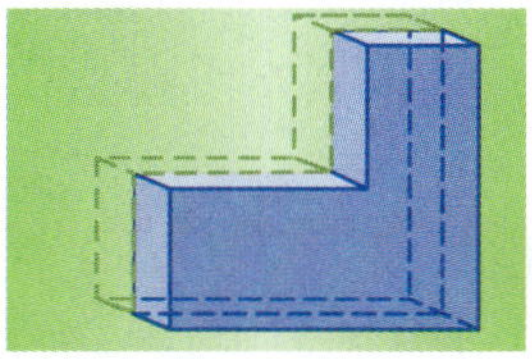

b. regular pentagonal prism

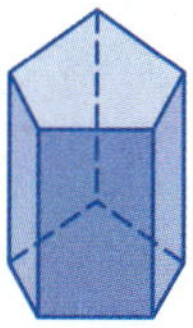

both plane symmetry and axis symmetry

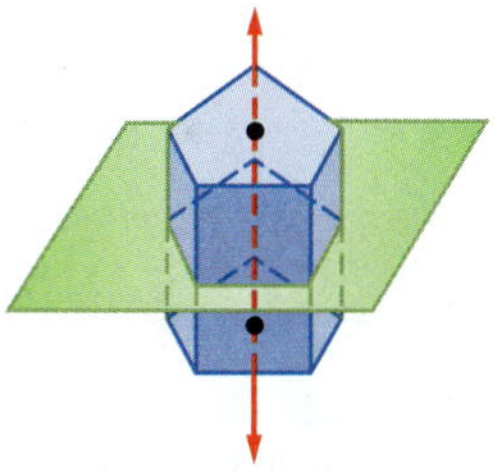

Guided Practice

SPORTS **State whether each piece of sports equipment appears to have *plane* symmetry, *axis* symmetry, *both*, or *neither* (ignoring the equipment's stitching or markings).**

3A.

3B.

3C.

3D.

Real-World Link

Aerodynamically designed to spin after it is thrown, a football's shape is that of a prolate spheroid. This means that its axis of symmetry is longer than its other axes.

Source: *Complete Idiot's Guide to Football*

Personal Tutor glencoe.com

Check Your Understanding

Example 1
p. 653

State whether the figure appears to have line symmetry. Write *yes* or *no*. If so, copy the figure, draw all lines of symmetry, and state their number.

1.
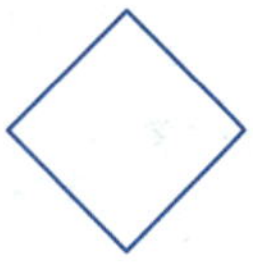

2.
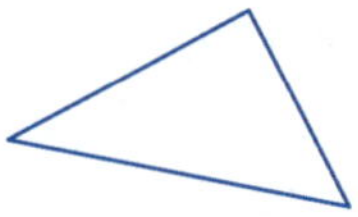

3.
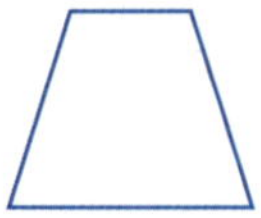

Example 2
p. 654

State whether the figure has rotational symmetry. Write *yes* or *no*. If so, copy the figure, locate the center of symmetry, and state the order and magnitude of symmetry.

4.
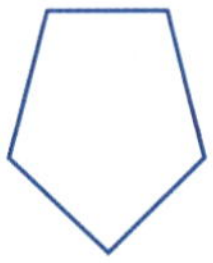

5.

6.

Examples 1 and 2
pp. 653–654

7. **U.S. CAPITOL** Completed in 1863, the dome is one of the most recent additions to the United States Capitol. It is supported by 36 iron ribs and has 108 windows, divided equally among three levels.

a. Excluding the spire of the dome, how many horizontal and vertical lines of symmetry does the dome appear to have?

b. Does the dome have rotational symmetry? If so, state the order and magnitude of symmetry.

Example 3
p. 655

8. State whether the figure has *plane* symmetry, *axis* symmetry, *both*, or *neither*.

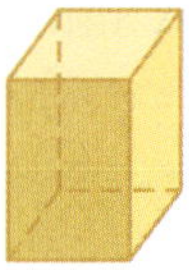

Practice and Problem Solving

= **Step-by-Step Solutions** begin on page R20.
Extra Practice begins on page 969.

Example 1
p. 653

State whether the figure appears to have line symmetry. Write *yes* or *no*. If so, copy the figure, draw all lines of symmetry, and state their number.

9.
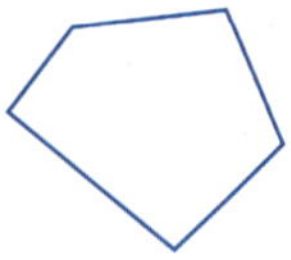

10.
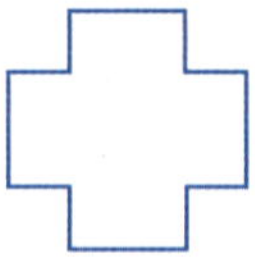

11.
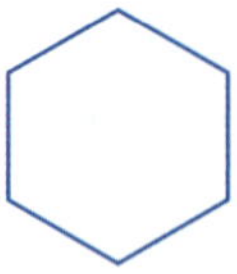

12.

13.
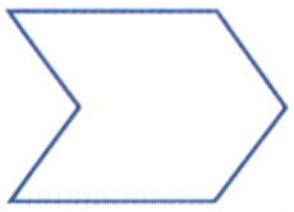

14.
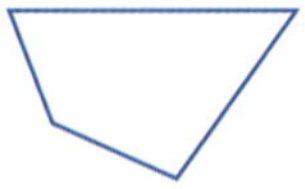

FLAGS **State whether each flag design appears to have line symmetry. Write *yes* or *no*. If so, copy the flag, draw all lines of symmetry, and state their number.**

15.

16.

17.

Example 2
p. 654

State whether the figure has rotational symmetry. Write *yes* or *no*. If so, copy the figure, locate the center of symmetry, and state the order and magnitude of symmetry.

18. 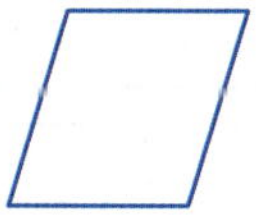

19. 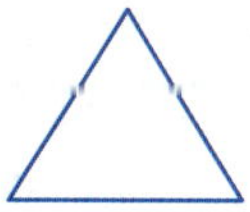

20. 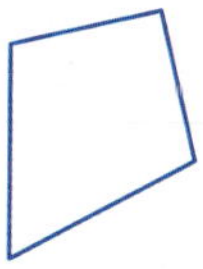

21. 

22. 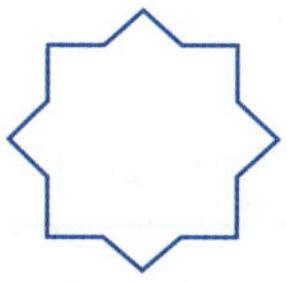

23. 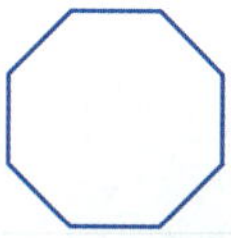

Example 3
p. 655

WHEELS State whether each wheel cover appears to have rotational symmetry. Write *yes* or *no*. If so, state the order and magnitude of symmetry.

24. 

25. 

26. 

Example 3
p. 655

State whether the figure has *plane* symmetry, *axis* symmetry, *both*, or *neither*.

27. 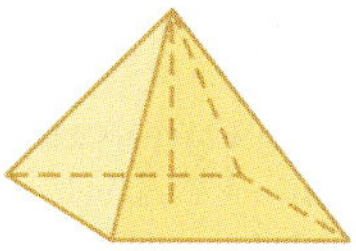

28. 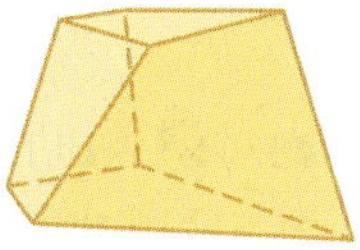

29. 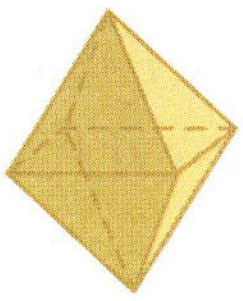

30. 

CONTAINERS Determine the number of horizontal and vertical planes of symmetry for each container shown below.

31. 

32. 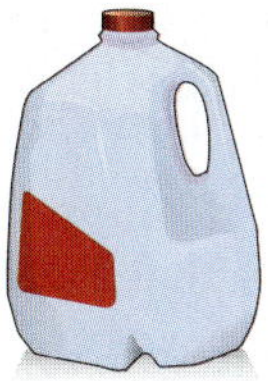

33. 

34. **PHOTOGRAPHY** Symmetry is an important component of photography. Photographers often use reflection in water to create symmetry in photos. The photo at the right is a long exposure shot of the Eiffel tower reflected in a pool.

a. Describe the two-dimensional symmetry created by the photo.

b. Is three-dimensional symmetry applicable? Explain your reasoning.

COORDINATE GEOMETRY Determine whether the figure with the given vertices has *line* symmetry and/or *rotational* symmetry.

35. $A(-4, 0), B(0, 4), C(4, 0), D(0, -4)$

36. $R(-3, 3), S(-3, -3), T(3, 3)$

37. $F(0, -4), G(-3, -2), H(-3, 2), J(0, 4), K(3, 2), L(3, -2)$

38. $W(-2, 3), X(-3, -3), Y(3, -3), Z(2, 3)$

ALGEBRA Graph the function and determine whether the graph has *line* and/or *rotational* symmetry. If so, state the order and magnitude of symmetry, and write the equations of any lines of symmetry.

39. $y = x$

40. $y = x^2 + 1$

41. $y = -x^3$

CRYSTALLOGRAPHY Determine whether the crystals below have *plane* symmetry and/or *axis* symmetry. If so, state the magnitude of symmetry.

42. cubic

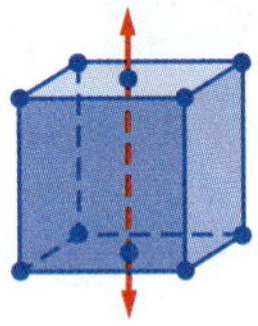

43. rhombohedral

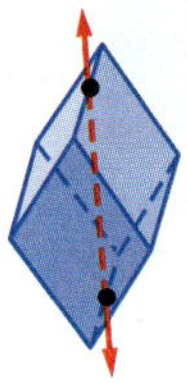

44. orthorhombic

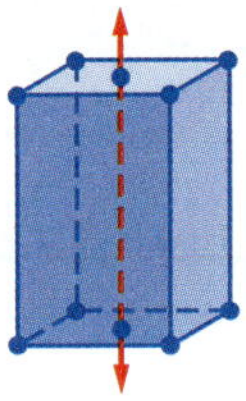

Real-World Link

The physical properties of a solid depend upon the arrangement of crystals. Diamonds, for example, have a cubic structure with very strong bonds that are tough to break. This is why diamonds are such hard materials.

Source: Virginia Tech Crystallography Laboratory

45. **MULTIPLE REPRESENTATIONS** In this problem, you will investigate rotational symmetry in regular polygons.
   a. **GEOMETRIC** Draw an equilateral triangle and determine its order of symmetry.
   b. **GEOMETRIC** Repeat the process in part **a** for a square, a regular pentagon, and a regular hexagon.
   c. **TABULAR** Tabulate the order of symmetry for each polygon.
   d. **VERBAL** Make a conjecture about the values of the order of symmetry for a regular polygon.

H.O.T. Problems

Use **H**igher-**O**rder **T**hinking Skills

46. **ERROR ANALYSIS** Jaime says that Figure A has only line symmetry, and Jewel says that Figure A has only rotational symmetry. Is either of them correct? Explain your reasoning.

Figure A

47. **CHALLENGE** A quadrilateral in the coordinate plane has exactly two lines of symmetry, $y = x - 1$ and $y = -x + 2$. Find a set of possible vertices for the figure. Graph the figure and the lines of symmetry.

48. **REASONING** A regular polyhedron has axis symmetry of order 3, but does not have plane symmetry. What is the figure? Explain.

49. **OPEN ENDED** Draw a figure that has line symmetry but not rotational symmetry. Explain.

50. **WRITING IN MATH** Compare and contrast line symmetry and rotational symmetry.

Standardized Test Practice

51. How many lines of symmetry can be drawn on the picture of the Canadian flag below?

A 0
B 1
C 2
D 4

52. GRIDDED RESPONSE What is the order of symmetry for the figure below?

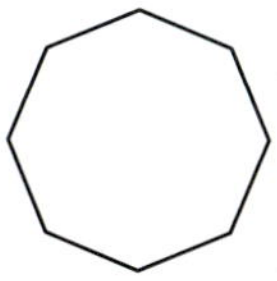

53. ALGEBRA A computer company ships computers in wooden crates that each weigh 45 pounds when empty. If each computer weighs no more than 13 pounds, which inequality *best* describes the total weight in pounds w of a crate of computers that contains c computers?

F $c \leq 13 + 45w$
G $c \geq 13 + 45w$
H $w \leq 13c + 45$
J $w \geq 13c + 45$

54. SAT/ACT What is the slope of the line determined by the linear equation $5x - 2y = 10$?

A -5
B $-\frac{5}{2}$
C $-\frac{2}{5}$
D $\frac{5}{2}$

Spiral Review

Triangle JKL has vertices $J(1, 5)$, $K(3, 1)$, and $L(5, 7)$. Graph $\triangle JKL$ and its image after the indicated transformation. (Lesson 9-4)

55. Translation: along $\langle -7, -1 \rangle$
Reflection: in x-axis

56. Translation: along $\langle 1, 2 \rangle$
Reflection: in y-axis

57. Quadrilateral $QRST$ is shown at the right. What is the image of point R after a rotation 180° counterclockwise about the origin? (Lesson 9-3)

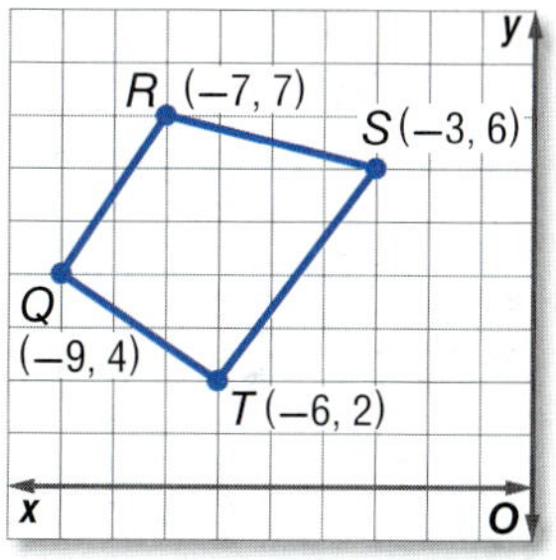

58. AMUSEMENT PARKS From the top of a roller coaster, 60 yards above the ground, a rider looks down and sees the merry-go-round and the Ferris wheel. If the angles of depression are 11° and 8° respectively, how far apart are the merry-go-round and the Ferris wheel? (Lesson 8-5)

Skills Review

Determine whether the dilation from Figure A to Figure B is an *enlargement* or a *reduction*. Then find the scale factor of the dilation. (Lesson 7-6)

59.

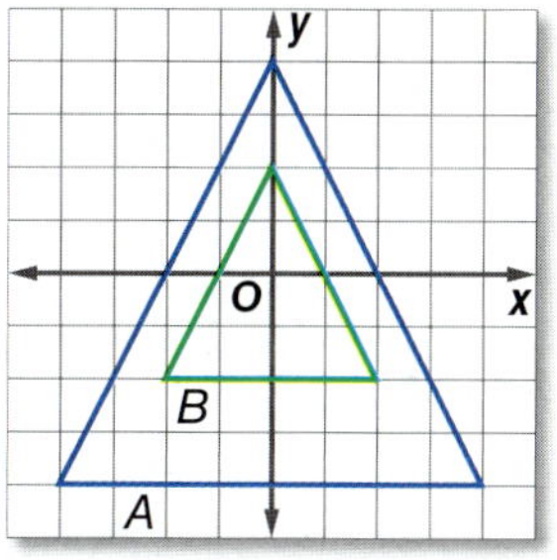

60.

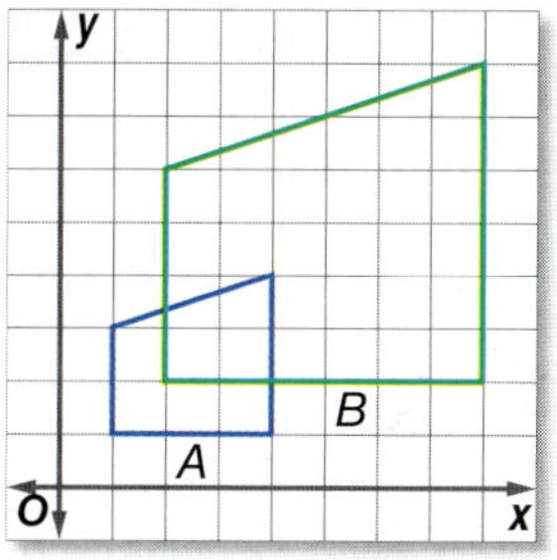

61.

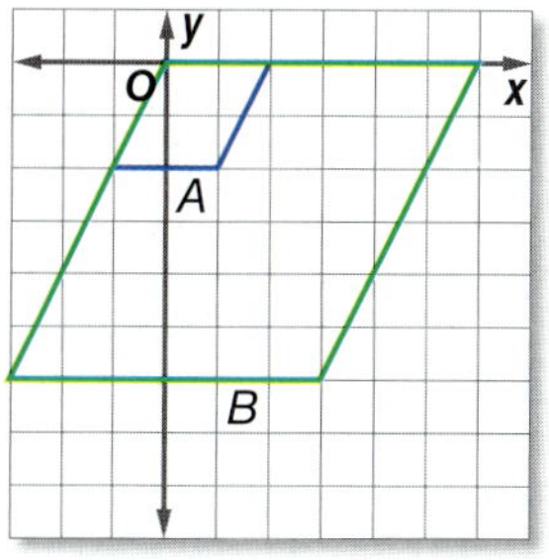

9-6 Dilations

Then

You identified dilations and verified them as similarity transformations. (Lesson 7-6)

Now

- Draw dilations.
- Draw dilations in the coordinate plane.

Math Online
glencoe.com

Why?

While many use digital cameras, some photographers still prefer traditional cameras and film to produce negatives. From these negatives, photographers can create scaled reproductions of an image.

Draw Dilations Recall from Lesson 7-6 that a dilation or *scaling* is a similarity transformation that enlarges or reduces a figure proportionally. Dilations are performed with respect to a *center* point and a *scale* factor.

Key Concept — Dilation

A dilation with center C and positive scale factor k, $k \neq 1$, maps a point P in a figure to its image such that

- if point P and C coincide, then the image and preimage are the same point, or
- if point P is not the center of dilation, then P' lies on $\overrightarrow{CP}$ and $CP' = k(CP)$.

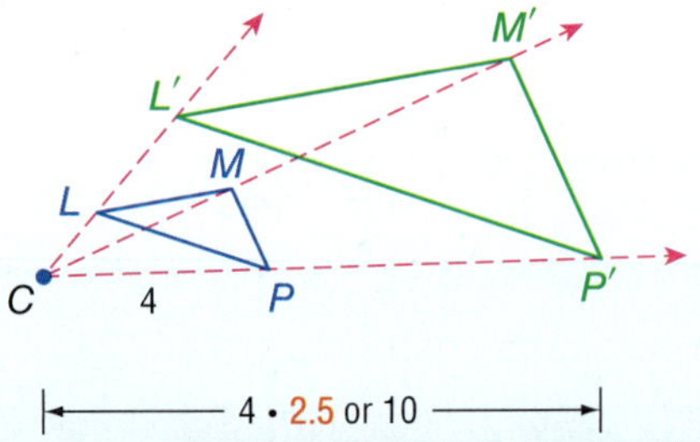

$\triangle L'M'P$ is the image of $\triangle LMP$ under a dilation with center C and scale factor of **2.5**.

EXAMPLE 1 Draw a Dilation

Copy $\triangle ABC$ and point D. Then use a ruler to draw the image of $\triangle ABC$ under a dilation with center D and scale factor $\frac{1}{2}$.

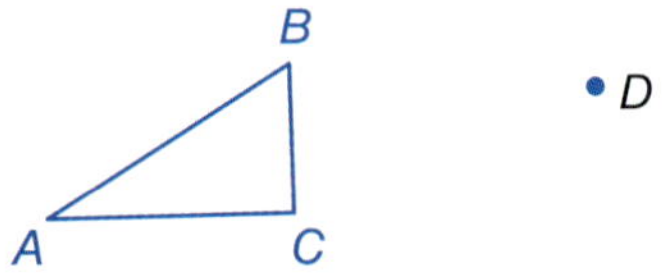

Step 1 Draw rays from D though each vertex.

Step 2 Locate A' on $\overrightarrow{DA}$ such that $DA' = \frac{1}{2}DA$.

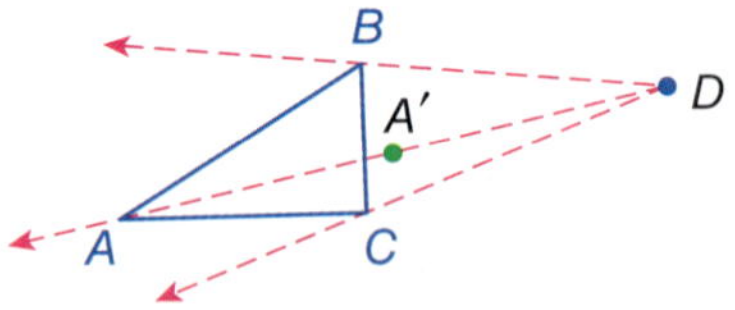

Step 3 Locate B' on $\overrightarrow{DB}$ and C' on $\overrightarrow{DC}$ in the same way. Then draw $\triangle A'B'C'$.

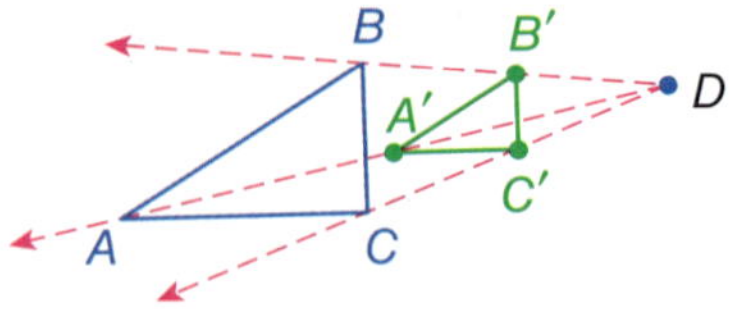

Guided Practice

Copy the figure and point J. Then use a ruler to draw the image of the figure under a dilation with center J and the scale factor k indicated.

1A. $k = \frac{3}{2}$

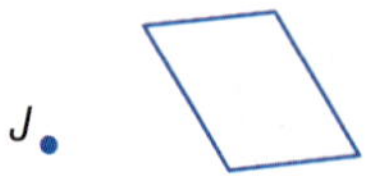

1B. $k = 0.75$

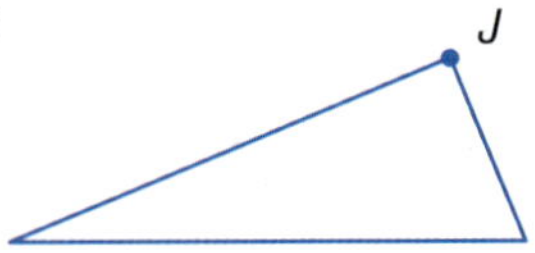

Personal Tutor glencoe.com

In Lesson 7-6, you also learned that if $k > 1$, then the dilation is an *enlargement*. If $0 < k < 1$, then the dilation is a *reduction*. Since $\frac{1}{2}$ is between 0 and 1, the dilation in Example 1 is a reduction.

A dilation with a scale factor of 1 is called an *isometry dilation*. It produces an image that coincides with the preimage. The two figures are congruent.

Real-World EXAMPLE 2 Find the Scale Factor of a Dilation

PHOTOGRAPHY **To create different-sized prints, you can adjust the distance between a film negative and the enlarged print by using a photographic enlarger. Suppose the distance between the light source C and the negative is 45 millimeters (CP). To what distance PP' should you adjust the enlarger to create a 22.75-centimeter wide print ($X'Y'$) from a 35-millimeter wide negative (XY)?**

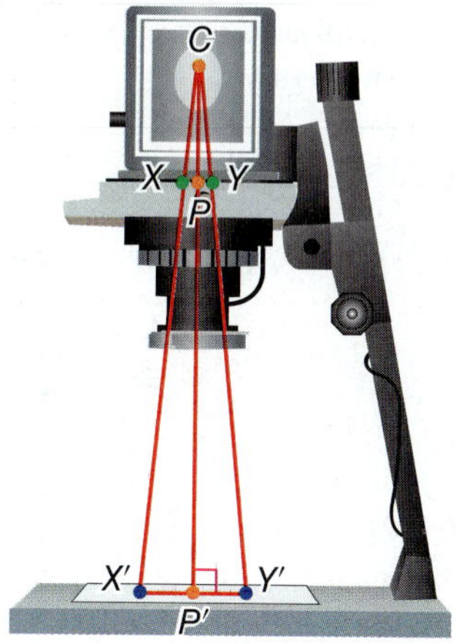

Understand This problem involves a dilation. The center of dilation is C, $XY = 35$ mm, $X'Y' = 22.75$ cm or 227.5 mm, and $CP = 45$ mm. You are asked to find PP'.

Plan Find the scale factor of the dilation from the preimage XY to the image $X'Y'$. Use the scale factor to find CP' and then use CP and CP' to find PP'.

Solve The scale factor k of the enlargement is the ratio of a length on the image to a corresponding length on the preimage.

$k = \frac{\text{image length}}{\text{preimage length}}$	**Scale factor of image**
$= \frac{X'Y'}{XY}$	**image = $X'Y'$, preimage = XY**
$= \frac{227.5}{35}$ or 6.5	**Divide.**

Use this scale factor of 6.5 to find CP'.

$CP' = k(CP)$	**Definition of dilation**
$= 6.5(45)$	**$k = 6.5$ and $CP = 45$**
$= 292.5$	**Multiply.**

Use CP' and CP to find PP'.

$CP + PP' = CP'$	**Segment Addition**
$45 + PP' = 292.5$	**$CP = 45$ and $CP' = 292.5$**
$PP' = 247.5$	**Subtract 45 from each side.**

So the enlarger should be adjusted so that the distance from the negative to the enlarged print (PP') is 247.5 millimeters or 24.75 centimeters.

Check Since the dilation is an enlargement, the scale factor should be greater than 1. Since $6.5 > 1$, the scale factor found is reasonable. ✓

Problem-Solving Tip

Use Estimation To prevent careless errors in your calculations, estimate the answer to a problem before solving. In Example 2, you can estimate the scale factor of the dilation to be about $\frac{240}{40}$ or 6. Then CP' would be about $6 \cdot 50$ or 300 and PP' about $300 - 50$ or 250 millimeters, which is 25 centimeters. A measure of 24.7 centimeters is close to this estimate, so the answer is reasonable.

Guided Practice

2. Determine whether the dilation from Figure Q to Q' is an *enlargement* or a *reduction*. Then find the scale factor of the dilation and x.

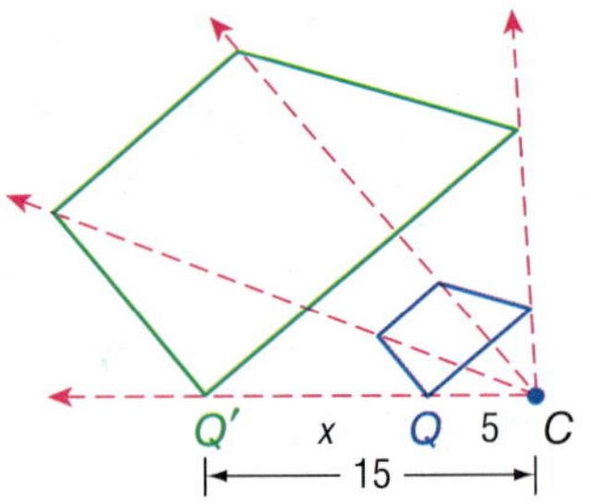

Personal Tutor glencoe.com

StudyTip

Negative Scale Factors Dilations can also have negative scale factors. You will investigate this type of dilation in Exercise 36.

Dilations in the Coordinate Plane You can use the following rules to find the image of a figure after a dilation centered at the origin.

Key Concept: Dilations in the Coordinate Plane

Words To find the coordinates of an image after a dilation centered at the origin, multiply the x- and y-coordinates of each point on the preimage by the scale factor of the dilation, k.

Symbols $(x, y) \rightarrow (kx, ky)$

Example

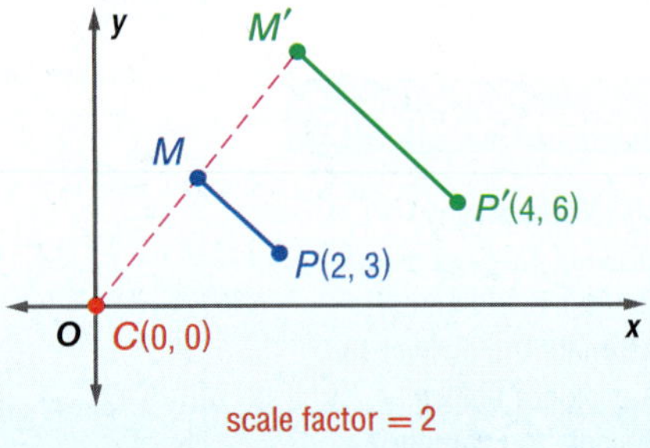

EXAMPLE 3 Dilations in the Coordinate Plane

Quadrilateral $JKLM$ has vertices $J(-2, 4)$, $K(-2, -2)$, $L(-4, -2)$, and $M(-4, 2)$. Graph the image of $JKLM$ after a dilation centered at the origin with a scale factor of 2.5.

Multiply the x- and y-coordinates of each vertex by the scale factor, 2.5.

$(x, y) \rightarrow (2.5x, 2.5y)$

$J(-2, 4) \rightarrow J'(-5, 10)$

$K(-2, -2) \rightarrow K'(-5, -5)$

$L(-4, -2) \rightarrow L'(-10, -5)$

$M(-4, 2) \rightarrow M'(-10, 5)$

Graph $JKLM$ and its image $J'K'L'M'$.

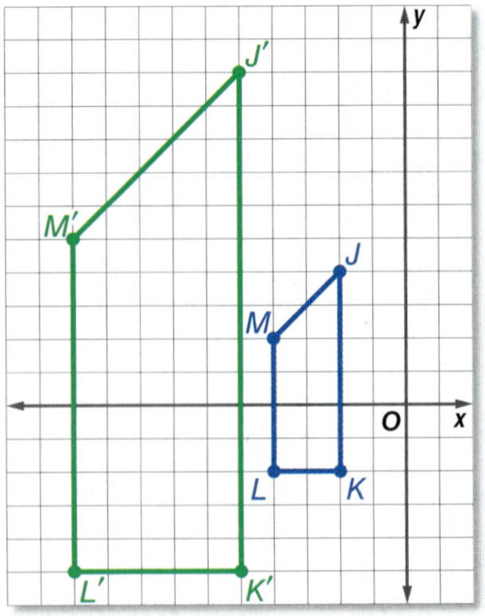

Guided Practice

Find the image of each polygon with the given vertices after a dilation centered at the origin with the given scale factor.

3A. $Q(0, 6)$, $R(-6, -3)$, $S(6, -3)$; $k = \frac{1}{3}$

3B. $A(2, 1)$, $B(0, 3)$, $C(-1, 2)$, $D(0, 1)$; $k = 2$

Personal Tutor glencoe.com

Check Your Understanding

Example 1 p. 660

Copy the figure and point M. Then use a ruler to draw the image of the figure under a dilation with center M and the scale factor k indicated.

1. $k = \frac{1}{4}$

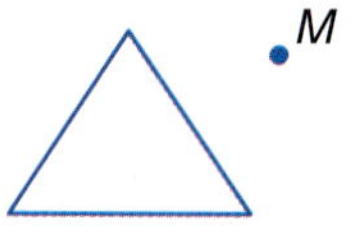

2. $k = 2$

Example 2 p. 661

3. Determine whether the dilation from Figure B to B' is an *enlargement* or a *reduction*. Then find the scale factor of the dilation and x.

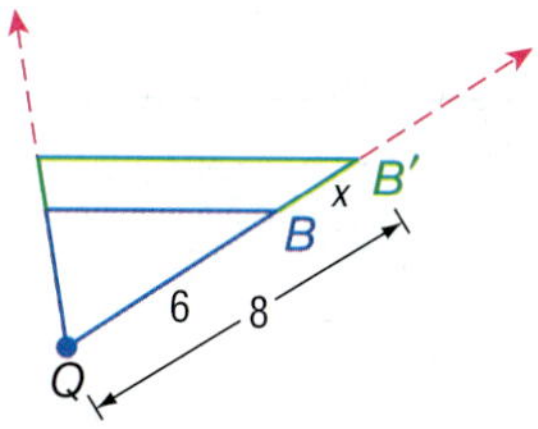

4. **BIOLOGY** Under a microscope, a single-celled organism 200 microns in length appears to be 50 millimeters long. If 1 millimeter = 1000 microns, what magnification setting (scale factor) was used? Explain your reasoning.

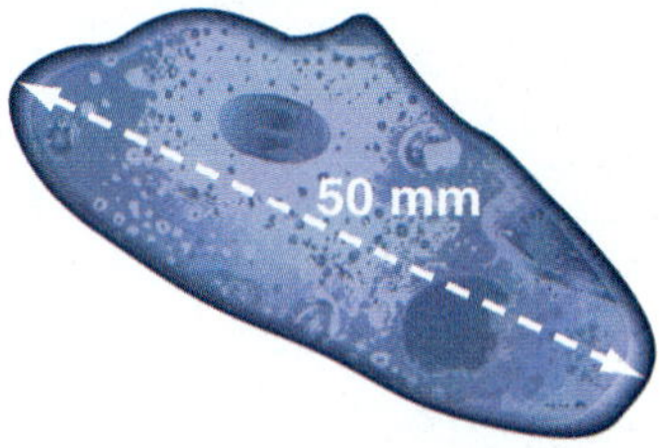

Example 3 p. 662

Graph the image of each polygon with the given vertices after a dilation centered at the origin with the given scale factor.

5. $W(0, 0), X(6, 6), Y(6, 0); k = 1.5$

6. $Q(-4, 4), R(-4, -4), S(4, -4), T(4, 4); k = \frac{1}{2}$

7. $A(-1, 4), B(2, 4), C(3, 2), D(-2, 2); k = 2$

8. $J(-2, 0), K(2, 4), L(8, 0), M(2, -4); k = \frac{3}{4}$

Practice and Problem Solving

= **Step-by-Step Solutions** begin on page R20.
Extra Practice begins on page 969.

Example 1 p. 660

Copy the figure and point S. Then use a ruler to draw the image of the figure under a dilation with center S and the scale factor k indicated.

9. $k = \frac{5}{2}$

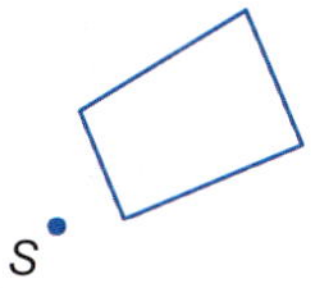

10. $k = 3$

11. $k = 0.8$

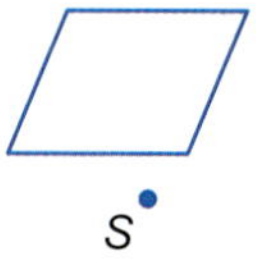

12. $k = \frac{1}{3}$

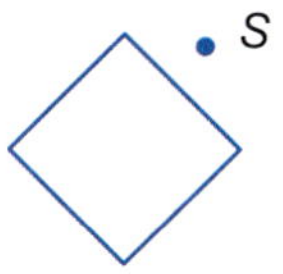

13. $k = 2.25$

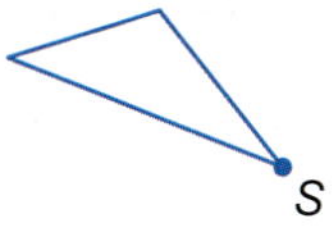

14. $k = \frac{7}{4}$

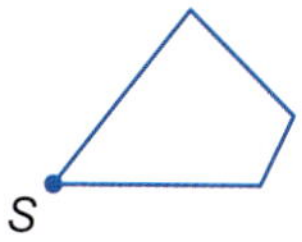

Example 2
p. 661

Determine whether the dilation from figure W to W' is an *enlargement* or a *reduction*. Then find the scale factor of the dilation and x.

15.

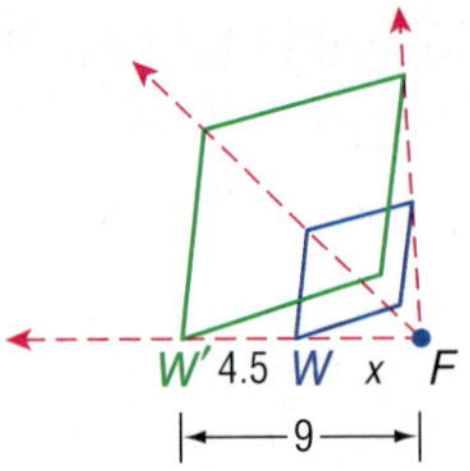

16.

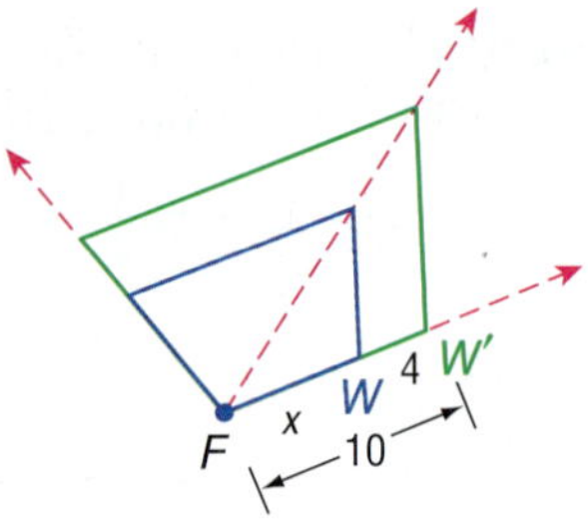

17.

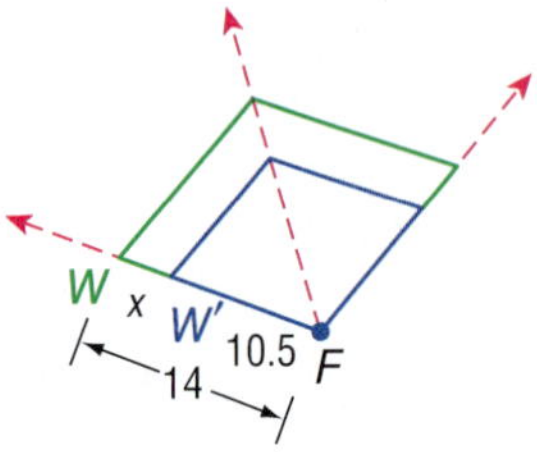

18.

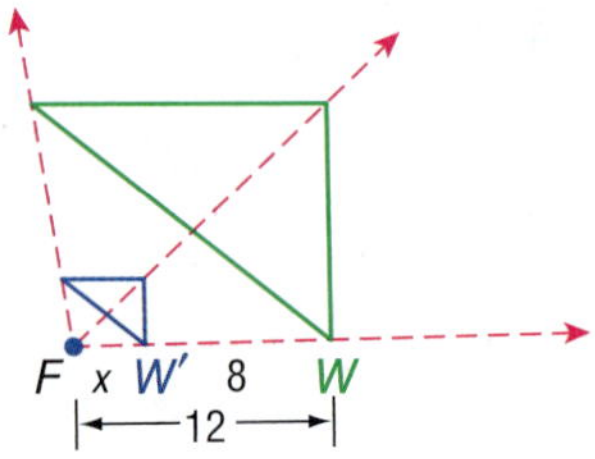

INSECTS When viewed under a microscope, each insect has the measurement given on the picture. Given the actual measure of each insect, what magnification was used? Explain your reasoning.

19.

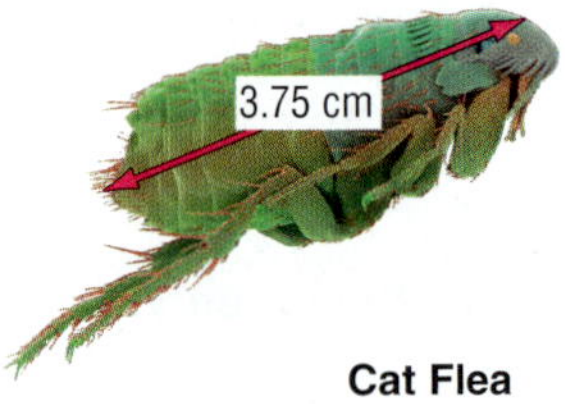

Cat Flea
Actual Length: 2.5 mm

20.

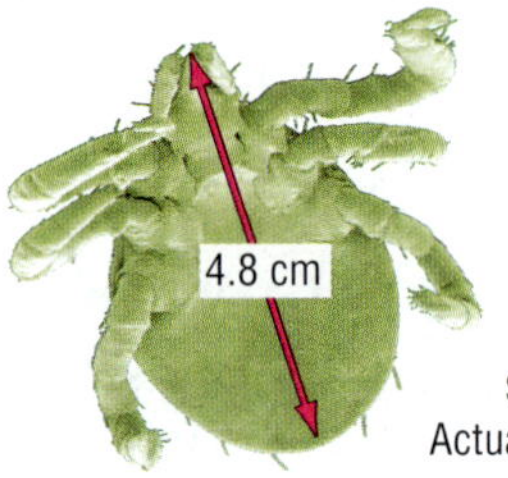

Spider Mite
Actual Length: 0.5 mm

Example 3
p. 662

Find the image of each polygon with the given vertices after a dilation centered at the origin with the given scale factor.

21. $J(-8, 0), K(-4, 4), L(-2, 0); k = 0.5$

22. $S(0, 0), T(-4, 0), V(-8, -8); k = 1.25$

23. $A(9, 9), B(3, 3), C(6, 0); k = \frac{1}{3}$

24. $D(4, 4), F(0, 0), G(8, 0); k = 0.75$

25. $M(-2, 0), P(0, 2), Q(2, 0), R(0, -2); k = 2.5$

26. $W(2, 2), X(2, 0), Y(0, 1), Z(1, 2); k = 3$

27. COORDINATE GEOMETRY Refer to the graph of $FGHJ$.

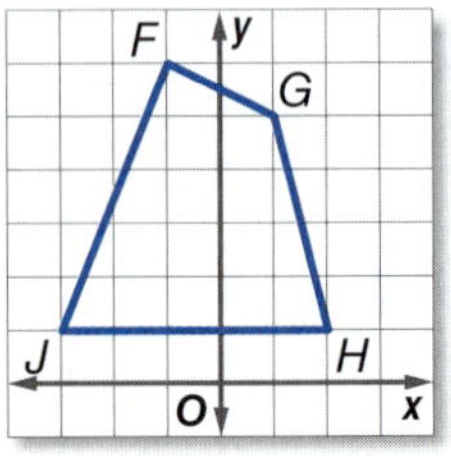

a. Dilate $FGHJ$ by a scale factor of $\frac{1}{2}$ centered at the origin, and then reflect the dilated image in the y-axis.

b. Complete the composition of transformations in part **a** in reverse order.

c. Does the order of the transformations affect the final image?

d. Will the order of a composition of a dilation and a reflection *always*, *sometimes*, or *never* affect the dilated image? Explain your reasoning.

Chuck Close, *Self Portrait*, 1997. Oil on canvas. The Museum of Modern Art, New York.

Real-World Link

Nearly all of Chuck Close's works use a grid as an underlying basis for the representation of an image.

Source: *Chuck Close: Process and Collaboration*

28. PHOTOGRAPHY AND ART To make a grid drawing in the style of Chuck Close, students overlay a $\frac{1}{4}$-inch grid on a 5-inch by 7-inch high contrast photo, overlay a $\frac{1}{2}$-inch grid on a 10-inch by 14-inch piece of drawing paper, and then sketch the image in each square of the photo to the corresponding square on the drawing paper.

a. What is the scale factor of the dilation?

b. To create an image that is 10 times as large as the original, what size grids are needed?

c. What would be the area of a grid drawing of a 5-inch by 7-inch photo that used 2-inch grids?

29. MEASUREMENT Determine whether the image shown is a dilation of *ABCD*. Explain your reasoning.

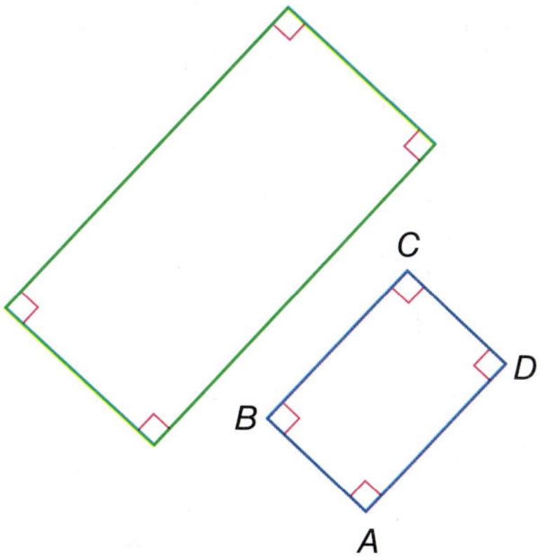

30. COORDINATE GEOMETRY *WXYZ* has vertices $W(6, 2)$, $X(3, 7)$, $Y(-1, 4)$, and $Z(4, -2)$.

a. Graph *WXYZ* and find the perimeter of the figure. Round to the nearest tenth.

b. Graph the image of *WXYZ* after a dilation of $\frac{1}{2}$ centered at the origin.

c. Find the perimeter of the dilated image. Round to the nearest tenth. How is the perimeter of the dilated image related to the perimeter of *WXYZ*?

31. CHANGING DIMENSIONS A three-dimensional figure can also undergo a dilation. Consider the rectangular prism shown.

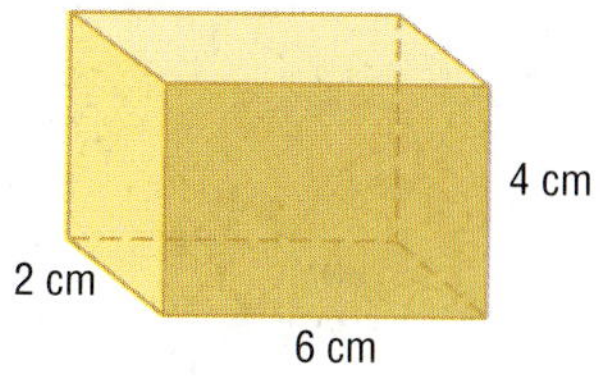

a. Find the surface area and volume of the prism.

b. Find the surface area and volume of the prism after a dilation with a scale factor of 2.

c. Find the surface area and volume of the prism after a dilation with a scale factor of $\frac{1}{2}$.

d. How many times as great is the surface area and volume of the image as the preimage after each dilation?

e. Make a conjecture as to the effect a dilation with a positive scale factor *r* would have on the surface area and volume of a prism.

32. COORDINATE GEOMETRY Refer to the graph of $\triangle DEF$.

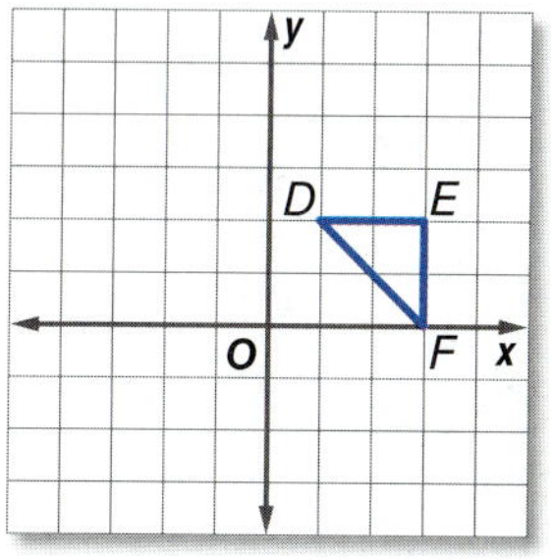

a. Graph the dilation of $\triangle DEF$ centered at point *D* with a scale factor of 3.

b. Describe the dilation as a composition of transformations including a dilation with a scale factor of 3 centered at the origin.

c. If a figure is dilated by a scale factor of 3 with a center of dilation (x, y), what composition of transformations, including a dilation with a scale factor of 3 centered at the origin, will produce the same final image?

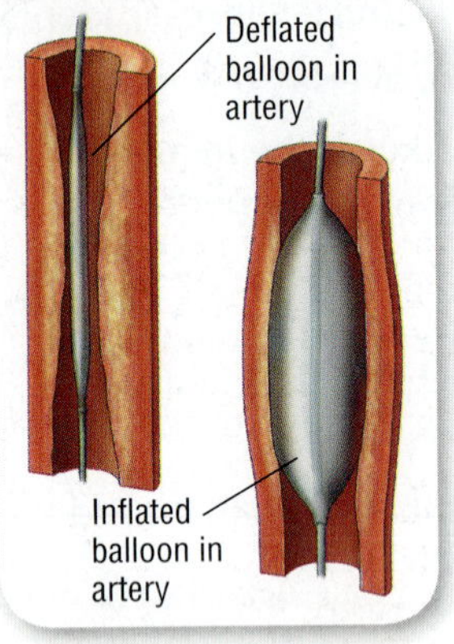

Real-World Link

If a coronary artery supplying blood to the heart is narrowed by an accumulation of cholesterol, it may be dilated with a balloon catheter–a small, hollow, flexible tube that has a balloon near the end of it.

Source: A.D.A.M. Healthcare Center

33. **HEALTH** Refer to the information at the left about coronary artery dilation.

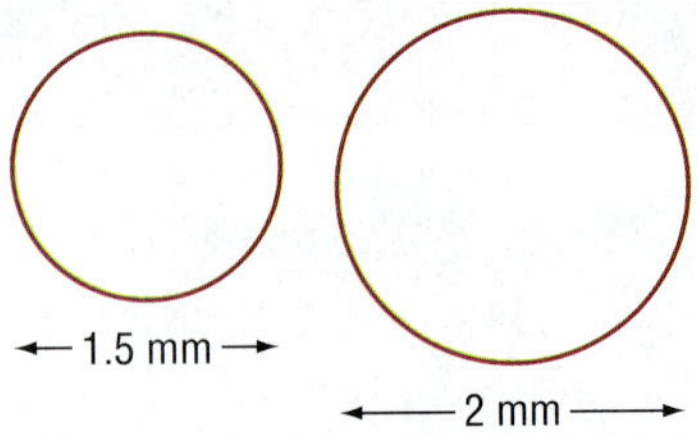

a. A surgeon inflates a balloon catheter in a patient's coronary artery, dilating the balloon as shown. Find the scale factor of this dilation.

b. Find the cross-sectional area of the balloon before and after the dilation.

Each figure shows a preimage and its image after a dilation centered at point *P*. Copy each figure, locate point *P*, and estimate the scale factor.

34.

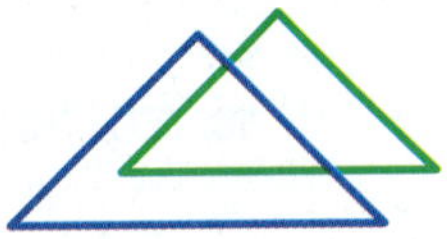

35.

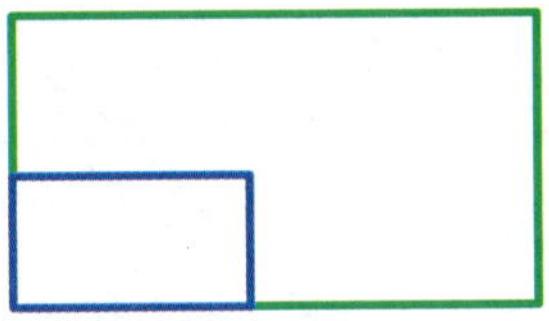

36. **MULTIPLE REPRESENTATIONS** In this problem, you will investigate dilations centered at the origin with negative scale factors.

a. **GEOMETRIC** Draw $\triangle ABC$ with points $A(-2, 0)$, $B(2, -4)$, and $C(4, 2)$. Then draw the image of $\triangle ABC$ after a dilation centered at the origin with a scale factor of -2.

b. **GEOMETRIC** Repeat the dilation with scale factors of $-\frac{1}{2}$ and -3.

c. **TABULAR** Record the coordinates for each dilation in a table.

d. **VERBAL** Make a conjecture about the function relationship for a dilation centered at the origin with a negative scale factor.

e. **ANALYTICAL** Write the function rule for a dilation centered at the origin with a scale factor of $-k$.

f. **VERBAL** Describe a dilation centered at the origin with a negative scale factor as a composition of transformations.

H.O.T. Problems

Use **H**igher-**O**rder **T**hinking Skills

37. **CHALLENGE** Find the equation for the dilated image of the line $y = 4x - 2$ if the dilation is centered at the origin with a scale factor of 1.5.

38. **WRITING IN MATH** Are parallel lines (parallelism) and collinear points (collinearity) preserved under all transformations? Explain.

39. **REASONING** Determine whether invariant points are *sometimes*, *always*, or *never* maintained for the transformations described below. If so, describe the invariant point(s). If not, explain why invariant points are not possible.

a. a dilation of $\triangle XYZ$ centered at the origin with a scale factor of 2

b. a dilation of $ABCD$ with a scale factor of 1

c. a rotation of $\overline{AB}$ 74° about B

d. a reflection of $\triangle MNP$ in the x-axis

e. a translation of $PQRS$ along $\langle 7, 3 \rangle$

40. **OPEN ENDED** Graph a triangle on the coordinate plane. Dilate the triangle so that the area of the dilation is four times the area of the original triangle. State the scale factor and center of your dilation.

41. **WRITING IN MATH** List the transformations that result in congruent figures, similar figures, and equal figures. Explain your reasoning.

Standardized Test Practice

42. EXTENDED RESPONSE Quadrilateral *PQRS* was dilated to form quadrilateral *WXYZ*.

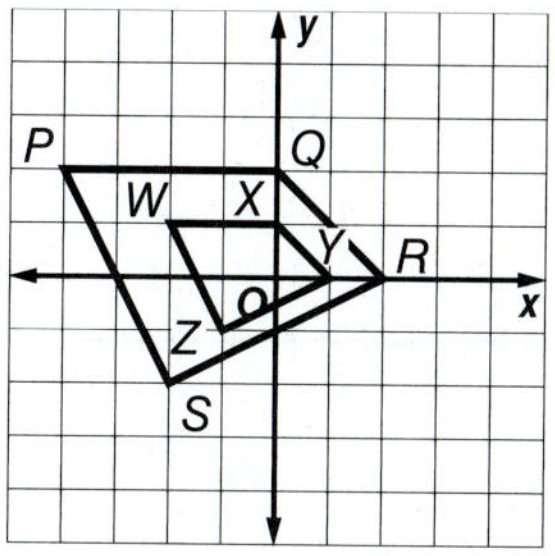

a. Is the dilation from *PQRS* to *WXYZ* an enlargement or reduction?

b. Which number *best* represents the scale factor for this dilation?

43. ALGEBRA How many ounces of pure water must a pharmacist add to 50 ounces of a 15% saline solution to make a solution that is 10% saline?

A 25   C 15
B 20   D 5

44. Tionna wants to replicate a painting in an art museum. The painting is 3 feet wide and 6 feet long. She decides on a dilation reduction factor of 0.25. What size paper should she use?

F 4 in. × 8 in.   H 8 in. × 16 in.
G 6 in. × 12 in.   J 10 in. × 20 in.

45. SAT/ACT For all x, $(x - 7)^2 = ?$

A $x^2 - 49$ C $x^2 - 14x - 49$
B $x^2 + 49$ D $x^2 - 14x + 49$

Spiral Review

State whether the figure appears to have line symmetry. Write *yes* or *no*. If so, copy the figure, draw all lines of symmetry, and state their number. (Lesson 9-5)

46.

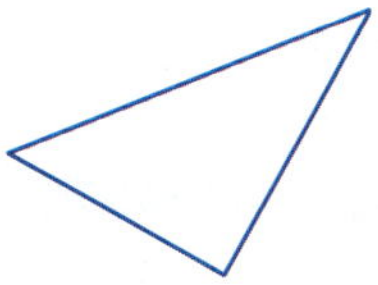

47.

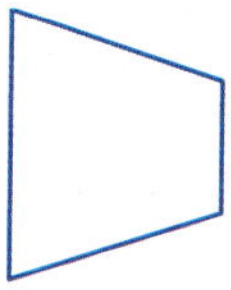

48.

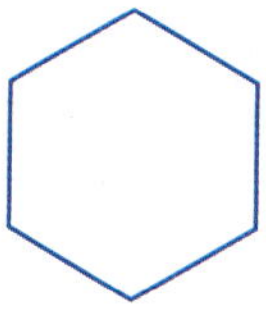

Describe the transformations that combined to map each figure. (Lesson 9-4)

49.

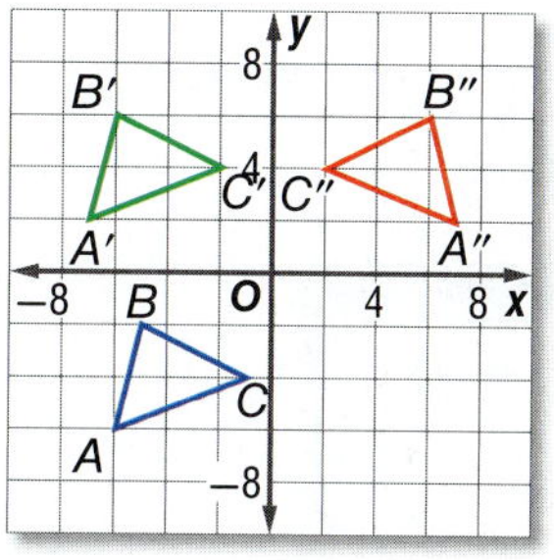

50.

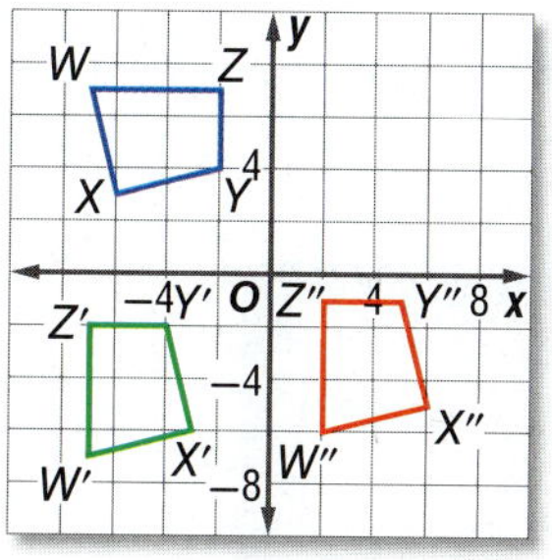

51. PAINTING A painter sets a ladder up to reach the bottom of a second-story window 16 feet above the ground. The base of the ladder is 12 feet from the house. While the painter mixes the paint, a neighbor's dog bumps the ladder, which moves the base 2 feet farther away from the house. How far up the side of the house does the ladder reach? (Lesson 8-2)

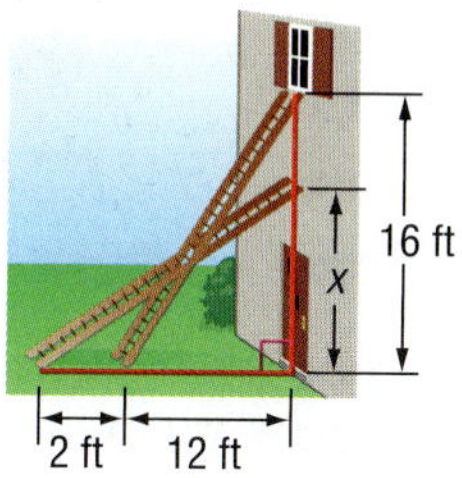

Skills Review

Find the value of x. (Lesson 0-5)

52. $58.9 = 2x$

53. $\frac{108.6}{\pi} = x$

54. $228.4 = \pi x$

55. $\frac{336.4}{x} = \pi$

EXTEND 9-6

Graphing Technology Lab
Transformations Using Matrices

Math Online glencoe.com
- Other Calculator Keystrokes
- Graphing Technology Personal Tutor

In Lesson 9-6, you dilated images by multiplying the x- and y-coordinates by a scale factor and graphing the resulting image. Transformations can also be performed by using a matrix, which is a rectangular group of numbers arranged in rows or columns. A matrix can represent a figure in the coordinate plane in which the first row corresponds to the x-coordinates of the figure and the second row corresponds to the y-coordinates, as shown below.

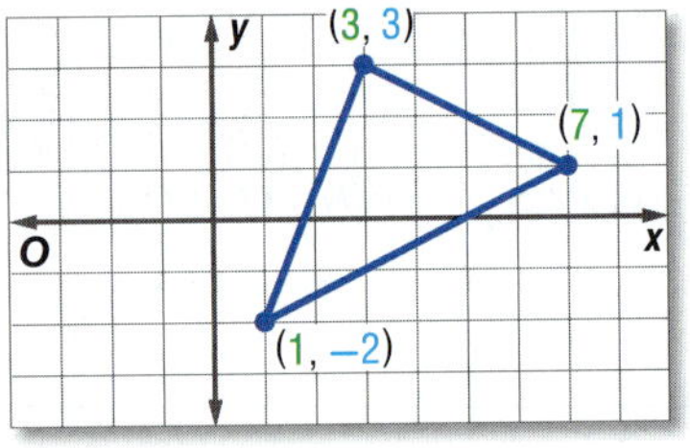

$$\begin{bmatrix} 3 & 7 & 1 \\ 3 & 1 & -2 \end{bmatrix} \begin{matrix} \leftarrow x\text{-coordinates} \\ \leftarrow y\text{-coordinates} \end{matrix}$$

In the following activities you will perform transformations by using matrix operations on a graphing calculator. Under the **Matrix** menu, select **EDIT**. Enter the values in the appropriate rows and columns as shown at the right.

ACTIVITY 1 Perform Transformations by Using Matrix Addition

Step 1 Enter a 2 × 3 matrix that represents a triangle with vertices (3, 2), (6, 6), and (1, 7) as [A].

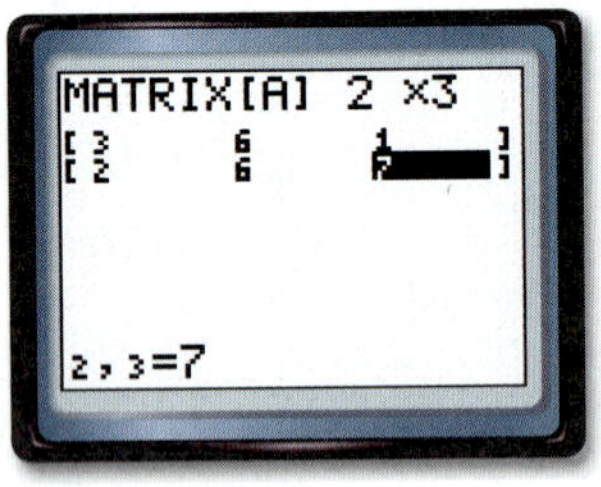

Step 2 Enter $\begin{bmatrix} 3 & 3 & 3 \\ 0 & 0 & 0 \end{bmatrix}$ into matrix [B] on your calculator. Then find [A] + [B].

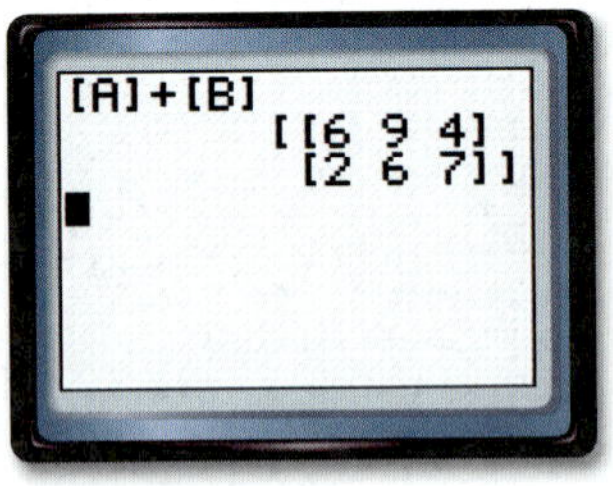

Analyze the Results

1. Graph the triangle from Step 1. Use the result from [A] + [B] to graph a second triangle. What kind of transformation do you observe?
2. Enter $\begin{bmatrix} 2 & 2 & 2 \\ -6 & -6 & -6 \end{bmatrix}$ into matrix [C]. Then find [A] + [C].
   Graph the triangle from Step 1 and the sum of [A] + [C]. Describe the type of transformation that occurs.
3. What matrix would you add to [A] in order to get $\begin{bmatrix} 7 & 10 & 5 \\ 5 & 9 & 10 \end{bmatrix}$?

ACTIVITY 2 Perform Transformations by Using Matrix Multiplication

Step 1 Enter a 2 × 3 matrix as [A] that represents a triangle with vertices (1, 1), (5, 2), and (4, 6).

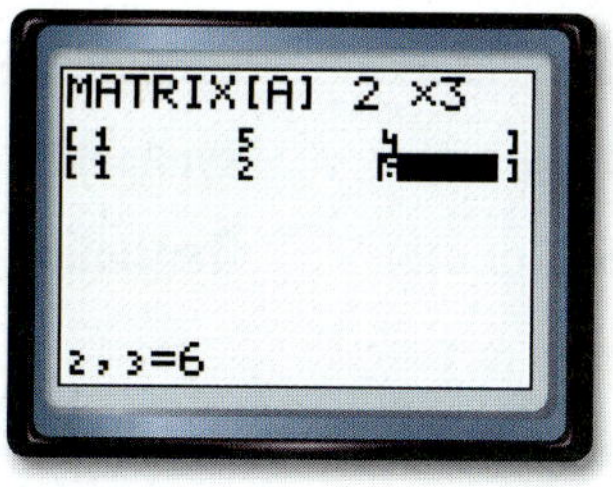

Step 2 Enter $\begin{bmatrix} 0 & -1 \\ 1 & 0 \end{bmatrix}$ into matrix [B] on your calculator. Then find [B] * [A].

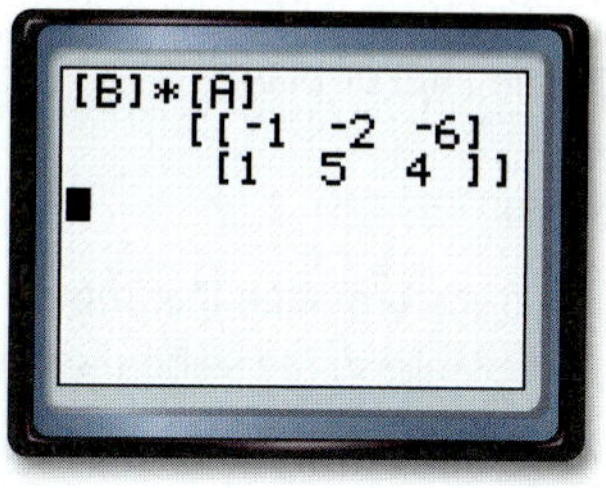

Analyze the Results

4. Graph the triangle from Step 1. Use the result from [B] * [A] to graph a second triangle. What kind of transformation do you observe?

5. Enter $\begin{bmatrix} 1 & 0 \\ 0 & -1 \end{bmatrix}$ into matrix [B]. Then find [B] * [A]. Graph the triangle from Step 1 and the product of [B] * [A]. Describe the type of transformation that occurs.

6. By what matrix would you multiply [A] in order to get $\begin{bmatrix} 1 & 2 & 6 \\ 1 & 5 & 4 \end{bmatrix}$?

ACTIVITY 3 Perform Transformations by Using Scalar Multiplication

Step 1 Enter a matrix as [A] that represents a triangle with vertices (2, 4), (6, 0), and (4, 6).

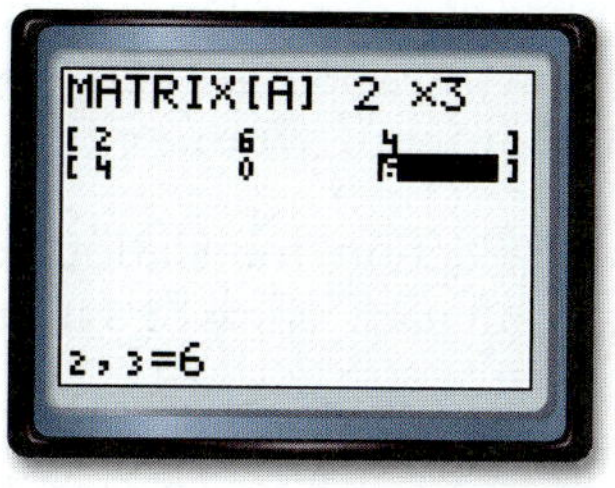

Step 2 Multiply [A] by 0.5 in your calculator.

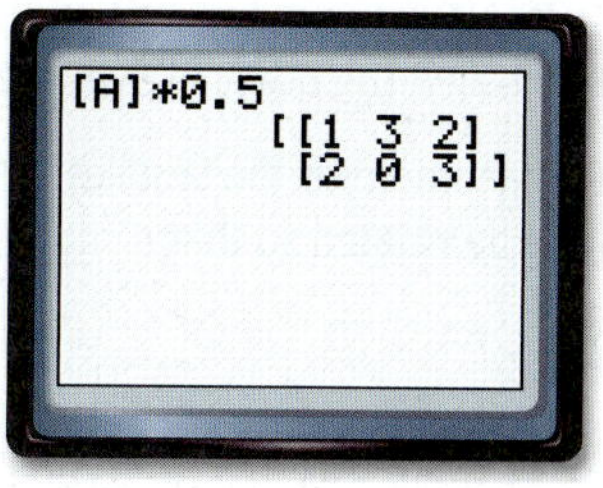

Analyze the Results

7. Graph the triangle from Step 1. Use the result from [A] * 0.5 to graph a second triangle. What kind of transformation do you observe?

8. Multiply [A] by 2. Graph the triangle from Step 1 and the product of [A] * 2. Describe the type of transformation that occurs.

9. **MAKE A CONJECTURE** What type of matrix operation would you use to perform each transformation?

a. rotation **b.** translation **c.** dilation **d.** reflection

CHAPTER 9 Study Guide and Review

Math Online glencoe.com
- STUDY TO GO
- Vocabulary Review

Chapter Summary

Key Concepts

Reflections (Lesson 9-1)

- A reflection is a transformation representing a flip of a figure over a point, line, or plane.

Translations (Lesson 9-2)

- A translation is a transformation that moves all points of a figure the same distance in the same direction.
- A translation maps each point to its image along a translation vector.

Rotations (Lesson 9-3)

- A rotation turns each point in a figure through the same angle about a fixed point.

Compositions of Transformations (Lesson 9-4)

- A translation can be represented as a composition of reflections in parallel lines and a rotation can be represented as a composition of reflections in intersecting lines.

Symmetry (Lesson 9-5)

- The line of symmetry in a figure is a line where the figure could be folded in half so that the two halves match exactly.
- The number of times a figure maps onto itself as it rotates from 0° to 360° is called the order of symmetry.
- The magnitude of symmetry is the smallest angle through which a figure can be rotated so that it maps onto itself.

Dilations (Lesson 9-6)

- Dilations enlarge or reduce figures proportionally.

FOLDABLES® Study Organizer

Be sure the Key Concepts are noted in your Foldable.

Key Vocabulary

angle of rotation (p. 632)
axis symmetry (p. 655)
center of rotation (p. 632)
composition of transformations (p. 641)
glide reflection (p. 641)
line of reflection (p. 615)
line of symmetry (p. 653)
line symmetry (p. 653)
magnitude of symmetry (p. 654)
order of symmetry (p. 654)
plane symmetry (p. 655)
rotational symmetry (p. 654)
symmetry (p. 653)
translation vector (p. 624)

Vocabulary Check

Choose the term that best completes each sentence.

1. When a transformation is applied to a figure, and then another transformation is applied to its image, this is a(n) (composition of transformations, order of symmetries).
2. If a figure is folded across a straight line and the halves match exactly, the fold line is called the (line of reflection, line of symmetry).
3. A (dilation, glide reflection) enlarges or reduces a figure proportionally.
4. The number of times a figure maps onto itself as it rotates from 0° to 360° is called the (magnitude of symmetry, order of symmetry).
5. A (line of reflection, translation vector) is the same distance from each point of a figure and its image.
6. A figure has (a center of rotation, symmetry) if it can be mapped onto itself by a rigid motion.
7. A glide reflection includes both a reflection and a (rotation, translation).
8. To rotate a point (90°, 180°) counterclockwise about the origin, multiply the y-coordinate by -1 and then interchange the x- and y-coordinates.
9. A (vector, reflection) is a congruence transformation.
10. A figure has (plane symmetry, rotational symmetry) if the figure can be mapped onto itself by a rotation between 0° and 360° about the center of the figure.

Lesson-by-Lesson Review

9-1 Reflections (pp. 613–621)

Graph each figure and its image under the given reflection.

11. rectangle *ABCD* with $A(2, -4)$, $B(4, -6)$, $C(7, -3)$, and $D(5, -1)$ in the *x*-axis

12. triangle *XYZ* with $X(-1, 1)$, $Y(-1, -2)$, and $Z(3, -3)$ in the *y*-axis

13. quadrilateral *QRST* with $Q(-4, -1)$, $R(-1, 2)$, $S(2, 2)$, and $T(0, -4)$ in the line $y = x$

14. **ART** Anita is making the two-piece sculpture shown for a memorial garden. In her design, one piece of the sculpture is a reflection of the other, to be placed beside a sidewalk that would be located along the line of reflection. Copy the figures and draw the line of reflection.

EXAMPLE 1

Graph $\triangle JKL$ with vertices $J(1, 4)$, $K(2, 1)$, and $L(6, 2)$ in the *x*-axis.

Multiply the *y*-coordinate of each vertex by -1.

$(x, y) \rightarrow (x, -y)$

$J(1, 4) \rightarrow J'(1, -4)$

$K(2, 1) \rightarrow K'(2, -1)$

$L(6, 2) \rightarrow L'(6, -2)$

Graph $\triangle JKL$ and its image $\triangle J'K'L'$.

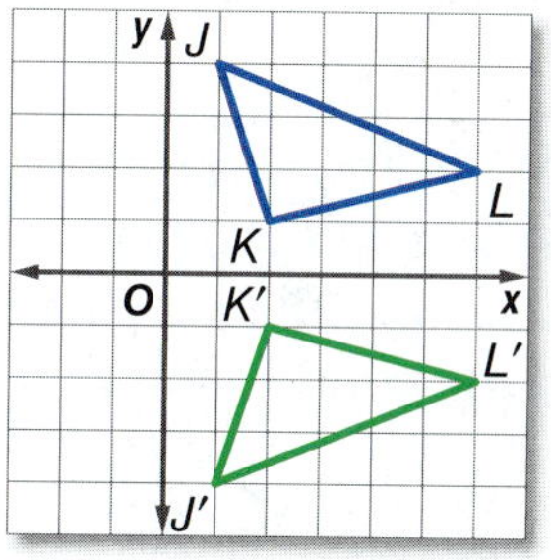

9-2 Translations (pp. 624–630)

15. Graph $\triangle ABC$ with vertices $A(0, -1)$, $B(2, 0)$, $C(3, -3)$ and its image along $\langle -5, 4 \rangle$.

16. Copy the figure and the given translation vector. Then draw the translation of the figure along the translation vector.

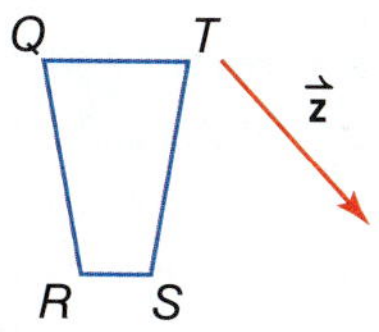

17. **DANCE** Five dancers are positioned onstage as shown. Dancers *B*, *F*, and *C* move along $\langle 0, -2 \rangle$, while dancer *A* moves along $\langle 5, -1 \rangle$. Draw the dancers' final positions.

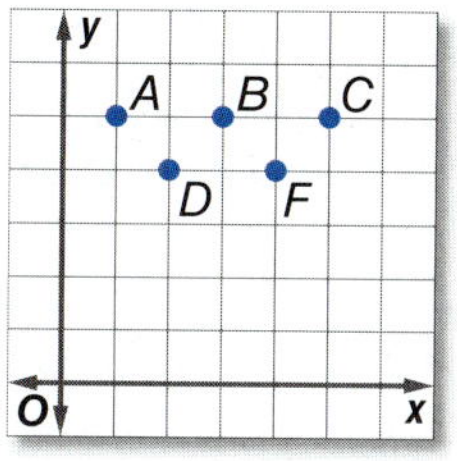

EXAMPLE 2

Graph $\triangle XYZ$ with vertices $X(2, 2)$, $Y(5, 5)$, $Z(5, 3)$ and its image along $\langle -3, -5 \rangle$.

The vector indicates a translation 3 units left and 5 units down.

$(x, y) \rightarrow (x - 3, y - 5)$

$X(2, 2) \rightarrow E'(-1, -3)$

$Y(5, 5) \rightarrow F'(2, 0)$

$Z(5, 3) \rightarrow G'(2, -2)$

Graph $\triangle XYZ$ and its image $\triangle X'Y'Z'$.

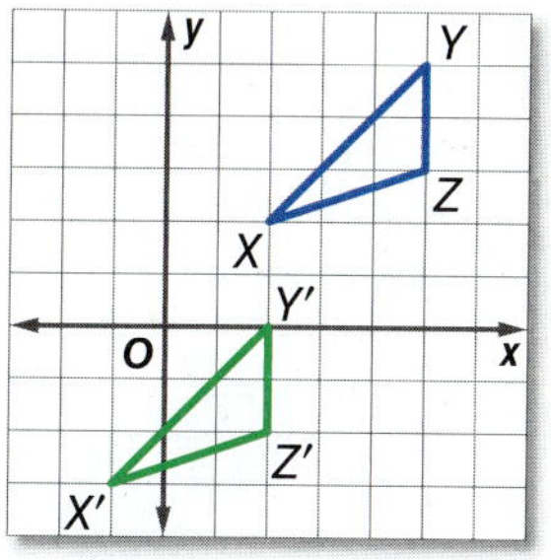

Study Guide and Review

9-3 Rotations (pp. 632–638)

18. Copy trapezoid *CDEF* and point *P*. Then use a protractor and ruler to draw a 50° rotation of *CDEF* about point *P*.

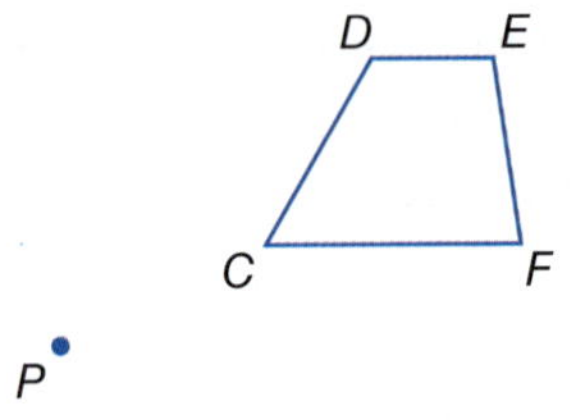

Graph each figure and its image after the specified rotation about the origin.

19. $\triangle MNO$ with vertices $M(-2, 2)$, $N(0, -2)$, $O(1, 0)$; 180°

20. $\triangle DGF$ with vertices $D(1, 2)$, $G(2, 3)$, $F(1, 3)$; 90°

Each figure shows a preimage and its image after a rotation about a point *P*. Copy each figure, locate point *P*, and find the angle of rotation.

21.

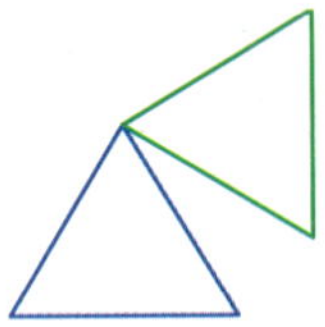

22.

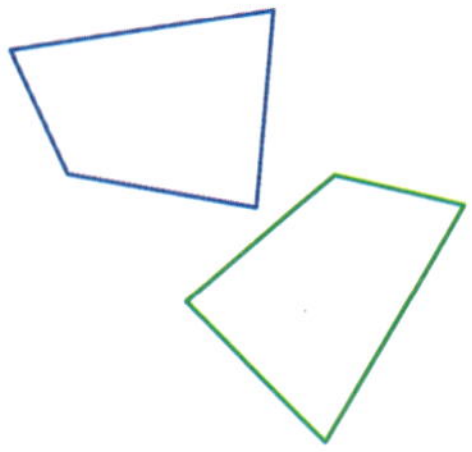

EXAMPLE 3

Triangle *ABC* has vertices $A(-4, 0)$, $B(-3, 4)$, and $C(-1, 1)$. Graph $\triangle ABC$ and its image after a rotation 270° about the origin.

One method to solve this is to combine a 180° rotation with a 90° rotation. Multiply the *x*- and *y*-coordinates of each vertex by -1.

(x, y)	→	$(-x, -y)$
$A(-4, 0)$	→	$A'(4, 0)$
$B(-3, 4)$	→	$B'(3, -4)$
$C(-1, 1)$	→	$C'(1, -1)$

Multiply the *y*-coordinate of each vertex by -1 and interchange.

$(-x, -y)$	→	$(-y, x)$
$A'(4, 0)$	→	$A'(0, -4)$
$B'(3, -4)$	→	$B'(-4, -3)$
$C'(1, -1)$	→	$C'(-1, -1)$

Graph $\triangle ABC$ and its image $\triangle A'B'C'$.

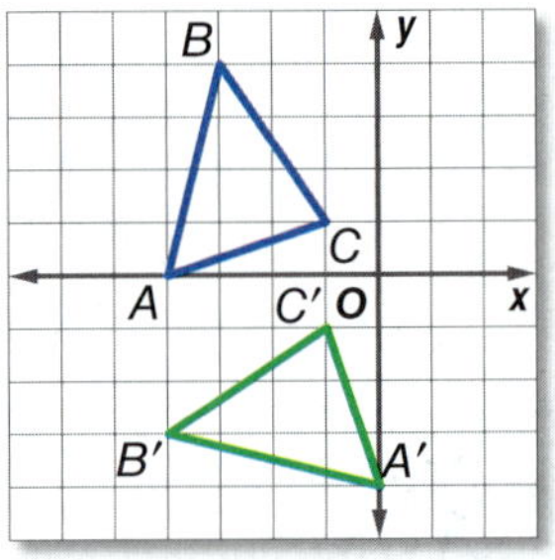

9-4 Compositions of Transformations (pp. 641–649)

Graph each figure with the given vertices and its image after the indicated transformation.

23. $\overline{CD}$: $C(3, 2)$ and $D(1, 4)$
Reflection: in $y = x$
Rotation: 270° about the origin.

24. $\overline{GH}$: $G(-2, -3)$ and $H(1, 1)$
Translation: along $\langle 4, 2\rangle$
Reflection: in the *x*-axis

25. **PATTERNS** Jeremy is creating a pattern for the border of a poster using a stencil. Describe the transformation combination that he used to create the pattern below.

EXAMPLE 4

The endpoints of $\overline{RS}$ are $R(4, 3)$ and $S(1, 1)$. Graph $\overline{RS}$ and its image after a translation along $\langle -5, -1\rangle$ and a rotation 180° about the origin.

Step 1 translation along $\langle -5, -1\rangle$

(x, y)	→	$(x - 5, y - 1)$
$R(4, 3)$	→	$R'(-1, 2)$
$S(1, 1)$	→	$S'(-4, 0)$

Step 2 rotation 180° about origin

(x, y)	→	$(-x, -y)$
$R'(-1, 2)$	→	$R''(1, -2)$
$S'(-4, 0)$	→	$S''(4, 0)$

(continued on the next page)

MIXED PROBLEM SOLVING
For mixed problem-solving practice, see page 1007.

26. Copy and reflect figure T in line ℓ and then line m. Then describe a single transformation that maps T onto T''.

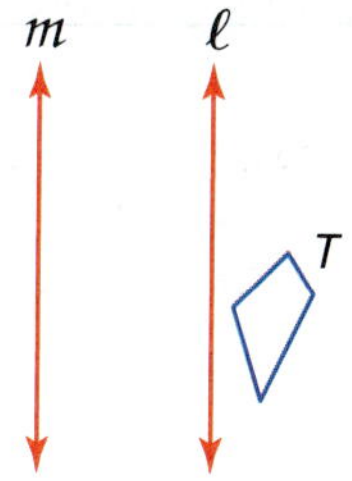

Step 3 Graph $\overline{RS}$ and its image $\overline{R''S''}$.

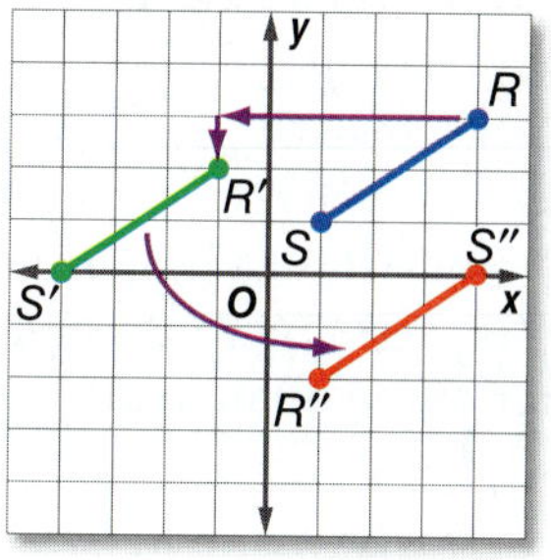

9-5 Symmetry (pp. 653–659)

State whether each figure appears to have line symmetry. Write *yes* or *no*. If so, copy the figure, draw all lines of symmetry, and state their number.

27.

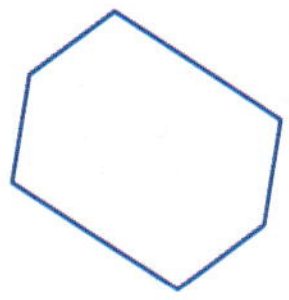

28.

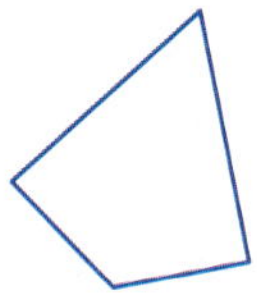

State whether each figure has rotational symmetry. Write *yes* or *no*. If so, copy the figure, locate the center of symmetry, and state the order and magnitude of symmetry.

29.

30.

31. KNITTING Amy is creating a pattern for a scarf she is knitting for her friend. How many lines of symmetry are there in the pattern?

EXAMPLE 5

State whether each figure has *plane* symmetry, *axis* symmetry, *both*, or *neither*.

a.

The light bulb has both plane and axis symmetry.

b.

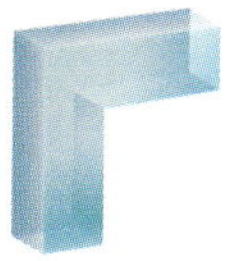

The prism has plane symmetry.

9-6 Dilations (pp. 660–667)

32. Copy the figure and point *S*. Then use a ruler to draw the image of the figure under a dilation with center *S* and scale factor $r = 1.25$.

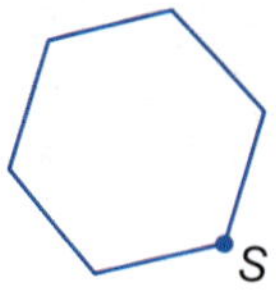

33. Determine whether the dilation from figure *W* to *W*′ is an *enlargement* or a *reduction*. Then find the scale factor of the dilation and *x*.

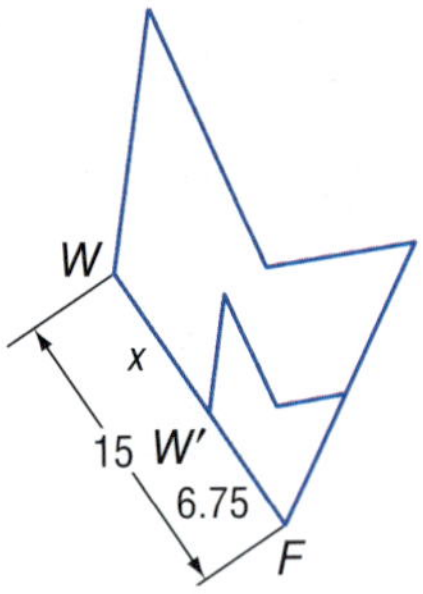

34. CLUBS The members of the Math Club use an overhead projector to make a poster. If the original image was 6 inches wide, and the image on the poster is 4 feet wide, what is the scale factor of the enlargement?

EXAMPLE 6

Square *ABCD* has vertices *A*(0, 0), *B*(0, 8), *C*(8, 8), and *D*(8, 0). Find the image of *ABCD* after a dilation centered at the origin with a scale factor of 0.5.

Multiply the *x*- and *y*-coordinates of each vertex by the scale factor, 0.5.

$(x, y) \rightarrow (0.5x, 0.5y)$

$A(0, 0) \rightarrow A'(0, 0)$

$B(0, 8) \rightarrow B'(0, 4)$

$C(8, 8) \rightarrow C'(4, 4)$

$D(8, 0) \rightarrow D'(4, 0)$

Graph *ABCD* and its image *A'B'C'D'*.

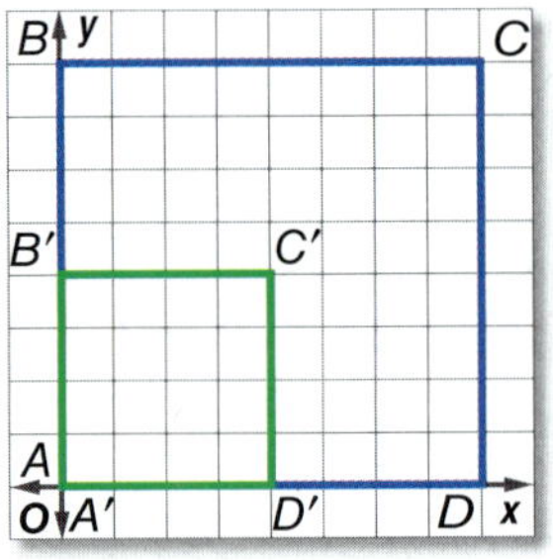

CHAPTER 9 Practice Test

Math Online glencoe.com
Practice Test

Copy the figure and the given line of reflection. Then draw the reflected image in this line using a ruler.

1.
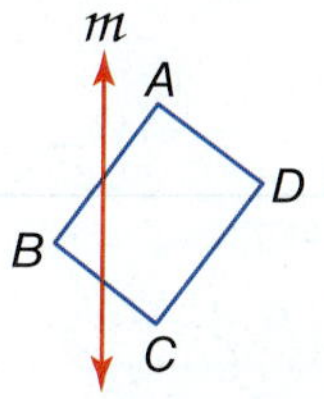

2.
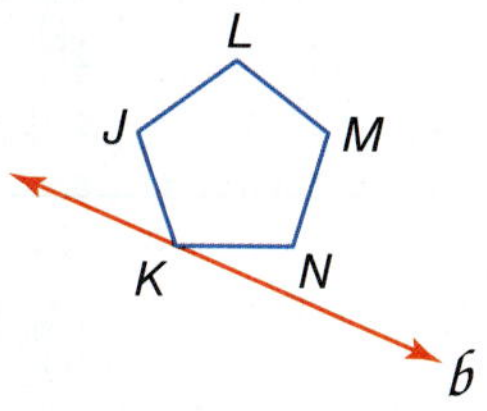

3. **PROJECTS** Eduardo wants to enlarge the picture below to 4 inches by 6 inches for a school project. If his school's copy machine can only enlarge up to 150% by whole number percents, find two whole number percents by which he can enlarge the piece and get as close to 4 inches by 6 inches or less.

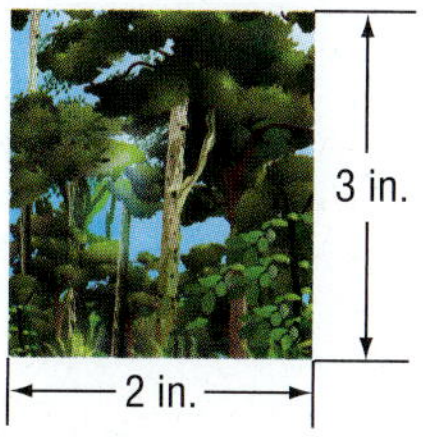

Copy the figure and point *M*. Then use a ruler to draw the image of the figure under a dilation with center *M* and the scale factor *r* indicated.

4. $r = 1.5$

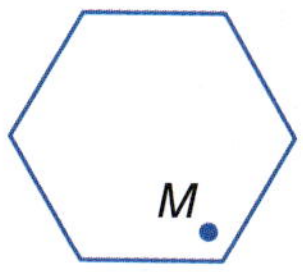

5. $r = \frac{1}{3}$

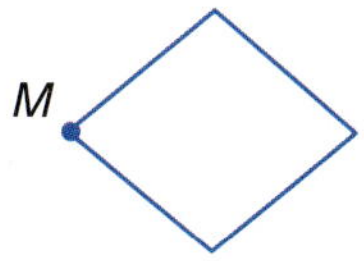

6. **PARKS** Isabel is on a ride at an amusement park that slides the rider to the right, and then rotates counterclockwise about its own center 60° every 2 seconds. How many seconds pass before Isabel returns to her starting position?

State whether each figure has *plane* symmetry, *axis* symmetry, *both*, or *neither*.

7.
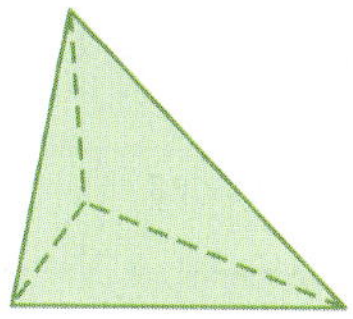

8.
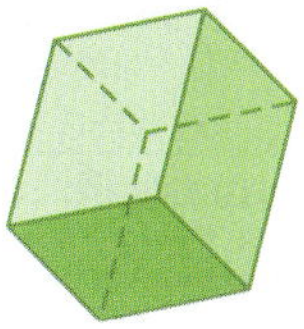

Graph each figure and its image under the given transformation.

9. $\square FGHJ$ with vertices $F(-1, -1)$, $G(-2, -4)$, $H(1, -4)$, and $J(2, -1)$ in the x-axis

10. $\triangle ABC$ with vertices $A(0, -1)$, $B(2, 0)$, $C(3, -3)$; $\langle -5, 4 \rangle$

11. quadrilateral $WXYZ$ with vertices $W(2, 3)$, $X(1, 1)$, $Y(3, 0)$, $Z(5, 2)$; 180° about the origin

Copy the figure and the given translation vector. Then draw the translation of the figure along the translation vector.

12.
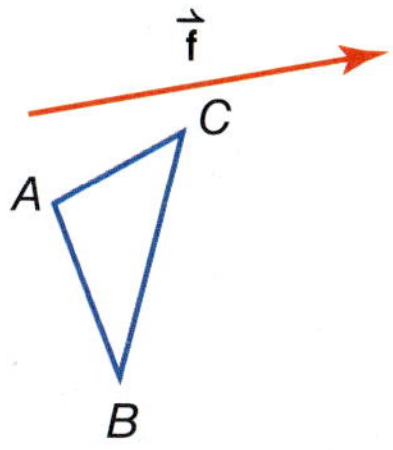

13.
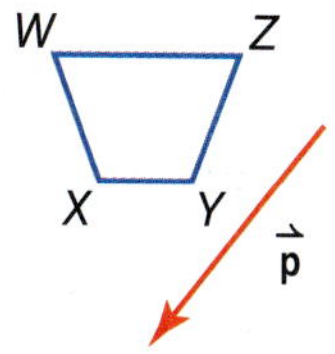

14. **ART** An artist's rendition of what Stonehenge, a famous archeological site in England, would have looked like before the stones fell or were removed, is shown below. What is the order and magnitude of symmetry for the outer ring?

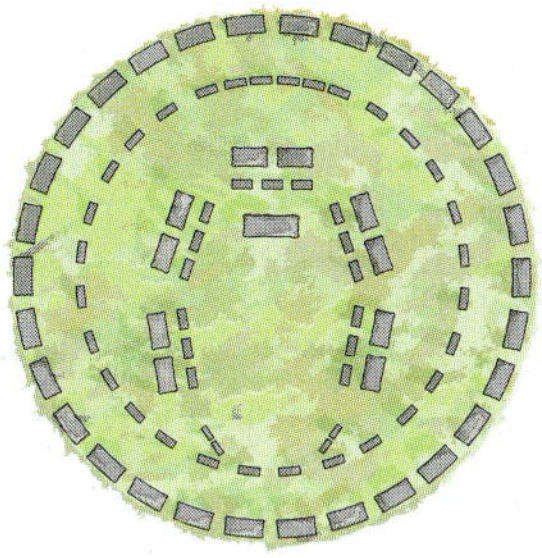

15. **MULTPLE CHOICE** What transformation or combination of transformations does the figure below represent?

A dilation
B glide reflection
C rotation
D translation

Work Backward

In most problems, a set of conditions or facts is given and you must find the end result. However, some problems give you the end result and ask you to find something that happened earlier in the process. To solve problems like this, you must work backward.

Strategies for Working Backward

Step 1

Look for keywords that indicate you will need to work backward to solve the problem.

Sample Keywords:

- What was the **original**...?
- What was the value **before**...?
- Where was the **starting** or **beginning**...?

Step 2

Undo the steps given in the problem statement to solve.

- List the sequence of steps from the beginning to the end result.
- Begin with the end result. Retrace the steps in reverse order.
- "Undo" each step using inverses to get back to the original value.

Step 3

Check your solution if time permits.

- Make sure your answer makes sense.
- Begin with your answer and follow the steps in the problem statement forward to see if you get the same end result.

EXAMPLE

Solve the problem below. Responses will be graded using the short-response scoring rubric shown.

Kelly is using a geometry software program to experiment with transformations on the coordinate grid. She began with a point and translated it 4 units up and 8 units left. Then she reflected the image across the x-axis. Finally, she dilated this new image by a scale factor of 0.5 with the origin to arrive at $(-1, -4)$. What were the original coordinates of the point?

Scoring Rubric

Criteria	Score
Full Credit: The answer is correct and a full explanation is provided that shows each step.	2
Partial Credit: • The answer is correct, but the explanation is incomplete. • The answer is incorrect, but the explanation is correct.	1
No Credit: Either an answer is not provided or the answer does not make sense.	0

Read the problem statement carefully. You are given a sequence of transformations of a point on a coordinate grid. You know the coordinates of the final image and are asked to find the original coordinates. Undo each transformation in reverse order to work backward and solve the problem.

Example of a 2-point response:

> original point → translation → reflection → dilation → end result
>
> Begin with the coordinates of the end result and work backward.
>
> Dilate by 2 to undo the dilation by 0.5:
>
> $(-1, -4) \rightarrow (-1 \times 2, -4 \times 2) = (-2, -8)$
>
> Reflect back across the x-axis to undo the reflection:
>
> $(-2, -8) \rightarrow (-2, 8)$
>
> Translate 4 units down and 8 units right to undo the translation:
>
> $(-2, 8) \rightarrow (-2 + 8, 8 - 4) = (6, 4)$
>
> The original coordinates of the point were (6, 4).

The steps, calculations, and reasoning are clearly stated. The student also arrives at the correct answer. So, this response is worth the full 2 points.

Exercises

Solve each problem. Show your work. Responses will be graded using the short-response scoring rubric given at the beginning of the lesson.

1. A flea landed on a coordinate grid. The flea hopped across the x-axis and then across the y-axis in the form of two consecutive reflections. Then it walked 9 units to the right and 4 units down. If the flea's final position was at $(4, -1)$, what point did it originally land on?

2. The coordinate grid below shows the final image when a point was rotated 90° clockwise about the origin, dilated by a scale factor of 2, and shifted 7 units right. What were the original coordinates?

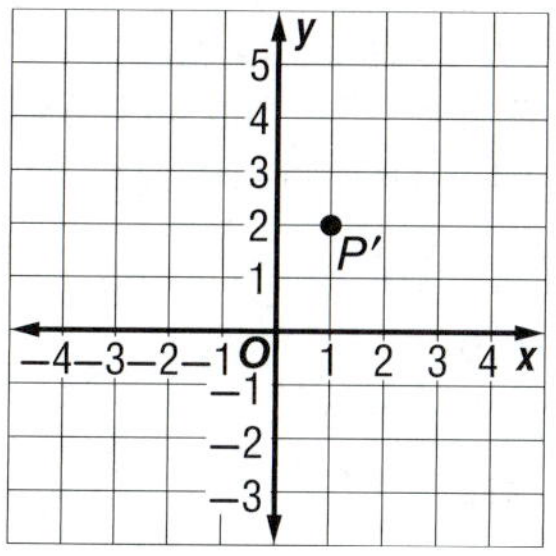

3. Figure $ABCD$ is an isosceles trapezoid.

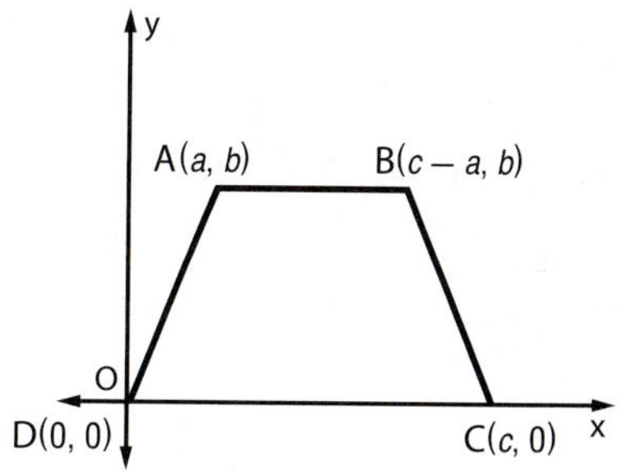

Which of the following are the coordinates of an endpoint of the median of $ABCD$?

A $\left(\frac{a+b}{2}, \frac{a+b}{2}\right)$

B $\left(\frac{2c-a}{2}, \frac{b}{2}\right)$

C $\left(\frac{c}{2}, 0\right)$

D $\left(\frac{c}{2}, b\right)$

4. If the measure of an interior angle of a regular polygon is 108, what type of polygon is it?

F octagon

G hexagon

H pentagon

J triangle

CHAPTER 9

Standardized Test Practice

Cumulative, Chapters 1 through 9

Multiple Choice

Read each question. Then fill in the correct answer on the answer document provided by your teacher or on a sheet of paper.

1. Point N has coordinates (4, −3). What will the coordinates of its image be after a reflection across the y-axis?

A $N'(-3, 4)$

B $N'(-4, 3)$

C $N'(4, 3)$

D $N'(-4, -3)$

2. Which pair of figures shows a reflection across the line followed by a translation up?

F

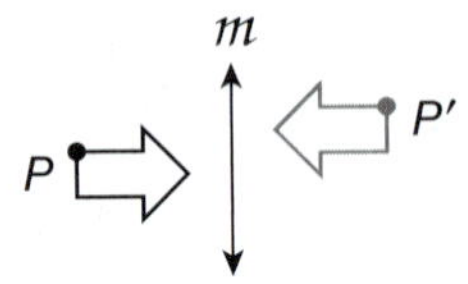

H

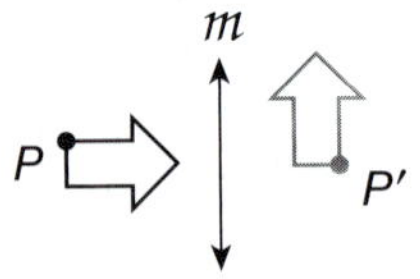

G

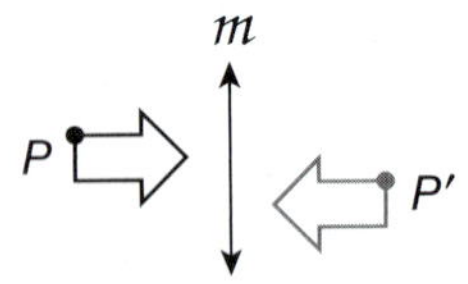

J

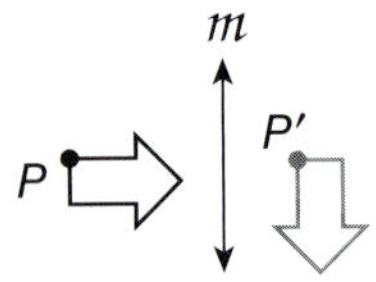

3. What is the angle of rotation that maps point T onto T' in the figure below?

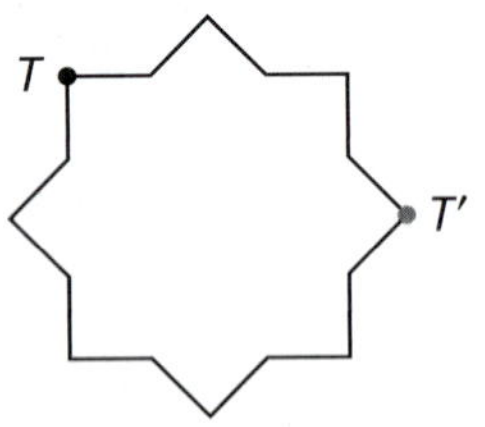

A 90°

B 120°

C 135°

D 145°

Test-Taking Tip

Question 3 How many points are there on the star? Divide 360° by this number to find the angle of rotation from one point to the next.

4. Given: $a \parallel b$

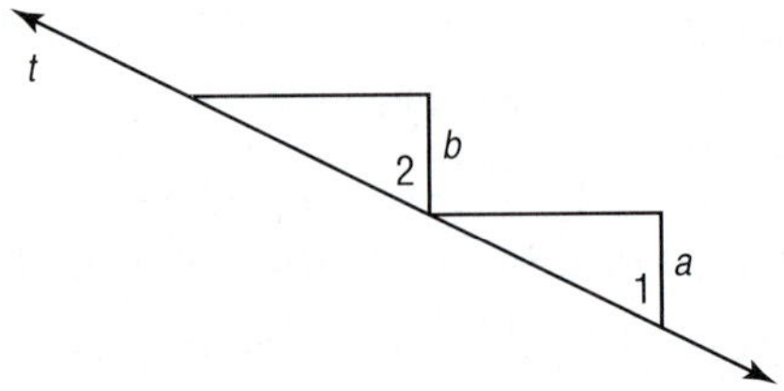

Which statement below justifies the conclusion that $\angle 1 \cong \angle 2$?

F If $a \parallel b$ and are cut by transversal t, then alternate exterior angles are congruent.

G If $a \parallel b$ and are cut by transversal t, then alternate interior angles are congruent.

H If $a \parallel b$ and are cut by transversal t, then corresponding angles are congruent.

J If $a \parallel b$ and are cut by transversal t, then vertical angles are congruent.

5. What is the geometric mean of 8 and 18?

A 9

B 10

C 11

D 12

6. Which of the following is a side length in isosceles triangle DEF?

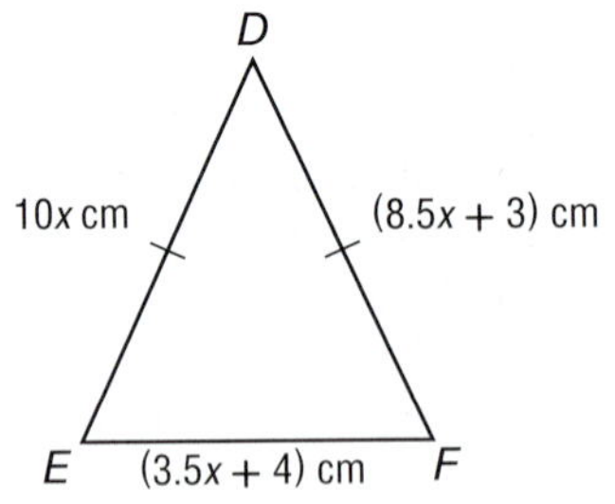

F 2 cm

G 8 cm

H 9 cm

J 11 cm

7. Which of the following has exactly two pairs of consecutive congruent sides?

A kite

B parallelogram

C rhombus

D trapezoid

Short Response/Gridded Response

Record your answers on the answer sheet provided by your teacher or on a sheet of paper.

8. State whether the figure has rotational symmetry. If so, copy the figure, locate the center, and state the order and magnitude of symmetry.

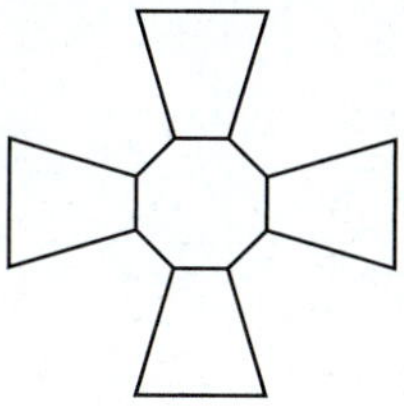

9. Dilate the figure shown on the coordinate grid by a scale factor of 1.5 centered at the origin.

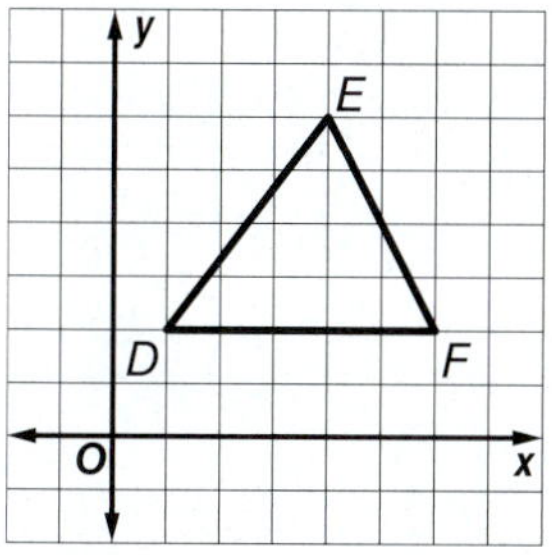

10. Complete the following statement.

According to the Angle Bisector Theorem, if a point is on the bisector of an angle, then it is _______ .

11. Regina left her office downtown and traveled 3 blocks west and 5 blocks north. Write a translation vector to describe her route.

12. What is the interior angle measure of the regular pentagon?

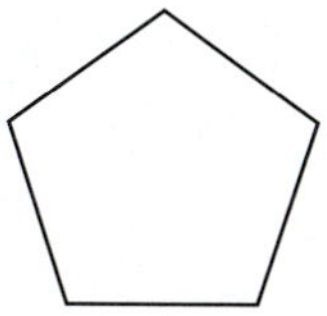

13. GRIDDED RESPONSE A group of 75 students were asked what types of movies they like to watch. The results are shown in the Venn diagram.

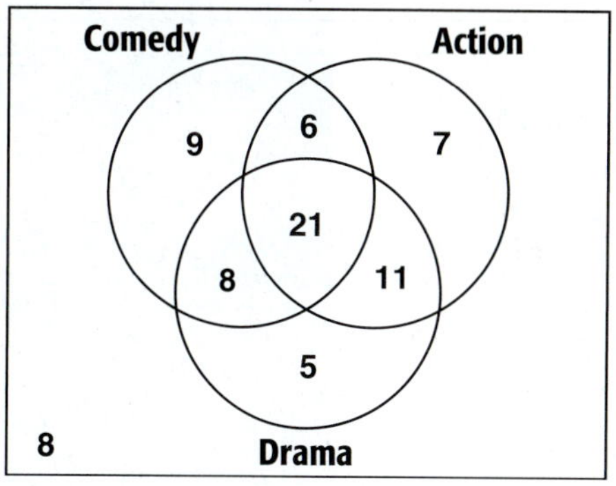

How many students said that they like to watch action and drama movies, but not comedy?

Extended Response

Record your answers on a sheet of paper. Show your work.

14. Rodrigo is making a scale model.

a. The actual length of the Golden Gate Bridge is about 9000 feet. If Rodrigo's model is 45 inches, what is the scale of his model?

b. How wide will Rodrigo's model of the bridge be if the actual width is 90 feet?

c. In Rodrigo's model, the tower will be 2.5 inches above the roadway. How high above the roadway is the actual tower?

Need Extra Help?

If you missed Question...	1	2	3	4	5	6	7	8	9	10	11	12	13	14
Go to Lesson or Page...	9-1	9-4	9-3	3-2	8-1	4-6	6-6	9-5	9-6	5-1	9-2	6-1	2-2	7-7

CHAPTER 10 Circles

Then

In Chapter 5, you learned about special segments and angle relationships in triangles.

Now

In Chapter 10, you will:

- Learn the relationships between central angles, arcs, and inscribed angles in a circle.
- Define and use secants and tangents.
- Use an equation to identify or describe a circle.

Why?

SCIENCE The actual shape of a rainbow is a complete circle. The portion of the circle that can be seen above the horizon is a special segment of a circle called an arc.

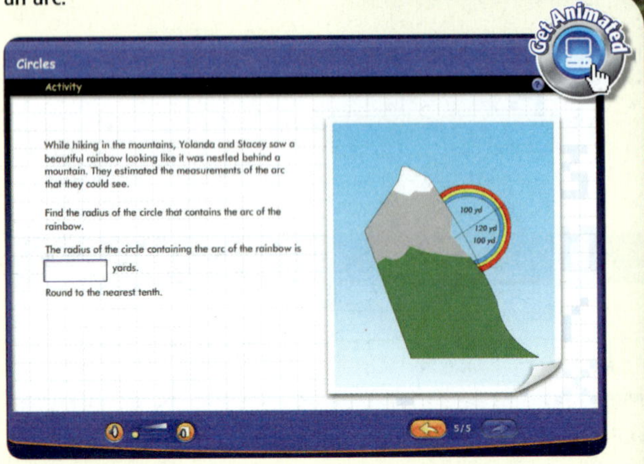

Get Ready for Chapter 10

Diagnose Readiness You have two options for checking Prerequisite Skills.

Text Option

Take the Quick Check below. Refer to the Quick Review for help.

*Quick*Check

Find the percent of the given number.
(Prerequisite Skill)

1. 26% of 500
2. 79% of 623
3. 19% of 82
4. 10% of 180
5. 92% of 90
6. 65% of 360
7. **TIPPING** A couple ate dinner at an Italian restaurant where their bill was \$32.50. If they want to leave an 18% tip, how much tip money should they leave?

8. Find x. Round to the nearest tenth. (Lesson 8-2)

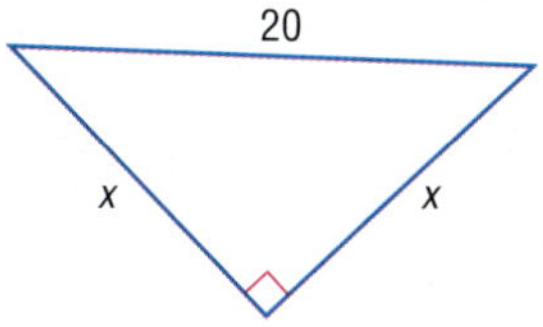

9. **CONSTRUCTION** Jennifer is putting a brace in a board, as shown at the right. Find the length of the board used for a brace.

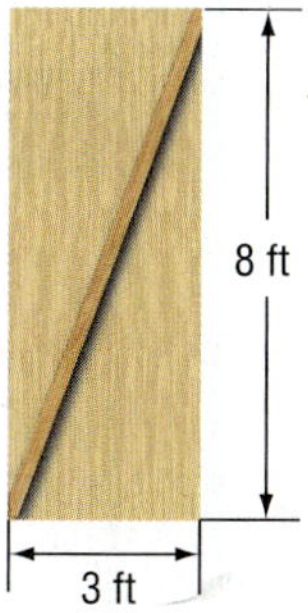

Solve each equation by using the Quadratic Formula. Round to the nearest tenth if necessary.
(Prerequisite Skill)

10. $5x^2 + 4x - 20 = 0$
11. $x^2 = x + 12$
12. **FIREWORKS** The Patriot Squad, a professional fireworks company, performed a show during a July 4th celebration. One of the rockets in the show failed to explode at a distance of $d = 80t - 16t^2$ from the ground. What length of time t did it take the rocket dud to hit the ground?

*Quick*Review

EXAMPLE 1

Find the percent of the given number.

$15\% \text{ of } 35 = (0.15)(35)$	Change the percent to a decimal.
$= 5.25$	Multiply.

So, 15% of 35 is 5.25.

EXAMPLE 2

Find x. Round to the nearest tenth.

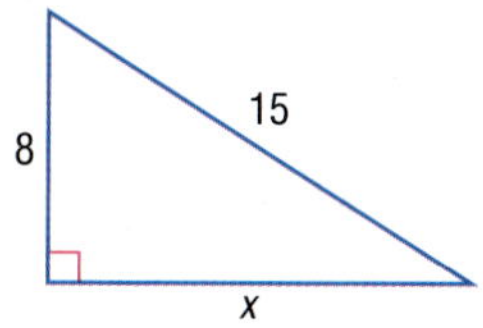

$a^2 + b^2 = c^2$	Pythagorean Theorem
$x^2 + 8^2 = 15^2$	Substitution
$x^2 + 64 = 225$	Simplify.
$x^2 = 161$	Subtract.
$x = \sqrt{161}$ or about 12.7	

EXAMPLE 3

Solve $x^2 + 3x - 40 = 0$ by using the Quadratic Formula. Round to the nearest tenth.

$x = \dfrac{-b \pm \sqrt{b^2 - 4ac}}{2a}$	Quadratic Formula
$= \dfrac{-3 \pm \sqrt{3^2 - 4(1)(-40)}}{2(1)}$	Substitution
$= \dfrac{-3 \pm \sqrt{169}}{2}$	Simplify.
$= 5 \text{ or } -8$	Simplify.

Online Option

Math Online Take a self-check Chapter Readiness Quiz at glencoe.com.

Get Started on Chapter 10

You will learn several new concepts, skills, and vocabulary terms as you study Chapter 10. To get ready, identify important terms and organize your resources. You may wish to refer to **Chapter 0** to review prerequisite skills.

FOLDABLES® Study Organizer

Circles Make this Foldable to help you organize your Chapter 10 notes on circles. Begin with nine sheets of paper.

1. **Trace** a 10-inch circle on each paper using a compass.

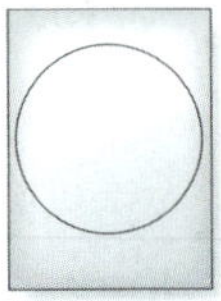

2. **Cut** out each of the circles.

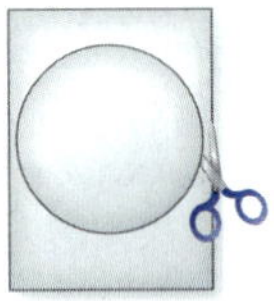

3. **Staple** an inch from the left side of the papers.

4. **Label** as shown.

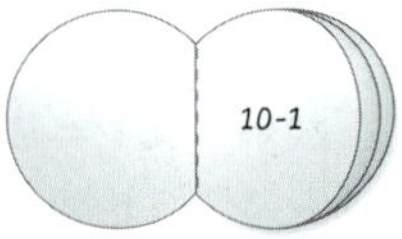

Math Online glencoe.com

- Study the chapter online
- Explore **Get Animated** activities
- Get extra help from your own **Personal Tutor**
- Use **Extra Examples** for additional help
- Take a **Self-Check Quiz**
- **Review Vocabulary** in fun ways

New Vocabulary

English		Español
circle	• p. 683 •	circulo
center	• p. 683 •	centro
radius	• p. 683 •	radio
chord	• p. 683 •	cuerda
diameter	• p. 683 •	diámetro
circumference	• p. 685 •	circunferencia
pi (π)	• p. 685 •	pi (π)
central angle	• p. 692 •	ángulo central
arc	• p. 692 •	arco
chord segment	• p. 736 •	segmento de cuerda

Review Vocabulary

coplanar • p. 5 • coplanar points that lie in the same plane

degree • p. 37 • grado $\frac{1}{360}$ of the circular rotation about a point

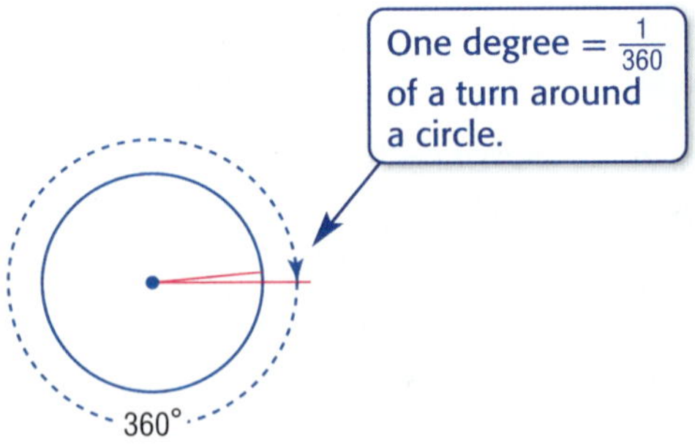

Multilingual eGlossary glencoe.com

10-1 Circles and Circumference

Then
You identified and used parts of parallelograms. (Lesson 6-2)

Now
- Identify and use parts of circles.
- Solve problems involving the circumference of a circle.

New Vocabulary
circle
center
radius
chord
diameter
congruent circles
concentric circles
circumference
pi (π)
inscribed
circumscribed

Math Online
glencoe.com

Why?

The maxAir ride shown speeds back and forth and rotates counterclockwise. At times, the riders are upside down 140 feet above the ground experiencing "airtime"—a feeling of weightlessness. The ride's width, or *diameter,* is 44 feet. You can find the distance that a rider travels in one rotation by using this measure.

Segments in Circles A **circle** is the locus or set of all points in a plane equidistant from a given point called the **center** of the circle.

Segments that intersect a circle have special names.

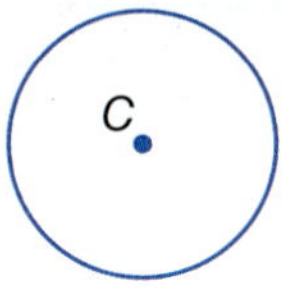

Circle C or $\odot C$

Key Concept — Special Segments in a Circle

A **radius** (plural radii) is a segment with endpoints at the center and on the circle.

Examples $\overline{CD}$, $\overline{CE}$, and $\overline{CF}$ are radii of $\odot C$.

A **chord** is a segment with endpoints on the circle.

Examples $\overline{AB}$ and $\overline{DE}$ are chords of $\odot C$.

A **diameter** of a circle is a chord that passes through the center and is made up of collinear radii.

Example $\overline{DE}$ is a diameter of $\odot C$. Diameter $\overline{DE}$ is made up of collinear radii $\overline{CD}$ and $\overline{CE}$.

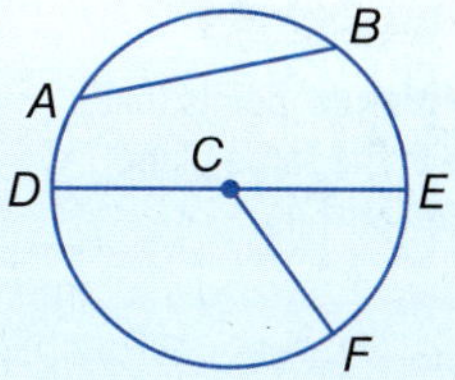

EXAMPLE 1 Identify Segments in a Circle

a. Name the circle and identify a radius.

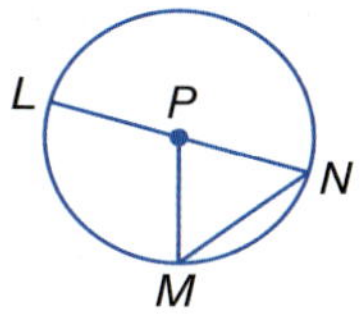

The circle has a center at P, so it is named circle P, or $\odot P$. Three radii are shown: $\overline{PL}$, $\overline{PN}$, and $\overline{PM}$.

b. Identify a chord and a diameter of the circle.

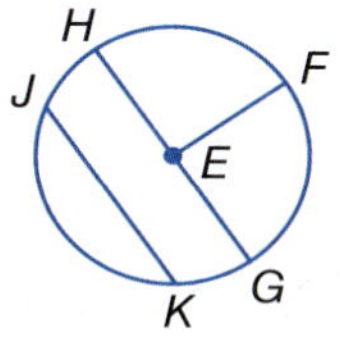

Two chords are shown: $\overline{JK}$ and $\overline{HG}$. $\overline{HG}$ goes through the center, so $\overline{HG}$ is a diameter.

Guided Practice

1. Name the circle, a radius, a chord, and a diameter of the circle.

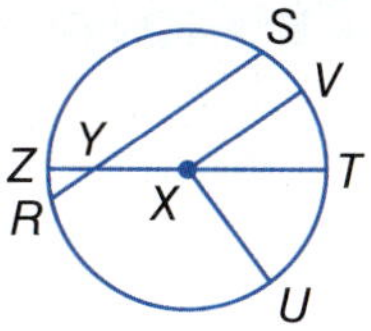

Personal Tutor glencoe.com

ReadingMath

Diameter and Radius The words *radius* and *diameter* are used to describe lengths as well as segments. Since a circle has many different radii and diameters, the phrases *the radius* and *the diameter* refer to lengths rather than segments.

By definition, the distance from the center of a circle to any point on the circle is always the same. Therefore, all radii r of a circle are congruent. Since a diameter d is composed of two radii, all diameters of a circle are also congruent.

Key Concept — Radius and Diameter Relationships

If a circle has radius r and diameter d, the following relationships are true.

Radius Formula $r = \frac{d}{2}$ or $r = \frac{1}{2}d$ **Diameter Formula** $d = 2r$

EXAMPLE 2 Find Radius and Diameter

If $QV = 8$ inches, what is the diameter of $\odot Q$?

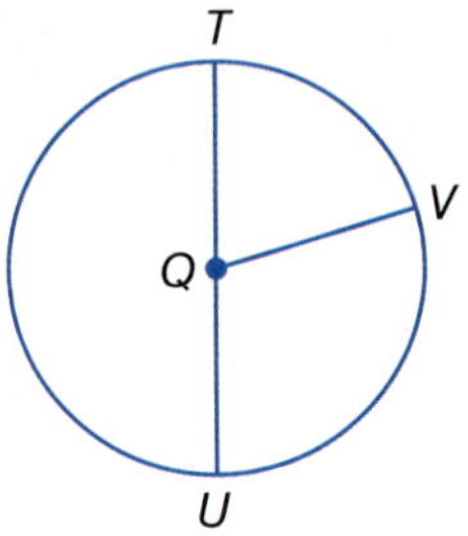

$d = 2r$ **Diameter Formula**

$= 2(8)$ or 16 **Substitute and simplify.**

The diameter of $\odot Q$ is 16 inches.

Guided Practice

2A. If $TU = 14$ feet, what is the radius of $\odot Q$?

2B. If $QT = 11$ meters, what is QU?

Personal Tutor glencoe.com

Review Vocabulary

coplanar points that lie in the same plane (Lesson 1-1)

As with other figures, pairs of circles can be congruent or share other special relationships.

Key Concept — Circle Pairs

Two circles are **congruent circles** if and only if they have congruent radii.

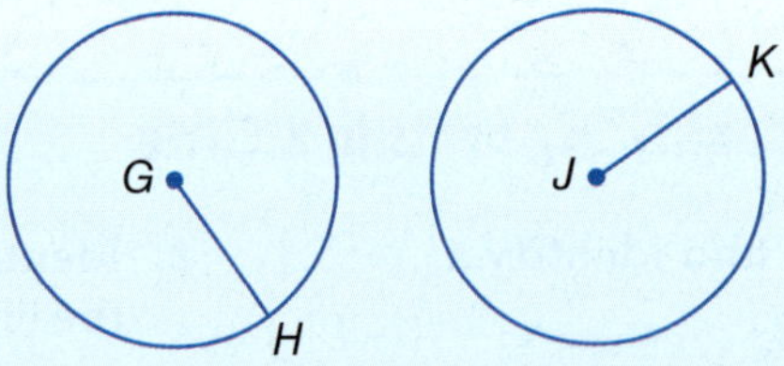

Example $\overline{GH} \cong \overline{JK}$, so $\odot G \cong \odot J$.

Concentric circles are coplanar circles that have the same center.

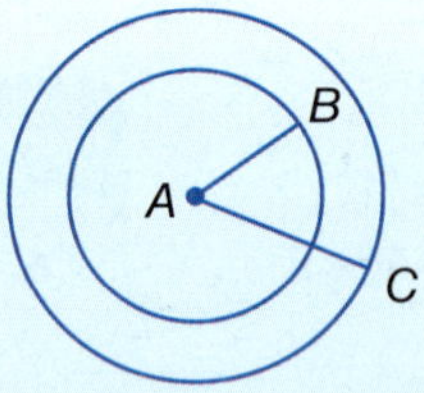

Example $\odot A$ with radius $\overline{AB}$ and $\odot A$ with radius $\overline{AC}$ are concentric.

Two circles can intersect in two different ways.

2 Points of Intersection	1 Point of Intersection	No Points of Intersection
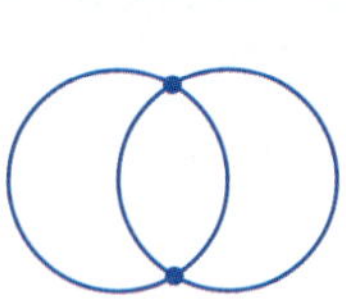		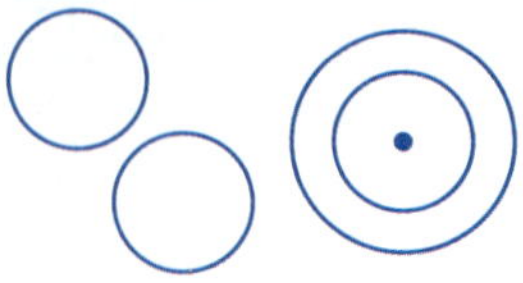

The segment connecting the centers of the two intersecting circles contains the radii of the two circles.

EXAMPLE 3 Find Measures in Intersecting Circles

The diameter of ⊙S is 30 units, the diameter of ⊙R is 20 units, and $DS = 9$ units. Find CD.

Since the diameter of ⊙S is 30, $CS = 15$.
$\overline{CD}$ is part of radius $\overline{CS}$.

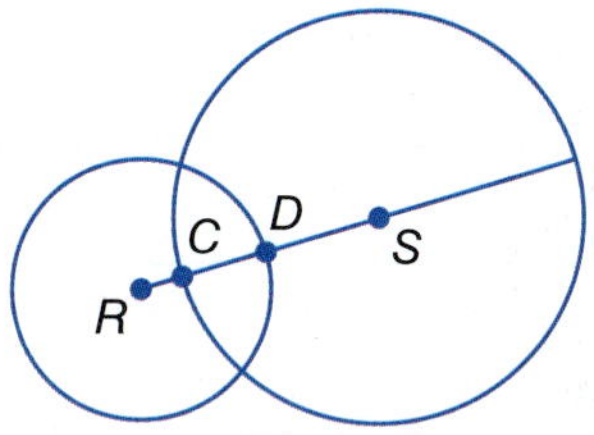

$CD + DS = CS$	**Segment Addition Postulate**
$CD + 9 = 15$	**Substitution**
$CD = 6$	**Subtract 9 from each side.**

Guided Practice

3. Use the diagram above to find RC.

Personal Tutor glencoe.com

Real-World Link

In 2005, Roger Federer and Andre Agassi played tennis on the helipad of the Burj Al Arab hotel in the United Arab Emirates. The helipad has a diameter of 79 feet and is nearly 700 feet high.

Source: Burj Al Arab, Emporis Buildings

Circumference The **circumference** of a circle is the distance around the circle. By definition, the ratio $\frac{C}{d}$ is an irrational number called **pi (π)**. Two formulas for circumference can be derived by using this definition.

$\frac{C}{d} = \pi$	**Definition of pi**
$C = \pi d$	**Multiply each side by *d*.**
$C = \pi(2r)$	**$d = 2r$**
$C = 2\pi r$	**Simplify.**

Key Concept

Circumference

Words If a circle has diameter d or radius r, the circumference C equals the diameter times pi or twice the radius times pi.

Symbols $C = \pi d$ or $C = 2\pi r$

Real-World EXAMPLE 4 Find Circumference

TENNIS Find the circumference of the helipad described at the left.

$C = \pi d$	**Circumference formula**
$= \pi(79)$	**Substitution**
$= 79\pi$	**Simplify.**
≈ 248.19	**Use a calculator.**

The circumference of the helipad is 79π feet or about 248.19 feet.

Guided Practice

Find the circumference of each circle described. Round to the nearest hundredth.

4A. radius = 2.5 centimeters

4B. diameter = 16 feet

Personal Tutor glencoe.com

These circumference formulas can also be used to determine the diameter and radius of a circle when the circumference is given.

StudyTip

Levels of Accuracy Since π is irrational, its value cannot be given as a terminating decimal. Using a value of 3 for π provides a quick estimate in calculations. Using a value of 3.14 or $\frac{22}{7}$ provides a closer approximation. For the most accurate approximation, use the π key on a calculator. Unless stated otherwise, assume that in this text, a calculator with a π key was used to generate answers.

EXAMPLE 5 Find Diameter and Radius

Find the diameter and radius of a circle to the nearest hundredth if the circumference of the circle is 106.4 millimeters.

$C = \pi d$	Circumference Formula	$r = \frac{1}{2}d$	Radius Formula
$106.4 = \pi d$	Substitution	$\approx \frac{1}{2}(33.87)$	$d \approx 33.87$
$\frac{106.4}{\pi} = d$	Divide each side by π.	≈ 16.94 mm	Use a calculator.
$33.87 \text{ mm} \approx d$	Use a calculator.		

Guided Practice

5. Find the diameter and radius of a circle to the nearest hundredth if the circumference of the circle is 77.8 centimeters.

Personal Tutor glencoe.com

A polygon is **inscribed** in a circle if all of its vertices lie on the circle. A circle is **circumscribed** about a polygon if it contains all the vertices of the polygon.

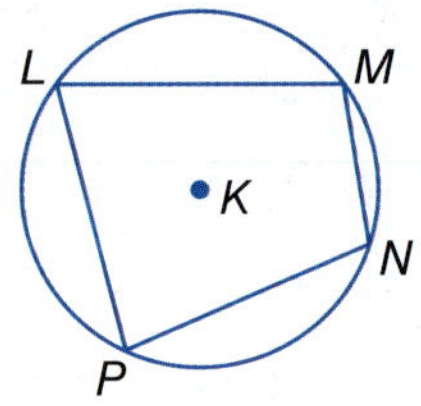

- Quadrilateral *LMNP* is *inscribed in* ⊙*K*.
- Circle *K* is *circumscribed about* quadrilateral *LMNP*.

STANDARDIZED TEST EXAMPLE 6

SHORT RESPONSE **A square with side length of 9 inches is inscribed in ⊙*J*. Find the exact circumference of ⊙*J*.**

Read the Test Item

You need to find the diameter of the circle and use it to calculate the circumference.

Solve the Test Item

First, draw a diagram. The diagonal of the square is the diameter of the circle and the hypotenuse of a right triangle.

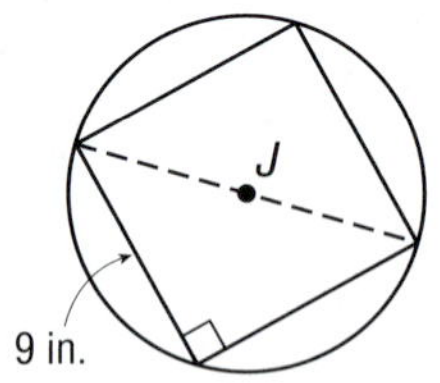

$a^2 + b^2 = c^2$	Pythagorean Theorem
$9^2 + 9^2 = c^2$	Substitution
$162 = c^2$	Simplify.
$9\sqrt{2} = c$	Take the positive square root of each side.

The diameter of the circle is $9\sqrt{2}$ inches.

Find the circumference in terms of π by substituting $9\sqrt{2}$ for d in $C = \pi d$. The exact circumference is $9\pi\sqrt{2}$ inches.

StudyTip

Circumcircle A *circumcircle* is a circle that passes through all of the vertices of a polygon.

Guided Practice

Find the exact circumference of each circle by using the given polygon.

6A. inscribed right triangle with legs 7 meters and 3 meters long

6B. circumscribed square with side 10 feet long

Personal Tutor glencoe.com

Check Your Understanding

Examples 1 and 2
pp. 683–684

For Exercises 1–4, refer to ⊙*N*.

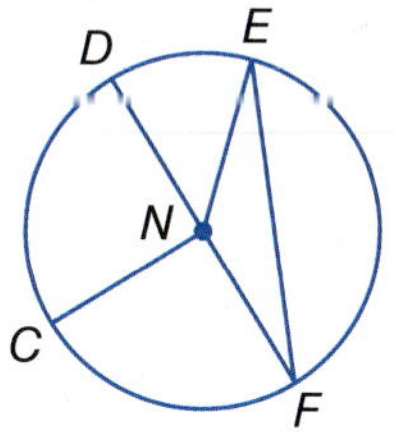

1. Name the circle.

2. Identify each.

a. a chord **b.** a diameter **c.** a radius

3. If $CN = 8$ centimeters, find DN.

4. If $EN = 13$ feet, what is the diameter of the circle?

Example 3
p. 685

The diameters of ⊙*A*, ⊙*B*, and ⊙*C* are 8 inches, 18 inches, and 11 inches, respectively. Find each measure.

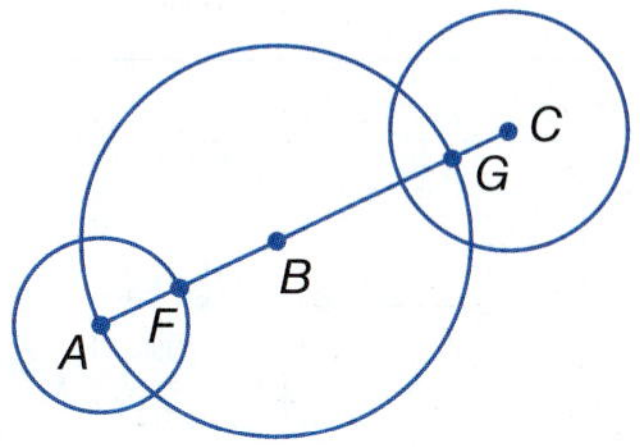

5. FG

6. FB

Example 4
p. 685

7. **RIDES** The circular ride described at the beginning of the lesson has a diameter of 44 feet. What are the radius and circumference of the ride? Round to the nearest hundredth, if necessary.

Example 5
p. 686

8. **POOLS** The circumference of the circular swimming pool shown is about 56.5 feet. What are the diameter and radius of the pool? Round to the nearest hundredth.

Example 6
p. 686

9. **SHORT RESPONSE** The right triangle shown is inscribed in ⊙*D*. Find the exact circumference of ⊙*D*.

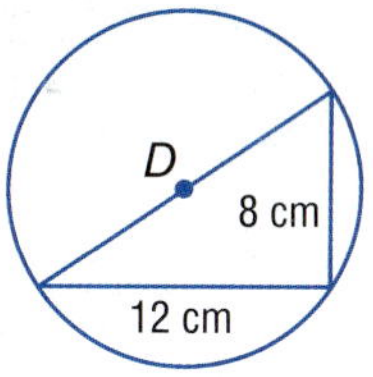

Practice and Problem Solving

● = **Step-by-Step Solutions** begin on page R20.
Extra Practice begins on page 969.

Examples 1 and 2
pp. 683–684

For Exercises 10–13, refer to ⊙*R*.

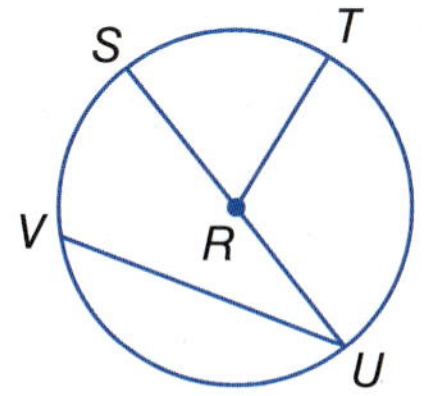

10. Name the center of the circle.

11. Identify a chord that is also a diameter.

12. Is $\overline{VU}$ a radius? Explain.

13. If $SU = 16.2$ centimeters, what is RT?

For Exercises 14–17, refer to ⊙*F*.

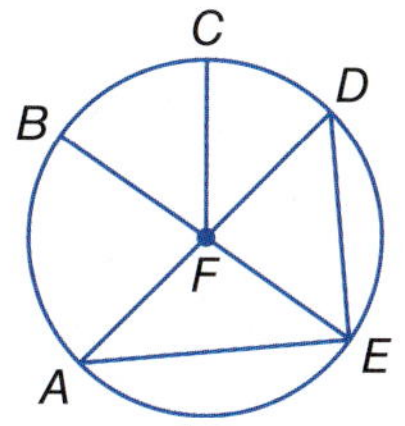

14. Identify a chord that is not a diameter.

15. If $CF = 14$ inches, what is the diameter of the circle?

16. Is $\overline{AF} \cong \overline{EF}$? Explain.

17. If $DA = 7.4$ centimeters, what is EF?

Example 3
p. 685

Circle J has a radius of 10 units, $\odot K$ has a radius of 8 units, and $BC = 5.4$ units. Find each measure.

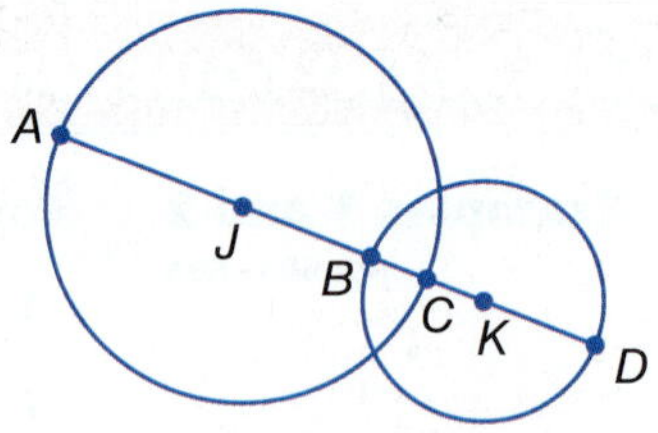

18. CK

19. AB

20. JK

21. AD

Example 4
p. 685

22. **PIZZA** Find the radius and circumference of the pizza shown. Round to the nearest hundredth, if necessary.

23. **BICYCLES** A bicycle has tires with a diameter of 26 inches. Find the radius and circumference of a tire. Round to the nearest hundredth, if necessary.

Example 5
p. 686

Find the diameter and radius of a circle with the given circumference. Round to the nearest hundredth.

24. $C = 18$ in.

25. $C = 124$ ft

26. $C = 375.3$ cm

27. $C = 2608.25$ m

Example 6
p. 686

Find the exact circumference of each circle by using the given inscribed or circumscribed polygon.

28.

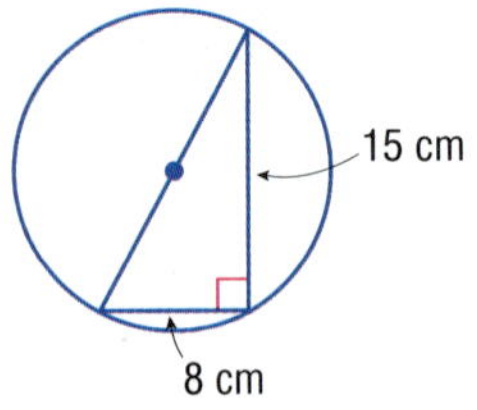

29.

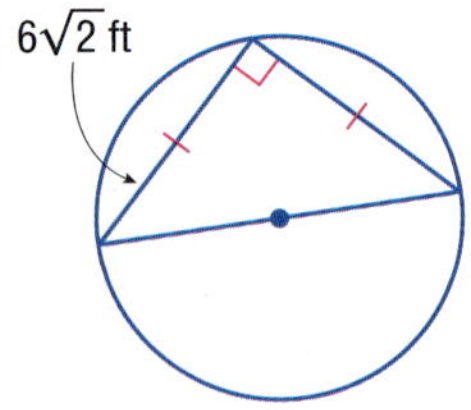

30.

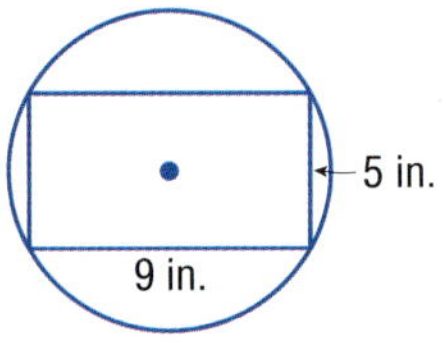

31.

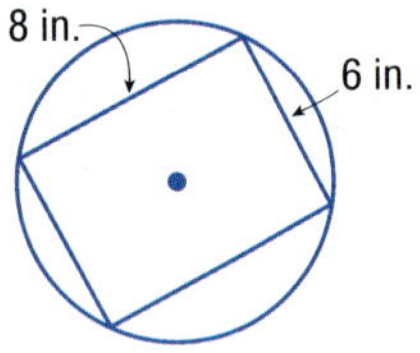

32.

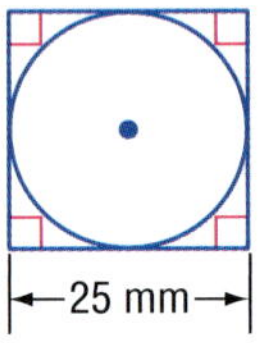

33.

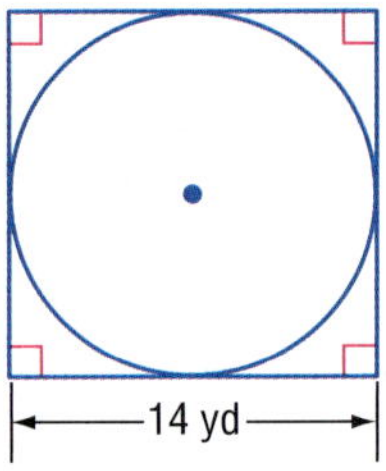

34. **DISC GOLF** Disc golf is similar to regular golf, except that a flying disc is used instead of a ball and clubs. For professional competitions, the maximum weight of a disc in grams is 8.3 times the diameter in centimeters. What is the maximum allowable weight for a disc with circumference 66.92 centimeters? Round to the nearest tenth.

35. **PATIOS** Mr. Martinez is going to build the patio shown.

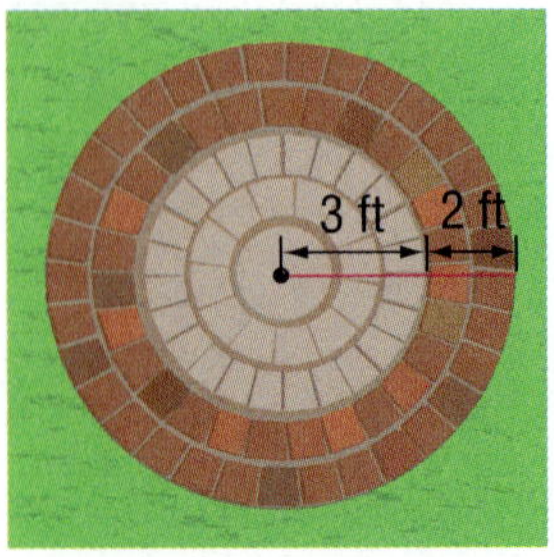

a. What is the patio's approximate circumference?

b. If Mr. Martinez changes the plans so that the inner circle has a circumference of approximately 25 feet, what should the radius of the circle be to the nearest foot?

The radius, diameter, or circumference of a circle is given. Find each missing measure to the nearest hundredth.

36. $d = 8\frac{1}{2}$ in., $r =$ __?__, $C =$ __?__

37. $r = 11\frac{2}{5}$ ft, $d =$ __?__, $C =$ __?__

38. $C = 35x$ cm, $d =$ __?__, $r =$ __?__

39. $r = \frac{x}{8}$, $d =$ _____, $C =$ __?__

Determine whether the circles in the figures below appear to be *congruent, concentric,* or *neither.*

40.

41.

42.

43. **HISTORY** The *Indian Shell Ring* on Hilton Head Island approximates a circle. If each unit on the coordinate grid represents 25 feet, how far would someone have to walk to go completely around the ring? Round to the nearest tenth.

44. **LANDSCAPING** A brick path is being installed around a circular pond. The pond has a circumference of 68 feet. The outer edge of the path is going to be 4 feet from the pond all the way around. What is the approximate circumference of the path? Round to the nearest hundredth.

45. **MULTIPLE REPRESENTATIONS** In this problem, you will explore changing dimensions in circles.

a. **GEOMETRIC** Use a compass to draw three circles in which the scale factor from each circle to the next larger circle is 1:2.

b. **TABULAR** Calculate the radius (to the nearest tenth) and circumference (to the nearest hundredth) of each circle. Record your results in a table.

c. **VERBAL** Explain why these three circles are geometrically similar.

d. **VERBAL** Make a conjecture about the ratio between the circumferences of two circles when the ratio between their radii is 2.

e. **ANALYTICAL** The scale factor from $\odot A$ to $\odot B$ is $\frac{b}{a}$. Write an equation relating the circumference (C_A) of $\odot A$ to the circumference (C_B) of $\odot B$.

f. **NUMERICAL** If the scale factor from $\odot A$ to $\odot B$ is $\frac{1}{3}$, and the circumference of $\odot A$ is 12 inches, what is the circumference of $\odot B$?

Real-World Link

A 150 pound person can burn 240 calories jogging 6 mph for 20 minutes, two times more than the same person would burn walking 4.5 mph for the same amount of time.

Source: Health Discovery

46. **EXERCISE** An indoor running track is shown.

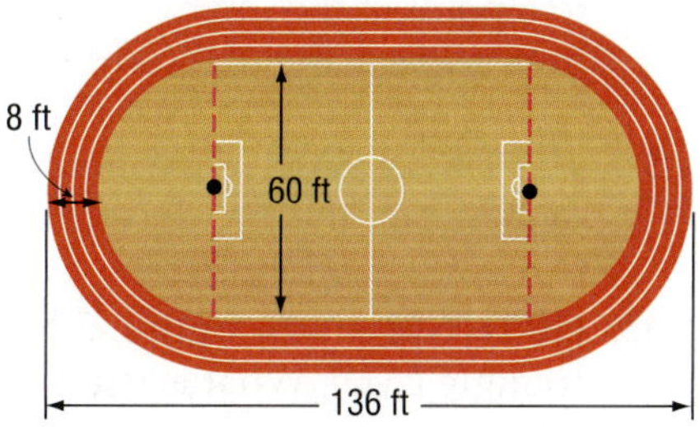

a. How much farther would someone running a lap on the outside lane travel than someone running a lap on the inside lane?

b. About how many times would a person have to run around the outside lane of the track in order to run a mile?

Real-World Link

You can use concentric circles when planning a vacation. Choose a place to stay where there are several things that you want to do within a 30-minute driving radius and spend most of your time in that circle. You will minimize your travel time by choosing the location for your lodging wisely.

Source: Slow Travel

47 **MAPS** The concentric circles on the map below show the areas that are 5, 10, 15, 20, 25, and 30 miles from downtown Phoenix.

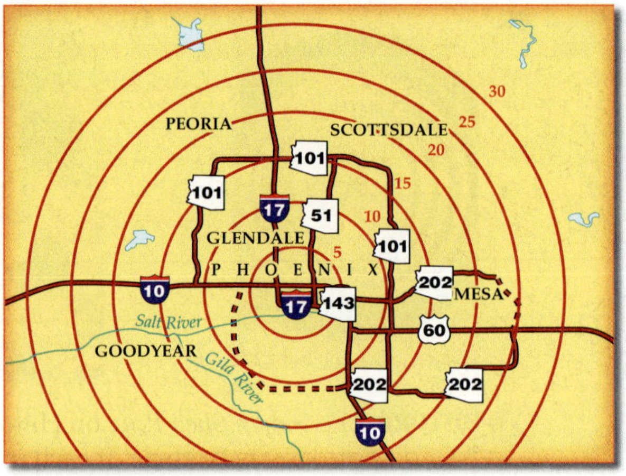

a. How much greater is the circumference of the outermost circle than the circumference of the center circle?

b. As the radii of the circles increase by 5 miles, by how much does the circumference increase?

H.O.T. Problems

Use **H**igher-**O**rder **T**hinking Skills

48. **OPEN ENDED** Draw a circle with a circumference between 8 and 12 centimeters. What is the radius of the circle?

49. **ERROR ANALYSIS** Cassie and Kendrick are each drawing a figure to represent the set of all points 4 centimeters from point *J*. Is either of them correct? Explain your reasoning.

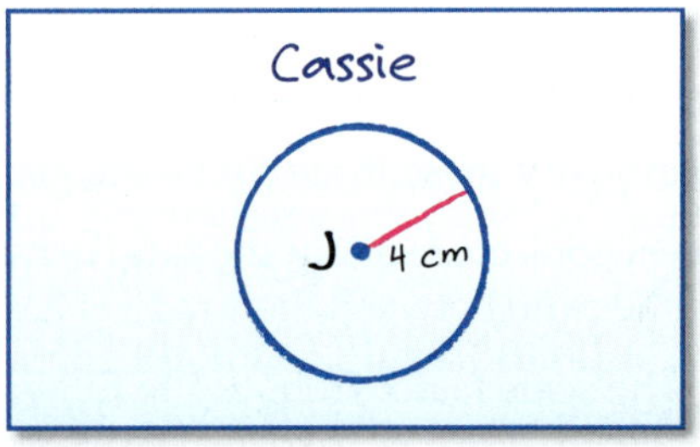

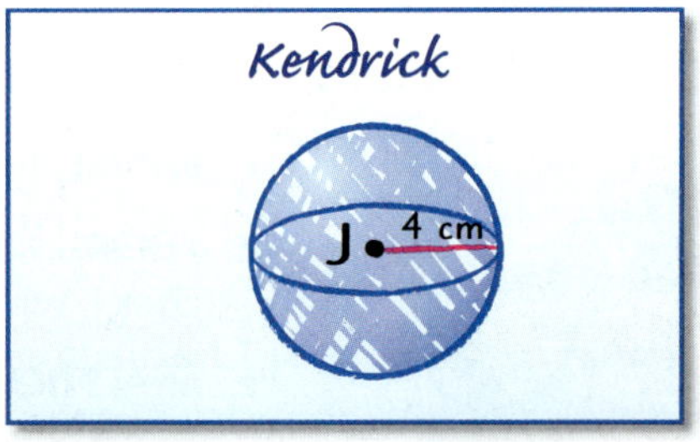

50. **CHALLENGE** The sum of the circumferences of circles *H*, *J*, and *K* shown at the right is 56π units. Find *KJ*.

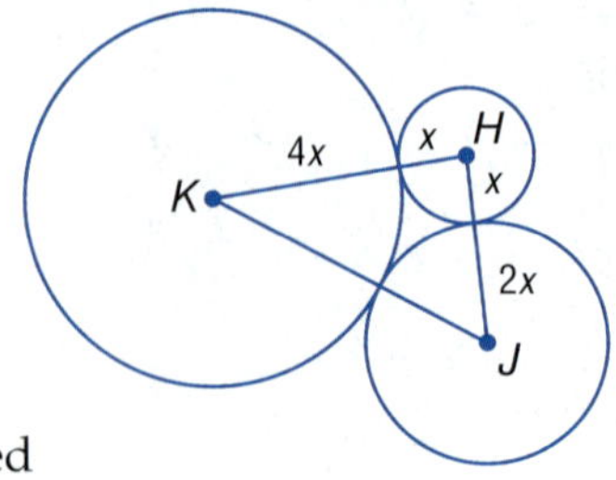

51. **REASONING** Is the distance from the center of a circle to a point in the interior of a circle *sometimes*, *always*, or *never* less than the radius of the circle? Explain.

52. **WRITING IN MATH** Give examples of how distance traveled depends on the circumference of a circle when used with vehicles.

53. **CHALLENGE** In the figure, $\odot P$ is inscribed in equilateral triangle *LMN*. What is the circumference of $\odot P$?

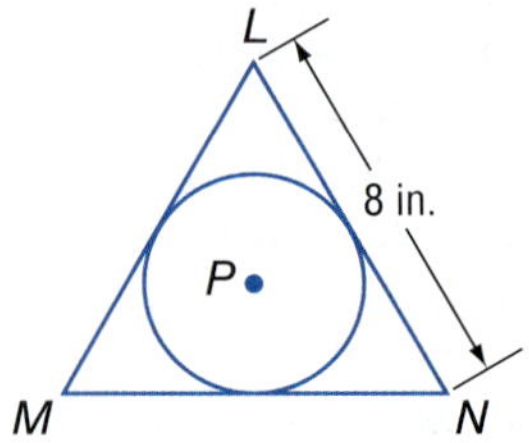

54. **WRITING IN MATH** Compare and contrast *congruent circles* and *concentric circles*.

Standardized Test Practice

55. GRIDDED RESPONSE What is the circumference of ⊙T? Round to the nearest tenth.

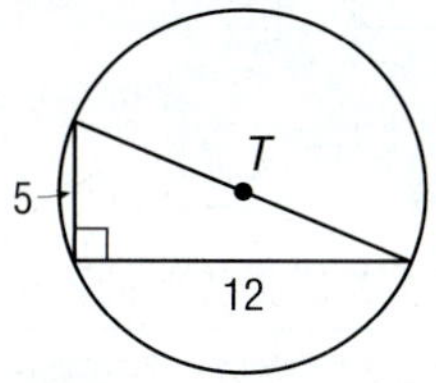

56. What is the radius of a table with a circumference of 10 feet?

A 1.6 ft **C** 3.2 ft
B 2.5 ft **D** 5 ft

57. ALGEBRA Bill is planning a circular vegetable garden with a fence around the border. If he can use up to 50 feet of fence, what radius can he use for the garden?

F 10 **H** 8
G 9 **J** 7

58. SAT/ACT What is the radius of a circle with an area of $\frac{\pi}{4}$ square units?

A 0.4 units **C** 2 units
B 0.5 units **D** 4 units

Spiral Review

Copy each figure and point B. Then use a ruler to draw the image of the figure under a dilation with center B and the scale factor r indicated. (Lesson 9-6)

59. $r = \frac{1}{5}$

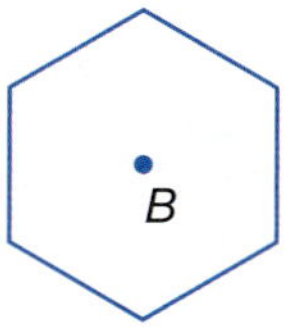

60. $r = \frac{2}{5}$

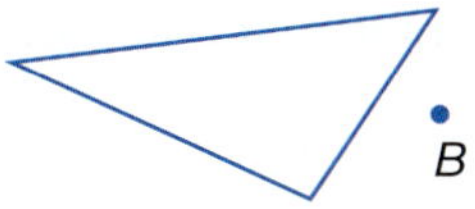

61. $r = 2$

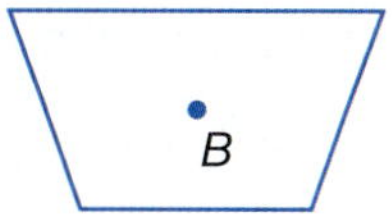

62. $r = 3$

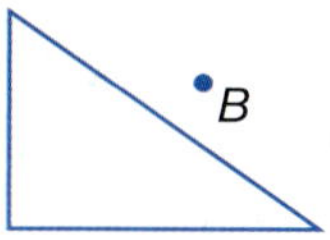

State whether each figure has rotational symmetry. If so, copy the figure, locate the center of symmetry, and state the order and magnitude of symmetry. (Lesson 9-5)

63.

64.

65.

66.

Determine the truth value of the following statement for each set of conditions. Explain your reasoning. (Lesson 2-2)

If you are over 18 years old, then you vote in all elections.

67. You are 19 years old and you vote.

68. You are 21 years old and do not vote.

Skills Review

Find x. (Lesson 1-5)

69.

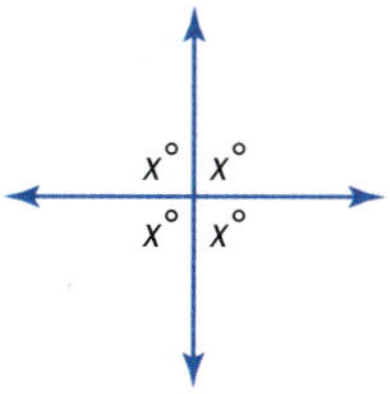

70.

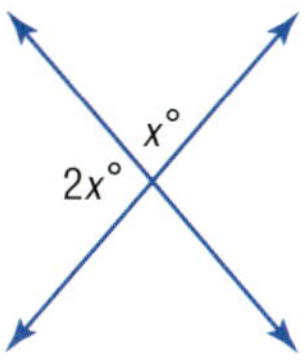

71.

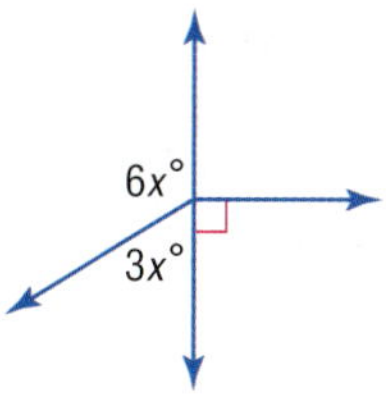

72.

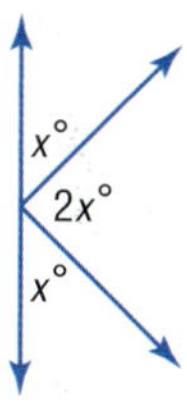

10-2 Measuring Angles and Arcs

Then

You measured angles and identified congruent angles. (Lesson 1-4)

Now

- Identify central angles, major arcs, minor arcs, and semicircles, and find their measures.
- Find arc lengths.

New Vocabulary

central angle
arc
minor arc
major arc
semicircle
congruent arcs
adjacent arcs
arc length

Math Online
glencoe.com

Why?

The thirteen stars of the Betsy Ross flag are arranged equidistant from each other and from a fixed point. The distance between consecutive stars varies depending on the size of the flag, but the measure of the central angle formed by the center of the circle and any two consecutive stars is always the same.

The Granger Collection, New York

Angles and Arcs A **central angle** of a circle is an angle with a vertex in the center of the circle. Its sides contain two radii of the circle. $\angle ABC$ is a central angle of $\odot B$.

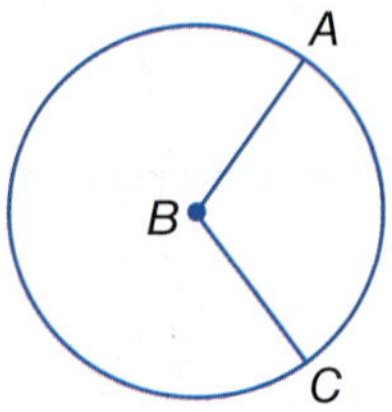

Recall from Lesson 1-4 that a *degree* is $\frac{1}{360}$ of the circular rotation about a point. This leads to the following relationship.

Key Concept — Sum of Central Angles

Words The sum of the measures of the central angles of a circle with no interior points in common is 360.

Example $m\angle 1 + m\angle 2 + m\angle 3 = 360$

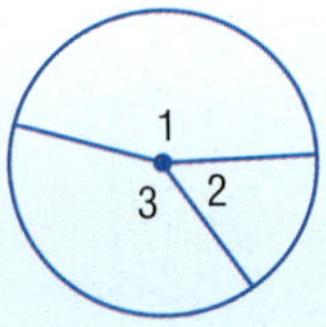

EXAMPLE 1 Find Measures of Central Angles

Find the value of *x*.

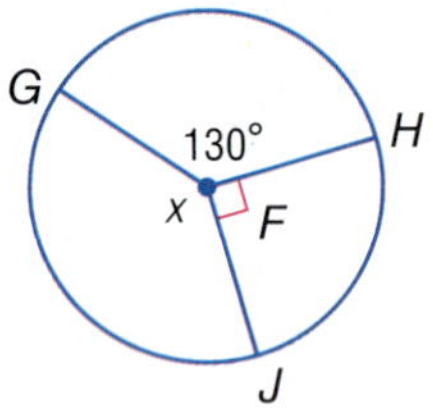

$m\angle GFH + m\angle HFJ + m\angle GFJ = 360$	**Sum of Central Angles**
$130 + 90 + m\angle GFJ = 360$	**Substitution**
$220 + m\angle GFJ = 360$	**Simplify.**
$m\angle GFJ = 140$	**Subtract 220 from each side.**

Guided Practice

1A.

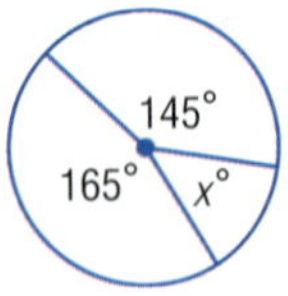

1B.

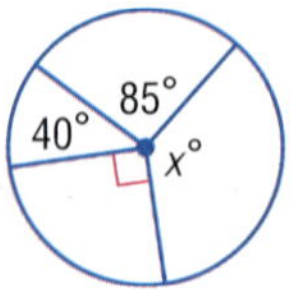

Personal Tutor glencoe.com

An **arc** is a portion of a circle defined by two endpoints. A central angle separates the circle into two arcs with measures related to the measure of the central angle.

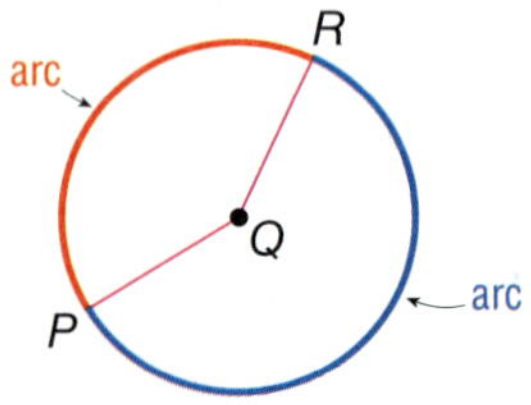

Key Concept Arcs and Arc Measure

Arc	Measure	
A **minor arc** is the shortest arc connecting two endpoints on a circle.	The measure of a minor arc is less than 180 and equal to the measure of its related central angle. $m\widehat{AB} = m\angle ACB = x$	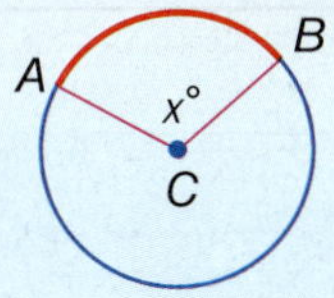
A **major arc** is the longest arc connecting two endpoints on a circle.	The measure of a major arc is greater than 180, and equal to 360 minus the measure of the minor arc with the same endpoints. $m\widehat{ADB} = 360 - m\widehat{AB} = 360 - x$	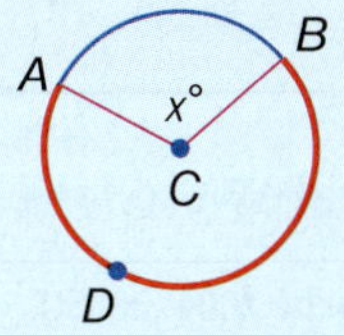
A **semicircle** is an arc with endpoints that lie on a diameter.	The measure of a semicircle is 180. $m\widehat{ADB} = 180$	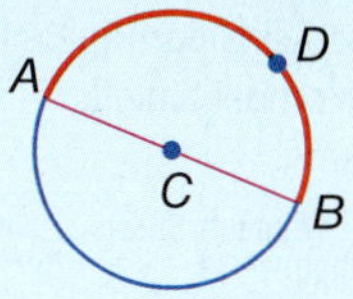

StudyTip

Naming Arcs Minor arcs are named by their endpoints. Major arcs and semicircles are named by their endpoints and another point on the arc that lies between these endpoints.

EXAMPLE 2 Classify Arcs and Find Arc Measures

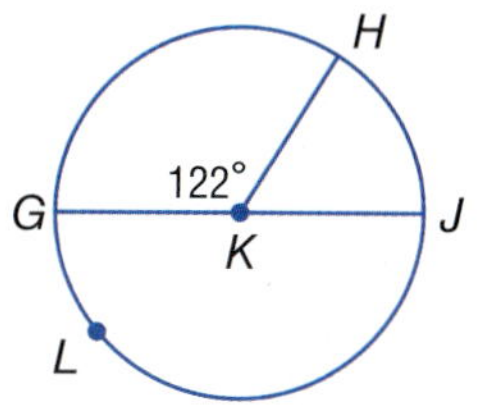

$\overline{GJ}$ is a diameter of ⊙K. Identify each arc as a *major arc*, *minor arc*, or *semicircle*. Then find its measure.

a. $m\widehat{GH}$

$\widehat{GH}$ is a minor arc, so $m\widehat{GH} = m\angle GKH$ or 122.

b. $m\widehat{GLH}$

$\widehat{GLH}$ is a major arc that shares the same endpoints as minor arc $\widehat{GH}$.

$m\widehat{GHL} = 360 - m\widehat{GH}$

$= 360 - 122$ or 238

c. $m\widehat{GLJ}$

$\widehat{GLJ}$ is a semicircle, so $m\widehat{GLJ} = 180$.

Math History Link

Euclid (c. 325–265 B.C.)
The 13 books of Euclid's *Elements* are influential works of science. In them, geometry and other branches of mathematics are logically developed. Book 3 of *Elements* is devoted to circles, arcs, and angles.

Guided Practice

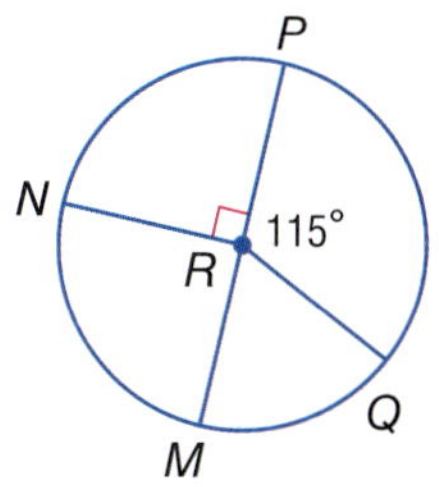

$\overline{PM}$ is a diameter of ⊙R. Identify each arc as a *major arc*, *minor arc*, or *semicircle*. Then find its measure.

2A. $\widehat{MQ}$ **2B.** $\widehat{MNP}$ **2C.** $\widehat{MNQ}$

Personal Tutor glencoe.com

Congruent arcs are arcs in the same or congruent circles that have the same measure.

Theorem 10.1

Words In the same circle or in congruent circles, two minor arcs are congruent if and only if their central angles are congruent.

Example If $\angle 1 \cong \angle 2$, then $\widehat{FG} \cong \widehat{HJ}$.
If $\widehat{FG} \cong \widehat{HJ}$, then $\angle 1 \cong \angle 2$.

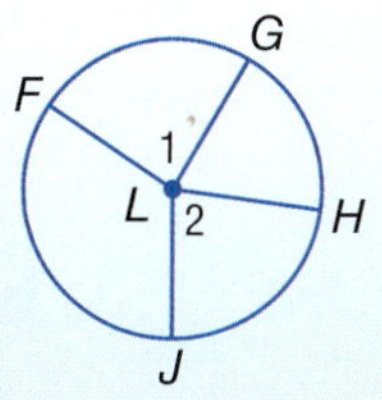

You will prove Theorem 10.1 in Exercise 52.

Real-World Link

During the 2004–2005 school year, approximately 3 million girls participated in high school athletic programs.

Source: Women's Sports Foundation

Real-World EXAMPLE 3 Find Arc Measures in Circle Graphs

SPORTS **Refer to the circle graph. Find each measure.**

Female Participation in Sports

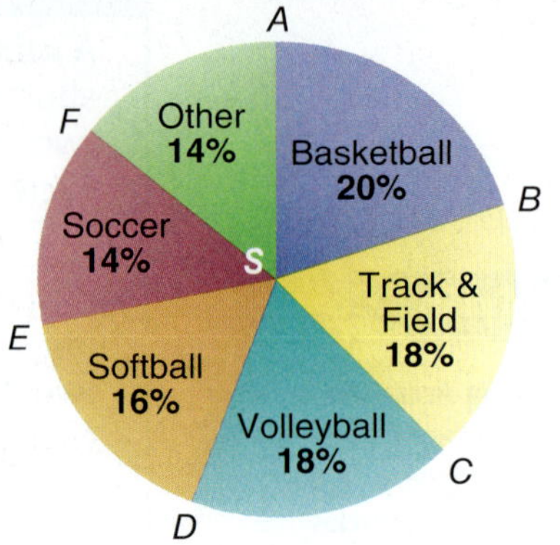

a. $m\widehat{CD}$

$\widehat{CD}$ is a minor arc. $m\widehat{CD} = m\angle CSD$

$\angle CSD$ represents 18% of the whole, or 18% of the circle.

$m\angle CSD = 0.18(360)$ **Find 18% of 360.**

$= 64.8$ **Simplify.**

b. $m\widehat{BC}$

The percents for volleyball and track and field are equal, so the central angles are congruent and the corresponding arcs are congruent.

$m\widehat{BC} = m\widehat{CD} = 64.8$

Guided Practice

3A. $m\widehat{EF}$ **3B.** $m\widehat{FA}$

Personal Tutor glencoe.com

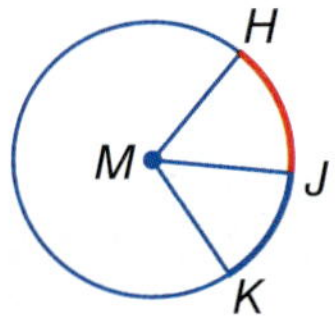

Adjacent arcs are arcs in a circle that have exactly one point in common. In $\odot M$, $\widehat{HJ}$ and $\widehat{JK}$ are adjacent arcs. As with adjacent angles, you can add the measures of adjacent arcs.

Postulate 10.1 Arc Addition Postulate

Words The measure of an arc formed by two adjacent arcs is the sum of the measures of the two arcs.

Example $m\widehat{XYZ} = m\widehat{XY} + m\widehat{YZ}$

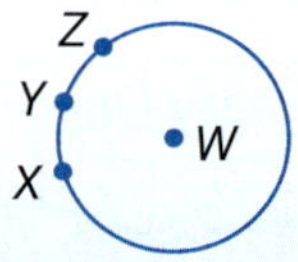

EXAMPLE 4 Use Arc Addition to Find Measures of Arcs

Find each measure in $\odot F$.

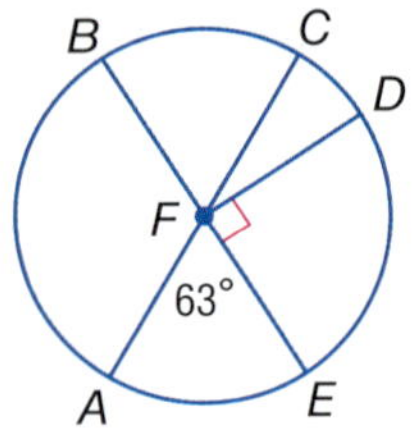

a. $m\widehat{AED}$

$m\widehat{AED} = m\widehat{AE} + m\widehat{ED}$ **Arc Addition Postulate**

$= m\angle AFE + m\angle EFD$ **$m\widehat{AE} = m\angle AFE$, $m\widehat{ED} = m\angle EFD$**

$= 63 + 90$ or 153 **Substitution**

b. $m\widehat{ADB}$

$m\widehat{ADB} = m\widehat{AE} + m\widehat{EDB}$ **Arc Addition Postulate**

$= 63 + 180$ or 243 **$\widehat{EDB}$ is a semicircle, so $m\widehat{EDB} = 180$.**

Guided Practice

4A. $m\widehat{CE}$ **4B.** $m\widehat{ABD}$

Personal Tutor glencoe.com

Arc Length Arc length is the distance between the endpoints along an arc measured in linear units. Since an arc is a portion of a circle, its length is a fraction of the circumference.

Watch Out!

Arc Length The length of an arc is given in linear units, such as centimeters. The measure of an arc is given in degrees.

Key Concept: Arc Length

Words The ratio of the **length of an arc ℓ** to the **circumference** of the circle is equal to the ratio of the **degree measure of the arc** to 360.

Proportion $\frac{\ell}{2\pi r} = \frac{x}{360}$ or

Equation $\ell = \frac{x}{360} \cdot 2\pi r$

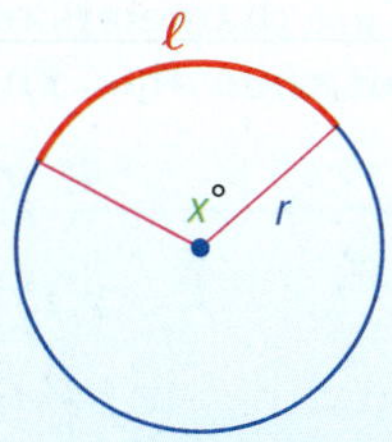

EXAMPLE 5 Find Arc Length

Find the length of $\widehat{ZY}$. Round to the nearest hundredth.

a.

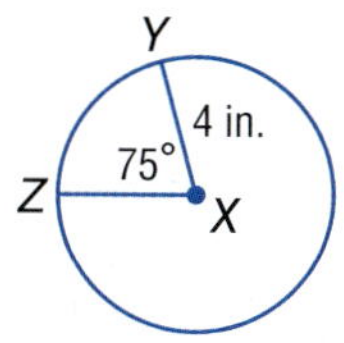

$\ell = \frac{x}{360} \cdot 2\pi r$	Arc Length Equation
$= \frac{75}{360} \cdot 2\pi(4)$	Substitution
≈ 5.24 in.	Use a calculator.

b.

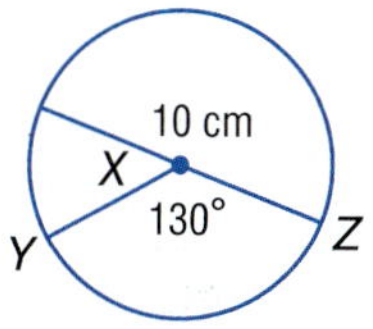

$\ell = \frac{x}{360} \cdot 2\pi r$	Arc Length Equation
$= \frac{130}{360} \cdot 2\pi(5)$	Substitution
≈ 11.34 cm	Use a calculator.

c.

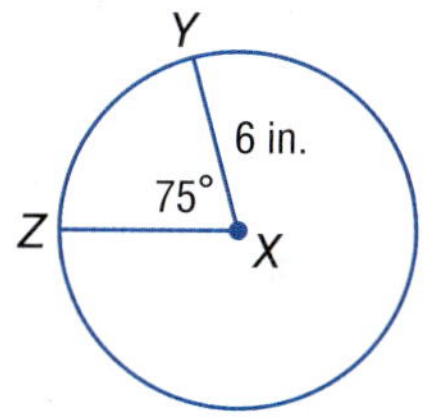

$\ell = \frac{x}{360} \cdot 2\pi r$	Arc Length Equation
$= \frac{75}{360} \cdot 2\pi(6)$	Substitution
≈ 7.85 in.	Use a calculator.

Notice that $\widehat{ZY}$ has the same measure, 75, in both Examples 5a and 5c. The arc lengths, however, are different. This is because they are in circles that have different radii.

StudyTip

Alternate Method The arc lengths in Examples 5a, 5b, and 5c could also have been calculated using the arc length proportion $\frac{\ell}{2\pi r} = \frac{x}{360}$.

Guided Practice

Find the length of $\widehat{AB}$. Round to the nearest hundredth.

5A.

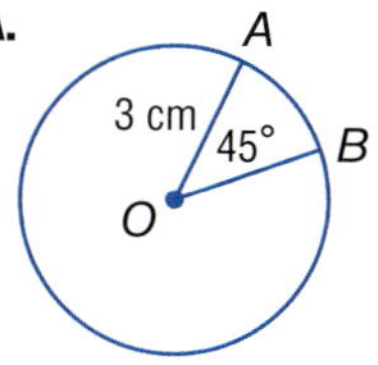

5B.

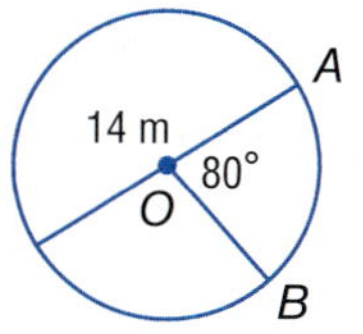

5C.

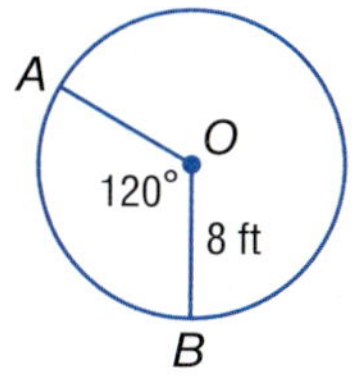

Personal Tutor glencoe.com

Check Your Understanding

Example 1 p. 692

Find the value of x.

1.

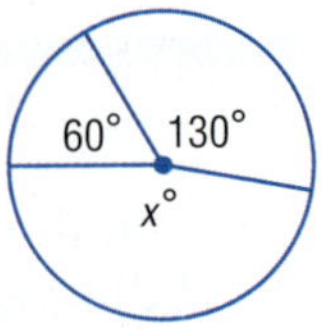

2.

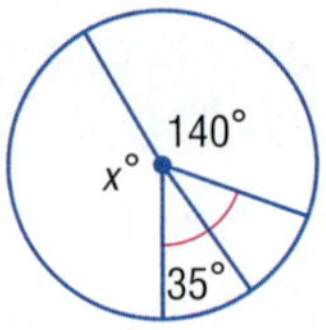

Example 2 p. 693

$\overline{HK}$ and $\overline{IG}$ are diameters of $\odot L$. Identify each arc as a *major arc*, *minor arc*, or *semicircle*. Then find its measure.

3. $m\widehat{IHJ}$ **4.** $m\widehat{HI}$ **5.** $m\widehat{HGK}$

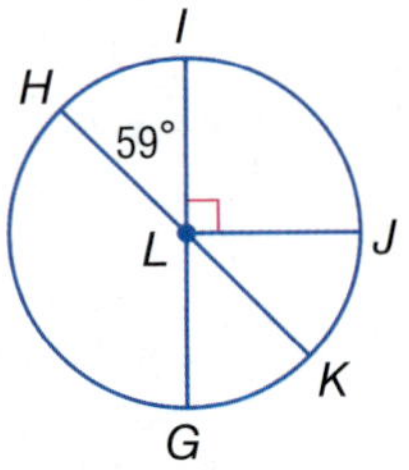

Example 3 p. 694

6. **RESTAURANTS** The graph shows the results of a survey taken by diners relating what is most important about the restaurants where they eat.

a. Find $m\widehat{AB}$.

b. Find $m\widehat{BC}$.

c. Describe the type of arc that the category Great Food represents.

What Diners Want

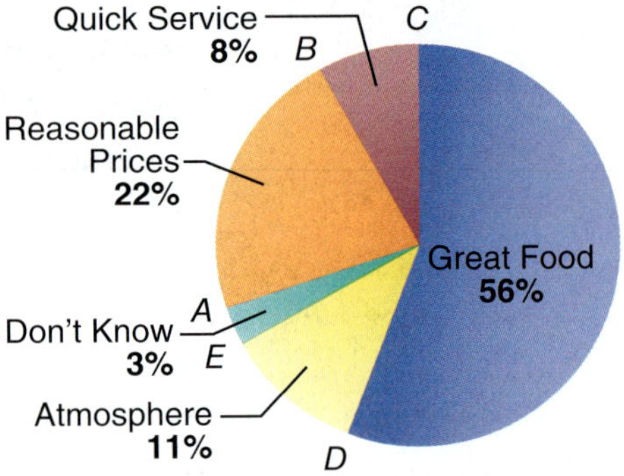

Source: *USA TODAY*

Example 4 p. 694

$\overline{QS}$ is a diameter of $\odot V$. Find each measure.

7. $m\widehat{STP}$

8. $m\widehat{QRT}$

9. $m\widehat{PQR}$

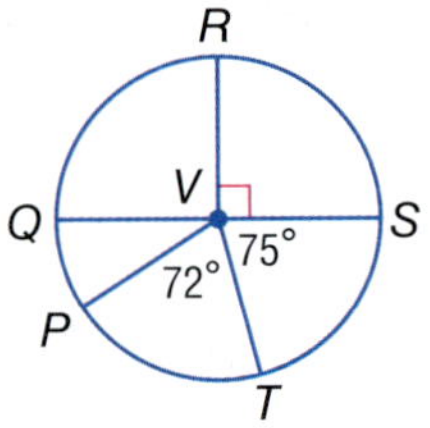

Example 5 p. 695

Find the length of $\widehat{JK}$. Round to the nearest hundredth.

10.

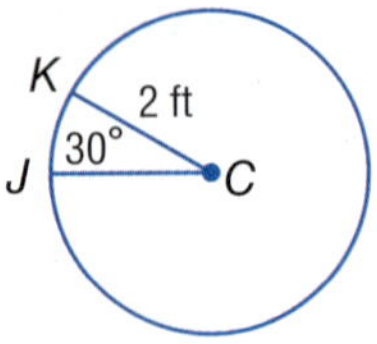

11.

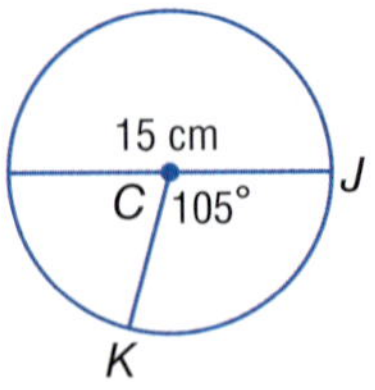

Practice and Problem Solving

● = **Step-by-Step Solutions** begin on page R20.
Extra Practice begins on page 969.

Example 1 p. 692

Find the value of x.

12.

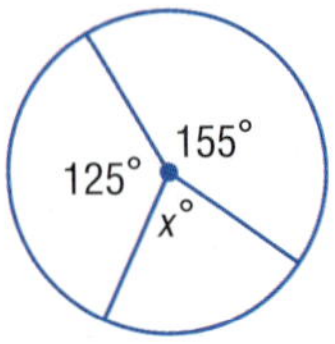

13

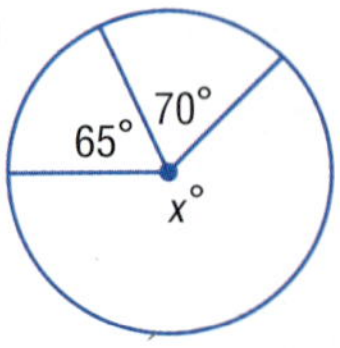

14.

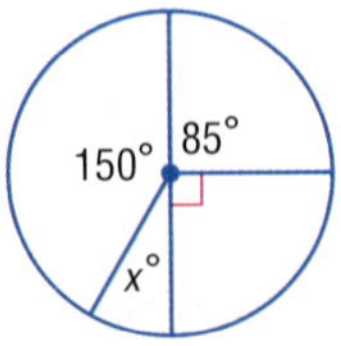

15.

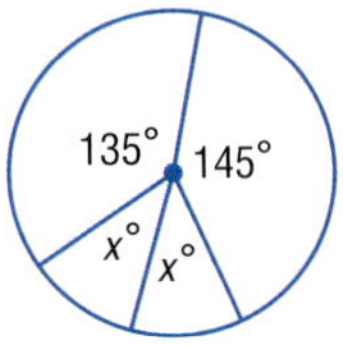

Example 2
p. 693

$\overline{AD}$ **and** $\overline{CG}$ **are diameters of** $\odot B$**. Identify each arc as a *major arc, minor arc,* or *semicircle.* Then find its measure.**

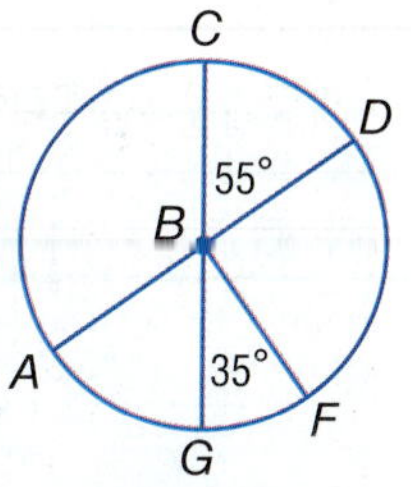

16. $m\widehat{CD}$ **17.** $m\widehat{AC}$ **18.** $m\widehat{CG}$

19. $m\widehat{CGD}$ **20.** $m\widehat{GCF}$ **21.** $m\widehat{ACD}$

22. $m\widehat{AG}$ **23.** $m\widehat{ACF}$

Example 3
p. 694

24. SHOPPING The graph shows the results of a survey in which teens were asked where the best place was to shop for clothes.

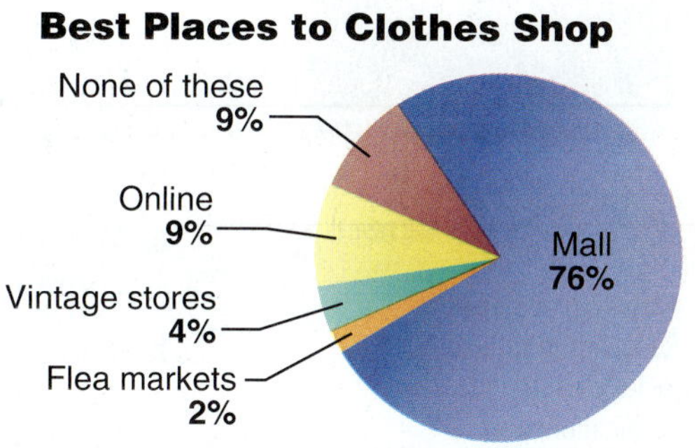

a. What would be the arc measures associated with the mall and vintage stores categories?

b. Describe the kinds of arcs associated with the first category and the last category.

c. Are there any congruent arcs in this graph? Explain.

Real-World Link

In a recent year, teenagers spent $33.1 billion on apparel and footwear.

Source: NPD Group

25. FOOD The table shows the results of a survey in which Americans were asked how long food could be on the floor and still be safe to eat.

Dropped Food	
Do you eat food dropped on the floor?	
Not safe to eat	78%
Three-second rule*	10%
Five-second rule*	8%
Ten-second rule*	4%

Source: American Diabetic Association
* The length of time the food is on the floor.

a. If you were to construct a circle graph of this information, what would be the arc measures associated with the first two categories?

b. Describe the kind of arcs associated with the first category and the last category.

c. Are there any congruent arcs in this graph? Explain.

Examples 2 and 4
pp. 693–694

ENTERTAINMENT Use the Ferris wheel shown to find each measure.

26. $m\widehat{FG}$ **27.** $m\widehat{JH}$

28. $m\widehat{JKF}$ **29.** $m\widehat{JFH}$

30. $m\widehat{GHF}$ **31.** $m\widehat{GHK}$

32. $m\widehat{HK}$ **33.** $m\widehat{JKG}$

34. $m\widehat{KFH}$ **35.** $m\widehat{HGF}$

Example 5
p. 695

Use $\odot P$ **to find the length of each arc. Round to the nearest hundredth.**

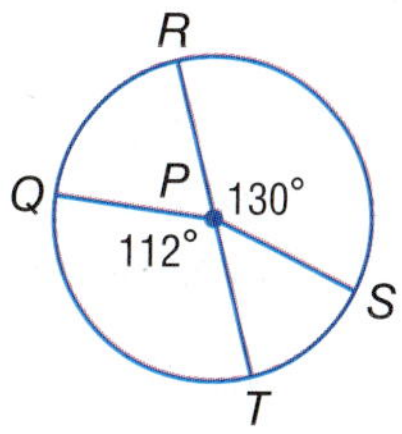

36. $\widehat{RS}$, if the radius is 2 inches

37 $\widehat{QT}$, if the diameter is 9 centimeters

38. $\widehat{QR}$, if $PS = 4$ millimeters

39. $\widehat{RS}$, if $RT = 15$ inches

40. $\widehat{QRS}$, if $RT = 11$ feet

41. $\widehat{RTS}$, if $PQ = 3$ meters

Real-World Career

Historical Researcher
Research in museums includes authentication, verification, and description of artifacts. Employment as a historical researcher requires a minimum of a bachelor's degree in history.

HISTORY The figure shows the stars in the Betsy Ross flag referenced at the beginning of the lesson.

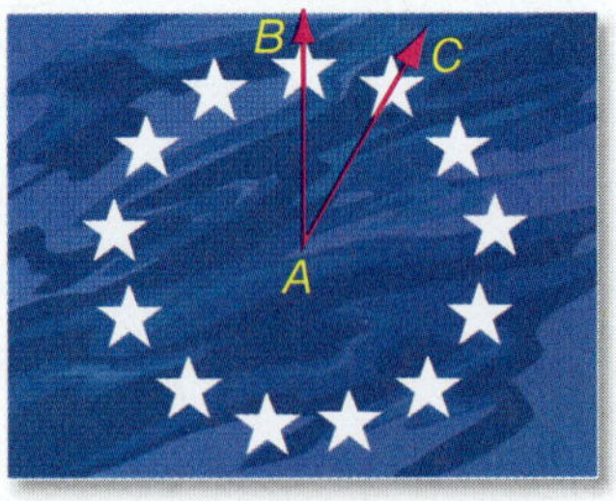

42. What is the measure of central angle A? Explain how you determined your answer.

43. If the diameter of the circle were doubled, what would be the effect on the arc length from the center of one star B to the next star C?

44. **FARMS** The *Pizza Farm* in Madera, California, is a circle divided into eight equal slices, as shown at the right. Each "slice" is used for growing or grazing pizza ingredients.

a. What is the total arc measure of the slices containing olives, tomatoes, and peppers?

b. The circle is 125 feet in diameter. What is the arc length of one slice? Round to the nearest hundredth.

Find each measure. Round each linear measure to the nearest hundredth and each arc measure to the nearest degree.

45. circumference of $\odot S$

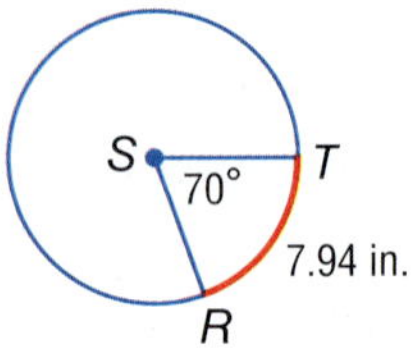

46. $m\widehat{CD}$

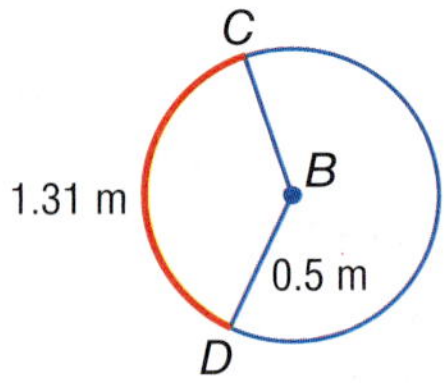

47. radius of $\odot K$

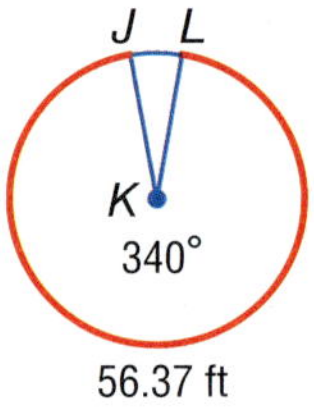

ALGEBRA **In $\odot C$, $m\angle HCG = 2x$ and $m\angle HCD = 6x + 28$. Find each measure.**

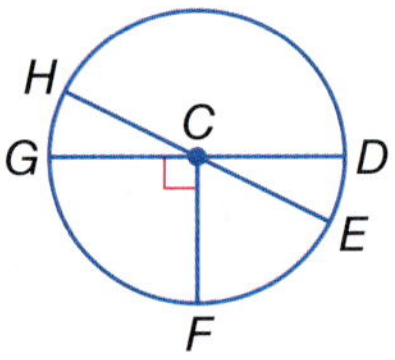

48. $m\widehat{EF}$

49. $m\widehat{HD}$

50. $m\widehat{HGF}$

51 **RIDES** A pirate ship ride follows a semi-circular path, as shown in the diagram.

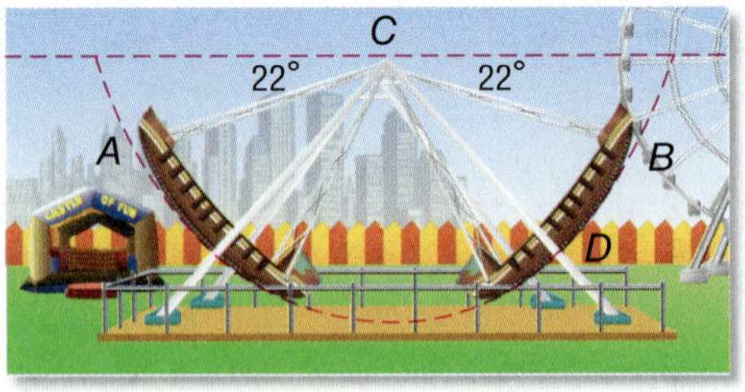

a. What is $m\widehat{AB}$?

b. If $CD = 62$ feet, what is the length of $\widehat{AB}$? Round to the nearest hundredth.

52. **PROOF** Write a two-column proof of Theorem 10.1.

Given: $\angle BAC \cong \angle DAE$

Prove: $\widehat{BC} \cong \widehat{DE}$

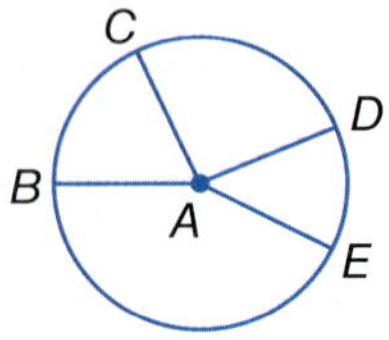

53. **COORDINATE GEOMETRY** In the graph, point M is located at the origin. Find each measure in $\odot M$. Round each linear measure to the nearest hundredth and each arc measure to the nearest tenth degree.

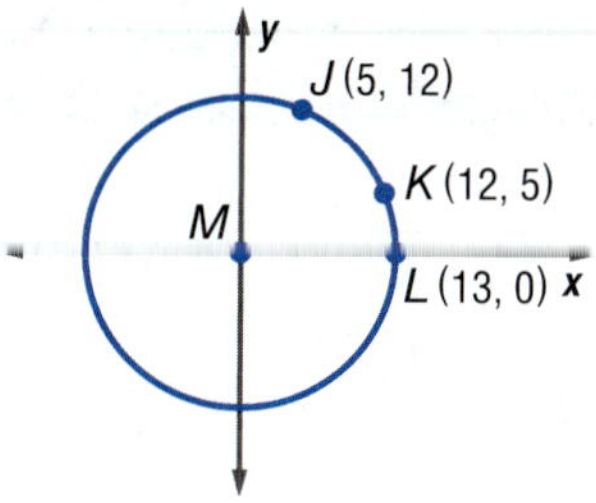

a. $m\widehat{JL}$ **b.** $m\widehat{KL}$ **c.** $m\widehat{JK}$

d. length of $\widehat{JL}$ **e.** length of $\widehat{JK}$

54. **MULTIPLE REPRESENTATIONS** In this problem, you will investigate the relationship between chords and arcs.

a. **GEOMETRIC** Draw a circle with congruent chords $\overline{AB}$ and $\overline{CD}$. Find the center of the circle. Repeat the process for two additional circles using different chord lengths.

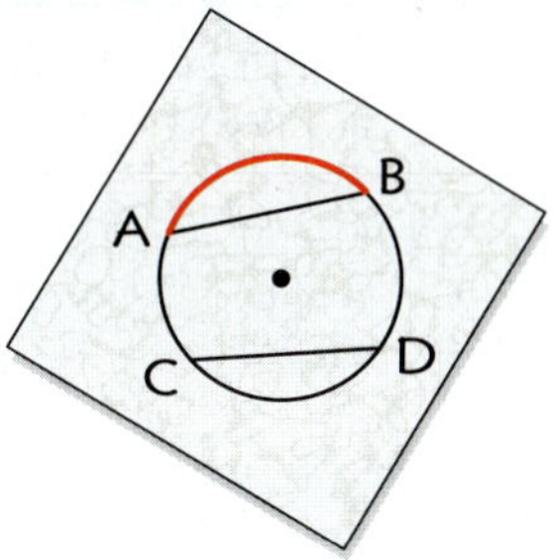

b. **CONCRETE** Cut a piece of patty paper larger than each circle and attach the patty paper at the center of each circle with a push pin. Trace the arc bounded by one of the segments on the patty paper. Rotate the patty paper around the push pin to compare the arc length that you traced to the arc formed by the second segment. Repeat this process for each circle.

c. **VERBAL** Make a conjecture about the relationship between the arcs bounded by congruent chords of a circle. Prove your conjecture.

H.O.T. Problems Use Higher-Order Thinking Skills

55. **ERROR ANALYSIS** Brody says that $\widehat{WX}$ and $\widehat{YZ}$ are congruent since their central angles have the same measure. Selena says they are not congruent. Is either of them correct? Explain your reasoning.

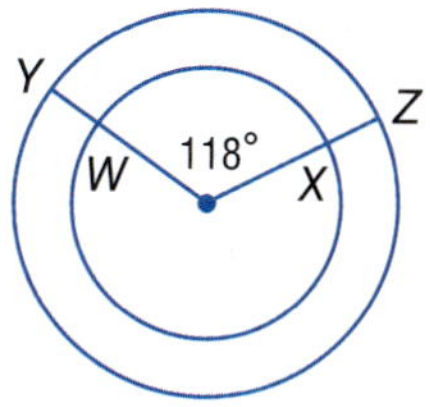

REASONING **Determine whether each statement is *sometimes, always,* or *never* true. Explain your reasoning.**

56. The measure of a minor arc is less than 180.

57. If a central angle is obtuse, its corresponding arc is a major arc.

58. The sum of the measures of adjacent arcs of a circle depends on the measure of the radius.

59. **CHALLENGE** The measures of $\widehat{LM}$, $\widehat{MN}$, and $\widehat{NL}$ are in the ratio 5:3:4. Find the measure of each arc.

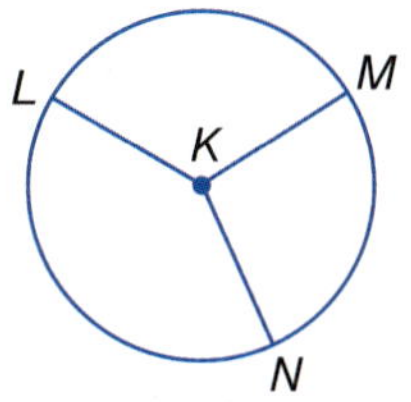

Watch Out!

Proportions
Remember that the order of the factors matters in proportions. For example, if the ratio of A to B to C is 1 to 2 to 3, $A:B:C = 1:2:3$.

60. **OPEN ENDED** Draw a circle and locate three points on the circle. Estimate the measures of the three nonoverlapping arcs that are formed. Then use a protractor to find the measure of each arc. Label your circle with the arc measures.

61. **CHALLENGE** The time shown on an analog clock is 8:10. What is the measure of the angle formed by the hands of the clock?

62. **WRITING IN MATH** Describe the three different types of arcs in a circle and the method for finding the measure of each one.

Standardized Test Practice

63. What is the value of x?

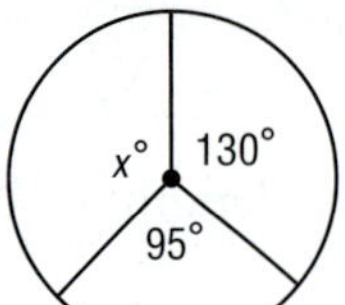

A 120

B 135

C 145

D 160

64. GRIDDED RESPONSE In $\odot B$, $m\angle LBM = 3x$ and $m\angle LBQ = 4x + 61$. What is the measure of $\angle PBQ$?

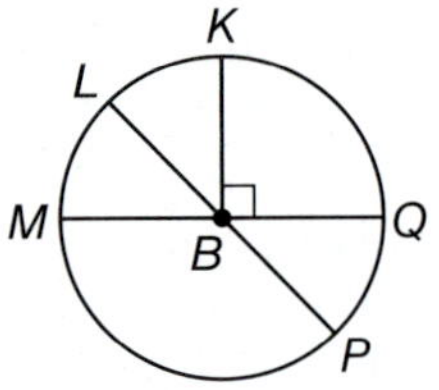

65. ALGEBRA A rectangle's width is represented by x and its length by y. Which expression best represents the area of the rectangle if the length and width are tripled?

F $3xy$

G $3(xy)^2$

H $9xy$

J $(xy)^3$

66. SAT/ACT What is the area of the shaded region if $r = 4$?

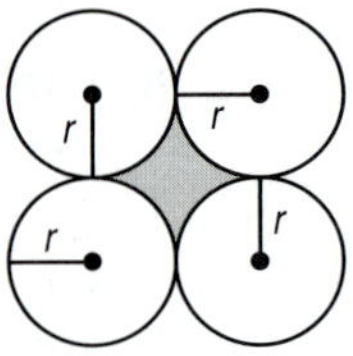

A $64 - 16\pi$

B $16 - 16\pi$

C $16 - 8\pi$

D $64 - 8\pi$

Spiral Review

Refer to $\odot J$. (Lesson 10-1)

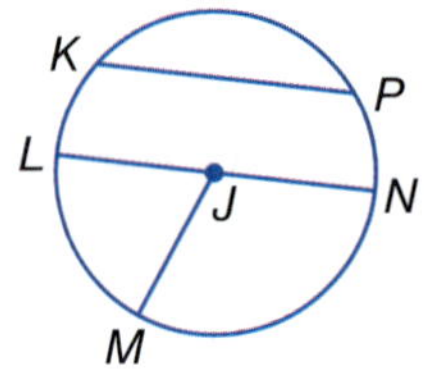

67. Name the center of the circle.

68. Identify a chord that is also a diameter.

69. If $LN = 12.4$, what is JM?

Find the image of each polygon with the given vertices after a dilation centered at the origin with the given scale factor. (Lesson 9-6)

70. $X(-1, 2)$, $Y(2, 1)$, $Z(-1, -2)$; $r = 3$

71. $A(-4, 4)$, $B(4, 4)$, $C(4, -4)$, $D(-4, -4)$; $r = 0.25$

72. BASEBALL The diagram shows some dimensions of Comiskey Park in Chicago, Illinois. $\overline{BD}$ is a segment from home plate to dead center field, and $\overline{AE}$ is a segment from the left field foul pole to the right field foul pole. If the center fielder is standing at C, how far is he from home plate? (Lesson 8-3)

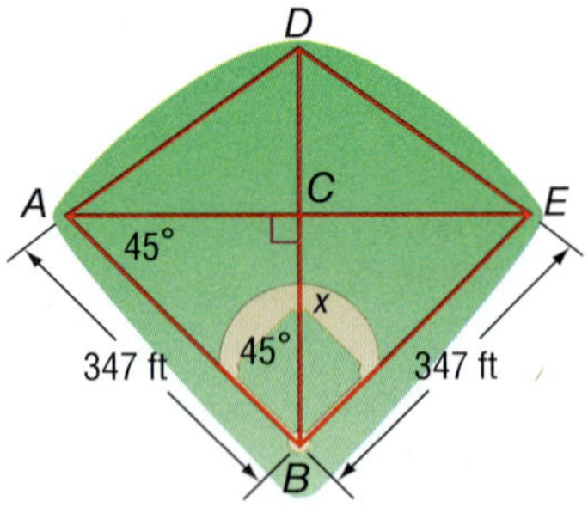

Find x, y, and z. (Lesson 8-1)

73.

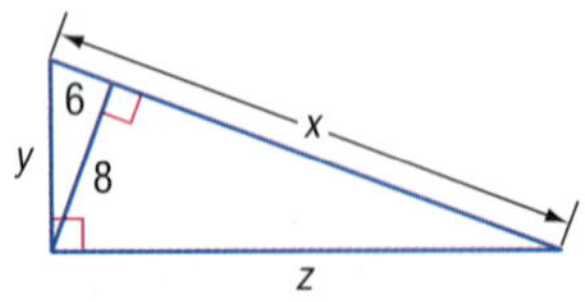

74.

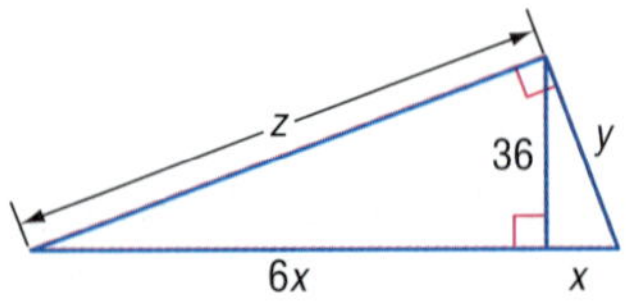

Skills Review

Find x. (Lesson 8-3)

75. $24^2 + x^2 = 26^2$

76. $x^2 + 5^2 = 13^2$

77. $30^2 + 35^2 = x^2$

10-3 Arcs and Chords

Then

You used the relationships between arcs and angles to find measures. (Lesson 10-2)

Now

- Recognize and use relationships between arcs and chords.
- Recognize and use relationships between arcs, chords, and diameters.

Math Online
glencoe.com

Why?

Embroidery hoops are used in sewing, quilting, and cross-stitching, as well as for embroidering. The endpoints of the cross-stitched snowflake shown are both the endpoints of a chord containing a side of the hexagon and the endpoints of an arc.

Arcs and Chords In Lesson 10-1, you learned that a *chord* is a segment with endpoints lying on a circle. If a chord is not a diameter, then its endpoints divide a circle into a major and a minor arc.

Theorem 10.2

Words In the same circle or in congruent circles, two minor arcs are congruent if and only if their corresponding chords are congruent.

Example $\overline{FG} \cong \overline{HJ}$ if and only if $\widehat{FG} \cong \widehat{HJ}$.

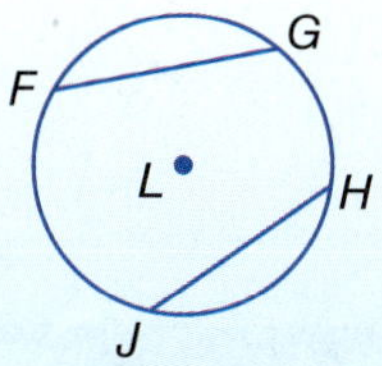

You will prove part 2 of Theorem 10.2 in Exercise 25.

Proof Theorem 10.2 (part 1)

Given: $\odot P$; $\widehat{QR} \cong \widehat{ST}$

Prove: $\overline{QR} \cong \overline{ST}$

Proof:

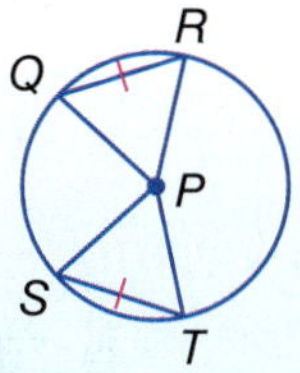

Statements	Reasons
1. $\odot P$; $\widehat{QR} \cong \widehat{ST}$	1. Given
2. $\angle QPR \cong \angle SPT$	2. If arcs are ≅, their corresponding central ∠s are ≅.
3. $\overline{QP} \cong \overline{PR} \cong \overline{SP} \cong \overline{PT}$	3. All radii of a circle are ≅.
4. $\triangle PQR \cong \triangle PST$	4. SAS
5. $\overline{QR} \cong \overline{ST}$	5. CPCTC

Real-World EXAMPLE 1 Use Congruent Chords to Find Arc Measure

CRAFTS In the embroidery hoop, $\overline{AB} \cong \overline{CD}$ and $m\widehat{AB} = 60$. Find $m\widehat{CD}$.

$\overline{AB}$ and $\overline{CD}$ are congruent chords, so the corresponding arcs $\widehat{AB}$ and $\widehat{CD}$ are congruent. $m\widehat{AB} = m\widehat{CD} = 60$

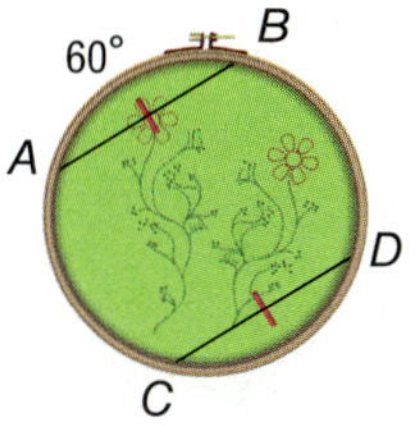

Guided Practice

1. If $m\widehat{AB} = 78$ in the embroidery hoop, find $m\widehat{CD}$.

Personal Tutor glencoe.com

EXAMPLE 2 Use Congruent Arcs to Find Chord Lengths

ALGEBRA In the figures, $\odot J \cong \odot K$ and $\widehat{MN} \cong \widehat{PQ}$. Find PQ.

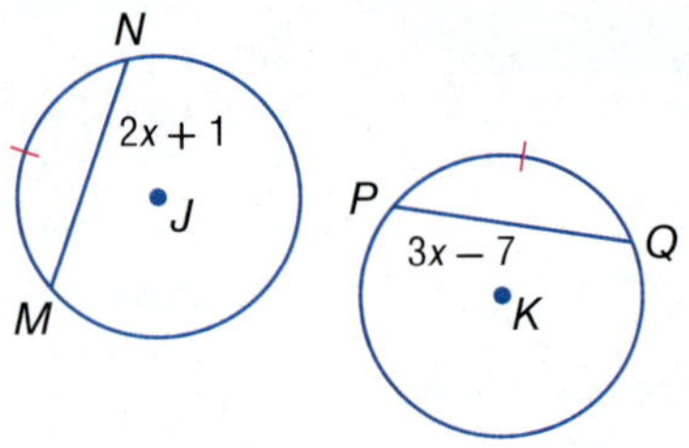

$\widehat{MN}$ and $\widehat{PQ}$ are congruent arcs in congruent circles, so the corresponding chords $\overline{MN}$ and $\overline{PQ}$ are congruent.

$MN = PQ$	**Definition of congruent segments**
$2x + 1 = 3x - 7$	**Substitution**
$8 = x$	**Simplify.**

So, $PQ = 3(8) - 7$ or 17.

Guided Practice

2. In $\odot W$, $\widehat{RS} \cong \widehat{TV}$. Find RS.

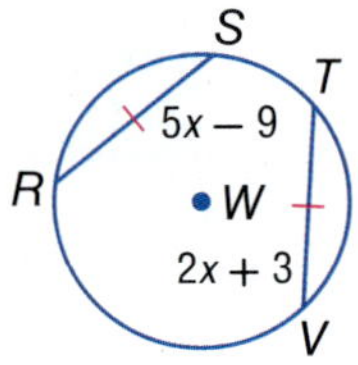

Personal Tutor glencoe.com

StudyTip

Arc Bisectors In the figure below, $\overline{FH}$ is an *arc bisector* of $\widehat{JHG}$.

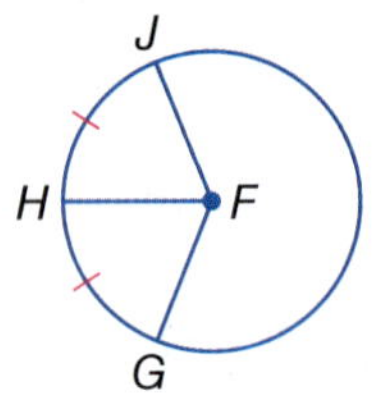

Bisecting Arcs and Chords If a line, segment, or ray divides an arc into two congruent arcs, then it *bisects* the arc.

Theorems

10.3 If a diameter (or radius) of a circle is perpendicular to a chord, then it bisects the chord and its arc.

Example If diameter $\overline{AB}$ is perpendicular to chord $\overline{XY}$, then $\overline{XZ} \cong \overline{ZY}$ and $\widehat{XB} \cong \widehat{BY}$.

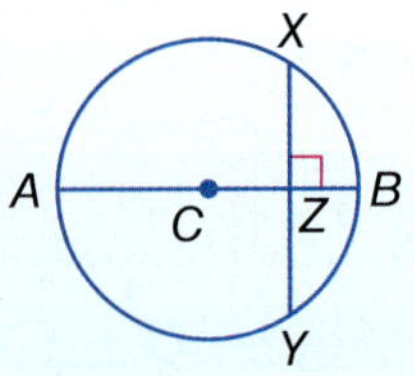

10.4 The perpendicular bisector of a chord is a diameter (or radius) of the circle.

Example If $\overline{AB}$ is a perpendicular bisector of chord $\overline{XY}$, then $\overline{AB}$ is a diameter of $\odot C$.

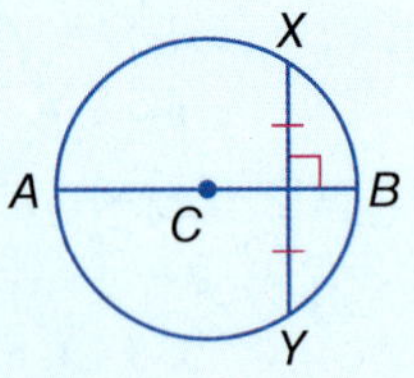

You will prove Theorems 10.3 and 10.4 in Exercises 26 and 28, respectively.

EXAMPLE 3 Use a Radius Perpendicular to a Chord

In $\odot S$, $m\widehat{PQR} = 98$. Find $m\widehat{PQ}$.

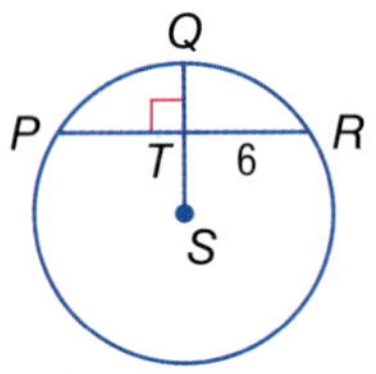

Radius $\overline{SQ}$ is perpendicular to chord $\overline{PR}$. So by Theorem 10.3, $\overline{SQ}$ bisects $\widehat{PQR}$. Therefore, $m\widehat{PQ} = m\widehat{QR}$. By substitution, $m\widehat{PQ} = \frac{98}{2}$ or 49.

Guided Practice

3. In $\odot S$, find PR.

Personal Tutor glencoe.com

Real-World Link

To make stained glass windows, glass is heated to a temperature of 2000 degrees, until it is the consistency of taffy. The colors are caused by the addition of metallic oxides.

Source: *Artistic Stained Glass* by Regg

Real-World EXAMPLE 4 Use a Diameter Perpendicular to a Chord

STAINED GLASS In the stained glass window, diameter $\overline{GH}$ is 30 inches long and chord $\overline{KM}$ is 22 inches long. Find JL.

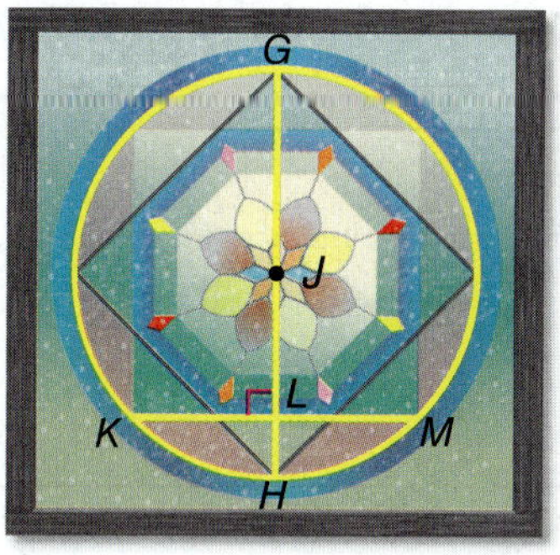

Step 1 Draw radius $\overline{JK}$.

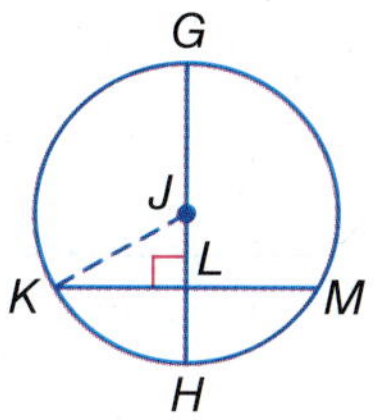

This forms right $\triangle JKL$.

Step 2 Find JK and KL.

Since $GH = 30$ inches, $JH = 15$ inches. All radii of a circle are congruent, so $JK = 15$ inches.

Since diameter $\overline{GH}$ is perpendicular to $\overline{KM}$, $\overline{GH}$ bisects chord $\overline{KM}$ by Theorem 10.3. So, $KL = \frac{1}{2}(22)$ or 11 inches.

Step 3 Use the Pythagorean Theorem to find JL.

$KL^2 + JL^2 = JK^2$	**Pythagorean Theorem**
$11^2 + JL^2 = 15^2$	**$KL = 11$ and $JK = 15$**
$121 + JL^2 = 225$	**Simplify.**
$JL^2 = 104$	**Subtract 121 from each side.**
$JL = \sqrt{104}$	**Take the positive square root of each side.**

So, JL is $\sqrt{104}$ or about 10.20 inches long.

Guided Practice

4. In $\odot R$, find TV. Round to the nearest hundredth.

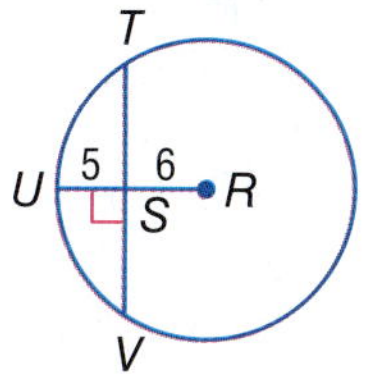

Personal Tutor glencoe.com

StudyTip

Drawing Segments You can add any known information to a figure to help you solve the problem. In Example 4, radius $\overline{JK}$ was drawn.

In addition to Theorem 10.2, you can use the following theorem to determine whether two chords in a circle are congruent.

Theorem 10.5

Words In the same circle or in congruent circles, two chords are congruent if and only if they are equidistant from the center.

Example $LX = LY$ if and only if $\overline{FG} \cong \overline{JH}$.

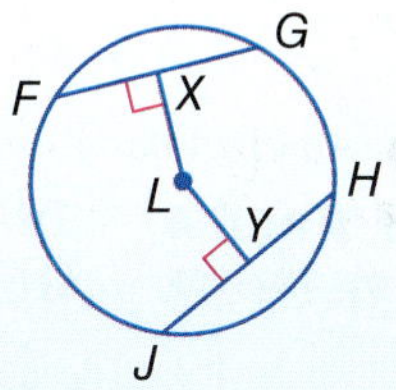

You will prove Theorem 10.5 in Exercises 29 and 30.

EXAMPLE 5 Chords Equidistant from Center

ALGEBRA In $\odot A$, $WX = XY = 22$. Find AB.

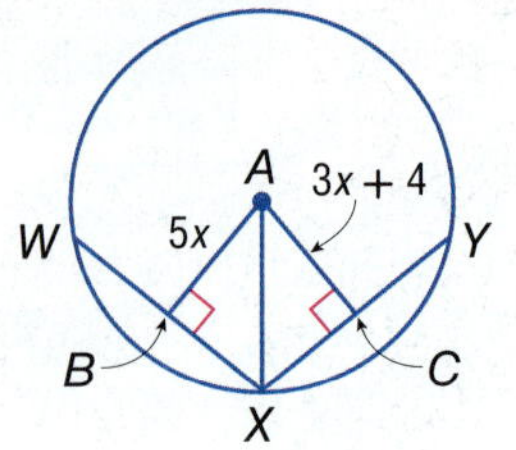

Since chords $\overline{WX}$ and $\overline{XY}$ are congruent, they are equidistant from A. So, $AB = AC$.

$AB = AC$

$5x = 3x + 4$ **Substitution**

$x = 2$ **Simplify.**

So, $AB = 5(2)$ or 10.

Guided Practice

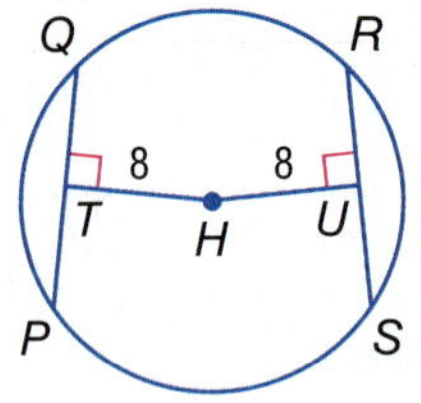

5. In $\odot H$, $PQ = 3x - 4$ and $RS = 14$. Find x.

Personal Tutor glencoe.com

You can use Theorem 10.5 to find the point equidistant from three noncollinear points.

Construction Circle Through Three Noncollinear Points

Step 1

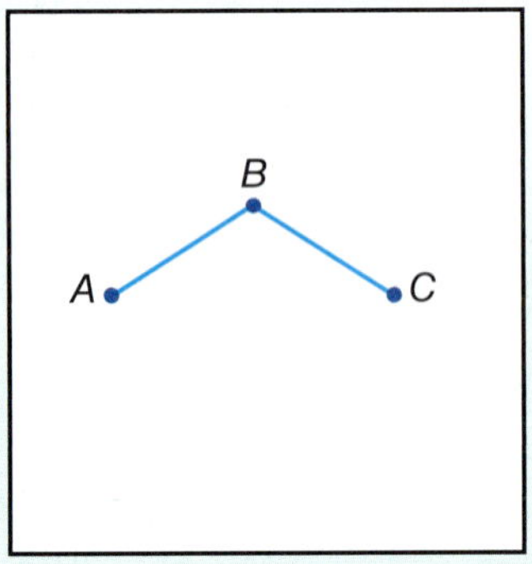

Draw three noncollinear points A, B, and C. Then draw segments $\overline{AB}$ and $\overline{BC}$.

Step 2

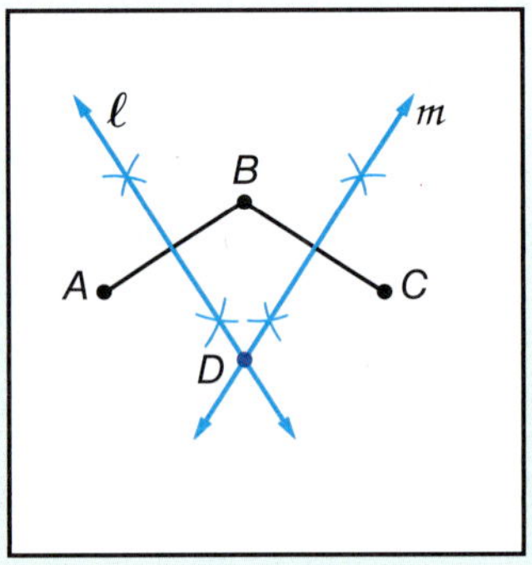

Construct the perpendicular bisectors ℓ and m of $\overline{AB}$ and $\overline{BC}$. Label the point of intersection D.

Step 3

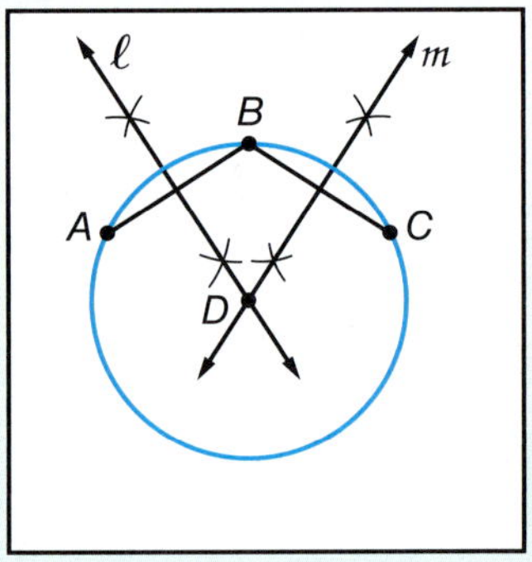

By Theorem 10.4, lines ℓ and m contain diameters of $\odot D$. With the compass at point D, draw a circle through points A, B, and C.

Check Your Understanding

Examples 1 and 2
pp. 701–702

ALGEBRA Find the value of x.

1.

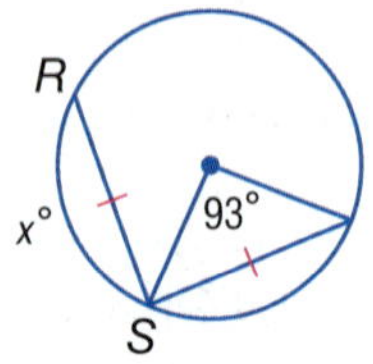

2.

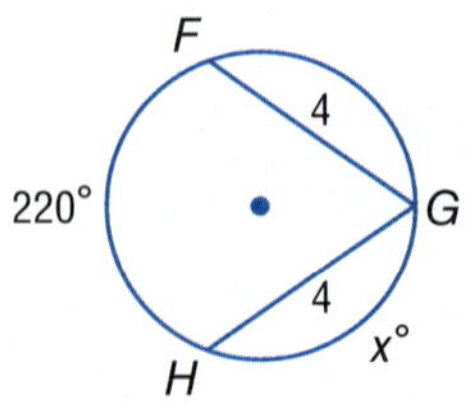

3.

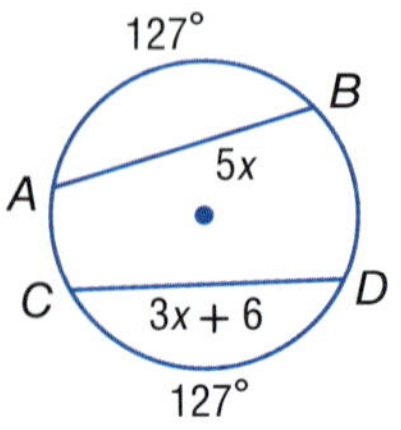

Examples 3 and 4
pp. 702–703

In $\odot P$, $JK = 10$ and $m\widehat{JLK} = 134$. Find each measure. Round to the nearest hundredth.

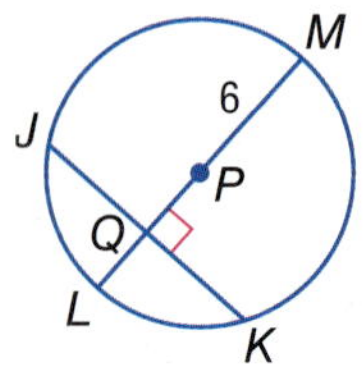

4. $m\widehat{JL}$

5. PQ

Example 5
p. 704

6. In $\odot J$, $GH = 9$, $KL = 4x + 1$. Find x.

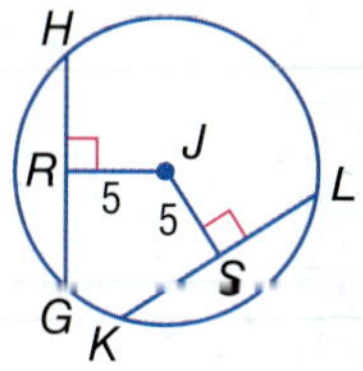

Practice and Problem Solving

● = **Step-by-Step Solutions** begin on page R20
Extra Practice begins on page 969.

Examples 1 and 2
pp. 701–702

ALGEBRA Find the value of x.

7.

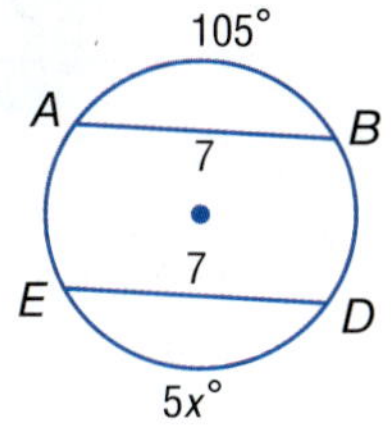

8.

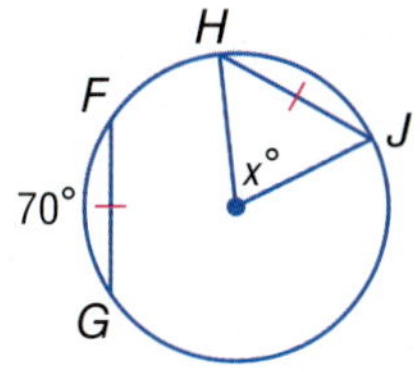

9.

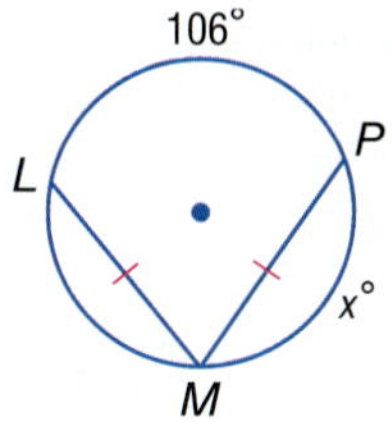

10.

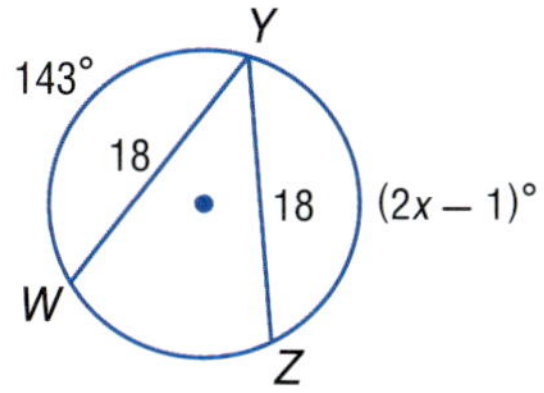

11.

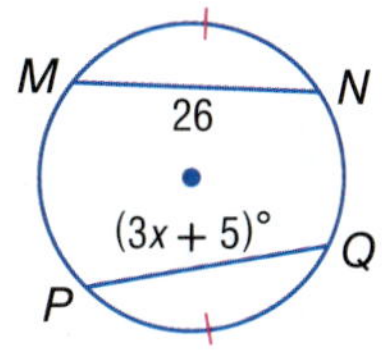

12.

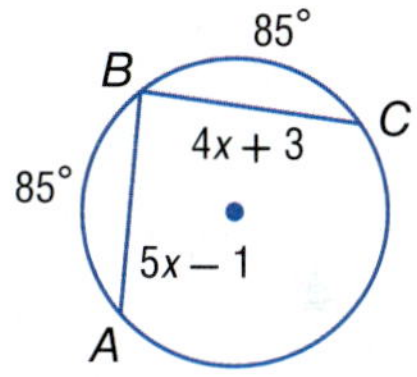

13 $\odot C \cong \odot D$

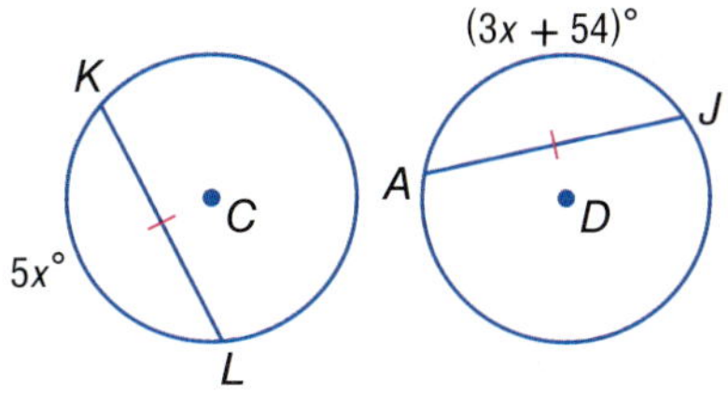

14. $\odot P \cong \odot Q$

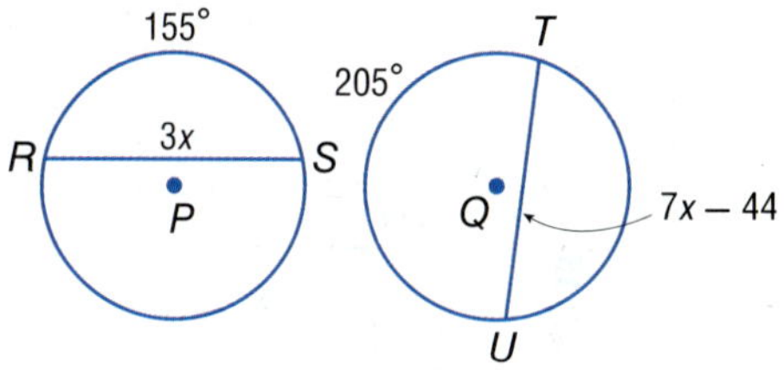

15. JEWELRY Angie is in a jewelry making class at her local arts center. She wants to make a pair of triangular earrings from a metal circle. She knows that $\widehat{AC}$ is 115°. If she wants to cut two equal parts off so that $\widehat{AB} = \widehat{BC}$, what is x?

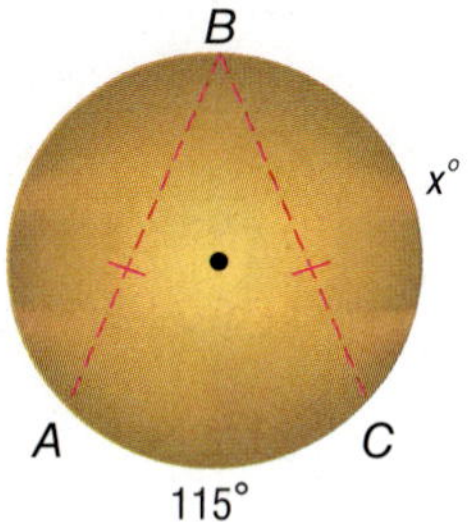

Examples 3 and 4
pp. 702–703

In $\odot A$, the radius is 14 and $CD = 22$. Find each measure. Round to the nearest hundredth, if necessary.

16. CE

17. EB

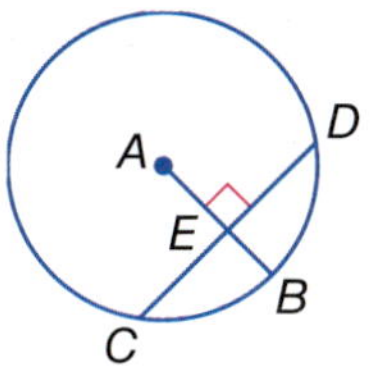

In $\odot H$, the diameter is 18, $LM = 12$, and $m\widehat{LM} = 84$. Find each measure. Round to the nearest hundredth, if necessary.

18. $m\widehat{LK}$

19. HP

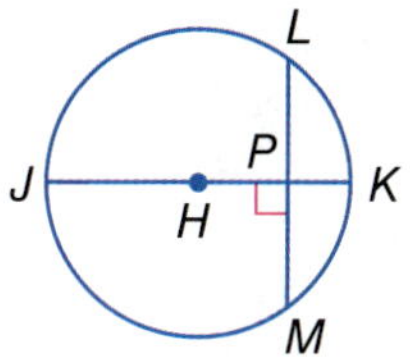

Real-World Link

A snowboard rail can be any nonsnow object that you slide the bottom of a snowboard across.

20. SNOWBOARDING The snowboarding rail shown is an arc of a circle in which $\overline{BD}$ is part of the diameter. If $\widehat{ABC}$ is about 32% of a complete circle, what is $m\widehat{AB}$?

21. ROADS The curved road at the right is part of $\odot C$, which has a radius of 88 feet. What is AB? Round to the nearest tenth.

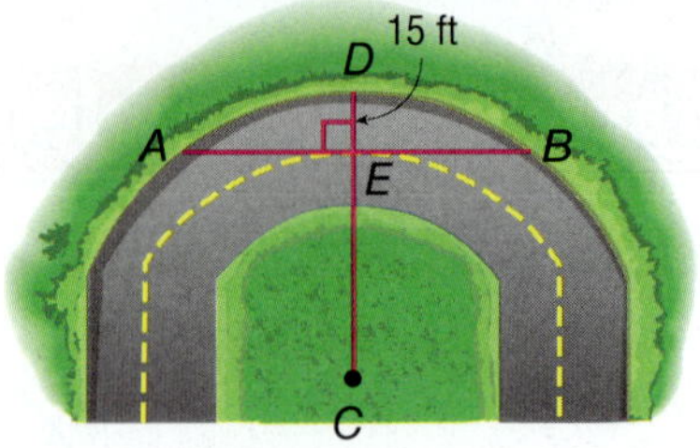

Example 5 p. 704

22. ALGEBRA In $\odot F$, $\overline{AB} \cong \overline{BC}$, $DF = 3x - 7$, and $FE = x + 9$. What is x?

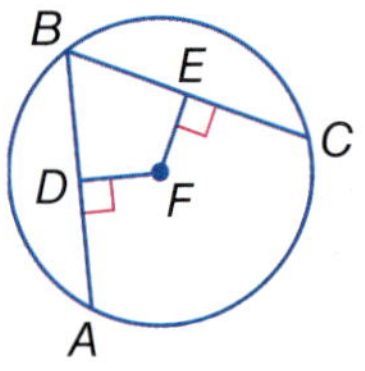

23. ALGEBRA In $\odot S$, $LM = 16$ and $PN = 4x$. What is x?

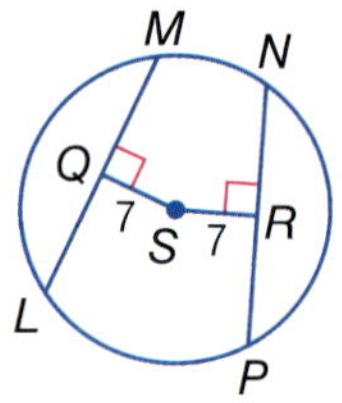

PROOF **Write a two-column proof.**

24. Given: $\odot P$, $\overline{KM} \perp \overline{JP}$
Prove: $\overline{JP}$ bisects $\overline{KM}$ and $\widehat{KM}$.

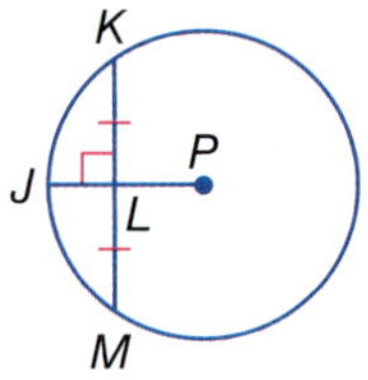

PROOF **Write the specified type of proof.**

25. paragraph proof of Theorem 10.2, part 2

Given: $\odot P$, $\overline{QR} \cong \overline{ST}$

Prove: $\widehat{QR} \cong \widehat{ST}$

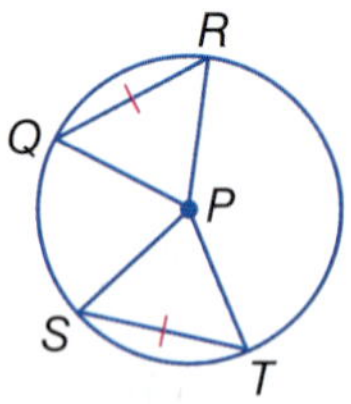

26. two-column proof of Theorem 10.3

Given: $\odot C$, $\overline{AB} \perp \overline{XY}$

Prove: $\overline{XZ} \cong \overline{YZ}$, $\widehat{XB} \cong \widehat{YB}$

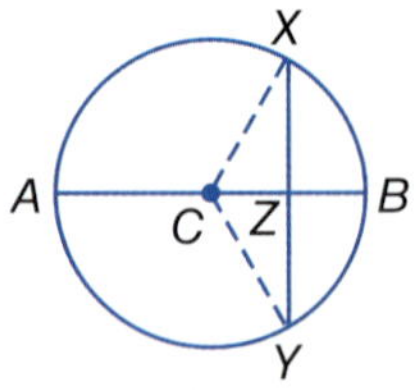

27. DESIGN Roberto is designing a logo for a friend's coffee shop according to the design at the right, where each chord is equal in length. What is the measure of each arc and the length of each chord?

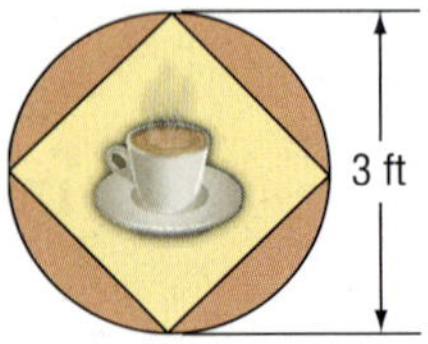

28. PROOF Write a two-column proof of Theorem 10.4.

PROOF **Write a two-column proof of the indicated part of Theorem 10.5.**

29. In a circle, if two chords are equidistant from the center, then they are congruent.

30. In a circle, if two chords are congruent, then they are equidistant from the center.

ALGEBRA **Find the value of x.**

31 $\overline{AB} \cong \overline{DF}$

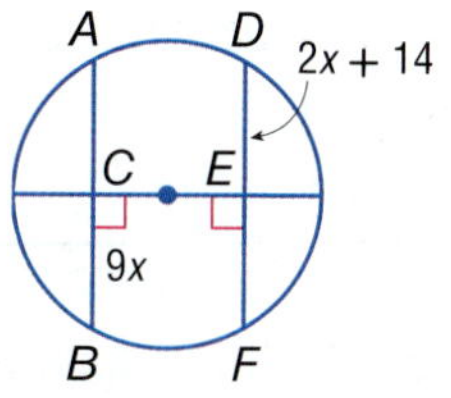

32. $\widehat{GJ} \cong \widehat{KH}$

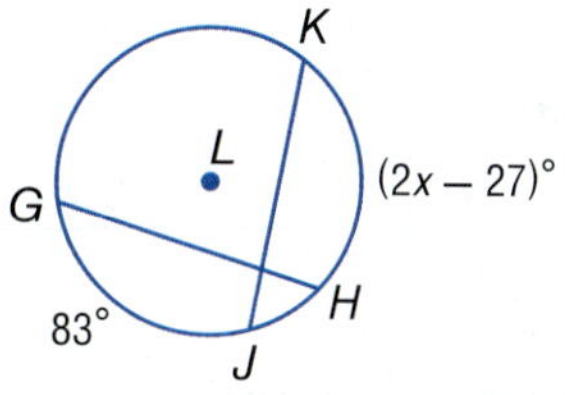

33. $\overline{TY} \cong \overline{WY}$

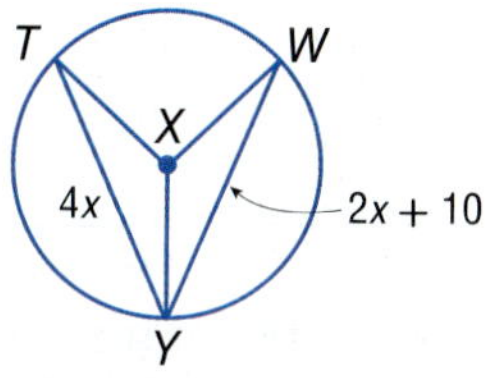

34. **ADVERTISING** A bookstore clerk wants to set up a display of new books. If there are three entrances into the store as shown in the figure at the right, where should the display be to get maximum exposure?

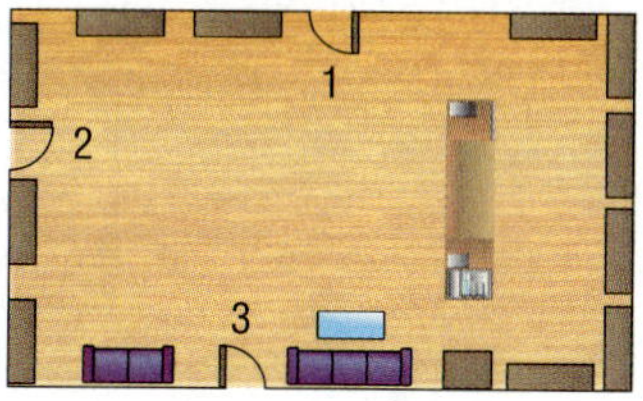

H.O.T. Problems Use Higher-Order Thinking Skills

35. **CHALLENGE** The common chord $\overline{AB}$ between $\odot P$ and $\odot Q$ is perpendicular to the segment connecting the centers of the circles. If $AB = 10$, what is the length of $\overline{PQ}$? Explain your reasoning.

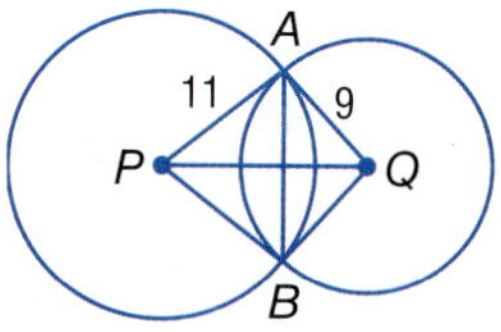

36. **REASONING** In a circle, $\overline{AB}$ is a diameter and $\overline{HG}$ is a chord that intersects $\overline{AB}$ at point X. Is it *sometimes*, *always*, or *never* true that $HX = GX$? Explain.

37. **CHALLENGE** Use a compass to draw a circle with chord $\overline{AB}$. Refer to this construction for the following problem.

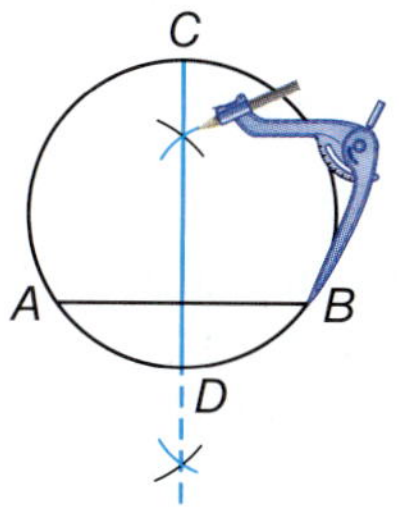

Step 1 Construct $\overline{CD}$, the perpendicular bisector of $\overline{AB}$.

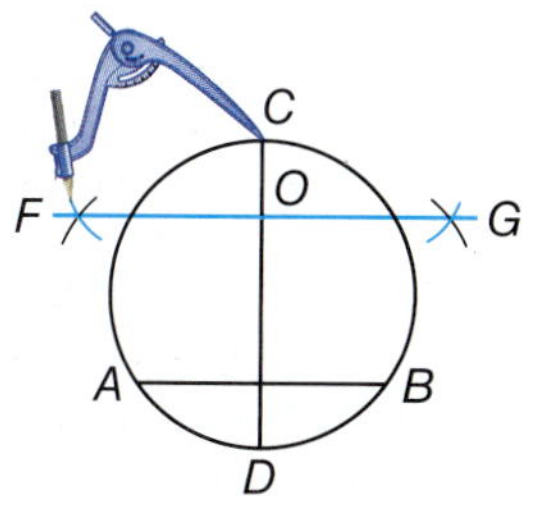

Step 2 Construct $\overline{FG}$, the perpendicular bisector of $\overline{CD}$. Label the point of intersection O.

a. Use an indirect proof to show that $\overline{CD}$ passes through the center of the circle by assuming that the center of the circle is *not* on $\overline{CD}$.

b. Prove that O is the center of the circle.

StudyTip

Indirect Proof Remember that an indirect proof is a proof by contradiction. You start an indirect proof by assuming that the conclusion is false.

38. **OPEN ENDED** Construct a circle and draw a chord. Measure the chord and the distance that the chord is from the center. Find the length of the radius.

39. **WRITING IN MATH** If the length of an arc in a circle is tripled, will the chord of the new arc be three times as long as the chord of the original arc? Draw a figure to support your conclusion.

Standardized Test Practice

40. If $CW = WF$ and $ED = 30$, what is DF?

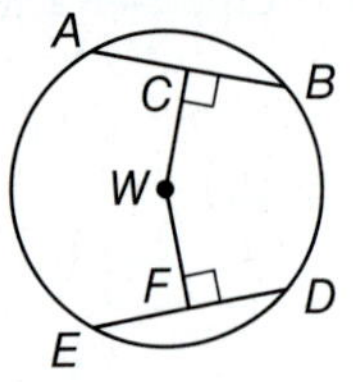

A 60

B 45

C 30

D 15

41. ALGEBRA Write the ratio of the area of the circle to the area of the square in simplest form.

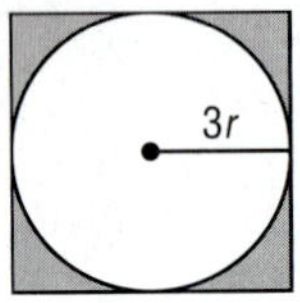

F $\frac{\pi}{4}$

G $\frac{\pi}{2}$

H $\frac{3\pi}{4}$

J π

42. SHORT RESPONSE The pipe shown is divided into five equal sections. How long is the pipe in feet (ft) and inches (in.)?

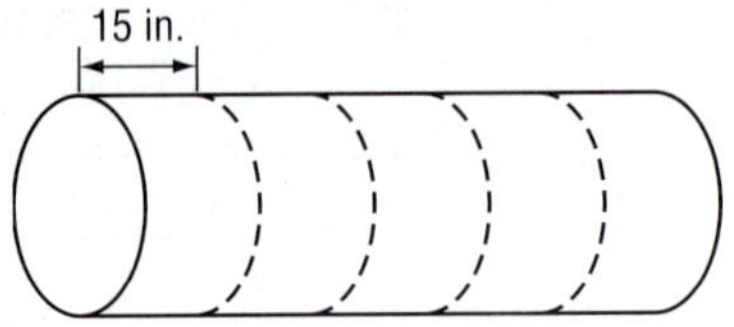

43. SAT/ACT Point B is the center of the circle and the coordinates of Point B are (3, 1). What is the area of the circle?

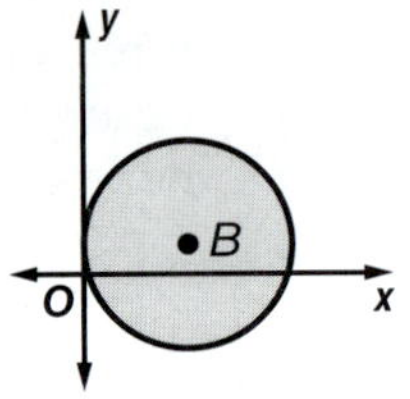

A π units^2

B 3π units^2

C 6π units^2

D 9π units^2

Spiral Review

Find x. (Lesson 10-2)

44.

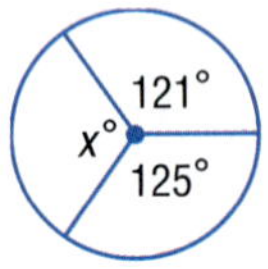

45.

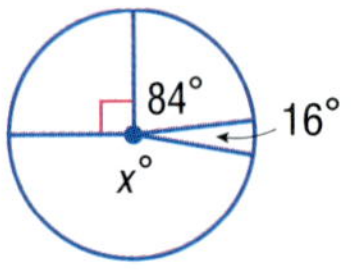

46.

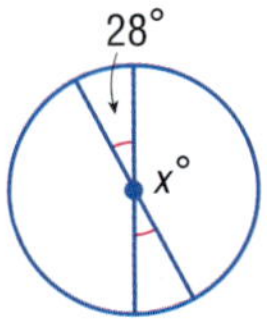

47. CRAFTS Ruby created a pattern to sew flowers onto a quilt by first drawing a regular pentagon that was 3.5 inches long on each side. Then she added a semicircle onto each side of the pentagon to create the appearance of five petals. How many inches of gold trim does she need to edge 10 flowers? Round to the nearest inch. (Lesson 10-1)

Determine whether each set of numbers can be the measures of the sides of a triangle. If so, classify the triangle as *acute*, *obtuse*, or *right*. Justify your answer. (Lesson 8-2)

48. 8, 15, 17

49. 20, 21, 31

50. 10, 16, 18

Skills Review

ALGEBRA Quadrilateral $WXYZ$ is a rhombus. Find each value or measure. (Lesson 6-5)

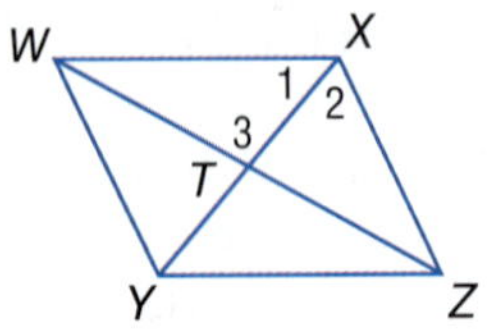

51. If $m\angle 3 = y^2 - 31$, find y.

52. If $m\angle XZY = 56$, find $m\angle YWZ$.

10-4 Inscribed Angles

Why?

The entrance to a school prom has a semicircular arch. Streamers are attached with one end at point A and the other end at point B. The middle of each streamer can then be attached to a different point P along the arch.

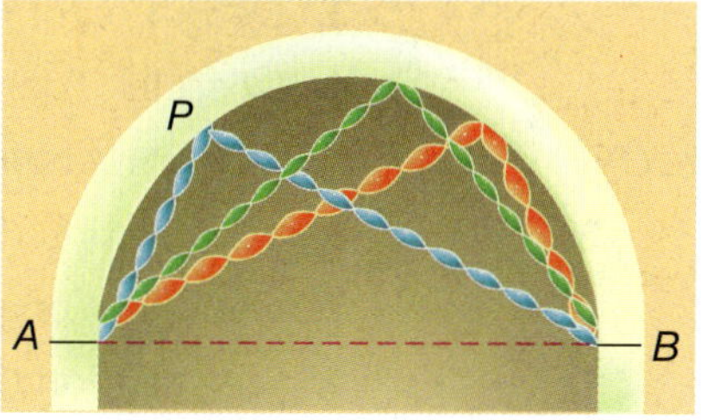

Then
You found measures of interior angles of polygons. (Lesson 6-1)

Now
- Find measures of inscribed angles.
- Find measures of angles of inscribed polygons.

New Vocabulary
inscribed angle
intercepted arc

Math Online
glencoe.com

Inscribed Angles Notice that the angle formed by each streamer appears to be congruent, no matter where point P is placed along the arch. An **inscribed angle** has a vertex on a circle and sides that contain chords of the circle. In $\odot C$, $\angle QRS$ is an inscribed angle.

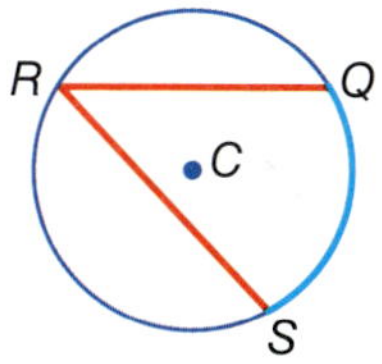

An **intercepted arc** has endpoints on the sides of an inscribed angle and lies in the interior of the inscribed angle. In $\odot C$, minor arc $\widehat{QS}$ is intercepted by $\angle QRS$.

There are three ways that an angle can be inscribed in a circle.

Case 1	Case 2	Case 3
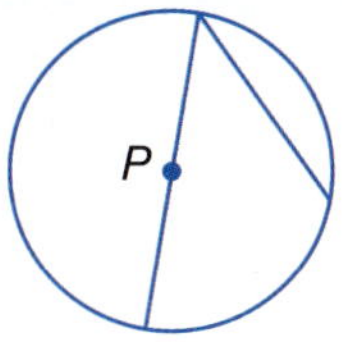	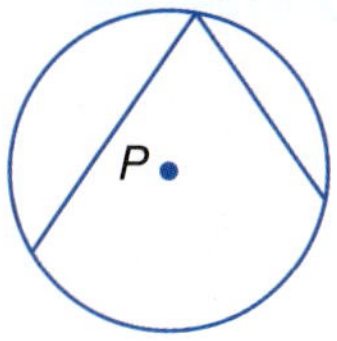	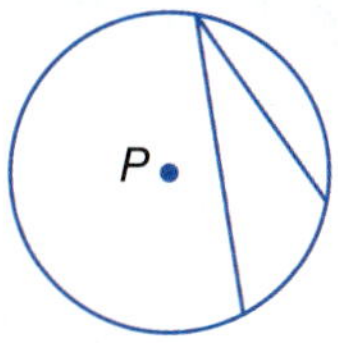
Center P is on a side of the inscribed angle.	Center P is inside the inscribed angle.	The center P is in the exterior of the inscribed angle.

In Case 1, the side of the angle is a diameter of the circle.

For each of these cases, the following theorem holds true.

Theorem 10.6 Inscribed Angle Theorem

Words If an angle is inscribed in a circle, then the measure of the angle equals one half the measure of its intercepted arc.

Example $m\angle 1 = \frac{1}{2}m\widehat{AB}$ and $m\widehat{AB} = 2m\angle 1$

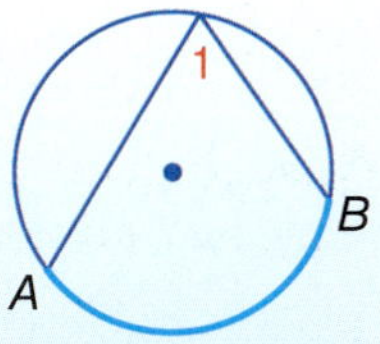

You will prove Cases 2 and 3 of the Inscribed Angle Theorem in Exercises 37 and 38.

Vocabulary Link

inscribed
Everyday Use: written on or in a surface, such as inscribing the inside of a ring with an inscription
Math Use: touching only the sides (or interior) of another figure

PROOF **Inscribed Angle Theorem (Case 1)**

Given: $\angle B$ is inscribed in $\odot P$.

Prove: $m\angle B = \frac{1}{2}m\widehat{AC}$

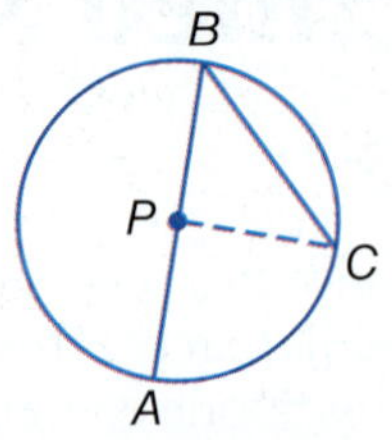

Proof:

Statements	Reasons
1. Draw an auxiliary radius $\overline{PC}$.	**1.** Two points determine a line.
2. $\overline{PB} \cong \overline{PC}$	**2.** All radii of a circle are ≅.
3. $\triangle PBC$ is isosceles.	**3.** Definition of isosceles triangle
4. $m\angle B = m\angle C$	**4.** Isosceles Triangle Theorem
5. $m\angle APC = m\angle B + m\angle C$	**5.** Exterior Angle Theorem
6. $m\angle APC = 2m\angle B$	**6.** Substitution (Steps 4, 5)
7. $m\widehat{AC} = m\angle APC$	**7.** Definition of arc measure
8. $m\widehat{AC} = 2m\angle B$	**8.** Substitution (Steps 6, 7)
9. $2m\angle B = m\widehat{AC}$	**9.** Symmetric Property of Equality
10. $m\angle B = \frac{1}{2}m\widehat{AC}$	**10.** Division Property of Equality

EXAMPLE 1 **Use Inscribed Angles to Find Measures**

Find each measure.

a. $m\angle P$

$$m\angle P = \frac{1}{2}m\widehat{MN}$$
$$= \frac{1}{2}(70) \text{ or } 35$$

b. $m\widehat{PO}$

$$m\widehat{PO} = 2m\angle N$$
$$= 2(56) \text{ or } 112$$

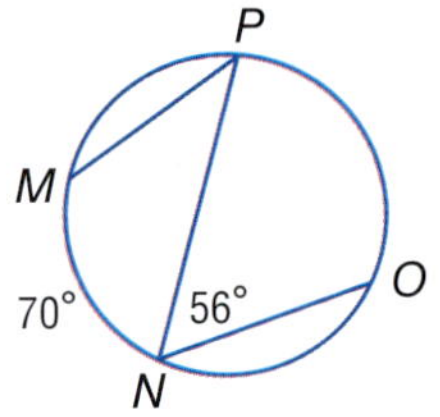

Guided Practice

Find each measure.

1A. $m\widehat{CF}$

1B. $m\angle C$

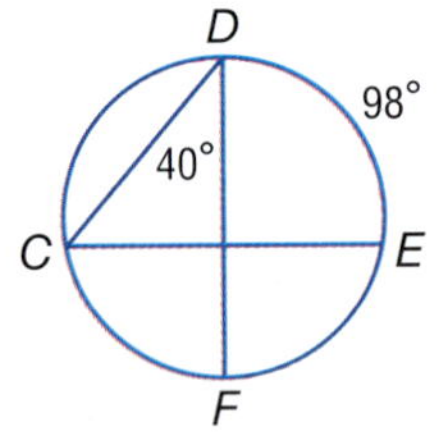

▶ **Personal Tutor** glencoe.com

Two inscribed angles that intercept the same arc of a circle are related.

Theorem 10.7

Words If two inscribed angles of a circle intercept the same arc or congruent arcs, then the angles are congruent.

Example $\angle B$ and $\angle C$ both intercept $\widehat{AD}$.
So, $\angle B \cong \angle C$.

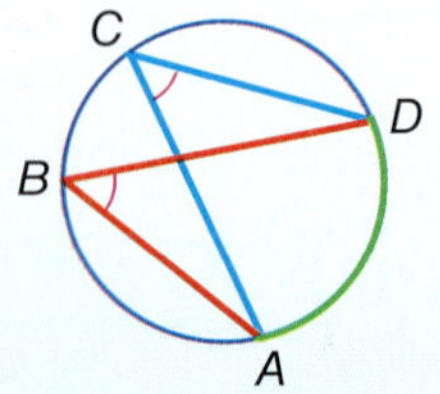

You will prove Theorem 10.7 in Exercise 39.

EXAMPLE 2 Use Inscribed Angles to Find Measures

ALGEBRA Find $m\angle T$.

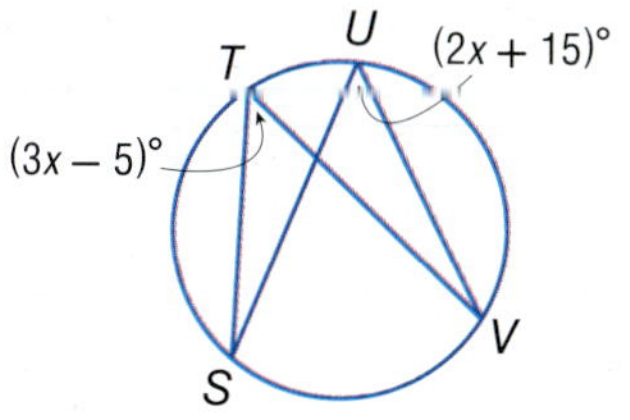

$\angle T \cong \angle U$	**$\angle T$ and $\angle U$ both intercept $\widehat{SV}$.**
$m\angle T = m\angle U$	**Definition of congruent angles**
$3x - 5 = 2x + 15$	**Substitution**
$x = 20$	**Simplify.**

So, $m\angle T = 3(20) - 5$ or 55.

Guided Practice

2. If $m\angle S = 3x$ and $m\angle V = (x + 16)$, find $m\angle S$.

Personal Tutor glencoe.com

EXAMPLE 3 Use Inscribed Angles in Proofs

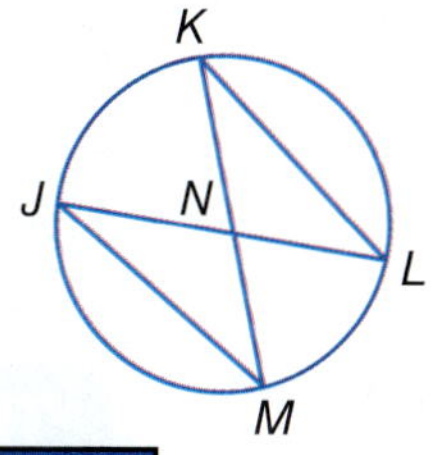

Write a two-column proof.

Given: $\widehat{JM} \cong \widehat{KL}$

Prove: $\triangle JMN \cong \triangle KLN$

Proof:

Statements	Reasons
1. $\widehat{JM} \cong \widehat{KL}$	**1.** Given
2. $\overline{JM} \cong \overline{KL}$	**2.** If minor arcs are $\cong$, their corresponding chords are $\cong$.
3. $\angle M$ intercepts $\widehat{JK}$. $\angle L$ intercepts $\widehat{JK}$.	**3.** Definition of intercepted arc
4. $\angle M \cong \angle L$	**4.** Inscribed $\angle$s of same arc are $\cong$.
5. $\angle JNM \cong \angle KNL$	**5.** Vertical $\angle$s are $\cong$.
6. $\triangle JMN \cong \triangle KLN$	**6.** AAS

Guided Practice

3. Given: $\widehat{QR} \cong \widehat{ST}$, $\widehat{PQ} \cong \widehat{PT}$

Prove: $\triangle PQR \cong \triangle PTS$

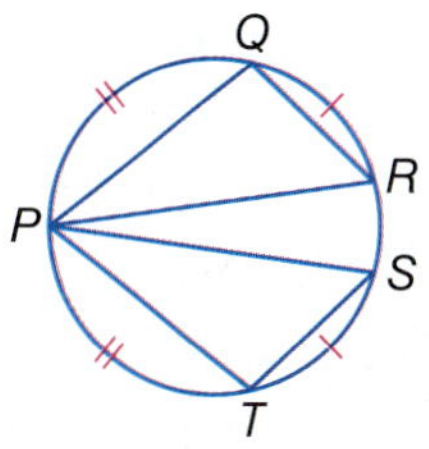

Personal Tutor glencoe.com

StudyTip

Inscribed Polygons Remember that for a polygon to be an inscribed polygon, *all* of its vertices must lie on the circle.

Angles of Inscribed Polygons Triangles and quadrilaterals that are inscribed in circles have special properties.

Theorem 10.8

Words An inscribed angle of a triangle intercepts a diameter or semicircle if and only if the angle is a right angle.

Example If $\widehat{FJH}$ is a semicircle, then $m\angle G = 90$.

If $m\angle G = 90$, then $\widehat{FJH}$ is a semicircle and $\overline{FH}$ is a diameter.

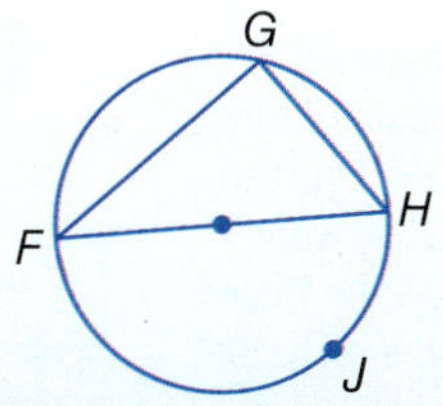

You will prove Theorem 10.8 in Exercise 40.

EXAMPLE 4 Find Angle Measures in Inscribed Triangles

ALGEBRA Find $m\angle F$.

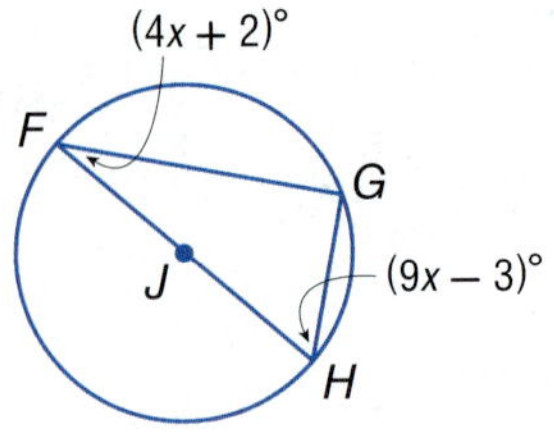

$\triangle FGH$ is a right triangle because $\angle G$ inscribes a semicircle.

$m\angle F + m\angle G + m\angle H = 180$	**Angle Sum Theorem**
$(4x + 2) + 90 + (9x - 3) = 180$	**Substitution**
$13x + 89 = 180$	**Simplify.**
$13x = 91$	**Subtract 89 from each side.**
$x = 7$	**Divide each side by 13.**

So, $m\angle F = 4(7) + 2$ or 30.

Guided Practice

4. If $m\angle F = 7x + 2$ and $m\angle H = 17x - 8$, find x.

Personal Tutor glencoe.com

While many different types of triangles, including right triangles, can be inscribed in a circle, only certain quadrilaterals can be inscribed in a circle.

StudyTip

Quadrilaterals Theorem 10.9 can be verified by considering that the arcs intercepted by opposite angles of an inscribed quadrilateral form a circle.

Theorem 10.9

Words If a quadrilateral is inscribed in a circle, then its opposite angles are supplementary.

Example If quadrilateral *KLMN* is inscribed in $\odot A$, then $\angle L$ and $\angle N$ are supplementary and $\angle K$ and $\angle M$ are supplementary.

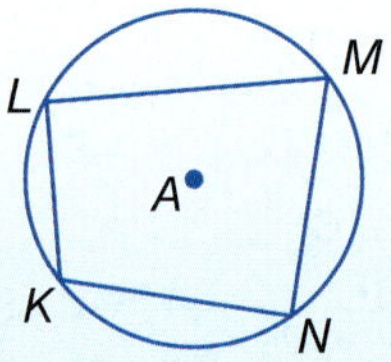

You will prove Theorem 10.9 in Exercise 31.

Real-World EXAMPLE 5 Find Angle Measures

JEWELRY The necklace charm shown uses a quadrilateral inscribed in a circle. Find $m\angle A$ and $m\angle B$.

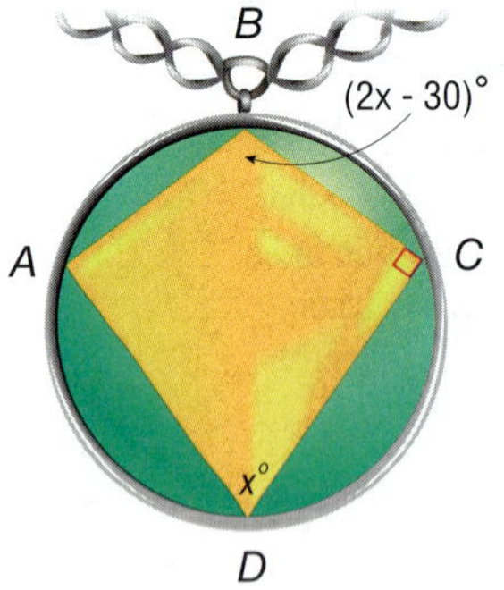

Since *ABCD* is inscribed in a circle, opposite angles are supplementary.

$m\angle A + m\angle C = 180$
$m\angle A + 90 = 180$
$m\angle A = 90$

$m\angle B + m\angle D = 180$
$(2x - 30) + x = 180$
$3x - 30 = 180$
$3x = 210$
$x = 70$

So, $m\angle A = 90$ and $m\angle B = 2(70) - 30$ or 110.

Guided Practice

5. Quadrilateral *WXYZ* is inscribed in $\odot V$. Find $m\angle X$ and $m\angle Y$.

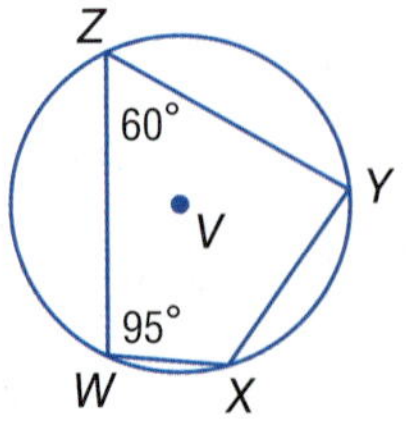

Personal Tutor glencoe.com

Real-World Link

Charms for jewelry first became popular during the age of the Egyptian Pharaohs. They were repopularized by Queen Victoria in the early 20th century and by Louis Vuitton in 2001.

Source: *My Mother's Charms*

Check Your Understanding

Example 1 p. 710

Find each measure.

1. $m\angle B$

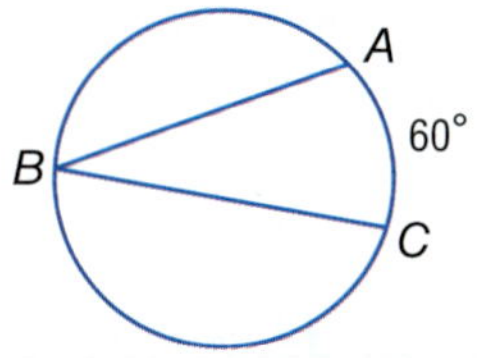

2. $m\widehat{RT}$

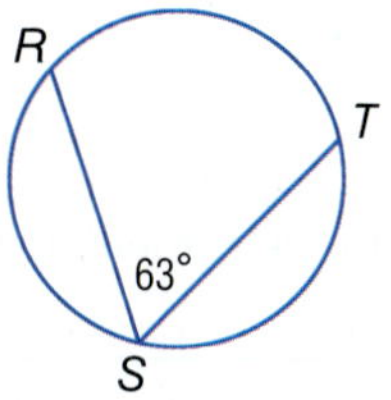

3. $m\widehat{WX}$

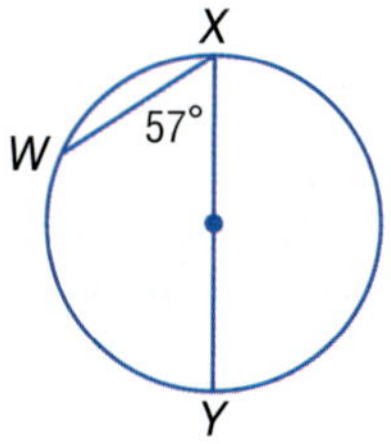

4. **SCIENCE** The diagram shows how light bends in a raindrop to make the colors of the rainbow. If $m\widehat{ST} = 144$, what is $m\angle R$?

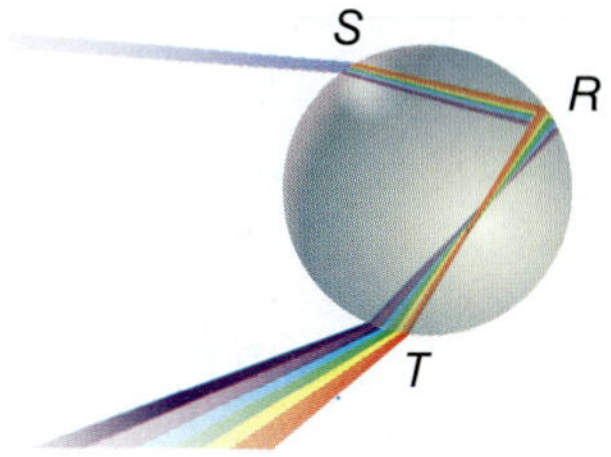

Example 2 p. 711

ALGEBRA Find each measure.

5. $m\angle H$

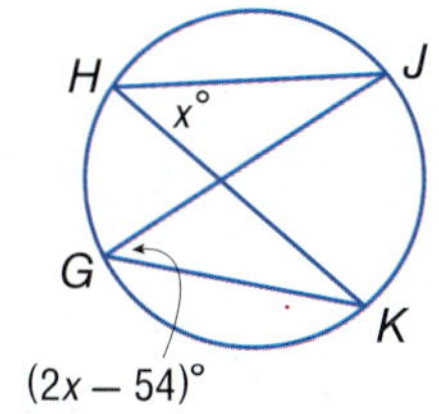

6. $m\angle B$

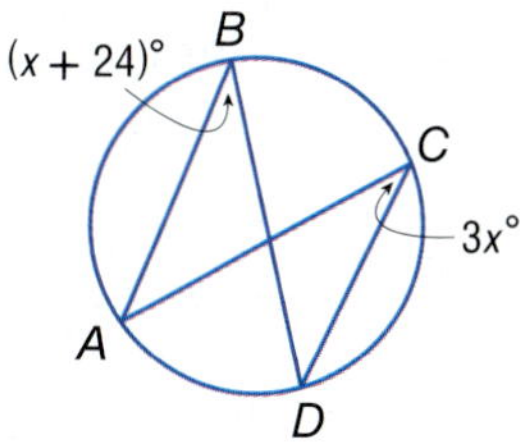

Example 3 p. 711

7. **PROOF** Write a two-column proof.

Given: $\overline{RT}$ bisects $\overline{SU}$.

Prove: $\triangle RVS \cong \triangle UVT$

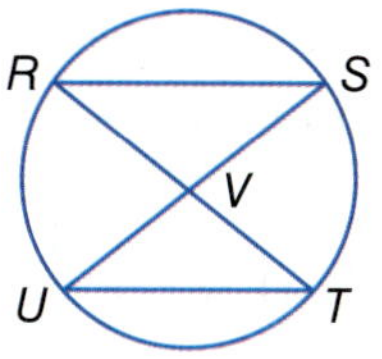

Examples 4 and 5 p. 712

ALGEBRA Find each value.

8. $m\angle R$

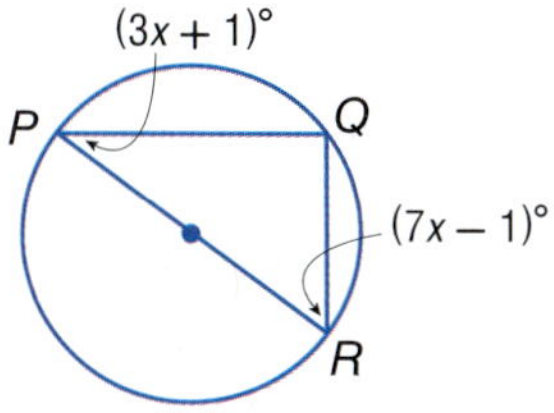

9. x

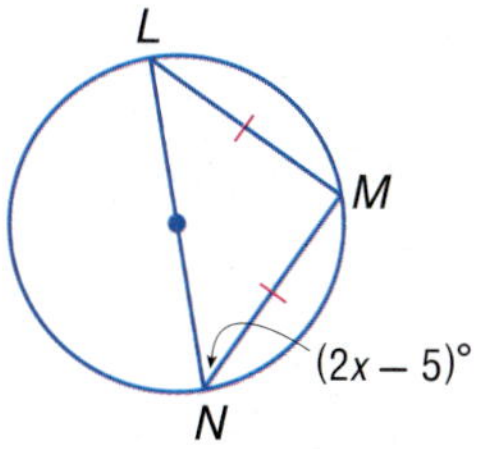

10. $m\angle C$ and $m\angle D$

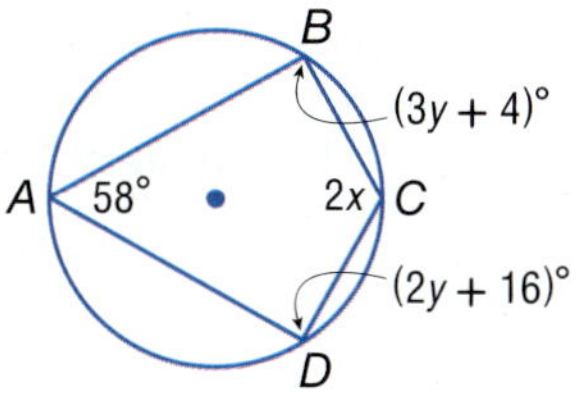

Practice and Problem Solving

● = **Step-by-Step Solutions** begin on page R20.
Extra Practice begins on page 969.

Example 1 p. 710

Find each measure.

11. $m\widehat{DH}$

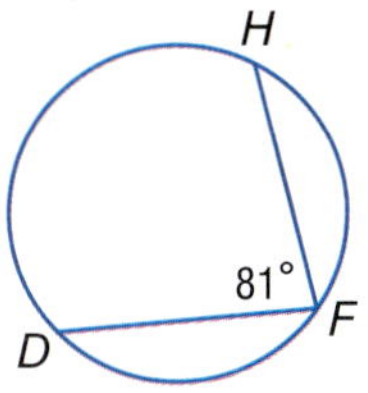

12. $m\angle K$

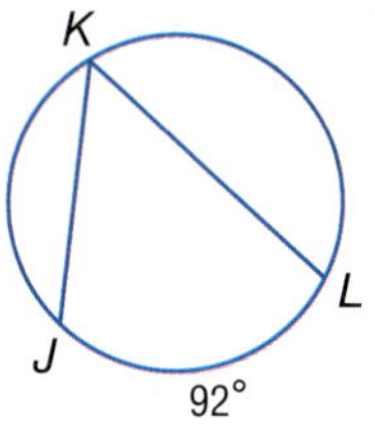

13 $m\angle P$

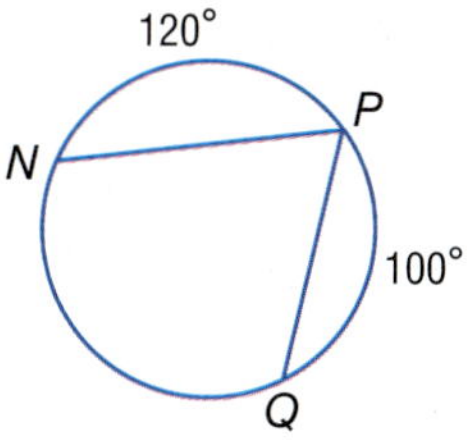

14. $m\widehat{AC}$

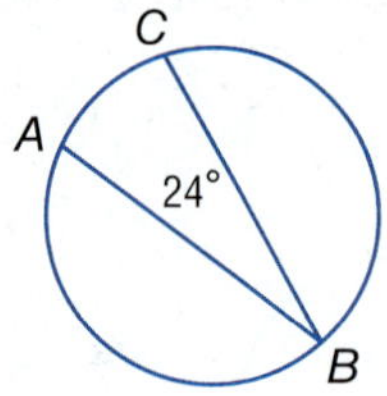

15. $m\widehat{GH}$

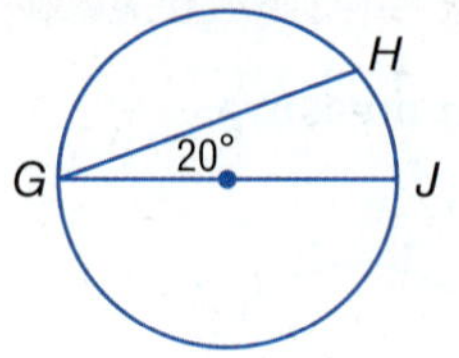

16. $m\angle S$

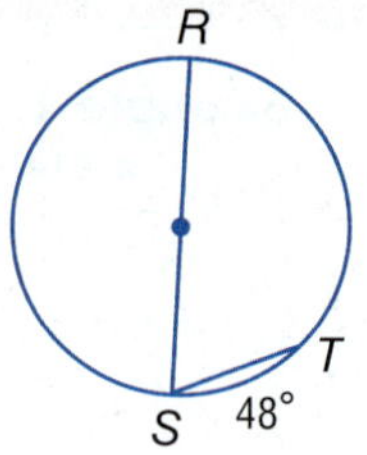

Example 2
p. 711

ALGEBRA **Find each measure.**

17. $m\angle R$

18. $m\angle S$

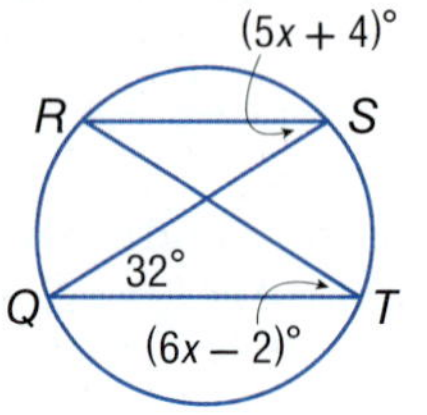

19. $m\angle A$

20. $m\angle C$

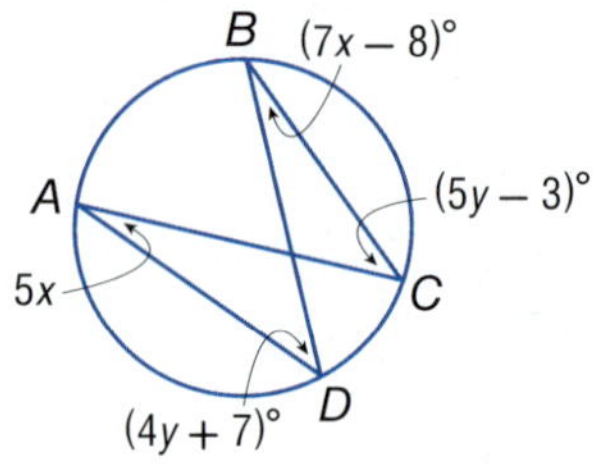

Example 3
p. 711

PROOF **Write the specified type of proof.**

21. paragraph proof

Given: $m\angle T = \frac{1}{2}m\angle S$

Prove: $m\widehat{TUR} = 2m\widehat{URS}$

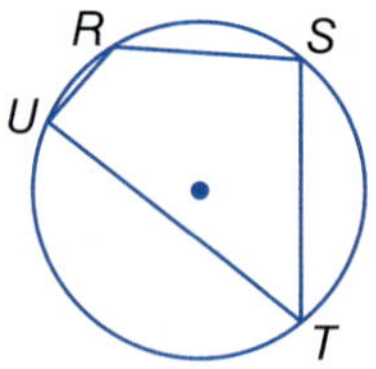

22. two-column proof

Given: $\odot C$

Prove: $\triangle KML \sim \triangle JMH$

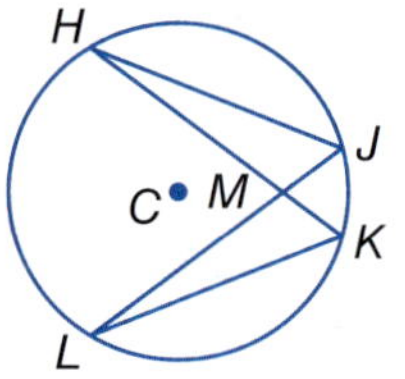

Example 4
p. 712

ALGEBRA **Find each value.**

23. x

24. $m\angle T$

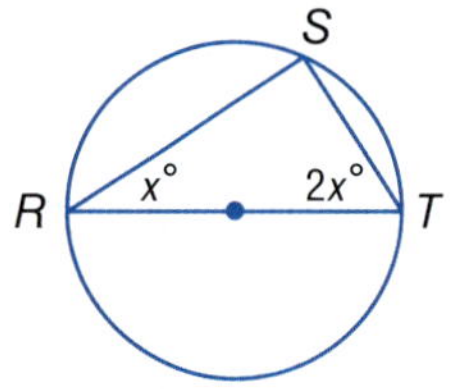

25. x

26. $m\angle C$

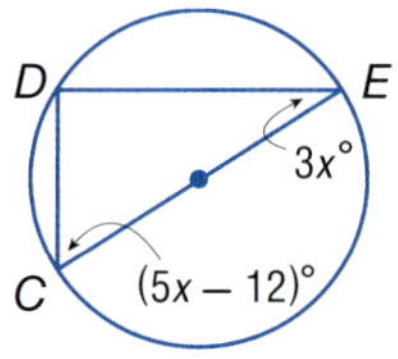

Example 5
p. 712

ALGEBRA **Find each measure.**

27. $m\angle T$

28. $m\angle Z$

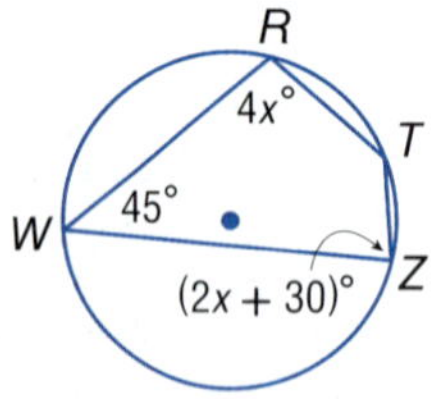

29. $m\angle H$

30. $m\angle G$

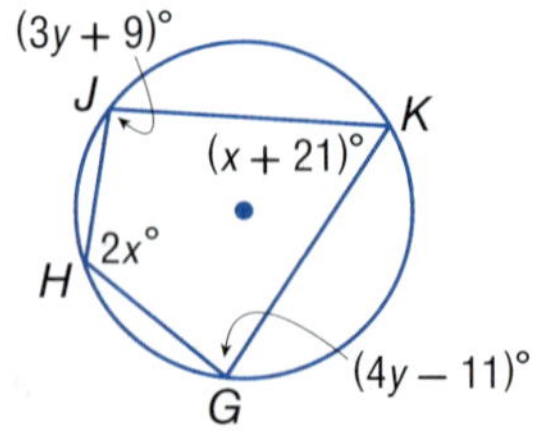

31. **PROOF** Write a paragraph proof for Theorem 10.9.

SIGNS **A stop sign in the shape of a regular octagon is inscribed in a circle. Find each measure.**

32. $m\widehat{NPQ}$

33 $m\angle RLQ$

34. $m\angle LRQ$

35. $m\angle LSR$

Real-World Link

String art can be used to form various geometric patterns.

36. **ART** Four different string art star patterns are shown. If all of the inscribed angles of each star shown are congruent, find the measure of each inscribed angle.

a. 

b. 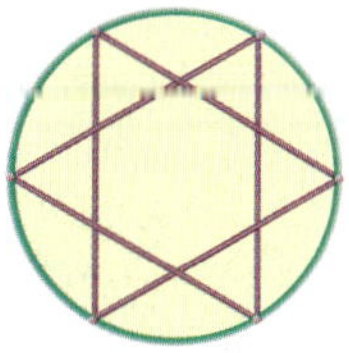

c. 

d. 

PROOF **Write a two-column proof for each case of Theorem 10.6.**

37. Case 2

Given: P lies inside $\angle ABC$. $\overline{BD}$ is a diameter.

Prove: $m\angle ABC = \frac{1}{2}m\widehat{AC}$

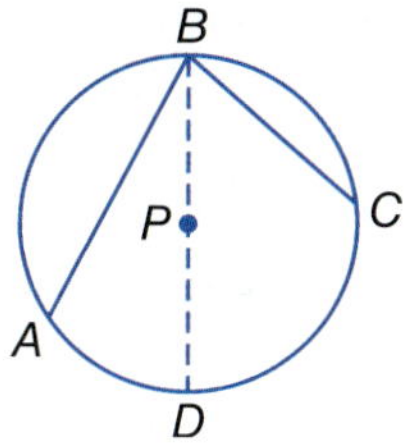

38. Case 3

Given: P lies outside $\angle ABC$. $\overline{BD}$ is a diameter.

Prove: $m\angle ABC = \frac{1}{2}m\widehat{AC}$

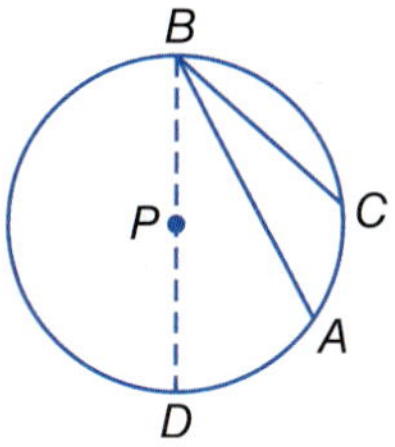

PROOF **Write the specified proof for each theorem.**

39. Theorem 10.7, two-column proof

40. Theorem 10.8, paragraph proof

41. **MULTIPLE REPRESENTATIONS** In this problem, you will investigate the relationship between the arcs of a circle that are cut by two parallel chords.

a. **GEOMETRIC** Use a compass to draw a circle with parallel chords $\overline{AB}$ and $\overline{CD}$. Connect points A and D by drawing segment $\overline{AD}$.

b. **NUMERICAL** Use a protractor to find $m\angle A$ and $m\angle D$. Then determine $m\widehat{AC}$ and $m\widehat{BD}$. What is true about these arcs? Explain.

c. **VERBAL** Draw another circle and repeat parts **a** and **b**. Make a conjecture about arcs of a circle that are cut by two parallel chords.

d. **ANALYTICAL** Use your conjecture to find $m\widehat{PR}$ and $m\widehat{QS}$ in the figure at the right. Verify by using inscribed angles to find the measures of the arcs.

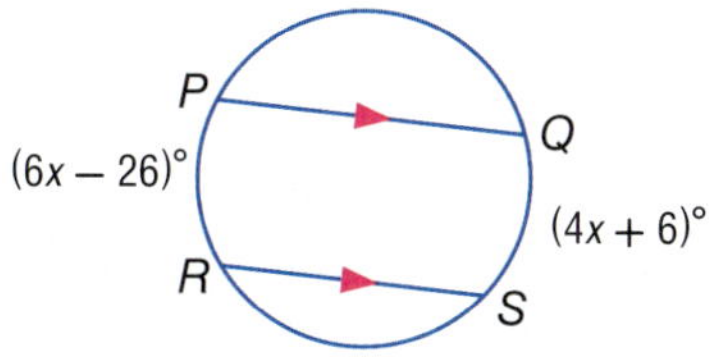

H.O.T. Problems

Use Higher-Order Thinking Skills

REASONING **Determine whether the quadrilateral can *always*, *sometimes*, or *never* be inscribed in a circle. Explain your reasoning.**

42. square **43.** rectangle **44.** parallelogram **45.** rhombus **46.** kite

47. **CHALLENGE** A square is inscribed in a circle. What is the ratio of the area of the circle to the area of the square?

48. **WRITING IN MATH** A 45°-45°-90° right triangle is inscribed in a circle. If the radius of the circle is given, explain how to find the lengths of the right triangle's legs.

49. **OPEN ENDED** Find and sketch a real-world logo with an inscribed polygon.

50. **WRITING IN MATH** Compare an d contrast inscribed angles and central angles of a circle. If they intercept the same arc, how are they related?

Standardized Test Practice

51. In the circle below, $m\widehat{AC} = 160$ and $m\angle BEC = 38$. What is $m\angle AEB$?

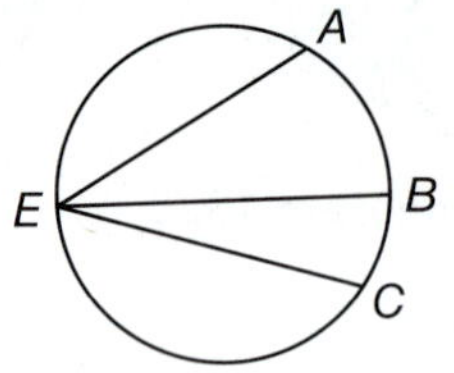

A 42 **C** 80
B 61 **D** 84

52. ALGEBRA Simplify $4(3x - 2)(2x + 4) + 3x^2 + 5x - 6$.

F $9x^2 + 3x - 14$ **H** $27x^2 + 37x - 38$
G $9x^2 + 13x - 14$ **J** $27x^2 + 27x - 26$

53. SHORT RESPONSE In the circle below, $\overline{AB}$ is a diameter, $AC = 8$ inches, and $BC = 15$ inches. Find the diameter, the radius, and the circumference of the circle.

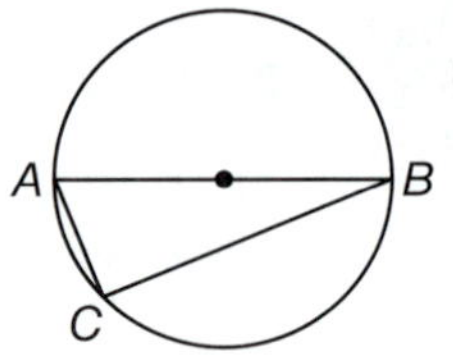

54. SAT/ACT The sum of three consecutive integers is −48. What is the least of the three integers?

A −15 **C** −17
B −16 **D** −18

Spiral Review

In $\odot M$, $FL = 24$, $HJ = 48$, and $m\widehat{HP} = 65$. Find each measure. (Lesson 10-3)

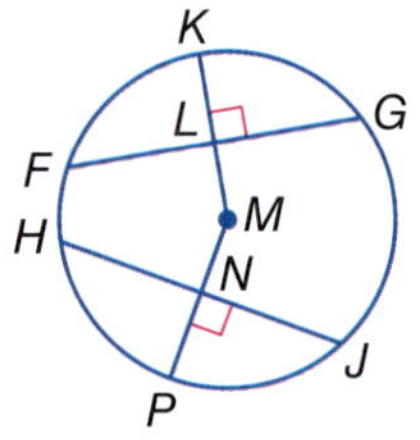

55. FG **56.** $m\widehat{PJ}$

57. NJ **58.** $m\widehat{HJ}$

Find x. (Lesson 10-2)

59.

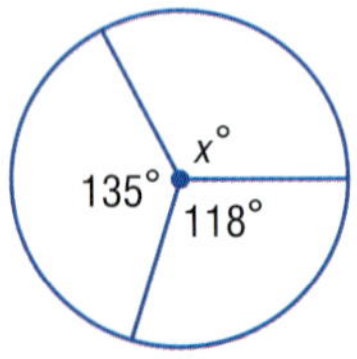

60.

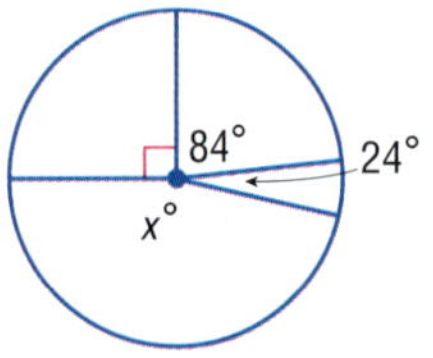

61.

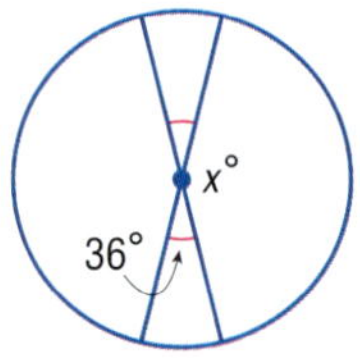

62. PHOTOGRAPHY In one of the first cameras invented, light entered an opening in the front. An image was reflected in the back of the camera, upside down, forming similar triangles. Suppose the image of the person on the back of the camera is 12 inches, the distance from the opening to the person is 7 feet, and the camera itself is 15 inches long. How tall is the person being photographed? (Lesson 7-3)

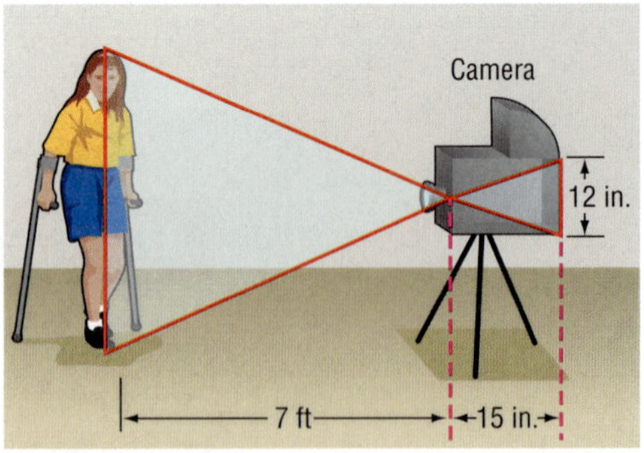

Skills Review

ALGEBRA Suppose B is the midpoint of $\overline{AC}$. Use the given information to find the missing measure. (Lesson 1-3)

63. $AB = 4x - 5$, $BC = 11 + 2x$, $AC = ?$

64. $AB = 6y - 14$, $BC = 10 - 2y$, $AC = ?$

65. $BC = 6 - 4m$, $AC = 8$, $m = ?$

66. $AB = 10s + 2$, $AC = 40$, $s = ?$

CHAPTER 10 Mid-Chapter Quiz

Lessons 10-1 through 10-4

For Exercises 1–4, refer to ⊙*A*. (Lesson 10-1)

1. Name the circle.
2. Name a diameter.
3. Name a chord that is not a diameter.

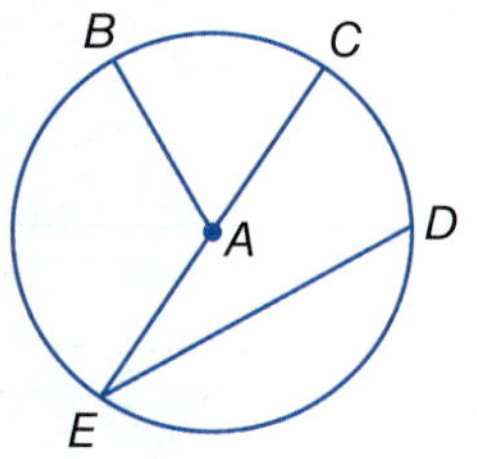

4. **BICYCLES** A bicycle has tires that are 24 inches in diameter. (Lesson 10-1)
   a. Find the circumference of one tire.
   b. How many inches does the tire travel after 100 rotations?

Find the diameter and radius of a circle with the given circumference. Round to the nearest hundredth. (Lesson 10-1)

5. $C = 23$ cm
6. $C = 78$ ft

7. **MULTIPLE CHOICE** Find the length of $\widehat{BC}$. (Lesson 10-2)

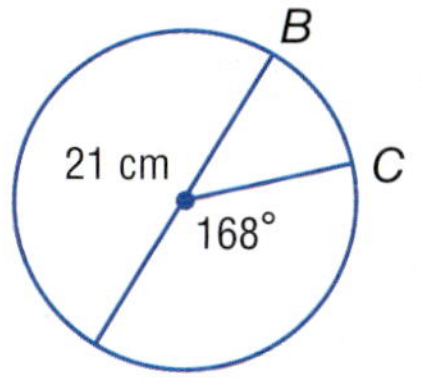

A 18°
B 2.20 cm
C 168°
D 30.79 cm

8. **MOVIES** The movie reel shown below has a diameter of 14.5 inches. (Lesson 10-2)

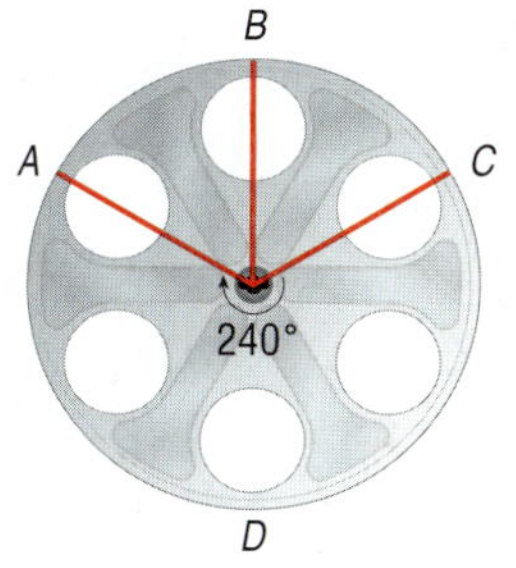

   a. Find $m\widehat{ADC}$.
   b. Find the length of $\widehat{ADC}$.

9. Find the value of x. (Lesson 10-3)

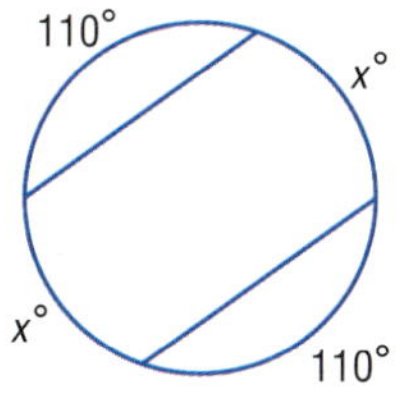

10. In ⊙B, $CE = 13.5$. Find BD. Round to the nearest hundredth. (Lesson 10-3)

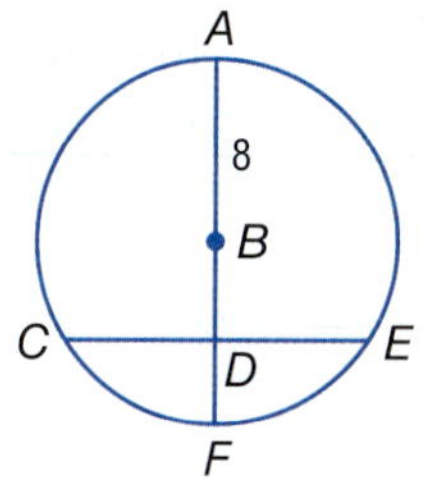

11. The two circles shown are congruent. Find x and the length of the chord. (Lesson 10-3)

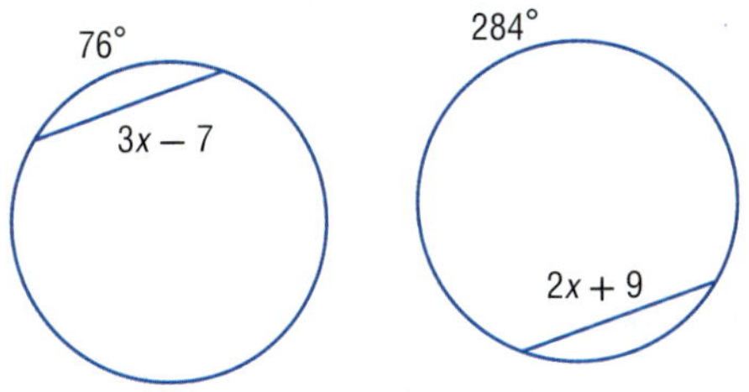

Find each measure. (Lesson 10-4)

12. $m\widehat{TU}$

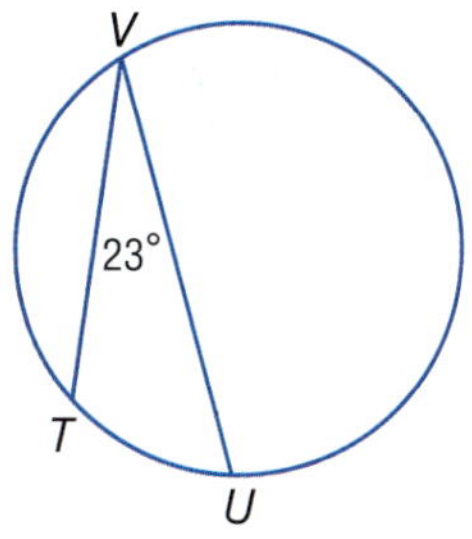

13. $m\angle A$

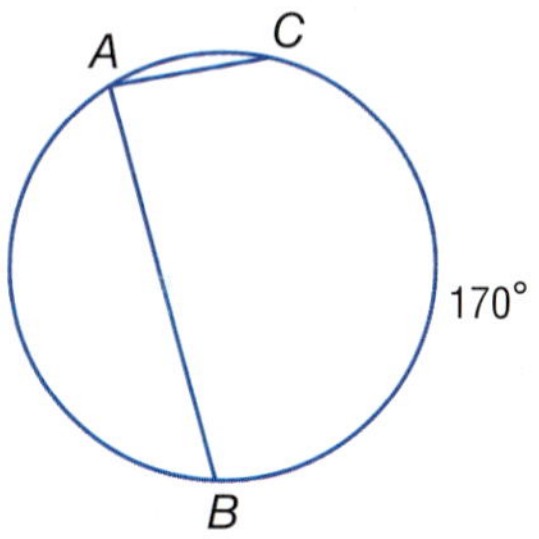

14. **MULTIPLE CHOICE** Find x. (Lesson 10-4)

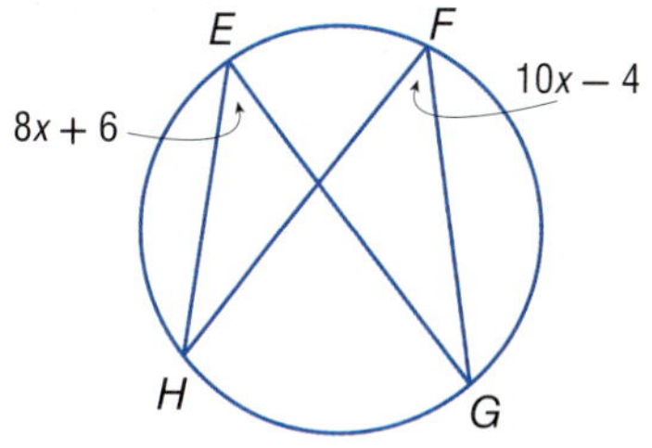

F 1.8
G 5
H 46
J 90

15. If a square with sides of 14 inches is inscribed in a circle, what is the diameter of the circle? (Lesson 10-4)

10-5 Tangents

Then
You used the Pythagorean Theorem to find side lengths of right triangles. (Lesson 8-2)

Now
- Use properties of tangents.
- Solve problems involving circumscribed polygons.

New Vocabulary
tangent
point of tangency
common tangent

Math Online
glencoe.com

Why?

The first bicycles were moved by pushing your feet on the ground. Modern bicycles use pedals, a chain, and gears. The chain loops around circular gears. The length of the chain between these gears is measured from the points of tangency.

Tangents A **tangent** is a line in the same plane as a circle that intersects the circle in exactly one point, called the **point of tangency**. $\overleftrightarrow{AB}$ is tangent to $\odot C$ at point A. $\overline{AB}$ and $\overrightarrow{AB}$ are also called tangents.

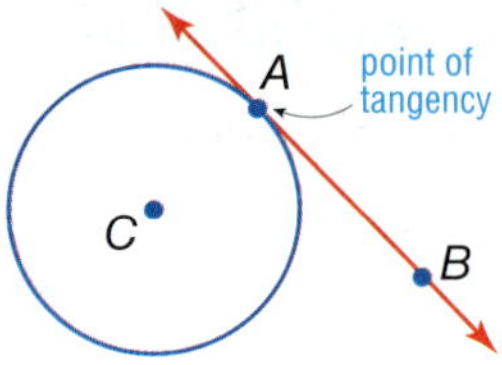

A **common tangent** is a line, ray, or segment that is tangent to two circles in the same plane. In each figure below, line ℓ is a common tangent of circles F and G.

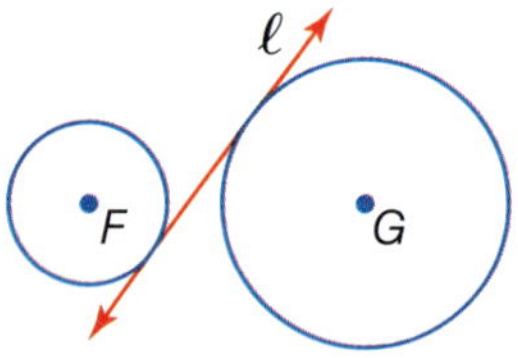

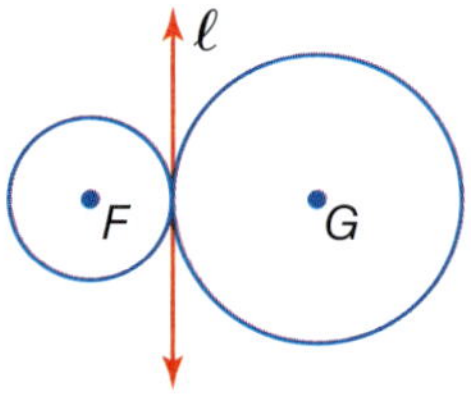

EXAMPLE 1 Identify Common Tangents

Copy each figure and draw the common tangents. If no common tangent exists, state *no common tangent*.

a.

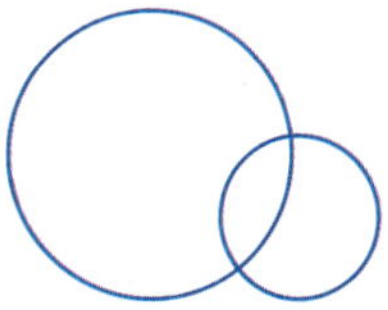

These circles have two common tangents.

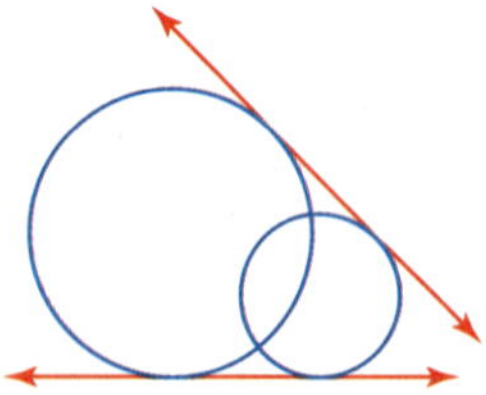

b.

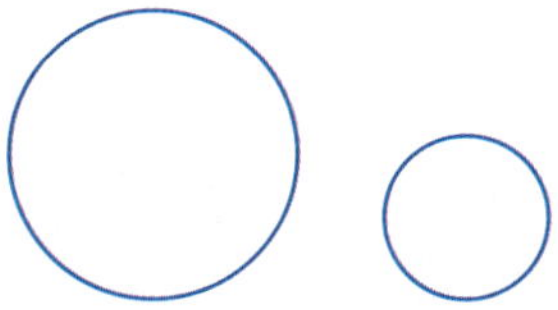

These circles have 4 common tangents.

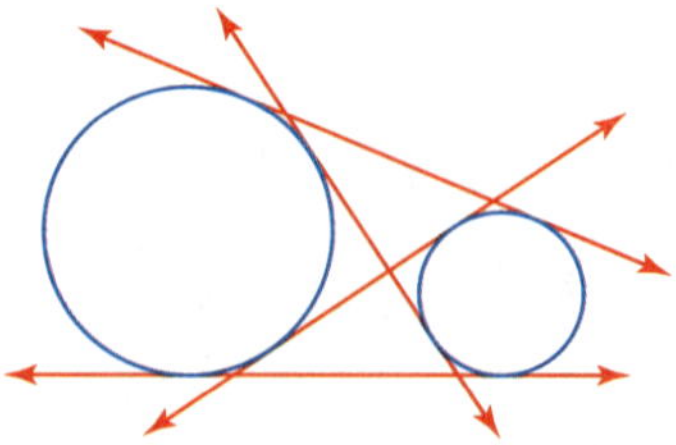

Guided Practice

1A.

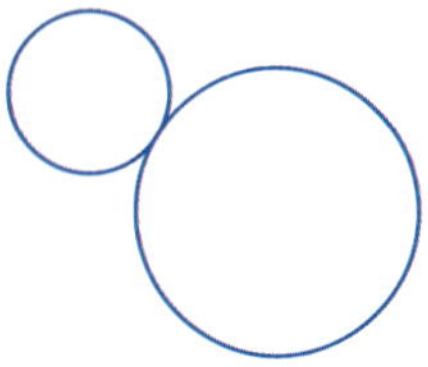

1B.

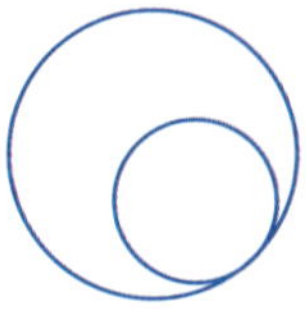

Personal Tutor glencoe.com

The shortest distance from a tangent to the center of a circle is the radius drawn to the point of tangency.

Theorem 10.10

Words In a plane, a line is tangent to a circle if and only if it is perpendicular to a radius drawn to the point of tangency.

Example Line ℓ is tangent to $\odot S$ if and only if $\ell \perp \overline{ST}$.

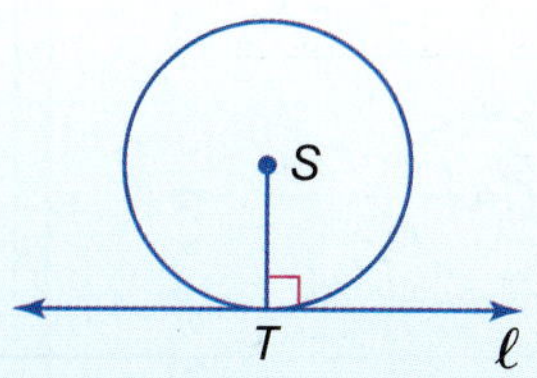

You will prove both parts of Theorem 10.10 in Exercises 32 and 33.

EXAMPLE 2 Identify a Tangent

$\overline{JL}$ is a radius of $\odot J$. Determine whether $\overline{KL}$ is tangent to $\odot J$. Justify your answer.

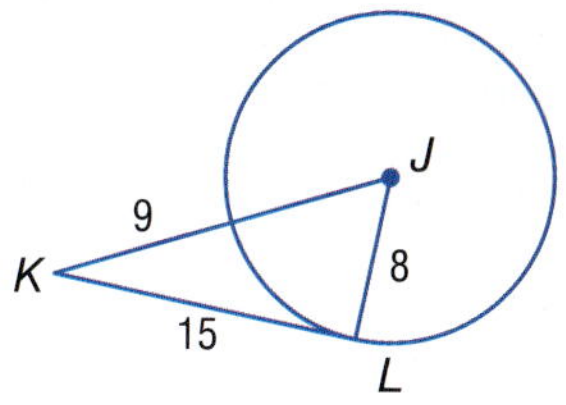

Test to see if $\triangle JKL$ is a right triangle.

$8^2 + 15^2 \stackrel{?}{=} (8 + 9)^2$ **Pythagorean Theorem**

$289 = 289$ ✓ **Simplify.**

$\triangle JKL$ is a right triangle with right angle JLK. So $\overline{KL}$ is perpendicular to radius $\overline{JL}$ at point L. Therefore, by Theorem 10.10, $\overline{KL}$ is tangent to $\odot J$.

Guided Practice

2. Determine whether $\overline{GH}$ is tangent to $\odot F$. Justify your answer.

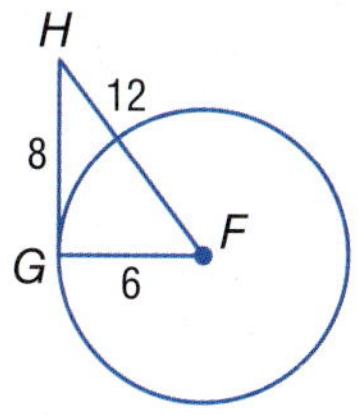

Personal Tutor glencoe.com

You can also use Theorem 10.10 to identify missing values.

EXAMPLE 3 Use a Tangent to Find Missing Values

$\overline{JH}$ is tangent to $\odot G$ at J. Find the value of x.

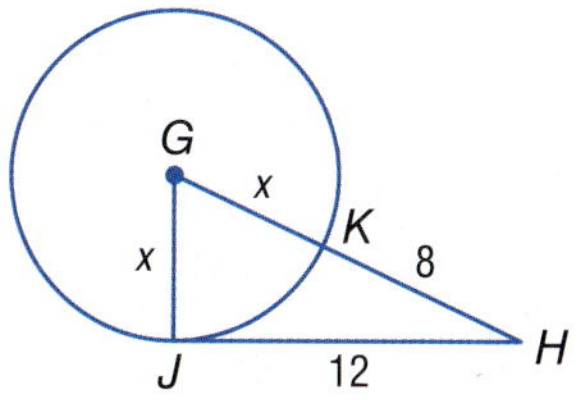

By Theorem 10.10, $\overline{JH} \perp \overline{GJ}$. So, $\triangle GHJ$ is a right triangle.

$GJ^2 + JH^2 = GH^2$ **Pythagorean Theorem**

$x^2 + 122 = (x + 8)^2$ **$GJ = x$, $JH = 12$, and $GH = x + 8$**

$x^2 + 144 = x^2 + 16x + 64$ **Multiply.**

$80 = 16x$ **Simplify.**

$5 = x$ **Divide each side by 16.**

Guided Practice

Find the value of x. Assume that segments that appear to be tangent are tangent.

3A.

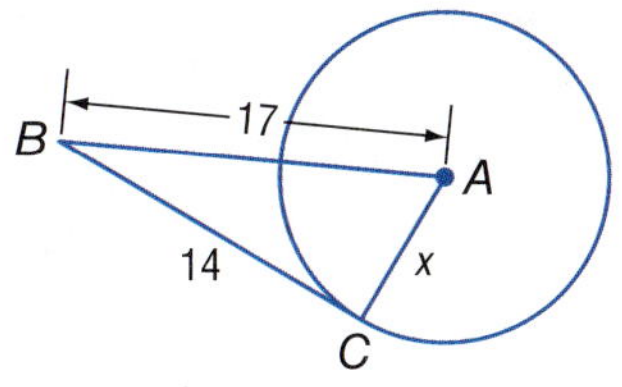

3B.

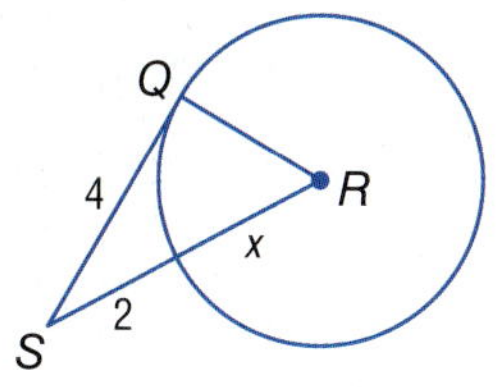

Personal Tutor glencoe.com

Problem-Solving Tip

Solve a Simpler Problem You can use the *solve a simpler problem* strategy by sketching and labeling the right triangles without the circles. A drawing of the triangle in Example 3 is shown below.

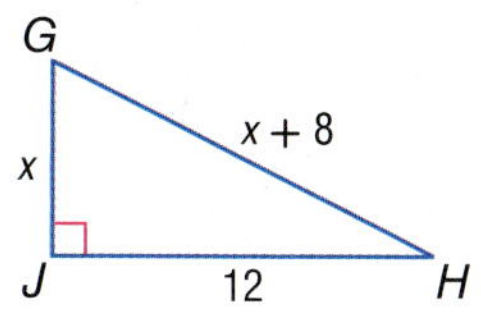

You can use Theorems 10.8 and 10.10 to construct a line tangent to a circle.

Construction — Line Tangent to a Circle Through an External Point

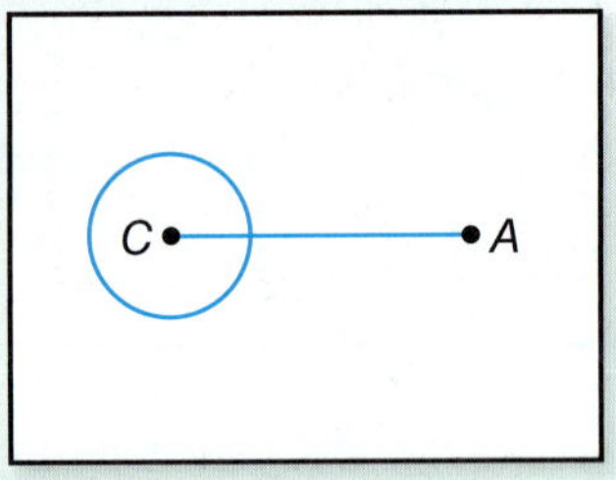

Step 1 Use a compass to draw circle C and a point A outside of circle C. Then draw $\overline{CA}$.

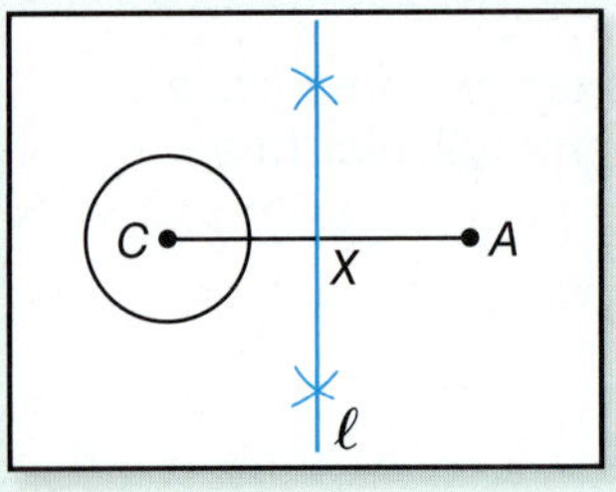

Step 2 Construct line ℓ, the perpendicular bisector of $\overline{CA}$. Label the point of intersection X.

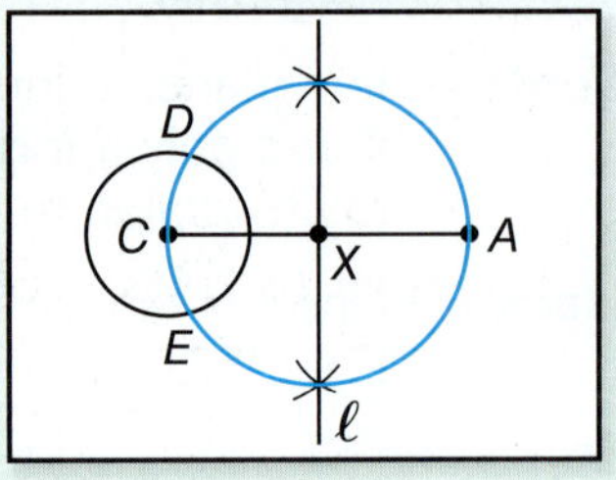

Step 3 Construct circle X with radius $\overline{XC}$. Label the points of intersection of the two circles D and E.

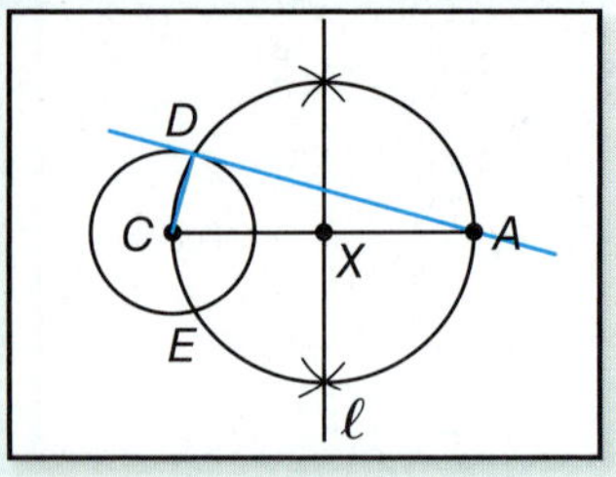

Step 4 Draw $\overleftrightarrow{AD}$ and $\overline{DC}$. $\triangle ADC$ is inscribed in a semicircle. So, $\angle ADC$ is a right angle and $\overleftrightarrow{AD}$ is tangent to $\odot C$.

You will justify this construction in Exercise 36 and construct a line tangent to a circle through a point on the circle in Exercise 34.

More than one line can be tangent to the same circle.

Theorem 10.11

Words If two segments from the same exterior point are tangent to a circle, then they are congruent.

Example If $\overline{AB}$ and $\overline{CB}$ are tangent to $\odot D$, then $\overline{AB} \cong \overline{CB}$.

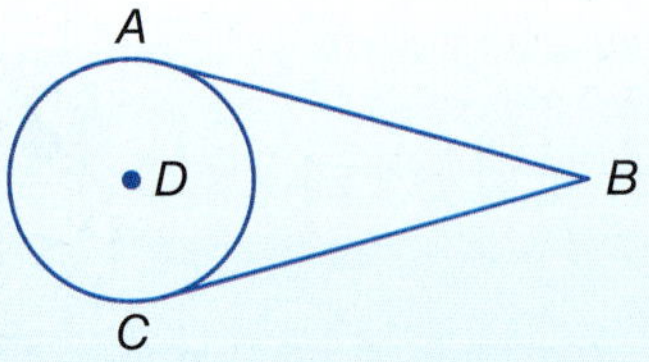

You will prove Theorem 10.11 in Exercise 28.

EXAMPLE 4 Use Congruent Tangents to Find Measures

ALGEBRA $\overline{AB}$ and $\overline{CB}$ are tangent to $\odot D$. Find the value of x.

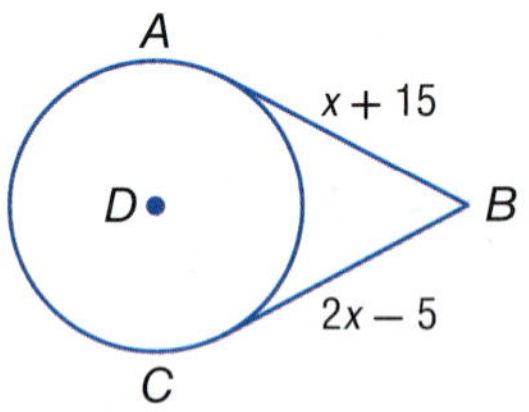

$AB = CB$	**Tangents from the same exterior point are congruent.**
$x + 15 = 2x - 5$	**Substitution**
$15 = x - 5$	**Subtract x from each side.**
$20 = x$	**Add 5 to each side.**

Guided Practice

ALGEBRA Find the value of x. Assume that segments that appear to be tangent are tangent.

4A.

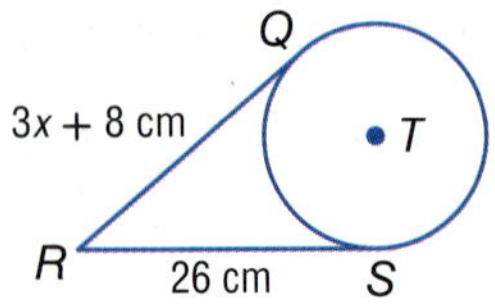

4B.

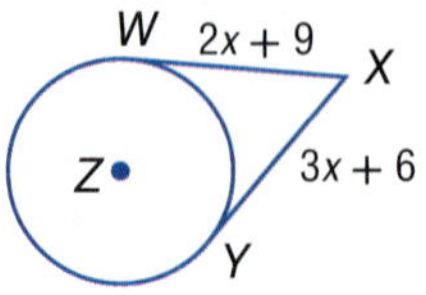

Personal Tutor glencoe.com

Circumscribed Polygons A polygon is circumscribed about a circle if every side of the polygon is tangent to the circle.

Circumscribed Polygons	Polygons Not Circumscribed
	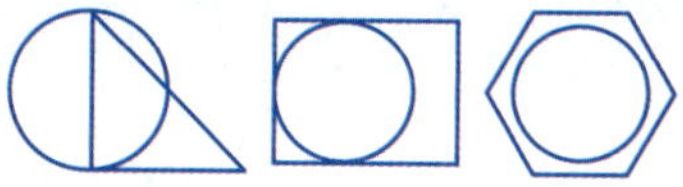

You can use Theorem 10.11 to find missing measures in circumscribed polygons.

Real-World EXAMPLE 5 Find Measures in Circumscribed Polygons

GRAPHIC DESIGN A graphic designer is giving directions to create a larger version of the triangular logo shown. If $\triangle ABC$ is circumscribed about $\odot G$, find the perimeter of $\triangle ABC$.

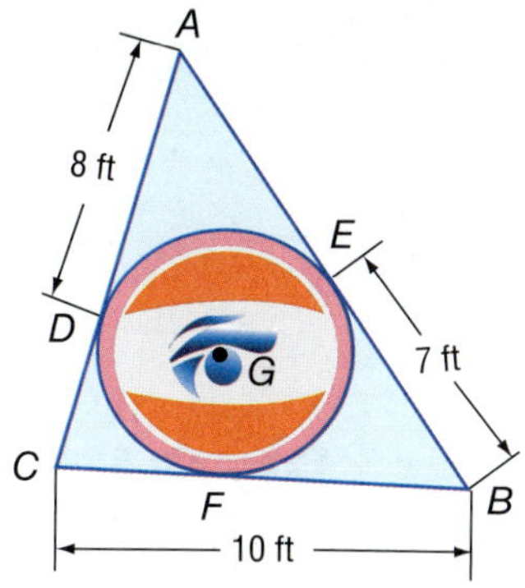

Step 1 Find the missing measures.

Since $\triangle ABC$ is circumscribed about $\odot G$, $\overline{AE}$ and $\overline{AD}$ are tangent to $\odot G$, as are $\overline{BE}$, $\overline{BF}$, $\overline{CF}$, and $\overline{CD}$. Therefore, $\overline{AE} \cong \overline{AD}$, $\overline{BF} \cong \overline{BE}$, and $\overline{CF} \cong \overline{CD}$.

So, $AE = AD = 8$ feet, $BF = BE = 7$ feet.

By Segment Addition, $CF + FB = CB$, so $CF = CB - FB = 10 - 7$ or 3 feet. So, $CD = CF = 3$ feet.

Step 2 Find the perimeter of $\triangle ABC$.

$$\text{perimeter} = AE + EB + BC + CD + DA$$
$$= 8 + 7 + 10 + 3 + 8 \text{ or } 36$$

So, the perimeter of $\triangle ABC$ is 36 feet.

> **Watch Out!**
>
> **Identifying Circumscribed Polygons** Just because the circle is tangent to one or more of the sides of a polygon does not mean that the polygon is circumscribed about the circle, as shown in the second set of figures.

Guided Practice

5. Quadrilateral $RSTU$ is circumscribed about $\odot J$. If the perimeter is 18 units, find x.

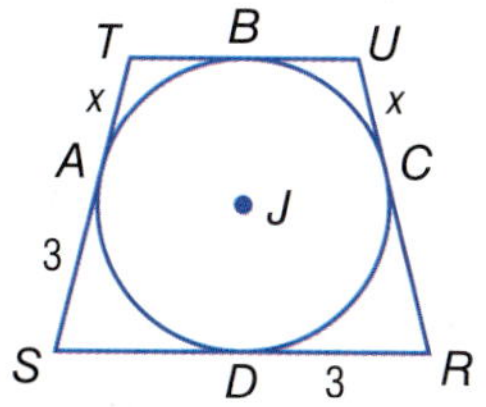

Personal Tutor glencoe.com

Check Your Understanding

Example 1 p. 718

1. Copy the figure shown, and draw the common tangents. If no common tangent exists, state *no common tangent*.

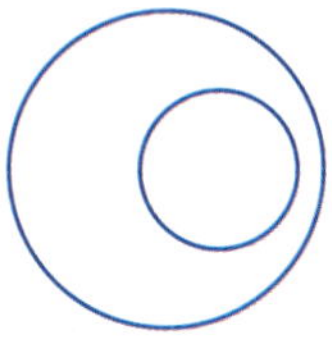

Example 2 p. 719

Determine whether $\overline{FG}$ is tangent to $\odot E$. Justify your answer.

2.

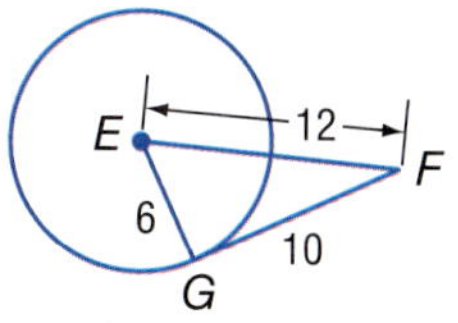

3

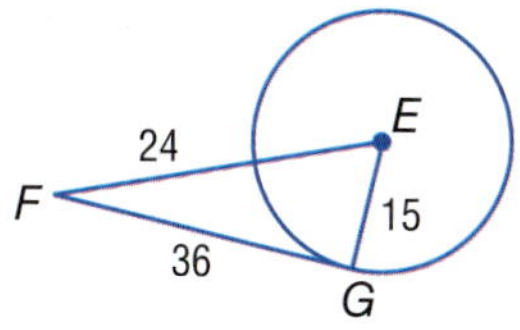

Examples 3 and 4
pp. 719–720

Find x. Assume that segments that appear to be tangent are tangent.

4. 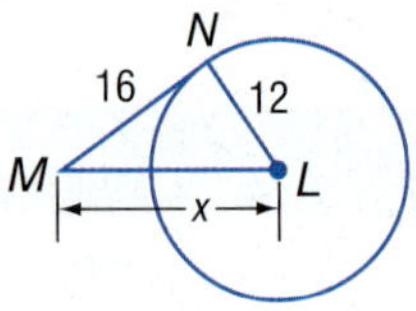

5. 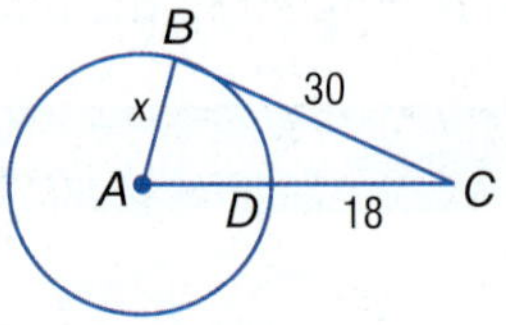

6. 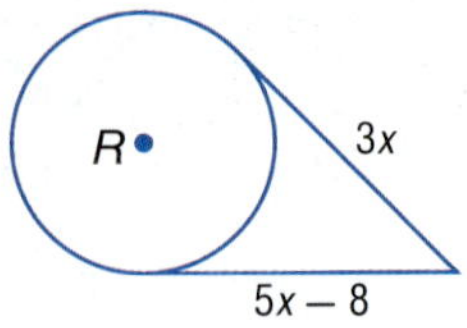

7. **LANDSCAPE ARCHITECT** A landscape architect is paving the two walking paths that are tangent to two approximately circular ponds as shown. The lengths given are in feet. Find the values of x and y. What is the total length of the walking paths?

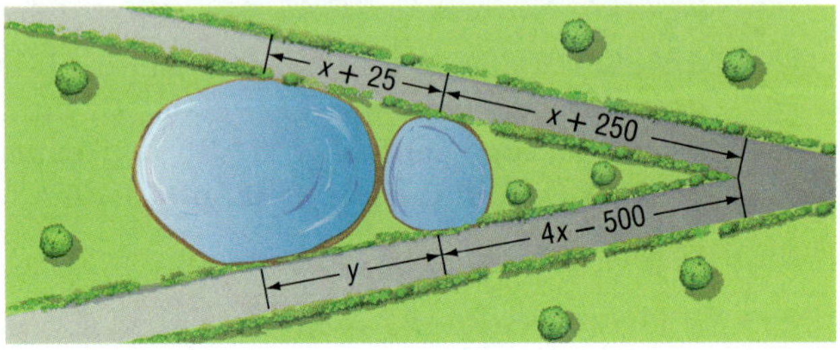

Example 5
p. 721

8. **ALGEBRA** Triangle JKL is circumscribed about $\odot R$.

   a. Find x.

   b. Find the perimeter of $\triangle JKL$.

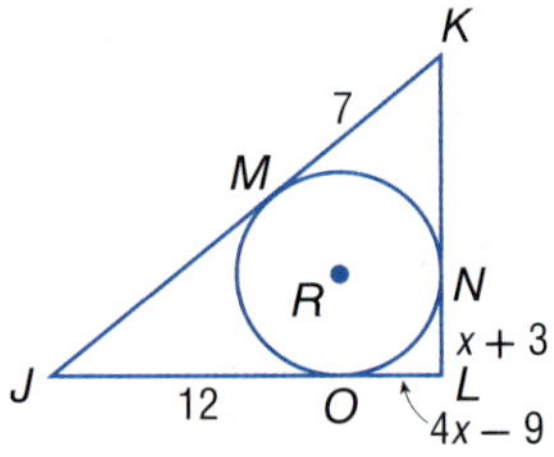

Practice and Problem Solving

= **Step-by-Step Solutions** begin on page R20.
Extra Practice begins on page 969.

Example 1
p. 718

Copy each figure and draw the common tangents. If no common tangent exists, state ***no common tangent.***

9. 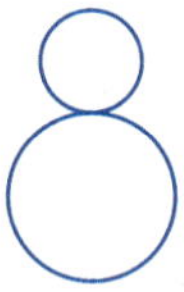

10. 

11. 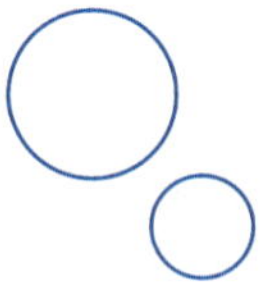

12. 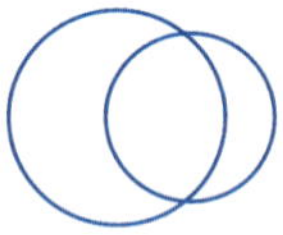

Example 2
p. 719

Determine whether each $\overline{XY}$ is tangent to the given circle. Justify your answer.

13. 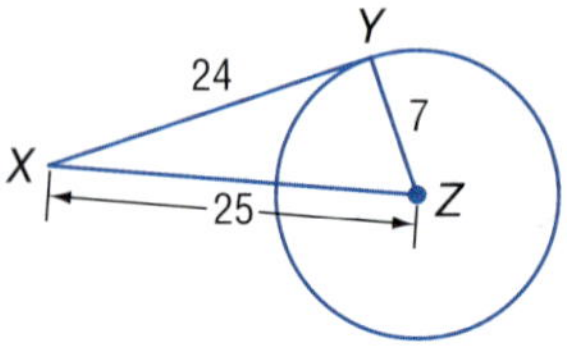

14. 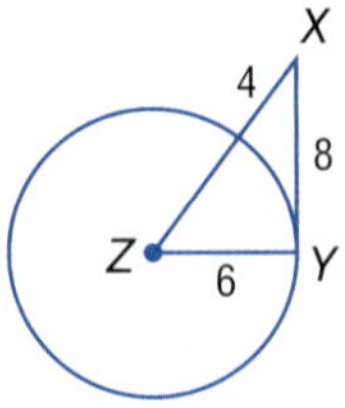

15. 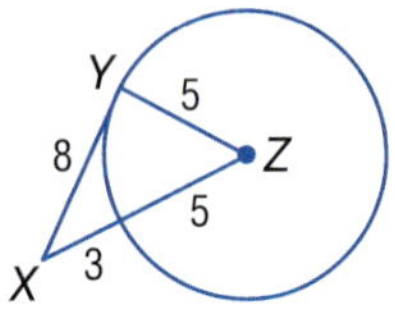

16. 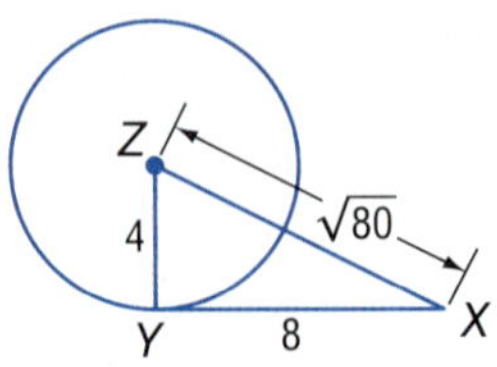

Examples 3 and 4
pp. 719–720

Find x. Assume that segments that appear to be tangent are tangent. Round to the nearest tenth if necessary.

17.
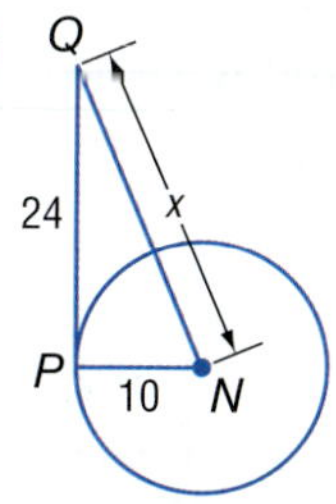

18.
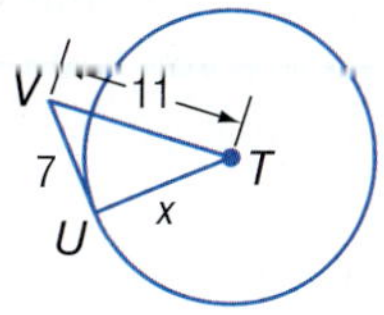

19.
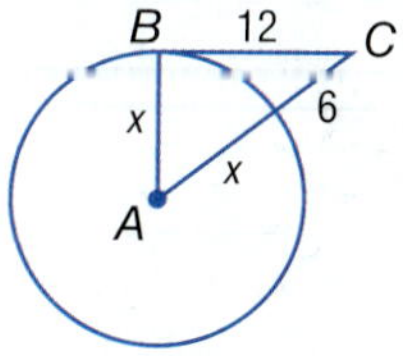

20.
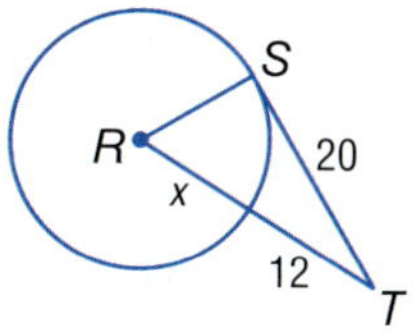

21.
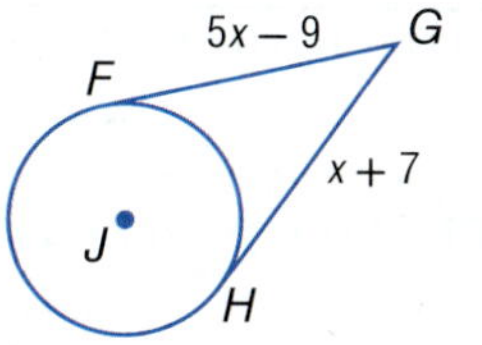

22.
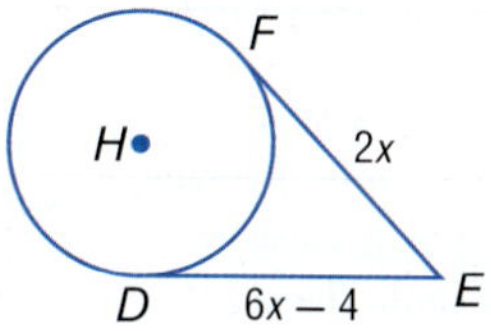

23. ARBORS In the arbor shown, $\overline{AC}$ and $\overline{BC}$ are tangents to $\odot D$. The radius of the circle is 26 inches and $EC = 20$ inches. Find each measure to the nearest hundredth.

a. AC

b. BC

Example 5
p. 721

Find the value of x. Then find the perimeter.

24.
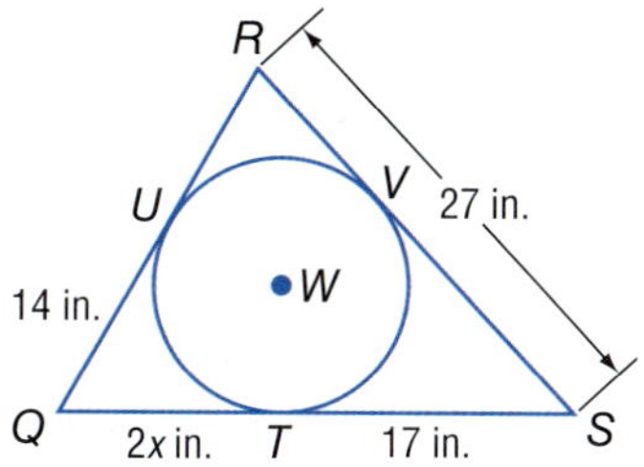

25.
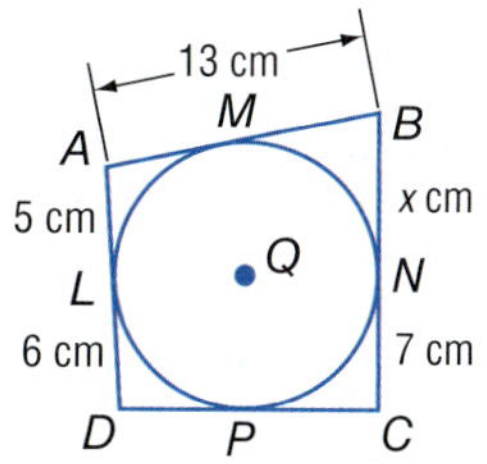

Find x to the nearest hundredth. Assume that segments that appear to be tangent are tangent.

26.
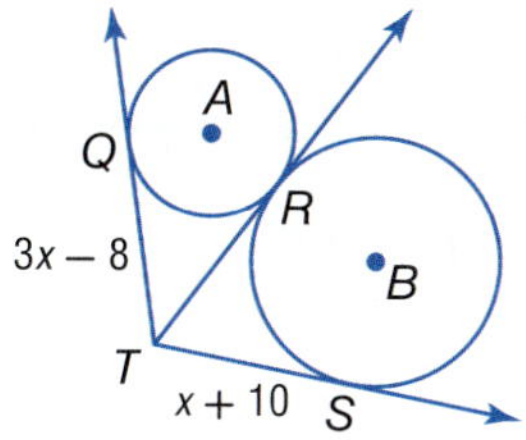

27.
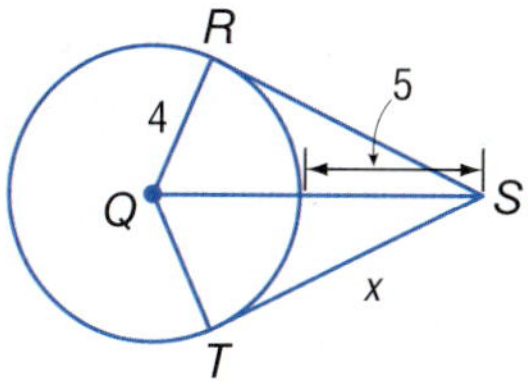

StudyTip

Identifying Tangents Do not assume that a segment that appears to be tangent to a circle is tangent, unless you are told to do so. The figure must either have a right angle symbol or include the measurements that confirm a right angle.

Write the specified type of proof.

28. two-column proof of Theorem 10.11

Given: $\overline{AC}$ is tangent to $\odot H$ at C.
$\overline{AB}$ is tangent to $\odot H$ at B.

Prove: $\overline{AC} \cong \overline{AB}$

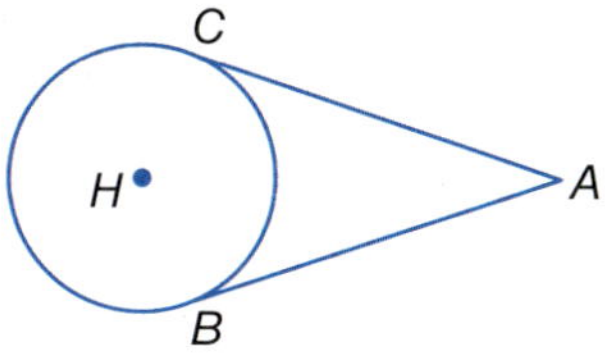

29. two-column proof

Given: Quadrilateral $ABCD$ is circumscribed about $\odot P$.

Prove: $AB + CD = AD + BC$

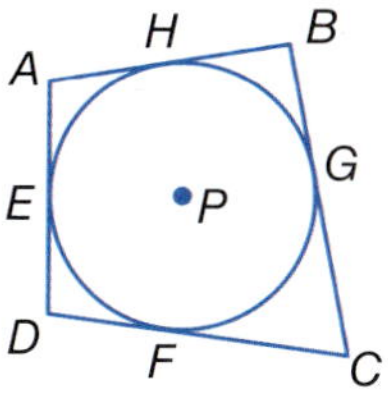

30. **SATELLITES** A satellite is 720 kilometers above Earth, which has a radius of 6360 kilometers. The region of Earth that is visible from the satellite is between the tangent lines $\overline{BA}$ and $\overline{BC}$. What is BA? Round to the nearest hundredth.

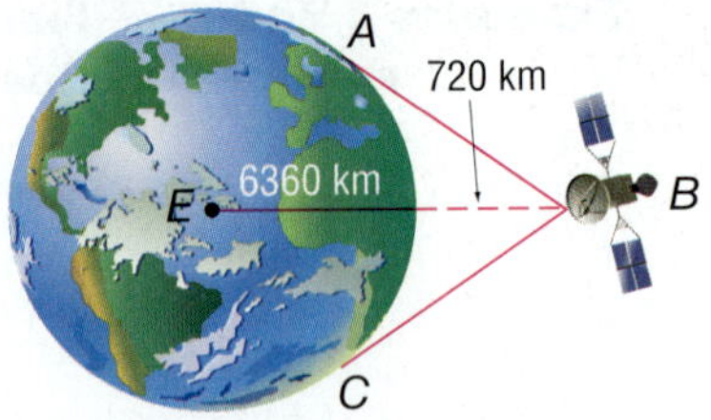

Real-World Link

There are more than 8000 large pieces of orbital debris, such as satellites, orbiting Earth. The objects travel approximately 8 kilometers per second. Objects more than 700 kilometers above Earth will remain in orbit for hundreds or even thousands of years.

Source: Tethers Unlimited, Inc.

31. **SPACE TRASH** *Orbital debris* refers to materials from space missions that still orbit Earth. In 2007, a 1400-pound ammonia tank was discarded from a space mission. Suppose the tank has an altitude of 435 miles. What is the distance from the tank to Earth's horizon? Assume that the radius of Earth is 4000 miles. Round to the nearest mile, and include a diagram of this situation with your answer.

32. **PROOF** Write an indirect proof to show that if a line is tangent to a circle, then it is perpendicular to a radius of the circle. (Part 1 of Theorem 10.10)

Given: ℓ is tangent to $\odot S$ at T; $\overline{ST}$ is a radius of $\odot S$.

Prove: $\ell \perp \overline{ST}$

(*Hint:* Assume ℓ is *not* $\perp$ to $\overline{ST}$.)

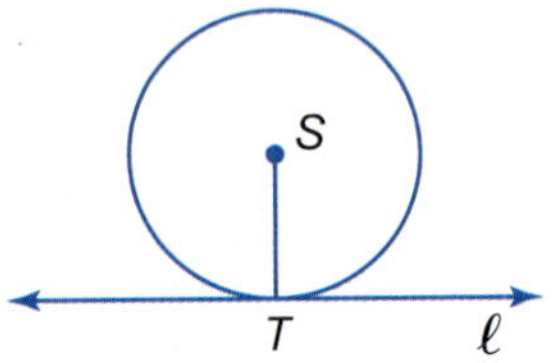

33. **PROOF** Write an indirect proof to show that if a line is perpendicular to the radius of a circle at its endpoint, then the line is a tangent of the circle. (Part 2 of Theorem 10.10)

Given: $\ell \perp \overline{ST}$; $\overline{ST}$ is a radius of $\odot S$.

Prove: ℓ is tangent to $\odot S$.

(*Hint:* Assume ℓ is *not* tangent to $\odot S$.)

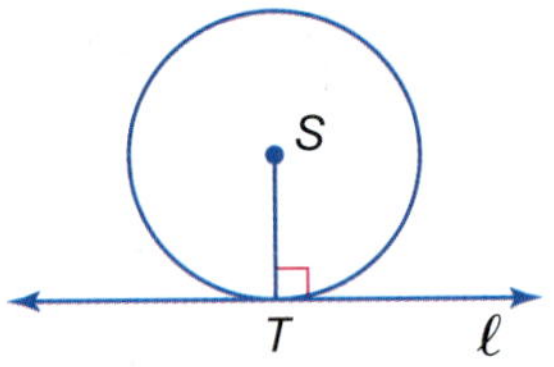

34. **CONSTRUCTION** Construct a line tangent to a circle through a point on the circle. Use a compass to draw $\odot A$. Choose a point P on the circle and draw $\overleftrightarrow{AP}$. Then construct a segment through point P perpendicular to $\overleftrightarrow{AP}$. Label the tangent line t. Explain and justify each step.

H.O.T. Problems

Use **H**igher-**O**rder **T**hinking Skills

35. **CHALLENGE** $\overline{PQ}$ is tangent to circles R and S. Find PQ. Explain your reasoning.

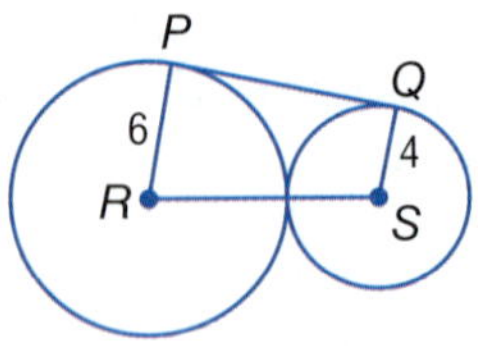

36. **WRITING IN MATH** Explain and justify each step in the construction on page 720.

37. **OPEN ENDED** Draw a circumscribed triangle and an inscribed triangle.

38. **REASONING** In the figure, $\overline{XY}$ and $\overline{XZ}$ are tangent to $\odot A$. $\overline{XZ}$ and $\overline{XW}$ are tangent to $\odot B$. Explain how segments $\overline{XY}$, $\overline{XZ}$, and $\overline{XW}$ can all be congruent if the circles have different radii.

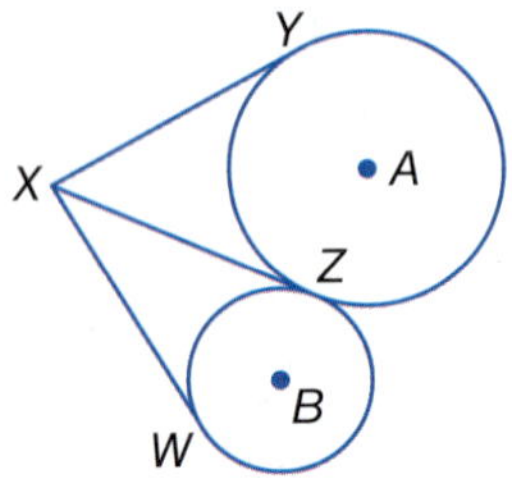

39. **WRITING IN MATH** How many tangents can be drawn from a point outside a circle, from a point on a circle, and from a point inside a circle? Explain your reasoning.

Standardized Test Practice

40. $\odot P$ has a radius of 10 centimeters, and $\overline{ED}$ is tangent to the circle at point D. F lies both on $\odot P$ and on segment $\overline{EP}$. If $ED = 24$ centimeters, what is the length of $\overline{EF}$?

A 10 cm
B 16 cm
C 21.8 cm
D 26 cm

41. SHORT RESPONSE A square is inscribed in a circle having a radius of 6 inches. Find the length of each side of the square.

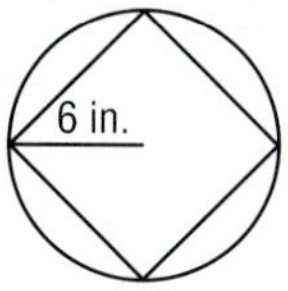

42. ALGEBRA Which of the following shows $25x^2 - 5x$ factored completely?

F $5x(x)$
G $5x(5x - 1)$
H $x(x - 5)$
J $x(5x - 1)$

43. SAT/ACT What is the perimeter of the triangle shown below?

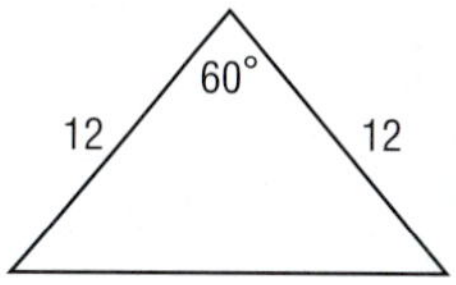

A 24 units
B 34.4 units
C 36 units
D 104 units

Spiral Review

Find each measure. (Lesson 10-4)

44. $m\widehat{JK}$

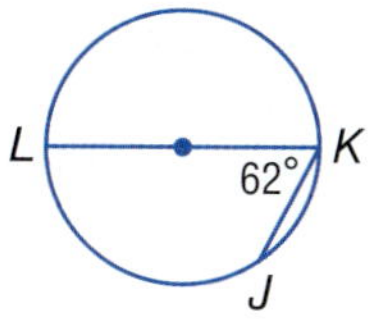

45. $m\angle B$

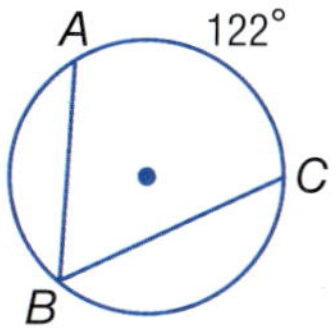

46. $m\widehat{VX}$

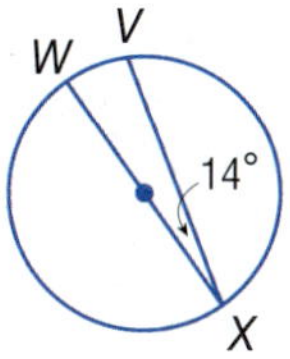

In $\odot F$, $GK = 14$ and $m\widehat{GHK} = 142$. Find each measure. Round to the nearest hundredth. (Lesson 10-3)

47. $m\widehat{GH}$

48. JK

49. $m\widehat{KM}$

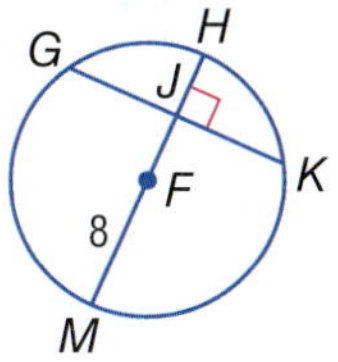

50. METEOROLOGY The altitude of the base of a cloud formation is called the *ceiling*. To find the ceiling one night, a meteorologist directed a spotlight vertically at the clouds. Using a theodolite, an optical instrument with a rotatable telescope, placed 83 meters from the spotlight and 1.5 meters above the ground, he found the angle of elevation to be 62.7°. How high was the ceiling? (Lesson 8-5)

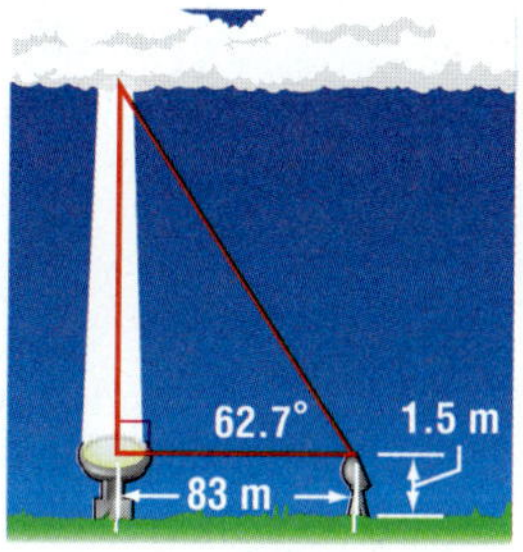

Determine whether the triangles are similar. If so, write a similarity statement. Explain your reasoning. (Lesson 7-3)

51.

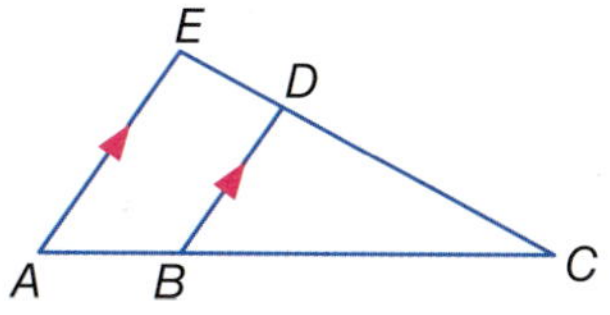

52.

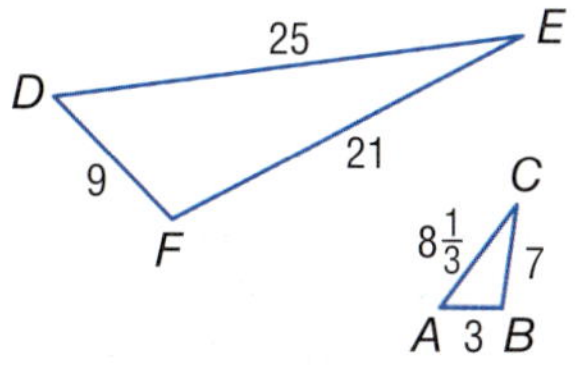

Skills Review

Solve each equation. (Lesson 0-5)

53. $15 = \frac{1}{2}[(360 - x) - 2x]$

54. $x + 12 = \frac{1}{2}[(180 - 120)]$

55. $x = \frac{1}{2}[(180 - 64)]$

EXTEND 10-5

Geometry Lab: Inscribed and Circumscribed Circles

In Lesson 5-1, you learned that there are special points of concurrency in a triangle. The *incenter* is the point at which the angle bisectors meet. It is equidistant from the sides of the triangle.

ACTIVITY 1 Construct a Circle Inscribed in a Triangle

Step 1

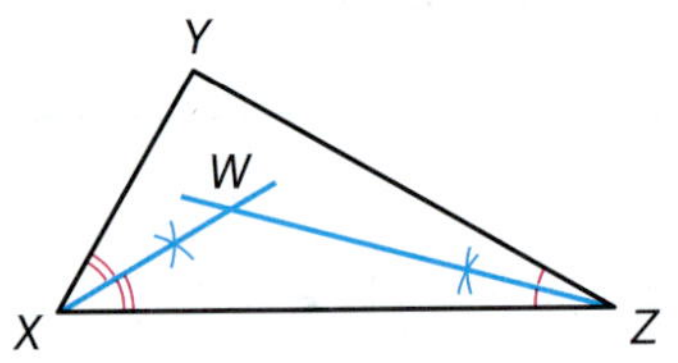

Draw a triangle XYZ and construct two angle bisectors of the triangle to locate the incenter W.

Step 2

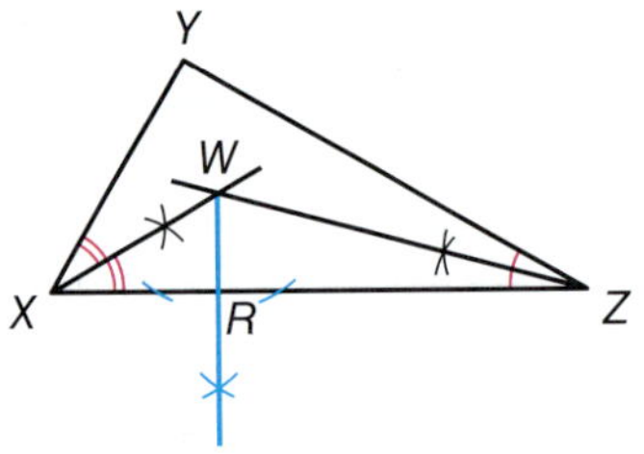

Construct a segment perpendicular to a side through the incenter. Label the intersection R.

Step 3

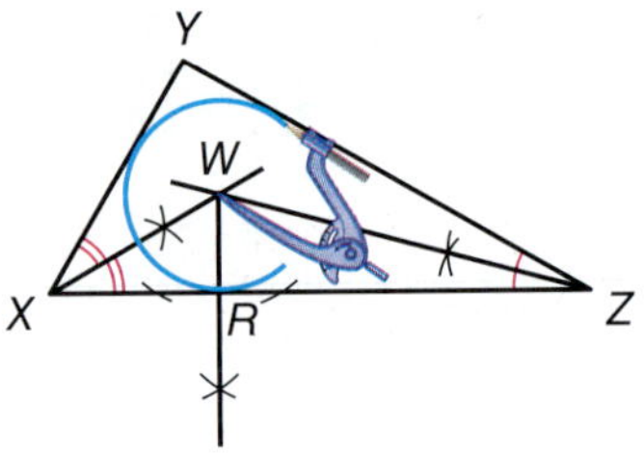

Set a compass of the length of $\overline{WR}$. Put the point of the compass on W and draw a circle with that radius.

ACTIVITY 2 Construct a Triangle Circumscribed About a Circle

Step 1

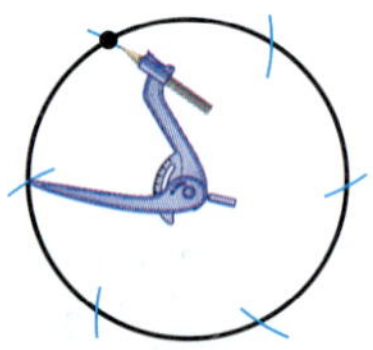

Construct a circle and draw a point. Use the same compass setting you used to construct the circle to construct an arc on the circle from the point. Continue as shown.

Step 2

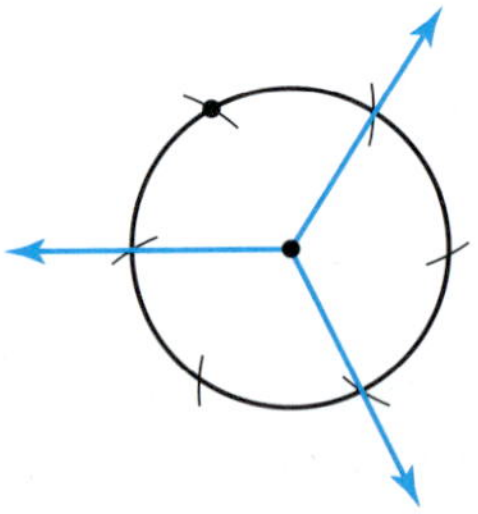

Draw rays from the center through every other arc.

Step 3

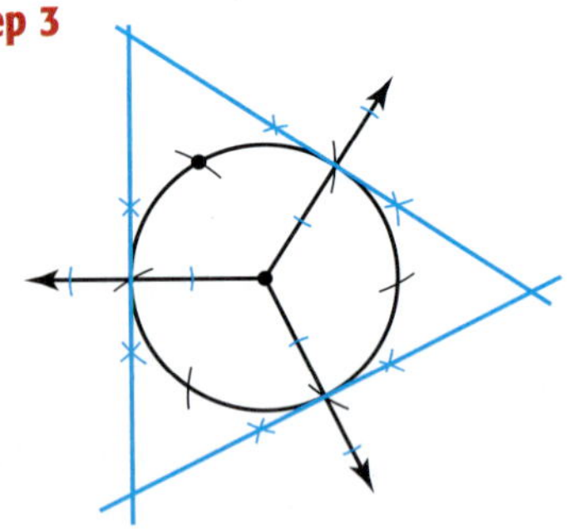

Construct a line perpendicular to each of the rays.

Model and Analyze

1. In Activity 1, why is it only necessary to construct the perpendicular to one side of the triangle?
2. Draw a right triangle and inscribe a circle in it.
3. **CHALLENGE** Circumscribe an obtuse triangle about a circle. Explain your reasoning.
4. **WRITING IN MATH** Why do you think the term *incenter* is a good term for the point it defines?

10-6 Secants, Tangents, and Angle Measures

Then
You found measures of segments formed by tangents to a circle.
(Lesson 10-5)

Now
- Find measures of angles formed by lines intersecting on or inside a circle.
- Find measures of angles formed by lines intersecting outside the circle.

New Vocabulary
secant

Math Online
glencoe.com

Why?

An average person's field of vision is about 180°. Most cameras have a much narrower viewing angle of between 20° and 50°. This viewing angle determines how much of a curved object a camera can capture on film.

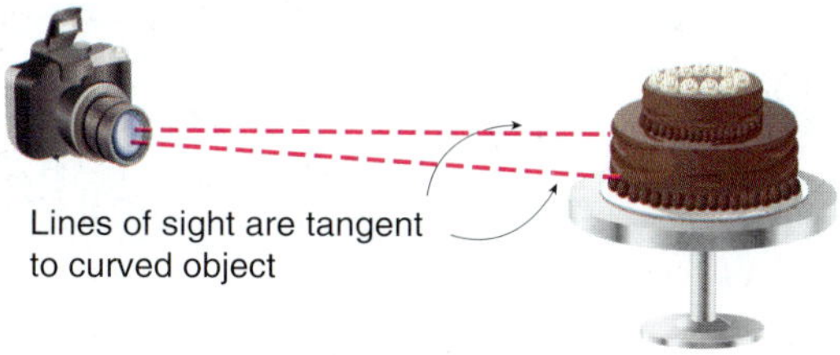

Intersections On or Inside a Circle A **secant** is a line that intersects a circle in exactly two points. Lines j and k are secants of $\odot C$.

When two secants intersect inside a circle, the angles formed are related to the arcs they intercept.

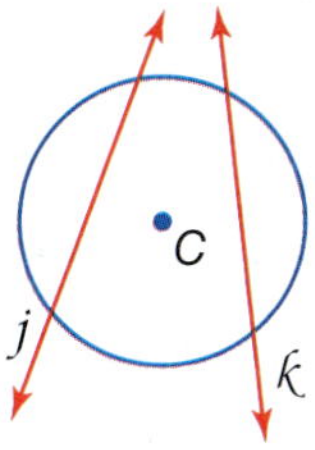

Theorem 10.12

Words If two secants or chords intersect in the interior of a circle, then the measure of an angle formed is one half the *sum* of the measure of the arcs intercepted by the angle and its vertical angle.

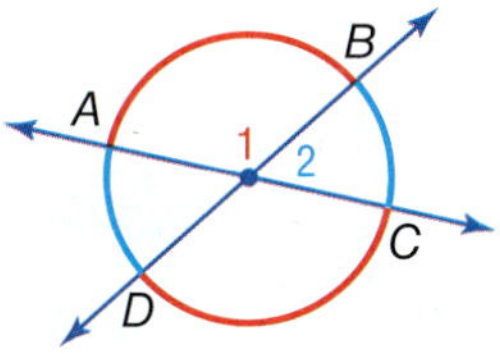

Example $m\angle 1 = \frac{1}{2}(m\overset{\frown}{AB} + m\overset{\frown}{CD})$ and $m\angle 2 = \frac{1}{2}(m\overset{\frown}{DA} + m\overset{\frown}{BC})$

Proof

Given: $\overleftrightarrow{HK}$ and $\overleftrightarrow{JL}$ intersect at M.

Prove: $m\angle 1 = \frac{1}{2}(m\overset{\frown}{JH} + m\overset{\frown}{LK})$

Proof:

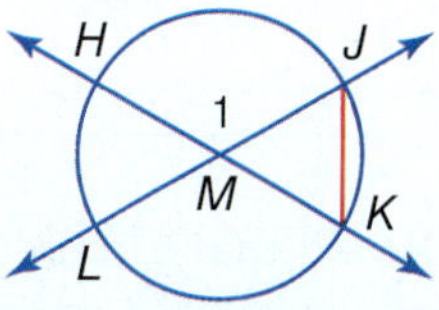

Statements	Reasons
1. $\overleftrightarrow{HK}$ and $\overleftrightarrow{JL}$ intersect at M.	**1.** Given
2. $m\angle 1 = m\angle MJK + m\angle MKJ$	**2.** Exterior Angle Theorem
3. $m\angle MJK = \frac{1}{2}m\angle\overset{\frown}{LK}$, $m\angle MKJ = \frac{1}{2}m\angle\overset{\frown}{JH}$	**3.** The measure of an inscribed $\angle$ equals half the measure of the intercepted arc.
4. $m\angle 1 = \frac{1}{2}m\angle\overset{\frown}{LK} + \frac{1}{2}m\angle\overset{\frown}{JH}$	**4.** Substitution
5. $m\angle 1 = \frac{1}{2}(m\overset{\frown}{JH} + m\overset{\frown}{LK})$	**5.** Distributive Property

EXAMPLE 1 Use Intersecting Chords or Secants

Find x.

a.

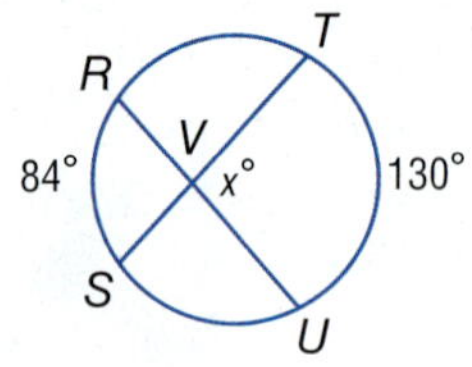

$m\angle TVU = \frac{1}{2}(m\widehat{RS} + m\widehat{TU})$ **Theorem 10.12**

$x = \frac{1}{2}(84 + 130)$ **Substitution**

$= \frac{1}{2}(214)$ or 107 **Simplify.**

b.

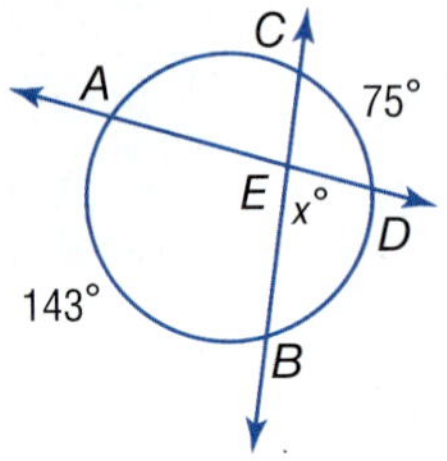

Step 1 Find $m\angle AEB$.

$m\angle AEB = \frac{1}{2}(m\widehat{AB} + m\widehat{CD})$ **Theorem 10.12**

$= \frac{1}{2}(143 + 75)$ **Substitution**

$= \frac{1}{2}(218)$ or 109 **Simplify.**

Step 2 Find x, the measure of $\angle DEB$.

$\angle AEB$ and $\angle DEB$ are supplementary angles.
So, $x = 180 - 109$ or 71.

c.

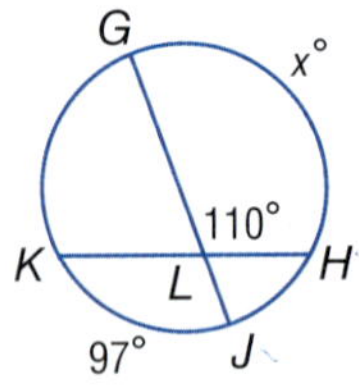

$m\angle GLH = \frac{1}{2}(m\widehat{GH} + m\widehat{KJ})$ **Theorem 10.12**

$110 = \frac{1}{2}(x + 97)$ **Substitution**

$220 = (x + 97)$ **Multiply each side by 2.**

$123 = x$ **Subtract 97 from each side.**

Guided Practice

1A.

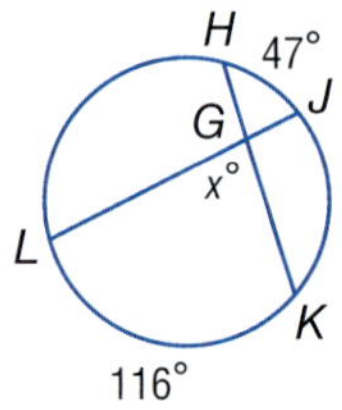

1B.

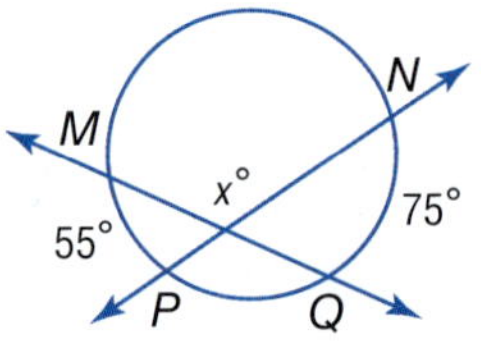

1C.

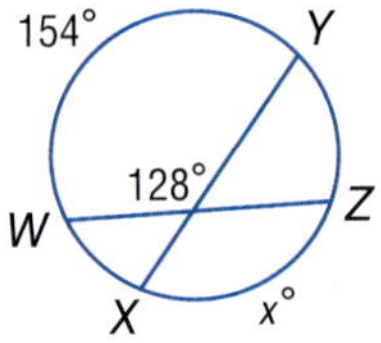

Personal Tutor glencoe.com

StudyTip

Alternative Method
In Example 1b, $m\angle DEB$ can also be found by first finding the sum of the measures of $\widehat{AC}$ and $\widehat{BD}$.

$m\widehat{AC} + m\widehat{BD}$
$= 360 - (m\widehat{AC} + m\widehat{CD})$
$= 360 - (143 + 75)$
$= 142$

$m\angle DEB$
$= \frac{1}{2}(m\widehat{AC} + m\widehat{BD})$
$= \frac{1}{2}(142)$ or 71

Recall that Theorem 10.6 states that the measure of an inscribed angle is half the measure of its intercepted arc. If one of the sides of this angle is tangent to the circle, this relationship still holds true.

Theorem 10.13

Words If a secant and a tangent intersect at the point of tangency, then the measure of each angle formed is one half the measure of its intercepted arc.

Example $m\angle 1 = \frac{1}{2}m\widehat{AB}$ and $m\angle 2 = \frac{1}{2}m\widehat{ACB}$

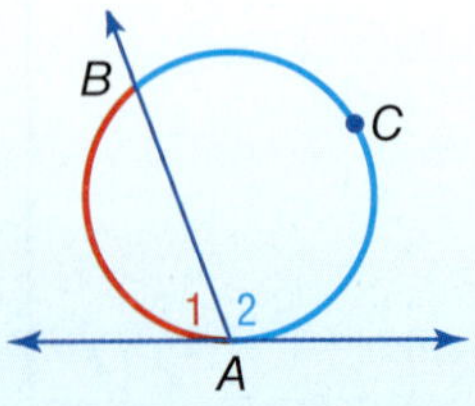

You will prove Theorem 10.13 in Exercise 33.

EXAMPLE 2 Use Intersecting Secants and Tangents

Find each measure.

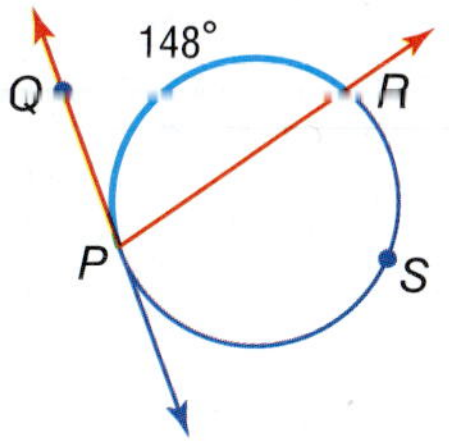

a. $m\angle QPR$

$m\angle QPR = \frac{1}{2}m\widehat{PR}$ — **Theorem 10.13**

$= \frac{1}{2}(148)$ or 74 — **Substitute and simplify.**

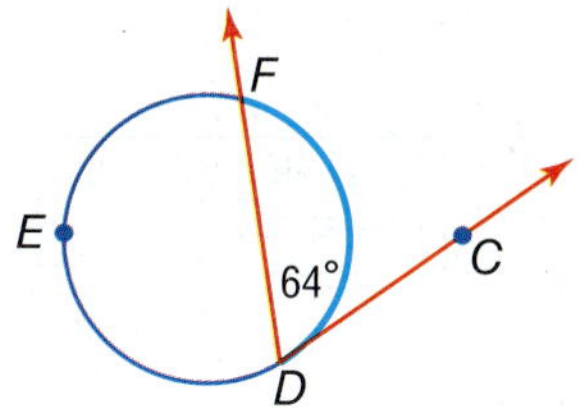

b. $m\widehat{DEF}$

$m\angle CDF = \frac{1}{2}m\widehat{FD}$ — **Theorem 10.13**

$64 = \frac{1}{2}m\widehat{FD}$ — **Substitution**

$128 = m\widehat{FD}$ — **Multiply each side by 2.**

$m\widehat{DEF} = 360 - m\widehat{FD} = 360 - 128$ or 232

Guided Practice

2A. Find $m\widehat{JLK}$.

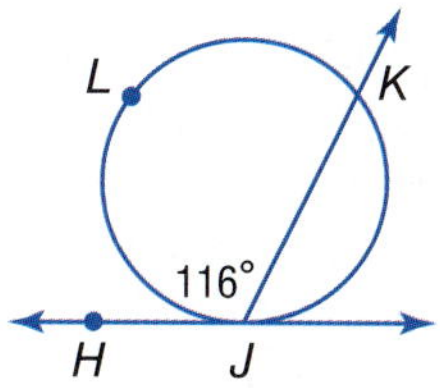

2B. Find $m\angle RQS$ if $m\widehat{QTS} = 238$.

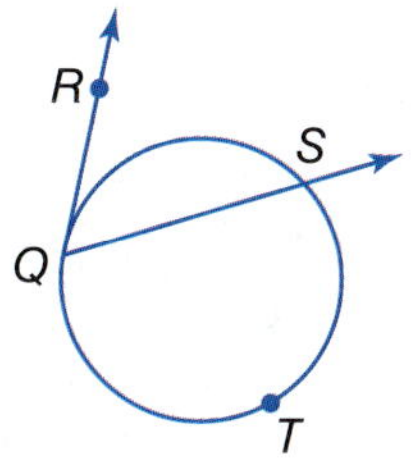

Personal Tutor glencoe.com

StudyTip

Absolute Value The measure of each $\angle A$ can also be expressed as half the absolute value of the difference of the arc measure. In this way, the order of the arc measures does not affect the outcome of the calculation.

Intersections Outside a Circle Secants and tangents can also meet outside a circle. The measure of the angle formed also involves half of the measures of the arcs they intercept.

Theorem 10.14

Words If two secants, a secant and a tangent, or two tangents intersect in the exterior of a circle, then the measure of the angle formed is one half the *difference* of the measures of the intercepted arcs.

Examples

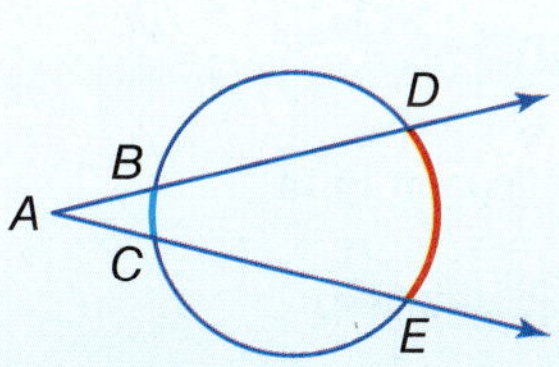

Two Secants

$m\angle A = \frac{1}{2}(m\widehat{DE} - m\widehat{BC})$

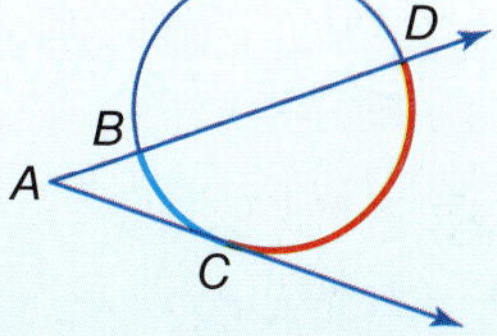

Secant-Tangent

$m\angle A = \frac{1}{2}(m\widehat{DC} - m\widehat{BC})$

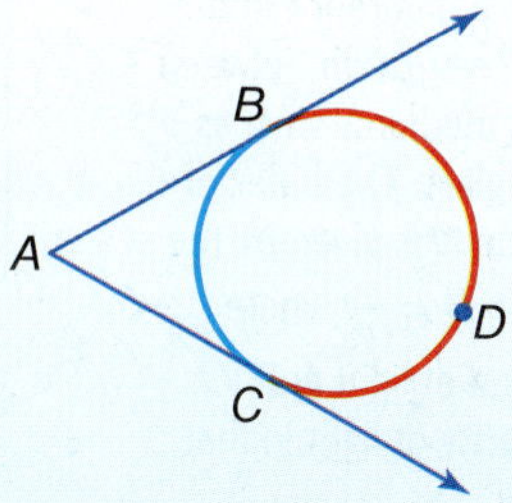

Two Tangents

$m\angle A = \frac{1}{2}(m\widehat{BDC} - m\widehat{BC})$

You will prove Theorem 10.14 in Exercises 30–32.

EXAMPLE 3 Use Tangents and Secants that Intersect Outside a Circle

Find each measure.

a. $m\angle L$

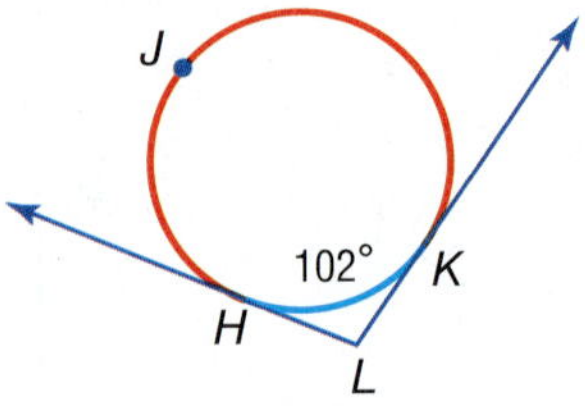

$m\angle L = \frac{1}{2}(m\widehat{HJK} - m\widehat{HK})$ **Theorem 10.14**

$= \frac{1}{2}[(360 - 102) - 102]$ **Substitution**

$= \frac{1}{2}(258 - 102)$ or 78 **Simplify.**

b. $m\widehat{CD}$

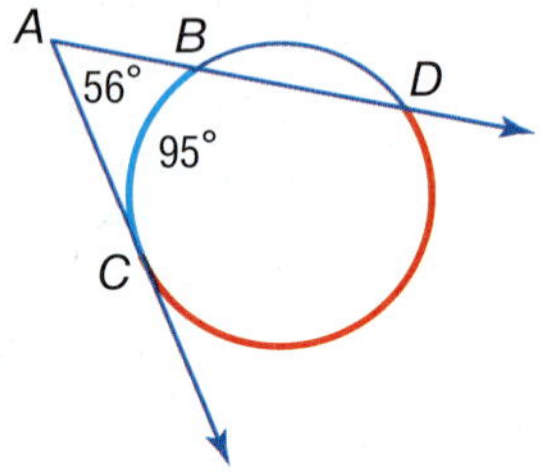

$m\angle A = \frac{1}{2}(m\widehat{CD} - m\widehat{BC})$ **Theorem 10.14**

$56 = \frac{1}{2}(m\widehat{CD} - 95)$ **Substitution**

$112 = m\widehat{CD} - 95$ **Multiply each side by 2.**

$207 = m\widehat{CD}$ **Add 95 to each side.**

Guided Practice

3A. $m\angle S$

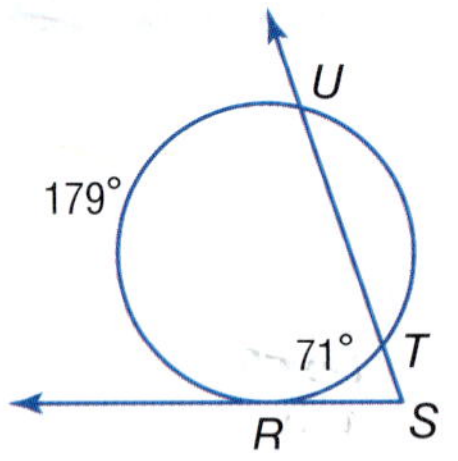

3B. $m\widehat{XZ}$

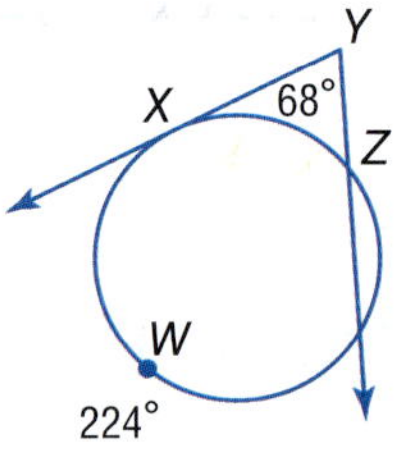

Personal Tutor glencoe.com

You can apply the properties of intersecting secants to solve real-world problems.

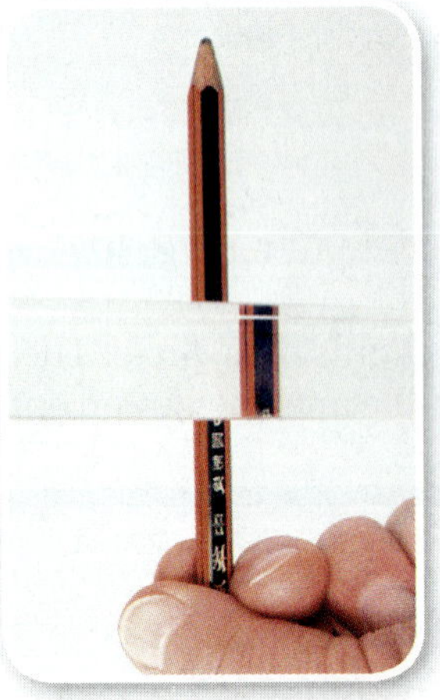

Real-World Link

There is a difference in the *index of refraction* between the two mediums such as air and glass. The index of refraction N is given by the equation $N = \frac{c}{V}$, where c is the speed of light and V is the velocity of light in that material.

Source: Microscopy Resource Center

Real-World EXAMPLE 4 Apply Properties of Intersecting Secants

SCIENCE The diagram shows the path of a light ray as it hits a drop of water. The ray is bent, or *refracted*, at points A, B, and C. If $m\widehat{AC} = 128$ and $m\widehat{XBY} = 84$, what is $m\angle D$?

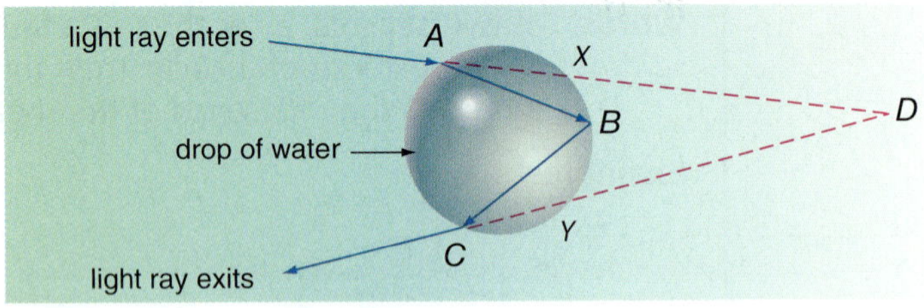

$m\angle D = \frac{1}{2}(m\widehat{AC} - m\widehat{XBY})$ **Theorem 10.14**

$= \frac{1}{2}(128 - 84)$ **Substitution**

$= \frac{1}{2}(44)$ or 22 **Simplify.**

Guided Practice

4. Find the value of x.

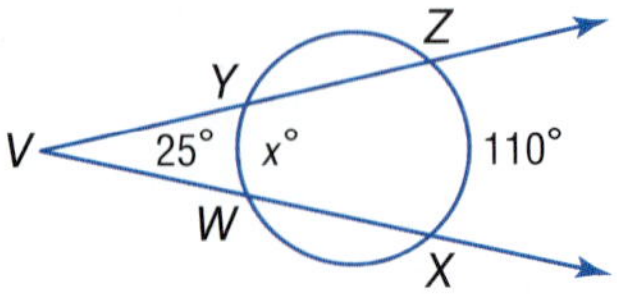

Personal Tutor glencoe.com

Concept Summary

Circle and Angle Relationships

Vertex of Angle	Model(s)	Angle Measure
on the circle	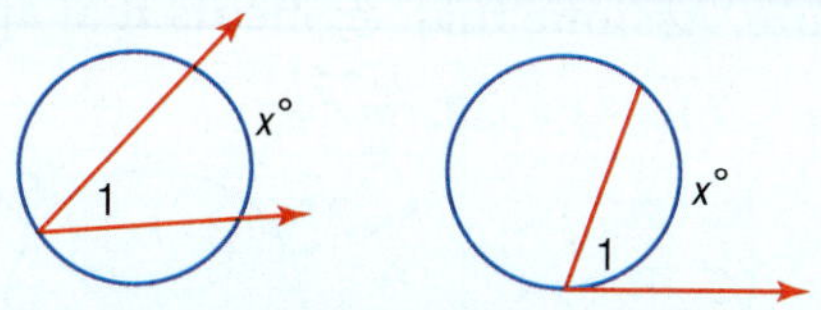	one half the measure of the intercepted arc $m\angle 1 = \frac{1}{2}x$
inside the circle	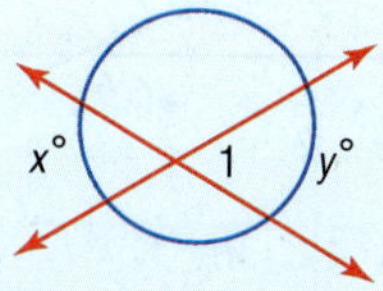	one half the measure of the sum of the intercepted arc $m\angle 1 = \frac{1}{2}(x + y)$
outside the circle	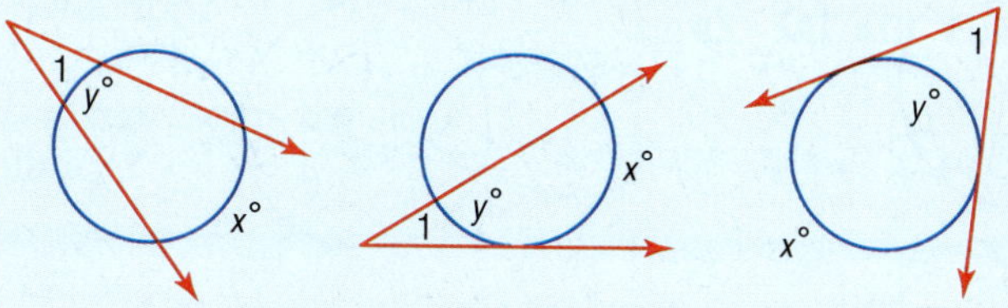	one half the measure of the difference of the intercepted arcs $m\angle 1 = \frac{1}{2}(x - y)$

Check Your Understanding

Examples 1 and 2
pp. 728–729

Find each measure. Assume that segments that appear to be tangent are tangent.

1. $m\angle 1$

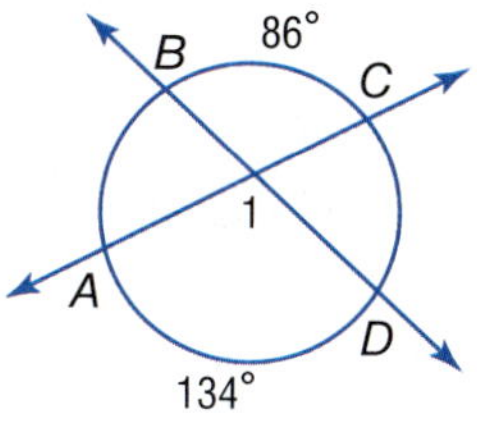

2. $m\widehat{TS}$

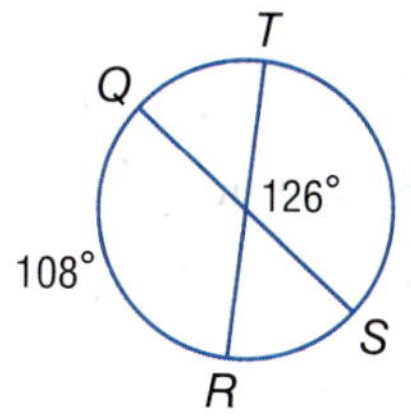

3. $m\angle 2$

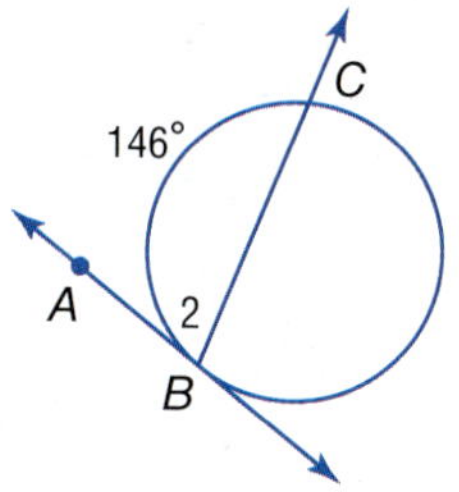

Examples 3 and 4
p. 730

4. $m\angle H$

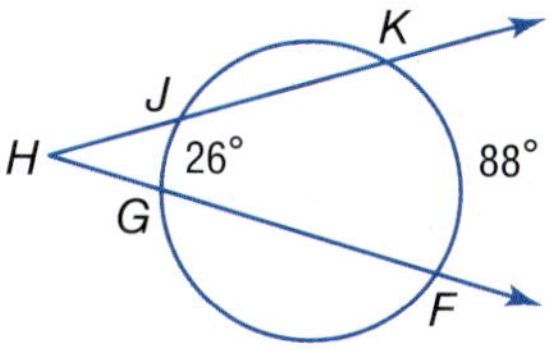

5. $m\widehat{QTS}$

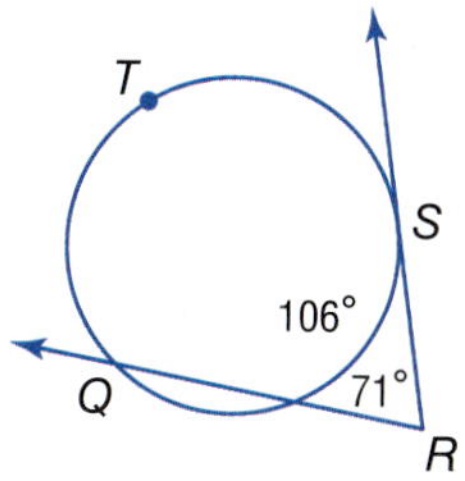

6. $m\widehat{LP}$

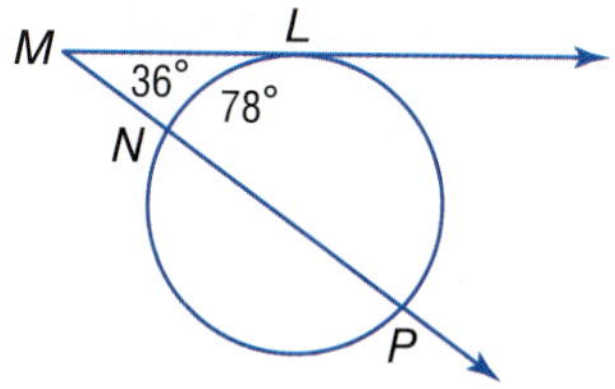

7 **STUNTS** A ramp is attached to the first of several barrels that have been strapped together for a circus motorcycle stunt as shown. What is the measure of the angle the ramp makes with the ground?

Practice and Problem Solving

● = **Step-by-Step Solutions** begin on page R20.
Extra Practice begins on page 969.

Examples 1 and 2
pp. 728–729

Find each measure. Assume that segments that appear to be tangent are tangent.

8. $m\angle 3$

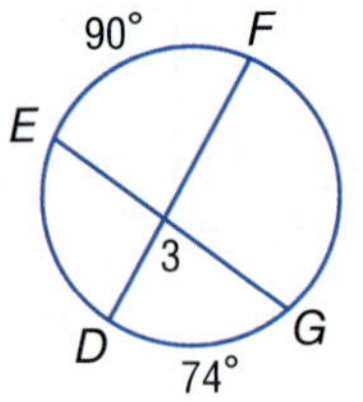

9. $m\angle 4$

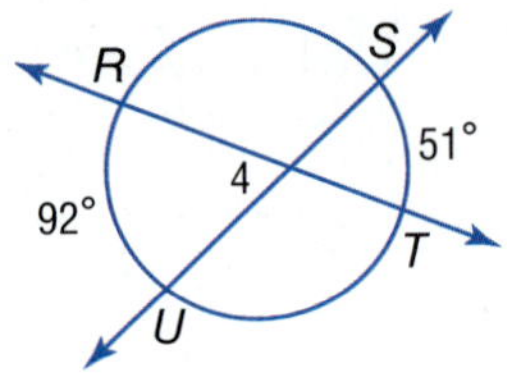

10. $m\angle JMK$

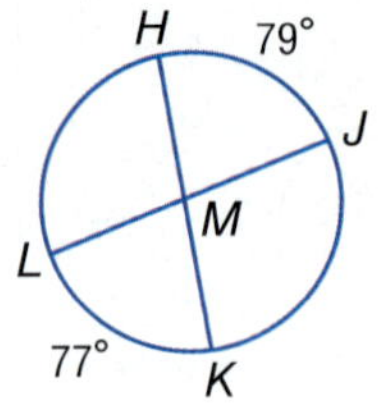

11. $m\widehat{RQ}$

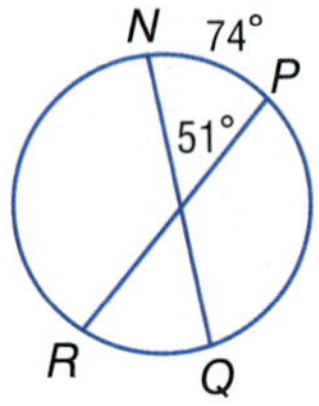

12. $m\angle K$

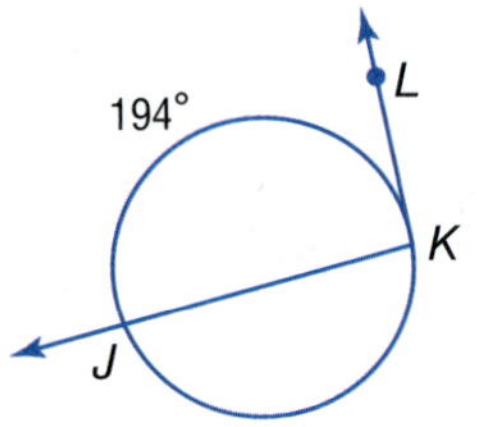

13. $m\widehat{PM}$

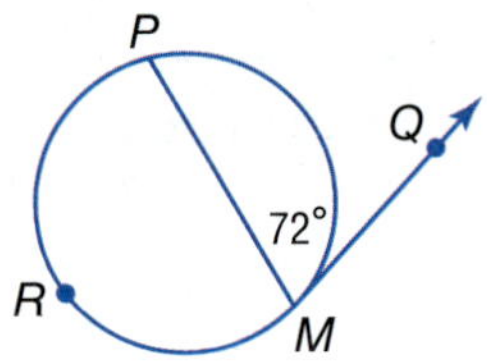

14. $m\angle ABD$

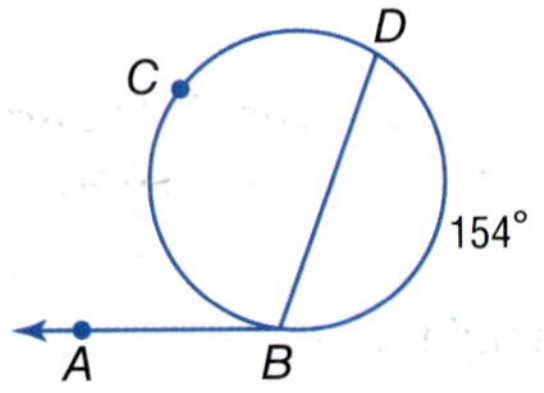

15. $m\angle DAB$

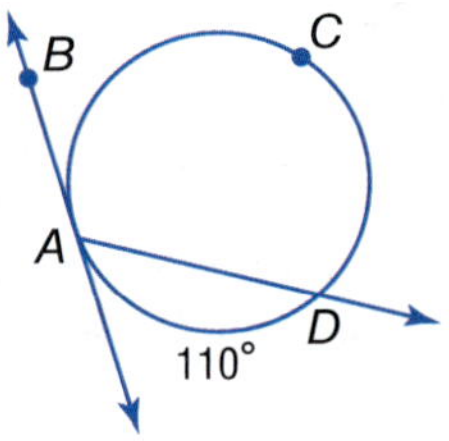

16. $m\widehat{GJF}$

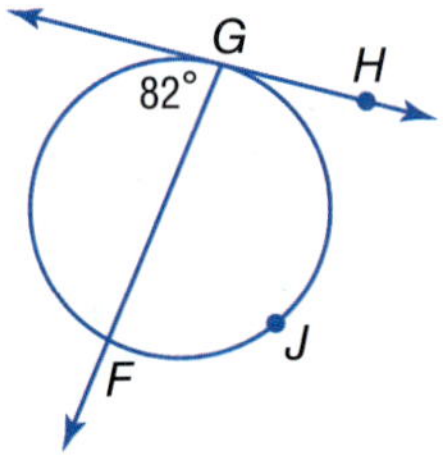

17. **SPORTS** The multi-sport field shown includes a softball field and a soccer field. If $m\widehat{ABC} = 200$, find each measure.

a. $m\angle ACE$

b. $m\angle ADC$

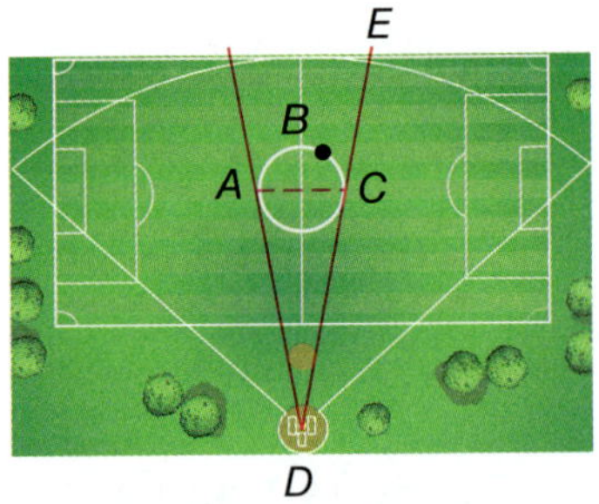

Examples 3 and 4
p. 730

Find each measure.

18. $m\angle A$

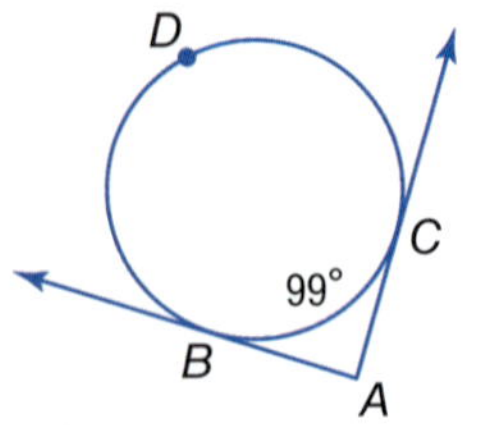

19. $m\angle W$

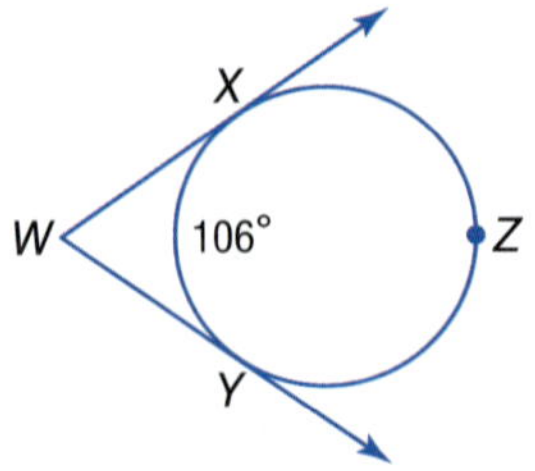

20. $m\widehat{JM}$

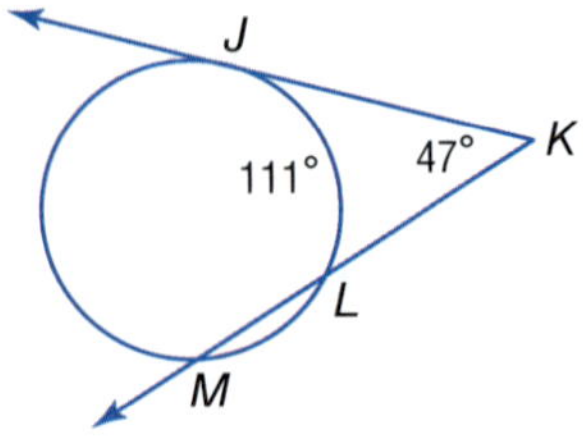

21. $m\widehat{XY}$

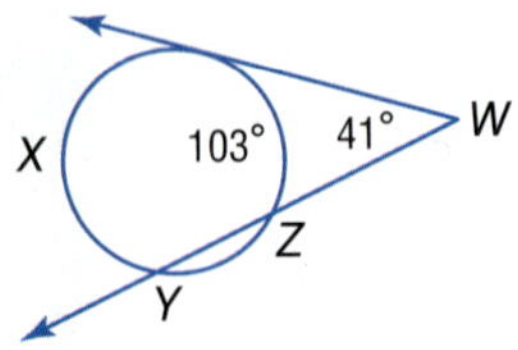

22. $m\angle R$

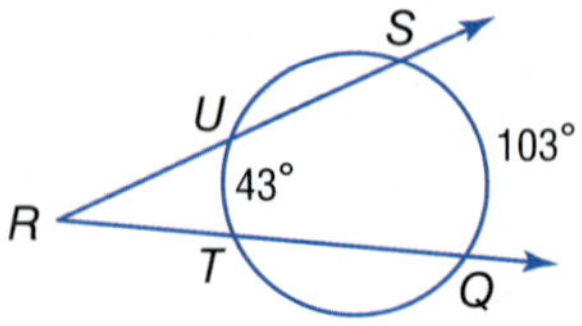

23. $m\widehat{SU}$

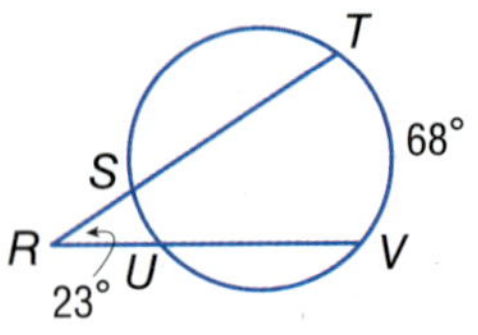

Real-World Career

Astronautical Engineer
Astronautical engineers are engineers that work with spacecraft. They may specialize in areas such as structural design or navigation and control.

Astronautical engineers must earn a bachelor's degree. A master's degree is favorable for those working in research positions.

24. JEWELRY In the circular necklace shown, A and B are tangent points. If $x = 260$, what is y?

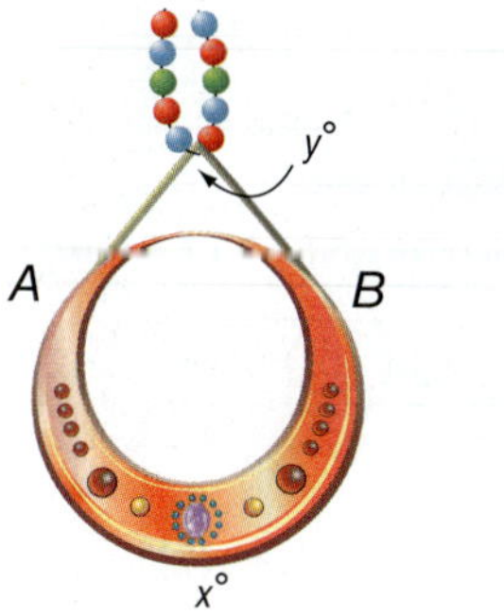

25. SPACE A satellite orbits above Earth's equator. Find x, the measure of the planet's arc, that is visible to the satellite.

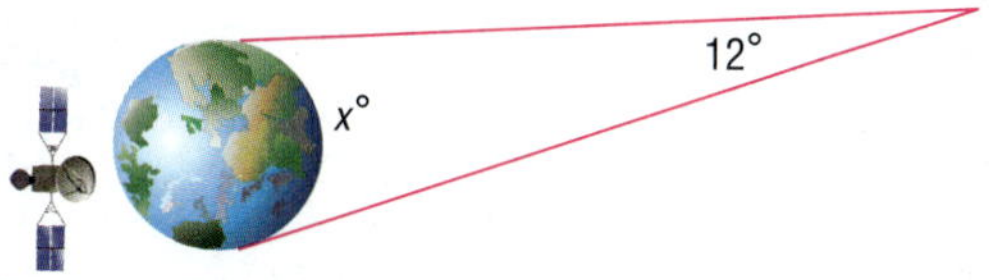

ALGEBRA **Find the value of x.**

26.

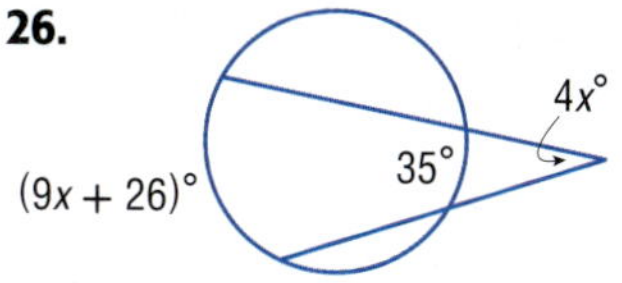

27

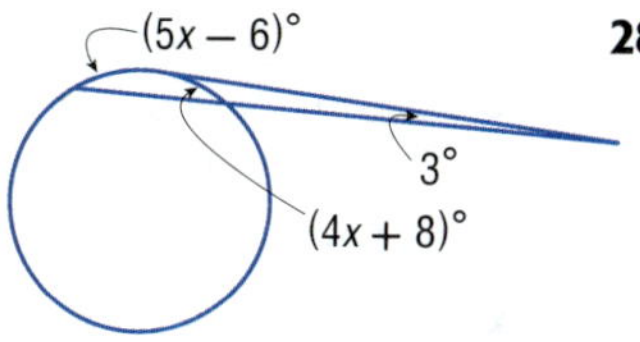

28.

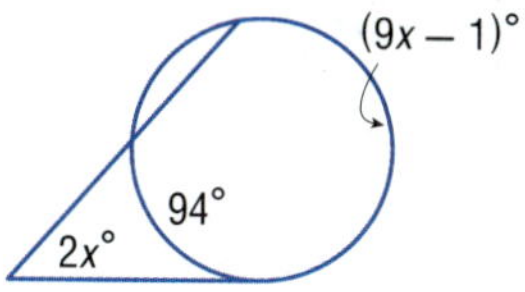

29. PHOTOGRAPHY A photographer frames a carousel in his camera shot as shown so that the lines of sight form tangents to the carousel.

a. If the camera's viewing angle is 35°, what is the arc measure of the carousel that appears in the shot?

b. If you want to capture an arc measure of 150° in the photograph, what viewing angle should be used?

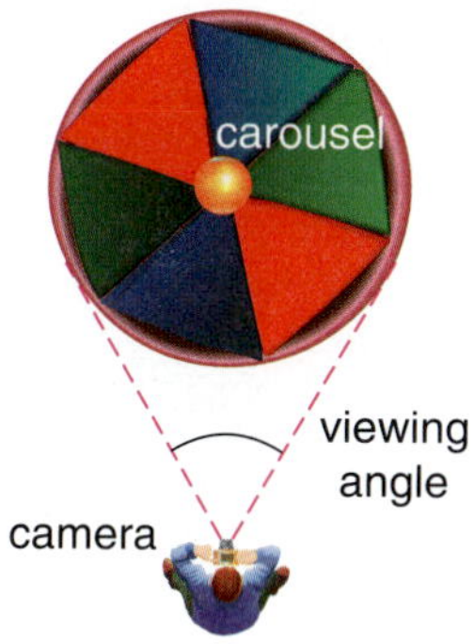

PROOF **For each case of Theorem 10.14, write a two-column proof.**

30. Case 1

Given: secants $\overrightarrow{AD}$ and $\overrightarrow{AE}$

Prove: $m\angle A = \frac{1}{2}(m\widehat{DE} - m\widehat{BC})$

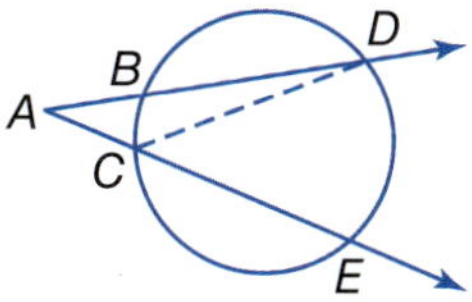

31. Case 2

Given: tangent $\overrightarrow{FM}$ and secant $\overrightarrow{FL}$

Prove: $m\angle F = \frac{1}{2}(m\widehat{LH} - m\widehat{GH})$

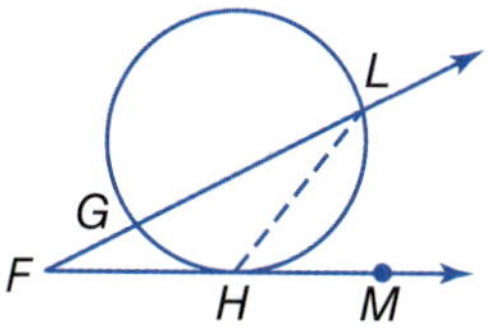

32. Case 3

Given: tangents $\overrightarrow{RS}$ and $\overrightarrow{RV}$

Prove: $m\angle R = \frac{1}{2}(m\widehat{SWT} - m\widehat{ST})$

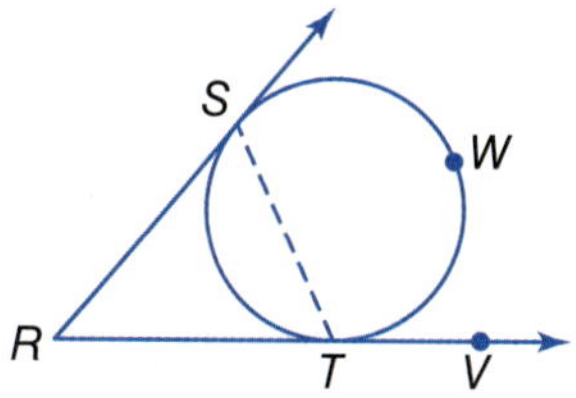

33. PROOF Write a paragraph proof of Theorem 10.13.

Given: tangent $\overleftrightarrow{AB}$ and secant $\overline{AC}$, $\angle CAB$ is obtuse

Prove: $m\angle CAB = \frac{1}{2}m\widehat{CDA}$

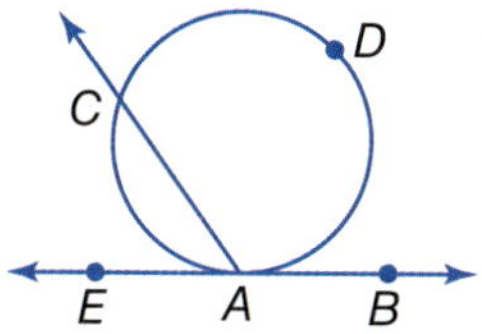

Real-World Link

Optical illusions were first used by the Greeks. They built their temples so that the roof was slanted. This gave the illusion that the temple was actually standing straight. They also made the columns bulge so that from a distance they would look perfectly proportioned.

Source: Odyssey Adventures

34. OPTICAL ILLUSION The design shown gives the optical illusion of rotating circles. $\overline{BC}$ is a diameter of $\odot Q$. If $m\angle A = 26$ and $m\widehat{CE} = 67$, what is $m\widehat{DE}$?

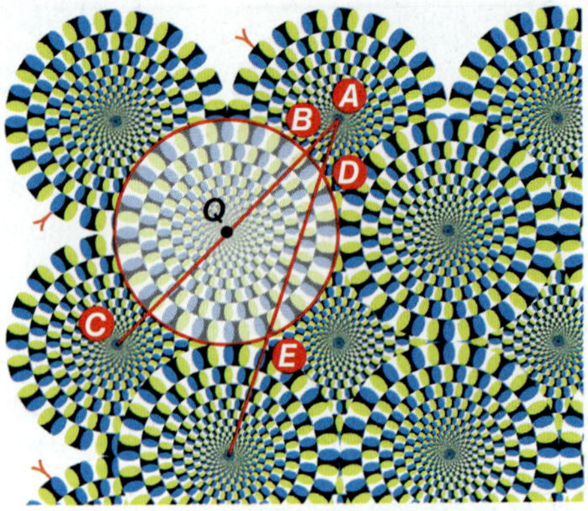

35. MULTIPLE REPRESENTATIONS In this problem, you will explore the relationship between Theorems 10.12 and 10.6.

a. GEOMETRIC Copy the figure shown. Then draw three successive figures in which the position of point D moves closer to point C, but points A, B, and C remain fixed.

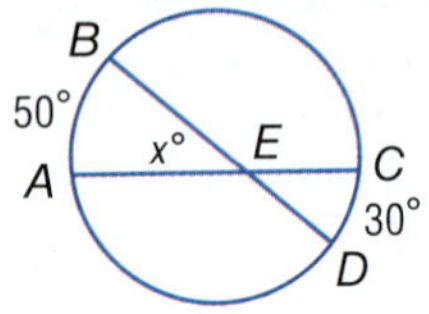

b. TABULAR Estimate the measure of $\widehat{CD}$ for each successive circle, recording the measures of $\widehat{AB}$ and $\widehat{CD}$ in a table. Then calculate and record the value of x for each circle.

c. VERBAL Describe the relationship between $m\widehat{AB}$ and the value of x as $m\widehat{CD}$ approaches zero. What type of angle does $\angle AEB$ become when $m\widehat{CD} = 0$?

d. ANALYTICAL Write an algebraic proof to show the relationship between Theorems 10.11 and 10.5 described in part **c**.

H.O.T. Problems

Use **H**igher-**O**rder **T**hinking Skills

36. WRITING IN MATH Explain how to find the measure of an angle formed by a secant and a tangent that intersect outside a circle.

37. CHALLENGE The circles below are concentric. What is x?

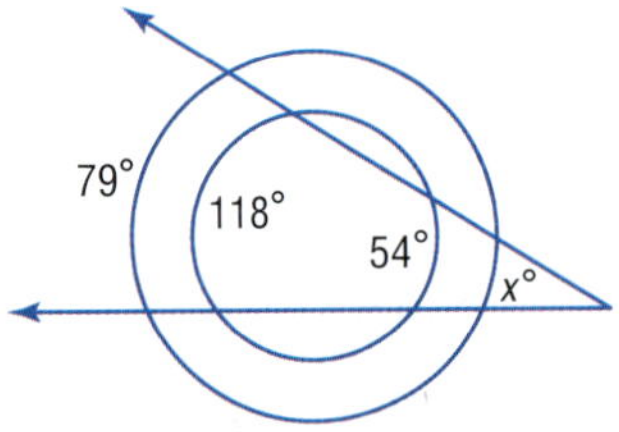

38. REASONING Isosceles $\triangle ABC$ is inscribed in $\odot D$. What can you conclude about $m\widehat{AB}$ and $m\widehat{BC}$? Explain.

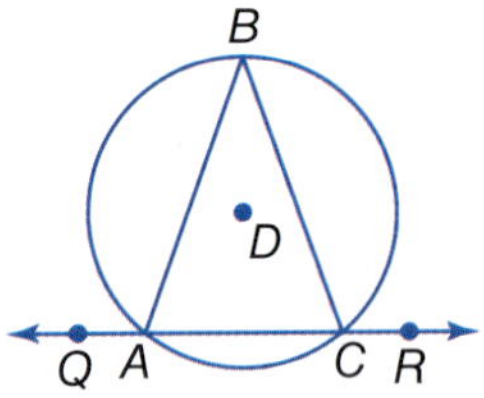

39. CHALLENGE In the figure, $\overline{JK}$ is a diameter and $\overrightarrow{GH}$ is a tangent.

a. Describe the range of possible values for $m\angle G$. Explain.

b. If $m\angle G = 34$, find the measures of minor arcs HJ and KH. Explain.

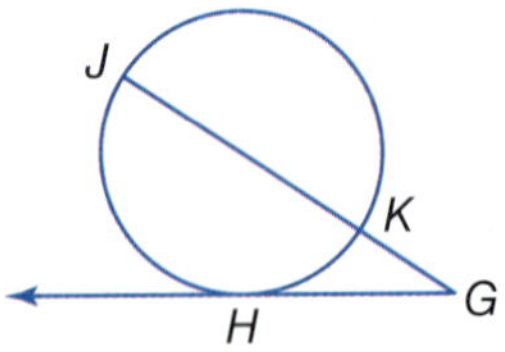

40. OPEN ENDED Draw a circle and two tangents that intersect outside the circle. Use a protractor to measure the angle that is formed. Find the measures of the minor and major arcs formed. Explain your reasoning.

41. WRITING IN MATH A circle is inscribed within $\triangle PQR$. If $m\angle P = 50$ and $m\angle Q = 60$, describe how to find the measures of the three minor arcs formed by the points of tangency.

Standardized Test Practice

42. What is the value of x if $m\widehat{NR} = 62$ and $m\widehat{NP} = 108$?

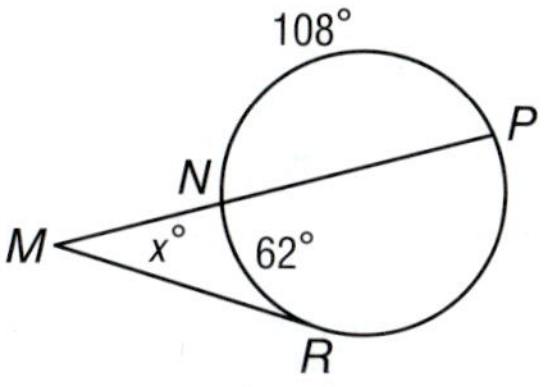

A 23°   C 64°
B 31°   D 128°

43. ALGEBRA Points $A(-4, 8)$ and $B(6, 2)$ are both on circle C, and $\overline{AB}$ is a diameter. What are the coordinates of C?

F (2, 10)   H (5, −3)
G (10, −6)   J (1, 5)

44. GRIDDED RESPONSE If $m\angle AED = 95$ and $m\widehat{AD} = 120$, what is $m\angle BAC$?

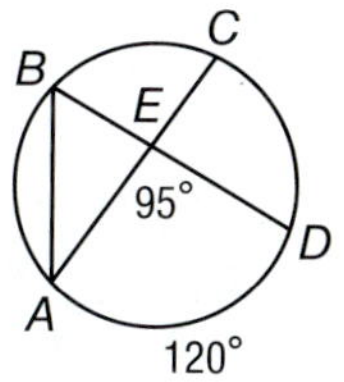

45. SAT/ACT If the circumference of the circle below is 16π, what is the total area of the shaded regions?

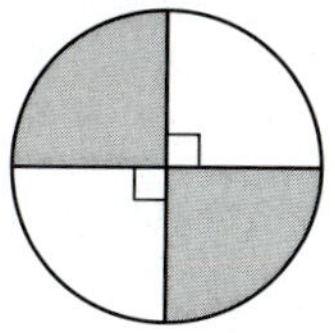

A 64π B 32π C 12π D 8π

Spiral Review

Find x. Assume that segments that appear to be tangent are tangent. (Lesson 10-5)

46.

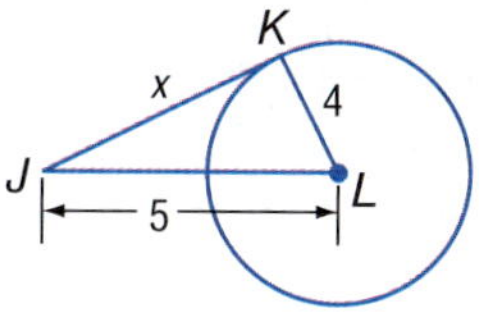

47.

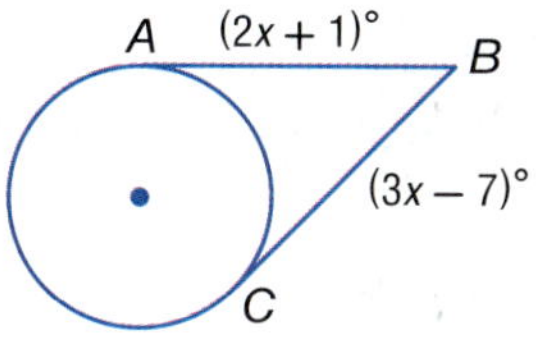

48.

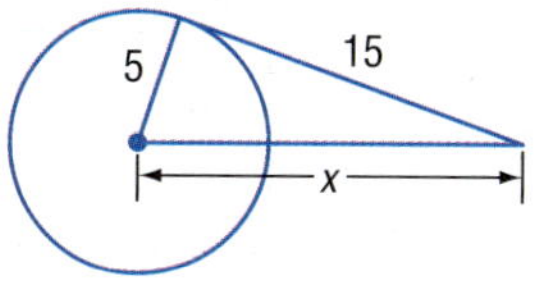

49. PROOF Write a two-column proof. (Lesson 10-4)

Given: $\widehat{MHT}$ is a semicircle; $\overline{RH} \perp \overline{TM}$.

Prove: $\frac{TR}{RH} = \frac{TH}{HM}$

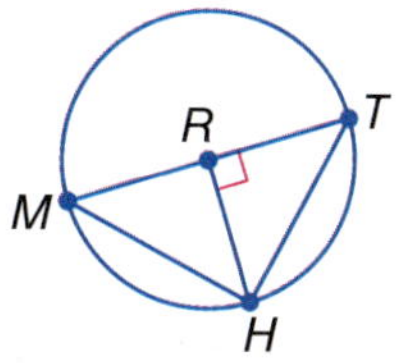

50. REMODELING The diagram at the right shows the floor plan of Trent's kitchen. Each square on the diagram represents a 3-foot by 3-foot area. While remodeling his kitchen, Trent moved his refrigerator from square A to square B. Describe one possible combination of transformations that could be used to make this move. (Lesson 9-4)

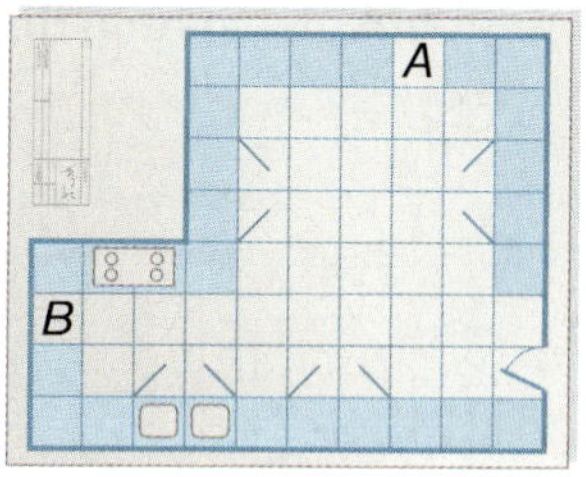

COORDINATE GEOMETRY Find the measure of each angle to the nearest tenth of a degree by using the Distance Formula and an inverse trigonometric ratio. (Lesson 8-4)

51. $\angle C$ in triangle BCD with vertices $B(-1, -5)$, $C(-6, -5)$, and $D(-1, 2)$

52. $\angle X$ in right triangle XYZ with vertices $X(2, 2)$, $Y(2, -2)$, and $Z(7, -2)$

Skills Review

Solve each equation. (Concepts and Skills, Lesson 4)

53. $x^2 + 13x = -36$

54. $x^2 - 6x = -9$

55. $3x^2 + 15x = 0$

56. $28 = x^2 + 3x$

57. $x^2 + 12x + 36 = 0$

58. $x^2 + 5x = -\frac{25}{4}$

10-7 Special Segments in a Circle

Why?

A large circular cake is cut lengthwise instead of into wedges to serve more people for a party. Only a small portion of the original cake remains. Using the geometry of circles, you can determine the diameter of the original cake.

Then

You found measures of diagonals that intersect in the interior of a parallelogram. (Lesson 6-2)

Now

- Find measures of segments that intersect in the interior of a circle.
- Find measures of segments that intersect in the exterior of a circle.

New Vocabulary

chord segment
secant segment
external secant segment
tangent segment

Math Online
glencoe.com

Segments Intersecting Inside a Circle When two chords intersect inside a circle, each chord is divided into two segments, called **chord segments**. In the figure, chord $\overline{AC}$ is divided into segments $\overline{AB}$ and $\overline{BC}$. Likewise, chord $\overline{ED}$ is divided into segments $\overline{EB}$ and $\overline{BD}$.

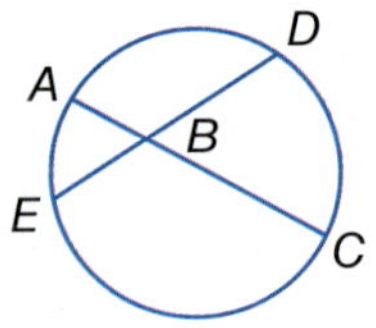

The following theorem describes the relationship among four segments formed by the intersection of two chords in the interior of a circle.

Theorem 10.15 Segments of Chords Theorem

Words If two chords intersect in a circle, then the products of the lengths of the chord segments are equal.

Example $AB \cdot BC = DB \cdot BE$

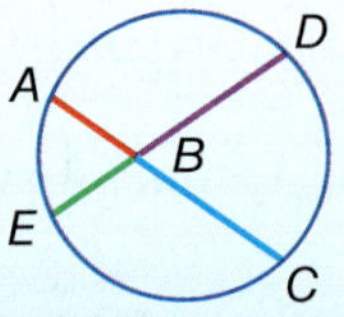

You will prove Theorem 10.15 in Exercise 23.

EXAMPLE 1 Use the Intersection of Two Chords

Find x.

a.

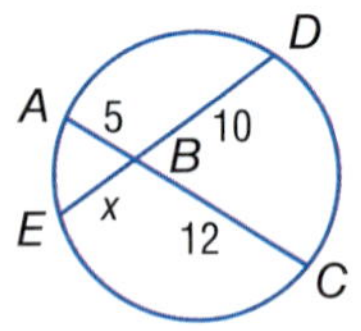

$AB \cdot BC = EB \cdot BD$	Theorem 10.15
$5 \cdot 12 = x \cdot 10$	Substitution
$60 = 10x$	Multiply.
$6 = x$	Divide each side by 10.

b.

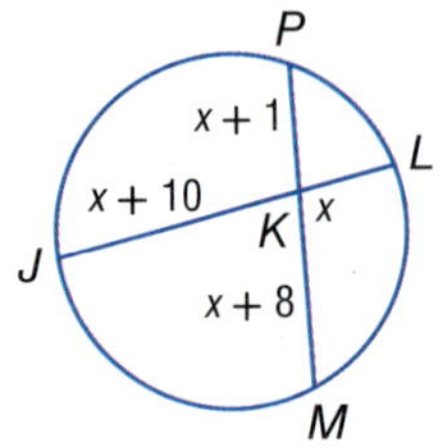

$JK \cdot KL = PK \cdot KM$	Theorem 10.15
$(x + 10) \cdot x = (x + 1)(x + 8)$	Substitution
$x^2 + 10x = x^2 + 9x + 8$	Multiply.
$10x = 9x + 8$	Subtract x^2 from each side.
$x = 8$	Subtract $9x$ from each side.

Guided Practice

1A.

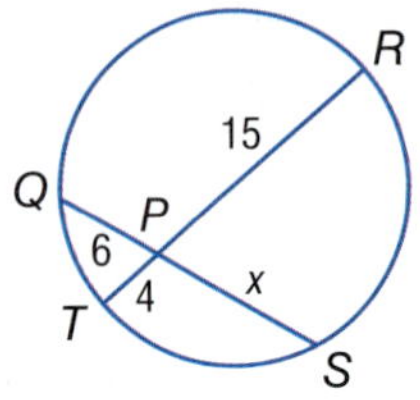

1B.

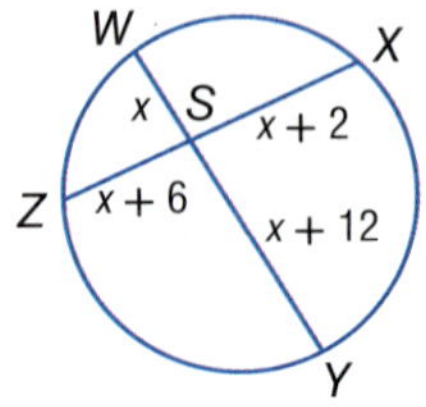

Personal Tutor glencoe.com

Real-World Link

The lower the Sun is to the horizon, the more of a rainbow you can see. At sunset, you could see a full semicircle of a rainbow with the top of the arch 42 degrees above the horizon.

Source: The National Center for Atmospheric Research

Real-World EXAMPLE 2 Find Measures of Segments in Circles

SCIENCE The true shape of a rainbow is a complete circle. However, we see only the arc of the circle that appears above Earth's horizon. What is the radius of the circle containing the arc of the rainbow shown?

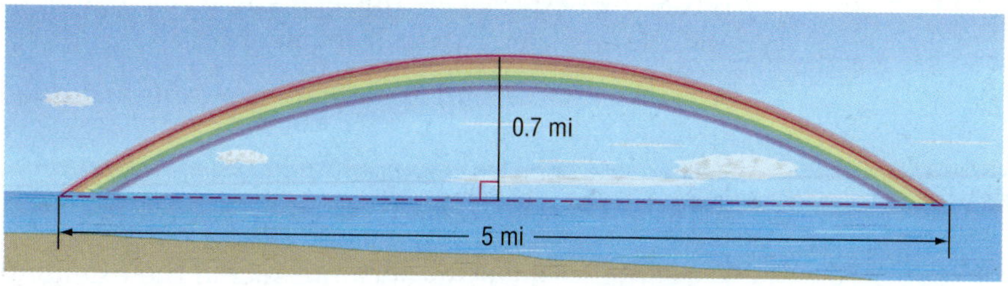

Understand You know that the rainbow's arc is part of a whole circle. $\overline{AC}$ is a chord of this circle, and $\overline{DB}$ is a perpendicular bisector of $\overline{AC}$.

Plan Draw a model. Since it bisects chord $\overline{AC}$, $\overline{DE}$ is a diameter of the circle. Use the products of the lengths of the intersecting chords to find the length of the diameter.

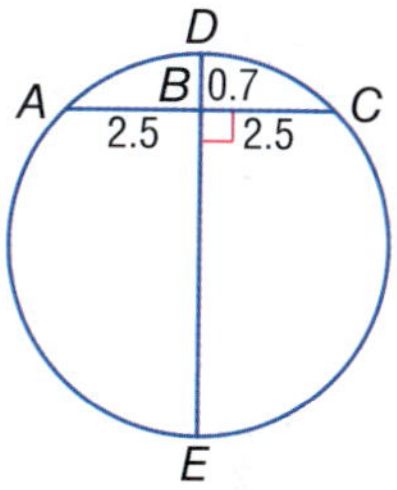

Solve

$AB \cdot BC = DB \cdot BE$	**Theorem 10.15**
$2.5 \cdot 2.5 = 0.7 \cdot BE$	**Substitution**
$6.25 = 0.7BE$	**Multiply.**
$8.9 \approx BE$	**Divide each side by 0.7.**

$DE = DB + BE$	**Segment Addition Postulate**
$= 0.7 + 8.9$	**Substitution**
$= 9.6$	**Add.**

Since the diameter of the circle is about 9.6 miles, the radius is about 9.6 ÷ 2 or 4.8 miles.

Check Use the Pythagorean Theorem to check the triangle in the circle formed by the radius, the chord, and part of the diameter.

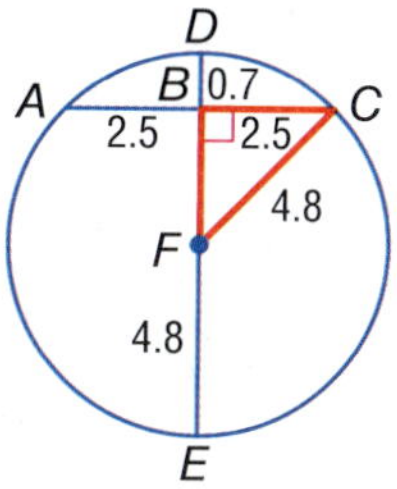

$DB + BF = DF$	**Segment Addition Postulate**
$0.7 + BF = 4.8$	**Substitution**
$BF = 4.1$	**Subtract 0.7 from each side.**

$BF^2 + BC^2 = CF^2$	**Pythagorean Theorem**
$4.1^2 + 2.5^2 \stackrel{?}{=} 4.8^2$	**Substitution**
$23.06 \approx 23.04$ ✓	**Simplify.**

Problem-Solving Tip

Make a Drawing When solving word problems involving circles, it is helpful to make a drawing and label all parts of the circle that are known. Use a variable to label the unknown measure.

Guided Practice

2. **ASTRODOME** The highest point, or apex, of the Astrodome is 208 feet high, and the diameter of the circle containing the arc is 710 feet. How long is the stadium from one side to the other?

Personal Tutor glencoe.com

Segments Intersecting Outside a Circle A **secant segment** is a segment of a secant line that has exactly one endpoint on the circle. In the figure, $\overline{AC}$, $\overline{AB}$, $\overline{AE}$ and $\overline{AD}$ are secant segments.

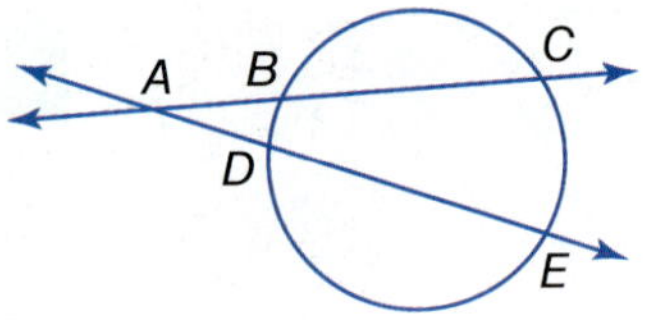

A secant segment that lies in the exterior of the circle is called an **external secant segment**. In the figure, $\overline{AB}$ and $\overline{AD}$ are external secant segments.

A special relationship exists among secants and external secant segments.

StudyTip

Simplify the Theorem Each side of the equation in Theorem 10.16 is the product of the lengths of the exterior part and the whole segment.

Theorem 10.16 Secant Segments Theorem

Words If two secants intersect in the exterior of a circle, then the product of the measures of one secant segment and its external secant segment is equal to the product of the measures of the other secant and its external secant segment.

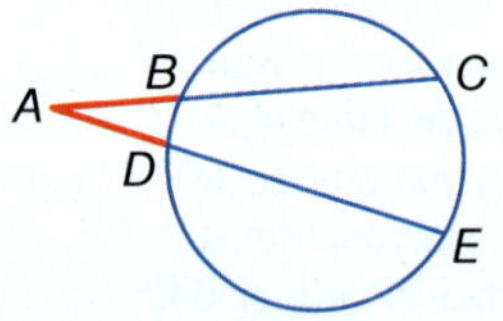

Example $AC \cdot \mathbf{AB} = AE \cdot \mathbf{AD}$

You will prove Theorem 10.16 in Exercise 24.

Watch Out!

Use the Correct Equation Be sure to multiply the length of the secant segment by the length of the external secant segment. Do not multiply the length of the internal secant segment, or chord, by the length of the external secant segment.

EXAMPLE 3 Use the Intersection of Two Secants

Find x.

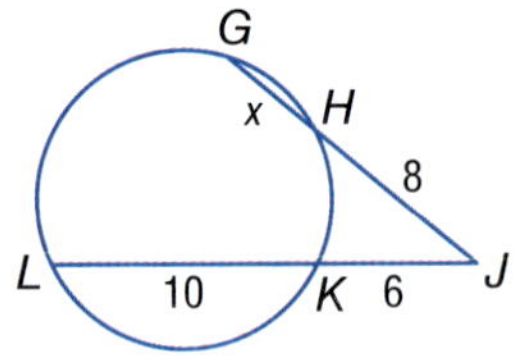

$JG \cdot JH = JL \cdot JK$	Theorem 10.16
$(x + 8)8 = (10 + 6)6$	Substitution
$8x + 64 = 96$	Multiply.
$8x = 32$	Subtract 64 from each side.
$x = 4$	Divide each side by 8.

Guided Practice

3A.

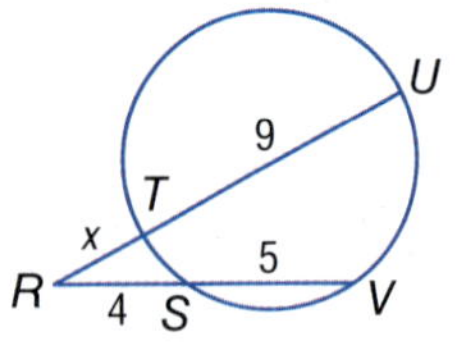

3B.

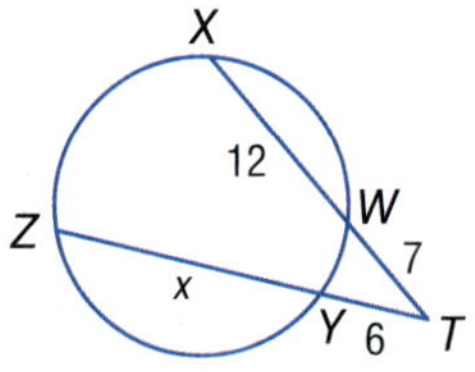

Personal Tutor glencoe.com

An equation similar to the one in Theorem 10.16 can be used when a secant and a tangent intersect outside a circle. In this case, the **tangent segment**, or segment of a tangent with one endpoint on the circle, is both the exterior and whole segment.

Theorem 10.17

Words If a tangent and a secant intersect in the exterior of a circle, then the square of the measure of the tangent is equal to the product of the measures of the secant and its external secant segment.

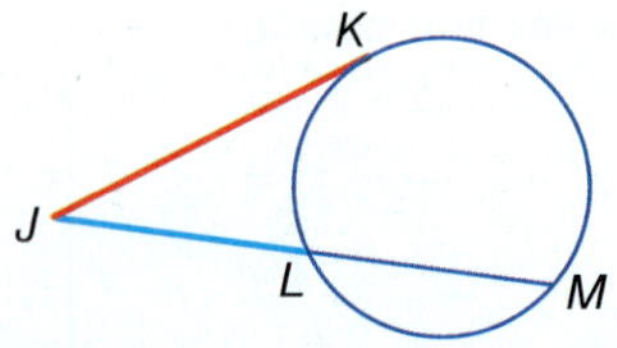

Example $\mathbf{JK}^2 = \mathbf{JL} \cdot JM$

You will prove Theorem 10.17 in Exercise 25.

EXAMPLE 4 Use the Intersection of a Secant and a Tangent

$\overline{PQ}$ is tangent to the circle. Find x. Round to the nearest tenth.

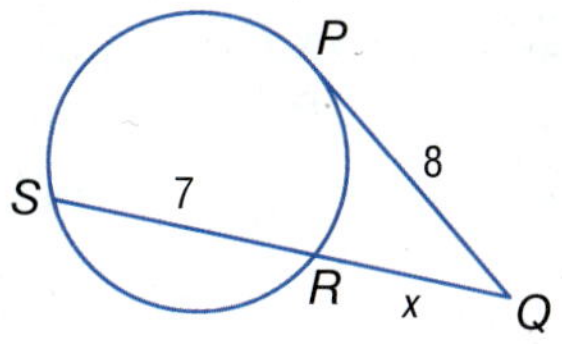

$PQ^2 = QR \cdot QS$	**Theorem 10.17**
$8^2 = x(x + 7)$	**Substitution**
$64 = x^2 + 7x$	**Multiply.**
$0 = x^2 + 7x - 64$	**Subtract 64 from each side.**

Since the expression is not factorable, use the Quadratic Formula.

$x = \dfrac{-b \pm \sqrt{b^2 - 4ac}}{2a}$	**Quadratic Formula**
$= \dfrac{-7 \pm \sqrt{7^2 - 4(1)(-64)}}{2(1)}$	**$a = 1$, $b = 7$, and $c = -64$**
$= \dfrac{-7 \pm \sqrt{305}}{2}$	**Simplify.**
≈ 5.2 or -12.2	**Use a calculator.**

Since lengths cannot be negative, the value of x is about 5.2.

Guided Practice

4. $\overline{AB}$ is tangent to the circle. Find x. Round to the nearest tenth.

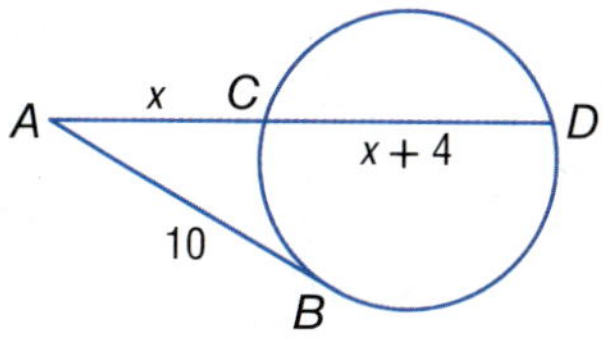

Personal Tutor glencoe.com

Check Your Understanding

Examples 1, 3, and 4
pp. 736, 738, 739

Find x. Assume that segments that appear to be tangent are tangent.

1.

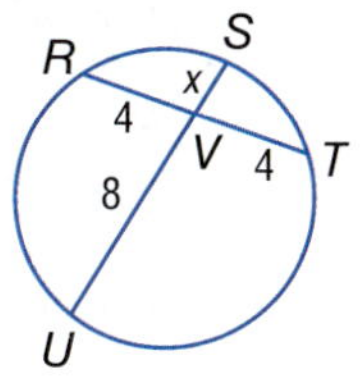

2.

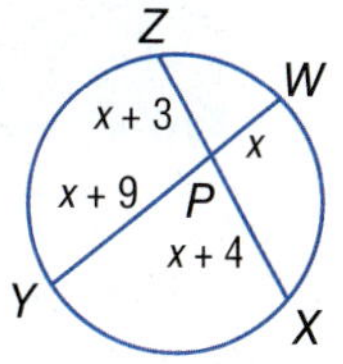

3.

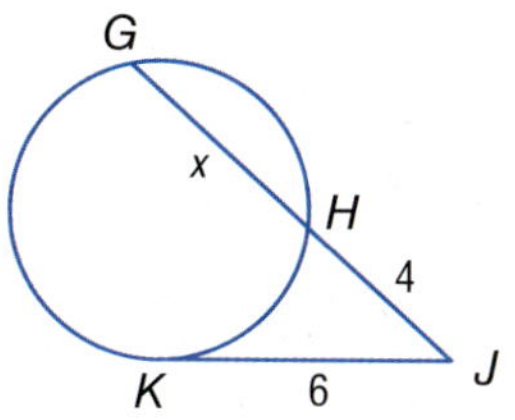

4.

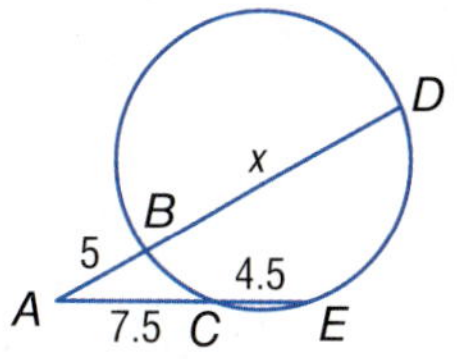

Example 2
p. 737

5 **SCIENCE** A piece of broken pottery found at an archaeological site is shown. $\overline{QS}$ lies on a diameter of the circle. What was the circumference of the original pottery? Round to the nearest hundredth.

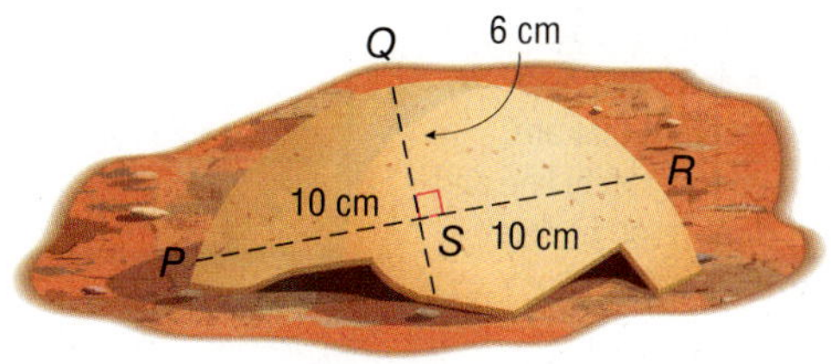

Practice and Problem Solving

= **Step-by-Step Solutions** begin on page R20.
Extra Practice begins on page 969.

Examples 1, 3, and 4
pp. 736, 738–739

Find x to the nearest tenth. Assume that segments that appear to be tangent are tangent.

6.

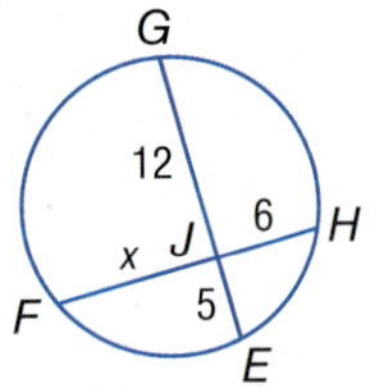

7.

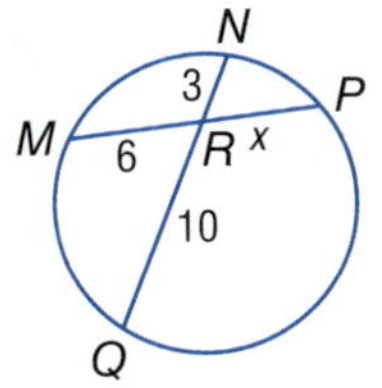

8.

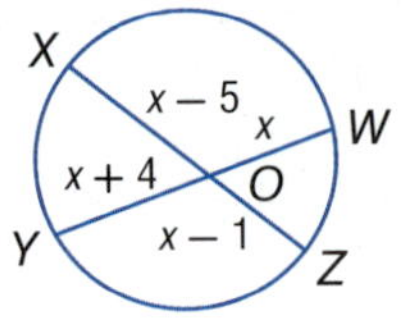

9.

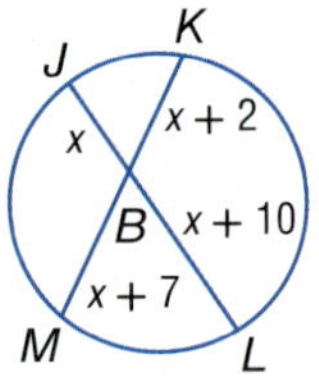

10.

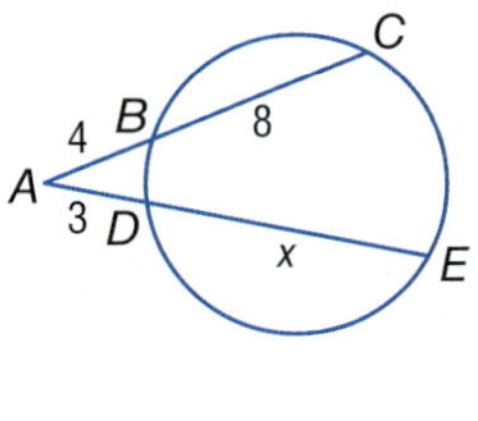

11.

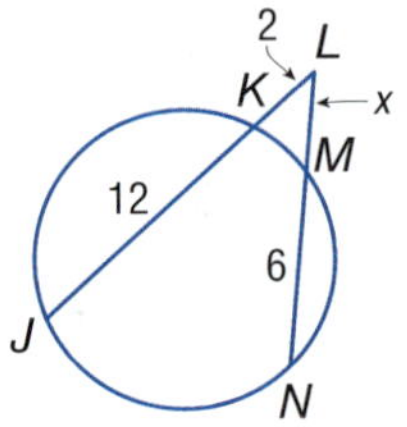

12.

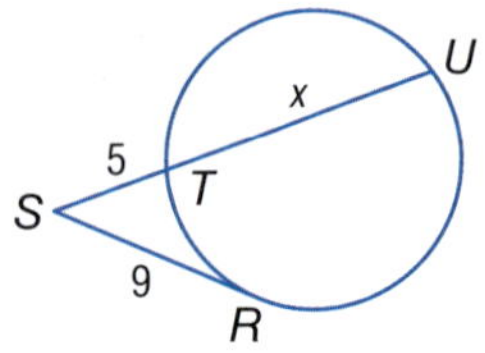

13

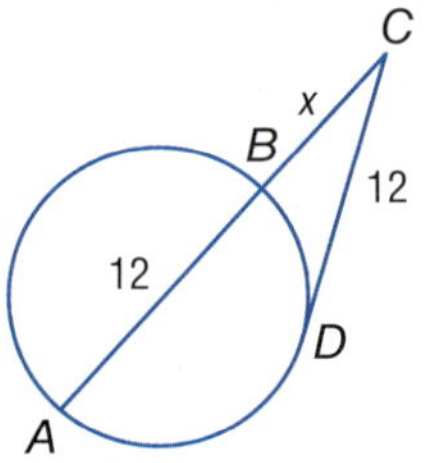

Example 2
p. 737

14. BRIDGES What is the diameter of the circle containing the arc of the Sydney Harbour Bridge shown? Round to the nearest tenth.

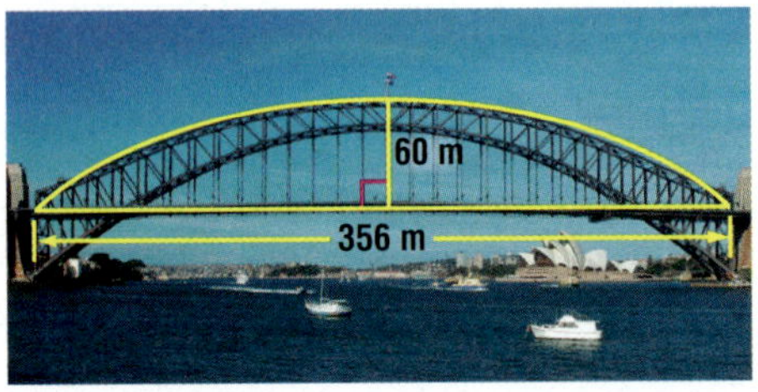

15. CAKES Sierra is serving cake at a party. If the dimensions of the remaining cake are shown below, what was the original diameter of the cake?

Real-World Link

Tours are conducted over the Sydney Harbour Bridge arch. Climbers wear overalls that are clipped to a safety line and can climb 1500 meters over the arch.

Source: TravelOnline

Find each variable to the nearest tenth. Assume that segments that appear to be tangent are tangent.

16.

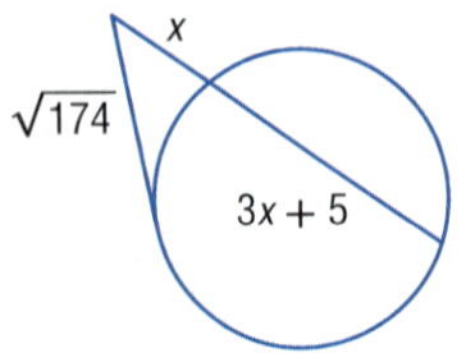

17.

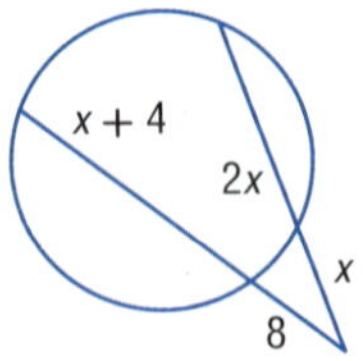

18.

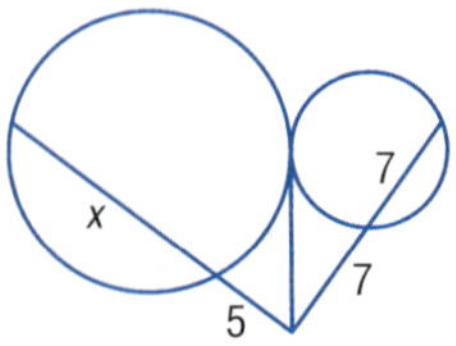

19.

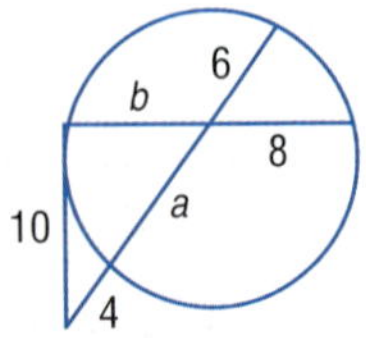

20.

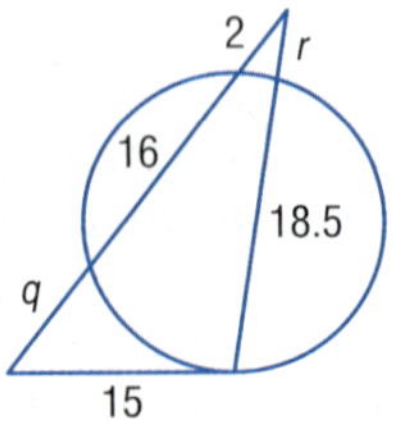

21.

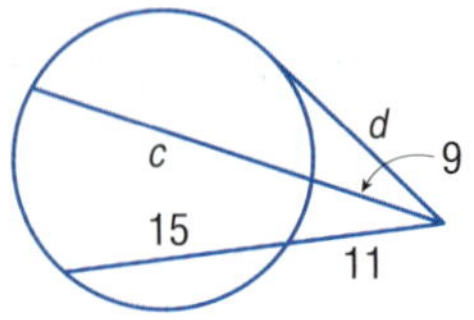

Real-World Link

The General Sherman Tree in Sequoia National Park is the largest tree in the world by volume. Its volume is estimated to be around 53,000 cubic feet.

Source: National Park Service

22. INDIRECT MEASUREMENT Gwendolyn is standing 16 feet from a giant sequoia tree and Chet is standing next to the tree, as shown. The distance between Gwendolyn and Chet is 27 feet. Draw a diagram of this situation, and then find the diameter of the tree.

PROOF **Prove each theorem.**

23. two-column proof of Theorem 10.15

Given: $\overline{AC}$ and $\overline{DE}$ intersect at B.

Prove: $AB \cdot BC = EB \cdot BD$

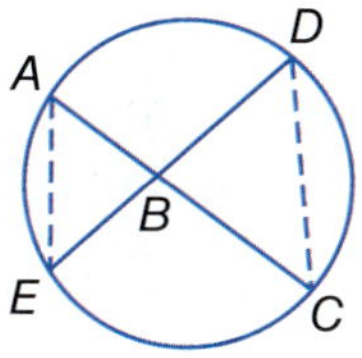

24. paragraph proof of Theorem 10.16

Given: Secants $\overline{AC}$ and $\overline{AE}$

Prove: $AB \cdot AC = AD \cdot AE$

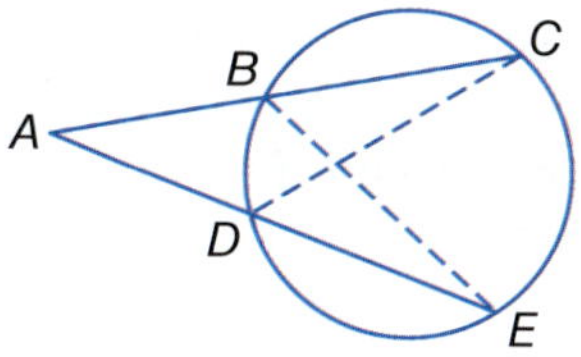

25. two-column proof of Theorem 10.17

Given: tangent $\overline{JK}$, secant $\overline{JM}$

Prove: $JK^2 = JL \cdot JM$

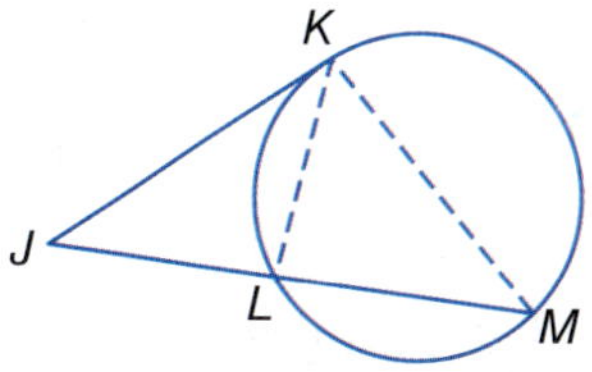

H.O.T. Problems Use Higher-Order Thinking Skills

26. ERROR ANALYSIS Tiffany and Jun are finding the value of x in the figure at the right. Tiffany wrote $3(5) = 2x$, and Jun wrote $3(8) = 2(2 + x)$. Is either of them correct? Explain your reasoning.

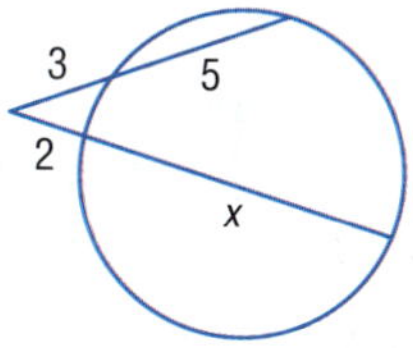

27. WRITING IN MATH Compare and contrast the methods for finding measures of segments when two secants intersect in the exterior of a circle and when a secant and a tangent intersect in the exterior of a circle.

28. CHALLENGE In the figure, a line tangent to circle M and a secant line intersect at R. Find a. Show the steps that you used.

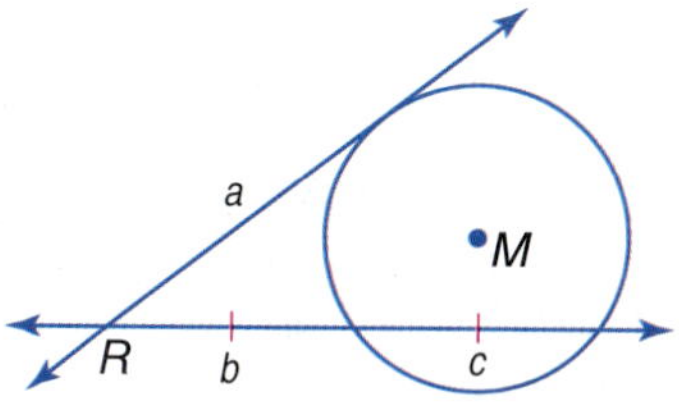

29. REASONING When two chords intersect at the center of a circle, are the measures of the intercepting arcs *sometimes*, *always*, or *never* equal to each other?

30. OPEN ENDED Investigate Theorem 10.17 by drawing and labeling a circle that has a secant and a tangent intersecting outside the circle. Measure and label the two parts of the secant segment to the nearest tenth of a centimeter. Use an equation to find the measure of the tangent segment. Verify your answer by measuring the segment.

31. WRITING IN MATH Describe the relationship among segments in a circle when two secants intersect inside a circle.

Standardized Test Practice

32. $\overline{TV}$ is tangent to the circle, and R and S are points on the circle. What is the value of x to the nearest tenth?

A 7.6
B 6.4
C 5.7
D 4.8

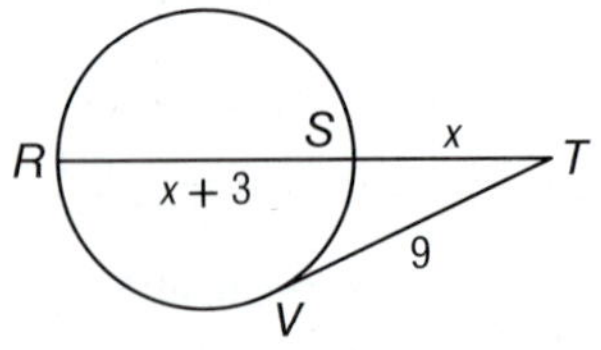

33. ALGEBRA A department store has all of its jewelry discounted 40%. It is having a sale that says you receive an additional 20% off the already discounted price. How much will you pay for a ring with an original price of $200?

F $80
G $96
H $120
J $140

34. EXTENDED RESPONSE The degree measures of minor arc $\widehat{AC}$ and major arc $\widehat{AC}$ are x and y, respectively.

a. If $m\angle ABC = 70°$, write two equations relating x and y.

b. Find x and y.

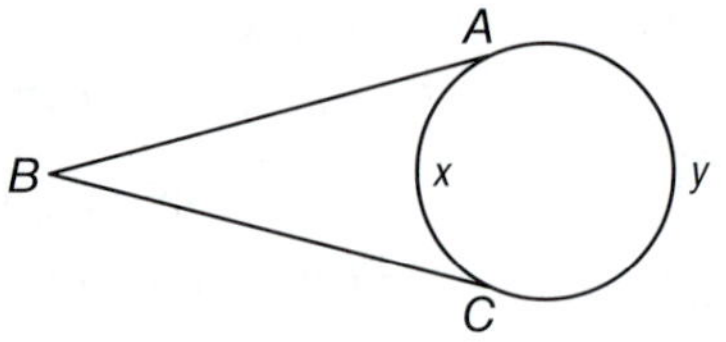

35. SAT/ACT During the first two weeks of summer vacation, Antonia earned $100 per week. During the next six weeks, she earned $150 per week. What was her average weekly pay?

A $112.50
B $125
C $135
D $137.50

Spiral Review

36. WEAVING Once yarn is woven from wool fibers, it is often dyed and then threaded along a path of pulleys to dry. One set of pulleys is shown. Note that the yarn appears to intersect itself at C, but in reality it does not. Use the information from the diagram to find $m\widehat{BH}$. (Lesson 10-6)

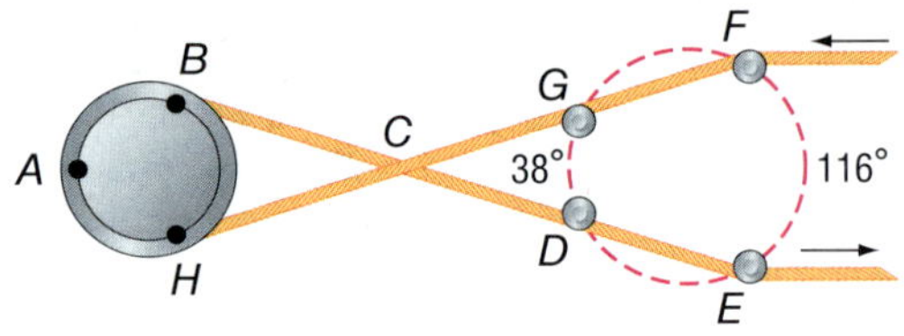

Copy the figure shown and draw the common tangents. If no common tangent exists, state *no common tangent*. (Lesson 10-5)

COORDINATE GEOMETRY Graph each figure and its image along the given vector. (Lesson 9-2)

41. $\triangle KLM$ with vertices $K(5, -2)$, $L(-3, -1)$, and $M(0, 5)$; $\langle -3, -4 \rangle$

42. quadrilateral $PQRS$ with vertices $P(1, 4)$, $Q(-1, 4)$, $R(-2, -4)$, and $S(2, -4)$; $\langle -5, 3 \rangle$

43. $\triangle EFG$ with vertices $E(0, -4)$, $F(-4, -4)$, and $G(0, 2)$; $\langle 2, -1 \rangle$

Skills Review

Write an equation in slope-intercept form of the line having the given slope and y-intercept. (Lesson 3-4)

44. m: 3, y-intercept: -4

45. m: 2, $(0, 8)$

46. m: $\frac{5}{8}$, $(0, -6)$

47. m: $\frac{2}{9}$, y-intercept: $\frac{1}{3}$

48. m: -1, b: -3

49. m: $-\frac{1}{12}$, b: 1

Graphing Technology Lab
Equations of Circles

Math Online glencoe.com
- Other Calculator Keystrokes
- Graphing Technology Personal Tutor

You can use TI-Nspire™ or TI-Nspire™ CAS technology to explore equations of circles.

ACTIVITY Graph a Circle

Step 1 Open a new **Graphs and Geometry** page.

Step 2 Select **Circle** from the **Shapes** menu and draw a circle so that the center does not lie on either axis.

Step 3 Select **Coordinates and Equations** from the **Tools** menu to display the equation for the circle.

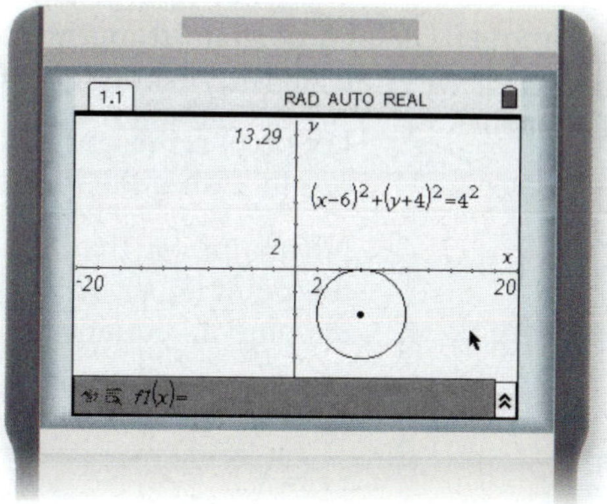

Step 4 Use your cursor to select the center of the circle. Click and drag the circle, and observe how the equation changes. Be sure to observe what happens to the equation when you place the center at the origin.

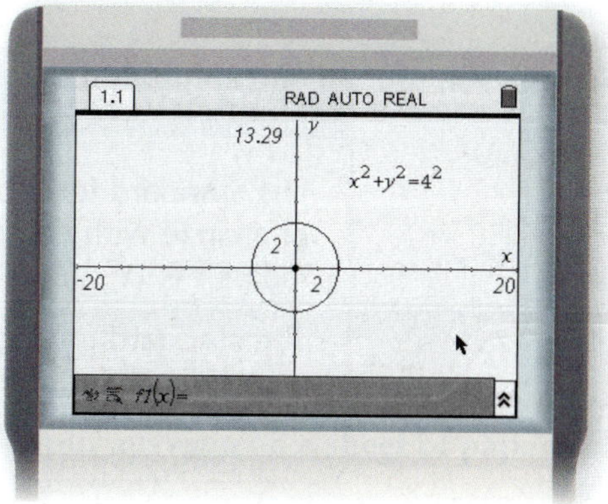

Step 5 Move the circle as you did in Step 4 to place the center at the origin. Select the circle using your cursor, and make the circle larger and smaller. Observe the effect on the equation.

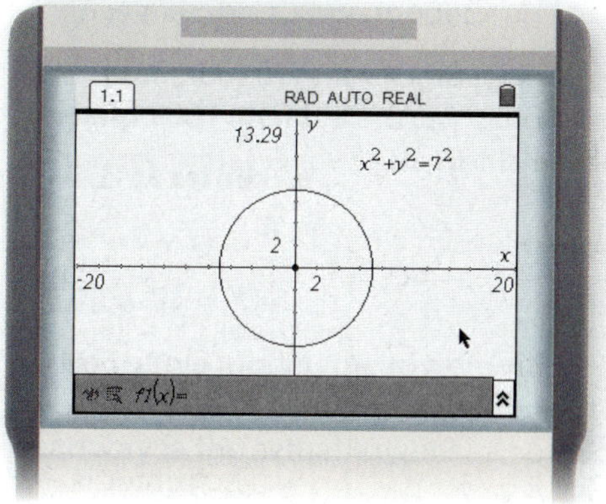

Analyze the Results

1. How does the equation for the circle change as you move its center?
2. How does the equation for the circle change as you increase and decrease its radius?
3. What do you think the equation for a circle centered at the origin with a radius of 4 would be?
4. What do you think the equation for a circle centered at the point (h, k) with a radius of r would be?

10-8 Equations of Circles

Then
You wrote equations of lines using information about their graphs. (Lesson 3-4)

Now
- Write the equation of a circle.
- Graph a circle on the coordinate plane.

New Vocabulary
compound locus

Math Online
glencoe.com

Why?

Telecommunications towers emit radio signals that are used to transmit cellular calls. Each tower covers a circular area, and towers are arranged so that a signal is available at any location in the coverage area.

Equation of a Circle Since all points on a circle are equidistant from the center, you can find an equation of a circle by using the Distance Formula.

Let (x, y) represent a point on a circle centered at the origin. Using the Pythagorean Theorem, $x^2 + y^2 = r^2$.

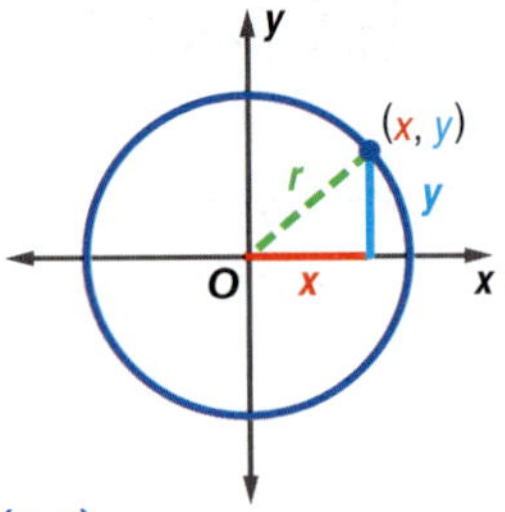

Now suppose that the center is not at the origin, but at the point (h, k). You can use the Distance Formula to develop an equation for the circle.

$d = \sqrt{(x_2 - x_1)^2 + (y_2 - y_1)^2}$ **Distance Formula**

$r = \sqrt{(x - h)^2 + (y - k)^2}$ **$d = r$, $(x_1, y_1) = (h, k)$, $(x_2, y_2) = (x, y)$**

$r^2 = (x - h)^2 + (y - k)^2$ **Square each side.**

Key Concept: Standard Form, Equation of a Circle

The standard form of the equation of a circle with center at (h, k) and radius r is $(x - h)^2 + (y - k)^2 = r^2$.

The standard form of the equation of a circle is also called the *center-radius* form.

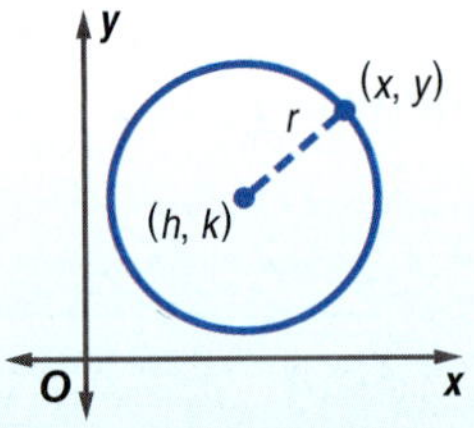

EXAMPLE 1 Write an Equation Using the Center and Radius

Write the equation of each circle.

a. center at (1, −8), radius 7

$(x - h)^2 + (y - k)^2 = r^2$ **Equation of a circle**

$(x - 1)^2 + [y - (-8)]^2 = 7^2$ **$(h, k) = (1, -8)$, $r = 7$**

$(x - 1)^2 + (y + 8)^2 = 49$ **Simplify.**

b. the circle graphed at the right

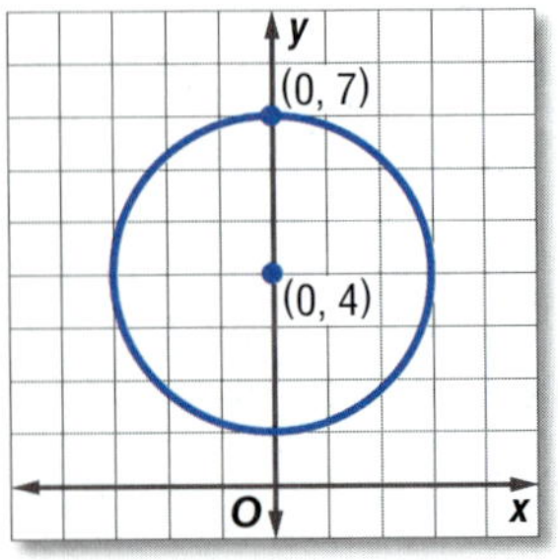

The center is at (0, 4) and the radius is 3.

$(x - h)^2 + (y - k)^2 = r^2$ **Equation of a circle**

$(x - 0)^2 + (y - 4)^2 = 3^2$ **$(h, k) = (0, 4)$, $r = 3$**

$x^2 + (y - 4)^2 = 9$ **Simplify.**

Guided Practice

1A. center at origin, radius $\sqrt{10}$

1B. center at $(4, -1)$, diameter 8

Personal Tutor glencoe.com

EXAMPLE 2 Write an Equation Using the Center and a Point

Write the equation of each circle.

a. center at (−2, 4), passes through (−6, 7)

Step 1 Find the distance between the points to determine the radius.

$r = \sqrt{(x_2 - x_1)^2 + (y_2 - y_1)^2}$ **Distance Formula**

$= \sqrt{[-6 - (-2)]^2 + (7 - 4)^2}$ **$(x_1, y_1) = (-2, 4)$ and $(x_2, y_2) = (-6, 7)$**

$= \sqrt{25}$ or 5 **Simplify.**

Step 2 Write the equation using $h = -2$, $k = 4$, and $r = 5$.

$(x - h)^2 + (y - k)^2 = r^2$ **Equation of a circle**

$[x - (-2)]^2 + (y - 4)^2 = 5^2$ **$h = -2$, $k = 4$, and $r = 5$**

$(x + 2)^2 + (y - 4)^2 = 25$ **Simplify.**

StudyTip

Radical Form In Example 2b, you can leave the radius in radical form because in the equation of a circle, the radius is squared.

b. the circle graphed at the right

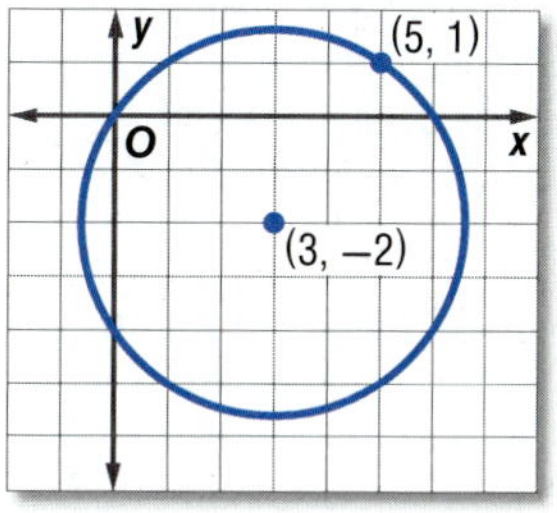

Step 1 Find the distance between the points to determine the radius.

$r = \sqrt{(x_2 - x_1)^2 + (y_2 - y_1)^2}$ **Distance Formula**

$= \sqrt{(5 - 3)^2 + [1 - (-2)]^2}$ **Substitute.**

$= \sqrt{13}$ **Simplify.**

Step 2 Write the equation using $h = 3$, $k = -2$, and $r = \sqrt{13}$.

$(x - h)^2 + (y - k)^2 = r^2$ **Equation of a circle**

$(x - 3)^2 + [y - (-2)]^2 = (\sqrt{13})^2$ **$h = 3$, $k = -2$, and $r = \sqrt{13}$**

$(x - 3)^2 + (y + 2)^2 = 13$ **Simplify.**

Guided Practice

2A. center at (5, 4), passes through (−3, 4)

2B. center at (−3, −5), passes through (0, 0)

Personal Tutor glencoe.com

Graph Circles You can use the equation of a circle to graph it on a coordinate plane.

EXAMPLE 3 Graph a Circle

The equation of a circle is $(x - 4)^2 + (y + 1)^2 = 9$. State the center and the radius. Then graph the equation.

Rewrite $(x - 4)^2 + (y + 1)^2 = 9$ to find the center and the radius.

$(x - 4)^2 + [y - (-1)]^2 = 3^2$

$\uparrow \qquad \uparrow \qquad \uparrow$

$(x - h)^2 + (y - k)^2 = r^2$

So, $h = 4$, $k = -1$, and $r = 3$. The center is at (4, −1), and the radius is 3.

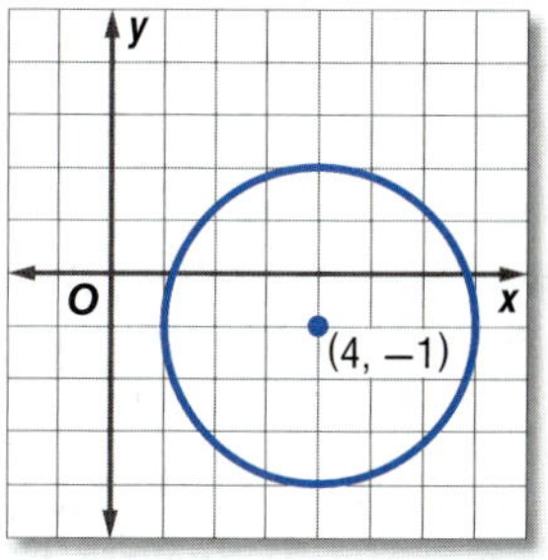

StudyTip

Euclid's Postulates You learned about three of Euclid's postulates in Lesson 3-5. Another of Euclid's postulates was that a circle can be drawn using any point as the center with any given radius.

Guided Practice

For each circle with the given equation, state the coordinates of the center and the measure of the radius. Then graph the equation.

3A. $x^2 + y^2 = 4$

3B. $(x + 4)^2 + (y - 7)^2 = 25$

Personal Tutor glencoe.com

Real-World Link

About 1000 tornadoes are reported across the United States each year. The most violent tornadoes have wind speeds of 250 mph or more. Damage paths can be a mile wide and 50 miles long.

Source: National Oceanic & Atmospheric Administration

Real-World EXAMPLE 4 Use Three Points to Write an Equation

TORNADOES Three tornado sirens are placed strategically on a circle around a town so they can be heard by all. Write the equation of the circle on which they are placed if the coordinates of the sirens are $A(-8, 3)$, $B(-4, 7)$, and $C(-4, -1)$.

Understand You are given three points that lie on a circle.

Plan Graph $\triangle ABC$ and construct the perpendicular bisectors of two sides to locate the center of the circle. Find the radius and then use the center and radius to write an equation.

Solve Construct the perpendicular bisectors of two sides. The center appears to be at $(-4, 3)$. The radius is 4.

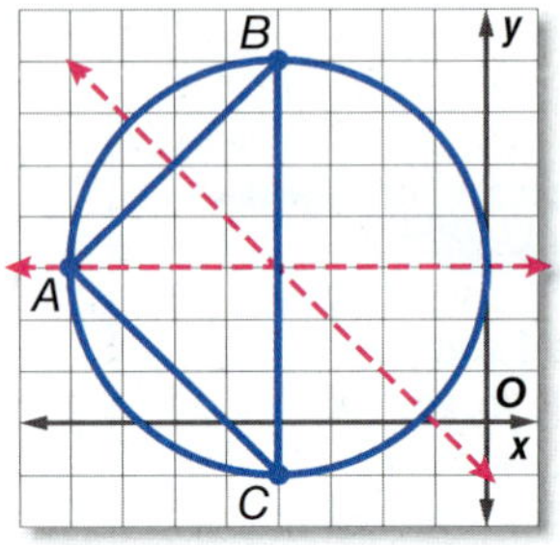

Write an equation.

$$(x - h)^2 + (y - k)^2 = r^2$$

$$[x - (-4)]^2 + (y - 3)^2 = 4^2$$

$$(x + 4)^2 + (y - 3)^2 = 16$$

Check Verify the center by finding the equations of the two bisectors and solving the system of equations. Verify the radius by finding the distance between the center and another point on the circle. ✔

Guided Practice

4. Write an equation of a circle that contains $R(1, 2)$, $S(-3, 4)$, and $T(-5, 0)$.

Personal Tutor glencoe.com

Check Your Understanding

Examples 1 and 2 pp. 744–745

Write the equation of each circle.

1. center at $(9, 0)$, radius 5

2. center at $(3, 1)$, diameter 14

3. center at origin, passes through $(2, 2)$

4. center at $(-5, 3)$, passes through $(1, -4)$

5.

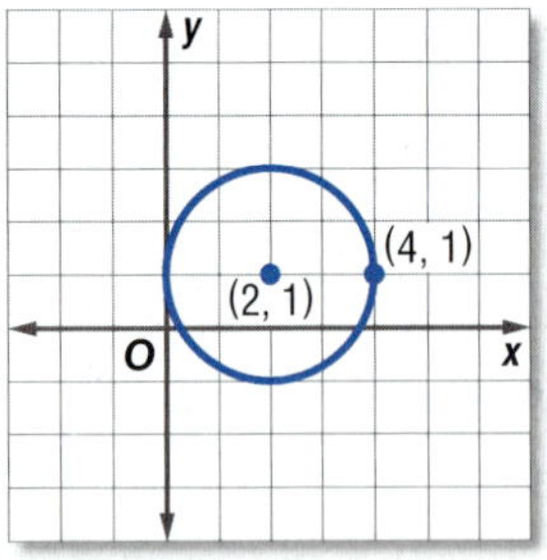

6.

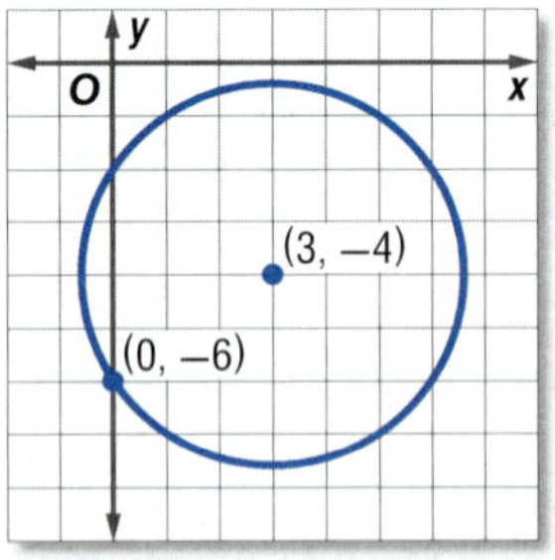

Example 3 p. 745

For each circle with the given equation, state the coordinates of the center and the measure of the radius. Then graph the equation.

7. $(x - 3)^2 + (y + 2)^2 = 16$

8. $x^2 + (y + 1)^2 = 4$

Example 4 p. 746

9. **RADIOS** Three radio towers are modeled by the points $R(4, 5)$, $S(8, 1)$, and $T(-4, 1)$. Determine the location of another tower equidistant from all three towers, and write an equation for the circle.

10. **COMMUNICATION** Three cell phone towers can be modeled by the points $X(6, 0)$, $Y(8, 4)$, and $Z(3, 9)$. Determine the location of another cell phone tower equidistant from the other three, and write an equation for the circle.

Practice and Problem Solving

= **Step-by-Step Solutions** begin on page R20.
Extra Practice begins on page 969.

Examples 1 and 2
pp. 744–745

Write the equation of each circle.

11. center at origin, radius 4

12. center at (6, 1), radius 7

13. center at (−2, 0), diameter 16

14. center at (8, −9), radius $\sqrt{11}$

15. center at (−3, 6), passes through (0, 6)

16. center at (1, −2), passes through (3, −4)

17.

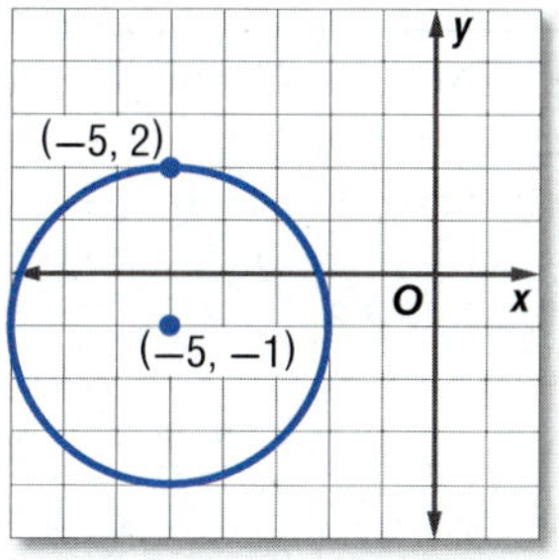

18.

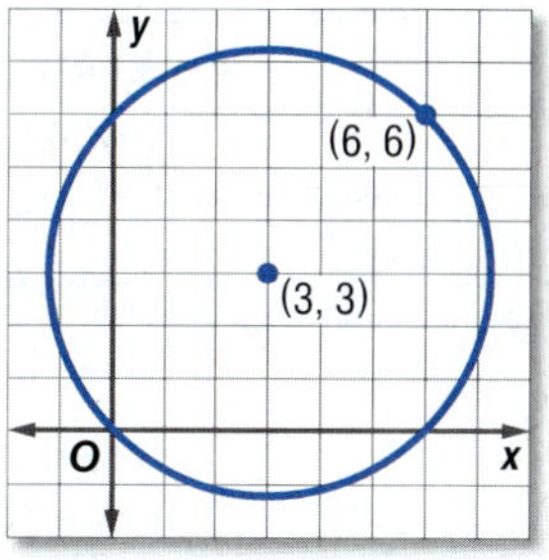

19. **WEATHER** A Doppler radar screen shows concentric rings around a storm. If the center of the radar screen is the origin and each ring is 15 miles farther from the center, what is the equation of the third ring?

20. **GARDENING** A sprinkler waters a circular area that has a diameter of 10 feet. The sprinkler is located 20 feet north of the house. If the house is located at the origin, what is the equation for the circle of area that is watered?

Example 3
p. 745

For each circle with the given equation, state the coordinates of the center and the measure of the radius. Then graph the equation.

21. $x^2 + y^2 = 36$

22. $(x - 2)^2 + (y - 1)^2 = 4$

23. $(x + 4)^2 + (y - 2)^2 = 16$

24. $(x - 8)^2 + y^2 = 64$

Example 4
p. 746

Write an equation of a circle that contains each set of points. Then graph the circle.

25. $A(1, 6), B(5, 6), C(5, 0)$

26. $F(3, -3), G(3, 1), H(7, 1)$

Write the equation of each circle.

27. a circle with a diameter having endpoints at (0, 4) and (6, −4)

28. a circle with $d = 22$ and a center translated 13 units left and 6 units up from the origin

29. **MODEL ROCKETS** Different-sized engines will launch model rockets to different altitudes. The higher a rocket goes, the larger the circle of possible landing sites becomes. Under normal wind conditions, the landing radius is three times the altitude of the rocket.

a. Write the equation of the landing circle for a rocket that travels 300 feet in the air.

b. What would be the radius of the landing circle for a rocket that travels 1000 feet in the air? Assume the center of the circle is at the origin.

30. **SKYDIVING** Three of the skydivers in the circular formation shown have approximate coordinates of $G(13, -2)$, $H(-1, -2)$, and $J(6, -9)$.

a. What are the approximate coordinates of the center skydiver?

b. If each unit represents 1 foot, what is the diameter of the skydiving formation?

31. **DELIVERY** Pizza and Subs offers free delivery within 6 miles of the restaurant. The restaurant is located 4 miles west and 5 miles north of Consuela's house.

a. Write and graph an equation to represent this situation if Consuela's house is at the origin of the coordinate system.

b. Describe what the graph represents. Can Consuela get free delivery if she orders pizza from Pizza and Subs? Explain how you know.

32. **MULTIPLE REPRESENTATIONS** In this problem, you will investigate a compound locus for a pair of points. A **compound locus** satisfies more than one distinct set of conditions.

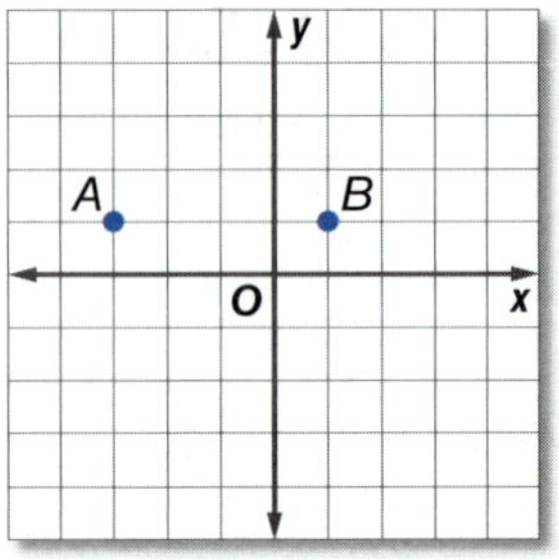

a. **TABULAR** Choose two points A and B in the coordinate plane. Locate 5 coordinates from the locus of points equidistant from A and B.

b. **GRAPHICAL** Represent this same locus of points by using a graph.

c. **VERBAL** Describe the locus of all points equidistant from a pair of points.

d. **GRAPHICAL** Using your graph from part **b**, determine and graph the locus of all points in a plane that are a distance of AB from B.

e. **VERBAL** Describe the locus of all points in a plane equidistant from a single point. Then describe the locus of all points that are both equidistant from A and B and are a distance of AB from B. How is the compound locus represented graphically?

StudyTip

Coordinate Plane Remember that when you have problems that use the coordinate plane, use tools like the Distance, Midpoint, and Slope Formulas to solve problems and to check your solutions.

33. The equation of a circle is $x^2 + 6x + y^2 - 2y = 15$. What are the center and radius of the circle?

34. Write an equation of the circle that has a diameter with endpoints at $(-6, 15)$ and $(2, 7)$.

35. A circle with a diameter of 12 has its center in the second quadrant. The lines $y = -4$ and $x = 1$ are tangent to the circle. Write an equation of the circle.

H.O.T. Problems Use Higher-Order Thinking Skills

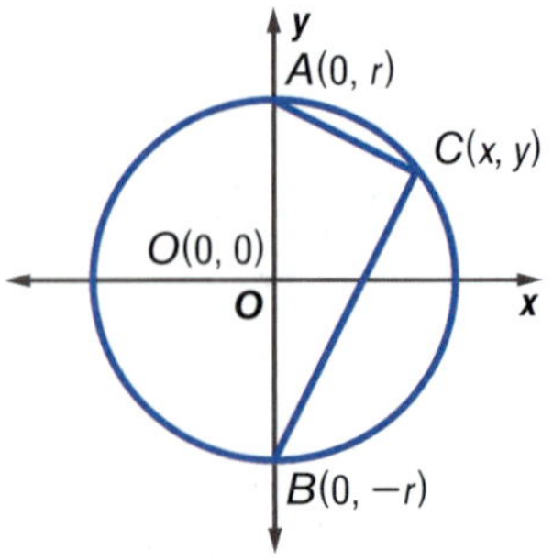

36. **CHALLENGE** Write a coordinate proof to show that if an inscribed angle intercepts the diameter of a circle, as shown, the angle is a right angle.

37. **REASONING** A circle has the equation $(x - 5)^2 + (y + 7)^2 = 16$. If the center of the circle is shifted 3 units right and 9 units up, what would be the equation of the new circle? Explain your reasoning.

38. **OPEN ENDED** Graph three noncollinear points on a coordinate plane. Draw a triangle by connecting the points. Then construct the circle that circumscribes it.

39. **WRITING IN MATH** Explain how the Distance Formula and the equation of a circle are related.

40. **WRITING IN MATH** Describe how the equation for a circle changes if the circle is translated a units to the right and b units down.

Standardized Test Practice

41. Which of the following is the equation of a circle with center (6, 5) that passes through (2, 8)?

A $(x - 6)^2 + (y - 5)^2 = 5^2$

B $(x - 5)^2 + (y - 6)^2 = 7^2$

C $(x + 6)^2 + (y + 5)^2 = 5^2$

D $(x - 2)^2 + (y - 8)^2 = 7^2$

42. ALGEBRA What are the solutions of $n^2 - 4n = 21$?

F 3, 7 **H** −3, 7

G 3, −7 **J** −3, −7

43. SHORT RESPONSE Solve: $5(x - 4) = 16$.

Step 1: $5x - 4 = 16$

Step 2: $5x = 20$

Step 3: $x = 4$

Which is the first incorrect step in the solution shown above?

44. SAT/ACT The center of ⊙F has coordinates (−4, 0). If the circle has a radius of 4, which point lies on ⊙F?

A (4, 0) **C** (4, 3)

B (0, 4) **D** (−4, 4)

Spiral Review

Find x. (Lesson 10-7)

45.

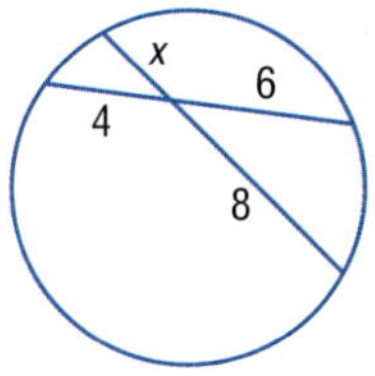

46.

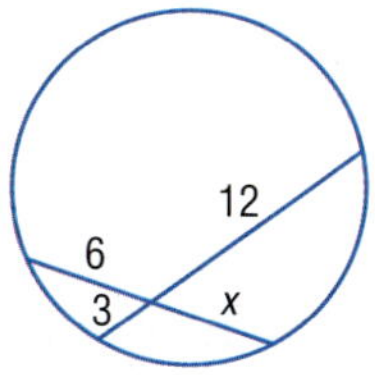

47.

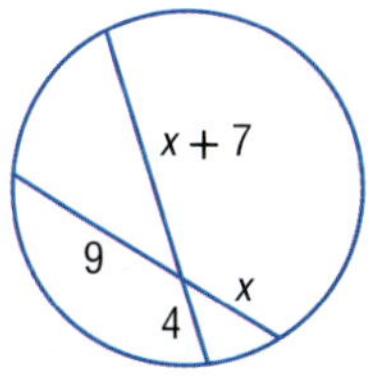

Find each measure. (Lesson 10-6)

48. $m\angle C$

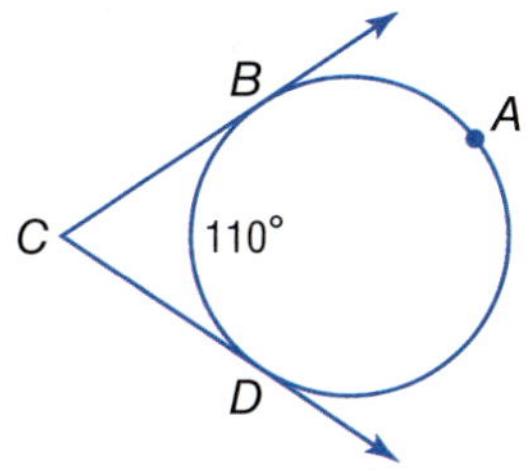

49. $m\angle K$

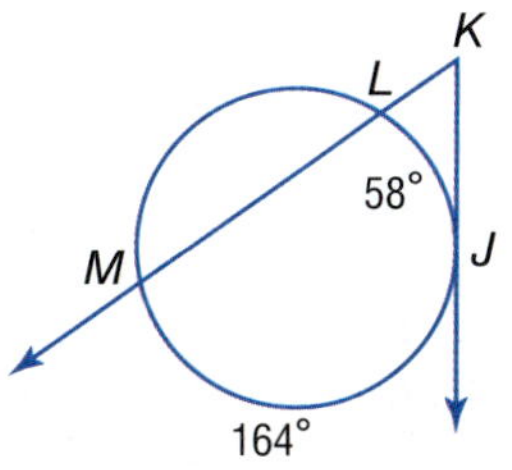

50. $m\widehat{YZ}$

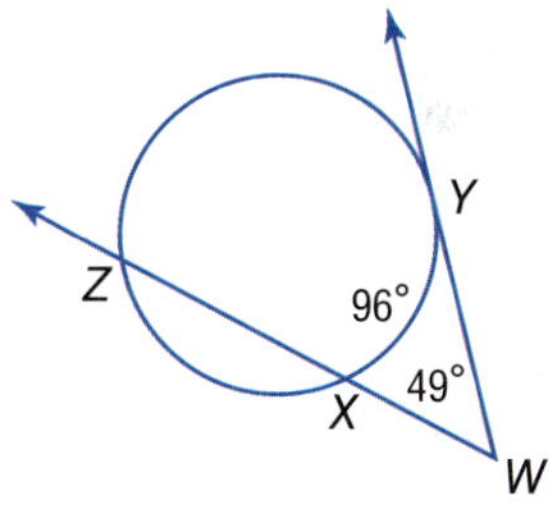

51. STREETS The neighborhood where Vincent lives has roundabouts where certain streets meet. If Vincent rides his bike once around the very edge of the grassy circle, how many feet will he have ridden? (Lesson 10-1)

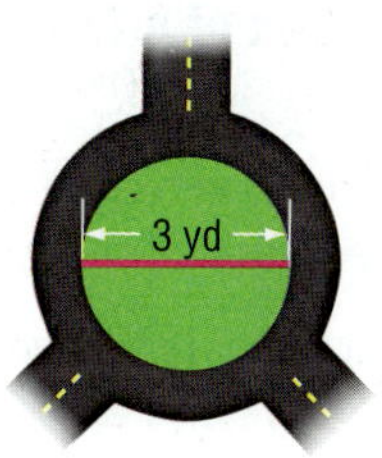

Skills Review

Find the perimeter and area of each figure. (Lesson 1-6)

52.

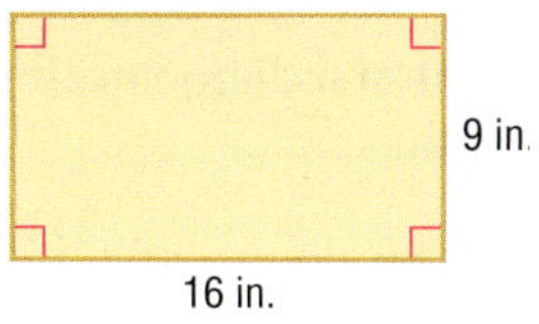

53.

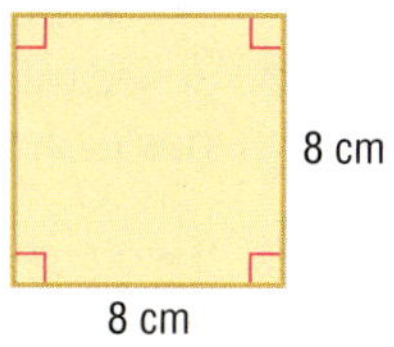

54.

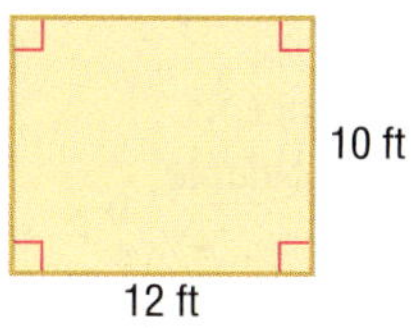

CHAPTER 10

Study Guide and Review

Math Online glencoe.com
- STUDY TO GO
- Vocabulary Review

Chapter Summary

Key Concepts

Circles and Circumference (Lesson 10-1)

- The circumference of a circle is equal to πd or $2\pi r$.

Angles, Arcs, Chords, and Inscribed Angles (Lessons 10-2 to 10-4)

- The sum of the measures of the central angles of a circle is 360°.
- The length of an arc is proportional to the length of the circumference.
- Diameters perpendicular to chords bisect chords and intercepted arcs.
- The measure of an inscribed angle is half the measure of its intercepted arc.

Tangents, Secants, and Angle Measures (Lessons 10-5 and 10-6)

- A line that is tangent to a circle intersects the circle in exactly one point and is perpendicular to a radius.
- Two segments tangent to a circle from the same exterior point are congruent.
- The measure of an angle formed by two secant lines is half the positive difference of its intercepted arcs.
- The measure of an angle formed by a secant and tangent line is half its intercepted arc.

Special Segments and Equation of a Circle (Lessons 10-7 and 10-8)

- The lengths of intersecting chords in a circle can be found by using the products of the measures of the segments.
- The equation of a circle with center (h, k) and radius r is $(x - h)^2 - (y - k)^2 = r^2$.

FOLDABLES® Study Organizer

Be sure the Key Concepts are noted in your Foldable.

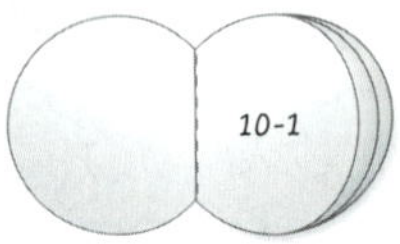

Key Vocabulary

adjacent arcs (p. 694)
arc (p. 692)
arc length (p. 695)
center (p. 683)
central angle (p. 692)
chord (p. 683)
chord segment (p. 736)
circle (p. 683)
circumference (p. 685)
circumscribed (p. 686)
common tangent (p. 718)
compound locus (p. 748)
concentric circles (p. 684)
congruent arcs (p. 693)
congruent circles (p. 684)
diameter (p. 683)
external segment (p. 738)
inscribed (p. 686)
inscribed angle (p. 709)
intercepted arc (p. 709)
major arc (p.693)
minor arc (p. 693)
pi (π) (p. 685)
point of tangency (p. 718)
radius (p. 683)
secant (p. 727)
secant segment (p. 738)
semicircle (p. 693)
tangent (p. 718)

Vocabulary Check

State whether each sentence is *true* or *false*. If *false*, replace the underlined word or phrase to make a true sentence.

1. Any segment with both endpoints on the circle is a <u>radius</u> of the circle.
2. A chord passing through the center of a circle is a <u>diameter</u>.
3. A <u>central angle</u> has the center as its vertex and its sides contain two radii of the circle.
4. An arc with a measure of less than 180° is a <u>major arc</u>.
5. An <u>intercepted arc</u> is an arc that has its endpoints on the sides of an inscribed angle and lies in the interior of the inscribed angle.
6. A <u>common tangent</u> is the point at which a line in the same plane as a circle intersects the circle.
7. A secant is a line that intersects a circle in exactly <u>one</u> point.
8. A secant segment is a segment of a <u>diameter</u> that has exactly one endpoint on the circle.
9. Two circles are <u>concentric</u> circles if and only if they have congruent radii.

Lesson-by-Lesson Review

10-1 Circles and Circumference (pp. 683–691)

For Exercises 10–12, refer to ⊙D.

10. Name the circle.

11. Name a radius.

12. Name a chord that is not a diameter.

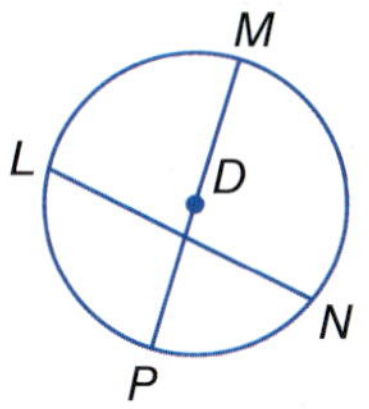

Find the diameter and radius of a circle with the given circumference. Round to the nearest hundredth.

13. $C = 43$ cm

14. $C = 26.7$ yd

15. $C = 108.5$ ft

16. $C = 225.9$ mm

EXAMPLE 1

Find the circumference of ⊙A.

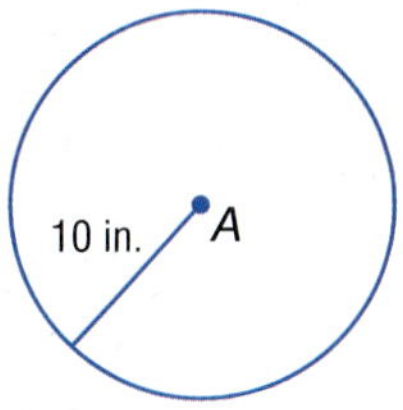

$C = 2\pi r$	**Circumference formula**
$= 2\pi(10)$	**Substitution**
≈ 62.83	**Use a calculator.**

The circumference of ⊙A is about 62.83 inches.

10-2 Measuring Angles and Arcs (pp. 692–700)

Find the value of x.

17.

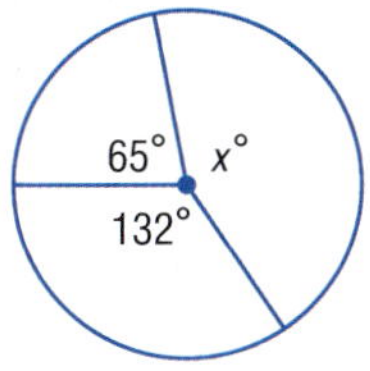

18.

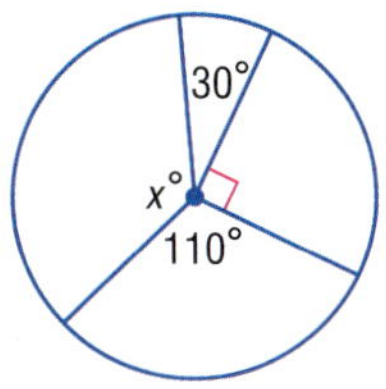

19. **MOVIES** The pie chart below represents the results of a survey taken by Mrs. Jameson regarding her students' favorite types of movies. Find each measure.

Mrs. Jameson's Students' Favorite Types of Movies

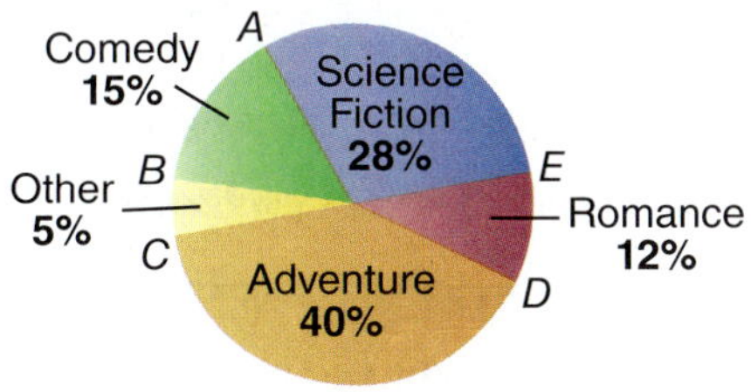

a. $m\widehat{AE}$

b. $m\widehat{BC}$

c. Describe the type of arc that the category Adventure represents.

EXAMPLE 2

Find the value of x.

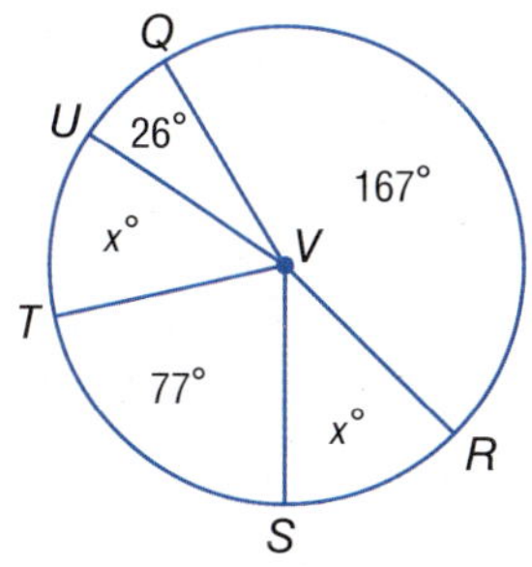

$m\angle QVR + m\angle RVS + m\angle SVT +$ $m\angle TVU + m\angle UVQ = 360°$	**Sum of Central Angles**
$167 + x + 77 + x + 26 = 360$	**Substitution**
$270 + 2x = 360$	**Simplify.**
$2x = 90$	**Subtract.**
$x = 45°$	**Divide.**

10-3 Arcs and Chords (pp. 701–708)

20. Find the value of x.

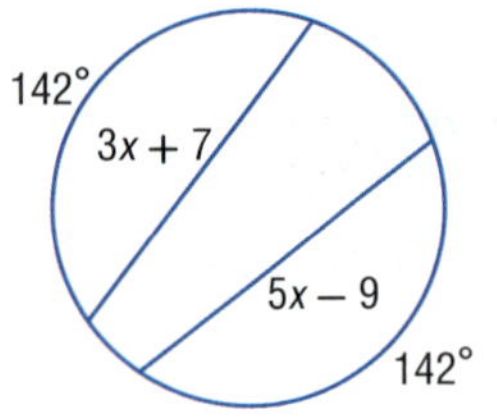

In ⊙K, $MN = 16$ and $m\widehat{MLN} = 98$. Find each measure. Round to the nearest hundredth.

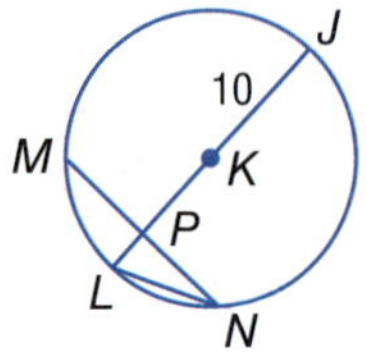

21. $m\widehat{NJ}$ **22.** LN

23. GARDENING The top of the trellis shown is an arc of a circle in which $\overline{CD}$ is part of the diameter. If $\widehat{ACB}$ is about 28% of a complete circle, what is $m\widehat{CB}$?

EXAMPLE 3

ALGEBRA In ⊙E, $EG = EF$. Find AB.

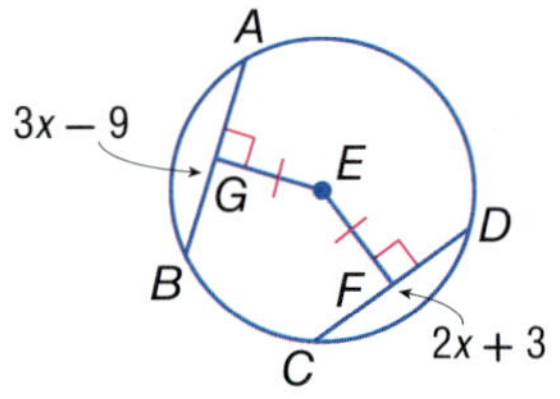

Since chords $\overline{EG}$ and $\overline{EF}$ are congruent, they are equidistant from E. So, $AB = CD$.

$AB = CD$	Theorem 10.5
$3x - 9 = 2x + 3$	Substitution
$3x = 2x + 12$	Add.
$x = 12$	Simplify.

So, $AB = 3(12) - 9$ or 27.

10-4 Inscribed Angles (pp. 709–716)

Find each measure.

24. $m\angle 1$

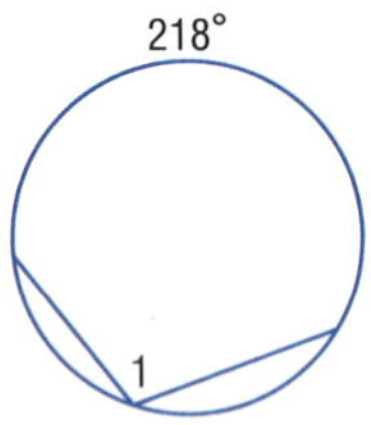

25. $m\widehat{GH}$

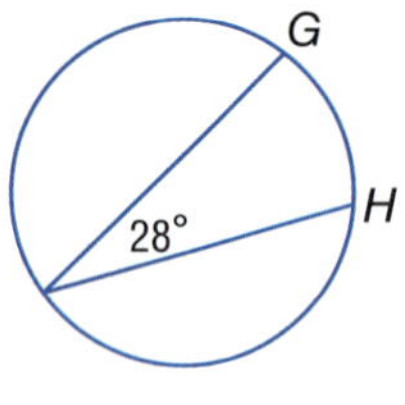

26. MARKETING In the logo at the right, $m\angle 1 = 42$. Find $m\angle 5$.

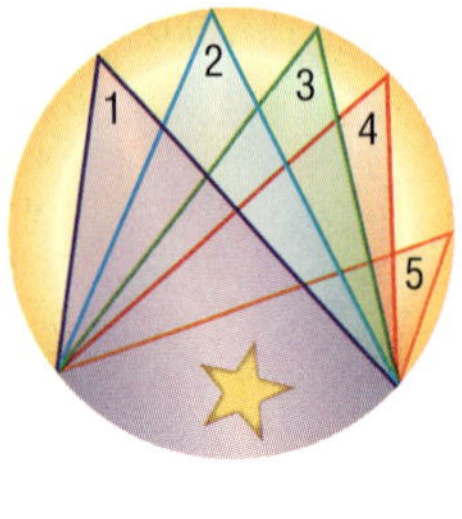

EXAMPLE 4

Find $m\angle D$ and $m\angle B$.

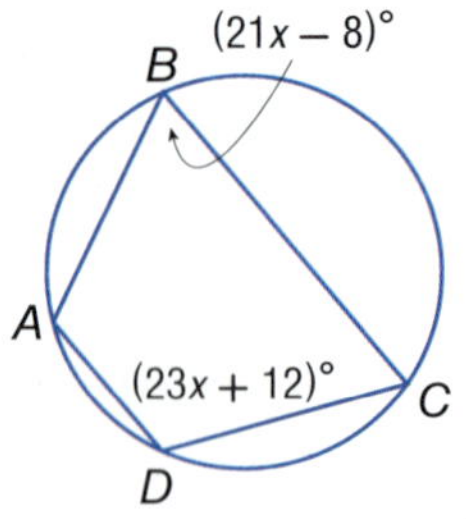

Since $ABCD$ is inscribed in a circle, opposite angles are supplementary.

$m\angle D + m\angle B = 180$	Definition of supplementary
$23x + 12 + 21x - 8 = 180$	Substitution
$44x + 4 = 180$	Simplify.
$44x = 176$	Subtract.
$x = 4$	Divide.

So, $m\angle D = 23(4) + 12$ or 104 and $m\angle B = 21(4) - 8$ or 76.

MIXED PROBLEM SOLVING
For mixed problem-solving practice, see page 1008.

10-5 Tangents (pp. 718–725)

27. SCIENCE FICTION In a story Todd is writing, instantaneous travel between a two-dimensional planet and its moon is possible when the time-traveler follows a tangent. Copy the figures below and draw all possible travel paths.

28. Find x and y. Assume that segments that appear to be tangent are tangent. Round to the nearest tenth if necessary.

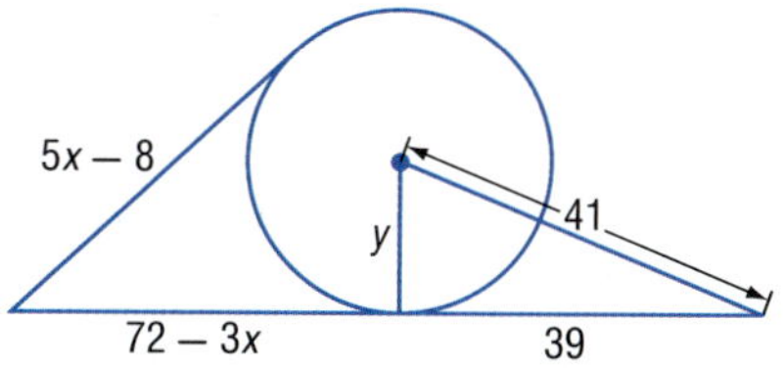

EXAMPLE 5

In the figure, $\overline{KL}$ is tangent to $\odot M$ at K. Find the value of x.

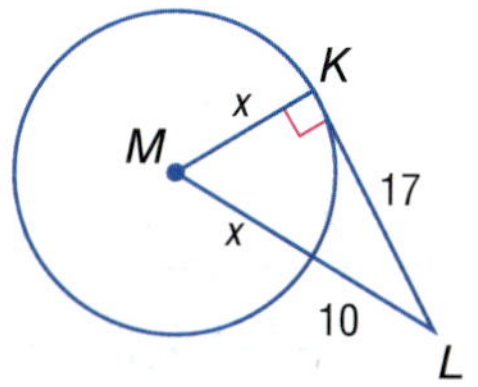

By Theorem 10.9, $\overline{MK} \perp \overline{KL}$. So, $\triangle MKL$ is a right triangle.

$KM^2 + KL^2 = ML^2$	Pythagorean Theorem
$x^2 + 17^2 = (x + 10)^2$	Substitution
$x^2 + 289 = x^2 + 20x + 100$	Multiply.
$289 = 20x + 100$	Simplify.
$189 = 20x$	Subtract.
$9.45 = x$	Divide.

10-6 Secants, Tangents, and Angle Measures (pp. 727–735)

Find each measure. Assume that segments that appear to be tangent are tangent.

29. $m\angle 1$

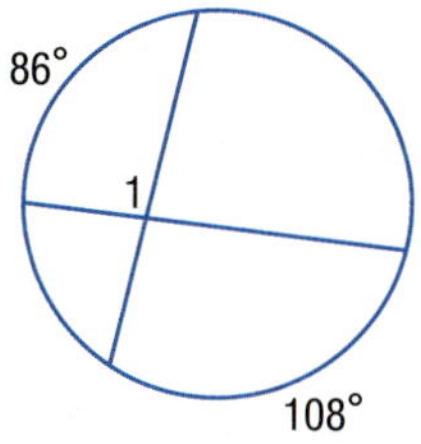

30. $m\widehat{AC}$

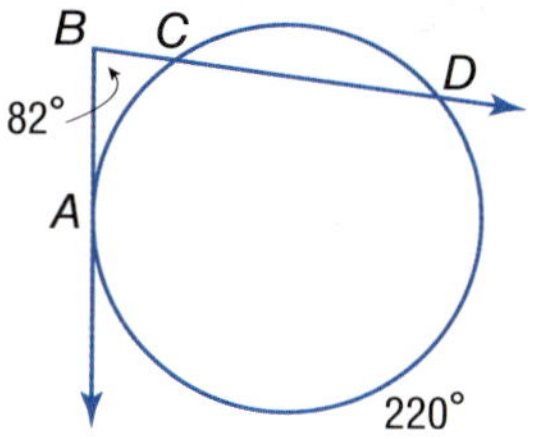

31. PHOTOGRAPHY Ahmed needs to take a close-up shot of an orange for his art class. He frames a shot of an orange as shown below, so that the lines of sight form tangents to the orange. If the measure of the camera's viewing angle is 34°, what is $m\widehat{ACB}$?

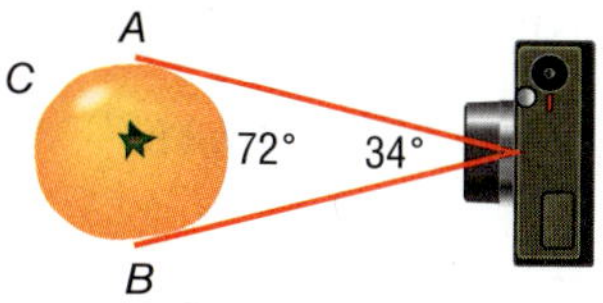

EXAMPLE 6

Find the value of x.

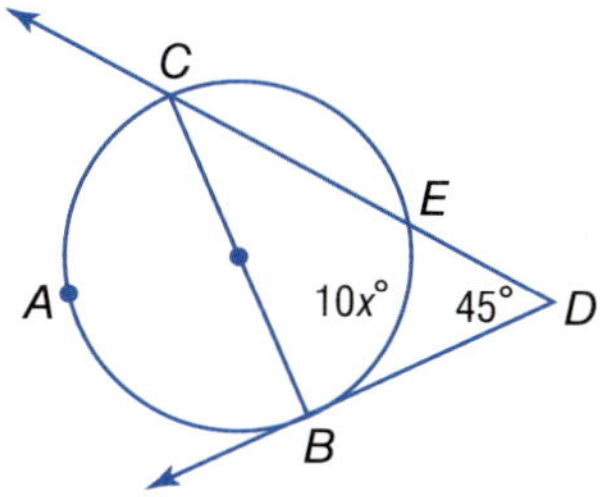

$\widehat{CAB}$ is a semicircle because $\overline{CB}$ is a diameter.

So, $m\widehat{CAB} = 180$.

$m\angle D = \frac{1}{2}(m\widehat{CB} - m\widehat{EB})$	Theorem 10.14
$45 = \frac{1}{2}(180 - 10x)$	Substitution
$90 = 180 - 10x$	Multiply.
$-90 = -10x$	Subtract.
$9 = x$	Divide.

CHAPTER 10 Study Guide and Review

10-7 Special Segments in a Circle (pp. 736–742)

Find x. Assume that segments that appear to be tangent are tangent.

32.

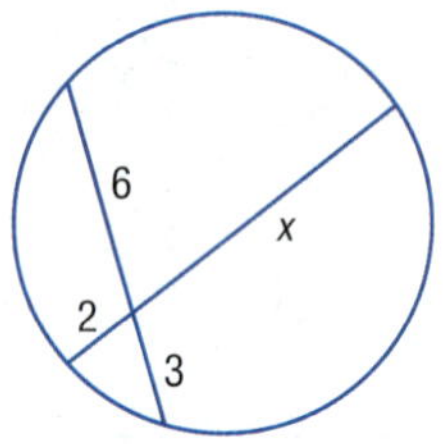

33.

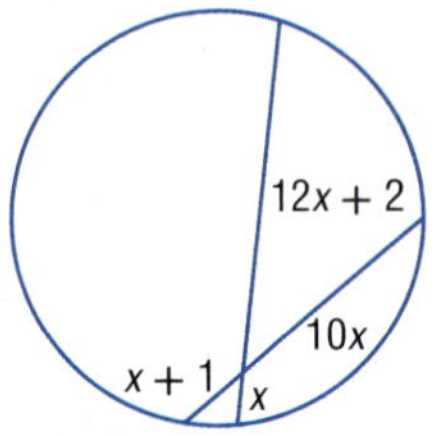

34. ARCHAEOLOGY While digging a hole to plant a tree, Henry found a piece of a broken saucer. What was the circumference of the original saucer? Round to the nearest hundredth.

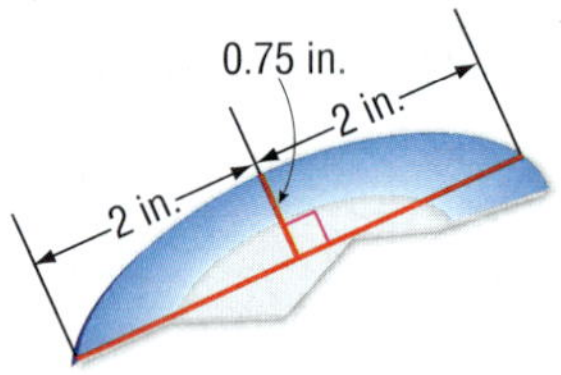

EXAMPLE 7

Find the diameter of circle M.

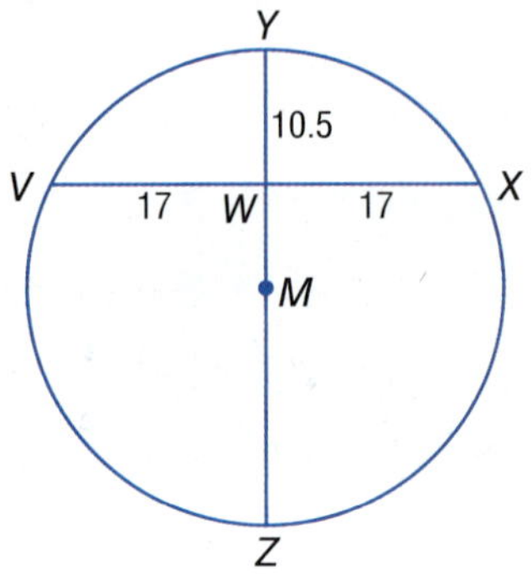

$VW \cdot WX = YW \cdot WZ$	**Theorem 10.14**
$17 \cdot 17 = 10.5 \cdot WZ$	**Substitution**
$289 = 10.5 \cdot WZ$	**Simplify.**
$27.5 \approx WZ$	**Divide each side by 10.5.**
$YZ = YW + WZ$	**Segment Addition Postulate**
$YZ = 10.5 + 27.5$	**Substitution**
$YZ = 38$	**Simplify.**

10-8 Equations of Circles (pp. 744–749)

Write the equation of each circle.

35. center at $(-2, 4)$, radius 5

36. center at $(1, 2)$, diameter 14

37. FIREWOOD In an outdoor training course, Kat learns a wood-chopping safety check that involves making a circle with her arm extended, to ensure she will not hit anything overhead as she chops. If her reach is 19 inches and the hatchet handle is 15 inches, what is the equation of Kat's safety circle?

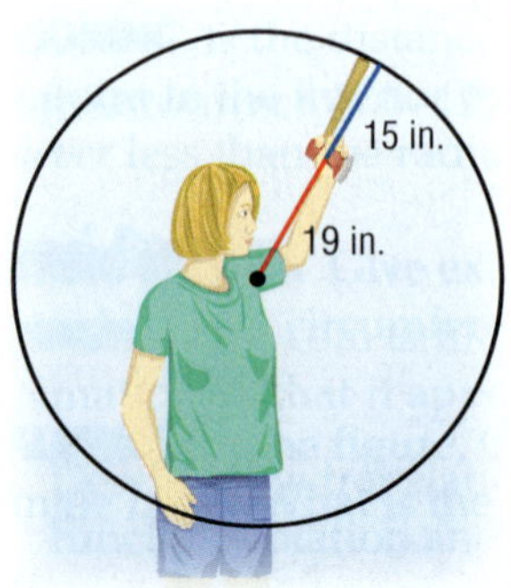

EXAMPLE 8

Write the equation of the circle graphed below.

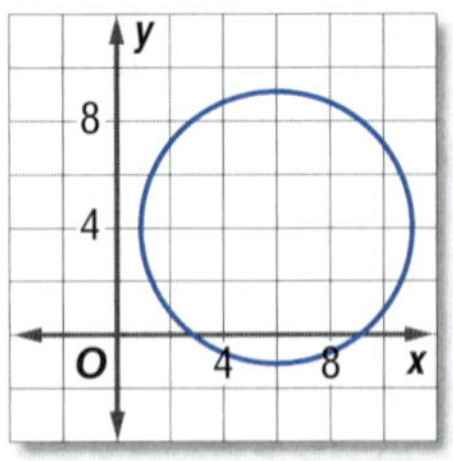

The center is at $(6, 4)$ and the radius is 5.

$(x - h)^2 + (y - k)^2 = r^2$	**Equation of a circle**
$(x - 6)^2 + (y - 4)^2 = 5^2$	**$(h, k) = (6, 4)$ and $r = 5$**
$(x - 6)^2 + (y - 4)^2 = 25$	**Simplify.**

CHAPTER 10 Practice Test

Math Online glencoe.com
Chapter Test

1. **POOLS** Amanda's family has a swimming pool that is 4 feet deep in their backyard. If the diameter of the pool is 25 feet, what is the circumference of the pool to the nearest foot?

2. Find the exact circumference of the circle below.

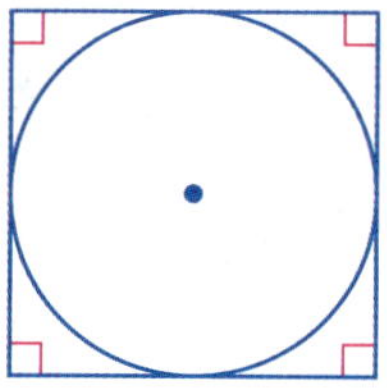

Find the value of x.

3.

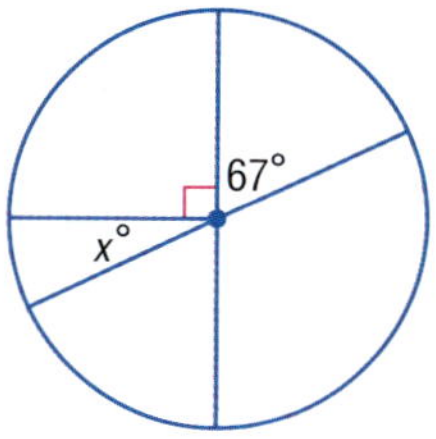

4.

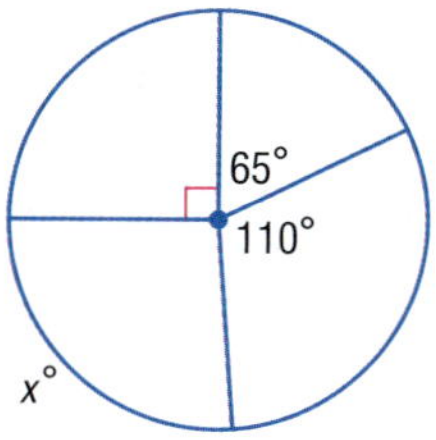

5.

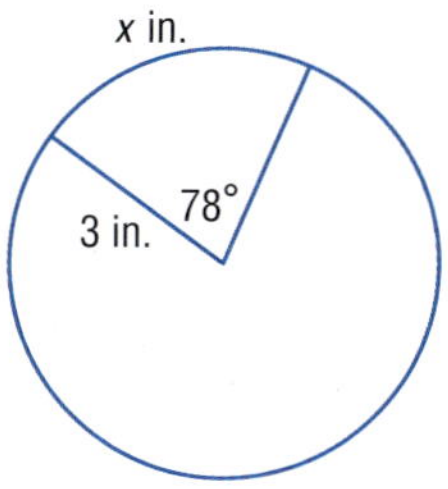

6.

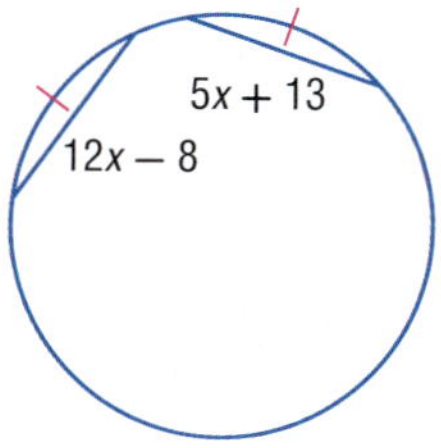

7. **MULTIPLE CHOICE** What is ED?

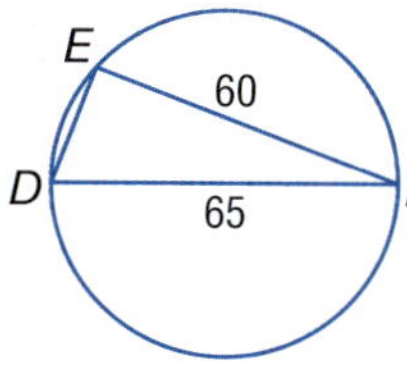

A 15

B 25

C 88.5

D not enough information

8. Find x if $\odot M \cong \odot N$.

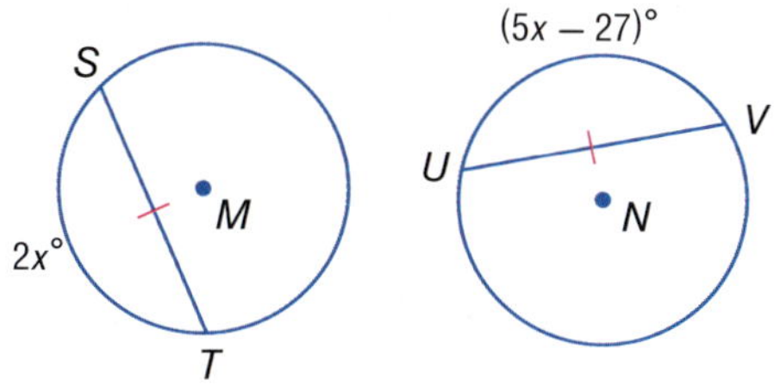

9. **MULTIPLE CHOICE** How many points are shared by concentric circles?

F 0

G 1

H 2

J infinite points

10. Determine whether $\overline{FG}$ is tangent to $\odot E$. Justify your answer.

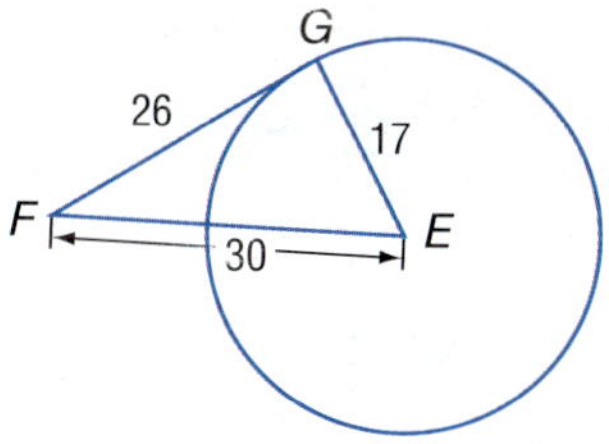

11. **MULTIPLE CHOICE** Which of the figures below shows a polygon circumscribed about a circle?

A

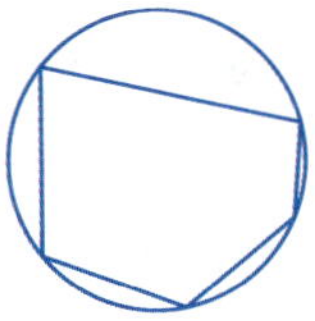

C

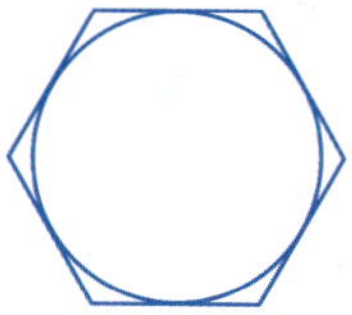

B

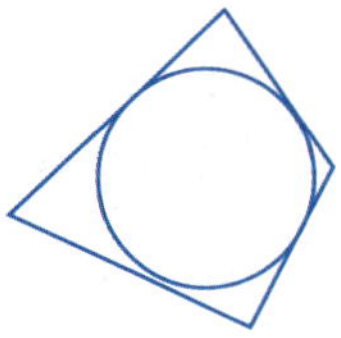

D

12. Find the perimeter of the triangle at the right. Assume that segments that appear to be tangent are tangent.

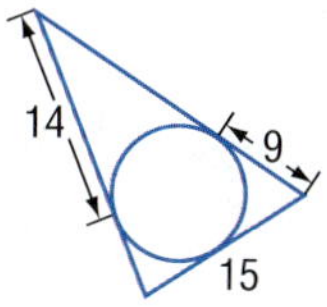

Find each measure.

13. $m\angle T$

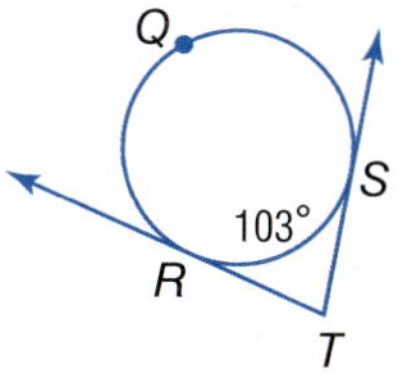

14. x

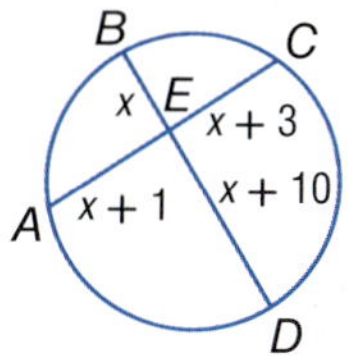

15. **FLOWERS** Hannah wants to encircle a tree trunk with a flower bed. If the center of the tree trunk is the origin and Hannah wants the flower bed to extend to 3 feet from the center of the tree, what is the equation that would represent the flower bed?

Properties of Circles

A circle is a unique shape in which the angles, arcs, and segments intersecting the circle have special properties and relationships. You should be able to identify the parts of a circle, write the equation of a circle, and solve for arc, angle, and segment measures in a circle.

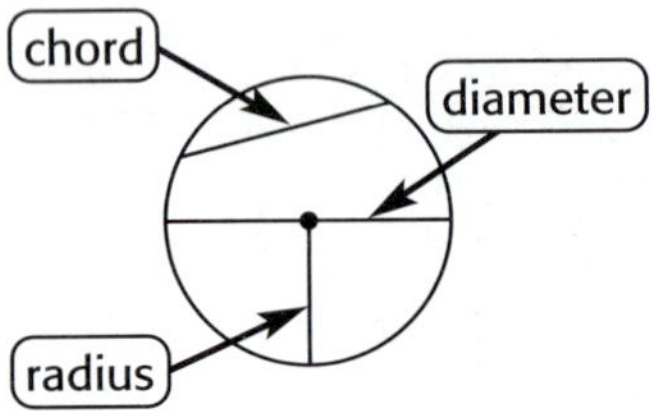

$r = \frac{1}{2}d$

$d = 2r$

$C = 2\pi r$ or πd

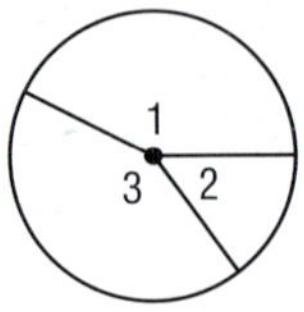

$m\angle 1 + m\angle 2 + m\angle 3 = 360$

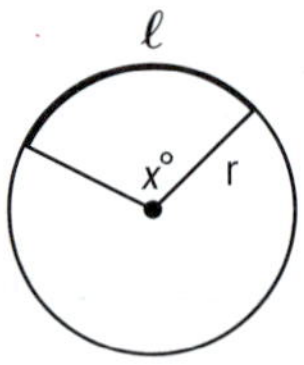

$\ell = \frac{x}{360} \cdot 2\pi r$

Strategies for Applying the Properties of Circles

Step 1

Review the parts of a circle and their relationships.

- Some key parts include: **radius, diameter, arc, chord, tangent, secant**
- Study the key theorems and the properties of circles as well as the relationships between the parts of a circle.

Step 2

Read the problem statement and study any figure you are given carefully.

- Determine what you are being asked to find.
- Fill in any information in the figure that you can.
- Determine which theorems or properties apply to the problem situation.

Step 3

Solve the problem and check your answer.

- Apply the theorems or properties to solve the problem.
- Check your answer to be sure it makes sense.

EXAMPLE

Read the problem. Identify what you need to know. Then use the information in the problem to solve.

Solve for x in the figure.

A 2 **C** 4

B 3 **D** 6

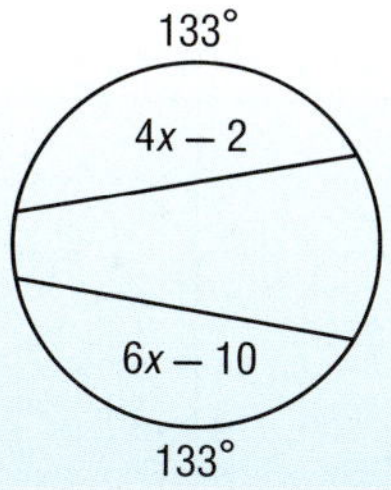

Read the problem statement and study the figure carefully. You are given a circle with two chords that correspond to congruent minor arcs. One important property of circles is that two chords are congruent if and only if their corresponding minor arcs are congruent. You can use this property to set up and solve an equation for x.

$4x - 2 = 6x - 10$	**Definition of Congruent Segments**
$4x - 6x = -10 + 2$	**Subtract.**
$-2x = -8$	**Simplify.**
$\frac{-2x}{-2} = \frac{-8}{-2}$	**Divide each side by –2.**
$x = 4$	**Simplify.**

So, the value of x is 4. The answer is C. You can check your answer by substituting 4 into each expression and making sure both chords have the same length.

Exercises

Read each problem. Identify what you need to know. Then use the information in the problem to solve.

1. Solve for x in the figure below.

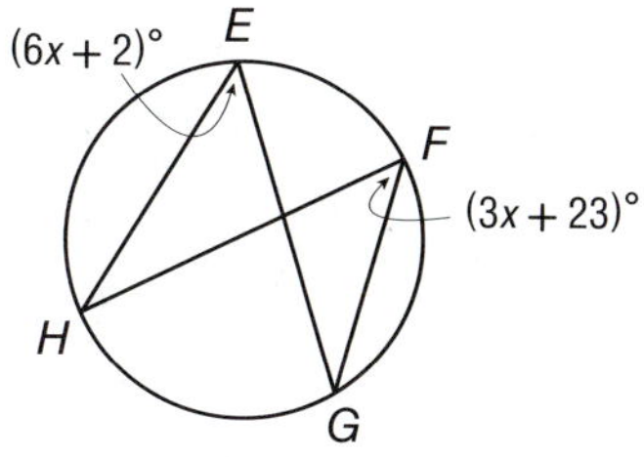

A 4 **C** 6

B 5 **D** 7

2. Triangle *RST* is circumscribed about the circle below. What is the perimeter of the triangle?

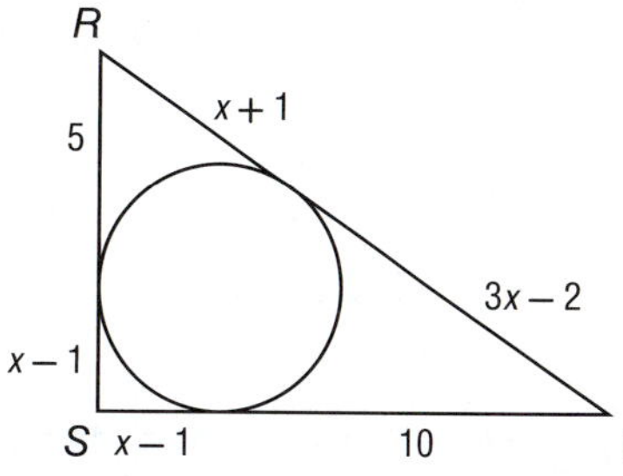

F 33 units **H** 37 units

G 36 units **J** 40 units

CHAPTER 10 Standardized Test Practice

Cumulative, Chapters 1 through 10

Multiple Choice

Read each question. Then fill in the correct answer on the answer document provided by your teacher or on a sheet of paper.

1. If $ABCD$ is a rhombus, and $m\angle ABC = 70°$, what is $m\angle 1$?

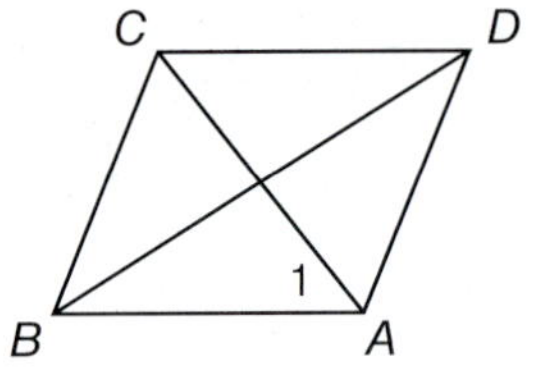

A 45° **C** 70°

B 55° **D** 125°

2. Karen argues that if you live in Greensboro, North Carolina, then you live in Guilford County. Which assumption would you need to make to form an indirect proof of this claim?

F Suppose someone lives in Guilford County, but not in Greensboro.

G Suppose someone lives in Greensboro, but not in Guilford County.

H Suppose someone lives in Greensboro and in Guilford County.

J Suppose someone lives in Guilford County and in Greensboro.

3. What is the value of x in the figure?

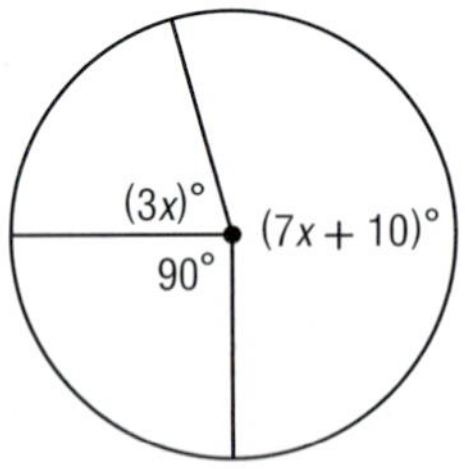

A 19 **C** 26

B 23 **D** 28

Test-Taking Tip

Question 3 Use the properties of circles to set up and solve an equation to find x.

4. Given $a \parallel b$, find $m\angle 1$.

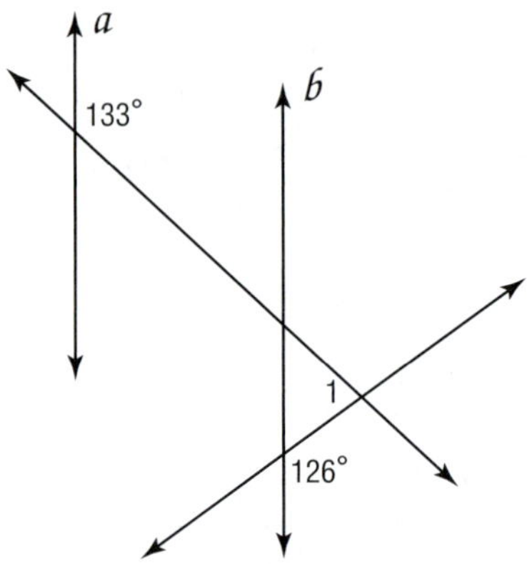

F 47°

G 54°

H 79°

J 101°

5. Which of the following conditions would *not* guarantee that a quadrilateral is a parallelogram?

A both pairs of opposite sides congruent

B both pairs of opposite angles congruent

C diagonals bisect each other

D one pair of opposite sides parallel

6. The ratio of the measures of the angles of the triangle below is 3:2:1. Which of the following is *not* an angle measure of the triangle?

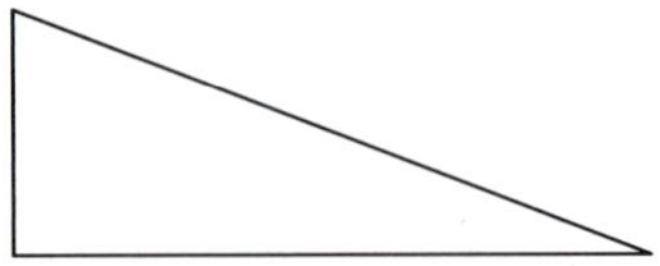

F 30°

G 45°

H 60°

J 90°

Short Response/Gridded Response

Record your answers on the answer sheet provided by your teacher or on a sheet of paper.

7. Does the figure shown have rotational symmetry? If so, give the order of symmetry.

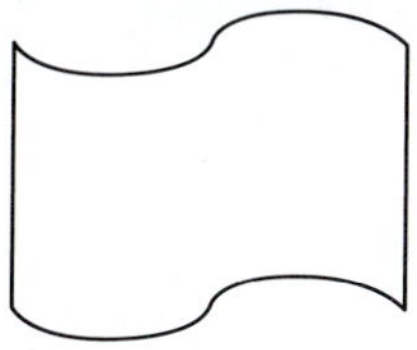

8. GRIDDED RESPONSE A square with 5-centimeter sides is inscribed in a circle. What is the circumference of the circle? Round your answer to the nearest tenth of a centimeter.

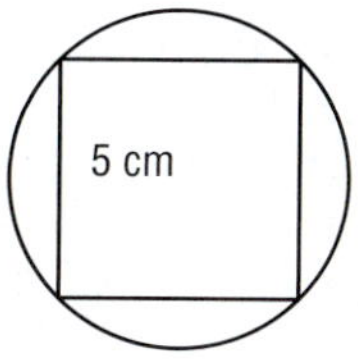

9. Solve for x in the figure. Show your work.

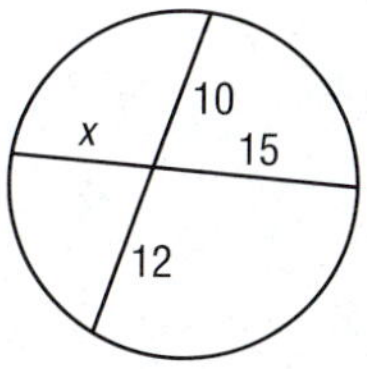

10. GRIDDED RESPONSE What is the perimeter of the right triangle below? Round your answer to the nearest tenth if necessary.

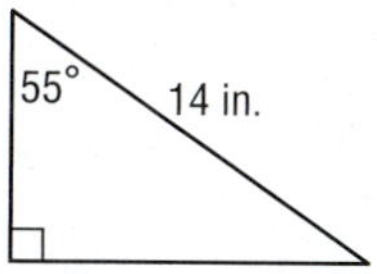

11. GRIDDED RESPONSE State the magnitude of rotational symmetry of the figure. Express your answer in degrees.

12. What is the length of $\overline{EF}$?

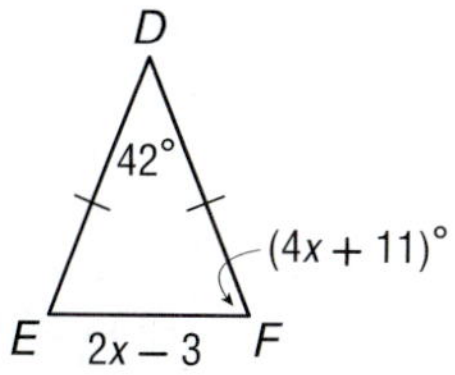

Extended Response

Record your answers on a sheet of paper. Show your work.

13. Use the circle shown to answer each question.

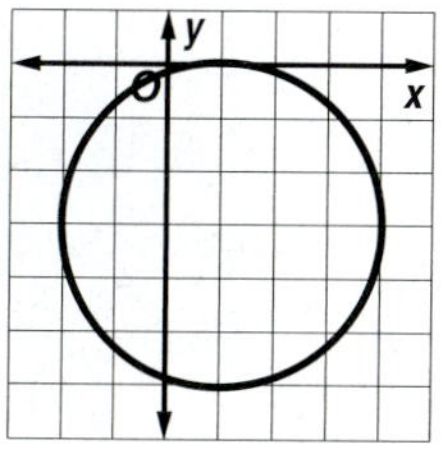

a. What is the center of the circle?

b. What is the radius of the circle?

c. Write an equation for the circle.

Need Extra Help?													
If you missed Question...	1	2	3	4	5	6	7	8	9	10	11	12	13
Go to Lesson or Page...	6-5	5-4	10-2	3-2	6-3	7-1	9-5	10-1	10-7	8-4	9-5	4-6	10-8

CHAPTER 11 Areas of Polygons and Circles

Then

In Chapter 10, you learned about circles and angles within circles.

Now

In Chapter 11, you will:

- Find areas of polygons.
- Solve problems involving areas and sectors of circles.
- Find scale factors using similar figures.

Why?

ART Artisans and craftsmen use area to determine the amount of raw materials that they will need for a project.

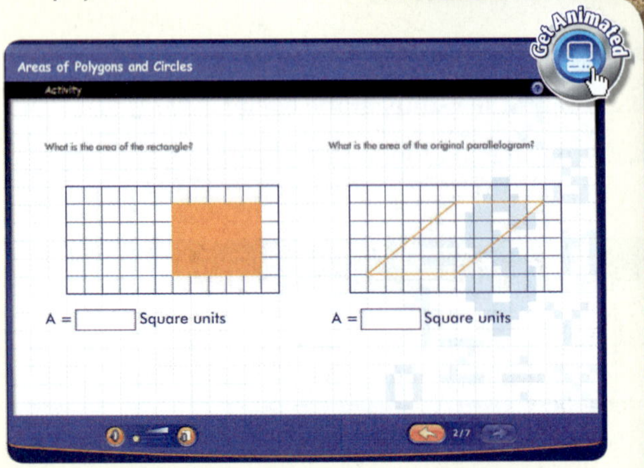

Get Ready for Chapter 11

Diagnose Readiness You have two options for checking Prerequisite Skills.

Text Option

Take the Quick Check below. Refer to the Quick Review for help.

*Quick*Check

The area and width of a rectangle are given. Find the length of the rectangle. (Lesson 1-6)

1. $A = 25, w = 5$

2. $A = 42, w = 6$

3. $A = 280, w = 14$

4. $A = 360, w = 60$

5. **GARDENS** Molly planted a garden with a length of 72 feet. If she bought enough fertilizer to cover 792 square feet, what width should she make the garden?

Evaluate each expression if $a = 9$, $b = 10$, $c = 12$, and $d = 13$. (Lesson 0-4)

6. $\frac{1}{2}a(b + c)$

7. $\frac{1}{2}(ab + cd)$

8. $\frac{1}{2}(a + bd)$

9. $\frac{1}{2}cd$

10. $\frac{1}{2}(ab + c)$

11. $\frac{1}{2}(a + d)$

Find h in each triangle. (Lesson 8-3)

12.

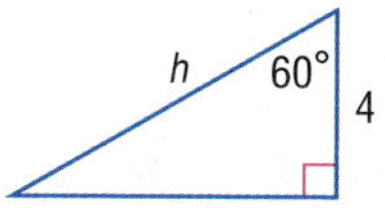

13.

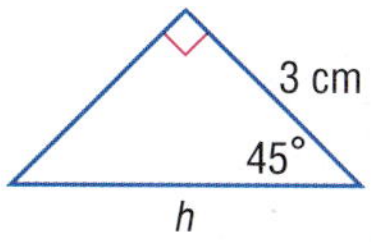

14. **LOOKOUT** The lookout on a pirate ship slides down a rope from the top of the mast 6 meters above the water. He can see the land at a 60° angle. How far does he slide?

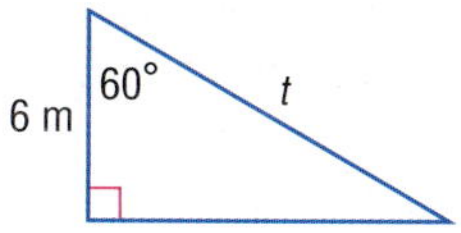

*Quick*Review

EXAMPLE 1

The area of a rectangle is 64 square units and the width is 4 units. Find the length.

$A = \ell w$	**Definition of Area**
$64 = \ell(4)$	**Substitution**
$16 = \ell$	**Divide each side by 4.**

The length is 16 units.

EXAMPLE 2

Evaluate $\frac{1}{2}x(2x + 3y)$ for $x = 4$ and $y = 12$.

$\frac{1}{2}x(2x + 3y) = \frac{1}{2}(4)[2(4) + 3(12)]$	**Substitution**
$= 2(8 + 36)$	**Simplify.**
$= 2(44)$	**Multiply.**
$= 88$	**Simplify.**

EXAMPLE 3

Find the value of h.

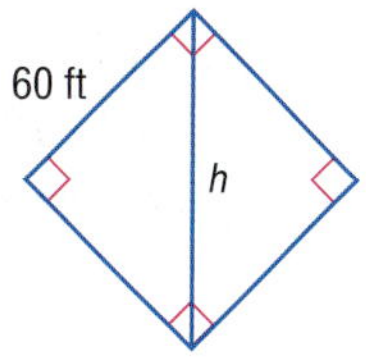

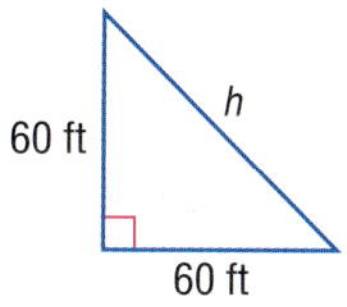

Since h is the hypotenuse of the triangle, the triangle can be redrawn as shown.

In a 45°-45°-90° triangle, the hypotenuse is $\sqrt{2}$ times the length of a leg.

$h = (\sqrt{2})60$

$= 84.85$

So, h is approximately 84.85 feet.

Online Option

Math Online Take a self-check Chapter Readiness Quiz at glencoe.com.

Get Started on Chapter 11

You will learn several new concepts, skills, and vocabulary terms as you study Chapter 11. To get ready, identify important terms and organize your resources. You may wish to refer to **Chapter 0** to review prerequisite skills.

FOLDABLES® Study Organizer

Areas of Polygons and Circles Make this Foldable to help you organize your Chapter 11 notes about areas of polygons and circles. Begin with three sheets of notebook paper.

1. **Stack** three sheets of paper and fold them in half, lengthwise.

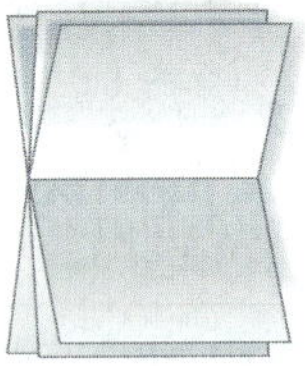

2. **Staple** the papers together one inch from the top fold.

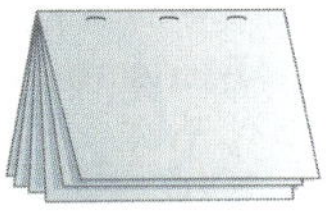

3. **Cut** the top sheet two inches from the top fold and each following sheet one inch longer than the previous sheet.

4. **Label** as shown.

Math Online glencoe.com

- Study the chapter online
- Explore **Get Animated** activities
- Get extra help from your own **Personal Tutor**
- Use **Extra Examples** for additional help
- Take a **Self-Check Quiz**
- **Review Vocabulary** in fun ways

New Vocabulary

English		Español
base of a parallelogram	• p. 763 •	base de un paralelogramo
height of a parallelogram	• p. 763 •	altura de un paralelogramo
base of a triangle	• p. 765 •	base de un triángulo
height of a triangle	• p. 765 •	altura de un triángulo
height of a trapezoid	• p. 773 •	altura de un trapecio
sector of a circle	• p. 783 •	sector de una círculo
center of a regular polygon	• p. 791 •	centro de un polígono regular
radius of a regular polygon	• p.791 •	radio de un polígono regular
apothem	• p. 791 •	apotema
central angle of a regular polygon	• p. 791 •	ángulo central de un polígono regular

Review Vocabulary

arc • p. 692 • arco a part of a circle that is defined by two endpoints

central angle • p. 692 • ángulo central an angle that intersects a circle in two points and has its vertex at the center of the circle

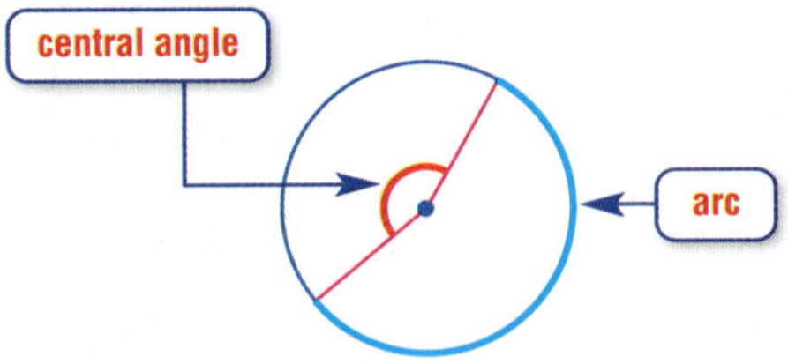

diagonal • p. 389 • diagonal a segment that connects nonconsecutive vertices of a polygon

Multilingual eGlossary glencoe.com

11-1 Areas of Parallelograms and Triangles

Then

You found areas of rectangles and squares. (Lesson 1-6)

Now

- Find perimeters and areas of parallelograms.
- Find perimeters and areas of triangles.

New Vocabulary

base of a parallelogram
height of a parallelogram
base of a triangle
height of a triangle

Math Online

glencoe.com

Why?

A tangram is an ancient Chinese puzzle that can be rearranged to form different images, such as the animals shown. The area of the puzzle, before and after being rearranged, remains the same. It is the sum of all the areas of its pieces.

Areas of Parallelograms In Lesson 6-2, you learned that a *parallelogram* is a quadrilateral with both pairs of opposite sides parallel. Any side of a parallelogram can be called the **base of a parallelogram**. The **height of a parallelogram** is the perpendicular distance between any two parallel bases.

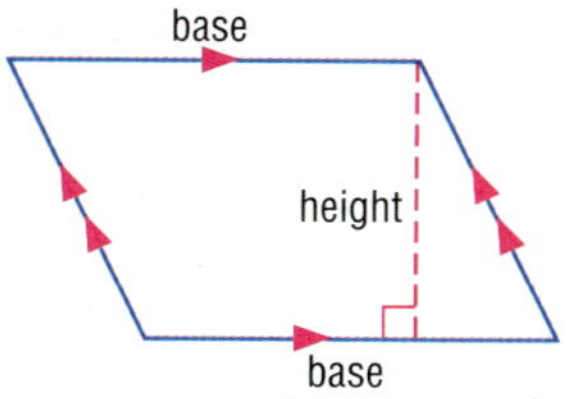

You can use the following postulate to develop the formula for the area of a parallelogram.

Postulate 11.1 **Area Addition Postulate**

The area of a region is the sum of the areas of its nonoverlapping parts.

In the figures below, a right triangle is cut off from one side of a parallelogram and translated to the other side as shown to form a rectangle with the same base and height.

Recall from Lesson 1-6 that the area of a rectangle is the product of its base and height. By the Area Addition Postulate, a parallelogram with base b and height h has the same area as a rectangle with base b and height h.

Key Concept **Area of a Parallelogram**

Words The area A of a parallelogram is the product of a base b and its corresponding height h.

Symbols $A = bh$

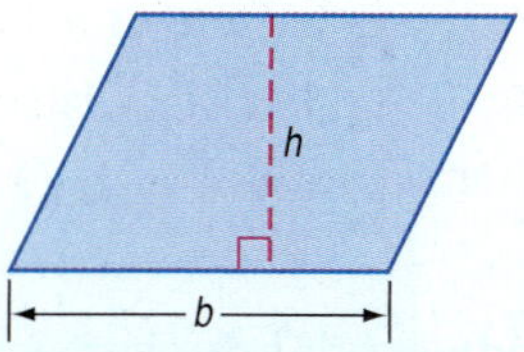

StudyTip

Heights of Figures The height of a figure can be measured by extending a base. In Example 1, the height of $\square ABCD$ that corresponds to base $\overline{DC}$ can be measured by extending $\overline{DC}$.

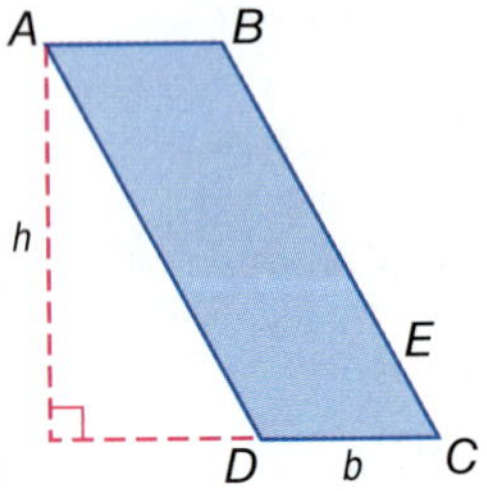

EXAMPLE 1 Perimeter and Area of a Parallelogram

Find the perimeter and area of $\square ABCD$.

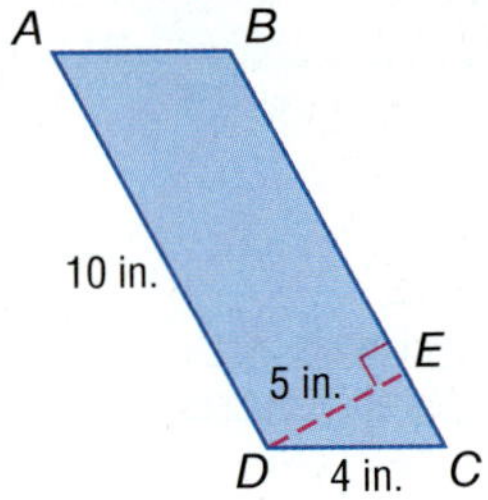

Perimeter

Since opposite sides of a parallelogram are congruent, $\overline{AB} \cong \overline{DC}$ and $\overline{BC} \cong \overline{AD}$. So $AB = 4$ inches and $BC = 10$ inches.

$$\text{Perimeter of } \square ABCD = AB + BC + DC + AD$$
$$= 4 + 10 + 4 + 10 \text{ or } 28 \text{ in.}$$

Area

The height given, DE, is 5 inches. $\overline{BC}$ is the base, which measures 10 inches.

$A = bh$ **Area of a parallelogram**

$= (10)(5)$ or 50 in^2 **$b = 10$ and $h = 5$**

Guided Practice

Find the perimeter and area of each parallelogram.

1A.

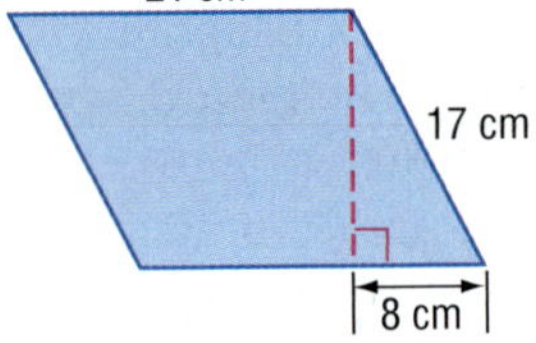

1B.

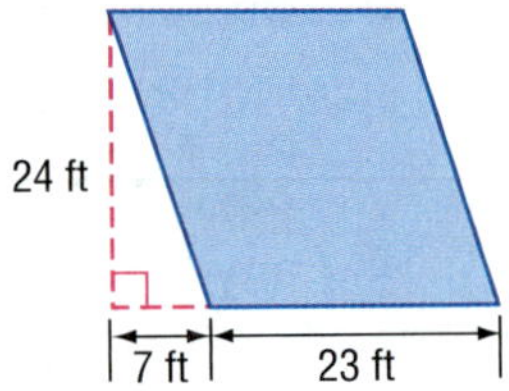

Personal Tutor glencoe.com

You may need to use trigonometry to find the area of a parallelogram.

EXAMPLE 2 Area of a Parallelogram

Find the area of $\square EFGH$.

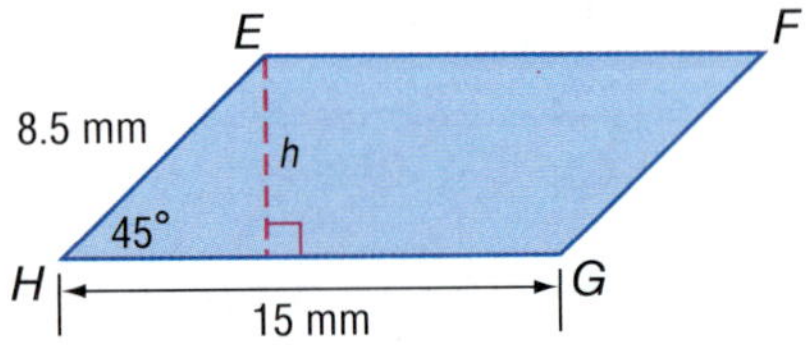

Step 1 Use a 45°-45°-90° triangle to find the height h of the parallelogram.

Recall that if the measure of the leg opposite the 45° angle is h, then the measure of the hypotenuse is $h\sqrt{2}$.

$h\sqrt{2} = 8.5$ **Substitute 8.5 for the measure of the hypotenuse.**

$h = \frac{8.5}{\sqrt{2}}$ or about 6 mm **Divide each side by $\sqrt{2}$.**

Step 2 Find the area.

$A \approx bh$ **Area of a parallelogram**

$\approx (15)(6)$ or 90 mm^2 **$b = 15$ and $h = 6$**

Guided Practice

Find the area of each parallelogram. Round to the nearest tenth if necessary.

2A.

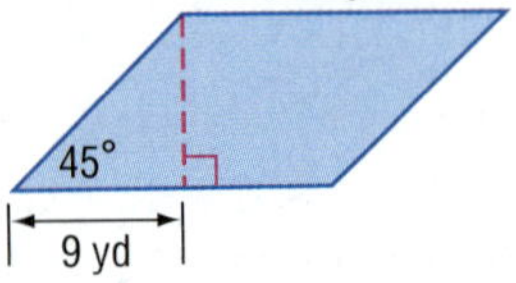

2B.

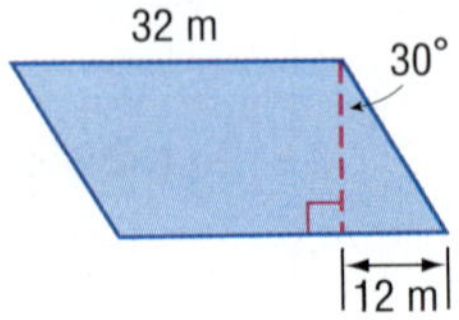

Personal Tutor glencoe.com

Watch Out!

Units of Length and Area Remember that perimeter is measured in linear units such as inches and centimeters. Area is measured in square units such as square feet and square millimeters.

Review Vocabulary

altitude of a triangle a segment from a vertex of a triangle to the line containing the opposite side and perpendicular to the line containing that side (Lesson 5-1)

Areas of Triangles Like the base of a parallelogram, the **base of a triangle** can be any side. The **height of a triangle** is the length of an altitude drawn to a given base.

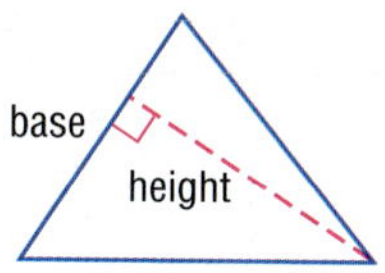

You can use the following postulate to develop the formula for the area of a triangle.

Postulate 11.2 **Area Congruence Postulate**

If two figures are congruent, then they have the same area.

In the figures below, a parallelogram is cut in half along a diagonal to form two congruent triangles with the same base and height.

By the Area Congruence Postulate, the two congruent triangles have the same area. So, one triangle with base b and height h has half the area of a parallelogram with base b and height h.

Key Concept **Area of a Triangle**

Words The area A of a triangle is one half the product of a base b and its corresponding height h.

Symbols $A = \frac{1}{2}bh$ or $A = \frac{bh}{2}$

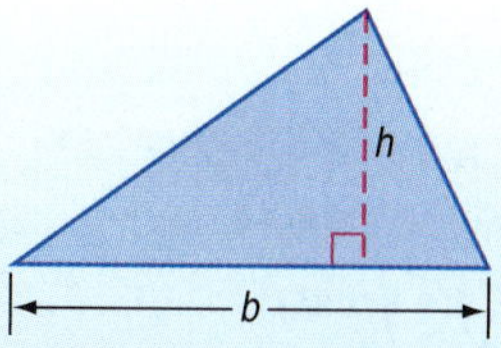

Real-World Link

Triangular gardens can serve as focal points in landscaping or simply result from intersecting walkways.

Real-World EXAMPLE 3 **Perimeter and Area of a Triangle**

GARDENING D'Andre needs enough mulch to cover the triangular garden shown and enough paving stones to border it. If one bag of mulch covers 12 square feet and one paving stone provides a 4-inch border, how many bags of mulch and how many stones does he need to buy?

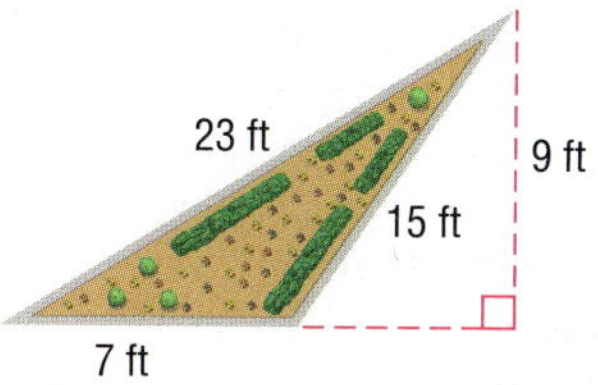

Step 1 Find the perimeter of the garden.

Perimeter of garden = 23 + 15 + 7 or 45 ft

Step 2 Find the area of the garden.

$A = \frac{1}{2}bh$ — Area of a triangle

$= \frac{1}{2}(7)(9)$ or 31.5 ft^2 — $b = 7$ and $h = 9$

Step 3 Use unit analysis to determine how many of each item are needed.

Bags of Mulch

$31.5\ \cancel{\text{ft}^2} \cdot \frac{1\text{ bag}}{12\ \cancel{\text{ft}^2}} = 2.625\text{ bags}$

Paving Stones

$45\ \cancel{\text{ft}} \cdot \frac{12\ \cancel{\text{in.}}}{1\ \cancel{\text{ft}}} \cdot \frac{1\text{ stone}}{4\ \cancel{\text{in.}}} = 135\text{ stones}$

Round the number of bags up so there is enough mulch. He will need 3 bags of mulch and 135 paving stones.

Guided Practice

Find the perimeter and area of each triangle.

3A.

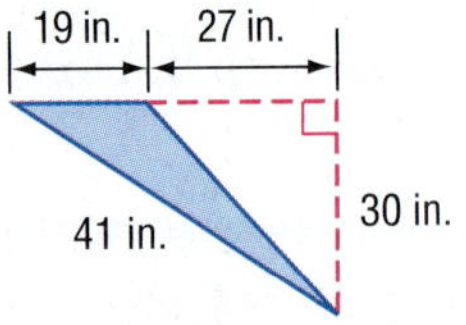

3B.

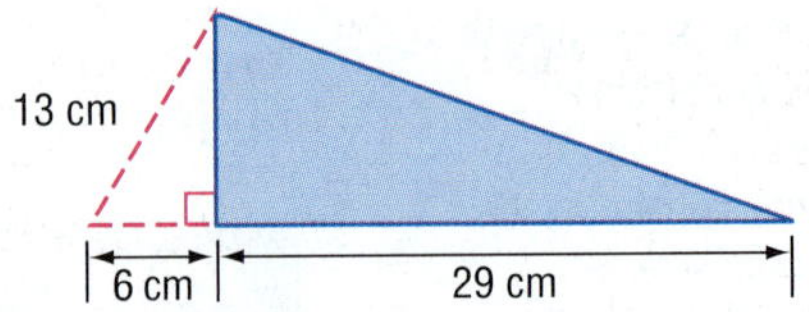

Personal Tutor glencoe.com

You can use algebra to solve for unknown measures in parallelograms and triangles.

EXAMPLE 4 Use Area to Find Missing Measures

ALGEBRA The height of a triangle is 5 centimeters more than its base. The area of the triangle is 52 square centimeters. Find the base and height.

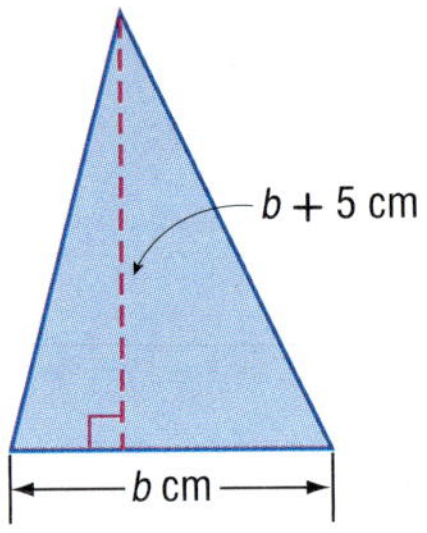

Step 1 Write expressions to represent each measure.

Let b represent the base of the triangle. Then the height is $b + 5$.

Step 2 Use the formula for the area of a triangle to find b.

$A = \frac{1}{2}bh$	**Area of a triangle**
$52 = \frac{1}{2}b(b + 5)$	**Replace A with 52 and h with $b + 5$.**
$104 = b(b + 5)$	**Multiply each side by 2.**
$104 = b^2 + 5b$	**Distributive Property**
$0 = b^2 + 5b - 104$	**Subtract 104 from each side.**
$0 = (b + 13)(b - 8)$	**Factor.**
$b + 13 = 0$ and $b - 8 = 0$	**Zero Product Property**
$b = -13$ $\quad$ $b = 8$	**Solve for b.**

Step 3 Use the expressions from Step 1 to find each measure.

Since a length cannot be negative, the base measures 8 centimeters and the height measures 8 + 5 or 13 centimeters.

StudyTip

Zero Product Property If the product of two factors is 0, then at least one of the factors must be 0.

Guided Practice

ALGEBRA Find x.

4A. $A = 148 \text{ m}^2$

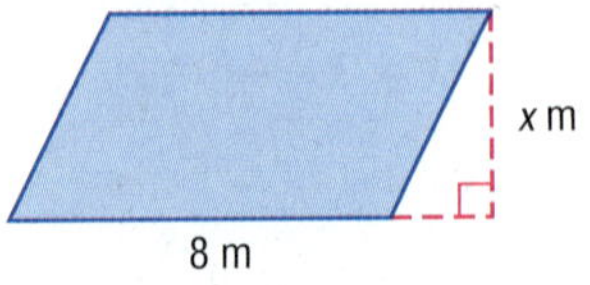

4B. $A = 357 \text{ in}^2$

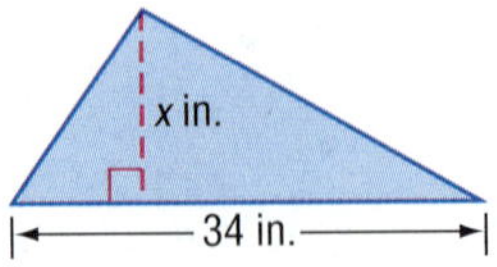

4C. ALGEBRA The base of a parallelogram is twice its height. If the area of the parallelogram is 72 square feet, find its base and height.

Personal Tutor glencoe.com

Check Your Understanding

Examples 1–3
pp. 764–765

Find the perimeter and area of each parallelogram or triangle. Round to the nearest tenth if necessary.

1.

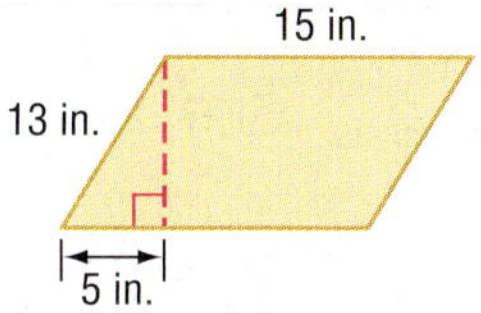

2.

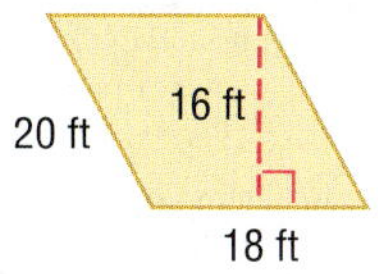

3.

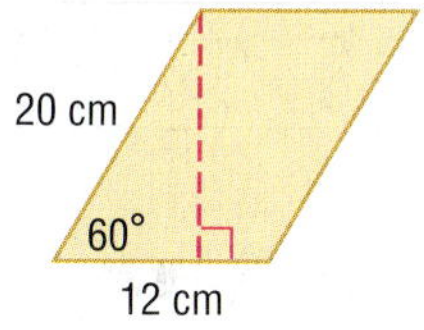

4.

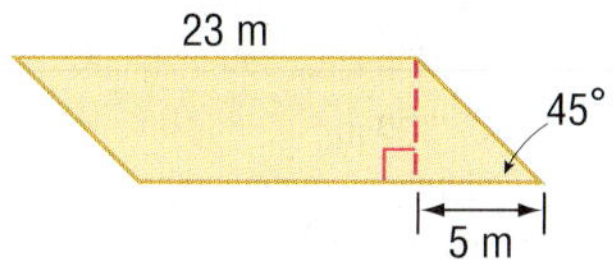

5.

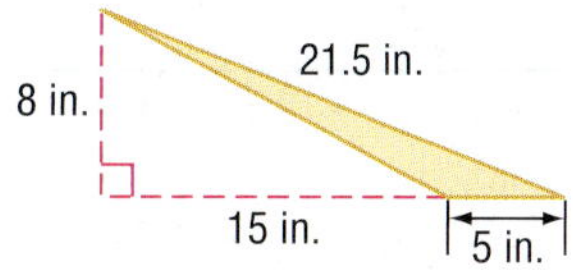

6.

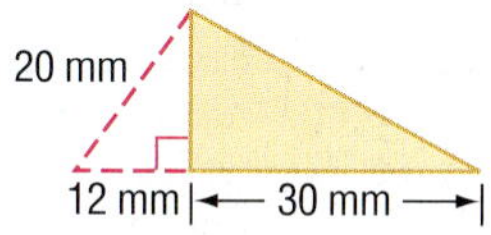

7. **CRAFTS** Marquez and Victoria are making pinwheels. Each pinwheel is composed of 4 triangles with the dimensions shown. Find the perimeter and area of one triangle.

Example 4
p. 766

Find x.

8. $A = 153 \text{ in}^2$

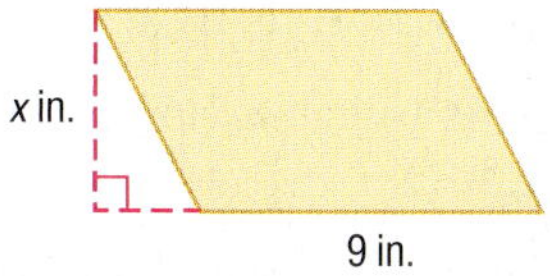

9. $A = 165 \text{ cm}^2$

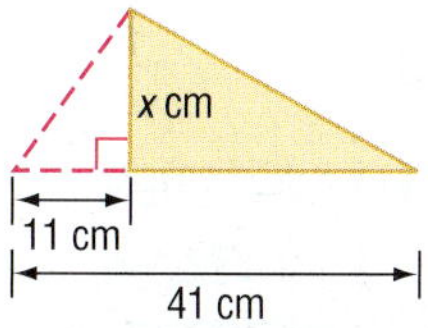

Practice and Problem Solving

= **Step-by-Step Solutions** begin on page R20.
Extra Practice begins on page 969.

Examples 1 and 3
pp. 764–765

Find the perimeter and area of each parallelogram or triangle. Round to the nearest tenth if necessary.

10.

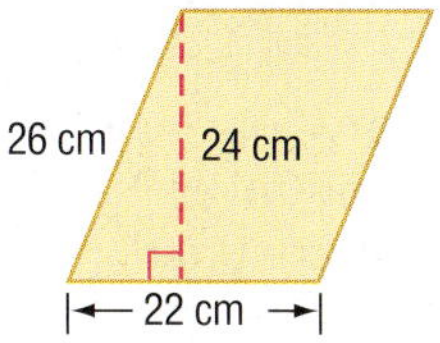

11

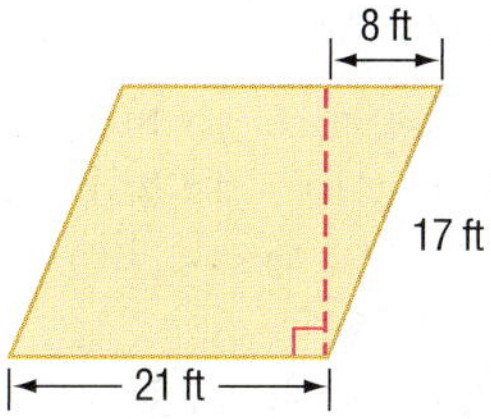

12.

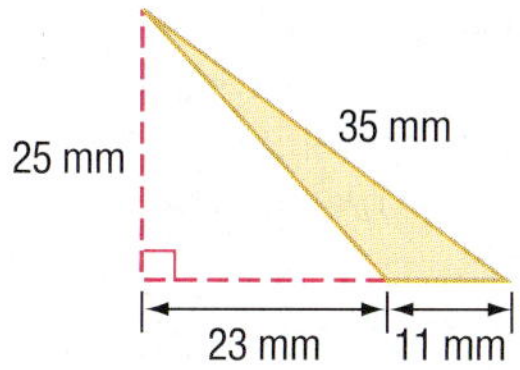

13.

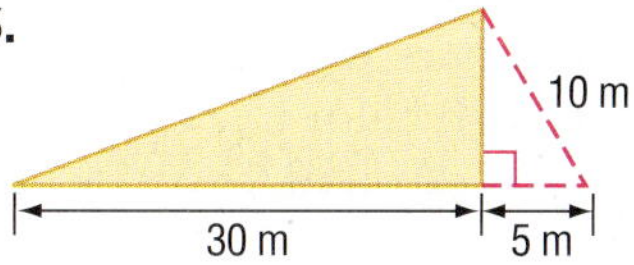

14.

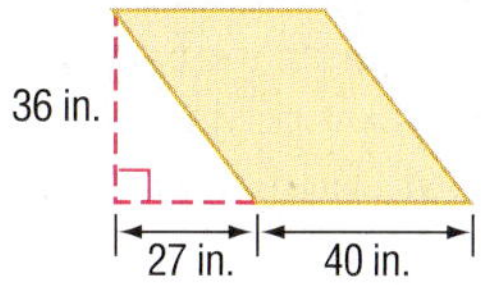

15.

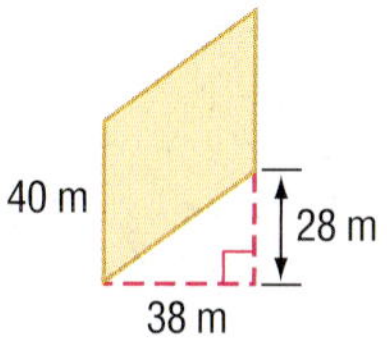

16. **TANGRAMS** The tangram shown is a 4-inch square.

a. Find the perimeter and area of the purple triangle. Round to the nearest tenth.

b. Find the perimeter and area of the blue parallelogram. Round to the nearest tenth.

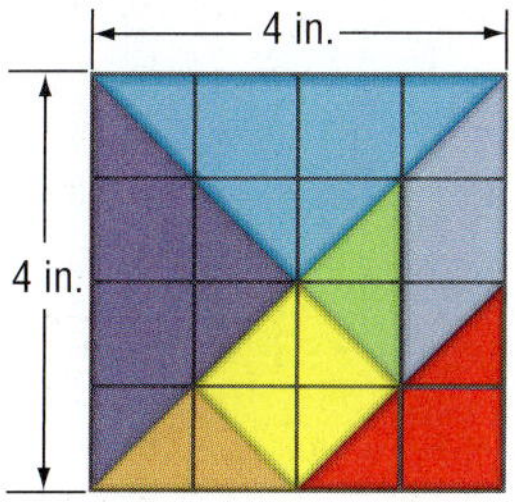

Example 2 p. 764 **Find the area of each parallelogram. Round to the nearest tenth if necessary.**

17.

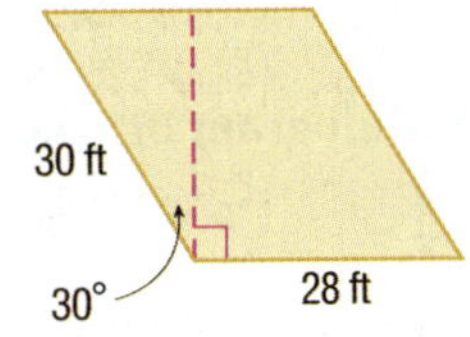

18.

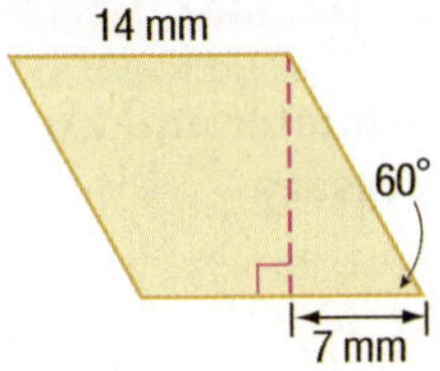

19.

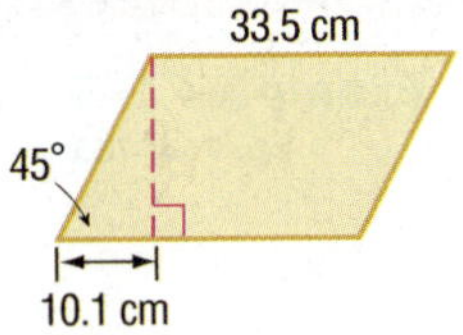

20.

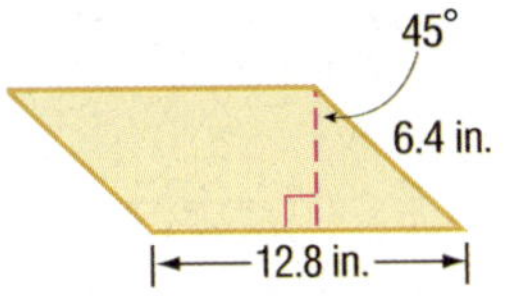

21.

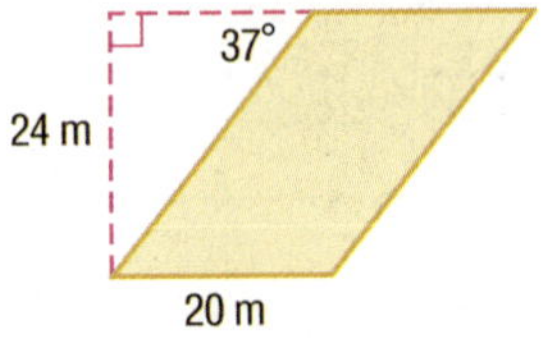

22.

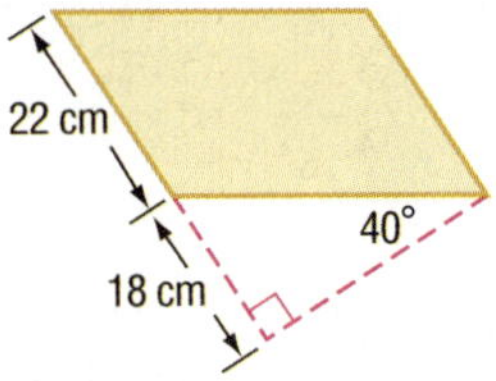

23 **WEATHER** Tornado watch areas are often shown on weather maps using parallelograms. What is the area of the region affected by the tornado watch shown? Round to the nearest square mile.

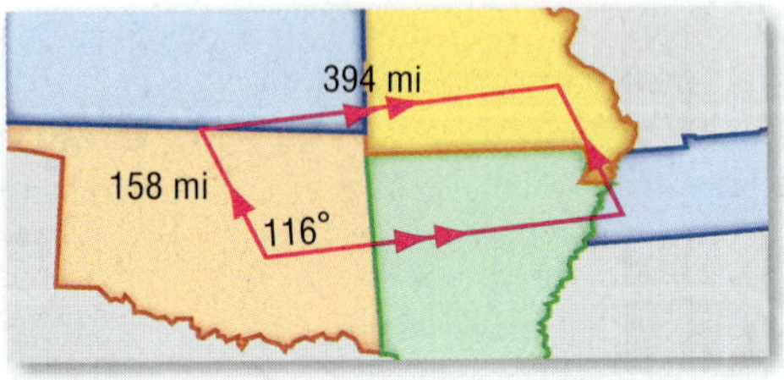

Example 4 p. 766

24. The height of a parallelogram is 4 millimeters more than its base. If the area of the parallelogram is 221 square millimeters, find its base and height.

25. The height of a parallelogram is one fourth of its base. If the area of the parallelogram is 36 square centimeters, find its base and height.

26. The base of a triangle is twice its height. If the area of the triangle is 49 square feet, find its base and height.

27. The height of a triangle is 3 meters less than its base. If the area of the triangle is 44 square meters, find its base and height.

Real-World Link

Many countries have triangular designs on their flags, including Jordan, Cuba, Zimbabwe, and The Bahamas.

Source: The Flag Institute

28. **FLAGS** Omar wants to make a replica of Guyana's national flag.

a. What is the area of the piece of fabric he will need for the red region? for the yellow region?

b. If the fabric costs $3.99 per square yard for each color and he buys exactly the amount of fabric he needs, how much will it cost to make the flag?

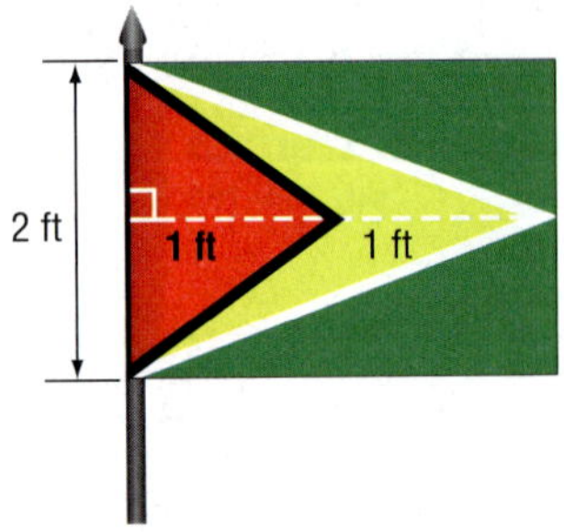

29. **DRAMA** Madison is in charge of the set design for her high school's rendition of *Romeo and Juliet*. One pint of paint covers 80 square feet. How many pints will she need of each color if the roof and tower each need 3 coats of paint?

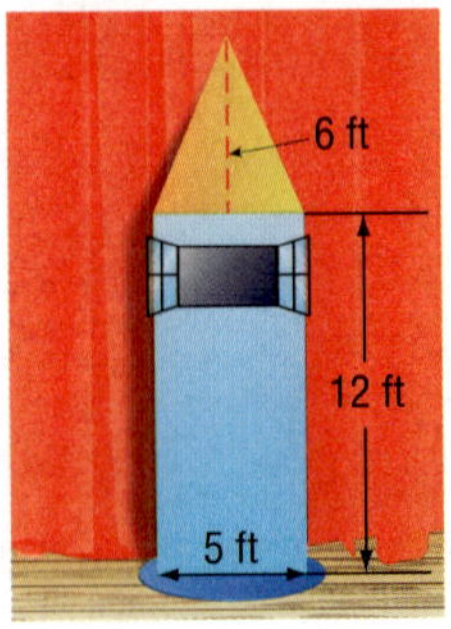

Find the area of each triangle.

30.

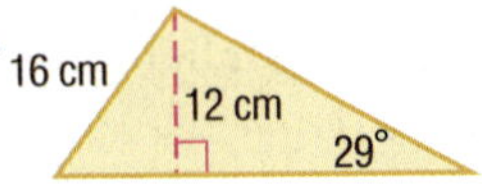

31.

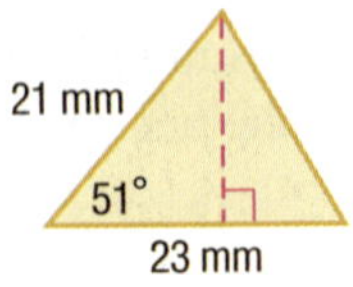

32.

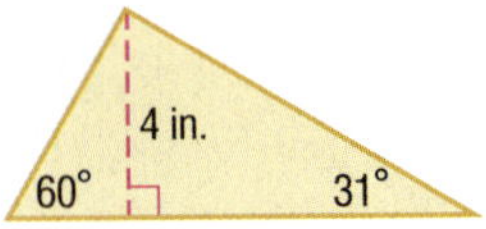

Math History Link

Heron of Alexandria
(c. 10–70 A.D.)

Heron was a mathematician and engineer in Roman Egypt. He developed a formula for finding the area of a triangle if the lengths of the sides are known.

COORDINATE GEOMETRY **Find the area of each figure. Explain the method that you used.**

33. ▱$ABCD$ with $A(4, 7)$, $B(2, 1)$, $C(8, 1)$, and $D(10, 7)$

34. $\triangle RST$ with $R(-8, -2)$, $S(-2, -2)$, and $T(-3, -7)$

35. **HERON'S FORMULA** Heron's Formula relates the lengths of the sides of a triangle to the area of the triangle. The formula is $A = \sqrt{s(s-a)(s-b)(s-c)}$, where s is the *semiperimeter*, or one half the perimeter, of the triangle and a, b, and c are the side lengths.

a. Use Heron's Formula to find the area of a triangle with side lengths 7, 10, and 4.

b. Show that the areas found for a 5-12-13 right triangle are the same using Heron's Formula and using the triangle area formula you learned earlier in this lesson.

36. **MULTIPLE REPRESENTATIONS** In this problem, you will investigate the relationship between the area and perimeter of a rectangle.

a. **ALGEBRAIC** A rectangle has a perimeter of 12 units. If the length of the rectangle is x and the width of the rectangle is y, write equations for the perimeter and area of the rectangle.

b. **TABULAR** Tabulate all possible whole-number values for the length and width of the rectangle, and find the area for each pair.

c. **GRAPHICAL** Graph the area of the rectangle with respect to its length.

d. **VERBAL** Describe how the area of the rectangle changes as its length changes.

e. **ANALYTICAL** For what whole-number values of length and width will the area be greatest? least? Explain your reasoning.

H.O.T. Problems

Use **H**igher-**O**rder **T**hinking Skills

37. **CHALLENGE** Find the area of $\triangle ABC$ graphed at the right. Explain your method.

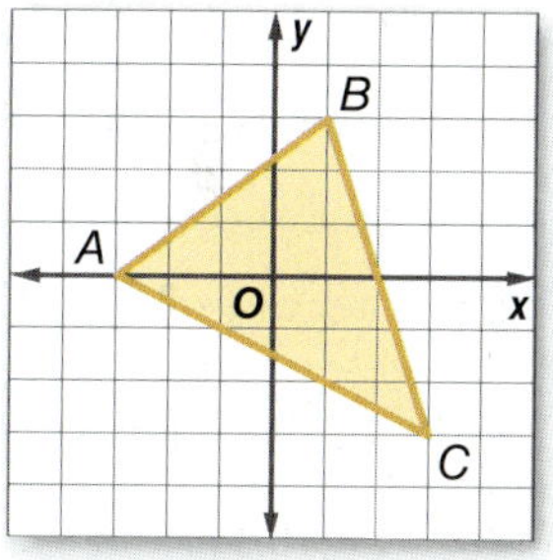

38. **REASONING** Will the perimeter of a nonrectangular parallelogram *always*, *sometimes*, or *never* be greater than the perimeter of a rectangle with the same area and the same height? Explain.

39. **WRITING IN MATH** Points J and L lie on line m. Point K lies on line p. If lines m and p are parallel, describe how the area of $\triangle JKL$ will change as K moves along line p.

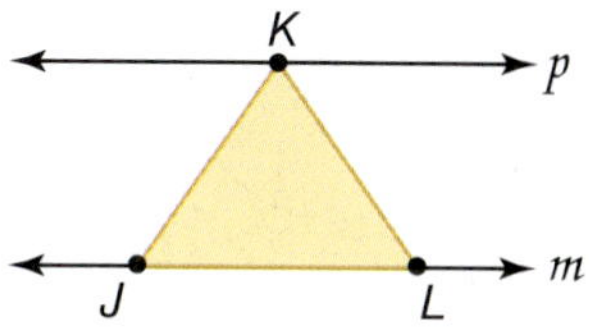

40. **OPEN ENDED** The area of a polygon is 35 square units. The height is 7 units. Draw three different triangles and three different parallelograms that meet these requirements. Label the base and height on each.

41. **WRITING IN MATH** Describe two different ways you could use measurement to find the area of parallelogram $PQRS$.

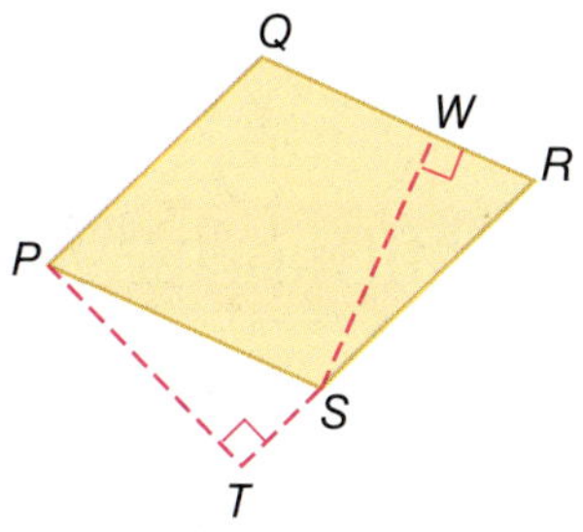

Standardized Test Practice

42. What is the area, in square units, of the parallelogram shown?

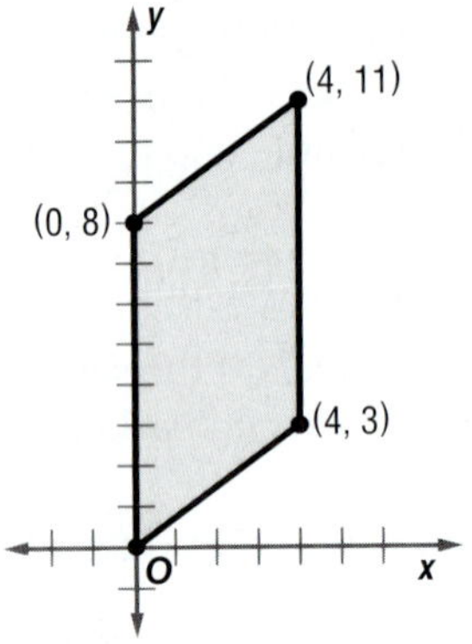

A 12
B 20
C 32
D 40

43. GRIDDED RESPONSE In parallelogram $ABCD$, $\overline{BD}$ and $\overline{AC}$ intersect at E. If $AE = 9$, $BE = 3x - 7$, and $DE = x + 5$, find x.

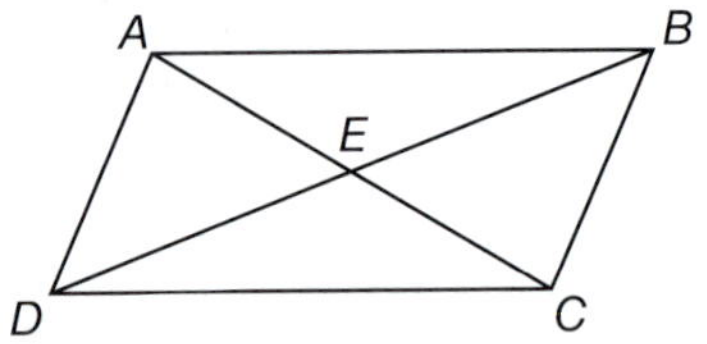

44. A wheelchair ramp is built that is 20 inches high and has a length of 12 feet as shown. What is the measure of the angle x that the ramp makes with the ground, to the *nearest* degree?

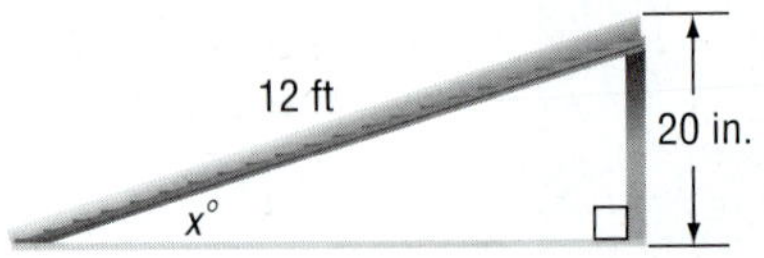

Note: Not drawn to scale.

F 8
G 16
H 37
J 53

45. SAT/ACT The formula for converting a Celsius temperature to a Fahrenheit temperature is $F = \frac{9}{5}C + 32$, where F is the temperature in degrees Fahrenheit and C is the temperature in degrees Celsius. Which of the following is the Celsius equivalent to a temperature of 86° Fahrenheit?

A 15.7° C
B 30° C
C 65.5° C
D 186.8° C

Spiral Review

Write the equation of each circle. (Lesson 10-8)

46. center at origin, $r = 3$

47. center at origin, $d = 12$

48. center at $(-3, -10)$, $d = 24$

49. center at $(1, -4)$, $r = \sqrt{17}$

Find x to the nearest tenth. Assume that segments that appear to be tangent are tangent. (Lesson 10-7)

50.

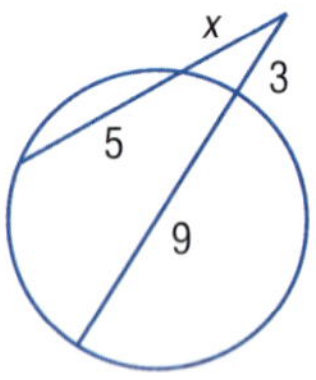

51.

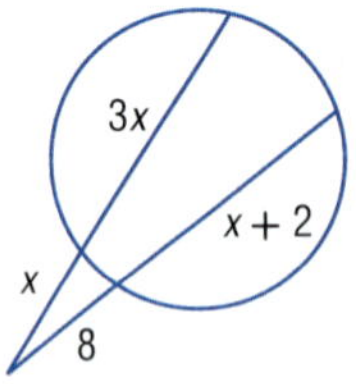

52.

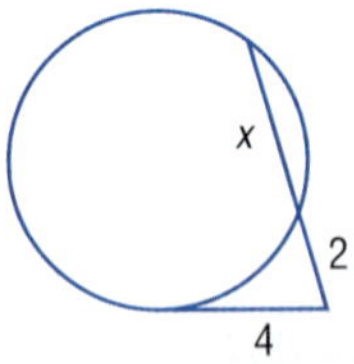

53. OPTIC ART Victor Vasarely created art in the *op art* style. This piece, *AMBIGU-B*, consists of multi-colored parallelograms. Describe one method to ensure that the shapes are parallelograms. (Lesson 6-3)

Skills Review

Evaluate each expression if $a = 2$, $b = 6$, and $c = 3$. (Lesson 0-4)

54. $\frac{1}{2}ac$

55. $\frac{1}{2}cb$

56. $\frac{1}{2}b(2a + c)$

57. $\frac{1}{2}c(b + a)$

58. $\frac{1}{2}a(2c + b)$

EXPLORE 11-2

Graphing Technology Lab

Areas of Trapezoids, Rhombi, and Kites

You can use the TI-Nspire™ or TI-Nspire™ CAS technology to explore special quadrilaterals.

ACTIVITY 1

Step 1 Open a new **Graphs and Geometry** page. Select **Show Grid** from the **View** so that points can be placed at integer coordinates.

Step 2 Select **Line** from the **Points and Lines** menu, and draw a line.

Step 3 Select **Parallel** from the **Construction** menu to draw a line parallel to your original line.

Step 4 Place an additional point on the parallel line you just constructed using **Point on** from the **Points and Lines** menu. Label the four points.

Step 1:

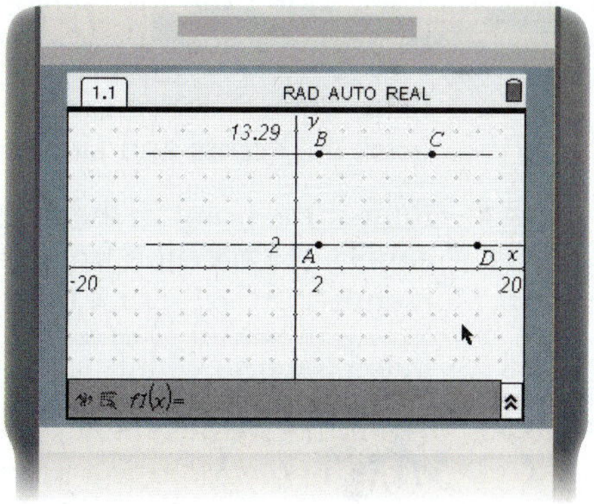

Step 5 From the **Shapes** menu, select **Polygon,** and draw a polygon using the four points you created. From the **Tools** menu, select **Attributes** and increase the line thickness of the polygon.

Step 5:

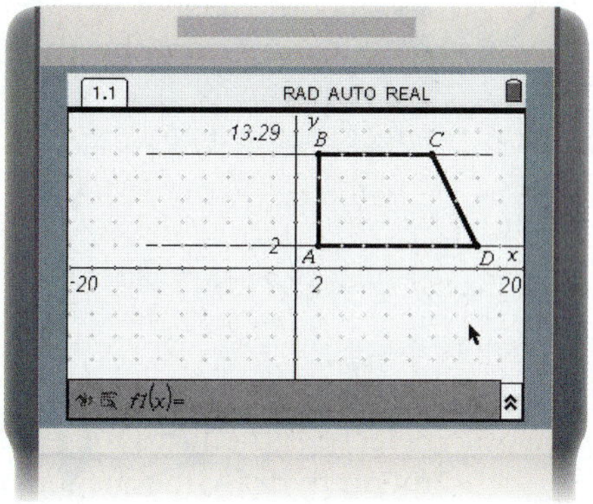

Step 6 Display the area of the polygon using the **Area** feature from the **Measurement** menu. Move each of the points and observe the effect on the area.

Step 6:

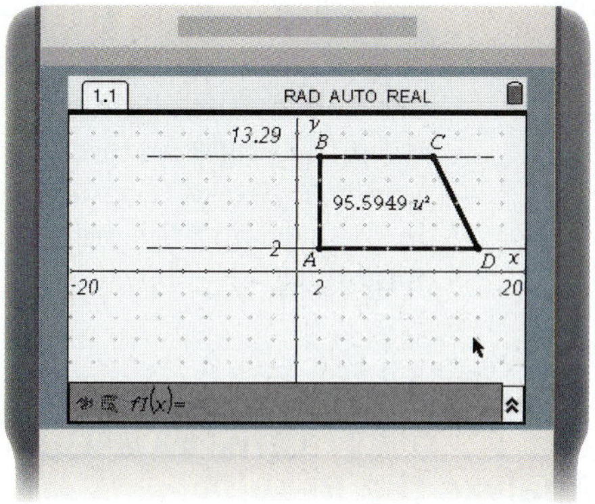

Analyze the Results

1. What type of quadrilateral is *ABCD*? Explain your reasoning.

2. **MAKE A CONJECTURE** Using the formulas you learned in Lesson 11-1, make a conjecture about the formula for the area of this type of quadrilateral if *BC* is b_1, *AD* is b_2, and *AB* is *h*. Explain.

(continued on the next page)

ACTIVITY 2

Step 1 Open a new **Graphs and Geometry** page. Select **Show Grid** from the **View** menu so that points can be placed at integer coordinates.

Step 2 Select **Line** from the **Points and Lines** menu, and draw a line.

Step 3 Place a point above the line by selecting **Point** from the **Points and Lines** menu.

Step 4 Reflect the point in the line from Step 2 by choosing **Reflection** from the **Transformation** menu.

Step 5 Label the four points as shown.

Step 6 From the **Shapes** menu, select **Polygon**, and draw a polygon using points *A*, *B*, *C*, and *D*.

Step 7 Display the area of the polygon using the **Area** feature from the **Measurement** menu. Move points *A*, *B*, and *C*, and observe the effect on the area.

Step 1:

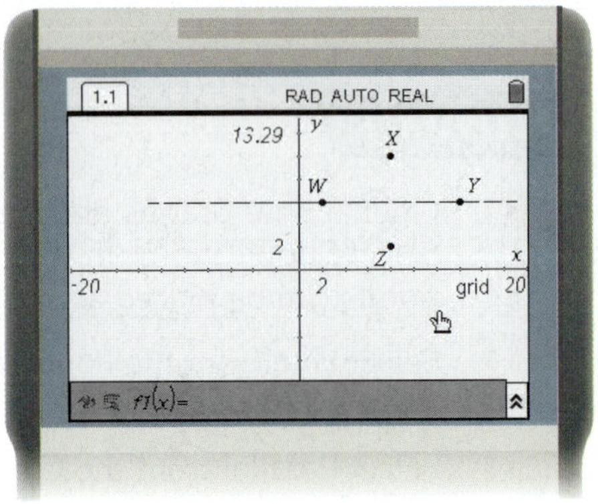

Step 6:

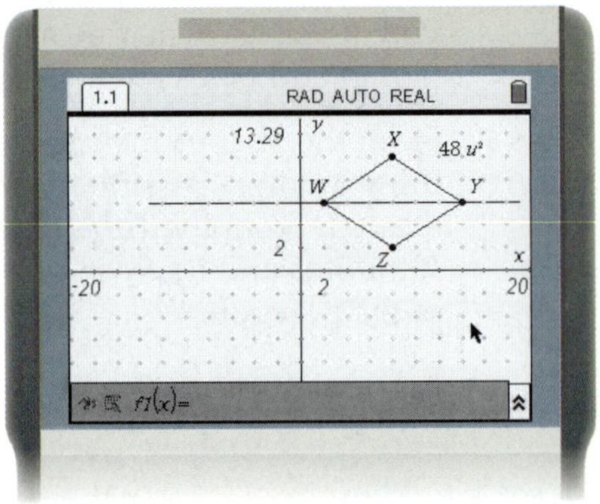

Step 8 Select **Segment** from the **Points and Lines** menu to draw the diagonals of *ABCD*.

Step 9 Display the lengths of the diagonals using the **Length** feature from the **Measurement** menu, and display the angle between the diagonals using the **Angle** feature. Continue to move points *A*, *B*, and *C*, and observe the effect on the area and the angle between the diagonals.

Step 8:

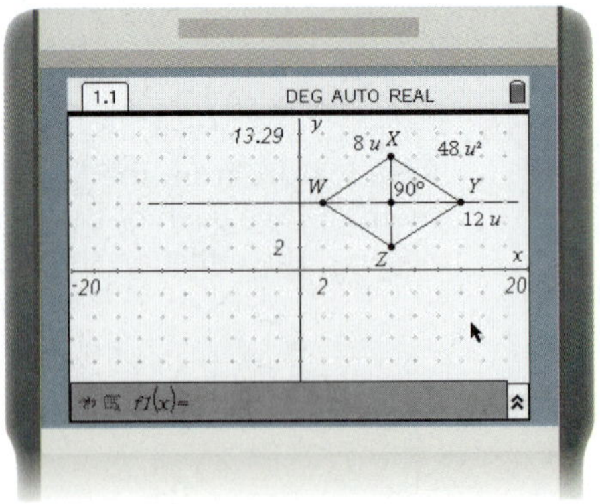

Analyze the Results

3. What type of quadrilateral is *WXYZ*? Explain your reasoning.

4. **MAKE A CONJECTURE** Using the formulas you learned in Lesson 11-1, develop a formula for the area of this type of quadrilateral. Let *WY* be d_1, and let *XZ* be d_2. Explain your reasoning.

5. **CHALLENGE** Construct a quadrilateral using two perpendicular lines and reflecting points on each as you did in Step 4 of Activity 2. What type of quadrilateral is formed? Does the formula for the area you developed in Exercise 4 apply?

11-2 Areas of Trapezoids, Rhombi, and Kites

Then

You found areas of triangles and parallelograms. (Lesson 11-1)

Now

- Find areas of trapezoids.
- Find areas of rhombi and kites.

New Vocabulary

height of a trapezoid

Math Online

glencoe.com

Why?

Brianna has turned her hobby of making designer handbags and totes into a small business. Among her designs is a trapezoid-shaped handbag. To estimate the amount of material needed to produce each handbag, she needs to calculate the area of a trapezoid.

Areas of Trapezoids In Lesson 6-6, you learned that a *trapezoid* is a quadrilateral with exactly one pair of parallel sides. These parallel sides are called *bases*. The **height of a trapezoid** is the perpendicular distance between its bases.

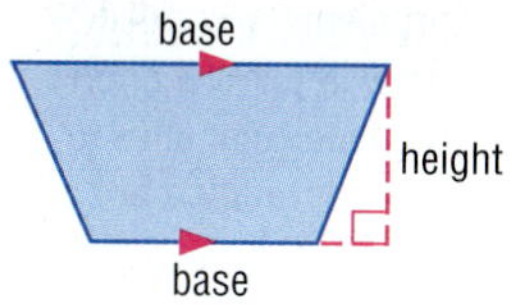

In the figure below, a glide reflection of the first trapezoid results in two congruent trapezoids that fit together to form a parallelogram.

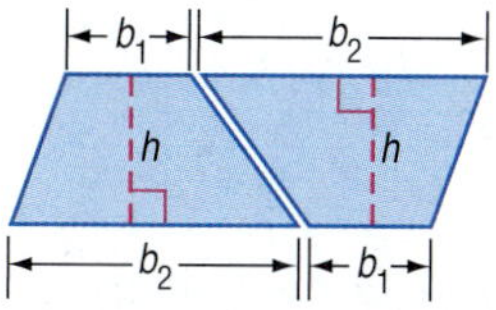

The area of the parallelogram is the product of the height h and the sum of the two bases, b_1 and b_2. The area of one trapezoid is one half the area of the parallelogram.

Key Concept Area of a Trapezoid

Words The area A of a trapezoid is one half the product of the height h and the sum of its bases, b_1 and b_2.

Symbols $A = \frac{1}{2}h(b_1 + b_2)$

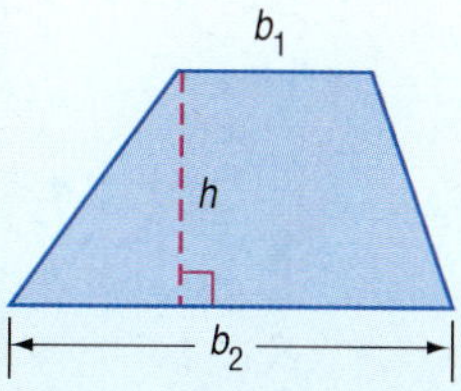

Real-World EXAMPLE 1 Area of a Trapezoid

CRAFTS One of Brianna's trapezoid-shaped totes is shown. Find the amount of material used to make the side shown.

$A = \frac{1}{2}h(b_1 + b_2)$ — Area of a trapezoid

$= \frac{1}{2}(30)(28 + 58)$ — $h = 30, b_1 = 28, b_2 = 58$

$= 1290 \text{ cm}^2$ — Simplify.

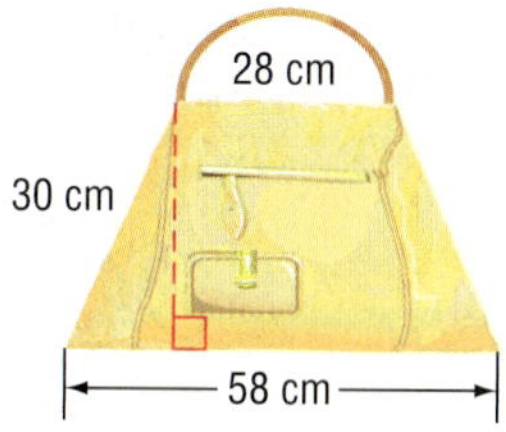

The tote requires 1290 square centimeters of material to make one side.

Guided Practice

1. **AUTOMOBILES** Find the area of glass used to make the windshield of a van shown at the right.

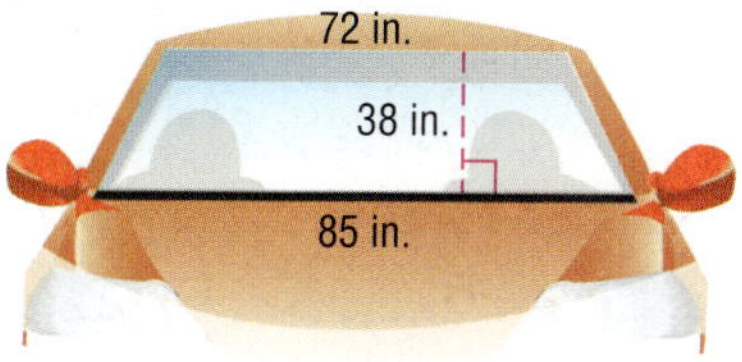

Personal Tutor glencoe.com

STANDARDIZED TEST EXAMPLE 2

OPEN ENDED **Emelia designed the pennant shown for her team. Find the area of the shaded portion of her team's pennant.**

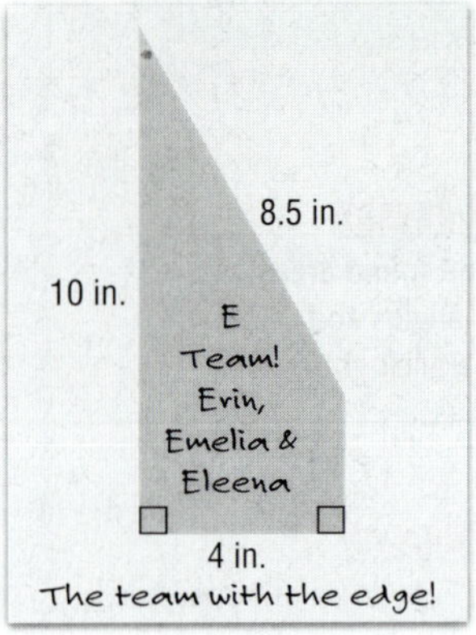

Read the Test Item

You are given a trapezoid with one base measuring 10 inches, a height of 4 inches, and a third side measuring 8.5 inches. To find the area of the trapezoid, first find the measure of the other base.

Solve the Test Item

Draw the segment shown to form a right triangle and a rectangle. The triangle has a hypotenuse of 8.5 inches and legs of 4 and ℓ inches. The rectangle has a length of 4 inches and a width of x inches.

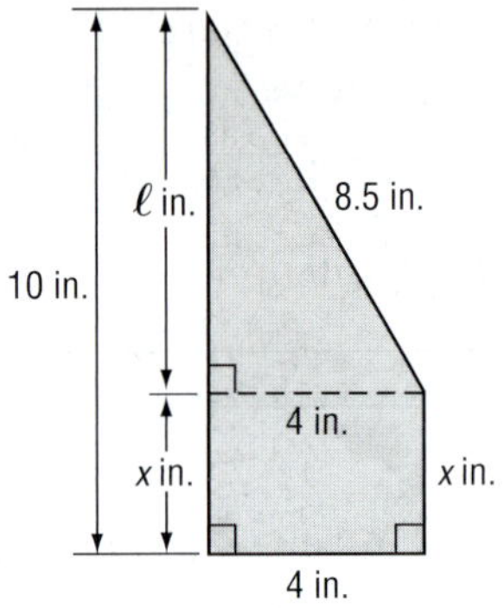

Use the Pythagorean Theorem to find ℓ.

$a^2 + b^2 = c^2$	**Pythagorean Theorem**
$\ell^2 + 4^2 = 8.5^2$	**$a = \ell$, $b = 4$, and $c = 8.5$**
$\ell^2 + 16 = 72.25$	**Simplify.**
$\ell^2 = 56.25$	**Subtract 16 from each side.**
$\ell = 7.5$	**Take the positive square root of each side.**

By Segment Addition, $\ell + x = 10$. So, $7.5 + x = 10$ and $x = 2.5$. The width of the rectangle is also the measure of the second base of the trapezoid.

$A = \frac{1}{2}h(b_1 + b_2)$	**Area of a trapezoid**
$= \frac{1}{2}(4)(10 + 2.5)$	**$h = 4$, $b_1 = 10$, and $b_2 = 2.5$**
$= 25$	**Simplify.**

So the pennant has an area of 25 square inches.

CHECK The area of the trapezoid is the sum of the areas of the right triangle and rectangle. The area of the triangle is $\frac{1}{2}(4)(7.5)$ or 15 square inches. The area of the rectangle is (4)(2.5) or 10 square inches. So the area of the trapezoid is 15 + 10 or 25 square inches. ✓

Test-Taking Tip

Separating Figures To solve some area problems, you need to draw in parallel and/or perpendicular lines to find information not provided.

Guided Practice

2. **OPEN ENDED** Owen designed the silver earrings shown that are shaped like isosceles trapezoids. What is the area of each earring?

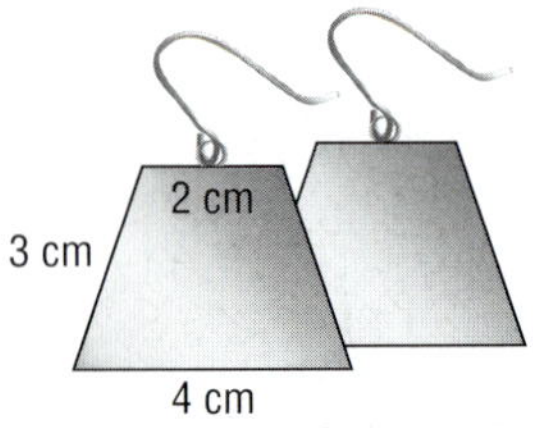

Personal Tutor glencoe.com

Areas of Rhombi and Kites Recall from Lessons 6-5 and 6-6 that a *rhombus* is a parallelogram with all four sides congruent and a *kite* is a quadrilateral with exactly two pairs of consecutive congruent sides.

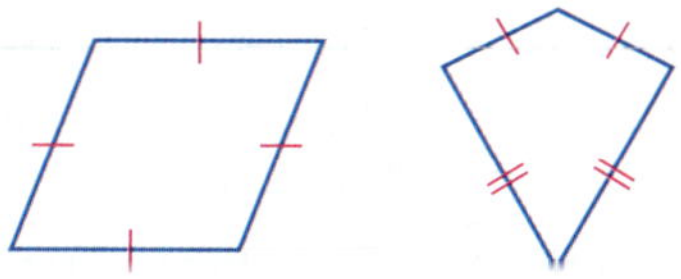
rhombus kite

The areas of rhombi and kites are related to the lengths of their diagonals.

Review Vocabulary

diagonal a segment that connects any two nonconsecutive vertices in a polygon (Lesson 6-1)

Key Concept — Area of a Rhombus or Kite

Words The area A of a rhombus or kite is one half the product of the lengths of its diagonals, d_1 and d_2.

Symbols $A = \frac{1}{2}d_1d_2$

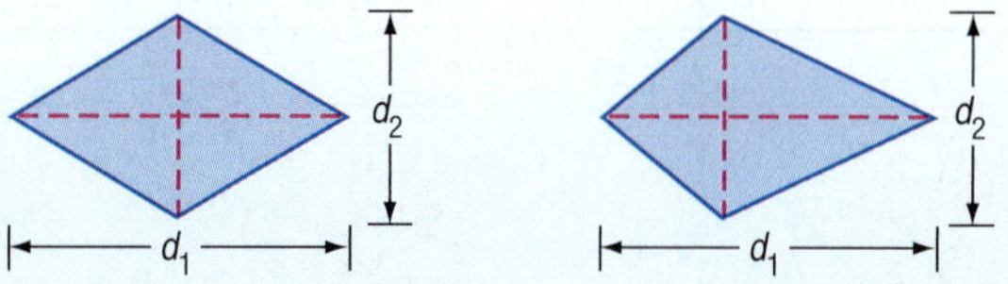

You will derive the formulas for the area of a rhombus and the area of a kite in Exercises 23 and 24.

EXAMPLE 3 Area of a Rhombus and a Kite

Find the area of each rhombus or kite.

a.

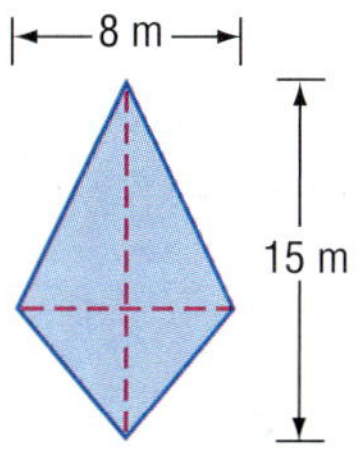

$A = \frac{1}{2}d_1d_2$ **Area of a kite**

$= \frac{1}{2}(8)(15)$ **$d_1 = 8$ and $d_2 = 15$**

$= 60 \text{ m}^2$ **Simplify.**

b.

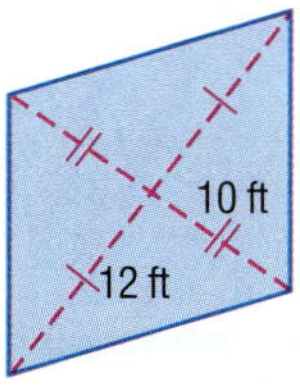

Step 1 Find the length of each diagonal.

Since the diagonals of a rhombus bisect each other, then lengths of the diagonals are 12 + 12 or 24 feet and 10 + 10 or 20 feet.

Step 2 Find the area of the rhombus.

$A = \frac{1}{2}d_1d_2$ **Area of a rhombus**

$= \frac{1}{2}(24)(20)$ **$d_1 = 24$ and $d_2 = 20$**

$= 240 \text{ ft}^2$ **Simplify.**

Guided Practice **Find the area of each rhombus or kite.**

3A.

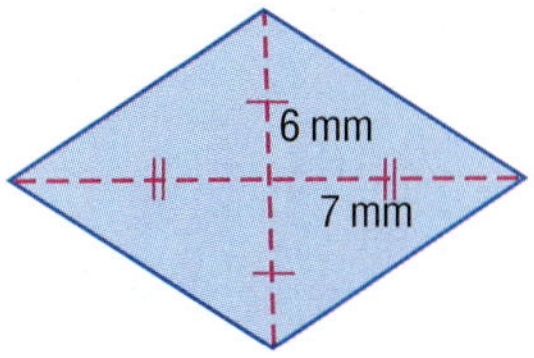

3B.

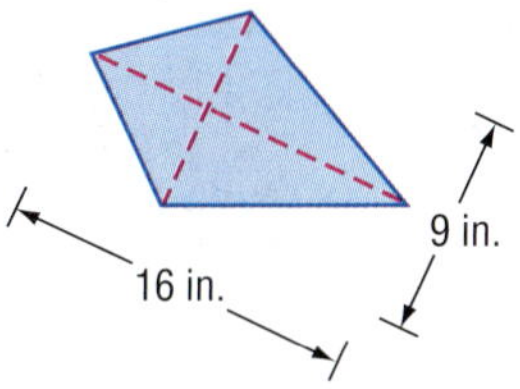

Personal Tutor glencoe.com

You can use algebra to solve for unknown measures in trapezoids, rhombi, and kites.

EXAMPLE 4 Use Area to Find Missing Measures

ALGEBRA **One diagonal of a rhombus is twice as long as the other diagonal. If the area of the rhombus is 169 square millimeters, what are the lengths of the diagonals?**

Step 1 Write an expression to represent each measure.

Let x represent the length of one diagonal. Then the length of the other diagonal is $2x$.

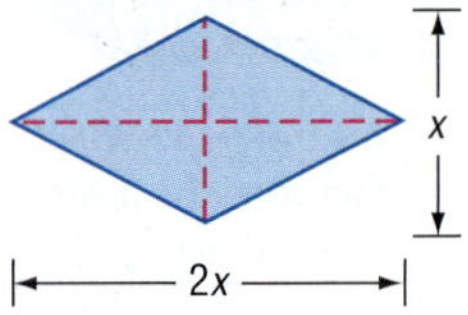

Step 2 Use the formula for the area of a rhombus to find x.

$A = \frac{1}{2}d_1d_2$	**Area of a rhombus**
$169 = \frac{1}{2}(x)(2x)$	**$A = 169$, $d_1 = x$, and $d_2 = 2x$**
$169 = x^2$	**Simplify.**
$13 = x$	**Take the positive square root of each side.**

So the lengths of the diagonals are 13 millimeters and 2(13) or 26 millimeters.

> **StudyTip**
>
> **Kites** Recall from Lesson 6-6 that the diagonals of kites are perpendicular.

Guided Practice

ALGEBRA Find x.

4A. $A = 92 \text{ in}^2$

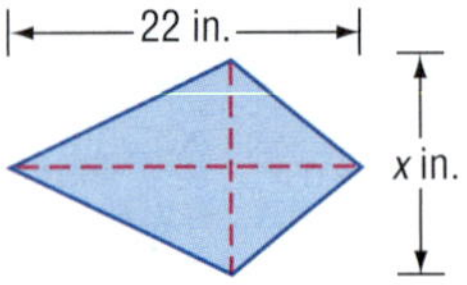

4B. $A = 177 \text{ cm}^2$

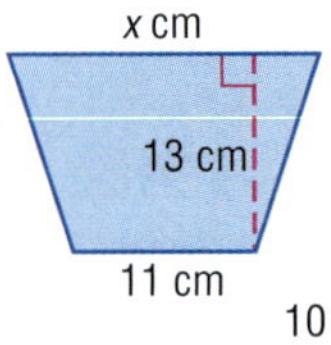

4C. **ALGEBRA** What is the area of the kite shown?

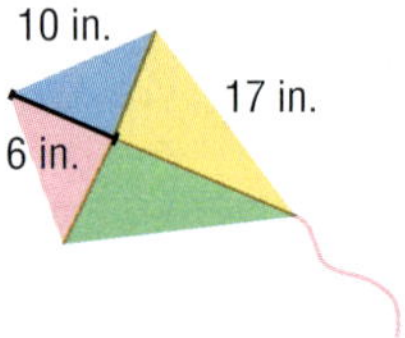

Personal Tutor glencoe.com

Concept Summary Areas of Polygons

Parallelogram	Triangles	Trapezoids	Rhombi and Kites
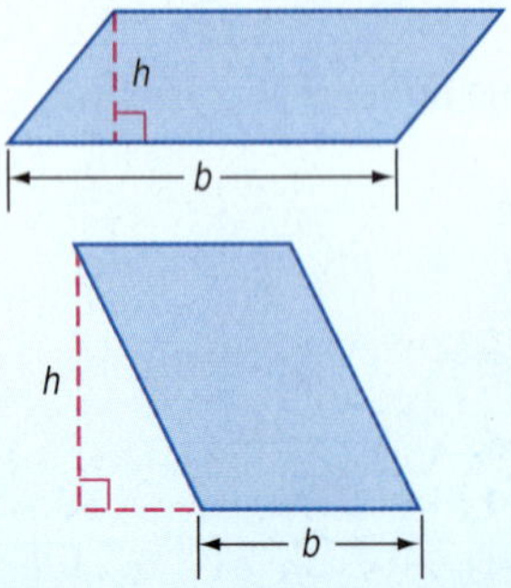	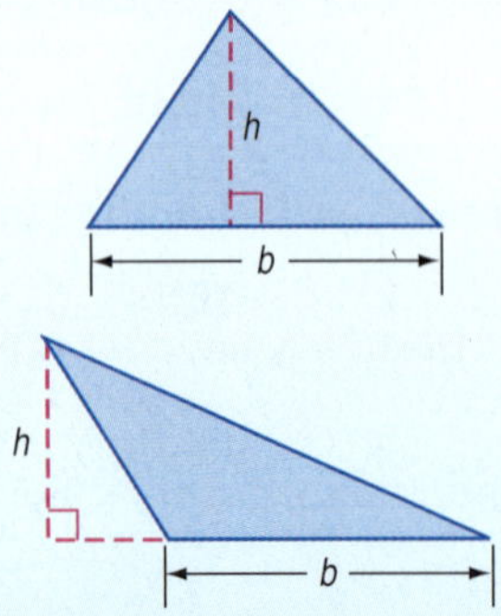	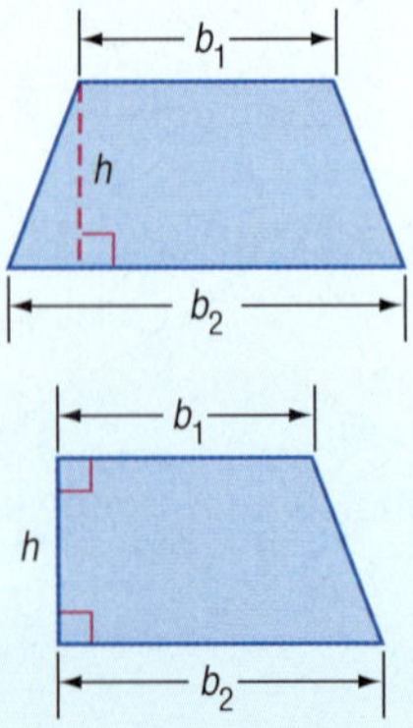	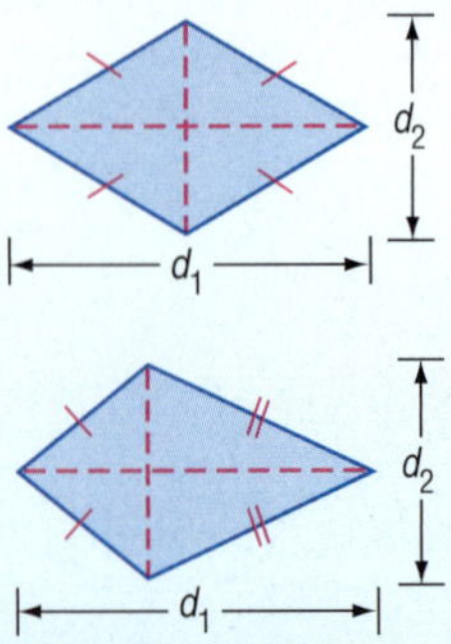
$A = bh$	$A = \frac{1}{2}bh$	$A = \frac{1}{2}h(b_1 + b_2)$	$A = \frac{1}{2}d_1d_2$

Check Your Understanding

Examples 1 and 3
pp. 773 and 775

Find the area of each trapezoid, rhombus, or kite.

1.

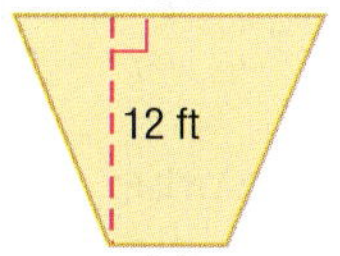

2.

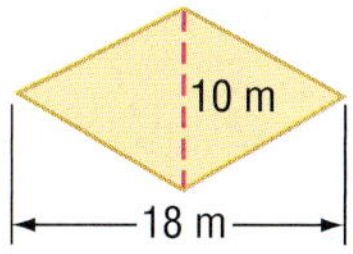

3.

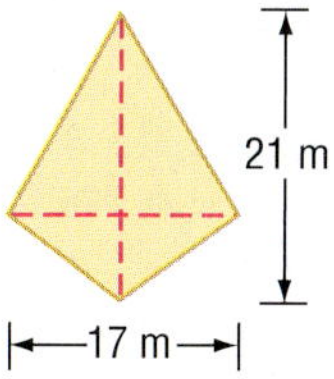

Example 2
p. 774

4. **OPEN ENDED** Suki is doing fashion design at 4-H Club. Her first project is to make a simple A-line skirt. How much fabric will she need according to the design at the right?

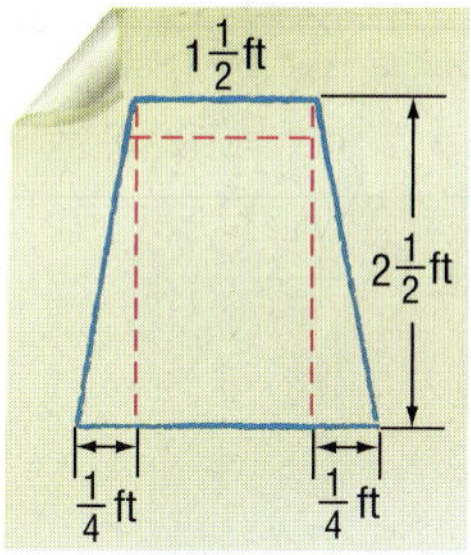

Example 4
p. 776

ALGEBRA **Find** x.

5. $A = 78 \text{ cm}^2$

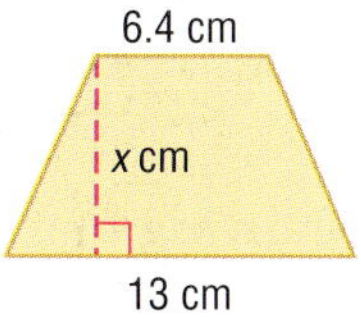

6. $A = 96 \text{ in}^2$

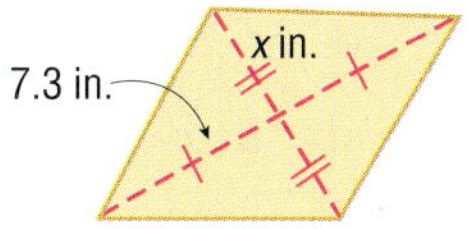

7. $A = 104 \text{ ft}^2$

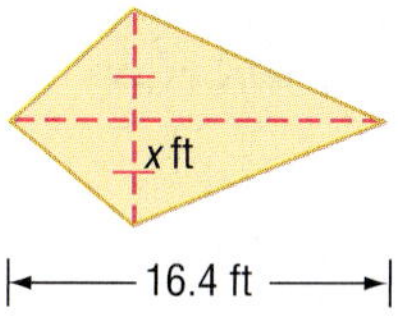

Practice and Problem Solving

= **Step-by-Step Solutions** begin on page R20.
Extra Practice begins on page 969.

Examples 1–3
pp. 773–775

Find the area of each trapezoid, rhombus, or kite.

8.

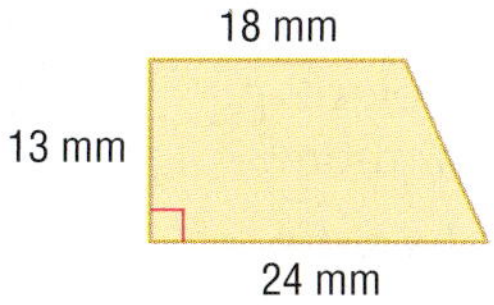

9.

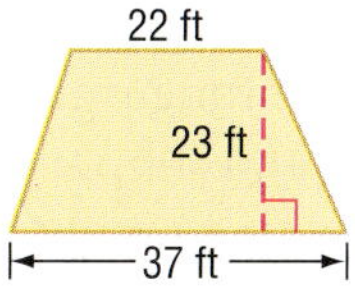

10.

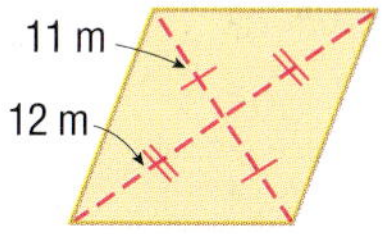

11.

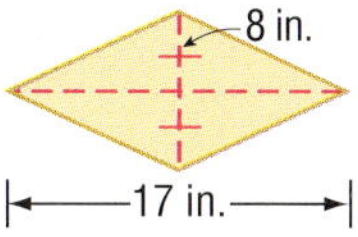

12.

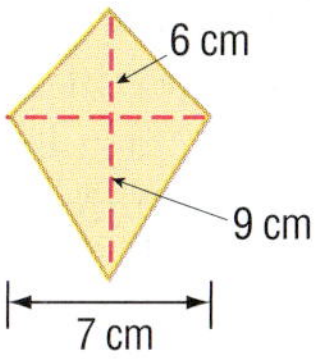

13.

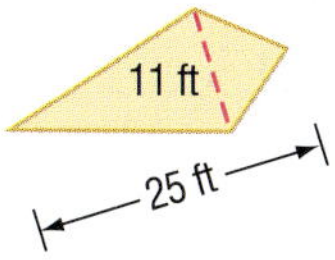

Real-World Link

Insects have compound eyes with many lenses. Because of this, the final image they see looks like a mosaic. Even so, the eye of a fly is six times more sensitive than the human eye.

Source: Microscopy-UK

MICROSCOPES **Find the area of the identified portion of each magnified image. Assume that the identified portion is either a trapezoid, rhombus, or kite. Measures are provided in microns.**

14. human skin

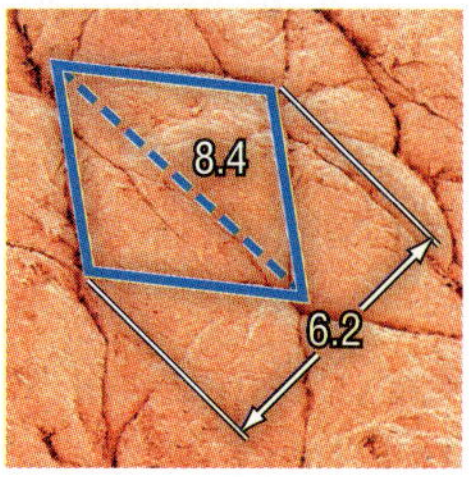

15 heartleaf plant

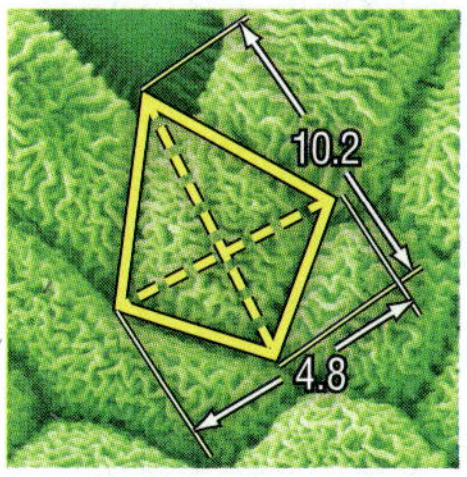

16. eye of a fly

17. **JOBS** Jimmy works on his neighbors' yards after school to earn extra money to buy a car. He is going to plant grass seed in Mr. Troyer's yard. What is the area of the yard?

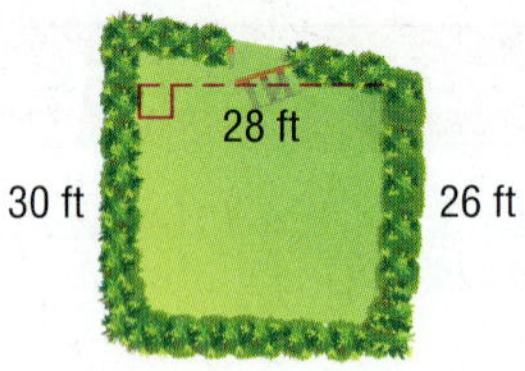

Example 4
p. 776

ALGEBRA **Find each missing length.**

18. One diagonal of a kite is twice as long as the other diagonal. If the area of the kite is 240 square inches, what are the lengths of the diagonals?

19. The area of a rhombus is 168 square centimeters. If one diagonal is three times as long as the other, what are the lengths of the diagonals?

20. A trapezoid has base lengths of 12 and 14 feet with an area of 322 square feet. What is the height of the trapezoid?

21. A trapezoid has a height of 8 meters, a base length of 12 meters, and an area of 64 square meters. What is the length of the other base?

Real-World Link

The National Honor Society was officially established in 1921. It was the first nationwide honor society. Its purpose is "To create enthusiasm for scholarship, to stimulate a desire to render service, to promote leadership, and to develop character in the students of secondary schools."

Source: National Honor Society

22. **HONORS** Estella has been asked to join an honor society at school. Before the first meeting, new members are asked to sand and stain the front side of a piece of wood in the shape of an isosceles trapezoid. What is the surface area that Estella will need to sand and stain?

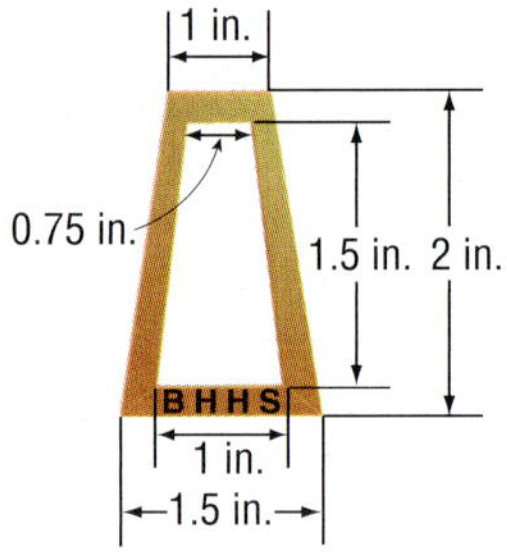

For each figure, provide a justification showing that $A = \frac{1}{2}d_1d_2$.

23.

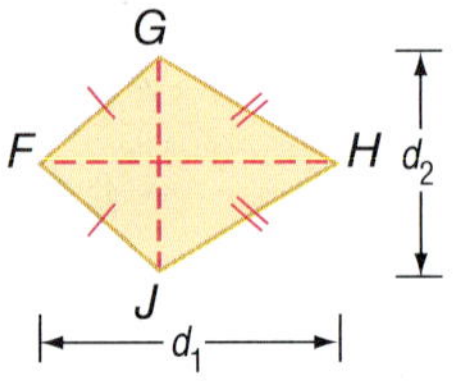

24.

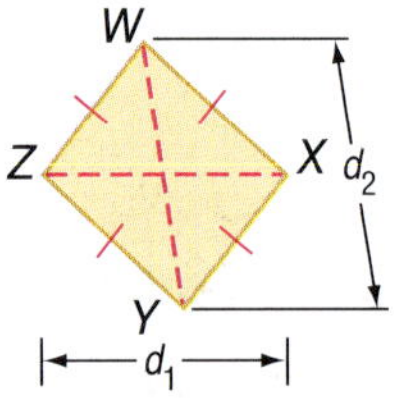

25. **CRAFTS** Ashanti is in a kite competition. The yellow, red, orange, green, and blue pieces of her kite design shown are congruent rhombi.

a. How much fabric of each color does she need to buy?

b. Competition rules require that the total area of each kite be no greater than 200 square inches. Does Ashanti's kite meet this requirement? Explain.

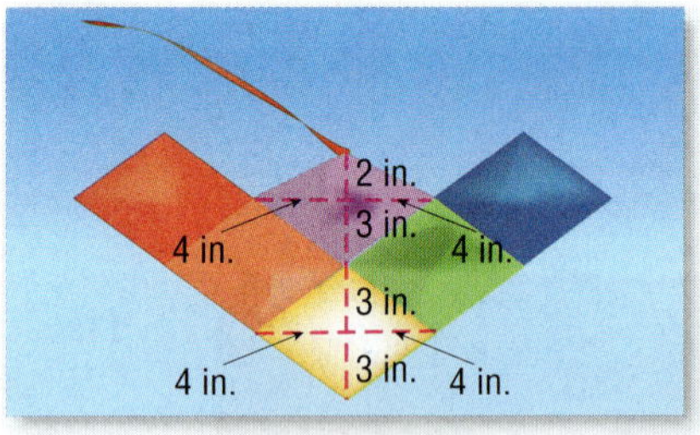

COORDINATE GEOMETRY **Find the area of each quadrilateral with the given vertices.**

26. $A(-8, 6)$, $B(-5, 8)$, $C(-2, 6)$, and $D(-5, 0)$

27. $W(3, 0)$, $X(0, 3)$, $Y(-3, 0)$, and $Z(0, -3)$

28. **METALS** When magnified in very powerful microscopes, some metals are composed of grains that have various polygonal shapes.

a. What is the area of figure 1 if the grain has a height of 4 microns and bases with lengths of 5 and 6 microns?

b. If figure 2 has perpendicular diagonal lengths of 3.8 microns and 4.9 microns, what is the area of the grain?

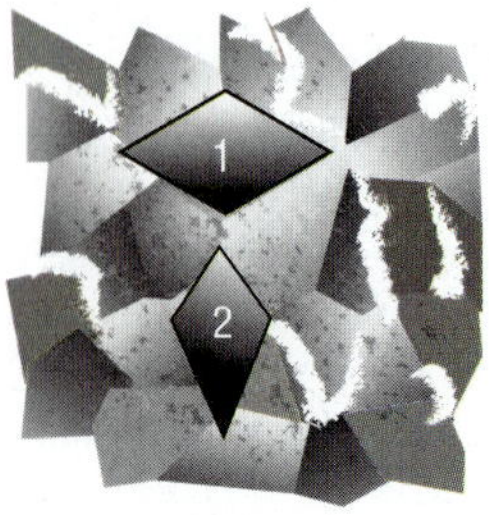

Real-World Link

James A. Garfield graduated from Williams College in Massachusetts in 1856. He planned to teach mathematics at Hiram College, but was elected into the Ohio Senate in 1859, and went on to become President of the United States in 1880.

Source: The Schiller Institute

29. **PROOF** The figure at the right is a trapezoid that consists of two congruent right triangles and an isosceles triangle. In 1876, James A. Garfield, the 20th president of the United States, discovered a proof of the Pythagorean Theorem using this diagram. Prove that $x^2 + y^2 = z^2$.

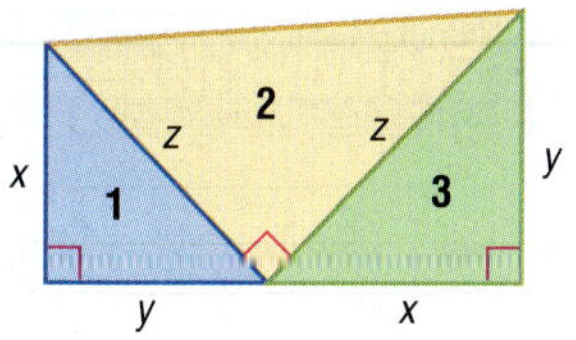

DIMENSIONAL ANALYSIS **Find the area of each figure. Convert each answer to square feet. Round to the nearest tenth.**

30.

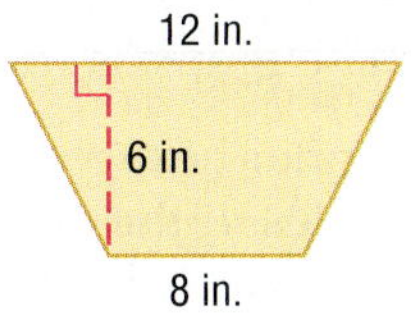

31.

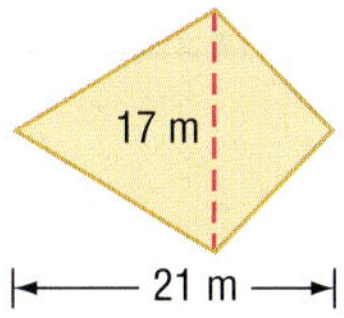

32.

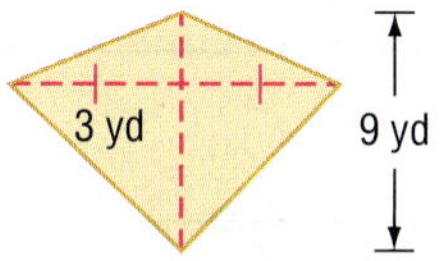

33. **MULTIPLE REPRESENTATIONS** In this problem, you will investigate perimeters of kites.

a. **GEOMETRIC** Draw a kite like the one shown if $x = 2$.

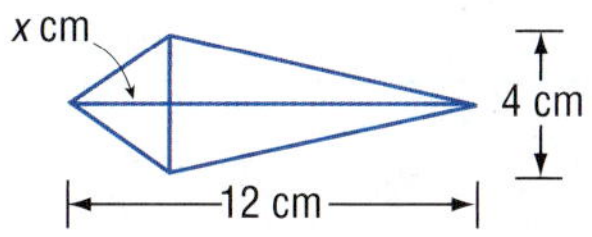

b. **GEOMETRIC** Repeat the process in part **a** for three x-values between 2 and 10 and for an x-value of 10.

c. **TABULAR** Measure and record in a table the perimeter of each kite, along with the x-value.

d. **GRAPHICAL** Graph the perimeter versus the x-value using the data from your table.

e. **ANALYTICAL** Make a conjecture about the value of x that will minimize the perimeter of the kite. What is the significance of this value?

H.O.T. Problems

Use **H**igher-**O**rder **T**hinking Skills

34. **ERROR ANALYSIS** Antonio and Madeline want to draw a trapezoid that has a height of 4 units and an area of 18 square units. Antonio says that only one trapezoid will meet the criteria. Madeline disagrees and thinks that she can draw several different trapezoids with a height of 4 units and an area of 18 square units. Is either of them correct? Explain your reasoning.

35. **CHALLENGE** Find x in parallelogram $ABCD$.

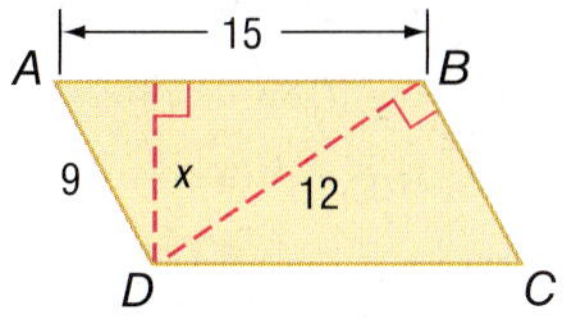

36. **OPEN ENDED** Draw a kite and a rhombus with an area of 6 square inches. Label and justify your drawings.

37. **REASONING** If the areas of two rhombi are equal, are the perimeters *sometimes*, *always*, or *never* equal? Explain.

38. **WRITING IN MATH** Use the formula for the area of a rhombus to explain why the formula $A = s^2$ for a square with side s is equivalent to the formula for the area of a rhombus.

Standardized Test Practice

39. The lengths of the bases of an isosceles trapezoid are shown below.

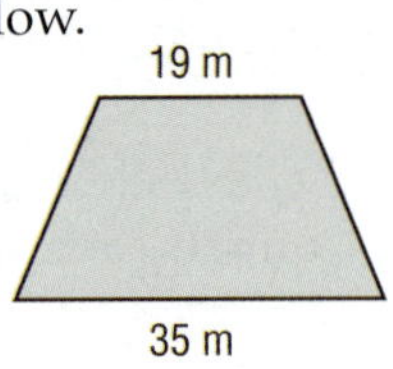

If the perimeter is 74 meters, what is its area?

A 162 m^2 **C** 332.5 m^2
B 270 m^2 **D** 342.25 m^2

40. SHORT RESPONSE One diagonal of a rhombus is three times as long as the other diagonal. If the area of the rhombus is 54 square millimeters, what are the lengths of the diagonals?

41. ALGEBRA What is the effect on the graph of the equation $y = \frac{1}{2}x$ when the equation is changed to $y = -2x$?

F The graph is moved 1 unit down.
G The graph is moved 1 unit up.
H The graph is rotated 45° about the origin.
J The graph is rotated 90° about the origin.

42. A regular hexagon is divided into 6 congruent triangles. If the perimeter of the hexagon is 48 centimeters, what is the height of each triangle?

A 4 cm **C** 8 cm
B $4\sqrt{3}$ cm **D** $8\sqrt{3}$ cm

Spiral Review

COORDINATE GEOMETRY **Find the area of each figure.** (Lesson 11-1)

43. $\triangle JKL$ with $J(-4, 3)$, $K(-9, -1)$, and $L(-4, -4)$

44. $\square RSTV$ with $R(-5, 7)$, $S(2, 7)$, $T(0, 2)$, and $V(-7, 2)$

45. WEATHER Meteorologists track severe storms using Doppler radar. A polar grid is used to measure distances as the storms progress. If the center of the radar screen is the origin and each ring is 10 miles farther from the center, what is the equation of the fourth ring? (Lesson 10-8)

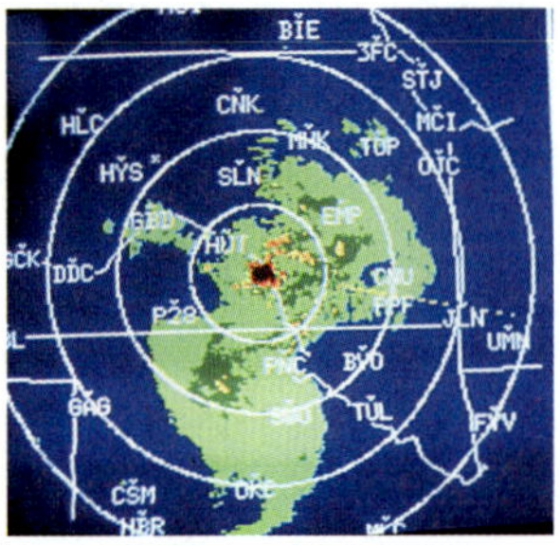

Find x and y. (Lesson 8-3)

46.

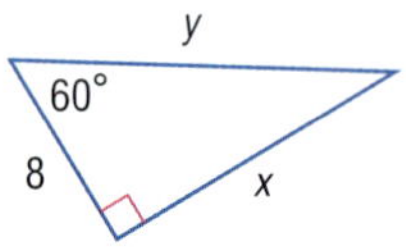

47.

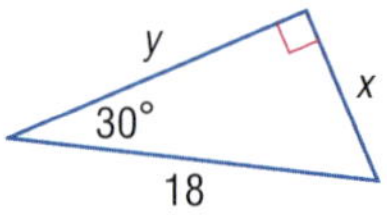

Use the Venn diagram to determine whether each statement is *always, sometimes,* or *never* true. (Lesson 6-5)

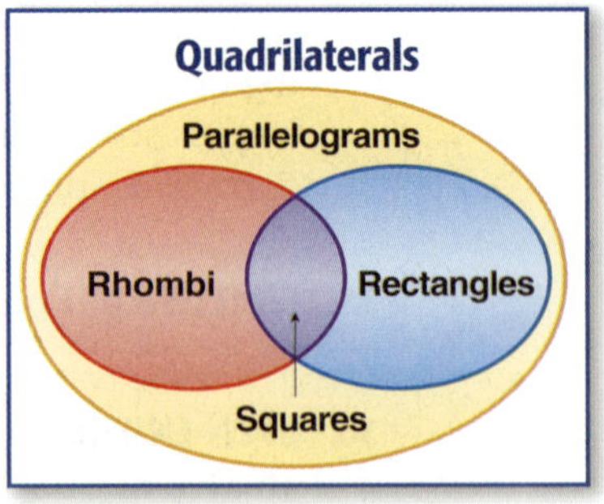

48. A parallelogram is a square.

49. A square is a rhombus.

50. A rectangle is a parallelogram.

51. A rhombus is a rectangle but not a square.

52. A rhombus is a square.

Skills Review

Find the circumference and area of each figure. Round to the nearest tenth. (Lesson 1-6)

53.

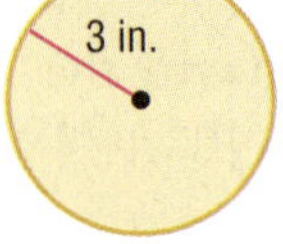

54.

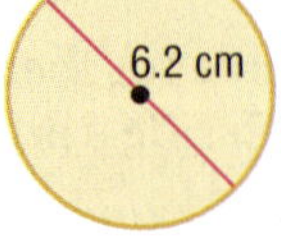

55.

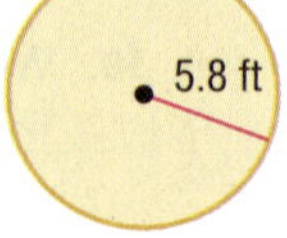

EXTEND 11-2

Spreadsheet Lab
Real Estate Property Values

Math Online glencoe.com
- Graphing Technology Personal Tutor

Objective
Use a spreadsheet to organize and compare descriptive data.

When a homeowner is determining the purchase price for their house, they often research the prices of similar properties in the neighborhood. By finding the sale prices of similar homes, a homeowner can estimate how much they would like to ask for their house.

ACTIVITY 1

Use the information below to organize a chart to compare four similar properties.

House A Property is in good condition. It was built in 1998 and has 1800 square feet. The house has 3 bedrooms, $2\frac{1}{2}$ baths and a 2-car garage. The lot is $\frac{1}{2}$ acre and the house has a full basement.

House B Property is in good condition. It was built in 1998 and has 1850 square feet. The house has 3 bedrooms, 2 baths and a 2-car garage. The lot is $\frac{1}{2}$ acre and the house has a full basement.

House C Property is in good condition. It was built in 1997 and has 1800 square feet. The house has 3 bedrooms, $2\frac{1}{2}$ baths and a 2-car garage. The lot is $\frac{1}{2}$ acre and the house has a full basement.

House D Property is in good condition. It was built in 1998 and has 1775 square feet. The house has 3 bedrooms, $2\frac{1}{2}$ baths and a 2-car garage. The lot is $\frac{1}{2}$ acre and the house has a full basement.

Design a spreadsheet to compare the houses using the information above.

Spreadsheet sample

◇	A	B	C	D	E	F
1			House A	House B	House C	House D
2	Property Condition					
3	Year Built					
4	Square Feet					
5	Bedrooms					
6	Baths					
7	Garage					
8	Lot size					
9	Basement					

Sheet 1 | Sheet 2 | Sheet 3

Model and Analyze

1. Fill in the spreadsheet using the descriptions of Houses A–D.
2. **MAKE A CONJECTURE** Which house do you think would have the highest selling price? the lowest selling price? Explain.
3. **WRITING IN MATH** Write a description of another house that would be similar to the listed houses. Add the data to the table.

11-3 Areas of Circles and Sectors

Why?

To determine whether a medium or large pizza is a better value, you can compare the cost per square inch. Divide the cost of each pizza by its area.

Then

You found the circumference of a circle. (Lesson 10-1)

Now

- Find areas of circles.
- Find areas of sectors of circles.

New Vocabulary

sector of a circle
segment of a circle

Math Online
glencoe.com

Areas of Circles In Lesson 10-1, you learned that the formula for the circumference C of a circle with radius r is given by $C = 2\pi r$. You can use this formula to develop the formula for the area of a circle.

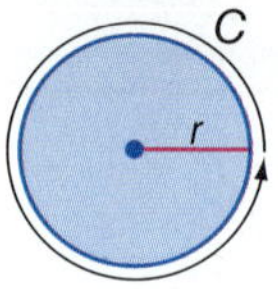

Below, a circle with radius r and circumference C has been divided into congruent pieces and then rearranged to form a figure that resembles a parallelogram.

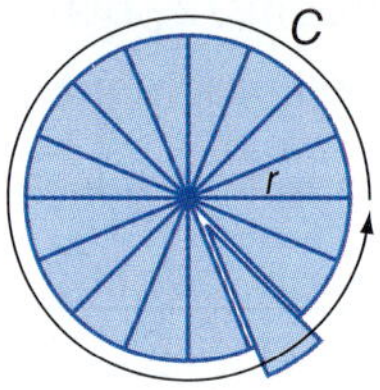

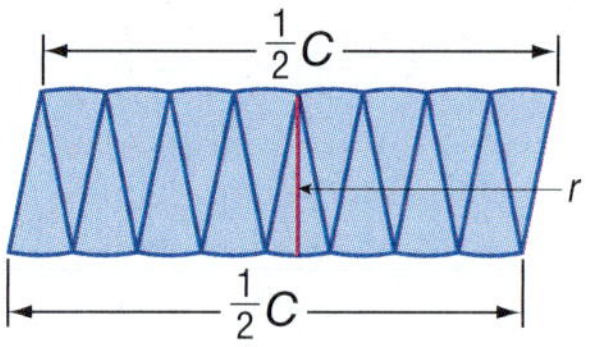

The base of the parallelogram is $\frac{1}{2}C$ and the height is r, so its area is $\frac{1}{2}C \cdot r$. Since $C = 2\pi r$, the area of the parallelogram is also $\frac{1}{2}(2\pi r)r$ or πr^2.

Key Concept — Area of a Circle

Words The area A of a circle is equal to π times the square of the radius r.

Symbols $A = \pi r^2$

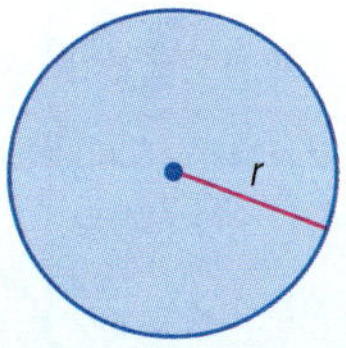

Real-World EXAMPLE 1 — Area of a Circle

SPORTS What is the area of the circular putting green shown to the nearest square foot?

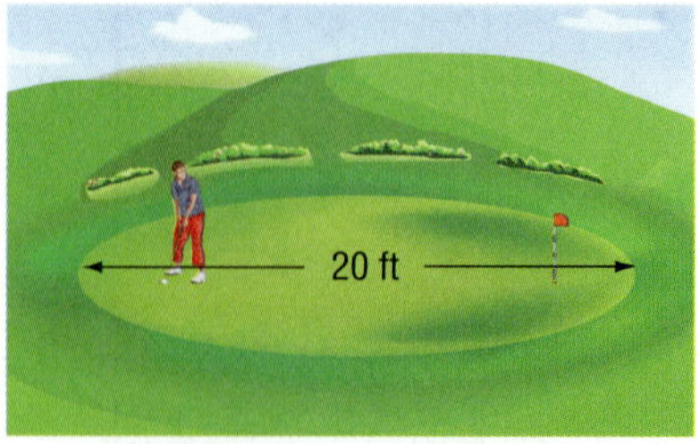

The diameter is 20 feet, so the radius is 10 feet.

$A = \pi r^2$	Area of a circle
$= \pi(10)^2$	$r = 10$
≈ 314	Use a calculator.

So, the area is about 314 square feet.

Guided Practice

1. **SPORTS** The archery target shown has a radius of 12 inches. What is the area of the target to the nearest square inch?

Personal Tutor glencoe.com

EXAMPLE 2 Use the Area of a Circle to Find a Missing Measure

ALGEBRA **Find the radius of a circle with an area of 95 square centimeters.**

$A = \pi r^2$	Area of a circle
$95 = \pi r^2$	$A = 95$
$\frac{95}{\pi} = r^2$	Divide each side by π.
$5.5 \approx r$	Use a calculator. Take the positive square root of each side.

The radius of the circle is about 5.5 centimeters.

Guided Practice

2. **ALGEBRA** The area of a circle is 196π square yards. Find the diameter.

Personal Tutor glencoe.com

Review Vocabulary

central angle an angle with a vertex in the center of a circle and with sides that contain two radii of the circle (Lesson 10-2)

arc a portion of a circle defined by two endpoints (Lesson 10-2)

Areas of Sectors A slice of a circular pizza is an example of a sector of a circle. A **sector of a circle** is a region of a circle bounded by a central angle and its intercepted arc. The formula for the area of a sector is similar to the formula for arc length.

Key Concept Area of a Sector

The ratio of the **area A of a sector** to the **area of the whole circle, πr^2**, is equal to the ratio of the **degree measure of the intercepted arc x** to 360.

Proportion: $\frac{A}{\pi r^2} = \frac{x}{360}$

Equation: $A = \frac{x}{360} \cdot \pi r^2$

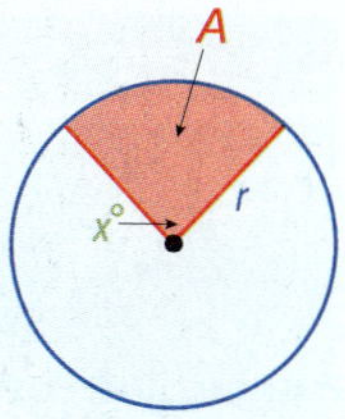

Real-World Link

About 3 billion pizzas are sold each year in the United States. That is equivalent to about 46 slices per person annually.

Source: ThinkQuest Library

Real-World EXAMPLE 3 Area of a Sector

PIZZA **A circular pizza has a diameter of 12 inches and is cut into 8 congruent slices. What is the area of one slice to the nearest hundredth?**

Step 1 Find the arc measure of a pizza slice.

Since the pizza is equally divided into 8 slices, each slice will have an arc measure of $360 \div 8$ or 45.

Step 2 Find the radius of the pizza. Use this measure to find the area of the sector, or slice.

The diameter is 12 inches, so the radius is 6 inches.

$A = \frac{x}{360} \cdot \pi r^2$	Area of a sector
$= \frac{45}{360} \cdot \pi(6)^2$	$x = 45$ and $r = 6$
≈ 14.14	Use a calculator.

So, the area of one slice of this pizza is about 14.14 square inches.

Guided Practice

Find the area of the shaded sector. Round to the nearest tenth.

3A.

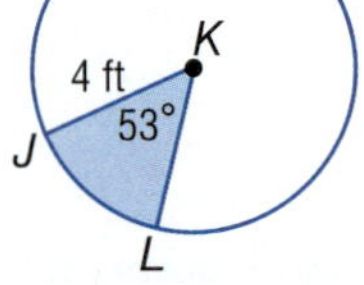

3B.

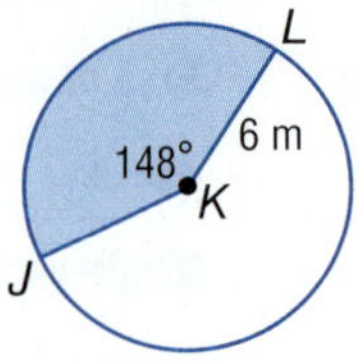

3C.

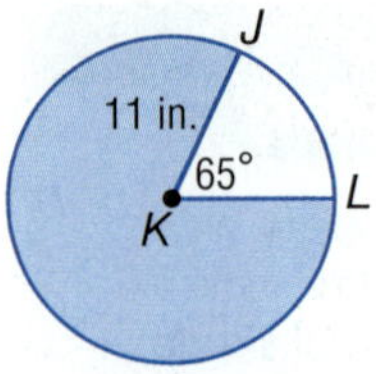

3D. **CRAFTS** The color wheel at the right is a tool that artists use to organize color schemes. If the diameter of the wheel is 10 inches and each of the 12 sections is congruent, find the approximate area covered by green hues.

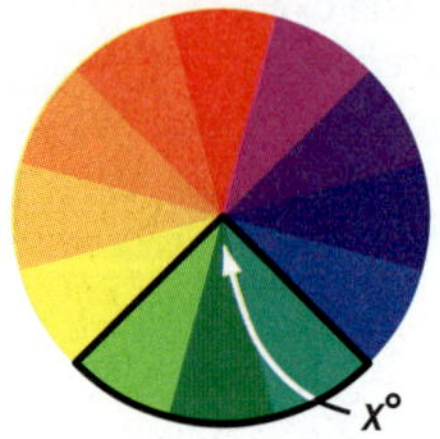

Personal Tutor glencoe.com

Check Your Understanding

Example 1 p. 782

CONSTRUCTION **Find the area of each circle. Round to the nearest tenth.**

1.

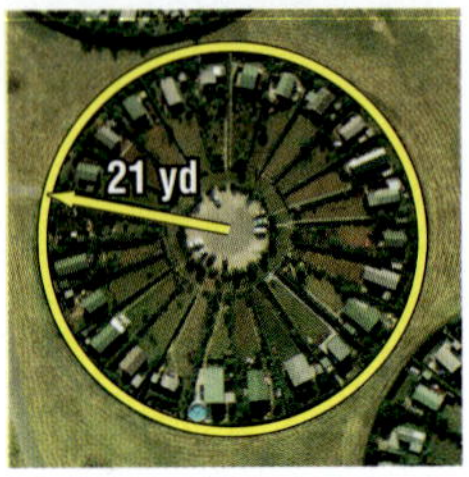

2.

Example 2 p. 783

Find the indicated measure. Round to the nearest tenth.

3. Find the diameter of a circle with an area of 74 square millimeters.

4. The area of circle is 88 square inches. Find the radius.

Example 3 p. 783

Find the area of each shaded sector. Round to the nearest tenth.

5.

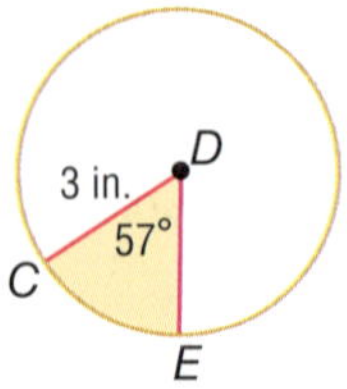

6.

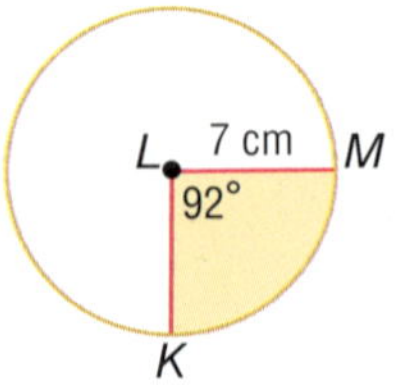

7. **BAKING** Chelsea is baking pies for a fundraiser at her school. She divides each pie into 6 equal slices.

a. What is the area, in square inches, for each slice of pie?

b. If each slice costs \$0.25 to make and she sells 8 pies at \$1.25 for each slice, how much money will she raise?

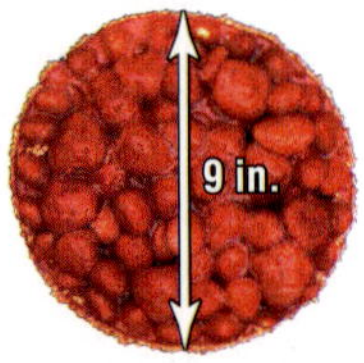

Practice and Problem Solving

● = **Step-by-Step Solutions** begin on page R20.
Extra Practice begins on page 969.

Example 1 p. 782

SPORTS **Find the area of each circle. Round to the nearest tenth.**

8. 

9. 

10. 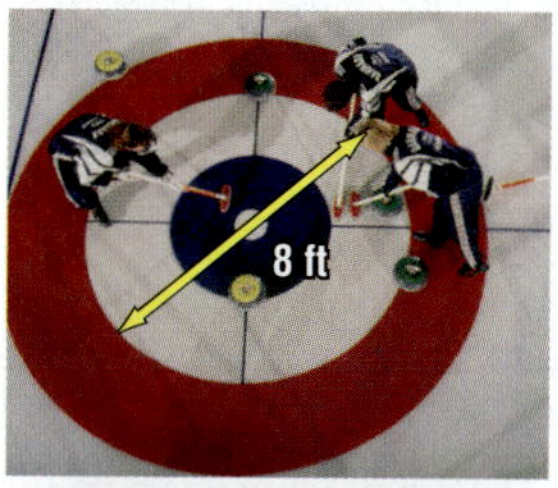

11. 

12. 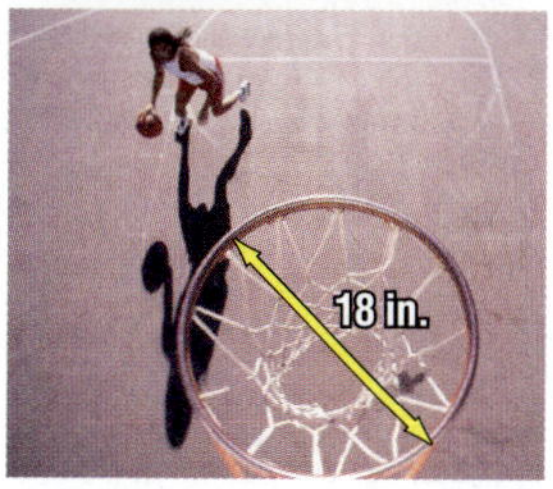

13. 

Example 2 p. 783

Find the indicated measure. Round to the nearest tenth, if necessary.

14. The area of a circle is 68 square centimeters. Find the diameter.

15. Find the diameter of a circle with an area of 94 square millimeters.

16. The area of circle is 112 square inches. Find the radius.

17. Find the radius of a circle with an area of 206 square feet.

Example 3 p. 783

Find the area of each shaded sector. Round to the nearest tenth, if necessary.

18. 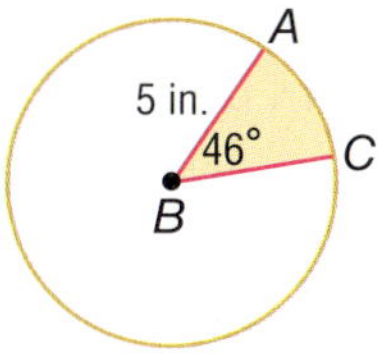

19. 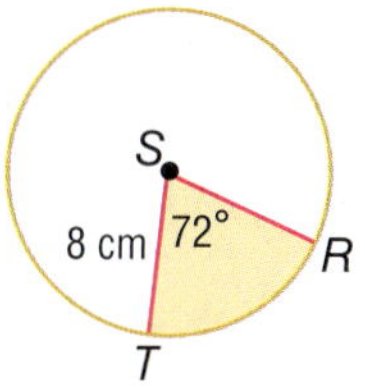

20. 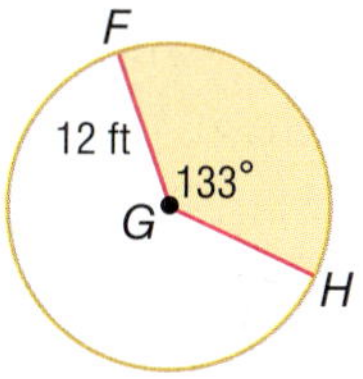

21. 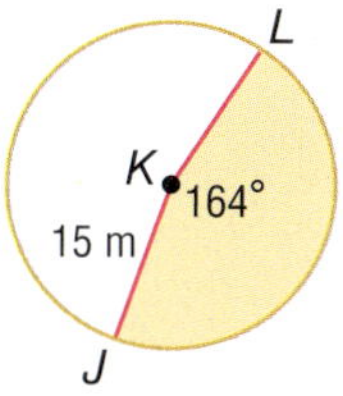

22. 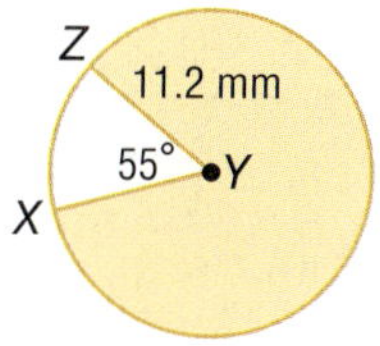

23. 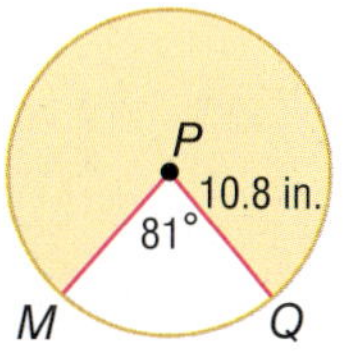

24. **CRAFTS** Aaliyah wants to make a multi-colored fabric clock as a gift for her friend's birthday. If she wants each of the four congruent sectors to be a different color, how much of each color of fabric will she need?

25. **JEWELRY** A jeweler makes a pair of earrings by cutting two 50° sectors from a silver disk.

a. Find the area of each sector.

b. If the weight of the silver disk is 2.3 grams, how many milligrams does the silver wedge for each earring weigh?

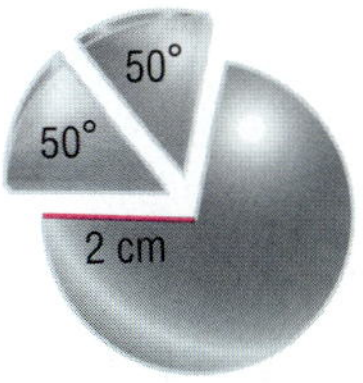

Real-World Link

The Bristlecone Pine is the world's oldest species of tree. It can reach an age of 4600 years.

Source: *Deep Creek Times*

26. PROM The table shows the results of a survey of students to determine their preference for a prom theme.

a. Create a circle graph with a diameter of 2 inches to represent these data.

b. Find the area of each theme's sector in your graph. Round to the nearest hundredth of an inch.

Theme	Percent
An Evening of Stars	11
Mardi Gras	32
Springtime in Paris	8
Night in Times Square	47
Undecided	2

ALGEBRA **The area A of each shaded region is given. Find x.**

27. $A = 66\text{ cm}^2$

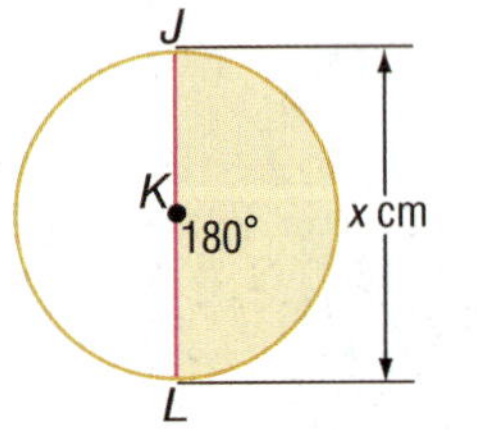

28. $A = 94\text{ in}^2$

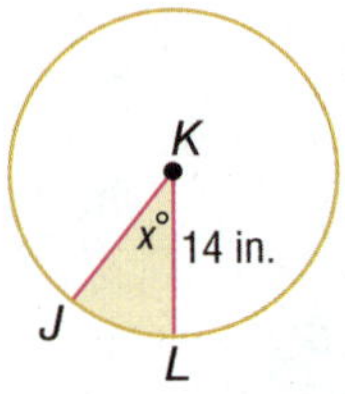

29. $A = 128\text{ ft}^2$

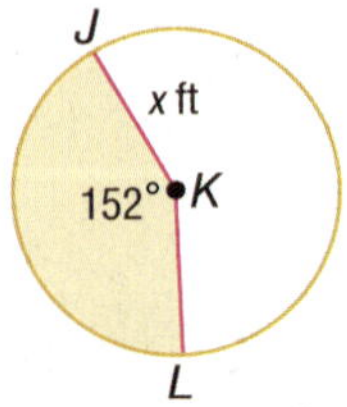

30. CRAFTS Luna is making tablecloths with the dimensions shown for a club banquet. Find the area of each tablecloth in square feet if each one is to just reach the floor.

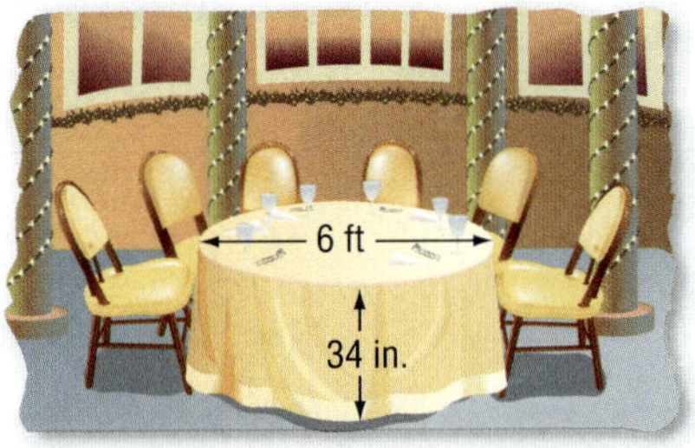

31. TREES The age of a living tree can be determined by multiplying the diameter of the tree by its growth factor, or rate of growth.

a. What is the diameter of a tree with a circumference of 2.5 feet?

b. If the growth factor of the tree is 4.5, what is the age of the tree?

Find the area of the shaded region. Round to the nearest tenth.

32.

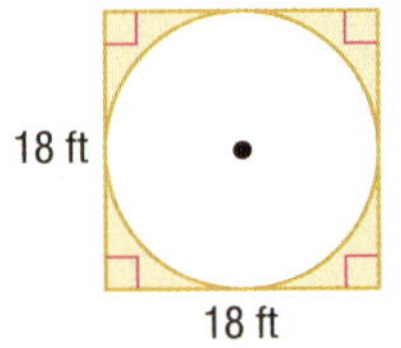

33.

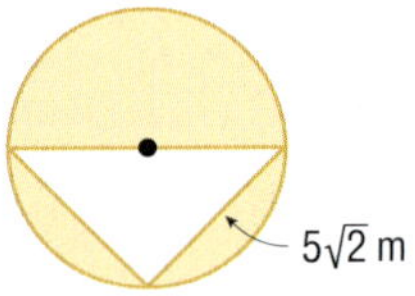

34.

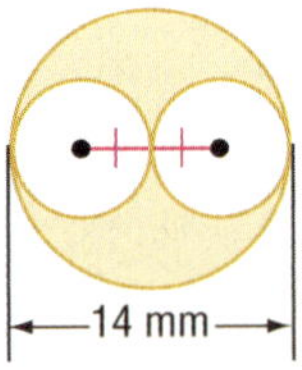

35.

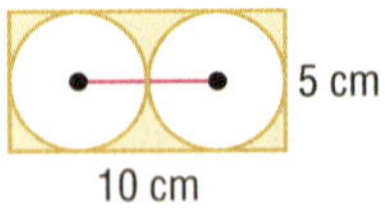

36.

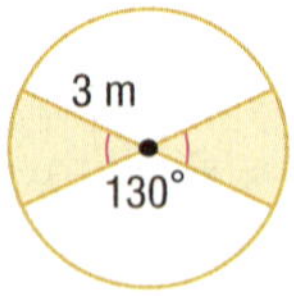

37.

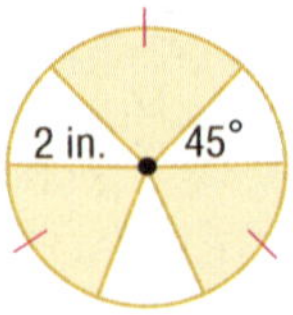

38. COORDINATE GEOMETRY What is the area of sector ABC shown on the graph?

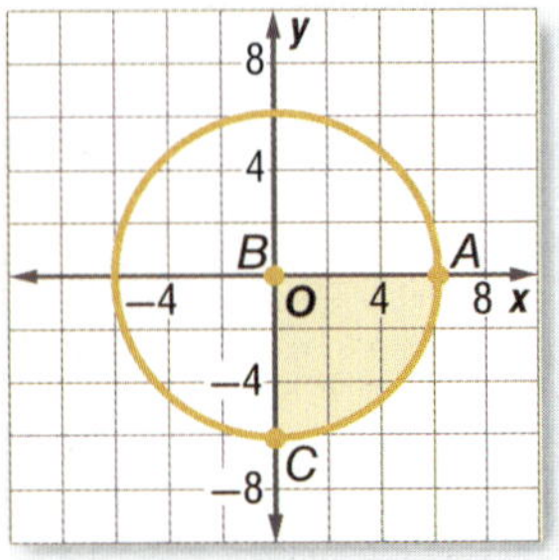

39. ALGEBRA The figure shown below is a sector of a circle. If the perimeter of the figure is 22 millimeters, find its area in square millimeters.

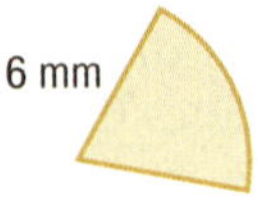

Find the area of each shaded region.

40.

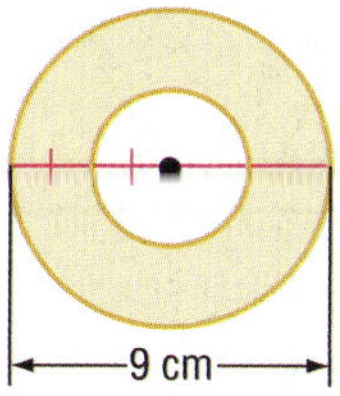

41

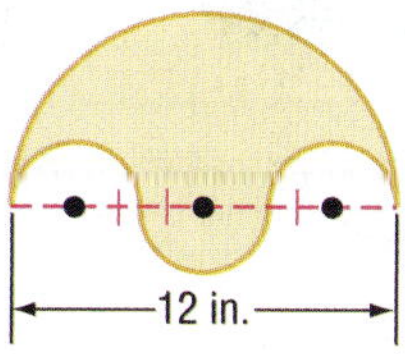

42.

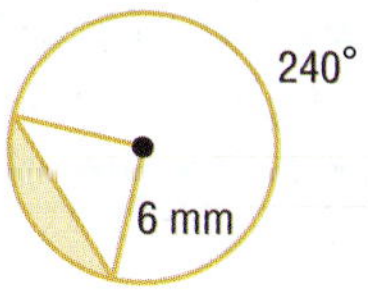

Problem-Solving Tip

Use Logical Reasoning The segments of circles become more difficult to draw as the central angle approaches 0° because they are smaller. Therefore, as the central angle decreases, logical reasoning tells you that the area of the segment should decrease.

43. **MULTIPLE REPRESENTATIONS** In this problem, you will investigate segments of circles. A **segment of a circle** is the region bounded by an arc and a chord.

a. **ALGEBRAIC** Write an equation for the area A of a segment of a circle with a radius r and a central angle of $x°$. (*Hint*: Use trigonometry to find the base and height of the triangle.)

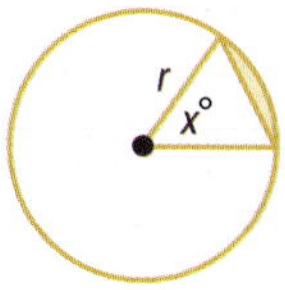

b. **TABULAR** Calculate and record in a table ten values of A for x-values ranging from 10 to 90 if r is 12 inches. Round to the nearest tenth.

c. **GRAPHICAL** Graph the data from your table with the x-values on the horizontal axis and the A-values on the vertical axis.

d. **ANALYTICAL** Use your graph to predict the value of A when x is 63. Then use the formula you generated in part **a** to calculate the value of A when x is 63. How do the values compare?

H.O.T. Problems

Use **H**igher-**O**rder **T**hinking Skills

44. **ERROR ANALYSIS** Kristen and Chase want to find the area of the shaded region in the circle shown. Is either of them correct? Explain your reasoning.

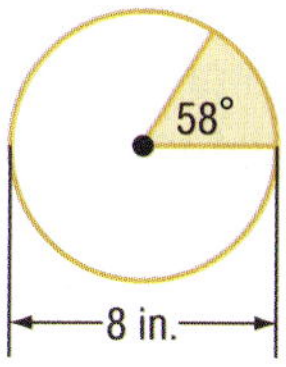

Kristen

$$A = \frac{x}{360} \cdot \pi r^2$$
$$= \frac{58}{360} \cdot \pi(8)^2$$
$$= 32.4 \text{ in}^2$$

Chase

$$A = \frac{x}{360} \cdot \pi r^2$$
$$= \frac{58}{360} \cdot \pi(4)^2$$
$$= 8.1 \text{ in}^2$$

45. **CHALLENGE** Find the area of the shaded region. Round to the nearest tenth.

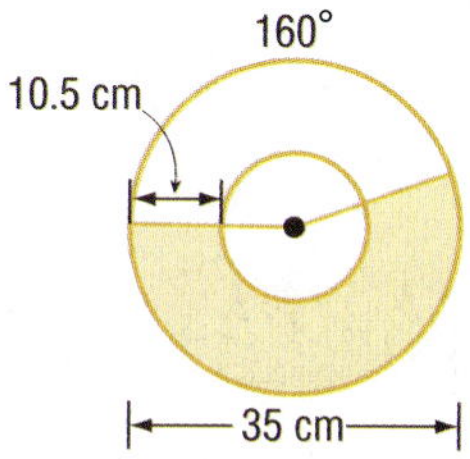

46. **REASONING** Refer to Exercise 43. Is the area of a sector of a circle *sometimes*, *always*, or *never* greater than the area of its corresponding segment?

47. **WRITING IN MATH** Describe two methods you could use to find the area of the shaded region of the circle. Which method do you think is more efficient? Explain your reasoning.

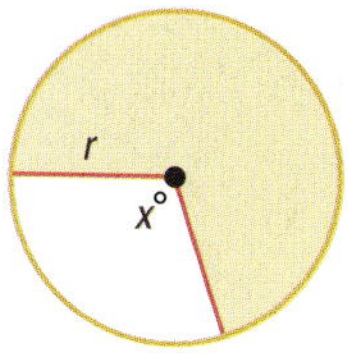

48. **OPEN ENDED** Draw a circle and a sector that makes up 60% of the circle. What is the central angle of the sector?

49. **WRITING IN MATH** If the radius of a circle doubles, will the measure of a sector of that circle double? Will it double if the arc measure of that sector doubles?

Standardized Test Practice

50. What is the area of the sector?

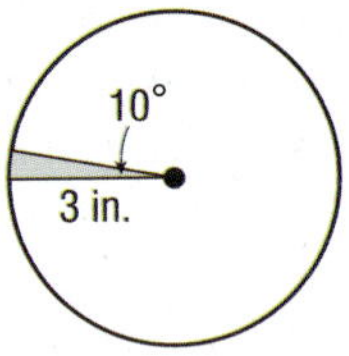

A $\frac{9\pi}{10}$ in^2

B $\frac{3\pi}{5}$ in^2

C $\frac{\pi}{4}$ in^2

D $\frac{\pi}{6}$ in^2

51. SHORT RESPONSE $\overleftrightarrow{MN}$ and $\overleftrightarrow{PQ}$ intersect at T. Find the value of x for which $m\angle MTQ = 2x + 5$ and $m\angle PTM = x + 7$. What are the degree measures of $\angle MTQ$ and $\angle PTM$?

52. ALGEBRA Raphael bowled 4 games and had a mean score of 130. He then bowled two more games with scores of 180 and 230. What was his mean score for all 6 games?

F 90

G 155

H 180

J 185

53. SAT/ACT The diagonals of rectangle $ABCD$ each have a length of 56 feet. If $m\angle BAC = 42°$, what is the length of $\overline{AB}$ to the nearest tenth of a foot?

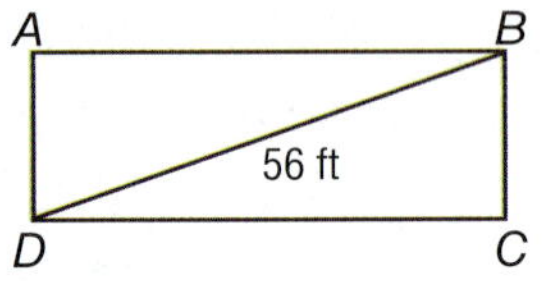

A 37.5

B 41.6

C 50.4

D 75.4

Spiral Review

Find each missing length. (Lesson 11-2)

54. One diagonal of a kite is half as long as the other diagonal. If the area of the kite is 188 square inches, what are the lengths of the diagonals?

55. The area of a rhombus is 175 square centimeters. If one diagonal is two times as long as the other, what are the lengths of the diagonals?

Find the area of each parallelogram. Round to the nearest tenth if necessary. (Lesson 11-1)

56.

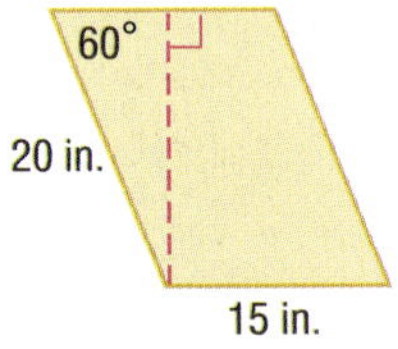

57.

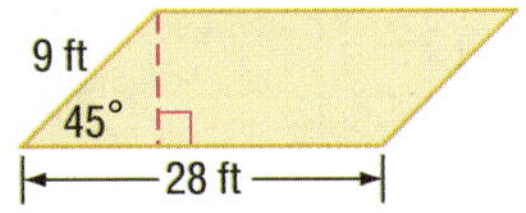

58.

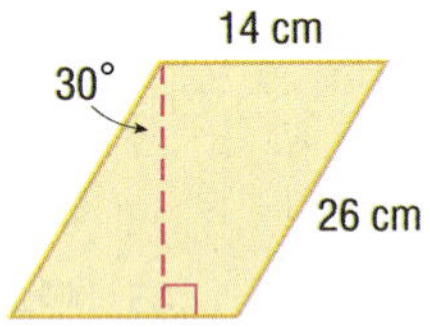

Skills Review

Find each measure. (Lesson 5-1)

59. XT

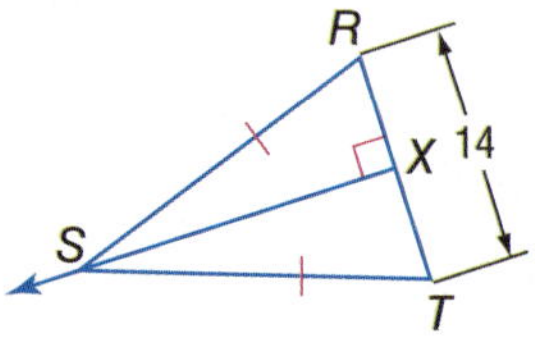

60. AC

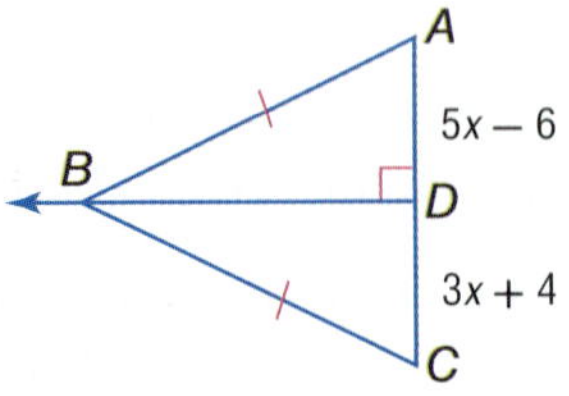

61. JK

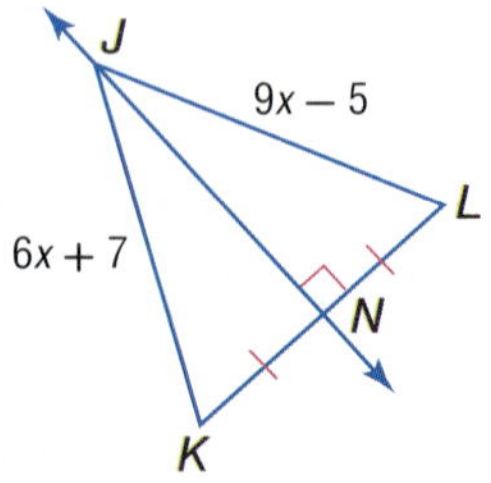

Mid-Chapter Quiz

Lessons 11-1 through 11-3

Find the perimeter and area of each parallelogram or triangle. Round to the nearest tenth if necessary. (Lesson 11-1)

1.
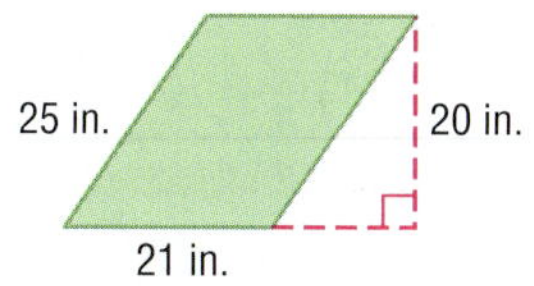

2.
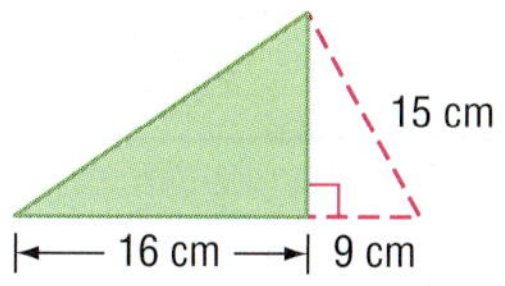

3.
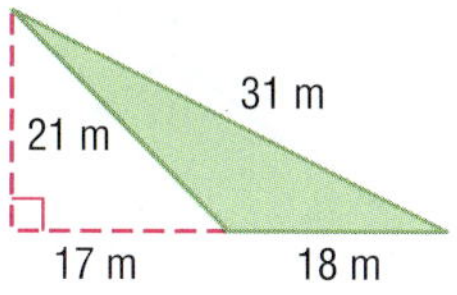

4.
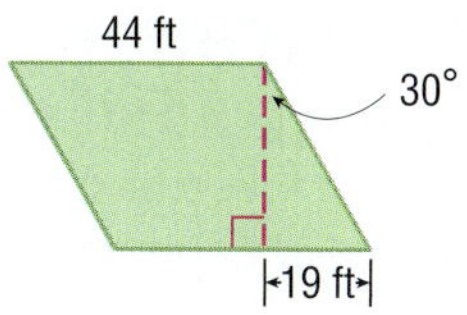

5. The height of a triangle is 8 inches more than its base. The area of the triangle is 104.5 square inches. Find the base and height. (Lesson 11-1)

6. **DESIGN** A plaque is made with a rhombus in the middle. If the diagonals of the rhombus measure 7 inches and 9 inches, how much space is available for engraving text onto the award? (Lesson 11-2)

7. **MULTIPLE CHOICE** The area of a kite is 4 square feet. If the tail is to be 3 times longer than the kite's long diagonal, and the short diagonal measures 2 feet, how long should the kite tail be? (Lesson 11-2)

A 4 feet
B 6 feet
C 7 feet
D 12 feet

Find the area of each trapezoid, rhombus, or kite. (Lesson 11-2)

8.
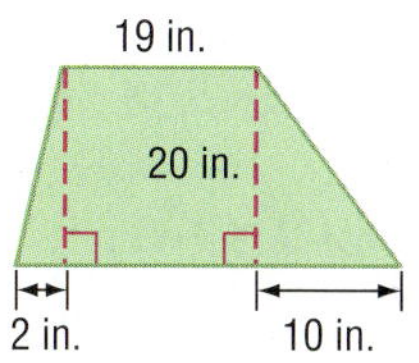

9.
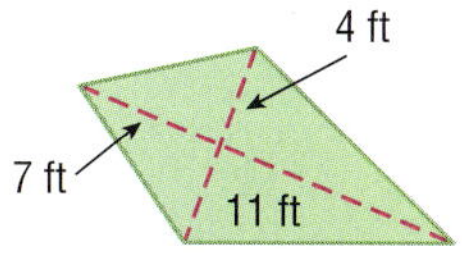

10.
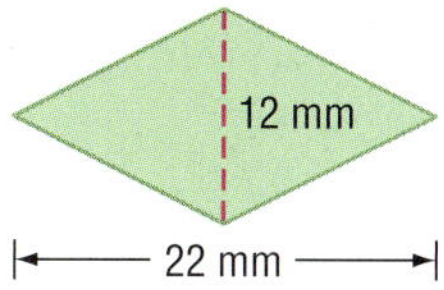

11.
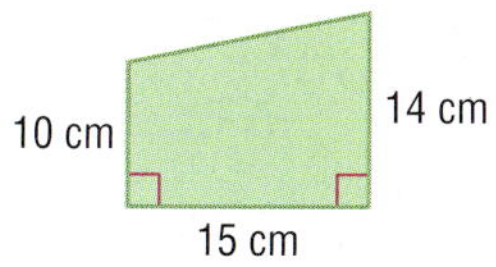

12. **ARCHAEOLOGY** The most predominant shape in Incan architecture is the trapezoid. The doorway pictured below is 3 feet wide at the top and 4 feet wide at the bottom. A person who is 5 feet 8 inches tall can barely pass through the doorway. How much fabric would be necessary to make a curtain for the doorway? (Lesson 11-2)

13. **ALGEBRA** A sector of a circle has a central angle measure of 30° and radius r. Write an expression for the perimeter of the sector in terms of r. (Lesson 11-3)

Find the area of each shaded sector. Round to the nearest tenth. (Lesson 11-3)

14.
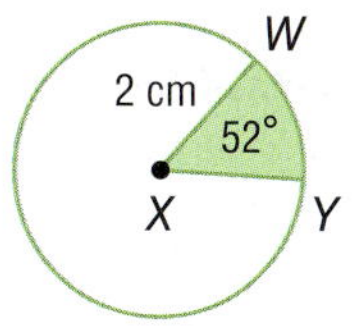

15.
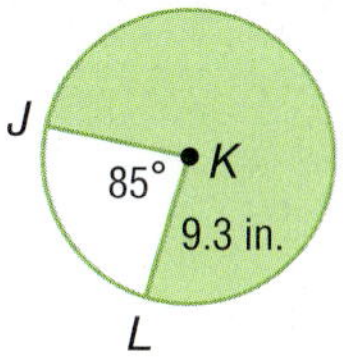

16.
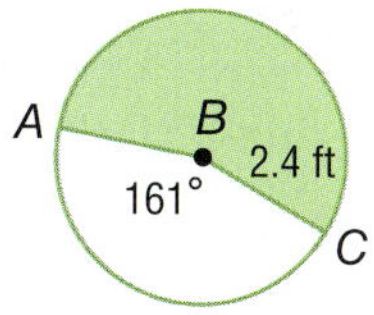

17.
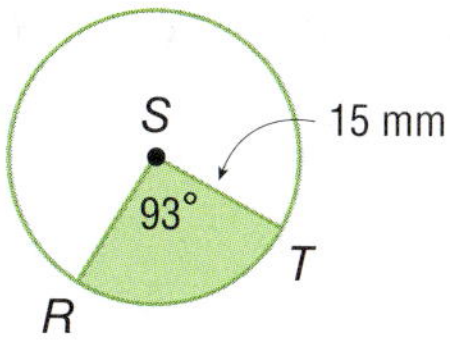

Find the indicated measure. Round to the nearest tenth. (Lesson 11-3)

18. The area of a circle is 52 square inches. Find the diameter.

19. Find the radius of a circle with an area of 104 square meters.

20. **FRUIT** The diameter of the orange slice shown is 9 centimeters. If each of the orange's 10 sections are congruent, find the approximate area covered by 8 sections. (Lesson 11-3)

EXPLORE **11-4**

Geometry Lab

Investigating Areas of Regular Polygons

Objective
Investigate the formula for the area of regular polygons.

The point in the interior of a regular polygon that is equidistant from all of the vertices is the *center* of the polygon. A segment from the center that is perpendicular to a side of the polygon is an **apothem**.

ACTIVITY

Step 1 Copy regular pentagon $ABCDE$ and its center O.

Step 2 Draw the apothem from O to side $\overline{AB}$ by constructing the perpendicular bisector of $\overline{AB}$. Label the apothem measure as a. Label the measure of $\overline{AB}$ as s.

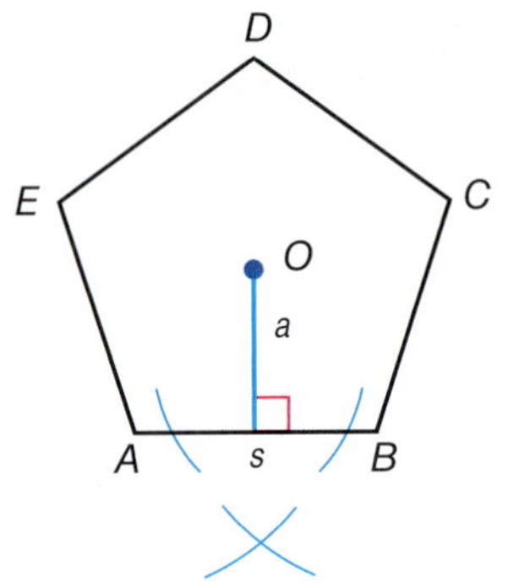

Step 3 Use a straightedge to draw $\overline{OA}$ and $\overline{OB}$.

Step 4 What measure in $\triangle AOB$ represents the base of the triangle? What measure represents the height?

Step 5 Find the area of $\triangle AOB$ in terms of s and a.

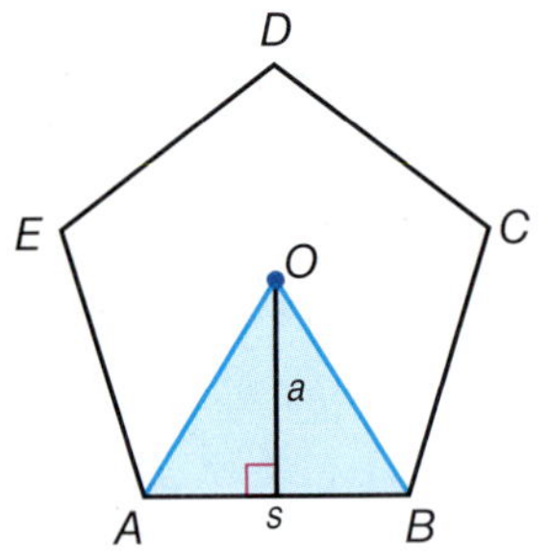

Step 6 Draw $\overline{OC}$, $\overline{OD}$, and $\overline{OE}$. What is true of the five small triangles formed?

Step 7 How do the areas of the five triangles compare?

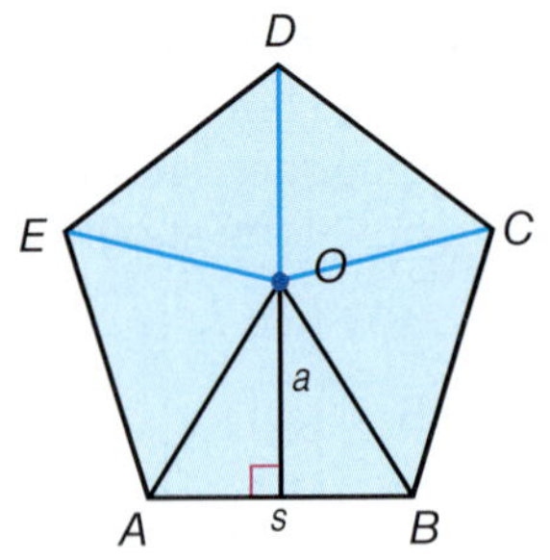

Analyze the Results

1. The area of a pentagon $ABCDE$ can be found by adding the areas of the given triangles that make up the pentagonal region.

$$A = \frac{1}{2}sa + \frac{1}{2}sa + \frac{1}{2}sa + \frac{1}{2}sa + \frac{1}{2}sa$$

$$A = \frac{1}{2}(sa + sa + sa + sa + sa) \text{ or } \frac{1}{2}(5sa)$$

What does $5s$ represent?

2. Write a formula for the area of a pentagon in terms of perimeter P.

11-4 Areas of Regular Polygons and Composite Figures

Then

You used inscribed and circumscribed figures and found the areas of circles. (Lessons 10-1 and 11-3)

Now

- Find areas of regular polygons.
- Find areas of composite figures.

New Vocabulary

center of a regular polygon
radius of a regular polygon
apothem
central angle of a regular polygon
composite figure

Math Online
glencoe.com

Why?

The top of the table shown is a regular hexagon. Notice that the top is composed of six congruent triangular sections. To find the area of the table top, you can find the sum of the areas of the sections.

Areas of Regular Polygons In the figure, a regular pentagon is *inscribed* in ⊙P, and ⊙P is *circumscribed* about the pentagon. The **center of a regular polygon** and the **radius of a regular polygon** are also the center and the radius of its circumscribed circle.

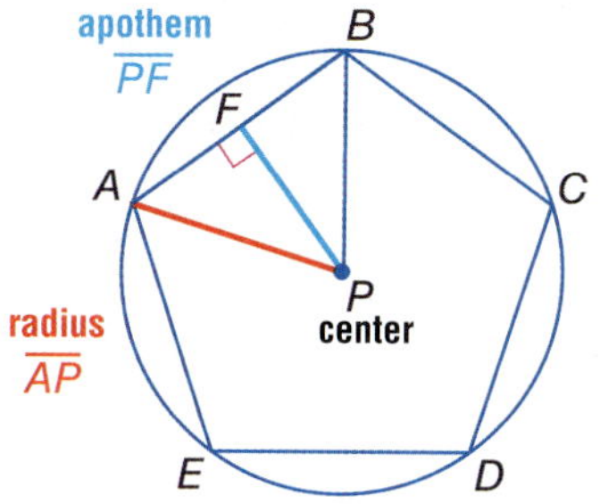

$\angle APB$ is a central angle of regular pentagon $ABCDE$.

A segment drawn from the center of a regular polygon perpendicular to a side of the polygon is called an **apothem**. Its length is the height of an isosceles triangle that has two radii as legs.

A **central angle of a regular polygon** has its vertex at the center of the polygon and its sides pass through consecutive vertices of the polygon. The measure of each central angle of a regular n-gon is $\frac{360}{n}$.

EXAMPLE 1 Identify Segments and Angles in Regular Polygons

Square $FGHJ$ is inscribed in ⊙K. Identify the center, a radius, an apothem, and a central angle of the polygon. Then find the measure of a central angle.

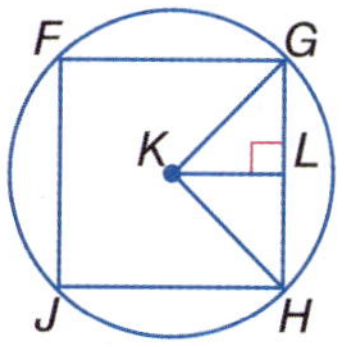

center: point K radius: $\overline{KG}$ or $\overline{KH}$

apothem: $\overline{KL}$ central angle: $\angle GKH$

A square is a regular polygon with 4 sides. Thus, the measure of each central angle of square $FGHJ$ is $\frac{360}{4}$ or 90.

Guided Practice

1. In the figure, regular hexagon $JKLMNP$ is inscribed in ⊙R. Identify the center, a radius, an apothem, and a central angle of the polygon. Then find the measure of a central angle.

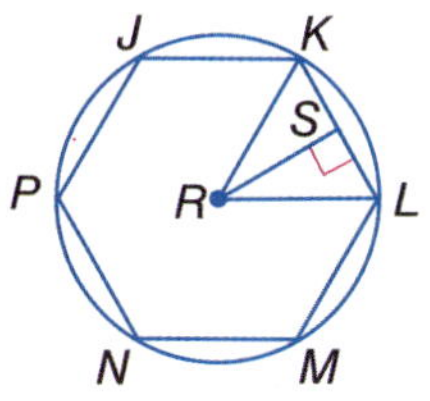

Personal Tutor glencoe.com

You can find the area of any regular n-gon by dividing the polygon into congruent isosceles triangles. This strategy is sometimes called *decomposing the polygon into triangles.*

Real-World EXAMPLE 2 Area of a Regular Polygon

ART **Kang created the stained glass window shown. The window is a regular octagon with a side length of 15 inches and an apothem of 18.1 inches. What is the area covered by the window?**

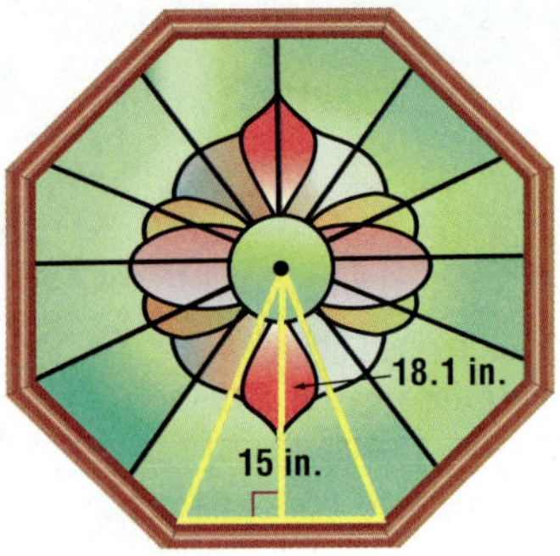

ReadingMath

Apothem Like the *radius* of a circle, the *apothem* of a polygon refers to the length of any apothem of the polygon.

Step 1 Divide the polygon into congruent isosceles triangles.

Since the polygon has 8 sides, the polygon can be divided into 8 congruent isosceles triangles, each with a base of 15 inches and a height of 18.1 inches.

Step 2 Find the area of one triangle.

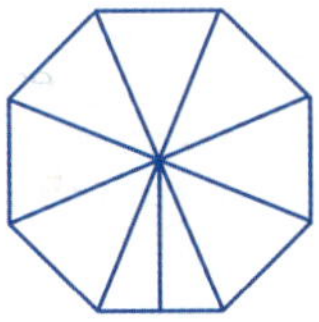

$A = \frac{1}{2}bh$ **Area of a triangle**

$= \frac{1}{2}(15)(18.1)$ **$b = 15$ and $h = 18.1$**

$= 135.75 \text{ in}^2$ **Simplify.**

Step 3 Multiply the area of one triangle by the total number of triangles.

Since there are 8 triangles, the area of the stained glass is $135.75 \cdot 8$ or 1086 square inches.

Guided Practice

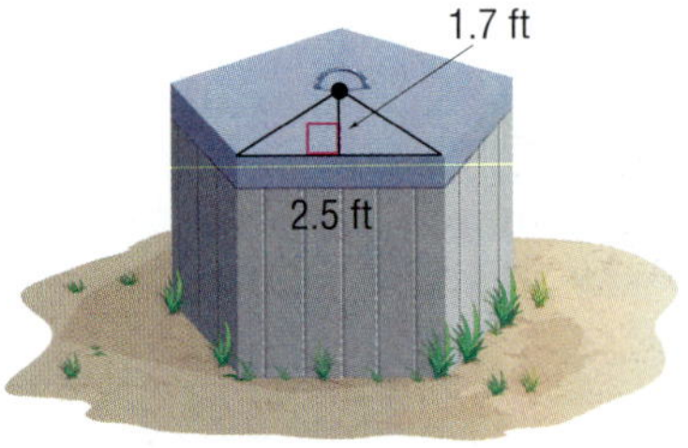

2. **HOT TUBS** The cover of the hot tub shown is a regular pentagon. If the side length is 2.5 feet and the apothem is 1.7 feet, find the area of the lid to the nearest tenth.

Watch Out!

Area of Regular Polygon This approach can only be applied to *regular* polygons.

Personal Tutor glencoe.com

From Example 2, we can develop a formula for the area of a regular n-gon with side length s and apothem a.

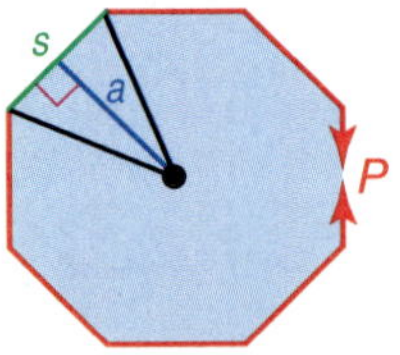

A = area of one triangle • number of triangles

$= \frac{1}{2}$ • base • height • number of triangles

$= \frac{1}{2} \cdot s \cdot a \cdot n$ **Base of triangle is s and height is a. The number of triangles is n.**

$= \frac{1}{2} \cdot a \cdot (n \cdot s)$ **Commutative and Associative Properties**

$= \frac{1}{2} \cdot a \cdot P$ **The perimeter P of the polygon is $n \cdot s$.**

Key Concept Area of a Regular Polygon

Words The area A of a regular n-gon with side length s is one half the product of the apothem a and perimeter P.

Symbols $A = \frac{1}{2}a(ns)$ or $A = \frac{1}{2}aP$

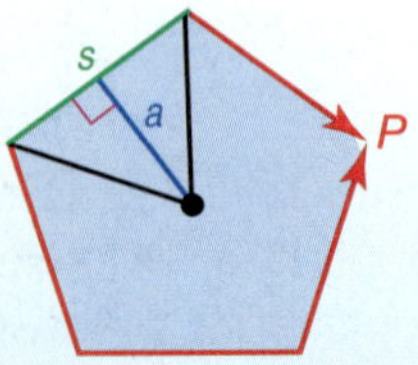

EXAMPLE 3 Use the Formula for the Area of a Regular Polygon

Find the area of each regular polygon. Round to the nearest tenth.

a. regular hexagon

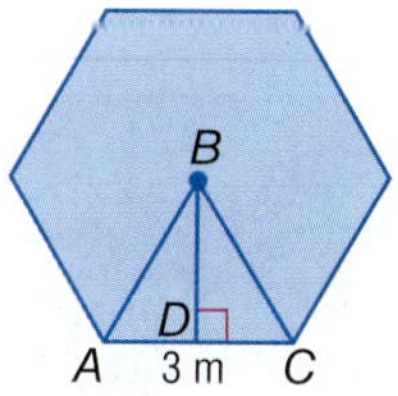

Step 1 Find the measure of a central angle.

A regular hexagon has 6 congruent central angles, so

$m\angle ABC = \frac{360}{6}$ or 60.

Step 2 Find the apothem.

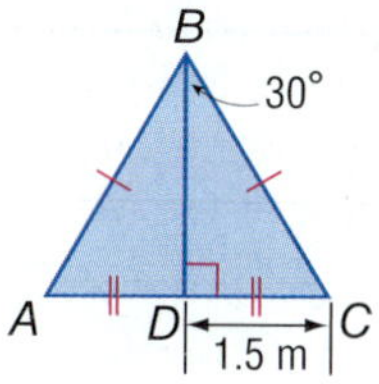

Apothem $\overline{BD}$ is the height of isosceles $\triangle ABC$. It bisects $\angle ABC$, so $m\angle DBC = 30$. It also bisects $\overline{AC}$, so $DC =$ 1.5 meters.

$\triangle BDC$ is a 30°-60°-90° triangle with a shorter leg that measures 1.5 meters, so $BD = 1.5\sqrt{3}$ meters.

Step 3 Use the apothem and side length to find the area.

$A = \frac{1}{2}aP$ **Area of a regular polygon**

$= \frac{1}{2}(1.5\sqrt{3})(18)$ **$a = 1.5\sqrt{3}$ and $P = 6(3)$ or 18**

$\approx 23.4 \text{ m}^2$ **Use a calculator.**

b. regular pentagon

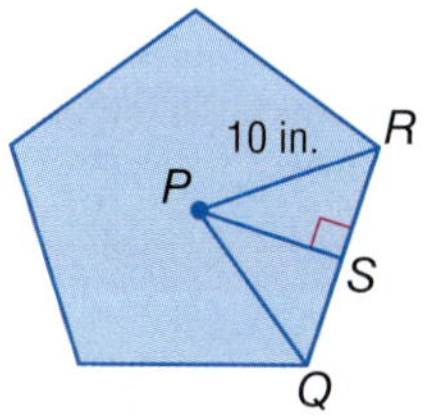

Step 1 A regular pentagon has 5 congruent central angles, so $m\angle QPR = \frac{360}{5}$ or 72.

Step 2 Apothem $\overline{PS}$ is the height of isosceles $\triangle RPQ$. It bisects $\angle RPQ$, so $m\angle RPS = 36$. Use trigonometric ratios to find the side length and apothem of the polygon.

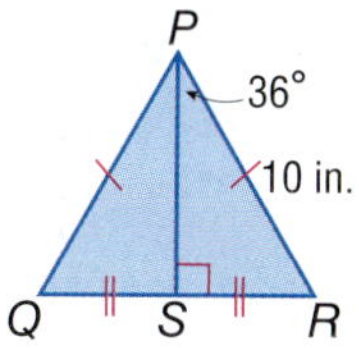

$\sin 36° = \frac{SR}{10}$ $\cos 36° = \frac{PS}{10}$

$10 \sin 36° = SR$ $10 \cos 36° = PS$

$QR = 2SR$ or $2(10 \sin 36°)$. So the pentagon's perimeter is $5 \cdot 2(10 \sin 36°)$ or $10(10 \sin 36°)$. The length of the apothem $\overline{PS}$ is $10 \cos 36°$.

Step 3 $A = \frac{1}{2}aP$ **Area of a regular polygon**

$= \frac{1}{2}(10 \cos 36°)[10(10 \sin 36°)]$ **$a = 10 \cos 36°$, $P = 10(10 \sin 36°)$**

$\approx 237.8 \text{ in}^2$ **Use a calculator.**

Guided Practice

3A.

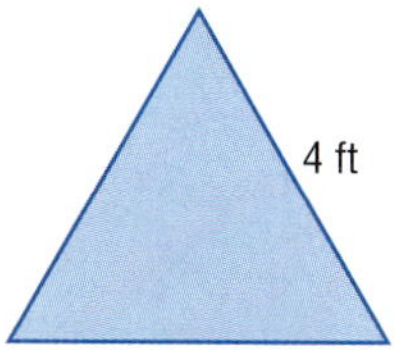

3B.

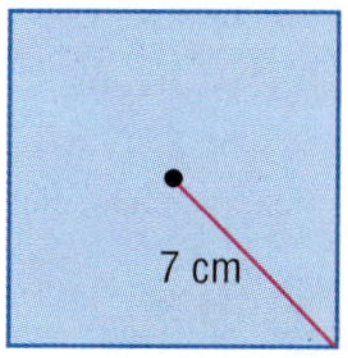

3C.

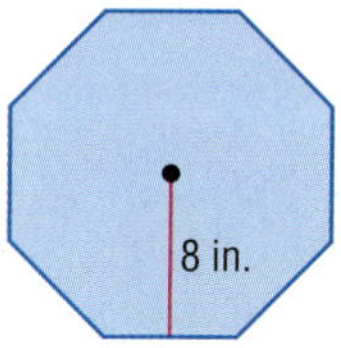

Personal Tutor glencoe.com

StudyTip

Altitude of an Isosceles Triangle The altitude of an isosceles triangle from its vertex to its base is also an angle bisector and median of the triangle.

Areas of Composite Figures A **composite figure** is a figure that can be separated into regions that are basic figures, such as triangles, rectangles, trapezoids, and circles. To find the area of a composite figure, find the area of each basic figure and then use the Area Addition Postulate.

Real-World Link

The first miniature golf course was built in Pinehurst, North Carolina, on a private estate owned by James Barber. There are currently between 5000 and 7500 miniature golf courses in the United States.

Source: Miniature Golf Association of the United States

EXAMPLE 4 Find the Area of a Composite Figure by Adding

MINIATURE GOLF The dimensions of a putting green at a miniature golf course are shown. How many square feet of carpet are needed to cover this green?

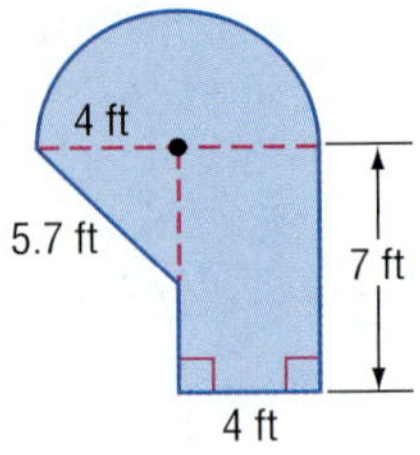

The area to be carpeted can be separated into a rectangle with a length of 4 feet and a width of 7 feet, a right triangle with a hypotenuse of 5.7 feet and a leg measuring 4 feet, and a semicircle with a radius of 4 feet.

Using the Pythagorean Theorem, the other leg of the right triangle is $\sqrt{5.7^2 - 4^2}$ or about 4.1 feet.

Area of green = **area of rectangle** + **area of triangle** + **area of semicircle.**

$$= \ell \cdot w + \frac{1}{2} \cdot b \cdot h + \frac{180}{360} \cdot \pi \cdot r^2$$

$$\approx 4 \cdot 7 + \frac{1}{2} \cdot 4 \cdot 4.1 + \frac{180}{360} \cdot \pi \cdot 4^2$$

$$\approx 28 + 8.2 + 8\pi \text{ or about } 61.3 \text{ ft}^2$$

So, about 62 square feet of carpet is needed.

Guided Practice

Find the area of each figure. Round to the nearest tenth if necessary.

4A.

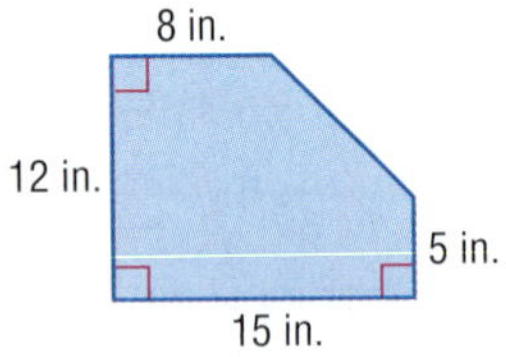

4B.

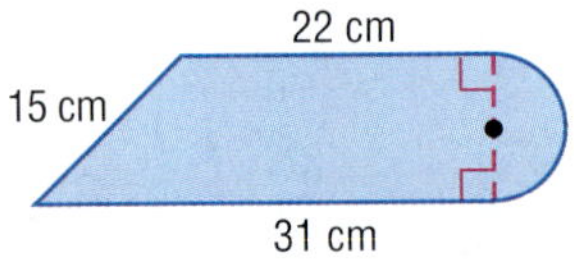

Personal Tutor glencoe.com

The areas of some figures can be found by subtracting the areas of basic figures.

EXAMPLE 5 Find the Area of a Composite Figure by Subtracting

Find the area of the figure. Round to the nearest tenth if necessary.

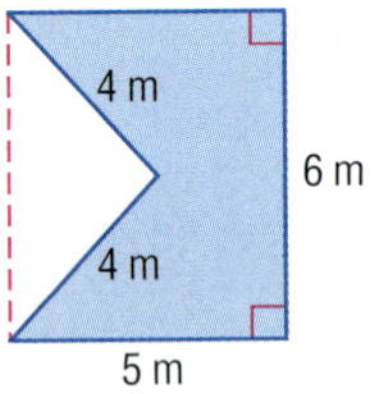

To find the area of the figure, subtract the area of the triangle from the area of the rectangle.

Using the Pythagorean Theorem, the height h of the triangle is $\sqrt{4^2 - 3^2}$ or $\sqrt{7}$ meters.

Area of figure = **Area of rectangle** − **Area of triangle**

$$= b \cdot h - \frac{1}{2}bh$$

$$= 5 \cdot 6 - \frac{1}{2}(6)(\sqrt{7})$$

$$\approx 30 - 7.9 \text{ or about } 22.1 \text{ m}^2$$

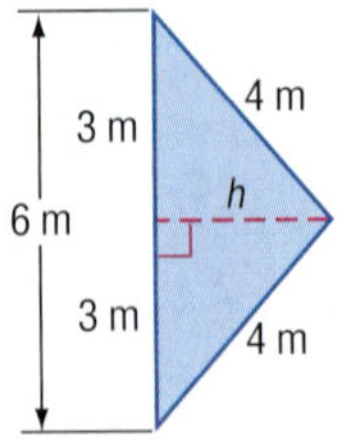

Guided Practice

5A.

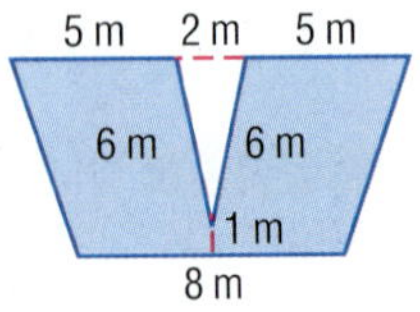

5B.

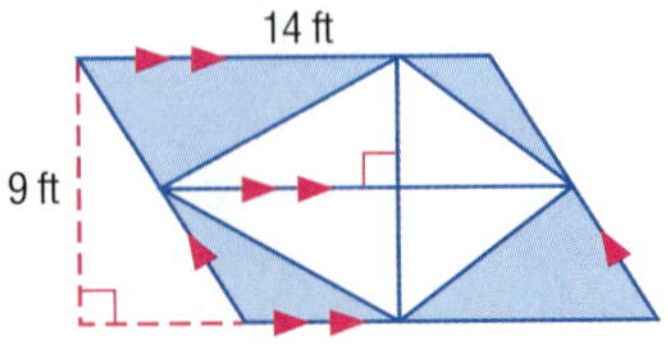

Personal Tutor glencoe.com

Check Your Understanding

Example 1
p. 791

1. In the figure, square $ABDC$ is inscribed in $\odot F$. Identify the center, a radius, an apothem, and a central angle of the polygon. Then find the measure of a central angle.

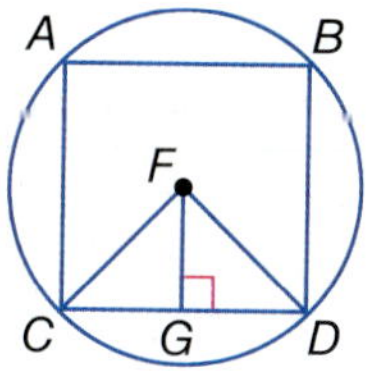

Examples 2 and 3
pp. 792–793

Find the area of each regular polygon. Round to the nearest tenth.

2.

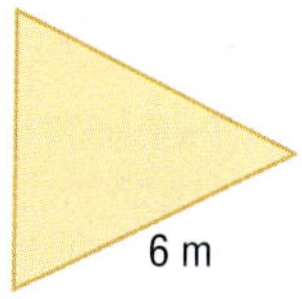

3.

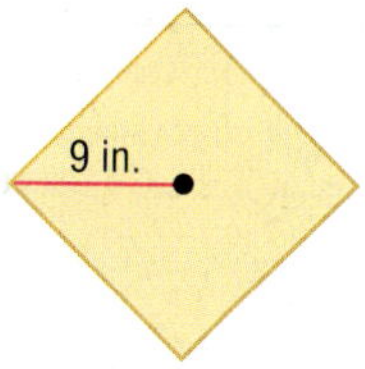

4. **POOLS** Kenton's job is to cover the community pool during fall and winter. Since the pool is in the shape of an octagon, he needs to find the area in order to have a custom cover made. If the pool has the dimensions shown at the right, what is the area of the pool?

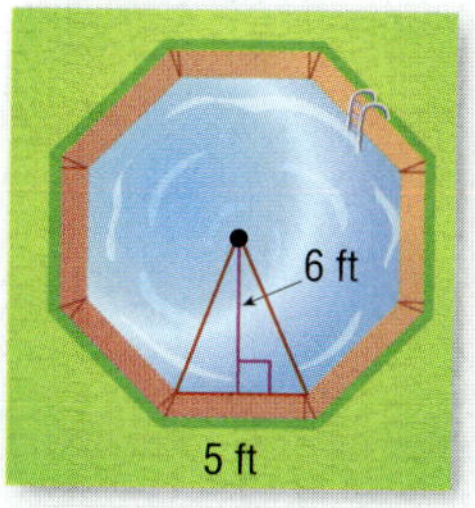

Examples 4 and 5
p. 794

Find the area of each figure. Round to the nearest tenth if necessary.

5.

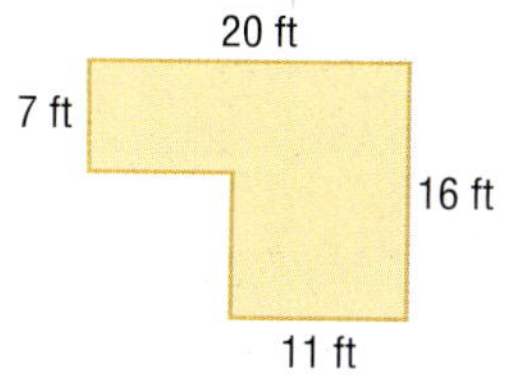

6.

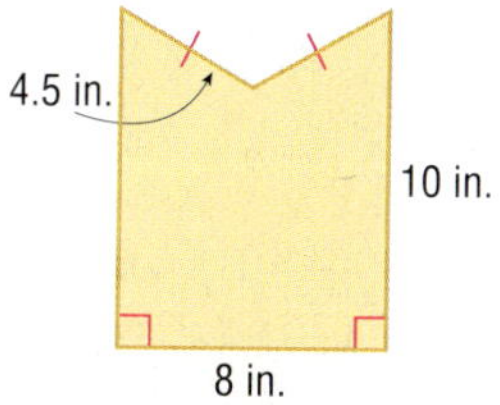

7 **BASKETBALL** The basketball court in Jeff's school is painted as shown.

a. What area of the court is blue? Round to the nearest square foot.

b. What area of the court is red? Round to the nearest square foot.

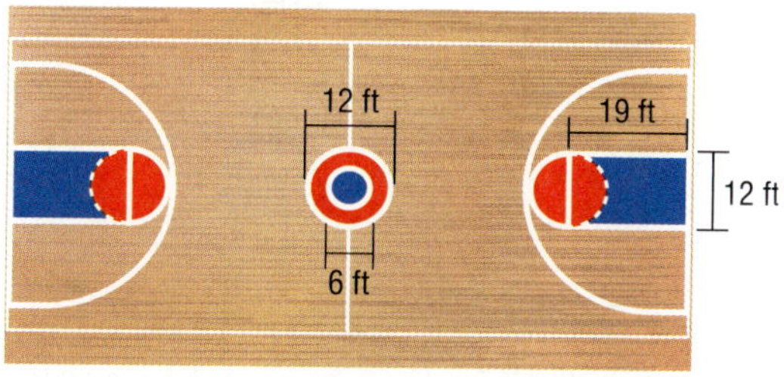

Note: Art not drawn to scale.

Practice and Problem Solving

= **Step-by-Step Solutions** begin on page R20.
Extra Practice begins on page 969.

Example 1
p. 791

In each figure, a regular polygon is inscribed in a circle. Identify the center, a radius, an apothem, and a central angle of each polygon. Then find the measure of a central angle.

8.

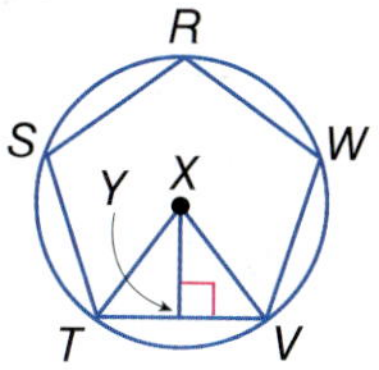

9.

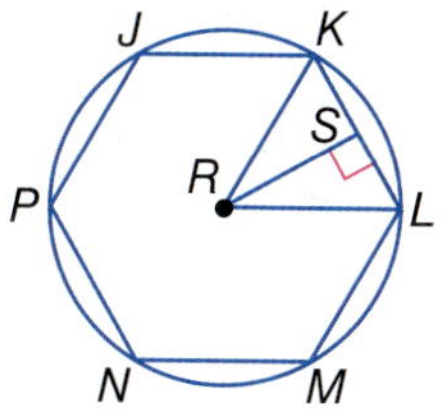

Examples 2 and 3
pp. 792–793

Find the area of each regular polygon. Round to the nearest tenth.

10.

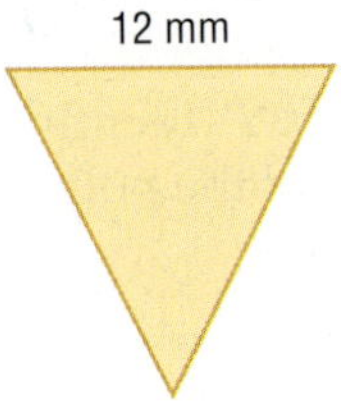

11.

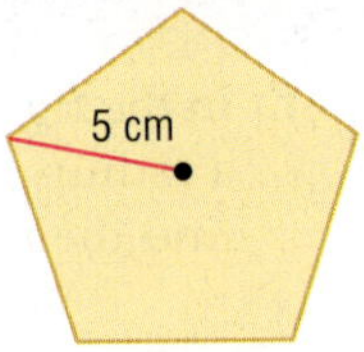

12.

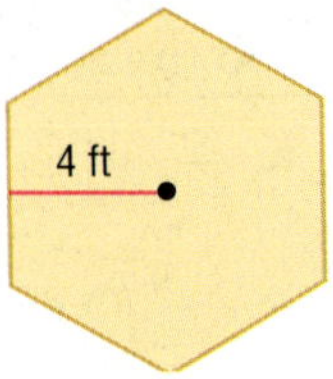

13.

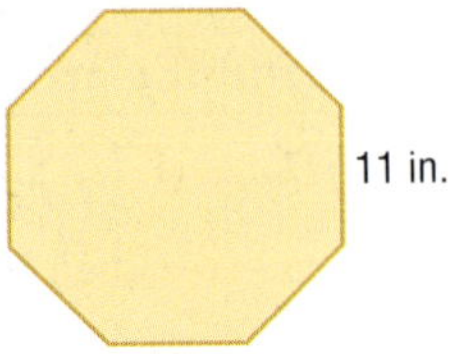

Example 4
p. 794

14. **CARPETING** Ignacio is helping his father install carpeting. Use the floor plan shown to find the area of floor to be carpeted.

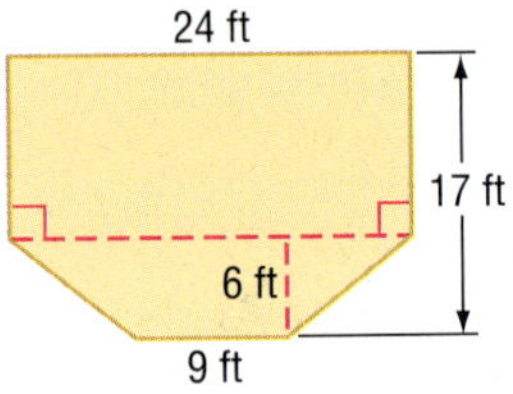

Examples 4 and 5
p. 794

Find the area of each figure. Round to the nearest tenth if necessary.

15

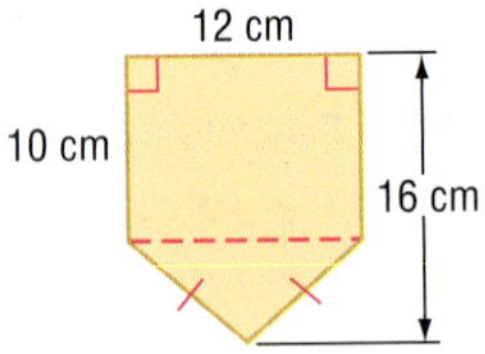

16.

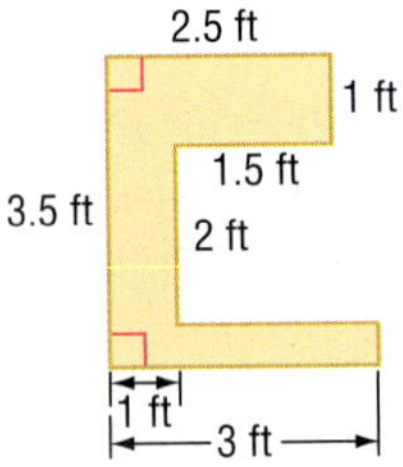

17.

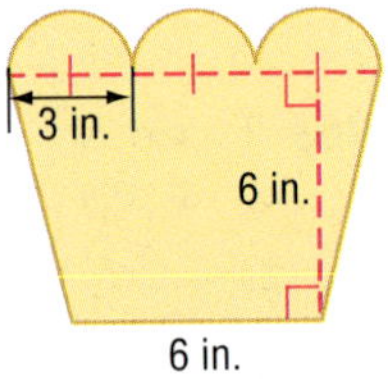

18.

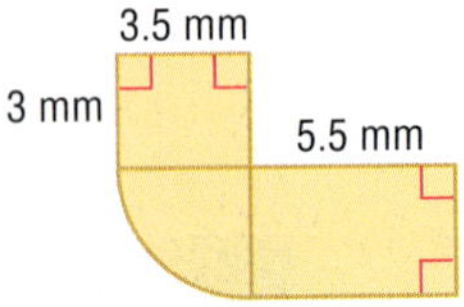

19.

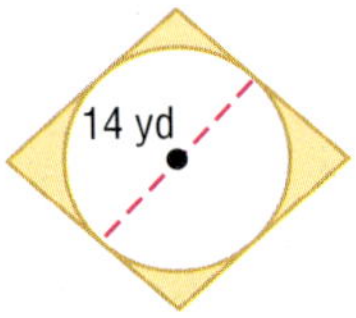

20.

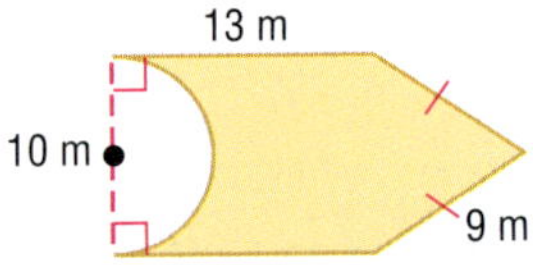

21. **CRAFTS** Latoya's greeting card company is making envelopes for the card from the pattern shown.

a. What area of paper is needed to make one envelope? Round to the nearest tenth.

b. If Latoya orders sheets of paper that are 2 feet by 4 feet, how many cards can she make per sheet?

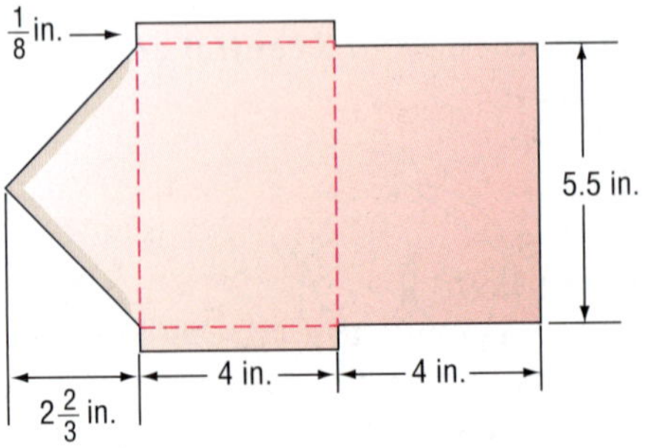

Find the area of each shaded region. Round to the nearest tenth.

22.

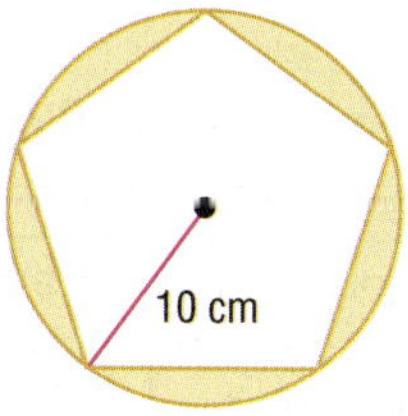

23.

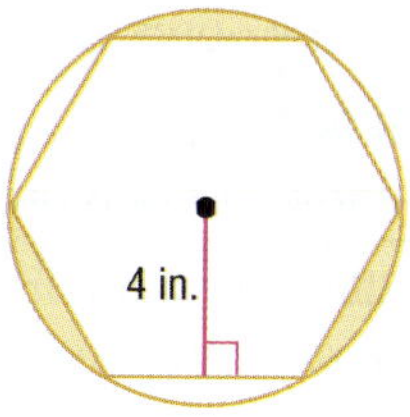

24.

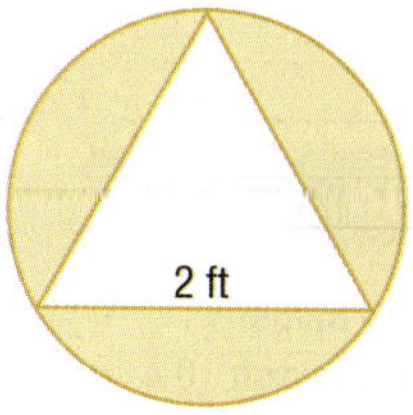

Real-World Link

Prices for apartments and houses are often compared using the price per square foot. Manhattan, New York, is one of the most expensive places to buy real estate, with an average price per square foot of $1013.

Source: Real Estate Board of New York

25. CONSTRUCTION Find the area of the bathroom floor in the apartment floor plan.

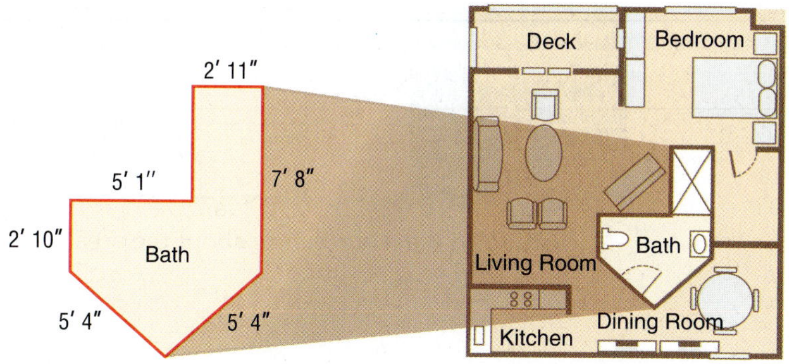

Use a compass and ruler to draw each figure. Then find the area of the figure. Round to the nearest tenth.

26. A rectangle that is 3 inches long and two inches wide shares each of its shorter sides with a semicircle with a radius of 1 inch.

27. A regular hexagon with a side length of 30 millimeters has a square that shares one of its sides removed from the interior of the hexagon.

28. An isosceles triangle with a base of 5 centimeters and a height of 2 centimeters is removed from a trapezoid that has one base of 6 centimeters, a second base of 4 centimeters, and a height of 2 centimeters.

Find the area of each shaded region. Round to the nearest tenth.

29

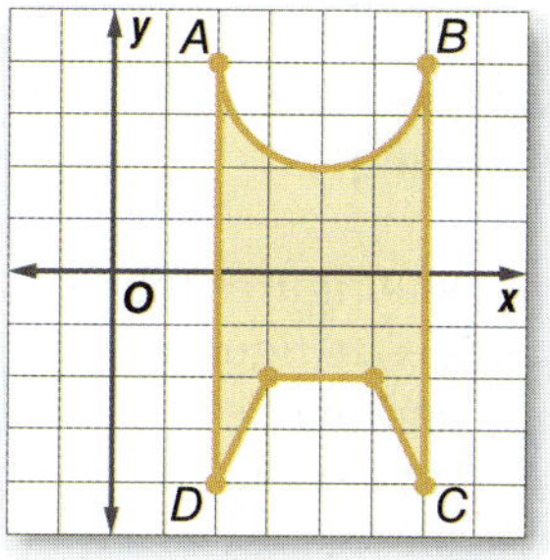

30.

31.

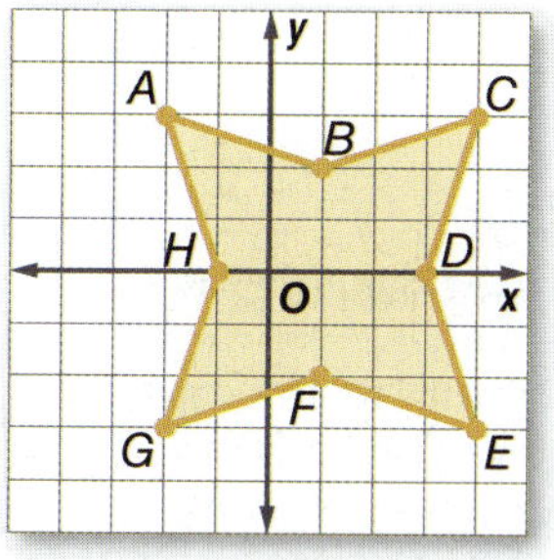

32. Find the total area of the shaded regions. Round to the nearest tenth.

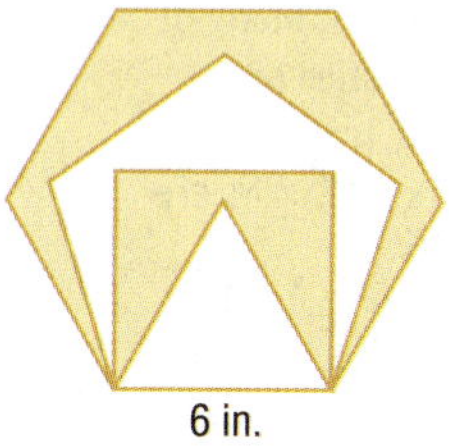

33. CHANGING DIMENSIONS Calculate the area of an equilateral triangle with a perimeter of 3 inches. Calculate the areas of a square, a regular pentagon, and a regular hexagon with perimeters of 3 inches. How does the area of a regular polygon with a fixed perimeter change as the number of sides increases?

34. **MULTIPLE REPRESENTATIONS** In this problem, you will investigate the areas of polygons inscribed in circles.

a. **GEOMETRIC** Draw a circle and inscribe a square. Repeat twice, inscribing a pentagon and a hexagon.

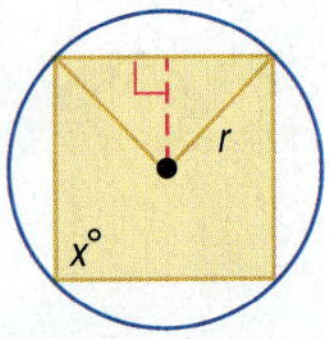

b. **ALGEBRAIC** Use the inscribed polygons from part a to develop a formula for the area of an inscribed polygon in terms of radius r, angle measure x, and number of sides n.

c. **TABULAR** Use the formula you developed in part **b** to complete the table below. The area you calculate will be in terms of r. Round to the nearest hundredth.

Number of Sides, n	4	5	6	8	10	20	50	100
Interior Angle Measure, x								
Area of Inscribed Polygon								

d. **VERBAL** Make a conjecture about the area of an inscribed polygon as the number of sides increases.

H.O.T. Problems Use Higher-Order Thinking Skills

35. **ERROR ANALYSIS** Chloe and Flavio want to find the area of the hexagon shown. Is either of them correct? Explain your reasoning.

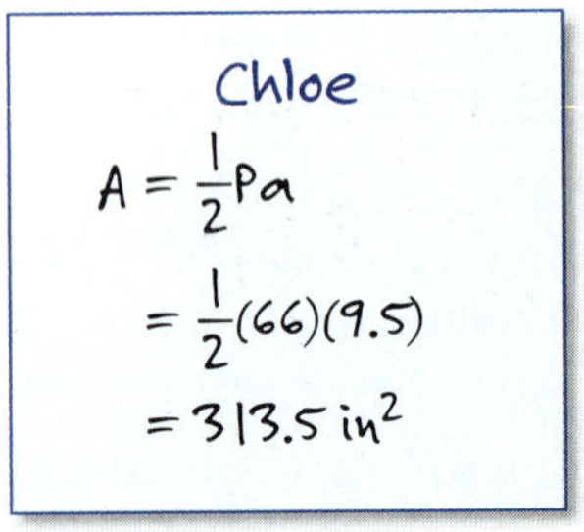

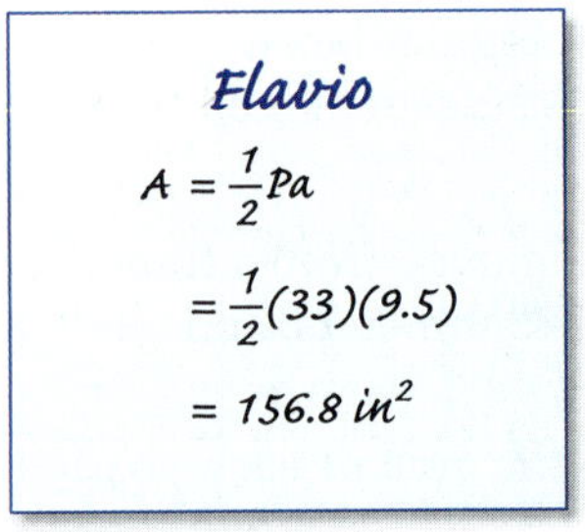

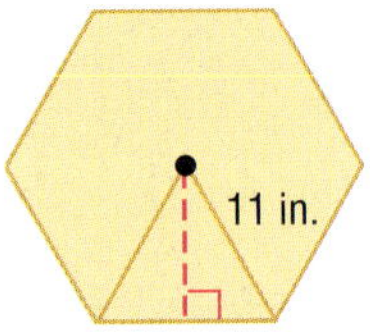

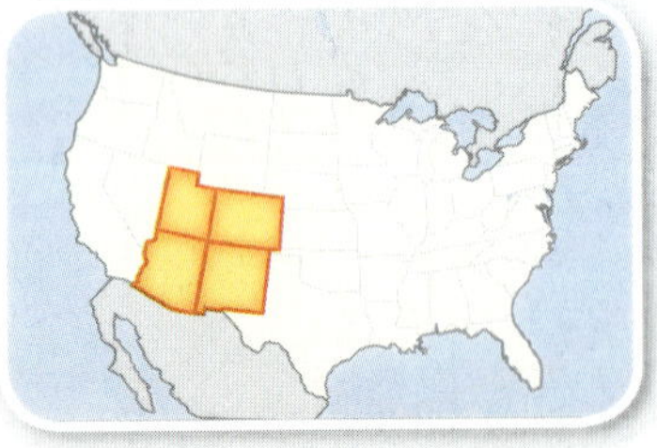

Real-World Link

Many of the western states are shaped like polygons. Nevada, Utah, Wyoming, Colorado, and New Mexico have particularly straight borders. Many of the eastern states are bordered by rivers, so some of their borders are not straight.

36. **CHALLENGE** Using the map shown, estimate the area of Nevada. Explain your reasoning.

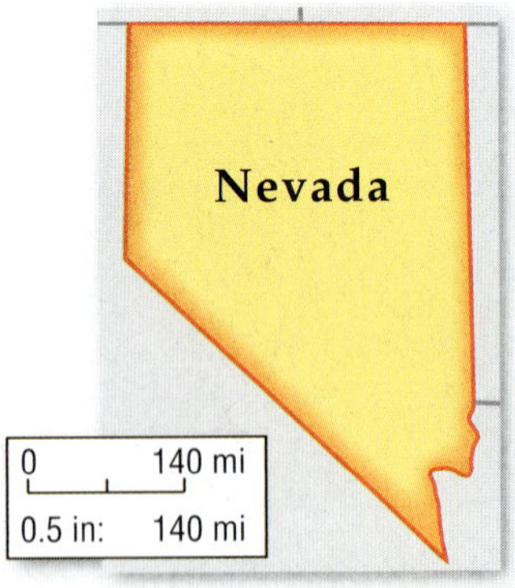

37. **OPEN ENDED** Draw a pair of composite figures that have the same area. Make one composite figure out of a rectangle and a trapezoid, and make the other composite figure out of a triangle and a rectangle. Show the area of each basic figure.

38. **REASONING** Is enough information provided to find the area of the shaded figure? If so, what is the area to the nearest square millimeter? If not, what additional information is needed? Explain.

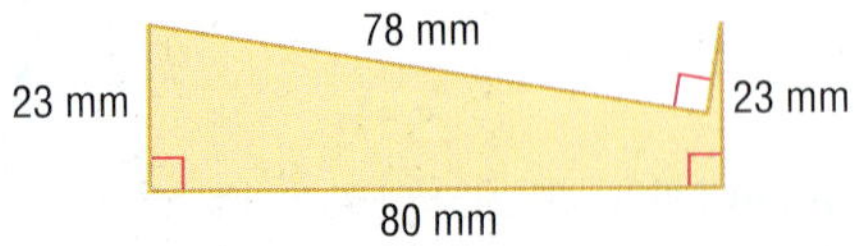

39. **WRITING IN MATH** Explain how to find the area of a regular octagon if you are only given the distance from the center of the figure to a vertex.

Standardized Test Practice

40. Which polynomial best represents the area of the regular pentagon shown below?

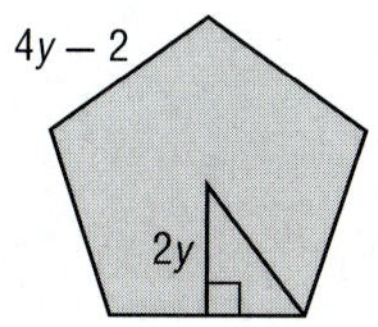

A $10y^2 - 5$
B $10y^2 + 5y$
C $20y^2 + 10$
D $20y^2 - 10y$

41. What is $27^{-\frac{2}{3}}$ in radical form?

F $\frac{1}{(\sqrt[3]{27})^2}$
G $(\sqrt[3]{27})^2$
H $\frac{1}{(\sqrt{27})^2}$
J $(\sqrt{27})^3$

42. **SHORT RESPONSE** Find the area of the shaded figure in square inches. Round to the nearest tenth.

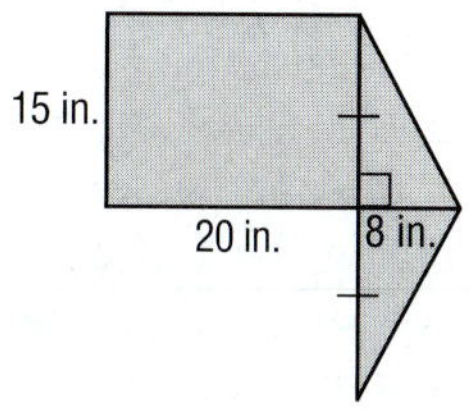

43. **SAT/ACT** If the $\cos\theta = -\frac{12}{13}$ and $\pi < \theta < \frac{3\pi}{2}$, what is the value of $\tan\theta$?

A $-\frac{13}{12}$
B $-\frac{5}{13}$
C $\frac{5}{12}$
D $\frac{12}{5}$

Spiral Review

Find the indicated measure. Round to the nearest tenth. (Lesson 11-3)

44. The area of a circle is 95 square feet. Find the radius.

45. Find the area of a circle whose radius is 9 centimeters.

46. The area of a circle is 256 square inches. Find the diameter.

47. Find the area of a circle whose diameter is 25 millimeters.

Find the area of each trapezoid, rhombus, or kite. (Lesson 11-2)

48.

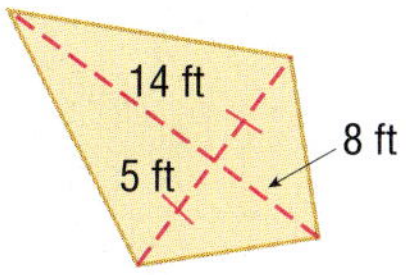

49.

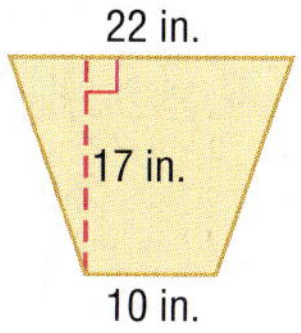

50.

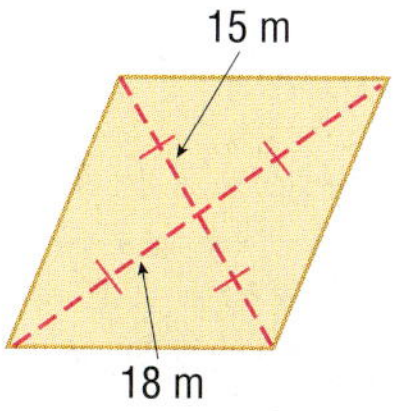

$\overline{EC}$ **and** $\overline{AB}$ **are diameters of ⊙*O*. Identify each arc as a *major arc*, *minor arc*, or *semicircle* of the circle. Then find its measure.** (Lesson 10-2)

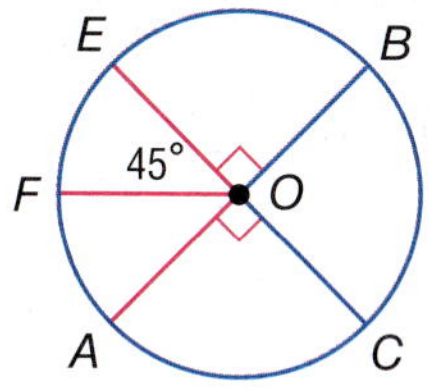

51. $m\widehat{ACB}$

52. $m\widehat{EB}$

53. $m\widehat{ACE}$

Skills Review

Each pair of polygons is similar. Find *x*. (Lesson 7-2)

54.

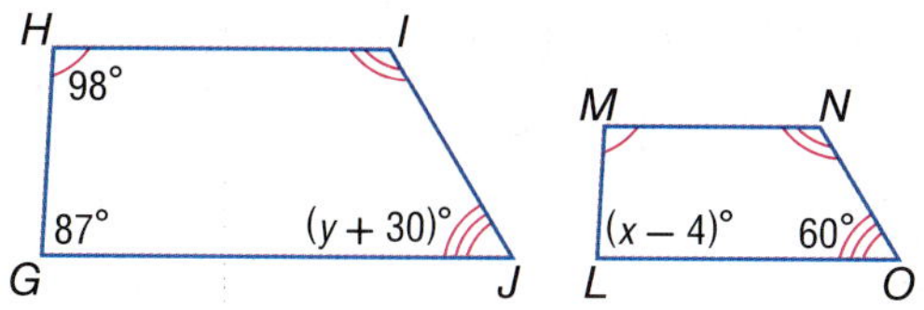

55.

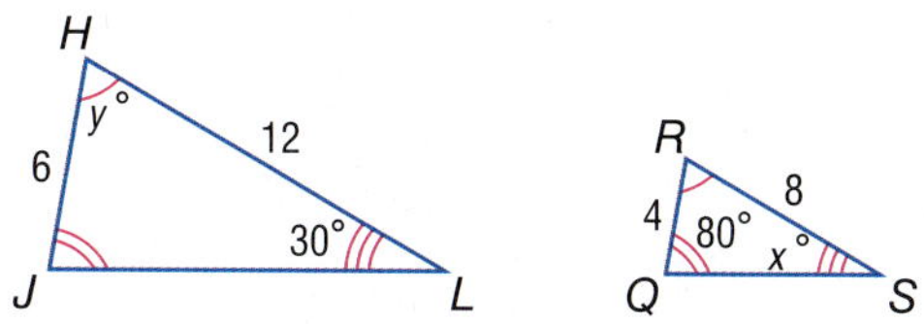

Geometry Lab
Areas of Irregular Figures

In Lesson 11-4, you found areas of composite figures by separating them into basic figures, finding the area of each basic figure, and finding the sum of the areas. You can also find the area of an irregular figure by using the same method. One way to approximate the shape of an irregular figure is to divide the figure into rectangles of equal width. In the branch of mathematics called calculus, this technique is called a *Reimann sum*, and is illustrated in the graph at the right.

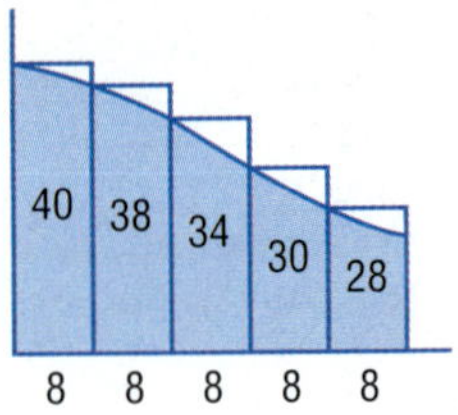

ACTIVITY 1 Estimate Area by Adding

Estimate the area of the figure.

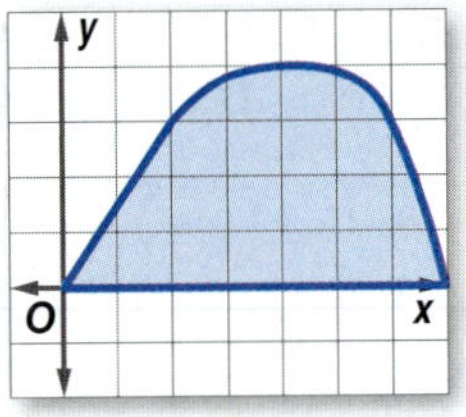

Step 1 Separate the irregular figure into rectangles of equal width.

Step 2 Find the area of each rectangle.

$A_1 = \ell w$ **Area of rectangle 1**

$= (1)(1)$ **$\ell = 1$ and $w = 1$**

$= 1$ **Simplify.**

Step 3 Add the areas of all of the rectangles.

$A_{\text{Total}} \approx A_1 + A_2 + A_3 + A_4 + A_5 + A_6 + A_7$

$\approx 1 + 3 + 4 + 4 + 4 + 4 + 3$ or 23 units^2

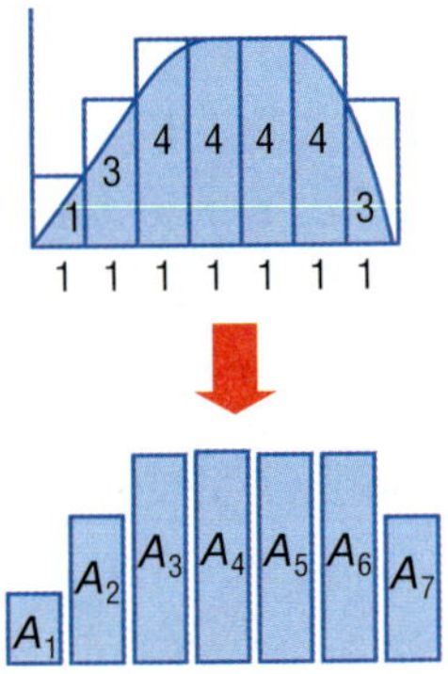

Step 1

ACTIVITY 2 Estimate Area by Counting

Estimate the area of the figure.

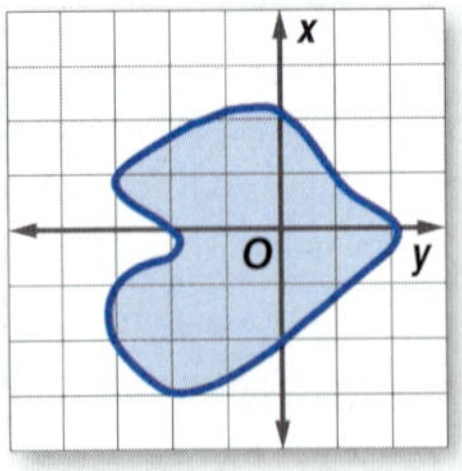

Step 1 Count the number of whole and partial squares inside the figure.

Step 2 Add the number of whole and partial squares.

$A \approx$ whole squares + partial squares

$\approx 12 + \frac{1}{2}(9)$ or 16.5 units^2

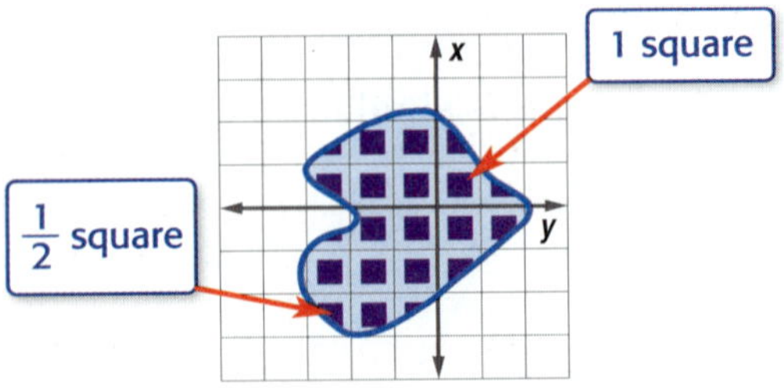

ACTIVITY 3 Estimate Area by Subtracting

Estimate the area of the figure.

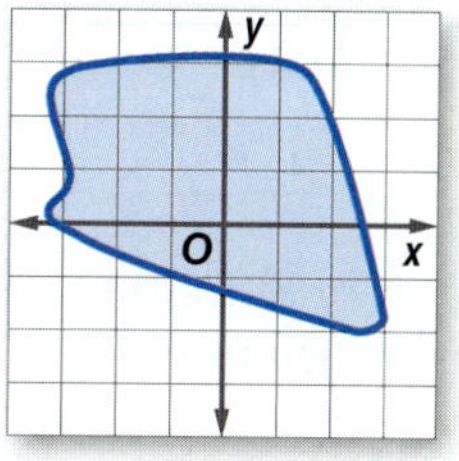

Step 1 Draw a rectangle around the irregular figure.

Step 2 Find the area of the rectangle.

$A_R = (6)(5)$ **$\ell = 6$ and $w = 5$**

$= 30$ **Simplify.**

Step 3 Find the areas of the unshaded triangles.

$A_1 \approx \frac{1}{2}(5)(2)$ **$b = 5$ and $h = 2$**

≈ 5 **Simplify.**

$A_2 \approx \frac{1}{2}(5)(1)$ **$b = 5$ and $h = 1$**

≈ 2.5 **Simplify.**

Step 4 Subtract the areas of the triangles from the area of the rectangle.

$A_{Total} \approx A_R - A_1 - A_2$ **Total area**

$\approx 30 - 5 - 2.5$ or 22.5 units^2 **Substitute.**

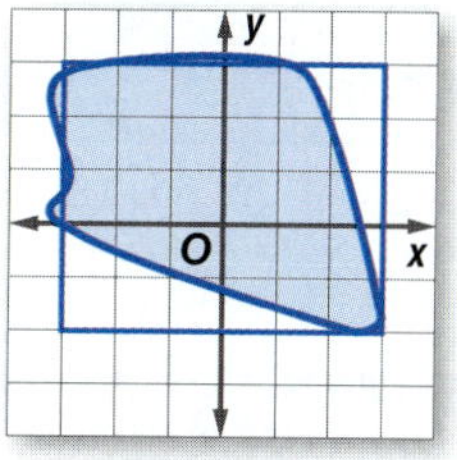

Step 1

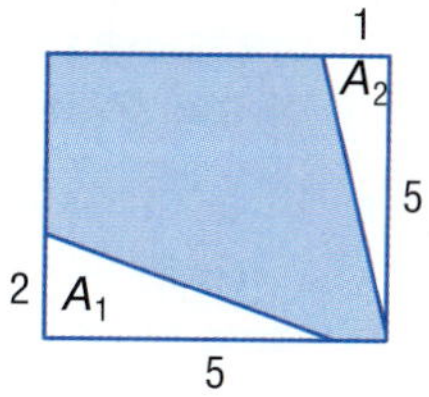

Step 2

Exercises

Estimate the area of each irregular figure.

1.

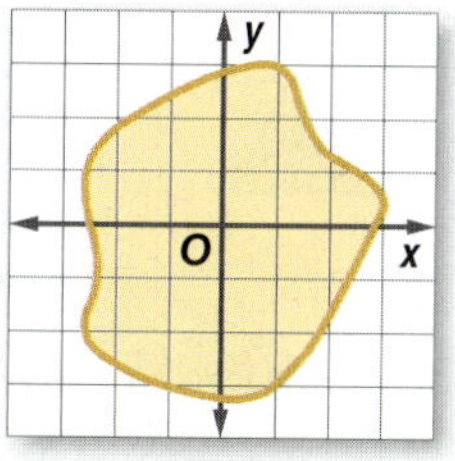

2.

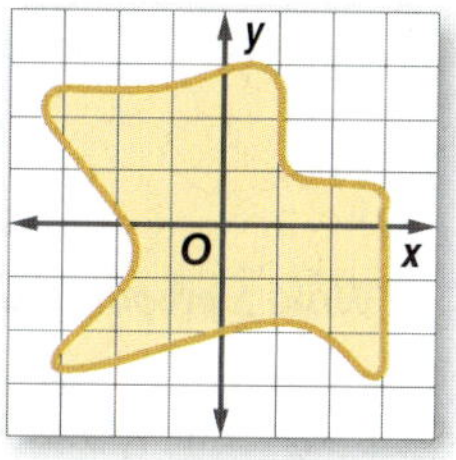

3.

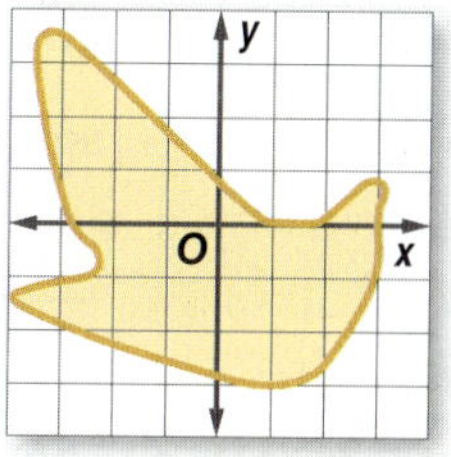

4. CHALLENGE The figure at the right is a graph of $y \leq \tan x$.

a. Estimate the shaded area in terms of units.

b. How would your estimate of the shaded area change if you used rectangles with a smaller width?

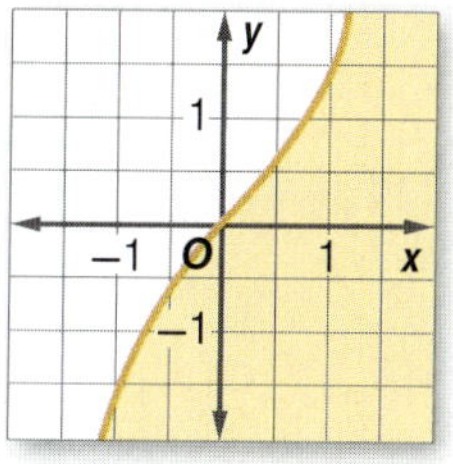

5. WRITING IN MATH Describe another method you could use to estimate the area of an irregular figure on the coordinate plane.

11-5 Areas of Similar Figures

Then

You used scale factors and proportions to solve problems involving the perimeters of similar figures. (Lesson 7-2)

Now

- Find areas of similar figures by using scale factors.
- Find scale factors or missing measures given the areas of similar figures.

Math Online
glencoe.com

Why?

Orlando and Mia like to make scale models of famous buildings. Since the base of a model is geometrically similar to the base of the actual building it represents, their areas are related.

Areas of Similar Figures In Lesson 7-2, you learned that if two polygons are similar, then their perimeters are proportional to the scale factor between them. The areas of two similar polygons share a different relationship.

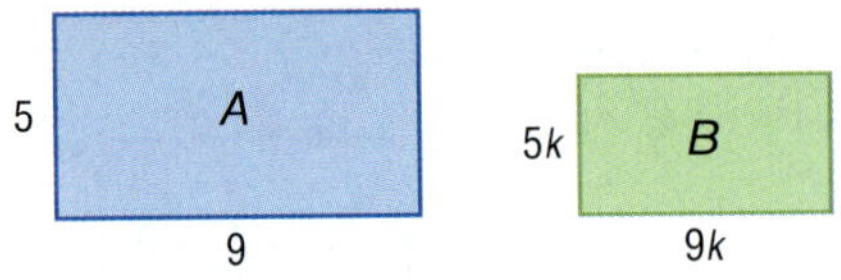

$$\frac{\text{perimeter of figure } B}{\text{perimeter of figure } A} = \frac{28k}{28} \text{ or } k$$

$$\frac{\text{area of figure } B}{\text{area of figure } A} = \frac{45k^2}{45} \text{ or } k^2$$

Theorem 11.1 Areas of Similar Polygons

Words If two polygons are similar, then their areas are proportional to the square of the scale factor between them.

Example If $ABCD \sim FGHJ$, then $\frac{\text{area of } FGHJ}{\text{area of } ABCD} = \left(\frac{FG}{AB}\right)^2$.

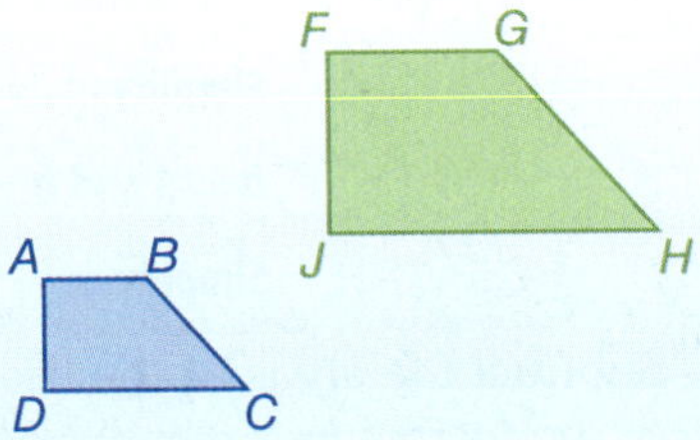

You will prove Theorem 11.1 for triangles in Exercise 22.

EXAMPLE 1 Find Areas of Similar Polygons

If $\triangle JKL \sim \triangle PQR$ and the area of $\triangle JKL$ is 30 square inches, find the area of $\triangle PQR$.

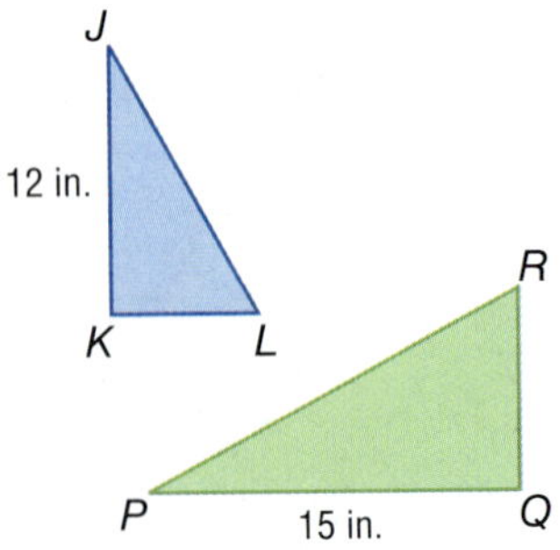

The scale factor between $\triangle PQR$ and $\triangle JKL$ is $\frac{15}{12}$ or $\frac{5}{4}$, so the ratio of their areas is $\left(\frac{5}{4}\right)^2$.

$\frac{\text{area of } \triangle PQR}{\text{area of } \triangle JKL} = \left(\frac{5}{4}\right)^2$ **Write a proportion.**

$\frac{\text{area of } \triangle PQR}{30} = \frac{25}{16}$ **Area of $\triangle JKL = 30$ and $\left(\frac{5}{4}\right)^2 = \frac{25}{16}$**

$\text{area of } \triangle PQR = \frac{25}{16} \cdot 30$ **Multiply each side by 30.**

$\text{area of } \triangle PQR = 46.875$ **Simplify.**

So the area of $\triangle PQR$ is about 46.9 square inches.

Guided Practice

For each pair of similar figures, find the area of the green figure.

1A.

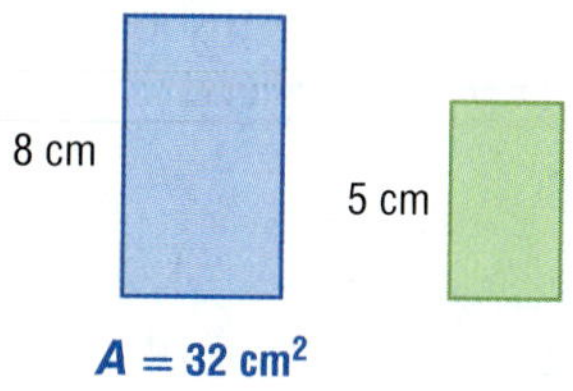

1B.

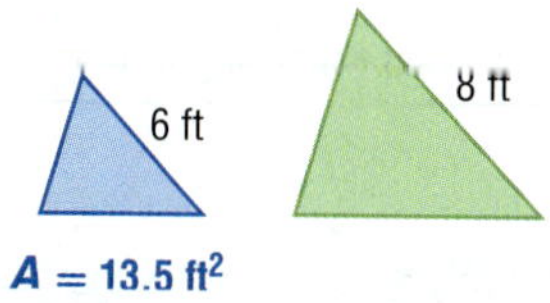

Personal Tutor glencoe.com

Scale Factors and Missing Measures in Similar Figures You can use the areas of similar figures to find the scale factor between them or a missing measure.

EXAMPLE 2 Use Areas of Similar Figures

The area of $\square ABCD$ is 150 square meters. The area of $\square FGHJ$ is 54 square meters. If $\square ABCD \sim \square FGHJ$, find the scale factor of $\square FGHJ$ to $\square ABCD$ and the value of x.

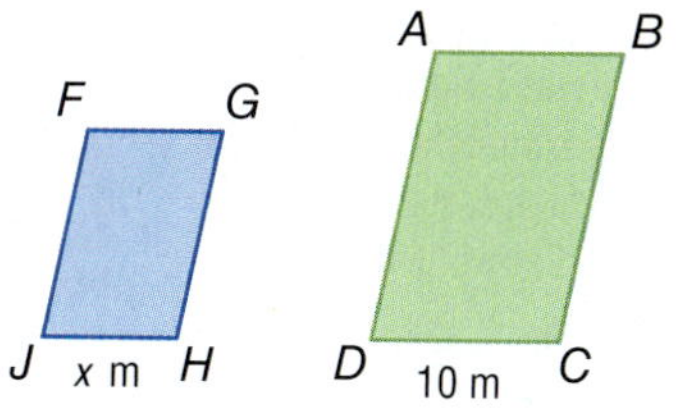

Let k be the scale factor between $\square FGHJ$ and $\square ABCD$.

$\frac{\text{area of } \square FGHJ}{\text{area of } \square ABCD} = k^2$ **Theorem 11.1**

$\frac{54}{150} = k^2$ **Substitution**

$\frac{9}{25} = k^2$ **Simplify.**

$\frac{3}{5} = k$ **Take the positive square root of each side.**

So the scale factor of $\square FGHJ$ to $\square ABCD$ is $\frac{3}{5}$. Use this scale factor to find the value of x.

$\frac{JH}{DC} = k$ **The ratio of corresponding lengths of similar polygons is equal to the scale factor between the polygons.**

$\frac{x}{10} = \frac{3}{5}$ **Substitution**

$x = \frac{3}{5} \cdot 10$ or 6 **Multiply each side by 10.**

CHECK Confirm that $\frac{JH}{DC}$ is equal to the scale factor.

$\frac{JH}{DC} = \frac{6}{10} = \frac{3}{5}$ ✓

Watch Out!

Writing Ratios When finding the ratio of the area of Figure A to the area of Figure B, be sure to write your ratio as $\frac{\text{area of figure A}}{\text{area of figure B}}$.

Guided Practice

For each pair of similar figures, use the given areas to find the scale factor of the blue to the green figure. Then find x.

2A.

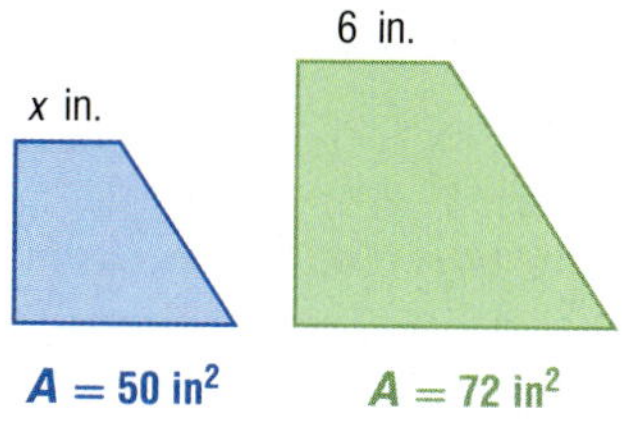

2B.

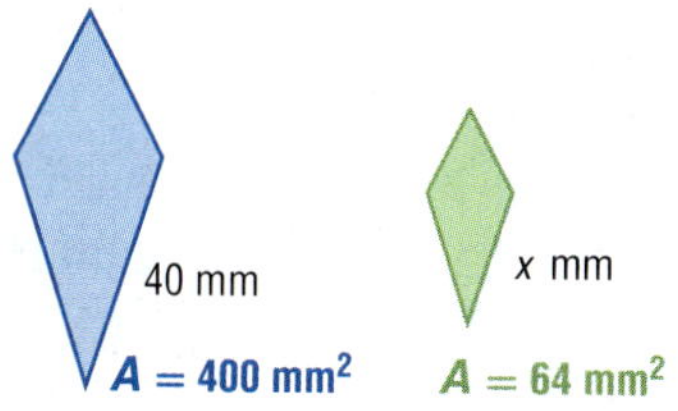

Personal Tutor glencoe.com

ReadingMath

Ratios Ratios can be written in different ways. For example, x to y, $x : y$, and $\frac{x}{y}$ are all representations of the ratio of x and y.

In Lesson 7-2, you learned that if all corresponding angles are congruent and all corresponding sides are proportional, then two polygons are similar. For this reason, all regular polygons with the same number of sides are similar.

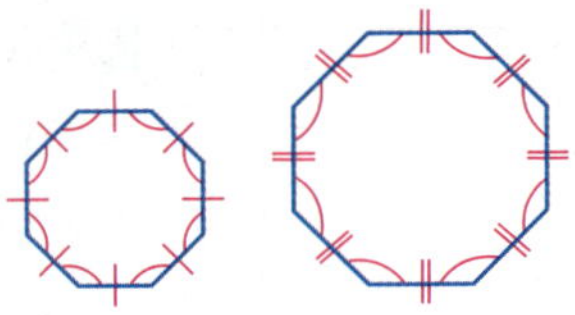

Real-World Link

The Pentagon building, including its center courtyard, occupies approximately 34 acres or 1,481,000 square feet of land. Each outer wall of the regular pentagonal building is 921 feet in length.

Source: U.S. Department of Defense

Real-World EXAMPLE 3 Scale Models

CRAFTS Use the information at the left. Orlando and Mia are making a scale model of the Pentagon. If the area of the base of their model is approximately 50 square inches, about how many times the length of each outer wall of the Pentagon is the length of the outer wall of the model?

Understand All regular pentagons are similar, so the base of the model is similar to the base of the Pentagon. You need to find the scale factor from the Pentagon to their model.

Plan The ratio of the areas of the bases of the two figures is equal to the square of the scale factor between them. Before comparing the two areas, write them so that they have the same units.

Solve Convert the area of the model's base to square feet.

$$50 \cancel{\text{in}^2} \cdot \frac{1 \text{ ft}^2}{144 \cancel{\text{in}^2}} \approx 0.3472 \text{ ft}^2$$

Next, write an equation using the ratio of the two areas in square inches. Let k represent the scale factor between the two bases.

$\frac{\text{area of model}}{\text{area of Pentagon}} = k^2$	**Theorem 11.1**
$\frac{0.3472 \text{ ft}^2}{1{,}481{,}000 \text{ ft}^2} \approx k^2$	**Substitution**
$2.34 \cdot 10^{-7} \approx k^2$	**Simplify using a calculator.**
$4.84 \cdot 10^{-4} \approx k$	**Take the positive square root of each side.**
$0.0005 \approx k$	**Write in standard form.**
$\frac{1}{2000} \approx k$	**Write as a simplified fraction.**

So the model's outer walls are about $\frac{1}{2000}$ the length of each outer wall of the Pentagon.

Check Multiply the area of the Pentagon's base by the square of this scale factor and compare to the given area of the model's base.

$$\frac{1{,}481{,}000 \cancel{\text{ft}^2}}{1} \cdot \frac{144 \text{ in}^2}{1 \cancel{\text{ft}^2}} \cdot \left(\frac{1}{2000}\right)^2 \approx 53 \text{ in}^2$$

This is close to the given area of 50 square inches, so our scale factor is reasonable. ✓

StudyTip

Similar Circles Since all circles have the same shape, all circles are similar. Therefore, the areas of two circles are also related by the square of the scale factor between them.

Guided Practice

3. CRAFTS Miyoki is crocheting two circles. The area of the larger circle is to be 2.5 times the size of the smaller. If the area of the smaller circle is about 50.2 square centimeters, what is the diameter of the larger?

Personal Tutor glencoe.com

Check Your Understanding

Example 1
p. 802

For each pair of similar figures, find the area of the green figure.

1.
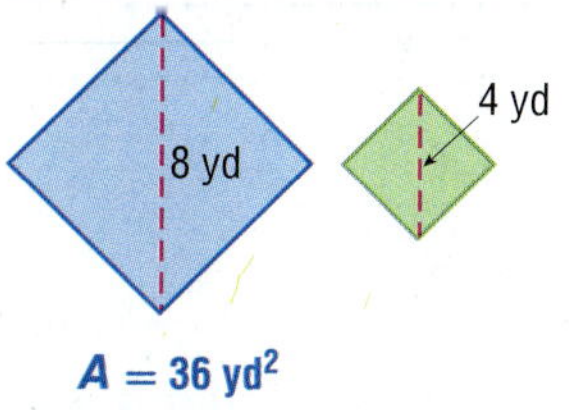

2.
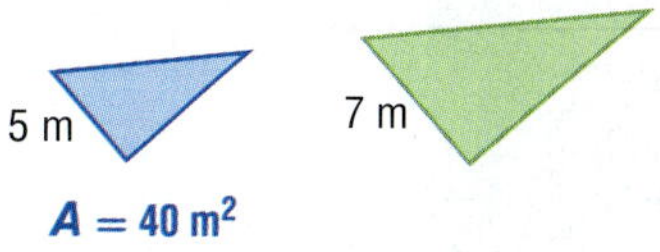

Example 2
p. 803

For each pair of similar figures, use the given areas to find the scale factor from the blue to the green figure. Then find *x*.

3.
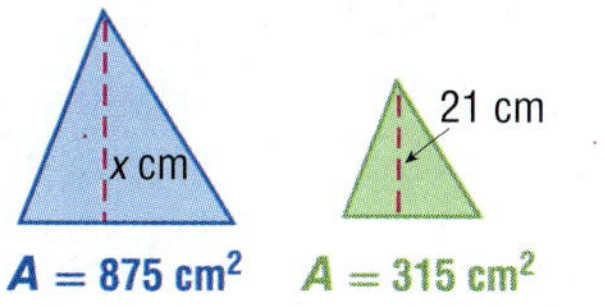

4.
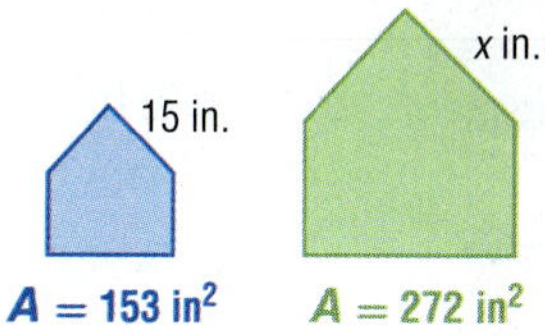

Example 3
p. 804

5. **MEMORIES** Zola has a picture frame that holds all of her school pictures. Each small opening is similar to the large opening in the center. If the center opening has an area of 33 square inches, what is the area of each small opening?

Practice and Problem Solving

● = **Step-by-Step Solutions** begin on page R20.
Extra Practice begins on page 969.

Example 1
p. 802

For each pair of similar figures, find the area of the green figure.

6.
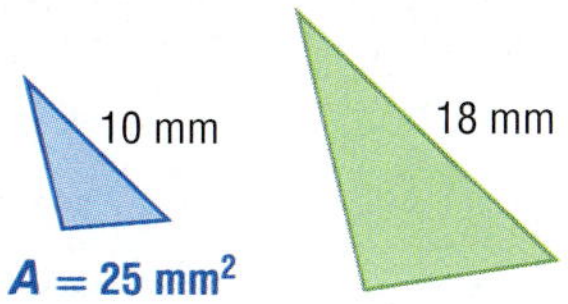

7.
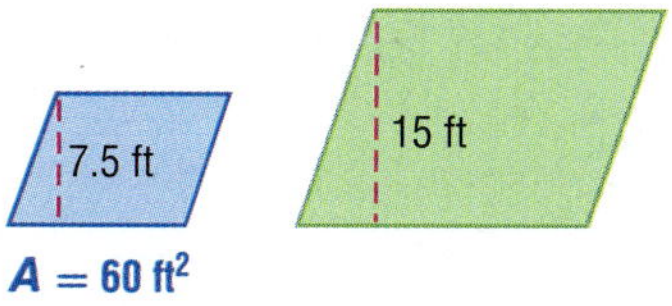

8.
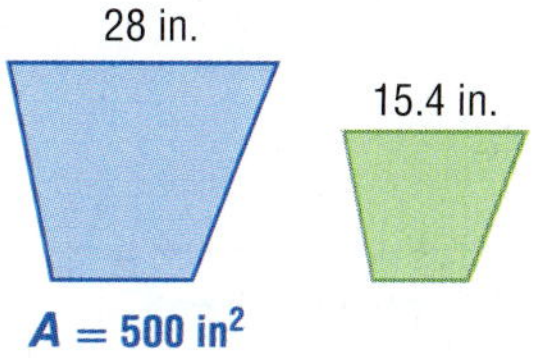

9.
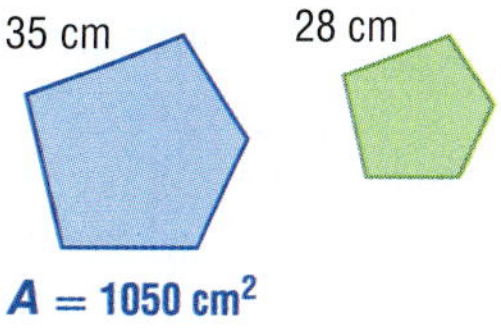

Example 2
p. 803

For each pair of similar figures, use the given areas to find the scale factor of the blue to the green figure. Then find *x*.

10.
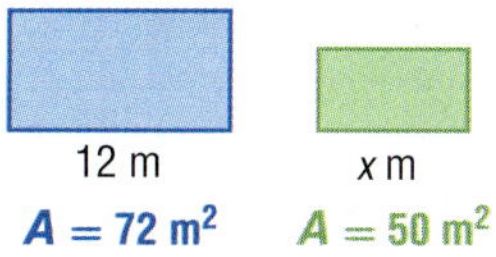

11.
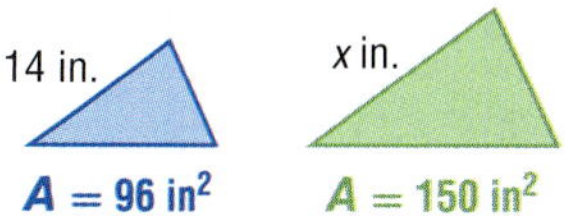

12.
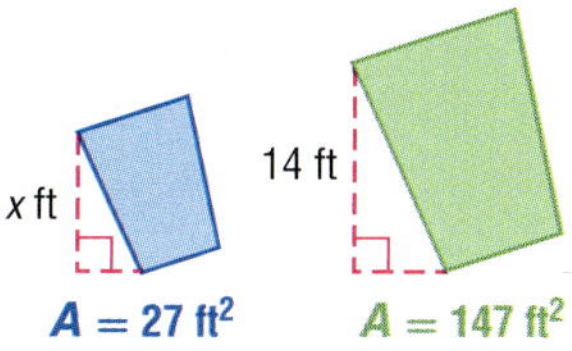

13.
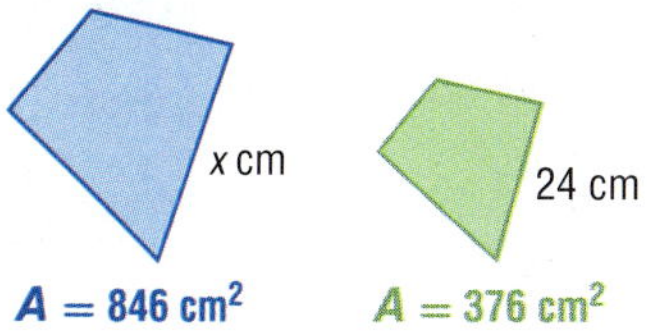

Example 3
p. 804

Real-World Career

Craft Artist

Craft artists create their art by hand to sell or exhibit. They work with a wide variety of materials including textiles, woods, metal, and ceramics.

Most artists receive some type of postsecondary training, and about 63% are self-employed. Craft artists make up about 3% of all artists.

14. CRAFTS Marina crafts unique trivets and other kitchenware. Each trivet is an equilateral triangle. The perimeter of the small trivet is 9 inches, and the perimeter of the large trivet is 12 inches. If the area of the small trivet is about 3.9 square inches, what is the approximate area of the large trivet?

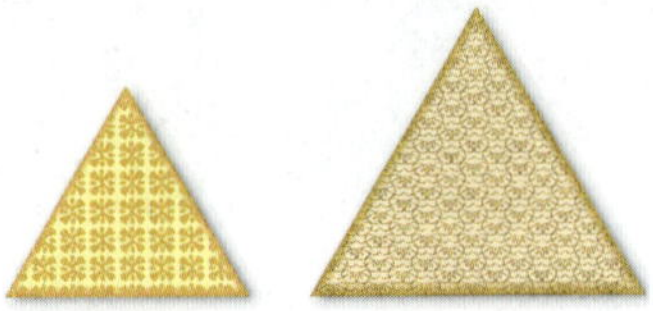

15. BAKING Kaitlyn wants to use one of two regular hexagonal cake pans for a recipe she is making. The side length of the larger pan is 4.5 inches, and the area of the base of the smaller pan is 41.6 square inches.

a. What is the side length of the smaller pan?

b. The recipe that Kaitlyn is using calls for a circular cake pan with an 8-inch diameter. Which pan should she choose? Explain your reasoning.

16. CHANGING DIMENSIONS A polygon has an area of 144 square meters.

a. If the area is doubled, how does each side length change?

b. How does each side length change if the area is tripled?

c. What is the change in each side length if the area is increased by a factor of x?

17. CHANGING DIMENSIONS A circle has a radius of 24 inches.

a. If the area is doubled, how does the radius change?

b. How does the radius change if the area is tripled?

c. What is the change in the radius if the area is increased by a factor of x?

18. HOME IMPROVEMENT Federico's family is putting hardwood floors in the two geometrically similar rooms shown. If the cost of flooring is constant and the flooring for the kitchen cost $2000, what will be the total flooring cost for the two rooms? Round to the nearest hundred dollars.

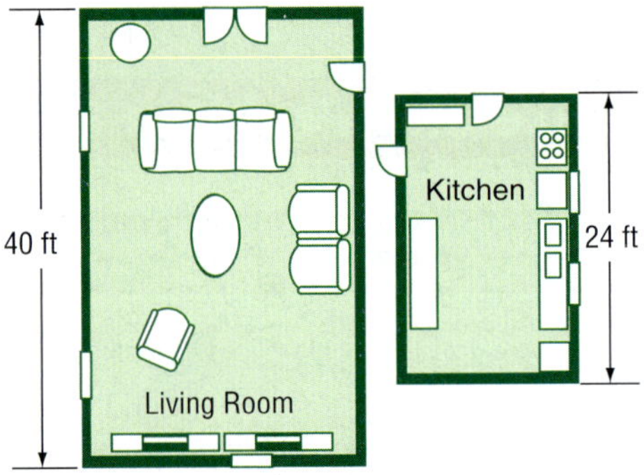

COORDINATE GEOMETRY **Find the area of each figure. Use the segment length given to find the area of a similar polygon.**

19 $J'L' = 3$

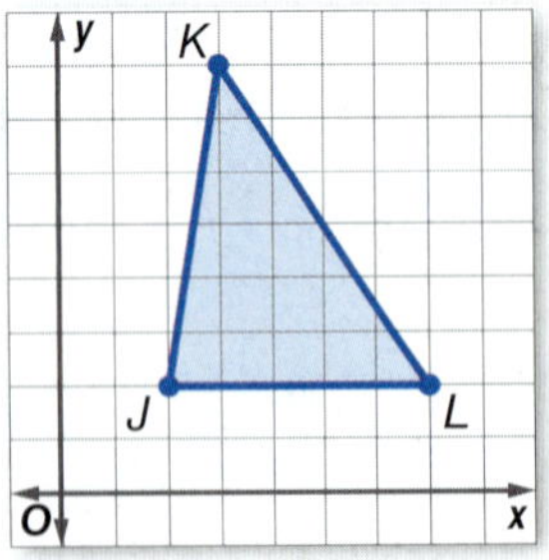

20. $W'X' = 8$

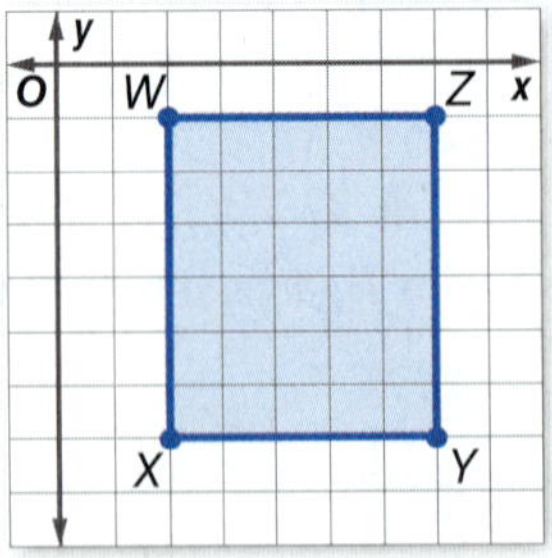

21. $B'C' = 5$

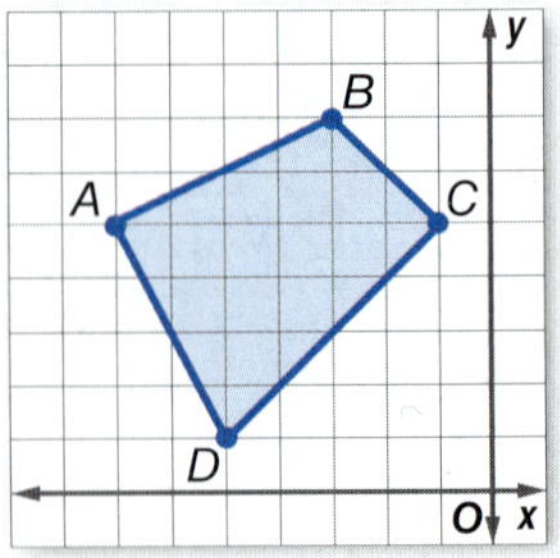

22. PROOF Write a paragraph proof.

Given: $\triangle ABC \sim \triangle XYZ$

Prove: $\dfrac{\text{area of } \triangle ABC}{\text{area of } \triangle XYZ} = \dfrac{a^2}{x^2}$

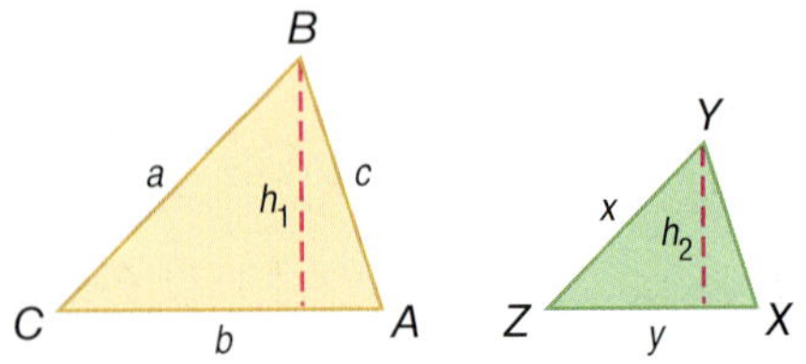

Real-World Link

Tennis is one of the most gender balanced sports, appealing almost equally to men and women.

Source: Nielsen Media Research

23. STATISTICS The graph shows the increase in high school tennis participation from 1995 to 2005.

a. Explain why the graph is misleading.

b. How could the graph be changed to more accurately represent the growth in high school tennis participation?

24. MULTIPLE REPRESENTATIONS In this problem, you will investigate changing dimensions proportionally in three-dimensional figures.

a. TABULAR Copy and complete the table below for each scale factor of a rectangular prism that is 2 inches by 3 inches by 5 inches.

Scale Factor	Length (in.)	Width (in.)	Height (in.)	Volume (in^3)	Ratio of Scaled Volume to Initial Volume
1	3	2	5		
2					
3					
4					
5					
10					

b. VERBAL Make a conjecture about the relationship between the scale factor and the ratio of the scaled volume to the initial volume.

c. GRAPHICAL Make a scatter plot of the scale factor and the ratio of the scaled volume to the initial volume using the STAT PLOT feature on your graphing calculator. Then use the STAT CALC feature to approximate the function represented by the graph.

d. ALGEBRAIC Write an algebraic expression for the ratio of the scaled volume to the initial volume in terms of scale factor k.

H.O.T. Problems

Use **H**igher-**O**rder **T**hinking Skills

25. ERROR ANALYSIS Violeta and Gavin are trying to come up with a formula that can be used to find the area of a circle with a radius r after it has been enlarged by a scale factor k. Is either of them correct? Explain your reasoning.

Violeta	Gavin
$A = k\pi r^2$	$A = \pi(r^2)^k$

26. CHALLENGE If you want the area of a polygon to be x% of its original area, by what scale factor should you multiply each side length?

27. REASONING A regular n-gon is enlarged, and the ratio of the area of the enlarged figure to the area of the original figure is R. Write an equation relating the perimeter of the enlarged figure to the perimeter of the original figure Q.

28. OPEN ENDED Draw a pair of similar figures with areas that have a ratio of 4:1. Explain.

29. WRITING IN MATH Explain how to find the area of an enlarged polygon if you know the area of the original polygon and the scale factor of the enlargement.

Standardized Test Practice

30. $\triangle ABC \sim \triangle PRT$, $AC = 15$ inches, $PT = 6$ inches, and the area of $\triangle PRT$ is 24 square inches. Find the area of $\triangle ABC$.

A 9.6 in^2
B 60 in^2
C 66.7 in^2
D 150 in^2

31. ALGEBRA Which of the following shows $2x^2 - 18xy - 72y^2$ factored completely?

F $(2x - 18y)(x + 4y)$
G $2(x - 9y)(x + 4y)$
H $(2x - 9y)(x + 4y)$
J $2(x - 12y)(x + 3y)$

32. EXTENDED RESPONSE The measures of two complementary angles are represented by $2x + 1$ and $5x - 9$.

a. Write an equation that represents the relationship between the two angles.

b. Find the degree measure of each angle.

33. SAT/ACT Which of the following are the values of x for which $(x + 5)(x - 4) = 10$?

A -5 and 4
B 5 and 6
C -6 and 5
D 6 and -5

Spiral Review

34. In the figure, square $WXYZ$ is inscribed in $\odot R$. Identify the center, a radius, an apothem, and a central angle of the polygon. Then find the measure of a central angle. (Lesson 11-4)

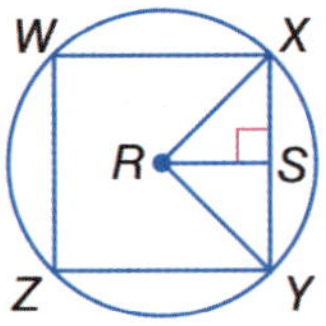

Find the area of the shaded region. Round to the nearest tenth. (Lesson 11-4)

35.

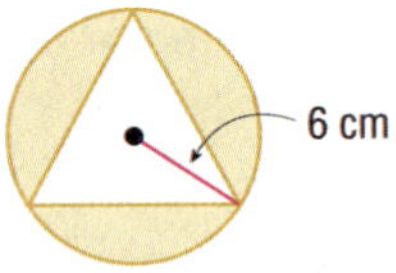

36.

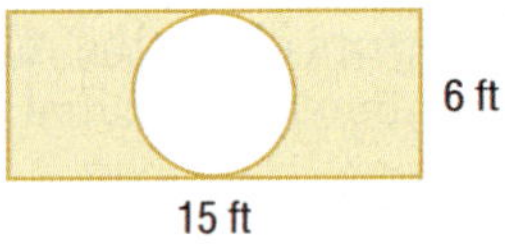

37.

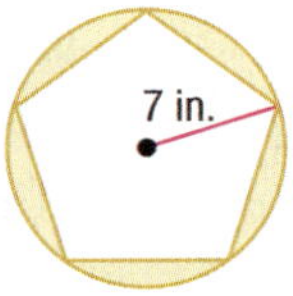

Find each measure. (Lesson 10-6)

38. $m\angle 5$

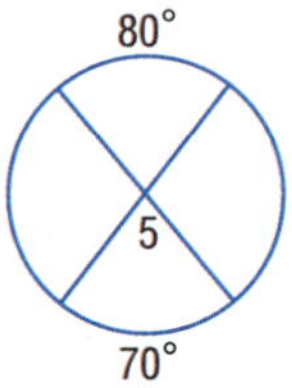

39. $m\angle 6$

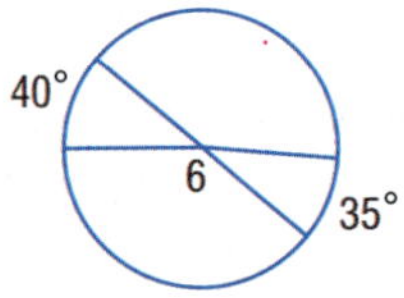

40. $m\angle 7$

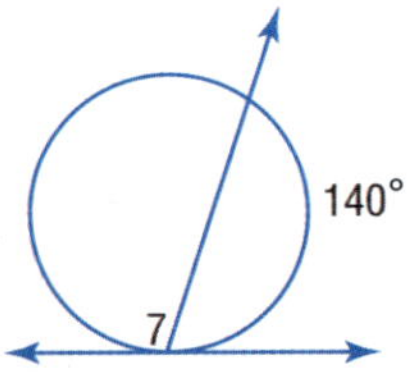

41. State whether the figure has *plane* symmetry, *axis* symmetry, *both*, or *neither*. (Lesson 9-5)

42. YEARBOOKS Tai resized a photograph that was 8 inches by 10 inches so that it would fit in a 4-inch by 4-inch area on a yearbook page. (Lesson 7-7)

a. Find the maximum dimensions of the reduced photograph.

b. What is the percent of reduction of the length?

Skills Review

Refer to the figure at the right to identify each of the following. (Lesson 3-1)

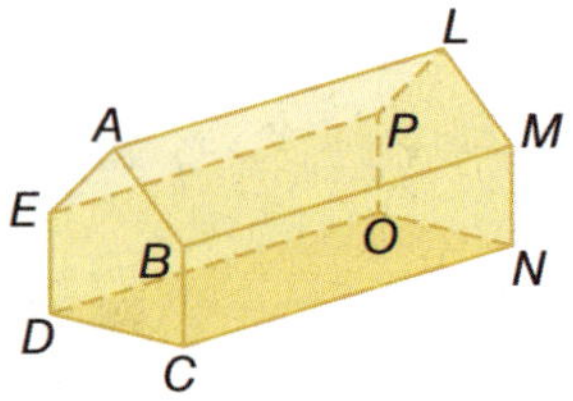

43. Name all segments parallel to $\overline{AE}$.

44. Name all planes intersecting plane BCN.

45. Name all segements skew to $\overline{DC}$.

Study Guide and Review

Math Online glencoe.com
- STUDY TO GO
- Vocabulary Review

Chapter Summary

Key Concepts

Parallelograms and Triangles (Lesson 11-1)

- The area A of a parallelogram is the product of a base b and its corresp onding height h. $A = bh$
- The area A of a triangle is one half the product of a base b and its corresponding height h. $A = \frac{1}{2}bh$ or $A = \frac{bh}{2}$

Trapezoids, Rhombi, and Kites (Lesson 11-2)

- The area A of a trapezoid is one half the product of the height h and the sum of its bases, b_1 and b_2. $A = \frac{1}{2}h(b_1 + b_2)$
- The area A of a rhombus or kite is one half the product of the lengths of its diagonals, d_1 and d_2. $A = \frac{1}{2}d_1d_2$

Areas of Circles and Sectors (Lesson 11-3)

- The area A of a circle is equal to π times the square of the radius r. $A = \pi r^2$
- The ratio of the area A of a sector to the area of the whole circle, πr^2, is equal to the ratio of the degree measure of the intercepted arc x to 360.
  Proportion: $\frac{A}{\pi r^2} = \frac{x}{360}$ Equation: $A = \frac{x}{360} \cdot \pi r^2$

Areas of Regular Polygons and Composite Figures (Lesson 11-4)

- The area A of a regular n-gon with side length s is one half the product of the apothem a and perimeter P. $A = \frac{1}{2}a(ns)$ or $A = \frac{1}{2}aP$

Areas of Similar Figures (Lesson 11-5)

- If two polygons are similar, then their areas are proportional to the square of the scale factor between them.
  If $ABCD \sim FGHJ$, then $\frac{\text{area of } FGHJ}{\text{area of } ABCD} = \left(\frac{FG}{AB}\right)^2$.

FOLDABLES® Study Organizer

Be sure the Key Concepts are noted in your Foldable.

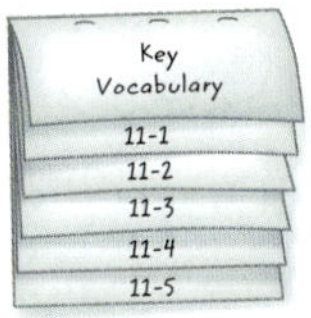

Key Vocabulary

- **base of a parallelogram** (p.763)
- **height of a parallelogram** (p.763)
- **base of a triangle** (p.765)
- **height of a triangle** (p.765)
- **height of a trapezoid** (p.773)
- **sector of a circle** (p.783)
- **center of a regular polygon** (p.791)
- **radius of a regular polygon** (p.791)
- **apothem of a polygon** (p.791)
- **central angle of a regular polygon** (p.791)
- **composite figure** (p.793)

Vocabulary Check

State whether each sentence is *true* or *false*. If *false*, replace the underlined term to make a true sentence.

1. The center of a trapezoid is the perpendicular distance between the bases.
2. A slice of pizza is a sector of a circle.
3. The center of a regular polygon is the distance from the middle to the circle circumscribed around the polygon.
4. The segment from the center of a square to the corner can be called the radius of the square.
5. A segment drawn perpendicular to a side of a regular polygon is called an apothem of the polygon.
6. The measure of each radial angle of a regular n-gon is $\frac{360}{n}$.
7. The apothem of a polygon is the perpendicular distance between any two parallel bases.
8. The height of a triangle is the length of an altitude drawn to a given base.
9. Any side of a parallelogram can be called the height of a parallelogram.
10. The center of a regular polygon is also the center of its circumscribed circle.

CHAPTER 11 Study Guide and Review

Lesson-by-Lesson Review

11-1 Areas of Parallelograms and Triangles (pp. 763–770)

Find the perimeter and area of each parallelogram or triangle. Round to the nearest tenth if necessary.

11.

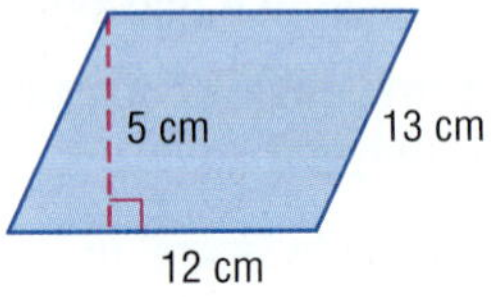

12.

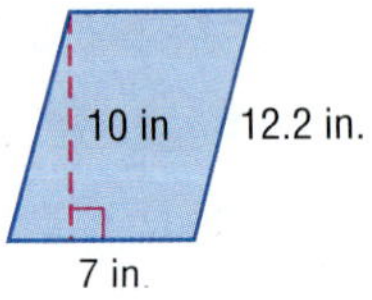

13.

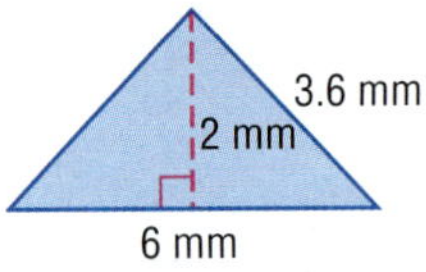

14.

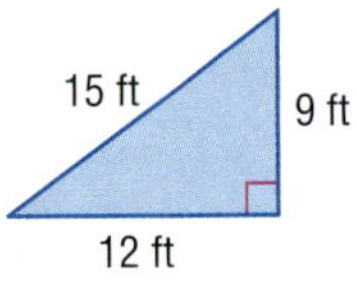

15. PAINTING Two of the walls of an attic in an A-frame house are triangular, each with a height of 12 feet and a width of 22 feet. How much paint is needed to paint one end of the attic?

EXAMPLE 1

Find the perimeter and area of $\square JKLM$.

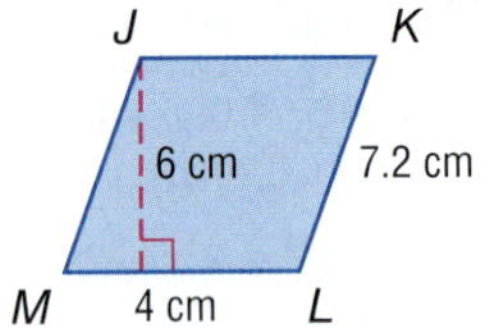

Perimeter

Perimeter of $\square JKLM = JK + KL + LM + JM$

$= 4 + 7.2 + 4 + 7.2$ or 22.4 cm

Area

$A = bh$ — **Area of a parallelogram**

$= (4)(6)$ or 24 cm^2 — $b = 4$ **and** $h = 6$

11-2 Areas of Trapezoids, Rhombi, and Kites (pp. 773–780)

Find the area of each trapezoid, rhombus, or kite.

16.

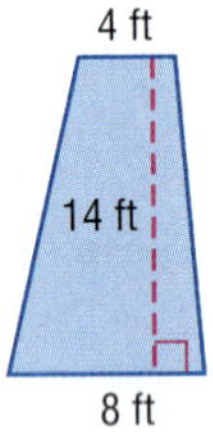

17.

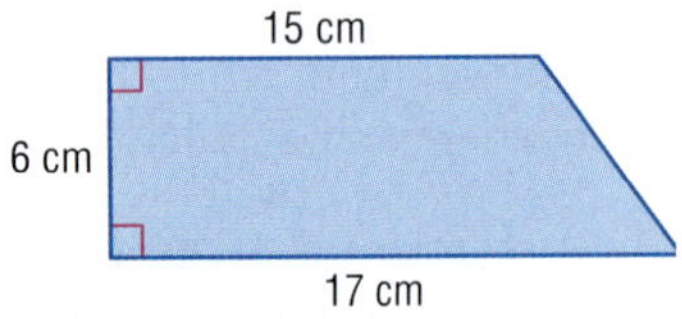

18.

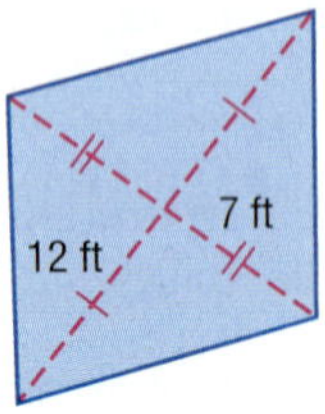

19.

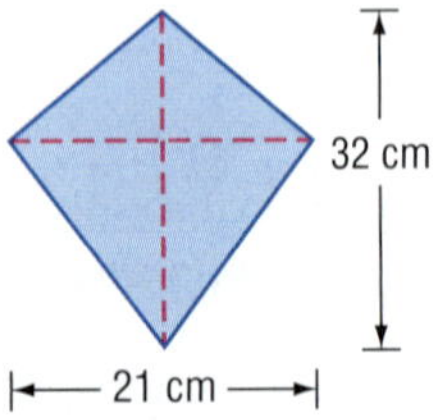

20. KITES Team Dragon's kite is 4 ft long and 3 ft across. How much fabric does it take to make their kite?

EXAMPLE 2

Find the area of each rhombus or kite.

a.

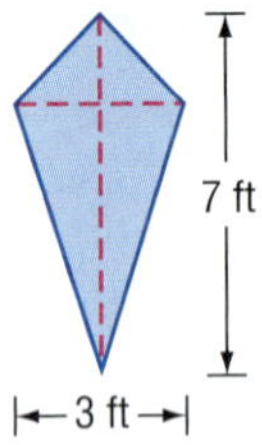

$A = \frac{1}{2}d_1d_2$ — **Area of a kite**

$= \frac{1}{2}(7)(3)$ — $d_1 = 7$ **and** $d_2 = 3$

$= 10.5\text{ ft}^2$ — **Simplify.**

b.

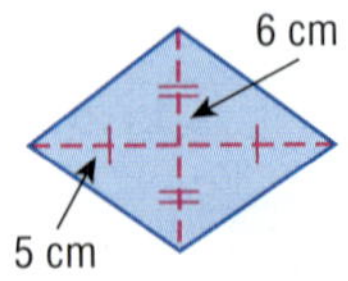

Since the diagonals of a rhombus bisect each other, the lengths of the diagonals are $6 + 6$ or 12 centimeters and $5 + 5$ or 10 centimeters.

$A = \frac{1}{2}d_1d_2$ — **Area of a rhombus**

$= \frac{1}{2}(10)(12)$ — $d_1 = 10$ **and** $d_2 = 12$

$= 60\text{ cm}^2$ — **Simplify.**

MIXED PROBLEM SOLVING
For mixed problem-solving practice, see page 1009.

11-3 Areas of Circles and Sectors (pp. 782–788)

Find the area of each shaded sector. Round to the nearest tenth.

21.

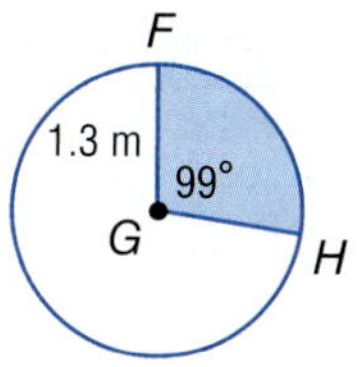

22.

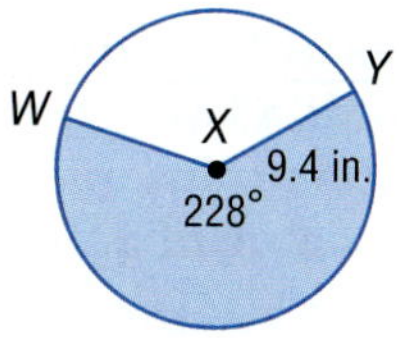

23. BICYCLES A bicycle tire decoration covers $\frac{1}{9}$ of the circle formed by the tire. If the tire has a diameter of 26 inches, what is the area of the decoration?

24. PIZZA Charlie and Kris ordered a 16-inch pizza and cut the pizza into 12 slices.

a. If Charlie ate 3 pieces, what area of the pizza did he eat?

b. If Kris ate 2 pieces, what area of the pizza did she eat?

c. What is the area of leftover pizza?

EXAMPLE 3

Find the area of the shaded sector. Round to the nearest tenth.

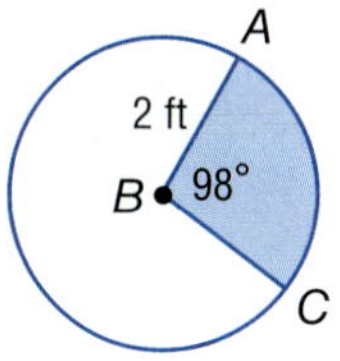

$A = \frac{x}{360} \cdot \pi r^2$ **Area of a sector**

$= \frac{98}{360} \cdot \pi(2)^2$ **Substitution**

$\approx 3.4 \text{ ft}^2$ **Simplify.**

11-4 Areas of Regular Polygons and Composite Figures (pp. 791–799)

Find the area of each regular polygon or composite figure. Round to the nearest tenth.

25.

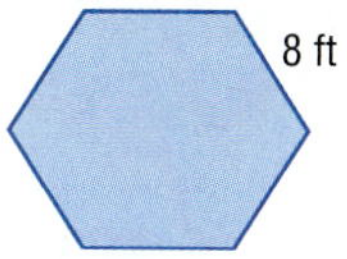

26.

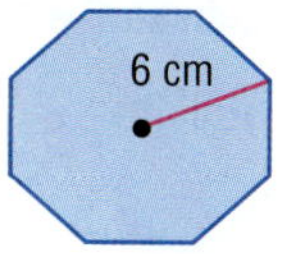

27.

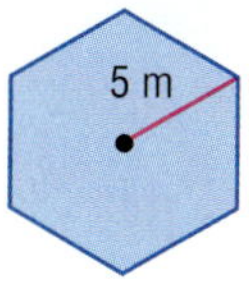

28.

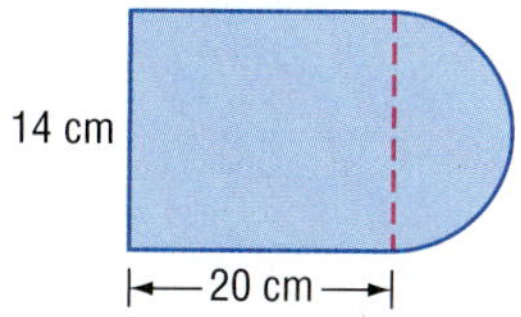

29. SIGNS Find the area of the stop sign below in square inches.

EXAMPLE 4

Find the area of the figure.

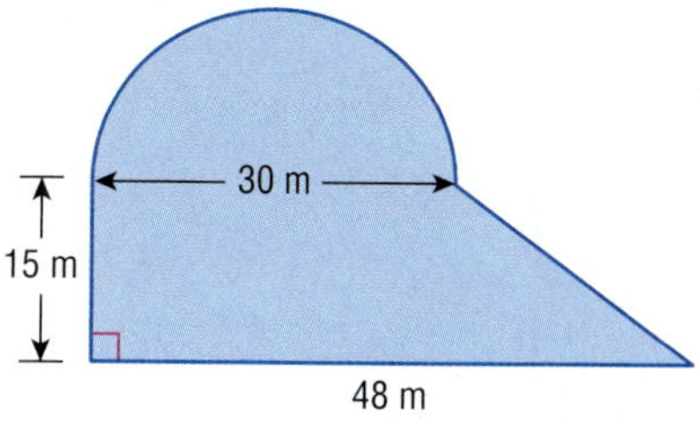

The composite shape is made up of a semicircle and a trapezoid.

Area = Area of semicircle + Area of trapezoid

$= \frac{180}{360} \cdot \pi \cdot r^2 + \frac{1}{2} \cdot h \cdot (b_1 + b_1)$

$\approx \frac{180}{360} \cdot \pi \cdot 15^2 + \frac{1}{2} \cdot 15 \cdot (30 + 48)$

$\approx 112.5\pi + 585$ or about 938.4 m^2

11-5 Areas of Similar Figures (pp. 802–808)

For each pair of similar figures, use the given areas to find the scale factor from the blue to the green figure. Then find x.

30.

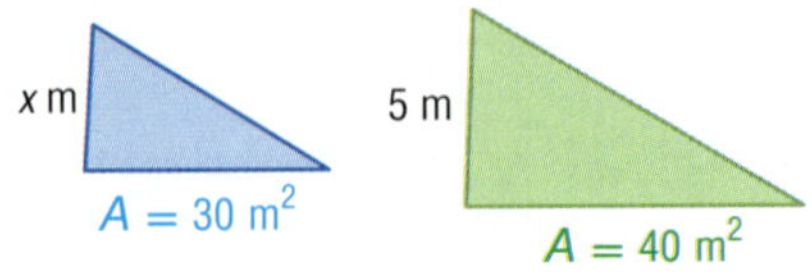

31.

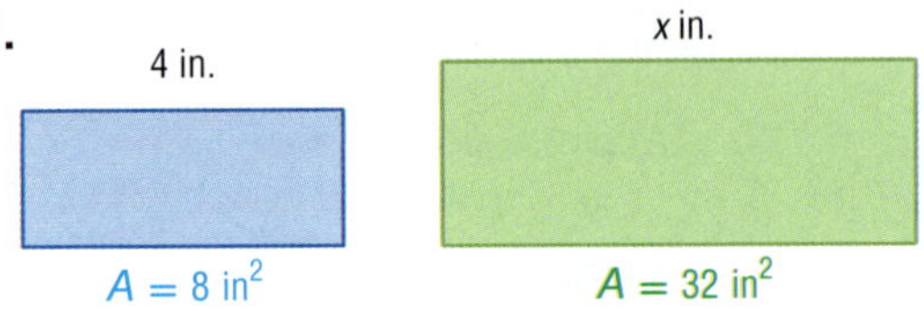

32.

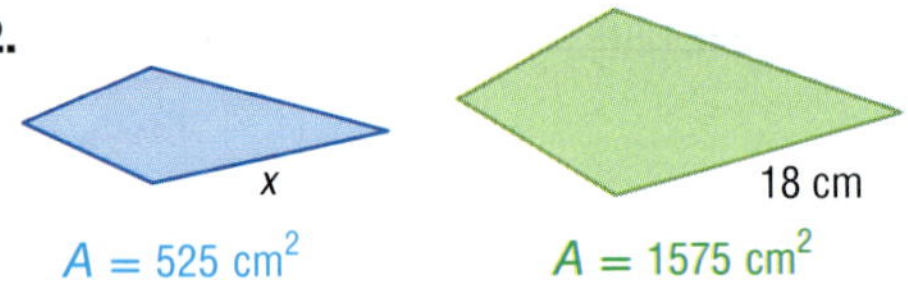

COORDINATE GEOMETRY **Find the area of each figure. Use the segment length given to find the area of a similar polygon.**

33. $R'S' = 3$

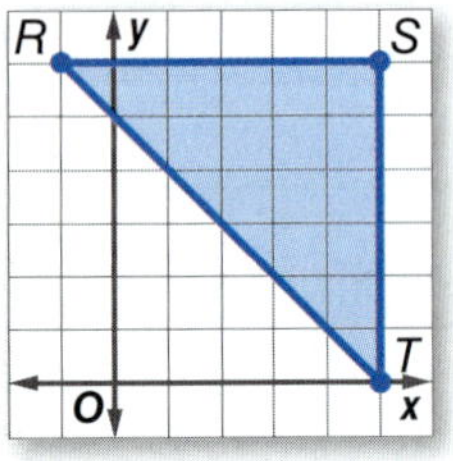

34. $K'L' = 15$

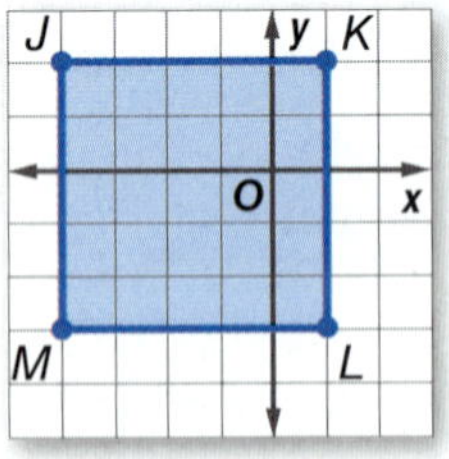

35. **LAND OWNERSHIP** Joshua's land is 600 square miles. A map of his land is 5 square feet. If one side of the map is 1.5 feet, how long is the corresponding side of the land?

EXAMPLE 5

The area of trapezoid *JKLM* is 138 square feet. The area of trapezoid *QRST* is 5.52 square feet. If trapezoid *JKLM* ~ trapezoid *QRST*, find the scale factor from trapezoid *JKLM* to trapezoid *QRST* and the value of x.

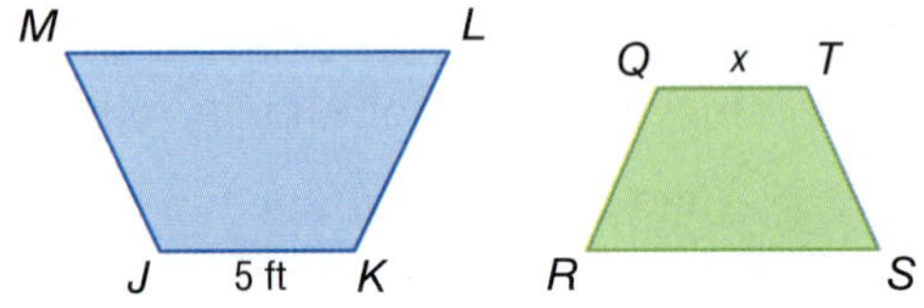

Let k be the scale factor between trapezoid *JKLM* and trapezoid *QRST*.

$\frac{\text{Area of trapezoid } JKLM}{\text{Area of trapezoid } QRST} = k^2$ **Theorem 11.1**

$\frac{138}{5.52} = k^2$ **Substitution**

$5 = k$ **Take the positive square root of each side.**

So, the scale factor from trapezoid *JKLM* to trapezoid *QRST* is 5. Use this scale factor to find the value of x.

$\frac{JK}{QT} = k$ **The ratio of corresponding lengths of similar polygons is equal to the scale factor between the polygons.**

$\frac{5}{x} = 5$ **Substitution**

$1 = x$ **Simplify.**

Practice Test

Math Online glencoe.com
Chapter Test

Find the area and perimeter of each figure. Round to the nearest tenth if necessary.

1.
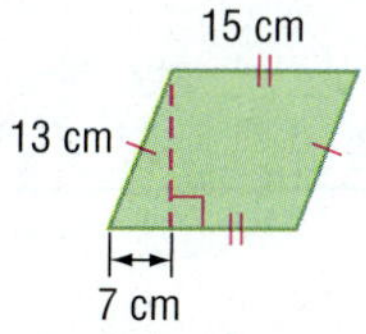

2.
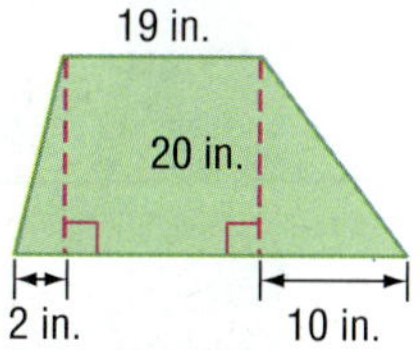

3.
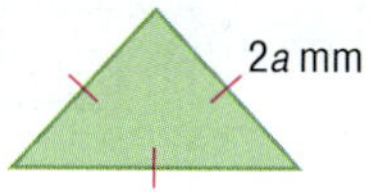

4.
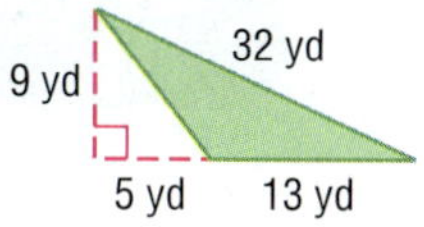

5. **ARCHAELOGY** The tile pattern shown was used in Pompeii for paving. If the diagonals of each rhombus are 2 and 3 inches, what area makes up each "cube" in the pattern?

Find the area of each figure. Round to the nearest tenth if necessary.

6.
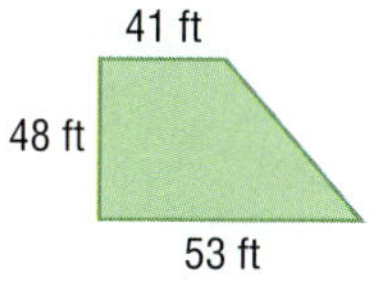

7.
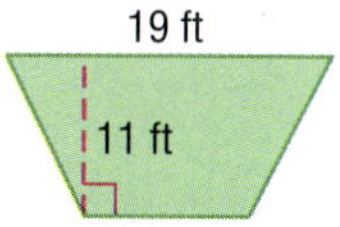

8.
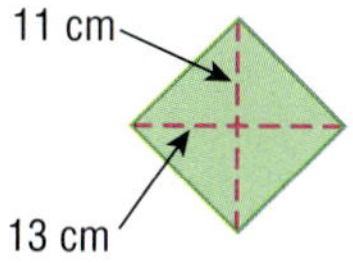

9.
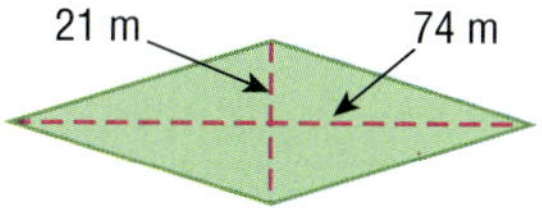

10. **GEMOLOGY** A gem is cut in a kite shape. It is 6.2 millimeters wide at its widest point and 5 millimeters long. What is the area?

11. **ALGEBRA** The area of a triangle is 16 square units. The base of the triangle is $x + 4$ and the height is x. Find x.

12. **ASTRONOMY** A large planetarium in the shape of a dome is being built. When it is complete, the base of the dome will have a circumference of 870 meters. How many square meters of land were required for this planetarium?

Find the area of each circle or sector. Round to the nearest tenth.

13.
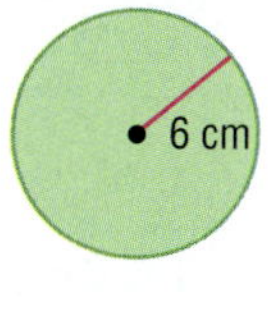

14.
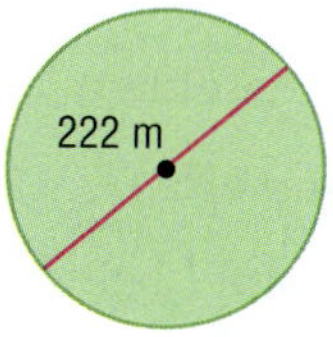

15.
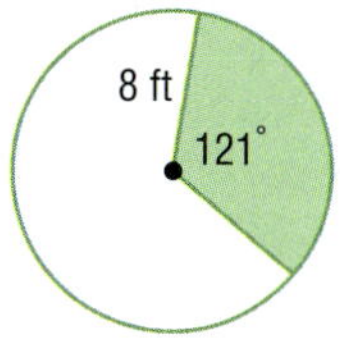

16.
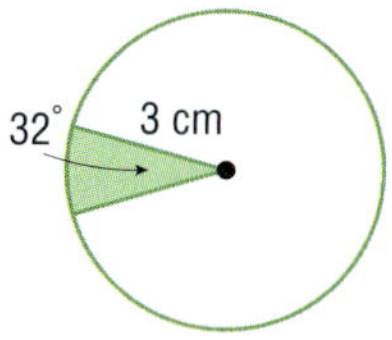

17. **MURALS** An artisan is creating a circular street mural for an art festival. The mural is going to be 50 feet wide.

a. Find the area of the mural to the nearest square foot.

b. One sector of the mural spans 38°. What is the area of this sector to the nearest square foot?

Find the perimeter and area of each figure. Round to the nearest tenth if necessary.

18.
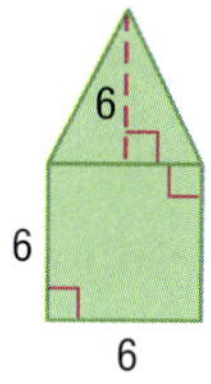

19.
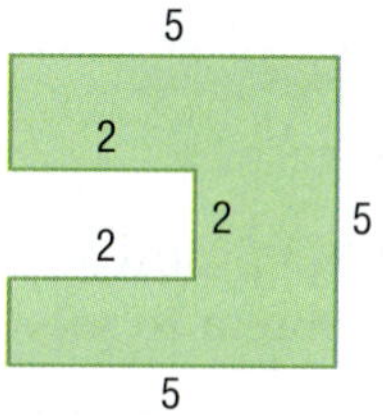

20. **BAKING** Todd wants to make a cheesecake for a birthday party. The recipe calls for a 9-inch diameter round pan. Todd only has square pans. He has an 8-inch square pan, a 9-inch square pan, and a 10-inch square pan. Which pan comes closest in area to the one that the recipe suggests?

CHAPTER 11

Preparing for Standardized Tests

Solve Multi-Step Problems

Some problems that you will encounter on standardized tests require you to solve multiple parts in order to come up with the final solution. Use this lesson to practice these types of problems.

Strategies for Solving Multi-Step Problems

Step 1

Read the problem statement carefully.

Ask yourself:

- What am I being asked to solve? What information is given?
- Are there any intermediate steps that need to be completed before I can solve the problem?

Step 2

Organize your approach.

- List the steps you will need to complete in order to solve the problem.
- Remember that there may be more than one possible way to solve the problem.

Step 3

Solve and check.

- Work as efficiently as possible to complete each step and solve.
- If time permits, check your answer.

EXAMPLE

Read the problem. Identify what you need to know. Then use the information in the problem to solve.

What is the area of the triangle? Round your answer to the nearest tenth if necessary.

A 137.4 m^2 **C** 170.5 m^2

B 161.3 m^2 **D** 186.9 m^2

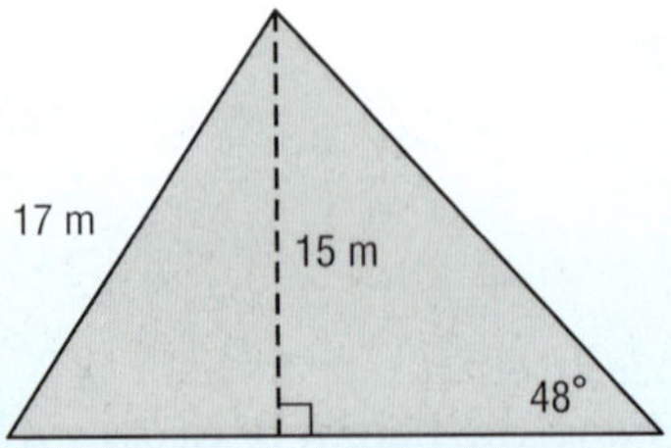

Read the problem statement and study the figure carefully. At first glance, the problem may appear fairly straightforward. Notice, however, that you must first find the base of the triangle before you can find its area. Organize an approach to solve the problem.

Step 1 Use the Pythagorean Theorem to find a.

Step 2 Use trigonometry to find b.

Step 3 Find the area of the triangle.

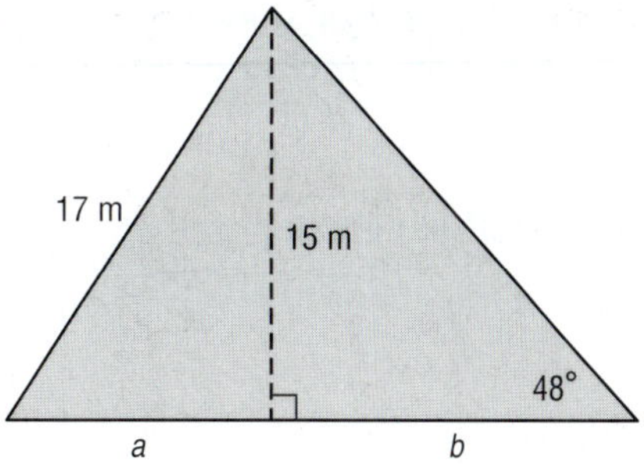

Step 1 Find a.	**Step 2** Find b.
$a^2 + 15^2 = 17^2$	$\tan 48° = \frac{15}{b}$
$a^2 = 289 - 225$	$b = \frac{15}{\tan 48°}$
$a^2 = 64$	$b \approx 13.506$
$a = 8$	

Step 3 Find the area of the triangle.

The base of the triangle is $a + b$ or about 21.506 meters.

$$A \approx \frac{1}{2}(21.506)(15)$$

$$\approx 161.3$$

So, the area of the triangle is about 161.3 square meters. The answer is B.

Exercises

Read the problem. Identify what you need to know. Then use the information in the problem to solve.

1. What is the area of the figure? Round to the nearest tenth.

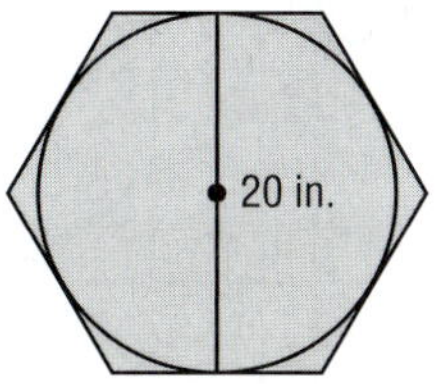

A 346.4 in^2

B 372.1 in^2

C 383.2 in^2

D 564.7 in^2

2. Kaleb is painting just the shaded part of the basketball key shown below. How much area will he need to cover? Round to the nearest tenth.

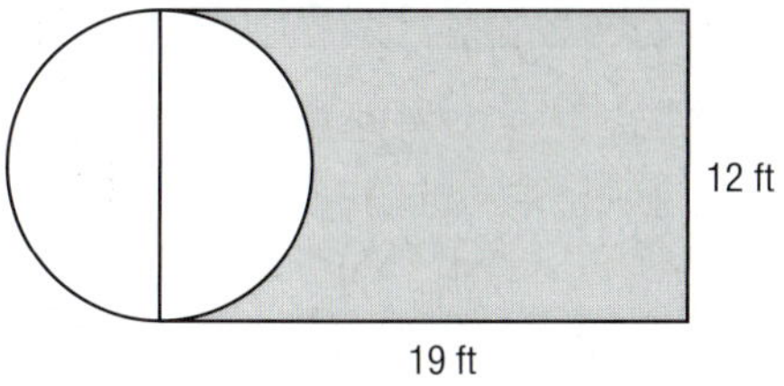

F 114.9 ft^2

G 142.4 ft^2

H 159.9 ft^2

J 171.5 ft^2

CHAPTER 11 Standardized Test Practice

Cumulative, Chapters 1 through 11

Multiple Choice

Read each question. Then fill in the correct answer on the answer document provided by your teacher or on a sheet of paper.

1. What is the value of x in the figure below?

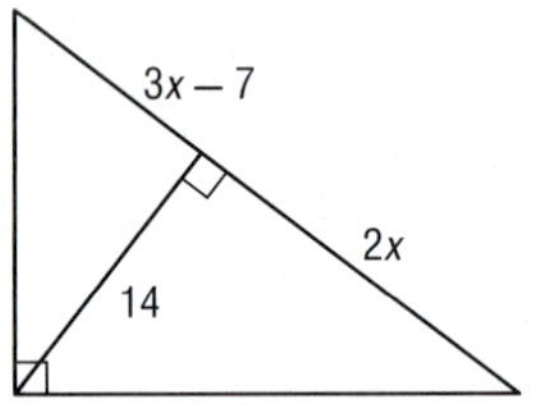

A 5

B 7

C 8

D 10

2. Which of the following is *not* a property of parallelograms?

F The opposite angles of a parallelogram are congruent.

G The opposite sides of a parallelogram are congruent.

H The consecutive angles of a parallelogram are supplementary.

J The consecutive angles of a parallelogram are complementary.

3. What is the area of the triangle below? Round your answer to the nearest tenth if necessary.

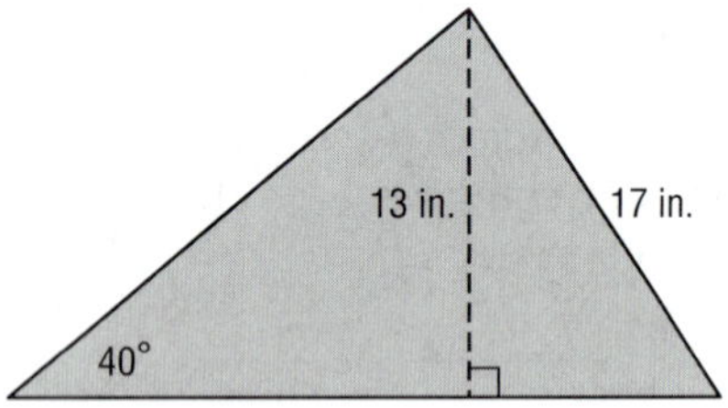

A 110.5 in^2

B 144.2 in^2

C 164.5 in^2

D 171.9 in^2

Test-Taking Tip

Question 3 Some problems require multiple steps to solve. You can use the Pythagorean Theorem and trigonometry to find the base of the triangle.

4. Given: $\overline{AC} \cong \overline{BD}$

$\overline{AC} \parallel \overline{BD}$

Prove: $\triangle ABC \cong \triangle DCB$

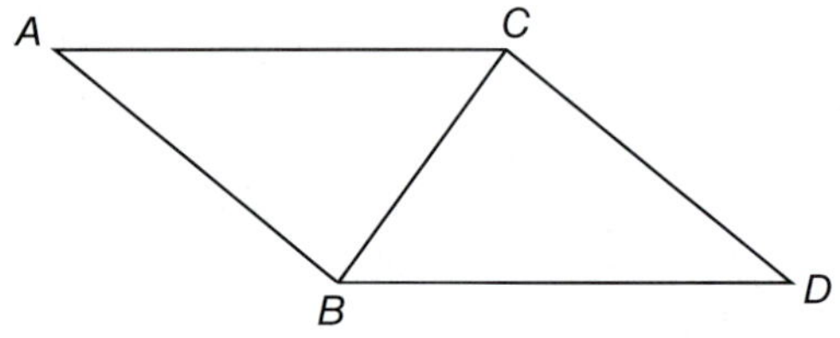

Statements	Reasons
1. $\overline{AC} \cong \overline{BD}$	**1.** Given
2. $\overline{AC} \parallel \overline{BD}$	**2.** Given
3. $\angle ACB \cong \angle DBC$	**3.** ___?___
4. $\overline{BC} \cong \overline{CB}$	**4.** Reflexive Property of Congruence
5. $\triangle ABC \cong \triangle DCB$	**5.** SAS

What is the missing line needed to complete the proof?

F Same side exterior angles are congruent.

G Vertical angles are congruent.

H Corresponding parts of congruent triangles are congruent.

J Alternate interior angles are congruent.

5. What is the slope of the line?

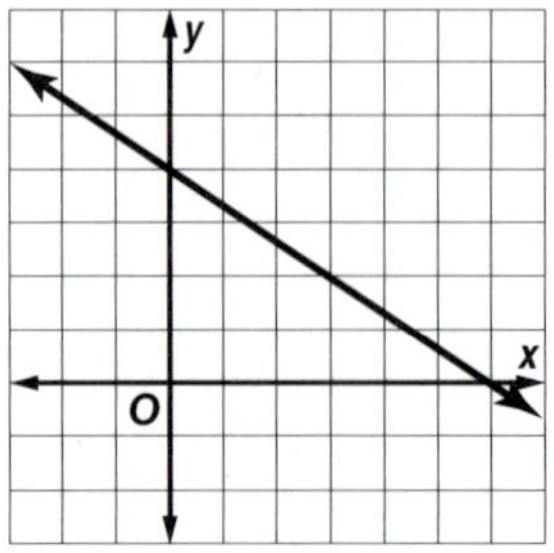

A $-\frac{2}{3}$

B $-\frac{3}{2}$

C $\frac{2}{3}$

D $\frac{3}{2}$

Short Response/Gridded Response

Record your answers on the answer sheet provided by your teacher or on a sheet of paper.

6. **GRIDDED RESPONSE** Suppose two similar rectangles have a scale factor of 3:5. The perimeter of the smaller rectangle is 21 millimeters. What is the perimeter of the larger rectangle? Express your answer in millimeters.

7. Copy the circles below on a sheet of paper and draw the common tangents, if any exist.

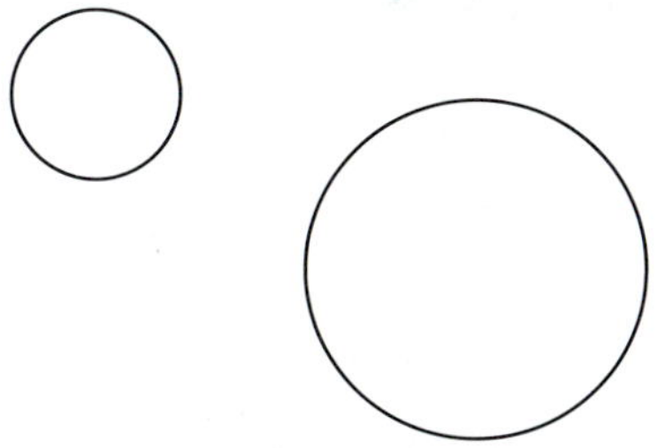

8. What is the contrapositive of the statement below?

> If a chord contains the center of a circle, then the chord is a diameter.

9. Copy the figure and point D. Then use a ruler to draw the image of the figure under a dilation with center D and a scale factor of 2.

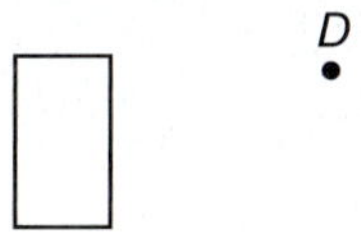

10. **GRIDDED RESPONSE** Solve for x in the figure below.

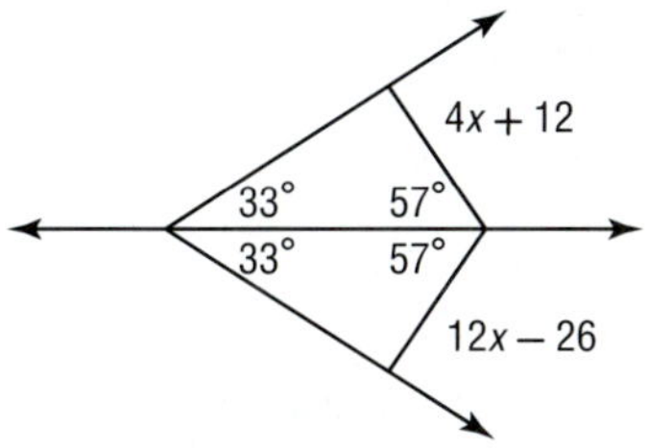

11. **GRIDDED RESPONSE** What is the area of the parallelogram below? Express your answer in square feet. Round to the nearest whole number if necessary.

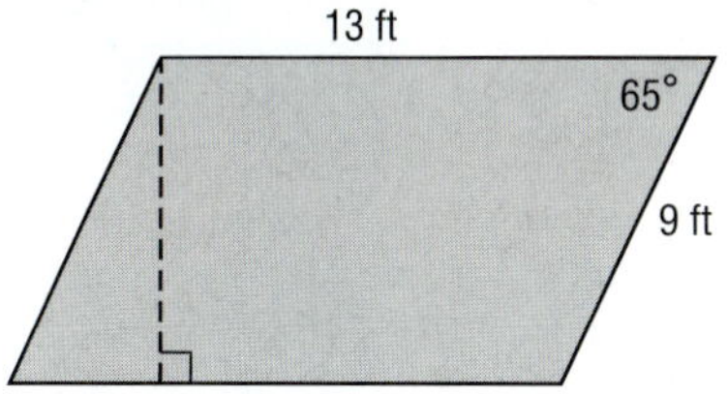

Extended Response

Record your answers on a sheet of paper. Show your work.

12. Use the figure below to answer each question.

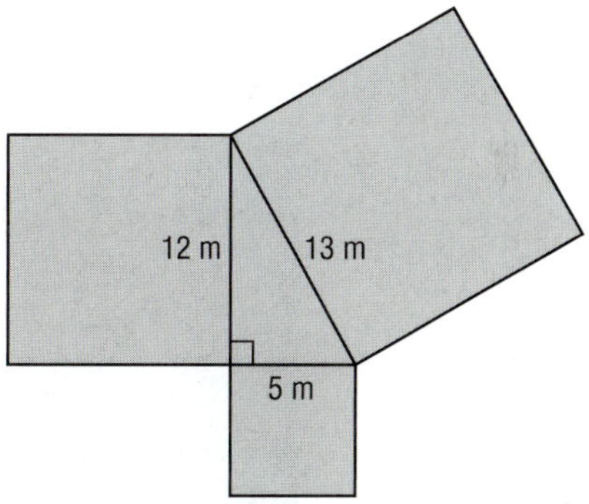

a. Find the area of each square and the area of the triangle.

b. What is the total area of the figure?

c. Explain how the areas of the squares model the Pythagorean Theorem.

Need Extra Help?												
If you missed Question...	1	2	3	4	5	6	7	8	9	10	11	12
Go to Lesson or Page...	8-1	6-2	11-1	4-4	3-3	7-2	10-5	2-3	9-6	5-1	11-1	11-4

CHAPTER 12 Extending Surface Area and Volume

Then

In Chapter 1, you identified and named three-dimensional figures and calculated surface area and volume for some common solids.

Now

In Chapter 12, you will:

- Find lateral areas, surface areas, and volumes of various solid figures.
- Investigate Euclidean and spherical geometries.
- Use properties of similar solids.

Why?

ARCHITECTURE

Architects use different types of solids to create designs that are both interesting and functional.

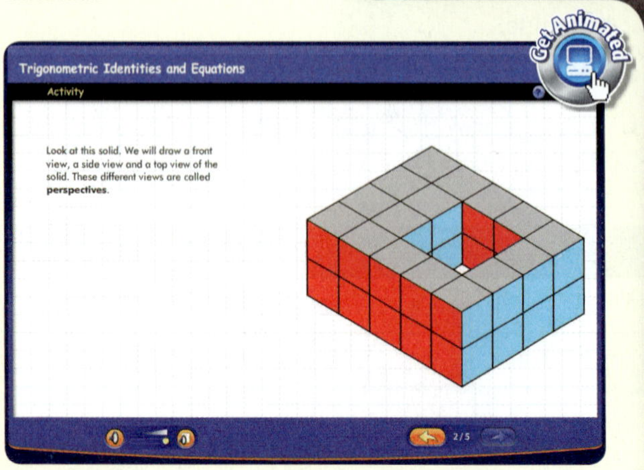

Get Ready for Chapter 12

Diagnose Readiness You have two options for checking Prerequisite Skills.

Text Option

Take the Quick Check below. Refer to the Quick Review for help.

QuickCheck

Determine whether each statement about the figure in Example 1 is *true, false,* or *cannot be determined.* (Lesson 1-1)

1. $\square ABCD$ lies in plane $\mathcal{M}$.
2. $\square CDHG$ lies in plane $\mathcal{N}$.
3. $\overline{AB}$ lies in plane $\mathcal{M}$.
4. $\overline{HG}$ lies in plane $\mathcal{N}$.
5. $\overline{AE}$ is $\perp$ to plane $\mathcal{M}$.
6. $\overline{DC} \parallel$ line ℓ.

Find the area of each figure. Round to the nearest tenth if necessary. (Lesson 11-2)

7. 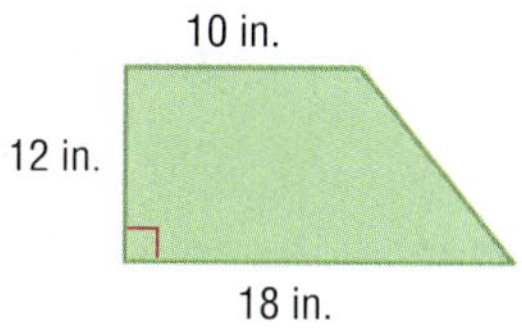

8. 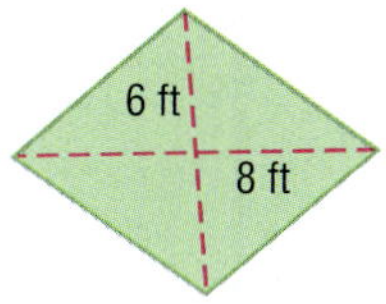

9. **CRAFTS** A seamstress wants to cover a kite frame with cloth. If the length of one diagonal is 16 inches and the other diagonal is 22 inches, find the area of the surface of the kite.

Find the value of the variable in each equation. (Prerequisite Skill)

10. $a^2 + 40^2 = 41^2$
11. $8^2 + b^2 = 17^2$
12. $a^2 + 6^2 = (7\sqrt{3})^2$

QuickReview

EXAMPLE 1

In the figure, $\overline{AD} \perp \ell$ and $ABCDEFGH$ is a cube. Determine whether plane $\mathcal{M} \perp$ plane $\mathcal{N}$.

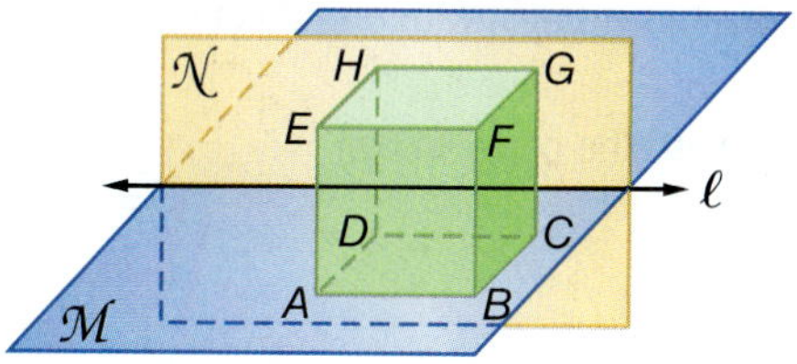

Plane $\mathcal{M} \perp$ plane $\mathcal{N}$ cannot be determined from the given information.

EXAMPLE 2

Find the area of the figure. Round to the nearest tenth if necessary.

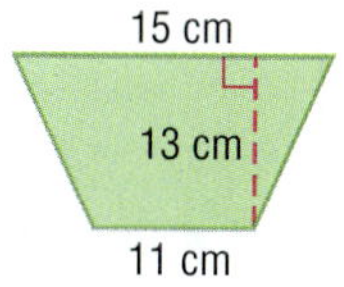

$A = \frac{1}{2}h(b_1 + b_2)$ **Area of a trapezoid**

$= \frac{1}{2}(13)(15 + 11)$ **Substitution**

$= \frac{1}{2}(13)(26)$ **Simplify.**

$= 169$ **Multiply.**

The area of the trapezoid is 169 cm^2.

EXAMPLE 3

Find the value of the variable in $8^2 + 7^2 = c^2$.

$c^2 = 8^2 + 7^2$ **Original equation**

$c^2 = 64 + 49$ **Evaluate the exponents.**

$c^2 = 113$ **Simplify.**

$c = \pm\sqrt{113}$ **Take the square root of each side.**

Online Option

Math Online Take a self-check Chapter Readiness Quiz at glencoe.com.

Get Started on Chapter 12

You will learn several new concepts, skills, and vocabulary terms as you study Chapter 12. To get ready, identify important terms and organize your resources. You may wish to refer to **Chapter 0** to review prerequisite skills.

FOLDABLES® Study Organizer

Surface Area and Volume Make this Foldable to help you organize your Chapter 12 notes about surface area and volume. Begin with one sheet of notebook paper.

1. **Fold** the paper in half.

2. **Fold** the paper again, two inches from the top.

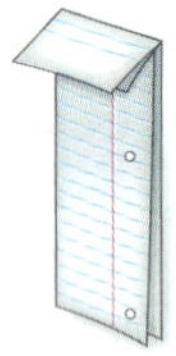

3. **Unfold** the paper.

4. **Label** as shown.

Math Online glencoe.com

- Study the chapter online
- Explore **Get Animated** activities
- Get extra help from your own **Personal Tutor**
- Use **Extra Examples** for additional help
- Take a **Self-Check Quiz**
- **Review Vocabulary** in fun ways

New Vocabulary

English		Español
right cylinder	• p. 821 •	cilindro recto
oblique cylinder	• p. 821 •	cilindro oblicuo
right prism	• p. 821 •	prisma recto
oblique prism	• p. 821 •	prisma oblicuo
isometric view	• p. 823 •	isométrico
cross section	• p. 824 •	sección transversal
lateral face	• p. 830 •	cara lateral
lateral edge	• p. 830 •	arista lateral
altitude	• p. 830 •	altura
lateral area	• p. 830 •	area lateral
axis	• p. 832 •	eje
regular pyramid	• p. 838 •	pirámide regular
slant height	• p. 838 •	altura oblicua
right cone	• p. 840 •	cono recto
oblique cone	• p. 840 •	cono oblicuo
great circle	• p. 865 •	círculo máximo
Euclidean geometry	• p. 873 •	geometría euclidea
spherical geometry	• p. 873 •	geometría esférica
similar solids	• p. 880 •	sólidos semejantes
congruent solids	• p. 880 •	sólidos congruentes

Review Vocabulary

regular polyhedron • p. 68 • poliedro regular a polyhedron in which all of the faces are regular congruent polygons

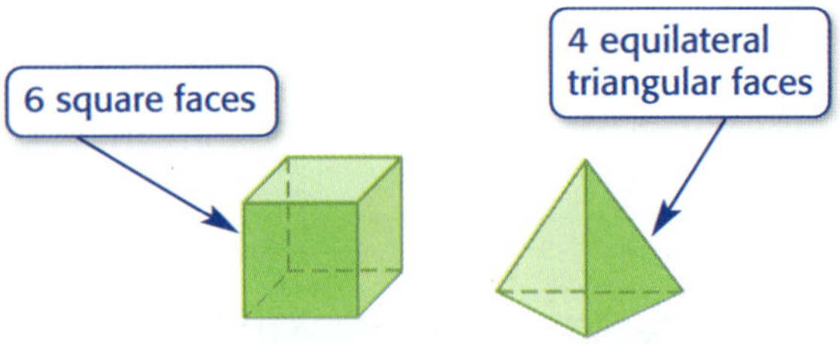

Multilingual eGlossary glencoe.com

EXPLORE

12-1 Geometry Lab
Right Solids and Oblique Solids

Objective

- Identify and draw right and oblique solids.
- Draw nets.

You know that in a right triangle, one of the sides is an altitude. However, in an obtuse triangle, the altitude is outside of the triangle. This same idea can be applied to solids.

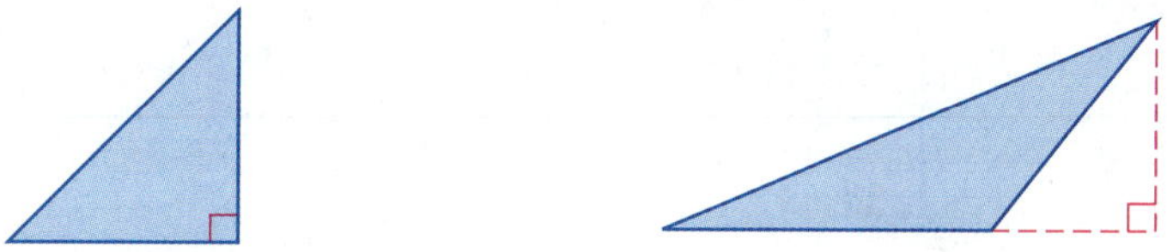

A prism with lateral edges that are also altitudes is called a **right prism**. If the lateral edges are not perpendicular to the bases, it is an **oblique prism**. Similarly, if the axis of a cylinder is also the altitude, then the cylinder is called a **right cylinder**. Otherwise, the cylinder is an **oblique cylinder**.

ACTIVITY 1 Right and Oblique Solids

Identify each figure as a right solid or an oblique solid.

a.

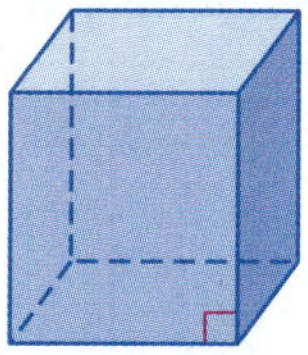

The altitude of the prism is the length of the lateral edge so it is a right solid.

b.

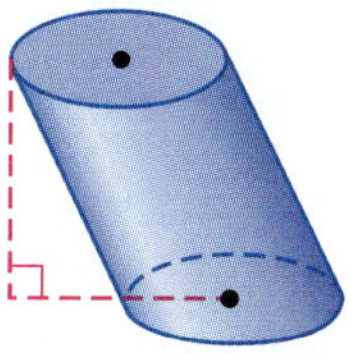

The altitude of the cylinder is not the length of the lateral edge so it is an oblique solid.

c. Draw a right cylinder and an oblique cylinder.

Step 1

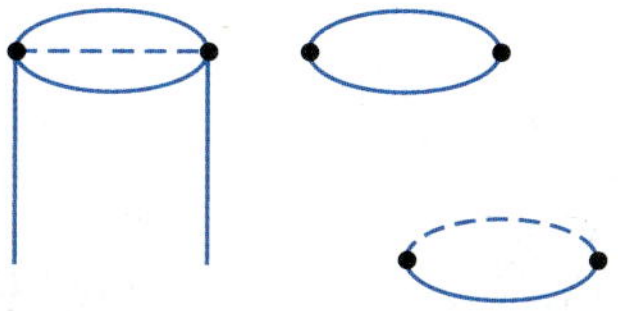

Draw two circles with points as shown above. Draw two congruent lines perpendicular to the existing dashed line. Draw a circle as shown above.

Step 2

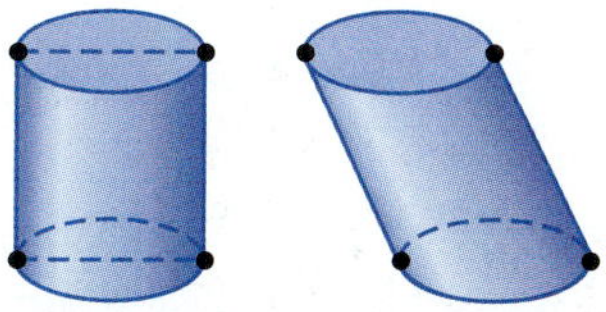

Create a circle with a dashed line that is perpendicular to the line created in Step 2. Connect the points as shown.

Exercises

Identify each figure as a right solid or an oblique solid.

1.

2.

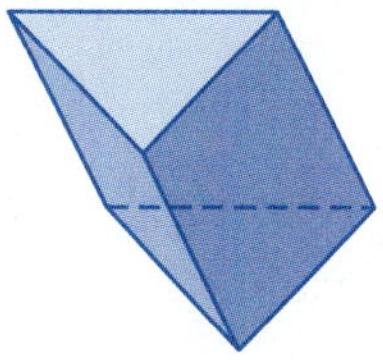

3.

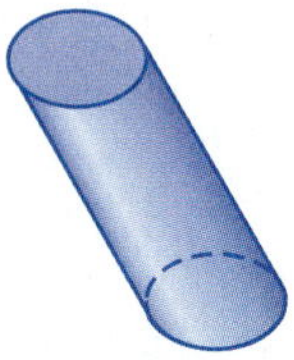

Sketch each solid. Describe the shapes of its bases and lateral faces.

4. oblique rectangular prism

5. right triangular prism

(continued on the next page)

Review Vocabulary

net two-dimensional pattern for a three-dimensional solid

ACTIVITY 2 Draw Nets

a.

Step 1

Identify the basic shapes which the cylinder contains. The bases of a cylinder are circles and the lateral surface of a cylinder is a rectangle.

Step 2

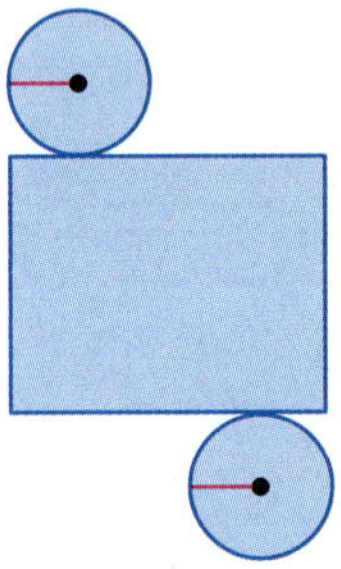

Use a net of the cylinder to show the relationship of the basic shapes which the cylinder contains.

b.

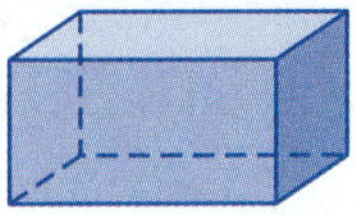

Step 1

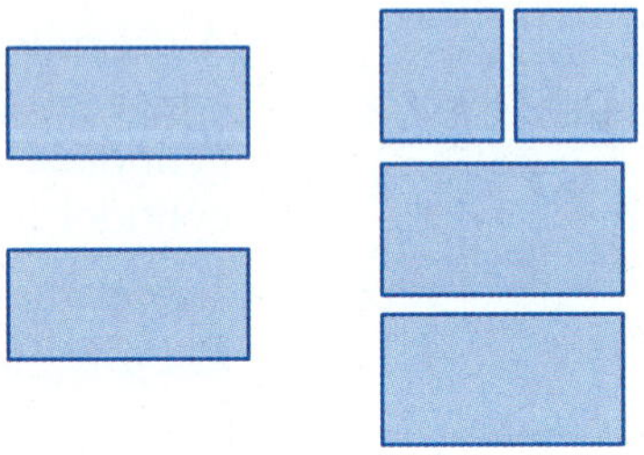

Identify the basic shapes the prism contains. The bases of this prism are rectangles and the lateral faces of this prism are rectangles.

Step 2

Use a net of the prism to show the relationship of the basic shapes which the prism contains.

Exercises

Draw the net of each figure.

6.

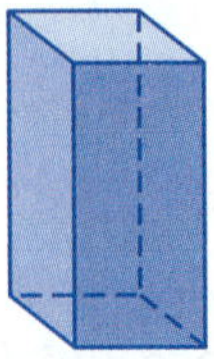

7.

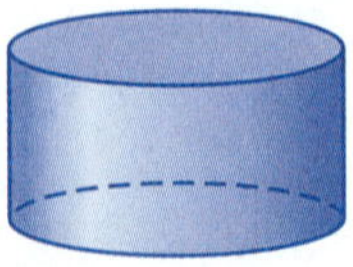

8.

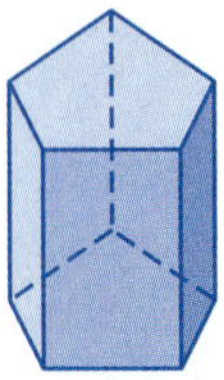

9. WRITING IN MATH Explain the difference between a right prism and an oblique prism.

12-1 Representations of Three-Dimensional Figures

Then

You identified parallel planes and intersecting planes in three-dimensional figures. (Lesson 3-1)

Now

- Draw isometric views of three-dimensional figures.
- Investigate cross sections of three-dimensional figures.

New Vocabulary

isometric view
cross section

Math Online
glencoe.com

Why?

Video game programmers use technology to make the gaming environments appear three-dimensional. As players move in the various video game worlds, objects are realistically shown from different perspectives.

Draw Isometric Views In video games, three-dimensional figures are represented on a two-dimensional screen. You can use isometric dot paper to draw **isometric views**, or corner views, of three-dimensional geometric solids on two-dimensional paper.

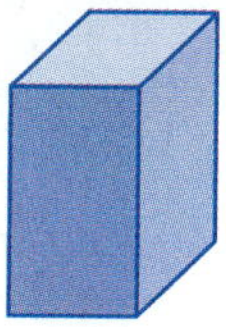

front view

isometric view

EXAMPLE 1 Use Dimensions of a Solid to Sketch a Solid

Use isometric dot paper to sketch a triangular prism 3 units high with two sides of the base that are 2 units long and 4 units long.

Step 1

Mark the corner of the solid. Draw 3 units down, 2 units to the left, and 4 units to the right. Then draw a triangle for the top of the solid.

Step 2

Draw segments 3 units down from each vertex for the vertical edges. Connect the appropriate vertices using a dashed line for the hidden edge.

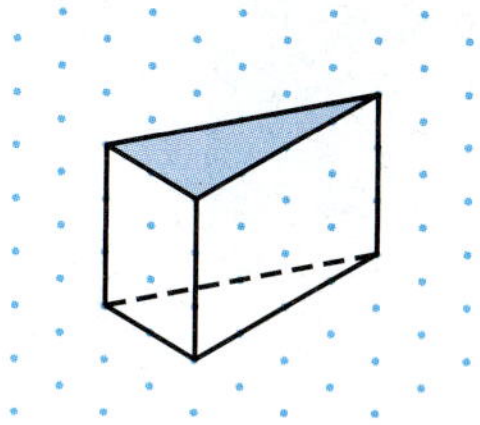

Guided Practice

1. Use isometric dot paper to sketch a rectangular prism 1 unit high, 5 units long, and 4 units wide.

Personal Tutor glencoe.com

Recall that an *orthographic drawing* shows the top, left, front, and right views of a solid. You can use an orthographic drawing to draw an isometric view of a three-dimensional figure.

The left, front, and right views of a cube are shown at the right.

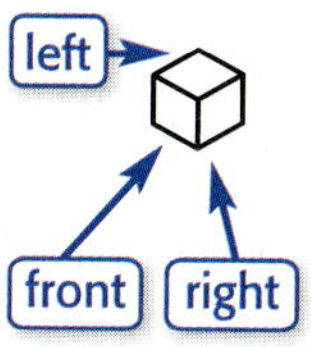

EXAMPLE 2 Use an Orthographic Drawing to Sketch a Solid

Use isometric dot paper and the orthographic drawing to sketch a solid.

- top view: There are two rows and two columns. The dark segments indicate that there are different heights.
- left view: The figure is 3 units high on the left.
- front view: The first column is 3 units high and the second column is 1 unit high.
- right view: The figure is 3 units high on the right. The dark segments indicate that there are breaks in the surface.

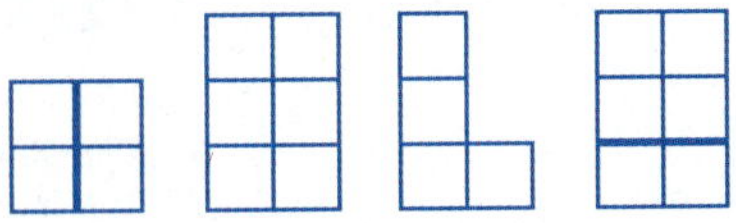

top view | left view | front view | right view

Connect the dots on the isometric dot paper to represent the edges of the solid. Shade the tops of each column.

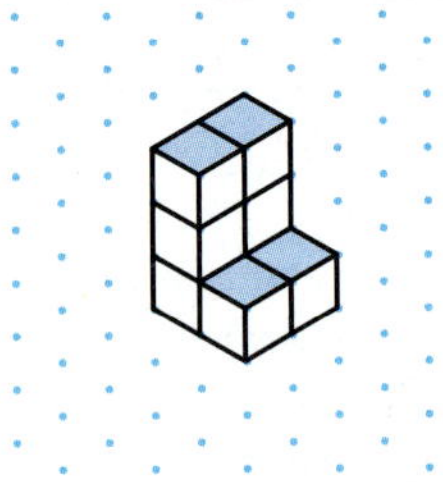

Guided Practice

2.

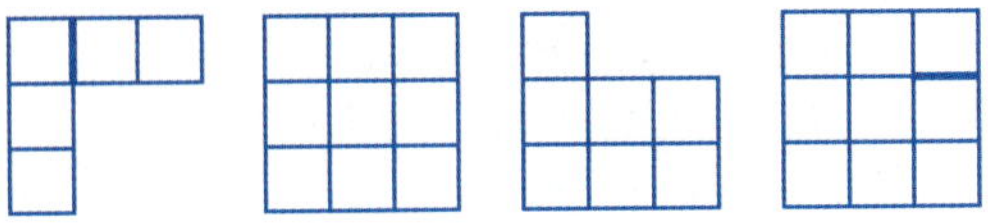

top view | left view | front view | right view

Personal Tutor glencoe.com

Investigate Cross Sections A **cross section** is the intersection of a solid and a plane. The shape of the cross section formed by the intersection of a plane and a three-dimensional figure depends on the angle of the plane.

Real-World Link

The largest pyramid ever constructed is about 63 miles from Mexico City. It is 177 feet tall and its base covers an area of nearly 45 acres.

Source: *Guinness World Records*

Real-World EXAMPLE 3 Identify Cross Sections of Solids

PYRAMIDS Scientists are able to use computers to study cross sections of ancient artifacts and structures. Determine the shape of each cross section of the pyramid below.

horizontal cut | angled cut | vertical cut

The horizontal cross section is a square. The angled cross section is a trapezoid. The vertical cross section is a triangle.

Guided Practice

3. **CAKES** Ramona has a cake pan shaped like half of a sphere, as shown at the right. Describe the shape of the cross sections of cakes baked in this pan if they are cut horizontally and vertically.

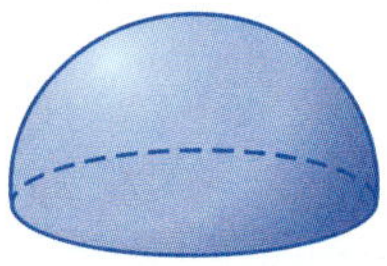

Personal Tutor glencoe.com

Check Your Understanding

Example 1 p. 823

Use isometric dot paper to sketch each prism.

1. triangular prism 2 units high, with two sides of the base that are 5 units long and 4 units long
2. rectangular prism 2 units high, 3 units wide, and 5 units long

Example 2 p. 824

Use isometric dot paper and each orthographic drawing to sketch a solid.

3.

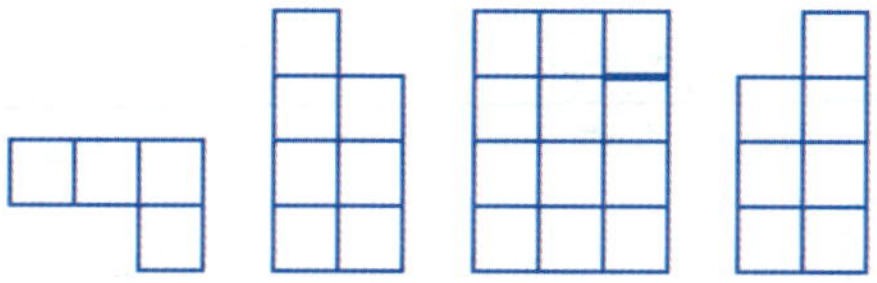

top view | left view | front view | right view

4.

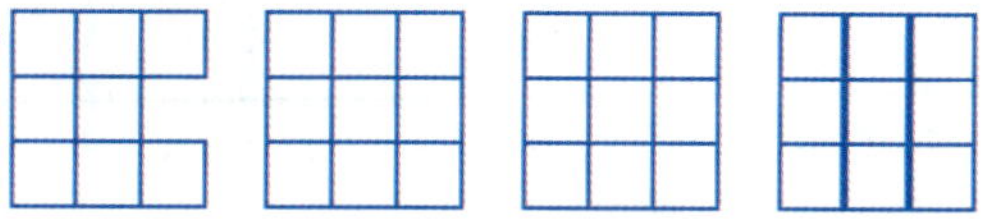

top view | left view | front view | right view

Example 3 p. 824

5. **FOOD** Describe how the cheese at the right can be sliced so that the slices form each shape.
   a. rectangle
   b. triangle
   c. trapezoid

Describe each cross section.

6.

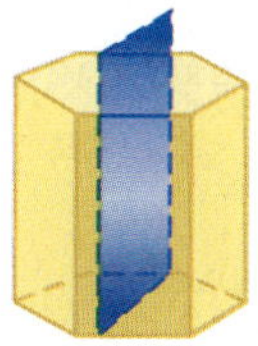

7.

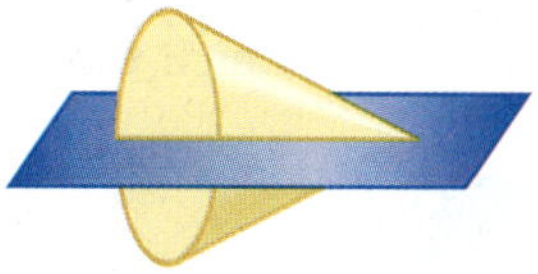

Practice and Problem Solving

= **Step-by-Step Solutions** begin on page R20.
Extra Practice begins on page 969.

Example 1 p. 823

Use isometric dot paper to sketch each prism.

8. cube 3 units on each edge
9. triangular prism 4 units high, with two sides of the base that are 1 unit long and 3 units long
10. triangular prism 4 units high, with two sides of the base that are 2 units long and 6 units long

Example 2 p. 824

Use isometric dot paper and each orthographic drawing to sketch a solid.

11.

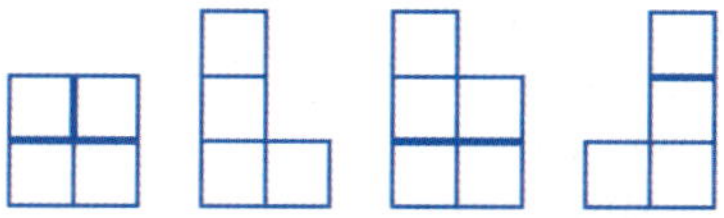

top view | left view | front view | right view

12.

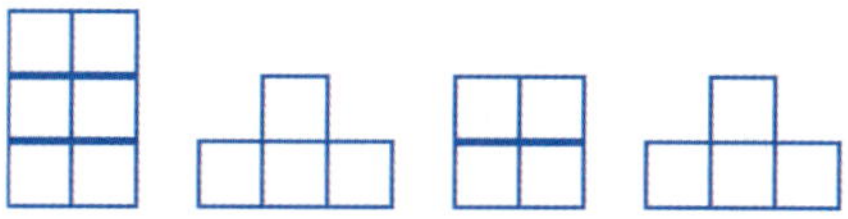

top view | left view | front view | right view

13.

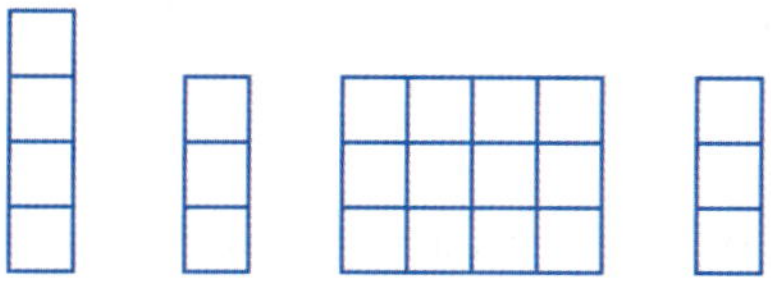

top view | left view | front view | right view

14.

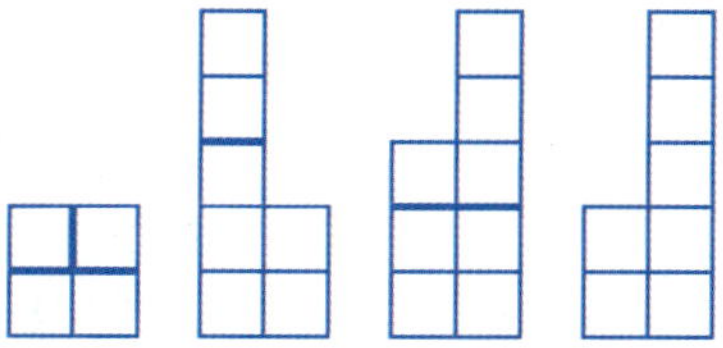

top view | left view | front view | right view

Example 3
p. 824

15. **ART** A piece of clay in the shape of a rectangular prism is cut in half as shown at the right.

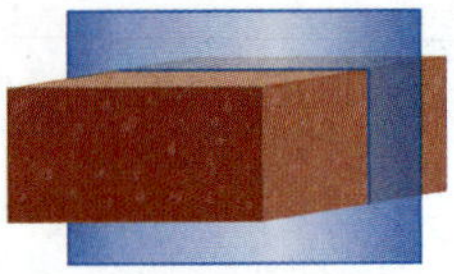

a. Describe the shape of the cross section.

b. Describe how the clay could be cut to make the cross section a triangle.

Describe each cross section.

16.

17.

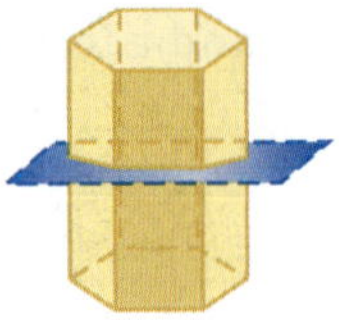

18.

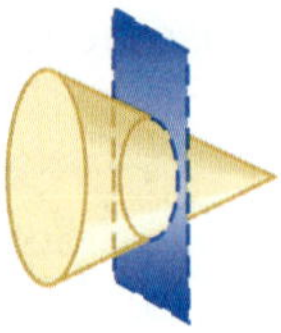

19.

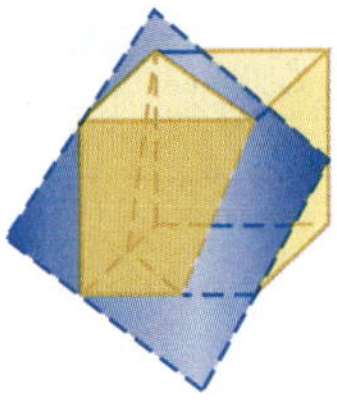

20. **ARCHITECTURE** Draw a top view, front view, and side view of the house at the right.

21. **COOKIES** Describe how to make a cut through a roll of cookie dough in the shape of a cylinder to make each shape.

a. circle **b.** rectangle **c.** oval

Review Vocabulary

regular polyhedron a polyhedron in which all of the faces are regular congruent polygons and all of the edges are congruent (Lesson 1-7)

Describe the net of each regular polyhedron.

22. octahedron **23.** tetrahedron **24.** dodecahedron

Sketch the cross section from a vertical slice of each figure.

25.

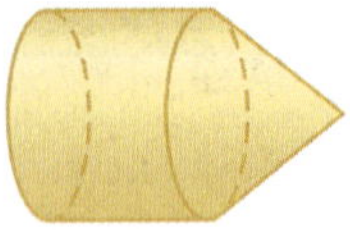

26.

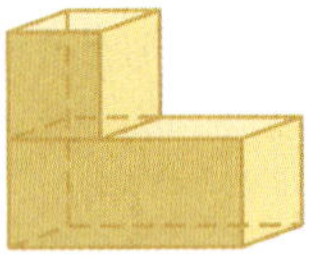

27.

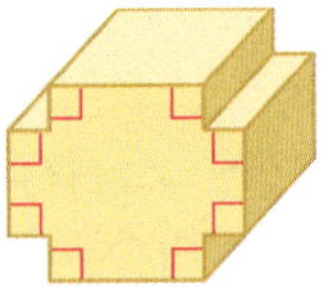

Real-World Link

The crystal shapes of minerals are grouped on the basis of geometric symmetry. The seven systems of three-dimensional patterns are shown below.

- cubic
- tetragonal
- hexagonal
- trigonal
- orthorhombic
- monoclinic
- triclinic

Source: Creetown Gem Rock Museum

28. **EARTH SCIENCE** Crystals are solids in which the atoms are arranged in regular geometrical patterns. Sketch a cross section from a horizontal slice of each crystal. Then describe the rotational symmetry about the vertical axis.

a. tetragonal **b.** hexagonal **c.** monoclinic

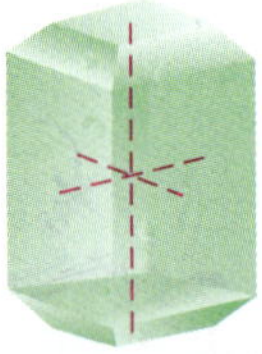

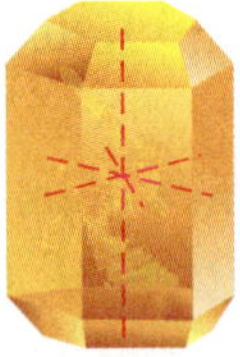

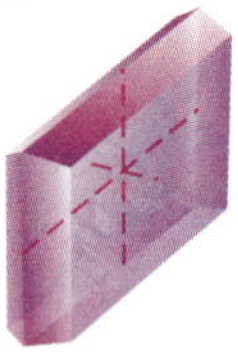

29. **ART** In a *perspective drawing*, a *vanishing point* is used to make the two-dimensional drawing appear three-dimensional. From one vanishing point, objects can be drawn from different points of view, as shown at the right.

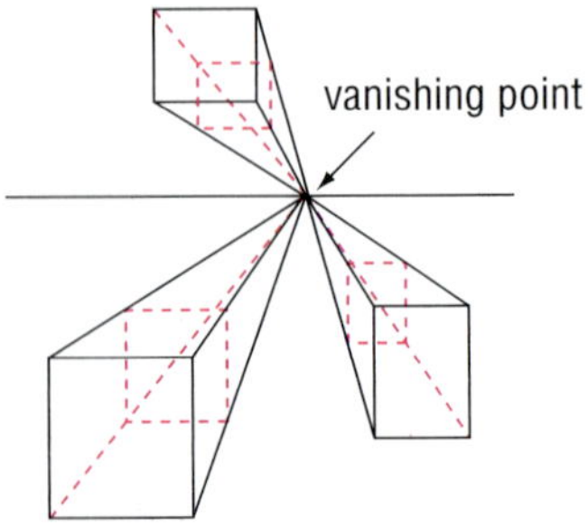

a. Draw a horizontal line and a vanishing point on the line. Draw a rectangle somewhere above the line and use the vanishing point to make a perspective drawing.

b. On the same drawing, draw a rectangle somewhere below the line and use the vanishing point to make a perspective drawing.

c. Describe the different views of the two drawings.

Draw the top, left, front, and right view of each solid.

30.

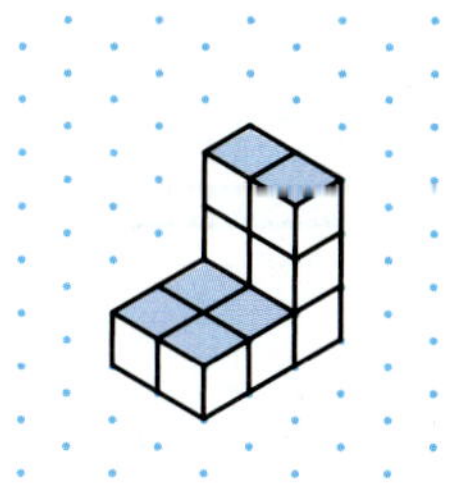

31.

32.

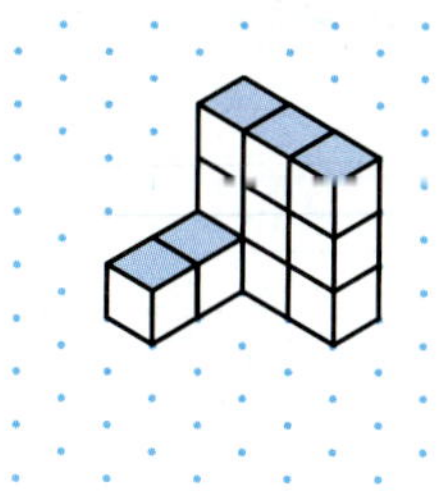

33. The top, front, and right views of a three-dimensional figure are shown at the right.

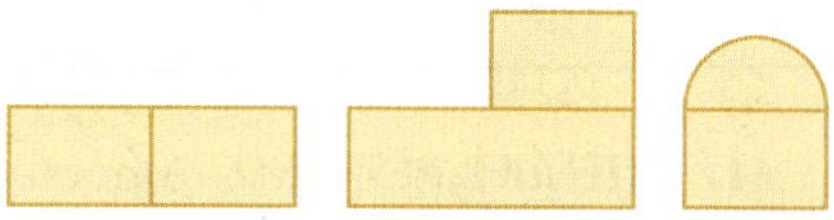

a. Make a sketch of the solid.

b. Describe two different ways that a rectangular cross section can be made.

c. Make a connection between the front and right views of the solid and cross sections of the solid.

34. **MULTIPLE REPRESENTATIONS** In this problem, you will investigate orthographic drawings.

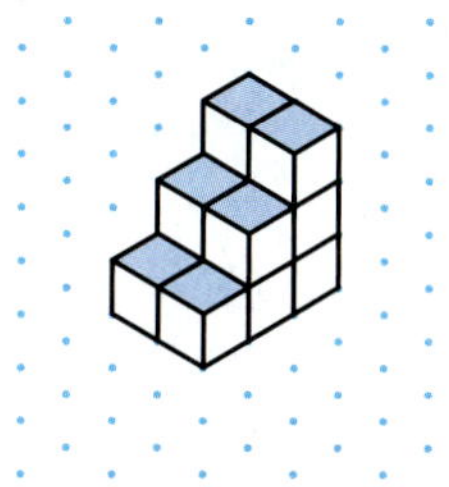

a. GEOMETRIC Create orthographic drawings of three different solids.

b. TABULAR Create a table that includes the number of cubes needed to construct the solid and the number of squares visible in the orthographic drawing.

c. VERBAL Is there a correlation between the number of cubes needed to construct a solid and the number of squares visible in the orthographic drawing? Explain.

H.O.T. Problems

Use **H**igher-**O**rder **T**hinking Skills

35. **CHALLENGE** The figure at the right is a cross section of a geometric solid. Describe the solid and how the cross section was made.

36. **REASONING** Determine whether the following statement is *true* or *false*. Explain your reasoning.

If the left, right, front, and back orthographic views of two objects are the same, then the objects are the same figure.

> **StudyTip**
>
> **Orthographic Views** Recall that all four orthographic views of a figure are required to define the figure.

37. **OPEN ENDED** Use isometric dot paper to draw a solid consisting of 12 cubic units. Then sketch the orthographic drawing for your solid.

38. **CHALLENGE** Draw the top view, front view, and side view of the solid figure at the right.

39. **WRITING IN MATH** A hexagonal pyramid is sliced through the vertex and the base so that the prism is separated into two congruent parts. Describe the cross section. Is there more than one way to separate the figure into two congruent parts? Will the shape of the cross section change? Explain.

Standardized Test Practice

40. Which polyhedron is represented by the net shown below?

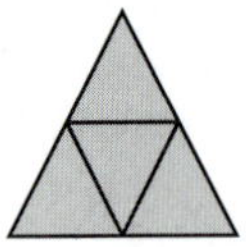

A cube
B octahedron
C triangular prism
D triangular pyramid

41. EXTENDED RESPONSE A homeowner wants to build a 3-foot-wide deck around his circular pool as shown below.

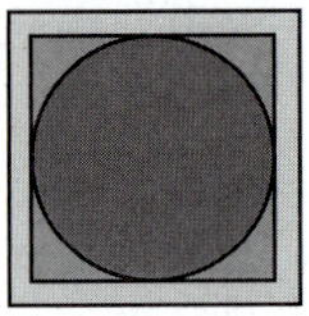

a. Find the outer perimeter of the deck to the nearest foot, if the circumference of the pool is about 81.64 feet.

b. What is the area of the top of the deck?

42. ALGEBRA Which inequality *best* describes the graph shown below?

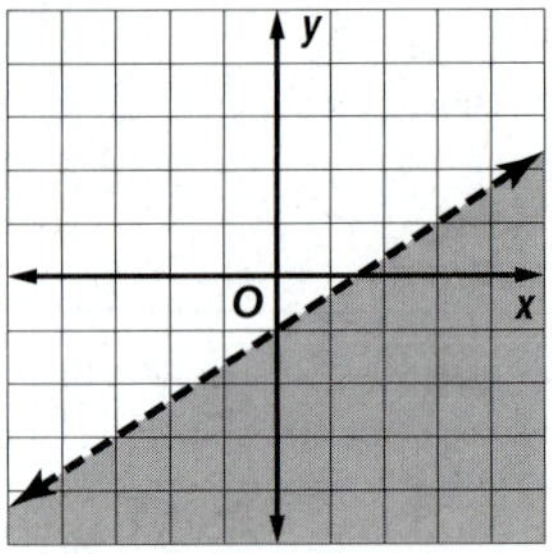

F $y < \frac{2}{3}x - 1$
G $y \leq \frac{2}{3}x - 1$
H $y > \frac{2}{3}x - 1$
J $y \geq \frac{2}{3}x - 1$

43. SAT/ACT Expand $(4\sqrt{5})^2$.

A 20
B $8\sqrt{5}$
C $16\sqrt{5}$
D 80

Spiral Review

For each pair of similar figures, use the given areas to find the scale factor from the blue to the green figure. Then find x. (Lesson 11-5)

44.

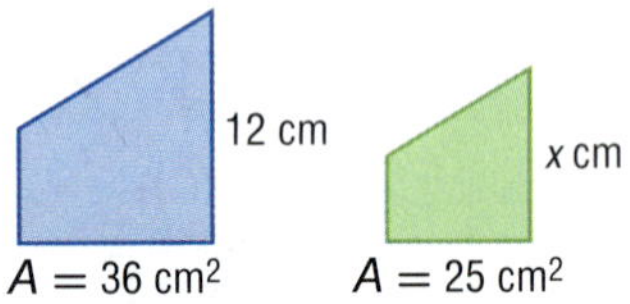

45.

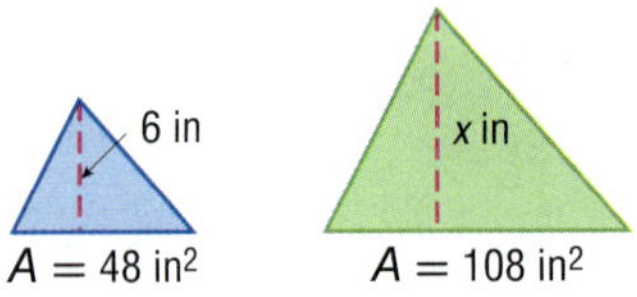

46.

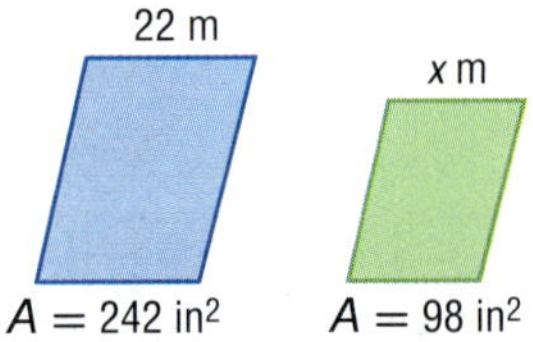

47. FURNITURE DESIGN Jenna wants to cover the cushions of her papasan chair with new fabric. There are seven congruent circular cushions with a diameter of 12 inches around a center cushion with a diameter of 20 inches. Find the area of fabric in square yards that she will need to cover both sides of the cushions. Allow an extra 3 inches of fabric around each cushion. (Lesson 11-4)

Skills Review

Find the perimeter or circumference and area of each figure. Round to the nearest tenth. (Lesson 1-6)

48.

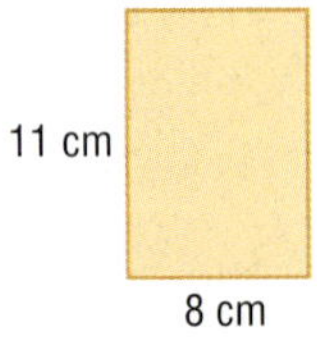

49.

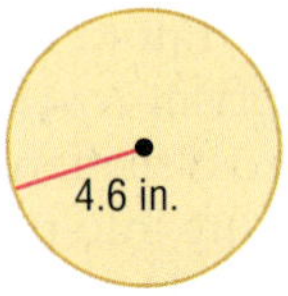

50.

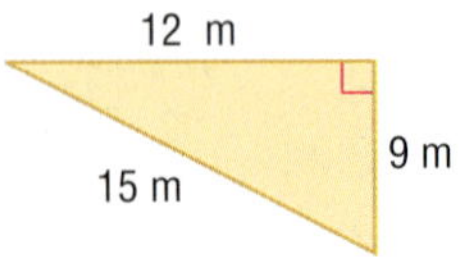

EXTEND
12-1

Geometry Lab
Topographic Maps

Objective
Interpret and draw topographic maps.

Maps are representations of Earth or some part of Earth. **Topographic maps** are representations of the three-dimensional surface of Earth on a two-dimensional piece of paper. In a topographic map, the *topography*, or shape of Earth's surface, is illustrated through the use of *contours*, which are imaginary lines that join locations with the same elevation.

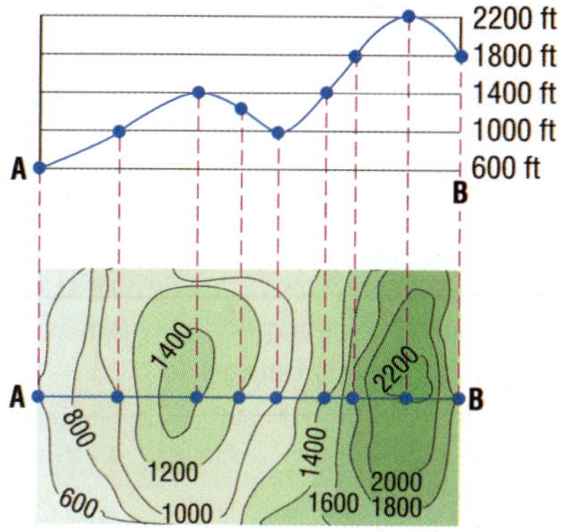

Some topographic maps show more than contours. These maps may include symbols that represent vegetation, rivers, and other landforms, as well as streets and buildings.

Follow these steps to read a topographic map.

- Thin lines represent contours. Since each contour is a line of equal elevation, they never cross. The closer together the contour lines, the steeper the slope.
- Contour lines form V shapes in valleys or riverbeds. The Vs point uphill.
- Most often, closed loops indicate that the surface slopes uphill on the inside and downhill on the outside. The innermost loop is the highest area.
- Pay attention to the colors. Blue represents water; green represents vegetation; red represents urban areas; black represents roads, trails, and railroads.
- The scale on a 1:24,000 map indicates that 1 inch equals 2000 feet.

ACTIVITY

Explore the Model

Use the topographic map above to answer these questions.

1. According to the scale, what is the vertical distance between each contour line?
2. What is the difference in height between the lowest and highest points?
3. What do you notice about the contour lines for the peaks of the hills?
4. Describe a steep slope on the topographic map. How do you know it is steep?
5. Explain how you would draw a topographic map given a side view of some hills.

Model and Analyze

6. Draw a topographic map similar to the map below for the side view of the hills from points *A* to *B*.

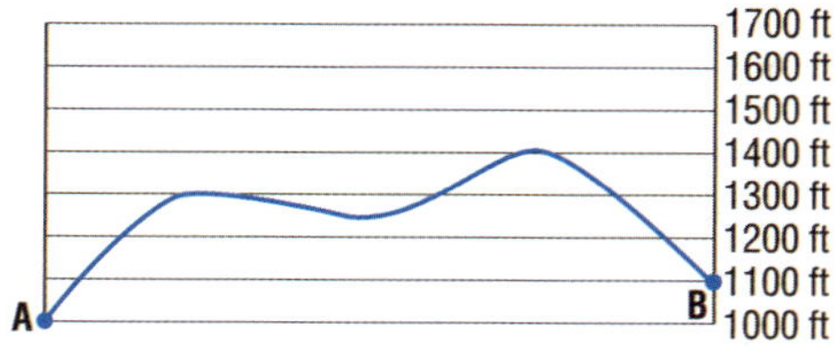

7. Draw a possible side view similar to the map below from points *A* to *B* of the hills from the topographic map. Measures are given in feet.

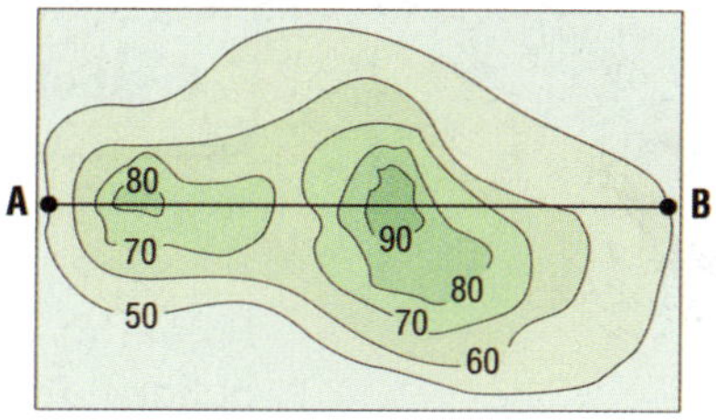

12-2 Surface Areas of Prisms and Cylinders

Then

You found areas of polygons. (Lesson 11-2)

Now

- Find lateral areas and surface areas of prisms.
- Find lateral areas and surface areas of cylinders.

New Vocabulary

lateral face
lateral edge
base edge
altitude
height
lateral area
axis
composite solid

Math Online
glencoe.com

Why?

Atlanta's Georgia Aquarium is the largest aquarium in the world, with more than 8 million gallons of water and more than 500 species from around the world. The aquarium has an underwater tunnel that is 100 feet long with 4574 square feet of viewing windows.

Lateral Areas and Surface Areas of Prisms In a solid figure, faces that are not bases are called **lateral faces**. Lateral faces intersect each other at the **lateral edges**, which are all parallel and congruent. The lateral faces intersect the base at the **base edges**. The **altitude** is a perpendicular segment that joins the planes of the bases. The **height** is the length of the altitude.

Recall that a prism is a polyhedron with two parallel congruent bases. In a right prism, the lateral edges are altitudes and the lateral faces are rectangles. In an oblique prism, the lateral edges are not perpendicular to the bases. At least one lateral face is not a rectangle.

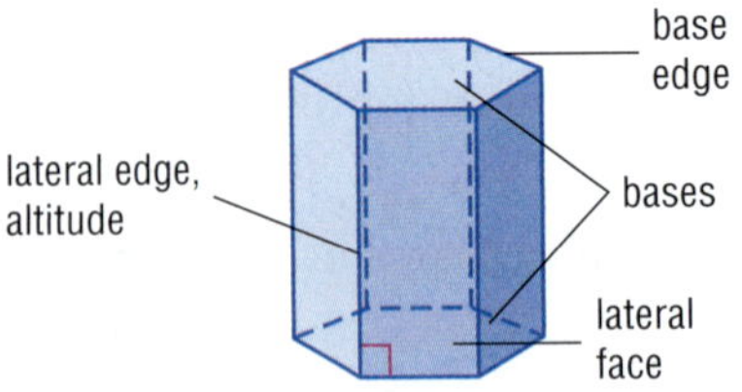

Right Prism

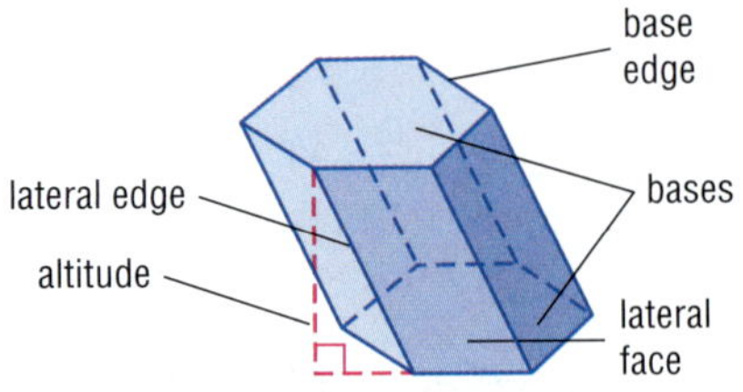

Oblique Prism

The **lateral area** L of a prism is the sum of the areas of the lateral faces. The net at the right shows how to find the lateral area of a prism.

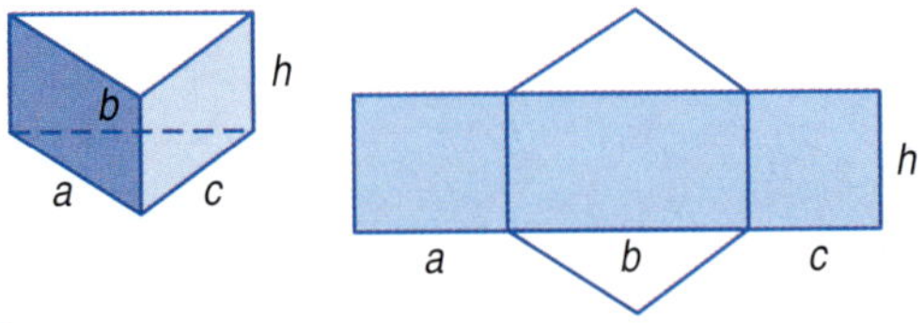

$L = a(h) + b(h) + c(h)$ — **Sum of areas of lateral faces**

$= (a + b + c)h$ — **Distributive Property**

$= Ph$ — $\boldsymbol{P = a + b + c}$

Key Concept: Lateral Area of a Prism

Words The lateral area L of a right prism is $L = Ph$, where h is the height of the prism and P is the perimeter of a base.

Symbols $L = Ph$

Model

From this point on, you can assume that solids in the text are right solids. If a solid is oblique, it will be clearly stated.

EXAMPLE 1 Lateral Area of a Prism

Find the lateral area of the prism. Round to the nearest tenth.

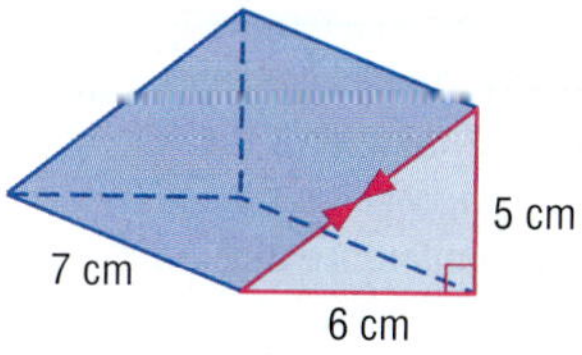

Step 1 Find the missing side length of the base.

$c^2 = 6^2 + 5^2$ — **Pythagorean Theorem**

$c^2 = 61$ — **Simplify.**

$c \approx 7.8$ — **Take the square root of each side.**

Step 2 Find the lateral area.

$L = Ph$ — **Lateral area of a prism**

$\approx (5 + 6 + 7.8)7$ — **Substitution**

≈ 131.6 — **Simplify.**

The lateral area is about 131.6 square centimeters.

Guided Practice

1. The length of each side of the base of a regular octagonal prism is 6 inches, and the height is 11 inches. Find the lateral area.

Personal Tutor glencoe.com

Watch Out!

Right Prisms The bases of a right prism are congruent, but the faces are not always congruent.

The surface area of a prism is the sum of the lateral area and the areas of the bases.

Key Concept Surface Area of a Prism

Words The surface area S of a right prism is $S = L + 2B$, where L is its lateral area and B is the area of a base.

Symbols $S = L + 2B$ or $S = Ph + 2B$

Model

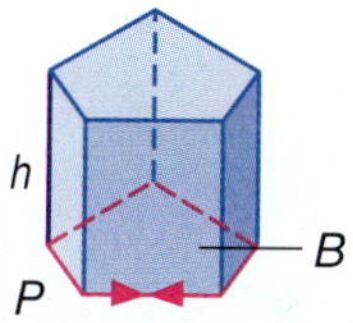

EXAMPLE 2 Surface Area of a Prism

Find the surface area of the rectangular prism.

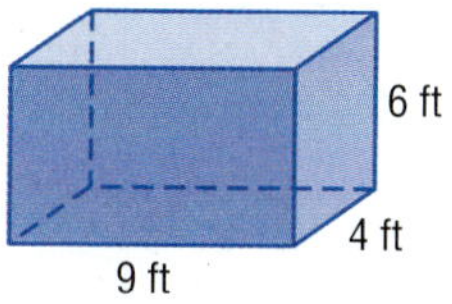

Use the 9-foot by 4-foot rectangle as the base.

$S = Ph + 2B$ — **Surface area of a prism**

$= (2 \cdot 9 + 2 \cdot 4)(6) + 2(9 \cdot 4)$ — **Substitution**

$= 228$ — **Simplify.**

The surface area of the prism is 228 square feet.

Guided Practice

2. Find the surface area of the triangular prism. Round to the nearest tenth.

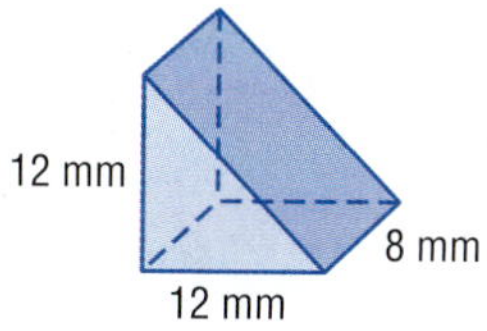

Personal Tutor glencoe.com

StudyTip

Alternate Method In Example 2, you can also use a 6-foot by 4-foot rectangle as the base. The height would be 9 feet.

StudyTip

Formulas An alternate formula for the lateral area of a cylinder is $L = \pi dh$, with πd as the circumference of a circle.

Lateral Areas and Surface Areas of Cylinders

The **axis** of a cylinder is the segment with endpoints that are centers of the circular bases. If the axis is also an altitude, then the cylinder is a right cylinder. If the axis is not an altitude, then the cylinder is an oblique cylinder.

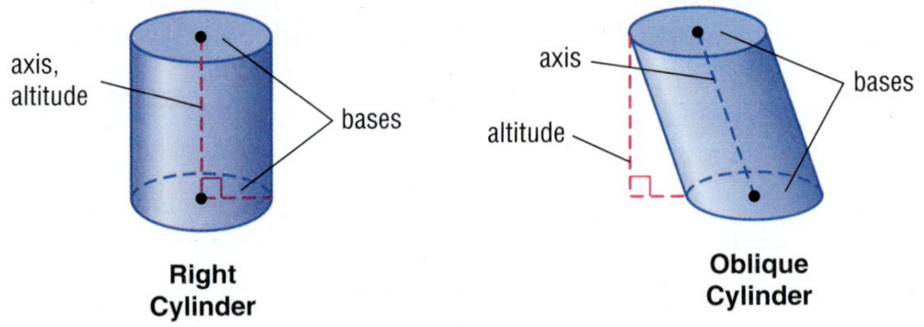

The lateral area of a cylinder is the area of the curved surface. Like a prism, the lateral area L equals Ph. Since the base is a circle, the perimeter is the circumference of the circle C. So, the lateral area is Ch or $2\pi rh$.

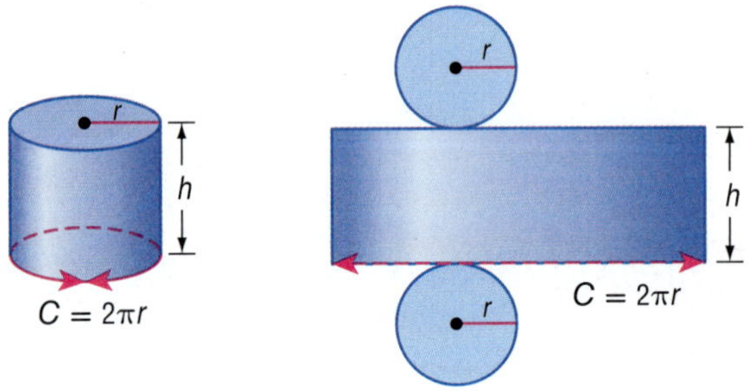

The surface area of a cylinder is the lateral area plus the areas of the bases.

Key Concept — Areas of a Cylinder

Words The lateral area L of a right cylinder is $L = 2\pi rh$, where r is the radius of a base and h is the height.

The surface area S of a right cylinder is $S = 2\pi rh + 2\pi r^2$, where r is the radius of a base and h is the height.

Model

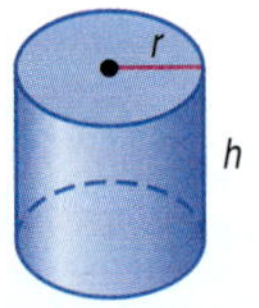

Symbols $L = 2\pi rh$

$S = L + 2B$ or $2\pi rh + 2\pi r^2$

EXAMPLE 3 Lateral Area and Surface Area of a Cylinder

Find the lateral area and the surface area of the cylinder. Round to the nearest tenth.

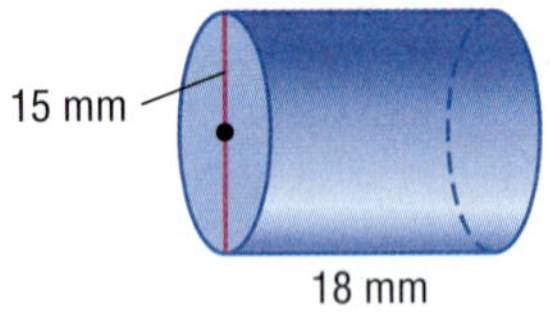

$L = 2\pi rh$ — **Lateral area of a cylinder**

$= 2\pi(7.5)(18)$ — **Replace r with 7.5 and h with 18.**

≈ 848.2 — **Use a calculator.**

$S = 2\pi rh + 2\pi r^2$ — **Surface area of a cylinder**

$\approx 848.2 + 2\pi(7.5)^2$ — **Replace $2\pi rh$ with 848.2 and r with 7.5.**

≈ 1201.6 — **Use a calculator.**

The lateral area is about 848.2 square millimeters, and the surface area is about 1201.6 square millimeters.

StudyTip

Estimation Before finding the lateral area of a cylinder, use mental math to estimate. To estimate, multiply the diameter by 3 (to approximate π) and then by the height of the cylinder.

Guided Practice

3A. $r = 5$ in., $h = 9$ in.

3B. $d = 6$ cm, $h = 4.8$ cm

Personal Tutor glencoe.com

Real-World EXAMPLE 4 Find Missing Dimensions

CRAFTS **Sheree used the rectangular piece of felt shown at the right to cover the curved surface of her cylindrical pencil holder. What is the radius of the pencil holder?**

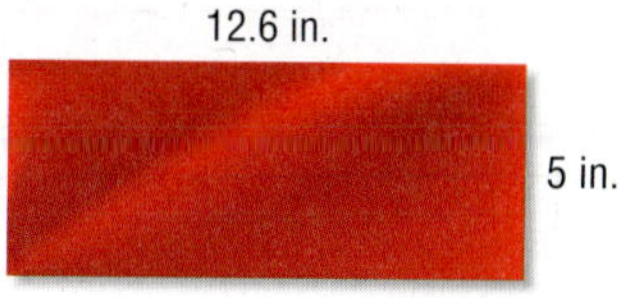

$L = 2\pi rh$	**Lateral area of a cylinder**
$63 = 2\pi r(5)$	**Replace L with $12.6 \cdot 5$ or 63 and h with 5.**
$63 = 10\pi r$	**Simplify.**
$2.0 \approx r$	**Divide each side by 10π.**

The radius of the pencil holder is about 2 inches.

Guided Practice

4. Find the diameter of a base of a cylinder if the surface area is 464π square centimeters and the height is 21 centimeters.

Personal Tutor glencoe.com

Check Your Understanding

Example 1 p. 831

1. Find the lateral area of the prism.

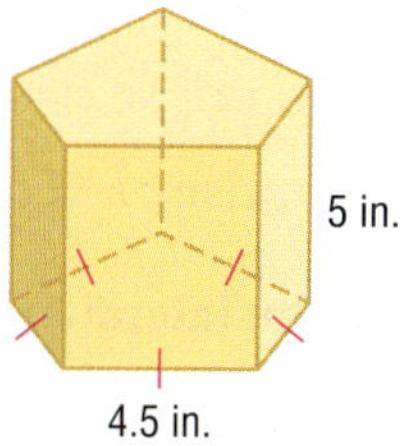

Examples 1 and 2 p. 831

Find the lateral area and surface area of each prism.

2.

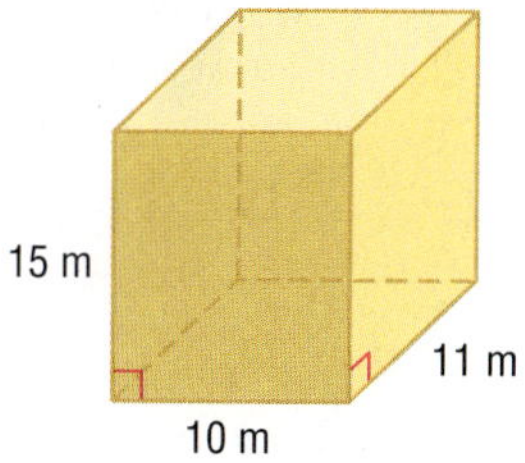

3.

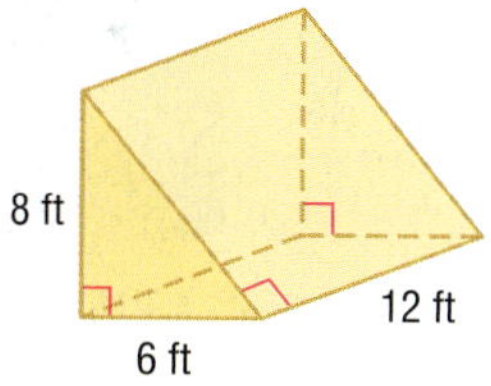

Example 3 p. 832

4. **CARS** Evan is buying new tire rims that are 14 inches in diameter and 6 inches wide. Determine the lateral area of each rim. Round to the nearest tenth.

Find the lateral area and surface area of each cylinder. Round to the nearest tenth.

5.

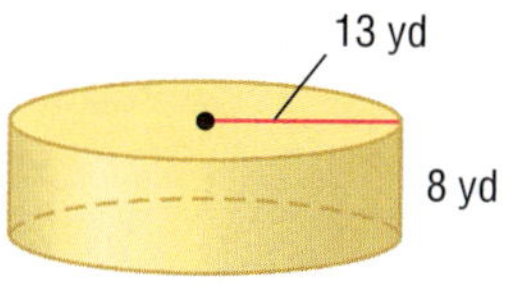

6.

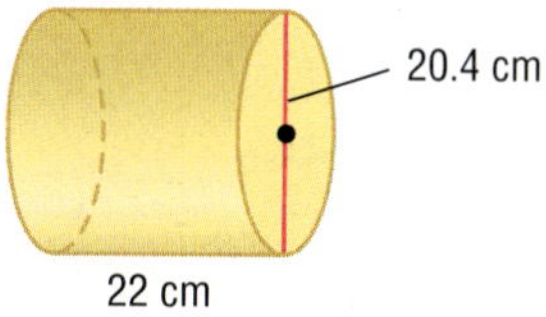

Example 4 p. 833

7. **FOOD** The can of soup at the right has a surface area of 286.3 square centimeters. What is the height of the can? Round to the nearest tenth.

8. The surface area of a cube is 294 square inches. Find the length of a lateral edge.

Practice and Problem Solving

● = **Step-by-Step Solutions** begin on page R20.
Extra Practice begins on page 969.

Examples 1 and 2 p. 831

Find the lateral area and surface area of each prism. Round to the nearest tenth if necessary.

9.
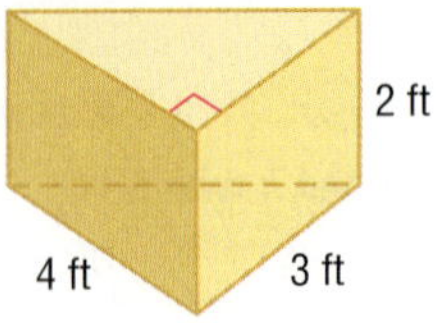

10.
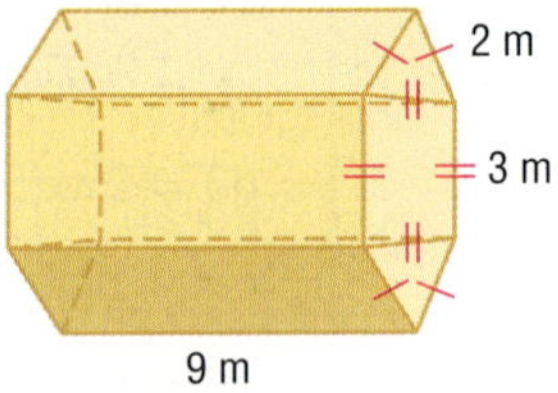

11.
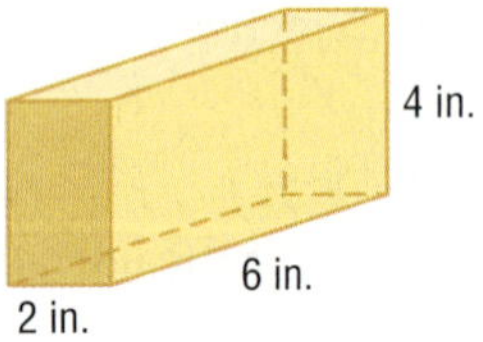

12.
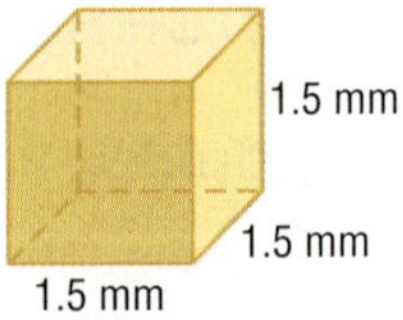

13.
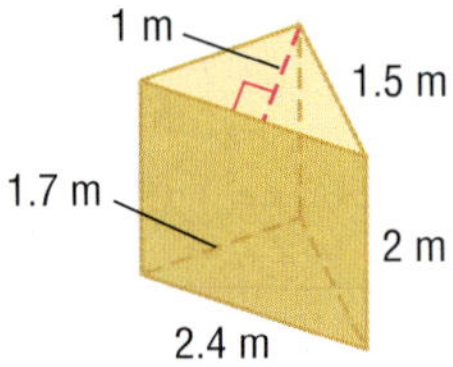

14.
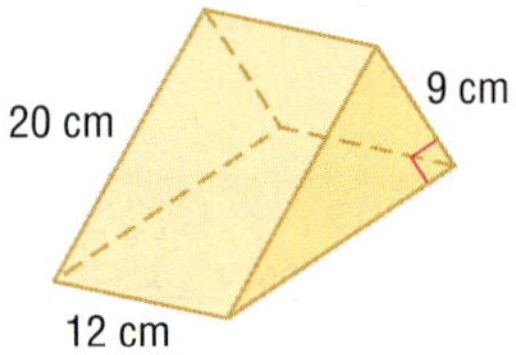

15. rectangular prism: $\ell = 25$ centimeters, $w = 18$ centimeters, $h = 12$ centimeters

16. triangular prism: $h = 6$ inches, right triangle base with legs 9 inches and 12 inches

Examples 1–3 pp. 831–832

CEREAL **Find the lateral area and the surface area of each cereal container. Round to the nearest tenth if necessary.**

17.

18.
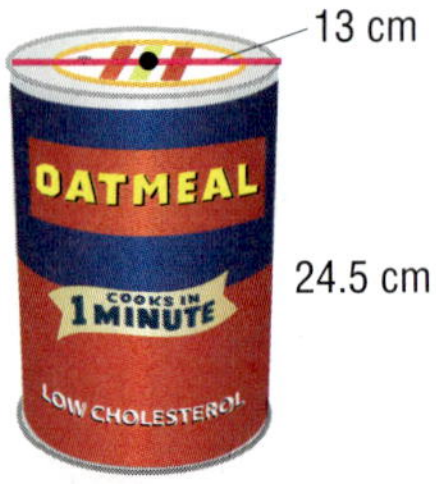

Example 3 p. 832

Find the lateral area and surface area of each cylinder. Round to the nearest tenth.

19.

20.
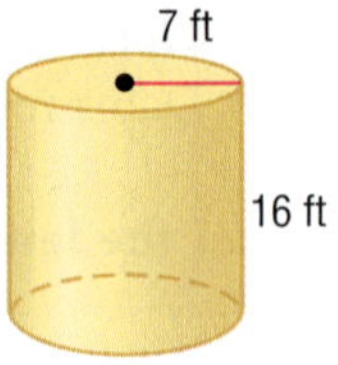

21.
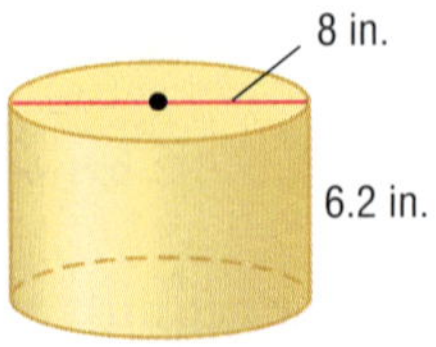

22.
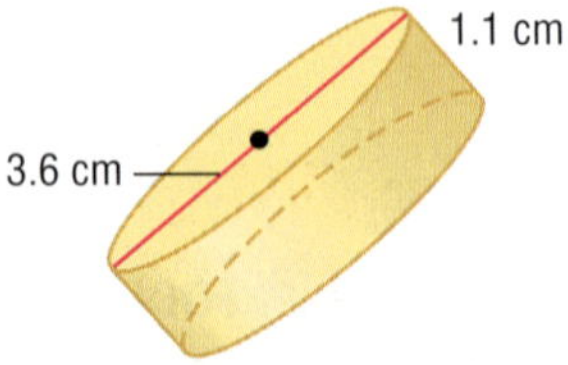

23. **WORLD RECORDS** The largest beverage can was a cylinder with height 4.67 meters and diameter 2.32 meters. What was the surface area of the can to the nearest tenth?

Example 4
p. 833

Use the given lateral area and the diagram to find the missing measure of each solid. Round to the nearest tenth if necessary.

24. $L = 48 \text{ in}^2$

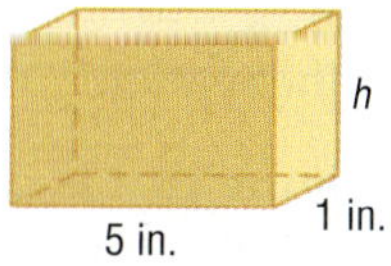

25. $L \approx 635.9 \text{ cm}^2$

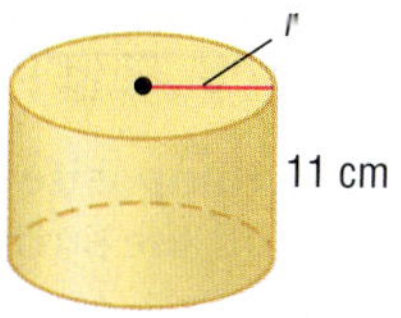

26. A right rectangular prism has a surface area of 1020 square inches, a length of 6 inches, and a width of 9 inches. Find the height.

27 A cylinder has a surface area of 256π square millimeters and a height of 8 millimeters. Find the diameter.

28. **MONUMENTS** The *monolith* shown at right mysteriously appeared overnight at Seattle, Washington's Manguson Park. It is a hollow rectangular prism 9 feet tall, 4 feet wide, and 1 foot deep.

a. Find the area in square feet of the structure's surfaces that lie above the ground.

b. Use dimensional analysis to find the area in square yards.

Real-World Link

The top-grossing movie genre from 1995 to 2007 was comedy. About 1260 comedies were made during these years, grossing more than $29 billion.

Source: Nash Information Services, LLC

29. **ENTERTAINMENT** The graphic shows the results of a survey in which people were asked where they like to watch movies.

a. Suppose the film can is a cylinder 12 inches in diameter. Explain how to find the surface area of the portion that represents people who prefer to watch movies at home.

b. If the film can is 3 inches tall, find the surface area of the portion in part **a**.

Find the lateral area and surface area of each oblique prism. Round to the nearest tenth if necessary.

30.

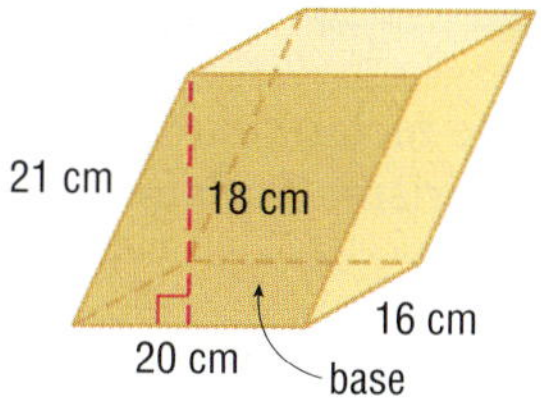

31.

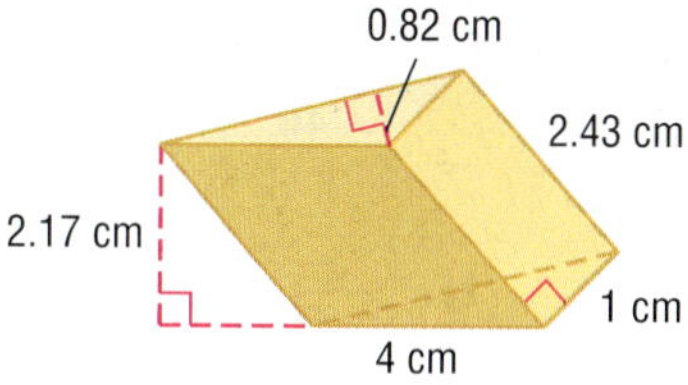

32. **LAMPS** The lamp shade is a cylinder of height 18 inches with a diameter of $6\frac{3}{4}$ inches.

a. What is the lateral area of the shade to the nearest tenth?

b. How does the lateral area change if the height is divided by 2?

33. Find the approximate surface area of a right hexagonal prism if the height is 9 centimeters and each base edge is 4 centimeters. (*Hint*: First, find the length of the apothem of the base.)

Review Vocabulary

area of a regular polygon $A = \frac{1}{2}Pa$, where P is the perimeter and a is the apothem (Lesson 11-4)

34. **MAIL** A mailer for posters is a triangular prism. The base is an equilateral triangle with sides that measure 6 inches.

a. Sketch a net for the prism.

b. Find the approximate surface area of the mailer.

A composite solid is a three-dimensional figure that is composed of simpler figures. Find the surface area of each composite solid. Round to the nearest tenth if necessary.

35.

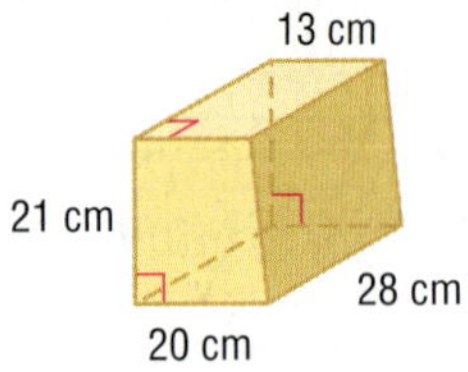

36.

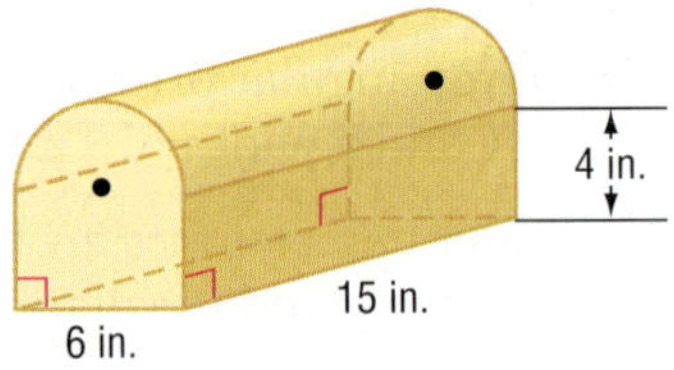

37.

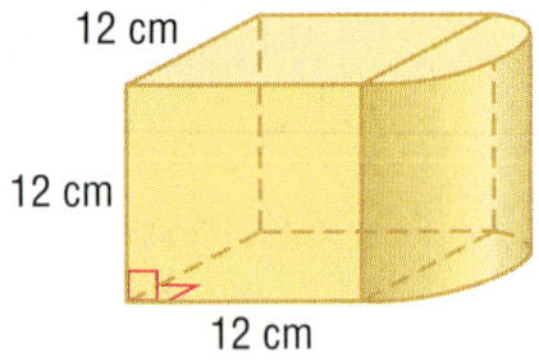

38. **MULTIPLE REPRESENTATIONS** In this problem, you will investigate the lateral area and surface area of a cylinder.

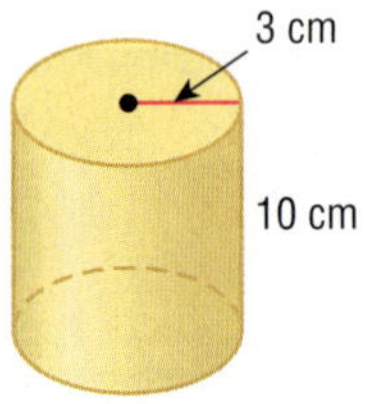

a. GEOMETRIC Sketch three different cylinders A, B, and C. Measure and label the radius and height.

b. TABULAR Create a table of the radius, height, lateral area, and surface area of cylinders A, B, and C. Write the areas in terms of π.

c. VERBAL If the radius is doubled, what effect does it have on the lateral area and the surface area of a cylinder? If the height is doubled, what effect does it have on the lateral area and the surface area of a cylinder?

StudyTip

Composite Solids To find the surface of a composite solid, it may be helpful to draw and label a net of the solid.

H.O.T. Problems

Use Higher-Order Thinking Skills

39. **ERROR ANALYSIS** Montell and Derek are finding the surface area of a cylinder with height 5 centimeters and radius 6 centimeters. Is either of them correct? Explain.

Montell	Derek
$S = \pi(6)^2 + \pi(6)(5)$	$S = 2\pi(6)^2 + 2\pi(6)(5)$
$= 36\pi + 30\pi$	$= 72\pi + 60\pi$
$= 66\pi \text{ cm}^2$	$= 132\pi \text{ cm}^2$

40. **WRITING IN MATH** Sketch an oblique rectangular prism, and describe the shapes that would be included in a net for the prism. Explain how the net is different from that of a right rectangular prism.

41. **REASONING** Compare and contrast finding the surface area of a prism and finding the surface area of a cylinder.

42. **OPEN ENDED** Give an example of two cylinders that have the same lateral area and different surface areas. Describe the lateral area and surface areas of each.

43. **CHALLENGE** A right prism has a height of h units and a base that is an equilateral triangle of side ℓ units. Find the general formula for the total surface area of the prism. Explain your reasoning.

44. **WRITING IN MATH** A rectangular prism and a triangular prism are the same height. The base of the triangular prism is an isosceles triangle, the altitude of which is the same as the height of the base of the rectangular prism. Compare the lateral areas of the prisms.

Standardized Test Practice

45. If the surface area of the right rectangular prism is 310 square centimeters, what is the measure of the height h of the prism?

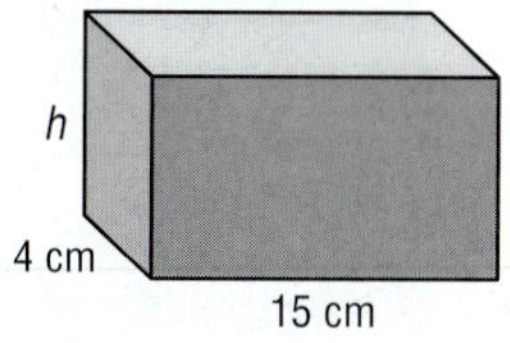

A 5 cm
B $5\frac{1}{6}$ cm
C 10
D $13\frac{3}{9}$ cm

46. SHORT RESPONSE A cylinder has a circumference of 16π inches and a height of 20 inches. What is the surface area of the cylinder in terms of π?

47. ALGEBRA The scores for a class on a 30-point math quiz are shown in the stem-and-leaf plot below. What was the mean score for this quiz?

Stem	Leaf
3	0 0
2	2 2 3 3 4 6 7 7 7 8 9
1	8 9

$2|8 = 28$

F 12
G 24
H 25
J 27

48. SAT/ACT What is the value of $f(-2)$ if $f(x) = x^3 + 4x^2 - 2x - 3$?

A -31
B $-\frac{9}{2}$
C 9
D 25

Spiral Review

Use isometric dot paper to sketch each prism. (Lesson 12-1)

49. rectangular prism 2 units high, 3 units long, and 2 units wide

50. triangular prism 2 units high with bases that are right triangles with legs 3 units and 4 units long

51. BAKING A bakery sells single-layer mini-cakes that are 3 inches in diameter for \$4 each. They also have a cake with a 9-inch diameter for \$15. If both cakes are the same thickness, which option gives you more cake for the money, nine mini-cakes or one 9-inch cake? Explain. (Lesson 11-5)

The diameters of ⊙*R*, ⊙*S*, and ⊙*T* are 10 inches, 14 inches, and 9 inches, respectively. Find each measure. (Lesson 10-1)

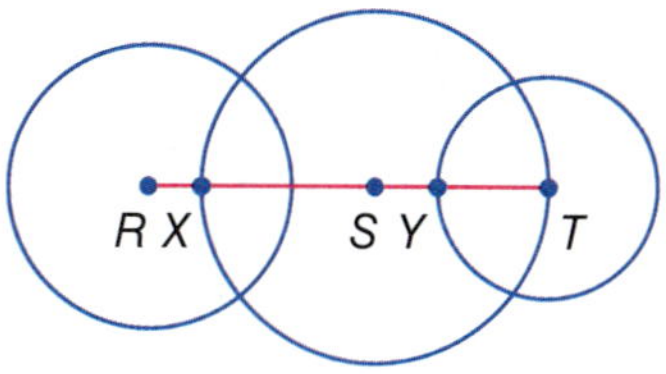

52. YX

53. SY

Skills Review

Find x. Round to the nearest tenth. (Lesson 8-4)

54.

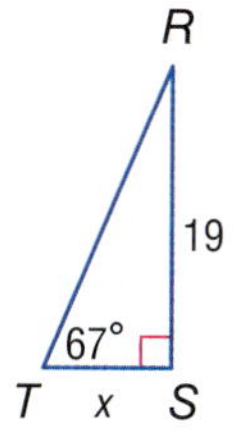

55.

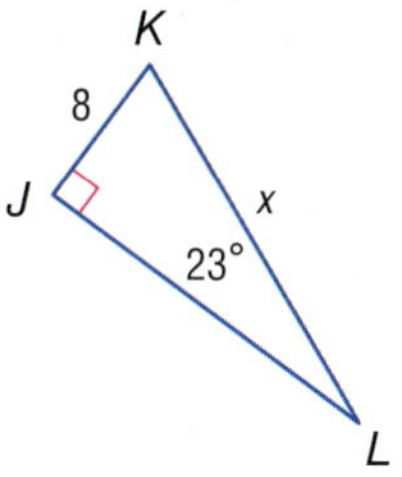

56.

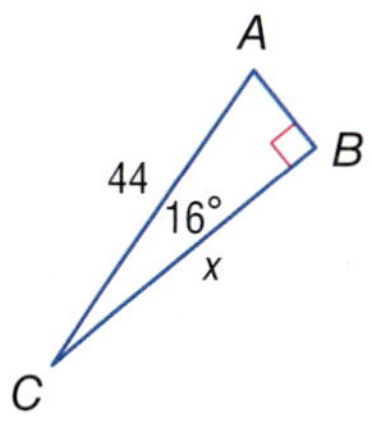

12-3 Surface Areas of Pyramids and Cones

Then

You found areas of regular polygons. (Lesson 11-3)

Now

- Find lateral areas and surface areas of pyramids.
- Find lateral areas and surface areas of cones.

New Vocabulary

regular pyramid
slant height
right cone
oblique cone

Math Online
glencoe.com

Why?

The Transamerica Pyramid in San Francisco, California, covers nearly one city block. Its unconventional design allows light and air to filter down to the streets around the building, unlike the more traditional rectangular prism skyscrapers.

Lateral Area and Surface Area of Pyramids The *lateral faces* of a pyramid intersect at a common point called the *vertex*. Two lateral faces intersect at a *lateral edge*. A lateral face and the base intersect at a *base edge*. The *altitude* is the segment from the vertex perpendicular to the base.

A **regular pyramid** has a base that is a regular polygon and the altitude has an endpoint at the center of the base. All the lateral edges are congruent and all the lateral faces are congruent isosceles triangles. The height of each lateral face is called the **slant height** ℓ of the pyramid.

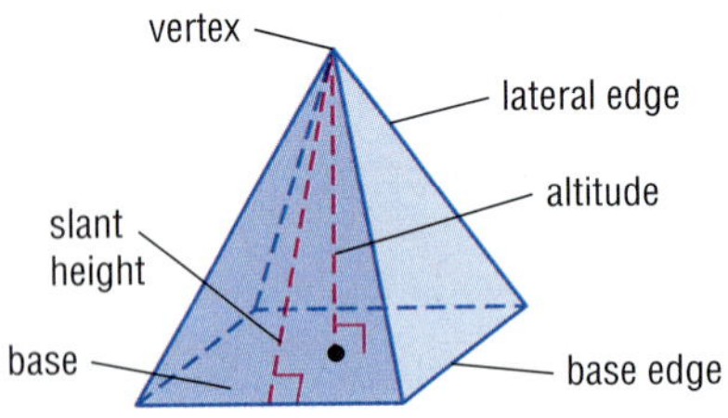

Regular Pyramid

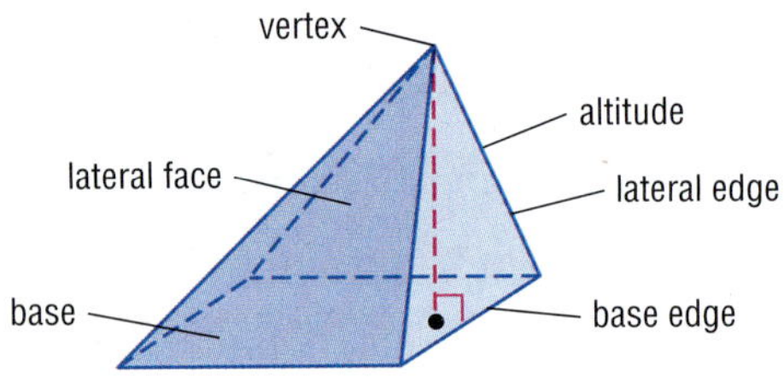

Nonregular Pyramid

The lateral area L of a regular pentagonal pyramid is the sum of the areas of all its congruent triangular faces as shown in the net at the right.

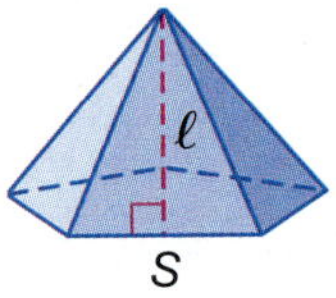

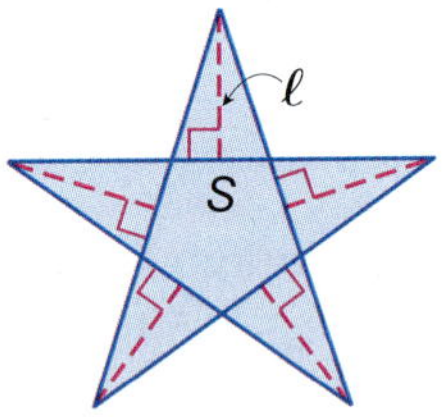

$$L = \frac{1}{2}s\ell + \frac{1}{2}s\ell + \frac{1}{2}s\ell + \frac{1}{2}s\ell + \frac{1}{2}s\ell \quad \text{Sum of the areas of the lateral faces}$$

$$= \frac{1}{2}\ell(s + s + s + s + s) \quad \text{Distributive Property}$$

$$= \frac{1}{2}P\ell \quad P = s + s + s + s + s$$

Key Concept **Lateral Area of a Regular Pyramid**

Words The lateral area L of a regular pyramid is $L = \frac{1}{2}P\ell$, where ℓ is the slant height and P is the perimeter of the base.

Symbols $L = \frac{1}{2}P\ell$

Model

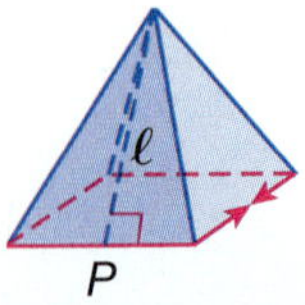

StudyTip

Alternative Method You can also find the lateral area of a pyramid by adding the areas of the congruent lateral faces.

area of one face: $\frac{1}{2}(4)(6) = 12 \text{ in}^2$

lateral area: $4 \cdot 12 = 48 \text{ in}^2$

EXAMPLE 1 Lateral Area of a Regular Pyramid

Find the lateral area of the square pyramid.

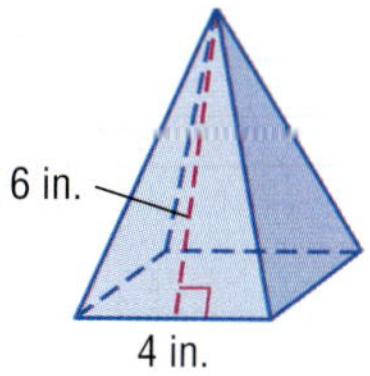

$L = \frac{1}{2}P\ell$	**Lateral area of a regular pyramid**
$= \frac{1}{2}(16)(6)$ or 48	$P = 4 \cdot 4$ **or 16,** $\ell = 6$

The lateral area is 48 square inches.

Guided Practice

1. Find the lateral area of a regular hexagonal pyramid with a base edge of 9 centimeters and a lateral height of 7 centimeters.

Personal Tutor glencoe.com

The surface area of a pyramid is the sum of the lateral area and the area of the base.

Key Concept Surface Area of a Regular Pyramid

Words The surface area S of a regular pyramid is $S = \frac{1}{2}P\ell + B$, where P is the perimeter of the base, ℓ is the slant height, and B is the area of the base.

Model

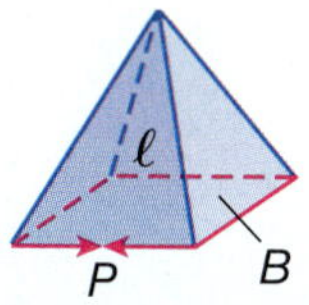

Symbols $S = \frac{1}{2}P\ell + B$

EXAMPLE 2 Surface Area of a Square Pyramid

Find the surface area of the square pyramid to the nearest tenth.

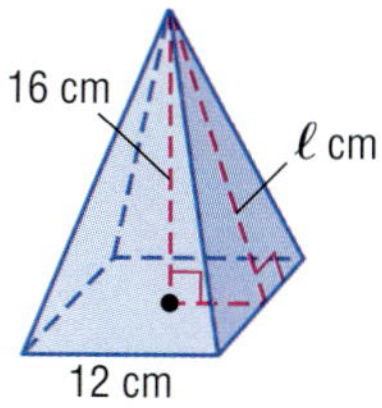

Step 1 Find the slant height.

$c^2 = a^2 + b^2$	**Pythagorean Theorem**
$\ell^2 = 16^2 + 6^2$	$a = 16, b = 6$, **and** $c = \ell$
$\ell = \sqrt{292}$	**Simplify.**

Step 2 Find the perimeter and area of the base.

$P = 4 \cdot 12$ or 48 cm $\qquad A = 12^2$ or 144 cm^2

Step 3 Find the surface area of the pyramid.

$S = \frac{1}{2}P\ell + B$	**Surface area of a regular pyramid**
$= \frac{1}{2}(48)\sqrt{292} + 144$	$P = 48, \ell = \sqrt{292}$, **and** $B = 144$
≈ 554.1	**Use a calculator.**

The surface area of the pyramid is about 554.1 square centimeters.

StudyTip

Making Connections The surface area of a pyramid equals $L + B$, not $L + 2B$, because a pyramid has only one base.

Guided Practice

2A.

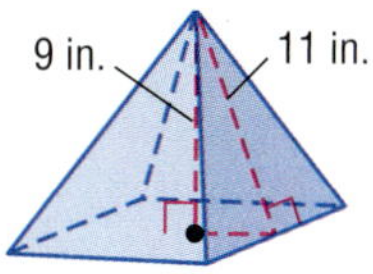

2B.

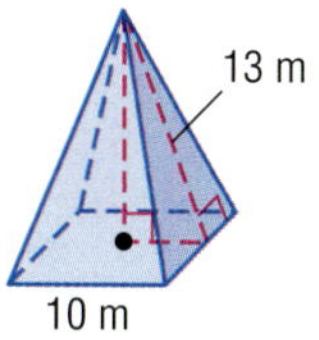

Personal Tutor glencoe.com

EXAMPLE 3 Surface Area of a Regular Pyramid

Find the surface area of the regular pyramid. Round to the nearest tenth.

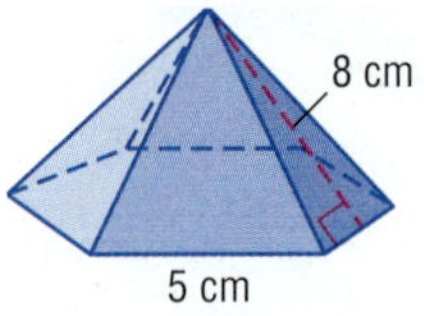

Step 1 Find the perimeter of the base.

$P = 6 \cdot 5$ or 30 cm

Step 2 Find the length of the apothem and the area of the base.

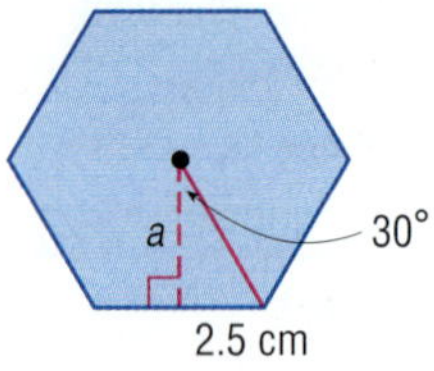

A central angle of the pentagon is $\frac{360°}{6}$ or 60°, so the angle formed in the triangle at the right is 30°.

$\tan 30° = \frac{2.5}{a}$ **Write a trigonometric ratio to find the apothem *a*.**

$a = \frac{2.5}{\tan 30°}$ **Solve for *a*.**

≈ 4.3 **Use a calculator.**

$A = \frac{1}{2}Pa$ **Area of a regular polygon**

$\approx \frac{1}{2}(30)(4.3)$ **Replace *P* with 30 and *a* with 4.3.**

≈ 64.5 **Multiply.**

So, the area of the base *B* is approximately 64.5 square centimeters.

Step 3 Find the surface area of the pyramid.

$S = \frac{1}{2}P\ell + B$ **Surface area of a regular pyramid**

$= \frac{1}{2}(30)(8) + 64.5$ **$P = 30$, $\ell = 8$, and $B \approx 64.5$**

≈ 184.5 **Simplify.**

The surface area of the pyramid is about 184.5 square centimeters.

Guided Practice

3A.

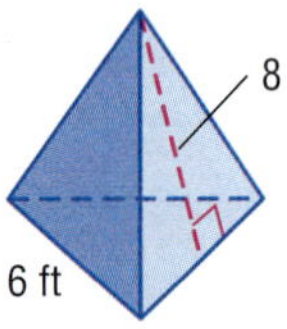

3B.

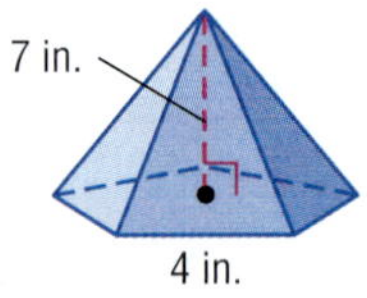

Personal Tutor glencoe.com

Lateral Area and Surface Area of Cones Recall that a cone has a circular base and a vertex. The axis of a cone is the segment with endpoints at the vertex and the center of the base. If the axis is also the altitude, then the cone is a **right cone**. If the axis is not the altitude, then the cone is an **oblique cone**.

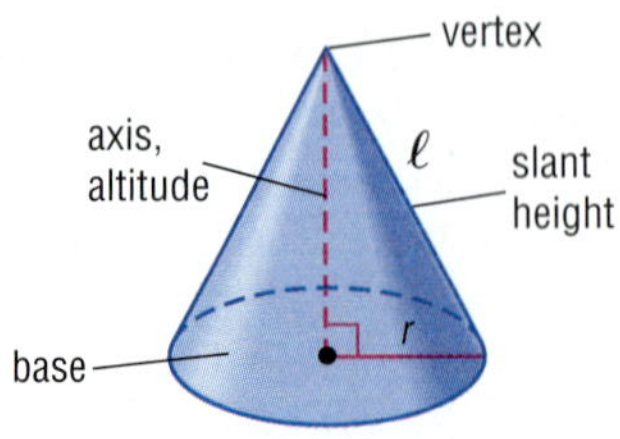

Right Cone

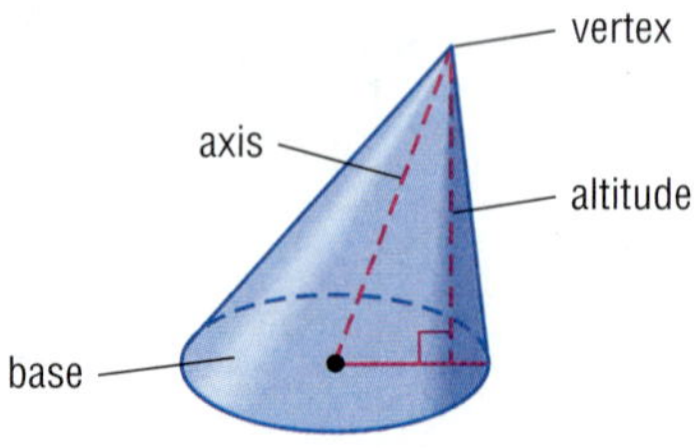

Oblique Cone

Review Vocabulary

trigonometric ratios

$\sin A = \frac{\text{opp}}{\text{hyp}}$

$\cos A = \frac{\text{adj}}{\text{hyp}}$

$\tan A = \frac{\text{opp}}{\text{adj}}$

(Lesson 8-4)

The net for a cone is shown at the right. The circle with radius r is the base of the cone. It has a circumference of $2\pi r$ and an area of πr^2. The sector with radius ℓ is the lateral surface of the cone. Its arc measure is $2\pi r$. You can use a proportion to find its area.

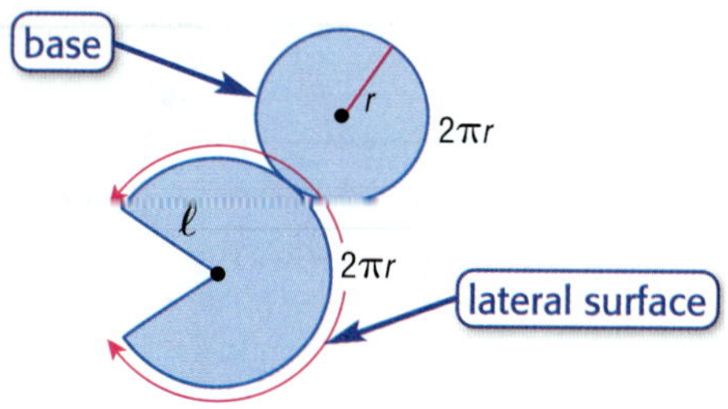

$$\frac{\text{area of sector}}{\text{area of circle}} = \frac{\text{measure of arc}}{\text{circumference of circle}}$$

$$\frac{\text{area of sector}}{\pi \ell^2} = \frac{2\pi r}{2\pi \ell}$$

$$\text{area of sector} = \pi \ell^2 \cdot \frac{2\pi r}{2\pi \ell} \text{ or } \pi r \ell$$

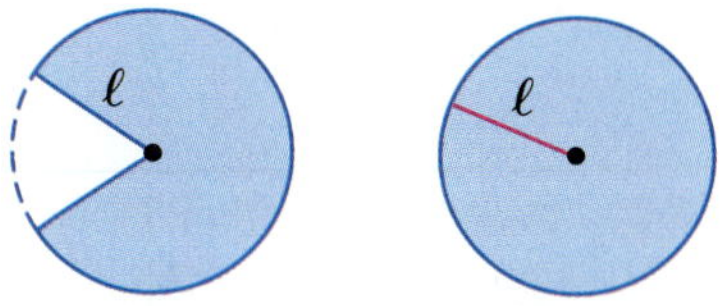

sector | **circle that contains the sector**

> **StudyTip**
>
> **Making Connections** Like a pyramid, the lateral area of a right circular cone L equals $\frac{1}{2}P\ell$. Since the base is a circle, the perimeter is the circumference of the base C. So, the lateral area is $\frac{1}{2}C\ell$.
>
> $L = \frac{1}{2}C\ell$
>
> $= \frac{1}{2}(2\pi r)\ell$
>
> $= \pi r \ell$

Key Concept — Lateral and Surface Area of a Cone

Words The lateral area L of a right circular cone is $L = \pi r \ell$, where r is the radius of the base and ℓ is the slant height.

Model

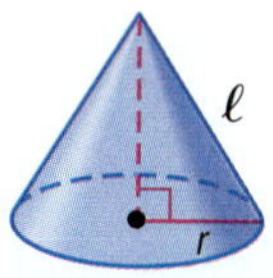

The surface area S of a right circular cone is $S = \pi r \ell + \pi r^2$, where r is the radius of the base and ℓ is the slant height.

Symbols $L = \pi r \ell$ $\quad S = \pi r \ell + \pi r^2$

Real-World EXAMPLE 4 — Lateral Area of a Cone

ARCHITECTURE The conical slate roof at the right has a height of 16 feet and a radius of 12 feet. Find the lateral area.

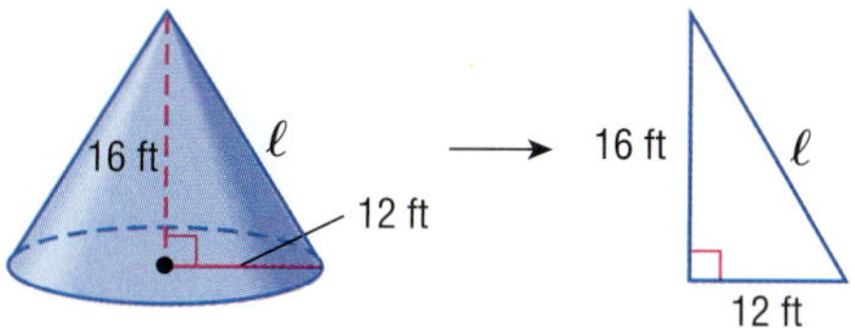

Step 1 Find the slant height ℓ.

$\ell^2 = 16^2 + 12^2$	**Pythagorean Theorem**
$\ell^2 = 400$	**Simplify.**
$\ell = 20$	**Take the positive square root of each side.**

Step 2 Find the lateral area L.

Estimate $L \approx 3 \cdot 12 \cdot 20$ **or 720** ft^2

$L = \pi r \ell$	**Lateral area of a cone**
$= \pi(12)(20)$	$r = 12$ **and** $\ell = 20$
≈ 754	**Use a calculator.**

The lateral area of the conical roof is about 754 square feet. The answer is reasonable compared to the estimate.

Guided Practice

4. ICE CREAM A waffle cone is $5\frac{1}{2}$ inches tall and the diameter of the base is $2\frac{1}{2}$ inches. Find the lateral area of the cone. Round to the nearest tenth.

Personal Tutor glencoe.com

EXAMPLE 5 Surface Area of a Cone

Find the surface area of a cone with a diameter of 14.8 centimeters and a slant height of 15 centimeters.

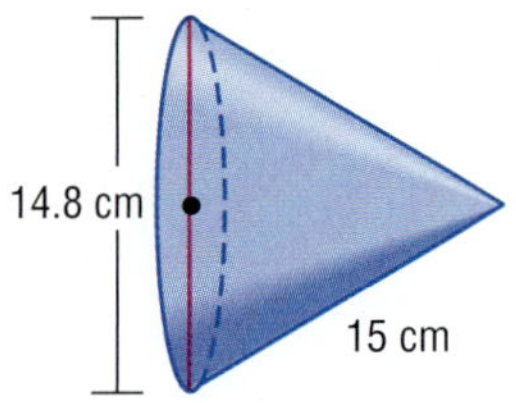

Estimate: $S \approx 3 \cdot 7 \cdot 20 + 3 \cdot 50$ or 570 cm^2

$S = \pi r\ell + \pi r^2$	**Surface area of a cone**
$= \pi(7.4)(15) + \pi(7.4)^2$	$r = 7.4$ **and** $\ell = 15$
≈ 520.8	**Use a calculator.**

The surface area of the cone is about 520.8 square centimeters. This is close to the estimate, so the answer is reasonable.

Guided Practice

Find the surface area of each cone. Round to the nearest tenth.

5A.

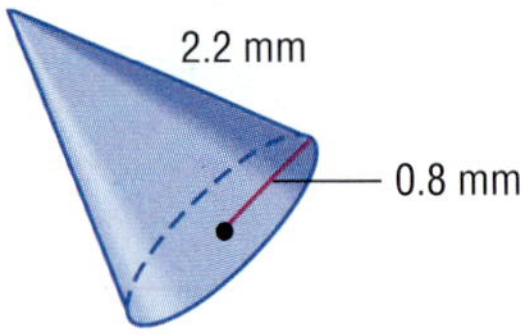

5B.

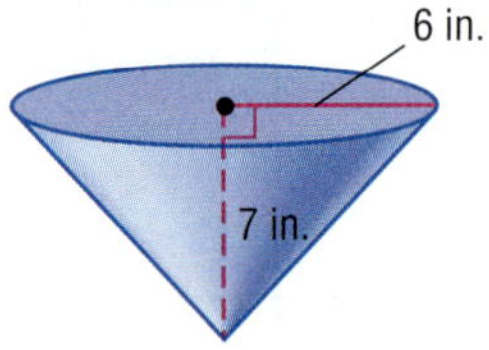

Personal Tutor glencoe.com

> **Watch Out!**
>
> **Bases** The bases of right prisms and right pyramids are not always regular polygons.

The formulas for lateral and surface area are summarized below.

Concept Summary

Lateral and Surface Areas of Solids

Solid	Model	Lateral Area	Surface Area
prism	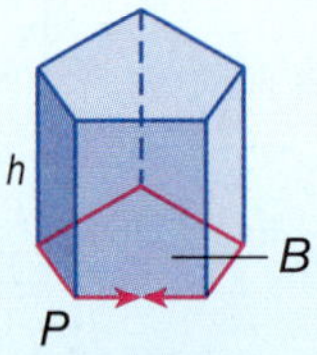	$L = Ph$	$S = L + 2B$ or $S = Ph + 2B$
cylinder	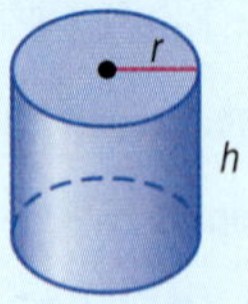	$L = 2\pi rh$	$S = L + 2B$ or $S = 2\pi rh + 2\pi r^2$
pyramid	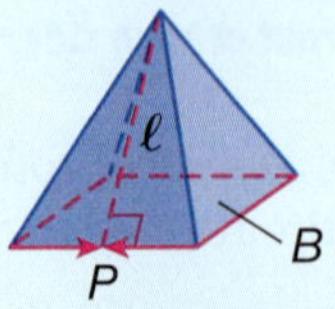	$L = \frac{1}{2}P\ell$	$S = \frac{1}{2}P\ell + B$
cone	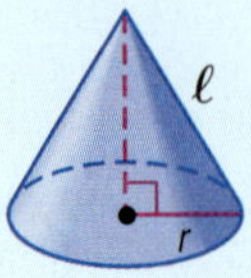	$L = \pi r\ell$	$S = \pi r\ell + \pi r^2$

Check Your Understanding

Examples 1–3
pp. 839–840

Find the lateral area and surface area of each regular pyramid. Round to the nearest tenth if necessary.

1.

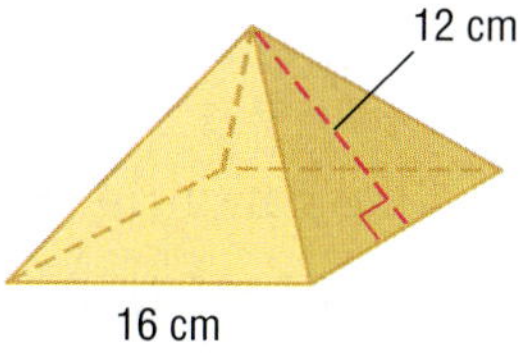

2.

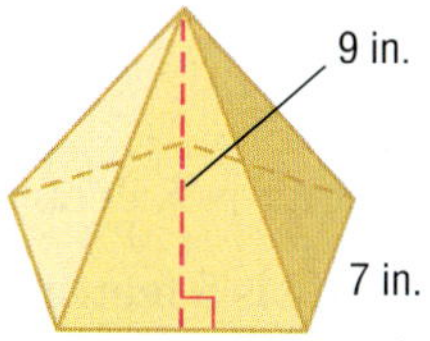

3.

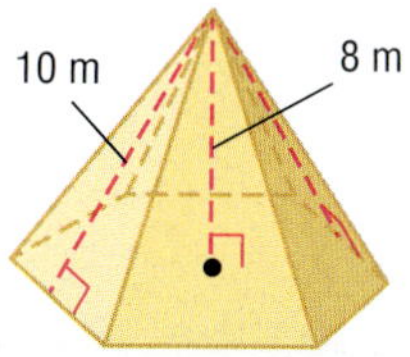

Examples 4 and 5
pp. 841–842

4. TENTS A conical tent is shown at the right. Round answers to the nearest tenth.

a. Find the lateral area of the tent and describe what it represents.

b. Find the surface area of the tent and describe what it represents.

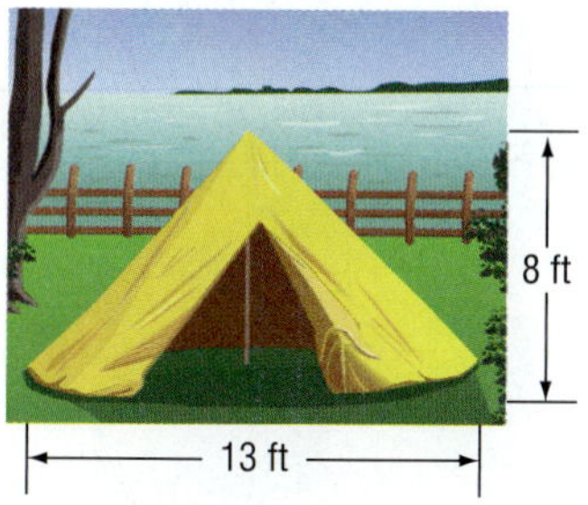

Find the lateral area and surface area of each cone. Round to the nearest tenth.

5.

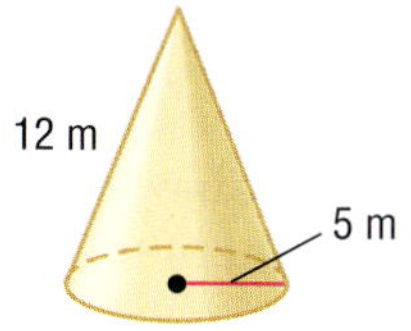

6.

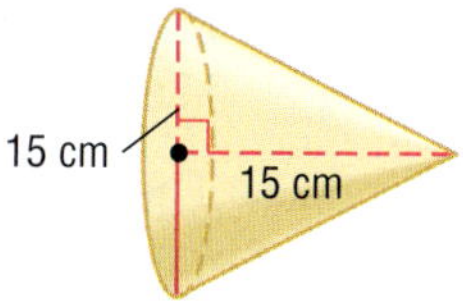

Practice and Problem Solving

= **Step-by-Step Solutions** begin on page R20.
Extra Practice begins on page 969.

Examples 1–3
pp. 839–840

Find the lateral area and surface area of each regular pyramid. Round to the nearest tenth if necessary.

7

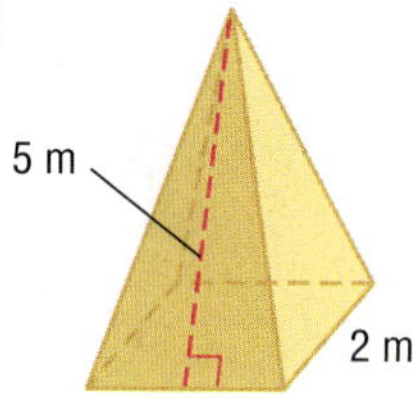

8.

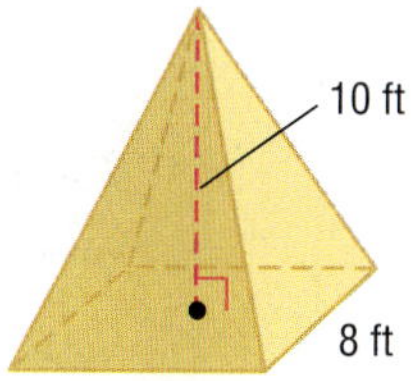

9.

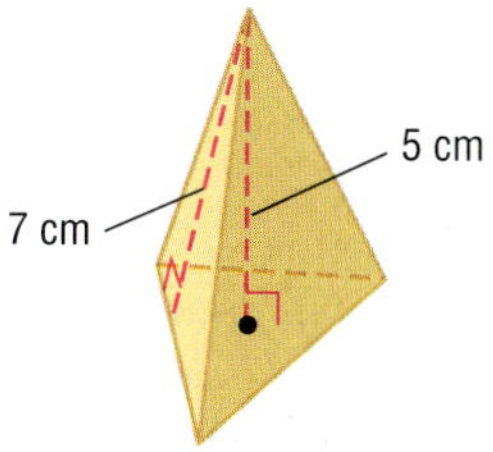

10.

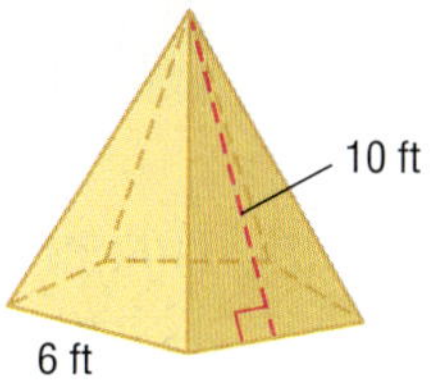

StudyTip

Draw a Diagram
When solving word problems involving solids, it is helpful to draw a figure and label the known parts. Use a variable to label the measure or measures that you need to find.

11. square pyramid with an altitude of 12 inches and a slant height of 18 inches

12. hexagonal pyramid with a base edge of 6 millimeters and a slant height of 9 millimeters

13. ARCHITECTURE Find the lateral area of a pyramid-shaped building that has a slant height of 210 feet and a square base 332 feet by 332 feet.

Examples 4 and 5
pp. 841–842

Find the lateral area and surface area of each cone. Round to the nearest tenth.

14.

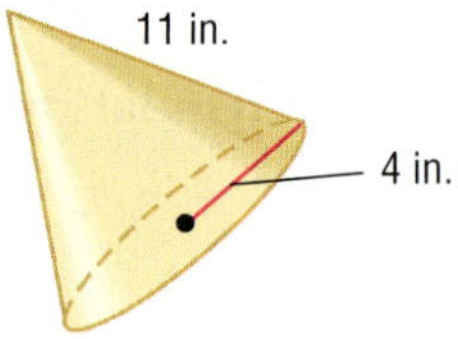

15.

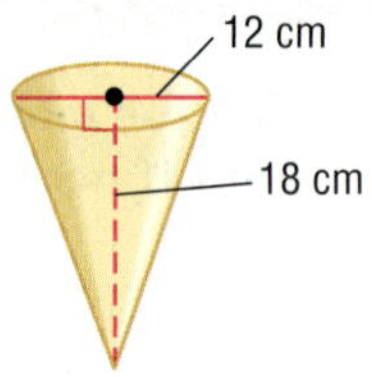

16. The diameter is 3.4 centimeters, and the slant height is 6.5 centimeters.

17. The altitude is 5 feet, and the slant height is $9\frac{1}{2}$ feet.

18. **MOUNTAINS** A conical mountain has a radius of 1.6 kilometers and a height of 0.5 kilometer. What is the lateral area of the mountain?

19 **HISTORY** Archaeologists recently discovered a 1500-year-old pyramid in Mexico City. The square pyramid measures 165 yards on each side and once stood 20 yards tall. What was the original lateral area of the pyramid?

Real-World Link

A volcano is a conical mountain built around a vent that connects with reservoirs of molten rock below Earth's surface. About 500 active volcanoes are known on Earth, not counting those that lie beneath the sea.

Source: United States Geological Survey

20. Describe two polyhedrons that have 7 faces.

21. What is the sum of the number of faces, vertices, and edges of an octagonal pyramid?

22. **TEPEES** The dimensions of two canvas tepees are shown in the table at the right. Not including the floors, approximately how much more canvas is used to make Tepee B than Tepee A?

Tepee	Diameter (ft)	Height (ft)
A	14	6
B	20	9

23. The surface area of a square pyramid is 24 square millimeters and the base area is 4 square millimeters. What is the slant height of the pyramid?

24. The surface area of a cone is 18π square inches and the radius of the base is 3 inches. What is the slant height of the cone?

25. The surface area of a triangular pyramid is 532 square centimeters, and the base is 24 centimeters wide with a hypotenuse of 25 centimeters. What is the slant height of the pyramid?

26. Find the lateral area of the tent to the nearest tenth.

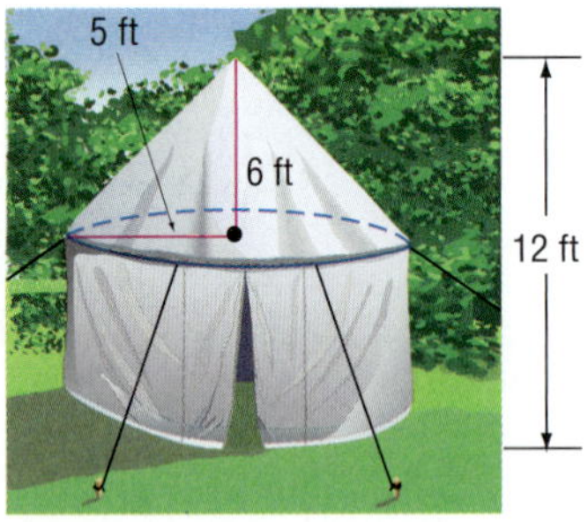

27. Find the surface area of the tank. Write in terms of π.

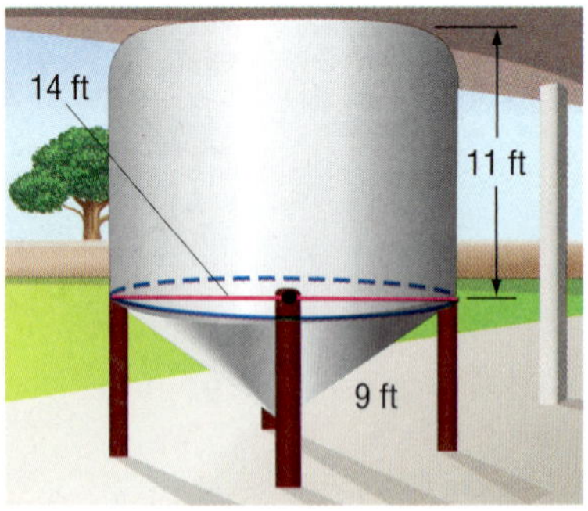

28. **CHANGING DIMENSIONS** A cone has a radius of 6 centimeters and a slant height of 12 centimeters. Describe how each change affects the surface area of the cone.

a. The radius and the slant height are doubled.

b. The radius and the slant height are divided by 3.

29. A solid has the net shown at the right.

a. Describe the solid.

b. Make a sketch of the solid.

Sketch each solid and a net that represents the solid.

30. hexagonal pyramid

31. rectangular pyramid

32. PETS A *frustum* is the part of a solid that remains after the top portion has been cut by a plane parallel to the base. The ferret tent shown at the right is a frustum of a regular pyramid.

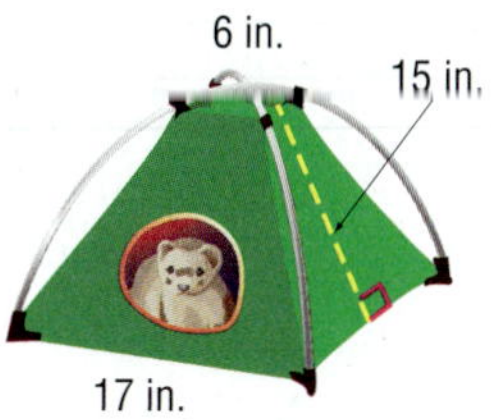

a. Describe the faces of the solid.

b. Find the lateral area and surface area of the frustum formed by the tent.

c. Another pet tent is made by cutting the top half off of a pyramid with a height of 12 centimeters, slant height of 20 centimeters and square base with side lengths of 32 centimeters. Find the surface area of the frustum.

Real-World Link

The average ferret adds 30%–40% of its body weight in the fall and loses this fat the following spring. The average life span of a ferret is 9 to 10 years.

Source: Animal Health Center

Find the surface area of each composite solid. Round to the nearest tenth.

33.

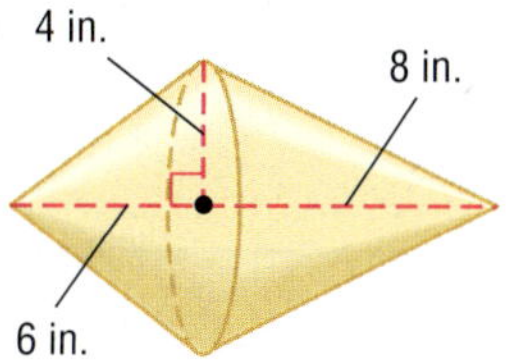

34.

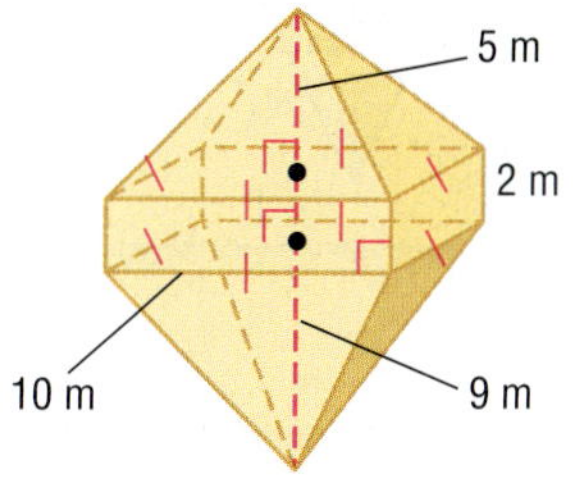

35. MULTIPLE REPRESENTATIONS In this problem, you will investigate the lateral and surface area of a square pyramid with a base edge of 3 units.

a. GEOMETRIC Sketch the pyramid on isometric dot paper.

b. TABULAR Make a table showing the lateral areas of the pyramid for a slant heights of 3 units and 9 units.

c. VERBAL Describe what happens to the lateral area of the pyramid if the slant height is tripled.

d. ANALYTICAL Make a conjecture about how the lateral area of a square pyramid is affected if both the slant height and the base edge are tripled. Then test your conjecture.

H.O.T. Problems

Use Higher-Order Thinking Skills

36. WRITING IN MATH Explain why an oblique cone does not have a slant height.

37. REASONING Classify the following statements as *sometimes*, *always*, or *never* true. Justify your reasoning.

The surface area of a cone of radius r and height h is less than the surface area of a cylinder of radius r and height h.

38. REASONING A cone and a pyramid have the same surface area. If the areas of their bases are also equal, do they have the same slant height as well? Explain.

39. OPEN ENDED Describe a pyramid that has a total surface area of 100 square units.

40. CHALLENGE Determine whether the following statement is *true* or *false*. Explain your reasoning.

A regular polygonal pyramid and a cone both have height h units and base perimeter P units. Therefore, they have the same total surface area.

41. WRITING IN MATH Describe how to find the surface area of a regular polygonal pyramid with an n-gon base, height h units, and an apothem of a units.

Standardized Test Practice

42. The top of a gazebo in a park is in the shape of a regular pentagonal pyramid. Each side of the pentagon is 10 feet long. If the slant height of the roof is about 6.9 feet, what is the lateral roof area?

A 34.5 ft^2 C 172.5 ft^2
B 50 ft^2 D 250 ft^2

43. SHORT RESPONSE What is the surface area of a cone with the dimensions shown?

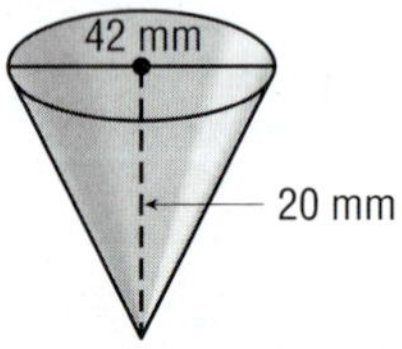

44. ALGEBRA Yu-Jun's craft store sells 3 handmade barrettes for \$9.99. Which expression can be used to find the total cost C of x barrettes?

F $C = \frac{9.99}{x}$ H $C = 3.33x$
G $C = 9.99x$ J $C = \frac{x}{3.33}$

45. SAT/ACT What is the slope of a line perpendicular to the line with equation $2x + 3y = 9$?

A $-\frac{3}{2}$ C $\frac{2}{3}$
B $-\frac{2}{3}$ D $\frac{3}{2}$

Spiral Review

46. Find the surface area of a cylinder with a diameter of 18 cm and a height of 12 cm. (Lesson 12-2)

Draw the back view and corner view of a figure given each orthographic drawing. (Lesson 12-1)

47.

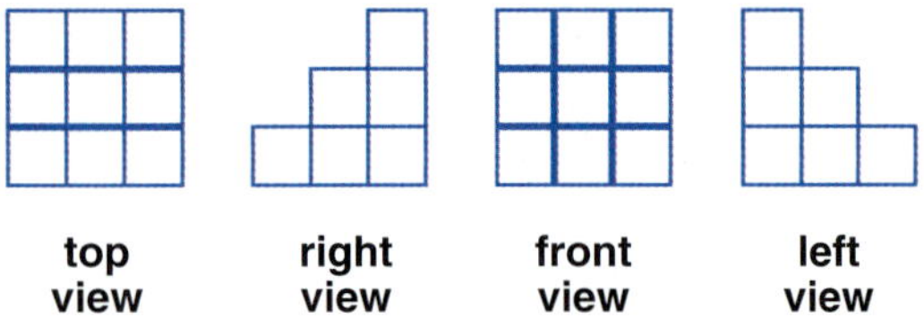

48.

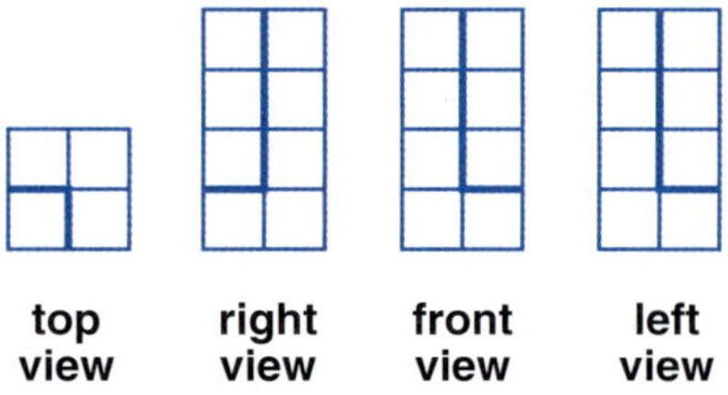

Graph each figure and its image in the given line. (Lesson 9-1)

***J*(2, 4), *K*(4, 0), *L*(7, 3)**

49. $\triangle JKL$; $x = 2$

50. $\triangle JKL$; $y = 1$

***Q*(4, 8), *R*(1, 6), *S*(2, 1), *T*(5, 5)**

51. $QRST$; $y = -1$

52. $QRST$; $x = 4$

***A*(−2, 6), *B*(−2, 1), *C*(3, 1), *D*(3, 4)**

53. $ABCD$; $x = 1$

54. $ABCD$; $y = -2$

Skills Review

Find the perimeter and area of each parallelogram, triangle, or composite figure. Round to the nearest tenth. (Lesson 11-1)

55.

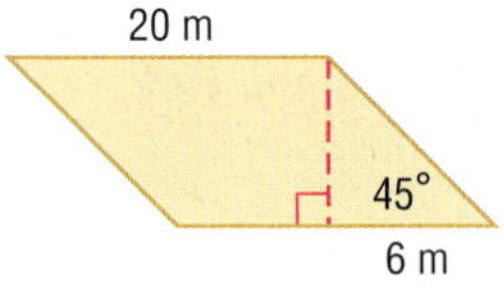

56.

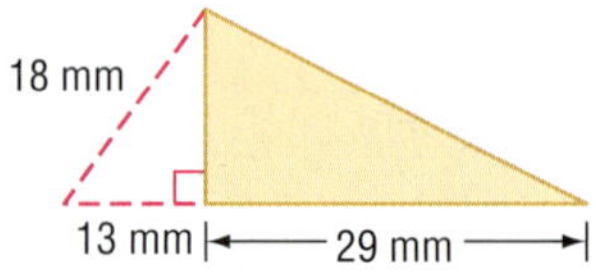

57.

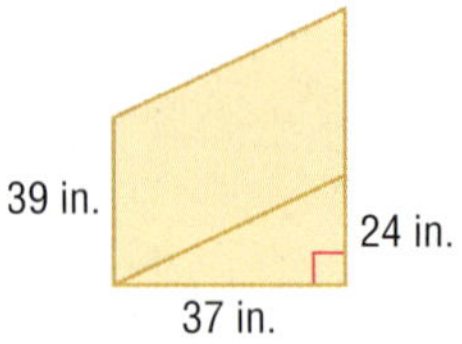

12-4 Volumes of Prisms and Cylinders

Why?

A Galileo thermometer uses floating glass spheres in a cylinder to show the temperature. The lowest floating sphere indicates the temperature. The amount of liquid in each thermometer depends on the height and radius of the cylinder.

Then
You found surface areas of prisms and cylinders. (Lesson 12-2)

Now
- Find volumes of prisms.
- Find volumes of cylinders.

Math Online
glencoe.com

Volume of Prisms In Galileo's thermometer, the amount of liquid contained in each sphere depends on its size. Recall that the *volume* of a solid is the measure of the amount of space the solid encloses. Volume is measured in cubic units.

The rectangular prism at the right has 6 • 4 or 24 cubic units in the bottom layer. Since there are two layers, the total volume is 24 • 2 or 48 cubic units.

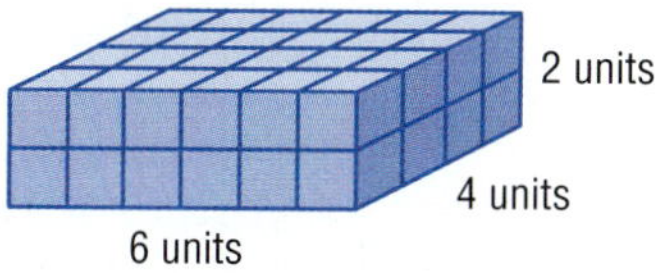

Key Concept: Volume of a Prism

Words The volume V of a prism is $V = Bh$, where B is the area of a base and h is the height of the prism.

Symbols $V = Bh$

Model

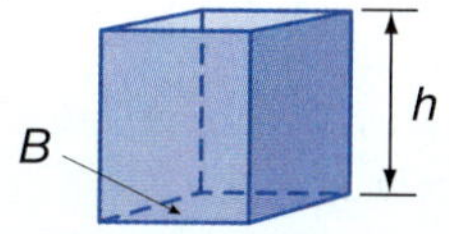

EXAMPLE 1 Volume of a Prism

Find the volume of the prism.

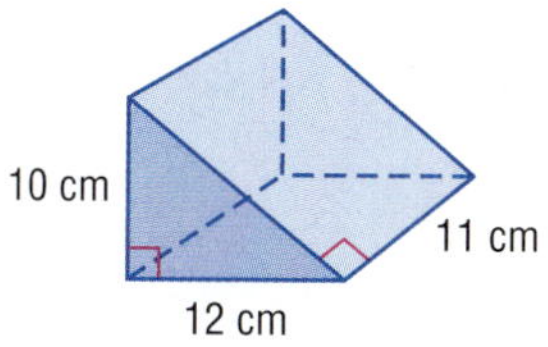

Step 1 Find the area of the base B.

$B = \frac{1}{2}bh$ **Area of a triangle**

$= \frac{1}{2}(12)(10)$ or 60 **$b = 12$ and $h = 10$**

Step 2 Find the volume of the prism.

$V = Bh$ **Volume of a prism**

$= 60(11)$ or 660 **$B = 60$ and $h = 11$**

The volume of the prism is 660 cubic centimeters.

Guided Practice

1A.

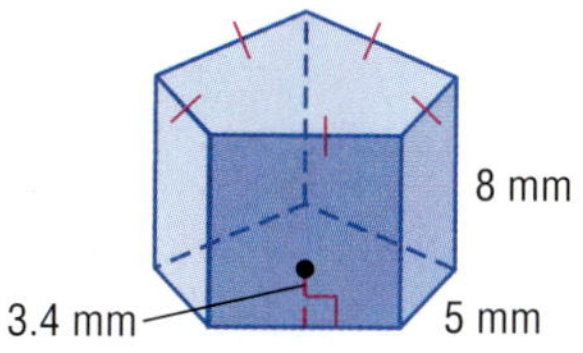

1B.

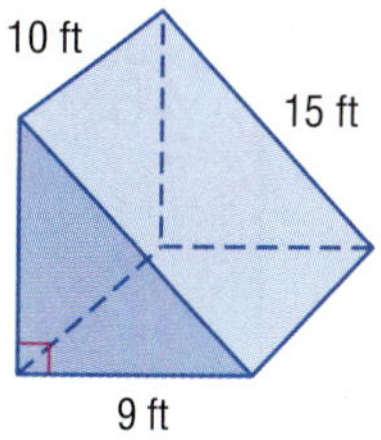

Personal Tutor glencoe.com

Volume of Cylinders Like a prism, the volume of a cylinder is the product of the area of the base and the height.

Key Concept — Volume of a Cylinder

Words The volume V of a cylinder is $V = Bh$ or $V = \pi r^2 h$, where B is the area of the base, h is the height of the cylinder, and r is the radius of the base.

Symbols $V = Bh$ or $V = \pi r^2 h$

Model

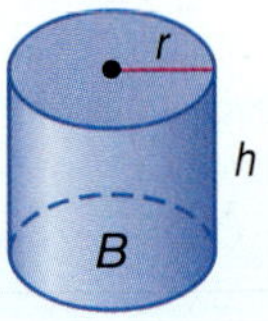

EXAMPLE 2 Volume of a Cylinder

Find the volume of the cylinder at the right.

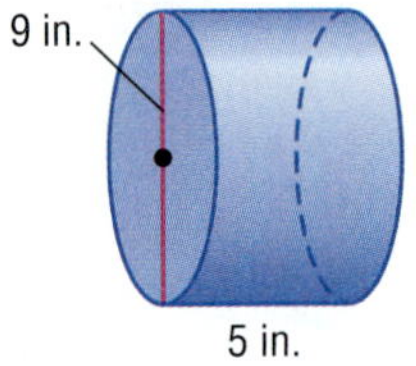

Estimate $V \approx 3 \cdot 5^2 \cdot 5$ or 375 in^3

$V = \pi r^2 h$	Volume of a cylinder
$= \pi(4.5)^2(5)$	$r = 4.5$ and $h = 5$
≈ 318.1	Use a calculator.

The volume of the cylinder is about 318.1 cubic inches. This is fairly close to the estimate, so the answer is reasonable.

Guided Practice

Find the volume of each cylinder. Round to the nearest tenth.

2A.

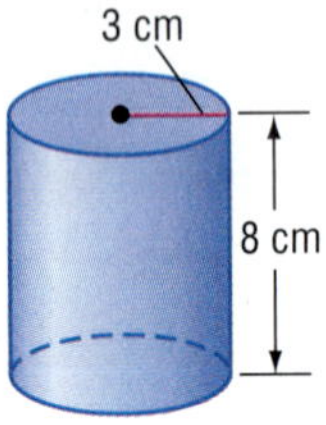

2B.

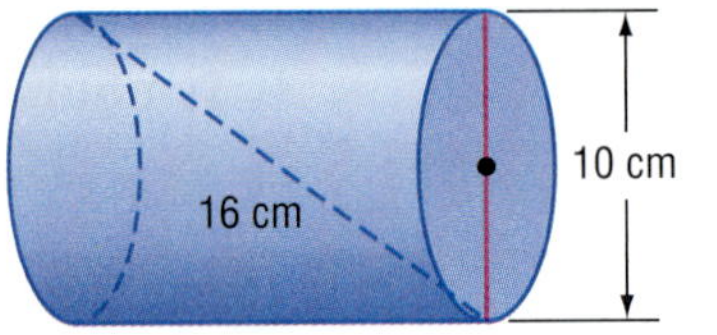

Personal Tutor glencoe.com

The first group of books at the right represents a right prism. The second group represents an oblique prism. Both groups have the same number of books. If all the books are the same size, then the volume of both groups is the same.

This demonstrates the following principle, which applies to all solids.

Key Concept — Cavalieri's Principle

Words If two solids have the same height h and the same cross-sectional area B at every level, then they have the same volume.

Models

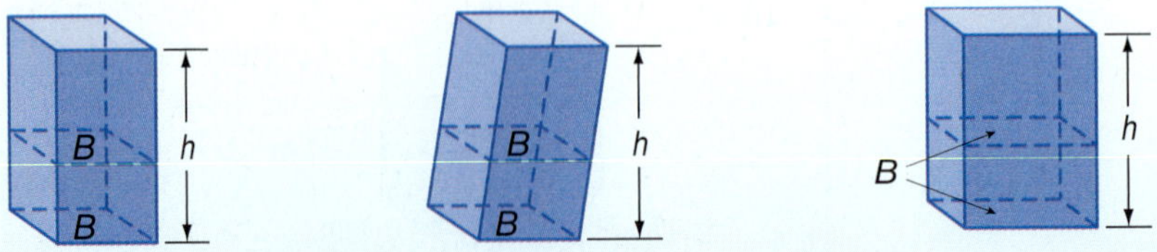

These prisms all have a volume of Bh.

Watch Out!

Cross-Sectional Area For solids with the same height to have the same volume, their cross-sections must have the same area. The cross sections of the different solids do not have to be congruent polygons.

ReadingMath

Height Recall that the height of a prism is the length of the altitude. The height may or may not be equal to the length of a side.

EXAMPLE 3 Volume of an Oblique Solid

Find the volume of an oblique hexagonal prism if the height is 6.4 centimeters and the base area is 17.3 square centimeters.

$V = Bh$ — **Volume of a prism**

$= 17.3(6.4)$ — $B = 17.3$ **and** $h = 6.4$

$= 110.72$ — **Simplify.**

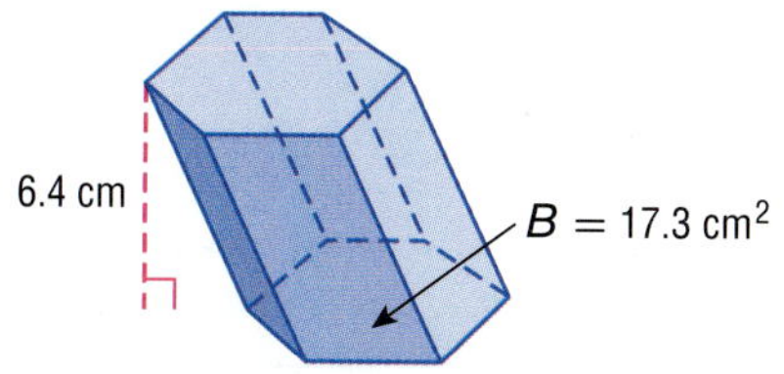

The volume is 110.72 cubic centimeters.

Guided Practice

3. Find the volume of an oblique cylinder that has a radius of 5 feet and a height of 3 feet. Round to the nearest tenth.

Personal Tutor glencoe.com

STANDARDIZED TEST EXAMPLE 4

Prisms A and B have the same length and width, but different heights. If the volume of Prism B is 150 cubic inches greater than the volume of Prism A, what is the length of each prism?

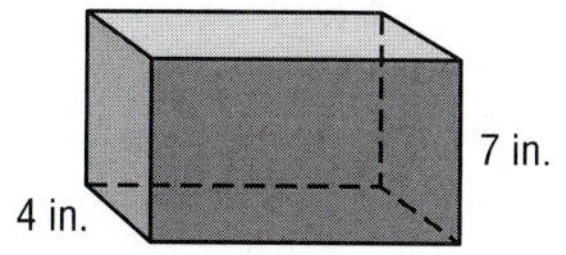

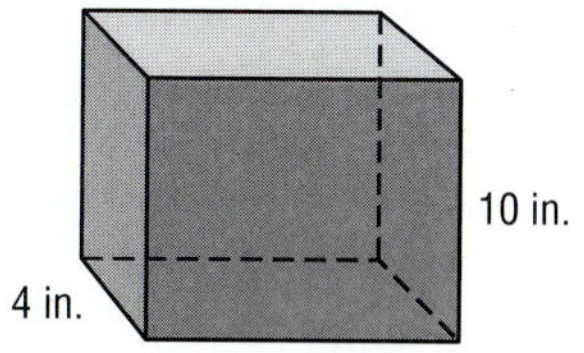

A 10 in. **B** $11\frac{1}{2}$ in. **C** 12 in. **D** $12\frac{1}{2}$ in.

Read the Test Item

You know two dimensions of each solid and that the difference between their volumes is 150 cubic inches.

Solve the Test Item

Volume of Prism B − Volume of Prism A = 150 — **Write an equation.**

$4\ell \cdot 10 - 4\ell \cdot 7 = 150$ — **Use** $V = Bh$.

$12\ell = 150$ — **Simplify.**

$\ell = 12\frac{1}{2}$ — **Divide each side by 12.**

The length of each prism is $12\frac{1}{2}$ inches. The correct answer is D.

Test-TakingTip

Writing Equations By choosing a variable to represent the unknown and setting up an equation, you make it easier to check for mistakes as you work out a problem.

Guided Practice

4. The containers at the right are filled with popcorn. About how many times as much popcorn does the larger container hold?

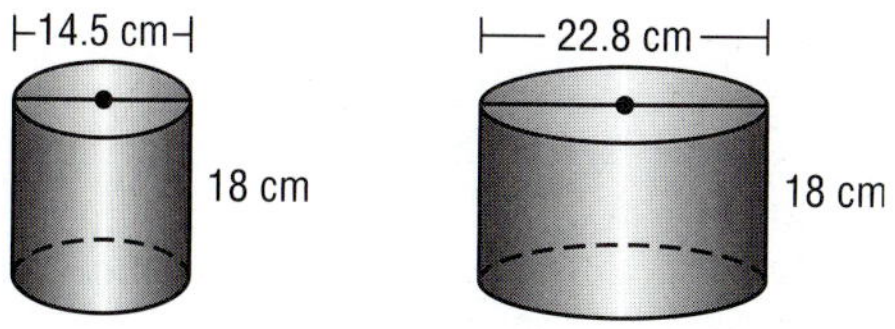

F 1.6 times as much

H 3.3 times as much

G 2.5 times as much

J 5.0 times as much

Personal Tutor glencoe.com

Check Your Understanding

Examples 1 and 3
pp. 847 and 849

Find the volume of each prism.

1.

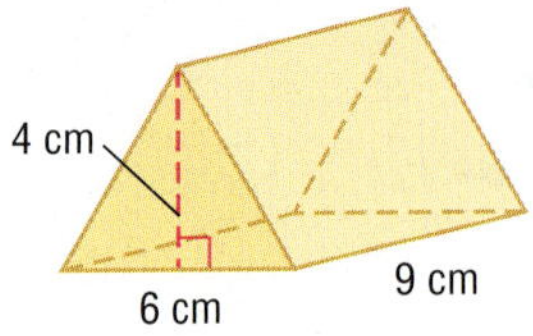

2.

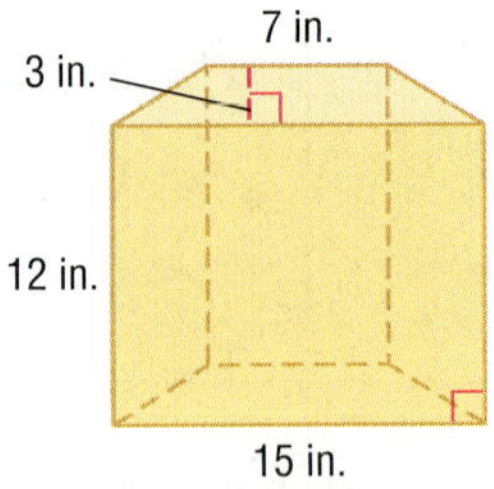

3. the oblique rectangular prism shown at the right

4. an oblique pentagonal prism with a base area of 42 square centimeters and a height of 5.2 centimeters

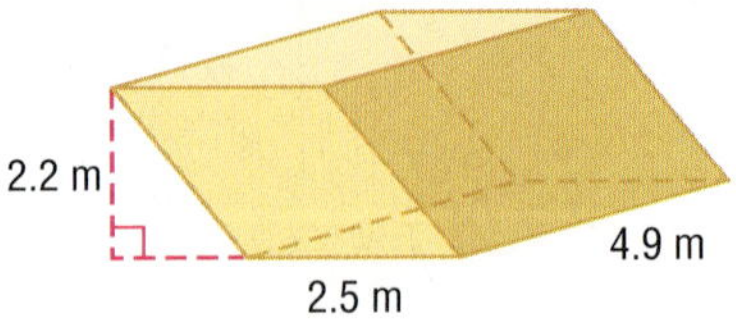

Examples 2–4
pp. 848–849

Find the volume of each cylinder. Round to the nearest tenth.

5.

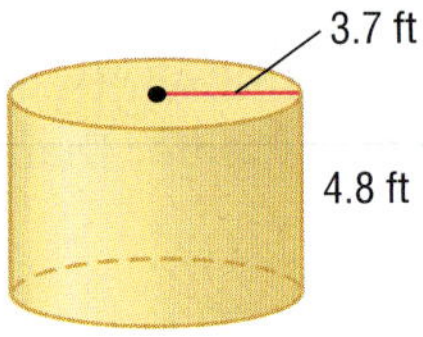

6.

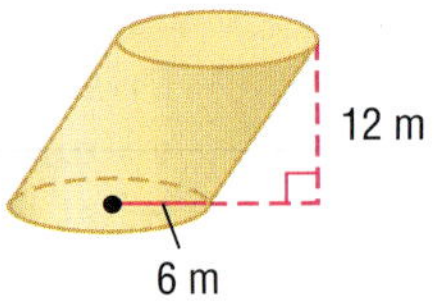

7. a cylinder with a diameter of 16 centimeters and a height of 5.1 centimeters

8. a cylinder with a radius of 4.2 inches and a height of 7.4 inches

9. **MULTIPLE CHOICE** A rectangular lap pool measures 80 feet long by 20 feet wide. If it needs to be filled to four feet deep and each cubic foot holds 7.5 gallons, how many gallons will it take to fill the lap pool?

A 4000 **B** 6400 **C** 30,000 **D** 48,000

Practice and Problem Solving

= **Step-by-Step Solutions** begin on page R20.
Extra Practice begins on page 969.

Examples 1 and 3
pp. 847 and 849

Find the volume of each prism.

10.

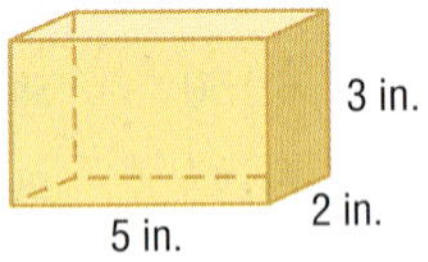

11.

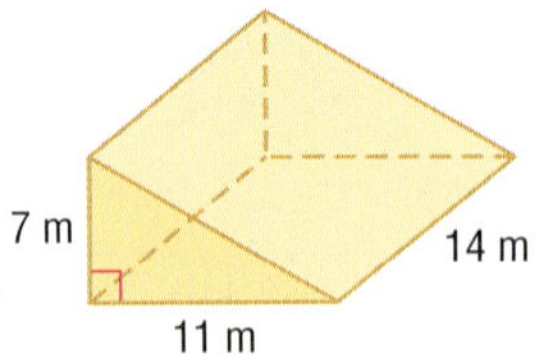

12.

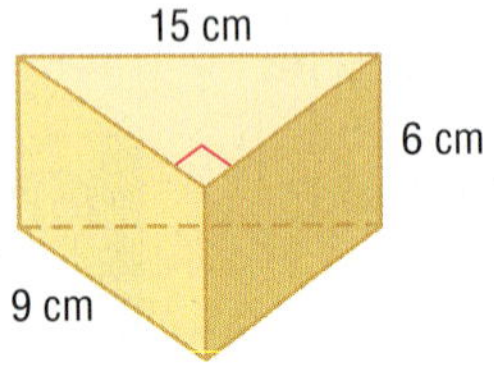

13.

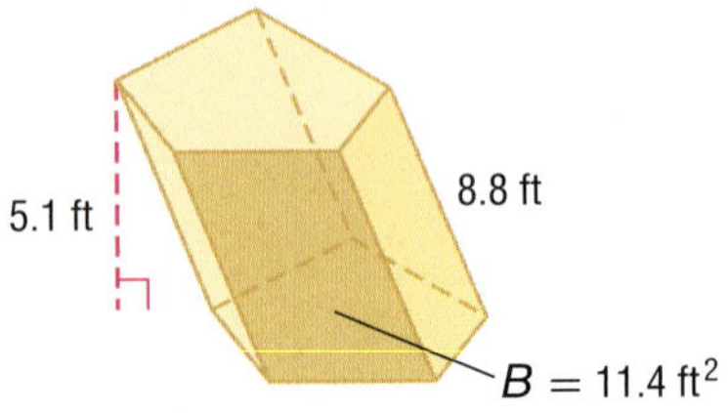

14. an oblique hexagonal prism with a height of 15 centimeters and with a base area of 136 square centimeters

15. a square prism with a base edge of 9.5 inches and a height of 17 inches

Examples 2 and 3
pp. 848–849

Find the volume of each cylinder. Round to the nearest tenth.

16.

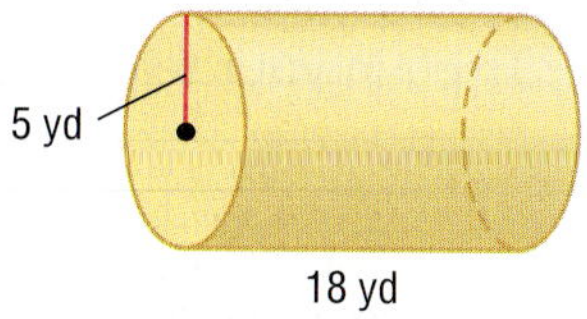

17

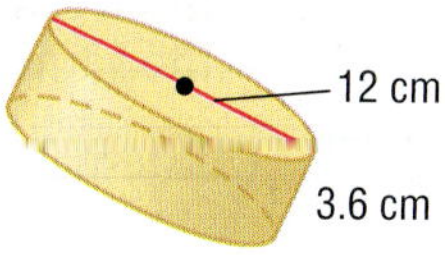

18.

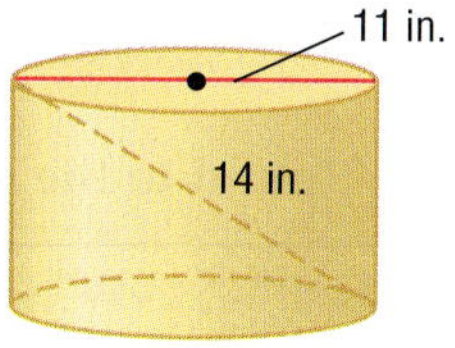

19.

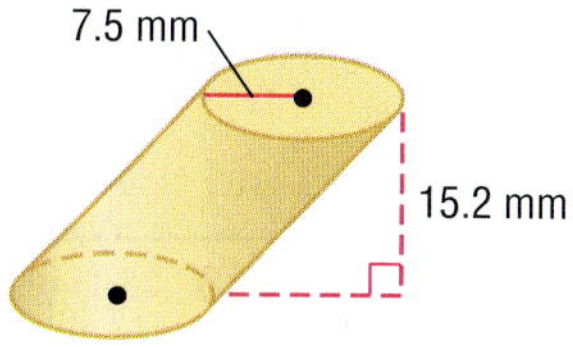

Example 4
p. 849

20. **PLANTER** A planter is in the shape of a rectangular prism 18 inches long, $14\frac{1}{2}$ inches deep, and 12 inches high. What is the volume of potting soil in the planter if the planter is filled to $1\frac{1}{2}$ inches below the top?

21. **SHIPPING** A box 18 centimeters by 9 centimeters by 15 centimeters is being used to ship two cylindrical candles. Each candle has a diameter of 9 centimeters and a height of 15 centimeters, as shown at the right. What is the volume of the empty space in the box?

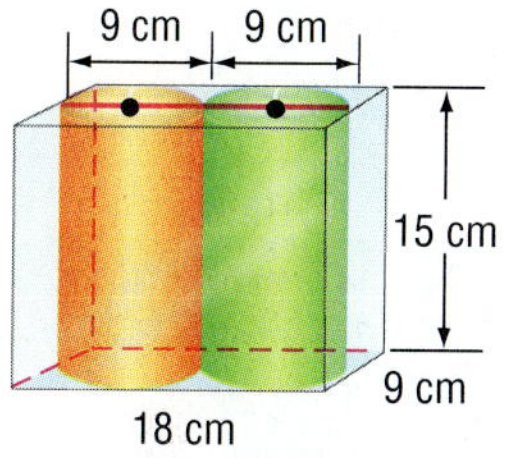

Real-World Link

Professional artists use between 10 and 5000 tons of sand to create sand scultpures.

Source: The Sand Sculpture Company

22. **SANDCASTLES** In a sandcastle competition, contestants are allowed to use only water, shovels, and 10 cubic feet of sand. To transport the correct amount of sand, they want to create cylinders that are 2 feet tall to hold enough sand for one contestant. What should the diameter of the cylinders be?

Find the volume of the solid formed by each net.

23.

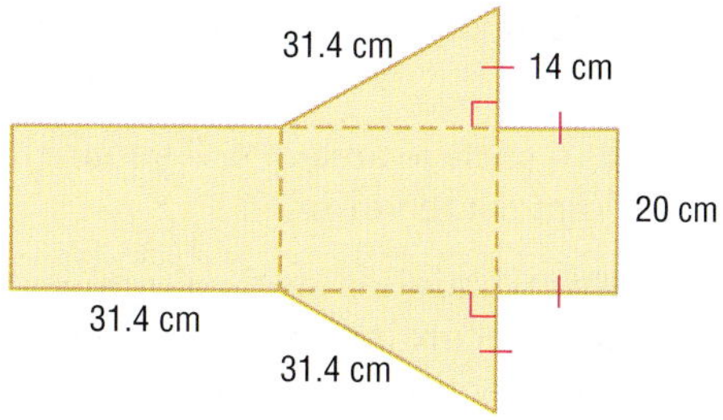

24.

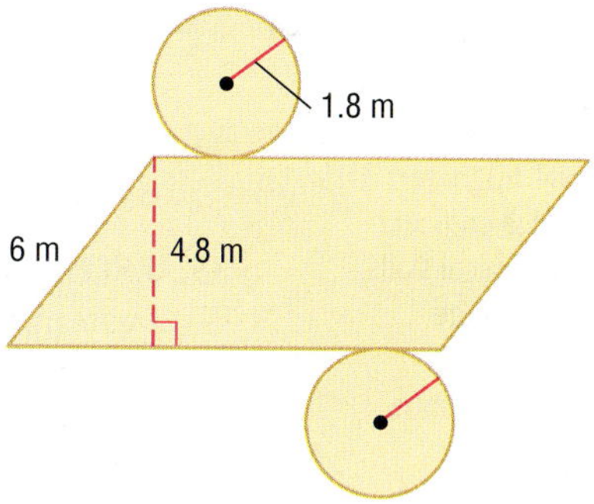

25. **FOOD** A cylindrical can of baked potato chips has a height of 27 centimeters and a radius of 4 centimeters. A new can is advertised as being 30% larger than the regular can. If both cans have the same radius, what is the height of the larger can?

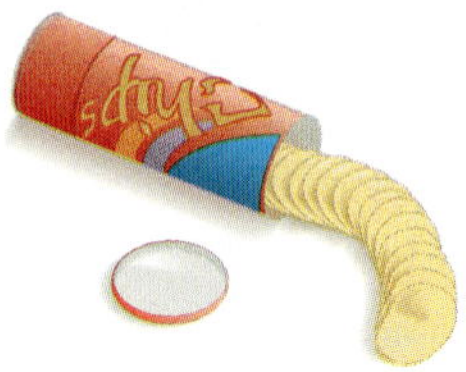

26. **CHANGING DIMENSIONS** A cylinder has a radius of 5 centimeters and a height of 8 centimeters. Describe how each change affects the volume of the cylinder.

a. The height is tripled.

b. The radius is tripled.

c. Both the radius and the height are tripled.

d. The dimensions are exchanged.

27. The prisms described below have the same height as the prism shown at the right. Which of the three prisms has the same volume as this prism? Explain your reasoning.

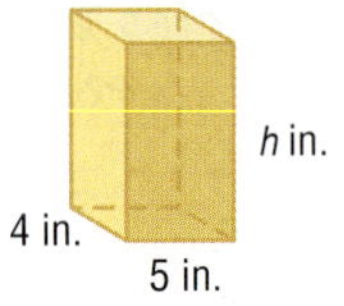

Prism A: The base is a right triangle with legs 8 inches and 5 inches.

Prism B: The base is a square with side lengths of 4.5 inches.

Prism C: The base is a hexagon with side lengths of 3 inches.

Find the volume of each composite solid. Round to the nearest tenth if necessary.

28.

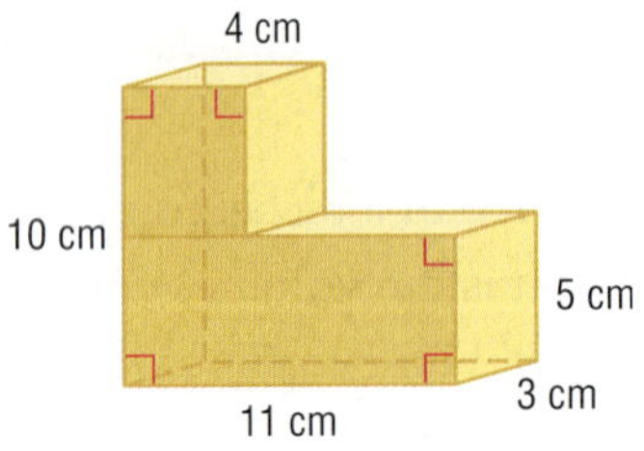

29.

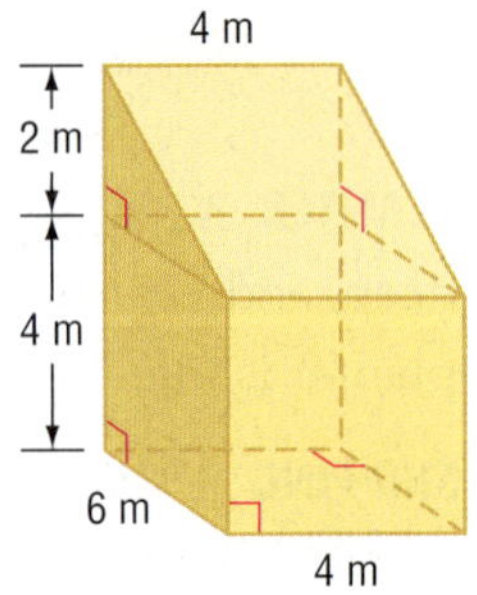

30.

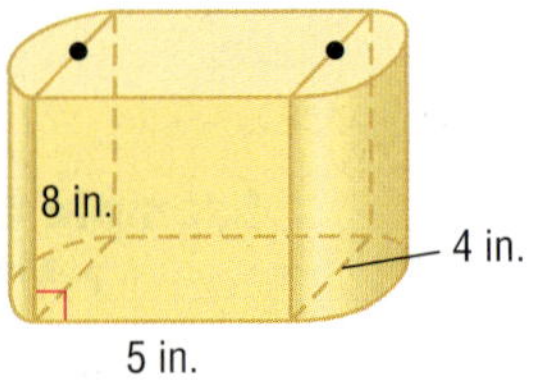

31 **MANUFACTURING** A can 12 centimeters tall fits into a rubberized cylindrical holder that is 11.5 centimeters tall, including 1 centimeter for the thickness of the base of the holder.The thickness of the rim of the holder is 1 centimeter. What is the volume of the rubberized material that makes up the holder?

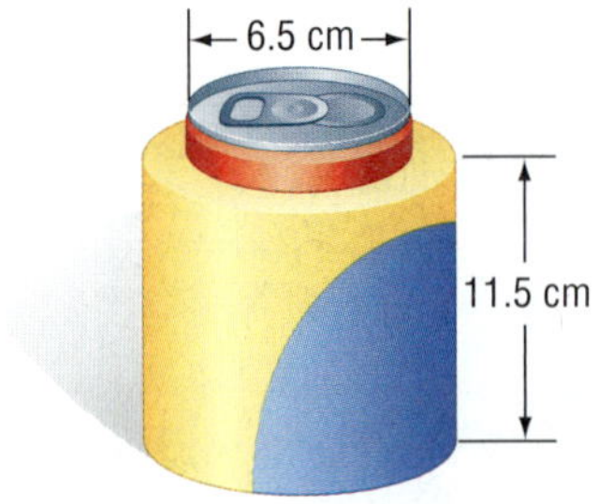

Find each measure to the nearest tenth.

32. A cylindrical can has a volume of 363 cubic centimeters. The diameter of the can is 9 centimeters. What is the height?

33. A cylinder has a surface area of 144π square inches and a height of 6 inches. What is the volume?

34. A rectangular prism has a surface area of 432 square inches, a height of 6 inches, and a width of 12 inches. What is the volume?

Real-World Career

Architectural Engineer
An architectural engineer applies the technical skills of engineering to the design, construction, operation, maintenance, and renovation of buildings.

Architectural engineers are required to have a bachelor's degree in engineering along with specialized coursework.

35. **ARCHITECTURE** A cylindrical stainless steel column is used to hide a ventilation system in a new building. According to the specifications, the diameter of the column can be between 30 centimeters and 95 centimeters. The height is to be 500 centimeters. What is the difference in volume between the largest and smallest possible column? Round to the nearest tenth cubic centimeter.

36. **SWIMMING POOLS** The base of a rectangular swimming pool is sloped so one end of the pool is 6 feet deep and the other end is 3 feet deep, as shown in the figure. If the width is 15 feet, find the volume of water it takes to fill the pool.

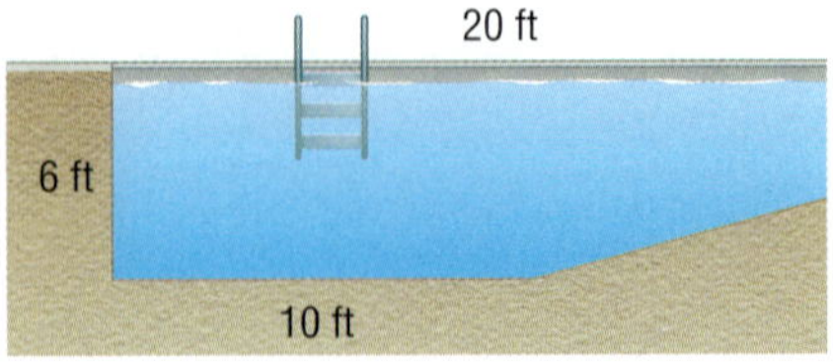

37. **CHANGING DIMENSIONS** A soy milk company is planning a promotion in which the volume of soy milk in each container will be increased by 25%. The company wants the base of the container to stay the same. What will be the height of the new containers?

38. **MEASUREMENT** Find a real prism or cylinder. Measure its dimensions to the nearest tenth of a centimeter and find the volume.

39 Find the volume of the regular pentagonal prism at the right by dividing it into five equal triangular prisms. Describe the base area and height of each triangular prism.

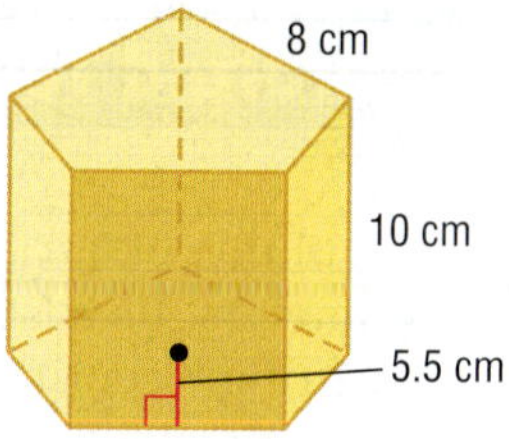

40. PATIOS Mr. Thomas is planning to remove an old patio and install a new rectangular concrete patio 20 feet long, 12 feet wide, and 4 inches thick. One contractor bid $2225 for the project. A second contractor bid $500 per cubic yard for the new patio and $700 for removal of the old patio. Which is the less expensive option? Explain.

Problem-SolvingTip

Make a Model When solving problems involving volume of solids, one way to help you visualize the problem is to make a model of the solid.

41. MULTIPLE REPRESENTATIONS In this problem, you will investigate right and oblique cylinders.

a. GEOMETRIC Draw a right cylinder and an oblique cylinder with a height of 10 meters and a diameter of 6 meters.

b. VERBAL A square prism has a height of 10 meters and a base edge of 6 meters. Is its volume *greater than*, *less than*, or *equal to* the volume of the cylinder? Explain.

c. ANALYTICAL Describe which change affects the volume of the cylinder more: multiplying the height by x or multiplying the radius by x. Explain.

H.O.T. Problems Use Higher-Order Thinking Skills

42. ERROR ANALYSIS Francisco and Valerie each calculated the volume of an equilateral triangular prism with an apothem of 4 units and height of 5 units. Is either of them correct? Explain your reasoning.

Francisco

$$\begin{aligned} V &= Bh \\ &= \frac{1}{2}aP \cdot h \\ &= \frac{1}{2}(4)(24\sqrt{3}) \cdot 5 \\ &= 240\sqrt{3} \text{ cubic units} \end{aligned}$$

Valerie

$$\begin{aligned} V &= Bh \\ &= \frac{\sqrt{3}}{2}s^2 \cdot h \\ &= \frac{\sqrt{3}}{2}(4\sqrt{3})^2 \cdot 5 \\ &= 120\sqrt{3} \text{ cubic units} \end{aligned}$$

43. CHALLENGE A cylindrical can at the right is used to fill a container with liquid. It takes three full cans to fill the container. Describe possible dimensions of the container if it is each of the following shapes.

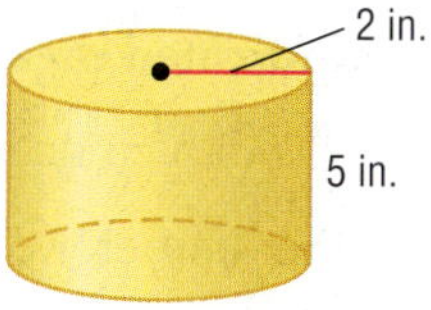

a. rectangular prism

b. square prism

c. triangular prism with a right triangle as the base

44. WRITING IN MATH Explain how two prisms with different heights can have the same volume. Give an example.

45. OPEN ENDED Draw and label a prism that has a volume of 50 cubic centimeters.

46. REASONING Determine whether the following statement is *true* or *false*. Explain.

Two cylinders with the same height and the same lateral area must have the same volume.

47. WRITING IN MATH How are the formulas for the volume of a prism and the volume of a cylinder similar? How are they different?

Standardized Test Practice

48. The volume of a triangular prism is 1380 cubic centimeters. Its base is a right triangle with legs measuring 8 centimeters and 15 centimeters. What is the height of the prism?

A 34.5 cm **C** 17 cm
B 23 cm **D** 11.5 cm

49. A cylindrical tank used for oil storage has a height that is half the length of its radius. If the volume of the tank is 1,122,360 ft^3, what is the tank's radius?

F 89.4 ft **H** 280.9 ft
G 178.8 ft **J** 561.8 ft

50. SHORT RESPONSE What is the ratio of the area of the circle to the area of the square?

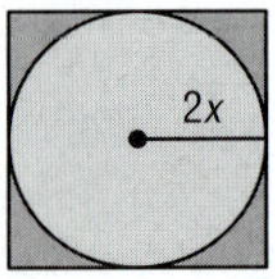

51. SAT/ACT A county proposes to enact a new 0.5% property tax. What would be the additional tax amount for a landowner whose property has a taxable value of $85,000?

A $4.25 **C** $425
B $170 **D** $42,500

Spiral Review

Find the lateral area and surface area of each regular pyramid. Round to the nearest tenth if necessary. (Lesson 12-3)

52.

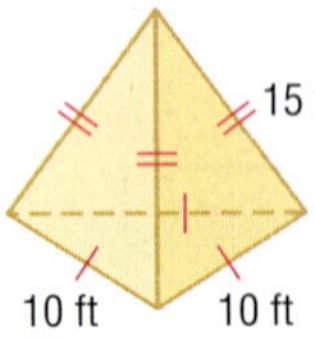

53.

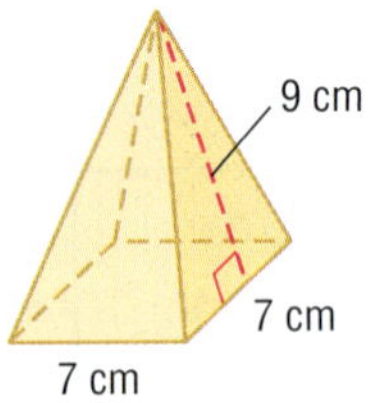

54.

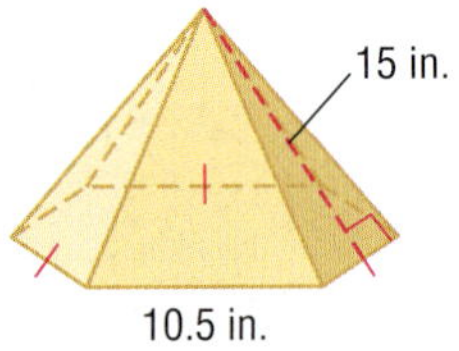

55. BAKING Many baking pans are given a special nonstick coating. A rectangular cake pan is 9 inches by 13 inches by 2 inches deep. What is the area of the inside of the pan that needs to be coated? (Lesson 12-2)

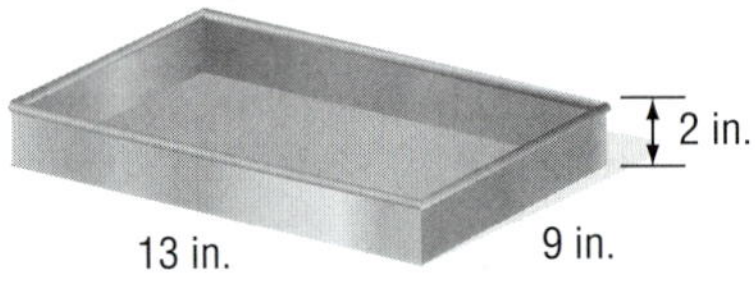

Find the indicated measure. Round to the nearest tenth. (Lesson 11-3)

56. The area of a circle is 54 square meters. Find the diameter.

57. Find the diameter of a circle with an area of 102 square centimeters.

58. The area of a circle is 191 square feet. Find the radius.

59. Find the radius of a circle with an area of 271 square inches.

Skills Review

Find the area of each trapezoid, rhombus, or kite. (Lesson 11-2)

60.

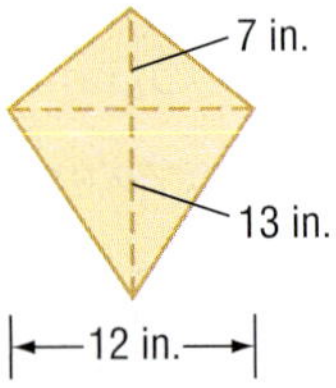

61.

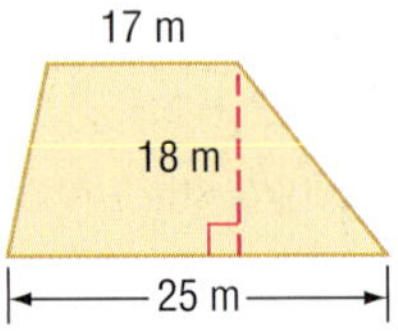

62.

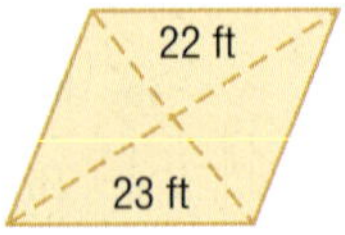

EXTEND
12-4
Graphing Technology Lab
Changing Dimensions

Math Online glencoe.com
- Other Calculator Keystrokes
- Graphing Technology Personal Tutor

You can use TI-Nspire™ or TI-Nspire™ CAS technology to investigate how changes in dimension affect the surface area and volume of a rectangular prism.

ACTIVITY

Step 1 Open a new **Lists and Spreadsheet** page.

Step 2 Move the cursor to the space beside the letter in each column and label the columns ℓ for length, w for width, h for height, sa for surface area, and v for volume.

Step 3 Insert the values for length, width, and height shown into the table.

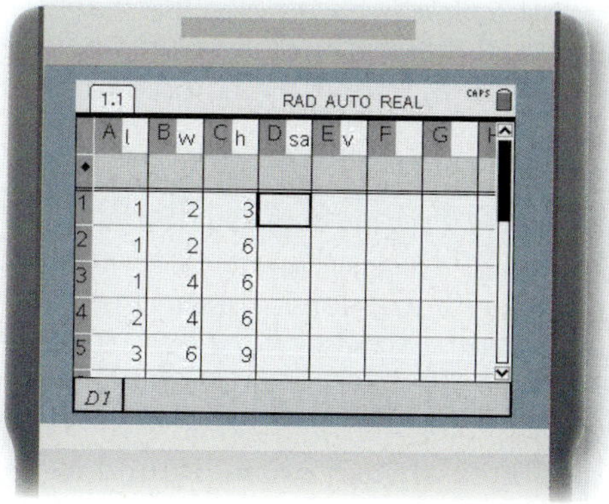

Step 4 Enter the formula for the surface area in terms of cells A1, B1, and C1 in cell D1.

Step 5 Enter the formula for the volume in terms of cells A1, B1, and C1 in cell E1.

Step 6 Select **Fill Down** from the **Data** menu with D1 highlighted and scroll down to fill in the surface areas for the other prisms. Repeat the process for volume.

Step 7 Add additional values and observe the effect on surface area and volume as one or more of the dimensions changes.

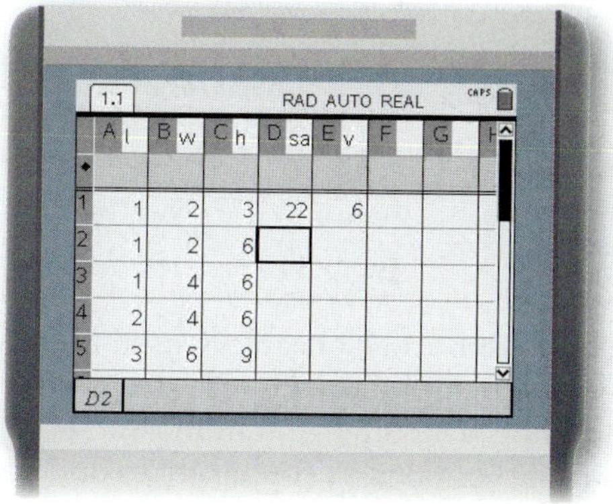

Analyze the Results

1. How does the surface area change when one of the dimensions is doubled? two of the dimensions? all three of the dimensions?
2. How does the volume change when one of the dimensions is doubled? two of the dimensions? all three of the dimensions?
3. How does the surface area change when all three of the dimensions are tripled?
4. How does the volume change when all three of the dimensions are tripled?
5. **MAKE A CONJECTURE** If the dimensions of a prism are all multiplied by a factor of 5, what do you think the ratio of the new surface area to the original surface area will be? the ratio of the new volume to the original volume? Explain.
6. **CHALLENGE** Write an expression for the ratio of the surface areas and the ratio of the volumes if all three of the dimensions of a prism are increased by a scale factor of k. Explain.

CHAPTER 12

Mid-Chapter Quiz

Lessons 12-1 through 12-4

1. Describe how to use isometric dot paper to sketch the following figure. (Lesson 12-1)

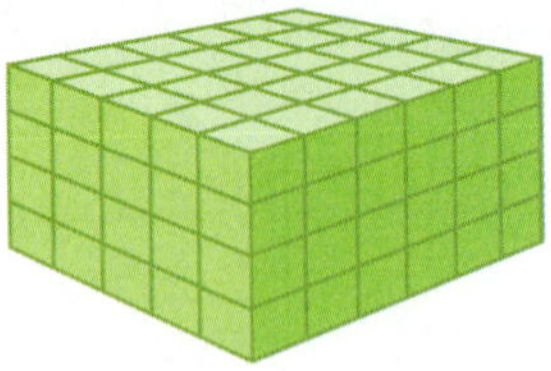

2. Use isometric dot paper to sketch a rectangular prism 2 units high, 3 units long, and 6 units wide. (Lesson 12-1)

3. Use isometric dot paper to sketch a triangular prism 5 units high, with two sides of the base that are 4 units long and 3 units long. (Lesson 12-1)

Find the lateral area of each prism. Round to the nearest tenth if necessary. (Lesson 12-2)

4.

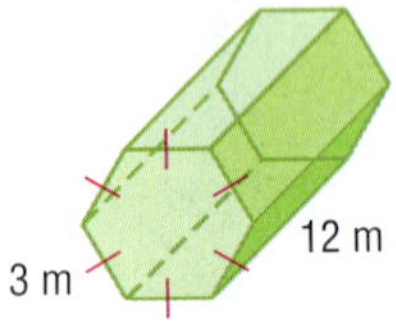

5.

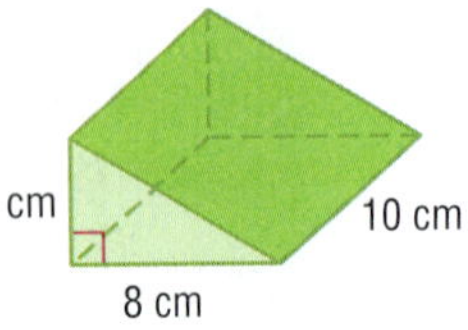

6. **MULTIPLE CHOICE** Coaxial cable is used to transmit long-distance telephone calls, cable television programming, and other communications. A typical coaxial cable contains 22 copper tubes and has a diameter of 3 inches. What is the approximate lateral area of a coaxial cable that is 500 feet long? (Lesson 12-2)

A. 16.4 ft^2 **C.** 294.5 ft^2

B. 196.3 ft^2 **D.** 392.7 ft^2

Find the lateral area and surface area of each cylinder. Round to the nearest tenth if necessary. (Lesson 12-2

7.

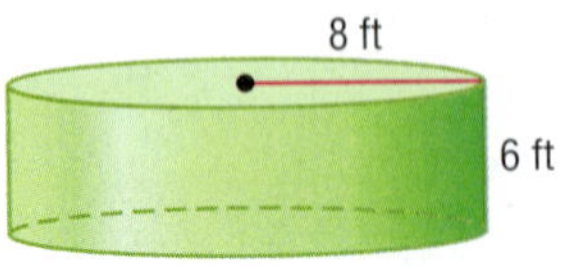

8.

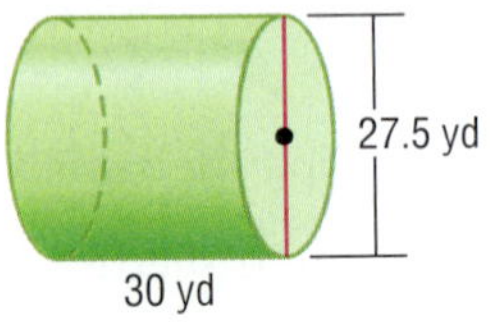

9.

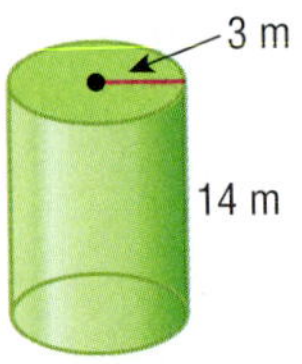

10.

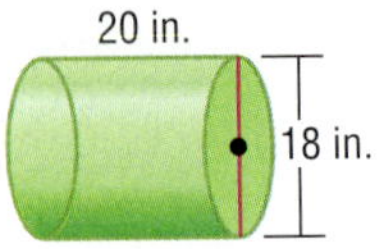

11. **COLLECTIONS** Soledad collects unique salt-and-pepper shakers. She inherited a pair of tetrahedral shakers from her mother. (Lesson 12-3)

   a. Each edge of a shaker measures 3 centimeters. Make a sketch of one shaker.

   b. Find the total surface area of one shaker.

Find the surface area of each regular pyramid or cone. Round to the nearest tenth if necessary. (Lesson 12-3)

12.

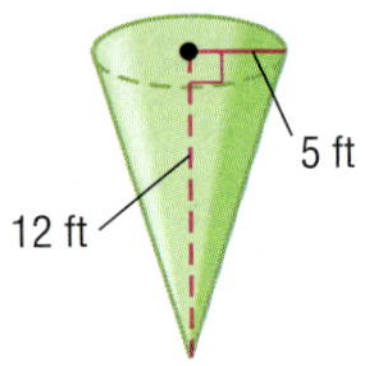

13.

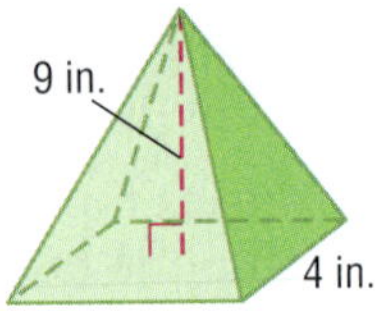

Find the volume of each prism or cylinder. Round to the nearest tenth if necessary. (Lesson 12-4)

14.

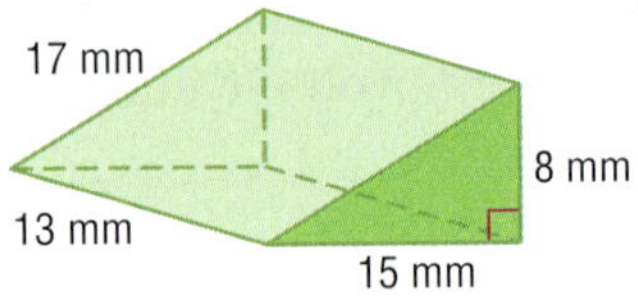

15.

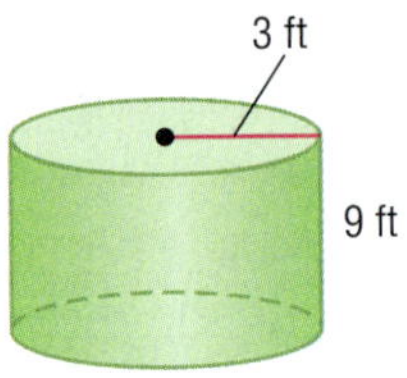

16.

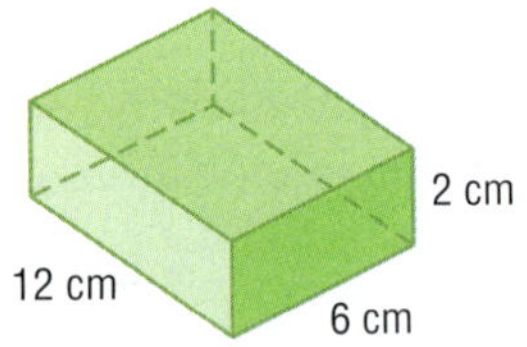

17.

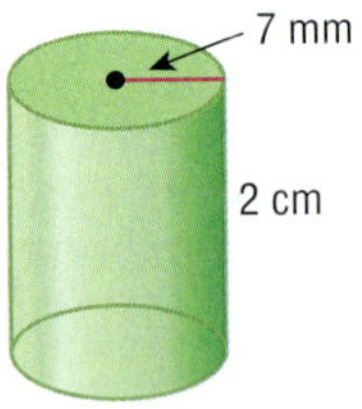

18.

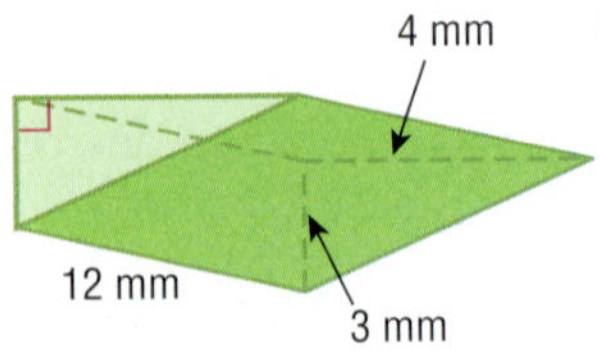

19.

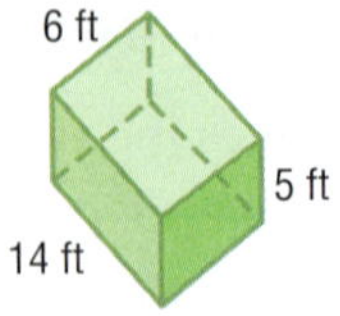

20. **METEOROLOGY** The TIROS weather satellites were a series of weather satellites that carried television and infrared cameras and were covered by solar cells. If the cylinder-shaped body of a TIROS had a diameter of 42 inches and a height of 19 inches, what was the volume available for carrying instruments and cameras? Round to the nearest tenth. (Lesson 12-4)

12-5 Volumes of Pyramids and Cones

Why?

Marta is making a clay model of an alum crystal for science class. The crystal shape is a composite of two congruent rectangular pyramids. The base of each pyramid is to be 1 inch by $1\frac{1}{2}$ inches and the total height of the model is 4 inches. Why is knowing the volume helpful in this situation?

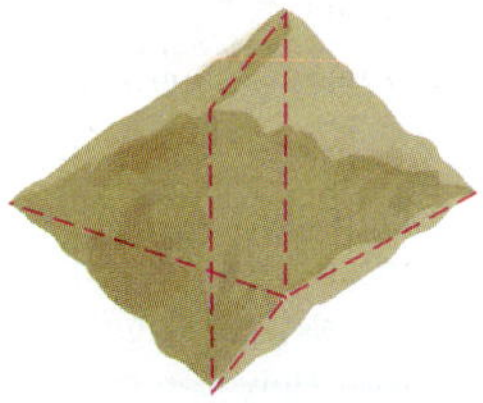

Then
You found surface areas of pyramids and cones. (Lesson 12-3)

Now
- Find volumes of pyramids.
- Find volumes of cones.

Math Online
glencoe.com

Volume of Pyramids The pyramid and prism at the right have the same base and height. You can see that the volume of the pyramid is less than the volume of the prism. In fact, the volume of the pyramid is one third the volume of the prism. If base areas and heights are the same, this is true for regular pyramids and nonregular pyramids.

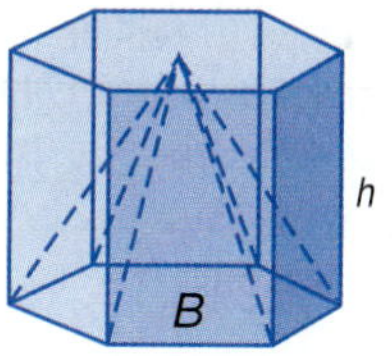

Key Concept — Volume of a Pyramid

Words The volume of a pyramid is $V = \frac{1}{3}Bh$, where B is the area of the base and h is the height of the pyramid.

Models

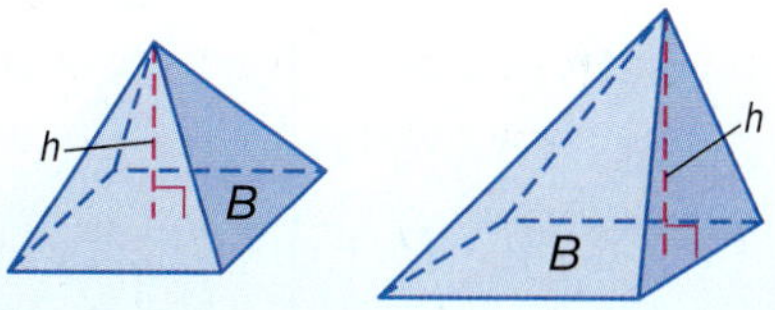

Symbols $V = \frac{1}{3}Bh$

EXAMPLE 1 Volume of a Pyramid

Find the volume of the pyramid.

$V = \frac{1}{3}Bh$ — **Volume of a pyramid**

$= \frac{1}{3}(9.5 \cdot 8)(9)$ — **$B = 9.5 \cdot 8$ and $h = 9$**

$= 228$ — **Simplify.**

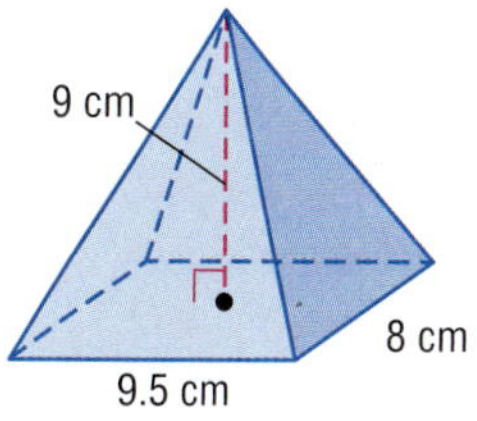

The volume of the pyramid is 228 cubic centimeters.

Guided Practice

Find the volume of each pyramid. Round to the nearest tenth if necessary.

1A.

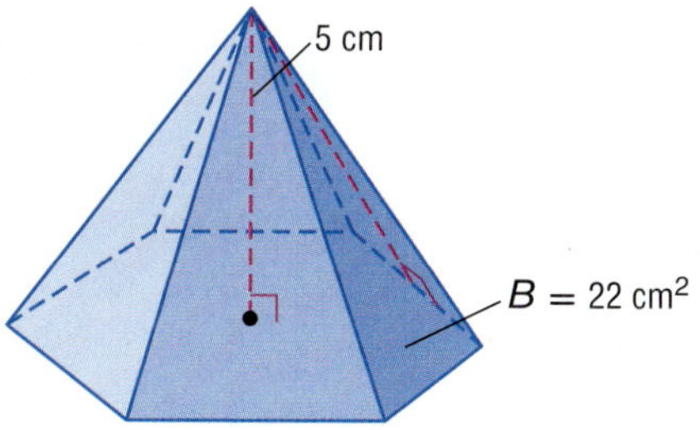

1B.

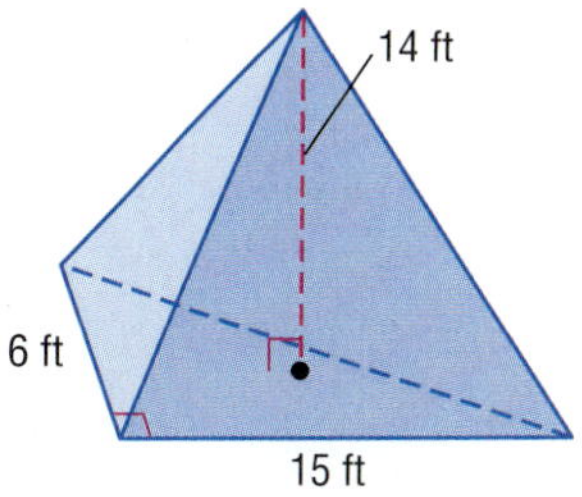

Personal Tutor glencoe.com

Volume of Cones If the heights of a cone and a cylinder are equal and the areas of the bases are equal, then the volume of the cylinder is three times the volume of the cone.

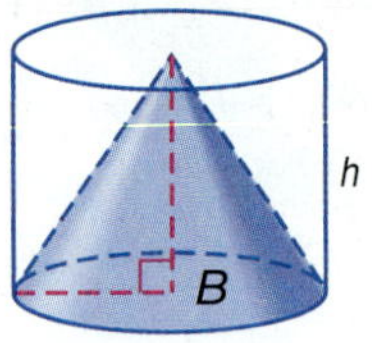

Watch Out!

Volumes of Cones The formula for the surface area of a cone only applies to right cones. However, the formula for volume applies to oblique cones as well as right cones.

Key Concept Volume of a Cone

Words The volume of a circular cone is $V = \frac{1}{3}Bh$, or $V = \frac{1}{3}\pi r^2 h$, where B is the area of the base, h is the height of the cone, and r is the radius of the base.

Models

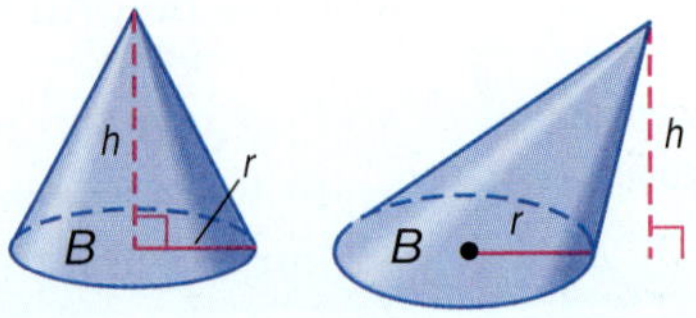

Symbols $V = \frac{1}{3}Bh$ or $V = \frac{1}{3}\pi r^2 h$

EXAMPLE 2 Volume of a Cone

Find the volume of each cone. Round to the nearest tenth.

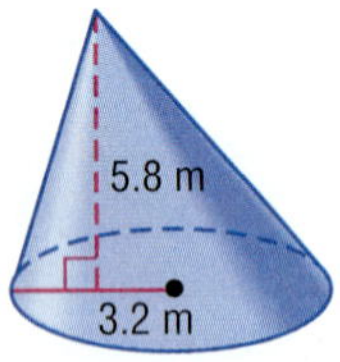

a. $V = \frac{1}{3}\pi r^2 h$ **Volume of a cone**

$\approx \frac{1}{3}\pi(3.2)^2(5.8)$ **$r = 3.2$ and $h = 5.8$**

≈ 62.2 **Use a calculator.**

The volume of the cone is approximately 62.2 cubic meters.

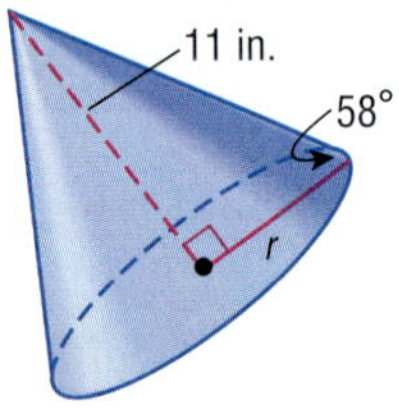

b. Step 1 Use trigonometry to find the radius.

$\tan 58° = \frac{11}{r}$ **$\tan \theta = \frac{\text{opp}}{\text{adj}}$**

$r = \frac{11}{\tan 58°}$ **Solve for r.**

$r \approx 6.9$ **Use a calculator.**

Step 2 Find the volume.

$V = \frac{1}{3}\pi r^2 h$ **Volume of a cone**

$\approx \frac{1}{3}\pi(6.9)^2(11)$ **$r \approx 6.9$ and $h = 11$**

≈ 548.4 **Use a calculator.**

The volume of the cone is approximately 548.4 cubic inches.

Guided Practice

2A.

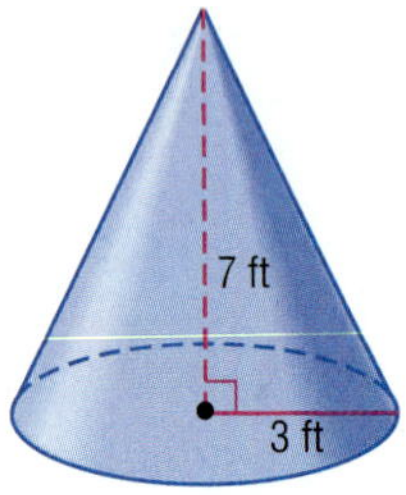

2B.

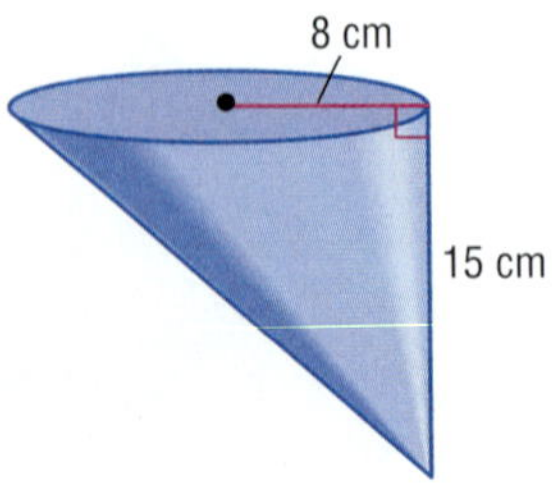

2C.

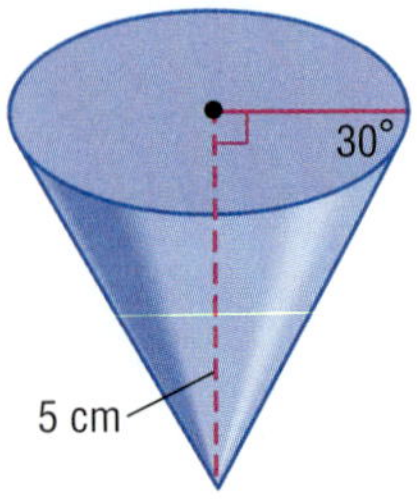

Personal Tutor glencoe.com

Real-World Link

The Washington Monument is the largest masonry structure in the world. By law, no other building in D.C. is allowed to be taller than the 555-foot-tall structure.

Source: Enchanted Learning

Real-World EXAMPLE 3 Find Real-World Volumes

ARCHITECTURE **At the top of the Washington Monument is a small square pyramid, called a *pyramidion*. This pyramid has a height of 55.5 feet with base edges of approximately 34.5 feet. What is the volume of the pyramidion? Round to the nearest tenth.**

Sketch and label the pyramid.

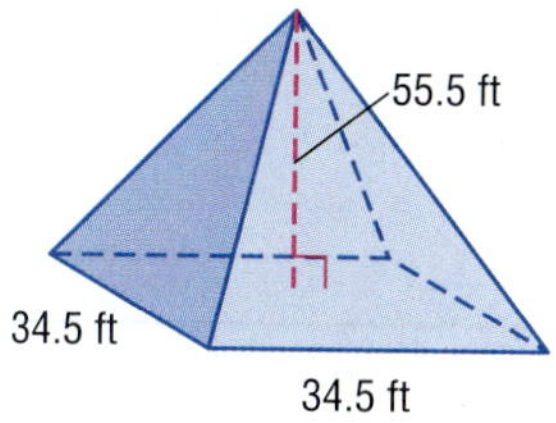

$V = \frac{1}{3}Bh$ — **Volume of a pyramid**

$= \frac{1}{3}(34.5 \cdot 34.5)(55.5)$ — **$B = 34.5 \cdot 34.5$, $h = 55.5$**

$\approx 22{,}019.6$ — **Simplify.**

The volume of the pyramidion atop the Washington Monument is about 22,019.6 cubic feet.

Guided Practice

3. **ARCHAEOLOGY** A pyramidion that was discovered in Saqqara, Egypt, in 1992 has a rectangular base 53 centimeters by 37 centimeters. It is 46 centimeters high. What is the volume of this pyramidion? Round to the nearest tenth.

Personal Tutor glencoe.com

The formulas for the volumes of solids are summarized below.

Concept Summary

Volumes of Solids

Solid	Model	Volume
prism	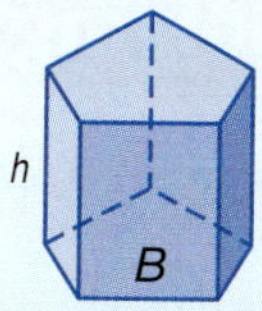	$V = Bh$
cylinder	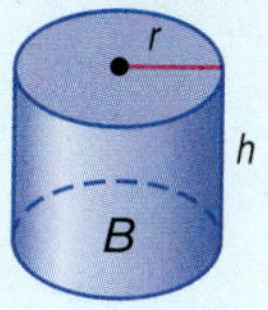	$V = Bh$ or $V = \pi r^2 h$
pyramid	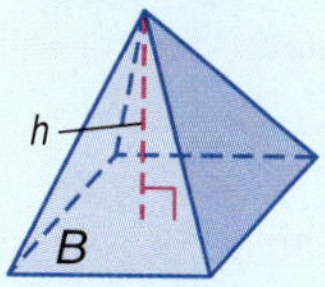	$V = \frac{1}{3}Bh$
cone	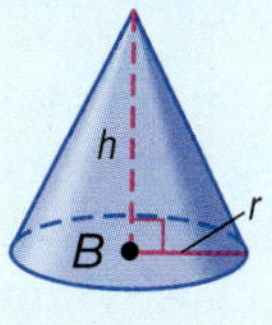	$V = \frac{1}{3}Bh$ or $V = \frac{1}{3}\pi r^2 h$

Check Your Understanding

Example 1 p. 857

Find the volume of each pyramid.

1.
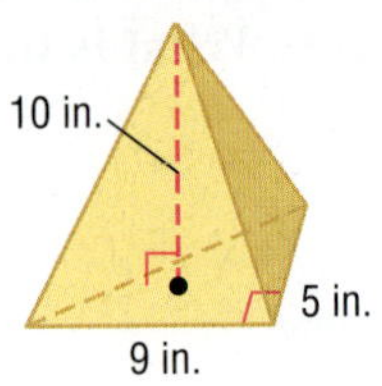

2.
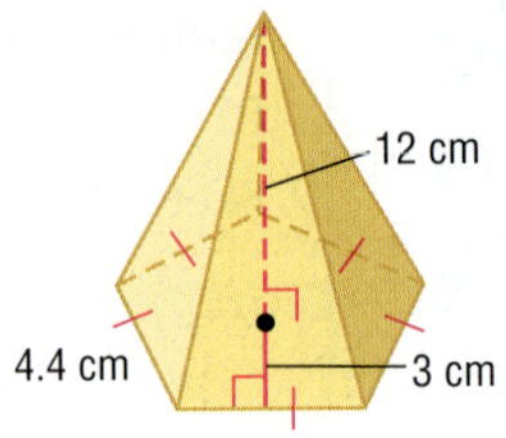

3. a rectangular pyramid with a height of 5.2 meters and a base 8 meters by 4.5 meters

4. a square pyramid with a height of 14 meters and a base with 8-meter side lengths

Example 2 p. 858

Find the volume of each cone. Round to the nearest tenth.

5.
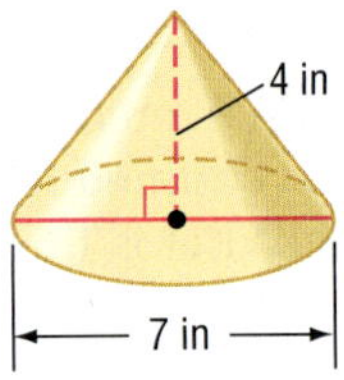

6.
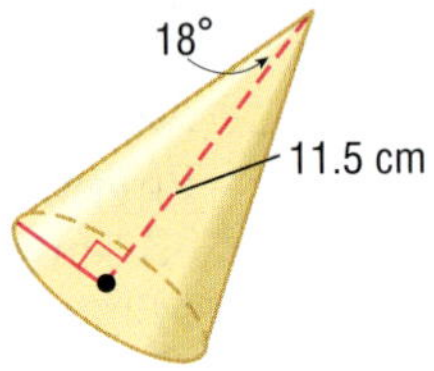

7. an oblique cone with a height of 10.5 millimeters and a radius of 1.6 millimeters

8. a cone with a slant height of 25 meters and a radius of 15 meters

Example 3 p. 859

9. **MUSEUMS** The sky dome of the National Corvette Museum in Bowling Green, Kentucky, is a conical building. If the height is 100 feet and the area of the base is about 15,400 square feet, find the volume of air that the heating and cooling systems would have to accommodate. Round to the nearest tenth.

Practice and Problem Solving

● = **Step-by-Step Solutions** begin on page R20.
Extra Practice begins on page 969.

Example 1 p. 857

Find the volume of each pyramid. Round to the nearest tenth if necessary.

10.
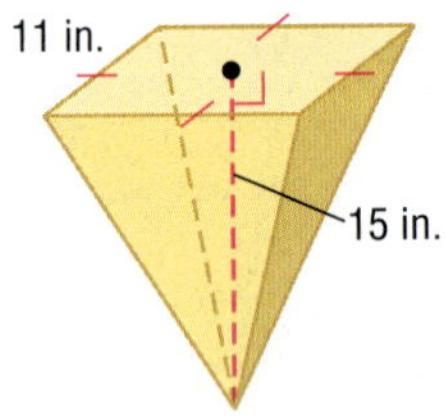

11.
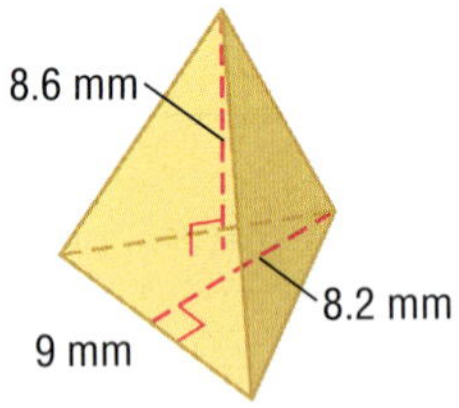

12.
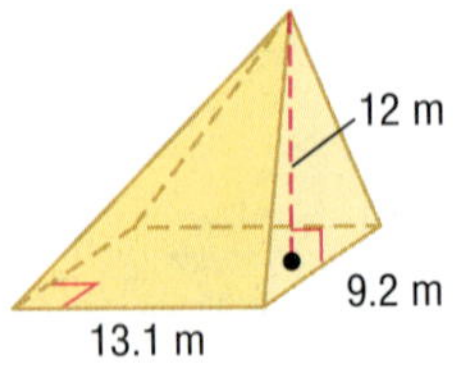

13.
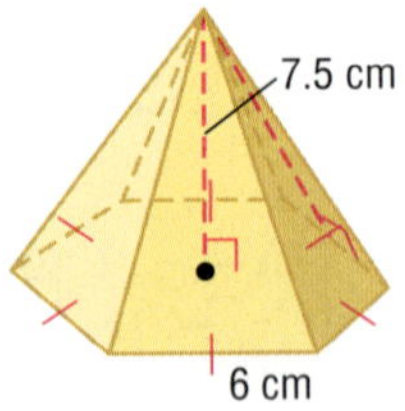

14. a pentagonal pyramid with a base area of 590 square feet and an altitude of 7 feet

15. a triangular pyramid with a height of 4.8 centimeters and a right triangle base with a leg 5 centimeters and hypotenuse 10.2 centimeters

16. A triangular pyramid with a right triangle base with a leg 8 centimeters and hypotenuse 10 centimeters has a volume of 144 cubic centimeters. Find the height.

Example 2 p. 858

Find the volume of each cone. Round to the nearest tenth.

17.

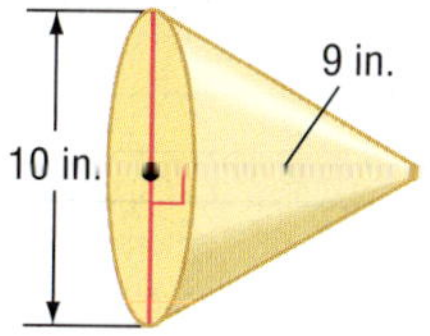

18.

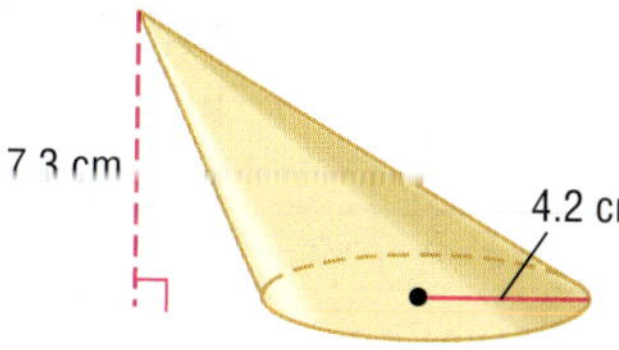

19.

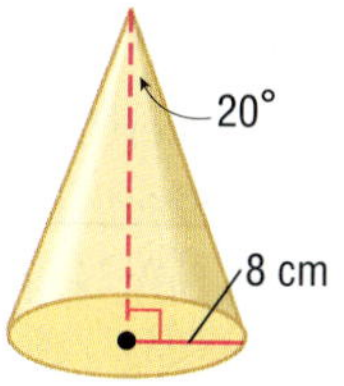

20.

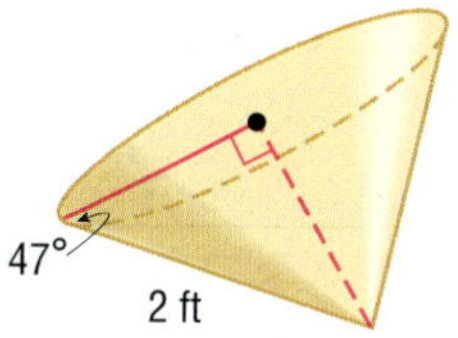

21. an oblique cone with a diameter of 16 inches and an altitude of 16 inches

22. a right cone with a slant height of 5.6 centimeters and a radius of 1 centimeter

Example 3 p. 859

23 **SNACKS** Approximately how many cubic centimeters of roasted peanuts will completely fill a paper cone that is 14 centimeters high and has a base diameter of 8 centimeters? Round to the nearest tenth.

Real-World Link

The 32-story Pyramid Arena in Memphis, Tennessee, is a sports and entertainment facility that has 21,000 seats. It is taller than the Statue of Liberty.

Source: The Pyramid

24. **ARCHITECTURE** The Pyramid Arena in Memphis, Tennessee, is the third largest pyramid in the world. It is approximately 350 feet tall, and its square base is 600 feet wide. Find the volume of this pyramid.

25. **GARDENING** A greenhouse is a regular octagonal pyramid with a height of 5 feet. The base has side lengths of 2 feet. What is the volume of the greenhouse?

Find the volume of each solid. Round to the nearest tenth.

26.

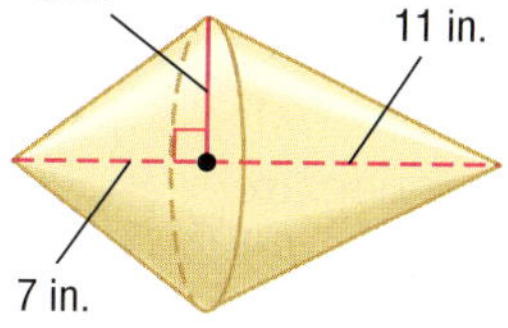

27.

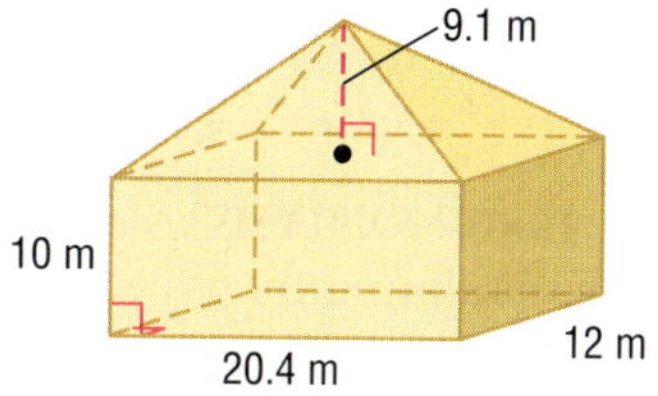

28.

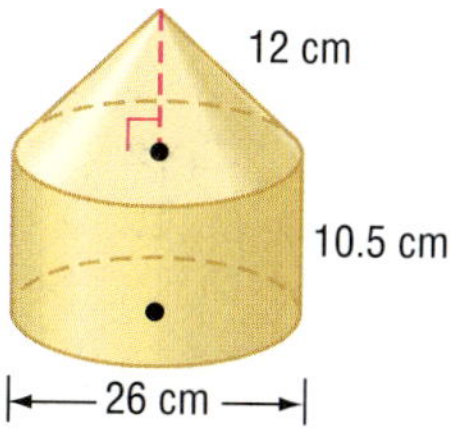

29. **GARDENING** The flowerpot shown is a frustum of an inverted cone. The volume of the frustum is the volume of the larger cone minus the volume of the smaller cone removed to create the frustum.

a. Use the similar triangles in the figure to find x.

b. What is the volume of the flowerpot?

b. If the potting soil is sold in 3-cubic-foot bags, how many planters could be filled with one bag of potting soil?

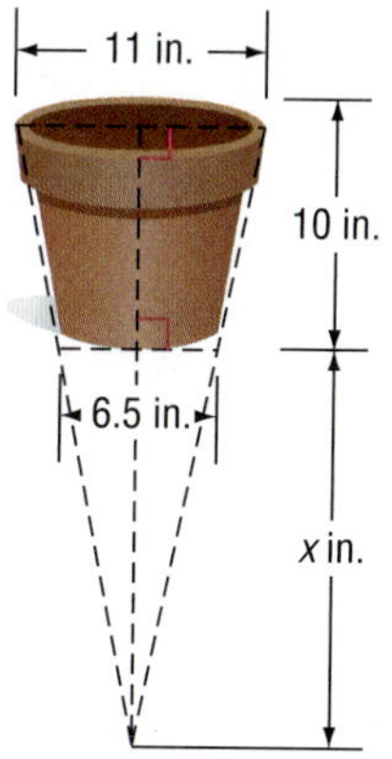

30. **SCIENCE** Refer to page 857. Determine the volume of the model. Explain why knowing the volume is helpful in this situation.

31. **CHANGING DIMENSIONS** A cone has a radius of 4 centimeters and a height of 9 centimeters. Describe how each change affects the volume of the cone.

a. The height is doubled.

b. The radius is doubled.

c. Both the radius and the height are doubled.

Find each measure. Round to the nearest tenth if necessary.

32. A square pyramid has a volume of 862.5 cubic centimeters and a height of 11.5 centimeters. Find the side length of the base.

33 The volume of a cone is 196π cubic inches and the height is 12 inches. What is the diameter?

34. The lateral area of a cone is 71.6 square millimeters and the slant height is 6 millimeters. What is the volume of the cone?

StudyTip

Draw a Diagram When solving problems involving volumes of solids, it is helpful to draw and label a diagram when no diagram is provided.

35. **MULTIPLE REPRESENTATIONS** In this problem, you will investigate rectangular pyramids.

a. **GEOMETRIC** Draw two pyramids with different bases that have a height of 10 centimeters and a base area of 24 square centimeters.

b. **VERBAL** What is true about the volumes of the two pyramids that you drew? Explain.

c. **ANALYTICAL** Explain how multiplying the base area and/or the height of the pyramid by 5 affects the volume of the pyramid.

H.O.T. Problems

Use **H**igher-**O**rder **T**hinking Skills

36. **CHALLENGE** Determine whether the following statement is *always*, *sometimes*, or *never* true. Justify your reasoning.

The volume of a cone with radius r and height h equals the volume of a prism with height h.

37. **ERROR ANALYSIS** Alexandra and Cornelio are calculating the volume of the cone at the right. Is either of them correct? Explain your answer.

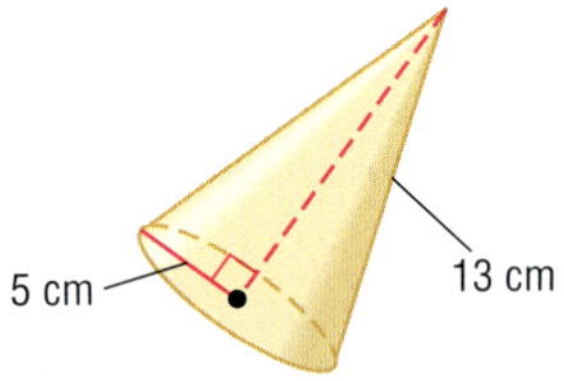

Alexandra

$V = \frac{1}{3}Bh$

$= \frac{1}{3}\pi(5^2)(13)$

$\approx 340.3 \text{ cm}^3$

Cornelio

$5^2 + 12^2 = 13^2$

$V = \frac{1}{3}Bh$

$= \frac{1}{3}\pi(5^2)(12)$

$\approx 314.2 \text{ cm}^3$

38. **REASONING** A cone has a volume of 568 cubic centimeters. What is the volume of a cylinder that has the same radius and height as the cone? Explain your reasoning.

39. **OPEN ENDED** Give an example of a pyramid and a prism that have the same base and the same volume. Explain your reasoning.

40. **WRITING IN MATH** Compare and contrast finding volumes of pyramids and cones with finding volumes of prisms and cylinders.

Standardized Test Practice

41. A conical sand toy has the dimensions as shown below. How many cubic centimeters of sand will it hold when it is filled to the top?

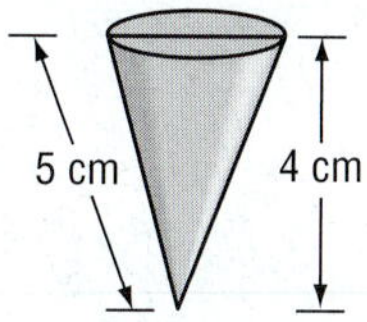

A 12π

B 15π

C $\frac{80}{3}\pi$

D $\frac{100}{3}\pi$

42. SHORT RESPONSE Brooke is buying a tent that is in the shape of a rectangular pyramid. The base is 6 feet by 8 feet. If the tent holds 88 cubic feet of air, how tall is the tent's center pole?

43. PROBABILITY A spinner has sections colored red, blue, orange, and green. The table below shows the results of several spins. What is the experimental probability of the spinner landing on orange?

F $\frac{1}{4}$

G $\frac{1}{5}$

H $\frac{9}{25}$

J $\frac{1}{2}$

Color	Frequency
red	6
blue	4
orange	5
green	10

44. SAT/ACT For all $x \neq -2$ or 0, $\frac{x^2 - 2x - 8}{x^2 + 2x} = ?$

A -8

B $x - 4$

C $\frac{-x - 4}{x}$

D $\frac{x - 4}{x}$

Spiral Review

Find the volume of each prism. (Lesson 12-4)

45.

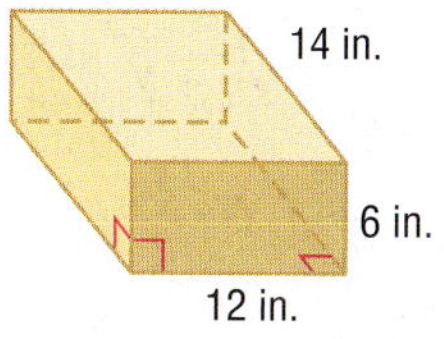

46.

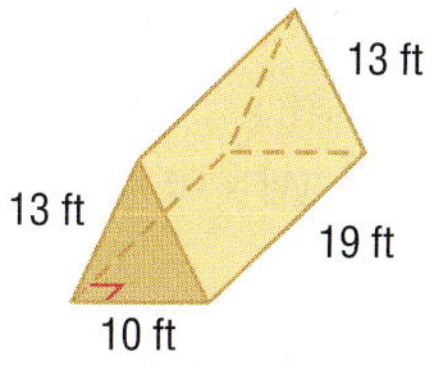

47.

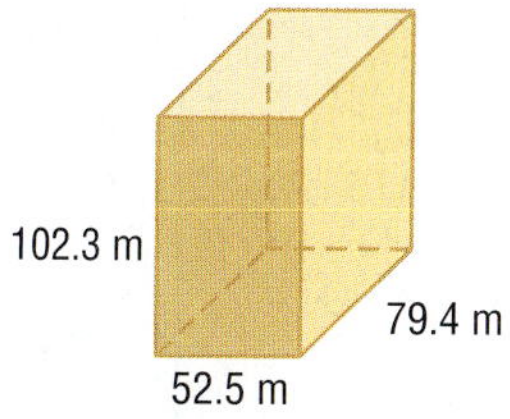

48. FARMING The picture shows a combination hopper cone and bin used by farmers to store grain after harvest. The cone at the bottom of the bin allows the grain to be emptied more easily. Use the dimensions in the diagram to find the entire surface area of the bin with a conical top and bottom. Write the exact answer and the answer rounded to the nearest square foot. (Lesson 12-3)

Skills Review

Find the area of each shaded region. Round to the nearest tenth. (Lesson 11-4)

49.

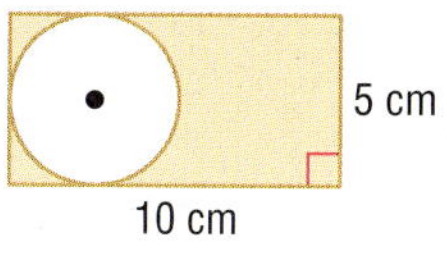

50.

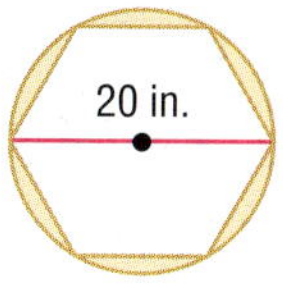

51.

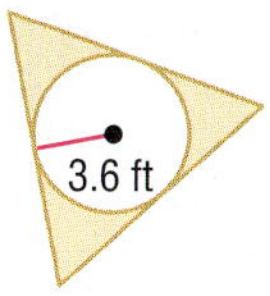

52.

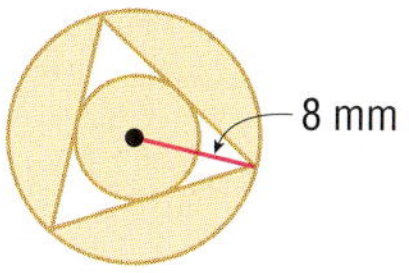

12-6 Surface Areas and Volumes of Spheres

Then

You found surface areas of prisms and cylinders. (Lesson 12-2)

Now

- Find surface areas of spheres.
- Find volumes of spheres.

New Vocabulary

great circle
pole
hemisphere

Math Online
glencoe.com

Why?

When you blow bubbles, soapy liquid surrounds a volume of air. Because of surface tension, the liquid maintains a shape that minimizes the surface area surrounding the air. The shape that minimizes surface area per unit of volume is a sphere.

Surface Area of Spheres Recall that a *sphere* is the locus of all points in space that are a given distance from a given point called the *center* of the sphere.

- A *radius* of a sphere is a segment from the center to a point on the sphere.
- A *chord* of a sphere is a segment that connects any two points on the sphere.
- A *diameter* of a sphere is a chord that contains the center.
- A *tangent* to a sphere is a line that intersects the sphere in exactly one point.

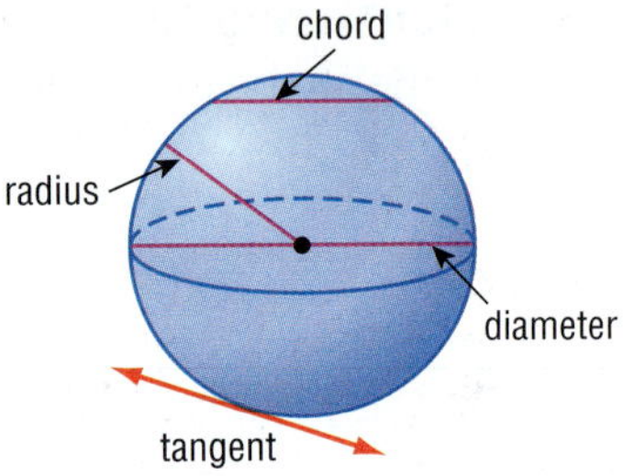

Key Concept — Surface Area of a Sphere

Words The surface area S of a sphere is $S = 4\pi r^2$, where r is the radius.

Symbols $S = 4\pi r^2$

Model

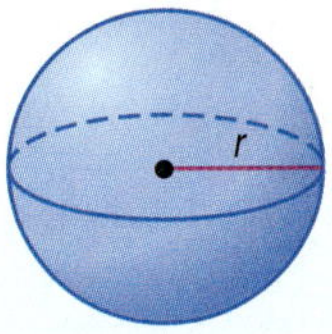

EXAMPLE 1 Surface Area of a Sphere

Find the surface area of the sphere. Round to the nearest tenth.

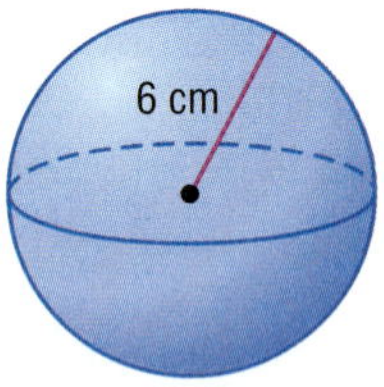

$S = 4\pi r^2$ — **Surface area of a sphere**

$= 4\pi(6)^2$ — **Replace r with 6.**

≈ 452.4 — **Use a calculator.**

The surface area is about 452.4 square centimeters.

Guided Practice

Find the surface area of each sphere. Round to the nearest tenth.

1A.

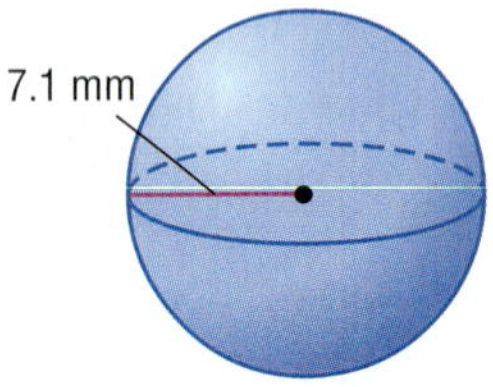

1B.

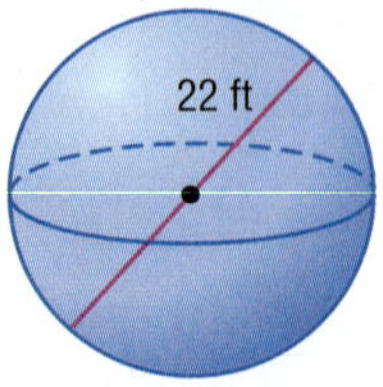

Personal Tutor glencoe.com

StudyTip

Great Circles A sphere has an infinite number of great circles.

A plane can intersect a sphere in a point or in a circle. If the circle contains the center of the sphere, the intersection is called a **great circle**. The endpoints of a diameter of a great circle are called **poles**.

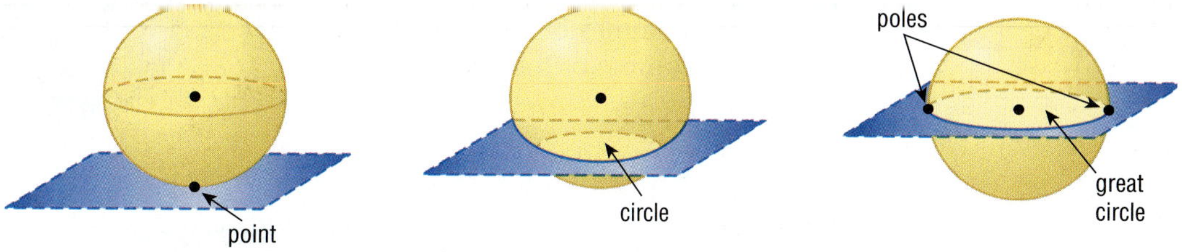

Since a great circle has the same center as the sphere and its radii are also radii of the sphere, it is the largest circle that can be drawn on a sphere. A great circle separates a sphere into two congruent halves, called **hemispheres**.

EXAMPLE 2 Use Great Circles to Find Surface Area

a. Find the surface area of the hemisphere.

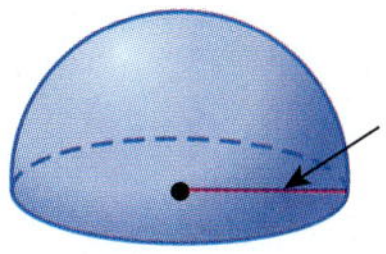

Find half the area of a sphere with a radius of 2.8 centimeters. Then add the area of the great circle.

$S = \frac{1}{2}(4\pi r^2) + \pi r^2$ **Surface area of a hemisphere**

$= \frac{1}{2}[4\pi(2.8)^2] + \pi(2.8)^2$ **Replace *r* with 2.8.**

$\approx 73.9 \text{ cm}^2$ **Use a calculator.**

Watch Out!

Area of Hemisphere When finding the surface area of a hemisphere, do not forget to include the area of the great circle.

b. Find the surface area of a sphere if the circumference of the great circle is 5π meters.

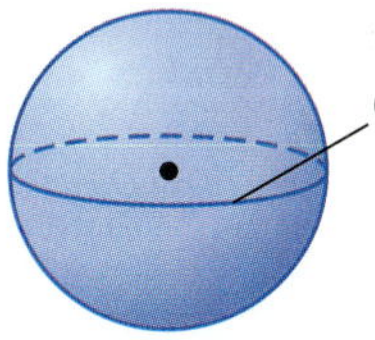

First, find the radius. The circumference of a great circle is $2\pi r$. So, $2\pi r = 5\pi$ or $r = 2.5$.

$S = 4\pi r^2$ **Surface area of a sphere**

$= 4\pi(2.5)^2$ **Replace *r* with 2.5.**

$\approx 78.5 \text{ m}^2$ **Use a calculator.**

c. Find the surface area of a sphere if the area of the great circle is approximately 130 square inches.

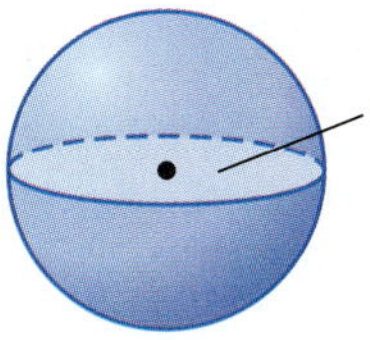

First, find the radius. The area of a great circle is πr^2. So, $\pi r^2 = 130$ or $r \approx 6.4$.

$S = 4\pi r^2$ **Surface area of a sphere**

$\approx 4\pi(6.4)^2$ **Replace *r* with 6.4.**

$\approx 514.7 \text{ in}^2$ **Use a calculator.**

Guided Practice

Find the surface area of each figure. Round to the nearest tenth if necessary.

2A. sphere: circumference of great circle = 16.2π ft

2B. hemisphere: area of great circle ≈ 94 mm^2

2C. hemisphere: circumference of great circle = 36π cm

Personal Tutor glencoe.com

Volume of Spheres Suppose a sphere with radius r contains infinitely many pyramids with vertices at the center of the sphere. Each pyramid has height r and base area B. The sum of the volumes of all the pyramids equals the volume of the sphere.

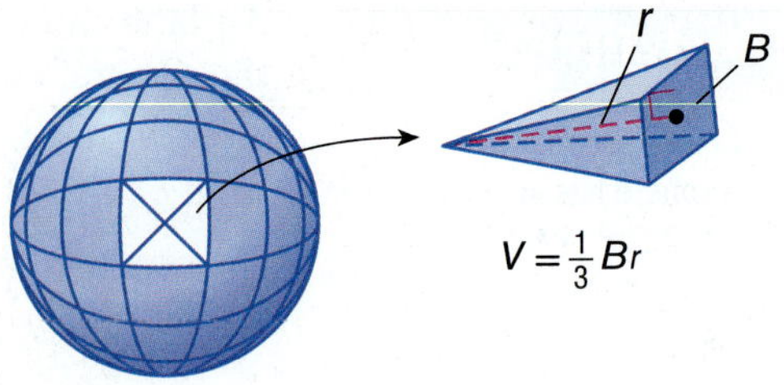

$V = \frac{1}{3}B_1r_1 + \frac{1}{3}B_2r_2 + \ldots + \frac{1}{3}B_nr_n$	**Sum of volumes of pyramids**
$= \frac{1}{3}r(B_1 + B_2 + \ldots + B_n)$	**Distributive Property**
$= \frac{1}{3}r(4\pi r^2)$	**The sum of the pyramid base areas equals the surface area of the sphere.**
$= \frac{4}{3}\pi r^3$	**Simplify.**

Key Concept Volume of a Sphere

Words The volume V of a sphere is $V = \frac{4}{3}\pi r^3$, where r is the radius of the sphere.

Symbols $V = \frac{4}{3}\pi r^3$

Model

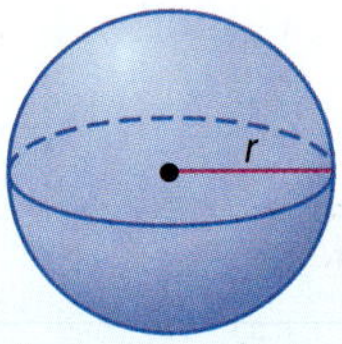

EXAMPLE 3 Volumes of Spheres and Hemispheres

Find the volume of each sphere or hemisphere. Round to the nearest tenth.

a. a hemisphere with a radius of 6 meters

Estimate: $V \approx \frac{1}{2} \cdot \frac{4}{3} \cdot 3 \cdot 6^3$ **or 432** m^3

$V = \frac{1}{2}\left(\frac{4}{3}\pi r^3\right)$	**Volume of a hemisphere**
$= \frac{2}{3}\pi(6)^3$	**Replace r with 6.**
$\approx 452.4 \text{ m}^3$	**Use a calculator.**

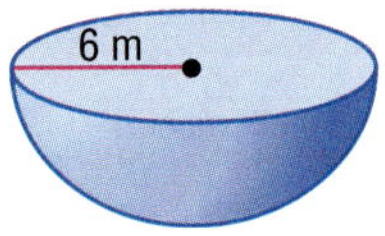

b. a sphere with a great circle circumference of 18π centimeters

Step 1 Find the radius of the sphere.

$C = 2\pi r$	**Circumference of a circle**
$18\pi = 2\pi r$	**Replace C with 18π.**
$r = 9$	**Solve for r.**

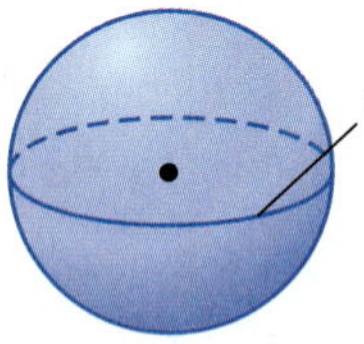

Step 2 Find the volume.

$V = \frac{4}{3}\pi r^3$	**Volume of a sphere**
$= \frac{4}{3}\pi(9)^3$	**Replace r with 9.**
$\approx 3053.6 \text{ cm}^3$	**Use a calculator.**

Guided Practice

3A. sphere: diameter = 7.4 in.

3B. hemisphere: area of great circle ≈ 249 mm^2

Personal Tutor glencoe.com

StudyTip

Units Remember to use the correct units when giving your answers. As with other solids, the surface area of a sphere is measured in square units, and volume is measured in cubic units.

Real-World Link

The University of North Carolina has won the greatest number of national championships in women's soccer since the first tournament in 1982. As of 2007, they have won 16 times.

Source: Fact Monster

Real-World EXAMPLE 4 Solve Problems Involving Solids

SOCCER **The soccer ball globe at the right was constructed for the 2006 World Cup soccer tournament. It takes up 47,916π cubic feet of space. Assume that the globe is a sphere. What is the circumference of the globe?**

Understand You know that the volume of the globe is 47,916π cubic feet. The circumference of the globe is the circumference of the great circle.

Plan First use the volume formula to find the radius. Then find the circumference of the great circle.

Solve

$V = \frac{4}{3}\pi r^3$ — **Volume of a sphere**

$47{,}916\pi = \frac{4}{3}\pi r^3$ — **Replace V with 47,916π.**

$35{,}937 = r^3$ — **Divide each side by $\frac{4}{3}\pi$.**

Use a calculator to find $\sqrt[3]{35{,}937}$.

35937 [^] [(] 1 [÷] 3 [)] [ENTER] 33

The radius of the globe is 33 feet. So, the circumference is $2\pi r = 2\pi(33)$ or approximately 207.3 feet.

Check You can work backward to check the solution.

If $C \approx 207.3$, then $r \approx 33$. If $r \approx 33$, then $V \approx \frac{4}{3}\pi \cdot 33^3$ or about 47,917π cubic feet. The solution is correct. ✓

Guided Practice

4. **BALLOONS** Ren inflates a spherical balloon to a circumference of about 14 inches. He then adds more air to the balloon until the circumference is about 18 inches. What volume of air was added to the balloon?

Personal Tutor glencoe.com

Concept Summary

Circles and Spheres

Figure	circle		sphere	
Model	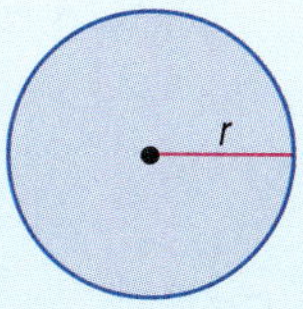		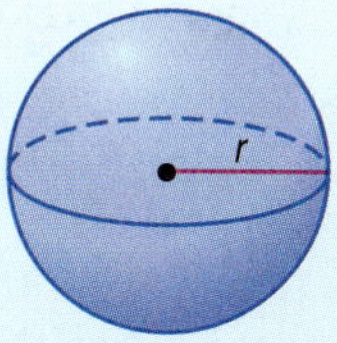	
Measure	circumference	area	surface area	volume
Formula	$C = 2\pi r$	$A = \pi r^2$	$S = 4\pi r^2$	$V = \frac{4}{3}\pi r^3$

Check Your Understanding

Examples 1 and 2
pp. 864–865

Find the surface area of each sphere or hemisphere. Round to the nearest tenth.

1.

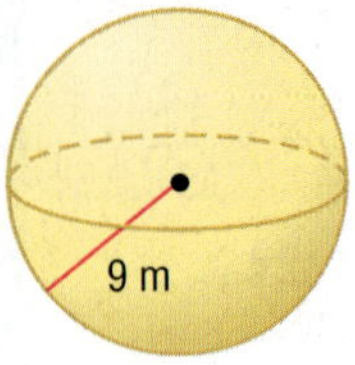

2.

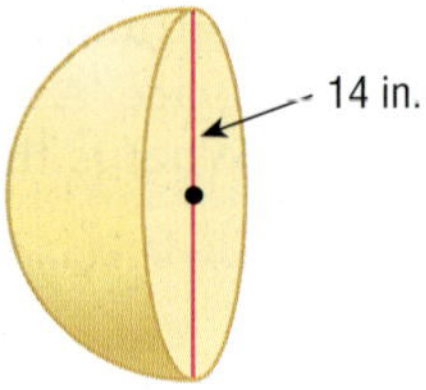

3. sphere: area of great circle = 36π yd^2

4. hemisphere: circumference of great circle ≈ 26 cm

Example 3
p. 866

Find the volume of each sphere or hemisphere. Round to the nearest tenth.

5. sphere: radius = 10 ft

6. hemisphere: diameter = 16 cm

7. hemisphere: circumference of great circle = 24π m

8. sphere: area of great circle = 55π in^2

Example 4
p. 867

9. **BASKETBALL** Basketballs used in professional games must have a circumference of $29\frac{1}{2}$ inches. What is the surface area of a basketball used in a professional game?

Practice and Problem Solving

= **Step-by-Step Solutions** begin on page R20.
Extra Practice begins on page 969.

Examples 1 and 2
pp. 864–865

Find the surface area of each sphere or hemisphere. Round to the nearest tenth.

10.

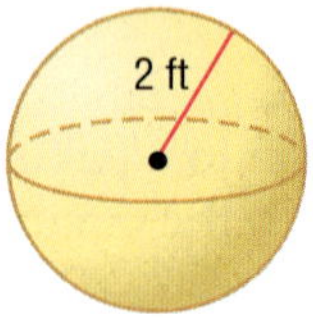

11.

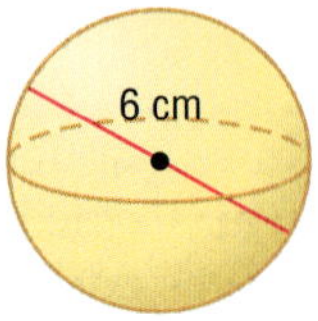

12.

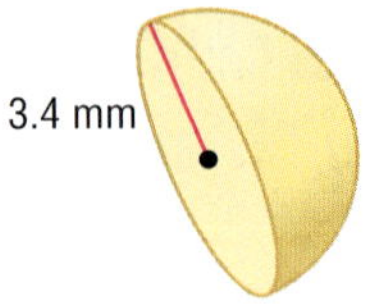

13.

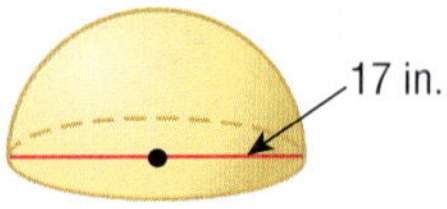

14. sphere: circumference of great circle = 2π cm

15. sphere: area of great circle ≈ 32 ft^2

16. hemisphere: area of great circle ≈ 40 in^2

17. hemisphere: circumference of great circle = 15π mm

Example 3
p. 866

Find the volume of each sphere or hemisphere. Round to the nearest tenth.

18.

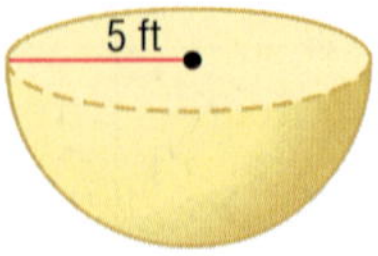

19

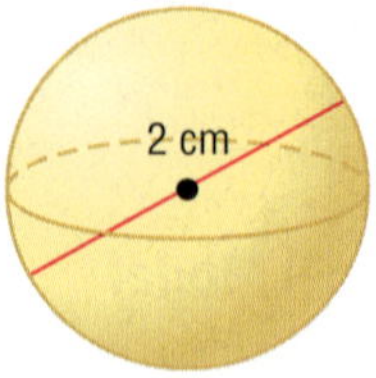

20. sphere: radius = 1.4 yd

21. hemisphere: diameter = 21.8 cm

22. sphere: area of great circle = 49π m^2

23. sphere: circumference of great circle ≈ 22 in.

24. hemisphere: circumference of great circle ≈ 18 ft

25. hemisphere: area of great circle ≈ 35 m^2

Example 4
p. 867

26. FISH A *puffer fish* is able to "puff up" when threatened by gulping water and inflating its body. The puffer fish at the right is approximately a sphere with a diameter of 5 inches. Its surface area when inflated is about 1.5 times its normal surface area. What is the surface area of the fish when it is *not* puffed up?

27. ARCHITECTURE The Reunion Tower in Dallas, Texas, is topped by a spherical dome that has a surface area of approximately 13,924π square feet. What is the volume of the dome? Round to the nearest tenth.

Real-World Link

Spherical tree houses weigh approximately 500 kilograms and can be as high as 35 meters off the ground.

Source: Free Spirit Spheres

28. TREE HOUSE The spherical tree house, or *tree sphere,* shown at the right has a diameter of 10.5 feet. Its volume is 1.8 times the volume of the first tree sphere that was built. What was the diameter of the first tree sphere? Round to the nearest foot.

Find the surface area and the volume of each solid. Round to the nearest tenth.

29.

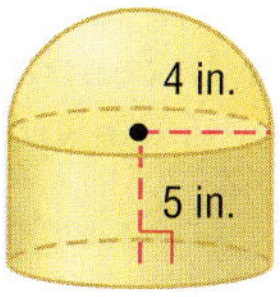

30.

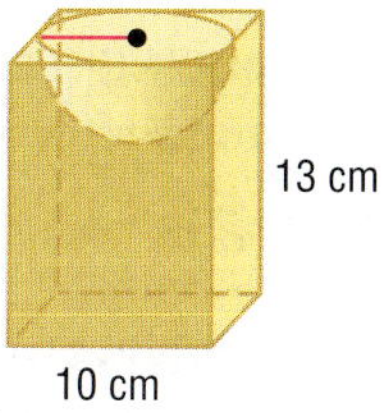

31. TOYS The spinning top at the right is a composite of a cone and a hemisphere.

a. Find the surface area and the volume of the top. Round to the nearest tenth.

b. If the manufacturer of the top makes another model with dimensions that are one-half of the dimensions of this top, what are its surface area and volume?

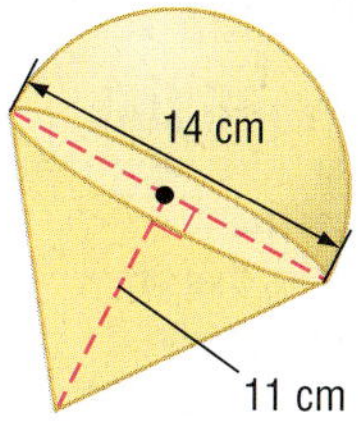

32. BALLOONS A spherical helium-filled balloon with a diameter of 30 centimeters can lift a 14-gram object. Find the size of a balloon that could lift a person who weighs 65 kilograms. Round to the nearest tenth.

Use sphere S to name each of the following.

33. a chord

34. a radius

35. a diameter

36. a tangent

37. a great circle

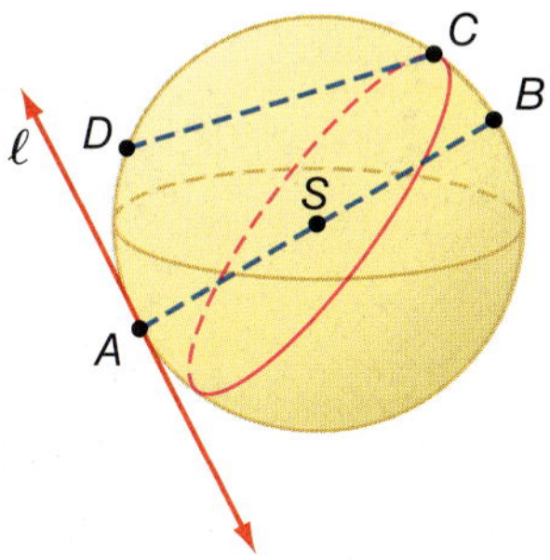

38. DIMENSIONAL ANALYSIS Which has greater volume: a sphere with a radius of 2.3 yards or a cylinder with a radius of 4 feet and height of 8 feet?

39. FOOD Suppose the orange in the graphic is a sphere with a radius of 4 centimeters. Round to the nearest tenth.

a. What is the volume of the portion of the sphere that represents orange production in California?

b. What is the surface area of the portion of the sphere that represents orange production in Florida?

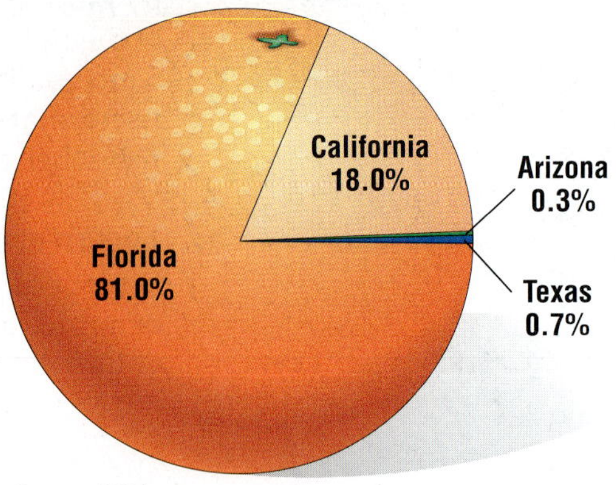

Source: USDA

Describe the number of vertical planes and the number of horizontal planes that produce reflection symmetry in each solid. Then describe the angles of rotation that produce rotation symmetry in each solid.

40.

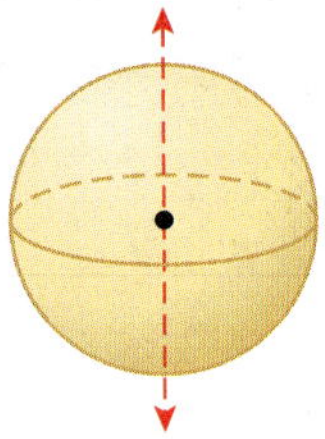

41.

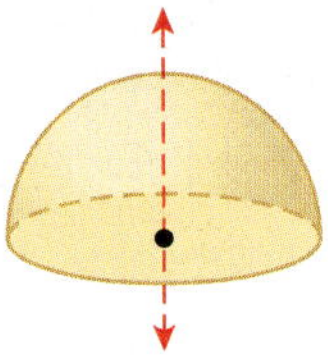

CHANGING DIMENSIONS **A sphere has a radius of 12 centimeters. Describe how each change affects the surface area and the volume of the sphere.**

42. The radius is multiplied by 4.

43. The radius is divided by 3.

44. NAVIGATIONAL COORDINATES *Latitude lines* circle Earth in an east-west direction.

a. Describe the longest latitude line. What is the geometric name for the circle created by this line?

b. The circumference of Earth is about 25,000 miles. Describe how you can use this to find the surface area of Earth. Then find the surface area.

c. If two planes are equidistant from the center of a sphere and intersect the sphere, then the circles of intersection are congruent. Describe how this property can be applied to latitude lines.

Real-World Link

Latitude lines are imaginary lines that are marked every twenty degrees north and south of the equator.

One degree is equivalent to about 111 kilometers.

Source: Visit and Learn

H.O.T. Problems

Use **H**igher-**O**rder **T**hinking Skills

45. CHALLENGE A cube has a volume of 216 cubic inches. Find the volume of a sphere that is circumscribed about the cube. Round to the nearest tenth.

46. REASONING Determine whether the following statement is *true* or *false*. If true, explain your reasoning. If false, provide a counterexample.

If a sphere has radius r, there exists a cone with radius r having the same volume.

47. OPEN ENDED Sketch a sphere showing two examples of great circles. Sketch another sphere showing two examples of circles formed by planes intersecting the sphere that are *not* great circles.

48. WRITING IN MATH Write a ratio comparing the volume of a sphere with radius r to the volume of a cylinder with radius r and height $2r$. Then describe what the ratio means.

Standardized Test Practice

49. GRIDDED RESPONSE What is the volume of the hemisphere shown below in cubic meters?

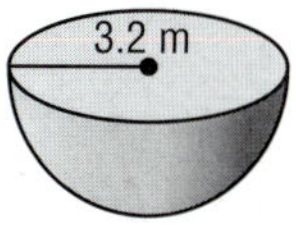

50. ALGEBRA What is the solution set of $3z + 4 < 6 + 7z$?

A $\{z|z > -0.5\}$ C $\{z|z < -0.5\}$

B $\{z|z > -2\}$ D $\{z|z < -2\}$

51. If the area of the great circle of a sphere is 33 ft^2, what is the surface area of the sphere?

F 42 ft^2 H 132 ft^2

G 117 ft^2 J 264 ft^2

52. SAT/ACT If a line ℓ is a perpendicular bisector of segment *AB* at *E*, how many points on line ℓ are the same distance from point *A* as from point *B*?

A none C two

B one D all points

Spiral Review

Find the volume of each pyramid. Round to the nearest tenth if necessary. (Lesson 12-5)

53.

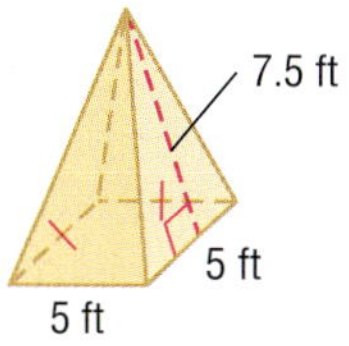

54.

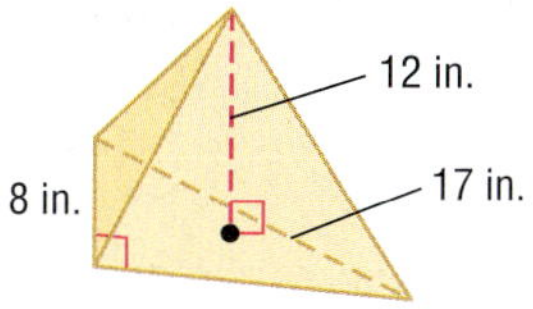

55.

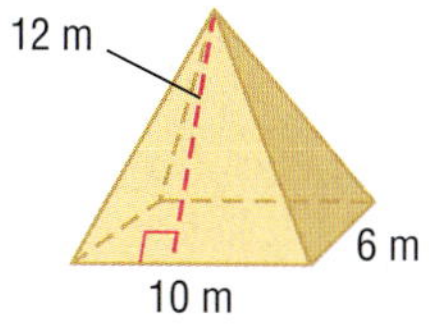

56. ENGINEERING The base of an oil drilling platform is made up of 24 concrete cylindrical cells. Twenty of the cells are used for oil storage. The pillars that support the platform deck rest on the four other cells. Find the total volume of the storage cells. (Lesson 12-4)

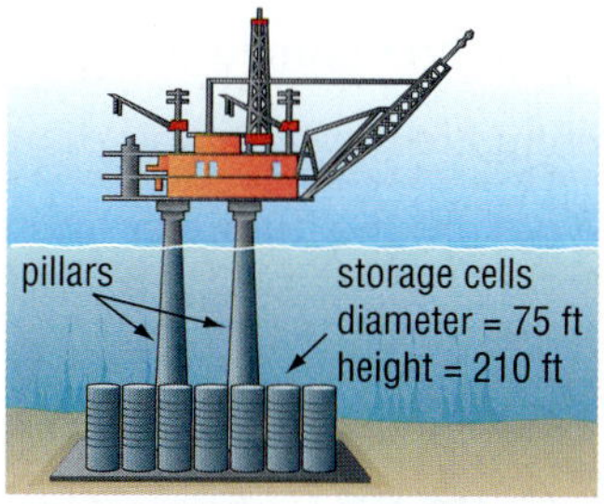

Find the area of each shaded region. Round to the nearest tenth. (Lesson 11-4)

57.

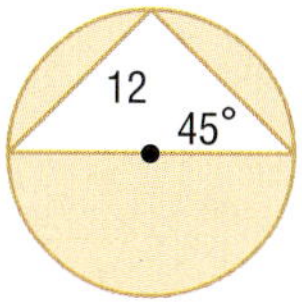

58.

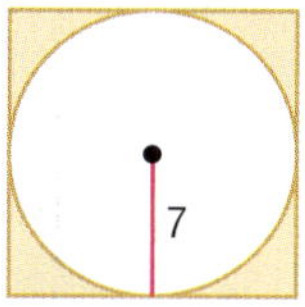

59.

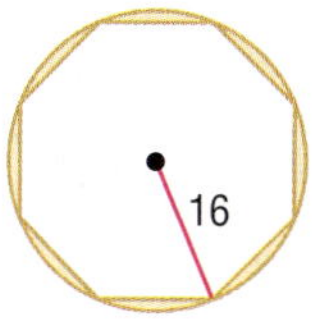

COORDINATE GEOMETRY Find the area of each figure. (Lesson 11-1)

60. ▱*WXYZ* with *W*(0, 0), *X*(4, 0), *Y*(5, 5), and *Z*(1, 5)

61. △*ABC* with *A*(2, −3), *B*(−5, −3), and *C*(−1, 3)

Skills Review

Refer to the figure. (Lesson 1-1)

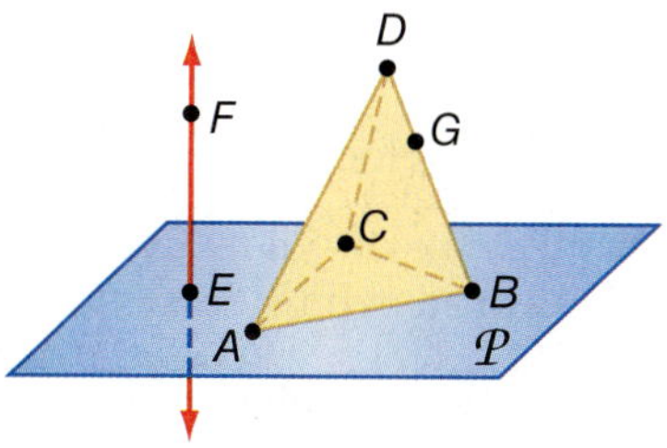

62. How many planes appear in this figure?

63. Name three points that are collinear.

64. Are points *G*, *A*, *B*, and *E* coplanar? Explain.

65. At what point do $\overleftrightarrow{EF}$ and $\overleftrightarrow{AB}$ intersect?

EXTEND
12-6

Geometry Lab
Locus and Spheres

Spheres are defined in terms of a locus of points in space. The definition of a sphere is the set of all points that are a given distance from a given point.

ACTIVITY 1 Locus of Points a Given Distance from Endpoints

Find the locus of all points that are equidistant from a segment.

Collect the Data

- Draw a given line segment with endpoints J and K

- Create a set of points that are equidistant from the segment.

Analyze

1. Draw a figure and describe the locus of points in space that are 8 units from a segment that is 30 units long.

2. What three-dimensional shapes form the figure?

3. What are the radii and diameters of each hemisphere?

4. What are the diameter and the height of the cylinder?

ACTIVITY 2 Spheres That Intersect

Find the locus of all points that are equidistant from the centers of two intersecting spheres with the same radius.

Collect the Data

- Draw a line segment.

- Draw congruent overlapping spheres, with the centers at the endpoints of the given line segment.

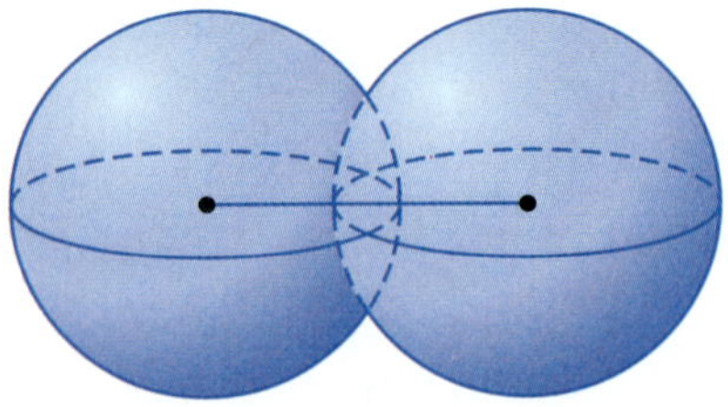

Analyze

5. What is the shape of the intersection of the upper hemispheres?

6. Can this be described as a locus of points in space or on a plane? Explain.

7. Describe the intersection as a locus.

8. **FIREWORKS** What is the locus of points that describes how particles from a fireworks explosion will disperse in an explosion at 400 feet above ground level if the expected distance a particle could travel is 200 feet?

12-7

Spherical Geometry

Then

You identified basic properties of spheres. (Lesson 12-6)

Now

- Describe sets of points on a sphere.
- Compare and contrast Euclidean and spherical geometries.

New Vocabulary

Euclidean geometry
spherical geometry
non-Euclidean geometry

Math Online
glencoe.com

Why?

Since Earth has a curved instead of a flat surface, the shortest path between two points on Earth is described by an arc of a great circle instead of a straight line.

Geometry on a Sphere In this text, we have studied **Euclidean geometry**, either in the plane or in space. In plane Euclidean geometry, a *plane* is a flat surface made up of points that extend infinitely in all directions. In **spherical geometry**, or geometry on a sphere, a plane is the surface of a sphere.

Lines are also defined differently in spherical geometry.

Key Concept — Lines in Plane and Spherical Geometry

Plane Euclidean Geometry

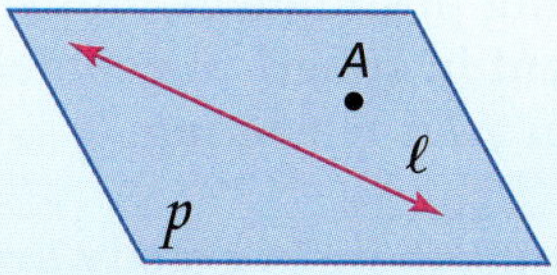

Plane $\mathcal{P}$ contains line ℓ and point A not on line ℓ.

Spherical Geometry

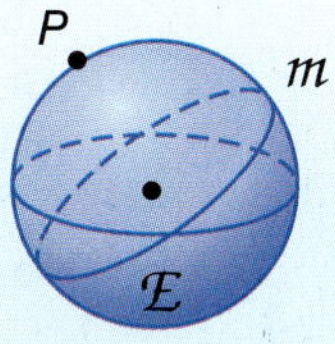

Sphere $\mathcal{E}$ contains great circle m and point P not on m. Great circle m is a line on sphere $\mathcal{E}$.

EXAMPLE 1 Describe Sets of Points on a Sphere

Name each of the following on sphere $\mathcal{F}$.

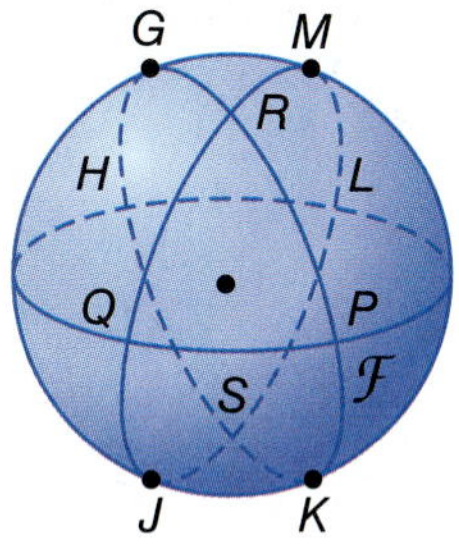

a. two lines containing point R

$\overleftrightarrow{GP}$ and $\overleftrightarrow{MQ}$ are lines on sphere $\mathcal{F}$ that contain point R.

b. a segment containing point K

$\overline{PS}$ is a segment on sphere $\mathcal{F}$ that contains point K.

c. a triangle

$\triangle RQP$ is a triangle on sphere $\mathcal{F}$.

Guided Practice

Name each of the following on sphere $\mathcal{F}$ above.

1A. two lines containing point P

1B. a segment containing point Q

1C. a triangle

Personal Tutor glencoe.com

Real-World EXAMPLE 2 Identify Lines in Spherical Geometry

ENTERTAINMENT Determine whether figure m on the mirror ball shown is a line in spherical geometry.

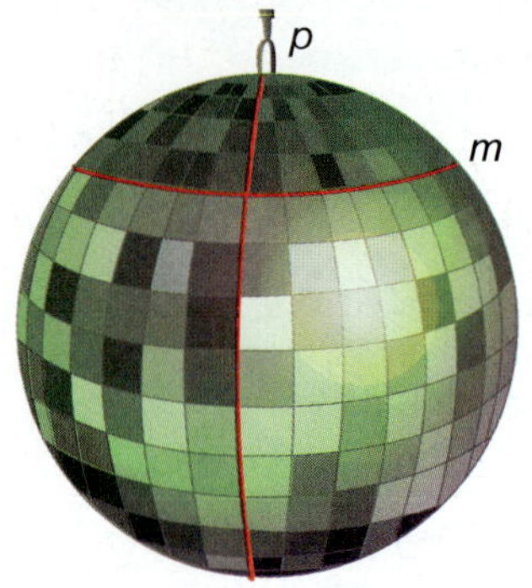

Notice that figure m does not go through the poles of the sphere. Therefore figure m is not a great circle and so not a line in spherical geometry.

Guided Practice

2. Determine whether figure p on the mirror ball shown is a line in spherical geometry.

Personal Tutor glencoe.com

StudyTip

Elliptical Geometry *Spherical geometry* is a subcategory of *elliptical geometry*.

Comparing Euclidean and Spherical Geometries While some postulates and properties of Euclidean geometry are true in spherical geometry, others are not, or are true only under certain circumstances.

EXAMPLE 3 Compare Plane Euclidean and Spherical Geometries

Tell whether the following postulate or property of plane Euclidean geometry has a corresponding statement in spherical geometry. If so, write the corresponding statement. If not, explain your reasoning.

a. Through any two points, there is exactly one line.

In the figure, notice that there is more than one great circle (line) through polar points A and B. However, there is only one great circle through nonpolar points C and D.

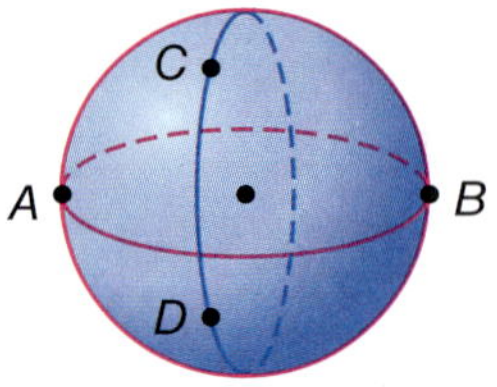

Therefore, a corresponding statement is that through any two nonpolar points, there is exactly one great circle (line).

b. If given a line and a point not on the line, then there exists exactly one line through the point that is parallel to the given line.

In the figure, notice that every great circle (line) containing point A will intersect line ℓ. Thus there exists no great circle through point A that is parallel to line ℓ.

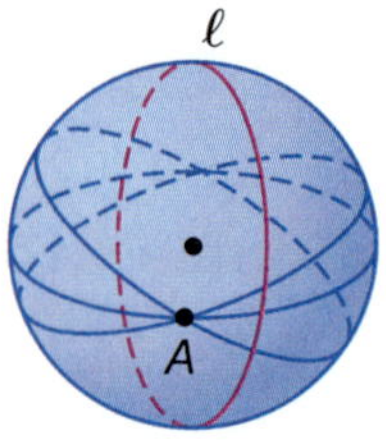

Guided Practice

3A. A line segment is the shortest path between two points.

3B. Through any two points, there is exactly one segment.

Personal Tutor glencoe.com

StudyTip

Finite Geometries *Planar networks* are another type of non-Euclidean geometry. You will learn about planar networks in Extend Lesson 13-5.

A **non-Euclidean geometry** is a geometry in which at least one of the postulates from Euclidean geometry fails. Notice in Example 3b that the Parallel Postulate does not hold true on a sphere. Lines, or great circles, cannot be parallel in spherical geometry. Therefore, spherical geometry is non-Euclidean.

Check Your Understanding

Example 1
p. 873

Name each of the following on sphere $\mathcal{B}$.

1. two lines containing point Q
2. a segment containing point L
3. a triangle
4. two segments on the same great circle

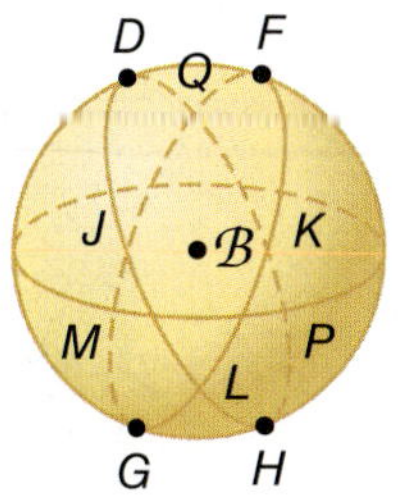

Example 2
p. 874

SPORTS Determine whether figure $\mathcal{X}$ on each of the spheres shown is a line in spherical geometry.

5.

6.

Example 3
p. 874

Tell whether the following postulate or property of plane Euclidean geometry has a corresponding statement in spherical geometry. If so, write the corresponding statement. If not, explain your reasoning.

7. The points on any line or line segment can be put into one-to-one correspondence with real numbers.
8. Perpendicular lines intersect at one point.

Practice and Problem Solving

● = **Step-by-Step Solutions** begin on page R20.
Extra Practice begins on page 969.

Example 1
p. 873

Name two lines containing point M, a segment containing point S, and a triangle in each of the following spheres.

9.

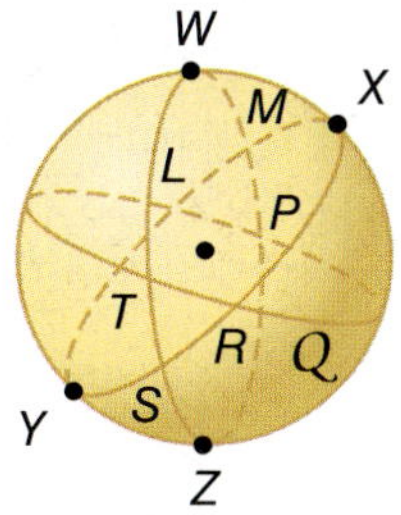

10.

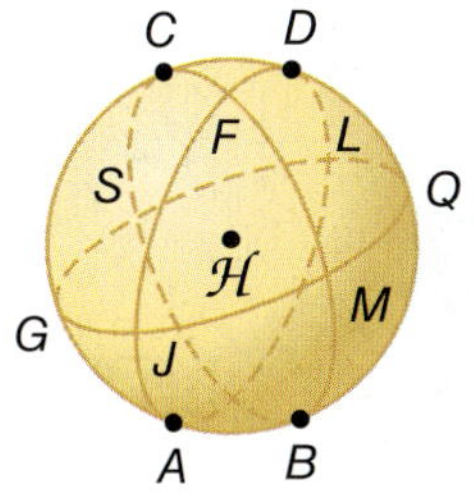

11. **SOCCER** Name each of the following on the soccer ball shown.
   a. two lines containing point B
   b. a segment containing point F
   c. a triangle
   d. a segment containing point C
   e. a line
   f. two lines containing point A

Example 2
p. 874

ARCHITECTURE Determine whether figure *w* on each of the spheres shown is a line in spherical geometry.

12.

13.

14. GEOGRAPHY Lines of latitude and longitude are used to describe positions on the Earth's surface. By convention, lines of longitude divide Earth vertically, while lines of latitude divide it horizontally.

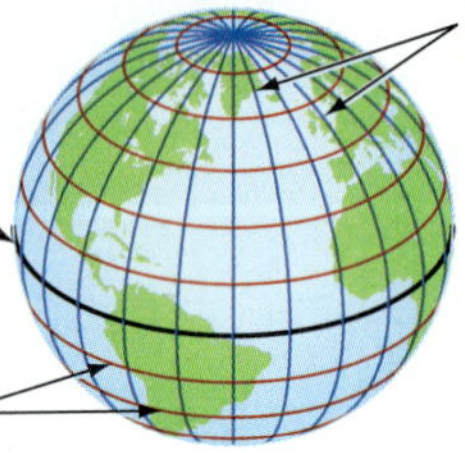

a. Are lines of longitude great circles? Explain.

b. Are lines of latitude great circles? Explain.

Example 3
p. 874

Tell whether the following postulate or property of plane Euclidean geometry has a corresponding statement in spherical geometry. If so, write the corresponding statement. If not, explain your reasoning.

15. A line goes on infinitely in two directions.

16. Perpendicular lines form four 90° angles.

17. If three points are collinear, exactly one is between the other two.

18. If M is the midpoint of $\overline{AB}$, then $\overline{AM} \cong \overline{MB}$.

On a sphere, there are two distances that can be measured between two points. Use each figure and the information given to determine the distance between points *J* and *K* on each sphere. Round to the nearest tenth. Justify your answer.

19.

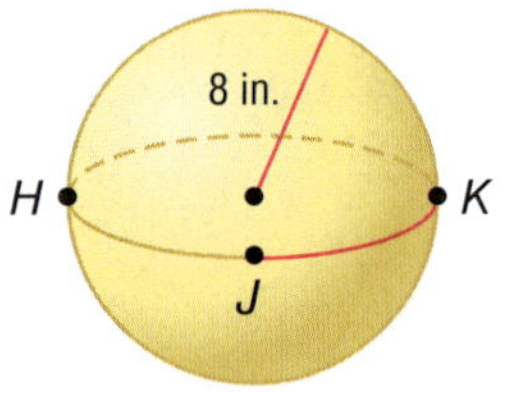

$m\widehat{JK} = 100$

20.

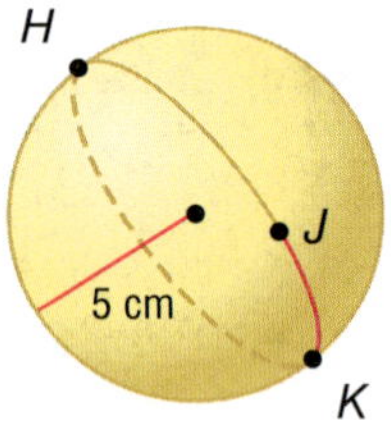

$m\widehat{JK} = 60$

21. GEOGRAPHY The location of Phoenix, Arizona, is 112° W longitude, 33.4° N latitude, and the location of Helena, Montana, is 112° W longitude, 46.6° N latitude. West indicates the location in terms of the prime meridian, and north indicates the location in terms of the equator. The mean radius of Earth is about 3963 miles.

a. Estimate the distance between Phoenix and Helena. Explain your reasoning.

b. Is there another way to express the distance between these two cities? Explain.

c. Can the distance between Washington, D.C., and London, England, which lie on approximately the same lines of latitude, be calculated in the same way? Explain your reasoning.

d. How many other locations are there that are the same distance from Phoenix, Arizona as Helena, Montana is? Explain.

Math History Link

Georg F.B. Riemann (1826–1866)
Spherical geometry is sometimes called *Riemann geometry*, after Georg Reimann, a German mathematician responsible for the Riemannian Postulate, which states that through a point not on a line, there are no lines parallel to the given line.

22. **MULTIPLE REPRESENTATIONS** In this problem, you will investigate triangles in spherical geometry.

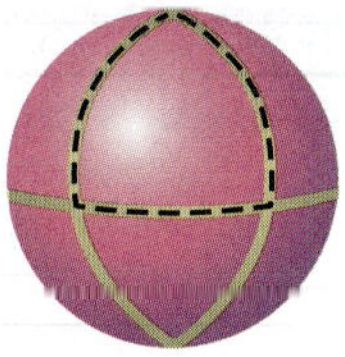

a. **CONCRETE** Use masking tape on a ball to mark three great circles. At least one of the three great circles should go through different poles than the other two. The great circles will form a triangle. Use a protractor to estimate the measure of each angle of the triangle.

b. **TABULAR** Tabulate the measure of each angle of the triangle formed. Remove the tape and repeat the process two times so that you have tabulated the measure of three different triangles. Record the sum of the measures of each triangle.

c. **VERBAL** Make a conjecture about the sum of the measures of a triangle in spherical geometry.

23. **QUADRILATERALS** Consider quadrilateral $ABCD$ on sphere $\mathcal{P}$. Note that it has four sides with $\overline{DC} \perp \overline{CB}$, $\overline{AB} \perp \overline{CB}$, and $\overline{DC} \cong \overline{AB}$.

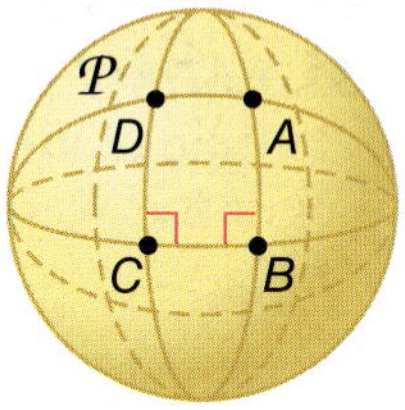

a. Is $\overline{CD} \perp \overline{DA}$? Explain your reasoning.

b. How does DA compare to CB?

c. Can a rectangle, as defined in Euclidean geometry, exist in non-Euclidean geometry? Explain your reasoning.

H.O.T. Problems

Use **H**igher-**O**rder **T**hinking Skills

24. **WRITING IN MATH** Compare and contrast Euclidean and spherical geometries. Be sure to include a discussion of planes and lines in both geometries.

25. **CHALLENGE** Geometries can be defined on curved surfaces other than spheres. Another type of non-Euclidean geometry is *hyperbolic geometry*. This geometry is defined on a curved saddle-like surface. Compare the sum of the angle measures of a triangle in hyperbolic, spherical, and Euclidean geometries.

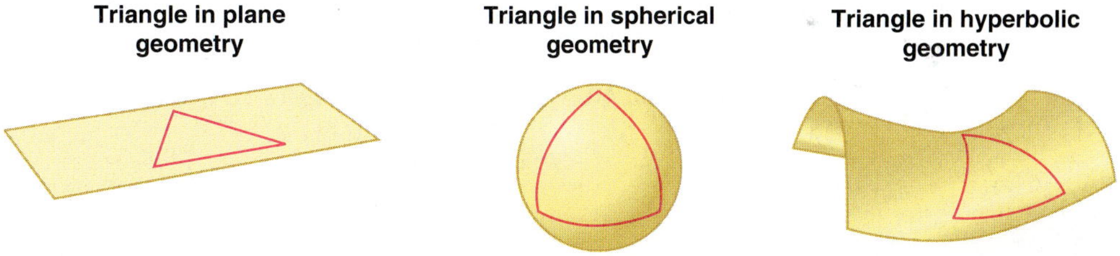

26. **OPEN ENDED** Sketch a sphere with three points so that two of the points lie on a great circle and two of the points do not lie on a great circle.

27. **REASONING** A *small circle* of a sphere intersects at least two points, but does not go through opposite poles. Points A and B lie on a small circle of sphere Q. Will two small circles *sometimes*, *always*, or *never* be parallel? Draw a sketch and explain your reasoning.

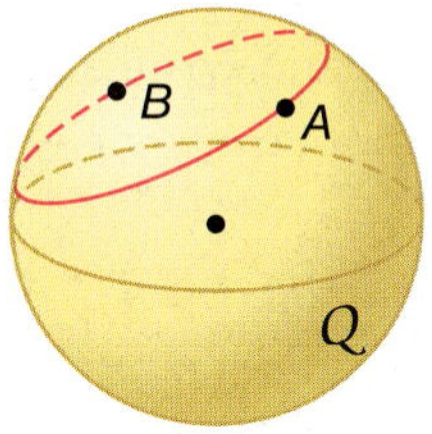

28. **WRITING IN MATH** Do similar or congruent triangles exist in spherical geometry? Explain your reasoning.

29. **REASONING** Is the statement *Spherical geometry is a subset of Euclidean geometry* true or false? Explain your reasoning.

30. **REASONING** Two planes are equidistant from the center of a sphere and intersect the sphere. What is true of the circles? Are they lines in spherical geometry? Explain.

Real-World Link

The Poincaré disc is a map of the entire hyperbolic plane. In *Circle Limit III* (1959), shown above, M.C. Escher created a tessellation of fish based on a hyperbolic tessellation of squares and triangles in the Poincaré disc.

Source: Wolfram MathWorld

Standardized Test Practice

31. Which of the following postulates or properties of spherical geometry is false?

A The shortest path between two points on a circle is an arc.
B If three points are collinear, any of the three points lies between the other two.
C A great circle is infinite and never returns to its original starting point.
D Perpendicular great circles intersect at two points.

32. SAT/ACT A car travels 50 miles due north in 1 hour and 120 miles due west in 2 hours. What is the average speed of the car?

F 50 mph **H** 60 mph
G 55 mph **J** none of the above

33. SHORT RESPONSE Name a line in sphere $\mathcal{P}$ that contains point D.

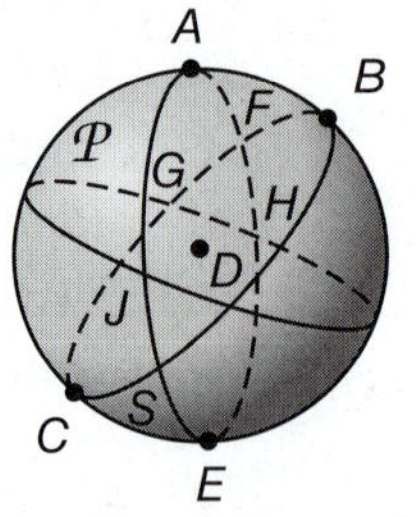

34. ALGEBRA The ratio of males to females in a classroom is 3:5. How many females are in the room if the total number of students is 32?

A 12 **C** 51
B 20 **D** 53

Spiral Review

Find the volume of each sphere or hemisphere. Round to the nearest tenth. (Lesson 12-6)

35. sphere: area of great circle = 98.5 m^2

36. sphere: circumference of great circle ≈ 23.1 in.

37. hemisphere: circumference of great circle ≈ 50.3 cm

38. hemisphere: area of great circle ≈ 3416 ft^2

Find the volume of each cone. Round to the nearest tenth. (Lesson 12-5)

39.

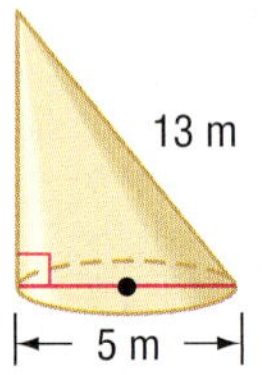

40.

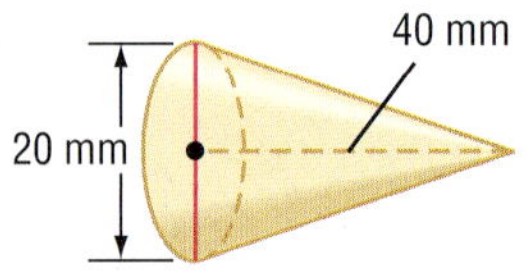

41.

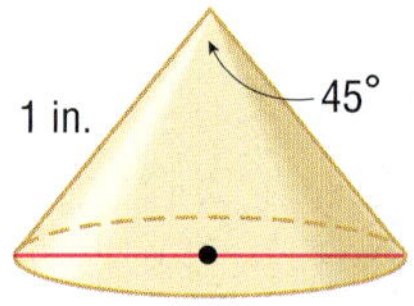

42. RADIOS Three radio towers are modeled by the points $A(-3, 4)$, $B(9, 4)$, and $C(-3, -12)$. Determine the location of another tower equidistant from all three towers, and write an equation for the circle which all three points lie on. (Lesson 10-8)

Skills Review

For each pair of similar figures, find the area of the green figure. (Lesson 11-5)

43.

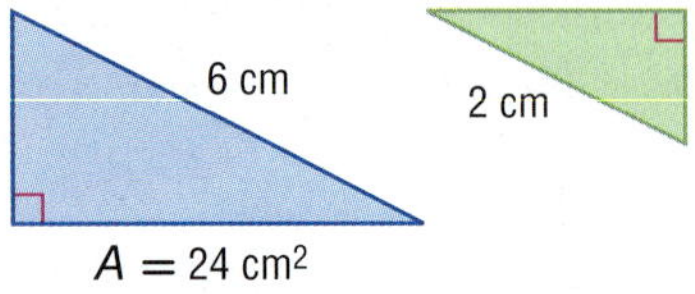

44.

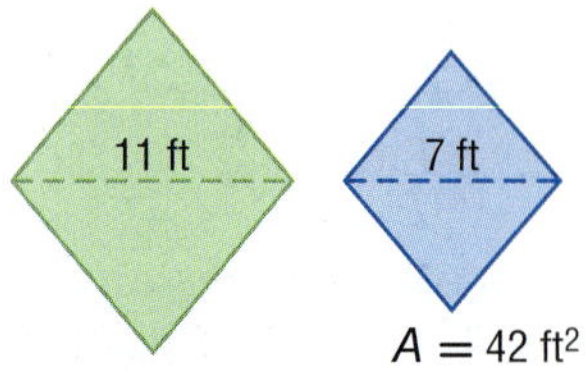

45.

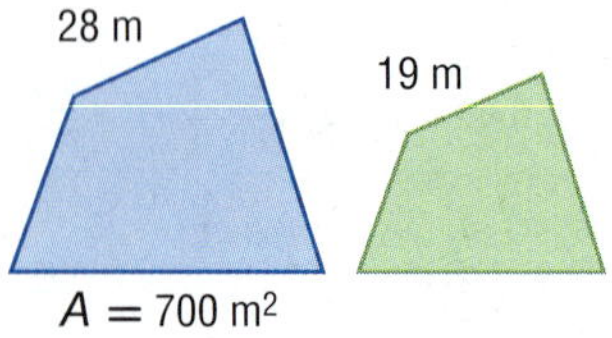

EXTEND

12-7 Geometry Lab Navigational Coordinates

Objective
Use a latitude and longitude measure to identify the hemispheres on which the location lies and estimate the location of a city using a globe or map.

A grid system of imaginary lines on Earth is used for locating places and navigation. Imaginary vertical lines drawn around the Earth through the North and South Poles are called **meridians** and determine the measure of **longitude**. Imaginary horizontal lines parallel to the equator are called **parallels** and determine the measure of **latitude**.

The basic units for measurements are degrees, minutes, and seconds. 1 degree (°) = 60 minutes (′), and 60 minutes = 60 seconds (″).

	Location of 0°	Direction	Maximum Degrees
Latitude (parallels)	equator	In northern hemisphere, all are degrees north. In southern hemisphere, all are degrees south.	180° at international dateline
Longitude (meridians)	Prime Meridian through Greenwich, England	In eastern hemisphere, all are degrees east. In western hemisphere, all are degrees west.	90° at each pole

ACTIVITY Investigate Latitude and Longitude

The table shows the latitude and longitude of three cities.

City	Latitude	Longitude
A	37°59′N	84°28′W
B	34°55′S	138°36′E
C	64°4′N	21°58′W

1. In which hemisphere is each city located?
2. Use a globe or map to name each city.
3. Earth is approximately a sphere with a radius of 3960 miles. The equator and all meridians are great circles. The circumference of a great circle is equal to the length of the equator or any meridian. Find the length of a great circle on Earth in miles.
4. Notice that the distance between each line of latitude is about the same. The distance from the equator to the North Pole is $\frac{1}{4}$ of the circumference of Earth, and each degree of latitude is $\frac{1}{90}$ of that distance. Estimate the distance between one pair of latitude lines in miles.

Analyze

The table shows the latitude and longitude of three cities.

City	Latitude	Longitude
F	1°28′S	48°29′W
G	13°45′N	100°30′E
H	41°17′S	174°47′E

1. Name the hemisphere in which each city is located.
2. Use a globe or map to name each city.
3. Find the approximate distance between meridians at latitude of about 22° N. The direct distance between the two cities below is about 1646 miles.

Calcutta, India	22°34′N	88°24′E
Hong Kong, China	22°20′N	114°11′E

12-8 Congruent and Similar Solids

Then
You compared surface areas and volumes of spheres. (Lesson 12-6)

Now
- Identify congruent or similar solids.
- Use properties of similar solids.

New Vocabulary
similar solids
congruent solids

Math Online
glencoe.com

Why?

The gemstones at the right are cut in exactly the same shape, but their sizes are different. Their shapes are *similar.*

Identify Congruent or Similar Solids

Similar solids have exactly the same shape but not necessarily the same size. All spheres are similar and all cubes are similar.

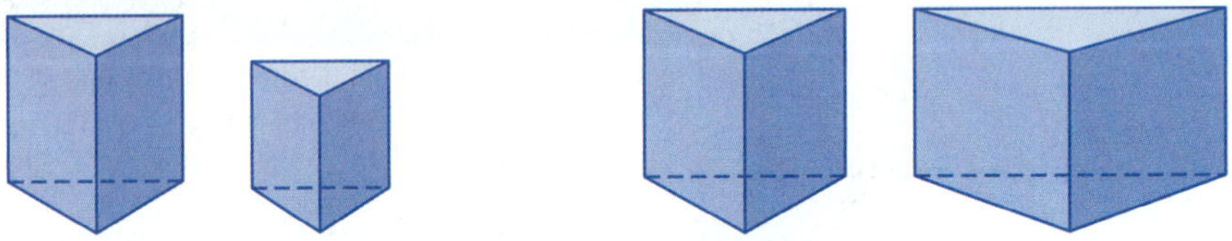

In similar solids, the corresponding linear measures, such as height and radius, have equal ratios. The common ratio is called the *scale factor.* If two similar solids are polyhedrons, their corresponding faces are similar.

Key Concept Similar Solids

Words Two solids are similar if they have the same shape and the ratios of their corresponding linear measures are equal.

Models

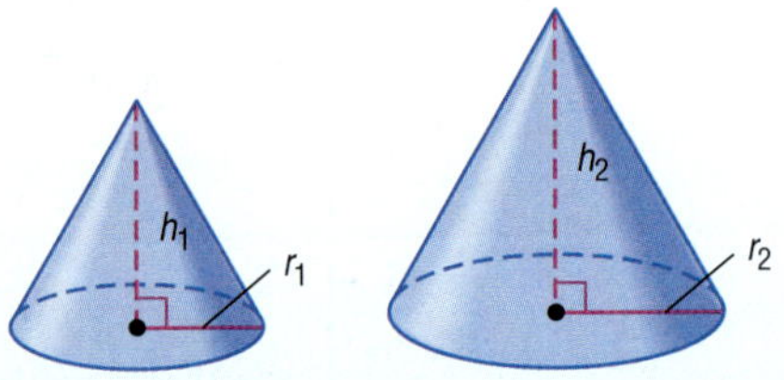

$$\frac{h_1}{h_2} = \frac{r_1}{r_2}$$

Congruent solids have exactly the same shape and the same size. Congruent solids are similar solids that have a scale factor of 1:1.

Key Concept Congruent Solids

Words
Two solids are congruent if they have the following characteristics.
- Corresponding angles are congruent.
- Corresponding edges are congruent.
- Corresponding faces are congruent.
- Volumes are equal.

Models

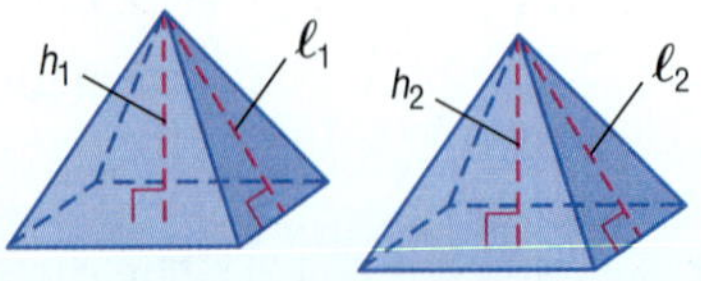

$$\frac{h_1}{h_2} = \frac{\ell_1}{\ell_2} = 1$$

EXAMPLE 1 Identify Similar and Congruent Solids

Determine whether each pair of solids is *similar, congruent,* or *neither*. If the solids are similar, state the scale factor.

a. the square pyramids

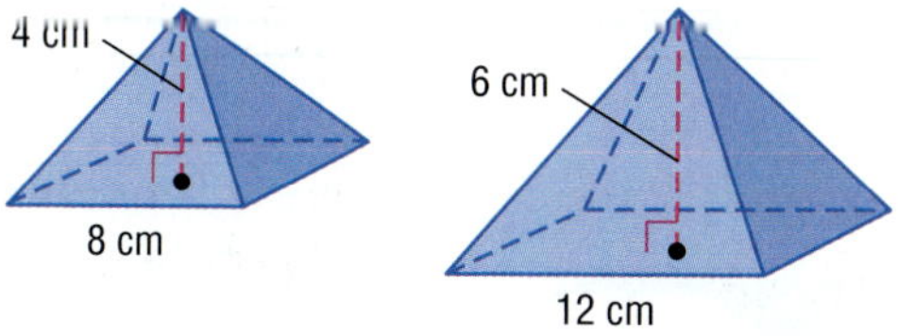

ratio of heights: $\frac{4}{6} = \frac{2}{3}$

ratio of base edges: $\frac{8}{12} = \frac{2}{3}$

The ratios of the corresponding measures are equal, so the pyramids are similar. The scale factor is 2:3. Since the scale factor is not 1:1, the solids are not congruent.

b. the rectangular prisms

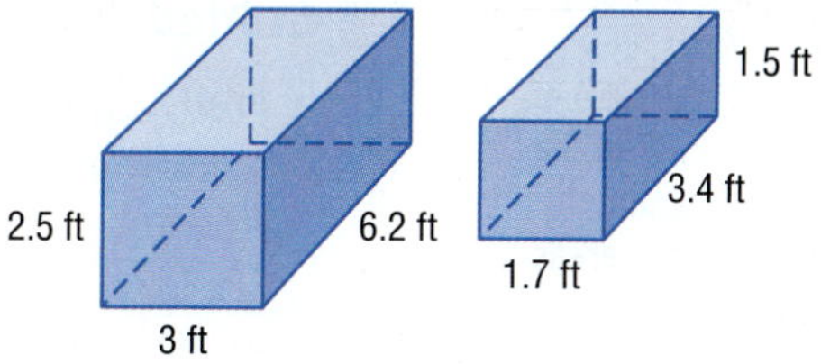

ratio of widths: $\frac{3}{1.7} \approx 1.76$

ratio of lengths: $\frac{6.2}{3.4} \approx 1.82$

ratio of heights: $\frac{2.5}{1.5} \approx 1.67$

Since the ratios of corresponding measures are not equal, the prisms are neither congruent nor similar.

StudyTip

Similar and Congruent Solids If two solids are similar, then their corresponding linear measures are proportional. If two solids are congruent, then their corresponding linear measures are equal.

Guided Practice

1A.

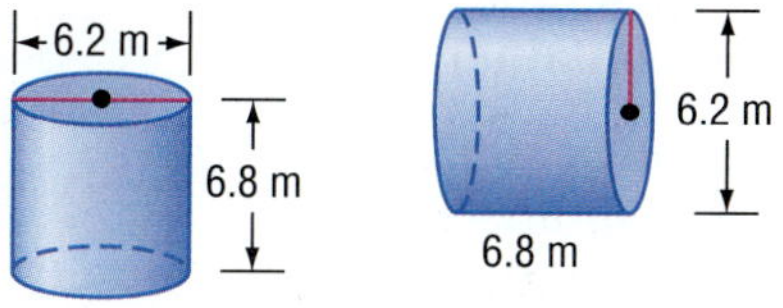

1B.

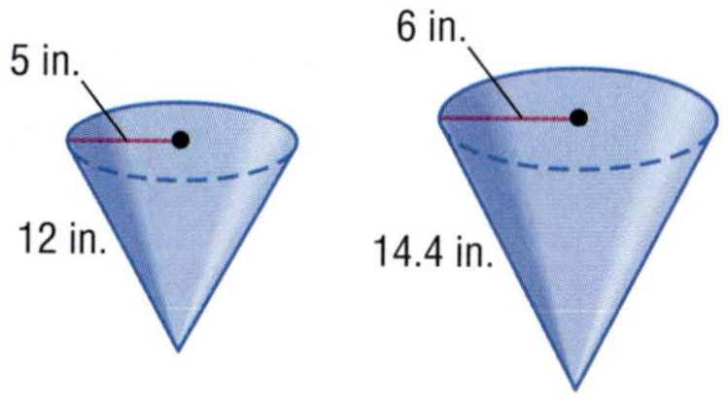

Personal Tutor glencoe.com

Properties of Congruent and Similar Solids

The cubes at the right are similar solids with a scale factor of 3:2.

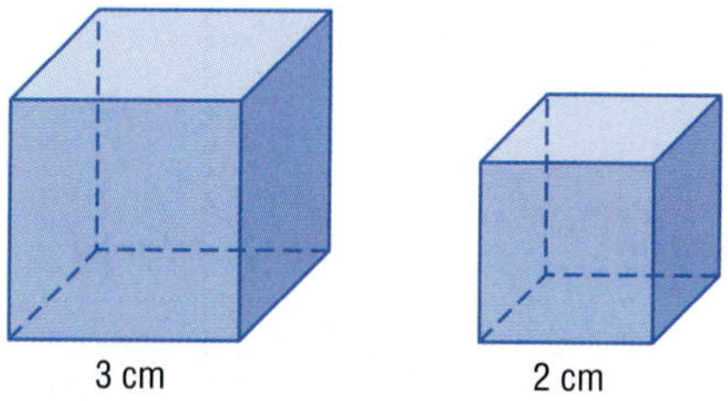

ratio of surface areas: 54:24 or 9:4

ratio of volumes: 27:8

Notice that the ratio of surface areas, 9:4, can be written as $3^2:2^2$. The ratio of volumes, 27:8, can be written as $3^3:2^3$. This suggests the following theorem.

Watch Out!

Theorem 12.1 Figures must be similar in order for Theorem 12.1 to apply.

Theorem 12.1

Words If two similar solids have a scale factor of $a:b$, then the surface areas have a ratio of $a^2:b^2$, and the volumes have a ratio of $a^3:b^3$.

Examples

scale factor	2:3
ratio of surface areas	4:9
ratio of volumes	8:27

Models

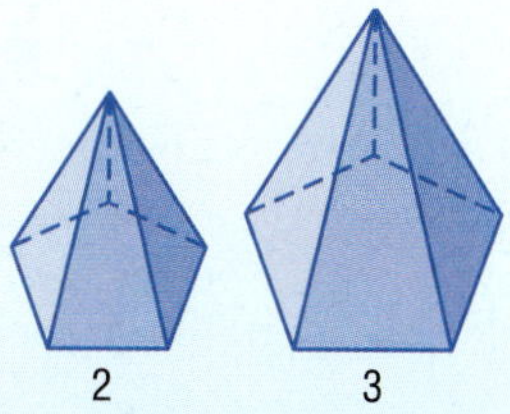

EXAMPLE 2 Use Similar Solids to Write Ratios

Two similar cones have radii of 10 millimeters and 15 millimeters. What is the ratio of the surface area of the small cone to the surface area of the large cone?

First, find the scale factor.

$\frac{\text{radius of small cone}}{\text{radius of large cone}} = \frac{10}{15} \text{ or } \frac{2}{3}$ **Write a ratio comparing the radii.**

The scale factor is $\frac{2}{3}$.

$\frac{a^2}{b^2} = \frac{2^2}{3^2} \text{ or } \frac{4}{9}$ **If the scale factor is $\frac{a}{b}$, then the ratio of surface areas is $\frac{a^2}{b^2}$.**

So, the ratio of the surface areas is 4:9.

StudyTip

Similar Solids and Area If two solids are similar, then the ratio of any corresponding areas is $a^2 : b^2$. In Example 2, the ratio of the lateral areas of the cones is 4:25, and the ratio of the base areas of the cones is 4:25.

Guided Practice

2. Two similar prisms have surface areas of 98 square centimeters and 18 square centimeters. What is the ratio of the height of the large prism to the height of the small prism?

Personal Tutor glencoe.com

Many real-world objects can be modeled by similar solids.

Real-World EXAMPLE 3 Use Similar Solids to Find Unknown Values

CONTAINERS The containers at the right are similar cylinders. Find the height *h* of the smaller container.

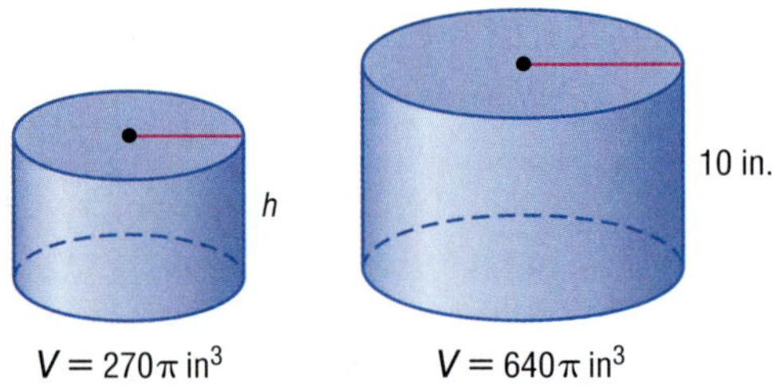

Understand You know the height of the larger container and the volumes of both containers.

Plan Use Theorem 12.1 to write a ratio comparing the volumes. Then find the scale factor and use it to find *h*.

Solve

$\frac{\text{volume of small container}}{\text{volume of large container}} = \frac{270\pi}{640\pi}$ **Write a ratio comparing volumes.**

$= \frac{27}{64}$ **Simplify.**

$= \frac{3^3}{4^3}$ **Write as $\frac{a^3}{b^3}$.**

The scale factor is 3:4.

Ratio of heights → $\frac{h}{10} = \frac{3}{4}$ **← Scale factor**

$h \cdot 4 = 10 \cdot 3$ **Find the cross products.**

$h = 7.5$ **Solve for *h*.**

So, the height of the smaller container is 7.5 inches.

Check Since $\frac{7.5}{10} = 0.75 = \frac{3}{4}$, the solution is correct. ✓

Guided Practice

3. VOLLEYBALL A regulation volleyball has a circumference of about 66 centimeters. The ratio of the surface area of that ball to the surface area of a children's ball is approximately 1.6:1. What is the circumference of the children's ball? Round to the nearest centimeter.

Personal Tutor glencoe.com

Check Your Understanding

Example 1 p. 881

Determine whether each pair of solids is *similar*, *congruent*, or *neither*. If the solids are similar, state the scale factor.

1.

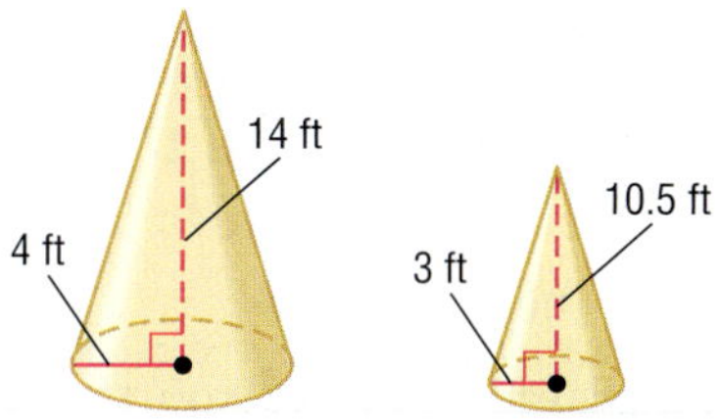

2.

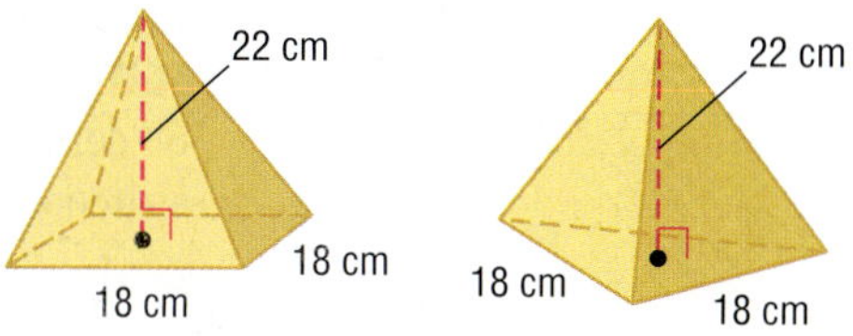

Example 2 p. 882

3. Two similar cylinders have radii of 15 inches and 6 inches. What is the ratio of the surface area of the small cylinder to the surface area of the large cylinder?

4. Two spheres have volumes of 36π cubic centimeters and 288π cubic centimeters. What is the ratio of the radius of the small sphere to the radius of the large sphere?

Example 3 p. 882

5. **EXERCISE BALLS** A company sells two different sizes of exercise balls. The ratio of the diameters is 15:11. If the diameter of the smaller ball is 55 centimeters, what is the volume of the larger ball? Round to the nearest tenth.

Practice and Problem Solving

● = **Step-by-Step Solutions** begin on page R20.
Extra Practice begins on page 969.

Example 1 p. 881

Determine whether each pair of solids is *similar*, *congruent*, or *neither*. If the solids are similar, state the scale factor.

6.

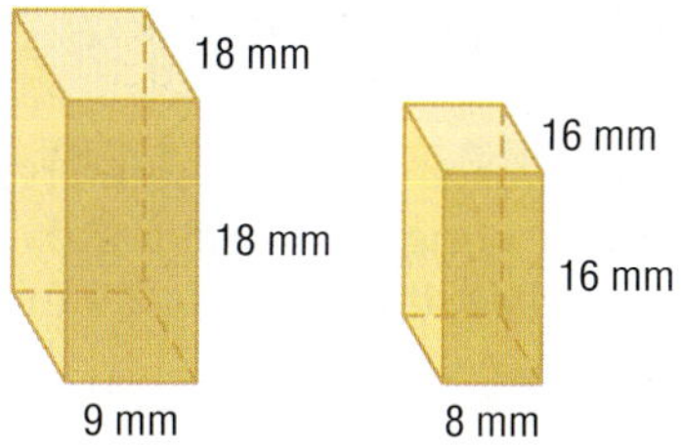

7.

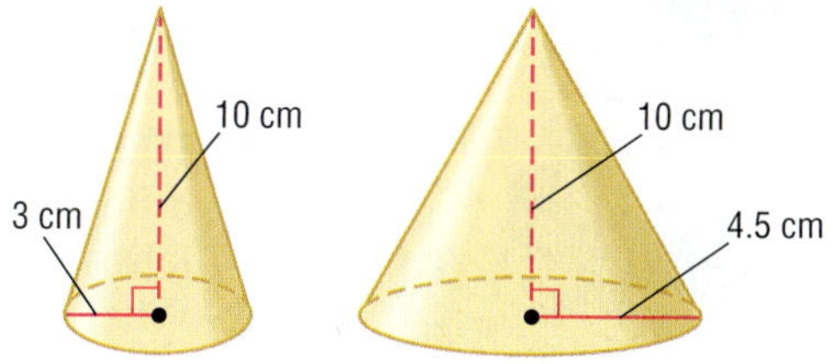

8.

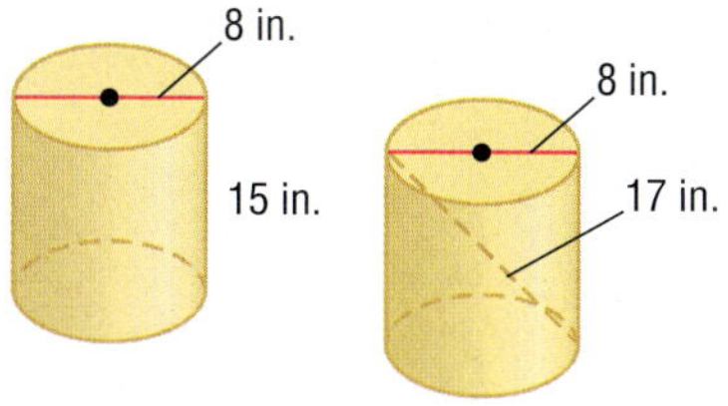

9.

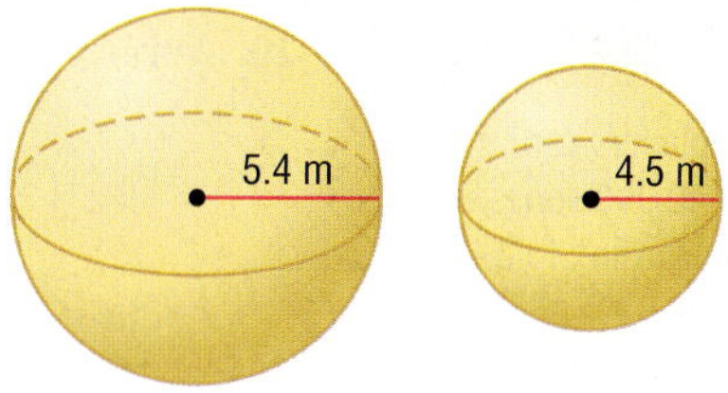

Example 2 p. 882

10. Two similar pyramids have slant heights of 6 inches and 12 inches. What is the ratio of the surface area of the small pyramid to the surface area of the large pyramid?

11. Two similar cylinders have heights of 35 meters and 25 meters. What is the ratio of the volume of the large cylinder to the volume of the small cylinder?

12. Two spheres have surface areas of 100π square centimeters and 16π square centimeters. What is the ratio of the volume of the large sphere to the volume of the small sphere?

13. Two similar hexagonal prisms have volumes of 250 cubic feet and 2 cubic feet. What is the ratio of the height of the large cylinder to the height of the small cylinder?

14. **DIMENSIONAL ANALYSIS** Two rectangular prisms are similar. The height of the first prism is 6 yards and the height of the other prism is 9 feet. If the volume of the first prism is 810 cubic yards, what is the volume of the other prism?

Example 3 p. 882

15. **FOOD** A small cylindrical can of tuna has a radius of 4 centimeters and a height of 3.8 centimeters. A larger and similar can of tuna has a radius of 5.2 centimeters.

    a. What is the scale factor of the cylinders?

    b. What is the volume of the larger can? Round to the nearest tenth.

16. **SUITCASES** Two suitcases are similar rectangular prisms. The smaller suitcase is 68 centimeters long, 47 centimeters wide, and 27 centimeters deep. The larger suitcase is 85 centimeters long.

    a. What is the scale factor of the prisms?

    b. What is the volume of the larger suitcase? Round to the nearest tenth.

17. **SCULPTURE** The sculpture shown at the right is a scale model of a cornet. If the sculpture is 26 feet long and a standard cornet is 14 inches long, what is the scale factor of the sculpture to a standard cornet?

18. The pyramids shown are congruent.

    a. What is the perimeter of the base of Pyramid A?

    b. What is the area of the base of Pyramid B?

    c. What is the volume of Pyramid B?

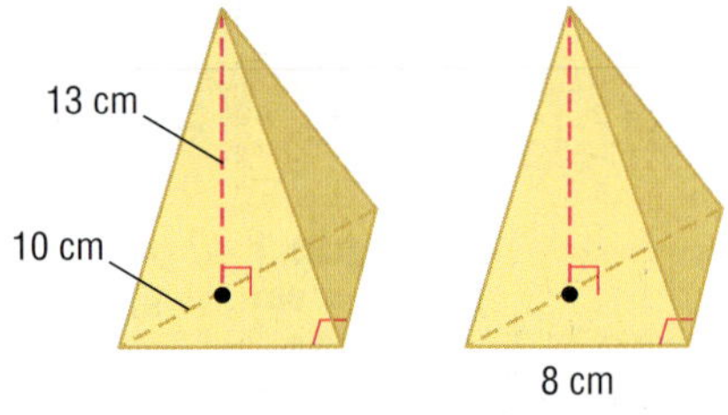

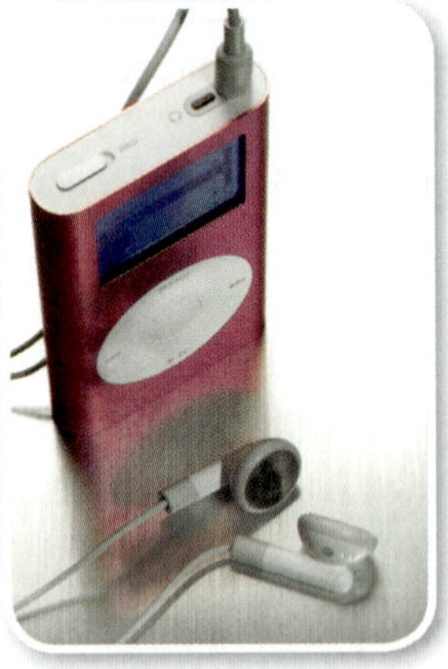

Real-World Link

Many MP3 players can store up to 40,000 songs. However, according to a recent survey, 90% of consumers have fewer than 1000 songs on their MP3 players.

Source: Jupiter Research

19. **TECHNOLOGY** Jalissa and Mateo each have the same type of MP3 player, but in different colors. The players are congruent rectangular prisms. The volume of Jalissa's player is 4.92 cubic inches, the width is 2.4 inches, and the depth is 0.5 inch. What is the height of Mateo's player?

Each pair of solids below is similar.

20. What is the surface area of the smaller solid shown below?

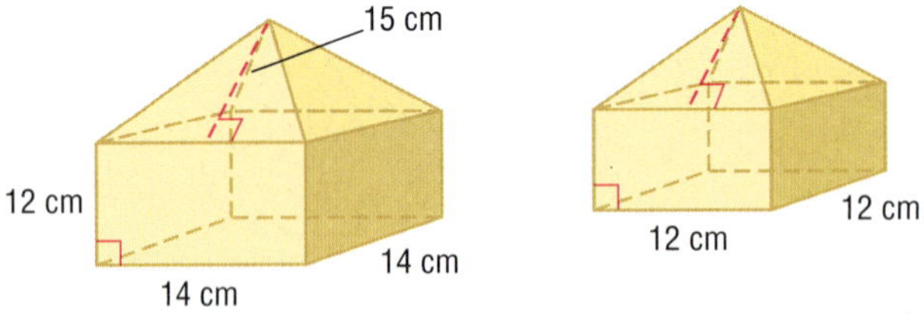

21. What is the volume of the larger solid shown below?

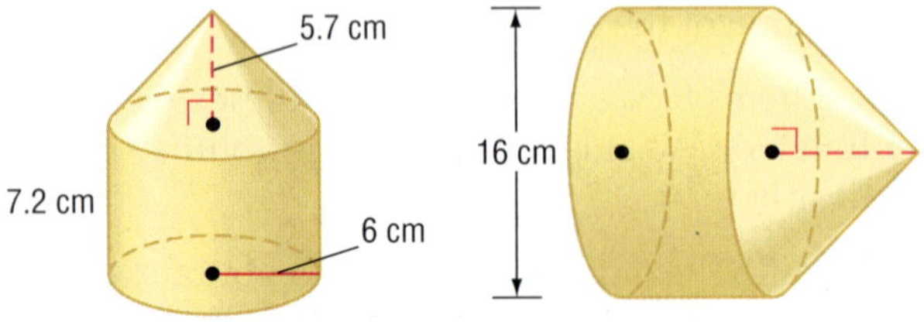

22. **DIMENSIONAL ANALYSIS** Two cylinders are similar. The height of the first cylinder is 23 cm and the height of the other cylinder is 8 in. If the volume of the first cylinder is 552π cm^3, what is the volume of the other prism? Use 2.54 cm = 1 in.

StudyTip

Check Solutions After finding missing measures of similar solids, work backward to check your solutions.

23. **DIMENSIONAL ANALYSIS** Two spheres are similar. The radius of the first sphere is 10 feet. The volume of the other sphere is 0.9 cubic meters. Use 2.54 cm = 1 in. to determine the scale factor from the first sphere to the second.

24. **ALGEBRA** Two similar cones have volumes of 343π cubic centimeters and 512π cubic centimeters. The height of each cone is equal to 3 times its radius. Find the radius and height of both cones.

25. **TENTS** Two tents are in the shape of hemispheres, with circular floors. The ratio of their floor areas is 9:12.25. If the diameter of the smaller tent is 6 feet, what is the volume of the larger tent? Round to the nearest tenth.

26. **MULTIPLE REPRESENTATIONS** In this problem, you will investigate similarity. The heights of two similar cylinders are in the ratio 2 to 3. The lateral area of the larger cylinder is 162π square centimeters, and the diameter of the smaller cylinder is 8 centimeters.

    a. **VERBAL** What is the height of the larger cylinder? Explain your method.

    b. **GEOMETRIC** Sketch and label the two cylinders.

    c. **ANALYTICAL** How many times as great is the volume of the larger cylinder as the volume of the smaller cylinder?

H.O.T. Problems Use Higher-Order Thinking Skills

27. **ERROR ANALYSIS** Cylinder X has a diameter of 20 centimeters and a height of 11 centimeters. Cylinder Y has a radius of 30 centimeters and is similar to Cylinder X. Did Laura or Paloma correctly find the height of Cylinder Y? Explain your reasoning.

Laura

Cylinder X: radius 10, height 11

Cylinder Y: radius 30, height a

$\frac{10}{30} = \frac{11}{a}$, so $a = 33$.

Paloma

Cylinder X: diameter 20, height 11

Cylinder Y: diameter 20, height a

$\frac{20}{20} = \frac{11}{a}$, so $a = 11$.

28. **CHALLENGE** The ratio of the volume of Cylinder A to the volume of Cylinder B is 1:5. Cylinder A is similar to Cylinder C with a scale factor of 1:2 and Cylinder B is similar to Cylinder D with a scale factor of 1:3. What is the ratio of the volume of Cylinder C to the volume of Cylinder D? Explain your reasoning.

29. **WRITING IN MATH** Explain how the surface areas and volumes of the similar prisms shown at the right are related.

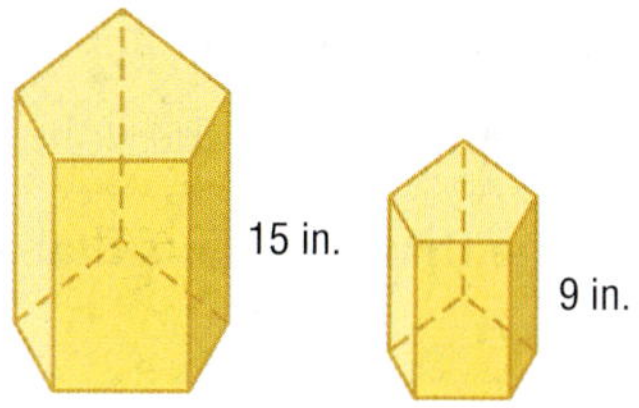

30. **OPEN ENDED** Describe two nonsimilar triangular pyramids with similar bases.

31. **REASONING** Plane $\mathcal{P}$ is parallel to the base of cone C, and the volume of the cone above the plane is $\frac{1}{8}$ of the volume of the cone below the plane. Find the height of cone C.

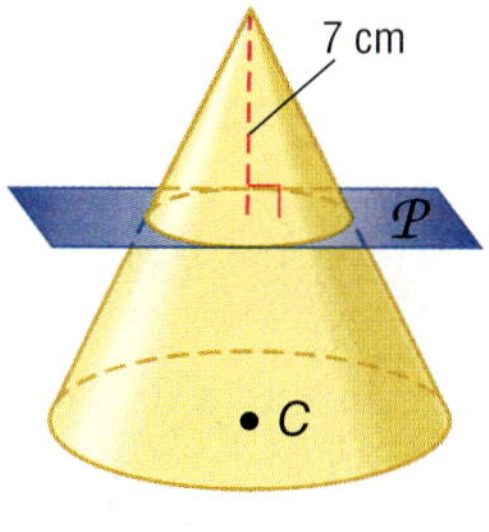

32. **WRITING IN MATH** Explain why all spheres are similar.

Standardized Test Practice

33. Two similar spheres have radii of 20π meters and 6π meters. What is the ratio of the surface area of the large sphere to the surface area of the small sphere?

A $\frac{100}{9}$ B $\frac{100}{3}$ C $\frac{10}{3}$ D $\frac{10}{9}$

34. What is the scale factor of the similar figures?

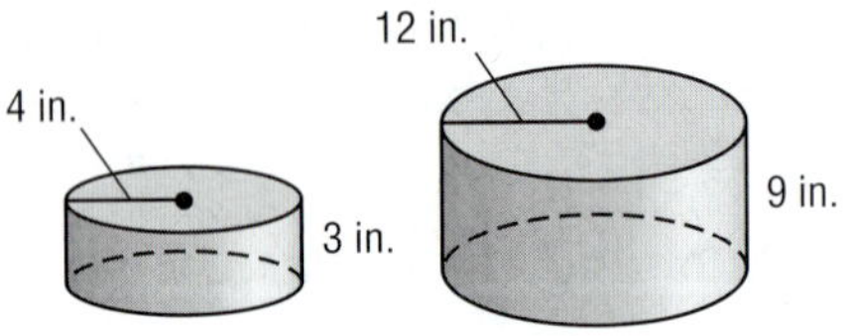

F 0.25 H 0.5
G 0.33 J 0.75

35. SHORT RESPONSE Point *A* and point *B* represent the locations of Timothy's and Quincy's houses. If each unit on the map represents one kilometer, how far apart are the two houses?

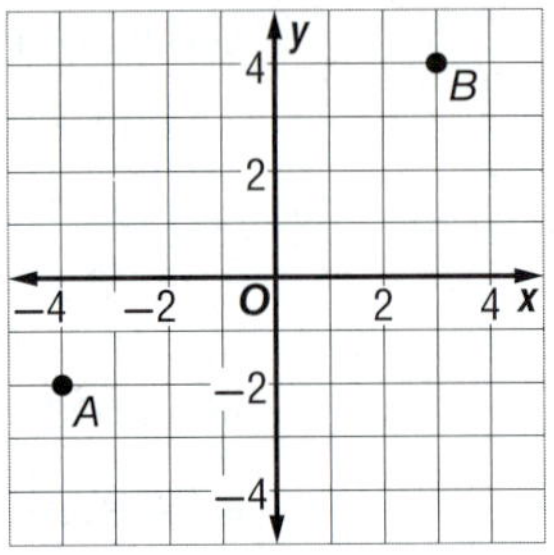

36. SAT/ACT If $\frac{x+2}{3} = \frac{(x+2)^2}{15}$, what is one possible value of x?

A 0 B 1 C 2 D 3

Spiral Review

Determine whether figure χ on each of the spheres shown is a line in spherical geometry. (Lesson 12-7)

37.

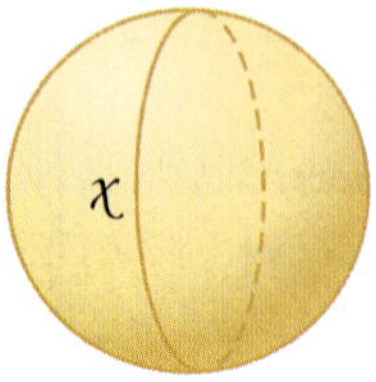

38.

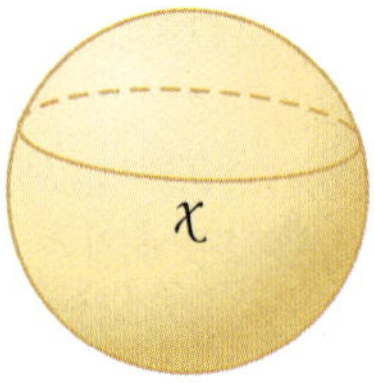

39.

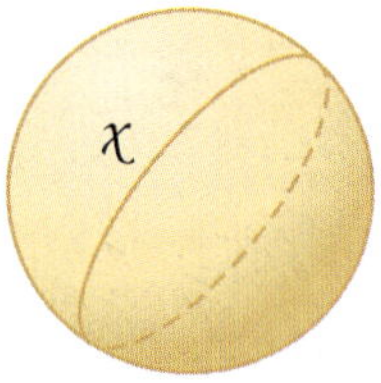

40. ENTERTAINMENT Some people think that the Spaceship Earth geosphere at Epcot in Disney World in Orlando, Florida, resembles a golf ball. The building is a sphere measuring 165 feet in diameter. A typical golf ball has a diameter of approximately 1.5 inches. (Lesson 12-6)

a. Find the volume of Spaceship Earth to the nearest cubic foot.

b. Find the volume of a golf ball to the nearest tenth.

c. What is the scale factor that compares Spaceship Earth to a golf ball?

d. What is the ratio of the volumes of Spaceship Earth to a golf ball?

Find x. Assume that segments that appear to be tangent are tangent. (Lesson 10-7)

41.

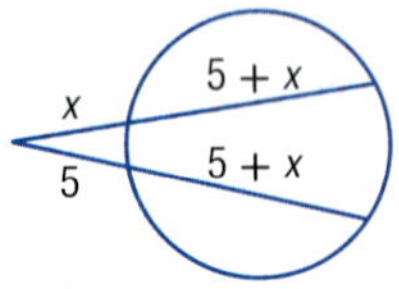

42.

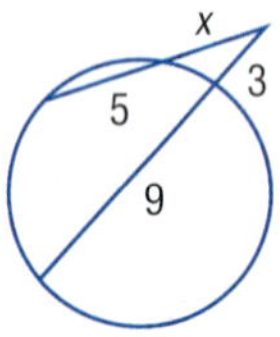

43.

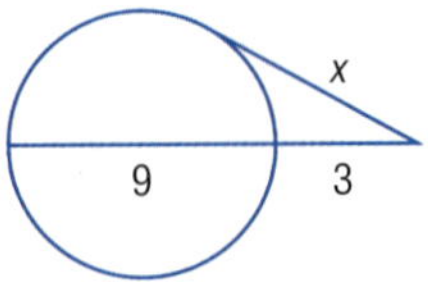

Skills Review

Express each fraction as a decimal to the nearest hundredth.

44. $\frac{8}{13}$ **45.** $\frac{17}{54}$ **46.** $\frac{11}{78}$ **47.** $\frac{43}{46}$

Math Online glencoe.com
- STUDY TO GO
- Vocabulary Review

Chapter Summary

Key Concepts

Representations of Three-Dimensional Figures (Lesson 12-1)

- Solids can be classified by bases, faces, edges, and vertices.

Surface Areas of Prisms and Cylinders (Lesson 12-2)

- Lateral surface area of a right prism: $L = Ph$
- Lateral surface area of a right cylinder: $L = 2\pi rh$

Surface Areas of Pyramids and Cones (Lesson 12-3)

- Lateral surface area of a pyramid: $L = \frac{1}{2}P\ell$
- Lateral surface area of a right cone: $L = \pi r\ell$

Volumes of Prisms and Cylinders (Lesson 12-4)

- Volume of prism or cylinder: $V = Bh$

Volumes of Pyramids and Cones (Lesson 12-5)

- Volume of a pyramid: $V = \frac{1}{3}Bh$
- Volume of a cone: $V = \frac{1}{3}\pi r^2 h$

Surface Areas and Volumes of Spheres (Lesson 12-6)

- Surface area of a sphere: $S = 4\pi r^2$
- Volume of a sphere: $V = \frac{4}{3}\pi r^3$

Congruent and Similar Solids (Lesson 12-8)

- Similar solids have the same shape, but not necessarily the same size.
- Congruent solids are similar solids with a scale factor of 1.

FOLDABLES® Study Organizer

Be sure the Key Concepts are noted in your Foldable.

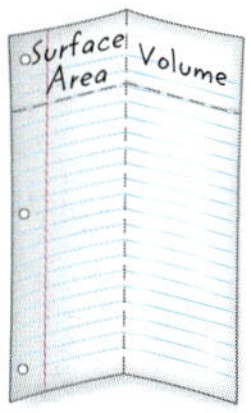

Key Vocabulary

altitude (p. 830)
axis (p. 832)
base edges (p. 830)
congruent solid (p. 880)
cross section (p. 824)
composite solid (p. 836)
Euclidean geometry (p. 873)
great circle (p. 865)
isometric view (p. 823)
lateral area (p. 830)
lateral edge (p. 830)
lateral face (p. 830)
non-Euclidean geometry (p. 874)
oblique cone (p. 840)
oblique cylinder (p. 821)
oblique prism (p. 821)
regular pyramid (p. 838)
right cone (p. 840)
right cylinder (p. 821)
right prism (p. 821)
segment of a circle (p. 787)
similar solids (p. 880)
slant height (p. 838)
spherical geometry (p. 873)
topographic map (p. 829)

Vocabulary Check

State whether each sentence is *true* or *false*. If *false*, replace the underlined term to make a true sentence.

1. Euclidean geometry deals with a system of points, great circles (lines), and spheres (planes).
2. Similar solids have exactly the same shape, but not necessarily the same size.
3. A right prism has an axis that is also an altitude.
4. The isometric view is when an object is viewed from a corner.
5. The perpendicular distance from the base of a geometric figure to the opposite vertex, parallel side, or parallel surface is the altitude.
6. Rotation symmetry is also called mirror symmetry.
7. The intersection of two adjacent lateral faces is the lateral edge.
8. Euclidean geometry refers to geometrical systems that are not in accordance with the Parallel Postulate.
9. A composite solid is a three-dimensional figure that is composed of simpler figures.
10. The slant height is the height of each lateral face of a pyramid or cone.

CHAPTER 12 Study Guide and Review

Lesson-by-Lesson Review

12-1 Representations of Three-Dimensional Figures (pp. 823–828)

Describe each cross section.

11.

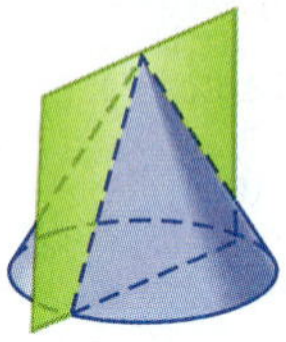

12.

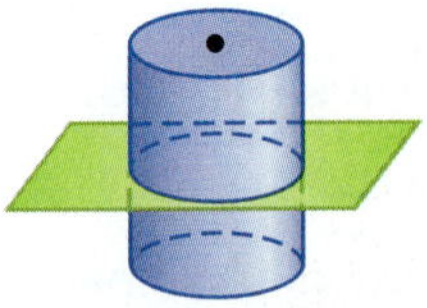

13. CAKE The cake shown is cut in half vertically. Describe the cross section of the cake.

EXAMPLE 1

Describe the vertical and horizontal cross sections of the figure shown below.

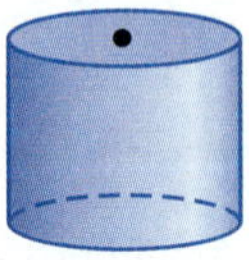

The vertical cross section is a rectangle.
The horizontal cross section is a circle.

12-2 Surface Areas of Prisms and Cylinders (pp. 830–837)

Find the lateral area and surface area of each prism. Round to the nearest tenth if necessary.

14.

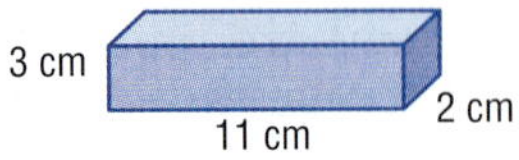

15.

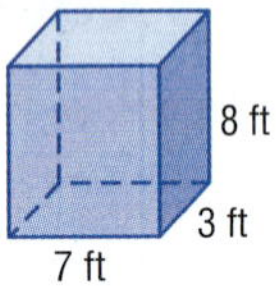

Find the lateral area and surface area of each cylinder. Round to the nearest tenth.

16.

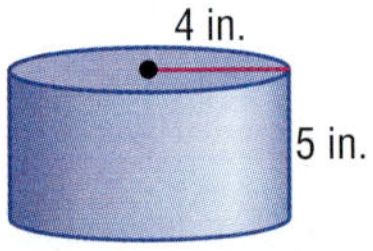

17.

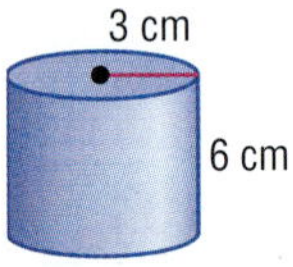

EXAMPLE 2

Find the surface area of the rectangular prism.

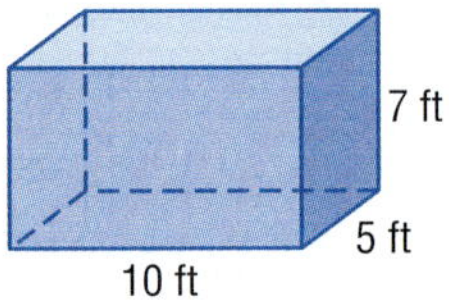

Use the 10-foot by 5-foot rectangle as the base.

$S = Ph + 2B$	**Surface area of a prism**
$= (2 \cdot 10 + 2 \cdot 5)(7) + 2(10 \cdot 5)$	**Substitution**
$= 310$	**Simplify.**

The surface area is 310 square feet.

12-3 Surface Areas of Pyramids and Cones (pp. 838–846)

Find the lateral area and the surface area of each regular pyramid. Round to the nearest tenth.

18.

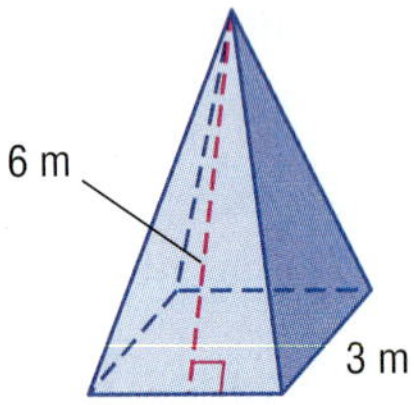

19.

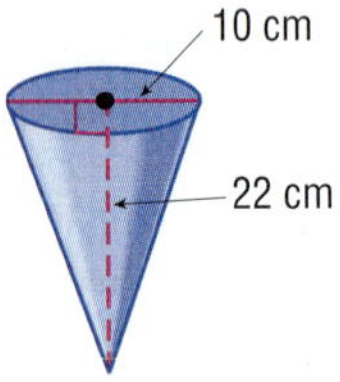

EXAMPLE 3

Find the surface area of the square pyramid. Round to the nearest tenth.

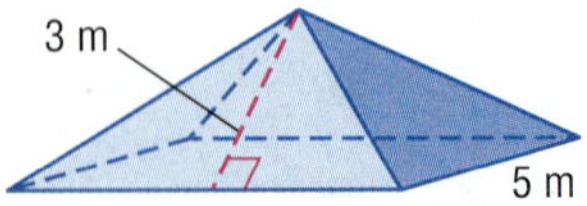

$S = \frac{1}{2}P\ell + B$	**Surface area of a regular pyramid**
$= \frac{1}{2}(4 \cdot 5)3 + 5 \cdot 5$	**$P = 4 \cdot 5$ or 20, $\ell = 3$, $B = 4 \cdot 5$**
$= 55$	**Simplify.**

The surface area is 55 square feet.

MIXED PROBLEM SOLVING
For mixed problem-solving practice, see page 1010.

12-4 Volumes of Prisms and Cylinders (pp. 847–854)

20. The volume of a cylinder is 770 cm^2. It has a height of 5 cm. Find its radius.

21. Find the volume of the triangular prism.

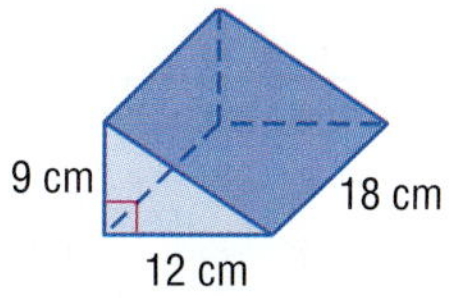

22. **TRAILERS** A semi-truck trailer is basically a rectangular prism. A typical height for the inside of these trailers is 108 inches. If the trailer is 8 feet wide and 20 feet long, what is the volume of the trailer?

EXAMPLE 4

Find the volume of the cylinder.

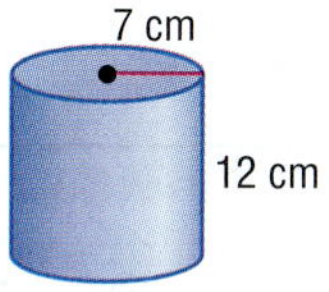

$V = \pi r^2 h$	**Volume of a cylinder**
$= \pi(7)^2(12)$	**$r = 7$ and $h = 12$**
≈ 1847.5	**Use a calculator.**

The volume is approximately 1847.5 cubic centimeters.

12-5 Volumes of Pyramids and Cones (pp. 857–863)

23. Find the volume of a cone that has a radius of 1 cm and a height of 3.4 cm.

24. Find the volume of the regular pyramid.

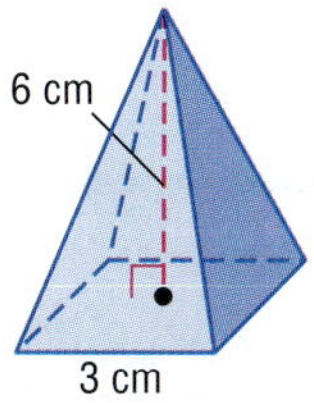

25. **ARCHITECTURE** The Great Pyramid measures 756 feet on each side of the base and the height is 481 feet. Find the volume of the pyramid.

EXAMPLE 5

Find the volume of the pyramid.

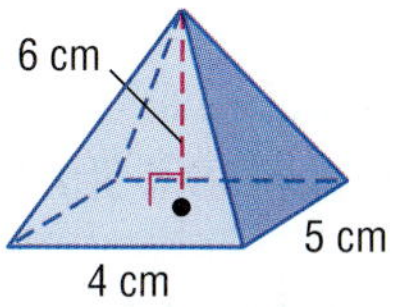

$V = \frac{1}{3}Bh$	**Volume of a pyramid**
$= \frac{1}{3}(4 \cdot 5)(6)$	**$B = 4 \cdot 8$ and $h = 6$**
$= 40$	**Simplify.**

The volume is 40 cubic centimeters.

12-6 Surface Area and Volume of Spheres (pp. 864–871)

Find the surface area of each figure.

26.

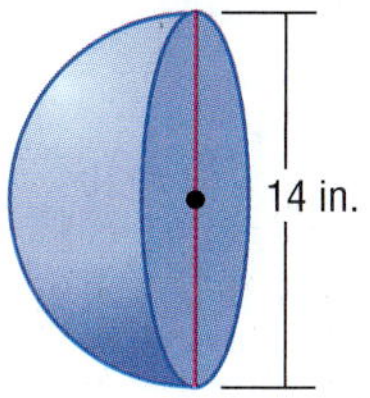

27.

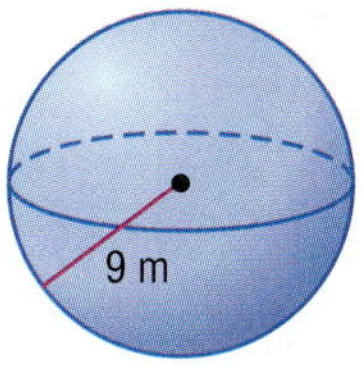

Find the volume of each sphere or hemisphere. Round to the nearest tenth.

28. hemisphere: circumference of great circle = 24π m

29. sphere: area of great circle = 55π in^2

30. **CONSTRUCTION** Cement is poured into a hemisphere that is 6 cm across. What is the volume of cement used?

EXAMPLE 6

Find the surface area and volume of the sphere. Round to the nearest tenth.

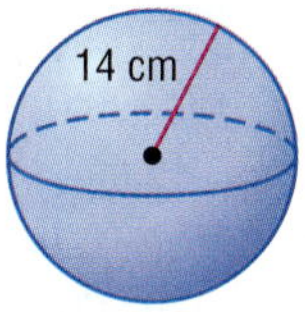

$S = 4\pi r^2$	**Surface area of a sphere**
$= 4\pi(14)^2$	**Substitute.**
≈ 2463	**Use a calculator.**

The surface area is about 2463 square centimeters.

$V = \frac{4}{3}\pi r^3$	**Volume of a sphere**
$= \frac{4}{3}\pi(14)^3$	**Replace r with 9.**
$\approx 11{,}494$ cm^3	**Use a calculator.**

The volume is about 11,494 cubic centimeters.

CHAPTER 12 Study Guide and Review

12-7 Spherical Geometry (pp. 873–878)

Name each of the following on sphere A.

31. two lines containing point C

32. a segment containing point H

33. a triangle containing point B

34. two lines containing point L

35. a segment containing point J

36. a triangle containing point K

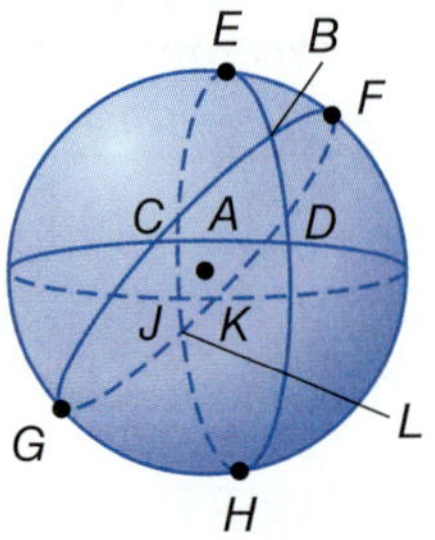

37. MARBLES Determine whether figure y on the sphere shown is a line in spherical geometry.

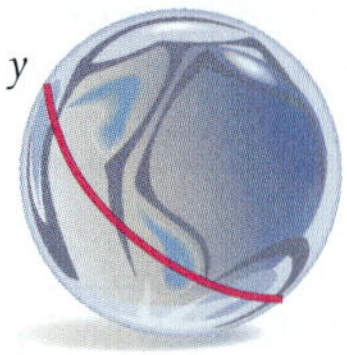

EXAMPLE 8

Name each of the following on sphere A.

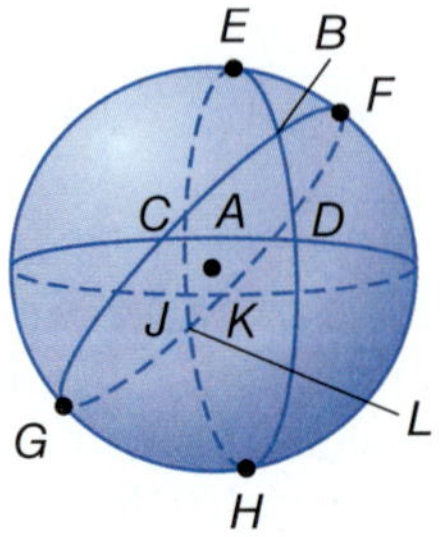

a. two lines containing point D
$\overleftrightarrow{EH}$, $\overleftrightarrow{CK}$

b. a segment containing point E
$\overline{DJ}$

12-8 Congruent and Similar Solids (pp. 880–886)

Determine whether each pair of solids is *similar*, *congruent*, or *neither*. If the solids are similar, state the scale factor.

38.

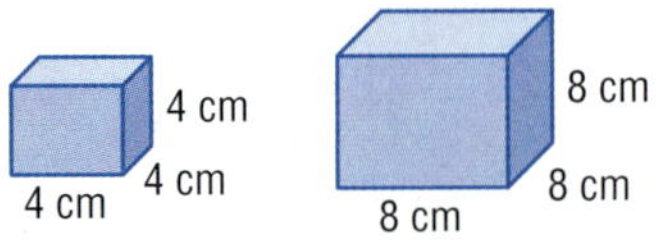

39.

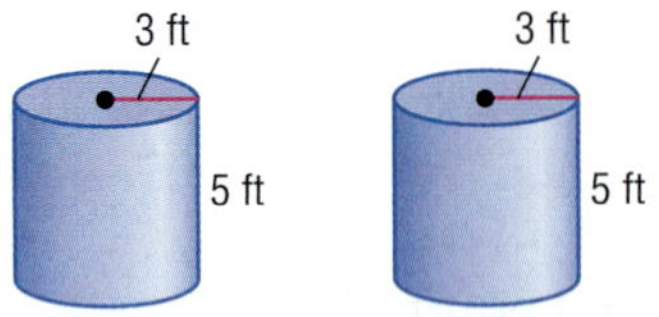

40.

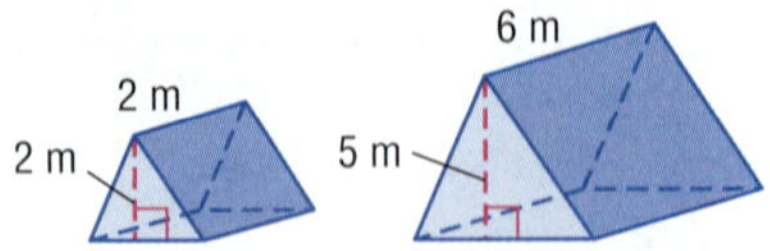

41.

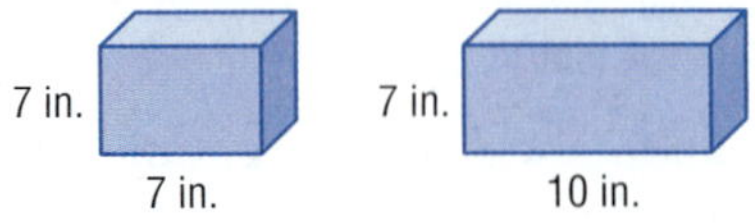

42. MODELS A collector's model car is scaled so that 1 inch on the model equals $5\frac{3}{4}$ feet on the actual car. If the model is $\frac{4}{5}$ inches high, how high is the actual car?

EXAMPLE 9

Determine whether each pair of solids is similar, congruent, or neither. If the solids are similar, state the scale factor.

a.

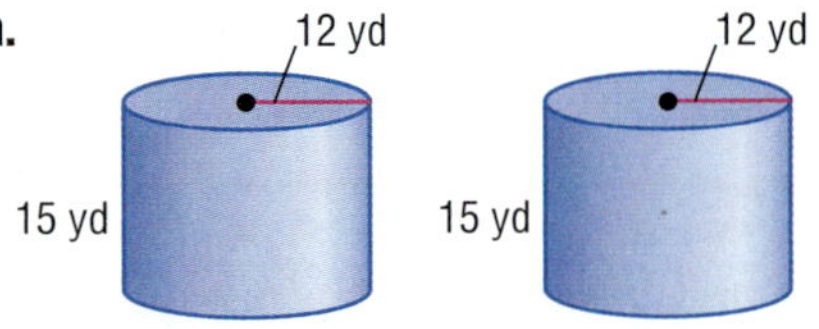

The ratios of the corresponding measures are equal and the scale factor is 1:1, so the solids are congruent.

b.

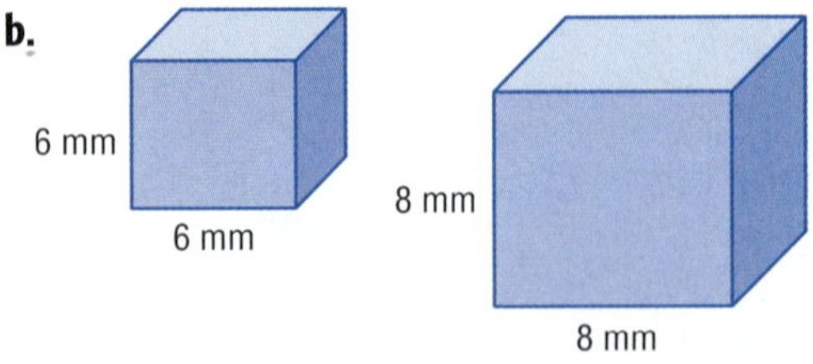

ratio of widths: $\frac{6}{8} = 0.75$

ratio of heights: $\frac{6}{8} = 0.75$

The ratios of the corresponding measures are equal, so the cubes are similar. The scale factor is 3:4. Since the scale factor is not 1:1, the solids are not congruent.

Math Online glencoe.com
Chapter Test

1. Use isometric dot paper and the orthographic drawings to sketch the solid.

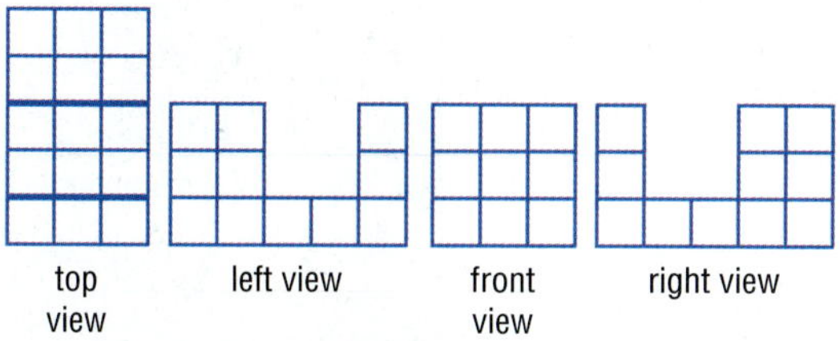

2. Describe the cross section.

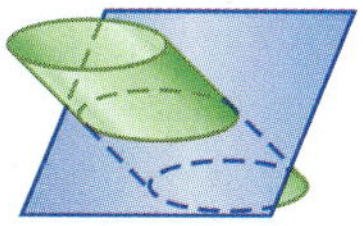

3. Find the surface area of the tent model. Round to the nearest tenth if necessary.

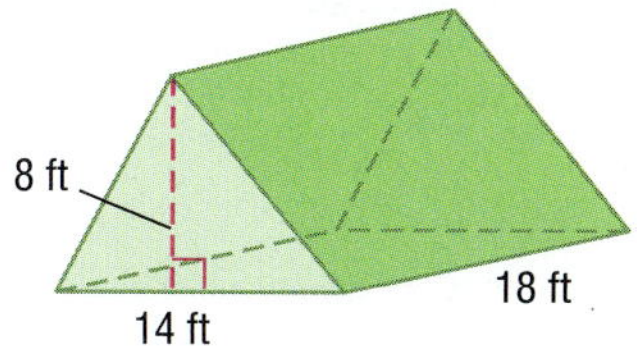

4. **CANDLES** A circular pillar candle is 2.8 inches wide and 6 inches tall. What are the lateral area and surface area of the candle? Round to the nearest tenth if necessary.

5. **TEA** A tea bag is shaped like a regular square pyramid. Each edge of the base is 4 centimeters, and the slant height is 5 centimeters. What is the surface area of the tea bag in square centimeters? Round to the nearest tenth if necessary.

6. **BEEHIVE** Estimate the lateral area and surface area of the Turkish beehive room. Round to the nearest tenth if necessary.

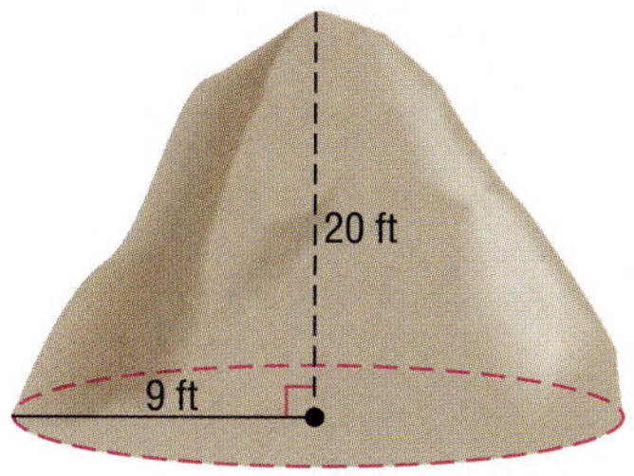

7. Find the volume of the candle in Exercise 4. Round to the nearest tenth if necessary.

8. Find the volume of the tea bag in Exercise 5. Round to the nearest tenth if necessary.

9. **EARTH** Earth's radius is approximately 6400 kilometers. What are the surface area and volume of the Earth? Round to the nearest tenth if necessary.

10. **SOFTBALL** A regulation softball has a circumference of 12 inches. What is the volume of the softball?

Name each of the following on sphere $\mathcal{A}$.

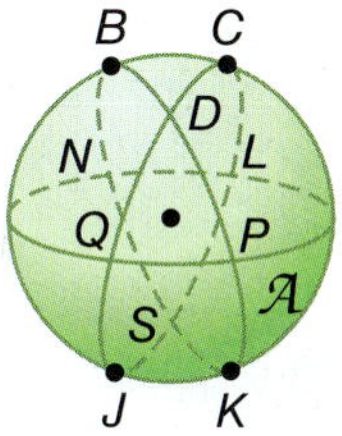

11. two lines containing point S

12. a segment containing point L

13. a triangle

14. two lines containing point D

15. a segment containing point P

16. Are these two cubes *similar*, *congruent*, or *neither*? Explain your reasoning.

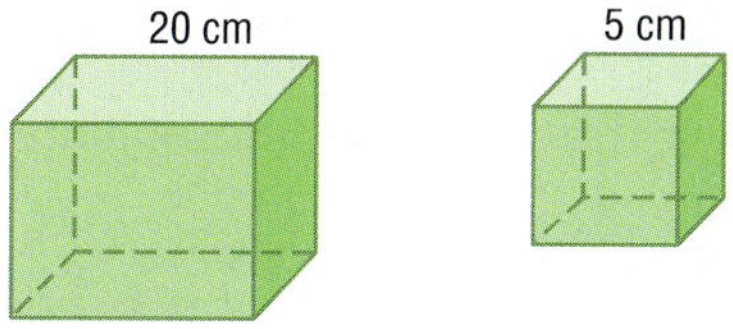

17. Two similar cylinders have heights of 75 centimeters and 25 centimeters. What is the ratio of the volume of the large cylinder to the volume of the small cylinder?

18. **BAKING** Two spherical pieces of cookie dough have radii of 3 centimeters and 5 centimeters, respectively. The pieces are combined to form one large spherical piece of dough. What is the approximate radius of the new sphere of dough? Round to the nearest tenth.

19. **ALGEBRA** A rectangular prism has a base with side lengths x and $x + 3$ and height $2x$. Find the surface area and volume of the prism.

20. **TRANSPORTATION** The traffic cone is 19 inches tall and has a radius of 5 inches.

a. Find the lateral area.

b. Find the surface area.

CHAPTER 12 Preparing for Standardized Tests

Make a Drawing

Making a drawing can be a very helpful way for you to visualize how to solve a problem. Sketch your drawings on scrap paper or in your test booklet (if allowed). Do not make any marks on your answer sheet other than your answers.

Strategies for Making a Drawing

Step 1

Read the problem statement carefully.

Ask yourself:

- What am I being asked to solve? What information is given?
- Would making a drawing help me visualize how to solve the problem?

Step 2

Sketch and label your drawing.

- Make your drawing as clear and accurate as possible.
- Label the drawing carefully. Be sure to include all of the information given in the problem statement.
- Fill in your drawing with information that can be gained from intermediate calculations.

EXAMPLE

Solve the problem below. Responses will be graded using the short-response scoring rubric shown.

A regular pyramid has a square base with 10-centimeter sides and a height of 12 centimeters. What is the total surface area of the pyramid? Round to the nearest tenth if necessary.

Scoring Rubric	
Criteria	**Score**
Full Credit: The answer is correct and a full explanation is provided that shows each step.	2
Partial Credit: • The answer is correct, but the explanation is incomplete. • The answer is incorrect, but the explanation is correct.	1

Read the problem statement carefully. You are given the dimensions of a square pyramid and asked to find the surface area. Sketching a drawing may help you visualize the problem and how to solve it.

Example of a 2-point response:

Use the Pythagorean Theorem to find the slant height, ℓ.

$\ell^2 = 5^2 + 12^2$

$\ell^2 = 169$

$\ell = 13$

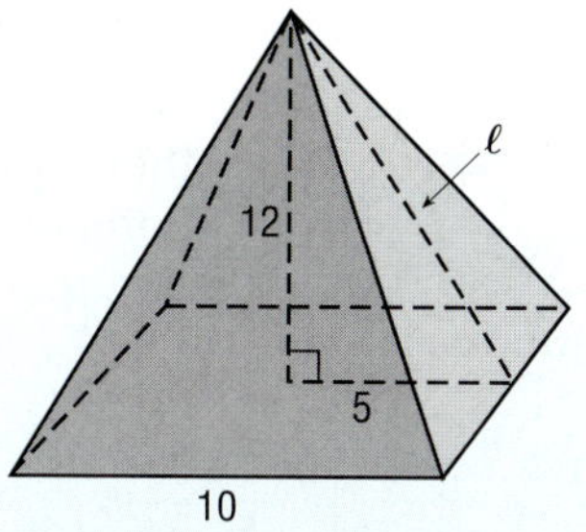

Find the lateral area.

$L = \frac{1}{2}P\ell$

$= \frac{1}{2}(40)(13)$

$= 260$

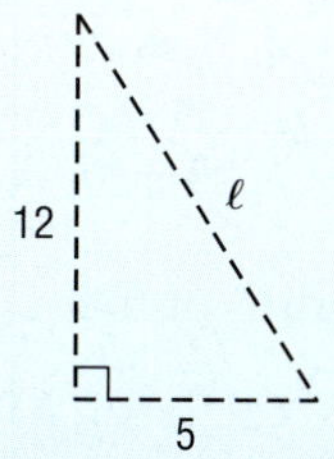

Add the area of the square base.

$S = 260 + 100$ or 360

The total surface area is 360 square centimeters.

The steps, calculations, and reasoning are clearly stated. The student also arrives at the correct answer. So, this response is worth the full 2 points.

Exercises

Solve each problem. Show your work. Responses will be graded using the short-response scoring rubric given at the beginning of the lesson.

1. A right circular cone has a slant height that is twice its radius. The lateral area of the cone is about 569 square millimeters. What is the radius of the cone? Round to the nearest whole millimeter.

2. From a single point in her yard, Marti measures and marks distances of 18 feet and 30 feet with stakes for two sides of her garden. How far apart should the two stakes be if the garden is to be rectangular shaped?

3. A passing boat is 310 feet from the base of a lighthouse. The angle of depression from the top of the lighthouse is 24°. What is the height of the lighthouse to the nearest tenth of a foot?

4. A regular hexagon is inscribed in a circle with a diameter of 12 centimeters. What is the exact area of the hexagon?

5. Luther is building a model rocket for a science fair project. He attaches a nosecone to a cylindrical body to form the rocket's fuselage. The rocket has a diameter of 4 inches and a total height (including the nosecone) of 2 feet 5 inches. The nosecone is 7 inches tall. What is the volume of the rocket? Give your answer rounded to the nearest tenth cubic inch.

6. Terry wants to measure the height of the top of the backboard of his basketball hoop. At 4:00, the shadow of a 4-foot fence post is 20 inches long, and the shadow of the backboard is 65 inches long. What is the height of the top of the backboard?

CHAPTER 12 Standardized Test Practice

Cumulative, Chapters 1 through 12

Multiple Choice

Read each question. Then fill in the correct answer on the answer document provided by your teacher or on a sheet of paper.

1. The Great Pyramid of Giza in Egypt originally had a height of about 148 meters. The base of the pyramid was a square with 230-meter sides. What was the original volume of the pyramid? Round to the nearest whole number.

A 1,786,503 m^3

B 2,609,733 m^3

C 104,128,752 m^3

D 122,716,907 m^3

2. If $\overline{HK}$ is tangent to circle O, what is the radius of the circle?

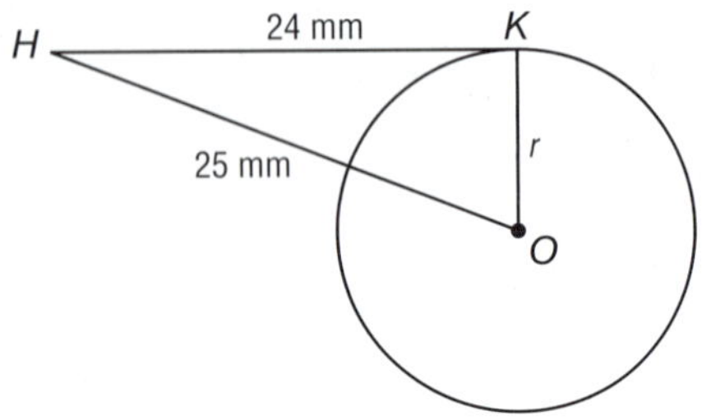

F 7 mm

G 8 mm

H 9 mm

J 10 mm

3. What is the sum of the interior angles of the figure?

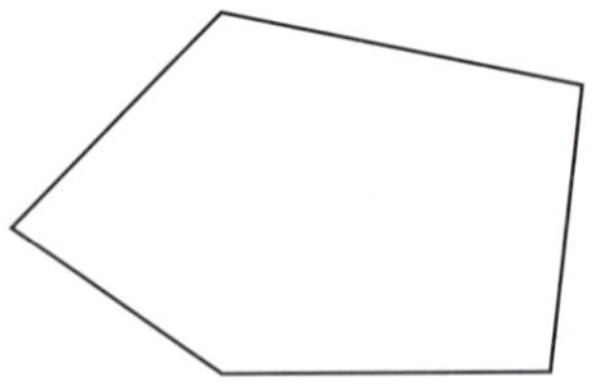

A 450°

B 540°

C 630°

D 720°

Test-Taking Tip

Question 1 You can eliminate some unreasonable answers by estimating first. Choices C and D are too large.

4. Eddie conducted a random survey of 50 students and found that 14 of them spend more than 2 hours each night doing homework. If there are 421 students at Eddie's school, predict how many of them spend more than 2 hours each night doing homework.

F 118

G 124

H 125

J 131

5. $\overline{RS}$ represents the height of Mount Mitchell, the highest point in the state of North Carolina. If $TU = 5013$ feet, $UV = 6684$ feet, and $TV = 8355$ feet, use the ASA Theorem to find the height of Mount Mitchell.

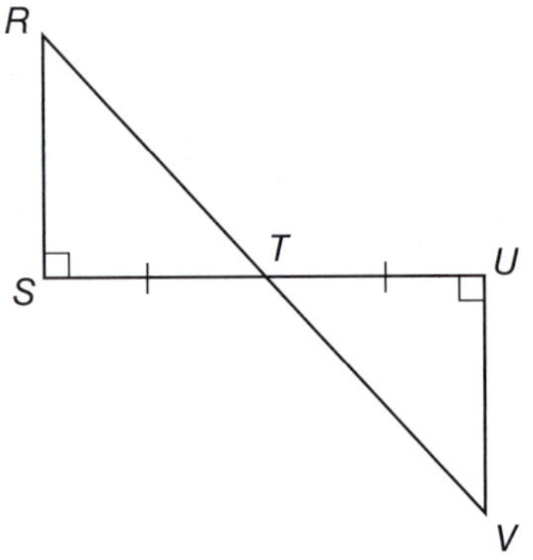

A 5013 ft

B 6684 ft

C 7154 ft

D 8355 ft

6. Triangle DEF is shown below.

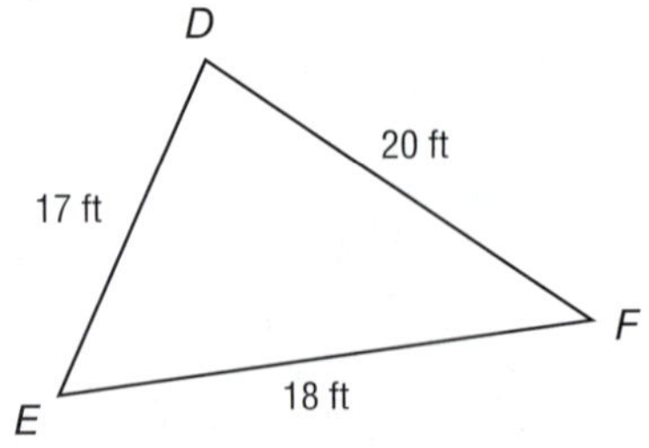

Which statement about this triangle is true?

F $m\angle F > m\angle D$

G $m\angle E > m\angle F$

H $m\angle D < m\angle F$

J $m\angle E < m\angle D$

Short Response/Gridded Response

Record your answers on the answer sheet provided by your teacher or on a sheet of paper.

7. Suppose the length of one diagonal of a kite is three times the length of the other diagonal. If the area of the kite is 96 square inches, what are the lengths of the diagonals? Show your work.

8. Copy the figure and point Y. Then use a protractor and ruler to draw the rotation of the figure 75° clockwise about point Y.

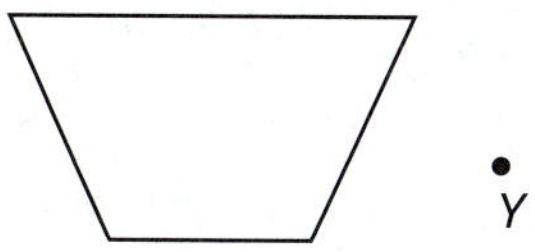

9. GRIDDED RESPONSE What is the perimeter of the isosceles triangle to the nearest tenth of a centimeter?

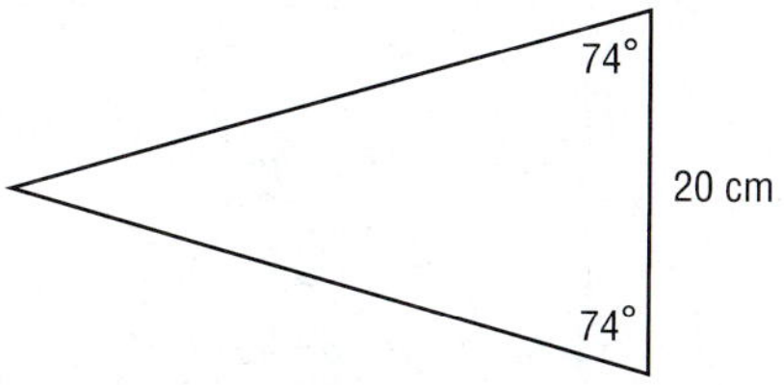

10. Determine whether the following statement is *sometimes*, *always*, or *never* true. Explain.

> The orthocenter of a right triangle is located at the vertex of the right angle.

11. GRIDDED RESPONSE **Given:** $c \parallel d$

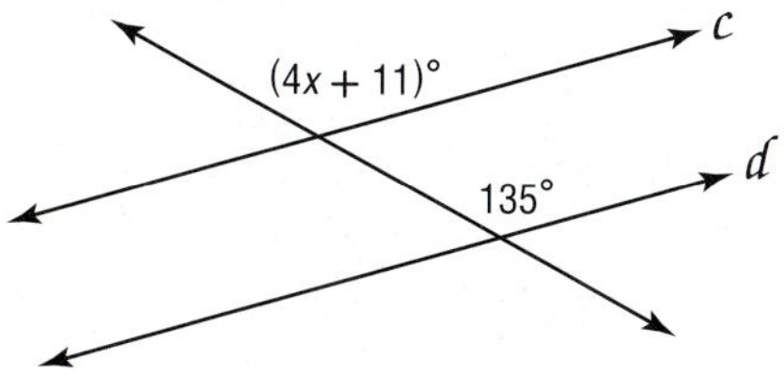

What is the value of x in the figure?

12. What is the lateral area of the square pyramid below? Round to the nearest tenth if necessary. Show your work.

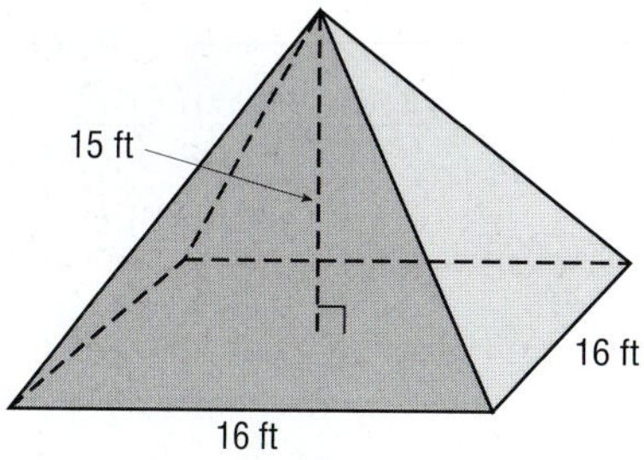

Extended Response

Record your answers on a sheet of paper. Show your work.

13. The two prisms below are similar figures.

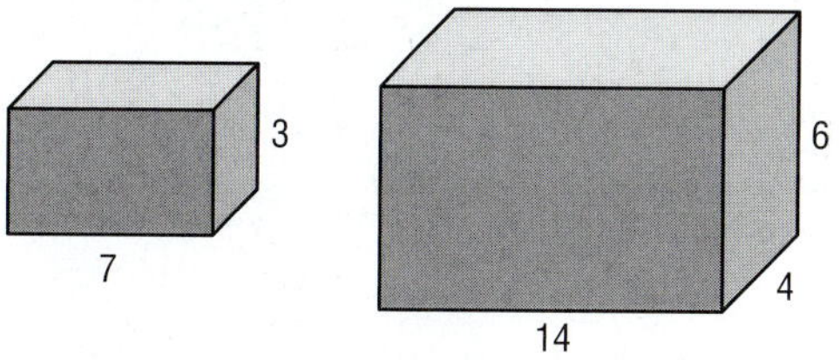

a. What is the scale factor from the smaller prism to the larger one?

b. What are the volumes of the prisms?

c. How many times as great is the volume of the larger prism as the smaller prism?

d. Suppose a solid figure has a volume of 40 cubic units. If its dimensions are scaled by a factor of 1.5, what will the volume of the new figure be?

Need Extra Help?

If you missed Question...	1	2	3	4	5	6	7	8	9	10	11	12	13
Go to Lesson or Page...	12-5	10-5	6-1	7-1	5-3	5-3	11-2	9-3	8-4	5-2	3-2	12-3	12-8

CHAPTER 13 Probability and Measurement

Then

In Chapter 0, you learned about experiments, outcomes, and events. You also found probabilities of simple events.

Now

In Chapter 13, you will:

- Represent sample spaces.
- Use permutations and combinations with probability.
- Find probabilities by using length and area.
- Find probabilities of compound events.

Why?

GAMES Probability can be used to predict the likelihood of different outcomes of the games that we play.

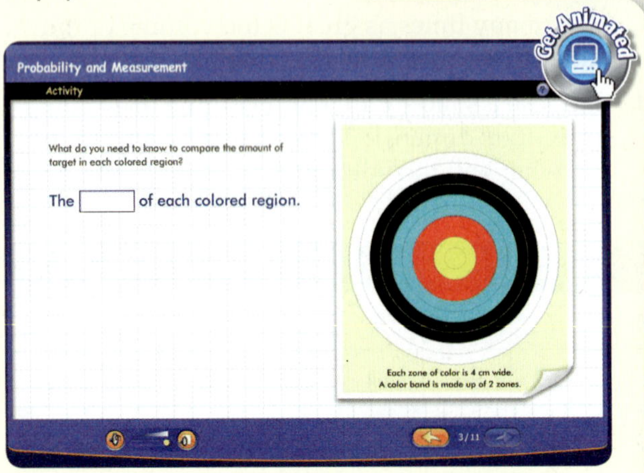

Get Ready for Chapter 13

Diagnose Readiness You have two options for checking Prerequisite Skills.

Text Option

Take the Quick Check below. Refer to the Quick Review for help.

*Quick*Check

Simplify. (Prerequisite Skill)

1. $\frac{1}{2} + \frac{3}{8}$ **2.** $\frac{7}{9} + \frac{2}{6}$ **3.** $\frac{2}{5} + \frac{7}{8}$

4. $\frac{2}{9} \cdot \frac{4}{8}$ **5.** $\frac{3}{7} \cdot \frac{21}{24}$ **6.** $\frac{3}{10} \cdot \frac{2}{9}$

7. SOCCER A soccer team has a 4.5-gallon cooler of water during games. How many 4-ounce cups can the team drink per game?

A die is rolled. Find the probability of each outcome. (Lesson 0-3)

8. P(greater than 1) **9.** P(odd)

10. P(less than 2) **11.** P(1 or 6)

12. GAMES Two friends are playing a game with a 20-sided die that has all of the letters of the alphabet except for Q, U, V, X, Y, and Z. What is the probability that the die will land on a vowel?

The table shows the results of an experiment in which a spinner numbered 1–4 was spun. (Lesson 0-3)

Outcome	Tally	Frequency
1	|||	3
2	𝍸 ||	7
3	𝍸 |	6
4	||||	4

13. What is the experimental probability that the spinner will land on a 4?

14. What is the experimental probability that the spinner will land on an odd number?

15. What is the experimental probability that the spinner will land on an even number?

*Quick*Review

EXAMPLE 1

Simplify $\frac{6}{9} \cdot \frac{1}{2}$.

$\frac{6}{9} \cdot \frac{1}{2} = \frac{6 \cdot 1}{9 \cdot 2}$ **Multiply the numerators and denominators.**

$= \frac{6}{18}$ or $\frac{1}{3}$ **Simplify.**

EXAMPLE 2

Suppose a die is rolled. What is the probability of rolling less than a five?

$P(\text{less than } 5) = \frac{\text{number of favorable outcomes}}{\text{number of possible outcomes}}$

$= \frac{4}{6}$ or $\frac{2}{3}$

The probability of rolling less than a five is $\frac{2}{3}$ or 67%.

EXAMPLE 3

A die was rolled. Find the experimental probability of rolling a 5.

Outcome	Tally	Frequency
1	||||	4
2	𝍸 ||	7
3	𝍸 |||	8
4	||||	4
5	||	2
6	𝍸	5

$P(5) = \frac{\text{number of times 5 occurs}}{\text{total number of outcomes}}$ or $\frac{2}{30}$

The experimental probability for getting a 5 is $\frac{2}{30}$ or 7%.

Online Option

Math Online Take a self-check Chapter Readiness Quiz at glencoe.com.

Get Started on Chapter 13

You will learn several new concepts, skills, and vocabulary terms as you study Chapter 13. To get ready, identify important terms and organize your resources. You may wish to refer to **Chapter 0** to review prerequisite skills.

FOLDABLES® Study Organizer

Probability and Measurement Make this Foldable to help you organize your Chapter 13 notes about probability. Begin with one sheet of paper.

1. **Fold** a sheet of paper lengthwise.

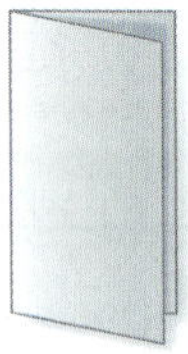

2. **Fold** in half two more times.

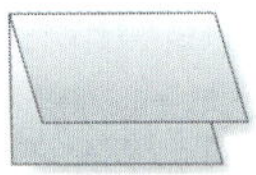

3. **Cut** along each fold on the left column.

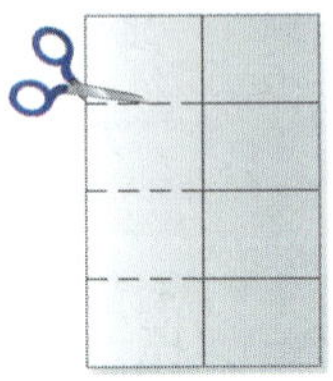

4. **Label** as shown.

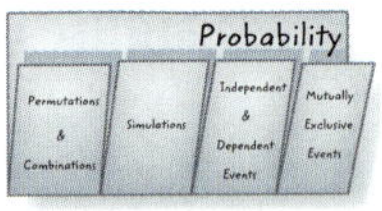

New Vocabulary

English		Español
sample space	• p. 899 •	espacio muestral
tree diagram	• p. 899 •	diagrama de árbol
permutation	• p. 906 •	permutación
factorial	• p. 906 •	factorial
circular permutation	• p. 909 •	permutación circular
combination	• p. 910 •	combinación
geometric probability	• p. 915 •	probabilidad geométrica
probability model	• p. 923 •	modelo de la probabilidad
simulation	• p. 923 •	simulación
random variable	• p. 925 •	variable aleatoria
expected value	• p. 925 •	valor previsto
compound events	• p. 931 •	eventos compuestos
independent events	• p. 931 •	eventos independientes
dependent events	• p. 931 •	eventos dependientes
conditional probability	• p. 933 •	probabilidad condicional
probability tree	• p. 933 •	árbol de la probabilidad
mutually exclusive	• p. 938 •	mutuamente exclusivos
complement	• p. 941 •	complemento

Review Vocabulary

event • p. P8 • evento one or more outcomes of an experiment

experiment • p. P8 • experimento a situation involving chance such as flipping a coin or rolling a die

Multilingual eGlossary glencoe.com

Math Online glencoe.com

- Study the chapter online
- Explore **Get Animated** activities
- Get extra help from your own **Personal Tutor**
- Use **Extra Examples** for additional help
- Take a **Self-Check Quiz**
- **Review Vocabulary** in fun ways

13-1 Representing Sample Spaces

Then

You calculated experimental probability. (Lesson 0-3)

Now

- Use lists, tables, and tree diagrams to represent sample spaces.
- Use the Fundamental Counting Principle to count outcomes.

New Vocabulary

sample space
tree diagram
two-stage experiment
multi-stage experiment
Fundamental Counting Principle

Math Online
glencoe.com

Why?

In a football game, a referee tosses a fair coin to determine which team will take possession of the football first. The coin can land on heads or tails.

Represent a Sample Space

You have learned the following about experiments, outcomes, and events.

Definition	Example
An *experiment* is a situation involving chance that leads to results called *outcomes*.	In the situation above, the experiment is tossing the coin.
An *outcome* is the result of a single performance or *trial* of an experiment.	The possible outcomes are landing on heads or tails.
An *event* is one or more outcomes of an experiment.	One event of this experiment is the coin landing on tails.

The **sample space** of an experiment is the set of all possible outcomes. You can represent a sample space by using an organized list, a table, or a **tree diagram**.

EXAMPLE 1 Represent a Sample Space

A coin is tossed twice. Represent the sample space for this experiment by making an organized list, a table, and a tree diagram.

For each coin toss, there are two possible outcomes, heads H or tails T.

Organized List

Pair each possible outcome from the first toss with the possible outcomes from the second toss.

H, H   T, T

H, T   T, H

Table

List the outcomes of the first toss in the left column and those of the second toss in the top row.

Outcomes	Heads	Tails
Heads	H, H	H, T
Tails	T, H	T, T

Tree Diagram

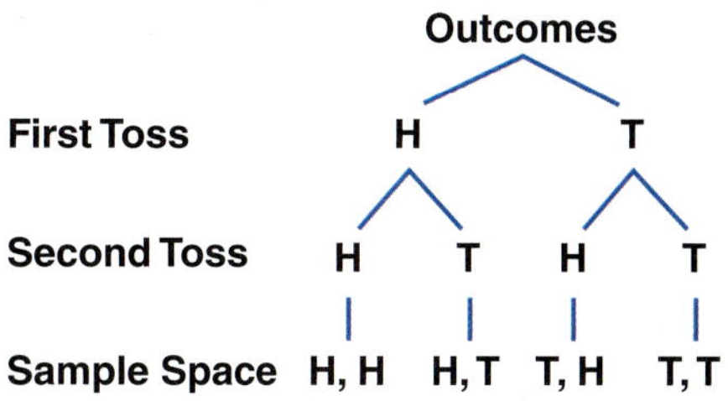

Guided Practice

1. A coin is tossed and then a number cube is rolled. Represent the sample space for this experiment by making an organized list, a table, and a tree diagram.

Personal Tutor glencoe.com

The experiment in Example 1 is an example of a **two-stage experiment**, which is an experiment with two stages or events. Experiments with more than two stages are called **multi-stage experiments**.

Real-World EXAMPLE 2 Multi-Stage Tree Diagrams

HAMBURGERS To take a hamburger order, Keandra asks each customer the questions from the script shown. Draw a tree diagram to represent the sample space for hamburger orders.

The sample space is the result of four stages.

- Burger size (K, R, or L)
- Cheese (C or NC)
- Tomato (T or NT)
- Pickles (P or NP)

Draw a tree diagram with four stages.

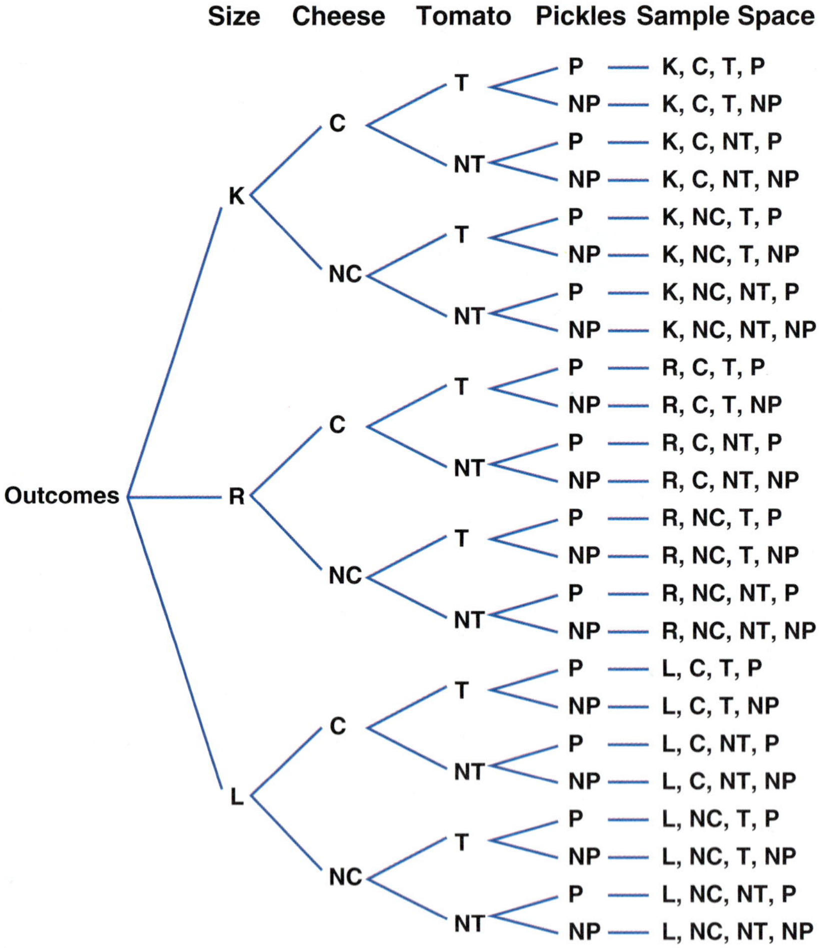

Guided Practice

2. **MUSIC** Yoki can choose a small MP3 player with a 4- or 8-gigabyte hard drive in black, teal, sage, or red. He can also get a clip and/or a dock to go with it. Make a tree diagram to represent the sample space for this situation.

Personal Tutor glencoe.com

Watch Out!

Hidden Stages
The words *and/or* in the third question for Example 2 suggest an additional stage in the ordering process. By making separate stages for choosing with or without tomato and with or without pickles, you allow for the possibility of choosing *both* tomato and pickles.

ReadingMath

Tree Diagram Notation
Choose notation for outcomes in your tree diagrams that will eliminate confusion. In Example 2, *C* stands for *cheese*, while *NC* stands for *no cheese*. Likewise, *NT* and *NP* stand for *no tomato* and *no pickles*, respectively.

Fundamental Counting Principle For some two-stage or multi-stage experiments, listing the entire sample space may not be practical or necessary. To find the *number* of possible outcomes, you can use the **Fundamental Counting Principle**.

Key Concept Fundamental Counting Principle

Words The number of possible outcomes in a sample space can be found by multiplying the number of possible outcomes from each stage or event.

Symbols In a k-stage experiment, let

n_1 = the number of possible outcomes for the first stage.

n_2 = the number of possible outcomes for the second stage after the first stage has occurred.

$\vdots$

n_k = the number of possible outcomes for the kth stage after the first $k - 1$ stages have occurred.

Then the total possible outcomes of this k-stage experiment is

$$n_1 \cdot n_2 \cdot n_3 \cdot \ldots \cdot n_k.$$

StudyTip

Multiplication Rule The Fundamental Counting Principle is sometimes called the *Multiplication Rule for Counting* or the *Counting Principle.*

Real-World EXAMPLE 3 Use the Fundamental Counting Principle

CLASS RINGS Haley has selected a size and overall style for her class ring. Now she must choose from the ring options shown. How many different rings could Haley create in her chosen style and size?

Ring Options	Number of Choices
metals	10
finishes	2
stone colors	12
stone cuts	5
side 1 activity logos	20
side 2 activity logos	20
band styles	2

Use the Fundamental Counting Principle.

metals		finishes		stone colors		stone cuts		side 1 logos		side 2 logos		band styles		possible outcomes
10	×	2	×	12	×	5	×	20	×	20	×	2	=	960,000

So Haley could select 960,000 different rings.

Real-World Link

More than 95 percent of high school students order a traditional ring style, which includes the name of the school, a stone, and the graduation year.

Source: *Fort Worth Star-Telegram*

Guided Practice

3. Find the number of possible outcomes for each situation.

A. The answer sheet shown is completed.

B. A die is rolled four times.

C. **SHOES** A pair of women's shoes comes in whole sizes 5 through 11 in red, navy, brown, or black. They can be leather or suede and come in three different widths.

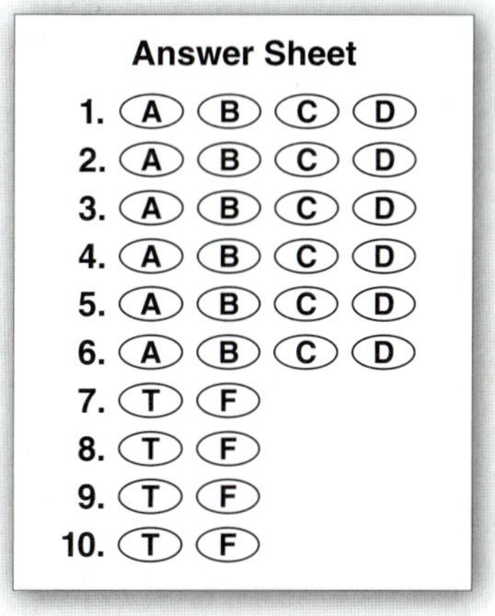

Personal Tutor glencoe.com

Check Your Understanding

Example 1 p. 899 **Represent the sample space for each experiment by making an organized list, a table, and a tree diagram.**

1. For each at bat, a player can either get on base or make an out. Suppose a player bats twice.

2. Quinton sold the most tickets in his school for the annual Autumn Festival. As a reward, he gets to choose twice from a grab bag with tickets that say "free juice" or "free notebook."

Example 2 p. 900

3. **TUXEDOS** Patrick is renting a prom tuxedo from the catalog shown. Draw a tree diagram to represent the sample space for this situation.

Example 3 p. 901 **Find the number of possible outcomes for each situation.**

4. Marcos is buying a cell phone and must choose a plan. Assume one of each is chosen.

Cell Phone Options	Number of Choices
phone style	15
minutes package	5
Internet access	3
text messaging	4
insurance	2

5. Desirée is creating a new menu for her restaurant. Assume one of each item is ordered.

Menu Titles	Number of Choices
Appetizer	8
Soup	4
Salad	6
Entree	12
Dessert	9

Practice and Problem Solving

● = **Step-by-Step Solutions** begin on page R20.
Extra Practice begins on page 969.

Example 1 p. 899 **Represent the sample space for each experiment by making an organized list, a table, and a tree diagram.**

6. Gina is a junior and has a choice for the next two years of either playing volleyball or basketball during the winter quarter.

7. Two different history classes in New York City are taking a trip to either the Smithsonian or the Museum of Natural History.

8. Simeon has an opportunity to travel abroad as a foreign exchange student during each of his last two years of college. He can choose between Ecuador or Italy.

9. A new club is formed, and a meeting time must be chosen. The possible meeting times are Monday or Thursday at 5:00 or 6:00 P.M.

10. An exam with multiple versions has exercises with triangles. In the first exercise, there is an obtuse triangle or an acute triangle. In the second exercise, there is an isosceles triangle or a scalene triangle.

11. PAINTING In an art class, students are working on two projects where they can use one of two different types of paints for each project. Represent the sample space for this experiment by making an organized list, a table, and a tree diagram.

Example 2 p. 900

Draw a tree diagram to represent the sample space for each situation.

12. BURRITOS At a burrito stand, customers have the choice of beans, pork, or chicken with rice or no rice, and cheese and/or salsa.

13. TRANSPORTATION Blake is buying a car and has a choice of sedan, truck, or van with leather or fabric interior, and a CD player and/or sunroof.

14. TREATS Ping and her friends go to a frozen yogurt parlor which has a sign like the one at the right. Draw a tree diagram for all possible combinations of cones with peanuts and/or sprinkles.

Example 3 p. 901

In Exercises 15–18, find the number of possible outcomes for each situation.

15 In the Junior Student Council elections, there are 3 people running for secretary, 4 people running for treasurer, 5 people running for vice president, and 2 people running for class president.

16. When signing up for classes during his first semester of college, Frederico has 4 class spots to fill with a choice of 4 literature classes, 2 math classes, 6 history classes, and 3 film classes.

17. Niecy is choosing one each of 6 colleges, 5 majors, 2 minors, and 4 clubs.

18. Evita works at a restaurant where she has to wear a white blouse, black pants or skirt, and black shoes. She has 5 blouses, 4 pants, 3 skirts, and 6 pairs of black shoes.

19. ART For an art class assignment, Mr. Green gives students their choice of two quadrilaterals to use as a base. One must have sides of equal length, and the other must have at least one set of parallel sides. Represent the sample space by making an organized list, a table, and a tree diagram.

20. BREAKFAST A hotel restaurant serves omelets with a choice of vegetables, ham, or sausage that come with a side of hash browns, grits, or toast.

a. How many different outcomes of omelet and one side are there if a vegetable omelet comes with just one vegetable?

b. Find the number of possible outcomes for a vegetable omelet if you can get any or all vegetables on any omelet.

21. COMPOSITE FIGURES Carlito is calculating the area of the composite figure at the right. In how many different ways can he do this? List the ways.

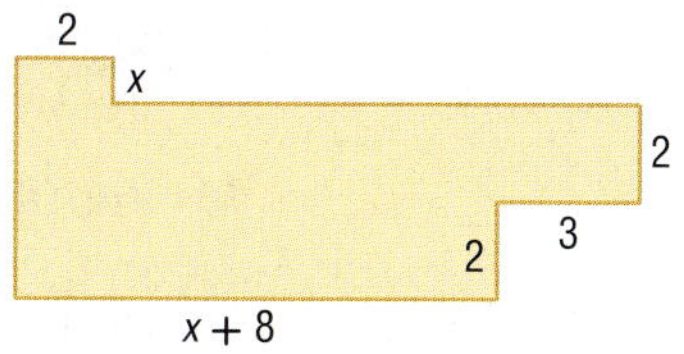

StudyTip

Replacing Objects
If you choose an object from a group of objects without replacing it, the number of objects in the group changes, so the number of possible outcomes changes.

Real-World Link

Bikes are less likely to be stolen in a well-lit location with other bicycles. It is best to secure the bike to a fixed object, such as a permanent bike rack or parking meter.

Source: National Bike Registry

22. **TRANSPORTATION** Miranda got a new bicycle lock that has a four-number combination. Each number in the combination is from 0 to 9.

a. How many combinations are possible if there are no restrictions on the number of times Miranda can use each number?

b. How many combinations are possible if Miranda can use each number only once? Explain.

23 **GAMES** Cody and Monette are playing a board game in which you roll two dice per turn.

a. In one turn, how many outcomes result in a sum of 8?

b. How many outcomes in one turn result in an odd sum?

24. **MULTIPLE REPRESENTATIONS** In this problem, you will investigate a sequence of events. In the first stage of a two-stage experiment, you spin Spinner 1 below. If the result is red, you flip a coin. If the result is yellow, you roll a die. If the result is green, you roll a number cube. If the result is blue, you spin Spinner 2.

a. **GEOMETRIC** Draw a tree diagram to represent the sample space for the experiment.

b. **LOGICAL** Draw a Venn diagram to represent the possible outcomes of the experiment.

c. **ANALYTICAL** How many possible outcomes are there?

d. **VERBAL** Could you use the Fundamental Counting Principle to determine the number of outcomes? Explain.

H.O.T. Problems Use Higher-Order Thinking Skills

25. **CHALLENGE** A box contains n different objects. If you remove three objects from the box, one at a time, without putting the previous object back, how many possible outcomes exist? Explain your reasoning.

26. **OPEN ENDED** Sometimes a tree diagram for an experiment is not symmetrical. Describe a two-stage experiment where the tree diagram is asymmetrical. Include a sketch of the tree diagram. Explain.

27. **WRITING IN MATH** Explain why it is not possible to represent the sample space for a multi-stage experiment by using a table.

28. **REASONING** Determine if the following statement is *sometimes*, *always*, or *never* true. Explain your reasoning.

When an outcome falls outside the sample space, it is a failure.

29. **REASONING** A multistage experiment has n possible outcomes at each stage. If the experiment is performed with k stages, write an equation for the total number of possible outcomes P. Explain.

30. **WRITING IN MATH** Explain when it is necessary to show all of the possible outcomes of an experiment by using a tree diagram and when using the Fundamental Counting Principle is sufficient.

Standardized Test Practice

31. PROBABILITY Alejandra can invite two friends to go out to dinner with her for her birthday. If she is choosing among four of her friends, how many possible outcomes are there?

A 4
B 6
C 8
D 9

32. SHORT RESPONSE What is the volume of the triangular prism shown below?

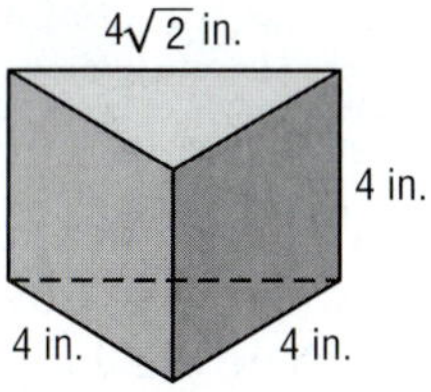

33. Brad's password must be five digits long, use the numbers 0–9, and the digits must not repeat. What is the maximum number of different passwords that Brad can have?

F 15,120
G 30,240
H 59,049
J 100,000

34. SAT/ACT A pizza shop offers 3 types of crust, 5 vegetable toppings, and 4 meat toppings. How many different pizzas could be ordered by choosing 1 crust, 1 vegetable topping, and 1 meat topping?

A 12
B 35
C 60
D infinite

Spiral Review

35. ARCHITECTURE To encourage recycling, the people of Rome, Italy, built a model of Basilica di San Pietro from empty beverage cans. The model was built to a 1:5 scale and was a rectangular prism that measured 26 meters high, 49 meters wide, and 93 meters long. Find the dimensions of the actual Basilica di San Pietro. (Lesson 12-8)

Using spherical geometry, name each of the following on sphere $\mathcal{W}$. (Lesson 12-7)

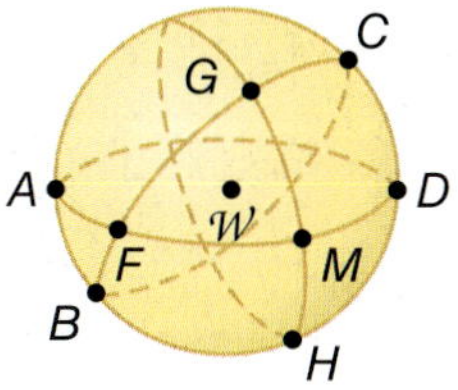

36. two lines containing point F

37. a segment containing point G

38. a triangle

Find the lateral area and surface area of each cylinder. Round to the nearest tenth. (Lesson 12-2)

39.

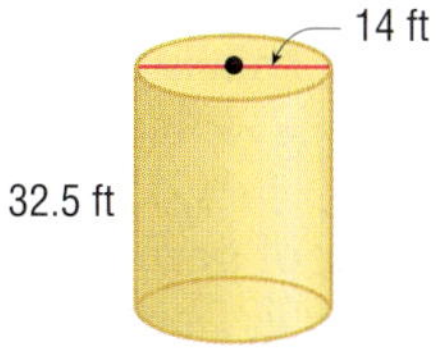

40.

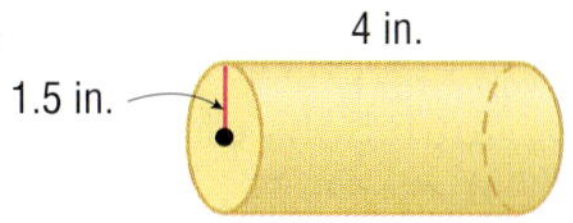

41.

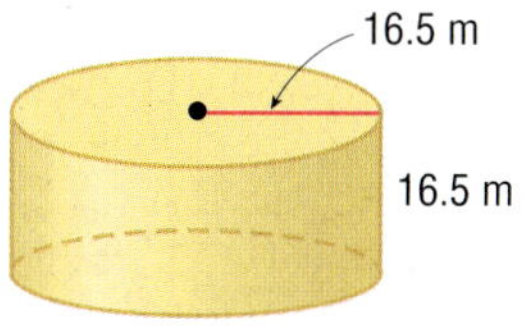

42. TELECOMMUNICATIONS The signal from a tower follows a ray that has its endpoint on the tower and is tangent to Earth. Suppose a tower is located at sea level as shown. Determine the measure of the arc intercepted by the two tangents. (Lesson 10-6)

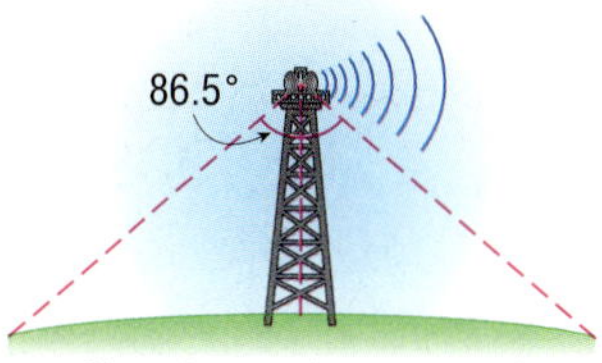

Note: Art not drawn to scale

COORDINATE GEOMETRY **Determine whether the figure with the given vertices has *line* symmetry and/or *rotational* symmetry.** (Lesson 9-5)

43. $Q(2, 2), R(7, 2), S(6, 6), T(3, 6)$

44. $J(-2, 2), K(-5, -1), L(-2, -4), M(1, -1)$

Skills Review

Find each quotient.

45. $\frac{5^2}{2}$

46. $\frac{3^3}{3 \cdot 2}$

47. $\frac{2^4 \cdot 6}{8}$

48. $\frac{2^3 \cdot 12}{6}$

49. $\frac{4^4 \cdot 3}{24}$

13-2

Probability with Permutations and Combinations

Then

You used the Fundamental Counting Principle. (Lesson 13-1)

Now

- Use permutations with probability.
- Use combinations with probability.

New Vocabulary

permutation
factorial
circular permutation
combination

Math Online
glencoe.com

Why?

Lina, Troy, Davian, and Mary are being positioned for a photograph. There are 4 choices for who can stand on the far left, leaving 3 choices for who can stand in the second position. For the third position, just 2 choices remain, and for the last position just 1 is possible.

Probability Using Permutations A **permutation** is an arrangement of objects in which order is important. One permutation of the four friends above is Troy, Davian, Mary, and then Lina. Using the Fundamental Counting Principle, there are $4 \cdot 3 \cdot 2 \cdot 1$ or 24 possible ordered arrangements of the friends.

The expression $4 \cdot 3 \cdot 2 \cdot 1$ used to calculate the number of permutations of these four friends can be written as 4!, which is read *4 factorial.*

Key Concept — **Factorial**

Words The **factorial** of a positive integer n, written $n!$, is the product of the positive integers less than or equal to n.

Symbols $n! = n \cdot (n-1) \cdot (n-2) \cdot \ldots \cdot 2 \cdot 1$, where $0! = 1$

EXAMPLE 1 Probability and Permutations of *n* Objects

SPORTS **Chanise and Renee are members of the lacrosse team. If the 20 girls on the team are each assigned a jersey number from 1 to 20 at random, what is the probability that Chanise's jersey number will be 1 and Renee's will be 2?**

Step 1 Find the number of possible outcomes in the sample space. This is the number of permutations of the 20 girls' names, or 20!.

Step 2 Find the number of favorable outcomes. This is the number of permutations of the other girls' names given that Chanise's jersey number is 1 and Renee's is 2: $(20-2)!$ or 18!.

Step 3 Calculate the probability.

$P(\text{Chanise 1, Renee 2}) = \frac{18!}{20!}$ ← number of favorable outcomes / ← number of possible outcomes

$= \frac{\overset{1}{\cancel{18!}}}{20 \cdot 19 \cdot \underset{1}{\cancel{18!}}}$ Expand 20! and divide out common factors.

$= \frac{1}{380}$ Simplify.

Guided Practice

1. **PHOTOGRAPHY** In the opening paragraph, what is the probability that Troy is chosen to stand on the far left and Davian on the far right for the photograph?

Personal Tutor glencoe.com

In the opening paragraph, suppose 6 friends were available, but the photographer wanted only 4 people in the picture. Using the Fundamental Counting Principle, the number of permutations of 4 friends taken from a group of 6 friends is 6 • 5 • 4 • 3 or 360.

Another way of describing this situation is the number of permutations of 6 friends taken 4 at a time, denoted ${}_6P_4$. This number can also be computed using factorials.

$${}_6P_4 = 6 \cdot 5 \cdot 4 \cdot 3 = \frac{6 \cdot 5 \cdot 4 \cdot 3 \cdot \mathbf{2} \cdot \mathbf{1}}{\mathbf{2} \cdot \mathbf{1}} = \frac{6!}{2!} = \frac{6!}{(6-4)!}$$

This suggests the following formula.

ReadingMath

Distinct The phrase *distinct objects* means that the objects are distinguishable as being different in some way.

Key Concept

Permutations

Symbols The number of permutations of n distinct objects taken r at a time is denoted by ${}_nP_r$ and given by ${}_nP_r = \frac{n!}{(n-r)!}$.

Example The number of permutations of 5 objects taken 2 at a time is ${}_5P_2 = \frac{5!}{(5-2)!} = \frac{5 \cdot 4 \cdot \cancel{3!}}{\cancel{3!}}$ or 20.

EXAMPLE 2 Probability and ${}_nP_r$

A class is divided into teams each made up of 15 students. Each team is directed to select team members to be officers. If Sam, Valencia, and Deshane are on a team, and the positions are decided at random, what is the probability that they are selected as president, vice president, and secretary, respectively?

Step 1 Since choosing officers is a way of ranking team members, order in this situation is important. The number of possible outcomes in the sample space is the number of permutations of 15 people taken 3 at a time, ${}_{15}P_3$.

$${}_{15}P_3 = \frac{15!}{(15-3)!} = \frac{15 \cdot 14 \cdot 13 \cdot \cancel{12!}}{\cancel{12!}} \text{ or } 2730$$

Step 2 The number of favorable outcomes is the number of permutations of the 3 students in their specific positions. This is 1!, or 1.

Step 3 So the probability of Sam, Valencia, and Deshane being selected as the three officers is $\frac{1}{2730}$.

StudyTip

Randomness When outcomes are decided at random, they are equally likely to occur and their probabilities can be calculated using permutations and combinations.

Guided Practice

2. A student identification card consists of 4 digits selected from 10 possible digits from 0 to 9. Digits cannot be repeated.

A. How many possible identification numbers are there?

B. Find the probability that a randomly generated card has the exact number 4213.

Personal Tutor glencoe.com

In a game, you must try to create a word using randomly selected letter tiles. Suppose you select the tiles shown. If you consider the letters O and O to be distinct, then there are 5! or 120 permutations of these letters.

Four of these possible arrangements are listed below.

POOLS POOLS SPOOL SPOOL

Notice that unless the Os are colored, several of these arrangements would look the same. Since there are 2 Os that can be arranged in 2! or 2 ways, the number of permutations of the letters O, P, O, L, and S can be written as $\frac{5!}{2!}$.

Key Concept — Permutations with Repetition

The number of distinguishable permutations of n objects in which one object is repeated r_1 times, another is repeated r_2 times, and so on, is

$$\frac{n!}{r_1! \cdot r_2! \cdot \ldots \cdot r_k!}.$$

Real-World Link

The two longest words in the Oxford Dictionary with no repeating letters are *uncopyrightable* and *dermatoglyphics*.

Source: Oxford Dictionaries

EXAMPLE 3 Probability and Permutations with Repetition

GAME SHOW On a game show, you are given the following letters and asked to unscramble them to name a U.S. river. If you selected a permutation of these letters at random, what is the probability that they would spell the correct answer of MISSISSIPPI?

Step 1 There is a total of 11 letters. Of these letters, I occurs 4 times, S occurs 4 times, and P occurs 2 times. So, the number of distinguishable permutations of these letters is

$$\frac{11!}{4! \cdot 4! \cdot 2!} = \frac{39{,}916{,}800}{1152} \text{ or } 34{,}650. \quad \textbf{Use a calculator.}$$

Step 2 There is only 1 favorable arrangement—MISSISSIPPI.

Step 3 The probability that a permutation of these letters selected at random spells Mississippi is $\frac{1}{34{,}650}$.

Guided Practice

3. TELEPHONE NUMBERS What is the probability that a 7-digit telephone number with the digits 5, 1, 6, 5, 2, 1, and 5 is the number 550-5211?

Personal Tutor glencoe.com

So far, you have been studying objects that are arranged in *linear* order. Notice that when the spices below are arranged in a line, shifting each spice one position to the right produces a different permutation—curry is now first instead of salt. There are 5! distinct permutations of these spices.

In a **circular permutation**, objects are arranged in a circle or loop. Consider the arrangements of these spices when placed on a turntable. Notice that rotating the turntable clockwise one position does *not* produce a different permutation—the order of the spices relative to each other remains unchanged.

Since 5 rotations of the turntable will produce the same permutation, the number of distinct permutations on the turntable is $\frac{1}{5}$ of the total number of arrangements when the spices are placed in a line.

$$\frac{1}{5} \cdot 5! = \frac{5 \cdot 4!}{5} \text{ or } 4!, \text{ which is } (5 - 1)!$$

StudyTip

Turning the Circle Over If the circular object looks the same when it is turned over, such as a plain key ring, then the number of permutations must be divided by 2.

Key Concept

Circular Permutations

The number of distinguishable permutations of n objects arranged in a circle with no fixed reference point is

$$\frac{n!}{n} \text{ or } (n - 1)!.$$

If the n objects are arranged relative to a fixed reference point, then the arrangements are treated as linear, making the number of permutations $n!$.

EXAMPLE 4 Probability and Circular Permutations

Find the indicated probability. Explain your reasoning.

a. JEWELRY If the 6 charms on the bracelet shown are arranged at random, what is the probability that the arrangement shown is produced?

Since there is no fixed reference point, this is a circular permutation. So there are $(6 - 1)!$ or $5!$ distinguishable permutations of the charms. Thus the probability that the exact arrangement shown is produced is $\frac{1}{5!}$ or $\frac{1}{120}$.

b. DINING You are seating a party of 4 people at a round table. One of the chairs around this table is next to a window. If the diners are seated at random, what is the probability that the person paying the bill is seated next to the window?

Since the people are seated around a table with a fixed reference point, this is a linear permutation. So there are $4!$ or 24 ways in which the people can be seated around the table. The number of favorable outcomes is the number of permutations of the other 3 diners given that the person paying the bill sits next to the window, $3!$ or 6.

So, the probability that the person paying the bill is seated next to the window is $\frac{6}{24}$ or $\frac{1}{4}$.

Guided Practice

4. **FOOTBALL** A team's 11 football players huddle together before a play.

   A. What is the probability that the fullback stands to the right of the quarterback if the team huddles together at random? Explain your reasoning.

   B. If a referee stands directly behind the huddle, what is the probability that the referee stands directly behind the halfback? Explain your reasoning.

Personal Tutor glencoe.com

Probability Using Combinations A **combination** is an arrangement of objects in which order is *not* important. Suppose you need to pack 3 of your 8 different pairs of socks for a trip. The order in which the socks are chosen does not matter, so the 3! or 6 groups of socks shown below would *not* be considered different. So, you would use combinations to determine the number of possible different sock choices.

A combination of n objects taken r at a time, or ${}_nC_r$, is calculated by dividing the number of permutations ${}_nP_r$ by the number of arrangements containing the same elements, $r!$.

StudyTip

Permutations and Combinations Use permutations when the order of an arrangement of objects is important and combinations when order is not important.

Key Concept — Combinations

Symbols The number of combinations of n distinct objects taken r at a time is denoted by ${}_nC_r$ and is given by ${}_nC_r = \frac{n!}{(n-r)!\,r!}$.

Example The number of combinations of 8 objects taken 3 at a time is

$${}_8C_3 = \frac{8!}{(8-3)!\,3!} = \frac{8!}{5!3!} = \frac{8 \cdot 7 \cdot \cancel{6} \cdot \cancel{5!}}{\cancel{5!} \cdot \cancel{6}} \text{ or } 56.$$

EXAMPLE 5 Probability and ${}_nC_r$

INVITATIONS For her birthday, Monica can invite 6 of her 20 friends to join her at a theme park. If she chooses to invite friends at random, what is the probability that friends Tessa, Guido, Brendan, Faith, Charlotte, and Rhianna are chosen?

Step 1 Since the order in which the friends are chosen does not matter, the number of possible outcomes in the sample space is the number of combinations of 20 people taken 6 at a time, ${}_{20}C_6$.

$${}_{20}C_6 = \frac{20!}{(20-6)!\,6!} = \frac{\cancel{20} \cdot 19 \cdot \cancel{18} \cdot 17 \cdot \overset{8}{\cancel{16}} \cdot 15 \cdot \cancel{14!}}{\cancel{14!} \cdot \cancel{6} \cdot \cancel{5} \cdot \cancel{4} \cdot \cancel{3} \cdot \cancel{2}} \text{ or } 38{,}760$$

Step 2 There is only 1 favorable outcome—that the six students listed above are chosen. The order in which they are chosen is not important.

Step 3 So the probability of these six friends being chosen is $\frac{1}{38{,}760}$.

Guided Practice

5. **GEOMETRY** If three points are randomly chosen from those named on the rectangle shown, what is the probability that they all lie on the same line segment?

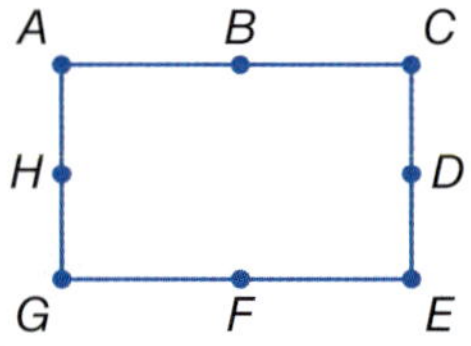

Personal Tutor glencoe.com

Check Your Understanding

Example 1 p. 906

1. **GEOMETRY** On an exam, you are asked to list the names of the polygons shown below in order of the number of sides, starting with the least number. If you choose two shapes at random, what is the probability that they are in the correct order?

Example 2 p. 907

2. **PLAYS** A high school performs a production of *A Raisin in the Sun* with each freshman English class of 18 students. If the three members of the crew are decided at random, what is the probability that Chase is selected for lighting, Jaden is selected for props, and Emelina for spotlighting?

Example 3 p. 908

3. **DRIVING** What is the probability that a license plate using the letters C, F, and F and numbers 3, 3, 3, and 1 will be CFF3133?

Example 4 p. 909

4. **CHEMISTRY** In chemistry lab, you need to test six samples that are randomly arranged on a circular tray.

   a. What is the probability that the arrangement shown at the right is produced?

   b. What is the probability that test tube 2 will be in the top middle position?

Example 5 p. 910

5. **DRAWINGS** Fifteen boys and fifteen girls entered a drawing for four free movie tickets. What is the probability that all four tickets were won by girls?

Practice and Problem Solving

● = **Step-by-Step Solutions** begin on page R20.
Extra Practice begins on page 969.

Example 1 p. 906

6. **CONCERTS** Nia and Chad are going to a concert with their high school's key club. If they choose a seat on the row below at random, what is the probability that Chad will be in seat C11 and Nia will be in C12?

7. **FAIRS** Alfonso and Colin each bought one raffle ticket at the state fair. If 50 tickets were randomly sold, what is the probability that Alfonso got ticket 14 and Colin got ticket 23?

Example 2 p. 907

8. **GYMNASTICS** The table shows the finalists for a floor exercises competition. The order in which they will perform will be chosen randomly.

   a. What is the probability that Cecilia, Annie, and Kimi are the first 3 gymnasts to perform, in any order?

   b. What is the probability that Cecilia is first, Annie is second, and Kimi is third?

Floor Exercises Finalists
Eliza Hernandez
Kimi Kanazawa
Cecilia Long
Annie Montgomery
Shenice Malone
Caroline Smith
Jessica Watson

9. **JOBS** A store randomly assigns their employees work identification numbers to track productivity. Each number consists of 5 digits ranging from 1–9. If the digits cannot repeat, find the probability that a randomly generated number is 25938.

10. **GROUPS** Two people are chosen randomly from a group of ten. What is the probability that Jimmy was selected first and George second?

Example 3 p. 908

11. **MAGNETS** Santiago bought some letter magnets that he can arrange to form words on his fridge. If he randomly selected a permutation of the letters shown below, what is the probability that they would form the word BASKETBALL?

12. **ZIP CODES** What is the probability that a zip code randomly generated from among the digits 3, 7, 3, 9, 5, 7, 2, and 3 is the number 39372?

Example 4 p. 909

13. **GROUPS** Keith is randomly arranging desks into circles for group activities. If there are 7 desks in his circle, what is the probability that Keith will be in the desk closest to the door?

14. **AMUSEMENT PARKS** Sylvie is at an amusement park with her friends. They go on a ride that has bucket seats in a circle. If there are 8 seats, what is the probability that Sylvie will be in the seat furthest from the entrance to the ride?

Example 5 p. 910

15. **PHOTOGRAPHY** If you are randomly placing 24 photos in a photo album and you can place four photos on the first page, what is the probability that you choose the photos at the right?

16. **ROAD TRIPS** Rita is going on a road trip across the U.S. She needs to choose from 15 cities where she will stay for one night. If she randomly pulls 3 city brochures from a pile of 15, what is the probability that she chooses Austin, Cheyenne, and Savannah?

17. **GAMES** Use the figure below. Assume that the balls are aligned at random.

a. What is the probability that in a row of 8 pool balls, the solid 2 and striped 11 would be first and second from the left?

b. What is the probability that if the 8 pool balls were mixed up at random, they would end up in the order shown?

c. What is the probability that in a row of seven balls, with three 8 balls, three 9 balls, and one 6 ball, the three 8 balls would be to the left of the 6 ball and the three 9 balls would be on the right?

d. If the balls were randomly rearranged and formed a circle, what is the probability that the 6 ball is next to the 7 ball?

18. How many lines are determined by 10 randomly selected points, no 3 of which are collinear? Explain your calculation.

19. Suppose 7 points on a circle are chosen at random, as shown at the right.

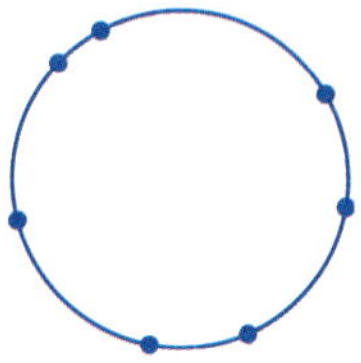

a. Using the letters A through E, how many ways can the points on the circle be named?

b. If one point on the circle is fixed, how many arrangements are possible?

Real-World Link

The Flying Horses Carousel, in Watch Hill, Rhode Island, is the oldest operating platform carousel in the nation. It was built in 1876.

Source: Martha's Vineyard Preservation Trust

20. RIDES A carousel has 7 horses and one bench seat that will hold two people. One of the horses does not move up or down.

a. How many ways can the seats on the carousel be randomly filled by 9 people?

b. If the carousel is filled randomly, what is the probability that you and your friend will end up in the bench seat?

c. If 6 of the 9 people randomly filling the carousel are under the age of 8, what is the probability that a person under the age of 8 will end up on the horse that does not move up or down?

21. LICENSES A camera positioned above a traffic light photographs cars that fail to stop at a red light. In one unclear photograph, the officer could see that the first letter was a Q, the second letter was an M or an N and the third letter was a B, P, or D. The first number was a 0, but the last two numbers were illegible. How many possible license plates fit this description?

22. MULTIPLE REPRESENTATIONS In this problem, you will investigate permutations.

a. NUMERICAL Randomly select three digits from 0 to 9. Find the possible permutations of the three integers.

b. TABULAR Repeat part **a** for four additional sets of three integers. You will use some digits more than once. Copy and complete the table below.

Integers	Permutations	Average of Permutations	$\frac{\text{Average of Permutations}}{37}$
1, 4, 7	147, 174, 417, 471, 714, 741	444	12

c. VERBAL Make a conjecture about the value of the average of the permutations of three digits between 0 and 9.

d. SYMBOLIC If the three digits are x, y, and z, is it possible to write an equation for the average A of the permutations of the digits? If so, write the equation. If not, explain why not.

H.O.T. Problems

Use **H**igher-**O**rder **T**hinking Skills

23. REASONING In a circular permutation with a fixed reference point, if n distinct objects are arranged in n positions, is the probability that a particular object will be in the nth position *sometimes, always,* or *never* $\frac{1}{n}$? Explain.

24. CHALLENGE A student claimed that permutations and combinations were related by $r! \cdot {}_nC_r = {}_nP_r$. Use algebra to show that this is true. Then explain why ${}_nC_r$ and ${}_nP_r$ differ by the factor $r!$.

25. OPEN ENDED Describe a situation in which the probability is given by $\frac{1}{{}_7C_3}$.

26. REASONING Is the following statement *sometimes, always,* or *never* true? Explain.

$${}_nP_r = {}_nC_r$$

27. PROOF Prove that ${}_nC_{n-r} = {}_nC_r$.

28. WRITING IN MATH Compare and contrast permutations and combinations.

Standardized Test Practice

29. PROBABILITY Four members of the pep band, two girls and two boys, always stand in a row when they sing. What is the probability that a girl will be at each end of the row if they line up in random order?

A $\frac{1}{24}$ **C** $\frac{1}{6}$

B $\frac{1}{12}$ **D** $\frac{1}{2}$

30. SHORT RESPONSE If you randomly select a permutation of the letters shown below, what is the probability that they would spell GEOMETRY?

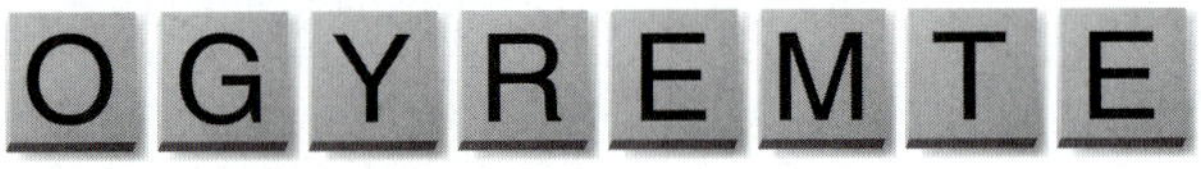

31. ALGEBRA Student Council sells soft drinks at basketball games and makes $1.50 from each. If they pay $75 to rent the concession stand, how many soft drinks would they have to sell to make $250 profit?

F 116 **H** 167

G 117 **J** 217

32. SAT/ACT The ratio of 12:9 is equal to the ratio of $\frac{1}{3}$ to

A 2 **C** 1

B $\frac{5}{4}$ **D** $\frac{1}{4}$

Spiral Review

33. SHOPPING A women's coat comes in sizes 4, 6, 8, or 10 in black, brown, ivory, and cinnamon. How many different coats could be selected? (Lesson 13-1)

34. Two similar prisms have surface areas of 256 square inches and 324 square inches. What is the ratio of the height of the small prism to the height of the large prism? (Lesson 12-8)

Find x. Round to the nearest tenth, if necessary. (Lesson 10-7)

35.

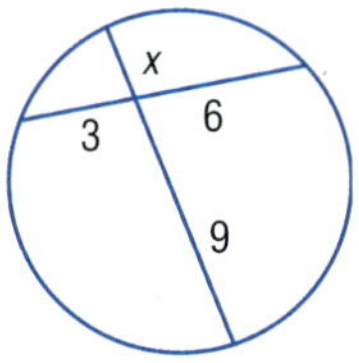

36.

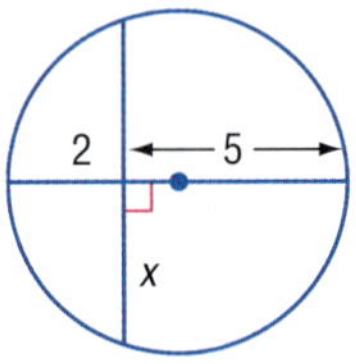

37.

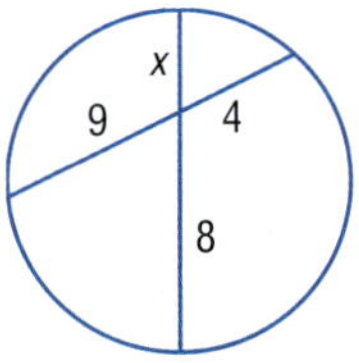

38. CHESS The bishop shown in square f8 can only move diagonally along dark squares. If the bishop is in c1 after two moves, describe the translation. (Lesson 9-2)

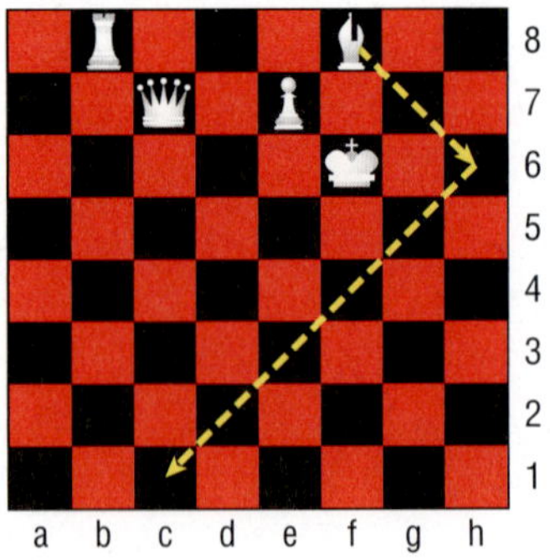

Skills Review

Use the number line to find each measure. (Lesson 1-3)

39. DF

40. AE

41. EF

42. BD

43. AC

44. CF

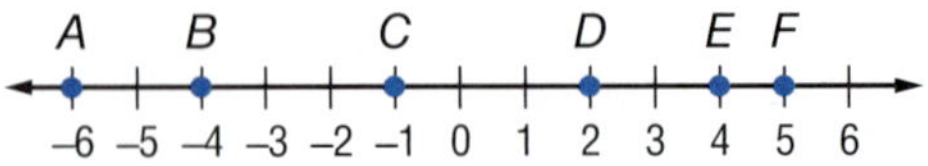

13-3 Geometric Probability

Then
You found probabilities of simple events. (Lesson 0-3)

Now
- Find probabilities by using length.
- Find probabilities by using area.

New Vocabulary
geometric probability

Math Online
glencoe.com

Why?

The object of the popular carnival game shown is to collect points by rolling a ball up an incline and into one of several circular target areas. The point value of each area is assigned based on the probability of a person landing a ball in that area.

Probability with Length The probability of winning the carnival game depends on the area of the target. Probability that involves a geometric measure such as length or area is called **geometric probability**.

Key Concept — Length Probability Ratio

Words If a line segment (1) contains another segment (2) and a point on segment (1) is chosen at random, then the probability that the point is on segment (2) is $\frac{\text{length of segment (2)}}{\text{length of segment (1)}}$.

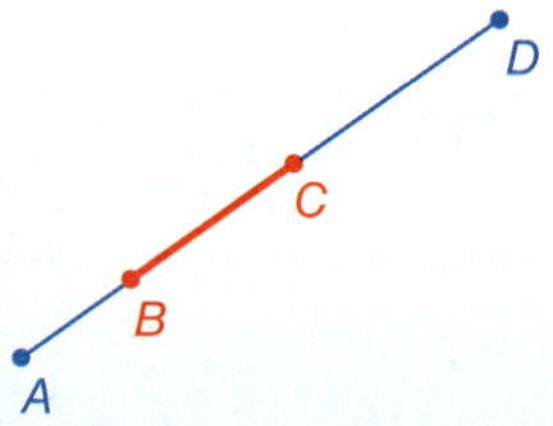

Example If a point E on $\overline{AD}$ is chosen at random, then $P(E \text{ is on } \overline{BC}) = \frac{BC}{AD}$.

EXAMPLE 1 Use Lengths to Find Geometric Probability

Point X is chosen at random on $\overline{JM}$. Find the probability that X is on $\overline{KL}$.

$P(X \text{ is on } \overline{KL}) = \frac{KL}{JM}$ — **Length probability ratio**

$= \frac{7}{14}$ — **$KL = 7$ and $JM = 3 + 7 + 4$ or 14**

$= \frac{1}{2}$, 0.5, or 50% — **Simplify.**

Guided Practice

Point X is chosen at random on $\overline{JM}$. Find the probability of each event.

1A. $P(X \text{ is on } \overline{LM})$

1B. $P(X \text{ is on } \overline{KM})$

Personal Tutor glencoe.com

Geometric probability can be used in many real-world situations that involve an infinite number of outcomes.

Real-World Link

A Chicago Transit Authority train arrives or departs a station like Addison on the Red Line every 15 minutes.

Source: Chicago Transit Authority

Real-World EXAMPLE 2 Model Real-World Probabilities

TRANSPORTATION Use the information at the left. Assuming that you arrive at Addison on the Red Line at a random time, what is the probability that you will have to wait 5 or more minutes for a train?

We can use a number line to model this situation. Since the trains arrive every 15 minutes, the next train will arrive in 15 minutes or less. On the number line below, the event of waiting 5 or more minutes is modeled by $\overline{BD}$.

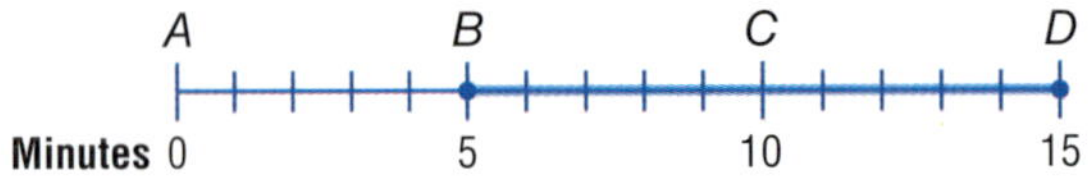

Find the probability of this event.

$P(\text{waiting 5 or more minutes}) = \frac{BD}{AD}$ **Length probability ratio**

$= \frac{10}{15}$ or $\frac{2}{3}$ **$BD = 10$ and $AD = 15$**

So, the probability of waiting 5 or more minutes for the next train is $\frac{2}{3}$ or about 67%.

Guided Practice

2. TEA Iced tea at a cafeteria-style restaurant is made in 8-gallon containers. Once the level gets below 2 gallons, the flavor of the tea becomes weak.

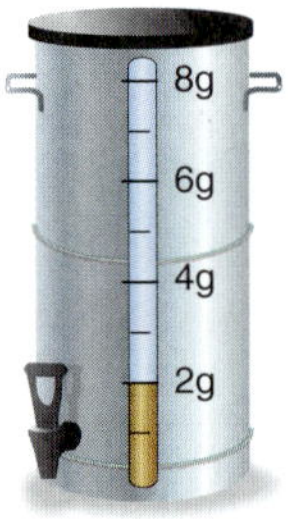

A. What is the probability that when someone tries to pour a glass of tea from the container, it is below 2 gallons?

B. What is the probability that the amount of tea in the container at any time is between 2 and 3 gallons?

Personal Tutor glencoe.com

Probability with Area Geometric probability can also involve area. The ratio for calculating geometric probability involving area is shown below.

Key Concept

Area Probability Ratio

Words If a region A contains a region B and a point E in region A is chosen at random, then the probability that point E is in region B is $\frac{\text{area of region } B}{\text{area of region } A}$.

Example If a point E is chosen at random in rectangle A, then $P(\text{point } E \text{ is in circle } B) = \frac{\text{area of circle } B}{\text{area of rectangle } A}$.

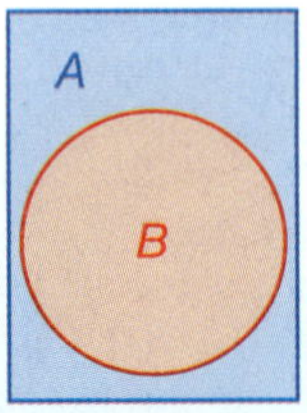

When determining geometric probabilities with targets, we assume

- that the object lands within the target area, and
- it is equally likely that the object will land anywhere in the region.

Real-World Link

Champion accuracy skydivers routinely land less than two inches away from the center of a target.

Source: *SkyDiving News*

Real-World EXAMPLE 3 Use Area to Find Geometric Probability

SKYDIVING Suppose a skydiver must land on a target of three concentric circles. If the diameter of the center circle is 2 yards and the circles are spaced 1 yard apart, what is the probability that the skydiver will land in the red circle?

You need to find the ratio of the area of the red circle to the area of the entire target. The radius of the red circle is 1 yard, while the radius of the entire target is 1 + 1 + 1 or 3 yards.

$$P(\text{skydiver lands in red circle}) = \frac{\text{area of red circle}}{\text{area of target}} \qquad \textbf{Area probability ratio}$$

$$= \frac{\pi(1)^2}{\pi(3)^2} \qquad A = \pi r^2$$

$$= \frac{\pi}{9\pi} \text{ or } \frac{1}{9} \qquad \textbf{Simplify.}$$

The probability that the skydiver will land in the red circle is $\frac{1}{9}$ or about 11%.

Guided Practice

3. SKYDIVING Find each probability using the example above.

A. P(skydiver lands in the blue region)

B. P(skydiver lands in white region)

Personal Tutor glencoe.com

You can also use an angle measure to find geometric probability. The ratio of the area of a sector of a circle to the area of the entire circle is the same as the ratio of the sector's central angle to 360. **You will prove this in Exercise 27.**

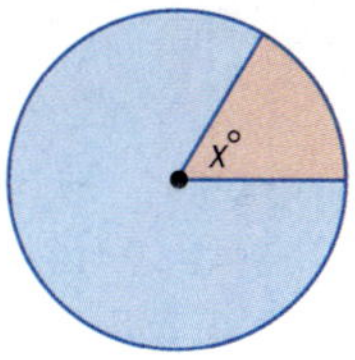

StudyTip

Use Estimation In Example 4b, the area of the purple sector is a little less than $\frac{1}{3}$ or 33% of the spinner. Therefore, an answer of 29% is reasonable.

EXAMPLE 4 Use Angle Measures to Find Geometric Probability

Use the spinner to find each probability.

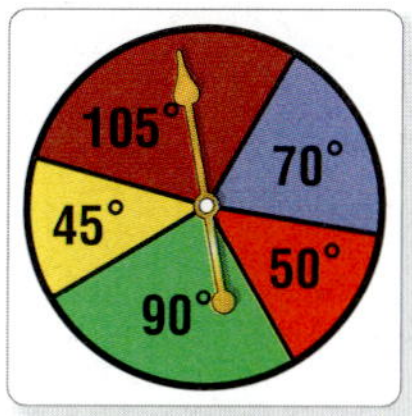

a. P(pointer landing on yellow)

The angle measure of the yellow region is **45**.

P(pointer landing on yellow) $= \frac{45}{360}$ or 12.5%

b. P(pointer landing on purple)

The angle measure of the purple region is **105**.

P(pointer landing on purple) $= \frac{105}{360}$ or about 29%

c. P(pointer landing on neither red nor blue)

The combined angle measures of the red and blue region are **50 + 70** or **120**.

P(pointer landing on neither red nor blue) $= \frac{360 - 120}{360}$ or about 67%

Guided Practice

4A. P(pointer landing on blue)

4B. P(pointer not landing on green)

Personal Tutor glencoe.com

Check Your Understanding

Example 1 p. 915

Point X is chosen at random on $\overline{AD}$. Find the probability of each event.

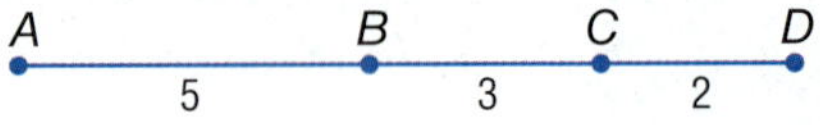

1. $P(X \text{ is on } \overline{BD})$
2. $P(X \text{ is on } \overline{BC})$

Example 2 p. 916

3. **CARDS** In a game of cards, 43 cards are used, including one joker. Four players are each dealt 10 cards and the rest are put in a pile. If Greg doesn't have the joker, what is the probability that either his partner or the pile have the joker?

Examples 3 and 4 p. 917

4. **ARCHERY** An archer aims at a target that is 122 centimeters in diameter with 10 concentric circles whose diameters decrease by 12.2 centimeters as they get closer to the center. Find the probability that the archer will hit the center.

5. **NAVIGATION** A camper lost in the woods points his compass in a random direction. Find the probability that the camper is heading in the N to NE direction.

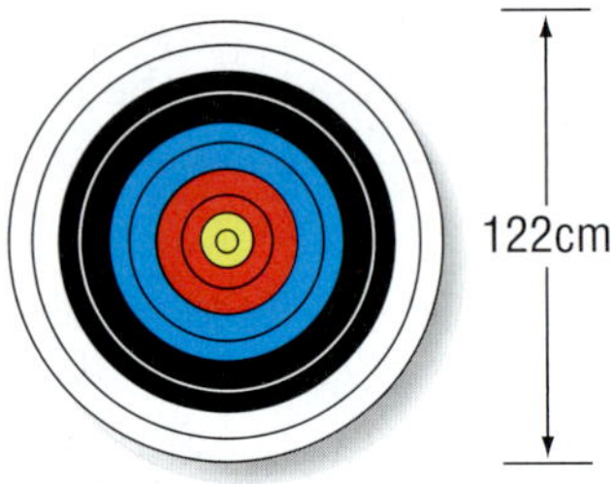

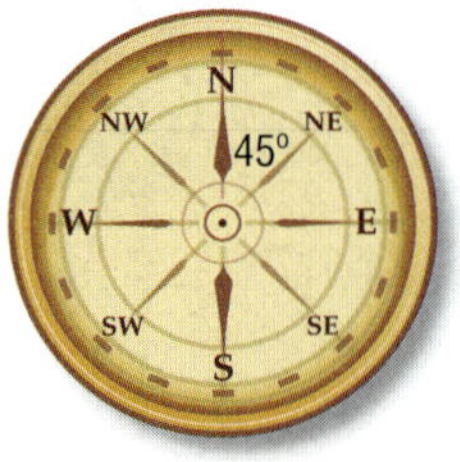

Practice and Problem Solving

● = **Step-by-Step Solutions** begin on page R20.
Extra Practice begins on page 969.

Example 1 p. 915

Point X is chosen at random on $\overline{FK}$. Find the probability of each event.

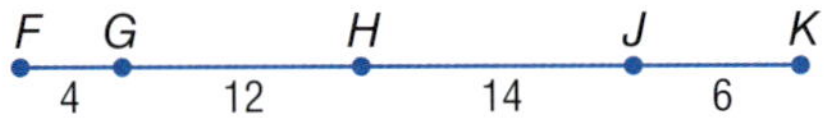

6. $P(X \text{ is on } \overline{FH})$
7. $P(X \text{ is on } \overline{GJ})$
8. $P(X \text{ is on } \overline{HK})$
9. $P(X \text{ is on } \overline{FG})$

10. **BIRDS** Four birds are sitting on a telephone wire. What is the probability that a fifth bird landing at a randomly selected point on the wire will sit at some point between birds 3 and 4?

Example 2 p. 916

11. **TELEVISION** Julio is watching television and sees an ad for a CD that he knows his friend wants for her birthday. If the ad replays at a random time in each 3-hour interval, what is the probability that he will see the ad again during his favorite 30-minute sitcom the next day?

Example 3 p. 917

Find the probability that a point chosen at random lies in the shaded region.

12.

13.

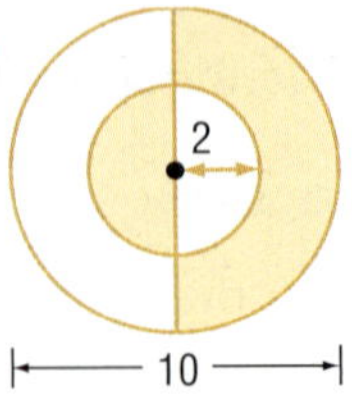

14.

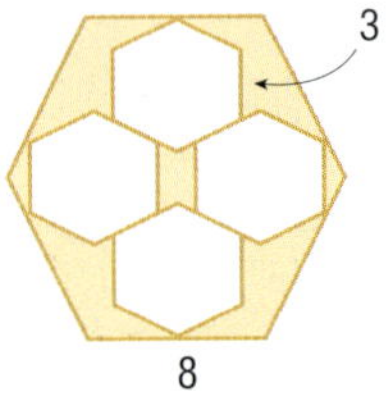

Example 4
p. 917

Use the spinner to find each probability. If the spinner lands on a line it is spun again.

15. P(pointer landing on yellow)

16. P(pointer landing on blue)

17. P(pointer not landing on green)

18. P(pointer landing on red)

19. P(pointer landing on neither red nor yellow)

Describe an event with a 33% probability for each model.

20.

21.

22.

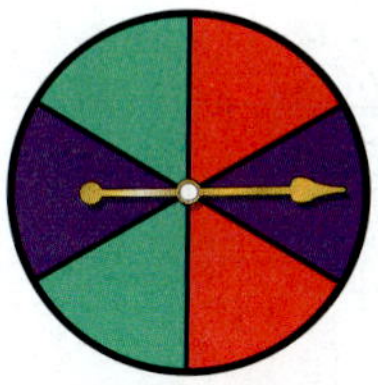

Real-World Link

Most traffic signals are red on top, yellow in the center, and green on the bottom so that people who are color blind can tell the color of the light based on position.

Source: Toledo-Bend

Find the probability that a point chosen at random lies in the shaded region.

23.

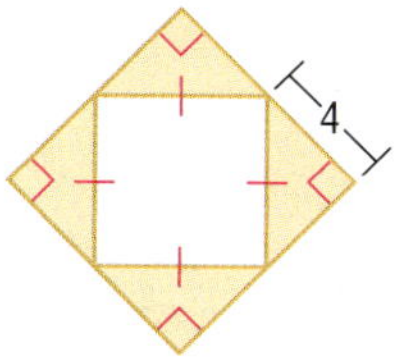

24.

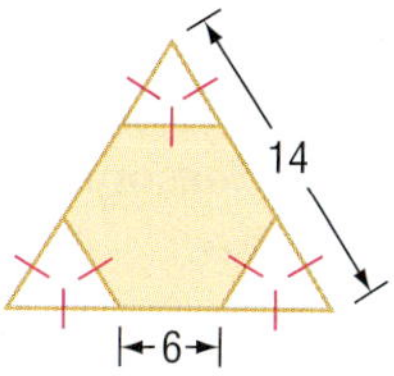

25.

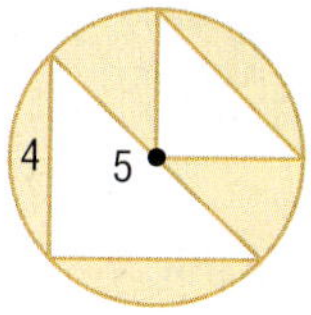

26. FARMING The layout for a farm is shown with each square representing a plot. Estimate the area of each field to answer each question.

a. What is the approximate combined area of the spinach and corn fields?

b. Find the probability that a randomly chosen plot is used to grow soybeans.

27. ALGEBRA Prove that the probability that a randomly chosen point in the circle will lie in the shaded region is equal to $\frac{x}{360}$.

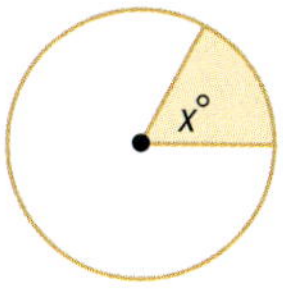

28. COORDINATE GEOMETRY If a point is chosen at random in the coordinate grid shown at the right, find each probability. Round to the nearest hundredth.

a. P(point inside the circle)

b. P(point inside the trapezoid)

c. P(point inside the trapezoid, square, or circle)

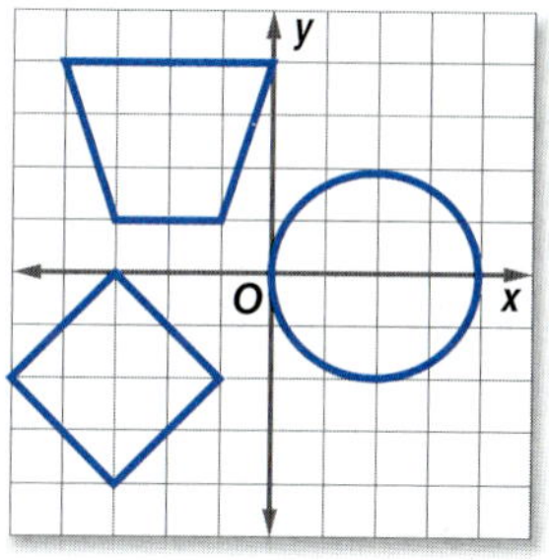

Find the probability that a point chosen at random lies in a shaded region.

29.

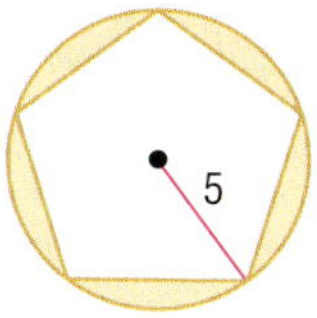

30.

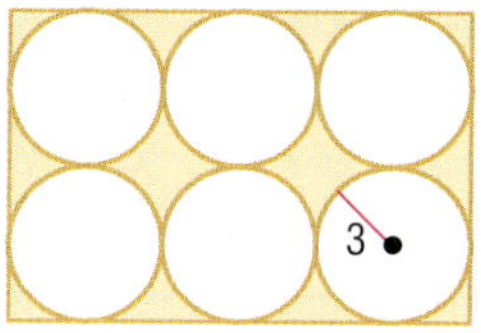

31.

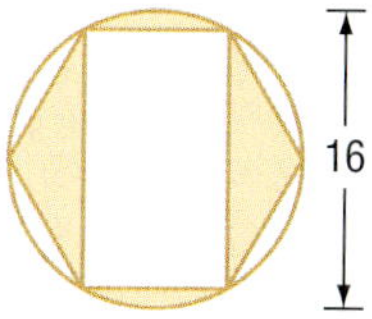

StudyTip

Systems of Inequalities Recall from Algebra 1 that a system of inequalities includes all of the points (x, y) that satisfy all of the inequalities in the system.

Real-World Link

The five "polar bear nations" in which polar bears are found include the U.S. (Alaska), Canada, Russia, Denmark (Greenland), and Norway. Polar bears do not live in the southern hemisphere.

Source: Polar Bears International

32. **COORDINATE GEOMETRY** Consider a system of inequalities, $1 \leq x \leq 6$, $y \leq x$, and $y \geq 1$. If a point (x, y) in the system is chosen at random, what is the probability that $(x - 1)^2 + (y - 1)^2 \geq 16$?

33. **VOLUME** The polar bear exhibit at a local zoo has a pool with the side profile shown. If the pool is 20 feet wide, what is the probability that a bear that is equally likely to swim anywhere in the pool will be in the incline region?

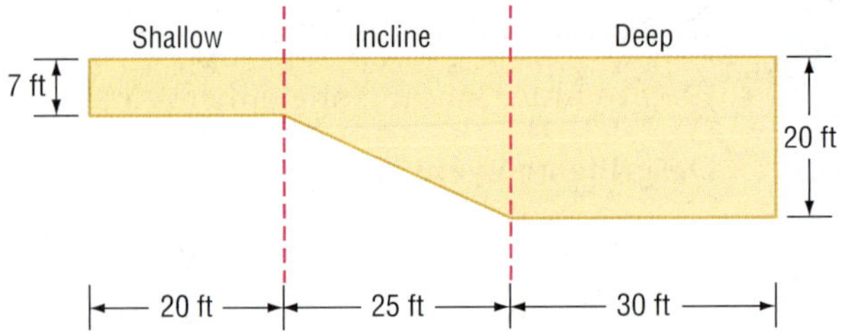

H.O.T. Problems Use Higher-Order Thinking Skills

34. **ERROR ANALYSIS** Carter and Meleah want to find the probability that a randomly chosen point within the circle will lie in the shaded region. Is either of them correct? Explain your reasoning.

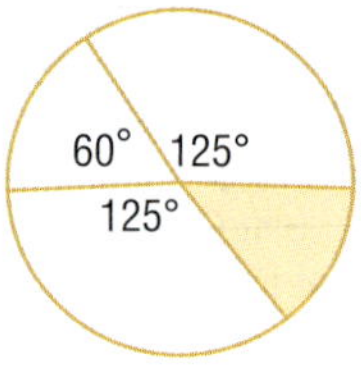

Carter

$p = \frac{\text{shaded sector}}{\text{total}}$
$= \frac{50}{360}$
$\approx 13.9\%$

Meleah

$p = \frac{\text{shaded sector}}{\text{total}}$
$= \frac{60}{360}$
$\approx 16.7\%$

35. **CHALLENGE** Find the probability that a point chosen at random would lie in the shaded area of the figure. Round to the nearest tenth of a percent.

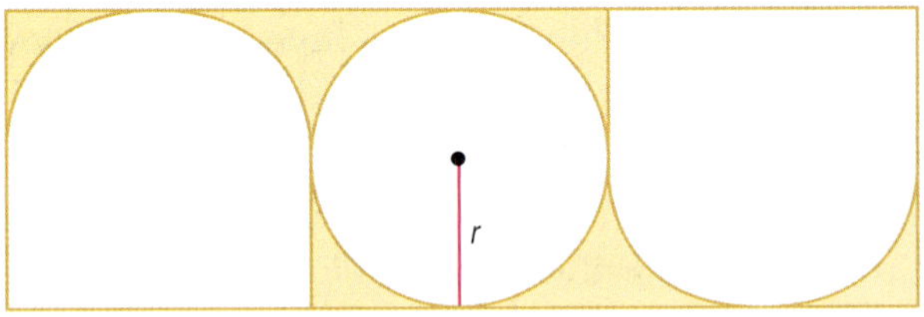

36. **REASONING** An isosceles triangle has a perimeter of 32 centimeters. If the lengths of the sides of the triangle are integers, what is the probability that the area of the triangle is exactly 48 square centimeters? Explain.

37. **REASONING** Explain why an athletic event, such as throwing a pitch, tossing a serve, or shooting a basket, is not a random event in the same sense as tossing a coin or drawing a card from a deck.

38. **OPEN ENDED** Represent a probability of 20% using three different geometric figures.

39. **WRITING IN MATH** Explain why the probability of a randomly chosen point falling in the shaded region of either of the squares shown is the same.

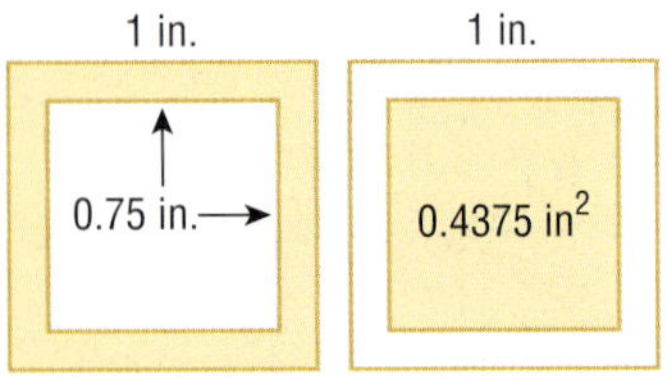

Standardized Test Practice

40. PROBABILITY A circle with radius 3 is contained in a square with side length 9. What is the probability that a randomly chosen point in the interior of the square will also lie in the interior of the circle?

A $\frac{1}{9}$ C $\frac{1}{3}$

B $\frac{\pi}{9}$ D $\frac{9}{\pi}$

41. ALGEBRA The area of Miki's room is $x^2 + 8x + 12$ square feet. A gallon of paint will cover an area of $x^2 + 6x + 8$ square feet. Which expression gives the number of gallons of paint that Miki will need to buy to paint her room?

F $\frac{x+6}{x+4}$ H $\frac{x+4}{x+6}$

G $\frac{x-4}{x-6}$ J $\frac{x-4}{x+6}$

42. EXTENDED RESPONSE The spinner is divided into 8 equal sections.

a. If the arrow lands on a number, what is the probability that it will land on 3?

b. If the arrow lands on a number, what is the probability that it will land on an odd number?

43. SAT/ACT A box contains 7 blue marbles, 6 red marbles, 2 white marbles, and 3 black marbles. If one marble is chosen at random, what is the probability that it will be red?

A 0.11 C 0.33

B 0.17 D 0.39

Spiral Review

44. PROM Four friends are sitting at a table together at the prom. What is the probability that a particular one of them will sit in the chair closest to the dance floor? (Lesson 13-2)

Represent the sample space for each experiment by making an organized list, a table, and a tree diagram. (Lesson 13-1)

45. Tito has a choice of taking music lessons for the next two years and playing drums or guitar.

46. Denise can buy a pair of shoes in either flats or heels in black or navy blue.

STAINED GLASS In the stained glass window design, all of the small arcs around the circle are congruent. Suppose the center of the circle is point *O*. (Lesson 10-4)

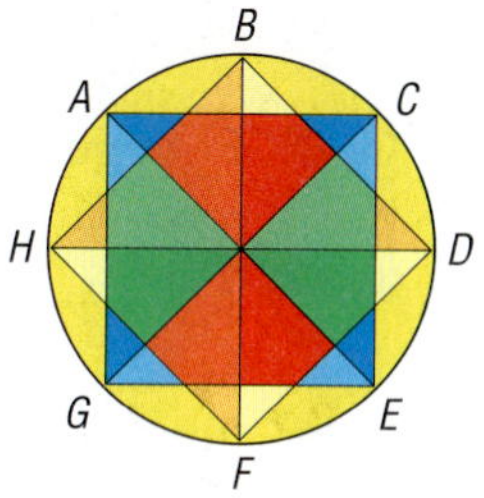

47. What is the measure of each of the small arcs?

48. What kind of figure is $\triangle AOC$? Explain.

49. What kind of figure is quadrilateral *BDFH*? Explain.

50. What kind of figure is quadrilateral *ACEG*? Explain.

Skills Review

Find the area of the shaded region. Round to the nearest tenth. (Lesson 11-3)

51.

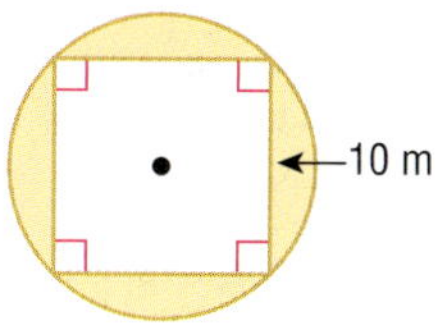

52.

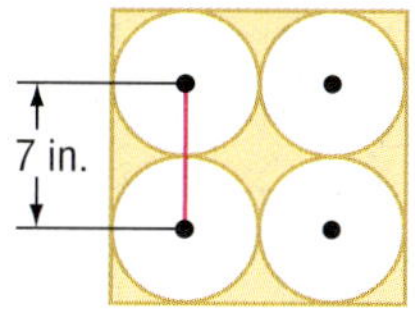

53.

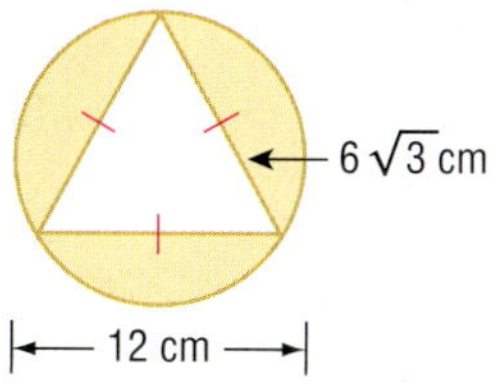

CHAPTER 13 Mid-Chapter Quiz

Lessons 13-1 through 13-3

1. **LUNCH** A deli has a lunch special, which consists of a sandwich, soup, dessert, and a drink for $4.99. The choices are in the table below. (Lesson 13-1)

Sandwich	Soup	Dessert	Drink
chicken salad	tomato	cookie	tea
ham	chicken noodle	pie	coffee
tuna	vegetable		cola
roast beef			diet cola
			milk

   a. How many different lunches can be created from the items shown in the table?

   b. If a soup and two desserts were added, how many different lunches could be created?

2. **FLAGS** How many different signals can be made with 5 flags from 8 flags of different colors? (Lesson 13-1)

3. **CLOTHING** Marcy has six colors of shirts: red, blue, yellow, green, pink, and orange. She has each color in short-sleeved and long-sleeved styles. Represent the sample space for Marcy's shirt choices by making an organized list, a table, and a tree diagram. (Lesson 13-1)

4. **SPELLING** A bag contains one tile for each letter of the word TRAINS. If you selected a permutation of these letters at random, what is the probability that they would spell TRAINS? (Lesson 13-2)

5. **CHANGE** Augusto has 3 pockets and 4 coins. In how many ways can he put the coins in his pockets? (Lesson 13-2)

6. **COINS** Ten coins are tossed simultaneously. In how many of the outcomes will the third coin turn up a head? (Lesson 13-2)

7. Find the probability that a point chosen at random lies in the shaded region. (Lesson 13-3)

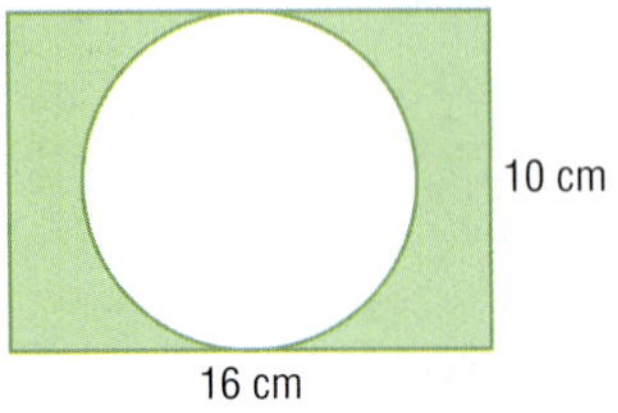

8. **CIRCUS** A 320 meter long tightrope is suspended between two poles. Assume that the line has an equal chance of breaking anywhere along its length. (Lesson 13-3)

   a. Determine the probability that a break will occur in the first 50 meters of the tightrope.

   b. Determine the probability that the break will occur within 20 meters of a pole.

Point A is chosen at random on $\overline{BE}$. Find the probability of each event. (Lesson 13-3)

9. $P(A \text{ is on } \overline{CD})$

10. $P(A \text{ is on } \overline{BD})$

11. $P(A \text{ is on } \overline{CE})$

12. $P(A \text{ is on } \overline{DE})$

Use the spinner to find each probability. If the spinner lands on a line, it is spun again. (Lesson 13-3)

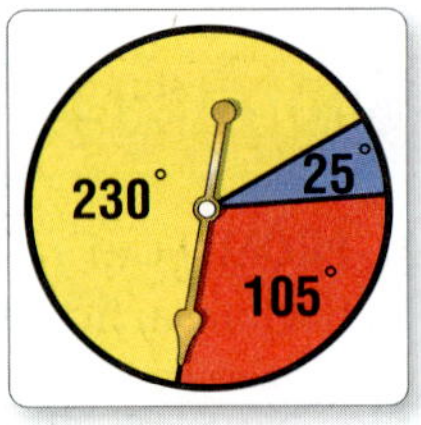

13. P(pointer landing on yellow)

14. P(pointer landing on blue)

15. P(pointer landing on red)

16. **GAMES** At a carnival, the object of a game is to throw a dart at the board and hit region III. (Lesson 13-3)

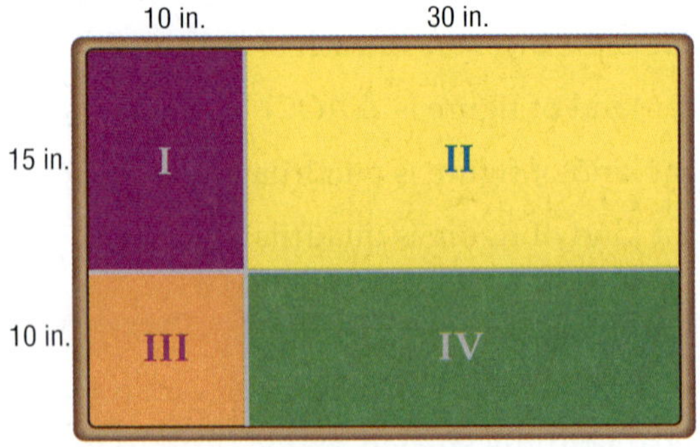

   a. What is the probability that it hits region I?

   b. What is the probability that it hits region II?

   c. What is the probability that it hits region III?

   d. What is the probability that it hits region IV?

13-4 Simulations

Why?

Based on practice, Allen knows that he makes 70% of his free throws. He wants to use this information to predict the number of free throws he is likely to make in games.

Then
You found probabilities by using geometric measures. (Lesson 13-3)

Now
- Design simulations to estimate probabilities.
- Summarize data from simulations.

New Vocabulary
probability model
simulation
random variable
expected value
Law of Large Numbers

Math Online
glencoe.com

Design a Simulation A **probability model** is a mathematical model used to match a random phenomenon. A **simulation** is the use of a probability model to recreate a situation again and again so that the likelihood of various outcomes can be estimated. To design a simulation, use the following steps.

Key Concept — Designing a Simulation

Step 1 Determine each possible outcome and its theoretical probability.

Step 2 State any assumptions.

Step 3 Describe an appropriate probability model for the situation.

Step 4 Define what a trial is for the situation and state the number of trials to be conducted.

An appropriate probability model has the same probabilities as the situation you are trying to predict. Geometric models are common probability models.

EXAMPLE 1 Design a Simulation by Using a Geometric Model

BASKETBALL **Allen made 70% of his free throws last season. Design a simulation that can be used to estimate the probability that he will make his next free throw this season.**

Step 1

Possible Outcomes		Theoretical Probability
• Allen makes a free throw.	→	70%
• Allen misses a free throw.	→	(100 − 70)% or 30%

Step 2 Our simulation will consist of 40 trials.

Step 3 One device that could be used is a spinner divided into two sectors, one containing 70% of the spinner's area and the other 30%. To create such a spinner, find the measure of the central angle of each sector.

Make Free Throw	Miss Free Throw
70% of 360° = 252°	30% of 360° = 108°

Step 4 A trial, one spin of the spinner, will represent shooting one free throw. A successful trial will be a made free throw and a failed trial will be a missed free throw. The simulation will consist of 40 trials.

Guided Practice

1. **RESTAURANTS** A restaurant attaches game pieces to its large drink cups, awarding a prize to anyone who collects all 6 game pieces. Design a simulation using a geometric model that can be used to estimate how many large drinks a person needs to buy to collect all 6 game pieces.

Personal Tutor glencoe.com

Problem-SolvingTip

Use a Simulation Simulations often provide a safe and efficient problem-solving strategy in situations that otherwise may be costly, dangerous, or impossible to solve using theoretical techniques. Simulations should involve data that are easier to obtain than the actual data you are modeling.

In addition to geometric models, simulations can also be conducted using dice, coin tosses, random number tables, and random number generators, such as those available on graphing calculators.

EXAMPLE 2 Design a Simulation by Using Random Numbers

EYE COLOR A survey of East High School students found that 40% had brown eyes, 30% had hazel eyes, 20% had blue eyes, and 10% had green eyes. Design a simulation that can be used to estimate the probability that a randomly chosen East High student will have one of these eye colors.

Step 1

Possible Outcomes		Theoretical Probability
Brown eyes	→	40%
Hazel eyes	→	30%
Blue eyes	→	20%
Green eyes	→	10%

Step 2 We assume that a student's eye color will fall into one of these four categories.

Step 3 Use the random number generator on your calculator. Assign the ten integers 0–9 to accurately represent the probability data. The actual numbers chosen to represent the outcomes do not matter.

Outcome	Represented by
Brown eyes	0, 1, 2, 3
Hazel eyes	4, 5, 6
Blue eyes	7, 8
Green eyes	9

Step 4 A trial will represent selecting a student at random and recording his or her eye color. The simulation will consist of 20 trials.

StudyTip

Random Number Generator To generate a set of random integers on a graphing calculator, press MATH and select randInt(under the PRB menu. Then enter the beginning and ending integer values for your range and the number of integers you want in each trial.

Guided Practice

2. **SOCCER** Last season, Yao made 18% of his free kicks. Design a simulation using a random number generator that can be used to estimate the probability that he will make his next free kick.

Personal Tutor glencoe.com

Summarize Data from a Simulation After designing a simulation, you will need to conduct the simulation and report the results. Include both numerical and graphical summaries of the simulation data, as well as an estimate of the probability of the desired outcome.

Real-World Link

Mark Price holds the record for the highest career free-throw percentage in the NBA at 90.4%.

Source: National Basketball Association

EXAMPLE 3 Conduct and Summarize Data from a Simulation

BASKETBALL Refer to the simulation in Example 1. Conduct the simulation and report the results using appropriate numerical and graphical summaries.

Make a frequency table and record the results after spinning the spinner 40 times.

Outcome	Tally	Frequency
Make Free Throw	𝍸 𝍸 𝍸 𝍸 𝍸 |	26
Miss Free Throw	𝍸 𝍸 ||||	14
Total		40

Based on the simulation data, calculate the probability that Allen will make his next free throw.

$$\frac{\text{number of made free throws}}{\text{number of free throws attempted}} = \frac{26}{40} \text{ or } 0.65$$ **This is an *experimental probability*.**

The probability that Allen makes his next free throw is 0.65 or 65%. Notice that this is close to the theoretical probability, 70%. So, the experimental probability of his missing the next free throw is 1 − 0.65 or 35%.

Make a bar graph of these results.

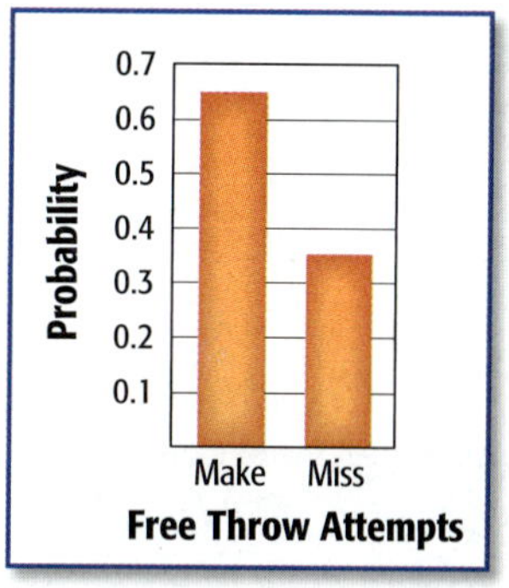

Guided Practice

3. EYE COLOR Use a graphing calculator to conduct the simulation in Example 2. Then report the results using appropriate numerical and graphical summaries.

Personal Tutor glencoe.com

A **random variable** is a variable that can assume a set of values, each with fixed probabilities. For example, in the experiment of rolling two dice, the random variable *X* can represent the sums of the potential outcomes on the dice. The table shows some of the *X*-values assigned to outcomes from this experiment.

Sum of Outcomes of Rolling Two Dice	
Outcome	***X*-Value**
(1, 1)	2
(1, 2)	3
(2, 1)	3
(4, 5)	9
(6, 6)	12

Expected value, also known as mathematical expectation, is the average value of a random variable that one *expects* after repeating an experiment or simulation a theoretically infinite number of times. To find the expected value $E(X)$ of a random variable *X*, follow these steps.

Review Vocabulary

experimental probability what *actually* occurs when a probability experiment is conducted

theoretical probability what *should* occur when a probability experiment is conducted (Lesson 0-3)

Key Concept Calculating Expected Value

Step 1 Multiply the value of *X* by its probability of occurring.

Step 2 Repeat Step 1 for all possible values of *X*.

Step 3 Find the sum of the results.

Since it is an average, an expected value does not have to be equal to a possible value of the random variable.

EXAMPLE 4 Calculate Expected Value

DARTS **Suppose a dart is thrown at the dartboard. The radius of the center circle is 1 centimeter and each successive circle has a radius 4 centimeters greater than the previous circle. The point value for each region is shown.**

a. Let the random variable Y represent the point value assigned to a region on the dartboard. Calculate the expected value $E(Y)$ from each throw.

First calculate the geometric probability of landing in each region.

$$\text{Region } 5 = \frac{\pi(1)^2}{\pi(1+4+4+4+4)^2} = \frac{1}{289} \qquad \text{Region } 4 = \frac{\pi(4+1)^2 - \pi(1)^2}{\pi(17)^2} = \frac{24}{289}$$

$$\text{Region } 3 = \frac{\pi(4+5)^2 - \pi(5)^2}{\pi(17)^2} = \frac{56}{289} \qquad \text{Region } 2 = \frac{\pi(4+9)^2 - \pi(9)^2}{\pi(17)^2} = \frac{88}{289}$$

$$\text{Region } 1 = \frac{\pi(4+13)^2 - \pi(13)^2}{\pi(17)^2} = \frac{120}{289}$$

$$E(Y) = 1 \cdot \frac{120}{289} + 2 \cdot \frac{88}{289} + 3 \cdot \frac{56}{289} + 4 \cdot \frac{24}{289} + 5 \cdot \frac{1}{289} \text{ or about } 1.96$$

The expected value of each throw is about 1.96.

b. Design a simulation to estimate the average value, or the average of the results of your simulation, of this game. How does this value compare with the expected value you found in part a?

Assign the integers 0–289 to accurately represent the probability data.

Region 1 = integers 1–120
Region 2 = integers 121–208
Region 3 = integers 209–264
Region 4 = integers 265–288
Region 5 = integer 289

Use a graphing calculator to generate 50 trials of random integers from 1 to 289. Record the results in a frequency table. Then calculate the average value of the outcomes.

Outcome	Frequency
Region 1	16
Region 2	13
Region 3	13
Region 4	8
Region 5	0

$$\text{average value} = 1 \cdot \frac{16}{50} + 2 \cdot \frac{13}{50} + 3 \cdot \frac{13}{50} + 4 \cdot \frac{8}{50} + 5 \cdot \frac{0}{50} = 2.26$$

The average value 2.26 is greater than the expected value 1.96.

Guided Practice

4. DICE If two dice are rolled, let the random variable X represent the sum of the potential outcomes.

A. Find the expected value $E(X)$.

B. Design and run a simulation to estimate the average value of this experiment. How does this value compare with the expected value you found in part A?

Personal Tutor glencoe.com

The difference in the average value from the simulation and the expected value in Example 4 illustrates the **Law of Large Numbers**: as the number of trials of a random process increases, the average value will approach the expected value.

StudyTip

Geometric Probability Remember that when determining geometric probabilities with targets, we assume that the object lands within the target area, and that it is equally likely that the object will land anywhere in the region.

The Granger Collection, New York

Math History Link

Jakob Bernoulli (1654–1705)

Bernoulli was a Swiss mathematician. It seemed obvious to him that the more observations made of a given situation, the better one would be able to predict future outcomes. He provided scientific proof of his Law of Large Numbers in his work *Ars Conjectandi* (Art of Conjecturing), published in 1713.

Check Your Understanding

Examples 1 and 3
pp. 923, 925

1. GRADES Clara got an A on 80% of her first semester Biology quizzes. Design and conduct a simulation using a geometric model to estimate the probability that she will get an A on a second semester Biology quiz. Report the results using appropriate numerical and graphical summaries.

Examples 2 and 3
pp. 924–925

2. FITNESS The table shows the percent of members participating in four classes offered at a gym. Design and conduct a simulation to estimate the probability that a new gym member will take each class. Report the results using appropriate numerical and graphical summaries.

Class	Sign-Up %
tae kwon do	45%
yoga	30%
swimming	15%
kick-boxing	10%

Example 4
p. 926

3. CARNIVAL GAMES The object of the game shown is to accumulate points by using a dart to pop the balloons. Assume that each dart will hit a balloon.

a. Calculate the expected value from each throw.

b. Design a simulation and estimate the average value of this game.

c. How do the expected value and average value compare?

Practice and Problem Solving

= **Step-by-Step Solutions** begin on page R20.
Extra Practice begins on page 969.

Examples 1 and 3
pp. 923, 925

Design and conduct a simulation using a geometric probability model. Then report the results using appropriate numerical and graphical summaries.

4. BOWLING Bridget is a member of the bowling club at her school. Last season she bowled a strike 60% of the time.

5. VIDEO GAMES Ian works at a video game store. Last year he sold 95% of the new-release video games.

6. MUSIC Kadisha is listening to a CD with her CD player set on the random mode. There are 10 songs on the CD.

7. BOARD GAMES Pilar is playing a board game with eight different categories, each with questions that must be answered correctly in order to win.

Examples 2 and 3
pp. 924–925

Design and conduct a simulation using a random number generator. Then report the results using appropriate numerical and graphical summaries.

8. MOVIES A movie theater reviewed sales from the previous year to determine which genre of movie sold the most tickets. The results are shown at the right.

Genre	Ticket %
Drama	40%
Mystery	30%
Comedy	25%
Action	5%

9. BASEBALL According to a baseball player's on-base percentages, he gets a single 60% of the time, a double 25% of the time, a triple 10% of the time, and a home run 5% of the time.

10. VACATION According to a survey done by a travel agency, 45% of their clients went on vacation to Europe, 25% went to Asia, 15% went to South America, 10% went to Africa, and 5% went to Australia.

Example 4
p. 926

11. TRANSPORTATION A car dealership's analysis indicated that 35% of the customers purchased a blue car, 30% purchased a red car, 15% purchased a white car, 15% purchased a black car, and 5% purchased any other color.

DARTBOARDS The dimensions of each dartboard below are given in inches. There is only one shot per game. Calculate the expected value of each dart game. Then design a simulation to estimate each game's average value. Compare the average and expected values. In each figure, ■ = 25, □ = 50, and ■ = 100 points.

12.

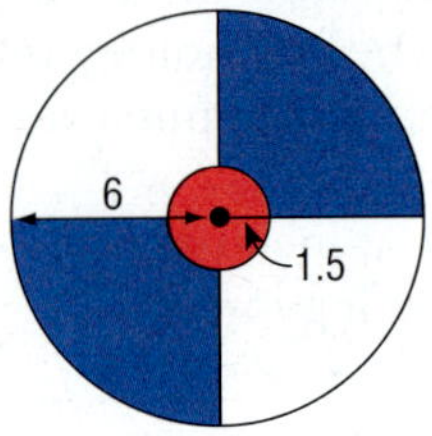

13.

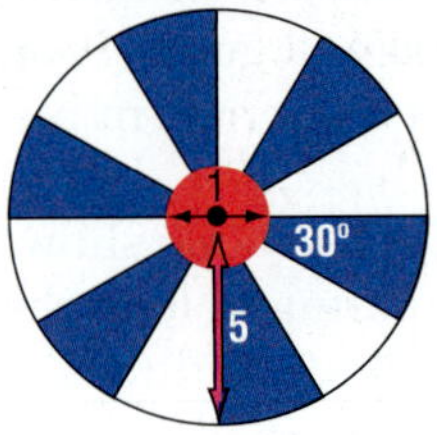

14.

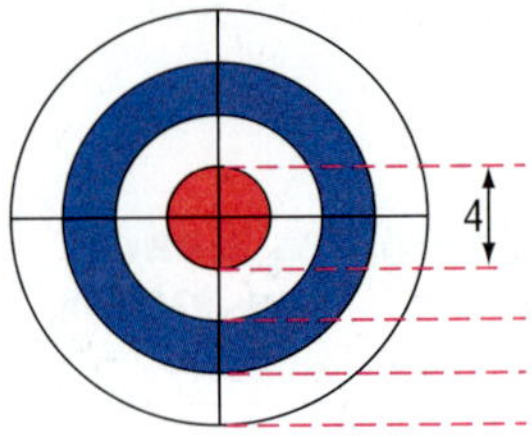

15. **CARDS** You are playing a team card game where a team can get 0 points, 1 point, or 3 points for a hand. The probability of your team getting 1 point for a hand is 60% and of getting 3 points for a hand is 5%.

a. Calculate your team's expected value for a hand.

b. Design a simulation and estimate your team's average value per hand.

c. Compare the values for parts **a** and **b**.

Real-World Career

Statisticians
Statisticians collect statistical data for various subject areas, including sports and games. They use computer software to analyze, interpret, and summarize the data. Most statisticians have a master's degree.

16. **CARNIVAL GAMES** The object of the game shown is to score points by rolling a ball up an incline into regions with different values. The probability that Susana will get 100 points in a roll is 20%, 200 points is 55%, 300 points is 20%, and 400 points is 5%.

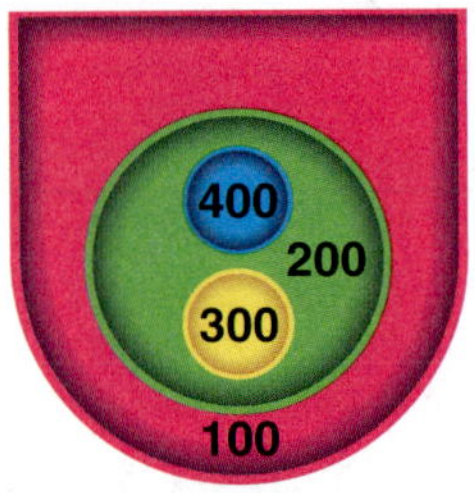

a. Calculate the expected value for each roll.

b. Design a simulation and estimate Susana's average value for this game.

c. Compare the values for parts **a** and **b**.

17. **BASEBALL** A baseball pitcher wants to track which areas of the strike zone his pitches have a higher probability of reaching. He divides the strike zone into six boxes as shown.

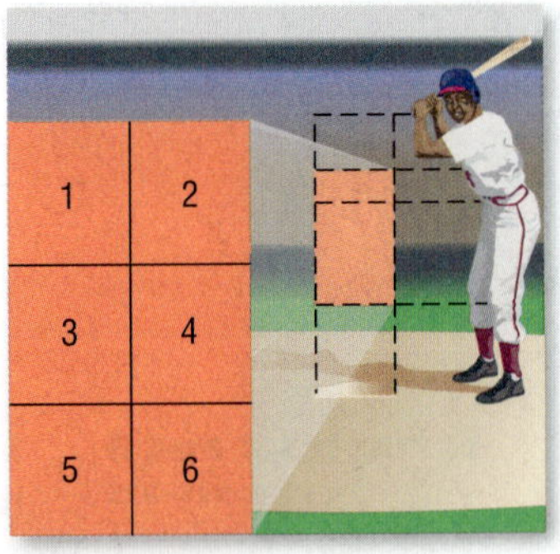

a. If the pitches are equally likely to hit each box, what is the probability that he will throw a strike in each box?

b. Design a simulation to estimate the probability of throwing a strike in each box.

c. Compare the values for parts **a** and **b**.

18. **BASKETBALL** Cynthia used her statistics from last season to design a simulation using a random number generator to predict what would happen the first time she got possession of the ball.

Integer Values	Points Scored	Frequency
1–14	0	31
15	1	0
16–28	2	17
29–30	3	2

a. Based on the frequency table, what did she assume was the theoretical probability that she would score two points in a possession?

b. What is Cynthia's average value for a possession? her expected value?

c. Would you expect the simulated data to be different? If so, explain how. If not, explain why.

19. **MULTIPLE REPRESENTATIONS** In this problem, you will investigate expected value.

a. **CONCRETE** Roll two dice 20 times and record the sum of each roll.

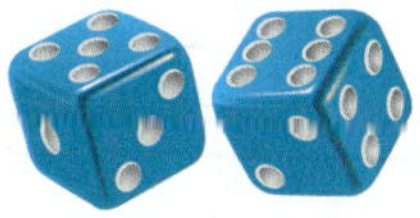

b. **NUMERICAL** Use the random number generator on a calculator to generate 20 pairs of integers between 1 and 6. Record the sum of each pair.

c. **TABULAR** Copy and complete the table below using your results from parts **a** and **b**.

Trial	Sum of Die Roll	Sum of Output from Random Number Generator
1		
2		
...		
20		

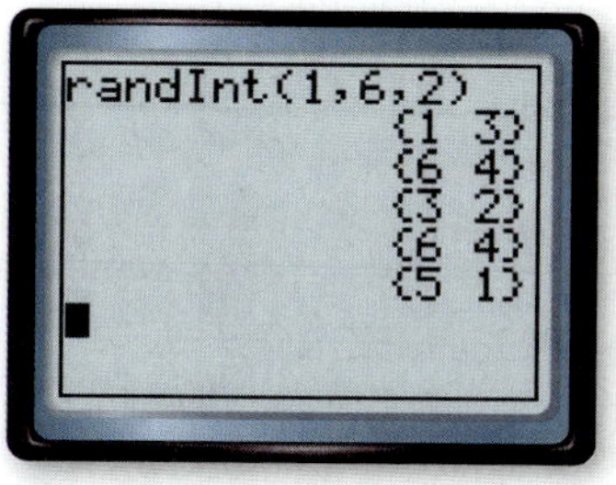

d. **GRAPHICAL** Use a bar graph to graph the number of times each possible sum occurred in the first 5 rolls. Repeat the process for the first 10 rolls and then all 20 outcomes.

e. **VERBAL** How does the shape of the bar graph change with each additional trial?

f. **GRAPHICAL** Graph the number of times each possible sum occurred with the random number generator as a bar graph.

g. **VERBAL** How do the graphs of the die trial and the random number trial compare?

h. **ANALYTICAL** Based on the graphs, what do you think the expected value of each experiment would be? Explain your reasoning.

Real-World Link

Computer-generated random numbers are actually called *pseudorandom numbers* because they are chosen from the same group of numbers or *seed*. Pseudorandom number generators are useful for applications such as programming movement in video games or a system that randomly selects music tracks.

H.O.T. Problems

Use Higher-Order Thinking Skills

20. **REASONING** An experiment has three equally likely outcomes *A*, *B*, and *C*. Is it possible to use the spinner shown in a simulation to predict the probability of outcome *C*? Explain your reasoning.

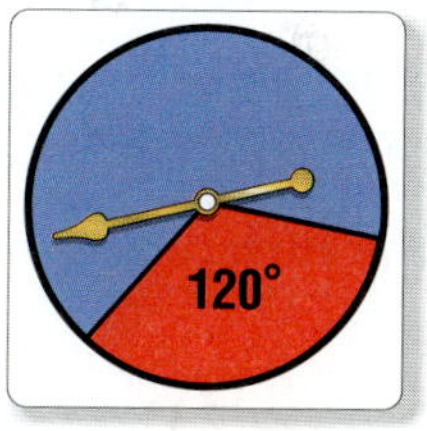

21. **REASONING** Can tossing a coin *sometimes*, *always*, or *never* be used to simulate an experiment with two possible outcomes? Explain.

22. **CHALLENGE** In an experiment, five coins are flipped simultaneously.

a. Design a simulation that can be used to predict the probability that exactly 3 of the coins will be tails.

b. Could the same simulation be used to predict the probability that at least 3 of the coins will be tails? Explain your reasoning.

c. Run your simulation. What is the experimental probability that exactly three of the coins will be tails?

23. **REASONING** When designing a simulation where darts are thrown at targets, what assumptions need to be made and why are they needed?

24. **OPEN ENDED** Describe an experiment in which the expected value is not a possible outcome. Explain.

25. **WRITING IN MATH** Summarize the process of designing and conducting a simulation.

Standardized Test Practice

26. PROBABILITY Kaya tosses three coins at the same time and repeats the process 9 more times. Her results are shown below where H represents heads and T represents tails. Based on Kaya's data, what is the probability that at least one of the group of 3 coins will land with heads up?

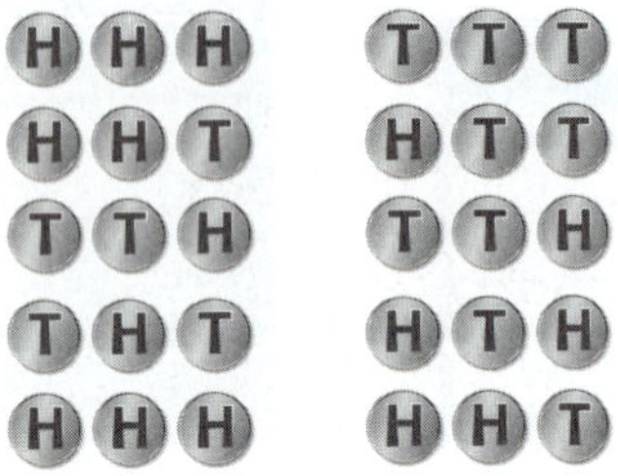

A 0.1 **B** 0.2 **C** 0.3 **D** 0.9

27. ALGEBRA Paul collects comic books. He has 20 books in his collection, and he adds 3 per month. In how many months will he have a total of 44 books in his collection?

F 5 **G** 6 **H** 8 **J** 15

28. SHORT RESPONSE Alberto designed a simulation to determine how many times a player would roll a number higher than 4 on a die in a board game with 5 rolls. The table below shows his results for 50 trials. What is the probability that a player will roll a number higher than 4 two or more times in 5 rolls?

Number of Rolls Greater Than 4	Frequency
0	8
1	15
2	18
3	9
4	0
5	0

29. SAT/ACT If a jar contains 150 peanuts and 60 cashews, what is the probability that a nut selected from the jar at random will be a cashew?

A 0.25 **B** 0.29 **C** 0.4 **D** 0.71

Spiral Review

Point X is chosen at random on $\overline{QT}$. Find the probability of each event. (Lesson 13-3)

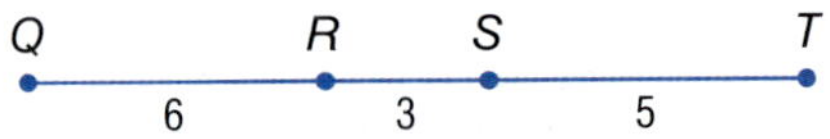

30. $P(X \text{ is on } \overline{QS})$

31. $P(X \text{ is on } \overline{RT})$

32. BOOKS Paige is choosing between 10 books at the library. What is the probability that she chooses 3 particular books to check out from the 10 initial books? (Lesson 13-2)

Find the surface area of each figure. Round to the nearest tenth. (Lesson 12-6)

33.

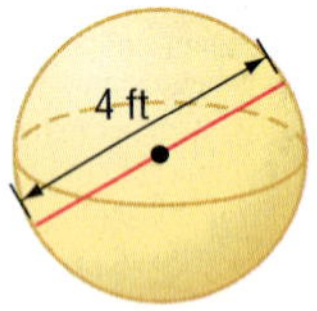

34.

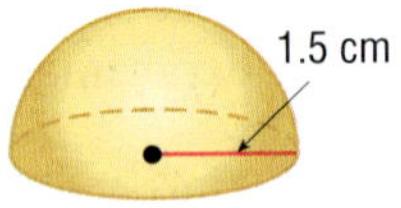

35.

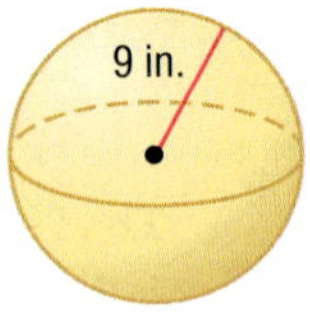

Skills Review

36. RECREATION A group of 150 students was asked what they like to do during their free time. (Lesson 2-2)

a. How many students like going to the movies or shopping?

b. Which activity was mentioned by 37 students?

c. How many students did *not* say they like movies?

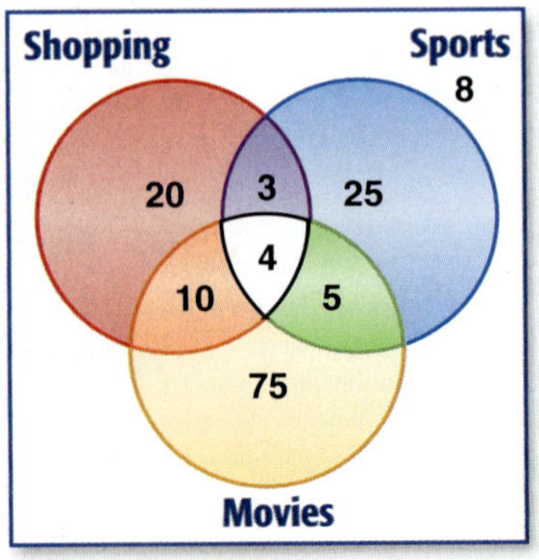

13-5 Probabilities of Independent and Dependent Events

Then

You found simple probabilities. (Lesson 0-3)

Now

- Find probabilities of independent and dependent events.
- Find probabilities of events given the occurrence of other events.

New Vocabulary

compound event
independent events
dependent events
conditional probability
probability tree

Math Online

glencoe.com

Why?

The 18 students in Mrs. Turner's chemistry class are drawing names to determine who will give his or her presentation first. James is hoping to be chosen first and his friend Arturo wants to be second.

Independent and Dependent Events

A **compound event** or *composite event* consists of two or more simple events. In the example above, James and Arturo being chosen to give their presentations first is a compound event. It consists of the event that Michael is chosen and the event that Arturo is chosen.

Compound events can be independent or dependent.

- Events A and B are **independent events** if the probability that A occurs does not affect the probability that B occurs.
- Events A and B are **dependent events** if the probability that A occurs in some way changes the probability that B occurs.

Consider choosing objects from a group of objects. If you replace the object each time, choosing additional objects are independent events. If you do not replace the object each time, choosing additional objects are dependent events.

EXAMPLE 1 Identify Independent and Dependent Events

Determine whether the events are *independent* or *dependent*. Explain your reasoning.

a. One coin is tossed, and then a second coin is tossed.

The outcome of the first coin toss in no way changes the probability of the outcome of the second coin toss. Therefore, these two events are *independent.*

b. In the class presentation example above, one student's name is chosen and not replaced, and then a second name is chosen.

After the first person is chosen, his or her name is removed and cannot be selected again. This affects the probability of the second person being chosen, since the sample space is reduced by one name. Therefore, these two events are *dependent.*

c. Wednesday's lottery numbers and Saturday's lottery numbers.

The numbers for one drawing have no bearing on the next drawing. Therefore, these two events are *independent*.

Guided Practice

1A. A card is selected from a deck of cards and put back. Then a second card is selected.

1B. Andrea selects a shirt from her closet to wear on Monday and then a different shirt to wear on Tuesday.

Personal Tutor glencoe.com

Suppose a coin is tossed and the spinner shown is spun. The sample space for this experiment is {(H, B), (H, R), (H, G), (T, B), (T, R), (T, G)}.

Using the sample space, the probability of the compound event of the coin landing on heads and the spinner on green is $P(\text{H and G}) = \frac{1}{6}$.

Notice that this same probability can be found by multiplying the probabilities of each simple event.

$$P(\text{H}) = \frac{1}{2} \qquad P(\text{G}) = \frac{1}{3} \qquad P(\text{H and G}) = \frac{1}{2} \cdot \frac{1}{3} \text{ or } \frac{1}{6}$$

This example illustrates the first of two Multiplication Rules for Probability.

ReadingMath

and The word *and* is a key word indicating to multiply probabilities.

Key Concept

Probability of Two Independent Events

Words The probability that two independent events both occur is the product of the probabilities of each individual event.

Symbols If two events A and B are independent, then

$$P(A \text{ and } B) = P(A) \cdot P(B).$$

This rule can be extended to any number of events.

StudyTip

Use an Area Model You can also use the area model shown below to calculate the probability that both slips are blue. The blue region represents the probability of drawing two successive blue slips. The area of this region is $\frac{9}{64}$ of the entire model.

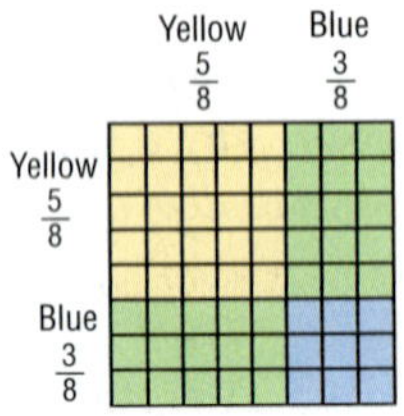

Real-World EXAMPLE 2 Probability of Independent Events

TRANSPORTATION Marisol and her friends are going to a concert. They put the slips of paper shown into a bag. If a person draws a yellow slip, he or she will ride in the van to the concert. A blue slip means he or she rides in the car.

Suppose Marisol draws a slip. Not liking the outcome, she puts it back and draws a second time. What is the probability that on each draw her slip is blue?

These events are independent since Marisol replaced the slip that she removed. Let B represent a blue slip and Y a yellow slip.

	Draw 1		Draw 2	
$P(B \text{ and } B) =$	$P(B)$	$\cdot$	$P(B)$	**Probability of independent events**
$=$	$\frac{3}{8}$	$\cdot$	$\frac{3}{8}$ or $\frac{9}{64}$	$P(B) = \frac{3}{8}$

So, the probability of Marisol drawing two blue slips is $\frac{9}{64}$ or about 14%.

Guided Practice

Find each probability.

2A. A coin is tossed and a die is rolled. What is the probability that the coin lands heads up and the number rolled is a 6?

2B. Suppose you toss a coin four times. What is the probability of getting four tails?

Personal Tutor glencoe.com

The second of the Multiplication Rules of Probability addresses the probability of two dependent events.

Key Concept — Probability of Two Dependent Events

Words The probability that two dependent events both occur is the product of the probability that the first event occurs and the probability that the second event occurs *after* the first event has already occurred.

Symbols If two events A and B are dependent, then

$$P(A \text{ and } B) = P(A) \cdot P(B|A).$$

This rule can be extended to any number of events.

Watch Out!

Conditional Notation The "|" symbol in the notation $P(B|A)$ should not be interpreted as a division symbol.

The notation $P(B|A)$ is read *the probability that event B occurs given that event A has already occurred.* This is called **conditional probability**.

EXAMPLE 3 Probability of Dependent Events

TRANSPORTATION Refer to Example 2. Suppose Marisol draws a slip and does not put it back. Then her friend Christian draws a slip. What is the probability that both friends draw a yellow slip?

These events are dependent since Marisol does not replace the slip that she removed.

$P(Y \text{ and } Y) = P(Y) \cdot P(Y|Y)$ — **Probability of dependent events**

$= \frac{5}{8} \cdot \frac{4}{7} \text{ or } \frac{5}{14}$ — **After the first yellow slip is chosen, 7 total slips remain, and 4 of those are yellow.**

So, the probability that both friends draw yellow slips is $\frac{5}{14}$ or about 36%.

CHECK You can use a tree diagram with probabilities, called a **probability tree**, to verify this result. Calculate the probability of each simple event at the first stage and each conditional probability at the second stage. Then multiply along each branch to find the probability of each outcome.

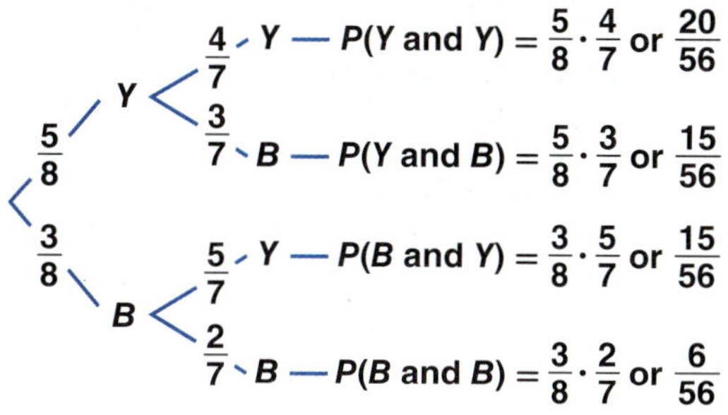

The sum of the probabilities should be 1.

$\frac{20}{56} + \frac{15}{56} + \frac{15}{56} + \frac{6}{56} = \frac{56}{56}$ or 1 ✓

Real-World Link

A recent study found that with three or more teenage passengers, 85% of fatal crashes of passenger vehicles driven by teens involved driver error, almost 50% involved speeding, and almost 70% involved a single vehicle.

Source: National Safety Council

Guided Practice

3. Three cards are selected from a standard deck of 52 cards. What is the probability that all three cards are diamonds if neither the first nor the second card is replaced?

Personal Tutor glencoe.com

Conditional Probabilities In addition to its use in finding the probability of two or more dependent events, conditional probability can be used when additional information is known about an event.

Suppose a die is rolled and it is known that the number rolled is odd. What is the probability that the number rolled is a 5?

There are only three odd numbers that can be rolled, so our sample space is reduced from {1, 2, 3, 4, 5, 6} to {1, 3, 5}. So, the probability that the number rolled is a 5 is $P(5|\text{odd}) = \frac{1}{3}$.

STANDARDIZED TEST EXAMPLE 4

Ms. Fuentes' class is holding a debate. The 8 students participating randomly draw cards numbered with consecutive integers from 1 to 8.

- **Students who draw odd numbers will be on the Proposition Team.**
- **Students who draw even numbers will be on the Opposition Team.**

If Jonathan is on the Opposition Team, what is the probability that he drew the number 2?

A $\frac{1}{8}$ **B** $\frac{1}{4}$ **C** $\frac{3}{8}$ **D** $\frac{1}{2}$

Read the Test Item

Since Jonathan is on the Opposition Team, he must have drawn an even number. So you need to find the probability that the number drawn was 2 given that the number drawn was even. This is a conditional probability problem.

Solve the Test Item

Let A be the event that an even number is drawn.
Let B be the event that the number 2 is drawn.

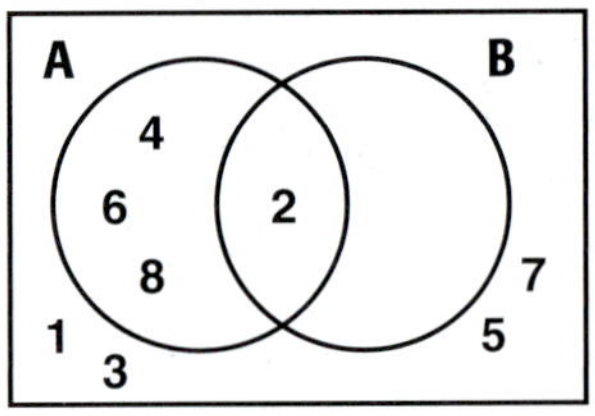

Draw a Venn diagram to represent this situation. There are only four even numbers in the sample space, and only one out of these numbers is a 2. Therefore, the $P(B|A) = \frac{1}{4}$. The answer is B.

Guided Practice

4. When two dice are rolled, what is the probability that one die is a 4, given that the sum of the two die is 9?

F $\frac{1}{6}$ **G** $\frac{1}{4}$ **H** $\frac{1}{3}$ **J** $\frac{1}{2}$

Personal Tutor glencoe.com

ReadingMath

Conditional Probability $P(5|\text{odd})$ is read *the probability that the number rolled is a 5 given that the number rolled is odd.*

Test-TakingTip

Use a Venn Diagram Use a Venn diagram to help you visualize the relationship between the outcomes of two events.

Since conditional probability reduces the sample space, the Venn diagram in Example 4 can be simplified as shown, with the intersection of the two events representing those outcomes in A and B. This suggests that $P(B|A) = \frac{P(A \text{ and } B)}{P(A)}$.

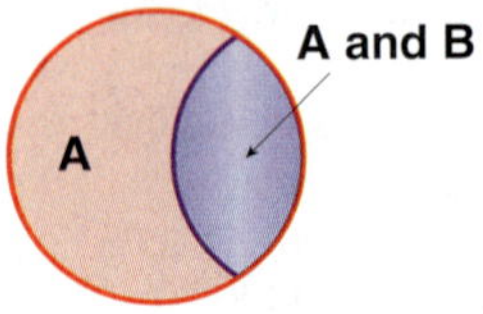

Key Concept Conditional Probability

The conditional probability of B given A is $P(B|A) = \frac{P(A \text{ and } B)}{P(A)}$, where $P(A) \neq 0$.

Check Your Understanding

Example 1
p. 931

Determine whether the events are *independent* or *dependent*. Explain.

1. Jeremy took the SAT on Saturday and scored 1350. The following week he took the ACT and scored 23.

2. Alita's basketball team is in the final four. If they win, they will play in the championship game.

Example 2
p. 932

3. **CARDS** A card is randomly chosen from a deck of 52 cards, replaced, and a second card is chosen. What is the probability of choosing both of the cards shown at the right?

Example 3
p. 933

4. **TRANSPORTATION** Isaiah is getting on the bus after work. It costs $0.50 to ride the bus to his house. If he has 3 quarters, 5 dimes, and 2 nickels in his pocket, find the probability that he will randomly pull out two quarters in a row. Assume that the events are equally likely to occur.

Example 4
p. 934

5. **GRIDDED RESPONSE** Every Saturday, 10 friends play dodgeball at a local park. To pick teams, they randomly draw cards with consecutive integers from 1 to 10. Odd numbers are on Team A, and even numbers are Team B. What is the probability that a player on Team B has drawn the number 10?

Practice and Problem Solving

● = **Step-by-Step Solutions** begin on page R20.
Extra Practice begins on page 969.

Examples 1–3
pp. 931–933

Determine whether the events are *independent* or *dependent*. Then find the probability.

6. In a game, you roll an even number on a die and then spin a spinner numbered 1 through 5 and get an odd number.

7. An ace is drawn, without replacement, from a deck of 52 cards. Then, a second ace is drawn.

8. In a bag of 3 green and 4 blue marbles, a blue marble is drawn and not replaced. Then, a second blue marble is drawn.

9. You roll two dice and get a 5 each time.

10. **GAMES** In a game, the spinner at the right is spun and a coin is tossed. What is the probability of getting an even number on the spinner and the coin landing on tails?

11. **GIFTS** Tisha's class is having a gift exchange. Tisha will draw first and her friend Brandi second. If there are 18 students participating, what is the probability that Brandi and Tisha draw each other's names?

Real-World Link

A recent study found that at the start of the summer, 60% of consumers had no plans to take a vacation over the next six months.

Source: *The New York Times*

12. **VACATION** A work survey found that 8 out of every 10 employees went on vacation last summer. If 3 employees' names are randomly chosen, with replacement, what is the probability that all 3 employees went on vacation last summer?

13. **CAMPAIGNS** The table shows the number of each color of Student Council campaign buttons Clemente has to give away. If given away at random, what is the probability that the first and second buttons given away are both red?

Button Color	Amount
blue	20
white	15
red	25
black	10

Example 4
p. 934

14. A red marble is selected at random from a bag of 2 blue and 9 red marbles and not replaced. What is the probability that a second marble selected will be red?

15. A die is rolled. If the number rolled is greater than 2, find the probability that it is a 6.

16. A quadrilateral has a perimeter of 12 and all of the side lengths are odd integers. What is the probability that the quadrilateral is a rhombus?

17. A spinner numbered 1 through 12 is spun. Find the probability that the number spun is an 11 given that the number spun was an odd number.

18. **CLASSES** The probability that a student takes geometry and French at Satomi's school is 0.064. The probability that a student takes French is 0.45. What is the probability that a student takes geometry if the student takes French?

19. **TECHNOLOGY** At Bell High School, 43% of the students own a CD player and 28% own a CD player and an MP3 player. What is the probability that a student owns an MP3 player if he or she also owns a CD player?

20. **PROOF** Use the formula for the probability of two dependent events $P(A \text{ and } B)$ to derive the conditional probability formula for $P(B|A)$.

21. **TENNIS** Refer to the information at the left. Kelly's first serve percentage is 40%, while her second serve percentage is 70%.
    a. Draw a probability tree that shows each outcome.
    b. What is the probability that Kelly will double fault?
    c. Design a simulation using a random number generator that can be used to estimate the probability that Kelly double faults on her next serve.

Real-World Link

A double fault in tennis is when the serving player fails to land their serve "in" without stepping on or over the service line in two chances.

22. **VACATION** A random survey was conducted to determine where families vacationed. The results indicated that $P(B) = 0.6$, $P(B \cap M) = 0.2$, and the probability that a family did not vacation at either destination is 0.1.

    a. What is the probability that a family vacations in the mountains?
    b. What is the probability that a family that has already visited the beach will also visit the mountains?

H.O.T. Problems

Use **H**igher-**O**rder **T**hinking Skills

23. **ERROR ANALYSIS** Wendy and Allison want to find the probability of A given B when $P(A) = 0.3$ and $P(B) = 0.3$. Events A and B are independent. Is either of them correct? Explain your reasoning.

Wendy	Allison
Since we don't know P(A and B), we cannot find P(A|B).	Since A and B are independent, P(A|B) = P(A).

24. **CHALLENGE** There are n objects in a bag. The probability of drawing object A and then object B without replacement is 2.4%. What is the value of n? Explain.

25. **REASONING** Is $P(A \text{ and } B) = P(B \text{ and } A)$ *sometimes, always,* or *never* true when A and B are independent? Explain your reasoning.

26. **OPEN ENDED** Describe a pair of independent events and a pair of dependent events. Explain your reasoning.

27. **WRITING IN MATH** Explain why the sum of the probabilities of the outcomes on a probability tree must be 1.

Standardized Test Practice

28. PROBABILITY Shannon will be assigned at random to 1 of 6 P.E. classes throughout the day and 1 of 3 lunch times. What is the probability that she will be in the second P.E. class and the first lunch?

A $\frac{1}{18}$ B $\frac{1}{9}$ C $\frac{1}{6}$ D $\frac{1}{2}$

29. ALGEBRA Tameron downloaded 2 videos and 7 songs to his MP3 player for \$10.91. Jake downloaded 3 videos and 4 songs for \$9.93. What is the cost of each video?

F \$0.99
G \$1.21
H \$1.42
J \$1.99

30. GRIDDED RESPONSE A bag contains 7 red jelly beans, 11 yellow jelly beans, and 13 green jelly beans. Victoro picks two jelly beans from the bag without looking. What is the probability that Victoro picks a green one and then a red one? Write the probability as a percent rounded to the nearest tenth.

31. SAT/ACT If the probability that it will snow on Tuesday is $\frac{4}{13}$, then what is the probability that it will *not* snow?

A $\frac{4}{9}$ B $\frac{9}{13}$ C $\frac{13}{9}$ D $\frac{13}{4}$

Spiral Review

32. SOFTBALL Zoe struck out during 10% of her at bats last season. Design and conduct a simulation to estimate the probability that she will strike out at her next at bat this season. (Lesson 13-4)

Use the spinner to find each probability. The spinner is spun again if it stops on a line. (Lesson 13-3)

33. P(pointer landing on red)

34. P(pointer landing on blue)

35. P(pointer landing on green)

36. P(pointer landing on yellow)

Determine whether each pair of solids is *similar, congruent,* or *neither.* If the solids are similar, state the scale factor. (Lesson 12-8)

37.

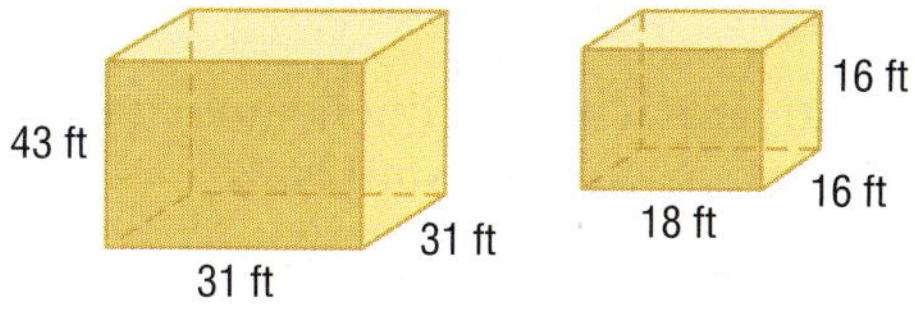

38.

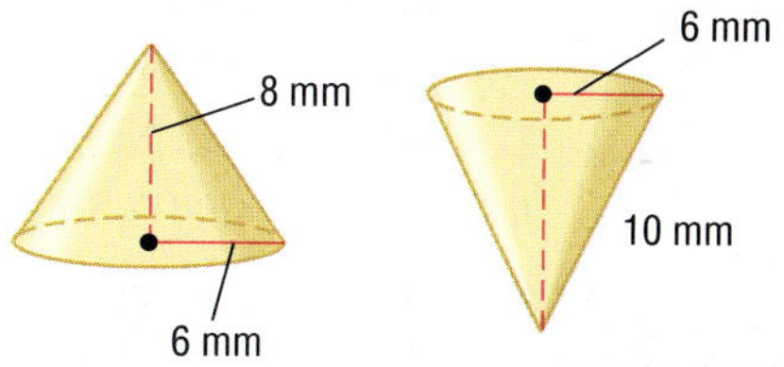

39. FIREWORKS Fireworks are shot from a barge on a river. There is an explosion circle inside which all of the fireworks will explode. Spectators sit outside a safety circle 800 feet from the center of the fireworks display. (Lesson 10-1)

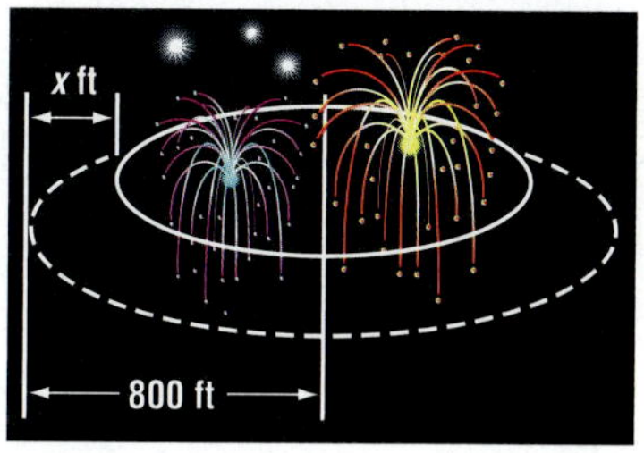

a. Find the approximate circumference of the safety circle.

b. If the safety circle is 200 to 300 feet farther from the center than the explosion circle, find the range of values for the radius of the explosion circle.

c. Find the least and maximum circumferences of the explosion circle to the nearest foot.

Skills Check

Find the number of possible outcomes for each situation. (Lesson 13-1)

40. Blanca chooses from 5 different flavors of ice cream and 3 different toppings.

41. Perry chooses from 6 colors and 2 seat designs for his new mountain bike.

42. A rectangle has a perimeter of 12 and integer side lengths.

43. Three number cubes are rolled simultaneously.

13-6

Probabilities of Mutually Exclusive Events

Then
You found probabilities of independent and dependent events. (Lesson 13-5)

Now
- Find probabilities of events that are mutually exclusive and events that are not mutually exclusive.
- Find probabilities of complements.

New Vocabulary
mutually exclusive events
complement

Math Online
glencoe.com

Why?
At Wayside High School, freshmen, sophomores, juniors, and seniors can all run for Student Council president. Dominic wants either a junior or a senior candidate to win the election. Trayvon wants either a sophomore or a female to win, but says, "If the winner is sophomore Katina Smith, I'll be thrilled!"

Mutually Exclusive Events In Lesson 13-4, you examined probabilities involving the intersection of two or more events. In this lesson, you will examine probabilities involving the union of two or more events.

$P(A \text{ and } B)$ ↑ **Indicates an intersection of two sample spaces.**

$P(A \text{ or } B)$ ↑ **Indicates a union of two sample spaces.**

To find the probability that one event occurs *or* another event occurs, you must know how the two events are related. If the two events cannot happen at the same time, they are said to be **mutually exclusive**. That is, the two events have no outcomes in common.

Real-World EXAMPLE 1 Identify Mutually Exclusive Events

ELECTIONS **Refer to the application above. Determine whether the events are *mutually exclusive* or *not mutually exclusive*. Explain your reasoning.**

a. a junior winning the election or a senior winning the election

These events are mutually exclusive. There are no common outcomes—a student cannot be both a junior and a senior.

b. a sophomore winning the election or a female winning the election

These events are not mutually exclusive. A female student who is a sophomore is an outcome that both events have in common.

c. drawing an ace or a club from a standard deck of cards.

Since the ace of clubs represents both events, they are not mutually exclusive.

Guided Practice

Determine whether the events are *mutually exclusive* or *not mutually exclusive*. Explain your reasoning.

1A. selecting a number at random from the integers from 1 to 100 and getting a number divisible by 5 or a number divisible by 10

1B. drawing a card from a standard deck and getting a 5 or a heart

1C. getting a sum of 6 or 7 when two dice are rolled

Personal Tutor glencoe.com

One way of finding the probability of two mutually exclusive events occurring is to examine their sample space.

When a die is rolled, what is the probability of getting a 3 or a 4? From the Venn diagram, you can see that there are two outcomes that satisfy this condition, 3 and 4. So,

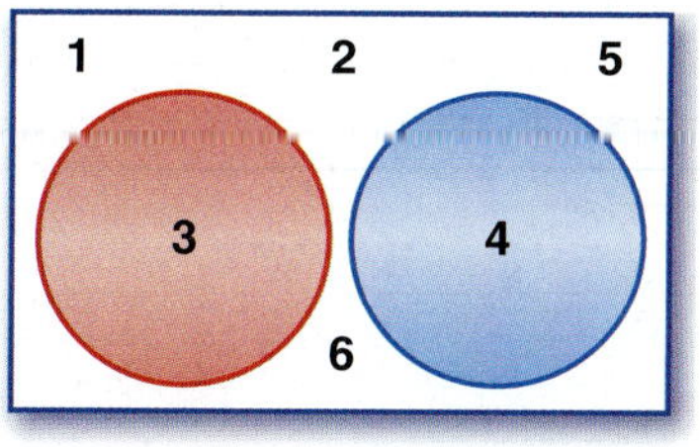

$$P(3 \text{ and } 4) = \frac{2}{6} \text{ or } \frac{1}{3}.$$

Notice that this same probability can be found by adding the probabilities of each simple event.

$P(3) = \frac{1}{6}$ $\quad P(4) = \frac{1}{6}$ $\quad P(3 \text{ and } 4) = \frac{1}{6} + \frac{1}{6} = \frac{2}{6} \text{ or } \frac{1}{3}$

This example illustrates the first of two Addition Rules for Probability.

ReadingMath

Or The word *or* is a key word indicating that at least one of the events occurs. *P*(*A* or *B*) is read as *the probability that A occurs or that B occurs*.

Key Concept

Probability of Mutually Exclusive Events

Words If two events *A* and *B* are mutually exclusive, then the probability that *A* or *B* occurs is the sum of the probabilities of each individual event.

Example If two events *A* or *B* are mutually exclusive, then $P(A \text{ or } B) = P(A) + P(B)$.

This rule can be extended to any number of events.

Real-World EXAMPLE 2 Mutually Exclusive Events

MUSIC Ramiro makes a playlist that consists of songs from three different albums by his favorite artist. If he lets his MP3 player select the songs from this list at random, what is the probability that the first song played is from Album 1 or Album 2?

Ramiro's Playlist

Album	Number of Songs
1	10
2	12
3	13

These are mutually exclusive events, since the songs selected cannot be from both Album 1 and Album 2.

Let event $A1$ represent selecting a song from Album 1.
Let event $A2$ represent selecting a song from Album 2.
There are a total of $10 + 12 + 13$ or 35 songs.

$P(A1 \text{ or } A2) = P(A1) + P(A2)$ **Probability of mutually exclusive events**

$= \frac{10}{35} + \frac{12}{35}$ **$P(A1) = \frac{10}{35}$ and $P(A2) = \frac{12}{35}$**

$= \frac{22}{35}$ **Add.**

So, the probability that the first song played is from Album 1 or Album 2 is $\frac{22}{35}$ or about 63%.

Real-World Link

An online survey in 2007 revealed that 79% of high school students own an MP3 player.

Source: Apple

Guided Practice

2A. Two dice are rolled. What is the probability that doubles are rolled or that the sum is 9?

2B. CARNIVAL GAMES If you win the ring toss game at a certain carnival, you receive a stuffed animal. If the stuffed animal is selected at random from among 15 puppies, 16 kittens, 14 frogs, 25 snakes, and 10 unicorns, what is the probability that a winner receives a puppy, a kitten, or a unicorn?

Personal Tutor glencoe.com

When a die is rolled, what is the probability of getting a number greater than 2 or an even number? From the Venn diagram, you can see that there are 5 numbers that are either greater than 2 or are an even number: 2, 3, 4, 5, and 6. So,

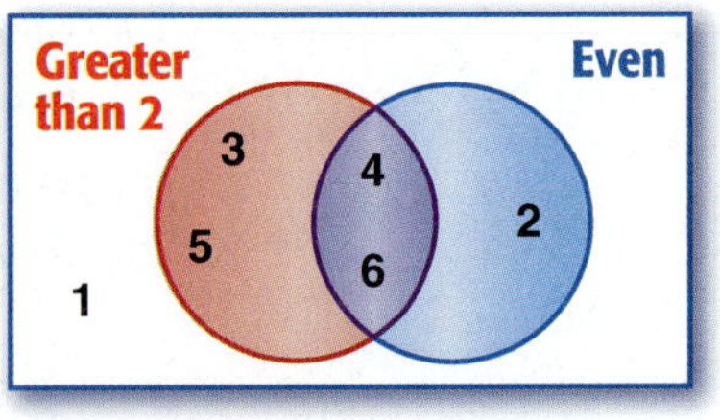

$$P(\text{greater than 2 or even}) = \frac{5}{6}.$$

Since it is possible to roll a number that is greater than 2 *and* an even number, these events are not mutually exclusive. Consider the probabilities of each individual event.

$$P(\text{greater than 2}) = \frac{4}{6} \qquad P(\text{even}) = \frac{3}{6}$$

If these probabilities were added, the probability of two outcomes, 4 and 6, would be counted twice—once for being numbers greater than 2 and once for being even numbers. You must subtract the probability of these common outcomes.

$$P(\text{greater than 2 or even}) = P(\text{greater than 2}) + P(\text{even}) - P(\text{greater than 2 and even})$$
$$= \frac{4}{6} + \frac{3}{6} - \frac{2}{6} \text{ or } \frac{5}{6}$$

This leads to the second of the Addition Rules for Probability.

Key Concept

Probability of Events That Are Not Mutually Exclusive

Words If two events *A* and *B* are not mutually exclusive, then the probability that *A* or *B* occurs is the sum of their individual probabilities minus the probability that both *A* and *B* occur.

Symbols If two events *A* and *B* are not mutually exclusive, then

$$P(A \text{ or } B) = P(A) + P(B) - P(A \text{ and } B).$$

Real-World Link

Juried art shows are shows in which artists are called to submit pieces and a panel of judges decides which art will be shown. They originated in the early 1800s to exhibit the work of current artists and educate the public.

Source: Humanities Web

Real-World EXAMPLE 3 Events That Are Not Mutually Exclusive

ART The table shows the number and type of paintings Namiko has created. If she randomly selects a painting to submit to an art contest, what is the probability that she selects a portrait or an oil painting?

Namiko's Paintings

Media	Still Life	Portrait	Landscape
watercolor	4	5	3
oil	1	3	2
acrylic	3	2	1
pastel	1	0	5

Since some of Namiko's paintings are both portraits and oil paintings, these events are not mutually exclusive. Use the rule for two events that are not mutually exclusive. The total number of paintings from which to choose is 30.

$$P(\text{oil or portrait}) = P(\text{oil}) + P(\text{portrait}) - P(\text{oil and portrait})$$
$$= \frac{1+3+2}{30} + \frac{5+3+2+0}{30} - \frac{3}{30} \quad \textbf{Substitution}$$
$$= \frac{6}{30} + \frac{10}{30} - \frac{3}{30} \text{ or } \frac{13}{30} \quad \textbf{Simplify.}$$

The probability that Namiko selects a portrait or an oil painting is $\frac{13}{30}$ or about 43%.

Guided Practice

3. What is the probability of drawing a king or a diamond from a standard deck of 52 cards?

Personal Tutor glencoe.com

Probabilities of Complements The complement of an event A consists of all the outcomes in the sample space that are not included as outcomes of event A.

When a die is rolled, the probability of getting a 4 is $\frac{1}{6}$. What is the probability of *not* getting a 4? There are 5 possible outcomes for this event: 1, 2, 3, 5, or 6. So, $P(\text{not } 4) = \frac{5}{6}$. Notice that this probability is also $1 - \frac{1}{6}$ or $1 - P(4)$.

Key Concept

Probability of the Complement of an Event

Words The probability that an event will not occur is equal to 1 minus the probability that the event will occur.

Symbols For an event A, $P(\text{not } A) = 1 - P(A)$.

ReadingMath

Complement The complement of event A can also be noted as A^C.

EXAMPLE 4 Complementary Events

RAFFLE Francisca bought 20 raffle tickets, hoping to win the $100 gift card to her favorite clothing store. If a total of 300 raffle tickets were sold, what is the probability that Francisca will not win the gift card?

Let event A represent selecting one of Francisca's tickets. Then find the probability of the complement of A.

$P(\text{not } A) = 1 - P(A)$ **Probability of a complement**

$= 1 - \frac{20}{300}$ **Substitution**

$= \frac{280}{300}$ or $\frac{14}{15}$ **Subtract and simplify.**

The probability that one of Francisca's tickets *will not* be selected is $\frac{14}{15}$ or about 93%.

Guided Practice

4. If the chance of rain is 70%, what is the probability that it will not rain?

Personal Tutor glencoe.com

Concept Summary Probability Rules

Types of Events	Words	Probability Rule
Independent Events	The outcome of a first event *does not affect* the outcome of the second event.	If two events A and B are independent, then $P(A \text{ and } B) = P(A) \cdot P(B)$.
Dependent Events	The outcome of a first event *does affect* the outcome of the other event.	If two events A and B are dependent, then $P(A \text{ and } B) = P(A) \cdot P(B \mid A)$.
Conditional	Additional information is known about the probability of an event.	The conditional probability of A given B is $P(A \mid B) = \frac{P(A \text{ and } B)}{P(B)}$.
Mutually Exclusive Events	Events *do not share* common outcomes.	If two events A or B are mutually exclusive, then $P(A \text{ or } B) = P(A) + P(B)$.
Not Mutually Exclusive Events	Events *do share* common outcomes.	If two events A and B are not mutually exclusive, then $P(A \text{ or } B) = P(A) + P(B) - P(A \text{ and } B)$.
Complementary Events	The outcomes of one event consist of all the outcomes in the sample space that are not outcomes of the other event.	For an event A, $P(\text{not } A) = 1 - P(A)$.

Real-World Link

About 81% of American motorists and their right-front passengers use a seat belt.

Source: National Highway Traffic Safety Administration

Real-World EXAMPLE 5 Identify and Use Probability Rules

SEAT BELTS Refer to the information at the left. Suppose two people are chosen at random from a group of 100 American motorists and passengers. If this group mirrors the population, what is the probability that at least one of them does not wear a seat belt?

Understand You know that 81% of Americans *do use* a seat belt. The phrase *at least one* means *one or more*. So, you need to find the probability that either

- the first person chosen does not use a seat belt *or*
- the second person chosen does not use a seat belt *or*
- both people chosen do not use a seat belt.

Plan The complement of the event described above is the event that both people chosen *do use* a seat belt. Find the probability of this event, and then find the probability of its complement.

Let event A represent choosing a person who does use a seat belt.

Let event B represent choosing a person who does use a seat belt after the first person has already been chosen.

These are two dependent events, since the outcome of the first event affects the probability of the outcome of the second event.

Solve

$P(A \text{ and } B) = P(A) \cdot P(B|A)$ — **Probability of dependent events**

$= \frac{81}{100} \cdot \frac{80}{99}$ — $P(A) = \frac{0.81(100)}{100}$ **or** $\frac{81}{100}$

$= \frac{6480}{9900} \text{ or } \frac{36}{55}$ — **Multiply.**

$P[\text{not } (A \text{ and } B)] = 1 - P(A \text{ and } B)$ — **Probability of a complement**

$= 1 - \frac{36}{55}$ — **Substitution**

$= \frac{19}{55}$ — **Subtract.**

So, the probability that at least one of the passengers does not use a seat belt is $\frac{19}{55}$ or about 35%.

Check Use logical reasoning to check the reasonableness of your answer.

The probability that one person chosen out of 100 *does not* wear his or her seat belt is (100 − 81)% or 19%. The probability that two people chosen out of 100 wear their seat belt should be greater than 19%. Since 35% > 19%, the answer is reasonable.

StudyTip

Key Probability Words When determining what type of probability you are dealing with in a situation, look for key words and correctly interpret their meaning.

and → independent or dependent events
or → mutually exclusive or not mutually exclusive
not → complementary events
and then → conditional
at least n → *n* or more
at most n → *n* or less

Guided Practice

5. CELL PHONES According to an online poll, 35% of American motorists routinely use their cell phones while driving. Three people are chosen at random from a group of 100 motorists. What is the probability that

A. at least two of them use their cell phone while driving?

B. no more than one use their cell phone while driving?

Personal Tutor glencoe.com

Check Your Understanding

Example 1 p. 938

Determine whether the events are *mutually exclusive* or *not mutually exclusive*. Explain your reasoning.

1. drawing a card from a standard deck and getting a jack or a club

2. adopting a cat or a dog

Example 2 p. 939

3. **JOBS** Adelaide is the employee of the month at her job. Her reward is to select at random from 4 gift cards, 6 coffee mugs, 7 DVDs, 10 CDs, and 3 gift baskets. What is the probability that an employee receives a gift card, coffee mug, or CD?

Example 3 p. 940

4. **CLUBS** According to the table, what is the probability that a student in a club is a junior or on the debate team?

Club	Soph.	Junior	Senior
Key	12	14	8
Debate	2	6	3
Math	7	4	5
French	11	15	13

Example 4 p. 941

Determine the probability of each event.

5. If you have a 2 in 10 chance of bowling a spare, what is the probability of missing the spare?

6. If the chance of living in a particular dorm is 75%, what is the probability of living in another dorm?

Example 5 p. 942

7. **PROM** In Armando's senior class of 100 students, 91% went to the senior prom. If two people are chosen at random from the entire class, what is the probability that at least one of them did not go to prom?

Practice and Problem Solving

= **Step-by-Step Solutions** begin on page R20.
Extra Practice begins on page 969.

Examples 1–3 pp. 938–940

Determine whether the events are *mutually exclusive* or *not mutually exclusive*. Then find the probability. Round to the nearest tenth of a percent, if necessary.

8. drawing a card from a standard deck and getting a jack or a six

9. rolling a pair of dice and getting doubles or a sum of 8

10. selecting a number at random from integers 1 to 20 and getting an even number or a number divisible by 3

11. tossing a coin and getting heads or tails

12. drawing an ace or a heart from a standard deck of 52 cards

13. rolling a pair of dice and getting a sum of either 6 or 10

14. **SPORTS** The table includes all of the programs offered at a sports complex and the number of participants aged 14–16. What is the probability that a player is 14 or plays basketball?

Graceland Sports Complex			
Age	Soccer	Baseball	Basketball
14	28	36	42
15	30	26	33
16	35	41	29

15. **GIFTS** An exchange student is moving back to Italy, and her homeroom class wants to get her a going away present. The teacher takes a survey of the class of 32 students and finds that 10 people chose a card, 12 chose a T-shirt, 6 chose a video, and 4 chose a bracelet. If the teacher randomly selects the present, what is the probability that the exchange student will get a card or a bracelet?

Example 4
p. 941

Determine the probability of each event.

16. rolling a pair of dice and not getting a 3

17. drawing a card from a standard deck and not getting a diamond

18. flipping a coin and not landing on heads

19. spinning a spinner numbered 1–8 and not landing on 5

20. **RAFFLE** Namid bought 20 raffle tickets. If a total of 500 raffle tickets were sold, what is the probability that Namid will not win the raffle?

21 **JOBS** Of young workers aged 18 to 25, 71% are paid by the hour. If two people are randomly chosen out of a group of 100 young workers, what is the probability that exactly one is paid by the hour?

Example 5
p. 942

22. **RECYCLING** Refer to the information at the left. If two Americans are chosen randomly from a group of 50, what is the probability that at most one of them recycles?

Real-World Link

Up from just 6% in 1960, 31% of Americans currently recycle.

Source: Energy Information Administration

CARDS **Suppose you pull a card from a standard 52-card deck. Find the probability of each event.**

23. The card is a 4.

24. The card is red.

25. The card is a face card.

26. The card is not a face card.

27. **MUSIC** A school carried out a survey of 265 students to see which types of music students would want played at a school dance. The results are shown in the Venn Diagram. Find each probability.

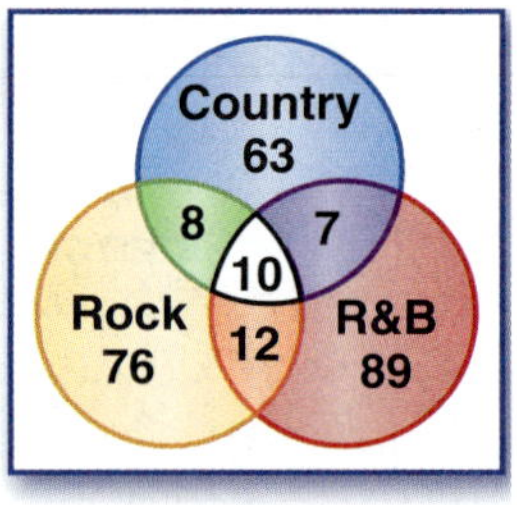

a. P(country or R&B)

b. P(rock and country or R&B and rock)

c. P(R&B but not rock)

d. P(all three)

H.O.T. Problems Use Higher-Order Thinking Skills

28. **ERROR ANALYSIS** Tetsuya and Mason want to determine the probability that a red marble will be chosen out of a bag of 4 red, 7 blue, 5 green, and 2 purple marbles. Is either of them correct? Explain your reasoning.

Tetsuya	Mason
$P(R) = \frac{4}{17}$	$P(R) = 1 - \frac{4}{18}$

29. **CHALLENGE** You roll 3 dice. What is the probability that the outcome of at least two of the dice will be less than or equal to 4? Explain your reasoning.

REASONING **Determine whether the following are mutually exclusive. Explain.**

30. choosing a quadrilateral that is a square and a quadrilateral that is a rectangle

31. choosing a triangle that is equilateral and a triangle that is equiangular

32. choosing a complex number and choosing a natural number

33. **OPEN ENDED** Describe a pair of events that are mutually exclusive and a pair of events that are not mutually exclusive.

34. **WRITING IN MATH** Explain why the sum of the probabilities of two mutually exclusive events is not always 1.

Standardized Test Practice

35. PROBABILITY Customers at a new salon can win prizes during opening day. The table shows the type and number of prizes. What is the probability that the first customer wins a manicure or a massage?

Prize	Number
manicure	10
pedicure	6
massage	3
facial	1

A 0.075
B 0.35
C 0.5
D 0.65

36. SHORT RESPONSE A cube numbered 1 through 6 is shown.

If the cube is rolled once, what is the probability that a number less than 3 or an odd number shows on the top face of the cube?

37. ALGEBRA What will happen to the slope of line p if it is shifted so that the y-intercept stays the same and the x-intercept approaches the origin?

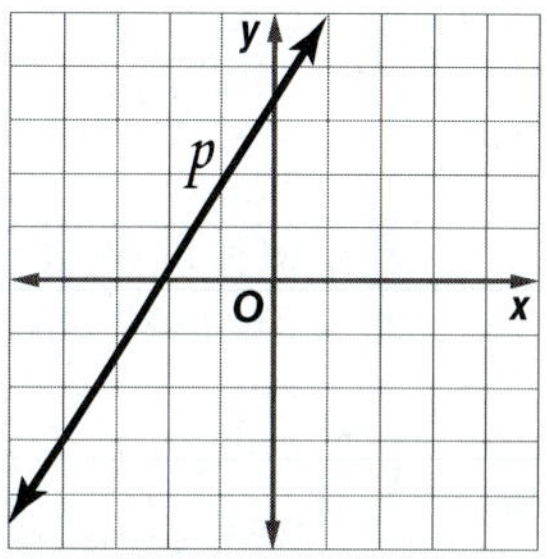

F The slope will become negative.
G The slope will become zero.
H The slope will decrease.
J The slope will increase.

38. SAT/ACT The probability of choosing a peppermint from a certain bag of candy is 0.25, and the probability of choosing a chocolate is 0.3. The bag contains 60 pieces of candy, and the only types of candy in the bag are peppermint, chocolate, and butterscotch. How many butterscotch candies are in the bag?

A 25 **B** 27 **C** 30 **D** 33

Spiral Review

Determine whether the events are *independent* or *dependent*. Then find the probability. (Lesson 13-5)

39. A king is drawn, without replacement, from a standard deck of 52 cards. Then, a second king is drawn.

40. You roll a die and get a 2. You roll another die and get a 3.

41. SPORTS A survey at a high school found that 15% of the athletes at the school play only volleyball, 20% play only soccer, 30% play only basketball, and 35% play only football. Design a simulation that can be used to estimate the probability that an athlete will play each of these sports. (Lesson 13-4)

Copy the figure and point P. Then use a ruler to draw the image of the figure under a dilation with center P and the scale factor r indicated. (Lesson 9-6)

42. $r = \frac{1}{2}$

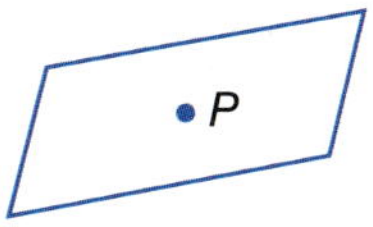

43. $r = 3$

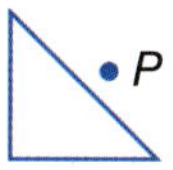

44. $r = \frac{1}{5}$

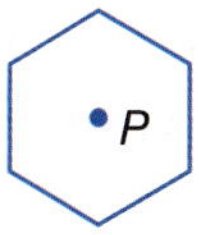

EXTEND 13-6

Geometry Lab: Graph Theory

Objective
To apply physical models, graphs, and networks to develop solutions in applied contexts.

Standard Course of Study

MBC.D.1.2 Use critical path analysis and weighted digraphs to optimally schedule large projects that are comprised of many smaller tasks.

Mathematical structures can be used to model relationships in a set. The study of these graphs is called *graph theory*. These **vertex-edge graphs** are not like graphs that can be seen on a coordinate plane. Each graph, also called a **network**, is a collection of vertices, called **nodes**, and segments called **edges** that connect the nodes.

The bus route in the figure is an example of a network. The school, each stop, and the garage are nodes in the network. The connecting streets, such as Long Street, are edges.

This is an example of a **traceable network** because all of the nodes are connected, and each edge is used once in the network.

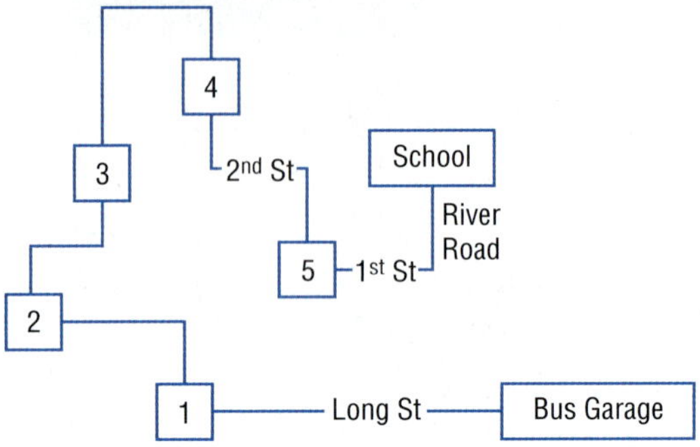

ACTIVITY 1

The graph represents the streets on Ava's newspaper route. To complete her route as quickly as possible, how can Ava ride her bike down each street only once?

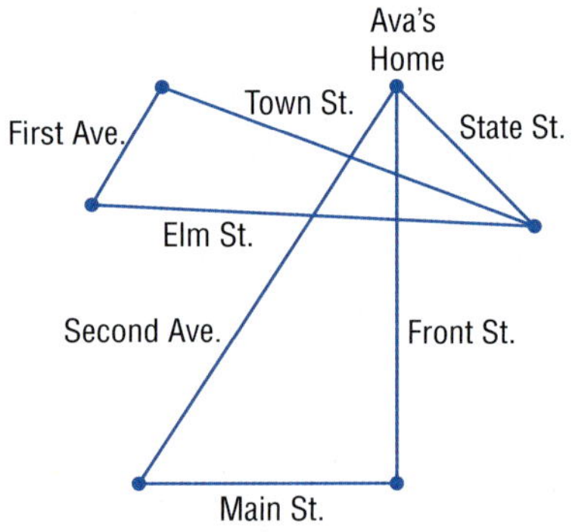

Step 1 Copy the graph onto your paper.

Step 2 Beginning at Ava's home, trace over her route without lifting your pencil. Remember to trace each edge only once.

Step 3 Describe Ava's route.

Analyze

1. Is there more than one traceable route that begins at Ava's house? If so, how many?
2. If it does not matter where Ava starts, how many traceable routes are possible?

Is each graph traceable? Write *yes* or *no*. Explain your reasoning.

3. 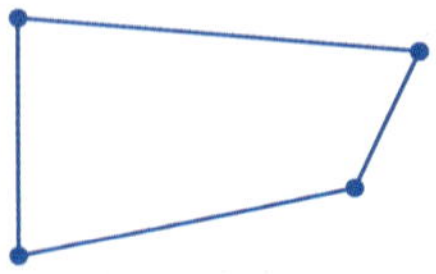

4. 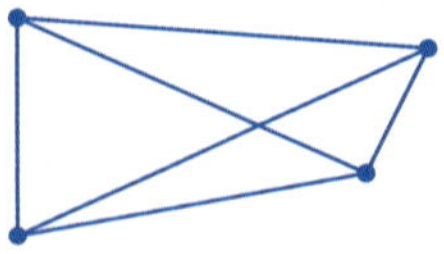

5. 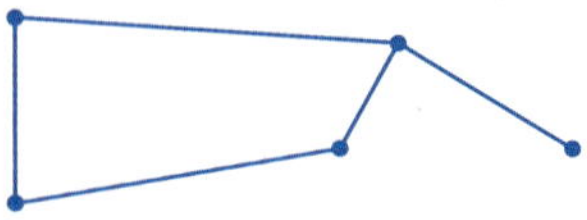

6. The campus for Centerburgh High School has five buildings built around the edge of a circular courtyard. There is a sidewalk between each pair of buildings.
   a. Draw a graph of the campus. Is the graph traceable?
   b. Suppose there are no sidewalks between pairs of adjacent buildings. Is it possible to reach all five buildings without walking down any sidewalk more than once?
7. **REASONING** Write a rule for determining whether a graph is traceable.

In a network, routes from one vertex to another are also called *paths*. **Weighted vertex-edge graphs** are graphs in which a value, or **weight**, is assigned to each edge. The **weight of a path** is the sum of the weights of the edges along the path. The **efficient route** is the path with the minimum weight.

ACTIVITY 2

The edges of the network have different weights. Find the efficient route from node A to node B.

Step 1 Find all of the possible paths from node A to node B. Label each path with the letters of the nodes along the path.

Step 2 Trace each path and add the weights of each edge. The path with the least weight is the efficient route: A-U-X-Y-Z-B. The weight is 54.

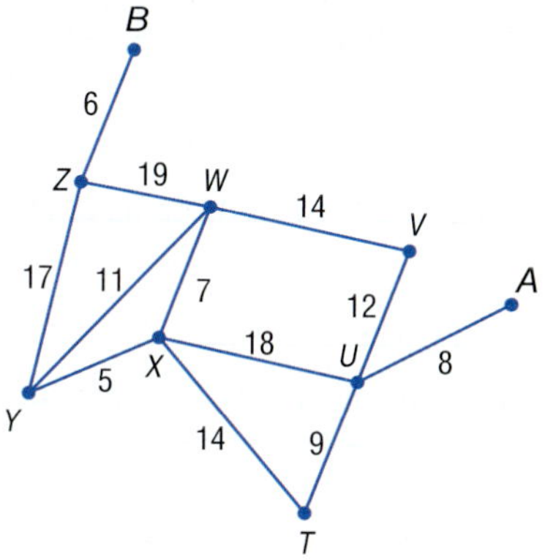

StudyTip

Weights of Edges Pay attention to the weights when determining the efficient route. It may not necessarily be the path with the fewest edges.

Math History Link

Leonhard Euler (1707–1783) Euler introduced graph theory in 1736 in a paper titled *Seven Bridges of Konigsburg,* a famous solved mathematics problem inspired by an actual place and situation. Also, Euler's formula relating the number of edges, vertices, and faces of a convex polyhedron is the origin of graph theory.

Model and Analyze

8. What is the longest path from A to B that does not cover any edges more than once?

Determine the efficient route from A to B for each network.

9.

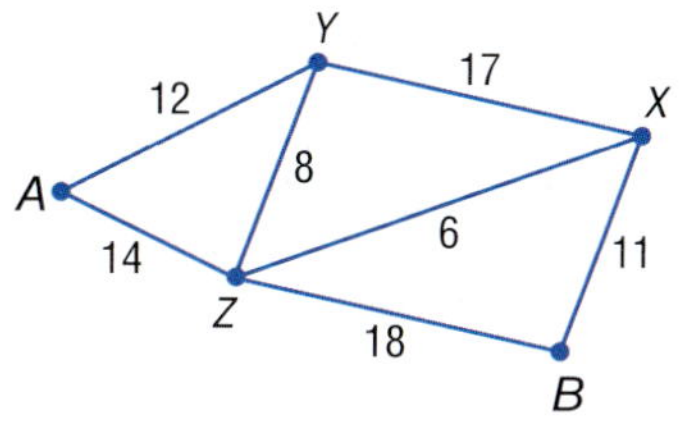

10.

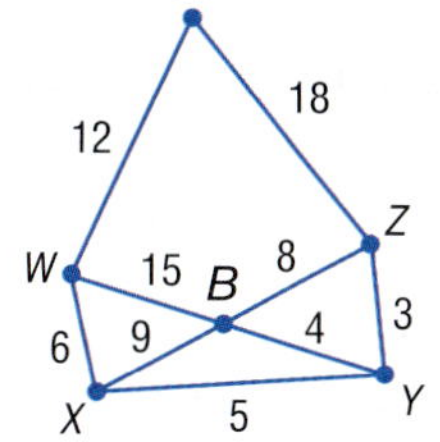

11.

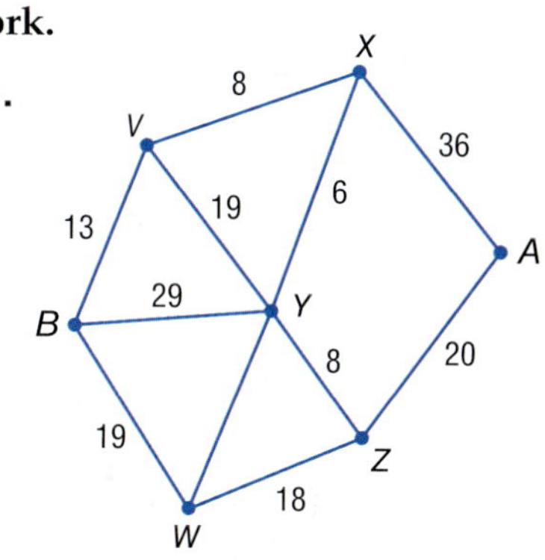

12. OPEN ENDED Create a network with 8 nodes and an efficient route with a value of 25.

13. WRITING IN MATH Explain your method for determining the efficient route of a network.

14. TRAVEL Use the network below to find each efficient route.

a. from Phoenix to New York

b. from Seattle to Atlanta

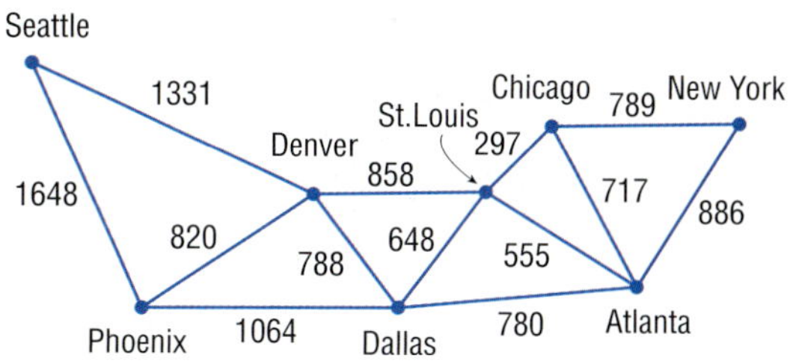

15. *Six Degrees of Separation* is a well-known example of graph theory. In this case, each person is a node and people are linked by an edge when they know each other.

a. Make a graph of the situation. Directly connect yourself to three other people that you know personally. This represents the first degree of separation.

b. Expand the graph to show the first three degrees of separation. Name a person who is within 3 degrees of you, and list the path.

Study Guide and Review

Math Online glencoe.com
- STUDY TO GO
- Vocabulary Review

Chapter Summary

Key Concepts

Representing Sample Spaces (Lesson 13-1)

- The sample space of an experiment is the set of all possible outcomes.
- A sample space can be determined by using an organized list, a table, or a tree diagram.

Permutations and Combinations (Lesson 13-2)

- In a permutation, order is important.
  $_nP_r = \frac{n!}{(n-r)!}$
- In a combination, order is not important.
  $_nC_r = \frac{n!}{(n-r)!r!}$

Geometric Probability (Lesson 13-3)

- If a region A contains a region B and a point E in region A is chosen at random, then the probability that point E is in region B is $\frac{\text{area of region } B}{\text{area of region } A}$.

Simulations (Lesson 13-4)

- A simulation uses a probability model to recreate a situation again and again so that the likelihood of various outcomes can be estimated.

Probabilities of Compound Events (Lessons 13-5 and 13-6)

- If event A does not affect the outcome of event B, then the events are independent and $P(A \text{ and } B) = P(A) \cdot P(B)$.
- If two events A and B are dependent, then $P(A \text{ and } B) = P(A) \cdot P(B|A)$.
- If two events A and B cannot happen at the same time, they are mutually exclusive and $P(A \text{ or } B) = P(A) + P(B)$.
- If two events A and B are not mutually exclusive, then $P(A \text{ or } B) = P(A) + P(B) - P(A \text{ and } B)$.

FOLDABLES® Study Organizer

Be sure the Key Concepts are noted in your Foldable.

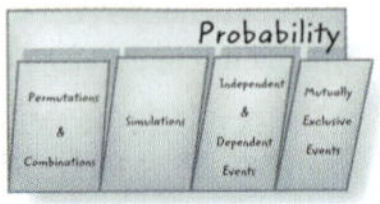

Key Vocabulary

- **circular permutation** (p. 909)
- **combination** (p. 910)
- **complement** (p. 941)
- **compound events** (p. 931)
- **conditional probability** (p. 933)
- **dependent events** (p. 931)
- **expected value** (p. 925)
- **factorial** (p. 906)
- **Fundamental Counting Principle** (p. 901)
- **geometric probability** (p. 915)
- **independent events** (p. 931)
- **mutually exclusive events** (p. 938)
- **permutation** (p. 906)
- **probability model** (p. 923)
- **probability tree** (p. 933)
- **random variable** (p. 925)
- **sample space** (p. 899)
- **simulation** (p. 923)
- **tree diagram** (p. 899)

Vocabulary Check

State whether each sentence is *true* or *false*. If *false*, replace the underlined term to make a true sentence.

1. A tree diagram uses line segments to display possible outcomes.
2. A permutation is an arrangement of objects in which order is NOT important.
3. Determining the arrangement of people around a circular table would require circular permutation.
4. Tossing a coin and then tossing another coin is an example of dependent events.
5. Geometric probability involves a geometric measure such as length or area.
6. $6! = 6 \cdot 5 \cdot 4 \cdot 3 \cdot 2 \cdot 1$, is an example of a factorial.
7. The set of all possible outcomes is the sample space.
8. Combining a coin toss and a roll of a die makes a simple event.
9. Grant flipped a coin 200 times to create a probability tree of the experiment.
10. Drawing two socks out of a drawer without replacing them are examples of mutually exclusive events.

MIXED PROBLEM SOLVING
For mixed problem-solving practice, see page 1011.

Lesson-by-Lesson Review

13-1 Representing Sample Spaces (pp. 899–905)

MBC.S.3.1

11. POPCORN A movie theater sells small (S), medium (M), and large (L) size popcorn with the choice of no butter (NB), butter (B), and extra butter (EB). Represent the sample space for popcorn orders by making an organized list, a table, and a tree diagram.

12. SHOES A pair of men's shoes comes in whole sizes 5 through 13 in navy, brown, or black. How many different pairs could be selected?

EXAMPLE 1

Three coins are tossed. Represent the sample space for this experiment by making an organized list.

Pair each possible outcome from the first toss with the possible outcomes from the second toss and third toss.

HHH, HHT, HTH, HTT, THH, THT, TTH, TTT

13-2 Probability with Permutations and Combinations (pp. 906–914)

13. DINING Three boys and three girls go out to eat together. The restaurant only has round tables. Fred does not want any girl next to him and Gena does not want any boy next to her. How many arrangements are possible?

14. DANCE The dance committee consisted of 10 students. The committee will select three officers at random. What is the probability that Alice, David, and Carlene are selected?

15. COMPETITION From 32 students, 4 are to be randomly chosen for an academic challenge team. In how many ways can this be done?

EXAMPLE 2

For a party, Lucita needs to seat four people at a round table. How many combinations are possible?

Since there is no fixed reference point, this is a circular permutation.

$P_n = (n - 1)!$ **Formula for circular permutation**

$P_4 = (4 - 1)!$ ***n* = 4**

$= 3!$ or 6 **Simplify.**

So, there are 6 ways for Casey to seat four people at a round table.

13-3 Geometric Probability (pp. 915–921)

MBC.S.1.1.

16. GAMES Measurements for a beanbag game are shown. What is the probability of each event?

a. P(hole)

b. P(no hole)

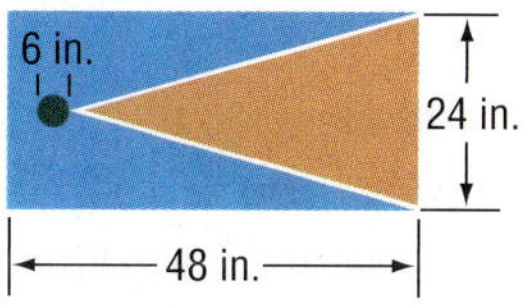

17. POOL Morgan, Phil, Callie, and Tyreese are sitting on the side of a pool in that order. Morgan is 2 feet from Phil. Phil is 4 feet from Callie. Callie is 3 feet from Tyreese. Oscar joins them.

a. Find the probability that Oscar sits between Morgan and Phil.

b. Find the probability that Oscar sits between Phil and Tyreese.

EXAMPLE 3

A carnival game is shown.

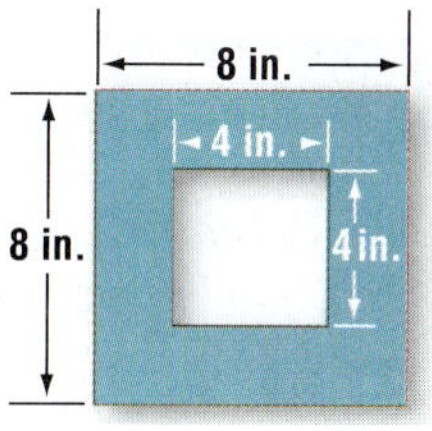

a. If Khianna threw 10 beanbags at the board, what is the probability that the beanbag went in the hole?

Area of hole $= 4 \cdot 4 = 16$

Area of board $= (8 \cdot 8) - 16 = 64 - 16$ or 48

$P(\text{hole}) = \frac{16}{64}$ or about 25%

b. What is the probability that the beanbag did not go in the hole?

$P(\text{no hole}) = \frac{48}{64}$ or about 75%

13-4 Simulations (pp. 923–930)

For each of the following, describe how you would use a geometric probability model to design a simulation.

18. **POLO** Max scores 35% of the goals his team earns in each water polo match.

19. **BOOKS** According to a survey, people buy 30% of their books in October, November, and December, 22% during January, February, and March, 23% during April, May, and June, and 25% during July, August, and September.

20. **OIL** The United States consumes 17.3 million barrels of oil a day. 63% is used for transportation, 4.9% is used to generate electricity, 7.8% is used for heating and cooking, and 24.3% is used for industrial processes.

EXAMPLE 4

Darius made 75% of his field goal kicks last season. Design a simulation that can be used to estimate the probability that he will make his next field goal kick this season.

Use a spinner that is divided into 2 sectors. Make one sector red containing 75% of the spinner's area and the other blue containing 25% of the spinner's area.

Spin the spinner 50 times. Each spin represents kicking a field goal. A successful trial will be a made field goal, and a failed trial will be a missed field goal.

13-5 Probabilities of Independent and Dependent Events (pp. 931–937)

21. **MARBLES** A box contains 3 white marbles and 4 black marbles. What is the probability of drawing 2 black marbles and 1 white marble in a row without replacing any marbles?

22. **CARDS** Two cards are randomly chosen from a standard deck of cards with replacement. What is the probability of successfully drawing, in order, a three and then a queen?

23. **PIZZA** A nationwide survey found that 72% of people in the United States like pizza. If 3 people are randomly selected, what is the probability that all three like pizza?

EXAMPLE 5

A bag contains 3 red, 2 white, and 6 blue marbles. What is the probability of drawing, in order, 2 red and 1 blue marble without replacement?

Since the marbles are not being replaced, the events are dependent events.

$$P(\text{red, red, blue}) = P(\text{red}) \cdot P(\text{red}) \cdot P(\text{blue})$$
$$= \frac{3}{11} \cdot \frac{2}{10} \cdot \frac{6}{9}$$
$$= \frac{2}{55} \text{ or about } 3.6\%$$

13-6 Probabilities of Mutually Exclusive Events (pp. 938–945)

24. **ROLLING DICE** Two dice are rolled. What is the probability that the sum of the numbers is 7 or 11?

25. **CARDS** A card is drawn from a deck of cards. Find the probability of drawing a 10 or a diamond.

26. **RAFFLE** A bag contains 40 raffle tickets numbered 1 through 40.
    a. What is the probability that a ticket chosen is an even number or less than 5?
    b. What is the probability that a ticket chosen is greater than 30 or less than 10?

EXAMPLE 6

Two dice are rolled. What is the probability that the sum is 5 or doubles are rolled?

These are mutually exclusive events because the sum of doubles can never equal 5.

$$P(\text{sum is 5 or doubles}) = P(\text{sum is 5}) + P(\text{doubles})$$
$$= \frac{4}{36} + \frac{6}{36}$$
$$= \frac{5}{18} \text{ or about } 27.8\%$$

CHAPTER 13

Practice Test

Math Online glencoe.com
Chapter Test

Point X is chosen at random on $\overline{AE}$. Find the probability of each event.

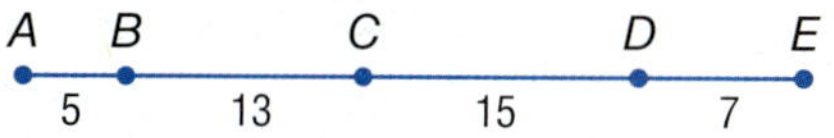

1. $P(X \text{ is on } \overline{AC})$
2. $P(X \text{ is on } \overline{CD})$
3. **BASEBALL** A baseball team fields 9 players. How many possible batting orders are there for the 9 players?
4. **TRAVEL** A traveling salesperson needs to visit four cities in her territory. How many distinct itineraries are there for visiting each city once?

Represent the sample space for each experiment by making an organized list, a table, and a tree diagram.

5. A box has 1 red ball, 1 green ball, and 1 blue ball. Two balls are drawn from the box one after the other, without replacement.
6. Shinsuke wants to adopt a pet and goes to his local humane society to find a dog or cat. While he is there, he decides to adopt two pets.
7. **ENGINEERING** An engineer is analyzing three factors that affect the quality of semiconductors: temperature, humidity, and material selection. There are 6 possible temperature settings, 4 possible humidity settings, and 6 choices of materials. How many combinations of settings are there?
8. **SPELLING** How many distinguishable ways are there to arrange the letters in the word "bubble"?
9. **PAINTBALL** Cordell is shooting a paintball gun at the target. What is the probability that he will shoot the shaded region?

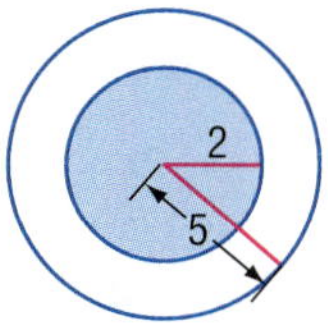

10. **PHONES** What is the probability that a phone number using the numbers 7, 7, 7, 2, 2, 2, and 6 will be 622-2777?
11. **TICKETS** Fifteen people entered the drawing at the right. What is the probability that Jodi, Dan, and Pilar all won the tickets?

Determine whether the events are *independent* or *dependent*. Then find the probability.

12. A deck of cards has 5 yellow, 5 pink, and 5 orange cards. Two cards are chosen from the deck with replacement. Find P(the first card is pink and the second card is pink).
13. There are 6 green, 2 red, 2 brown, 4 navy, and 2 purple marbles in a hat. Sadie picks 2 marbles from the hat without replacement. What is the probability that the first marble is brown and the second marble is not purple?

Use the spinner to find each probability. If the spinner lands on a line, it is spun again.

14. P(pointer landing on purple)
15. P(pointer landing on red)
16. P(pointer not landing on yellow)
17. **FOOTBALL** According to a football team's offensive success rate, the team punts 40% of the time, kicks a field goal 30% of the time, loses possession 5% of the time, and scores a touchdown 25% of the time. Design a simulation using a random number generator. Report the results using appropriate numerical and graphical summaries.

Determine whether the events are *mutually exclusive* or *not mutually exclusive*. Explain your reasoning.

18. a person owning a car and a truck
19. rolling a pair of dice and getting a sum of 7 and 6 on the face of one die
20. a playing card being both a spade and a club
21. **GRADES** This quarter, Todd earned As in his classes 45% of the time. Design and conduct a simulation using a geometric probability model. Then report the results using appropriate numerical and graphical summaries.

CHAPTER 13

Preparing for Standardized Tests

Organize Data

Sometimes you may be given a set of data that you need to analyze in order to solve items on a standardized test. Use this section to practice organizing data and to help you solve problems.

Strategies for Organizing Data

Step 1

When you are given a problem statement containing data, consider:

- **making a list** of the data.
- **using a table** to organize the data.
- **using a data display** (such as a *bar graph, Venn diagram, circle graph, line graph, box-and-whisker plot*, etc.) to organize the data.

Step 2

Organize the data.

- Create your table, list, or data display.
- If possible, fill in any missing values that can be found by intermediate computations.

Step 3

Analyze the data to solve the problem.

- Reread the problem statement to determine what you are being asked to solve.
- Use the properties of geometry and algebra to work with the organized data and solve the problem.
- If time permits, go back and check your answer.

EXAMPLE

Read the problem. Identify what you need to know. Then use the information in the problem to solve.

Of the students who speak a foreign language at Marie's school, 18 speak Spanish, 14 speak French, and 16 speak German. There are 8 students who only speak Spanish, 7 who speak only German, 3 who speak Spanish and French, 2 who speak French and German, and 4 who speak all three languages. If a student is selected at random, what is the probability that he or she speaks Spanish or German, but not French?

A $\frac{9}{16}$ B $\frac{2}{5}$ C $\frac{5}{18}$ D $\frac{7}{12}$

Read the problem carefully. The data is difficult to analyze as it is presented. Use a Venn diagram to organize the data and solve the problem.

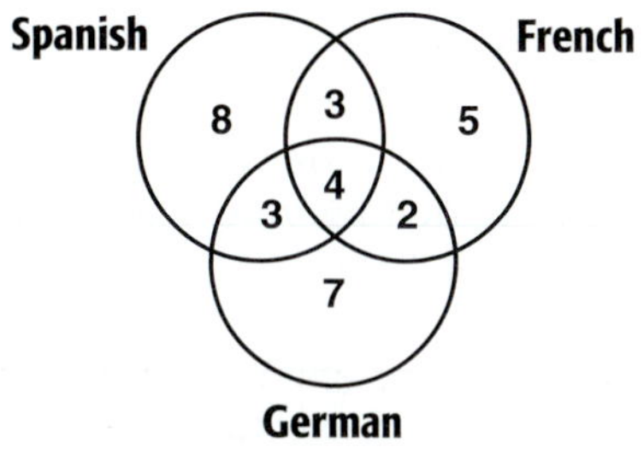

Step 1 Draw three circles, each representing a language.

Step 2 Fill in the data given in the problem statement.

Step 3 Fill in the missing values. For example, you know that 18 students speak Spanish and 14 students speak French.

$18 - 8 - 3 - 4 = 3$ (Spanish and German)
$14 - 3 - 4 - 2 = 5$ (only French)

Step 4 Solve the problem. You are asked to find the probability that a randomly selected student speaks Spanish or German, but not French. From the Venn diagram, you can see that there are 32 total students. Of these, $8 + 3 + 7$, or 18 students speak Spanish or German, but not French. So, the probability is $\frac{18}{32}$ or $\frac{9}{16}$. So, the correct answer is A.

Exercises

Read the problem. Identify what you need to know. Then organize the data to solve the problem.

1. Alana has the letter tiles A, H, M, and T in a bag. If she selects a permutation of the tiles at random, what is the probability she will spell the word MATH?

A $\frac{3}{50}$
B $\frac{1}{24}$
C $\frac{1}{12}$
D $\frac{1}{4}$

2. The table below shows the number of freshmen, sophomores, juniors, and seniors involved in basketball, soccer, and volleyball. What is the probability that a randomly selected student is a junior or plays volleyball?

Sport	Fr	So	Jr	Sr
Basketball	7	6	5	6
Soccer	6	4	8	7
Volleyball	9	2	4	6

F $\frac{4}{21}$
G $\frac{5}{21}$
H $\frac{5}{17}$
J $\frac{17}{35}$

3. Find the probability that a point chosen at random lies in the shaded region.

A 0.22
B 0.25
C 0.28
D 0.32

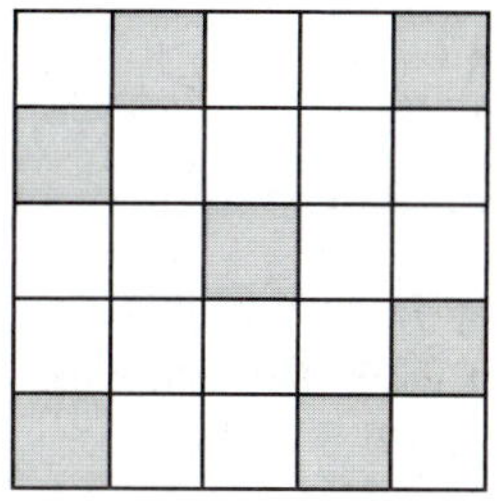

4. There are 10 sophomore, 8 junior, and 9 senior members in student council. Each member is assigned to help plan one school activity during the year. There are 4 sophomores working on the field day and 6 working on the pep rally. Of the juniors, 2 are working on the field day and 5 are working on the school dance. There are 2 seniors working on the pep rally. If each activity has a total of 9 students helping to plan it, what is the probability that a randomly selected student council member is a junior or is working on the field day?

F $\frac{1}{5}$
G $\frac{4}{18}$
H $\frac{5}{9}$
J $\frac{2}{3}$

CHAPTER 13

Standardized Test Practice

Cumulative, Chapters 1 through 13

Multiple Choice

Read each question. Then fill in the correct answer on the answer document provided by your teacher or on a sheet of paper.

1. A machine is making steel washers by cutting out 10-millimeter circular disks from 34-millimeter circular disks as shown below. What is the area of each washer to the nearest tenth?

A 75.4 mm^2

B 829.4 mm^2

C 986.5 mm^2

D 3317.5 mm^2

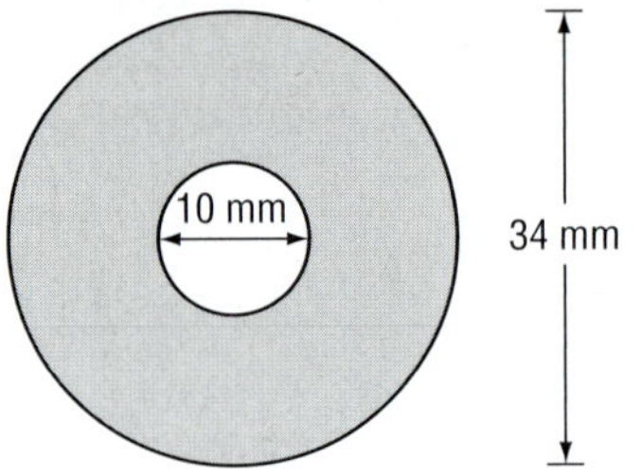

2. How much paper is needed to make the drinking cup below? Round to the nearest tenth.

F 73.4 cm^2

G 70.7 cm^2

H 67.9 cm^2

J 58.8 cm^2

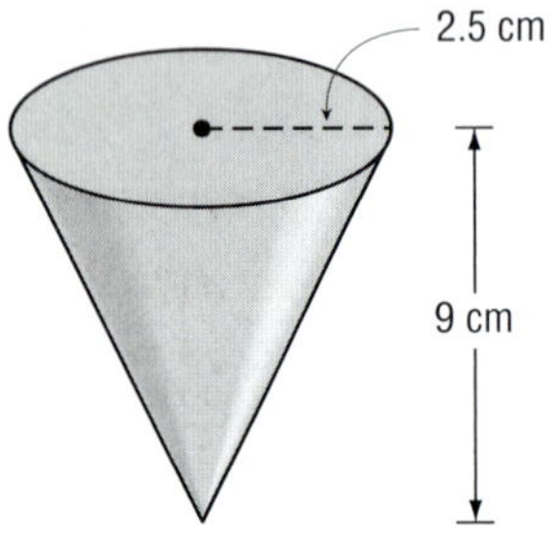

3. Which of the following properties of real numbers justifies the statement below?

If $3x - 2 = 7x + 12$, then
$3x - 2 + 2 = 7x + 12 + 4$.

A Addition Property of Equality

B Reflection Property of Equality

C Subtraction Property of Equality

D Symmetric Property of Equality

Test-TakingTip

Question 4 What is the probability of rolling doubles with two number cubes? Multiply this by the number of trials.

4. What is the expected number of times Clarence will roll doubles with two number cubes in 90 trials? (Doubles occur when both number cubes show the same number in a trial.)

F 6 G 9 H 10 J 15

5. The Venn diagram shows the states in the U.S. in which the population is greater than 10,000,000 and the population density is greater than 200 people per square mile. Which statement is false?

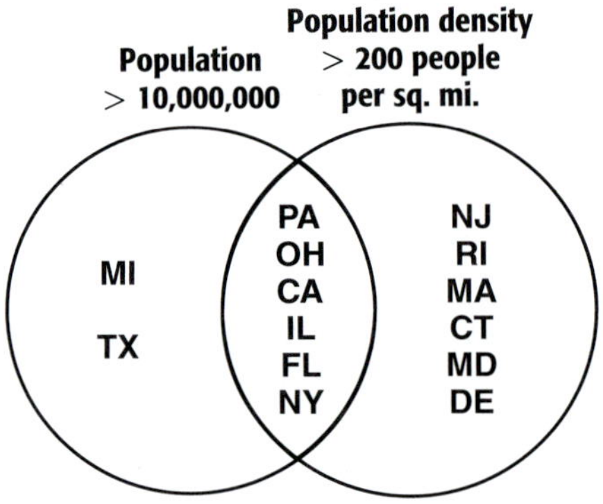

A In California (CA), the population is greater than 10,000,000, and the density is greater than 200.

B In Maryland (MD), the population is greater than 10,000,000 or the density is greater than 200.

C 14 states have a density greater than 200.

D 8 states have a population greater than 10,000,000.

6. Which of the following correctly shows the relationship between the angle measures of triangle *RST*?

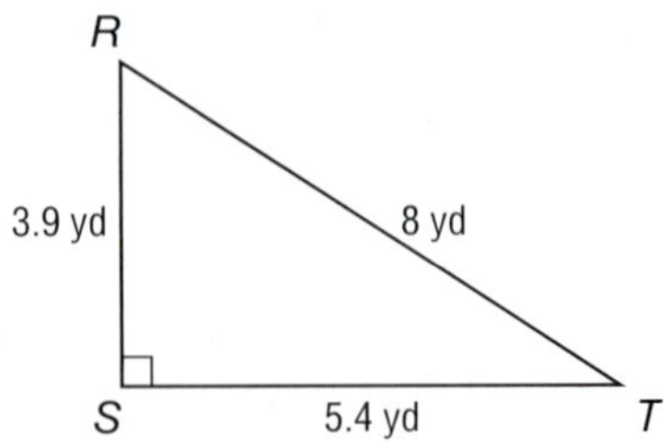

F $m\angle S < m\angle R < m\angle T$

G $m\angle T < m\angle S < m\angle R$

H $m\angle R < m\angle S < m\angle T$

J $m\angle T < m\angle R < m\angle S$

Short Response/Gridded Response

Record your answers on the answer sheet provided by your teacher or on a sheet of paper.

7. GRIDDED RESPONSE What is $m\angle S$ in the figure below? Express your answer in degrees.

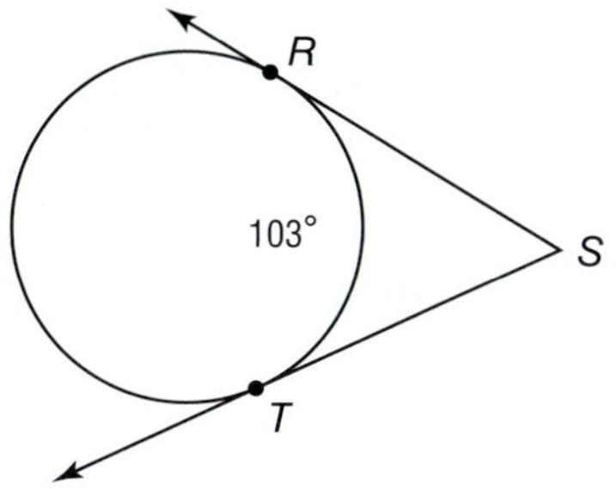

8. Does the figure have rotational symmetry? If so, give the order of symmetry.

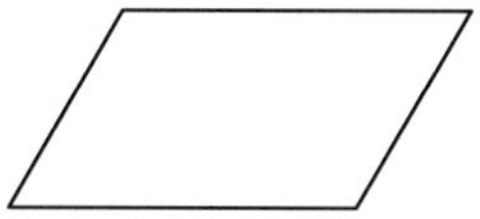

9. GRIDDED RESPONSE Segment AD bisects $\angle CAB$ in the triangle below. What is the value of x?

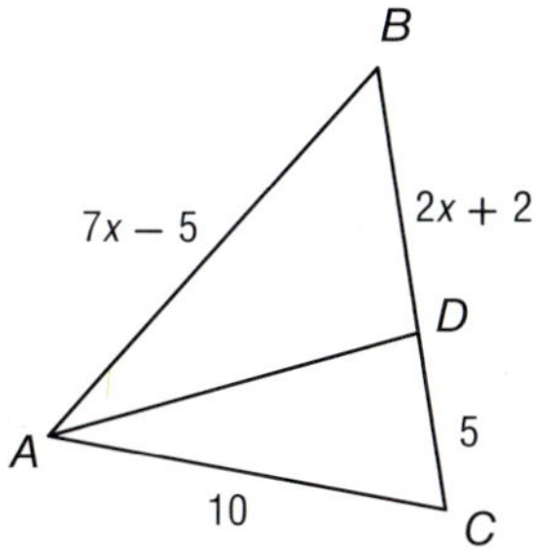

10. GRIDDED RESPONSE Armando leans an 18-foot ladder against the side of his house to clean out the gutters. The base of the ladder is 5 feet from the wall. How high up the side of the house does the ladder reach? Express your answer in feet, rounded to the nearest tenth.

11. Solve for x in the triangle below.

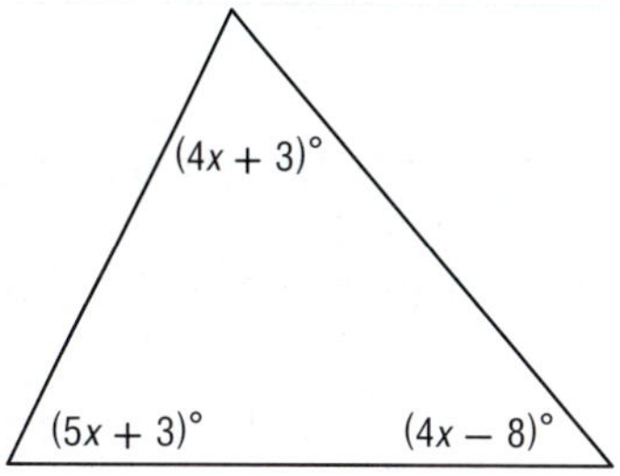

12. What effect does doubling the dimensions of the rectangle below have on its area and perimeter?

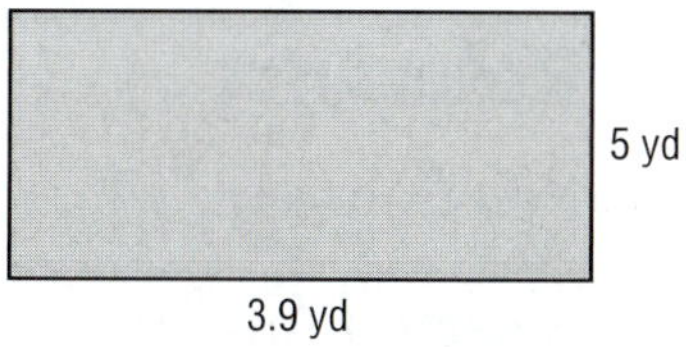

Extended Response

Record your answers on a sheet of paper. Show your work.

13. A bag contains 3 red chips, 5 green chips, 2 yellow chips, 4 brown chips, and 6 purple chips. One chip is chosen at random, the color noted, and the chip returned to the bag.

a. Suppose two trials of this experiment are conducted. Are the events independent or dependent? Explain.

b. What is the probability that both chips are purple?

c. What is the probability that the first chip is green and the second is brown?

Need Extra Help?

If you missed Question...	1	2	3	4	5	6	7	8	9	10	11	12	13
Go to Lesson or Page...	11-3	12-3	2-6	13-4	2-2	5-3	10-6	9-5	5-1	8-2	4-2	1-6	13-5

Student Handbook

Built-In Workbooks

Reference

How to Use the Student Handbook

The Student Handbook is the additional skill and reference material found at the end of the text. This Handbook can help you answer these questions.

What If I Need Problem-Solving Practice?

You have probably used several different problem solving strategies in previous math courses. The **Problem Solving Handbook** section provides examples and problems for refreshing your skills at using various strategies.

What If I Need More Practice?

You, or your teacher, may decide that working through some additional problems would be helpful. The **Extra Practice** section provides these problems for each lesson so you have ample opportunity to practice new skills.

What If I Have Trouble with Word Problems?

The **Mixed Problem Solving** portion of the book provides additional word problems that use the skills presented in each lesson. These problems give you real-world situations where math can be applied.

What If I Forget What I Learned Last Year?

Use the **Concepts and Skills Bank** section to refresh your memory about things you have learned in other math classes. Here's a list of the topics covered in your book.

1. Solving for a Specific Variable
2. Multiplying Polynomials
3. Dividing Polynomials
4. Factoring to Solve Equations
5. Linear Quadratic Systems
6. Mean, Median, and Mode
7. Displaying Data

What If I Need to Review a Concept?

The **Key Concepts** section lists all of the important concepts that were highlighted in your text along with the pages where they appear.

What If I Need to Check a Homework Answer?

The answers to odd-numbered problems are included in **Selected Answers and Solutions**. Check your answers to make sure you understand how to solve all of the assigned problems.

What If I Forget a Vocabulary Word?

The **English-Spanish Glossary** provides a list of important or difficult words used throughout the textbook. It provides a definition in English and Spanish as well as the page number(s) where the word can be found.

What If I Need to Find Something Quickly?

The **Index** alphabetically lists the subjects covered throughout the entire textbook and the pages on which each subject can be found.

What If I Forget a Formula?

Inside the back cover of your math book is a list of **Formulas and Symbols** that are used in the book.

Problem-Solving Handbook

Problem-Solving Strategy: Look for a Pattern

There are many problem-solving strategies in mathematics. One of the most common is to **look for a pattern**. In order to use this strategy, analyze the first few numbers in a pattern and identify a rule that is used to go from the first number in the pattern to the second, and then to the third, and so on. Then use the rule to extend the pattern and find a solution.

EXAMPLE

A function passes through the points shown. List the coordinates of each point. Describe the pattern in the coordinates and predict the next point in the pattern in the positive direction.

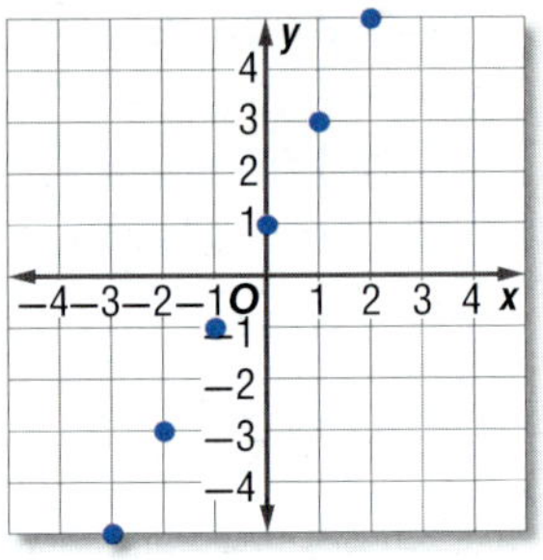

Step 1 List the coordinates of the points shown on the graph.
$(-3, -5)$, $(-2, -3)$, $(-1, -1)$, $(0, 1)$, $(1, 3)$, and $(2, 5)$

Step 2 Identify the pattern in the x-coordinates and the y-coordinates.

As the x-coordinates increase by 1, the y-coordinates increase by 2 each time. So in the positive direction, the next point would be immediately following $(2, 5)$.

The next point is $(2 + 1, 5 + 2)$ or $(3, 7)$.

Practice

Solve each problem by looking for a pattern.

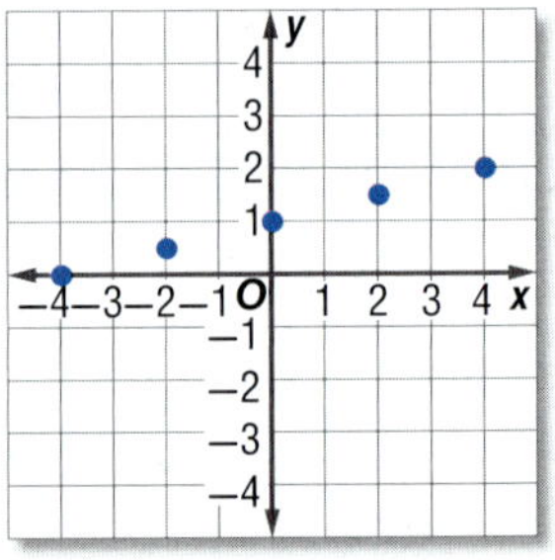

1. A function passes through the points shown.
   a. Describe the pattern in the coordinates, and predict the next point in the pattern in the positive direction.
   b. Predict the next point in the pattern in the negative direction.
2. Two workers can make two chairs in two days. How many chairs can 8 workers working at the same rate make in 20 days?
3. The sum of the measures of the angles of a triangle is 180°.
   a. Use the sum of the angles of the triangles to determine the sum of the measures of the angles of each polygon by drawing all the diagonals from one vertex for a quadrilateral, a pentagon and a hexagon.
   b. Write a rule to find the sum of the measures of the angles of an n-gon.

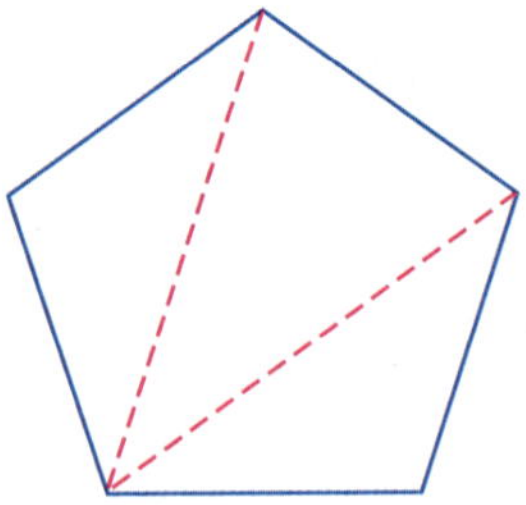

4. Courtney travels south on her bicycle riding 8 miles per hour. One hour later, her friend Horacio starts riding his bicycle from the same location. If he travels south at 10 miles per hour, how long will it take him to catch Courtney?
5. What is the perimeter of the twelfth figure?

Figure 1 Perimeter = 6 | **Figure 2** Perimeter = 8 | **Figure 3** Perimeter = 10

6. A basketball bounces back 0.6 of its height on every bounce. If the basketball is dropped from 200 feet, how high does it bounce on the fifth bounce? Round to the nearest tenth.

Problem-Solving Strategy: Create a Table

One strategy for solving problems is to **create a table**. A table allows you to organize information in an understandable way.

Real-World EXAMPLE

A fruit machine accepts dollars, and each piece of fruit costs 65 cents. If the machine gives only nickels, dimes, and quarters, what combinations of those coins are possible as change for a dollar?

The machine will give back \$1.00 − \$0.65 or 35 cents in change in a combination of nickels, dimes, and quarters.

Make a table showing different combinations of nickels, dimes, and quarters that total 35 cents. Organize the table by starting with the combinations that include the most quarters.

quarters	dimes	nickels
1	1	0
1	0	2
0	3	1
0	2	3
0	1	5
0	0	7

The total for each combination of coins is 35 cents. There are 6 combinations possible.

Practice

Solve each problem by creating a table.

1. How many ways can you make change for a half-dollar using only nickels, dimes, and quarters?

2. A penny, a nickel, a dime, and a quarter are in a purse. How many amounts of money are possible if you grab two coins at random?

3. Laura, Josie, and Marcus ate lunch together at the cafeteria. Each had a different item: a peanut butter sandwich, a hamburger, and a peanut butter and jelly sandwich. Josie ate the sandwich that had a bun. Laura does not like jelly. Marcus sat between the student eating the peanut butter sandwich and the student allergic to peanuts. Which student had which sandwich?

4. Aria asked her friends whether they used wrapping paper, gift bags, recycled paper, or no wrapping for birthday presents. Create a table to show how many students preferred each method of the 24 students she asked. Then predict how many would choose each method if 120 students were asked.

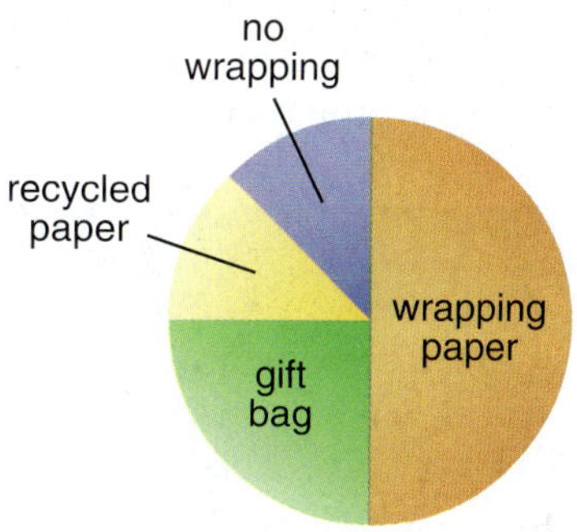

5. At Midas High School, students are selling popcorn at a football game. Each small bag of popcorn is \$1.25. Each large bag of popcorn is \$2.25. Create a table to show the purchase price of five bags of popcorn with every possible combination of large and/or small bags.

6. The equation for a semicircle is $y = \sqrt{16 - x^2}$. Create a table to show five ordered pairs with x-coordinates belonging to the set $\{-4, -2, 0, 2, 4\}$.

Problem-Solving Strategy: Make a Chart

Data presented in a problem can be organized by **making a chart**. This problem-solving strategy allows you to see patterns and relationships among data.

Real-World EXAMPLE

It takes an average driver 1.5 seconds to begin braking after they see an obstruction. The driver can safely decelerate a car or light truck with good tires on a dry street surface at the rate of about 15 feet per second (fps). The distance a vehicle will travel while braking can be found by multiplying the initial velocity by the deceleration time and dividing by 2. Find the stopping distance for a car traveling at 45, 55, 65, and 75 miles per hour.

Step 1 Make a chart that includes the given information and the information to be found. Before setting up your chart, think about how speed, distance, and time are related.

Since the distance is listed in feet, find the rate of speed it takes to stop in feet per second. To convert from miles per hour to feet per second, multiply by a conversion factor of $\frac{5280 \text{ ft}}{3600 \text{ s}}$. Then find the deceleration time by dividing the initial velocity by 15 fps.

Initial Velocity (mph)	Initial Velocity (fps)	Deceleration Time (s)	Distance Traveled before Braking (ft)	Distance Traveled while Braking (ft)	Total Stopping Distance (ft)
45	66	4.4			
55	80.7	5.38			
65	95.3	6.35			
75	110	7.33			

Step 2 To find the distance traveled before braking, multiply the initial velocity by the reaction time, 1.5 seconds. Then find the distance traveled while braking. Add to find the total stopping distance.

Initial Velocity (mph)	Initial Velocity (fps)	Deceleration Time (s)	Distance Traveled before Braking (ft)	Distance Traveled while Braking (ft)	Total Stopping Distance (ft)
45	66	4.4	99	145.2	244.2
55	80.7	5.38	121.05	217.08	338.13
65	95.3	6.35	142.95	302.58	445.53
75	110	7.33	165	403.15	568.15

Practice

Solve each problem by making a chart.

1. As the length of a square doubles, the area increases by a scale factor. Using the squares in the diagram, make a chart of each length and each area. Then find the scale factor.

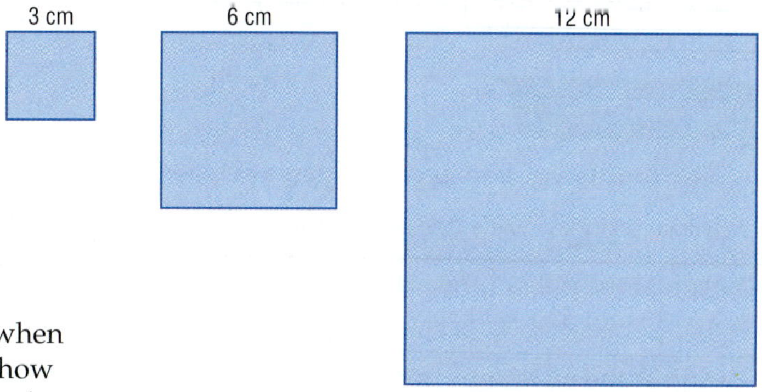

2. Given $f(x) = 2x + 3$ and $g(x) = -x - 3$, use a table to find $f(x) - g(x)$ for all positive integers less than or equal to 6.

3. The chart shows at which point drivers determine when they are going to fill the gas tank. Make a chart to show how many people of 200 surveyed would be expected to have each response.

Source: Bruskin/Goldring for Exxon

4. The following table shows the official state reptile in each state. Make a tally chart that shows how many states have turtles (tortoise, terrapin), snakes, alligators, lizards, toads, or none. Find the ratio of states with alligators as their official state reptile compared to the states with no official state reptile.

AL	red-bellied turtle	LA	American alligator	OH	black racer
AK	none	ME	none	OK	collared lizard
AZ	ridge-nosed rattlesnake	MD	diamondback terrapin	OR	none
AR	none	MA	garter snake	PA	none
CA	desert tortoise	MI	painted turtle	RI	none
CO	none	MN	Blanding's turtle	SC	loggerhead turtle
CT	none	MS	American alligator	SD	none
DE	none	MO	three-toed turtle	TN	eastern box turtle
FL	American alligator	MT	none	TX	horned lizard
GA	gopher tortoise	NE	none	UT	none
HI	none	NV	desert tortoise	VT	none
ID	none	NH	none	VA	none
IL	painted turtle	NJ	none	WA	none
IN	none	NM	whiptail lizard	WV	none
IA	none	NY	snapping turtle	WI	none
KS	ornate box turtle	NC	eastern box turtle	WY	horned toad
KY	none	ND	none		

Problem-Solving Strategy: Guess-and-Check

To solve some problems, you can make a reasonable guess and then check it in the problem. You can then use the results to improve your guess until you find the solution. This strategy is called **guess-and-check**.

EXAMPLE

The product of two even integers is close to 1000.

Make a guess. Let's try 24 and 26. → $24 \times 26 = 624$ ← This product is too low.

Adjust the guess upward.
Try 30 and 32. → $30 \times 32 = 960$ ← This product is still too low.

Adjust the guess upward again.
Try 34 and 36. → $34 \times 36 = 1224$ ← This product is too high.

Try between 30 and 34.
Try 32 and 34. → $32 \times 34 = 1088$ ← This is the correct product.

The integers are 32 and 34.

Practice

Solve each problem by using the guess-and-check strategy.

1. The product of two consecutive odd integers is 783. What are the integers?
2. This year, Mr. Jefferson's age is a multiple of 11. Next year, his age will be a multiple of 9. How old is Mr. Jefferson this year?
3. Rafael is burning a CD for Selma. The CD will hold 35 minutes of music. Which songs should he select from the list to record the maximum time on the CD without going over?

Song	A	B	C	D	E	F	G	H	I	J
Time	5 min 4 s	9 min 10 s	4 min 12 s	3 min 9 s	3 min 44 s	4 min 30 s	5 min 0 s	7 min 21 s	4 min 33 s	5 min 58 s

4. The Science Club sold candy bars and soft pretzels to raise money for an animal shelter. They raised a total of \$62.75. They made \$0.25 profit on each candy bar and \$0.30 profit on each pretzel sold. How many of each did they sell?
5. Lacie bought four items at a school book fair. The prices of the books are shown in the table. Which four items did Lacie buy if she spent \$25.20?
6. Jerrica has 8 coins, all dimes, quarters and pennies. If she has \$0.89, how many of each coin does she have?

Book	Price ($)
comic book	2.10
graphic novel	5.00
fiction book	5.90
atlas	22.40
calendar	16.00

7. One angle of a triangle is shown. Find the other two angles if their product is 1,216.

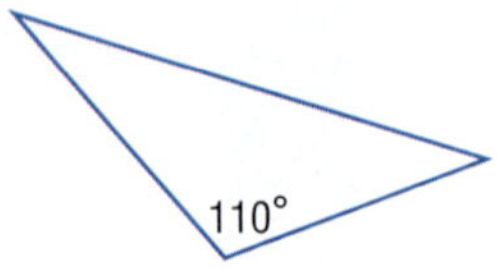

8. Anita sold tickets to the school musical. She had 2 bills worth \$75 for the tickets she sold. If all the money was in \$5 bills, \$10 bills, and \$20 bills, how many of each bill did she have?

Problem-Solving Strategy: Work Backward

On most problems, a set of conditions or facts is given and an end result must be found. However, some problems start with the result and ask for something that happened earlier. The strategy of **working backward** can be used to solve problems like this. To use this strategy, start with the end result and *undo* each step.

Real-World EXAMPLE

Kendrick spent half of the money he had this morning on lunch. After lunch, he loaned his friend a dollar. Now he has $1.50. How much money did Kendrick have this morning?

Start with the end result, $1.50, and work backward to find the amount Kendrick had this morning.

Kendrick now has $1.50 ⟶	$1.50
Undo the $1.00 he loaned to his friend. ⟶	+ $1.00
	$2.50
Undo the half he spent for lunch. ⟶	× 2
	$5.00

Add $1.00 to undo giving his friend $1.00.

Multiply by 2 to undo spending half the original amount.

Kendrick had $5.00 this morning.

CHECK Kendrick started with $5.00. If he spent half of that, or $2.50, on lunch, and loaned his friend $1.00, he would have $1.50 left. This matches the amount stated in the problem, so the solution is correct.

Practice

Solve each problem by working backward.

1. Tia used half of her allowance to buy a ticket to the class play. Then she spent $0.75 for an ice cream cone. Now she has $2.25 left. How much is her allowance?
2. Lawanda put $15 of her paycheck in savings. Then she spent one half of what was left on clothes. She paid $24 for a concert ticket and later spent one half of what was then left on a book. When she got home, she had $14 left. What was the amount of Lawanda's paycheck?
3. Mr. and Mrs. Delgado each own an equal number of shares of a stock. Mr. Delgado sells one-third of his shares for $2700. What was the total value of Mr. and Mrs. Delgado's stock before the sale?
4. A certain point was reflected in the x-axis, and then moved up three units and two units to the left. The final coordinates of the point are (3, 5). What were the original coordinates?

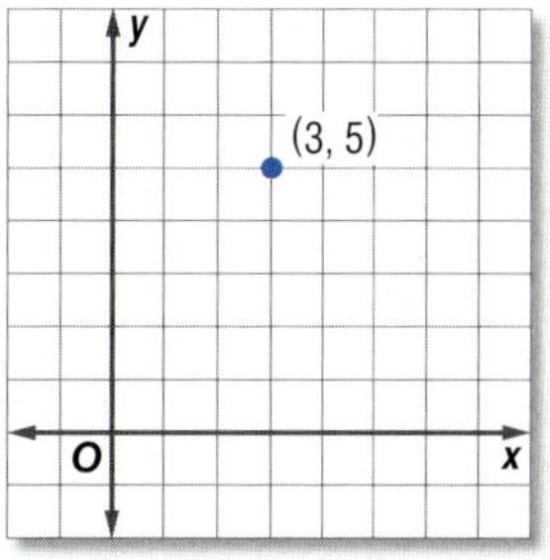

5. A certain bacteria doubles its population every 12 hours. After 3 full days, there are 1600 bacteria in a culture. How many bacteria were there at the beginning of the first day?
6. To catch a 7:30 A.M. bus, Don needs 30 minutes to get dressed, 30 minutes for breakfast, and 5 minutes to walk to the bus stop. What time should he wake up?
7. If a lizard in its cage weighs 23 pounds, find the weight of the lizard if the cage weighs 19 pounds and the sand weighs 2 pounds.

Problem-Solving Strategy: Solve a Simpler Problem

One of the strategies you can use to solve a problem is to **solve a simpler problem**. To use this strategy, first solve a simpler or more familiar case of the problem. Then use the same concept and relationships to solve the original problem.

EXAMPLE

Find the sum of the numbers 1 through 500.

Consider a simpler problem. Find the sum of the numbers 1 through 10. Notice that you can group the addends into partial sums as shown below.

$1 + 2 + 3 + 4 + 5 + 6 + 7 + 8 + 9 + 10 = 55$

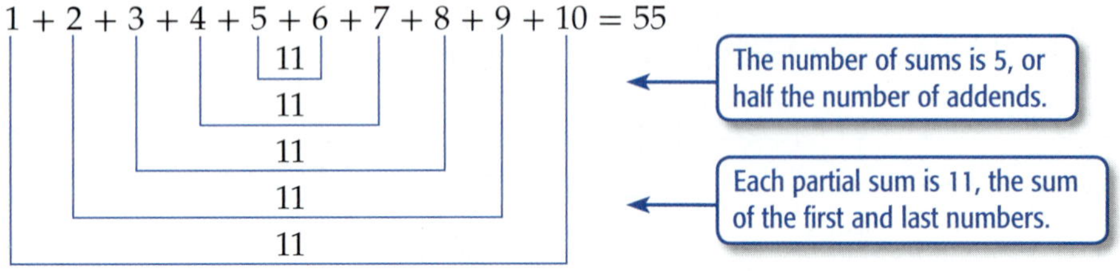

The sum is 5×11 or 55.

Use the same concepts to find the sum of the numbers 1 through 500.

$$1 + 2 + 3 + \ldots + 499 + 500 = 250 \times 501$$
$$= 125{,}250$$

Multiply half the number of addends, 250, by the sum of the first and last numbers, 501.

Practice

Solve each problem by solving a simpler problem.

1. Find the number of squares of any size in the game board shown at the right.
2. Find the sum of the whole numbers through 1000.
3. How many links are needed to join 30 pieces of chain into one long chain?
4. Three people can pick six baskets of apples in one hour. How many baskets of apples can 2 people pick in one-half hour?
5. Find the number of triangles of any size in the figure at the left.
6. A shirt shop has 112 orders for T-shirt designs. Three designers can make 2 shirts in 2 hours. How many designers are needed to complete the orders in 8 hours?
7. Find the area of the composite figure at the right with the given measures.
8. Stamps for postcards cost $0.24, and stamps for first-class letters cost $0.41. Diego wants to send postcards and letters to 10 friends. If he has $3.50 for stamps, how many postcards and how many letters can he send?

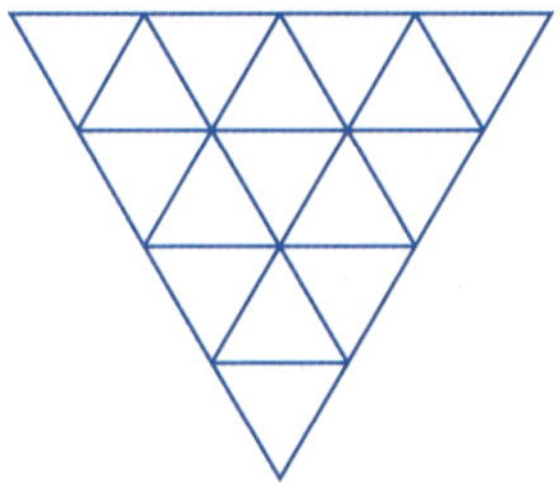

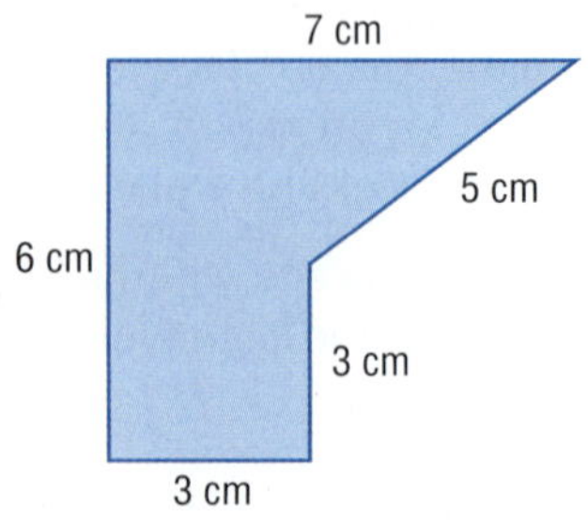

Problem-Solving Strategy: Draw a Diagram

Another strategy for solving problems is to **draw a diagram**. There will be times when a sketch or diagram will give you a better picture of how to tackle a mathematics problem. Adding details like units, labels, and numbers to the drawing or sketch can help you make decisions on how to solve the problem.

Real-World EXAMPLE

Imani is trying to determine the number of 9-inch tiles needed to cover her patio. The rectangular patio measures 8-feet by 10-feet. What is the minimum number of 9-inch tiles Imani should purchase?

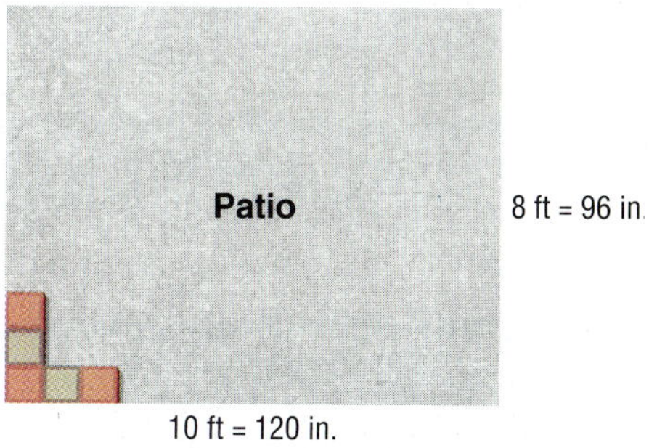

First, draw a diagram of the situation. Express the measurement of the patio in inches.

If each tile is 9 square inches, the minimum number of tiles for the width of the patio is $96 \div 9 \approx 10.7$ or 11 tiles.

The minimum number of tiles for the length of the patio is $120 \div 9 \approx 13.3$ or 14 tiles.

So, the minimum number of tiles Imani needs to cover the patio is 11×14 or 154 tiles.

Practice

Solve each problem by drawing a diagram.

1. Kevin was hired to paint a mural on a wall that measures 5 feet by 20 feet. Starting from the center of the wall, he will paint a square that measures 3 feet. The dimensions of the next square will be .5 times greater than and centered on the previous square. How many squares can Kevin paint on the wall?
2. Find the number of line segments that can be drawn between any two vertices of an octagon.

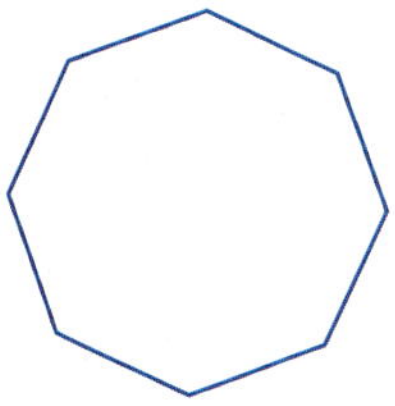

3. A 500-gallon water tank is being filled with water. Eighty gallons of water are in the tank after 4 minutes. How long will it take to fill the tank?
4. It takes 42 minutes to cut a 2-inch by 4-inch piece of wood into 7 equally-sized pieces. How long will it take to cut a similar 2-inch by 4-inch piece into 4 equally-sized pieces?
5. There are 7 people in a meeting. If each person shakes hands with all the other people, how many handshakes take place?
6. How many different teams of 3 players can be chosen from 8 players?
7. Nitarren is trying to decide how many 9-inch diameter pies will fit on her dessert table that measures 4 feet by 2 feet. How many pies could Nitarren fit on the table?
8. Which of these shapes can be made by slicing a cone with a plane—a circle, an oval, a rectangle, a square, a triangle?

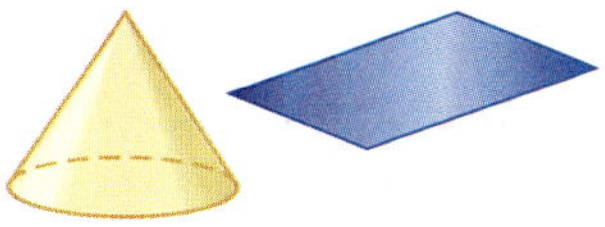

9. Two lines can share no points or one point. How many points can two circles share?

Problem-Solving Strategy: Use Estimation

When you need to make a decision on the basis of inexact information, a common strategy is to **use estimation**. Estimation is often used when an exact answer is not required or when mental math is used rather than a calculator or paper and pencil. You should use estimation to determine if your answer is reasonable.

Real-World EXAMPLE

In a recent year, 51 million international visitors travelled to the United States. Given the information in the table, estimate what percentage of the visitors were from Japan.

Source: Travel Industry Association

Step 1 Determine about how many international visitors came to the United States.

51 million or 51,000,000 would be an easier number to work with if it was rounded to only one digit in the ten millions place, so round 51,000,000 to 50,000,000.

Step 2 Determine from the chart the number of visitors who were from Japan.

According to the chart, 5,000,000 visitors to the United States were from Japan.

Step 3 To determine the percentage, divide.

$$\frac{\text{number of visitors from Japan}}{\text{total number of visitors}} = \frac{5{,}000{,}000}{50{,}000{,}000}$$
$$= 0.1 \text{ or } 10\%$$

About 10% of the visitors to the United States were from Japan.

Practice

Solve each problem by using estimation.

1. If Café Mocha charges $1.98 for each cup of coffee, about how much money did Café Mocha earn in March?

2. The length of Fun Center's go-kart track is 843 feet. If Nadia circled the track 9 times, about how many feet did she travel?

Coffee of the Month Sales

Month	Number of Cups Sold
January	850
February	765
March	587
April	500
May	387

3. In 2007, 82% of all Americans had at least one e-mail account. Of the 301,139,947 Americans, about how many checked their e-mail daily?

4. Sarah solves the following problem. Her answer is 8 doses. Use estimation to determine whether or not Sarah's answer is correct. Explain your reasoning.

How many $\frac{4}{5}$-ounce doses of medicine are in a 10-ounce jar?

Source: UCLA Center for Communication Policy

Problem-Solving Strategy: Eliminate Unnecessary Information

A useful problem-solving strategy is to learn how to **eliminate unnecessary information**. If there is a diagram, it is important to determine if all or some of the information is necessary to find a solution.

Real-World EXAMPLE

Twila is making a quilt that shows a house repeating on each block of the quilt. Which information is unnecessary to find the area of the white door of the house?

The dimensions needed to find the area of the door are the length and width of the door. So, the dimensions of $2\frac{1}{2}$ inches by $5\frac{1}{2}$ inches are needed. The other dimensions, such as the width of the window or the size of the square, are unnecessary.

Practice

1. Miranda is making a mosaic with the tiles at the right. Which information is not needed to determine the area that each tile will cover?

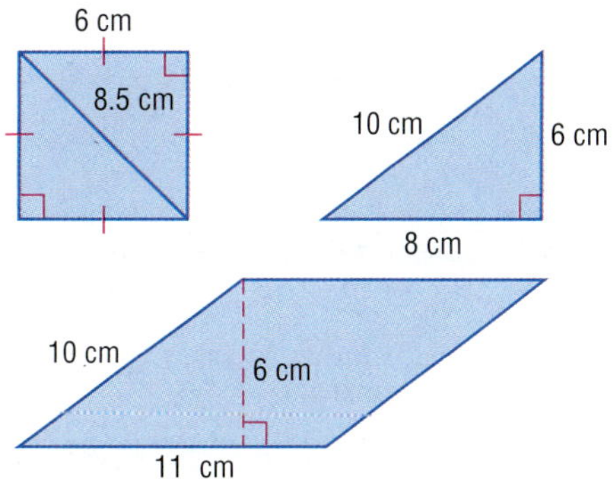

2. Which information in the table below is not necessary to find the difference in height between the tallest building and the 4th tallest building?

Rank	Building, City	Year	Stories	Height	
				m	ft
1.	Taipei 101, Taipei, Taiwan	2004	101	508	1667
2.	Petronas Tower 1, Kuala Lumpur, Malaysia	1998	88	452	1483
3.	Petronas Tower 2, Kuala Lumpur, Malaysia	1998	88	452	1483
4.	Sears Tower, Chicago	1974	110	442	1451
5.	Jin Mao Building, Shanghai	1999	88	421	1381

Source: Council on Tall Buildings and Urban Habitat, 2006

3. The Gemini North telescope was placed in Mauna Kea, Hawaii, in the year 2000. The Gemini South telescope was placed in Cerro Pachon, Chili, in the year 2001. Each of the twin telescopes are 8.1 meters in diameter. What information is not necessary to find the circumference of the base of the telescopes?

Problem-Solving Strategy: Write an Equation

A natural outcome of recognizing mathematical patterns and organizing data is to **write an equation**. Look at a set of data or read a word problem to determine which values are constants and which values vary. Figure out the dependent and independent variables in order to write an equation to reflect the given situation.

Real-World EXAMPLE

For every $10 gift card sold, the theater department at Wallace High School earns $1.25. Write an equation to represent the amount raised based on the number of cards sold.

Step 1 Make a table of data to represent the number of cards sold and the amount of money raised.

Number of Cards	Amount of Money ($)
1	1.25
2	2.50
3	3.75
4	5.00

Step 2 Find the value that varies.

The value that varies is the number of cards sold and the amount raised based on the number of cards sold.

Step 3 Find the value that is constant for each card sold.

The value that is constant for each card sold is the amount of money raised per card, $1.25. This is the slope of the line.

Step 4 Write the equation that shows how the total amount changes based on the number of cards sold.

The equation is $y = 1.25x$.

Practice

Solve each problem by writing an equation.

1. The area of a rectangular parking lot is to be 20,000 square feet. The table shows some possible dimensions for the lot. Write an equation that can be used to find the width of the parking lot for any given length.

Length (ft)	Width (ft)
20	1000
50	400
100	200
200	100
400	50

2. Iceland spent approximately 8.8% of its Gross Domestic Product on public health expenditures in 2006, the highest percentage of all countries. If Iceland had a GDP of $11,380,000,000 in 2006, write an equation to show how much money was spent on public health expenditures.

3. When Jonathan was 100 miles from home, he stopped for a break. From this point on, he will travel 50 miles each hour. The graph shows his distance from home, with the rest stop as the starting point. Write an equation that shows the distance from home given the hours traveled.

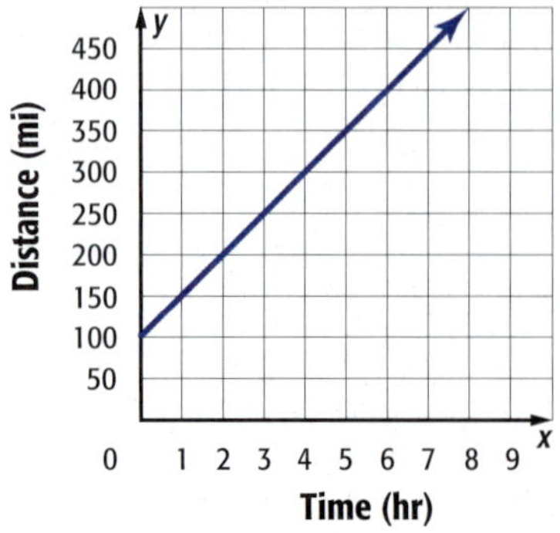

Extra Practice

Lesson 1-1 Points, Lines, and Planes (pp. 5–12)

Refer to the figure.

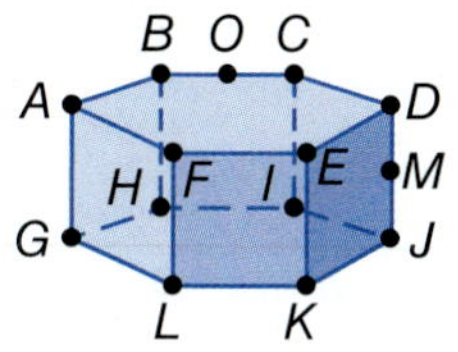

1. How many planes are shown in the figure?
2. Name three collinear points.
3. Name all planes that contain point G.
4. Name the intersection of plane ABD and plane DJK.
5. Name two planes that do not intersect.
6. Name a plane that contains $\overleftrightarrow{FK}$ and $\overleftrightarrow{EL}$.

Draw and label a figure for each relationship.

7. Line a intersects planes $\mathcal{A}$, $\mathcal{B}$, and $\mathcal{C}$ at three distinct points.
8. Planes $\mathcal{X}$ and $\mathcal{Z}$ intersect in line m. Line b intersects the two planes in two distinct points.

Lesson 1-2 Linear Measure (pp. 14–21)

Find the measurement of each segment. Assume that each figure is not drawn to scale.

1. $\overline{AB}$

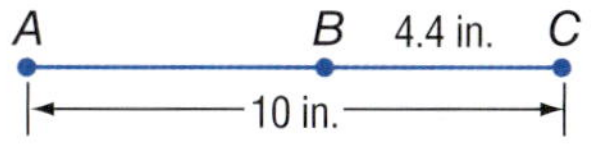

2. $\overline{JL}$

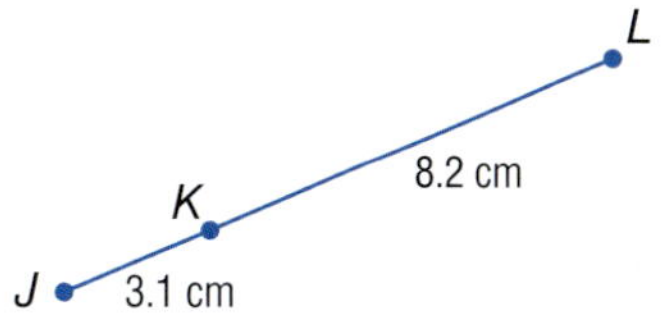

Find the value of the variable and BC if B is between A and C.

3. $AB = 4x$, $BC = 5x$; $AB = 16$
4. $AB = 17$, $BC = 3m$, $AC = 32$
5. $AB = 9a$, $BC = 12a$, $AC = 42$
6. $AB = 25$, $BC = 3b$, $AC = 7b + 13$

Lesson 1-3 Distance and Midpoints (pp. 25–35)

Find the distance between each pair of points.

1. $A(0, 0)$, $B(-3, 4)$
2. $C(-1, 2)$, $N(5, 10)$
3. $X(-6, -2)$, $Z(6, 3)$
4. $M(-5, -8)$, $O(3, 7)$
5. $T(-10, 2)$, $R(6, -10)$
6. $F(5, -6)$, $N(-5, 6)$

Find the coordinates of the midpoint of a segment with the given endpoints.

7. $A(0, 0)$, $D(-2, -8)$
8. $D(-4, -3)$, $E(2, 2)$
9. $K(-4, -5)$, $M(5, 4)$
10. $R(-10, 5)$, $S(8, 4)$
11. $B(2.8, -3.4)$, $Z(1.2, 5.6)$
12. $D(-6.2, 7)$, $K(3.4, -4.8)$

Find the coordinates of the missing endpoint if B is the midpoint of $\overline{AC}$.

13. $C(0, 0)$, $B(5, -6)$
14. $C(-7, -4)$, $B(3, 5)$
15. $C(8, -4)$, $B(-10, 2)$
16. $C(6, 8)$, $B(-3, 5)$
17. $C(6, -8)$, $B(3, -4)$
18. $C(-2, -4)$, $B(0, 5)$

Lesson 1-4 Angle Measure (pp. 36–44)

For Exercises 1–14, use the figure at the right. Name the vertex of each angle.

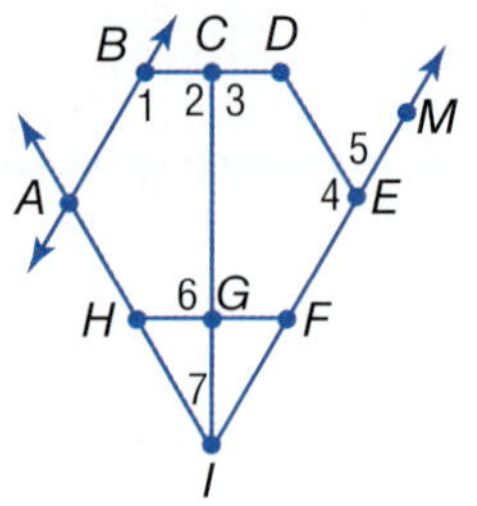

1. ∠1
2. ∠4
3. ∠6
4. ∠7

Name the sides of each angle.

5. ∠*AIE*
6. ∠4
7. ∠6
8. ∠*AHF*

Write another name for each angle.

9. ∠3
10. ∠*DEF*
11. ∠2

Classify each angle as *right, acute,* or *obtuse*. Then use a protractor to measure the angle to the nearest degree.

12. ∠*ABC*
13. ∠*CGF*
14. ∠*HIF*

Lesson 1-5 Angle Relationships (pp. 46–54)

Name an angle pair that satisfies each condition.

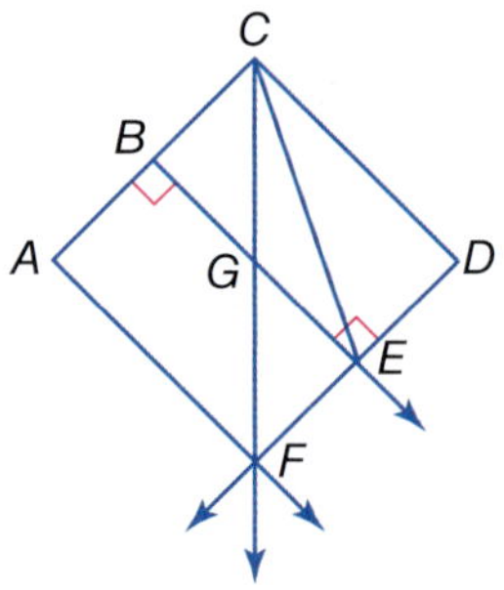

1. two acute vertical angles
2. two obtuse vertical angles
3. a pair of complementary adjacent angles
4. a pair of supplementary adjacent angles
5. a pair of congruent supplementary adjacent angles
6. **ALGEBRA** The measure of ∠1 is nine less than the measure of ∠2. If ∠1 and ∠2 form a linear pair, what are their measures?
7. **ALGEBRA** The measure of the complement of an angle is 17 more than the measure of the angle. Find the measure of each angle.

Lesson 1-6 Two-Dimensional Figures (pp. 56–64)

Name each polygon by its number of sides. Then classify it as *convex* or *concave* and *regular* or *irregular*.

1.
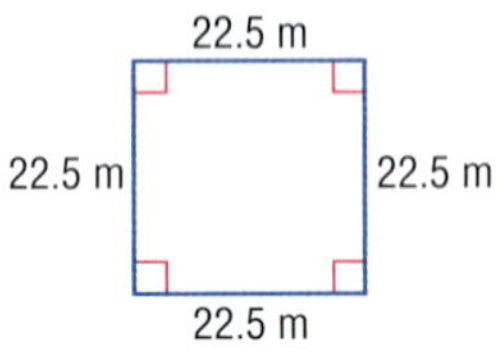

2.
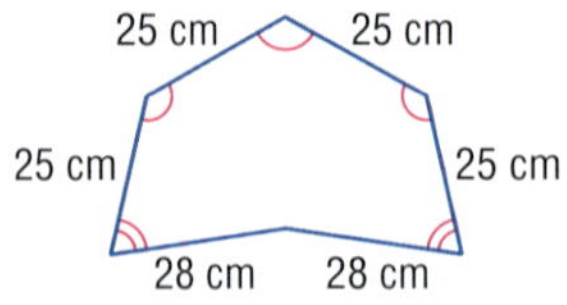

3.
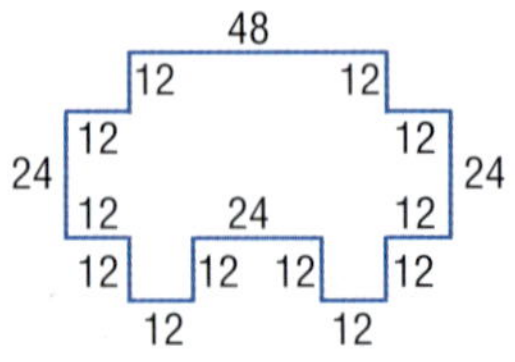

COORDINATE GEOMETRY **Graph each figure with the given vertices and identify the figure. Then find the perimeter and area of each figure.**

4. $X(3, 3)$, $Y(-2, 1)$, $Z(3, 1)$
5. $A(1, 6)$, $B(1, 2)$, $C(3, 2)$
6. $J(-3, 5)$, $K(2, 5)$, $L(2, -3)$, $M(-3, -3)$
7. $P(-3, 3)$, $Q(4, 3)$, $R(3, -1)$, $S(-4, -1)$

Lesson 1-7 Three-Dimensional Figures (pp. 67–74)

Determine whether the solid is a polyhedron. Then identify each solid. If it is a polyhedron, name the bases, faces, edges, and vertices.

1.

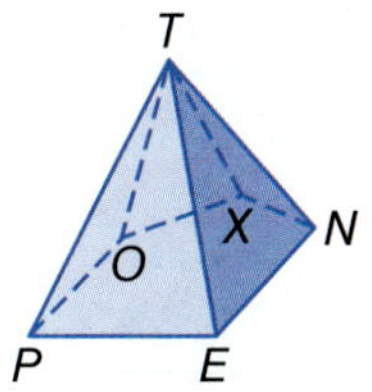

2.

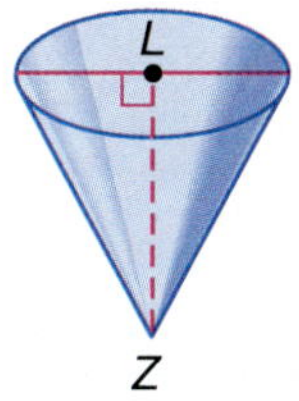

3.

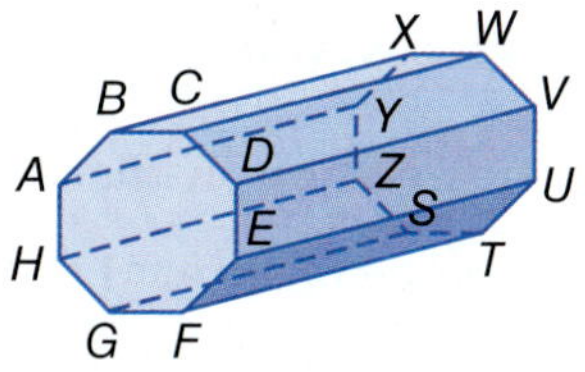

Find the surface area and volume of each solid to the nearest tenth.

4.

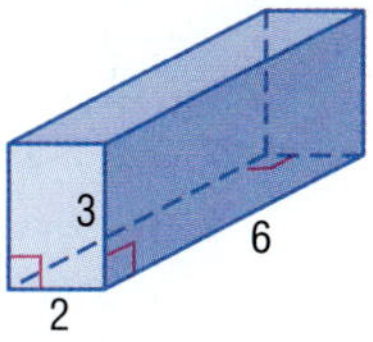

5.

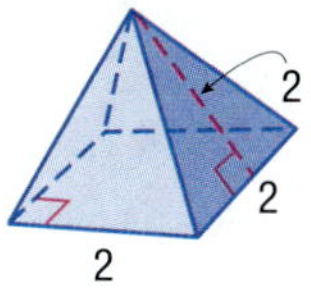

6.

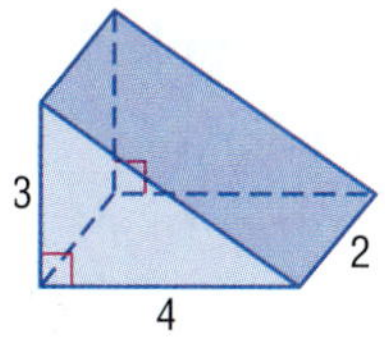

Lesson 2-1 Inductive Reasoning and Conjecture (pp. 89–96)

Make a conjecture about each value or geometric relationship.

1. Lines j and k are parallel.

2. $A(-1, -7)$, $B(4, -7)$, $C(4, -3)$, $D(-1, -3)$

3. $\overline{AB}$ bisects $\overline{CD}$ at K.

4. $\overrightarrow{SR}$ is an angle bisector of $\angle TSU$.

Write a conjecture that describes the pattern in each sequence. Then use your conjecture to find the next item in the sequence.

5.

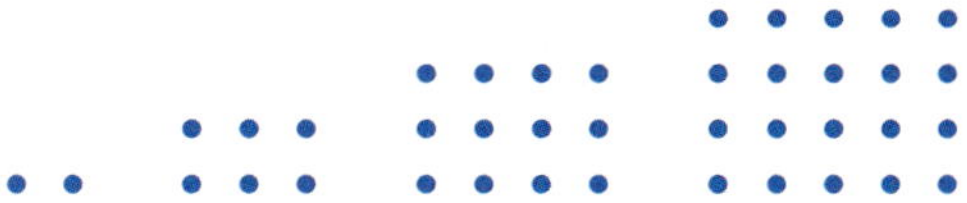

6.

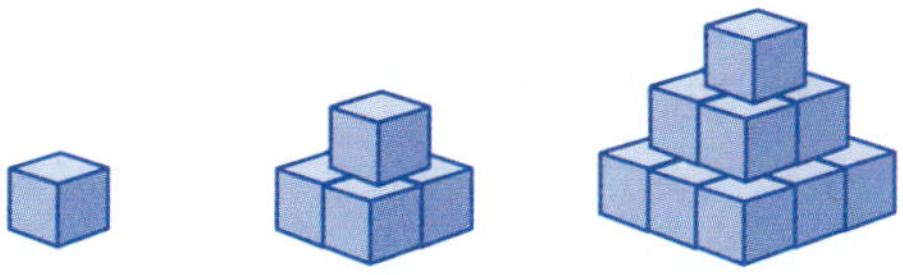

Lesson 2-2 Logic (pp. 97–104)

Use the following statements to write a compound statement for each conjunction and disjunction. Then find its truth value. Explain your reasoning.

p: $(-3)^2 = 9$

q: A robin is a fish.

r: An acute angle measures less than 90°.

1. p and q

2. p or q

3. p and r

4. p or r

5. $\sim p$ or q

6. p or $\sim r$

7. $q \wedge r$

8. $(p \wedge q) \vee r$

9. $\sim p \vee \sim r$

Copy and complete each truth table.

10.

p	q	$\sim q$	$p \vee \sim q$
T			
T			
F			
F			

11.

p	q	$\sim p$	$\sim q$	$\sim p \vee \sim q$
T	T			
T	F			
F	T			
F	F			

Lesson 2-3 Conditional Statements (pp. 105–113)

Identify the hypothesis and conclusion of each conditional statement.

1. If no sides of a triangle are equal, then it is a scalene triangle.
2. If it rains today, you will be wearing your raincoat.

Write each statement in if-then form.

3. The sum of the measures of two supplementary angles is 180.
4. A triangle with two congruent sides is an isosceles triangle.

Write the converse, inverse, and contrapositive of each conditional statement. Determine whether each related conditional is *true* or *false*. If a statement is false, find a counterexample.

5. All triangles are polygons.
6. If two angles are congruent angles, then they have the same measure.

Lesson 2-4 Deductive Reasoning (pp. 115–123)

Determine whether each conclusion is based on *inductive* or *deductive* reasoning.

1. At Thomas Jefferson High School, if you are present at school every day during a grading period, you get a perfect attendance pin. Nina has been at school every day during the grading period; therefore she will receive a perfect attendance pin.
2. Every Thursday Mike's neighbor washes his car. Today is Thursday, so Mike concludes his neighbor will wash his car.

Determine whether the stated conclusion is valid based on the information given. If not, write *invalid*. Explain your reasoning.

3. **Given:** If it rains, then the field will be muddy. If the field is muddy, then the game will be cancelled.

   Conclusion: If it rains, the game will be cancelled.

4. **Given:** If you read a book, then you enjoy reading. If you enjoy reading, then you have read *To Kill a Mockingbird*.

   Conclusion: If you read a book, then you have read *To Kill a Mockingbird*.

Lesson 2-5 Postulates and Paragraph Proofs (pp. 125–132)

Determine whether each statement is *always*, *sometimes*, or *never true*. Explain.

1. $\overleftrightarrow{RS}$ is perpendicular to $\overleftrightarrow{PS}$.
2. Three points will lie on one line.
3. Points B and C are in plane $\mathcal{K}$. A line perpendicular to line BC is in plane $\mathcal{K}$.

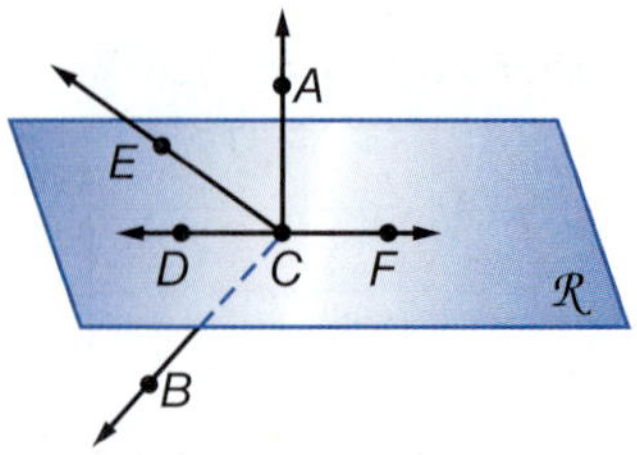

In the figure at the right, $\overleftrightarrow{EC}$ and $\overleftrightarrow{CD}$ are in plane $\mathcal{R}$, and F is on $\overleftrightarrow{CD}$. State the postulate that can be used to show each statement is true.

4. $\overleftrightarrow{DF}$ lies in plane $\mathcal{R}$.
5. E and C are collinear.
6. D, F, and E are coplanar.
7. E and F are collinear.

Lesson 2-6 Algebraic Proof (pp. 134–141)

State the property that justifies each statement.

1. If $x - 5 = 6$, then $x = 11$.

2. If $a - b = r$, then $r = a - b$.

3. If $AB = CD$ and $CD = EF$, then $AB = EF$.

4. If $HJ + 5 = 20$, then $HJ = 15$.

5. Complete the following proof.

Given: $\frac{5x - 1}{8} = 3$

Prove: $x = 5$

Proof:

Statements	Reasons
a. ?	**a.** Given
b. ?	**b.** Multiplication Prop.
c. $5x - 1 = 24$	**c.** ?
d. $5x = 25$	**d.** ?
e. ?	**e.** Division Property

Lesson 2-7 Proving Segment Relationships (pp. 142–148)

PROOF **Prove the following.**

1. If $\overline{AB} \cong \overline{AC}$ and $\overline{PC} \cong \overline{QB}$, then $\overline{AP} \cong \overline{AQ}$.

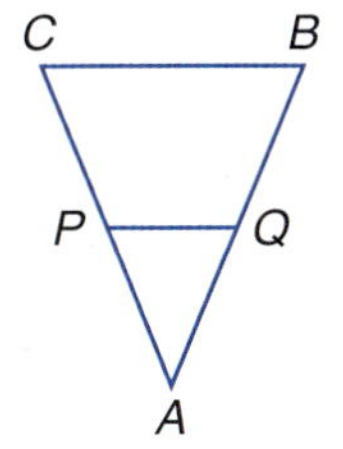

2. If $\overline{LM} \cong \overline{PN}$ and $\overline{XM} \cong \overline{XN}$, then $\overline{LX} \cong \overline{PX}$.

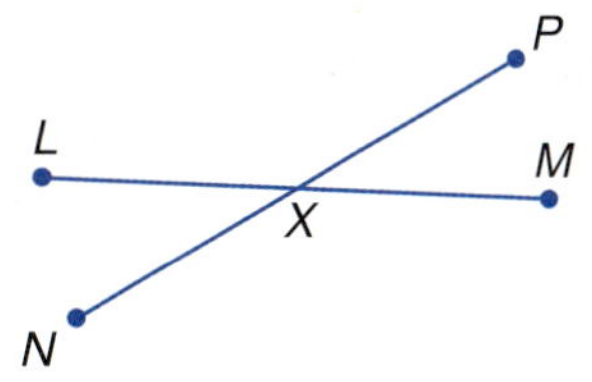

Lesson 2-8 Proving Angle Relationships (pp. 149–157)

Find the measure of each numbered angle, and name the theorems that justify your work.

1. $m\angle 9 = 141 + x$

$m\angle 10 = 25 + x$

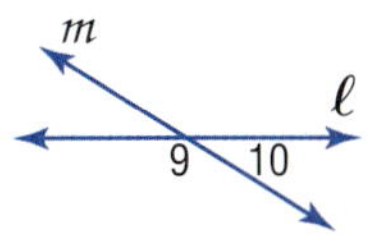

2. $m\angle 11 = x + 40$

$m\angle 12 = x + 10$

$m\angle 13 = 3x + 30$

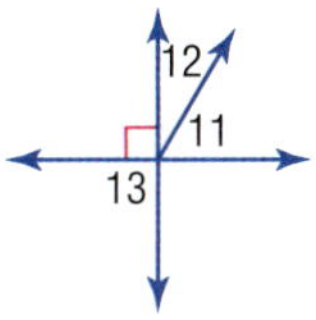

3. $m\angle 14 = x + 25$

$m\angle 15 = 4x + 50$

$m\angle 16 = x + 45$

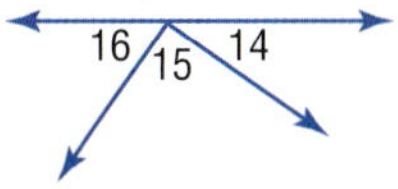

4. **PROOF** Write a two column proof.

Given: $\overrightarrow{VX}$ bisects $\angle WVY$.

$\overrightarrow{VY}$ bisects $\angle XVZ$.

Prove: $\angle WVX \cong \angle YVZ$.

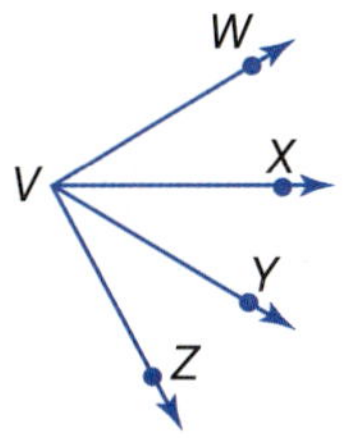

Lesson 3-1 Parallel Lines and Transversals (pp. 171–176)

Refer to the figure to identify each of the following.

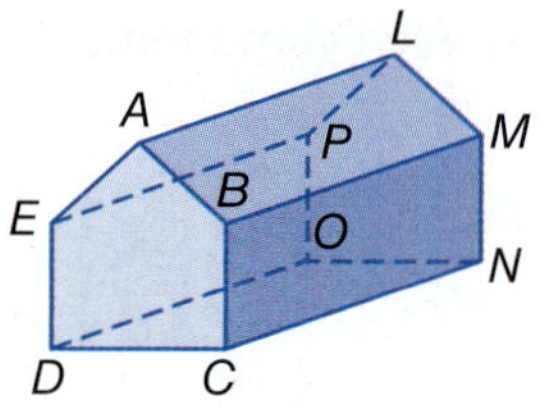

1. all segments parallel to $\overline{AE}$
2. all planes intersecting plane *BCN*
3. all segments skew to $\overline{DC}$

Identify the transversal connecting each pair of angles. Then classify the relationship between each pair of angles as *alternate interior, alternate exterior, corresponding,* or *consecutive interior* angles.

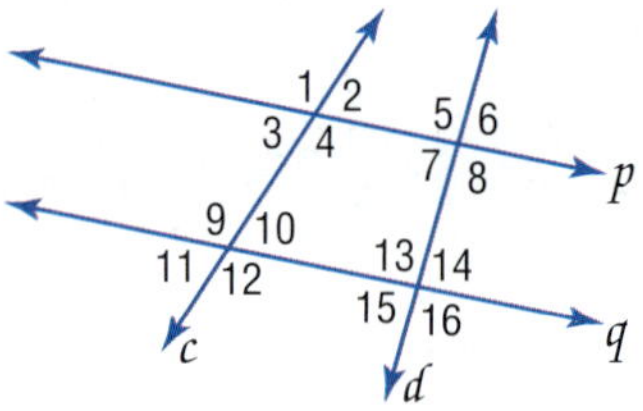

4. ∠2 and ∠5
5. ∠9 and ∠13
6. ∠12 and ∠13
7. ∠3 and ∠6

Lesson 3-2 Angles and Parallel Lines (pp. 178–184)

In the figure, $m\angle 3 = 43$. Find the measure of each angle.

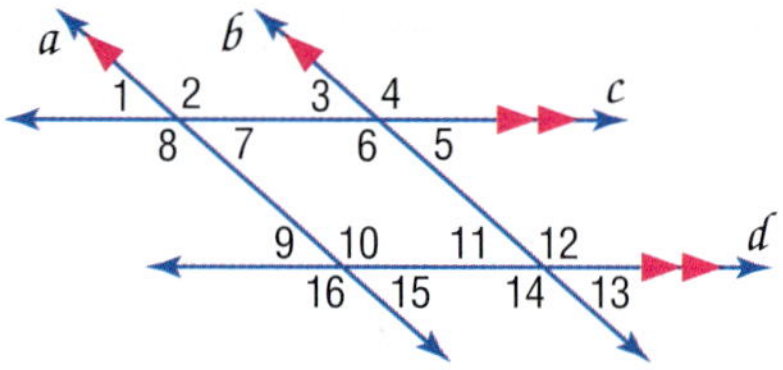

1. ∠2
2. ∠7
3. ∠10
4. ∠11
5. ∠6
6. ∠4

Find the value of the variable(s) in each figure. Explain your reasoning.

7.

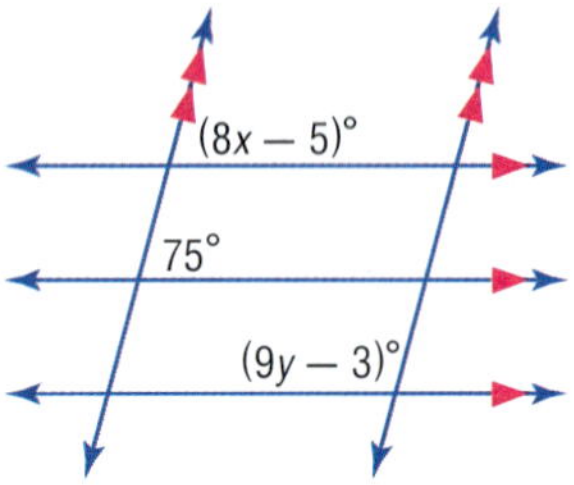

8.

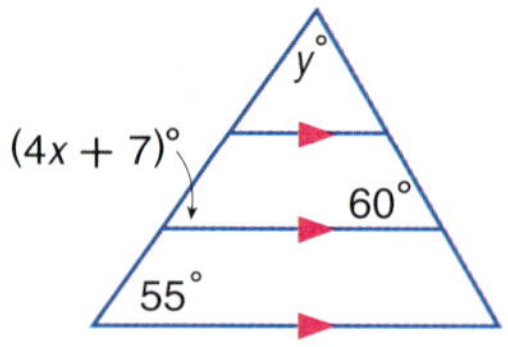

Lesson 3-3 Slopes of Lines (pp. 186–194)

Find the slope of each line.

1.

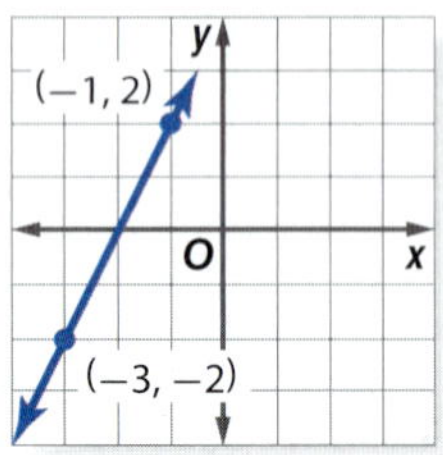

2.

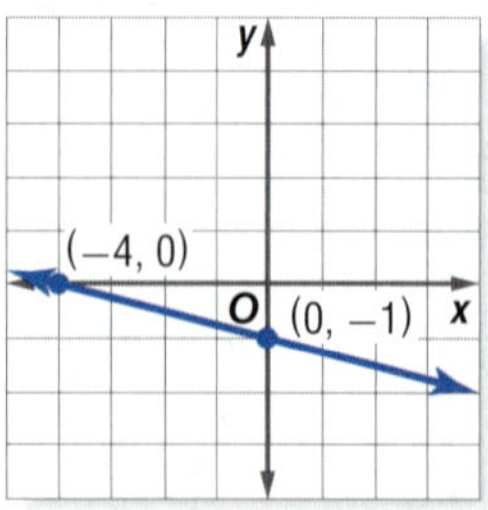

3.

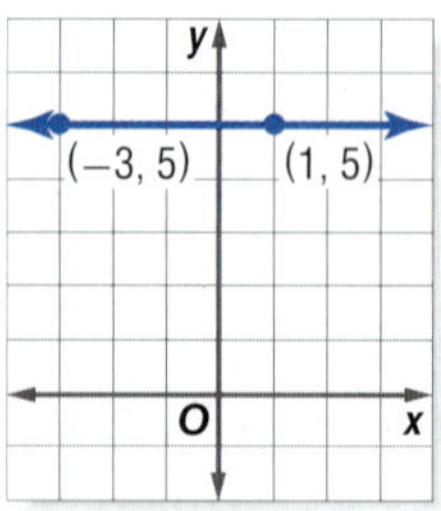

Determine whether $\overleftrightarrow{RS}$ and $\overleftrightarrow{TU}$ are *parallel, perpendicular,* or *neither.* Graph each line to verify your answer.

4. *R*(3, 5), *S*(5, 6), *T*(−2, 0), *U*(4, 3)
5. *R*(5, 11), *S*(2, 2), *T*(−1, 0), *U*(2, 1)
6. *R*(−1, 4), *S*(−3, 7), *T*(5, −1), *U*(8, 1)
7. *R*(−2, 5), *S*(−4, 1), *T*(3, 3), *U*(1, 5)

Lesson 3-4 Equations of Lines (pp. 196–203)

Write an equation in slope-intercept form of the line having the given slope and y-intercept. Then graph the line.

1. $m = 1$, y-intercept: -5 **2.** $m = -\frac{1}{2}$, y-intercept: $\frac{1}{2}$ **3.** $m = 3$, $b = -\frac{1}{4}$

Write an equation in point-slope form of the line having the given slope that contains the given point. Then graph the line.

4. $m = 3$, $(-2, 4)$ **5.** $m = -4$, $(0, 3)$ **6.** $m = \frac{2}{3}$, $(5, -7)$

Write an equation in slope-intercept form for each line.

7. p **8.** q

9. r **10.** s

11. parallel to line q, contains $(2, -5)$

12. perpendicular to line p, contains $(0, 0)$

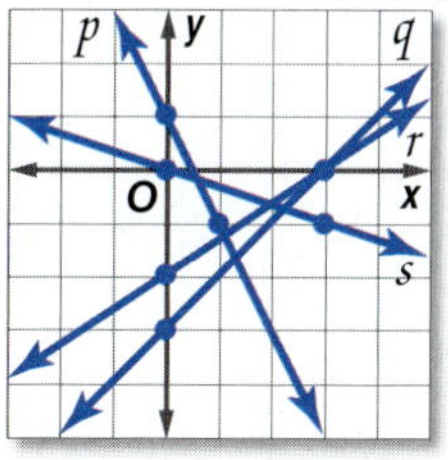

Lesson 3-5 Proving Lines Parallel (pp. 205–212)

Given the following information, determine which lines, if any, are parallel. State the postulate or theorem that justifies your answer.

1. $\angle 9 \cong \angle 16$ **2.** $\angle 10 \cong \angle 16$

3. $\angle 12 \cong \angle 13$ **4.** $m\angle 12 + m\angle 14 = 180$

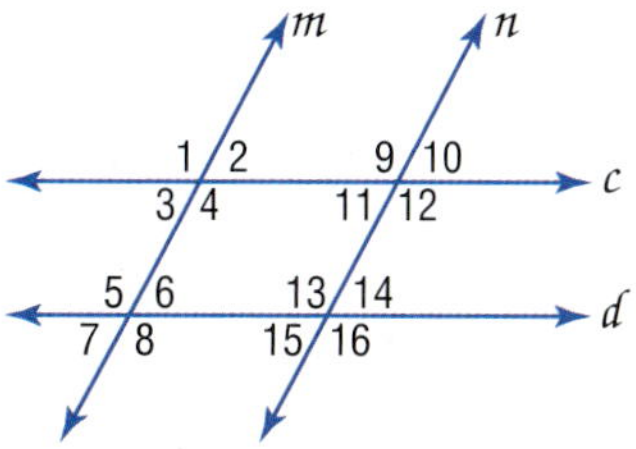

Find x so that $r \parallel s$. Identify the postulate or theorem you used.

5.

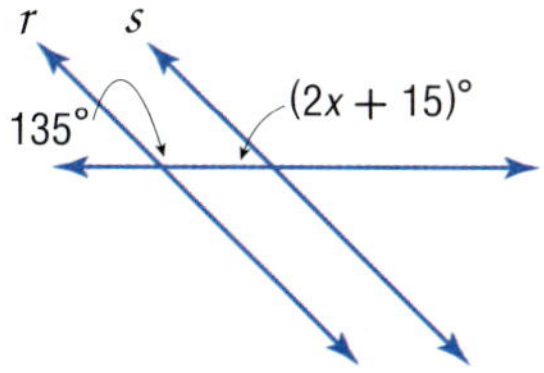

6.

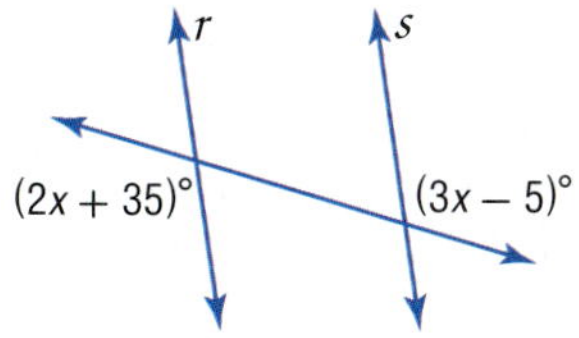

7.

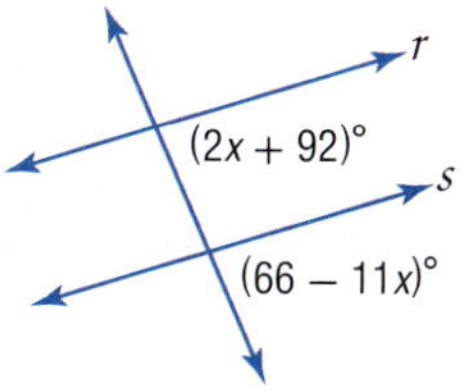

Lesson 3-6 Perpendiculars and Distance (pp. 213–222)

Copy each figure. Construct the segment that represents the distance indicated.

1. P to $\overleftrightarrow{RS}$

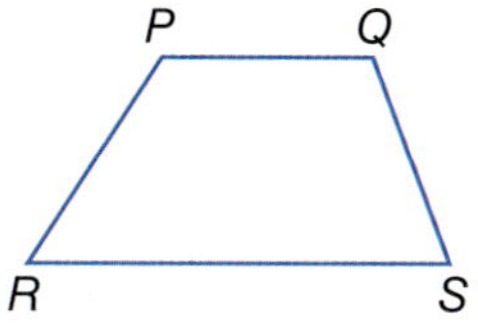

2. J to $\overleftrightarrow{KL}$

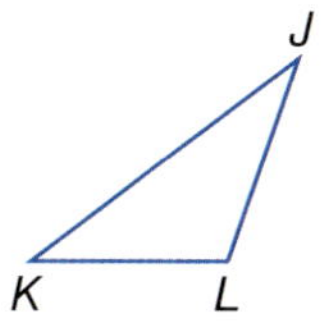

3. B to $\overleftrightarrow{FE}$

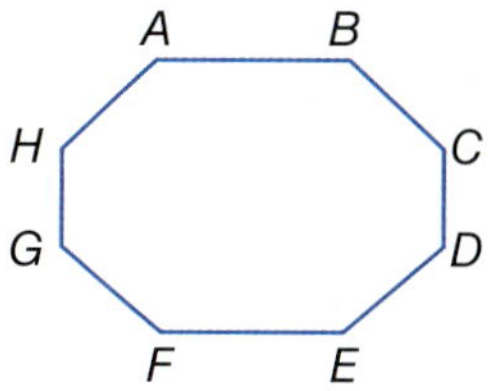

Find the distance between each pair of parallel lines with the given equations.

4. $y = \frac{2}{3}x - 2$
$y = \frac{2}{3}x + \frac{1}{2}$

5. $y = 2x + 4$
$y - 2x = -5$

6. $x + 4y = -6$
$x + 4y = 4$

COORDINATE GEOMETRY **Find the distance from P to ℓ.**

7. Line ℓ contains points $(0, 4)$ and $(-4, 0)$. Point P has coordinates $(2, -1)$.

8. Line ℓ contains points $(3, -2)$ and $(0, 2)$. Point P has coordinates $(-2.5, 3)$.

Lesson 4-1 Classifying Triangles (pp. 235–242)

Classify each triangle as *acute, equiangular, obtuse,* or *right*.

1.

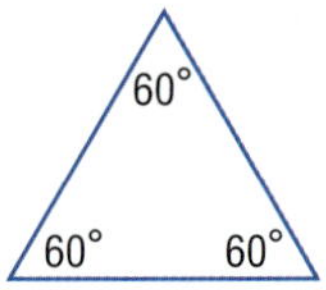

2.

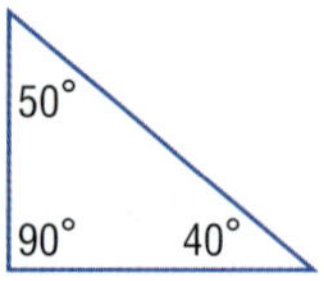

3.

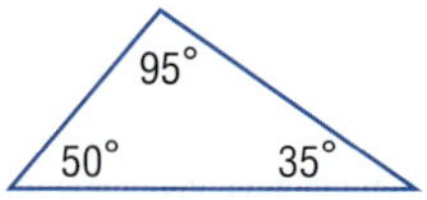

If point Q is the midpoint of $\overline{JL}$, and R is the midpoint of $\overline{JK}$, classify each triangle as *equilateral, isosceles,* or *scalene*.

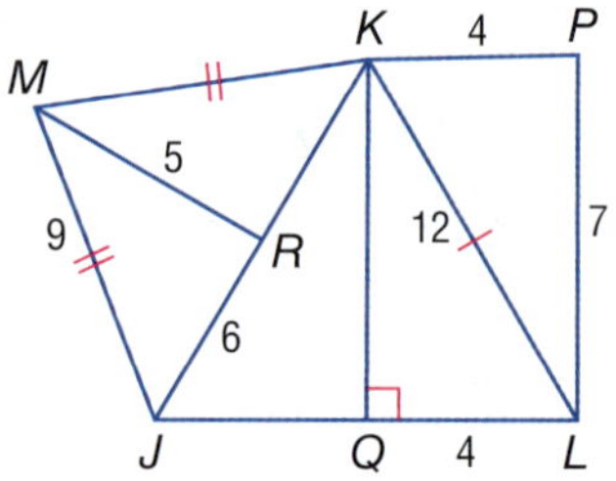

4. $\triangle KPL$

5. $\triangle QKL$

6. $\triangle JMK$

7. $\triangle KJQ$

8. $\triangle LKJ$

9. $\triangle MRK$

Lesson 4-2 Angles of Triangles (pp. 244–252)

Find each measure.

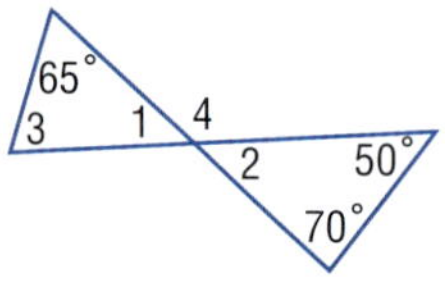

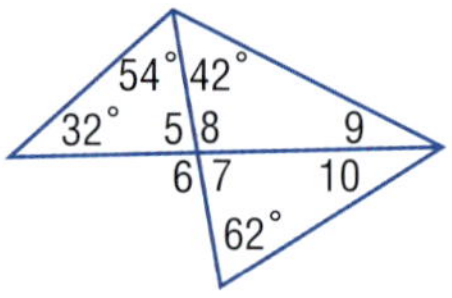

1. $m\angle 1$

2. $m\angle 2$

3. $m\angle 3$

4. $m\angle 4$

5. $m\angle 5$

6. $m\angle 6$

7. $m\angle 7$

8. $m\angle 8$

9. $m\angle 9$

10. $m\angle 10$

Lesson 4-3 Congruent Triangles (pp. 253–261)

Show that polygons are congruent by identifying all congruent corresponding parts. Then write a congruence statement.

1.

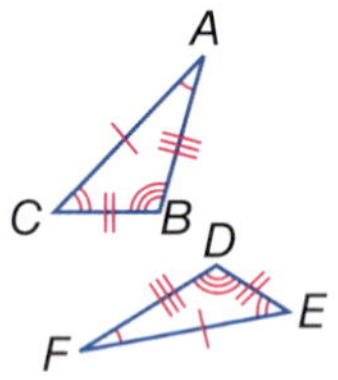

2.

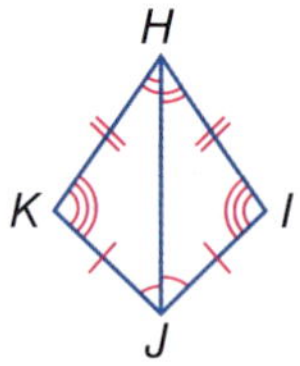

3.

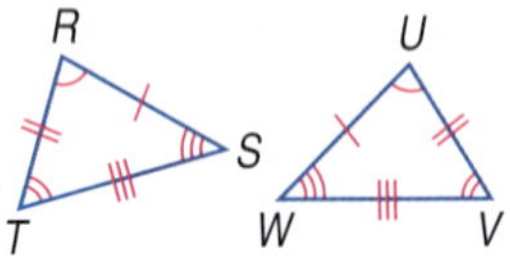

4.

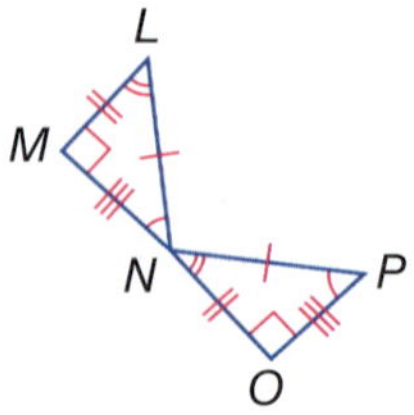

5. Write a two-column proof.

Given: $\triangle ANG \cong \triangle NGA$
$\triangle NGA \cong \triangle GAN$

Prove: $\triangle AGN$ is equilateral and equiangular.

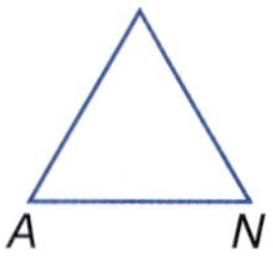

Lesson 4-4 Proving Congruence: SSS, SAS (pp. 262–270)

Determine whether $\triangle RST \cong \triangle JKL$. Explain.

1. $R(-6, 2)$, $S(-4, 4)$, $T(-2, 2)$, $J(6, -2)$, $K(4, -4)$, $L(2, -2)$
2. $R(-6, 3)$, $S(-4, 7)$, $T(-2, 3)$, $J(2, 3)$, $K(5, 7)$, $L(6, 3)$

PROOF **Write a two-column proof.**

3. **Given:** $\triangle GWN$ is equilateral.
$\overline{WS} \cong \overline{WI}$
$\angle SWG \cong \angle IWN$
Prove: $\triangle SWG \cong \triangle IWN$

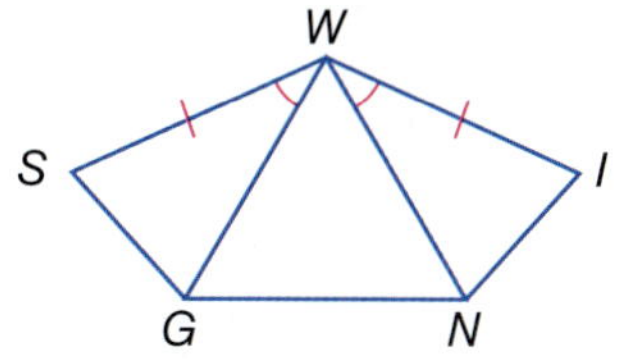

4. **Given:** $\triangle ANM \cong \triangle ANI$
$\overline{DI} \cong \overline{OM}$
$\overline{ND} \cong \overline{NO}$
Prove: $\triangle DIN \cong \triangle OMN$

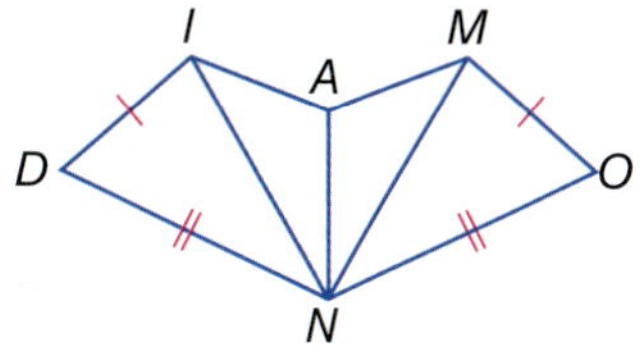

Lesson 4-5 Proving Triangles Congruent–ASA, AAS (pp. 273–280)

PROOF **Write a paragraph proof.**

1. **Given:** $\triangle TEN$ is isosceles with base $\overline{TN}$.
$\angle 1 \cong \angle 4$, $\angle T \cong \angle N$
Prove: $\triangle TEC \cong \triangle NEA$

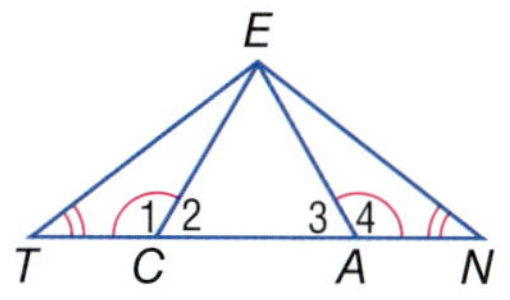

2. **Given:** $\angle S \cong \angle W$
$\overline{SY} \cong \overline{YW}$
Prove: $\overline{ST} \cong \overline{WV}$

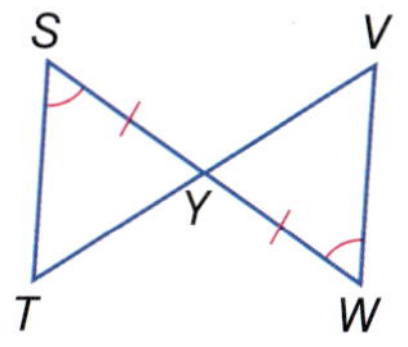

PROOF **Write a flow proof.**

3. **Given:** $\angle 1 \cong \angle 2$, $\angle 3 \cong \angle 4$
Prove: $\overline{PT} \cong \overline{LX}$

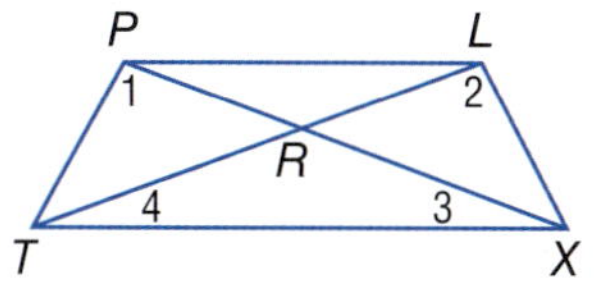

4. **Given:** $\overline{FP} \parallel \overline{ML}$, $\overline{FL} \parallel \overline{MP}$
Prove: $\overline{MP} \cong \overline{FL}$

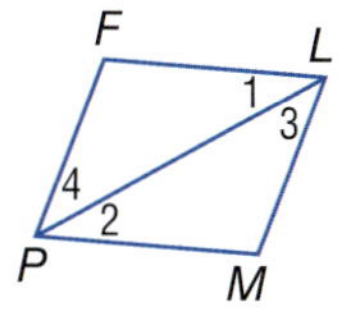

Lesson 4-6 Isosceles and Equilateral Triangles (pp. 283–291)

Refer to the figure at the right.

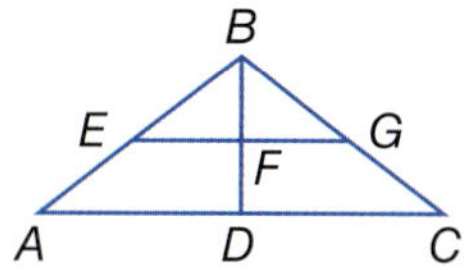

1. If $\overline{AD} \cong \overline{BD}$, name two congruent angles.
2. If $\overline{BF} \cong \overline{FG}$, name two congruent angles.
3. If $\overline{BE} \cong \overline{BG}$, name two congruent angles.
4. If $\angle FBE \cong \angle FEB$, name two congruent segments.
5. If $\angle BCA \cong \angle BAC$, name two congruent segments.
6. If $\angle DBC \cong \angle BCD$, name two congruent segments.

Lesson 4-7 Congruence Transformations (pp. 294–300)

Identify the type of congruence transformation shown as a *reflection, translation,* or *rotation*.

1.

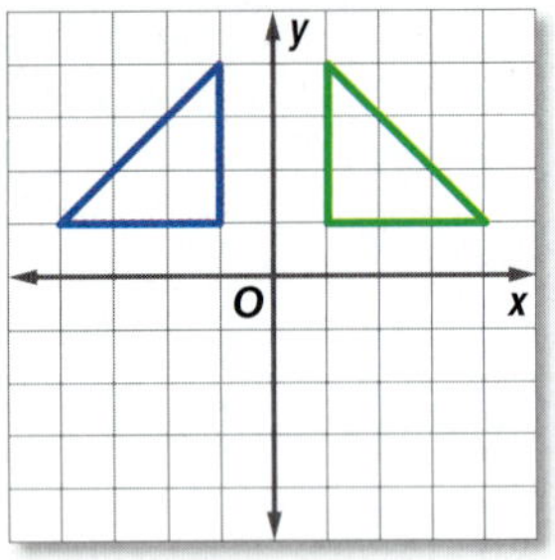

2.

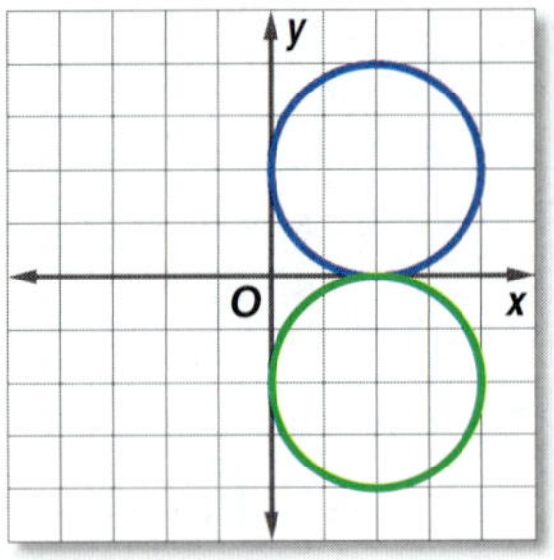

3.

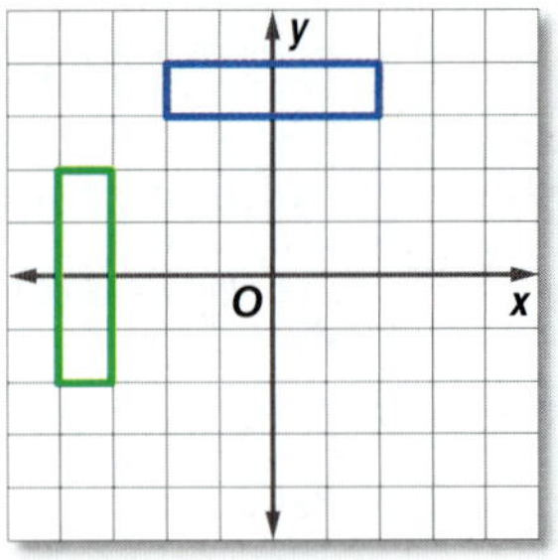

4.

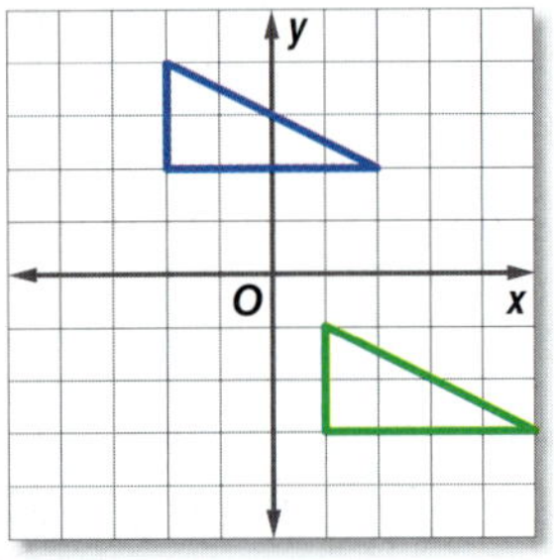

5.

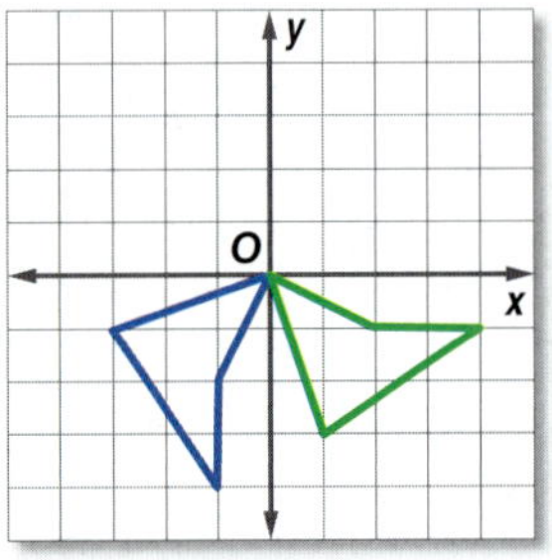

6.

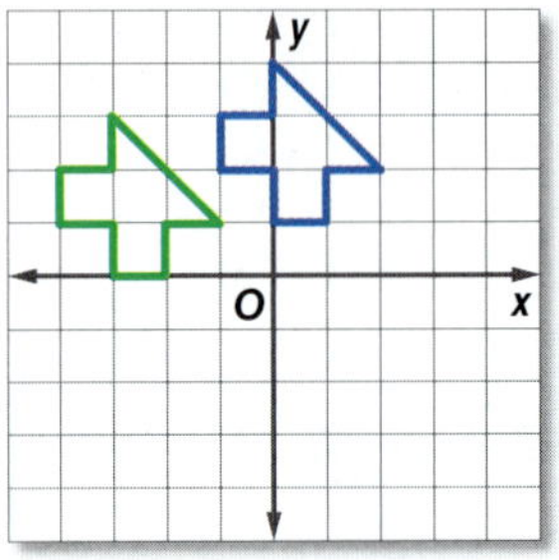

Lesson 4-8 Triangles and Coordinate Proof (pp. 301–307)

Position and label each triangle on the coordinate plane.

1. isosceles $\triangle ABC$ with base $\overline{BC}$ that is r units long

2. equilateral $\triangle XYZ$ with sides $4b$ units long

Name the missing coordinate(s) of each triangle.

3.

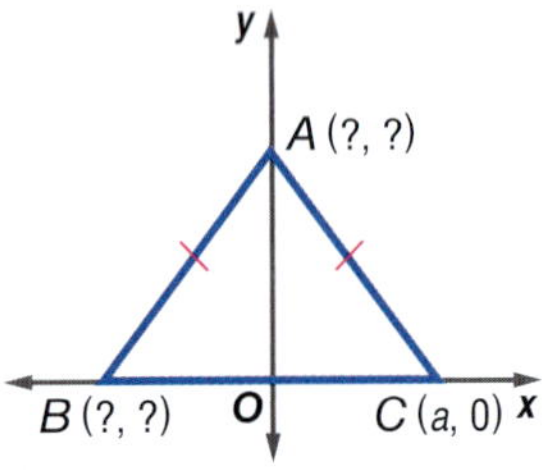

4.

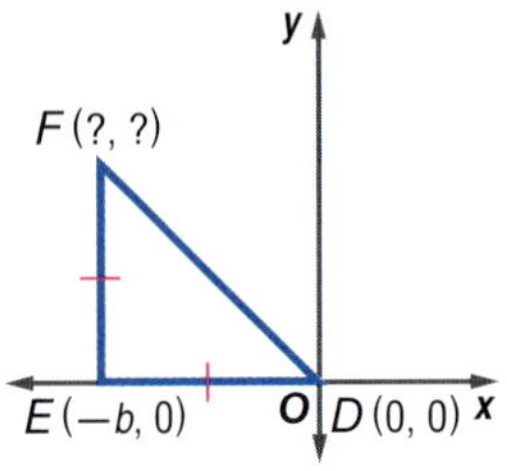

5.

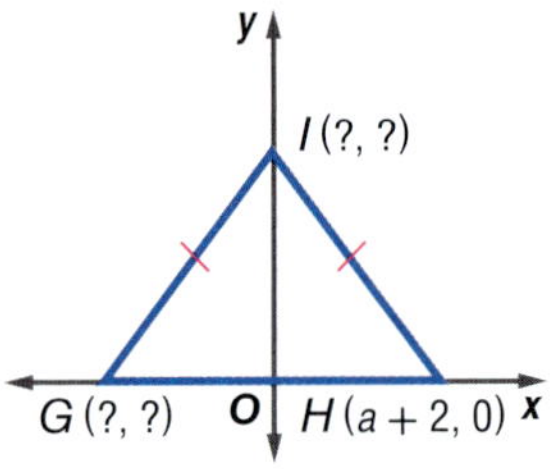

Lesson 5-1 Bisectors of Triangles (pp. 322–331)

Find each measure.

1. JL

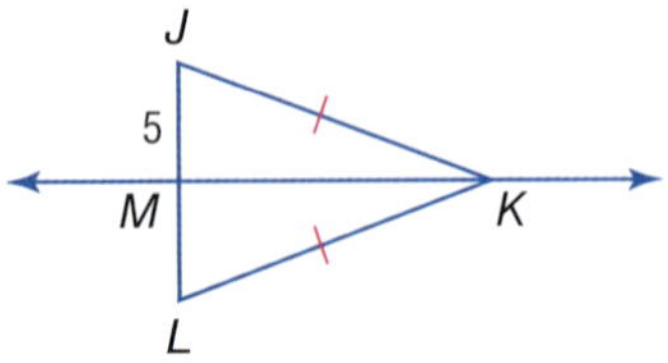

2. XY

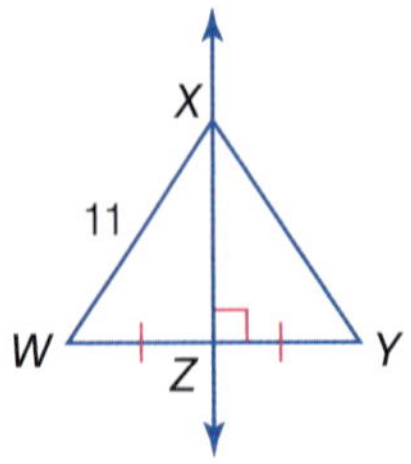

3. BC

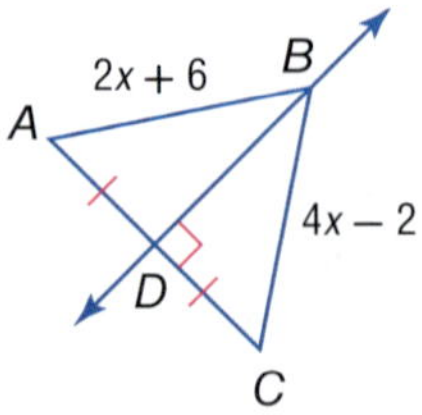

4. Find LC if M is the incenter of $\triangle ABC$.

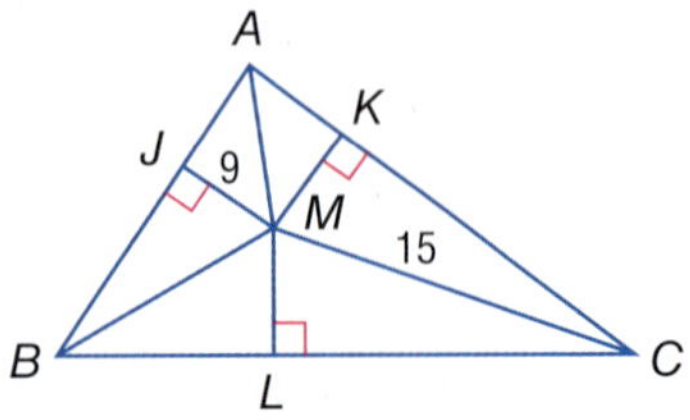

Lesson 5-2 Medians and Altitudes of Triangles (pp. 333–341)

In $\triangle XYZ$, Q is the centroid, $ZQ = 10$, and $XA = 12$.

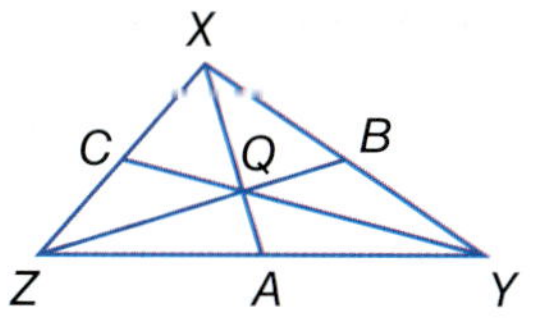

1. XQ
2. QA
3. BZ

State whether each sentence is *always, sometimes,* or *never* true.

4. The circumcenter and incenter of a triangle are the same point.
5. The three altitudes of a triangle intersect at a point inside the triangle.
6. In an equilateral triangle, the circumcenter, incenter, and centroid are the same point.
7. The incenter is inside of a triangle.

Lesson 5-3 Inequalities in One Triangle (pp. 342–349)

Use the figure at the right to determine the relationship between the measures of the given angles.

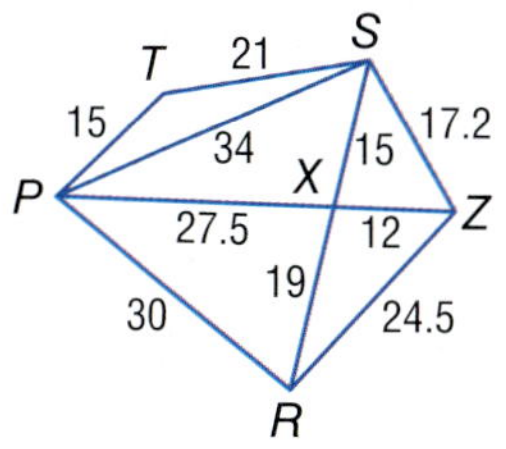

1. $\angle TPS$, $\angle TSP$
2. $\angle PRZ$, $\angle ZPR$
3. $\angle SPZ$, $\angle SZP$
4. $\angle SPR$, $\angle SRP$

List the angles and sides of each triangle in order from smallest to largest.

5.

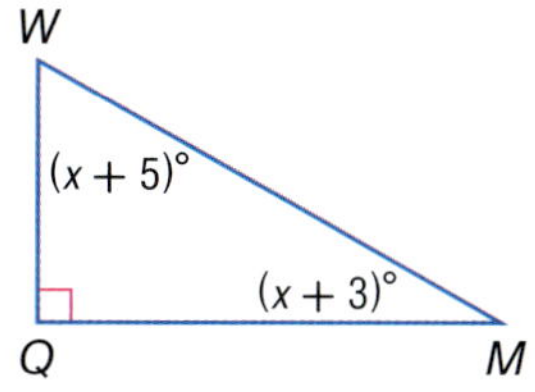

6.

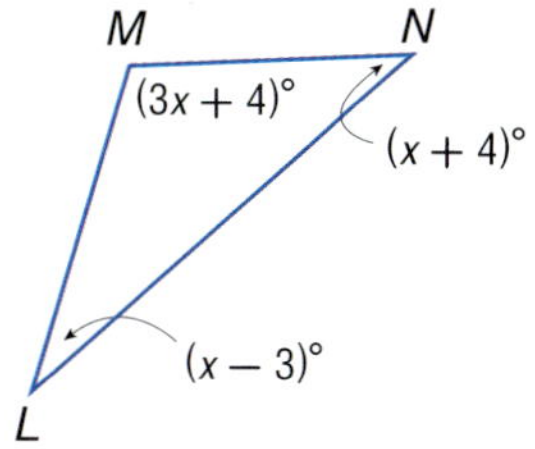

Lesson 5-4 Indirect Proof (pp. 351–358)

State the assumption you would make to start an indirect proof of each statement.

1. $\angle ABC \cong \angle XYZ$
2. $\overrightarrow{RS}$ bisects $\angle ARC$.
3. An angle bisector of an equilateral triangle is also a median.

Write an indirect proof of each statement.

4. **Given:** $\angle AOY \cong \angle AOX$; $\overline{XO} \not\cong \overline{YO}$

   Prove: $\overrightarrow{AO}$ is not the angle bisector of $\angle XAY$.

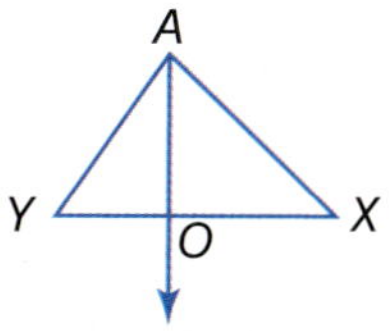

5. **Given:** $\triangle RUN$

   Prove: There can be no more than one right angle in $\triangle RUN$.

Lesson 5-5 The Triangle Inequality (pp. 360–366)

Is it possible to form a triangle with the given side lengths? If not, explain why not.

1. 2, 2, 6 **2.** 2, 3, 4 **3.** 6, 8, 10 **4.** 1, 1, 2

5. 15, 20, 30 **6.** 1, 3, 5 **7.** 2.5, 3.5, 6.5 **8.** 0.3, 0.4, 0.5

Find the range for the measure of the third side of a triangle given the measures of two sides.

9. 6 in. and 10 in. **10.** 2 cm and 5 cm **11.** 20 ft and 12 ft **12.** 8 m and 8 m

13. 18 km and 36 km **14.** 32 yd and 34 yd **15.** 2 mm and 29 mm **16.** 80 in. and 25 in.

Write a two-column proof.

17. Given: $RS = RT$
Prove: $UV + VS > UT$

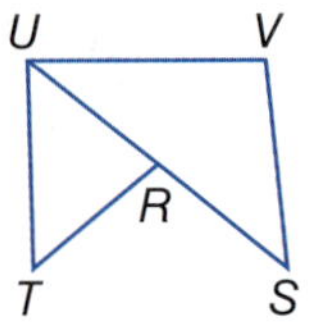

18. Given: quadrilateral $ABCD$
Prove: $AD + CD + AB > BC$

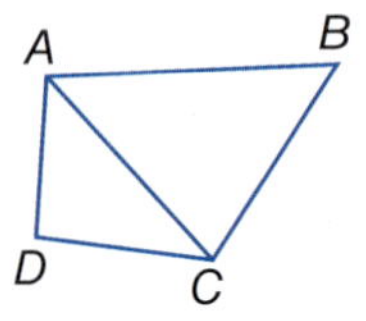

Lesson 5-6 Inequalities in Two Triangles (pp. 367–376)

Compare the given measures.

1. XZ and OZ

2. $m\angle ZIO$ and $m\angle ZUX$

3. $m\angle AEZ$ and $m\angle AZE$

4. IO and AE

5. $m\angle AZE$ and $m\angle IZO$

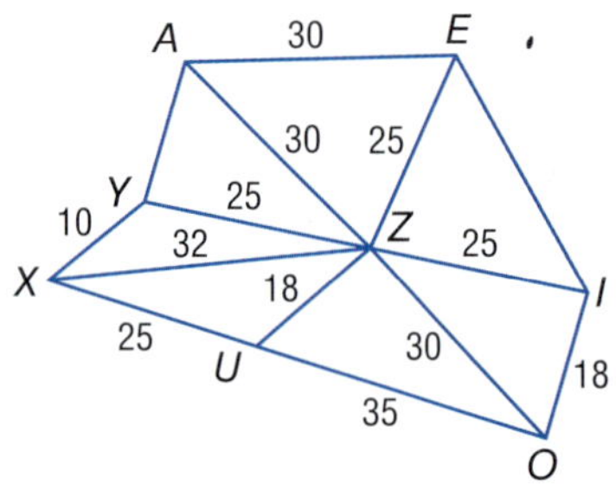

Write an inequality for the range of values for x.

6.

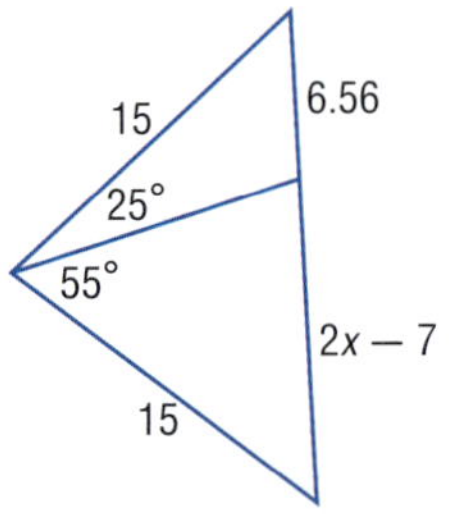

7.

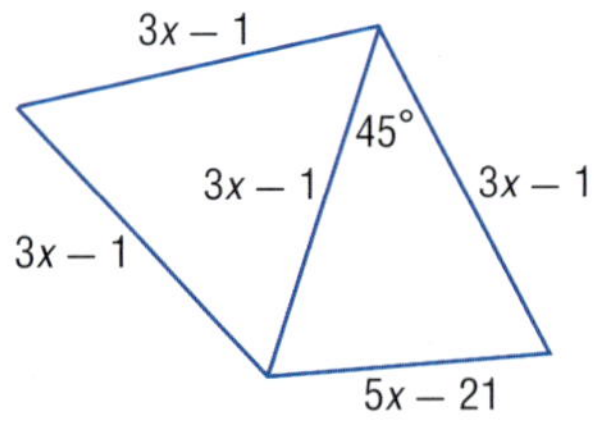

Lesson 6-1 Angles of Polygons (pp. 389–397)

Find the sum of the measures of the interior angles of each convex polygon.

1. 25-gon **2.** 30-gon **3.** 22-gon

4. 17-gon **5.** $5a$-gon **6.** b-gon

The measure of an interior angle of a regular polygon is given. Find the number of sides in each polygon.

7. 156 **8.** 168 **9.** 162

Find the measures of an exterior angle and an interior angle given the number of sides of a regular polygon. Round to the nearest tenth, if necessary.

10. 15 **11.** 13 **12.** 42

Lesson 6-2 Parallelograms (pp. 399–407)

Use ▱$RSTU$ to find each measure.

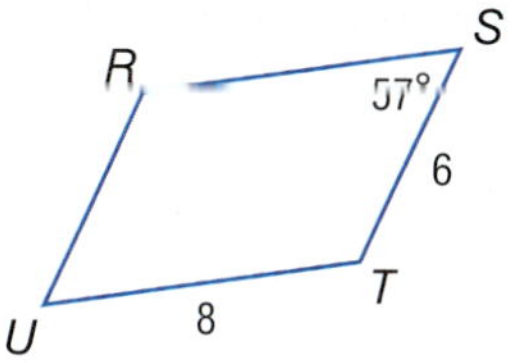

1. $m\angle TUR$
2. RU
3. SR
4. $m\angle UTS$

COORDINATE GEOMETRY **Find the coordinates of the intersection of the diagonals of ▱$ABCD$ with the given vertices.**

5. $A(0, 5)$, $B(6, 3)$, $C(6, -3)$, $D(0, -1)$
6. $A(-1, 2)$, $B(3, 2)$, $C(1, -4)$, $D(-3, -4)$

Lesson 6-3 Tests for Parallelograms (pp. 409–417)

Determine whether each quadrilateral is a parallelogram. Justify your answer.

1. 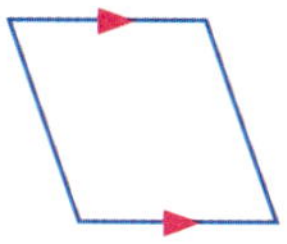

2. 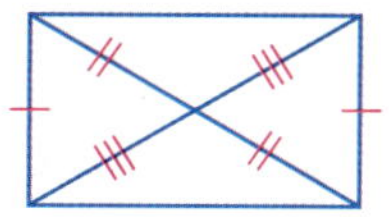

3. 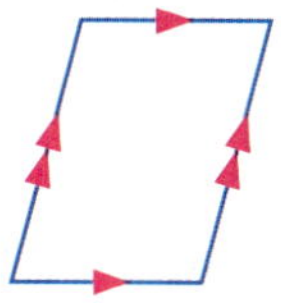

ALGEBRA **Find x and y so that each quadrilateral is a parallelogram.**

4. 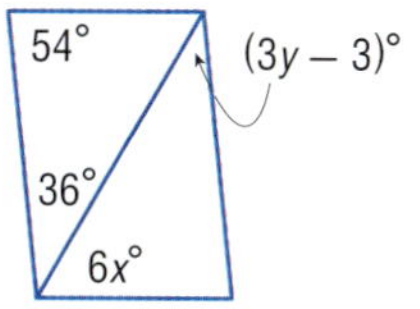

5. 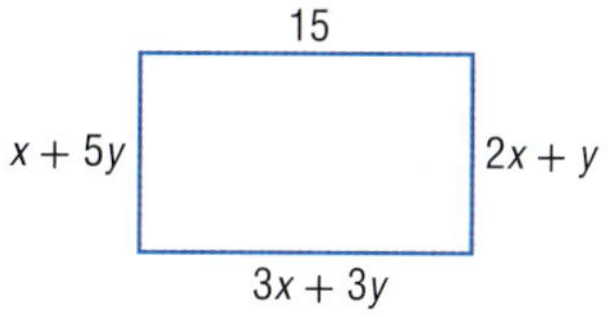

6. 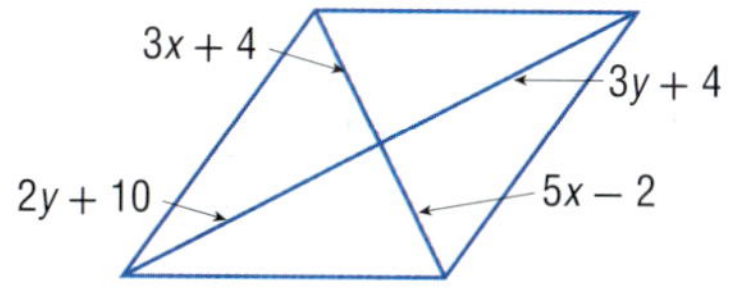

COORDINATE GEOMETRY **Graph each quadrilateral with the given vertices. Determine whether the figure is a parallelogram. Justify your answer with the method indicated.**

7. $L(-3, 2)$, $M(5, 2)$, $N(3, -6)$, $O(-5, -6)$; Slope Formula
8. $W(-5, 6)$, $X(2, 5)$, $Y(-3, -4)$, $Z(-8, -2)$; Distance Formula
9. $Q(-5, 4)$, $R(0, 6)$, $S(3, -1)$, $T(-2, -3)$; Midpoint Formula
10. $G(-5, 0)$, $H(-13, 5)$, $I(-10, 9)$, $J(-2, 4)$; Distance and Slope Formulas

Lesson 6-4 Rectangles (pp. 419–425)

Quadrilateral $LMNO$ is a rectangle. Find each measure if $m\angle 5 = 38$.

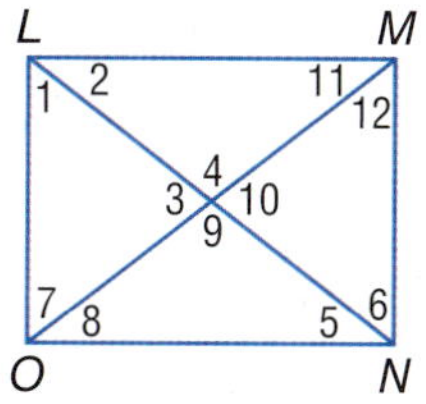

1. $m\angle 1$
2. $m\angle 2$
3. $m\angle 3$
4. $m\angle 4$
5. $m\angle 6$
6. $m\angle 7$

ALGEBRA **Quadrilateral $QRST$ is a rectangle.**

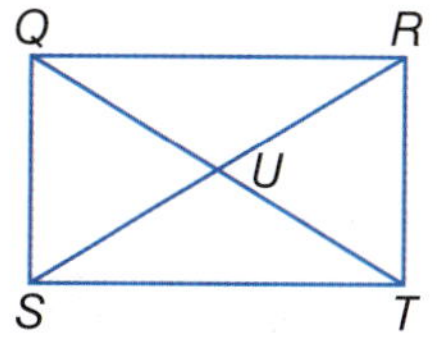

7. If $QU = 2x + 3$ and $UT = 4x - 9$, find SU.
8. If $RU = 3x - 6$ and $UT = x + 9$, find RS.
9. If $m\angle STQ = 5x + 3$ and $m\angle RTQ = x + 3$, find x.
10. If $m\angle SRQ = x^2 + 6$ and $m\angle RST = 36 - x$, find $m\angle SRT$.

Extra Practice

Lesson 6-5 Rhombi and Squares (pp. 426–434)

Quadrilateral $QRST$ is a rhombus. If $m\angle QRS = m\angle TSR - 40$ and $TS = 15$; find each measure.

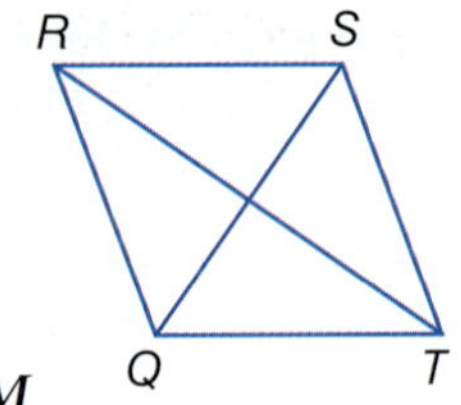

1. $m\angle TSQ$
2. $m\angle QRS$
3. $m\angle SRT$
4. QR

COORDINATE GEOMETRY **Given each set of vertices, determine whether $\square JKLM$ is a *rhombus*, a *rectangle*, or a *square*. List all that apply. Explain.**

5. $J(1, 10), K(-4, 0), L(7, 2), M(12, 12)$
6. $J(-7, 3), K(-2, 3), L(1, 7), M(-4, 7)$
7. $J(1, 5), K(6, 5), L(6, 10), M(1, 10)$
8. $J(-2, -1), K(-4, 3), L(1, 5), M(3, 1)$

Lesson 6-6 Kites and Trapezoids (pp. 435–444)

COORDINATE GEOMETRY **For each quadrilateral with the given vertices, verify that the quadrilateral is a trapezoid and determine whether the figure is an isosceles trapezoid.**

1. $A(0, 9), B(3, 4), C(-5, 4), D(-2, 9)$
2. $Q(1, 4), R(4, 6), S(10, 7), T(1, 1)$
3. $L(1, 2), M(4, -1), N(3, -5), O(-3, 1)$
4. $W(1, -2), X(3, -1), Y(7, -2), Z(1, -5)$

5. For trapezoid $ABCD$, E and F are midpoints of the legs. Find CD.

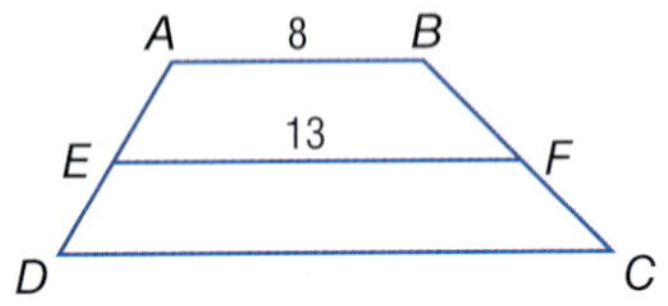

6. For isosceles trapezoid $QRST$, find the length of the median, $m\angle S$, and $m\angle R$.

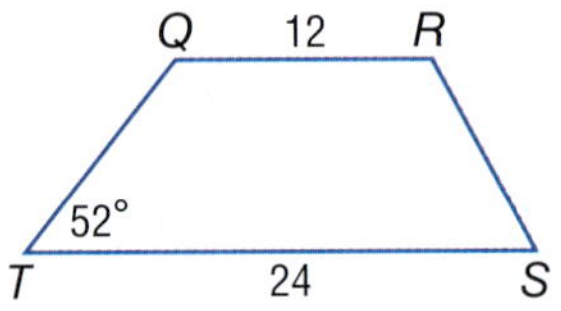

Lesson 7-1 Ratios and Proportions (pp. 457–463)

1. **ARCHITECTURE** The ratio of the height of a model of a house to the actual house is 1:63. If the width of the model is 16 inches, find the width of the actual house in feet.

2. **CONSTRUCTION** A 64-inch long board is divided into lengths in the ratio 2:3. What are the two lengths into which the board is divided?

Solve each proportion.

3. $\frac{x+4}{26} = -\frac{1}{3}$
4. $\frac{3x+1}{14} = \frac{5}{7}$
5. $\frac{x-3}{4} = \frac{x+1}{5}$
6. $\frac{2x+2}{2x-1} = \frac{1}{3}$

7. The ratio of the measures of the three sides of a triangle is 9:6:5. Its perimeter is 100 inches. Find the measure of each side.

8. The ratio of the measures of the three sides of a triangle is $\frac{1}{2} : \frac{1}{3} : \frac{1}{5}$. Its perimeter is 6.2 centimeters. Find the measure of each side.

Lesson 7-2 Similar Polygons (pp. 465–473)

List all pairs of congruent angles, and write a proportion that relates the corresponding sides for each pair of similar polygons.

1. $\triangle ABC \sim \triangle XYZ$

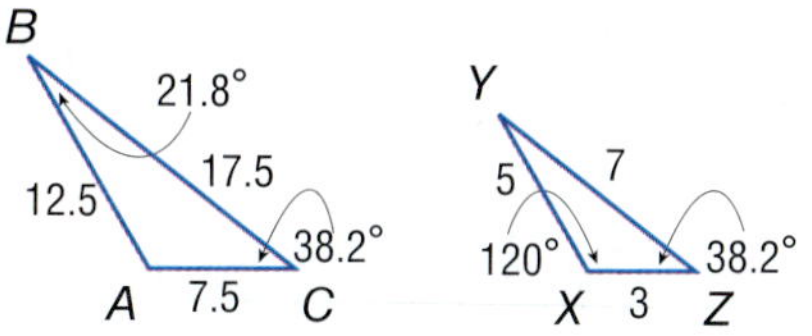

2. $RSTU \sim VWXY$

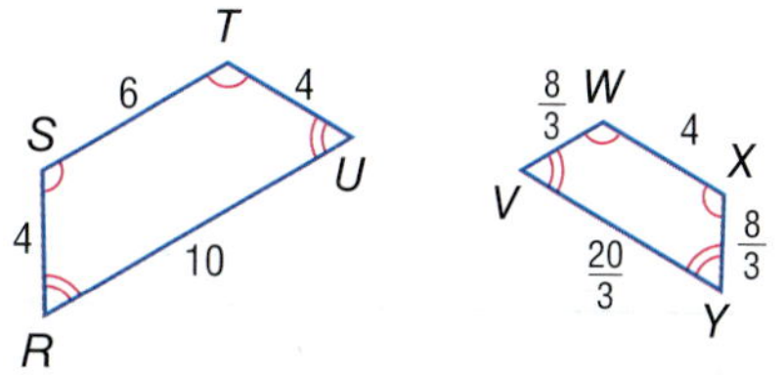

3. Two similar rectangles have a scale factor of 6:3. The perimeter of the small rectangle is 20 meters. Find the perimeter of the large rectangle.

4. Two similar squares have a scale factor of 2:5. The perimeter of the large square is 45 inches. Find the perimeter of the small square.

Lesson 7-3 Similar Triangles (pp. 474–483)

Determine whether the triangles are similar. If so, write a similarity statement. Explain your reasoning.

1.

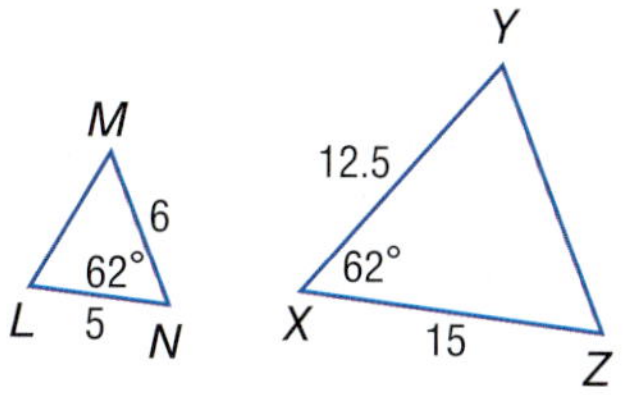

2.

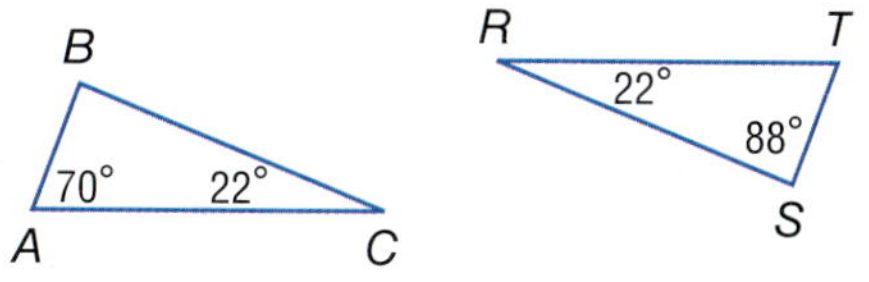

ALGEBRA **Identify the similar triangles. Find each measure.**

3. RT, SV

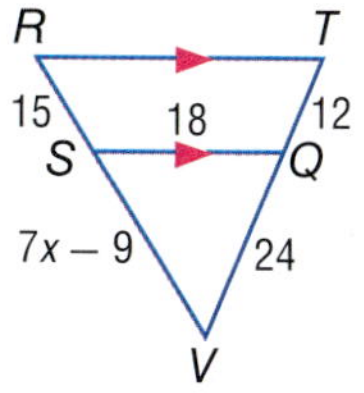

4. PN, MN

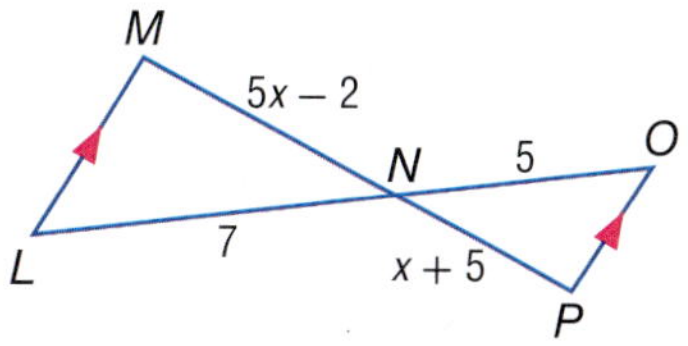

Lesson 7-4 Parallel Lines and Proportional Parts (pp. 484–493)

1. If $HI = 28$, $LH = 21$, and $LK = 8$, find IJ.

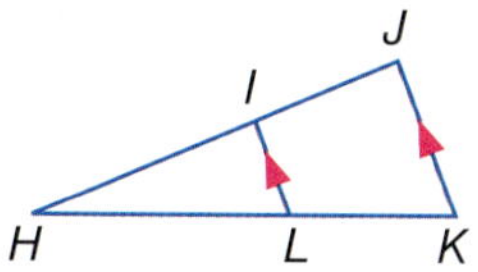

2. Find x, AD, DR, and QR if $AU = 15$, $QU = 25$, $AD = 3x + 6$, $DR = 8x - 2$, and $UD = 15$.

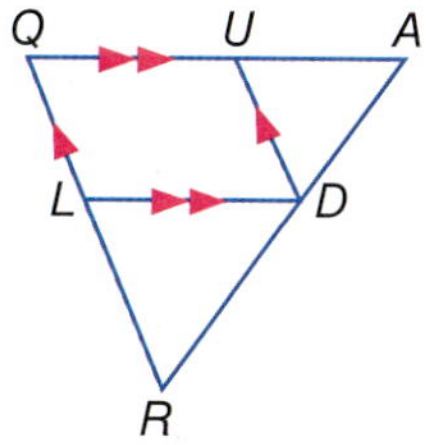

Determine the value of x so that $\overline{XY} \parallel \overline{LM}$.

3. $XL = 3$, $YM = 5$, $LD = 9$, $MD = x + 3$

4. $YM = 3$, $LD = 3x + 1$, $XL = 4$, $MD = x + 7$

5. $MD = 5x - 6$, $YM = 3$, $LD = 5x + 1$, $XL = 5$

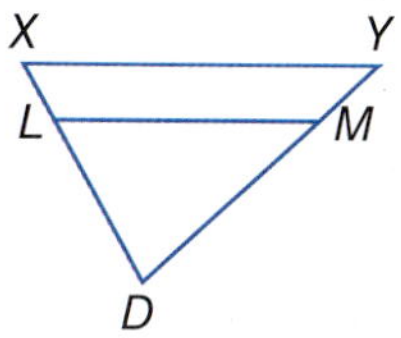

Extra Practice

Lesson 7-5 Parts of Similar Triangles (pp. 495–502)

Find x.

1.

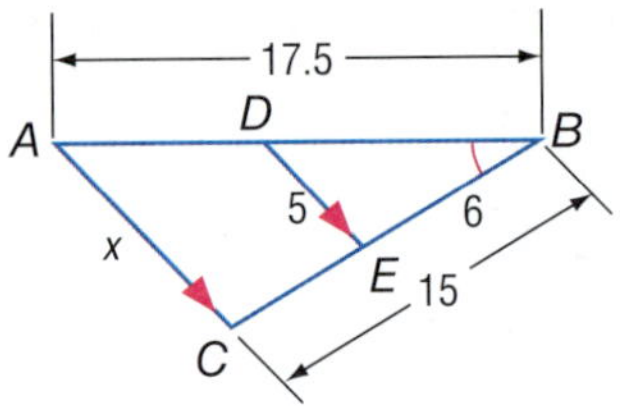

2.

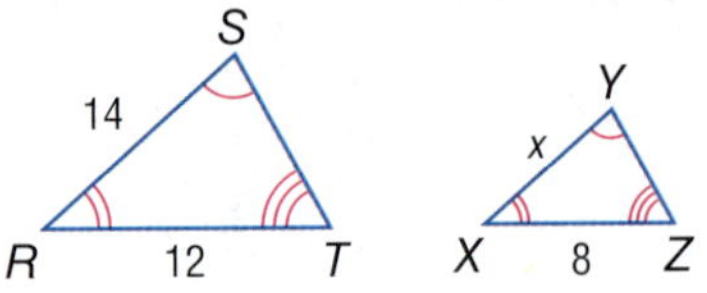

3. **PROOF** Write a paragraph proof.

Given: $\triangle ABC \sim \triangle PQR$

$\overline{BD}$ is an altitude of $\triangle ABC$.

$\overline{QS}$ is an altitude of $\triangle PQR$.

Prove: $\frac{QP}{BA} = \frac{QS}{BD}$

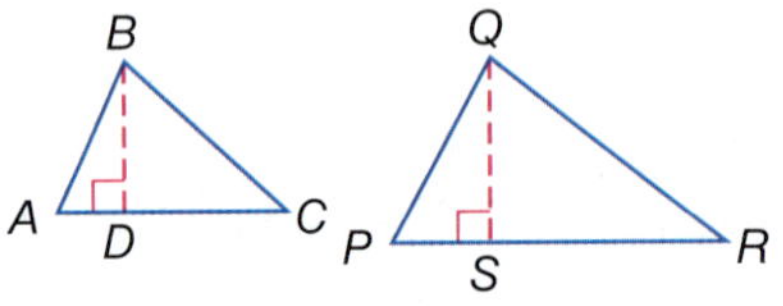

Lesson 7-6 Similarity Transformations (pp. 505–511)

Determine whether the dilation from Figure A to Figure B is an *enlargement* or a *reduction*. Then find the scale factor of the dilation.

1.

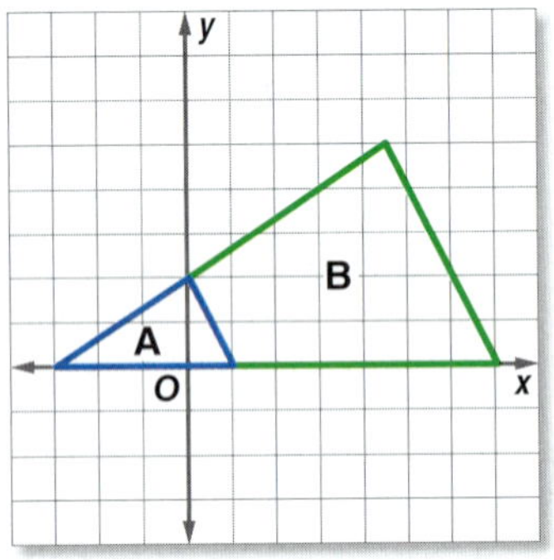

2.

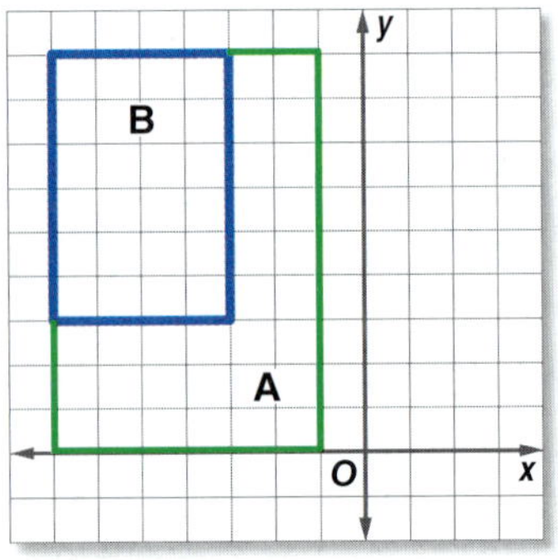

Graph the original figure and its dilated image. Then verify that the dilation is a similarity transformation.

3. $A(0, 0)$, $B(0, 4)$, $C(4, 0)$; $E(0, 2)$, $F(2, 0)$

4. $L(0, 3)$, $M(-6, 6)$, $N(-6, 0)$; $Q(-3, 4.5)$, $P(-3, 1.5)$

Lesson 7-7 Scale Drawings and Models (pp. 512–517)

1. **MODELS** If the length of a scale model of a ship is 9 inches, and the actual length of the ship is 150 feet, what is the scale factor of the model?

2. **MAPS** The scale on a map is 0.5 inches : 25 miles. What is the actual distance between two cities that are 2.4 inches apart on the map?

3. **SOLAR SYSTEM** In creating an accurate scale model of our solar system, Daniel placed Earth 1 foot from the sun. The actual distance from earth to the Sun is 93,000,000 miles. If the actual distance from Pluto to the Sun is 3,695,950,000 miles, how far from the Sun would Lana need to place Pluto in her model?

4. **CRAFTS** Lucia is building a corkboard that is 45 inches tall and 63 inches wide. She wants to build a smaller corkboard with a similar shape for the kitchen. Choose an appropriate scale for the corkboard.

Lesson 8-1 Geometric Mean (pp. 531–539)

Find the geometric mean between each pair of numbers.

1. 8 and 12
2. 15 and 20
3. 1 and 2
4. 4 and 16
5. $3\sqrt{2}$ and $6\sqrt{2}$
6. $\frac{1}{2}$ and 10
7. $\frac{3}{8}$ and $\frac{1}{2}$
8. $\frac{\sqrt{2}}{2}$ and $\frac{3\sqrt{2}}{2}$
9. $\frac{1}{10}$ and $\frac{7}{10}$

Find *x*, *y*, and *z*.

10.

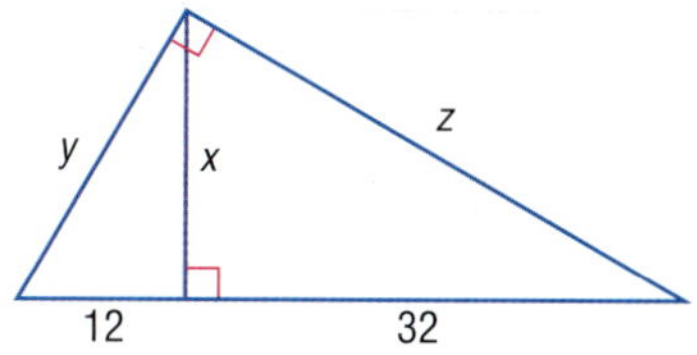

11.

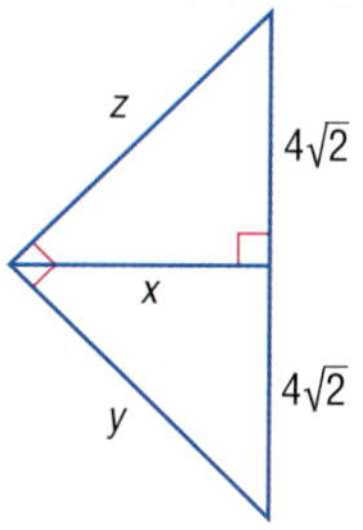

12.

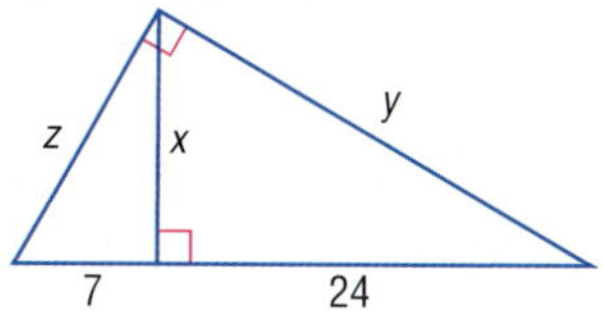

Lesson 8-2 The Pythagorean Theorem and Its Converse (pp. 541–549)

Find *x*.

1.

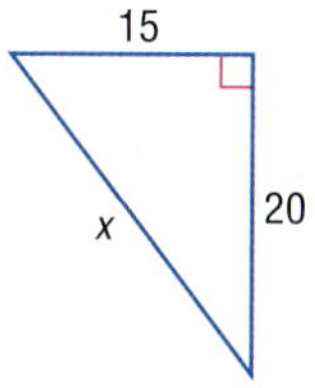

2.

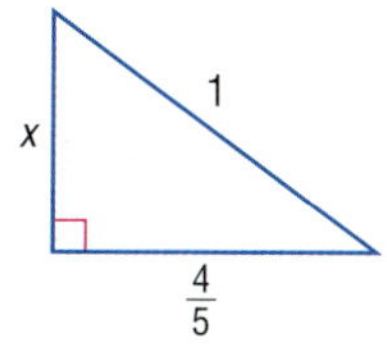

3.

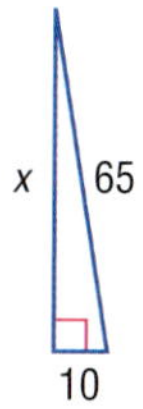

4.

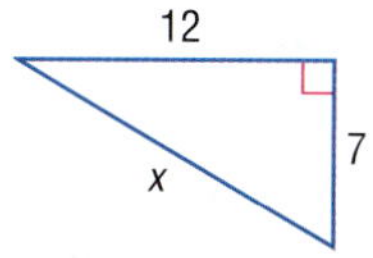

Determine whether each set of numbers can be the measures of the sides of a triangle. If so, classify the triangle as *acute*, *obtuse*, or *right*. Justify your answer.

5. 1, 1, 2
6. 21, 28, 35
7. 3, 5, 7
8. 2, 5, 7
9. 24, 45, 51
10. $\frac{1}{3}, \frac{5}{3}, \frac{\sqrt{26}}{3}$
11. $\frac{6}{11}, \frac{8}{11}, \frac{10}{11}$
12. $\frac{1}{2}, \frac{1}{2}, 1$
13. $\frac{\sqrt{6}}{3}, \frac{\sqrt{10}}{5}, \frac{\sqrt{240}}{15}$

Lesson 8-3 Special Right Triangles (pp. 552–560)

Find *x* and *y*.

1.

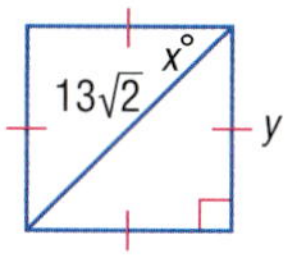

2.

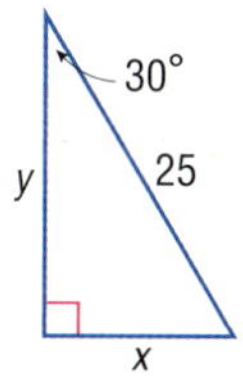

3.

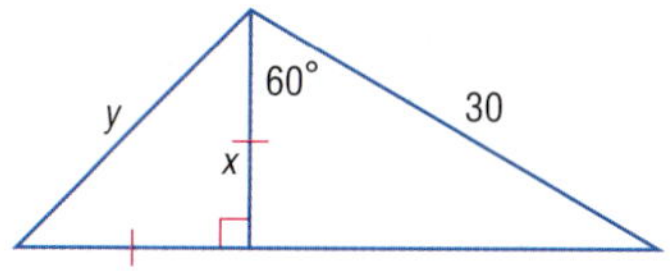

4.

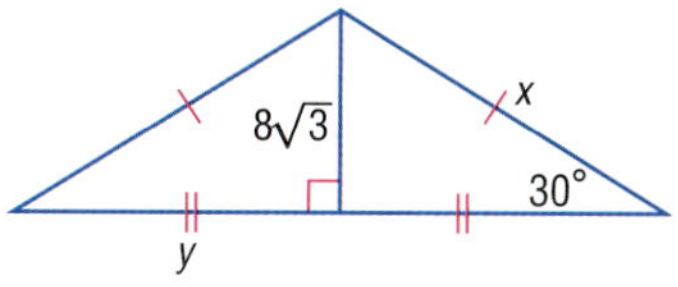

5.

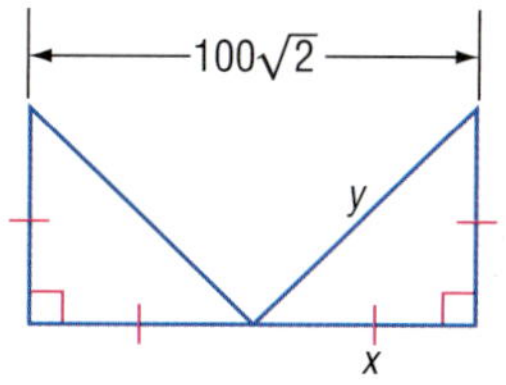

6.

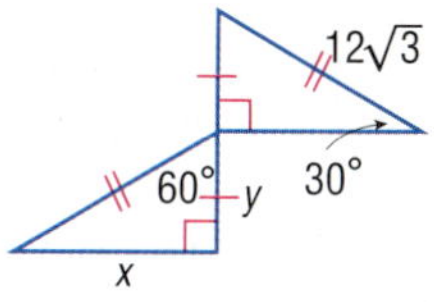

Lesson 8-4 Trigonometry (pp. 562–571)

Express each ratio as a fraction and as a decimal to the nearest hundredth.

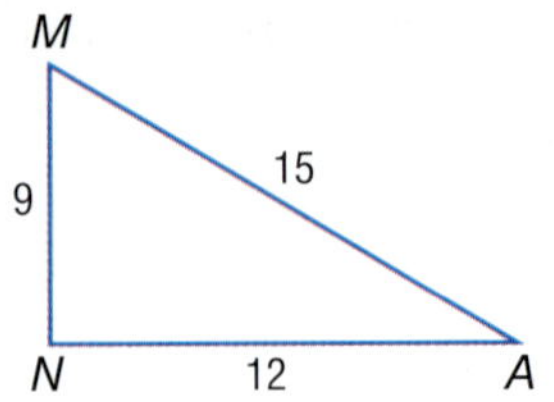

1. sin M
2. cos M
3. cos A
4. tan A
5. sin A
6. tan M

Find x. Round to the nearest tenth.

7. 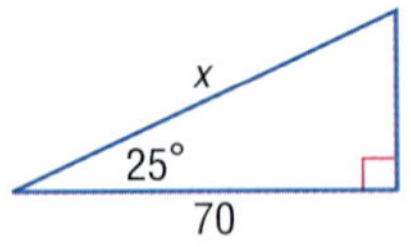

8. 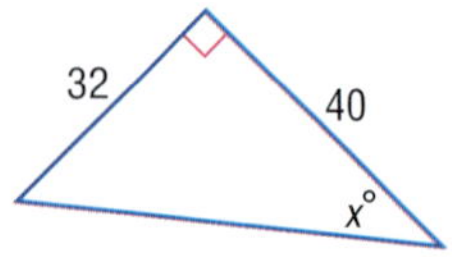

9. 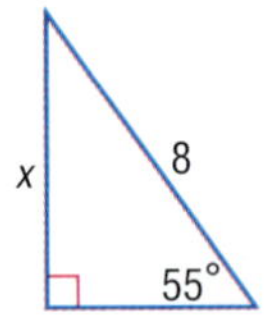

Lesson 8-5 Angles of Elevation and Depression (pp. 574–581)

1. **COMMUNICATIONS** A house is located below a hill that has a satellite dish. If $MN = 450$ feet and $RN = 120$ feet, what is the measure of the angle of elevation to the top of the hill?

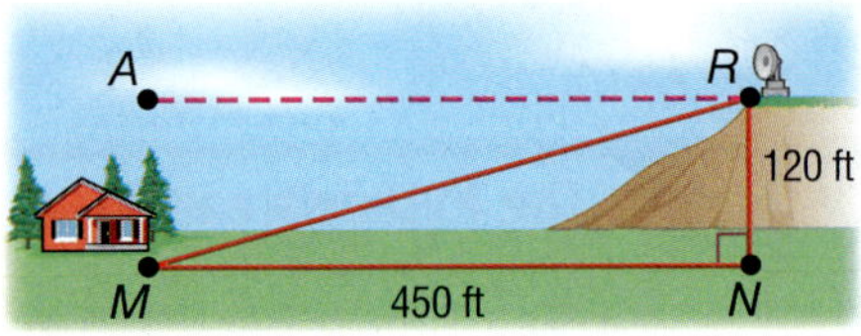

2. **AMUSEMENT PARKS** Elena is at the top of the Mighty Screamer roller coaster. Her friend Bryn is at the bottom of the coaster waiting for the next ride. If the angle of depression from Elena to Bryn is 26° and OL is 75 feet, what is the distance from L to C?

3. **SKIING** Mitchell is at the top of the Bridger Peak ski run. His brother Scott is looking up from the ski lodge at I. If the angle of elevation from Scott to Mitchell is 13° and the distance from K to I is 2000 ft, what is the length of the ski run SI?

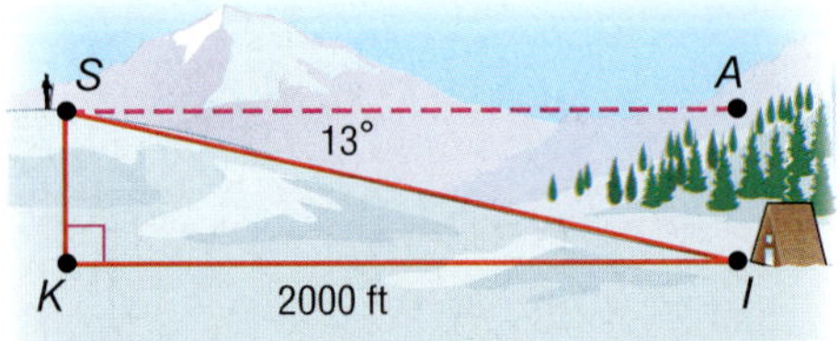

Lesson 8-6 The Law of Sines and Law of Cosines (pp. 582–591)

Solve each triangle. Round angle measures to the nearest degree and side measures to the nearest tenth.

1. 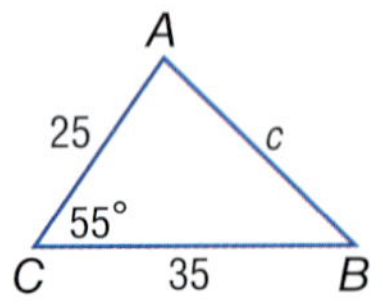

2. 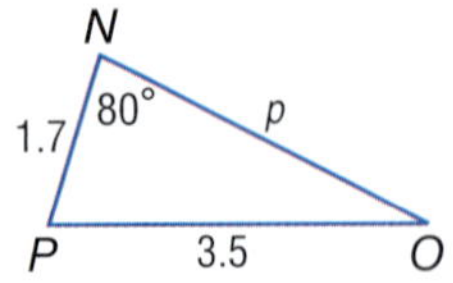

3. 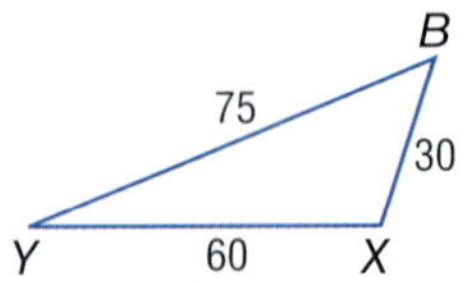

4. Solve $\triangle ABC$ if $m\angle C = 62$, $a = 28.5$, $m\angle B = 33$
5. Solve $\triangle XYZ$ if $y = 3.6$, $z = 3.7$, $m\angle Z = 55$
6. Solve $\triangle RST$ if $m\angle S = 35$, $m\angle R = 65$, $t = 50$
7. Solve $\triangle JKL$ if $m\angle J = 122$, $m\angle L = 15$, $j = 33.2$

Lesson 8-7 Vectors (pp. 593–600)

Write the component form of each vector.

1. 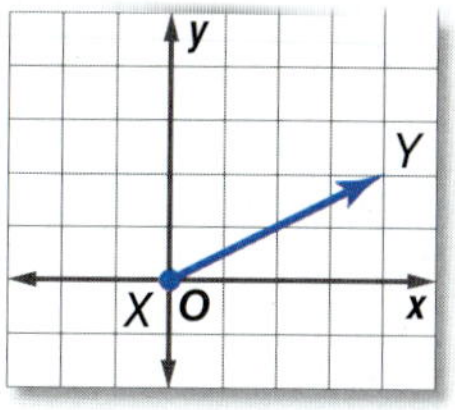

2. 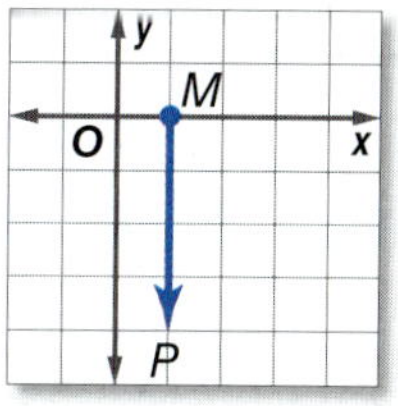

Find the magnitude and direction of each vector.

3. $\overrightarrow{RS}$: $R(-2, -1)$ and $S(3, -1)$
4. $\overrightarrow{ST}$: $S(-2, -4)$ and $T(-6, 3)$
5. $\overrightarrow{PQ}$: $P(6, 1)$ and $Q(-3, 6)$
6. $\overrightarrow{EF}$: $E(9, -2)$ and $F(9, -6)$

Copy the vectors to find each sum or difference.

7. $\vec{m} + \vec{q}$ 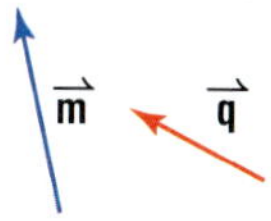

8. $\vec{\ell} - \vec{o}$ 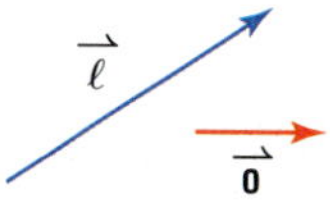

Lesson 9-1 Reflections (pp. 615–623)

COORDINATE GEOMETRY **Graph each figure and its image under the given reflection.**

1. $\triangle ABN$ with vertices $A(2, 2)$, $B(3, -2)$, and $N(-3, -1)$ in the x-axis
2. rectangle $BARN$ with vertices $B(3, 3)$, $A(3, -4)$, $R(-1, -4)$, and $N(-1, 3)$ in the line $y = x$
3. trapezoid $ZOID$ with vertices $Z(2, 3)$, $O(2, -4)$, $I(-3, -3)$, and $D(-3, 1)$ in the origin
4. $\triangle PQR$ with vertices $P(-2, 1)$, $Q(2, -2)$, and $R(-3, -4)$ in the y-axis
5. square $BDFH$ with vertices $B(-4, 4)$, $D(-1, 4)$, $F(-1, 1)$, and $H(-4, 1)$ in the origin
6. quadrilateral $QUAD$ with vertices $Q(1, 3)$, $U(3, 1)$, $A(-1, 0)$, and $D(-3, 4)$ in the line $y = -1$
7. $\triangle CAB$ with vertices $C(0, 4)$, $A(1, -3)$, and $B(-4, 0)$ in the line $x = -2$

Lesson 9-2 Translations (pp. 624–630)

Copy the figure and the given translation vector. Then draw the translation of the figure along the translation vector.

1. 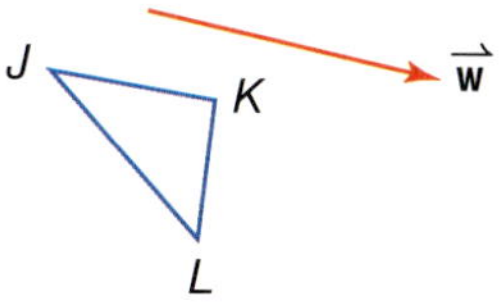

2. 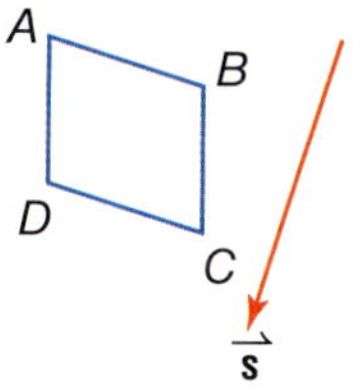

3. 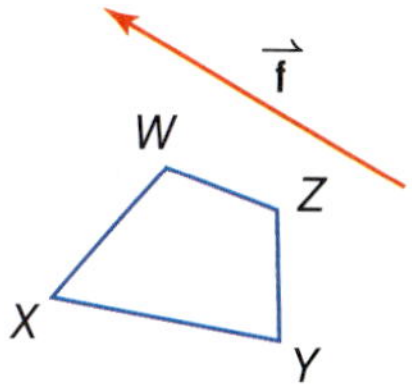

Graph each figure and its image along the given vector.

4. $\overline{LM}$ with endpoints $L(2, 3)$ and $M(-4, 1)$; $\langle 2, 1\rangle$
5. $\triangle DEF$ with vertices $D(1, 2)$, $E(-2, 1)$, and $F(-3, -1)$; $\langle -1, -3\rangle$
6. quadrilateral $WXYZ$ with vertices $W(1, 1)$, $X(-2, 3)$, $Y(-3, -2)$, and $Z(2, -2)$; $\langle 1, -1\rangle$
7. pentagon $ABCDE$ with vertices $A(1, 3)$, $B(-1, 1)$, $C(-1, -2)$, $D(3, -2)$, and $E(3, 1)$; $\langle -2, 3\rangle$
8. $\triangle RST$ with vertices $R(-4, 3)$, $S(-2, -3)$, and $T(2, -1)$; $\langle 3, -2\rangle$

Lesson 9-3 Rotations (pp. 632–638)

Copy each polygon and point *K*. Then use a protractor and ruler to draw the specified rotation of each figure about point *K*.

1. 60°

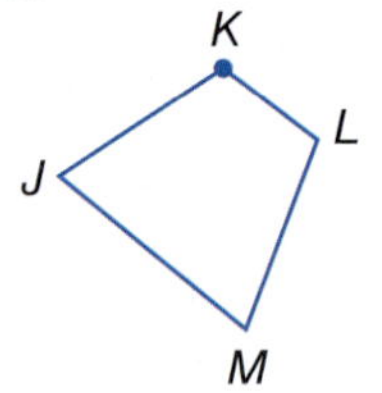

2. 90°

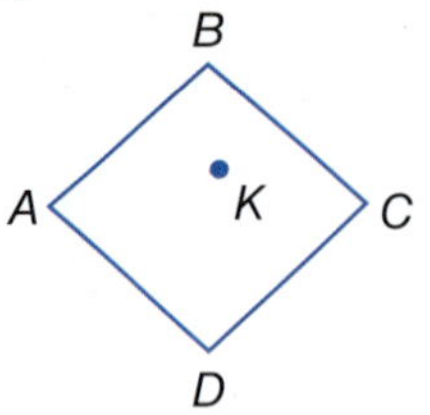

3. 120°

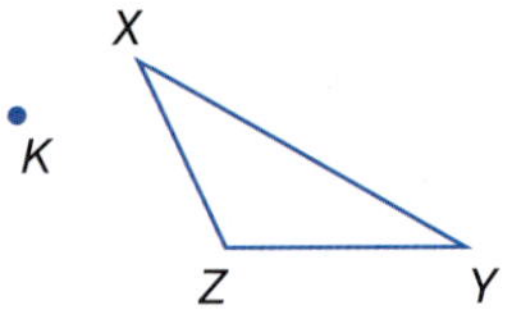

COORDINATE GEOMETRY **Graph each figure and its image after the specified rotation about the origin.**

4. $\triangle KLM$ with vertices $K(4, 2)$, $L(1, 3)$, and $M(2, 1)$; 90°

5. $\triangle FGH$ with vertices $F(-3, -3)$, $G(2, -4)$, and $H(-1, -1)$; 180°

Lesson 9-4 Compositions of Transformations (pp. 641–649)

Graph each figure with the given vertices and its image after the indicated transformation.

1. $\overline{MN}$: $M(2, 2)$ and $N(3, 5)$
Reflection: in x-axis; **Rotation:** 180° about the origin

2. $\overline{JK}$: $J(-1, 2)$ and $K(2, -2)$
Translation: along $\langle -3, -3 \rangle$; **Reflection:** in $y = x$

3. $\triangle ABC$: $A(-3, -2)$, $B(-7, -2)$, $C(-5, -4)$
Translation: along $\langle -1, -2 \rangle$; **Reflection:** in y-axis

4. $\triangle MNP$: $M(3, 4)$, $N(-1, 2)$, $P(1, 5)$
Translation: along $\langle 0, 3 \rangle$; **Reflection:** in $y = x$

Copy and reflect figure *L* in line *m* and then line *q*. Then describe a single transformation that maps *L* onto *L″*.

5.

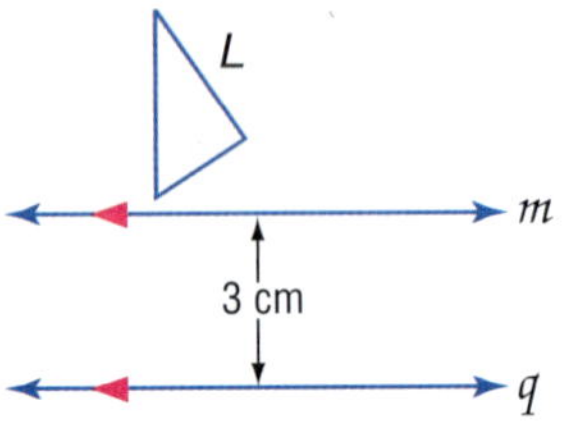

6.

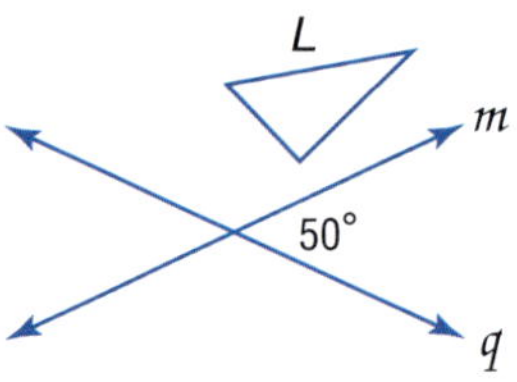

Lesson 9-5 Symmetry (pp. 653–659)

State whether the figure appears to have line symmetry. Write *yes* or *no*. If so, copy the figure, draw all lines of symmetry, and state their number.

1.

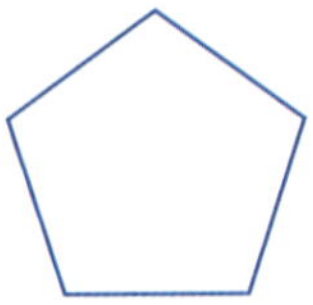

2.

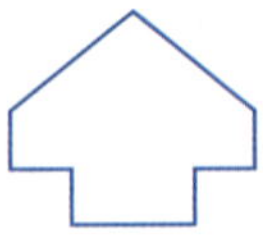

3.

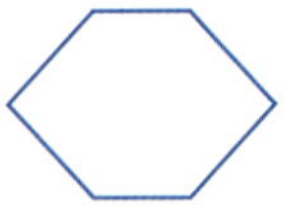

4.

State whether the figure has rotational symmetry. Write *yes* or *no*. If so, copy the figure, locate the center of symmetry, and state the order and magnitude of symmetry.

5.

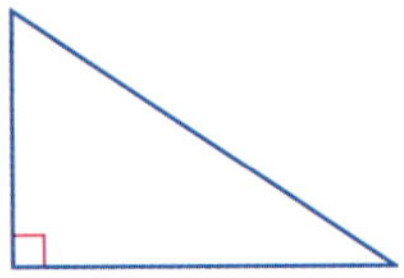

6.

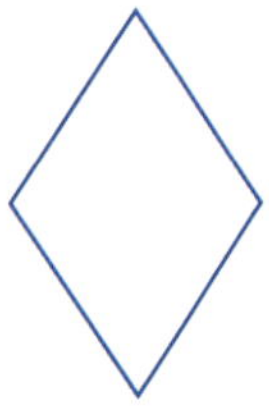

7.

8.

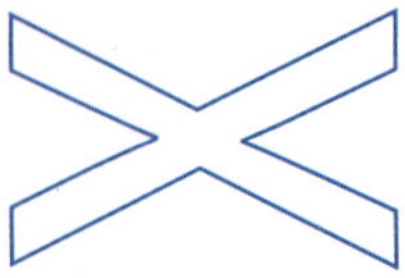

Lesson 9-6 Dilations (pp. 660–667)

Copy the figure and point *P*. Then use a ruler to draw the image of the figure under a dilation with center *P* and the scale factor *r* indicated.

1. $r = 2$

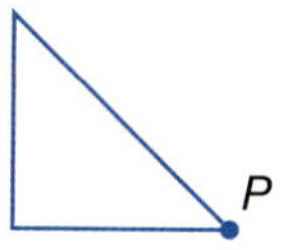

2. $r = \frac{1}{3}$

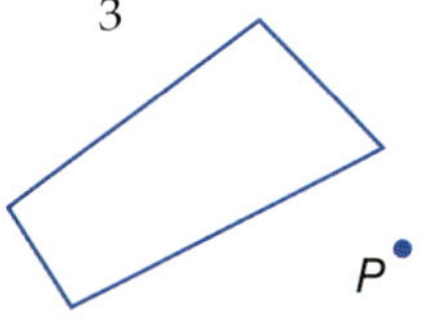

3. $r = 1.5$

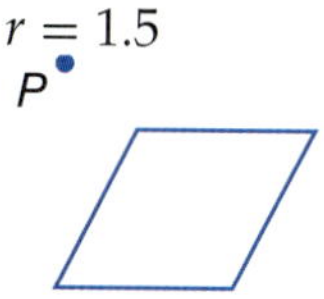

Find the image of each polygon with the given vertices after a dilation centered at the origin with the given scale factor.

4. $R(2, 3), S(0, 0), T(2, 0); r = 3$

5. $W(-2, 6), X(4, 6), Y(4, 2), Z(-2, 2); r = \frac{1}{2}$

6. $J(-3, 1), K(1, 5), L(5, 1), M(1, -3); r = 0.25$

7. $A(1, 2), B(3, 2), C(4, 0), D(0, 0); r = 2.5$

Lesson 10-1 Circles and Circumference (pp. 683–691)

Find the exact circumference of each circle by using the given inscribed or circumscribed polygon.

1.

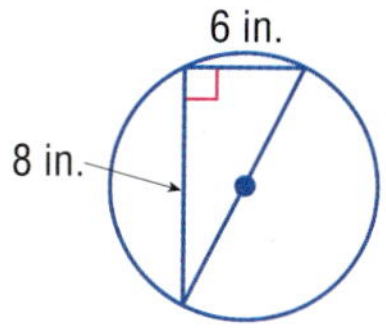

2.

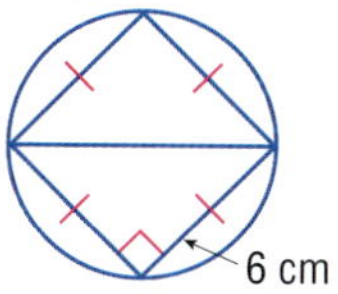

3.

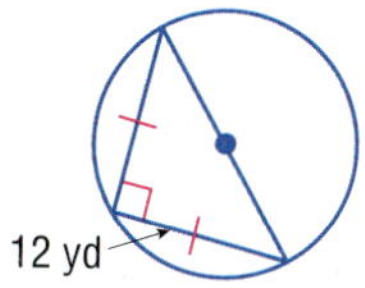

4.

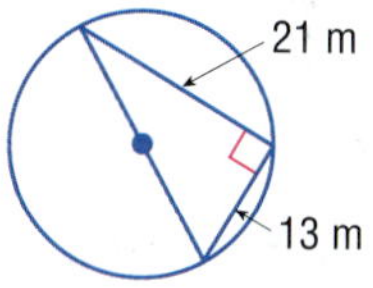

The radius, diameter, or circumference of a circle is given. Find each missing measure to the nearest hundredth if necessary.

5. $r = 18$ in., $d =$ __?__, $C =$ __?__

6. $d = 34.2$ ft, $r =$ __?__, $C =$ __?__

7. $C = 12\pi$ m, $r =$ __?__, $d =$ __?__

8. $C = 84.8$ mi, $r =$ __?__, $d =$ __?__

9. $d = 8.7$ cm, $r =$ __?__, $C =$ __?__

10. $r = 3b$ in., $d =$ __?__, $C =$ __?__

Lesson 10-2 Measuring Angles and Arcs (pp. 692–700)

Find *x*.

1.

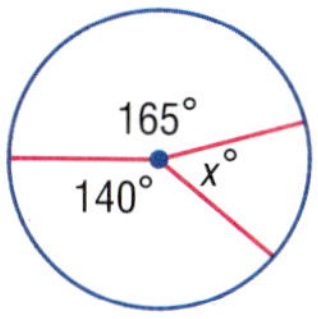

2.

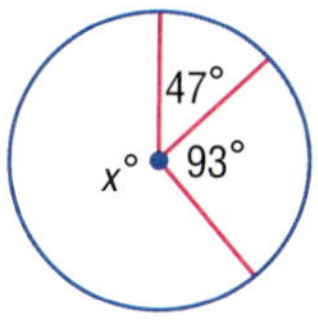

3.

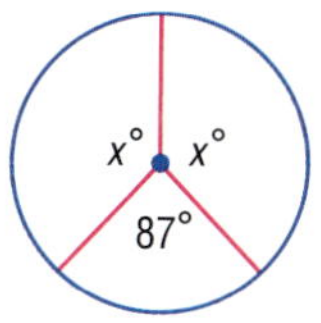

4.

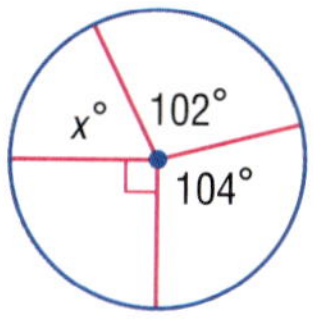

$\overline{WS}$, $\overline{VR}$, and $\overline{QT}$ are diameters of $\odot X$. Identify each arc as a *major arc*, *minor arc*, or *semicircle*. Then find its measure.

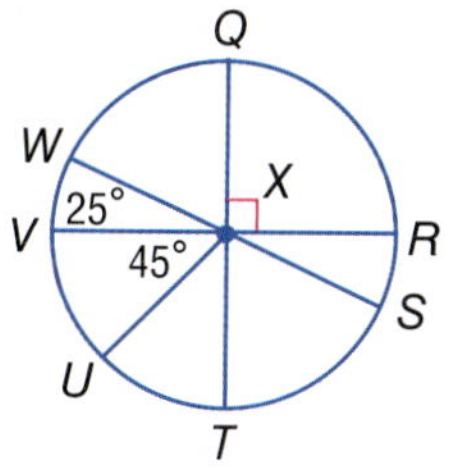

5. $m\widehat{QR}$

6. $m\widehat{QW}$

7. $m\widehat{TU}$

8. $m\widehat{WRV}$

9. $m\widehat{SV}$

10. $m\widehat{QRT}$

Lesson 10-3 Arcs and Chords (pp. 701–708)

In $\odot S$, $HJ = 22$, $LG = 18$, $m\widehat{IJ} = 35$, and $m\widehat{LM} = 30$. Find each measure.

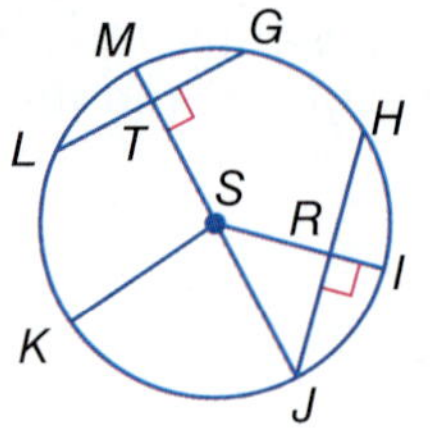

1. HR
2. RJ
3. LT
4. TG
5. $m\widehat{HJ}$
6. $m\widehat{LG}$
7. $m\widehat{MG}$
8. $m\widehat{HI}$

In $\odot R$, $CR = RF$, and $ED = 30$. Find each measure.

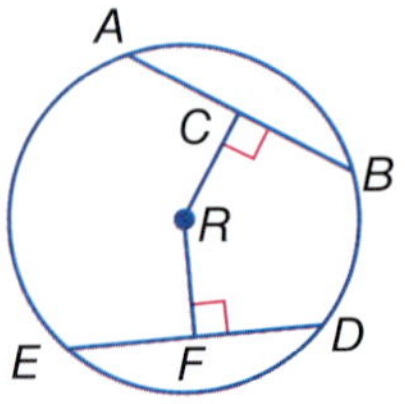

9. AB
10. EF
11. DF
12. BC

Lesson 10-4 Inscribed Angles (pp. 709–716)

Find each measure.

1. $m\angle S$

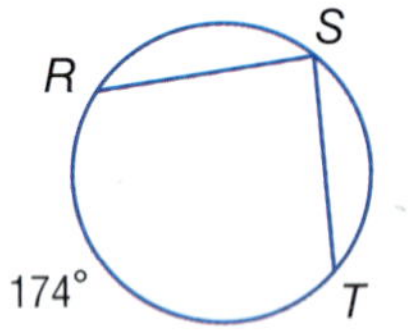

2. $m\widehat{AC}$

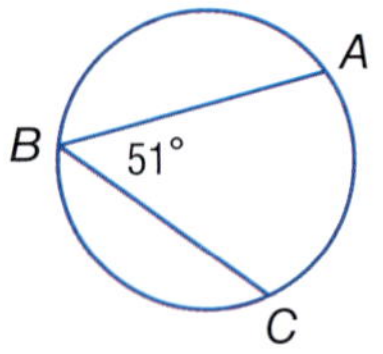

3. $m\angle K$

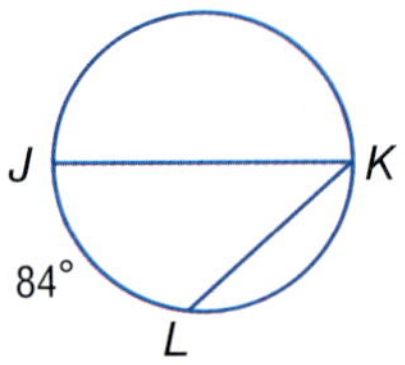

ALGEBRA Find each value.

4. x
5. $m\angle F$

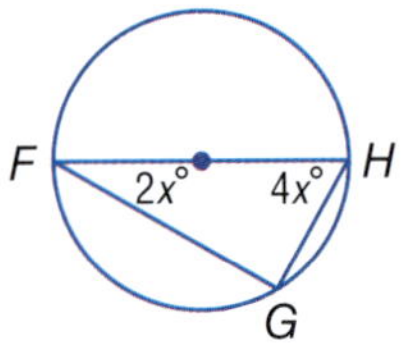

6. x
7. $m\angle S$

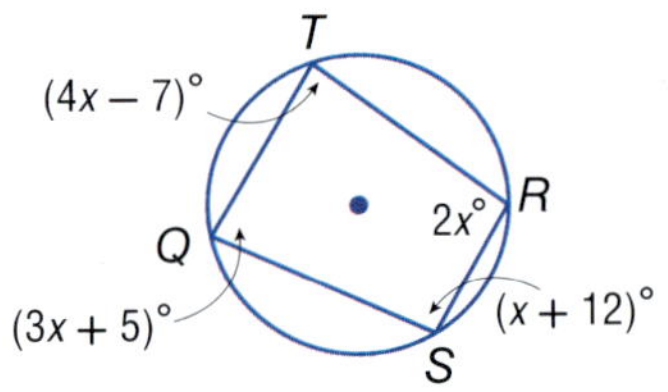

Lesson 10-5 Tangents (pp. 718–725)

Determine whether each segment is tangent to the given circle. Justify your answer.

1.

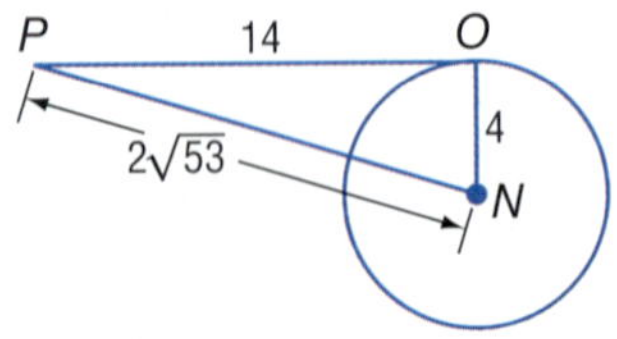

2.

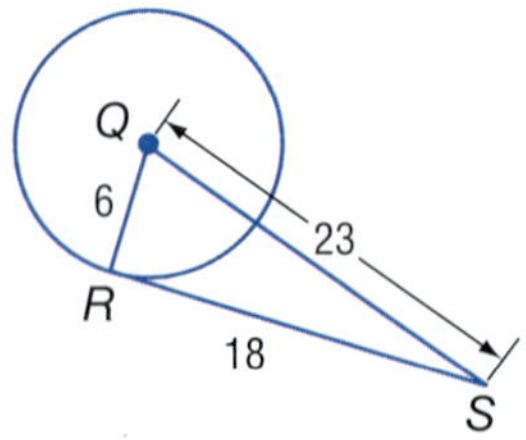

Find x. Assume that segments that appear to be tangent are tangent.

3.

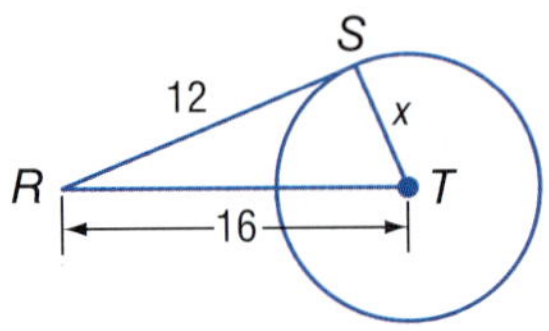

4.

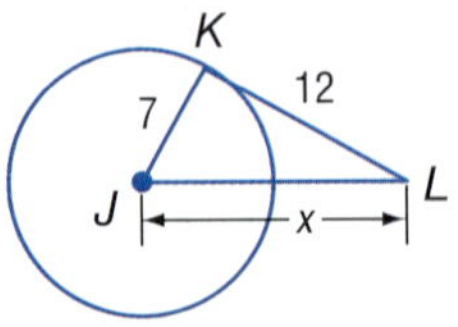

5.

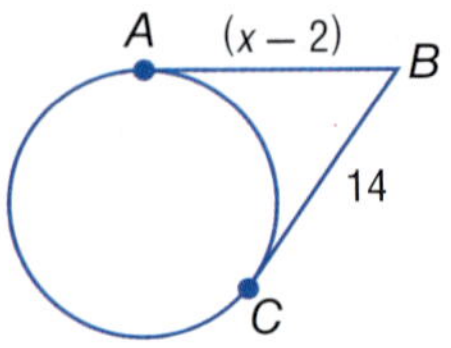

Lesson 10-6 Secants, Tangents, and Angle Measures (pp. 727–735)

Find each measure. Assume that segments that appear to be tangent are tangent.

1. $m\angle 3$

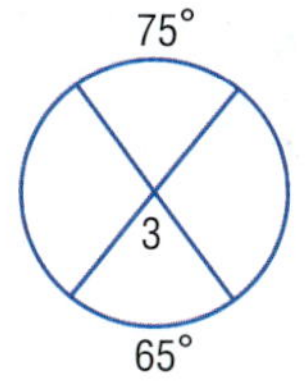

2. $m\angle 1$

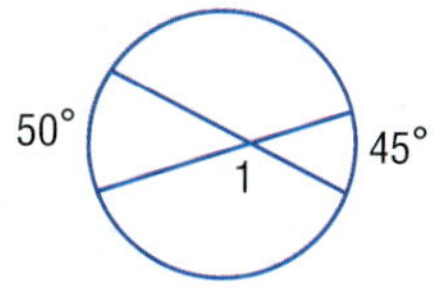

3. $m\angle 9$

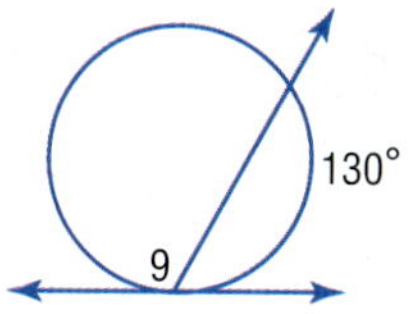

Find x. Assume that any segment that appears to be tangent is tangent.

4. $m\angle Y$

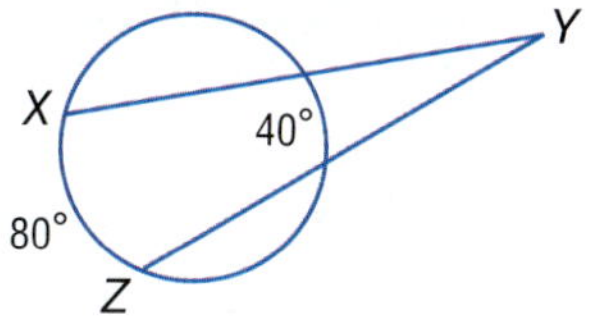

5. $m\angle J$

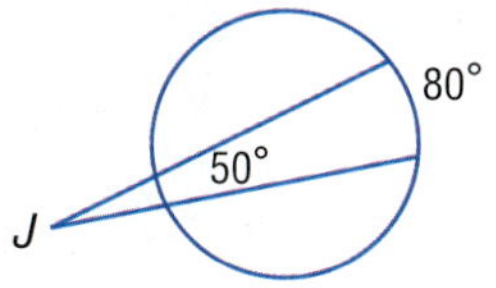

6. $m\widehat{AD}$

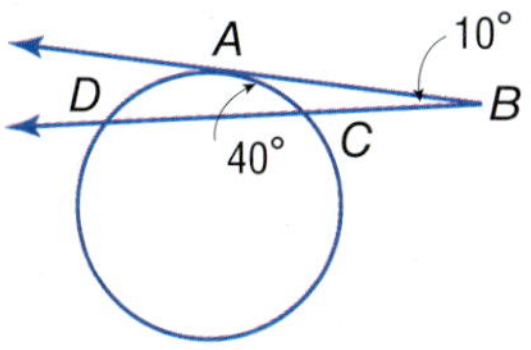

Lesson 10-7 Special Segments in a Circle (pp. 736–742)

Find x.

1.

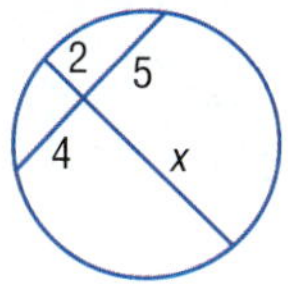

2.

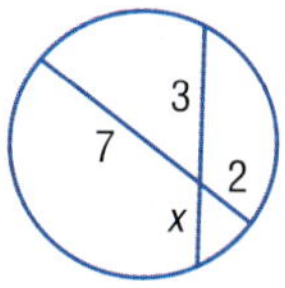

3.

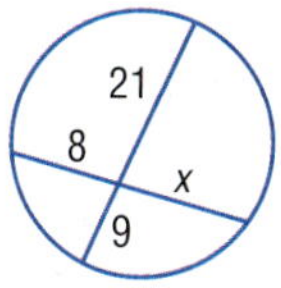

Find each variable to the nearest tenth. Assume that segments that appear to be tangent are tangent.

4.

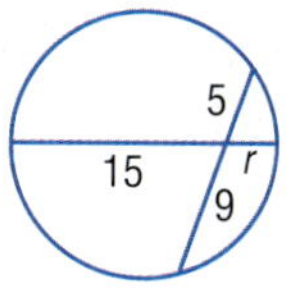

5.

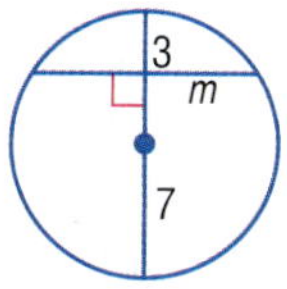

6.

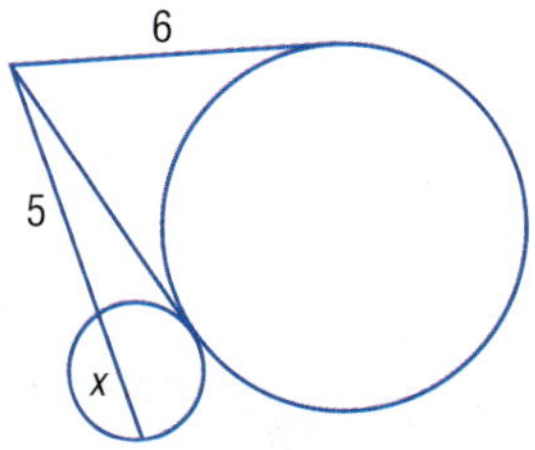

Lesson 10-8 Equations of Circles (pp. 744–749)

Write the equation of each circle.

1. center at $(1, -2)$, $r = 2$

2. center at origin, $r = 4$

3. center at $(-3, -4)$, $r = \sqrt{11}$

4. center at $(3, -1)$, $d = 6$

5. center at $(6, 12)$, $r = 7$

6. center at $(4, 0)$, $d = 8$

7. center at $(6, -6)$, $d = 22$

8. center at $(-5, 1)$, $d = 2$

For each circle with the given equation, state the coordinates of the center and the measure of the radius. Then graph the equation.

9. $x^2 + y^2 = 25$

10. $x^2 + y^2 - 3 = 1$

11. $(x - 3)^2 + (y + 1)^2 = 9$

12. $(x - 1)^2 + (y - 4)^2 = 1$

Extra Practice

Lesson 11-1 Areas of Parallelograms and Triangles (pp. 763–770)

Find the perimeter and area of each parallelogram or triangle. Round to the nearest tenth, or triangle if necessary.

1.

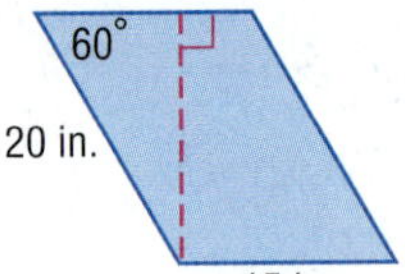

2.

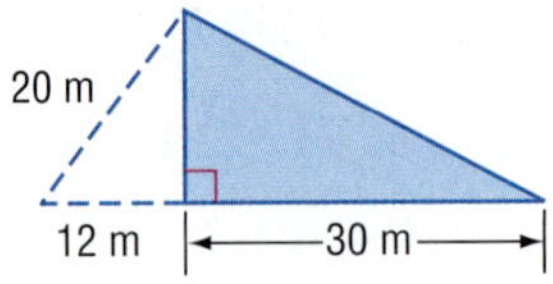

3.

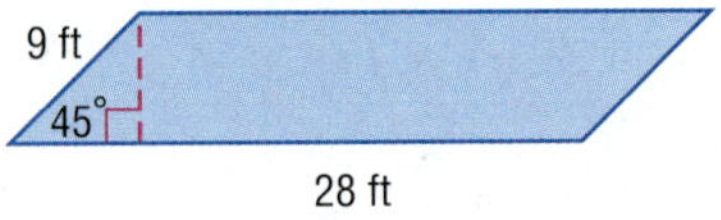

Find x.

4. $A = 144 \text{ in}^2$

5. $A = 720 \text{ in}^2$

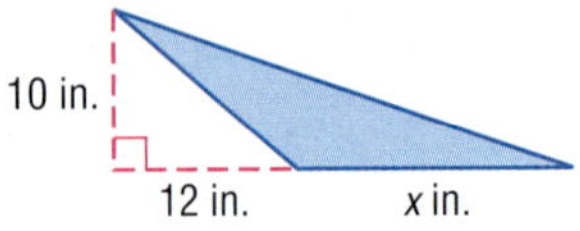

6. $A = 780 \text{ cm}^2$

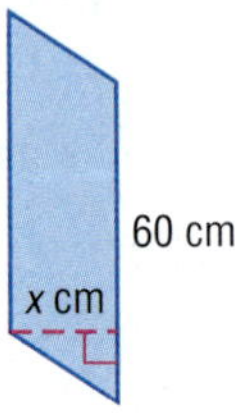

7. $A = 315 \text{ mm}^2$

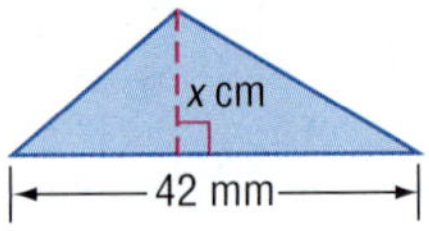

Lesson 11-2 Areas of Trapezoids, Rhombi, and Kites (pp. 773–780)

Find the area of each trapezoid, rhombus, or kite.

1.

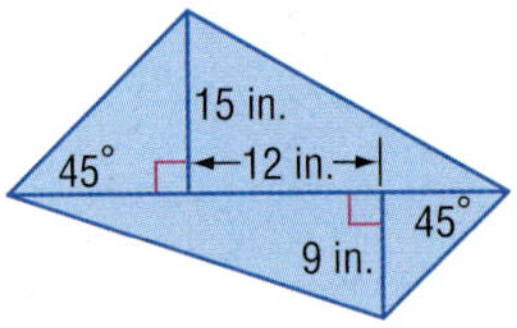

2.

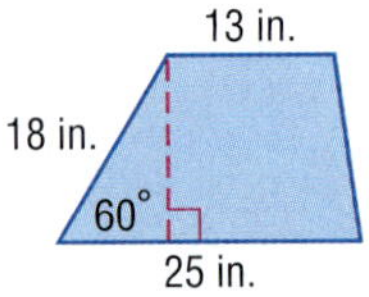

3.

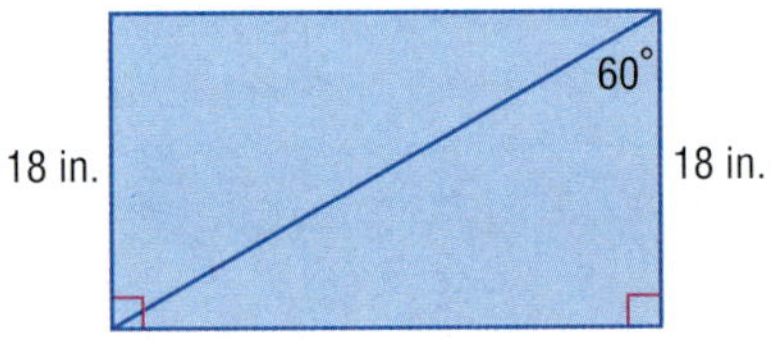

Find x.

4. $A = 325.68 \text{ mm}^2$

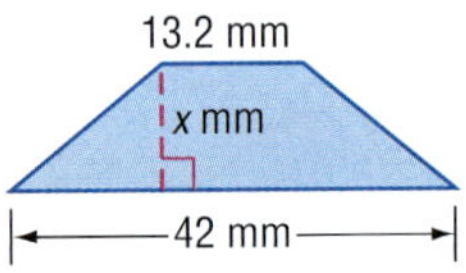

5. $A = 206.8 \text{ cm}^2$

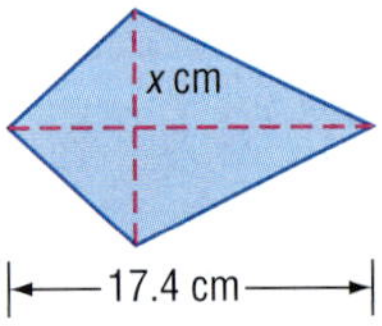

6. $A = 315 \text{ in}^2$

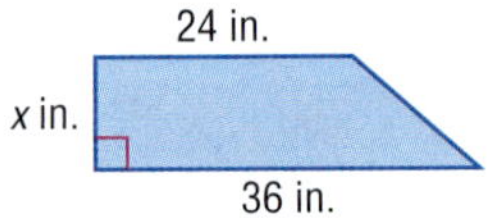

7. $A = 104.31 \text{ m}^2$

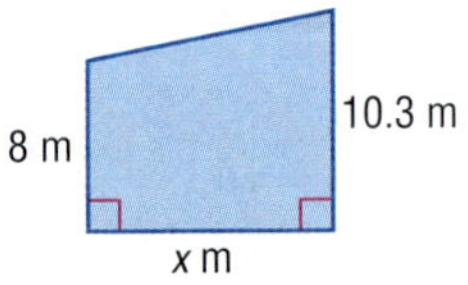

Lesson 11-3 Areas of Circles and Sectors (pp. 782–788)

Find the area of each shaded sector. Round to the nearest tenth.

1.

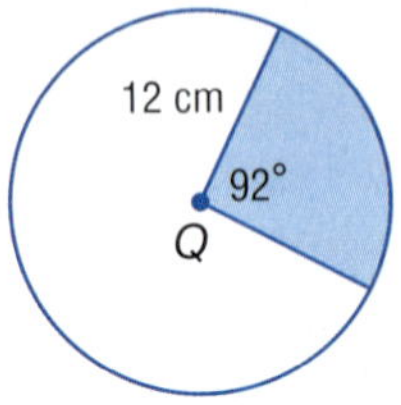

2.

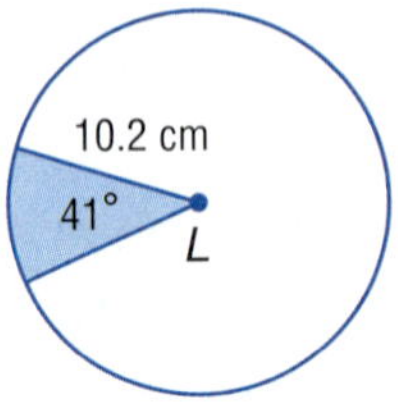

3.

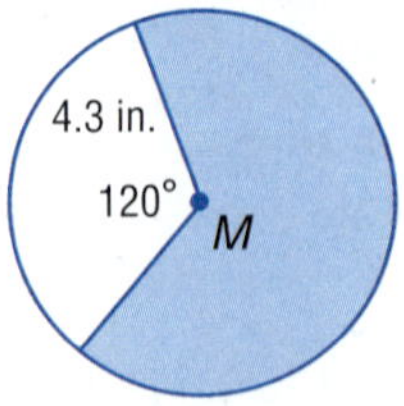

Find the area of each shaded region. Round to the nearest tenth.

4.

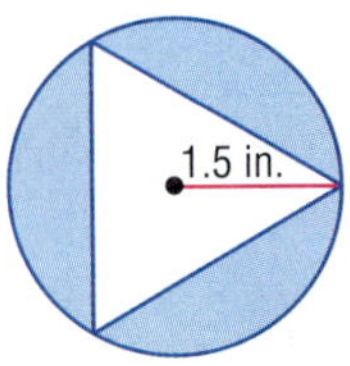

5.

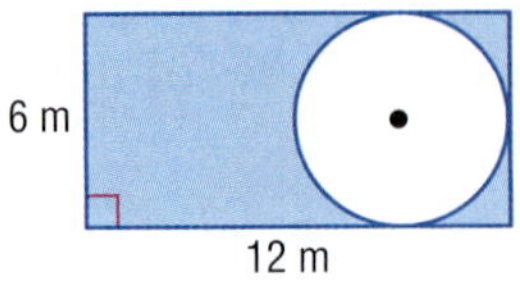

6.

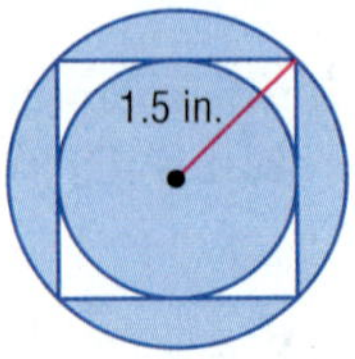

Extra Practice

Lesson 11-4 Areas of Regular Polygons and Composite Figures (pp. 791–799)

Find the area of each figure. Round to the nearest tenth, if necessary.

1.

2.

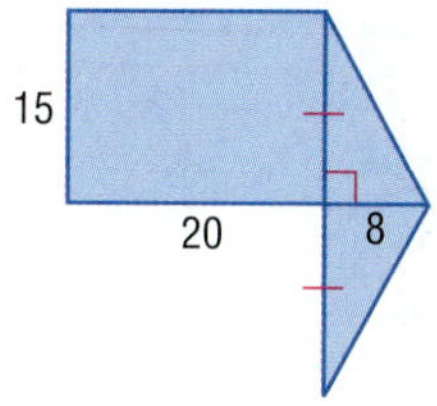

3.

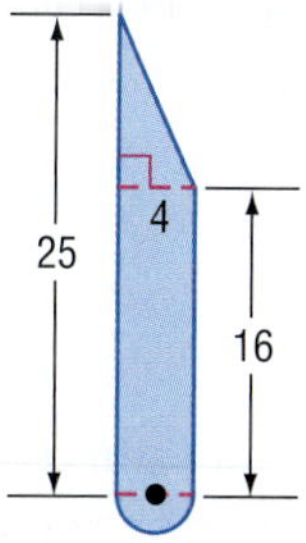

Find the area of each shaded region. Round to the nearest tenth.

4.

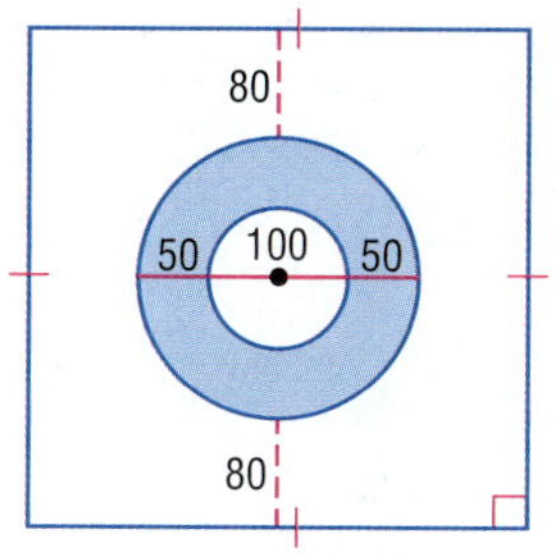

5.

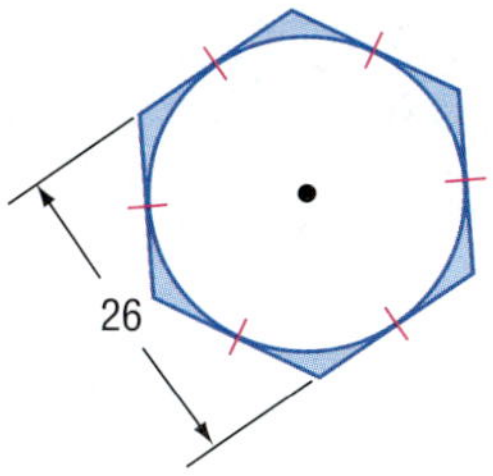

6.

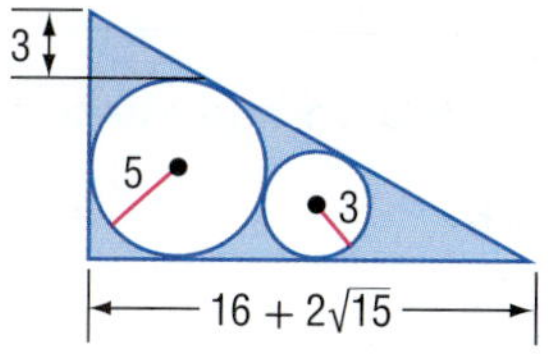

Lesson 11-5 Areas of Similar Figures (pp. 802–808)

For each pair of similar figures, find the area of the green figure.

1.

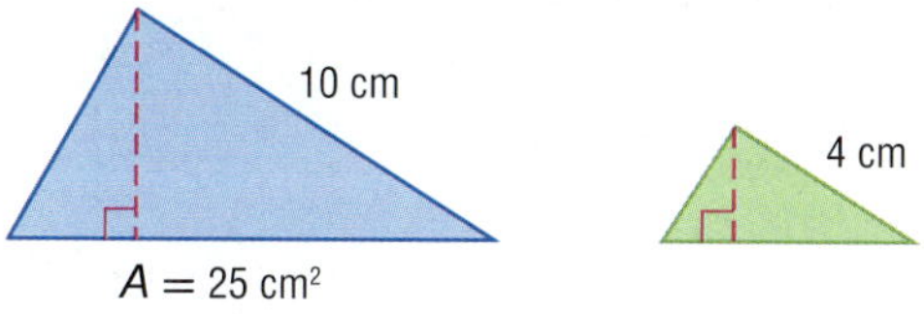

2.

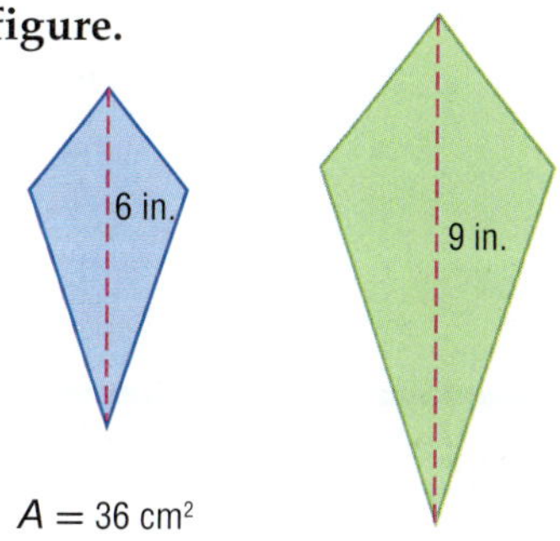

For each pair of similar figures, use the given areas to find the scale factor of the blue to the green figure. Then find x.

3.

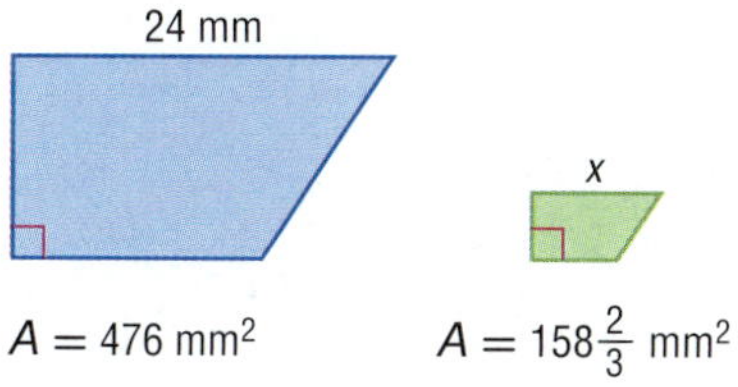

4.

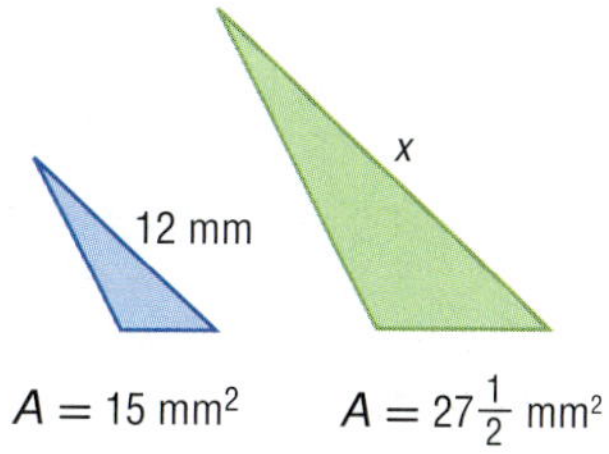

Lesson 12-1 Representations of Three-Dimensional Figures (pp. 823–828)

Use isometric dot paper and each orthographic drawing to sketch a solid.

1.

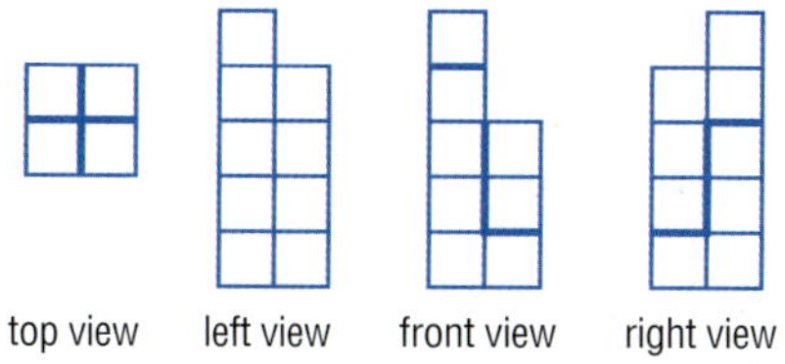

2.

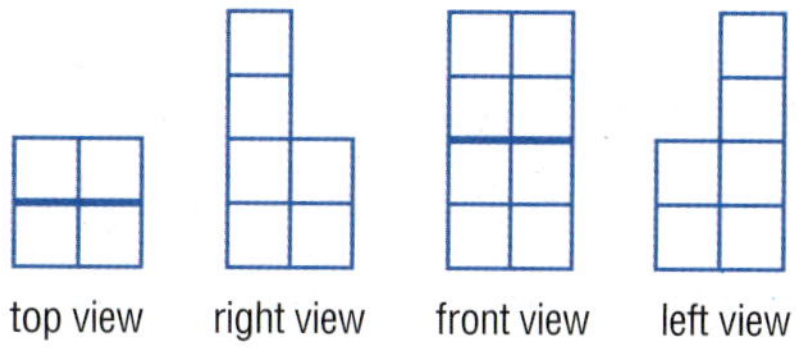

Use isometric dot paper to sketch each prism.

3. rectangular prism 6 units high, 6 units long, and 3 units wide

4. triangular prism 2 units high, with bases that are right triangles with legs 3 units and 7 units long

Lesson 12-2 Surface Areas of Prisms and Cylinders (pp. 830–837)

Find the lateral area and the surface area of each prism. Round to the nearest tenth if necessary.

1.

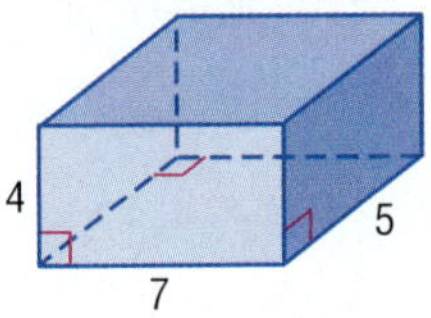

2.

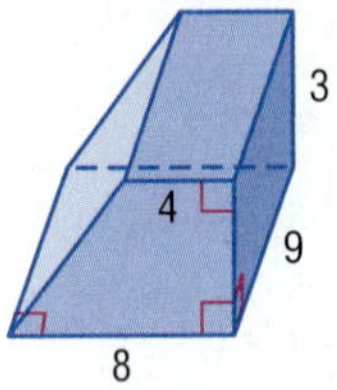

3.

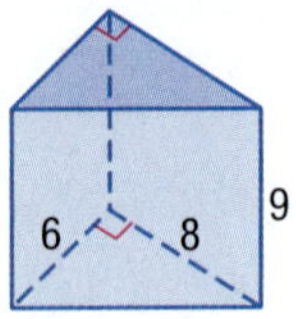

4. The surface area of a right triangular prism is 228 square inches. The base is a right triangle with legs measuring 6 inches and 8 inches. Find the height of the prism.

Find the lateral area and surface area of each cylinder. Round to the nearest tenth.

5.

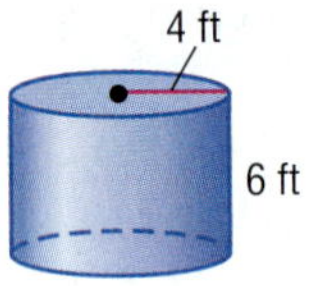

6.

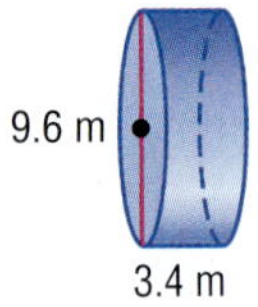

7.

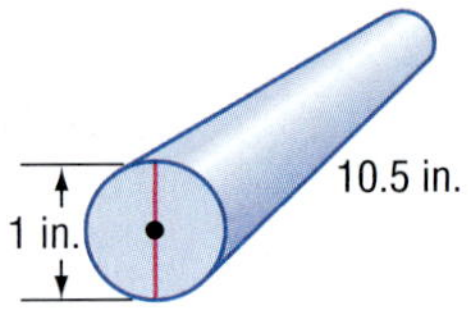

8.

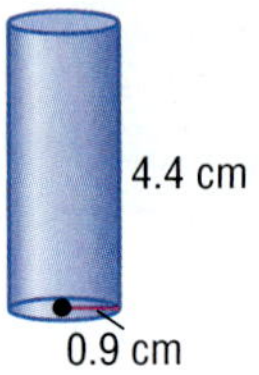

Find the lateral area and surface area of a cylinder with the given dimensions. Round to the nearest tenth.

9. $r = 2$ ft, $h = 3.5$ ft

10. $d = 15$ in., $h = 20$ in.

11. $r = 3.7$ m, $h = 6.2$ m

12. $d = 19$ mm, $h = 32$ mm

Lesson 12-3 Surface Areas of Pyramids and Cones (pp. 838–846)

Find the lateral area and surface area of each regular pyramid. Round to the nearest tenth if necessary.

1.

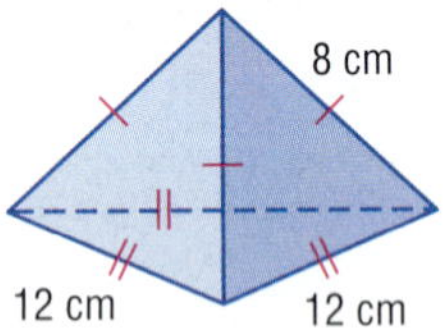

2.

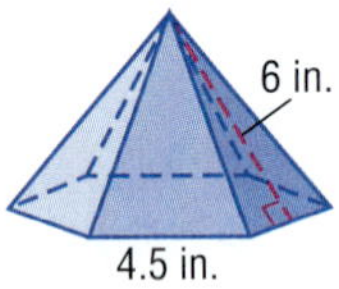

3.

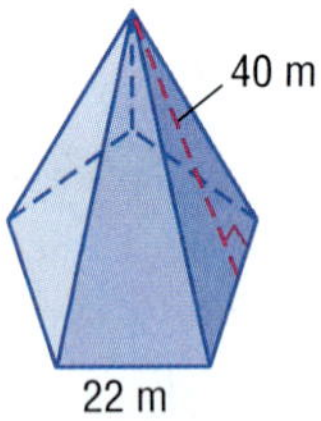

4.

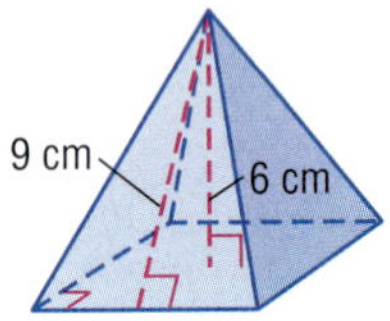

5.

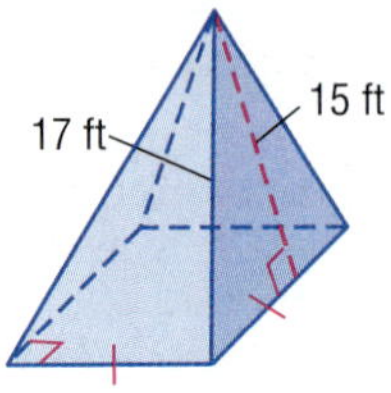

6.

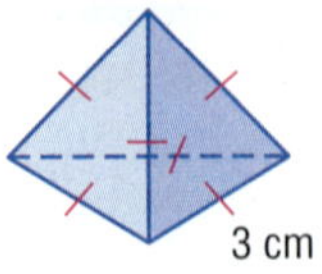

Find the lateral area and surface area of each cone. Round to the nearest tenth.

7.

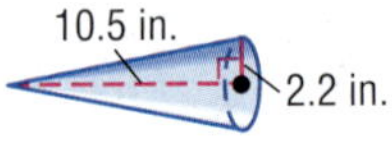

8.

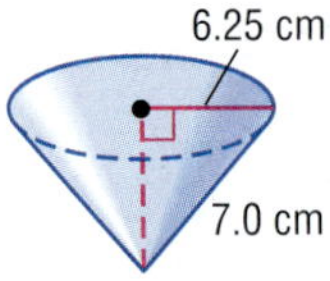

9.

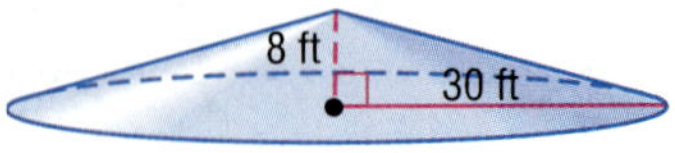

Lesson 12-4 Volumes of Prisms and Cylinders (pp. 847–854)

Find the volume of each prism or cylinder. Round to the nearest tenth if necessary.

1.

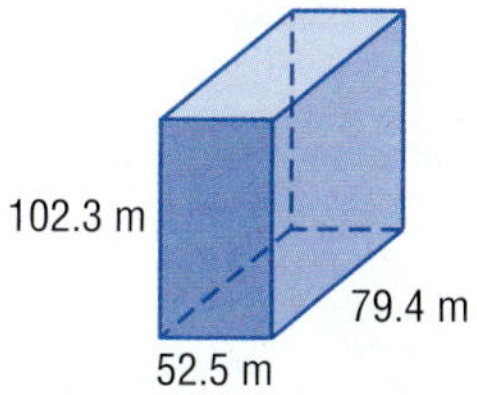

2.

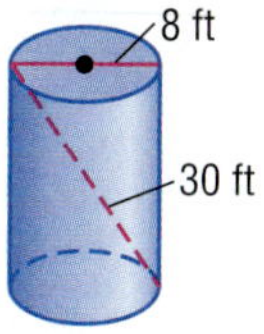

3.

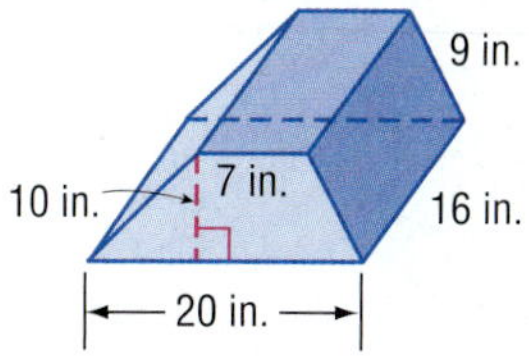

COMPOSITE SOLIDS Find the volume of each composite solid to the nearest tenth if necessary.

4.

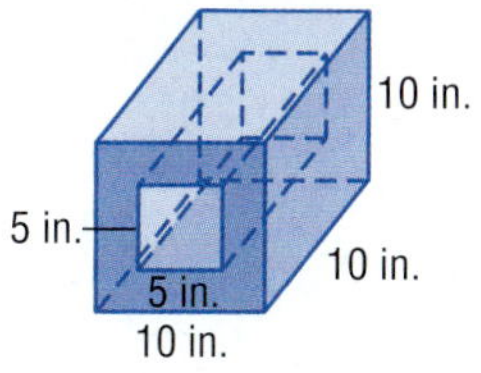

5.

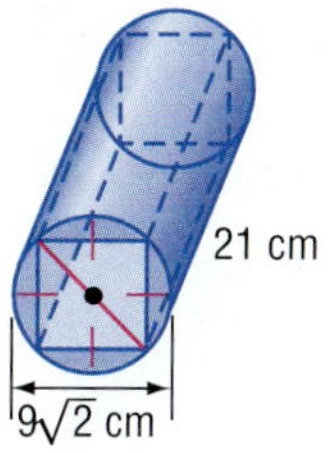

6.

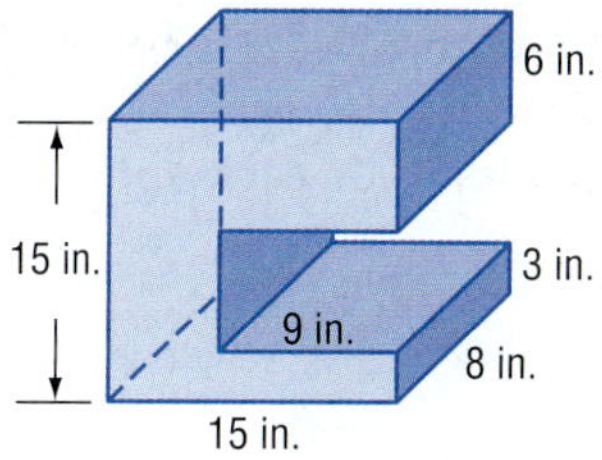

7. a cylinder with a radius of 6.8 centimeters and a height of 10.2 centimeters

8. a cylinder with a diameter of 13.6 yards and a height of 2.4 yards

9. a prism with a base area of 11.3 square inches and a height of 7.1 inches

10. a prism with a height of $8\sqrt{3}$ feet and a base area of $12\sqrt{3}$ feet

11. a cylinder with a diameter of 31.4 millimeters and a height of 12.9 millimeters

Lesson 12-5 Volumes of Pyramids and Cones (pp. 857–863)

Find the volume of each pyramid or cone. Round to the nearest tenth if necessary.

1.

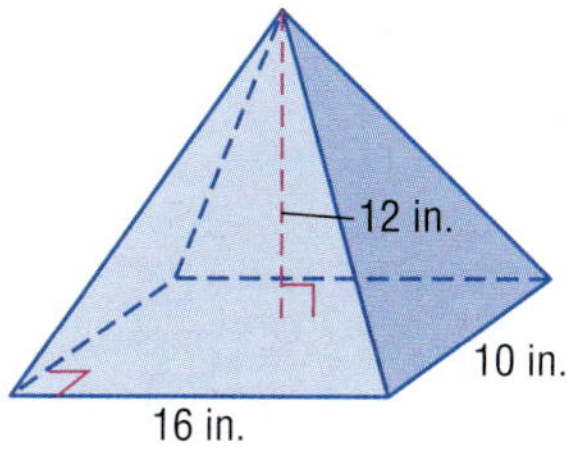

2.

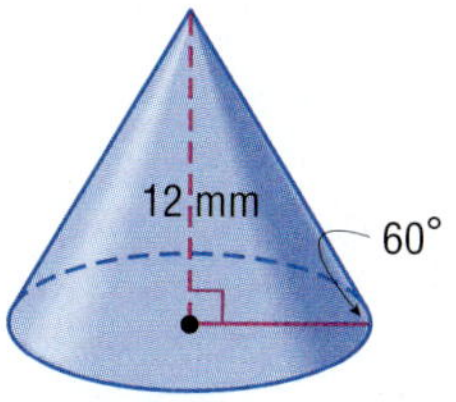

3.

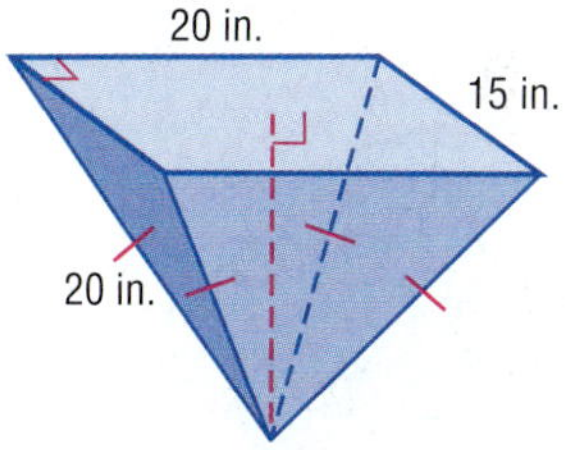

4.

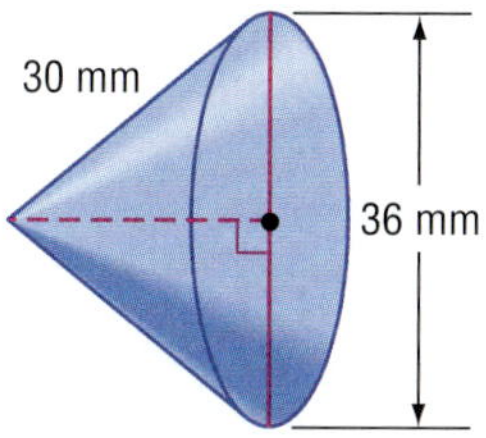

5.

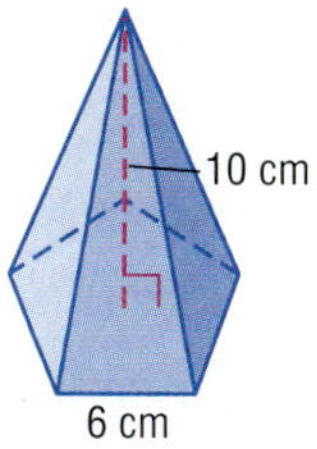

6.

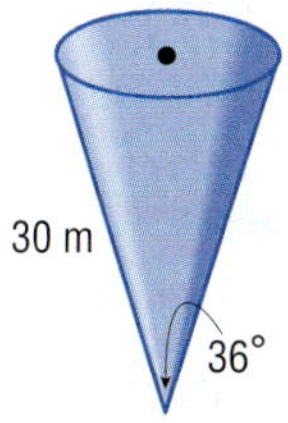

7. a pyramid with a height of 4.7 meters and a base 5 meters by 7 meters

8. a cone with a height of 11 inches and a radius of 6.5 inches

9. a pyramid with a base area of 29.4 square yards and a height of 25.8 yards

10. a cone with a slant height of 42 centimeters and a radius of 25.7 centimeters

11. a cone with a diameter of 46.3 feet and a height of 17.2 feet

Lesson 12-6 Surface Areas and Volumes of Spheres (pp. 864–871)

Find the surface area of each sphere or hemisphere. Round to the nearest tenth.

1.

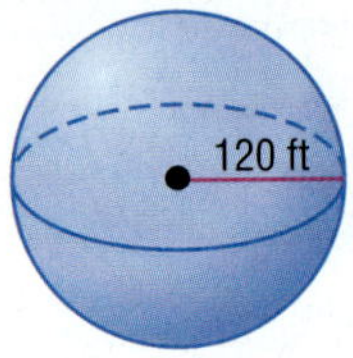

2.

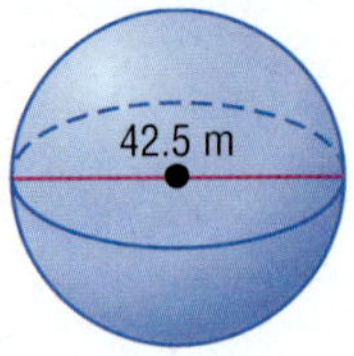

3.

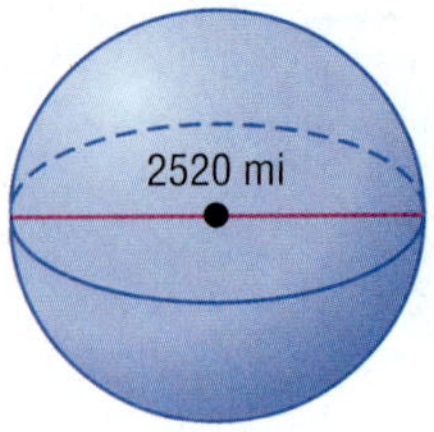

4. hemisphere: circumference of great circle ≈ 40.8 in.

5. sphere: circumference of great circle ≈ 30.2 ft

6. sphere: area of great circle ≈ 814.3 m^2

7. hemisphere: area of great circle ≈ 227 km^2

Find the volume of each sphere or hemisphere. Round to the nearest tenth.

8.

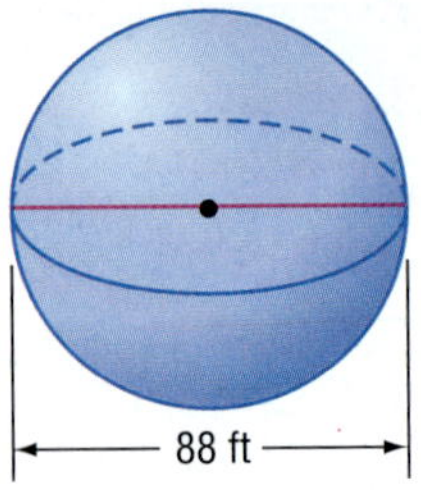

9. $C = 4$ m

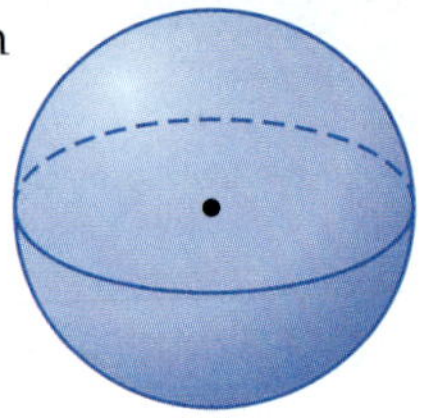

10.

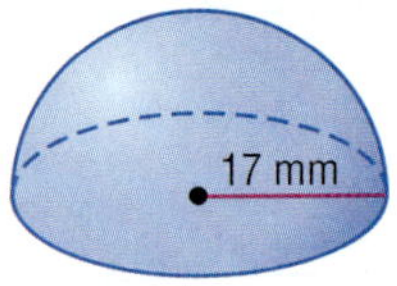

11. sphere: diameter = 3 cm

12. hemisphere: radius = $7\sqrt{2}$ m

13. hemisphere: diameter = 90 ft

14. sphere: radius = 0.5 in.

Lesson 12-7 Spherical Geometry (pp. 873–878)

Name two lines containing point *Q*, a segment containing point *L*, and a triangle in each of the following spheres.

1.

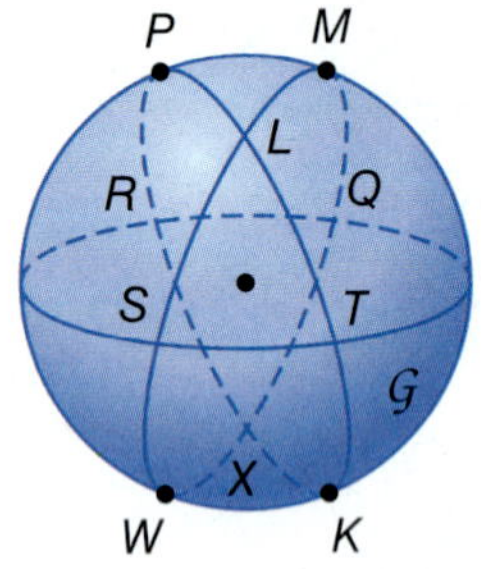

2.

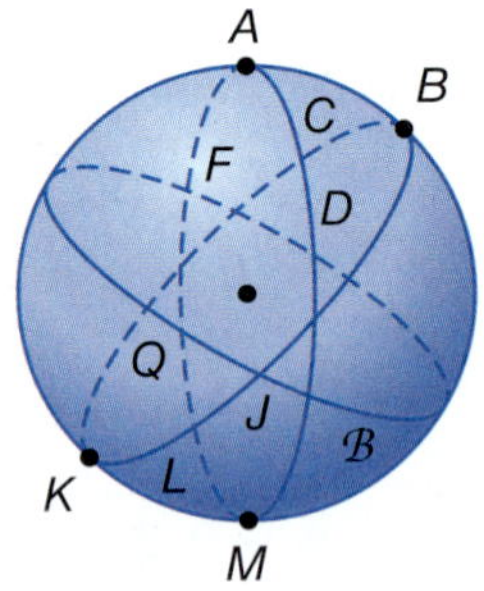

Lesson 12-8 Congruent and Similar Solids (pp. 880–886)

Determine whether each pair of solids is *similar*, *congruent*, or *neither*. If the solids are similar, state the scale factor.

1.

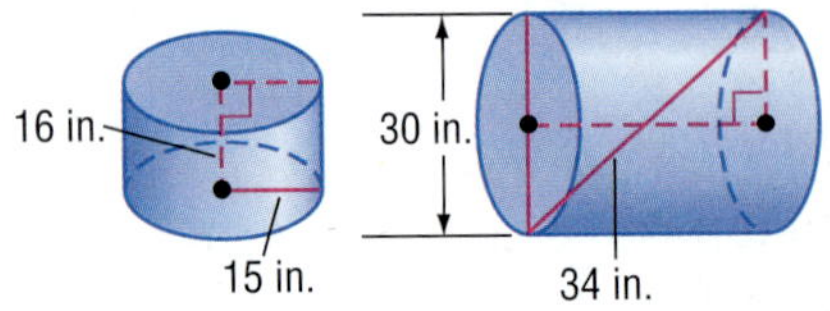

2.

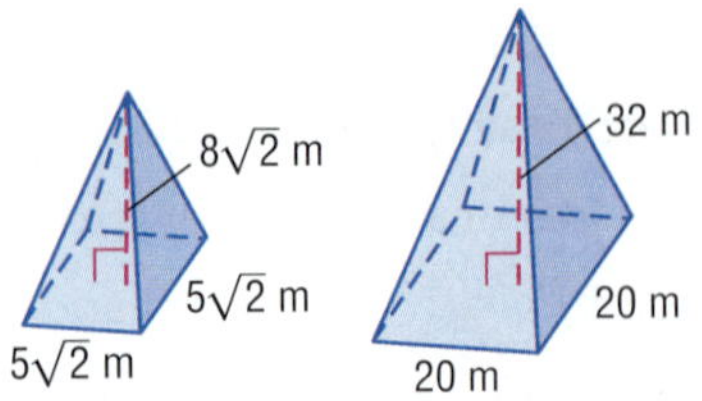

Lesson 13-1 Representing Sample Spaces (pp. 899–905)

Represent the sample space for each experiment using an organized list, a table, and a tree diagram.

1. Sean wants to purchase two pets. He can purchase a turtle, a lizard, or a snake.
2. Beatriz has the option of doing conservation work in Equador or Peru this year. Next year she can work in the other country.
3. The student government meetings need to have an assigned time. The possible meeting times are Wednesday or Thursday at 3:30 P.M. or 4:30 P.M.

Find the number of possible outcomes for each situation.

4. Janet is buying a new computer.

Computer Options	Number of Choices
color	7
screen size	5
memory	3
speed	4

5. Kareem is ordering dinner from a menu at a restaurant.

Menu Items	Number of Choices
drink	12
appetizer	7
entree	10
dessert	4

Lesson 13-2 Permutations and Combinations (pp. 906–914)

1. **BOOKS** Sixteen different books are placed on a shelf at random. What is the probability that *Great Expectations* will be shelved first and *The Doll House* will be shelved second?
2. **SALES** A company uses a program to randomly generate telephone numbers to make sales calls. What is the probability that a phone number randomly generated from among the digits 5, 3, 5, 1, 2, 5, and 3 is 555-2331?
3. **GEOMETRY** If four points are randomly chosen from those named on the triangle shown, what is the probability that they lie on a vertex?

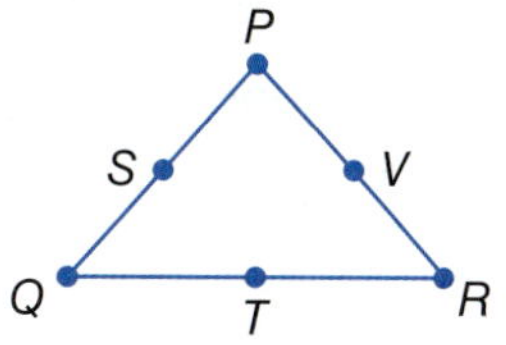

Lesson 13-3 Geometric Probability (pp. 915–921)

Point X is chosen at random on $\overline{AQ}$. Find the probability of each event.

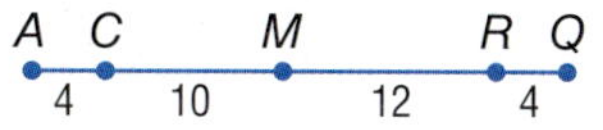

1. $P(X \text{ is on } \overline{AM}.)$
2. $P(X \text{ is on } \overline{CQ}.)$
3. $P(X \text{ is on } \overline{RQ}.)$
4. $P(X \text{ is on } \overline{MR}.)$

Use the spinner to find each probability.

5. *P*(pointer landing on orange)
6. *P*(pointer landing on green)
7. *P*(pointer landing on red)
8. *P*(pointer landing on blue)

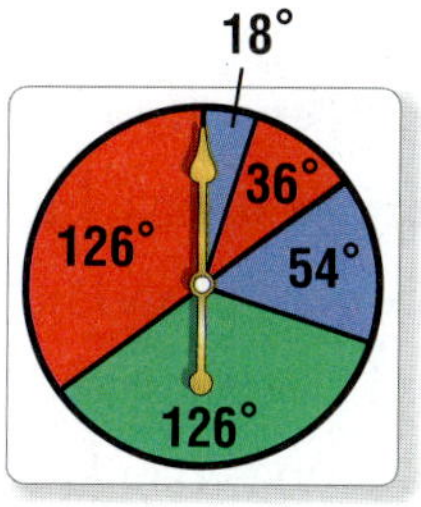

Lesson 13-4 Simulations (pp. 923–930)

Design and conduct a simulation using a geometric probability model. Then report the results using appropriate numerical and graphical summaries.

1. **MUSC** Marcia works at a music store. Last year she sold 45% of the used CDs.
2. **TELEVISION** A commercial plays on a television station 3 out of 5 weekdays.
3. **SOCCER** Todd is on the soccer team at his school. Last season he made at least one goal in 75% of the games.
4. **GAMES** Camilla is playing a board game with a spinner numbered 1 through 5. The table at the right shows the percentage of times she spun each number.

Number	Spinning Results
1	25%
2	20%
3	15%
4	30%
5	10%

Extra Practice

Lesson 13-5 Probabilities of Independent and Dependent Events (pp. 931–937)

Determine whether the events are *independent* or *dependent*. Then find the probability.

1. In a bag of 6 red and 8 blue marbles, a red marble is drawn and not replaced. Then, a blue marble is drawn.
2. You roll two dice and get an odd number each time.
3. An even-numbered card is drawn, without replacement, from a deck of 52 cards. Then, an odd-numbered card is drawn.
4. A spinner numbered 1 through 10 is spun. Find the probability that the number spun is an eight, given that the number spun was an even number.
5. A six-sided die is rolled. If the number rolled is greater than 3, find the probability that it is a 5.

Lesson 13-6 Probabilities of Mutually Exclusive Events (pp. 938–945)

Determine whether the events are *mutually exclusive* or *not mutually exclusive*. Explain your reasoning.

1. rolling a pair of dice and getting two odds or a sum of ten
2. drawing a three or a five from a standard deck of 52 cards
3. selecting a number at random from integers 1 to 50 and getting an odd number or a number divisible by 5

Determine the probability of each event.

4. rolling two dice and not getting a three each time
5. drawing a card from a standard deck and not getting a heart or a diamond
6. spinning a spinner numbered 1 through 10 and not landing on a two
7. flipping two coins and not landing on tails either time

Mixed Problem Solving

Chapter 1 Tools of Geometry (pp. 2–85)

LIGHTHOUSES For Exercises 1–3, name the geometric terms modeled by the objects in the picture. (Lesson 1-1)

1. the light bulb
2. the light going out over the water
3. the surface of the water

4. **DISTANCE** The ice cream shop is between the school and the library on Main Street. It is 500 yards from the school to the library. The ice cream shop is $\frac{3}{10}$ of the way from the school to the library. How far is it from the school to the ice cream shop? (Lesson 1-2)

5. **TRAVEL** The triangle with vertices at Los Angeles, California, Las Vegas, Nevada, and Phoenix, Arizona, is roughly a right triangle. Use the triangle to find the distance from Phoenix to Los Angeles. (Lesson 1-3)

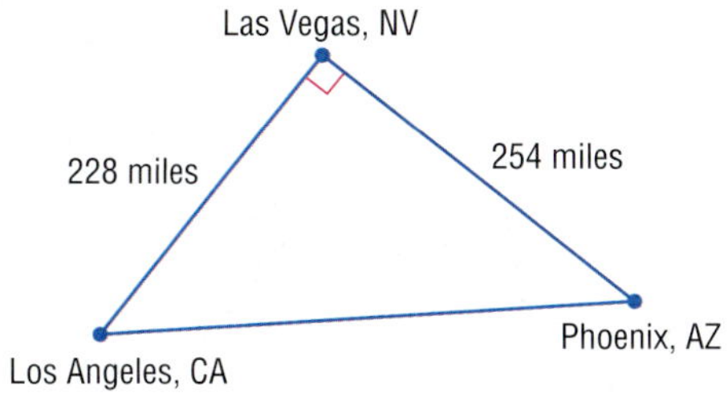

6. **ORIENTEERING** Delilah is using the compass shown below to find directions. (Lesson 1-4)

   a. If she is traveling ENE, the direction that bisects the angle made between E and NE, what is her heading in degrees?

   b. What direction is she traveling if her heading is 225°?

7. **POOL** As a ball hits the side of a pool table, its path is reflected. The angle at which the ball strikes the side is the *angle of incidence*. The angle that the path is reflected is the *angle of reflection*. The angle of incidence and the angle of reflection are congruent. In the diagram below, if $\angle EAC = 96°$, find the angle of reflection and $\angle CAB$. (Lesson 1-5)

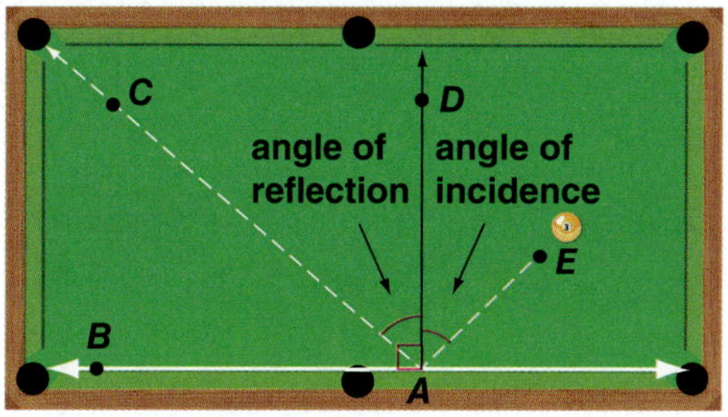

8. **PIZZA** If you divide the cost of a pizza by the area of the pizza, you can determine the cost per square inch of pizza. If the pizzas below cost the same, which is less expensive per square inch? (Lesson 1-6)

9. **GIFTS** Jazmin needs to wrap the gift box shown below. If the wrapping paper at the store is sold in increments of 1 square foot, how many square feet of paper will she need to buy? (Lesson 1-7)

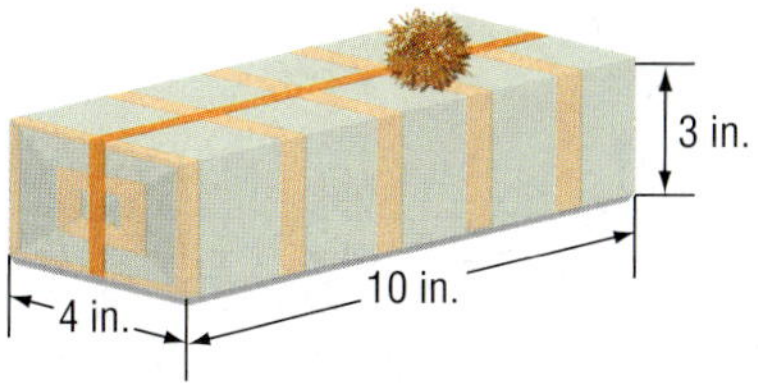

10. **FISH TANKS** Ben purchased a cylindrical fish tank. The diameter of the base is 8 inches, and it is 12 inches tall. What volume of water will fill the tank? (Lesson 1-7)

1. **PATTERNS** Write a conjecture that describes the pattern in the following sequence. Then use your conjecture to draw the next item in the sequence. (Lesson 2-1)

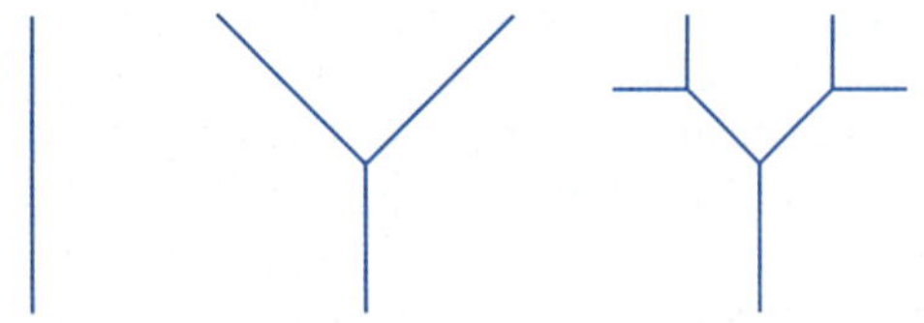

2. **AFTER GRADUATION** The Venn diagram shows the number of seniors that plan to attend college, to work full time, or do neither next fall. (Lesson 2-2)

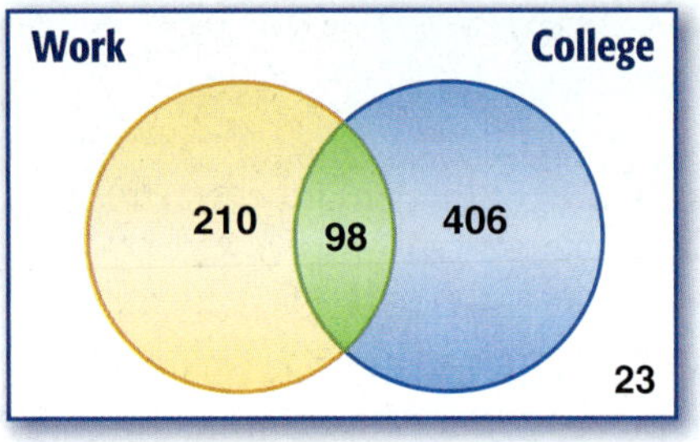

a. How many students plan to work or attend college?

b. How many students plan to work full time while attending college next fall?

c. What does the 23 located in the region outside both the work and college regions represent?

d. How many students plan to go to college next fall?

3. **FORESTRY** In different regions of the country, different variations of trees dominate the landscape. Write the three conditionals in if-then form. (Lesson 2-3)

a. In Colorado, aspen trees cover high areas of the mountains.

b. In Florida, cypress trees rise from swamps.

c. In Vermont, maple trees are prevalent.

4. **MUSIC** Draw a valid conclusion from the given statements, if possible. Then state whether your conclusion was drawn using the Law of Detachment or the Law of Syllogism. If no valid conclusion can be drawn, write *no valid conclusion* and explain your reasoning. (Lesson 2-4)

Given: If you have an MP3 player, then you can listen to your favorite song.

Jill is listening to her favorite song.

5. **FIND THE ERROR** Tino and Marie are making a sculpture for a new garden area outside their school. They decide to place a triangular sheet of corrugated metal on top of three poles. Tino says for the pieces to all meet, at least two of the poles have to be the same height. Marie claims that the poles can all be different heights, and the sheet of metal will still be able to rest on all three poles. Is either of them correct? Use a postulate to explain your reasoning. (Lesson 2-5)

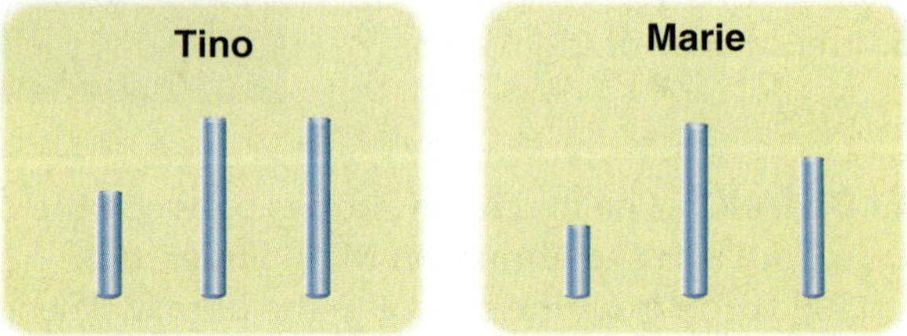

6. **SPEED** The distance formula is $d = rt$, where d is distance, r is rate (or speed), and t is time. (Lesson 2-6)

a. Prove that if the distance and rate are known, then the formula $t = \frac{d}{r}$ gives the time of the trip.

b. Write a proof to show that when the distance is constant and the time is halved, the rate is doubled.

7. **QUILTING** The cross quilt block can be used along with solid fabric squares to make the following pattern. (Lesson 2-7)

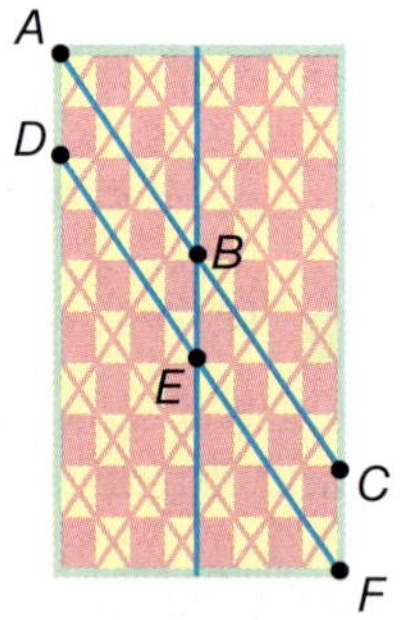

$\overline{AC}$ is congruent to $\overline{DF}$. $\overline{AB}$ is congruent to $\overline{EF}$. Prove that $\overline{BC}$ is congruent to $\overline{DE}$.

8. **RIVERS** Tributaries of rivers sometimes form a linear pair of angles when they meet the main river. The Yellowstone River forms the linear pair $\angle 1$ and $\angle 2$ with the Missouri River. If $m\angle 1$ is 28, find $m\angle 2$. (Lesson 2-8)

HIGH VOLTAGE TOWERS
Classify the relationship between each pair of angles. (Lesson 3-1)

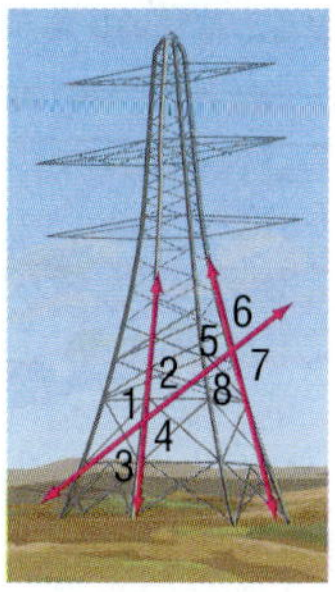

1. $\angle 2$ and $\angle 5$
2. $\angle 1$ and $\angle 7$
3. $\angle 6$ and $\angle 2$
4. $\angle 4$ and $\angle 5$

5. **ART** This box was drawn using a vanishing point. Although the illusion is that the face of the box is a rectangle, it is not. $\overleftrightarrow{AC}$ and $\overleftrightarrow{BD}$ are parallel. If $m\angle 2 = 118$, find $m\angle 1$. (Lesson 3-2)

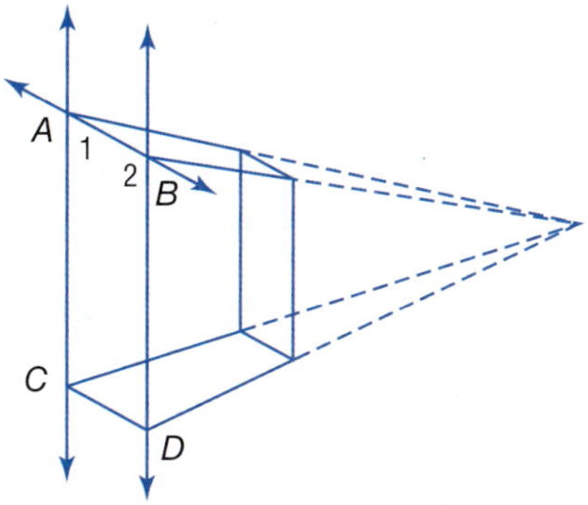

6. **INLINE SKATING** The map for the local mixed-use recreation trail where Isaac goes to inline skate has a 1320-foot section with a 5% grade. (Lesson 3-3)
   a. Express a 5% grade as a fraction in simplest terms.
   b. As Isaac travels 100 feet horizontally, how much does his position change vertically? Use the figure below to help you decide.

7. **MOVIES** Cristina can either join a movie club that charges \$5 per month plus \$1 per movie, or she can rent movies for the regular price of \$2 per movie. (Lesson 3-4)
   a. Write and graph two equations that show the amount of money spent based on the number of movies rented.
   b. If Cristina generally watches 8 movies a month, should she join the club? Explain your answer.
   c. If Cristina joins the movie club for one year and rents 75 movies throughout the year, how much money will she have spent?

8. **ESPALIER** Espalier is the technique of pruning and training a plant to grow flat against a wall. Is it possible to prove that any of the branches are parallel? If so, explain how. If not, explain why not. (Lesson 3-5)

9. **UTILITIES** Housing developers often locate the shortest distance from a house to the water main so that a minimum of pipe is required to connect the house to the water supply. Copy the diagram, and draw a possible location for the pipe. (Lesson 3-5)

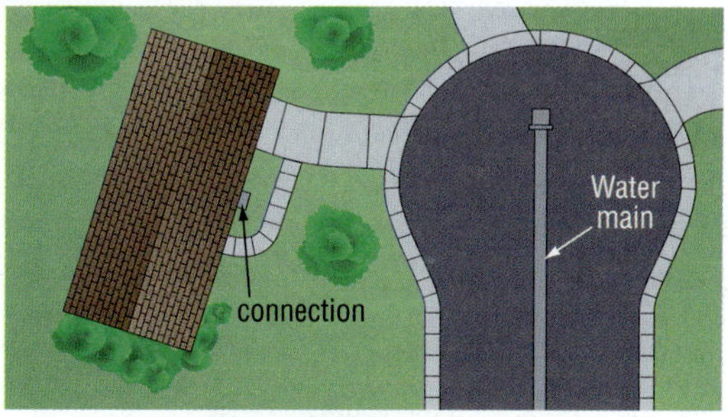

10. **HIKING** While hiking, Micah went off the trail to take a picture from a cliff. Copy the diagram below, and construct the segment that represents the shortest route Micah can take back to the marked trail. (Lesson 3-6)

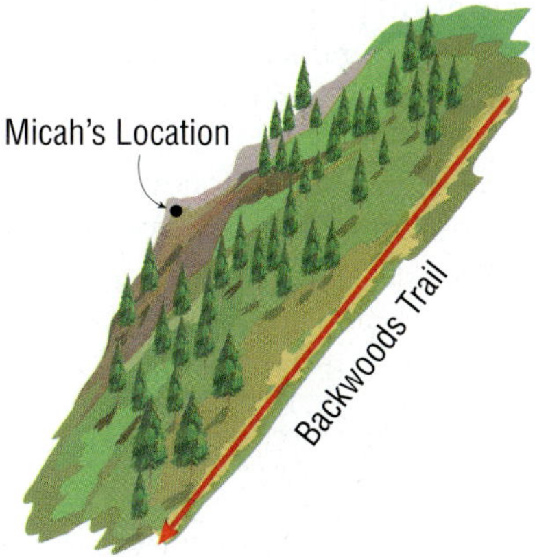

1. **LOGOS** Jenny and Alec are working on a logo for the school newspaper in the shape of an isosceles triangle. Jenny wants to make the legs twice as long as the base. Alec wants to make them 12 more than half the base. They decide to compromise and do one leg each way. If it is still an isosceles triangle, how long are the legs and how long is the base? (Alec and Jenny are working in centimeters.) (Lesson 4-1)

2. **BICYCLING** In the bicycle shown, $m\angle C = 60$ and $m\angle D = 105$. Find the measure of $\angle B$. (Lesson 4-2)

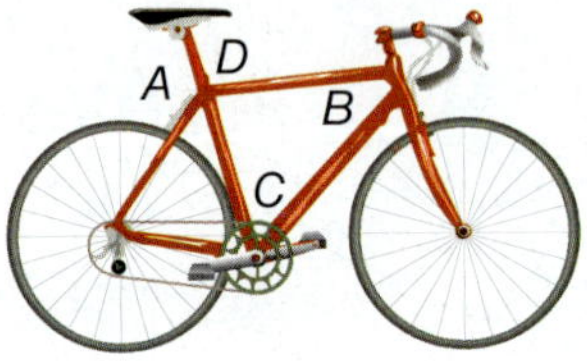

3. **STATUE OF LIBERTY** On the Statue of Liberty's crown, $\triangle STU$ is congruent to $\triangle LBY$. Name three pairs of congruent segments and three pairs of congruent angles. (Lesson 4-3)

4. **BRIDGES** Kai-Yo knows that the sides of a bridge, $\overline{AC}$ and $\overline{BD}$, are parallel and have the same length. She needs to be certain that the far side of the bridge is the same width as the near side to ensure that she can tow her wagon across the bridge. Write a paragraph proof to reassure Kai-Yo that $\overline{CD}$ has the same length as $\overline{AB}$. (Lesson 4-4)

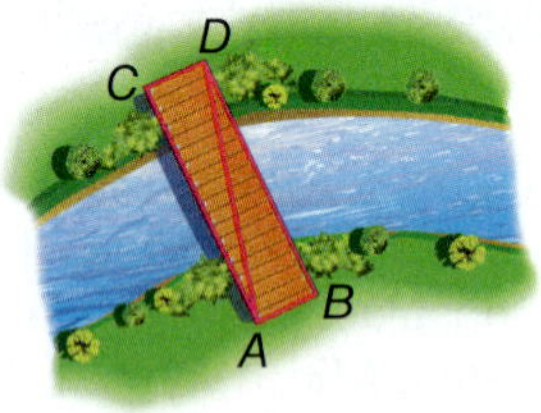

5. **STUNTS** Demetrius is a stunt rider for a traveling bike show. Setting up for the show, he sees a creek with high banks and a clear area on both sides. He knows that with his trick bike and special ramps, he can jump 59 feet. After choosing his landing spot, he measures off the following triangles. Can he make the jump? Explain your reasoning. (Lesson 4-5)

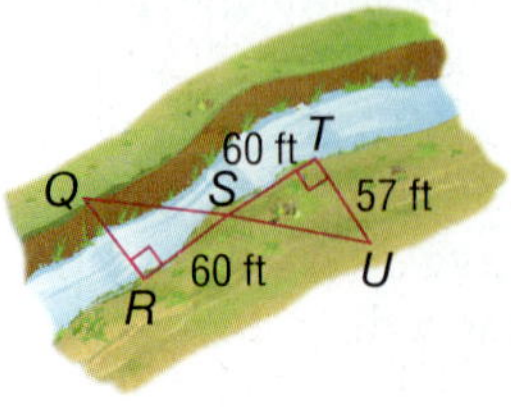

6. **CONSUMER SCIENCES** Lily's class is studying a unit on hospitality. She is in charge of setting the table decoratively for an Asian meal. If she folds the napkin on each plate into the shape of an isosceles triangle, and then places chopsticks on top of the napkin so that they also form an isosceles triangle, what angle will each chopstick make with the edge of the napkin? (Lesson 4-6)

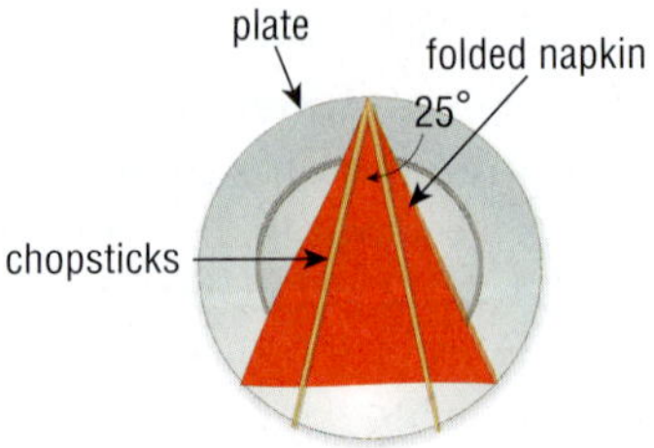

7. **YEARBOOKS** Takeisha is working on a page layout for the yearbook. She has placed a triangular picture on the page. Viewing the page as a coordinate plane, she decides to reflect the shape of the picture in the y-axis and then in the x-axis to place another triangular picture on the page. Verify that the triangles are congruent after two reflections. (Lesson 4-7)

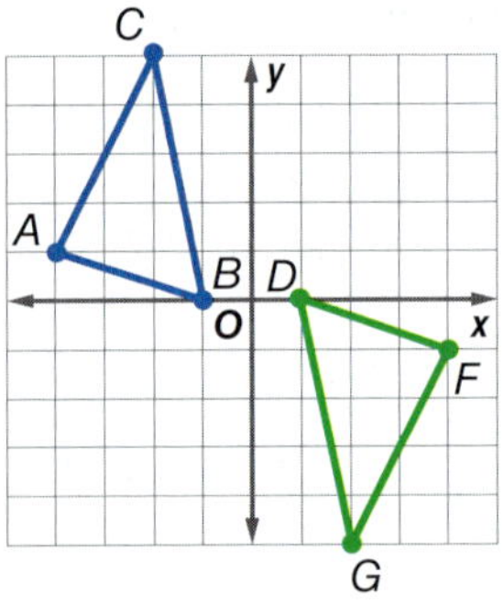

8. **HIKING** Petra and Joe are hiking. Petra hikes 300 feet east of the camp and then hikes 500 feet north. Joe hikes 500 feet west of the camp and then 300 feet north. Prove that Joe, Petra, and the camp form a right triangle. (Lesson 4-8)

9. **STEEPLECHASE** Write a coordinate proof to prove that $\triangle ABD \cong \triangle FBD$ in the hurdle shown. (Lesson 4-8)

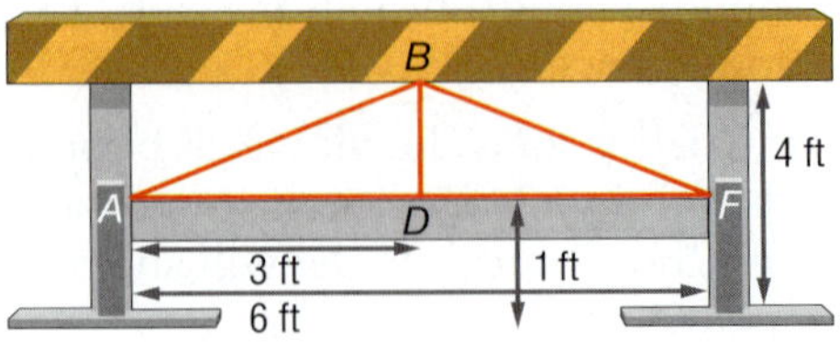

1. **OFFICE DESIGN** The copy machine C, filing cabinet F, and supply shelves S are positioned in an office as shown. Copy the diagram, and find the location for the center of a work table so that it is the same distance from all three points. (Lesson 5-1)

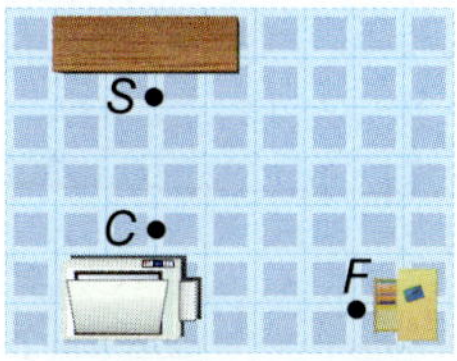

2. **FOUNTAINS** The Garden Club decides to donate a fountain to the town park. It will fit within a triangle formed by three walkways. Copy the diagram and find the best location for the center of the fountain so it will spray the largest circle possible without spraying any of the walkways. (Lesson 5-1)

3. **LIGHTING** Santos is designing a modern-style lamp with a triangular top, to deflect the light downward. He has laid out the triangle design on a coordinate grid. If the deflector is to balance on a single point, find the coordinates of that point. (Lesson 5-2)

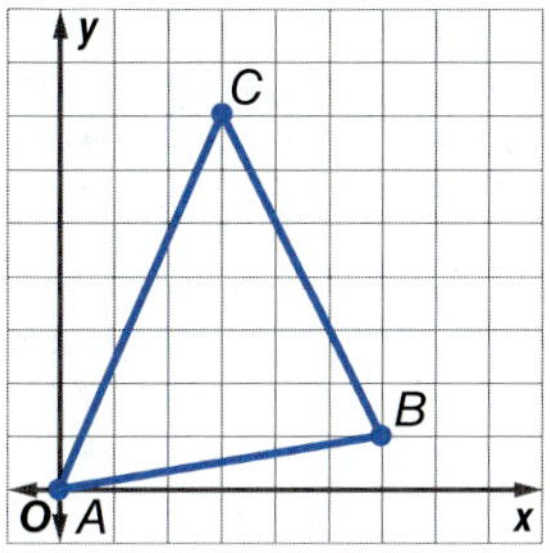

4. **MAP SENSE** Three cities lie in a triangle roughly in the positions shown. Which two cities are the farthest apart? (Lesson 5-3)

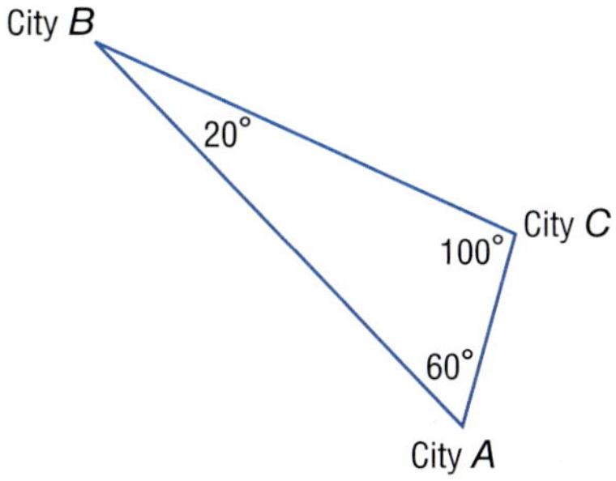

5. **TREEHOUSES** Mrs. Sanchez is constructing the framework for part of a treehouse for her daughter. She plans to install braces at the ends of a certain floor support, as shown. Which brace will be longer—the brace attached to A or B? (Lesson 5-3)

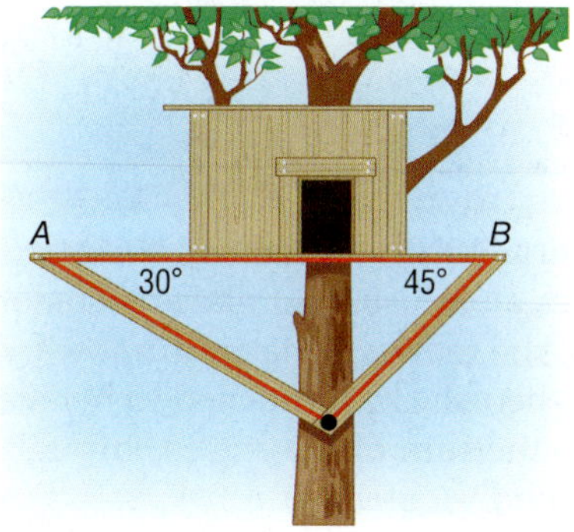

6. **WORK HABITS** Christa worked over 6 hours today, taking only two short breaks. Use indirect reasoning to prove that she worked longer than 2 hours during at least one work session. (Lesson 5-4)

7. **CARPENTRY** Felicia is making a triangular frame. She has two pieces of wood, one of which is 15 inches and the other is 10 inches. If she would like to nail together a frame without cutting the wood, within what range of lengths would a third piece of wood need to lie? (Lesson 5-5)

8. **CRAFTS** Andrew is learning to make stained glass windows. He would like to make a triangle with three pieces of trim. The lengths of the pieces are 3 inches, 6 inches, and 10 inches. Can he make a triangle with these three pieces? Explain your reasoning. (Lesson 5-5)

9. **SPORTS** Wesley and Nadia are playing tetherball. The photo below shows them at two different points in the game. Who was standing closer to the pole? Explain. (Lesson 5-6)

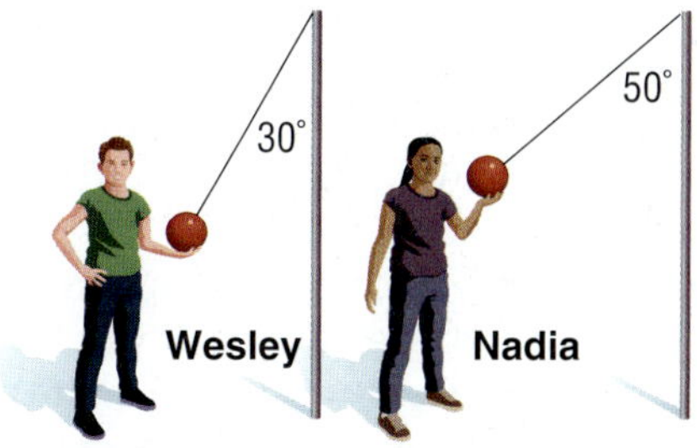

10. **SNOWSHOEING** Two groups of friends leave the same cabin to go snowshoeing. Group A goes 2.5 miles due north and then turns 79° east of north and travels 2.5 miles further. Group B goes 2.5 miles due south and then turns 86° east of south and travels 2.5 miles. Who is now closer to the cabin? Use a diagram to explain your reasoning. (Lesson 5-6)

1. **OUTDOOR DRAMA** The shape of the stage of an amphitheater is that of a heptagon. What is the sum of the measures of the interior angles of the stage? (Lesson 6-1)

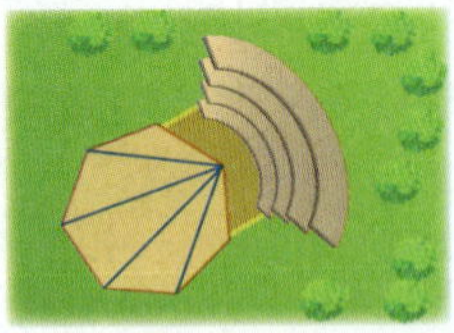

2. **MARCHING BAND** A member of a marching band marches a path shaped like an octagon. At each corner, she makes a sharp turn toward the next side. When she has returned to her starting point, what is the sum of the angles through which she has turned? (Lesson 6-1)

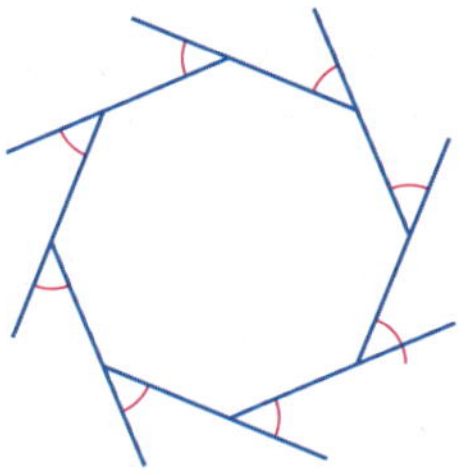

3. **AUTO REPAIR** A scissor jack is used to lift part of a car to make repairs. $ABCD$ is a parallelogram. As the jack is cranked up, $m\angle A$ and $m\angle C$ increase. What must happen to $m\angle B$ and $m\angle D$? Draw a picture to show the change. (Lesson 6-2)

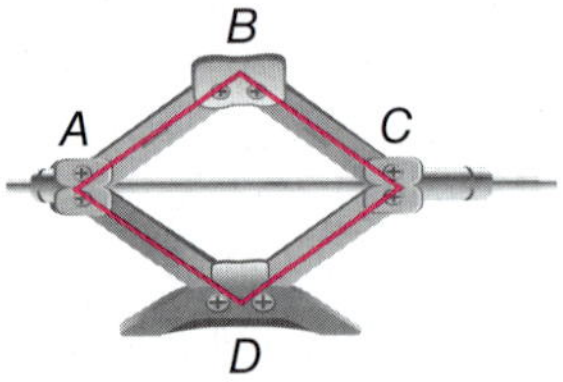

4. **CONSTRUCTION** Che is laying out wooden forms into which he will pour concrete to make a patio. The layout is to be several parallelograms laid together. If he has two 3-foot pieces and two 4-foot pieces of framing, can he be sure that he is making a parallelogram? Explain. (Lesson 6-3)

5. **DECORATING** Skye decorated a bulletin board using ribbon and large buttons. If you know that $AC = BD$ and $m\angle CAB = m\angle DBE$, write a two-column proof to show that quadrilateral $ACDB$ is a parallelogram. (Lesson 6-3)

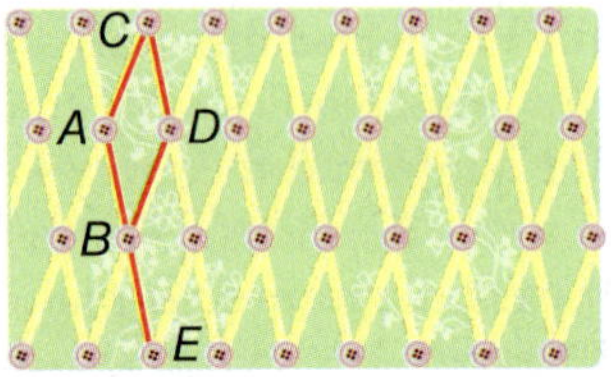

6. **MINIATURE GOLF** A section of a miniature golf course will be a rectangle 10 feet long and 6 feet wide. Suppose the contractor marked the boundaries with the corners at A, B, C, and D. The contractor measured BD and AC and found that $AC > BD$. Describe how to move stakes L and K to make $ABCD$ a rectangle. (Lesson 6-4)

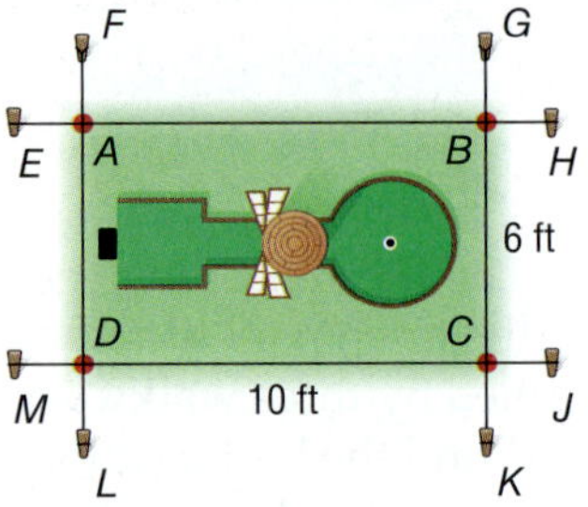

7. **WINTER** Four boys are standing such that their positions are the vertices of a rectangle. They are throwing snowballs. (Lesson 6-4)

a. If Carl is 100 feet from Ben, how far is Abe from Dan?

b. If Abe and Ben throw snowballs along the diagonals at the same time and the snowballs collide in mid air, how far had the snowballs traveled at the time of the collision?

8. **ART** The figure below appears to be a cube, but is actually made up of three rhombi. Can the figure be translated such that the image of $\overline{EF}$ lines up with $\overline{BC}$? Explain your reasoning. Draw a sketch to support your explanation. (Lesson 6-5)

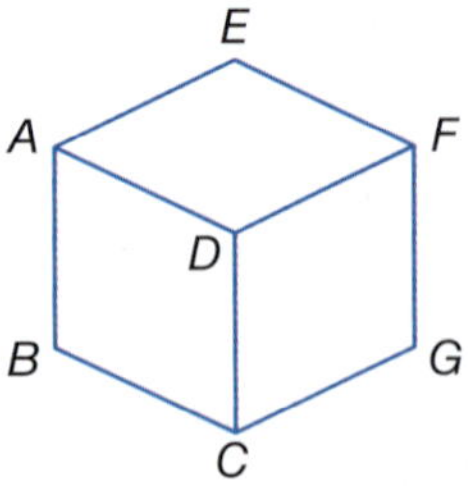

9. **JEWELRY** Rosamaria is making a jade pendant in the shape of a trapezoid. She is putting a piece of pewter trim on both bases and across the midsegment of the trapezoid. What is the total length of the pewter trim that she will use? (Lesson 6-6)

1. **AUTOMOBILES** Turell drove 326 miles and used 11.9 gallons of gas. What is his approximate gas mileage, expressed as miles per gallon? (Lesson 7-1)

2. **PHOTO EDITING** Kylie is making a computer presentation for computer class. She wants to use this photo of a frog resized to fill the screen. Determine whether the rectangular images are similar. If so, write the similarity statement and scale factor. Explain your reasoning. (Lesson 7-2)

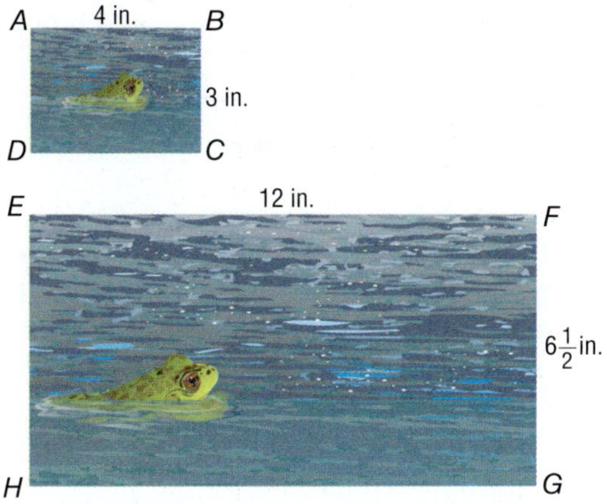

3. **HORTICULTURE** Macy needs to determine the height of a tree in her yard. On a sunny afternoon, Macy's shadow is 2 feet 6 inches long. If she is 4 feet 9 inches tall, and the shadow of the tree is 8 feet long, how tall is the tree to the nearest foot? (Lesson 7-3)

4. **ICE SKATING** Hiro is choreographing a skating routine. He wants to use two similar triangles that cover a portion of the rink. Does his plan show similar triangles? Explain your reasoning. (Lesson 7-3)

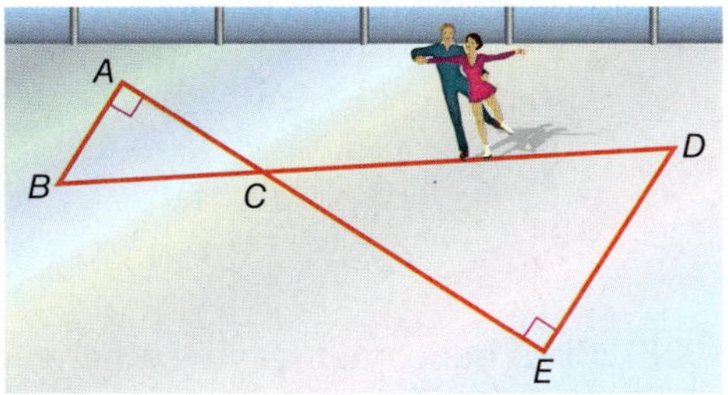

5. **SIGNS** The design of a neon sign for a restaurant is shown below. The designer is using three parallel grey lines to help lay out the letters evenly. Find x to the nearest tenth of a foot. (Lesson 7-4)

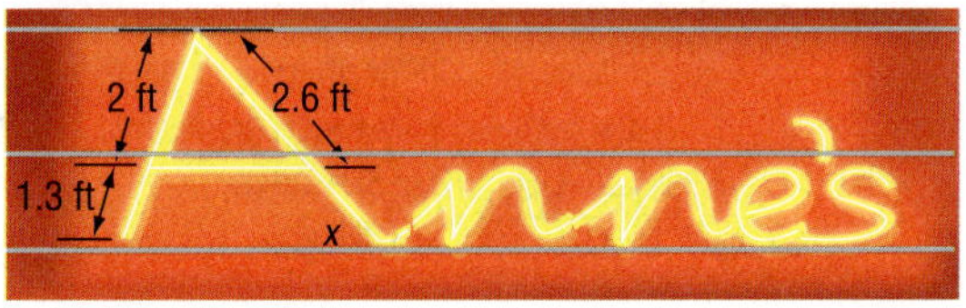

6. **HOUSES** In an A-frame house, the roof slopes to the ground. Find the width x of the second floor. (Lesson 7-4)

7. **VISION** A tree that is 20 feet tall forms a retinal image that is 9 millimeters tall. If $\triangle QRV \sim \triangle TRS$ and the distance from the pupil to the retina is 25 millimeters, how far away from your pupil is the tree? (Lesson 7-5)

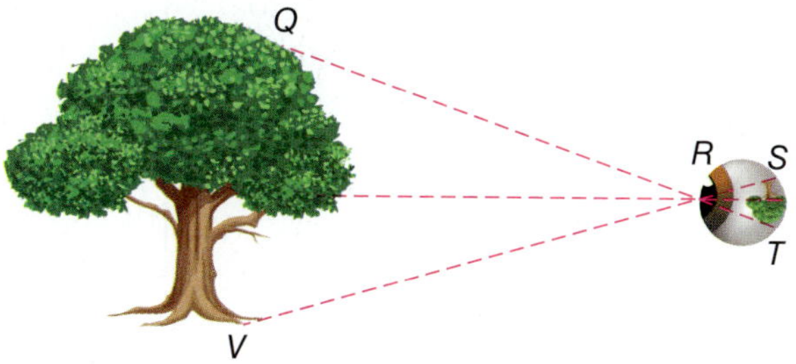

8. **BANNER ART** The National Honor Society is having a fundraiser selling houseplants. Jim and Edmundo are in charge of making a banner by tracing the brochure onto a transparency and using the overhead projector to project the image onto the poster paper. If the width of the pot in the brochure is 4 inches and it is $2\frac{1}{2}$ feet wide on the banner, what is the scale factor of the dilation? (Lesson 7-6)

9. **BASKETS** The Longaberger Basket Company's home office is a scale model of one of the baskets they make. The scale factor is 600 : 1. (Lesson 7-7)

a. At the roof, the building is 208 feet long by 142 feet wide. What are the dimensions of the top of the basket after which it is modeled?

b. If the bottom of an actual basket is 3.8 inches by 2.5 inches, what are the dimensions around the bottom of the building?

10. **MAPS** The scale on a map is 0.3 inch: 100 miles. The distance between Portland, Maine, and Portland, Oregon, is 9.6 inches on the map. Find the actual distance between the two cities. (Lesson 7-7)

1. **NATURE** Clare is using a book to sight the top and bottom of a cliff across the road from her school. She knows she is exactly 75 feet from the base of the cliff. If her eye level is 5 feet from the ground, find the height of the cliff. (*Note*: The picture is not to scale.) (Lesson 8-1)

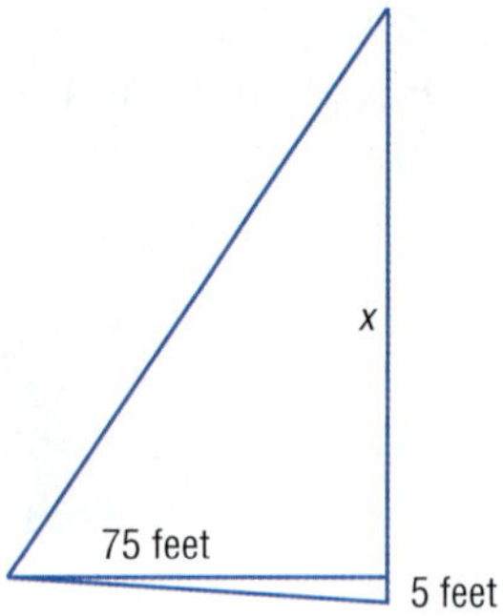

2. **CONSTRUCTION** Gerardo is repairing his shed. He wants to add a diagonal piece as shown by the dashed line. How long should it be? (Lesson 8-2)

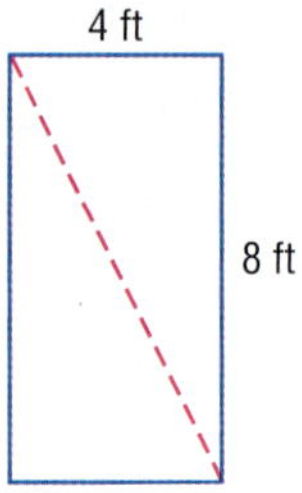

3. **HOME DECORATING** The Coles want to make an 8-foot-tall wreath for a decoration for their house. They plan to build a hexagonal frame and wrap it with garland. For the entire frame to be 8 feet tall, what should be the length of each side to the nearest inch? (*Hint:* A hexagon can be made up of six equilateral triangles.) (Lesson 8-3)

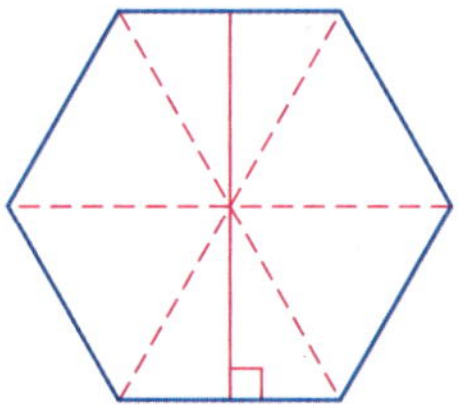

4. **SURVIVAL SKILLS** Mariah is putting up a shelter to sleep in while backpacking, by tying a rope between two trees, laying a tarp across it, and staking it down as shown. The open end of the shelter makes an equilateral triangle, and the rope is 3 feet from the ground. How long is each side of the open end? (Lesson 8-3)

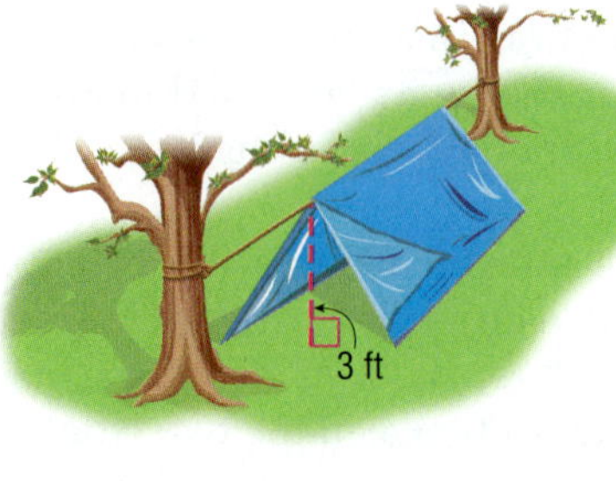

5. **ACCESSIBILITY** A ramp into a local department store raises at an angle of 5°. If you travel 3 meters on the ramp, how far have you traveled vertically? (Lesson 8-4)

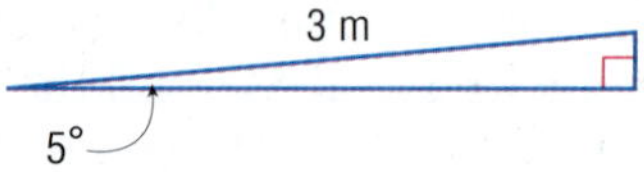

6. **SPACE FLIGHT** A space shuttle is directed towards the Moon, but drifts 0.8° from its calculated path. If the distance from Earth to the Moon is 240,000 miles, by how many miles will the space shuttle miss its intended target? (Lesson 8-4)

7. **HALF TIME** At a football game half-time show, the drum major throws his baton 52 feet straight up in the air. What is the angle of elevation for the peak of the toss for a person sitting 60 feet away, 20 feet up in the bleachers? (Lesson 8-5)

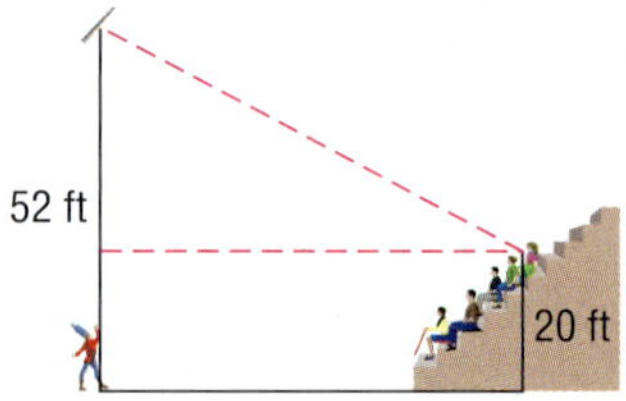

8. **BRIDGES** This old postcard shows a picture of the Y-Bridge in Zanesville, Ohio. Using the measurements shown, determine the length of a bridge that would connect point *B* directly to point *C*. (Lesson 8-6)

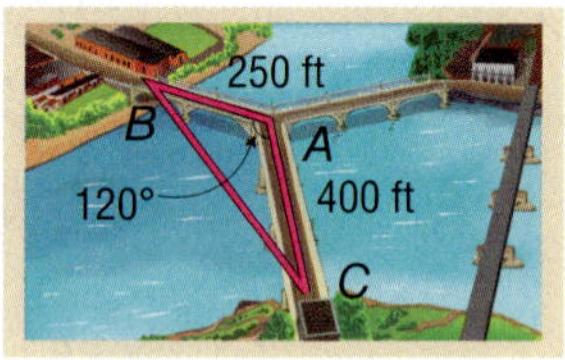

9. **HELICOPTERS** A helicopter leaving a hangar travels 50 meters north, turns and travels 200 meters east, and then lifts off 300 meters into the air. What is the total displacement of the helicopter? (Lesson 8-7)

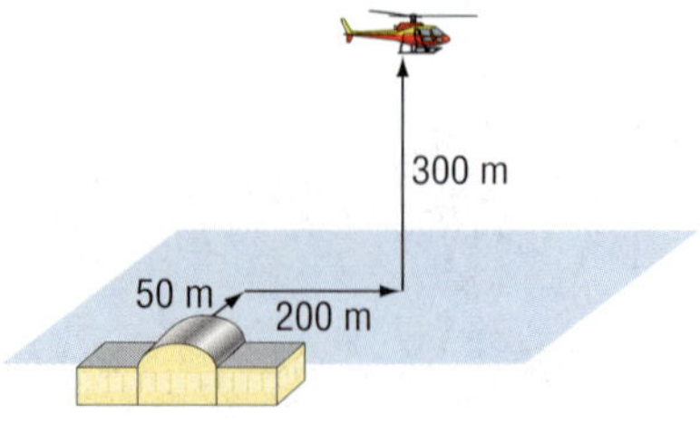

1. **CAMPING** Jin is camping with a friend. They plan to hike to see the rock bridge one day, then back to camp, and hike to the falls on the next day. Where along the trail should they place their camp in order to minimize the distance they must hike? (Lesson 9-1)

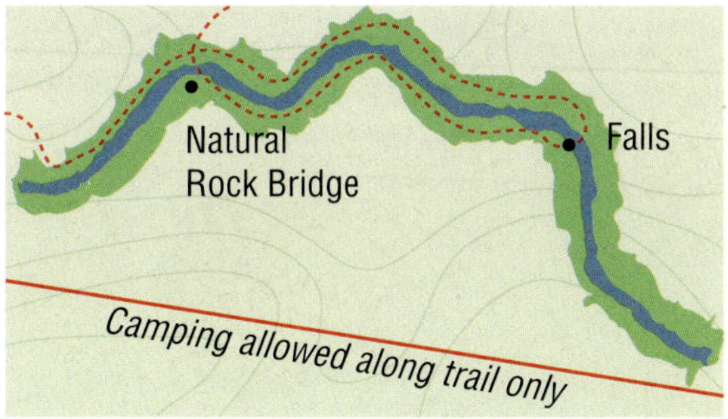

2. **ANTS** Twelve ants are walking on a mirror. Each ant has 6 legs. How many legs can be seen during this journey? (Lesson 9-1)

3. **MAPS** Jailene's house and several places she visits are shown on the grid. (Lesson 9-2)

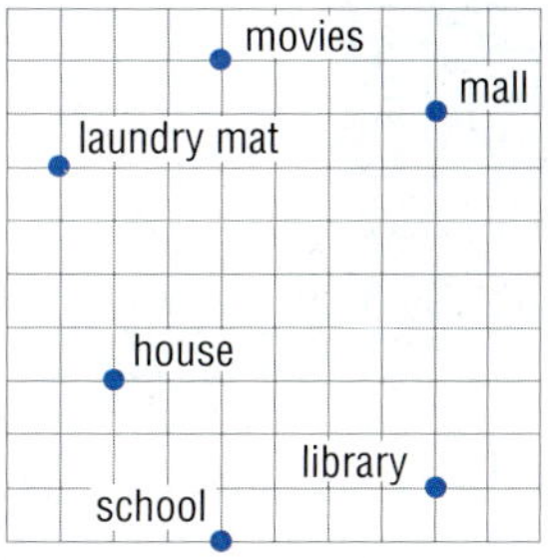

a. If she leaves home and travels 2 blocks south and 6 blocks east, what is her new location?

b. Use words to describe two possible translations that will take Jailene home from the mall.

4. **WEATHER** A windsock is used at the airport to indicate the direction of the wind. If it is pointing south and rotates 135° counterclockwise, what is the new wind direction? (Lesson 9-3)

5. **QUILTING** Aurelia laid out the pattern below to use in making a quilt. What combination of transformations did she use to create the pattern? (Lesson 9-4)

6. **ANIMATION** Computers are often used to create animation. Describe the transformation combination used to animate the star shown below. (Lesson 9-4)

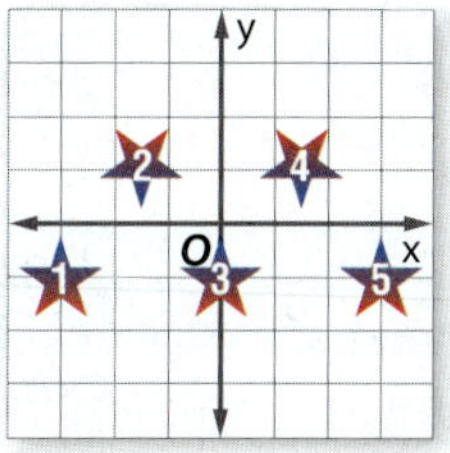

7. **MERRY-GO-ROUND** State the order and magnitude of symmetry for the merry-go-round shown, if the rails are evenly spaced. (Lesson 9-5)

8. **LANDMARKS** The Washington Monument is an obelisk built in 1884 in honor of George Washington. State whether the monument has plane symmetry, axis symmetry, or neither. (Lesson 9-5)

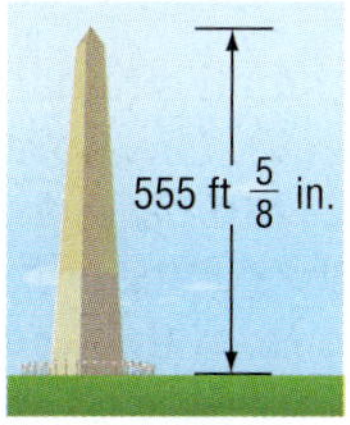

9. **AMUSEMENT PARKS** The Kings Island amusement park near Cincinnati, Ohio, has a $\frac{1}{3}$ scale replica of the Eiffel Tower. If it stands 331 feet 6 inches tall, how tall is the Eiffel tower in Paris? (Lesson 9-6)

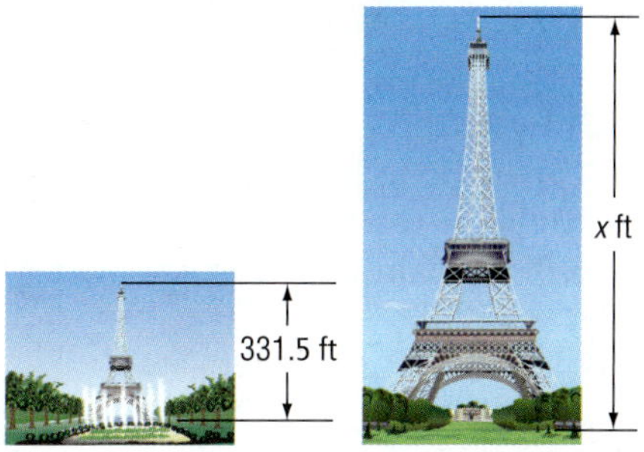

10. **MODELS** Rey is building a model using a scale factor of $r = 200$. Is he more likely building a model of a train or a germ? Explain your reasoning. (Lesson 9-6)

1. **ARCHITECTURE** The Houston Astrodome is 710 feet in diameter. If you walk all the way around the building, how far have you walked? Express your answer to the nearest foot. (Lesson 10-1)

2. **CLOCKS** While watching the pendulum of a clock, Thomas noticed that it swung from 26 minutes after the hour to 34 minutes after the hour. If the point at which the pendulum is attached to the clock is the center of the circle that makes up the face of the clock, identify the arc made by the pendulum as a *major arc*, *minor arc*, or *semicircle* of a circle. Then find its measure. (Lesson 10-2)

3. **LANDSCAPING** A small circular park has a diameter of 175 feet. There are three trees on the edge of the park with paths between them. There is a fountain at the center of the park. What is the shortest distance from the fountain to the path between Tree *A* and Tree *B*? (Lesson 10-3)

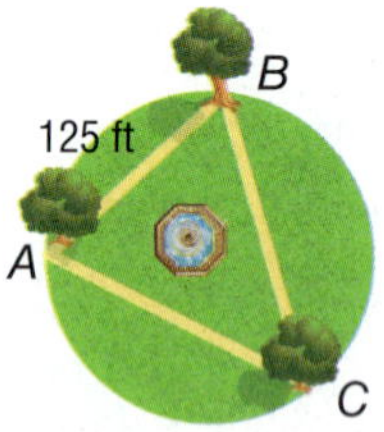

4. **TROPICS** Two people move their line of sight through an angle to look from the hut to the palm tree. What is the relationship between the angles at *A* and *B*? Explain your answer. (Lesson 10-4)

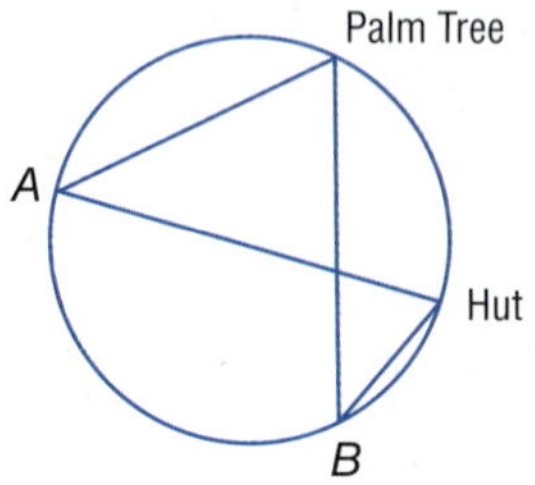

5. **ALIENS** The rays shown are tangent to Zadeek's planet Dwook at Island *A* and Island *B*. Zadeek is one sixth of his favorite number of miles from Island *A* and 1100 less than twice his favorite number of miles from Island *B*. What is Zadeek's favorite number of miles? (Lesson 10-5)

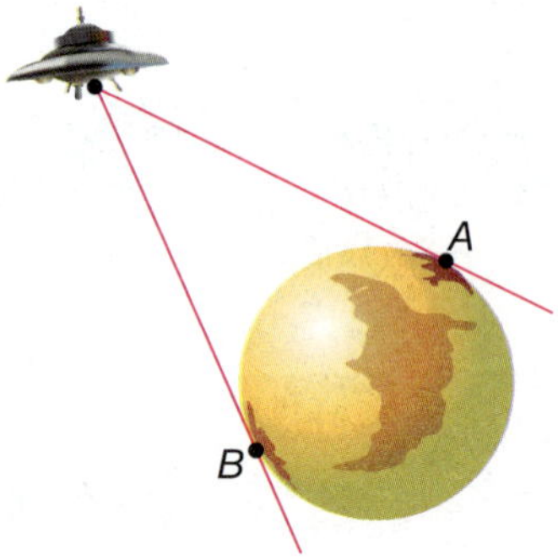

6. **PHOTOGRAPHY** Travis frames a barrel in a camera shot as shown so that the lines of sight form tangents to the barrel. If the camera's viewing angle is 22°, what is the arc measure of the barrel that appears in the shot? (Lesson 10-6)

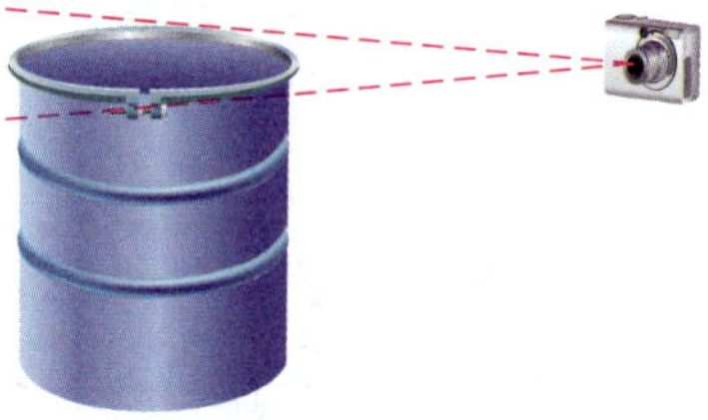

7. **CULVERTS** What is the diameter of the circle containing the arc of the culvert shown? Round to the nearest tenth. (Lesson 10-7)

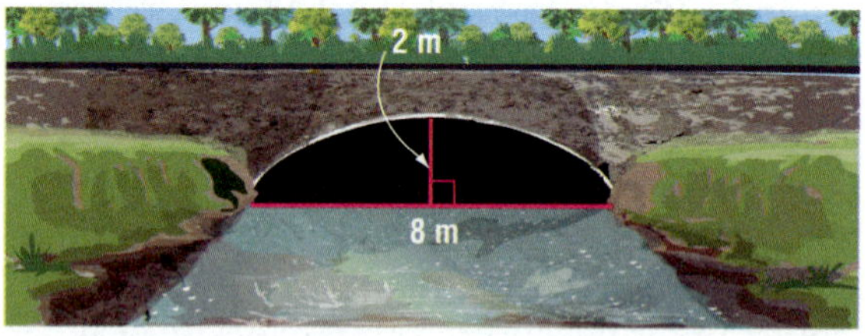

8. **DRILL TEAM** Estrella was planning a drill team routine on graph paper. She was placing the girls in a circle, and having them march in to join hands at the center of the circle. Three of the girls in the circular formation have approximate coordinates of *R*(10, 12), *S*(9, 5), and *T*(3, 13). (Lesson 10-8)

   a. What are the approximate coordinates of the center where they will join hands?

   b. What is the equation of the circle?

   c. If each unit is 3 feet, how far will each girl travel to reach the center of the circle?

1. **PARKING LOTS** A new restaurant has a 48-foot by 16 foot rectangle in which to make parking spaces. The owners want each space to be longer than 16 feet, so they slant the spaces, making three congruent parallelograms with two triangular areas left over. (Lesson 11-1)

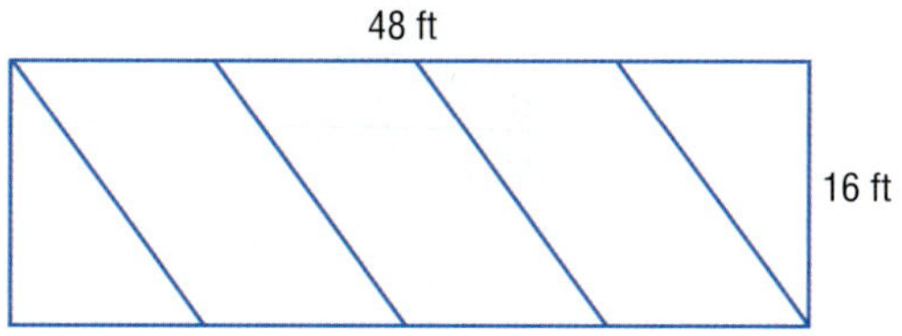

   a. Find the perimeter and area of each parking space.

   b. Find the perimeter and area of each of the triangles.

   c. The restaurant owners want to plant ground cover shrubs on the triangles. If each shrub will cover at most one square yard, how many shrubs do they need to completely cover the two triangles?

2. **ROADS** Find the perpendicular distance between the longer two sides of the crosswalk shown. (Lesson 11-1)

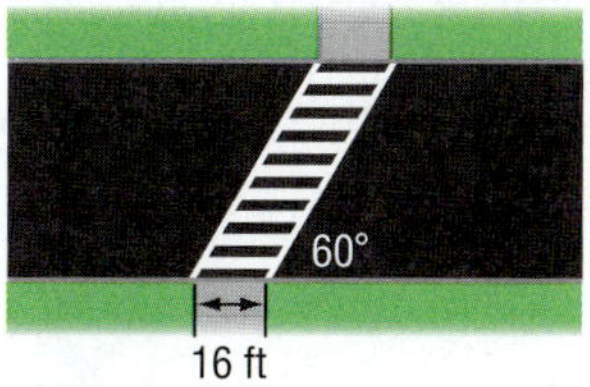

3. **MURALS** Joni is designing a mural for the side of a building. The wall is 15 feet high and 50 feet long. If she covers the area with a kite as shown, what is the area of the kite? (Lesson 11-2)

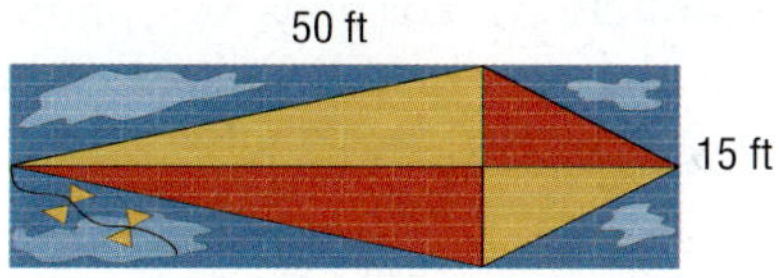

4. **CLOCKS** The diameter of the blue circle of the clock shown is 10 inches. The radius of the green circle is 7 inches. (Lesson 11-3)

   a. Find the area of the blue and pink sector.

   b. Find the area of the green sector.

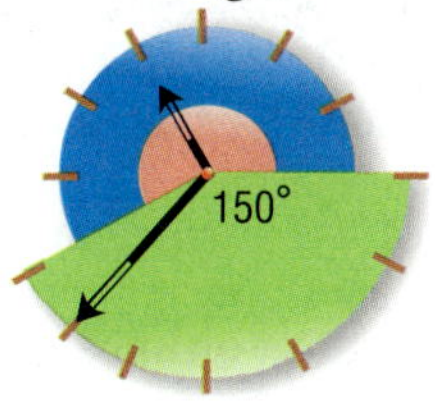

5. **GAZEBOS** Antoinette has a gazebo with a circular floor with a 10-foot diameter. The ceiling is a regular heptagon that could be inscribed in the circle. Find the area of the ceiling. (Lesson 11-4)

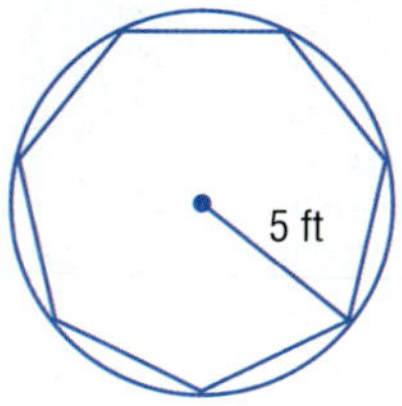

6. **REAL ESTATE** Pete's house is on an irregularly shaped lot as shown below. What is the area of the lot to the nearest square foot? (Lesson 11-4)

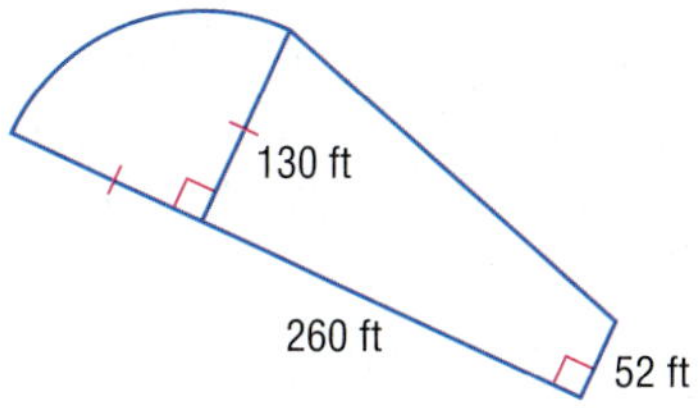

7. **CHOREOGRAPHY** Ayako is choreographing a dance number for the drama club's spring musical. If she can use all of the semicircular stage, the rectangular apron, and the triangular protrusion, how many square yards does she have to work with? Express your answer to the nearest square yard. (Lesson 11-4)

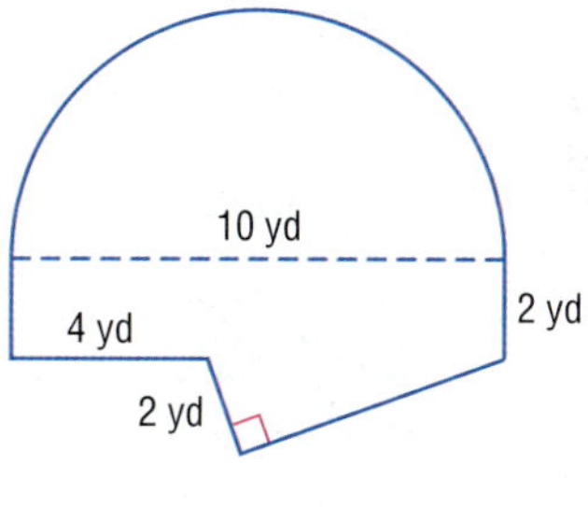

8. **MODELS** Anna is making a model of her house. The area of the living space of the model is 70 square inches. If the area of the actual living space is 1750 square feet, how many feet are represented by each inch of the model? (Lesson 11-5)

9. **CRAFTS** Kate is pasting pictures into a scrapbook over the decorative papers shown below. If the area of the larger triangle is 36 square inches, what area of paper does she need for the smaller triangle? (Lesson 11-5)

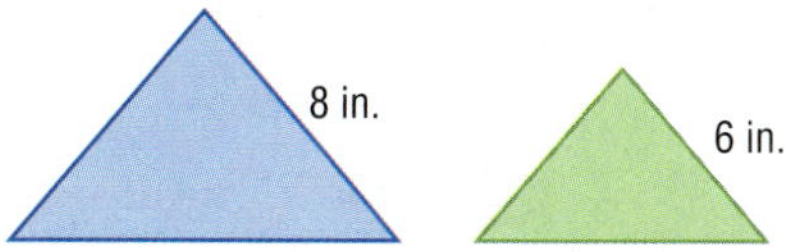

1. **SPEAKERS** The top and front views of a speaker for a stereo system are shown. (Lesson 12-1)

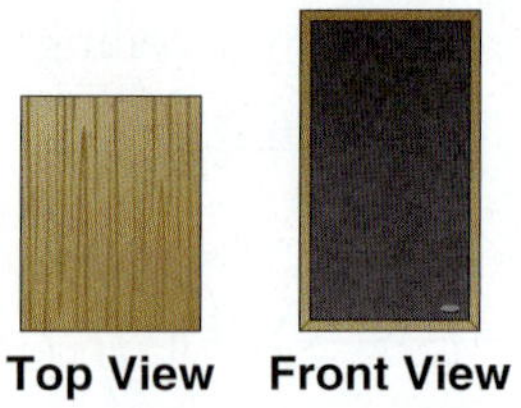

a. Is it possible to determine the shape of the speaker? Explain.

b. Describe possible shapes for the speaker. Draw the left and right views of one of the possible shapes.

2. **MANUFACTURING** An office has recycling barrels that are cylindrical with cardboard sides and plastic lids and bases. Each barrel is 3 feet tall, with a diameter of 30 inches. How many square feet of cardboard are used to make each barrel? (Lesson 12-2)

3. **WELDING** Ofelia is learning to weld by making a wind chime out of three identical square pyramids. The bottom of each pyramid is open. How much metal does Ofelia need to make all three pyramids if they are 5 centimeters wide at the bottom and 6 centimeters tall? (Lesson 12-3)

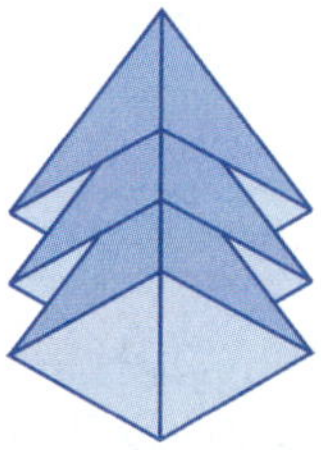

4. **HYDRATION** Tyra has a water bottle that is 3 inches in diameter and 8 inches tall. It has a cylindrical insert that is frozen and inserted into the bottle to keep the water cool. If the insert has a diameter of $\frac{3}{4}$ inch and a height of 6 inches, what is the volume of water that the bottle holds with the insert in place? (Lesson 12-4)

5. **DIRT** A mound of dirt is roughly cone shaped, 6 feet high, and 8 feet in diameter. What is the approximate volume of the mound? Express your answer in cubic yards. (Lesson 12-5)

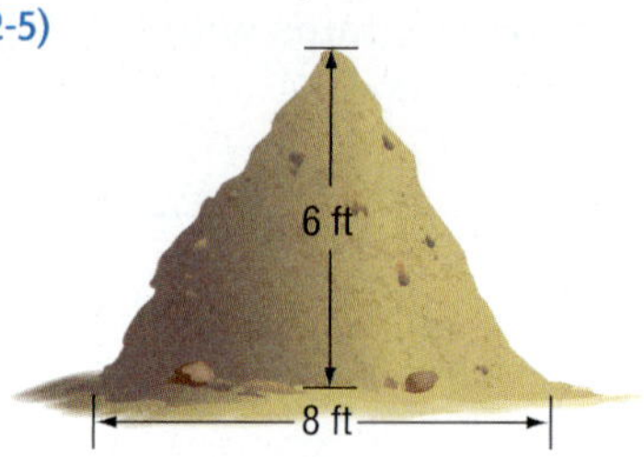

6. **BALLS** The diameter of a tennis ball is 2.5 inches. Dustin has another ball with a diameter three times that of the tennis ball. Express all answers to the nearest tenth. (Lesson 12-6)

a. What is the surface area of the tennis ball?

b. What is the surface area of Dustin's ball?

c. How many times as large is the volume of Dustin's ball as the tennis ball?

d. Find the volume of each ball.

7. **ART** Determine whether figure *S* on the sculpture shown is a line in spherical geometry. (Lesson 12-7)

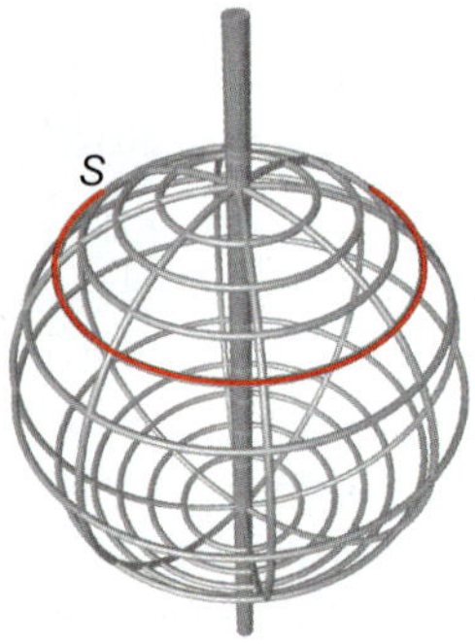

8. **ENTERTAINMENT** Mirror balls are spheres that are covered with reflective tiles. One ball has a diameter of 4 inches, and another has a diameter of 20 inches. (Lesson 12-8)

a. Find the scale factor of the two spheres.

b. Find the ratio of the surface areas of the spheres.

c. Find the ratio of the volumes of the two spheres.

1. **UNIFORMS** Patriot High School is designing a new uniform for the soccer team. The school colors are red and blue, but they always use white in their uniforms as well. They want to use two of the colors in one of two specific areas on the jersey, and the third color for the shorts. Represent the sample space for uniform color choices using a tree diagram. (Lesson 13-1)

2. **MODELS** Daevon bought a kit for building a model light sword. He must choose one item from each category shown. How many different model light swords could Daevon create? (Lesson 13-1)

Light Sword Parts	Number of Choices
hilt	3
colored light filter	4
blades	2
crystals	5

3. **SCHOOL UNIFORMS** The students at Celeste's school wear uniforms, but they still have many choices. They may wear either long-sleeved or short-sleeved button-down shirts or polo-style shirts. The shirt colors are blue, yellow, hunter green, white, or red. They may also wear either navy or khaki pants. In addition to pants, the girls may choose between a plaid skirt and jumper. How many different combinations of a uniform top and bottom can a girl choose to wear? Explain your reasoning. (Lesson 13-2)

4. **GAMES** For a game at a party, Natasha hid a treasure box full of candy somewhere on her family's small farm. If the entire farm is 20 acres, what is the probability the treasure box is hidden in the yard? (Lesson 13-3)

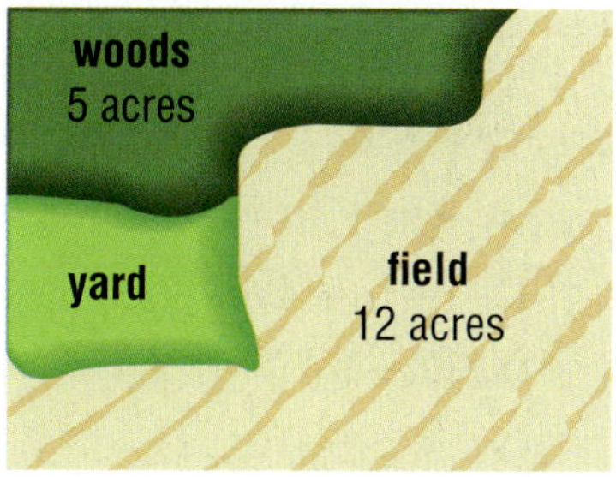

5. **GAMES** A square game board has black and white stripes of equal width, as shown. What is the chance that a dart that strikes the board will land on a white stripe? (Lesson 13-3)

6. **NEWSPAPERS** Sunee delivers newspapers to 75 houses. She throws the papers toward the front door of each house. Last week, she threw 80% of the papers onto the houses' front porches. Design a simulation that can be used to estimate the probability that she will get the next paper she delivers onto the front porch of the house. (Lesson 13-4)

7. **MUSIC** Vince can preset up to 12 radio stations on his car radio. The table shows how many stations he has preset for each of the given types of music. If he pushes a preset button at random twice because he didn't like the song playing on the first station, what is the probability that he will choose a classic rock station followed by the jazz station? (Lesson 13-5)

Type of Music	Number of Stations
hip-hop	1
rock	2
country	3
classic rock	4
jazz	1
not set	1

8. **GAMES** If you have drawn and kept a 4 and a 7 from a standard deck of cards, what is the chance that you will draw another 7 next? (Lesson 13-6)

9. **DOGS** The table shows the ages and genders of the dogs at the animal shelter. What is the probability that a dog is a female or over 5 years old? (Lesson 13-6)

Age	Male	Female
under 1 year	6	5
1–5 years	8	7
6–10 years	4	6
over 10 years	3	5

Concepts and Skills Bank

1 Solving for a Specific Variable

When an equation contains more than one variable, it is often useful to solve the equation for one of the variables.

EXAMPLE 1

Solve $3x - 4y = 7$ for y.

$3x - 4y = 7$	**Original equation**
$3x - 4y - 3x = 7 - 3x$	**Subtract $3x$ from each side.**
$-4y = 7 - 3x$	**Simplify.**
$\frac{-4y}{-4} = \frac{7 - 3x}{-4}$	**Divide each side by -4.**
$y = \frac{7 - 3x}{-4}$ or $\frac{3x - 7}{4}$	**Simplify.**

The value of y is $\frac{3x - 7}{4}$.

You can use the Distributive Property to isolate the variable for which you are solving.

EXAMPLE 2

Solve $2m - t = km + 5$ for m.

$2m - t = km + 5$	**Original equation**
$2m - t - km = km + 5 - km$	**Subtract km from each side.**
$2m - t - km = 5$	**Simplify.**
$2m - t - km + t = 5 + t$	**Add t to each side.**
$2m - km = 5 + t$	**Simplify.**
$m(2 - k) = 5 + t$	**Use the Distributive Property.**
$\frac{m(2 - k)}{2 - k} = \frac{5 + t}{2 - k}$	**Divide each side by $2 - k$.**
$m = \frac{5 + t}{2 - k}$	**Simplify.**

The value of m is $\frac{5 + t}{2 - k}$. Because division by 0 is undefined, $2 - k \neq 0$ or $k \neq 2$.

Exercises

Solve for the indicated variable.

1. $2a + 3b = 6$, for b

2. $3x + 5y = -15$, for y

3. $y = mx + b$, for m

4. $cd - 4 = 5c$, for d

5. $P = 2w + 2\ell$, for w

6. $-6 + 7j = 9k$, for j

7. $\frac{gh - 4}{2} = 6$, for g

8. $\frac{2pt + 7t}{3} = -2$, for t

9. $\frac{2}{3}m + 4p = -2$, for m

10. $-3ac - 4bd = ac$, for a

11. $1 + jk = -2 + j$, for k

12. $-2x + 4y = 3(x - 1)$, for x

13. $4j(p - 2) = 2p$, for p

14. $a(2 - b) = 4$, for b

15. $d(c + 5) = -3$, for c

16. $V = \frac{1}{3}\pi r^2 h$, for r

17. $a^2 + b^2 = c^2$, for a

18. $A = \frac{1}{2}bh$, for h

2 Multiplying Polynomials

The **Product of Powers** rule states that for any number a and all integers m and p, $a^m \cdot a^p = a^{m+p}$.

EXAMPLE 1

Simplify each expression.

a. $(4p^5)(p^4)$

$$\begin{aligned}(4p^5)(p^4) &= (4)(1)(p^5 \cdot p^4)\\ &= (4)(1)(p^{5+4})\\ &= 4p^9\end{aligned}$$

b. $(3yz^5)(-9y^2z^2)$

$$\begin{aligned}(3yz^5)(-9y^2z^2) &= (3)(-9)(y \cdot y^2)(z^5 \cdot z^2)\\ &= -27(y^{1+2})(z^{5+2})\\ &= -27y^3z^7\end{aligned}$$

The Distributive Property can be used to multiply a monomial by a polynomial.

EXAMPLE 2

Simplify $3x^3(-4x^2 + x - 5)$.

$3x^3(-4x^2 + x - 5) = 3x^3(-4x^2) + 3x^3(x) - 3x^3(5)$ **Distributive Property**

$= -12x^5 + 3x^4 - 15x^3$ **Multiply.**

To find the power of a power, multiply the exponents. This is called the **Power of a Power** rule.

EXAMPLE 3

Simplify each expression.

a. $(-3x^2y^4)^3$

$$\begin{aligned}(-3x^2y^4)^3 &= (-3)^3(x^2)^3(y^4)^3\\ &= -27x^6y^{12}\end{aligned}$$

b. $(xy)^3(-2x^4)^2$

$$\begin{aligned}(xy)^3(-2x^4)^2 &= x^3y^3(-2)^2(x^4)^2\\ &= x^3y^3(4)x^8\\ &= 4x^3 \cdot x^8 \cdot y^3\\ &= 4x^{11}y^3\end{aligned}$$

To multiply two binomials, find the sum of the products of

F the *First* terms,
O the *Outer* terms,
I the *Inner* terms, and
L the *Last* terms.

EXAMPLE 4

Find $(2x - 3)(x + 1)$.

F O I L

$(2x - 3)(x + 1) = (2x)(x) + (2x)(1) + (-3)(x) + (-3)(1)$ **FOIL method**

$= 2x^2 + 2x - 3x - 3$ **Multiply.**

$= 2x^2 - x - 3$ **Combine like terms.**

Concepts and Skills Bank

The Distributive Property can be used to multiply any two polynomials.

EXAMPLE 5

Find $(3x - 2)(2x^2 + 7x - 4)$.

$(3x - 2)(2x^2 + 7x - 4) = 3x(2x^2 + 7x - 4) - 2(2x^2 + 7x - 4)$ **Distributive Property**

$= 6x^3 + 21x^2 - 12x - 4x^2 - 14x + 8$ **Distributive Property**

$= 6x^3 + 17x^2 - 26x + 8$ **Combine like terms.**

Three special products are $(a + b)^2 = a^2 + 2ab + b^2$,
$(a - b)^2 = a^2 - 2ab + b^2$, and
$(a + b)(a - b) = a^2 - b^2$.

EXAMPLE 6

Find each product.

a. $(2x - z)^2$

$(a - b)^2 = a^2 - 2ab + b^2$ **Square of a difference**

$(2x - z)^2 = (2x)^2 - 2(2x)(z) + (z)^2$ **$a = 2x$ and $b = z$**

$= 4x^2 - 4xz + z^2$ **Simplify.**

b. $(3x + 7)(3x - 7)$

$(a + b)(a - b) = a^2 - b^2$ **Product of sum and difference**

$(3x + 7)(3x - 7) = (3x)^2 - (7)^2$ **$a = 3x$ and $b = 7$**

$= 9x^2 - 49$ **Simplify.**

Exercises

Find each product.

1. $(3q^2)(q^5)$
2. $(5m)(4m^3)$
3. $\left(\frac{9}{2}c\right)(8c^5)$
4. $(n^6)(10n^2)$
5. $(fg^8)(15f^2g)$
6. $(6j^4k^4)(j^2k)$
7. $(2ab^3)(4a^2b^2)$
8. $\left(\frac{8}{5}x^3y\right)\left(4x^3y^2\right)$
9. $-2q^2(q^2 + 3)$
10. $5p(p - 18)$
11. $15c(-3c^2 + 2c + 5)$
12. $8x(-4x^2 - x + 11)$
13. $4m^2(-2m^2 + 7m - 5)$
14. $8y^2(5y^3 - 2y + 1)$
15. $\left(\frac{3}{2}m^3p^2\right)^2$
16. $(-2c^3d^2)^2$
17. $(-5wx^5)^3$
18. $(6a^5b)^3$
19. $(k^2\ell)^3(13k^2)^2$
20. $(-5w^3x^2)^2(2w^5)^2$
21. $(-7y^3z^2)(4y^2)^4$
22. $\left(\frac{1}{2}p^2t^2\right)^2\left(4pt^3\right)^3$
23. $(m - 1)(m - 4)$
24. $(k - 7)(k - 2)$
25. $(x - 3)(x + 4)$
26. $(a + 3)(a - 6)$
27. $(5d + 3)(d - 4)$
28. $(q + 2)(3q + 5)$
29. $(2q + 3)(5q + 2)$
30. $(2a - 3)(2a - 5)$
31. $(d + 1)(d - 1)$
32. $(4a - 3)(4a + 3)$
33. $(c - 5)^2$
34. $(3f - g)^2$
35. $(2r - 5)^2$
36. $\left(t + \frac{8}{3}\right)^2$
37. $(x + 4)(x^2 - 5x - 2)$
38. $(x - 2)(x^2 + 3x - 7)$
39. $(3b - 2)(3b^2 + b + 1)$
40. $(2j + 7)(j^2 - 2j + 4)$

3 Dividing Polynomials

The **Quotient of Powers** rule states that for any nonzero number a and all integers m and p, $\frac{a^m}{a^p} = a^{m-p}$.

To find the power of a quotient, find the power of the numerator and the power of the denominator.

EXAMPLE 1

Simplify.

a. $\frac{x^5y^8}{-xy^3}$

$\frac{x^5y^8}{-xy^3} = \left(\frac{x^5}{-x}\right)\left(\frac{y^8}{y^3}\right)$ — **Group powers that have the same base.**

$= -(x^{5-1})(y^{8-3})$ — **Quotient of powers**

$= -x^4y^5$ — **Simplify.**

b. $\frac{w^{-2}x^4}{2w^{-5}}$

$\frac{w^{-2}x^4}{2w^{-5}} = \frac{1}{2}\left(\frac{w^{-2}}{w^{-5}}\right)x^4$ — **Group powers that have the same base.**

$= \frac{1}{2}(w^{-2-(-5)})x^4$ — **Quotient of powers**

$= \frac{1}{2}w^3x^4$ — **Simplify.**

You can divide a polynomial by a monomial by separating the terms of the numerator.

EXAMPLE 2

Simplify $\frac{15x^3 - 3x^2 + 12x}{3x}$.

$\frac{15x^3 - 3x^2 + 12x}{3x} = \frac{15x^3}{3x} - \frac{3x^2}{3x} + \frac{12x}{3x}$ — **Divide each term by $3x$.**

$= 5x^2 - x + 4$ — **Simplify.**

Division can sometimes be performed using factoring.

EXAMPLE 3

Find $(n^2 - 8n - 9) \div (n - 9)$.

$(n^2 - 8n - 9) \div (n - 9) = \frac{n^2 - 8n - 9}{(n-9)}$ — **Write as a rational expression.**

$= \frac{(n-9)(n+1)}{(n-9)}$ — **Factor the numerator.**

$= \frac{\cancel{(n-9)}(n+1)}{\cancel{(n-9)}}$ — **Divide by the GCF.**

$= n + 1$ — **Simplify.**

Concepts and Skills Bank

EXAMPLE 4

Find $(n^3 - 4n^2 + 12) \div (n - 3)$.

In this case, there is no n term, so you must rename the dividend using 0 as the coefficient of the missing term.

$(n^3 - 4n^2 + 12) \div (n - 3) = (n^3 - 4n^2 + 0n + 12) \div (n - 3)$

Divide the first term of the dividend, n^3, by the first term of the divisor, n.

$$
\begin{array}{r l}
n^2 - n - 3 & \\
n - 3\overline{)n^3 - 4n^2 + 0n + 12} & \\
(-)\ \underline{n^3 - 3n^2} & \textbf{Multiply } n^2 \textbf{ and } n - 3. \\
-n^2 + 0n & \textbf{Subtract and bring down } 0n. \\
(-)\ \underline{-n^2 + 3n} & \textbf{Multiply } -n \textbf{ and } n - 3. \\
-3n + 12 & \textbf{Subtract and bring down 12.} \\
(-)\ \underline{-3n + 9} & \textbf{Multiply } -3 \textbf{ and } n - 3. \\
3 & \textbf{Subtract.}
\end{array}
$$

Therefore, $(n^3 - 4n^2 + 12) \div (n - 3) = n^2 - n - 3 + \frac{3}{n - 3}$. Since the quotient has a nonzero remainder, $n - 3$ is not a factor of $n^3 - 4n^2 + 12$.

Exercises

Find each quotient.

1. $\frac{a^2c^2}{2a}$

2. $\frac{5q^5r^3}{q^2r^2}$

3. $\frac{b^2d^5}{8b^{-2}d^3}$

4. $\frac{5p^{-3}x}{2p^{-7}}$

5. $\frac{3r^{-3}n^2t^4}{2r^2nt^{-3}}$

6. $\frac{3x^3y^{-1}z^5}{xyz^2}$

7. $\left(\frac{w^4}{6}\right)^3$

8. $\left(\frac{-3q^2}{5}\right)^3$

9. $\left(\frac{-2y^2}{7}\right)^2$

10. $\left(\frac{5m^2}{3}\right)^4$

11. $\frac{4z^2 - 16z - 36}{4z}$

12. $(5d^2 + 8d - 20) \div (10d)$

13. $(p^3 - 12p^2 + 3p + 8) \div (4p)$

14. $(b^3 + 4b^2 + 10) \div (2b)$

15. $\frac{a^3 - 6a^2 + 4a - 3}{a^2}$

16. $\frac{8x^2y - 10xy^2 + 6x^3}{2x^2}$

17. $\frac{c^2 - 2c - 8}{c - 4}$

18. $(r^2 + 9r + 20) \div (r + 5)$

19. $(t^2 - 7t + 12) \div (t - 3)$

20. $(c^2 + 3c - 54) \div (c + 9)$

21. $(2q^2 - 9q - 5) \div (q - 5)$

22. $\frac{3z^2 - 2z - 5}{z + 1}$

23. $\frac{(m^3 + 3m^2 - 5m + 1)}{m - 1}$

24. $(d^3 - 2d^2 + 4d + 24) \div (d + 2)$

25. $(2j^3 + 5j + 26) \div (j + 2)$

26. $\frac{2x^3 + 3x^2 - 176}{x - 4}$

27. $(x^2 + 6x - 3) \div (x + 4)$

28. $\frac{h^3 + 2h^2 - 6h + 1}{h - 2}$

4 Factoring to Solve Equations

Some polynomials can be factored using the Distributive Property.

EXAMPLE 1

Factor $5t^2 + 15t$.

Find the greatest common factor (GCF) of $5t^2$ and $15t$.

$5t^2 = 5 \cdot t \cdot t$, $15t = 3 \cdot 5 \cdot t$ — **GCF: $5 \cdot t$ or $5t$**

$5t^2 + 15t = 5t(t) + 5t(3)$ — **Rewrite each term using the GCF.**

$= 5t(t + 3)$ — **Distributive Property**

To factor polynomials of the form $x^2 + bx + c$, find two integers m and p so that $mp = c$ and $m + p = b$. Then write $x^2 + bx + c$ using the pattern $(x + m)(x + p)$.

To factor polynomials of the form $ax^2 + bx + c$, find two integers m and p with a product equal to ac and with a sum equal to b. Write $ax^2 + bx + c$ using the pattern $ax^2 + mx + px + c$. Then factor by grouping.

EXAMPLE 2

Factor each polynomial.

a. $x^2 - 8x + 15$

b is negative and c is positive.

In this equation, b is -8 and c is 15. This means that $m + p$ is negative and mp is positive. So m and p must both be negative.

Factors of 15	Sum of Factors
−1, −15	−16
−3, −5	−8

$x^2 - 8x + 15 = (x + m)(x + p)$ — **The correct factors are −3 and −5.**

$= (x - 3)(x - 5)$ — **Write the pattern; $m = -3$ and $p = -5$**

b. $5x^2 - 19x - 4$

b is negative and c is negative.

In this equation, a is 5, b is -19, and c is -4. Find two numbers with a product of -20 and with a sum of -19.

Factors of −20	Sum of Factors
−2, 10	8
2, −10	−8
−1, 20	19
1, −20	−19

$5x^2 - 19x - 4 = 5x^2 + mx + px - 4$

$= 5x^2 + x + (-20)x - 4$

$= (5x^2 + x) - (20x + 4)$

$= x(5x + 1) - 4(5x + 1)$ — **Factor the GCF from each group.**

$= (x - 4)(5x + 1)$ — **Distributive Property**

Here are some special products.

Perfect Square Trinomials

$a^2 + 2ab + b^2 = (a + b)(a + b) = (a + b)^2$

$a^2 - 2ab + b^2 = (a - b)(a - b) = (a - b)^2$

Difference of Squares

$a^2 - b^2 = (a + b)(a - b)$

Concepts and Skills Bank

EXAMPLE 3

Factor each polynomial.

a. $9x^2 + 6x + 1$ ← The first and last terms are perfect squares, and the middle term is equal to $2(3x)(1)$.

$$9x^2 + 6x + 1 = (3x)^2 + 2(3x)(1) + 1^2 \quad \text{Write as } a^2 + 2ab + b^2.$$
$$= (3x + 1)^2 \quad \text{Factor using the pattern.}$$

b. $x^2 - 9 = 0$ ← This is a difference of squares.

$$x^2 - 9 = x^2 - (3)^2 \quad \text{Write in the form } a^2 - b^2.$$
$$= (x - 3)(x + 3) \quad \text{Factor the difference of squares.}$$

The binomial $x - a$ is a factor of the polynomial $f(x)$ if and only if $f(a) = 0$. Since 0 times any number is equal to zero, this implies that we can use factoring to solve equations.

EXAMPLE 4

Solve $x^2 - 5x + 4 = 0$ by factoring.

$x^2 - 5x + 4 = 0$ — Original equation

$(x - 1)(x - 4) = 0$ — Factor the polynomial.

$x - 1 = 0$ or $x - 4 = 0$ — Zero Product Property

$x = 1$ $\quad$ $x = 4$

Exercises

Factor each polynomial.

1. $u^2 - 12u$
2. $w^2 + 4w$
3. $7j^2 - 28j$
4. $2g^2 + 24g$
5. $6x^2 + 2x$
6. $5t^2 - 30t$
7. $z^2 + 10z + 21$
8. $n^2 + 8n + 15$
9. $h^2 + 8h + 12$
10. $x^2 + 14x + 48$
11. $m^2 + 6m - 7$
12. $b^2 + 2b - 24$
13. $q^2 - 9q + 18$
14. $p^2 - 5p + 6$
15. $a^2 - 3a - 4$
16. $k^2 - 4k - 32$
17. $n^2 - 7n - 44$
18. $y^2 - 3y - 88$
19. $3z^2 + 4z - 4$
20. $2y^2 + 9y - 5$
21. $5x^2 + 7x + 2$
22. $3c^2 + 11c - 4$
23. $6r^2 - 5r + 1$
24. $8a^2 + 15a - 2$
25. $w^2 - \frac{9}{4}$
26. $c^2 - 64$
27. $r^2 + 14r + 49$
28. $b^2 + 18b + 81$
29. $j^2 - 12j + 36$
30. $4t^2 - 25$

Solve each equation by factoring.

31. $10r^2 - 35r = 0$
32. $4x^2 + 20x = 0$
33. $k^2 + 13k + 36 = 0$
34. $w^2 - 8w + 12 = 0$
35. $c^2 - 5c - 14 = 0$
36. $z^2 - z - 42 = 0$
37. $2y^2 - 5y - 12 = 0$
38. $3b^2 - 4b - 15 = 0$
39. $t^2 + 12t + 32 = 0$
40. $u^2 + 5u + \frac{25}{4} = 0$
41. $q^2 - 8q + 16 = 0$
42. $a^2 - 6a + 9 = 0$

Concepts and Skills Bank

5 Linear Quadratic Equations

If the graphs of a system of equations are a parabola and a line, the system may have zero, one, or two solutions. Some of the possible situations are shown below.

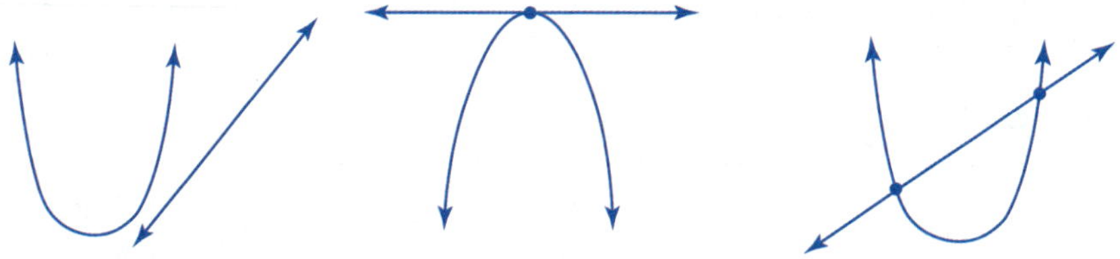

EXAMPLE 1

Solve the system of equations.

$\mathbf{x^2 - 4y^2 = 9}$

$\mathbf{4y - x = 3}$

Use substitution to solve the system. First rewrite $4y - x = 3$ as $x = 4y - 3$.

$x^2 - 4y^2 = 9$	**First equation in the system**
$(4y - 3)^2 - 4y^2 = 9$	**Substitute $4y - 3$ for x.**
$12y^2 - 24y = 0$	**Simplify.**
$y^2 - 2y = 0$	**Divide each side by 12.**
$y(y - 2) = 0$	**Factor.**
$y = 0$ or $y - 2 = 0$	**Zero Product Property**
$y = 2$	**Solve for y.**

Now solve for x.

$x = 4y - 3$	**Equation for x in terms of y**	$x = 4y - 3$
$= 4(0) - 3$	**Substitute the y-values.**	$= 4(2) - 3$
$= -3$	**Simplify.**	$= 5$

The solutions of the system are $(-3, 0)$ and $(5, 2)$. Based on the graph, these solutions are reasonable.

Exercises

Find the exact solution(s) of each system of equations.

1. $y = x + 6$, $y = x^2$
2. $y = x + 2$, $y = x^2$
3. $y = 4x - 3$, $y = x^2$
4. $y = 3x - 4$, $y = x^2$
5. $y = 2x^2$, $y = x - 1$
6. $y = 2x^2$, $y = x + 3$
7. $y = 3$, $y^2 + 16 = x^2$
8. $y^2 = x^2 - 8$, $y = -1$
9. $y^2 = x^2 + 12$, $y = -4$
10. $y^2 - 20 = x^2$, $y = 6$
11. $y = x^2 - 4$, $y = -4$
12. $y = 6$, $y = x^2 + 6$
13. $y = -x + 4$, $y^2 + x^2 = 2$
14. $y = x + 5$, $y^2 + x^2 = 4$
15. $4x^2 = 25 - y^2$, $2x + y = -1$

6 Mean, Median, and Mode

Mean, median, and mode are measures of central tendency that are often used to represent a set of data.

- To find the **mean**, find the sum of the data and divide by the number of items in the data set. (The mean is often called the average.)
- To find the **median**, arrange the data in numerical order. The median is the middle number. If there is an even number of data, the median is the mean of the two middle numbers.
- The **mode** is the number (or numbers) that appears most often in a set of data. If no item appears most often, the set has no mode.

EXAMPLE 1

Michelle is saving to buy a car. She saved $200 in June, $300 in July, $400 in August, and $150 in September. What was her mean (or average) monthly savings?

mean = sum of monthly savings/number of months

$$= \frac{\$200 + \$300 + \$400 + \$150}{4}$$

$$= \frac{\$1050}{4} \text{ or } \$262.50$$

Michelle's mean monthly savings was $262.50.

EXAMPLE 2

Find the median of the data.

To find the median, order the numbers from least to greatest. The median is in the middle. The two middle numbers are 3.7 and 4.1.

$$\frac{3.7 + 4.1}{2} = 3.9$$

There is an even number of data. Find the mean of the middle two.

Peter's Best Running Times

Week	Minutes to Run a Mile
1	4.5
2	3.7
3	4.1
4	4.1
5	3.6
6	3.4

EXAMPLE 3

GOLF Four players tied for first in the 2001 PGA Tour Championship. The scores for each player for each round are shown in the table below. What is the mode score?

Player	Round 1	Round 2	Round 3	Round 4
Mike Weir	68	66	68	68
David Toms	73	66	64	67
Sergio Garcia	69	67	66	68
Ernie Els	69	68	65	68

Source: ESPN

The mode is the score that occurred most often. Since the score of 68 occurred 6 times, it is the mode of these data.

The **range** of a set of data is the difference between the greatest and the least values of the set. It describes how a set of data varies.

EXAMPLE 4

Find the range of the data. **{6, 11, 18, 4, 9, 15, 6, 3}**

The greatest value is 18 and the least value is 3. So, the range is 18 − 3 or 15.

Exercises

Find the mean, median, mode, and range for each set of data. Round to the nearest tenth if necessary.

1. {2, 8, 12, 13, 15}
2. {66, 78, 78, 64, 34, 88}
3. {87, 95, 84, 89, 100, 82}
4. {99, 100, 85, 96, 94, 99}
5. {9.9, 9.9, 10, 9.9, 8.8, 9.5, 9.5}
6. {501, 503, 502, 502, 502, 504, 503, 503}
7. {7, 19, 15, 13, 11, 17, 9}
8. {6, 12, 21, 43, 1, 3, 13, 8}
9. {0.8, 0.04, 0.9, 1.1, 0.25}
10. $\left\{2\frac{1}{2}, 1\frac{7}{8}, 2\frac{5}{8}, 2\frac{3}{4}, 2\frac{1}{8}\right\}$

11. **CHARITY** The table shows the amounts collected by classes at Jackson High School. Find the mean, median, mode, and range of the data.

Amounts Collected for Charity			
Class	Amount	Class	Amount
A	$150	E	$10
B	$300	F	$25
C	$55	G	$200
D	$40	H	$100

12. **SCHOOL** The table shows Pilar's grades in chemistry class for the semester. Find her mean, median, and mode scores, and the range of her scores.

Chemistry Grades	
Assignment	Grade (out of 100)
Homework	100
Electron Project	98
Test I	87
Atomic Mass Project	95
Test II	88
Phase Change Project	90
Test III	95

13. **WEATHER** The table shows the precipitation for the month of July in Cape Hatteras, North Carolina, in various years. Find the mean, median, mode, and range of the data.

July Precipitation in Cape Hatteras, North Carolina												
Year	1990	1991	1992	1993	1994	1995	1996	1997	1998	1999	2000	2001
Inches	4.22	8.58	5.28	2.03	3.93	1.08	9.54	4.94	10.85	2.66	6.04	3.26

Source: National Climatic Data Center

14. **SCHOOL** Kaitlyn's scores on her first five algebra tests are 88, 90, 91, 89, and 92. What test score must Kaitlyn earn on the sixth test so that her mean score will be at least 90?
15. **GOLF** Colin's average for three rounds of golf is 94. What is the highest score he can receive for the fourth round to have an average (mean) of 92?
16. **SCHOOL** Mika has a mean score of 21 on his first four Spanish quizzes. If each quiz is worth 25 points, what is the highest possible mean score he can have after the fifth quiz?
17. **SCHOOL** To earn a grade of B in math, Latisha must have an average (mean) score of at least 84 on five math tests. Her scores on the first three tests are 85, 89, and 82. What is the lowest total score that Latisha must have on the last two tests to earn a B test average?

7 Displaying Data

There are various techniques that can be used when organizing and displaying data. A **bar graph** can be used to compare various categories of data by showing each as a bar whose length is related to the frequency of the event.

EXAMPLE 1

SURVEYS The table at the right shows the results of a survey of student's favorite types of pizza. Make a bar graph to display the data.

Outcome	Tally	Frequency
cheese	卌 卌 ||||	14
pepperoni	卌 卌 卌 卌 |	21
sausage	卌 卌	10
vegetable	卌	5

Step 1 Draw and label a horizontal axis and a vertical axis. Add a title to the top of the graph.

Step 2 Draw a bar to represent each pizza. The vertical scale is the number of students who chose each pizza. The horizontal scale identifies the pizza choice.

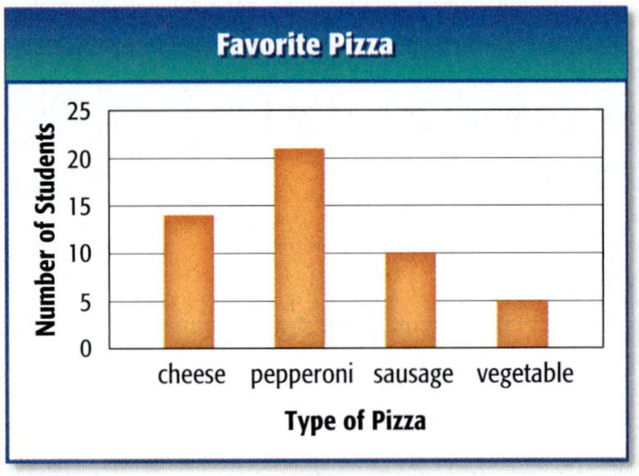

A **line graph** is another method that can be used to display data. Line graphs usually show how data changes over a period of time.

EXAMPLE 2

POPULATION The change in population from 1900 to 2000 in the United States is shown in the table below. Make a line graph of the data.

Year	Population (millions)	Year	Population (millions)
1900	76.1	1960	180.7
1920	106.5	1980	227.2
1940	132.1	2000	281.4

Step 1 Draw and label a horizontal and vertical axis. Add a title to the top of the graph.

Step 2 Plot each data point.

Step 3 Draw a line connecting each point, as shown.

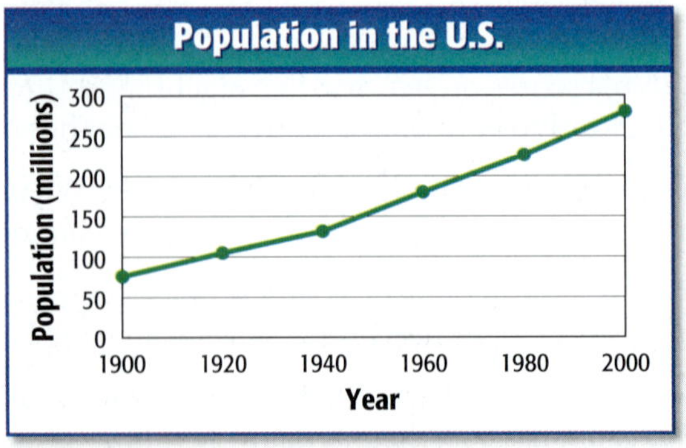

A **circle graph** can be used to compare parts of a data set to the whole set of data. Each sector of the circle represents a portion of the data set and the entire circle represents all of the data within that set.

EXAMPLE 3

POPULATION The regions of a city's population are distributed as shown. Make a circle graph to represent the data.

Region	NE	SE	Central	NW	SW
Percent	18	14	33	16	19

Step 1 Convert each of the percentages into degrees by using a proportion. Round to the nearest whole degree. Remember that a full circle is 360°.

Region	NE	SE	Central	NW	SW
Percent	18	14	33	16	19
Degrees	65	50	119	58	68

Step 2 Use a compass to draw a circle. Then draw a radius.

Step 3 Use a protractor to draw a 65° angle. This sector represents the NE Region. Repeat for the remaining regions.

Step 4 Label each sector of the graph with the region and percent. Title the graph.

Population

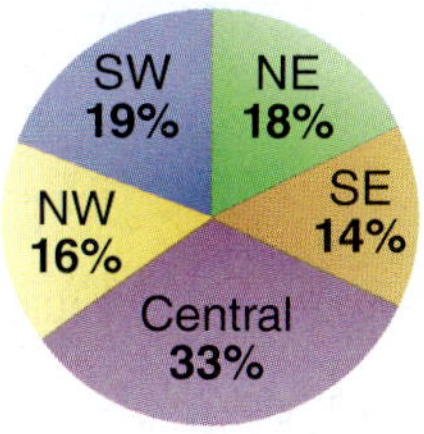

Exercises

1. **LANGUAGES** The results of a foreign language survey are shown in the table at the right. Make a bar graph to display the data.

Outcome	Tally	Frequency
French	||||/ ||||/ ||||/ ||||/		21
German	||||/ ||||/ ||	12
Japanese	||||/ ||||/	10
Spanish	||||/ ||||/ ||||/ ||||/ |||	23

2. **RECYCLING** The amount of paper recycled (in millions of tons) in the United States from 1997 to 2006 is shown in the table at the right. Make a line graph of the data.

Year	Recycled Paper	Year	Recycled Paper
1997	44	2002	47.6
1998	45.1	2003	49.3
1999	46.8	2004	50.3
2000	47.3	2005	51.3
2001	47	2006	53.5

Make a circle graph to represent the data in each exercise.

3. **CELL PHONES** A survey was conducted about which phone service company was used. The results are shown in the table.

Company	Talk to Me	On the Go	L.O.L.	Tomorrow	Voice Age
Results (%)	23	32	17	16	12

4. **PEP RALLY** Out of 700 students, 102 freshmen, 187 sophomores, 189 juniors, and 222 seniors attended the bonfire.

5. **MATH CLASSES** Mr. Blackwell has 62 Algebra I students, 85 Geometry students, 28 Algebra II students, and 21 Precalculus students.

Key Concepts

Tools of Geometry — Chapter 1

Undefined Terms (p. 5)

A point is a location. It has neither shape nor size.

A line is made up of points and has no thickness or width.

A plane is a flat surface made up of points that extend infinitely in all directions.

Betweenness of Points (p. 15)

Point M is between points P and Q if and only if P, Q, and M are collinear and $PM + MQ = PQ$.

Congruent Segments (p. 16)

Congruent segments have the same measure.

Distance Formula (on a Number Line) (p. 25)

The distance between two points is the absolute value of the difference between their coordinates.

Distance Formula (in Coordinate Plane) (p. 26)

If P has coordinates (x_1, y_1) and Q has coordinates (x_2, y_2), then

$$PQ = \sqrt{(x_2 - x_1)^2 + (y_2 - y_1)^2}.$$

Midpoint Formula (on a Number Line) (p. 27)

If $\overline{AB}$ has endpoints x_1 and x_2 on a number line, then the midpoint M of $\overline{AB}$ has coordinate $\frac{x_1 + x_2}{2}$.

Midpoint Formula (in Coordinate Plane) (p. 27)

If $\overline{PQ}$ has endpoints $P(x_1, y_1)$ and $Q(x_2, y_2)$ in the coordinate plane, then the midpoint M of $\overline{PQ}$ has coordinate $M\left(\frac{x_1 + x_2}{2}, \frac{y_1 + y_2}{2}\right)$.

Special Angle Pairs (p. 46)

Adjacent angles are two angles that lie in the same plane and have a common vertex and a common side, but no common interior points.

A linear pair is a pair of adjacent angles with noncommon sides that are opposite rays.

Vertical angles are two nonadjacent angles formed by two intersecting lines.

Angle Pair Relationships (p. 47)

Vertical angles are congruent.

Complementary angles are two angles with measures that have a sum of 90.

Supplementary angles are two angles with measures that have a sum of 180.

The angles in a linear pair are supplementary.

Perpendicular Lines (p. 48)

Perpendicular lines intersect to form four right angles.

Perpendicular lines intersect to form congruent adjacent angles.

Segments and rays can be perpendicular to lines or to other line segments and rays.

A right angle symbol indicates that two lines are perpendicular.

Polygons (p. 56)

A polygon is a closed figure formed by a finite number of coplanar segments called *sides* that have a common endpoint, are noncollinear, and intersect exactly two other sides at their endpoints.

The vertex of each angle is a vertex of the polygon.

A polygon is named by the letters of its vertices, written in order of consecutive vertices.

Types of Solids (p. 67)

Polyhedrons

A prism has two parallel congruent faces called bases connected by parallelogram faces.

A pyramid has a polygonal base and three or more triangular faces that meet at a common vertex.

Not Polyhedrons

A cylinder has congruent parallel circular bases connected by a curved surface.

A cone has a circular base connected by a curved surface to a single vertex.

A sphere is a set of points in space that are the same distance from a given point. A sphere has no faces, edges, or vertices.

Platonic Solids (p. 68)

A tetrahedron has four equilateral triangular faces.

A hexahedron or cube has six square faces.

An octahedron has eight equilateral triangular faces.

A dodecahedron has twelve regular pentagonal faces.

An icosahedron has twenty equilateral triangular faces.

Reasoning and Proof — Chapter 2

Negation, Conjunction, Disjunction (p. 98)

A negation is a statement that has the opposite meaning and truth value of an original statement.

A conjunction is a compound statement formed by joining two or more statements using the word *and*.

A disjunction is a compound statement formed by joining two or more statements using the word *or*.

Conditional Statement (p. 105)

An if-then statement is of the form *if p, then q*.

The hypothesis of a conditional statement is the phrase immediately following the word *if*.

The conclusion of a conditional statement is the phrase immediately following the word *then*.

Related Conditionals (p. 107)

A conditional statement is a statement that can be written in the form *if p, then q*.

The converse is formed by exchanging the hypothesis and the conclusion of the conditional.

The inverse is formed by negating both the hypothesis and the conclusion of the conditional.

The contrapositive is formed by negating both the hypothesis and the conclusion of the converse of the conditional.

Logically Equivalent Statements (p. 108)
A conditional and its contrapositive are logically equivalent.

The converse and inverse of a conditional are logically equivalent.

Biconditional Statement (p. 114)
A biconditional statement is the conjunction of a conditional and its converse.

Law of Detachment (p. 116)
If $p \rightarrow q$ is a true statement and p is true, then q is true.

Law of Syllogism (p. 117)
If $p \rightarrow q$ and $q \rightarrow r$ are true statements, then $p \rightarrow r$ is a true statement.

Postulate 2.1 (p. 125) Through any two points, there is exactly one line.

Postulate 2.2 (p. 125) Through any three noncollinear points, there is exactly one plane.

Postulate 2.3 (p. 125) A line contains at least two points.

Postulate 2.4 (p. 125) A plane contains at least three noncollinear points.

Postulate 2.5 (p. 125) If two points lie in a plane, then the entire line containing those points lies in that plane.

Postulate 2.6 (p. 125) If two lines intersect, then their intersection is exactly one point.

Postulate 2.7 (p. 125) If two planes intersect, then their intersection is a line.

Theorem 2.1 Midpoint (p. 127)
If M is the midpoint of $\overline{AB}$, then $\overline{AM} \cong \overline{MB}$.

Properties of Real Numbers (p. 134)

Addition Property of Equality	If $a = b$, then $a + c = b + c$.
Subtraction Property of Equality	If $a = b$, then $a - c = b - c$.
Multiplication Property of Equality	If $a = b$, then $a \cdot c = b \cdot c$.
Division Property of Equality	If $a = b$ and $c \neq 0$, then, $\frac{a}{c} = \frac{b}{c}$.
Reflexive Property of Equality	$a = a$
Symmetric Property of Equality	If $a = b$, then $b = a$.
Transitive Property of Equality	If $a = b$ and $b = c$, then $a = c$.
Substitution Property of Equality	If $a = b$, then a may be replaced by b in any equation or expression.
Distributive Property	$a(b + c) = ab + ac$

Postulate 2.8 Ruler (p. 142)
The points on any line or line segment can be put into one-to-one correspondence with real numbers.

Postulate 2.9 Segment Addition (p. 142)
If A, B, and C are collinear, then point B is between A and C if and only if $AB + BC = AC$.

Theorem 2.2 Properties of Segment Congruence (p. 143)

Reflexive Property of Congruence	$\overline{AB} \cong \overline{AB}$
Symmetric Property of Congruence	If $\overline{AB} \cong \overline{CD}$, then $\overline{CD} \cong \overline{AB}$.
Transitive Property of Congruence	If $\overline{AB} \cong \overline{CD}$ and $\overline{CD} \cong \overline{EF}$, then $\overline{AB} \cong \overline{EF}$.

Postulate 2.10 Protractor (p. 149)
Given any angle, the measure can be put into one-to-one correspondence with real numbers between 0 and 180.

Postulate 2.11 Angle Addition (p. 149)
D is in the interior of $\angle ABC$ if and only if $m\angle ABD + m\angle DBC = m\angle ABC$.

Theorem 2.3 Supplement (p. 150)
If two angles form a linear pair, then they are supplementary angles.

Theorem 2.4 Complement (p. 150)
If the noncommon sides of two adjacent angles form a right angle, then the angles are complementary angles.

Theorem 2.5 Properties of Angle Congruence (p. 151)

Reflexive Property of Congruence	$\angle 1 \cong \angle 1$
Symmetric Property of Congruence	If $\angle 1 \cong \angle 2$, then $\angle 2 \cong \angle 1$.
Transitive Property of Congruence	If $\angle 1 \cong \angle 2$ and $\angle 2 \cong \angle 3$, then $\angle 1 \cong \angle 3$.

Theorem 2.6 Congruent Supplements (p. 151)
Angles supplementary to the same angle or to congruent angles are congruent.
Abbreviation: $\angle$s suppl. to same $\angle$ or $\cong$ $\angle$s are $\cong$.

Theorem 2.7 Congruent Complements (p. 151)
Angles complementary to the same angle or to congruent angles are congruent.
Abbreviation: $\angle$s compl. to same $\angle$ or $\cong$ $\angle$s are $\cong$.

Theorem 2.8 Vertical Angles (p. 152)
If two angles are vertical angles, then they are congruent.
Abbreviation: Vert. $\angle$s are $\cong$.

Theorem 2.9 (p. 153) Perpendicular lines intersect to form four right angles.

Theorem 2.10 (p. 153) All right angles are congruent.

Theorem 2.11 (p. 153) Perpendicular lines form congruent adjacent angles.

Theorem 2.12 (p. 153) If two angles are congruent and supplementary, then each angle is a right angle.

Theorem 2.13 (p. 153) If two congruent angles form a linear pair, then they are right angles.

Parallel and Perpendicular Lines — Chapter 3

Parallel and Skew (p. 171)
Parallel lines are coplanar lines that do not intersect.
Skew lines are lines that do not intersect and are not coplanar.
Parallel planes are planes that do not intersect.

Angle Pairs Formed by Two Lines and a Transversal (p. 172)
Interior angles lie in the region between lines that are intersected by a transversal.
Exterior angles lie in the two regions that are not between lines that are intersected by a transversal.
Consecutive interior angles are interior angles that lie on the same side of a transversal.
Alternate interior angles are nonadjacent interior angles that lie on opposite sides of a transversal.
Alternate exterior angles are nonadjacent exterior angles that lie on opposite sides of a transversal.
Corresponding angles lie on the same side of a transversal and on the same side of two lines that are intersected by the transversal.

Postulate 3.1 Corresponding Angles (p. 178)
If two parallel lines are cut by a transversal, then each pair of corresponding angles is congruent.

Theorem 3.1 Alternate Interior Angles Theorem (p. 179)
If two parallel lines are cut by a transversal, then each pair of alternate interior angles is congruent.

Theorem 3.2 Consecutive Interior Angles Theorem (p. 179)
If two parallel lines are cut by a transversal, then each pair of consecutive interior angles is supplementary.

Theorem 3.3 Alternate Exterior Angles Theorem (p. 179)
If two parallel lines are cut by a transversal, then each pair of alternate exterior angles is congruent.

Theorem 3.4 Perpendicular Transversal (p. 180)
In a plane, if a line is perpendicular to one of two parallel lines, then it is perpendicular to the other.

Slope of a Line (p. 186)
In a coordinate plane, the slope of a line is the ratio of the change along the y-axis to the change along the x-axis between any two points on the line.

Postulate 3.2 Slopes of Parallel Lines (p. 189)
Two nonvertical lines have the same slope if and only if they are parallel. All vertical lines are parallel.

Postulate 3.3 Slopes of Perpendicular Lines (p. 189)
Two nonvertical lines are perpendicular if and only if the product of their slopes is -1. Vertical and horizontal lines are perpendicular.

Forms of the Equation of a Nonvertical Line (p. 196)
The slope-intercept form of a linear equation is $y = mx + b$, where m is the slope of the line and b is the y-intercept.

The point-slope form of a linear equation is $y - y_1 = m(x - x_1)$, where (x_1, y_1) is any point on the line and m is the slope of the line.

Equations of Horizontal and Vertical Lines (p. 198)
The equation of a horizontal line is $y = b$, where b is the y-intercept of the line.

The equation of a vertical line is $x = a$, where a is the x-intercept of the line.

Postulate 3.4 Converse of Corresponding Angles (p. 205)
If two parallel lines are cut by a transversal so that corresponding angles are congruent, then the lines are parallel.

Postulate 3.5 Parallel (p. 206)
If given a line and a point not on the line, then there exists exactly one line through the point that is parallel to the given line.

Theorem 3.5 Alternate Exterior Angles Converse (p. 206)
If two lines in a plane are cut by a transversal so that a pair of alternate exterior angles is congruent, then the two lines are parallel.

Theorem 3.6 Consecutive Interior Angles Converse (p. 206)
If two lines in a plane are cut by a transversal so that a pair of consecutive interior angles is supplementary, then the lines are parallel.

Theorem 3.7 Alternate Interior Angles Converse (p. 206)
If two lines in a plane are cut by a transversal so that a pair of alternate interior angles is congruent, then the lines are parallel.

Theorem 3.8 Perpendicular Transversal Converse (p. 206)
In a plane, if two lines are perpendicular to the same line, then they are parallel.

Postulate 3.6 Perpendicular (p. 213)
If given a line and a point not on the line, then there exists exactly one line through the point that is perpendicular to the given line.

Distance Between a Point and a Line (p. 213)
The distance from a line to a point not on the line is the length of the segment perpendicular to the line from the point.

Distance Between Parallel Lines (p. 216)
The distance between two parallel lines is the perpendicular distance between one of the lines and any point on the other line.

Theorem 3.9 Two Lines Equidistant from a Third (p. 216)
In a plane, if two lines are each equidistant from a third line, then the two lines are parallel to each other.

Chapter 4
Congruent Triangles

Classifications of Triangles by Angles (p. 235)

An acute triangle has three acute angles.

An equiangular triangle has three congruent acute angles.

An obtuse triangle has one obtuse angle.

A right triangle has one right angle.

Classifications of Triangles by Sides (p. 236)

An equilateral triangle has three congruent sides.

An isosceles triangle has at least two congruent sides.

A scalene triangle has no congruent sides.

Theorem 4.1 Triangle Angle-Sum (p. 244)
The sum of the measures of the angles of a triangle is 180.

Theorem 4.2 Exterior Angle (p. 246)
The measure of an exterior angle of a triangle is equal to the sum of the measures of the two remote interior angles.

Corollary 4.1 (p. 247) The acute angles of a right triangle are complementary.

Corollary 4.2 (p. 247) There can be at most one right or obtuse angle in a triangle.

Theorem 4.3 Third Angles (p. 255)
If two angles of one triangle are congruent to two angles of a second triangle, then the third angles of the triangles are congruent.

Theorem 4.4 Properties of Triangle Congruence (p. 256)

Reflexive Property of Triangle Congruence $\triangle ABC \cong \triangle ABC$

Symmetric Property of Triangle Congruence If $\triangle ABC \cong \triangle EFG$, then $\triangle EFG \cong \triangle ABC$.

Transitive Property of Triangle Congruence If $\triangle ABC \cong \triangle EFG$ and $\triangle EFG \cong \triangle JKL$, then $\triangle ABC \cong \triangle JKL$.

Postulate 4.1 Side-Side-Side (SSS) Congruence (p. 262)
If three sides of one triangle are congruent to three sides of a second triangle, then the triangles are congruent.

Postulate 4.2 Side-Angle-Side (SAS) Congruence (p. 264)
If two sides and the included angle of one triangle are congruent to two sides and the included angle of a second triangle, then the triangles are congruent.

Postulate 4.3 Angle-Side-Angle (ASA) Congruence (p. 273)
If two angles and the included side of one triangle are congruent to two angles and the included side of another triangle, then the triangles are congruent.

Theorem 4.5 Angle-Angle-Side (AAS) Congruence (p. 274)
If two angles and a nonincluded side of one triangle are congruent to the corresponding two angles and side of a second triangle, then the two triangles are congruent.

Theorem 4.6 Leg-Leg (LL) Congruence (p. 282)
If the legs of one right triangle are congruent to the corresponding legs of another right triangle, then the triangles are congruent.

Theorem 4.7 Hypotenuse-Angle (HA) Congruence (p. 282)
If the hypotenuse and acute angle of one right triangle are congruent to the hypotenuse and corresponding acute angle of another right triangle, then the two triangles are congruent.

Theorem 4.8 Leg-Angle (LA) Congruence (p. 282)
If one leg and an acute angle of one right triangle are congruent to the corresponding leg and acute angle of another right triangle, then the triangles are congruent.

Theorem 4.9 Hypotenuse-Leg (HL) Congruence (p. 282)
If the hypotenuse and a leg of one right triangle are congruent to the hypotenuse and corresponding leg of another right triangle, then the triangles are congruent.

Theorem 4.10 Isosceles Triangle Theorem (p. 283)
If two sides of a triangle are congruent, then the angles opposite those sides are congruent.

Theorem 4.11 Converse of Isosceles Triangle Theorem (p. 283)
If two angles of a triangle are congruent, then the sides opposite those angles are congruent.

Corollary 4.3 (p. 284) A triangle is equilateral if and only if it is equiangular.

Corollary 4.4 (p. 284) Each angle of an equilateral triangle measures 60.

Reflections, Translations, and Rotations (p. 294)
A reflection or *flip* is a transformation over a line called the *line of reflection*. Each point of the preimage and its image are the same distance from the line of reflection.

A translation or *slide* is a transformation that moves all points of the original figure the same distance in the same direction.

A rotation or *turn* is a transformation around a fixed point called the *center of rotation*, through a specific angle, and in a specific direction. Each point of the original figure and its image are the same distance from the center.

Relationships in Triangles — Chapter 5

Theorem 5.1 Perpendicular Bisector Theorem (p. 322)
If a point is on the perpendicular bisector of a segment, then it is equidistant from the endpoints of the segment.

Theorem 5.2 Converse of the Perpendicular Bisector Theorem (p. 322)
If a point is equidistant from the endpoints of a segment, then it is on the perpendicular bisector of the segment.

Theorem 5.3 Circumcenter Theorem (p. 323)
The perpendicular bisectors of a triangle intersect at a point called the *circumcenter* that is equidistant from the vertices of the triangle.

Theorem 5.4 Angle Bisector Theorem (p. 325)
If a point is on the bisector of an angle, then it is equidistant from the sides of the angle.

Theorem 5.5 Converse of the Angle Bisector Theorem (p. 325)
If a point in the interior of an angle is equidistant from the sides of the angle, then it is on the bisector of the angle.

Theorem 5.6 Incenter Theorem (p. 326)
The angle bisectors of a triangle intersect at a point called the *incenter* that is equidistant from each side of the triangle.

Theorem 5.7 Centroid Theorem (p. 333)
The medians of a triangle intersect at a point called the centroid that is two thirds of the distance from each vertex to the midpoint of the opposite side.

Definition of Inequality (p. 342)
For any real numbers a and b, $a > b$ if and only if there is a positive number c such that $a = b + c$.

Properties of Inequality for Real Numbers (p. 342)

Comparison Property of Inequality	$a < b$, $a = b$, or $a > b$
Transitive Property of Inequality	1. If $a < b$ and $b < c$, then $a < c$. 2. If $a > b$ and $b > c$, then $a > c$.
Addition Property of Inequality	1. If $a > b$, then $a + c > b + c$. 2. If $a < b$, then $a + c < b + c$.
Subtraction Property of Inequality	1. If $a > b$, then $a - c > b - c$. 2. If $a < b$, then $a - c < b - c$.

Theorem 5.8 Exterior Angle Inequality Theorem (p. 343)
The measure of an exterior angle of a triangle is greater than the measure of either of its corresponding remote interior angles.

Theorem 5.9 (p. 344) If one side of a triangle is longer than another side, then the angle opposite the longer side has a greater measure than the angle opposite the shorter side.

Theorem 5.10 (p. 344) If one angle of a triangle has a greater measure than another angle, then the side opposite the greater angle is longer than the side opposite the lesser angle.

Theorem 5.11 Triangle Inequality Theorem (p. 360)
The sum of the lengths of any two sides of a triangle must be greater than the length of the third side.

Theorem 5.13 Hinge Theorem (p. 367)
If two sides of a triangle are congruent to two sides of another triangle, and the included angle of the first is larger than the included angle of the second, then the third side of the first triangle is longer than the third side of the second triangle.

Theorem 5.14 Converse of the Hinge Theorem (p. 367)
If two sides of a triangle are congruent to two sides of another triangle, and the third side in the first is longer than the third side in the second, then the included angle of the first triangle is greater than the included angle in the second triangle.

Quadrilaterals — Chapter 6

Theorem 6.1 Polygon Interior Angles Sum (p. 389)
The sum of the interior angle measures of an n-sided convex polygon is $(n - 2) \cdot 180$.

Theorem 6.2 Polygon Exterior Angles Sum (p. 392)
The sum of the exterior angle measures of a convex polygon, one angle at each vertex, is 360.

Theorem 6.3 (p. 399) If a quadrilateral is a parallelogram, then its opposite sides are congruent.

Theorem 6.4 (p. 399) If a quadrilateral is a parallelogram, then its opposite angles are congruent.

Theorem 6.5 (p. 399) If a quadrilateral is a parallelogram, then its consecutive angles are supplementary.

Theorem 6.6 (p. 399) If a parallelogram has one right angle, then it has four right angles.

Theorem 6.7 (p. 401) If a quadrilateral is a parallelogram, then its diagonals bisect each other.

Theorem 6.8 (p. 401) If a quadrilateral is a parallelogram, then each diagonal separates the parallelogram into two congruent triangles.

Theorem 6.9 (p. 409) If both pairs of opposite sides of a quadrilateral are congruent, then the quadrilateral is a parallelogram.

Abbreviation: If both pairs of opp. sides are $\cong$, then quad. is a ▱.

Theorem 6.10 (p. 409) If both pairs of opposite angles of a quadrilateral are congruent, then the quadrilateral is a parallelogram.

Abbreviation: If both pairs of opp. $\angle$s are $\cong$, then quad. is a ▱.

Theorem 6.11 (p. 409) If the diagonals of a quadrilateral bisect each other, then the quadrilateral is a parallelogram.

Abbreviation: If diag. bisect each other, then quad. is a ▱.

Theorem 6.12 (p. 409) If one pair of opposite sides of a quadrilateral is both parallel and congruent, then the quadrilateral is a parallelogram.

Abbreviation: If one pair of opp. sides is $\cong$ and $||$, then the quad. is a ▱.

Theorem 6.13 Diagonals of a Rectangle (p. 419)
If a parallelogram is a rectangle, then its diagonals are congruent.

Theorem 6.14 (p. 420) If the diagonals of a parallelogram are congruent, then the parallelogram is a rectangle.

Theorem 6.15 (p. 426) If a parallelogram is a rhombus, then its diagonals are perpendicular.

Theorem 6.16 (p. 426) If a parallelogram is a rhombus, then each diagonal bisects a pair of opposite angles.

Theorem 6.17 (p. 428) If the diagonals of a parallelogram are perpendicular, then the parallelogram is a rhombus.

Theorem 6.18 (p. 428) If one diagonal of a parallelogram bisects a pair of opposite angles, then the parallelogram is a rhombus.

Theorem 6.19 (p. 428) If one pair of consecutive sides of a parallelogram are congruent, the parallelogram is a rhombus.

Theorem 6.20 (p. 428) If a quadrilateral is both a rectangle and a rhombus, then it is a square.

Theorem 6.21 (p. 435) If a trapezoid is isosceles, then each pair of base angles are congruent.

Theorem 6.22 (p. 435) If a trapezoid has one pair of congruent base angles, then it is an isosceles trapezoid.

Theorem 6.23 (p. 435) A trapezoid is isosceles if and only if its diagonals are congruent.

Theorem 6.24 Trapezoid Midsegment Theorem (p. 437)
The midsegment of a trapezoid is parallel to each base and its measure is one half the sum of the lengths of the bases.

Theorem 6.25 (p. 439) If a quadrilateral is a kite, then its diagonals are perpendicular.

Theorem 6.26 (p. 439) If a quadrilateral is a kite, then exactly one pair of opposite angles is congruent.

Cross Products Property (p. 458)
In a proportion, the product of the extremes equals the product of the means.

Equivalent Proportions (p. 459)
Equivalent proportions have identical cross products.

Similar Polygons (p. 465)
Two polygons are similar if and only if their corresponding angles are congruent and corresponding side lengths are proportional.

Theorem 7.1 Perimeters of Similar Polygons (p. 467)
If two polygons are similar, then their perimeters are proportional to the scale factor between them.

Postulate 7.1 Angle-Angle (AA) Similarity (p. 474)
If two angles of one triangle are congruent to two angles of another triangle, then the triangles are similar.

Theorem 7.2 Side-Side-Side (SSS) Similarity (p. 475)
If the corresponding side lengths of two triangles are proportional, then the triangles are similar.

Theorem 7.3 Side-Angle-Side (SAS) Similarity (p. 475)
If the lengths of two sides of one triangle are proportional to the lengths of two corresponding sides of another triangle and the included angles are congruent, then the triangles are similar.

Theorem 7.4 Properties of Similarity (p. 477)
Reflexive Property of Similarity $\triangle ABC \sim \triangle ABC$

Symmetric Property of Similarity If $\triangle ABC \sim \triangle DEF$, then $\triangle DEF \sim \triangle ABC$.

Transitive Property of Similarity If $\triangle ABC \sim \triangle DEF$ and $\triangle DEF \sim \triangle XYZ$, then $\triangle ABC \sim \triangle XYZ$.

Theorem 7.5 Triangle Proportionality Theorem (p. 484)
If a line is parallel to one side of a triangle and intersects the other two sides, then it divides the sides into segments of proportional lengths.

Theorem 7.6 Converse of the Triangle Proportionality Theorem (p. 485)
If a line intersects two sides of a triangle and separates the sides into proportional corresponding segments, then the line is parallel to the third side of the triangle.

Theorem 7.7 Triangle Midsegment Theorem (p. 485)
A midsegment of a triangle is parallel to one side of the triangle, and its length is one half the length of that side.

Corollary 7.1 Proportional Parts of Parallel Lines (p. 486)
If three or more parallel lines intersect two transversals, then they cut off the transversals proportionally.

Corollary 7.2 Congruent Parts of Parallel Lines (p. 487)
If three or more parallel lines cut off congruent segments on one transversal, then they cut off congruent segments on every transversal.

Theorem 7.8 (p. 495) If two triangles are similar, the lengths of corresponding altitudes are proportional to the lengths of corresponding sides.

Abbreviation: $\sim \triangle$s have corr. altitudes proportional to corr. sides.

Theorem 7.9 (p. 495) If two triangles are similar, the lengths of corresponding angle bisectors are proportional to the lengths of corresponding sides.

Abbreviation: $\sim \triangle$s have corr. $\angle$ bisectors proportional to corr. sides.

Theorem 7.10 (p. 495) If two triangles are similar, the lengths of corresponding medians are proportional to the lengths of corresponding sides.

Abbreviation: ~ △s have corr. medians proportional to corr. sides.

Theorem 7.11 Triangle Angle Bisector (p. 498)
An angle bisector in a triangle separates the opposite side into two segments that are proportional to the lengths of the other two sides.

Types of Dilations (p. 505)
A dilation with a scale factor greater than 1 produces an enlargement, or an image that is larger than the original figure.

A dilation with a scale factor between 0 and 1 produces a reduction, an image that is smaller than the original figure.

Right Triangles and Trigonometry — Chapter 8

Geometric Mean (p. 531)
The geometric mean of two positive numbers a and b is the number x such that $\frac{a}{x} = \frac{x}{b}$. So $x^2 = ab$ and $x = \sqrt{ab}$.

Theorem 8.1 (p. 532) If the altitude is drawn to the hypotenuse of a right triangle, then the two triangles formed are similar to the original triangle and to each other.

Theorem 8.2 Geometric Mean (Altitude) Theorem (p. 533) The altitude drawn to the hypotenuse of a right triangle separates the hypotenuse into two segments. The length of this altitude is the geometric mean between the lengths of these two segments.

Theorem 8.3 Geometric Mean (Leg) Theorem (p. 533) The altitude drawn to the hypotenuse of a right triangle separates the hypotenuse into two segments. The length of a leg of this triangle is the geometric mean between the length of the hypotenuse and the segment of the hypotenuse adjacent to that leg.

Theorem 8.4 Pythagorean Theorem (p. 541)
In a right triangle, the sum of the squares of the lengths of the legs is equal to the square of the length of the hypotenuse.

Common Pythagorean Triples (p. 542)
A pythagorean triple is a group of three nonzero whole numbers a, b, and c, such that $a^2 + b^2 = c^2$.

Theorem 8.5 Converse of the Pythagorean Theorem (p. 544)
If the sum of the squares of the lengths of the shortest sides of a triangle is equal to the square of the length of the longest side, then the triangle is a right triangle.

Theorem 8.6 (p. 544) If the square of the length of the longest side of a triangle is less than the sum of the squares of the lengths of the other two sides, then the triangle is an acute triangle.

Theorem 8.7 (p. 544) If the square of the length of the longest side of a triangle is greater than the sum of the squares of the lengths of the other two sides, then the triangle is an obtuse triangle.

Distance and Midpoint Formulas in Space (p. 550)
If A has coordinates $A(x_1, y_1, z_1)$ and B has coordinates $B(x_2, y_2, z_2)$, then $AB = \sqrt{(x_2 - x_1)^2 + (y_2 - y_1)^2 + (z_2 - z_1)^2}$. The midpoint of $\overline{AB}$ has coordinates $\left(\frac{x_1 + x_2}{2}, \frac{y_1 + y_2}{2}, \frac{z_1 + z_2}{2}\right)$.

Theorem 8.8 45°-45°-90° Triangle Theorem (p. 552)

In a 45°-45°-90° triangle, the legs ℓ are congruent and the length of the hypotenuse h is $\sqrt{2}$ times the length of a leg.

Theorem 8.9 30°-60°-90° Triangle Theorem (p. 554)

In a 30°-60°-90° triangle, the length of the hypotenuse h is 2 times the length of the shorter leg s, and the longer leg ℓ is $\sqrt{3}$ times the length of the shorter leg.

Trigonometric Ratios (p. 562)

If $\triangle ABC$ is a right triangle with acute $\angle A$, then the sine of $\angle A$ (written sin A) is the ratio of the length of the leg opposite $\angle A$ (opp) to the length of the hypotenuse (hyp).

If $\triangle ABC$ is a right triangle with acute $\angle A$, then the cosine of $\angle A$ (written cos A) is the ratio of the length of the leg adjacent $\angle A$ (adj) to the length of the hypotenuse (hyp).

If $\triangle ABC$ is a right triangle with acute $\angle A$, then the tangent of $\angle A$ (written tan A) is the ratio of the length of the leg opposite $\angle A$ (opp) to the length of the leg adjacent $\angle A$ (adj).

Inverse Trigonometric Ratios (p. 565)

If $\angle A$ is an acute angle and the sine of A is x, then the inverse sine of x is the measure of $\angle A$.

If $\angle A$ is an acute angle and the cosine of A is x, then the inverse cosine of x is the measure of $\angle A$.

If $\angle A$ is an acute angle and the tangent of A is x, then the inverse tangent of x is the measure of $\angle A$.

Reciprocal Trigonometric Ratios (p. 572)

The cosecant of $\angle A$ (written csc A) is the reciprocal of sin A.

The secant of $\angle A$ (written sec A) is the reciprocal of cos A.

The cotangent of $\angle A$ (written cot A) is the reciprocal of tan A.

Theorem 8.10 Law of Sines (p. 582)

If $\triangle ABC$ has lengths a, b, and c, representing the lengths of the sides opposite the angles with measures A, B, and C, then $\frac{\sin A}{a} = \frac{\sin B}{b} = \frac{\sin C}{c}$.

Theorem 8.11 Law of Cosines (p. 583)

If $\triangle ABC$ has lengths a, b, and c, representing the lengths of the sides opposite the angles with measures A, B, and C, then $a^2 = b^2 + c^2 - 2bc \cos A$, $b^2 = a^2 + c^2 - 2ac \cos B$, and $c^2 = a^2 + b^2 - 2ab \cos C$.

Equal, Parallel, and Opposite Vectors (p. 594)

Equal Vectors Two vectors are equal if and only if they have the same magnitude and direction.

Parallel Vectors Two vectors are parallel if and only if they have the same or opposite direction.

Opposite Vectors Two vectors are opposites if they have the same magnitude and opposite directions.

Vector Addition (p. 595)

Parallelogram Method Place both vectors at the same initial point and complete the parallelogram. The resultant is the diagonal of the parallelogram.

Triangle Method Place the initial point of the second vector at the terminal point of the first. The resultant connects the initial point of the first vector and the terminal point of the second.

Reflection in a Line (p. 615)

A reflection in a line maps a point to its image such that if the point is on the line, then the image and preimage are the same point, or if the point does not lie on the line, the line is the perpendicular bisector of the segment joining the two points.

Reflection in the x- or y-axis (p. 617)

Reflections in the x-axis To reflect a point in the x-axis, multiply its y-coordinate by -1.

Reflections in the y-axis To reflect a point in the y-axis, multiply its x-coordinate by -1.

Reflection in Line $y = x$ (p. 618)

To reflect a point in the line $y = x$, interchange the x- and y-coordinates.

Translation (p. 624)

A translation maps each point to its image along a vector, called the translation vector, such that each segment joining a point and its image has the same length as the vector, and this segment is also parallel to the vector.

Translation in the Coordinate Plane (p. 625)

To translate a point along vector $\langle a, b \rangle$, add a to the x-coordinate and b to the y-coordinate.

Rotation (p. 632)

A rotation about a fixed point, called the center of rotation, through an angle of $x°$ maps a point to its image such that if the point is the center of rotation, then the image and preimage are the same point, or if the point is not the center of rotation, then the image and preimage are the same distance from the center of rotation and the measure of the angle of rotation formed by the preimage, center of rotation, and image points is x.

Rotations in the Coordinate Plane (p. 633)

90° Rotation To rotate a point 90° counterclockwise about the origin, multiply the y-coordinate by -1 and then interchange the x- and y-coordinates.

180° Rotation To rotate a point 180° counterclockwise about the origin, multiply the x- and y-coordinates by -1.

270° Rotation To rotate a point 270° counterclockwise about the origin, multiply the x-coordinate by -1 and then interchange the x- and y-coordinates.

Theorem 9.1 Composition of Isometries (p. 642)

The composition of two (or more) isometries is an isometry.

Theorem 9.2 Reflections in Parallel Lines (p. 642)

The composition of two reflections in parallel lines can be described by a translation vector that is perpendicular to the two lines, and twice the distance between the two lines.

Theorem 9.3 Reflections in Intersecting Lines (p. 643)

The composition of two reflections in intersecting lines can be described by a rotation about the point where the lines intersect and through an angle that is twice the measure of the acute or right angle formed by the lines.

Line Symmetry (p. 653)

A figure in the plane has line symmetry (or *reflection symmetry*) if the figure can be mapped onto itself by a reflection in a line, called a line of symmetry (or *axis of symmetry*).

Rotational Symmetry (p. 654)

A figure in the plane has rotational symmetry (or *radial symmetry*) if the figure can be mapped onto itself by a rotation between 0° and 360° about the center of the figure, called the center of symmetry (or *point of symmetry*).

Three-Dimensional Symmetries (p. 655)

Plane Symmetry A three dimensional figure has plane symmetry if the figure can be mapped onto itself by a reflection in a plane.

Axis Symmetry A three dimensional figure has axis symmetry if the figure can be mapped onto itself by a rotation between 0° and 360° in a line.

Dilation (p. 660)

A dilation with center C and positive scale factor r, $r \neq 1$, maps a point P in a figure to its image such that if point P and C coincide, then the image and preimage are the same point, or if point P is not the center of dilation, then P' lies on $\overrightarrow{CP}$ and $CP' = r(CP)$.

Dilations in the Coordinate Plane (p. 662)

To find the coordinates of an image after a dilation centered at the origin, multiply the x- and y-coordinates of each point on the preimage by the scale factor of the dilation, r.

Chapter 10 Circles

Special Segments in a Circle (p. 683)

A radius (plural radii) is a segment with endpoints at the center and on the circle.

A chord is a segment with endpoints on the circle.

A diameter of a circle is a chord that passes through the center and is made up of collinear radii.

Circle Pairs (p. 684)

Two circles are congruent circles if and only if they have congruent radii.

Concentric circles are coplanar circles that have the same center.

Circumference (p. 685)

If a circle has diameter d or radius r, the circumference C equals the diameter times pi or twice the radius times pi.

Sum of Central Angles (p. 692)

The sum of the measures of the central angles of a circle with no interior points in common is 360.

Arcs and Arc Measure (p. 693)

The measure of a minor arc is less than 180° and equal to the measure of its related central angle.

The measure of a major arc is greater than 180° and equal to 360° minus the measure of the minor arc with the same endpoints.

The measure of a semicircle is 180°.

Theorem 10.1 (p. 693) In the same circle or in congruent circles, two minor arcs are congruent if and only if their central angles are congruent.

Postulate 10.1 (p. 694) The measure of an arc formed by two adjacent arcs is the sum of the measures of the two arcs.

Arc Length (p. 695)

The ratio of the length of an arc ℓ to the circumference of the circle is equal to the ratio of the degree measure of the arc to 360.

Theorem 10.2 (p. 701) In the same circle or in congruent circles, two minor arcs are congruent if and only if their corresponding chords are congruent.

Theorem 10.3 (p. 702) If a diameter (or radius) of a circle is perpendicular to a chord, then it bisects the chord and its arc.

Theorem 10.4 (p. 702) The perpendicular bisector of a chord is a diameter (or radius) of the circle.

Theorem 10.5 (p. 703) In the same circle or in congruent circles, two chords are congruent if and only if they are equidistant from the center.

Theorem 10.6 Inscribed Angle Theorem (p. 709)
If an angle is inscribed in a circle, then the measure of the angle equals one half the measure of its intercepted arc.

Theorem 10.7 (p. 710) If two inscribed angles of a circle intercept the same arc or congruent arcs, then the angles are congruent.

Theorem 10.8 (p. 711) An inscribed angle of a triangle intercepts a diameter or semicircle if and only if the angle is a right angle.

Theorem 10.9 (p. 712) If a quadrilateral is inscribed in a circle, then its opposite angles are supplementary.

Theorem 10.10 (p. 719) In a plane, a line is tangent to a circle if and only if it is perpendicular to a radius drawn to the point of tangency.

Theorem 10.11 Tangent to a Circle (p. 720)
If two segments from the same exterior point are tangent to a circle, then they are congruent.

Theorem 10.12 (p. 727) If two secants or chords intersect in the interior of a circle, then the measure of an angle formed is one half the *sum* of the measure of the arcs intercepted by the angle and its vertical angle.

Theorem 10.13 (p. 728) If a secant and a tangent intersect at the point of tangency, then the measure of each angle formed is one half the measure of its intercepted arc.

Theorem 10.14 (p. 729) If two secants, a secant and a tangent, or two tangents intersect in the exterior of a circle, then the measure of the angle formed is one half the *difference* of the measures of the intercepted arcs.

Theorem 10.15 Segments of Chords Theorem (p. 736)
If two chords intersect in a circle, then the products of the lengths of the chord segments are equal.

Theorem 10.16 Secant Segments Theorem (p. 738)
If two secants intersect in the exterior of a circle, then the product of the measures of one secant segment and its external secant segment is equal to the product of the measures of the other secant and its external secant segment.

Theorem 10.17 (p. 738) If a tangent and a secant intersect in the exterior of a circle, then the square of the measure of the tangent is equal to the product of the measures of the secant and its external secant segment.

Standard Form of the Equation of a Circle (p. 744)
The standard form of the equation of a circle with center at (h, k) and radius r is $(x - h)^2 + (y - k)^2 = r^2$.

Areas of Polygons and Circles — Chapter 11

Area of a Parallelogram (p. 763)
The area A of a parallelogram is the product of a base b and its corresponding height h.

Postulate 11.1 Area Addition Postulate (p. 763)
The area of a region is the sum of the areas of its nonoverlapping parts.

Postulate 11.2 Area Congruence Postulate (p. 765)
If two figures are congruent, then they have the same area.

Area of a Triangle (p. 765)
The area A of a triangle is one half the product of a base b and its corresponding height h.

Area of a Trapezoid (p. 773)
The area A of a trapezoid is one half the product of the height h and the sum of its bases, b_1 and b_2.

Area of a Rhombus or Kite (p. 775)
The area A of a rhombus or kite is one half the product of the lengths of its diagonals, d_1 and d_2.

Area of a Circle (p. 782)
The area A of a circle is equal to π times the square of the radius r.

Area of a Sector (p. 783)
The ratio of the area A of a sector to the area of the whole circle, πr^2, is equal to the ratio of the degree measure of the intercepted arc x to 360.

Area of a Regular Polygon (p. 792)
The area A of a regular n-gon with side length s is one half the product of the apothem a and perimeter P.

Theorem 11.1 Areas of Similar Polygons (p. 802)
If two polygons are similar, then their areas are proportional to the square of the scale factor between them.

Extending Surface Area and Volume — Chapter 12

Lateral Area of a Prism (p. 830)
The lateral area L of a right prism is $L = Ph$, where h is the height of the prism and P is the perimeter of each base.

Surface Area of a Prism (p. 831)
The surface area S of a right prism is $S = L + 2B$, where L is its lateral area and B is the area of each base.

Areas of a Cylinder (p. 832)
Lateral Area The lateral area L of a right cylinder is $L = 2\pi rh$, where r is the radius of a base and h is the height.

Surface Area of a Cylinder The surface area S of a right cylinder is $S = 2\pi rh + 2\pi r^2$, where r is the radius of a base and h is the height.

Lateral Area of a Regular Pyramid (p. 838)
The lateral area L of a regular pyramid is $L = \frac{1}{2}P\ell$, where ℓ is the slant height and P is the perimeter of the base.

Surface Area of a Regular Pyramid (p. 839)
The surface area S of a regular pyramid is $S = \frac{1}{2}P\ell + B$, where P is the perimeter of the base, ℓ is the slant height, and B is the area of the base.

Lateral Area of a Cone (p. 841)
The lateral area L of a right circular cone is $L = \pi r\ell$, where r is the radius of the base and ℓ is the slant height.

Surface Area of a Cone (p. 841)
The surface area S of a right circular cone is $S = \pi r\ell + \pi r^2$, where r is the radius of the base and ℓ is the slant height.

Volume of a Prism (p. 847)
The volume V of a prism is $V = Bh$, where B is the area of a base and h is the height of the prism.

Volume of a Cylinder (p. 848)
The volume V of a cylinder is $V = Bh$ or $V = \pi r^2 h$, where B is the area of the base, h is the height of the cylinder, and r is the radius of the base.

Cavalieri's Principle (p. 848)
If two solids have the same height h and the same cross-sectional area B at every level, then they have the same volume.

Volume of a Pyramid (p. 857)
The volume of a pyramid is $V = \frac{1}{3}Bh$, where B is the area of the base and h is the height of the pyramid.

Volume of a Cone (p. 858)
The volume of a circular cone is $V = \frac{1}{3}Bh$ or $V = \frac{1}{3}\pi r^2 h$, where B is the area of the base, h is the height of the cone, and r is the radius of the base.

Surface Area of a Sphere (p. 864)
The surface area S of a sphere is $S = 4\pi r^2$, where r is the radius.

Volume of a Sphere (p. 866)
The volume V of a sphere is $V = \frac{4}{3}\pi r^3$, where r is the radius of the sphere.

Theorem 12.1 (p. 881) If two similar solids have a scale factor of $a:b$, then the surface areas have a ratio of $a^2:b^2$, and the volumes have a ratio of $a^3:b^3$.

Probability and Measurement — Chapter 13

Fundamental Counting Principle (p. 901)
All possible outcomes in a sample space can be found by multiplying the number of possible outcomes from each stage or event.

Factorial (p. 906)
The factorial of a positive integer n, written $n!$, is the product of the integers less than or equal to n.

Permutations (p. 907)
The number of permutations of n distinct objects taken r at a time is denoted by ${}_nP_r$ and given by ${}_nP_r = \frac{n!}{(n-r)!}$.

Permutations with Repetition (p. 908)
The number of distinguishable permutations of n objects in which one object is repeated r_1 times, another is repeated r_2 times, and so on, is

$$\frac{n!}{r_1! \cdot r_2! \cdot \ldots \cdot r_k}.$$

Circular Permutations (p. 909)
The number of distinguishable permutations of n objects arranged in a circle with no fixed reference point is

$$\frac{n!}{n} \text{ or } (n-1)!.$$

Combinations (p. 910)
The number of combinations of n distinct objects taken r at a time is denoted by ${}_nC_r$ and is given by ${}_nC_r = \frac{n!}{(n-r)!r!}$.

Length Probability Ratio (p. 915)
If a line segment (1) contains another segment (2) and a point on segment (1) is chosen at random, then the probability that the point is on segment (2) is

$$\frac{\text{length of segment (2)}}{\text{length of segment (1)}}.$$

Key Concepts

Area Probability Ratio (p. 916)

If a region A contains a region B and a point E in region A is chosen at random, then the probability that point E is in region B is $\frac{\text{area of region } B}{\text{area of region } A}$.

Probability of Two Independent Events (p. 932)

The probability that two independent events both occur is the product of the probabilities of each individual event.

Probability of Two Dependent Events (p. 933)

The probability that two dependent events both occur is the product of the probability that the first event occurs and the probability that the second event occurs *after* the first event has already occurred.

Conditional Probability (p. 934)

The conditional probability of B given A is $P(B|A) = \frac{P(A \text{ and } B)}{P(A)}$, where $P(A) \neq 0$.

Probability of Mutually Exclusive Events (p. 939)

If two events A and B are mutually exclusive, then the probability that A or B occurs is the sum of the probabilities of each individual event.

Probability of Events That Are Not Mutually Exclusive (p. 940)

If two events A and B are not mutually exclusive, then the probability that A or B occurs is the sum of their individual probabilities minus the probability that both A and B occur.

Probability of the Complement of an Event (p. 941)

The probability that an event will not occur is equal to 1 minus the probability that the event will occur.

For Homework Help, go to Hotmath.com
Complete, step-by-step solutions of most odd-numbered exercises are provided free of charge.

Chapter 0 Preparing for Geometry

Page P5 Lesson 0-1

1. cm **3.** kg **5.** mL **7.** 10 **9.** 10,000 **11.** 0.18 **13.** 2.5 **15.** 24 **17.** 0.370 **19.** 4 **21.** 5 **23.** 16 **25.** 208 **27.** 9050

Page P7 Lesson 0-2

1. 20 **3.** 12.1 **5.** 16 **7.** 12 **9.** 5.4 **11.** 22.47 **13.** 1.125 **15.** 5.4 **17.** 15 **19.** 367.9 g **21.** 735.8 g

Page P9 Lesson 0-3

1. $\frac{1}{3}$ or 33% **3.** $\frac{2}{3}$ or 67% **5.** $\frac{1}{3}$ or 33% **7.** $\frac{13}{28}$ or about 46% **9.** $\frac{11}{14}$ or about 79% **11.** $\frac{9}{70}$ or about 13% **13.** $\frac{9}{28}$ or about 32% **15.** $\frac{1}{28}$ or about 3.6% **17.** $\frac{13}{28}$ or about 46% **19.** $\frac{13}{14}$ or about 93% **21.** $\frac{1}{10}$ or 10%; $\frac{1}{8}$ or 12.5% **23.** $\frac{2}{5}$ or 40%; $\frac{3}{8}$ or 37.5%

Page P10 Lesson 0-4

1. 3 **3.** −2 **5.** −1 **7.** −26 **9.** 26 **11.** 15

Page P12 Lesson 0-5

1. −8 **3.** 15 **5.** −72 **7.** $-\frac{15}{2}$ **9.** $\frac{7}{2}$ **11.** −15 **13.** −7 **15.** −7 **17.** −1 **19.** 60 **21.** −4 **23.** 4 **25.** 15 **27.** 21 **29.** −2 **31.** $-\frac{29}{2}$ **33.** −6 **35.** 1

Page P14 Lesson 0-6

1. $\{x|x < 13\}$ **3.** $\{y|y < 5\}$ **5.** $\{t|t > -42\}$ **7.** $\{d|d \leq 4\}$ **9.** $\{k|k \geq -3\}$ **11.** $\{z|z < -2\}$ **13.** $\{m|m < 29\}$ **15.** $\{b|b \geq -16\}$ **17.** $\{z|z > -2\}$ **19.** $\{b|b \leq 10\}$ **21.** $\{q|q \geq 2\}$ **23.** $\left\{w|w \geq -\frac{7}{3}\right\}$

Page P16 Lesson 0-7

1. (−2, 3) **3.** (2, 2) **5.** (−3, 1) **7.** (4, 1) **9.** (−1, −1) **11.** (3, 0) **13.** (2, −4) **15.** (−4, 2) **17.** none **19.** IV **21.** I **23.** III

25.

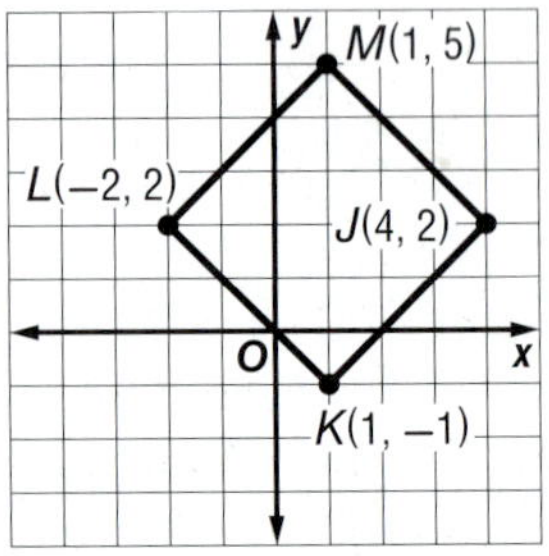

27.

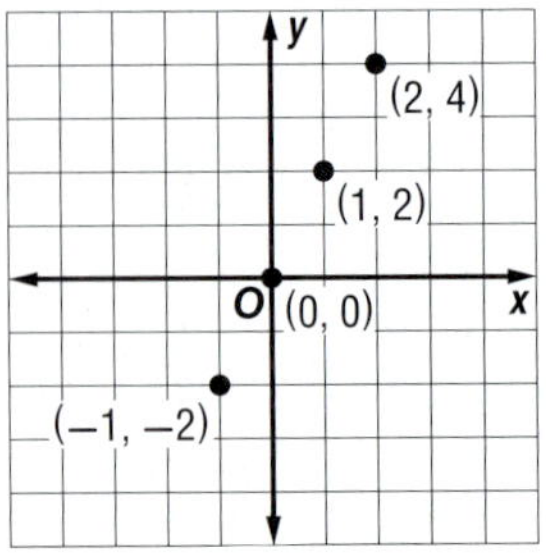

29.

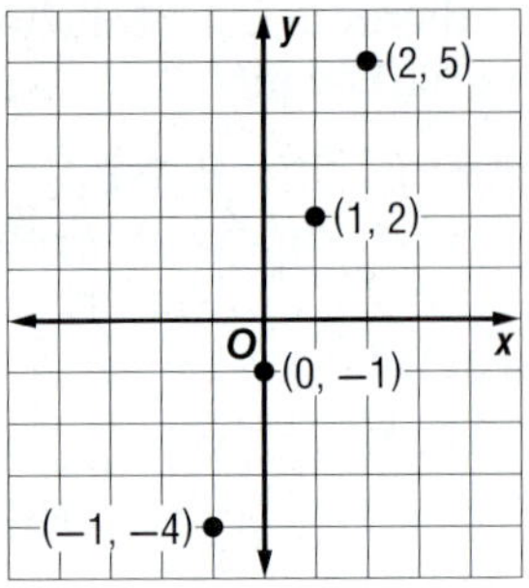

Page P18 Lesson 0-8

1. (2, 0) **3.** no solution **5.** (2, −5) **7.** $\left(-\frac{4}{3}, 3\right)$ **9.** (4, 1) **11.** elimination, no solution **13.** elimination or substitution, (3, 0) **15.** elimination or substitution, (−6, 4)

Page P20 Lesson 0-9

1. $4\sqrt{2}$ **3.** $10\sqrt{5}$ **5.** 6 **7.** $7x|y^3|\sqrt{2x}$ **9.** $\frac{9}{7}$ **11.** $\frac{3\sqrt{14}}{4}$ **13.** $\frac{p\sqrt{30p}}{9}$ **15.** $\frac{20 + 8\sqrt{3}}{13}$ **17.** $\frac{\sqrt{3}}{4}$ **19.** $\frac{6\sqrt{5} + 3\sqrt{10}}{2}$

Chapter 1 Tools of Geometry

Page 3 Chapter 1 Get Ready

1.

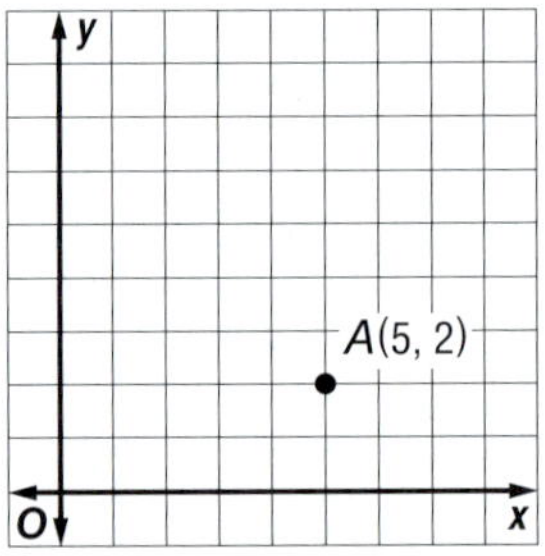

3.

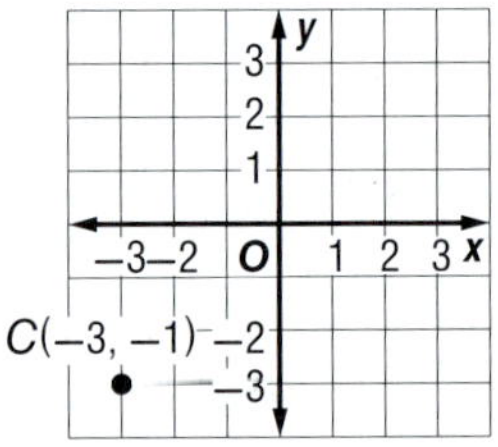

5. e5 **7.** $6\frac{29}{36}$ **9.** $5\frac{2}{15}$ **11.** 81 **13.** 153 **15.** 6

Pages 5–12 Lesson 1-1

1. Sample answer: *m* **3.** $\mathcal{B}$ **5.** plane
7. Sample answer:

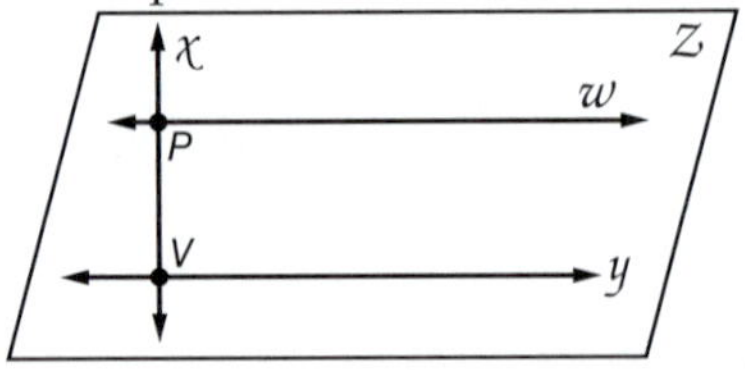

9. Sample answer: *A*, *H*, and *B* **11.** Yes; points *B*, *D*, and *F* lie in plane *BDF*.

13. Sample answer: *n* and *q* **15.** *R*

17 Points *A*, *B*, and *C* are contained in plane $\mathcal{R}$. Since point *P* is not contained in plane $\mathcal{R}$, it is not coplanar with points *A*, *B*, and *C*.

19. points *A* and *P* **21.** Yes; line *n* intersects line *q* when the lines are extended. **23.** intersecting lines

25. two planes intersecting in a line **27.** point
29. line **31.** intersecting planes
33. Sample answer:

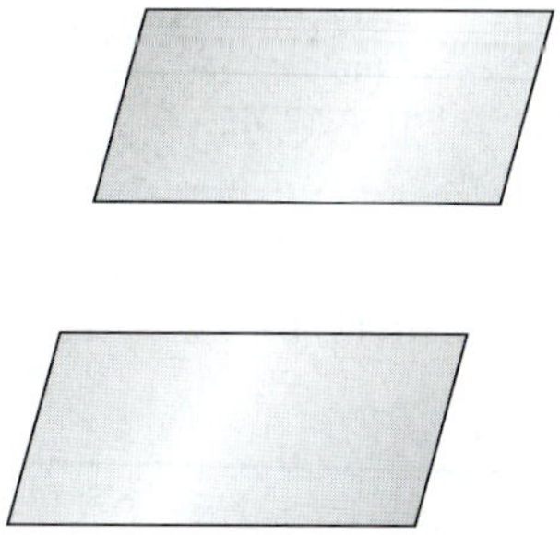

35. Sample answer:

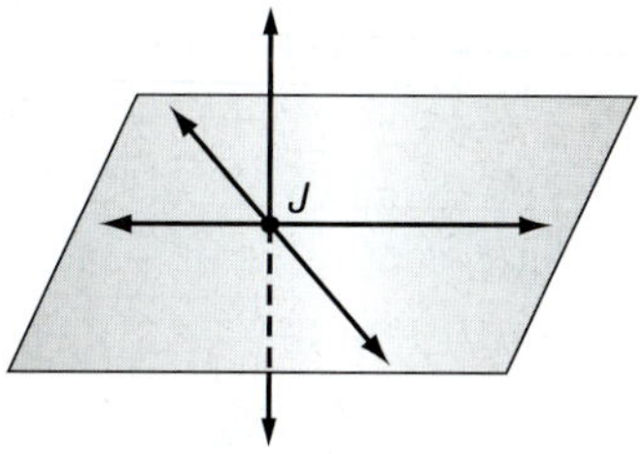

37. Sample answer:

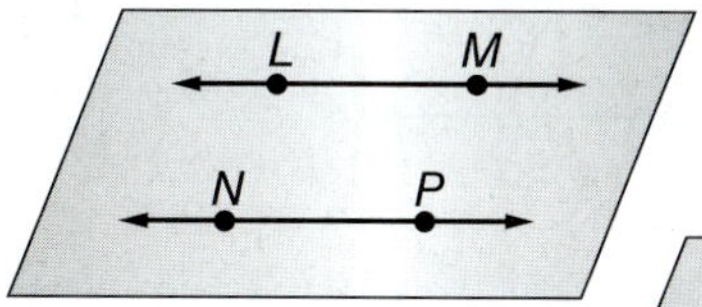

39. Sample answer:

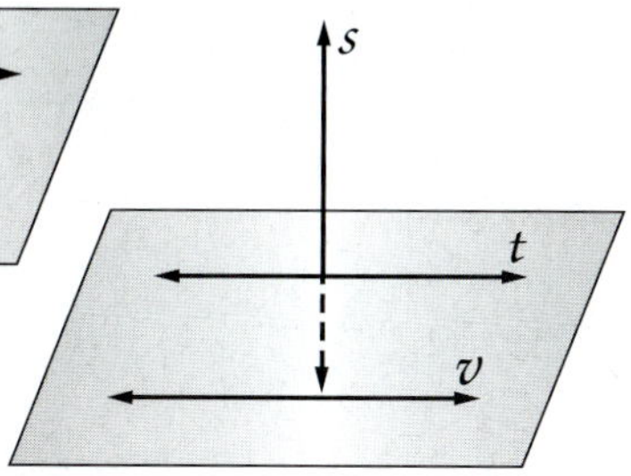

41. edges **43.** Sample answer: M and N
45 The planes appear to be parallel. Since they do not have any points in common, they do not intersect.
47. No; V does not lie in the same plane.
49 **a.** The intersection between the signs and the pole is represented by a point. **b.** The two planes intersect in a line.

51a.

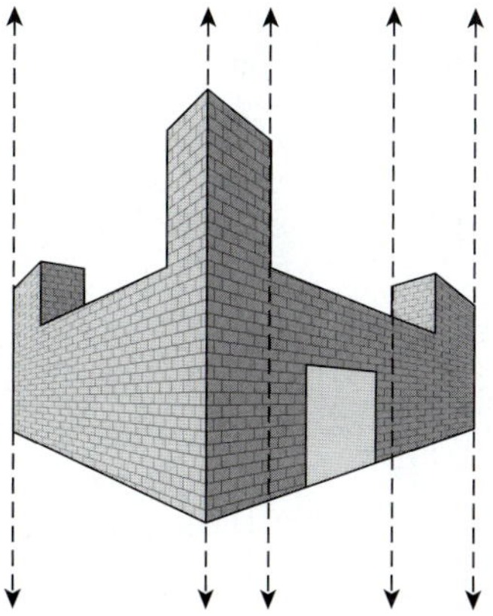

51b.

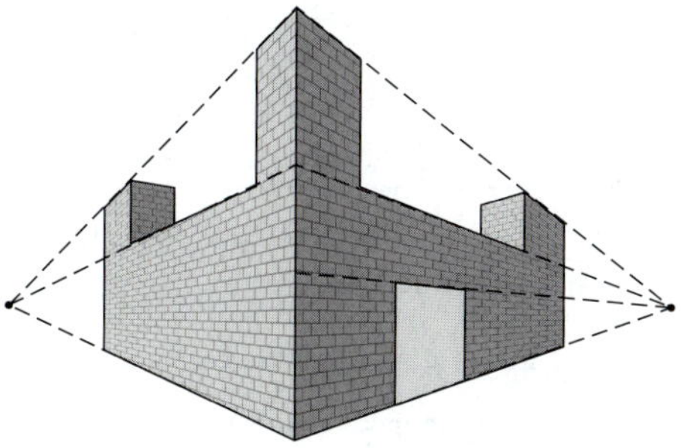

51c. Sample answer: They get closer together.
53. Sample answer: The airplanes are in different horizontal planes.

55 **a.** There are four ways to choose three points: FGH, FGK, GHK, and FHK. Only one way, FGH, has three points collinear. So, the probability is $\frac{1}{4}$.
b. There is exactly one plane through any three noncollinear points and infinitely many planes through three collinear points. Therefore, the probability that the three points chosen are coplanar is 1.

57. Sample answer:

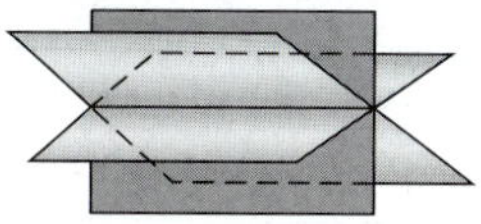

59. 4 **61.** Sample answer: A table is a finite plane. It is not possible to have a real-life object that is an infinite plane because all real-life objects have boundaries.

63. H **65.** B **67.** $6\sqrt{7}$ **69.** $\frac{\sqrt{2}}{2}$ **71.** $\frac{|a^3|\sqrt{6}}{9}$
73. $4\sqrt{15} - 8\sqrt{3}$

75.

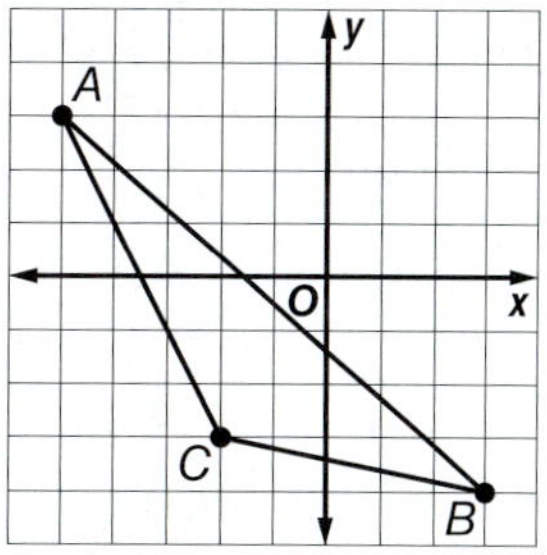

77. 425.2 g **79.** 1.25 kg
81. > **83.** < **85.** >

Pages 14–21 Lesson 1-2

1. 5.7 cm or 57 mm **3.** $1\frac{7}{8}$ in. **5.** 3.8 in.

7

$BD = 4x$	Given
$12 = 4x$	Substitution
$\frac{12}{4} = \frac{4x}{4}$	Divide each side by 4.
$3 = x$	Simplify.
$BC = 2x$	Given
$= 2(3)$	$x = 3$
$= 6$	Simplify.

9. $\overline{AG} \cong \overline{FG}$, $\overline{BG} \cong \overline{EG}$, $\overline{CG} \cong \overline{DG}$ **11.** 38 mm
13. $\frac{15}{16}$ in. **15.** 1.1 cm **17.** 1.5 in. **19.** 4.2 cm **21.** $c = 18$; $YZ = 72$ **23.** $a = 4$; $YZ = 20$ **25.** $n = 4\frac{1}{3}$; $YZ = 1\frac{2}{3}$
27 Yes; $KJ = 4$ in. and $HL = 4$ in. Since the segments have the same measure, they are congruent.
29. no **31.** yes
33 Sample answer: All the segments that have one slash are congruent: $\overline{AB} \cong \overline{BC} \cong \overline{CD} \cong \overline{DE} \cong \overline{DG} \cong \overline{BG} \cong \overline{CG}$, $\overline{AC} \cong \overline{EC}$. All segments with two slashes are congruent: $\overline{AH} \cong \overline{HG} \cong \overline{GF} \cong \overline{FE}$, $\overline{AG} \cong \overline{HF} \cong \overline{GE}$. All segments with three slashes are congruent: $\overline{BH} \cong \overline{DF}$.
35. Sample answer: $\overline{BD} \cong \overline{CE}$; $\overline{BD} \cong \overline{PQ}$; $\overline{YZ} \cong \overline{JK}$; $\overline{PQ} \cong \overline{RS}$; $\overline{GK} \cong \overline{KL}$ **37.** If point B is between points A and C, and you know AB and BC, add AB and BC to find AC. If you know AB and AC, subtract AB from

AC to find BC. **39.** $JK = 12$, $KL = 16$ **41.** Units of measure are used to differentiate between size and distance, as well as for precision. An advantage is that the standard of measure of a cubit is always available. A disadvantage is that a cubit would vary in length depending on whose arm was measured.
43. D **45.** D **47.** Sample answer: plane CDF
49. points C, B, and F **51a.** about 2.1 s
51b. about 9.7 in. **53.** $\{p|p > 9\}$ **55.** $\{x|x \leq -13\}$
57. 12 **59.** 5.5 **61.** $\sqrt{185}$

Pages 25–35 Lesson 1-3

1. 8

3 $AB = \sqrt{(x_2 - x_1)^2 + (y_2 - y_1)^2}$ Distance Formula

$= \sqrt{(2-4)^2 + (-3-9)^2}$ $(x_1, y_1) = (4, 9)$ and $(x_2, y_2) = (2, -3)$

$= \sqrt{(-2)^2 + (-12)^2}$ Subtract.

$= \sqrt{4 + 144}$ or $\sqrt{148}$ Simplify.

The distance between the time capsules is $\sqrt{148}$ or about 12.2 units.

5. $\sqrt{58}$ or about 7.6 units **7.** -3 **9.** $(4, -5.5)$

11 Let G be (x_1, y_1) and J be (x_2, y_2) in the Midpoint Formula.

$F\left(\frac{x_1 + 6}{2}, \frac{y_1 + (-2)}{2}\right) = F(1, 3.5)$ $(x_2, y_2) = (6, -2)$

Write two equations to find the coordinates of G.

$\frac{x_1 + 6}{2} = 1$ Midpoint Formula $\frac{y_1 + (-2)}{2} = 3.5$

$x_1 + 6 = 2$ Multiply each side by 2. $y_1 + (-2) = 7$

$x_1 = -4$ Simplify. $y_1 = 9$

The coordinates of G are $(-4, 9)$.

13. 5 **15.** 9 **17.** 12 **19.** $\sqrt{89}$ or about 9.4 units
21. $\sqrt{58}$ or about 7.6 units **23.** $\sqrt{208}$ or about 14.4 units **25.** $\sqrt{65}$ or about 8.1 units **27.** $\sqrt{53}$ or about 7.3 units **29.** $\sqrt{18}$ or about 4.2 units
31. 4.5 mi **33.** 6 **35.** -4.5 **37.** 3

39 $M\left(\frac{x_1 + x_2}{2}, \frac{y_1 + y_2}{2}\right)$ Midpoint Formula

$= M\left(\frac{22 + 15}{2}, \frac{4 + 7}{2}\right)$ $(x_1, y_1) = (22, 4)$ and $(x_2, y_2) = (15, 7)$

$= M\left(\frac{37}{2}, \frac{11}{2}\right)$ Simplify.

$= M(18.5, 5.5)$ Simplify.

The coordinates of the midpoint are $(18.5, 5.5)$.

41. $(-6.5, -3)$ **43.** $(-4.2, -10.4)$ **45.** $\left(-\frac{1}{2}, \frac{1}{2}\right)$ **47.** $A(1, 6)$
49. $C(16, -4)$ **51.** $C(-12, 13.25)$ **53.** 58 **55.** 4.5

57 **a.** If the center of the court is the origin, the player in the first is half of the length of the court or 47 feet to the right and half of the width of the court or 25 feet down. Since you go to the right, the x-coordinate is positive, and since you go down, the y-coordinate is negative. The ordered pair is $(47, -25)$. **b.** The distance that the ball travels is the distance between $(0, 0)$ and $(47, -25)$.

$d = \sqrt{(x_2 - x_1)^2 + (y_2 - y_1)^2}$ Distance Formula

$d = \sqrt{(47 - 0)^2 + (-25 - 0)^2}$ $(x_1, y_1) = (0, 0)$, $(x_2, y_2) = (47, -25)$

$d = \sqrt{2834}$ Simplify.

$d \approx 53.2$ Use a calculator.

The distance between the two players is about 53.2 feet.

59. =AVERAGE(B2:D2) **61.** $(-5, 0)$, $(7, 0)$
63. $\left(-1\frac{1}{2}, -1\right)$ **65.** ± 5

67 **a.** Sample answer:

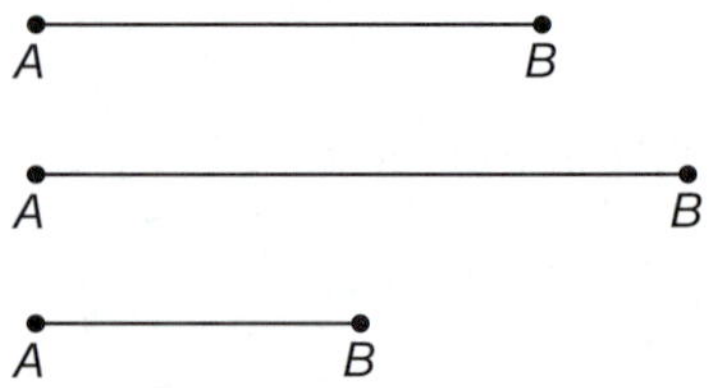

b. Sample answer:

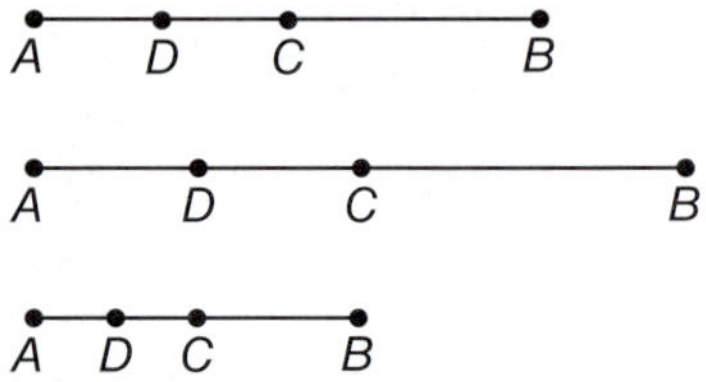

c. Sample answer:

line	AB (cm)	AC (cm)	AD (cm)
1	4	2	1
2	6	3	1.5
3	3	1.5	0.75

d. $AC = \frac{1}{2}AB$ Definition of midpoint

$AC = \frac{1}{2}x$ $AB = x$

$AD = \frac{1}{2}AC$ Definition of midpoint

$AD = \frac{1}{2}\left(\frac{1}{2}x\right)$ Substitution

$AD = \frac{1}{4}x$ Simplify.

e. Look for a pattern.

Number of Midpoints	Length of Smallest Segment
1	$\frac{1}{2}x$
2	$\frac{1}{2} \cdot \frac{1}{2}x = \frac{1}{2(2)}x$
3	$\frac{1}{2} \cdot \frac{1}{2(2)}x = \frac{1}{2(3)}x$
4	$\frac{1}{2} \cdot \frac{1}{2(3)}x = \frac{1}{2(4)}x$
n	$\frac{1}{2n}x$

Sample answer: If n midpoints are found, then the smallest segment will have a measure of $\frac{1}{2n}x$.

69. Sample answer: Sometimes; only when the segment lies on the x- or y-axis.

71. Sample answer:

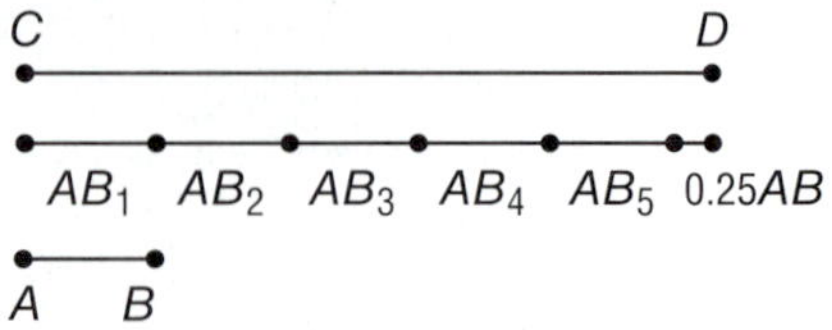

Draw $\overline{AB}$. Next, draw a construction line and place point C on it. From point C, strike 6 arcs in succession of length AB. On the sixth $\overline{AB}$ length, perform a segment bisector two times to create a $\frac{1}{4}AB$ length. Label the endpoint D. **73.** C **75.** B **77.** $2\frac{1}{8}$ in.

79.

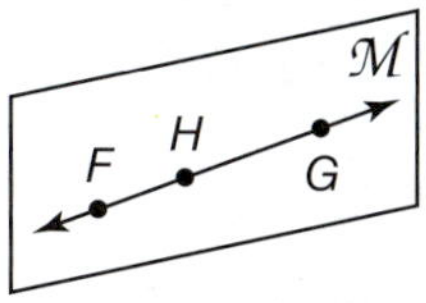

81. $4x + 38 \leq 75$; 9.25 lb or less **83.** 15.5 **85.** $2\frac{1}{3}$ **87.** 4

Pages 36–44 Lesson 1-4

1. U **3.** $\angle XYU$, $\angle UYX$ **5.** acute; 40 **7.** right; 90 **9.** 156 **11a.** 45; When joined together, the angles form a right angle, which measures 90. If the two angles that form this right angle are congruent, then the measure of each angle is $90 \div 2$ or 45. The angle of the cut is an acute angle. **11b.** The joint is the angle bisector of the frame angle. **13.** P **15.** M **17.** $\overrightarrow{NV}$, $\overrightarrow{NM}$ **19.** $\overrightarrow{RP}$, $\overrightarrow{RQ}$ **21.** $\angle TPQ$ **23.** $\angle TPN$, $\angle NPT$, $\angle TPM$, $\angle MPT$ **25.** $\angle 4$ **27.** S, Q

29 Sample answer: $\angle MPR$ and $\angle PRQ$ share points P and R.

31. 90, right **33.** 45, acute **35.** 135, obtuse

37

$m\angle ABE = m\angle EBF$	Definition of $\cong$ $\angle$s
$2n + 7 = 4n - 13$	Substitution
$7 = 2n - 13$	Subtract $2n$ from each side.
$20 = 2n$	Add 13 to each side.
$10 = n$	Divide each side by 2.

$m\angle ABE = 2n + 7$	Given
$= 2(10) + 7$	$n = 10$
$= 20 + 7$ or 27	Simplify.

39. 16 **41.** 47 **43a.** about 50 **43b.** about 140 **43c.** about 20 **43d.** 0

45. acute

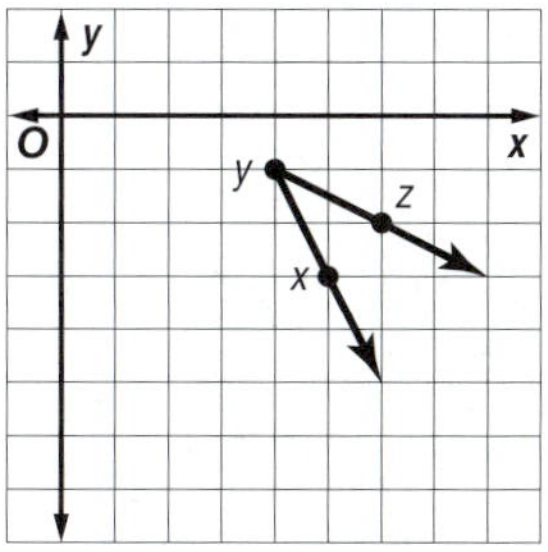

47 **a.** $m\angle 1 \approx 110$; since $110 > 90$, the angle is obtuse. **b.** $m\angle 2 \approx 85$; since $85 < 90$, the angle is acute. **c.** about 15; If the original path of the light is extended, the measure of the angle the original path makes with the refracted path represents the number of degrees the path of the light changed. The sum of the measure of this angle and the measure of $\angle 3$ is 180. The measure of $\angle 3$ is $360 - (110 + 85)$ or 165, so the measure of the angle the original path makes with the refracted path is $180 - 165$ or 15. **49.** The two angles formed are acute angles. Sample answer: With my compass at point A, I drew an arc in the interior of the angle. With the same compass setting, I drew an arc from point C that intersected the arc from point A. From the vertex, I drew $\overrightarrow{BD}$. I used the same compass setting to draw the intersecting arcs, so $\overrightarrow{BD}$ bisects $\angle ABC$ or the measurement of $\angle ABD$ and $\angle DBC$ are the same. Therefore, $\overrightarrow{BD}$ bisects $\angle ABC$. **51.** Sometimes; sample answer: For example, if you add an angle measure of 4 and an angle measure of 6, you will have an angle measure of 10, which is still acute. But if you add angles with measure of 50 and 60, you will have an obtuse angle with a measure of 110. **53.** Sample answer: To measure an acute angle, you can fold the corner of the paper so that the edges meet. This would bisect the angle, allowing you to determine whether the angle was between 0° and 45° or between 45° and 90°. If the paper is folded two more times in the same manner and cut off this corner of the paper, the fold lines would form the increments of a homemade protractor that starts at 0° on one side and progresses in $90 \div 8$ or 11.25° increments, ending at the adjacent side, which would indicate 90°. You can estimate halfway between each fold line, which would give you an accuracy of $11.25° \div 2$ or about 6°. The actual measure of the angle shown is 52°. An estimate between 46° and 58° would be acceptable. **55.** Sample answer: Leticia's survey does not represent the entire student body because she did not take a random sample; she only took a sample of students from one major.
57. D **59.** 8.25 **61.** 15.81 **63.** 10.07 **65.** $x = 11$; $ST = 22$ **67.** 4.5 **69.** 56 **71.** 14.75 **73.** 24.8 **75.** $17\frac{1}{3}$

Pages 46–54 Lesson 1-5

1. $\angle ZVY$, $\angle WVU$ **3a.** vertical **3b.** 15

5 If $x \perp y$, then $m\angle 2 = 90$ and $m\angle 3 = 90$.

$m\angle 2 = 3a - 27$	Given
$90 = 3a - 27$	Substitution
$117 = 3a$	Add 27 to each side.
$39 = a$	Divide each side by 3.

$m\angle 3 = 2b + 14$	Given
$90 = 2b + 14$	Substitution
$76 = 2b$	Subtract 14 from each side.
$38 = b$	Divide each side by 3.

7. Yes: they share a common side and vertex, so they are adjacent. Since $m\angle EDB + m\angle BDA + m\angle ADC = 90$, $\angle EDB$ and $\angle BDA$ cannot be complementary or supplementary. **9.** Sample answer: $\angle BFC$, $\angle DFE$ **11.** $\angle FDG$, $\angle GDE$ **13.** Sample answer: $\angle CBF$, $\angle ABF$ **15.** $\angle GDE$ **17.** $\angle CAE$ **19.** 65

21

$2x + 25 = 3x - 10$	Vertical $\angle$s are $\cong$ and have equal measures.
$25 = x - 10$	Subtract $2x$ from each side.
$35 = x$	Add 10 to each side.

$3x - 10 + y = 180$	Def. of supplementary $\angle$s
$3(35) - 10 + y = 180$	Substitution
$105 - 10 + y = 180$	Multiply.
$95 + y = 180$	Simplify.
$y = 85$	Subtract 95 from each side.

23. $x = 48$; $y = 21$ **25.** $m\angle F = 63$; $m\angle E = 117$ **27.** 40

29 If $\angle KNM$ is a right angle, then $m\angle KNM = 90$.

$m\angle KNL + m\angle LNM = m\angle KNM$	Sum of parts = whole
$6x - 4 + 4x + 24 = 90$	Substitution
$10x + 20 = 90$	Combine like terms.
$10x = 70$	Subtract 20 from each side.
$x = 7$	Divide each side by 10.

31. 92 **33.** 53; 37 **35.** $a = 8; b = 54$ **37.** Yes; the angles form a linear pair. **39.** No; the measures of each angle are unknown. **41.** No; the angles are not adjacent. **43.** Sample answer: $\angle 1$ and $\angle 3$

45 $\angle 1$ and $\angle 3$ are vertical angles, so they are congruent; $m\angle 3 = m\angle 1 = 110$. $\angle 1$ and $\angle 4$ are a linear pair, so they are supplementary.

$m\angle 4 + m\angle 1 = 180$	Def. of supplementary $\triangle$
$m\angle 4 + 110 = 180$	Substitution
$m\angle 4 = 70$	Subtract 110 from each side.

47. Sample answer: Yes; if the wings are not rotated at all, then all of the angles are right angles, which are neither acute nor obtuse. **49.** Yes; angles that are right or obtuse do not have complements because their measures are greater than or equal to 90. **51a.** Line a is perpendicular to plane $\mathcal{P}$. **51b.** Line m is in plane $\mathcal{P}$. **51c.** Any plane containing line a is perpendicular to plane $\mathcal{P}$. **53.** B **55.** J **57.** 125, obtuse **59.** 90, right **61.** $\left(-3\frac{1}{2}, 1\right)$ **63.** 81.5 cm **65.** $\overline{FG} \cong \overline{HJ} \cong \overline{JK} \cong \overline{FL}$, $\overline{GH} \cong \overline{LK}$; $\angle F \cong \angle J$, $\angle G \cong \angle H \cong \angle K \cong \angle L$ **67.** $\overline{WX} \cong \overline{XY} \cong \overline{YZ} \cong \overline{ZW}$; $\angle W \cong \angle Y$, $\angle X \cong \angle Z$

Pages 56–64 Lesson 1-6

1. pentagon; concave; irregular **3.** octagon; regular **5.** hexagon; irregular **7.** ≈ 40.2 cm; ≈ 128.7 cm^2 **9.** C **11.** triangle; convex; regular

13 The polygon has 8 sides, so it is an octagon. All of the lines containing the sides of the polygon will pass through the interior of the octagon, so it is concave. Since the polygon is not convex, it is irregular.

15. 11-gon; concave; irregular **17.** 7.8 m; ≈ 3.1 m^2 **19.** 26 in.; 42.3 in^2

21

$c^2 = a^2 + b^2$	Pythagorean Theorem
$c^2 = 6.5^2 + 4.5^2$	$a = 6.5, b = 4.5$
$c^2 = 62.5$	Simplify.
$c \approx 7.9$	Simplify.

$P = a + b + c$	Perimeter of a triangle
$\approx 6.5 + 4.5 + 7.9$	Substitution
≈ 18.9 cm	Simplify.

$A = \frac{1}{2}bh$	Area of a triangle
$= \frac{1}{2}(4.5)(6.5)$	Substitution
≈ 14.6 cm^2	Simplify.

23. ≈ 2.55 in.
25. triangle; $P = 5 + \sqrt{32} + \sqrt{17}$ or about 14.78 units; $A = 10$ units^2

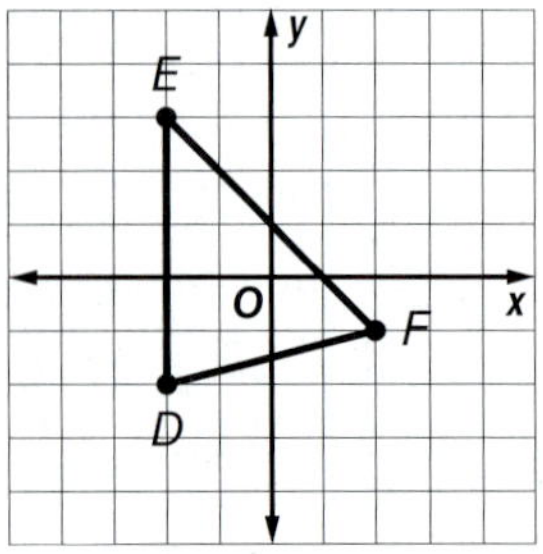

27. quadrilateral or square; $P = 20$ units; $A = 25$ units^2

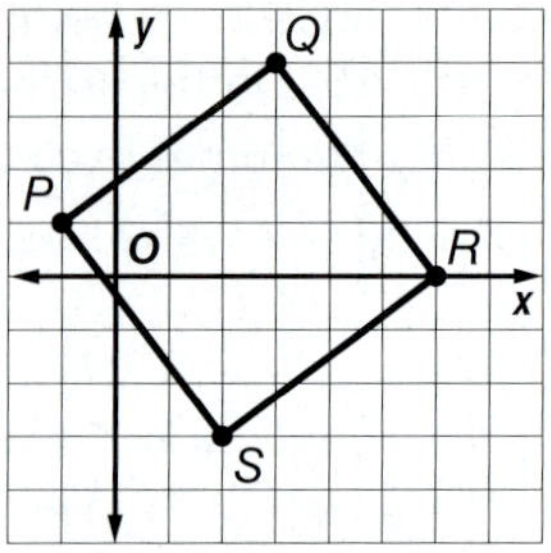

29a. 14 ft **29b.** 12 ft^2 **29c.** The perimeter doubles; the area quadruples. The perimeter of a rectangle with dimensions 6 ft and 8 ft is 28 ft, which is twice the perimeter of the original figure since $2 \cdot 14$ ft $= 28$ ft. The area of a rectangle with dimensions 6 ft and 8 ft is 48 ft^2, which is four times the area of the original figure, since $4 \cdot 12$ ft$^2 = 48$ ft^2. **29d.** The perimeter is halved; the area is divided by 4. The perimeter of a rectangle with dimensions 1.5 ft and 2 ft is 7 ft, which is half the perimeter of the original figure, since $\frac{1}{2} \cdot 14$ ft $=$ 7 ft. The area of a rectangle with dimensions 1.5 ft and 2 ft is 3 ft^2, which is $\frac{1}{4}$ the area of the original figure, since $\frac{1}{4} \cdot 12$ ft$^2 = 3$ ft^2. **31.** 60 yd, 6 yd

33 **a.**

$C = \pi d$	Circumference
$= \pi(8)$	$d = 8$
≈ 25.1	Simplify.

$C = \pi d$	Circumference
$= \pi(10)$	$d = 10$
≈ 31.4	Simplify.

minimum circumference: 25.1 in.; maximum circumference: 31.4 in.

b.

$A = \pi r^2$	Area of a circle
$= \pi(4)^2$	$r = 4$
≈ 50.3	Simplify.

$A = \pi r^2$	Circumference
$- \pi(5)^2$	$r = 5$
≈ 78.5	Simplify.

minimum area: 50.3 in^2; maximum area: 78.5 in^2

35. 21.2 m **37.** $2\pi\sqrt{32}$ or about 35.5 units **39.** $12\sqrt{6}$ or about 29.4 in. **41.** 12.5 cm **43a.** 128 ft **43b.** about 69% **45.** 290.93 units^2 **47.** Sample answer: The pentagon is convex, since no points of the lines drawn on the edges are in the interior. The pentagon is regular since all of the angles and sides were constructed with the same measurement, making them congruent to each other.

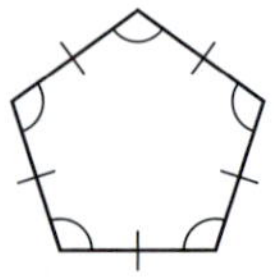

49. Sample answer: If a convex polygon is equiangular but not also equilateral, then it is not a regular polygon. Likewise, if a polygon is equiangular and equilateral, but not convex, then it is not a regular polygon. **51.** F **53.** B **55.** No; we do not know anything about these measures. **57.** Yes; they form a linear pair. **59.** elimination; $x = -3, y = -1$ **61.** substitution; $x = -4, y = -2.5$ **63.** 24 **65.** 169.6

Pages 67–74 Lesson 1-7

1. not a polyhedron; cylinder

3 $T = PH + 2B$ Surface area of a prism
$= (14)(3) + 2(12)$ $P = 14$ cm, $h = 3$ cm, $B = 12$ cm^2
$= 66$ cm^2 Simplify.

$V = BH$ Volume of a prism
$= (12)(3)$ $B = 12$ cm^2, $h = 3$ cm
$= 36$ cm^3 Simplify.

5a. ≈ 27.2 in^3 **5b.** 13.6π or about 42.7 in^2
7. rectangular pyramid; a polyhedron **9.** rectangular prism; a polyhedron **11.** cylinder; not a polyhedron **13.** not a polyhedron; cone **15.** not a polyhedron; sphere **17.** a polyhedron; pentagonal pyramid; base: *JHGFD*; faces: $\overline{JHGFD}$, $\triangle JEH$, $\triangle HEG$, $\triangle GEF$, $\triangle FED$, $\triangle EDJ$; edges: $\overline{HG}$, $\overline{GF}$, $\overline{FD}$, $\overline{DJ}$, $\overline{JH}$, $\overline{EJ}$, $\overline{EH}$, $\overline{EG}$, $\overline{EF}$, $\overline{ED}$; vertices: *J*, *H*, *G*, *F*, *D*, *E* **19.** 121.5 m^2; 91.1 m^3

21 $T = PH + 2B$ Surface area of a prism
$= (24)(5) + 2(24)$ $P = 24$ cm, $h = 5$ cm, $B = 24$ cm^2
$= 168$ cm^2 Simplify.

$V = BH$ Volume of a prism
$= (24)(5)$ $B = 24$ cm^2, $h = 5$ cm
$= 120$ cm^3 Simplify.

23. 150π or about 471.2 mm^2; 250π or about 785.4 mm^3

25 **a.** $V = \pi r^2 h$ Volume of a cylinder
$= \pi\left(7\frac{3}{4}\right)^2\left(11\frac{3}{4}\right)$ $r = 7\frac{3}{4}$ in., $h = 11\frac{3}{4}$ in.
≈ 2217.1 in^3 Simplify.

b. $T = 2\pi rh + 2\pi r^2$ Surface area of a cylinder
$= 2\pi\left(7\frac{3}{4}\right)\left(11\frac{3}{4}\right) + 2\pi\left(7\frac{3}{4}\right)^2$ $r = 7\frac{3}{4}$ in., $h = 11\frac{3}{4}$ in.
≈ 949.5 in^2 Simplify.

27. 3 in. **29.** 1212 in^2; 1776 in^3 **31a.** 96 in^2
31b. 113.1 in^2 **31c.** prism: 2 cans; cylinder: 3 cans
31d. 2.18 in.; if the height is 10 in., then the surface area of the rectangular cake is 152 in^2. To find the radius of a cylindrical cake with the same height, solve the equation $152 = \pi r^2 + 20\pi r$. The solutions are $r = -22.18$ or $r = 2.18$. Using a radius of 2.18 in. gives surface area of about 152 in^2.

33 $1 \text{ ft}^3 = (12 \text{ in.})^3 = 1728 \text{ in}^3$
$4320 \text{ in}^3 \cdot \frac{1 \text{ ft}^3}{1728 \text{ in}^3} = 2.5 \text{ ft}^3$

35. The volume of the original prism is 4752 cm^3. The volume of the new prism is 38,016 cm^3. The volume increased by a factor of 8 when each dimension was doubled. **37.** Neither; sample answer: the surface area is twice the sum of the areas of the top, front, and left side of the prism or $2(5 \cdot 3 + 5 \cdot 4 + 3 \cdot 4)$, which is 94 in^2. **39a.** cone **39b.** cylinder
41. 27 mm^3 **43.** 55.2 **45.** F **47.** quadrilateral; convex; regular **49.** dodecagon; concave; irregular
51. 10 **53.** The intersection of a plane and a line not in the plane is a point. **55.** Two lines intersect in one point.

57.

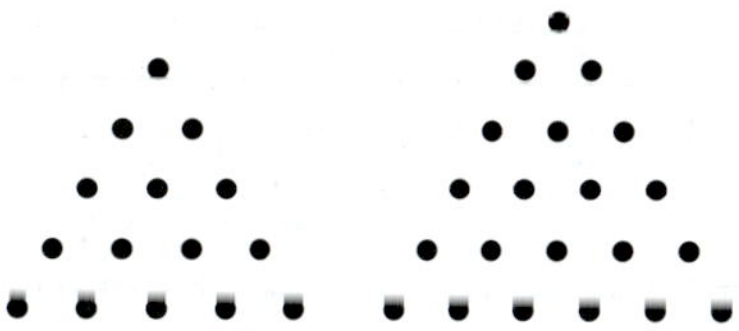

59.

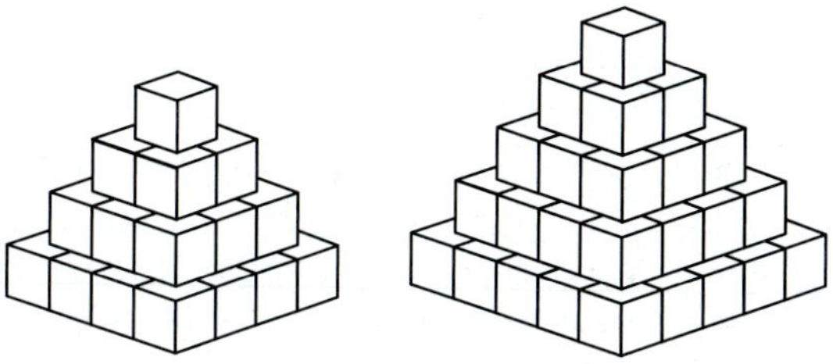

Pages 76–80 Chapter 1 Study Guide and Review

1. plane **3.** perpendicular **5.** point *P* **7.** point *W*
9. line **11.** $x = 6$, $XP = 27$ **13.** yes **15.** 1.5 mi **17.** 10
19. $(16, -6.5)$ **21.** $(-27, 16)$ **23.** *G* **25.** $\overrightarrow{CA}$ and $\overrightarrow{CH}$
27. Sample answer: $\angle A$ and $\angle B$ are right, $\angle E$ and $\angle C$ are obtuse, and $\angle D$ is acute. **29.** Sample answer: $\angle QWP$ and $\angle XWV$ **31.** 66 **33.** dodecagon, concave, irregular **35.** Option 1 = 12,000 ft^2, Option 2 = 12,100 ft^2, Option 3 $\approx$ 15,393.8 ft^2. Option 3 provides the greatest area. **37.** hexagonal prism. Bases: *ABCDEF* and *GHJKLM*; Faces: $\square ABHG$, $\square BCJH$, $\square CDKJ$, $\square DELK$, $\square EFML$, $\square FAGM$; Edges: $\overline{AB}$, $\overline{BC}$, $\overline{CD}$, $\overline{DE}$, $\overline{EF}$, $\overline{FA}$, $\overline{GH}$, $\overline{HJ}$, $\overline{JK}$, $\overline{KL}$, $\overline{LM}$, $\overline{MG}$, $\overline{AG}$, $\overline{BH}$, $\overline{CJ}$, $\overline{DK}$, $\overline{EL}$, $\overline{FM}$; Vertices: *A*, *B*, *C*, *D*, *E*, *F*, *G*, *H*, *J*, *K*, *L*, *M*
39. 384 in^2; 384 in^3 **41.** 72 m^2, 36 m^3 **43.** ≈ 23.6 in^2, ≈ 7.1 in^3

Chapter 2 Reasoning and Proof

Page 87 Chapter 2 Get Ready

1. 31 **3.** 14 **5.** 12 **7.** $x^2 + 3$ **9.** -7 **11.** 10.8
13. $4x = 52$; \$13 **15.** $\angle CXD$, $\angle DXE$ **17.** 38

Pages 89–96 Lesson 2-1

1. Each cost is \$2.25 more than the previous cost; \$11.25.
3. In each figure, the shading moves to the next point clockwise.

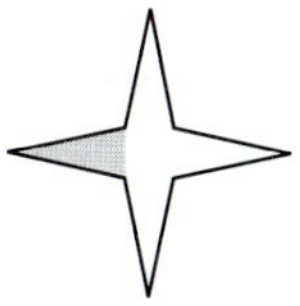

5 3, 3, 6, 9, 15

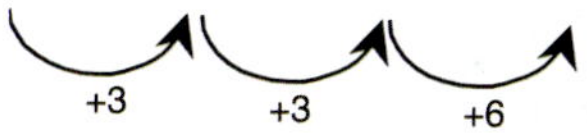

Beginning with the third element, each element in the pattern is the sum of the previous two elements. So, the next element will be 15 + 9 or 24.
7. The product of two even numbers is an even number.
9. The set of points in a plane equidistant from point *A* is a circle.

11a.

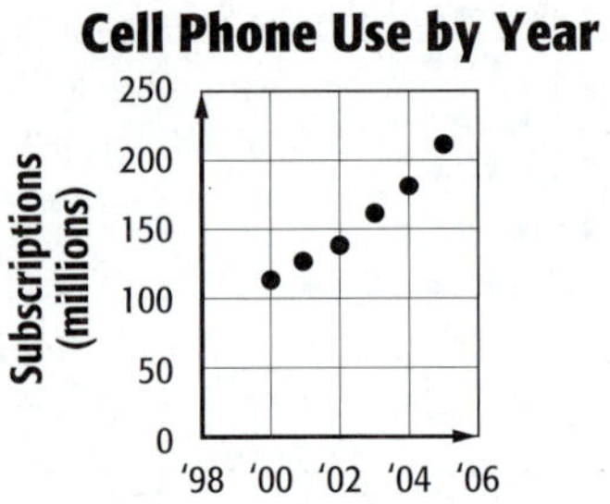

11b. Sample answer: About 392,000,000 Americans will use cell phones in 2010.

13

If a ray intersects a segment at its midpoint and forms adjacent angles that are not right angles, then the ray is not perpendicular to the segment. **15.** Each element in the pattern is three more than the previous element; 18.

17. Each element has an additional two as part of the number; 22222. **19.** Each element is one half the previous element; $\frac{1}{16}$. **21.** Each percentage is 7% less than the previous percentage; 79%. **23.** Each meeting is two months after the previous meeting; July.

25.

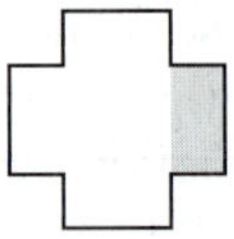

In each figure, the shading moves to the next area of the figure counter clockwise.

27.

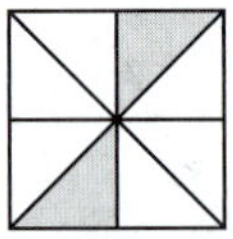

The shading of the lower triangle in the upper right quadrant of the first figure moves clockwise through each set of triangles from one figure to the next.

29. Sample answer: It is drier in the west and hotter in the south than other parts of the country, so less water would be readily available.

31 First, list examples: $1 \cdot 3 = 3$, $3 \cdot 5 = 15$, $7 \cdot 9 = 63$, $11 \cdot 11 = 121$. All the products are odd numbers. So, a conjecture about the product of two odd numbers is that the product is an odd number.

33. They are equal. **35.** The points equidistant from A and B form the perpendicular bisector of $\overline{AB}$. **37.** The area of the rectangle is two times the area of the square.

39a.

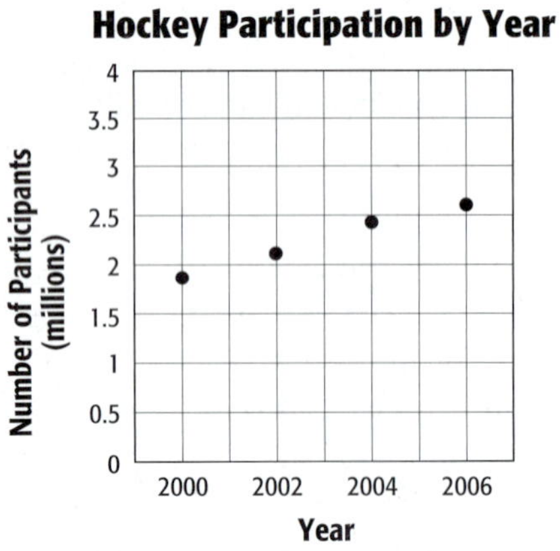

39b. Sample answer: More people over the age 7 will play hockey in the future. The number of people playing hockey increases each year, so the graph suggests that even more people will play hockey in subsequent years. **41.** False; sample answer: Suppose $x = 2$, then $-x = -2$.

43. False; sample answer:

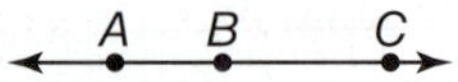

45. False; sample answer: The length could be 4 m and the width could be 5 m.

47 **a.** For each even number from 10 to 20, write the number as the sum of two primes. Sample answer: $10 = 5 + 5$, $12 = 5 + 7$, $14 = 7 + 7$, $16 = 5 + 11$, $18 = 7 + 11$, $20 = 7 + 13$ **b.** The number 3 can be written as $0 + 3$ and as $1 + 2$. Since neither 0 nor 1 is a prime number, 3 cannot be written as the sum of two primes. So, the conjecture is false.

49a.

Age (y)	Height (in.)
3	33
5	42
7	49
9	53

49b.

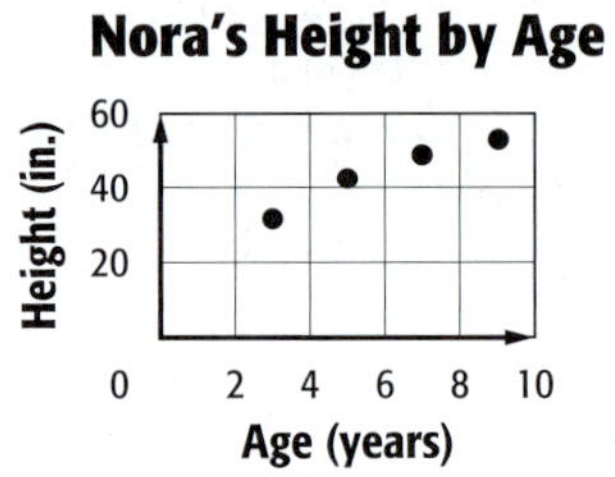

49c. Sample answer: 61 in. **49d.** Sample answer: No; Nora's growth will start to slow down and eventually stop. **51.** Jack; 2 is an even prime number.

53. Sample answer: False; if the two points create a straight angle that includes the third point, then the conjecture is true. If the two points do not create a straight angle with the third point, then the conjecture is false. **55.** B **57.** G **59.** 132 m^2; 60 m^3 **61.** 54 cm^2; 27 cm^3 **63.** 26.69 **65.** plane **67.** 18 **69.** 5, 6

Pages 97–104 Lesson 2-2

1. A week has seven days, and there are 60 minutes in an hour. p and r is true, because p is true and r is true.

3 $q \vee r$: There are 20 hours in a day, or there are 60 minutes in an hour. A disjunction is true if at least one of the statements is true. So, $q \vee r$ is true because r is true. It does not matter that q is false.

5. A week has seven days, or there are 60 minutes in an hour. $p \vee r$ is true, because p is true and r is true.

7.

p	q	$\sim q$	$p \vee \sim q$
T	T	F	T
T	F	T	T
F	T	F	F
F	F	T	T

9.

p	q	$\sim p$	$\sim q$	$\sim p \vee \sim q$
T	T	F	F	F
T	F	F	T	T
F	T	T	F	T
F	F	T	T	T

11. $\overrightarrow{DB}$ is the angle bisector of $\angle ADC$, and $\overline{AD} \cong \overline{DC}$. p and r is true because p is true and r is true.

13. $\overline{AD} \cong \overline{DC}$ or $\overrightarrow{DB}$ is not the angle bisector of $\angle ADC$. r or $\sim p$ is true because r is true and $\sim p$ is false.

15. $\overrightarrow{DB}$ is not the angle bisector of $\angle ADC$, or $\overline{AD} \not\cong \overline{DC}$. $\sim p$ or $\sim r$ is false because $\sim p$ is false and $\sim r$ is false. **17.** Springfield is the capital of Illinois, and Illinois shares a border with Kentucky. $p \wedge r$ is true because p is true and r is true. **19.** Illinois does not share a border with Kentucky, or Illinois is to the west of Missouri. $\sim r \vee s$ is false because $\sim r$ is false and s is false. **21.** Springfield is not the capital of Illinois, and Illinois does not share a border with Kentucky. $\sim p \wedge \sim r$ is false because $\sim p$ is false and $\sim r$ is false.

23.

p	q	$\sim p$	$\sim p \wedge q$
T	T	F	F
T	F	F	F
F	T	T	T
F	F	T	F

25.

p	r	$p \wedge r$
T	T	T
T	F	F
F	T	F
F	F	F

27.

p	r	$p \vee r$
T	T	T
T	F	T
F	T	T
F	F	F

29.

p	r	$\sim p$	$\sim p \wedge r$
T	T	F	F
T	F	F	F
F	T	T	T
F	F	T	F

31 **a.** The students who dive are represented by the intersection of the two sets and the nonintersecting portion of the Dive region. So, there are 3 + 4 or 7 students who dive. **b.** The students who participate in swimming or diving or both are represented by the union of the sets. There are 19 + 3 + 4 or 26 students who swim, dive, or do both. **c.** The students who swim and dive are represented by the intersection of the two sets. There are 3 students who both swim and dive.

33a. 50 **33b.** 40 **33c.** 110 **33d.** 20 **33e.** These teens do not use any of the listed electronics.

35 Make columns with the headings p, q, $\sim q$, r, $\sim q \vee r$, and $p \wedge (\sim q \vee r)$. List the possible combinations of truth values for p, q, and r. Use the truth values of q to find the truth values of $\sim q$. Use the truth values for each part of $\sim q \vee r$ to find the truth value of the compound statement. Then use the truth values for each part of $p \wedge (\sim q \vee r)$ to find the truth value of the compound statement.

p	q	$\sim q$	r	$\sim q \vee r$	$p \wedge (\sim q \vee r)$
T	T	F	T	T	T
T	F	T	T	T	T
T	T	F	F	F	F
T	F	T	F	T	T
F	T	F	T	T	F
F	F	T	T	T	F
F	T	F	F	F	F
F	F	T	F	T	F

If p and r are true, and q is true or false, then $p \wedge (\sim q \vee r)$ is true.

37.

p	q	$\sim q$	r	$\sim r$	$\sim q \wedge \sim r$	$p \vee (\sim q \wedge \sim r)$
T	T	F	T	F	F	T
T	F	T	T	F	F	T
T	T	F	F	T	F	T
T	F	T	F	T	T	T
F	T	F	T	F	F	F
F	F	T	T	F	F	F
F	T	F	F	T	F	F
F	F	T	F	T	T	T

true

39.

p	$\sim p$	q	r	$\sim r$	$(\sim p \vee q)$	$(\sim p \vee q) \vee \sim r$
T	F	T	T	F	T	T
T	F	F	T	F	F	F
T	F	T	F	T	T	T
T	F	F	F	T	F	T
F	T	T	T	F	T	T
F	T	F	T	F	T	T
F	T	T	F	T	T	T
F	T	F	F	T	T	T

If r is true or false, then $(\sim p \vee q) \vee \sim r$ is true.

41. Never; integers are rational numbers, not irrational. **43.** There exists at least one square that is not a rectangle. **45.** No students have classes in C-wing. **47.** Every segment has a midpoint. **49.** Sample answer: A triangle has three sides, and a square has four sides. Both are true, so the compound statement is true. **51.** 22 in^2; The area of a triangle is $\frac{1}{2}bh$. The base of the triangle is 11 inches and the height of the triangle is 4 inches, so the area of the triangle is $\frac{1}{2}(11)(4)$ or 22 in^2. **53.** A **55.** triangular prism; bases: $\triangle MNO$, $\triangle PQR$; faces: $\triangle MNO$, $\triangle PQR$, $OMPR$, $ONQR$, $PQNM$; edges: $\overline{MN}$, $\overline{NO}$, $\overline{OM}$, $\overline{PQ}$, $\overline{QR}$, $\overline{PR}$, $\overline{NQ}$, $\overline{MP}$, $\overline{OR}$; vertices: M, N, O, P, Q, and R **57.** triangular pyramid; base: $\triangle HJK$; faces: $\triangle HJK$, $\triangle HLK$, $\triangle KLJ$, $\triangle HLJ$; edges: $\overline{HK}$, $\overline{KJ}$, $\overline{HJ}$, $\overline{HL}$, $\overline{KL}$, $\overline{JL}$; vertices: H, K, J, and L **59.** −1 **61.** −7 **63.** 25 **65.** 14 **67.** 10

Pages 105–113 Lesson 2-3

1. H: today is Friday; C: tomorrow is Saturday
3. H: two angles are supplementary; C: the sum of the measures of the angles is 180

5 **hypothesis:** You are sixteen years old.
conclusion: You are eligible to drive.
statement in if-then form: If you are sixteen years old, then you are eligible to drive.

7. If the angle is acute, then its measure is between 0 and 90. **9a.** If moisture in the air condenses and falls, then it rains. **9b.** If a cumulonimbus cloud has supercooled moisture, then hail forms. **9c.** If the temperature is freezing in all or most of the atmosphere, then precipitation falls as snow.

Selected Answers and Solutions

11. False; Charlotte, Michigan; The hypothesis of the conditional is true, but the conclusion is false. The counterexample shows that the conditional statement is false. **13.** False; the animal could be a leopard. The hypothesis of the conditional is true, but the conclusion is false. This counterexample shows that the conditional statement is false. **15.** True; the hypothesis is false, since pigs cannot fly. A conditional with a false hypothesis is always true, so this conditional statement is true. **17.** If a number is a whole number, then it is an integer. Converse: If a number is an integer, then it is a whole number. False; Sample answer: −3. Inverse: If a number is not a whole number, then it is not an integer. False: Sample answer: −3. Contrapositive: If a number is not an integer, then it is not a whole number; true. **19.** H: you lead; C: I will follow **21.** H: two angles are vertical; C: they are congruent **23.** H: there is no struggle; C: there is no progress **25.** H: a convex polygon has five sides; C: it is a pentagon **27.** If you were at the party, then you received a gift. **29.** If a figure is a circle, then the area is πr^2. **31.** If an angle is right, then the angle measures 90 degrees. **33.** If the museum is the Andy Warhol Museum, then most of the collection is Andy Warhol's artwork.

35 To show that a conditional is false, you need only to find one counterexample. 9 is an odd number, but not divisible by 5. The hypothesis of the conditional is true, but the conclusion is false. So, this counterexample shows that the conditional statement is false.

37. False; the angle drawn is an acute angle whose measure is not 45. The hypothesis of the conditional is true, but the conclusion is false. This counterexample shows that the conditional statement is false.

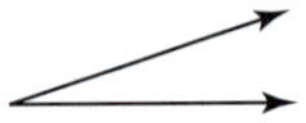

39. True; when this hypothesis is true, the conclusion is also true, since an angle and its complement's sum is 90. So, the conditional statement is true. **41.** True; the hypothesis is false, since red and blue paint make purple paint. A conditional with a false hypothesis is always true, so this conditional statement is true. **43.** False; the animal could be a falcon. The hypothesis of the conditional is true, but the conclusion is false. This counterexample shows that the conditional statement is false.
45. False; these lines intersect, but do not form right angles. The hypothesis of the conditional is true, but the conclusion is false. This counterexample shows that the conditional statement is false.

47. Converse: If you live in Illinois, then you live in Chicago. False: You can live in Springfield. Inverse: If you do not live in Chicago, then you do not live in Illinois. False: You can live in Springfield. Contrapositive: If you do not live in Illinois, then you do not live in Chicago; true. **49.** Converse: If two angles are congruent, then they have the same measure; true. Inverse: If two angles do not have the same measure, then the angles are not congruent; true. Contrapositive: If two angles are not congruent, then they do not have the same measure; true. **51.** If segments are congruent, then they have the same length. Converse: If segments have the same length, then they are congruent; true. Inverse: If segments are not congruent, then they do not have the same length; true. Contrapositive: If segments do not have the same length, then they are not congruent; true. **53.** If an animal has stripes, then it is a zebra; false: a zebra duiker has stripes.

55 The inverse is formed by negating both the hypothesis and the conclusion of the conditional. Inverse: If an animal does not have stripes, then it is not a zebra. This is a true statement.

57a. Sample answer: If a compound is an acid, it contains hydrogen. If a compound is a base, it contains hydroxide. If a compound is a hydrocarbon, it contains only hydrogen and carbon. **57b.** Sample answer: If a compound contains hydrogen, it is an acid. False; a hydrocarbon contains hydrogen. If a compound contains hydroxide, it is a base; true. If a compound contains only hydrogen and carbon, it is a hydrocarbon; true.

59 The blue area of the Venn diagram includes nonlinear functions but not quadratic functions. So, if a function is nonlinear, it may or may not be a quadratic function. Therefore, the conditional is false.

61. True; the deciduous area and the evergreen area have no common areas, so a deciduous tree cannot be evergreen. **63.** Sample answer: Kiri; when the hypothesis of a conditional is false, the conditional is always true. **65.** True; since the conclusion is false, the converse of the statement must be true. The converse and inverse are logically equivalent, so the inverse is also true. **67.** The hypothesis q of the inverse statement is *I received a detention*. The conclusion p of the inverse statement is *I did not arrive at school on time*. So the conditional A is $p \rightarrow q$: If I did not arrive at school on time, then I recieved a detention. So the converse of statement A is $\sim p \rightarrow \sim q$: If I did arrive at school on time, then I did not receive a detention. The contrapositve of Statement A is $\sim q \rightarrow \sim p$: If I did not receive a detention, then I arrived at school on time. **69.** A **71.** 0.00462

73.

p	q	p and q
T	T	T
T	F	F
F	T	F
F	F	F

75.

p	q	$\sim p$	$\sim p \wedge q$
T	T	F	F
T	F	F	F
F	T	T	T
F	F	T	F

77. H, J, and K are noncollinear.

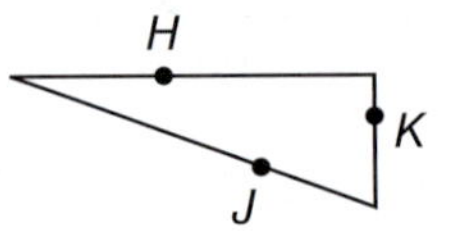

79. R, S, and T are collinear. **81.** $\overline{BC} \cong \overline{CD}$, $\overline{BE} \cong \overline{ED}$, $\overline{BA} \cong \overline{DA}$ **83.** about 9000 kg **85.** Divide each side by 8. **87.** Multiply each side by 3.

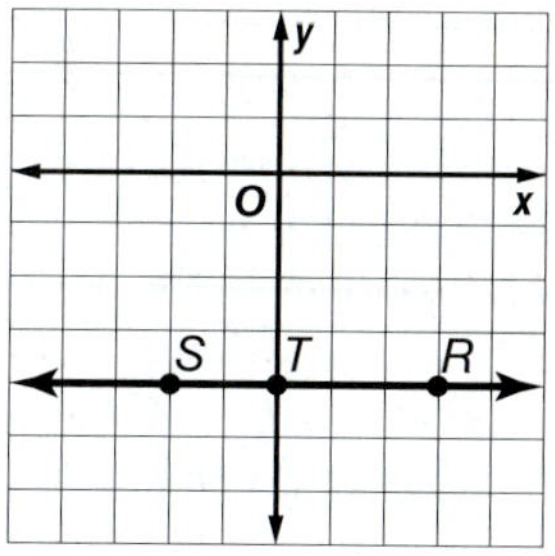

Pages 115–123 Lesson 2-4

1 Olivia is basing her conclusion on facts provided to her by her high school, not on a pattern of observations, so she is using deductive reasoning.
3. valid; Law of Detachment **5.** Invalid: Bayview could be inside or outside the public beach's circle.

Beaches

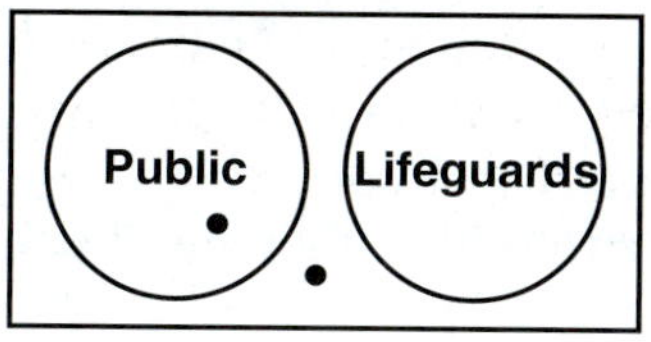

7. C **9.** No valid conclusion; ∠1 and ∠2 do not have to be vertical in order to be congruent. **11.** inductive reasoning **13.** deductive reasoning **15.** inductive reasoning

17 The given statement *Figure ABCD has four right angles* satisfies the conclusion of the true conditional. However, having a true conditional and a true conclusion does not make the hypothesis true. The figure could be a rectangle. So, the conclusion is invalid.
19. Invalid; your battery could be dead because it was old. **21.** valid; Law of Detachment
23. Valid; Monday is outside of the days when the temperature drops below 32°F, so it cannot be inside the days when it snows circle either, so the conclusion is valid.

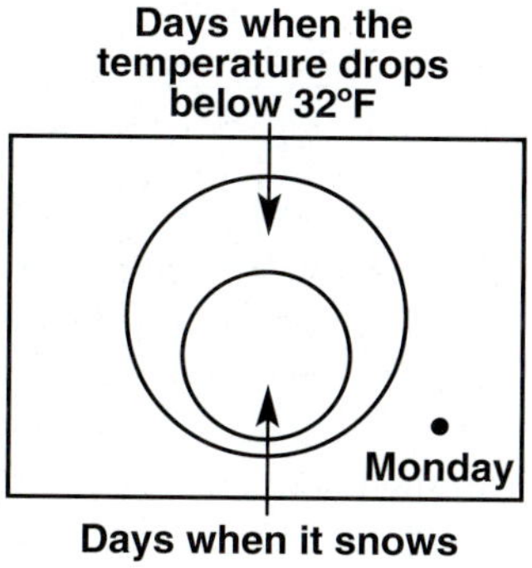

25. Invalid; Sabrina could be inside just the nurses' circle or inside the intersection of the circles, so the conclusion is invalid.

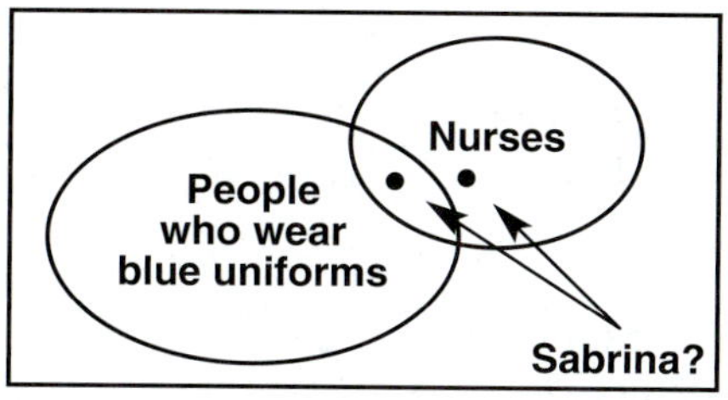

27 The given statement *Ms. Rodriguez has just purchased a vehicle that has four-wheel drive* satisfies the conclusion of the true conditional. However, having a true conditional and a true conclusion does not make the hypothesis true. Ms. Rodriguez's car might be in the Four-wheel-drive section of the diagram that is not a sport-utility vehicle. So, the conclusion is invalid.

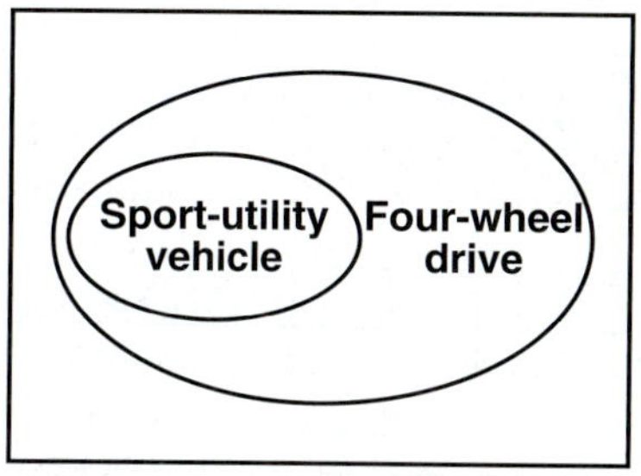

29. no valid conclusion **31.** no valid conclusion **33.** If two lines are not parallel, then they intersect in a point. **35.** Figure *ABCD* has all sides congruent; Law of Detachment.

37 You can reword the first given statement: If you are a ballet dancer, then you like classical music. Statement (1): If you are a ballet dancer, then you like classical music. Statement (2): If you like classical music, then you enjoy the opera. Since the conclusion of Statement (1) is the hypothesis of Statement (2), you can apply the Law of Syllogism. A valid conclusion: If you are a ballet dancer, then you enjoy the opera.
39. No valid conclusion; knowing a conclusion is true, does not imply the hypothesis will be true.

41a. **Florida Marlins Hits vs. At Bats**

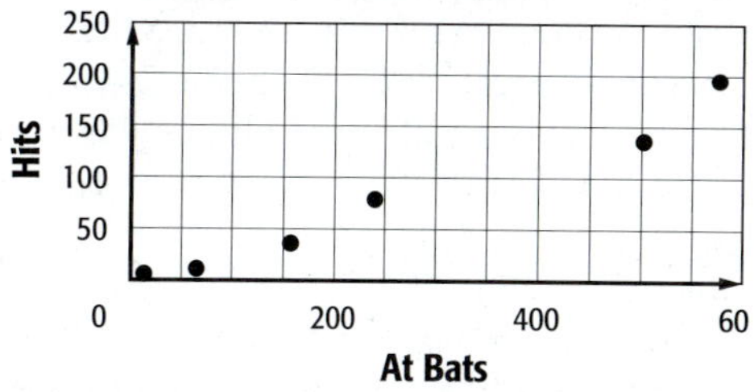

41b. Sample answer: About 91; inductive; a pattern was used to reach the conclusion. **41c.** Sample answer: The player with 240 at bats got more hits; deductive; the facts provided in the table were used to reach the conclusion. **43.** Law of Detachment: $[(p \rightarrow q) \wedge p] \rightarrow q$; Law of Syllogism: $[(p \rightarrow q) \wedge (q \rightarrow r)] \rightarrow (p \rightarrow r)$
45. Jonah's statement can be restated as, "Jonah is in group B and Janeka is in group B." In order for this compound statement to be true, both parts of the statement must be true. If Jonah was in group A, he would not be able to say that he is in group B, since students in group A must always tell the truth. Therefore the statement that Jonah is in group B is true. For the compound statement to be false, the statement that Janeka is in group B must be false. Therefore, Jonah is in group B and Janeka is in group A. **47.** D **49.** $\frac{26}{11}$
51a. If you live in Hawaii or Arizona then you do not observe Daylight Savings Time.

51b. If you do not observe Daylight Savings Time, then you live in Hawaii or Arizona; true.

53.

p	$\sim p$	q	$\sim q$	$\sim p$ or $\sim q$
T	F	T	F	F
T	F	F	T	T
F	T	T	F	T
F	T	F	T	T

55.

y	$\sim y$	z	$\sim y$ or z
T	F	T	T
T	F	F	F
F	T	T	T
F	T	F	T

57. 18 **59.** Yes; the symbol denotes that $\angle DAB$ is a right angle. **61.** Yes; the sum of their measures is $m\angle ADC$, which is 90. **63.** No; we do not know $m\angle ABC$.

Pages 125–132 Lesson 2-5

1. The left side and front side have a common edge line r. Planes $\mathcal{P}$ and $\mathcal{Q}$ only intersect along line r. Postulate 2.7, which states if two planes intersect, then their intersection is a line. **3.** The front bottom edge of the figure is line n which contains points D, C, and E. Postulate 2.3, which states a line contains at least two points. **5.** Points D and E, which are on line n, lie in plane $\mathcal{Q}$. Postulate 2.5, which states that if two points lie in a plane, then the entire line containing those points lies in that plane.

7 Postulate 2.7 states that if two planes intersect, then their intersection is a line. However, if three planes intersect, then their intersection may be a line or a point. So, the statement is sometimes true.

9. Always; Postulate 2.1 states through any two points, there is exactly one line. **11.** Postulate 2.3; a line contains at least two points. **13.** Postulate 2.4; a plane contains at least three noncollinear points. **15.** Since C is the midpoint of $\overline{AE}$ and $\overline{DB}$, $CA = CE = \frac{1}{2}AE$ and $CD = CB = \frac{1}{2}DB$ by the definition of midpoint. We are given $\overline{AE} \cong \overline{DB}$, so $AE = DB$ by the definition of congruent segments. By the multiplication property, $\frac{1}{2}DB = \frac{1}{2}AE$. So, by substitution, $AC = CB$. **17.** The edges of the sides of the bottom layer of the cake intersect. Plane $\mathcal{P}$ and $\mathcal{Q}$ of this cake intersect only once in line m. Postulate 2.7; if two planes intersect, then their intersection is a line. **19.** The top edge of the bottom layer of the cake is a straight line n. Points C, D, and K lie along this edge, so they lie along line n. Postulate 2.3; a line contains at least two points.

21 The bottom right part of the cake is a side. The side contains points K, E, F, and G and forms a plane. Postulate 2.2, which states that through any three noncollinear points, there is exactly one plane, shows that this is true.

23. The top edges of the bottom layer form intersecting lines. Lines h and g of this cake intersect only once at point J. Postulate 2.6; if two lines intersect, then their intersection is exactly one point. **25.** Never; Postulate 2.1 states through any two points, there is exactly one line. **27.** Always; Postulate 2.5 states if two points lie in a plane, then the entire line containing those points lies in that plane. **29.** Sometimes; the points must be noncollinear.

31. Given: L is the midpoint of $\overline{JK}$. $\overline{JK}$ intersects $\overline{MK}$ at K. $\overline{MK} \cong \overline{JL}$

Prove: $\overline{LK} \cong \overline{MK}$

Proof: We are given that L is the midpoint of $\overline{JK}$ and $\overline{MK} \cong \overline{JL}$. By the Midpoint Theorem, $\overline{JL} \cong \overline{LK}$. By the Transitive Property of Equality, $\overline{LK} \cong \overline{MK}$.

33a. Southside Blvd.; sample answer: Since there is a line between any two points, and Southside Blvd. is the line between point A and point B, it is the shortest route between the two. **33b.** I-295

35 Points E, F, and G lie along the same line. Postulate 2.3 states that a line contains at least two points.

37. Postulate 2.1; through any two points, there is exactly one line. **39.** Postulate 2.4; a plane contains at least three noncollinear points. **41.** Postulate 2.7; if two planes intersect, then their intersection is a line.

43a.

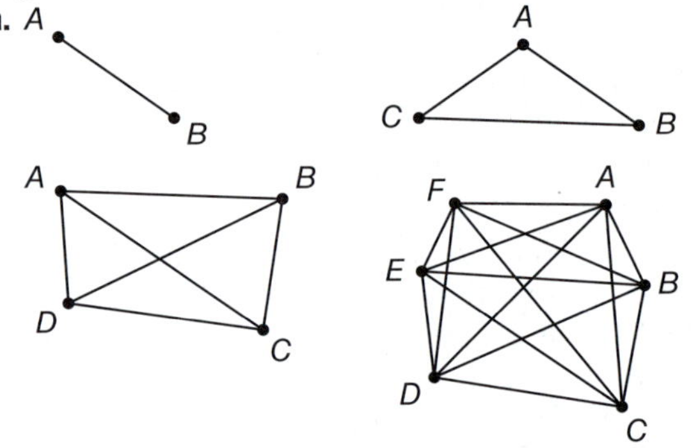

43b.

Number of Computers	Number of Connections
2	1
3	3
4	6
5	10
6	15

43c. $n - 1$

43d. $\frac{n(n-1)}{2}$

45. Lisa is correct. Sample answer: The proof should begin with the given, which is that $\overline{AB}$ is congruent to $\overline{BD}$ and A, B, and D are collinear. Therefore, Lisa began the proof correctly. **47a.** Plane $\mathcal{Q}$ is perpendicular to plane $\mathcal{P}$. **47b.** Line a is perpendicular to plane $\mathcal{P}$. **49.** Sometimes; three coplanar lines may have 0, 1, 2, or 3 points of intersection, as shown in the figures below.

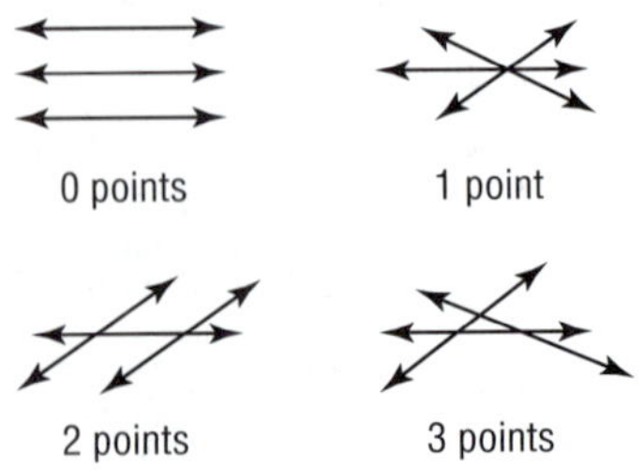

Postulates 2.1–2.5 were used. Through points A and B there is exactly one line, n, satisfying Postulate 2.1. For the noncollinear points A, B, and C there is exactly one plane, $\mathcal{P}$, satisfying Postulate 2.2. Line n contains points A and B,

satisfying Postulate 2.3. Plane $\mathcal{P}$ contains the noncollinear points A, B, and C, satisfying Postulate 2.4. Line n, containing points A and B, lies entirely in plane $\mathcal{P}$, satisfying Postulate 2.5. **51.** A **53.** H **55.** no conclusion **57.** If people are happy, then they rarely correct their faults. **59.** True; M is on $\overline{AB}$ and $AM + MB = AB$, so $p \wedge q$ is true. **61.** 19 m; 20 m of edging **63.** 5.5 **65.** 2, −2

Pages 134–141 Lesson 2-6

1. Trans. Prop. **3.** Sym. Prop.

5. Given: $\frac{y+2}{3} = 3$

Prove: $y = 7$

Proof:

Statements (Reasons)

a. $\frac{y+2}{3} = 3$ (Given)

b. $3\left(\frac{y+2}{3}\right) = 3(3)$ (Mult. Prop.)

c. $y + 2 = 9$ (Subs.)

d. $y = 7$ (Subt. Prop.)

7. Given: $\overline{AB} \cong \overline{CD}$

Prove: $x = 7$

Proof:

Statements (Reasons)

1. $\overline{AB} \cong \overline{CD}$ (Given)

2. $AB = CD$ (Def. of congruent segments)

3. $4x - 6 = 22$ (Subs. Prop.)

4. $4x = 28$ (Add. Prop.)

5. $x = 7$ (Div. Prop.)

9. Subt. Prop.

11

$4x - 5 = x + 12$	Original equation
$4x - 5 + 5 = x + 12 + 5$	Addition Property of Equality
$4x = x + 17$	Simplify.

If $4x - 5 = x + 12$, then $4x = x + 17$ by the Addition Property of Equality.

13. Dist. Prop. **15.** Trans. Prop.

17. Given: $\frac{8-3x}{4} = 32$

Prove: $x = -40$

Proof:

Statements (Reasons)

a. $\frac{8-3x}{4} = 32$ (Given)

b. $4\left(\frac{8-3x}{4}\right) = 4(32)$ (Mult. Prop.)

c. $8 - 3x = 128$ (Subs.)

d. $-3x = 120$ (Subt. Prop.)

e. $x = -40$ (Div. Prop.)

19. Given: $-\frac{1}{3}n = 12$

Prove: $n = -36$

Proof:

Statements (Reasons)

1. $-\frac{1}{3}n = 12$ (Given)

2. $-3\left(-\frac{1}{3}n\right) = -3(12)$ (Mult. Prop.)

3. $n = -36$ (Subs.)

21 a. Use properties of equality to justify each step in solving the equation for a.

Given: $d = vt + \frac{1}{2}at^2$

Prove: $a = \frac{2d - 2vt}{t^2}$

Proof:

Statements (Reasons)

1. $d = vt + \frac{1}{2}at^2$ (Given)

2. $d - vt = vt - vt + \frac{1}{2}at^2$ (Subtraction Property)

3. $d - vt = \frac{1}{2}at^2$ (Substitution)

4. $2(d - vt) = 2\left(\frac{1}{2}at^2\right)$ (Multiplication Property)

5. $2(d - vt) = at^2$ (Substitution)

6. $2d - 2vt = at^2$ (Distributive Property)

7. $\frac{2d - 2vt}{t^2} = \frac{at^2}{t^2}$ (Division Property)

8. $\frac{2d - 2vt}{t^2} = a$ (Substitution)

9. $a = \frac{2d - 2vt}{t^2}$ (Symmetric Property)

b. $a = \frac{2d - 2vt}{t^2}$	Given
$= \frac{2(2850) - 2(50)(30)}{30^2}$	$d = 2850, t = 30, v = 50$
$= 3$	Simplify.

The acceleration of the object is 3 ft/s^2. The Substitution Property of Equality justifies this calculation.

23. Given: $\overline{DF} \cong \overline{EG}$

Prove: $x = 10$

Proof:

Statements (Reasons)

1. $\overline{DF} \cong \overline{EG}$ (Given)

2. $DF = EG$ (Def. of $\cong$ segs)

3. $11 = 2x - 9$ (Subs.)

4. $20 = 2x$ (Add. Prop.)

5. $10 = x$ (Div. Prop.)

6. $x = 10$ (Symm. Prop.)

25 Use properties of equality and the definition of congruent angles to justify each step in proving $x = 100$.

Given: $\angle Y \cong \angle Z$

Prove: $x = 100$

Proof:

Statements (Reasons)

1. $\angle Y \cong \angle Z$ (Given)

2. $m\angle Y = m\angle Z$ (Definition of $\cong$ ⦞s)

3. $x + 10 = 2x - 90$ (Substitution)

4. $10 = x - 90$ (Subtraction Property)

5. $100 = x$ (Addition Property)

6. $x = 100$ (Symmetric Property)

27a. Given: $V = \frac{P}{I}$

Prove: $\frac{V}{2} = \frac{P}{2I}$

Proof:

Statements (Reasons)

1. $V = \frac{P}{I}$ (Given)

Selected Answers and Solutions

2. $\frac{1}{2} \cdot V = \frac{1}{2} \cdot \frac{P}{I}$ (Mult. Prop.)
3. $\frac{V}{2} = \frac{P}{2I}$ (Mult. Prop.)

27b. Given: $V = \frac{P}{I}$
Prove: $2V = \frac{2P}{I}$
Proof:
Statements (Reasons)
1. $V = \frac{P}{I}$ (Given)
2. $2 \cdot V = 2 \cdot \frac{P}{I}$ (Mult. Prop)
3. $2V = \frac{2P}{I}$ (Mult. Prop.)

29. Given: $c^2 = a^2 + b^2$
Prove: $a = \sqrt{c^2 - b^2}$
Proof:
Statements (Reasons)
1. $a^2 + b^2 = c^2$ (Given)
2. $a^2 + b^2 - b^2 = c^2 - b^2$ (Subt. Prop.)
3. $a^2 = c^2 - b^2$ (Subs.)
4. $a = \pm\sqrt{c^2 - b^2}$ (Sq. Root Prop.)
5. $a = \sqrt{c^2 - b^2}$ (Length cannot be negative.)

31 The relation "is taller than" is not an equivalence relation because it fails the Reflexive and Symmetric properties. You cannot be taller than yourself (reflexive); if you are taller than your friend, then it does not imply that your friend is taller than you (symmetric).

33. The relation "$\neq$" is not an equivalence relation because it fails the Reflexive Property, since $a \neq a$ is not true. **35.** The relation "$\approx$" is not an equivalence relation because it fails the Reflexive Property, since $a \approx a$ is not true.

37. Given: $AP = 2x + 3$, $PB = \frac{3x+1}{2}$, $AB = 10.5$
Prove: $\frac{AP}{AB} = \frac{2}{3}$

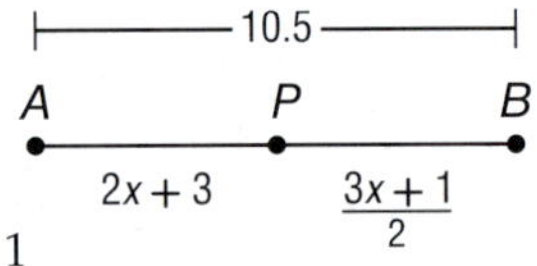

Proof:
Statements (Reasons)
1. $AP = 2x + 3$, $PB = \frac{3x+1}{2}$, $AB = 10.5$ (Given)
2. $AP + PB = AB$ (Def. of a segment)
3. $2x + 3 + \frac{3x+1}{2} = 10.5$ (Subt.)
4. $2 \cdot \left(2x + 3 + \frac{3x+1}{2}\right) = 2 \cdot 10.5$ (Mult. Prop.)
5. $2 \cdot \left(2x + 3 + \frac{3x+1}{2}\right) = 21$ (Subs. Prop.)
6. $2 \cdot 2x + 2 \cdot 3 + 2 \cdot \frac{3x+1}{2} = 21$ (Dist. Prop.)
7. $4x + 6 + 3x + 1 = 21$ (Mult. Prop.)
8. $7x + 7 = 21$ (Add. Prop.)
9. $7x + 7 - 7 = 21 - 7$ (Subt. Prop.)
10. $7x = 14$ (Subs.)
11. $x = 2$ (Div. Prop.)
12. $AP = 2(2) + 3$ (Subs.)
13. $AP = 4 + 3$ (Mult. Prop.)
14. $AP = 7$ (Add. Prop.)
15. $\frac{AP}{AB} = \frac{7}{10.5}$ (Subs.)
16. $\frac{AP}{AB} = 0.\overline{6}$ (Div. Prop.)
17. $\frac{2}{3} = 0.\overline{6}$ (Div. Prop.)
18. $\frac{AP}{AB} = \frac{2}{3}$ (Trans. Prop.)

39. Sometimes; sample answer: If $a^2 = 1$ and $a = 1$, then $b = \sqrt{1}$ or 1. The statement is also true if $a = -1$, then $b = 1$. If $b = 1$, then $\sqrt{b} = 1$ since the square root of a number is nonnegative. Therefore, the statement is sometimes true. **41.** An informal or paragraph proof is a kind of proof in which the steps are written out in complete sentences, in paragraph form. This type of proof is identical in content, but different in form, from a two-column proof, in which the statements (conclusions) are listed in one column, and the reasons for why each statement is true are listed in another column. **43.** 83° **45.** D **47.** Never; the sum of two supplementary angles is 180°, so two obtuse angles can never be supplementary. **49.** Yes, by the Law of Detachment. **51.** (4, −3) **53.** (1, 2) **55.** (−1, −1) **57.** 2.4 cm

Pages 142–148 *Lesson 2-7*

1. Given: $\overline{LK} \cong \overline{NM}$, $\overline{KJ} \cong \overline{MJ}$
Prove: $\overline{LJ} \cong \overline{NJ}$
Proof:
Statements (Reasons)
a. $\overline{LK} \cong \overline{NM}$, $\overline{KJ} \cong \overline{MJ}$ (Given)
b. $LK = NM$, $KJ = MJ$ (Def. of $\cong$ segs.)
c. $LK + KJ = NM + MJ$ (Add. Prop.)
d. $LJ = LK + KJ$; $NJ = NM + MJ$ (Seg. Add. Post.)
e. $LJ = NJ$ (Subs.)
f. $\overline{LJ} \cong \overline{NJ}$ (Def. of $\cong$ segs.)

3 Use the definition of congruent segments and the Substitution Property of Equality.
Given: $\overline{AR} \cong \overline{CR}$; $\overline{DR} \cong \overline{BR}$
Prove: $AR + DR \cong CR + BR$
Proof:
Statements (Reasons)
1. $\overline{AR} \cong \overline{CR}$, $\overline{DR} \cong \overline{BR}$ (Given)
2. $AR = CR$, $DR = BR$ (Definition of $\cong$ segments)
3. $AR + DR = AR + BR$ (Addition Property)
4. $AR + DR = CR + BR$ (Substitution Property)

5. Given: $\overline{AB} \cong \overline{CD}$, $AB + CD = EF$
Prove: $2AB = EF$
Proof:
Statements (Reasons)
1. $\overline{AB} \cong \overline{CD}$, $AB + CD = EF$ (Given)
2. $AB = CD$ (Def. of $\cong$ segs.)
3. $AB + AB = EF$ (Subs.)
4. $2AB = EF$ (Subs. Prop.)

7 Use the Reflexive Property of Equality and the definition of congruent segments.
Given: $\overline{AB}$
Prove: $\overline{AB} \cong \overline{AB}$

Proof:

Statements (Reasons)

1. $\overline{AB}$ (Given)
2. $AB = AB$ (Reflexive Property)
3. $\overline{AB} \cong \overline{AB}$ (Definition of $\cong$ segments)

9. Given: $\overline{SC} \cong \overline{HR}$ and $\overline{HR} \cong \overline{AB}$

Prove: $\overline{SC} \cong \overline{AB}$

Proof:

Statements (Reasons)

1. $\overline{SC} \cong \overline{HR}$ and $\overline{HR} \cong \overline{AB}$ (Given)
2. $SC = HR$ and $HR = AB$ (Def. of $\cong$ segs.)
3. $SC = AB$ (Trans. Prop.)
4. $\overline{SC} \cong \overline{AB}$ (Def. of $\cong$ segs.)

11. Given: E is the midpoint of $\overline{DF}$ and $\overline{CD} \cong \overline{FG}$.

Prove: $\overline{CE} \cong \overline{EG}$

Proof:

Statements (Reasons)

1. E is the midpoint of $\overline{DF}$ and $\overline{CD} \cong \overline{FG}$. (Given)
2. $DE = EF$ (Def. of midpoint)
3. $CD = FG$ (Def. of $\cong$ segs.)
4. $CD + DE = EF + FG$ (Add. Prop.)
5. $CE = CD + DE$ and $EG = EF + FG$ (Seg. Add. Post.)
6. $CE = EG$ (Subs.)
7. $\overline{CE} \cong \overline{EG}$ (Def. of $\cong$ segs.)

13a. Given: $\overline{AC} \cong \overline{GI}$, $\overline{FE} \cong \overline{LK}$, $AC + CF + FE = GI + IL + LK$

Prove: $\overline{CF} \cong \overline{IL}$

Proof:

Statements (Reasons)

1. $\overline{AC} \cong \overline{GI}$, $\overline{FE} \cong \overline{LK}$, $AC + CF + FE = GI + IL + LK$ (Given)
2. $AC + CF + FE = AC + IL + LK$ (Subs.)
3. $AC - AC + CF + FE = AC - AC + IL + LK$ (Subt. Prop.)
4. $CF + FE = IL + LK$ (Subs. Prop.)
5. $CF + FE = IL + FE$ (Subs.)
6. $CF + FE - FE = IL + FE - FE$ (Subt. Prop.)
7. $CF = IL$ (Subs. Prop.)
8. $\overline{CF} \cong \overline{IL}$ (Def. of $\cong$ segs.)

13b. Sample answer: I measured $\overline{CF}$ and $\overline{IL}$, and both were 1.5 inches long, so the two segments are congruent.

15 a. Use the definition of midpoint and the segment addition postulate to prove.

Given: $\overline{SH} \cong \overline{TF}$; P is the midpoint of $\overline{SH}$ and $\overline{TF}$.

Prove: $\overline{SP} \cong \overline{TP}$

Proof:

Statements (Reasons)

1. $\overline{SH} \cong \overline{TF}$, P is the midpoint of $\overline{SH}$, P is the midpoint of $\overline{TF}$. (Given)
2. $SH = TF$ (Definition of $\cong$ segments)
3. $SP = PH$, $TP = PF$ (Definition of midpoint)
4. $SH = SP + PH$, $TF = TP + PF$ (Segment Addition Postulate)
5. $SP + PH = TP + PF$ (Substitution)
6. $SP + SP = TP + TP$ (Substitution)
7. $2SP = 2TP$ (Substitution)
8. $SP = TP$ (Division Property)
9. $\overline{SP} \cong \overline{TP}$ (Definition of $\cong$ segments)

b.

$SP = \frac{1}{2}SH$	Definition of midpoint
$= \frac{1}{2}(127.3)$ or 63.54	Substitution

Since $\overline{TF} \cong \overline{SH}$, then $FP = SP = 63.54$.

Since $\triangle SPF$ is a right triangle, use the Pythagorean Theorem to find SF, the distance from first base to second base.

$c^2 = a^2 + b^2$	Pythagorean Theorem
$SF^2 = SP^2 + FP^2$	Substitution
$SF^2 = 63.54^2 + 63.54^2$	Substitution
$SF^2 \approx 8074.6632$	Simplify.
$SF \approx 90$	Take the positive square root of each side.

The distance from first base to second base is about 90 feet.

17. Neither; Since $\overline{AB} \cong \overline{CD}$ and $\overline{CD} \cong \overline{BF}$, then $\overline{AB} \cong \overline{BF}$ by the Transitive Property of Congruence.

19. No; congruence refers to segments. Segments cannot be added, only the measures of segments.

21.

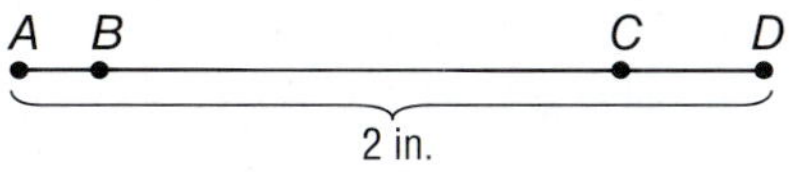

23. D **25.** 18

27. Given: $AC = DF$, $AB = DE$

Prove: $BC = EF$

Proof:

Statements (Reasons)

1. $AC = DF$, $AB = DE$ (Given)
2. $AC = AB + BC$; $DF = DE + EF$ (Seg. Add. Post.)
3. $AB + BC = DE + EF$ (Subs.)
4. $BC = EF$ (Subt. Prop.)

29. 60, 30, 90, 60, 120, 60 **31.** $9\sqrt{2}$ **33.** $3y^4\sqrt{5x}$ **35.** 8

Pages 149–157 Lesson 2-8

1 $m\angle 1 = 90$ because $\angle 1$ is a right angle.

$m\angle 2 + m\angle 3 = 90$	Complement Theorem
$26 + m\angle 3 = 90$	$m\angle 2 = 26$
$26 + m\angle 3 - 26 = 90 - 26$	Subtraction Property
$m\angle 3 = 64$	Substitution

3. $m\angle 4 = 114$, $m\angle 5 = 66$; Suppl. Thm.

5. Given: $\angle 2 \cong \angle 6$

Prove: $\angle 4 \cong \angle 8$

Proof:

Statements (Reasons)

1. $\angle 2 \cong \angle 6$ (Given)
2. $m\angle 2 + m\angle 4 = 180$, $m\angle 6 + m\angle 8 = 180$ (Suppl. Thm.)
3. $m\angle 2 + m\angle 8 = 180$ (Subs.)
4. $m\angle 2 - m\angle 2 + m\angle 4 = 180 - m\angle 2$, $m\angle 2 - m\angle 2 + m\angle 8 = 180 - m\angle 2$ (Subt. Prop.)
5. $m\angle 4 = 180 - m\angle 2$, $m\angle 8 = 180 - m\angle 2$ (Subt. Prop.)
6. $m\angle 4 = m\angle 8$ (Subs.)
7. $\angle 4 \cong \angle 8$ (Def. $\cong$ ∡)

7. Given: $\angle 4 \cong \angle 7$

Prove: $\angle 5 \cong \angle 6$

Proof:

Statements (Reasons)

1. $\angle 4 \cong \angle 7$ (Given)
2. $\angle 4 \cong \angle 5$ and $\angle 6 \cong \angle 7$ (Vert. $\angle$s Thm.)
3. $\angle 7 \cong \angle 5$ (Subs.)
4. $\angle 5 \cong \angle 6$ (Subs.)

9. $m\angle 3 = 62$, $m\angle 1 = m\angle 4 = 45$ ($\cong$ Comp. and Suppl. Thm.) **11.** $m\angle 9 = 156$, $m\angle 10 = 24$ ($\cong$ Suppl. Thm.)

13

$m\angle 6 + m\angle 7 = 180$	Supplement Thm.
$2x - 21 + 3x - 34 = 180$	Substitution
$5x - 55 = 180$	Substitution
$5x - 55 + 55 = 180 + 55$	Addition Property
$5x = 235$	Substitution
$\frac{5x}{5} = \frac{235}{5}$	Division Property
$x = 47$	Substitution

$m\angle 6 = 2x - 21$	Given
$m\angle 6 = 2(47) - 21$ or 73	Substitution

$m\angle 7 = 3x - 34$	Given
$m\angle 7 = 3(47) - 34$ or 107	Substitution

$\angle 8 \cong \angle 6$	Vertical Angles Theorem
$m\angle 8 = m\angle 6$	Definition of $\cong$ $\angle$s
$= 73$	Substitution

15. Given: $\angle 5 \cong \angle 6$

Prove: $\angle 4$ and $\angle 6$ are supplementary.

Proof:

Statements (Reasons)

1. $\angle 5 \cong \angle 6$ (Given)
2. $m\angle 5 = m\angle 6$ (Def. of $\cong$ $\angle$s)
3. $\angle 4$ and $\angle 5$ are supplementary. (Def. of linear pairs)
4. $m\angle 4 + m\angle 5 = 180$ (Def. of $\angle$s)
5. $m\angle 4 + m\angle 6 = 180$ (Subs.)
6. $\angle 4$ and $\angle 6$ are supplementary. (Def. of $\angle$s)

17. Given: $\angle ABC$ is a right angle.

Prove: $\angle 1$ and $\angle 2$ are complementary angles.

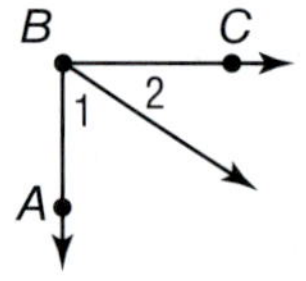

Proof:

Statements (Reasons)

1. $\angle ABC$ is a right angle. (Given)
2. $m\angle ABC = 90$ (Def. of rt. $\angle$s)
3. $m\angle ABC = m\angle 1 + m\angle 2$ ($\angle$ Add. Post.)
4. $90 = m\angle 1 + m\angle 2$ (Subst.)
5. $\angle 1$ and $\angle 2$ are complementary angles. (Def. of comp. $\angle$s)

19. Given: $\angle 1 \cong \angle 2$, $\angle 2 \cong \angle 3$

Prove: $\angle 1 \cong \angle 3$

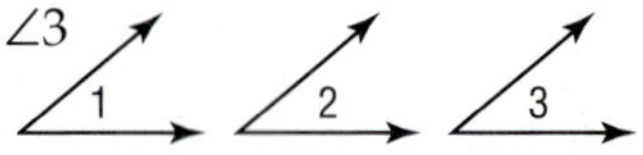

Proof:

Statements (Reasons)

1. $\angle 1 \cong \angle 2$, $\angle 2 \cong \angle 3$ (Given)
2. $m\angle 1 = m\angle 2$, $m\angle 2 = m\angle 3$ (Def. of $\cong$ $\angle$s)
3. $m\angle 1 = m\angle 3$ (Trans. Prop.)
4. $\angle 1 \cong \angle 3$ (Def. of $\cong$ $\angle$s)

21. Given: $\angle 1 \cong \angle 4$

Prove: $\angle 2 \cong \angle 3$

Proof:

Statements (Reasons)

1. $\angle 1 \cong \angle 4$ (Given)
2. $\angle 1 \cong \angle 2$, $\angle 3 \cong \angle 4$ (Vert. $\angle$s are $\cong$.)
3. $\angle 1 \cong \angle 3$ (Trans. Prop.)
4. $\angle 2 \cong \angle 3$ (Subs.)

23. Given: $\angle 1$ and $\angle 2$ are rt. $\angle$s.

Prove: $\angle 1 \cong \angle 2$

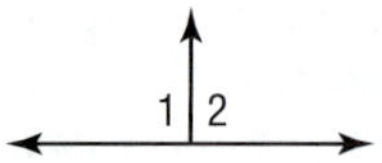

Proof:

Statements (Reasons)

1. $\angle 1$ and $\angle 2$ are rt. $\angle$s. (Given)
2. $m\angle 1 = 90$, $m\angle 2 = 90$ (Def. of rt. $\angle$s)
3. $m\angle 1 = m\angle 2$ (Subs.)
4. $\angle 1 \cong \angle 2$ (Def. of $\cong$ $\angle$s)

25. Given: $\angle 1 \cong \angle 2$, $\angle 1$ and $\angle 2$ are supplementary.

Prove: $\angle 1$ and $\angle 2$ are rt. $\angle$s.

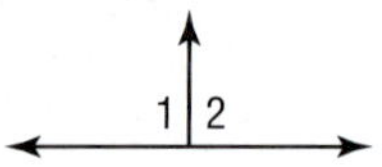

Proof:

Statements (Reasons)

1. $\angle 1 \cong \angle 2$, $\angle 1$ and $\angle 2$ are supplementary. (Given)
2. $m\angle 1 + m\angle 2 = 180$ (Def. of supp. $\angle$s)
3. $m\angle 1 = m\angle 2$ (Def. of $\cong$ $\angle$s)
4. $m\angle 1 + m\angle 1 = 180$ (Subs.)
5. $2(m\angle 1) = 180$ (Subs.)
6. $m\angle 1 = 90$ (Div. Prop.)
7. $m\angle 2 = 90$ (Subs. (steps 3, 6))
8. $\angle 1$ and $\angle 2$ are rt. $\angle$s. (Def. of rt. $\angle$s)

27. Since the path of the pendulum forms a right angle, $\angle ABC$ is a right angle, or measures 90. $\overrightarrow{BR}$ divides $\angle ABC$ into $\angle ABR$ and $\angle CBR$. By the Angle Addition Postulate, $m\angle ABR + m\angle CBR = m\angle ABC$, and, using substitution, $m\angle ABR + m\angle CBR = 90$. Substituting again, $m\angle 1 + m\angle 2 = 90$. We are given that $m\angle 1$ is 45°, so, substituting, $45 + m\angle 2 = 90$. Using the Subtraction Property, $45 - 45 + m\angle 2 = 90 - 45$, or $m\angle 2 = 45$. Since $m\angle 1$ and $m\angle 2$ are equal, $\overrightarrow{BR}$ is the bisector of $\angle ABC$ by the definition of angle bisector.

29 To prove lines ℓ and m are perpendicular, show that $\angle 1$, $\angle 3$, and $\angle 4$ are right $\angle$s.

Given: $\angle 2$ is a right angle.

Prove: $\ell \perp m$

Proof:

Statements (Reasons)

1. $\angle 2$ is a right angle. (Given)
2. $m\angle 2 = 90$ (Definition of a rt. $\angle$)
3. $\angle 2 \cong \angle 3$ (Vert. $\angle$s are $\cong$.)
4. $m\angle 3 = 90$ (Substitution)
5. $m\angle 1 + m\angle 2 = 180$ (Supplement Theorem)
6. $m\angle 1 + 90 = 180$ (Substitution)
7. $m\angle 1 + 90 - 90 = 180 - 90$ (Subtraction Property)
8. $m\angle 1 = 90$ (Substitution)
9. $\angle 1 \cong \angle 4$ (Vertical $\angle$s are $\cong$.)
10. $\angle 4 \cong \angle 1$ (Symmetric Property)
11. $m\angle 4 = m\angle 1$ (Definition of $\cong$ $\angle$s)
12. $m\angle 4 = 90$ (Substitution)
13. $\ell \perp m$ ($\perp$ lines intersect to form four rt. $\angle$s.)

31. Given: $\overrightarrow{XZ}$ bisects $\angle WXY$, and $m\angle WXZ = 45$.
Prove: $\angle WXY$ is a right angle.
Proof:
<u>Statements (Reasons)</u>
1. $\overrightarrow{XZ}$ bisects $\angle WXY$ and $m\angle WXZ = 45$. (Given)
2. $\angle WXZ \cong \angle ZXY$ (Def. of $\angle$ bisector)
3. $m\angle WXZ = m\angle ZXY$ (Def. of $\cong$ $\triangle$)
4. $m\angle ZXY = 45$ (Subs.)
5. $m\angle WXY = m\angle WXZ + m\angle ZXY$ ($\angle$ Add. Post.)
6. $m\angle WXY = 45 + 45$ (Subs.)
7. $m\angle WXY = 90$ (Subs.)
8. $\angle WXY$ is a right angle. (Def. of rt. $\angle$)

33. Each of these theorems uses the words "or to congruent angles" indicating that this case of the theorem must also be proven true. The other proofs only addressed the "to the same angle" case of the theorem.

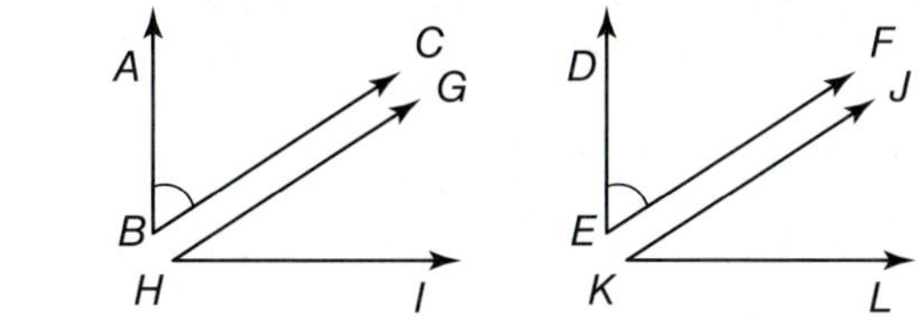

Given: $\angle ABC \cong \angle DEF$, $\angle GHI$ is complementary to $\angle ABC$, $\angle JKL$ is complementary to $\angle DEF$.
Prove: $\angle GHI \cong \angle JKL$
Proof:
<u>Statements (Reasons)</u>
1. $\angle ABC \cong \angle DEF$, $\angle GHI$ is complementary to $\angle ABC$, $\angle JKL$ is complementary to $\angle DEF$. (Given)
2. $m\angle ABC + m\angle GHI = 90$, $\angle DEF + \angle JKL = 90$ (Def. of compl. $\triangle$)
3. $m\angle ABC + m\angle JKL = 90$ (Subs.)
4. $90 = m\angle ABC + m\angle JKL$ (Symm. Prop.)
5. $m\angle ABC + m\angle GHI = m\angle ABC + m\angle JKL$ (Trans. Prop.)
6. $m\angle ABC - m\angle ABC + m\angle GHI = m\angle ABC - m\angle ABC + m\angle JKL$ (Subt. Prop.)
7. $m\angle GHI = m\angle JKL$ (Subs. Prop.)
8. $\angle GHI \cong \angle JKL$ (Def. of $\cong$ $\triangle$)

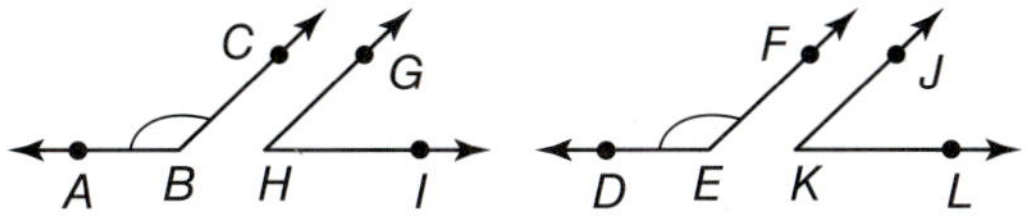

Given: $\angle ABC \cong \angle DEF$, $\angle GHI$ is supplementary to $\angle ABC$, $\angle JKL$ is supplementary to $\angle DEF$.
Prove: $\angle GHI \cong \angle JKL$
Proof:
<u>Statements (Reasons)</u>
1. $\angle ABC \cong \angle DEF$, $\angle GHI$ is supplementary to $\angle ABC$, $\angle JKL$ is supplementary to $\angle DEF$. (Given)
2. $m\angle ABC + m\angle GHI = 180$, $m\angle DEF + m\angle JKL = 180$ (Def. of compl. $\triangle$)
3. $m\angle ABC + m\angle JKL = 180$ (Subs.)
4. $180 = m\angle ABC + m\angle JKL$ (Symm. Property)
5. $m\angle ABC + m\angle GHI = m\angle ABC + m\angle JKL$ (Trans. Prop.)
6. $m\angle ABC - m\angle ABC + m\angle GHI = m\angle ABC - m\angle ABC + m\angle JKL$ (Subt. Prop.)
7. $m\angle GHI = m\angle JKL$ (Subs. Prop.)
8. $\angle GHI \cong \angle JKL$ (Def. of $\cong$ $\triangle$)

35. Sample answer: Since protractors have the scale for both acute and obtuse angles along the top, the supplement is the measure of the given angle on the other scale. **37.** A **39.** B **41.** Subtratcion Prop. **43.** Substitution **45.** true **47.** true **49.** line n **51.** point W **53.** Yes; it intersects both m and n when all three lines are extended.

Pages 158–162 Chapter 2 Study Guide and Review

1. false; theorem **3.** true **5.** true **7.** true **9.** false; negation **11.** false; two nonadjacent supplementary angles **13.** Sample answer: Dogs or other pets may threaten or chase wildlife that might not be present in his local park. **15.** A plane contains at least three noncollinear points and the sum of the measures of two complementary angles is not 180°; true. **17a.** 18 **17b.** 14 **17c.** 22 **19.** true **21.** $PQRS$ is a parallelogram. **23.** valid; Law of Detachment **25.** Sometimes; if the three points are collinear, they will be contained in multiple planes, but if they are noncollinear, they will be contained in only one plane. **27.** Sometimes; if the angles are adjacent, they will form a right angle, but if they are not adjacent, they will not. **29.** Symmetric **31.** Distributive Property **33.** Transitive Property

35. <u>Statements (Reasons)</u>
1. $PQ = RS$, $PQ = 5x + 9$, $RS = x - 31$ (Given)
2. $5x + 9 = x - 31$ (Subs. Prop.)
3. $4x = 9 = -31$ (Subt. Prop.)
4. $4x = -40$ (Subt. Prop.)
5. $x = -10$ (Div. Prop.)

37. <u>Statements (Reasons)</u>
1. x is the midpoint of $\overline{WX}$ and $\overline{VZ}$. (Given)
2. $\overline{WX} \cong \overline{YX}$, $\overline{VX} \cong \overline{ZX}$ (Def. of midpoint)
3. $WX = YX$, $VX = ZX$ (Def. of $\cong$)
4. $VX = VW = WX$, $ZX = ZY + YX$ (Seg. Add. Post)
5. $VW + WX = ZY + YX$ (Subs.)
6. $VW = ZY$ (Sut. Prop.)

39. Segment Addition Postulate **41.** 127

43. <u>Statements (Reasons)</u>
1. $\angle 1 \cong \angle 4$, $\angle 2 \cong \angle 3$ (Given)
2. $m\angle 1 = m\angle 4$, $m\angle 2 = m\angle 3$ (Def. of $\cong$)
3. $m\angle 1 + m\angle 2 = m\angle 3 + m\angle 4$ (Add. Prop.)
4. $m\angle 1 + m\angle 2 = m\angle AFC$, $m\angle 3 + m\angle 4 = m\angle EFC$ ($\angle$ Add. Post.)
5. $m\angle AFC = m\angle EFC$ (Subs.)
6. $\angle AFC = \angle EFC$ (Def. of $\cong$)

Chapter 3 Parallel and Perpendicular Lines

Page 169 Chapter 3 Get Ready

1. 4 **3.** Yes, points C and D lie in plane CBD. **5.** 113 **7.** 90 **9.** -1 **11.** $\frac{1}{3}$

Pages 171–176 Lesson 3-1

1. *TUV* **3.** $\overline{YX}$, $\overline{TU}$, $\overline{ZW}$

5 Angle ∠1 and ∠8 are nonadjacent exterior angles that lie on opposite sides of the transversal. So, they are alternate exterior angles.

7. alternate interior **9.** line *n*; corresponding **11.** line *m*; consecutive interior **13.** $\overline{CL}$, $\overline{EN}$, $\overline{BK}$, $\overline{AJ}$

15 The segments that do not intersect $\overline{BC}$ and are not in the same plane as $\overline{BC}$ are skew to $\overline{BC}$. Any of the following are skew to $\overline{BC}$: $\overline{EN}$, $\overline{AJ}$, $\overline{DM}$, $\overline{NM}$, $\overline{NJ}$, $\overline{JK}$, or $\overline{ML}$.

17. $\overline{KL}$, $\overline{CL}$, $\overline{BK}$, $\overline{ML}$, $\overline{DM}$, $\overline{NM}$, $\overline{KJ}$ **19.** $\overline{JK}$ **21.** line *s*; corresponding **23.** line *t*; alternate interior **25.** line *t*; alternate exterior **27.** line *t*; consecutive interior **29.** line *s*; alternate exterior **31.** line *b*; vertical **33.** line *c*; alternate interior **35.** line *f*; corresponding **37a.** Sample answer: Since the lines are coplanar and they cannot touch, they are parallel. **37b.** Line *q* is a transversal of lines *p* and *m*. **39.** skew **41.** parallel **43.** intersecting

45 **a.** The treads, or the upper horizontal parts of the stairs, are parallel. **b.** The treads of the two steps at the top of the incline lie in the same plane. So, they are coplanar. **c.** The treads of the steps on the incline of the escalator do not intersect and are not parallel to the treads of the steps on the bottom of the conveyor. So, they are skew.

47a.

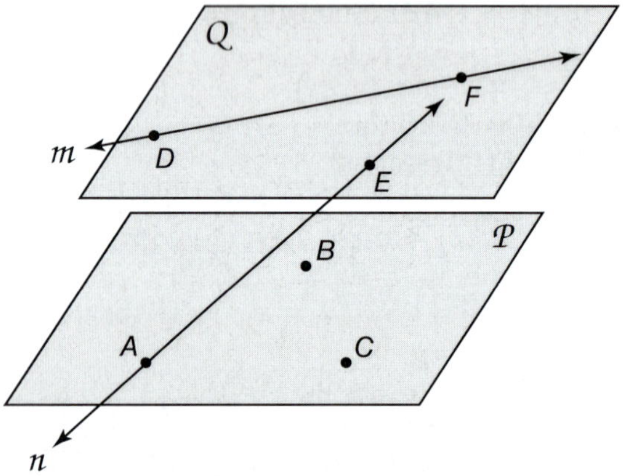

47b. parallel **47c.** skew **49.** Sometimes; $\overleftrightarrow{AB}$ intersects $\overleftrightarrow{EF}$ depending on where the planes intersect. **51.** B **53.** (0, 4), (−6, 0) **55.** $m\angle 9 = 86$, $m\angle 10 = 94$ **57.** $m\angle 19 = 140$, $m\angle 20 = 40$ **59.** 3.75 **61.** 90 **63.** 45

Pages 178–184 Lesson 3-2

1. 94; Corresponding Angle Postulate **3.** 86; Corresponding Angle Postulate and Supplement Angle Theorem **5.** 79; Vertical Angle Theorem, Consecutive Interior Angles Theorem **7.** $m\angle 2 = 93$, $m\angle 3 = 87$, $m\angle 4 = 87$ **9.** $x = 114$ by the Alternate Exterior Angles Theorem **11.** 22; Corresponding Angles Postulate **13.** 158; Def. of Supplementary Angles **15.** 18; Corresponding Angles Postulate **17.** 162; Supplement Angles Theorem **19.** 18; Alternate Exterior Angles Postulate

21 If the radiation rays form parallel lines, then ∠1 and ∠3 are corresponding angles. So, according to the Corresponding Angles Postulate, ∠1 and ∠3 are congruent.

23. Supplementary; since ∠3 and ∠4 are a linear pair, they are supplementary. ∠4 and ∠5 are congruent because they are alternate exterior angles, so ∠3 is supplementary to ∠4.

25

$3x - 15 = 105$	Corresponding Angles Postulate
$3x = 120$	Add 15 to each side.
$x = 40$	Divide each side by 3.

$(3x - 15) + (y + 25) = 180$	Supplement Theorem
$105 + y + 25 = 180$	Substitution
$y + 130 = 180$	Simplify.
$y = 50$	Subtract 130 from each side.

So, $x = 40$ by the Corresponding Angles Postulate; $y = 50$ by the Supplement Theorem.

27. $x = 42$ by the Consecutive Interior Angles Theorem; $y = 14$ by the Consecutive Interior Angles Theorem **29.** $x = 60$ by the Consecutive Interior Angles Theorem; $y = 10$ by the Supplement Theorem **31.** Congruent; Alternate Interior Angles **33.** Congruent; vertical angles are congruent.

35. Given: $\ell \parallel m$
Prove: $\angle 1 \cong \angle 8$
$\angle 2 \cong \angle 7$

Proof:

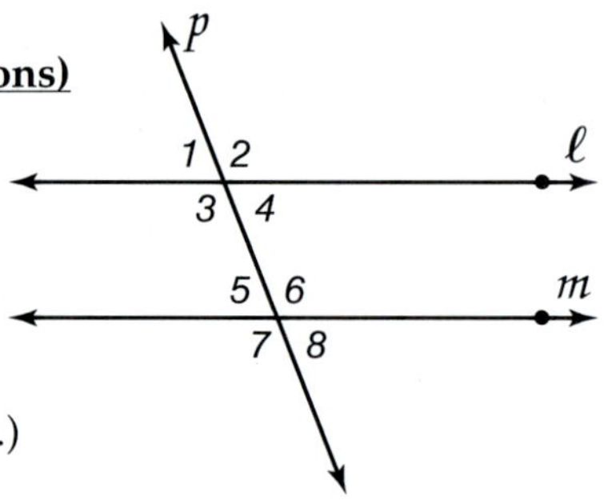

<u>**Statements (Reasons)**</u>
1. $\ell \parallel m$ (Given)
2. $\angle 1 \cong \angle 5$, $\angle 2 \cong \angle 6$ (Corr. ∠s Post.)
3. $\angle 5 \cong \angle 8$, $\angle 6 \cong \angle 7$ (Vertical ∠s Thm.)
4. $\angle 1 \cong \angle 8$, $\angle 2 \cong \angle 7$ (Trans. Prop.)

37. Given: $m \parallel n$, $t \perp m$
Prove: $t \perp n$
Proof:

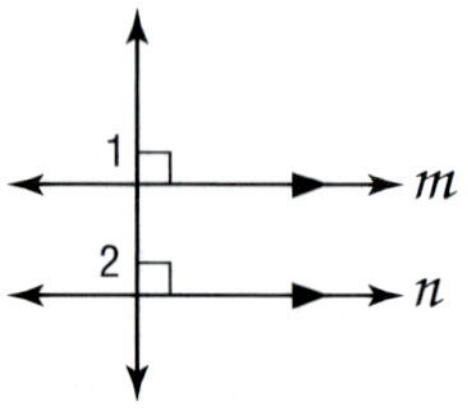

<u>**Statements (Reasons)**</u>
1. $m \parallel n$, $t \perp m$ (Given)
2. ∠1 is a right angle. (Def. of ⊥)
3. $m\angle 1 = 90$ (Def. of rt. ∠s)
4. $\angle 1 \cong \angle 2$ (Corr. ∠s Post.)
5. $m\angle 1 = m\angle 2$ (Def. of ≅ ∠s)
6. $m\angle 2 = 90$ (Subs.)
7. ∠2 is a right angle. (Def. of rt. ∠s)
8. $t \perp n$ (Def. of ⊥ lines)

39 Draw a line parallel to the two given lines and label angles 1, 2, 3, and 4. So, $x = m\angle 2 + m\angle 3$.

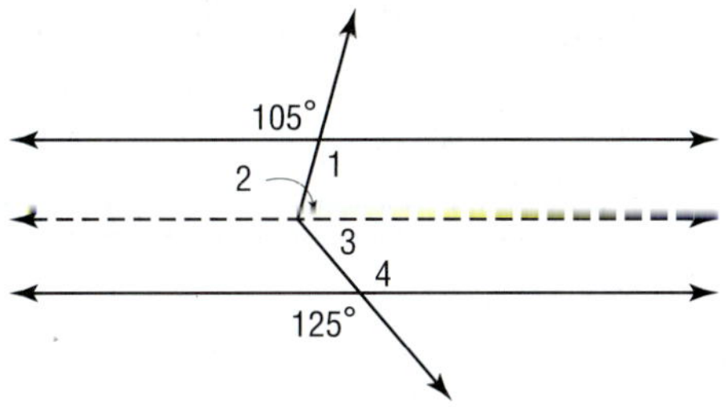

$m\angle 1 = 105$ because vertical angles are congruent and their measures are equal. $\angle 1$ and $\angle 2$ are supplementary by the Consecutive Interior Angles Theorem.

$m\angle 1 + m\angle 2 = 180$	Definition of supplementary angles
$105 + m\angle 2 = 180$	Substitution
$m\angle 2 = 75$	Subtract 105 from each side.

$m\angle 4 = 125$ because vertical angles are congruent and their measures are equal. $\angle 3$ and $\angle 4$ are supplementary by the Consecutive Interior Angles Theorem.

$m\angle 3 + m\angle 4 = 180$	Definition of supplementary angles
$m\angle 3 + 125 = 180$	Substitution
$m\angle 3 = 55$	Subtract 125 from each side.

$m\angle 2 + m\angle 3 = x$	Angle Addition Postulate
$75 + 55 = x$	Substitution
$130 = x$	Simplify.

41a. Sample answer for m and n:

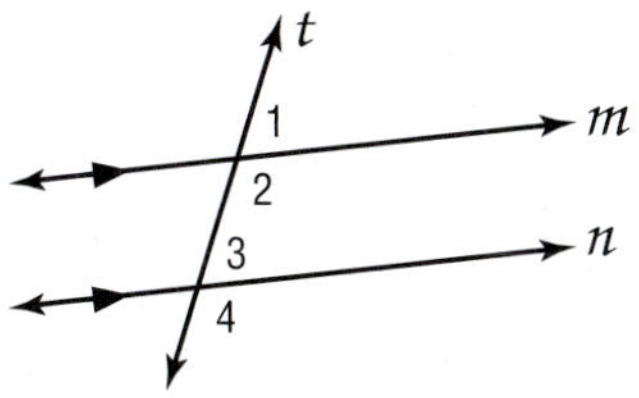

41b. Sample answer:

$m\angle 1$	$m\angle 2$	$m\angle 3$	$m\angle 4$
60	120	60	120
45	135	45	135
70	110	70	110
90	90	90	90
25	155	25	155
30	150	30	150

41c. Sample answer: Angles on the exterior of a pair of parallel lines located on the same side of the transversal are supplementary. **41d.** Inductive; a pattern was used to make a conjecture.

41e. Given: parallel lines m and n cut by transversal t
Prove: $\angle 1$ and $\angle 4$ are supplementary.

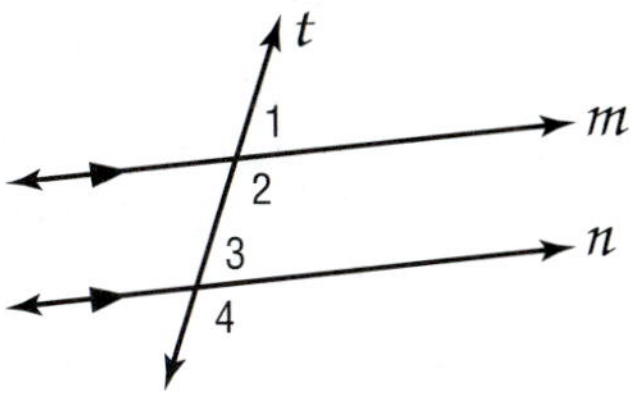

Proof:

Statements (Reasons)

1. Lines m and n are parallel and cut by transversal t. (Given)
2. $m\angle 1 + m\angle 2 = 180$ (Suppl. Thm.)
3. $\angle 2 \cong \angle 4$ (Corr. $\angle$s are $\cong$.)
4. $m\angle 2 = m\angle 4$ (Def. of congruence)
5. $m\angle 1 + m\angle 4 = 180$ (Subs.)
6. $\angle 1$ and $\angle 4$ are supplementary. (Definition of supplementary angles)

43. In both theorems, a pair of angles is formed when two parallel lines are cut by a transversal. However, in the Alternate Interior Angles Theorem, each pair of alternate interior angles that is formed are congruent, whereas in the Consecutive Interior Angles Theorem, each pair of angles formed is supplementary.
45. $x = 171$ or $x = 155$; $y = 3$ or $y = 5$ **47.** C **49.** I and II **51.** Skew lines; the planes are flying in different directions and at different altitudes.

53. $m\angle 6 = 43, m\angle 7 = 90$ **55.** 15 **57.** $\frac{7}{3}$ **59.** 1
61. $-\frac{3}{5}$

Pages 186–194 Lesson 3-3

1. -1 **3.** $\frac{6}{5}$

5

slope of $\overleftrightarrow{WX} = \frac{y_2 - y_1}{x_2 - x_1}$	Slope formula
$= \frac{5-4}{4-2}$	$(x_1, y_1) = (2, 4), (x_2, y_2) = (4, 5)$
$= \frac{1}{2}$	Simplify.
slope of $\overleftrightarrow{YZ} = \frac{y_2 - y_1}{x_2 - x_1}$	Slope formula
$= \frac{-7-1}{8-4}$	$(x_1, y_1) = (4, 1)$, $(x_2, y_2) = (8, -7)$
$= \frac{-8}{4}$ or -2	Simplify.
$\frac{1}{2}(-2) = -1$	Product of slopes

Since the product of the slopes is -1, $\overleftrightarrow{WX}$ is perpendicular to $\overleftrightarrow{YZ}$.

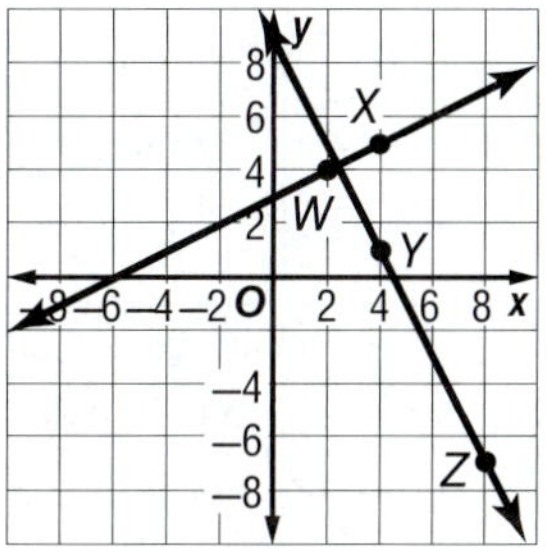

7. parallel

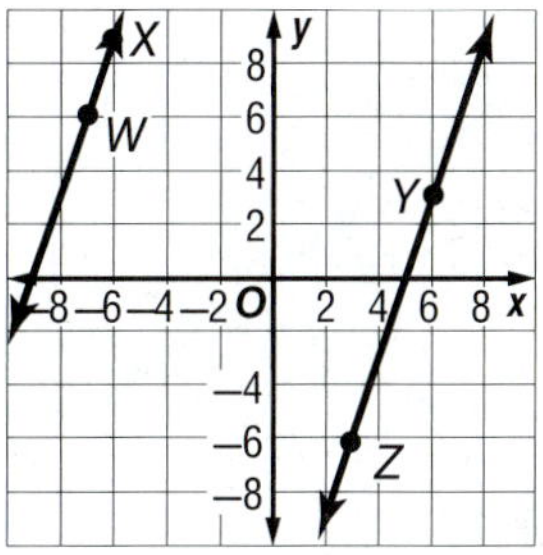

9.

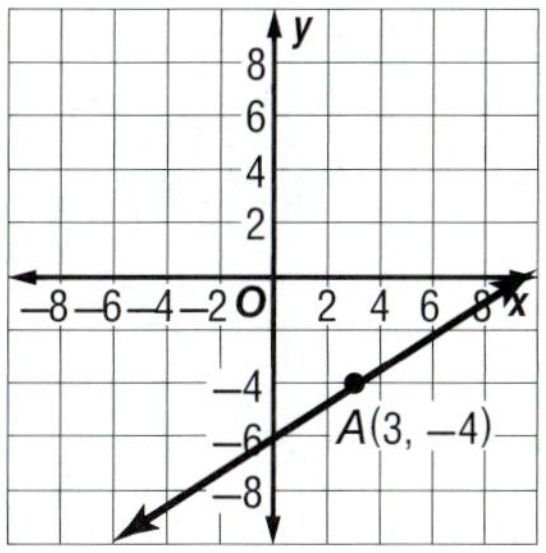

11.

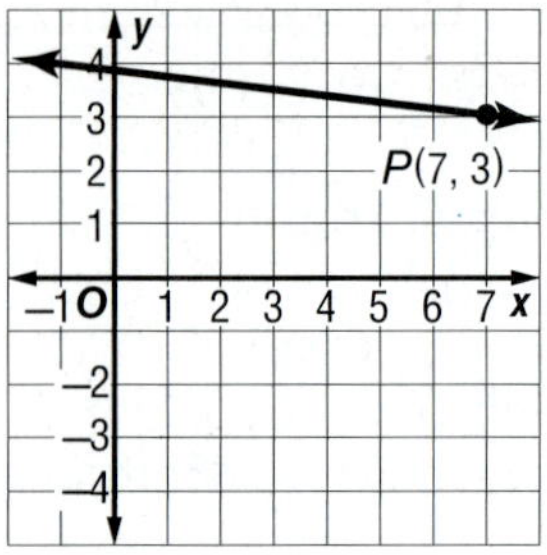

13 Substitute $(-3, 1)$ for (x_1, y_1) and $(4, -2)$ for (x_2, y_2).

$m = \dfrac{y_2 - y_1}{x_2 - x_1}$	Slope formula
$= \dfrac{-2 - 1}{4 - (-3)}$	Substitution
$= -\dfrac{3}{7}$	Simplify.

15. 8 **17.** undefined **19.** 1 **21.** 0 **23.** undefined

25. $-\dfrac{1}{6}$

27a.

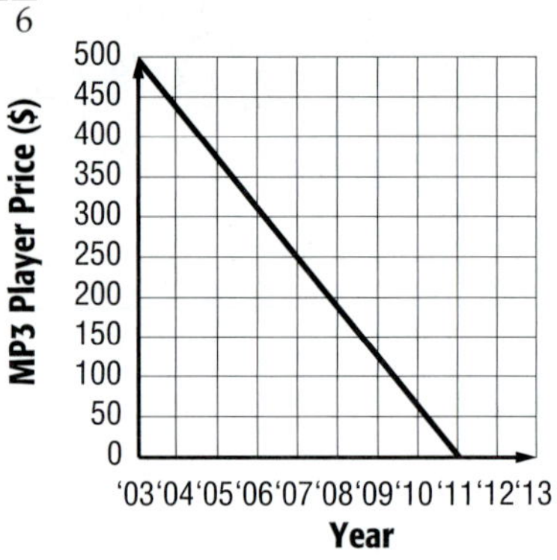

27b. $62.25

27c. $63.25

29. parallel

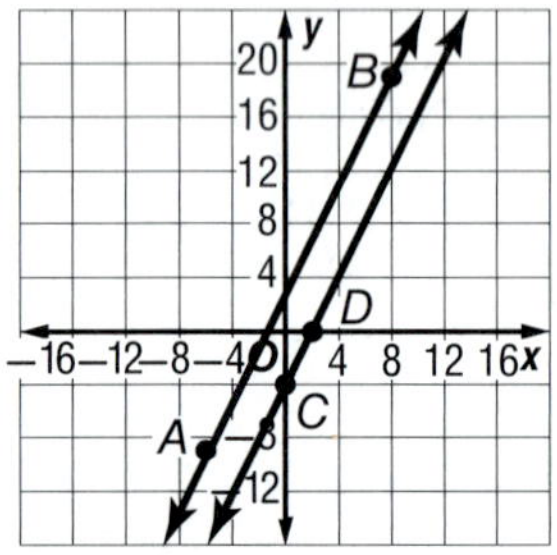

31. perpendicular

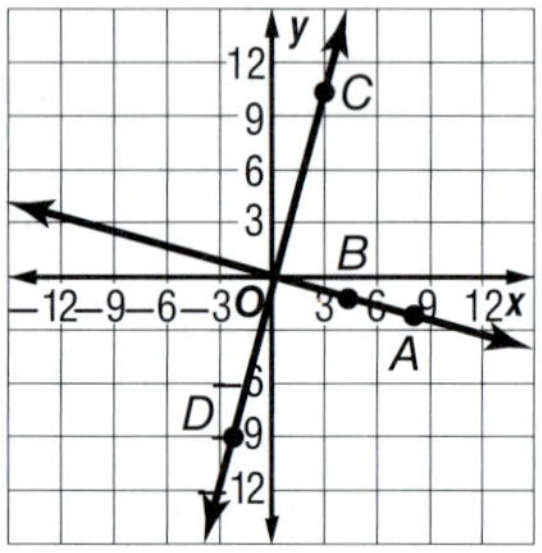

33. neither

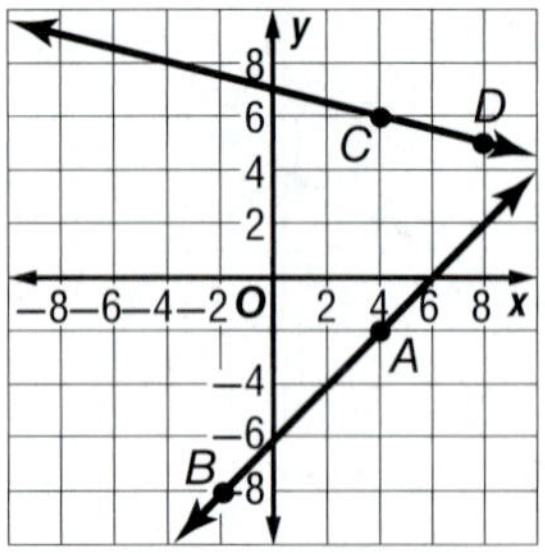

35.

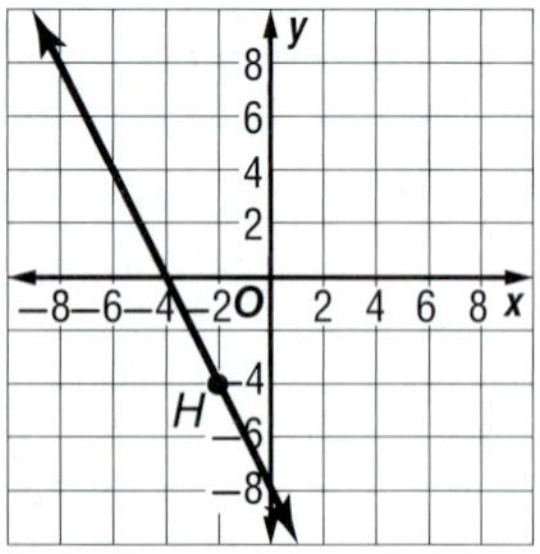

37.

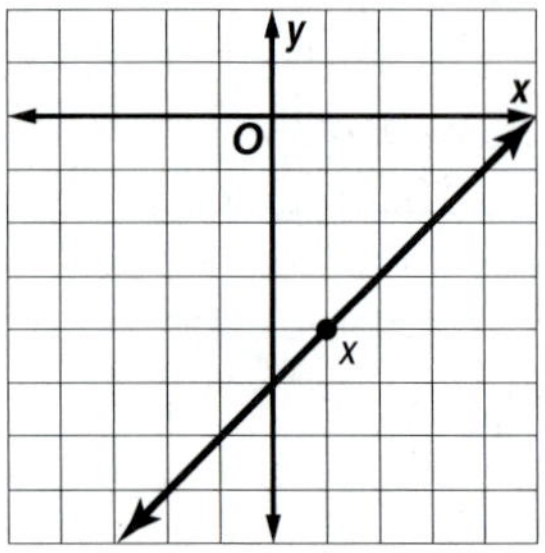

39.

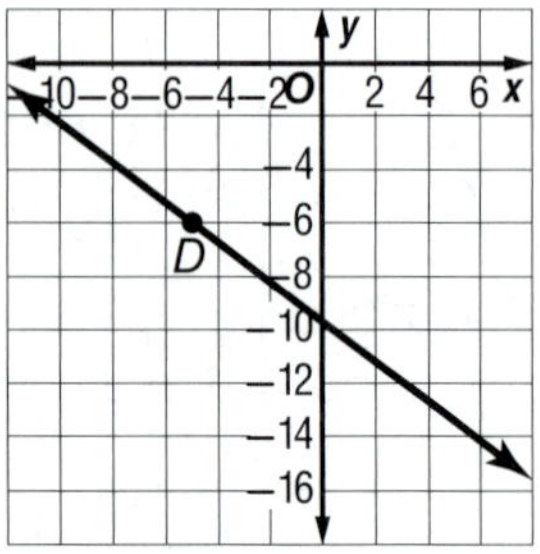

41. Line 2

43

slope of Line 1 $= \dfrac{y_2 - y_1}{x_2 - x_1}$	Slope formula
$= \dfrac{0 - (-4)}{7 - (-9)}$	$(x_1, y_1) = (-9, -4)$, $(x_2, y_2) = (7, 0)$
$= \dfrac{4}{16}$	Subtract.
$= \dfrac{1}{4}$	Simplify.

slope of Line 2 $= \dfrac{y_2 - y_1}{x_2 - x_1}$	Slope formula
$= \dfrac{4 - 1}{7 - 0}$	$(x_1, y_1) = (0, 1)$, $(x_2, y_2) = (7, 4)$
$= \dfrac{3}{7}$	Subtract.

Since $\frac{1}{4} < \frac{3}{7}$, the slope of Line 2 is steeper than the slope of Line 1.

45a. the bald eagle

45b.

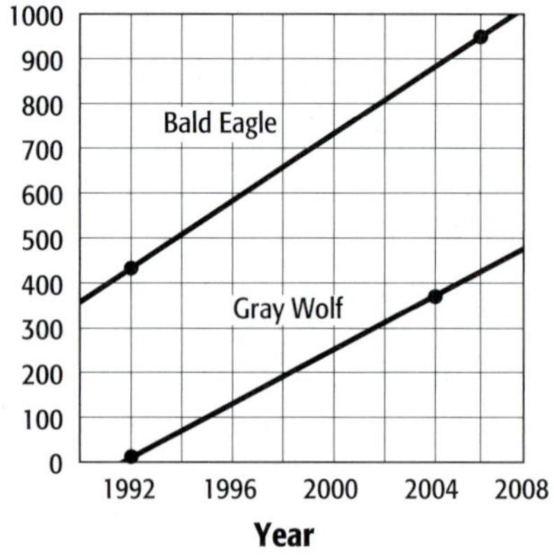

45c. 1189 bald eagles; 494 gray wolves

47. $y = -8$

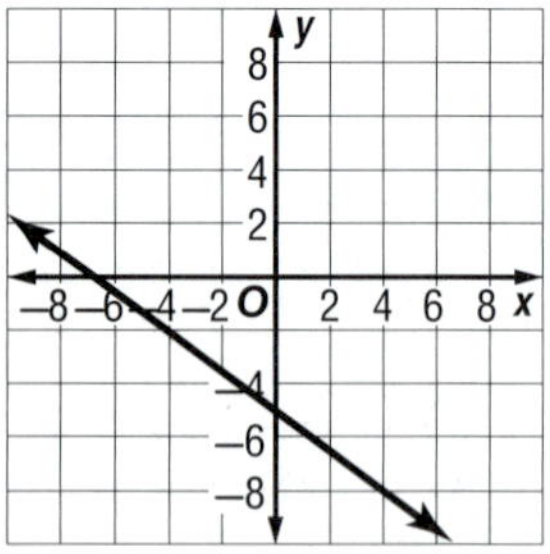

49. $y = 0$

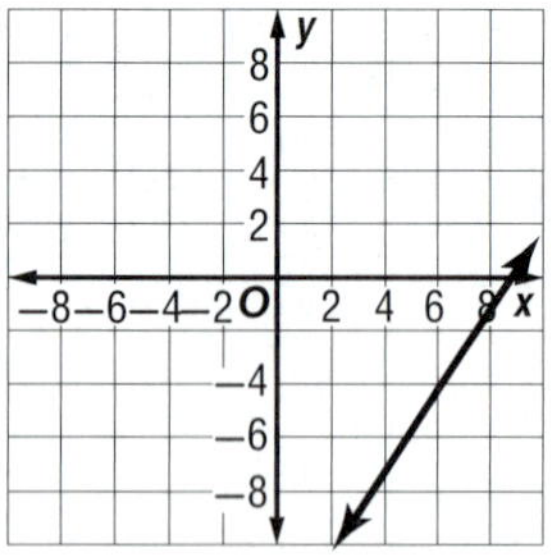

51 a.

Time (hours)	Distance Walking (miles)	Distance Riding Bikes (miles)
0	0	0
1	3.5	10
2	7	20
3	10.5	30
4	14	40

Multiply the time spent walking by 3.5 and the time spent riding by 10 to find the distances travelled.

b. Use the table from part **a** to create a graph with time on the horizontal axis and distance on the vertical axis.

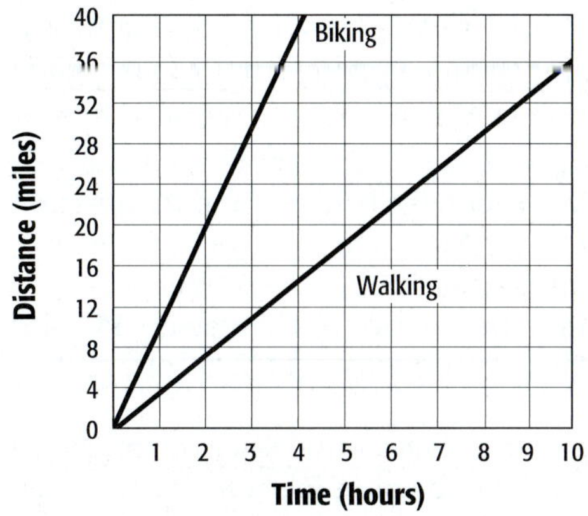

c. their speed **d.** Sample answer: Yes, they can make it if they ride their bikes. If they walk, it takes over two hours to go eight miles, so they wouldn't be home in time and they wouldn't get to spend any time in the store. If they ride their bikes, they can travel there in 24 minutes. If they spend 30 minutes in the store and spend 24 minutes riding home, the total amount of time they will use is 24 + 30 + 24 = 78 minutes, which is 1 hour and 18 minutes.

53. Terrell; Hale subtracted the x-coordinates in the wrong order. **55.** The Sears Tower has a vertical or undefined and the Leaning Tower of Pisa has a positive slope. **57.** Sample answer: (4, −3) and (5, −5) lie along the same line as point X and Y. The slope between all of the points is −2. To find additional points, you can take any point on the line and subtract 2 from the y-coordinate and add 1 to the x-coordinate. **59.** 2:3 **61.** C **63.** 123 **65.** 57 **67.** *ABC, ABQ, PQR, CDS, APU, DET* **69.** valid **71.** 6 **73.** $y = -2x + 3$

Pages 196–203 *Lesson 3-4*

1. $y = 4x - 3$

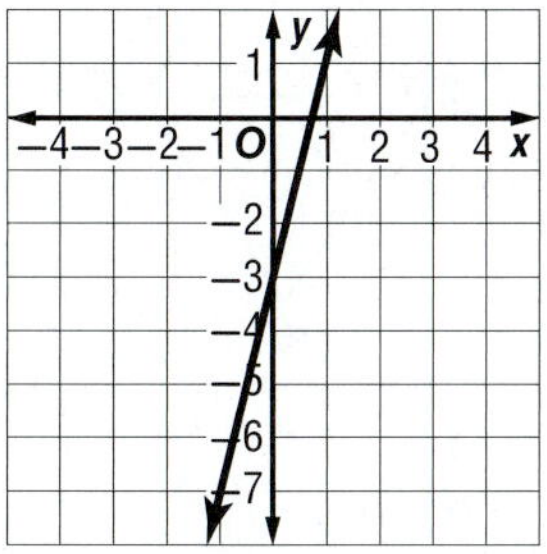

3. $y = -\frac{2}{3}x + 5$

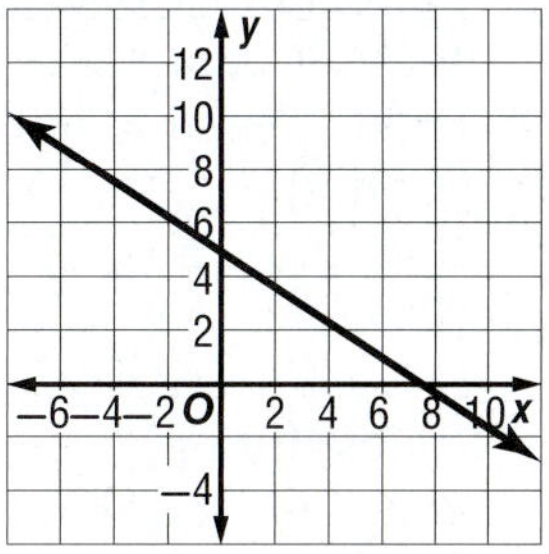

5. $y + 3 = \frac{1}{4}(x + 2)$

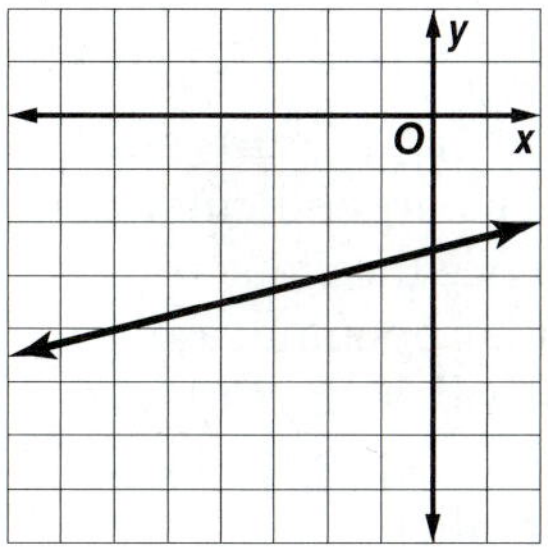

7. $y = \frac{5}{4}x - 1$

9. $y = \frac{9}{7}x - \frac{19}{7}$

11. $y = 4x + 9$

13. $y = -5x - 2$

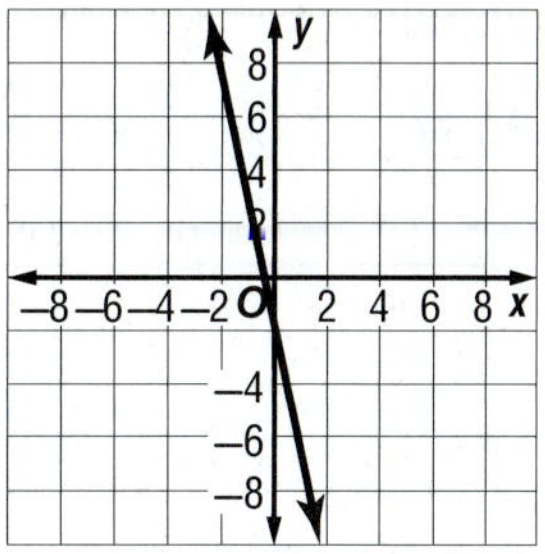

15. $y = 9x + 2$

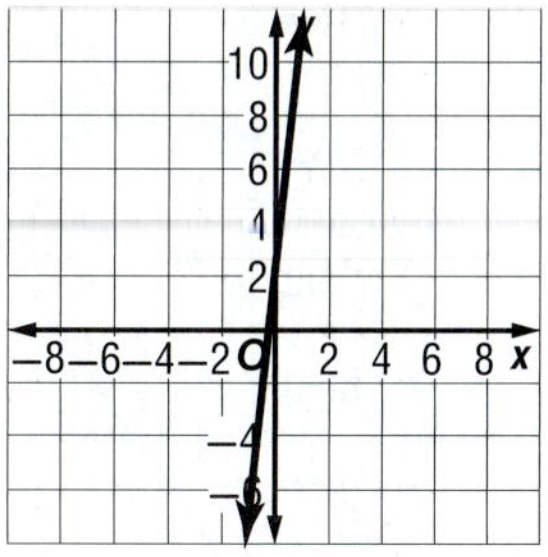

17. $y = -\frac{3}{4}x + 4$

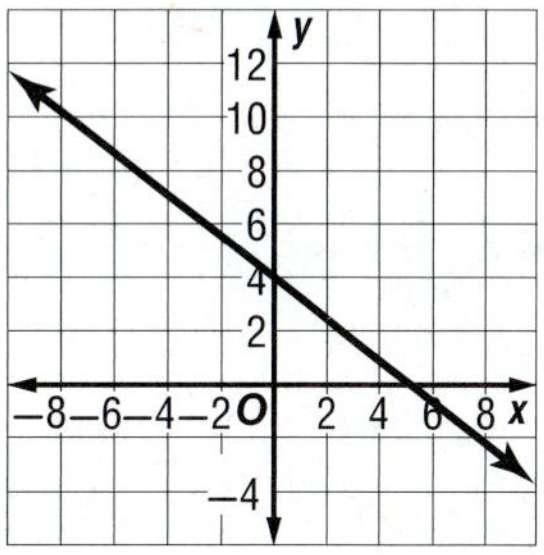

19

$y - y_1 = m(x - x_1)$	Point-slope form
$y - 11 = 2(x - 3)$	$m = 2$, $(x_1, y_1) = (3, 11)$
$y = 2x + 5$	Simplify.

Graph the given point (3, 11). Use the slope 2 or $\frac{2}{1}$ to find another point 2 units up and 1 unit to the right.

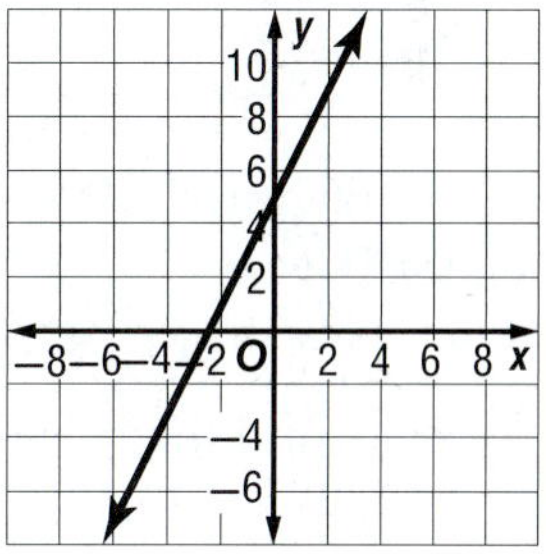

21. $y - 9 = -7(x - 1)$

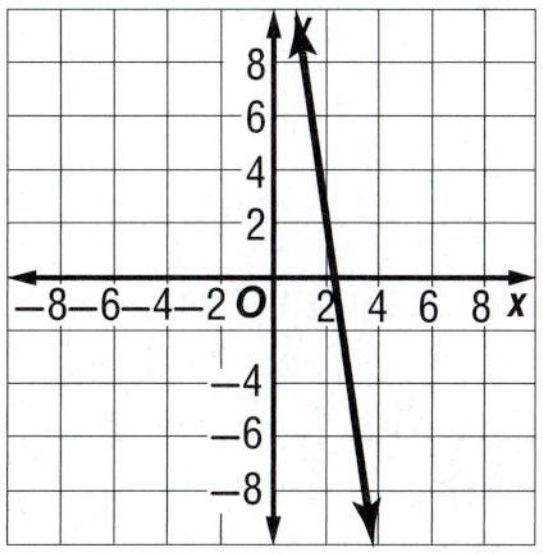

23. $y + 6 = -\frac{4}{5}(x + 3)$

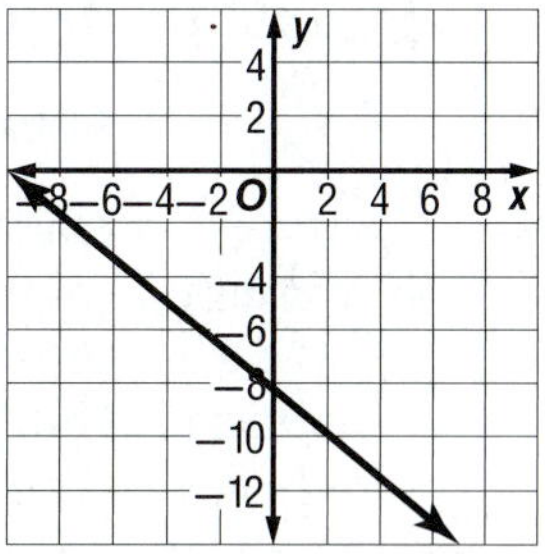

25. $y = -4$ **27.** $x = -3$ **29.** $y = -\frac{3}{4}x + 3$

31. $y = -\frac{10}{3}x + \frac{38}{3}$ **33.** $y = \frac{3}{2}x - \frac{1}{2}$ **35.** $y = \frac{2}{3}x - 2$

37. $y = -2x - 18$ **39.** $y = -\frac{2}{3}x + 6$

41 **a.** For each person, the cost increases \$5.50. So, the rate of change, or slope, is 5.5. The y-intercept represents the cost when there are 0 people, or \$400. Let y represent the total cost and x represent the number of people who attend the party.

$y = mx + b$	Slope-intercept form
$y = 5.5x + 400$	$m = 5.5$, $b = 400$

b. Let the x-axis represent the number of people and the y-axis represent the total cost.

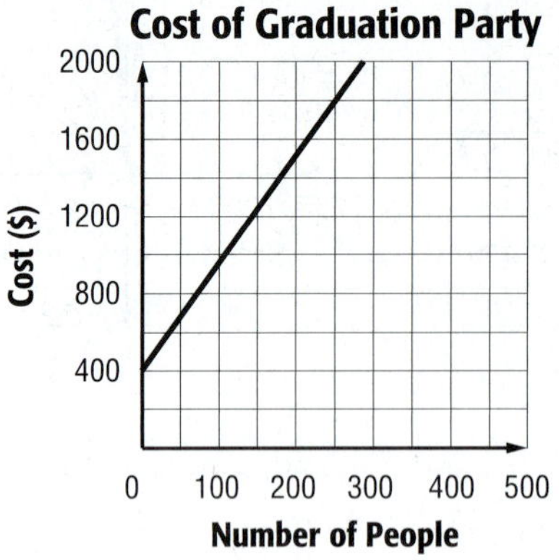

c. If $\frac{2}{3}$ of the class attends, then $\frac{2}{3} \cdot 285$ or 190 people attend.

$y = 5.5x + 400$	Write the equation.
$= 5.5(190) + 400$	$x = 190$
$= 1045 + 400$	Multiply.
$= 1445$	Simplify.

The party will cost $1445.

d.

$y = 5.5x + 400$	Write the equation.
$2000 = 5.5x + 400$	$y = 2000$
$1600 = 5.5x$	Subtract.
$y = 290.9$	Simplify.

If they raise $2000, 290 people can attend the party.

43. p **45.** $n, f,$ or r **47.** perpendicular **49.** neither

51 First, find the slope of the line through (3, 2) and (−7, 2)

$m = \frac{y_2 - y_1}{x_2 - x_1}$	Slope formula
$= \frac{2-2}{-7-3}$	$(x_1, y_1) = (3, 2), (x_2, y_2) = (-7, 2)$
$= \frac{0}{-10}$	Subtract.
$= 0$	Simplify.

Since the slope of the line is 0, it is a horizontal line. A line is perpendicular to a horizontal line if it is vertical, so find the vertical line through the point (−8, 12). The line through the point (−8, 12) perpendicular to the line through (3, 2) and (−7, 2) is $x = -8$.

53. $C = 15(x - 1) + 40$ or $C = 15x + 25$ **55.** 14 **57.** Sample answer: $x = -3, y = -7; y = 2x - 1, y = -\frac{1}{2}x - \frac{17}{2}$ **59.** Sample answer: When given the slope and y-intercept, the slope-intercept form is easier to use. When given two points, the point-slope form is easier to use. When given the slope and a point, the point-slope form is easier to use.
61. H **63.** B **65.** 2 **67.** $x = 3, y \approx 26.33$
69. Gas-O-Rama is also a quarter mile from Lacy's home; the two gas stations are half a mile apart.
71. consecutive interior **73.** alternate exterior

Pages 205–212 *Lesson 3-5*

1. $j \parallel k$; Converse of Corresponding Angles Postulate

3 Angle 3 and ∠10 are alternate exterior angles of lines ℓ and m. Since $\angle 3 \cong \angle 10$, $\ell \parallel m$ by the Alternate Exterior Angles Converse Theorem.

5. 20 **7.** Sample answer: Yes; since the alternate exterior angles are congruent, the backrest and footrest are parallel. **9.** $u \parallel v$; Alternate Exterior ∡ Converse
11. $r \parallel s$; Consecutive Interior ∡ Converse
13. $u \parallel v$; Alternate Interior ∡ Converse **15.** $r \parallel s$; Corresponding ∡ Converse **17.** 22; Corr. ∡ Post.

19 The angles are consecutive interior angles. For lines m and n to be parallel, consecutive interior angles must be supplementary, according to the Consecutive Interior Angles Converse Theorem.

$(7x - 2) + (10 - 3x) = 180$	Definition of supp. ∡
$4x + 8 = 180$	Simplify.
$4x = 172$	Subtract 8 from each side.
$x = 43$	Divide each side by 4.

21. 36; Alt. Ext. ∡ Thm.
23a. ∠1 and ∠2 are supplementary. **23b.** Def. of linear pair **23c.** ∠2 and ∠3 are supplementary.
23d. Trans. Prop. **23e.** Converse of Corr. ∡ Post.

25. Given: $\angle 1 \cong \angle 3$, $\overline{AC} \parallel \overline{BD}$
Prove: $\overline{AB} \parallel \overline{CD}$

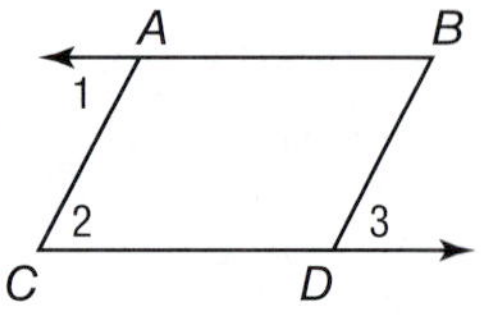

Proof:
Statements (Reasons)
1. $\angle 1 \cong \angle 3$, $\overline{AC} \parallel \overline{BD}$ (Given)
2. $\angle 2 \cong \angle 3$ (Corr. ∡ Post.)
3. $\angle 1 \cong \angle 2$ (Trans. Prop.)
4. $\overline{AB} \parallel \overline{CD}$ (If alternate ∡ are ≅, then lines are ∥.)

27. Given: $\angle ABC \cong \angle ADC$, $m\angle A + m\angle B = 180$
Prove: $\overline{AB} \parallel \overline{CD}$

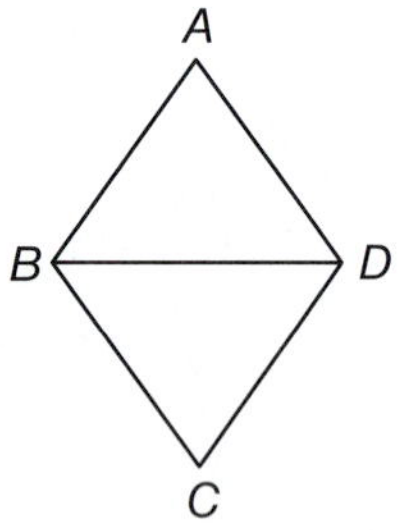

Proof:
Statements (Reasons)
1. $\angle ABC \cong \angle ADC$, $m\angle A + m\angle B = 180$ (Given)
2. $m\angle ABC \cong m\angle ADC$ (Def. of ≅ ∡)
3. $m\angle A + m\angle ADC = 180$ (Substitution)
4. ∠A and ∠ADC are supplementary. (Def. of supplementary ∡)
5. $\overline{AB} \parallel \overline{CD}$ (If consec. int. ∡ are supplementary, then lines are ∥.)

29 The Converse of the Perpendicular Transversal Theorem states that two lines perpendicular to the same line are parallel. Since the slots, or the bottom of each rectangular opening, are perpendicular to each of the sides, the slots are parallel. Since any pair of slots is perpendicular to the sides, they are also parallel.

31. **Given:** $\angle 1 \cong \angle 2$
Prove: $\ell \parallel m$
Proof:

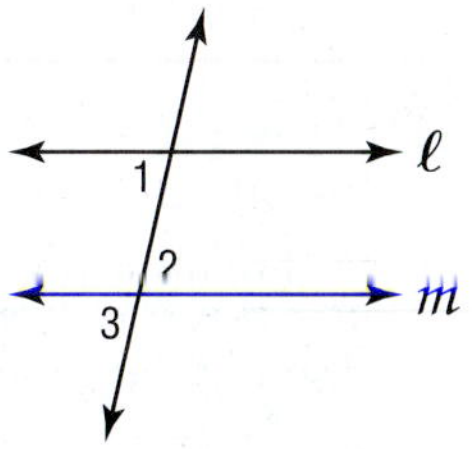

__Statements (Reasons)__

1. $\angle 1 \cong \angle 2$ (Given)
2. $\angle 2 \cong \angle 3$ (Vertical ⦞)
3. $\angle 1 \cong \angle 3$ (Transitive Prop.)
4. $\ell \parallel m$ (If corr ⦞ are ≅, then lines are ∥.)

33. $r \parallel s$; Sample answer: The corresponding angles are congruent. Since the angles are equal, the lines are parallel. **35.** $r \parallel s$; Sample answer: The alternate exterior angles are congruent. Since the angles are equal, the lines are parallel. **37.** Daniela; $\angle 1$ and $\angle 2$ are alternate interior angles for $\overline{WX}$ and $\overline{YZ}$, so if alternate interior angles are congruent, then the lines are parallel.

39a.

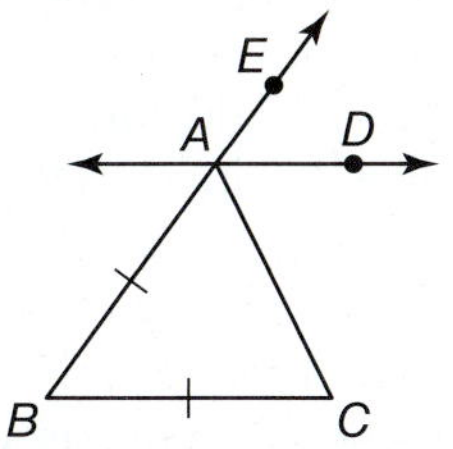

39b. Sample answer: Using a straightedge, the lines are equidistant. So they are parallel. **39c.** Sample answer: $\overleftrightarrow{AB}$ is a transversal for $\overleftrightarrow{BC}$ and $\overleftrightarrow{AD}$. $\angle ABC$ was copied to construct $\angle EAD$. So, $\angle ABC \cong \angle EAD$. $\angle ABC$ and $\angle EAD$ are corresponding angles, so by the converse of corresponding angles postulate, $\overleftrightarrow{AD} \parallel \overleftrightarrow{BC}$. **41a.** We know that $m\angle 1 + m\angle 2 = 180$. Since $\angle 2$ and $\angle 3$ are linear pairs, $m\angle 2 + m\angle 3 = 180$. By substitution, $m\angle 1 + m\angle 2 = m\angle 2 + m\angle 3$. By subtracting $m\angle 2$ from both sides we get $m\angle 1 = m\angle 3$. $\angle 1 \cong \angle 3$, by the definition of congruent angles. Therefore, $a \parallel c$ since the corresponding angles are congruent. **41b.** We know that $a \parallel c$ and $m\angle 1 + m\angle 3 = 180$. Since $\angle 1$ and $\angle 3$ are corresponding angles, they are congruent and their measures are equal. By substitution, $m\angle 3 + m\angle 3 = 180$ or $2m\angle 3 = 180$. By dividing both sides by 2, we get $m\angle 3 = 90$. Therefore, $t \perp c$ since they form a right angle. **43.** Sample answer: In order for supplementary angles to be both supplementary and congruent, they must be 90°, because they must sum to 180° and their angle measures must be equal. That means that the transversal must be perpendicular to the parallel lines.

45. G

47. H

49. $y = \frac{4}{5}x - 9$

51. 6 hours

53.

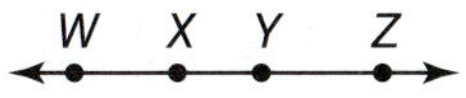

55. 8.6 m; 3.5 m^2 **57.** 10, 8.3

Pages 213–222 Lesson 3-6

1.

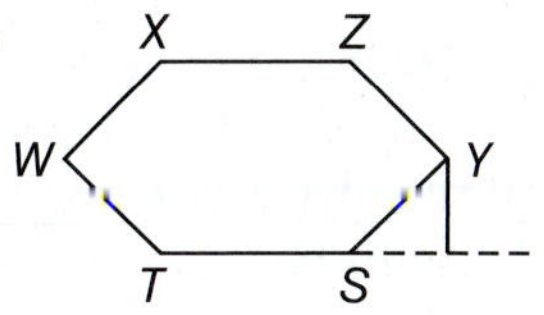

3. The formation should be that of two parallel lines that are also parallel to the 50-yard line; the band members have formed two lines that are equidistant from the 50-yard line, so by Theorem 3.9, the two lines formed are parallel. **5.** $\sqrt{10}$ units

7 Let ℓ represent $y = -2x + 4$ and let m represent $y = -2x + 14$. The slope of the lines is -2. Write an equation for line p. The slope of p is the opposite reciprocal of -2, or $\frac{1}{2}$. Use the y-intercept of line ℓ, (0, 4), as one of the endpoints of the perpendicular segment.

$(y - y_1) = m(x - x_1)$	Point-slope form
$(y - 4) = \frac{1}{2}(x - 0)$	$(x_1, y_1) = (0, 4), m = \frac{1}{2}$
$y - 4 = \frac{1}{2}x$	Simplify.
$y = \frac{1}{2}x + 4$	Add 4 to each side.

Use a system of equations to find the point of intersection of lines m and p.
Equation for m: $y = -2x + 14$
Equation for p: $y = \frac{1}{2}x + 4$

$y = -2x + 14$	Equation for m
$\frac{1}{2}x + 4 = -2x + 14$	Use Equation p to substitute $\frac{1}{2}x + 4$ for y.
$\frac{5}{2}x + 4 = 14$	Add $2x$ to each side.
$\frac{5}{2}x = 10$	Subtract 4 from each side.
$x = 4$	Multiply each side by $\frac{2}{5}$.
$y = \frac{1}{2}x + 4$	Equation for p
$y = \frac{1}{2}(4) + 4$	Substitute 4 for x.
$y = 6$	Simplify.

The point of intersection is (4, 6). Use the Distance Formula to find the distance between (0, 4) and (4, 6).

$d = \sqrt{(x_2 - x_1)^2 + (y_2 - y_1)^2}$	Distance Formula
$= \sqrt{(4 - 0)^2 + (6 - 4)^2}$	$x_1 = 0, y_1 = 4,$ $x_2 = 4, y_2 = 6$
$= \sqrt{20}$ or $2\sqrt{5}$	Simplify.

The distance between the lines is $2\sqrt{5}$ units.

9.

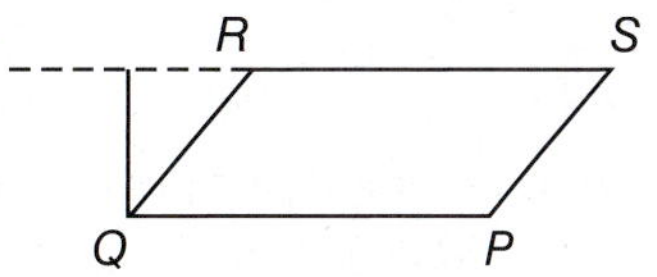

11.

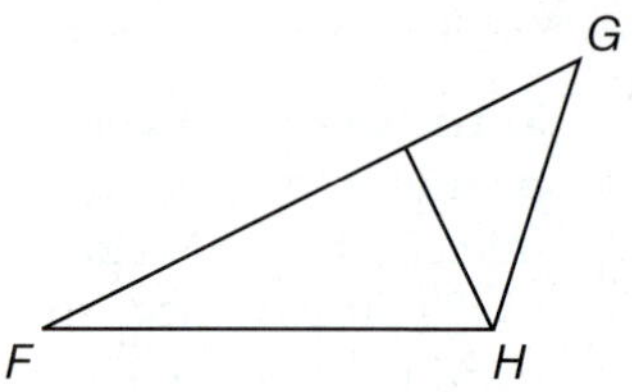

13. No; a driveway perpendicular to the road would be the shortest. The angle the driveway makes with the road is less than 90°, so it is not the shortest possible driveway.

15 Find the slope and y-intercept of line ℓ and write the equation for the line.

$m = \frac{y_2 - y_1}{x_2 - x_1} = \frac{4 - (-3)}{7 - 0}$ or 1

Since ℓ contains $(0, -3)$, the y-intercept is -3. So, the equation for line ℓ is $y = 1x + (-3)$ or $y = x - 3$. The slope of a line perpendicular to ℓ, line w, is -1. Write the equation of line w through $(4, 3)$ with slope -1.

$y = mx + b$	Slope-intercept form
$3 = -1(4) + b$	$m = -1, (x, y) = (4, 3)$
$3 = -4 + b$	Simplify.
$7 = b$	Add 4 to each side.

So, the equation for line w is $y = -x + 7$. Solve the system of equations to determine the point of intersection.

line ℓ:	$y = \quad x - 3$	
line w:	$(+)\ y = -x + 7$	
	$2y = \quad 4$	Add the two equations.
	$y = \quad 2$	Divide each side by 2.

Solve for x.

$y = x - 3$	Equation for line ℓ
$2 = x - 3$	$y = 2$
$5 = x$	Add 3 to each side.

The point of intersection is $(5, 2)$. Let this be point Q. Use the Distance Formula to determine the distance between $P(4, 3)$ and $Q(5, 2)$.

$d = \sqrt{(x_2 - x_1)^2 + (y_2 - y_1)^2}$	Distance Formula
$= \sqrt{(5 - 4)^2 + (2 - 3)^2}$	$x_1 = 4, y_1 = 3,$ $x_2 = 5, y_2 = 2$
$= \sqrt{2}$	Simplify.

The distance between the lines is $\sqrt{2}$ units.

17. 6 units **19.** $\sqrt{10}$ units **21.** 6 units **23.** $\sqrt{26}$ units **25.** 21 units **27.** $4\sqrt{17}$ units **29.** $\sqrt{14.76}$ units **31.** 5 units **33.** 6 units

35 He can conclude that the right and left sides of the bulletin board are not parallel, since the perpendicular distance between one line and any point on the other line must be equal anywhere on the lines for the two lines to be parallel. In this case, the length of the top of the bulletin board is not equal to the length of the bottom of the bulletin board.

39b. Place point C any place on line m. The area of the triangle is $\frac{1}{2}$ the height of the triangle times the length of the base of the triangle. The numbers stay constant regardless of the location of C on line m. **39c.** 16.5 in^2

41 To find out if the lines will intersect, determine if they are parallel. If they are parallel the perpendicular distance between the two lines at any point will be equal. Use a ruler to measure the distance between points A and C and points B and D. $AC = 1.2$ centimeters and $BD = 1.35$ centimeters. The lines are not parallel, so they will intersect. Shenequa is correct.

43. $a = \pm 1$; $y = \frac{1}{2}x + 6$ and $y = \frac{1}{2}x + \frac{7}{2}$, or $y = -\frac{1}{2}x + 6$ and $y = -\frac{1}{2}x + \frac{7}{2}$

45a.

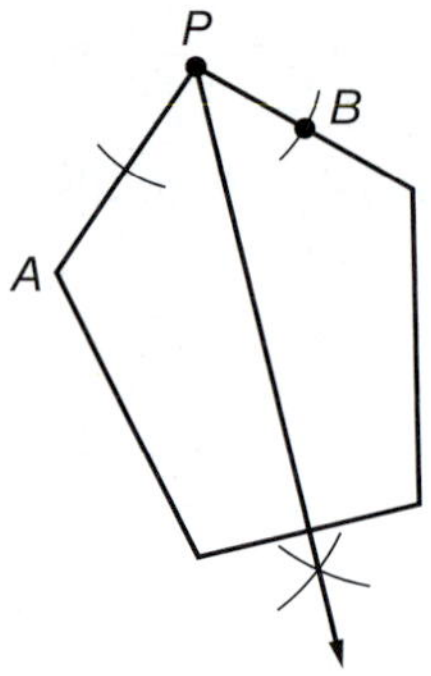

45b. Sample answer: Using a protractor, the measurement of the constructed angle is equal to 90. So, the line constructed from vertex P is perpendicular to the nonadjacent side chosen. **45c.** Sample answer: The same compass setting was used to construct points A and B. Then the same compass setting was used to construct the perpendicular line to the side chosen. Since the compass setting was equidistant in both steps, a perpendicular line was constructed.

47. Sample answer: First the line perpendicular to the pair of parallel lines is found. Then the point of intersection is found between the perpendicular line and the other line not used in the first step. Last, the Distance Formula is used to determine the distance between the pair of intersection points. This value is the distance between the pair of parallel lines. **49.** H **51.** G **53.** $y + 1 = \frac{1}{4}(x - 3)$

55. $y - 3 = -(x + 2)$

57. Given: $AB = BC$

Prove: $AC = 2BC$

Proof:

<u>**Statements (Reasons)**</u>

1. $AB = BC$ (Given)
2. $AC = AB + BC$ (Seg. Add. Post.)
3. $AC = BC + BC$ (Substitution)
4. $AC = 2BC$ (Substitution)

59. Sample answer: Robin ⊥ Cardinal; Bluebird divides two of the angles formed by Robin and Cardinal into pairs of complementary angles.

61. 5 **63.** 13 **65.** 5

Pages 213–222 Chapter 3 Study Guide and Review

1. false; parallel **3.** true **5.** true **7.** false; congruent **9.** corresponding **11.** alternate exterior **13.** skew lines **15.** 57; $\angle 5 \cong \angle 13$ by Corr. ∡ Post. and 13 and 14 form a linear pair **17.** 123; $\angle 11 \cong \angle 5$ by Alt. Int. ∡ Thm. and $\angle 5 \cong \angle 1$ by Alt. Ext. ∡ Thm. **19.** 57; $\angle 1 \cong \angle 3$ by Corr. ∡ Post. and $\angle 3$ and $\angle 6$ form a linear pair.

21. perpendicular **23.** parallel

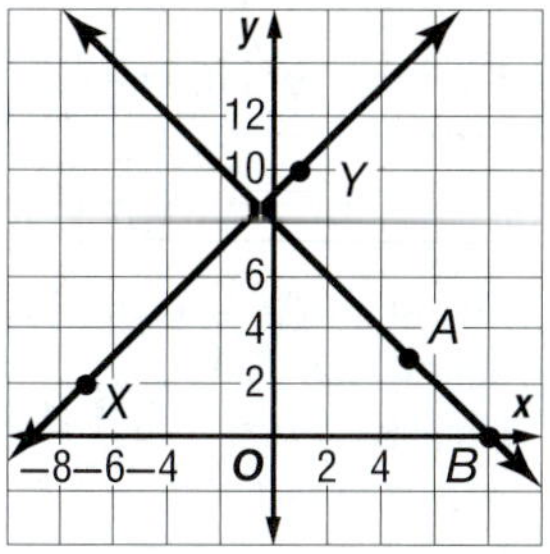

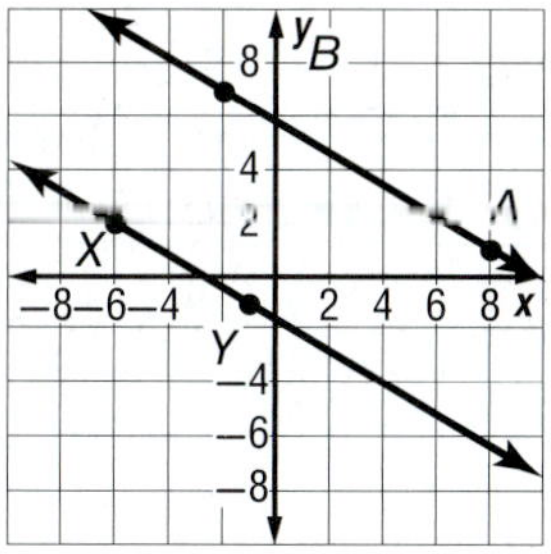

25.

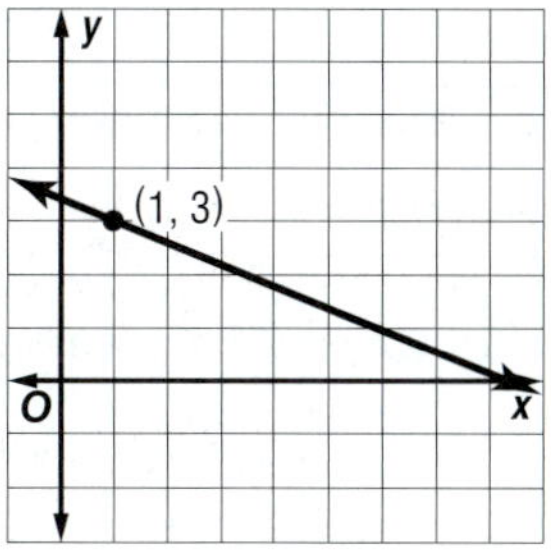

27. $y + 9 = 2(x - 4)$
29. $y = 5x - 3$
31. $y = -\frac{2}{3}x + 10$
33. $C = 20h + 50$
35. none
37. $v \parallel z$; Alternate Exterior Angles Converse Thm.

39. 135 **41.**

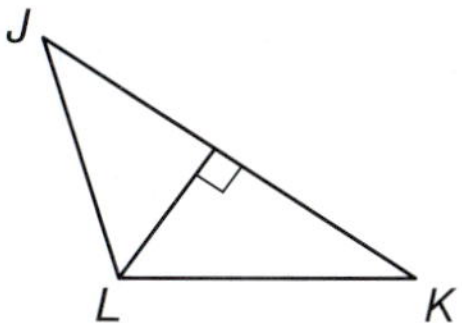

Chapter 4 Congruent Triangles

Page 233 Chapter 4 Get Ready

1. right **3.** obtuse **5.** 84°; Alternate Exterior Angles **7.** 10.8 **9.** 18.0 **11.** 144.2 mi

Pages 235–242 Lesson 4-1

1. right **3.** equiangular **5.** obtuse; $\angle BDC > 90°$ **7.** isosceles

9 By the definition of midpoint, $FK = KH$.

$FK + KH = FH$	Segment Addition Postulate
$KH + KH = FH$	Substitution
$2KH = FH$	Simplify.
$2(2.5) = FH$	Substitution
$5 = FH$	Simplify.

Since $\overline{GH} \cong \overline{FG}$, $GH = FG$ or 5. Since $GH = FG = FH = 5$, the triangle has three sides with the same measure. Therefore, the triangle has three congruent sides, so it is equilateral.

11. scalene **13.** $x = 5$, $QR = RS = QS = 25$ **15.** obtuse **17.** right **19.** acute **21.** obtuse **23.** acute **25.** right **27.** equilateral **29.** scalene **31.** scalene **33.** scalene **35.** equilateral

37 Since $\triangle FGH$ is equilateral, $FG = GH$.

$FG = GH$	Given
$3x + 10 = 9x - 8$	Substitution
$10 = 6x - 8$	Subtract $3x$ from each side.
$18 = 6x$	Add 8 to each side.
$3 = x$	Divide each side by 3.
$FG = 3x + 10$	Given
$= 3(3) + 10$ or 19	$x = 3$

$FG = GH = HF = 19$

39 Because the base of the prism formed is an equilateral triangle, the mirror tile must be cut into three strips of equal width. Since the original tile is a 12-inch square, each strip will be 12 inches long by 12 ÷ 3 or 4 inches wide.

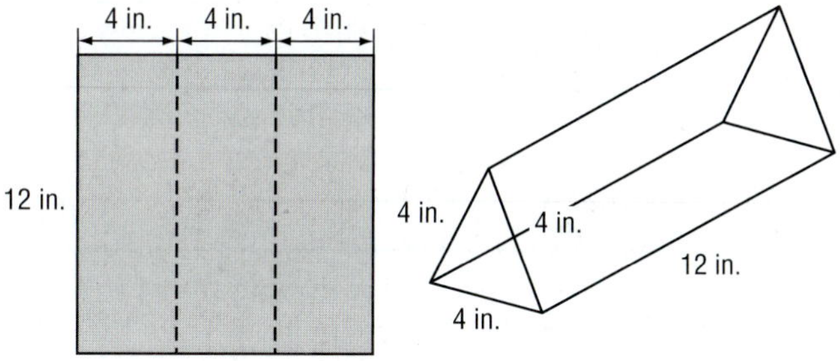

41. isosceles obtuse **43.** scalene; $XZ = 3\sqrt{5}$, $XY = \sqrt{113}$, $YZ = 2\sqrt{26}$ **45.** isosceles; $XZ = 2$, $XY = 2\sqrt{2}$, $YZ = 2$

47. Given: $m\angle ADC = 120$
Prove: $\triangle DBC$ is acute.
Proof: $\angle ADB$ and $\angle BDC$ form a linear pair. $\angle ADB$ and $\angle BDC$ are supplementary because if two angles form a linear pair, then they are supplementary. So, $m\angle ADB + m\angle BDC = 180$. We know $m\angle ADC = 120$, so by substitution, $120 + m\angle BDC = 180$. Subtract to find that $m\angle BDC = 60$. We already know that $\angle B$ is acute because $\triangle ABC$ is acute. $\angle BCD$ must also be acute because $\angle C$ is acute and $m\angle C = m\angle ACD + m\angle BCD$. $\triangle DBC$ is acute by definition.

49. $x = 15$; $FG = 35$, $GH = 35$, $HF = 35$ **51.** $x = 3$; $MN = 13$, $NP = 13$, $PM = 11$

53.

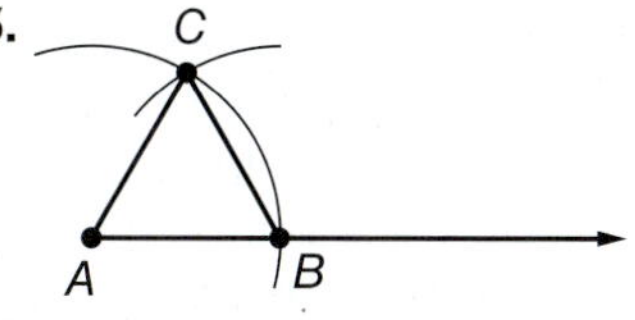

Sample answer: In $\triangle ABC$, $AB = BC = AC = 1.3$ cm. Since all sides have the same length, they are all congruent. Therefore the triangle is equilateral. $\triangle ABC$ was constructed using AB as the length of each side. Since the arc for each segment is the same, the triangle is equilateral.

55 **a.** Sample answer: acute isosceles

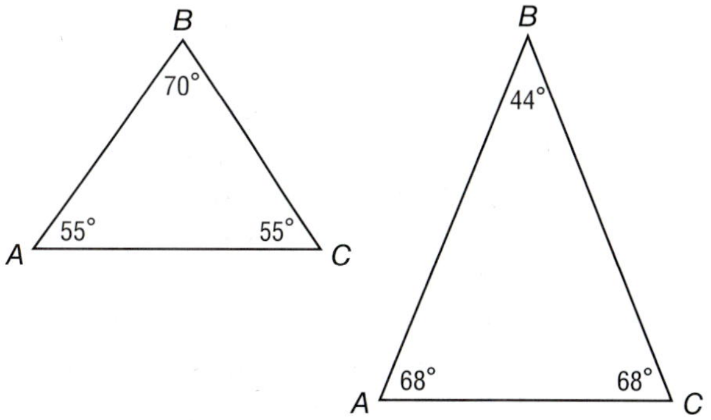

right isosceles

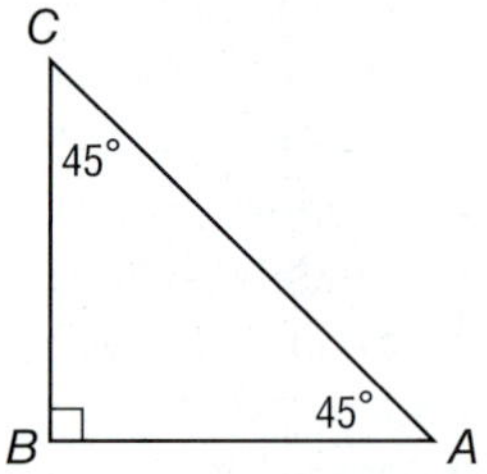

obtuse isosceles

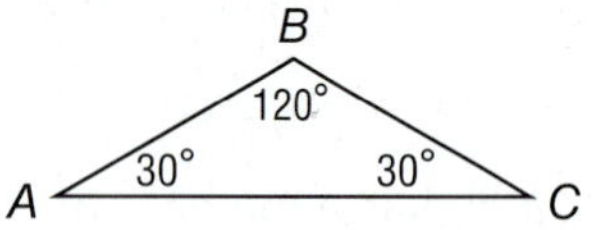

b.

$m\angle A$	$m\angle C$	$m\angle B$	Sum of Angle Measures
55	55	70	180
68	68	44	180
45	45	90	180
30	30	120	180

c. Sample answer: In an isosceles triangle, the angles opposite the congruent sides have the same measure. The sum of the angle measures of an isosceles triangle is 180. **d.** If the measures of the angles opposite the congruent sides of an isosceles triangle have the same measure, then if one angle measures x, then the other angle also measures x. If the sum of the measures of the angles of an isosceles triangle is 180, the measure of the third angle is $180 - (x + x)$ or $180 - 2x$. So, the other two angles measure x and $180 - 2x$.

57. Never; all equiangular triangles have three 60° angles, so they do not have a 90° angle. Therefore they cannot be right triangles. **59.** Never; all equilateral triangles are also equiangular, which means all of the angles are 60°. A right triangle has one 90° angle.

61. Sample answer:

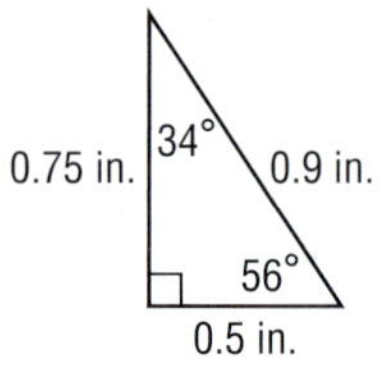

63. Not possible; all equilateral triangles have three acute angles. **65.** A **67.** 13.5 **69.** 7 **71.** $2\sqrt{5}$ **73.** Two lines in a plane that are perpendicular to the same line are parallel. **75.** H: you are a teenager; C: you are at least 13 years old

77. H: you have a driver's license; C: you are at least 16 years old **79.** Plane AEB intersects with plane $\mathcal{N}$ in $\overleftrightarrow{AB}$. **81.** Points D, C, and B lie in plane $\mathcal{N}$, but point E does not lie in plane $\mathcal{N}$. Thus, they are not coplanar. **83.** cons. int. **85.** alt. ext.

Pages 244–252 Lesson 4-2

1. 58 **3.** 80 **5.** 49 **7.** 78 **9.** 61 **11.** 151 **13.** 30

15
$m\angle L + m\angle M + m\angle 2 = 180$ Triangle Angle-Sum Theorem
$31 + 90 + m\angle 2 = 180$ Substitution
$121 + m\angle 2 = 180$ Simplify.
$m\angle 2 = 59$ Subtract 121 from each side.

$\angle 1$ and $\angle 2$ are congruent vertical angles. So, $m\angle 1 = 59$.

$m\angle 1 + m\angle 3 + m\angle P = 180$ Triangle Angle-Sum Theorem
$59 + m\angle 3 + 22 = 180$ Substitution
$81 + m\angle 3 = 180$ Simplify.
$m\angle 3 = 99$ Subtract 118 from each side.

$m\angle 1 = 59, m\angle 2 = 59, m\angle 3 = 99$

17. 79 **19.** 21

21
$m\angle A + m\angle B = 148$ Exterior Angle Theorem
$(2x - 15) + (x - 5) = 148$ Substitution
$3x - 20 = 148$ Simplify.
$3x = 168$ Add 20 to each side.
$x = 56$ Divide each side by 2.

So, $m\angle ABC = 56 - 5$ or 51.

23. 78 **25.** 39 **27.** 55 **29.** 35 **31.** $x = 30$; 30, 60

33 In $\triangle ABC$, $\angle B$ and $\angle C$ are congruent, so $m\angle B = m\angle C$.

$m\angle A = 3(m\angle B)$ $m\angle A$ is to be 3 times $m\angle B$.
$m\angle A + m\angle B + m\angle C = 180$ Triangle Angle-Sum Theorem
$3(m\angle B) + m\angle B + m\angle B = 180$ Substitution
$5(m\angle B) = 180$ Simplify.
$m\angle B = 36$ Divide each side by 5.

$m\angle C = m\angle B = 36$ $m\angle A = 3m\angle B = 3(36)$ or 108

35. Given: $\triangle MNO$; $\angle M$ is a right angle.
Prove: There can be at most one right angle in a triangle.
Proof: In $\triangle MNO$, M is a right angle. $m\angle M + m\angle N + m\angle O = 180$. $m\angle M = 90$, so $m\angle N + m\angle O = 90$. If N were a right angle, then $m\angle O = 0$. But that is impossible, so there cannot be two right angles in a triangle.
Given: $\triangle PQR$; $\angle P$ is obtuse.
Prove: There can be at most one obtuse angle in a triangle.
Proof: In $\triangle PQR$, $\angle P$ is obtuse. So $m\angle P > 90$. $m\angle P + m\angle Q + m\angle R = 180$. It must be that $m\angle Q + m\angle R < 90$. So, $\angle Q$ and $\angle R$ must be acute.

37. $m\angle 1 = 65, m\angle 2 = 20, m\angle 3 = 95, m\angle 4 = 40, m\angle 5 = 110, m\angle 6 = 45, m\angle 7 = 70, m\angle 8 = 65$ **39.** 67°, 23° **41.** $z < 23$; Sample answer: Since the sum of the measures of the angles of a triangle is 180 and $m\angle X = 157$, $157 + m\angle Y + m\angle Z = 180$, so $m\angle Y + m\angle Z = 23$. If $m\angle Y$ was 0, then $m\angle Z$ would equal 23. But since an angle must have a measure greater than 0, $m\angle Z$ must be less than 23, so $z < 23$.

43 Use the Triangle Angle-Sum Theorem and the Addition Property to prove the statement is true.
Given: $RSTUV$ is a pentagon.
Prove: $m\angle S + m\angle STU + m\angle TUV + m\angle V + m\angle VRS = 540$

Proof:

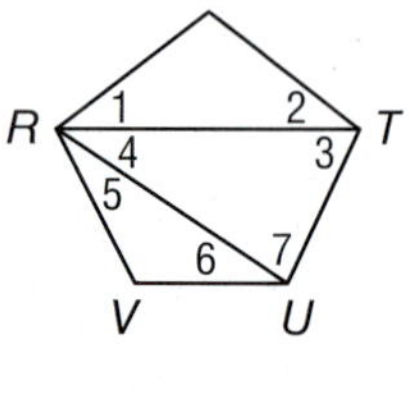

Statements (Reasons)
1. $RSTUV$ is a pentagon. (Given)
2. $m\angle S + m\angle 1 + m\angle 2 = 180$; $m\angle 3 + m\angle 4 + m\angle 7 = 180$; $m\angle 6 + m\angle V + m\angle 5 = 180$

3. $m\angle S + m\angle 1 + m\angle 2 + m\angle 3 + m\angle 4 + m\angle 7 + m\angle 6 + m\angle V + m\angle 5 = 540$ (Addition Property) **4.** $m\angle VRS = m\angle 1 + m\angle 4 + m\angle 5$; $m\angle TUV = m\angle 7 + m\angle 6$; $m\angle STU = m\angle 2 + m\angle 3$ ($\angle$ Addition) **5.** $m\angle S + m\angle STU + m\angle TUV + m\angle V + m\angle VRS = 540$ (Substitution)

45a. Sample answer:

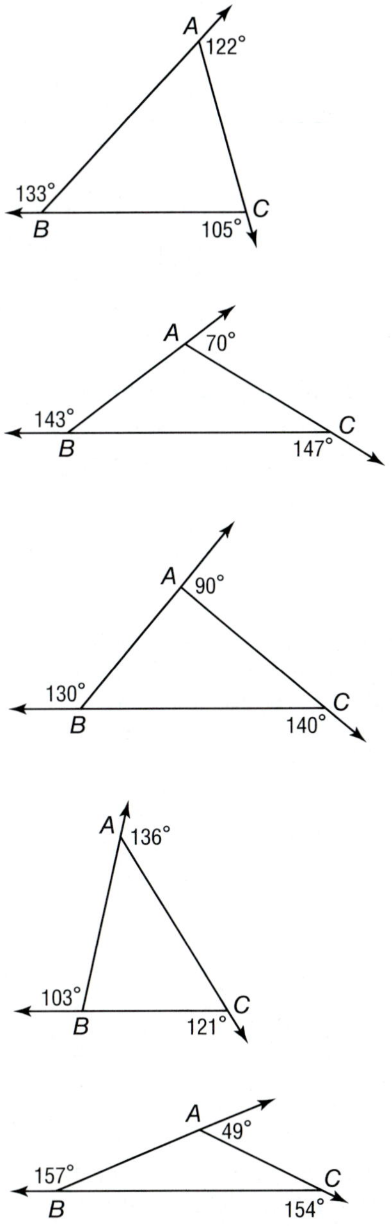

45b. Sample answer:

$m\angle 1$	$m\angle 2$	$m\angle 3$	Sum of Angle Measures
122	105	133	360
70	147	143	360
90	140	130	360
136	121	103	360
49	154	157	360

45c. Sample answer: The sum of the measures of the exterior angles of a triangle is 360.

45d. $m\angle 1 + m\angle 2 + m\angle 3 = 360$

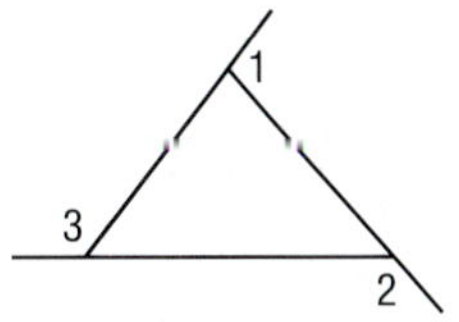

45e. The Exterior Angle Theorem tells us that $m\angle 3 = m\angle BAC + m\angle BCA$, $m\angle 2 = m\angle BAC + m\angle CBA$, $m\angle 1 = m\angle CBA + m\angle BCA$. Through substitution, $m\angle 1 + m\angle 2 + m\angle 3 = m\angle CBA + m\angle BCA + m\angle BAC + m\angle CBA + m\angle BAC + m\angle BCA$. This can be simplified to $m\angle 1 + m\angle 2 + m\angle 3 = 2m\angle CBA + 2m\angle BCA + 2m\angle BAC$. The Distributive Property can be applied and gives $m\angle 1 + m\angle 2 + m\angle 3 = 2(m\angle CBA + m\angle BCA + m\angle BAC)$. The Triangle Angle-Sum Theorem tells us that $m\angle CBA + m\angle BCA + m\angle BAC = 180$. Through substitution we have $m\angle 1 + m\angle 2 + m\angle 3 = 2(180) = 360$.
47. The measure of $\angle a$ is the supplement of the exterior angle with measure 110, so $m\angle a = 180 - 110$ or 70. Because the angles with measures b and c are congruent, $b = c$. Using the Exterior Angle Theorem, $b + c = 110$. By substitution, $b + b = 110$, so $2b = 110$ and $b = 55$. Because $b = c$, $c = 55$. **49.** $y = 13, z = 14$
51. Sample answer: Since an exterior angle is acute, the adjacent angle must be obtuse. Since another exterior angle is right, the adjacent angle must be right. A triangle cannot contain both a right and an obtuse angle because it would be more than 180 degrees. Therefore, a triangle cannot have an obtuse, acute, and a right exterior angle. **53.** 100°, 115°, 145°
55. D **57.** obtuse **59.** $\sqrt{26}$ units
61. 

63. Multiplication Property **65.** Addition Property
67. Substitution Property

Pages 253–261 Lesson 4-3

1. $\angle Y \cong \angle S$, $\angle X \cong \angle R$, $\angle XZY \cong \angle RZS$, $\overline{YX} \cong \overline{SR}$, $\overline{YZ} \cong \overline{SZ}$, $\overline{XZ} \cong \overline{RZ}$; $\triangle YXZ \cong \triangle SRZ$ **3.** $\frac{1}{2}$ in.; Sample answer: The nut is congruent to the opening for the $\frac{1}{2}$-in. socket.

5

$\angle M \cong \angle R$	CPCTC
$m\angle M = m\angle R$	Definition of congruence
$y + 10 = 2y - 40$	Substitution
$10 = y - 40$	Subtract y from each side.
$50 = y$	Add 40 to each side.

7. 16; $\angle N$ corresponds to $\angle X$. By the Third Angles Theorem, $m\angle N = 64$, so $4x = 64$. **9.** $\angle X \cong \angle A$, $\angle Y \cong \angle B$, $\angle Z \cong \angle C$, $\overline{XY} \cong \overline{AB}$, $\overline{XZ} \cong \overline{AC}$, $\overline{YZ} \cong \overline{BC}$; $\triangle XYZ \cong \triangle ABC$ **11.** $\angle R \cong \angle J$, $\angle T \cong \angle K$, $\angle S \cong \angle L$, $\overline{RT} \cong \overline{JK}$, $\overline{TS} \cong \overline{KL}$, $\overline{RS} \cong \overline{JL}$; $\triangle RTS \cong \triangle JKL$ **13.** 20

15

$\overline{ED} \cong \overline{UT}$	CPCTC
$ED = UT$	Definition of congruence
$3z + 10 = z + 16$	Substitution
$2z + 10 = 16$	Subtract z from each side.
$2z = 6$	Subtract 10 from each side.
$z = 3$	Divide each side by 3.

17a. $\triangle ABC \cong \triangle MNO$; $\triangle DEF \cong \triangle PQR$ **17b.** $\overline{AB} \cong \overline{MN}$, $\overline{BC} \cong \overline{NO}$, $\overline{AC} \cong \overline{MO}$, $\overline{DE} \cong \overline{PQ}$, $\overline{EF} \cong \overline{QR}$, $\overline{DF} \cong \overline{PR}$ **17c.** $\angle A \cong \angle M$, $\angle B \cong \angle N$, $\angle C \cong \angle O$, $\angle D \cong \angle P$, $\angle E \cong \angle Q$, $\angle F \cong \angle R$

19 $148 + 18 + a = 180$ Triangle Angle-Sum Theorem
$166 + a = 180$ Simplify.
$a = 14$ Subtract 166 from each side.

If two angles of one triangle are congruent to two angles of another triangle, then the third angles of the triangles are congruent. So, $3x + y = 14$ and $5x - y = 18$. Solve the system of equations.

$3x + y = 14$
$(+)\ 5x - y = 18$
$8x = 32$ Add the equations.
$x = 4$ Divide each side by 8.

$3x + y = 14$ Original equation
$3(4) + y = 14$ $x = 4$
$12 + y = 14$ Simplify.
$y = 2$ Subtract 12 from each side.

21. Given: $\angle A \cong \angle D$
$\angle B \cong \angle E$
Prove: $\angle C \cong \angle F$
Proof:

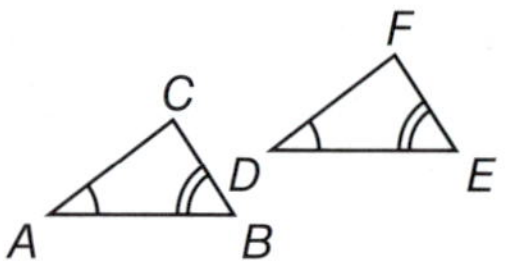

Statements (Reasons)
1. $\angle A \cong \angle D$, $\angle B \cong \angle E$ (Given)
2. $m\angle A = m\angle D$, $m\angle B = m\angle E$ (Def. of $\cong$)
3. $m\angle A + m\angle B + m\angle C = 180$, $m\angle D + m\angle E + m\angle F = 180$ ($\angle$ Sum Theorem)
4. $m\angle A + m\angle B + m\angle C = m\angle D + m\angle E + m\angle F$ (Trans. Prop.)
5. $m\angle D + m\angle E + m\angle C = m\angle D + m\angle E + m\angle F$ (Subst.)
6. $m\angle C = m\angle F$ (Subt. Prop.)
7. $\angle C \cong \angle F$ (Def. of $\cong$ $\angle$s)

23. Given: $\overline{BD}$ bisects $\angle B$.
$\overline{BD} \perp \overline{AC}$
Prove: $\angle A \cong \angle C$
Proof:
Statements (Reasons)
1. $\overline{BD}$ bisects $\angle B$, $\overline{BD} \perp \overline{AC}$. (Given)
2. $\angle ABD \cong \angle DBC$ (Def. of angle bisector)
3. $\angle ADB$ and $\angle BDC$ are right angles. ($\perp$ lines form rt. $\angle$s.)
4. $\angle ADB \cong \angle BDC$ (All rt. $\angle$s are $\cong$.)
5. $\angle A \cong \angle C$ (Third $\angle$ Thm.)

25. Sample answer: Both of the punched flowers are congruent to the flower on the stamp, because it was used to create the images. According to the Transitive Property of Polygon Congruence, the two stamped images are congruent to each other because they are both congruent to the flowers on the punch.

27. Given: $\triangle DEF$
Prove: $\triangle DEF \cong \triangle DEF$

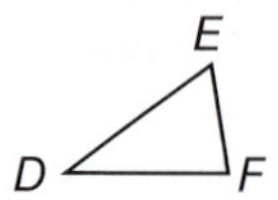

Proof:

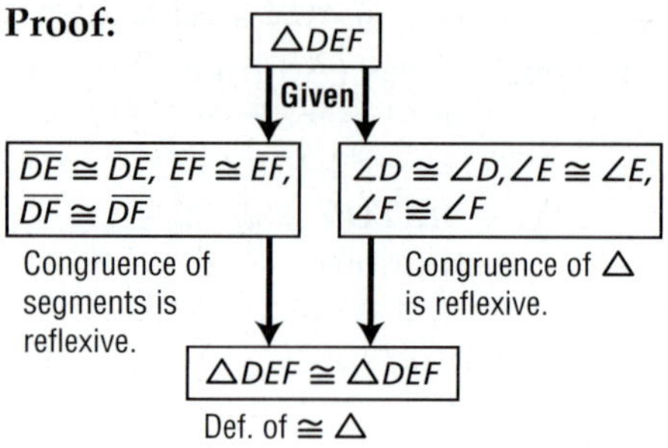

29.

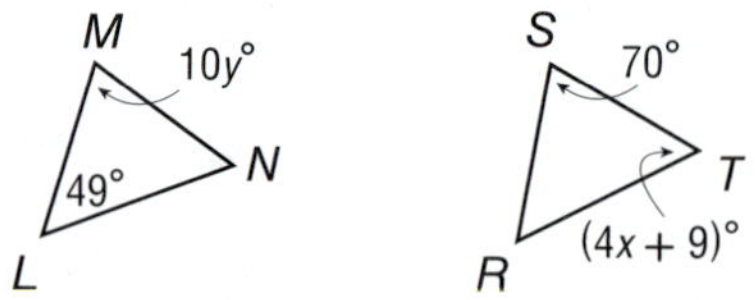

$x = 13$; $y = 7$

31 **a.** All the longer sides of the triangles are congruent and all the shorter sides are congruent. Sample answer: $\overline{AB} \cong \overline{CB}$, $\overline{AB} \cong \overline{DE}$, $\overline{AB} \cong \overline{FE}$, $\overline{CB} \cong \overline{DE}$, $\overline{CB} \cong \overline{FE}$, $\overline{DE} \cong \overline{FE}$, $\overline{AC} \cong \overline{DF}$ **b.** If the area is a square, then each of the four sides measures $\sqrt{100}$ or 10 feet. So, the perimeter of the square is 4(10) or 40 ft. The pennant string will need to be 40 ft long. **c.** Each pennant and the distance to the next pennant is 6 in. or 0.5 ft. So, the number of pennants is $40 \div 0.5$ or 80.

33a. If two triangles are congruent, then their areas are equal. **33b.** If the areas of a pair of triangles are equal, then the triangles are congruent; false; If one triangle has a base of 2 and a height of 6 and a second triangle has a base of 3 and a height of 4, then their areas are equal, but they are not congruent. **33c.** No; sample answer: Any pair of equilateral triangles that have the same base also have the same height, so it is not possible to draw a pair of equilateral triangles with the same area that are not congruent.
33d. yes;
sample answer:

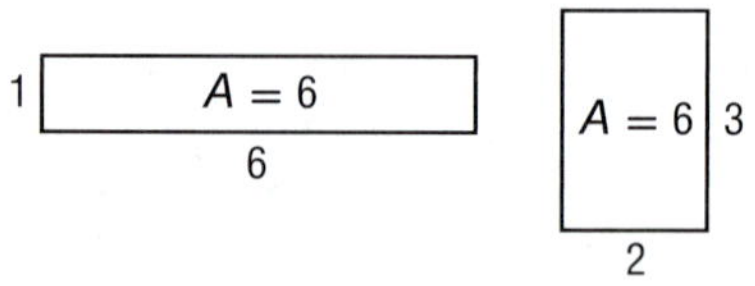

33e. No; any pair of squares that have the same area have the same side length, which is the square root of the area. If their areas are equal, they are congruent. **33f.** Regular n-gons; If two regular n-gons are congruent, then they have the same area. All regular n-gons have the same shape, but may have different sizes. If two regular n-gons have the same area, then they not only have the same shape but also the same size. Therefore, they are congruent. **35.** diameter, radius, or circumerence; Sample answer: Two circles are the same size if they have the same diameter, radius, or circumference, so she can determine if the hoops are congruent if she measures any of them. **37.** Both; Sample answer: $\angle A$ corresponds with $\angle Y$, $\angle B$ corresponds with $\angle X$, and $\angle C$ corresponds with $\angle Z$. $\triangle CAB$ is the same triangle as $\triangle ABC$ and $\triangle ZXY$ is the same triangle as $\triangle XYZ$. **39.** $x = 16$, $y = 8$

41. False; $\angle A \cong \angle X$, $\angle B \cong \angle Y$, $\angle C \cong \angle Z$, but corresponding sides are not congruent.

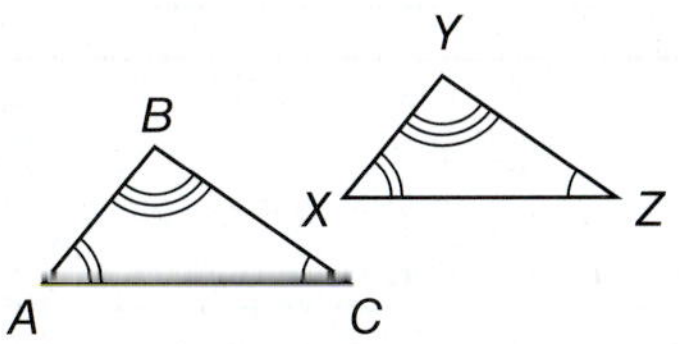

43. Sometimes; Equilateral triangles will be congruent if one pair of corresponding sides are congruent. **45.** 5 **47.** B **49.** 59 **51.** $JK = 2\sqrt{146}$, $KL = \sqrt{290}$, $JL = \sqrt{146}$; scalene **53.** $JK = 5$, $KL = 5\sqrt{2}$, $JL = 5$; isosceles **55.** always **57.** complementary angles

Pages 262–270 Lesson 4-4

1a. two

1b. Given: $AB = CD$; $DA = BC$

Prove: $\triangle ABC \cong \triangle CDA$

Proof:

Statements (Reasons)

1. $AB = CD$; $DA = BC$ (Given)
2. $\overline{AB} \cong \overline{CD}$, $\overline{BC} \cong \overline{DA}$ (Def. of $\cong$)
3. $\overline{AC} \cong \overline{CA}$ (Reflex. Prop. $\cong$)
4. $\triangle ABC \cong \triangle CDA$ (SSS)

1c. Sample answer: Since all of the squares in the pattern are congruent, all of the triangles that form the squares are congruent. The bases form two lines and the legs are transversals. The corresponding angles are congruent, so the lines formed by the bases are parallel.

3 Sample answer: We are given that $\overline{LP} \cong \overline{NO}$ and $\angle LPM \cong \angle NOM$. Since $\triangle MOP$ is equilateral, $\overline{MO} \cong \overline{MP}$ by the definition of an equilateral triangle. So, two sides and the included angle of $\triangle LMP$ are congruent to two sides and the included angle of $\triangle NMO$. Therefore, $\triangle LMP$ is congruent to $\triangle NMO$ by the Side-Angle-Side Congruence Postulate.

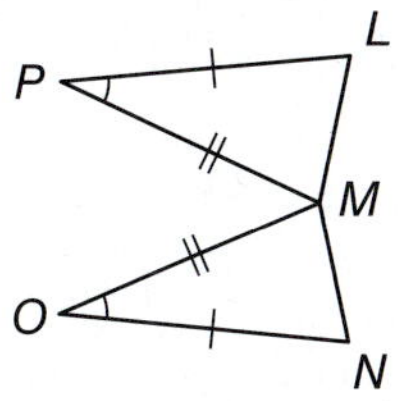

5. Given: $\overline{QR} \cong \overline{SR}$ and $\overline{ST} \cong \overline{QT}$

Prove: $\triangle QRT \cong \triangle SRT$

Proof: We know that $\overline{QR} \cong \overline{SR}$ and $\overline{ST} \cong \overline{QT}$. $\overline{RT} \cong \overline{RT}$ by the Reflexive Property. Since $\overline{QR} \cong \overline{SR}$, $\overline{ST} \cong \overline{QT}$, and $\overline{RT} \cong \overline{RT}$, $\triangle QRT \cong \triangle SRT$ by SSS.

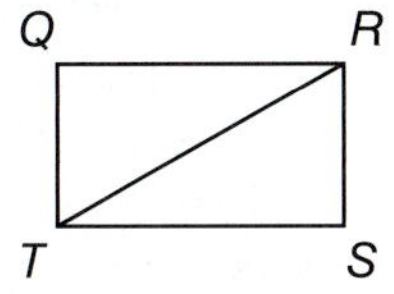

7. Given: $\overline{AB} \cong \overline{ED}$, $\angle ABC$ and $\angle EDC$ are right angles, and C is the midpoint of $\overline{BD}$.

Prove: $\triangle ABC \cong \triangle EDC$

Proof:

Statements (Reasons)

1. $\overline{AB} \cong \overline{ED}$, $\angle ABC$ and $\angle EDC$ are right angles, and C is the midpoint of $\overline{BD}$. (Given)
2. $\angle ABC \cong \angle EDC$ (All rt. ∠s $\cong$)
3. $\overline{BC} \cong \overline{CD}$ (Midpoint Thm.)
4. $\overline{CD} \cong \overline{DC}$ (Reflex. Prop. $\cong$)
5. $\overline{BC} \cong \overline{DC}$ (Trans. Prop.)
6. $\triangle ABC \cong \triangle EDC$ (SAS)

9 Use $d = \sqrt{(x_2 - x_1)^2 + (y_2 - y_1)^2}$ to find the lengths of the sides of $\triangle MNO$.

$MN = \sqrt{(-1 - 0)^2 + [-4 - (-1)]^2}$ $(x_1, y_1) = (0, -1)$, $(x_2, y_2) = (-1, -4)$

$= \sqrt{1 + 9}$ or $\sqrt{10}$ Simplify.

$NO = \sqrt{[-4 - (-1)]^2 + [-3 - (-4)]^2}$ $(x_1, y_1) = (-1, -4)$, $(x_2, y_2) = (-4, -3)$

$= \sqrt{9 + 1}$ or $\sqrt{10}$ Simplify.

$MO = \sqrt{(-4 - 0)^2 + [-3 - (-1)]^2}$ $(x_1, y_1) = (0, -1)$, $(x_2, y_2) = (-4, -3)$

$= \sqrt{16 + 4}$ or $\sqrt{20}$ Simplify.

Find the lengths of the sides of $\triangle QRS$.

$QR = \sqrt{(4 - 3)^2 + [-4 - (-3)]^2}$ $(x_1, y_1) = (3, -3)$, $(x_2, y_2) = (4, -4)$

$= \sqrt{1 + 1}$ or $\sqrt{2}$ Simplify.

$RS = \sqrt{(3 - 4)^2 + [3 - (-4)]^2}$ $(x_1, y_1) = (4, -4)$, $(x_2, y_2) = (3, 3)$

$= \sqrt{1 + 49}$ or $\sqrt{50}$ Simplify.

$QS = \sqrt{(3 - 3)^2 + [3 - (-3)]^2}$ $(x_1, y_1) = (3, -3)$, $(x_2, y_2) = (3, 3)$

$= \sqrt{0 + 36}$ or 6 Simplify.

$MN = \sqrt{10}$, $NO = \sqrt{10}$, $MO = \sqrt{20}$, $QR = \sqrt{2}$, $RS = \sqrt{50}$, and $QS = 6$. The corresponding sides are not congruent, so the triangles are not congruent.

11. $MN = \sqrt{10}$, $NO = \sqrt{10}$, $MO = \sqrt{20}$, $QR = \sqrt{10}$, $RS = \sqrt{10}$, and $QS = \sqrt{20}$. Each pair of corresponding sides has the same measure, so they are congruent. $\triangle MNO \cong \triangle QRS$ by SSS.

13. Given: R is the midpoint of $\overline{QS}$ and $\overline{PT}$.

Prove: $\triangle PRQ \cong \triangle TRS$

Proof: Since R is the midpoint of $\overline{QS}$ and $\overline{PT}$, $\overline{PR} \cong \overline{RT}$ and $\overline{RQ} \cong \overline{RS}$ (by definition of a midpoint). $\angle PRQ \cong \angle TRS$ by the Vertical Angles Theorem.

So, $\triangle PRQ \cong \triangle TRS$ by SSS.

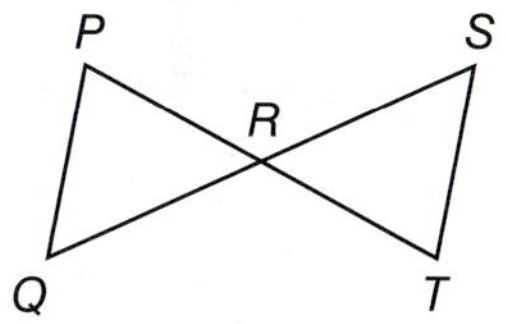

15. Given: $\triangle XYZ$ is equilateral. $\overline{WY}$ bisects $\angle Y$.

Prove: $\overline{XW} \cong \overline{ZW}$

Proof: We know that $\overline{WY}$ bisects $\angle Y$, so $\angle XYW \cong \angle ZYW$. Also, $\overline{YW} \cong \overline{YW}$ by the Reflexive Property. Since $\triangle XYZ$ is equilateral it is a special type of isosceles triangle, so $\overline{XY} \cong \overline{ZY}$. By the Side-Angle-Side Congruence Postulate, $\triangle XYW \cong \triangle ZYW$. By CPCTC, $\overline{XW} \cong \overline{ZW}$.

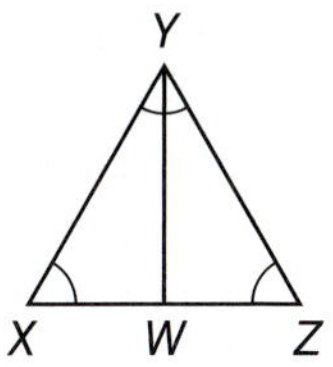

17 The triangles have two pairs of congruent sides. Since triangles cannot be proven congruent using only two sides, it is not possible to prove congruence.

19. SAS

21. Given: $\overline{MJ} \cong \overline{ML}$;
K is the midpoint of $\overline{JL}$.
Prove: $\triangle MJK \cong \triangle MLK$

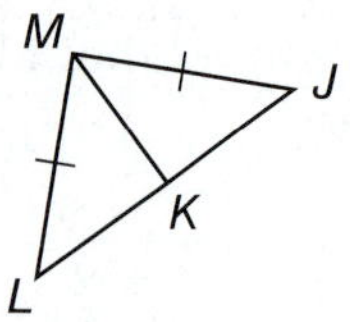

Proof:

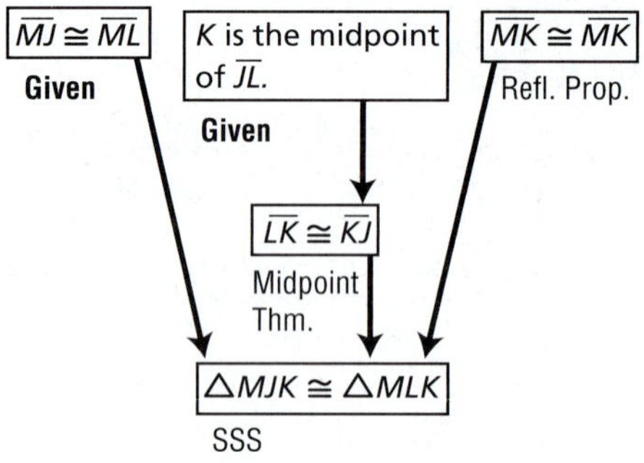

23a. Given: $\overline{TS} \cong \overline{SF} \cong \overline{FR} \cong \overline{RT}$; $\angle TSF$, $\angle SFR$, $\angle FRT$, and $\angle RTS$ are right angles.
Prove: $\overline{RS} \cong \overline{TF}$

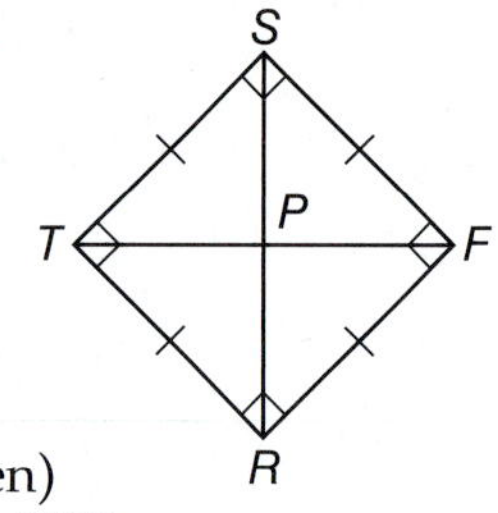

Proof:

<u>**Statements (Reasons)**</u>

1. $\overline{TS} \cong \overline{SF} \cong \overline{FR} \cong \overline{RT}$ (Given)
2. $\angle TSF$, $\angle SFR$, $\angle FRT$, and $\angle RTS$ are right angles. (Given)
3. $\angle STR \cong \angle TRF$ (All rt. $\angle$s are $\cong$.)
4. $\triangle STR \cong \triangle TRF$ (SAS)
5. $\overline{RS} \cong \overline{TF}$ (CPCTC)

23b. Given: $\overline{TS} \cong \overline{SF} \cong \overline{FH} \cong \overline{HT}$; $\angle TSF$, $\angle SFH$, $\angle FHT$, and $\angle HTS$ are right angles.
Prove: $\angle SRT \cong \angle SRF$

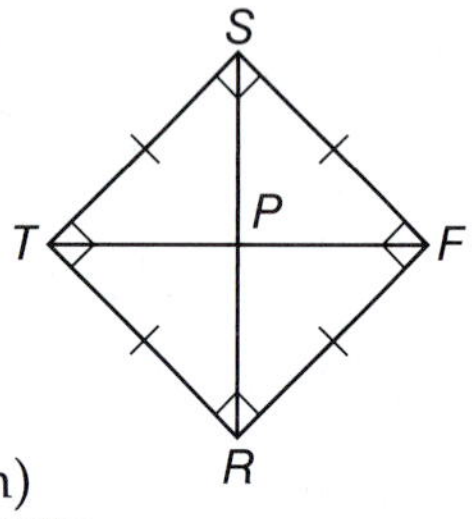

Proof:

<u>**Statements (Reasons)**</u>

1. $\overline{TS} \cong \overline{SF} \cong \overline{FR} \cong \overline{RT}$ (Given)
2. $\angle TSF$, $\angle SFR$, $\angle FRT$, and $\angle RTS$ are right angles. (Given)
3. $\angle STR \cong \angle SFR$ (All rt. $\angle$s are $\cong$.)
4. $\triangle STR \cong \triangle SFR$ (SAS)
5. $\angle SRT \cong \angle SRF$ (CPCTC)

25. Given: $\triangle EAB \cong \triangle DCB$
Prove: $\triangle EAD \cong \triangle DCE$

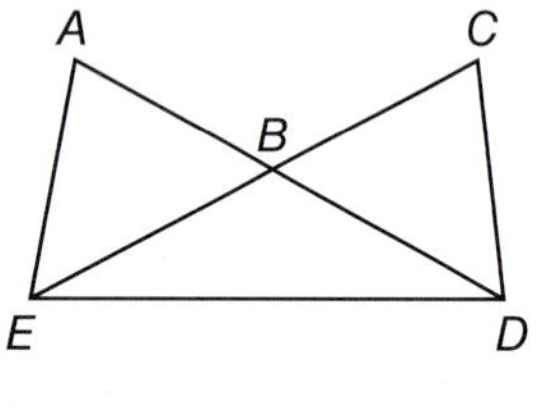

Proof:

<u>**Statements (Reasons)**</u>

1. $\triangle EAB \cong \triangle DCB$ (Given)
2. $\overline{EA} \cong \overline{DC}$ (CPCTC)
3. $\overline{ED} \cong \overline{DE}$ (Reflex. Prop.)
4. $\overline{AB} \cong \overline{CB}$ (CPCTC)
5. $\overline{DB} \cong \overline{EB}$ (CPCTC)
6. $AB = CB$, $DB = EB$ (Def. $\cong$ segments)
7. $AB + DB = CB + EB$ (Add. Prop. =)
8. $AD = AB + DB$, $CE = CB + EB$ (Seg. addition)
9. $AD = CE$ (Subst. Prop. =)
10. $\overline{AD} \cong \overline{CE}$ (Def. $\cong$ segments)
11. $\triangle EAD \cong \triangle DCE$ (SSS)

27 By the Reflexive Property, $\overline{WX} \cong \overline{WX}$. $\angle WXZ \cong \angle WXY$ since all right angles are congruent.

$\angle WXZ \cong \angle WXY$	All right angles are congruent.
$m\angle WXZ = m\angle WXY$	Definition of congruence
$90 = 20y + 10$	Substitution
$80 = 20y$	Subtract 10 from each side.
$y = 4$	Divide each side by 2.

$\overline{XY} \cong \overline{XZ}$	CPCTC
$XY = XZ$	Definition of congruence
$3y + 7 = 19$	Substitution
$3y = 12$	Subtract 7 from each side.
$y = 4$	Divide each side by 3.

If $y = 4$, then by SAS, the triangles are congruent.

29a. Sample answer: Method 1: You could use the Distance Formula to find the length of each of the sides, and then use the Side-Side-Side Congruence Postulate to prove the triangles congruent. Method 2: You could find the slopes of $\overline{ZX}$ and $\overline{WY}$ to prove that they are perpendicular and that $\angle WYZ$ and $\angle WYX$ are both right angles. You can use the Distance Formula to prove that $\overline{XY}$ is congruent to $\overline{ZY}$. Since the triangles share the leg $\overline{WY}$, you can use the Side-Angle-Side Congruence Postulate; Sample answer: I think that method 2 is more efficient, because you only have two steps instead of three.

29b. Sample answer: Yes; the slope of $\overline{WY}$ is -1 and the slope of $\overline{ZX}$ is 1, and -1 and 1 are opposite reciprocals, so $\overline{WY}$ is perpendicular to $\overline{ZX}$. Since they are perpendicular, $\angle WYZ$ and $\angle WYX$ are both 90°. Using the Distance Formula, the length of $\overline{ZY}$ is $\sqrt{(4-1)^2 + (5-2)^2}$ or $3\sqrt{2}$, and the length of $\overline{XY}$ is $\sqrt{(7-4)^2 + (8-5)^2}$ or $3\sqrt{2}$. Since $\overline{WY}$ is congruent to $\overline{WY}$, $\triangle WYZ$ is congruent to $\triangle WYX$ by the Side-Angle-Side Congruence Postulate. **31.** Shada; for SAS the angle must be the included angle and here it is not included. **33.** Case 1: You know the hypotenuses are congruent and two corresponding legs are congruent. Then the Pythagorean Theorem says that the other legs are congruent so the triangles are congruent by SSS. Case 2: You know the legs are congruent and the right angles are congruent, then the triangles are congruent by SAS. **35.** F **37.** C **39.** 18 **41.** $y = -\frac{1}{5}x - 4$ **43.** $y = x + 3$ **45.** False; a 16-year-old could be a freshman, sophomore, junior, or senior. The hypothesis of the conditional is true, but the conclusion is false. This counterexample shows that the conditional statement is false. **47.** Trans. Prop. **49.** Substitution

Pages 273–280 Lesson 4-5

1. Given: $\overline{CB}$ bisects $\angle ABD$ and $\angle ACD$.
Prove: $\triangle ABC \cong \triangle DBC$

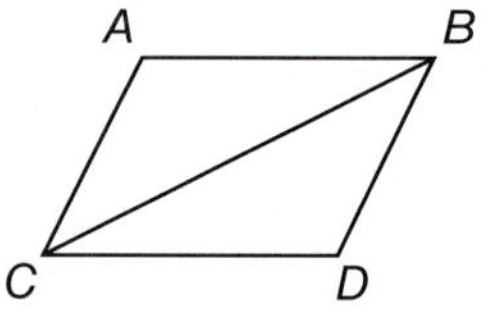

Proof:

Statements (Reasons)

1. $\overline{CB}$ bisects $\angle ABD$ and $\angle ACD$. (Given)
2. $\angle ABC \cong \angle DBC$ (Def. of $\angle$ bisector)
3. $\overline{BC} \cong \overline{BC}$ (Refl. prop.)
4. $\angle ACB \cong \angle DCB$ (Def. of $\angle$ bisector)
5. $\triangle ABC \cong \triangle DBC$ (ASA)

3. Given: $\angle K \cong \angle M$, $\overline{JK} \cong \overline{JM}$, $\overline{JL}$ bisects $\angle KLM$.

Prove: $\triangle JKL \cong \triangle JML$

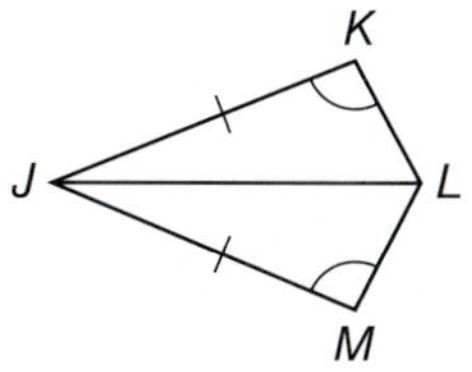

Proof: We are given $\angle K \cong \angle M$, $\overline{JK} \cong \overline{JM}$, and $\overline{JL}$ bisects $\angle KLM$. Since $\overline{JL}$ bisects $\angle KLM$, we know $\angle KLJ \cong \angle MLJ$. So, $\triangle JKL \cong \triangle JML$ is congruent by the AAS Congruence Theorem.

5 **a.** We know $\angle BAE$ and $\angle DCE$ are congruent because they are both right angles. $\overline{AE}$ is congruent to $\overline{EC}$ by the Midpoint Theorem. From the Vertical Angles Theorem, $\angle DEC \cong \angle BEA$. So, two angles and the included side of $\triangle DCE$ are congruent to two angles and the included side of $\triangle BAE$. By ASA, the surveyor knows that $\triangle DCE \cong \triangle BAE$. By CPCTC, $\overline{DC} \cong \overline{AB}$, so the surveyor can measure $\overline{DC}$ and know the distance between A and B.

b.

$\triangle DCE \cong \triangle BAE$	SAS
$\overline{DC} \cong \overline{AB}$	CPCTC
$DC = AB$	Definition of congruence
$550 = AB$	Substitution

So, by the definition of congruence, $AB = 550$ m.

7. Given: $\angle W \cong \angle Y$, $\overline{WZ} \cong \overline{YZ}$, $\overline{XZ}$ bisects $\angle WZY$.

Prove: $\triangle XWZ \cong \triangle XYZ$

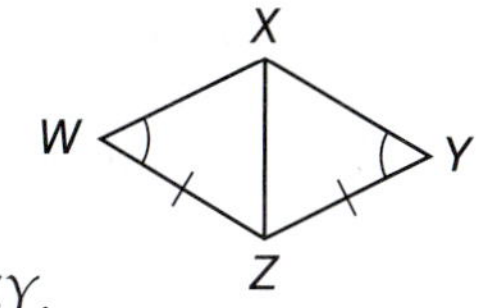

Proof: It is given that $\angle W \cong \angle Y$, $\overline{WZ} \cong \overline{YZ}$, and $\overline{XZ}$ bisects $\angle WZY$. By the definition of angle bisector, $\angle WZX \cong \angle YZX$. The Angle-Side-Angle Congruence Postulate tells us that $\triangle XWZ \cong \triangle XYZ$.

9 Use the Alternate Interior Angle Theorem and AAS to prove the triangles congruent.

Given: V is the midpoint of $\overline{YW}$; $\overline{UY} \parallel \overline{XW}$.

Prove: $\triangle UVY \cong \triangle XVW$

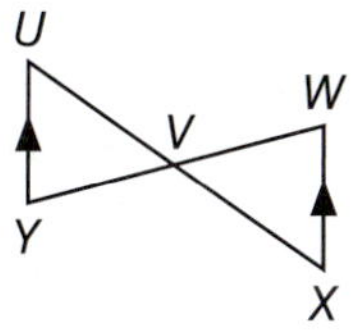

Proof:

Statements (Reasons)

1. V is the midpoint of $\overline{YW}$; $\overline{UY} \parallel \overline{XW}$. (Given)
2. $\overline{YV} \cong \overline{VW}$ (Midpoint Theorem)
3. $\angle VWX \cong \angle VYU$ (Alt. Int. $\angle$ Thm.)
4. $\angle VUY \cong \angle VXW$ (Alt. Int. $\angle$ Thm.)
5. $\triangle UVY \cong \triangle XVW$ (AAS)

11. Given: $\angle A$ and $\angle C$ are right angles. $\angle ABE \cong \angle CBD$, $\overline{AE} \cong \overline{CD}$

Prove: $\overline{BE} \cong \overline{BD}$

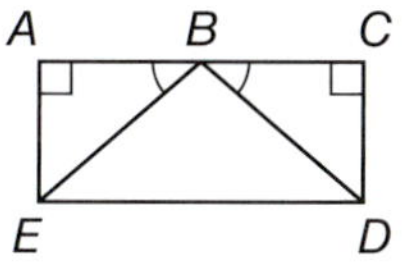

Proof:

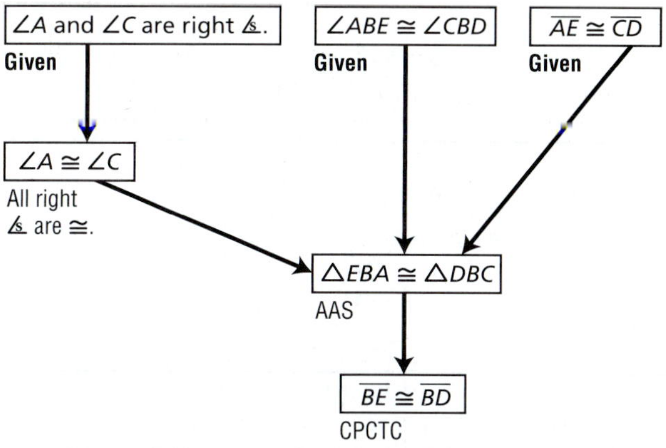

13a. $\angle HJK \cong \angle GFK$ since all right angles are congruent. We are given that $\overline{JK} \cong \overline{KF}$. $\angle HKJ$ and $\angle FKG$ are vertical angles, so $\angle HKJ \cong \angle FKG$ by the Vertical Angles Theorem. By ASA, $\triangle HJK \cong \triangle GFK$, so $\overline{FG} \cong \overline{HJ}$ by CPCTC. **13b.** No; $HJ = 1350$ m, so $FG = 1350$ m. If the regatta is to be 1500 m, the lake is not long enough, since $1350 < 1500$.

15 If the triangles are congruent, $HJ = QJ$. Solve for y.

$HJ = QJ$	CPCTC
$9 = 2y - 1$	$HJ = 9$, $QJ = 2y - 1$
$y = 5$	Simplify.

17. Given: $\overline{AE} \perp \overline{DE}$, $\overline{EA} \perp \overline{AB}$, C is the midpoint of $\overline{AE}$.

Prove: $\overline{CD} \cong \overline{CB}$

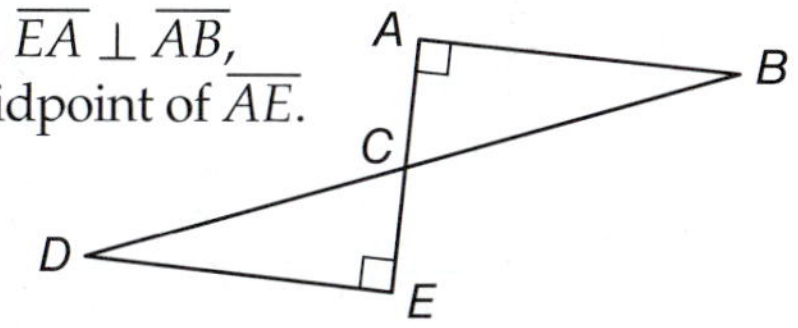

Proof: We are given that $\overline{AE}$ is perpendicular to $\overline{DE}$, $\overline{EA}$ is perpendicular to $\overline{AB}$, and C is the midpoint of $\overline{AE}$. Since $\overline{AE}$ is perpendicular to $\overline{DE}$, $m\angle CED = 90$. Since $\overline{EA}$ is perpendicular to $\overline{AB}$, $m\angle BAC = 90$. $\angle CED \cong \angle BAC$ because all right angles are congruent. $\overline{AC} \cong \overline{CE}$ from the Midpt. Thm. $\angle ECD \cong \angle ACB$ because they are vertical angles. Angle-Side-Angle gives us that $\triangle CED \cong \triangle CAB$. $\overline{CD} \cong \overline{CB}$ because corresponding parts of congruent triangles are congruent.

19. Given: $\angle K \cong \angle M$, $\overline{KP} \perp \overline{PR}$, $\overline{MR} \perp \overline{PR}$

Prove: $\angle KPL \cong \angle MNL$

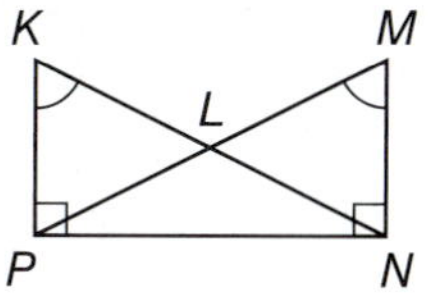

Proof:

Statements (Reasons)

1. $\angle K \cong \angle M$, $\overline{KP} \perp \overline{PR}$, $\overline{MR} \perp \overline{PR}$ (Given)
2. $\angle KPR$ and $\angle MRP$ are both right angles. (Def. of $\perp$)
3. $\angle KPR \cong \angle MRP$ (All rt. $\angle$s are congruent.)
4. $\overline{PR} \cong \overline{PR}$ (Refl. Prop.)
5. $\triangle KPR \cong \triangle MRP$ (AAS)
6. $\overline{KP} \cong \overline{MR}$ (CPCTC)
7. $\angle KLP \cong \angle MLR$ (Vertical angles are $\cong$.)
8. $\triangle KLP \cong \triangle MLR$ (AAS)
9. $\angle KPL \cong \angle MRL$ (CPCTC)

21 Since $m\angle ACB = m\angle ADB = 44$, and $m\angle CBA = m\angle DBA = 68$, then $\angle ACB \cong \angle ADB$ and $\angle CBA \cong \angle DBA$ by the definition of congruence. $\overline{AB} \cong \overline{AB}$

by the Reflexive Property, so $\triangle ACB \cong \triangle ADB$ by AAS. Then $\overline{AC} \cong \overline{AD}$ by CPCTC. Since $AC = AD$ by the definition of congruence, the two seat stays are the same length.

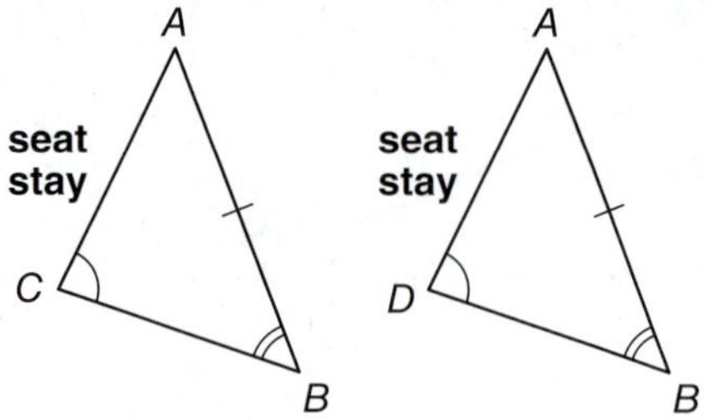

23. Tyrone; Lorenzo showed that all three corresponding angles were congruent, but AAA is not a proof of triangle congruence.

25.

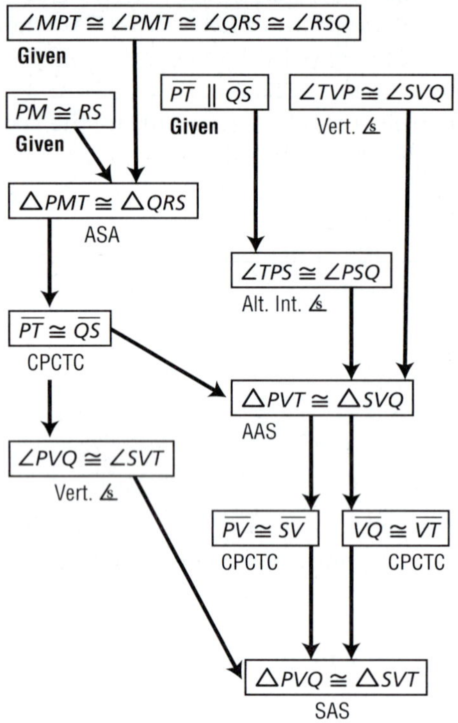

27. B **29.** J **31.** $AB = \sqrt{125}$, $BC = \sqrt{221}$, $AC = \sqrt{226}$, $XY = \sqrt{125}$, $YZ = \sqrt{221}$, $XZ = \sqrt{226}$. The corresponding sides have the same measure and are congruent. $\triangle ABC \cong \triangle XYZ$ by SSS.

33. $x = 19$; $y = 3$

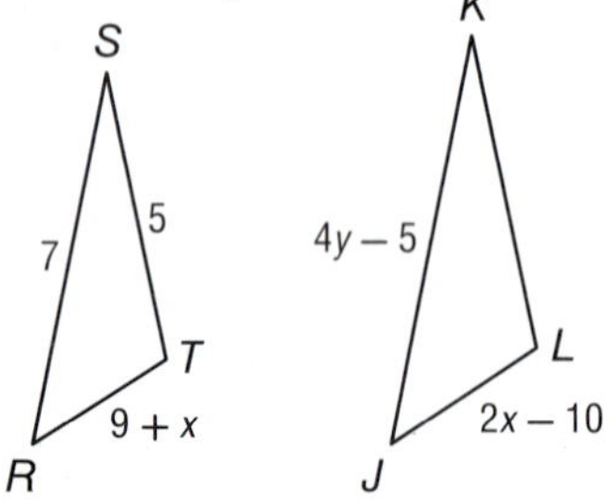

35.

p	q	$\sim p$	$\sim p \vee q$
F	T	T	T
T	T	F	T
F	F	T	T
T	F	F	F

37. Given: $\angle 2 \cong \angle 1$
$\angle 1 \cong \angle 3$
Prove: $\overline{AB} \parallel \overline{DE}$

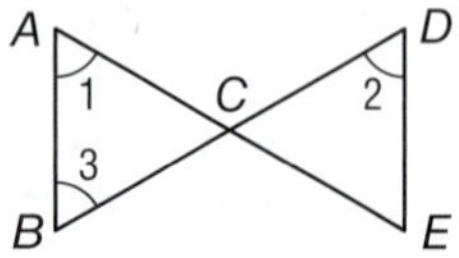

Statements (Reasons)
1. $\angle 2 \cong \angle 1$, $\angle 1 \cong \angle 3$ (Given)
2. $\angle 2 \cong \angle 3$ (Trans. Prop.)
3. $\overline{AB} \parallel \overline{DE}$ (If alt. int. ∠s are ≅, lines are ∥.)

Pages 283–291 Lesson 4-6

1. $\angle BAC$ and $\angle BCA$ **3.** 12 **5.** 12
7. Given: $\triangle ABC$ is isosceles; $\overline{EB}$ bisects $\angle ABC$.
Prove: $\triangle ABE \cong \triangle CBE$
Proof:
Statements (Reasons)
1. $\triangle ABC$ is isosceles; $\overline{EB}$ bisects $\angle ABC$. (Given)
2. $\overline{AB} \cong \overline{BC}$ (Def. of isosceles)
3. $\angle ABE \cong \angle CBE$ (Def. of ∠ bisector)
4. $\overline{BE} \cong \overline{BE}$ (Refl. Prop.)
5. $\triangle ABE \cong \triangle CBE$ (SAS)

9 $\angle ABE$ is opposite $\overline{AE}$ and $\angle AEB$ is opposite $\overline{AB}$. Since $\overline{AE} \cong \overline{AB}$, $\angle ABE \cong \angle AEB$.
11. $\angle ACD$ and $\angle ADC$ **13.** $\overline{BF}$ and $\overline{BC}$ **15.** 60 **17.** 4

19 The triangle is equiangular, so it is also equilateral. All the sides are congruent.

$2x + 11 = 6x - 9$	Definition of congruence
$11 = 4x - 9$	Subtract 2x from each side.
$20 = 4x$	Add 9 to each side.
$5 = x$	Divide each side by 4.

21. $x = 11$, $y = 11$
23. Given: $\triangle HJM$ is an isosceles triangle and $\triangle HKL$ is an equilateral triangle. $\angle JKH$, $\angle HKL$ and $\angle HLK$, $\angle MLH$ are supplementary.

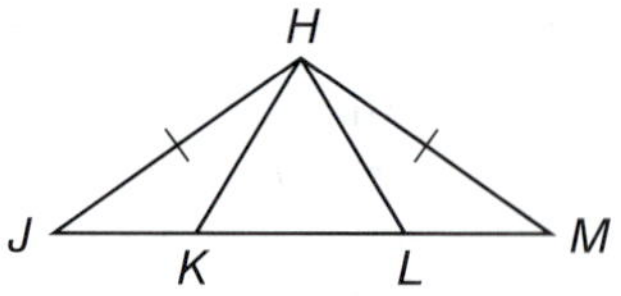

Prove: $\angle JHK \cong \angle MHL$
Proof: We are given that $\triangle HJM$ is an isosceles triangle and $\triangle HKL$ is an equilateral triangle, $\angle JKH$ and $\angle HKL$ are supplementary and $\angle HLK$ and $\angle MLH$ are supplementary. From the Isosceles Triangle Theorem, we know that $\angle HJK \cong \angle HML$. Since $\triangle HKL$ is an equilateral triangle, we know $\angle HLK \cong \angle LKH \cong \angle KHL$ and $\overline{HL} \cong \overline{KL} \cong \overline{HK}$. $\angle JKH$, $\angle HKL$ and $\angle HLK$, $\angle MLH$ are supplementary, and $\angle HKL \cong \angle HLK$, we know $\angle JKH \cong \angle MLH$ by the Congruent Supplements Theorem. By AAS, $\triangle JHK \cong \triangle MLH$. By CPCTC, $\angle JHK \cong \angle MHL$.

25a. 65°; Since $\triangle ABC$ is an isosceles, $\angle ABC \cong \angle ACB$, so $180 - 50 = 130$ and $\frac{130}{2}$ or 65.
25b. Given: $\overline{BE} \cong \overline{CD}$
Prove: $\triangle AED$ is isosceles.
Proof:
Statements (Reasons)
1. $\overline{AB} \cong \overline{AC}$, $\overline{BE} \cong \overline{CD}$ (Given)
2. $AB = AC$, $BE = CD$ (Def. of congruence)
3. $AB + BE = AE$, $AC + CD = AD$ (Seg. Add. Post.)
4. $AB + BE = AC + CD$ (Add. Prop. of Eq.)
5. $AE = AD$ (Subst.)
6. $\overline{AE} \cong \overline{AD}$ (Def. of congruence)
7. $\triangle AED$ is isosceles. (Def. of isosceles)

25c. Given: $\overline{BC} \parallel \overline{ED}$ and $\overline{ED} \cong \overline{AD}$
Prove: $\triangle ADE$ is equilateral.
Proof:

Statements (Reasons)
1. $\overline{AB} \cong \overline{AC}$ (Given)
$\overline{BC} \parallel \overline{ED}$ and $\overline{ED} \cong \overline{AD}$
2. $\angle ABC \cong \angle ACB$ (Isos. △ Thm.)
3. $m\angle ABC = m\angle ACB$ (Def. of ≅)
4. $\angle ABC \cong \angle AED$, $\angle ACB \cong \angle ADE$ (Corr. ∠s)
5. $m\angle ABC = m\angle AED$, $m\angle ACB = m\angle ADE$ (Def. of ≅)
6. $m\angle AED = m\angle ACB$ (Subst.)
7. $m\angle AED = m\angle ADE$ (Subst.)
8. $\angle AED \cong \angle ADE$ (Def. of ≅)
9. $\overline{AD} \cong \overline{AE}$ (Conv. of Isos. △ Thm.)
10. $\triangle ADE$ is equilateral. (Def. of equilateral △)

25d. One pair of congruent corresponding sides and one pair of congruent corresponding angles; since we know that the triangle is isosceles, if one leg is congruent to a leg of $\triangle ABC$, then you know that both pairs of legs are congruent. Because the base angles of an isosceles triangle are congruent, if you know that one pair of angles are congruent, you know that all pairs of angles are congruent. Therefore, with one pair on congruent corresponding side and one pair of congruent corresponding angles, the triangles can be proved congruent using either ASA or SAS.

27.

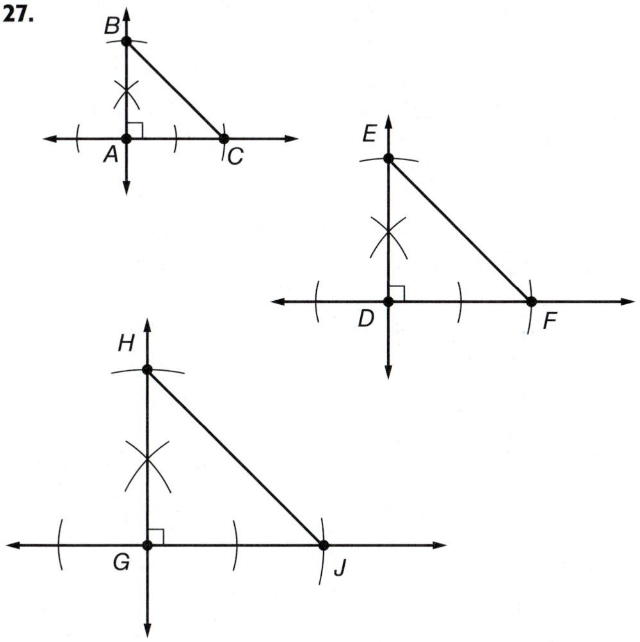

Sample answer: I constructed a pair of perpendicular segments and then used the same compass setting to mark points equidistant from their intersection. I measured both legs for each triangle. Since $AB = AC = 1.3$ cm, $DE = DF = 1.9$ cm, and $GH = GJ = 2.3$ cm, the triangles are isosceles. I used a protractor to confirm that $\angle A$, $\angle D$, and $\angle G$ are all right angles.

29 Since $\overline{AD} \cong \overline{CD}$, base angles CAD and ACD are congruent by the Isosceles Triangle Theorem. So, $m\angle CAD = m\angle ACD$.

$m\angle CAD + m\angle ACD + m\angle D = 180$	Triangle Sum Theorem
$m\angle CAD + m\angle CAD + 92 = 180$	Substitution
$2m\angle CAD + 92 = 180$	Simplify.
$2m\angle CAD = 88$	Subtract 92 from each side.
$m\angle CAD = 44$	Simplify.

31. 136

33. Given: Each triangle is isosceles, $\overline{BG} \cong \overline{HC}$, $\overline{HD} \cong \overline{JF}$, $\angle G \cong \angle H$, and $\angle H \cong \angle J$.
Prove: The distance from B to F is three times the distance from D to F.
Proof:
Statements (Reasons)
1. Each triangle is isosceles, $\overline{BG} \cong \overline{HC}$, $\overline{HD} \cong \overline{JF}$, $\angle G \cong \angle H$, and $\angle H \cong \angle J$. (Given)
2. $\angle G \cong \angle J$ (Trans. Prop.)
3. $\overline{BG} \cong \overline{CG}$, $\overline{HC} \cong \overline{HD}$, $\overline{JD} \cong \overline{JF}$ (Def. of Isosceles)
4. $\overline{BG} \cong \overline{JD}$ (Trans. Prop.)
5. $\overline{HC} \cong \overline{JD}$ (Trans. Prop.)
6. $\overline{CG} \cong \overline{JF}$ (Trans. Prop.)
7. $\triangle BCG \cong \triangle CDH \cong \triangle DFJ$ (SAS)
8. $\overline{BC} \cong \overline{CD} \cong \overline{DF}$ (CPCTC)
9. $BC = CD = DF$ (Def. of congruence)
10. $BC + CD + DF = BF$ (Seg. Add. Post.)
11. $DF + DF + DF = BF$ (Subst.)
12. $3DF = BF$ (Addition)

35. Case I

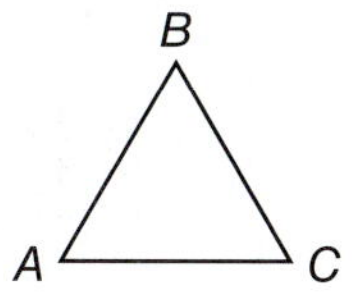

Given: $\triangle ABC$ is an equilateral triangle.
Prove: $\triangle ABC$ is an equiangular triangle.
Proof:
Statements (Reasons)
1. $\triangle ABC$ is an equilateral triangle. (Given)
2. $\overline{AB} \cong \overline{AC} \cong \overline{BC}$ (Def. of equilateral △)
3. $\angle A \cong \angle B \cong \angle C$ (Isosceles △ Th.)
4. $\triangle ABC$ is an equiangular triangle. (Def. of equilateral)

Case II

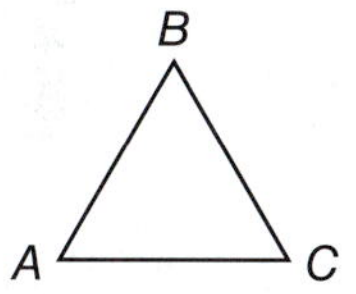

Given: $\triangle ABC$ is an equiangular triangle.
Prove: $\triangle ABC$ is an equilateral triangle.
Proof:
Statements (Reasons)
1. $\triangle ABC$ is an equiangular triangle. (Given)
2. $\angle A \cong \angle B \cong \angle C$ (Def. of equiangular △)
3. $\overline{AB} \cong \overline{AC} \cong \overline{BC}$ (If 2 ∠s of a △ are ≅ then the sides opp. those ∠s are ≅.)
4. $\triangle ABC$ is an equilateral triangle. (Def. of equilateral)

37. Given: $\triangle ABC$, $\angle A \cong \angle C$
Prove: $\overline{AB} \cong \overline{CB}$

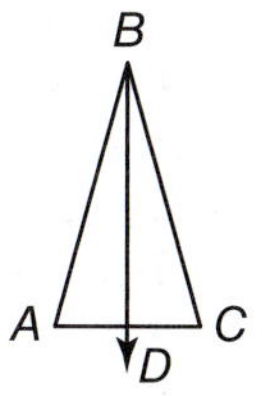

Proof:
Statements (Reasons)
1. Let $\overrightarrow{BD}$ bisect $\angle ABC$. (Protractor Post.)
2. $\angle ABD \cong \angle CBD$ (Def. of ∠ bisector)
3. $\angle A \cong \angle C$ (Given)
4. $\overline{BD} \cong \overline{BD}$ (Refl. Prop.)
5. $\triangle ABD \cong \triangle CBD$ (AAS)
6. $\overline{AB} \cong \overline{CB}$ (CPCTC)

39. 14

41

$m\angle LPM + m\angle LPQ = 180$	Supplement Theorem
$(3x - 55) + (2x + 10) = 180$	Substitution
$5x - 45 = 180$	Simplify.
$5x = 225$	Add 45 to each side.
$x = 45$	Divide each side by 45.

$m\angle LPM = 3x - 55$	Given
$= 3(45) - 55$	Substitution
$= 135 - 55$ or 80	Simplify.

Because $\overline{LM} \cong \overline{LP}$, base angles LMP and LPM are congruent by the Isosceles Triangle Theorem. So, $m\angle LMP = m\angle LPM = 80$.

43. 80

45. Given: $\triangle WJZ$ is equilateral, and $\angle ZWP \cong \angle WJM \cong \angle JZL$.

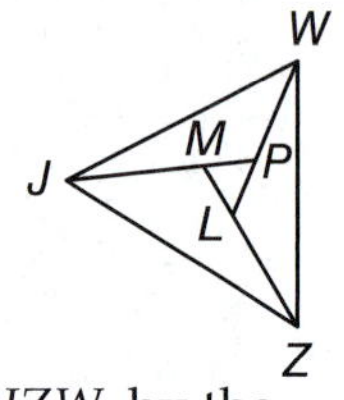

Prove: $\overline{WP} \cong \overline{ZL} \cong \overline{JM}$

Proof: We know that $\triangle WJZ$ is equilateral, since an equilateral $\triangle$ is equiangular, $\angle ZWJ \cong \angle WJZ \cong \angle JZW$. So, $m\angle ZWJ = m\angle WJZ = m\angle JZW$, by the definition of congruence. Since $\angle ZWP \cong \angle WJM \cong \angle JZL$, $m\angle ZWP = m\angle WJM = m\angle JZL$, by the definition of congruence. By the Angle Addition Postulate, $m\angle ZWJ = m\angle ZWP + m\angle PWJ$, $m\angle WJZ = m\angle WJM + m\angle MJZ$, $m\angle JZW = m\angle JZL + m\angle LZW$. By substitution, $m\angle ZWP + m\angle PWJ = m\angle WJM + m\angle MJZ = m\angle JZL + m\angle LZW$. Again by substitution, $m\angle ZWP + m\angle PWJ = m\angle ZWP + m\angle PJZ = m\angle ZWP + m\angle LZW$. By the Subtraction Property, $m\angle PWJ = m\angle PJZ = m\angle LZW$. By the definition of congruence, $\angle PWJ \cong \angle PJZ \cong \angle LZW$. So, by ASA, $\triangle WZL \cong \triangle ZJM \cong \triangle JWP$. By CPCTC, $\overline{WP} \cong \overline{ZL} \cong \overline{JM}$.

47. Never; the measure of the vertex angle will be $180 - 2$(measure of the base angle) so if the base angles are integers, then 2(measure of the base angle) will be even and $180 - 2$(measure of the base angle) will be even. **49.** It is not possible because a triangle cannot have more than one obtuse angle. **51.** The sum of the measures of the angles must be 180 and the base angles have the same measure, so the measure of the vertex angle will be equal to $180 - 2$(measure of the base angle). **53.** 185 **55.** D
57. $SU = \sqrt{17}$, $TU = \sqrt{2}$, $ST = 5$, $XZ = \sqrt{29}$, $YZ = 2$, $XY = 5$; the corresponding sides are not congruent; the triangles are not congruent.

59. Given: $AC = BD$

Prove: $AB = CD$

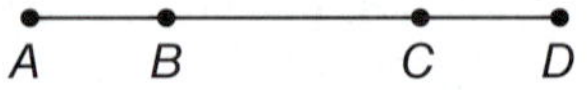

1. $AC = BD$ (Given)
2. $AC = AB + BC$, $BD = BC + CD$ (Seg. Add. Post.)
3. $AB + BC = BC + CD$ (Subst.)
4. $\overline{BC} \cong \overline{BC}$ (Reflexive)
5. $BC = BC$ (Def. of $\cong$ Segs.)
6. $AB = CD$ (Subt. Prop.)

61. Add. Prop. **63.** Trans. Prop. **65.** A, K, B or B, J, C

67. Given: $\angle ACB \cong \angle ABC$

Prove: $\angle XCA \cong \angle YBA$

__Statements (Reasons)__

1. $\angle ACB \cong \angle ABC$ (Given)
2. $\angle XCA$ and $\angle ACB$ are a linear pair. $\angle ABC$ and $\angle ABY$ are a linear pair. (Def. of linear pair)
3. $\angle XCA$, $\angle ABC$ and $\angle ABC$, $\angle ABY$ are supplementary. (Suppl. Thm.)
4. $\angle XCA \cong \angle YBA$ (△ suppl. to $\cong$ △ are $\cong$.)

Pages 294–300 Lesson 4-7

1. translation **3.** reflection **5.** $\triangle LKJ$ is a reflection of $\triangle XYZ$. $XY = 7$, $YZ = 8$, $XZ = \sqrt{113}$, $KJ = 8$, $LJ = \sqrt{113}$, $LK = 7$. $\triangle XYZ \cong \triangle LKJ$ by SSS. **7.** reflection

9 The image in green can be found by translating the blue figure to the right and up, by reflecting the figure in the line $y = -x$, or by rotating the figure 180° in either direction.

11. rotation **13.** translation **15.** rotation

17

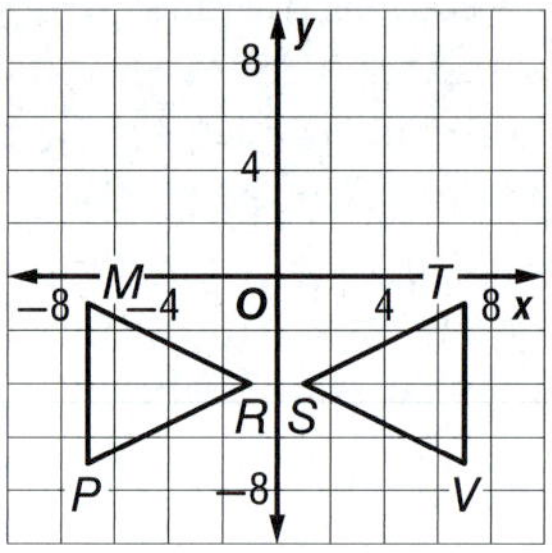

Use $d = \sqrt{(x_2 - x_1)^2 + (y_2 - y_1)^2}$ to find the lengths of the sides of $\triangle MPR$.

$MP = \sqrt{[-7 - (-7)]^2 + [-7 - (-1)]^2}$	$(x_1, y_1) = (-7, -1)$, $(x_2, y_2) = (-7, -7)$
$= \sqrt{0 + 36}$ or 6	Simplify.
$PR = \sqrt{[-1 - (-7)]^2 + [-4 - (-7)]^2}$	$(x_1, y_1) = (-7, -7)$, $(x_2, y_2) = (-1, -4)$
$= \sqrt{36 + 9}$ or $\sqrt{45}$	Simplify.
$MR = \sqrt{[-1 - (-7)]^2 + [-4 - (-1)]^2}$	$(x_1, y_1) = (-7, -1)$, $(x_2, y_2) = (-1, -4)$
$= \sqrt{36 + 9}$ or $\sqrt{45}$	Simplify.

Find the lengths of the sides of $\triangle TVS$.

$ST = \sqrt{(7 - 1)^2 + [-1 - (-4)]^2}$	$(x_1, y_1) = (1, -4)$, $(x_2, y_2) = (7, -1)$
$= \sqrt{36 + 9}$ or $\sqrt{45}$	Simplify.
$TV = \sqrt{(7 - 7)^2 + [-7 - (-1)]^2}$	$(x_1, y_1) = (7, -1)$, $(x_2, y_2) = (7, -7)$
$= \sqrt{0 + 36}$ or 6	Simplify.
$SV = \sqrt{(7 - 1)^2 + [-7 - (-4)]^2}$	$(x_1, y_1) = (1, -4)$, $(x_2, y_2) = (7, -7)$
$= \sqrt{36 + 9}$ or $\sqrt{45}$	Simplify.

In $\triangle MPR$, $MP = 6$, $PR = \sqrt{45}$, and $MR = \sqrt{45}$. In $\triangle TVS$, $ST = \sqrt{45}$, $TV = 6$, $SV = \sqrt{45}$. $\triangle MPR \cong \triangle TVS$ by SSS. $\triangle TVS$ is a reflection of $\triangle MPR$.

19. 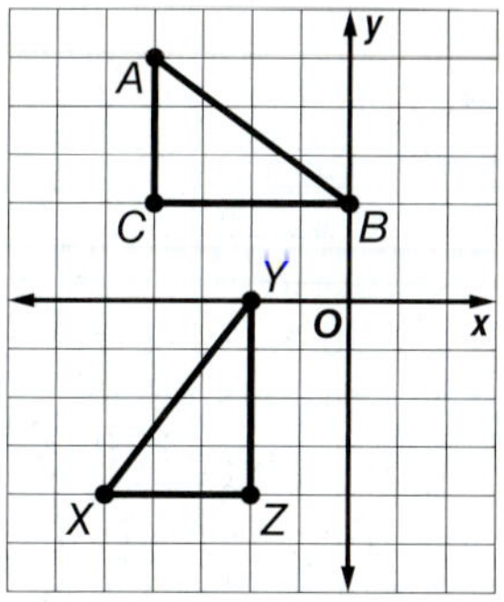

$\triangle XYZ$ is a rotation of $\triangle ABC$. $AB = 5$, $BC = 4$, $AC = 3$, $XY = 5$, $YZ = 4$, $XZ = 3$. Since $AB = XY$, $BC = YZ$, and $AC = XZ$, $\overline{AB} \cong \overline{XY}$, $\overline{BC} \cong \overline{YZ}$, and $\overline{AC} \cong \overline{XZ}$, $\triangle ABC \cong \triangle XYZ$ by SSS.

21. reflection and rotation

23. reflection

25. rotation

27. Rotation; the knob is the center of rotation.

29 **a.** Tionne used the stencil on one side, then flipped it and used the other side, then flipped it again to create the third flower in the design. She could have also used the stencil, then turned it to create the second flower, and turned it again to create the third flower. So, she could have used reflections or rotations. **b.** Tionne used the stamp, then turned it to create the second flower, and turned it again to create the third flower. So, she used rotations.

31a. translation, reflection **31b.** Sample answer: The triangles must be either isosceles or equilateral. When triangles are isosceles or equilateral, they have a line of symmetry, so reflections result in the same figure. **33.** Sample answer: A person looking in a mirror sees a reflection of himself or herself.

35. Sample answer: A faucet handle rotates when you turn the water on. **37.** no; 75% **39.** J **41.** 4 **43.** 10 **45.** yes; Law of Detachment **47.** (8, 9.5) **49.** (−7.5, 3.5) **51.** (1, 9.5)

Pages 301–307 Lesson 4-8

1. 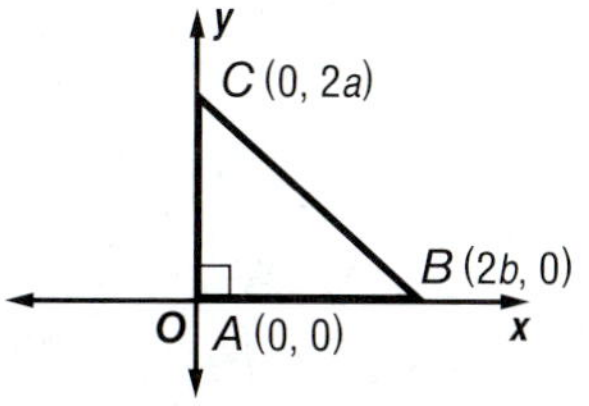

3. $T(2a, 0)$

5. $DC = \sqrt{[-a-(-a)]^2 + (b-0)^2}$ or b

$GH = \sqrt{(a-a)^2 + (b-0)^2}$ or b

Since $DC = GH$, $\overline{DC} \cong \overline{GH}$.

$DF = \sqrt{(0-a)^2 + \left(\frac{b}{2} - b\right)^2}$ or $\sqrt{a^2 + \frac{b^2}{4}}$

$GF = \sqrt{(a-0)^2 + \left(b - \frac{b}{2}\right)^2}$ or $\sqrt{a^2 + \frac{b^2}{4}}$

$CF = \sqrt{(0-a)^2 + \left(\frac{b}{2} - 0\right)^2}$ or $\sqrt{a^2 + \frac{b^2}{4}}$

$HF = \sqrt{(a-0)^2 + \left(0 - \frac{b}{2}\right)^2}$ or $\sqrt{a^2 + \frac{b^2}{4}}$

Since $DF = GF = CF = HF$, $\overline{DF} \cong \overline{GF} \cong \overline{CF} \cong \overline{HF}$. $\triangle FGH \cong \triangle FDC$ by SSS.

7. 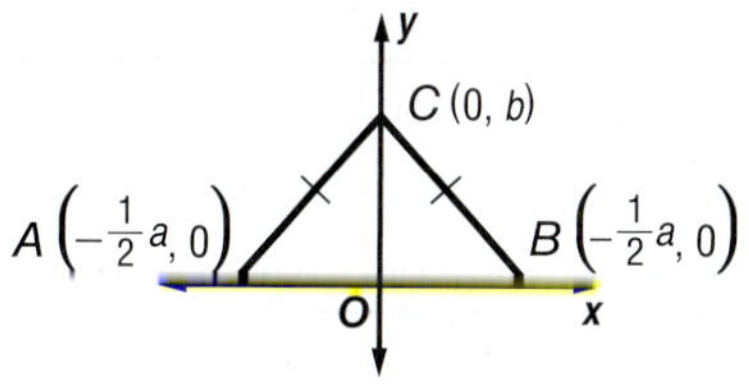

9 Since this is a right triangle, each of the legs can be located on an axis. Placing the right angle of the triangle, $\angle T$, at the origin will allow the two legs to be along the x- and y-axes. Position the triangle in the first quadrant. Since R is on the y-axis, its x-coordinate is 0. Its y-coordinate is $3a$ because the leg is $3a$ units long. Since S is on the x-axis, its y-coordinate is 0. Its x-coordinate is $3a$ because the leg is $3a$ units long.

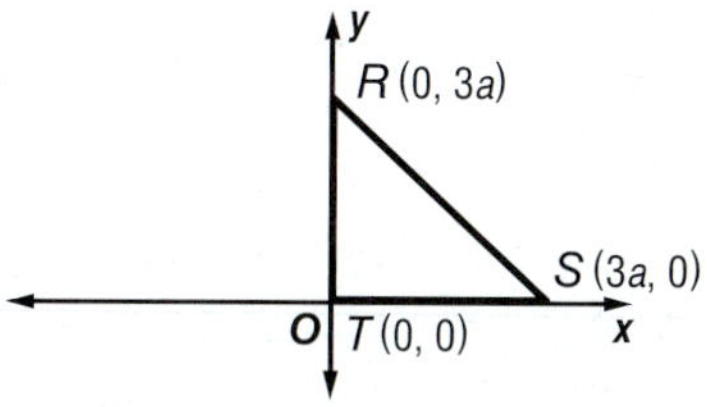

11. 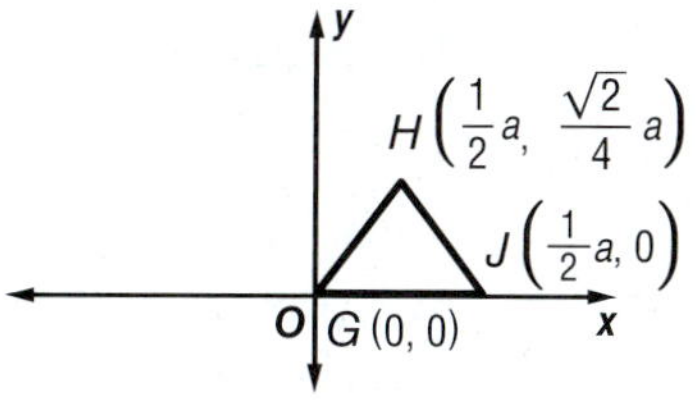

13. $C(a, a)$, $Y(a, 0)$

15 Vertex N is positioned at the origin. So, its coordinates are (0, 0). Vertex L is on the x-axis, so its y-coordinate is 0. The coordinates of vertex L are $(3a, 0)$. $\triangle NJL$ is isosceles, so the x-coordinate of J is located halfway between 0 and $3a$, or $1.5a$. The coordinates of vertex J are $(1.5a, b)$. So, the vertices are $N(0, 0)$, $J(1.5a, b)$, $L(3a, 0)$.

17. $H(2b, 2b\sqrt{3})$, $N(0, 0)$, $D(4b, 0)$

19. Given: Isosceles $\triangle ABC$ with $\overline{AC} \cong \overline{BC}$; R and S are midpoints of legs $\overline{AC}$ and $\overline{BC}$.

Prove: $\overline{AS} \cong \overline{BR}$

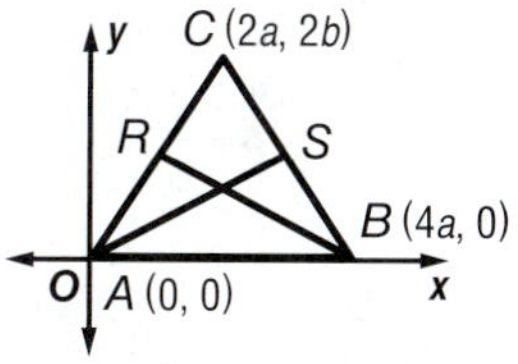

Proof:

The coordinates of S are $\left(\frac{2a + 4a}{2}, \frac{2b + 0}{2}\right)$ or $(3a, b)$.

The coordinates of R are $\left(\frac{2a + 0}{2}, \frac{2b + 0}{2}\right)$ or (a, b).

$AS = \sqrt{(3a - 0)^2 + (b - 0)^2}$ or $\sqrt{9a^2 + b^2}$

$BR = \sqrt{(4a - a)^2 + (0 - b)^2}$ or $\sqrt{9a^2 + b^2}$

Since $AS = BR$, $\overline{AS} \cong \overline{BR}$.

21. Given: Right $\triangle ABC$ with right $\angle BAC$; P is the midpoint of $\overline{BC}$.

Prove: $AP = \frac{1}{2}BC$

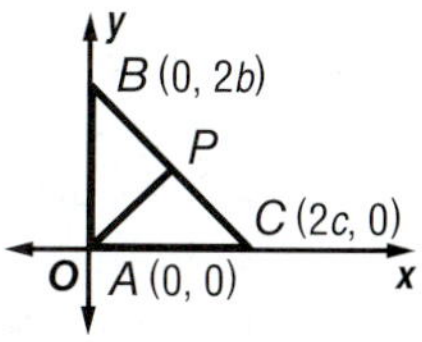

Proof:

Midpoint P is $\left(\frac{0 + 2c}{2}, \frac{2b + 0}{2}\right)$ or (c, b).

$AP = \sqrt{(c - 0)^2 + (b - 0)^2}$ or $\sqrt{c^2 + b^2}$

Selected Answers and Solutions

$BC = \sqrt{(c - 0)^2 + (0 - 2b)^2} = \sqrt{4c^2 + 4b^2}$ or $2\sqrt{c^2 + b^2}$

$\frac{1}{2}BC = \sqrt{c^2 + b^2}$

So, $AP = \frac{1}{2}BC$.

23. The distance between Raleigh and Durham is about 0.2 units between Raleigh and Chapel Hill is about 4.03 units, and between Durham and Chapel Hill is about 3.88 units. Since none of these distances are the same, the Research Triangle is scalene.

25. slope of $\overline{XY} = 1$, slope of $\overline{YZ} = -1$, slope of $\overline{ZX} = 0$; since the slopes of two sides of $\triangle XYZ$ are negative reciprocals, $\triangle XYZ$ is a right triangle.

27. The slope between the tents is $-\frac{4}{3}$. The slope between the ranger's station and the tent located at (12, 9) is $\frac{3}{4}$. Since $-\frac{4}{3} \cdot \frac{3}{4} = -1$, the triangle formed by the tents and ranger's station is a right triangle.

29 a.

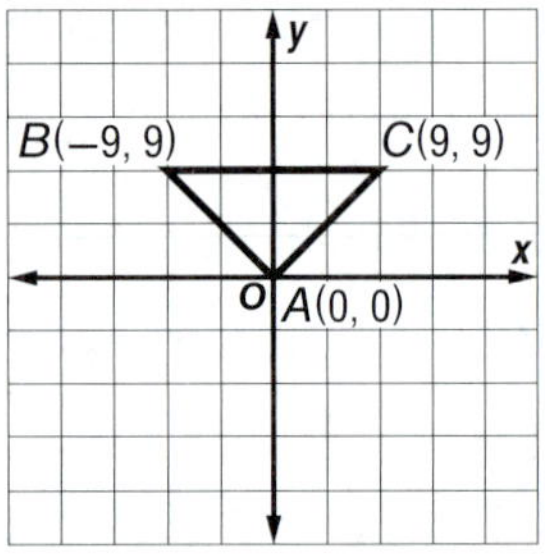

The equation of the line along which the first vehicle lies is $y = x$. The slope is 1 because the vehicle travels the same number of units north as it does east of the origin and the y-intercept is 0. The equation of the line along which the second vehicle lies is $y = -x$. The slope is -1 because the vehicle travels the same number of units north as it does west of the origin and the y-intercept is 0.

b. The paths taken by both the first and second vehicles are 300 yards long. Therefore, the paths are congruent. If two sides of a triangle are congruent, then the triangle is isosceles. You can also write a coordinate proof to prove the triangle formed is isosceles.

Given: $\triangle ABC$

Prove: $\triangle ABC$ is an isosceles right triangle.

Proof: By the Distance Formula,

$AB = \sqrt{(-a - 0)^2 + (a - 0)^2}$ or $\sqrt{2a^2}$ and

$AC = \sqrt{(a - 0)^2 + (a - 0)^2}$ or $\sqrt{2a^2}$. So, $AB = AC$

and $\overline{AB} \cong \overline{AC}$. The triangle is isosceles. By the Slope Formula, the slope of $\overline{AB}$ is $\frac{a}{-a}$ or -1 and the slope of $\overline{AC}$ is $\frac{a}{a}$ or 1. Since the slopes are negative reciprocals, the sides of the triangle are perpendicular and therefore form a right angle. So, $\triangle ABC$ is an isosceles right triangle.

c. The paths taken by the first two vehicles form the hypotenuse of isosceles right triangles.

$a^2 + b^2 = c^2$	Pythagorean Theorem
$a^2 + a^2 = 300^2$	$b = a$
$2a^2 = 90{,}000$	Simplify.
$a^2 = 45{,}000$	Divide each side by 2.
$a = 150\sqrt{2}$	Take the positive square root of each side.

First vehicle: $(a, a) = (150\sqrt{2}, 150\sqrt{2})$; second vehicle: $(-a, a) = (-150\sqrt{2}, 150\sqrt{2})$
The third vehicle travels due north and therefore, remains on the y-axis; third vehicle: (0, 212)

d. The y-coordinates of the first two vehicles are $150\sqrt{2} \approx 212.13$, while the y-coordinate of the third vehicle is 212. Since all three vehicles have approximately the same y-coordinate, they are approximately collinear. The midpoint between the first and second vehicles is

$$\left(\frac{150\sqrt{2} + (-150\sqrt{2})}{2}, \frac{150\sqrt{2} + 150\sqrt{2}}{2}\right)$$

or approximately (0, 212.13). This is the approximate location of the third vehicle.

31. $(a, 0)$ **33.** Sample answer: $(4a, 0)$

35. Given: $\triangle ABC$ with coordinates $A(0, 0)$, $B(a, b)$, and $C(c, d)$ and $\triangle DEF$ with coordinates $D(0 + n, 0 + m)$, $E(a + n, b + m)$, and $F(c + n, d + m)$

Prove: $\triangle DEF \cong \triangle ABC$

Proof:

$AB = \sqrt{(a - 0)^2 + (b - 0)^2}$ or $\sqrt{a^2 + b^2}$

$DE = \sqrt{[a + n - (0 + n)]^2 + [b + m - (0 + m)]^2}$ or $\sqrt{a^2 + b^2}$

Since $AB = DE$, $\overline{AB} \cong \overline{DE}$.

$BC = \sqrt{(c - a)^2 + (d - b)^2}$ or $\sqrt{c^2 - 2ac + a^2 + d^2 - 2bd + b^2}$

$EF = \sqrt{[c + n - (a + n)]^2 + [d + m - (b + m)]^2}$ or $\sqrt{c^2 - 2ac + a^2 + d^2 - 2bd + b^2}$

Since $BC = EF$, $\overline{BC} \cong \overline{EF}$.

$CA = \sqrt{(c - 0)^2 + (d - 0)^2}$ or $\sqrt{c^2 + d^2}$

$FD = \sqrt{[0 + n - (c + n)]^2 + [0 + m - (d + m)]^2}$ or $\sqrt{c^2 + d^2}$

Since $CA = FD$, $\overline{CA} \cong \overline{FD}$.

Therefore, $\triangle DEF \cong \triangle ABC$ by the SSS Postulate.

37a. Using the origin as a vertex of the triangle makes calculations easier because the coordinates are (0, 0).
37b. Placing at least one side of the triangle on the x- or y-axis makes it easier to calculate the length of the side since one of the coordinates will be 0.
37c. Keeping a triangle within the first quadrant makes all of the coordinates positive, and makes the calculations easier. **39.** D **41.** B **43.** reflection, translation, or rotation **45.** $\angle TSR \cong \angle TRS$
47. Sample answer: $\triangle RQV \cong \triangle SQV$ **49.** 2 **51.** 3.6

Pages 308–312 Chapter 4 Study Guide & Review

1. true **3.** true **5.** false; base **7.** true **9.** false;

coordinate proof **11.** obtuse **13.** right **15.** $x = 6$, $JK = KL = JL = 24$ **17.** 70 **19.** 82 **21.** $\angle D \cong \angle J$, $\angle A \cong \angle F$, $\angle C \cong \angle H$, $\angle B \cong \angle G$, $\overline{AB} \cong \overline{FG}$, $\overline{BC} \cong \overline{HG}$, $\overline{DC} \cong \overline{JH}$, $\overline{DA} \cong \overline{JF}$; polygon $ABCD \cong$ polygon $FGHJ$
23. $\triangle BFG \cong \triangle CGH \cong \triangle DHE \cong \triangle AEF$, $\triangle EFG \cong \triangle FGH \cong \triangle GHE \cong \triangle HEF$ **25.** No, the corresp. sides of the 2 △s are not ≅. **27.** not possible

29. Given: $\overline{AB} \parallel \overline{DC}$, $\overline{AB} \cong \overline{DC}$
Prove: $\triangle ABE \cong \triangle CDE$
Proof:
Statements (Reasons)
1. $\overline{AB} \parallel \overline{DC}$ (Given)
2. $\angle A \cong \angle DCE$ (Alt. Int. ∠s Thm.)
3. $\overline{AB} \parallel \overline{DC}$ (Given)
4. $\angle ABE \cong \angle D$ (Alt. Int. ∠s Thm.)
5. $\triangle ABE \cong \triangle CDE$ (ASA)

31. 3
33. 77.5
35. reflection
37. rotation

39.

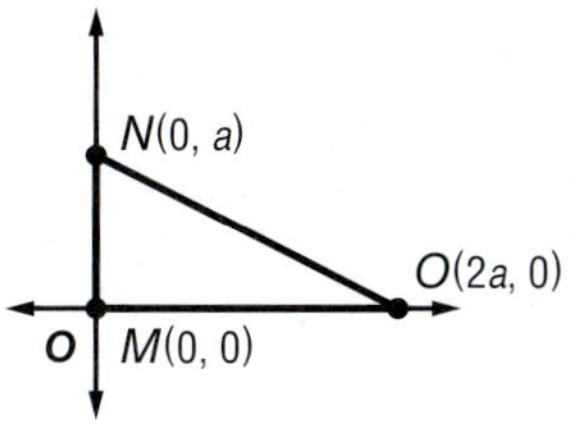

41. Given: $\triangle DSH$ with vertices $D(8, 28)$, $S(0, 0)$, and $H(19, 7)$
Prove: $\triangle DSH$ is scalene.
Proof:
Statements (Reasons)
1. $D(8, 28)$, $S(0, 0)$, and $H(19, 7)$ (Given)
2. $DS = \sqrt{(8-0)^2 + (28-0)^2}$ or $\sqrt{848}$ (Distance Formula)
3. $SH = \sqrt{(19-0)^2 + (7-0)^2}$ or $\sqrt{410}$ (Distance Formula)
4. $DH = \sqrt{(8-19)^2 + (28-7)^2}$ or $\sqrt{562}$ (Distance Formula)
5. $\triangle DSH$ is scalene (Definition of scalene)

Chapter 5 Relationships in Triangles

Page 319 Chapter 5 Get Ready

1. 9 **3.** 10 ft
5. $JK = KL = LM = MJ$

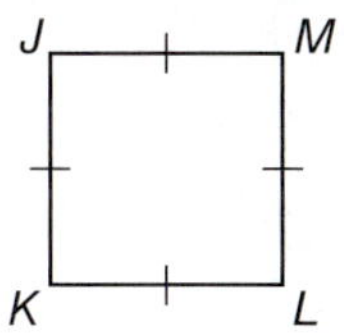

7. Sometimes; the conjecture is true when E is between D and F, otherwise it is false. **9.** $-6 > x$
11. $x > 41$

Pages 322–331 Lesson 5-1

1. 12 **3.** 15 **5.** 8 **7.** 12

9 $\overrightarrow{MP}$ is the perpendicular bisector of $\overline{LN}$.

$LP = NP$	Perpendicular Bisector Theorem
$2x - 4 = x + 5$	Substitution
$x - 4 = 5$	Subtract x from each side.
$x = 9$	Add 4 to each side.

$NP = 9 + 5$ or 14

11. 6 **13.** 4 **15.**

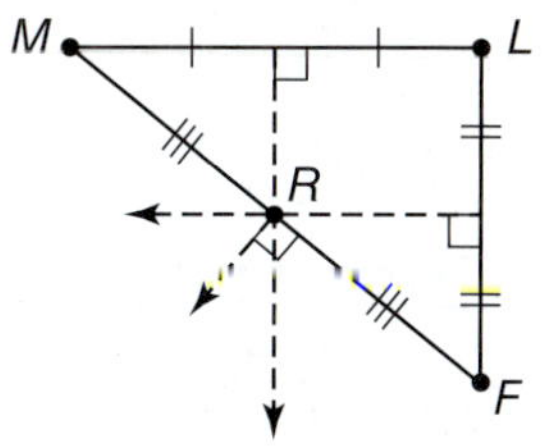

17. $\overline{CD}$, $\overline{BD}$
19. $\overline{BH}$ **21.** 11

23 Since $\overline{QM} \perp \overrightarrow{NM}$, $\overline{QP} \perp \overrightarrow{NP}$, and $QM = QP$, Q is equidistant from the sides of $\angle PNM$. By the Converse of the Angle Bisector Theorem, $\overrightarrow{NQ}$ bisects $\angle PNM$.

$\angle PNQ \cong \angle QNM$	Definition of angle bisector
$m\angle PNQ = m\angle QNM$	Definition of congruent angles
$4x - 8 = 3x + 5$	Substitution
$x - 8 = 5$	Subtract $3x$ from each side.
$x = 13$	Add 8 to each side.

$m\angle PNM = m\angle PNQ + m\angle QNM$	Angle Addition Postulate
$= (4x - 8) + (3x + 5)$	Substitution
$= 7x - 3$	Simplify.
$= 7(13) - 3$ or 88	$x = 13$

25. 42° **27.** 7.1 **29.** 33°

31 Sketch the table and draw the three angle bisectors of the triangle.

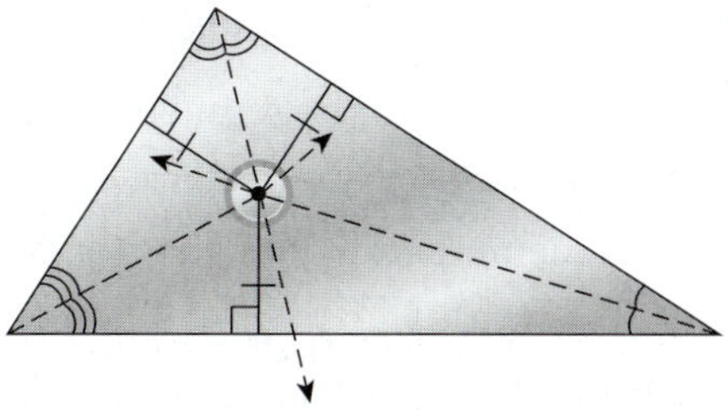

Find the point of concurrency of the angle bisectors of the triangle, the incenter. This point is equidistant from each side of the triangle. So, the centerpiece should be placed at the incenter.

33. No; we need to know whether the perpendicular segments are congruent. **35.** No; we need to know whether the hypotenuses of the triangles are congruent.

37. Given: $\overline{CA} \cong \overline{CB}$, $\overline{AD} \cong \overline{BD}$
Prove: C and D are on the perpendicular bisector of $\overline{AB}$.
Proof:

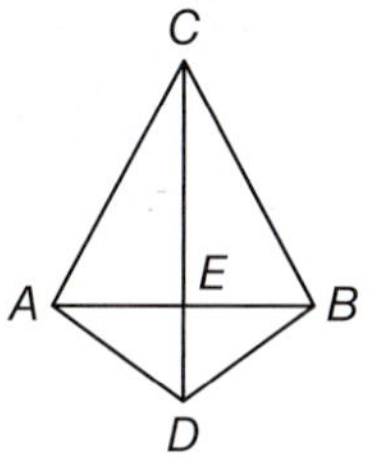

Statements (Reasons)
1. $\overline{CA} \cong \overline{CB}$, $\overline{AD} \cong \overline{BD}$ (Given)
2. $\overline{CD} \cong \overline{CD}$ (Congruence of segments is reflexive.)
3. $\triangle ACD \cong \triangle BCD$ (SSS)
4. $\angle ACD \cong \angle BCD$ (CPCTC)
5. $\overline{CE} \cong \overline{CE}$ (Congruence of segments is reflexive.)
6. $\triangle CEA \cong \triangle CEB$ (SAS)
7. $\overline{AE} \cong \overline{BE}$ (CPCTC)
8. E is the midpoint of $\overline{AB}$. (Def. of midpoint)
9. $\angle CEA \cong \angle CEB$ (CPCTC)
10. $\angle CEA$ and $\angle CEB$ form a linear pair. (Def. of linear pair)
11. $\angle CEA$ and $\angle CEB$ are supplementary. (Supplement Theorem)

12. $m\angle CEA + m\angle CEB = 180$ (Def. of supplementary)
13. $m\angle CEA + m\angle CEA = 180$ (Substitution Prop.)
14. $2m\angle CEA = 180$ (Substitution Prop.)
15. $m\angle CEA = 90$ (Division Prop.)
16. $\angle CEA$ and $\angle CEB$ are rt. $\angle$. (Def. of rt. $\angle$)
17. $\overline{CD} \perp \overline{AB}$ (Def. of $\perp$)
18. $\overline{CD}$ is the perpendicular bisector of $\overline{AB}$. (Def. of $\perp$ bisector)
19. C and D are on the perpendicular bisector of $\overline{AB}$. (Def. of point on a line)

39. Given: $\overline{CD}$ is the $\perp$ bisector of $\overline{AB}$. E is a point on $\overline{CD}$.
Prove: $EA = EB$

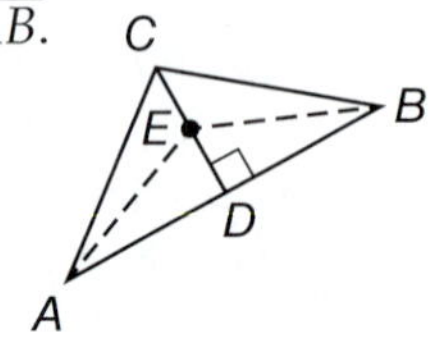

Proof: $\overline{CD}$ is the $\perp$ bisector of $\overline{AB}$. By definition of $\perp$ bisector, D is the midpoint of $\overline{AB}$. Thus, $\overline{AD} \cong \overline{BD}$ by the Midpoint Theorem. $\angle CDA$ and $\angle CDB$ are right angles by the definition of perpendicular. Since all right angles are congruent, $\angle CDA \cong \angle CDB$. Since E is a point on $\overline{CD}$, $\angle EDA$ and $\angle EDB$ are right angles and are congruent. By the Reflexive Property, $\overline{ED} \cong \overline{ED}$. Thus, $\triangle EDA \cong \triangle EDB$ by SAS. $\overline{EA} \cong \overline{EB}$ because CPCTC, and by definition of congruence, $EA = EB$.

41. $y = -\frac{7}{2}x + \frac{15}{4}$; The perpendicular bisector bisects the segment at the midpoint of the segment. The midpoint is $\left(\frac{1}{2}, 2\right)$. The slope of the given segment is $\frac{2}{7}$, so the slope of the perpendicular bisector is $-\frac{7}{2}$.

43. Given: $\overrightarrow{PX}$ bisects $\angle QPR$. $\overline{XY} \perp \overline{PQ}$ and $\overline{XZ} \perp \overline{PR}$
Prove: $\overline{XY} \cong \overline{XZ}$

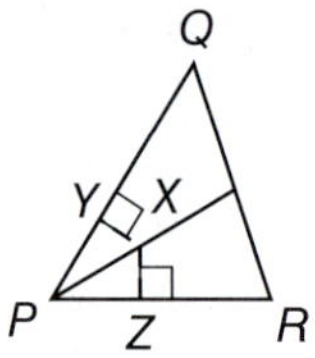

Proof:
Statements (Reasons)
1. $\overrightarrow{PX}$ bisects $\angle QPR$, $\overline{XY} \perp \overline{PQ}$, and $\overline{XZ} \perp \overline{PR}$. (Given)
2. $\angle YPX \cong \angle ZPX$ (Definition of angle bisector)
3. $\angle PYX$ and $\angle PZX$ are right angles. (Definition of perpendicular)
4. $\angle PYX \cong \angle PZX$ (Right angles are congruent.)
5. $\overline{PX} \cong \overline{PX}$ (Reflexive Property)
6. $\triangle PYX \cong \triangle PZX$ (AAS)
7. $\overline{XY} \cong \overline{XZ}$ (CPCTC)

45 The circumcenter is the point where the perpendicular bisectors of a triangle intersect. You can find the circumcenter by locating the point of intersection of two of the perpendicular bisectors.

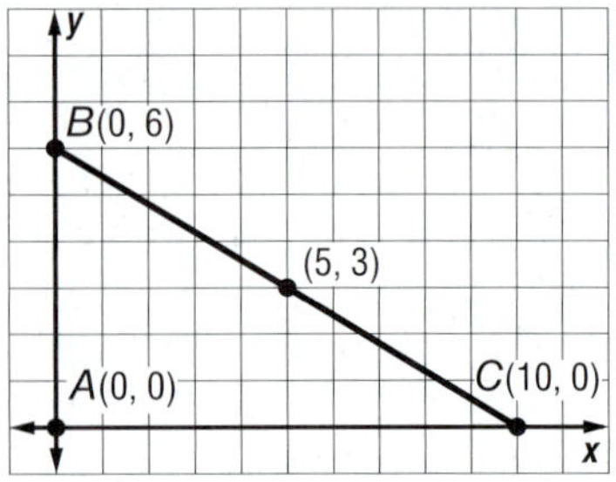

The equation of the perpendicular bisector of $\overline{AB}$ is $y = 3$. The equation of the perpendicular bisector of $\overline{AC}$ is $x = 5$. These lines intersect at (5, 3). The circumcenter is located at (5, 3).

47. a plane perpendicular to the plane in which $\overline{CD}$ lies and bisecting $\overline{CD}$

49. Sample answer:

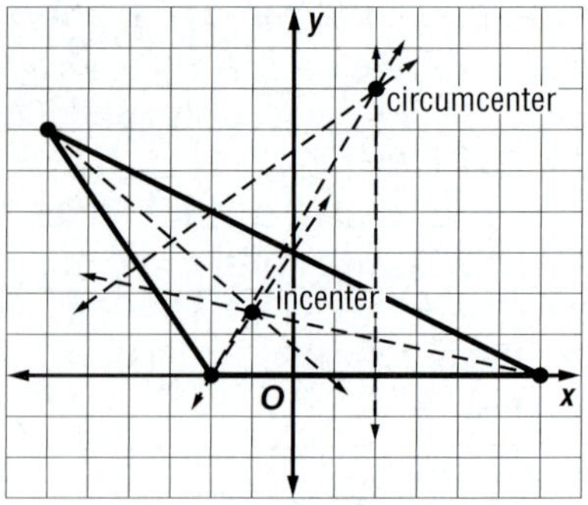

51. always
Given: $\triangle ABC$ is isosceles with legs $\overline{AB}$ and $\overline{BC}$; $\overline{BD}$ is the $\perp$ bisector of $\overline{AC}$.
Prove: $\overline{BD}$ is the angle bisector of $\angle ABC$.

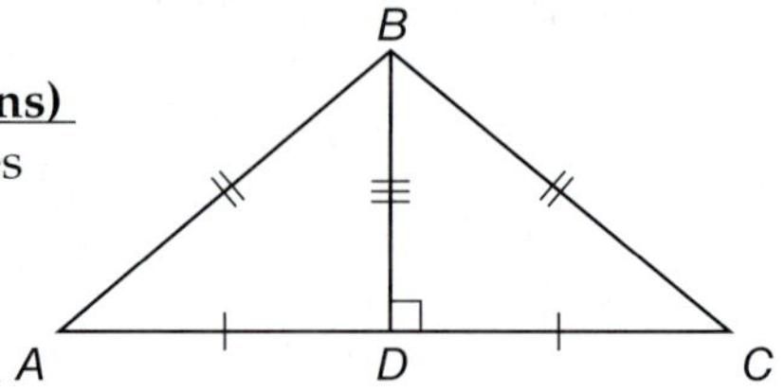

Proof:
Statements (Reasons)
1. $\triangle ABC$ is isosceles with legs $\overline{AB}$ and $\overline{BC}$. (Given)
2. $\overline{AB} \cong \overline{BC}$ (Def. of isosceles $\triangle$)
3. $\overline{BD}$ is the $\perp$ bisector of $\overline{AC}$. (Given)
4. D is the midpoint of $\overline{AC}$. (Def. of segment bisector)
5. $\overline{AD} \cong \overline{DC}$ (Def. of midpoint)
6. $\overline{BD} \cong \overline{BD}$ (Reflexive Property)
7. $\triangle ABD \cong \triangle CBD$ (SSS)
8. $\angle ABD \cong \angle CBD$ (CPCTC)
9. $\overline{BD}$ is the angle bisector of $\angle ABC$. (Def. of $\angle$ bisector)

53. Given: Plane $\mathcal{Z}$ is an angle bisector of $\angle KJH$. $\overline{KJ} \cong \overline{HJ}$
Prove: $\overline{MH} \cong \overline{MK}$
Proof:
Statements (Reasons)
1. Plane $\mathcal{Z}$ is an angle bisector of $\angle KJH$; $\overline{KJ} \cong \overline{HJ}$ (Given)
2. $\angle KJM \cong \angle HJM$ (Definition of angle bisector)
3. $\overline{JM} \cong \overline{JM}$ (Reflexive Property)
4. $\triangle KJM \cong \triangle HJM$ (SAS)
5. $\overline{MH} \cong \overline{MK}$ (CPCTC)

55. A **57.** J **59.** $L(a, b)$ **61.** $S(-2b, 0)$ and $R(0, c)$

63. $\triangle JKL$ is a translation of $\triangle XYZ$; $JK = 2$, $KL = 4$, $JL = \sqrt{20}$, $XY = 2$, $YZ = 4$, $XZ = \sqrt{20}$. $\triangle JKL \cong \triangle XYZ$ by SSS.

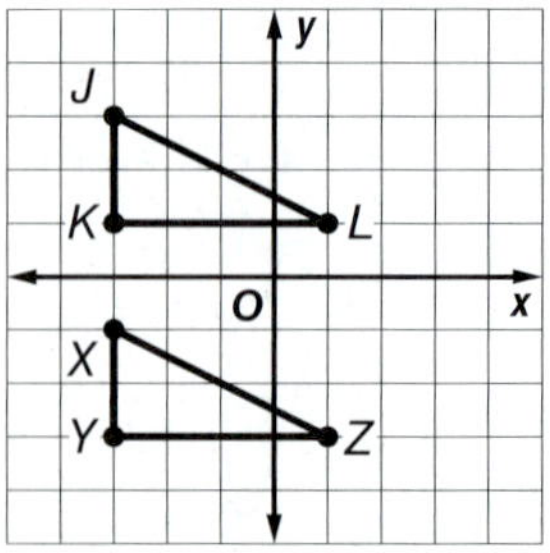

65. $\sqrt{5}$
67. $m = 42t + 450$; \$1164

69. Given: $\triangle MLP$ is isosceles. N is the midpoint of $\overline{MP}$.
Prove: $\overline{LN} \perp \overline{MP}$

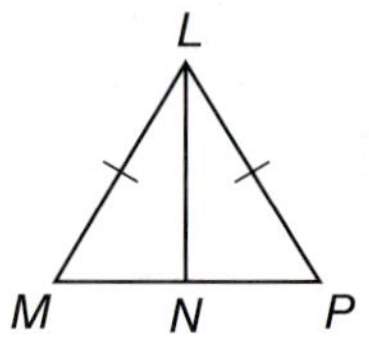

Proof:
Statements (Reasons)
1. $\triangle MLP$ is isosceles. (Given)
2. $\overline{ML} \cong \overline{PL}$ (Definition of isosceles $\triangle$)
3. $\angle M \cong \angle P$ (Isosceles $\triangle$ Th.)

4. N is the midpoint of $\overline{MP}$. (Given)
5. $\overline{MN} \cong \overline{PN}$ (Def. of midpoint)
6. $\triangle MNL \cong \triangle PNL$ (SAS)
7. $\angle LNM \cong \angle LNP$ (CPCTC)
8. $m\angle LNM = m\angle LNP$ (Def. of $\cong$ ∠s)
9. $\angle LNM$ and $\angle LNP$ are a linear pair. (Def. of a linear pair)
10. $m\angle LNM + m\angle LNP = 180$ (Sum of measure of linear pair of ∠s $= 180$)
11. $2m\angle LNM = 180$ (Substitution)
12. $m\angle LNM = 90$ (Division)
13. $\angle LNM$ is a right angle. (Def. of rt. $\angle$)
14. $\overline{LN} \perp \overline{MP}$ (Def. of $\perp$)

Pages 333–341 Lesson 5-2

1

$PC = \frac{2}{3}FC$	Centroid Theorem
$PC = \frac{2}{3}(PF + PC)$	Segment Addition and Substitution
$PC = \frac{2}{3}(6 + PC)$	$PF = 6$
$PC = 4 + \frac{2}{3}PC$	Distributive Property
$\frac{1}{3}PC = 4$	Subtract $\frac{2}{3}PC$ from each side.
$PC = 12$	Multiply each side by 3.

3. (5, 6) **5.** 4.5 **7.** 13.5 **9.** 6 **11.** (3, 6)

13 The centroid is the point of balance for a triangle. Use the Midpoint Theorem to find the midpoint M of the side with endpoints at (0, 8) and (6, 4). The centroid is two-thirds the distance from the opposite vertex to that midpoint.

$$M\left(\frac{0+6}{2}, \frac{8+4}{2}\right) = M(3, 6)$$

The distance from $M(3, 6)$ to the point at (3, 0) is $6 - 0$ or 6 units. If P is the centroid of the triangle, then $P = \frac{2}{3}(6)$ or 4 units up from the point at (3, 0). The coordinates of P are $(3, 0 + 4)$ or (3, 4).

15. (−4, −4) **17.** median **19.** median **21.** 3 **23.** $\frac{1}{2}$

25

$\overline{AC} \cong \overline{DC}$	Definition of median
$AC = DC$	Definition of congruence
$4x - 3 = 2x + 9$	Substitution
$2x - 3 = 9$	Subtract $2x$ from each side.
$2x = 12$	Add 3 to each side.
$x = 6$	Divide each side by 2.

$m\angle ECA = 15x + 2$	Given
$= 15(6) + 2$	$x = 6$
$= 90 + 2$ or 92	Simplify.

$\overline{EC}$ is not an altitude of $\triangle AED$ because $m\angle ECA = 92$. If $\overline{EC}$ were an altitude, then $m\angle ECA$ must be 90.

27. altitude **29.** median

31. Given: $\triangle XYZ$ is isosceles. $\overline{WY}$ bisects $\angle Y$.
Prove: $\overline{WY}$ is a median.
Proof: Since $\triangle XYZ$ is isosceles, $\overline{XY} \cong \overline{ZY}$. By the definition of angle bisector, $\angle XYW \cong \angle ZYW$. $\overline{YW} \cong \overline{YW}$ by the Reflexive Property. So, by SAS, $\triangle XYW \cong \triangle ZYW$. By CPCTC, $\overline{XW} \cong \overline{ZW}$. By the definition of a midpoint, W is the midpoint of $\overline{XZ}$. By the definition of a median, $\overline{WY}$ is a median.

33a.

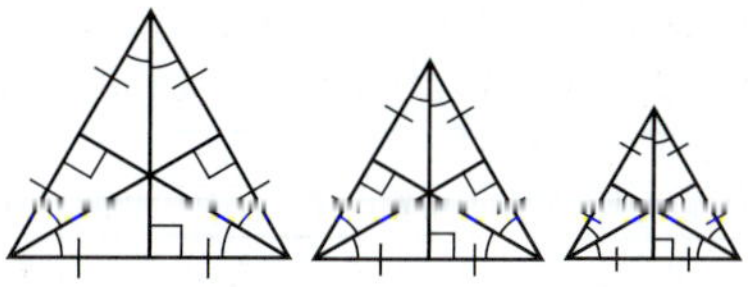

33b. Sample answer: The four points of concurrency of an equilateral triangle are all the same point.

33c.

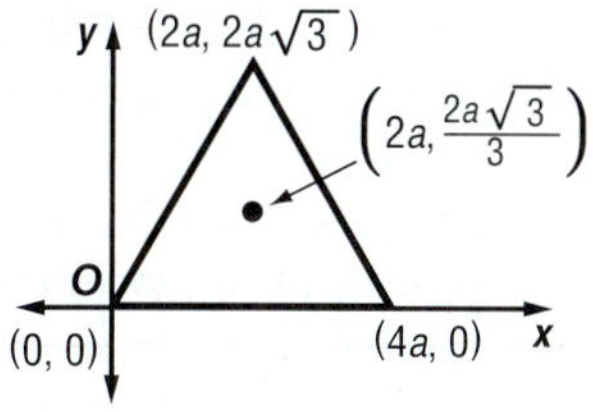

35. 7

37 Sample answer: Kareem is correct. According to the Centroid Theorem, $AP = \frac{2}{3}AD$. The segment lengths are transposed.

39. $\left(1, \frac{5}{3}\right)$; Sample answer: I found the midpoint of $\overline{AC}$ and used it to find the equation for the line that contains point B and the midpoint of $\overline{AC}$, $y = \frac{10}{3}x - \frac{5}{3}$. I also found the midpoint of $\overline{BC}$ and the equation for the line between point A and the midpoint of $\overline{BC}$, $y = -\frac{1}{3}x + 2$. I solved the system of two equations for x and y to get the coordinates of the centroid, $\left(1, \frac{5}{3}\right)$.

41. $2\sqrt{13}$ **43.** Sample answer: Each median divides the triangle into two smaller triangles of equal area, so the triangle can be balanced along any one of those lines. To balance the triangle on one point, you need to find the point where these three balance lines intersect. The balancing point for a rectangle is the intersection of the segments connecting the midpoints of the opposite sides, since each segment connecting these midpoints of a pair of opposite sides divides the rectangle into two parts with equal area. **45.** 3 **47.** B **49.** 5

51.

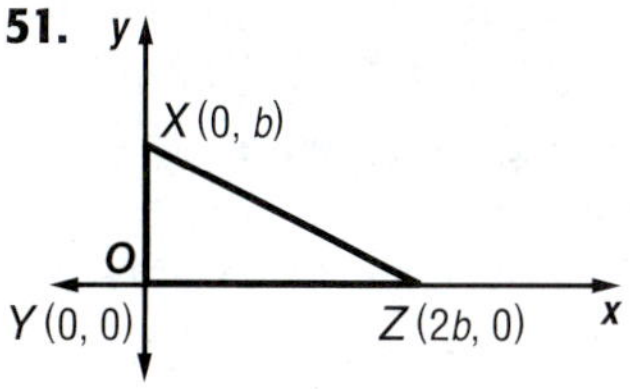

53. neither

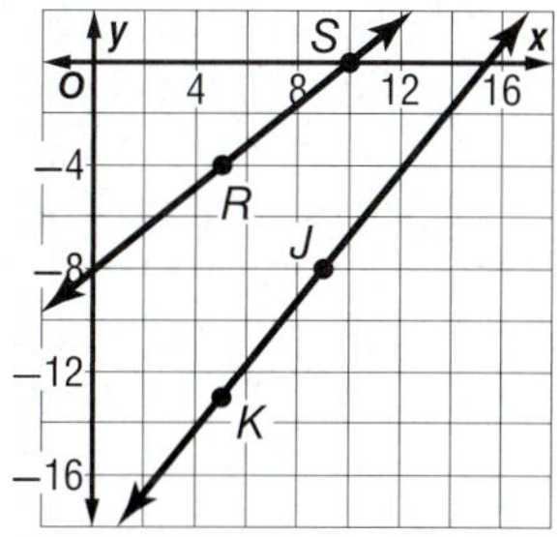

55. Because the lines are perpendicular, the angles formed are right angles. All right angles are congruent. Therefore, $\angle 1$ is congruent to $\angle 2$.

Pages 342–349 Lesson 5-3

1. $\angle 1, \angle 2$ **3.** $\angle 4$ **5.** $\angle A, \angle C, \angle B$; $\overline{BC}, \overline{AB}, \overline{AC}$ **7.** $\overline{BC}$; Sample answer: Since the angle across from segment $\overline{BC}$ is larger than the angle across from $\overline{AC}$, $\overline{BC}$ is longer. **9.** $\angle 1, \angle 2$ **11.** $\angle 1, \angle 3, \angle 6, \angle 7$ **13.** $\angle 5, \angle 9$

15 The sides from shortest to longest are $\overline{RT}, \overline{RS}, \overline{ST}$. The angles opposite these sides are $\angle S$, $\angle T$, and $\angle R$, respectively. So the angles from smallest to largest are $\angle S$, $\angle T$, and $\angle R$.

17. $\angle L, \angle P, \angle M; \overline{PM}, \overline{ML}, \overline{PL}$ **19.** $\angle C, \angle D, \angle E; \overline{DE}, \overline{CE}, \overline{CD}$

21 To use Theorem 5.9, first show that the measure of the angle opposite $\overline{YZ}$ is greater than the measure of the angle opposite $\overline{XZ}$. If $m\angle X = 90$, then $m\angle Y + m\angle Z = 90$, so $m\angle Y < 90$ by the definition of inequality. So $m\angle X > m\angle Y$. According to Theorem 5.9, if $m\angle X > m\angle Y$, then the length of the side opposite $\angle X$ must be greater than the length of the side opposite $\angle Y$. Since $\overline{YZ}$ is opposite $\angle X$, and $\overline{XZ}$ is opposite $\angle Y$, then $YZ > XZ$. So YZ, the length of the top surface of the ramp, must be greater than the length of the ramp.

23. $\angle P, \angle Q, \angle M; \overline{MQ}, \overline{PM}, \overline{PQ}$ **25.** $\angle 2$ **27.** $\angle 3$ **29.** $\angle 8$ **31.** $\angle BCF > \angle CFB$ **33.** $\angle DBF < \angle BFD$ **35.** $RP > MP$ **37.** $SM > RQ$

39 Use the Distance Formula $d = \sqrt{(x_2 - x_1)^2 + (y_2 - y_1)^2}$ to find the lengths of the sides.

$AB = \sqrt{[-2 - (-4)]^2 + (1 - 6)^2}$	$x_1 = -4, x_2 = -2, y_1 = 6, y_2 = 1$
$= \sqrt{29}$	Simplify.
≈ 5.4	Use a calculator.
$BC = \sqrt{[5 - (-2)]^2 + (6 - 1)^2}$	$x_1 = -2, x_2 = 5, y_1 = 1, y_2 = 6$
$= \sqrt{74}$	Simplify.
≈ 8.6	Use a calculator.
$AC = \sqrt{[5 - (-4)]^2 + (6 - 6)^2}$	$x_1 = -4, x_2 = 5, y_1 = 6, y_2 = 6$
$= \sqrt{81}$ or 9	Simplify.

Since $AB < BC < AC$, $\angle C < \angle A < \angle B$. The angles in order from smallest to largest are $\angle C, \angle A, \angle B$.

41. AB, BC, AC, CD, BD; In $\triangle ABC$, $AB < BC < AC$ and in $\triangle BCD$, $BC < CD < BD$. By the figure $AC < CD$, so $BC < AC < CD$. **43.** Sample answer: $\angle R$ is an exterior angle to $\triangle PQR$, so by the Exterior Angle Inequality, $m\angle R$ must be greater than $m\angle Q$. The markings indicate that $\angle R \cong \angle Q$, indicating that $m\angle R = m\angle Q$. This is a contradiction of the Exterior Angle Inequality Theorem, so the markings are incorrect. **45.** Sample answer: 10; $m\angle C > m\angle B$, so if $AB > AC$, Theorem 5.10 is satisfied. Since $10 > 6$, $AB > AC$. **47.** $m\angle 1, m\angle 2 = m\angle 5, m\angle 4, m\angle 6, m\angle 3$; Sample answer: Since the side opposite $\angle 5$ is the smallest side in that triangle and $m\angle 2 = m\angle 5$, we know that $m\angle 4$ and $m\angle 6$ are both greater than $m\angle 2$ and $m\angle 5$. Since the side opposite $\angle 2$ is longer than the side opposite $\angle 1$, we know that $m\angle 1$ is less than $m\angle 2$ and $m\angle 5$. Since $m\angle 1$ is less than $m\angle 4$, we know that $m\angle 3$ is greater than $m\angle 6$ **49.** D **51a.** $t = 2.5h + 198$

51b. 6 **51c.** \$180 **53.** 9 **55.** $y = -5x + 7$; The perpendicular bisector bisects the segment at the midpoint of the segment. The midpoint is $\left(\frac{1}{2}, \frac{9}{2}\right)$. The slope of the given segment is $\frac{1}{5}$, so the slope of the perpendicular bisector is -5.

57. Given: T is the midpoint of $\overline{SQ}$.
$\overline{SR} \cong \overline{QR}$

Prove: $\triangle SRT \cong \triangle QRT$

Proof:

Statements (Reasons)

1. T is the midpoint of SQ. (Given)
2. $\overline{ST} \cong \overline{TQ}$ (Def. of midpoint)
3. $\overline{SR} \cong \overline{QR}$ (Given)
4. $\overline{RT} \cong \overline{RT}$ (Reflexive Prop.)
5. $\triangle SRT \cong \triangle QRT$ (SSS)

59. false **61.** true

Pages 351–358 Lesson 5-4

1. $\overline{AB} \not\cong \overline{CD}$

3 The conclusion of the conditional statement is $x < 6$. If $x < 6$ is false, then x must be greater than or equal to 6. The negation of the conclusion is $x \geq 6$.

5. Given: $2x + 3 < 7$

Prove: $x < 2$

Indirect Proof: Step 1

Assume that $x > 2$ or $x = 2$ is true.

Step 2

x	2	3	4	5	6
$2x + 3$	7	9	11	13	15

When $x > 2$, $2x + 3 > 7$ and when $x = 2$, $2x + 3 = 7$.

Step 3 In both cases, the assumption leads to the contradiction of the given information that $2x + 3 < 7$. Therefore, the assumption that $x \geq 2$ must be false, so the original conclusion that $x < 2$ must be true.

7. Use a = average or $\frac{\text{number of points scored}}{\text{number of games played}}$.

Proof:

Indirect Proof: Step 1 Assume that Christina's average points per game was greater than or equal to 3, $a \geq 3$.

Step 2	**CASE 1**	**CASE 2**
	$a = 3$	$a > 3$
	$3 \stackrel{?}{=} \frac{13}{6}$	$\frac{13}{6} \stackrel{?}{>} 3$
	$3 \neq 2.2$	$2.2 \not> 3$

Step 3 The conclusions are false, so the assumption must be false. Therefore, Christina's average points per game was less than 3.

9. Given: $\triangle ABC$ is a right triangle; $\angle C$ is a right angle.

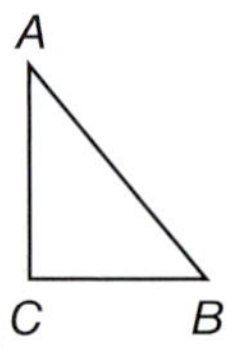

Prove: $AB > BC$ and $AB > AC$

Indirect Proof: Step 1 Assume that the hypotenuse of a right triangle is $\angle C$ is a longest side. That is, $AB < BC$ or $AB < AC$.

Step 2 If $AB < BC$, then $m\angle C < m\angle A$. Since $m\angle C = 90$, $m\angle A > 90$. So, $m\angle C + m\angle A > 180$.

By the same reasoning, $m\angle C + m\angle B > 180$.

Step 3 Both relationships contradict the fact that the sum of the measures of the angles of a triangle equals 180. Therefore, the hypotenuse must be the longest side of a right triangle.

11. $x \leq 8$ **13.** The lines are not parallel. **15.** The triangle is equiangular.

17 To write an indirect proof, first identify the conclusion. Find the negation of the conclusion and assume that it is true. Make a table of values to show that the negation of the conclusion is false. Since the assumption leads to a contradiction, you can conclude that the original conclusion must be true.

Given: $2x - 7 > -11$

Prove: $x > -2$

Indirect Proof: Step 1 The negation of $x > -2$ is $x \leq -2$. So, assume that $x \leq -2$ is true.

Step 2 Make a table with several possibilities for x assuming $x < -2$ or $x = -2$.

x	−6	−5	−4	−3	−2
$2x - 7$	−19	−17	−15	−13	−11

When $x < -2$, $2x - 7 < -11$ and when $x = -2$, $2x - 7 = -11$.

Step 3 In both cases, the assumption leads to the contradiction of the given information that $2x - 7 > -11$. Therefore, the assumption that $x \leq -2$ must be false, so the original conclusion that $x > -2$ must be true.

19. Given: $-3x + 4 < 7$

Prove: $x > -1$

Indirect Proof: Step 1 Assume that $x \leq -1$ is true.

Step 2 When $x < -1$, $-3x + 4 > 7$ and when $x = -1$, $-3x + 4 = 7$.

Step 3 In both cases, the assumption leads to the contradiction of the given information that $-3x + 4 > 7$. Therefore, the assumption that $x \leq -1$ must be false, so the original conclusion that $x > -1$ must be true.

x	−5	−4	−3	−2	−1
$-3x + 4$	19	16	13	10	7

21. Let the cost of one game be x and the other be y.

Given: $x + y > 80$

Prove: $x > 40$ or $y > 40$

Indirect Proof: Step 1 Assume that $x \leq 40$ and $y \leq 40$.

Step 2 If $x \leq 40$ and $y \leq 40$, then $x + y \leq 40 + 40$ or $x + y \leq 80$. This is a contradiction because we know that $x + y > 80$.

Step 3 Since the assumption that $x \leq 40$ and $y \leq 40$ leads to a contradiction of a known fact, the assumption must be false. Therefore, the conclusion that $x > 40$ or $y > 40$ must be true. Thus, at least one of the games had to cost more than \$40.

23. Given: xy is an odd integer.

Prove: x and y are odd integers.

Indirect Proof: Step 1 Assume that x and y are not both odd integers. That is, assume that either x or y is an even integer.

Step 2 You only need to show that the assumption that x is an even integer leads to a contradiction, since the argument for y is an even integer follows the same reasoning. So, assume that x is an even integer and y is an odd integer. This means that $x = 2k$ for some integer k and $y = 2m + 1$ for some integer m.

$xy = (2k)(2m + 1)$ Subst. of assumption

$= 4km + 2k$ Dist. Prop.

$= 2(km + k)$ Dist. Prop.

Since k and m are integers, $km + k$ is also an integer. Let p represent the integer $km + k$. So xy can be represented by $2p$, where p is an integer. This means that xy is an even integer, but this contradicts the given that xy is an odd integer.

Step 3 Since the assumption that x is an even integer and y is an odd integer leads to a contradiction of the given, the original conclusion that x and y are both odd integers must be true.

25. Given: x is an odd number.

Prove: x is not divisible by 4.

Indirect Proof: Step 1 Assume x is divisible by 4. In other words, 4 is a factor of x.

Step 2 Let $x = 4n$, for some integer n.

$x = 2(2n)$

So, 2 is a factor of x which means x is an even number, but this contradicts the given information.

Step 3 Since the assumption that x is divisible by 4 leads to a contradiction of the given, the original conclusion x is not divisible by 4 must be true.

27. Given: $XZ > YZ$

Prove: $\angle X \not\cong \angle Y$

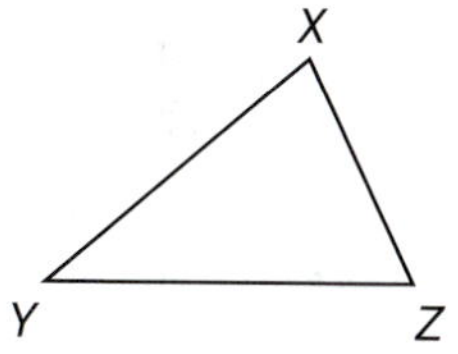

Indirect Proof: Step 1 Assume that $\angle X \cong \angle Y$.

Step 2 $\overline{XZ} \cong \overline{YZ}$ by the converse of the isosceles △ theorem.

Step 3 This contradicts the given information that $XZ > YZ$. Therefore, the assumption $\angle X \cong \angle Y$ must be false, so the original conclusion $\angle X \not\cong \angle Y$ must be true.

29. Given: $\triangle ABC$ is isosceles.

Prove: Neither of the base angles is a right angle.

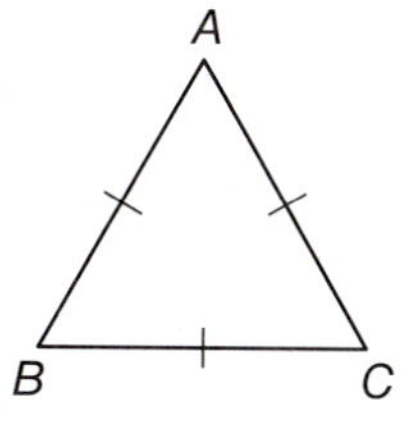

Indirect Proof: Step 1 Assume that $\angle B$ is a right angle.

Step 2 By the Isosceles △ Theorem, $\angle C$ is also a right angle.

Step 3 This contradicts the fact that a triangle can have no more than one right angle. Therefore, the assumption that $\angle B$ is a right angle must be false, so the original conclusion neither of the base angles is a right angle must be true.

31. Given: $m\angle A > m\angle ABC$
Prove: $BC > AC$

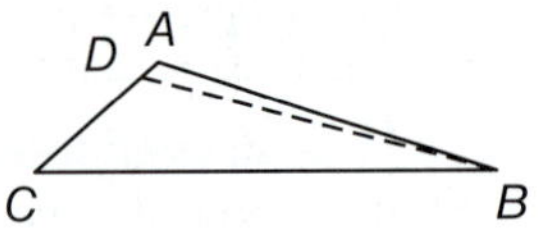

Proof:
Assume $BC \not> AC$. By the Comparison Property, $BC = AC$ or $BC < AC$.
Case 1: If $BC = AC$, then $\angle ABC \cong \angle A$ by the Isosceles Triangle Theorem. (If two sides of a triangle are congruent, then the angles opposite those sides are congruent.) But, $\angle ABC \cong \angle A$ contradicts the given statement that $m\angle A > m\angle ABC$. So, $BC \neq AC$.
Case 2: If $BC < AC$, then there must be a point D between A and C so that $\overline{DC} \cong \overline{BC}$. Draw the auxiliary segment $\overline{BD}$. Since $DC = BC$, by the Isosceles Triangle Theorem $\angle BDC \cong \angle DBC$. Now $\angle BDC$ is an exterior angle of $\triangle BAD$ and by the Exterior Angles Inequality Theorem (the measure of an exterior angle of a triangle is greater than the measure of either corresponding remote interior angle) $m\angle BDC > m\angle A$. By the Angle Addition Postulate, $m\angle ABC = m\angle ABD + m\angle DBC$. Then by the definition of inequality, $m\angle ABC > m\angle DBC$. By Substitution and the Transitive Property of Inequality, $m\angle ABC > m\angle A$. But this contradicts the given statement that $m\angle A > m\angle ABC$. In both cases, a contradiction was found, and hence our assumption must have been false. Therefore, $BC > AC$.

33. We know that the other team scored 3 points, and Katsu thinks that they made a three point shot. We also know that a player can score 3 points by making a basket and a foul shot.
Step 1 Assume that a player for the other team made a two-point basket and a foul shot.
Step 2 The other team's score before Katsu left was 26, so their score after a two-point basket and a foul shot would be 26 + 3 or 29.
Step 3 The score is correct when we assume that the other team made a two-point basket and a foul shot, so Katsu's assumption may not be correct. The other team could have made a three-point basket or a two-point basket and a foul shot.

35 a. To write an indirect proof, first identify the conclusion. Find the negation of the conclusion and assume that it is true. Then use the data to prove that the assumption is false.
Step 1 50% is half, and the statement says more than half of the teens polled said that they recycle, so assume that less than 50% recycle.
Step 2 The data shows that 51% of teens said that they recycle, and 51% > 50%, so the number of teens that recycle is not less than half.
Step 3 This contradicts the data given. Therefore, the assumption is false, and the conclusion more than half of the teens polled said they recycle must be true.
b. According to the data, 23% of 400 teenagers polled said that they participate in Earth Day. Verify that 23% of 400 is 92.

$400 \cdot 23\% \stackrel{?}{=} 92$
$400 \cdot 0.23 \stackrel{?}{=} 92$
$92 = 92$

37 **Given:** $\overline{AB} \perp$ line p
Prove: $\overline{AB}$ is the shortest segment from A to line p.
Indirect Proof: Step 1
Assume $\overline{AB}$ is not the shortest segment from A to p.

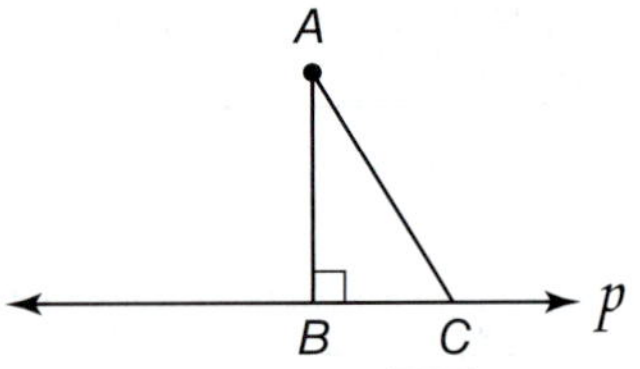

Step 2 Since $\overline{AB}$ is not the shortest segment from A to p, there is a point C such that $\overline{AC}$ is the shortest distance. $\triangle ABC$ is a right triangle with hypotenuse $\overline{AC}$, the longest side of $\triangle ABC$ since it is across from the largest angle in $\triangle ABC$ by the Angle-Side Relationships in Triangles Theorem.
Step 3 This contradicts the fact that $\overline{AC}$ is the shortest side. Therefore, the assumption is false, and the conclusion, $\overline{AB}$ is the shortest side, must be true.

39a. $n^3 + 3$ **39b.** Sample answer:

n	$n^3 + 3$
2	11
3	30
10	1003
11	1334
24	13,827
25	15,628
100	1,000,003
101	1,030,304
526	145,531,579
527	146,363,186

39c. Sample answer: When $n^3 + 3$ is even, n is odd.
39d. Indirect Proof: Step 1
Assume that n is even. Let $n = 2k$, where k is some integer.

Step 2

$n^3 + 3 = (2k)^3 + 3$	Substitute assumption
$= 8k^3 + 3$	Simplify.
$= (8k^3 + 2) + 1$	Replace 3 with 2 + 1 and group the first two terms.
$= 2(4k^3 + 1) + 1$	Distributive Property

Since k is an integer, $4k^3 + 1$ is also an integer. Therefore, $n^3 + 3$ is odd.
Step 3 This contradicts the given information that $n^3 + 3$ is even. Therefore, the assumption is false, so the conclusion that n is odd must be true.

41. Sample answer: $\triangle ABC$ is scalene.
Given: $\triangle ABC$; $AB \neq BC$; $BC \neq AC$; $AB \neq AC$
Prove: $\triangle ABC$ is scalene.

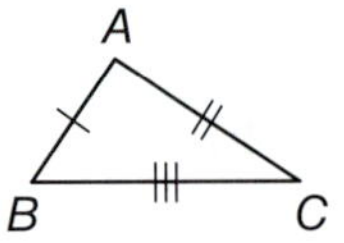

Indirect Proof:
Step 1 Assume that $\triangle ABC$ is not scalene.
Case 1: $\triangle ABC$ is isosceles.
Step 2 If $\triangle ABC$ is isosceles, then $AB = BC$, $BC = AC$, or $AB = AC$.
Step 3 This contradicts the given information, so $\triangle ABC$ is not isosceles.

Case 2: $\triangle ABC$ is equilateral.
In order for a triangle to be equilateral, it must also be isosceles, and Case 1 proved that $\triangle ABC$ is not isosceles. Thus, $\triangle ABC$ is not equilateral. Therefore, $\triangle ABC$ is scalene. **43.** Neither; sample answer: Since the hypothesis is true when the conclusion is false, the statement is false. **45.** $y = 2x - 7$ **47.** J

49. Given: $\overline{RQ}$ bisects $\angle SRT$.
Prove: $m\angle SQR > m\angle SRQ$

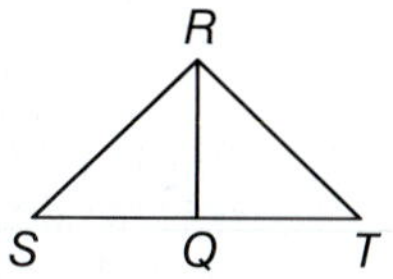

Proof:
Statements (Reasons)
1. $\overline{RQ}$ bisects $\angle SRT$. (Given)
2. $\angle SRQ \cong \angle QRT$ (Def. of bisector)
3. $m\angle QRS = m\angle QRT$ (Def. of $\cong$ ∡)
4. $m\angle SQR = m\angle T + m\angle QRT$ (Exterior Angle Theorem)
5. $m\angle SQR > m\angle QRT$ (Def. of Inequality)
6. $m\angle SQR > m\angle SRQ$ (Substitution)

51. $\left(1\frac{2}{5}, 2\frac{3}{5}\right)$ **53.** 64 **55.** $\sqrt{5} \approx 2.2$ **57.** true **59.** false

Pages 360–366 ***Lesson 5-5***

1 Yes; check each inequality.

$5 + 7 \overset{?}{>} 10$ $\quad 5 + 10 \overset{?}{>} 7$ $\quad 7 + 10 \overset{?}{>} 5$
$12 > 10$ ✓ $\quad 15 > 7$ ✓ $\quad 17 > 5$ ✓

Since the sum of each pair of side lengths is greater than the third side length, sides with lengths 5 cm, 7 cm, and 10 cm will form a triangle.

3. yes

5. Given: $\overline{XW} \cong \overline{YW}$
Prove: $YZ + ZW > XW$

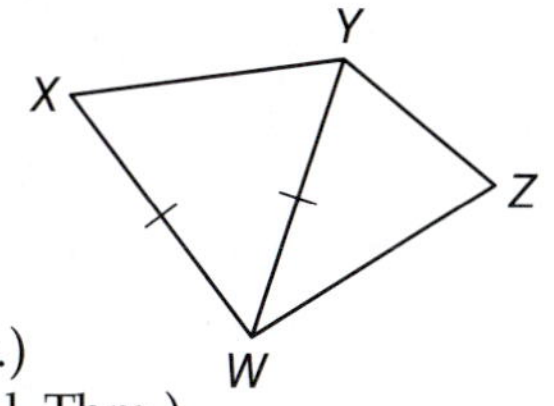

Proof:
Statements (Reasons)
1. $\overline{XW} \cong \overline{YW}$ (Given)
2. $XW = YW$ (Def. of $\cong$ segs.)
3. $YZ + ZW > YW$ ($\triangle$ Inequal. Thm.)
4. $YZ + ZW > XW$ (Subst.)

7. yes **9.** no; $2.1 + 4.2 \not> 7.9$ **11.** yes **13.** 6 m $< n <$ 16 m **15.** 5.4 in. $< n <$ 13 in. **17.** $5\frac{1}{3}$ yd $< n <$ 10 yd

19. Given: $\overline{JL} \cong \overline{LM}$
Prove: $KJ + KL > LM$

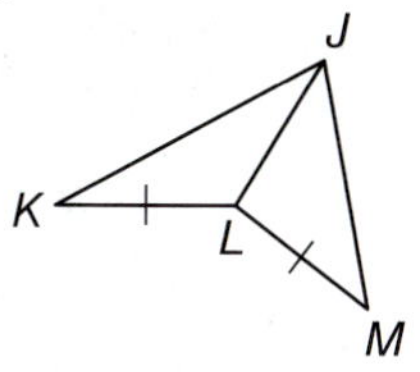

Proof:
Statements (Reasons)
1. $\overline{JL} \cong \overline{LM}$ (Given)
2. $JL = LM$ (Def. of $\cong$ segments)
3. $KJ + KL > JL$ ($\triangle$ Inequal. Thm.)
4. $KJ + KL > LM$ (Subt.)

21
$$XY + YZ > XZ$$
$$(4x - 1) + (2x + 7) > x + 13$$
$$6x + 6 > x + 13$$
$$5x > 7$$
$$x > \frac{7}{5}$$

$$XY + XZ > YZ$$
$$(4x - 1) + (x + 13) > 2x + 7$$
$$5x + 12 > 2x + 7$$
$$3x > -5$$
$$x > -\frac{5}{3}$$

$$YZ + XZ > XY$$
$$(2x + 7) + (x + 13) > 4x - 1$$
$$3x + 20 > 4x - 1$$
$$21 > x$$

Since x must be greater than $\frac{7}{5}$, greater than $-\frac{5}{3}$, and less than 21, possible values of x are $\frac{7}{5} < x < 21$.

23. Given: $\triangle ABC$
Prove: $AC + BC > AB$

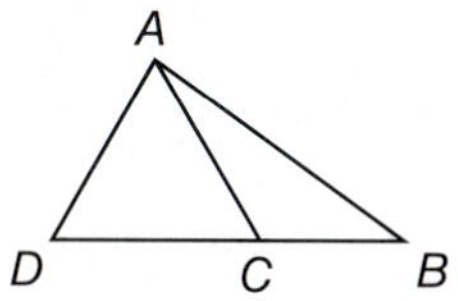

Proof:
Statements (Reasons)
1. Construct $\overline{CD}$ so that C is between B and D and $\overline{CD} \cong \overline{AC}$. (Ruler Post.)
2. $CD = AC$ (Def. of $\cong$)
3. $\angle CAD \cong \angle ADC$ (Isos. $\triangle$ Thm.)
4. $m\angle CAD = m\angle ADC$ (Def. of $\cong$ ∡)
5. $m\angle BAC + m\angle CAD = m\angle BAD$ ($\angle$ Add. Post.)
6. $m\angle BAC + m\angle ADC = m\angle BAD$ (Subst.)
7. $m\angle ADC < m\angle BAD$ (Def. of inequality)
8. $AB < BD$ (Angle-Side Relationships in Triangles)
9. $BD = BC + CD$ (Seg. Add. Post.)
10. $AB < BC + CD$ (Subst.)
11. $AB < BC + AC$ (Subst. (Steps 2, 10))

25. $2 < x < 10$ **27.** $1 < x < 11$ **29.** $x > 0$ **31.** Yes; sample answer: The measurements on the drawing do not form a triangle. According to the Triangle Inequality Theorem, the sum of the lengths of any two sides of a triangle is greater than the length of the third side. The lengths in the drawing are 1 ft, $3\frac{7}{8}$ ft, and $6\frac{3}{4}$ ft. Since $1 + 3\frac{7}{8} \not> 6\frac{3}{4}$, the triangle is impossible. They should recalculate their measurements before they cut the wood.

33 A triangle 3 feet by 4 feet by x feet is formed. The length of the third side x must be less than the sum of the lengths of the other two sides. So, $x < 3 + 4$ or $x < 7$. Since the awning drapes 6 inches or 0.5 feet over the front, the total length should be less than $7 + 0.5$ or 7.5 feet. She should buy no more than 7.5 feet.

35. Yes; $\sqrt{99} \approx 9.9$ since $\sqrt{100} = 10$, $\sqrt{48} \approx 6.9$ since $\sqrt{49} = 7$, and $\sqrt{65} \approx 8.1$ since $\sqrt{64} = 8$. So, $9.9 + 6.9 > 8.1$, $6.9 + 8.1 > 9.9$, and $8.1 + 9.9 > 6.9$.
37. no; $\sqrt{122} \approx 11.1$ since $\sqrt{121} = 11$, $\sqrt{5} \approx 2.1$ since $\sqrt{4} = 2$, and $\sqrt{26} \approx 5.1$ since $\sqrt{25} = 5$. So, $2.1 + 5.1 \not> 11.1$.

39 Use the Distance Formula $d = \sqrt{(x_2 - x_1)^2 + (y_2 - y_1)^2}$ to find the lengths of the sides.

$FG = \sqrt{[3 - (-4)]^2 + (-3 - 3)^2}$ $\quad x_1 = -4, x_2 = 3, y_1 = 3, y_2 = -3$
$= \sqrt{85}$ $\quad$ Simplify.
≈ 9.2 $\quad$ Use a calculator.
$GH = \sqrt{(4 - 3)^2 + [6 - (-3)]^2}$ $\quad x_1 = 3, x_2 = 4, y_1 = -3, y_2 = 6$

$= \sqrt{82}$	Simplify.
≈ 9.1	Use a calculator.
$FH = \sqrt{[4 - (-4)]^2 + (6 - 3)^2}$	$x_1 = -4, x_2 = 4,$ $y_1 = 3, y_2 = 6$
$= \sqrt{73}$	Simplify.
≈ 8.5	Use a calculator.

$FG + GH \overset{?}{>} FH$	$FG + FH \overset{?}{>} GH$	$GH + FH \overset{?}{>} FG$
$9.2 + 9.1 \overset{?}{>} 8.5$	$9.2 + 8.5 \overset{?}{>} 9.1$	$9.1 + 8.5 \overset{?}{>} 9.2$
$18.3 > 8.5$	$17.7 > 9.1$	$17.6 > 9.2$

Since $FG + GH > FH$, $FG + FH > GH$, and $GH + FH > FG$, the coordinates are the vertices of a triangle.

41. yes; $QR + QS > RS$, $QR + RS > QS$, and $QS + RS > QR$ **43.** The perimeter is greater than 36 and less than 64. Sample answer: From the diagram we know that $\overline{AC} \cong \overline{EC}$ and $\overline{DC} \cong \overline{BC}$, and $\angle ACB \cong \angle ECD$ because vertical angles are congruent, so $\triangle ACB \cong \triangle ECD$. Using the Triangle Inequality Theorem, the minimum value of AC and EC is 2 and the maximum value is 16. Therefore, the minimum value of the perimeter is greater than $2(2 + 7 + 9)$ or 36, and the maximum value of the perimeter is less than $2(16 + 7 + 9)$ or 64. **45.** Sample answer: The Triangle Inequality Theorem states that the sum of the lengths of two sides of a triangle is always greater than the length of the third side of a triangle, so you set up three inequalities. For example, for a triangle with side lengths a, b, and c, you set up $a + b > c$, $a + c > b$, and $b + c > a$. Usually, one of the inequalities results in a negative number and is not necessary to find the minimum and maximum values for the unknown side. The other two inequalities give the value that the side must be greater than and the value that the side must be less than.

47.

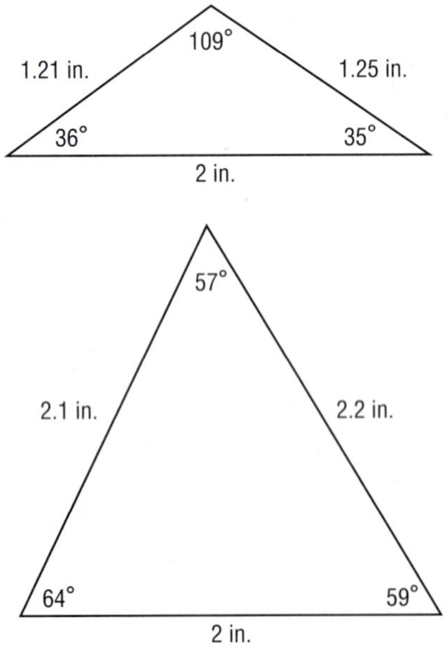

49. B **51.** H **53.** $y > 6$ or $y < 6$ **55.** 132 mi $< d <$ 618 mi **57.** 15; Alt. Ext. ∠s Thm. **59.** $x = \frac{4}{3} \approx 1.3$; $JK = 4$ **61.** $x = 2$; $JK = KL = JL = 14$ **63.** $x = 7$; $SR = RT = 24$, $ST = 19$

Pages 367–376 Lesson 5-6

1. $m\angle ACB > m\angle GDE$

3 $\overline{QR} \cong \overline{SR}$, $\overline{TR} \cong \overline{TR}$, and $m\angle QRT < m\angle SRT$. By the Hinge Theorem, $QT < ST$.

5a. $\overline{AB} \cong \overline{DE}$, $\overline{AC} \cong \overline{DF}$ **5b.** $\angle D$; Sample answer: Since $EF > BC$, according to the Hinge Theorem, $m\angle D > m\angle A$ **7.** $\frac{5}{3} < x < B$

9. Given: $\overline{AD} \cong \overline{CB}$, $DC < AB$
Prove: $m\angle CBD < m\angle ADB$

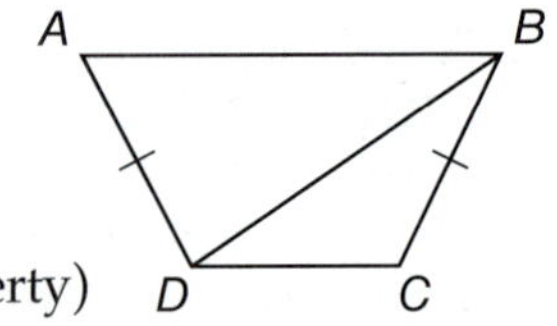

Statements (Reasons)
1. $\overline{AD} \cong \overline{CB}$ (Given)
2. $\overline{DB} \cong \overline{DB}$ (Reflexive Property)
3. $DC < AB$ (Given)
4. $m\angle CBD < m\angle ADB$ (SSS Inequality)

11. $m\angle MLP < m\angle TSR$

13 $\overline{TU} \cong \overline{VU}$, $\overline{WU} \cong \overline{WU}$, and $WT < WV$. By the Converse of the Hinge Theorem, $m\angle TUW > m\angle VUW$.

15. $JK > HJ$ **17.** $2 < x < 6$

19 Two sides of one triangle are congruent to two sides of the other triangle and the third side of the first triangle is less than the third side of the second triangle. So, by the Converse of the Hinge Theorem, the included angle of the first triangle is less than the included angle of the second triangle.

$0 < x + 20 < 41$	Converse of Hinge Theorem
$-20 < x < 21$	Subtract 20 from each.

The range of values containing x is $-20 < x < 21$.

21. $\overline{RS}$; sample answer: The height of the crane is the same and the length of the crane arm is fixed, so according to the Hinge Theorem, the side opposite the smaller angle is shorter. Since $29° < 52°$, $RS < MN$.

23. Given: $\overline{LK} \cong \overline{JK}$, $\overline{RL} \cong \overline{RJ}$, K is the midpoint of $\overline{QS}$, $m\angle SKL > m\angle QKJ$
Prove: $RS > QR$

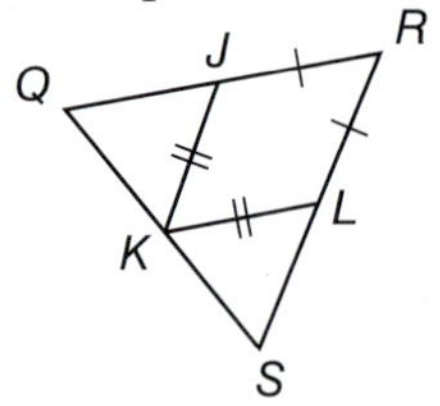

Statements (Reasons)
1. $\overline{LK} \cong \overline{JK}$, $\overline{RL} \cong \overline{RJ}$, K is the midpoint of $\overline{QS}$, $m\angle SKL > m\angle QKJ$ (Given)
2. $SK = QK$ (Def. of midpoint)
3. $SL > QJ$ (Hinge Thm.)
4. $RL = RJ$ (Def. of ≅ segs.)
5. $SL + RL > RL + RJ$ (Add. Prop.)
6. $SL + RL > QJ + RJ$ (Sub.)
7. $RS = SL + RL$, $QR = QJ + RJ$ (Seg. Add. Post.)
8. $RS > QR$ (Subst.)

25. Given: $\overline{XU} \cong \overline{VW}$, $VW > XW$, $\overline{XU} \parallel \overline{VW}$
Prove: $m\angle XZU > m\angle UZV$

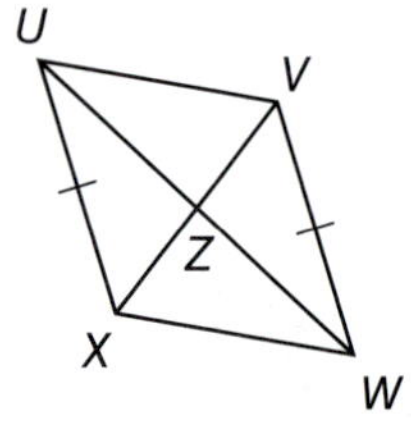

Statements (Reasons)
1. $\overline{XU} \cong \overline{VW}$, $\overline{XU} \parallel \overline{VW}$ (Given)
2. $\angle UXV \cong \angle XVW$, $\angle XUW \cong \angle UWV$ (Alt. Int. ∠s Thm.)
3. $\triangle XZU \cong \triangle VZW$ (ASA)
4. $\overline{XZ} \cong \overline{VZ}$ (CPCTC)
5. $\overline{WZ} \cong \overline{WZ}$ (Refl. Prop.)
6. $VW > XW$ (Given)

7. $m\angle VZW > m\angle XZW$ (Converse of Hinge Thm.)
8. $\angle VZW \cong \angle XZU$, $\angle XZW \cong \angle VZU$ (Vert. ≜ are ≅.)
9. $m\angle VZW = m\angle XZU$, $m\angle XZW = m\angle VZU$ (Def. of ≅ ≜)
10. $m\angle XZU > m\angle UZV$ (Subst.)

27 **a.** Sample answer: Use a ruler to measure the distance from her shoulder to her fist for each position. The distance is 1.6 cm for Position 1 and 2 cm for Position 2. Therefore, the distance from her shoulder to her fist is greater in Position 2. **b.** Sample answer: In each position, a triangle formed. The distance from her shoulder to her elbow and from her elbow to her wrist is the same in both triangles. Using the measurements in part **a** and the Converse of the Hinge Theorem, you know that the measure of the angle opposite the larger side is larger, so the angle formed by Anica's elbow is greater in Position 2.

29. Given: $\overline{PR} \cong \overline{PQ}$, $SQ > SR$
Prove: $m\angle 1 < m\angle 2$

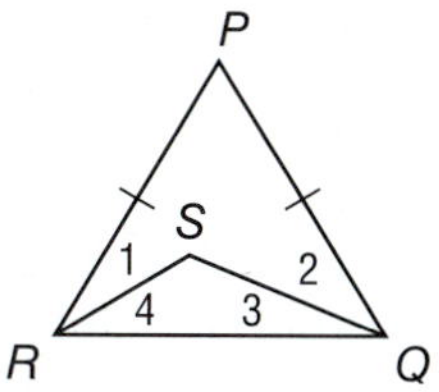

Statements (Reasons)
1. $\overline{PR} \cong \overline{PQ}$ (Given)
2. $\angle PRQ \cong \angle PQR$ (Isos. △ Thm.)
3. $m\angle PRQ = m\angle 1 + m\angle 4$, $m\angle PQR = 2 + m\angle 3$ (Angle Add. Post.)
4. $m\angle PRQ = m\angle PQR$ (Def. of ≅ ≜)
5. $m\angle 1 + m\angle 4 = m\angle 2 + m\angle 3$ (Subst.)
6. $SQ > SR$ (Given)
7. $m\angle 4 > m\angle 3$ (∠-Side Relationship Thm.)
8. $m\angle 4 = m\angle 3 + x$ (Def. of inequality)
9. $m\angle 1 + m\angle 4 - m\angle 4 = m\angle 2 + m\angle 3 - (m\angle 3 + x)$ (Subt. Prop.)
10. $m\angle 1 = m\angle 2 - x$ (Subst.)
11. $m\angle 1 + x = m\angle 2$ (Add. Prop.)
12. $m\angle 1 < m\angle 2$ (Def. of inequality)

31. $CB < AB$ **33.** $m\angle BGC < m\angle BAF$

35 $\overline{WZ} \cong \overline{YZ}$, $\overline{ZU} \cong \overline{ZU}$, and $m\angle WZU > m\angle YZU$. By the Hinge Theorem, $WU > YU$.

37a.

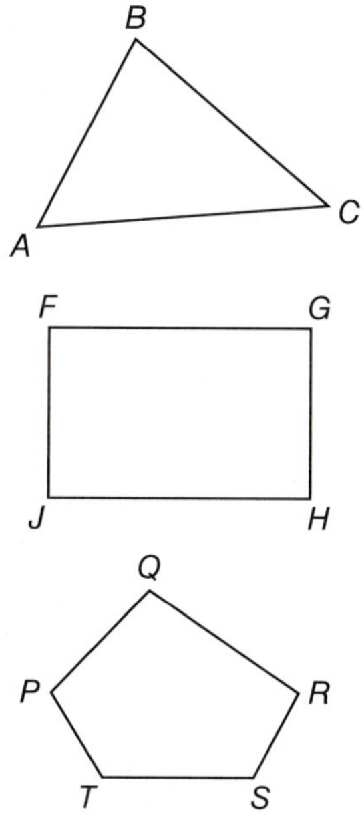

37b. ∠ measures: 59, 76, 45; 90, 90, 90, 90; 105, 100, 96, 116, 123; Sum of ≜: 180, 360, 540
37c. Sample answer: The sum of the angles of the polygon is equal to 180 times two less than the number of sides of the polygon.
37d. Inductive; sample answer: Since I used a pattern to determine the relationship, the reasoning I used was inductive.
37e. $(n - 2)180$

39.

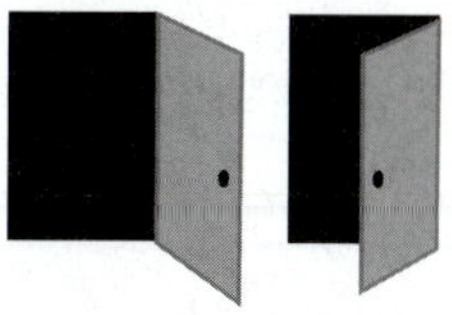

A door; as the door opens, the door opening increases as the angle made by the hinge increases. As the door closes, the door opening decreases as the angle made by the hinge decreases. This is similar to the side opposite the angle in a triangle, because as the side opposite an angle increases the measure of the angle also increases. As the side decreases, the angle also decreases. **41.** Never; from the Converse of the Hinge Theorem, $\angle ADB < \angle BDC$. $\angle ADB < \angle BDC$ form a linear pair. So, $m\angle ADB + m\angle BDC = 180$. Since, $m\angle BDC > m\angle ADB$, $m\angle BDC$ must be greater than 90 and $m\angle ADB$ must be smaller than 90. So, by the definition of obtuse and acute angles, $m\angle BDC$ is always obtuse and $m\angle ADB$ is always acute. **43.** $2.8 < x < 12$ **45.** F **47.** $1.2 \text{ cm} < n < 7.6 \text{ cm}$ **49.** $6 \text{ m} < n < 12 \text{ m}$ **51.** $x = 8$

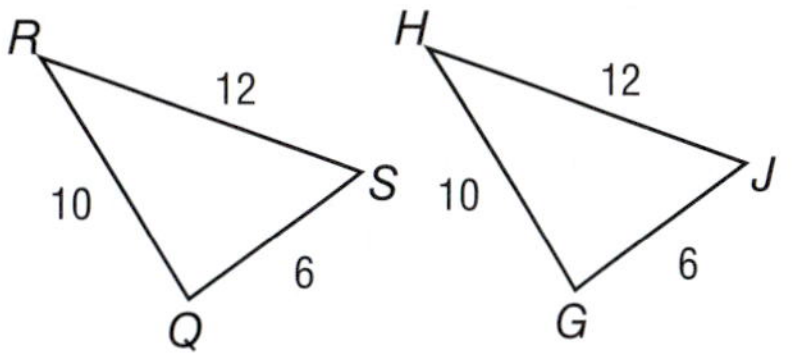

53. A **55.** $\angle 3$, $\angle ACB$ **57.** $x = 66$ by the Consecutive Interior Angles Theorem; $y = 35$ by the Consecutive Interior Angles Theorem

Pages 377–380 *Chapter 5* *Study Guide and Review*

1. false; orthocenter **3.** true **5.** false; median **7.** false; false **9.** false; the vertex opposite that side **11.** 5 **13.** 34 **15.** (2, 3) **17.** $\angle S$, $\angle R$, $\angle T$; $\overline{RT}$, $\overline{TS}$, $\overline{SR}$ **19.** The shorter path is for Sarah to get Irene and then go to Anna's house. **21.** $\triangle FGH$ is not congruent to $\triangle MNO$. **23.** $y \geq 4$
25. Let the cost of one DVD be x, and the cost of the other DVD be y.
Given: $x + y > 50$
Prove: $x > 25$ or $y > 25$
Indirect proof: Step 1 Assume that $x \leq 25$ and $y \leq 25$.
Step 2 If $x \leq 25$ and $y \leq 25$, then $x + y \leq 25 + 25$, or $x + y \leq 50$. This is a contradiction because we know that $x + y > 50$.
Step 3 Since the assumption that $x \leq 25$ and $y \leq 25$ leads to a contradiction of a known fact, the assumption must be false. Therefore, the conclusion that $x > 25$ or $y > 25$ must be true. Thus, at least one DVD had to be over $25.
27. no; $3 + 4 < 8$ **29.** Let x be the length of the third side. $6.5 \text{ cm} < x < 14.5 \text{ cm}$ **31.** $m\angle ABC > m\angle DEF$ **33.** Rose

Chapter 6 Quadrilaterals

Page 387 *Chapter 6* *Get Ready*

1. 150 **3.** 54 **5.** 137 **7.** $x = 1$, $WX = XY = YW = 9$
9. Des Moines to Phoenix = 1153 mi, Des Moines to Atlanta = 738 mi, Phoenix to Atlanta = 1591 mi

Pages 389–397 Lesson 6-1

1. 1440 **3.** $m\angle X = 36$, $m\angle Y = 72$, $m\angle Z = 144$, $m\angle W = 108$

5 $(n - 2) \cdot 180 = (16 - 2) \cdot 180$ $n = 6$
$= 14 \cdot 180$ or 2520 Simplify.

The sum of the interior angle measures is 2520. So, the measure of one interior angle is $2520 \div 16$ or 157.5.

7. 36 **9.** 68 **11.** 45 **13.** 3240 **15.** 5400

17 $(n - 2) \cdot 180 = (4 - 2) \cdot 180$ $n = 4$
$= 2 \cdot 180$ or 360 Simplify.

The sum of the interior angle measures is 360.

$360 = m\angle J + m\angle K + m\angle L + m\angle M$	Sum of interior angle measures
$360 = (3x - 6) + (x + 10) + x + (2x - 8)$	Substitution
$360 = 7x - 4$	Combine like terms.
$364 = 7x$	Add 4 to each side.
$52 = x$	Simplify.

$m\angle J = 3x - 6 = 3(52) - 6$ or 150 $m\angle K = x + 10 = 52 + 10$ or 62

$m\angle L = x = 52$ $m\angle M = 2x - 8 = 2(52) - 8$ or 96

19. $m\angle U = 60$, $m\angle V = 193$, $m\angle W = 76$, $m\angle Y = 68$, $m\angle Z = 143$ **21.** 150 **23.** 144 **25a.** 720 **25b.** Yes, 120; sample answer: Since the measures of the sides of the hexagon are equal, it is regular and the measures of the angles are equal. That means each angle is $720 \div 6$ or 120. **27.** 4 **29.** 15

31
$$21 + 42 + 29 + (x + 14) + x + (x - 10) + (x - 20) = 360$$
$$4x + 76 = 360$$
$$4x = 284$$
$$x = 71$$

33. 37 **35.** 72 **37.** 24 **39.** 51.4, 128.6 **41.** 25.7, 154.3 **43.** Consider the sum of the measures of the exterior angles N for an n-gon.

N = sum of measures of linear pairs − sum of measures of interior angles
$= 180n - 180(n - 2)$
$= 180n - 180n + 360$
$= 360$

So, the sum of the exterior angle measures is 360 for any convex polygon. **45.** 105, 110, 120, 130, 135, 140, 160, 170, 180, 190

47 **a.** $60 \text{ ft} \div 8 = 7.5 \text{ ft}$ Perimeter ÷ number of sides
b. $(n - 2) \cdot 180 = (8 - 2) \cdot 180$ $n = 8$
$= 6 \cdot 180$ or 1080 Simplify.

Sample answer: The sum of the interior angle measures is 1080. So, the measure of each angle of a regular octagon is $1080 \div 8$ or 135. So if each side of the board makes up half of the angle, each one measures $135 \div 2$ or 67.5.

49. Liam; by the Exterior Angle Sum Theorem, the sum of the measures of any convex polygon is 360.

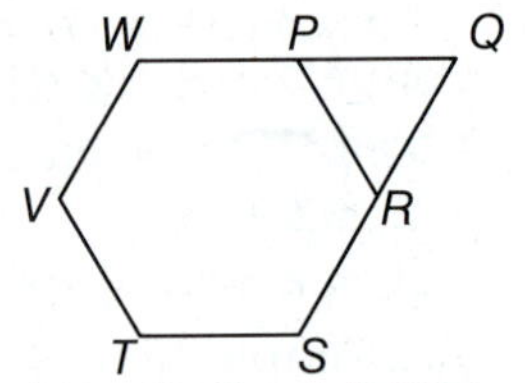

51. Always; by the Exterior Angle Sum Theorem, $m\angle QPR = 60$ and $m\angle QRP = 60$. Since the interior angle measures of a triangle add up to 180, the measure of $\angle PQR = 180 - m\angle QPR - m\angle QRP = 180 - 60 - 60 = 60$. So, $\triangle PQR$ is an equilateral triangle. **53.** The Interior Angle Sum Theorem is derived from the pattern between the number of sides in a polygon and the number of triangles. The formula is the product of the sum of the measures of the angles in a triangle, 180, and the number of triangles in the polygon. **55.** 72 **57** C **59.** $ML < JM$ **61.** 3 **63.** $\angle E \cong \angle G$; $\angle EFH \cong \angle GHF$; $\angle EHF \cong \angle GFH$; $\overline{EF} \cong \overline{GH}$; $\overline{EH} \cong \overline{GF}$; $\overline{FH} \cong \overline{FH}$; $\triangle EFH \cong \triangle GHF$ **65.** $\angle 1$ and $\angle 5$, $\angle 4$ and $\angle 6$, $\angle 2$ and $\angle 8$, $\angle 3$ and $\angle 7$

Pages 399–407 Lesson 6-2

1a. 148 **1b.** 125 **1c.** 4 **3.** 15 **5.** $w = 5$, $b = 4$

7. Given: ▱$ABCD$, $\angle A$ is a right angle.
Prove: $\angle B$, $\angle C$, and $\angle D$ are right angles. (Theorem 6.6)
Proof: By definition of a parallelogram, $\overline{AB} \parallel \overline{CD}$. Since $\angle A$ is a right angle, $\overline{AC} \perp \overline{AB}$. By the Perpendicular Transversal Theorem, $\overline{AC} \perp \overline{CD}$. $\angle C$ is a right angle, because perpendicular lines form a right angle. $\angle B \cong \angle C$ and $\angle A \cong \angle D$ because opposite angles in a parallelogram are congruent. $\angle C$ and $\angle D$ are right angles, since all right angles are congruent.

9

$m\angle R + m\angle Q = 180$	Consecutive angles are supplementary.
$m\angle R + 128 = 180$	Substitution
$m\angle R = 52$	Subtract 128 from each side.

11. 5

13

a. $\overline{JH} \cong \overline{FG}$	Opposite sides are congruent.
$JH = FG$	Definition of congruence
$= 1$ in.	Substitution
b. $\overline{GH} \cong \overline{FJ}$	Opposite sides are congruent.
$GH = FJ$	Definition of congruence
$= \frac{3}{4}$ in.	Substitution
c. $\angle JFG \cong \angle JHG$	Opposite angles are congruent.
$m\angle JFG = m\angle JHG$	Definition of congruence
$= 62$	Substitution
d. $m\angle FJH + m\angle JHG = 180$	Consecutive angles are supplementary.
$m\angle FJH + 62 = 180$	Substitution
$m\angle FJH = 118$	Subtract 62 from each side.

15. $a = 7$, $b = 11$ **17.** $x = 5$, $y = 17$ **19.** $x = 58$, $y = 63.5$ **21.** (2.5, 2.5)

23. Given: $WXTV$ and $ZYVT$ are parallelograms.
Prove: $\overline{WX} \cong \overline{ZY}$

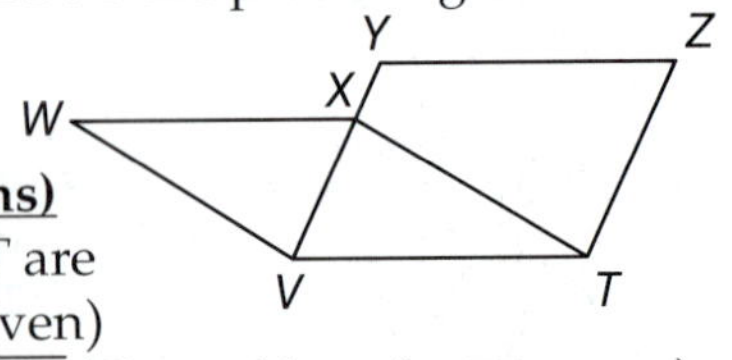

Proof:

Statements (Reasons)

1. $WXTV$ and $ZYVT$ are parallelograms. (Given)
2. $\overline{WX} \cong \overline{VT}$, $\overline{VT} \cong \overline{YZ}$ (Opp. sides of a ▱ are ≅.)
3. $\overline{WX} \cong \overline{ZY}$ (Trans. Prop.)

25. Given: $\triangle ACD \cong \triangle CAB$
Prove: $\overline{DP} \cong \overline{PB}$

Proof:
Statements (Reasons)
1. $\triangle ACD \cong \triangle CAB$ (Given)
2. $\angle ACD \cong \angle CAB$ (CPCTC)
3. $\angle DPC \cong \angle BPA$ (Vert. ∠s are ≅.)
4. $\overline{AB} \cong \overline{CD}$ (CPCTC)
5. $\triangle ABP \cong \triangle CDP$ (AAS)
6. $\overline{DP} \cong \overline{PB}$ (Diag. of a ▱ bisect each other.)

27. Given: ▱$WXYZ$
Prove: $\triangle WXZ \cong \triangle YZX$ (Theorem 6.8)
Proof:

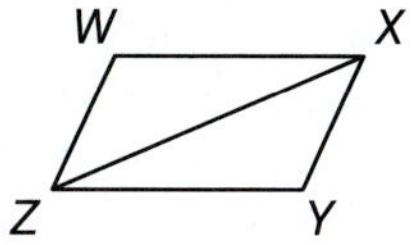

Statements (Reasons)
1. ▱$WXYZ$ (Given)
2. $\overline{WX} \cong \overline{ZY}$, $\overline{WZ} \cong \overline{XY}$ (Opp. sides of a ▱ are ≅.)
3. $\angle ZWX \cong \angle XYZ$ (Opp. ∠s of a ▱ are ≅.)
4. $\triangle WXZ \cong \triangle YZX$ (SAS)

29. Given: $ACDE$ is a parallellogram.
Prove: $\overline{EC}$ bisects $\overline{AD}$. (Theorem 6.7)

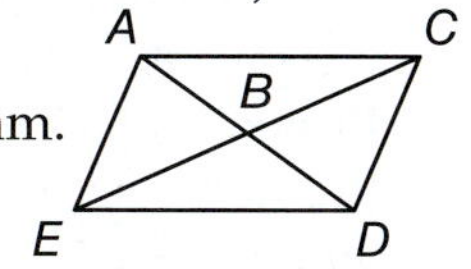

Proof: It is given that $ACDE$ is a parallelogram. Since opposite sides of a parallelogram are congruent, $\overline{EA} \cong \overline{DC}$. By definition of a parallelogram, $\overline{EA} \parallel \overline{DC}$. $\angle AEB \cong \angle DCB$ and $\angle EAB \cong \angle CDB$ because alternate interior angles are congruent. $\triangle EBA \cong \triangle CBD$ by ASA. $\overline{EB} \cong \overline{BC}$ and $\overline{AB} \cong \overline{BD}$ by CPCTC. By the definition of segment bisector, $\overline{EC}$ bisects $\overline{AD}$ and $\overline{AD}$ bisects $\overline{EC}$.

31. 3

33 $\angle AFB$ and $\angle BFC$ form a linear pair.

$\angle AFB + \angle BFC = 180$	Supplement Theorem
$\angle AFB + 49 = 180$	Substitution
$\angle AFB = 131$	Subtract 49 from each side.

35. 29 **37.** (0, −1); Sample answer: Opposite sides of a parallelogram are parallel. Since the slope of $\overline{BC} = \frac{-6}{2}$, the slope of $\overline{AD}$ must also be $\frac{-6}{2}$. To locate vertex D, start from vertex A and move down 6 and right 2.

39 First, use properties involving opposite angles and opposite sides of a parallelogram to help prove that $\triangle YUZ$ and $\triangle VXW$ are right angles. Then use the Hypotenuse-Angle Congruence Theorem to prove the triangles are congruent.
Given: ▱$YWVZ$, $\overline{VX} \perp \overline{WY}$, $\overline{YU} \perp \overline{VZ}$
Prove: $\triangle YUZ \cong \triangle VXW$
Proof:

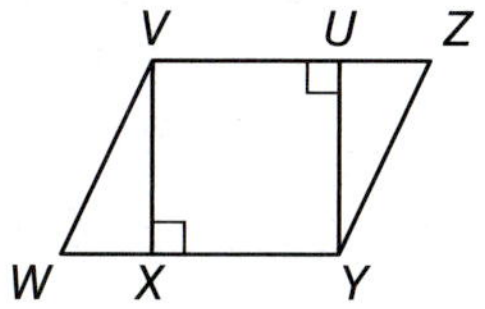

Statements (Reasons)
1. ▱$YWVZ$, $\overline{VX} \perp \overline{WY}$, $\overline{YU} \perp \overline{VZ}$ (Given)
2. $\angle Z \cong \angle W$ (Opp. ∠s of a ▱ are ≅.)
3. $\overline{WV} \cong \overline{ZY}$ (Opp. sides of a ▱ are ≅.)
4. $\angle VXW$ and $\angle YUZ$ are rt. ∠s. (⊥ lines form four rt. ∠s.)
5. $\triangle VXW$ and $\triangle YUZ$ are rt. △s. (Def. of rt. △s)
6. $\triangle YUZ \cong \triangle VXW$ (HA)

41. 7

43.

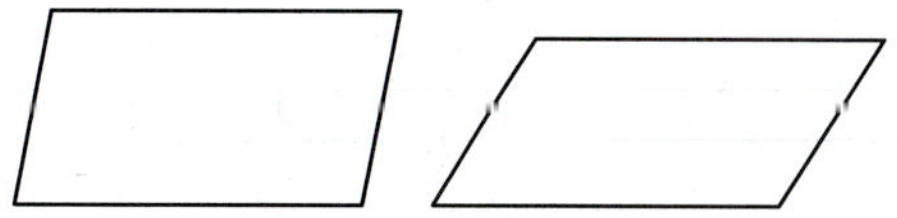

45. Sample answer: In a parallelogram, the opposite sides and angles are congruent. Two consecutive angles in a parallelogram are supplementary. If one angle of a parallelogram is right, then all the angles are right. The diagonals of a parallelogram bisect each other. **47.** 13 **49.** B **51.** 9 **53.** 18 **55.** 100
57. not a polyhedron; cylinder; bases: circles S and T
59. not a polyhedron; cone; base: circle B, vertex: A
61. diagonal; $-\frac{4}{5}$

Pages 409–417 Lesson 6-3

1. Yes; each pair of opposite angles are congruent.
3. $AP = CP$, $BP = DP$; sample answer: If the diagonals of a quadrilateral bisect each other, then the quadrilateral is a parallelogram, so if $AP = CP$ and $BP = DP$, then the string forms a parallelogram.

5 If both pairs of opposite sides are congruent, then the quadrilateral is a parallelogram.

$2x + 3 = x + 7$	Congruent sides have equal measures.
$x + 3 = 7$	Subtract x from each side.
$x = 4$	Subtract 3 from each side.

$3y - 5 = y + 11$	Congruent sides have equal measures.
$2y - 5 = 11$	Subtract y from each side.
$2y = 16$	Add 5 to each side.
$y = 8$	Divide each side by 2.

7

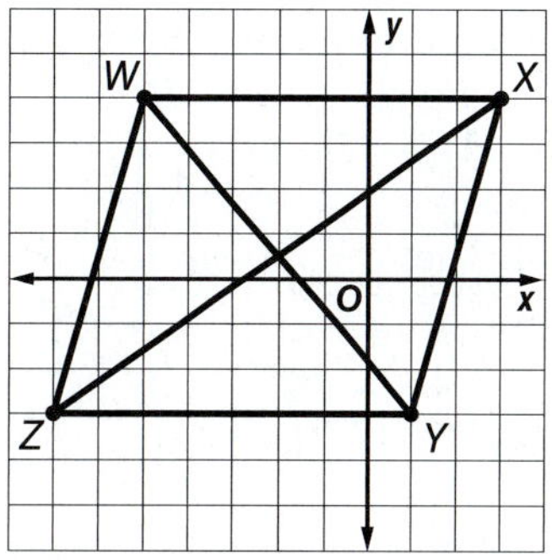

If the diagonals bisect each other, then it is a parallelogram.

midpoint of $\overline{WY}$: $\left(\frac{-5+1}{2}, \frac{4+(-3)}{2}\right) = \left(-2, \frac{1}{2}\right)$

midpoint of $\overline{XZ}$: $\left(\frac{3+(-7)}{2}, \frac{4+(-3)}{2}\right) = \left(-2, \frac{1}{2}\right)$

The midpoint of $\overline{WY}$ and $\overline{XZ}$ is $\left(-2, \frac{1}{2}\right)$. Since the diagonals bisect each other, $WXYZ$ is a parallelogram.

9. Yes; both pairs of opposite sides are congruent.
11. No; none of the tests for parallelograms are fulfilled. **13.** Yes; the diagonals bisect each other.

15.

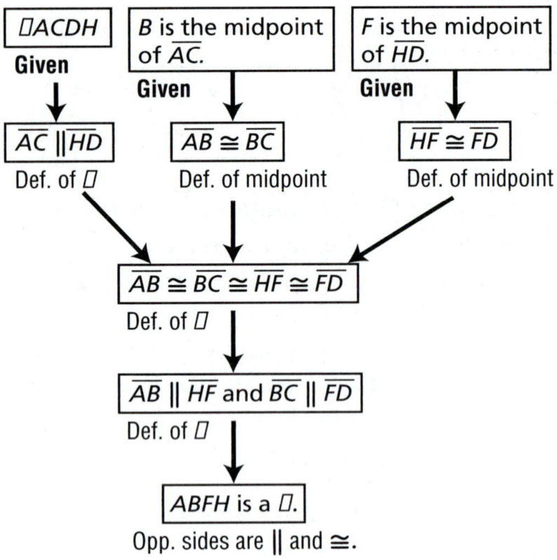

17. Given: $ABEF$ is a parallelogram; $BCDE$ is a parallelogram.
Prove: $ACDF$ is a parallelogram.

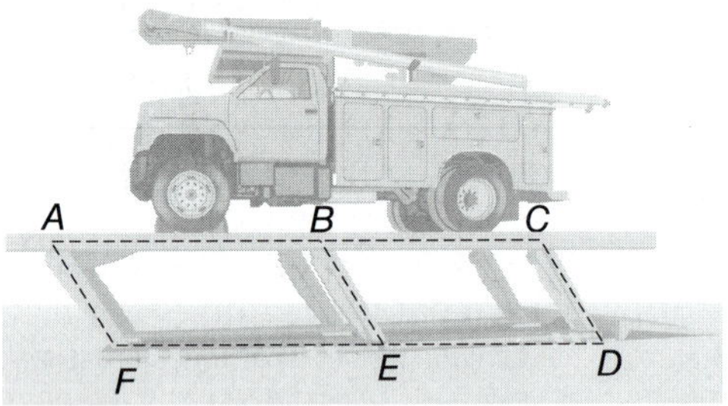

Proof:
Statements (Reasons)
1. $ABEF$ is a parallelogram; $BCDE$ is a parallelogram. (Given)
2. $\overline{AF} \cong \overline{BE}$, $\overline{BE} \cong \overline{CD}$, $\overline{AF} \parallel \overline{BE}$, $\overline{BE} \parallel \overline{CD}$ (Def. of ▱)
3. $\overline{AF} \cong \overline{CD}$, $\overline{AF} \parallel \overline{CD}$ (Trans. Prop.)
4. $ACDF$ is a parallelogram. (If one pair of opp. sides is ≅ and ∥, then the quad. is a ▱.)

19. $x = 8, y = 9$ **21.** $x = 11, y = 7$ **23.** $x = 4, y = 3$

25. No; both pairs of opposite sides must be congruent. The distance between K and L is $\sqrt{53}$. The distance between L and M is $\sqrt{37}$. The distance between M and J is $\sqrt{50}$. The distance between J and K is $\sqrt{26}$. Since, both pairs of opposite sides are not congruent, $JKLM$ is not a parallelogram.

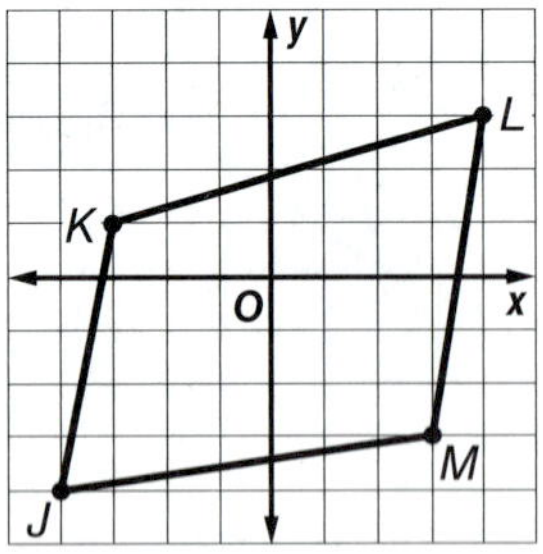

27. Yes; a pair of opposite sides must be parallel and congruent. Slope of $\overline{QR} = \frac{7}{2} =$ slope of $\overline{ST}$, so $\overline{QR} \parallel \overline{ST}$. The distance between Q and $R = \sqrt{53} =$ distance between S and T, so $\overline{QR} \cong \overline{ST}$. So, $QRST$ is a parallelogram.

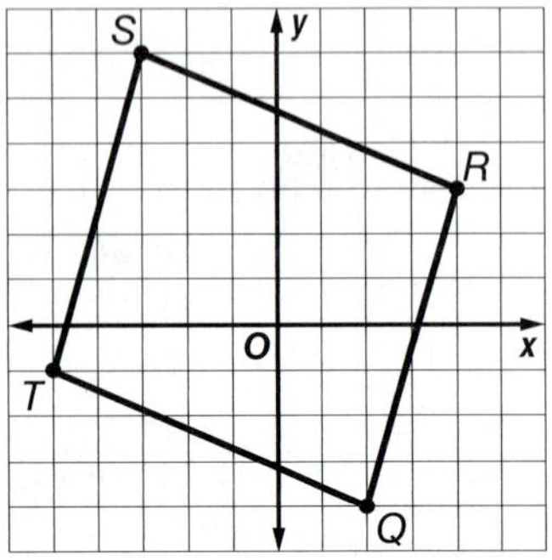

29. Given: $ABCD$ is a parallelogram. $\angle A$ is a right angle.
Prove: $\angle B$, $\angle C$, and $\angle D$ are right angles.

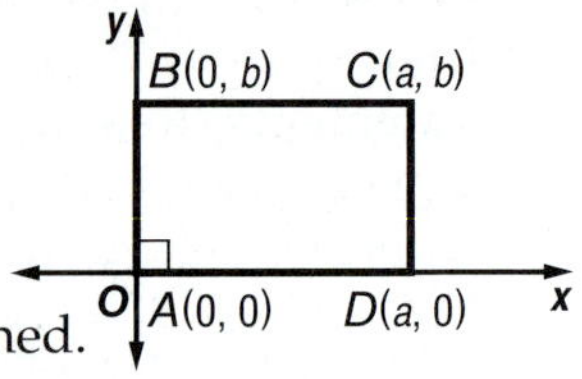

Proof:
slope of $\overline{BC} = \left(\frac{b-b}{a-0}\right)$ or 0
The slope of $\overline{CD}$ is undefined.
slope of $\overline{AD} = \left(\frac{0-0}{a-0}\right)$ or 0
The slope of $\overline{AB}$ is undefined.
Therefore, $\overline{BC} \perp \overline{CD}$, $\overline{CD} \perp \overline{AD}$, and $\overline{AB} \perp \overline{BC}$. So, $\angle B$, $\angle C$, and $\angle D$ are right angles.

31 **a.** Use the Segment Addition Postulate to rewrite AC and CF each as the sum of two measures. Then use substitution, the Subtraction Property, and the Transitive Property to show that $\overline{BC} \cong \overline{DE}$. By proving that $BCDE$ is a parallelogram, you can prove that $\overline{BE} \parallel \overline{CD}$.
Given: $\overline{AC} \cong \overline{CF}$, $\overline{AB} \cong \overline{CD} \cong \overline{BE}$, and $\overline{DF} \cong \overline{DE}$
Prove: $\overline{BE} \parallel \overline{CD}$
Proof: We are given that $\overline{AC} \cong \overline{CF}$, $\overline{AB} \cong \overline{CD} \cong \overline{BE}$, and $\overline{DF} \cong \overline{DE}$. $AC = CF$ by the definition of congruence. $AC = AB + BC$ and $CF = CD + DF$ by the Segment Addition Postulate and $AB + BC = CD + DF$ by substitution. Using substitution again, $AB + BC = AB + DF$, and $BC = DF$ by the Subtraction Property. $\overline{BC} \cong \overline{DF}$ by the definition of congruence, and $\overline{BC} \cong \overline{DE}$ by the Transitive Property. If both pairs of opposite sides of a quadrilateral are congruent, then the quadrilateral is a parallelogram, so $BCDE$ is a parallelogram. By the definition of a parallelogram, $\overline{BE} \parallel \overline{CD}$.

b. Let x = width of the copy. If $AB = 12$, then $CD = BE = 12$. So, $CF = 12 + 8$ or 20.

$\frac{CF}{BE} = \frac{\text{width of copy}}{\text{width of original}}$	
$\frac{20}{12} = \frac{x}{5.5}$	Substitution
$20(5.5) = 12x$	Cross multiply.
$110 = 12x$	Simplify.
$9.2 \approx x$	Divide each side by 12.

The width of the original object is about 9.2 in.

33. Given: $\overline{AB} \cong \overline{DC}$, $\overline{AB} \parallel \overline{DC}$
Prove: $ABCD$ is a parallelogram.

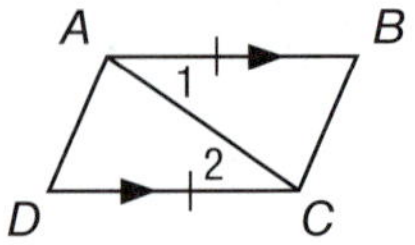

Proof:
Statements (Reasons)
1. $\overline{AB} \cong \overline{DC}$, $\overline{AB} \parallel \overline{DC}$ (Given)
2. Draw $\overline{AC}$. (Two points determine a line.)
3. $\angle 1 \cong \angle 2$ (If two lines are ∥, then alt. int. ∠s are ≅.)
4. $\overline{AC} \cong \overline{AC}$ (Refl. Prop.)
5. $\triangle ABC \cong \triangle CDA$ (SAS)
6. $\overline{AD} \cong \overline{BC}$ (CPCTC)
7. $ABCD$ is a parallelogram. (If both pairs of opp. sides are ≅, then the quad. is ▱.)

35. Opposite sides of a parallelogram are parallel and congruent. Since the slope of $\overline{AB}$ is 0, the slope of $\overline{DC}$ must be 0. The y-coordinate of vertex D must be c. The length of $\overline{AB}$ is $(a + b) - 0$ or $a + b$. So the length of $\overline{DC}$

must be $a + b$. If the x-coordinate of D is $-b$ and the x-coordinate of C is a, then the length of $\overline{DC}$ is $a - (-b)$ or $a + b$. So, the coordinates are $C(a, c)$ and $D(-b, c)$.

37 Sample answer: Since the two vertical rails are both perpendicular to the ground, he knows that they are parallel to each other. If he measures the distance between the two rails at the top of the steps and at the bottom of the steps, and they are equal, then one pair of sides of the quadrilateral formed by the handrails is both parallel and congruent, so the quadrilateral is a parallelogram. Since the quadrilateral is a parallelogram, the two hand rails are parallel by definition.

39a. Sample answer:

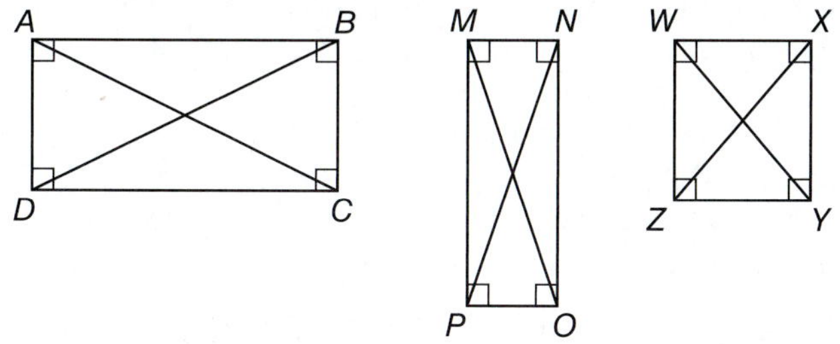

39b. Sample answer:

Rectangle	Side	Length
ABCD	$\overline{AC}$	3.3 cm
	$\overline{BD}$	3.3 cm
MNOP	$\overline{MO}$	2.8 cm
	$\overline{NP}$	2.8 cm
WXYZ	$\overline{WY}$	2.0 cm
	$\overline{XZ}$	2.0 cm

39c. Sample answer: The diagonals of a rectangle are congruent.

41. Sample answer: The theorems are converses of each other. The hypothesis of Theorem 6-3 is "a figure is a parallelogram," and the hypothesis of 6-9 is "both pairs of opposite sides of a quadrilateral are congruent." The conclusion of Theorem 6-3 is "opposite sides are congruent," and the conclusion of 6-9 is "the quadrilateral is a parallelogram."

43.

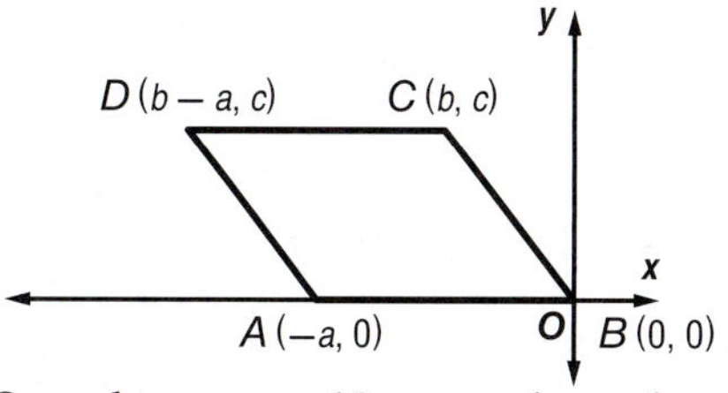

45. Sample answer: You can show that: both pairs of opposite sides are congruent or parallel, both pairs of opposite angles are congruent, diagonals bisect each other, or one pair of opposite sides is both congruent and parallel. **47.** 4 **49.** D **51.** (4.5, 1.5) **53.** 35

55. Given: $P + W > 2$ (P is time spent in the pool; W is time spent lifting weights.)

Prove: $P > 1$ or $W > 1$

Proof:

Step 1: Assume $P \leq 1$ and $W \leq 1$.

Step 2: $P + W \leq 2$

Step 3: This contradicts the given statement. Therefore he did at least one of these activities for more than an hour.

57.

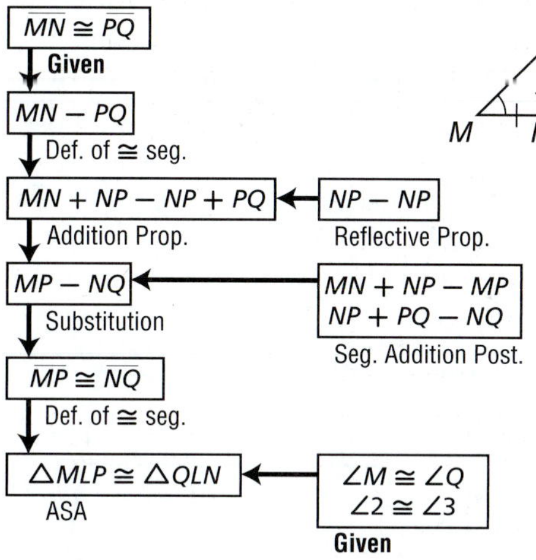

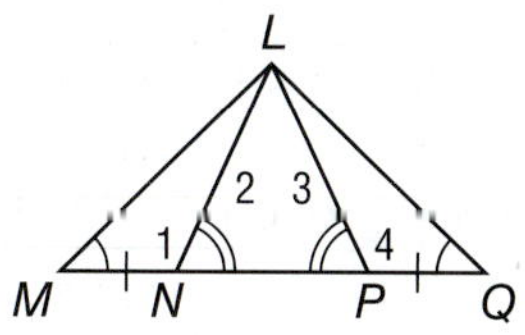

59. not perpendicular

Pages 419–425 Lesson 6-4

1. 7 ft **3.** 33.5 **5.** 11

7. Given: $ABDE$ is a rectangle; $\overline{BC} \cong \overline{DC}$.

Prove: $\overline{AC} \cong \overline{EC}$

Proof:

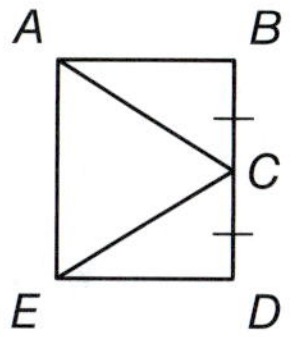

Statements (Reasons)

1. $ABDE$ is a rectangle; $\overline{BC} \cong \overline{DC}$. (Given)
2. $ABDE$ is a parallelogram. (Def. of rectangle)
3. $\overline{AB} \cong \overline{DE}$ (Opp. sides of a ▱ are ≅.)
4. $\angle B$ and $\angle D$ are right angles. (Def. of rectangle)
5. $\angle B \cong \angle D$ (All rt ∠s are ≅.)
6. $\triangle ABC \cong \triangle EDC$ (SAS)
7. $\overline{AC} \cong \overline{EC}$ (CPCTC)

9. Yes; $AB = 5 = CD$ and $BC = 8 = AD$. So, $ABCD$ is a parallelogram. $BD = \sqrt{89} = AC$, so the diagonals are congruent. Thus, $ABCD$ is a rectangle.

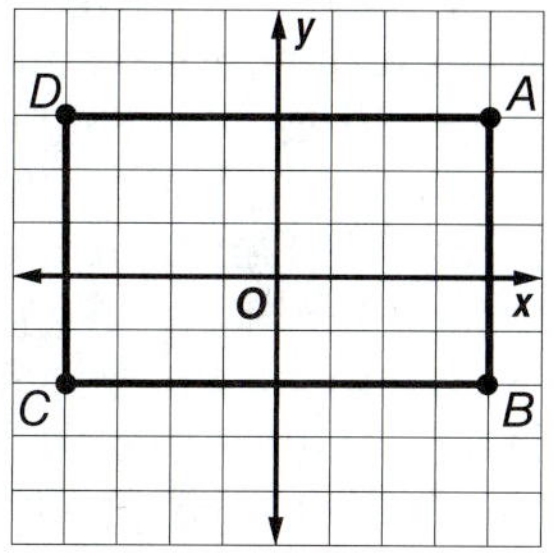

11 Since all angles in a rectangle are right angles, $\angle ABC$ is a right triangle.

$AC^2 + AB^2 = BC^2$	Pythagorean Theorem
$2^2 + 6^2 = BC^2$	Substitution
$40 = BC^2$	Simplify.
$6.3 \approx BC$	Take the positive square root of each side.

$BC \approx 6.3$ ft

13. 25 **15.** 43 **17.** 38 **19.** 46

21. Given: $QTVW$ is a rectangle; $\overline{QR} \cong \overline{ST}$.

Prove: $\triangle SWQ \cong \triangle RVT$

Proof:

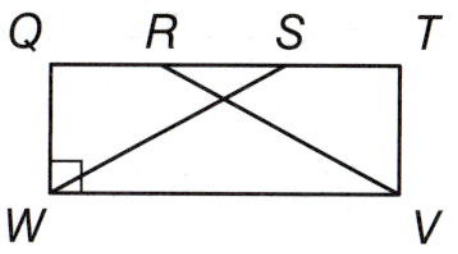

Statements (Reasons)

1. $QTVW$ is a rectangle; $\overline{QR} \cong \overline{ST}$. (Given)
2. $QTVW$ is a parallelogram. (Def. of rectangle)
3. $\overline{WQ} \cong \overline{VT}$ (Opp sides of a ▱ are ≅.)
4. $\angle Q$ and $\angle T$ are right angles. (Def. of rectangle)
5. $\angle Q \cong \angle T$ (All rt ∠s are ≅.)

6. $QR = ST$ (Def. of ≅ segs.)
7. $\overline{RS} \cong \overline{RS}$ (Refl. Prop.)
8. $RS = RS$ (Def. of ≅ segs.)
9. $QR + RS = RS + ST$ (Add. prop.)
10. $QS = QR + RS$, $RT = RS + ST$ (Seg. Add. Post.)
11. $QS = RT$ (Subst.)
12. $\overline{QS} \cong \overline{RT}$ (Def. of ≅ segs.)
13. $\triangle SWQ \cong \triangle RVT$ (SAS)

23. No; $JK = \sqrt{65} = LM$, $KL = \sqrt{37} = MJ$, so $JKLM$ is a parallelogram. $KM = \sqrt{106}$; $JL = \sqrt{98}$. $KM \neq JL$, so the diagonals are not congruent. Thus, $JKLM$ is not a rectangle.

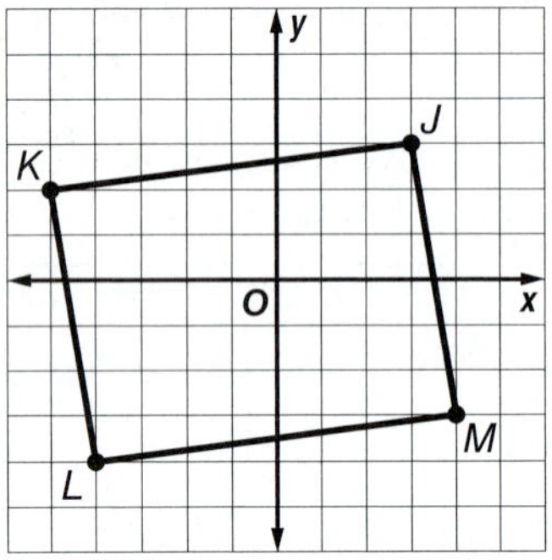

25. No; slope of $\overline{GH} = \frac{1}{8}$ = slope of $\overline{JK}$ and slope of $\overline{HJ} = -6$ = slope of $\overline{KG} = -6$. So, $GHJK$ is a parallelogram. The product of the slopes of consecutive sides $\neq -1$, so the consecutive sides are not perpendicular. Thus, $GHJK$ is not a rectangle. **27.** 40

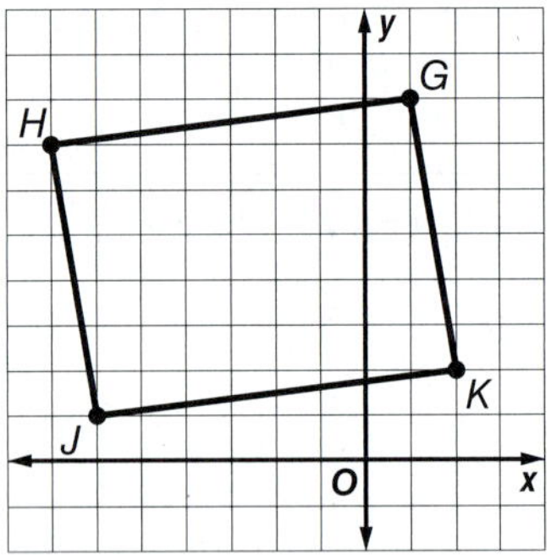

29 A rectangle has four right angles, so $m\angle CDB = 90$. Since a rectangle is a parallelogram, opposite sides are parallel. Alternate interior angles of parallel lines are congruent. So, $\angle 3 \cong \angle 2$ and $m\angle 3 = m\angle 2 = 40$. $m\angle 4 = 90 - 40$ or 50. Since diagonals of a rectangle are congruent and bisect each other, the triangle with angles 4, 5, and 6 is isosceles with $m\angle 6 = m\angle 4$.

$m\angle 4 + m\angle 5 + m\angle 6 = 180$	Triangle Angle-Sum Theorem
$m\angle 4 + m\angle 5 + m\angle 4 = 180$	$m\angle 6 = m\angle 4$
$50 + m\angle 5 + 50 = 180$	Substitution
$100 + m\angle 5 = 180$	Simplify.
$m\angle 5 = 80$	Subtract 100 from each side.

31. 100

33. Given: $WXYZ$ is a rectangle with diagonals $\overline{WY}$ and $\overline{XZ}$.
Prove: $\overline{WY} \cong \overline{XZ}$

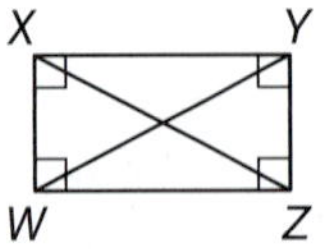

Proof:

Statements (Reasons)
1. $WXYZ$ is a rectangle with diagonals $\overline{WY}$ and $\overline{XZ}$. (Given)
2. $\overline{WX} \cong \overline{ZY}$ (Opp. sides of a ▱ are ≅.)
3. $\overline{WZ} \cong \overline{WZ}$ (Refl. Prop.)
4. $\angle XWZ$ and $\angle YZW$ are right angles. (Def. of ▭)
5. $\angle XWZ \cong \angle YZW$ (All right ∠s are ≅.)
6. $\triangle XWZ \cong \triangle YZW$ (SAS)
7. $\overline{WY} \cong \overline{XZ}$ (CPCTC)

35. $ABCD$ is a parallelogram, and $\angle B$ is a right angle. Since $ABCD$ is a parallelogram and has one right angle, then it has four right angles. So by the definition of a rectangle, $ABCD$ is a rectangle.

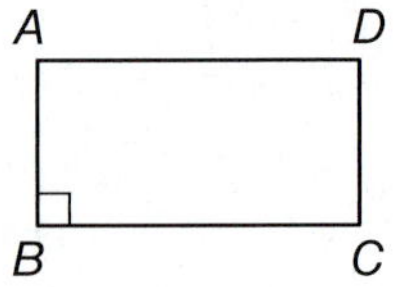

37. Sample answer: Using a protractor, $m\angle P = m\angle Q = 90$. The same compass setting was used to locate points R and S so they are the same distance from points P and Q, respectively. RQ and SP are equal so the diagonals are congruent. Thus, $PQSR$ is a rectangle. **39.** 5 **41.** No; sample answer: Both pairs of opposite sides are congruent, so the sign is a parallelogram, but no measure is given that can be used to prove that it is a rectangle.

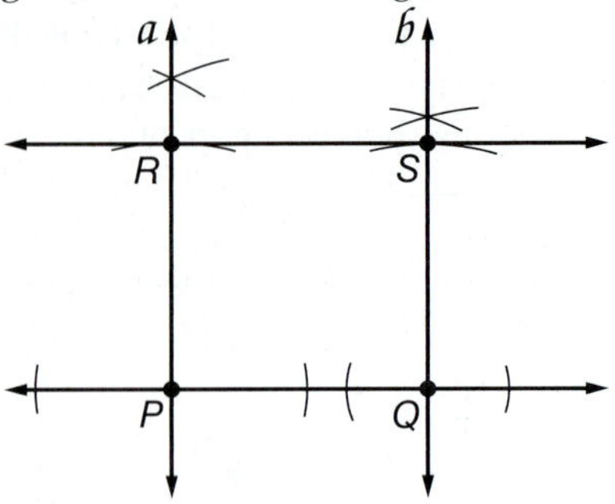

43 Draw ▱ $ABCD$ on the coordinate plane. Find the slopes of $\overline{AD}$ and $\overline{AB}$ and show that the lines are perpendicular.

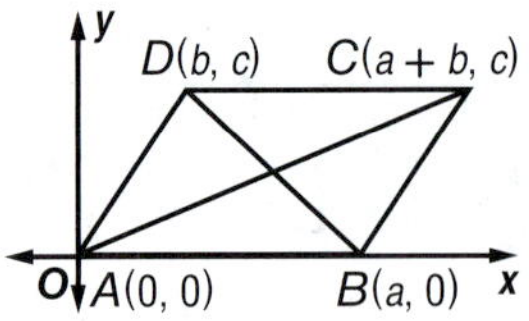

Given: ▱$ABCD$ and $\overline{AC} \cong \overline{BD}$
Prove: ▱$ABCD$ is a rectangle.
Proof:

$AC = \sqrt{(a + b - 0)^2 + (c - 0)^2}$

$BD = \sqrt{(b - a)^2 + (c - 0)^2}$

But $AC = BD$ and

$$\sqrt{(a + b - 0)^2 + (c - 0)^2} = \sqrt{(b - a)^2 + (c - 0)^2}.$$
$$(a + b - 0)^2 + (c - 0)^2 = (b - a)^2 + (c - 0)^2$$
$$(a + b)^2 + c^2 = (b - a)^2 + c^2$$
$$a^2 + 2ab + b^2 + c^2 = b^2 - 2ab + a^2 + c^2$$
$$2ab = -2ab$$
$$4ab = 0$$
$$a = 0 \text{ or } b = 0$$

Because A and B are different points, $a \neq 0$. Then $b = 0$. The slope of $\overline{AD}$ is undefined and the slope of $\overline{AB} = 0$. Thus, $\overline{AD} \perp \overline{AB}$. $\angle DAB$ is a right angle and $ABCD$ is a rectangle.

45. $x = 6$, $y = -10$ **47.** 6 **49.** Sample answer: All rectangles are parallelograms because, by definition, both pairs of opposite sides are parallel. Parallelograms with right angles are rectangles, so some parallelograms are rectangles, but others with non-right angles are not. **51.** J **53.** B **55.** $x = 8$, $y = 22$ **57.** 31.2 cm **59.** $\overline{AH}$ and $\overline{AJ}$ **61.** $\angle AJK$ and $\angle AKJ$ **63.** $\sqrt{101}$

Pages 426–434 Lesson 6-5

1. 32

3. Given: $ABCD$ is a rhombus with diagonal $\overline{DB}$.
Prove: $\overline{AP} \cong \overline{CP}$
Proof:

__Statements (Reasons)__
1. $ABCD$ is a rhombus with diagonal $\overline{DB}$. (Given)
2. $\angle ABP \cong \angle CBP$ (Diag. of rhombus bisects $\angle$)
3. $\overline{PB} \cong \overline{PB}$ (Refl. Prop.)
4. $\overline{AB} \cong \overline{CB}$ (Def. of rhombus)
5. $\triangle APB \cong \triangle CPB$ (SAS)
6. $\overline{AP} \cong \overline{CP}$ (CPCTC)

5. Rectangle, rhombus, square; consecutive sides are perpendicular, all sides are congruent. **7.** 14 **9.** 28

11

$m\angle ABC + m\angle BCD = 180$	Consecutive $\angle$s are supp.
$(2x - 7) + (2x + 3) = 180$	Substitution
$4x - 4 = 180$	Simplify.
$4x = 184$	Add 4 to each side.
$x = 46$	Divide each side by 4.

$m\angle BCD = 2x + 3$	Given
$= 2(46) + 3$ or 95	Substitution

$\angle DAB \cong \angle BCD$	Opposite angles are congruent.
$m\angle DAB = m\angle BCD$	Definition of congruence
$= 95$	Substitution

13. Given: $\overline{WZ} \parallel \overline{XY}$, $\overline{WX} \parallel \overline{ZY}$, $\overline{WZ} \cong \overline{ZY}$
Prove: $WXYZ$ is a rhombus.
Proof:

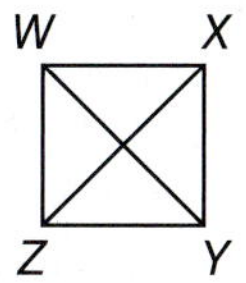

__Statements (Reasons)__
1. $\overline{WZ} \parallel \overline{XY}$, $\overline{WX} \parallel \overline{ZY}$, $\overline{WZ} \cong \overline{ZY}$ (Given)
2. $WXYZ$ is a ▱. (Both pairs of opp. sides are $\parallel$.)
3. $WXYZ$ is a rhombus. (If one pair of consecutive sides of a ▱ are $\cong$, the ▱ is a rhombus.)

15. Given: $JKQP$ is a square. $\overline{ML}$ bisects $\overline{JP}$ and $\overline{KQ}$.
Prove: $JKLM$ is a parallelogram.
Proof:

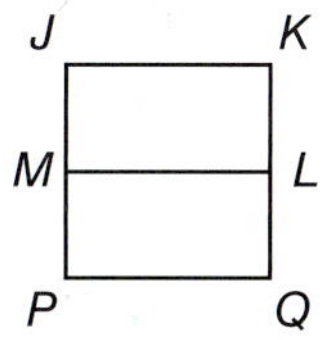

__Statements (Reasons)__
1. $JKQP$ is a square. $\overline{ML}$ bisects $\overline{JP}$ and $\overline{KQ}$. (Given)
2. $JKQP$ is a parallelogram. (All squares are parallelograms.)
3. $\overline{JM} \parallel \overline{KL}$ (Def. of ▱)
4. $\overline{JP} \cong \overline{KQ}$ (Def. of ▱)
5. $JP = KQ$ (Def of $\cong$ segs.)
6. $JM = MP$, $KL = LQ$ (Def. of bisects)
7. $JP = JM + MP$, $KQ = KL + LQ$ (Seg. Add Post.)
8. $JP = 2JM$, $KQ = 2KL$ (Subst.)
9. $2JM = 2KL$ (Subst.)
10. $JM = KL$ (Division Prop.)
11. $\overline{KL} \cong \overline{JM}$ (Def. of $\cong$ segs.)
12. $JKLM$ is a parallelogram. (If one pair of opp. sides is $\cong$ and $\parallel$, then the quad. is a ▱.)

17. Rhombus; Sample answer: The measure of angle formed between the two streets is 29, and vertical angles are congruent, so the measure of one angle of the quadrilateral is 29. Since the crosswalks are the same length, the sides of the quadrilateral are congruent. Therefore, they form a rhombus. **19.** Rhombus; the diagonals are perpendicular. **21.** None; the diagonals are not congruent or perpendicular.

23 The diagonals of a rhombus are perpendicular, so $\triangle ABP$ is a right triangle.

$AP^2 + PB^2 = AB^2$	Pythagorean Theorem
$AP^2 + 12^2 = 15^2$	Substitution
$AP^2 + 144 = 225$	Simplify.
$AP^2 = 81$	Subtract 144 from each side.
$AP = 9$	Take the square root of each side.

25. 24 **27.** 6 **29.** 90 **31.** square **33.** rectangle

35. Given: $ABCD$ is a parallelogram; $\overline{AC} \perp \overline{BD}$.
Prove: $ABCD$ is a rhombus.

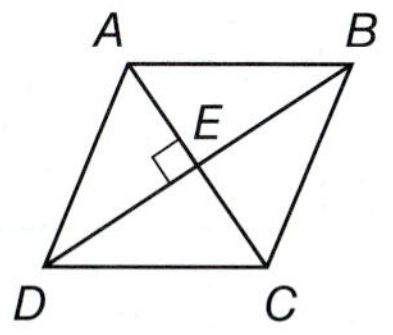

Proof: We are given that $ABCD$ is a parallelogram. The diagonals of a parallelogram bisect each other, so $\overline{AE} \cong \overline{EC}$. $\overline{BE} \cong \overline{BE}$ because congruence of segments is reflexive. We are also given that $\overline{AC} \perp \overline{BD}$. Thus, $\angle AEB$ and $\angle BEC$ are right angles by the definition of perpendicular lines. Then $\angle AEB \cong \angle BEC$ because all right angles are congruent. Therefore, $\triangle AEB \cong \triangle CEB$ by SAS. $\overline{AB} \cong \overline{CB}$ by CPCTC. Opposite sides of parallelograms are congruent, so $\overline{AB} \cong \overline{CD}$ and $\overline{BC} \cong \overline{AD}$. Then since congruence of segments is transitive, $\overline{AB} \cong \overline{CD} \cong \overline{BC} \cong \overline{AD}$. All four sides of $ABCD$ are congruent, so $ABCD$ is a rhombus by definition.

37. Given: $ABCD$ is a parallelogram; $\overline{AB} \cong \overline{BC}$.
Prove: $ABCD$ is a rhombus.

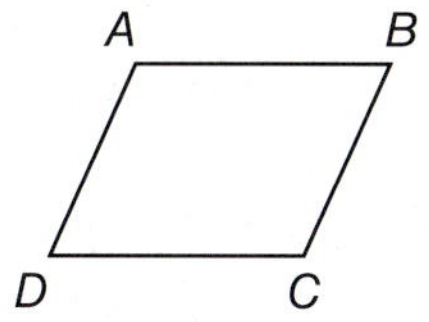

Proof: Opposite sides of a parallelogram are congruent, so $\overline{BC} \cong \overline{AD}$ and $\overline{AB} \cong \overline{CD}$. We are given that $\overline{AB} \cong \overline{BC}$. So, by the Transitive Property, $\overline{BC} \cong \overline{CD}$. So, $\overline{BC} \cong \overline{CD} \cong \overline{AB} \cong \overline{AD}$. Thus, $ABCD$ is a rhombus by definition.

39. Sample answer: If the diagonals of a parallelogram are perpendicular to each other, then the parallelogram is a rhombus.

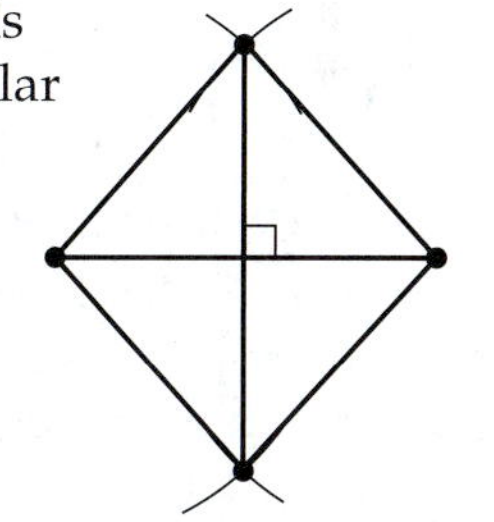

41. Given: $ABCD$ is a square.
Prove: $\overline{AC} \perp \overline{DB}$

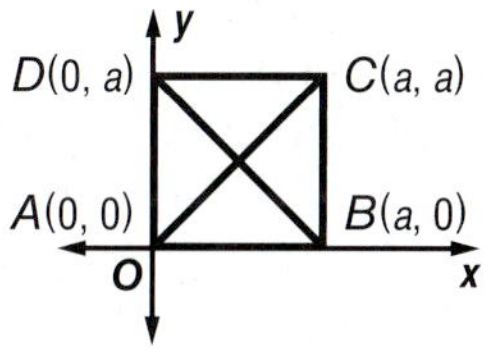

Proof:
slope of $\overline{DB} = \dfrac{0 - a}{a - 0}$ or -1
slope of $\overline{AC} = \dfrac{0 - a}{0 - a}$ or 1
The slope of $\overline{AC}$ is the negative reciprocal of the slope of $\overline{DB}$, so they are perpendicular.

43 Use the properties of exterior angles to classify the quadrilaterals. Sample answer: Since the octagons are regular, each side is congruent, and the quadrilaterals share common sides with the octagons, so the quadrilaterals are either rhombuses or squares. The vertices of the quadrilaterals are formed by the exterior angles of the sides of the octagons adjacent to the vertices. The sum of the measures of the exterior angles of a polygon is always 360 and since a regular octagon has 8 congruent exterior angles, each one measures 360 ÷ 8 or 45. As shown in the diagram, each angle of the quadrilaterals in the pattern measures 45 + 45 or 90. Therefore, the quadrilateral is a square.

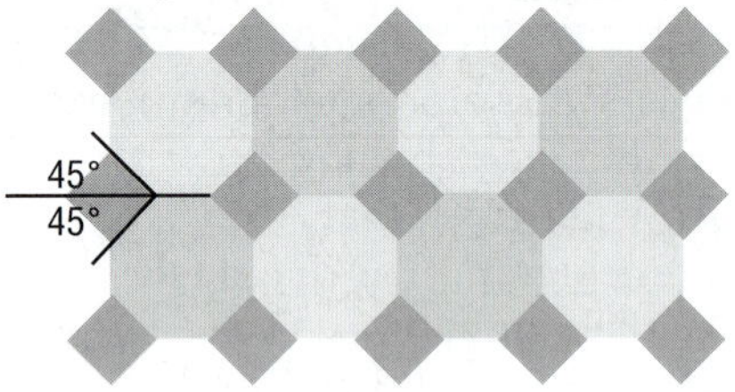

45a.

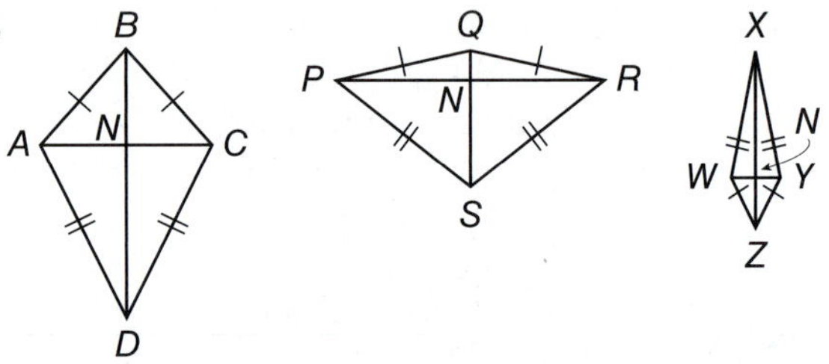

45b.

Figure	Distance from *N* to each vertex along shorter diagonal		Distance from *N* to each vertex along longer diagonal	
ABCD	0.8 cm	0.8 cm	0.9 cm	1.5 cm
PQRS	1.2 cm	1.2 cm	0.3 cm	0.9 cm
WXYZ	0.2 cm	0.2 cm	1.1 cm	0.4 cm

45c. Sample answer: The shorter diagonal of a kite is bisected by the longer diagonal. **47.** True; sample answer: A rectangle is a quadrilateral with four right angles and a square is both a rectangle and a rhombus, so a square is always a rectangle. Converse: If a quadrilateral is a rectangle then it is a square. False; sample answer: A rectangle is a quadrilateral with four right angles. It is not necessarily a rhombus, so it is not necessarily a square. Inverse: If a quadrilateral is not a square, then it is not a rectangle. False; sample answer: A quadrilateral that has four right angles and two pairs of congruent sides is not a square, but it is a rectangle. Contrapositive: If a quadrilateral is not a rectangle, then it is not a square. True; sample answer: If a quadrilateral is not a rectangle, it is also not a square by definition. **49.** Sample answer: (0, 0), (6, 0), (0, 6), (6, 6) ; the diagonals are perpendicular, and any four points on the lines equidistant from the intersection of the lines will be the vertices of a square. **51.** B **53.** H **55.** 52 **57.** 38 **59.** Yes; both pairs of opposite sides are congruent. **61.** No; the Triangle Inequality Theorem states that the sum of the lengths of any two sides of a triangle must be greater than the length of the third side. Since 22 + 23 = 45, the sides of Monifa's backyard cannot be 22 ft, 23 ft and 45 ft. **63.** 2 **65.** $\frac{5}{4}$

Pages 435–444 Lesson 6-6

1. 101 **3.** $\overline{BC} \parallel \overline{AD}$, $\overline{AB} \nparallel \overline{CD}$; *ABCD* is a trapezoid. **5.** 1.2 **7.** 70 **9.** 70

11

$\overline{XZ} \cong \overline{YW}$	Diagonals are congruent.
$XZ = YW$	Definition of congruence.
$18 = YW$	Substitution
$18 = YP + PW$	Segment Addition Postulate
$18 = 3 + PW$	Substitution
$15 = PW$	Subtract 3 from each side.

13. $\overline{JK} \parallel \overline{LM}$, $\overline{KL} \nparallel \overline{JM}$; *JKLM* is a trapezoid, but not isosceles since $KL = \sqrt{26}$ and $JM = 5$. **15.** $\overline{XY} \parallel \overline{WZ}$, $\overline{WX} \nparallel \overline{YZ}$; *WXYZ* is a trapezoid, but not isosceles since $XZ = \sqrt{74}$ and $WY = \sqrt{68}$. **17.** 10 **19.** 8 **21.** 17

23 The G key represents the midsegment of the trapezoid. Use the Trapezoid Midsegment Theorem to find the length of the key.

length of G key $= \frac{1}{2}$(length of C key + length of D key)

$= \frac{1}{2}(6 + 1.8)$ Substitution

$= \frac{1}{2}(7.8)$ or 3.9 Simplify.

The length of the G key is 3.9 in.

25. $\sqrt{20}$ **27.** 75

29. Given: *ABCD* is a trapezoid; $\angle D \cong \angle C$.

Prove: Trapezoid *ABCD* is isosceles.

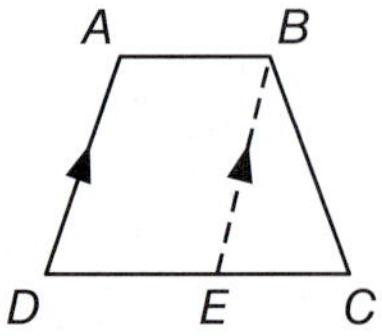

Proof: By the Parallel Postulate, we can draw the auxillary line $\overline{EB} \parallel \overline{AD}$. $\angle D \cong \angle BEC$, by the Corr. ∠s Thm. We are given that $\angle D \cong \angle C$, so by the Trans. Prop, $\angle BEC \cong \angle C$. So, $\triangle EBC$ is isosceles and $\overline{EB} \cong \overline{BC}$. From the definition of a trapezoid, $\overline{AB} \parallel \overline{DE}$. Since both pairs of opposite sides are parallel, *ABED* is a parallelogram. So, $\overline{AD} \cong \overline{EB}$. By the Transitive Property, $\overline{BC} \cong \overline{AD}$. Thus, *ABCD* is an isosceles trapezoid.

31. Given: *ABCD* is a kite with $\overline{AB} \cong \overline{BC}$ and $\overline{AD} \cong \overline{DC}$.

Prove: $\overline{BD} \perp \overline{AC}$

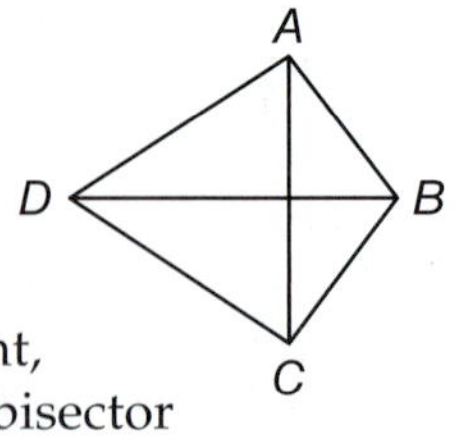

Proof: We know that $\overline{AB} \cong \overline{BC}$ and $\overline{AD} \cong \overline{DC}$. So, *B* and *D* are both equidistant from *A* and *C*. If a point is equidistant from the endpoints of a segment, then it is on the perpendicular bisector of the segment. The line that contains *B* and *D* is the perpendicular bisector of $\overline{AC}$, since only one line exists through two points. Thus, $\overline{BD} \perp \overline{AC}$.

33. Given: *ABCD* is a trapezoid with median $\overline{EF}$.

Prove: $\overline{EF} \cong \overline{AB}$ and $\overline{EF} \cong \overline{DC}$ and $EF = \frac{1}{2}(AB + DC)$

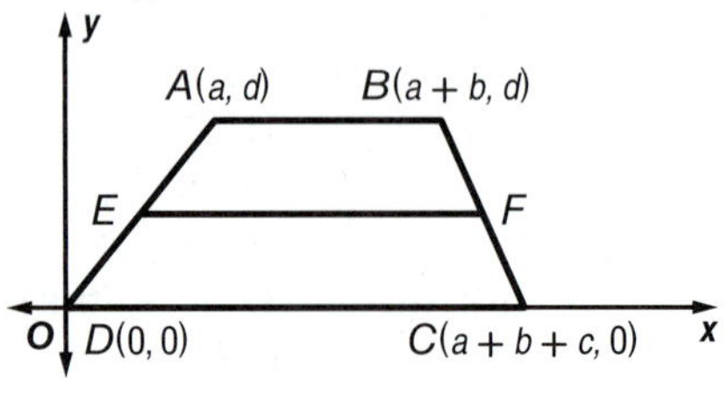

Proof:
By the definition of the median of a trapezoid, E is the midpoint of $\overline{AD}$ and F is the midpoint of $\overline{BC}$.

Midpoint E is $\left(\frac{a+0}{2}, \frac{d+0}{2}\right)$ or $\left(\frac{a}{2}, \frac{d}{2}\right)$.

Midpoint F is $\left(\frac{a+b+a+b+c}{2}, \frac{d+0}{2}\right)$ or $\left(\frac{2a+2b+c}{2}, \frac{d}{2}\right)$.

The slope of $\overline{AB} = 0$, the slope of $\overline{EF} = 0$, and the slope of $\overline{DC} = 0$. Thus, $\overline{EF} \parallel \overline{AB}$ and $\overline{EF} \parallel \overline{DC}$.

$$AB = \sqrt{[(a+b)-a]^2 + (d-d)^2} = \sqrt{b^2} \text{ or } b$$

$$DC = \sqrt{[(a+b+c)-0]^2 + (0-0)^2}$$
$$= \sqrt{(a+b+c)^2} \text{ or } a+b+c$$

$$EF = \sqrt{\left(\frac{2a+2b+c-a}{2}\right)^2 + \left(\frac{d}{2}-\frac{d}{2}\right)^2}$$
$$= \sqrt{\left(\frac{a+2b+c}{2}\right)^2} \text{ or } \frac{a+2b+c}{2}$$

$$\frac{1}{2}(AB+DC) = \frac{1}{2}[b+(a+b+c)]$$
$$= \frac{1}{2}(a+2b+c)$$
$$= \frac{a+2b+c}{2}$$
$$= EF$$

Thus, $\frac{1}{2}(AB+DC) = EF$.

35. 15 **37.** 28 ft **39.** 70 **41.** 2 **43.** 20 **45.** 10 in. **47.** 105

49 $\angle ZWX \cong \angle ZYX$ because one pair of opposite sides of a kite is congruent and $\angle ZWX \cong \angle ZYX$. So, $m\angle ZYX = m\angle ZWX = 10x$.

$m\angle ZWX + m\angle WXY + m\angle ZYX + m\angle WZY = 360$	Polygon Interior Angles Sum Theorem
$10x + 120 + 10x + 4x = 360$	Substitution
$24x + 120 = 360$	Simplify.
$24x = 240$	Subtract 120 from each side.
$x = 10$	Divide each side by 24.

So, $m\angle ZYX = 10x$ or 100.

51. Given: $ABCD$ is an isosceles trapezoid.
Prove: $\angle DAC \cong \angle CBD$

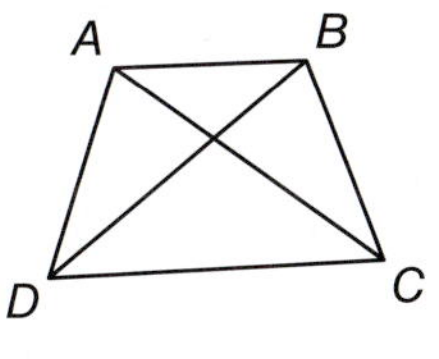

Proof:
<u>Statements (Reasons)</u>
1. $ABCD$ is an isosceles trapezoid. (Given)
2. $\overline{AD} \cong \overline{BC}$ (Def. of isos. trap.)
3. $\overline{DC} \cong \overline{DC}$ (Refl. Prop.)
4. $\overline{AC} \cong \overline{BD}$ (Diags. of isos. trap. are ≅.)
5. $\triangle ADC \cong \triangle BCD$ (SSS)
6. $\angle DAC \cong \angle CBD$ (CPCTC)

53. Sometimes; opp ∠s are supplementary in an isosceles trapezoid. **55.** Always; by def. a square is a quadrilateral with 4 rt. ∠s and 4 ≅ sides. Since by def., a rhombus is a quadrilateral with 4 ≅ sides, a square is always a rhombus. **57.** Sometimes; only if the parallelogram has 4 rt. ∠s and/or congruent diagonals, is it a rectangle.

59.

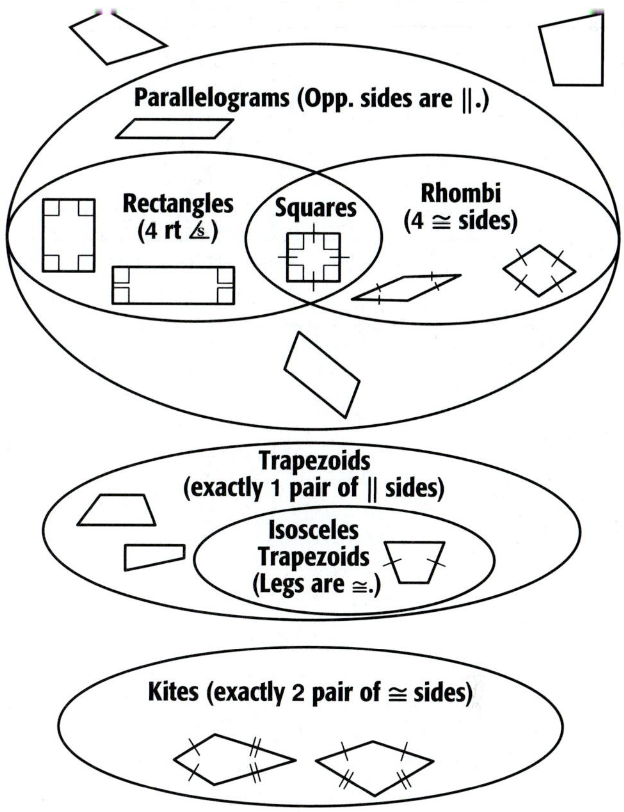

61 slope of $\overline{WX} = \frac{4-4}{3-(-3)} = 0$

slope of $\overline{XY} = \frac{3-4}{5-3} = -\frac{1}{2}$

slope of $\overline{YZ} = \frac{1-3}{-5-5} = \frac{1}{5}$

slope of $\overline{WZ} = \frac{1-4}{-5-(-3)} = \frac{3}{2}$

Since the figure has no parallel sides, it is just a quadrilateral.

63. Given: isosceles trapezoid $ABCD$ with $\overline{AD} \cong \overline{BC}$
Prove: $\overline{BD} \cong \overline{AC}$

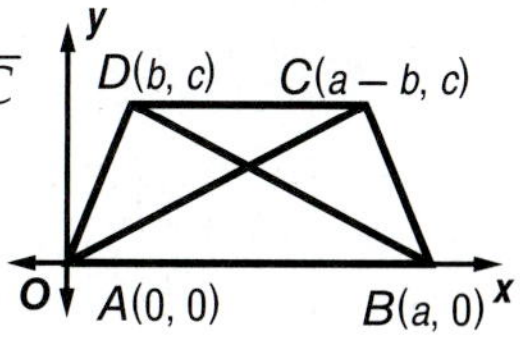

Proof:
$DB = \sqrt{(a-b)^2 + (0-c)^2}$ or $\sqrt{(a-b)^2 + c^2}$

$AC = \sqrt{[(a-b)-0]^2 + (c-0)^2}$ or $\sqrt{(a-b)^2 + c^2}$

$BD = AC$ and $\overline{BD} \cong \overline{AC}$

65. Belinda; $m\angle D = m\angle B$. So $m\angle A + m\angle B + m\angle C + m\angle D = 360$ or $m\angle A + 100 + 45 + 100 = 360$. So, $m\angle A = 115$. **67.** Never; a square has all 4 sides ≅, while a kite does not have any opposite sides congruent. **69.** A quadrilateral must have exactly one pair of sides parallel to be a trapezoid. If the legs are congruent, then the trapezoid is an isosceles trapezoid. If a quadrilateral has exactly two pairs of consecutive congruent sides with the opposite sides not congruent, the quadrilateral is a kite. A trapezoid and a kite both have four sides. In a trapezoid and isosceles trapezoid, both have exactly one pair of parallel sides. **71.** 76 **73.** B **75.** 18 **77.** 9

79. No; slope of $\overline{JK} = \frac{1}{8} =$ slope of $\overline{LM}$ and slope of $\overline{KL} = -6 =$ slope of $\overline{MJ}$. So, $JKLM$ is a parallelogram. The product of the slopes of consecutive sides $\neq -1$, so the consecutive sides are not perpendicular. Thus, $JKLM$ is not a rectangle.

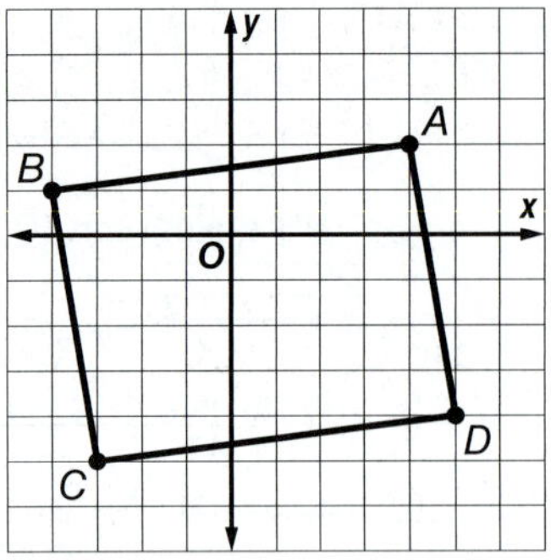

81. Given: $\angle CMF \cong \angle EMF$, $\angle CFM \cong \angle EFM$

Prove: $\triangle DMC \cong \triangle DME$

Proof:

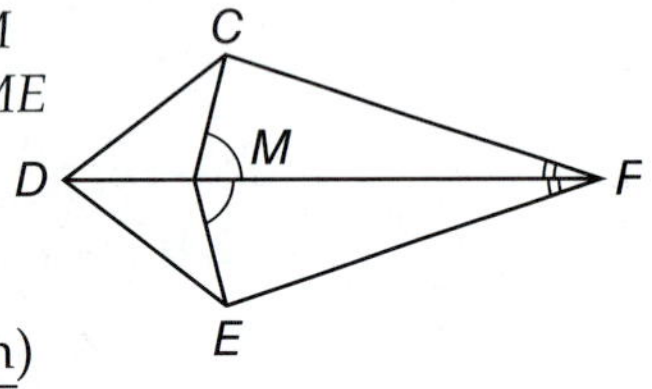

<u>Statements (Reasons)</u>

1. $\angle CMF \cong \angle EMF$, $\angle CFM \cong \angle EFM$ (Given)
2. $\overline{MF} \cong \overline{MF}$, $\overline{DM} \cong \overline{MF}$ (Reflexive Property)
3. $\triangle CMF \cong \triangle EMF$ (ASA)
4. $\overline{CM} \cong \overline{EM}$ (CPCTC)
5. $\angle DMC$ and $\angle CMF$ are supplementary and $\angle DME$ and $\angle EMF$ are supplementary. (Supplement Th.)
6. $\angle DMC \cong \angle DME$ (⦣ suppl. to $\cong$ ⦣ are $\cong$.)
7. $\triangle DMC \cong \triangle DME$ (SAS)

83. 1

Pages 445–448 Chapter 6 Study Guide and Review

1. false, both pairs of base angles **3.** false, diagonal **5.** true **7.** false, is always **9.** true **11.** 1440 **13.** 720 **15.** 26 **17.** 18 **19.** 115 **21.** $x = 37, y = 6$ **23.** yes, Theorem 6.11

25. Given: ▱$ABCD$, $\overline{AE} \cong \overline{CF}$

Prove: Quadrilateral $EBFD$ is a parallelogram

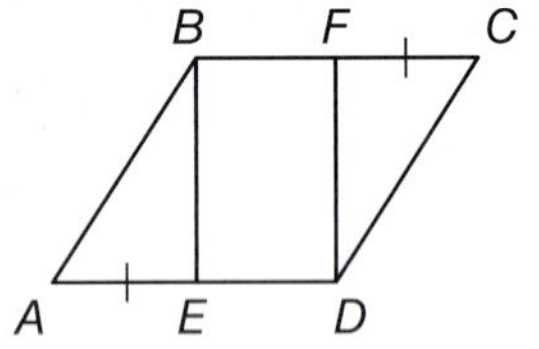

<u>Statements (Reasons)</u>

1. $ABCD$, $\overline{AE} \cong \overline{CF}$ (Given)
2. $AE = CF$ (Def. of $\cong$ segs)
3. $\overline{BC} \cong \overline{AD}$ (Opp. sides of a ▱ are $\cong$)
4. $BC = AD$ (Def. of $\cong$ segs)
5. $BC = BF + CF$, $AD = AE + ED$ (Seg. Add. Post.)
6. $BF + CF = AE + ED$ (Subst.)
7. $BF + AE = AE + ED$ (Subst.)
8. $BF = ED$ (Subt. Prop.)
9. $\overline{BF} \cong \overline{ED}$ (Def. of $\cong$ segs)
10. $\overline{BF} \parallel \overline{ED}$ (Def. of ▱)
11. Quadrilateral $EBFD$ is a parallelogram. (If one pair of opposite sides is parallel and congruent then it is a parallelogram.)

27. $x = 5, y = 12$ **29.** 33 **31.** 64 **33.** 6 **35.** 55 **37.** 35 **39.** Rectangle, rhombus, square; all sides are $\cong$, consecutive sides are $\perp$. **41.** 19.2 **43a.** Sample answer: The legs of the trapezoids are part of the diagonals of the square. The diagonals of a square bisect opposite angles, so each base angle of a trapezoid measures 45°. One pair of sides is parallel and the base angles are congruent. **43b.** $16 + 8\sqrt{2} \approx 27.3$ in.

Chapter 7 Proportions and Similarity

Page 455 Chapter 7 Get Ready

1. 4 or –4 **3.** –37 **5.** 64 **7.** 64.5

Pages 457–463 Lesson 7-1

1. 23:50 **3.** 30, 75, 60 **5.** 16 **7.** 8 **9.** 18 **11.** Movie A; 1:2

13 The ratio of the sides can be written as $9x:7x:5x$.

$9x + 7x + 5x = 191.1$	Perimeter of a triangle
$21x = 191.1$	Combine like terms.
$x = 9.1$	Divide each side by 21.

The measures of the sides are 9(9.1) or 81.9 inches, 7(9.1) or 63.7 inches, and 5(9.1) or 45.5 inches.

15. 2.2 ft **17.** 54, 108, 18 **19.** 75, 60, 45 **21.** $\frac{15}{8}$ **23.** 3 **25.** 8 **27.** 3

29

$\frac{7}{500} = \frac{x}{350}$	← 13- to 17-year-old vegetarians ← 13- to 17-year olds
$7 \cdot 350 = 500 \cdot x$	Cross Products Property
$2450 = 500x$	Simplify.
$4.9 = x$	Divide each side by 500.

About 5 13- to 17-year-olds would be vegetarian.

31. 3, –3.5 **33.** 12.9, –0.2 **35.** 2541 in^2 **37.** 48, 96, 144, 72 **39a.** Yes; the ratio of sides is 1.617; 22.83 ft. **39b.** No: the HDTV aspect ratio is 1.77778 and the fullscreen aspect ratio is 1.33333. Neither television set is a golden rectangle since the ratios of the lengths to the widths are not the golden ratio.

41. Given: $\frac{a}{b} = \frac{c}{d}$, $b \neq 0, d \neq 0$

Prove: $ad = bc$

Proof:

<u>Statements (Reasons)</u>

1. $\frac{a}{b} = \frac{c}{d}$ (Given)
2. $(bd)\frac{a}{b} = (bd)\frac{c}{d}$ (Mult. Prop.)
3. $da = bc$ (Subs. Prop.)
4. $ad = bc$ (Comm. Prop.)

43 **a.**

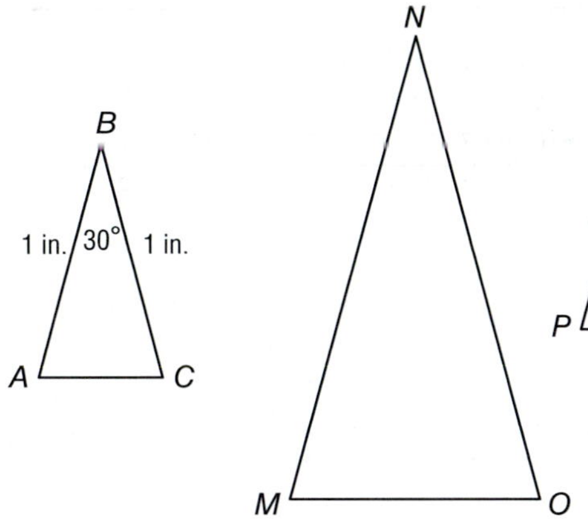

b.

Triangle	*ABC*	*MNO*	*PQR*
Leg length	1 in.	2 in.	0.5 in.
Perimeter	2.5 in.	5 in.	1.25 in.

c. Sample answer: When the vertex angle of an isosceles triangle is held constant and the leg length is increased or decreased by a factor, the perimeter of the triangle increases or decreases by the same factor.

45. 5:2 **47.** $\frac{2}{3} = \frac{5}{7.5}$; You need to multiply $\frac{2}{3}$ by a factor of 2.5 to get $\frac{5}{7.5}$. The factor of the other three proportions is 2.8. **49.** Both are written with fractions. A ratio compares two quantities with division, while a proportion equates two ratios. First, ratios are used to write proportions. Then the cross products are used to solve the proportion. **51.** F **53.** D **55.** 2.5 **57.** Since a square is a parallelogram, the diagonals bisect each other. Since a square is a rhombus, the diagonals are congruent. Therefore, the distance from first base to third base is equal to the distance between home plate and second base. Thus, the distance from home plate to the center of the infield is 127 ft $3\frac{3}{8}$ in. divided by 2 or 63 ft $7\frac{11}{16}$ in. This distance is longer than the distance from home plate to the pitcher's mound so the pitcher's mound is not located in the center of the field. It is about 3 feet closer to home. **59.** $-\frac{4}{7} < x < \frac{136}{7}$ **61.** $\angle 1, \angle 4, \angle 11$ **63.** $\angle 2, \angle 6, \angle 9, \angle 4$

65. Given: $\triangle ABC \cong \triangle DEF$

$\triangle DEF \cong \triangle GHI$

Prove: $\triangle ABC \cong \triangle GHI$

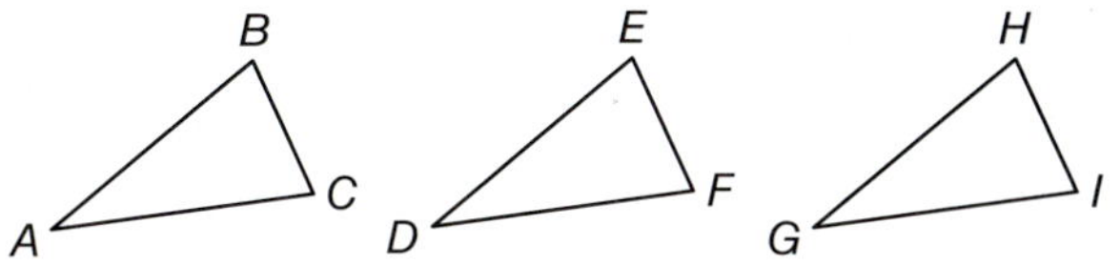

Proof:

You are given that $\triangle ABC \cong \triangle DEF$. Because corresponding parts of congruent triangles are congruent, $\angle A \cong \angle D$, $\angle B \cong \angle E$, $\angle C \cong \angle F$, $\overline{AB} \cong \overline{DE}$, $\overline{BC} \cong \overline{EF}$, and $\overline{AC} \cong \overline{DF}$. You are also given that $\triangle DEF \cong \triangle GHI$. So $\angle D \cong \angle G$, $\angle E \cong \angle H$, $\angle F \cong \angle I$, $\overline{DE} \cong \overline{GH}$, $\overline{EF} \cong \overline{HI}$, and $\overline{DF} \cong \overline{GI}$ by CPCTC. Therefore, $\angle A \cong \angle G$, $\angle B \cong \angle H$, $\angle C \cong \angle I$, $\overline{AB} \cong \overline{GH}$, $\overline{BC} \cong \overline{HI}$, and $\overline{AC} \cong \overline{GI}$ because congruence of angles and segments is transitive. Thus, $\triangle ABC \cong \triangle GHI$ by the defintion of congruent triangles.

Pages 465–473 Lesson 7-2

1 Using the similarity statement, $\angle A$ and $\angle Z$ are corresponding angles, $\angle B$ and $\angle Y$ are corresponding angles, and $\angle C$ and $\angle X$ are corresponding angles. So, $\angle A \cong \angle Z$, $\angle B \cong \angle Y$, and $\angle C \cong \angle X$. $\overline{AC}$ and $\overline{ZX}$ are corresponding sides, $\overline{BC}$ and $\overline{YX}$ are corresponding sides, and $\overline{AB}$ and $\overline{ZY}$ are corresponding sides. So, $\frac{AC}{ZX} = \frac{BC}{YX} = \frac{AB}{ZY}$.

3. no; $\frac{NQ}{WZ} \neq \frac{QR}{WX}$ **5.** 6 **7.** 22 ft **9.** $\angle J \cong \angle P$, $\angle F \cong \angle S$, $\angle M \cong \angle T$; $\angle H \cong \angle Q$, $\frac{PQ}{JH} = \frac{TS}{MF} = \frac{SQ}{FH} = \frac{TP}{MJ}$

11. $\angle D \cong \angle K$, $\angle F \cong \angle M$, $\angle G \cong \angle J$; $\frac{DF}{KM} = \frac{FG}{MJ} = \frac{GD}{JK}$

13 Two sides and the included angle of $\triangle LTK$ are congruent to two sides and the included angle of $\triangle MTK$, so $\triangle LTK \cong \triangle MTK$. By CPCTC, the triangles have all corresponding angles congruent and all corresponding sides congruent. So, $\triangle LTK \sim \triangle MTK$, with a scale factor of 1.

15. no; $\frac{AD}{WM} \neq \frac{DK}{ML}$ **17.** Yes; sample answer: The ratio of the longer dimensions of the screens is approximately 1.1 and the ratio of the shorter dimensions of the screens is approximately 1.1.

19

$\frac{SB}{JH} = \frac{BP}{HT}$	Similarity proportion
$\frac{2}{3} = \frac{x+3}{2x+2}$	$SB = 2$, $JH = 3$, $BP = x + 3$, $HT = 2x + 2$
$2(2x + 2) = 3(x + 3)$	Cross Products Property
$4x + 4 = 3x + 9$	Distributive Property
$x + 4 = 9$	Subtract $3x$ from each side.
$x = 5$	Subtract 4 from each side.

21. 3 **23.** 10.8 **25.** 18.9 **27.** 40 m **29.** $\angle A \cong \angle V$, $\angle B \cong \angle X$, $\angle D \cong \angle Z$, $\angle F \cong \angle T$; $\frac{AB}{VX} = \frac{BD}{XZ} = \frac{DF}{ZT} = \frac{FA}{TV} = 2$ **31.** $\overline{AC}, \overline{AD}$ **33.** $\angle ABH, \angle ADF$

35

$\angle D \cong \angle P$	Corresponding angles of similar polygons are congruent.
$m\angle D = m\angle P$	Definition of congruence
$x + 34 = 97$	Substitution
$x = 63$	Subtract 34 from each side.

$\angle C \cong \angle R$	Corresponding angles of similar polygons are congruent.
$m\angle C = m\angle R$	Definition of congruence
$83 = 3y - 13$	Substitution
$96 = 3y$	Add 13 to each side.
$32 = y$	Divide each side by 3.

37. 52 in. by 37 in. **39.** no; $\frac{BC}{XY} \neq \frac{AB}{WX}$ **41.** Never; sample answer: Parallelograms have both pairs of opposite sides parallel. Trapezoids have exactly one pair of parallel legs. Therefore, the two figures cannot be similar because they can never be the same type of figure. **43.** Sometimes; sample answer: If corresponding angles are congruent and corresponding sides are proportional, two isosceles triangles are similar. **45.** Always; sample answer: Equilateral triangles always have three 60° angles, so the angles of one equilateral triangle are always congruent to the angles of a second equilateral triangle. The three sides of an equilateral triangle are always congruent, so the ratio of each pair of legs of one triangle to a second triangle will always be the same. Therefore, a pair of equilateral triangles is always similar.

47 Let ℓ = new length and w = new width.

$\frac{2\frac{1}{3}}{\ell} = \frac{2}{3}$	Use the lengths and scale factor to write a proportion.
$2\frac{1}{3} \cdot 3 = \ell \cdot 2$	Cross Products Property
$\frac{7}{3} \cdot 3 = 2\ell$	Write $2\frac{1}{3}$ as $\frac{7}{3}$.
$7 = 2\ell$	Multiply.
$\frac{7}{2} = \ell$	Divide each side by 7.

The new length is $\frac{7}{2}$ inches or $3\frac{1}{2}$ inches.

$\frac{1\frac{2}{3}}{w} = \frac{2}{3}$	Use the widths and scale factor to write a proportion.
$1\frac{2}{3} \cdot 3 = w \cdot 2$	Cross Products Property
$\frac{5}{3} \cdot 3 = 2w$	Write $1\frac{2}{3}$ as $\frac{5}{3}$.
$5 = 2w$	Multiply.
$\frac{5}{2} = w$	Divide each side by 2.

The new width is $\frac{5}{2}$ inches or $2\frac{1}{2}$ inches.

49a. $\frac{a}{3a} = \frac{b}{3b} = \frac{c}{3c} = \frac{a+b+c}{3(a+b+c)} = \frac{1}{3}$

49b. No; the sides are no longer proportional. **51.** 4

53. Sample answer:

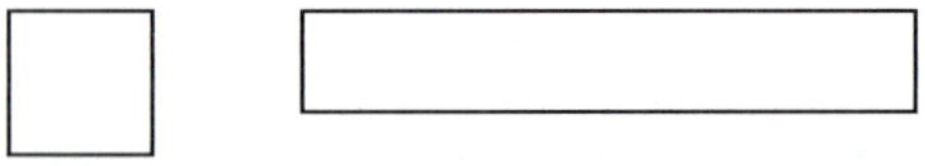

55. Sample answer: Two figures are congruent if they are the same size and shape. In congruent figures, corresponding angles are congruent and the corresponding sides are congruent. When two figures are similar, corresponding angles are congruent and corresponding sides are proportional. Figures that are congruent are also similar, since corresponding angles are congruent and corresponding sides are proportional. Two figures that are similar are only congruent if the ratio of their corresponding side lengths is 1. If figures are equal, then they are the same figure. **57.** G **59.** C

61. Given: E and C are midpoints of $\overline{AD}$ and $\overline{DB}$. $\overline{AD} \cong \overline{DB}$; $\angle A \cong \angle 1$

Prove: $ABCE$ is an isosceles trapezoid.

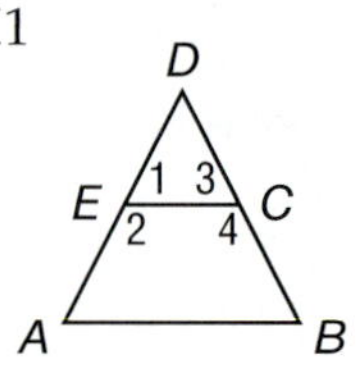

Proof:

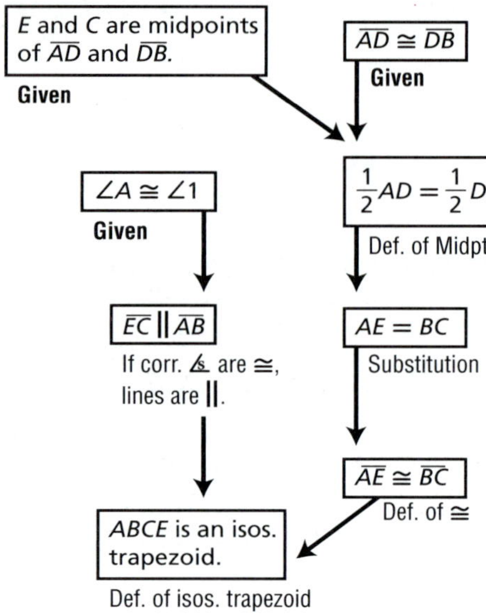

63. $x \leq 4$ **65.** The angle bisector of the vertex angle of an isosceles triangle is not an altitude of the triangle. **67.** 128 **69.** 68 **71.** $x = 2$, $RT = 8$, $RS = 8$

Pages 474–483 Lesson 7-3

1. Yes; $\triangle YXZ \sim \triangle VWZ$ by AA Similarity. **3.** No; corresponding sides are not proportional. **5.** C **7.** $\triangle QVS \sim \triangle RTS$; 20 **9.** Yes; $\triangle XUZ \sim \triangle WUY$ by SSS Similarity.

11 $\frac{CB}{DB} = \frac{10}{6}$ or $\frac{5}{3}$ and $\frac{BA}{BF} = \frac{9+6}{9} = \frac{15}{9}$ or $\frac{5}{3}$. $m\angle CBA = m\angle DBF = 38$, so $\angle CBA \cong \angle DBF$. Since the length of the sides that include the congruent angles are proportional, $\triangle CBA \sim \triangle DBF$ by SAS Similarity.

13. No; not enough information to determine. If $JH = 3$ or $WY = 24$, then $\triangle JHK \sim \triangle XWY$ by SSS Similarity. **15.** Yes; sample answer: Since $\overline{AB} \cong \overline{EB}$ and $\overline{CB} \cong \overline{DB}$, $\frac{AB}{CB} = \frac{EB}{DB}$. $\angle ABE \cong \angle CBD$ because vertical angles are congruent. Therefore, $\triangle ABE \sim \triangle CBD$ by SAS Similarity.

17 Since $\overline{RS} \parallel \overline{PT}$, $\angle QRS \cong \angle QPT$ and $\angle QSR \cong \angle QTP$ because they are corresponding angles. By AA Similarity, $\triangle QRS \sim \triangle QPT$.

$\frac{RS}{PT} = \frac{QS}{QT}$	Definition of similar polygons
$\frac{12}{16} = \frac{x}{20}$	$RS = 12$, $PT = 16$, $QS = x$, $QT = 20$
$12 \cdot 20 = 16 \cdot x$	Cross Products Property
$240 = 16x$	Simplify.
$15 = x$	Divide each side by 16.

Since $QS + ST = 20$ and $QS = 15$, then $ST = 5$.

19. $\triangle HJK \sim \triangle NQP$; 15, 10 **21.** $\triangle GHJ \sim \triangle GDH$; 14, 20
23. about 12.8 ft
25. Given: $\angle B \cong \angle E$, $\overline{QP} \parallel \overline{BC}$; $\overline{QP} \cong \overline{EF}$, $\frac{AB}{DE} = \frac{BC}{EF}$
Prove: $\triangle ABC \sim \triangle DEF$

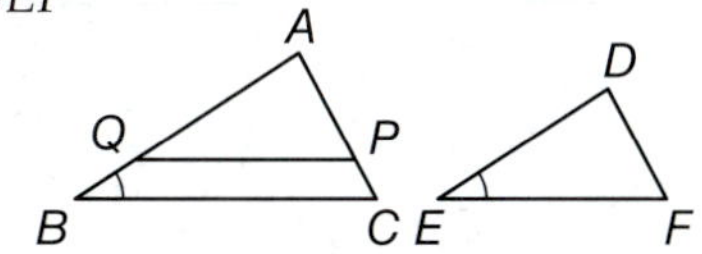

Proof:
Statements (Reasons)

1. $\angle B \cong \angle E$, $\overline{QP} \parallel \overline{BC}$; $\overline{QP} \cong \overline{EF}$, $\frac{AB}{DE} = \frac{BC}{EF}$ (Given)
2. $\angle APQ \cong \angle C$, $\angle AQP \cong \angle B$ (Corr. ∠s Post.)
3. $\angle AQP \cong \angle E$ (Trans. Prop.)
4. $\triangle ABC \sim \triangle AQP$ (AA Similarity)
5. $\frac{AB}{AQ} = \frac{BC}{QP}$ (Def. of ~ △s)
6. $AB \cdot QP = AQ \cdot BC$; $AB \cdot EF = DE \cdot BC$ (Cross products)
7. $QP = EF$ (Def. of ≅ segs.)
8. $AB \cdot EF = AQ \cdot BC$ (Subst.)
9. $AQ \cdot BC = DE \cdot BC$ (Subst.)
10. $AQ = DE$ (Div. Prop.)
11. $\overline{AQ} \cong \overline{DE}$ (Def. of ≅ segs.)
12. $\triangle AQP \cong \triangle DEF$ (SAS)
13. $\angle APQ \cong \angle F$ (CPCTC)
14. $\angle C \cong \angle F$ (Trans. Prop.)
15. $\triangle ABC \sim \triangle DEF$ (AA Similarity)

27. Given: $\triangle XYZ$ and $\triangle ABC$ are right triangles; $\frac{XY}{AB} = \frac{YZ}{BC}$.
Prove: $\triangle YXZ \sim \triangle BAC$

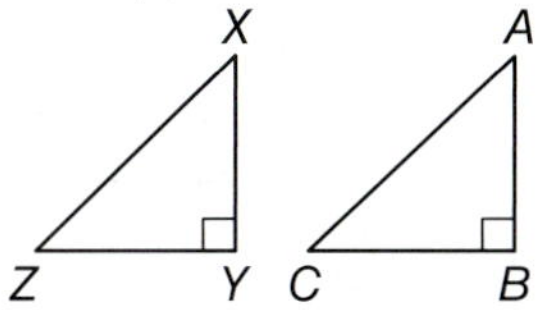

Proof:
Statements (Reasons)

1. $\triangle XYZ$ and $\triangle ABC$ are right triangles. (Given)
2. $\angle XYZ$ and $\angle ABC$ are right angles. (Def. of rt. △)
3. $\angle XYZ \cong \angle ABC$ (All rt. ∠s are ≅.)
4. $\frac{XY}{AB} = \frac{YZ}{BC}$ (Given)
5. $\triangle YXZ \sim \triangle BAC$ (SAS Similarity)

29. $26\frac{2}{3}$ in.

31 Use $d = \sqrt{(x_2 - x_1)^2 + (y_2 - y_1)^2}$ to find the side lengths of $\triangle XYZ$ and $\triangle WYV$.

$XY = \sqrt{[5 - (-1)]^2 + [3 - (-9)]^2} = 6\sqrt{5}$
$(x_1, y_1) = (-1, -9), (x_2, y_2) = (5, 3)$

$YZ = \sqrt{(-1 - 5)^2 + (6 - 3)^2} = 3\sqrt{5}$
$(x_1, y_1) = (5, 3), (x_2, y_2) = (-1, 6)$

$XZ = \sqrt{[-1 - (-1)]^2 + [6 - (-9)]^2} = 15$
$(x_1, y_1) = (-1, -9), (x_2, y_2) = (-1, 6)$

$WY = \sqrt{(5 - 1)^2 + [3 - (-5)]^2} = 2\sqrt{5}$
$(x_1, y_1) = (1, -5), (x_2, y_2) = (5, 3)$

$YV = \sqrt{(1 - 5)^2 + (5 - 3)^2} = 2\sqrt{5}$
$(x_1, y_1) = (5, 3), (x_2, y_2) = (1, 5)$

$WV = \sqrt{(1 - 1)^2 + [5 - (-5)]^2} = 10$
$(x_1, y_1) = (1, -5), (x_2, y_2) = (1, 5)$

Since the perimeter of $\triangle XYZ = 6\sqrt{5} + 3\sqrt{5} + 15$ or $9\sqrt{5} + 15$ and the perimeter of $\triangle WYV = 4\sqrt{5} + 2\sqrt{5} + 10$ or $6\sqrt{5} + 10$, the ratio of the perimeters is $\frac{9\sqrt{5} + 15}{6\sqrt{5} + 10} = \frac{3(3\sqrt{5} + 5)}{2(3\sqrt{5} + 5)}$ or $\frac{3}{2}$.

33. $\angle C \cong \angle C'$, since all right angles are congruent. Line ℓ is a transversal of parallel segments $\overline{BC}$ and $\overline{B'C'}$, so $\angle ABC \cong \angle A'B'C'$ since corresponding angles of parallel lines are congruent. Therefore, by AA Similarity, $\triangle ABC \sim \triangle A'B'C'$. So $\frac{BC}{AC}$, the slope of line ℓ through points A and B, is equal to $\frac{B'C'}{A'C'}$, the slope of line ℓ through points A' and B'.

35 Since $\angle CAB \cong \angle CAB$ by the Reflexive Property and $\angle ABC \sim \angle ADE$ by the definition of right angles, $\triangle ABC \sim \triangle ADE$ by the AA Similarity Postulate.

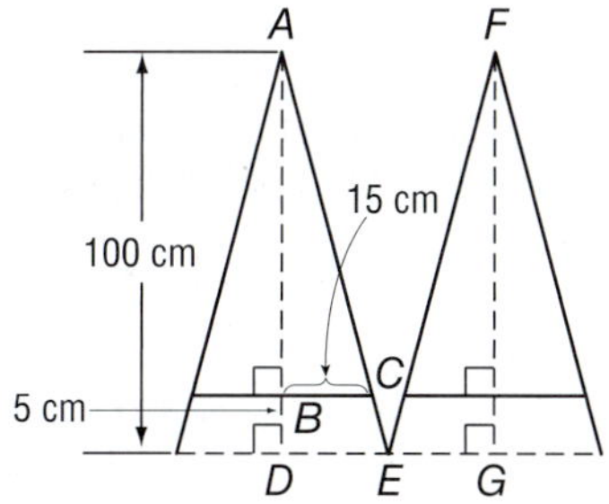

$\frac{AB}{BC} = \frac{AD}{DE}$		Definition of similar polygons
$\frac{100}{15} = \frac{100 + 5}{DE}$		Substitution
$100 \cdot DE = 105 \cdot 15$		Cross Products Property
$100DE = 1575$		Simplify.
$DE = 15.75$		Divide each side by 100.

Since the △s formed by the laser sources are ≅, $\overline{DE} \cong \overline{GE}$. So, $DG = 15.75 + 15.75$ or 31.5. Since $\overline{AD}$ and $\overline{FG}$ are ||, $AF = DG$. So, $AF = 31.5$. The laser sources should be placed 31.5 cm apart.

37. Sample answer: The AA Similarity Postulate, SSS Similarity Theorem, and SAS Similarity Theorem are all tests that can be used to determine whether two triangles are similar. The AA Similarity Postulate is used when two pairs of congruent angles on two triangles are given. The SSS Similarity Theorem is used when the corresponding side lengths of two triangles are given. The SAS Similarity Theorem is used when two proportional side lengths and the included angle on two triangles are given. **39.** 6
41. Sample answer: Choose a side of the original triangle and measure it. Draw a segment that is twice as long as the one you measured. Measure the angles between the side that you measured on the original triangle and the other two sides. Construct angles congruent to the two angles on the original triangle from your segment. Extend the two new segments until they meet. The new triangle will be similar to the first triangle and twice as large. **43a.** $\frac{6}{x - 2} = \frac{4}{5}$
43b. 9.5; 7.5 **45.** B **47.** $\angle X \cong \angle R$, $\angle W \cong \angle Q$, $\angle Y \cong \angle S$; $\angle Z \cong \angle T$, $\frac{WX}{QR} = \frac{ZY}{TS} = \frac{WZ}{QT} = \frac{XY}{RS}$
49. 12 **51.** 52.3 **53.** Sample answer: If one pair of

opposite sides are congruent and parallel, the quadrilateral is a parallelogram. **55.** not possible

57. Given: $r \parallel t$; $\angle 5 \cong \angle 6$
Prove: $\ell \parallel m$

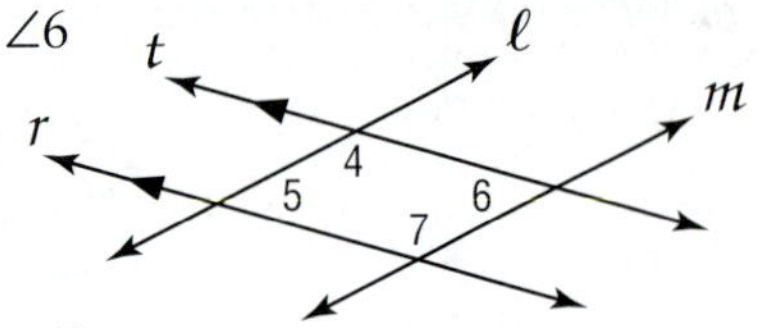

Proof:
Statements (Reasons)
1. $r \parallel t$; $\angle 5 \cong \angle 6$ (Given)
2. $\angle 4$ and $\angle 5$ are supplementary. (Consecutive Interior Angles Theorem)
3. $m\angle 4 + m\angle 5 = 180$ (Def. of Suppl. ∡)
4. $m\angle 5 = m\angle 6$ (Def. of ≅ ∡)
5. $m\angle 4 + m\angle 6 = 180$ (Substitution)
6. $\angle 4$ and $\angle 6$ are supplementary. (Def. of Suppl. ∡)
7. $\ell \parallel m$ (If cons. int. ∡ are suppl., then lines are ∥.)

Pages 484–493 Lesson 7-4

1. 10 **3.** Yes; $\frac{AD}{DC} = \frac{BE}{EC} = \frac{2}{3}$, so $\overline{DE} \parallel \overline{AB}$. **5.** 11
7. 2360.3 ft **9.** $x = 20$; $y = 2$

11 If $AB = 12$ and $AC = 16$, then $BC = 4$.

$\frac{AB}{BC} = \frac{AE}{ED}$	Triangle Proportionality Theorem
$\frac{12}{4} = \frac{AE}{5}$	Substitute.
$12 \cdot 5 = 4 \cdot AE$	Cross Products Property
$60 = 4AE$	Multiply.
$15 = AE$	Divide each side by 4.

13. 10 **15.** yes; $\frac{ZV}{VX} = \frac{WY}{YX} = \frac{11}{5}$ **17.** no; $\frac{ZV}{VX} \neq \frac{WY}{YX}$

19

$m\angle PHM + m\angle PHJ + m\angle JHL = 180$	Definition of a straight angle
$44 + m\angle PHJ + 76 = 180$	Substitution
$120 + m\angle PHJ = 180$	Simplify.
$m\angle PHJ = 60$	Subtract 120 from each side.

By the Triangle Midsegment Theorem, $\overline{PH} \parallel \overline{KL}$.

$\angle PHJ \cong \angle JHL$	Alternate Interior Angles Theorem
$m\angle PHJ = m\angle JHL$	Definition of congruence
$60 = x$	Substitution

21. 1.35 **23.** 1.2 in. **25.** $x = 18$; $y = 3$ **27.** $x = 48$; $y = 15$

29. Given: $\overleftrightarrow{AD} \parallel \overleftrightarrow{BE} \parallel \overleftrightarrow{CF}$, $\overline{AB} \cong \overline{BC}$
Prove: $\overline{DE} \cong \overline{EF}$

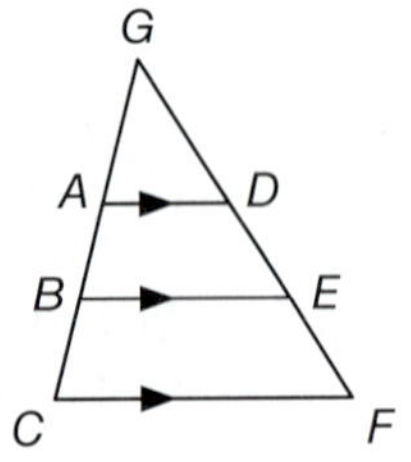

Proof:
From Corollary 7.1, $\frac{AB}{BC} = \frac{DE}{EF}$.
Since $\overline{AB} \cong \overline{BC}$, $AB = BC$ by definition of congruence.
Therefore, $\frac{AB}{BC} = 1$.
By substitution, $1 = \frac{DE}{EF}$. Thus, $DE = EF$. By definition of congruence, $\overline{DE} \cong \overline{EF}$.

31. Given: $\frac{DB}{AD} = \frac{EC}{AE}$
Prove: $\overline{DE} \parallel \overline{BC}$

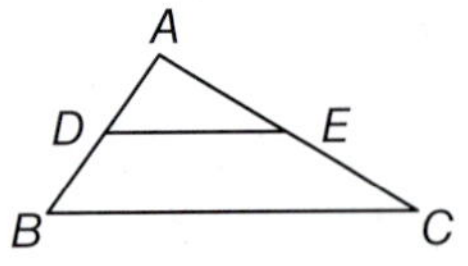

Proof:
Statements (Reasons)
1. $\frac{DB}{AD} = \frac{EC}{AE}$ (Given)
2. $\frac{AD}{AD} + \frac{DB}{AD} = \frac{AE}{AE} + \frac{EC}{AE}$ (Add. Prop.)
3. $\frac{AD + DB}{AD} = \frac{AE + EC}{AE}$ (Subst.)
4. $AB = AD + DB$, $AC = AE + EC$ (Seg. Add. Post.)
5. $\frac{AB}{AD} = \frac{AC}{AE}$ (Subst.)
6. $\angle A \cong \angle A$ (Refl. Prop.)
7. $\triangle ADE \cong \triangle ABC$ (SAS Similarity)
8. $\angle ADE \cong \angle ABC$ (Def. of ~ polygons)
9. $\overline{DE} \parallel \overline{BC}$ (If corr. ∡ are ≅, then the lines are ∥.)

33. 9

35 If $CA = 10$ and $CD = 2$, then $DA = 8$.

$\frac{CE}{EB} = \frac{CD}{DA}$	Triangle Proportionality Theorem
$\frac{t-2}{t+1} = \frac{2}{8}$	Substitute.
$(t-2)(8) = (t+1)(2)$	Cross Products Property
$8t - 16 = 2t + 2$	Distributive Property
$6t - 16 = 2$	Subtract $2t$ from each side.
$6t = 18$	Add 16 to each side.
$t = 3$	Divide each side by 3.

If $t = 3$, then $CE = 3 - 2$ or 1.

37. 8, 7.5
39. $\triangle ABC \sim \triangle ADE$ SAS Similarity **41.** 6

$\frac{AD}{AB} = \frac{DE}{BC}$	Def. ~ △s
$\frac{40}{100} = \frac{DE}{BC}$	Substitution
$\frac{2}{5} = \frac{DE}{BC}$	Simplify.
$\frac{2}{5}BC = DE$	Multiply.

43 All the triangles are isosceles. Segment EH is the midsegment of triangle ABC. Therefore, segment EH is half of the length of AC, which is $35 \div 2$ or 17.5 feet. Similarly, FG is the midsegment of triangle BEH, so $FG = 17.5 \div 2$ or 8.75 feet. To find DJ, use the vertical altitude which is 12 feet. Let the altitude from B to the segment AC meet the segment DJ at K. Find BC using the Pythagorean Theorem.

$BC^2 = BK^2 + KC^2$
$BC^2 = 12^2 + 17.5^2$
$BC = \sqrt{12^2 + 17.5^2}$
$BC \approx 21.22$ in.

Since the width of each piece of siding is the same, $BJ = \frac{3}{4}BC$, which is about $\frac{3}{4}(21.22)$ or 15.92 in.
Now, use the Triangle Proportionality Theorem.

$\frac{AC}{BC} = \frac{DJ}{BJ}$
$\frac{35}{21.22} = \frac{DJ}{15.92}$
$21.22(DJ) = (15.92)(35)$
$21.22(DJ) = 557.2$
$DJ \approx 26.25$ in.

45. Sample answer:

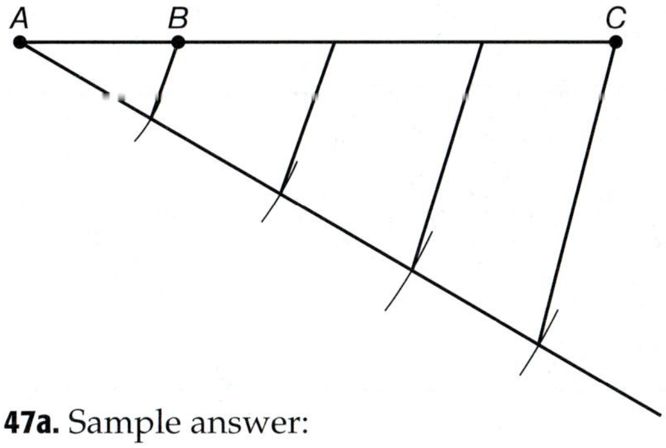

47a. Sample answer:

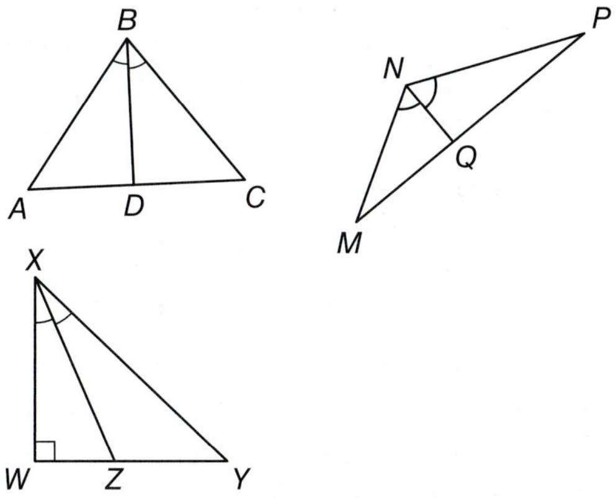

47b.

Triangle	Length		Ratio	
ABC	*AD*	1.1 cm	$\frac{AD}{CD}$	1.0
	CD	1.1 cm		
	AB	2.0 cm	$\frac{AB}{CB}$	1.0
	CB	2.0 cm		
MNP	*MQ*	1.4 cm	$\frac{MQ}{PQ}$	0.8
	PQ	1.7 cm		
	MN	1.6 cm	$\frac{MN}{PN}$	0.8
	PN	2.0 cm		
WXY	*WZ*	0.8 cm	$\frac{WZ}{YZ}$	0.7
	YZ	1.2 cm		
	WX	2.0 cm	$\frac{WX}{YX}$	0.7
	YX	2.9 cm		

47c. Sample answer: The proportion of the segments created by the angle bisector of a triangle is equal to the proportion of their respective consecutive sides.

49. Always; sample answer: $\overline{FH}$ is a midsegment. Let $BC = x$, then $FH = \frac{1}{2}x$. *FHCB* is a trapezoid, so $DE = \frac{1}{2}(BC + FH) = \frac{1}{2}\left(x + \frac{1}{2}x\right) = \frac{1}{2}x + \frac{1}{4}x = \frac{3}{4}x$. Therefore, $DE = \frac{3}{4}BC$.

51.

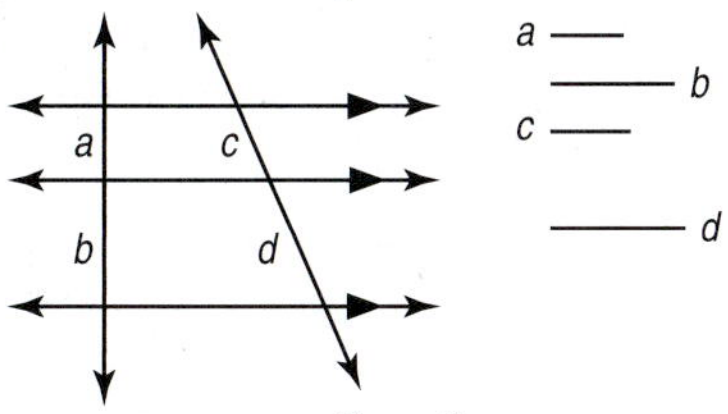

By Corollary 7.1, $\frac{a}{b} = \frac{c}{d}$.

53. 8 **55.** G **57.** $\triangle ABE \sim \triangle CDE$ by AA Similarity; 6.25

59. $\triangle WZT \sim \triangle WXY$ by AA Similarity; 7.5

61. $\overline{QR} \parallel \overline{TS}$, $\overline{QT} \nparallel \overline{RS}$; *QRST* is an isosceles trapezoid since $RS = \sqrt{26} = QT$. **63.** 6 **65.** 56 **67.** $\frac{2}{3}$

69. 2.1 **71.** 8.7

Pages 495–502 Lesson 7-5

1 The triangles are similar by AA Similarity.

$\frac{x}{10} = \frac{12}{15}$	$\sim\triangle$s have corr. medians proportional to the corr. sides.
$x \cdot 15 = 10 \cdot 12$	Cross Products Property
$15x = 120$	Simplify.
$x = 8$	Divide each side by 15.

3. 35.7 ft **5.** 20 **7.** 8.5 **9.** 18

11

$\frac{15}{27} = \frac{28 - b}{b}$	Triangle Angle Bisector Theorem
$15 \cdot b = 27(28 - b)$	Cross Products Property
$15b = 756 - 27b$	Multiply.
$42b = 756$	Add 27*b* to each side.
$b = 18$	Divide each side by 42.

13. 15

15

$\frac{AB}{JK} = \frac{AD}{JM}$	$\sim\triangle$s have corr. altitudes proportional to the corr. sides.
$\frac{9}{21} = \frac{4x - 8}{5x + 3}$	Substitute.
$9(5x + 3) = 21(4x - 8)$	Cross Products Property
$45x + 27 = 84x - 168$	Distributive Property
$27 = 39x - 168$	Subtract 45*x* from each side.
$195 = 39x$	Add 168 to each side.
$5 = x$	Divide each side by 39.

17. 4

19. Given: $\triangle ABC \sim \triangle RST$
$\overline{AD}$ is a median of $\triangle ABC$.
$\overline{RU}$ is a median of $\triangle RST$.

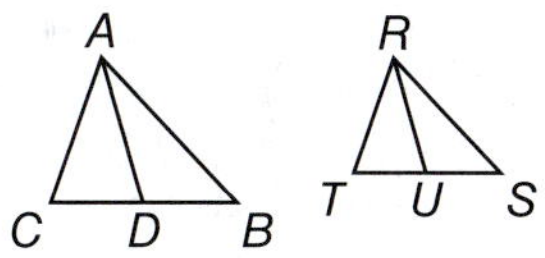

Prove: $\frac{AD}{RU} = \frac{AB}{RS}$

Proof:

Statements (Reasons)

1. $\triangle ABC \sim \triangle RST$; $\overline{AD}$ is a median of $\triangle ABC$; $\overline{RU}$ is a median of $\triangle RST$. (Given)
2. $CD = DB$; $TU = US$ (Def. of median)
3. $\frac{AB}{RS} = \frac{CB}{TS}$ (Def. of $\sim\triangle$s)
4. $CB = CD + DB$; $TS = TU + US$ (Seg. Add. Post.)
5. $\frac{AB}{RS} = \frac{CD + DB}{TU + US}$ (Subst.)
6. $\frac{AB}{RS} = \frac{DB + DB}{US + US}$ or $\frac{2(DB)}{2(US)}$ (Subst.)
7. $\frac{AB}{RS} = \frac{DB}{US}$ (Subst.)
8. $\angle B \cong \angle S$ (Def. of $\sim\triangle$s)
9. $\triangle ABD \sim \triangle RSU$ (SAS Similarity)
10. $\frac{AD}{RU} = \frac{AB}{RS}$ (Def. of $\sim\triangle$s)

21. 3 **23.** 70

25. Given: $\overline{CD}$ bisects $\angle ACB$.
By construction, $\overline{AE} \parallel \overline{CD}$.

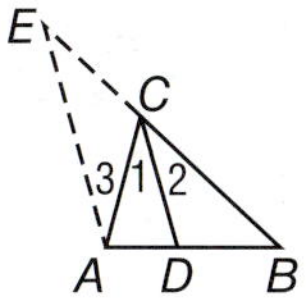

Prove: $\frac{AD}{DB} = \frac{AC}{BC}$

Proof:

Statements (Reasons)

1. $\overline{CD}$ bisects $\angle ACB$; By construction, $\overline{AE} \parallel \overline{CD}$. (Given)
2. $\frac{AD}{DB} = \frac{EC}{BC}$ ($\triangle$ Prop. Thm.)
3. $\angle 1 \cong \angle 2$ (Def. of $\angle$ Bisector)
4. $\angle 3 \cong \angle 1$ (Alt. Int. $\angle$ Thm.)
5. $\angle 2 \cong \angle E$ (Corr. $\angle$ Post.)
6. $\angle 3 \cong \angle E$ (Trans. Prop.)
7. $\overline{EC} \cong \overline{AC}$ (Converse of Isos. $\triangle$ Thm.)
8. $EC = AC$ (Def. of $\cong$ segs.)
9. $\frac{AD}{DB} = \frac{AC}{BC}$ (Subst.)

27. Given: $\triangle QTS \sim \triangle XWZ$, $\overline{TR}$, $\overline{WY}$ are $\angle$ bisectors.

Prove: $\frac{TR}{WY} = \frac{QT}{XW}$

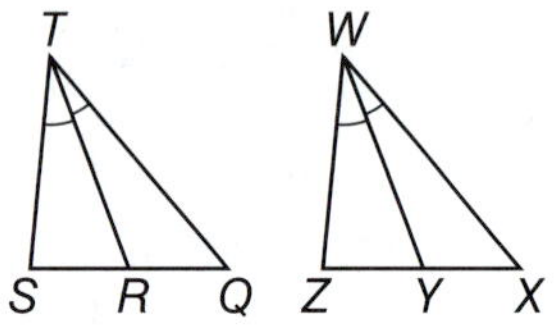

Proof:

Statements (Reasons)

1. $\triangle STQ \sim \triangle ZWX$, $\overline{TR}$ and $\overline{WY}$ are angle bisectors. (Given)
2. $\angle STQ \cong \angle ZWX$, $\angle Q \cong \angle X$ (Def of $\sim \triangle$s)
3. $\angle STR \cong \angle QTR$, $\angle ZWY \cong \angle XWY$ (Def. $\angle$ bisector)
4. $m\angle STQ = m\angle STR + m\angle QTR$, $m\angle ZWX = m\angle ZWY + m\angle XWY$ ($\angle$ Add. Post.)
5. $m\angle STQ = 2m\angle QTR$, $m\angle ZWX = 2m\angle XWY$ (Subst.)
6. $2m\angle QTR = 2m\angle XWY$ (Subst.)
7. $m\angle QTR = m\angle XWY$ (Div. Prop.)
8. $\triangle QTR \sim \triangle XWY$ (AA Similarity)
9. $\frac{TR}{WY} = \frac{QT}{XW}$ (Def. of $\sim \triangle$s)

29 Since the segment from Trevor to Ricardo is an angle bisector, the segments from Ricardo to Craig and from Ricardo to Eli are proportional to the segments from Trevor to Craig and from Trevor to Eli. Since Craig is closer to Trevor than Eli is, Craig is also closer to Ricardo than Eli is. So, Craig will reach Ricardo first.

31. Chun; by the Angle Bisector Theorem, the correct proportion is $\frac{5}{8} = \frac{15}{x}$. **33.** $PS = 18.4$, $RS = 24$

35. Both theorems have a segment that bisects an angle and has proportionate ratios. The Triangle Angle Bisector Theorem pertains to one triangle, while Theorem 7.9 pertains to similar triangles. Unlike the Triangle Angle Bisector Theorem, which separates the opposite side into segments that have the same ratio as the other two sides, Theorem 7.9 relates the angle bisector to the measures of the sides. **37.** 2.2 **39.** B

41. $x = 2$; $y = 3$ **43.** $KP = 5$, $KM = 15$, $MR = 13\frac{1}{3}$, $ML = 20$, $MN = 12$, $PR = 16\frac{2}{3}$

45. Given: $\overline{EF} \cong \overline{HF}$
G is the midpoint of $\overline{EH}$.

Prove: $\triangle EFG \cong \triangle HFG$

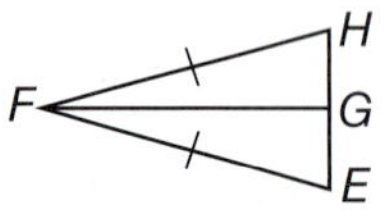

Proof:

Statements (Reasons)

1. $\overline{EF} \cong \overline{HF}$; G is the midpoint of $\overline{EH}$. (Given)
2. $\overline{EG} \cong \overline{GH}$ (Def. of midpoint)
3. $\overline{FG} \cong \overline{FG}$ (Reflexive Prop.)
4. $\triangle EFG \cong \triangle HFG$ (SSS)

47. 5 **49.** $\sqrt{137} \approx 11.7$ **51.** $\sqrt{340} \approx 18.4$

Pages 505–511 Lesson 7-6

1. enlargement; 2

3 The ratio comparing the widths is $\frac{152.5}{27} \approx 5.6$.

The ratio comparing the lengths is $\frac{274}{78} \approx 3.5$.

Sample answer: Since $\frac{152.5}{27} \neq \frac{274}{78}$, a table tennis table is not a dilation of a tennis court.

5. $\frac{RJ}{KJ} = \frac{SJ}{LJ} = \frac{RS}{KL} = \frac{1}{2}$, so $\triangle RSJ \sim \triangle KLJ$ by SSS Similarity. **7.** reduction; $\frac{1}{2}$ **9.** enlargement; 2

11. reduction **13.** No; sample answer: Since $\frac{1.2}{2.5} \neq \frac{1.25}{3}$, the design and the actual tattoo are not proportional. Therefore, the tattoo is not a dilation of the design.

15.

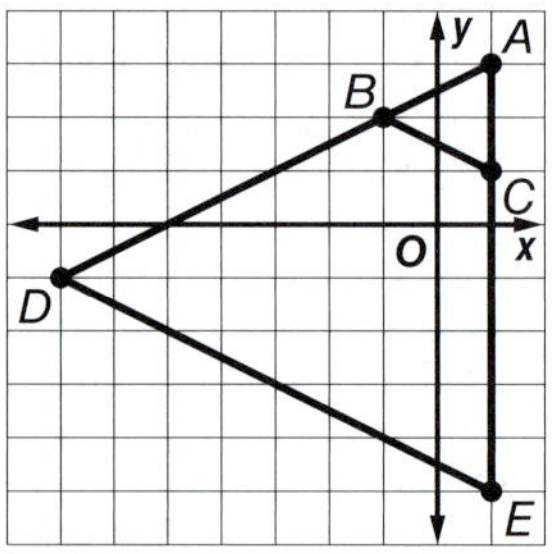

$\angle A \cong \angle A$ and $\frac{AB}{AD} = \frac{AC}{AE} = \frac{1}{4}$, so $\triangle ABC \sim \triangle ADE$ by SAS Similarity.

17.

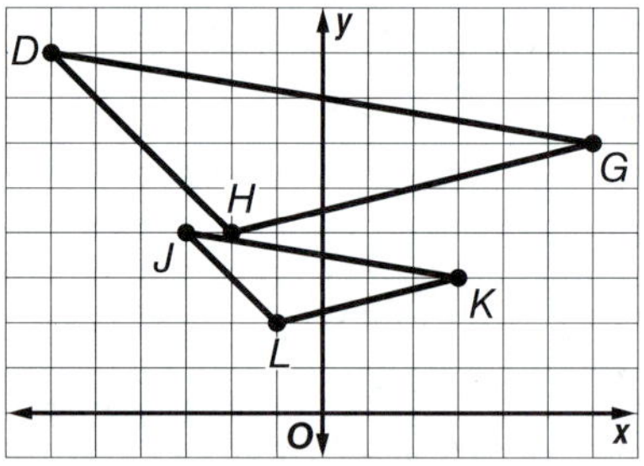

$\frac{JK}{DG} = \frac{KL}{GH} = \frac{JL}{DH} = \frac{1}{2}$, so $\triangle JKL \sim \triangle DGH$ by SSS Similarity.

19

$\frac{AC}{AB} = \frac{AZ}{AY}$	Definition of similar polygons
$\frac{4}{AB} = \frac{12}{6}$	$AC = 4$, $AZ = 12$, $AY = 6$
$4 \cdot 6 = AB \cdot 12$	Cross Products Property
$24 = 12AB$	Simplify.
$2 = AB$	Divide each side by 12.

Since $AB = 2$, the coordinates of B are (0, –2).

21 **a.** Sample answer: $\triangle ABC$: $A(0, 0)$, $B(1, 3)$, $C(4, 2)$ Draw points D and E so that $AD = 2AB$ and $AE = 2AC$. Connect the points to form $\triangle ADE$: $A(0, 0)$, $D(2, 6)$, $E(8, 4)$.

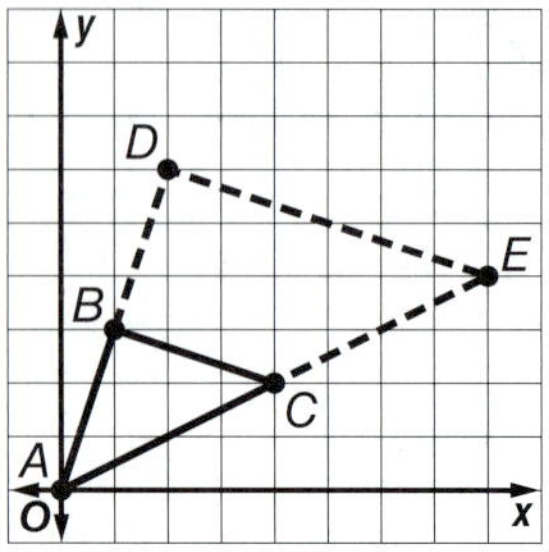

b. Sample answer: $\triangle MNP$: $M(0, 0)$, $N(1, 2)$, $P(3, 1)$ Draw points Q and R so that $MQ = 3MN$ and $MR = 3MP$. Connect the points to form $\triangle MQR$: $M(0, 0)$, $Q(3, 6)$, $R(9, 3)$.

$\triangle TWX$: $T(0, 0)$, $W(0, 7)$, $X(4, 0)$

Draw points Y and Z so that $TY = 0.5TW$ and $TZ = 0.5TX$. Connect the points to form $\triangle TYZ$: $T(0, 0)$, $Y(0, 3.5)$, $Z(2, 0)$.

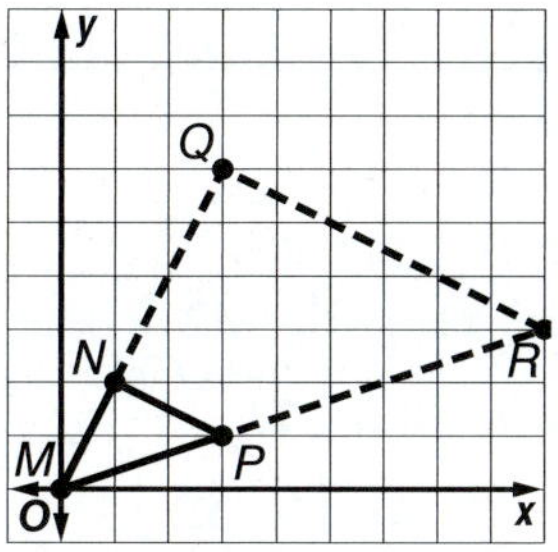

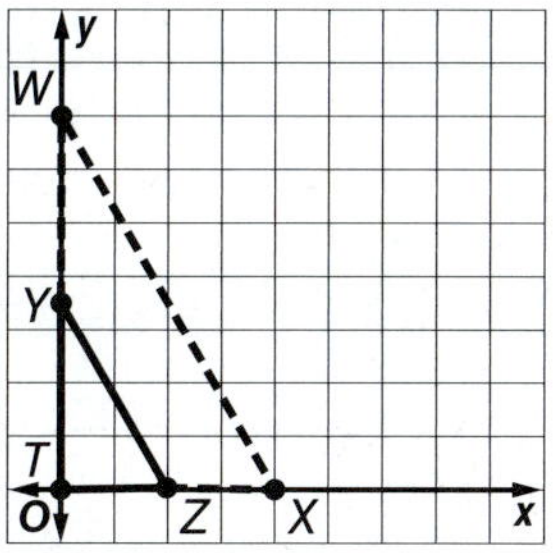

c.

Coordinates											
$\triangle ABC$		$\triangle ADE$		$\triangle MNP$		$\triangle MQR$		$\triangle TWX$		$\triangle TYZ$	
A	(0, 0)	A	(0, 0)	M	(0, 0)	M	(0, 0)	T	(0, 0)	T	(0, 0)
B	(1, 3)	D	(2, 6)	N	(1, 2)	Q	(3, 6)	W	(0, 7)	Y	(0, 3.5)
C	(4, 2)	E	(8, 4)	P	(3, 1)	R	(9, 3)	X	(4, 0)	Z	(2, 0)

d. Sample answer: Multiply the coordinates of the given triangle by the scale factor to get the coordinates of the dilated triangle.

$\triangle ADE$: $A(0 \cdot 2, 0 \cdot 2)$ or $A(0, 0)$, $D(1 \cdot 2, 3 \cdot 2)$ or $D(2, 6)$, $E(4 \cdot 2, 2 \cdot 2)$ or $E(8, 4)$

$\triangle MQR$: $M(0 \cdot 3, 0 \cdot 3)$ or $M(0, 0)$, $Q(1 \cdot 3, 2 \cdot 3)$ or $Q(3, 6)$, $R(3 \cdot 3, 1 \cdot 3)$ or $R(9, 3)$

$\triangle TYZ$: $T(0 \cdot 0.5, 0 \cdot 0.5)$ or $T(0, 0)$, $Y(0 \cdot 0.5, 7 \cdot 0.5)$ or $Y(0, 3.5)$, $Z(4 \cdot 0.5, 0 \cdot 0.5)$ or $Z(2, 0)$

23. No; sample answer: Since the x-coordinates are multiplied by 3 and the y-coordinates are multiplied by 2, $\triangle XYZ$ is 3 times as wide and only 2 times as tall as $\triangle PQR$. Therefore, the transformation is not a dilation. **25.** Sample answer: Architectural plans are reductions. **27.** Sample answer: If a transformation is an enlargement, the lengths of the transformed object will be greater than the original object, so the scale factor will be greater than 1. If a transformation is a reduction, the lengths of the transformed object will be less than the original object, so the scale factor will be less than 1, but greater than 0. If the transformation is a congruence transformation, the scale factor is 1, because the lengths of the transformed object are equal to the lengths of the original object. **29.** $\frac{1}{2}$ **31.** C

33. yes; $\frac{AC}{BD} = \frac{DE}{CE} = \frac{4}{3}$ **35.** no; $\frac{AB}{CD} \neq \frac{AE}{CE}$ **37.** 117

39. Given: $\triangle ABC$
S is the midpoint of $\overline{AC}$.
T is the midpoint of $\overline{BC}$.
Prove: $\overline{ST} \parallel \overline{AB}$

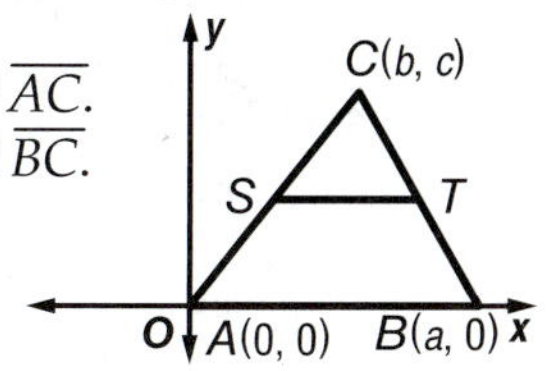

Proof:

Midpoint S is $\left(\frac{b+0}{2}, \frac{c+0}{2}\right)$ or $\left(\frac{b}{2}, \frac{c}{2}\right)$.

Midpoint T is $\left(\frac{a+b}{2}, \frac{0+c}{2}\right)$ or $\left(\frac{a+b}{2}, \frac{c}{2}\right)$.

Slope of $\overline{ST} = \dfrac{\frac{c}{2} - \frac{c}{2}}{\frac{a+b}{2} - \frac{b}{2}}, = \dfrac{0}{\frac{a}{2}}$ or 0.

Slope of $\overline{AB} = \frac{0-0}{a-0} = \frac{0}{a}$ or 0.

$\overline{ST}$ and $\overline{AB}$ have the same slope so, $\overline{ST} \parallel \overline{AB}$.

41. 8 **43.** 0.003 **45.** 0.17

Pages 512–517 Lesson 7-7

1. about 117 mi **3a.** 6 in.:50 ft **3b.** $\frac{1}{100}$ **5.** 380 km **7.** 173 km

9 a. $\frac{\text{replica height}}{\text{statue height}} = \frac{10 \text{ in.}}{10 \text{ ft}} = \frac{1 \text{ in.}}{1 \text{ ft}}$
The scale of the replica is 1 inch:1 foot.

b. $\frac{10 \text{ in.}}{10 \text{ ft}} = \frac{10 \text{ in.}}{10 \text{ ft}} \cdot \frac{1 \text{ ft}}{12 \text{ in.}} = \frac{1}{12}$
The scale factor is 1:12. So, the replica is $\frac{1}{12}$ times as tall as the actual sculpture.

11. Sample answer: 1 in. = 12 ft

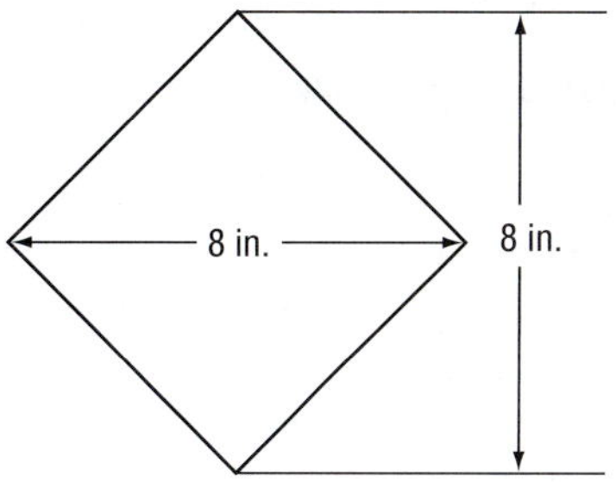

13. about 2.7 h or 2 h and 42 min **15.** 1.61 km

17 a. Scale Model to Rocket

$\frac{1 \text{ in.}}{12 \text{ ft}} = \frac{7 \text{ in.}}{x \text{ ft}}$	Write a proportion.
$1 \cdot x = 12 \cdot 7$	Cross Product Property
$x = 84$	Simplify.

The height of the rocket is 84 feet.

b. Since 12 ft = 144 in., $\frac{1 \text{ in.}}{12 \text{ ft}} = \frac{1 \text{ in.}}{144 \text{ in.}}$.

Scale Model to Rocket

$\frac{1 \text{ in.}}{144 \text{ in.}} = \frac{x \text{ in.}}{70 \text{ in.}}$	Write a proportion.
$1 \cdot 70 = 144 \cdot x$	Cross Product Property
$70 = 144x$	Simplify.
$0.5 \approx x$	Divide each side by 144.

The diameter of the model is about 0.5 inch.

19 Scale Actual Tower to Replica

$\frac{3}{1} = \frac{986 \text{ ft}}{x \text{ ft}}$	Write a proportion.
$3 \cdot x = 1 \cdot 986$	Cross Product Property
$3x = 986$	Simplify.
$x \approx 329$	Divide each side by 3.

The height of the ride is about 329 feet.

21. Felix; sample answer: The ratio of the actual high school to the replica is $\frac{75}{1.5}$ or 50:1. **23.** The first drawing will be larger. The second drawing will be $\frac{1}{6}$ the size of the first drawing, so the scale factor is 1:6. **25.** Both can be written as ratios comparing lengths. A scale factor must have the same unit of measure for both measurements. **27.** D **29.** B **31.** 12 **33.** 17.5 **35.** 29.3 **37.** 3.5 **39.** 10.5 **41.** 8 **43.** $JK = \sqrt{10}$, $KL = \sqrt{10}$, $JL = \sqrt{20}$, $XY = \sqrt{10}$, $YZ = \sqrt{10}$, and $XZ = \sqrt{20}$. Each pair of corresponding sides has the same measure so they are congruent. $\triangle JKL \cong \triangle XYZ$ by SSS. **45.** 8 **47.** 48 **49.** $3\sqrt{77}$

Pages 518–522 Chapter 7 Study Guide and Review

1. j **3.** g **5.** d **7.** b **9.** 49 **11.** 10 or −10 **13.** 120 in. **15.** No, the polygons are not similar because the corresponding sides are not proportional. **17.** 16.5 **19.** Yes, $\triangle ABE \sim \triangle ADC$ by the SAS ~ Thm. **21.** No, the triangles are not similar because not all corresponding angles are congruent. **23.** 34.2 feet **25.** 22.5 **27.** 6 **29.** 633 mi **31.** enlargement; 2 **33.** 10.5 inches **35.** 95.8 mi

Chapter 8 Right Angles and Trigonometry

Page 529 Chapter 8 Get Ready

1. $4\sqrt{7}$ **3.** $10\sqrt{3}$ **5.** $\frac{3}{4}$ **7.** 10 **9.** $x = 17.2$ in., 68.8 in. of trim

11.

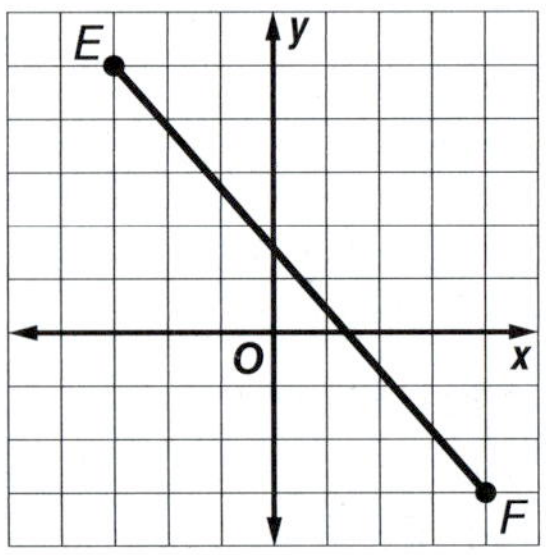

Pages 531–539 Lesson 8-1

1. 10 **3.** $10\sqrt{6}$ or 24.5 **5.** $x = 6$; $y = 3\sqrt{5} \approx 6.7$; $z = 6\sqrt{5} \approx 13.4$ **7.** 18 ft 11 in.

9

$x = \sqrt{ab}$	Definition of geometric mean
$= \sqrt{16 \cdot 25}$	$a = 16$ and $b = 25$
$= \sqrt{(4 \cdot 4) \cdot (5 \cdot 5)}$	Factor.
$= 4 \cdot 5$ or 20	Simplify.

11. $12\sqrt{6} \approx 29.4$ **13.** $3\sqrt{3} \approx 5.2$ **15.** $\triangle WXY \sim \triangle XZY \sim \triangle XZW$ **17.** $\triangle HGF \sim \triangle HIG \sim \triangle GIF$

19

$17 = \sqrt{6 \cdot (y - 6)}$	Geometric Mean (Altitude) Theorem
$289 = 6 \cdot (y - 6)$	Square each side.
$\frac{289}{6} = y - 6$	Divide each side by 6.
$54\frac{1}{6} = y$	Add 6 to each side.
$54.2 \approx y$	Write as a decimal.

$x = \sqrt{6 \cdot y}$	Geometric Mean (Leg) Theorem
$= \sqrt{6 \cdot 54\frac{1}{6}}$	$y = 54\frac{1}{6}$
$= \sqrt{325}$	Multiply.
$= 5\sqrt{13}$	Simplify.
≈ 18.0	Use a calculator.

$z = \sqrt{(y - 6) \cdot y}$	Geometric Mean (Leg) Theorem
$= \sqrt{\left(54\frac{1}{6} - 6\right) \cdot 54\frac{1}{6}}$	$y = 54\frac{1}{6}$
$= \sqrt{48\frac{1}{6} \cdot 54\frac{1}{6}}$	Subtract.
≈ 51.1	Use a calculator.

21. $x = 4.7$; $y = 1.8$; $z = 13.1$ **23.** $x = 24\sqrt{2} \approx 33.9$; $y = 8\sqrt{2} \approx 11.3$; $z = 32$ **25.** 161.8 ft **27.** $\frac{\sqrt{30}}{7}$ or 0.8 **29.** $x = \frac{3\sqrt{3}}{2} \approx 2.6$; $y = \frac{3}{2}$; $z = 3$ **31.** 11 **33.** 3.7 ft **35.** 5 **37.** 4

39. Given: $\angle PQR$ is a right angle. $\overline{QS}$ is an altitude of $\triangle PQR$.

Prove: $\triangle PSQ \sim \triangle PQR$
$\triangle PQR \sim \triangle QSR$
$\triangle PSQ \sim \triangle QSR$

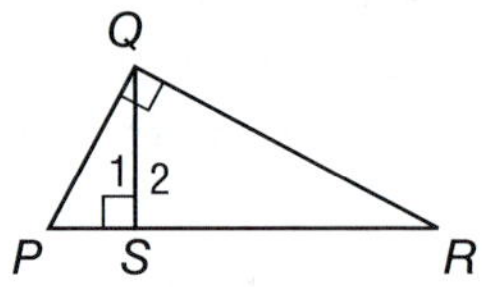

Proof:

Statements (Reasons)

1. $\angle PQR$ is a right angle. $\overline{QS}$ is an altitude of $\triangle PQR$. (Given)
2. $\overline{QS} \perp \overline{RP}$ (Definition of altitude)
3. $\angle 1$ and $\angle 2$ are right angles. (Definition of perpendicular lines)
4. $\angle 1 \cong \angle PQR$, $\angle 2 \cong \angle PQR$ (All right ∠s are ≅.)
5. $\angle P \cong \angle P$, $\angle R \cong \angle R$ (Congruence of angles is reflexive.)
6. $\triangle PSQ \sim \triangle PQR$, $\triangle PQR \sim \triangle QSR$ (AA Similarity Statements 4 and 5)
7. $\triangle PSQ \sim \triangle QSR$ (Similarity of triangles is transitive.)

41. Given: $\angle ADC$ is a right angle. $\overline{DB}$ is an altitude of $\triangle ADC$.

Prove: $\frac{AB}{AD} = \frac{AD}{AC}$
$\frac{BC}{DC} = \frac{DC}{AC}$

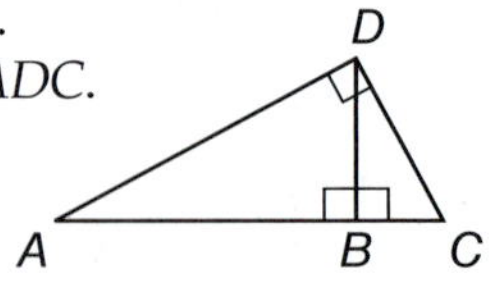

Proof:

Statements (Reasons)

1. $\angle ADC$ is a right angle. $\overline{DB}$ is an altitude of $\triangle ADC$. (Given)
2. $\triangle ADC$ is a right triangle. (Definition of right triangle)
3. $\triangle ABD \sim \triangle ADC$, $\triangle DBC \sim \triangle ADC$ (If the altitude is drawn from the vertex of the rt. ∠ to the hypotenuse of a rt. △, then the 2 △s formed are similar to the given △ and to each other.)
4. $\frac{AB}{AD} = \frac{AD}{AC}$, $\frac{BC}{DC} = \frac{DC}{AC}$ (Def. of similar triangles)

43

$x = \sqrt{ab}$	Definition of geometric mean
$= \sqrt{7 \cdot 12}$	$a = 7$ and $b = 12$

$= \sqrt{84}$ Multiply.

≈ 9 Simplify.

The average rate of return is about 9%.

45 Sample answer: The geometric mean of two consecutive integers is $\sqrt{x(x+1)}$ and the average of two consecutive integers is $\frac{x+(x+1)}{2}$.

$$\sqrt{x(x+1)} \stackrel{?}{=} \frac{x+(x+1)}{2}$$
$$\sqrt{x^2+x} \stackrel{?}{=} \frac{2x+1}{2}$$
$$\sqrt{x^2+x} \stackrel{?}{=} x+\frac{1}{2}$$
$$x^2+x \stackrel{?}{=} \left(x+\frac{1}{2}\right)^2$$
$$x^2+x \stackrel{?}{=} x^2+x+\frac{1}{4}$$
$$0 \neq \frac{1}{4}$$

If you set the two expressions equal to each other, the equation has no solution. So, the statement is never true.

47. Sometimes; sample answer: When the product of the two integers is a perfect square, the geometric mean will be a positive integer. **49.** Neither; sample answer: On the similar triangles created by the altitude, the leg that is x units long on the smaller triangle corresponds with the leg that is 8 units long on the larger triangle, so the correct proportion is $\frac{4}{x} = \frac{x}{8}$ and x is about 5.7. **51.** Sample answer: 9 and 4, 8 and 8; In order for two whole numbers to result in a whole-number geometric mean, their product must be a perfect square. **53.** Sample answer: Both the arithmetic and the geometric mean calculate a value between two given numbers. The arithmetic mean of two numbers a and b is $\frac{a+b}{2}$, and the geometric mean of two numbers a and b is $\sqrt{ab}$. The two means will be equal when $a = b$.

Justification:

$$\frac{a+b}{2} = \sqrt{ab}$$
$$\left(\frac{a+b}{2}\right)^2 = ab$$
$$\frac{(a+b)^2}{4} = ab$$
$$(a+b)^2 = 4ab$$
$$a^2+2ab+b^2 = 4ab$$
$$a^2-2ab+b^2 = 0$$
$$(a-b)^2 = 0$$
$$a-b = 0$$
$$a = b$$

55. 10 **57.** C

59. $\frac{AB}{DE} = \frac{BC}{EF} = \frac{AC}{DF} = 3$, so $\triangle ABC \sim \triangle DEF$ by SSS Similarity.

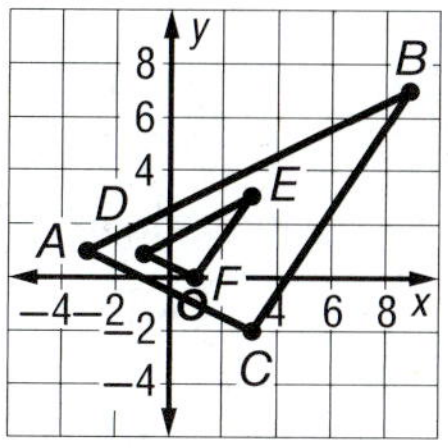

61. $\angle M \cong \angle Q$ and $\frac{PM}{SQ} = \frac{MN}{QR} = \frac{1}{2}$, so $\triangle MNP \sim \triangle QRS$ by SAS Similarity.

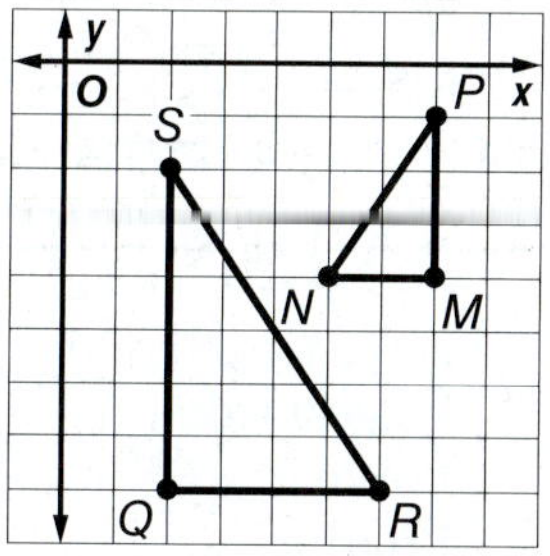

63. octagon **65.** $x = 34, y = \pm 5$ **67.** hexagonal pyramid; base: $ABCDEF$, faces: $ABCDEF$, AGF, FGE, EGD, DGC, CGB, BGA; edges: $\overline{AF}$, $\overline{FE}$, $\overline{ED}$, $\overline{DC}$, $\overline{CB}$, $\overline{BA}$, $\overline{AG}$, $\overline{FG}$, $\overline{EG}$, $\overline{DG}$, $\overline{CG}$, and $\overline{BG}$; vertices: A, B, C, D, E, F, and G **69.** cone; base: circle Q; vertex: P

71. $\frac{16\sqrt{3}}{3}$ **73.** $\frac{3\sqrt{55}}{11}$

Pages 541–549 Lesson 8-2

1. 12

3 The side opposite the right angle is the hypotenuse, so $c = 16$.

$a^2 + b^2 = c^2$	Pythagorean Theorem
$4^2 + x^2 = 16^2$	$a = 4$ and $b = x$
$16 + x^2 = 256$	Simplify.
$x^2 = 240$	Subtract 16 from each side.
$x = \sqrt{240}$	Take the positive square root of each side.
$x = 4\sqrt{15}$	Simplify.
$x \approx 15.5$	Use a calculator.

5. D **7.** yes; obtuse
$26^2 \stackrel{?}{=} 16^2 + 18^2$
$676 > 256 + 324$
9. 20 **11.** $\sqrt{21} \approx 4.6$

13. $\frac{\sqrt{10}}{5} \approx 0.6$

15 16 and 30 are both multiples of 2: $16 = 2 \cdot 8$ and $30 = 2 \cdot 15$. Since 8, 15, 17 is a Pythagorean triple, the missing hypotenuse is $2 \cdot 17$ or 34.

17. 70 **19.** about 3 ft

21. yes; obtuse
$21^2 \stackrel{?}{=} 7^2 + 15^2$
$441 > 49 + 225$

23. yes; right
$20.5^2 \stackrel{?}{=} 4.5^2 + 20^2$
$420.25 = 20.25 + 400$

25. yes; acute
$7.6^2 \stackrel{?}{=} 4.2^2 + 6.4^2$
$57.76 < 17.64 + 40.96$

27. 15 **29.** $4\sqrt{6} \approx 9.8$

31. acute; $XY = \sqrt{29}$, $YZ = \sqrt{20}$, $XZ = \sqrt{13}$; $\sqrt{29}^2 + \sqrt{20}^2 > \sqrt{13}^2$ **33.** right; $XY = 6$, $YZ = 10$, $XZ = 8$; $6^2 + 8^2 = 10^2$

35. Given: $\triangle ABC$ with sides of measure a, b, and c, where $c^2 = a^2 + b^2$

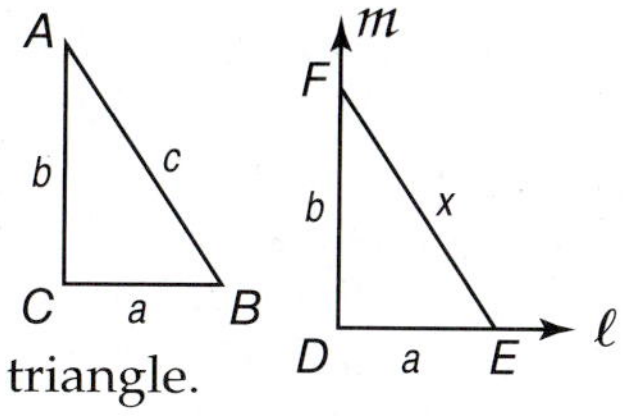

Prove: $\triangle ABC$ is a right triangle.

Proof:

Draw $\overline{DE}$ on line ℓ with measure equal to a. At D, draw line $m \perp \overline{DE}$. Locate point F on m so that $DF = b$. Draw $\overline{FE}$ and call its measure x. Because $\triangle FED$ is a right triangle, $a^2 + b^2 = x^2$. But $a^2 +$

$b^2 = c^2$, so $x^2 = c^2$ or $x = c$. Thus, $\triangle ABC \cong \triangle FED$ by SSS. This means $\angle C \cong \angle D$. Therefore, $\angle C$ must be a right angle, making $\triangle ABC$ a right triangle.

37. Given: In $\triangle ABC$, $c^2 > a^2 + b^2$ where c is the length of the longest side.

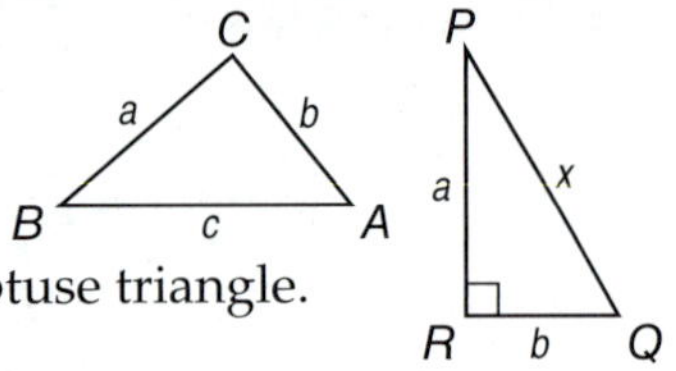

Prove: $\triangle ABC$ is an obtuse triangle.

Proof:

Statements (Reasons)

1. In $\triangle ABC$, $c^2 > a^2 + b^2$ where c is the length of the longest side. In $\triangle PQR$, $\angle R$ is a right angle. (Given)
2. $a^2 + b^2 = x^2$ (Pythagorean Theorem)
3. $c^2 > x^2$ (Substitution Property)
4. $c > x$ (A property of square roots)
5. $m\angle R = 90°$ (Definition of a right angle)
6. $m\angle C > m\angle R$ (Converse of the Hinge Theorem)
7. $m\angle C > 90°$ (Substitution Property of Equality)
8. $\angle C$ is an obtuse angle. (Definition of an obtuse angle)
9. $\triangle ABC$ is an obtuse triangle. (Definition of an obtuse triangle)

39. $P = 36$ units; $A = 60$ square units^2 **41.** 15

43 **Scale** **Width to Length**

$\frac{16}{9} = \frac{41 \text{ in.}}{x \text{ in.}}$ Write a proportion.

$16 \cdot x = 9 \cdot 41$ Cross Product Property

$16x = 369$ Simplify.

$x = \frac{369}{16}$ Divide each side by 16.

The length of the television is about 23 inches.

$a^2 + b^2 = c^2$ Pythagorean Theorem

$\left(\frac{369}{16}\right)^2 + 41^2 = c^2$ $a = \frac{369}{16}$ and $b = 41$

$\sqrt{\left(\frac{369}{16}\right)^2 + 41^2} = c$ Take the positive square root of each side.

$47.0 \approx c$ Use a calculator.

The screen size is about 47 inches.

45 The side opposite the right angle is the hypotenuse, so $c = x$.

$a^2 + b^2 = c^2$ Pythagorean Theorem

$8^2 + (x - 4)^2 = x^2$ $a = 8$ and $b = x - 4$

$64 + x^2 - 8x + 16 = x^2$ Find 8^2 and $(x - 4)^2$.

$-8x + 80 = 0$ Simplify.

$80 = 8x$ Add $8x$ to each side.

$10 = x$ Divide each side by 8.

47. $\frac{1}{2}$ **49.** 5.4 **51.** Right; sample answer: If you double or halve the side lengths, all three sides of the new triangles are proportional to the sides of the original triangle. Using the Side-Side-Side Similarity Theorem, you know that both of the new triangles are similar to the original triangle, so they are both right.

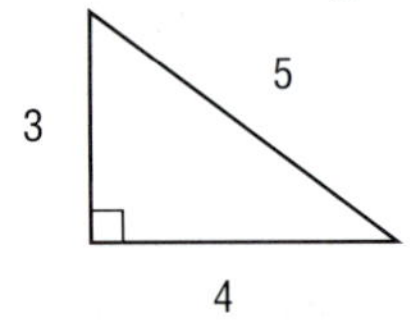

53. D **55.** 250 units **57.** 6 **59.** $6\sqrt{5} \approx 13.4$

61. 1 in. = 2 ft; 6 in. × 4 in. **63.** yes; AA

65. Given: $\overline{FG} \perp \ell$; $\overline{FG}$ is any nonperpendicular segment from F to ℓ.

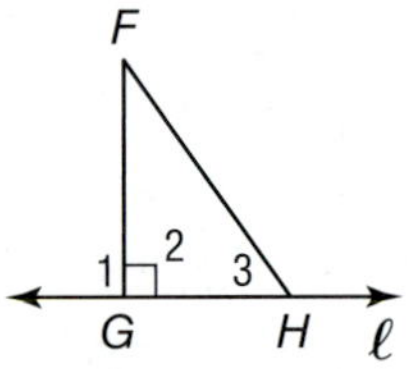

Prove: $FH > FG$

Proof:

Statements (Reasons)

1. $\overline{FG} \perp \ell$ (Given)
2. $\angle 1$ and $\angle 2$ are right angles. ($\perp$ lines form right angles.)
3. $\angle 1 \cong \angle 2$ (All right angles are congruent.)
4. $m\angle 1 = m\angle 2$ (Definition of congruent angles)
5. $m\angle 1 > m\angle 3$ (Exterior Angle Inequality Thm.)
6. $m\angle 2 > m\angle 3$ (Substitution Property.)
7. $FH > FG$ (If an $\angle$ of a $\triangle$ is > a second $\angle$, then the side opposite the greater $\angle$ is longer than the side opposite the lesser $\angle$.)

67. 50 **69.** 40 **71.** $\frac{7\sqrt{5}}{5}$ **73.** $2\sqrt{3}$ **75.** 2

Pages 552–560 Lesson 8-3

1. $5\sqrt{2}$ **3.** 22 **5.** $x = 14$; $y = 7\sqrt{3}$ **7.** Yes; sample answer: The height of the triangle is about $3\frac{1}{2}$ in., so since the diameter of the roll of art is less than the diameter of the opening, it will fit. **9.** $\frac{15\sqrt{2}}{2}$ or $7.5\sqrt{2}$

11 In a 45°-45°-90° triangle, the length of the hypotenuse is $\sqrt{2}$ times the length of a leg.

$h = x\sqrt{2}$ Theorem 8.8

$= 18\sqrt{3} \cdot \sqrt{2}$ Substitution

$= 18\sqrt{6}$ $\sqrt{3} \cdot \sqrt{2} = \sqrt{6}$

13. $20\sqrt{2}$ **15.** $\frac{11\sqrt{2}}{2}$ **17.** $8\sqrt{2}$ or 11.3 cm **19.** $x = 10$; $y = 20$ **21.** $x = \frac{17\sqrt{3}}{2}$; $y = \frac{17}{2}$ **23.** $x = \frac{14\sqrt{3}}{3}$; $y = \frac{28\sqrt{3}}{3}$ **25.** $16\sqrt{3}$ or 27.7 ft **27.** 22.6 ft

29 In a 45°-45°-90° triangle, the length of the hypotenuse is 2 times the length of a leg.

$h = x\sqrt{2}$ Theorem 8.8

$6 = x\sqrt{2}$ Substitution

$\frac{6}{\sqrt{2}} = x$ Divide each side by $\sqrt{2}$.

$\frac{6}{\sqrt{2}} \cdot \frac{\sqrt{2}}{\sqrt{2}} = x$ Rationalize the denominator.

$\frac{6 \cdot \sqrt{2}}{\sqrt{2} \cdot \sqrt{2}} = x$ Multiply.

$\frac{6\sqrt{2}}{2} = x$ $\sqrt{2} \cdot \sqrt{2} = 2$

$3\sqrt{2} = x$ Simplify.

$h = x\sqrt{2}$ Theorem 8.8

$y = 6\sqrt{2}$ Substitution

31. $x = 5$; $y = 10$ **33.** $x = 45$; $y = 12\sqrt{2}$

35 In a 30°-60°-90° triangle, the length of the hypotenuse is 2 times the length of the shorter leg.

$h = 2s$ Theorem 8.9

$= 2(25)$ or 50 Substitution

The zip line's length is about 50 feet.

37. $x = 9\sqrt{2}$; $y = 6\sqrt{3}$; $z = 12\sqrt{3}$
39. 7.5 ft; 10.6 ft; 13.0 ft **41.** (6, 9) **43.** (4, −2)

45 a.

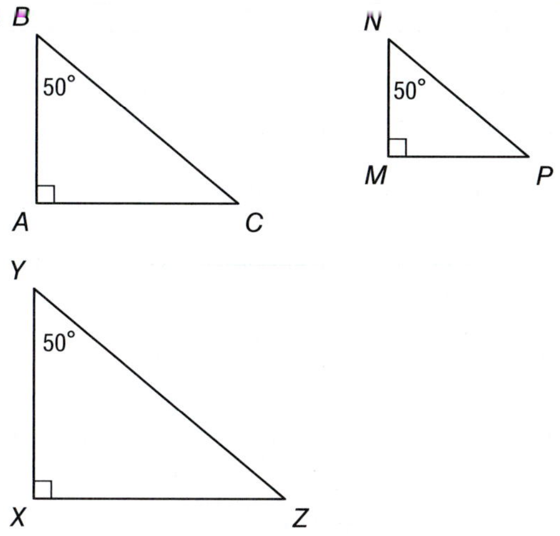

b. Measure the sides of the triangles to the nearest tenth of a centimeter. Find the ratios to the nearest tenth of a centimeter. Sample answer:

Triangle	Length				Ratio	
ABC	*AC*	2.4 cm	*BC*	3.2 cm	$\frac{BC}{AC}$	1.3
MNP	*MP*	1.7 cm	*NP*	2.2 cm	$\frac{NP}{MP}$	1.3
XYZ	*XZ*	3.0 cm	*YZ*	3.9 cm	$\frac{YZ}{XZ}$	1.3

c. Sample answer: In a right triangle with a 50° angle, the ratio of the leg opposite the 50° angle to the hypotenuse will always be the same, 1.3.

47. Sample answer: Let ℓ represent the length. $\ell^2 + w^2 = (2w)^2$; $\ell^2 = 3w^2$; $\ell = w\sqrt{3}$.

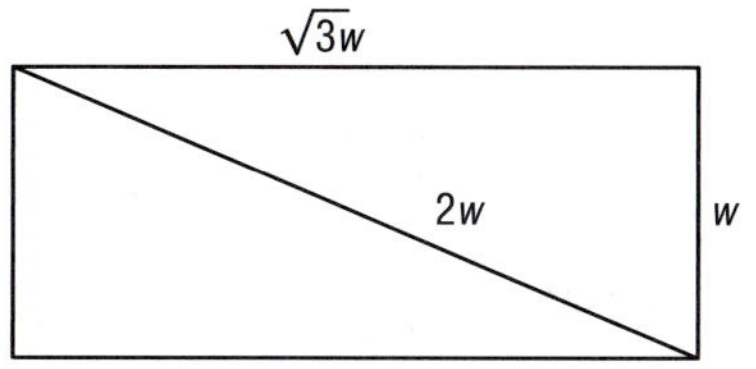

49. 37.9 **51.** C **53.** (−13, −3) **55.** 15 ft **57.** $x \approx 16.9$, $y \approx 22.6$, $z \approx 25.4$ **59.** 36, 90, 54 **61.** 45, 63, 72 **63.** ∠1, ∠4, ∠11 **65.** ∠2, ∠6, ∠9, ∠8, ∠7 **67.** 12.0

Pages 562–571 Lesson 8-4

1. $\frac{16}{20} = 0.80$ **3.** $\frac{12}{20} = 0.60$ **5.** $\frac{16}{20} = 0.80$ **7.** $\frac{\sqrt{3}}{2} \approx 0.87$ **9.** 27.44 **11.** about 1.2 ft **13.** 44.4° **15.** $RS \approx 6.7$; $m\angle R \approx 42$; $m\angle T \approx 48$

17

$$\sin J = \frac{\text{opp}}{\text{hyp}} = \frac{56}{65} \approx 0.86 \qquad \cos J = \frac{\text{adj}}{\text{hyp}} = \frac{33}{65} \approx 0.51$$

$$\tan J = \frac{\text{opp}}{\text{adj}} = \frac{56}{33} \approx 1.70 \qquad \sin L = \frac{\text{opp}}{\text{hyp}} = \frac{33}{65} \approx 0.51$$

$$\cos L = \frac{\text{adj}}{\text{hyp}} = \frac{56}{65} \approx 0.86 \qquad \tan L = \frac{\text{opp}}{\text{adj}} = \frac{33}{56} \approx 0.59$$

19. $\frac{84}{85} = 0.99$; $\frac{13}{85} = 0.15$; $\frac{84}{13} = 6.46$; $\frac{13}{85} = 0.15$; $\frac{84}{85} = 0.99$; $\frac{13}{84} = 0.15$ **21.** $\frac{\sqrt{3}}{2} = 0.87$; $\frac{2\sqrt{2}}{4\sqrt{2}} = 0.50$; $\sqrt{3} = 1.73$; $\frac{2\sqrt{2}}{4\sqrt{2}} = 0.50$; $\frac{\sqrt{3}}{2} = 0.87$; $\frac{\sqrt{3}}{3} = 0.58$ **23.** $\frac{\sqrt{3}}{2} \approx 0.87$ **25.** $\frac{1}{2}$ or 0.5 **27.** $\frac{1}{2}$ or 0.5 **29.** 28.7 **31.** 57.2 **33.** 17.4

35 Let $m\angle A = 55$ and let x be the height of the roller coaster.

$\sin A = \frac{\text{opp}}{\text{hyp}}$	Definition of sine ratio
$\sin 55 = \frac{x}{98}$	Substitution
$98 \cdot \sin 55 = x$	Multiply each side by 98.
$80 \approx x$	Use a calculator.

The height of the roller coaster is about 80 feet.

37. 61.4° **39.** 28.5° **41.** 21.8° **43.** $WX = 15.1$; $XZ = 9.8$; $m\angle W = 33$ **45.** $ST = 30.6$; $m\angle R = 58$; $m\angle T = 32$

47 $JL = \sqrt{[-2-(-2)]^2 + [4-(-3)]^2} = 7$

$KJ = \sqrt{[-2-(-7)]^2 + [-3-(-3)]^2} = 5$

$\tan K = \frac{\text{opp}}{\text{adj}}$	Definition of tangent ratio
$= \frac{7}{5}$	Substitution
$m\angle K = \tan^{-1}\left(\frac{7}{5}\right) \approx 54.5°$	Use a calculator.

49. 51.3° **51.** 13.83 in.; 7.50 in^2 **53.** 8.73 ft; 3.40 ft^2 **55.** 0.92

57 The triangle is isosceles, so two sides measure 32 and the two smaller triangles each have a side that measures x. Let $m\angle A = 54$.

$\cos A = \frac{\text{adj}}{\text{hyp}}$	Definition of cosine ratio
$\cos 54 = \frac{x}{32}$	Substitution
$32 \cdot \cos 54 = x$	Multiply each side by 32.
$18.8 \approx x$	Simplify.
$\sin A = \frac{\text{opp}}{\text{hyp}}$	Definition of sine ratio
$\sin 54 = \frac{y}{32}$	Substitution
$32 \cdot \sin 54 = y$	Multiply each side by 32.
$25.9 \approx y$	Simplify.

59. $x = 9.2$; $y = 11.7$

61a.

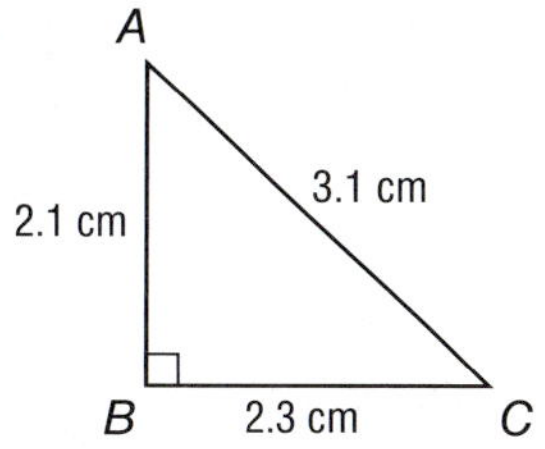

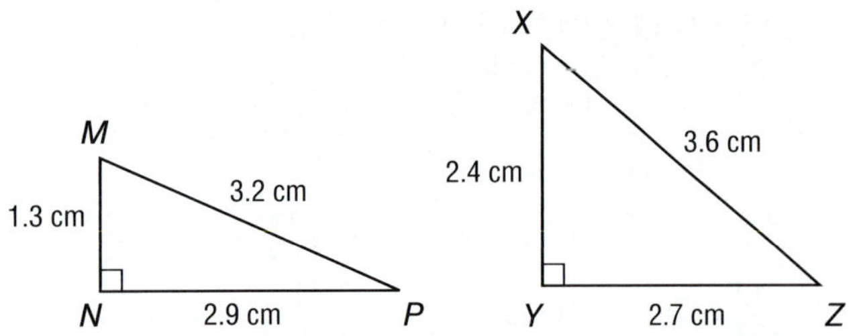

61b. Sample answer:

Triangle	Trigonometric Ratios				Sum of Ratios Squared	
ABC	cos *A*	0.677	sin *A*	0.742	$(\cos A)^2 + (\sin A)^2$	1
	cos *C*	0.742	sin *C*	0.677	$(\cos C)^2 + (\sin C)^2$	1
MNP	cos *M*	0.406	sin *M*	0.906	$(\cos M)^2 + (\sin M)^2$	1
	cos *P*	0.906	sin *P*	0.406	$(\cos P)^2 + (\sin P)^2$	1
XYZ	cos *X*	0.667	sin *X*	0.75	$(\cos X)^2 + (\sin X)^2$	1
	cos *Z*	0.75	sin *Z*	0.667	$(\cos Z)^2 + (\sin Z)^2$	1

61c. Sample answer: The sum of the cosine squared and the sine squared of an acute angle of a right triangle is 1. **61d.** $(\sin x)^2 + (\cos x)^2 = 1$

61e. Sample answer:

$(\sin A)^2 + (\cos A)^2 \stackrel{?}{=} 1$ Conjecture

$\left(\frac{y}{r}\right)^2 + \left(\frac{x}{r}\right)^2 \stackrel{?}{=} 1$ $\sin A = \frac{y}{r}, \cos A = \frac{x}{r}$

$\frac{y^2}{r^2} + \frac{x^2}{r^2} \stackrel{?}{=} 1$ Simplify.

$\frac{y^2 + x^2}{r^2} \stackrel{?}{=} 1$ Combine fractions with like denominators.

$\frac{r^2}{r^2} \stackrel{?}{=} 1$ Pythagorean Theorem

$1 = 1$ Simplify.

63. Sample answer: Yes; since the values of sine and cosine are both calculated by dividing one of the legs of a right triangle by the hypotenuse, and the hypotenuse is always the longest side of a right triangle, the values will always be less than 1. You will always be dividing the smaller number by the larger number. **65.** Sample answer: To find the measure of an acute angle of a right triangle, you can find the ratio of the leg opposite the angle to the hypotenuse and use a calculator to find the inverse sine of the ratio, you can find the ratio of the leg adjacent to the angle to the hypotenuse and use a calculator to find the inverse cosine of the ratio, or you can find the ratio of the leg opposite the angle to the leg adjacent to the angle and use a calculator to find the inverse tangent of the ratio. **67.** H **69.** D **71.** $x = 7\sqrt{2}; y = 14$

73. yes; right
$17^2 \stackrel{?}{=} 8^2 + 15^2$
$289 = 64 + 225$

75. yes; obtuse
$35^2 \stackrel{?}{=} 30^2 + 13^2$
$1225 > 900 + 169$

77. no; $8.6 > 3.2 + 5.3$ **79.** $5\frac{7}{15}$ h or 5 h 28 min
81. $x = 1, y = \frac{3}{2}$ **83.** 260 **85.** 18.9 **87.** 157.1

Pages 574–581 Lesson 8-5

1. 27.5 ft **3.** 14.2 ft

5 Make a sketch.

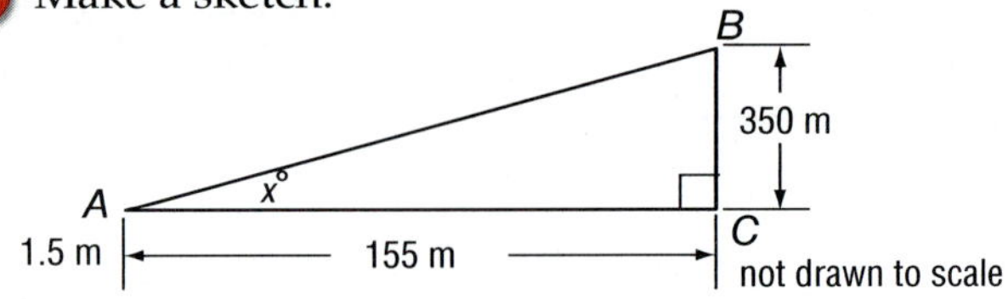

$\tan A = \frac{BC}{AC}$ $\quad$ tan = $\frac{\text{opposite}}{\text{adjacent}}$

$\tan x° = \frac{348.5}{155}$ $\quad$ $m\angle A = x$, $BC = 350 - 1.5$ or 348.5, $AC = 155$

$x = \tan^{-1}\left(\frac{348.5}{155}\right)$ $\quad$ Solve for *x*.

$x \approx 66.0$ $\quad$ Use a calculator.

The angle of elevation is about 66°.

7. 14.8°

9 Make a sketch.

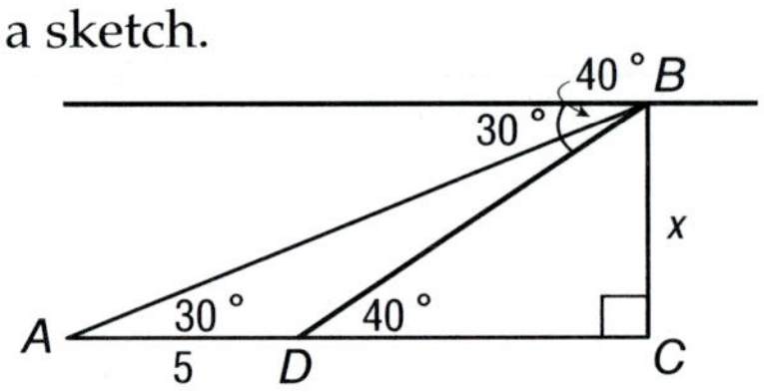

$\tan 30 = \frac{x}{5} + DC$ $\quad$ tan = $\frac{\text{opposite}}{\text{adjacent}}$

$(5 + DC) \tan 30 = x$ $\quad$ Solve for *x*.

$\tan 40 = \frac{x}{DC}$ $\quad$ tan = $\frac{\text{opposite}}{\text{adjacent}}$

$DC \tan 40 = x$ $\quad$ Solve for *x*.

$DC \tan 40 = (5 + DC) \tan 30$ $\quad$ Substitution

$DC \tan 40 = 5 \tan 30 + DC \tan 30$ $\quad$ Distributive Property

$DC \tan 40 - DC \tan 30 = 5 \tan 30$ $\quad$ Subtract *DC* tan 30 from each side.

$DC(\tan 40 - \tan 30) = 5 \tan 30$ $\quad$ Factor *DC*.

$DC = \frac{5 \tan 30}{\tan 40 - \tan 30}$ $\quad$ Divide each side by tan 40 − tan 30.

$DC \approx 11.0$ $\quad$ Use a calculator.

$\tan A = \frac{BC}{AC}$ $\quad$ tan = $\frac{\text{opposite}}{\text{adjacent}}$

$\tan 30 = \frac{x}{16.0}$ $\quad$ $A = 30$, $BC = x$, $AC = 5 + 11.0$ or 16.0

$16.0 \tan 30 = x$ $\quad$ Multiply each side by 16.0.

$9.3 \approx x$ $\quad$ Use a calculator.

The platform is about 9.3 feet high.

11. about 1309 ft **13.** 16.6° **15.** 240.2 ft

17 Make a sketch.

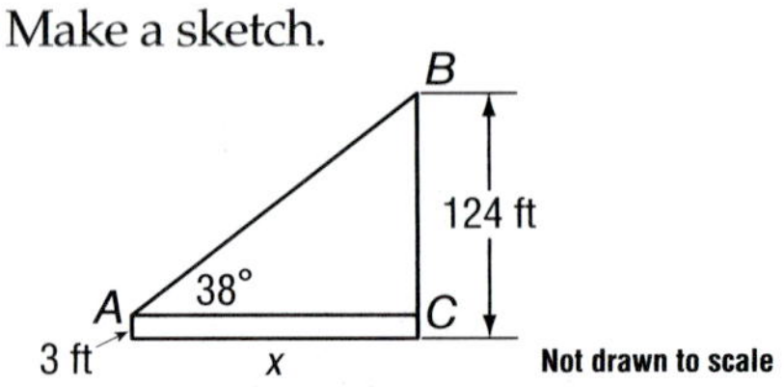

$\tan A = \frac{BC}{AC}$ — $\tan = \frac{\text{opposite}}{\text{adjacent}}$

$\tan 38° = \frac{121}{x}$ — $m\angle A = 38$, $BC = 124 - 3$ or 121, $AC = x$

$x = \frac{121}{\tan 38°}$ — Solve for x.

$x \approx 154.9$ — Use a calculator.

You should place the tripod about 154.9 feet from the monument.

19a. 74.8° **19b.** 110.1 m

21 Make two sketches.

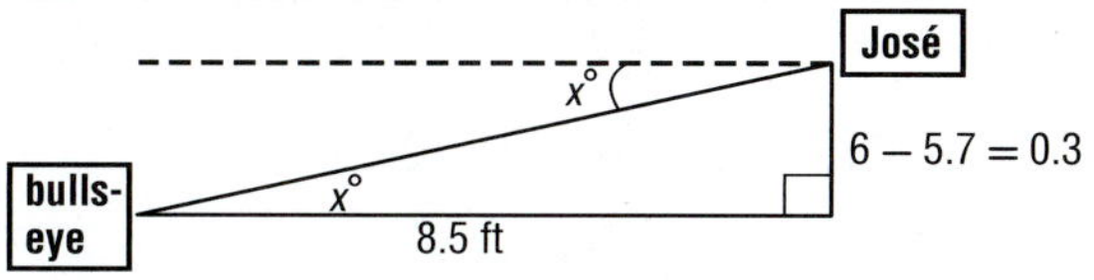

$\tan x° = \frac{0.3}{8.5}$ — $\tan = \frac{\text{opp}}{\text{adj}}$

$x = \tan^{-1}\left(\frac{0.3}{8.5}\right)$ — Solve for x.

$x \approx 2.02$ — Use a calculator.

José throws at an angle of depression of 2.02°.

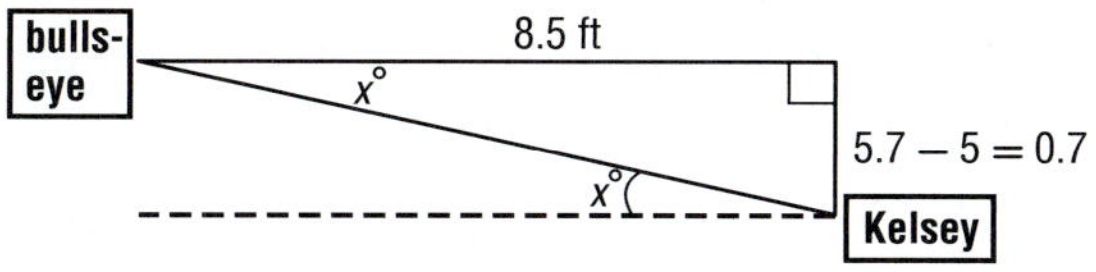

$\tan x° = \frac{0.7}{8.5}$ — $\tan = \frac{\text{opp}}{\text{adj}}$

$x = \tan^{-1}\left(\frac{0.7}{8.5}\right)$ — Solve for x.

$x \approx 4.71$ — Use a calculator.

Kelsey throws at an angle of elevation of 4.71°.

23. Rodrigo; sample answer: Since your horizontal line of sight is parallel to the other person's horizontal line of sight, the angles of elevation and depression are congruent according to the Alternate Interior Angles Theorem. **25.** True; sample answer: As a person moves closer to an object, the horizontal distance decreases, but the height of the object is constant. The tangent ratio will increase, and therefore the measure of the angle also increases. **27.** Sample answer: If you sight something with a 45° angle of elevation, you don't have to use trigonometry to determine the height of the object. Since the legs of a 45°-45°-90° are congruent, the height of the object will be the same as your horizontal distance from the object. **29.** 6500 ft **31.** B

33. $\frac{20}{15} = 1.33$ **35.** $\frac{15}{20} = 0.75$ **37.** $\frac{20}{25} = 0.80$

39. Given: $\overline{CD}$ bisects $\angle ACB$. By construction, $\overline{AE} \parallel \overline{CD}$.

Prove: $\frac{AD}{DB} = \frac{AC}{BC}$

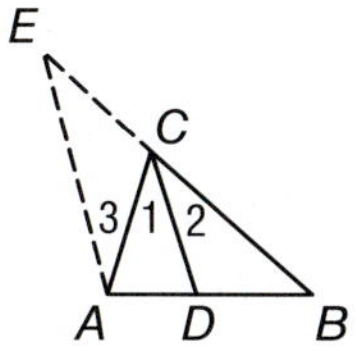

Proof:

Statements (Reasons)

1. $\overline{CD}$ bisects $\angle ACB$. By construction, $\overline{AE} \parallel \overline{CD}$. (Given)
2. $\frac{AD}{DB} = \frac{EC}{BC}$ (Triangle Proportionality Theorem)
3. $\angle 1 \cong \angle 2$ (Definition of Angle Bisector)
4. $\angle 3 \cong \angle 1$ (Alternate Interior Angle Theorem)
5. $\angle 2 \cong \angle E$ (Corresponding Angle Postulate)
6. $\angle 3 \cong \angle E$ (Transitive Prop.)
7. $\overline{EC} \cong \overline{AC}$ (Isosceles △ Th.)
8. $EC = AC$ (Def. of congruent segments)
9. $\frac{AD}{DB} = \frac{AC}{BC}$ (Substitution)

41. (7, 4) **43.** (−5, 6) **45.** 2 **47.** 2.1

Pages 582–591 Lesson 8-6

1. 6.1

3

$\frac{\sin A}{a} = \frac{\sin B}{b}$ — Law of Sines

$\frac{\sin 60°}{70} = \frac{\sin x°}{73}$ — $m\angle A = 60$, $a = 70$, $m\angle B = x$, $b = 73$

$73 \sin 60° = 70 \sin x°$ — Cross Products Property

$\frac{73 \sin 60°}{70} = \sin x°$ — Divide each side by 70.

$\sin^{-1}\left(\frac{73 \sin 60°}{70}\right) = x$ — Use the inverse sine ratio.

$x \approx 65$ — Use a calculator.

5. 8.3 **7.** 47.1 ft **9.** $m\angle M \approx 27$, $m\angle N \approx 42$, $MP \approx 35.8$ **11.** $m\angle D \approx 73$, $m\angle E \approx 62$, $m\angle F \approx 45$ **13.** 4.1 **15.** 22.8 **17.** 45 **19.** 29 **21.** 2.8 in.

23

$a^2 = b^2 + c^2 - 2bc \cos A$ — Law of Cosines

$x^2 = 1.2^2 + 3.0^2 - 2(1.2)(3.0) \cos 123°$ — Substitution

$x^2 = 10.44 - 7.2 \cos 123°$ — Simplify.

$x = \sqrt{10.44 - 7.2 \cos 123°}$ — Take the square root of each side.

$x \approx 3.8$ — Use a calculator.

25. 98 **27.** 112 **29.** 126.2 ft **31.** $m\angle B = 34$, $AB \approx 9.5$, $CA \approx 6.7$

33

$j^2 = k^2 + \ell^2 - 2k\ell \cos J$ — Law of Cosines

$29.7^2 = 30.0^2 + 24.6^2 - 2(30.0)(24.6) \cos J$ — $j = 29.7$, $k = 30.0$, $\ell = 24.6$

$882.09 = 1505.16 - 1476 \cos J$ — Simplify.

$-623.07 = -1476 \cos J$ — Subtract 1505.16 from each side.

$\frac{-623.07}{-1476} = \cos J$ — Divide each side by −1476.

$m\angle J = \cos^{-1}\left(\frac{623.07}{1476}\right)$ — Use the inverse cosine ratio.

$m\angle J \approx 65$ — Use a calculator.

$\frac{\sin J}{j} = \frac{\sin K}{k}$ — Law of Sines

$\frac{\sin 65°}{29.7} = \frac{\sin K}{30.0}$ — $m\angle J \approx 65$, $j = 29.7$, and $k = 30.0$

$30.0 \sin 65° = 29.7 \sin K$ — Cross Products Property

$\frac{30.0 \sin 65°}{29.7} = \sin K$ — Divide each side by 29.7.

$m\angle K = \sin^{-1}\left(\frac{30.0 \sin 65°}{29.7}\right)$ — Use the inverse sine ratio.

$m\angle K \approx 66$ — Use a calculator.

By the Triangle Angle Sum Theorem, $m\angle L \approx 180 - (65 + 66)$ or 49.

35. $m\angle G \approx 75$, $m\angle J \approx 71$, $HJ \approx 20.3$ **37.** $m\angle P \approx 35$, $m\angle R \approx 75$, $RP \approx 14.6$ **39.** $m\angle C \approx 23$, $m\angle D \approx 67$,

$m\angle E \approx 90$ **41.** $m\angle A = 35, AB \approx 7.5, BC \approx 9.8$
43. 82.4 ft **45a.** Def. of sine **45b.** Mult. Prop.
45c. Subs. **45d.** Div. Prop. **47.** 24.3

49

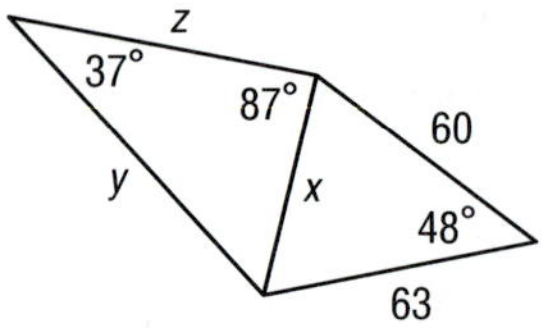

$a^2 = b^2 + c^2 - 2bc \cos A$	Law of Cosines
$x^2 = 60^2 + 63^2 - 2(60)(63) \cos 48°$	Substitution
$x^2 = 7569 - 7560 \cos 48°$	Simplify.
$x = \sqrt{7569 - 7560 \cos 48°}$	Take the square root of each side.
$x \approx 50.1$	Use a calculator.
$\frac{\sin A}{a} = \frac{\sin B}{b}$	Law of Sines
$\frac{\sin 37°}{50.1} = \frac{\sin 87°}{y}$	$m\angle A = 37°, a = x = 50.1, m\angle B = 87°, b = y$
$y \sin 37° = 50.1 \sin 87°$	Cross Products Property
$y = \frac{50.1 \sin 87°}{\sin 37°}$	Divide each side by sin 37°.
$y \approx 83.1$	Use a calculator.

By the Triangle Angle Sum Theorem, the angle opposite side z measures $180 - (37 + 87)$ or 56.

$\frac{\sin A}{a} = \frac{\sin B}{b}$	Law of Sines
$\frac{\sin 37°}{50.1} = \frac{\sin 56°}{z}$	$m\angle A = 37°, a = x = 50.1, m\angle B = 87°, b = z$
$z \sin 37° = 50.1 \sin 56°$	Cross Products Property
$z = \frac{50.1 \sin 56°}{\sin 37°}$	Divide each side by sin 37°.
$z \approx 69.0$	Use a calculator.

$$\text{perimeter} = 60 + 63 + y + z$$
$$\approx 60 + 63 + 83.1 + 69.0 \text{ or } 275.1$$

51

$a^2 = b^2 + c^2 - 2bc \cos A$	Law of Cosines
$x^2 = 8^2 + 6^2 - 2(8)(6) \cos 71.8°$	Substitution
$x^2 = 100 - 96 \cos 71.8°$	Simplify.
$x = \sqrt{100 - 96 \cos 71.8°}$	Take the square root of each side.
$x \approx 8.4$ in.	Use a calculator.

53a.

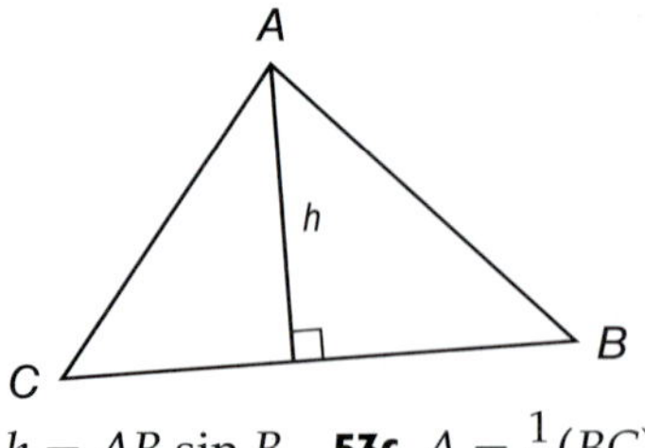

53b. $h = AB \sin B$ **53c.** $A = \frac{1}{2}(BC)(AB \sin B)$

53d. 57.2 units^2 **53e.** $A = \frac{1}{2}(BC)(CA \sin C)$ **55.** 68.9°

57a. Sample answer:

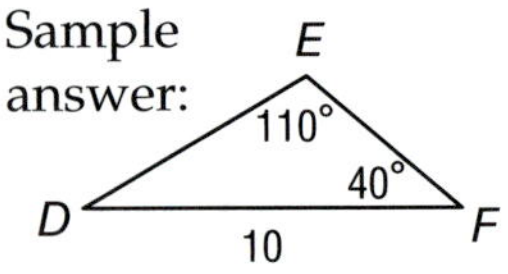

57b. Sample answer:

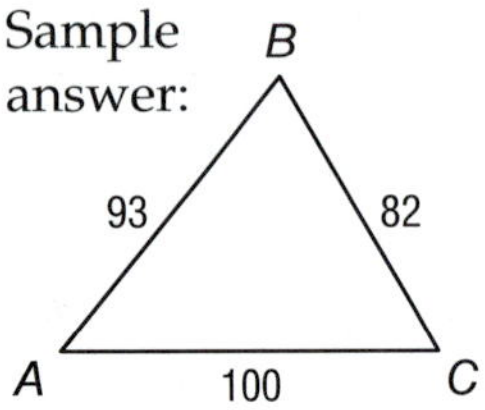

59. A **61.** 329.7 **63.** 269.6 ft **65.** 45 **67.** Never; sample answer: Since an equilateral triangle has three congruent sides and a scalene triangle has three non-congruent sides, the ratios of the three pairs of sides can never be equal. Therefore, an equilateral triangle and a scalene triangle can never be similar.
69. $Q(a, a), P(a, 0)$ **71.** $\sqrt{80} \approx 8.9$ **73.** $\sqrt{72} \approx 85$

Pages 593–600 Lesson 8-7

1. $\langle 5, 5 \rangle$ **3.** 6 units; 0°
5.

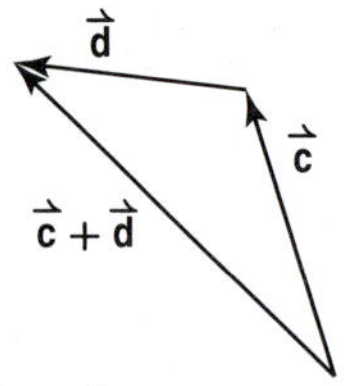

7. The velocity is 109.9 m/h at an angle of 7.3° with the ground. **9.** $\langle 5, 0 \rangle$

11

$\overrightarrow{KJ} = \langle x_2 - x_1, y_2 - y_1 \rangle$	Component form of vector
$= \langle -7 - (-1), -5 - (-2) \rangle$	$x_1 = -1, x_2 = -7, y_1 = -2, y_2 = -5$
$= \langle -6, -3 \rangle$	Simplify.

13. $\langle -3, -6 \rangle$ **15.** 6 units, 270° **17.** $5\sqrt{2} \approx 7.1$ units, 45°
19. 5 units, 53.1°

21.

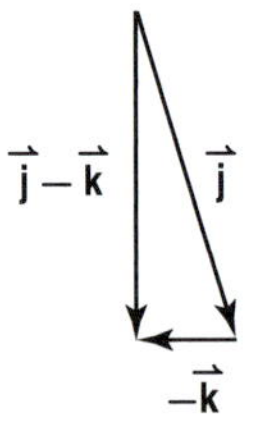

23.

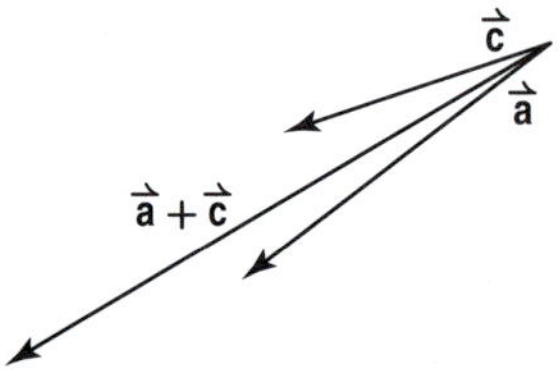

25.

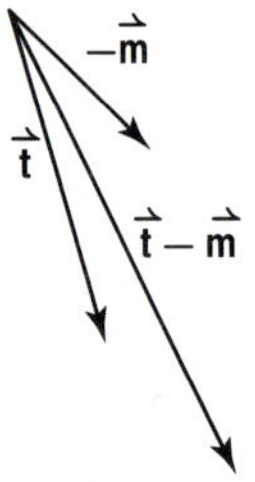

27a. 7.8 m/s, 4.4°
27b. about 12.8 seconds
29. $\langle -2, -10 \rangle$
31. $\langle a + c, b + d \rangle$
33. $\langle -p, 2r \rangle$

35

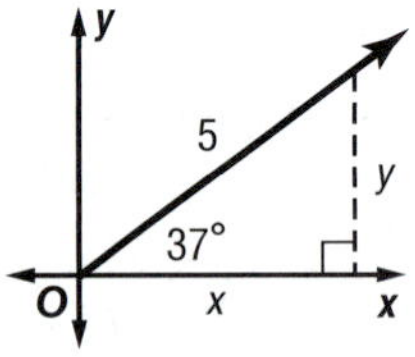

$\cos 37° = \frac{x}{5}$	$\cos = \frac{\text{adj}}{\text{hyp}}$
$5 \cos 37 = y$	Multiply each side by 5.
$4.0 \approx x$	Simplify.
$\sin 37° = \frac{y}{5}$	$\sin = \frac{\text{opp}}{\text{hyp}}$
$5 \sin 37 = y$	Multiply each side by 5.
$3.0 \approx y$	Simplify.

The vector in component form is $\langle 4, 3 \rangle$.
37. $\langle -1.7, 6.8 \rangle$

39 If the initial location of the firework is at (0, 0), then when it has traveled 70 feet horizontally and 400 feet vertically, it is at (70, 400).

$\langle x_2 - x_1, y_2 - y_1 \rangle = \langle 70 - 0, 400 - 0 \rangle$	Component form of vector
$= \langle 70, 400 \rangle$	Simplify.

41a. Sample answer:

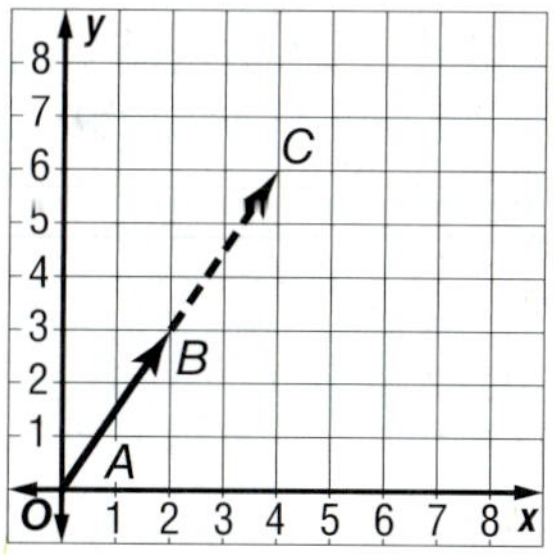

41b. Sample answer:

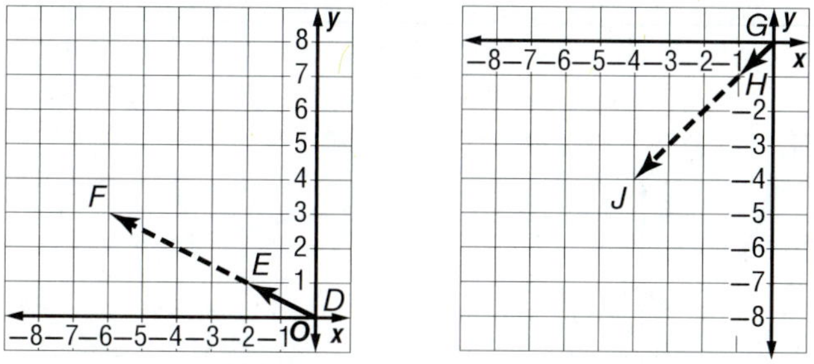

41c. Sample answer:

Coordinates				Scalar Quantity
B	(2, 3)	*C*	(4, 6)	2
E	(−2, 1)	*F*	(−6, 3)	3
H	(−1, −1)	*J*	(−4, −4)	4

41d. Sample answer: The coordinates of the final point of the vector that has been multiplied by a scalar quantity are the coordinates of the final point of the initial vector multiplied by the scalar quantity.

41e. $k(x, y)$ or (kx, ky) **43.** $\sqrt{20} \approx 4.5$ m/s, 26.6°

45a.

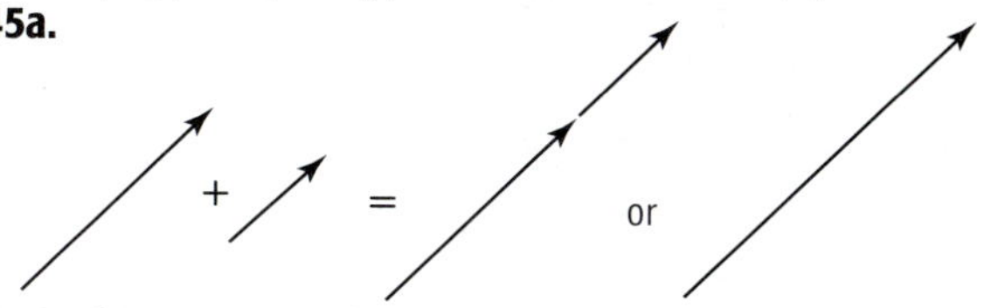

Sample answer: The sum of the two vectors is also parallel to both of the original vectors.

45b.

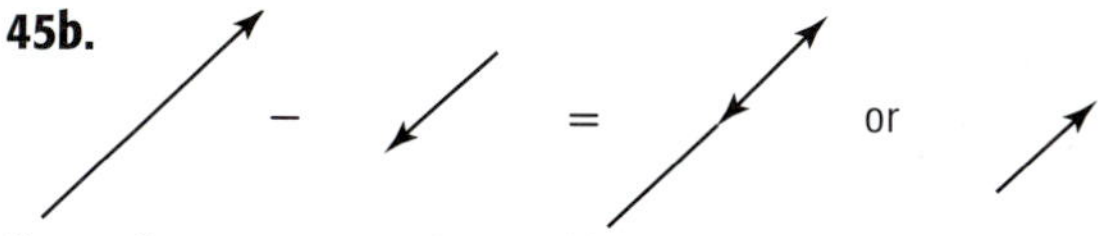

Sample answer: The difference of the two vectors is also parallel to both of the original vectors.

47a. $\langle 2, 3\rangle + \langle 5, -1\rangle = \langle 7, 2\rangle$ **47b.** $\sqrt{7^2 + 2^2} = \sqrt{53}$ or 7.3 mi **49.** J **51.** 33.1 **53.** 117.6 **55.** 60 **57.** 120 **59.** 60 **61.** $\triangle 1 \cong \triangle 10$, $\triangle 2 \cong \triangle 9$, $\triangle 3 \cong \triangle 8$, $\triangle 4 \cong \triangle 7$, $\triangle 5 \cong \triangle 6$ **63.** $\triangle$s 1, 5, 6, and 11, $\triangle$s 3, 8, 10, and 12, $\triangle$s 2, 4, 7, and 9

Pages 518–522 Chapter 8 Study Guide and Review

1. false, geometric **3.** false, sum. **5.** true **7.** true **9.** false, Law of Cosines **11.** 6 **13.** $\frac{8}{3}$ **15.** 50 ft **17.** $9\sqrt{3} \approx 15.6$

19. yes; acute

$16^2 \stackrel{?}{=} 13^2 + 15^2$

$256 < 169 + 225$

21. 18.4 m **23.** $x = 4\sqrt{2}$, $y = 45°$ **25.** $\frac{5}{13}$, 0.38 **27.** $\frac{12}{13}$, 0.92 **29.** $\frac{5}{12}$, 0.42 **31.** 32.2 **33.** 63.4° and 26.6° **35.** 86.6 feet **37.** 33.2 **39.** $\langle -6, -5\rangle$ **41.** $\langle -8, 1\rangle$

Chapter 9 Tranformations and Symmetry

Page 613 Chapter 9 Get Ready

1. rotation **3.** translation **5.** $\langle -16, -28\rangle$ **7.** reduction; $\frac{1}{2}$

Pages 615–623 Lesson 9-1

1.

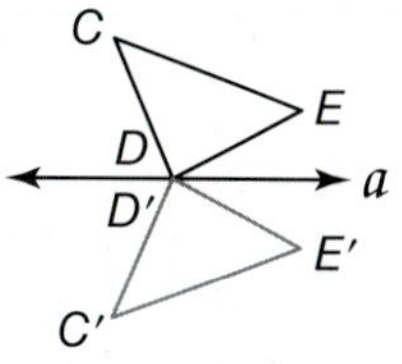

3.

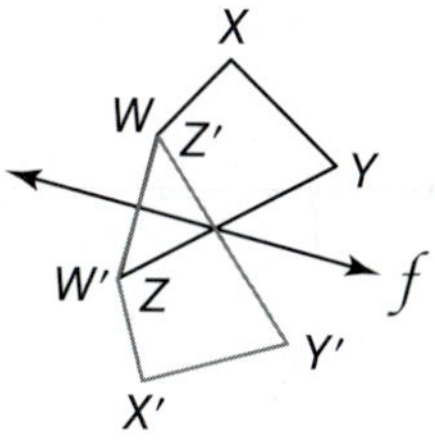

5.

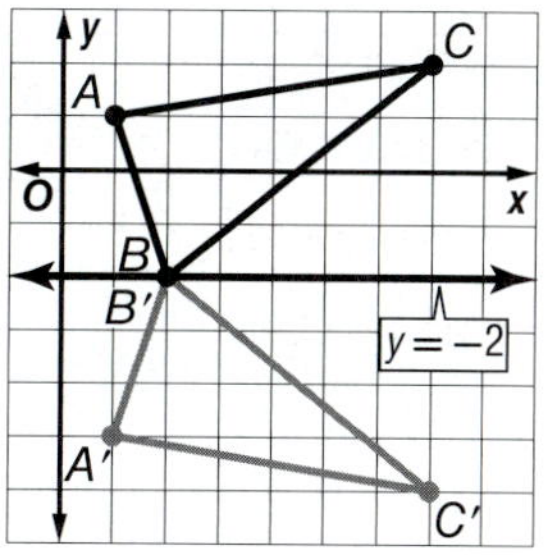

7.

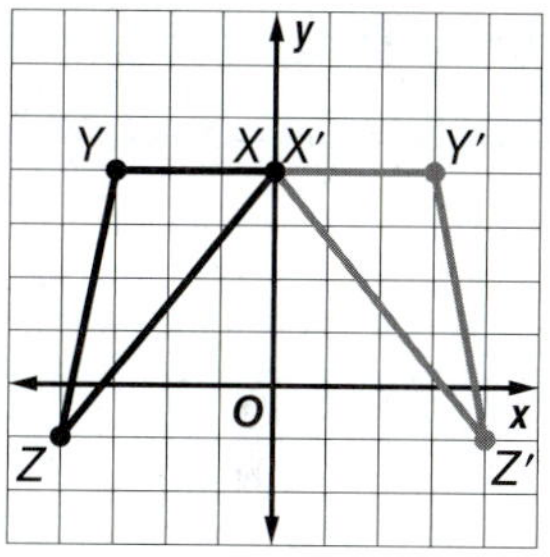

9.

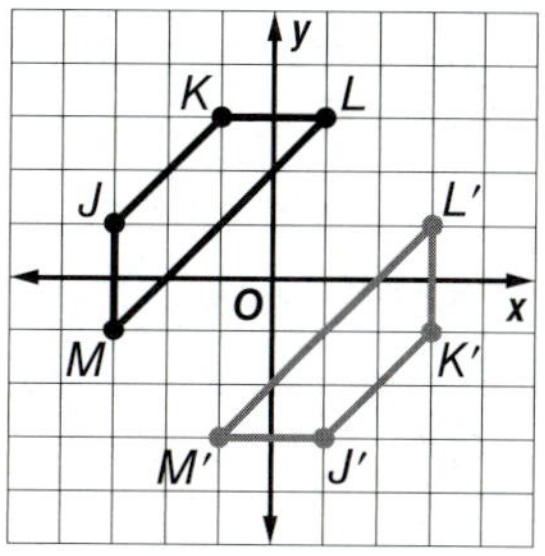

11.

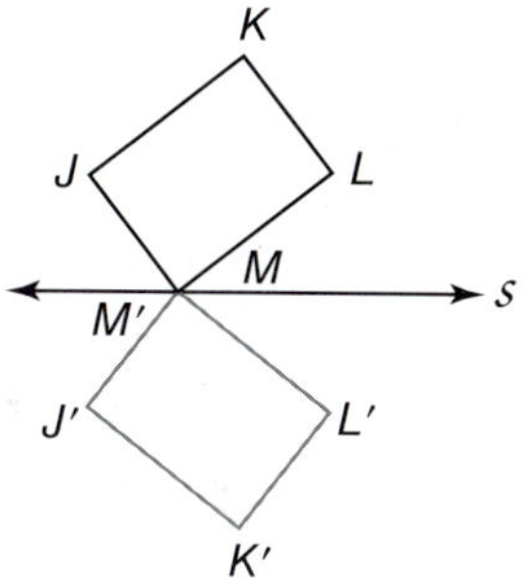

13 Draw a line through each vertex that is perpendicular to line *t*. Measure the distance from point *A* to line *t*. Then locate *A*′ the same distance from line *t* on the opposite side. Repeat to locate points *B*′, *C*′, and *D*′. Then connect vertices *A*′, *B*′, *C*′ and *D*′ to form the reflected image.

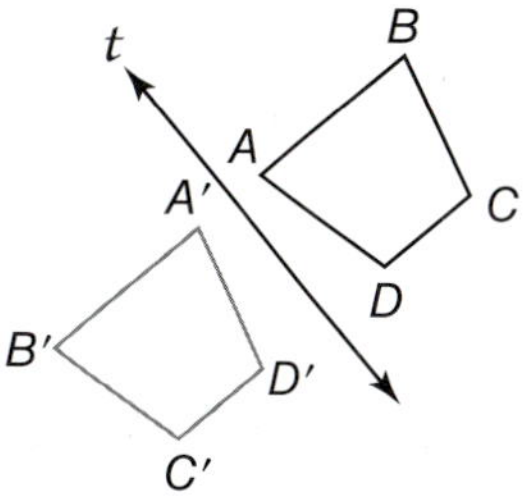

15.

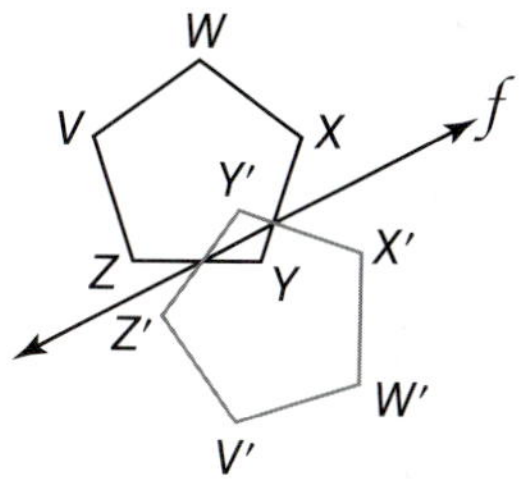

17.

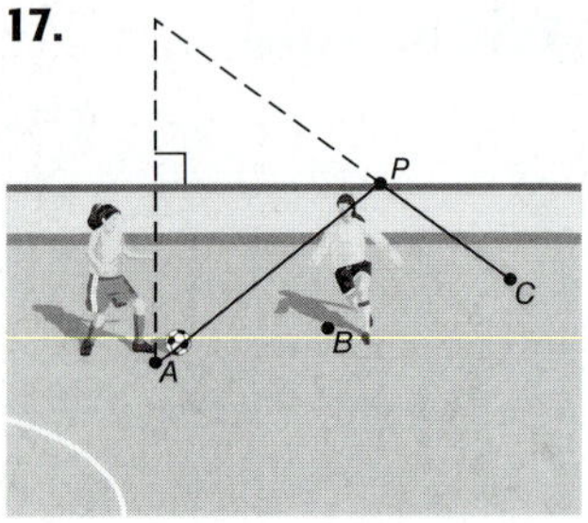

19.

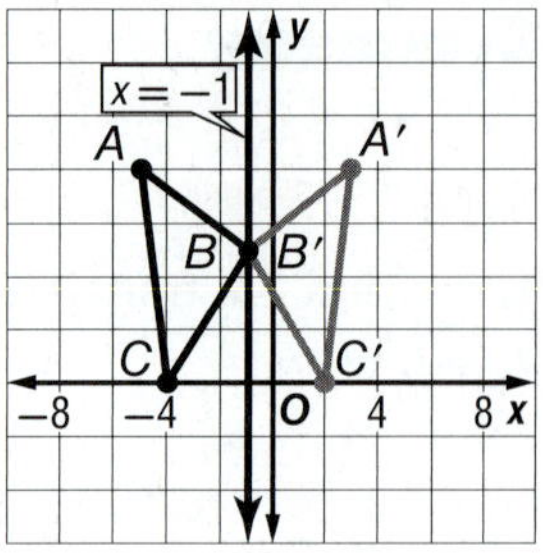

21.

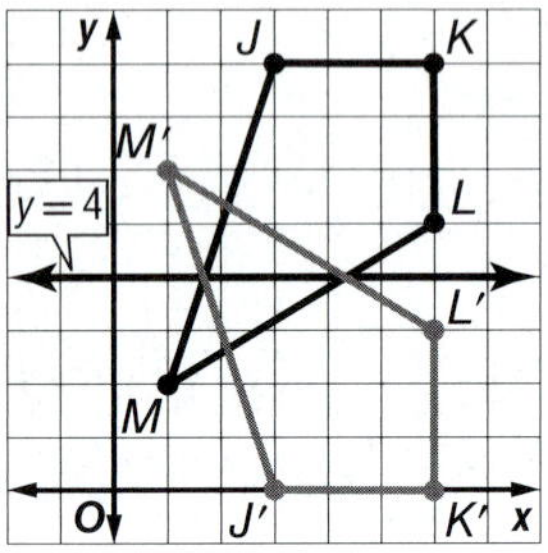

23.

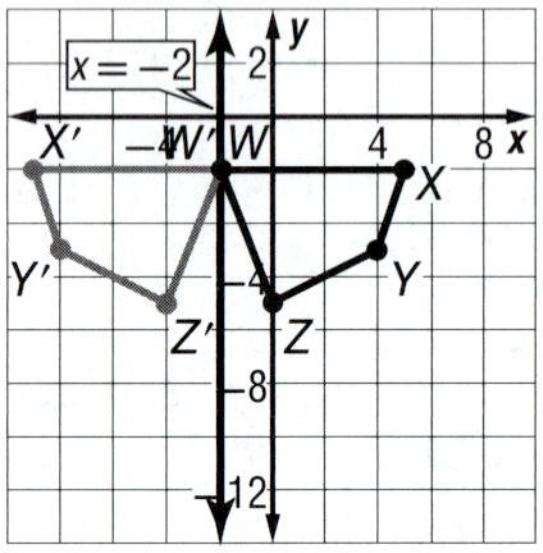

25 Multiply the x-coordinate of each vertex by -1.

(x, y)	→	$(-x, y)$
$J(-4, 6)$	→	$J'(4, 6)$
$K(0, 6)$	→	$K'(0, 6)$
$L(0, 2)$	→	$L'(0, 2)$
$M(-4, 2)$	→	$M'(4, 2)$

Graph $JKLM$ and its image $J'K'L'M'$

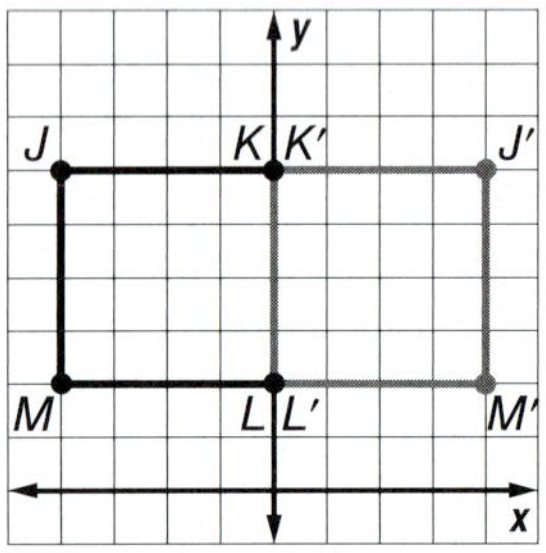

27.

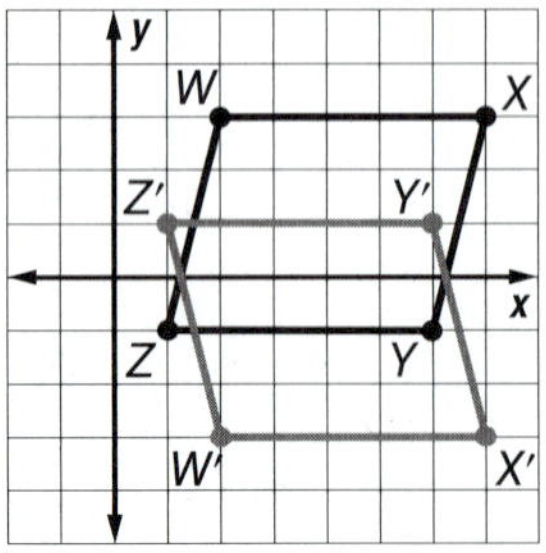

29.

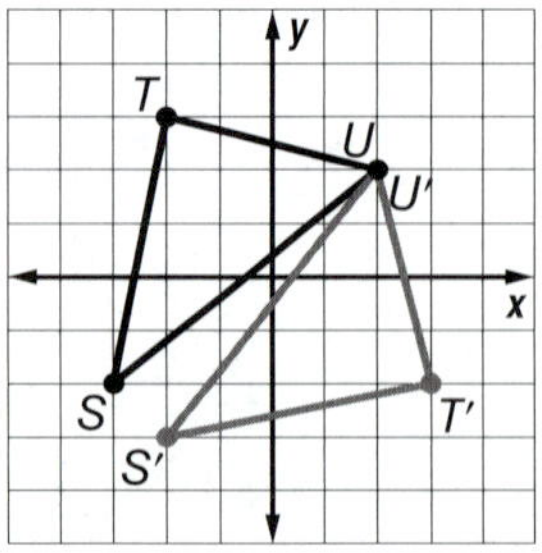

31.

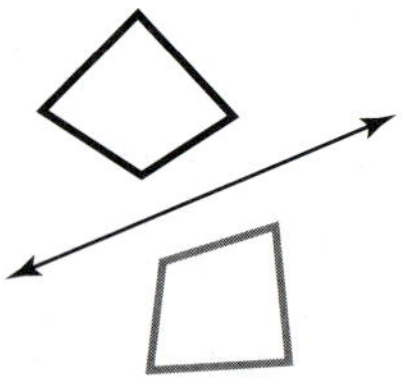

33.

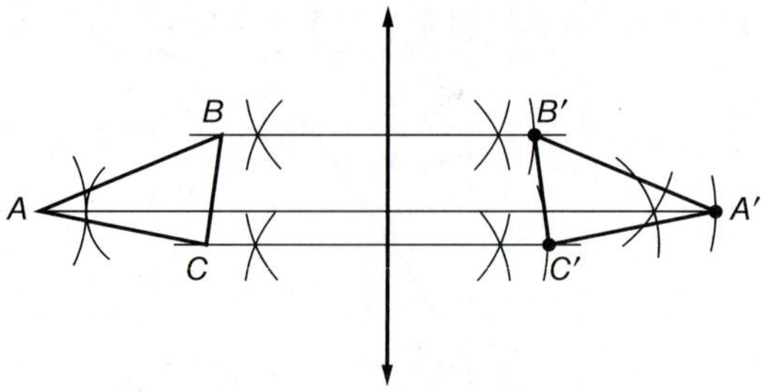

35 **a.** The reflections of the zebras are shown in the water. So, the water separates the zebras and their reflections. **b.** The water is a flat surface that extends in all directions. It represents a finite plane.

37.

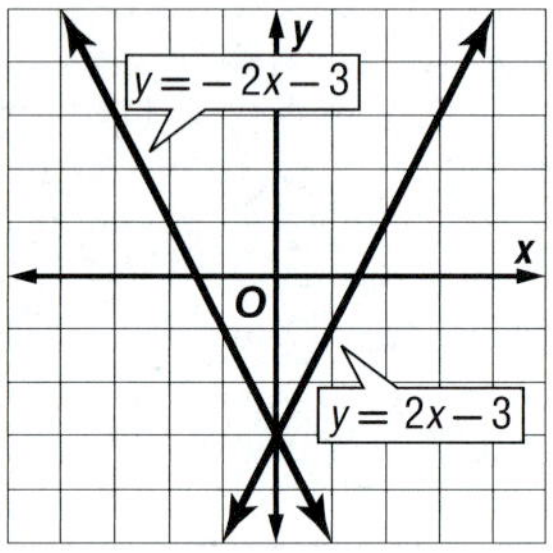

39.

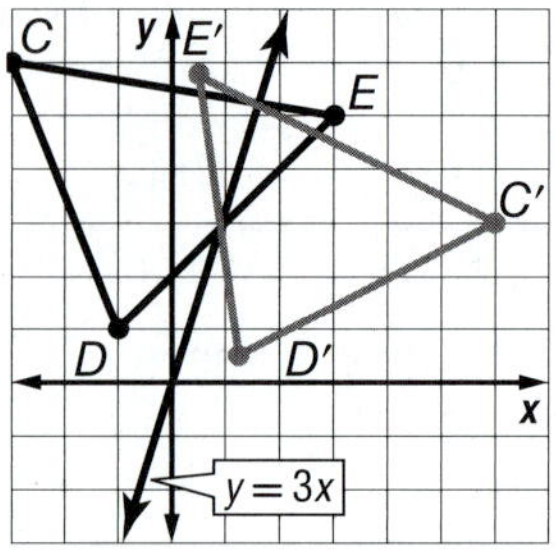

41 In the reflection, the x-coordinates stay the same, but the y-coordinates are opposites. The equation of the reflected image is $-y = \frac{1}{2}x^2$ or $y = -\frac{1}{2}x^2$.

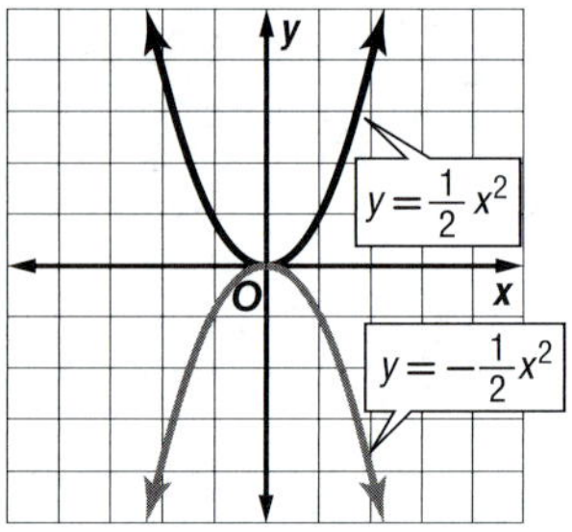

43.

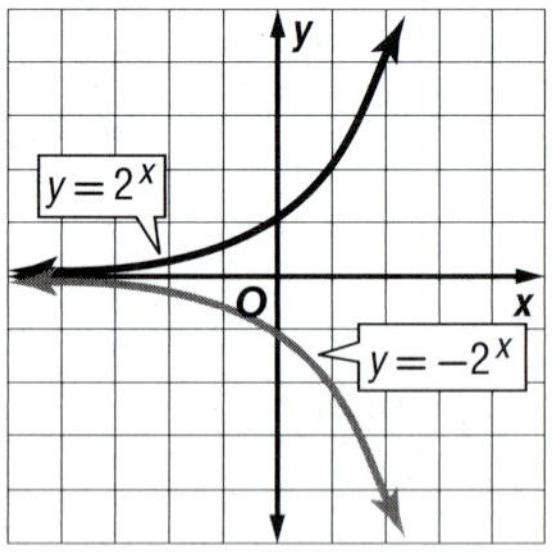

45. Jamil; sample answer: When you reflect a point across the x-axis, the reflected point is in the same place horizontally, but not vertically. When (2, 3) is reflected across the x-axis, the coordinates of the reflected point are (2, −3) since it is in the same location horizontally, but the other side of the x-axis vertically. **47.** (a, b) **49.** The slope of the line connecting the two points is $\frac{3}{5}$. The Midpoint Formula can be used to find the midpoint between the two points, which is $\left(\frac{3}{2}, \frac{3}{2}\right)$. Using the point-slope form, the equation of the line is $y = -\frac{5}{3}x + 4$. (The slope of the bisector is $-\frac{5}{3}$ because it is the negative reciprocal of the slope $\frac{3}{5}$.) **51.** Construct

P, Q, R collinear with Q between P and R. Draw line ℓ, then construct perpendicular lines from P, Q, and R to line ℓ. Show equidistance or similarity of slope.
53. B **55.** B **57.** $\langle 3, 2\rangle$ **59.** about 536 ft **61.** $AB > FD$
63. $m\angle FBA > m\angle DBF$ **65.** $2\sqrt{13} \approx 7.2$, 146.3°
67. $2\sqrt{122} \approx 22.1$, 275.2°

Pages 624–630 Lesson 9-2

1.

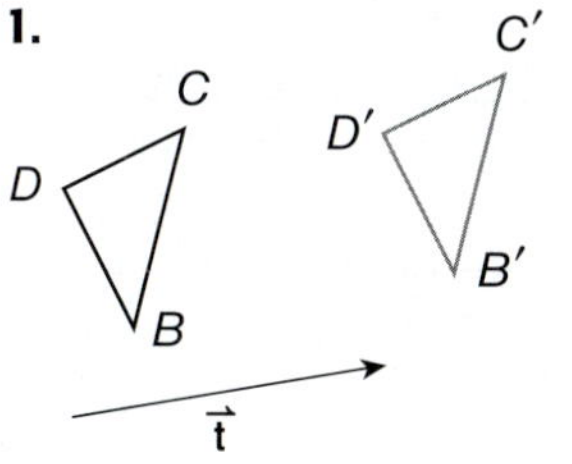

3.

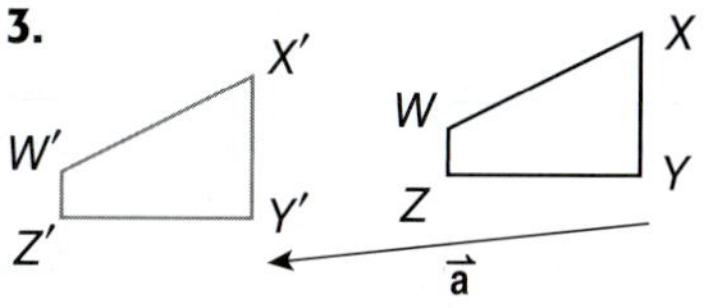

5.

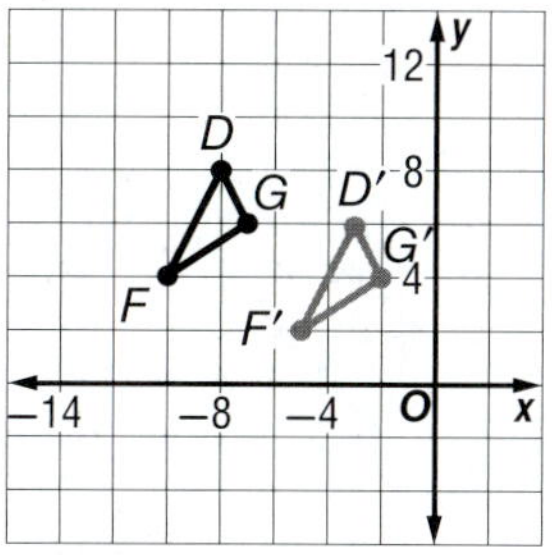

7. $(x, y) \rightarrow (x + 3, y - 5)$

9.

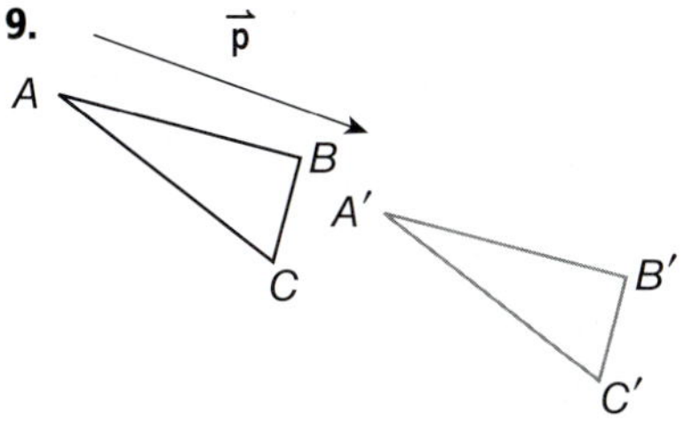

11.

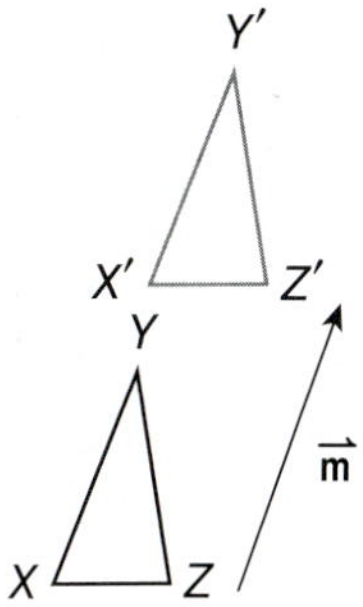

13.

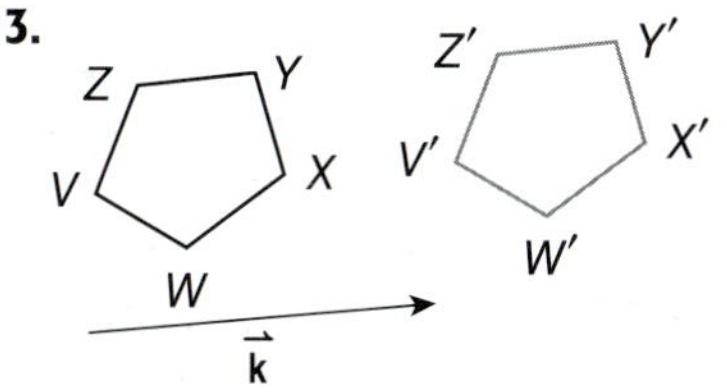

15 The vector indicates a translation 2 units left and 5 units up.

(x, y)	$\rightarrow$	$(x - 2, y + 5)$
$M(4, -5)$	$\rightarrow$	$M'(2, 0)$
$N(5, -8)$	$\rightarrow$	$N'(3, -3)$
$P(8, -6)$	$\rightarrow$	$P'(6, -1)$

Graph $\triangle MNP$ and its image $\triangle M'N'P'$.

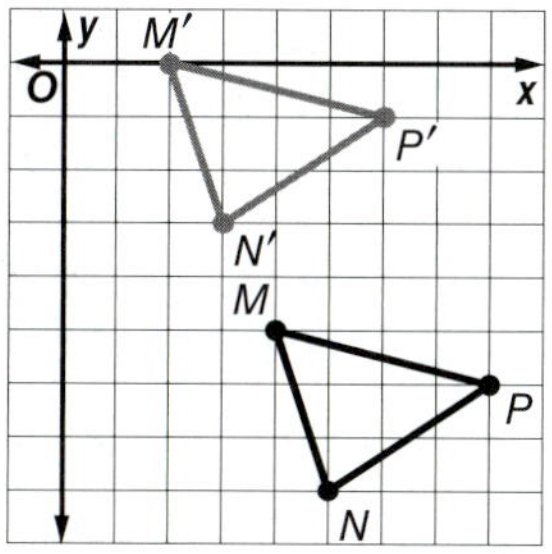

17.

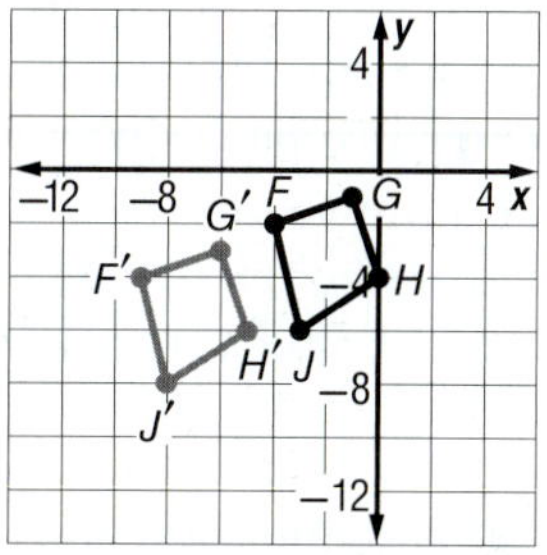

19.

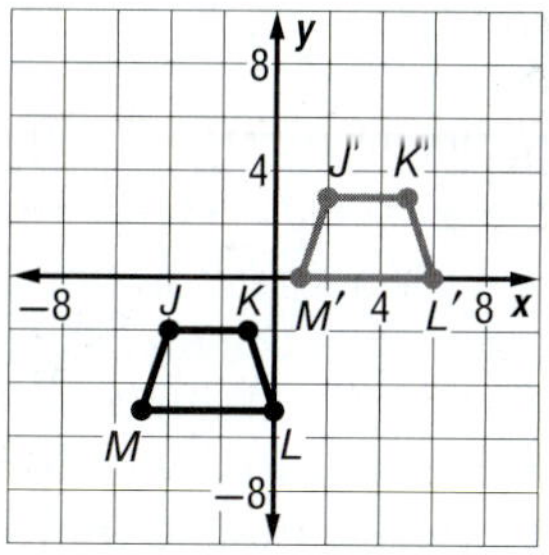

21 -12 represents a horizontal change of 12 yards left and 17 represents a vertical change of 17 yards up. A vector that indicates a translation 12 units left and 17 units up is $\langle -12, 17\rangle$.

23. $\langle 3, -5\rangle$ **25.** They move to the right 13 seats and back one row; $\langle 13, -1\rangle$.

27 The translation vector $\langle -2, 0\rangle$ represents a horizontal shift 2 units left. The equation of the reflected image is $y = -[x - (-2)]^3$ or $y = -(x + 2)^3$.

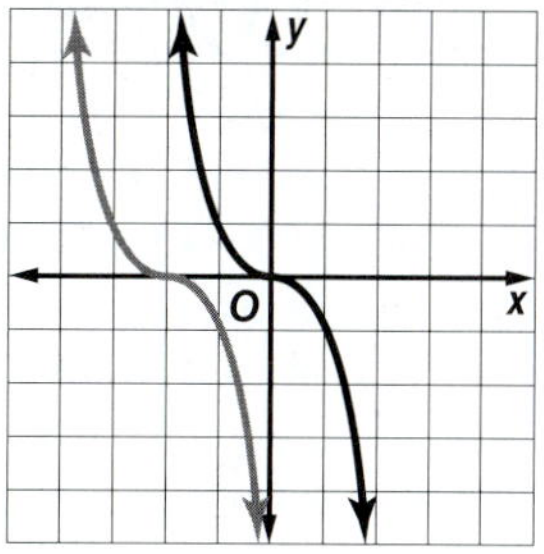

29a.

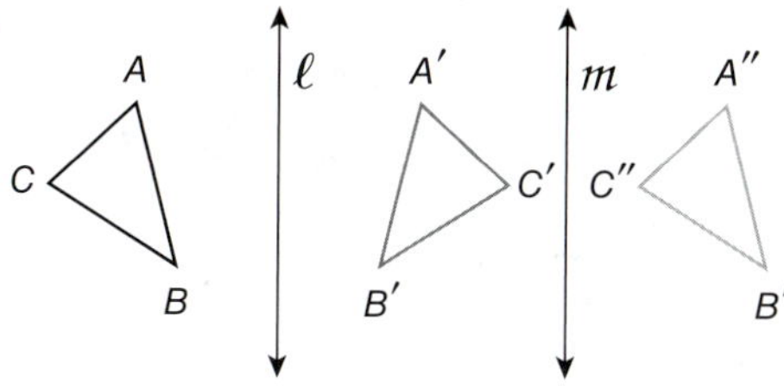

29b.

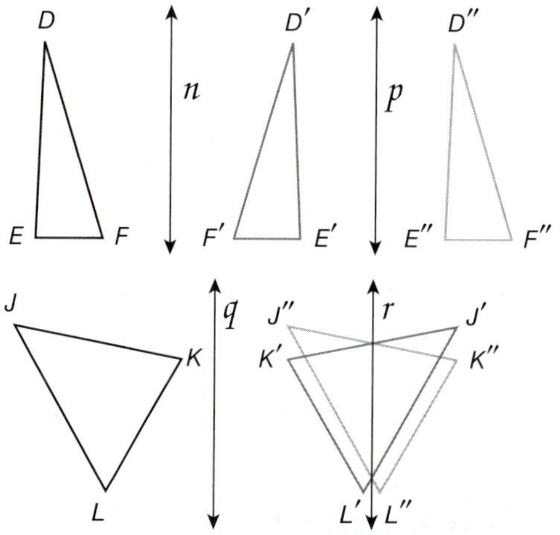

29c.

Distance Between Corresponding Points (cm)		Distance Between Parallel Lines (cm)	
A and A'', B and B'', C and C''	4.4	ℓ and m	2.2
D and D'', E and E'', F and F''	5.6	n and p	2.8
J and J'', K and K'', L and L''	2.8	q and r	1.4

Selected Answers and Solutions

29d. Sample answer: The composition of two reflections in vertical lines can be described by a horizontal translation that is twice the distance between the two vertical lines. **31.** $y = m(x - a) + 2b$; $2b - ma$ **33.** Sample answer: Both vector notation and function notation describe the distance a figure is translated in the horizontal and vertical directions. Vector notation does not give a rule in terms of initial location, but function notation does. For example, the translation a units to the right and b units up from the point (x, y) would be written $\langle a, b\rangle$ in vector notation and $(x + a, y + b)$ in function notation. **35.** D **37.** F

39.

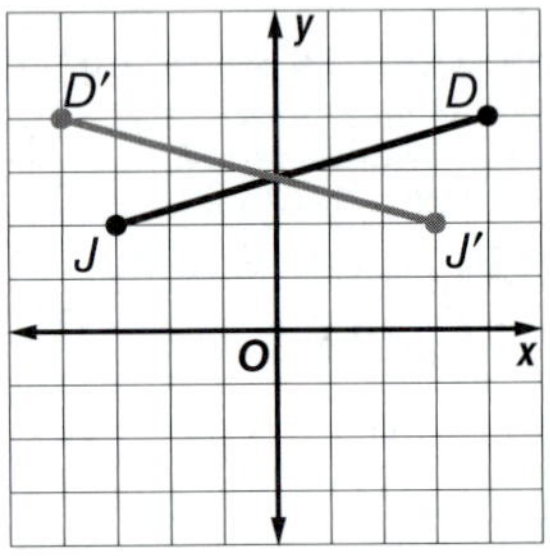

41.

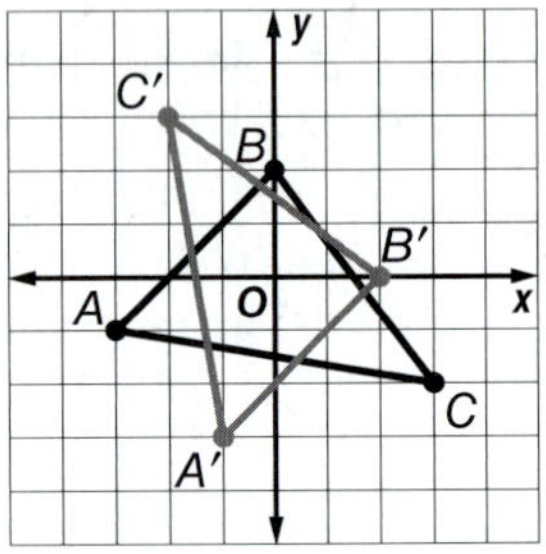

43. $\vec{c} - \vec{d}$

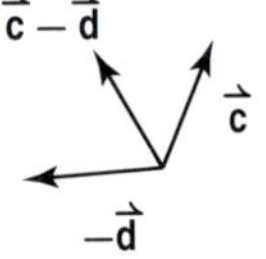

45.

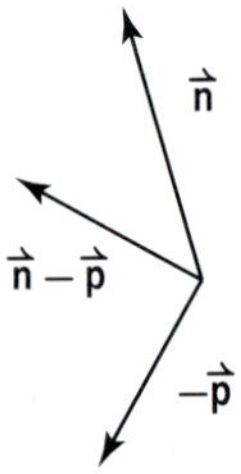

47. 100 **49.** 80
51. obtuse; 110
53. obtuse; 140

Pages 632–638 Lesson 9-3

1.

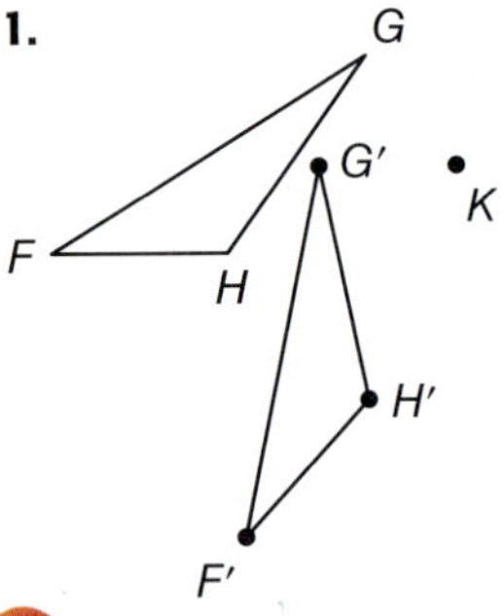

3 Multiply the x- and y-coordinates by -1.

(x, y)	→	$(-x, -y)$
$D(-2, 6)$	→	$D'(2, -6)$
$F(2, 8)$	→	$F'(-2, -8)$
$G(2, 3)$	→	$G'(-2, -3)$

Graph $\triangle DFG$ and its image $\triangle D'F'G'$.

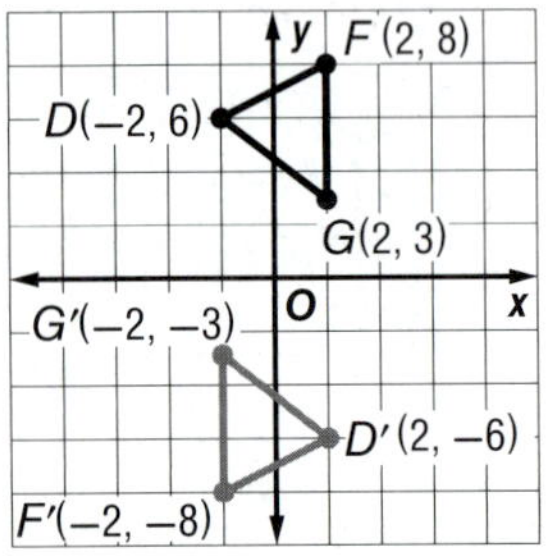

5.

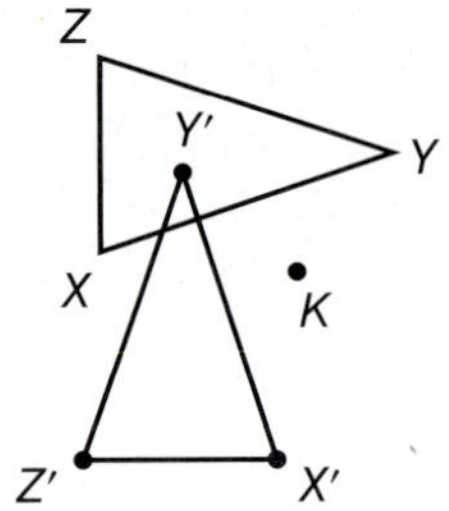

7.

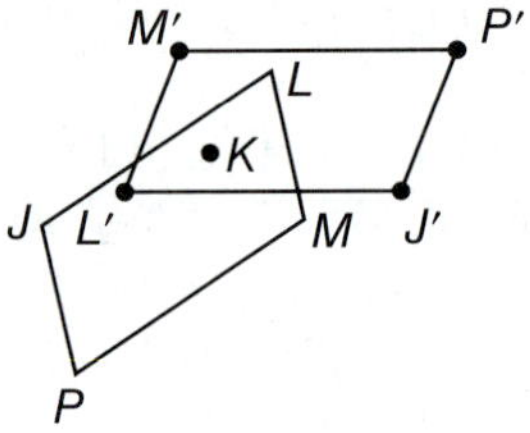

9.

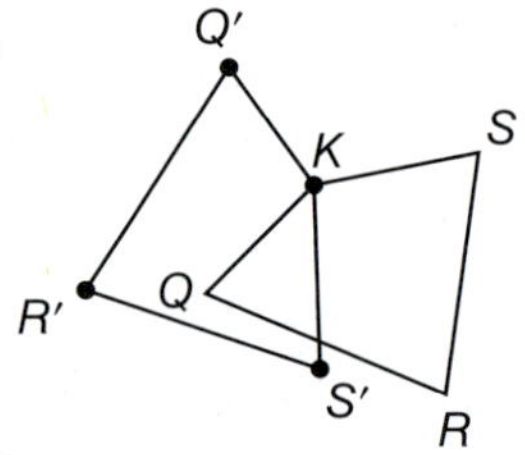

11. 120°; 360° ÷ 6 petals = 60° per petal. Two petal turns is 2 • 60° or 120°.
13. 154.2°; 360° ÷ 7 petals = 51.4° per petal. Three petal turns is 3 • 51.4° or 154.2°.

15.

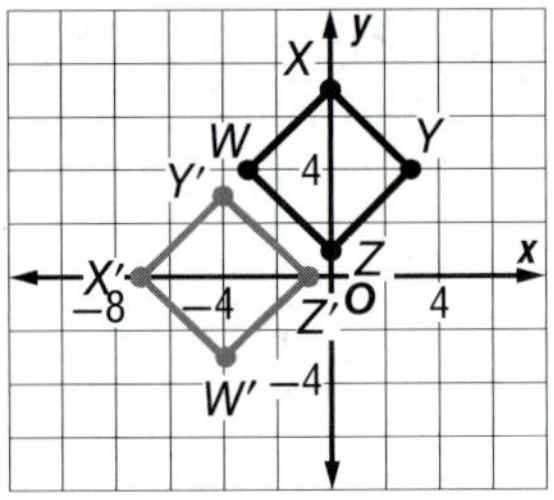

17.

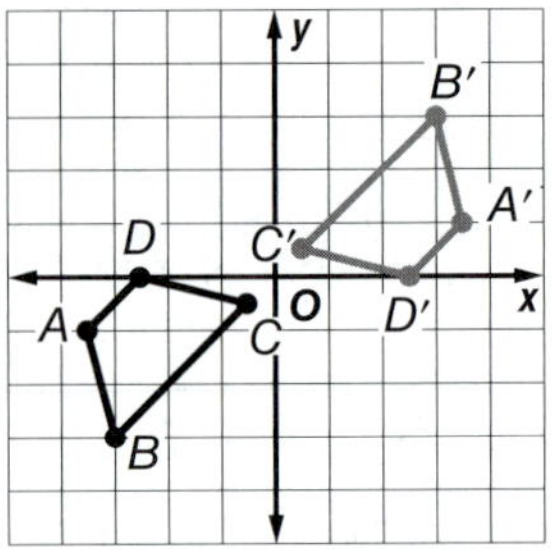

19.

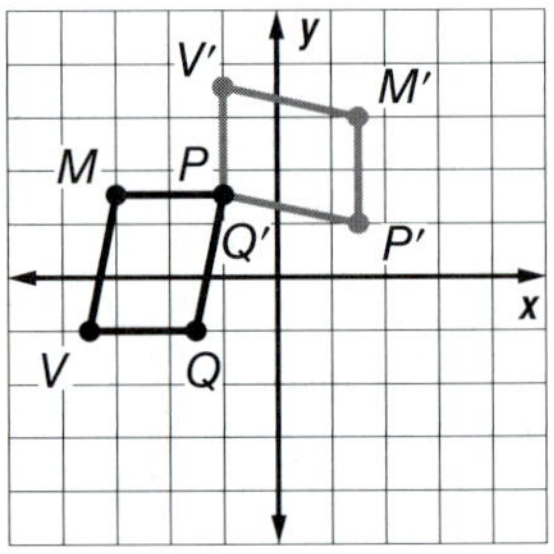

21a. 10°
21b. about 1.7 seconds
23. 125°

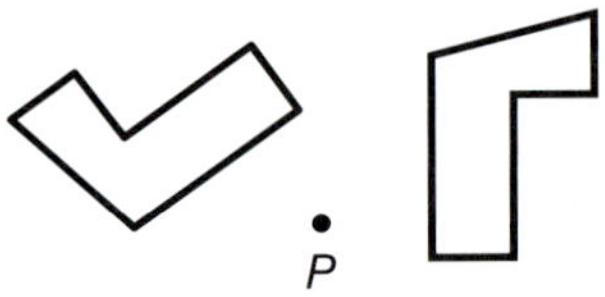

25. $y = -x + 2$; parallel **27.** $y = -x - 2$; collinear
29. x-intercept: $y = 2x + 4$; y-intercept: $y = 2x + 4$

31 After 31 seconds, the ride will have rotated 31 • 0.25 or 7.75 times. So, she will be in the bottom position of the ride, with the car having the center at (0, −4). After 31 seconds, the car will have rotated 31 • 0.5 or 15.5 times. So, she will be in the position directly across from her starting position. Her position after 31 seconds would be (2, −4).

33 **a.**

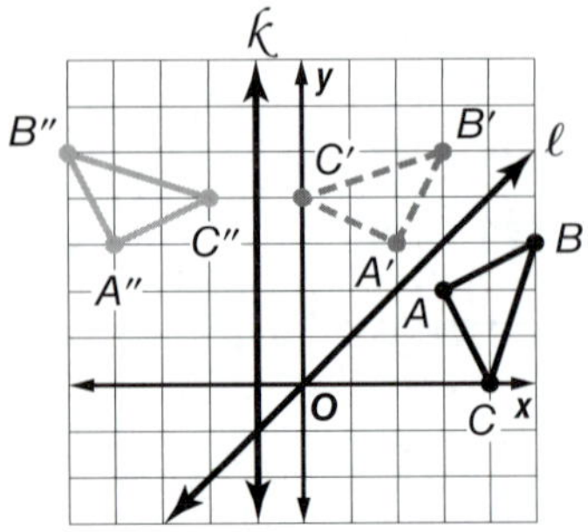

b. Sample answer: Let the x-axis be line n and the y-axis be line p. Draw $\triangle DEF$ with vertices $D(-4, 1)$, $E(-2, 3)$, and $F(0, 1)$. Reflect $\triangle DEF$ in the line n to get $\triangle D'E'F'$ with vertices $D'(-4, -1)$,

$E'(-2, -3)$, and $F'(0, -1)$. Reflect $\triangle D'E'F'$ in the line p to get $\triangle D''E''F''$ with vertices $D''(4, -1)$, $E''(2, -3)$, and $F''(0, -1)$.

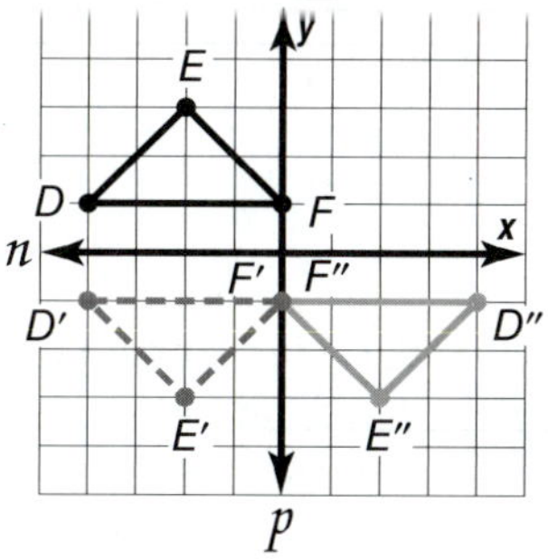

Sample answer: Let the line $y = -x$ be line q and the line $y = 0.5$ be line r. Draw $\triangle MNP$ with vertices $M(0, -4)$, $N(1, 2)$, and $P(3, -5)$. Reflect $\triangle MNP$ in the line q to get $\triangle M'N'P'$ with vertices $M'(4, 0)$, $N'(2, -1)$, and $P'(5, -3)$. Reflect $\triangle M'N'P'$ in the line r to get $\triangle M''N''P''$ with vertices $M''(4, 1)$, $N''(2, 2)$, and $P''(5, 4)$.

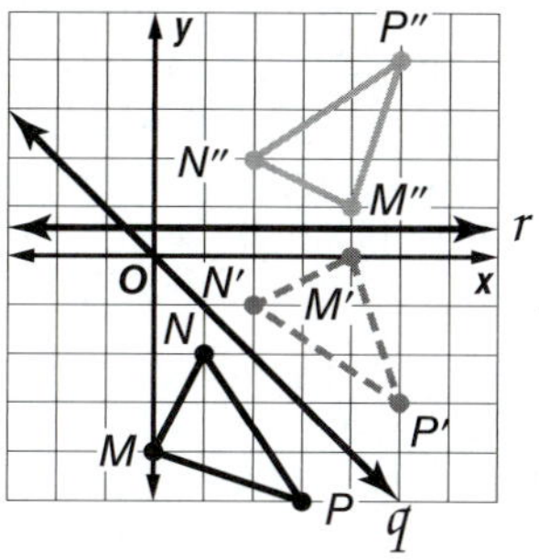

c.

Angle of Rotation Between Figures		Angle Between Intersecting Lines	
$\triangle ABC$ and $\triangle A''B''C''$	90°	ℓ and m	45°
$\triangle DEF$ and $\triangle D''E''F''$	180°	n and p	90°
$\triangle MNP$ and $\triangle M''N''P''$	90°	q and r	45°

d. Sample answer: The measure of the angle of rotation about the point where the lines intersect is twice the measure of the angle between the two intersecting lines.

35. Sample answer: $(-1, 2)$; Since $\triangle CC'P$ is isosceles and the vertex angle of the triangle is formed by the angle of rotation, both $m\angle PCC'$ and $m\angle PC'C$ are 40° because the base angles of isosceles triangles are congruent. When you construct a 40° angle with a vertex at C and a 40° angle with a vertex at C', the intersection of the rays forming the two angles intersect at the point of rotation, or $(-1, 2)$.

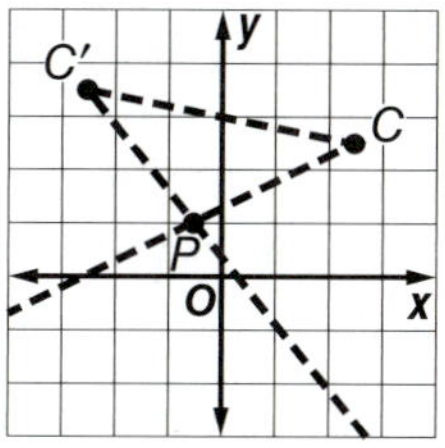

37. No; sample answer: When a figure is reflected about the x-axis, the x-coordinates of the transformed figure remain the same, and the y-coordinates are negated. When a figure is rotated 180° about the origin, both the x- and y-coordinates are negated. Therefore, the transformations are not equivalent. **39.** D **41.** J

43. 50 mi

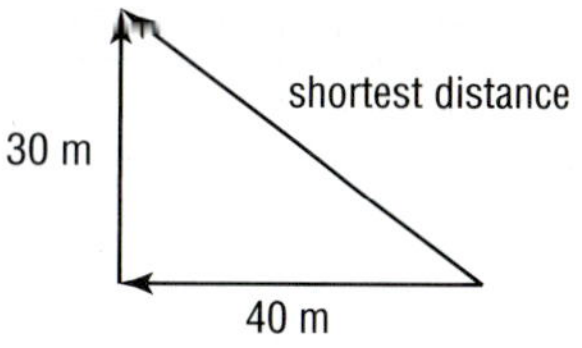

45.

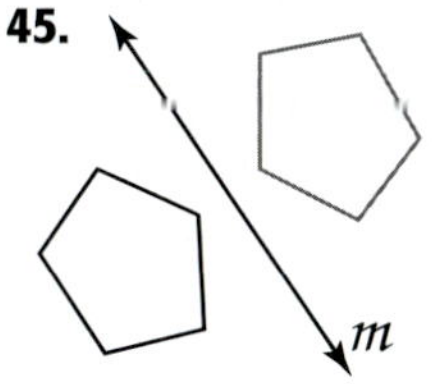

47. reflection **49.** rotation or reflection

Pages 641–649 Lesson 9-4

1.

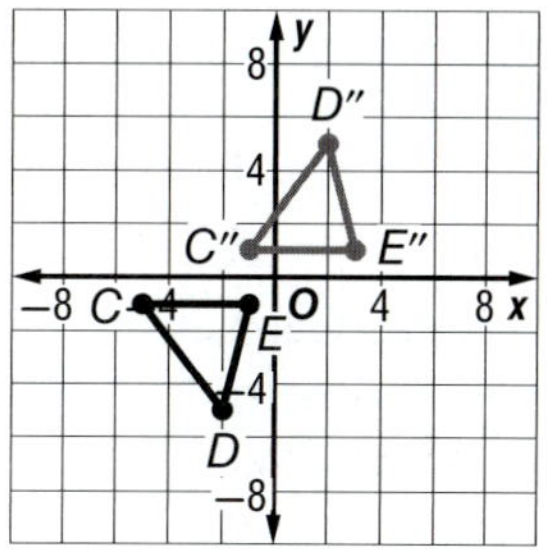

3.

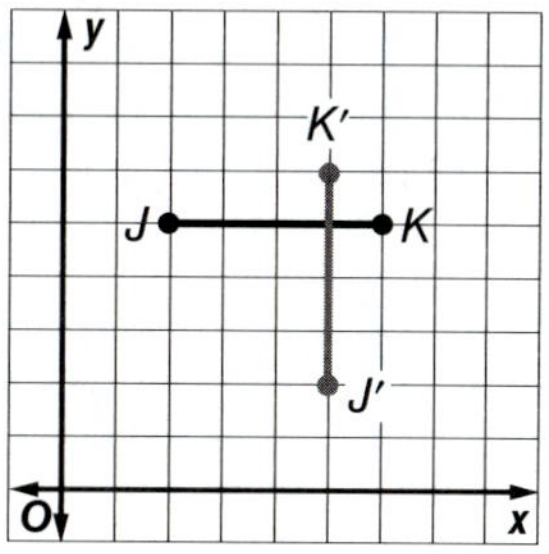

5.

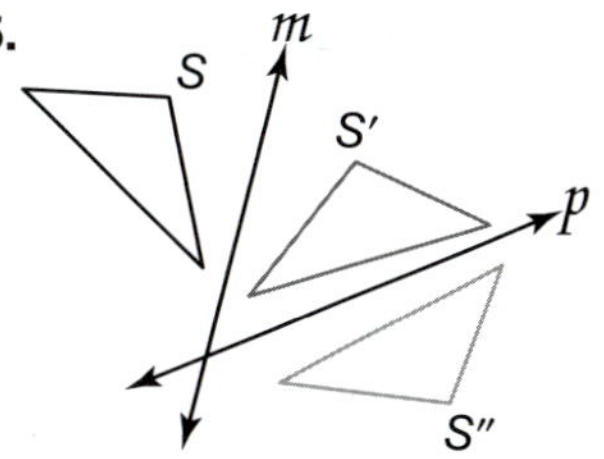

rotation clockwise 100° about the point where lines m and p intersect.

7 translation along $\langle 2, 0\rangle$ reflection in x-axis

(x, y)	$\rightarrow$	$(x + 2, y)$	(x, y)	$\rightarrow$	$(x, -y)$
$R(1, -4)$	$\rightarrow$	$R'(3, -4)$	$R'(3, -4)$	$\rightarrow$	$R''(3, 4)$
$S(6, -4)$	$\rightarrow$	$S'(8, -4)$	$S'(8, -4)$	$\rightarrow$	$S''(8, 4)$
$T(5, -1)$	$\rightarrow$	$T'(7, -1)$	$T'(7, -1)$	$\rightarrow$	$T''(7, 1)$

Graph $\triangle RST$ and its image $\triangle R''S''T''$.

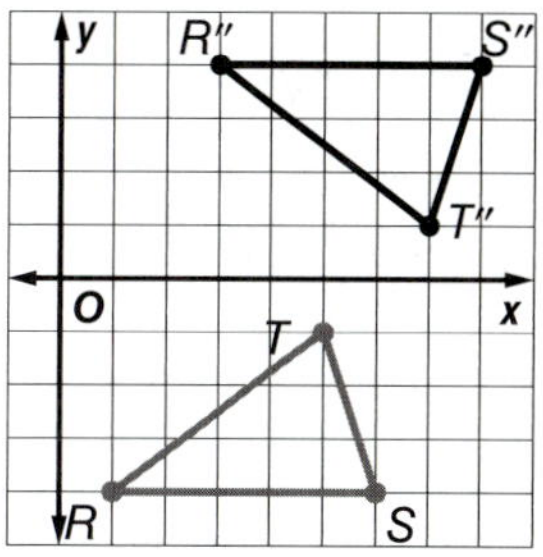

9.

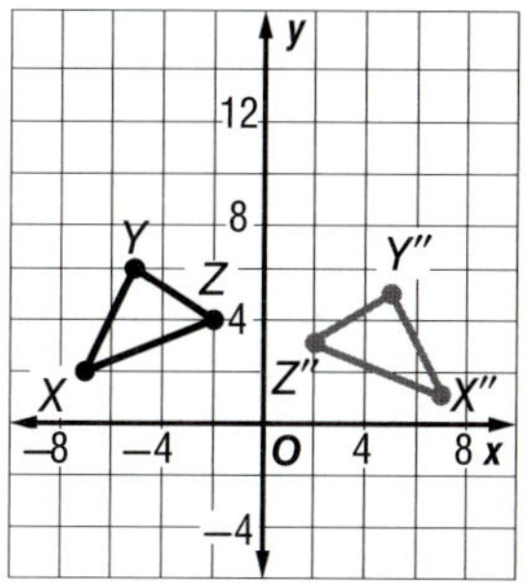

11.

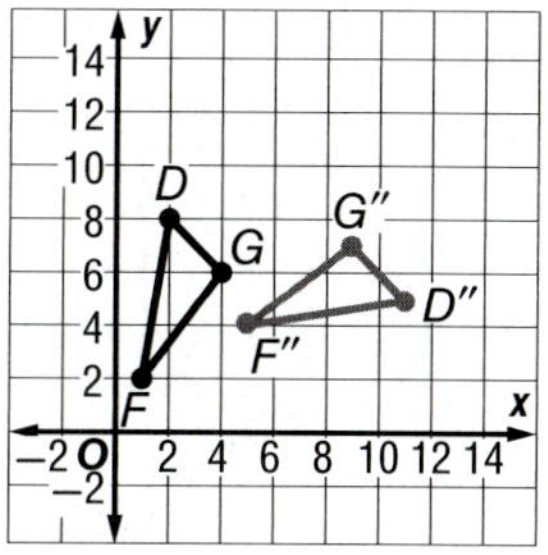

13.

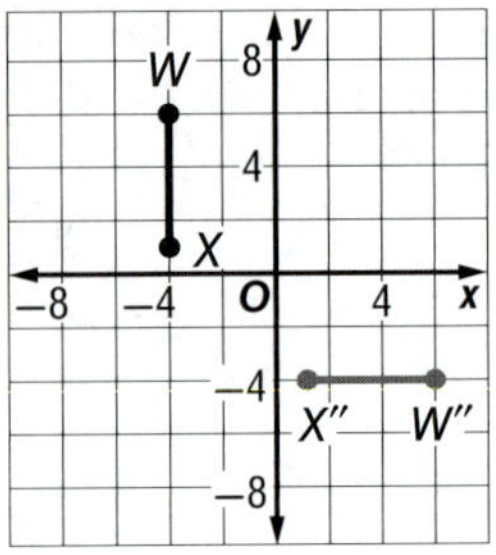

15.

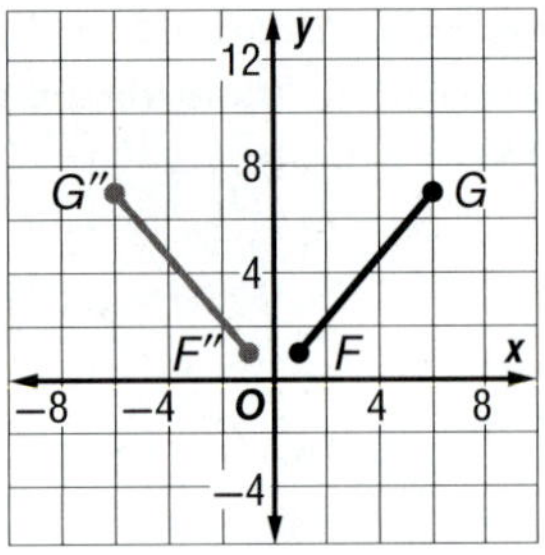

17.

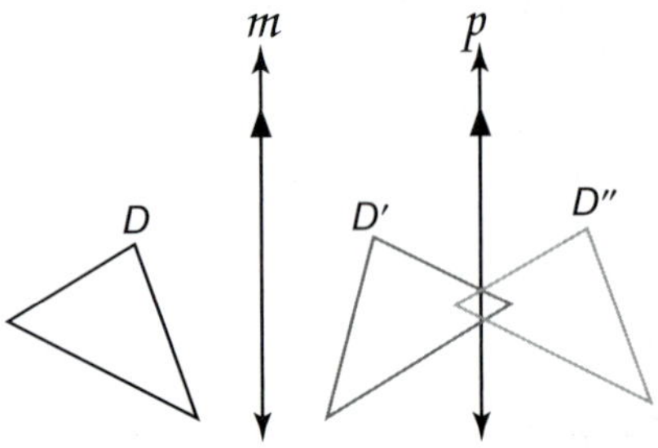

horizontal translation 4 cm to the right.

19.

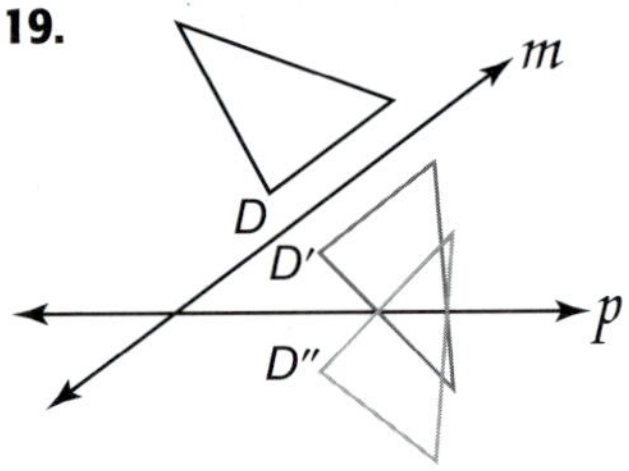

vertical translation 2.4 in. down; 70° rotation about the point where lines m and p intersect

21. translation **23.** rotation

25 rotation 90° about origin reflection in x-axis

rotation 90° about origin			reflection in x-axis		
(x, y)	→	$(-y, x)$	(x, y)	→	$(x, -y)$
$(-2, -5)$	→	$(5, -2)$	$(5, -2)$	→	$(5, 2)$
$(0, 1)$	→	$(-1, 0)$	$(-1, 0)$	→	$(-1, 0)$

Graph line y' through points at $(5, -2)$ and $(-1, 0)$.
Graph line y'' through points at $(5, 2)$ and $(-1, 0)$.

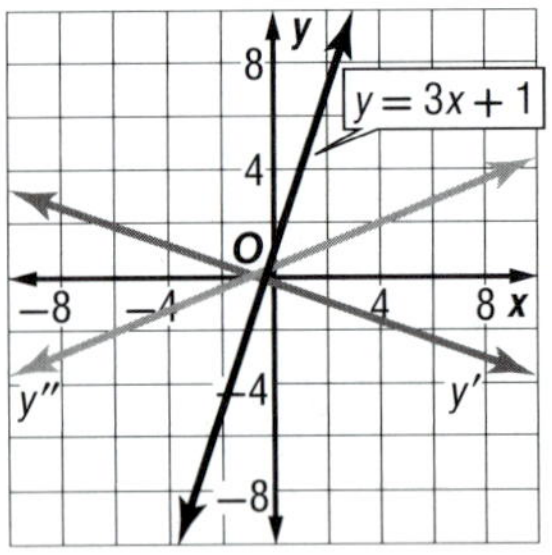

27. $A''(3, 1)$, $B''(2, 3)$, $C''(1, 0)$

29 A translation occurs from Step 1 to Step 2. A rotation 90° counterclockwise occurs from Step 2 to Step 3. So, the transformations are a translation and a 90° rotation.

31. $(x + 5.5, y)$ reflected in the line that separates the left prints from the right prints **33.** double reflection **35.** rotation 180° about the origin and reflection in the x-axis

37. Given: Lines ℓ and m intersect at point P. A is any point not on ℓ or m.

Prove: a. If you reflect point A in line m, and then reflect its image A' in line ℓ, A'' is the image of A after a rotation about point P.

b. $m\angle APA'' = 2(m\angle SPR)$

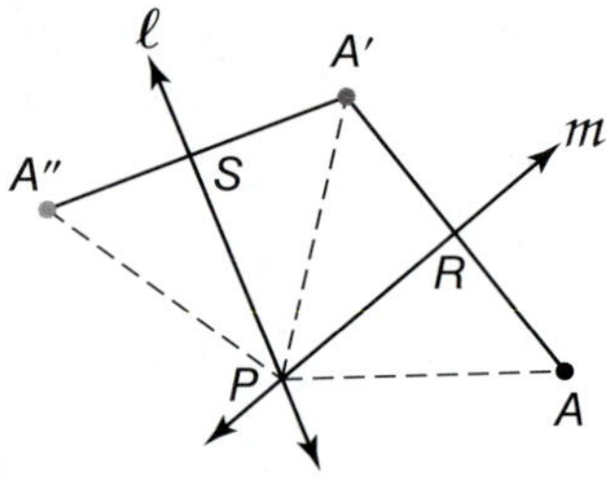

Proof: We are given that line ℓ and line m intersect at point P and that A is not on line ℓ or line m. Reflect A over line m to A' and reflect A' over line ℓ to A''. By the definition of reflection, line m is the perpendicular bisector of $\overline{AA'}$ at R, and line ℓ is the perpendicular bisector of $\overline{A'A''}$ at S. $\overline{AR} \cong \overline{A'R}$ and $\overline{A'S} \cong \overline{A''S}$ by the definition of a perpendicular bisector. Through any two points there is exactly one line, so we can draw auxiliary segments $\overline{AP}$, $\overline{A'P}$, and $\overline{A''P}$. $\angle ARP$, $\angle A'RP$, $\angle A'SP$ and $\angle A''SP$ are right angles by the definition of perpendicular bisectors. $\overline{RP} \cong \overline{RP}$ and $\overline{SP} \cong \overline{SP}$ by the Reflexive Property. $\triangle ARP \cong \triangle A'RP$ and $\triangle A'SP \cong \triangle A''SP$ by the SAS Congruence Postulate. Using CPCTC, $\overline{AP} \cong \overline{A'P}$ and $\overline{A'P} \cong \overline{A''P}$, and $\overline{AP} \cong \overline{A''P}$ by the Transitive Property. By the definition of a rotation, A'' is the image of A after a rotation about point P. Also using CPCTC, $\angle APR \cong \angle A'PR$ and $\angle A'PS \cong \angle A''PS$. By the definition of congruence, $m\angle APR = m\angle A'PR$ and $m\angle A'PS = m\angle A''PS$. $m\angle APR + m\angle A'PR + m\angle A'PS + m\angle A''PS = m\angle APA''$ and $m\angle A'PS + m\angle A'PR = m\angle SPR$ by the Angle Addition Postulate. $m\angle A'PR + m\angle A'PR + m\angle A'PS + m\angle A'PS = m\angle APA''$ by Substitution, which simplifies to $2(m\angle A'PR + m\angle A'PS) = m\angle APA''$. By Substitution, $2(m\angle SPR) = m\angle APA''$.

39. Sample answer: No; there are not invariant points in a glide reflection because all of the points are translated along a vector. Perhaps for compositions of transformations, there may be invariant points when a figure is rotated and reflected, rotated twice, or reflected twice. **41.** Yes; sample answer: If a segment with endpoints (a, b) and (c, d) is to be reflected about the x-axis, the coordinates of the endpoints of the reflected image are $(a, -b)$ and $(c, -d)$. If the segment is then reflected about the line $y = x$, the coordinates of the endpoints of the final image are $(-b, a)$ and $(-d, c)$. If the original image is first reflected about $y = x$, the coordinates of the endpoints of the reflected image are (b, a) and (d, c). If the segment is then reflected about the x-axis, the coordinates of the endpoints of the final image are $(b, -a)$ and $(d, -c)$.

43 Sample answer: When two rotations are performed on a single image, the order of the rotations does not affect the final image when the two rotations are centered at the same point. For example, if $\triangle ABC$ is rotated 45° clockwise about the origin and then rotated 60° clockwise about the origin, $\triangle A''B''C''$ is the same as if the figure were first rotated 60° clockwise about the origin and then rotated 45° clockwise about the origin. If $\triangle ABC$ is rotated 45° clockwise about the origin and then rotated 60° clockwise about $P(2, 3)$, $\triangle A''B''C''$ is different than if the figure were first rotated 60° clockwise about $P(2, 3)$ and then rotated 15° clockwise about the origin. So, the order of the rotations sometimes affects the location of the final image.

45. A **47.** H

49.

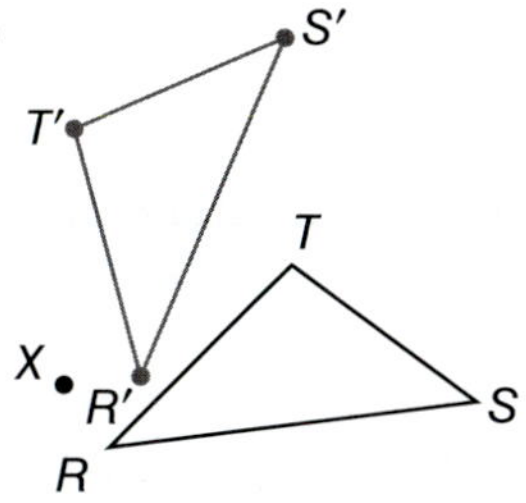

51.

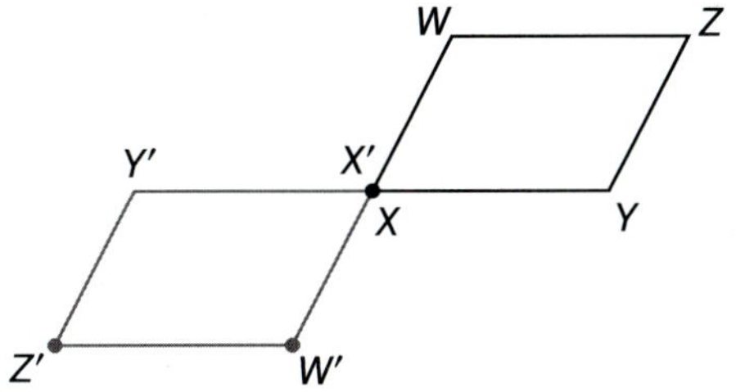

53.

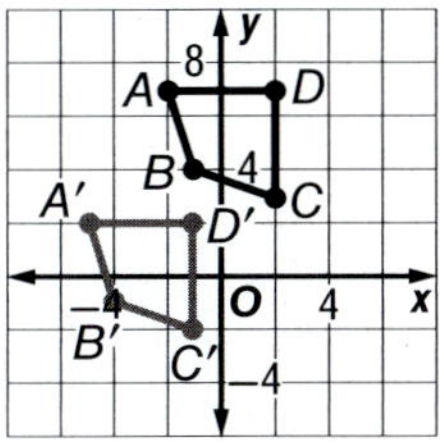

55.

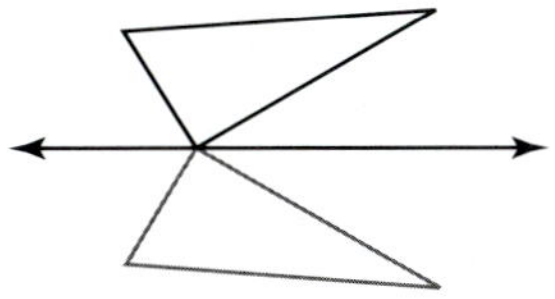

57.

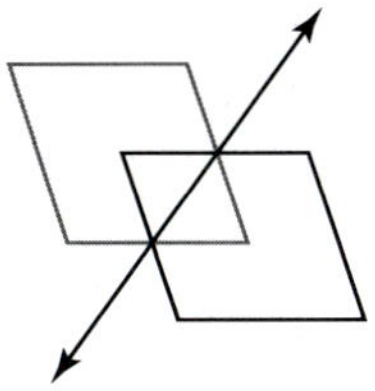

Pages 653–659 Lesson 9-5

1. yes; 4

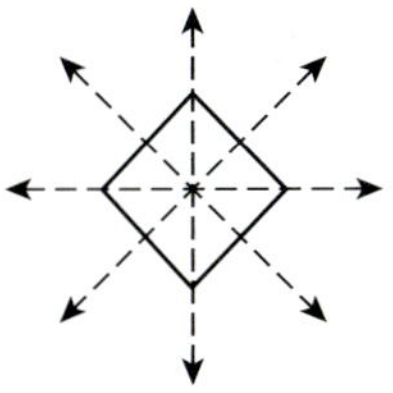

3. yes; 1

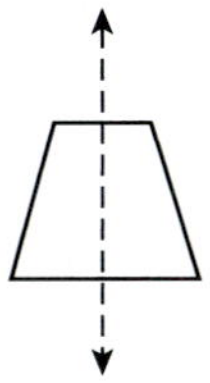

5. yes; 2; 180°

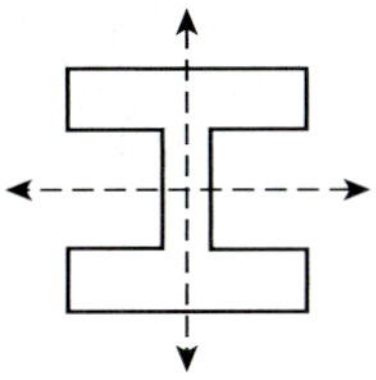

7 **a.** There is no line in which the dome can be reflected onto itself horizontally. So, there are no horizontal lines of symmetry. The dome has 36 • 2 or 72 vertical lines of symmetry. **b.** The dome has order 36 symmetry and magnitude 360° ÷ 36 or 10°.

9. no **11.** yes; 6

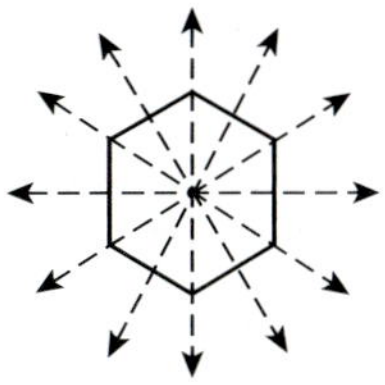

13. yes; 1

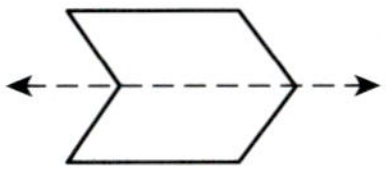

15. no **17.** yes; 1

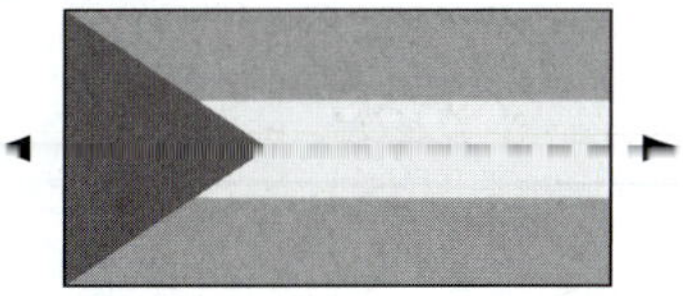

19. yes; 3; 120°

21 No; there is no rotation between 0° and 360° that maps the shape onto itself. So, the figure does not have rotational symmetry.

23. yes; 8; 45° **25.** yes; 8; 45° **27.** both **29.** both

31. no horizontal, infinitely many vertical

33. 1 horizontal, infinitely many vertical

35 Figure $ABCD$ is a square. It has 4 lines of symmetry and has order 4 symmetry and magnitude 360° ÷ 4 or 90°. So, it has line symmetry and rotational symmetry.

37. line and rotational

39. rotational; 2; 180°; line symmetry; $y = -x$

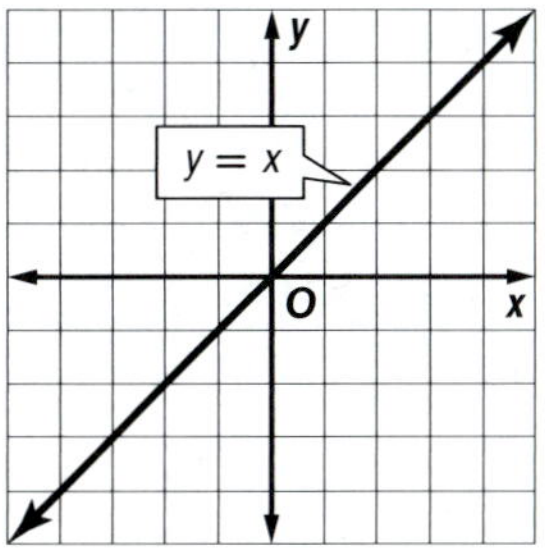

41. rotational; 2; 180°

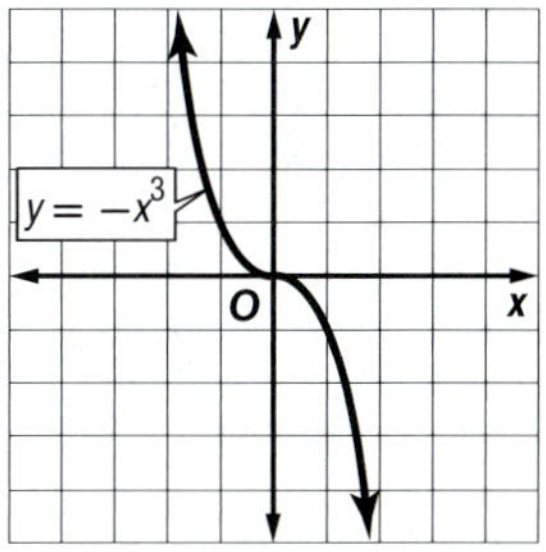

43. plane and axis; 180

45a. order of symmetry = 3

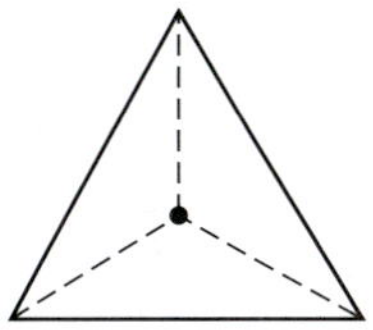

45b. order of symmetry = 4

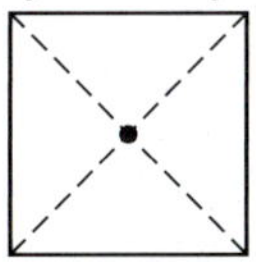

order of symmetry = 5

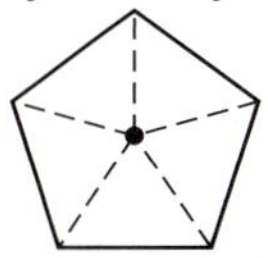

order of symmetry = 6

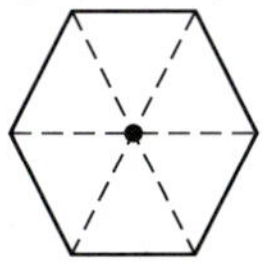

45c.

Polygon	Order of Symmetry
equilateral triangle	3
square	4
regular pentagon	5
regular hexagon	6

45d. Sample answer: The order of symmetry for a regular polygon is the number of sides of the polygon.

47. Sample answer: (–1, 0), (2, 3), (4, 1), and (1, –2)

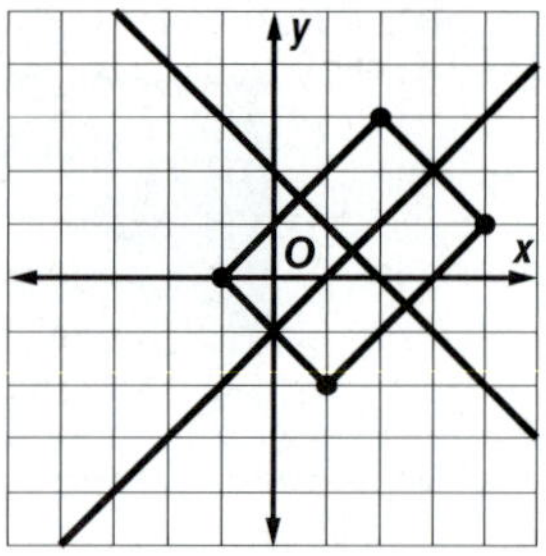

49.

Sample answer: An isosceles triangle has line symmetry from the vertex angle to the base of the triangle, but it does not have rotational symmetry because it cannot be rotated from 0° to 360° and map onto itself. **51.** B **53.** H

55.

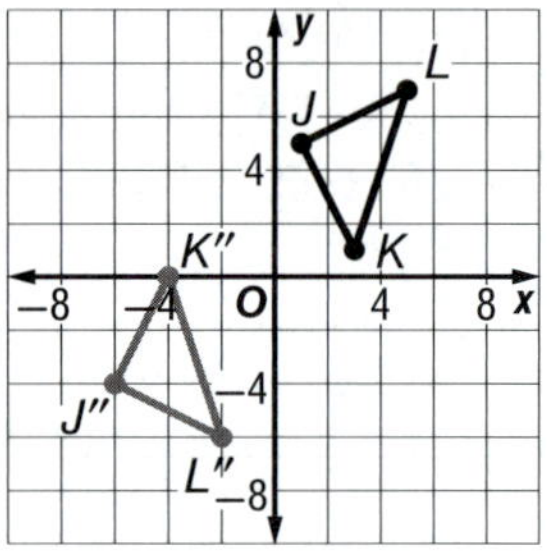

57. (7, –7)

59. reduction; $\frac{1}{2}$

61. enlargement; 3

Pages 660–667 Lesson 9–6

1.

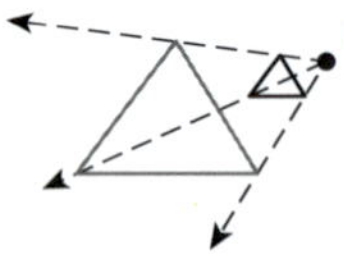

3 The figure increases in size from B to B', so it is an enlargement.

$$\frac{\text{image length}}{\text{preimage length}} = \frac{QB'}{QB}$$
$$= \frac{8}{6} \text{ or } \frac{4}{3}$$

$$QB + BB' = QB'$$
$$6 + x = 8$$
$$x = 2$$

5.

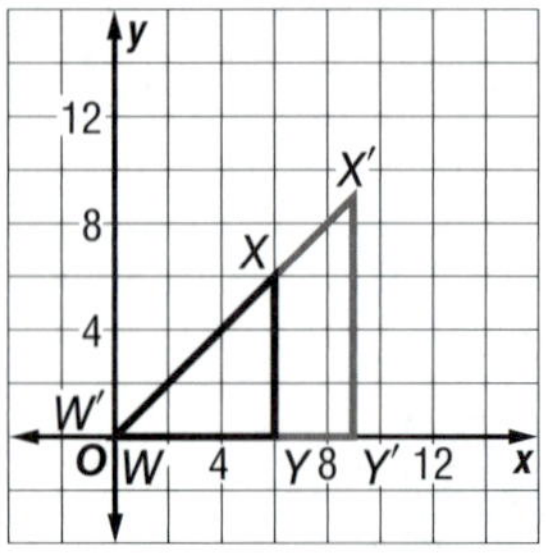

7.

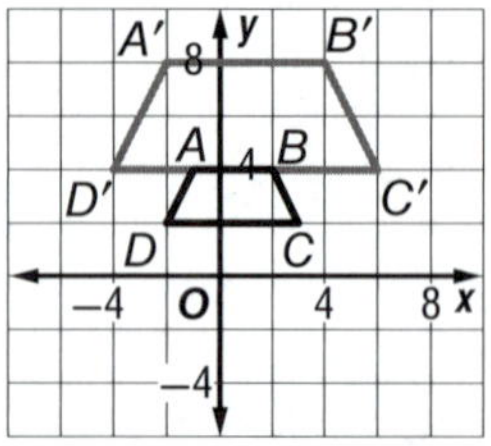

9.

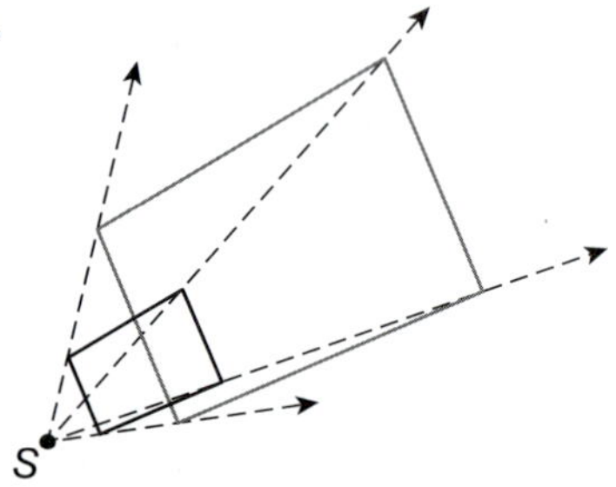

11.

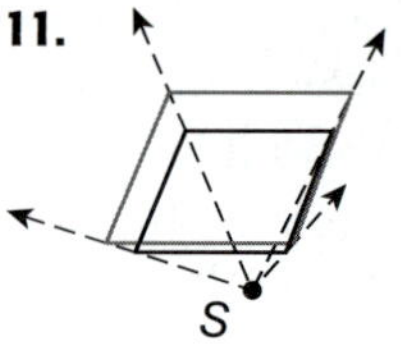

13.

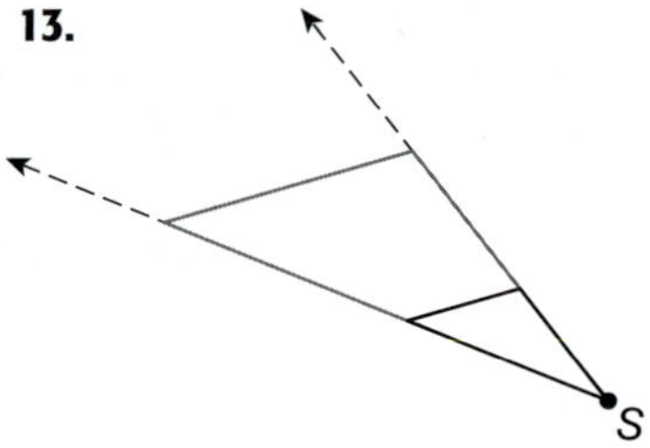

15. enlargement; 2; 4.5 **17.** reduction; $\frac{3}{4}$; 3.5

19. 15×; The insect's image length in millimeters is 3.75 • 10 or 37.5 mm. The scale factor of the dilation is $\frac{37.5}{2.5}$ or 15.

21 Multiply the x- and y-coordinates of each vertex by the scale factor, 0.5.

(x, y)	→	$(0.5x, 0.5y)$
$J(-8, 0)$	→	$J'(-4, 0)$
$K(-4, 4)$	→	$K'(-2, 2)$
$L(-2, 0)$	→	$L'(-1, 0)$

Graph JKL and its image $J'K'L'$.

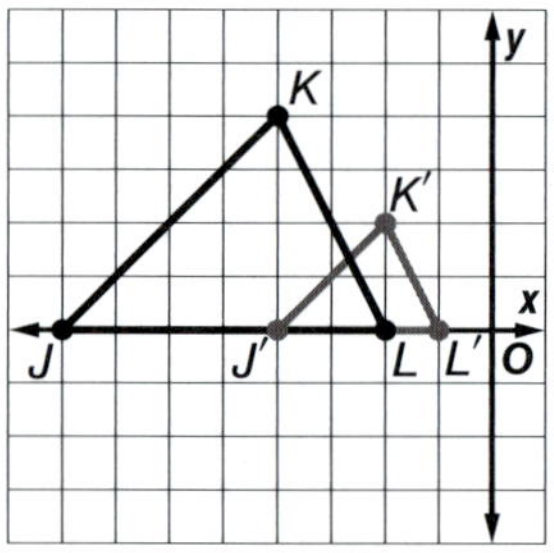

23.

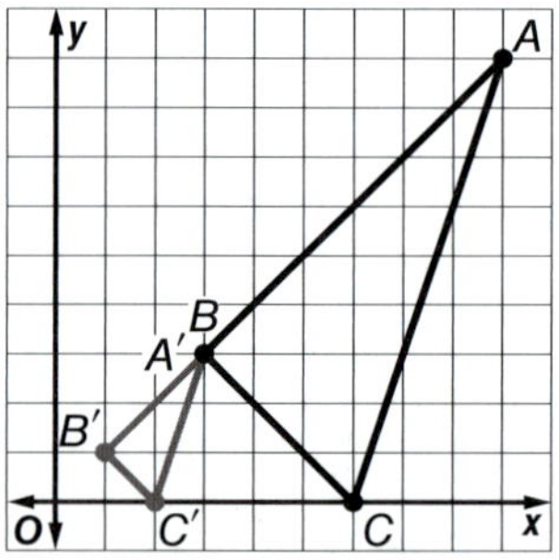

25.

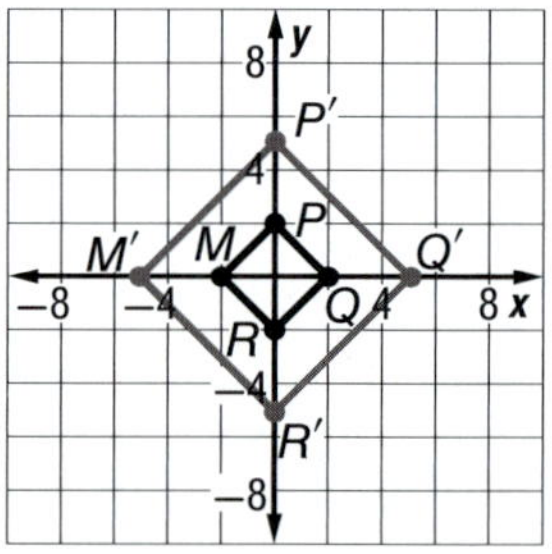

27a.

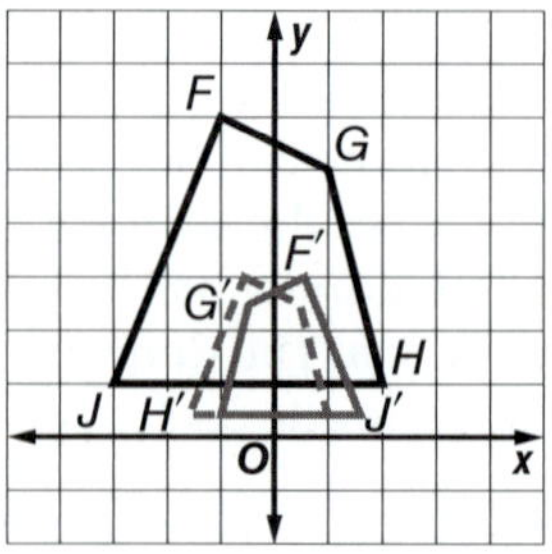

27b.

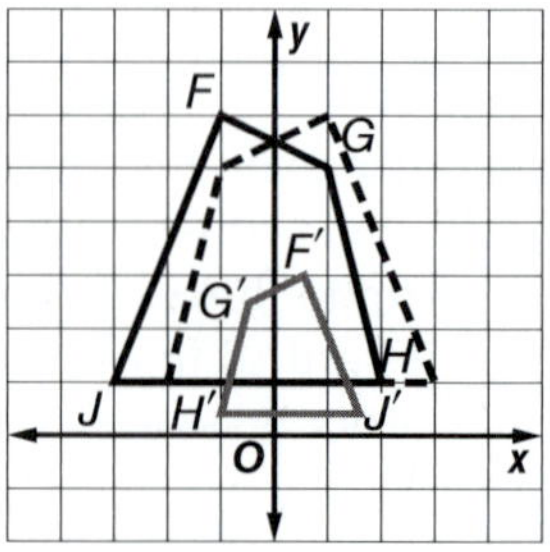

27c. no **27d.** Sometimes; sample answer: For the order of a composition of a dilation centered at the origin and a reflection to be unimportant, the line of reflection must contain the origin, or must be of the form $y = mx$.
29. No; sample answer: The measures of the sides of the rectangles are not proportional, so they are not similar and cannot be a dilation.

31 **a.** $T = Ph + 2B$ Surface area of a prism
$= (16)(4) + 2(12)$ $P = 16$ cm, $h = 4$ cm, $B = 12$ cm^2
$= 88$ cm^2 Simplify.

$V = Bh$ Volume of a prism
$= (12)(4)$ $B = 12$ cm^2, $h = 4$ cm
$= 48$ cm^3 Simplify.

b. Multiply the dimensions by the scale factor 2: length $= 6 \cdot 2$ or 12 cm, width $= 2 \cdot 2$ or 4 cm, height $= 4 \cdot 2$ or 8 cm.

$T = Ph + 2B$ Surface area of a prism
$= (32)(8) + 2(48)$ $P = 32$ cm, $h = 8$ cm, $B = 48$ cm^2
$= 352$ cm^2 Simplify.

$V = Bh$ Volume of a prism
$= (48)(8)$ $B = 48$ cm^2, $h = 8$ cm
$= 384$ cm^3 Simplify.

c. Multiply the dimensions by the scale factor $\frac{1}{2}$: length $= 6 \cdot \frac{1}{2}$ or 3 cm, width $= 2 \cdot \frac{1}{2}$ or 1 cm, height $= 4 \cdot \frac{1}{2}$ or 2 cm.

$T = Ph + 2B$ Surface area of a prism
$= (8)(2) + 2(3)$ $P = 8$ cm, $h = 2$ cm, $B = 3$ cm^2
$= 22$ cm^2 Simplify.

$V = Bh$ Volume of a prism
$= (3)(2)$ $B = 3$ cm^2, $h = 2$ cm
$= 6$ cm^3 Simplify.

d. surface area of preimage: 88 cm^2
surface area of image with scale factor 2: 352 cm^2 or $(88 \cdot 4)$ cm^2
surface area of image with scale factor $\frac{1}{2}$: 22 cm^2 or $\left(88 \cdot \frac{1}{4}\right)$ cm^2
The surface area is 4 times greater after dilation with scale factor 2, $\frac{1}{4}$ as great after dilation with scale factor $\frac{1}{2}$.

volume of preimage: 48 cm^3
volume of image with scale factor 2: 384 cm^3 or $(48 \cdot 8)$ cm^3
volume of image with scale factor $\frac{1}{2}$: 6 cm^3 or $\left(48 \cdot \frac{1}{8}\right)$ cm^2
The volume is 8 times greater after dilation with scale factor 2; $\frac{1}{8}$ as great after dilation with scale factor $\frac{1}{2}$.

e. The surface area of the preimage would be multiplied by r^2. The volume of the preimage would be multiplied by r^3.

33 **a.** $k = \dfrac{\text{diameter of image}}{\text{diameter of preimage}}$
$= \dfrac{2 \text{ mm}}{1.5 \text{ mm}}$
$= \dfrac{2}{1\frac{1}{2}}$
$= 2 \cdot \frac{2}{3}$
$= \frac{4}{3}$ or $1\frac{1}{3}$

b. $A = \pi r^2$ Area of a circle
$= \pi(0.75)^2$ $r = 1.5 \div 2$ or 0.75
≈ 1.77 mm^2 Use a calculator.

$A = \pi r^2$ Area of a circle
$= \pi(1)^2$ $r = 2 \div 2$ or 1
≈ 3.14 mm^2 Use a calculator.

35. $\frac{11}{5}$

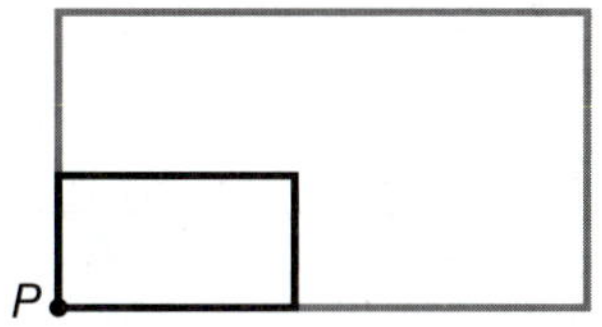

37. $y = 4x - 3$

39a. Sometimes; sample answer: If one of the vertices of the triangle is located at the origin, then that vertex would remain invariant under the dilation. If none of the vertices are located at the origin, then no points will remain invariant under the dilation. **39b.** Always; sample answer: Since a dilation of 1 maps an image onto itself, all four vertices will remain invariant under the dilation. **39c.** Always; sample answer: Since the rotation is centered at B, point B will always remain invariant under the rotation. **39d.** Sometimes: sample answer: If one of the vertices is on the x-axis, then that point will remain invariant under reflection. If two vertices are on the x-axis, then the two vertices located on the x-axis will remain invariant under reflection. **39e.** Never; when a figure is translated, all points move an equal distance. Therefore, no points can remain invariant under translation. **41.** Sample answer: Translations, reflections, and rotations produce congruent figures because the sides and angles of the preimage are congruent to the corresponding sides and angles of the image. Dilations produce similar figures, because the angles of the preimage and the image are congruent and the sides of the preimage are proportional to the corresponding sides of the image. A dilation with a scale factor of 1 produces an equal figure because the image is mapped onto its corresponding parts in the preimage. **43.** A **45.** D

47. yes; 1

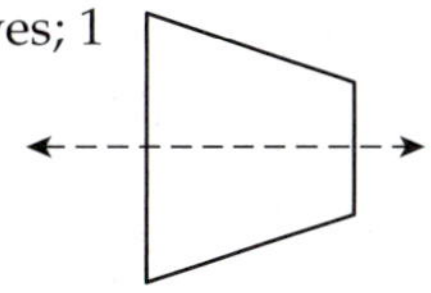

49. translation along $\langle -1, 8 \rangle$ and reflection in the y-axis
51. $2\sqrt{51}$ ft ≈ 14.3 ft
53. 34.6 **55.** 107.1

Pages 670–674 Chapter 9 Study Guide and Review

1. composition of transformations **3.** dilation
5. line of reflection **7.** translation **9.** reflection

11.

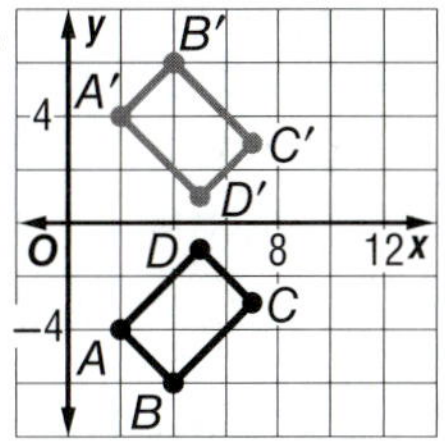

13.

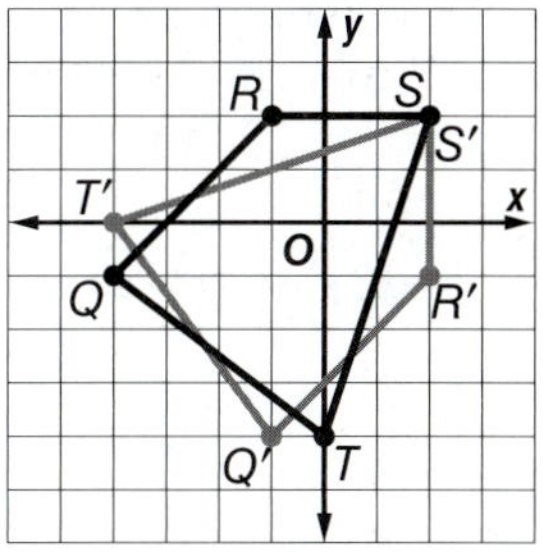

15.
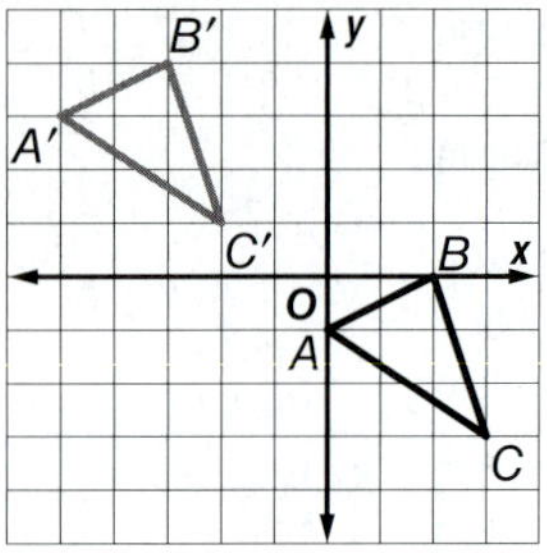

17.
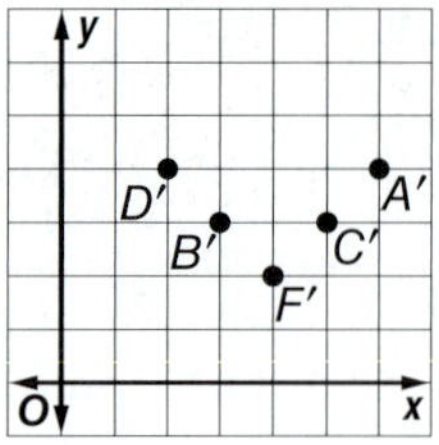

19.
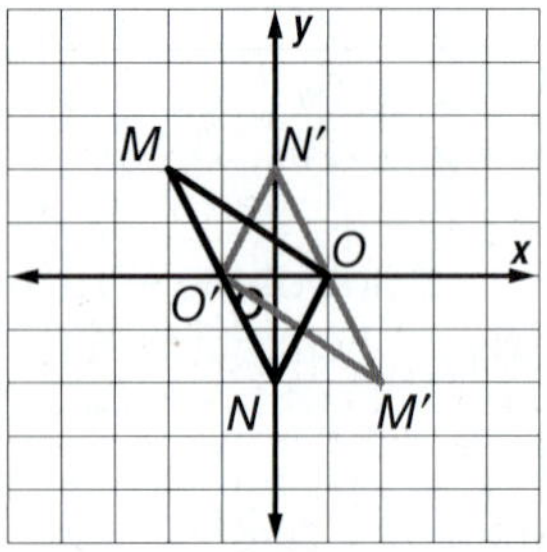

21. 90°
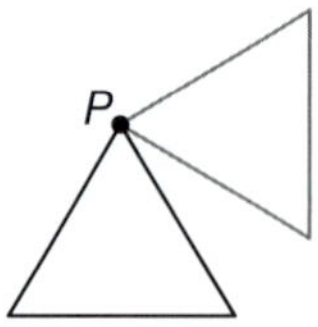

23.
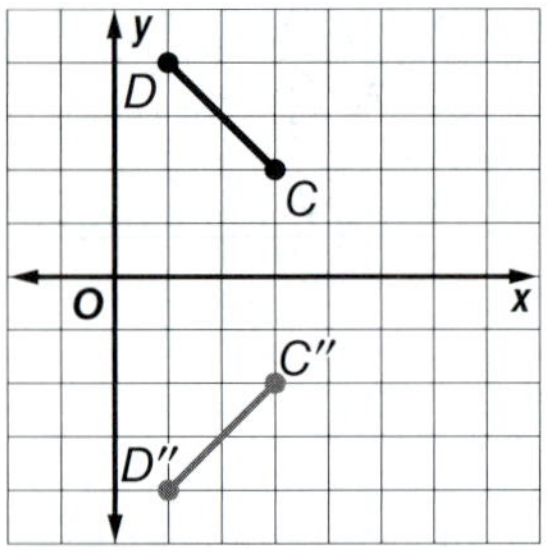

25. Sample answer: translation right and down, translation of result right and up.

27. yes; 2
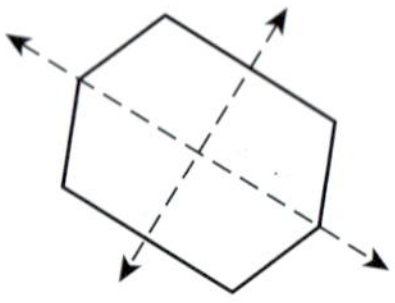

29. yes; 4; 90°
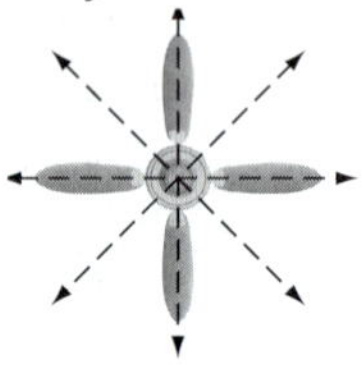

31. 4 33. reduction; 8.25; 0.45

Chapter 10 Circles

Page 681 Chapter 10 Get Ready

1. 130 3. 15.58 5. 82.8 7. $5.85 9. 8.5 ft
11. −3, 4

Pages 683–691 Lesson 10-1

1. ⊙N 3. 8 cm 5. 14 in. 7. 22 ft; 138.23 ft
9. $4\pi\sqrt{13}$ cm 11. $\overline{SU}$ 13. 8.1 cm

15 $d = 2r$ Diameter formula
$= 2(14)$ or 28 in. Substitute and simplify.

17. 3.7 cm 19. 14.6 21. 30.6 23. 13 in.; 81.68 in.
25. 39.47 ft; 19.74 ft 27. 830.23 m; 415.12 m

29
$a^2 + b^2 = c^2$ Pythagorean Theorem
$(6\sqrt{2})^2 + (6\sqrt{2})^2 = c^2$ Substitution
$144 = c^2$ Simplify.
$12 = c$ Take the positive square root of each side.

The diameter is 12π feet.

$C = \pi d$ Circumference formula
$= \pi(12)$ Substitution
$= 12\pi$ ft Simplify.

31. 10π in. 33. 14π yd 35a. 31.42 ft 35b. 4 ft
37. 22.80 ft; 71.63 ft 39. $0.25x$; $0.79x$ 41. neither

43 The radius is 3 units, or $3 \cdot 25 = 75$ feet.
$C = 2\pi r$ Circumference formula
$= 2\pi(75)$ Substitution
$= 150\pi$ Simplify.
≈ 471.2 ft Use a calculator.

45a. Sample answer:
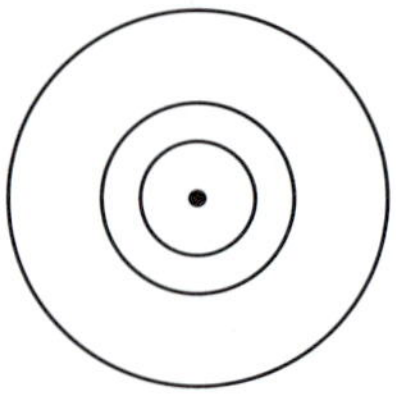

45b.

Circle Radius (cm)	Circumference (cm)
0.5	3.14
1	6.28
2	12.57

45c. They all have the same shape—circular.
45d. The ratio of their circumferences is also 2.
45e. $(C_B) = \frac{b}{a}(C_A)$ 45f. 4 in.

47 a. $C = 2\pi r$ Circumference formula
$= 2\pi(30)$ Substitution
$= 60\pi$ Simplify.

$C = 2\pi r$ Circumference formula
$= 2\pi(5)$ Substitution
$= 10\pi$ Simplify.

$60\pi - 10\pi = 50\pi \approx 157.1$ mi

b. If $r = 5$, $C = 10\pi$; if $r = 10$, $C = 20\pi$; if $r = 15$, $C = 30\pi$, and so on. So, as r increases by 5, C increases by 10π or by about 31.42 miles.

49. Both; Cassie's set of points is equidistant from J in two-dimensional space. Kendrick's set of points are also equidistant from point J, but in three-dimensional space.
51. Always; a radius is a segment drawn between the center of the circle and a point on the circle. A segment drawn from the center to a point inside the circle will always have a length less than the radius of the circle.

53. $\frac{8\pi}{\sqrt{3}}$ or $\frac{8\pi\sqrt{3}}{3}$ 55. 40.8 57. J 59.

61.
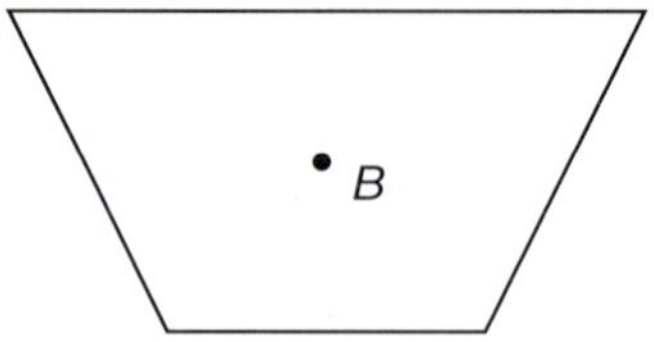

63. no 65. no

67. True; sample answer: Since the hypothesis is true and the conclusion is true, then the statement is true for the conditions. 69. 90 71. 20

Pages 692–700 Lesson 10-2

1. 170 3. major arc; 270 5. semicircle; 180 7. 147
9. 123 11. 13.74 cm

13 $65 + 70 + x = 360$ Sum of central angles
$135 + x = 360$ Simplify.
$x = 225$ Subtract 135 from each side.

15. 40 **17.** minor arc; 125 **19.** major arc; 305 **21.** semicircle; 180 **23.** major arc; 270 **25a.** 280.8; 36 **25b.** major arc; minor arc **25c.** No; no categories share the same percentage of the circle. **27.** 60 **29.** 300 **31.** 180 **33.** 220 **35.** 120

37 $\ell = \frac{x}{360} \cdot 2\pi r$ Arc length equation

$= \frac{112}{360} \cdot 2\pi(4.5)$ Substitution.

≈ 8.80 cm Use a calculator.

39. 17.02 in. **41.** 12.04 m **43.** The length of the arc would double. **45.** 40.84 in. **47.** 9.50 ft **49.** 142

51 **a.** $m\overset{\frown}{AB} = m\angle ACB$ $\overset{\frown}{AB}$ is a minor arc.

$= 180 - (22 + 22)$ Angle Addition Postulate

$= 180 - 44$ or 136 Simplify.

b. $\ell = \frac{x}{360} \cdot 2\pi r$ Arc length equation

$= \frac{136}{360} \cdot 2\pi(62)$ Substitution.

≈ 147.17 ft Use a calculator.

53 **a.**

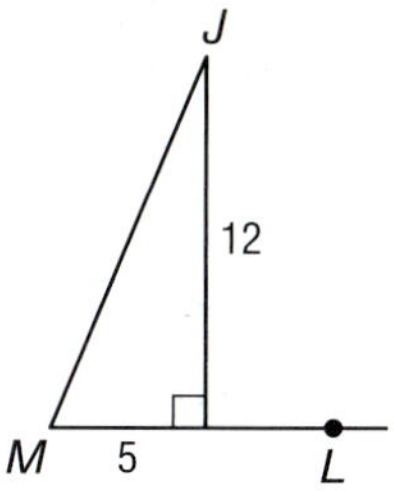

$\tan \angle JML = \frac{12}{5}$

$m\angle JML = \tan^{-1}\left(\frac{12}{5}\right)$

$\approx 67.4°$

$m\overset{\frown}{JL} = m\angle JML \approx 67.4°$

b.

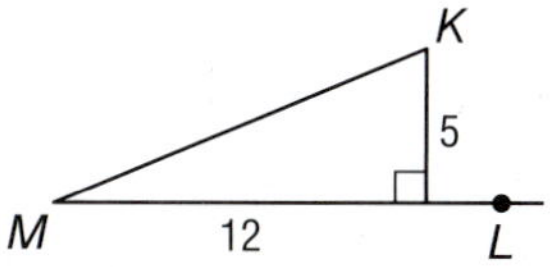

$\tan \angle KML = \frac{5}{12}$

$m\angle KML = \tan^{-1}\left(\frac{5}{12}\right)$

$\approx 22.6°$

$m\overset{\frown}{KL} = m\angle KML \approx 22.6°$

c. $m\angle JMK = m\angle JML - m\angle KML$

$\approx 67.4 - 22.6$

$\approx 44.8°$

$m\overset{\frown}{JK} = m\angle JMK \approx 44.8°$

d. $r = \sqrt{(x_2 - x_1)^2 + (y_2 - y_1)^2}$ Distance Formula

$= \sqrt{(5 - 0)^2 + (12 - 0)^2}$ $(x_1, y_1) = (0, 0)$ and $(x_2, y_2) = (5, 12)$

$= 13$ Simplify.

$\ell = \frac{x}{360} \cdot 2\pi r$ Arc length equation

$= \frac{67.4}{360} \cdot 2\pi(13)$ Substitution

≈ 15.29 units Use a calculator.

e. $\ell = \frac{x}{360} \cdot 2\pi r$ Arc length equation

$= \frac{44.8}{360} \cdot 2\pi(13)$ Substitution

≈ 10.16 units Use a calculator.

55. Selena; the circles are not congruent because they do not have congruent radii. So, the arcs are not congruent. **57.** Never; obtuse angles intersect arcs between 90° and 180° **59.** $m\overset{\frown}{LM} = 150$, $m\overset{\frown}{MN} = 90$, $m\overset{\frown}{NL} = 120$ **61.** 175 **63.** B **65.** H **67.** J **69.** 6.2

71.

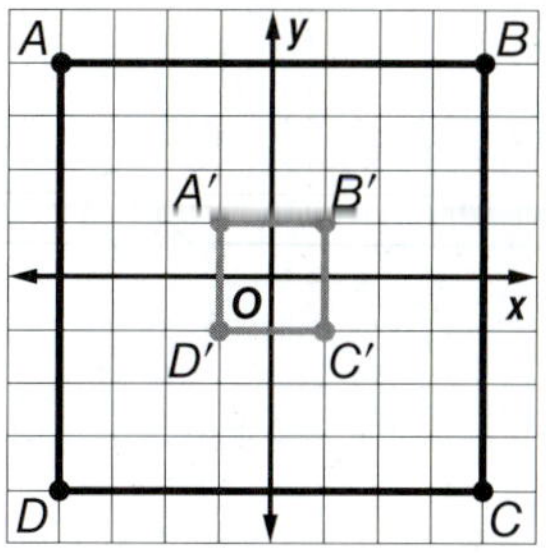

73. $x = \frac{50}{3}$; $y = 10$; $z = \frac{40}{3}$ **75.** 10, −10 **77.** 46.1, −46.1

Pages 701–708 Lesson 10-3

1 $\overset{\frown}{ST}$ is a minor arc, so $m\overset{\frown}{ST} = 93$. $\overline{RS}$ and $\overline{ST}$ are congruent chords, so the corresponding arcs $\overset{\frown}{RS}$ and $\overset{\frown}{ST}$ are congruent.

$\overset{\frown}{RS} \cong \overset{\frown}{ST}$ Corresponding arcs are congruent.

$m\overset{\frown}{RS} = m\overset{\frown}{ST}$ Definition of congruent arcs

$x = 93$ Substitution

3. 3 **5.** 3.32 **7.** 21 **9.** 127 **11.** 7

13 $\overline{KL}$ and $\overline{AJ}$ are congruent chords in congruent circles, so the corresponding arcs $\overset{\frown}{KL}$ and $\overset{\frown}{AJ}$ are congruent.

$\overset{\frown}{KL} \cong \overset{\frown}{AJ}$ Corresponding arcs are congruent.

$m\overset{\frown}{KL} = m\overset{\frown}{AJ}$ Definition of congruent arcs

$5x = 3x + 54$ Substitution

$2x = 54$ Subtract $3x$ from each side.

$x = 27$ Divide each side by 2.

15. 122.5° **17.** 5.34 **19.** 6.71

21 $DE + EC = DC$ Segment Addition Postulate

$15 + EC = 88$ Substitution

$EC = 73$ Subtract 15 from each side.

$EC^2 + EB^2 = CB^2$ Pythagorean Theorem

$73^2 + EB^2 = 88^2$ Substitution

$EB^2 = 2415$ Subtract 73^2 from each side.

$EB \approx 49.14$ Take the positive square root of each side.

$EB = \frac{1}{2}AB$ $\overline{DC} \perp \overline{AB}$, so $\overline{DC}$ bisects $\overline{AB}$.

$2EB = AB$ Multiply each side by 2.

$2(49.14) \approx AB$ Substitution

$98.3 \approx AB$ Simplify.

23. 4

25. Proof:

Because all radii are congruent, $\overline{QP} \cong \overline{PR} \cong \overline{SP} \cong \overline{PT}$. You are given that $\overline{QR} \cong \overline{ST}$, so $\triangle PQR \cong \triangle PST$ by SSS. Thus, $\angle QPR \cong \angle SPT$ by CPCTC. Since the central angles have the same measure, their intercepted arcs have the same measure and are therefore congruent.

Thus, $\overset{\frown}{QR} \cong \overset{\frown}{ST}$.

27. Each arc is 90°, and each chord is 2.12 ft.

29. Given: $\odot L$, $\overline{LX} \perp \overline{FG}$, $\overline{LY} \perp \overline{JH}$, $\overline{LX} \cong \overline{LY}$

Prove: $\overline{FG} \cong \overline{JH}$

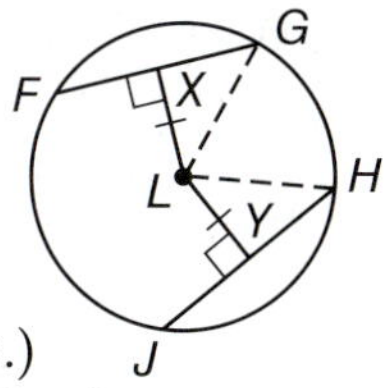

Proof:

Statements (Reasons)

1. $\overline{LG} \cong \overline{LH}$ (All radii of a ⊙ are ≅.)
2. $\overline{LX} \perp \overline{FG}$, $\overline{LY} \perp \overline{JH}$, $\overline{LX} \cong \overline{LY}$ (Given)
3. $\angle LXG$ and $\angle LYH$ are right ∡. (Def. of ⊥ lines)

4. $\triangle XGL \cong \triangle YHL$ (HL)
5. $\overline{XG} \cong \overline{YH}$ (CPCTC)
6. $XG = YH$ (Def. of $\cong$ segments)
7. $2(XG) = 2(YH)$ (Multiplication Property)
8. $\overline{LX}$ bisects $\overline{FG}$; $\overline{LY}$ bisects $\overline{JH}$. (A radius $\perp$ to a chord bisects the chord.)
9. $FG = 2(XG)$, $JH = 2(YH)$ (Def. of seg. bisector)
10. $FG = JH$ (Substitution)
11. $\overline{FG} \cong \overline{JH}$ (Def. of $\cong$ segments)

31 Since $\overline{AB} \perp \overline{CE}$ and $\overline{DF} \perp \overline{CE}$, $\overline{CE}$ bisects $\overline{AB}$ and $\overline{DF}$.
Since $\overline{AB} \cong \overline{DF}$, $AB = DF$.

$CB = \frac{1}{2}AB$	Definition of bisector
$CB = \frac{1}{2}DF$	Substitution
$CB = DE$	Definition of bisector
$9x = 2x + 14$	Substitution
$7x = 14$	Subtract $2x$ from each side.
$x = 2$	Divide each side by 7.

33. 5 **35.** About 17.3; P and Q are equidistant from the endpoints of $\overline{AB}$ so they both lie on the perpendicular bisector of $\overline{AB}$, so $\overline{PQ}$ is the perpendicular bisector of $\overline{AB}$. Hence, both segments of $\overline{AB}$ are 5. Since $\overline{PS}$ is perpendicular to chord $\overline{AB}$, $\angle PSA$ is a right angle. So, $\triangle PSA$ is a right triangle. By the Pythagorean Theorem, $PS = \sqrt{(PA)^2 - (AS)^2}$. By substitution, $PS = \sqrt{11^2 - 5^2}$ or $\sqrt{96}$. Similarly, $\triangle ASQ$ is a right triangle with $SQ = \sqrt{(AQ)^2 - (AS)^2} = \sqrt{9^2 - 5^2}$ or $\sqrt{56}$. Since $PQ = PS + SQ$, $PQ = \sqrt{96} + \sqrt{56}$ or about 17.3.

37a. Given: $\overline{CD}$ is the perpendicular bisector of chord $\overline{AB}$ in $\odot X$.
Prove: $\overline{CD}$ contains point X.
Proof:

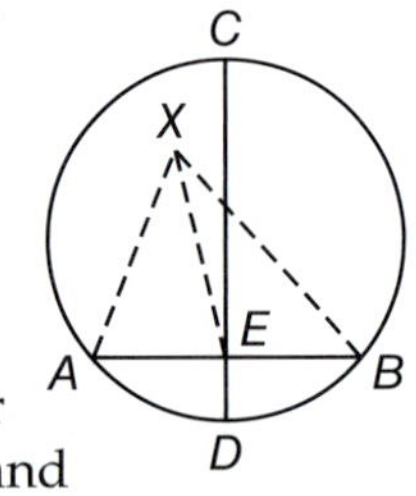

Suppose X is not on $\overline{CD}$. Draw $\overline{XE}$ and radii $\overline{XA}$ and $\overline{XB}$. Since $\overline{CD}$ is the perpendicular bisector of $\overline{AB}$, E is the midpoint of $\overline{AB}$ and $\overline{AE} \cong \overline{EB}$. Also, $\overline{XA} \cong \overline{XB}$, since all radii of a $\odot$ are $\cong$. $\overline{XE} \cong \overline{XE}$ by the Reflexive Property. So, $\triangle AXE \cong \triangle BXE$. By CPCTC, $\angle XEA \cong \angle XEB$. Since $\angle XEA$ and $\angle XEB$ are congruent adjacent angles that make up $\angle AEB$, $\overline{XE} \perp \overline{AB}$. Then $\overline{XE}$ is the perpendicular bisector of $\overline{AB}$. But $\overline{CD}$ is also the perpendicular bisector of $\overline{AB}$. This contradicts the uniqueness of a perpendicular bisector of a segment. Thus, the assumption is false, and center X must be on $\overline{CD}$.

37b. Given: In $\odot X$, X is on $\overline{CD}$ and $\overline{FG}$ bisects $\overline{CD}$ at O.
Prove: Point O is point X.
Proof:

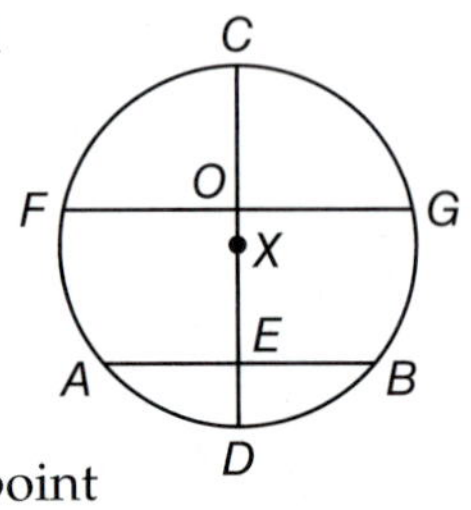

Since point X is on $\overline{CD}$ and C and D are on $\odot X$, $\overline{CD}$ is a diameter of $\odot X$. Since $\overline{FG}$ bisects $\overline{CD}$ at O, O is the midpoint of $\overline{CD}$. Since the midpoint of a diameter is the center of a circle, O, is the center of the circle. Therefore, point O is point X.

39. No; sample answer: In a circle with a radius of 12, a 60° arc would include a chord of 12. If the arc is tripled to 180°, the included chord is 24, which is not a triple of 12.

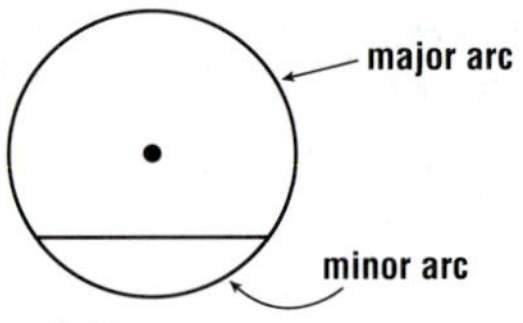

41. F **43.** D **45.** 170 **47.** 275 in.
49. yes; obtuse
$31^2 \stackrel{?}{=} 20^2 + 21^2$
$961 > 400 + 441$
51. ± 11

Pages 701–708 Lesson 10-4

1. 30 **3.** 66 **5.** 54
7. Given: $\overline{RT}$ bisects $\overline{SU}$.
Prove: $\triangle RVS \cong \triangle UVT$
Proof:

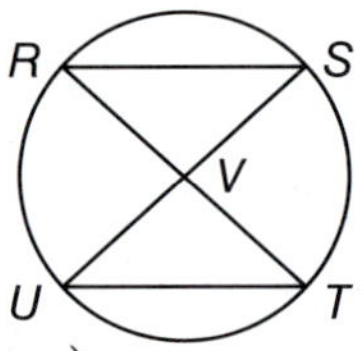

<u>**Statements (Reasons)**</u>
1. $\overline{RT}$ bisects $\overline{SU}$. (Given)
2. $\overline{SV} \cong \overline{VU}$ (Def. of segment bisector)
3. $\angle SRT$ intercepts $\widehat{ST}$. $\angle SUT$ intercepts $\widehat{ST}$. (Def. of intercepted arc)
4. $\angle SRT \cong \angle SUT$ (Inscribed $\angle$s of same arc are $\cong$.)
5. $\angle RVS \cong \angle UVT$ (Vertical $\angle$s are $\cong$.)
6. $\triangle RVS \cong \triangle UVT$ (AAS)

9. 25 **11.** 162

13

$m\widehat{NP} + m\widehat{PQ} + m\widehat{QN} = 360$	Arc Addition Theorem
$120 + 100 + m\widehat{QN} = 360$	Substitution
$220 + m\widehat{QN} = 360$	Simplify.
$m\angle\widehat{QN} = 140$	Subtract 220 from each side.
$m\angle P = \frac{1}{2}m\widehat{QN}$	$\angle P$ intercepts $\widehat{QN}$.
$= \frac{1}{2}(140)$ or 70	Substitution

15. 140 **17.** 32 **19.** 20
21. Given: $m\angle T = \frac{1}{2}m\angle S$
Prove: $m\widehat{TUR} = 2m\widehat{URS}$
Proof:

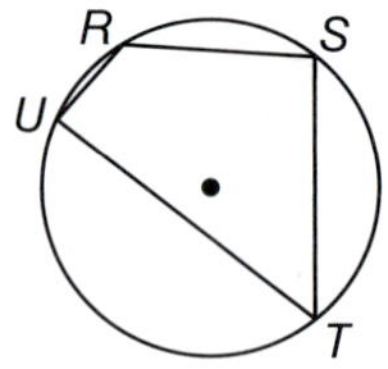

$m\angle T = \frac{1}{2}m\angle S$ means that $m\angle S = 2m\angle T$. Since $m\angle S = \frac{1}{2}m\widehat{TUR}$ and $m\angle T = \frac{1}{2}m\widehat{URS}$, the equation becomes $\frac{1}{2}m\widehat{TUR} = 2\left(\frac{1}{2}m\widehat{URS}\right)$. Multiplying each side of the equation by 2 results in $m\widehat{TUR} = 2m\widehat{URS}$.

23. 30 **25.** 12.75 **27.** 135 **29.** 106
31. Given: Quadrilateral $ABCD$ is inscribed in $\odot O$.
Prove: $\angle A$ and $\angle C$ are supplementary. $\angle B$ and $\angle D$ are supplementary.
Proof: By arc addition and the definitions of arc measure and the sum of central angles,

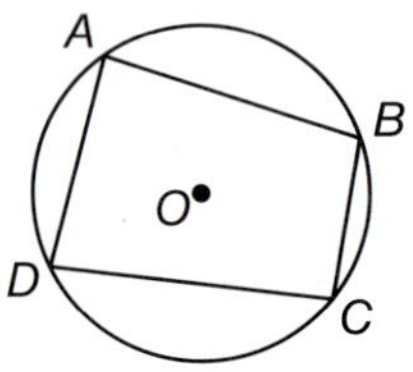

$m\widehat{DCB} + m\widehat{DAB} = 360$. Since

$m\angle C = \frac{1}{2}m\widehat{DAB}$ and

$m\angle A = \frac{1}{2}m\widehat{DCB}$, $m\angle C + m\angle A = \frac{1}{2}(m\widehat{DCB} + m\widehat{DAB})$, but $m\widehat{DCB} + m\widehat{DAB} = 360$, so $m\angle C + m\angle A = \frac{1}{2}(360)$ or 180. This makes $\angle C$ and $\angle A$ supplementary. Because the sum of the measures of the interior angles of a quadrilateral is 360, $m\angle A + m\angle C + m\angle B + m\angle D = 360$. But $m\angle A + m\angle C = 180$, so $m\angle B + m\angle D = 180$, making them supplementary also.

33 Since all the sides of the sign are congruent, all the corresponding arcs are congruent.

$8m\widehat{QR} = 360$, so $m\widehat{QR} = \frac{360}{8}$ or 45.

$m\angle RLQ = \frac{1}{2}m\widehat{QR}$

$= \frac{1}{2}(45)$ or 22.5

35. 135

37. Proof:

Statements (Reasons)

1. $m\angle ABC = m\angle ABD + m\angle DBC$ ($\angle$ Addition Th.)
2. $m\widehat{ADC} = m\widehat{AD} + m\widehat{DC}$ (Arc Addition Theorem)
3. $\frac{1}{2}m\widehat{ADC} = \frac{1}{2}m\widehat{AD} + \frac{1}{2}m\widehat{DC}$ (Multiplication Property)
4. $m\angle ABD = \frac{1}{2}m\widehat{AD}$, $m\angle DBC = \frac{1}{2}m\widehat{DC}$ (The measure of an inscribed $\angle$ whose side is a diameter is half the measure of the intercepted arc (Case 1).)
5. $\frac{1}{2}m\widehat{ADC} = m\angle ABD + m\angle DBC$ (Substitution (Steps 3, 4))
6. $\frac{1}{2}m\widehat{ADC} = m\angle ABC$ (Substitution (Steps 5, 1))

39 Use the Inscribed Angle Theorem to find the measures of $\angle FAE$ and $\angle CBD$. Then use the definition of congruent arcs and the Multiplication Property of Equality to help prove that the angles are congruent.

Given: $\angle FAE$ and $\angle CBD$ are inscribed; $\widehat{DC} \cong \widehat{EF}$

Prove: $\angle FAE = \angle CBD$

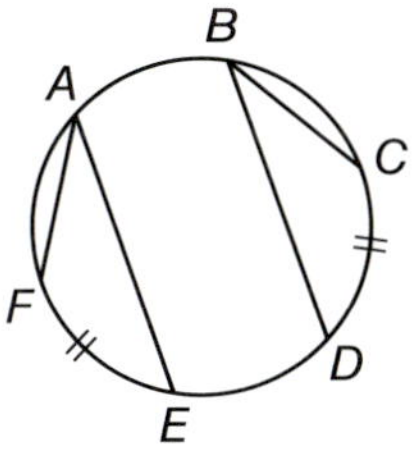

Proof:

Statements (Reasons)

1. $\angle FAE$ and $\angle CBD$ are inscribed; $\widehat{EF} \cong \widehat{DC}$ (Given)
2. $m\angle FAE = \frac{1}{2}m\widehat{EF}$; $m\angle CBD = \frac{1}{2}m\widehat{DC}$ (Measure of an inscribed $\angle$ = half measure of intercepted arc.)
3. $m\widehat{EF} = m\widehat{DC}$ (Def. of $\cong$ arcs)
4. $\frac{1}{2}m\widehat{EF} = \frac{1}{2}m\widehat{DC}$ (Mult. Prop.)
5. $m\angle FAE = m\angle CBD$ (Substitution)
6. $\angle FAE \cong \angle CBD$ (Def. of $\cong$ $\angle$s)

41a.

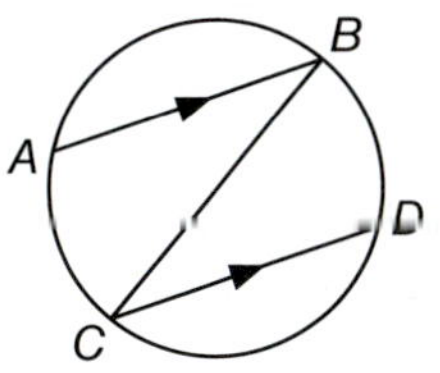

41b. Sample answer: $m\angle A = 30$, $m\angle D = 30$; $m\widehat{AC} = 60$, $m\widehat{BD} = 60$; The arcs are congruent because they have equal measures. **41c.** Sample answer: In a circle, two parallel chords cut congruent arcs. **41d.** 70; 70 **43.** Always; rectangles have right angles at each vertex, therefore opposite angles will be inscribed in a semicircle. **45.** Sometimes; a rhombus can be inscribed in a circle as long as it is a square. Since the opposite angles of rhombi that are not squares are not supplementary, they can not be inscribed in a circle. **47.** $\frac{\pi}{2}$ **49.** See students' work. **51.** A **53.** $d = 17$ in., $r = 8.5$ in., $C = 17\pi$ or about 53.4 in. **55.** 48 **57.** 24 **59.** 107 **61.** 144 **63.** 54 **65.** $\frac{1}{2}$

Pages 718–725 Lesson 10-5

1. no common tangent

3 $FG^2 + GE^2 \stackrel{?}{=} FE^2$

$36^2 + 15^2 \stackrel{?}{=} (24 + 15)^2$

$1521 = 1521$

$\triangle EFG$ is a right triangle with right angle EGF. So $\overline{FG}$ is perpendicular to radius $\overline{EG}$ at point G. Therefore, by Theorem 10.10, $\overline{FG}$ is tangent to $\odot E$.

5. 16 **7.** $x = 250$; $y = 275$; 1550 ft

9.

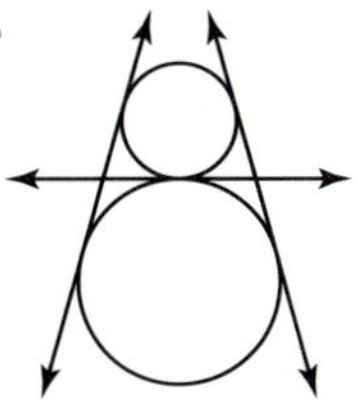

11.

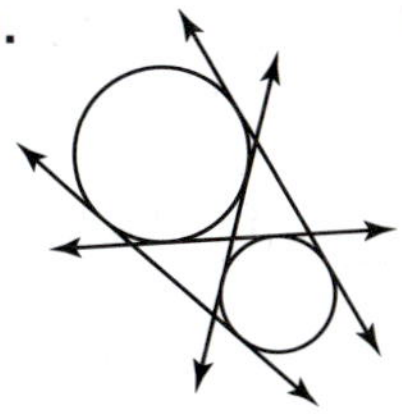

13. yes; 625 = 625

15 $XY^2 + YZ^2 \stackrel{?}{=} XZ^2$

$8^2 + 5^2 \stackrel{?}{=} (3 + 5)^2$

$89 \neq 64$

Since $\triangle XYZ$ is not a right triangle, $\overline{XY}$ is not perpendicular to radius $\overline{YZ}$. So, $\overline{XY}$ is not tangent to $\odot Z$.

17 $\overrightarrow{QP}$ is tangent to $\odot N$ at P. So, $\overrightarrow{QP} \perp \overline{PN}$ and $\triangle PQN$ is a right triangle.

$QP^2 + PN^2 = QN^2$	Pythagorean Theorem
$24^2 + 10^2 = x^2$	$QP = 24$, $PN = 10$, and $QN = x$
$576 + 100 = x^2$	Multiply.
$676 = x^2$	Simplify.
$26 = x$	Take the positive square root of each side.

19. 9 **21.** 4 **23a.** 37.95 in. **23b.** 37.95 in. **25.** 8; 52 cm **27.** 8.06

29. Given: Quadrilateral $ABCD$ is circumscribed about $\odot P$.

Prove: $AB + CD = AD + BC$

Statements (Reasons)

1. Quadrilateral $ABCD$ is circumscribed about $\odot P$. (Given)

2. Sides $\overline{AB}$, $\overline{BC}$, $\overline{CD}$, and $\overline{DA}$ are tangent to $\odot P$ at points H, G, F, and E, respectively. (Def. of circumscribed)
3. $\overline{EA} \cong \overline{AH}$; $\overline{HB} \cong \overline{BG}$; $\overline{GC} \cong \overline{CF}$; $\overline{FD} \cong \overline{DE}$ (Two segments tangent to a circle from the same exterior point are $\cong$.)
4. $AB = AH + HB$, $BC = BG + GC$, $CD = CF + FD$, $DA = DE + EA$ (Segment Addition)
5. $AB + CD = AH + HB + CF + FD$; $DA + BC = DE + EA + BG + GC$ (Substitution)
6. $AB + CD = AH + BG + GC + FD$; $DA + BC = FD + AH + BG + GC$ (Substitution)
7. $AB + CD = FD + AH + BG + GC$ (Commutative Prop. of Add.)
8. $AB + CD = DA + BC$ (Substitution)

31

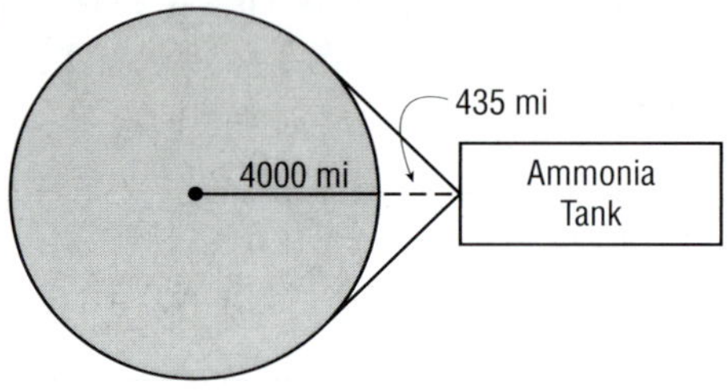

$4000^2 + x^2 = (4000 + 435)^2$	Pythagorean Theorem
$x^2 = 3{,}669{,}225$	Subtract 4000^2 from each side.
$x \approx 1916$ mi	Take the positive square root of each side.

33. Proof: Assume that ℓ is not tangent to $\odot S$. Since ℓ intersects $\odot S$ at T, it must intersect the circle in another place. Call this point Q. Then $ST = SQ$. But if $\overline{ST} \perp \ell$, then $\overline{ST}$ must be the shortest segment from S to ℓ. Since T and Q are two different points on ℓ, this is a contradiction. Therefore, ℓ is tangent to $\odot S$.

35. Sample answer: Using the Pythagorean Theorem, $2^2 + x^2 = 10^2$, so $x \approx 9.8$. Since $PQSR$ is a rectangle, $PQ = x = 9.8$.

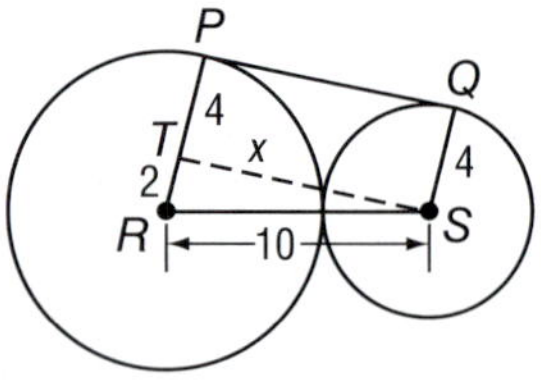

37. Sample answer:

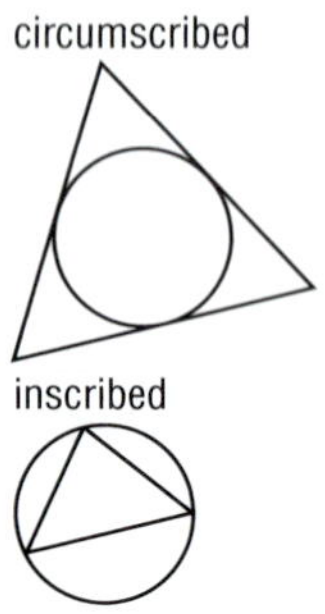

39. From a point outside the circle, two tangents can be drawn. From a point on the circle, one tangent can be drawn. From a point inside the circle, no tangents can be drawn because a line would intersect the circle in two points.
41. $6\sqrt{2}$ or about 8.5 in.

43. C **45.** 61 **47.** 71 **49.** 109 **51.** Yes; $\triangle AEC \sim \triangle BDC$ by AA Similarity. **53.** 110 **55.** 58

Pages 727–735 Lesson 10-6

1. 110 **3.** 73 **5.** 248

7 Draw and label a diagram.

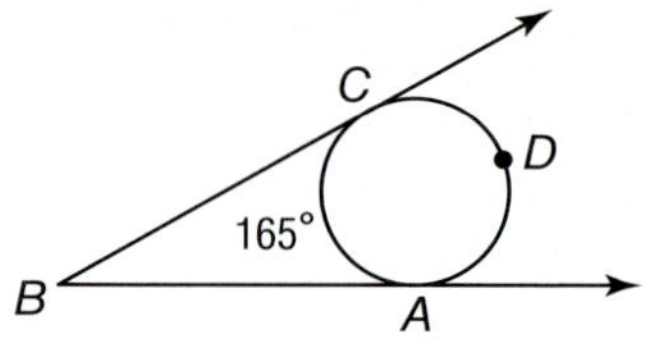

$m\angle B = \frac{1}{2}(m\widehat{CDA} - m\widehat{CA})$	Theorem 10.14
$= \frac{1}{2}[(360 - 165) - 165]$	Substitution
$= \frac{1}{2}(195 - 165)$	Subtract.
$= \frac{1}{2}(30)$ or 15	Simplify.

9. 71.5

11

$51 = \frac{1}{2}(m\widehat{RQ} + m\widehat{NP})$	Theorem 10.12
$51 = \frac{1}{2}(m\widehat{RQ} + 74)$	Substitution
$102 = m\widehat{RQ} + 74$	Multiply each side by 2.
$28 = m\widehat{RQ}$	Subtract 74 from each side.

13. 144 **15.** 125 **17a.** 100 **17b.** 20 **19.** 74 **21.** 185
23. 22 **25.** 168

27

$3 = \frac{1}{2}[(5x - 6) - (4x + 8)]$	Theorem 10.14
$6 = (5x - 6) - (4x + 8)$	Multiply each side by 2.
$6 = x - 14$	Simplify.
$20 = x$	Add 14 to each side.

29a. 145 **29b.** 30

31. <u>Statements (Reasons)</u>
1. $\overrightarrow{FM}$ is a tangent to the circle and $\overrightarrow{FL}$ is a secant to the circle. (Given)
2. $m\angle FLH = \frac{1}{2}m\widehat{HG}$, $m\angle LHM = \frac{1}{2}m\widehat{LH}$ (The meas. of an inscribed $\angle = \frac{1}{2}$ the measure of its intercepted arc.)
3. $m\angle LHM = m\angle FLH + m\angle F$ (Exterior $\triangle$ Th.)
4. $\frac{1}{2}m\widehat{LH} = \frac{1}{2}m\widehat{HG} + m\angle F$ (Substitution)
5. $\frac{1}{2}m\widehat{LH} - \frac{1}{2}m\widehat{HG} = m\angle F$ (Subtraction Prop.)
6. $\frac{1}{2}(m\widehat{LH} - m\widehat{HG}) = m\angle F$ (Distributive Prop.)

33. Proof:
$\angle CAB$ and $\angle CAE$ form a linear pair, so $m\angle CAB + m\angle CAE = 180$. Since $\angle CAB$ is obtuse, $\angle CAE$ is acute and Case 1 applies, so $m\angle CAE = \frac{1}{2}m\widehat{CA}$. $m\widehat{CA} + m\widehat{CDA} = 360$, so $\frac{1}{2}m\widehat{CA} + \frac{1}{2}m\widehat{CDA} = 180$ by Mult. Prop., and $m\angle CAE + \frac{1}{2}m\widehat{CDA} = 180$ by substitution. By the Transitive Prop., $m\angle CAB + m\angle CAE = m\angle CAE + \frac{1}{2}m\widehat{CDA}$, so by Subtraction Prop., $m\angle CAB = \frac{1}{2}m\widehat{CDA}$

35 **a.** Sample answer:

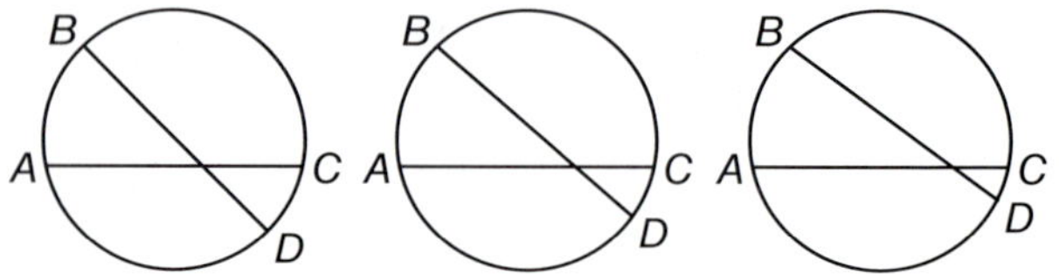

b. Sample answer:

Arc	Circle 1	Circle 2	Circle 3
$\overset{\frown}{CD}$	25	15	5
$\overset{\frown}{AB}$	50	50	50
x	37.5	32.5	27.5

c. As the measure of $\overset{\frown}{CD}$ gets closer to 0, the measure of x approaches half of $m\overset{\frown}{AB}$; $\angle AEB$ becomes an inscribed angle.
d. Theorem 10.12 states that if two chords intersect in the interior of a circle, then the measure of an angle formed is one half the sum of the measure of the arcs intercepted by the angle and its vertical angle. Use this theorem to write an equation relating x, $m\overset{\frown}{AB}$, and $m\overset{\frown}{CD}$. Then let $m\overset{\frown}{CD} = 0$ and simplify. The result is Theorem 10.6, the Inscribed Angle Theorem.

$x = \frac{1}{2}(m\overset{\frown}{AB} + m\overset{\frown}{CD})$

$x = \frac{1}{2}(m\overset{\frown}{AB} + 0)$

$x = \frac{1}{2}m\overset{\frown}{AB}$

37. 15
39a. $m\angle G \leq 90$; $m\angle G < 90$ for all values except when $\overrightarrow{JG} \perp \overrightarrow{GH}$ at G, then $m\angle G = 90$. **39b.** $m\overset{\frown}{KH} = 56$; $m\overset{\frown}{HJ} = 124$; Because a diameter is involved the intercepted arcs measure $(180 - x)$ and x degrees. Hence, solving $\frac{180 - x - x}{2} = 34$ leads to the answer.
41. Sample answer: Using Theorem 10.14, $60° = \frac{1}{2}[(360° - x) - x]$ or $120°$; repeat for $50°$ to get $130°$. The third arc can be found by adding $50°$ and $60°$ and subtracting from $360°$ to get $110°$. **43.** J **45.** B **47.** 8
49. Given: $\overset{\frown}{MHT}$ is a semicircle.
$\overline{RH} \perp \overline{TM}$.
Prove: $\frac{TR}{RH} = \frac{TH}{HM}$

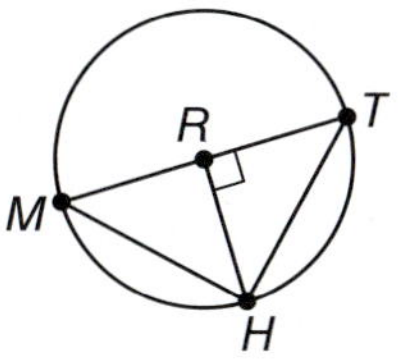

Proof:
Statements (Reasons)
1. MHT is a semicircle; $\overline{RH} \perp \overline{TM}$. (Given)
2. $\angle THM$ is a right angle. (If an inscribed $\angle$ intercepts a semicircle, the $\angle$ is a rt. $\angle$.)
3. $\angle TRH$ is a right angle. (Def. of $\perp$ lines)
4. $\angle THM \cong \angle TRH$ (All rt. angles are $\cong$.)
5. $\angle T \cong \angle T$ (Reflexive Prop.)
6. $\triangle TRH \sim \triangle THM$ (AA Sim.)
7. $\frac{TR}{RH} = \frac{TH}{HM}$ (Def. of $\sim \triangle$s)

51. 54.5° **53.** −4, −9 **55.** 0, −5 **57.** −6

Pages 736–742 Lesson 10-7

1. 2 **3.** 5
5 Let T be the endpoint of $\overline{QT}$, the diameter that passes through point S.

$PS \cdot SR = QS \cdot ST$	Theorem 10.15
$10 \cdot 10 = 6 \cdot ST$	Substitution
$100 = 6ST$	Simplify.
$\frac{50}{3} = ST$	Divide each side by 6.

So, the diameter of the circle is $6 + \frac{50}{3}$ or $\frac{68}{3}$ centimeters.

$C = \pi d$	Circumference formula
$= \pi\left(\frac{68}{3}\right)$	Substitution
≈ 71.21 cm	Use a calculator.

7. 5 **9.** 14 **11.** 3.1
13

$CD^2 = CB \cdot CA$	Theorem 10.17
$12^2 = x \cdot (x + 12)$	Substitution
$144 = x^2 + 12x$	Simplify.
$0 = x^2 + 12x - 144$	Subtract 144 from each side.
$x = \frac{-b \pm \sqrt{b^2 - 4ac}}{2a}$	Quadratic Formula
$= \frac{-12 \pm \sqrt{12^2 - 4(1)(-144)}}{2(1)}$	$a = 1, b = 12, c = -144$
$= \frac{-12 \pm \sqrt{720}}{2}$	Disregard the negative solution.
≈ 7.4	Use a calculator.

15. 13 in. **17.** 7.1 **19.** $a \approx 15$; $b \approx 11.3$ **21.** $c \approx 22.8$; $d \approx 16.9$
23 Inscribed angles that intercept the same arc are congruent. Use this theorem to find two pairs of congruent angles. Then use AA Similarity to show that two triangles in the figure are similar. Finally, use the definition of similar triangles to write a proportion. Find the cross products.
Proof:
Statements (Reasons)
1. $\overline{AC}$ and $\overline{DE}$ intersect at B. (Given)
2. $\angle A \cong \angle D$, $\angle E \cong \angle C$ (Inscribed $\angle$s that intercept the same arc are $\cong$.)
3. $\triangle ABE \sim \triangle DBC$ (AA Similarity)
4. $\frac{AB}{BD} = \frac{EB}{BC}$ (Def. of $\sim \triangle$s)
5. $AB \cdot BC = EB \cdot BD$ (Cross products)

25. Proof:
Statements (Reasons)
1. tangent $\overline{JK}$ and secant $\overline{JM}$ (Given)
2. $m\angle KML = \frac{1}{2}m\overset{\frown}{KL}$ (The measure of an inscribed $\angle$ equals half the measure of its intercept arc.)
3. $m\angle JKL = \frac{1}{2}m\overset{\frown}{KL}$ (The measure of an $\angle$ formed by a secant and a tangent = half the measure of its intercepted arc.)
4. $m\angle KML = m\angle JKL$ (Substitution)
5. $\angle KML \cong m\angle JKL$ (Definition of $\cong$ $\angle$s)
6. $\angle J \cong \angle J$ (Reflexive Property)
7. $\triangle JMK \sim \triangle JKL$ (AA Similarity)
8. $\frac{JK}{JL} = \frac{JM}{JK}$ (Definition of $\sim \triangle$s)
9. $JK^2 \cdot BC = JL \cdot JM$ (Cross products)

27. Sample answer: When two secants intersect in the exterior of a circle, the product equation equates the product of the exterior segment measure and the whole segment measure for each secant. When a secant and a tangent intersect, the product involving the tangent segment becomes (measure of tangent segments)^2 because the exterior segments and the whole segments are the same segment.
29. Sometimes; they are equal when the chords are perpendicular. **31.** Sample answer: The product of the parts on one intersecting chord equals the product of the parts of the other chord. **33.** G **35.** D

37.

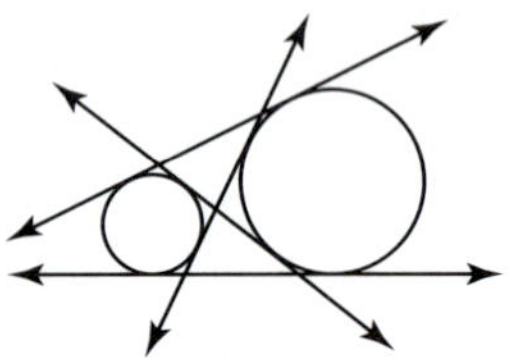

39.

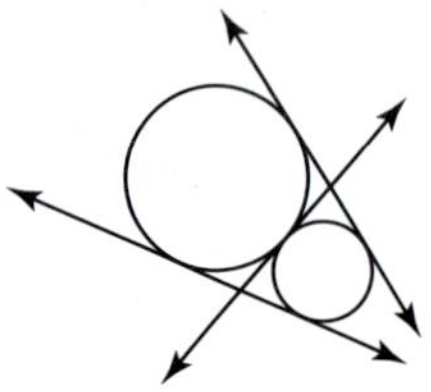

41.

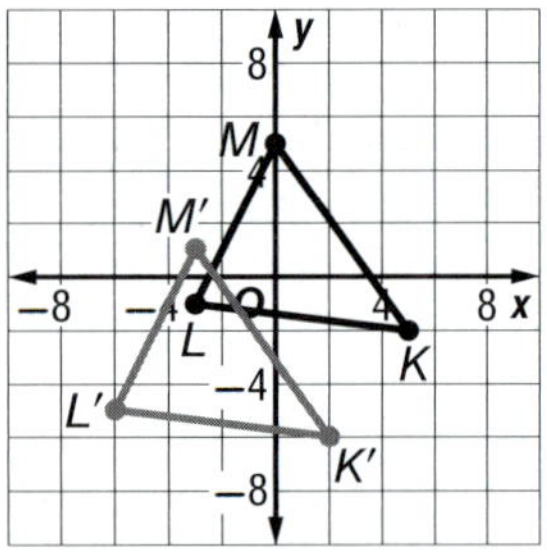

43.

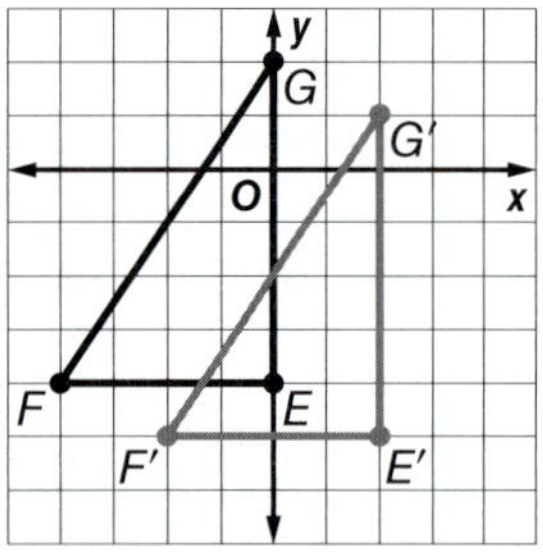

45. $y = 2x + 8$ **47.** $y = \frac{2}{9}x + \frac{1}{3}$ **49.** $y = -\frac{1}{12}x + 1$

Pages 744–749 Lesson 10-8

1. $(x - 9)^2 + y^2 = 25$

3

$$r = \sqrt{(x_2 - x_1)^2 + (y_2 - y_1)^2} \quad \text{Distance Formula}$$
$$= \sqrt{(2 - 0)^2 + (2 - 0)^2} \quad (x_1, y_1) = (0, 0) \text{ and } (x_2, y_2) = (2, 2)$$
$$= \sqrt{8} \quad \text{Simplify.}$$
$$(x - h)^2 + (y - k)^2 = r^2 \quad \text{Equation of a circle}$$
$$(x - 0)^2 + (y - 0)^2 = (\sqrt{8})^2 \quad h = 0, k = 0, \text{ and } r = \sqrt{8}$$
$$x^2 + y^2 = 8 \quad \text{Simplify.}$$

5. $(x - 2)^2 + (y - 1)^2 = 4$
7. $(3, -2)$; 4

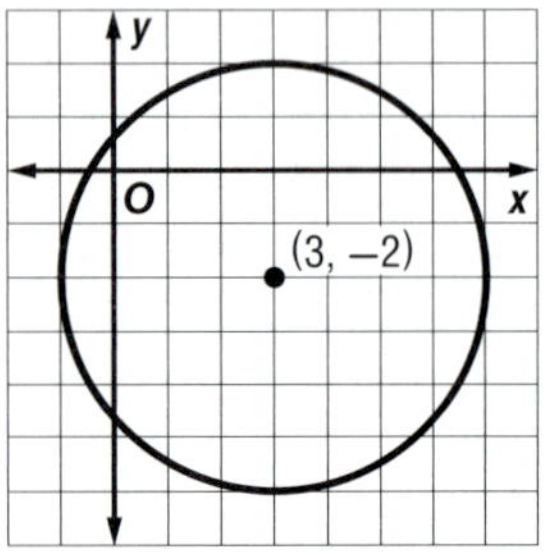

9. $(2, -1)$; $(x - 2)^2 + (y + 1)^2 = 40$
11. $x^2 + y^2 = 16$
13. $(x + 2)^2 + y^2 = 64$
15. $(x + 3)^2 + (y - 6)^2 = 9$
17. $(x + 5)^2 + (y + 1)^2 = 9$

19 The third ring has a radius of 15 + 15 + 15 or 45 miles.

$$(x - h)^2 + (y - k)^2 = r^2 \quad \text{Equation of a circle}$$
$$(x - 0)^2 + (y - 0)^2 = 45^2 \quad h = 0, k = 0, \text{ and } r = 45$$
$$x^2 + y^2 = 2025 \quad \text{Simplify.}$$

21. $(0, 0)$; 6

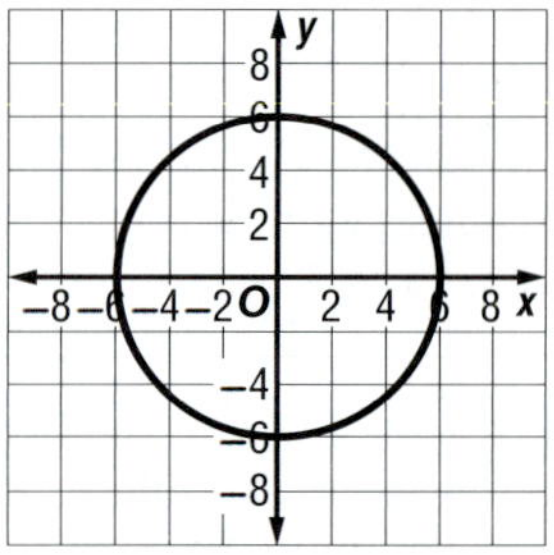

23. $(-4, 2)$; 4

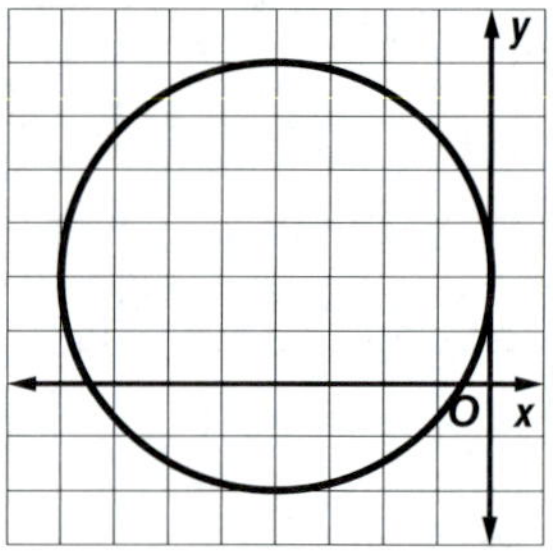

25. $(x - 3)^2 + (y - 3)^2 = 13$

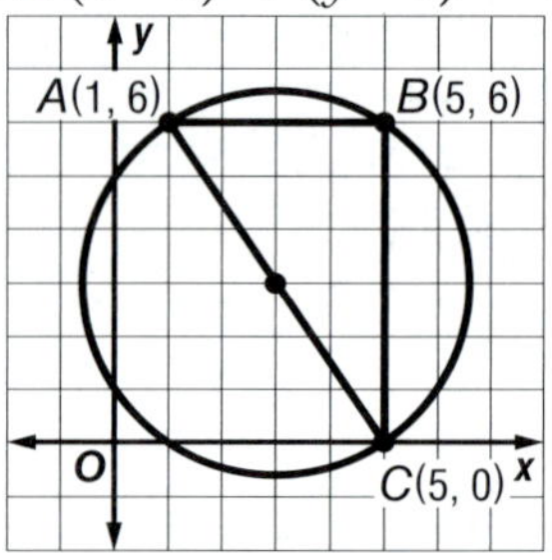

27. $(x - 3)^2 + y^2 = 25$
29a. $x^2 + y^2 = 810{,}000$
29b. 3000 ft

31a. $(x + 4)^2 + (y - 5)^2 = 36$

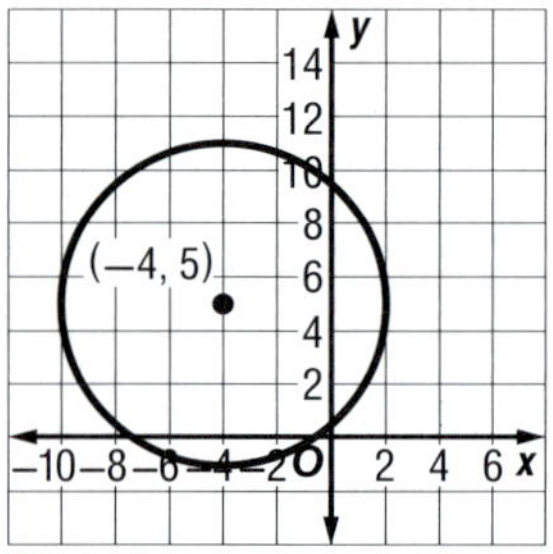

31b. The circle represents the boundary of the delivery region. All homes within the circle get free delivery. Consuela's home at (0, 0) is located outside the circle, so she cannot get free delivery.

33 Write the equation $x^2 + 6x + y^2 - 2y = 15$ in the form $(x - h)^2 + (y - k)^2 = r^2$.

$$x^2 + 6x + y^2 - 2y = 15 \quad \text{Original equation}$$
$$(x^2 + 6x + ?) + (y^2 - 2y + ?) = 15 \quad \text{Associative Property}$$
$$(x^2 + 6x + 9) + (y^2 - 2y + 1) = 15 + 9 + 1 \quad \text{Add 9 and 1 to each side.}$$
$$(x + 3)^2 + (y - 1)^2 = 25 \quad \text{Simplify.}$$

The center is at $(-3, 1)$ and the radius is 5.

35. $(x + 5)^2 + (y - 2)^2 = 36$ **37.** $(x - 8)^2 + (y - 2)^2 = 16$; the first circle has its center at $(5, -7)$. If the circle is shifted 3 units right and 9 units up, the new center is at $(8, 2)$, so the new equation becomes $(x - 8)^2 + (y - 2)^2 = 16$. **39.** Sample answer: A circle is the locus of all points in a plane (coordinate plane) a given distance (the radius) from a given point (the center) The equation of a circle is derived using the Distance Formula by using the given point and the radius.
41. A **43.** Step 1 **45.** 3 **47.** 5.6 **49.** 53 **51.** 157 ft
53. 32 cm; 64 cm^2

Pages 750–754 Chapter 10 Study Guide and Review

1. false, chord **3.** true **5.** true **7.** false, two **9.** false, congruent **11.** $\overline{DM}$ or $\overline{DP}$ **13.** 13.69 cm; 6.84 cm **15.** 34.54 ft; 17.27 ft **17.** 163 **19a.** 100.8 **19b.** 18 **19c.** minor arc **21.** 131 **23.** 50.4 **25.** 56

27.

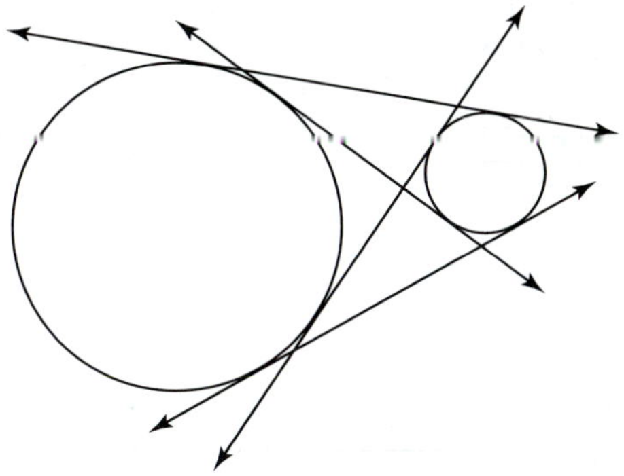

29. 97 **31.** 146 **33.** 4 **35.** $(x + 2)^2 + (y - 4)^2 = 25$ **37.** The radius of the circle is 19 + 15 or 34 inches, and (h, k) is (0, 0). Therefore the equation is $(x - 0)^2 + (y - 0)^2 = 34^2$ or $x^2 + y^2 = 34^2$.

Chapter 11 Areas of Polygons and Circles

Page 761 Chapter 11 Get Ready

1. 5 **3.** 20 **5.** 11 ft **7.** 123 **9.** 78 **11.** 11 **13.** $3\sqrt{2}$ cm

Pages 763–770 Lesson 11-1

1. 56 in., 180 in^2 **3.** 64 cm, 207.8 cm^2 **5.** 43.5 in., 20 in^2 **7.** 28.5 in., 33.8 in^2 **9.** 11 cm

11 Perimeter = 21 + 17 + 21 + 17 or 76 ft

Use the Pythagorean Theorem to find the height.

$8^2 + h^2 = 17^2$ Pythagorean Theorem

$64 + h^2 = 289$ Simplify.

$h^2 = 225$ Subtract 64 from each side.

$h = 15$ Take the positive square root of each side.

$A = bh$ Area of a parallelogram

$= 21(15)$ or 315 ft^2 $b = 21$ and $h = 15$

13. 69.9 m, 129.9 m^2 **15.** 174.4 m, 1520 m^2 **17.** 727.5 ft^2 **19.** 338.4 cm^2 **21.** 480 m^2

23

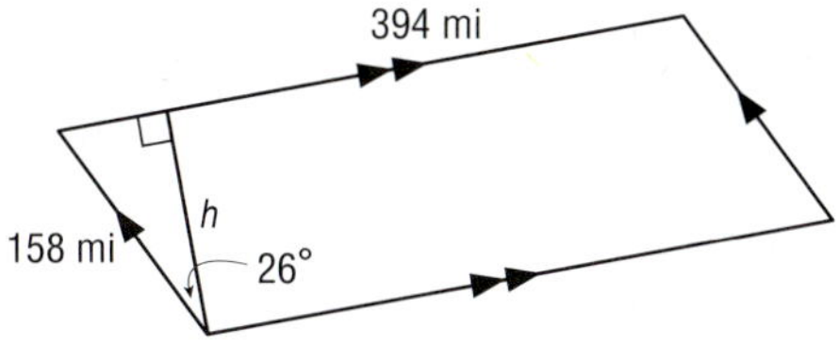

$\cos 26° = \frac{h}{158}$ $\cos = \frac{\text{adjacent}}{\text{hypotenuse}}$

$158 \cos 26° = h$ Multiply each side by 158.

$142 \approx h$ Use a calculator.

$A = bh$ Area of a parallelogram

$\approx 394(142)$ or 55,948 mi^2 $b = 394$ and $h = 142$

25. $b = 12$ cm; $h = 3$ cm **27.** $b = 11$ m; $h = 8$ m **29.** 1 pint dark blue, 3 pints of light blue **31.** 187.7 mm^2

33 Graph the parallelogram then measure the length of the base and the height and calculate the area.

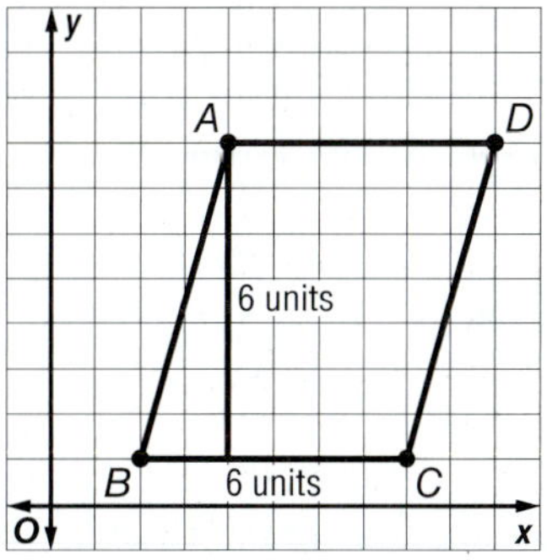

$A = bh$ Area of a parallelogram

$\approx 6(6)$ or 36 units^2 $b = 6$ and $h = 6$

35a. 10.9 units^2

35b. $\sqrt{s(s - a)(s - b)(s - c)} \stackrel{?}{=} \frac{1}{2}bh$

$\sqrt{15(15 - 5)(15 - 12)(15 - 13)} \stackrel{?}{=} \frac{1}{2}(5)(12)$

$\sqrt{15(10)(3)(2)} \stackrel{?}{=} 30$

$\sqrt{900} \stackrel{?}{=} 30$

$30 = 30$

37. 15 units^2; Sample answer: I inscribed the triangle in a 6-by-6 square. I found the area of the square and subtracted the areas of the three right triangles inside the square that were positioned around the given triangle. The area of the given triangle is the difference, or 15 units^2. **39.** Sample answer: The area will not change as K moves along line p. Since lines m and p are parallel, the perpendicular distance between them is constant. That means that no matter where K is on line p, the perpendicular distance to line p, or the height of the triangle, is always the same. Since point J and L are not moving, the distance between them, or the length of the base, is constant. Since the height of the triangle and the base of the triangle are both constant, the area will always be the same. **41.** Sample answer: To find the area of the parallelogram, you can measure the height $\overline{PT}$ and then measure one of the bases $\overline{PQ}$ or $\overline{SR}$ and multiply the height by the base to get the area. You can also measure the height $\overline{SW}$ and measure one of the bases $\overline{QR}$ or $\overline{PS}$ and then multiply the height by the base to get the area. It doesn't matter which side you choose to use as the base, as long as you use the height that is perpendicular to that base to calculate the area. **43.** 6 **45.** B **47.** $x^2 + y^2 = 36$ **49.** $(x - 1)^2 + (y + 4)^2 = 17$ **51.** 5.6 **53.** Sample answer: if each pair of opposite sides are parallel, the quadrilateral is a parallelogram. **55.** 9 **57.** 12

Pages 773–780 Lesson 11-2

1. 132 ft^2 **3.** 178.5 mm^2 **5.** 8 cm **7.** 6.3 ft **9.** 678.5 ft^2 **11.** 136 in^2 **13.** 137.5 ft^2

15 $A = \frac{1}{2}d_1d_2$ Area of a kite

$= \frac{1}{2}(4.8)(10.2)$ $d_1 = 4.8$ and $d_2 = 10.2$

$= 24.48$ Simplify.

The area is about 24.5 square microns.

17. 784 ft^2

19 Let x represent the length of one diagonal. Then the length of the other diagonal is $3x$.

$A = \frac{1}{2}d_1d_2$ Area of a rhombus

$168 = \frac{1}{2}(x)(3x)$ $A = 168$, $d_1 = x$, and $d_2 = 3x$

$168 = \frac{3}{2}x^2$ Simplify.

$112 = x^2$ Multiply each side by $\frac{2}{3}$.

$\sqrt{112} = x$ Take the positive square root of each side.

So the lengths of the diagonals are $\sqrt{112}$ or about 10.6 centimeters and $3(\sqrt{112})$ or about 31.7 centimeters.

21. 4 m **23.** The area of $\triangle HJF = \frac{1}{2}\left(\frac{1}{2}d_1\right)d_2$ and the area of $\triangle HGF = \frac{1}{2}\left(\frac{1}{2}d_1\right)d_2$. Therefore, the area of $\triangle HJF = \frac{1}{4}d_1d_2$, and the area of $\triangle HGF = \frac{1}{4}d_1d_2$. The area of kite $FGHJ$ is equal to the area of $\triangle HJF$ + the area of $\triangle HGF$ or $\frac{1}{4}d_1d_2 + \frac{1}{4}d_1d_2$. After simplification, the area of kite $FGHJ$ is equal to $\frac{1}{2}d_1d_2$. **25a.** 24 in^2 each of yellow, red, orange, green, and blue; 20 in^2 of purple **25b.** Yes; her kite has an area of 140 in^2, which is less than 200 in^2. **27.** 18 sq. units **29.** The area of a trapezoid is $\frac{1}{2}h(b_1 + b_2)$. So, $A = \frac{1}{2}(x + y)(x + y)$ or $\frac{1}{2}(x^2 + xy + y^2)$. The area of $\triangle 1 = \frac{1}{2}(y)(x)$, $\triangle 2 = \frac{1}{2}(z)(z)$, and $\triangle 3 = \frac{1}{2}(x)(y)$. The area of $\triangle 1 + \triangle 2 + \triangle 3 = \frac{1}{2}xy + \frac{1}{2}z^2 + \frac{1}{2}xy$. Set the area of the trapezoid equal to the combined areas of the triangles to get $\frac{1}{2}(x^2 + 2xy + y^2) = \frac{1}{2}xy + \frac{1}{2}z^2 + \frac{1}{2}xy$. Multiply by 2 on each side: $x^2 + 2xy + y^2 = 2xy + z^2$. When simplified, $x^2 + y^2 = z^2$.

31 $A = \frac{1}{2}d_1d_2$ Area of a kite

$= \frac{1}{2}(17)(21)$ $d_1 = 17$ and $d_2 = 21$

$= 178.5\text{ m}^2$ Simplify.

$178.5\text{ m}^2 \cdot \frac{10.764\text{ ft}^2}{1\text{ m}^2} \approx 1921.4\text{ ft}^2$

33a.

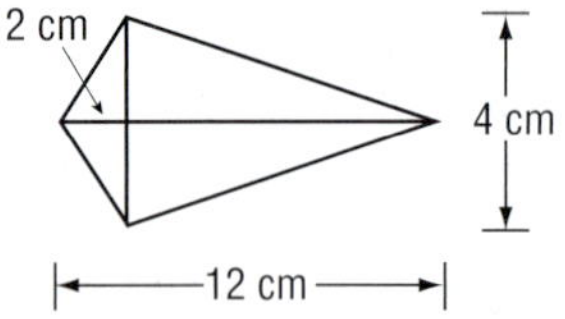

33b.

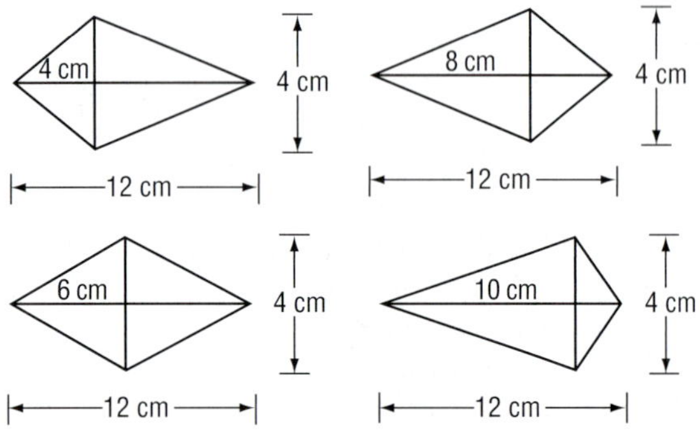

33c.

x	P
2 cm	28.0 cm
4 cm	27.2 cm
6 cm	26.9 cm
8 cm	27.2 cm
10 cm	28.0 cm

33d.

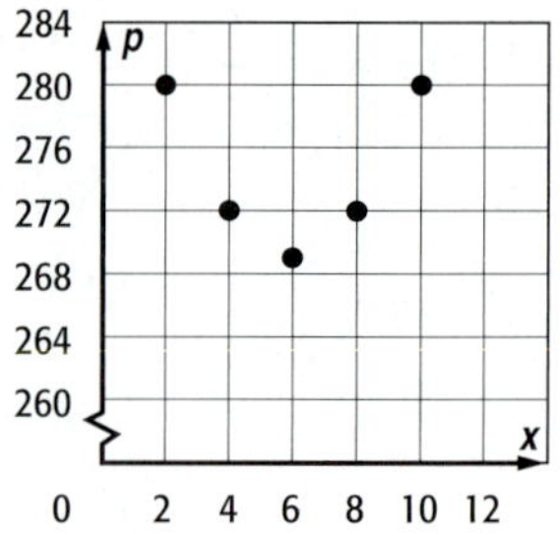

33e. Sample answer: Based on the graph, the perimeter will be minimized when $x = 6$. This value is significant because when $x = 6$, the figure is a rhombus. **35.** 7.2 **37.** Sometimes; sample answer: If the areas are equal, it means that the products of the diagonals are equal. The only time that the perimeters will be equal is when the diagonals are also equal, or when the two rhombi are congruent. **39.** A **41.** J **43.** 17.5 units^2 **45.** $x^2 + y^2 = 1600$ **47.** $x = 9$; $y = 9\sqrt{3}$ **49.** always **51.** never **53.** 18.8 in.; 28.3 in^2 **55.** 36.4 ft; 105.7 ft^2

Pages 782–788 Lesson 11-3

1. 1385.4 yd^2

3 $A = \pi r^2$ Area of a circle

$74 = \pi r^2$ $A = 74$

$23.55 \approx r^2$ Divide each side by π.

$4.85 \approx r$ Take the positive square root of each side.

So, the diameter is $2 \cdot 4.85$ or about 9.7 millimeters.

5. 4.5 in^2 **7a.** 10.6 in^2 **7b.** \$48 **9.** 78.5 yd^2 **11.** 14.2 in^2 **13.** 78.5 ft^2 **15.** 10.9 mm **17.** 8.1 ft

19 $A = \frac{x}{360} \cdot \pi r^2$ Area of a sector

$= \frac{72}{360} \cdot \pi(8)^2$ $x = 72$ and $r = 8$

$\approx 40.2\text{ cm}^2$ Use a calculator.

21. 322 m^2 **23.** 284 in^2 **25a.** 1.7 cm^2 each **25b.** about 319.4 mg **27.** 13 **29.** 9.8

31 **a.** $C = \pi d$ Circumference of a circle

$2.5 = \pi d$ $C = 2.5$

$0.8\text{ ft} \approx d$ Divide each side by π.

b. age = diameter · growth factor

= 0.8 · 4.5 or 3.6 yr

33. 53.5 m^2 **35.** 10.7 cm^2 **37.** 7.9 in^2 **39.** 30 mm^2

41 The area equals the area of the large semicircle with a radius of 6 in. plus the area of a small semicircle minus 2 times the area of a small semicircle. The radius of each small semicircles is 2 in.

$A = \frac{1}{2}\pi(6)^2 + \frac{1}{2}\pi(2)^2 - 2\left[\frac{1}{2}\pi(2)^2\right]$

$= 18\pi + 2\pi - 4\pi$

$= 16\pi$

$\approx 50.3\text{ in}^2$

43a. $A = \frac{x\pi r^2}{360} - r^2\left[\sin\left(\frac{x}{2}\right)\cos\left(\frac{x}{2}\right)\right]$

43b.

x	A
10	0.1
20	0.5
30	1.7
40	4.0
45	5.6
50	7.7
60	13.0
70	20.3
80	29.6
90	41.1

43c.

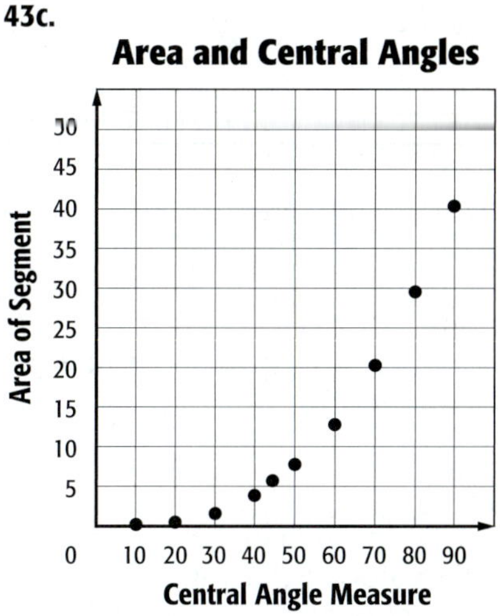

43d. Sample answer: From the graph, it looks like the area would be about 15.5 when x is 63°. Using the formula, the area is 15.0 when x is 63°. The values are very close because I used the formula to create the graph. **45.** 925.5 cm^2 **47.** Sample answer: You can find the shaded area of the circle by subtracting x from 360° and using the resulting measure in the formula for the area of a sector. You could also find the shaded area by finding the area of the entire circle, finding the area of the unshaded sector using the formula for the area of a sector, and subtracting the area of the unshaded sector from the area of the entire circle. The method in which you find the ratio of the area of a sector to the area of the whole circle is more efficient. It requires less steps, is faster, and there is a lower probability for error. **49.** Sample answer: If the radius of the circle doubles, the area will not double. If the radius of the circle doubles, the area will be four times as great. Since the radius is squared, if you multiply the radius by 2, you multiply the area by 2^2, or 4. If the arc length of a sector is doubled, the area of the sector is doubled. Since the arc length is not raised to a power, if the arc length is doubled, the area would also be twice as large. **51.** $x = 56$; $m\angle MTQ = 117$; $m\angle PTM = 63$ **53.** B **55.** 13.2 cm, 26.4 cm **57.** 178.2 in^2 **59.** 7 **61.** 31

Pages 791–799 Lesson 11-4

1. center: point F, radius: $\overline{FD}$, apothem: $\overline{FG}$, central angle: $\angle CFD$, 90° **3.** 162 in^2 **5.** 239 ft^2

7 a. The blue area equals the area of the center circle with a radius of 3 ft plus 2 times the quantity of the area a rectangle 19 ft by 12 ft minus the area of a semicircle with a radius of 6 ft.

Area
= Area of circle + 2 · Area of rectangle − Area of semicircle

$= \pi r^2 + 2 \cdot \left(\ell w - \frac{1}{2}\pi r^2\right)$

$= \pi(3)^2 + 2\left[(19(12) - \frac{1}{2}\pi(6)^2\right]$

$= 9\pi + 2(228 - 18\pi)$

$= 9\pi + 456 - 36\pi$

$= 456 - 27\pi$

≈ 371 ft^2

b. The red area equals the area of the center circle with a radius of 6 ft minus the center circle with a radius of 3 ft plus 2 times the area of a circle with a radius of 6 ft.

Area
= Area of large circle − Area of small circle + 2 · Area of circle

$= \pi r^2 - \pi r^2 + 2 \cdot \pi r^2$

$= \pi(6)^2 - \pi(3)^2 + 2\pi(6)^2$

$= 36\pi - 9\pi + 72\pi$

$= 99\pi$

≈ 311 ft^2

9. center: point R, radius: $\overline{RL}$, apothem: $\overline{RS}$, central angle: $\angle KRL$, 60° **11.** 59.4 cm^2 **13.** 584.2 in^2

15 The figure can be separated into a rectangle with a length of 12 cm and a width of 10 cm and a triangle with a base of 12 cm and a height of 16 cm − 10 cm or 6 cm.

Area of figure = Area of rectangle + Area of triangle

$= \ell w + \frac{1}{2}bh$

$= 12(10) + \frac{1}{2}(12)(6)$

$= 120 + 36$ or 156 cm^2

17. 55.6 in^2 **19.** 42.1 yd^2 **21a.** 52.3 in^2 **21b.** 16
23. 5.8 in^2 **25.** 63.417 ft^2
27. 1438.3 mm^2

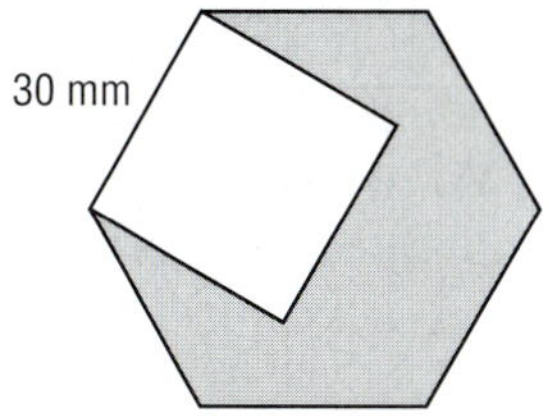

29 To find the area of the shaded region, find the area of the rectangle 8 units by 4 units minus the area of the semicircle with a radius of 2 units minus the area of the trapezoid with bases 4 units and 2 units and height 2 units.

Area of figure
= Area of rectangle − Area of semicircle − Area of trapezoid

$= \ell w - \frac{1}{2}\pi r^2 - \frac{1}{2}h(b_1 + b_2)$

$= 8(4) - \frac{1}{2}\pi(2)^2 - \frac{1}{2}(2)(4 + 2)$

$= 32 - 2\pi - 6$

$= 26 - 2\pi$

≈ 19.7 units^2

31. 24 units^2 **33.** 0.43 in^2; 0.56 in^2; 0.62 in^2; 0.65 in^2; Sample answer: When the perimeter of a regular polygon is constant, as the number of sides increases, the area of the polygon increases.
35. Chloe; sample answer: The measure of each angle of a regular hexagon is 120°, so the segments from the center to each vertex form 60° angles. The triangles formed by the segments from the center to each vertex are equilateral, so each side of the

hexagon is 11 in. The perimeter of the hexagon is 66 in. Using trigonometry, the length of the apothem is about 9.5 in. Putting the values into the formula for the area of a regular polygon and simplifying, the area is about 313.5 in^2.

37. Sample answer:

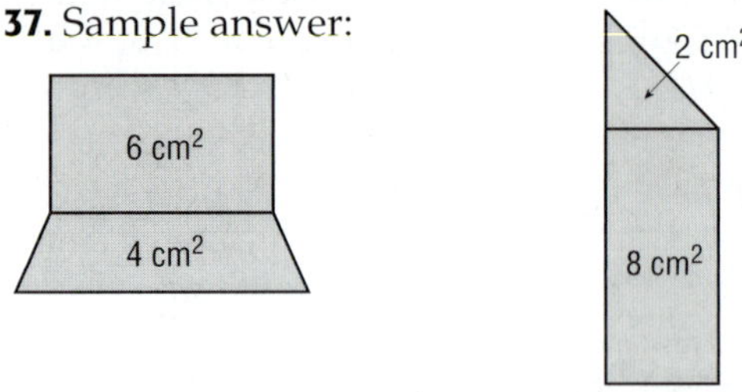

39. Sample answer: If you draw an apothem to the side of the octagon to form a right triangle with the side of the figure and the segment from the center to the vertex, you can use the triangle to find the apothem and the perimeter. Since the octagon is a regular octagon, the measure of each vertex is 135°, so the angle formed by the side of the figure and the segment from the center to the vertex is 67.5°. Using the sine ratio, the length of the apothem is the radius multiplied by sin 67.5. Using the cosine ratio, one half of the length of a side is the radius multiplied by cos 67.5, so the side is twice the radius multiplied by cos 67.5. Find the perimeter by multiplying the length of the side by 8. Use the formula $A = \frac{1}{2}Pa$ with the perimeter and apothem lengths you calculated. **41.** F **43.** C **45.** 254.5 cm^2 **47.** 490.9 mm^2 **49.** 272 in^2 **51.** semicircle; 180 **53.** major arc; 270 **55.** 30

Pages 802–808 Lesson 11-5

1. 9 yd^2 **3.** $\frac{5}{3}$; 35 **5.** 5.28 in^2

7 The scale factor between the parallelograms is $\frac{7.5}{15}$ or $\frac{1}{2}$, so the ratio of their areas is $\left(\frac{1}{2}\right)^2$ or $\frac{1}{4}$.

$$\frac{\text{area of small figure}}{\text{area of large figure}} = \frac{1}{4} \quad \text{Write a proportion.}$$

$$\frac{60}{\text{area of large figure}} = \frac{1}{4} \quad \text{Substitution}$$

$$60 \cdot 4 = \text{area of large figure} \cdot 1 \quad \text{Cross multiply.}$$

$$240 = \text{area of large figure} \quad \text{Simplify.}$$

So the area of the large parallelogram is 240 ft^2.

9. 672 cm^2 **11.** $\frac{4}{5}$; 17.5 **13.** $\frac{3}{2}$; 36 **15a.** 4 in. **15b.** Larger; sample answer: The area of a circular pie pan with an 8 in. diameter is about 50 in^2. The area of the larger pan is 52.6 in^2, and the area of the smaller pan is 41.6 in^2. The area of the larger pan is closer to the area of the circle, so Kaitlyn should choose the larger pan to make the recipe. **17a.** If the area is doubled, the radius changes from 24 in. to 33.9 in. **17b.** If the area is tripled, the radius changes from 24 in. to 41.6 in. **17c.** If the area changes by a factor of x, then the radius changes from 24 in. to $24\sqrt{x}$ in.

19 Area of $\triangle JKL = \frac{1}{2}bh$

$$= \frac{1}{2}(5)(6) \text{ or } 15 \text{ square units}$$

The scale factor between the triangles is $\frac{5}{3}$, so the ratio of their areas is $\left(\frac{5}{3}\right)^2$ or $\frac{25}{9}$.

$$\frac{\text{area of } \triangle JKL}{\text{area of } \triangle J'K'L'} = \frac{25}{9} \quad \text{Write a proportion.}$$

$$\frac{15}{\triangle J'K'L'} = \frac{25}{9} \quad \text{Area of } \triangle JKL = 15$$

$$15 \cdot 9 = \text{area of } \triangle J'K'L' \cdot 25 \quad \text{Cross multiply.}$$

$$5.4 = \text{area of } \triangle J'K'L' \quad \text{Divide each side by 25.}$$

So the area of $\triangle J'K'L'$ is 5.4 units^2.

21. area of $ABCD = 18$; area of $A'B'C'D' \approx 56.2$

23 **a.** Sample answer: The graph is misleading because the tennis balls used to illustrate the number of participants are circles. When the height of the tennis ball increases, the width of the tennis ball also increases. For example, the height of the tennis ball representing 1995 is about 2.6 and the height of the tennis ball representing 2000 is about 3. So, the rate of increase in the heights is $\frac{3 - 2.6}{2000 - 1995}$ or about 8%. The area of the circle representing 1995 is $\pi(1.3)^2$ and the area of the circle representing 2000 is $\pi(1.5)^2$. So, the rate of increase in the areas is $\frac{2.25\pi - 1.69\pi}{2000 - 1995}$ or about 35%. The area of the tennis ball increases at a greater rate than the height of the tennis ball, so it looks like the number of participants in high school tennis is increasing more than it actually is.

b. Sample answer: If you use a figure with a constant width to represent the participation in each year and only change the height, the graph would not be misleading. For example, use rectangles of equal width and height that varies.

25. Neither; sample answer: In order to find the area of the enlarged circle, you can multiply the radius by the scale factor and substitute it into the area formula, or you can multiply the area formula by the scale factor squared. The formula for the area of the enlargment is $A = \pi(kr)^2$ or $A = k^2\pi r^2$. **27.** $P_{enlarged} = Q\sqrt{R}$ **29.** Sample answer: If you know the area of the original polygon and the scale factor of the enlargement, you can find the area of the enlarged polygon by multiplying the original area by the scale factor squared. **31.** J **33.** C **35.** 66.3 cm^2 **37.** 37.4 in^2 **39.** 142.5 **41.** both **43.** $\overline{LP}$ **45.** $\overline{BM}, \overline{AL}, \overline{EP}, \overline{OP}, \overline{PL}, \overline{LM}, \overline{MN}$

Pages 809–812 Chapter 11 Study Guide and Review

1. false; height **3.** false; radius **5.** true **7.** false; height of a parallelogram **9.** false; base **11.** $P = 50$ cm; $A = 60$ cm^2 **13.** $P = 13.2$ mm; $A = 6$ mm^2 **15.** 132 in^2 **17.** 96 cm^2 **19.** 336 cm^2 **21.** 1.5 m^2 **23.** 59 in^2 **25.** 166.3 ft^2 **27.** 65.0 m^2 **29.** 696 in^2 **31.** $\frac{1}{2}$; 8

33. area of $\triangle RST = 18$ square units; area of $\triangle R'S'T' = 4.5$ square units **35.** 16.4 miles

Chapter 12 Extending Surface Area and Volume

Page 819 Chapter 12 Get Ready

1. true **3.** true **5.** true **7.** 168 in^2 **9.** 176 in^2 **11.** ± 15

Pages 823–828 Lesson 12-1

1. Sample answer:

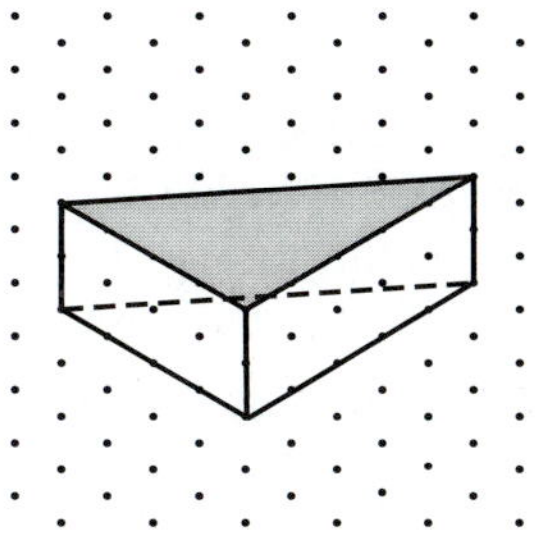

3.

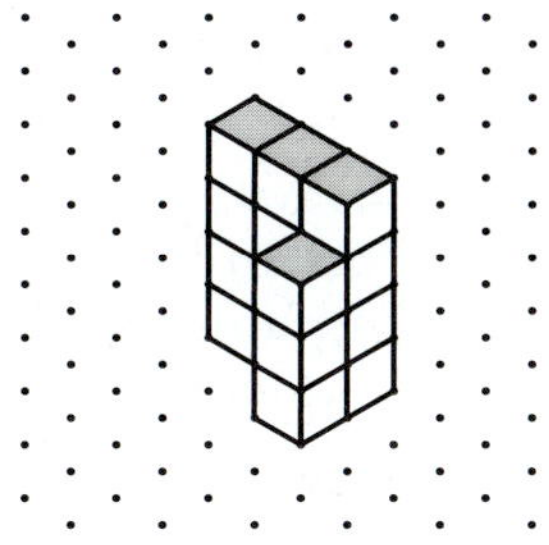

5a. slice vertically **5b.** slice horizontally **5c.** slice at an angle **7.** triangle

9 Sample answer: First mark the corner of the solid. Then draw 4 units down, 1 unit to the left, and 3 units to the right. Draw a triangle for the top of the solid. Draw segments 4 units down from each vertex for the vertical edges. Connect the appropriate vertices. Use a dashed line for the hidden edge.

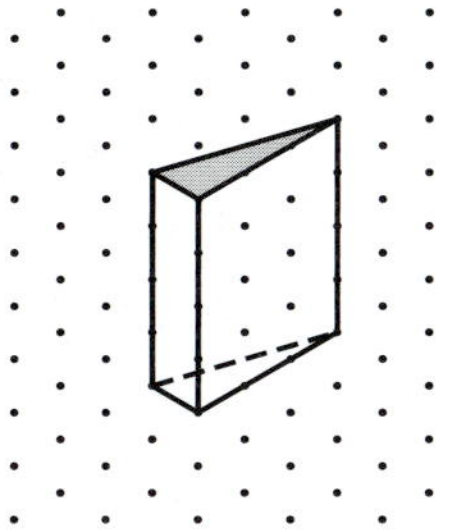

11.

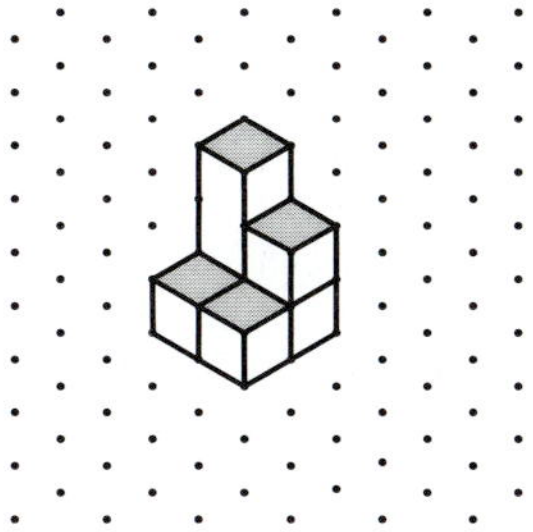

13.

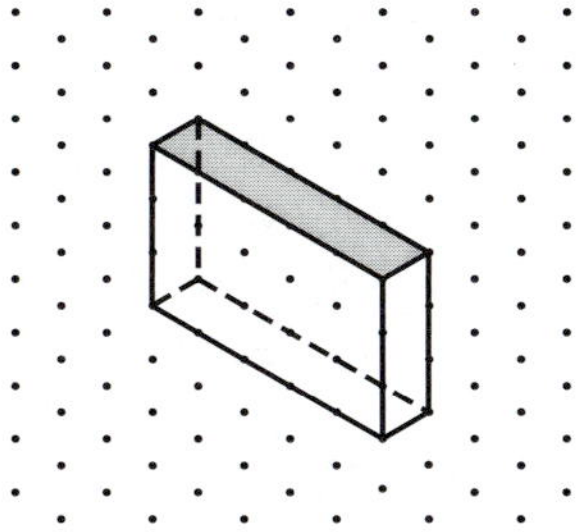

15 **a.** The cross section is a four-sided figure with opposite sides parallel and congruent and with all angles right angles. So, the cross section is a rectangle.
b. To make the cross section of a triangle, cut off the corner of the clay through one of the vertices to the opposite face of the figure.

17. hexagon **19.** trapezoid
21a. Make a vertical cut. **21b.** Make a horizontal cut.
21c. Make an angled cut. **23.** 4 equilateral triangles
25. Sample answer:

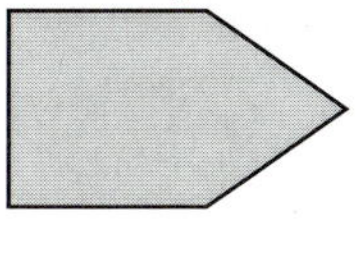

27. Sample answer:

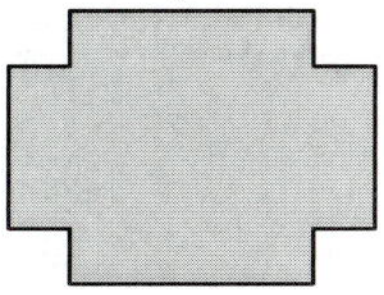

29a–b. Sample answer:

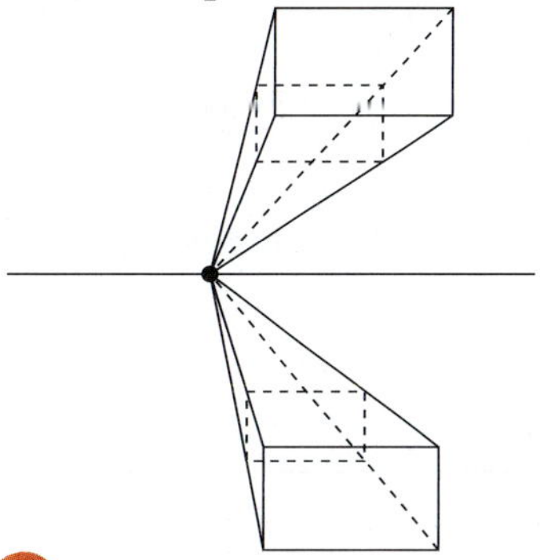

29c. Sample answer: The first drawing shows a view of the object from the bottom. The second drawing shows a view of the object from the top.

31 Top view: There are two rows and 3 columns when viewed from above. Use dark segments to indicate that there are different heights. Left view: The figure is 3 units high and 2 units wide. Front view: The first column is 3 units high, the second is 2 units high, and the third is 1 unit high. Right view: The figure is 3 units high and 2 units wide. Use dark segments to indicate that there are two breaks in this surface.

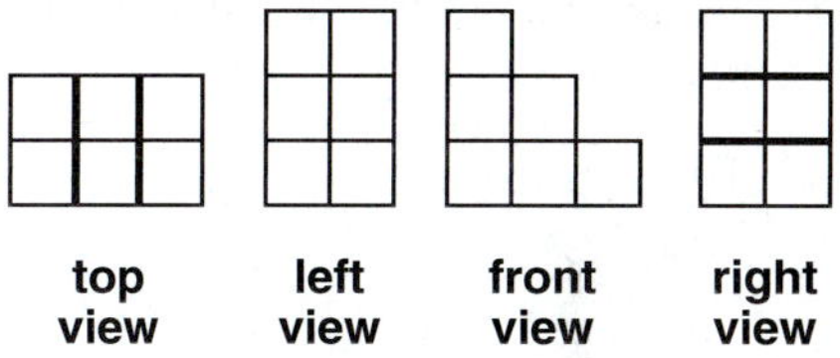

33a. Sample answer:

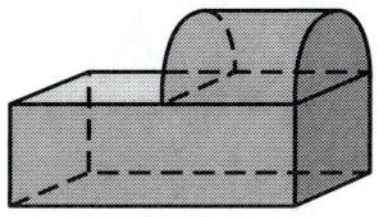

33b. Make a horizontal cut through the bottom part of the figure or make a vertical cut through the left side of the figure.

33c. The front view of the solid is the cross section when a vertical cut is made lengthwise. The right view of the solid is the cross section when a vertical cut is made through the right side of the figure. **35.** A cone is sliced at an angle.

37. Sample answer:

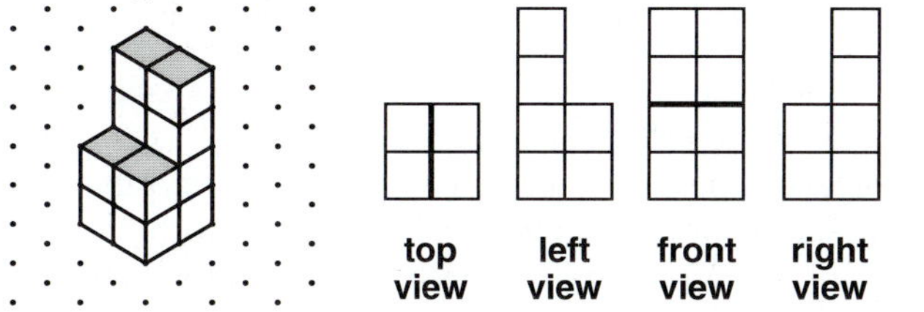

39. The cross section is a triangle. There are six different ways to slice the pyramid so that two equal parts are formed because the figure has six planes of symmetry. In each case, the cross section is an isosceles triangle. Only the side lengths of the triangles change. **41a.** inner side of deck = circumference of pool $= 81.64 \div \pi \approx 26$ ft; outer side of deck $= 26 + 3 + 3 = 32$ ft; outer perimeter of deck $= 4 \times 32 = 128$ ft **41b.** area of deck $= (2 \times 3 \times 32) + (2 \times 3 \times 26) = 348$ square feet **43.** D **45.** $\frac{3}{2}$; 9 **47.** about 3.6 yd^2 **49.** 28.9 in.; 66.5 in^2

Pages 830–837 Lesson 12-2

1. 112.5 in^2 **3.** $L = 288\ \text{ft}^2$; $S = 336\ \text{ft}^2$ **5.** $L \approx 653.5\ \text{yd}^2$; $S \approx 1715.3\ \text{yd}^2$

7 $S = 2\pi rh + 2\pi r^2$ Surface area of a cylinder
$286.3 = 2\pi(3.4)h + 2\pi(3.4)^2$ Replace S with 286.3 and r with 3.4.
$286.3 \approx 21.4h + 72.6$ Use a calculator to simplify.
$213.7 \approx 21.4h$ Subtract 72.6 from each side.
$10.0 \approx h$ Divide each side by 21.4.
The height of the can is about 10.0 cm.

9 Find the missing side length of the base.
$c^2 = 4^2 + 3^2$ Pythagorean Theorem
$c^2 = 25$ Simplify.
$c = 5$ Take the square root of each side.

$L = Ph$ Lateral area of a prism
$= (4 + 3 + 5)2$ Substitution
$= 24$ Simplify.
The lateral area is 24 ft^2.

$S = Ph + 2B$ Surface area of a prism
$= (4 + 3 + 5)(2) + 2\left(\frac{1}{2} \cdot 4 \cdot 3\right)$ Substitution
$= 24 + 12$ or 36 Simplify.
The surface area is 36 ft^2.

11. Sample answer: $L = 64$ in^2; $S = 88$ in^2 **13.** $L = 11.2$ m^2; $S = 13.6$ m^2 **15.** $L = 1032$ cm^2, $S = 1932$ cm^2 (18 × 25 base); $L = 1332$ cm^2; $S = 1932$ cm^2 (25 × 12 base); $L = 1500$ cm^2; $S = 1932$ cm^2 (18 × 12 base) **17.** $L = 1484.8$ cm^2; $S = 1745.2$ cm^2 **19.** $L \approx 282.7$ mm^2; $S \approx 339.3$ mm^2 **21.** $L \approx 155.8$ in^2; $S \approx 256.4$ in^2 **23.** 42.5 m^2 **25.** $r = 9.2$ cm

27 $S = 2\pi rh + 2\pi r^2$ Surface area of a cylinder
$256\pi = 2\pi r(8) + 2\pi r^2$ Replace S with 256π and h with 8.
$128 = 8r + r^2$ Divide each side by 2π.
$0 = r^2 + 8r - 128$ Subtract 128 from each side.
$0 = (r + 16)(r - 8)$ Factor.
$r + 16 = 0$ or $r - 8 = 0$ Zero Product Property
$r = -16$ $r = 8$
Since the radius cannot be negative, $r = 8$. So, the diameter is $8 \cdot 2$ or 16 millimeters.

29a. First find the area of the sector. Then find the lateral area of the cylinder, which is the product of 73% of the circumference of the base and the height of the cylinder. Last, add the areas of the two rectangles formed when a portion is cut.
29b. 283.7 in^2 **31.** $L = 20.8$ cm^2; $S = 24.5$ cm^2
33. about 299.1 cm^2

35 The composite figure has trapezoid bases. The trapezoids have bases 20 cm and 13 cm and a height of 21 cm. To find the length of the fourth side of the trapezoid x, use the Pythagorean Theorem.

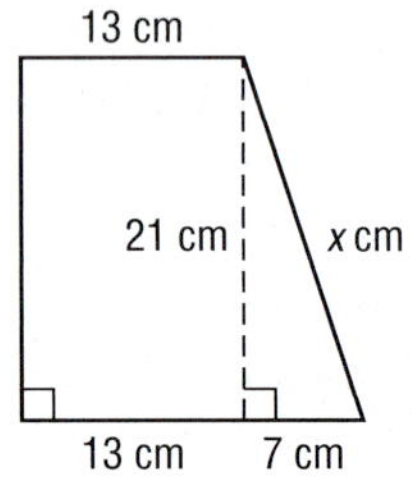

$x^2 = 21^2 + 7^2$ Pythagorean Theorem
$x^2 = 490$ Simplify.
$x \approx 22.136$ Take the square root of each side.

$S = Ph + 2B$ Surface area of a prism
$\approx (21 + 13 + 22.136 + 20)(28) + 2\left[\frac{1}{2}(21)(20 + 13)\right]$ Substitution
$\approx 2131.8 + 693$ or 2824.8 cm^2 Simplify.

37. 1059.3 cm^2 **39.** Derek; sample answer: $S = 2\pi r^2 + 2\pi rh$, so the surface area of the cylinder is $2\pi(6)^2 + 2\pi(6)(5)$ or 132π cm^2. **41.** To find the surface area of any solid figure, find the area of the base (or bases) and add to the area of the sides of the figure. The faces and bases of a rectangular prism are rectangles. Since the bases of a cylinder are circles, the "side" of a cylinder is a rectangle. **43.** $\frac{\sqrt{3}}{2}\ell^2 + 3\ell h$; the area of an equilateral triangle of side ℓ is $\frac{\sqrt{3}}{4}\ell^2$ and the perimeter of the triangle is 3ℓ. So, the total surface area is $\frac{\sqrt{3}}{2}\ell^2 + 3\ell h$. **45.** A **47.** H

49.

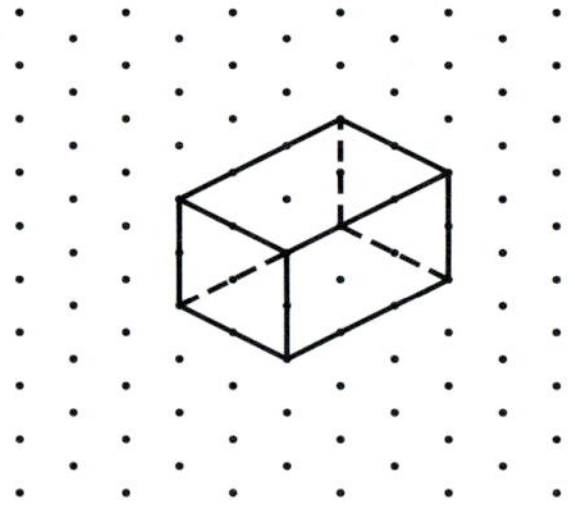

51. The volumes are equal. Nine minicakes are the same size as one 9-inch cake, but nine minicakes cost 9($4) or $36 while the 9-inch cake is only $15, so the 9-inch cake is a better buy.
53. 2.5 in. **55.** 20.5

Pages 838–846 Lesson 12-3

1. $L = 384$ cm^2; $S = 640$ cm^2 **3.** $L \approx 207.8$ m^2; $S \approx 332.6$ m^2 **5.** $L \approx 188.5$ m^2; $S \approx 267.0$ m^2

7 $L = \frac{1}{2}P\ell$ Lateral area of a regular pyramid
$= \frac{1}{2}(8)(5)$ $P = 2 \cdot 4$ or 8, $\ell = 5$
$= 20$ m^2 Simplify.

$S = \frac{1}{2}P\ell + B$ Surface area of a regular pyramid
$= 20 + 4$ $\frac{1}{2}P\ell = 20, B = 4$
$= 24$ m^2 Simplify.

9. $L \approx 178.2$ cm^2; $S \approx 302.9$ cm^2 **11.** $L \approx 966.0$ in^2; $S \approx 1686.0$ in^2 **13.** 139,440 ft^2 **15.** $L \approx 357.6$ cm^2; $S \approx 470.7$ cm^2 **17.** $L \approx 241.1$ ft^2; $S \approx 446.1$ ft^2

19 Use the Pythagorean Theorem to find the slant height ℓ.

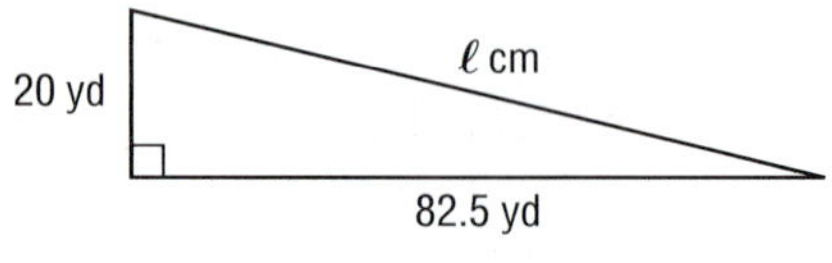

$c^2 = a^2 + b^2$ Pythagorean Theorem
$\ell^2 = 20^2 + 82.5^2$ $a = 20, b = 82.5, c = \ell$

$\ell^2 = 7206.25$ Simplify.

$\ell = \sqrt{7206.25}$ Take the square root of each side.

$L = \frac{1}{2}P\ell$ Lateral area of a regular pyramid

$= \frac{1}{2}(660)(\sqrt{7206.25})$ $P = 154 \cdot 4$ or 660, $\ell = 6$

$\approx 28{,}013.6\ \text{yd}^2$ Use a calculator.

21. 34 **23.** 5 mm **25.** 16 cm **27.** $266\pi\ \text{ft}^2$
29a. nonregular pyramid with a square base
29b. Sample answer:

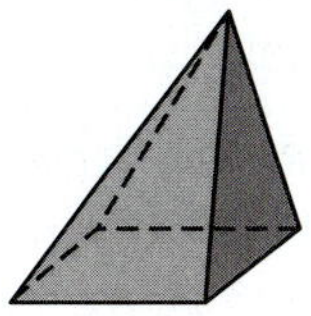

31. Sample answer:

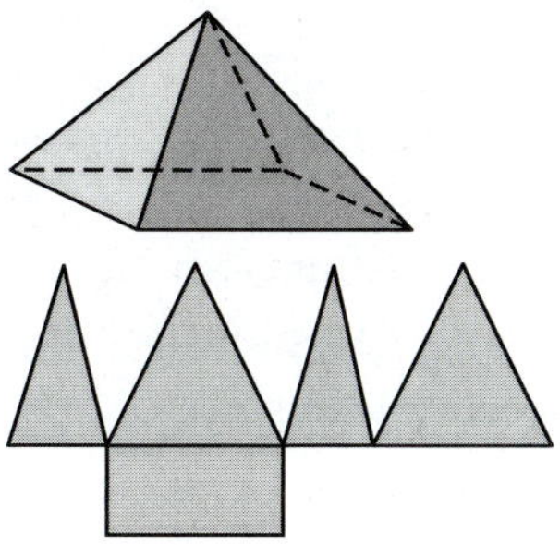

33 The surface area of the figure is the sum of the lateral areas of the cones. Find the slant heights, ℓ_1 and ℓ_2, of the cones.

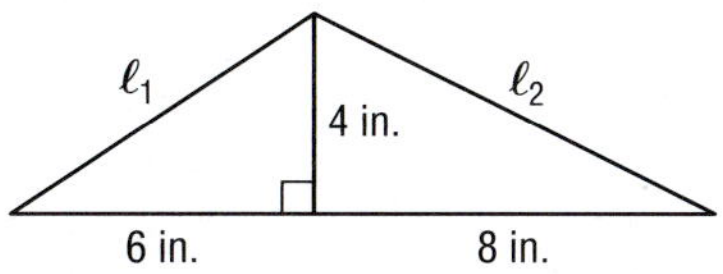

$\ell_1^2 = 4^2 + 6^2$ $\ell_2^2 = 4^2 + 8^2$
$\ell_1^2 = 52$ $\ell_2^2 = 80$
$\ell_1 = \sqrt{52}$ $\ell_2 = \sqrt{80}$

$S =$ Lateral area of Cone 1 + Lateral area of Cone 2
$= \pi r\ell_1 + \pi r\ell_2$
$= \pi(4)(\sqrt{52}) + \pi(4)(\sqrt{80})$
$\approx 203.0\ \text{in}^2$

35a. Sample answer:

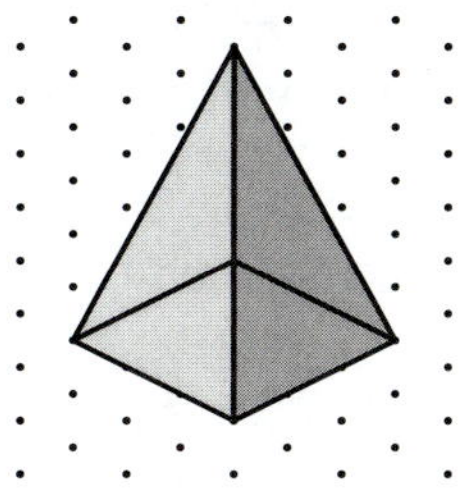

35b.

Slant Height (units)	Lateral Area (units)^2
1	6
3	18
9	54

35c. The lateral area is tripled. **35d.** The lateral area is multiplied by 3^2 or 9. **37.** Always; if the heights and radii are the same, the surface area of the cylinder will be greater since it has two circular bases and additional lateral area. **39.** Sample answer: a square pyramid with a base edge of 5 units and a slant height of 7.5 units
41. Use the apothem, the height, and the Pythagorean Theorem to find the slant height ℓ of the pyramid. Then use the central angle of the n-gon and the apothem to find the length of one side of the n-gon. Then find the perimeter. Finally, use $S = \frac{1}{2}P\ell + B$ to find the surface area. The area of the base B is $\frac{1}{2}Pa$. **43.** $3297\ \text{mm}^2$ **45.** D

47.

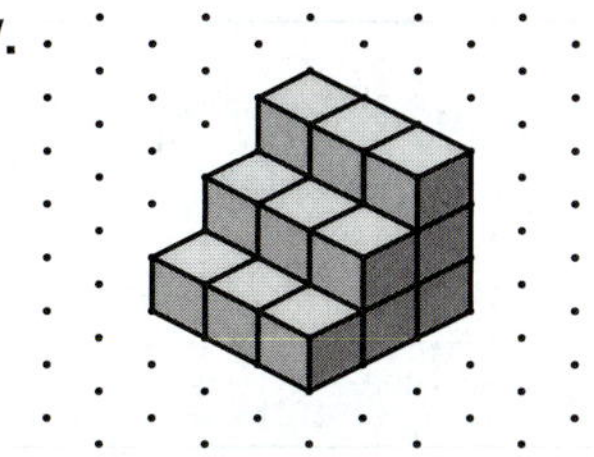

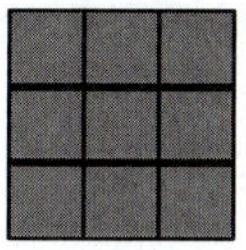

back view

49.

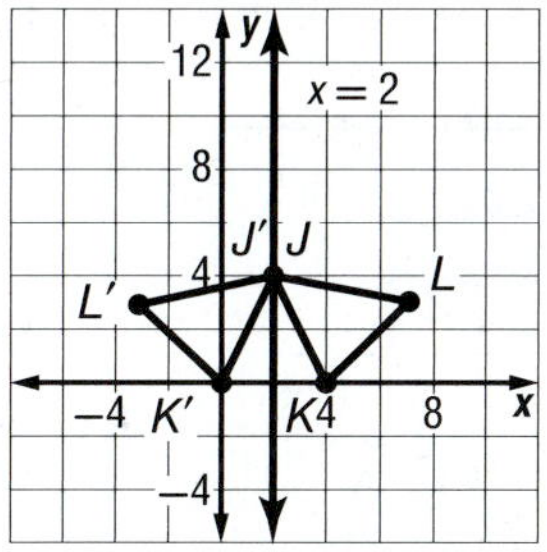

51.

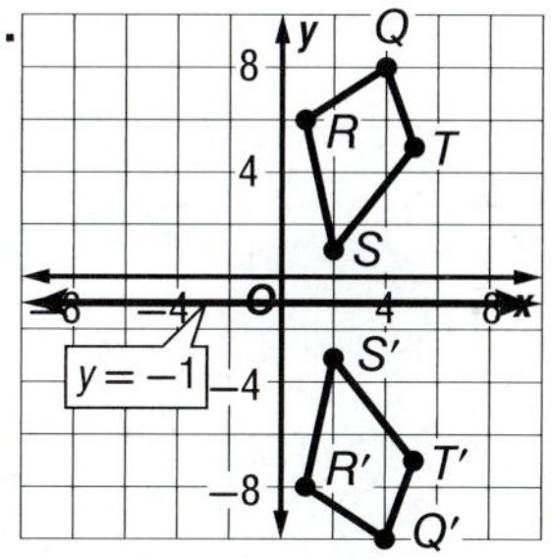

53.

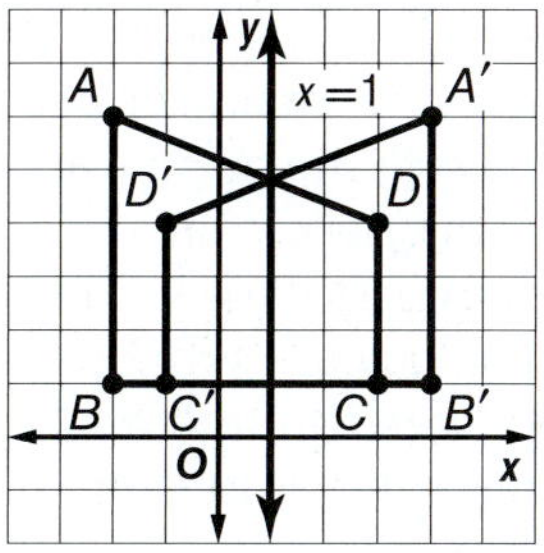

55. 57 m, $120\ \text{m}^2$
57. 166.2 in., $1443\ \text{in}^2$

Pages 847–854 Lesson 12-4

1. $108\ \text{cm}^3$ **3.** $26.95\ \text{m}^3$ **5.** $206.4\ \text{ft}^3$ **7.** $1025.4\ \text{cm}^3$ **9.** D

11 $V = Bh$ Volume of a prism
$= 38.5(14)$ $B = \frac{1}{2}(11)(7)$ or 38.5, $h = 14$
$= 539\ \text{m}^3$ Simplify.

13. $58.14\ \text{ft}^3$ **15.** $1534.25\ \text{in}^3$

17 $V = \pi r^2 h$ Volume of a cylinder
$= \pi(6)^2(3.6)$ Replace r with $12 \div 2$ or 6 and h with 3.6.
$\approx 407.2\ \text{cm}^3$ Use a calculator.

19. $2686.1\ \text{mm}^3$ **21.** $521.5\ \text{cm}^3$ **23.** $3934.9\ \text{cm}^3$
25. 35.1 cm **27.** Prism A; the area of the base is $20\ \text{in}^2$, which equals area of the base of the rectangular prism. Two prisms have equal volumes if their height is the same and their bases have equal areas. **29.** $120\ \text{m}^3$

31 Find the volume of the cylinder with a height of 10.5 cm and a radius of $8.5 \div 2$ or 4.25 cm. Subtract from that the volume of the cylinder with a height of 10.5 cm and a radius of $6.5 \div 2$ or 3.25 cm. Add the volume of the bottom cylinder that has a height of 1 cm and a radius of $6.5 \div 2$ or 3.25 cm.

Volume = Volume of large cylinder −
Volume of small cylinder +
Volume of bottom
$= \pi r_1^2 h_1 - \pi r_2^2 h_1 + \pi r_2^2 h_2$

$= \pi(4.25)^2(10.5) - \pi(3.25)^2(10.5) + \pi(3.25)^2(1)$ $r_1 = 4.25, r_2 = 3.25, h_1 = 10.5, h_2 = 1$

$\approx 280.6 \text{ cm}^3$ Use a calculator.

33. 678.6 in^3 **35.** 3,190,680.0 cm^3 **37.** $11\frac{1}{4}$ in.

39 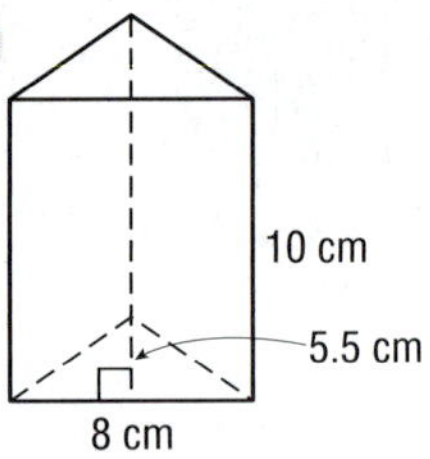

Each triangular prism has a base area of $\frac{1}{2}(8)(5.5)$ or 22 cm^2 and a height of 10 cm. The volume of each triangular prism is 22 • 10 or 220 cm^3. So, the volume of five triangular prisms is 220 • 5 or 1100 cm^3.

41a. 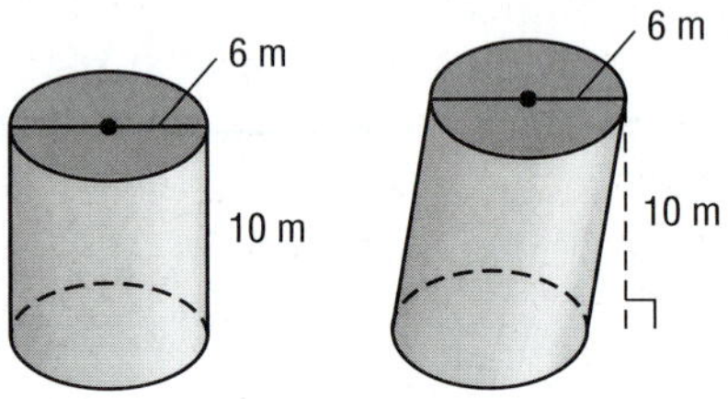

41b. Greater than; a square with a side length of 6 m has an area of 36 m^2. A circle with a diameter of 6 m has an area of 9π or 28.3 m^2. Since the heights are the same, the volume of the square prism is greater. **41c.** Multiplying the radius by x; since the volume is represented by $\pi r^2 h$, multiplying the height by x makes the volume x times greater. Multiplying the radius by x makes the volume x^2 times greater. **43.** Sample answer: The can holds $\pi(2)^2(5)$ or 20π in^3 of liquid. Therefore, the container holds 60π in^3. **43a.** base 3 in. by 5 in., height 4π in. **43b.** base 5 in. per side, height $\frac{12}{5}\pi$ in. **43c.** base with legs measuring 3 in. and 4 in., height 10π in.

45. Sample answer:

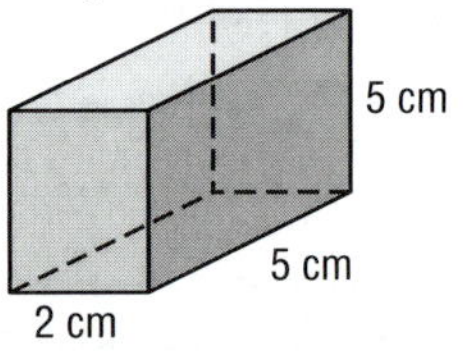

47. Both formulas involve multiplying the area of the base by the height. The base of a prism is a polygon, so the expression representing the area varies, depending on the type of polygon it is. The base of a cylinder is a circle, so its area is πr^2. **49.** F **51.** C **53.** 126 cm^2; 175 cm^2 **55.** 205 in^2 **57.** 11.4 cm **59.** 9.3 in. **61.** 378 m^2

Pages 857–863 Lesson 12-5

1. 75 in^3 **3.** 62.4 m^3 **5.** 51.3 in^3 **7.** 28.1 mm^3 **9.** 513,333.3 ft^3

11 $V = \frac{1}{3}Bh$ Volume of a pyramid

$= \frac{1}{3}(36.9)(8.6)$ $B = \frac{1}{2} \cdot 9 \cdot 8.2$ or 36.9, $h = 8.6$

$\approx 105.8 \text{ cm}^3$ Simplify.

13. 233.8 cm^3 **15.** 35.6 cm^3 **17.** 235.6 in^3 **19.** 1473.1 cm^3 **21.** 1072.3 in^3

23 $V = \frac{1}{3}\pi r^2 h$ Volume of a cone

$= \frac{1}{3}\pi(4)^2(14)$ Replace r with $\frac{8}{2}$ or 4 and h with 14.

$\approx 234.6 \text{ cm}^3$ Use a calculator.

25. 32.2 ft^3 **27.** 3190.6 m^3 **29a.** $\frac{130}{9}$ **29b.** $\approx$ 614.6 in^3 or 0.4 ft^3 **29c.** 7 flowerpots **31a.** The volume is doubled. **31b.** The volume is multiplied by 2^2 or 4. **31c.** The volume is multiplied by 2^3 or 8.

33 $V = \frac{1}{3}\pi r^2 h$ Volume of a cone

$196\pi = \frac{1}{3}\pi r^2(12)$ Replace V with 196π and h with 12.

$196\pi = 4\pi r^2$ Simplify.

$49 = r^2$ Divide each side by 4π.

$7 = r$ Take the square root of each side.

The radius of the cone is 7 inches, so the diameter is 7 • 2 or 14 inches.

35a. Sample answer:

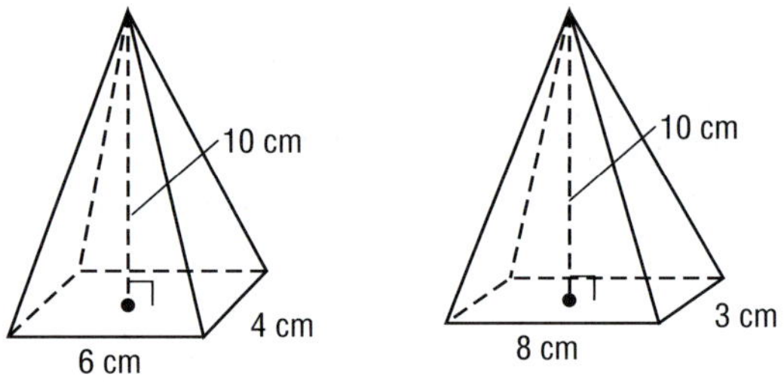

35b. The volumes are the same. The volume of a pyramid equals one third times the base area times the height. So, if the base areas of two pyramids are equal and their heights are equal, then their volumes are equal. **35c.** If the base area is multiplied by 5, the volume is multiplied by 5. If the height is multiplied by 5, the volume is multiplied by 5. If both the base area and the height are multiplied by 5, the volume is multiplied by 5 • 5 or 25. **37.** Cornelio; Alexandra incorrectly used the slant height. **39.** Sample answer: A square pyramid with a base area of 16 and a height of 12, a prism with a square base of area 16 and height of 4; if a pyramid and prism have the same base, then in order to have the same volume, the height of the pyramid must be 3 times as great as the height of the prism. **41.** A **43.** G **45.** 1008.0 in^3 **47.** 426,437.6 m^3 **49.** 30.4 cm^2 **51.** 26.6 ft^2

Pages 864–871 Lesson 12-6

1. 1017.9 m^2 **3.** 452.4 yd^2 **5.** 4188.8 ft^3 **7.** 3619.1 m^3 **9.** 277.0 in^2 **11.** 113.1 cm^2 **13.** 680.9 in^2 **15.** 128 ft^2 **17.** 530.1 mm^2

19 $V = \frac{4}{3}\pi r^3$ Volume of a sphere

$= \frac{4}{3}\pi(1)^3$ $r = \frac{2}{2}$ or 1

$\approx 4.2 \text{ cm}^3$ Use a calculator.

21. 2712.3 cm^3 **23.** 179.8 in^3 **25.** 77.9 m^3 **27.** 860,289.5 ft^3

29 Surface area $= \frac{1}{2} \cdot$ Area of sphere + Lateral area of cylinder + Area of circle

$$= \frac{1}{2}(4\pi r^2) + 2\pi rh + \pi r^2$$
$$= \frac{1}{2}(4\pi)(4)^2 + 2\pi(4)(5) + \pi(4)^2$$
$$\approx 276.5 \text{ in}^2$$

Volume = Volume of hemisphere + Volume of cylinder

$$= \frac{1}{2}\left(\frac{4}{3}\pi r^3\right) + \pi r^2 h$$
$$= \frac{1}{2}\left(\frac{4}{3}\pi \cdot 4^3\right) + \pi(4)^2(5)$$
$$\approx 385.4 \text{ in}^3$$

31a. 594.6 cm^2; 1282.8 cm^3 **31b.** 148.7 cm^2; 160.4 cm^3 **33.** $\overline{DC}$ **35.** $\overline{AB}$ **37.** $\odot S$

39 **a.** $V = \frac{4}{3}\pi r^3$ Volume of a sphere

$= \frac{4}{3}\pi(4)^3$ $r = \frac{2}{2}$ or 1

≈ 268.1 cm^3 Use a calculator.

Since the portion of the sphere that represents orange production in California is 18.0% of the total sphere, its volume is 0.18 • 268.1 or about 48.3 cm^3.

b. $S = 4\pi r^2$ Surface area of a sphere

$\approx 4\pi(4)^2$ Replace r with 2.5.

≈ 201.1 cm^2 Use a calculator.

Since the portion of the sphere that represents orange production in Florida is 81.0% of the total sphere, its surface area is 0.81 • 201.1 or about 162.9 cm^2.

41. There is an infinite number of vertical planes that produce reflection symmetry. There are no horizontal planes that produce reflection symmetry. There is an infinite number of angles of rotation. **43.** The surface area is divided by 3^2 or 9. The volume is divided by 3^3 or 27. **45.** 587.7 in^3

47.

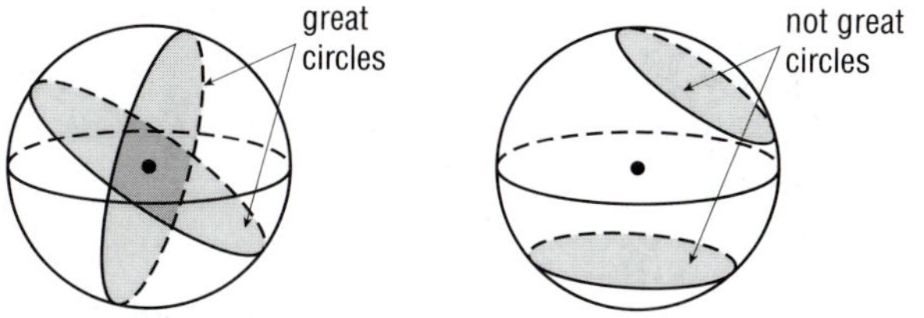

49. 68.6 **51.** H **53.** 58.9 ft^3 **55.** 207.8 m^3 **57.** 154.2 units^2 **59.** 80.2 units^2 **61.** 21 units^2 **63.** D, B, and G **65.** $\overleftrightarrow{EF}$ and $\overleftrightarrow{AB}$ do not intersect. $\overleftrightarrow{AB}$ lies in plane $\mathcal{P}$, but only E lies in plane $\mathcal{P}$.

Pages 873–878 Lesson 12-7

1. $\overleftrightarrow{DP}$, $\overleftrightarrow{FM}$ **3.** $\triangle JKL$, $\triangle QMP$

5 Figure x does not go through the poles of the sphere. So, figure x is not a great circle and so not a line in spherical geometry.

7. The points on any great circle or arc of a great circle can be put into one-to-one correspondence with real numbers. **9.** Sample answers: $\overleftrightarrow{WZ}$ and $\overleftrightarrow{XY}$, $\overline{RY}$ or $\overline{TZ}$, $\triangle RST$ or $\triangle MPL$ **11a.** $\overleftrightarrow{AD}$ and $\overleftrightarrow{FC}$ **11b.** Sample answers: $\overline{BG}$ and $\overline{AH}$ **11c.** Sample answers: $\triangle BCD$ and $\triangle ABF$ **11d.** $\overline{QD}$ and $\overline{BL}$ **11e.** $\overleftrightarrow{MJ}$ **11f.** $\overleftrightarrow{MB}$ and $\overleftrightarrow{KF}$ **13.** no

15 Every great circle (line) is finite and returns to its original starting point. Thus there exists no great circle that goes on infinitely in two directions.

17. Yes; If three points are collinear, any one of the three points is between the other two. **19.** 14.0 in.; since 100 degrees is $\frac{5}{18}$ of 360 degrees, $\frac{5}{18} \times$ circumference of the great circle ≈ 14.0. **21a.** about 913 mi; the cities are 13.2° apart on the same great circle, so $\frac{13.2}{360} \times 2\pi \times 3963$ give the distance between them. **21b.** Yes; sample answer: Since the cities lie on a great circle, the distance between the cities can be expressed as the major arc or the minor arc. The sum of the two values is the circumference of Earth. **21c.** No; sample answer: Since lines of latitude do not go through opposite poles of the sphere, they are not great circles. Therefore, the distance cannot be calculated in the same way. **21d.** Sample answer: Infinite locations. If Phoenix were a point on the sphere, then there are infinite points that are equidistant from that point.

23 **a.** No; if $\overline{CD}$ were perpendicular to $\overline{DA}$, then $\overline{DA}$ would be parallel to $\overline{CB}$. This is not possible, since there are no parallel lines in spherical geometry. **b.** $DA < CB$ because $\overline{CB}$ appears to lie on a great circle. **c.** No; since there are no parallel lines in spherical geometry, the sides of a figure cannot be parallel. So, a rectangle, as defined in Euclidean geometry, cannot exist in non-Euclidean geometry.

25. Sample answer: In plane geometry, the sum of the measures of the angles of a triangle is 180. In spherical geometry, the sum of the measures of the angles of a triangle is greater than 180. In hyperbolic geometry, the sum of the measures of the angles of a triangle is less than 180. **27.** Sometimes; sample answer: Since small circles cannot go through opposite poles, it is possible for them to be parallel, such as lines of latitude. It is also possible for them to intersect when two small circles can be drawn through three points, where they have one point in common and two points that occur on one small circle and not the other.

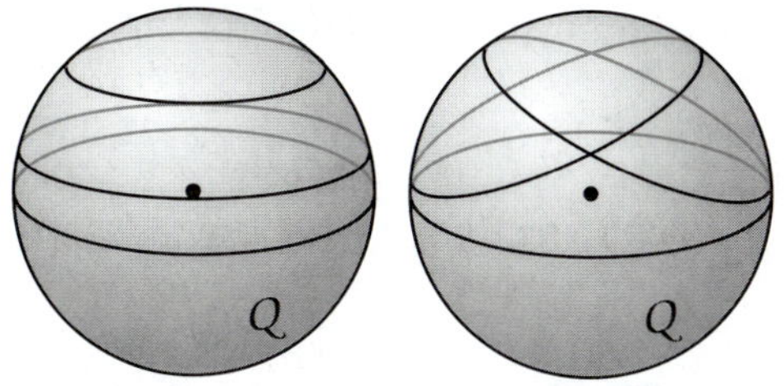

29. False; sample answer; Spherical geometry is non-Euclidean, so it cannot be a subset of Euclidean geometry. **31.** C **33.** Sample answer: $\overline{BC}$ **35.** 735.4 m^3 **37.** 1074.5 cm^3 **39.** 78.5 m^3 **41.** 0.1 in^3 **43.** 2.7 cm^2 **45.** 322.3 m^2

Selected Answers and Solutions

Pages 880–886 Lesson 12-8

1. similar; 4:3 **3.** 4:25 **5.** 220,893.2 cm^3
7. neither **9.** similar; 6:5

11 $\frac{\text{height of large cylinder}}{\text{height of small cylinder}} = \frac{35}{25} \text{ or } \frac{7}{5}$

The scale factor is $\frac{7}{5}$. If the scale factor is $\frac{a}{b}$, then the ratio of volumes is $\frac{a^3}{b^3} = \frac{7^3}{5^3}$ or $\frac{343}{125}$. So, the ratio of the volumes is 343:125.

13. 5:1 **15a.** 10:13 **15b.** 419.6 cm^3

17 scale factor $= \frac{26 \text{ ft}}{14 \text{ in.}}$ Write a ratio comparing the lengths.
$= \frac{312 \text{ in.}}{14 \text{ in.}}$ 26 ft = 26 • 12 or 312 in.
$= \frac{156}{7}$ Simplify.

The scale factor is 156:7.

19. 4.1 in. **21.** 2439.6 cm^3 **23.** about 5.08 to 1

25 $\frac{\text{area of smaller tent}}{\text{area of larger tent}} = \frac{9}{12.25}$ Write a ratio comparing the floor areas.
$= \frac{3^2}{3.5^2}$ Write as $\frac{a^2}{b^2}$.

The scale factor is 3:3.5.

ratio of diameters → $\frac{6}{d} = \frac{3}{3.5}$ ← scale factor
$6 \cdot 3.5 = d \cdot 3$ Find the cross products.
$7 = d$ Solve for *d*.

So, the diameter of the larger tent is 7 feet.

$V = \frac{1}{2}\left(\frac{4}{3}\pi r^3\right)$ Volume of a hemisphere
$= \frac{1}{2}\left(\frac{4}{3}\pi \cdot 3.5^3\right)$ Radius $= \frac{7}{2}$ or 3.5
≈ 89.8 Use a calculator.

The volume of the larger tent is about 89.8 ft^3.

27. Laura; because she compared corresponding parts of the similar figures. Paloma incorrectly compared the diameter of X to the radius of Y. **29.** Since the scale factor is 15:9 or 5:3, the ratio of the surface areas is 25:9 and the ratio of the volumes is 125:27. So, the surface area of the larger prism is $\frac{25}{9}$ or about 2.8 times the surface area of the smaller prism. The volume of the larger prism is $\frac{125}{27}$ or about 4.6 times the volume the smaller prism. **31.** 14 cm **33.** A **35.** $\sqrt{85} \approx 9.2$ km **37.** yes **39.** yes **41.** 5 **43.** 6 **45.** 0.31 **47.** 0.93

Pages 887–890 Chapter 12 Study Guide and Review

1. false, Spherical geometry **3.** false, right cone **5.** true **7.** true **9.** true **11.** triangle **13.** rectangle **15.** 160 ft^2; 202 ft^2 **17.** 113.1 cm^2; 169.6 cm^2 **19.** 354.4 cm^2; 432.9 cm^2 **21.** 972 cm^3 **23.** 3.6 cm^3 **25.** 91,636,242 ft^3 **27.** 1017.9 m^2 **29.** 1708.6 in^3 **31.** $\overleftrightarrow{FG}$, $\overleftrightarrow{DJ}$ **33.** $\triangle CBD$ **35.** $\overline{KC}$ **37.** no **39.** congruent **41.** neither

Chapter 13 Probability and Measurement

Page 897 Chapter 13 Get Ready

1. $\frac{7}{8}$ **3.** $1\frac{11}{40}$ **5.** $\frac{3}{8}$ **7.** 144 **9.** $\frac{1}{2}$ or 50% **11.** $\frac{1}{3}$ or 33% **13.** $\frac{1}{5}$ or 20% **15.** $\frac{11}{20}$ or 55%

Pages 899–905 Lesson 13-1

1. S, S O, O
S, O O, S

Outcomes	Safe	Out
Safe	S, S	S, O
Out	O, S	O, O

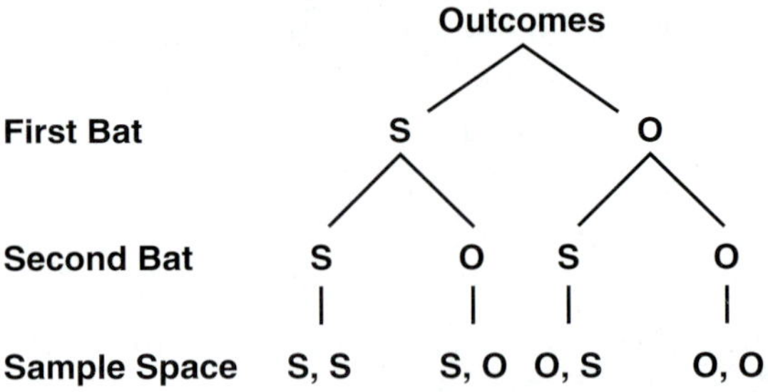

3.

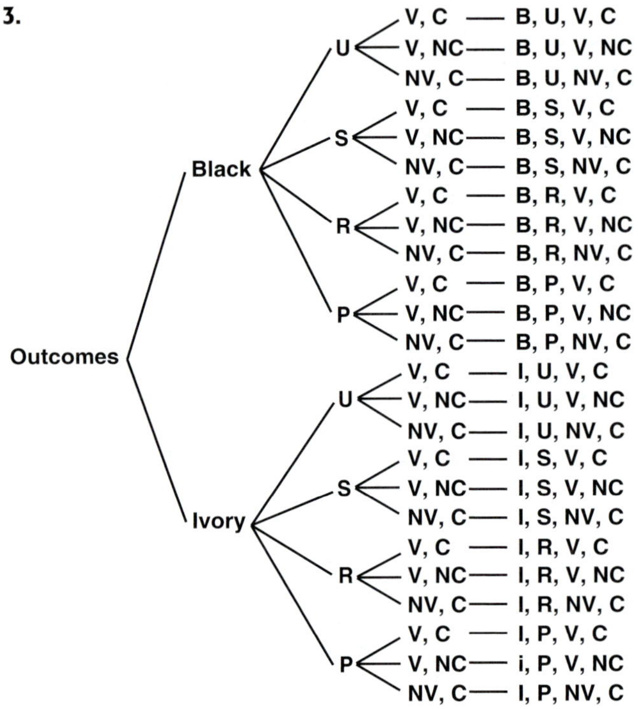

5 Possible Outcomes
= Appetizers × Soups × Salads × Entrees × Desserts
= 8 × 4 × 6 × 12 × 9 or 20,736

7. S, S N, N
S, N N, S

Outcomes	Smithsonian	Natural
Smithsonian	S, S	S, N
Natural	N, S	N, N

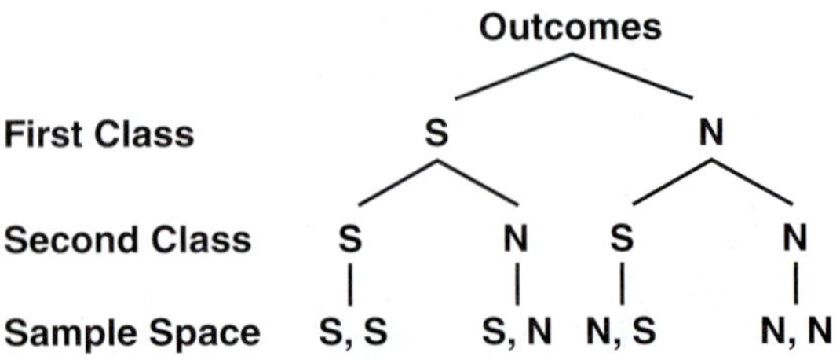

9. M, 5 T, 5
M, 6 T, 6

Outcomes	5	6
Monday	M, 5	M, 6
Thursday	T, 5	T, 6

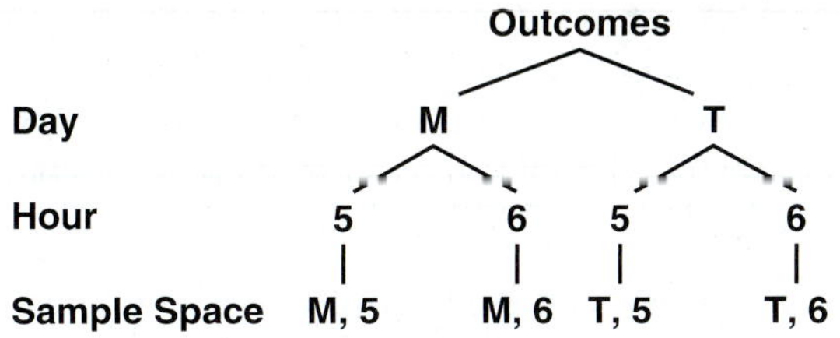

11. O, O A, A
O, A A, O

Outcomes	Oil	Acrylic
Oil	O, O	O, A
Acrylic	A, O	A, A

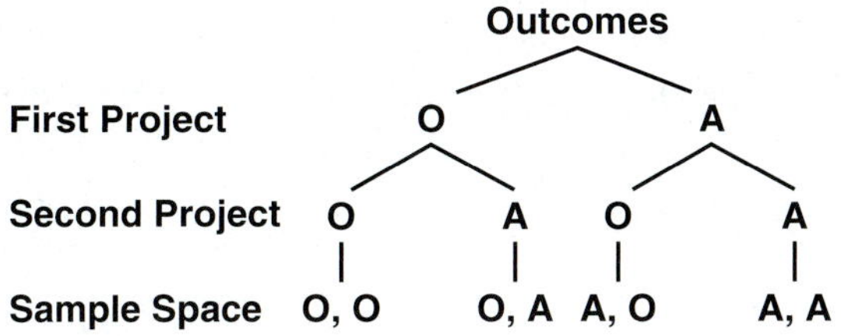

13. S = sedan, T = truck, V = van, L = leather, F = fabric, P = CD player, NP = no CD player, R = sunroof, NR = no sunroof

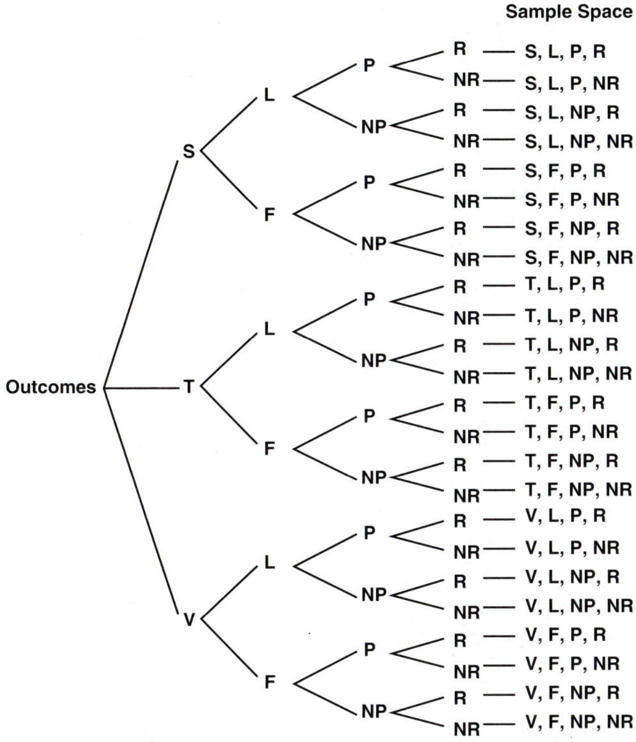

15 Possible Outcomes = Secretary × Treasurer × Vice President × President
= 3 × 4 × 5 × 2 or 120

17. 240 **19.** H = rhombus, P = parallelogram, R = rectangle, S = square, T = trapezoid; H, P; H, R; H, S; H, T; H, H; S, P; S, R; S, S; S, T; S, H

Outcomes	Rhombus	Square
Parallelogram	H, P	S, P
Rectangle	H, R	S, R
Square	H, S	S, S
Trapezoid	H, T	S, T
Rhombus	H, H	S, H

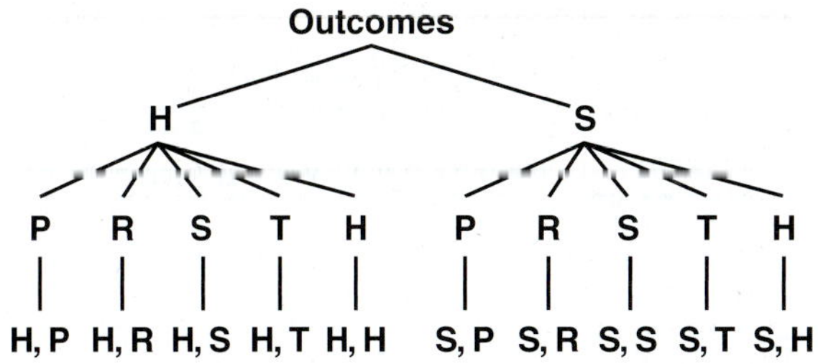

21. 6 different ways:
$4(x + 6) + 2(3) + 2(x + 4)$;
$2(x + 11) + 2(x + 8) + 2(x)$;
$2(x + 4) + 2(x + 9) + 2(x + 6)$;
$2(x) + 2(3) + 4(x + 8)$;
$2(x) + 2(x + 8) + 2(3) + 2(x + 8)$;
$2(x) + 2(3) + 2(4) + 2(x + 6) + 2(x + 6)$

23 **a.** The rolls that result in a sum of 8 are 2 and 6, 3 and 5, 4 and 4, 5 and 3, 6 and 2. So, there are 5 outcomes.
b. The rolls that result in an odd sum are shown.

1, 2	1, 4	1, 6
2, 1	2, 3	2, 5
3, 2	3, 4	3, 6
4, 1	4, 3	4, 5
5, 2	5, 4	5, 6
6, 1	6, 3	6, 5

So, there are 18 outcomes.

25. $n^3 - 3n^2 + 2n$; Sample answer: There are n objects in the box when you remove the first object, so after you remove one object, there are $n - 1$ possible outcomes. After you remove the second object, there are $n - 2$ possible outcomes. The number of possible outcomes is the product of the number of outcomes of each experiment or $n(n - 1)(n - 2)$.
27. Sample answer: You can list the possible outcomes for one stage of an experiment in the columns and the possible outcomes for the other stage of the experiment in the rows. Since a table is two dimensional, it would be impossible to list the possible outcomes for three or more stages of an experiment. Therefore, tables can only be used to represent the sample space for a two-stage experiment.
29. $P = n^k$; Sample answer: The total number of possible outcomes is the product of the number of outcomes for each of the stages 1 through k. Since there are k stages, you are multiplying n by itself k times which is n^k. **31.** B **33.** G **35.** 130 m high, 245 m wide, and 465 m long **37.** $\overline{FC}$ **39.** 1429.4 ft^2, 1737.3 ft^2 **41.** 1710.6 m^2, 3421.2 m^2 **43.** line **45.** 12.5 **47.** 12 **49.** 32

Pages 906–914 *Lesson 13-2*

1. $\frac{1}{20}$ or 0.05 **3.** $\frac{1}{420}$ **5.** $\frac{13}{261}$ or about 5%

7 The number of possible outcomes is 50!. The number of favorable outcomes is (50 − 2)! or 48!.

P(Alfonso 14, Colin 23)

$= \frac{48!}{50!}$ — Number of favorable outcomes / Number of possible outcomes

$= \frac{\cancel{48!}^{1}}{50 \cdot 49 \cdot \cancel{48!}_{1}}$ — Expand 48! and divide out common factors.

$= \frac{1}{2450}$ — Simplify.

9. $\frac{1}{15{,}120}$

11 There is a total of 10 letters. Of these letters, B occurs 2 times, A occurs 2 times, and L occurs 2 times. So, the number of distinguishable permutations of these letters is

$\frac{10!}{2! \cdot 2! \cdot 2!} = \frac{3{,}628{,}800}{8}$ or 453,600 — Use a calculator.

There is only 1 favorable arrangement—BASKETBALL. So, the probability that a permutation of these letters selected at random spells basketball is $\frac{1}{453{,}600}$.

13. $\frac{1}{7}$ **15.** $\frac{1}{10{,}626}$ **17a.** $\frac{1}{56}$ **17b.** $\frac{1}{40{,}320}$ **17c.** $\frac{1}{140}$ **17d.** $\frac{2}{7}$ **19a.** 720 **19b.** 5040

21 Find the number of ways to choose the second letter times the number of ways to choose the third letter times the number of ways to choose the last two numbers.

possible license plates $= {}_2C_1 \cdot {}_3C_1 \cdot {}_{10}C_1 \cdot {}_{10}C_1$

$= 2 \cdot 3 \cdot 10 \cdot 10$ or 600

23. Always; sample answer: The number of favorable outcomes is $(n-1)!$ and the number of possible outcomes is $n!$. Therefore, the probability is $\frac{1}{n}$.
25. Sample answer: A bag contains seven marbles that are red, orange, yellow, green, blue, purple, and black. The probability that the orange, blue, and black marbles will be chosen if three marbles are drawn at random can be calculated using a combination.

27.

$$C(n, n-r) \stackrel{?}{=} C(n, r)$$

$$\frac{n!}{[n-(n-r)]!(n-r)!} \stackrel{?}{=} \frac{n!}{(n-r)!r!}$$

$$\frac{n!}{r!(n-r)!} \stackrel{?}{=} \frac{n!}{(n-r)!r!}$$

$$\frac{n!}{(n-r)!r!} = \frac{n!}{(n-r)!r!}$$

29. C **31.** J **33.** 16 **35.** 2 **37.** 4.5 **39.** 3 **41.** 1 **43.** 5

Pages 915–921 Lesson 13-3

1. $\frac{1}{2}$, 0.5, or 50% **3.** $\frac{13}{33}$, 0.39, or about 39% **5.** $\frac{1}{8}$, 0.125, or 12.5% **7.** $\frac{13}{18}$, 0.72, or 72% **9.** $\frac{1}{9}$, 0.11, or 11% **11.** $\frac{1}{6}$, 0.17, or about 17%

13 You need to find the ratio of the area of the shaded region to the area of the entire region. The area of shaded region equals the area of the large semicircle minus the area of the small semicircle plus the area of the small semicircle. So, the area of the shaded region equals the area of the large semicircle. Since the area of the large semicircle equals half the total area, P(landing in shaded region) $= \frac{1}{2}$, 0.5, or 50%.

15 P(pointer landing on yellow) $= \frac{44}{360}$ or about 12.2%

17. 69.4% **19.** 62.2% **21.** Sample answer: a point between 10 and 20 **23.** $\frac{1}{2}$, 0.5, or 50%
25. 53.5% **27.** Sample answer: The probability that a randomly chosen point will lie in the shaded region is ratio of the area of the sector to the area of the circle.

P(randomly chosen point lies in sector) $= \frac{\text{area of sector}}{\text{area of circle}}$

P(randomly chosen point lies in sector) $= \frac{\frac{x}{360} \cdot \pi r^2}{\pi r^2}$

P(randomly chosen point lies in sector) $= \frac{x}{360}$

29. 0.24 or 24% **31.** 0.33 or 33%

33 volume of shallow region $= Bh = (7 \cdot 20) \cdot 20$ or 2800 ft^3

volume of incline region $= Bh = \frac{1}{2}(25)(7 + 20) \cdot 20$ or 6750 ft^3

volume of deep region $= Bh = (20 \cdot 30) \cdot 20$ or 12,000 ft^3

P(bear swims in the incline region)

$= \frac{\text{volume of incline region}}{\text{volume of pool}}$

$= \frac{6750}{2800 + 6750 + 12{,}000}$

≈ 0.31 or 31%

35. 14.3% **37.** Sample answer: Athletic events should not be considered random because there are factors involved, such as pressure and ability that have an impact on the success of the event. **39.** Sample answer: The probability of a randomly chosen point lying in the shaded region of the square on the left is found by subtracting the area of the unshaded square from the area of the larger square and finding the ratio of the difference of the areas to the area of the larger square. The probability is $\frac{1^2 - 0.75^2}{1^2}$ or 43.75%. The probability of a randomly chosen point lying in the shaded region of the square on the left is the ratio of the area of the shaded square to the area of the larger square, which is $\frac{0.4375}{1}$ or 43.75%. Therefore, the probability of a randomly chosen point lying in the shaded area of either square is the same.
41. F **43.** C
45. D, D G, G
D, G G, D

Outcomes	Drums	Guitar
Drums	D, D	D, G
Guitar	G, D	G, G

Selected Answers and Solutions

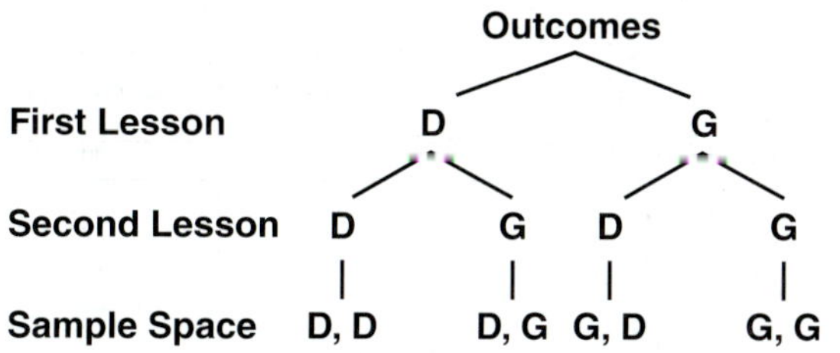

47. 45 **49.** Square; each angle intercepts a semicircle, making them 90° angles. Each side is a chord of congruent arcs, so the chords are congruent. **51.** 57.1 m^2 **53.** 66.3 cm^2

Pages 923–930 Lesson 13-4

1. Sample answer: Use a spinner that is divided into two sectors, one containing 80% or 288° and the other containing 20% or 72°. Do 20 trials and record the results in a frequency table.

Outcome	Frequency
A	17
below an A	3
Total	20

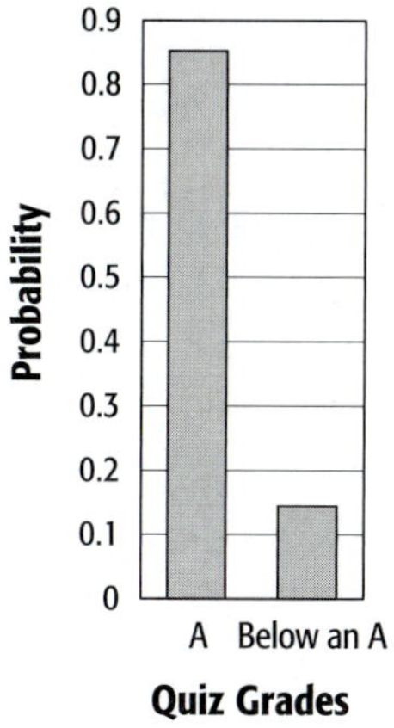

The probability of Clara getting an A on her next quiz is 0.85. The probability of earning any other grade is 1 − 0.85 or 0.15. **3a.** 36 **3b.** Sample answer: Use a random number generator to generate integers 1 through 25 where 1–16 represents 25 points, 17–24 represents 50 points, and 25 represents 100 points. Do 50 trials and record the results in a frequency table.

Outcome	Frequency
25	29
50	21
100	0

The average value is 35.5. **3c.** Sample answer: The expected value and average value are very close.

5 Sample answer: Use a spinner that is divided into two sectors, one containing 95% or 342° and the other containing 5% or 18°.
Sale: 95% of 360 = 342°
No Sale: 5% of 360 = 18°

Do 50 trials and record the results in a frequency table.

Outcome	Frequency
sale	46
no sale	4
Total	50

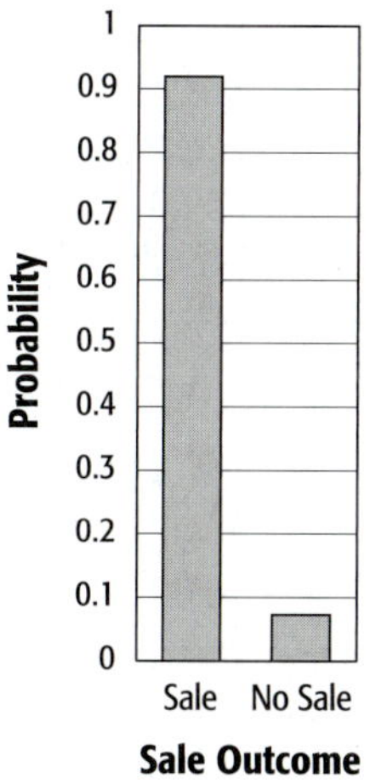

Based on the simulation, the experimental probability of Ian selling a game is 0.92. The experimental probability of not selling a game is 1 − 0.92 or 0.08.

7. Sample answer: Use a spinner that is divided into 8 equal sectors, each 45°. Do 50 trials and record the results in a frequency table.

Outcome	Frequency
Category 1	3
Category 2	3
Category 3	6
Category 4	13
Category 5	4
Category 6	9
Category 7	7
Category 8	5
Total	50

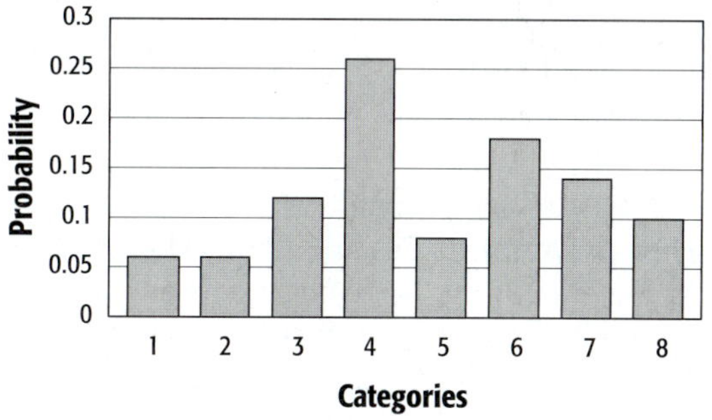

The probability of landing on Categories 1 and 2 is 0.06, Category 3 is 0.12, Category 4 is 0.26, Category 5 is 0.08, Category 6 is 0.18, Category 7 is 0.14, and Category 8 is 0.1.

9. Sample answer: Use a random number generator to generate integers 1 through 20, where 1–12 represents a single, 13–17 represents a double, 18–19 represents a triple, and 20 represents a home run. Do 20 trials and record the results in a frequency table.

Outcome	Frequency
single	13
double	4
triple	2
home run	1
Total	20

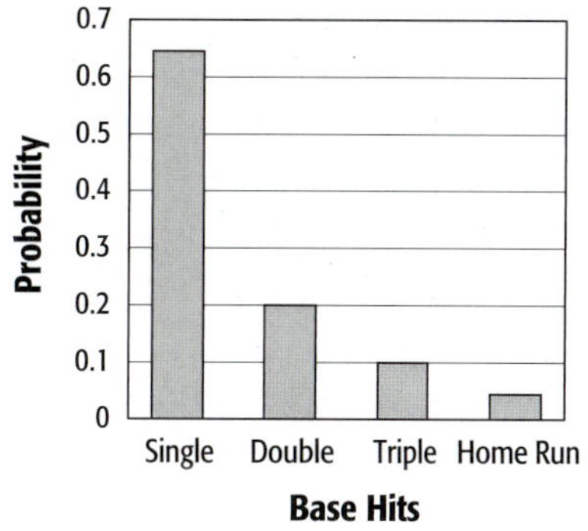

The probability of the baseball player hitting a single is 0.65, a double is 0.2, a triple is 0.1, and a home run is 0.05.

11. Sample answer: Use a random number generator to generate integers 1 through 20, where 1–7 represents blue, 8–13 represents red, 14–16 represents white, 17–19 represents black, and 20 represents all other colors. Do 50 trials and record the results in a frequency table.

Outcome	Frequency
blue	17
red	14
black	7
white	10
other	2
Total	50

The probability of a customer buying a blue car is 0.34, buying a red car is 0.28, buying a black car is 0.14, buying a white car is 0.2, and any other color is 0.04.

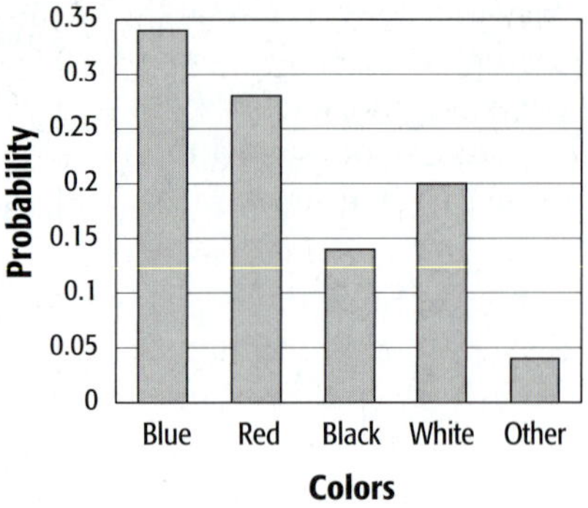

13 Calculate the geometric probability of landing on each color.

$$P(\text{red}) = \frac{\text{area of red}}{\text{area of circle}} = \frac{\pi(0.5)^2}{\pi(5)^2} = 0.01$$

The blue and white areas are equal. Their total area is the area of the large circle minus the area of the red circle or $\pi(5)^2 - \pi(0.5)^2$ or about 77.8 square units. So, the blue area and the white area are each about 38.9 square units.

$$P(\text{blue}) = \frac{\text{area of blue}}{\text{area of circle}} = \frac{38.9}{\pi(5)^2} \approx 0.495$$

$P(\text{white}) = P(\text{blue}) \approx 0.495$

$E(Y) = 25 \cdot 0.495 + 50 \cdot 0.495 + 0.01 \cdot 100$

$E(Y) = 38.125$

Sample answer: Assign the integers 1–1000 to accurately represent the probability data. Blue = integers 1–495, White = integers 496–990, Red = integers 991–1000. Use the calculator to generate 500 trials of random integers from 1 to 1000. Record the results in a frequency table.

Outcome	Frequency
red	0
blue	29
white	21
total	50

Then calculate the average value of the outcomes.

$$\frac{29}{50} \cdot 25 + \frac{21}{50} \cdot 50 + \frac{0}{50} \cdot 100 = 35.5$$

Average value = 35.5; the expected value is greater than the average value.

15a. 0.75 **15b.** Sample answer: Use a random number generator to generate integers 1 through 20, where 1–7 represents 0 points, 8–19 represents 1 point, and 20 represents 3 points. Do 50 trials and record the results in a frequency table; average value = 0.76.

Outcome	Frequency
0	16
1	32
3	2

15c. Sample answer: the two values are almost equal.

17a. There is a $\frac{1}{6}$ or 16.7% probability of throwing a strike in each box.

17b. Sample answer:

Strike Area	Accuracy (%)
1	15
2	17
3	19
4	22
5	19
6	8
Total	100

17c. Sample answer: Some of the values are higher or lower, but most are very close to 16.7%.

19a. Sample answer:

Sum of Die Roll
9
10
6
6
7
9
5
9
5
7
6
5
7
3
9
7
6
7
8
7

19b. Sample answer:

Sum of Output from Random Number Generator
4
10
5
10
6
7
12
3
7
4
7
9
3
6
4
11
5
7
5
3

19c. Sample answer:

Trial	Sum of Die Roll	Sum of Output from Random Number Generator
1	9	4
2	10	10
3	6	5
4	6	10
5	7	6
6	9	7
7	5	12
8	9	3
9	5	7
10	7	4
11	6	7
12	5	9
13	7	3
14	3	6
15	9	4
16	7	11
17	6	5
18	7	7
19	8	5
20	7	3

19d. Sample answer:

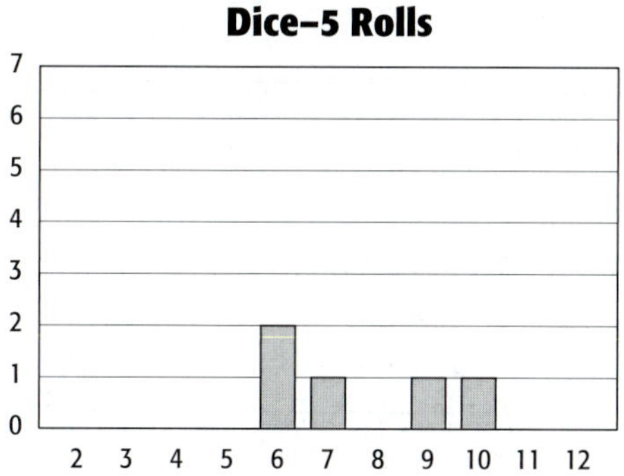

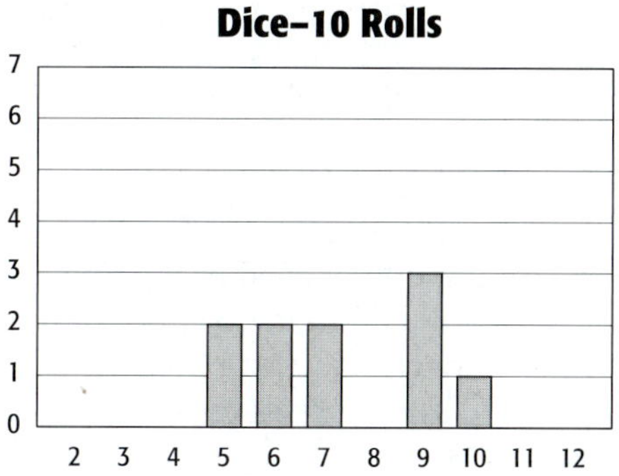

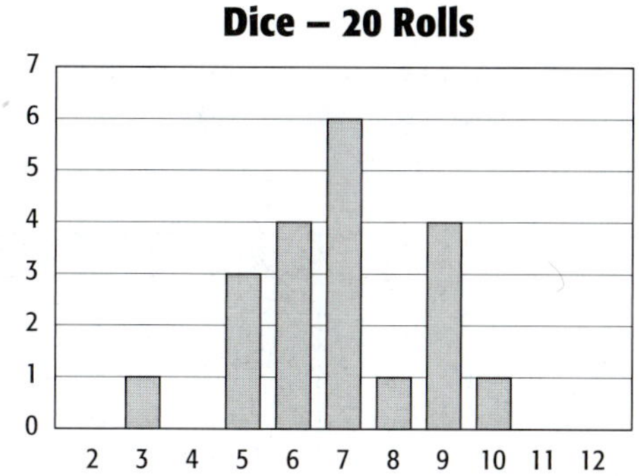

19e. Sample answer: The bar graph has more data points at the middle sums as more trials are added.
19f. Sample answer:

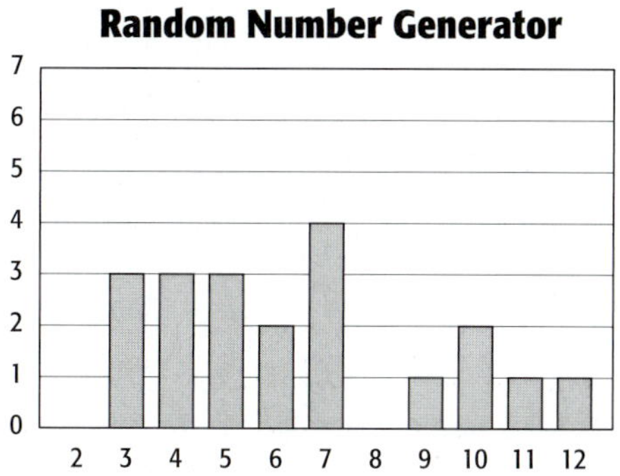

19g. Sample answer: They both have the most data points at the middle sums. **19h.** Sample answer: The expected value in both experiments is 7 because it is the sum that occurs most frequently.

21 Sometimes; sample answer: Flipping a coin can be used to simulate and experiment with two possible outcomes when both of the outcomes are equally likely. For example, if there is an equal number of boys and girls in a class, then flipping a coin can be used to simulate choosing one person from the class. If the probabilities of the occurrence of the two outcomes are different, flipping a coin is not an appropriate simulation. For example, if 55% of the class is girls and 45% is boys, then flipping a coin cannot be used to simulate choosing one person from the class.
23. Sample answer: We assume that the object lands within the target area, and that it is equally likely that the object will land anywhere in the region. These are needed because in the real world it will not be equally likely to land anywhere in the region.
25. Sample answer: To design a simulation, first you have to determine all of the possible outcomes and the theoretical probability associated with each of those outcomes. When you know each outcome and the probability of each outcome, you should state any assumptions you are making. Next, you choose the model that you want to use in your simulation, define each outcome in terms of your model, and determine how many trials you will do. When your simulation is planned, you run the number of trials that you have determined will provide a good set of data, recording the data for each trial. Finally, you analyze the data to determine the experimental probability of one or more of the outcomes.

27. H **29.** B **31.** $\frac{4}{7}$, 0.57, or 57% **33.** 50.3 ft^2
35. 1017.9 in^2

Pages 931–937 Lesson 13-5

1. The outcome of Jeremy taking the SAT in no way changes the probability of the outcome of his ACT test. Therefore, these two events are *independent*.

3. $\frac{1}{2704}$ or 3.7×10^{-4} **5.** $\frac{1}{5}$ or 0.20

7 These events are dependent since the card is not replaced.

$P(A \text{ and } A) = P(A) \cdot P(A\|A)$	Probability of dependent events
$= \frac{4}{52} \cdot \frac{3}{51}$	After first ace is drawn, 51 cards remain, and 3 of those are aces.
$= \frac{12}{2652}$ or $\frac{1}{221}$	Multiply.

The probability is $\frac{1}{221}$ or about 0.5%.

9. independent; $\frac{1}{36}$ or about 3% **11.** $\frac{1}{306}$ or about 0.3%
13. $\frac{20}{161}$ or about 12% **15.** $\frac{1}{4}$ or 25% **17.** $\frac{1}{6}$ or 17%

19 $P(\text{own MP3 player} \mid \text{own CD player})$

$$= \frac{P(\text{own MP3 player and CD player})}{P(\text{own CD player})}$$

$$= \frac{0.28}{0.43}$$

$$\approx 0.65$$

21a.

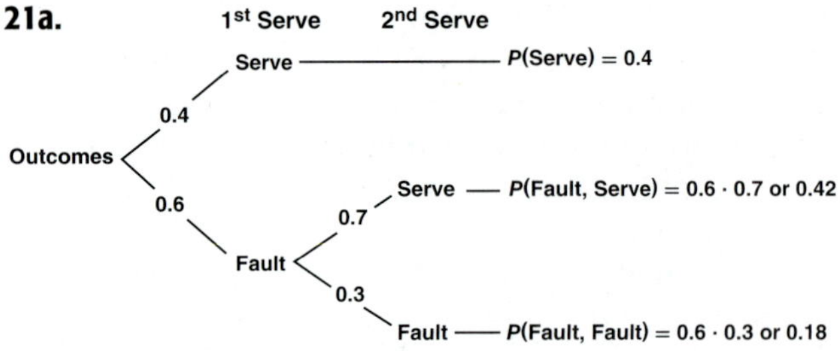

21b. 0.18 or 18% **21c.** Sample answer: I would use a random number generator to generate integers 1 through 50. The integers 1–9 will represent a double fault, and the integers 10–50 will represent the other possible outcomes. The simulation will consist of 50 trials. **23.** Allison; sample answer: Since the events are independent, $P(A \text{ and } B) = P(A) \cdot P(B)$. Substituting for $P(A \text{ and } B)$ in the formula $P(A|B) = \frac{P(A \text{ and } B)}{P(B)}$, $P(A|B) = \frac{P(A) \cdot P(B)}{P(B)}$ or $P(A)$.

25. Always; sample answer: $P(A \text{ and } B)$ and $P(B \text{ and } A)$ are both the product of $P(A)$ and $P(B)$. Since they are independent, the probability of the second event is not changed by the first event.

27. Sample answer: A probability tree shows all of the possible outcomes of a compound event that occur in the sample space. The probability of each outcome is the proportion of the outcome to the whole sample space. The whole sample space is represented by 1, so the sum of the probabilities of all of the outcomes must equal 1. **29.** J **31.** B **33.** 0.25 **35.** 0.07 **37.** neither **39a.** 5026.5 ft **39b.** 500–600 ft **39c.** 3142 ft; 3770 ft **41.** 12 **43.** 216

Pages 938–945 Lesson 13-6

1. not mutually exclusive **3.** $\frac{2}{3}$ or about 67%

5. The probability of missing the spare is $\frac{8}{10}$ or 80%.

7. 17.3%

9 Since rolling two fours is both getting doubles and getting a sum of 8, the events are not mutually exclusive.

$P(\text{doubles or a sum of 8})$

$= P(\text{doubles}) + P(\text{a sum of 8}) - P(\text{doubles and a sum of 8})$

$= \frac{6}{36} + \frac{5}{36} - \frac{1}{36}$

$= \frac{10}{36}$ or about 27.8%

11. mutually exclusive; 100% **13.** mutually exclusive; $\frac{2}{9}$ or about 22.2% **15.** $\frac{7}{16}$ or about 43.8% **17.** $\frac{3}{4}$ or about 75% **19.** $\frac{7}{8}$ or about 87.5%

21 Find the probability that the first worker is paid by the hour plus the probability that that the second worker is paid by the hour. The probability that a worker is not paid by the hour is 1 − 0.71 or 0.29.

$P(\text{first paid}) + P(\text{second paid})$

$= P(\text{first paid}) \cdot P(\text{second not paid}) + P(\text{second paid}) \cdot P(\text{first not paid})$

$= 0.71(0.29) + 0.71(0.29)$

$\approx 0.21 + 0.21$ or 0.42

The probability is about 0.42 or 42%.

23. $\frac{1}{13}$ or 7.7% **25.** $\frac{3}{13}$ or 23.1%

27a. 71.3% **27b.** 11.3% **27c.** 36.2% **27d.** 3.8% **29.** 0.74; sample answer: There are three outcomes in which the values of two or more of the dice are less than or equal to 4 and one outcome where the values of all three of the dice are less than or equal to 4. You have to find the probability of each of the four scenarios and add them together. **31.** Not mutually exclusive; sample answer: If a triangle is equilateral, it is also equiangular. The two can never be mutually exclusive. **33.** Sample answer: If you pull a card from a deck, it can be either a 3 or a 5. The two events are mutually exclusive. If you pull a card from a deck, it can be a 3 and it can be red. The two events are not mutually exclusive. **35.** D **37.** J **39.** dependent; $\frac{1}{221}$ or 0.5%

41. Sample answer: Use a random number generator to generate integers 1 through 20 in which 1–7 represent football, 8–13 represent basketball, 14–17 represent soccer, and 18–20 represent volleyball. Do 20 trials, and record the results in a frequency table.

Outcome	Frequency
football	7
basketball	6
soccer	5
volleyball	2
Total	20

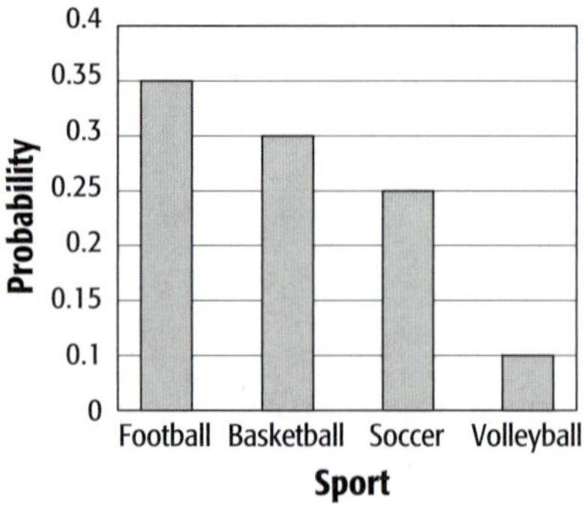

The probability that an athlete plays only football is 0.35, only basketball is 0.30, only soccer is 0.25, and only volleyball is 0.1.

43.

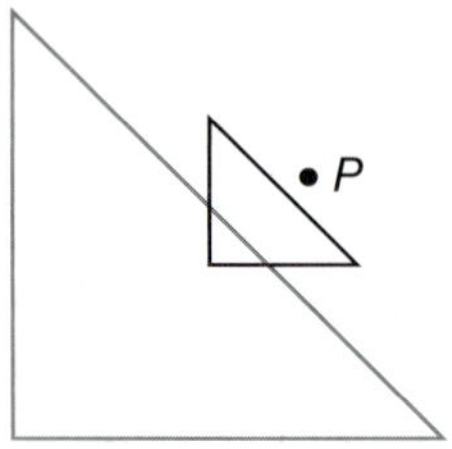

1. true **3.** true **5.** true **7.** true **9.** false, simulation
11. S, NB; S, B; S, EB; M, NB; M, B; M, EB; L, NB; L, LB; L, EB

Outcomes	No Butter	Butter	Extra Butter
Small	S, NB	S, B	S, EB
Medium	N, NB	M, B	M, EB
Large	L, NB	L, B	L, EB

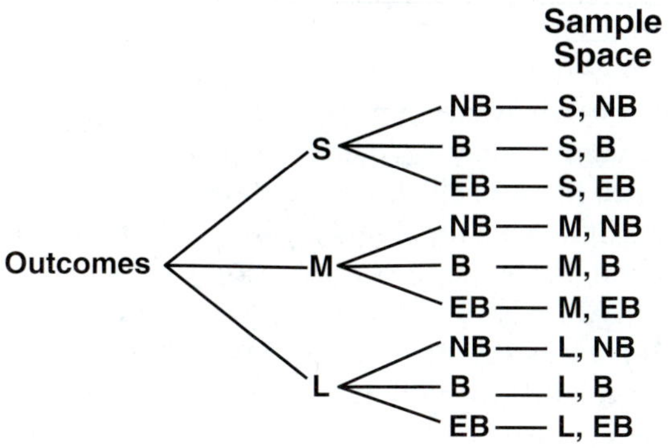

13. 4 **15.** 35,960 **17a.** $\frac{2}{9}$ **17b.** $\frac{7}{9}$ **19.** Sample answer: Use a spinner that is divided into 4 sectors, 108°, 79.2°, 82.8°, and 90°. Perform 50 trials and record the results in a frequency table. The results can be used to determine the probability of when a particular book will be purchased. **21.** $\frac{6}{35}$ **23.** 37% **25.** $\frac{4}{13}$

Photo Credits

Cover (batter)moodboard/CORBIS, (stadium)Kevin Reece/Icon SMI/CORBIS, (baseball)C Squared Studios/Getty Images, (bkrd)Chad Baker/Getty Images; **2–3** Alaska Stock LLC/Alamy Images; **5** Brand X Pictures/JupiterImages; **6** Comstock/SuperStock; **10** (l)Artiga Photo/CORBIS, (r)Thinkstock/JupiterImages; **11** Don Mason/Getty Images; **14** (t)HIP/Art Resource, NY, (b)Science Museum/SSPL/The Image Works; **17** Rick Rickman/NewSport/CORBIS; **18** (t)Tracy Kahn/CORBIS, (b)Stockbyte; **19** (l)moodboard/CORBIS, (r)Coinery/Alamy Images; **23** The Steve Bicknell Style Library/Alamy Images; **24** TNT Magazine/Alamy Images; **31** Tyler Stableford/Getty Images; **34** A. Huber/U. Starke/zefa/CORBIS; **35** (l)Joseph T. Collins/Photo Researchers, (r)Redmond Durrell/Alamy Images; **36** Michael Newman/PhotoEdit; **42** Dimitri Vervits/ImageState; **43** (l)Richard Wahlstrom/Stone/Getty Images, (r)Chesh/Alamy Images; **46** Zia Soleil/Iconica/Getty Images; **47** Mark Karrass/CORBIS; **53** NASA; **55** Ed-Imaging; **56** JupiterImages/Brand X/Alamy Images; **63** Philippe Hays/Alamy Images; **67** Digital Vision/Alamy Images; **68** Richard Nowitz/National Geographic Image Collection; **70** (bkgd)Mark Ransom Photography, (candles)Nichola Evans/Iconica/Getty Images; **71** (tl)FoodPhotography Eising/Stockfood, (tc)Image Source Black/Alamy Images, (tr)David Wasserman/JupiterImages, (cl)Lauri Rotko/Getty Images, (c)Cone 6 Productions/Photodisc/Getty Images, (cr)Tetra Images/Getty Images; **72** Courtesy of Arango Inc.; **74** (l)PhotoDisc/Getty Images, (c)Mark Ransom Photography, (r)Robert Landau/CORBIS; **82** Will & Deni McIntyre/CORBIS; **86–87** Frank Krahmer/Masterfile; **89** Michael Newman/PhotoEdit; **91** image100/CORBIS; **93** Stockbyte/Getty Images; **94** Ryan McVay/Getty Images; **95** The Granger Collection, New York; **102** Jose Luis Pelaez, Inc./Blend Images/CORBIS; **103 108** Masterfile; **110** Karen Hardy/Sygma/CORBIS; **111** Tom McHugh/Photo Researchers; **114** Troy Wayrynen/NewSport/CORBIS; **115** Dorling Kindersley/Getty Images; **117** Kevin Schafer/CORBIS; **120** Entertainment Software Rating Boards; **121** Reuters/CORBIS; **124** Marc Romanelli/PhotoDisc/Getty Images; **125** Jim Sugar/Science Faction; **126** Jose Fuste Raga/CORBIS; **130** David Schmidt/Masterfile; **131** (t)Greg Ryan/Alamy Images, (b)Masterfile Royalty Free; **134** Rick Gomez/Masterfile; **135** Jim Craigmyle/CORBIS; **138** Berkeley Formula SAE; **139** Peter Arnold/Digital Vision/Getty Images; **140** CORBIS; **142** Mary Kate Denny/PhotoEdit; **144** Nancy Sheehan/PhotoEdit; **146** Barros & Barros/Stone/Getty Images; **147** Diamond Images/Getty Images; **149** Danita Delimont/Alamy Images; **152** Joseph Sohm/Visions of America/CORBIS; **155** Breck P. Kent/Animals Animals; **156** Joseph Sohm/Visions of America/CORBIS; **164** Image Source Pink/SuperStock; **168–169** Jeremy Woodhouse/Masterfile; **171** Phil Schermeister/CORBIS; **173** Steve Crise/CORBIS; **174** (l)Robert Llewellyn/CORBIS, (r)Gerard Rancinan/Sygma/CORBIS; **175** Comstock/SuperStock; **178** age fotostock/SuperStock; **179** David Jay Zimmerman/CORBIS; **181** Creatas/SuperStock; **182** Garret Anglin/Getty Images; **185** Horizons Companies; **186** Uli Weismeier/zefa/CORBIS; **188** JupiterImages/Creatas/Alamy Images; **192** Joe Robbins/Getty Images; **193** (l)Rick Gomez/CORBIS, (c)image100/CORBIS, (r)Image Source/CORBIS; **198** Roger Viollet/The Image Works; **201** Stockbyte/PunchStock; **202** Richard Hutchings/PhotoEdit; **205** William Hamilton/SuperStock; **208** (t)Abode/Beateworks/CORBIS, (b)Robert Michael/CORBIS; **210** (t)Danita Delimont/Alamy Images, (c)Design Pics Inc./Alamy Images, (b)John Edward Linden/Arcaid/CORBIS; **214** Rachel Epstein/PhotoEdit; **220** Bonnie Kamin/PhotoEdit; **232–233** Alan Becker/Getty Images; **235** Robert Stahl/Getty Images; **236** (l)fStop/SuperStock, (r)DK Limited/CORBIS; **238** (l)Brad Simmons/Beateworks/CORBIS, (c)Natalie Tepper/Arcaid/CORBIS, (r)Ashley Cooper/CORBIS; **239** (l)Tiziana and Gianne Baldizzone/CORBIS, (c)Brand X Pictures/PunchStock, (r)Todd Gipstein/Getty Images; **240** (l)Shigeru Tanaka/Getty Images, (r)Purestock/SuperStock; **241** Emmanuel Dunand/AFP/Getty Images; **243** Ed-Imaging; **245** Kitt Cooper-Smith/Alamy Images; **247** (l)Hill Street Studios/Stock This Way/CORBIS, (r)Ingram Publishing/SuperStock; **248** (t)Andreas Herpens/iStockphoto, (c)Peter Adams/Digital Vision/Getty Images, (bl)Masterfile, (br)Jonelle Weaver/Getty Images; **250** (l)The Garden Picture Library/Alamy Images, (r)Jan Stromme/Getty Images; **253** Klaus Hackenberg/zefa/CORBIS; **254** The Granger Collection, New York; **255** f1 online/Alamy Images; **258** (t)Gary Corbett/Alamy Images, (b)Longview/Getty Images; **259** (l)Joseph Pobereskin/Getty Images, (r)Rudy Sulgan/CORBIS; **260** Art Vandalay/Getty Images; **265** Richard Southall/Arcaid/CORBIS; **267** Tony Arruza/CORBIS; **268** moodboard/CORBIS; **273** Ben Blankenburg/CORBIS; **278** (t)Robert Michael/CORBIS, (b)Arcaid/CORBIS; **279** UpperCut Images/SuperStock; **283** Richard Cummins/CORBIS; **286** (l)Gary Williams/epa/CORBIS, (r)B.S.P.I./CORBIS; **288** SW Productions/Brand X/CORBIS; **289** (l)Stefan Schuetz/zefa/CORBIS, (r)Jean-Yves Bruel/Masterfile; **290** Stockbyte/SuperStock; **294** Parque/zefa/CORBIS; **295** Chuck Franklin/Alamy Images; **297** (l)Grand Tour/CORBIS, (r)Richard Cummins/CORBIS; **298** (tl)moodboard/CORBIS, (tr)David Clapp/Arcaid/CORBIS, (cl)Geostock/Getty Images, (c)George D. Lepp/CORBIS, (cr)Ron Chapple/CORBIS, (bl)Kristy-

Anne Glubish/Design Pics/CORBIS, (bc)Michael S. Lewis/CORBIS, (br)Kelly Mooney Photography/CORBIS; **301** Martyn Goddard/CORBIS; **306** Ben Blankenburg/CORBIS; **314** Jose Luis Pelaez, Inc./CORBIS; **318–319** Graham Henderson/Elizabeth Whiting & Associates/CORBIS; **321** Ed-Imaging; **322** Dan Forer/Beateworks/CORBIS; **324** Masterfile Royalty Free; **329** Stuart O'Sullivan/Getty Images; **332** Ed-Imaging; **333** James Randklev/Stone/Getty Images; **338** (cl)moodboard/CORBIS, (cr)Igor Zhorov/iStockphoto, (bl)Kelly-Mooney Photography/CORBIS, (br)Blaine Harrington III/CORBIS; **339** Mark Ransom Photography; **340** The Granger Collection, New York; **342** age fotostock/SuperStock; **345** SW Productions/Brand X/CORBIS; **347** Todd Bigelow/Aurora/Getty Images; **351** Ian Shaw/Alamy Images; **352** Stockbyte/Getty Images; **356** Tami Chappell/Reuters/CORBIS; **362** CORBIS/PunchStock; **364** Reuters/CORBIS; **369** Richard Hamilton Smith/CORBIS; **374** Rick Gomez/CORBIS; **375** RNHRD NHS Trust/Getty Images; **382** Tetra Images/CORBIS; **386–387** image100/age fotostock; **389** Masterfile; **391** PhotoSpin, Inc./Alamy Images; **393** Alan Schein Photography/CORBIS; **395** Stephen Dunn/Getty Images; **396** Fred Dufour/AFP/Getty Images; **400** moodboard/SuperStock; **403** Adrian Muttitt/Alamy Images; **404** Mario Tama/Getty Images; **405** Richard Cummins/SuperStock; **406** Louie Psihoyos/CORBIS; **410** Michael Mahovlich/Masterfile; **413** Erich Lessing/Art Resource, NY; **415** SSPL/The Image Works; **416** Supapixx/Alamy Images; **419** Bob Daemmrich/PhotoEdit; **420** AP Photo/Ames Tribune, Nirmalendu Majumdar/AP Images; **421** Jim West/Alamy; **423** Marc Serota/Getty Images; **424** Mark Ransom Photography; **426** SuperStock, Inc./SuperStock; **429** Waltraud Grubitzsch/epa/CORBIS; **432** (l)PhotoDisc/SuperStock, (c)Jon Hicks/CORBIS, (r)Atlantide Phototravel/CORBIS; **433** Mosaic of Dionysus Riding a Leopard c.180 AD/House of Masks, Delos, Greece/The Bridgeman Art Library; **435** AM Corporation/Alamy Images; **436** K-Photos/Alamy Images; **439** Guy Grenier/Masterfile; **441** Blue Line Pictures/Getty Images; **442** age fotostock/SuperStock; **450** Comstock/JupiterImages; **454–455** Andrew Hasson/Alamy Images; **457** (TV)David Samuel Robbins/Getty Images, (flowers)Steven P. Lynch, (b)John Capella/Sports Imagery/Getty Images; **459** Rob Melnychuk/Brand X/CORBIS; **461** Masterfile Royalty Free; **462** John Kelly/Getty Images; **465** (l)David Young-Wolff/Alamy Images, (r)D. Hurst/Alamy Images; **466** Jim Craigmyle/CORBIS; **471** PhotoDisc/SuperStock; **472** (l)JJ/Getty Images, (r)Masterfile Royalty Free; **481** Andrew D. Bernstein/NBAE/Getty Images; **482** CORBIS/SuperStock; **491** PoodlesRock/CORBIS; **497** Brand X Pictures/PunchStock; **501** Dennis MacDonald/age fotostock; **506** CORBIS/SuperStock; **509** (l)C Squared Studios/Getty Images, (r)PhotoDisc/Getty Images; **510** Cathy Melloan/PhotoEdit; **512** David Rogers; **513** (l)Timothy Hursley/SuperStock, (r)Construction and photo by architectural-models.com; **514** Wm. Baker/GhostWorx Images/Alamy Images; **515** AP Photo/Robert F. Bukaty; **516** Don Despain/www.rekindlephotos.com/Alamy Images; **528–529** Kwame Zikomo/SuperStock; **531** Masterfile Royalty Free; **534** Jeff Greenberg/PhotoEdit; **537** (tl)Alan Schein Photography/CORBIS, (tc)Image Source/SuperStock, (tr)Richard Cummins/SuperStock, (b)David Noble Photography/Alamy Images; **547** (cl)Alec Pytlowany/Masterfile, (TV)David Samuel Robbins/Getty Images, (Football Game)Kevin C. Cox/Getty Images; **548** Nancy Sheehan/PhotoEdit; **558** RubberBall/SuperStock; **559** Tim Mantoani/Masterfile; **562** Tom Bean/CORBIS; **564** J. A. Kraulis/Masterfile; **569** Rodolfo Arpia/Alamy Images; **574** Stock Connection Distribution/Alamy Images; **576** Compassionate Eye Foundation/Colorblind Images/Getty Images; **578** Image Plan/CORBIS; **579** (t)Visual Arts Library (London)/Alamy Images, (cl)WildCountry/CORBIS, (cr)Andrew Brown/Ecoscene/CORBIS; **580** ThinkStock/SuperStock; **582** Ben Mangor/SuperStock; **584** Bettmann/CORBIS; **586** Maurizio Borgese/Hemis/CORBIS; **587** (t)Roger Harris/Photo Researchers, Inc., (c)Roger Ressmeyer/CORBIS; **588** Bettmann/CORBIS; **589** Juan Mabromata/AFP/Getty Images; **590** (l)Louie Psihoyos/CORBIS, (r)Rose Hartman/CORBIS; **593** Walter Lockwood/CORBIS; **596** Chris Birck/NBAE/Getty Images; **599** AP Photo/Brian Bohannon; **612–613** Werner Bollmann/age fotostock; **615** age fotostock/SuperStock; **621** (l)Jason Molyneaux/Masterfile, (r)Masterfile Royalty Free; **622** Science Museum/SSPL/The Image Works; **624** DreamWorks Distribution LLC/Special Anti-Pesto Still (Aardma/Bureau L.A. Collection/CORBIS); **626** Scott Boehm/Getty Images; **628** Otto Greule/Allsport/Getty Images; **631** Ed-Imaging; **632** age fotostock/SuperStock; **635** (l)Brian Elliot/Alamy Images, (c)JupiterImages/Thinkstock/Alamy Images, (r)Pick and Mix Images/Alamy Images; **636** (l)Renaud Visage/Getty Images, (r)Creatas/SuperStock; **637** Image Source/CORBIS; **638** Stella Snead/Bruce Coleman, Inc.; **641** age fotostock/SuperStock; **644** DIOMEDIA/Alamy Images; **646** PoodlesRock/CORBIS; **647** Juniors Bildarchiv/Alamy Images; **653** (tl)Brand X Pictures/PunchStock, (tr)Jochen Tack/Alamy Images, (cl bl)Stockbyte/age fotostock, (c)Don Hammond/Design Pics/CORBIS, (cr)Siede Preis/Getty Images, (bc)Don Hammond/Design Pics/CORBIS; **654** (l)Mark Ransom Photography, (cl)JupiterImages/Thinkstock/Alamy Images, (cr)Organica/Alamy Images, (r)CuboImages srl/Alamy Images; **655** (l)age fotostock/SuperStock, (cl)D. Hurst/Alamy Images, (cr)Image Source/PunchStock, (r b)Radlund & Associates/Getty Images; **656** Hisham F. Ibrahim/Getty Images; **657** (tl)Car Culture/CORBIS, (tc)Mike Rinnan/Alamy Images, (tr)Klaus Hackenberg/zefa/CORBIS, (b)dubassy/Alamy Images; **658** Digital Vision/Alamy Images; **660** John Lund/Getty Images; **664** (l)Visuals

Unlimited/CORBIS, (r)Mediscan/CORBIS; **665** Digital Image © The Museum of Modern Art/Licensed by SCALA/Art Resource, ® Chuck Close, courtesy of PaceWilderstein, New York; **666** Visuals Unlimited/CORBIS; **676** Jose Luis Pelaez, Inc./CORBIS; **680–681** Charles O'Rear/CORBIS; **683** M Stock/Alamy Images; **685** epa/CORBIS; **689** (tl tc)Garry Gay/Alamy Images, (tr)Gary Roebuck/Alamy Images, (b)UpperCut Images/Alamy Images; **690** Digital Vision Ltd./SuperStock; **691** (l)Siede Pries/PhotoDisc, (c)Spike Mafford/PhotoDisc, (r)Phillip Hayson/Photo Researchers; **692–693** The Granger Collection, New York; **694** Aflo Foto Agency/Alamy Images; **697** Jose Luis Pelaez, Inc./Getty Images; **698** (l)imagebroker/Alamy Images; (r)Darren Schmall; **703** (l)Dan Lim/Masterfile, (r)Darren Schmall; **706** Buzz Pictures/Alamy Images; **709** moodboard/Alamy Images; **712** Steve Gorton/Getty Images; **715** Larry Hamill; **718** David Young-Wolff/PhotoEdit; **724** Detlev van Ravenswaay/Photo Researchers, Inc.; **727** Bob Daemmrich/PhotoEdit; **730** sciencephotos/Alamy Images; **733** Brownie Harris/CORBIS; **734** (l)blickwinkel/Alamy Images, (r)Courtesy Akiyoshi Kitaoka; **737** (t)Daniel Dempster Photography/Alamy Images, (b)Henry Diltz/CORBIS; **740** (l)Daryl Benson/Masterfile, (r)Ilya Genki/Alamy Images; **741** Pixtal/SuperStock; **744** Gloria H. Chomica/Masterfile; **746** Gene & Karen Rhoden/Visuals Unlimited; **760–761** Jim West/PhotoEdit; **765** Skyscan Photolibrary/Alamy Images; **769** Johnathan Smith, Cordaiy Photo Library Ltd./CORBIS; **773** Juice Images/age fotostock; **777** (cl)Susumu Nishinaga/Photo Researchers, Inc., (bl bc)Micro Discovery/CORBIS, (br)Visuals Unlimited/CORBIS; **778** Joe Atlas/age fotostock; **779** Bettmann/CORBIS; **780** NOAA; **781** Steven Puetzer/Masterfile; **782** (t)Lois Ellen Frank/CORBIS, (b)Lawrence M. Sawyer/Getty Images; **783** Michael Newman/PhotoEdit; **784** (cl)Imageplus/CORBIS, (cr)Tom Nebbia/CORBIS, (b)Olaf Hirschberg/zefa/CORBIS; **785** (tl)Pete Saloutos/zefa/CORBIS, (tc)Arctic-Images/CORBIS, (tr)Dimitri Iundt/TempSport/CORBIS, (cl)Royalty-Free/CORBIS, (c)ML Sinibaldi/CORBIS, (cr)Duomo/CORBIS; **786** Don Mason/CORBIS; **791** Lindsey Stock/Alamy Images; **794** Brand X/SuperStock; **797** Art Kowalsky/Alamy Images; **802** Tim Garcha/zefa/CORBIS; **804** Digital Vision/Getty Images; **806** Masterfile Royalty Free; **807** Ron Dahlquist/SuperStock; **814** Jose Luis Pelaez, Inc./CORBIS; **818–819** age fotostock/SuperStock; **823** AP Photo/Nam Y. Huh; **824** E & E Image Library/HIP/The Image Works; **826** Charles D. Winters/Photo Researchers, Inc.; **828** Christie's Images; **830** Streeter Lecka/Getty Images; **835** (l)altrendo images/Getty Images, (r)AP Photo/Cheryl Hatch; **838** age fotostock/SuperStock; **841** Medioimages/PhotoDisc/Getty Images; **844** Michele Falzone/JAI/CORBIS; **845** Juniors Bildarchiv/age fotostock; **847** age fotostock/SuperStock; **851** David Burton/Alamy Images; **852** Javier Larrea/age fotostock; **859** Brand X Pictures/PunchStock; **861** Terry Smith Images/Alamy Images; **863** Courtesy M-K Distributors, Conrad M; **864** Rommel/Masterfile; **867** (l)AP Photo/Sara D. Davis, (r)Miguel Villagran/dpa/CORBIS; **869** (t)Masterfile Royalty Free, (b)Tom Chudleigh; (c)ArkReligion.com/Alamy Images, **870** Koji Kitagawa/SuperStock; **873** Bloomimage/CORBIS; **875** (cl)Hugh Threlfall/Alamy Images, (cr)D. Hurst/Alamy Images, (b)Randy Faris/CORBIS; **876** (tc)Atlantide Phototravel/CORBIS, (tr)Datacraft/age fotostock, (b)The Granger Collection, New York; **877** M.C. Escher "Circle Limit II"®2008 The M.C. Escher Company-Holland. All rights reserved. www.mcescher.com<http://www.mcescher.com/>; **880** Jewelry specialist/Alamy Images; **884** (l)Steve Broulis/SuperStock, (r)Baron Photography & Image Finders, Inc.; **886** F. Stuart Westmorland/Photo Researchers; **892** Mary Kate Denny/PhotoEdit; **896–897** Hill Street Studios/Blend Images/CORBIS; **899** Robert Michael/CORBIS; **901** VStock/Alamy Images; **904** Masterfile Royalty Free; **906** Digital Vision/Alamy Images; **908** SuperStock, Inc./SuperStock; **913** Lynda Schemansky/age fotostock; **915** Gabe Palmer/Alamy Images; **916** Ingram Publishing/SuperStock; **917** Horizon International Images Limited/Alamy Images; **919** Royalty-Free/CORBIS; **920** age fotostock/SuperStock; **923** VStock LLC/age fotostock; **925** Andrew D. Bernstein/NBAE/Getty Images; **926** The Granger Collection, New York; **928** Jiang Jin/SuperStock; **929** David Young-Wolff/PhotoEdit; **931** CORBIS; **933** Masterfile Royalty Free; **935** Teo Lannie/Getty Images; **936** Oliver Lang/AFP/Getty Images; **938** Image Source Pink/Alamy Images; **939** Philip Lee Harvey/Getty Images; **940** David Young-Wolff/PhotoEdit; **942** Digital Vision/Alamy Images; **944** David Young-Wolff/PhotoEdit; **947** The Granger Collection, New York; **952** Spencer Grant/PhotoEdit; **956** Eclipse Studios; **1005** David R. Frazier Photolibrary, Inc./Alamy Images; **1008** Bettmann/CORBIS; **1010** Aaron Haupt; **1021** Jim West/Alamy Images.

Glossary/Glosario

Math Online

A mathematics multilingual eGlossary is available at **glencoe.com**. The glossary includes the following languages:

Arabic	Haitian Creole	Tagalog
Bengali	Hmong	Urdu
Brazilian Portuguese	Korean	Vietnamese
Chinese	Russian	
English	Spanish	

Cómo usar el glosario en español:

1. Busca el término en inglés que desees encontrar.
2. El término en español, junto con la definición, se encuentran en la columna de la derecha.

English / Español

A

acute angle (p. 38) An angle with a degree measure less than 90.

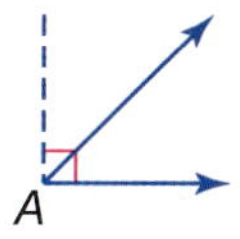

$0 < m\angle A < 90$

ángulo agudo Ángulo cuya medida en grados es menos de 90.

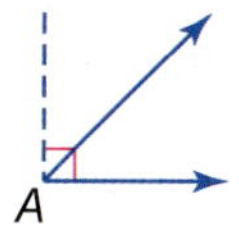

$0 < m\angle A < 90$

acute triangle (p. 235) A triangle in which all of the angles are acute angles.

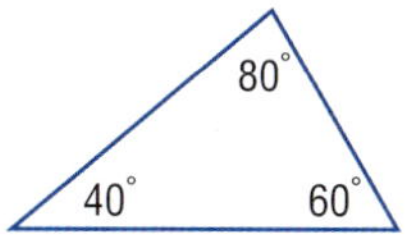

three acute angles

triángulo acutángulo Triángulo cuyos ángulos son todos agudos.

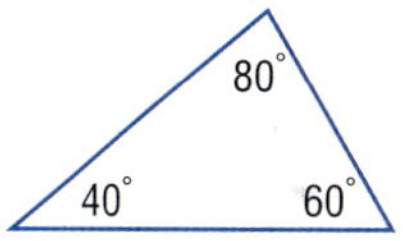

tres ángulos agudos

adjacent angles (p. 46) Two angles that lie in the same plane, have a common vertex and a common side, but no common interior points.

ángulos adyacentes Dos ángulos que yacen sobre el mismo plano, tienen el mismo vértice y un lado en común, pero ningún punto interior en común.

adjacent arcs (p. 694) Arcs in a circle that have exactly one point in common.

arcos adyacentes Arcos en un círculo que tienen un solo punto en común.

algebraic proof (p. 134) A proof that is made up of a series of algebraic statements. The properties of equality provide justification for many statements in algebraic proofs.

demostración algebraica Demostración que se realiza con una serie de enunciados algebraicos. Las propiedades de la igualdad proveen justificación para muchas enunciados en demostraciones algebraicas.

alternate exterior angles (p. 132) In the figure, transversal t intersects lines ℓ and m. $\angle 5$ and $\angle 3$, and $\angle 6$ and $\angle 4$ are alternate exterior angles.

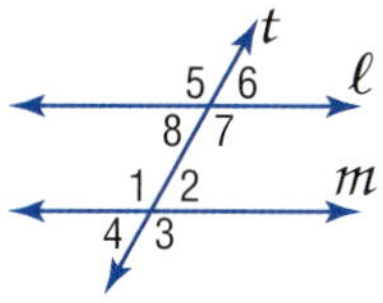

ángulos alternos externos En la figura, la transversal t interseca las rectas ℓ y m. $\angle 5$ y $\angle 3$; y $\angle 6$ y $\angle 4$ son ángulos alternos externos.

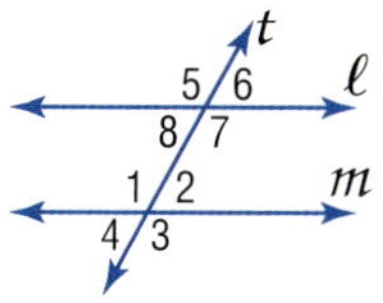

alternate interior angles (p. 172) In the figure at the bottom of page R123, transversal t intersects lines ℓ and m. $\angle 1$ and $\angle 7$, and $\angle 2$ and $\angle 8$ are alternate interior angles.

altitude **1.** (p. 333) In a triangle, a segment from a vertex of the triangle to the line containing the opposite side and perpendicular to that side. **2.** (p. 830) In a prism or cylinder, a segment perpendicular to the bases with an endpoint in each plane. **3.** (p. 830) In a pyramid or cone, the segment that has the vertex as one endpoint and is perpendicular to the base. **4.** (p. 706) In a parallelogram, any segment perpendicular to the bases, with endpoints on each base.

ambiguous case of the Law of Sines (p. 592) Given the measures of two sides and a nonincluded angle, there exist two possible triangles.

angle (p. 36) The intersection of two noncollinear rays at a common endpoint. The rays are called *sides* and the common endpoint is called the *vertex*.

angle bisector (p. 39) A ray that divides an angle into two congruent angles.

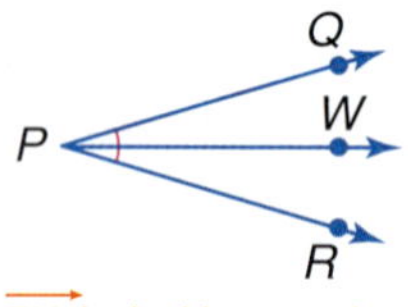

$\overrightarrow{PW}$ is the bisector of $\angle P$.

angle of depression (p. 574) The angle between the line of sight and the horizontal when an observer looks downward.

angle of elevation (p. 574) The angle between the line of sight and the horizontal when an observer looks upward.

angle of rotation (p. 630) The angle through which a preimage is rotated to form the image.

apothem (p. 791) A segment that is drawn from the center of a regular polygon perpendicular to a side of the polygon.

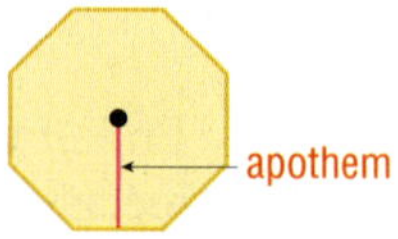

arc (p. 692) A part of a circle that is defined by two endpoints.

area (p. 58) The number of square units needed to cover a surface.

ángulos alternos internos En la figura anterior, la transversal t interseca las rectas ℓ y m. $\angle 1$ y $\angle 7$; y $\angle 2$ y $\angle 8$ son ángulos alternos internos.

altura **1.** En un triángulo, segmento trazado desde uno de los vértices del triángulo hasta el lado opuesto y que es perpendicular a dicho lado. **2.** En un prisma o un cilindro, segmento perpendicular a las bases con un extremo en cada plano. **3.** En una pirámide o un cono, segmento que tiene un extremo en el vértice y que es perpendicular a la base. **4.** En un paralelogramo, cualquier segmento de recta perpendicular a las bases y con extremos en cada una de las bases.

caso ambiguo de la ley de los senos Dadas las medidas de dos lados y de un ángulo no incluido, existen dos triángulos posibles.

ángulo Intersección de dos rayos no colineales en un extremo común. Los rayos se llaman *lados* y el punto común se llama *vértice*.

bisectriz de un ángulo Rayo que divide un ángulo en dos ángulos congruentes.

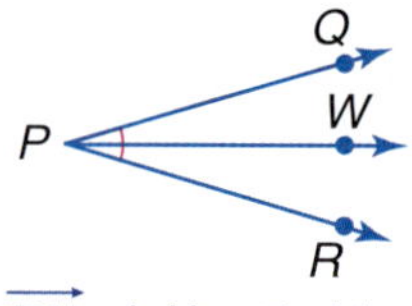

$\overrightarrow{PW}$ es la bisectriz del $\angle P$.

ángulo de depresión Ángulo formado por la horizontal y la línea de visión de un observador que mira hacia abajo.

ángulo de elevación Ángulo formado por la horizontal y la línea de visión de un observador que mira hacia arriba.

ángulo de rotación Ángulo a través del cual se rota una preimagen para formar la imagen.

apotema Segmento trazado desde el centro de un polígono regular hasta uno de sus lados y que es perpendicular a dicho lado.

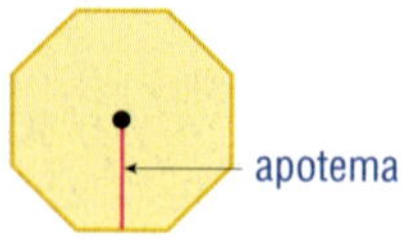

arco Parte de un círculo definida por dos extremos.

área Número de unidades cuadradas necesarias para cubrir una superficie.

auxiliary line (p. 244) An extra line or segment drawn in a figure to help complete a proof.

línea auxiliar Recta o segmento de recta adicional que se traza en una figura para ayudar a completar una demostración.

axiom (p. 125) A statement that is accepted as true.

axioma Enunciado que se acepta como verdadero.

axis **1.** (p. 693) In a cylinder, the segment with endpoints that are the centers of the bases. **2.** (p. 711) In a cone, the segment with endpoints that are the vertex and the center of the base.

eje **1.** En un cilindro, el segmento cuyos extremos son el centro de las bases. **2.** En un cono, el segmento cuyos extremos son el vértice y el centro de la base.

axis of symmetry (p. 655) Symmetry in a three-dimensional figure that occurs if the figure can be mapped onto itself by a rotation between 0° and 360° in a line.

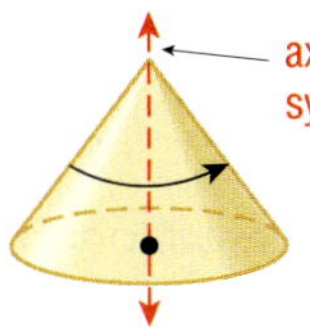

eje de simetría Simetría que ocurre en una figura tridimensional si la figura se puede aplicar sobre sí misma, mientras se gira entre 0° y 360° sobre una recta.

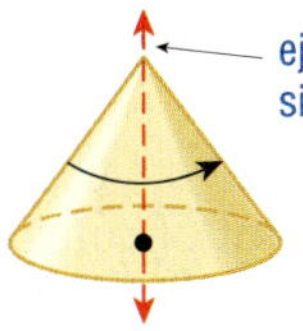

B

base angle of an isosceles triangle (p. 283) See *isosceles triangle.*

ángulo de la base de un triángulo isósceles Ver *triángulo isósceles.*

base edges (p. 830) The intersection of the lateral faces and bases in a solid figure.

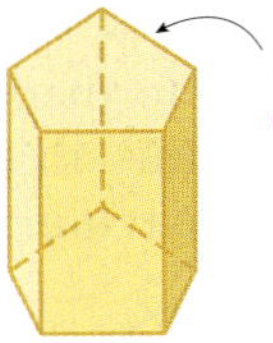

aristas de las bases Intersección de las bases con las caras laterales en una figura sólida.

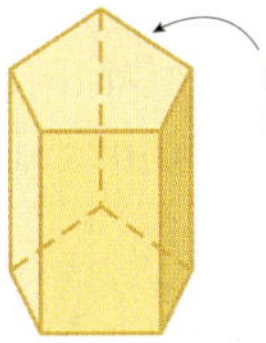

base of parallelogram (p. 763) Any side of a parallelogram.

base de un paralelogramo Cualquier lado de un paralelogramo.

base of a polyhedron (p. 67) The two parallel congruent faces of a polyhedron.

base de un poliedro Las dos caras paralelas y congruentes de un poliedro.

between (p. 15) For any two points A and B on a line, there is another point C between A and B if and only if A, B, and C are collinear and $AC + CB = AB$.

entre Para cualquier par de puntos A y B de una recta, existe otro punto C ubicado entre A y B si y sólo si A, B y C son colineales y $AC + CB = AB$.

betweenness of points (p. 15) See *between.*

intermediación de puntos Ver *entre.*

biconditional (p. 114) The conjunction of a conditional statement and its converse.

bicondicional Conjunción entre un enunciado condicional y su recíproco.

C

center of circle (p. 683) The central point where radii form a locus of points called a circle.

centro de un círculo Punto central desde el cual los radios forman un lugar geométrico de puntos llamado círculo.

center of dilation (p. 505) The center point from which dilations are performed.

center of rotation (p. 630) A fixed point around which shapes move in a circular motion to a new position.

center of symmetry (p. 653) See *point of symmetry*.

central angle (p. 692) An angle that intersects a circle in two points and has its vertex at the center of the circle.

central angle of regular polygon (p. 791) An angle that has its vertex at the center of a polygon and with sides that pass through consecutive vertices of the polygon.

centroid (p. 333) The point of concurrency of the medians of a triangle.

chord **1.** (p. 683) For a given circle, a segment with endpoints that are on the circle. **2.** (p. 711) For a given sphere, a segment with endpoints that are on the sphere.

chord segments (p. 737) Segments that form when two chords intersect inside a circle.

circle (p. 683) The locus of all points in a plane equidistant from a given point called the *center* of the circle.

P is the center of the circle.

circular permutation (p. 908) A permutation of objects that are arranged in a circle or loop.

circumcenter (p. 323) The point of concurrency of the perpendicular bisectors of a triangle.

circumference (pp. 58, 685) The distance around a circle.

circumscribed (p. 685) A circle is circumscribed about a polygon if the circle contains all the vertices of the polygon.

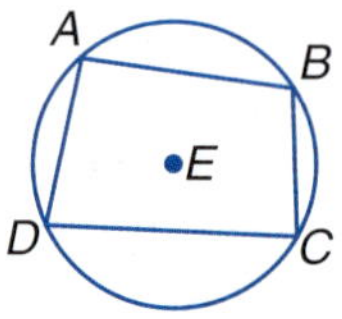

⊙*E* is circumscribed about quadrilateral *ABCD*.

centro de la homotecia Punto fijo en torno al cual se realizan las homotecias.

centro de rotación Punto fijo alrededor del cual gira una figura hasta alcanzar una posición dada.

centro de simetría Vea *punto de simetría*.

ángulo central Ángulo que interseca un círculo en dos puntos y cuyo vértice está en el centro del círculo.

ángulo central de un polígono Ángulo cuyo vértice está en el centro de un polígono y cuyos lados pasan por vértices consecutivos del polígono.

baricentro Punto de intersección de las medianas de un triángulo.

cuerda **1.** Para cualquier círculo, segmento cuyos extremos están en el círculo. **2.** Para cualquier esfera, segmento cuyos extremos están en la esfera.

segmentos de cuerda Segmentos que se forman cuando dos cuerdas se intersecan dentro de un círculo.

círculo Lugar geométrico formado por todos los puntos en un plano, equidistantes de un punto dado llamado *centro* del círculo.

P es el centro del círculo.

permutación circular Permutación de objetos que se arreglan en un círculo o un bucle.

circuncentro Punto de intersección de las mediatrices de un triángulo.

circunferencia Distancia alrededor de un círculo.

circunscrito Un polígono está circunscrito a un círculo si todos sus vértices están contenidos en el círculo.

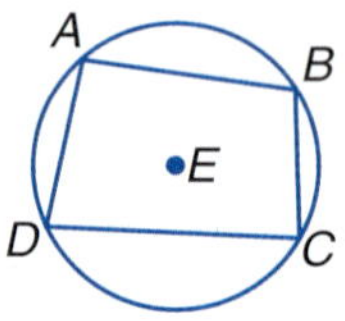

⊙*E* está circunscrito al cuadrilátero *ABCD*.

collinear (p. 5) Points that lie on the same line.

P, *Q*, and *R* are collinear.

common tangent (p. 718) A line or segment that is tangent to two circles in the same plane.

complement (p. 941) The complement of an event *A* consists of all the outcomes in the sample space that are not included as outcomes of event *A*.

complementary angles (p. 47) Two angles with measures that have a sum of 90.

component form (p. 593) A vector expressed as an ordered pair, ⟨change in *x*, change in *y*⟩.

composite figure (p. 793) A figure that can be separated into regions that are basic figures.

composite solid (p. 836) A three-dimensional figure that is composed of simpler figures.

composition of reflections (p. 639) Successive reflections in parallel lines.

composition of transformations (p. 641) The resulting transformation when a transformation is applied to a figure and then another transformation is applied to its image.

compound event (p. 931) An event that consists of two or more simple events.

compound locus (p. 748) A locus or set of points which satisfies more than one condition.

compound statement (p. 97) A statement formed by joining two or more statements.

concave polygon (p. 56) A polygon for which there is a line containing a side of the polygon that also contains a point in the interior of the polygon.

concentric circles (p. 684) Coplanar circles with the same center.

conclusion (p. 105) In a conditional statement, the statement that immediately follows the word *then*.

concurrent lines (p. 323) Three or more lines that intersect at a common point.

conditional statement (p. 105) A statement that can be written in *if-then* form.

colineal Puntos que yacen sobre la misma recta.

P, *Q* y *R* son colineales.

tangente común Recta o segmento de recta tangente a dos círculos en el mismo plano.

complemento El complemento de un evento *A* consiste en todos los resultados en el espacio muestral que no se incluyen como resultados del evento *A*.

ángulos complementarios Dos ángulos cuyas medidas suman 90°.

componente Vector expresado en forma de par ordenado, ⟨cambio en *x*, cambio en *y*⟩.

figura compuesta Figura que se puede separar en regiones con formas de figuras básicas.

solido compuesto Figura tridimensional formada por figuras más simples.

composición de reflexiones Reflexiones sucesivas en rectas paralelas.

composición de transformaciones Transformación que resulta cuando se aplica una transformación a una figura y luego se le aplica otra transformación a su imagen.

evento compuesto Evento que consiste en dos o más eventos simples.

lugar geométrico compuesto Conjunto de puntos que satisfacen más de una condición.

enunciado compuesto Enunciado formado por la unión de dos o más enunciados.

polígono cóncavo Polígono para el cual existe una recta que contiene un lado del polígono y un punto en el interior del polígono.

círculos concéntricos Círculos coplanarios con el mismo centro.

conclusión Parte de un enunciado condicional que está escrito justo después de la palabra *entonces*.

rectas concurrentes Tres o más rectas que se intersecan en un punto común.

enunciado condicional Enunciado escrito en la forma *si-entonces*.

cone (p. 67) A solid with a circular base, a vertex not contained in the same plane as the base, and a lateral surface area composed of all points in the segments connecting the vertex to the edge of the base.

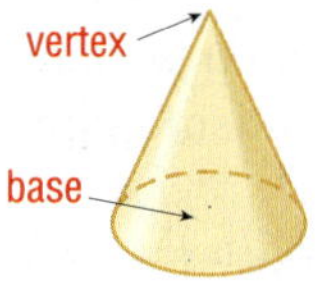

congruence transformations (p. 294) A mapping for which a geometric figure and its image are congruent.

congruent (p. 17) Having the same measure.

congruent arcs (p. 693) Arcs in the same circle or in congruent circles that have the same measure.

congruent circles (p. 684) Two circles that have congruent radii.

congruent polygons (p. 253) Polygons in which all matching parts are congruent.

congruent solids (p. 880) Two solids with the same shape, size and scale factor of 1:1.

conjecture (p. 89) An educated guess based on known information.

conjunction (p. 97) A compound statement formed by joining two or more statements with the word *and*.

consecutive interior angles (p. 172) In the figure, transversal t intersects lines ℓ and m. There are two pairs of consecutive interior angles: $\angle 8$ and $\angle 1$, and $\angle 7$ and $\angle 2$.

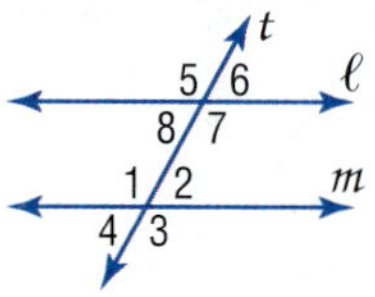

cono Sólido de base circular cuyo vértice no yace en el mismo plano que la base y cuya área de superficie lateral está formada por todos los puntos en los segmentos que conectan el vértice con el borde de la base.

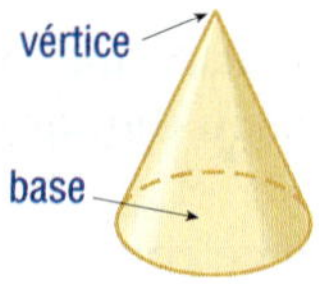

transformaciones de congruencia Aplicación en la cual una figura geométrica y su imagen son congruentes.

congruente Que tienen la misma medida.

arcos congruentes Arcos que tienen la misma medida y que pertenecen al mismo círculo o a círculos congruentes.

círculos congruentes Dos círculos que tienen radios congruentes.

polígonos congruentes Polígonos cuyas partes correspondientes son todas congruentes.

sólidos congruentes Dos sólidos con la misma forma, tamaño y factor de escala de 1:1.

conjetura Juicio basado en información conocida.

conjunción Enunciado compuesto que se obtiene al unir dos o más enunciados con la palabra *y*.

ángulos internos consecutivos En la figura, la transversal t interseca las rectas ℓ y m. La figura presenta dos pares de ángulos internos consecutivos: $\angle 8$ y $\angle 1$; y $\angle 7$ y $\angle 2$.

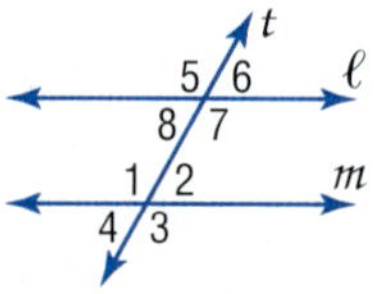

construction (p. 17) A method of creating geometric figures without the benefit of measuring tools. Generally, only a pencil, straightedge, and compass are used.

contrapositive (p. 107) The statement formed by negating both the hypothesis and conclusion of the converse of a conditional statement.

converse (p. 107) The statement formed by exchanging the hypothesis and conclusion of a conditional statement.

convex polygon (p. 56) A polygon for which there is no line that contains both a side of the polygon and a point in the interior of the polygon.

coordinate proofs (p. 301) Proofs that use figures in the coordinate plane and algebra to prove geometric concepts.

coplanar (p. 5) Points that lie in the same plane.

corner view (p. 680) The view from a corner of a three-dimensional figure, also called the *perspective view*.

corollary (p. 247) A statement that can be easily proved using a theorem is called a corollary of that theorem.

corresponding angles (p. 172) In the figure, transversal t intersects lines ℓ and m. There are four pairs of corresponding angles: $\angle 5$ and $\angle 1$, $\angle 8$ and $\angle 4$, $\angle 6$ and $\angle 2$, and $\angle 7$ and $\angle 3$.

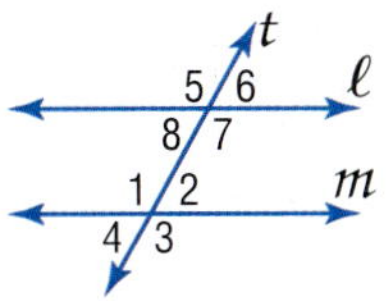

corresponding parts (p. 253) Matching parts of congruent polygons.

cosecant (p. 572) The reciprocal of the sine of an angle in a right triangle.

cosine (p. 562) For an acute angle of a right triangle, the ratio of the measure of the leg adjacent to the acute angle to the measure of the hypotenuse.

cotangent (p. 572) The ratio of the adjacent to the opposite side of a right triangle.

counterexample (p. 92) An example used to show that a given statement is not always true.

construcción Método para dibujar figuras geométricas sin el uso de instrumentos de medición. En general, sólo requiere de un lápiz, una regla y un compas.

antítesis Enunciado formado por la negación tanto de la hipótesis como de la conclusión del recíproco de un enunciado condicional.

recíproco Enunciado que se obtiene al intercambiar la hipótesis y la conclusión de un enunciado condicional dado.

polígono convexo Polígono para el cual no existe recta alguna que contenga un lado del polígono y un punto en el interior del polígono.

demostraciones en coordenadas Demostraciones que usan figuras en el plano de coordenadas y álgebra para demostrar conceptos geométricos.

coplanar Puntos que yacen en el mismo plano.

vista de esquina Vista desde una de las esquinas de una figura tridimensional. También se conoce como *vista en perspectiva*.

corolario Un enunciado que se puede demostrar fácilmente usando un teorema se conoce como corolario de dicho teorema.

ángulos correspondientes En la figura, la transversal t interseca las rectas ℓ y m. La figura muestra cuatro pares de ángulos correspondientes: $\angle 5$ y $\angle 1$, $\angle 8$ y $\angle 4$, $\angle 6$ y $\angle 2$; y $\angle 7$ y $\angle 3$.

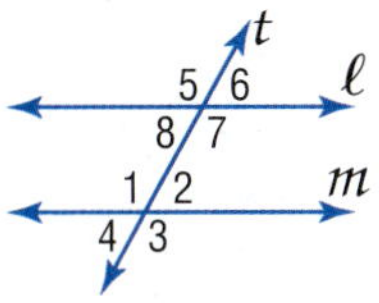

partes correspondientes Partes que coinciden de polígonos congruentes.

cosecante Recíproco del seno de un ángulo en un triángulo rectángulo.

coseno Para cualquier ángulo agudo de un triángulo rectángulo, razón de la medida del cateto adyacente al ángulo agudo a la medida de la hipotenusa.

cotangente Razón de la medida del cateto adyacente a la medida del cateto opuesto de un triángulo rectángulo.

contraejemplo Ejemplo que se usa para demostrar que un enunciado dado no siempre es verdadero.

cross products (p. 458) In the proportion $\frac{a}{b} = \frac{c}{d}$, where $b \neq 0$ and $d \neq 0$, the cross products are ad and bc. The proportion is true if and only if the cross products are equal.

productos cruzados En la proporción $\frac{a}{b} = \frac{c}{d}$, donde $b \neq 0$ y $d \neq 0$, los productos cruzados son ad y bc. La proporción es verdadera si y sólo si los productos cruzados son iguales.

cross section (p. 824) The intersection of a solid and a plane.

sección transversal Intersección de un sólido con un plano.

cylinder (p. 67) A figure with bases that are formed by congruent circles in parallel planes.

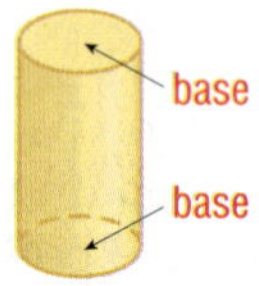

cilindro Figura cuyas bases son círculos congruentes ubicados en planos paralelos.

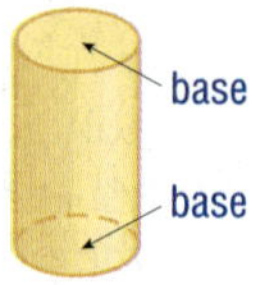

D

deductive argument (p. 127) A proof formed by a group of algebraic steps used to solve a problem.

argumento deductivo Demostración que consta de un conjunto de pasos algebraicos que se usan para resolver un problema.

deductive reasoning (p. 115) A system of reasoning that uses facts, rules, definitions, or properties to reach logical conclusions.

razonamiento deductivo Sistema de razonamiento que emplea hechos, reglas, definiciones o propiedades para obtener conclusiones lógicas.

degree (p. 37) A unit of measure used in measuring angles and arcs. An arc of a circle with a measure of 1° is $\frac{1}{360}$ of the entire circle.

grado Unidad de medida que se usa para medir ángulos y arcos. El arco de un círculo que mide 1° equivale a $\frac{1}{360}$ del círculo completo.

diagonal (p. 389) In a polygon, a segment that connects nonconsecutive vertices of the polygon.

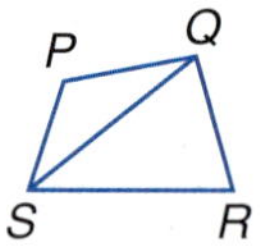

$\overline{SQ}$ is a diagonal.

diagonal Recta que conecta vértices no consecutivos de un polígono.

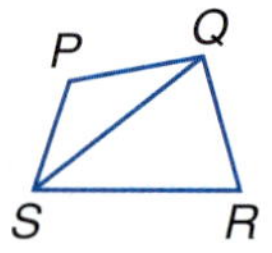

$\overline{SQ}$ es una diagonal.

diameter **1.** (p. 683) In a circle, a chord that passes through the center of the circle. **2.** (p. 711) In a sphere, a segment that contains the center of the sphere, and has endpoints that are on the sphere.

diámetro **1.** En un círculo, cuerda que pasa por el centro. **2.** En una esfera, segmento que incluye el centro de la esfera y cuyos extremos están ubicados en la esfera.

dilation (p. 505) A transformation determined by a center point C and a scale factor k. When $k > 0$, the image P' of P is the point on $\overrightarrow{CP}$ such that $CP' = |k| \cdot CP$. When $k < 0$, the image P' of P is the point on the ray opposite $\overrightarrow{CP}$ such that $CP' = k \cdot CP$.

homotecia Transformación determinada por un punto central C y un factor de escala k. Cuando $k > 0$, la imagen P' de P es el punto en $\overrightarrow{CP}$ tal que $CP' = |k| \cdot CP$. Cuando $k < 0$, la imagen P' de P es el punto en el rayo opuesto $\overrightarrow{CP}$ tal que $CP' = k \cdot CP$.

direct isometry (p. 516) An isometry in which the image of a figure is found by moving the figure intact within the plane.

isometría directa Isometría en la cual se obtiene la imagen de una figura, al mover la figura intacta dentro del plano.

direction (p. 593) The measure of the angle that a vector forms with the positive x-axis or any other horizontal line.

disjunction (p. 98) A compound statement formed by joining two or more statements with the word *or*.

distance between two points (p. 25) The length of the segment between two points.

E

edge (p. 946) A line that connects two nodes in a network.

edge of a polyhedron (p. 67) A line segment where the faces of a polyhedron intersect.

enlargement (p. 505) An image that is larger that the original figure.

equal vectors (p. 594) Vectors that have the same magnitude and direction.

equiangular polygon (p. 57) A polygon with all congruent angles.

equiangular triangle (p. 235) A triangle with all angles congruent.

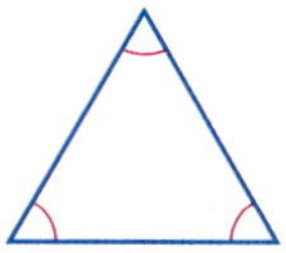

equidistant (p. 216) The distance between two lines measured along a perpendicular line is always the same.

equilateral polygon (p. 57) A polygon with all congruent sides.

equilateral triangle (p. 236) A triangle with all sides congruent.

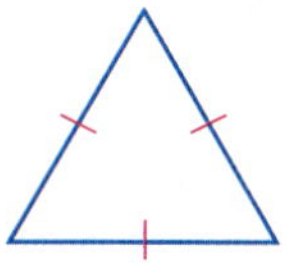

Euclidean geometry (p. 873) A geometrical system in which a plane is a flat surface made up of points that extend infinitely in all directions.

dirección Medida del ángulo que forma un vector con el eje x positivo o con cualquier otra recta horizontal.

disyunción Enunciado compuesto que se forma al unir dos o más enunciados con la palabra *o*.

distancia entre dos puntos Longitud del segmento entre dos puntos.

arista Recta que conecta dos nodos en una red.

arista de un poliedro Segmento de recta donde se intersecan las caras de un poliedro.

ampliación Imagen que es más grande que la figura original.

vectores iguales Vectores con la misma magnitud y dirección.

polígono equiangular Polígono cuyos ángulos son todos congruentes.

triángulo equiangular Triángulo cuyos ángulos son todos congruentes.

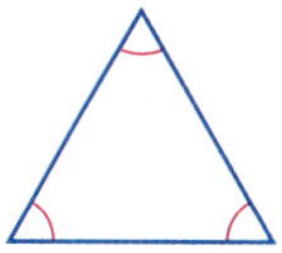

equidistante La distancia entre dos rectas que siempre permanece constante cuando se mide a lo largo de una perpendicular.

polígono equilátero Polígono cuyos lados son todos congruentes.

triángulo equilátero Triángulo cuyos lados son todos congruentes.

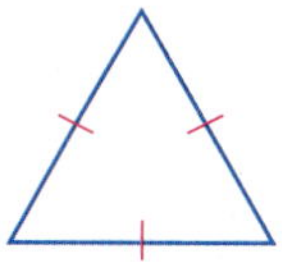

geometría euclidiana Sistema geométrico en el cual un plano es una superficie plana formada por puntos que se extienden infinitamente en todas las direcciones.

expected value (p. 925) Also *mathematical expectation*, is the average value of a random variable that one *expects* after repeating an experiment or simulation an infinite number of times.

extended ratios (p. 457) Ratios that are used to compare three or more quantities.

exterior (p. 36) A point is in the exterior of an angle if it is neither on the angle nor in the interior of the angle.

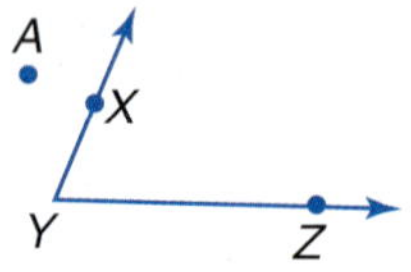

A is in the exterior of $\angle XYZ$.

exterior angle (p. 246) An angle formed by one side of a triangle and the extension of another side.

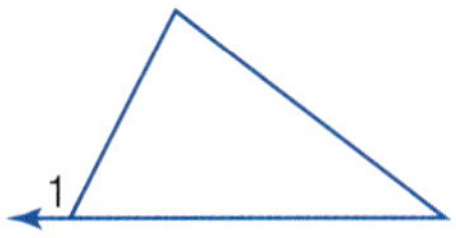

$\angle 1$ is an exterior angle.

external secant segment (p. 737) A secant segment that lies in the exterior of the circle.

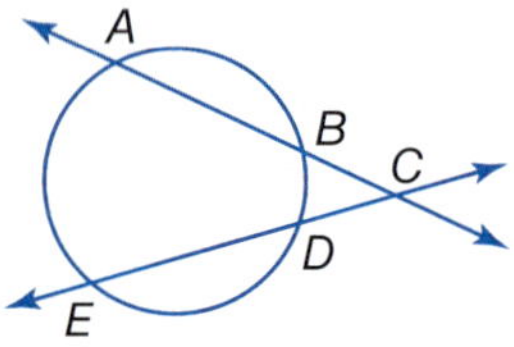

$\overline{BC}$ and $\overline{CD}$ are external secant segments

external segment (p. 738) A segment that lies in the exterior of a circle.

extremes (p. 458) In $\frac{a}{b} = \frac{c}{d}$, the numbers *a* and *d*.

valor esperado También *expectativa matemática*, el valor promedio de una variable aleatoria que uno *espera* después de repetir un experimento o un simulacro un número infinito de veces.

razones extendidas Razones que se utilizan para comparar tres o más cantidades.

exterior Un punto yace en el exterior de un ángulo si no se ubica ni en el ángulo ni en el interior del ángulo.

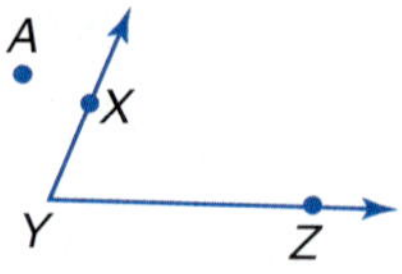

A está en el exterior del $\angle XYZ$.

ángulo externo Ángulo formado por un lado de un triángulo y la prolongación de otro de sus lados.

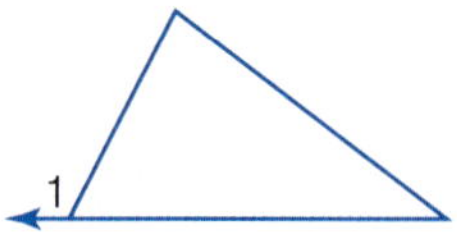

$\angle 1$ es un ángulo externo.

segmento secante externo Segmento secante que yace en el exterior del círculo.

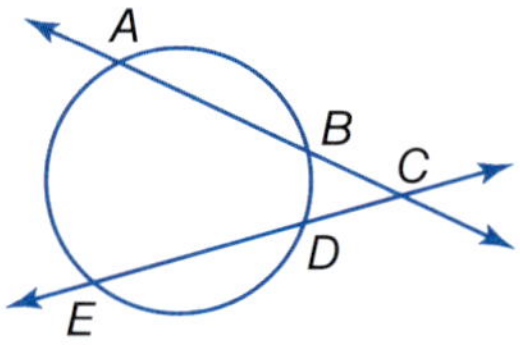

$\overline{BC}$ y $\overline{CD}$ son segmentos secantes externos

segmento externo Segmento que yace en el exterior de un círculo.

extremos Los números *a* y *d* en $\frac{a}{b} = \frac{c}{d}$.

F

face of a polyhedron (p. 67) A flat surface of a polyhedron.

factorial (p. 906) The product of the integers less than or equal to a positive integer *n*, written as *n*!

finite plane (p. 10) A plane that has boundaries or does not extend indefinitely.

cara de un poliedro Superficie plana de un poliedro.

factorial Producto de los enteros menores o iguales a un número positivo *n*, escrito *n*!, escrito como *n!*

plano finito Plano que tiene límites o que no se extiende indefinidamente.

flow proof (p. 246) A proof that organizes statements in logical order, starting with the given statements. Each statement is written in a box with the reason verifying the statement written below the box. Arrows are used to indicate the order of the statements.

demostración de flujo Demostración que organiza los enunciados en orden lógico, comenzando con los enunciados dados. Cada enunciado se escribe en una casilla y debajo de cada casilla se escribe el argumento que verifica dicho enunciado. El orden de los enunciados se indica con flechas.

formal proof (p. 135) A two-column proof containing statements and reasons.

demostración formal Demostración en dos columnas que contiene enunciados y razonamientos.

fractal (p. 423) A figure generated by repeating a special sequence of steps infinitely often. Fractals often exhibit self-similarity.

fractal Figura que se obtiene mediante la repetición infinita de una sucesión particular de pasos. Los fractales a menudo exhiben autosemejanza.

frustum (p. 704) The part of a solid that remains after the top portion has been cut by a plane parallel to the base.

tronco Parte de un sólido que queda después de que la parte superior ha sido cortada por un plano paralelo a la base.

Fundamental Counting Principle (p. 901) A method used to determine the number of possible outcomes in a sample space by multiplying the number of possible outcomes from each stage or event.

principio fundamental de contar Método para determinar el número de resultados posibles en un espacio muestral multiplicando el número de resultados posibles de cada etapa o evento.

G

geometric mean (p. 531) For any positive numbers a and b, the positive number x such that $\frac{a}{x} = \frac{x}{b}$.

media geométrica Para todo número positivo a y b, existe un número positivo x tal que $\frac{a}{x} = \frac{x}{b}$.

geometric probability (p. 915) Using the principles of length and area to find the probability of an event.

probabilidad geométrica Uso de los principios de longitud y área para calcular la probabilidad de un evento.

glide reflection (p. 639) The composition of a translation followed by a reflection in a line parallel to the translation vector.

reflexión de deslizamiento Composición de una traslación seguida por una reflexión en una recta paralela al vector de la traslación.

great circle (p. 864) A circle formed when a plane intersects a sphere with its center at the center of the sphere.

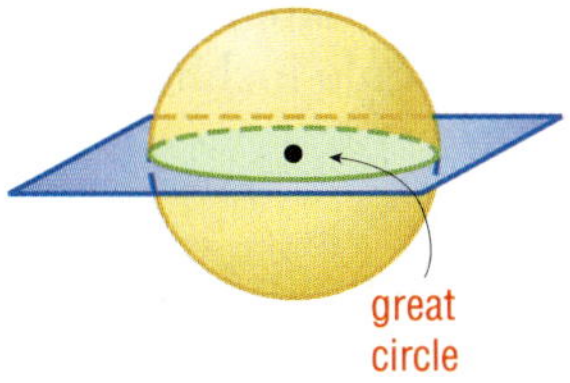

círculo mayor Círculo que se forma cuando un plano interseca una esfera y cuyo centro es el mismo que el centro de la esfera.

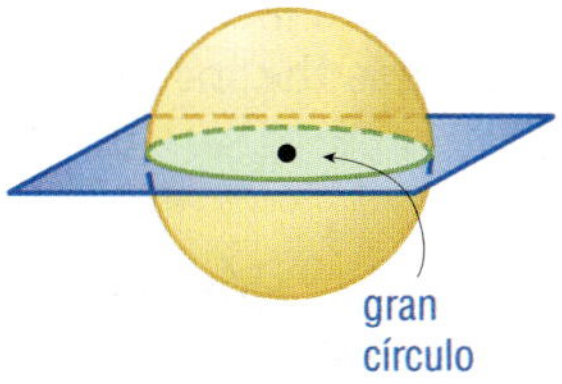

H

height of a parallelogram (p. 763) The length of an altitude of a parallelogram.

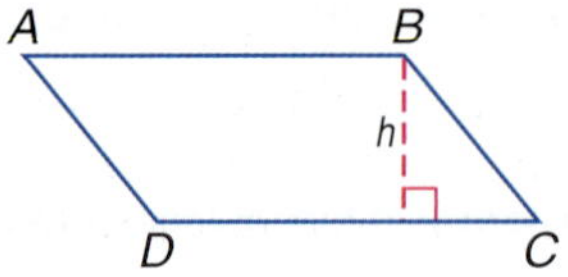

h is the height of parallelogram *ABCD*.

height of a trapezoid (p. 773) The perpendicular distance between the bases of a trapezoid.

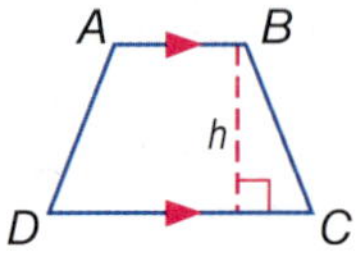

h is the height of trapezoid *ABCD*.

height of a triangle (p. 764) The length of an altitude drawn to a given base of a triangle.

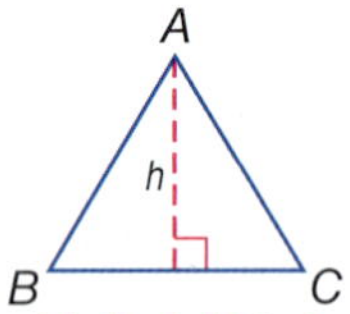

h is the height of triangle *ABC*.

hemisphere (p. 864) One of the two congruent parts into which a great circle separates a sphere.

hypothesis (p. 105) In a conditional statement, the statement that immediately follows the word *if*.

altura de un paralelogramo Longitud del segmento perpendicular que va desde la base hasta el vértice opuesto a ella.

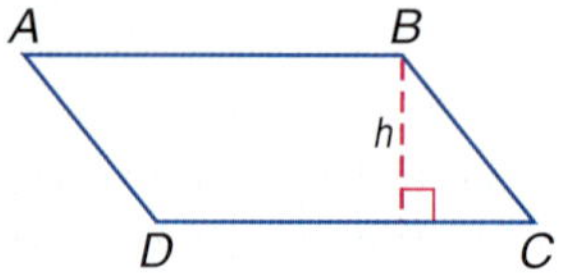

h es la altura del paralelogramo *ABCD*.

altura de un trapecio Distancia de la perpendicular entre las bases de un trapecio.

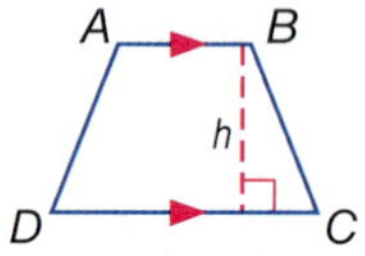

h es la altura del trapecio *ABCD*.

altura de un triángulo Longitud de una altura trazada a una base dada de un triángulo.

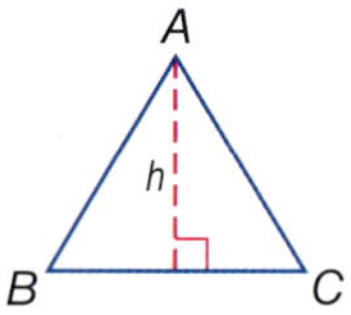

h es la altura del triángulo *ABC*.

hemisferio Una de las dos partes congruentes en las cuales un círculo mayor divide una esfera.

hipótesis Enunciado escrito inmediatamente después de la palabra *si* en un enunciado condicional.

I

if-then statement (p. 105) A compound statement of the form "if *p*, then *q*," where *p* and *q* are statements.

image (p. 294) A figure that results from the transformation of a geometric figure.

incenter (p. 326) The point of concurrency of the angle bisectors of a triangle.

included angle (p. 264) In a triangle, the angle formed by two sides is the included angle for those two sides.

included side (p. 273) The side of a triangle that is a side of each of two angles.

enunciado si-entonces Enunciado compuesto de la forma "si *p*, entonces *q*," donde *p* y *q* son enunciados.

imagen Figura que resulta de la transformación de una figura geométrica.

incentro Punto de intersección de las bisectrices interiores de un triángulo.

ángulo incluido En un triángulo, el ángulo formado por dos lados es el ángulo incluido de esos dos lados.

lado incluido Lado de un triángulo común a dos de sus ángulos.

indirect isometry (p. 516) An isometry that cannot be performed by maintaining the orientation of the points, as in a direct isometry.

indirect proof (p. 351) In an indirect proof, one assumes that the statement to be proved is false. One then uses logical reasoning to deduce that a statement contradicts a postulate, theorem, or one of the assumptions. Once a contradiction is obtained, one concludes that the statement assumed false must in fact be true.

indirect reasoning (p. 351) Reasoning that assumes that the conclusion is false and then shows that this assumption leads to a contradiction of the hypothesis like a postulate, theorem, or corollary. Then, since the assumption has been proved false, the conclusion must be true.

inductive reasoning (p. 89) Reasoning that uses a number of specific examples to arrive at a plausible generalization or prediction. Conclusions arrived at by inductive reasoning lack the logical certainty of those arrived at by deductive reasoning.

informal proof (p. 127) A paragraph proof.

inscribed (p. 686) A polygon is inscribed in a circle if each of its vertices lie on the circle.

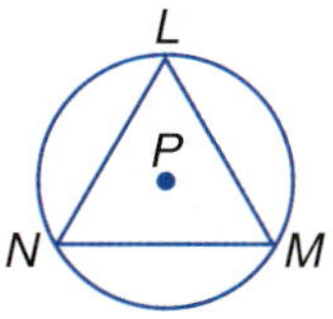

$\triangle LMN$ is inscribed in $\odot P$.

inscribed angle (p. 709) An angle that has a vertex on a circle and sides that contain chords of the circle.

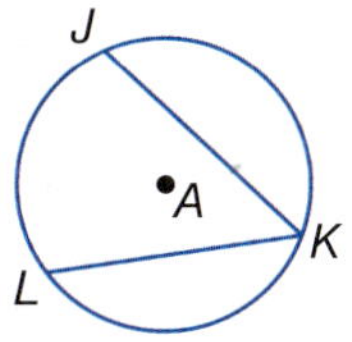

In $\odot A$, $\angle JKL$ is an inscribed angle.

intercepted (p. 709) An angle intercepts an arc if and only if each of the following conditions are met.

1. The endpoints of the arc lie on the angle.
2. All points of the arc except the endpoints are in the interior of the circle.
3. Each side of the angle has an endpoint of the arc.

isometría indirecta Tipo de isometría que no se puede obtener manteniendo la orientación de los puntos, como ocurre con la isometría directa.

demostración indirecta En una demostración indirecta, se supone que el enunciado a demostrar es falso. Después, se deduce lógicamente que existe un enunciado que contradice un postulado, un teoarema o una de las conjeturas. Una vez hallada una contradicción, se concluye que el enunciado que se suponía falso debe ser, en realidad, verdadero.

razonamiento indirecto Razonamiento en que primero se supone que la conclusión es falsa y luego se demuestra que esta conjetura lleva a una contradicción de la hipótesis como un postulado, un teorema o un corolario. Finalmente, como se ha demostrado que la conjetura es falsa, la conclusión debe ser verdadera.

razonamiento inductivo Razonamiento que usa varios ejemplos específicos para lograr una generalización o una predicción pausible. Las conclusiones obtenidas por razonamiento inductivo carecen de la certeza lógica de aquellas obtenidas por razonamiento deductivo.

demostración informal Demostración en forma de párrafo.

inscrito Un polígono está inscrito en un círculo si todos sus vértices yacen en el círculo.

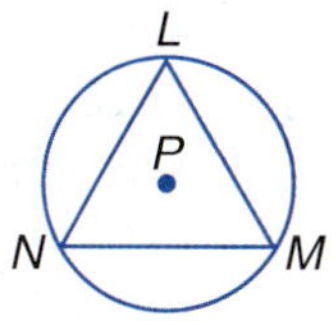

$\triangle LMN$ está inscrito en $\odot P$.

ángulo inscrito Ángulo cuyo vértice está en un círculo y cuyos lados contienen cuerdas del círculo.

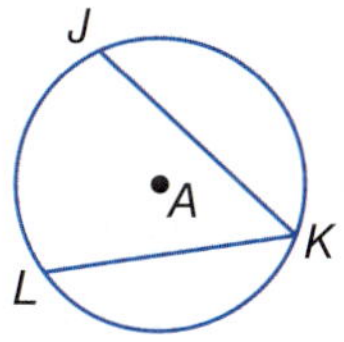

En $\odot A$, $\angle JKL$ es un ángulo inscrito.

intersecado Un ángulo interseca un arco si y sólo si se cumple cada una de las siguientes condiciones.

1. Los extremos del arco yacen en el ángulo.
2. Todos los puntos del arco, excepto los extremos, yacen en el interior del círculo.
3. Cada lado del ángulo tiene un extremo del arco.

intercepted arc (p. 709) An arc that has endpoints on the sides of an inscribed angle and lies in the interior of the inscribed angle.

interior (p. 36) A point is in the interior of an angle if it does not lie on the angle itself and it lies on a segment with endpoints that are on the sides of the angle.

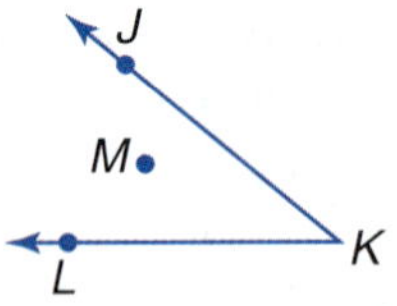

M is in the interior of $\angle JKL$.

interior angles (p. 172) Angles that lie between two transversals that intersect the same line.

intersection (p. 6) A set of points common to two or more geometric figures.

inverse (p. 107) The statement formed by negating both the hypothesis and conclusion of a conditional statement.

inverse cosine (p. 565) The inverse function of cosine, or $\cos^{-1}$. If the cosine of an acute $\angle A$ is equal to x, then $\cos^{-1} x$ is equal to the measure of $\angle A$.

inverse sine (p. 565) The inverse function of sine, or $\sin^{-1}$. If the sine of an acute $\angle A$ is equal to x, then $\sin^{-1} x$ is equal to the measure of $\angle A$.

inverse tangent (p. 565) The inverse function of tangent, or $\tan^{-1}$. If the tangent of an acute $\angle A$ is equal to x, then $\tan^{-1} x$ is equal to the measure of $\angle A$.

irrational number (p. 26) A number that cannot be expressed as a terminating or repeating decimal.

irregular figure (p. 800) A polygon with sides and angles that are not all congruent.

isometric view (p. 823) Corner views of three-dimensional objects on two-dimensional paper.

isometry (p. 294) A mapping for which the original figure and its image are congruent.

isosceles trapezoid (p. 435) A trapezoid in which the legs are congruent, both pairs of base angles are congruent, and the diagonals are congruent.

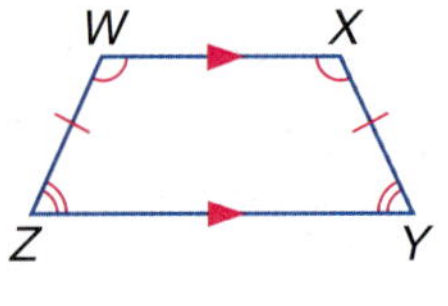

arco intersecado Arco cuyos extremos están en los lados de un ángulo inscrito y que yace en el interior del ángulo.

interior Un punto se encuentra en el interior de un ángulo si no yace en el ángulo como tal y está en un segmento cuyos extremos están en los lados del ángulo.

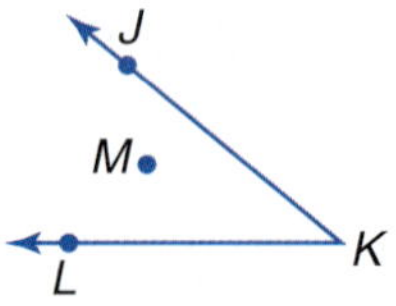

M está en el interior del $\angle JKL$.

ángulos interiores Ángulos que yacen entre dos transversales que interscan la misma recta.

intersección Conjunto de puntos comunes a dos o más figuras geométricas.

inverso Enunciado que se obtiene al negar tanto la hipótesis como la conclusión de un enunciado condicional.

inverso del coseno Función inversa del coseno, o $\cos^{-1}$. Si el coseno de un $\angle A$ agudo es igual a x, entonces $\cos^{-1} x$ es igual a la medida del $\angle A$.

inverso del seno Función inversa del seno o $\sin^{-1}$. Si el seno de un $\angle A$ agudo es igual a x, entonces $\sin^{-1} x$ es igual a la medida del $\angle A$.

inverso de la tangente Función inversa de la tangente o $\tan^{-1}$. Si la tangente de un $\angle A$ agudo es igual a x, entonces $\tan^{-1} x$ es igual a la medida del $\angle A$.

número irracional Número que no se puede expresar como un decimal terminal o periódico.

figura irregular Polígono cuyos lados y ángulos no son todos congruentes.

vista isométrica Vistas de las esquinas de sólidos geométricos tridimensionales sobre un papel bidimensional.

isometría Aplicación en la cual la figura original y su imagen son congruentes.

trapecio isósceles Trapecio cuyos catetos son congruentes, ambos pares de ángulos de las bases son congruentes y las diagonales son congruentes.

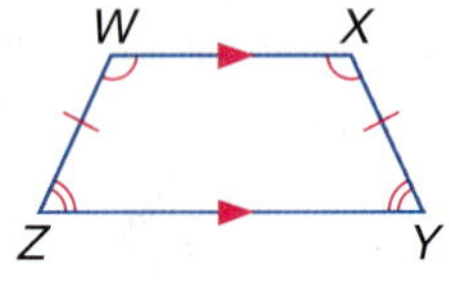

isosceles triangle (p. 236) A triangle with at least two sides congruent. The congruent sides are called *legs*. The angles opposite the legs are *base angles*. The angle formed by the two legs is the *vertex angle*. The side opposite the vertex angle is the *base*.

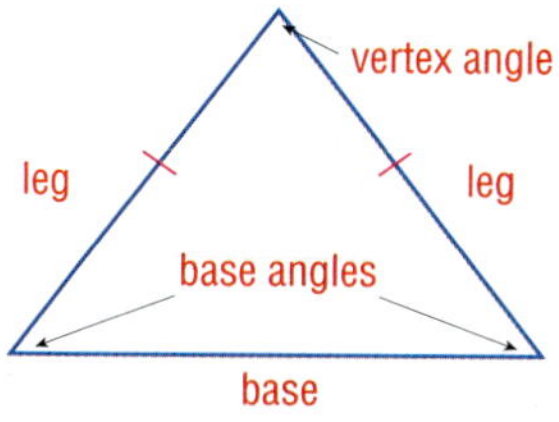

iteration (p. 423) A process of repeating the same procedure over and over again.

triángulo isósceles Triángulo que tiene por lo menos dos lados congruentes. Los lados congruentes se llaman *catetos*. Los ángulos opuestos a los catetos son los *ángulos de la base*. El ángulo formado por los dos catetos es el *ángulo del vértice*. El lado opuesto al ángulo del vértice es la *base*.

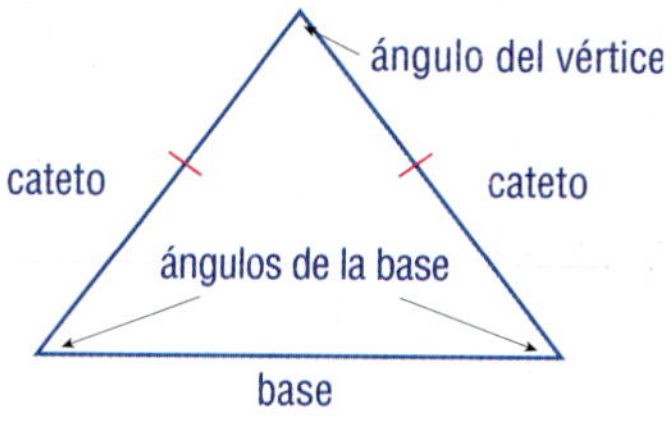

iteración Proceso de repetir el mismo procedimiento una y otra vez.

K

kite (p. 438) A quadrilateral with exactly two distinct pairs of adjacent congruent sides.

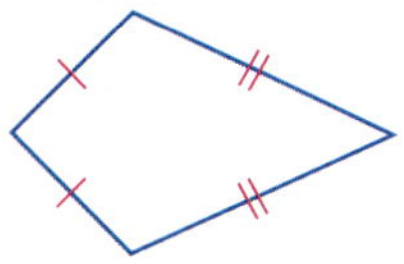

cometa Cuadrilátero que tiene exactamente dos pares diferentes de lados congruentes y adyacentes.

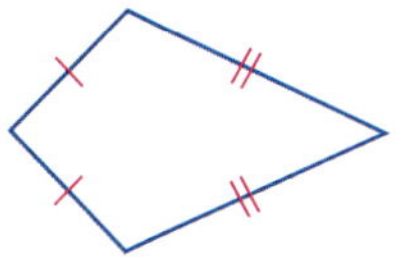

L

lateral area (p. 830) For prisms, pyramids, cylinders, and cones, the area of the figure, not including the bases.

lateral edges **1.** (p. 830) In a prism, the intersection of two adjacent lateral faces. **2.** (p. 686) In a pyramid, lateral edges are the edges of the lateral faces that join the vertex to vertices of the base.

lateral faces **1.** (p. 830) In a prism, the faces that are not bases. **2.** (p. 699) In a pyramid, faces that intersect at the vertex.

latitude (p. 879) A measure of distance north or south of the equator.

Law of Cosines (p. 583) Let $\triangle ABC$ be any triangle with a, b, and c representing the measures of sides opposite the angles with measures A, B, and C respectively. Then the following equations are true.

$a^2 = b^2 + c^2 - 2bc \cos A$

$b^2 = a^2 + c^2 - 2ac \cos B$

$c^2 = a^2 + b^2 - 2ab \cos C$

área lateral En prismas, pirámides, cilindros y conos, es el área de la figura, sin incluir el área de las bases.

aristas laterales **1.** En un prisma, la intersección de dos caras laterales adyacentes. **2.** En una pirámide, las aristas de las caras laterales que unen el vértice de la pirámide con los vértices de la base.

caras laterales **1.** En un prisma, las caras que no forman las bases. **2.** En una pirámide, las caras que se intersecan en el vértice.

latitud Medida de la distancia al norte o al sur del ecuador.

ley de los cosenos Sea $\triangle ABC$ cualquier triángulo donde a, b y c son las medidas de los lados opuestos a los ángulos que miden A, B y C respectivamente. Entonces las siguientes ecuaciones son verdaderas.

$a^2 = b^2 + c^2 - 2bc \cos A$

$b^2 = a^2 + c^2 - 2ac \cos B$

$c^2 = a^2 + b^2 - 2ab \cos C$

Law of Detachment (p. 115) If $p \rightarrow q$ is a true conditional and p is true, then q is also true.

Law of Large Numbers (p. 925) Law that states that as the number of trials of a random process increases, the average value will approach the expected value.

Law of Sines (p. 582) Let $\triangle ABC$ be any triangle with a, b, and c representing the measures of sides opposite the angles with measures A, B, and C respectively. Then, $\frac{\sin A}{a} = \frac{\sin B}{b} = \frac{\sin C}{c}$.

Law of Syllogism (p. 117) If $p \rightarrow q$ and $q \rightarrow r$ are true conditionals, then $p \rightarrow r$ is also true.

legs of a right triangle (p. 26) The shorter sides of a right triangle.

legs of a trapezoid (p. 435) The nonparallel sides of a trapezoid.

legs of an isosceles triangle (p. 283) The two congruent sides of an isosceles triangle.

line (p. 6) A basic undefined term of geometry. A line is made up of points and has no thickness or width. In a figure, a line is shown with an arrowhead at each end. Lines are usually named by lowercase script letters or by writing capital letters for two points on the line, with a double arrow over the pair of letters.

line of reflection (p. 613) A line through a figure that separates the figure into two mirror images.

line of symmetry (p. 653) A line that can be drawn through a plane figure so that the figure on one side is the reflection image of the figure on the opposite side.

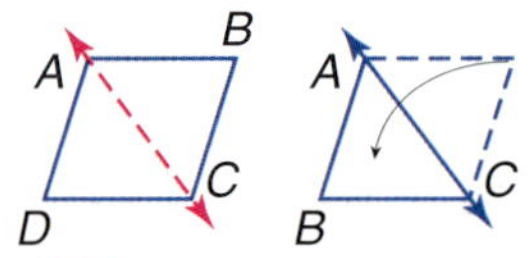

$\overleftrightarrow{AC}$ is a line of symmetry.

line segment (p. 14) A measurable part of a line that consists of two points, called endpoints, and all of the points between them.

ley de indiferencia Si $p \rightarrow q$ es un enunciado condicional verdadero y p es verdadero, entonces q también es verdadero.

ley de los grandes números Ley que establece que a medida que aumenta el número de ensayos de un proceso aleatorio, el valor promedio se aproximará al valor esperado.

ley de los senos Sea $\triangle ABC$ cualquier triángulo donde a, b y c representan las medidas de los lados opuestos a los ángulos que miden A, B y C respectivamente. Entonces, $\frac{\text{sen } A}{a} = \frac{\text{sen } B}{b} = \frac{\text{sen } C}{c}$.

ley del silogismo Si $p \rightarrow q$ y $q \rightarrow r$ son enunciados condicionales verdaderos, entonces $p \rightarrow r$ también es verdadero.

catetos de un triángulo rectángulo Lados más cortos de un triángulo rectángulo.

catetos de un trapecio Los lados no paralelos de un trapecio.

catetos de un triángulo isósceles Las dos lados congruentes de un triángulo isósceles.

recta Término geométrico básico no definido. Una recta está formada por puntos y carece de grosor o ancho. En una figura, una recta se representa con una flecha en cada extremo. Generalmente se designan con letras minúsculas o con las dos letras mayúsculas de dos puntos sobre la recta y una flecha doble sobre el par de letras.

línea de reflexión Línea que divide una figura en dos imágenes especulares.

eje de simetría Recta que se traza a través de una figura plana, de modo que un lado de la figura es la imagen reflejada del lado opuesto.

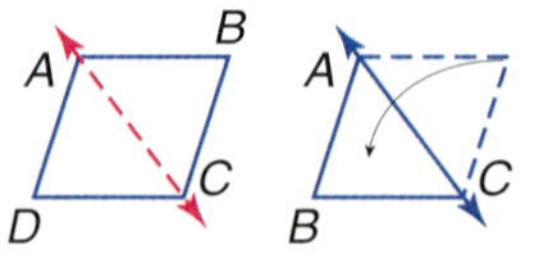

$\overleftrightarrow{AC}$ es un eje de simetría.

segmento de recta Sección medible de una recta que consta de dos puntos llamados extremos y todos los puntos entre ellos.

linear pair (p. 46) A pair of adjacent angles whose noncommon sides are opposite rays.

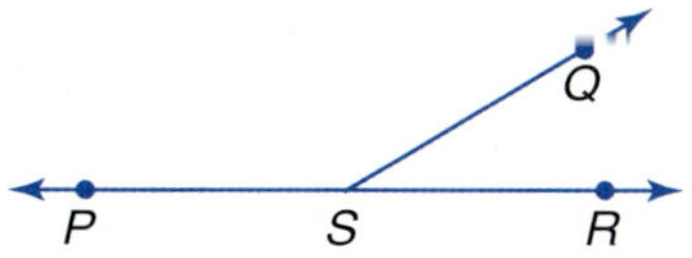

$\angle PSQ$ and $\angle QSR$ are a linear pair.

par lineal Par de ángulos adyacentes cuyos lados no comunes forman rayos opuestos.

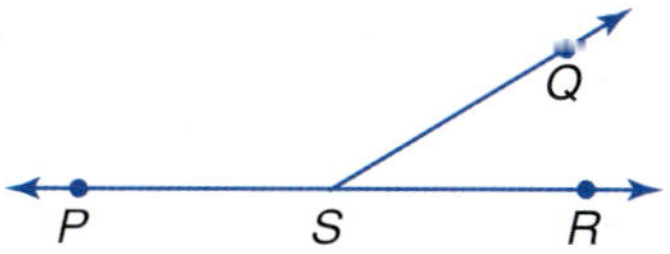

$\angle PSQ$ y $\angle QSR$ forman un par lineal.

locus (p. 11) The set of points that satisfy a given condition.

lugar geométrico Conjunto de puntos que satisfacen una condición dada.

logically equivalent (p. 108) Statements that have the same truth values.

lógicamente equivalentes Enunciados que poseen los mismos valores verdaderos.

longitude (p. 879) A measure of distance east or west of the prime meridian.

longitud Medida de la distancia al este o al oeste del primer meridiano.

M

magnitude (p. 593) The length of a vector.

magnitud Longitud de un vector.

major arc (p. 563) An arc with a measure greater than 180.
$\widehat{ACB}$ is a major arc.

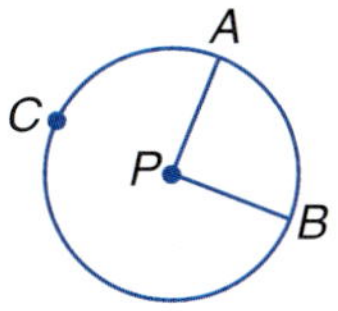

arco mayor Arco que mide más de 180°.
$\widehat{ACB}$ es un arco mayor.

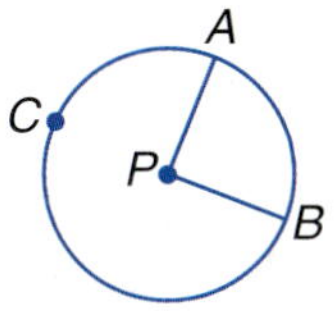

means (p. 458) In $\frac{a}{b} = \frac{c}{d}$, the numbers b and c.

medias Los números b y c en la proporción $\frac{a}{b} = \frac{c}{d}$.

median **1.** (p. 333) In a triangle, a line segment with endpoints that are a vertex of a triangle and the midpoint of the side opposite the vertex.
2. (p. 358) In a trapezoid, the segment that joins the midpoints of the legs.

mediana **1.** En un triángulo, segmento de recta cuyos extremos son un vértice del triángulo y el punto medio del lado opuesto a dicho vértice. **2.** En un trapecio, segmento que une los puntos medios de los catetos.

meridians (p. 879) Imaginary vertical lines drawn around the Earth through the North and South Poles.

meridianos Líneas verticales imaginarias dibujadas alrededor de la Tierra que van del polo norte al polo sur.

midpoint (p. 27) The point on a segment exactly halfway between the endpoints of the segment.

punto medio Punto en un segmento que yace exactamente en la mitad, entre los extremos del segmento.

midsegment (p. 437) A segment with endpoints that are the midpoints of two sides of a triangle.

segmento medio Segmento cuyos extremos son los puntos medios de dos lados de un triángulo.

midsegment of trapezoid (p. 437) A segment that connects the legs of a trapezoid.

segmento medio de un trapecio Segmento que conecta los puntos medios de los catetos de un trapecio.

midsegment of triangle (p. 485) A segment with endpoints that are the midpoints of two sides of a triangle.

segmento medio de un triángulo Segmento cuyos extremos son los puntos medios de dos lados de un triángulo.

minor arc (p. 693) An arc with a measure less than 180. $\widehat{AB}$ is a minor arc.

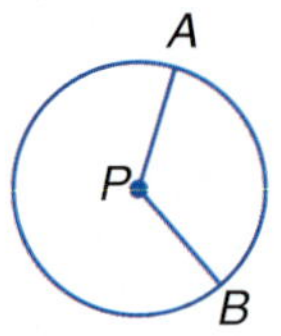

arco menor Arco que mide menos de 180°. $\widehat{AB}$ es un arco menor.

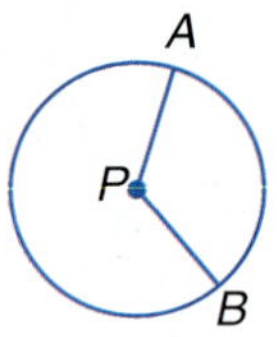

multi-stage experiments (p. 899) Experiments with more than two stages.

experimentos multietápicos Experimentos con más de dos etapas.

mutually exclusive (p. 938) Two events that have no outcomes in common.

mutuamente exclusivos Eventos que no tienen resultados en común.

N

negation (p. 97) If a statement is represented by *p*, then *not p* is the negation of the statement.

negación Si *p* representa un enunciado, entonces *no p* es la negación del enunciado.

net (p. 75) A two-dimensional figure that when folded forms the surfaces of a three-dimensional object.

red Figura bidimensional que al ser plegada forma las superficies de un objeto tridimensional.

network (p. 946) A graph of interconnected vertices.

red Gráfica de vértices interconectados.

***n*-gon** (p. 57) A polygon with *n* sides.

***en*ágono** Polígono con *n* lados.

node (p. 946) A collection of vertices.

nodo Colección de vértices.

non-Euclidean geometry (p. 874) The study of geometrical systems that are not in accordance with the Parallel Postulate of Euclidean geometry.

geometría no euclidiana El estudio de sistemas geométricos que no satisfacen el postulado de las paralelas de la geometría euclidiana.

O

oblique cone (p. 840) A cone that is not a right cone.

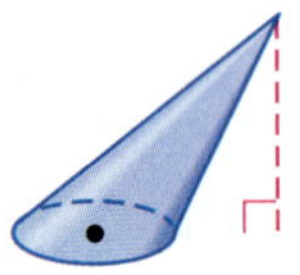

cono oblicuo Cono que no es un cono recto.

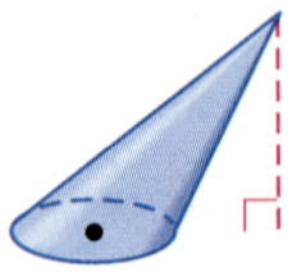

oblique cylinder (p. 831) A cylinder that is not a right cylinder.

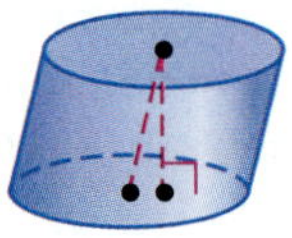

cilindro oblicuo Cilindro que no es un cilindro recto.

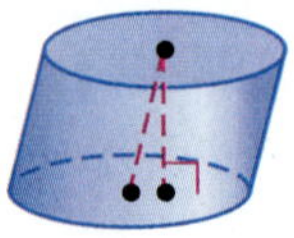

Glossary/Glosario

oblique prism (p. 830) A prism in which the lateral edges are not perpendicular to the bases.

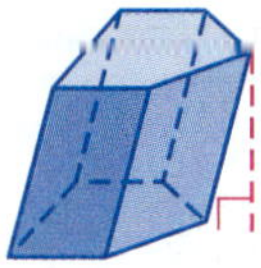

prisma oblicuo Prisma cuyas aristas laterales no son perpendiculares a las bases.

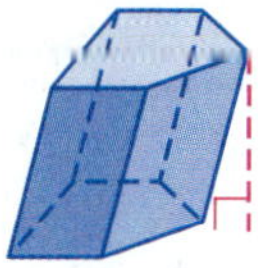

obtuse angle (p. 38) An angle with degree measure greater than 90 and less than 180.

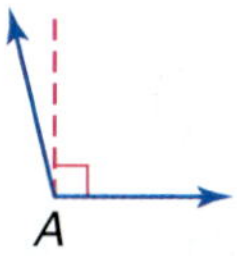

$90 < m\angle A < 180$

ángulo obtuso Ángulo que mide más de 90° y menos de 180°.

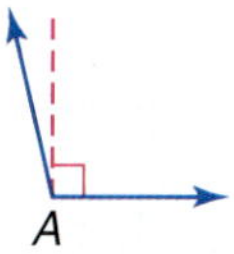

$90 < m\angle A < 180$

obtuse triangle (p. 235) A triangle with an obtuse angle.

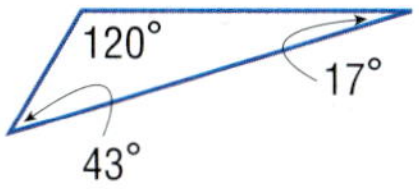

one obtuse angle

triángulo obtusángulo Triángulo con un ángulo obtuso.

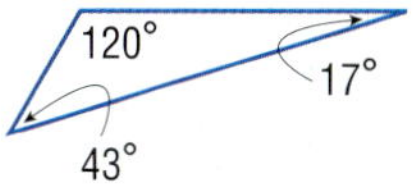

un ángulo obtuso

opposite rays (p. 36) Two rays $\overrightarrow{BA}$ and $\overrightarrow{BC}$ such that B is between A and C.

rayos opuestos Dos rayos $\overrightarrow{BA}$ y $\overrightarrow{BC}$ donde B está entre A y C.

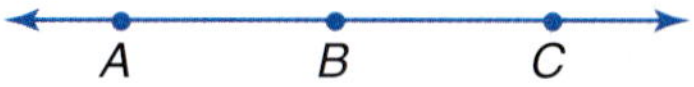

order of symmetry (p. 652) The number of times a figure can map onto itself as it rotates from 0° to 360°.

orden de simetría Número de veces que una figura se puede aplicar sobre sí misma mientras gira de 0° a 360°.

ordered triple (p. 550) Three numbers given in a specific order used to locate points in space.

triple ordenado Tres números dados en un orden específico que sirven para ubicar puntos en el espacio.

orthocenter (p. 335) The point of concurrency of the altitudes of a triangle.

ortocentro Punto de intersección de las alturas de un triángulo.

orthographic drawing (p. 75) The two-dimensional top view, left view, front view, and right view of a three-dimensional object.

proyección ortogonal Vista bidimensional superior, del lado izquierdo, frontal y del lado derecho de un objeto tridimensional.

P

paragraph proof (p. 127) An informal proof written in the form of a paragraph that explains why a conjecture for a given situation is true.

demostración en párrafo Demostración informal escrita en párrafo que explica por qué una conjetura para una situación dada es verdadera.

parallel lines (p. 171) Coplanar lines that do not intersect.

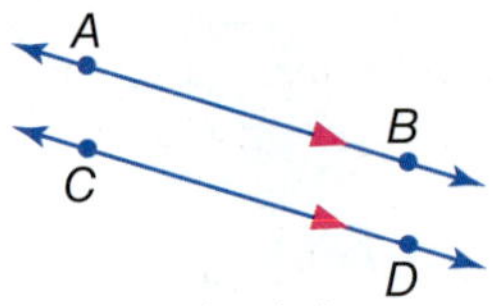

$\overleftrightarrow{AB} \parallel \overleftrightarrow{CD}$

parallel planes (p. 171) Planes that do not intersect.

parallel vectors (p. 594) Vectors that have the same or opposite direction.

parallelogram (pp. 322, 399) A quadrilateral with parallel opposite sides. Any side of a parallelogram may be called a *base*.

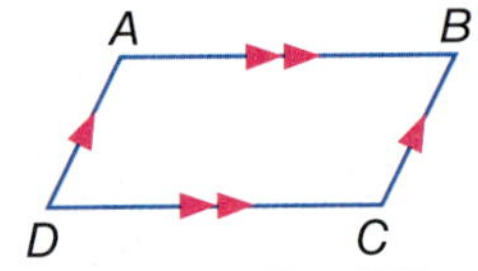

$\overline{AB} \parallel \overline{DC}; \overline{AD} \parallel \overline{BC}$

parallelogram method (p. 595) A method used to find the resultant of two vectors in which you place the vectors at the same initial point, complete a parallelogram, and draw the diagonal.

parallels (p. 879) Imaginary horizontal lines parallel to the equator.

perimeter (p. 58) The sum of the lengths of the sides of a polygon.

permutation (p. 906) An arrangement of objects in which order is important.

perpendicular bisector (p. 322) In a triangle, a line, segment, or ray that passes through the midpoint of a side and is perpendicular to that side.

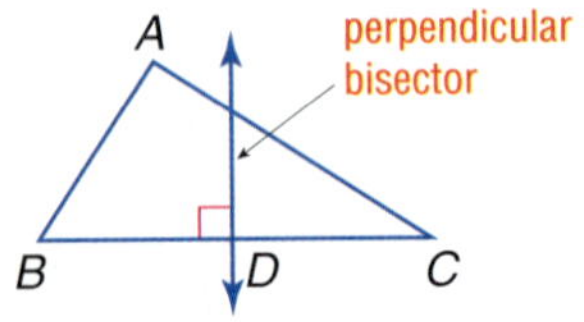

D is the midpoint of $\overline{BC}$.

perpendicular lines (p. 48) Lines that form right angles.

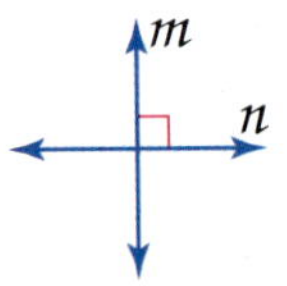

line $m \perp$ line n

rectas paralelas Rectas coplanares que no se intersecan.

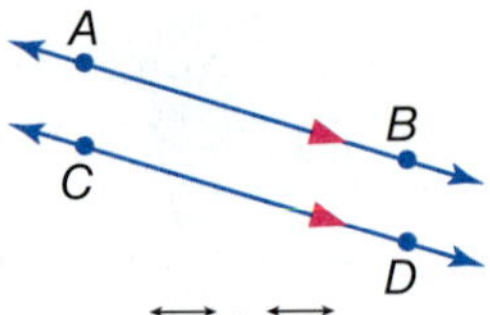

$\overleftrightarrow{AB} \parallel \overleftrightarrow{CD}$

planos paralelos Planos que no se intersecan.

vectores paralelos Vectores con la misma dirección o dirección opuesta.

paralelogramo Cuadrilátero cuyos lados opuestos son paralelos y cuya *base* puede ser cualquiera de sus lados.

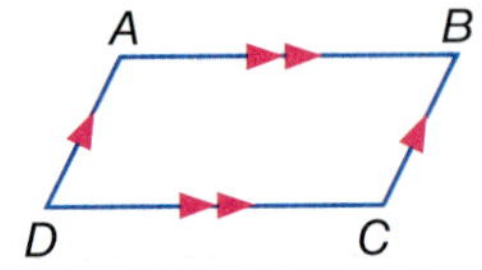

$\overline{AB} \parallel \overline{DC}; \overline{AD} \parallel \overline{BC}$

método del paralelogramo Método que se usa para hallar la resultante de dos vectores en que se dibujan los vectores con el mismo punto de origen, se completa un paralelogramo y se traza la diagonal.

paralelos Rectas horizontales imaginarias paralelas al ecuador.

perímetro Suma de la longitud de los lados de un polígono.

permutación Disposición de objetos en la cual el orden es importante.

mediatriz Recta, segmento de recta o rayo perpendicular que corta un lado del triángulo en su punto medio.

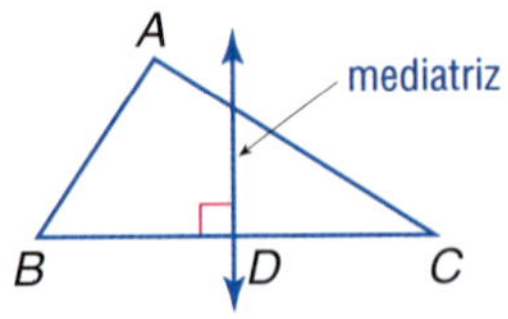

D es el punto medio de $\overline{BC}$.

rectas perpendiculares Rectas que forman ángulos rectos.

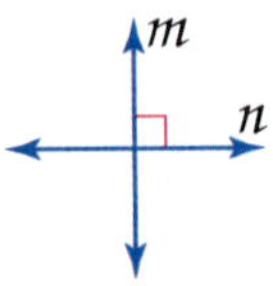

recta $m \perp$ recta n

perspective view (p. 680) The view of a three-dimensional figure from the corner.

pi (π) (p. 685) An irrational number represented by the ratio of the circumference of a circle to the diameter of the circle.

plane (p. 5) A basic undefined term of geometry. A plane is a flat surface made up of points that has no depth and extends indefinitely in all directions. In a figure, a plane is often represented by a shaded, slanted four-sided figure. Planes are usually named by a capital script letter or by three noncollinear points on the plane.

plane Euclidean geometry (p. 188) Geometry based on Euclid's axioms dealing with a system of points, lines, and planes.

plane symmetry (p. 653) Symmetry in a three-dimensional figure that occurs if the figure can be mapped onto itself by a reflection in a plane.

Platonic solids (p. 67) The five regular polyhedra: tetrahedron, hexahedron, octahedron, dodecahedron, or icosahedron.

point (p. 5) A basic undefined term of geometry. A point is a location. In a figure, points are represented by a dot. Points are named by capital letters.

point of concurrency (p. 323) The point of intersection of concurrent lines.

point of symmetry (p. 652) The common point of reflection for all points of a figure.

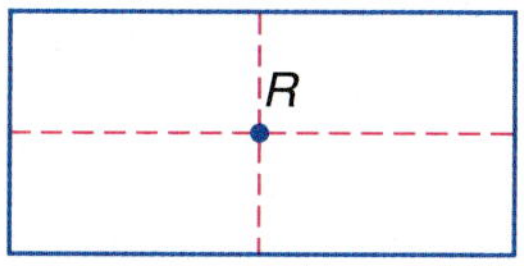

R is a point of symmetry.

point of tangency (p. 718) For a line that intersects a circle in only one point, the point at which they intersect.

point-slope form (p. 196) An equation of the form $y - y_1 = m(x - x_1)$, where (x_1, y_1) are the coordinates of any point on the line and m is the slope of the line.

poles (p. 864) The endpoints of the diameter of a great circle.

vista en perspectiva Vista de una figura tridimensional desde una de sus esquinas.

pi (π) Número irracional representado por la razón de la circunferencia de un círculo al diámetro del mismo.

plano Término geométrico básico no definido. Superficie plana sin espesor formada por puntos y que se extiende hasta el infinito en todas direcciones. En una figura, los planos a menudo se representan con una figura inclinada y sombreada y se designan con una letra mayúscula o con tres puntos no colineales del plano.

geometría del plano euclidiano Geometría basada en los axiomas de Euclides, los cuales abarcan un sistema de puntos, rectas y planos.

simetría plana Simetría de una figura tridimensional que ocurre si la figura se puede aplicar sobre sí misma mediante una reflexión en el plano.

sólidos platónicos Los cinco poliedros regulares siguientes: tetraedro, hexaedro, octaedro, dodecaedro e icosaedro.

punto Término geométrico básico no definido. Un punto representa un lugar o ubicación. En una figura, se representan con una marca puntual y se designan con letras mayúsculas.

punto de concurrencia Punto de intersección de rectas concurrentes.

punto de simetría El punto común de reflexión de todos los puntos de una figura.

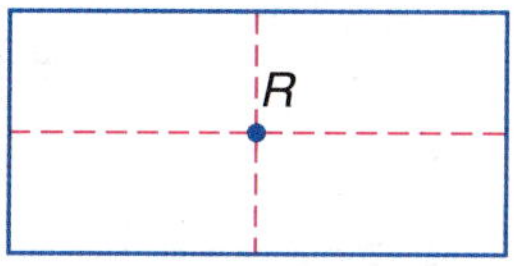

R es un punto de simetría.

punto de tangencia Punto de intersección de una recta en un círculo en un solo punto.

forma punto-pendiente Ecuación de la forma $y - y_1 = m(x - x_1)$, donde (x_1, y_1) representan las coordenadas de un punto cualquiera sobre la recta y m representa la pendiente de la recta.

polos Los extremos del diámetro de un círculo mayor.

polygon (p. 56) A closed figure formed by a finite number of coplanar segments called *sides* such that the following conditions are met:

1. The sides that have a common endpoint are noncollinear.
2. Each side intersects exactly two other sides, but only at their endpoints, called the *vertices*.

polyhedrons (p. 67) Closed three-dimensional figures made up of flat polygonal regions. The flat regions formed by the polygons and their interiors are called *faces*. Pairs of faces intersect in segments called *edges*. Points where three or more edges intersect are called *vertices*.

postulate (p. 125) A statement that describes a fundamental relationship between the basic terms of geometry. Postulates are accepted as true without proof.

preimage (p. 294) The graph of an object before a transformation.

prism (p. 67) A solid with the following characteristics:

1. Two faces, called *bases*, are formed by congruent polygons that lie in parallel planes.
2. The faces that are not bases, called *lateral faces*, are formed by parallelograms.
3. The intersections of two adjacent lateral faces are called *lateral edges* and are parallel segments.

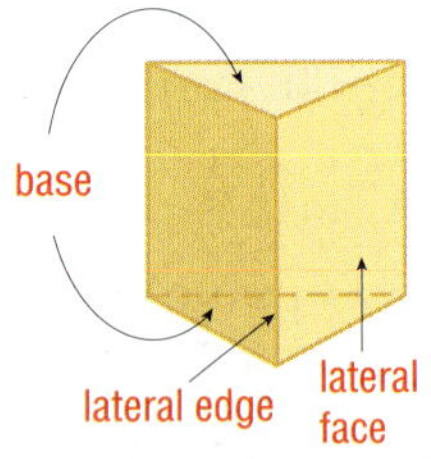

triangular prism

probability model (p. 923) A mathematical model used to match a random phenomenon.

proof (p. 126) A logical argument in which each statement you make is supported by a statement that is accepted as true.

proof by contradiction (p. 351) An indirect proof in which one assumes that the statement to be proved is false. One then uses logical reasoning to deduce a statement that contradicts a postulate, theorem, or one of the assumptions. Once a contradiction is obtained, one concludes that the statement assumed false must in fact be true.

polígono Figura cerrada formada por un número finito de segmentos coplanares llamados *lados*, tal que satisface las siguientes condiciones:

1. Los lados que tienen un extremo común son no colineales.
2. Cada lado interseca exactamente dos lados más, pero sólo en sus extremos, llamados *vértices*.

poliedros Figuras tridimensionales cerradas formadas por regiones poligonales planas. Las regiones planas definidas por un polígono y sus interiores se llaman *caras*. Cada intersección entre dos caras se llama *arista*. Los puntos donde se intersecan tres o más aristas se llaman *vértices*.

postulado Enunciado que describe una relación fundamental entre los términos geométricos básicos. Los postulados se aceptan como verdaderos sin necesidad de demostración.

preimagen Gráfica de una figura antes de una transformación.

prisma Sólido con las siguientes características:

1. Dos caras, llamadas *bases*, formadas por polígonos congruentes que yacen en planos paralelos.
2. Las caras que no son bases, llamadas *caras laterales*, son paralelogramos.
3. Las intersecciones de dos caras laterales adyacentes se llaman *aristas laterales* y son segmentos paralelos.

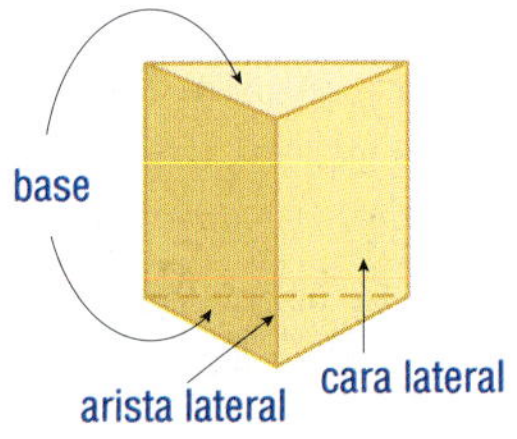

prisma triangular

modelo de probabilidad Modelo matemático que se usa para relacionar un fenómeno aleatorio.

demostración Argumento lógico en el cual cada enunciado que se hace está respaldado por un enunciado que se acepta como verdadero.

demostración por contradicción Demostración indirecta en la cual se supone que el enunciado a demostrarse es falso. Luego, se usa el razonamiento lógico para inferir un enunciado que contradiga el postulado, teorema o una de las conjeturas. Una vez que se obtiene una contradicción, se concluye que el enunciado que se supuso falso es, en realidad, verdadero.

proportion (p. 458) An equation of the form $\frac{a}{b} = \frac{c}{d}$ that states that two ratios are equal.

pyramid (p. 67) A solid with the following characteristics:

1. All of the faces, except one face, intersect at a point called the *vertex*.
2. The face that does not contain the vertex is called the *base* and is a polygonal region.
3. The faces meeting at the vertex are called *lateral faces* and are triangular regions.

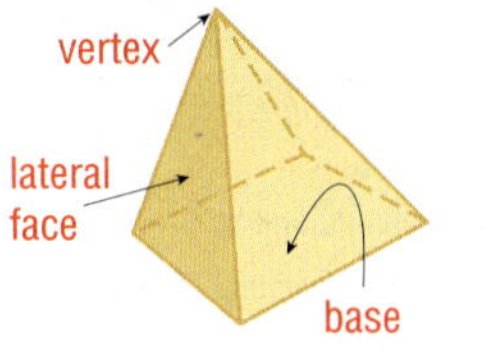

rectangular pyramid

Pythagorean triple (p. 541) A group of three whole numbers that satisfies the equation $a^2 + b^2 = c^2$, where c is the greatest number.

proporción Ecuación de la forma $\frac{a}{b} = \frac{c}{d}$ que establece que dos razones son iguales.

pirámide Sólido con las siguientes características:

1. Todas las caras, excepto una, se intersecan en un punto llamado *vértice*.
2. La cara sin el vértice se llama *base* y es una región poligonal.
3. Las caras que se encuentran en los vértices se llaman *caras laterales* y son regiones triangulares.

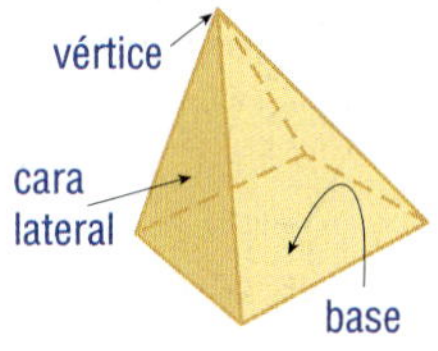

pirámide rectangular

triple pitagórico Grupo de tres números enteros que satisfacen la ecuación $a^2 + b^2 = c^2$, donde c es el número mayor.

R

radius **1.** (p. 655) In a circle, any segment with endpoints that are the center of the circle and a point on the circle. **2.** (p. 711) In a sphere, any segment with endpoints that are the center and a point on the sphere.

radius of a regular polygon (p. 791) The radius of a circle circumscribed about a polygon.

random variable (p. 925) A variable that can assume a set of values, each with fixed probabilities.

rate of change (p. 187) Describes how a quantity is changing over time.

ratio (p. 457) A comparison of two quantities using division.

ray (p. 36) $\overrightarrow{PQ}$ is a ray if it is the set of points consisting of $\overline{PQ}$ and all points S for which Q is between P and S.

rectangle (p. 419) A quadrilateral with four right angles.

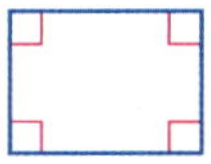

radio **1.** En un círculo, cualquier segmento cuyos extremos son el centro y un punto del círculo. **2.** En una esfera, cualquier segmento cuyos extremos son el centro y un punto de la esfera.

radio de un polígono regular Radio de un círculo circunscrito alrededor de un polígono.

variable aleatoria Variable que puede tomar un conjunto de valores, cada uno con probabilidades fijas.

tasa de cambio Describe cómo cambia una cantidad a través del tiempo.

razón Comparación de dos cantidades mediante división.

rayo $\overrightarrow{PQ}$ es un rayo si es el conjunto de puntos formado por $\overline{PQ}$ y todos los puntos S para los cuales Q se ubica entre P y S.

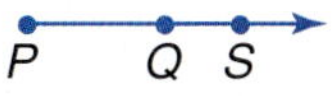

rectángulo Cuadrilátero con cuatro ángulos rectos.

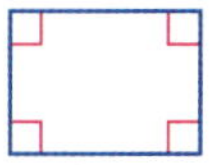

reduction (p. 505) An image that is smaller than the original figure.

reflection (p. 294) A transformation representing a flip of the figure over a point, line, or plane.

regular polygon (p. 57) A convex polygon in which all of the sides are congruent and all of the angles are congruent.

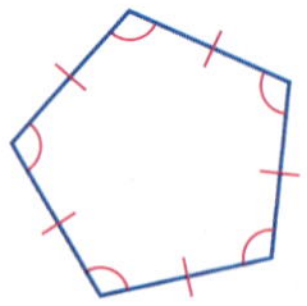

regular pentagon

regular polyhedron (p. 68) A polyhedron in which all of the faces are regular congruent polygons.

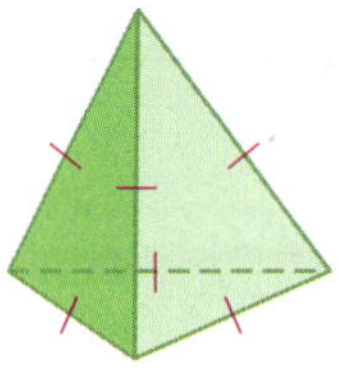

regular prism (p. 60) A right prism with bases that are regular polygons.

regular pyramid (p. 838) A pyramid with a base that is a regular polygon.

regular tessellation (p. 648) A tessellation formed by only one type of regular polygon.

related conditionals (p. 107) Statements that are based on a given conditional statement.

remote interior angles (p. 246) The angles of a triangle that are not adjacent to a given exterior angle.

resultant (p. 594) The sum of two vectors.

rhombus (p. 426) A quadrilateral with all four sides congruent.

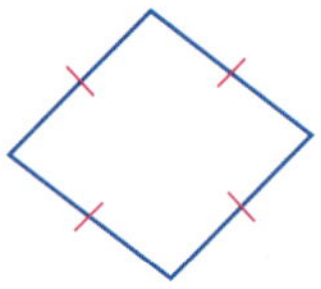

reducción Imagen más pequeña que la figura original.

reflexión Transformación en la cual una figura se "voltea" a través de un punto, una recta o un plano.

polígono regular Polígono convexo cuyos lados y ángulos son congruentes.

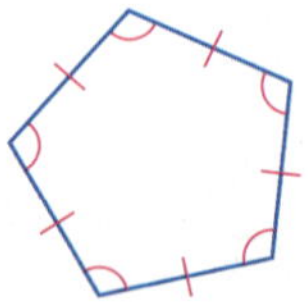

pentágono regular

poliedro regular Poliedro cuyas caras son polígonos regulares congruentes.

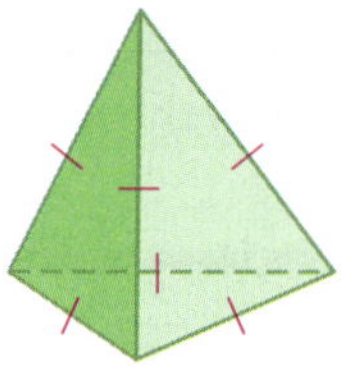

prisma regular Prisma recto cuyas bases son polígonos regulares.

pirámide regular Pirámide cuya base es un polígono regular.

teselado regular Teselado formado por un solo tipo de polígono regular.

condicionales relacionados Enunciados que se basan en un enunciado condicional dado.

ángulos internos no adyacentes Ángulos de un triángulo que no son adyacentes a un ángulo externo dado.

resultante Suma de dos vectores.

rombo Cuadrilátero con cuatro lados congruentes.

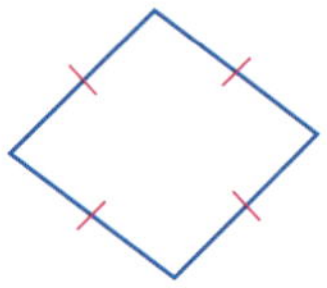

right angle (p. 32) An angle with a degree measure of 90.

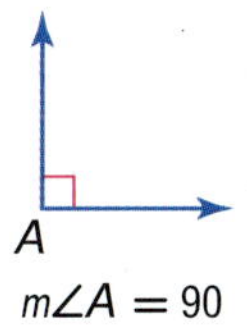

ángulo recto Ángulo que mide 90°.

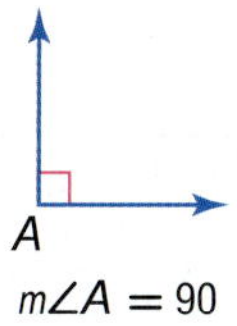

right cone (p. 840) A cone with an axis that is also an altitude.

cono recto Cono cuyo eje es también su altura.

right cylinder (p. 832) A cylinder with an axis that is also an altitude.

cilindro recto Cilindro cuyo eje es también su altura.

right prism (p. 830) A prism with lateral edges that are also altitudes.

prisma recto Prisma cuyas aristas laterales también son su altura.

right triangle (p. 235) A triangle with a right angle. The side opposite the right angle is called the *hypotenuse*. The other two sides are called *legs*.

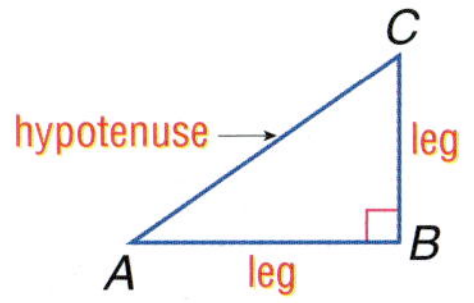

triángulo rectángulo Triángulo con un ángulo recto. El lado opuesto al ángulo recto se conoce como *hipotenusa*. Los otros dos lados se llaman *catetos*.

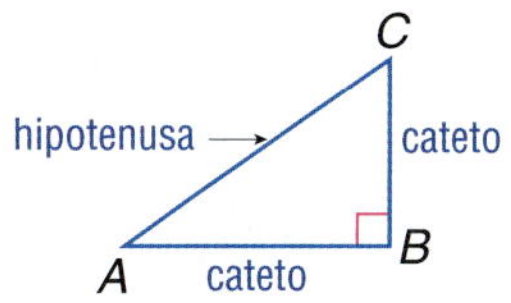

rotation (p. 294) A transformation that turns every point of a preimage through a specified angle and direction about a fixed point, called the *center of rotation*.

rotación Transformación en la cual se hace girar cada punto de la preimagen a través de un ángulo y una dirección determinadas alrededor de un punto llamado *centro de rotación*.

rotational symmetry (p. 652) If a figure can be rotated less than 360° about a point so that the image and the preimage are indistinguishable, the figure has rotational symmetry.

simetría rotacional Si una imagen se puede girar menos de 360° alrededor de un punto, de modo que la imagen y la preimagen sean idénticas, entonces la figura tiene simetría rotacional.

S

sample space (p. 899) The set of all possible outcomes of an experiment.

espacio muestral El conjunto de todos los resultados posibles de un experimento.

scalar (p. 536) A constant multiplied by a vector.

escalar Constante multiplicada por un vector.

scalar multiplication (p. 536) Multiplication of a vector by a scalar.

multiplicación escalar Multiplicación de un vector por un escalar.

scale factor (p. 466) The ratio of the lengths of two corresponding sides of two similar polygons or two similar solids.

factor de escala Razón entre las longitudes de dos lados correspondientes de dos polígonos o sólidos semejantes.

scale factor of dilation (p. 505) The ratio of a length on an image to a corresponding length on the preimage.

factor de escala de homotecia Razón de una longitud en la imagen a una longitud correspondiente en la preimagen.

scalene triangle (p. 236) A triangle with no two sides congruent.

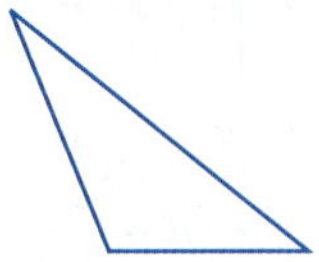

secant (p. 572) Any line that intersects a circle in exactly two points.

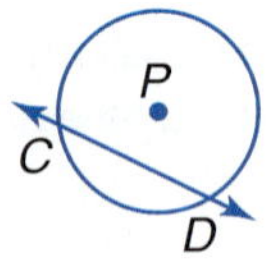

$\overleftrightarrow{CD}$ is a secant of $\odot P$.

secant segment (p. 738) A segment of a secant line that has exactly one endpoint on the circle.

sector of a circle (p. 783) A region of a circle bounded by a central angle and its intercepted arc.

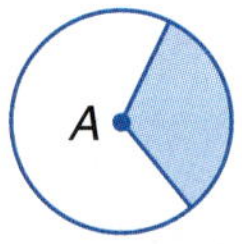

The shaded region is a sector of $\odot A$.

segment (p. 14) See *line segment*.

segment bisector (p. 29) A segment, line, or plane that intersects a segment at its midpoint.

segment of a circle (p. 667) The region of a circle bounded by an arc and a chord.

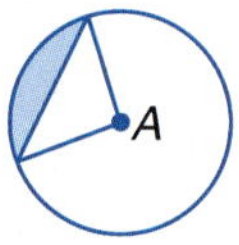

The shaded region is a segment of $\odot A$.

self-similar (p. 423) If any parts of a fractal image are replicas of the entire image, the image is self-similar.

semicircle (p. 693) An arc that measures 180.

semi-regular tessellation (p. 520) A uniform tessellation formed using two or more regular polygons.

sides of an angle (p. 36) The rays of an angle.

triángulo escaleno Triángulo que no tiene dos lados congruentes.

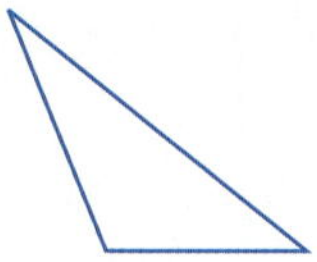

secante Cualquier recta que interseca un círculo exactamente en dos puntos.

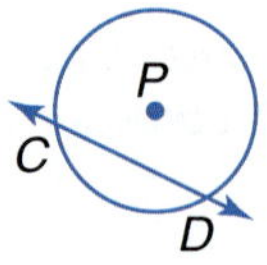

$\overleftrightarrow{CD}$ es una secante de $\odot P$.

segmento secante Segmento de una recta secante que tiene exactamente un extremo en el círculo.

sector circular Región de un círculo limitada por un ángulo central y su arco de intersección.

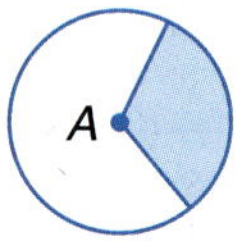

La región sombreada es un sector de $\odot A$.

segmento Ver *segmento de recta*.

mediatriz Segmento, recta o plano que interseca un segmento en su punto medio.

segmento de un círculo Región de un círculo limitada por un arco y una cuerda.

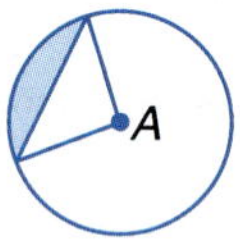

La región sombreada es un segmento de $\odot A$.

autosemejante Si cualquier parte de una imagen fractal es una réplica de la imagen completa, entonces la imagen es autosemejante.

semicírculo Arco que mide 180°.

teselado semiregular Teselado uniforme compuesto por dos o más polígonos regulares.

lados de un ángulo Los rayos de un ángulo.

Sierpinski Triangle (p. 423) A self-similar fractal described by Waclaw Sierpinski. The figure was named for him.

similar solids (p. 880) Solids that have exactly the same shape, but not necessarily the same size.

similarity ratio (p. 466) The scale factor between two similar polygons

similarity transformation (p. 505) When a figure and its transformation image are similar.

simulation (p. 923) A probability model used to recreate a situation again and again so the likelihood of various outcomes can be estimated.

sine (p. 562) For an acute angle of a right triangle, the ratio of the measure of the leg opposite the acute angle to the measure of the hypotenuse.

skew lines (p. 171) Lines that do not intersect and are not coplanar.

slant height (p. 838) The height of the lateral side of a pyramid or cone.

slope (p. 186) For a (nonvertical) line containing two points (x_1, y_1) and (x_2, y_2), the number m given by the formula $m = \frac{y_2 - y_1}{x_2 - x_1}$ where $x_2 \neq x_1$.

slope-intercept form (p. 196) A linear equation of the form $y = mx + b$. The graph of such an equation has slope m and y-intercept b.

solving a triangle (p. 472) Finding the measures of all of the angles and sides of a triangle.

space (p. 7) A boundless three-dimensional set of all points.

sphere (p. 67) In space, the set of all points that are a given distance from a given point, called the *center*.

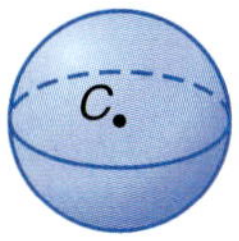

C is the center of the sphere.

triángulo de Sierpinski Fractal autosemejante descrito por el matemático Waclaw Sierpinski. La figura se nombró en su honor.

sólidos semejantes Sólidos que tienen exactamente la misma forma, pero no necesariamente el mismo tamaño.

razón de semejanza Factor de escala entre dos polígonos semejantes.

transformación de semejanza Cuando una figura y su imagen transformada son semejantes.

simulacro Modelo de probabilidad que se usa para reconstruir una situación repetidas veces y así poder estimar la probabilidad de varios resultados.

seno Para un ángulo agudo de un triángulo rectángulo, razón entre la medida del cateto opuesto al ángulo agudo a la medida de la hipotenusa.

rectas alabeadas Rectas que no se intersecan y que no son coplanares.

altura oblicua Altura de la cara lateral de una pirámide o un cono.

pendiente Para una recta (no vertical) que contiene dos puntos (x_1, y_1) y (x_2, y_2), el número m viene dado por la fórmula $m = \frac{y_2 - y_1}{x_2 - x_1}$ donde $x_2 \neq x_1$.

forma pendiente-intersección Ecuación lineal de la forma $y = mx + b$, donde la pendiente es m y la intersección y es b.

resolver un triángulo Calcular las medidas de todos los ángulos y todos los lados de un triángulo.

espacio Conjunto tridimensional no acotado de todos los puntos.

esfera En el espacio, conjunto de todos los puntos a cierta distancia de un punto dado llamado *centro*.

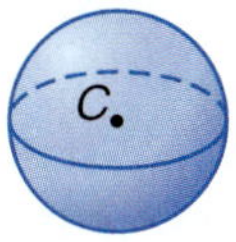

C es el centro de la esfera.

spherical geometry (p. 873) The branch of geometry that deals with a system of points, great circles (lines), and spheres (planes).

square (p. 426) A quadrilateral with four right angles and four congruent sides.

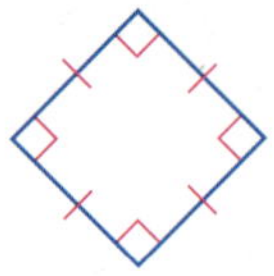

standard position (p. 593) When the initial point of a vector is at the origin.

statement (p. 97) Any sentence that is either true or false, but not both.

supplementary angles (p. 47) Two angles with measures that have a sum of 180.

surface area (p. 69) The sum of the areas of all faces and side surfaces of a three-dimensional figure.

symmetry (p. 651) A figure has symmetry if there exists a rigid motion—reflection, translation, rotation, or glide reflection—that maps the figure onto itself.

geometría esférica Rama de la geometría que estudia los sistemas de puntos, los círculos mayores (rectas) y las esferas (planos).

cuadrado Cuadrilátero con cuatro ángulos rectos y cuatro lados congruentes.

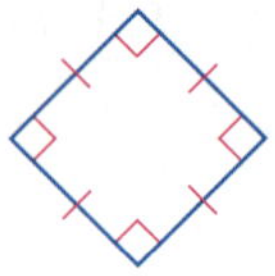

posición estándar Cuando el punto inicial de un vector es el origen.

enunciado Cualquier suposición que puede ser falsa o verdadera, pero no ambas.

ángulos suplementarios Dos ángulos cuya suma es igual a 180°.

área de superficie Suma de las áreas de todas las caras y superficies laterales de una figura tridimensional.

simetría Una figura tiene simetría si existe un movimiento rígido (reflexión, traslación, rotación o reflexión con deslizamiento) que aplica la figura sobre sí misma.

T

tangent **1.** (p. 562) For an acute angle of a right triangle, the ratio of the measure of the leg opposite the acute angle to the measure of the leg adjacent to the acute angle. **2.** (p. 587) A line in the plane of a circle that intersects the circle in exactly one point. The point of intersection is called the *point of tangency*. **3.** (p. 711) A line that intersects a sphere in exactly one point.

tangent segment (p. 738) A segment of a tangent with one endpoint on a circle that is both the exterior and whole segment.

tessellation (p. 648) A pattern that covers a plane by transforming the same figure or set of figures so that there are no overlapping or empty spaces.

theorem (p. 127) A statement or conjecture that can be proven true by undefined terms, definitions, and postulates.

topographic map (p. 829) A representation of a three-dimensional surface on a flat piece of paper.

tangente **1.** Para un ángulo agudo de un triángulo rectángulo, razón de la medida del cateto opuesto al ángulo agudo a la medida del cateto adyacente al ángulo agudo. **2.** Recta en el plano de un círculo que interseca el círculo en exactamente un punto. El punto de intersección se conoce como *punto de tangencia*. **3.** Recta que interseca una esfera en exactamente un punto.

segmento de la tangente Segmento de una tangente con un extremo en un círculo que es tanto el segmento externo como el segmento completo.

teselado Patrón con que se cubre un plano aplicando la misma figura o conjunto de figuras, sin que haya traslapes ni espacios vacíos.

teorema Enunciado o conjetura que se puede demostrar como verdadero mediante términos geométricos básicos, definiciones y postulados.

mapa topográfico Representación de una superficie tridimensional sobre una hoja de papel.

traceable network (p. 946) A network in which all of the nodes are connected and each edge is used once when the network is used.

transformation (p. 505) In a plane, a mapping for which each point has exactly one image point and each image point has exactly one preimage point.

translation (p. 294) A transformation that moves all points of a figure the same distance in the same direction.

translation vector (p. 622) The vector in which a translation maps each point to its image.

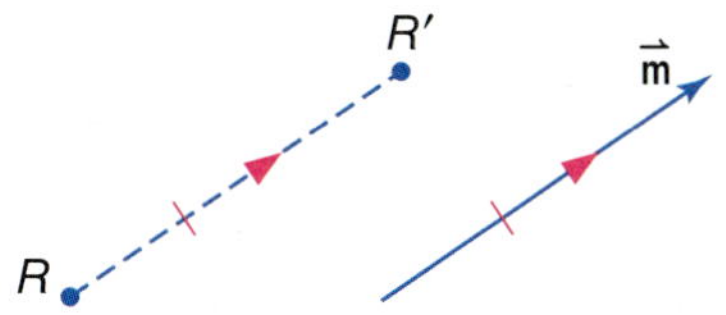

Point *R′*, is a translation of point *R* along translation vector *m*.

transversal (p. 172) A line that intersects two or more lines in a plane at different points.

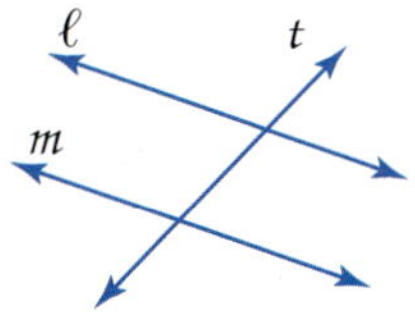

Line *t* is a transversal.

trapezoid (p. 435) A quadrilateral with exactly one pair of parallel sides. The parallel sides of a trapezoid are called *bases*. The nonparallel sides are called *legs*. The pairs of angles with their vertices at the endpoints of the same base are called *base angles*.

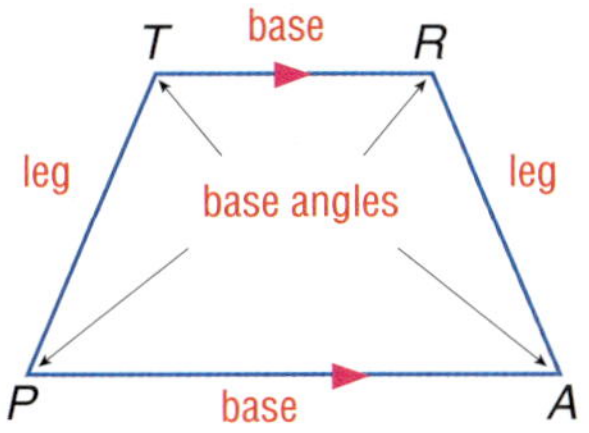

tree diagram (p. 899) An organized table of line segments (branches) which shows possible experiment outcomes.

triangle method (p. 595) A method used to find the resultant of two vectors in which the second vector is connected to the terminal point of the first and the resultant is drawn from the initial point of the first vector to the terminal point of the second vector.

red detectable Red en la cual todos los nodos están conectados y cada arista se utiliza una vez al usarse la red.

transformación En un plano, aplicación para la cual cada punto del plano tiene un único punto de la imagen y cada punto de la imagen tiene un único punto de la preimagen.

traslación Transformación que mueve todos los puntos de una figura la misma distancia, en la misma dirección.

vector de traslación Vector en el cual una traslación aplica cada punto a su imagen.

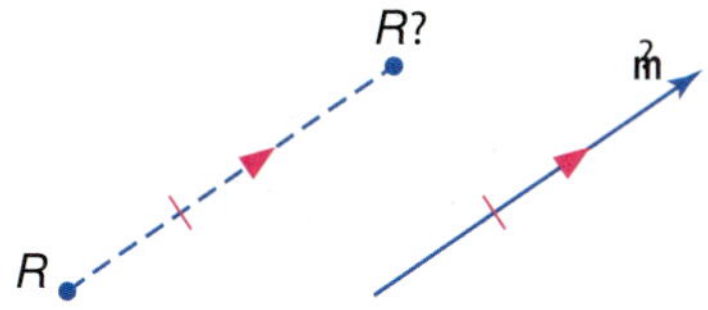

El punto *R*?, es la traslación del punto *R* a lo largo del vector *m* de traslación.

transversal Recta que interseca dos o más rectas en diferentes puntos del mismo plano.

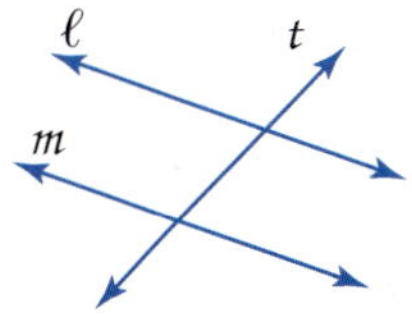

La recta *t* es una transversal.

trapecio Cuadrilátero con sólo un par de lados paralelos. Los lados paralelos del trapecio se llaman *bases*. Los lados no paralelos se llaman *catetos*. Los pares de ángulos cuyos vértices coinciden en los extremos de la misma base son los *ángulos de la base*.

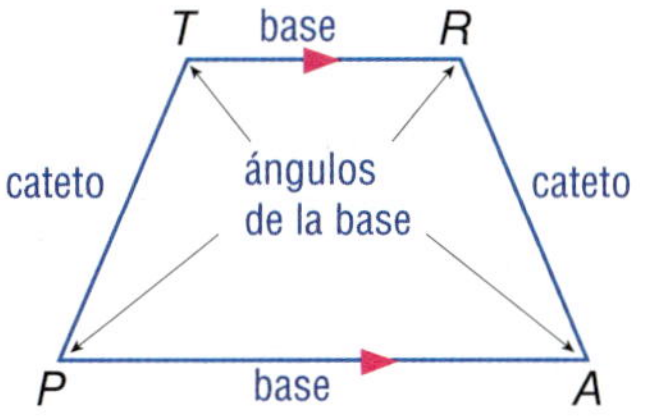

diagrama de árbol Tabla organizada de segmentos de recta (ramas) que muestra los resultados posibles de un experimento.

método del triángulo Método para calcular la resultante de dos vectores en el cual el segundo vector está conectado al extremo del primero y la resultante se traza desde el punto inicial del primer vector al extremo del segundo vector.

trigonometric ratio (p. 562) A ratio of the lengths of sides of a right triangle.

razón trigonométrica Razón de las longitudes de los lados de un triángulo rectángulo.

trigonometry (p. 562) The study of the properties of triangles and trigonometric functions and their applications.

trigonometría Estudio de las propiedades de los triángulos, de las funciones trigonométricas y sus aplicaciones.

truth table (p. 99) A table used as a convenient method for organizing the truth values of statements.

tabla de validez Tabla que se utiliza para organizar de una manera conveniente los valores verdaderos de los enunciados.

truth value (p. 97) The truth or falsity of a statement.

valor de verdad Condición de un enunciado de ser verdadero o falso.

two-column proof (p. 135) A formal proof that contains statements and reasons organized in two columns. Each step is called a *statement*, and the properties that justify each step are called *reasons*.

demostración de dos columnas Demostración formal que contiene enunciados y razones organizados en dos columnas. Cada paso se llama *enunciado* y las propiedades que lo justifican son las *razones*.

two-stage experiment (p. 899) An experiment with two stages or events.

experimento de dos pasos Experimento que consta de dos pasos o eventos.

U

undefined term (p. 5) Words, usually readily understood, that are not formally explained by means of more basic words and concepts. The basic undefined terms of geometry are point, line, and plane.

término geométrico básico no definido Palabras que por lo general se entienden fácilmente y que no se explican formalmente mediante palabras o conceptos más básicos. Los términos geométricos básicos no definidos son el punto, la recta y el plano.

uniform tessellations (p. 648) Tessellations containing the same arrangement of shapes and angles at each vertex.

teselados uniformes Teselados con el mismo patrón de formas y ángulos en cada vértice.

V

vector (p. 593) A directed segment representing a quantity that has both magnitude, or length, and direction.

vector Segmento dirigido que representa una cantidad, la cual posee tanto magnitud o longitud como dirección.

vertex angle of an isosceles triangle (p. 283) See *isosceles triangle*.

ángulo del vértice de un triángulo isósceles Ver *triángulo isósceles*.

vertex of an angle (pp. 36, 67) The common endpoint of an angle.

vértice de un ángulo Extremo común de un ángulo.

vertex of a polyhedron (p. 60) The intersection of three edges of a polyhedron.

vértice de un poliedro Intersección de las aristas de un poliedro.

vertex of a polygon (p. 56) The vertex of each angle of a polygon.

vértice de un polígono Vértice de cada ángulo de un polígono.

vertical angles (p. 46) Two nonadjacent angles formed by two intersecting lines.

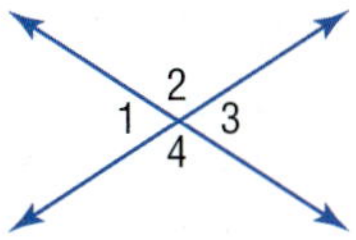

$\angle 1$ and $\angle 3$ are vertical angles.
$\angle 2$ and $\angle 4$ are vertical angles.

ángulos opuestos por el vértice Dos ángulos no adyacentes formados por dos rectas que se intersecan.

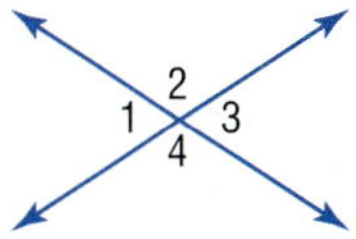

$\angle 1$ y $\angle 3$ son ángulos opuestos por el vértice.
$\angle 2$ y $\angle 4$ son ángulos opuestos por el vértice.

volume (p. 69) A measure of the amount of space enclosed by a three-dimensional figure.

volumen Medida de la cantidad de espacio que contiene una figura tridimensional.

Index

Index

Index

Index

D

E

Index

Index

H

I

Index

Index

L

M

N

O

P

Q

Index

R

S

Index

Index

T

Index

U

V

W

X

Y

Z

Formulas

Coordinate Geometry

Slope	$m = \frac{y_2 - y_1}{x_2 - x_1}$
Distance on a number line:	$d = \lvert a - b \rvert$
Distance on a coordinate plane:	$d = \sqrt{(x_2 - x_1)^2 + (y_2 - y_1)^2}$
Distance in space:	$d = \sqrt{(x_2 - x_1)^2 + (y_2 - y_1)^2 + (z_2 - z_1)^2}$
Distance arc length:	$\ell = \frac{x}{360} \cdot 2\pi r$
Midpoint on a number line:	$M = \frac{a + b}{2}$
Midpoint on a coordinate plane:	$M = \left(\frac{x_1 + x_2}{2}, \frac{y_1 + y_2}{2}\right)$
Midpoint in space:	$M = \left(\frac{x_1 + x_2}{2}, \frac{y_1 + y_2}{2}, \frac{z_1 + z_2}{2}\right)$

Perimeter and Circumference

square	$P = 4s$	**rectangle**	$P = 2\ell + 2w$	**circle**	$C = 2\pi r$ or $C = \pi d$

Area

square	$A = s^2$	**triangle**	$A = \frac{1}{2}bh$
rectangle	$A = \ell w$ or $A = bh$	**regular polygon**	$A = \frac{1}{2}Pa$
parallelogram	$A = bh$	**circle**	$A = \pi r^2$
trapezoid	$A = \frac{1}{2}h(b_1 + b_2)$	**sector of a circle**	$A = \frac{x}{360} \cdot \pi r^2$
rhombus	$A = \frac{1}{2}d_1 d_2$ or $A = bh$		

Lateral Surface Area

prism	$L = Ph$	**pyramid**	$L = \frac{1}{2}P\ell$
cylinder	$L = 2\pi rh$	**cone**	$L = \pi r\ell$

Total Surface Area

prism	$T = Ph + 2B$	**cone**	$T = \pi r\ell + \pi r^2$
cylinder	$T = 2\pi rh + 2\pi r^2$	**sphere**	$T = 4\pi r^2$
pyramid	$T = \frac{1}{2}P\ell + B$		

Volume

cube	$V = s^3$	**pyramid**	$V = \frac{1}{3}Bh$
rectangular prism	$V = \ell wh$	**cone**	$V = \frac{1}{3}\pi r^2 h$
prism	$V = Bh$	**sphere**	$V = \frac{4}{3}\pi r^3$
cylinder	$V = \pi r^2 h$		

Equations for Figures on a Coordinate Plane

slope-intercept form of a line	$y = mx + b$	**circle**	$(x - h)^2 + (y - k)^2 = r^2$
point-slope form of a line	$y - y_1 = m(x - x_1)$		

Trigonometry

Law of Sines	$\frac{\sin A}{a} = \frac{\sin B}{b} = \frac{\sin C}{c}$	**Law of Cosines**	$a^2 = b^2 + c^2 - 2bc \cos A$ $b^2 = a^2 + c^2 - 2ac \cos B$ $c^2 = a^2 + b^2 - 2ab \cos C$
Pythagorean Theorem	$a^2 + b^2 = c^2$		

Symbols

$\neq$	is not equal to	$p \leftrightarrow q$	biconditional statement, p if and only if q	A'	the image of preimage A
$\approx$	is approximately equal to	$\parallel$	is parallel to	$\rightarrow$	is mapped onto
$\cong$	is congruent to	$\nparallel$	is not parallel to	$\odot A$	circle with center A
$\sim$	is similar to	$\perp$	is perpendicular to	π	pi
$\angle$, $\measuredangle$	angle, angles	$\triangle$	triangle	$\widehat{AB}$	minor arc with endpoints A and B
$m\angle A$	degree measure of $\angle A$	$>, \geq$	is greater than, is greater than or equal to	$\widehat{ABC}$	major arc with endpoints A and C
$^\circ$	degree	$<, \leq$	is less than, is less than or equal to	$m\widehat{AB}$	degree measure of arc AB
$\overleftrightarrow{AB}$	line containing points A and B	$\square$	parallelogram	$f(x)$	f of x, the value of f at x
$\overline{AB}$	segment with endpoints A and B	n-gon	polygon with n sides	!	factorial
$\overrightarrow{AB}$	ray with endpoint A containing B	$a:b$	ratio of a to b	${}_nP_r$	permutation of n objects taken r at a time
AB	measure of $\overline{AB}$, distance between points A and B	(x, y)	ordered pair	${}_nC_r$	combination of n objects taken r at a time
$\sim p$	negation of p, not p	(x, y, z)	ordered triple	$P(A)$	probability of A
$p \wedge q$	conjunction of p and q	$\sin x$	sine of x	$P(A\|B)$	the probability of A given that B has already occurred
$p \vee q$	disjunction of p and q	$\cos x$	cosine of x		
$p \longrightarrow q$	conditional statement, if p then q	$\tan x$	tangent of x		
		$\vec{\mathbf{a}}$	vector **a**		
		$\overrightarrow{AB}$	vector from A to B		
		$\lvert\overrightarrow{AB}\rvert$	magnitude of the vector from A to B		

Measures

Metric	Customary
Length	
1 kilometer (km) = 1000 meters (m)	1 mile (mi) = 1760 yards (yd)
1 meter = 100 centimeters (cm)	1 mile = 5280 feet (ft)
1 centimeter = 10 millimeters (mm)	1 yard = 3 feet
	1 yard = 36 inches (in.)
	1 foot = 12 inches
Volume and Capacity	
1 liter (L) = 1000 milliliters (mL)	1 gallon (gal) = 4 quarts (qt)
1 kiloliter (kL) = 1000 liters	1 gallon = 128 fluid ounces (fl oz)
	1 quart = 2 pints (pt)
	1 pint = 2 cups (c)
	1 cup = 8 fluid ounces
Weight and Mass	
1 kilogram (kg) = 1000 grams (g)	1 ton (T) = 2000 pounds (lb)
1 gram = 1000 milligrams (mg)	1 pound = 16 ounces (oz)
1 metric ton (t) = 1000 kilograms	